AF333896

THE PALEOBIOLOGY
of
PLANT PROTISTS

THE PALEOBIOLOGY
of
PLANT PROTISTS

Helen Tappan

University of California, Los Angeles

W. H. Freeman and Company
San Francisco

Sponsoring Editor: John H. Staples; *Project Editor:* Patricia Brewer; *Copyeditor:* Sean Cotter; *Designer and Illustration Coordinator:* Brenn Lea Pearson; *Production Coordinator:* Fran Mitchell; *Compositor:* Bi-Comp, Inc.; *Printer and Binder:* Arcata Book Group.

Library of Congress Cataloging in Publication Data
Tappan, Helen Niña, 1917–
   The paleobiology of plant protists.

   Includes bibliographies and index.
   1. Micropaleontology. 2. Paleobotany.
3. Unicellular organisms. I. Title.
QE719.T36   560   80-14675
ISBN 0-7167-1109-5

CONTENTS IN BRIEF

List of Tables — xv

Preface — xvii

Abbreviations — xxi

Introduction — 1

Chapter 1 — 9
The Procaryotes: Bacteria and Blue-Greens

Chapter 2 — 107
The Rhodophyta

Chapter 3 — 148
Acritarcha or Hystrichophyta

Chapter 4 — 225
Dinoflagellates

Chapter 5 — 463
Ebridians

Chapter 6 — 490
Xanthophyta and Chrysophyta

Chapter 7 — 535
Silicoflagellates

Chapter 8 — 567
Diatoms

Chapter 9 — 678
Haptophyta, Coccolithophores, and Other Calcareous Nannoplankton

Chapter 10 — 804
Green Algae: Prasinophyta, Chlorophyta, and Euglenophyta

Chapter 11 — 913
Charophytes and Umbellinaceans

Glossary — 964

Index — 979

CONTENTS IN DETAIL

List of Tables   xv

Preface   xvii

Abbreviations   xxi

Introduction   1
   References 8

**Chapter 1   The Procaryotes:
Bacteria and Blue-Greens**   9
Primitive Life in the Fossil Record   10
The Bacteria   10
  The Cell 11
    The Capsule or Sheath 13
    Cell Wall 13
    The Gram Reaction 14
    Nucleoid or Genophore 14
    Flagella 15
    Cysts and Endospores 16
  Reproduction 16
    Asexual Reproduction 17
    Sexual Reproduction 17
      Conjugation 17
      Transformation 17
      Transduction 18
  Metabolism 18
    Bacterial Photosynthesis 19
    Chemotrophic Metabolism 20
      Carbon Cycle 20
      Nitrogen Cycle 22
      Sulfur Cycle 22
      Phosphorus Cycle 22
  Distribution 23
  Bacteria as Geologic Agents 24

    Iron-Depositing Bacteria 24
    Elemental Sulfur Deposits 27
    Petroleum 27
    Limestone Deposition 28
    Hot Spring Stromatolites 28
  Evolution 28
  Fossil Record 31
    Precambrian Fossils 32
    Phanerozoic Fossils 35
  Classification 38
Cyanophyta or Blue-Green Algae   43
  The Cell 43
    The Wall 43
    The Mucilage Sheath 43
    Nucleoplasm 44
    Pigments, Thylakoids, and Storage
    Products 45
    Vesicles, Granules, and
    Inclusions 46
  Growth and Reproduction 46
    Filaments, or Trichomes, and
    Hormogonia 47
      False and True Branching 47
    Sheets and Packets of Cells 48
    Akinetes, or Resting Spores, and
    Endospores 48
    Heterocysts 49
    Mutagenesis and Genetic
    Recombination 51
  Motility 52
  Metabolism 53
    Photosynthesis 53
    Nitrogen Fixation 53
    Mineral Requirements 54
    pH Limitations 54

Oxygen Limitations 55
External Metabolites 55
Calcium Deposition 55
Iron and Sulfur Deposits 55
Distribution 57
Endolithic and Soil Algae 57
Fresh Water 57
Hot Springs 58
Algal Mats and Stromatolites 58
Calcium Dissolution 59
Planktonic Cyanophytes 62
Endosymbionts 62
Fossil Record 63
Stromatolites in the Fossil
Record 63
Calcareous Tubes 65
Fossil Boring and Perforating
Algae 65
Nonmineralized Remains 66
Cyanophyte Evolution 80
Classification 80
Prochlorophyta, the Green Procaryotes 84
Origin of the Eucaryotes 84
The Evolutionary or Autogenous
Origin of Eucaryotes 85
Endosymbiotic Origin of
Eucaryotes 89
Nature of the Host Symbiont 90
Nucleus 90
Plastids and Mitochondria as
Endosymbionts 91
Origin of the Eucaryotic
9+2 Flagella 93
The Precambrian Eucaryote
Controversy 93
Time of Appearance of the
Eucaryote Cell 95
References 96

Chapter 2 *The Rhodophyta* 107
The Living and Fossil Rhodophyta 107
The Organism 107
Cytoplasm 108
Nucleus 108
Pigments, Thylakoids, and
Plastids 108
Storage Products 109
Iodine, Bromine, and
Chlorine 109
The Wall and Sheath 110

Calcification 110
Pit Connections 112
Specialized Cells 113
Growth and Reproduction 113
Morphologic Phases of the
Bangiophycidae 113
Unicellular Rhodophyta 113
Gliding Movement in the
Porphyridiaceae 113
Multicellular
Bangiophycidae 114
Spore Production 116
Morphologic Phases of the
Florideophycidae 116
The Gametophyte 117
The Carposporophyte 123
Tetrasporophytes 123
Life Histories 124
Ecology and Distribution 126
Rhodoliths 127
Geologic Occurrence 128
Evolution 136
Classification 138
References 143

**Chapter 3 *Acritarcha or
Hystrichophyta* 148**
Acritarch Morphology 149
The Vesicle 149
Size 149
Symmetry 149
Clusters or "Colonies" 149
The Wall 152
Wall Structure 152
Wall Ultrastructure 152
Wall Composition 153
Preservation 154
Surface Sculpture 154
Processes 160
Solid Processes 160
Trabeculae 160
Tubular Processes 160
Filamentous Processes 171
Tubular Vellum 171
Excystment Openings 173
Terminal or Lateral Rupture 173
Epityche 176
Equatorial Rupture or Median
Split 176
Cryptopylome 176

Pylome or Cyclopyle   178
Pseudopylome   181
Systematic Position of the Acritarchs   **181**
Occurrence   **185**
Geologic Distribution   185
Evolutionary Trends   197
Contamination and Reworking   199
Paleoecology   199
Temperature   199
Proximity to Shore   200
Depth of Water   202
Salinity   204
Bioturbation of Sediments   204
Classification   **205**
Suprageneric Categories Proposed
for the Acritarcha   206
Acritarch Genera   208
References   **212**

Chapter 4   *Dinoflagellates*   **225**
The Cell   **229**
The Protoplast   231
Cytoplasm   231
Nucleus   231
Plastids and Pigments   237
Oil Drops   240
Pyrenoids   240
Eyespots and Ocelli   242
Pusules   244
Trichocysts and
Nematocysts   244
Flagella   247
Tentacle or Prod   250
Internal Skeleton   250
The Amphiesma   256
The Wall   256
Outer or Plasma
Membrane   257
Vesicular Layer   257
Thecal Layer   258
Pellicular Layer   261
Chemical Composition   261
Plate Patterns (Tabulation) of the
Theca   261
Tabulation of the
Peridiniales   261
Tabulation of the
Prorocentrales   265
Tabulation of the
Dinophysiales   269
Lists   271

The Cyst   **271**
Background of Cyst Studies   272
Nature of the Cyst   275
Cyst Wall Composition   278
Cyst Wall Structure   281
Morphology of the Resting
Cyst   284
Major Cyst Types   284
Reflected Tabulation
(Paratabulation or
Pseudotabulation)   287
Pandasutural Zone   288
Horns, Processes, and
Crests   288
Cyst Wall Surface   290
Flagellar Markings   290
Inclusions Within the Cyst   293
Archeopyle and
Operculum   293
Systematic Relationships of Cysts
and Thecae   299
Physiology   **302**
Movement   302
Flagellar Action   302
Cell Asymmetry   304
Rate of Movement   306
Phototaxis and Mechanism of
Orientation   306
Nutrition   308
Autotrophs, Auxotrophs, and
Myxotrophs   308
Holozoic Dinoflagellates   308
Saprophytes   309
Parasites   309
Host Symbionts   311
Endosymbionts   313
Bioluminescence   316
Toxins   317
Reproduction   319
Vegetative Reproduction   319
Rates of Division   319
Simple Binary or Multiple
Fission   319
Ecdysis   320
Thecal Partition   323
Chain Formation   325
Growth   326
Exuviation and Autotomy   327
Sexual Reproduction   329
Nuclear Cyclosis and
Knot-Stage   330

Isogamy 330
Heterogamy 334
Diploid Vegetative Cell 335
Summary 337

Ecology 338
Population Dynamics 338
Productivity 338
Blooms and Red Tides 339
Habitat 342
Marine 342
Physical Parameters 343
Salinity 344
Temperature 344
Nutrient Requirements 345
Distribution 345
Cold Atlantic Waters 345
Cold Pacific Waters 345
Tropical Oceans 346
Interstitial Flora 347
Tidepools 348
Fresh Water 348

The Fossil Record 349
Historical Background 349
Geologic History 351
Paleozoic 352
Mesozoic 353
Cenozoic 358
Nonmarine Fossil
Dinoflagellates 364

Classification 364
Phylogenetic Relationships 364
Classification of Living
Dinoflagellates 368
Classification of Fossil and Living
Cysts 383

References 435

Chapter 5   Ebridians 463
The Living Organism 464
The Protoplast 464
The Skeleton 465
Skeletal Ornamentation 465
The Lorica 469

Reproduction 472

Occurrence 472
Ecology 472
Geologic Occurrence 474
Evolutionary Trends 476

Classification 477

References 487

***Chapter 6   Xanthophyta and
Chrysophyta*** 490
**Division Xanthophyta, Class
Xanthophyceae** 491
The Cell 491
Motile Cell or Zoospores 491
Plastids and Pigments 493
Cell Wall 493
Reproduction 494
Occurrence 496

Class Eustigmatophyceae 496
The Cell 497

Classification of the Xanthophyta 499

Division Chrysophyta 499
The Cell 499
Cell Wall or Lorica 499
Cytology 501
Flagella 501
Choanoflagellate Collar 504
Nutrition 504
Reproduction 504
Cysts 506
Occurrence 510
Habitat 510
Geologic Record 510
Classification 518

References 530

Chapter 7   Silicoflagellates 535
The Living Organism 535
The Protoplast 535
The Skeleton 538

Reproduction 542

Occurrence 546
Ecology 546
Physical Environment 546
Geologic Record 547
Abundance and Silica
Deposition 547
Biostratigraphy 549
Paleotemperature
Interpretation 552

Evolutionary Trends 553

Classification 554

References 561

Chapter 8   Diatoms 567
The Diatom Cell 568
The Protoplast 568
Nucleus 568
Plastids 571
Pigments 571

Pyrenoids 571
Extracellular Sheaths, Stalks, and
Matrix 572
The Siliceous Frustule 572
Wall Composition 572
Silica Deposition 573
Frustule Morphology 573
Shape and Symmetry 576
The Girdle 576
Raphe System 577
Fine Structure of the Wall 581
Areolae, Pseudoloculi, and
Alveoli 581
Elevations, Horns, and
Folds 585
Hyaline Fields 585
Keels, Costae, Craticula, and
Collars 587
Internal Ducts and
Stigma 587
Ocelli, Pseudocelli, and
Pseudonodules 587
Processes 591

Physiology 595
Vegetative Reproduction 595
Sexual Reproduction and Auxospore
Formation 596
Sexual Reproduction in the
Centric Diatoms 597
Male Gamete Formation 597
Female Gamete 598
Fertilization and Auxospore
Formation 598
Autogamy 599
Sexual Reproduction in the
Pennate Diatoms 599
Statospores or Resting Cysts 600
Intraspecific Variation 603
Size Range 603
Life History Variations 604
Other Morphologic
Variations 605
Habitat 605
Substrate Attachment and
Colonies 607
Planktonic Species and Flotational
Adaptations 607
Movement 610
Nature of Movement 610
Mechanism of Movement 610
Symbiosis and

Commensalism 613
Parasitism and Grazers 613
Nutrition 613
Heterotrophy 613
Photosynthesis and
Productivity 614
Ecologic Factors 615
Light 615
Temperature 616
Salinity 616
Hydrogen Ion 616
Silicon 617
Nitrogen 618
Phosphorus 618
Sulfur 619
Boron 619
Barium 619
Manganese 619
External Metabolites 619
Vitamins 620
Distribution 620
Present Occurrence 620
Marine Habitat 620
Plankton 621
Benthos 621
Geographic Distribution in the
Seas 622
The Marine Succession 625
Freshwater Habitat 626
Diatomaceous Oozes and
Diatomites 628
Distribution of Diatomaceous
Sediments 628
Effect of Selective Solution 632
Origin and Accumulation of
Diatomaceous Sediments 634
Diagenesis and Preservation of
Diatom Silica 636
Geologic Occurrence 636
Marine Record 637
Freshwater Record 641
Paleotemperatures 650
Evolution of the Diatoms 652
Classification 654
References 659

**Chapter 9 Haptophyta,
Coccolithophores, and Other
Calcareous Nannoplankton** 678
The Living Organism 679
Nucleus 679

Plastids, Pigments, and
Pyrenoids 681
Other Organelles and Cytoplasmic
Inclusions 681
Flagella and Haptonema 683
  Cell Movement and Haptonemal
  Function 683
Cell Covering 684
  Scales 684
  Coccoliths 684
    Organic Matrix of the
    Coccolith 684
    Relationship of Coccolith
    Formation and
    Photosynthesis 687
    Cytology of Coccolith
    Formation 688
    Coccolith Mineralogy 692
    Coccolith Morphology and
    Crystal Structure 694
      Holococcoliths 694
      Heterococcoliths 695
      Ortholithae 706
      Micrococcoliths and
      Macrococcoliths 706
      Malformed Coccoliths 710
Reproduction and the Life Cycle 710
  Coccolith-Covered Motile and
  Nonmotile Pelagic Stages 710
  Coccolith-Covered Nonmotile Stage;
  Naked or Scale-Covered Motile
  Stage 712
  Coccolith-Covered Motile Stage;
  Benthic Filamentous Nonmotile
  Stage 714
  Coccolith-Covered Motile Stage;
  Cysts and Palmella Stage 715
  Scale-Covered Motile Stage; Coccoid
  or Palmelloid Nonmotile Stage 715
Ecology and Distribution 715
  Physical Parameters 715
    Depth of Water 715
    Temperature 719
    Salinity 721
    Calcium, Magnesium, Strontium,
    and Iron 723
    Light 723
    Oxygen 724
    Nitrogen 724
    Vitamins 724
    Heterotrophy 725

Fossil Calcareous Nannoplankton 725
  Background Studies 725
    Problems in Studying Fossil
    Coccoliths 726
  Geologic Occurrence 727
    Paleozoic Calcareous
    Nannoplankton 727
    Mesozoic Calcareous
    Nannoplankton 728
      Nannoconids 728
      Schizospheres 729
      Stomiosphaeraceae 733
      Coccoliths 735
    Cenozoic Calcareous
    Nannoplankton 745
      Braarudosphaeraceae 745
      Thoracosphaeraceae 747
      Ceratolithaceae 750
      Discoasters 751
      Coccoliths 752
  Coccolith Oozes and Chalks 761
  Paleoecology 769
**Evolution of the Calcareous
Nannoplankton** 771
Classification 776
References 785

**Chapter 10 Green Algae:
Prasinophyta, Chlorophyta, and
Euglenophyta** 804
The Prasinophyta 804
  The Cell 805
  Reproduction 806
  Cyst or Phycoma 807
    Development of the Cyst 808
    Cyst Wall Structure and
    Composition 809
  Occurrence 812
    Present Distribution 812
    Geologic Occurrence 813
  Classification 816
The Chlorophyta 825
  The Organism 825
    Pigments, Plastids, and Storage
    Products 825
    Nucleus 826
    Flagella 826
    Cell Wall 826
    Calcium Carbonate, Silica, and Cell
    Inclusions 828
  Reproduction 828

Asexual Reproduction 828
Sexual Reproduction 829
Rock-Forming Chlorophyta 831
Life History and Distribution 833
Orders Chlamydomonadales,
Volvocales, and
Tetrasporales 834
Order Chlorococcales 836
Chlorococcaceae 837
Chlorellaceae 838
Hydrodictyaceae 838
Coelastraceae 839
Botryococcaceae 841
Oocystaceae 841
Scenedesmaceae 842
Class Codiolophyceae 843
Ulotricaceae 843
Microsporaceae 845
Chaetophoraceae 845
Trentepohliaceae 845
Ulvaceae 845
Orders Cladophorales, Codiales,
Caulerpales, and Derbesiales 846
Order Siphonocladales 852
Order Receptaculitales 852
Order Dasycladales 860
Fossil Dasyclads 864
Distribution of Dasyclads 881
Order Oedogoniales 882
Order Zygnematales 882
Zygnemataceae 882
The Desmids 884
Classification 887
Division Euglenophyta **895**
The Euglenid Cell 895
Nutrition 896
Reproduction 896
Occurrence 896
Fossil Euglenids 896
Classification 897
References **897**

Chapter 11   Charophytes and
Umbellinaceans **913**
The Living Organism **915**
Reproduction 915
The Antheridium 917
The Sporophydium 919
The Oosporangium and
Gyrogonite or Lime Shell 919
Calcification 919
Size and Shape of the Lime
Shell 923
The Basal Pole 924
The Apex 924
Oospore Membrane 926
Utricle 926
Germination 927
Ecology 927
Depth 927
Salinity 930
Oxygenation 930
Lime 930
Hydrogen Ion 930
Fossil Charophytes **931**
Geologic Occurrence 931
Umbellinaceans 932
Paleozoic Charophytes 937
Mesozoic Charophytes 941
Cenozoic Charophytes 951
General Evolutionary Trends 951
Charophyte Classification and
Systematic Relationships **954**
References **957**

Glossary **964**

Index **979**

LIST OF TABLES

Table 1.1   Bacterial respiration   **20**

Table 1.2   Classification of the Kingdom Procaryota   **39**

Table 1.3   Classification of the Division Cyanophyta   **81**

Table 1.4   Classification of the Division Prochlorophyta   **84**

Table 2.1   Classification of the Division Rhodophyta   **138**

Table 4.1   Mean number of chromosomes reported in some dinoflagellate nuclei   **233**

Table 4.2   Pigments identified by thin-layer chromatography in selected dinoflagellates   **239**

Table 4.3   Comparison of thecate cell and resting cyst of living dinoflagellates   **273**

Table 4.4   Motile cell and separately named equivalent resting cyst of some living species   **299**

Table 4.5   Biostratigraphic zonation based on dinoflagellate assemblages   **361**

Table 4.6   Classification of living dinoflagellates, with representative genera   **368**

Table 4.7   Classification of fossil dinoflagellates (cysts and spicules)   **386**

Table 5.1   Classification of the Ebriophyceae   **480**

Table 6.1   Classification of the Xanthophyta   **498**

Table 6.2   Plankton cells per cubic centimeter of seawater   **510**

Table 6.3   Classification of the Chrysophyta   **518**

Table 6.4   Cyst genera and families of the Chrysophyceae   **528**

Table 7.1   Surface water temperatures as indicated by the *Dictyocha/ Distephanus* ratios   **552**

Table 7.2   Classification of the silicoflagellates   **554**

Table 8.1   Progressive decrease in mean size of diatom population with successive cell divisions, according to Macdonald-Pfitzer Hypothesis   **595**

Table 8.2   Rate of sinking in the water column   **609**

Table 8.3   Upper Cretaceous (Late Campanian and Maastrichtian) diatom zones of Hajós (1975)   **639**

Table 8.4   Cenozoic diatom biostratigraphy, largely from data of the Deep Sea Drilling Program   **644**

Table 8.5   Classification of the diatoms   **655**

Table 9.1   Comparison of coccolithophore cell number per liter, salinity, and temperature at Utsira, in the southern part of the west coast of Norway, 1945   **717**

Table 9.2   Mesozoic biostratigraphic zonation based on calcareous nannoplankton   **748**

Table 9.3   Paleogene biostratigraphic zonation based on calcareous nannoplankton   **762**

Table 9.4   Neogene biostratigraphic zonation

             based on calcareous nannoplankton   766

Table 9.5   Paleogene latitudinal zonation of coccolithophores   771

Table 9.6   Classification of the Haptophyta and other calcareous nannoplankton   776

Table 10.1   Classification of the Prasinophyta   818

Table 10.2   Classification of the Chlorophyta   888

Table 10.3   Classification of the Euglenophyta   898

Table 11.1   Charophyta polarity indices   924

Table 11.2   Types of sculpture of the oosporangial membrane   927

Table 11.3   Classification of the charophytes   955

PREFACE

A man will turn over half a library to make one book.

Boswell's *Life of Dr. Johnson*

The study of microscopic fossil organisms, both animal and plant, is the subject of micropaleontology. Like other branches of science, micropaleontology overlaps and interacts with many disciplines. In order to determine what information about any particular group of protists is available, the micropaleontologist must become familiar with much of the primary literature in phycology, protozoology, bacteriology, microbiology, micropaleontology, botany, zoology, and paleontology, as well as various specialized journals in microscopy, cytology, ultrastructure, biochemistry, ecology, evolution, and oceanography.

In recent years, micropaleontologists have increasingly emphasized the study of various photosynthetic protists, in particular, such phytoplankton groups as the coccolithophores, dinoflagellates, and diatoms, in addition to the better known foraminifera and radiolaria, or the metazoan microfossils such as the ostracods and conodonts. The paleontologist is most concerned with those organisms producing preservable skeletal or covering materials, whether mineralized or of resistant organic compounds. In contrast, the biologist has generally tended to emphasize those living organisms that lack hard parts or resistant walls and whose cell contents can be studied without difficulty. As a result the basic information concerning the particular living organisms of interest to the paleontologist may be widely scattered. Teaching courses in micropaleontology in the absence of a comprehensive modern summary of pertinent biologic information and the similar absence of an up-to-date summary of the fossil record provided the impetus that led to the preparation of this volume. During the 10 years of intensive work that it has required, much new material has become available, so that the resultant volume is larger and more comprehensive than had originally been planned. Some of the material may be skimmed in an introductory course, while the advanced student may use the extensive reference lists of each chapter as a base for more detailed study of specialized topics.

The purpose of the present volume is to compile and summarize the available knowledge concerning the fossil and living plant protists, the procaryotes and eucaryotic algae; to indicate their probable evolutionary relationships, as well as their value for biostratigraphy; and to show their fluctuations in diversity and abundance through geologic time, and their changing marine, freshwater, or terrestrial distribution in response to differing environmental conditions, as a basis for interpreting the nature of the paleoenvironments. Although thus intended to serve the needs of the micropaleontologist, it is hoped that a summary of the paleontologic information will also be useful for biologists and geologists, and provide a basis for additional cross-disciplinary studies that will further advance our knowledge of the microscopic biosphere.

TAXONOMY

This volume is not intended as a systematic treatise; not all living or fossil genera are listed in the classifications included in each chapter, and no detailed references are given for the taxa, although author and dates are cited. Some groups with limited fossil records have only a very abbreviated classification appended. Authorship is cited for species at the first place each appears in the volume, and is not repeated where the same taxon is mentioned later. The first reference to a species herein may be found by reference to the Index.

An attempt has been made to utilize the correct (current) name for each taxon, wherever cited. Thus, zonal names in the biostratigraphic charts have been modified to reflect recent taxonomic changes, and illustrations from previous publications appear under the newer names, reflecting later proposed genera or determinations of synonymy.

A very few taxonomic transfers or name corrections have been made herein: *Duvernaysphaera tenuimarginata* (ex *Veliferites*), in the Prasinophyta; *Umbellina lineata* (ex *Umbella*), in the umbellinaceans; Subfamily Cribrosphaerelloideae (pro Cribrosphaeroideae), Family Lithostromatiaceae (pro Lithostromationidae, Lithostromationaceae), Family Acanthoicaceae (pro Acanthoiceae), and Order Podorhabdales (pro Podorhabdinales), all in the haptophyte coccolithophores; and Hermesinaceae (pro Hermesinidae), in the ebriophycean Pyrrhophyta.

ACKNOWLEDGMENTS

This volume would not have been possible without the valuable assistance and encouragement at all stages provided by my husband and frequent collaborator, Alfred R. Loeblich, Jr. He critically read the manuscript, helped proofread every word in typescript and in proof, provided invaluable criticism, suggestions, references, and many illustrations, both with the light microscope and the SEM, and placed in proper form for the bibliography the references that I had compiled on cards. Various systematic indices that he prepared (either alone or jointly with me or others) also were invaluable in updating taxonomy.

Because of the large number of illustrations, acknowledgments for those that have been previously published are keyed by author and date to the References of the various chapters. My gratitude is herewith expressed both to the authors and to the publishers of each.

In addition, I am grateful for original figures supplied by the following colleagues: Hans M. Bolli, SEM of *Pithonella;* J. P. Bujak, photos of Triassic dinoflagellates; Vivienne Cassie, SEM of *Melosira;* R. E. Dickerson, diagram of bacterial phylogeny; Warren S. Drugg, SEM and light photos of Tertiary dinoflagellates; the late Ralph Emerson, photo of parasitic dinoflagellate on euphausid; W. R. Evitt, light photos and SEM of dinoflagellates; D. Fütterer, SEM of calciodinellaceans and thoracosphaeraceans; Hans Gocht, SEM of dinoflagellates; Claude Greuet, diagrams of dinoflagellate ocellar and trichocyst structures; Merton E. Hill III, light photos and SEM of Cretaceous coccolithophores; Robert W. Holmes, SEM of diatoms; Susumu Honjo, light photos and SEM of modern coccolithophores; Lazar Jerković, SEM of diatoms; Wolfhart Langer, thin section and SEM figures of *Sycidium;* Sigurd Locker, drawings of silicoflagellates; Alfred R. Loeblich III, for many SEM, TEM, and light photos of dinoflagellates and coccolithophores, and for sketches from living material of other algae, from which figures were drafted; Andrew McIntyre, TEM of living coccolithophores; David J. McIntyre, photos of Cretaceous dinoflagellates; B. L.

Mamet, photos of Umbellinaceae; S. B. Manum, photos of fossil prasinophytes and dinoflagellates; the late V. P. Maslov, photos of charophytes from the USSR; Eduardo Musacchio, photos of Cretaceous charophytes from Argentina; Denise Noël, SEM of surface of coccolith limestone; Sarah T. Pierce (Damassa), photos of the prasinophyte *Schizophacus;* A. A. Prashnowsky, TEM of Precambrian bacteria; the late James M. Schopf, TEM of Pennsylvanian bacteria; J. W. Schopf, SEM, TEM, and transmitted light photos of Precambrian algae and bacteria; Eiji Takahashi, TEM of *Mallomonas;* Mario Trejo H., photos of *Nannoconus* rosettes from Mexico; John H. Wall, photos of Cretaceous acritarchs and dinoflagellates; Graeme J. Wilson, photos of New Zealand *Wetzeliella.* New light photos, SEM, and drawings prepared by Alfred Loeblich, Jr. or myself are not individually indicated.

H. W. Johansen kindly identified some of the calcareous red algae that I have collected and illustrated herein.

Figures were drafted by Martha Matthews, Vicki Doyle Jones (those on dinoflagellates), or myself.

The quality of this volume has been greatly enhanced by the meticulous editing and careful cross-checking of references, tables, figures, and glossary by Sean Cotter, manuscript editor, to whom I am most grateful.

Finally, as suggested by the quotation heading this preface, and indicated as well by the extensive references appended to the various chapters, I am grateful to innumerable co-workers in phycology and micropaleontology for exchange of reprints and other valuable publications, and to the many librarians in the Geology Library, Biomedical Library, and University Research Library at the University of California, Los Angeles, whose work also was increased as I "turned over half a library" in the preparation of this volume.

April 1980   *Helen Tappan*

ABBREVIATIONS

Abbreviations for journal titles in the References for each chapter are those of the *World List of Scientific Periodicals* (London: Butterworths).

A few additional abbreviations are used through the text, as follows:

Å	Angstrom; 0.1 nm, or 10^{-10} m
ATP	adenosine triphosphate
CCD	carbonate compensation depth
DNA	deoxyribonucleic acid
DSDP	Deep Sea Drilling Program
G+C	Guanine + Cytosine (organic bases in DNA, varying in percent)
ICBN	International Code of Botanical Nomenclature
ICZN	International Code of Zoological Nomenclature
l	liter
MTOC	Microtubule organizing center
μm	0.001 mm
nm	0.001 μm; 10 Å
pg	picograms; 10^{-12} g
RNA	ribonucleic acid
rRNA	ribosomal RNA
SEM	scanning electron microscope (or microphotograph)
syn.	synonyms, in classification tables
TEM	transmission electron microscope (or micrograph)
tRNA	transfer RNA
‰	parts per thousand

THE PALEOBIOLOGY
of
PLANT PROTISTS

INTRODUCTION

To see clearly something that no one has ever examined before, whether one is using a hand lens or an electron microscope, is a creative intellectual act; and to realize how a new observation is related to previous knowledge is an act of insight of the kind on which all advance in scientific knowledge is based.

R. E. Holttum, 1961

*P*erhaps in no other branch of paleontology have so many advances occurred in the last quarter century as in micropaleontology; in part because of the availability of new tools such as the electron microscope; in part because of the impetus provided most recently by the Deep Sea Drilling Program (DSDP), through which a wealth of new material has become available for study by many specialists, allowing an unprecedented compilation of the results; and in part because of the realization that the astronomically abundant microfossils provided information not only for age dating, but also as to paleotemperatures, oceanic circulation, depth of water and distance from shore, and the origin and early evolution of life. Furthermore, because of the position of most of these organisms at the base of the aquatic food chains, their fluctuations in abundance and diversity through geologic time resulted in selection pressures that modified the course of evolution of the entire biosphere.

The organization of this volume into chapters is on a systematic basis, largely because few specialists work with all these groups, and because methods of study vary along similar lines. Thus different techniques are required for a study of fossil dinoflagellates from those used for the calcareous nannoplankton or the siliceous diatoms and silicoflagellates. Discussion of evolutionary relationships also are facilitated by a prior discussion of the morphologic and biochemical features of each group. Biostratigraphic zonations based on the various types of fossils are included, as are discussions of other ways that the fossil record has been utilized in interpretation of the geologic past.

The plant protists considered herein include the Procaryota or Monera (bacteria and blue-green algae) and most of the eucaryotic algae

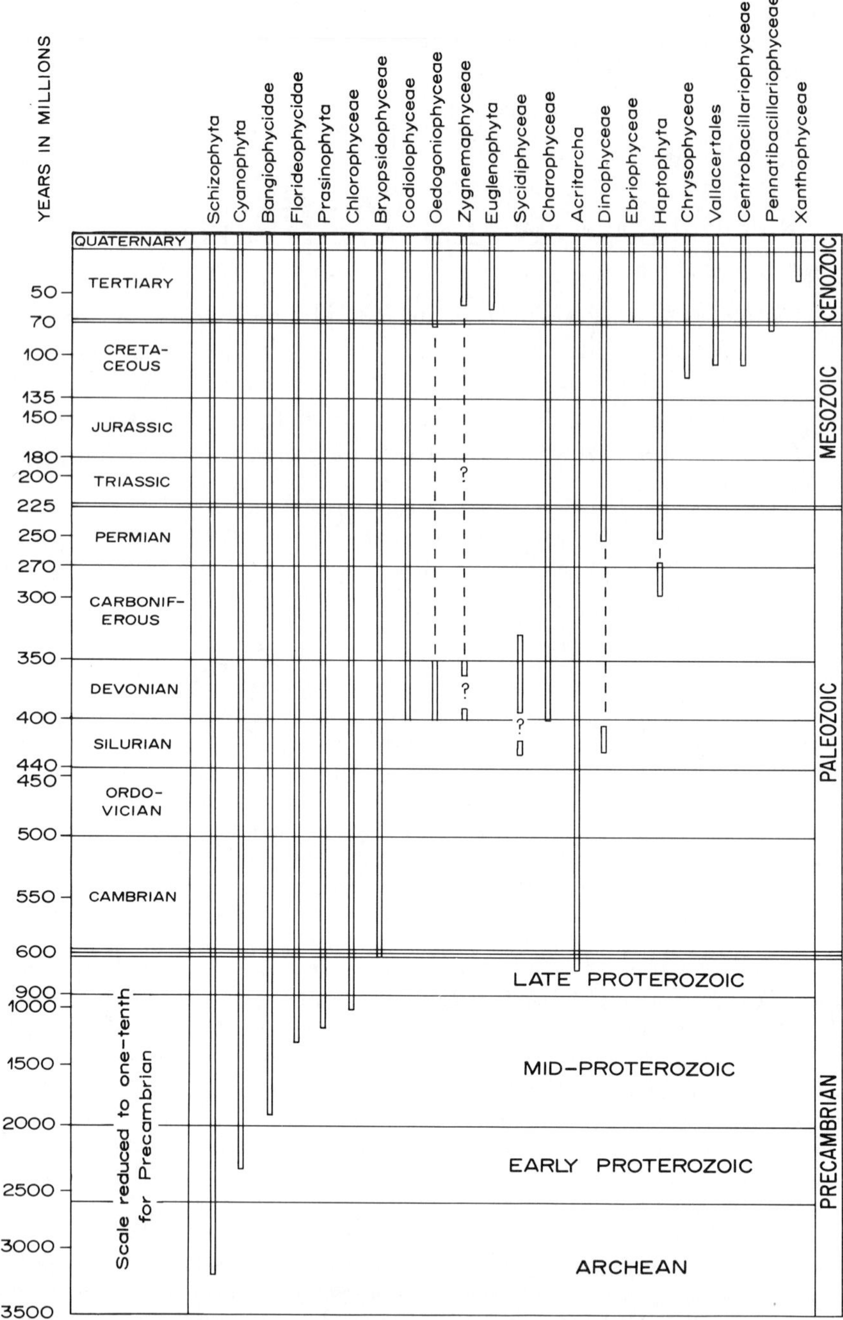

Figure 1
Geologic occurrence of major groups of the plant protists, mostly shown to
the level of classes. Both the extent of the geologic record and the diversity of
organisms involved are demonstrated.

and their obvious relatives. Not included are the strictly animal protists, such as the Rhizopodea and Ciliatea (Protozoa). These plant protists have been termed Protoctista, but the latter term originally excluded the multicellular red, green, and brown algae, hence is not strictly applicable to the groups here covered. Although emphasis is placed on those groups with a good fossil record (see Figure 1), related forms also are discussed to aid in understanding their evolutionary relationships. Various microfossils of uncertain relationship are included, and reasons are given for their inclusion. A few groups of uncertain affinity that are omitted may eventually be found to belong with the plant protists, such as the Paleozoic calcispheres (variously regarded as coccoid algae or as primitive protozoans) or Chitinozoa (now regarded by some as referrable to the fungi, but also placed with protozoans or with various metazoans). Other modern algae that are not included, in view of the improbability of any fossil record, are the cryptomonads and chloromonads. Although a few compression fossils are known of the brown algae, the group is not discussed, nor are the fungi covered herein.

After these various omissions, some 13 divisions (or phyla) are included, in approximate order of evolutionary advancement. The procaryotes of Chapter 1 are the Bacteriophyta, Cyanophyta, and Prochlorophyta; 10 eucaryotic divisions include the Rhodophyta (Chapter 2), which are most primitive in many regards, and Pyrrhophyta (Chapters 4, 5), whose nuclear and other features attest to a primitive condition; the Acritarcha (Chapter 3) have a long geologic record, and at least some may belong to an unknown but extinct group; the Xanthophyta (Chapter 6) here include both the Xanthophyceae and Eustigmatophyceae, although they may not be closely related; the Chrysophyta (Chapters 6, 7) here include the Bikosoecophyceae, Chrysophyceae, and Craspedomonadophyceae, although some place the last group in the protozoans. The Bacillariophyta (diatoms, Chapter 8), and Haptophyta (coccolithophores, Chapter 9) have long been regarded as chrysophytes, but differ greatly in the nature of their vegetative cells. Green algal groups are generally agreed to have

diverged early from the brown-pigmented divisions; the Precambrian record of the Prasinophyta (Chapter 10) supports the morphological evidence of the primitive nature of this division, which differs in many ways from the other green pigmented algae, Chlorophyta, Euglenophyta (both in Chapter 10), and Charophyta (Chapter 11).

As knowledge of any group of organisms increases, there is a tendency toward more and more detailed classification, as a means of summarizing and ordering this knowledge and for making predictions; the hierarchical classifications also include more and more lower taxa. Perhaps one of the most obvious results of the explosion of knowledge of the plant protists in all its aspects is the similar taxonomic explosion in described genera and species. The accompanying figures, showing the cumulative increase in generic taxa assigned to acritarchs (Figure 2), fossil dinoflagellates (Figure 3), fossil prasinophytes (Figure 4), and to the living and fossil Ebriophyceae (Figure 5), silicoflagellates (Figure 6), and calcareous nannoplankton (Figure 7), indicate graphically the rapid development of these aspects of micropaleontology. Similarly rapid increases in knowledge of living taxa have resulted from a combination of new techniques for laboratory culture, new techniques for preparation of material, new instrumentation for study, and new biochemical techniques for determination of genetic similarities. Much of this biological information is summarized herein, particularly as it applies to evolutionary interrelationships of the higher categories, life histories, ecology, and present distribution, those features of particular relevance to the study and interpretation of the fossil record.

Because of their abundance, the various plant protists have been important rock builders; the rise of the coccolithophores in the Mesozoic resulted in the accumulation of extensive oceanic chalks and limestones; shallow water reef deposits were formed at many times in the past by the benthic calcareous red and green algae; diatoms and other siliceous microplankton similarly provided much of the opaline silica for the diatomites and cherts of the Cenozoic. The abundance of organic matter

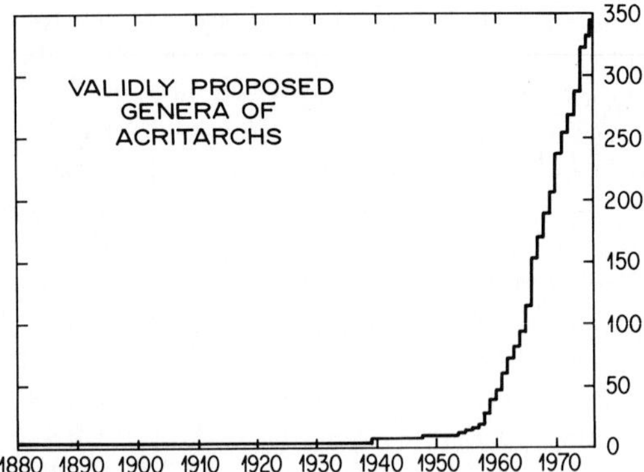

Figure 2
Cumulative total of genera proposed for organisms referrable to the acritarchs, showing most of the taxa to have been described within the past two decades.

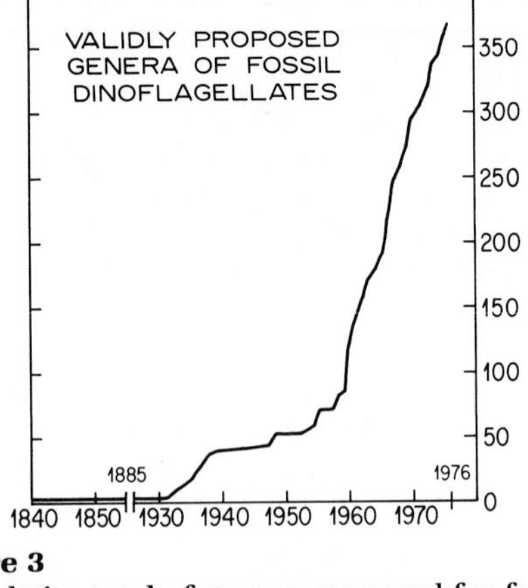

Figure 3
Cumulative total of genera proposed for fossil dinoflagellates (hence based on the encysted stage or endoskeletal remains).

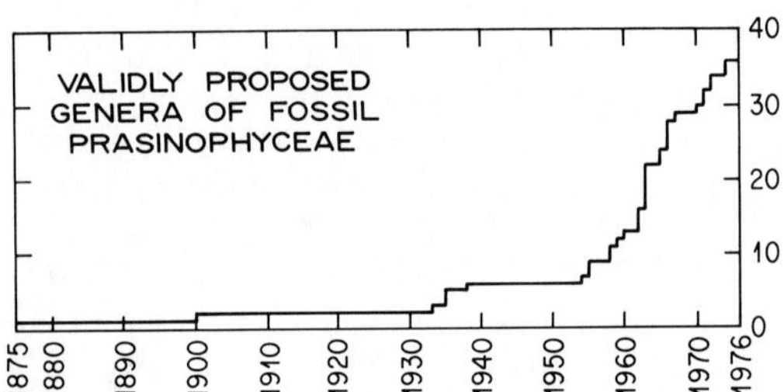

Figure 4
Cumulative total of genera proposed for fossil Prasinophyta (Prasinophyceae), known only from the resistant phycoma stage.

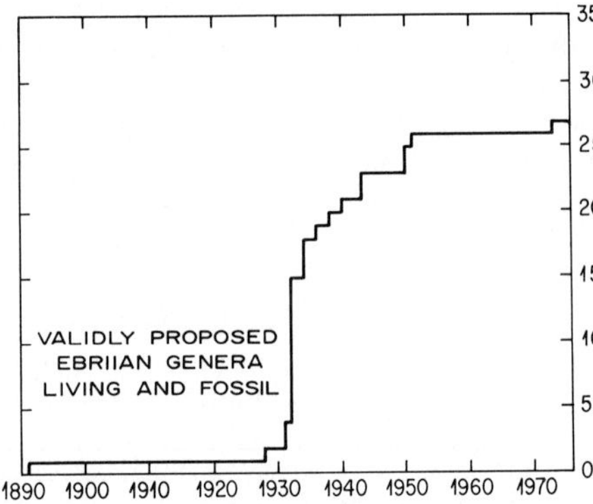

Figure 5
Cumulative total of genera proposed for the living and fossil Ebriophyceae. Unlike most of the other groups considered here, most of the taxonomic work was done on these during the 1930s and 1940s, with little major change subsequently.

produced by these protists, both as cell material and as excreted organic compounds, supplied the dispersed organic carbon in sediments and the hydrocarbons of petroleum deposits and algal coals, and allowed the accumulation of atmospheric oxygen, meanwhile recording the process of photosynthetic carbon-fixation in their isotopic fractionation of the reduced carbon compounds (Tappan, 1968; Tappan and Loeblich, 1971).

The plant protists have shown a general diversification at the higher taxonomic levels. Species and genera also diversified rapidly at certain periods of the past, underwent rapid biotic turnover (rapid replacement of earlier taxa by newly evolved ones) at other periods,

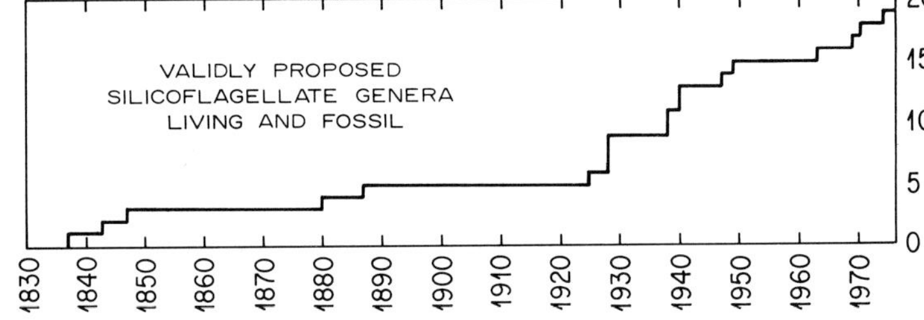

Figure 6
Cumulative total of genera proposed for living and fossil silicoflagellates (Vallacertales).

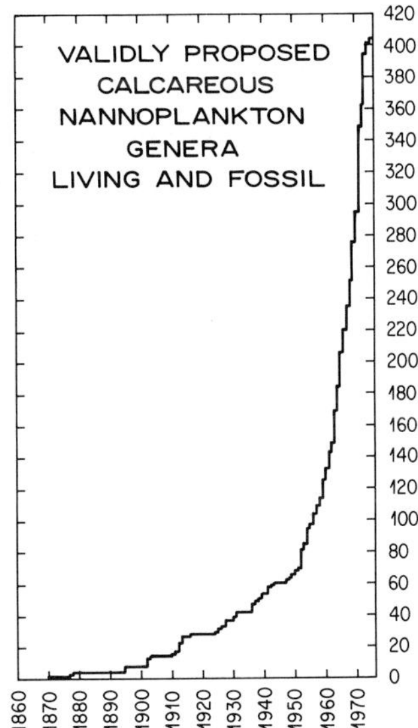

Figure 7
Cumulative total of genera proposed for living and fossil calcareous nannoplankton (coccolithophorids, discoasters, and probably related forms), showing a major increase over the past two decades.

and at still other times suffered a decrease in diversity as extinctions outnumbered newly evolved replacements (Tappan and Loeblich, 1972). These fluctuations in diversity appear to be related to changing oceanographic conditions (circulation and upwelling) and the effectiveness of nutrient recycling (of phosphates, nitrates, vitamins, etc.; Bramlette, 1965).

As seas encroached onto the continents, nutrient supplies from land favored extensive phytoplankton growth as well as that of the benthic algae; but as the land areas continued to decrease in area and elevation, the nutrient influx slowed. The progressively more equable maritime climates depressed oceanic circulation and upwelling, and phytoplankton cells that sank below the photic zone could not return to the upper water levels but were buried on the sea floor. The dispersed carbon and other organic remains buried in the sediments thus trapped the nutrients that no longer could be recycled. The nutrient depletion resulted in heavy selection pressures among the remaining phytoplankton, and many became extinct, particularly the oceanic types. The rate of accumulation of biogenic ooze also dropped precipitously, resulting in worldwide disconformities, even in the deep sea, as at the Cretaceous-Tertiary boundary (Tappan and Loeblich, 1973).

Even at the times of widespread reduction in diversity (extinctions) a few highly adaptable species persisted, commonly in near-shore habitats. Examples of these persistent forms are the *Tasmanites* and related prasinophytes that survived the Devonian acritarch extinctions. Similar persistent or "disaster" taxa include *Braarudosphaera* and *Thoracosphaera,* which survived the major extinctions of calcareous nannoplankton at the end of the Mesozoic, and the dinoflagellates *Deflandrea* and *Spiniferites,* which similarly survived the Cretaceous crisis to persist into the Paleocene.

The successive periods of evolution, diversification, and extinction of various plant protists that form the base of the oceanic food chains not only changed the levels of primary productivity but variously resulted in concurrent diversification of the coexisting animal taxa, produced strong selective pressures for improvements in food gathering techniques or adaptation to a different food source such as detritus, or resulted in the selective extinction of the least efficient (sessile suspension feeders, such as bryozoa, corals, brachiopods, and stalked echinoderms) or those comprising the higher trophic levels (Tappan, 1970).

The evolutionary changes in the ecosystem resulting from variations in primary productivity resemble the successional stages of modern ecosystems that similarly reflect changes in energy flow over a shorter time period (see Figure 8; Tappan, 1971). The early stage of succession is characterized by only a few, highly adaptable species, such as the previously mentioned "disaster" species. These species generally are small, of simple morphology, and characterized by short life cycles and rapid growth, numerous individuals, and high variability within a species. In the global succession, they evolved rapidly and were rapidly replaced by others. In later successional stages, diversity increased, as did equability of diversity; some taxa attained larger sizes and greater morphologic complexity; and life cycles became longer and more complex and food chains more varied and interconnected, involving detritus-feeders as well as filter-feeders, grazers, predators, and parasites. Such complex ecosystems also were more susceptible to disruption, as occurred when nutrient depletion attained crisis levels. The wave of extinctions that followed such a disruption might have long-lasting effects, for unlike stages of a modern ecosystem, which depend on immigrants to effect the successional changes, the geologic ecosystems could only become reestablished by the slower process of evolution of new taxa to succeed the earlier ones. The Paleozoic extinctions left a Triassic global ecosystem of very low diversity, and the marine extinctions of the Cretaceous similarly were followed by the low diversity of the Danian and slow evolutionary expansion of the later Paleocene.

A knowledge of the diversity and abundance of the plant protists at various times in the geologic past thus aids in understanding the evolutionary changes in the remainder of the biota. To date, no paleoecologic study has yet related the evolutionary changes in the entire megafauna to those occurring in the associated microfauna and microflora. Perhaps the present summary of the status of the plant protists may aid the development of more complete paleoecologic studies in the future.

In a review of the status of protistan micropaleontology about a decade ago, Tappan and Loeblich (1968) mentioned some of the problems involved, such as their simultaneous consideration both as animals and as plants, hence involving both the botanical and zoological nomenclatural rules. This difficulty has now been resolved in that most of the fossil photosynthetic protists are regarded under the botanical code. A decade ago, stratigraphical and taxonomic studies had scarcely begun for the organic-walled plant protist fossils, but already dinoflagellates, acritarchs, and prasinophytes have been shown to be stratigraphically useful. The observed dearth of knowledge of the encysted stage of living taxa has been remedied in large part by the studies of living dinoflagellates, in which cysts are shown commonly to represent the zygote produced during sexual reproduction, and the similar life history studies involving the phycoma stage of the prasinophytes.

The taxonomic problems due to the difficulty of correlating studies of calcareous nannoplankton with the light microscope (including the use of crossed nicols under the polarizing microscope) to those made with the TEM has been resolved by new techniques whereby the same specimen can be viewed with both types of instrumentation; the orientation problem that characterized the use of carbon replicas and the TEM has been nullified by use of the SEM. The necessity for life cycle studies of the coccolithophores also has been partially met, such studies indicating morphologically distinct generations, one of which might be benthic, nonmotile, and even filamentous. Similar free-living haptophytes that lacked coccoliths but

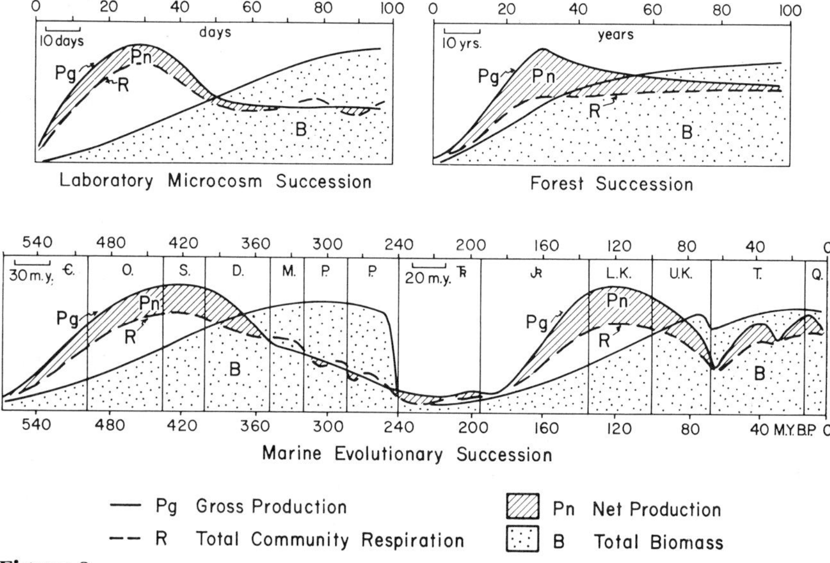

Figure 8

Patterns of ecologic succession. Ecologic succession in all types of ecosystems, regardless of size or scale, shows a similar pattern of interrelated primary production, total community utilization of the productivity (indicated as respiration), and total biomass. In each system, both gross production and respiration increase rapidly at first and later decrease as the biomass enlarges. Net productivity (excess of production over consumption) also decreases, as better utilization results from the more complex food webs of the later stages of succession. The hierarchical nature of ecosystems is indicated by the small scale and short duration (a matter of days) in a laboratory culture succession, as compared to a forest succession (occupying about a century). As productivity and rate of energy flow also provide strong evolutionary pressures, the global evolutionary pattern also can be compared to an ecologic succession on a much larger scale, but as in the case of other ecologic successions, the delicately balanced climax stages may be susceptible to disruption. Such disruption on a global scale is reflected in the major periods of extinction, with much slower recovery of the previous level of diversity. From Tappan, 1971.

which have organic scales strongly resembling coccoliths suggest that evolutionary diversification may have occurred in this group prior to the development of mineralized plates on the scale cover; studies of the living cells led to the separation of the Haptophyta from the Chrysophyta. The dearth of reliable stratigraphic data for the siliceous microfossils (ebrians, chrysophyte cysts, silicoflagellates, diatoms) has been overcome by the wealth of such documentation by the DSDP. Predictions were made (Tappan and Loeblich, 1968) that micropaleontology was evolving from an almost exclusively biostratigraphic emphasis to include studies of ecology, evolution, and interspecific relationships; and that precise stratigraphic, taxonomic, and morphologic documentation would allow utilization of the many varied microfossils in interpretation of paleogeography, studies of past climates, oceanic current patterns, and even ecosystem structure and relationships. That these predictions also have been fulfilled over the past decade will be amply demonstrated in the remainder of this volume.

REFERENCES

Bramlette, M. N., Massive extinctions in biota at the end of Mesozoic time. *Science,* v. 148, p. 1696 – 1699, 1 fig., 1965.

Holttum, R. E., Plant taxonomy as a scientific discipline. *Advanc. Sci.,* v. 18(73), p. 1 – 9, 1961.

Tappan, Helen, Primary production, isotopes, extinctions and the atmosphere. *Palaeogeogr. Palaeoclimat. Palaeoecol.,* v. 4, p. 187 – 210, 1 fig., 1968.

Tappan, Helen, Phytoplankton abundance and Late Paleozoic extinctions: a reply. *Palaeogeogr. Palaeoclimat. Palaeoecol.,* v. 8, p. 56 – 66, 1970.

Tappan, Helen, Microplankton, ecological succession, and evolution. *Proc. N. Am. Paleont. Conv. Chicago, Sept. 1969,* pt. H, p. 1058 – 1103, figs. 1 – 8, 1971.

Tappan, Helen, and A. R. Loeblich, Jr., Protista other than Foraminiferida, p. 1342 – 1346. *In* R. C. Moore et al., Developments, trends and outlooks in paleontology. *J. Paleont.,* v. 42, p. 1327 – 1377, 1968.

Tappan, Helen, and A. R. Loeblich, Jr., Geobiologic implications of fossil phytoplankton evolution and time-space distribution, p. 247 – 340, 1 pl., 12 figs. *In* R. Kosanke and A. T. Cross, *Symposium on palynology of the Late Cretaceous and Early Tertiary.* Spec. Pap. Geol. Soc. Am., no. 127, 1971.

Tappan, Helen, and A. R. Loeblich, Jr., Fluctuating rates of protistan evolution, diversification and extinction. *24th Int. Geol. Congr., Sec. 7 Paleont.,* p. 205 – 213, figs. 1 – 6, 1972.

Tappan, Helen, and A. R. Loeblich, Jr., Evolution of the oceanic plankton. *Earth-Sci. Rev.,* v. 9, p. 207 – 240, figs. 1 – 9, 1973.

THE PROCARYOTES: Bacteria and Blue-Greens

There are perhaps as many, and probably more, different species of bacteria in the sea than there are on land, and future studies may reveal that marine bacteria are as important as those which live on land or in fresh water environments.

C. E. Zobell, 1946

The smallest and least differentiated living organisms, generally believed to be most similar to the first primitive organisms on the earth, are the **procaryote** Schizophyta or Protophyta, commonly known as the bacteria and blue-green algae. Tiny organisms approaching the lower limits resolvable with the compound light microscope, they have a very simple morphology and have been grouped together because of their similar lack of nuclei, plastids, and sexual reproduction (Stanier and van Niel, 1941). Recent studies have shown that these negative characters require modification, but that there are positive similarities. A nuclear membrane is lacking, a single "chromosome" is present, and there are no vacuoles or membrane-bound organelles such as mitochondria and plastids;

neither cytoplasmic streaming nor amoeboid movement occurs. The mucilaginous sheath surrounding one or several blue-green cells and the capsule present in many bacteria have a similar structure of extremely fine fibrils. In addition to the slime layer or capsule, the bacteria and blue-greens have a cell wall of similar microstructure, "nuclear" division occurs by fission of the nucleoplasm rather than by mitotic division, and cell division is characterized by centripetal growth of a wall in some bacteria. Both include representatives that can fix atmospheric nitrogen. Of limited morphologic variety, the procaryotes may be biochemically complex and have become adapted to the widest range of environments of any organisms.

PRIMITIVE LIFE IN THE FOSSIL RECORD

Long a favorite subject for speculation, the origin of life has been discussed recently from the varied viewpoints of experimental chemistry, the biochemistry of biogenic materials, the geochemistry of the primitive oceans and atmosphere and of meteorites, the micropaleontologic record, historical geology, radiometric dating, and astronomy.

From the stabilization of the earth's crust about 4.5 billion years ago to the beginning of the Cambrian about 600 million years ago, the long period of Precambrian time has a very sparse fossil record of ancient life. Strangely, although unicellular organisms are generally regarded as more primitive, early studies concentrated on a search for Precambrian invertebrate megafossils, whether shelled or soft-bodied, or for their tracks or trails. Most reported fossils proved to be of questionable biogenicity, or their Precambrian age was in doubt (Cloud, 1968). The supposedly primitive organism *Eozoon*, the subject of much discussion during the last half of the nineteenth century, was shown to be only an inorganic precipitate, and many so-called microfossils proved to be inorganic crystals and spherulites.

During the prolonged yet unrewarding search for unquestionable megafossils, some fossil remains were found. Among those generally accepted as of biogenic origin are the abundant calcareous stromatolitic lamellar structures, similar to those formed by modern blue-green algae. In addition, a few Precambrian microfossils were described as much as 50 years ago, and many of the sedimentary limestones, graphites, and iron ores had been recognized as biogenic, but the micropaleontologic study of the Precambrian saw its real beginning in the mid-1960s. Examination of thin sections and acid-digested cherts and shales at high magnifications and with both light and electron microscopes has demonstrated the presence of a diverse assemblage of bacteria and blue-green algae in rocks as old as 3.1 billion years, together with probable red and green algae and fungi in the later Precambrian.

Features that are both ubiquitous and basic are probably primitive in either biological evolution or the evolution of biochemical systems. Identical structures and processes are less likely to evolve independently in different organisms than to be preserved in lineages that have since diversified. Thus relationships of the various groups of protists are hypothesized on the basis of the joint possession of characters, or their absence, whether these are morphologic or biochemical. Bacteria and blue-green algae share many such features and lack many that characterize all other organisms, plant or animal. Both have a lengthy fossil record and appear most similar to the probable earliest life on earth of any living organisms. However, their differences in morphology and biochemistry are of major importance, particularly the nature of their photosynthesis; hence they are discussed separately.

THE BACTERIA

The main function of each individual bacterial cell, as for any organism, is growth and eventual replication. There is no tissue differentiation, hence no division of function. The ratio of surface area to volume is highest in the smallest organisms, hence bacterial chemical and metabolic activities are particularly efficient. Many of these activities comprise fundamental links in the geochemical processes that allow the presence and continuation of other forms of life and have considerable geochemical and evolutionary significance. Among these important bacterial processes are the fixation of atmospheric nitrogen, and the decomposition of organic matter to release fixed carbon as a requisite step in the carbon cycle.

Some bacteria are photosynthetic, their pigments localized in simple rounded particles or lamellae that lack the surrounding membrane of the complex plastids of other algae and green plants. Bacterial photosynthesis is less efficient than that of green plants, or even of the blue-green algae, and does not produce oxygen.

As they have no mineralized skeleton, bacteria would at first glance seem very unlikely candidates for fossilization, and a record of their presence in past geologic periods might appear improbable. However, like most plants, bacteria have a relatively firm cell wall, and fossil bacteria and blue-green algae have been recorded from many sedimentary rocks, ranging from those Precambrian strata that have been most thoroughly searched for traces of life to very recent sediments. Living bacteria are classified on the basis of physiology, cultural characteristics, and pathogenicity, rather than morphology; hence species, genera, and families of bacteria may not be equivalent to such taxa of organisms with a greater variety of recognizable morphological features. The photosynthetic bacteria include cocci, bacilli, and spirilla, and the lactic acid bacteria also may occur as cocci, rods, or filaments. These and other groups probably represent ecological-physiological communities (e.g., the sulfur bacteria), but from the economic, medical, or geochemical view such groupings are convenient as common names. Relationships of the fossil representatives are difficult to determine, as morphologically they may appear identical to living species, and the biochemical techniques of identification are not available for fossils. Even here, the occurrence in particular rock facies or within fossilized plant tissues or bone may suggest an affinity to the modern sulfur bacteria, iron bacteria, or pathogens.

Living bacteria probably are descendants of the same very early ancestor as the present animal and plant kingdoms, the three modern groups having evolved in different directions. The physiological, ecological, and biochemical evolution of the procaryotes is equally important as the morphologic and cytologic differentiation of the eucaryotes.

That the bacteria are not really simple can be demonstrated by their potential genetic variation. An average bacterium, such as *Pseudomonas,* has a sequence of about 4×10^6 nucleotide pairs on its single "chromosome," which would theoretically allow some $10^{24,000,000}$ different combinations. Because of the double-stranded nature of DNA (deoxyribonucleic acid), only half of these are possible, and while many others probably would not be viable, the num-

ber still possible must be astronomically high. Other bacteria range from 3×10^5 nucleotide pairs (*Mycoplasma*) on the chromosome to about 5×10^6 (*Escherichia*), or some 25 percent more than in *Pseudomonas.* Living bacteria number some 10,000 species, each of which may have many varieties and strains; however, the total number of individual cells on the earth is only about 10^{27}, a very minor fraction of the number of different forms theoretically possible. Experiments with mutation rates in vertebrates suggest that an average change of 10 nucleotide pairs per year since the Permian produced the diversification of birds and mammals from the reptiles. At similar rates, a bacterial chromosome would undergo a complete change of its base pairs every million years. Although their mutation rates may differ, these figures suggest that the bacterial flora even of two successive geological periods would be quite distinct (De Ley, 1968).

The Cell

The simplest and smallest bacteria are the true bacteria or Eubacteriales (Schizosporales herein). Morphologically they are the least complex of any living organism visible at the limits of the compound light microscope. Other bacteria variously resemble protozoa, algae, fungi, or slime molds.

Three cell forms are characteristic of the true bacteria (see Figure 1.1): (1) the **coccus,** an ellipsoidal or spherical cell, averaging 0.5 to 1.0 μm in diameter; (2) the **bacillus** or rod, a cylinder or elongate ellipsoid about 0.5 μm in diameter and 1.5 μm in length, but rarely up to 10 times that length; and (3) the **vibrio,** a slightly larger elongate curved rod of a half turn, which is termed a spire or **spirillum** when it consists of several helices. The latter may be 1.5 μm in diameter and up to 13 or 14 μm in length.

Cell increase is normally by transverse fission, but the fission may not result in complete separation, and various types of cell aggregates are formed. Successive cell divisions of cocci in the same plane may result in long chains of cells, those in two planes at right angles form

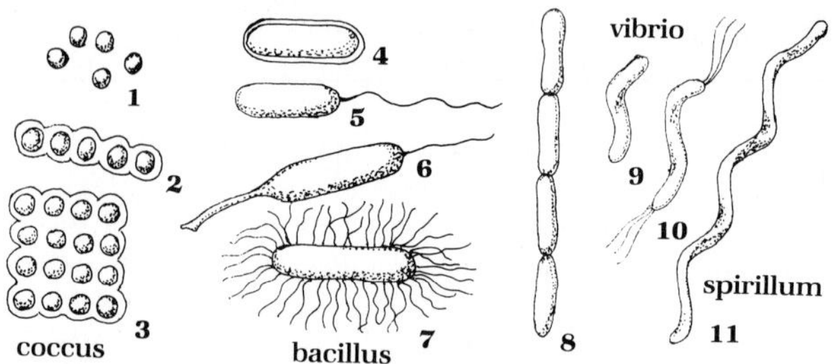

Figure 1.1

Bacterial cell form. **1.** Single coccus. **2.** Chain of cocci with mucilage sheath. **3.** Sheet of cocci, held together by the enclosing slime. **4.** Single bacillus with capsule. **5.** Bacillus with a single polar flagellum, or monotrichate. **6.** Bacillus with mucilage stalk for attachment, and single flagellum at opposite pole. **7.** Bacillus with many flagella, or peritrichate. **8.** Chain of bacilli. **9.** Single vibrio. **10.** Vibrio with tuft of flagella at each pole, or lophotrichate. **11.** Spirillum. 1−3, 8, approximately ×1500; 4−7, approximately ×3500; 9−11, approximately ×500.

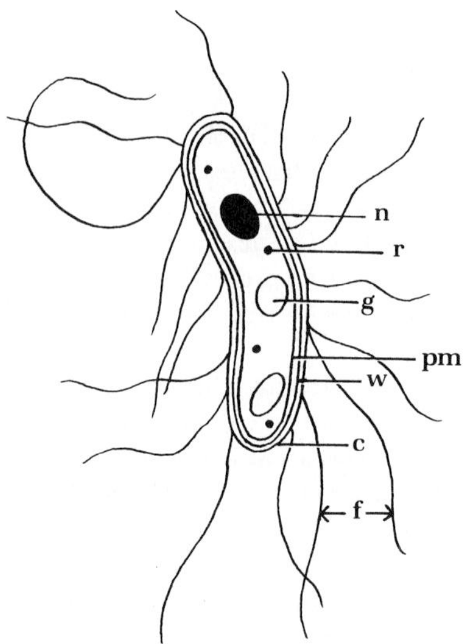

Figure 1.2

Cell organization in bacteria. n, nuclear material or genophore; r, ribosomes; g, granules of fat, sulfur, etc.; pm, plasma membrane; w, wall; c, capsule of slime; f, flagella.

tetrads or sheets of cells, and successive divisions in three planes produce regular cubical packets of eight cells termed **sarcinae** (singular form, **sarcina**). Rod-shaped and helical cells only divide at right angles to their long axis, thus incomplete separation may form chains of cells. Some chains (**trichomes**) have readily distinguishable transverse divisions, whereas others (**filaments**) have a continuous cell wall with no evidence of multicellularity except after special staining.

The protoplasm is chemically complex. The **nucleoid** (a small particle somewhat resembling a nucleus), cytoplasm, and cell wall contain a wide variety of organic compounds, largely protein, RNA (ribonucleic acid), carbohydrates, and lipids. The nuclear material is concentrated in the central part of the cell (see Figure 1.2), although there is no well-defined membrane-bounded nucleus. The remainder of the cell consists of uniform granular cytoplasm, with some small particles of protein and ribonucleic acid, the **ribosomes** that are the site of protein synthesis. Granules of fat, protein, or starch or globules of minerals such as sulfur or calcium carbonate also may be present.

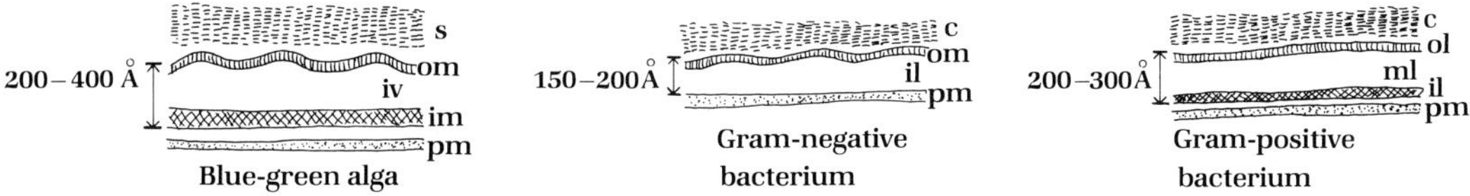

Figure 1.3

Comparison of cell walls of bacteria and blue-green algae, showing innermost plasma membrane (pm), wall (between brackets), and external slime layer forming the sheath (s) of the algae and the capsule (c) of the bacteria. The cyanophyte wall consists of the commonly crenulated outer membrane (om), the inner investment (iv), and inner membrane (im), best differentiated in electron micrographs. Gram-negative bacteria have an outer membrane (om), and an intermediate layer (il) in the wall, and gram-positive ones have outer (ol), middle (ml), and inner (il) layers, although the outer layer is not crenulated. Redrawn with slight modification from Echlin and Morris, 1965, *Biol. Rev.,* v. 40, p. 143–187; used with permission of Cambridge University Press.

The Capsule or Sheath

The cell is enclosed by three layers (see Figure 1.3), the inner or cytoplasmic membrane, the cell wall (which may have more than one layer), and an outermost covering of slime. The innermost, flexible, easily deformed and differentially permeable cytoplasmic membrane is composed of lipids and lipoproteins. A rigid cell wall of distinctive composition surrounds the membrane. The outermost layer of slime varies in thickness and composition with the environmental conditions or from species to species. A thick, gummy, and mucilaginous slime layer around a single cell is termed a **capsule;** that surrounding a filament is termed the **sheath.** Eubacterial capsules contain different chemical substances, ranging from a complex of polysaccharides to a simple polypeptide of a single amino acid. One species has a slime layer of cellulose, forming a loose fibrillar network. Separate strains of a single species may have capsules of several different strain-specific polysaccharides.

In habitats with a paucity of dissolved nutrients, the slime layer may form a **stalk.** The simplest stalks are for the attachment of a single cell; others may be much ramified or form a rosette with a cell at the end of each of the radially arranged stalks. Stalks may be very elongate and slender, or may be twisted. They may be wholly organic, organic with an iron-encrusted surface, or composed entirely of ferric hydroxide. An example of the latter is *Gallionella,* which grows only in iron-bearing waters, either fresh or marine. Other species deposit manganese compounds on the slime surface.

Cell Wall

A cell wall is characteristic of the plant kingdom. In higher algae and plants it is composed of cellulose, a polysaccharide or polymer of glucose. The cell wall of fungi consists of chitin, a polymer of an amino sugar. In contrast, the unique bacterial cell wall consists of a mucocomplex substance of two amino sugars (one of these, muramic acid, was known only in the bacteria until recently identified in blue-green algae), and three or four amino acids, and may also contain lipids and proteins. Two or three layers in the wall may have different compositions. Gram-negative species (see succeeding section on the Gram reaction) have very thin walls only about 10 nm in thickness. The thin wall is seen in electron microscopy to consist of two electron-opaque layers separated by a less dense layer. The wall consists of 5 to 10 percent mureins (mucopeptide) and 20 to 30 percent lipids, with varied amounts of protein, lipoprotein, and lipopolysaccharide, the latter component having important endotoxic and antigen properties (Glauert and Thornley, 1969). Gram-positive species have thicker walls (20 to 35 nm) that appear three-layered and resemble those of blue-green algae. They contain no lipids, have a much smaller number of amino acids, but have two amino sugars; mucopeptides range from 50 to as much as 90 percent of the dry weight (Echlin and Morris, 1965).

Amino acids are important in the formation of proteins, and all amino acids known to occur in proteins are found in the bacterial wall. In higher organisms, the amino acids occur in the cytoplasm and nucleus, but not as major constituents of the wall.

The rigid cell wall is not deformed by an osmotic pressure of up to 300 pounds per square inch, and its function appears to be wholly mechanical, giving the cell its shape. If the cell wall is removed by laboratory procedures, the naked protoplast will become spherical; it will continue to carry out all metabolic activities, although the semipermeable cytoplasmic membrane then will be very sensitive to osmotic changes.

In electron microscopy the different layers of the wall may have different and distinctive surface patterns. Excluding the plasma membrane at the inside and the slime layers at the outside, the other wall layers have patterns ranging from the wavy, undulating surface of the Gram-negative *Escherichia coli* (Migula) Castellani & Chalmers, to a linear pattern in *Caulobacter* and hexagonal patterns in *Spirillum* and *Azotobacter*. Some *Bacillus* have a hexagonal pattern and others have a tetragonal one; *Lampropedia* has an outer echinate layer held by fine fibers to an inner punctate membrane, both layers showing different hexagonal surface patterns. The extent of such surface features in the bacteria is not yet known, as many are delicate and readily lost in the process of preparation for microscopy (Glauert and Thornley, 1969).

Bacterial respiratory enzymes are found in the plasma membrane, whereas in other organisms these enzymes occur in special organelles, the **mitochondria**, within the cell. Intrusive tubules or lamellae may arise from the plasma membrane of some species; the peripherally arranged chromatophores of the purple nonsulfur bacteria arise in this way. Similar invagination of portions of the wall and plasma membrane, containing the amino acids, enzymes, and so forth, may have given rise to such cell organelles of the higher protists as the mitochondria and plastids, although an alternative theory suggests that they arose by endosymbiosis of entire procaryote cells.

The Gram Reaction

A distinctive cytological character of the eubacteria that has proven most useful in their identification is the differential reaction to a staining procedure first developed by a Danish physician, and named the **Gram stain** in his honor. A heat-fixed smear of the bacteria is first stained with a solution of crystal violet and a dilute solution of iodine. All cells stain blue. The cells are next treated with an organic solvent, alcohol or acetone, which results in decolorizing the **Gram-negative** bacteria; **Gram-positive** bacteria retain the violet color. The division into these two major groups correlates well with many other characteristics of the eubacteria. All polarly flagellated ones are Gram-negative, their wall contains a high proportion of lipids, and the cells are more resistant to antibiotics. Gram-positive bacteria include all spore-forming ones, and the wall is more complex, although containing no lipids, and provides a barrier to the decolorizing agent, thereby preventing the leaching of the dye from the cytoplasm.

Nucleoid or Genophore

In previous interpretations, bacteria have been regarded as completely lacking a nucleus (the original basis for the Kingdom Monera), or at the opposite extreme the entire cell has been regarded as a naked nucleus. Nuclear material has been described as diffused throughout the cell, restricted to a central body although not limited by a nuclear membrane, or occurring in one or more masses. The lack of agreement was due partly to the minute size of the organism, the absence of a bounding nuclear membrane, and the lack of a mitotic division in which distinct chromosomes could be observed.

With cytochemical studies involving staining, a single structureless chromatin granule is faintly distinguishable from the surrounding cytoplasm. More than one is present only if the cell is undergoing division. Because of the rapid rate of reproduction, "nuclear" division may be somewhat in advance of cell separation, and an elongated cell may appear to have two or three such structures. Differing in many ways from the eucaryotic nucleus, this structure of the bacteria is termed a **genophore** or **nucleoid** rather than a chromosome or nucleus.

In *Escherichia coli,* the most intensively studied species of bacteria, the genes are arranged linearly in a single linkage group, equivalent to a single "chromosome" (Cairns, 1963; Bleecken et al., 1966). Autoradiography has shown it to be a skein of double-stranded DNA fibrils, forming a closed loop or "circular chromosome" about 700 to 900 μm in length and packed in a loosely defined "nuclear" region less than 0.1 μm^3 in volume. In replication, the two strands separate, each of the daughter cells taking one of the original strands and forming a new one.

Flagella

Motile bacteria may have one or more flagella, but bacterial flagella are quite unlike those of other organisms. The flagella and cilia of higher protists and all other organisms have an identical structure, consisting in section of a ring of nine paired fibrils surrounding an additional central pair, all attached to a basal plate. In contrast, bacterial flagella appear structureless in section, apparently consisting of two or three very fine protein microfibrils of about 10 to 30 nm (0.01 to 0.03 μm; see Figure 1.4). As a protein helix is about 1 nm (0.001 μm), a bacterial flagellum contains very few molecules. The flagella are composed largely of the single protein flagellin, and are demonstrated by radioautography of labelled material to lengthen by the addition of subunits of flagellin at their distal end (Emerson et al., 1970).

Because of their very small size, the flagella are scarcely visible in the light microscope

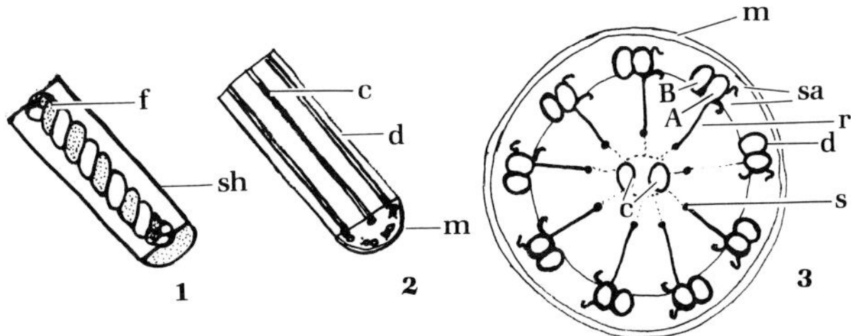

Figure 1.4
Flagellar structure. **1.** Diagram of bacterial flagellum, with two double fibrils (f) twisted together (one of the double fibrils is stippled) and surrounded by a sheath. **2,3.** Structure of flagella and cilia of eucaryotes, with a peripheral ring of nine paired fibrils (A, B) forming doublets (d); A-fibrils with side arms (sa); paired fibrils surround two central unpaired fibrils (c); radial spokes (r) lead from the A-B doublets to secondary fibrils (s) near the center; bounding membrane (m). The wide occurrence of the 9+2 structured flagella and cilia in nearly all eucaryotic plants and animals suggests a common origin.

without a heavily coating stain that merely indicates the gross outline. Somewhat more detail is apparent in phase contrast, but most of the information concerning bacterial flagella is from electron microscopy. Some flagella attach at a small granule within the cell membrane and wall, as they remain attached to the protoplast from which the cell wall has been removed. Flagella of other bacteria are reportedly derived from the cell wall, held together as a fibrous layer around the wall, or separating and peeling off as flagella. Such specimens with most numerous flagella have a reduced fibrous layer.

Bacterial cells may have 1 to 4 flagella, up to 50 to 100 per cell. Not all flagella are formed at once, and old cells commonly have more flagella than newly formed ones. The free-swimming bacteria may be: (1) **monotrichate**, with only one flagellum usually at one pole; (2) **amphitrichate**, with one or more flagella at each pole (although this may represent cells in the process of division); (3) **lophotrichate**, with a tuft of flagella at each pole; (4) **peritrichate**, with many flagella over the lateral surface; or (5) **atrichate**, lacking flagella and moving only by Brownian movement or current action.

The helical motion of the flagella may allow the cell to attain a velocity of about 50 μm per second, but such motility is possible only in a liquid medium. When grown on agar plates, even normally motile types form stationary colonies. The protein of the flagella is very similar to that of muscle, and flagellar movement may be related to muscle movement. Isolated muscle fibers contract when treated with ATP (adenosine triphosphate), and bacterial cells apparently generate ATP to cause flagellar motion. Formation of the ATP is related to the process of respiration. When normally aerobic bacteria are examined under a coverslip whose edges are sealed to prevent access of additional air, active movement of the flagella ceases when the original dissolved oxygen is exhausted. However, when *Spirillum volutans* Ehrenberg was squeezed between a slide and cover glass to restrict the flagellar rotation, the cell body rotated. The appearance of helical waves thus is not due to contraction or bending of the flagella, but to the simple rotation of rigidly held flagella (Mussill and Jarosch, 1972).

Cysts and Endospores

Resting stages of bacteria may be cysts formed of the entire cell, or **endospores** in which only part of the original cell is utilized. The cysts of the Myxobacterales, known as **microcysts**, are small spheres with thick walls capable of withstanding desiccation and able to germinate even after long periods of time. They are formed on specialized fruiting structures. Eubacteria may also form simple cysts, each from a single cell, that are somewhat less resistant than the endospores.

The highly resistant endospores, characteristic of the true bacterial family Bacillaceae, may withstand several hours in boiling water. This resting stage represents a form of reproduction without multiplication. When placed in conditions that favor spore production, the internal structure of a *Bacillus* cell is modified to become a **sporangium**. A round opaque area is isolated within the cell, separated from the remaining cytoplasm by a ring of less dense material. The spore coat then forms around this. An endospore may form in the center of a cell or at one end, but does not comprise the entire cell. The completed spherical or ellipsoidal endospore may have a larger diameter than the cell itself, causing the cell wall to bulge locally (see Figure 1.5). The water content of the spores is low, and they are dense and highly refractive. When conditions again become favorable, the endospore germinates to produce a single cell.

Spore-forming bacteria characteristically are soil-dwelling, although a few are pathogenic in insects or in higher animals (e.g., those causing tetanus and botulism). Aerobic, facultative anaerobic, and strictly anaerobic spore-forming types are known. One coccus species forms endospores, but all others are formed by bacilli.

Reproduction

The simplest organisms without much cell differentiation commonly have a dominant **haploid** phase in their life cycle, and a **diploid** phase (with doubled number of chromosomes) is rare. The larger and more complex or-

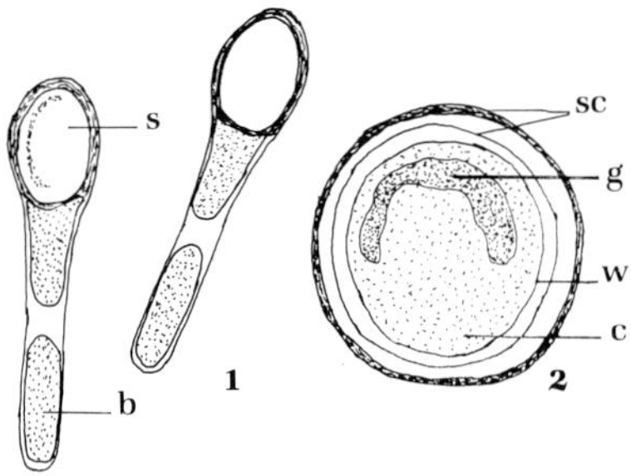

Figure 1.5
Formation of endospores by bacteria. **1.** Two cells of *Clostridium* showing the rodlike bacillus (b), enlarged at one end to form a round to ovate spore (s); approximately ×4100. **2.** Structure of an endospore showing sc, spore coats; g, the genophore or nucleoid; w, spore wall; and c, cytoplasm of the endospore; approximately ×33,000.

ganisms are favored by a diploid cycle, as the cells are then genetically more stable. Mutations are immediately expressed in haploid cells, whereas a mutation may be recessive in a diploid organism, and is then expressed only when present in genetic material received from both parents. Diploidy thus reduces variability and gives stability, the introduction of sexual reproduction allowing genetic recombination to be of relatively greater importance. Haploid organisms, especially those with a rapid growth rate, are more susceptible to the effects of mutations, and may thus respond more rapidly to changing environments by the processes of mutation and selection.

Asexual Reproduction

Asexual reproduction in bacteria may be, in its simplest form, transverse or binary fission. The cell elongates and the nuclear material divides before cell division. Gram-negative bacteria then develop a simple median constriction that finally results in a complete separation without forming a cross wall. In contrast, a ring of wall material forms in the Gram-positive bacteria and by centripetal growth forms a complete cross wall. Only the plasma membrane and

inner wall layer are involved in the division. If the median wall is not completed, the cells may remain connected to produce chains or clusters of cells, according to the plane of cell division. In other forms, the two newly formed cells may completely separate. Bacterial cells may undergo fission every 20 to 30 minutes, and increase in numbers until their waste products accumulate to limit further growth or until some essential nutrient is exhausted. Growth and multiplication is normally very rapid in the early stages, leveling off at the maximum, then declining rapidly.

Sexual Reproduction

Three methods of obtaining new genetic combinations have been discovered in the bacteria: conjugation, transformation, and transduction, each allowing recombination of genetic material for trial in new environments.

Conjugation. In the process of **conjugation**, two cells become temporarily joined at their ends for the transferral of genetic material from one (the donor cell) to the other (recipient). The donor cell dies after transfer of its genetic material. The chromosomal material is always inserted in the same order (as determined by genetic studies of *Escherichia coli*), and commonly the cells separate after only a part of the genetic material of the donor cell is transferred. The recipient thus becomes only partially diploid. In the subsequent recombination, a part of the new genetic material replaces a corresponding part of the chromosome of the recipient cell. The unincorporated genetic material from both original cells is then discarded, leaving the recipient cell again in the haploid phase.

Transformation. **Transformation** is an unusual form of recombination observed in some laboratory strains but not known certainly to occur in nature. In this process, some molecules of the DNA that carries the genetic information of the cell move from cell to cell of a strain of bacteria through the solution in which they are growing. In this form of sexual reproduction, only the isolated molecules move from cell to cell. In all cases known to date, the genetic material obtained by transformation allows recovery of an ability that had been lost by mutation (e.g., ability to produce certain molecules, such as the slime layer characteristic of a particular bacterial strain). Transformation introduces DNA from a donor strain to reendow the mutant bacterium with the ability to synthesize the necessary enzymes.

Transduction. Transduction is another strange type of genetic transfer, in which the genetic material is transferred from one bacterium to another by a **bacteriophage**, a filterable virus that infects the bacteria as an intracellular parasite. One of the phages that attacks *E. coli* has a prismatic head, probably consisting of a single DNA molecule, and a tail which is hexagonal in section and consists of a few protein molecules (see Figure 1.6). The phage develops in a host bacterial cell, and meanwhile incorporates a small amount of the host genetic material, perhaps in exchange for some of its own. The infected cell may show no apparent change for 15 to 60 minutes, then suddenly undergoes lysis (disintegrates), releasing a large quantity of new phage particles. Each phage then moves to a new host, into which it injects both its own genetic complement and that obtained from the former host cell. The new host cell then becomes a partial diploid, as in the processes of conjugation or transformation. A partially diploid culture may be maintained or the cell may segregate a haploid chromosome, discarding the excess genetic material. Some bacterial properties are known to result from the phage genetic material, being totally absent from uninfected strains. The close relationship of the bacteria and viruses (particularly the phages) is indicated by the ability of the genetic material of one to replace that of the other. Genes controlling a given host property may occur only in the bacterial cell, only in the phage, or in both.

In any of the above methods, complete genetic transfer is unlikely, and in transformation

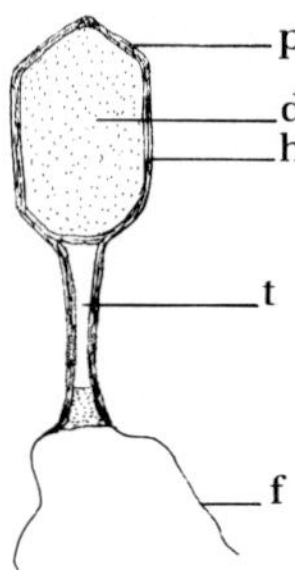

Figure 1.6
A bacteriophage particle, showing prismatic head (h) and narrow tail (t); head consisting (probably) of a single molecule of DNA (d) with an external protein coat (p); tail hollow, but distally plugged and with attachment fibers (f) by which it attaches to the host cell. Approximately ×145,000.

and transduction only small bits of genetic material are transferred.

Metabolism

Two major metabolic processes concern all organisms: the transformation of matter into the various molecular and macromolecular components of the cell, and obtaining a supply of energy. These metabolic aims may be met in different ways. Classically organisms were divided into two types on the basis of nutritional methods, the **autotrophs** (plants) obtaining energy phototrophically from sunlight, and the **heterotrophs** (particularly the animals) using only organic compounds. The separation lacks definite boundaries, however, as many phototrophic species also require exogenous organic compounds. The type of energy source has also been used as a basis for division. The phototrophs are the photosynthetic organisms, and the chemotrophs (or chemosynthetic forms) use dark chemical oxidation-reduction reactions. Separation may also be based on the nature of the substance oxidized, the lithotrophs utilizing inorganic substances and the organotrophs oxidizing organic matter for their sole source of energy. Organotrophic organisms include all animals, the fungi, and the majority of the bacteria. Lithotrophic metabolism is known only in the bacteria that for a source of energy may oxidize sulfur, oxidize ammonium salts to nitrites, or oxidize nitrites to nitrates, meanwhile obtaining carbon from CO_2 by chemosynthesis, for a source of both food and energy.

Bacterial Photosynthesis

A few bacteria are photosynthetic, containing green and purple pigments in spherical or lamellar vesicles that lack a surrounding membrane and are thus unlike the complex plastids of eucaryotic organisms. They appear to arise as invaginations from the plasma membrane. The pigments of the bacteria are termed bacteriochlorophylls. Green bacteria possess bacteriochlorophyll *c* (formerly chlorobium chlorophyll 660) and *d* (formerly chlorobium chlorophyll 650), and yellow carotenoids. Purple bacteria have the bacteriochlorophylls *a* and *b*, and red, purple, and yellow carotenoids. Because different pigments absorb light of different wave lengths, the possession of more than one photosynthetically active pigment allows a more efficient use of the sunlight. Algae possessing different pigments are also particularly characteristic of separate depths in the ocean, depending on the depth to which the various wave lengths of light will penetrate. The green, blue, brown, and red algae successively occupy greater depths in the oceans.

The different kinds of chlorophyll all are porphyrin compounds of similar structure, consisting of a complex organic molecule containing nitrogen and magnesium atoms. The simplest known chlorophyll, protochlorophyll, is found in plants grown in the dark. When light is available, the protochlorophyll is converted to true chlorophyll. Chlorophyll *a* is formed by the addition of two hydrogen atoms. Bacteriochlorophyll is similar, but has four more hydrogen atoms and one more oxygen atom than does protochlorophyll, as well as a slightly different arrangement of other atoms. Even this slight difference has far-reaching consequences, as the various forms of chlorophyll react to light of different wave lengths, and as a

result vary in efficiency. For each mole of glucose $(C_6H_{12}O_6)$ produced by green plant photosynthesis, 675,000 calories of energy are stored. Bacterial photosynthesis stores much less energy.

Photosynthesis in all organisms is reducible to the equation:

$$CO_2 + 2H_2A \xrightarrow{\text{light}} \overset{\text{cell material}}{(CH_2O)} + H_2O + 2A$$

where H_2A is any oxidizable compound and A the corresponding oxidation (dehydrogenation) product. In green plants, oxygen substitutes for A, thus the hydrogen donor is water, and the process produces free oxygen. When oxygen is substituted in the preceding equation, and water subtracted from each side, the result is:

$$CO_2 + 2H_2O \xrightarrow{\text{light}} (CH_2O) + H_2O + 2O \quad \text{or}$$

$$CO_2 + H_2O \longrightarrow (CH_2O) + O_2$$

In bacterial photosynthesis, various substances may substitute for A, but never oxygen, hence the process is always anaerobic and never results in the production of oxygen. If sulfur supplies the oxidizable material, water and free sulfur result:

$$CO_2 + 2H_2S \xrightarrow{\text{light}} (CH_2O) + H_2O + 2S$$

The sulfur purple and green bacteria are obligate anaerobes, and cannot live in the presence of oxygen. They only require light, CO_2, and inorganic sulfur compounds, such as hydrogen sulfide. The nonsulfur purple bacteria used organic compounds instead of sulfur, and can grow in the presence of oxygen (facultative anaerobes). They could thus be termed photoorganotrophic.

Photosynthesis is thus a process of deriving nutrients from inorganic compounds and sunlight. However, many photosynthetic organisms require certain additional compounds as well as carbon dioxide from their environment. Some algae may require an external source of certain vitamins, and some nonphotosynthetic bacteria (the lactic acid bacteria) must obtain carbon from another organic source, rather than from the atmosphere. Possibly a genetic mutation in the past resulted in the loss of enzymes capable of synthesizing these necessary compounds, so that the modern organisms are dependent on an external source.

Chemotrophic Metabolism

Photosynthesizing bacteria represent only a small fraction of the Schizophyta; chemotrophic ones are most abundant, and are important geochemical agents.

Whether the oxidation is of organic compounds (chemoorganotrophic) or inorganic ones (chemolithotrophic), the process is similar. In either form of oxidation electrons are removed from an atom or molecule. Thus iron is oxidized from Fe^{2+} to Fe^{3+} by the removal of one electron. Organic compounds are oxidized by the removal of hydrogen atoms (dehydrogenation). Neither electrons nor hydrogen atoms can accumulate as such, hence oxidation of one substance requires the simultaneous reduction of another, to which the excess atoms or electrons are attached. Both the oxidant and reductant must be present in sufficient quantities in the environment, and bacteria have been able to make use of nearly all possible combinations. If the oxidant is organic, the process is termed fermentation or mineralization; if an inorganic oxidant is used, such as sulfates that are reduced to sulfides, or nitrates that are reduced to nitrogen gas, ammonia, or nitrous oxide, the process is termed respiration (see Table 1.1).

Bacterial respiration is accompanied by a complex series of chemical reactions, each of which is catalyzed by a specific enzyme.

Carbon Cycle. The chemical elements in the biosphere must be constantly recycled from the living to nonliving state. Organotrophic bacteria

Table 1.1
Bacterial respiration.

Reductant	Oxidant	Product	Organism
H_2	O_2	H_2O	hydrogen bacteria (*Hydrogenomonas*)
H_2	SO_4^{2-}	$H_2O + S^{2-}$	*Desulfovibrio* (sulfate-reducing bacteria)
Organic compounds (e.g., CH_4)	O_2	$CO_2 + H_2O$	many bacteria (e.g., *Methanomonas*, methane-oxidizing); also other plants, animals
NH_3	O_2	$NO_2^{2-} + H_2O$	*Nitrosomonas* — nitrifying bacteria
NO_2^{2-}	O_2	$NO_3^- + H_2O$	*Nitrobacter*
Organic compounds	NO_3	$N_2 + CO_2$	denitrifying bacteria (*Pseudomonas*)
Fe^{2+}	O_2	Fe^{3+}	*Thiobacillus* (iron-depositing bacteria)
S^{2-}	O_2	$SO_4^{2-} + H_2O$	*Thiobacillus* (sulfur-oxidizing bacteria)

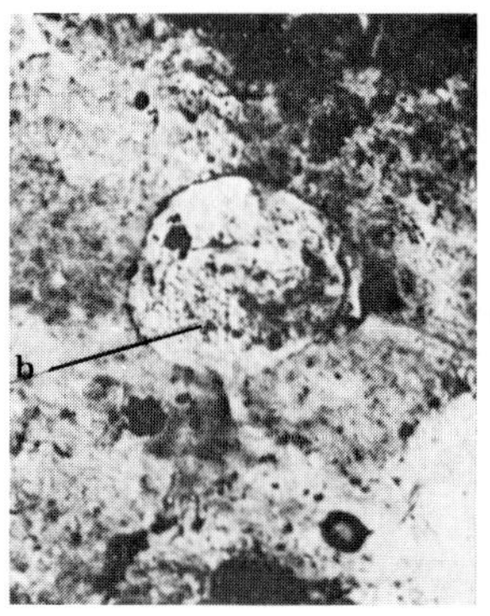

1

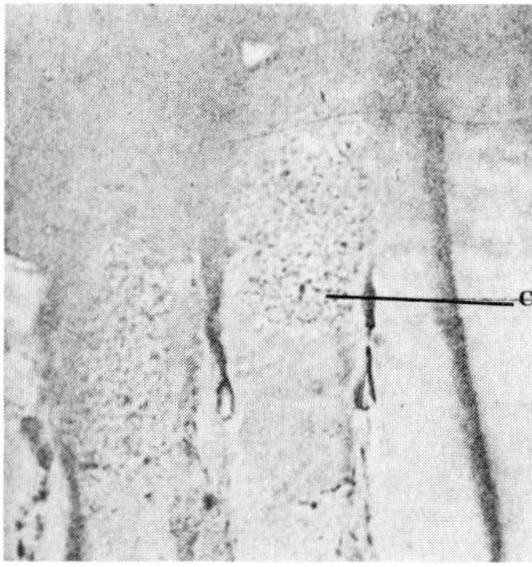

2

Figure 1.7
Micrococcus lignitum Renault &
Roche, lower Eocene, France.
1. Plant macrospore with surface
covered by many bacterial cells
(b), ×310. **2.** Longitudinal section
of wood showing two tracheids
with many bacteria; c, cross wall
that is similarly covered with
bacterial cells; ×1200. From Renault and Roche, 1898.

that decompose previously formed organic compounds are an essential link in the carbon cycle, returning the organically fixed carbon to the atmosphere as CO_2. The decomposition of organic compounds also liberates nitrogen as ammonia, phosphorus as inorganic phosphate, and sulfur as hydrogen sulfide. Some species of bacteria, either as a saprophyte or parasite, is able to oxidize every naturally occurring organic compound. Some of these decomposers are fossilized with the organic remains, whether of plants or animals (see Figure 1.7).

The organotrophic bacteria also fix carbon in their own cells, but this is of minor amount as compared to the activity of "green" plants, which annually fix about 30 billion tons of carbon, or an average of about 1000 tons per second. A large proportion of the fixation of carbon occurs in the oceans, which presently contain about 2600 billion tons of organic matter, in true solution, in colloidal solution, in suspension in the living organisms, and in organic detritus. Distribution is uneven, the greatest quantity being in shallow near-shore waters and in the upper layers of water. The greatest quantity is in true solution, however, and the proportion in solution increases with depth, although the total lessens. The greatest amount of suspended organic matter is within the upper 100 meters, where 90 percent of the plankton is concentrated. Bacterial decomposition of all this organic matter is continuous at all depths, much of it taking place within the water column. Some is consumed by other organisms, but if

not wholly decomposed, is further acted on by the bacteria. Organic matter and dead organisms ranging from 0.05 to 0.2 mm in size are largely decomposed in the upper 1500 meters of water, and only their skeletal material is deposited at the sea floor. Material ranging from 0.2 mm to 1.0 mm in size is profoundly decomposed by 2000 meters, and only particles originally greater than one millimeter in size reach the sea floor without decomposition. Mineralization by animals (plankton feeders) reconverts about 5 percent of the annual carbon production of the plants, and bacterial activity is responsible for the remaining 95 percent. The upper 25 meters of water contain up to 30,000 to 40,000 bacterial cells per cubic centimeter, compared to an average of 1500 to 4500 phytoplankton cells. As obviously the consumers can never equal or exceed the producers, their relation is more accurately shown by biomass than cell numbers. The total phytoplankton biomass averages about twice that of the much smaller bacterial cells.

Although bacteria are important consumers of organic matter, they may also be utilized as food by other organisms in their larval stages, or even as adults (e.g., Protozoa, plankton, and deposit-feeding invertebrates). Bacterial action may occur at many places within the complex food chains of the oceans, resulting in reconstitution and chemical change of most organic matter before its incorporation in the bottom sediments.

Bacteria are also concentrated at the

sediment-water interface on the ocean bottom, with up to 10 million individuals present in a gram of sediment, including about 89 percent bacilli, 8 percent cocci, and the remainder spores. These bacteria continue the processes of organic decomposition until the material is too deeply buried.

Nitrogen Cycle. Although oxygen is the oxidant source for most organisms, including many bacteria, other bacteria have anaerobic respiration. Three types occur: the fermenting anaerobes that oxidize organic matter, the denitrifying or nitrate-reducing bacteria, and others that may grow either by fermenting processes or aerobic respiration (facultative anaerobes). Bacteria are important in nearly all steps of the nitrogen cycle, beginning with nitrogen fixation from the atmosphere. Nitrogen-fixing bacteria may be symbionts (as in the roots of legumes) or free-living in soil. *Azotobacter* is an aerobic form, *Clostridium* anaerobic. A few photosynthetic bacteria, such as the sulfate-reducing *Desulfovibrio*, also fix atmospheric nitrogen and are thus able to live solely on air and water. The nitrogenous compounds fixed by the bacteria are incorporated into complex organic compounds by animals and plants. These are released to the soil as waste products or at death of the organism. Bacterial and fungal action then results in decomposition of the remains, and the release of ammonia, NH_3. Most nitrogen in living matter is reduced as amino nitrogen in proteins. It is first converted to ammonia, and then oxidized by other bacteria (nitrifying bacteria) as a source of energy; *Nitrosomonas* oxidizes the ammonia to nitrite:

$$NH_3 + 1\tfrac{1}{2}O_2 \rightarrow NO_2{}^{2-} + H^+ + H_2O + energy$$

Others (e.g., *Nitrobacter*) oxidize the nitrites to nitrates, which is a better source of nitrogen for plants than the more reduced forms:

$$NO_2{}^{2-} + \tfrac{1}{2}O_2 \rightarrow NO_3{}^- + energy$$

In contrast to the nitrogen-fixing and nitrifying bacteria, the denitrifying bacteria reverse the above processes, changing nitrates to nitrites, nitrites to ammonia, and ammonia to free nitrogen. Their action depletes the soil nitrogen, which must then again be fixed by the nitrogen-fixing bacteria or blue-green algae.

Sulfur Cycle. Most biological sulfur is in a reduced state. Microorganisms convert the organic sulfur to inorganic sulfide, commonly hydrogen sulfide. Others oxidize the hydrogen sulfide to free sulfur (e.g., *Beggiatoa*) and the cells may become stuffed with tiny sulfur granules:

$$H_2S + \tfrac{1}{2}O_2 \rightarrow S^0 + H_2O$$

Another group of bacteria, the autotrophic sulfur bacteria such as *Thiobacillus thiooxidans* Waksman & Joffe, utilize free sulfur and produce sulfuric acid:

$$S^0 + H_2O + 1\tfrac{1}{2}O_2 \rightarrow \underset{\substack{\text{sulfuric}\\\text{acid}}}{H_2SO_4} + energy$$

Although the sulfur cycle is not as critical as the carbon or nitrogen cycles, because there is a sufficient supply of sulfates, the action of the sulfur bacteria is geochemically important. Bacterial processes may have played a major role in the formation of many elemental sulfur deposits, as in the cap rock of Gulf Coast salt domes.

Phosphorus Cycle. The presence of phosphate may be critical for phytoplankton that require a source of this compound. Bacteria are active in the regeneration of nutrients in the sea from the organic compounds in the animal and plant remains. However, phytoplankton and small zooplankton species may directly release nitrogen and phosphorus into the sea, sometimes in quantities equal to their total body content over periods ranging from only about 10 minutes to a day. In contrast, bacterial action on decaying organic matter may even deplete available phosphorus, as the bacterial cytoplasm contains a higher proportion of phosphorus than does the decaying matter. The added phosphorus requirement is absorbed by the bacteria from that dissolved in the seawater, reducing the supply. Any excess phosphorus may be stored in bacterial cytoplasm as well (Johannes, 1968). Thus geologic periods of increased marine bacterial activity, for example in the Permian, may have adversely affected the nutrient supply available for phytoplankton and the dependent food web (Tappan, 1970).

Distribution

Bacteria occur in the atmosphere to a height of four miles, and in the water and on the sea floor three miles below sea level. They exist in hot springs and are found below 0°C in Antarctic ice. Some are adapted to a normal atmosphere, and others thrive in the absence of oxygen. There is a similar range of tolerance to the hydrogen ion concentration, some species flourishing at a pH near 0, and others at a pH of 11. Although they can survive either high temperature or high acidity, they cannot survive both simultaneously. In neutral or alkaline hot springs they can occur up to the boiling point of water (92 to 100°C, depending on the altitude), but in acid waters, at a pH of 2 to 3, the upper temperature limit is 75 to 80°C (Brock and Darland, 1970). Many bacteria are free-living in a wide variety of habitats, but others are symbiotic, parasitic, or pathogenic in various host animals and plants.

Bacteria that normally live at a pressure of one atmosphere can tolerate pressures of 3000 to 6000 atmospheres in the laboratory, and as the highest pressure in the ocean depth approximates about 1000 atmospheres, pressure can have little limiting effect on bacteria. Limiting factors in the sea are predation (by protozoans, filter-feeding and deposit-feeding animals, etc.), food supply or concentration of nutrients, light, temperature, sedimentation rate (bacteria being carried downward in the water column by attachment to particulate matter), inhibitory compounds produced by other bacteria or phytoplankton, and the presence of bacteriophages (near shore; but this may not occur in the open sea).

Bacteria are the most widely distributed of all living organisms, both because they can withstand great physical extremes of temperature, moisture, acidity, and salinity, and because they are adapted to a wide variety of energy sources, ranging from the decomposition of all natural organic compounds to chemical oxidation of nonorganic compounds, and photosynthesis. As discussed earlier, bacteria decompose organic matter, oxidize organic and inorganic compounds, and reduce others, and fix various atmospheric elements. In a large natural area, all these actions may proceed simultaneously, the dominant process varying with the availability of the different compounds. For example, in an aerobic environment, the dominant bacteria will vary with the supply of carbon dioxide, sulfates, and nitrates. Microenvironments are exceedingly important because of the very small size of the bacteria. A single fragment of a decaying organism may represent one type of environment, with other quite different habitats in the immediately adjacent mud, soil, or water. Aquatic habitats may also be distinctly stratified. As different photosynthetic pigments absorb light of differing wavelengths, the purple sulfur bacteria that absorb light into the infrared region commonly are found in large dense masses beneath the green and blue-green algae, which do not have the same light-absorbing potential. Similarly, the green sulfur bacteria absorb light of a wavelength intermediate between those used by the purple bacteria and the other algae.

The nonphotosynthetic sulfur bacteria that utilize hydrogen sulfide also require oxygen, hence occur densely where oxygenated water borders on that containing H_2S. If the quantity of hydrogen sulfide is low, the bacteria will be near the bottom, moving up in the water column as the concentration increases. In the Black Sea, a layer of sulfur bacteria occurs at a depth of about 200 meters, and muds below this depth are black.

Bacteria in soils far outnumber all other soil-inhabiting organisms. They may number from 1×10^3 to 10×10^9 per gram, the usual range being between 1×10^6 and 10×10^6. Most soil bacteria occur in the upper 15 to 30 cm, one hectare of soil to 30 cm depth containing about 22.4×10^{10} cells.

In clear lakes and streams, few bacteria are present, and these predominantly are aerobic oxidizing species. Sulfur bacteria, including the green and purple sulfur bacteria and the large colorless filamentous forms, occur in sulfur springs and sulfide environments, where they may deposit both sulfur and calcium carbonate. Iron-bearing springs, pipelines, and the regions near coal mines whose waters contain ferrous iron contain a flora of iron-depositing bacteria.

The various environmentally adapted bacteria are not always closely related. The sulfur bacteria include various members of the orders Rhodospirillales and Beggiatoales, and range morphologically from free motile cells to stalked single cells or filaments.

Bacteria as Geologic Agents

Bacterial decay of organic matter is the main source of energy in the early diagenesis of sediments. Bacteria are important agents in the sedimentary deposition of calcium carbonate, calcium phosphate, ferric hydroxide, and iron sulfide (pyrite or marcasite), in the formation of elemental sulfur and the later decomposition or weathering of rocks, and in the various chemical cycles previously mentioned. They undoubtedly play an active role in the formation of petroleum and also in the decomposition of certain petroleum compounds.

Iron-Depositing Bacteria

Ehrenberg (1836) first noted the importance of microorganisms in the formation of bog iron ores, and described the responsible organism, *Gallionella*, as a floating iron factory. He mistook the iron-encrusted twisted stalk of the bacterium for the organism and thought it to be a diatom, as he failed to observe the very tiny organism at the end of the stalk.

Present in all iron-bearing waters, the iron-precipitating bacteria include various unrelated forms, a few coccoid (*Siderocapsa*) and bacillar types (*Bacillus*) of the Order Schizosporales or Eubacteria (see Figure 1.8), and the more common filamentous sheathed threads (*Crenothrix, Sphaerotilus*) and elongate stalked forms (*Gallionella*) of the Sphaerotilales. Very similar-appearing fossils are found in sectioned pyrite and other iron-rich sediments from the Precambrian to the present (see Figure 1.9).

Iron deposition may occur in various ways: (1) Bacteria may produce hydrogen sulfide by the decomposition of sulfur-containing proteins, or by the reduction of free sulfur or its oxides. The hydrogen sulfide then reacts with ferrous sulfate to precipitate iron sulfide. (2) The sulfate reducers may reduce the sulfites, sulfates, and thiosulfates to sulfides, and form the iron sulfide directly. (3) If calcium is present, calcium sulfide is formed, reacts with carbon dioxide and water to form hydrogen sulfide, and the latter precipitates iron sulfide:

$$CaS + CO_2 + H_2O \rightarrow CaCO_3 + H_2S$$

Bog and lake iron ores contain bacterially deposited ferric hydroxide. The iron bacterium, formerly termed *Ferrobacillus* but now = *Thiobacillus,* obtains energy by converting ferrous iron to the ferric state, depositing it as ferric hydroxide. Other sedimentary iron ores, like those of the Precambrian of the Lake Superior region, are ferrous carbonates deposited in a reducing environment. Similar iron carbonates in Carboniferous rocks commonly are associated with coals.

Iron silicates, such as glauconite, occur in Cretaceous and lower Tertiary strata. Glauconite is common inside foraminifers and other microfossils, and probably results from the formation of sulfides by organic decay, the sulfides being oxidized to ferric hydroxide, free-

---

Figure 1.8
Bacteria, probably short chains of cocci, described as "segmented microbes" from Pennsylvanian pyrite, Ohio. **1.** TEM of carbon replica, ×11,000; bar = 2 μm. **2.** TEM of carbon replica of deeply etched surface, ×7000; bar = 3 μm. From J. M. Schopf et al., 1965, *Proc. Am. Phil. Soc.,* v. 109, p. 288—308; by permission of American Philosophical Society.

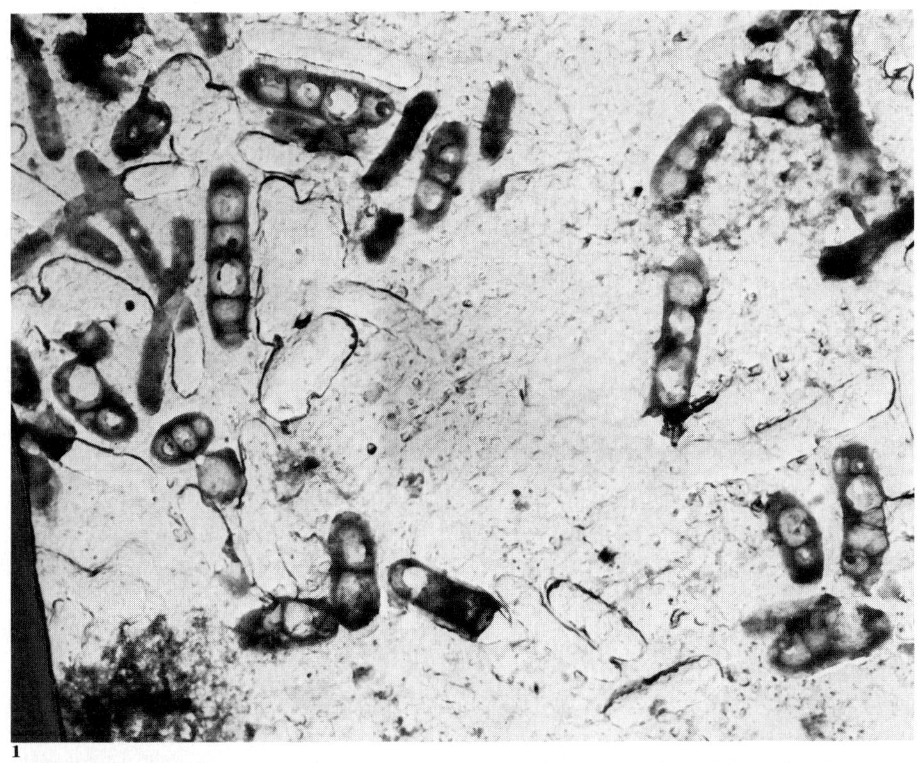

1

2

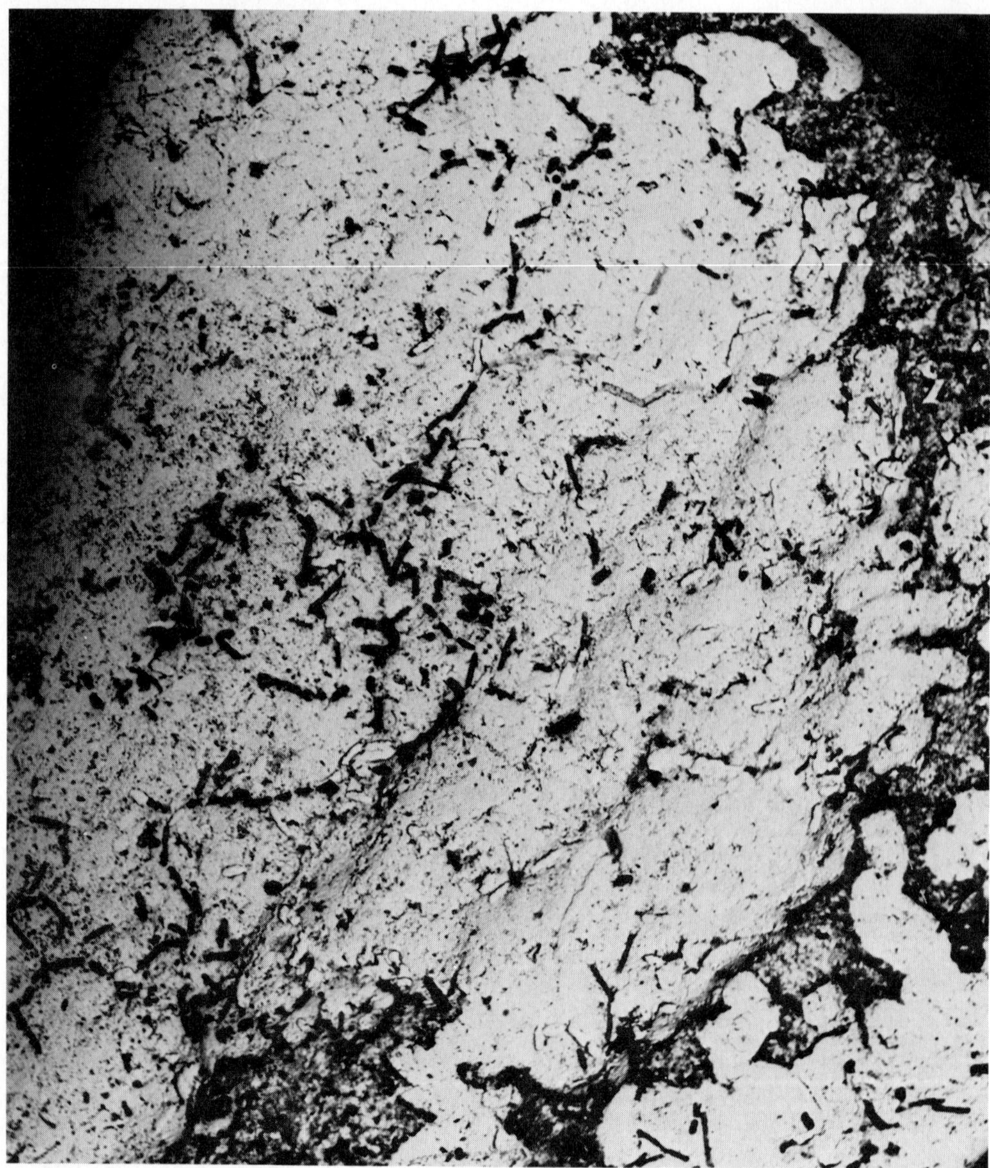

Figure 1.9
Sphaerotilus catenulatus J. M. Schopf et al., Pennsylvanian, Ohio. TEM of carbon replica of polished section of pyrite, showing chains of cells from lower left to upper center; mucilaginous sheath poorly preserved, pyrite-filled cells appear black, those shown as outline are molds; ×3520. From J. M. Schopf et al., 1965, *Proc. Am. Phil. Soc.*, v. 109, p. 288−308; by permission of American Philosophical Society.

ing sulfuric acid, and finally resulting in glauconite.

Elemental Sulfur Deposits

Sedimentary deposits include more than 90 percent of the world's sulfur reserves. Some of this sulfur may have been precipitated simultaneously with the deposition of the surrounding rocks, or syngenetically. Other deposits may be of secondary origin, or epigenetic.

Syngenetic sulfur may be produced directly in a reducing environment by oxidation of the hydrogen sulfide to free sulfur granules (e.g., by *Beggiatoa*) that sink to the bottom to accumulate in the calcareous sediments. Sulfur may also be formed by the reduction of sulfates to sulfides in the ooze, the sulfides then diffusing to the surface, where thiobacteria oxidize them to molecular sulfur.

Epigenetic sulfur deposits are commonly associated with oil deposits, as in the cap rock of salt domes. The crystalline sulfur fills fractures and replaces some of the porous cap rock. A nonbiological origin of the sulfur, from material in the cap, would require temperatures of 700 to 1000°C, but there is evidence that the salt domes never had temperatures over 100°C.

In oil-bearing strata near the surface, *Desulfovibrio desulfuricans* (Beijerinck) Kluyver & van Niel has been postulated to decompose the crude oils, releasing hydrogen sulfide. Sulfide waters rise through fractures in the rock to the contact with oxygenated waters near the surface, where the sulfides are converted to finely crystalline free sulfur, partly inorganically and partly by the action of thiobacteria, *Thiobacillus thioparus* Beijerinck. Similar reactions are thought to have been the source of the elemental sulfur in the porous cap rocks of salt domes, and the possible bacterial action on the petroleum and on the sulfates, gypsum, and anhydrite of the cap rock is thought to have resulted in the accumulation of the elemental sulfur deposits.

Petroleum

Petroleum is generally regarded as of biogenic origin. Bacteria undoubtedly played a part in its formation, by the gradual decomposition of organic matter, thereby increasing the proportion of carbon and hydrogen and decreasing the amounts of oxygen and nitrogen. In the anaerobic destruction of organic matter, small quantities of hydrocarbons are produced within the bacterial cell. These small droplets could gradually accumulate in the sediment, although an appreciable quantity would require much time for accumulation and transformation to petroleum. An estimated 87 barrels (13 tons) of hydrocarbon might be produced per square mile of ocean surface annually, if all the phytoplankton production were so converted. As obviously much of it is consumed as food, or by bacterial action, and thereby oxidized, only a small fraction of that amount could be available for accumulation.

Laboratory studies indicate that hydrocarbons may be biologically produced by various organisms or by bacterial decomposition of organic matter. Production may occur either aerobically or anaerobically, but preservation requires a reducing environment (anaerobic conditions) to prevent destruction of the hydrocarbons by bacterial oxidation processes. Abiologic transformation of organic matter would require much greater time and high temperatures. Thus, according to Oppenheimer (1965), a 1 percent conversion of organic matter in shale to bitumen requires 8.4×10^4 years at 100°C, or 6.7×10^7 years at 60°C. The porphyrins commonly present in crude oil indicate that it never attained a temperature over 230°C, although in laboratory experiments high temperatures might decrease the time of the conversion process.

Bacteria may play an important role in petroleum transformation and recovery, as well as in its original accumulation. Thus secondary oil extraction may be facilitated by bacterial action. Bacteria introduced into oil-bearing strata may increase limestone porosity by the formation of carbonic acid, or may increase the mobility of the oil either by the production of gases or by changing the high molecular hydrocarbons to more mobile ones. These desirable results are due to partial or limited molecular changes in decomposition of petroleum hydrocarbons. Although aerobic bacterial action on hydrocarbons is known to occur, the anaerobic action suggested to occur at depth has yet to be proven conclusively. The postulated changes may continue to a less desirable end result, from the human viewpoint, in which the oil may be destroyed by conversion to methane gas. Similar aerobic bacterial action is known to have a deleterious effect on asphalt surfaces.

Limestone Deposition

Early studies indicated a relation of some limestones to the decomposition of nitrogenous substances. A bacterium thought to be the agent in this process was described as *Bacterium calcis,* later transferred to the genus *Pseudomonas* as *P. calcis* (Drew) Kellerman & Smith. Drew (1914) found bacteria to be abundant in seawater, and his studies of bacterial cultures in calcium-enriched seawater suggested that bacteria were responsible for precipitation of the calcium carbonate. Kellerman (1915) found that bacterial denitrification and fermentation resulted in the formation of crystals and oolites of calcium carbonate in either marine or fresh water. Lipman (1924) showed that none of the bacteria could deposit lime from pure seawater, but that all species did if calcium salts were added to the medium as had been done by Drew. Later studies showed that bacterial action was effective if organic substances were present, however (LaLou, 1957). Glucose was added to a kilogram of black euxinic mud and seawater in an aquarium. Various bacteria in the mud acted on the glucose, releasing carbon dioxide and causing the solution of any calcium carbonate then in the mud. Action of the sulfate-reducing bacteria released hydrogen sulfide, of which part was fixed as FeS by iron bacteria and part was lost to the atmosphere. The increased pH caused a reaction of the magnesium and calcium ions with the CO_2-saturated medium to form bicarbonate. Where the CO_2 tension was reduced at the air-water interface, the carbonate crystallized in a film, first as calcite, later in aragonite spherules, and finally in a film of dolomite. The only requisites for carbonate deposition by bacteria were the presence of organic matter, a sufficiently high temperature, maximum light (hence favored in shallow water), and quiet and seldom-renewed water. The

process thus would be most effective in shallow lagoons. Although the bacteria did not directly deposit the calcium carbonate, they caused the physico-chemical changes that resulted in carbonate deposition. Meanwhile iron bacteria had precipitated pyrite in the cultures, perhaps explaining the common association of pyrite with oolitic and other limestones.

Hot Spring Stromatolites

As discussed later under the blue-green algae, **stromatolites** are laminar sedimentary deposits formed by sediment-trapping or depositing organic mats. Most modern and probably fossil ones are produced by blue-green algae, but those of hot springs, as at Yellowstone National Park, are formed largely by filamentous bacteria less than 1 μm in diameter (Brock, 1968, 1969; Doemel and Brock, 1974). The laminae result from differential migration of the bacteria in response to light intensity; the filamentous photosynthetic bacteria form a surface layer at night, but migrate below a layer of unicellular blue-greens (*Synechococcus*) during the day. The entrapped sediment is dominantly detrital silica from the siliceous sinter of the geyser basins, so that the stromatolites are primarily siliceous. Because the filamentous bacteria can produce stromatolites, the mere presence of these laminated structures in Precambrian rocks is not sufficient evidence of the presence of the oxygen-producing blue-green algae (Walter et al., 1972).

Evolution

Although bacteria appear to resemble the primitive organism, their limited morphology and small size have provided little evidence as to evolution within the group. In earlier studies of bacteria, the coccus was regarded as the most primitive form, but modern cocci are relatively advanced and require complex foods, such as secretions from the skin or glands of vertebrates. As few cocci are free-living in water or soil, they are probably not primitive. Others have suggested the spirillum as primitive, but the rodlike forms appear earliest as fossils. On the basis of molecular-biological experiments on DNA:RNA hybridization and various nutritional and physiological evidence, as is used to identify and classify living species, other theories and even phylogenetical trees have been proposed; but none are wholly satisfactory in detail as yet, in view of the absence of verifiable data from the various lines of evidence and from the fossil record.

Bacteria are regarded as lacking a nuclear membrane, but Bisset (1973) suggested that the cell membrane of the bacteria may in fact be a nuclear membrane. This inner membrane carries the attachment of the bacterial flagella and the nuclear system, structures that in eucaryotes are associated with the nuclear membrane. One of the other layers of the wall could have been derived from an original cell membrane. This theory would regard bacteria as secondarily simplified, particularly specialized in their small size, low volume-surface ratio, and rapid metabolism, and derived from more primitive, flagellate protists. However, the bacterial flagella and other characters are unlike those of other protists, and the wall of Gram-positive species is like that of blue-green algae.

Many bacteria are very similar to certain blue-green algae, and some may represent blue-green algae that have secondarily lost the pigment system. Thus, *Beggiatoa*, *Thiothrix*, and *Thioploca* may be **apochlorotic** forms of *Oscillatoria*, *Phormidium*, and *Schizothrix* (Stanier and van Niel, 1941), and are included in classifications of blue-green algae by some. Most bacteria do not appear to be secondarily derived, however.

Because their morphologic differences are limited, bacterial evolution is generally inferred from the biochemical character or metabolic capabilities of living taxa, rather than from evidence in the fossil record. Hall (1971) suggested that the nature of procaryote evolution was determined by the problem of obtaining energy in an (originally) anaerobic environment. Each major bacterial metabolic type probably arose as a biological invention in response to a requirement for some compound that was in short supply. Those organisms able to solve the problem could then invade new areas, increase in numbers, diversify, and provide the basis for the next step, as outlined below.

The earliest energy-generating system may have been similar to the simplest one now utilized, the substrate level phosphorylation. This process of fermentation produces 2 to 3 moles of ATP per mole of glucose, whereas the more efficient aerobic metabolism of other organisms produces 34 moles of ATP per mole of glucose. In fermentation, hydrogen is extracted at one stage and later returned to the products of the first reaction. This rearrangement into compounds at the same oxidation-reduction level produces no net oxidation of the substrate and wastes much of the energy as heat. Living organisms with this type of fermentative metabolism, such as *Clostridium* and the Lactobacillaceae, are mostly Gram-positive anaerobes, persisting in an environment that has remained stable through geologic time, although a few have become adapted to tolerate oxygen (see Figure 1.10).

The next evolutionary stage in metabolism was the linkage of phosphorylation to an electron transport, and primitively was probably similar to the first generation of ATP in aerobic respiration by transfer of electrons from a pyridine nucleotide to a flavoprotein. Some living anaerobic bacteria, such as *Clostridium* and *Streptococcus,* seem to do this.

An improvement in the system accompanied the use of inorganic electron acceptors; as the substrate and products did not have to be at the same net oxidation-reduction level, more energy could be obtained. The reduction of CO_2, sulfates, and nitrates was related to the appearance of the cytochromes, heme-containing electron-carrier proteins. The heme proteins may have originally evolved to decompose the toxic hydrogen peroxide resulting from ultraviolet bombardment of the surface waters, but could then be modified for electron-transport. Methane-producers are mostly Gram-negative, reduce CO_2 to methane, and produce relatively little energy. Sulfate reducers, also Gram-negative, are even less efficient, but obtain more ATP in reducing the sulfates to sulfides than do strictly fermentative species. As

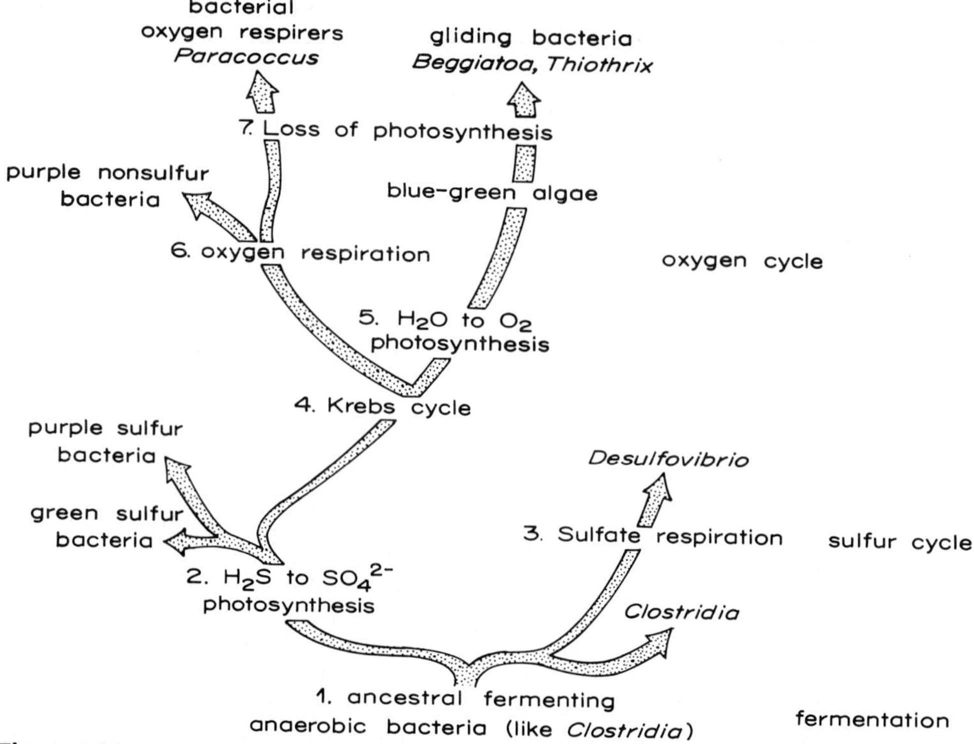

Figure 1.10

Evolutionary diversification of the procaryotes, modified from Dickerson et al., 1976, *J. Molec. Biol.*, v. 100, p. 473 — 491. Copyright by Academic Press, Inc. (London), Ltd., reproduced by permission. **1.** Fermenting anaerobic bacteria. **2.** Sulfate-producing photosynthesis by green and purple bacteria. **3.** Sulfate respiration completing the sulfur cycle. **4.** Evolution in sulfur bacteria of the Krebs cycle (citric acid cycle, an ATP-producing system in respiration) leads to use of alternate energy source by purple nonsulfur bacteria. **5.** Water-splitting photosynthesis by blue-green algae produces oxygen. **6.** Respiration by blue-green algae and purple nonsulfur bacteria utilizes oxygen released by photosynthesis, completing the oxygen cycle. **7.** Bacterial oxygen respirers like *Paracoccus* evolved as mutations impaired bacterial photosynthesis; a similar loss of photosynthesis by blue-green algae led to the gliding bacteria such as *Beggiatoa*.

they use waste products of the fermentators, they are widely distributed. The nitrate reducers, producing nitrite or reduced nitrogen compounds including gaseous nitrogen, can obtain more energy, approaching that of aerobic metabolism. Nitrate respiration probably preceded aerobic respiration, and is now used by various *Bacillus, Escherichia, Micrococcus, Pseudomonas, Spirilla, Staphylococcus,* and *Thiobacillus.*

Photosynthesis, obtaining energy from light, and using an organic substrate for growth were the next evolutionary advances. Probably some nitrate or sulfate reducer first obtained additional energy by trapping light photons as a heme protein underwent a minor mutation to produce chlorophyll. The photosynthetic organisms then could invade habitats with only sufficient carbon for cell building, and not for energy use as well. The eventual improvement of the system by the blue-green algae to use water as an electron donor provided a major advance in energy metabolism, and resulted in the production of free oxygen as a byproduct.

As oxygen became abundant, oxygen respiration arose, probably as a modification of nitrate respiration, as the systems have in common most components of the electron transport chain. Some aerobic bacteria grow best in somewhat lower oxygen levels, perhaps reflecting conditions when they first evolved. The autotrophic (photosynthetic) bacteria may have been more abundant at first, but as oxygen became more abundant in the atmosphere they were limited to special habitats at the interface of the organic matter or reduced minerals with the air.

Thus evidence from their metabolism suggests that Gram-positive bacteria are primitive, the Gram-negative trait and variations in wall structure appearing independently in sulfate reducers, phototrophs, and aerobes (Hall, 1971).

Another line of evidence as to bacterial evolution is based on biochemistry. The DNA of most organisms contains four kinds of organic bases, adenine, thymine, cytosine, and guanine, and the various kinds of DNA are characterized by their respective molar nucleotide composition, in practice represented as the percent G+C (guanine + cytosine). In plants the range is from about 37 to 49 percent G+C, vertebrates have only about 42 percent G+C, the range in invertebrates is higher, and in the protists varies some 30 percent, ranging in bacteria from 25 to 75 percent G+C. Organisms with many morphological, physiological, and biochemical features in common also show very similar percent G+C. These values have been obtained for a large variety of bacteria, those of different species within most genera being very close. Genera with similar percent G+C may be closely related; if the percentage is very different, there can be few nucleotide sequences in common and the organisms cannot be closely related evolutionarily. One suggested phylogenetic sequence of increasing morphological complexity also shows an increase in percent G+C, from *Corynebacterium* (48 to 59 percent) to *Arthrobacter* (63 to 65 percent G+C), *Mycobacterium* (60 to 70 percent), and the actinomycetes (68 to 78 percent), apparently corroborating the hypothesized sequence (De Ley, 1968).

The higher percent G+C may have had a strong selective value in bacterial evolution. Those species most exposed to sunlight appear to have the higher percent, perhaps adaptive to avoid thymine-specific ultraviolet radiation damage (Singer and Ames, 1970). Aerially reproducing species, aquatic ones, and those with carotenoid pigments have a higher percent G+C and lower thymine, whereas the non-photosynthetic obligate anaerobes and parasites, with no exposure to sunlight, tend to have a low guanine and cytosine content.

As bacteria may differ by some 95 percent of their chromosome, they are more evolutionarily diversified than mammals, for example, which share some 20 percent of their DNA. The diversification has been particularly marked in the Enterobacteriaceae, in which *Escherichia coli* differs as much from other genera of the family, such as *Proteus* and *Serratia*, as it does from the genera *Pseudomonas* and *Bacillus*, which represent distinct families and even orders (*Pseudomonas*). The free-living nitrogen-fixing bacteria also have widely diverse G+C values, suggesting that this may have been a very old characteristic, originating early and retained during the later evolutionary divergence. Nitrogen-fixing species are found in the Gram-negative aerobic Azotobacteraceae and Bacillaceae and the Gram-positive anaerobic spore-forming Rhizobiaceae (e.g., the symbiotic nitrogen-fixing *Rhizobium*) and in various photosynthetic bacteria (De Ley, 1968).

On the basis of characteristics of the 16S rRNA (ribosomal RNA), the methanogenic bacteria are regarded as a distinct group, only distantly related to other bacteria (Fox et al., 1977). Other bacteria are closer to blue-green algae in the common sequences of the rRNA than to the methanogens. Two separate groups were recognized: One included most *Methanobacterium*, the other *Methanosarcina* and *Methanospirillum*, the former being Gram-positive and the latter dominantly Gram-negative.

Fossil Record

Although there is abundant evidence throughout geologic time of bacterial activity similar to that at present, the actual fossil record is less well documented, both because of the very small size of the original organisms and because of the difficulty of preventing contamination.

Small organic remains without mineralized hard parts were only recently recognized to be commonly preserved. The advance of palynological studies has given added impetus to the use of similar techniques for the study of many of the smaller microfossils in addition to pollen and spores.

A more usual method for bacterial studies is by the preparation of very thin sections. Bacteria have been reported from such thin sections of limestone, chert, iron ore, pyrite, oil shale, coal, salt, bone, coprolites, and fossil wood. Bacteria have also been reported as fossils in oil, peat, and other carbonaceous deposits.

The use of higher magnifications for study, including electron microscopy, has aided in the recognition of fossil bacteria. Perfection of the necessary delicate techniques has allowed recognition of a preserved record in nearly all types and ages of sedimentary rocks, from the Precambrian onward. Although morphologically simple and thus difficult to define taxonomically, some of the more distinctive types of bacteria are useful in paleoenvironmental interpretation.

Precambrian Fossils

The scarcity of Precambrian megafossils (other than the algal stromatolites) in contrast to the diverse and abundant Cambrian record has led to a concentrated search for any record of Precambrian life. Since Walcott (1916) first reported fossil bacteria from the Algonkian Galla-tin Formation of Montana, a wide variety of microorganisms has been described from the Precambrian cherts.

Probably the oldest known specimens are those from the early Precambrian black cherts of the upper Onverwacht Swartkoppie Formation of South Africa. Radiometric dating of associated rocks indicates an age of approximately 3.2 billion years. *Eobacterium isolatum* Barghoorn & Schopf (see Figure 1.11) resembles modern eubacteria and consists of rodlike cells between 0.45 and 0.7 μm in length and 0.18 to 0.32 μm in diameter. Organic filamentous matter associated with the bacterial fossils may also be biogenic in origin.

Less recognizable but probable evidence of Precambrian life is the carbonaceous matter in the lower Precambrian Soudan Formation of Minnesota (2.7 to 3 billion years old), in graphitic-appearing layers and lenses. A variety of apparent bacteria, cocci, bacilli, and filaments were observed in the 2.15-billion-year-old Witwatersrand System of South Africa, some in electron micrographs of surface replicas (Schidlowski, 1966, 1970; Oberlies and Prashnowsky, 1968), although some may not be indigenous.

Ranging from small, ovoid, sporelike bodies to threadlike, septate, or radiating filaments, fossils in cherts of the 1.9-billion-year-old Gunflint Formation (middle Huronian) of Minnesota and Ontario (J. W. Schopf et al., 1965) suggest the presence of eubacteria (see Figure 1.12), thread bacteria, and blue-green algae (similar to the modern *Oscillatoria*). These organisms apparently were sealed in the gelatinous cherty mass while still growing (Cloud, 1965). Various types of preservation include the amber to

Figure 1.11
Precambrian procaryotes. **1.** Possible microspores, TEM of surface replica of 2.8-billion-year-old stromatolite from the Bulawayan Group of Rhodesia, ×5500, from Oberlies and Prashnowsky, 1968 (suggested by Cloud, 1976, to be modern hyphomycete fungal spore contaminants). **2—7.** *Eobacterium isolatum* in chert of the Fig Tree Series (age 3.1 billion years), South Africa; TEM, from Barghoorn and Schopf, 1966, *Science*, v. 152, p. 758—763; copyright 1966, American Association for the Advancement of Science; reproduced by permission. **2,3.** Specimens with rodlike organically preserved cell (electron dense, hence appearing white at lower part), and its imprint above, from which it was displaced in preparation. **4.** Imprint of more elongate cell, 0.75 μm long. **5.** Cell intersected by chalcedony grain boundary, showing its indigenous nature. **6,7.** Cross sections of the bacilli, showing cell wall (arrow in 6). Bar in each photo = 1 μm; 5 and 6 are the same magnification as 7.

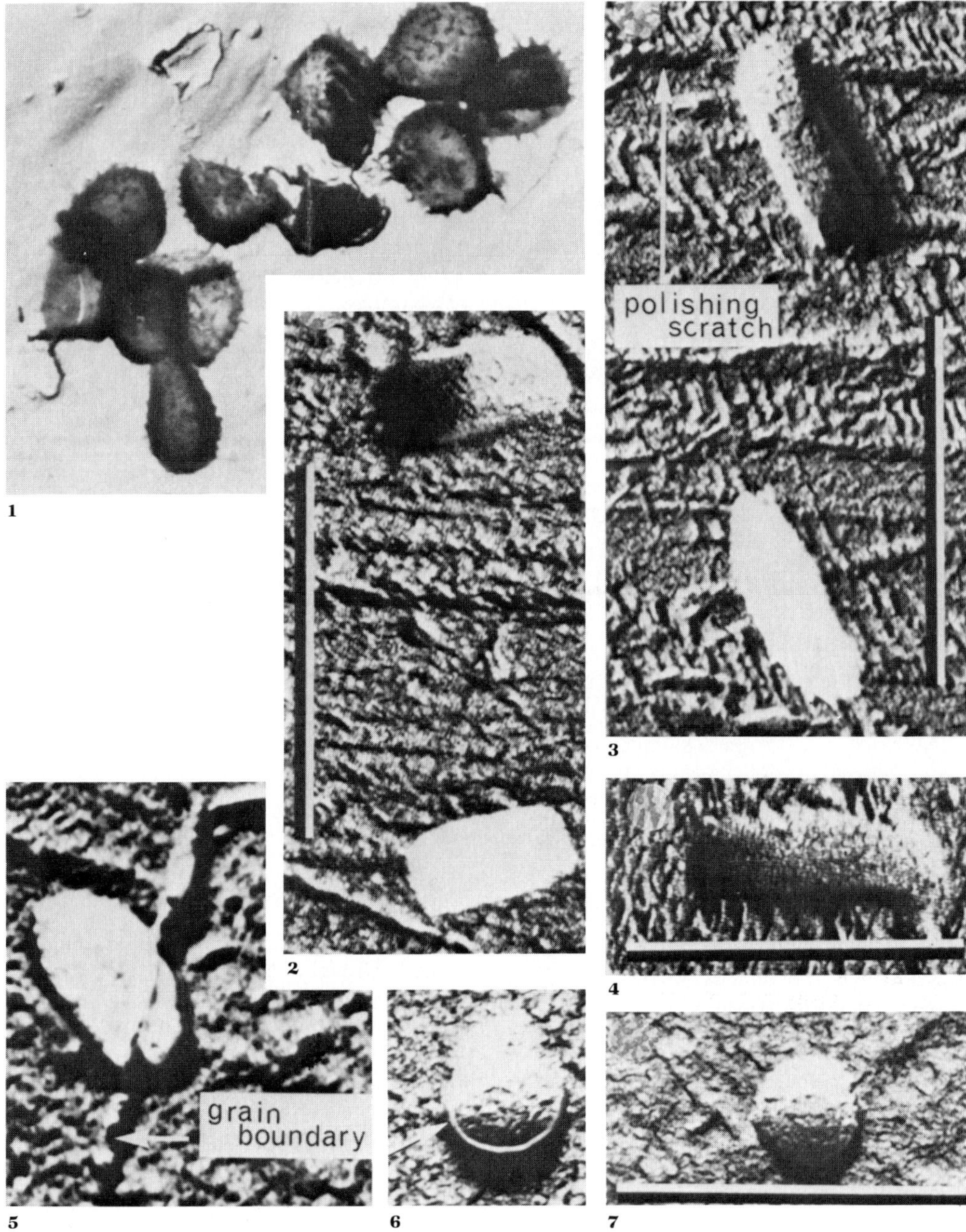

1

2

3

4

5

6

7

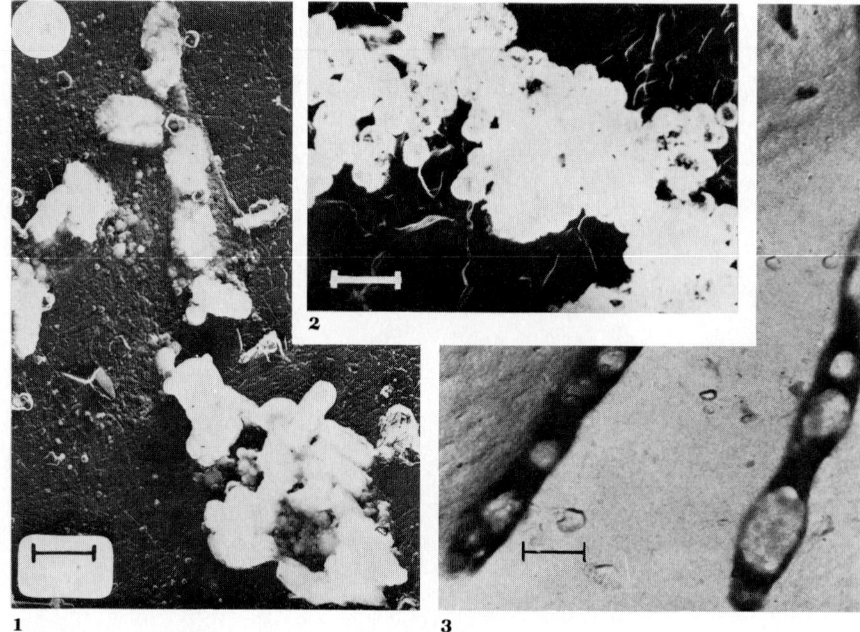

Figure 1.12

Precambrian bacteria. **1.** TEM of bacilli in carbon-platinum replica, ×19,700.
2. TEM of group of cocci, each about 0.35 μm in diameter, ×9200. 1,2, from the
Gunflint Formation (age 1.9 billion years), Ontario; from J. W. Schopf et al., 1965,
Science, v. 152, p. 758 – 763. Copyright 1966, American Association for the Advancement of Science; reproduced by permission. **3.** TEM of filamentous bacteria,
Witwatersrand Supergroup (age about 2.1 billion years), South Africa, ×10,000;
from Oberlies and Prashnowsky, 1968. Bar in each photo = 1 μm.

brown organic residues and films, filaments,
and globules best observed in chips or very thin
sections, as they tend to break up in the
palynological process of hydrofluoric dissolution. Crystals and aggregates of pyrite may outline or replace the organic filaments and
sporelike bodies that closely resemble the modern iron-depositing *Sphaerotilus* and *Gallionella.* Other filaments are preserved in the
chert matrix as carbonate or hematite, the latter particularly in the red jaspers. The hematite
may represent later oxidation of pyrite or a replacement of original organic matter.

An assemblage from the 2-billion-year-old
Kasegalik Formation, Belcher Islands, Canada,
included probable rodlike and filamentous bacteria and spheroidal unicells and trichomes that
may represent either bacteria or blue-green
algae (Hofmann, 1974, 1976).

In the glacial varves of the middle Precambrian Gowganda Formation of the Cobalt Series
(upper Huronian) of Ontario, fossilized bacterial cells, chains of cells, and filaments resembling actinomycete hyphae were obtained from
sectioned sulfide crystals. Modern Actinomycetales are largely nonmarine, living in
lake waters, sediments, and soils, suggesting
that the Precambrian ones may have had a similar habitat and probably were active in the formation of the sulfides in the Gowganda argillites
(Jackson, 1967). The Gunflint *Eoastrion* (see
Figure 1.13) also may be an actinoymycete.

Filamentous bacteria, unicellular and
filamentous blue-green algae, and possible red

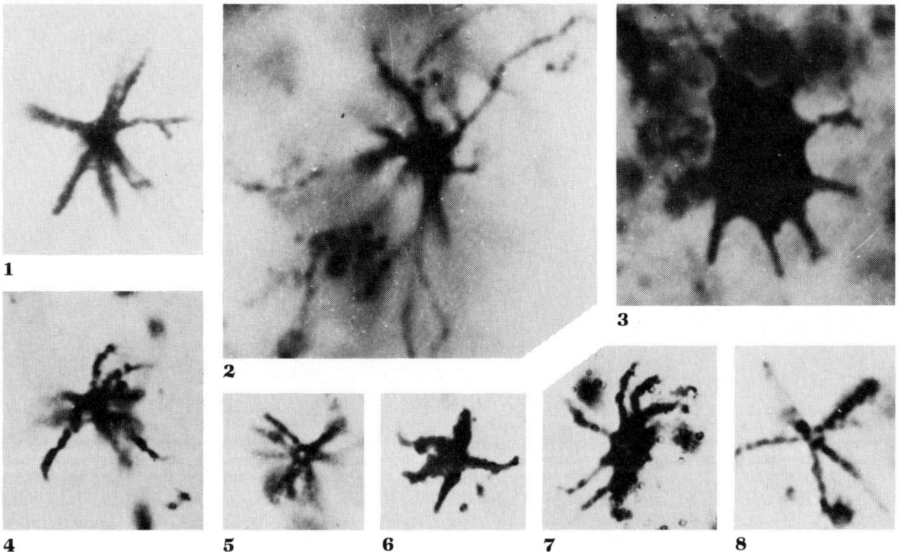

Figure 1.13
Possible actinomycetes from the 1.9-billion-year-old Gunflint Chert, Ontario,
Canada. 1–3. *Eoastrion bifurcatum* Barghoorn. **1**, ×1300; **2**, ×2384; **3**, ×1028.
4–8. *E. simplex* Barghoorn. 4,5, ×1748; 6,7, ×928; 8, ×2384. From Barghoorn and
Tyler, 1965b, *Science,* v. 147, p. 563–577. Copyright 1965, American Association for
the Advancement of Science; reproduced by permission.

or green algae have been described from the
1.5-billion-year old Barney Creek Formation,
McArthur Group of northern Australia (J.
Oehler, 1977). Most of the bacterial fossils are
pyritized.

The upper Precambrian (Keeweenawan)
Nonesuch Shale (about 1.1 billion years old)
contains crude oil and many recognizable hydrocarbons of biological origin, whose composition, including phytane and pristane, suggests
the degradation of chlorophyll (Meinschein et
al., 1964). The particular compounds present
also indicate that the material was indigenous
and not derived from younger sediments. The
Nonesuch Shale seems to have been deposited
in a shallow deltaic region, in a highly organic,
periodically reducing environment. A few
filamentous objects in sections are suggestive of
plant tissues.

Bacterial fossils were illustrated in Precambrian rocks from Finland about 1 billion years
old (Oberlies and Prashnowsky, 1968). Fila-

ments and cell aggregates of probable bacteria
and blue-green algae are present in the Precambrian quartzite and chert of central and
western Australia, and possibly simple green
and red algae are represented as well (Schopf,
1968).

Phanerozoic Fossils

The variety of rock types in which bacterial fossils occur was mentioned above. Their geologic
distribution is equally extensive. Fossil free-
living iron bacteria have been reported in iron
ores, bauxite, and pyrite from all ages, beginning in early Precambrian, and others have
been reported from sedimentary rocks of all
types, sandstone, shale, and limestone, through
a similar age range. They are characteristic in
cherts, the salt and sulfur of salt domes, coal,
and phosphates.

Pennsylvanian bacteria similar to modern

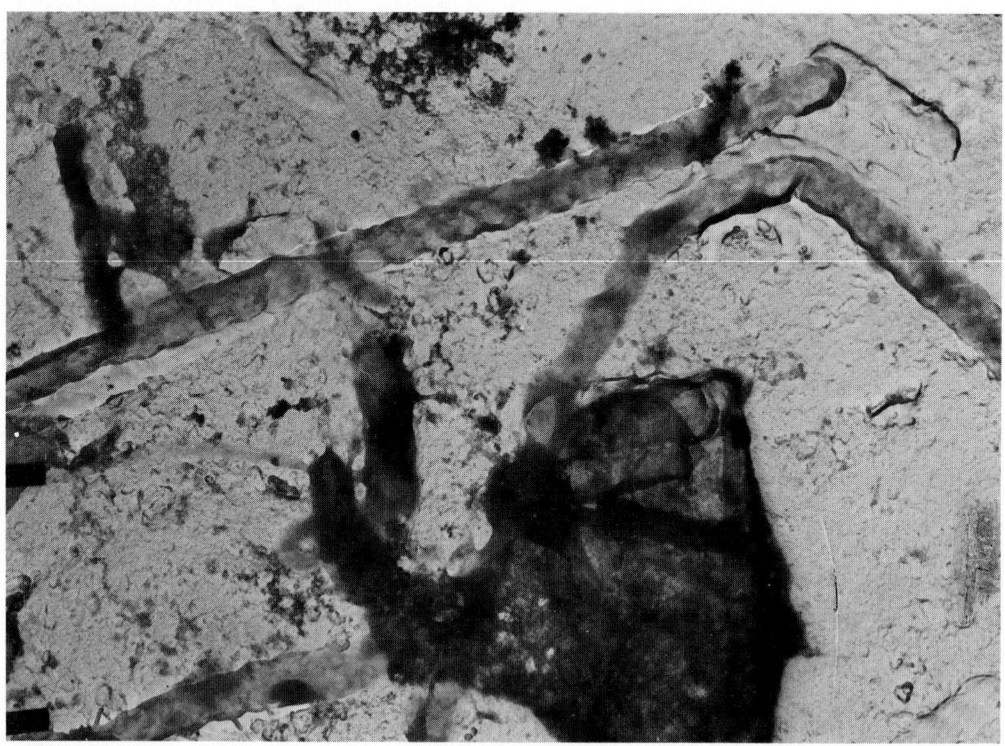

Figure 1.14
Gallionella pyritica J. M. Schopf et al., Pennsylvanian, Ohio. TEM of carbon replica, showing several spiral stalks similar to those of the modern iron bacteria; approximately × 15,000. From J. M. Schopf et al., 1965, *Proc. Am. Phil. Soc.*, v. 109, p. 288−308; by permission of American Philosophical Society.

Gallionella in appearance and habitat are preserved as ferrous sulfide (see Figure 1.14), although the living *Gallionella* deposits ferric hydroxide. The different composition may be due to later alteration.

In the aquatic environment, bacteria may play an important part of the food chain, providing food for various forms of benthic or planktonic protozoans, invertebrate larvae, and adult filter-feeding and deposit-feeding organisms. Occasionally, fossilized bacteria have been found in fecal pellets, for example in early Tertiary marine phosphates of Tunisia. These pellets are about 150 μm in diameter and contain fragments of diatoms, radiolaria, and foraminifera, within which are small rodlike bacteria 0.5 to 1 μm in diameter and from 1 to 3 μm in length, and some short chains of cells.

Coprolites also occur in nonmarine rocks, those in the upper Eocene Bridger Formation of Wyoming containing an abundance of silicified bacteria similar to modern intestinal saprophytes (Bradley, 1946). The coprolites are probably of both mammalian and reptilian origin, and contain desmids, bone fragments, fish scales, chrysophytes, and cyprid ostracods in addition to the bacteria. The assemblage suggests that the animals drank from a poorly drained acidic pond at the low water stage in a dry hot climate. The coprolites were buried by volcanic ash and preserved as calcium phosphate, and their undigested contents were silicified. Even parasitic and pathogenic bacteria have a fossil record, having been reported in plant tissues, fossil bone, and fossilized coprolites from Devonian to the present.

The results of bacterial action on fossil wood has been reported from Mississippian and later rocks (Tieghem, 1879; Renault, 1899), and silicified saprophytic bacteria occur in the fruits and sporangia of the Carboniferous fern *Pecopteris*. Similar bacteria-caused pathologic structures in fossil bone and teeth are widely reported, for example from the Devonian placoderm fish such as the antiarch *Bothriolepis* from Canada and the arthrodire *Coccosteus* of Scotland, in fragments of bone in Permian coprolites from Texas, and in Pliocene elephant tusks from Nebraska (Moodie, 1923).

Because of their small size, the absence of many diagnostic morphologic features, and the possibility of contamination in the field or laboratory, particular care is required to be certain of the true fossil nature of specimens, as to both their biogenicity and their indigenous character. For example, bacteria described from a Lower Cretaceous lake deposit in Nevada (Bradley, 1963) later were determined to be tiny inorganic spheres of fluorite produced during the acid maceration of the limestone (Bradley, 1968), and not biogenic in origin.

Viable bacteria have been reported to persist *in situ* in the ancient rocks of all ages, from the Precambrian (Dombrowski, 1968), the Zechstein (Permian) of Germany, to at least the Cretaceous. Thin sections of a Mississippian rock salt from Nova Scotia, carefully collected to avoid contamination, were found to contain bacteria and bacterial spores completely embedded in the salt, and five different living species were isolated and cultured from the salt (Dombrowski et al., 1965). They were regarded as "living fossils" that remained vital since the sediments were deposited; bacteria are known to remain viable for some time, and the preservation in dry salt was regarded as advantageous. Some were isolated from actual salt crystals and thus were believed to be proven indigenous.

Evidence against these bacteria being truly ancient was summarized by De Ley (1968). Accidental contamination by bacteria in handling is possible, and other workers have not been able to detect viable bacteria from similar samples (Bien and Schwarz, 1965). Ground water contains large numbers of bacteria, and the layers of Zechsteinsalze, from which the Permian bacteria were isolated, are in continuous contact with such ground water. The same bacterium can be isolated from the salt and from well water, and the Zechstein is readily permeable to foreign bacteria through microcapillaries. Many modern species of bacteria have been shown to penetrate rock; bacteria experimentally labelled with radioactive phosphorus, ^{32}P, were found to penetrate 14.25 inches of a 2-inch-diameter rock core of the Berea Sandstone at atmospheric pressures, as well as 3 inches of 1-inch-diameter cores of Lower Devonian and Mississippian limestones and Cretaceous sandstone, indicating the strong likelihood of contamination (Myers and McCready, 1966). The possibility of organisms remaining dormant for such extended periods also is refuted by both observational and experimental data. Spore-forming bacteria may remain viable in dry soil for about 300 years, but human remains and coprolites of 1000 to 3500 years old are sterile. Revival of a dormant cell requires the simultaneous presence of over 1000 enzymes and cofactors, many of which decompose readily even at $-20°C$ in the dry state. Even the more stable protein compounds, amino acids, decompose spontaneously over geologic time. Furthermore, *Pseudomonas halocrenaea* Dombrowski isolated from the Permian salt was found to have a nucleotide composition and sequence that was indistinguishable from the living *P. aeruginosa* on the basis of some 88 different phenotypic tests. In addition, the *Pseudomonas* that was isolated from the Permian salt was shown to be very pathogenic for mice, although such a pathogenicity could only have arisen after the development of the mammals. The living *P. aeruginosa* is a common inhabitant of soil and water, hence probably only recently infiltrated the salt deposits. Had it in fact been of Permian age, it could scarcely have changed at all by mutation in 200 million years, in contrast to the very high mutation rate of bacteria in laboratories and hospitals. In explanation of this rapid evolutionary rate of living bacteria, it is noted that the entire recorded history of man includes less than 100 generations, a number exceeded in a few days by *Escherichia coli* (Mandel, 1969).

Probably the earliest microorganisms in the Precambrian were heterotrophic, feeding on the abiotic organic compounds present in the seas. As chemoautotrophic species evolved, they served as a continuing food source for the heterotrophs. Photoautotrophic species appeared next, and probably gave rise to the oxygen-producing blue-green algae, perhaps 2 to 3 billion years ago. Nitrifying and denitrifying bacteria could only have arisen later, after the atmosphere became more oxygenic.

In the Precambrian, the intense ultraviolet irradiation probably resulted in rapid changes in the DNA base composition, in the way that artificial mutants are now obtained in the laboratory. Thus diversification probably was rapid, leading to the invasion of a variety of habitats. The symbiotic and pathogenic forms probably adapted from nonpathogenic stocks after the advent of the various host organisms. Cretaceous reptile teeth are known to have caries, and various other diseases of bones have been observed in fossils. In addition, various extinct plants, invertebrates, and vertebrates probably also had parasites and bacteria-caused diseases that became extinct with their hosts, only a few of which are recognizable in the fossil remains.

Classification

First thought to be small animals, these simple forms were later placed in the plant kingdom, the blue-green algae or Cyanophyta being related to the algae, and the bacteria or Schizomycetes (fission fungi) being placed with the fungi. Because of similarities in these two groups, and their differences from other animals and plants, the bacteria and blue-green algae were later placed in a separate kingdom, the Monera or Mychota, or in a separate division of the plant kingdom, the Schizophyta or Protophyta. The original basis for the kingdom Monera was the supposed absence of a nucleus, no longer a valid character, as nuclear material is present in the form of chromatin granules or microfibrils, but many important distinctions allow the continued separation of the bacteria from both the plant and animal kingdoms.

The first international microbiological congress in 1930 recommended preparation of a separate code of nomenclature for bacteria and viruses, which had previously been covered by the botanical code. The first approved code was published in 1948 (Buchanan and Breed) for the bacteria and viruses, and later revisions have appeared.

Distinctions between many of the major groups of protists are difficult to define, and certain flagellates and others are simultaneously regarded both as animals (protozoans) and as plants (algae). Similarly, certain of the Schizophyta have been regarded variously as bacteria or as colorless ''blue-green'' algae, hence may be covered under both botanical and bacteriological codes.

The present classification is based on *Bergey's Manual of Determinative Bacteriology* (Buchanan and Gibbons, 1974), but with some modifications. As used herein the Kingdom Procaryota includes the three divisions Cyanophyta (blue-green algae), Prochlorophyta (procaryotes with green algal pigmentation), and Bacteriophyta (bacteria). Other classifications place all photosynthetic procaryotes (blue-greens, red and green bacteria) together as a Division Photobacteria, with those that are indifferent to light combined in the Division Scotobacteria (nonphotosynthetic bacteria, rickettsias, and mycoplasmas). Most bacteria have a poor fossil record, and the difference in photosynthetic product of the bacteria and blue-green algae has been so important geologically, and to the evolution of life, that the photosynthetic procaryotes are here regarded as representing three divisions (see Table 1.2).

Includes Bacteriophyta (both Scotobacteria, or nonphotosynthetic bacteria, and the photosynthetic bacteria) and blue-green algae. Equivalent to Monera, Mychota, Schizophyta, Protophyta.

Division Bacteriophyta Steinecke 1931
Synonyms include Phylum Archezoa Haeckel 1894, Class Schizomycetes Naegeli 1857, Division Schizomycophyta Pascher 1931, Division Schizomycetae Stanier and van Niel 1941.

The photosynthetic bacteria probably are closer to the blue-green algae than to the other bacteria. Some heterotrophic bacteria probably were primitive and ancestral to photosynthetic ones, but many living nonphotosynthetic bacteria are highly evolved.

"Scotobacteria"
Procaryotes that are indifferent to light.

A. Order Myxobacterales Thaxter 1892
Gliding rods, 1.5 μm in diameter, forming colonies that may glide as a unit, typically embedded in a layer of slime; Gram-negative, chemoorganotrophs, aerobic, respiratory metabolism; reproduce by binary fission; may form complex fruiting bodies that produce resting cells (myxospores); saprophytes and parasites in soil and marine and freshwater animals and plants; decompose cellulose and starch.

1. Family Myxococcaceae Jahn 1924
Soil-dwelling; vegetative cells tapered, microcysts spherical or oval.
2. Family Archangiaceae Jahn 1924
Rodlike microcysts not formed in sporangia.
3. Family Cystobacteraceae McCurdy 1970
Rodlike microcysts formed in sporangia.
4. Family Polyangiaceae Jahn 1924
Vegetative cells with bluntly rounded ends; myxospores similar.

B. Order Beggiatoales Buchanan 1956
Syn.: Cytophagales Leadbetter 1974
Rods or spirilla, single or in trichomes; nonflagellate, gliding or creeping on solid surfaces; transverse fission; nonphotosynthetic, but may contain sulfur granules or calcium carbonate crystals in or on the cells; in soil and fresh and marine water; in sulfur springs, stagnant pools, and decomposing matter, or free-living. Permian to Holocene.

1. Family Cytophagaceae Stanier 1940
Chlamydospores formed by individual cells, not in cysts.
2. Family Beggiatoaceae Migula 1894
Colorless sulfur bacteria; aerobes, Gram-negative, cells in chains; metabolism respiratory; mixotrophs or chemoorganotrophs, deposit sulfur granules in presence of H_2S. *Beggiatoa* Trevisan 1842.
3. Family Simonsiellaceae Steed 1962
Occur in oral cavities of animals.
4. Family Leucotrichaceae Buchanan 1957
Aerobic, aquatic, nonmotile, colorless sulfur bacteria.
Leucothrix Oersted 1844.
5. Family Achromatiaceae Massart 1901
Gliding, colorless, spherical cells, may contain granules of sulfur or $CaCO_3$.
6. Family Pelonemataceae Skuja 1956

C. Order Sphaerotilales Copeland 1956
Syn.: Chlamydobacteriales Buchanan 1917; Desmobacteriales Pribram 1929
Sheathed bacteria, cell in trichomes, may be attached; may have motile swarm spores or nonmotile conidia. Motile forms have polar flagella. Sheath may contain ferric hydroxide or manganese oxide. Freshwater and marine. Precambrian to Holocene.

1. Family Sphaerotilaceae Pribram 1929.
Sheathed filaments with false branching. Cells may escape from sheath, become lophotrichous, and function as swarm spores.
?Ramacia J. Oehler 1977 (Precambrian); *Sphaerotilus* Kützing 1833 (Pennsylvanian, Holocene).
2. Family Leptotrichaceae Schröter 1866
Aquatic bacteria, producing stalks that may be elongate, the minute cells being terminal.
?Coleobacter J. Oehler 1977 (Precambrian); *Fer-*

(*continued*)

rimonilis J. Oehler 1977 (Precambrian); *Gallionella* Ehrenberg 1838 (iron-depositing; Pennsylvanian; Holocene); *Leptothrix* Kützing 1843.

3. Family Crenotrichaceae Hansgirg 1888
Iron bacteria, filamentous, with nonmotile conidia.
Crenothrix Cohn 1870.

D. Order Hyphomicrobiales Douglas 1957
Budding or appendaged bacteria; ovoid cells, commonly in aggregates, may be stalked; motile forms with one polar flagellum; reproduce by budding rather than fission. In mud and freshwater ponds and streams, ?marine; aerobic, heterotrophic, or chemoorganotrophic; *Metallogenium* is Mn-oxidizing. Precambrian, Holocene.

1. Family Hyphomicrobiaceae
Hyphomicrobium Stutzer & Hartleb 1898.
2. Family Caulobacteraceae Henrici & Johnson 1935
Caulobacter Henrici & Johnson 1935; ?*Eoastrion* Barghoorn 1965 (Precambrian); *Metallogenium* Perfil'ev & Gabe 1961 (?Precambrian; Holocene).

E. Order Spirochaetales Buchanan 1917
Spirilla from 3 to 500 μm in length; may have axial filament; nonflagellate, motile, move by whirling about the long axis; transverse fission; aerobic, anaerobic, or facultative; chemoheterotrophic; free-living in stagnant fresh or marine water; saprophytic or parasitic in mollusks and vertebrates; some insect-borne and pathogenic.

1. Family Spirochaetaceae Swellengrebel 1907
Borrelia Swellengrebel 1907; *Spirochaeta* Ehrenberg 1835; *Treponema* Schaudinn 1905.

F. Order Schizosporales Cohn 1872
Syn.: Eubacteria Schröter 1886; Haplobacteriacei Fischer 1895; Eubacteriales Buchanan 1917; Pseudomonadales Orla-Jensen 1921
Typical bacteria, cocci, bacilli, spirilla, not differentiated along axis; Gram-negative (families 1 — 14) or Gram-positive with few exceptions (families 15 — 22).
Precambrian to Holocene.

1. Family Spirillaceae Migula 1894
Spiral and curved bacteria (spirilla and vibrio forms), polar flagella; no spores; Gram-negative anaerobes and aerobes; chemoorganotrophs, require carbohydrates, reducing sulfates to H_2S.
Desulfovibrio Kluyver & van Niel 1936; *Spirillum* Ehrenberg 1832.
2. Family Pseudomonadaceae Winslow et al., 1917
Gram-negative rods and cocci, may be motile with one or tuft of polar flagella; aerobic, respiratory metabolism; chemoorganotrophs, heterotrophs, denitrifying.
Pseudomonas Migula 1894.
3. Family Azotobacteraceae Pribram 1933
Gram-negative rods, peritrichous flagella or nonmotile clusters in capsule; N-fixing, fixing at least 10 mg of atmospheric nitrogen per gram of carbohydrate consumed.
Azomonas Winogradsky 1938; *Azotobacter* Beijerinck 1901; *Beijerinckia* Derx 1950; *Derxia* Jensen et al., 1960.
4. Family Rhizobiaceae Conn 1938
Gram-negative rods, atrichous or peritrichous; in soils, parasitic on plants (blights, wilts), and symbiotic in legumes, producing nodules on the roots, and benefiting the host by nitrogen-fixation.
Rhizobium Frank 1889.
5. Family Methanomonadaceae Breed 1957
Syn.: Methylomonadaceae Leadbetter 1974
Gram-negative rods; aerobic, may oxidize methane, methyl, hydrogen, or carbon monoxide.
Methanomonas Orla-Jensen 1909.
6. Family Halobacteriaceae Gobbons 1974
Red-colored, requiring a high level of NaCl, and living in environments containing up to 25 percent salt.
7. Family Enterobacteriaceae Rahn 1937
Syn.: Bacteriaceae McNab 1877; Achromobacteriaceae Breed 1945.

Gram-negative facultatively anaerobic rods; mostly commensal in the gut of animals; some pathogenic; some soil-dwelling; sugar-fermenting. Eocene, Holocene.

Escherichia Castellani & Chalmers 1919 (reported in Eocene coprolites; living); *Klebsiella* Trevisan 1885; *Serratia* Bizio 1823 (produces extracellular red pigment); *Proteus* Hauser 1885.

8. Family Vibrionaceae Véron 1965

9. Family Bacteroidaceae Pribram 1933
Gram-negative anaerobic bacteria.

10. Family Neisseriaceae Prévot 1933
Gram-negative cocci and coccobacilli; non-flagellated, occur in pairs, chains, or masses; aerobic; obligate parasites causing disease.
?Lampropedia Schröter 1886.

11. Family Veillonellaceae Rogosa 1971
Gram-negative anaerobic cocci.

12. Family Nitrobacteraceae Buchanan 1917
Gram-negative, chemolithotrophic bacteria, cocci, bacilli, and spirilla forms, no endospores, may have subpolar or peritrichous flagella; aerobic, N-oxidizing; in soil and fresh and marine water.
Nitrobacter Winogradsky 1892; *Nitrosomonas* Winogradsky 1892.

13. Family Siderocapsaceae Pribram 1929
Gram-negative, chemolithotrophic bacteria, depositing iron and manganese oxides, in iron-containing waters.
Siderocapsa Molisch 1910.

14. Family Thiobacteriaceae Migula 1894
Gram-negative chemolithotrophic bacteria, oxidizing sulfur and reduced sulphur compounds; some also use ferrous compounds.
Thiobacillus Beijerinck 1904 (syn.: *Ferrobacillus*) deposits coating of ferric oxide; *Thiobacterium* Janke 1924.

15. Family Methanobacteriaceae Barker 1956
Anaerobic cocci and rods, motile or non-motile, Gram-negative or positive; no spores; produce methane by reduction of CO_2.
Methanobacterium Kluyver & van Niel 1936.

16. Family Micrococcaceae Pribram 1929
Gram-positive cocci, 0.5 to 3.5 μm in diameter, divide to form clusters or packets of cells; motile or nonmotile; chemoorganotrophs, respiration or fermentation, aerobic or facultatively anaerobic; in soil and fresh water, parasitic or saprophytic.
Micrococcus Cohn 1872 (nonpathogenic; Precambrian, Eocene, Holocene); *Staphylococcus* Rosenbach 1884 (pathogenic).

17. Family Streptococcaceae Deibel & Seeley 1974
Gram-positive cocci that form pairs, tetrads, or chains, mostly nonmotile; facultative anaerobes, fermentative metabolism, chemo-organotrophs producing lactic, acetic, and formic acids, ethanol and CO_2 from carbohydrates, saprophytes in milk involved in cheese-making; commensals and pathogens.
Streptococcus Rosenbach 1884.

18. Family Peptococcaceae Rogosa 1971
Gram-positive cocci, including those that digest cellulose in digestive tract of Ruminantia.

19. Family Bacillaceae Fischer 1895
Endospore-forming rods and rarely cocci, mostly Gram-positive; motile with lateral or peritrichous flagella, or nonmotile; endospores with central core, median layer, and outer spore coat; aerobic, facultative, or anaerobic; chemoorganotrophs, some fix atmospheric nitrogen; mostly saprophytic, but some important pathogens; range from desert soils to marine. Precambrian to Holocene.
Bacillus Cohn 1872; *Clostridium* Prazmowski 1880 (some N-fixing).
Precambrian genera: *Biocatenoides* Schopf 1968; *Eobacterium* Barghoorn & Schopf 1967; *Rhicnonema* Hofmann 1976.

20. Family Lactobacillaceae Winslow et al., 1917
Gram-positive rods, non-spore-forming; produce lactic acid.

21. Family Corynebacteriaceae Lehmann & Neumann 1907
Gram-positive rods, irregular outline or club-shaped; non-spore-forming; parasitic and pathogenic in plants and animals.
Arthrobacter Conn & Dimmick 1947; *Corynebacterium* Lehmann & Neumann 1896.

22. Family Propionibacteriaceae Delwiche 1957
Gram-positive rods, non-spore-forming; ferment sugars and lactic acid to produce pro-

(*continued*)

pionic and acetic acids and CO_2; commensal in ruminants.
Eubacterium Prévot 1938.

G. Order Actinomycetales Buchanan 1917

Gram-positive, elongate cells may branch as mycelia that resemble fungi but cell wall not of chitin or cellulose; hyphae usually 1.0 μm or less in diameter, up to maximum of 1.5 μm; may form spores in sporangia. In soil or fresh water or dairy products, symbiotic and N-fixing in plants, or parasitic in animals. Middle Precambrian to Holocene.

1. Family Actinomycetaceae Buchanan 1918
Actinomyces Harz 1877; *Archaeotrichion* Schopf 1968 (Precambrian).

2. Family Mycobacteriaceae Chester 1897
Mycobacterium Lehmann & Neumann 1896 (causes tuberculosis).

3. Family Frankiaceae Becking 1970

4. Family Actinoplanaceae Couch 1955

5. Family Dermatophilaceae Austwick 1958

6. Family Nocardiaceae Castellani & Chalmers 1919

7. Family Streptomycetaceae Waksman & Henrici 1943

Readily isolated from air to soil; source of streptomycin.
Streptomyces Waksman & Henrici 1943.

8. Family Micromonosporaceae Krasil'nikov 1938

H. Order Rickettsiales Gieszczkiewicz 1939

Gram-negative rods or cocci, bacterialike walls, no flagella; obligately intracellular in eucaryotic cells, parasitic or mutualistic, may be pathogenic (e.g., typhus), with insects as primary host; divide only within host cells.

I. Order Chlamydiales Storz & Page 1971

Obligate intracellular parasites, Gram-negative cocci.

J. Order Mycoplasmatales Freundt 1955

Syn.: Class Mollicutes Edward & Freundt 1967

Gram-negative, nonmotile, no endospores, single triple-layered membrane, no true cell wall; reproduction by fragmentation of filament into coccoid, filterable elements, down to 0.2 μm in diameter. Parasitic to pathogenic in mammals and birds; also saprophytic. May prove to be eubacteria that have lost ability to form cell wall, as by treatment with penicillin.
Mycoplasma Nowak 1929.

"Photobacteria"
Includes the photosynthetic bacteria and the procaryotic algae.

K. Order Rhodospirillales Pfennig & Trüper 1971

Syn.: Rhodobacteria Molisch 1907.

Photosynthetic, containing the pigments bacteriochlorophyll *a, b, c,* or *d,* and carotenoids; photosynthesis uses as oxidizable external electron donors the reduced sulfur compounds, molecular H_2, or organic compounds; does not produce oxygen; some N-fixing; mostly aquatic; Gram-negative, cells as coccus, rod, vibrio, or spirillum, 0.3 to 6 μm in diameter.

1. Family Rhodobacillaceae Copeland 1956

Syn.: Athiorhodaceae Molisch 1907; Rhodospirillaceae Pfennig & Trüper 1971

Anaerobic, non-sulfur-depositing purple and brown bacteria.
Rhodomicrobium Duchow & Douglas 1949.

2. Family Chromatiaceae (Migula) Bavendamm 1924

Syn.: Rhodobacteriaceae Migula 1895; Thiorhodaceae Molisch 1907

Purple sulfur bacteria, aerobic.
Chromatium Perty 1852; *Ectothiorhodospira* Pelsh 1936.

3. Family Chlorobiaceae Copeland 1956

Syn.: Chlorobacteriaceae Geitler & Pascher ex van Niel 1948

Anaerobic, green sulfur bacteria, nonmotile, producing irregular or regular gelatinous colonies.
Chlorobium Nadson 1906.

Division Cyanophyta

Blue-green algae, Cyanobacteria (see Table 1.3).

Division Prochlorophyta

Green-pigmented "blue-green" algae (see Table 1.4).

CYANOPHYTA OR BLUE-GREEN ALGAE

Blue-green algae, variously termed Cyanophyta, Myxophyceae, Cyanophyceae, and Cyanobacteria, are clearly separated from all other algae by their low state of cell differentiation. The cells are commonly from 2 to 4 μm in size, although some may be larger and visible to the naked eye. They may occur singly, as chains of cells (**trichomes**), or as a sheet of cells (**thallus**); **filaments** consist of one or more trichomes enveloped in a gelatinous sheath. The colonies of cells and filaments may form gelatinous, leathery, mealy, or stony masses, and may be free-floating or attached to a substrate.

The Cell

As was true of the bacteria, no membrane-limited organelles occur within the cell, although a variety of vesicles and inclusions may be present (see Figure 1.15).

The cytoplasm is surrounded by a plasma membrane (**plasmalemma**), from which folds may penetrate the cytoplasm. The infolding plasmalemma has been suggested to give rise to the new pigment-containing thylakoid membranes, by budding off vesicles that elongate into the thylakoids and then are distributed in the cell. However, the complex folding prevents positive proof of this origin of the thylakoids.

The Wall

External to the plasmalemma is a thin firm four-layered wall, similar in structure and composition to that of Gram-positive bacteria. It is composed of murein (also known as mucopeptide, peptidoglycan, and glycopeptide), together with carbohydrates, amino acids, and fatty acids, whereas eucaryotic plants have a main wall component of polysaccharide, particularly cellulose. The four wall layers were observed in electron microscopy, and numbered L_I, L_{II}, L_{III}, and L_{IV}. Each is about 10 nm in thickness. L_I is just outside the plasmalemma,

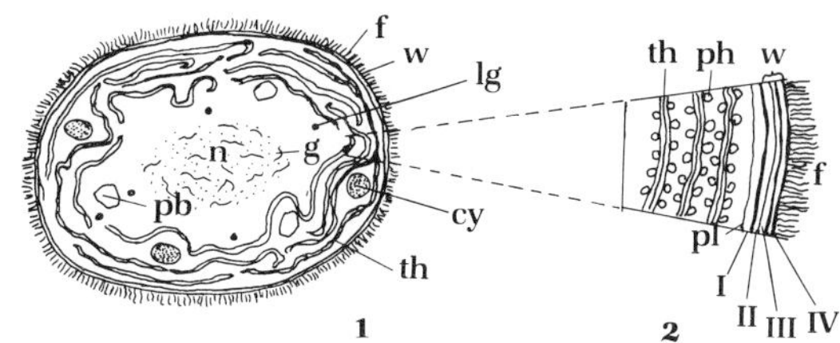

Figure 1.15
Diagram of sectioned unicellular blue-green alga, such as *Synechococcus,* based on information from electron microscopy. **1.** Entire cell, showing n, nucleoplasm; g, genophore of DNA fibrils with associated ribosomes; th, thylakoids; cy, cyanophycin granule (structured granule); lg, lipid granules; pb, polyhedral bodies (polyphosphate granules); w, wall; f, fibrous mucilage sheath; approximately ×7500. **2.** Enlargement of part of outer section, showing th, thylakoids; ph, phycobilin granules; pl, plasmalemma; w, wall of four layers (I, II, III, IV); exterior with f, fibers of mucilage; approximately ×25,000.

is electron transparent, and of variable thickness, hence has been suggested to be an artifact, and to represent only the space between the next layer and the plasmalemma. L_{II} is electron opaque, and L_{III} is electron transparent and composed of parallel fibrils, arranged to form a right-handed helix in *Oscillatoria princeps* Vaucher; contractions of these fibrils are regarded as the dominant factor in the gliding motility of the blue-greens, as discussed later. The outermost wall layer, L_{IV}, appears double tracked in section. All wall layers have a globular substructure.

The Mucilage Sheath

The wall is surrounded by a mucilage sheath, which may be almost amorphous and difficult to distinguish unless some form of dye is added to the medium; others have a mucilage sheath with very sharp outer boundary. Martin and Wyatt (1974) recognized three separate types of sheath in blue-green algae as seen in the light microscope: (1) The sheath may reflect the shape of the cells and be visible without staining. It may be adhesive and produce a tangled mass of cells and filaments, particularly in terrestrial species such as *Lyngbya.* (2) Other species, such as some *Nostoc,* seem to have broad amorphic slimy shrouds that evenly surround the cells or filament, becoming progressively less dense outward. Produced intermittently, it allows more dispersed growth, particularly in a liquid medium. (3) The third type is intermediate, consisting of slimy sheaths with a sharp outline, forming a cloudlike appearance. Most aquatic types, especially those of *Nostoc,* have a sheath of this form.

In electron microscopy the sheath is seen to be constructed of fibers held together in a mucilage. The orientation of the fibers may differ in different species, although the method of fixation may result in some variation, as seen in electron microscopy. The fibrils arise from the cell wall, but are not firmly attached, as the trichomes may turn within the sheath. Some *Lyngbya* have a three-layered sheath, the inner and outer layers consisting of long intertwining parallel fibrils, and the median layer consisting of disorganized fibrils. In some *Nostoc* and *Calothrix,* the fibrils form a reticulum (Lang, 1968).

The prominent sheath of the blue-green algae probably aids survival during desiccation, particularly of those species that do not produce resistant spores. Among aquatic species, mucilage production varies with the degree of oxygenation of the water. *Microcystis aeruginosa* Kützing produces more mucilage under turbulent conditions or with artificial aeration in laboratory culture, and produces less sheath material under low oxygen tension, as at the bottom of a lake (Whitton and Sinclair, 1975).

The mucilage is a complex polysaccharide, that of various species having one to three distinctive sugars (arabinose, galactose, glucose, mannose, rhamnose, xylose) and glucuronic acid. In culture, *Anabaena cylindrica* Lemmermann formed a complex of the slime envelope with iron, using 10 percent or more of the iron in the medium, the iron content giving a brown color to the sheath. In addition, the iron content may aid the preservation of the sheath in fossils, as the mucous envelopes appear to have better chance of preservation than the cell walls alone (Golubic and Hofmann, 1976).

Nucleoplasm

The nuclear region consists of a colorless central body termed the **nucleoplasm**, DNA-plasm, chromatin-apparatus, or **centroplasm**, equivalent to the bacterial genophore. No nuclear membrane or nucleoli are present, although this region stains in a manner like that of the nucleus of eucaryotes. Unlike the true nucleus, this central body may store reserve food materials. Except in the smallest cells, the nucleoplasm is not sharply differentiated and commonly interfingers with the peripheral chromatoplasm that contains the pigments (Wolk, 1973). When preparations of cells are fixed by Kellenberger fixative, the nucleoplasm is shown to consist of fine fibrils of an average diameter of 0.2 to 0.3 nm, as in the bacteria.

Unlike the mitosis of eucaryotes, cell division of blue-green algae involves only simple fission of the nuclear material. Fission of the nucleoplasm may precede cell division, so that cells of *Oscillatoria amoena* (Kützing) Gomont may contain from one to eight separate chromosomelike structures, depending on the position of the cell within the filament. Larger blue-green cells commonly have a number of these individual rodlike structures, suggested to indicate the presence of several complete genetic complements, or **genomes,** per cell. Such polyploidy may explain the ultraviolet resistance of the blue-greens and the difficulty of obtaining recessive mutants.

Agmenellum quadruplicatum (Meneghini) de Brébisson, a marine species, has a base composition and kinetic complexity very similar to the bacterium *Escherichia coli* (Roberts et al., 1977). The DNA G+C is 48 percent in *Agmenellum* and 50 percent in *E. coli.* A variety of experiments (buoyant density, thermal denaturation, and measurements of viscoelastic retardation time of the DNA) demonstrated that *Agmenellum* contains two or three copies of its genome, even at a slow rate of growth (20-hour generation time), probably as separate genome-sized molecules, thus substantiating the possible polyploidy that had been postulated.

About 5 percent of the cell DNA in *Agmenellum* was shown in electron microscopy to be in the form of closed circles like bacterial DNA (Roberts and Koths, 1976). The circles were in at least six size classes (the bacterium *E. coli* has up to five size classes), the smaller ones being most abundant. The function of the circular DNA or the possible existence of sequence homologies between the size classes is unknown.

Pigments, Thylakoids, and Storage Products

The distinctive pigments of the blue-green algae are not contained in a true membrane-limited plastid like that of eucaryotic plants. Instead, much of the cell is filled with **thylakoids**, lamellae composed of two membranes joined at the ends and enclosing an intrathylakoidal space of variable width, thus forming a closed, flattened sac. In cross section, the usual appearance is of two opaque lines separated by a transparent space, but in electron microscopy the thylakoids are seen to have a layer with a structure of globular particles from 5 to 7 nm in diameter, a lamellated homogeneous part, and a layer of 10 to 20 nm particles (Lang, 1968).

The Chroococcales have thylakoids in rows paralleling the longitudinal wall and close to the cell periphery, as do the nonspiral Oscillatoriaceae. In other blue-greens, the thylakoids may be highly distorted and distributed throughout the cell, as in many nostocaceans. The thylakoids are the site of photosynthesis in the cell, and contain the photosynthetic pigments. The amount of chlorophyll in the cell is directly proportional to the amount of thylakoid membrane present. As thylakoids also are present in cells grown in the dark (hence that do not photosynthesize), they probably also are the site of respiration (Fogg et al., 1973), serving as a mitochondrial equivalent. At the outer surfaces of the thylakoids are numerous small granules (about 35 nm in diameter), the **phycobilisomes**, which contain aggregates of the phycobiliprotein pigments that are characteristic of the blue-greens.

The common pigment of all green plants, chlorophyll *a*, is present in the Cyanophyta (in contrast to the bacteriochlorophylls of the other photosynthetic procaryotes). The chlorophyll *a* has different molecular forms: chlorophyll *a* 680, a long-wave form associated with photosystem I (producing ATP), and chlorophyll *a* 670, a short-wave form associated with the water-splitting and oxygen-producing photosystem II. In addition to the chlorophyll, there are four different carotenoid accessory pigments and three kinds of phycobiliprotein, the blue allophycocyanin and phycocyanin, and the red pigment phycoerythrin. The usual bluish-green color results from the combined phycocyanin and chlorophyll, but combinations of some of the other pigments may result in a variety of colors. The photosynthetic activities of the blue-green algae result in the production of oxygen as in other plants but unlike the bacteria, as well as the formation of sugars and glycogen. Glycogen is a carbohydrate food reserve commonly found in fungi and animals, but not in green plants which store food reserves as starch.

The fatty acids present in the cells of the blue-green algae well demonstrate the intermediate character of this group. All photosynthetic eucaryotes contain polyunsaturated fatty acids as components of the lipids in the plastids. In contrast, both photosynthetic bacteria and nonphotosynthetic ones have either saturated or monounsaturated fatty acids. The difference was once regarded as related to the oxygen production of the eucaryotes, but the blue-green algae refute this suggestion. Blue-greens appear to include two types. One resembles the bacteria in having very little (maximum of 4 percent) or no polyunsaturates, and consisting mostly of the rodlike unicellular species such as *Synechococcus.* Most filamentous species and a few unicellular ones have a high percentage of polyunsaturated fatty acids, ranging from 27 to 50 percent of the total fatty acids of the cell. This group includes species of *Anabaena, Calothrix, Chlorogloea,* and *Oscillatoria* (Kenyon and Stanier, 1970). The unicellular strains, probably including representatives of

Aphanocapsa, Chroococcus, and *Gloeocapsa,* showed both types of composition; however, the two groups not only differed in fatty acid composition, but also in the nature of the DNA base composition and the general cell size. Those with none or a low amount of polyunsaturated fatty acids had the largest cell size, ranging from 3 to 6.8 μm in diameter, and had a DNA percent G+C of 34.7 to 38.2. The other group had a high polyunsaturated composition, cell diameters of 2.2 to 3 μm, and a DNA percent G+C of 46.9 to 47.2. This group appears to be the only known blue-green algal representatives with a lipid composition and DNA base composition similar to that of the eucaryote plastid, hence is of interest in connection with the theory of the endosymbiotic origin of eucaryote plastids.

Although generally the more complex blue-green algae have the higher percent of polyunsaturated fatty acids, at least one exception is known. *Hapalosiphon* is a filamentous form, possesses akinetes and heterocysts, and shows true branching, yet has only monounsaturated fatty acids (Holton et al., 1968). This hot-springs alga was formerly suggested to be a relict of former generally warmer temperatures on earth, but may also be a more advanced form that has lost the capability of producing the polyunsaturates.

The thylakoids thus are involved in photosynthesis and respiration, the enzymes on these membranes combining the functions of such separate eucaryote organelles as the plastids, mitochondria, dictyosomes, and endoplasmic reticulum.

Vesicles, Granules, and Inclusions

Outside the central body, the remainder of the cell contains, in addition to the thylakoids, various protein granules and gas and oil vesicles.

The **cyanophycin granules,** also termed **structured granules,** are irregularly spherical or polyhedral in shape, colorless when unstained, but stain densely with osmium; peripheral in position, they serve for storage of nitrogen. Present in vegetative cells, they are particularly prominent in the resting cells, or akinetes.

Polyphosphate bodies are smaller, spherical, and colorless, but become black when stained for polyphosphates. They are located near the center of the cell and are active in storage of phosphorus.

Polyhedral bodies are angular in shape and show medium electron density. They are the carboxysomes, storing the enzyme that is essential in CO_2 fixation in photosynthesis. They are common in the area of the nucleoplasm.

Ribosomes are granules 10 to 15 nm in diameter, common in the cytoplasm away from the thylakoids, but particularly in the central region, near the DNA fibrils. They are believed to contain RNA (Fogg et al., 1973).

Gas vesicles are proteinaceous, tubular in shape, and may occur in large numbers. They disappear under pressure. As the membranes of these vacuoles are permeable to nitrogen, oxygen, and argon, they probably are not for gas storage, but may provide light shielding as well as aiding buoyancy. Lipoid droplets may also be present in the cells.

Growth and Reproduction

Neither gametes nor flagellated zoospores occur in the Cyanophyta. Cell division occurs by binary fission, the varying planes of division resulting in filaments, sheets, or packets of cells.

As might be expected in these primitive forms, the rate of reproduction is rapid, the maximum known rate being 12 doublings per day in the coccoid *Anacystis nidulans* (Richter in Wittrock & Nordstedt) Drouet & Daily. In conditions favorable for their growth, as in warm eutrophic lakes and ponds, dense water blooms may result. Species characteristic of such blooms include *Aphanizomenon flos-aquae* (Linnaeus) Ralfs and *Microcystis aeruginosa,* or the species that is an indicator for increased eutrophication, *Oscillatoria rubescens* De Candolle. Because gas vacuoles are present in the cells, many bloom-forming species are buoyant and may rise to the surface in great numbers, forming a scum or mat. The gas normally allows the cell to rise toward the surface, where increased light causes a reduction in the gas vacuoles, allowing the cell to

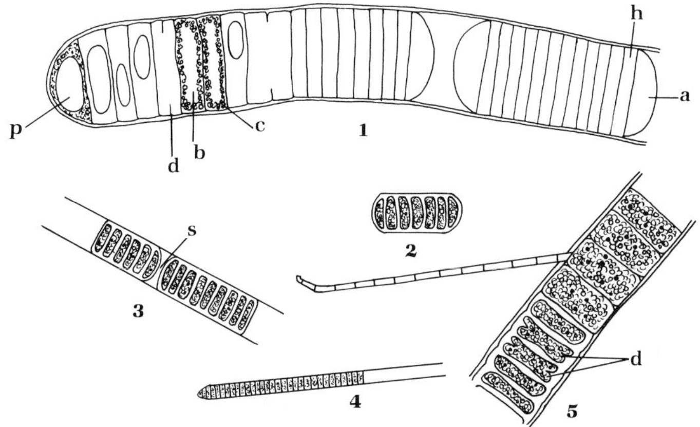

Figure 1.16
Oscillatoria sp. **1.** Parent filament at left; h, developing hormogonium; a, apical cell; p, pseudovacuole; d, division of cells in progress; b, central body; c, chromoplasm. **2.** Isolated hormogone. **3.** Part of filament showing s, separation disc. **4.** Terminal end of filament. **5.** Parts of filaments, one with enlarged cells, and some in process of division (d). 1,5 ×725; 2,3 ×325; 4, ×75.

sink, the continuous vertical migrations making available the nutrients from a greater water volume.

Surface blooms prevent the normal process of vertical migration and the algae are stranded at the surface where they may eventually die.

Filaments, or Trichomes, and Hormogonia

The trichome is a filament, or cylinder of cytoplasm, partitioned by cross walls, and may be encased in a mucilage sheath. Cell division occurs by median constriction of a cell, the two inner wall layers L_I and L_{II} growing inwards in ringlike fashion, pushing the thylakoids inward and gradually cutting across them as the new septum is completed. Commonly division occurs simultaneously in many or all cells of a trichome, and when growth is rapid, each cell in a filament may contain up to seven partly completed walls. Possibly the cross walls may not completely close at the center, leaving fine interconnections between adjacent cells, termed **plasmodesmata.**

Filaments reproduce vegetatively by fragmentation, either at random or at separation disks, which may be atrophied cells, into segments termed **hormogonia** (see Figure 1.16). The hormogonia, chief method of reproduction in the Nostocales and Stigonematales, are capable of movement. When they come to rest, they lengthen by repeated cell divisions to form a new filament. Certain motile forms, such as *Oscillatoria*, may be regarded as permanent hormogonia which do not become attached.

Soon after separation into hormogonia, the terminal cells of the broken end take on the spherical, conical, or tapered shape characteristic of the species.

False and True Branching. Although some filamentous cyanophytes (Nostocales) may not branch, **true branching** occurs in the Stigonematales. An intermediate cell of a trichome divides in a plane parallel to the axis of the trichome, and continues to divide to produce a branch, as in *Stigonema* or *Hapalosiphon* (see Figure 1.17).

In other cyanophytes, **false branching** occurs. The growth and cell division of a trichome may be more rapid than that of the sheath, causing it to buckle, break through the sheath,

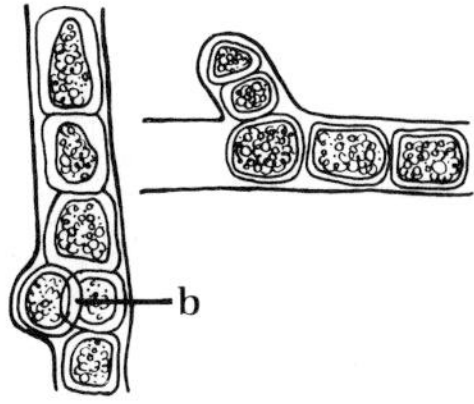

Figure 1.17
Hapalosiphon sp. Portions of filaments, showing formation of b, a true branch; ×650.

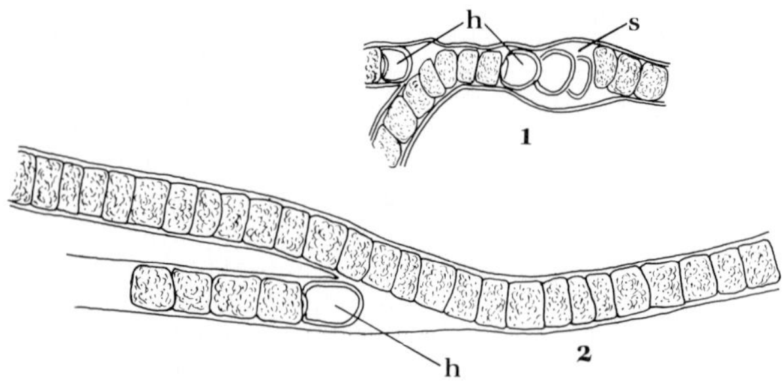

Figure 1.18
Tolypothrix sp. **1.** Filament with sheath (s) beginning to buckle at heterocysts (h), leading to false branching, ×550. **2.** False branching at heterocysts (h), ×650.

and finally break apart, the two ends continuing on as separate trichomes. In other species, particularly those with heterocysts, the filament may break adjacent to the heterocyst, and the free end project through the sheath to appear as a false branch, as in *Tolypothrix* (see Figure 1.18). A modification of the false branching in the Mastigocladaceae results in a V- or Y-like appearance. Here the filament growth is rapid, causing a buckling and breakage, with elongation of the two ends producing a V, or with one continuing to elongate, and appearing Y-like.

Sheets and Packets of Cells

Some blue-green algae remain as single unicells and do not produce filaments. When dividing, a new wall may begin to form in these before the previous division has completed separation, resulting in packets of cells. Depending on the succession of divisions, the result may be a flat sheet of cells, or three-dimensional groups, which may be held together by the mucilage sheaths after cell separation is completed (see Figure 1.19).

Division of the unicellular forms occurs by constriction of the wall as in the filaments, but in these all layers are involved, and the newly formed septum splits the two daughter cells apart.

Akinetes, or Resting Spores, and Endospores

Some cells develop much-thickened walls and become resting spores, or **akinetes** (see Figure 1.20). The spherical, oblong, or cylindrical akinetes are crowded with food reserves and have a firm, two-layered membrane. The outer layer may be sculptured. Some species produce akinetes under unfavorable conditions, such as a nitrogen deficiency, whereas others form akinetes upon attaining a certain size. The akinetes germinate to produce a new filament (see Figure 1.21).

Endospores are commonly formed by species that do not develop hormogonia. As they germinate immediately, they are not regarded as a resting stage comparable to the akinete. The cell that produces endospores be-

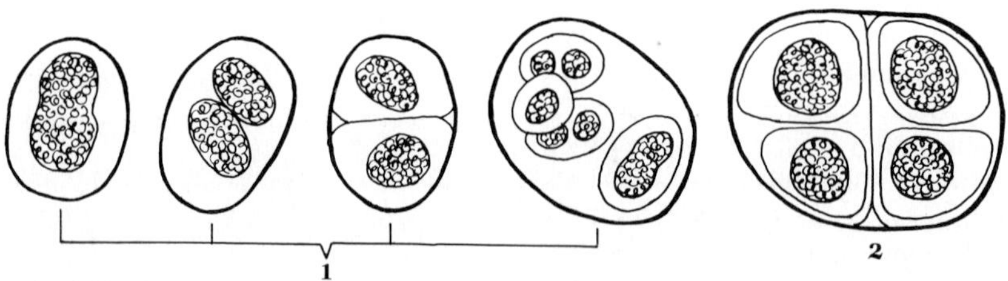

Figure 1.19
Chroococcus sp. **1.** Sequence of cell division, ×1000. **2.** Four-celled colony, ×2200.

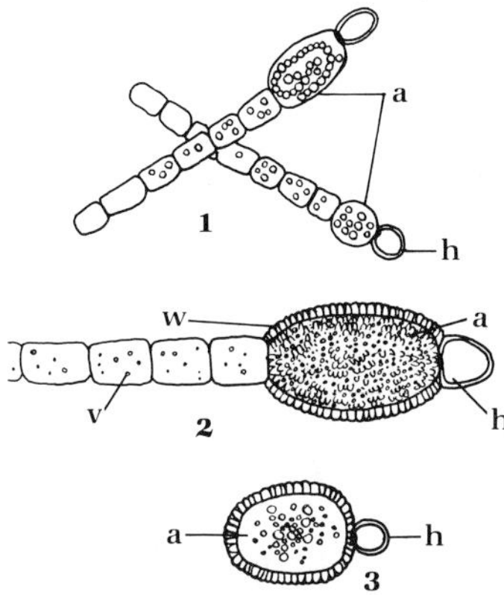

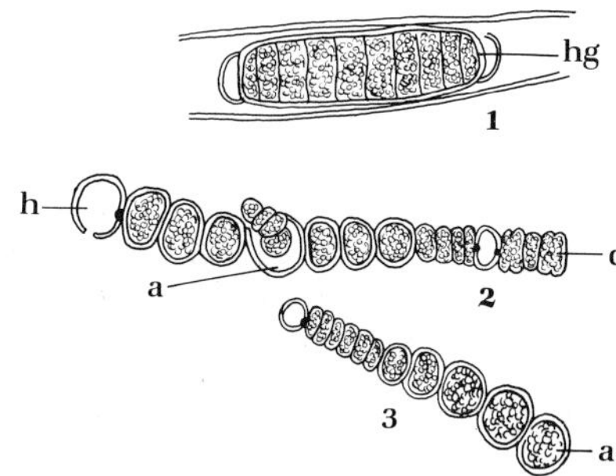

Figure 1.20
Cylindrospermum sp. **1.** Filaments with a, young akinetes; and h, terminal heterocyst; ×625. **2.** Terminal part, enlarged, ×750, showing v, vegetative cells; a, mature akinete with thick wall (w); and h, terminal heterocyst. **3.** a, mature akinete; and h, heterocyst; ×725.

Figure 1.21
1. *Lyngbya* sp., showing hg, the short hormogone, ×425. **2,3.** *Anabaena* sp. ×325. **2.** Showing h, heterocyst; a, akinete in process of germination; and d, dividing cells. **3.** Filament with a, akinete, and terminal heterocyst at opposite end.

comes somewhat enlarged, and its contents divide successively in three planes. The rounded spores are liberated and later germinate to form new colonies.

Heterocysts

The Nostocales (except the Oscillatoriaceae) and Stigonematales contain a distinctive structure, the **heterocyst** (see Figure 1.22), which arises from a somewhat enlarged vegetative cell, has a thick refractive two-layered wall, and rather hyaline, yellowish contents. A conspicuous pore is present at one or both poles, depending on whether the heterocyst is terminal or intercalary in position within the filament. They may be solitary or occur in series, and commonly are regularly spaced along the filaments. They contain no phycobilisomes and no polyglucan granules; the thylakoids are tightly packed, with reduced interlamellar space, and grade into the contorted lamellae near the poles, which may break down to form a polar plug in the pore channel (see Figure 1.23).

The function of the heterocysts was long uncertain; suggested functions include that of a storage organ; a specialized reproductive structure that is functionless except on rare occasions; some relationship with spore formation, as they commonly are associated with akinetes on the filaments; or a relationship with growth and cell division. The usual presence of heterocysts in nitrogen-fixing species, and many laboratory experiments to confirm this association, suggest that they are the site of nitrogen fixation (Fay et al., 1968; Ogawa and Carr, 1970). As the isolated heterocysts do not fix nitrogen, those within the filaments apparently obtain some benefit from adjacent vegetative cells, with which they appear to interconnect. Cultures of *Anabaena cylindrica* grown without combined nitrogen available have well-developed heterocysts, whereas those grown in the presence of combined nitrogen have only proheterocysts, recognizable but not fully developed. If these filaments are later transferred to a nitrogen-deficient medium, the heterocysts mature fully (Wilcox, 1970).

The yellowish color of the heterocysts results from the absence of bilin pigments and smaller amounts of chlorophyll *a* and carotenoids than in the usual vegetative cell. Sufficient chlorophyll *a* and β-carotene are present to have an active photosystem I (as in bacteria), but the absence of the phycobilisomes with the pigment phycocyanin prevents the action of photosystem II, hence there is no CO_2 fixation

Figure 1.22
Nostoc sp. **1.** Short filament with
terminal heterocyst. **2.** Part of a
squashed *Nostoc* ball with many
filaments. **3,4.** Mature akinetes. 2,
×150; others, ×645.

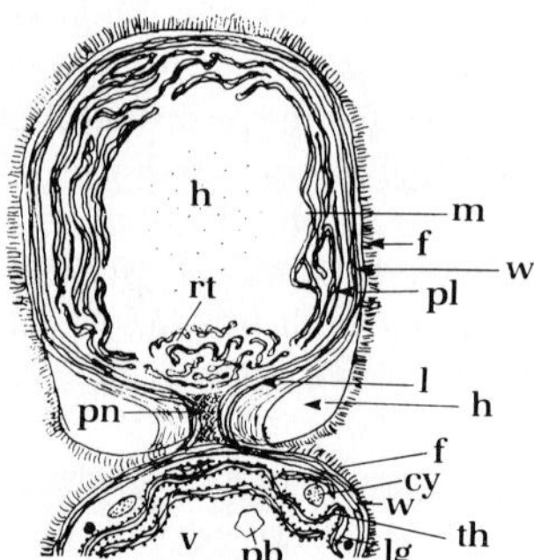

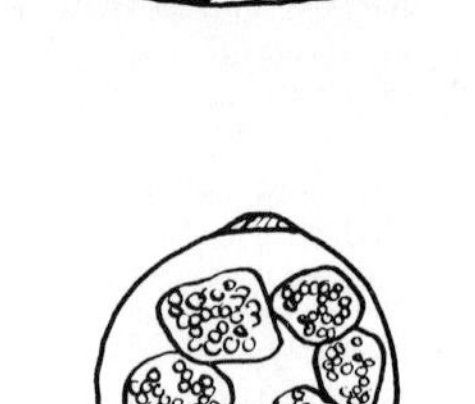

Figure 1.23
Diagrammatic view of a mature terminal
heterocyst (h), with a few scattered ribosome
granules; m, peripheral membrane system,
lacking phycobilisomes; rt, reticulate lamellae
near pore channel; pl, plasmalemma; w, four-
layered wall; f, fibrous mucilage; pn, osmiophilic polar nodule that plugs the pore
channel; l, laminated layer; h, homogeneous
layer. Heterocysts lack most of the inclusions
and granules such as occur in adjacent vegetative cell (v): pb, polyhedral body; lg, lipid
granule; cy, cyanophycin granule; th,
thylakoids with phycobilisomes; w, four-
layered wall; f, fibrous mucilage. Approximately ×7500.

or O_2 production (Fogg et al., 1973). The heterocysts thus obtain their required organic carbon from adjacent vegetative cells. If the adjacent cell to a terminal heterocyst develops into a mature spore, thus severing the intercellular connection or plasmodesmata, the heterocyst dies. If a terminal heterocyst dies, the adjacent cell develops into a heterocyst.

Mutagenesis and Genetic Recombination

Cyanophytes lack a true sexual reproduction, this absence having been regarded as the reason for their relatively slow evolutionary rate, at least as indicated by morphologic differentiation. Studies of the DNA in blue-greens indicate that they are polyploid, containing at least two or three copies of the genome and making them highly resistant to UV radiation or other causes of mutation, thereby also hindering studies of possible genetic recombination.

Nevertheless, some genetic transfer has been demonstrated experimentally in the cyanophytes. As might be expected, this is closer to such processes in bacteria than to those of eucaryotic organisms, and includes examples of both recombination and transformation. Cyanophage viruses are known to infect species of *Lyngbya, Phormidium,* and *Plectonema,* but as yet transduction like that of bacteria has not been recognized in the blue-greens.

Mutants largely have been restricted to varied resistance to antibiotics; whether the mutation is "nuclear" or extranuclear is uncertain. Mutations obtained from blue-green algae include streptomycin-resistant strains that are 100 to several 1000 times more resistant than the wild type (Bazin, 1968), as well as those resistant to other antibiotics. Morphologic mutants are less common, but a filamentous colony, with twisted, distorted, and exceptionally long filaments (up to 15 times as long as normal) was produced by *"Anacystis nidulans"* (= *Synechococcus elongatus* Nägeli). Other mutants had varied growth forms and modification of cell division. In treatment with a strong alkylating agent, the sheetlike *Agmenellum quadruplicatum* produced nonseptate filamentous mutants that did not revert to the coccoid form. Whether these variations were the result of a single mutation or a series of these is uncertain. Another type of mutant developed a blockage in the process of nitrate assimilation, hence required reduced nitrogen in the medium. Pigment mutants have been obtained whose absorption spectra differed from those of wild cells, but their interpretation is hindered by the wide normal variation in pigmentation with differing light intensity, light wavelength, or availability of nutrients. Possible mutants with UV resistance, non-nitrogen-fixing mutants, apochlorotic (colorless) strains, photosensitive strains able to grow only in the dark, and temperature sensitive mutants also have been obtained experimentally (see reviews in Houghton, 1972; van Baalen in Carr and Whitton, 1973).

Genetic transformation was reported by Shestakov and Khyen (1970) in mutants of *"A. nidulans"* that were resistant to either streptomycin or erythromycin and a filamentous mutant that was sensitive to both antibiotics. DNA was collected from the antibiotic-resistant algal mutants in culture by various chemical treatments and centrifugation. When this donor DNA was added to selected recipient wild type cells that were sensitive to the antibiotics, some recipient cells became resistant. The frequency of resistant cells was not as high as occurs in bacterial transformation, but was more than 10,000 times more frequent than would occur by spontaneous mutation to resistant strains. Erythromycin resistance was transformed on both streptomycin resistant and filamentous strains. As the transformation was low in the dark, but greatly enhanced in the light, the uptake of DNA may depend on the energy supplied by photosynthesis.

Genetic recombination also was shown to occur in *"A. nidulans"* by Kumar (1962), recombinants occurring in 1 of 100 million cells, roughly the same number as in recombinants of the bacterium *E. coli.* Similar recombination experiments were made with antibiotic resistant clones of *"A. nidulans"* by Bazin (1968). Those resistant to different antibiotics were mixed, and subcultures were obtained that were resistant to both, although at a low frequency. Recombination was also reported in another species, *Anabaena doliolum* Bharadwaja, in which two successive exposures to UV produced nonsporulating mutants (Singh, 1967). Progeny of mixtures of the two strains included some that could sporulate, with spores of two types, some like the parent and some like the wild type, suggesting a heterozygosity. The frequency of mutants was 1 in 27,500, as 16 nonsporulating colonies were obtained. After mixing, 20 of the 12,000 colonies sporulated; germination of the spores resulted in colonies that mostly did not sporulate, but three colonies did sporulate, indicating stable recombinants.

Anastomosis of filaments and cells has also been reported in *Nostoc,* although this has not been repeated experimentally. The evidence to date suggests that genetic recombination does occur in blue-green algae, although at a low frequency, and that favorable mutations in separate strains could be combined in this way.

Motility

Although no flagella are present at any stage of the life history of the blue-green algae, many species have a characteristic gliding motion. The gliding may progress in either direction, and may be accompanied by rotation, reversal of direction, flectional activity, and production of mucilage sheath. The rate of movement differs with the species, temperature, pH, viscosity of the medium, and light intensity, to a maximum of about 2.5 to 11.1 μm per second.

The mechanism of gliding has been suggested to be forcible sheath extrusion (similar to the slime production of motile pennate diatoms), but this could not explain the reversals of direction or the slippage of the sheath backward and forward over immobilized trichomes. Walsby (1968) suggested that the filament might be moved by an undulating surface, and that mucilage secretion was not the compelling force. In *Anabaena cylindrica* the abundant mucilage accumulated as transverse rings around cells that did not move. Time lapse photography showed that the rings were moved along the cell in either direction, hence the glid-

ing mechanism was spaced along the cell, not only at the ends. The region of secretion is restricted, but the gliding mechanism occurred along the entire cell length, suggesting that the filament used the mucilage as a substrate for movement, and that secretion was not the propelling mechanism.

Jarosch (1964) proposed that motility resulted from waves in helical fibrils on the cell surface; such fibrils in *Oscillatoria princeps* are arranged in parallel arrays, wrapped around the trichome in a right-handed helix with a 60° pitch (Halfen and Castenholz, 1970, 1971). The proteinaceous fibrils are about 5 to 8 nm in diameter, about 55 fibrils present in 1 μm (Halfen, 1973), and appear to correspond to a middle layer of the rigid cell wall. The fibrils are continuous over the surface of the trichome, irrespective of the cross walls of adjacent cells. Motility was suggested to result from the unidirectional traveling waves of bending of these fibrils acting against a solid substrate or the closely associated elastic sheath. The wavelike distortions demonstrated by electron microscopy suggest that several adjacent fibrils act together (Halfen, 1973).

Gliding movement is associated with a right-hand rotation of the trichome in relation to the direction of movement. When the rotation is impeded, progression ceases. If the sheath lies against a solid substrate, the trichome is propelled; if the sheath is suspended in liquid, the fibril movement causes the sheath to be displaced in a helical fashion. Reversal of movement results from a change in polarity and direction of wave propagation. *Oscillatoria animalis* Agardh ex Gomont was found to have similar fibrils, but these formed a 60° left-handed helix, and rotation of the trichome also is in a left-handed direction (Halfen and Castenholz, 1970, 1971).

Unicellular species and trichomes with more widely spaced cells could have such a fibrillar system over a single cell length, and species that do not rotate in gliding may have straight longitudinal fibers rather than helically wound ones.

Motility is dependent on at least minimum concentrations of calcium, as well as temperature limits slightly less than the limits required for cell growth. In some species it is also light dependent. Gliding may be either toward or away from the light, depending on the light wavelength and intensity.

Metabolism

The morphologically simple blue-green algae are highly diverse in their metabolic capabilities. Most are photosynthetic, although a few can grow in the dark, and some colorless species are known. Many are nitrogen-fixing, others produce vitamin B compounds, and some deposit calcium carbonate, iron, or sulfur granules.

As the nitrogen-fixing forms obtain nitrogen and carbon dioxide from the atmosphere, and require only light, water, and a few minor elements in addition, their nutritional requirements are the simplest of any organism.

Photosynthesis

Blue-green algal photosynthesis consists of two systems, the long wavelength Photosystem I and the short wavelength Photosystem II, in contrast to the bacterial photosynthesis, which has a single photoreaction. In blue-greens, energy is trapped by the accessory pigments phycocyanin and allophycocyanin, and probably also phycoerythrin (Photosystem II), and then transferred to the longer wavelength chlorophyll *a*, which produces the free oxygen by using water as the hydrogen donor in reducing CO_2 (Photosystem I). Blue-green photosynthesis is similar to that of eucaryotic plants and much advanced over that of the bacteria.

The primitive nature of the blue-greens is demonstrated by the fact that some may also have an anoxygenic photosynthesis. *Oscillatoria limnetica* Lemmerman can utilize sulfides (H_2S or Na_2S) as an electron donor for CO_2 photoassimilation, as do the photosynthetic bacteria, oxidizing them to elemental sulfur. It also can use water in the usual green plant photosynthesis. In nature, it grows in environments of high sulfide concentrations and is adapted to shift from oxygenic to anoxygenic photosynthesis as conditions fluctuate (Cohen et al., 1975a, b).

Nitrogen Fixation

Like some bacteria, some of the blue-green algae are nitrogen-fixing, nearly all being referrable to the Nostocales: *Anabaena, Cylindrospermum,* and *Nostoc* belong to the Nostocaceae, the planktonic marine *Trichodesmium* to the Oscillatoriaceae, and a few others belong to the Rivulariaceae and Scytonemataceae. *Chlorogloea,* a nitrogen-fixing form that is commonly unicellular, had been referred to the Chroococcales, but is known also to produce short filaments and to have heterocysts and hence is a nostocalean. *Mastigocladus* is also nitrogen-fixing, and here is placed in the Stigonematales, although a separate order has also been proposed for it.

As already noted, most nitrogen-fixing species have heterocysts, cells specialized as the site of nitrogen fixation. As the heterocysts lack chlorophyll *a,* allophycocyanin, and phycocyanin, the principal components of Photosystem II, they cannot produce oxygen. Photosystem I is active, however, and the ATP it generates promotes the reduction of nitrogen (Thomas, 1970). Nitrogen-fixation commonly occurs only in low oxygen environments, and the few nonheterocystous algae that are nitrogen-fixing do so under anaerobic conditions (Stewart, 1977).

Their ability to fix nitrogen allows blue-green algae to become dominant in lakes during the summer, after combined nitrogen has been depleted from the water by the spring blooms of green algae and diatoms. Heterocyst-bearing blue-greens appear first, and are later followed by other blue-greens that do not fix nitrogen. Even the nitrogen-fixing species require phosphates, and do not develop heterocysts in the absence of phosphate. In Lake Erie, two phytoplankton peaks (each about 900 cell/ml) occurred in 1967, the first in mid-June consisting of about 75 percent green algae, with the remaining 25 percent diatoms and non-heterocyst-bearing blue-greens. The second peak occurred in mid-August and consisted of 86 percent heterocyst-bearing *Aphanizomenon* and *Anabaena;* green algae and diatoms each represented less than 4 percent of the cells. Nitrate concentrations were highest in early June, but had decreased to nearly zero by the end of the green algal bloom (Ogawa and Carr, 1970). However, other factors also may be important in the dominance of blue-green algae, for if nutrients are extremely plentiful, the blue-greens generally surpass the green algae, probably as they are more efficient in obtaining CO_2 where this becomes limited by high productivity. Addition of CO_2, nitrogen, and phosphorus to a lake that had been dominated by the blue-green algae resulted in a shift to green algal dominance; either the addition of CO_2 or the reduction of the pH stimulated the chlorophytes (Shapiro, 1973).

Bacteria are presently important in nitrogen-fixation, particularly in the terrestrial environment, where they are symbiotic in legumes. However, nitrogen-fixing bacteria have very limited habitats, hence blue-green algae probably are more important on a worldwide scale, and were much more so in the pre-Cretaceous world, prior to the evolution of the angiosperms. They are particularly important in present freshwater lakes, and planktonic blue-greens also fix nitrogen in the seas.

Dichothrix fucicola Bornet & Flahault ex Bornet & Flahault is a heterocystous blue-green that is nitrogen-fixing in the open ocean. It was reported to be epiphytic on pelagic *Sargassum* in areas where the carbon to nitrogen ratios indicate nitrogen starvation (Carpenter, 1972). More common as a marine nitrogen-fixing alga is *Trichodesmium,* a planktonic form similar to (and possibly synonymous with) *Oscillatoria.* In the marine tropics its blooms frequently result in "red water." *Trichodesmium* lacks true heterocysts, but occurs in colonies of several hundred trichomes, with the centrally located trichomes being somewhat differentiated in having reduced pigmentation (as is true of heterocysts of other species). In experiments, they did not incorporate $^{14}CO_2$ in photosynthesis, hence produce no oxygen that would have deactivated the nitrogenase. If the central trichomes are exposed to oxygen by disruption of the colonies, nitrogen fixation ceases abruptly. Thus within the center of the large mass of trichomes, this oscillatoriacean can fix nitrogen without having heterocysts to protect the nitrogenase from oxygen. Wave turbulence would disrupt the masses of trichomes and restrict this activity, hence *Trichodesmium* blooms are in calm seas (Carpenter and Price, 1976).

The only other nonheterocystous nitrogen-fixing cyanophyte is the coccoid *Gloeocapsa*, whose thick mucilage sheath may serve to protect it from the inhibiting effect of oxygen, as do the thick-walled heterocysts of the filamentous species (Whitton and Sinclair, 1975).

Mineral Requirements

A few species have distinct mineral requirements. *Anabaena cylindrica*, commonly found in highly saline areas, is the first plant for which sodium has been found to be essential. The requirement is not merely an osmotic one, as various osmotically active substitutes were tried in cultures, but could not be utilized in place of sodium. Other blue-green algae also grow in sodium-containing media and appear to utilize sodium in place of potassium. Possibly the periodic blooms of the Cyanophyta in freshwater lakes in regions affected by civilization are related to sodium enrichment instead of the increased organic matter.

Like other algae, *Anabaena* and other blue-greens require traces of iron, molybdenum, copper, manganese, vanadium, and cobalt, the required amounts varying with the species. Copper concentrations that inhibit *Phormidium* will stimulate growth of *Oscillatoria.* For many, the limit between copper needs and toxicity is narrow.

pH Limitations

Nearly ubiquitous in their occurrence, the blue-greens are completely absent from highly acid waters below a pH of about 4, and show very little growth below pH 6. Alkaline conditions of pH 7 to 8.5 are optimal. The wide occurrence of blue-green algae in the Precambrian thus indicates that waters were probably not acidic, as had once been postulated in explanation of the absence of skeletal invertebrates.

Oxygen Limitations

While the blue-green algae produce oxygen in photosynthesis, they are not well adapted to high oxygen levels, and actually function best when oxygen is limited. Their tolerance of low oxygen allows masses of the colonies to overwinter on the bottom, whereas high oxygen levels are inhibitory, or even cause death of the cells by photo-oxidation. Species that can tolerate and survive high oxygen concentrations are characterized by an abundance of mucilage, which has been suggested as a defense mechanism (Whitton and Sinclair, 1975).

External Metabolites

Blue-green algae resemble bacteria in being important producers of organic compounds required by other plankton. Many synthesize vitamin B_{12} and may represent an important source of this vitamin in fresh water. Other important external metabolites include the polypeptides secreted by *Anabaena,* which render tricalcium phosphate more soluble and hence available for other organisms.

The small amounts of the external metabolites (vitamins, etc.) present in solution at any one time do not accurately indicate the production of these substances. Although released to the water in abundance, they are as rapidly utilized but can be identified by their accumulation in unialgal, bacteria-free cultures. Symbiotic algae in lichens supply not only the carbon source but also the nitrogen and thiamine required by the fungal member.

Calcium Deposition

Many freshwater blue-greens deposit lime, forming the travertine in lime-rich streams and around hot springs, and the calcareous lake balls of inland lakes. The lime is always deposited within the enveloping mucilage, never within the cell proper. Isolated crystals first appear in the slime, then clusters of crystals form and finally coalesce. This lime deposition may be related to their CO_2 fixation during photosynthesis, but other factors are also involved, as only certain species produce the calcareous deposits.

In a study of the travertine deposits of the Arbuckle Mountains, Oklahoma, Emig (1917) found three types of calcareous deposits; a banded travertine, a pisolitic form, and a cavernous type (see Figure 1.24). A variety of blue-green (*Oscillatoria, Lyngbya, Rivularia*), green (*Oedogonium, Trochiscia, Cladophora,* and *Haematococcus*), and xanthophyte (*Vaucheria*) algae were present in the travertine deposits at springs and at rapids and waterfalls in the streams. The banded travertine was least common, but some large masses were 30 to 90 cm thick. It contained only the blue-green *Lyngbya* and *Oscillatoria* and appeared to accumulate at the rate of about 1 mm per month.

The pisolitic travertine consisted of granules of 1 to 8 mm diameter that gradually enlarged to fuse in a homogeneous deposit. It included the same blue-greens, but *Haematococcus* also was present one year. The cavernous travertine is of less uniform structure and contains cavities where moss tufts occurred. *Vaucheria* seemed to be dominant, although a wide variety of blue-green and green algae also was present (Emig, 1917).

Iron and Sulfur Deposits

Similar in form and habitat to the bacteria, some blue-green algae (*Oscillatoria*) may contain sulfur granules, and the filamentous sulfur bacteria *Beggiatoa* and *Thiothrix* are sometimes included in the blue-green algae. Other species resemble the iron bacteria and deposit iron hydroxide in and upon the sheath (*Lyngbya, Anabaena, Chamaesiphon*), but rarely produce sedimentary iron deposits like those resulting from bacterial activity. *Paracapsa* is an exception that does form iron accumulations.

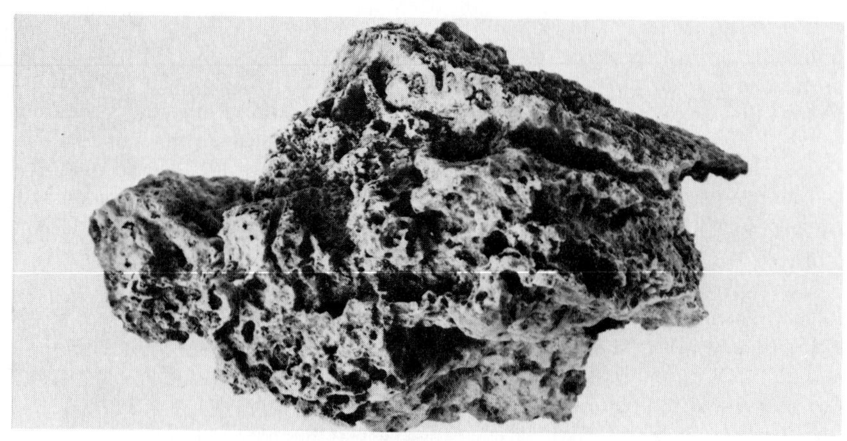

1

2

3

Distribution

The Cyanophyta are widely distributed. They may be aquatic, live on damp soil or rock, or occur within the soil to a depth of a meter. Aquatic forms include both benthic and planktonic species, largely in fresh water, but some occur in marine, intertidal, and salt marsh environments. They occur in hot springs, Antarctic lakes, soil crusts from the Antarctic desert, and other arid regions such as the Negev desert; some are symbionts in fungi, algae, bryophytes, cycads, or rhizopod protozoans; and they are one of the two partners of lichen associations.

With relatively simple nutritional requirements, they have persisted in a wide variety of habitats since early in the Precambrian. They are the first colonizers to appear on freshly eroded soil and on new volcanic surfaces or newly exposed coastal regions, both protecting the soil against further erosion and increasing its fertility. They are thus important pioneers ecologically, and geologically were probably the earliest colonizers of the land.

Endolithic and Soil Algae

Like the bacteria, a variety of blue-green algae are soil-dwelling or endolithic in extreme habitats. In the extremely dry Negev Desert of Israel, the soil algae may be **epedaphic** (living at the surface) or **endedaphic** (living within the soil); others are **hypolithic** (living on the under-surface of stones at the soil interface), **chasmolithic** (living in fissures or cracks in the rocks), or **endolithic** (within the rock itself). Both green and blue-green algae were present in each of these habitats where dew supplies the only moisture, which are as dry as any inhabited by lichens (Friedmann et al., 1967). Endedaphic algae included *Oscillatoria* and *Lyng-bya;* epedaphic ones (*Oscillatoria*) were present only after torrential winter rains. The greatest variety (*Chroococcus, Gloeocapsa, Lyngbya, Microcoleus, Nostoc, Schizothrix, Scytonema,* and *Tolypothrix*) were present under rocks or in fissures. The most extreme ecologic conditions on the present earth are those of the endolithic algae. Possibly they originally obtained access through cracks, later dispersal being by the slow growth of the algae. Algae were present between grains of sandstone, but may have bored into the limestones. Extensive layers of algae were produced a few millimeters below the surface, giving it a greenish appearance. The few species in this environment could not be cultured, but included tiny chroococcoid blue-greens and small unicellular green algae (Friedmann et al., 1967).

In another extreme habitat, that of the Antarctic deserts, extreme aridity is combined with low temperatures, but here also the alga *Gloeocapsa* was present in pores of the orthoquartzite, forming a greenish zone 2 to 3 mm thick, about 1 to 1.5 mm below the surface (Friedmann and Ocampo-Friedmann, 1976). Free-living *Nostoc* in Antarctica is sufficiently abundant in alkaline areas to form an "algal peat" 10 to 15 cm deep (Fogg et al., 1973).

In less extreme terrestrial habitats, the cyanophytes average from 5000 to 20,000 individuals per gram of soil. They are especially prevalent in moist, neutral to alkaline areas, in soil, on rocks, or on trees, and are among the earliest colonizers of new lava surfaces, being present on the Island of Surtsey off Iceland within five years after its volcanic origin (Fogg et al., 1973).

Fresh Water

Blue-green algae are particularly characteristic of fresh water, ranging in habitat from rapid

---

Figure 1.24
Travertine from Turner Falls on Honey Creek, Arbuckle Mountains, Oklahoma. **1.** Cavernous travertine, one-half natural size. **2,3.** Opposite sides of specimen showing banding due to the filamentous algae. Each beaded filament is a hollow calcareous tube, the beaded appearance reflecting cell divisions. Natural size.

mountain streams to stagnant ponds, where they occur on rocks, as epiphytes in shallow fresh water, or as plankton. Many prefer warm temperatures, as indicated by the flora of hot springs or the seasonal blooms of temperate species. Approximately 1300 species belonging to about 111 genera are recognized in fresh water (Bourrelly, 1970), although other systems place many of these in synonymy (Drouet, 1959, 1968, 1973; Drouet and Daily, 1956).

Aquatic species are completely absent from highly acidic environments (below about pH 4), grow well in neutral waters, but find optimal conditions in an alkaline environment rich in oxidizable organic compounds. They tolerate minimal concentrations of nitrogen and phosphorus.

A few species are toxic, particularly *Anabaena flos-aquae* (Linnaeus) de Brébisson, but also including some *Aphanizonemon* and *Microcystis.* Not all filaments of these species are toxic, hence some blooms may be toxic and others are not. In culture, toxin production was depressed when certain bacteria were present, particularly some of the Gram-negative Enterobacteriaceae (Carmichael and Gorham, 1977). The toxins may be lethal to fish and to other animals drinking the water. In addition, rice fields with heavy growths of *Anabaena* were found to be free of mosquito larvae, whereas other fields nearby that had no algae were infested with the insects. The algal toxins appeared to kill the larvae.

Hot Springs

Blue-green algae are more tolerant of high temperatures than are eucaryotic algae, occurring in neutral and alkaline hot springs up to temperatures of 73 to 75°C. In acidic waters, they tolerate temperatures only to 55 or 56°C, similar to the pH-limited temperature tolerances of bacteria, with which they may be associated in the hot springs (Brock and Darland, 1970). The only cyanophyte to live at such high temperatures is the unicellular *Synechococcus lividus* Copeland, which is associated with mats of filamentous bacteria in hot springs up to 62 to 75°C in the western United States, Japan, and

New Zealand. However, the optimal temperature for even these thermal strains is between 40 and 55°C.

Algal Mats and Stromatolites

In a littoral marine environment, certain Chroococcales and Nostocales may form conspicuous belts on the rocks. Other species of these, together with species of the Chamaesiphonales and Pleurocapsales, occur in shallow water or as pioneer elements in salt-marsh muds. Some Chamaesiphonales occur as marine epiphytes on other algae, and others are important in reef building. The algae form dense mats in which sediment particles are trapped, producing laminated organo-sedimentary deposits. Similar structures are found in sediments of all ages, and have been termed **stromatolites** (Kalkowsky, 1908); the sets of laminae or successive algal mats that form the stromatolite are **stromatoids** (Kalkowsky, 1908; Monty, 1977). Their form is controlled in part by the algal mat and in part by environmental factors. Generally, those in the intertidal area have been regarded as solely resulting from entrapped detrital sediment, whereas those of the supratidal to various subtidal depths may result in part from algal lime deposition.

Stromatolites in the Bahamas are produced by a nearly pure assemblage of the filamentous *Schizothrix calcicola* (Agardh) Gomont. The algal mats consist of filaments (of less than 1 μm diameter) held in a firm mucilage; this organic layer traps varied amounts of calcareous sedimentary particles. The laminated structure results from the noctidiurnal growth cycle; the organic lamina grows in the early part of the day, whereas the calcareous layer forms in the late day and night, the double layer formed daily being up to 600 μm thick (Monty, 1965). The same microstructure and general growth form of the stromatolites occurred from the low intertidal zone to 2 m below low tide in the shallow infralittoral zone. The laminations thus result from the periodicity of algal growth, which is most rapid in the early morning, slower in the afternoon, and ceases at night. In addition to the diurnal layering, growth fluctuates seasonally, with much of the annual accumulation taking place during July and August (Monty, 1965).

Additional blue-greens may be associated with *Schizothrix* in the algal mats. In the supratidal region, stromatolites of the *Scytonema* belt have a fine-grained matrix and clotty appearance resulting from the various heads of active lime precipitation, interspersed with calcified tubes, broken calcareous chips, and polygons. The terrestrial species *Scytonema myochrous* (Dillwyn) Agardh ex Bornet & Flahault is associated with *Schizothrix calcicola* in these supratidal mats, the relative dominance of the two depending on precipitation and varying from the rainy season to one of drought. If it is dry, *Scytonema* is present at the surface of the mat, and *Schizothrix* protected from desiccation at the base. In wet weather, *Schizothrix* migrates to the surface and thickens the mat (Monty, 1967).

The intertidal region is the *Rivularia* belt, where radiating filaments of *Rivularia* and some *Dichothrix* form mats to rounded heads, trapping fine sand and silt if carbonate is absent. The typical *Schizothrix* belt is from the low intertidal to infralittoral region. *Schizothrix* does not have strong bundles of filaments like those of *Rivularia* that trap detrital grains, but fine particles are agglutinated on its mucilaginous surface to produce laminated mats and domes (Monty, 1967).

Modern blue-green algae commonly are more conspicuous in fresh water than in the seas, and lacustrine algal biscuits occur widely; yet most stromatolites in the fossil record appear to be marine in origin. Nevertheless, modern marine stromatolites are forming in Australia, Florida, and the Bahamas, and some lacustrine ones are known as fossils. The absence of grazing invertebrates has been suggested as an explanation of their greater marine importance in the Precambrian (Garrett, 1970), as in some coastal regions various mollusks may restrict the growth of algal mats.

In contrast, a dense population of *Oscillatoria limosa* Agardh is found in the inner lagoon of the coral reef of Moorea Island in French Polynesia, where it is heavily grazed by holothurians. Large amounts of viable filaments are released among the fecal pellets of the holothurian, suggesting that grazing by this invertebrate may aid propagation and dispersal of the cyanophyte (Sournia, 1976).

An unusual annual or seasonal lamination is produced by algal mats in a lake near the shore of the Gulf of Aqaba. *Oscillatoria limnetica* mats here are beneath two dense layers of photosynthetic bacteria. In the intertidal zone and in shallow water, the summer algal mat consists of the coccoid blue-green algae *Aphanothece* and *Aphanocapsa* and some diatoms. In winter, a layer of the filamentous *Oscillatoria, Spirulina, Lyngbya,* and *Microcoleus,* together with bacteria, replaces the summer mat. The filamentous species are in a deeper layer, as shading for one or two days results in their movement up through the usual summer mat. The rate of sediment influx is very low and much of the organic matter decays, so that accumulation is extremely slow, the laminae thus being seasonal rather than diurnal (Krumbein and Cohen, 1977). In deeper water, a flocculose algal mat covers the bottom and actively photosynthesizes in winter, but disintegrates and floats to the surface in summer. As the bottom may be completely anaerobic, with low light intensity, the *Oscillatoria limnetica* algal bloom uses H_2S for photosystem II (like that of bacteria), producing sulfur granules on the filament surface instead of free oxygen. The species changes to the usual oxygenic photosynthesis if the H_2S concentration is decreased. Perhaps these simple cyanophytes surviving in special environments have retained the capability for either type of photosynthesis from an early period in their evolution (Krumbein and Cohen, 1977).

Calcium Dissolution

Many boring and perforating forms are known, principally among the Nostocales (*Plectonema, Kyrtuthrix*) and Pleurocapsales (*Hyella, Dalmatella*), with a few representing the Stigonematales (*Mastigocoleus*). Similar borings are produced by some fungi and red algae, and it is not always possible to determine the organism responsible. Fungal mycelia may be

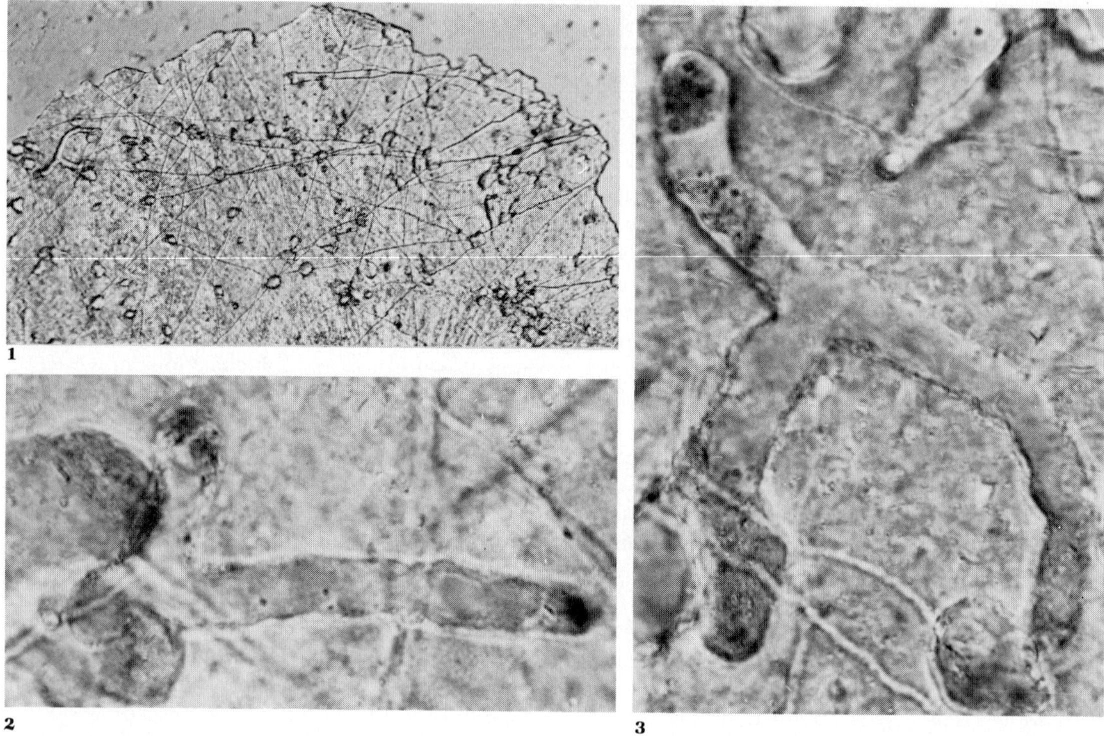

Figure 1.25
Perforating algae, in keel of foraminiferan *Laticarinina pauperata* (Parker & Jones), Recent, North Caribbean Sea, at 700 m, with phase contrast microscopy. **1.** Thin section of part of keel, showing perforations of algae and fungi, ×68. **2,3.** *Mastigocoleus testarum,* showing branching perforations, irregularly arranged, and possible heterocysts, ×730.

larger and commonly are branching, whereas the algal tubes are small, may be unbranched, may appear segmented, or may have globular terminations (see Figures 1.25 and 1.26).

Perforating algae may burrow into carbonate rock surfaces and are thought to play a part in the formation of coral atolls by destruction of the central part of the reefs. The mechanism of dissolution is not known certainly, although it may involve secretion of oxalic acid. Both freshwater and marine perforating forms are known. In the sea they may occur to a depth of 50 m, although they rarely are found at such depths. In the shallow carbonate shelf of Florida, and in the Bahamas, they are common, but the boring decreases abruptly at about 18.5 m (Swinchatt, 1969). Because of this depth restriction, algal borings in rocks and fossils have been suggested as a paleodepth indicator, suggesting depths of less than 36 m and, unless

Figure 1.26
Perforating algae in keel of foraminiferan *Laticarinina pauperata,* Recent, North Caribbean Sea, at 700 m, in phase contrast microscopy. **1.** Thin section of part of keel, showing perforations perpendicular to keel periphery, ×120. **2.** *Kyrtuthrix* sp. perforations perpendicular to surface and broader internally, ×1300. **3.** ?Perforating fungus, with much finer straight tubes extending from subspherical cavities. Similar straight tubes penetrate the algal tubes in part 2. ×1300.

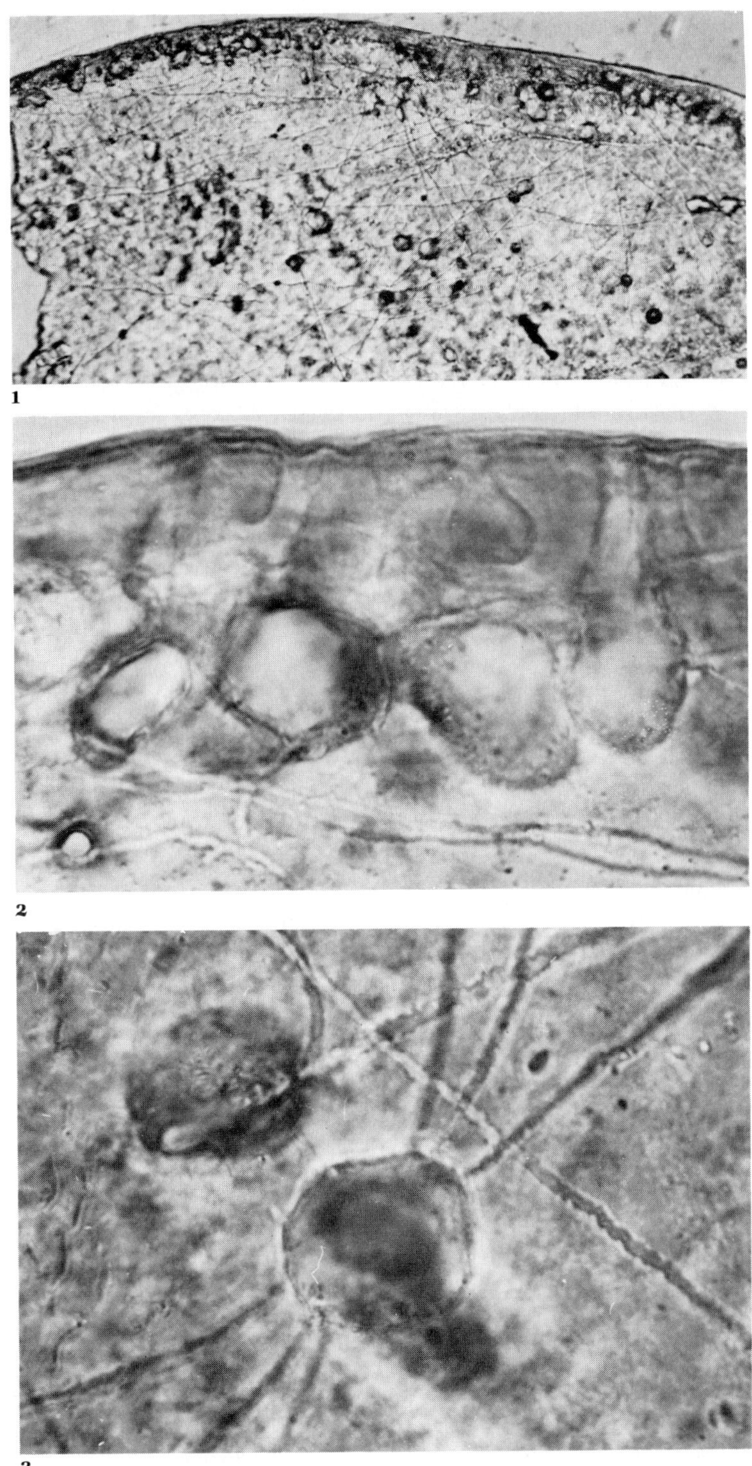

1
2
3

reworking had occurred, probably less than 18 m.

In the West Indies, blue-green algae that are endolithic borers include *Mastigocoleus testarum* Lagerheim, the most abundant species from the intertidal area to 30 m, and *Plectonema terebrans* Bornet & Flahault, second in importance over the same range, whereas *Hyella* sp. was patchy. *Calothrix* sp. was abundant from depths of 5 to 15 m, and *Scytonema* sp. occurred sporadically from the intertidal area to about 22 m (Perkins and Tsentas, 1976). Boring green algae were next in abundance, and red algae and fungi were present largely in the deeper water.

The limestone-perforating algae showed a tunnel direction and wall sculpture in controlled experiments that was determined by the crystal cleavage and twinning of the calcite. The dissolution was carried out by the terminal cells of the filaments, the space dissolved having the shape of a miniature crystal. The tunnel size was related to the size of the filaments, and different species showed characteristic patterns of branching (Golubic, 1969). The expanded terminal cells of *Hormatonema paulocellulare* Ercegović produced closely packed parallel tunnels of 8 to 15 μm in diameter, penetrating the crystal from 30 to 50 μm. *Hyella caespitosa* Bornet & Flahault formed laterally or dichotomously branched tufts; the tunnels were 4 to 6 μm in diameter and penetrated to 120 to 300 μm.

Planktonic Cyanophytes

Relatively few blue-green species are planktonic, although some Oscillatoriaceae are important in the tropical waters. The best known of these is *Trichodesmium* (or *Oscillatoria,* to which it is also referred), represented by at least four species in the area near Madagascar, where blooms of these nitrogen-fixing species cause red water (Sournia, 1968, 1970). Some 26 additional genera are present as well, including *Richelia* as an epiphyte on the diatom *Chaetoceros* and as an endosymbiont in the diatom *Rhizosolenia* (Sournia, 1970).

Like the unicellular blue-greens in hot springs, the freshwater planktonic species are invariably associated with bacteria. If atmospheric CO_2 is limited, addition to a culture of a carbon source assimilable by bacteria will enhance the algal growth. Similarly, in highly productive lakes, where rapid growth of the cyanophytes depletes CO_2, bacterial growth on dead organic matter returns CO_2 to the system, hence is probably important in the continuance of the algal blooms (Lange, 1970).

Although some blue-green algae are toxic and inhibiting to invertebrates and vertebrates, marine planktonic species may be important components of the food chain. Copepod larvae need a substrate for growth and support, and in culture appear to be attracted to the blue-green *Trichodesmium thiebautii* Gomont. In the Atlantic off northeastern South America, there is a positive correlation in occurrence of larvae of the copepod *Macrostella* with *Trichodesmium* (Calef and Grice, 1966).

Endosymbionts

Cyanophytic symbionts are present in protozoans, other algae, fungi, liverworts, and higher plants. Probably more symbionts belong to this group than to any other, possibly excepting the green algae. Because of the characteristic pigmentation the blue-green endosymbionts have been termed **cyanelles.** Some are very similar to free-living chroococcoid species, but others have become modified and lack the outer sheath and cell wall, retaining only the plasma membrane to separate the cytoplasm of symbiont and host. Those present as endosymbionts in the green alga *Glaucocystis* appear to have lost some metabolic functions as well, as they cannot grow and divide when removed from the host (Holm-Hansen, 1968). Within the host, *Glaucocystis nostochinearum* Itzigsohn, the blue-green plastidlike cells occur singly or in groups of two to four, their division coinciding with that of the host nucleus as it produces autospores. Each newly formed spore contains stellate masses of the symbiont (Fenwick, 1966). Although similar to the free-living genera *Aphanothece* or *Synechococcus*, their lack of a cell wall led to their description as a separate taxon, *Skujapelta.*

Similar blue-green symbionts in the cryptomonad *Cyanophora* also lack a cell wall; their pigments are typically cyanophycean, and when isolated from the host they develop cyanophycin granules. This symbiont has been described as *Cyanocyta.* Both *Glaucocystis* and *Cyanophora* also are sometimes regarded as red algae (Schnepf and Brown, 1971). A symbiont of various marine planktonic diatoms, *Richelia,* resembles the free-living *Anabaenopsis.*

The thecamoebian *Paulinella* always contains two large sausage-shaped specimens of the symbiotic endophyte *Synechococcus* (see Figure 1.27). At fission of the rhizopod, each of the two new daughter cells takes one of the endosymbiont cells, which then also soon divides in half. The close relationship of host and endosymbiont even led to the mistaken description of *Paulinella* and its endosymbiont as the "alga" *Cyanospira aeruginosa* Chodat.

Filamentous blue-greens also may be endosymbionts. Those of the liverworts, ferns and cycads, and a single angiosperm are Nostocales (especially *Anabaena* and *Nostoc*), and both the unicellular Chroococcales and the Nostocales may be symbiotic with fungi to form lichens.

The common occurrence of blue-green endosymbionts, as well as ultrastructural features, lend some credence to the suggestion that eucaryotic plastids may have arisen originally as endosymbiotic procaryotes.

Fossil Record

Stromatolites in the Fossil Record

Probably the most extensive and best-known record of the blue-green algae as fossils is that of the stromatolites, the laminated organosedimentary structures produced by algal mats that trap detrital sediments and may also deposit some calcium carbonate. Alternations of dark and light lamellae reflect the relative quantities of organic and mineral material. Similar structures were first recognized in modern lacustrine deposits, but marine stromatolites also are known in various regions, from the intertidal to subtidal depths. The globular lami-

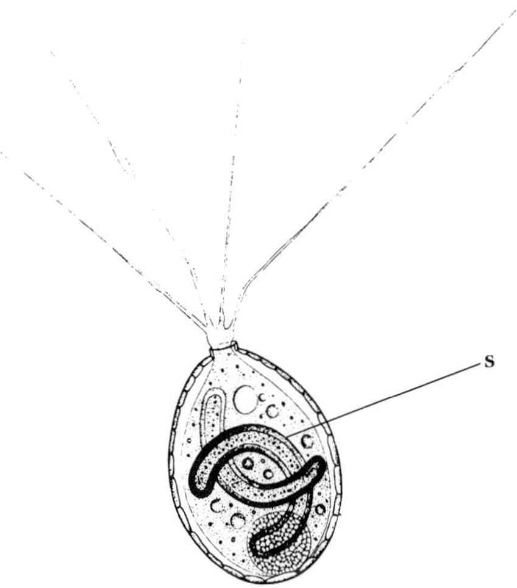

Figure 1.27
Synechococcus, a blue-green algal symbiont in the euglyphacean protozoan *Paulinella chromatophora.* Each protozoan has two of the horseshoe-shaped symbionts, binary fission of *Paulinella* being followed by division of the symbiont. ×1000. From Lauterborn, 1895.

nated "lake biscuits" or "lake balls" have a similar origin, but were rolled rather than stationary during growth. Fossil counterparts are termed **oncolites** (see Figure 1.28). Both stromatolites and oncolites occur in rocks of all ages, although particularly common in the Precambrian and early Paleozoic. Silicified stromatolites contain some of the oldest known fossil organisms, such as those of the 2.7-billion-year-old Bulawayan Group of Rhodesia and the 1.9- to 2.3-billion-year-old Pretoria Group, Transvaal Sequence of South Africa, and provide much of the fossil record of the Precambrian. They are most abundant through the upper Proterozoic rocks, gradually declining in importance through the lower Paleozoic, and are only locally common in the Mississippian, Permian, Triassic, and Tertiary.

Early studies of stromatolites regarded them as distinctive genera and species, but evidence from the modern examples indicates the importance of the sediment-trapping mechanisms, leading others to regard them as

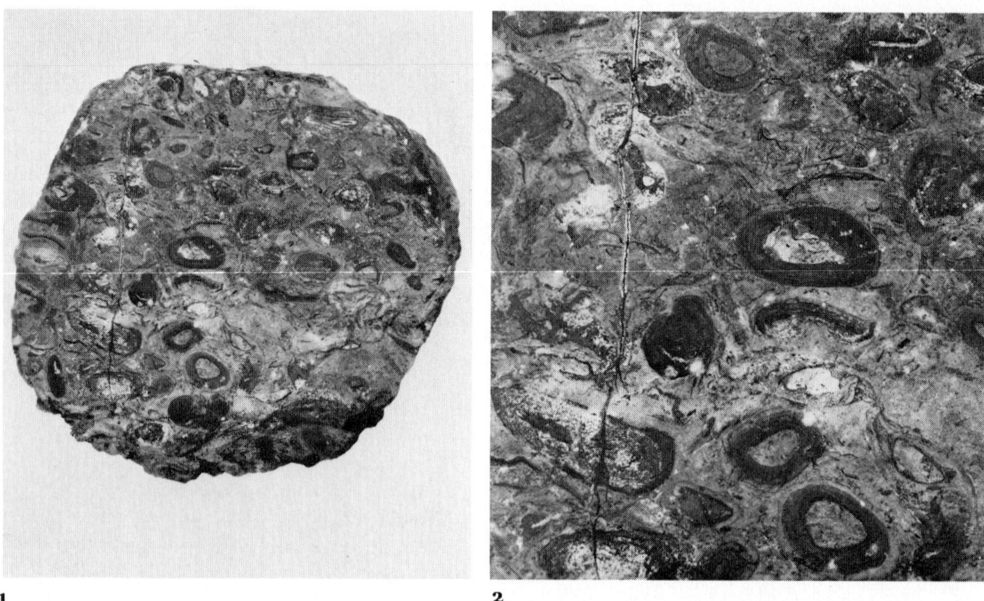

Figure 1.28
Oncolites, Lower Cambrian Chambless Limestone, Marble Mountains, California. **1.** Weathered surface, ×0.5; **2.** Enlarged, ×1.25.

sedimentary structures alone. Yet, if stromatolite morphology and lamination were controlled solely by the environment, as suggested by the sediment entrapment origin, similar forms should occur throughout geologic time. However, different forms appear to characterize different periods, suggesting that some of these changes may reflect changes in the algal mat assemblage or in the nature of the dominant species. Thus some stromatolites may be produced by a dominant species that controls the morphology, even if accompanied by other minor species. Others may reflect growth of an association of species in particular environments.

Three main types of structures are known in modern stromatolites, and their fossil counterparts probably represent similar environments. (1) The laterally linked hemispheroids, such as *Collenia*, develop in protected intertidal mud flats, where wave action is slight. (2) Discrete vertically stacked hemispheroids (*Cryptozoon*) form in exposed intertidal mud flats, where wave scour prevents growth of an algal mat between the stromatolites. (3) Discrete spheroids with concentric arrangement, or randomly stacked hemispheroids (*Oncolites*) form in low intertidal areas exposed to wave action. Wave action causes the partially aggregated material to roll, resulting in balls or spheres.

In addition to variations in gross morphology and the form and nature of laminae, a distinctive organic microstructure may include well-preserved cells and filaments. Only those stromatolites with both a distinctive form and well-preserved organic remains are recognized as separate taxa by some (e.g., Vologdin, 1962); others describe the morphological features such as size and shape, close or open spacing, presence and nature of branching, surface projections, and lamellar structure and convexity. Non-Linnaean Groups and Forms are described by some, particularly the Russian workers, who have found stromatolites useful for Precambrian stratigraphy (*e.g.,* Raaben, 1969, 1972; Krylov, 1975; also see Cloud and Semikhatov, 1969; Glaessner et al., 1969). Others regard the sedimentary control as more important, hence use only descriptive symbols and formulae, rather than names (Logan et al., 1964). As stromatolites, whether regarded as Groups and Forms or identified by symbols, do not refer to biological species, they are not further discussed herein. (More detail is available in such recent summaries as those of Cloud and Semikhatov, 1969; Hofmann, 1969a, 1969b, 1973, 1975; Komar, 1966; Krylov, 1975; Logan et al., 1960, 1964; Preiss, 1972, 1973, 1974; Raaben, 1969, 1972; Vologdin, 1962; Walter, 1976.)

Because the algae are photosynthetic, stromatolites are generally regarded as indicating shallow water, if not intertidal. However, some modern ones occur to greater depths, and in the Devonian of the Canning Basin, Australia, algal stromatolites occur in the reef, back-reef, and fore-reef facies, some probably having formed at depths of at least 45 m (Playford and Cockbain, 1969).

The growth of the stromatolites is directionally regulated by the light source (phototropism), hence is normally vertical. Observation of some with inclined growth led Vologdin (1963) to postulate a seasonal effect, with light strongest from the south. The direction of the strong heliotropism of certain colonies in the Precambrian of China was suggested by Vologdin to indicate polar migration. However, in Shark Bay, Australia, the low intertidal stromatolites were inclined westerly away from shore, and aligned perpendicular to the meridionally trending shoreline (Hofmann, 1973). Thus, phototropism is not the only cause of inclination of the stromatolite heads and its use for paleogeographic interpretation requires caution.

Calcareous Tubes

Somewhat similar to the oncolites are the algal balls consisting of a tangle of calcareous tubes rather than concentric laminations, commonly referred to *Girvanella* (see Figure 1.29). Origi-

nally described as a foraminiferan, they were later referred to the algae, and occur in rocks of Cambrian to Cretaceous age. The tubes are of regular diameter and may be closely packed and subparallel or more widely spaced and contorted. Branching occurs rarely. Although the systematic position is uncertain, a blue-green algal affinity is probable. *Girvanella* has been suggested to be a depth indicator in the Silurian of Sweden, where it is associated with tabulate corals and stromatoporoids. The presence of *Girvanella* suggests depth limits for some of the Silurian brachiopod communities as well (Lauritzen and Worsley, 1974), just as the presence of fossil algal borings also is indicative of relatively shallow depths. However, cyanophyte pigments like those of the red algae allow them to utilize light penetrating to greater depths than can green algae; use of *Girvanella* as a depth indicator thus may not be a reliable method (Riding, 1975).

Freshwater calcareous algae have been reported from the Upper Cretaceous and Tertiary (Freytet and Plaziat, 1966; Maslov, 1967; Rutte, 1954), although variously referred to the Cyanophyta or green algae.

Fossil Boring and Perforating Algae

Excavations in fossil shells have been recorded in Ordovician trilobites and brachiopods, referred to the genus *Palaeachlya* (see Figure 1.30) in Silurian and Devonian foraminifers, sponges, corals, and brachiopods (Duncan, 1876); in Permo-Carboniferous corals, bryozoa, brachiopods, crinoids, and mollusca; in Mesozoic and Cenozoic sponges, foraminifers (particularly the porcellaneous species), corals, molluscs, and calcareous algae; and as *Chaetophorites* (Silurian, Jurassic, and Tertiary) and *Palaeopede* (Devonian, Jurassic, and Cretaceous). Similar-appearing structures occur in modern and fossil fish scales; and such borings in fossil bone have been referred erroneously to the fungal genus *Mycelites* in Devonian and Carboniferous fish, Permian and Triassic amphibians, and Mesozoic and Tertiary fish and reptile bones.

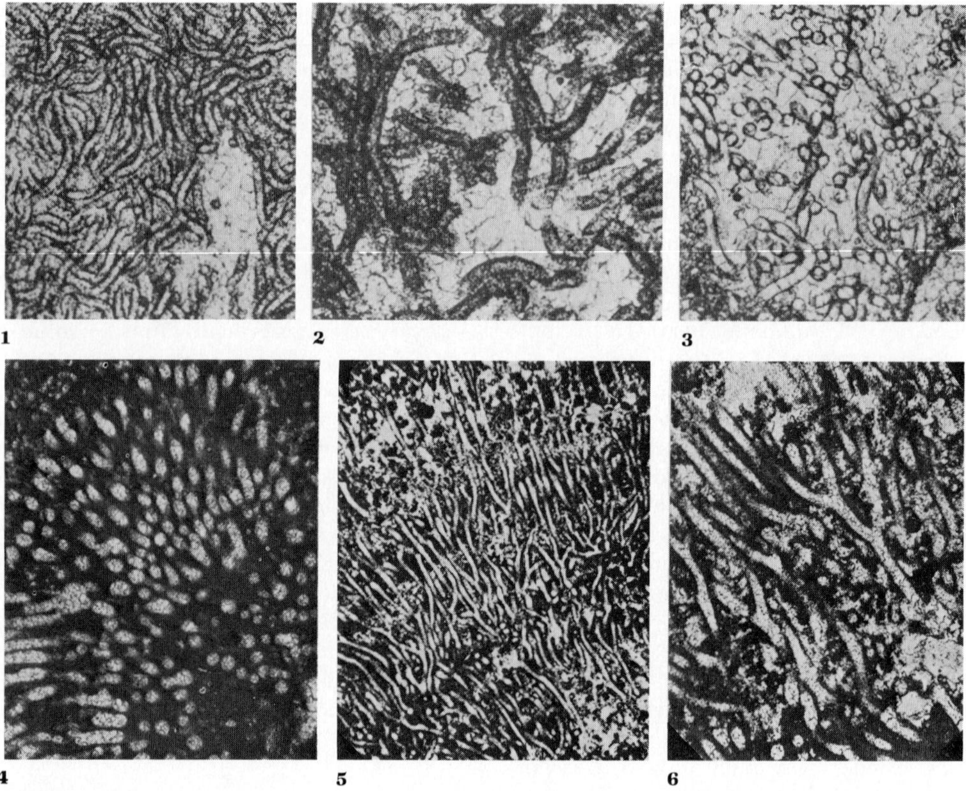

Figure 1.29

Calcite-depositing "Porostromata." **1 – 3**, *Girvanella problematica* Nicholson & Etheridge, Stinchar Limestone, Upper Ordovician, Girvan, Scotland; from Wood, 1957. **1.** Closely packed meandering tubes, ×70. **2.** ×96. **3.** Small aggregation, showing cross sections and possible cell walls dividing the tubes, especially in V-shaped tube at lower center of figure; ×70. **4.** *Garwoodia gregaria* (Nicholson) Wood, Mississippian, England; partially oblique section, showing diameter of threads in both longitudinal and transverse section; ×34, from Wood, 1941. **5,6.** *Ortonella furcata* Garwood, Mississippian, England; from Garwood, 1914. **5.** ×16. **6.** Enlarged to show dichotomously branching tubes, ×44.

Nonmineralized Remains

In addition to the blue-green algal fossil stromatolites, borings, and excavations, microscopic vegetal material found in some stromatolites and in bedded cherts has in part been referred to this group.

The stromatolites *Collenia* and *Cryptozoon* show only concentric laminae and lack a cellular or filamentous structure. However, stromatolites containing preserved unicellular microfossils are found from the lower Precambrian onward; for example, from the Algonkian Gallatin Formation of Montana. Filamentous microfossils are preserved in stromatolites and bedded cherts from the middle Precambrian (e.g., the Gunflint Formation of southern Ontario) and from many localities of late Precambrian age (such as the Bitter Springs Formation of central Australia); the

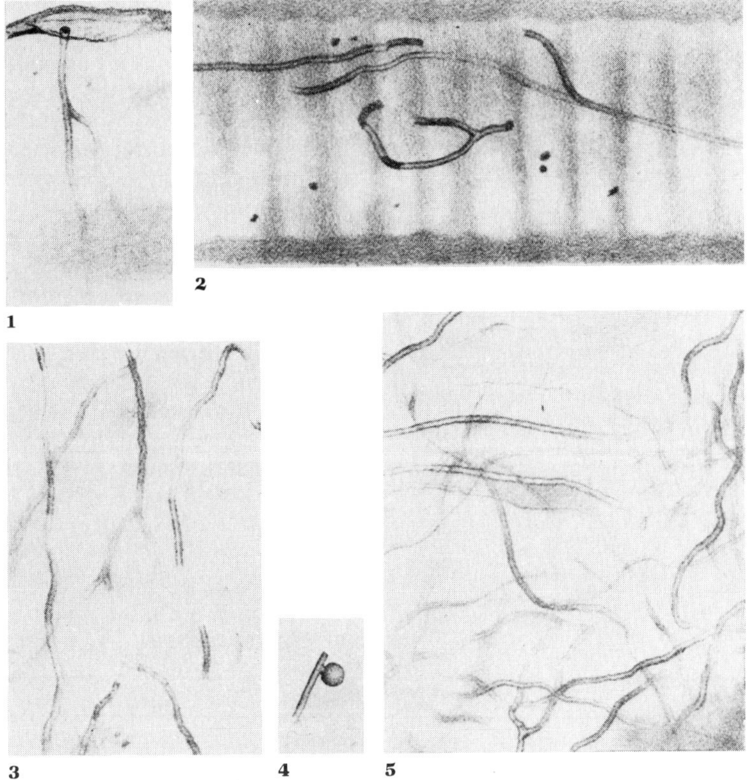

Figure 1.30
Perforating alga, *Palaeachlya perforans* Duncan. **1,3,4.** Boring in coral *Goniophyllum,* Upper Silurian, Canada. **2.** Perforations in shell of brachiopod embedded in the coral. **5.** Perforations in coral *Thamnastraea,* Miocene, Tasmania. All ×350, from Duncan, 1876.

Devonian of Scotland; Pennsylvanian and Permian of Kansas, Colorado, and New Mexico; Jurassic of France; and Tertiary and Pleistocene lake beds of Colorado, Utah, and Nevada.

In contrast to all other groups of fossils, more fossil genera and species of the blue-green algae have been described from the Precambrian than from the entire Phanerozoic. In part, this is a reflection of the long duration of this "age of the blue-green algae," but probably much of the explanation lies in the intensive search in recent years for Precambrian fossils and its concentration on bedded cherts, or silicified stromatolites, facies that are most likely to contain the cyanophytes, but also most likely to

allow their preservation. The few examples of similar Phanerozoic facies that have been studied, such as the Rhynie Chert of Scotland, suggest that additional blue-green fossils should be found in a similar concentration in younger strata.

The very simple morphology and small size of the coccoid blue-greens, and the possibility of contamination by fungal filaments in the field or laboratory, has led to doubt as to the validity of some of the described Precambrian taxa. The likelihood of postmortem degradation of filaments and sheaths also has led to questions as to the taxonomic limits of the fossil genera and species. Preservation of the sheath

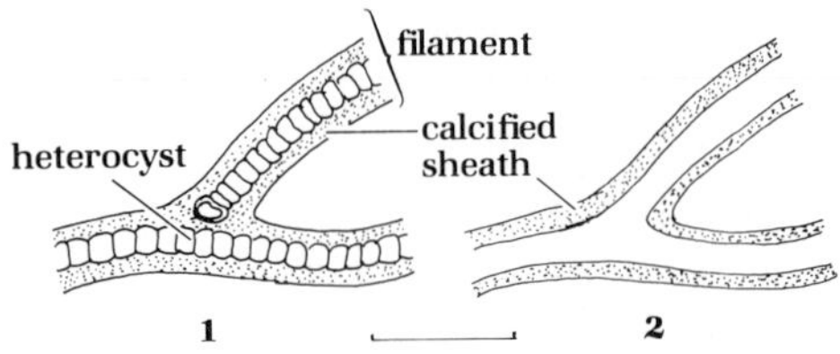

Figure 1.31
Diagrammatic longitudinal section of *Tolypo-thrix,* showing postmortem loss of morpho-logic detail in a calcified filamentous blue-green alga. **1.** Part of calcified living filament, show-ing false branching and heterocysts. **2.** Empty calcified sheath similar to many fossil speci-mens, lacking any indication of the type of branching, size of trichomes, or presence of heterocysts. Bar = 20 μm. Redrawn from Rid-ing, 1977.

without the enclosed filament or trichome also may obscure diagnostic characters (see Figure 1.31). Even allowing for these difficulties in pres-ervation and interpretation, and the resultant divergence of opinions (e.g., Schopf, 1968, 1974; Cloud, 1973; Knoll and Barghoorn, 1975; Schopf and Oehler, 1976; Hofmann, 1976; D. Oehler, 1976; Golubic and Barghoorn, 1977), a general evolutionary sequence seems apparent.

In the early Precambrian, the coccoid blue-green alga *Archaeosphaeroides barbertonensis* Schopf & Barghoorn (see Figure 1.32) occurs in the South African Fig Tree Series (Schopf and Barghoorn, 1967). Unicells and filaments have also been reported from the Fig Tree Series by Pflug (1967) and from the Onverwacht (Brooks et al., 1973); spheroids of 7 to 10 μm diameter occur in the Theespruit Formation (lower On-

Figure 1.32
Early Precambrian cyanophytes.
1. Spherical organic-walled microfossil in SEM, from acid maceration of chert in Onverwacht Series, Swaziland Sys-tem, 3.2 billion years old, from South Africa; bar = 1 μm. From Engel et al., 1968. **2,3.** *Archaeosphaeroides barber-tonensis* Schopf & Barghoorn, in thin sections of chert of Fig Tree Series, Swaziland System, South Africa, show-ing irregularly reticulate surface; bar = 10 μm in both photos. From Schopf and Barghoorn, 1967, *Science,* v. 156, p. 508 – 512. Copyright 1967, American Association for the Advancement of Science; reproduced by permission.

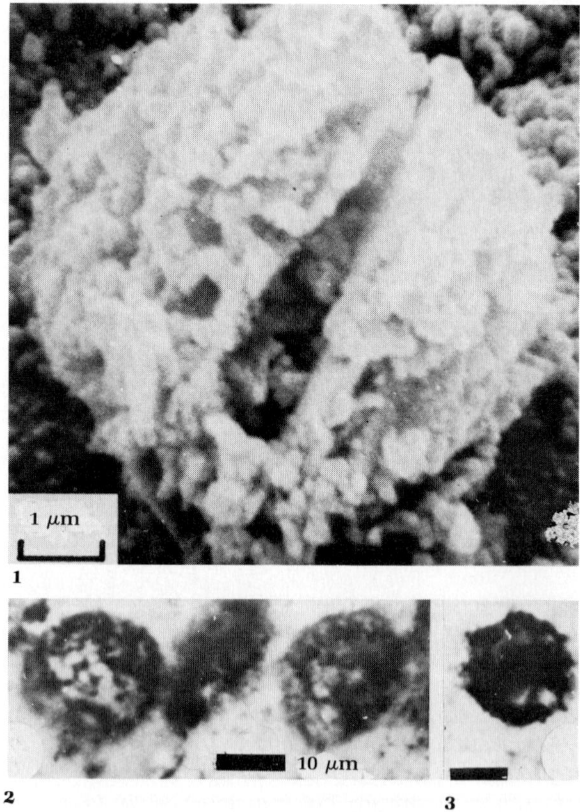

verwacht), and filaments 4 to 8 μm in diameter and up to 20 μm long (the filament illustrated by Brooks et al., 1973, would be 50 μm long according to the indicated scale, but the bar scale probably is mislabelled 5 μm instead of 2 μm) are present in the Swartkoppie Formation and the upper Onverwacht. These Fig Tree and Onverwacht specimens have been questioned as possible pseudofossils or contaminants (Schopf, 1974), but extracted organic matter from the rocks has a composition similar to sporopollenin (Brooks et al., 1973), offering some support for their biogenicity. However, the material reported from the lower Precambrian Soudan Formation (Cloud et al., 1965), probably does represent artifacts rather than fossils (Cloud, 1973).

Much better evidence is available for the early Proterozoic, from about 2.5 to 1.8 billion years ago, including the oldest widely occurring stromatolites; for example, in the 2.3-billion-year-old Transvaal Dolomite of South Africa, which contains filamentous algae resembling the modern *Oscillatoria* and *Rivularia* (MacGregor et al., 1974) and the nostocacean *Petraphera* (Nagy, 1974). This period was that of the wide occurrence of the banded iron formations (from 2.2 to 1.9 billion years ago), those alternating beds of chert and sedimentary iron deposits suggested by Cloud (1965) as indicating the gradual change to an oxygenic atmosphere of about 1.0 percent the present level. Redbeds are present from about 2 billion years. The best-known assemblage of this period of time is that of the Gunflint Formation of southern Ontario (see Figure 1.33), about 1.9 billion years old (Barghoorn and Tyler, 1965a, b; Cloud, 1965; Cloud and Hagen, 1965; Edhorn, 1973). This assemblage included representatives of the Chroococcaceae in *Chlamydomonopsis* and possibly *Cumulosphaera* (Edhorn, 1973) and *Huroniospora* (Barghoorn and Tyler, 1965b), although the latter may be a red alga, as discussed in Chapter 2. Many filamentous taxa also occur in the Gunflint, including the oscillatoriacean *Animikiea* and *Entosphaeroides* (Barghoorn and Tyler, 1965b) and *Palaeospiralis* and *Palaeospirulina* (Edhorn, 1973), the nostocacean *Gunflintia* (see Figure 1.34 and Barghoorn and Tyler, 1965b; Licari and Cloud, 1968; Cloud and Licari,

1972), the scytonematacean *Palaeoscytonema*, and the rivulariacean *Primorivularia* (both Edhorn, 1973). *Archaeorestis* (see Figure 1.35) appears to be nonseptate and siphonate, of irregular diameter, and showing branching. It probably represents a fossilized sheath, lacking the trichome, and may represent either true or false branching (compare Figure 1.31). A major diversification of the filamentous cyanophytes thus occurred during the middle Precambrian. The presence of taxa such as *Gunflintia* and *Petraphera* with heterocysts also indicates that atmospheric oxygen already had attained the levels requiring these specialized cells for protection from oxygen during nitrogen-fixation.

Another Gunflint taxon of uncertain relationship, possibly blue-green or bacterial, is *Kakabekia* (Barghoorn and Tyler, 1965b), characterized by a strange umbrellalike projection or mantle (see Figure 1.36). A similar living form, isolated from soil samples in Wales and described as *Kakabekia barghoorniana* Siegel and Siegel, provided little information as to the relationships of the Precambrian species, although suggesting that the fossils were reproductive bodies, possibly of blue-green algae & Siegel, provided little information as to the 1.8-billion-year-old Belcher Group of the Hudson Bay region of Canada (Hofmann, 1974, 1975, 1976; Hofmann and Jackson, 1969) includes an assemblage containing some similarities with the Gunflint, but others of aspect like the much younger Bitter Springs Formation of Australia (see Figure 1.37). The chroococcaceans *Eosynechococcus* and *Melasmatosphaera*, the entophysalidacean *Eoentophysalis*, and the oscillatoriacean *Rhicnonema* were new, and *Kakabekia* originally was described from the Gunflint, but in addition to these, species were referred to 12 genera previously described from the Bitter Springs Formation: *Archaeotrichion, Biocatenoides, Caryosphaeroides, Eomycetopsis, Eozygion, Glenobotrydion, Globophycus, Halythrix, Myxococcoides, Palaeoanacystis, Sphaerophycus,* and *Zosterosphaera.*

During the middle Proterozoic, from perhaps 1.7 to 1.25 billion years ago, diversification continued; a colonial form similar to the modern chroococcalean *Eucapsis* was described from the Paradise Creek Formation, Queensland, Aus-

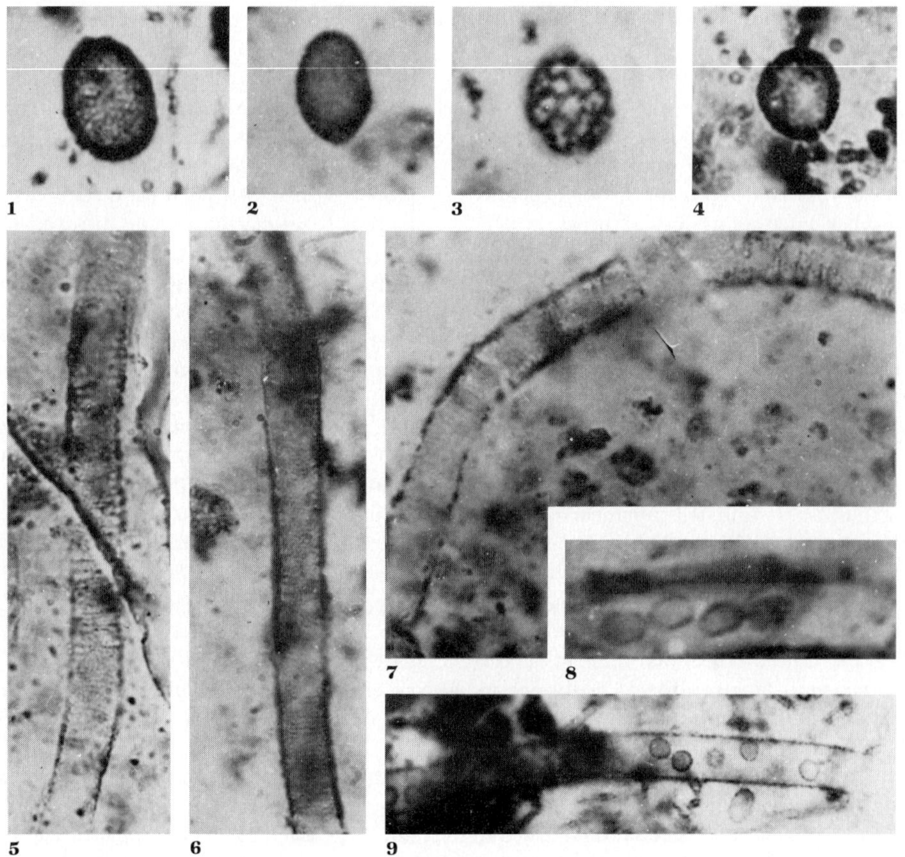

Figure 1.33

Chroococcaceae and Oscillatoriaceae from the 1.9-billion-year-old Gunflint Chert, Ontario, Canada. **1.** *Huroniospora microreticulata* Barghoorn, ×1412. **2.** *H. psilata* Barghoorn. **3,4.** *H. macroreticulata.* 2,3, ×2824; 4, ×1536. As noted, *Huroniospora* also has been referred to red algae or fungi. **5–7.** *Animikiea septata* Barghoorn, with transverse septa; 5, ×820; 6, ×800; 7, ×1200. **8,9.** *Entosphaeroides amplus* Barghoorn, filament containing sporelike bodies interpreted as endospores; 8, ×2048; 9, ×1200. All from Barghoorn and Tyler, 1965b, *Science,* v. 147, p. 563–577. Copyright 1965, American Association for the Advancement of Science; reproduced by permission.

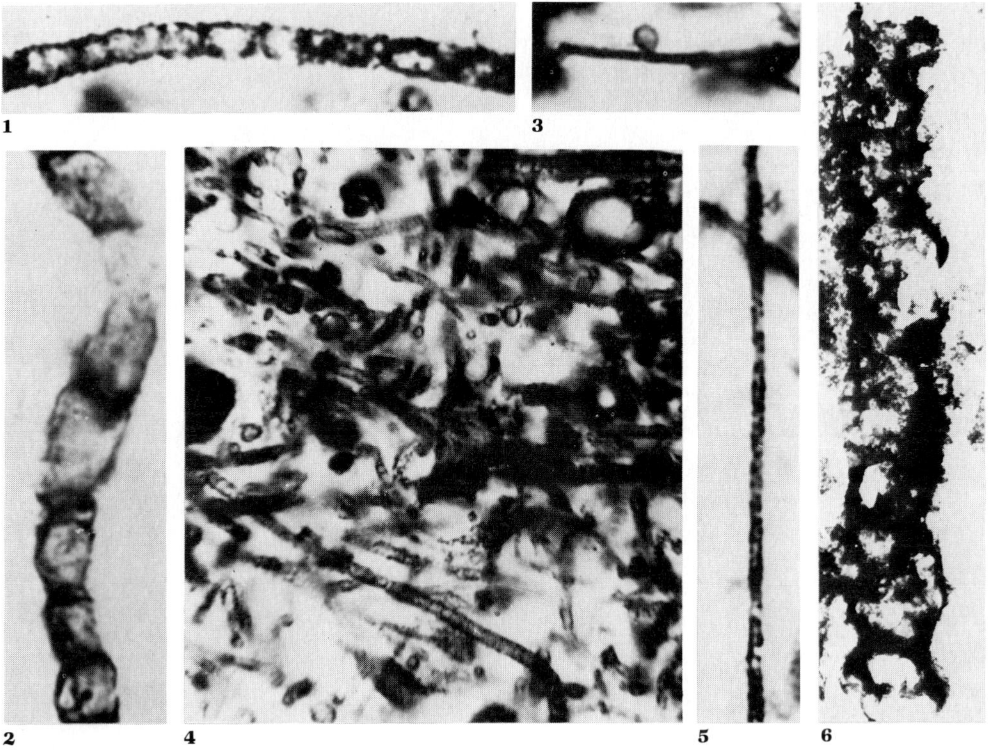

Figure 1.34

Nostocaceae from the Gunflint Chert, Ontario, Canada. **1,2.** *Gunflintia grandis* Barghoorn, septate filaments, ×2048. **3−6.** *G. minuta* Barghoorn, with narrower filaments; 3, ×884; 4, ×908; 5, ×1380. 1−5. In transmitted light of thin sections of chert, from Barghoorn and Tyler, 1965b, *Science,* v. 147, p. 563−577. Copyright 1965, American Association for the Advancement of Science; reproduced by permission. **6.** TEM of acid maceration, ×9200; photo from J. W. Schopf.

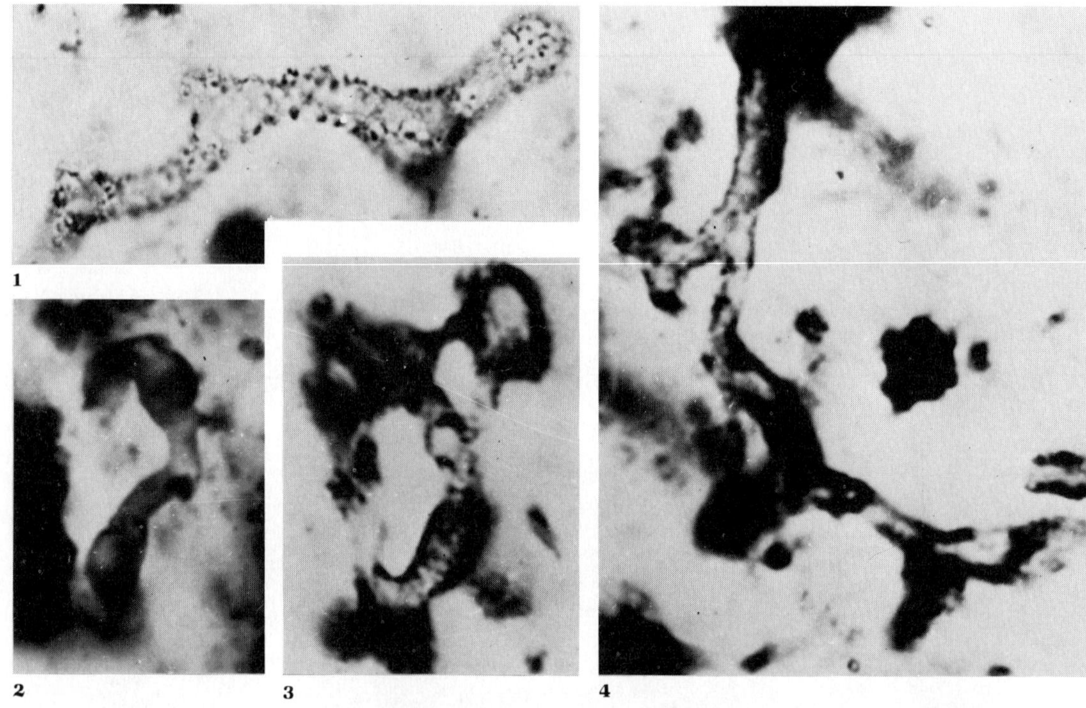

Figure 1.35

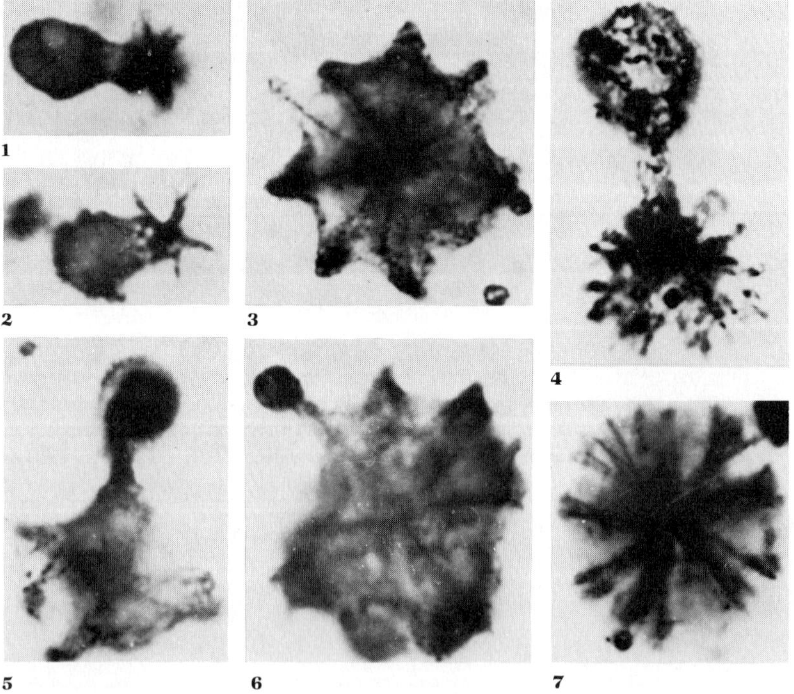

Figure 1.36

Figure 1.35 (*facing page, top*)
Archaeorestis schreiberensis Barghoorn, Gunflint Chert, Ontario, Canada; thin section of chert in transmitted light. Probably a cyanophyte sheath, showing branching but no partitions; nature of branching cannot be determined, as filament is not preserved. 1, ×1005; 2, ×505; 3,4, ×2415; from Barghoorn and Tyler, 1965b, *Science*, v. 147, p. 563−577. Copyright 1965, American Association for the Advancement of Science; reproduced by permission.

Figure 1.36 (*facing page, bottom*)
Kakabekia umbellata Barghoorn, thin sections of Gunflint Chert, from Ontario, Canada, in transmitted light. Proportions of bulblike structure and the umbrellalike crown vary, and are regarded as representing different stages in the ontogeny of the organism. 1−6, ×1900; 7, ×1000; from Barghoorn and Tyler, 1965b, *Science*, v. 147, p. 563−577. Copyright 1965, American Association for the Advancement of Science; reproduced by permission.

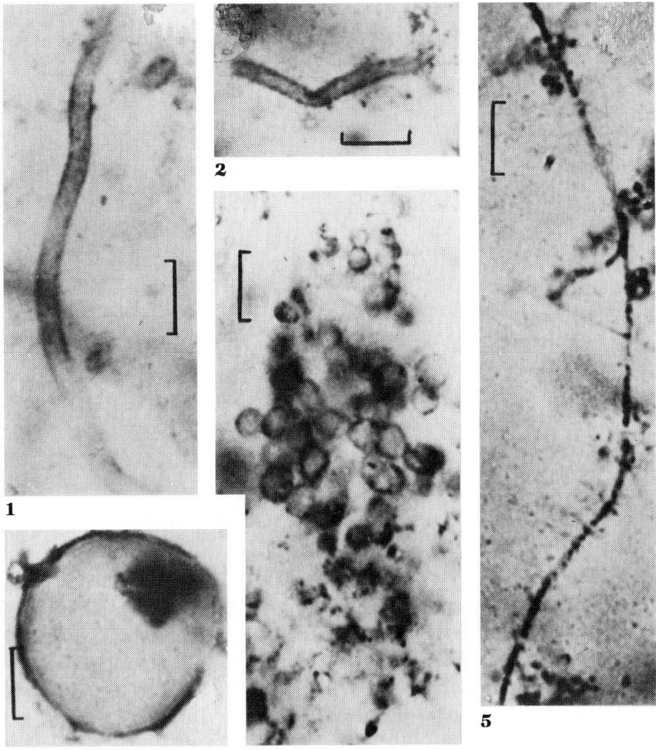

Figure 1.37
Belcher Group, 1.8 billion years old, Hudson Bay region, Canada. **1,2.** *Eomycetopsis filiformis* Schopf, originally regarded as fungal, but possibly a cyanophyte sheath. **3.** *Glenobotrydion majorinum* Schopf & Blacic, containing dark spot variously interpreted as a preserved organelle or as condensed and partly degraded cytoplasm. **4.** *Sphaerophycus parvum* Schopf. **5.** *Biocatenoides sphaerula* Schopf. These species are regarded as identical to those originally described from the much younger (900 million years old) Bitter Springs Formation of Australia. Bar = 10 μm in each photo. All from Hofmann and Jackson, 1969, *Can. J. Earth Sci.*, v. 6, p. 1137−1144; reproduced by permission of the National Research Council of Canada.

tralia, about 1.6 billion years old. Rectangular packets of cells consisted of four rows of four to eight hemispherical cells each, the size suggesting a procaryote (Licari et al., 1969), although similar growth habits occur in a variety of eucaryotic algae (Tappan, 1976).

In the Amelia Dolomite, McArthur Group, of Northern Territories, Australia, also some 1.6 billion years in age, unicells and colonial coccoid forms include *Sphaerophycus*, *Huroniospora*, *Myxococcoides*, and *Palaeoanacystis* (Muir, 1974), a mixture of older and younger taxa.

The 1.3-billion-year-old Beck Spring Dolomite, Pahrump Group, of California contains a large algal assemblage, including some probable eucaryotes, such as *Latisphaera*, *Maculosphaera*, and *Palaeosiphonella* (Cloud et al., 1969; Licari, 1978). However, *Bullasphaera*, compared to a chrysophycean cyst, may be an artifact (true chrysophycean cysts are unknown prior to the Late Cretaceous), and inner "bodies" or "spots" in some cells (*Latisphaera*, *Maculosphaera*) that were regarded as eucaryote organelles may represent only remnants of coagulated cytoplasm (Golubic and Barghoorn, 1977), as do similar structures widely reported in Precambrian microfossils.

Within the late Proterozoic, between 1.2 billion and 600 million years ago, undoubted eucaryotes include red and green algae, and even metazoans, but controversy remains as to how many of the fossil taxa are eucaryotes and how many are blue-green algae or even distinct taxa. For example, *Sphaerocongregus variabilis* Moorman was described from the 725- to 850-million-year-old Hector Formation, of Alberta, Canada, as a coccoid, planktonic, endospore-forming cyanophyte, commonly replaced by pyrite (Allison and Moorman, 1973; Moorman, 1974), but as discussed by Javor and Mountjoy (1976), *Sphaerocongregus* may be a blue-green alga, bacterium, eucaryote, or only diagenetically produced pseudofossils, including framboidal pyrite.

The Pleurocapsales have a Precambrian representative in the species *Palaeopleurocapsa wopfnerii* Knoll et al. (see Figure 1.38). In the 1.0-billion-year-old Skillogalee Dolomite of Australia, it occurs as filaments and sheaths up to 72 μm in length and 25 μm wide, and one large single tuft is visible to the naked eye. Branching was dichotomous (Knoll et al., 1975).

The most diverse assemblage in the Precambrian is that of the 900-million-year-old Bitter Springs Formation, Northern Territories, Australia (Schopf, 1968; Schopf and Barghoorn, 1969; Schopf and Blacic, 1971; D. Oehler, 1976). Single unicells and colonies and a variety of filamentous algae are present. Chroococcaceans (including those transferred here by Knoll and Barghoorn, 1975; and Hofmann, 1976; see Figure 1.39) are referred to *Bigeminococcus*, *Caryosphaeroides*, *Eozygion*, *Glenobotrydion*, *Globophycus*, *Gloeodiniopsis*, *Myxococcoides*, *Palaeoanacystis*, *Phanerosphaerops*, *Sphaerophycus*, and *Zosterosphaera*. Oscillatoriaceans of the Precambrian are *Calyptothrix*, *Cephalophytarion*, *Contortothrix*, *Cyanonema*, *Filiconstrictosus*, *Heliconema*, *Obconicophycus*, *Oscillatoriopsis*, *Palaeolyngbya*, *Partitiofilum*, *Siphonophycus*, and *Tenuofilum*. Nostocaceans (see Figure 1.40) include *Anabaenidium*, *Archaeonema*, *Nostocopsis*, and *Veteronostocale*; *Halythrix* represents the Scytonemataceae, and *Caudiculophycus* the Rivular-

---

Figure 1.38
Cyanophytes from the 1.0-billion-year-old Skillogalee Dolomite, South Australia. **1—4.** *Archaeonema longicellulare* Schopf, originally described from the 900-million-year-old Bitter Springs Formation. **1.** Septate filament. **2—4.** Enlarged cell in each filament regarded as a heterocyst. **5,6.** *Myxococcoides muricata* Schopf & Barghoorn. 1—6, from Schopf and Barghoorn, 1969; bar = 10 μm. **7—10.** *Palaeopleurocapsa wopfnerii* Knoll, Barghoorn & Golubić. **7.** Showing parallel dichotomous branching. **8.** Entire thallus. **9.** Showing trilamellar sheath. **10.** Showing degraded protoplasm in center, the cell outline indicated by the better preserved sheath. 7,9,10, ×960; 8, ×360; from Knoll et al., 1975.

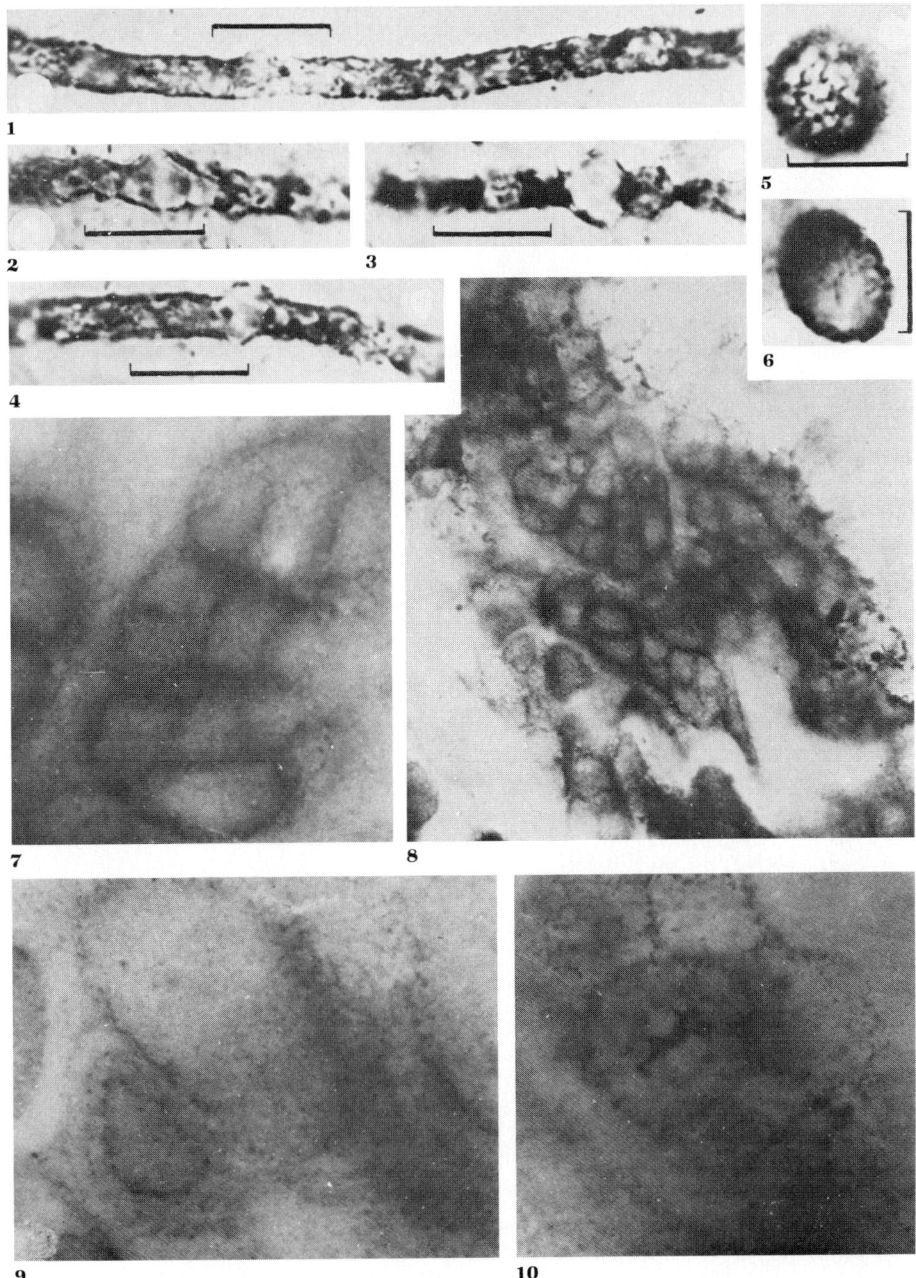

1

2

3

4

5

6

7

8

9

10

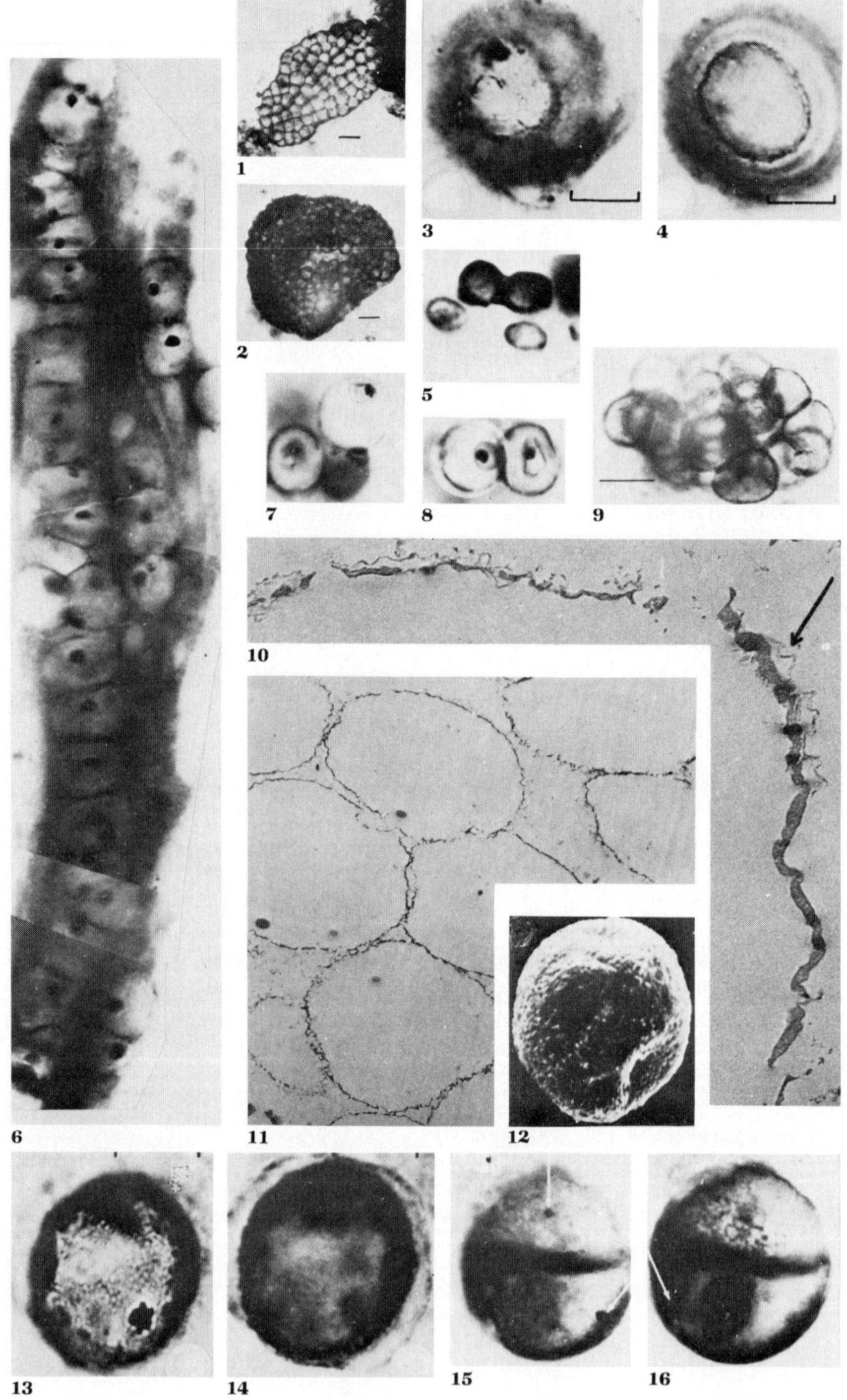

Figure 1.39
Chroococcaceans from the 900-million-year-old Bitter Springs Formation, central Australia.
Bar = 10 μm when given. **1,2.** *Palaeoanacystis vulgaris* Schopf, a colonial taxon. **3,4.** *Gloeodiniopsis lamellosa* Schopf, at two levels of focus, showing lamellar sheath. **5.** *Sphaerophycus parvum,* apparently in process of fission, $\times 1720$. **6—8.** *Glenobotrydion aenigmatis* Schopf. **6.** Sheath-enclosed colony. At inner wall surface of each cell lies a discoid body previously interpreted as pyrenoid, or a centrally located body regarded as a nuclear residue, hence as evidence of eucaryotic structure. Occurrence of similar forms in the older Belcher Group of Canada led to interpretation of the presumed organelles as degraded cytoplasm remnants. **7,8.** Small cell clusters, $\times 984$. **9—12.** *Myxococcoides minor* Schopf. **9.** Colony of globular cells with reticulate surfaces. **10.** TEM, ultra-thin section of acid-maceration-freed specimen, showing inner thick granular electron-dense layer and outer thin membranous layer (arrow), $\times 14,400$. **11.** TEM, ultra-thin section of colony, $\times 2400$. **12.** SEM of isolated cell, $\times 1600$. **13,14.** *Globophycus rugosum* Schopf, at two levels of focus. **15,16.** *Zosterosphaera tripunctata* Schopf, at two levels of focus, showing surficial pits or pores (arrows), $\times 1000$. 1—9, 13—16, from Schopf, 1968; 10—12, from Schopf, 1970.

---

iaceae. Even if some of these should prove synonymous, the variety remains impressive.

Many other reports of blue-green algae in the USSR are not as accurately dated radiometrically, but most are from upper Precambrian strata. They include the chroococcaceans *Echaninia, Gonamophyton, Murandavia, Praechroococcus, Uranovia,* and *Vesicophyton,* the entophysalidacean *Kareliana,* nostocacean *Dikimdinella,* and rivulariaceans *Aldanella, Azyrtalia, Jatuliana, Nelcanella,* and *Panomnimella,* as well as others referred to new families, such as *Protospira* and *Thysanoplanta.*

Among the post-Paleozoic occurrences of note are those of the Ordovician *Gloeocapsomorpha* (see Figure 1.41), which may be confused with the green alga *Botryococcus,* and *Schizothrichites* (see Figure 1.42; Starmach, 1963), and the flora of the Devonian Rhynie Chert of Scotland (Croft and George, 1959; Kidston and Lang, 1921). Occurrences in younger rocks have been in bituminous or calcareous deposits, as in the Mississippian (Renault, 1899; Maslov, 1956), Triassic (Bock, 1961; Fliche, 1905), Eocene (Bradley, 1970), Oligocene (Fliche, 1886; Rutte, 1953, 1954), and Miocene (Maslov, 1955). Perforating algae also are known from the early Paleozoic to the present, although not all can be definitely assigned to blue-green, green, or red algae or fungi.

Figure 1.40 (overleaf)
Bitter Springs Oscillatoriaceae and Rivulariaceae, Australia. **1.** *Heliconema australiense* Schopf, showing spiral growth habit. **2.** *Oscillatoriopsis obtusa* Schopf. **3.** *Siphonophycus kestron* Schopf. **4,5.** *Palaeolyngbya barghoorniana* Schopf, sheath-enclosed, some partial septation indicating vegetative division. **5.** Disrupted filament, showing cells in both side and end (arrow) views. **6—8.** *Cephalophytarion grande* Schopf with capitate end; left portion of filament shown in part 6 is shown diagrammatically in part 7. **8.** Two closely appressed filaments. **9,10.** *Caudiculophycus rivularioides* Schopf, a rivulariacean taxon with multicellular terminal hair. **11—14.** *Eomycetopsis robusta* Schopf, described as fungal, but regarded by others as possibly a cyanophyte sheath. **11.** Acid maceration, SEM. **12.** Grazing section in TEM, showing reticulate surface structure. **13.** Acid maceration. **14.** Longitudinal section, TEM, showing irregular thickness and absence of layering of the wall. 11,12 $\times 2800$; bar in all others = 10 μm. 1—10, 13, from Schopf, 1968; 11, 12, 14, from Schopf, 1970.

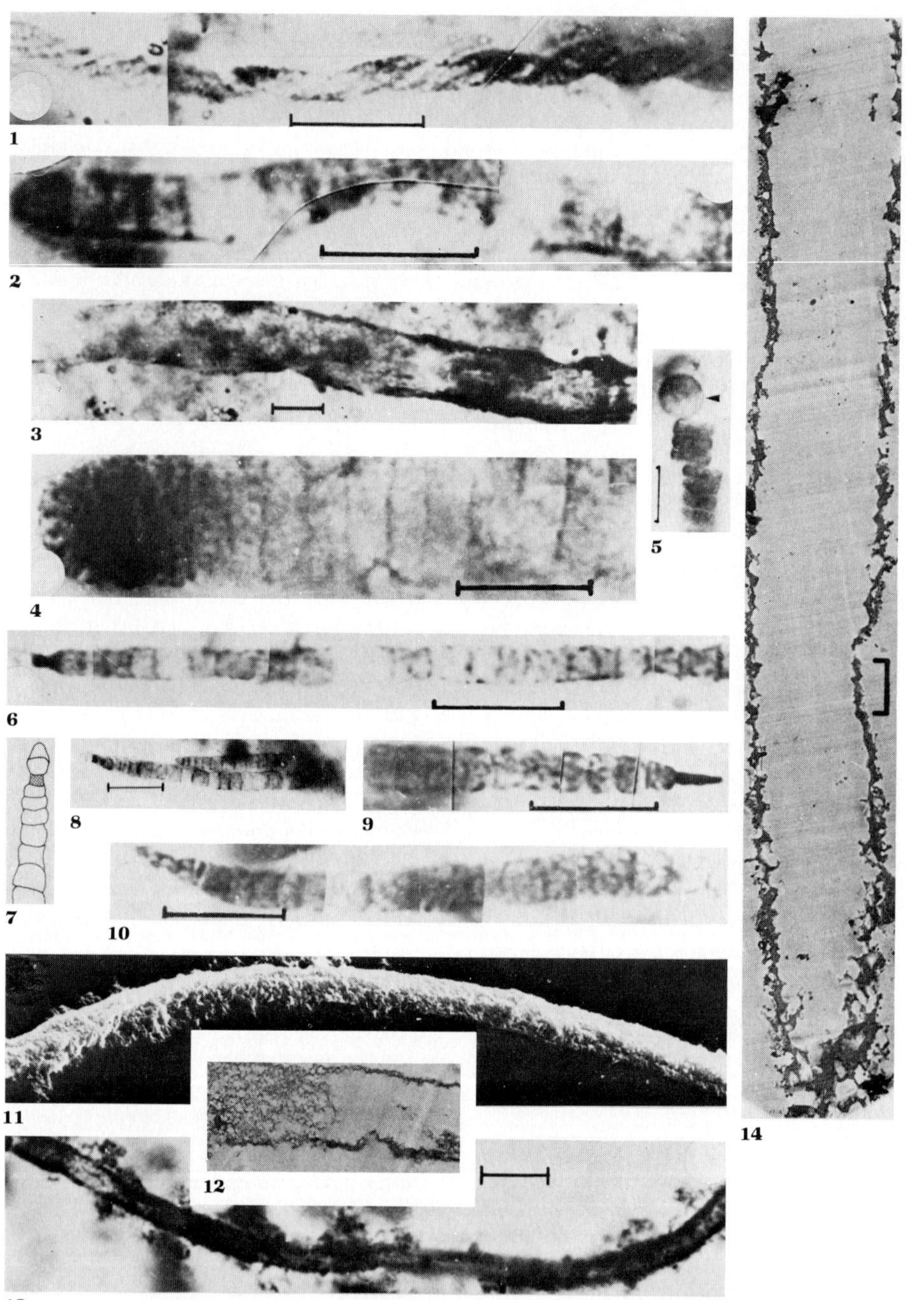

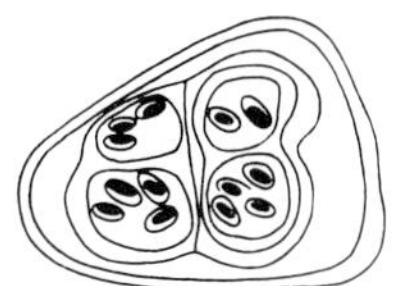

Figure 1.41
Gloeocapsomorpha prisca
Zalessky, Ordovician, Poland,
×355. From Adamczak, 1963.

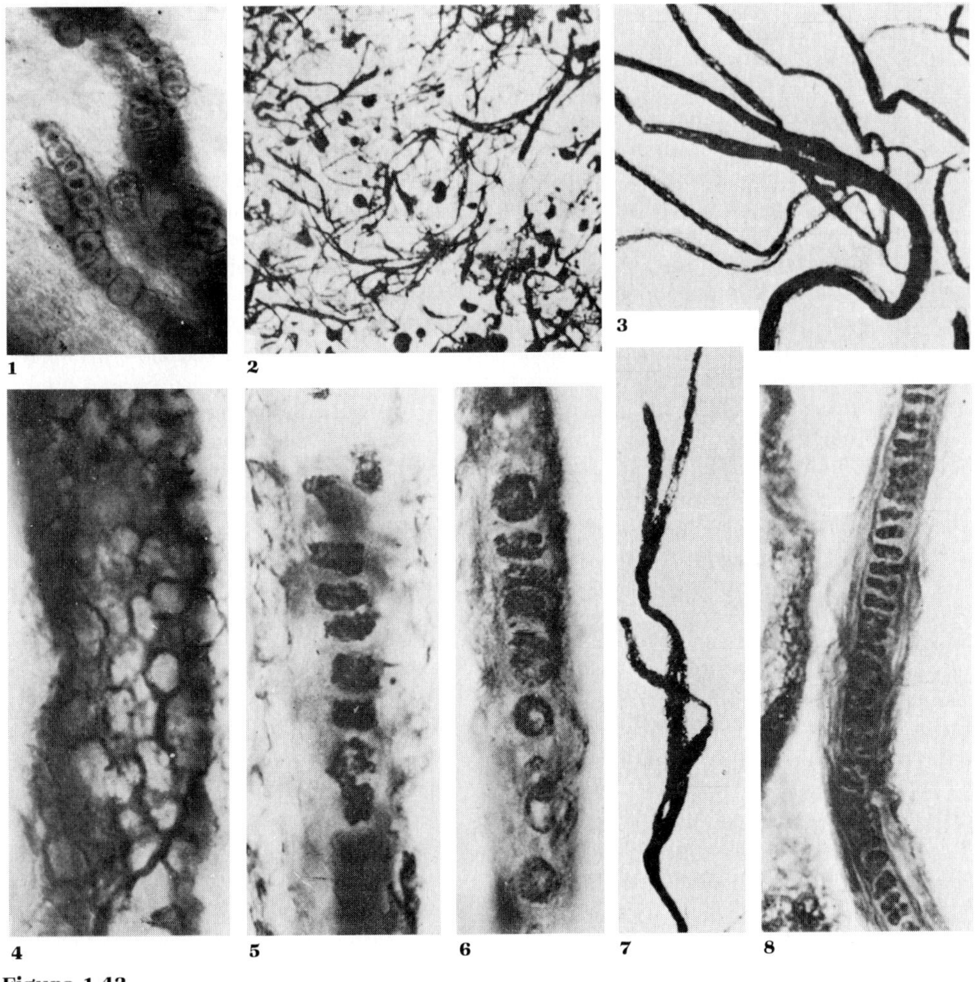

Figure 1.42
Early and middle Paleozoic Cyanophyta. **1.** *Kidstoniella fritschii* Croft & George, Rhynie Chert, Middle Devonian, Scotland, ×348; from Croft and George, 1959. **2 — 8.** *Schizothrichites ordoviciensis* Starmach, Lower Ordovician, Tremadocian, Poland, in thin sections of the chalcedony; from Starmach, 1963. **2.** Showing many filaments, ×28. **3.** Showing branching. **4,5.** Surface and optical section of a fragment, showing cellular divisions and surface reticulation. **6.** Trichome with loosened cells. **7.** Filament showing three branches, as in the living *Schizothrix*. **8.** Fragment, with some cells near top of figure apparently in process of division. 3,7, ×90; 4—6,8, ×960.

Cyanophyte Evolution

Their many similarities indicate that the bacteria were ancestral to the blue-green algae. The resemblance of the cyanophyte *Chroococcus* to the bacterium *Micrococcus,* and of *Eucapsis* to *Sarcina,* other than the pigments present in the blue-greens, led Stanier and van Niel (1941) to suggest that the Chroococcales evolved from the coccoid bacteria as a photosynthetic line paralleling at first the evolution of the purple bacteria, but also progressing farther. The similarity in wall composition and ultrastructure, and in the method of cell division of blue-green algae and Gram-positive bacteria (Echlin and Morris, 1965), provides additional evidence in support of this postulated ancestry.

The oldest known fossils in the South African Onverwacht probably include both bacteria and blue-green algae; the earliest life of some 3.5 billion years ago or more may have left no fossil record, either because it lacked a resistant composition, or because of the scarcity of suitable nonmetamorphosed Archaean rocks. Cells of both procaryote groups are protected by slime or mucus coatings or sheaths that trap sediment particles in the mats formed by the colonies. These stromatolitic mats provided protection against solar ultraviolet radiation (Margulis et al., 1976) prior to the development of an oxygenic atmosphere.

Both solitary cells and filamentous species appeared early. As mentioned previously, the similarity of some of the filamentous bacteria to the cyanophytes may result from the evolution of the bacteria such as *Beggiatoa, Thioploca,* and *Thiothrix* from the similar blue-green *Oscillatoria, Schizothrix,* and *Phormidium* by a secondary loss of their pigments. These are even regarded as apochlorotic cyanophytes in some modern classifications, although herein included with the bacteria.

Nitrogenous salts in solution absorb ultraviolet light, and the fixation of nitrogen by both bacteria and some blue-green algae may originally have provided protection from radiation as well as a source of cell nitrogen. The process of nitrate reduction also was a precursor to aerobic respiration, which evolved as oxygen accumulated in the atmosphere. The time of origin of blue-green algae, with their oxygen-producing photosynthesis, and the ensuing period of time before the atmosphere became oxygenic are still uncertain. Some regard the development of an ozone screen, at about 1 percent of the present level of atmospheric oxygen, to have occurred near the close of the Precambrian, providing a trigger for the evolution of the metazoa; others suggest it was present by about 2.0 billion years ago (Cloud, 1965, 1968; Schopf, 1974); or even within a few million years of the origin of the first blue-greens in the Archaean, perhaps 2.6 billion years ago (Dimroth and Kimberley, 1976). The presence of probable heterocysts in the fossil *Gunflintia* and in *Petraphera* supports the suggestion of an early oxygenic atmosphere, although others regard these as possibly degraded cells that burst within the sheath (Golubic and Barghoorn, 1977) rather than as specialized cells.

The simple Chroococcales probably were present in the early Precambrian, the colonial *Eucapsis*-like forms and filamentous Nostocales by middle Precambrian, and Pleurocapsales by late Precambrian. The perforating forms of the Stigonematales were represented by the middle Paleozoic (Silurian and Devonian).

The decrease in stromatolite diversity, loss of some morphologic types, and their generally smaller size after the Precambrian led to the suggestion that the presence of grazing invertebrates inhibited their development in the Phanerozoic. However, changes in the composition of the fossil communities may have affected the stromatolite morphology. The decline of the blue-greens may have been related to the decrease in atmospheric CO_2 and increase of oxygen, and to competition provided by the red and green algae as these evolved (Monty, 1973). Modern blue-greens are most abundant when other algae are restricted by one factor or another of the environment.

Classification

The history of classification of the blue-green algae was reviewed by Papenfuss (1955), who placed them as a Class Schizophyceae with the bacteria (Class Schizomyceteae) in the Division

Schizophyta. They also have been variously termed the Myxophyceae, Phycochromaceae, and Cyanobacteria. The classification is based on morphological features, cell form, type of cell division and type of resulting colony, and a few features of cell structure, such as the presence of inclusions, polar granules, gas vacuoles, or specialized cells such as akinetes and heterocysts. Because colony form may be influenced by environmental features, as shown by cultures, some specialists have drastically reduced the number of species, genera, and families recognized (Drouet and Daily, 1956; Drouet, 1968), but as certain of these supposed ecovariants occur together in nature, they may be true species. Blue-green taxonomy of the lower ranks thus is somewhat unstable; that of the higher categories given in Table 1.3 is generally accepted but may require modification as more biochemical data are obtained.

Table 1.3
Classification of the Division Cyanophyta.

Division Cyanophyta (Sachs 1874) Pascher 1931

I. Class Schizophyceae Cohn 1880

A. Order Chroococcales Wettstein 1924
Unicellular or colonial, colonies without polarity; reproduction by fission or endospores; no heterocysts, no hormogonia. Marine, freshwater, and terrestrial. Precambrian to Holocene.

1. Family Chroococcaceae Nägeli 1849
Abundacapsa Licari 1978 (Precambrian); *Africanosphaeroides* Boureau 1975 (Precambrian); *Aphanocapsa* Nägeli 1849; *Aphanocapsites* Maslov 1956 (Mississippian); *?Archaeosphaeroides* Schopf & Barghoorn 1967 (Precambrian); *Bigeminococcus* Schopf & Blacic 1971 (Precambrian); *Bisacculoides* J. Oehler 1977 (Precambrian); *?Caryosphaeroides* Schopf 1968 (Precambrian); *?Catinella* Pflug 1966 (Precambrian); *Chlamydomonopsis* Edhorn 1973 (Precambrian); *Chlorogloea* Wille 1900; *Chroococcus* Nägeli 1849 (Eocene, Holocene); *?Cumulosphaera* Edhorn 1973 (Precambrian); *Echaninia* Vologdin & Drozdova 1969 (Precambrian); *Eosynechococcus* Hofmann 1976 (Precambrian); *Eozygion* Schopf & Blacic 1971 (Precambrian); *Eucapsis* Clements & Shantz 1909 (Precambrian, Holocene); *?Glenobotrydion* Schopf 1968 (Precambrian); *?Globophycus* Schopf 1968 (Precambrian); *Gloeocapsa* Kützing 1843; *Gloeocapsomorpha* Zalessky 1917 (Ordovician to Jurassic); *Gloeodiniopsis* Schopf 1968 (Precambrian); *Gloioconis* Renault 1896 (Permian); *Gonamophyton* Vologdin & Drozdova 1964 (Precambrian); *?Huroniospora* Barghoorn 1965 (Precambrian; possibly a red alga); *Melasmatosphaera* Hofmann 1976 (Precambrian); *Merismopedia* Meyen 1839 (syn.: *Agmenellum* de Brébisson 1839); *Microcystis* Kützing 1833 (Devonian, Holocene); *?Millaria* Pflug 1966 (Precambrian); *Murandavia* Vologdin 1965 (Precambrian); *Myxococcoides* Schopf 1968 (Precambrian); *Nanococcus* J. Oehler 1977 (Precambrian); *Palaeoanacystis* Schopf 1968 (Precambrian); *Palaeomicrocystis* Korde 1955 (Cambrian); *Phanerosphaerops* Schopf & Blacic 1971 (Precambrian); *Praechrococcus* Vologdin) 1962 (Precambrian); *Sphaerophycus* Schopf 1968 (Precambrian); *Stipulella* Maslov 1956 (Mississippian); *Subtetrapedia* Renault 1899 (Mississippian); *Synechococcus* Nägeli 1849 (includes endosymbionts); *Uranovia* Korde 1958 (Precambrian, Cambrian); *Vesicophyton* Vologdin & Drozdova 1969 (Precambrian); *?Zosterosphaera* Schopf 1968 (Precambrian). *Kakabekia* Barghoorn 1965 and *Stellasphaeroides* Boureau 1975 (both Precambrian) may belong here.

2. Family Entophysalidaceae Geitler 1925
Bogutchania Korde in Vologdin & Korde 1965 (Ordovician); *Entophysalis* Kützing 1843; *Entophysalites* Larishchev 1952 (Jurassic); *Eoentophysalis* Hofmann 1976 (Precambrian); *Kareliana* Korde in Vologdin & Korde 1965 (Precambrian); *Paracapsa* Naumann 1924

(*continued*)

(iron-depositing); ?*Sphaerocongregus* Moorman 1974 (Precambrian); ?*Synsphaeroides* Boureau 1975 (Precambrian).

3. Family Tubiellaceae Elenkin 1934
Johannesbaptista J. de Toni 1934.

B. Order Chamaesiphonales Wettstein 1924
Unicellular or colonial; colonies with distinct polarity; reproduction by endospores or exospores; no heterocysts. Mostly marine, few in fresh water. Holocene.

1. Family Chamaesiphonaceae Borzi 1882
Chamaesiphon A. Braun & Grunow 1865 (iron-depositing).
2. Family Cyanophanaceae Geitler 1955
Cyanophanon Geitler 1955.
3. Family Clastidiaceae Drouet & Daily 1952
4. Family Siphononemataceae Geitler 1925
5. Family Pascherinemataceae Geitler 1942
Syn.: Endonemataceae Pascher 1929
Pascherinema J. de Toni 1936 (syn.: *Endonema* Pascher 1929 non Jussieu 1846).

C. Order Pleurocapsales Geitler 1925
Filamentous, reproduce by endospores; no heterocysts. Marine and freshwater. Precambrian to Holocene.

1. Family Chroococcidiaceae Geitler 1933
Myxosarcina Printz 1921.
2. Family Hyellaceae Borzi 1914
Syn.: Pleurocapsaceae Geitler 1925
Hyella Bornet & Flahault 1888 (perforating); *Oncobyrsella* Johnson 1937 (Oligocene); *Palaeopleurocapsa* Knoll, Barghoorn & Golubić 1975 (Precambrian); *Pleurocapsa* Thuret ex Hauck 1885.

D. Order Nostocales Geitler 1925
Filamentous, not separated into prostrate and upright filaments; multiplication by hormogonia (short, motile filaments); may lack or have heterocysts; spores and akinetes present. Mostly freshwater, some marine; Precambrian to Holocene.

1. Family Protospiraceae Vologdin 1969
Protospira Vologdin 1969 (Precambrian).
2. Family Thysanoplantaceae Vologdin & Titorenko 1966

Thysanoplanta Vologdin & Titorenko 1966 (Precambrian).

3. Family Oscillatoriaceae (S. F. Gray) Dumortier ex Kirchner 1898
Animikiea Barghoorn 1965 (Precambrian); *Archaeothrix* Kidston & Lang 1921 (Devonian); *Calyptothrix* Schopf 1968 (Precambrian); *Cephalophytarion* Schopf 1968 (Precambrian); *Contortothrix* Schopf 1968 (Precambrian); *Crinalium* Crow 1927 (Eocene; Holocene); *Cyanonema* Schopf 1968 (Precambrian); ?*Entosphaeroides* Barghoorn 1965 (Precambrian; possibly a cyanophycean sheath occupied by a chroococcacean); *Filiconstrictosus* Schopf & Blacic 1971 (Precambrian); *Heliconema* Schopf 1968 (Precambrian); *Lyngbya* C. A. Agardh 1824 (may be iron-depositing); *Marpolia* Walcott 1919 (Cambrian); *Obconicophycus* Schopf & Blacic 1971 (Precambrian); *Oscillatoria* Vaucher 1803; *Oscillatoriopsis* Schopf 1968 (Precambrian); *Oscillatoriites* Zalessky 1926 (Jurassic); ?*Palaeospiralis* Edhorn 1973 (Precambrian); *Palaeospirulina* Edhorn 1973 (Precambrian); *Palaeolyngbya* Schopf 1968 (Precambrian); *Partitiofilum* Schopf & Blacic 1971 (Precambrian); *Phormidium* Kützing 1843 (?Permian, Tertiary, Holocene); *Rhicnonema* Hofmann 1976 (Precambrian); *Schizothrichites* Starmach 1963 (Ordovician); *Schizothrix* Kützing 1843 (?Precambrian, Holocene); *Siphonophycus* Schopf 1968 (Precambrian); *Spirulina* Turpin 1827 (Eocene-Holocene); *Tenuofilum* Schopf 1968 (Precambrian); *Trichodesmium* Ehrenberg ex Gomont 1892 (marine plankton, N-fixing). *Palaeomicrocoleus* Korde in Vologdin & Korde 1965 (Precambrian) represents trails of ambient mineral grains originally figured by Gruner, and not an organism, according to Cloud (1965).

4. Family Gomontiellaceae Elenkin 1936
Gomontiella Teodoresco 1901.

5. Family Nostocaceae Dumortier 1829 ex Engler 1892
Syn.: Anabaenaceae Elenkin 1934
Anabaena Bory 1822 (Devonian, Holocene); *Anabaenidium* Schopf 1968 (Precambrian); *Aphanizomenon* Morren 1838; *Archaeonema* Schopf 1968 (Precambrian); *Beckspringia* Licari 1978 (Precambrian); ?*Confervites* Brogniart 1828

(Tertiary); *Cylindrospermum* Kützing 1843 (N-fixing); *Dikimdinella* Kolosov 1966 (Precambrian); *?Filamentella* Pflug 1965 (Precambrian); *Gunflintia* Barghoorn 1965 (Precambrian); *Morania* Walcott 1919 (Cambrian); *Nodularites* Maslov 1956 (Cambrian, U. Cretaceous); *Nostoc* Vaucher 1803 (Devonian, Tertiary, Holocene); *Nostocopsis* Schopf 1968 (Precambrian); *Palaeonostoc* Sastri, Venkatachala & Desikachary 1972 (Cambrian); *Petraphera* L. A. Nagy 1974 (Precambrian); *Veteronostocale* Schopf & Blacic 1971 (Precambrian); *Visheraia* Korde 1958 (Silurian, Devonian). The Devonian *Izhella* Antropov 1955 may be a parathuramminid foraminifer, a poorly preserved chroococcalean colony, or a nostocacean.

6. Family Scytonemataceae Kützing 1843 ex Bornet & Flahault 1886
Encrusta Daley 1975 (Oligocene); *Epivalvia* Daley 1975 (Oligocene); *Halythrix* Schopf 1968 (Precambrian); *Palaeoscytonema* Edhorn 1973 (Precambrian); *Plectonema* Thuret 1875 ex Gomont 1892 (perforating alga; nonheterocystous, N-fixing under anaerobic conditions); *Rhyniella* Croft & George 1959 (Devonian); *Scytonema* C. A. Agardh 1824 (Tertiary, Pleistocene, Holocene; perforating alga); *Tolypothrix* Kützing 1843.

7. Family Microchaetaceae Lemmermann 1907
Microchaete Thuret 1875 (nom. conserv. ICBN; non *Microchaete* Bentham 1845).

8. Family Rivulariaceae Kützing 1843 ex Bornet & Flahault 1886
Aldanella Kolosov 1966 (Precambrian); *Apophoretella* Elliott 1975 (Jurassic); *Azyrtalia* Vologdin & Drozdova 1969 (Precambrian); *Calothrix* Agardh ex Bornet & Flahault 1886; *Caudiculophycus* Schopf 1968 (Precambrian); *Dendractis* Reis 1923 (Tertiary); *Dichothrix* Zanardini 1858 (marine, nitrogen-fixing); *Dimorphostroma* Reis 1921 (Tertiary); *Globuloella* Korde 1958 (Cambrian); *Gloeotrichata* W. Bock 1961 (Triassic); *Gloeotrichia* J. Agardh 1842; *?Jatuliana* Korde in Vologdin & Korde 1965 (Precambrian); *Kyrtuthrix* Ercegović 1929 (marine, perforating alga); *Nelcanella* Vologdin & Drozdova 1964 (Precambrian); *Panomninella* Kolosov 1966 (Precambrian); *Primorivularia* Edhorn 1973 (Precambrian); *Protorivularia* Butin 1959 (Precambrian); *Rivularia* C. A. Agardh 1824 (Miocene-Holocene); *Rivu-*

larialithus Maslov 1955 (Miocene); *Rivularites* Fliche 1905 (Triassic); *Ternithrix* Reis 1928 (Miocene); *Zonotrichites* Bornemann 1886 (Silurian to Triassic). *Palaeorivularia* Korde 1958 is an artifact due to the growth of fibrous chalcedony, and not an organism.

E. Order Stigonematales Geitler 1925
Filamentous, with distinct prostrate and upright filaments; reproduce by hormogonia or rarely by akinetes; true branching; heterocysts present. Freshwater; Devonian to Holocene.

1. Family Stigonemataceae Hassal 1845
Hapalosiphon Nägeli in Kützing 1849 (Eocene; Holocene); *Kidstoniella* Croft & George 1959 (Devonian); *Langiella* Croft & George 1959 (Devonian); *Stigonema* C. A. Agardh 1824.
2. Family Capsosiraceae Geitler 1925
Syn.: Pulvinulariaceae Geitler 1925
3. Family Mastigocladopsidaceae Iyengar & Desikachary 1946
Mastigocladopsis Iyengar & Desikachary 1946.
4. Family Mastigocladaceae Geitler 1925
Iyengariella Desikachary 1953 (N-fixing); *Mastigocladus* Cohn 1862.
5. Family Borzinemataceae Geitler 1942
Syn.: Diplonemataceae (Borzi) Elenkin 1934
Borzinema J. de Toni 1936 (syn.: *Diplonema* Borzi 1917, non Ehrenberg 1833 nec Don 1837).
6. Family Nostochopsidaceae Geitler 1925
?Chaetophorites Fliche 1886 (Silurian, Jurassic, Oligocene; perforating); *Mastigocoleus* Lagerheim 1886 (perforating); *?Palaeachlya* Duncan 1876 (Silurian; perforating); *?Palaeopede* Etheridge 1899 (Devonian, Jurassic, Cretaceous; perforating).

Uncertain position. "Porostromata"
Calcified sheaths of blue-green algae, such as the living *Plectonema* and *Scytonema*, but for which details of the cells or trichomes cannot be distinguished as fossils. Some may be green or red algae. *?Bevocastria* Garwood 1931 (Mississippian); *Cayeuxia* Frollo 1938 (Jurassic); *?Garwoodia* Wood 1941 (Mississippian); *Girvanella* Nicholson & Etheridge 1880 (Cambrian to Cretaceous, Holocene); *?Hedstroemia* Rothpletz 1913 (Silurian); *?Mitcheldeania* Wethered 1886 (Mississippian); *?Ortonella* Garwood 1941 (Cambrian to Jurassic); *Sphaerocodium* Rothpletz 1890 (Triassic).

PROCHLOROPHYTA, THE GREEN PROCARYOTES

A major distinction between the blue-green algae and the eucaryotic algae is their possession of a single chlorophyll, chlorophyll *a*; living green algae and euglenids have chlorophylls *a* and *b*, chrysophytes and diatoms have *a* and *c*, and red algae, *a* and *d*. Blue-green algae also have bilin pigments that give their usual red or blue appearance. However, a few procaryotic algae with a distinctive green appearance have been reported. Examination by Lewin and Withers (1975) of the pigment complex in detail of one of these, referred to *Synechocystis*, showed the presence of chlorophyll *a* and *b*, β-carotene (which was some 65 percent of the yellow pigmentation), and three xanthophylls. No red or blue bilin pigments were present. This species was symbiotic with colonial ascidians in Baja California, Mexico. They also reported a similar procaryote lacking bilin pigments from Australia, and observed that certain algal mats from Yellowstone Park had been reported to contain chlorophyll *b*. True green algae are not known to withstand such high temperatures, hence they suggested that the hot-spring species probably was similar to the marine *Synechococcus*. Later detailed studies with electron microscopy showed the sectioned symbiont to have the typical procaryote ultrastructure. The 8- to 14-μm-diameter cell has an electron-transparent central area (genophore) with fine fibrils of DNA and a wall of four layers. Most of the cell interior is filled with many thylakoids with smooth surfaces and no projecting phycobilisomes (corresponding to their lack of bilin pigments); the thylakoids occur in aggregates of up to 20 double lamellae, mostly forming a broad peripheral band, but with a few crossing the cell interior. A few tiny (less than 1.2 μm diameter) spherical bodies with a crystalline substructure of hexagonal prisms resemble the polyhedral bodies of bacteria and blue-green algae (Lewin, 1975; Schulz-Baldes and Lewin, 1976). In view of these distinctive features, intermediate between the blue-green and green algae, the species *Prochloron didemni* (Lewin) Lewin was placed in a new

family, order, class, and the Division Prochlorophyta Lewin 1976 (see Table 1.4). Although presently monotypic, additional species probably have been misidentified as Cyanophyta, and will be reassigned after a study of their pigments. However, others would prefer redefinition of the Cyanophyta to include taxa of various pigmentation (Antia, 1977).

ORIGIN OF THE EUCARYOTES

Eucaryotic cells have distinct nuclei and mitochondria bounded by double membranes, and eucaryotic plants have similarly membrane-bounded plastids; most eucaryotes also have characteristic flagella or cilia with the 9+2 fibril structure at some stage of their life cycle and have distinctive methods of cell division. Their uniformity in structure as well as in their biochemical processes strongly indicates a common ancestry, and as all present life also is generally believed to be related, the eucaryotes must have evolved from the procaryotes; how this occurred is the subject of varied hypotheses.

As already discussed, the photosynthetic procaryotes probably evolved from heterotrophic

Table 1.4
Classification of the Division Prochlorophyta.

Division Prochlorophyta   Lewin 1976
Photosynthetic procaryotes that resemble blue-green algae in lacking membrane-bound organelles, but differ in possessing both chlorophyll *a* and *b*, and in lacking red or blue bilin pigments.

Class Prochlorophyceae Lewin 1977
Order Prochlorales Lewin 1977
    Family Prochloraceae Lewin 1977
    Unicellular prochlorophytes; multiply by binary division; may be extracellular symbionts of marine ascidians (*Prochloron*). Living.
    Prochloron Lewin 1977.

---

procaryotes. Similarly, cyanophytes probably evolved from the photosynthetic bacteria, although some have suggested the reverse evolutionary direction. The different pigment complex of the blue-green algae allowed the use of water as a hydrogen donor for carbon fixation in photosynthesis and produced oxygen as a by-product. As oxygen accumulated in the atmosphere, aerobic metabolism became necessary.

Oxygen respiration may have evolved from the nitrate respiration of bacteria, as the two processes have in common most components of the electron transport system (Hall, 1971). The new processes made available more energy, allowing the evolution of more elaborate organisms and requiring an increase in DNA. Eventually the segregation of the DNA into a distinct nucleus was advantageous, and as the cells increased in size, other components were segregated into separate organelles. Among these organelles are the mitochondria, which are essential for eucaryote respiration, as they contain the electron transport system that generates ATP by oxidative phosphorylation, in addition to many important enzymes. Other organelles are the plastids containing the pigments for photosynthesis, but which do not respire (Stanier and van Niel, 1962). Interestingly, both plastids and mitochondria contain some DNA that partly controls the growth and multiplication of these organelles, although they are in part controlled by genes of the cell nucleus. The important and major change from an undifferentiated procaryote cell to the eucaryote cell of larger size and with distinctive organelles occurred during the Precambrian, but as the evidence supplied by the fossil record is equivocal, both the time of first appearance and the way in which this evolutionary step occurred are conjectural. "Whether or not life originated more than once, it is certain that life possessing nuclei came into existence once only, by evolution from non-nucleate life. This conclusion is as certain as any which can be based on induction—it is established by the uniformity of the nucleus, in its structure and in its behavior, in mitosis, in sexual reproduction, and as the vehicle of Mendelian heredity, wherever it occurs" (Copeland, 1947, p. 342).

The development of the membrane-bounded nucleus, plastids, mitochondria, and other organelles in the primitive eucaryote classically was assumed to be endogenous, produced by differentiation of the cell itself, and variations on this model are discussed first.

The contrasting hypothesis suggesting that these organelles arose by endosymbiosis also is an old hypothesis that has been revived as including a series of symbioses.

The Evolutionary or Autogenous Origin of Eucaryotes

Briefly, the evolutionary hypothesis suggests that the organelles of the eucaryote arose by differentiation within the cell itself, hence are of autogenous origin, arising either by complex pseudopodial projections from the surface (see Figure 1.43) or by invaginations of the outer membranes to enclose parts of the cell within the cell membranes. The photosynthetic lamellae of the cyanophytes and bacteria appear to arise from the membrane in this way. Some of these complex membrane folds formed the endoplasmic reticulum, Golgi apparatus, and nuclear envelope, and some pinched off segments of the cytoplasm and areas of the dispersed DNA-containing genophore to become the mitochondria and plastids. In support of this suggestion is the fact that the organelle membranes are always paired. Similarly paired membranes form the extensive membrane system of the endoplasmic reticulum. The double membrane surrounding mitochondria has been found connected to the endoplasmic reticulum, and continuities have been observed between the endoplasmic reticulum, outer plastid membrane, outer mitochondrial membrane, cell membrane or plasmalemma, the outer nuclear membrane, and other intracellular structures (F. J. R. Taylor, 1974); Golgi bodies also are a part of the membrane system. In addition, photosynthetic lamellae in the bacterium *Ectothiorhodospira mobilis* Pelsh are stacked in a manner resembling the grana stacks of higher plant plastids, the individual stacks being surrounded by a double lamella

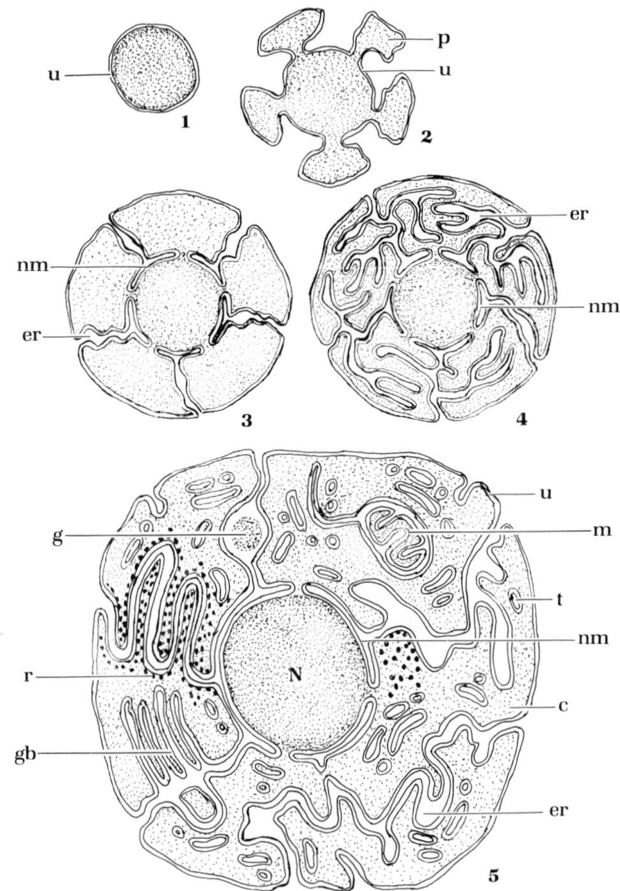

that belongs to the lamellar system. These lamellae are derived by invagination of the outer plasma membrane, in a way very similar to the proposed autogenous theories of plastid formation (Stubbe, 1971).

Robertson (1962) suggested that the cell organelles were produced by the development of projections from the cell surface, whereas Raff and Mahler (1972) and Chadefaud (1974) postulate invaginations from the surface. Raff and Mahler suggested that the protoeucaryote was a relatively large, advanced, aerobic cell, whose large size required an increase in respiratory membrane surface, obtained by invagination of the inner cell membrane (as in some modern bacteria), with membrane-bound vesicles later pinched off from the invaginations. As some of

the protein synthesis needed to be related to these organelles, DNA that was involved in this synthesis and located near the respiratory invaginations of the procaryote was pinched off with the organelles as they formed.

Chadefaud (1974) modified the invagination hypothesis as a two-stage process (see Figure 1.44), the first stage producing the respiratory crests and thylakoids, similar to those of modern procaryotes. Nuclear material became segregated in the center, although some remained in the peripheral regions. Later invaginations enclosing the respiratory crests and their associated DNA evolved into mitochondria, and the membrane invaginations around the thylakoids and associated DNA became the plastids. Additional membrane invaginations resulted in the endoplasmic reticulum, whose innermost ramifications formed the perforate nuclear envelope around the major central region of DNA to form the typical eucaryote nucleus.

Under the autogenous hypotheses, the protoeucaryote must have evolved from a photosynthetic procaryote (cyanophyte), as otherwise the unlikely multiple independent origins of chlorophyll *a*-based photosynthesis would be implied (F. J. R. Taylor, 1976). The protoeucaryote of the autogenous hypotheses has been termed the "Uralga" (Klein and Cronquist, 1967).

A modification of the autogenous theories was proposed to involve evolution of the eucaryote nucleus from a coenocytic (Goksøyr, 1967) or colonial procaryote (Nass, 1969). As the cytoplasm of the colony or coenocyte fused, the procaryote genophores combined to form the larger eucaryote nucleus. The similarity of dinoflagellate chromosomes (see Chapter 4) to those of bacteria (both lack histones) thus is the result of the primitive pooling of the procaryote DNA; eucaryote nuclei with histone-containing chromosomes evolved later. Mitochondria and plastids may then have arisen either autogenously or symbiotically.

Coenocytic or colonial associations may not have been necessary to produce the larger nucleus, however. Much later evolutionary change has involved polyploidy, the doubling of chromosomes without their later division among new daughter cells. A similar process as the cell size increased may have been the source of the larger nuclei.

A detailed explanation for the presence of DNA in the mitochondria was suggested by Reijnders (1975). The original aerobic protoeucaryote cell had multiple copies of its genome (as do many blue-green algae), which were trapped in separate invaginations of the inner layer of the wall. After developing a stable relationship between these compartmentalized DNAs and the relatively large cell, opposing selection pressures favored either the elimination of redundant information and centralized control with enslavement of the protomitochondrion by the protonucleus, or the maintenance of the status quo as in the ances-

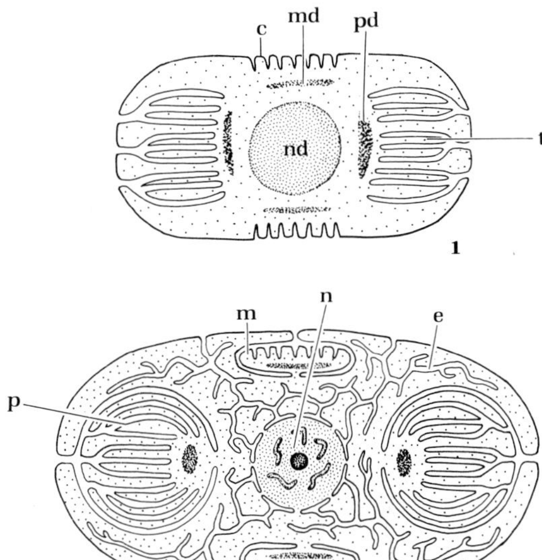

Figure 1.44
Hypothetical origin of the eucaryote cell by invagination of the outer membrane in two stages. **1.** First stage, in which invaginations produced the respiratory crests (c) similar to the cristae of the mitochondria; thylakoids (t) containing the chlorophyll pigments like those of the cyanophyceans form from the plasmalemma; segregation of most of the cell DNA produced the nuclear DNA (nd) at the center, but left some DNA peripherally associated with the respiratory crests to become the mitochondrial DNA (md), and with the thylakoids to become the plastidial DNA (pd). **2.** Secondary stage of invaginations in which thylakoids and their DNA are enclosed as the plastids (p), and the respiratory crests and associated DNA are enclosed to become the mitochondria (m); additional ramifying invaginations form the endoplasmic reticulum (e), which also produced the nuclear envelope surrounding the nucleus (n) with its chromosomes. Redrawn from Chadefaud, 1974.

tral cyanophyte because of the large cell size and to avoid the dangers of possible unfavorable mutations. Selection eventually favored a division of labor and metabolic specialization, and elimination of the redundant genes. The differences in mitochondrial genome size between animal and plant cells resulted from differential loss, and not from complicated insertion-excision processes of gene transfer as suggested for the endosymbiotic hypothesis.

Bogorad (1975) described an autogenous method of eucaryote evolution as the cluster-clone hypothesis. Gene clusters were formed in the procaryote cell, each cluster being a separate genome. Later a membrane developed around each cluster and the immediate cytoplasm, and from these structures the nucleus, mitochondria, and plastids eventually evolved; the gene-free space remained as the cell cytoplasm. Each separate gene-containing compartment divided and reproduced itself, producing clones of nuclei, plastids, and mitochondria, the entire complex evolving into the present variety of eucaryotic cells. The membrane enclosures may have developed in sequence, rather than at one time, the nucleus developing first, and other DNA included with the mitochondria and plastids as these evolved.

Evolution of a blue-green alga lacking a cell wall into a eucaryote was suggested by Cavalier-Smith (1975) to be the result of phagocytosis. The ancestor was regarded as an isolated cyanophyte, with oxygen-evolving photosynthesis and oxygen-based respiration, living in a shallow-water, detritus-rich environment. Perhaps associated bacteria produced enzymes that destroyed the peptidoglycan cell wall. The development of an ability to phagocytose the associated bacteria during the night, high latitude winters, or other dark environments would provide a selective advantage. Some modern algae may have such alternate nutritional methods. Eucaryote cell division occurs by a similar process of furrowing, involving a calcium-activated contractile actin-myosin-like microfilament system. Evolution of this system would have made possible both phagocytosis and amoeboid movement, and eventually cytoplasmic streaming and organelle movement. The increased cell size associated

with phagotrophy made advantageous a compartmentalization of the cell into regions specializing in particular functions, and the increase of membrane material provided a mechanism for the compartmentalization (see Figure 1.45). Chromosome segregation evolved in relation to the microtubule system, as procaryotelike attachment of DNA to the plasma membrane would not be possible with a highly mobile cell surface. Invagination of the area of attachment produced the nuclear membrane. Cytosis also provided a means of cell fusion of the wall-free cells, probably originally as a response to nutrient limitation, as fused cells would have a doubled internal food supply. Cavalier-Smith (1975) suggested that cell fusion probably evolved even before the development of the nuclear envelope, as many properties of the eucaryote chromosomes suggest a simultaneous evolution of mitosis and meiosis. A probable primitive sexuality also was postulated by Gabriel (1960). Increased size of a circular chromosome like that of bacteria would lead to breakage and eventual selection for linear chromosomes, with their advantage of allowing recombination by pairing more readily. The distinctive features of eucaryote chromosomes, mitosis and meiosis (and sexual reproduction), probably evolved together over a short period of time in the earliest stage of eucaryote evolution, as a direct consequence of cytosis (Cavalier-Smith, 1975). The protoalga was similar to a red alga; later acquisition of flagella produced the chromoamoeboflagellate. New cell walls were independently acquired by the red algae and by the flagellated algal lineages; loss of sexual reproduction as in the euglenids was secondary, as was loss of plastids in evolution of the protozoa and their animal descendants. Higher fungi may have evolved from red or green algae by loss of plastids, and lower fungi similarly arose from the brownish flagellates by loss of plastids.

An expanded autogenous sequence was proposed by F. J. R. Taylor (1976) as beginning with a procaryote having a wall, ribosomes in the matrix, and genome dispersed into several membrane associated areas; lamellae throughout the cell performed both photosynthesis and respiration, and polar granules may have been

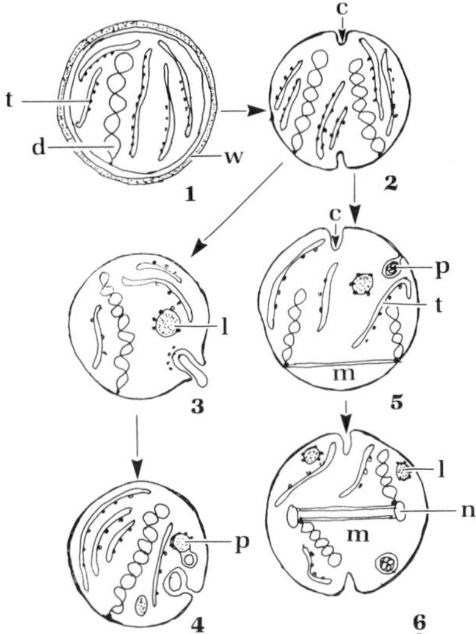

Figure 1.45
Origin of the eucaryotic cell by phagocytosis.
1. Ancestral cyanophyte with cell wall (w), circular chromosome of DNA (d) twisted and attached to cell membrane, and thylakoids (t) with ribosomes. **2.** Alga has lost the cell wall and developed an acto-myosin dependent cleavage (c). **3.** Protolysosomes (l) evolve from the thylakoids as digestive organelles that may partly or completely surround the prey before liberating the enzymes. **4.** Completely phagocytic form with protolysosomes fusing with the phagosomes (p). **5.** Evolution of spindle microtubules (m) aids separation of the daughter chromosomes in division so that each half retains one chromosome; chromosomes still attached to cell membrane. **6.** Attachment sites of the chromosomes engulfed to become the protonuclear envelope (n), with spindle now completely intracellular.
Simplified and redrawn from Cavalier-Smith, 1975.

primitive microtubule organizing centers (the MTOCs of Pickett-Heaps, 1969). Multiplication of one of the dispersed genophores formed a large central nucleoid; surrounded by the innermost lamellae, it formed the nucleus to which the MTOCs moved. Gradual membrane specialization produced thylakoids that were specialized for photosynthesis, and other membranes for respiration; still other membranes evolved as the endoplasmic reticulum and such derivatives as the nuclear envelope, Golgi bodies, and so forth. Following loss of the procaryotic cell wall, the onset of cytosis or engulfment produced vacuoles and the membrane system around the evolving organelles and associated genomes. Mitotic spindle microtubules appeared and size increased, as eucaryotes are generally about 10 times the average size of procaryotes. Red algae may have diverged at this stage. Later the MTOCs became centriolelike and gave rise to flagella.

Endosymbiotic Origin of Eucaryotes

Near the turn of the century, the origin of eucaryote plastids from symbiotic blue-green algae was proposed by Schimper (1883), Mereschkowsky (1905), and Famintzen (1907), and the origin of mitochondria from endosymbiotic bacteria (Altmann, 1890), but were given very little credence. In recent years, new lines of evidence have been used in support of the origin of these and other organelles by endosymbiosis.

Modern revivals of this theory range from those accepting an endosymbiotic origin for plastids only (Ris, 1961), to those regarding both plastids and mitochondria as endosymbiotic (Goksøyr, 1967; Raven, 1970), to the "Serial Symbiosis Hypothesis" (see Figure 1.46), in which not only plastids and mitochondria but also flagella and the nuclear spindle are regarded as resulting from symbioses (Sagan, 1967; Margulis, 1970, 1971).

The proposed sequence consisted of heterotrophic anaerobes, then phototrophic anaerobes followed by oxygen-producing phototrophs (blue-green algae) and, as the oxygen increased in the environment, by aerobic procaryotes. The first step in the evolution of the eucaryotes consisted of the ingestion of an aerobic bacterium (the protomitochondrion) into the cytoplasm of a heterotrophic anaerobe, and the change of this association to an obligate symbiosis (Sagan, 1967), producing an aerobic amoeboid organism. A similar procedure resulted in the production of flagella from a

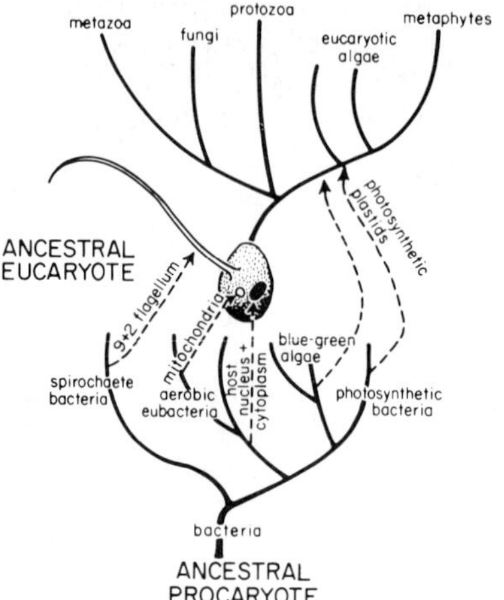

Figure 1.46
The origin of the eucaryote cell by serial symbiosis. The ancestral procaryote was a heterotrophic anaerobe, but diversified into aerobic eubacteria, photosynthetic bacteria, spirochaetes, and blue-green algae. In response to environmental changes, an anaerobe became the host cell in which endosymbiotic aerobic bacteria later became the protomitochondria. An endosymbiotic spirochaete became the flagellum of the primitive eucaryotic amoeboid protozoan, its descendents becoming the protozoa, fungi, and metazoa. Another symbiotic relationship between these eucaryote protozoans and endosymbiotic procaryote cyanophytes led to the permanent capture of the latter as plastids, giving rise to the various groups of algae and eventually to higher plants. From Tappan, 1974, modified and redrawn from Margulis, 1970.

motile endosymbiont such as a spirochaete bacterium. Later the chromosomal centromeres and centrioles were differentiated from the 9+2 flagellar basal bodies, making possible the development of mitosis. Finally, another symbiotic association of a blue-green alga with the new eucaryote cells evolved into the plastids of the eucaryotic algae and higher plants.

Nature of the Host Symbiont

The host cell for the symbiont is variously regarded as having been an anerobic heterotroph ingesting an aerobic bacterium that evolved into

a mitochondrion (Sagan, 1967), a coenocytic group of cells (Goksøyr, 1967), or an aerobic phagocyte that relied on peroxisomes for its respiratory metabolism, rather than a phosphorylating electron transport system (Duve, 1973). Biochemical evidence suggesting that the original host cell of the eucaryote was aerobic includes the presence of enzymes in the cytoplasm to protect cell contents from autooxidation; the requirement for products of aerobic eucaryotic metabolism by even the few known anaerobic eucaryotes; synthesis of steroids and unsaturated fatty acids only by aerobic pathways in the eucaryote cytoplasm; and cytochromes functioning in the cytoplasm that differ from those of the mitochondria (Raff and Mahler, 1972, 1973).

Nucleus

One difficulty with the endosymbiont hypothesis is that of explaining the origin of the nucleus and its double membrane (Cavalier-Smith, 1975). This is generally regarded as developed from the outer membrane as in the autogenous theories, either directly or indirectly from the endoplasmic reticulum. In contrast, the similarly double membrane of the plastids and mitochondria are postulated as being partly from the host (outermost of the two membranes) and partly from the symbiont (the inner organellar membrane).

Goksøyr (1967) suggested that the nucleus arose by the accumulation of the DNA as a polyploid from a coenocytic group of procaryotes, followed by development of a primitive mitotic mechanism as in procaryotes, and then development of the membrane. As mutations occurred in some of the repeated genomes, greater differentiation became possible.

However, a greater difficulty lies in the fact that most of the mitochondrial components are controlled by the nuclear genome, including parts of the cytochrome complex, which is assumed to be endosymbiont derived. Perhaps many of the genes were transferred from the symbiont protomitochondrion to the nucleus by conjugation or transformation, as in bacteria, although what possible advantage could have resulted in the gene migration is uncertain (Reijnders, 1975). The protomitochondrion originally may have had multiple copies of its genome (polyploid), some of which migrated to the nucleus; the later differential loss of redundant information resulted in some being retained in the mitochondrion and some being present in the nucleus (Bogorad, 1975).

Plastids and Mitochondria as Endosymbionts

Most discussions of the endosymbiont hypothesis of eucaryote origin concentrate on the similarity of plastids to entire cells of blue-green algae, and of mitochondria to aerobic bacteria. Both mitochondria and plastids of eucaryotes are similar in size (typically 1 to 5 μm) and structure to entire procaryotic cells, and their growth and division is semiautonomous, only partly controlled by the nuclear DNA. The biochemical processes of the mitochondria and plastids, in particular their protein-synthesizing system, also are similar to those of entire procaryote cells, and these organelles arise only by division of preexisting ones, never *de novo*. The plastids of eucaryotes are multilamellate, resembling the entire cyanophyte cell with its closed sacs or thylakoids bounded by a membrane about 7 to 8 nm thick, and containing chlorophyll *a* and accessory pigments. Isolated plastids can be maintained in culture, growing and dividing outside a cell (Ridley and Leech, 1970). Both plastids and mitochondria contain DNA as 2.5 nm fibrils that are not organized into chromosomes, as well as RNA. The RNA and DNA are distinct from that of the nucleus and replicate independently; like that of procaryotes, the DNA is histone-free and bound to membranes. A mitochondrion contains from two to six copies of its DNA in individual compartments, similar to a filamentous bacterium that replicated but failed to separate into distinct cells. Animal mitochondrial DNA is double stranded and circular like that of bacteria, whereas that of plants is in longer linear pieces, similar to that of the plastids (Cohen, 1970).

The bacterium *Paracoccus denitrificans* (Beijerinck) Davis shows many biochemical similarities to a eucaryote mitochondrion (John and Whatley, 1975), including some that had been previously regarded as sufficiently distinct to bring into question the symbiotic hypothesis (Raff and Mahler, 1972). Possibly such an aerobic bacterium endosymbiont became the protomitochondrion, utilizing fermentation products of the host (e.g., lactic acid) as do the respiratory endosymbiotic bacteria of the living giant amoeba *Pelomyxa*, which lacks its own mitochondria.

The amount of DNA in plastids is about 10 to 20 percent that of bacteria, but 10 to 20 times as much as that present in animal mitochondria (Cohen, 1973). If inherited from a symbiont, it has decreased in amount, and some of the plastid functions of the procaryote DNA have been taken over by the nucleus of the eucaryote cell. Similarly, the production of cytochrome *c*, an essential part of the mitochondrial respiratory process, is controlled by the cell nucleus (Hall, 1973).

The plastid DNA of *Euglena gracilis* Klebs has been hybridized in the laboratory with the ribosomal RNA from several species of blue-green algae and bacteria (Pigott and Carr, 1972), and showed a homology of 30 to 9 percent with the algae, up to 32 percent homology with *Nostoc muscorum* C. A. Agardh and *Chlorogloea fritschii* (now regarded as *Nostoc fritschii* (Mitra) Schwabe & El Ayouty), and a ribosomal RNA homology of 47 percent for *Gloeocapsa alpicola* (Lyngbye) Bornet. The last species also has a polyunsaturated fatty acid distribution most similar to that of the *Euglena* plastids. All bacteria tested showed significantly lower homologies, ranging from 1.5 percent in a *Pseudomonas* to 5.9 percent in the photosynthetic *Rhodomicrobium* (Pigott and Carr, 1972).

A study of the 16S ribosomal RNA of the *Euglena gracilis* plastid showed sequence similarities to the Bacillaceae (bacteria) and to *Anacystis* (blue-green alga), but an even closer relationship to the plastids of the red alga *Porphyridium*, sufficiently close to resemble members of a single genus or family of procaryotes (Buetow, 1976), in spite of the differences in pigments. The plastid rRNA was considered to be procaryotic because of the many sequences shared with certain blue-green algae.

Many bacteria and blue-green algae are

known to be endosymbiotic in other algae, protozoans, and higher plants, and some have become obligate symbionts, no longer able to grow and divide separately. Occasionally it is difficult to determine if a particular organism is a single individual or a symbiotic association. For example the rhizopod *Paulinella* with its blue-green symbiont *Synechococcus* also was described as a single-celled alga, *Cyanospira*, and the genera *Glaucocystis* and *Gloeochaete* are regarded either as apoplastidic green algae with blue-green symbionts (Geitler, 1923; Echlin, 1966) or as simple red algal cells (Schnepf and Brown, 1971).

If plastids arose as endosymbiotic blue-green algae, the earliest eucaryote may have resembled *Glaucocystis* or *Gloeochaete*, as the cyanophyte pigment composition is most like that of the red algae. The evolutionary changes in pigment composition then may have occurred within the eucaryote lineages. Another possibility is that the pigment diversification may have occurred within the procaryotes, and the green, brown, red, and yellow-brown algae evolved from separate symbioses. As the DNA base ratios of plastids range from 25 to 60 percent G+C (guanine and cytosine), an origin in various ways and from different procaryotes appears probable (D. L. Taylor, 1970; Fogg et al., 1973). Red algae and Cryptophyta obtained blue-green symbionts that became plastids, whereas the green Chlorophyta, Prasinophyta, and Euglenophyta may have had symbionts of a completely different group of procaryotes characterized by chlorophylls *a* and *b* (Raven, 1970), like the recently described Prochlorophyta (Lewin, 1976). A third hypothetical group of symbionts with chlorophylls *a* and *c* would have given rise to the brown-pigmented algae, Chrysophyta, Xanthophyta, Pyrrhophyta, and Phaeophyta.

However, although cyanophytes and prochlorophytes may be similar to the hypothetical protoplastids, there is no known procaryote with chlorophyll *a* and *c* nor any evidence that such ever existed. Furthermore, as all groups of eucaryotic algae have similar pyrenoids associated with the plastids, and some members of all groups also have an endoplasmic reticulum (Lee, 1972), the divergence

may well have occurred after the development of the primitive eucaryotes. Lee (1972) suggested that the first eucaryote was a colorless cryptophyte, which obtained blue-green symbionts to evolve into the modern Cryptophyta, from which the red algae arose by loss of flagella and the pigment chlorophyll *c*, noting that the loss of a nonessential structure might require a single mutation, in contrast to the many mutations, long period of time, and unlikely coincidence of acquiring similar or identical biochemical pathways or new structures by evolution. Chromophyte groups merely retained the flagella and chlorophyll *c* and diverged structurally, whereas in the green algae groups chlorophyll *c* was replaced by chlorophyll *b*. The major differences in the plastid RNA and that of the nucleus could be a result of the endosymbiosis, or because the nuclear genome evolved more rapidly than that of the mitochondria and plastids after these organelles were segregated by the evolutionary process (Buetow, 1976).

Arguments against the symbiotic origin of plastids include the fact that algal plastids are commonly of complex shape and structure, whereas those of higher plants are extremely simple, and the presence of some DNA in the nucleus that controls many aspects of these organelles. In addition, the previously mentioned bacterium *Ectothiorhodospira* has stacked photosynthetic lamellae very similar to those of higher plant plastids, surrounded by a double lamella that arises from the outer plasma membrane, much as suggested in the invagination hypothesis of plastid origin (Stubbe, 1971).

Although sequence data are available as yet for only about a dozen procaryotes and perhaps an equal number of eucaryotes (ranging from *Euglena* to yeast and from rye to mammals), analysis of the data concerning amino acid and nucleotide sequences (ferredoxin, 5S ribosomal RNA, *c*-type cytochromes, and azurin-plastocyanin) allowed construction of evolutionary trees (Schwartz and Dayhoff, 1978). Ferredoxin data placed the anaerobic heterotrophic bacteria (*Clostridium*) at the base of the tree, as most like the ancestral form. Photosynthetic anaerobes that use H_2S as an electron donor (*Chlorobium, Chromatium*) diverged early from the heterotrophic bacteria, and aerobic or facultatively aerobic forms (*Bacillus, Desulfovibrio*) diverged soon after. Data provided by the 5S ribosomal RNA sequences suggest that the cytoplasm of the green algae and higher plants differs greatly from that of blue-green algae, and were consistent with the plastids having evolved from endosymbiotic blue-greens. Data from the *c*-type cytochromes further supported the hypothesis that mitochondria arose from endosymbiotic forms similar to the photosynthetic Rhodospirillaceae, which may have lost their photosynthetic ability just prior to invading the host symbiont. The latter was considered to have been a facultatively aerobic procaryote that diverged from the ancestral type at about the same time as did *Bacillus* and *Escherichia*.

Origin of the Eucaryotic 9+2 Flagella

Spirochaete bacteria have been suggested as free-living counterparts of a motile endosymbiont that evolved into the flagellum of the eucaryote cell (Sagan, 1967; Margulis, 1970). The "9+2 homologue" then evolved into the centrioles and chromosomal centromeres of the mitotic apparatus. Some spirochaetes are known to be associated with living protozoa as ectosymbionts, and may even be responsible for the movement of the host; furthermore, living spirochaetes have an axial filament that is structurally equivalent to a bundle of bacterial flagella, with a set of fibrils helically wrapped around the cell and anchored at the poles (Stanier and van Niel, 1962). In addition, internal microtubules were reported in the parasitic spirochaete, *Pillotina calotermitidis* Hollande & Gharagozlou (Hollande and Gharagozlou, 1967, pl. 2, figs. 5—8; Margulis et al., 1978).

Nevertheless, the symbiotic origin of the flagella is questioned by some. Centrioles and basal bodies appear to be products of the microtubule organizing centers (MTOCs) that are ubiquitous in eucaryotes and essential in the formation of microtubules, in the process of mitosis, and for flagellar development and movement. The simplest MTOCs are electron-dense globules or discs in which the microtubules, centrioles, and basal bodies arise *de novo,* and not by fission of preexisting ones, as would be the case if they had evolved from a symbiotic spirochaete as a 9+2 homologue. In addition, no spirochaete is known to have a structure as complex as that of the cilia and flagella of eucaryotes, and the latter are not known to contain DNA as do the mitochondria and plastids.

The centriole is not an autonomous organelle, but only one form of MTOC (Pickett-Heaps, 1971). It is not active in formation of the mitotic spindle, but is assured of its own division by attachment to the spindle apparatus. In cell division, the spindle MTOC duplicates and splits, separating by elongation of microtubules between the two parts. The centrioles themselves are inert and are not even present in many plant cells. When present, both centrioles and centrosomes generally are associated with the nuclear envelope, and unlike mitochondria and plastids, they are not surrounded by a double membrane (Schnepf and Brown, 1971).

The Precambrian Eucaryote Controversy

As many fossil eucaryotes are present in latest Precambrian rocks, including representatives of a number of invertebrate phyla, the time of origin of the eucaryote also must be within the Precambrian. As described earlier, many algal and bacterial microfossils have been reported from the Precambrian, but the absence of pigments and so forth makes controversial their assignment among the modern algal divisions, or even the determination of their procaryote or eucaryote nature.

In the upper Precambrian 900-million-year-old Bitter Springs Formation, Schopf (1968) described some 30 species, most of which were assigned to various blue-green algal families, but of which some were regarded as probable green algae on the basis of inclusions within the fossil cells that were regarded as preserved nuclei, pyrenoids, or other organelles. However, similar inner bodies or spots, at first regarded as nuclei, plastids, or eyespots, were found by other workers in nearly every Precambrian assemblage (Cloud et al., 1969; Edhorn, 1973; Hofmann and Jackson, 1969; Licari, 1978; D. Oehler, 1976, 1977): the 1.0-billion-year Skillogallee Dolomite and 1.15-billion-year Blyth Dolomite of Australia; the 1.2-billion-year Dismal Lakes Group of Canada; the 1.3-billion-year Beck Spring Dolomite of California; the 1.5-billion-year Bungle Bungle Dolomite and 1.6-billion-year Amelia Dolomite of Australia; and the 1.8-billion-year Belcher Group and 1.9-billion-year Gunflint Formation of Canada (summarized by Schopf, 1974). Meanwhile, examination of modern blue-green algal mats and laboratory simulations indicated that identical structures could be produced by diagenetic changes (Awramik et al., 1972; Golubic and Hofmann, 1976; Golubic and Barghoorn, 1977; Knoll and Barghoorn, 1975; J. H. Oehler, 1976). During diagenesis the sheaths persisted long after the cell contents had completely disintegrated. Possibly as many as 12 of the species described from the Bitter Springs Formation may represent only varying stages of such degradation of a chroococcalean blue-green (Knoll and Barghoorn, 1975), although some may represent eucaryotes with preserved pyrenoids (D. Oehler, 1977).

Other methods for recognition of eucaryotes among Precambrian fossils were suggested after suspicion arose that the supposed fossilized organelles were merely degraded cytoplasmic contents of entire cells within the sheaths. Observing the nature of budding in Gunflint *Huroniospora,* unlike the usual binary fission of the modern blue-green algae, Darby (1974) suggested a fungal affinity for this taxon. However, fungi are not only eucaryotic, but are generally thought to have evolved from other eucaryotic algae by the loss of their plastids, implying an even earlier origin of eucaryotes. Tappan (1976) suggested a red algal affinity as more probable for both *Huroniospora* and *Eosphaera* of the Gunflint, regarding the red algae as the most primitive eucaryotes. Both the budding and apparent sheathlike cover of *Eosphaera* are paralleled in living unicellular Bangiophycidae.

A study of size ranges of Precambrian unicells also suggests eucaryote affinities for some of those in the late Precambrian (Schopf and D. Oehler, 1976). Modern cyanophytes range from 1 to 55 μm in diameter, but are generally about 4 μm and only rarely 15 μm or more. In contrast, green algae range from 1 to 350 μm in diameter, with over half being larger than 20 μm. Early Proterozoic unicells are less than 40 μm in diameter, whereas those of the late Proterozoic range up to 80 μm. Much larger taxa (acritarchs) range up to 250 μm in the late Precambrian of 850 million years ago in the USSR.

Size alone may be suggestive but is not an absolute criterion. As indicated in the later chapters, some very small species are present in every group of eucaryote algae. Furthermore, it is possible that the size categories are extensively partitioned by the wide variety of extant algal groups, whereas less diverse early representatives in the Precambrian may not have been as limited in size.

Other morphological criteria include the siphonous filaments with rare cross walls in the Beck Spring and Skillogalee, regarded as probable siphonaceous Chlorophyta or Xanthophyta (Schopf and D. Oehler, 1976); the presumed fungal hyphae with rare offset septa in the Bitter Springs; and the cell tetrads in the Bitter Springs and Amelia Dolomite that resemble the tetrads of some modern red algae and rare examples of green algae (Schopf and Blacic, 1971; J. H. Oehler et al., 1976), although regarded by others as possible accidental associations of blue-green unicells such as *Chroococcus* (Golubic and Barghoorn, 1977). The presence of a triradial impression on isolated cells and of tetrads still enclosed in a sheath suggests that they do not represent blue-green algae, whose sheaths would prevent the accidental close packing to form a tetrad as an artifact, and would also prevent the development of such triradiate impressions on isolated cells.

Possible megascopic eucaryotic algae also have been described from the 1.3-billion-year-old Belt Supergroup of Montana (Walter et al., 1976); because of their size, they were regarded not as cyanophytes, but as possible green, red, or brown algae. If the Gunflint *Eosphaera* and *Huroniospora* are unicellular red algae, these filaments and thalli may well represent the megascopic rhodophytes.

Time of Appearance of the Eucaryote Cell

Many lines of evidence show a relationship of the organelles of eucaryotes to the procaryote cells, but whether the eucaryotes evolved by symbiosis or by compartmentalization remains hypothetical.

Under either the evolutionary or the symbiotic theory, the eucaryotes probably have always been aerobic, inheriting this biosynthetic pathway from their procaryotic ancestor (Uzzell and Spolsky, 1973). The greatest evolutionary advance has occurred in the nuclear genome, as that of the mitochondria and plastids remains much more like that of the procaryotes. Similarly, certain ultrastructural features probably are retained primitive states.

If the primitive eucaryote was originally aerobic, its appearance must have followed the accumulation of an oxygenic atmosphere. The level of atmospheric oxygen that had accumulated prior to evolution of the eucaryote and the time elapsing are both uncertain. The first appearance of an oxygenic atmosphere thus has been estimated variously as occurring near the close of the Precambrian, as about 2 billion years ago at the time of the Gunflint Formation deposition, or perhaps as early as 2.6 billion years ago (Dimroth and Kimberley, 1976).

As discussed above, the affinity of many Precambrian taxa is uncertain, and from the available fossil evidence, some regard the earliest eucaryotes as present in the Gunflint Formation (1.9 billion years old), others place them in the Amelia Dolomite (1.6 billion years old), Belt Supergroup or Beck Spring (1.3 billion years old), or in the Bitter Springs (900 million years old), and others recognize as reliable no evidence of Precambrian eucaryotes older than the metazoans of the Ediacarian (Knoll and Barghoorn, 1975).

A biochemical technique utilizing the cytochrome c and tRNA (transfer RNA) of a wide variety of organisms (McLaughlin and Dayhoff, 1970, 1973) has been used to determine a possible time scale for the divergence of eucaryotes and procaryotes that is independent of the fossil record. The rate of sequence change in the tRNA is nearly identical in all organisms, hence

their degree of difference is an indication of their relative recency of divergence. In comparing the cytochrome sequences, the average number of amino acid differences was 40.5 between animals (various insects and vertebrates from fish to mammals) and plants (wheat), and 44.9 between the animals and fungi (baker's yeast, the basidiomycete *Candida,* and the mold *Neurospora*). The plant and fungi differences averaged 49.3, a greater number than that of differences between animals and plants or animals and fungi. Comparison of these sequences with procaryotes (the purple bacterium *Rhodospirillum*) showed differences of 66.1 between the animals and procaryotes, 69.0 between the plants and procaryotes, and 74.3 between the fungi and procaryotes. Thus there is greater similarity between the three eucaryote groups than between any of these and the procaryotes. The differences suggest that the divergence of eucaryotes and procaryotes was 2.6 times more distant in time than divergence of the three eucaryote groups. If the divergence of these eucaryote kingdoms is minimally placed at the 700-million-year-old Ediacarian, in which the first metazoan fossils are found, the eucaryotes and procaryotes diverged at least some 1.8 billion years ago. Obviously, some evolution of protozoans preceded the metazoans, although they have no Precambrian fossil record; if the 900-million-year-old Bitter Springs fungi are correctly assigned, however, the time of procaryote-eucaryote divergence may have been as early as 2.3 billion years ago.

A similar study of the 5S ribosomal RNA and cytochrome c was made by Kimura and Ohta (1973) for a variety of procaryotes and eucaryotes. Assuming that the annual rate of nucleotide substitutions for each macromolecule is constant, as shown by McLaughlin and Dayhoff (1970), the mutational distance from a fish to a mammal is about one-third that between a mammal and yeast cell (Kimura and Ohta, 1973). If the fish and tetrapod lineages diverged 400 million years ago, then the ancestors of mammals and yeasts (animals and fungi) must have diverged about 1.2 billion years ago, and the divergence between eucaryotes and procaryotes occurred about 1.8 to 2 billion years ago. This theoretical timetable is surprisingly close to the maximum suggested by the fossil record, and approximately that of the Gunflint Iron Formation of 1.9 billion years (Tappan, 1976). Because most of the Precambrian assemblages described to date are from stromatolites whose modern counterparts similarly are dominated by blue-green algae and bacteria, possibly better fossil evidence of the eucaryotes will be found in other sedimentary rocks such as the shales of the Belt Supergroup (Walter et al., 1976).

REFERENCES

Adamczak, Ewa, *Gloeocapsomorpha prisca* Zalessky (Sinice) z ordowickich Głazow Narzutowych Polski [*Gloeocapsomorpha prisca* Zalessky (Cyanophyceae) from Ordovician erratic boulders in Poland]. *Acta Palaeont. Pol.,* v. 8, p. 465–472, 1 pl., 1 text-fig., 1963.

Allison, C. W., and M. A. Moorman, Microbiota from the Late Proterozoic Tindir Group, Alaska. *Geology,* v. 1, p. 65–68, fig. 1, 1973.

Allsopp, A., Phylogenetic relationships of the Procaryota and the origin of the eucaryotic cell. *New Phytol.,* v. 68, p. 591–612, pl. 1, fig. 1, 1969.

Altmann, Richard, *Die Elementarorganismen und ihre Beziehungen zu den Zellen.* Leipzig: Veit and Co., 145 p., 21 pls., 2 figs., 1890.

Antia, N. J., A critical appraisal of Lewin's Prochlorophyta. *Br. Phycol. J.,* v. 12, p. 271–276, fig. 1, 1977.

Awramik, S. M., S. Golubić, and E. S. Barghoorn, Blue-green algal cell degradation and its implication for the fossil record. *Geol. Soc. Am., Abstr. Progm.,* v. 4, p. 438, 1972.

Barghoorn, E. S., W. G. Meinschein, and J. W. Schopf, Paleobiology of a Precambrian shale. *Science,* v. 148, p. 461–472, text-figs. 1–9, 1965.

Barghoorn, E. S., and J. W. Schopf, Microorganisms from the Late Precambrian of Central Australia. *Science,* v. 150, p. 337–339, text-figs. 1–8, 1965.

Barghoorn, E. S., and J. W. Schopf, Microorganisms three billion years old from the Precambrian of South Africa. *Science,* v. 152, p. 758–763, text-figs. 1–12, 1966.

Barghoorn, E. S., and S. A. Tyler, Microorganisms of Middle Precambrian age from the Animikie Series, Ontario, Canada, Chapter 3, p. 93–118, text-figs. 1–7. *In* G. Mamikunian and M. H. Briggs (Eds.), *Current aspects of exobiology.* Pasadena: California Institute of Technology, 1965a.

Barghoorn, E. S., and S. A. Tyler, Microorganisms from the Gunflint Chert. *Science,* v. 147, p. 563–577, text-figs. 1–10, 1965b.

Barton, H. M., and D. J. Jones, Electron microfossils. *Science,* v. 108, p. 745–746, 1 fig., 1948.

Bazin, M. J., Sexuality in a blue-green alga: genetic recombination in *Anacystis nidulans. Nature,* v. 218, p. 282–283, 1968.

Bergh, S. V., Fossilifierade svavelbakterier uti alunskiffern på Kinnekulle [Fossilized sulfur bacteria in aluminum schists at Kinnekulle, Sweden]. *Geol. För. Stockh. Förh.,* v. 50, p. 413–418, 5 figs., 1928.

Bertrand, C. E., Les coprolithes des Bernissart. *Mém. Mus. R. Hist. Nat. Belg.,* v. 1, p. 10–154, pls. 1–15, 1903.

Bien, Erika, and W. Schwartz, Geomikrobiologische Untersuchungen VI. Über das Vorkommen konservierter toter und lebender Bakterienzellen in Salzgesteinen. *Z. Allg. Mikrobiol.,* v. 5, p. 185–205, figs. 1–7, 1965.

Bisset, K. A., Do bacteria have a nuclear membrane? *Nature,* v. 241, p. 45, 1973.

Bleecken, S., G. Strohbach, and Eva Sarfert, Autoradiography of bacterial chromosomes. *Z. Allg. Mikrobiol.,* v. 6, p. 121–123, figs. 1, 2, 1966.

Bock, Wilhelm, New fresh water algae of the eastern American Triassic. *Proc. Pa. Acad. Sci.,* v. 35, p. 77–81, 3 figs., 1961.

Bogorad, Lawrence, Evolution of organelles and eukaryotic genomes. *Science,* v. 188, p. 891–898, figs. 1–7, 1975.

Bordovskiy, O. K., Sources of organic matter in marine basins. *Mar. Geol.,* v. 3, p. 5–31, 1965a.

Bordovskiy, O. K., Transformation of organic matter in bottom sediments and its early diagenesis. *Mar. Geol.,* v. 3, p. 83–114, 1965b.

Boureau, Edouard, Evolution paléocytologique des organismes précambriens des Richât de Mauritanie. *C. R. Hebd. Séanc. Acad. Sci., Paris,* v. 280, ser. D, p. 2321–2324, pls. 1, 2, 1975.

Bourrelly, P., *Les algues d'eau douce. Initiation à la systématique.* Tome III *Les algues bleues et rouges. Les eugléniens, peridiniens et cryptomonadines.* Paris: Éd. N. Boubée et Cie., 512 p., 134 pls., figs., 1970.

Bradley, W. H., Freshwater algae from the Green River Formation of Colorado. *Bull. Torrey Bot. Club,* v. 56, p. 421–428, pls. 22, 23, 1929.

Bradley, W. H., Coprolites from the Bridger Formation of Wyoming: their composition and microörganisms. *Am. J. Sci.,* v. 244, p. 215–239, pls. 1–4, 1946.

Bradley, W. H., Unmineralized fossil bacteria. *Science,* v. 141, p. 919–921, 4 figs., 1963.

Bradley, W. H., Unmineralized fossil bacteria: a retraction. *Science,* v. 160, p. 437, 1968.

Bradley, W. H., Eocene algae and plant hairs from the Green River Formation of Wyoming. *Am. J. Bot.,* v. 57, p. 782–785, figs. 1–10, 1970.

Brock, T. D., Taxonomic confusion concerning certain filamentous blue-green algae. *J. Phycol.,* v. 4, p. 178–179, 1968.

Brock, T. D., Vertical zonation in hot spring algal mats. *Phycologia,* v. 8, p. 201–205, figs. 1–5, 1969.

Brock, T. D., Lower pH limit for the existence of blue-green algae: evolutionary and ecological implications. *Science,* v. 179, p. 480–483, 1973.

Brock, T. D., and G. K. Darland, Limits of microbial existence: temperature and pH. *Science,* v. 169, p. 1316–1318, figs. 1, 2, 1970.

Brooks, J., M. D. Muir, and G. Shaw, Chemistry and morphology of Precambrian microorganisms. *Nature,* v. 244, p. 215–217, fig. 1, 1973.

Buchanan, R. E., and R. S. Breed (Eds.), International bacteriological code of nomenclature. *J. Bact.,* v. 55, p. 287–306, 1948.

Buchanan, R. E., and N. E. Gibbons (Eds.), *Bergey's Manual of Determinative Bacteriology,* 8th ed. Baltimore: Williams and Wilkins Co., xxvi + 1246 p., 1974.

Buetow, D. E., Phylogenetic origin of the chloroplast. *J. Protozool.,* v. 23, p. 41–47, figs. 1–3, 1976.

Burris, R. H., Biological nitrogen fixation. *A. Rev. Pl. Physiol.,* v. 17, p. 155–184, figs. 1, 2, 1966.

Cairns, John, The bacterial chromosome and its manner of replication as seen by autoradiography. *J. Molec. Biol.,* v. 6, p. 208–213, 1963.

Calef, G. W., and G. D. Grice, Relationship between the blue-green alga *Trichodesmium thiebautii* and the copepod *Macrosetella gracilis* in the plankton off northeastern South America. *Ecology,* v. 47, p. 855–856, 1 fig., 1966.

Carmichael, W. W., and P. R. Gorham, Factors influencing the toxicity and animal susceptibility of *Anabaena flos-aquae* (Cyanophyta) blooms. *J. Phycol.,* v. 13, p. 97–101, fig. 1, 1977.

Carpenter, E. J., Nitrogen fixation by a blue-green epiphyte on pelagic *Sargassum. Science,* v. 178, p. 1207–1209, 1972.

Carpenter, E. J., and C. C. Price IV, Marine *Oscillatoria* (*Trichodesmium*): explanation for aerobic nitrogen fixation without heterocysts. *Science,* v. 191, p. 1278–1280, figs. 1, 2, 1976.

Carr, N. G., and B. A. Whitton (Eds.), *The biology of blue-green algae.* Bot. Monogr., v. 9, Berkeley and Los Angeles: Univ. Calif. Press, x + 676 p., 1973.

Cavalier-Smith, T., The origin of nuclei and of eucaryotic cells. *Nature,* v. 256, p. 463–468, figs. 1–5, 1975.

Cayeux, L., Existence de bactéries dans les roches sédimentaires anciennes, sutures que les phosphates. *C. R. Somm. Séanc. Soc. Géol. Fr.,* fasc. 10, p. 115–116, 1937.

Chadefaud, Marius, Possibilité d'une origine non symbiotique de la cellule des Eucaryotes. *C. R. Hebd. Séanc. Acad. Sci., Paris,* v. 278, ser. D, p. 3079–3081, figs. 1, 2, 1974.

Cloud, P. E., Notes on stromatolites. *Am. J. Sci.,* v. 40, p. 363–379, pls. 1, 2, 1942.

Cloud, P. E., Significance of the Gunflint (Precambrian) microflora. *Science,* v. 148, p. 27–35, 4 figs., 1965.

Cloud, P. E., Pre-metazoan evolution and the origins of the Metazoa, p. 1–72. *In* E. T. Drake (Ed.), *Evolution and environment.* New Haven: Yale Univ. Press, 1968.

Cloud, P. E., Pseudofossils: a plea for caution. *Geology,* v. 1, p. 123–127, figs. 1–7A, 1973.

Cloud, P. E., Beginnings of biospheric evolution and the biogeochemical consequences. *Paleobiology,* v. 2, p. 351–387, pls. 1–5, 1976.

Cloud, P. E., J. W. Gruner, and H. Hagen, Carbonaceous rocks of the Soudan Iron Formation (Early Precambrian). *Science,* v. 148, p. 1713–1716, 1 fig., 1965.

Cloud, P. E., and H. Hagen, Electron microscopy of the Gunflint microflora: preliminary results. *Proc. Natn. Acad. Sci. U.S.A.,* v. 54, p. 1–8, 24 figs., 1965.

Cloud, P. E., and G. R. Licari, Ultrastructure and geologic relations of some two-aeon old nostocacean algae from northeastern Minnesota. *Am. J. Sci.,* v. 272, p. 138–149, pls. 1–3, 1972.

Cloud, P. E., G. R. Licari, L. A. Wright, and B. W. Troxel, Proterozoic eucaryotes from eastern California. *Proc. Natn. Acad. Sci. U.S.A.,* v. 62, p. 623–630, figs. 1–12, 1969.

Cloud, P. E., Mary Moorman, and David Pierce, Sporulation and ultrastructure in a Late Proterozoic cyanophyte: some implications for taxonomy and plant phylogeny. *Q. Rev. Biol.,* v. 50, p. 131–150, 20 figs., 1975.

Cloud, P. E., and M. A. Semikhatov, Proterozoic stromatolite zonation. *Am. J. Sci.,* v. 267, p. 1017–1061, pls. 1–7, figs. 1–15, 1969.

Cohen, S. S., Are/were mitochondria and chloroplasts microorganisms? *Am. Scient.,* v. 58, p. 281–289, 1970.

Cohen, S. S., Mitochondria and chloroplasts revisited. *Am. Scient.,* v. 61, p. 437–445, 1973.

Cohen, Yehuda, B. B. Jørgensen, Etana Padan, and Moshe Shilo, Sulphide-dependent anoxygenic photosynthesis in the cyanobacterium *Oscillatoria limnetica. Nature,* v. 257, p. 489–492, figs. 1–3, 1975a.

Cohen, Yehuda, Etana Padan, and Moshe Shilo, Facultative anoxygenic photosynthesis in the cyanobacterium *Oscillatoria limnetica. J. Bact.,* v. 123, p. 855 – 861, figs. 1 – 5, 1975b.

Colom, G., Las "algas perforantes." *Estudios Geol. Inst. Invest. Geol. Lucas Mallada,* v. 9, p. 353 – 368, pls. 35 – 42; 2 text figs., 1953.

Copeland, H. F., Progress report on basic classification. *Am. Nat.,* v. 81, p. 340 – 350, 1947.

Copeland, H. F., *The classification of lower organisms.* Palo Alto: Pacific Books, ix + 302 p., 45 figs., 1956.

Croft, W. N., and E. A. George, Blue-green algae from the Middle Devonian of Rhynie, Aberdeenshire. *Bull. Br. Mus. Nat. Hist., Geology,* v. 3, p. 339 – 353, pls. 41 – 44, 1959.

Daley, Brian, Shell encrusting algae from the Bembridge Marls (Lattorfian) of the Isle of Wight, Hampshire, England. *Revue Micropaléont.,* v. 17 (1974), p. 15 – 22, pls. 1 – 3, 1975.

Darby, D. G., Reproductive modes of *Huroniospora microreticulata* from cherts of the Precambrian Gunflint Iron-Formation. *Bull. Geol. Soc. Am.,* v. 85, p. 1595 – 1596, 1 fig., 1974.

De Ley, J., Molecular biology and bacterial phylogeny, p. 103 – 156, figs. 1 – 13. *In* T. Dobzhansky, M. K. Hecht, and W. C. Steere (Eds.), *Evolutionary biology,* v. 2. New York: Appleton-Century Crofts, 1968.

De Ley, J., Phylogeny of procaryotes. *Taxon,* v. 23, p. 291 – 300, figs. 1 – 4, 1974.

Desikachary, T. V., *Cyanophyta.* New Delhi: Indian Council of Agricultural Research, 686 p., 139 pls., 1959.

Dickerson, R. E., Russell Timkovich, and R. J. Almassy, The cytochrome fold and the evolution of bacterial energy metabolism. *J. Molec. Biol.,* v. 100, p. 473 – 491, figs. 1 – 8, 1976.

Dimroth, Erich, and M. M. Kimberley, Precambrian atmospheric oxygen: evidence in the sedimentary distributions of carbon, sulfur, uranium, and iron. *Can. J. Earth Sci.,* v. 13, p. 1161 – 1185, figs. 1 – 21, 1976.

Doemel, W. N., and T. D. Brock, Bacterial stromatolites: origin of laminations. *Science,* v. 184, p. 1083 – 1085, figs. 1 – 3, 1974.

Dombrowski, H., Des bactéries vivantes des dépôts de sel du Paléozoïque. Problèmes géologiques et minéralogiques. *Revue Géogr. Phys. Géol. Dyn.,* v. 10, p. 21 – 26, fig. 1, 1968.

Dombrowski, H., C. G. I. Friedlaender, R. Kühn, and D. H. Loring, Bacteriological investigations of Carboniferous rock salt from Pugwash, Nova Scotia. *Eclog. Geol. Helv.,* v. 58, p. 967 – 974, 4 figs., 1965.

Drew, G. H., On the precipitation of calcium carbonate in the sea by marine bacteria and on the action of denitrifying bacteria in tropical and temperature seas. *Papers Mar. Biol. Lab. Tortugas, Publs. Carnegie Instn.,* 182, p. 7 – 45, 1914.

Drouet, Francis, Myxophyceae, Chapter 5, p. 95 – 114. *In* W. T. Edmonton (Ed.), *Ward and Whipple's freshwater biology,* 2nd ed. New York: John Wiley and Sons, 1959.

Drouet, Francis, *Revision of the classification of the Oscillatoriaceae.* Monogr. Acad. Nat. Sci. Philad., v. 15, 370 p., 131 figs., 1968.

Drouet, Francis, *Revision of the Nostocaceae with cylindrical trichomes (formerly Scytonemataceae and Rivulariaceae).* New York: Hafner Press, 292 p., 83 figs. 1973.

Drouet, Francis, and W. A. Daily, *Revision of the coccoid Myxophyceae.* Butler Univ. Bot. Stud., v. 12, 218 p., 377 figs., 1956.

Duncan, P. M., On some unicellular algae parasitic within Silurian and Tertiary corals, with a notice on their presence in *Calceola sandalina* and other fossils. *Q. Jl. Geol. Soc. Lond.,* v. 32, p. 205 – 211, pl. 16, 1876.

Duve, C. de, Origin of mitochondria. *Science,* v. 182, p. 85, 1973.

Echlin, Patrick, The cyanophytic origin of higher plant chloroplasts. *Br. Phycol. Bull.,* v. 3, p. 150 – 151, 1966.

Echlin, Patrick, and I. Morris, The relationships between blue-green algae and bacteria. *Biol. Rev.,* v. 40, p. 143 – 187, figs. 1 – 4, 1965.

Edhorn, Anna-Stina, Further investigations of fossils from the Animikie, Thunder Bay, Ontario. *Proc. Geol. Ass. Can.,* v. 25, p. 37 – 66, pls. 1 – 10, 1973.

Eglinton, G., P. M. Scott, T. Belsky, A. L. Burlingame, and M. Calvin, Hydrocarbons of biological origin from a one-billion-year-old sediment. *Science,* v. 145, p. 263 – 264, 1964.

Ehlers, E. G., D. V. Stiles, and J. D. Berle, Fossil bacteria in pyrite. *Science,* v. 148, p. 1719 – 1721, 5 figs., 1965.

Ehrenberg, C. G., Vorläufige Mittheilungen ueber das wirklige Vorkommen fossiler Infusorien und ihre grosse Verbreitung. *Annln Phys.,* v. 38, p. 213 – 227, 1836.

Elliott, G. F., *Zonotrichites* (calcareous algae) from the Arabian Triassic. *Eclog. Geol. Helv.,* v. 57, p. 567 – 570, pls. 1, 2, 1964.

Elliott, G. F., Transported algae as indicators of different marine habitats in the English Middle Jurassic. *Palaeontology,* v. 18, p. 351 – 366, pls. 48 – 50, 1975.

Emerson, S. U., K. Tokuyasu, and M. I. Simon, Bacterial flagella: polarity of elongation. *Science,* v. 169, p. 190 – 192, figs. 1, 2, 1970.

Emig, W. H., The travertine deposits of the Arbuckle Mountains, Oklahoma, with reference to the plant agencies concerned in their formation. *Bull. Okla. Geol. Surv.,* 29, p. 1 – 76, figs. 1 – 4, 1917.

Engel, A. E. J., B. Nagy, L. A. Nagy, C. G. Engel, G. O. W. Kremp, and C. M. Drew, Alga-like forms in Onverwacht Series, South Africa: oldest recognized lifelike forms on earth. *Science,* v. 161, p. 1005–1008, 4 figs., 1968.

Etheridge, R., Jun., On two additional perforating bodies believed to be thallophytic cryptogams, from the Lower Palaeozoic rocks of N. S. Wales. *Rec. Aust. Mus.,* v. 3, p. 121–127, pl. 23, 1899.

Famintzin, A., Die Symbiose als Mittel der Synthese von Organismen. *Biol. Zbl.,* v. 27, p. 353–364, 1907.

Fannin, N. G. T., Stromatolites from the Middle Old Red Sandstone of Western Orkney. *Geol. Mag.,* v. 106, p. 77–88, pl. 7, figs. 1–5, 1969.

Fay, P., W. D. P. Stewart, A. E. Walsby, and G. E. Fogg, Is the heterocyst the site of nitrogen fixation in blue-green algae? *Nature,* v. 220, p. 810–812, figs. 1, 2, 1968.

Fenton, C. L., and M. A. Fenton, Belt Series of the north: stratigraphy, sedimentation, paleontology. *Bull. Geol. Soc. Am.,* v. 48, p. 1873–1970, 19 pls., 20 figs., 1937.

Fenwick, M. G. Some rare and interesting algae from Port Radium, N. W. T. Canada. *Trans. Am. Microsc. Soc.,* v. 85, p. 477–480, 1 pl., 1966.

Fliche, Paul, Note sur les flores Tertiaires des environs de Mulhouse. *Bull. Soc. Ind. Mulhouse,* v. 56, p. 348–362, 1886.

Fliche, Paul, Flore fossile du Trias en Lorraine et en Franche-Comté. *Bull. Séanc. Soc. Sci. Nancy,* v. 6, p. 1–66, pls. 1–5, 1905.

Fogg, G. E., W. D. P. Stewart, P. Fay, and A. E. Walsby, *The blue-green algae.* London and New York: Academic Press, viii + 460 p., figs., 1973.

Fox, G. E., L. J. Magrum, W. E. Balch, R. S. Wolfe, and C. R. Woese, Classification of methanogenic bacteria by 16S ribosomal RNA characterization. *Proc. Natn. Acad. Sci. U.S.A.,* v. 74, p. 4537–4541, 1 fig., 1977.

Freytet, Pierre, and J. C. Plaziat, Importance des constructions algaires dues à des cyanophycées dans les formations continentales du Crétacé supérieur et de l'Éocène du Languedoc. *Bull. Soc. Géol. Fr.,* sér. 7, v. 7 (1965), p. 679–694, pls. 25–27, text-figs. 1–3, text-pls. 1–3, 1966.

Friedmann, I., Y. Lipkin, and Roseli Ocampo-Paus, Desert algae of the Negev (Israel). *Phycologia,* v. 6, p. 185–200, figs. 1–17, 1967.

Friedmann, E. I., and R. Ocampo, Endolithic blue-green algae in the Dry Valleys: primary producers in the Antarctic desert ecosystem. *Science,* v. 193, p. 1247–1249, 1976.

Friedmann, E. I., and R. Ocampo-Friedmann, Endolithic blue-green algae in the Dry Valleys (Antarctica). *J. Phycol.* (Suppl.), v. 12, p. 20 (Abstr.), 1976.

Fritsch, F. E., *The structure and reproduction of the algae,* v. 2, *Foreword, Phaeophyceae, Rhodophyceae, Myxophyceae,* xiv + 939 p., 336 figs. Cambridge: University Press, 1952.

Gabriel, M. L., Primitive genetic mechanisms and the origin of chromosomes. *Am. Nat.,* v. 94, p. 257–269, figs. 1–3, 1960.

Garrett, Peter, Phanerozoic stromatolites: noncompetitive ecologic restriction by grazing and burrowing animals. *Science,* v. 169, p. 171–173, fig. 1, 1970.

Garwood, E. J., Some new rock-building organisms from the Lower Carboniferous beds of Westmoreland. *Geol. Mag.,* n.s., dec. 6, v. 1, p. 265–271, pls. 20, 21, 1914.

Geitler, L., Der Zellbau von *Glaucocystis nostochinearum* und *Gloeochaete Wittrockiana* und die Chromatophoren-Symbiose Theorie von Mereschkowsky. *Arch. Protistenk.,* v. 47, p. 1–24, pl. 1, figs. 1–8, 1923.

Glaessner, M. F., W. V. Preiss, and M. R. Walter, Precambrian columnar stromatolites in Australia: morphological and stratigraphic analysis. *Science,* v. 164, p. 1056–1059, figs. 1–13, 1969.

Glauert, A. M., and M. J. Thornley, The topography of the bacterial cell wall. *A. Rev. Microbiol.,* v. 23, p. 159–198, 1969.

Goksøyr, Jostein, Evolution of eucaryotic cells. *Nature,* v. 214, p. 1161, fig. 1, 1967.

Golubic, Stjepko, Distribution, taxonomy and boring patterns of marine endolithic algae. *Am. Zool.,* v. 9, p. 747–751, figs. 1–6, 1969.

Golubic, Stjepko, and E. S. Barghoorn, Interpretation of microbial fossils with special reference to the Precambrian, p. 1–14, figs. 1–5. *In* Erik Flügel (Ed.), *Fossil algae, recent results and developments.* Berlin, Heidelberg, New York: Springer-Verlag, 1977.

Golubic, Stjepko, and H. J. Hofmann, Comparison of Holocene and Mid-Precambrian Entophysalidaceae (Cyanophyta) in stromatolitic algal mats: cell division and degradation. *J. Paleont.,* v. 50, p. 1074–1082, pls. 1, 2, 1976.

Goswami, S. K., Occurrence of *Scytonema* sp. in the lignite of Kashmir Valley. *Curr. Sci.,* v. 24, p. 56, fig. 1, 1955.

Gross, J. D., and L. Caro, Genetic transfer in bacterial mating. *Science,* v. 150, p. 1679–1684, 1965.

Gruner, J. W., The origin of sedimentary iron formations. The Biwabik Formation of the Mesabi Range. *Econ. Geol.,* v. 17, p. 407–460, figs. 42–46, 1922.

Gunia, Tadeusz, Mikroflora Prekambryjskich wapieni okolicy Dusznik Zdroju (Sudety Środkowe) [Mikroflora of Pre-Cambrian limestones of the Duszniki Zdrój Region (the Central Sudeten)]. *Roczn. Pol. Tow. Geol.,* v. 44, p. 65–92, pls. 1–6, 1974.

Halfen, L. N., Gliding motility of *Oscillatoria:* ultrastructural and chemical characterization of the fibrillar layer. *J. Phycol.,* v. 9, p. 248−253, figs. 1−12, 1973.

Halfen, L. N., and R. W. Castenholz, Gliding in a blue-green alga: a possible mechanism. *Nature,* v. 225, p. 1163−1165, figs. 1−3, 1970.

Halfen, L. N., and R. W. Castenholz, Gliding motility in the blue-green alga *Oscillatoria princeps. J. Phycol.,* v. 7, p. 133−145, 18 figs., 1971.

Hall, J. B., Evolution of the procaryotes. *J. Theor. Biol.,* v. 30, p. 429−454, 1971.

Hall, J. B., The nature of the host in the origin of the eukaryote cell. *J. Theor. Biol.,* v. 38, p. 413−418, 1973.

Harder, E. C., Iron-depositing bacteria and their geologic relations. *Prof. Pap. U.S. Geol. Surv.,* 113, p. 1−89, pls. 1−12, text-figs. 1−14, 1919.

Hessland, I., Investigations of the Lower Ordovician of the Siljan District, Sweden. II. Lower Ordovician penetrative and enveloping algae from the Siljan District. *Bull. Geol. Instn. Univ. Upsala,* v. 33, p. 409−428, 10 pls., 2 text-figs., 1949.

Hocht, H., H. H. Martin, and O. Kandler, Zur Kenntnis der chemischen Zusammensetzung der Zellwand der Blaualgen. *Z. PflPhysiol.,* v. 53, p. 39−57, 1965.

Hofmann, H. J., Stromatolites from the Proterozoic Animikie and Sibley Groups, Ontario. *Geol. Surv. Pap. Can.* 68−69, v + 77 p., 22 pls., 9 figs., 1969a.

Hofmann, H. J., Attributes of stromatolites. *Geol. Surv. Pap. Can.* 69−39, 58 p., 22 figs., 1969b.

Hofmann, H. J., Stromatolites: characteristics and utility. *Earth-Science. Rev.,* v. 9, p. 339−373, figs. 1−12, 1973.

Hofmann, H. J., Mid-Precambrian prokaryotes (?) from the Belcher Islands, Canada. *Nature,* v. 249, p. 87−88, fig. 1, 1974.

Hofmann, H. J., Stratiform Precambrian stromatolites Belcher Islands, Canada: relations between silicified microfossils and microstructure. *Am. J. Sci.,* v. 275, p. 1121−1132, pls. 1−4, fig. 1, 1975.

Hofmann, H. J., Precambrian microflora, Belcher Islands, Canada: significance and systematics. *J. Paleont.,* v. 50, p. 1040−1073, pls. 1−9, figs. 1−7, 1976.

Hofmann, H. J., and G. D. Jackson, Precambrian (Aphebian) microfossils from Belcher Islands, Hudson Bay. *Can. J. Earth Sci.,* v. 6, p. 1137−1144, figs. 1−20, 1969.

Hollande, Andre, and Iran Gharagozlou, Morphologie infrastructurale de *Pillotina calotermitidis* nov. gen., nov. sp., spirochaetale de l'intestin de *Calotermes praecox. C. R. Hebd. Seanc. Acad. Sci., Paris,* ser. D, v. 265, p. 1309−1312, pls. 1−3, 1 fig., 1967.

Holm-Hansen, Osmund, Ecology, physiology and biochemistry of blue-green algae. *A. Rev. Microbiol.,* v. 22, p. 47−70, 1968.

Holton, R. W., H. H. Blecker, and T. S. Stevens, Fatty acids in blue-green algae: possible relation to phylogenetic position. *Science,* v. 160, p. 545−547, 1968.

Holwill, M. E. J., and R. E. Burge, A hydrodynamic study of the motility of flagellated bacteria. *Archs Biochem. Biophys.,* v. 101, p. 249−260, figs. 1−7, 1963.

Houghton, J. A., The genetics of blue-green algae. *Sci. Prog., Oxf.,* v. 60, p. 407−412, 1972.

Howe, M. A., The geologic importance of the lime-secreting algae, with a description of a new travertine-forming organism. *Prof. Pap. U.S. Geol. Surv.,* 170E, p. 57−65, pls. 19−23, 1932.

International Committee on Bacteriological Nomenclature, *International code of nomenclature of bacteria and viruses.* Ames, Iowa: Iowa State Univ. Press, 186 p., 1958.

Jackson, T. A., Fossil Actinomycetes in Middle Precambrian glacial varves. *Science,* v. 155, p. 1003−1005, 3 figs., 1967.

Jarosch, R., Gleitbewegung und Torsion von Oscillatorien. *Öst. Bot. Z.,* v. 111, p. 143−148, figs. 1−6, 1964.

Javor, B. J., and E. W. Mountjoy, Late Proterozoic microbiota of the Miette Group, southern British Columbia. *Geology,* v. 4, p. 111−119, 6 figs., 1976.

Johannes, R. E., Nutrient regeneration in lakes and oceans. *Adv. Microbiol. Sea,* v. 1, p. 203−213, fig. 1, 1968.

John, Philip, and F. R. Whatley, *Paracoccus denitrificans* and the evolutionary origin of the mitochondrion. *Nature,* v. 254, p. 495−498, figs. 1−3, 1975.

Jones, G. E., R. L. Starkey, H. W. Feely, and J. L. Kulp, Biological origin of native sulfur in salt domes of Texas and Louisiana. *Science,* v. 123, p. 1124−1125, 1956.

Kalkowsky, E., Oolith und Stromatolith in norddeutschen Buntsandstein. *Z. Dt. Geol. Ges.,* v. 60, p. 68−125, figs. 1−3, 1908.

Kellerman, K. F., Relation of bacteria to deposition of calcium carbonate. *Bull. Geol. Soc. Am.,* v. 26, p. 58, 1915.

Kenyon, C. N., and R. Y. Stanier, Possible evolutionary significance of polyunsaturated fatty acids in blue-green algae. *Nature,* v. 227, p. 1164−1165, fig. 1, 1970.

Kidston, Robert, and W. H. Lang, On Old Red Sandstone plants showing structure, from the Rhynie Chert Bed, Aberdeenshire, Pt. 5, The Thallophyta occurring in the Peat-Bed. *Trans. R. Soc. Edinb.,* v. 52, p. 855−902, pls. 1−10, 1921.

Kimura, Motoo, and Tomoko Ohta, Eukaryotes-prokaryotes divergence estimated by 5S ribosomal RNA sequences. *Nature, New Biol.,* v. 243, p. 199−200, 1973.

Kirchner, O., Schizophyceae, p. 45—92, figs. 48—62. *In* A. Engler and K. Prantl, *Die naturlichen Pflanzenfamilien*, pt. I 1a. Leipzig: W. Engelmann, 1900.

Klein, R. M., and A. Cronquist, A consideration of the evolutionary and taxonomic significance of some biochemical, micromorphological, and physiological characters in the thallophytes. *Q. Rev. Biol.*, v. 42, p. 105—296, figs. 1—26, 1967.

Knaysi, Georges, *Elements of bacterial cytology*, 2nd. ed. Ithaca: Comstock Publishing Co., Inc. 375 p., 49 pls., 123 figs. 1951.

Knoll, A. H., and E. S. Barghoorn, Precambrian eukaryotic organisms: a reassessment of the evidence. *Science*, v. 190, p. 52—54, fig. 1, 1975.

Knoll, A. H., and E. S. Barghoorn, Archean microfossils showing cell division from the Swaziland System of South Africa. *Science*, v. 198, p. 396—398, figs. 1, 2, 1977.

Knoll, A. H., E. S. Barghoorn, and Stjepko Golubić, *Paleopleurocapsa wopfnerii* gen. et sp. nov.: a Late Precambrian alga and its modern counterpart. *Proc. Natn. Acad. Sci. U.S.A.*, v. 72, p. 2488—2492, figs. 1, 2, 1975.

Kolosov, P. N., Novye vidy dokembriyskikh vodorosley basseyna reki Olekmy [New species of Precambrian algae from the Olekma River Basin]. *Dokl. Akad. Nauk SSSR*, v. 171, p. 978—980, fig. 1, 1966.

Komar, V. A., Stromatolity verkhnedokembriyskikh otlozheniy severa Sibirskoy Platformy i ikh stratigraficheskoe znachenie [Stromatolites of the Upper Precambrian deposits of the north Siberian Platform and their stratigraphic importance]. *Trudy Inst. Geol. Nauk, Mosk.*, vyp. 154, p. 1—122, pls. 1—20, figs. 1—25, 1966.

Korde, K. B., V. P. Maslov, and I. N. Krylova, Tip Cyanophyta (Schizophyceae) Sinezelenye vodorosli, p. 29—54. *In* Yu. A. Orlov (Ed.), *Osnovy Paleontologii*, v. 14, *Vodorosli, Mokhoobraznye, Psilofitovye, Plaunovidnye, Chlenistostebel'nye, Paporotniki.* Moscow: Akad. Nauk SSSR, 1963.

Krumbein, W. E., and Y. Cohen, Primary production, mat formation and lithification: contribution of oxygenic and facultative anoxygenic cyanobacteria, p. 37—56, figs. 1—19. *In* Erik Flügel (Ed.), *Fossil algae, recent results and developments.* Berlin, Heidelberg, New York: Springer-Verlag, 1977.

Krylov, I. N., Stromatolity Rifeya i Fanerozoya SSSR [Stromatolites of the Riphean and Phanerozoic of the USSR]. *Trudy Inst. Geol. Nauk, Mosk.*, vyp. 274, p. 1—241, pls. 1—20, figs. 1—78, 1975.

Kumar, H. D., Apparent genetic recombination in a blue-green alga. *Nature*, v. 196, p. 1121—1122, 1 fig., 1962.

Kuznetsov, S. I., M. V. Ivanov, and N. N. Lyalikova, *Introduction to geological microbiology*, Ed. C. H. Oppenheimer, Trans. P. T. Broneer. New York: McGraw Hill Book Co., 252 p., 64 figs., 1963.

LaLou, C., Studies on the bacterial precipitation of carbonates in sea water. *J. Sedim. Petrol.*, v. 27, p. 190—195, 1957.

Lang, N. J., The fine structure of blue-green algae. *A. Rev. Microbiol.*, v. 22, p. 15—46, figs. 1—25, 1968.

Lange, Willy, Cyanophyta-Bacteria systems: Effects of added carbon compounds or phosphate on algal growth at low nutrient concentrations. *J. Phycol.*, v. 6, p. 230—234, 1 fig., 1970.

Lauritzen, Ørnulf, and David Worsley, Algae as depth indicators in the Silurian of the Oslo region. *Lethaia*, v. 7, p. 157—161, figs. 1—3, 1974.

Lauterborn, Robert, Protozoenstudien, II. *Paulinella chromatophora* nov. gen. nov. spec., ein beschalter Rhizopode des Süsswassers mit blaugrünen chromatophorenartigen Einschlüssen. *Z. Wiss. Zool.*, v. 59, p. 537—544, pl. 30, 1895.

Lee, R. E., Origin of plastids and the phylogeny of algae. *Nature*, v. 237, p. 44—46, 2 figs., 1972.

Lewin, R. A., Microscopic algae associated with didemnid ascidians. *J. Phycol. (Suppl.)*, v. 11, p. 18 (Abstr.), 1975.

Lewin, R. A., Prochlorophyta as a proposed new division of algae. *Nature*, v. 261, p. 697—698, 1976.

Lewin, R. A., *Prochloron*, type genus of the Prochlorophyta. *Phycologia*, v. 16, p. 217, 1977.

Lewin, R. A., and N. W. Withers, Extraordinary pigment composition of a prokaryotic alga. *Nature*, v. 256, p. 735—737, fig. 1, 1975.

Licari, G. R., Biogeology of the Late Pre-Phanerozoic Beck Spring Dolomite of Eastern California. *J. Paleont.*, v. 52, p. 767—792, 3 pls., 2 figs., 1978.

Licari, G. R., and P. E. Cloud, Reproductive structures and taxonomic affinities of some nannofossils from the Gunflint Iron Formation. *Proc. Natn. Acad. Sci. U.S.A.*, v. 59, p. 1053—1060, figs. 1—20, 1968.

Licari, G. R., and P. E. Cloud, Prokaryotic algae associated with Australian Proterozoic stromatolites. *Proc. Natn. Acad. Sci. U.S.A.*, v. 69, p. 2500—2504, figs. 1—5, 1972.

Licari, G. R., P. E. Cloud, and W. D. Smith, A new chroococcacean alga from the Proterozoic of Queensland. *Proc. Natn. Acad. Sci. U.S.A.*, v. 62, p. 56—62, figs. 1—4, 1969.

Lipman, Charles B., A critical and experimental study of Drew's bacterial hypothesis on $CaCO_3$ precipitation in the sea. *Publs. Carnegie Instn.*, 340, *Dept. Mar. Biol.*, v. 19, p. 179—191, 1924.

Logan, B. W., R. Rezak, and R. N. Ginsburg, Classification and environmental significance of stromatolites. *Bull. Geol. Soc. Am.*, v. 71, p. 1918—1919, 1960.

Logan, B. W., R. Rezak, and R. N. Ginsburg, Classification and environmental significance of algal stromatolites. *J. Geol.*, v. 72, p. 68—83, pls. 1—4, figs. 1—5, 1964.

MacGregor, I. M., J. F. Truswell, and K. A. Eriksson, Filamentous algae from the 2300 m.y. old Transvaal Dolomite. *Nature,* v. 247, p. 538—540, figs. 1—5, 1974.

McLaughlin, P. J., and M. O. Dayhoff, Eukaryotes versus prokaryotes: an estimate of evolutionary distance. *Science,* v. 168, p. 1469—1471, figs. 1—3, 1970.

McLaughlin, P. J., and M. O. Dayhoff, Eukaryotic evolution: a view based on cytochrome *c* sequence data. *J. Molec. Evol.,* v. 2, p. 99—116, figs. 1—5, 1973.

Mandel, M., New approaches to bacterial taxonomy: perspective and prospects. *A. Rev. Microbiol.,* v. 23, p. 239—274, 1969.

Margulis, Lynn, *Origin of eucaryotic cells.* New Haven and London: Yale Univ. Press, 349 p., 1970.

Margulis, Lynn, The origin of plant and animal cells. *Am. Scient.,* v. 59, p. 230—235, 1971.

Margulis, Lynn, Leleng To, and David Chase, Microtubules in prokaryotes. *Science,* v. 200, p. 1118—1124, figs. 1—4, 1978.

Margulis, Lynn, J. C. G. Walker, and Mitchell Rambler, Reassessment of roles of oxygen and ultra-violet light in Precambrian evolution. *Nature,* v. 264, p. 620—624, 1976.

Marshall, C. G. A., J. W. May, and C. J. Perret, Fossil microorganisms: possible presence in Precambrian shield of western Australia: *Science,* v. 144, p. 290—292, 1 fig., 1964.

Martin, T. C., and J. T. Wyatt, Extracellular investments in blue-green algae with particular emphasis on the genus *Nostoc. J. Phycol.,* v. 10, p. 204—210, figs. 1—5, 1974.

Maslov, V. P., O novykh formakh Tretichnykh vodorosley [On new forms of Tertiary algae]. *Dokl. Akad. Nauk SSSR,* v. 103, p. 145—148, text-fig. 1, 1955.

Maslov, V. P., O novom iskopaemom semeystve Bagryanykh i dvukh novykh rodakh Sinezelenykh vodorosley Karbona [On a new fossil family of red algae and two new genera of blue-green algae from the Carboniferous]. *Dokl. Akad. Nauk SSSR,* v. 107, p. 151—154, figs. 1—4, 1956.

Maslov, V. P., Mikrokodii [Microcodieae]. *Paleont. Zh.,* no. 1, p. 100—109, figs. 1—6, 1967.

Meinschein, W. G., Soudan Formation: Organic extracts of Early Precambrian rocks. *Science,* v. 150, p. 601—605, 10 figs., 1965.

Meinschein, W. G., E. S. Barghoorn, and J. W. Schopf, Biological remnants in a Precambrian sediment. *Science,* v. 145, p. 262—263, 1964.

Mereschkowsky, C., Über Natur und Ursprung der Chromatophoren im Pflanzenreiche. *Biol. Zbl.,* v. 25, p. 593—604, 689—690, 1905.

Migula, W., Schizophyta: Schizomycetes, p. 1—33, figs. 1—47. *In* A. Engler and K. Prantl, *Die natürlichen Pflanzenfamilien,* pt. I 1a. Leipzig: W. Engelmann, 1900.

Monty, C. L. V., Recent algal stromatolites in the Windward Lagoon, Andros Island, Bahamas. *Annls. Soc. Géol. Belg.,* v. 88, p. B269—B276, pls. 1, 2, 1965.

Monty, C. L. V., Distribution and structure of Recent stromatolitic algal mats, Eastern Andros Island, Bahamas. *Annls. Soc. Géol. Belg.,* v. 90 (1966—1967), p. B55—B100, pls. 1—19, figs. 1—10, 1967.

Monty, C. L. V., Precambrian background and Phanerozoic history of stromatolitic communities, an overview. *Annls. Soc. Géol. Belg.,* v. 96, p. 585—624, 1973.

Monty, C. L. V., Evolving concepts in the nature and ecological significance of stromatolites, p. 15—35. *In* Erik Flügel (Ed.), *Fossil algae, recent results and developments.* Berlin, Heidelberg, and New York: Springer-Verlag, 1977.

Moodie, R. L., *Paleopathology: an introduction to the study of ancient evidences of disease.* Urbana: Univ. Illinois Press, 576 p., 117 pls., 49 figs., 1923.

Moorman, Mary, Microbiota of the Late Proterozoic Hector Formation, southwestern Alberta, Canada. *J. Paleont.,* v. 48, p. 524—539, pls. 1—3, figs. 1—3, 1974.

Muir, M. D., Microfossils from the Middle Precambrian McArthur Group, Northern Territory, Australia. *Origins Life,* v. 5, p. 105—118, figs. 1—36, 1974.

Müller, A., and W. Schwartz, Über das Vorkommen von Mikroorganismen in Salzlagerstätten. (Geomikrobiologische Untersuchungen III). *Z. Dt. Geol. Ges.,* v. 105, p. 789—802, 6 figs., 1955.

Mussill, Michael, and Robert Jarosch, Bacterial flagella rotate and do not contract. *Protoplasma,* v. 75, p. 465—469, figs. 1, 2, 1972.

Myers, G. E., and R. G. L. McCready, Bacteria can penetrate rock. *Can. J. Microbiol.,* v. 12, p. 477—484, 1966.

Nagy, L. A., Transvaal stromatolite: First evidence for the diversification of cells about 2.2 × 10^9 years ago. *Science,* v. 183, p. 514—516, fig. 1, 1974.

Nass, S., The significance of the structural and functional similarities of bacteria and mitochondria. *Int. Rev. Cytol.,* v. 25, p. 55—129, figs. 1—5, 1969.

Niel, C. B. van, Classification and taxonomy of the bacteria and blue-green algae, p. 89—114. *In* E. L. Kessel (Ed.), *A century of progress in the natural sciences, 1853-1953.* San Francisco: Calif. Acad. Sci., 1955.

Niel, C. B. van, and R. Y. Stanier, Bacteria, Chapter 3, p. 16—46. *In* W. T. Edmondson (Ed.), *Ward and Whipple's fresh water biology.* New York: John Wiley and Sons, 1959.

Oberlies, F., and A. A. Prashnowsky, Biogeochemische und elektronen-mikroskopische Untersuchung präkambrischer Gesteine. *Naturwissenschaften,* v. 55, p. 25—28, figs. 1—8, 1968.

Oehler, D. Z., Transmission electron microscopy of organic microfossils from the Late Precambrian Bitter Springs Formation of Australia: techniques and survey of preserved ultrastructure. *J. Paleont.,* v. 50, p. 90–106, pls. 1–4, fig. 1, 1976.

Oehler, D. Z., Pyrenoid-like structures in Late Precambrian algae from the Bitter Springs Formation of Australia. *J. Paleont.,* v. 51, p. 885–901, pls. 1–4, 1977.

Oehler, J. H., Experimental studies in Precambrian paleontology: structural and chemical changes in blue-green algae during simulated fossilization in synthetic chert. *Bull. Geol. Soc. Am.,* v. 87, p. 117–129, 14 figs., 1976.

Oehler, J. H., Microflora of the H. Y. C. Pyritic Shale Member of the Barney Creek Formation (McArthur Group), middle Proterozoic of northern Australia. *Alcheringa,* v. 1, p. 315–349, figs. 1–14, 1977.

Oehler, J. H., D. Z. Oehler, and M. D. Muir, On the significance of tetrahedral tetrads of Precambrian algal cells. *Origins Life,* v. 7, p. 259–267, figs. 1–4, 1976.

Ogawa, R. E., and J. F. Carr, The influence of nitrogen on heterocyst production in blue-green algae. *Limnol. Oceanogr.,* v. 14, p. 342–351, figs. 1–7, 1970.

Oppenheimer, C. H., Evidence of fossil bacteria in phosphate rocks. *Publs. Inst. Mar. Sci. Univ. Tex.,* v. 5, p. 156–159, figs. 1–2b, 1958.

Oppenheimer, C. H., Bacterial production of hydrocarbon-like materials. *Z. Allg. Mikrobiol.,* v. 5, p. 284–307, 1965.

Papenfuss, G. F., Classification of the algae, p. 115–224. *In* E. L. Kessel (Ed.), *A century of progress in the natural sciences, 1853–1953.* San Francisco: Calif. Acad. Sci., 1955.

Perkins, R. D., and C. I. Tsentas, Microbial infestation of carbonate substrates planted on the St. Croix Shelf, West Indies. *Bull. Geol. Soc. Am.,* v. 87, p. 1615–1628, 46 figs., 1976.

Pflug, H. D., Niedere Algen und ähnliche Kleinformen aus dem Algonkium der Belt-Serie. *Oberhess. Gesell. Natur-Heilkunde zu Giessen, N.F., Naturw. Abt.,* v. 33, p. 403–411, 1 pl., 1964.

Pflug, H. D., Organische Reste aus der Belt-Serie (Algonkium) von Nordamerika. *Paläont. Z.,* v. 39, p. 10–25, pls. 1–4, 1965.

Pflug, H. D., Structured organic remains from the Fig Tree Series (Precambrian) of the Barberton Mountain Land (South Africa). *Revue Palaeobot. Palynol.,* v. 5, p. 9–29, pls. 1–4, fig. 1, 1967.

Pia, J., Die vorzeitliche Spaltpilze und ihre Lebensspuren. *Paleobiologica,* v. 1, nos. 5–7, p. 457–474, 1928.

Pia, J., Die kalklösenden Thallophyten. *Arch. Hydrobiol.,* v. 31, p. 264–328; v. 32, p. 341–398, 1937.

Pickett-Heaps, J. D., The evolution of the mitotic apparatus: an attempt at comparative ultrastructural cytology in dividing plant cells. *Cytobios,* v. 1, p. 257–280, figs. 1–16, 1969.

Pickett-Heaps, J. D., The autonomy of the centriole: fact or fallacy. *Cytobios,* v. 3, p. 205–214, figs. 1, 2, 1971.

Pigott, G. H., and N. G. Carr, Homology between nucleic acids of blue-green algae and chloroplasts of *Euglena gracilis. Science,* v. 175, p. 1259–1261, 1972.

Playford, P. E., and A. E. Cockbain, Algal stromatolites: deep water forms in the Devonian of Western Australia. *Science,* v. 165, p. 1008–1010, 3 figs., 1969.

Plöchinger, Benno, Fossile Bakterien in den Tennengebirgs—Manganschiefern? *Mikroskopie,* v. 7, p. 197–201, 4 figs., 1952.

Pratje, O., Fossile kalkbohrende Algen (*Chaetophorites gomontoides*) in Liaskalken. *Zentbl. Miner. Geol. Paläont.,* p. 299–301, 3 figs., 1922.

Preiss, W. V., The systematics of South Australian Precambrian and Cambrian stromatolites. Part I. *Trans. R. Soc. S. Aust.,* v. 96, p. 67–100, 15 figs., 1972.

Preiss, W. V., The systematics of South Australian Precambrian and Cambrian stromatolites. Part II. *Trans. R. Soc. S. Aust.,* v. 97, p. 91–125, 1973.

Preiss, W. V., The systematics of South Australian Precambrian and Cambrian stromatolites. Part III. *Trans. R. Soc. S. Aust.,* v. 98, p. 185–208, 12 figs., 1974.

Provasoli, L., Nutrition and ecology of protozoa and algae. *A. Rev. Microbiol.,* v. 12, p. 279–308, 1958.

Raaben, M. E., Columnar stromatolites and Late Precambrian stratigraphy. *Am. J. Sci.,* v. 267, p. 1–18, figs. 1–6, 1969.

Raaben, M. E., Stromatolity Verkhnego Rifeya. Tungussidy i Kussiellidy [Stromatolites of the Upper Riphean, Tungussida and Kussiellida]. *Trudy Inst. Geol. Nauk,* Mosk., vyp. 217, p. 6–51, pls. 1–30, 1972.

Raff, R. A., and H. R. Mahler, The non symbiotic origin of mitochondria. *Science,* v. 177, p. 575–582, figs. 1–4, 1972.

Raff, R. A., and H. R. Mahler, [Reply to comments] Origin of mitochondria. *Science,* v. 180, p. 517, 1973.

Raven, P. H., A multiple origin for plastids and mitochondria. *Science,* v. 169, p. 641–646, 1970.

Reijnders, L., The origin of mitochondria. *J. Molec. Evol.,* v. 5, p. 167–176, 1975.

Renault, Bernard, Sur quelques micro-organismes des combustibles fossiles. *Bull. Soc. Ind. Minér. St.-Étienne,* v. 13, p. 865–1169, figs. 1–34; v. 14, p. 5–159, figs. 35–101, 1899.

Renault, Bernard, Du rôle de quelques bactériacées fossiles au point de vue géologique. *Int. Geol. Congr., 8, Paris, 1900,* v. 1, p. 646–663, pls. 7–9, 13 figs., 1901.

Renault, Bernard, and A. Roche, Étude sur la constitution des lignites et les organismes qu'ils renferment suivie d'une note préliminaire sur les schistes lignitifères de Menat et du Bois-d'Asson. *Bull. Soc. Hist. Nat. Autun,* v. 11, p. 201–239, pls. 11–13, 1898.

Riding, Robert, *Girvanella* and other algae as depth indicators. *Lethaia*, v. 8, p. 173–179, 1975.

Riding, Robert, Calcified *Plectonema* (Blue-green algae), a Recent example of *Girvanella* from Aldabra Atoll. *Palaeontology*, v. 20, p. 33–46, pl. 11, figs. 1–5, 1977.

Ridley, S. M., and R. M. Leech, Division of chloroplasts in an artificial environment. *Nature*, v. 227, p. 463–465, figs. 1–4, 1970.

Rippel, A., Fossile Mikroorganismen in einem permischen Salzlager. *Arch. Mikrobiol.*, v. 6, p. 350–358, 7 figs., 1935.

Ris, Hans, Annual invitation lecture. Ultrastructure and molecular organization of genetic systems. *Can. J. Gen. Cytol.*, v. 3, p. 95–120, 1961.

Roberts, T. M., L. C. Klotz, and A. R. Loeblich III, Characterization of a blue-green algal genome. *J. Molec. Biol.*, v. 110, p. 341–361, figs. 1–8, 1977.

Roberts, T. M., and K. E. Koths, The blue-green alga *Agmenellum quadruplicatum* contains covalently closed DNA circles. *Cell*, v. 9, p. 551–557, figs. 1–4, 1976.

Roberts, T. M., A. R. Loeblich III, and L. C. Klotz, Studies on the DNA of the blue-green alga *Agmenellum quadruplicatum. J. Phycol.*, v. 11 (Suppl.), p. 16 (Abstr.), 1975.

Robertson, J. D., The membrane of the living cell. *Scient. Am.*, v. 206(4), p. 64–72, figs., 1962.

Roux, W., Über eine im Knochen lebende Gruppe von Fadenpilzen (*Mycelites ossifragus*). *Z. Wiss. Zool.*, v. 45, p. 227–254, pl. 14, 1887.

Rutte, Erwin, Gesteinsbildende Algen aus dem Eozän von Kleinkems am Isteiner Klotz in Baden. *Neues Jb. Geol. Paläont. Mh.*, no. 11, p. 498–506, figs. 1–5, 1953.

Rutte, Erwin, Süsswasserkalke und Kalkalgenbildungen in der chattischen Unteren Süsswassermolasse von Hoppetenzell nördlich Stockach/Baden. *Geol. Jb.*, v. 69, p. 517–536, pls. 38, 39, figs. 1–4, 1954.

Sagan, Lynn, On the origin of mitosing cells. *J. Theor. Biol.*, v. 14, p. 225–274, figs. 1, 2, 1967.

Sarles, W. B., W. C. Frazier, J. B. Wilson, and S. G. Knight, *Microbiology, general and applied,* 2nd. ed. New York: Harper and Brothers, xiii + 491 p., 69 figs., 1956.

Sastri, V. V., B. S. Venkatachala, and T. V. Desikachary, A fossil nostocaceae from India, p. 159–160, pl. 1. *In* T. V. Desikachary, *Taxonomy and biology of blue-green algae.* Madras: Univ. Madras Centre for Advanced Study in Botany, Banglor Press, 1972.

Schidlowski, Manfred, Zellular strukturierte Elemente aus dem Präkambrium des Witwatersrand-Systems (Südafrika). *Z. Dt. Geol. Ges.*, v. 115 (1963), p. 783–786, pls. 20, 21, 1966.

Schidlowski, Manfred, Elektronenoptische Identifizierung zellartiger Mikrostrukturen aus dem Präkambrium des Witwatersrand-Systems (>2.15 Mrd. Jahre). *Paläont. Z.*, v. 44, p. 128–133, pls. 14–16, 1970.

Schimper, A. F. W., Ueber die Entwickelung der Chlorophyllkörner und Farbkörper. *Bot. Ztg.*, v. 41, p. 105–114, pl. 1, 1883.

Schnepf, Eberhard, and R. M. Brown, Jr., On relationships between endosymbiosis and the origin of plastids and mitochondria, p. 299–322. *In* J. Reinert and H. Ursprung (Eds.), *Origin and continuity of cell organelles* (v. 2 of *Results and problems in cell differentiation*). New York, Heidelberg, Berlin: Springer-Verlag, 1971.

Schopf, J. M., E. G. Ehlers, D. V. Stiles, and J. D. Birle, Fossil iron bacteria preserved in pyrite. *Proc. Am. Phil. Soc.*, v. 109, p. 288–308, 20 figs., 1965.

Schopf, J. W., Microflora of the Bitter Springs Formation, Late Precambrian, Central Australia. *J. Paleont.*, v. 42, p. 651–688, pls. 77–86, figs. 1–6, 1968.

Schopf, J. W., Electron microscopy of organically preserved Precambrian microorganisms. *J. Paleont.*, v. 44, p. 1–6, pls. 1, 2, 1970.

Schopf, J. W., Paleobiology of the Precambrian: The age of blue-green algae. *Evolut. Biol.*, v. 7, p. 1–43, figs. 1–10, 1974.

Schopf, J. W., Precambrian paleobiology: problems and perspectives. *A. Rev. Earth Planet. Sci.*, v. 3, p. 213–249, figs. 1–4, 1975.

Schopf, J. W., and E. S. Barghoorn, Alga-like fossils from the early Precambrian of South Africa. *Science*, v. 156, p. 508–512, 8 figs., 1967.

Schopf, J. W., and E. S. Barghoorn, Microorganisms from the Late Precambrian of South Australia. *J. Paleont.*, v. 43, p. 111–118, pls. 21, 22, 1969.

Schopf, J. W., E. S. Barghoorn, M. D. Maser, and R. O. Gordon, Electron microscopy of fossil bacteria two billion years old. *Science*, v. 149, p. 1365–1367, figs. 1–6, 1965.

Schopf, J. W., and J. M. Blacic, New microorganisms from the Bitter Springs Formation (Late Precambrian) of the north-central Amadeus Basin, Australia. *J. Paleont.*, v. 45, p. 925–960, pls. 105–113, 1971.

Schopf, J. W., and D. Z. Oehler, How old are the eukaryotes? *Science*, v. 193, p. 47–49, figs. 1–3, 1976.

Schulz-Baldes, M., and R. A. Lewin, Fine structure of *Synechocystis didemni* (Cyanophyta: Chroococcales). *Phycologia*, v. 15, p. 1–6, figs. 1–10, 1976.

Schwartz, R. M., and M. O. Dayhoff, Origins of prokaryotes, eukaryotes, mitochondria, and chloroplasts. *Science*, v. 199, p. 395–403, 6 figs., 1978.

Shapiro, Joseph, Blue-green algae: why they became dominant. *Science,* v. 179, p. 382—384, figs. 1, 2, 1973.

Shestakov, S. V., and N. T. Khyen, Evidence for genetic transformation in blue-green alga *Anacystis nidulans. Molecular General Genetics,* v. 107, p. 372—375, 1970.

Siegel, S. M., Karen Roberts, Henry Nathan, and Olive Daly, Living relative of the microfossil *Kakabekia. Science,* v. 156, p. 1231—1234, figs. 1—5, 1967.

Siegel, S. M., and B. Z. Siegel, A living organism morphologically comparable to the Precambrian genus *Kakabekia. Am. J. Bot.,* v. 55, p. 684—687, figs. 1—3, 1968.

Singer, C. E., and B. N. Ames, Sunlight ultraviolet and bacterial DNA base ratios. *Science,* v. 170, p. 822—826, fig. 1, 1970.

Singh, H. N., Genetic control of sporulation in the blue-green alga *Anabaena doliolum* Bharadwaja. *Planta,* v. 75, p. 33—38, fig. 1, 1967.

Sistrom, W. R., *Microbial life.* New York: Holt Rinehart and Winston, Inc., Modern biology series, 112 p., figs., 1962.

Smith, G. M., *Cryptogamic botany,* v. 1. New York: McGraw Hill Book Co., 545 p., 299 figs., 1938.

Smith, P. V., Jr., Studies on origin of petroleum: occurrence of hydrocarbons in Recent sediments. *Bull. Am. Ass. Petrol. Geol.,* v. 38, p. 377—404, 6 figs., 1954.

Sournia, Alain, La cyanophycée *Oscillatoria* (= *Trichodesmium*) dans le plancton marin: taxinomie, et observations dans le Canal de Mozambique. *Nova Hedwigia,* v. 15, p. 1—12, pls. 1, 2, 1968.

Sournia, Alain, Les cyanophycées dans le plancton marin. *Année Biol.,* v. 9, p. 63—76, figs. 1—9, 1970.

Sournia, Alain, Ecologie et productivité d'une Cyanophycée en milieu corallien: *Oscillatoria limosa* Agardh. *Phycologia,* v. 15, p. 363—366, 1976.

Stanier, R. Y., M. Doudoroff, and E. A. Adelberg, *The microbial world,* 2nd ed. Englewood Cliffs, N.J.: Prentice-Hall Inc., 753 p., figs., 1965.

Stanier, R. Y., and C. B. van Niel, The main outlines of bacterial classification. *J. Bact.,* v. 42, p. 437—466, figs. 1, 2, 1941.

Stanier, R. Y., and C. B. van Niel, The concept of a bacterium. *Arch. Mikrobiol.,* v. 42, p. 17—35, 1962.

Starmach, Karol, Blue-green algae from the Tremadocian of the Holy Cross Mountains (Poland). *Acta Palaeont. Pol.,* v. 8, p. 451—459, pls. 1, 2, fig. 1, 1963.

Stewart, W. D. P., A botanical ramble among the blue-green algae. *Br. phycol. J.,* v. 12, p. 89—115, figs. 1—36, 1977.

Stubbe, Wilfried, Origin and continuity of plastids, p. 65—81, figs. 1—6. *In* J. Reinert and H. Ursprung (Eds.), *Origin and continuity of cell organelles* (v. 2 of *Results and problems in cell differentiation*). New York, Heidelberg, Berlin: Springer-Verlag, 1971.

Swinchatt, J. P., Algal boring, a possible depth indicator in carbonate rocks and sediments. *Bull. Geol. Soc. Am.,* v. 80, p. 1391—1396, figs. 1, 2, 1969.

Tappan, Helen, Phytoplankton abundance and Late Paleozoic extinctions: a reply. *Palaeogeogr. Palaeoclimat. Palaeoecol.,* v. 8, p. 56—66, 1970.

Tappan, Helen, Molecular oxygen and evolution, p. 81—135, figs. 1—9. *In* O. Hayaishi (Ed.), *Molecular oxygen in biology: topics in molecular oxygen research.* Amsterdam, Oxford: North Holland Publishing Co., 1974.

Tappan, Helen, Possible eucaryotic algae (Bangiophycidae) among early Proterozoic microfossils. *Bull. Geol. Soc. Am.,* v. 87, p. 633—639, figs. 1—21, 1976.

Tasnádi-Kubacska, A., *Paläopathologie, Pathologie der vorzeitlichen Tiere.* Jena: Gustav Fischer Verlag, 269 p., 293 figs., 1962.

Taylor, D. L., Chloroplasts as symbiotic organelles. *Int. Rev. Cytol.,* v. 27, p. 29—64, figs. 1—5, 1970.

Taylor, F. J. R., Implications and extensions of the serial endosymbiosis theory of the origin of eukaryotes. *Taxon,* v. 23, p. 229—258, 1974.

Taylor, F. J. R., Autogenous theories for the origin of eukaryotes. *Taxon,* v. 25, p. 377—390, figs. 1, 2, 1976.

Thomas, Joseph, Absence of the pigments of Photosystem II of photosynthesis in heterocysts of blue-green algae. *Nature,* v. 228, p. 181—183, 1970.

Tieghem, Ph. van, Sur la fermentation butyrique (*Bacillus amylobacter*) à l'époque de la houille. *C. R. hebd. Séanc. Acad. Sci., Paris,* v. 89, p. 1102—1104, 1879.

Tilden, J. E., A phycological examination of fossil salt from three localities in the southern states. *Am. J. Sci.,* v. 219 (ser. 5, v. 19), p. 297—304, 1930.

Tyler, S. A., and E. S. Barghoorn, Occurrence of structurally preserved plants in Precambrian rocks of the Canadian Shield. *Science,* v. 119, p. 606—608, 4 figs., 1954.

Uzzell, Thomas, and Chris Spolsky, Origin of mitochondria. *Science,* v. 180, p. 516—517, 1973.

Vologdin, A. G., *Drevneyshie vodorosli SSSR* [*Ancient algae of the USSR*]. Moscow: Akad. Nauk SSSR, 656 p., 46 pls. 1962.

Vologdin, A. G., Stromatolity i fototropizm [Stromatolites and phototropism]. *Dokl. Akad. Nauk SSSR,* v. 151, p. 683—686, fig. 1, 1963.

Vologdin, A. G., Ostatki organizmov iz Ladozhskoy Serii Proterozoya Karelii [Remains of organisms from the Ladozh Series of the Proterozoic of Karelia]. *Dokl. Akad. Nauk SSSR,* v. 175, p. 1143—1146, figs. 1, 2, 1967.

Vologdin, A. G., Otkrytiye ostatkov organismov v ver-khney svite Krivorozhskoy Serii Dokembriya Ukrainy [A discovery of organic remains in the upper suite of the Krivorozh Series of the Ukrainian Precambrian]. *Dokl. Akad. Nauk SSSR,* v. 188, p. 446—449, figs. 1, 2, 1969a.

Vologdin, A. G., Sinezelenye vodorosli Kembriya iz Ushchel'ya Ulug-Shangan, Tuva [Cambrian blue-green algae from Ulug Shangan Gorge, Tuva]. *Dokl. Akad. Nauk SSSR,* v. 188, p. 1376—1379, figs. 1—4, 1969b.

Vologdin, A. G., Ostatki organizmov iz Shungitov Dokembriya Karelii [Organic remains from Shungites of the Precambrian of Karelia]. *Dokl. Akad. Nauk SSSR,* v. 193, p. 1163—1166, figs. 1—3, 1970.

Vologdin, A. G., and N. A. Drozdova, Neskol'ko vidov vodorosley iz Gonamskoy Svity Uchurskoy Serii Pro-terozoya Ayano-Mayskogo Rayona Dal'nego Vostoka [Some algal species from the Gonam Beds of the Uchur Series of the Proterozoic of the Ayano-Maysk Region of the Far East]. *Dokl. Akad. Nauk SSSR,* v. 159, p. 114—116, 1 fig., 1964a.

Vologdin, A. G., and N. A. Drozdova, Iskopaemaya sinezelenaya vodorosl' v pozdnedokembriyskikh ot-lozheniyakh Dal'nego Vostoka [Fossil blue-green algae from Late Precambrian deposits of the Far East]. *Dokl. Akad. Nauk SSSR,* v. 159, p. 576—578, fig. 1, 1964b.

Vologdin, A. G., and N. A. Drozdova, Vodorosli Sem. Gloeocapsaceae v osadkakh Dokembriya Bate-nevskogo Kryazha [Algae of the family Gloeocap-saceae in the Precambrian of Batenev Ridge]. *Dokl. Akad. Nauk SSSR,* v. 186, p. 1419—1421, fig. 1, 1969a.

Vologdin, A. G., and N. A. Drozdova, Novye sinezelenye vodorosli Dokembriyskogo vozrasta iz Batenevskogo Kryazha [New Precambrian blue-green algae from Batenev Ridge]. *Dokl. Akad. Nauk SSSR,* v. 187, p. 440—442, fig. 1, 1969b.

Vologdin, A. G., and N. A. Drozdova, Kotkrytiyu vod-orosley semeystva Rivulariaceae v pozdnem Dokem-brii [On a discovery of algae of the family Rivulariaceae in the Late Precambrian]. *Dokl. Akad. Nauk SSSR,* v. 187, p. 1162—1163, fig. 1, 1969c.

Vologdin, A. G., and K. B. Korde, Neskol'ko vidov drev-nikh Cyanophyta i ikh tsenozy [Several species of an-cient Cyanophyta and their coenoses]. *Dokl. Akad. Nauk SSSR,* v. 164, p. 429—432, fig. 1, 1965.

Vologdin, A. G., and T. N. Titorenko, Proterozoyskie vodorosli s reki Kurtun, yugo-zapadnoe Pribaykal'e [Proterozoic algae from Kurtun River, southwes-tern Pribaikal]. *Dokl. Akad. Nauk SSSR,* v. 166, p. 1436—1439, figs. 1—4, 1966.

Walcott, C. D., Evidences of primitive life. *Rep. Smithson. Instn.,* 1915, p. 235—255, 18 pls., 1916.

Walsby, A. E., Mucilage secretion and the movements of blue-green algae. *Protoplasma,* v. 65, p. 223—238, figs. 1—17, 1968.

Walsby, A. E., The extracellular products of *Anabaena cylindrica* Lemm. I. Isolation of a macromolecular pigment-peptide complex and other components. *Br. Phycol. J.,* v. 9, p. 371—381, fig. 1, 1974.

Walter, M. R., Stromatolites and the biostratigraphy of the Australian Precambrian and Cambrian. *Spec. Pap. Palaeontology,* no. 11, p. 1—190, pls. 1—34, figs. 1—55, 1972.

Walter, M. R., *Stromatolites.* Amsterdam, Oxford, New York: Elsevier Scient. Publ. Co., xii + 790 p., 1976.

Walter, M. R., John Bauld, and T. D. Brock, Siliceous algal and bacterial stromatolites in hot spring and geyser effluents of Yellowstone National Park. *Sci-ence,* v. 178, p. 402—405, figs. 1, 2, 1972.

Walter, M. R., J. H. Oehler, and D. Z. Oehler, Mega-scopic algae 1300 million years old from the Belt Supergroup, Montana: a reinterpretation of Walcott's *Helminthoidichnites. J. Paleont.,* v. 50, p. 872—881, pls. 1, 2, 1976.

Whitton, B. A., and C. Sinclair, Ecology of blue-green algae. *Sci. Prog., Oxf.,* v. 62, p. 429—446, figs. 1—3, 1975.

Wilcox, Michael, One-dimensional pattern found in blue-green algae. *Nature,* v. 228, p. 686—687, figs. 1, 2, 1970.

Wildon, D. C., and F. V. Mercer, The ultrastructure of the heterocyst and akinete of the blue-green algae. *Arch. Mikrobiol.,* v. 47, p. 19—31, figs. 1—15, 1963.

Wolk, C. P., Physiology and cytological chemistry of blue-green algae. *Bact. Rev.,* v. 37, p. 32—101, figs. 1—16, 1973.

Wood, Alan, The Lower Carboniferous calcareous algae *Mitcheldeania* Wethered and *Garwoodia,* gen. nov. *Proc. Geol. Ass.,* v. 52, p. 216—226, pls. 13—15, 1941.

Wood, Alan, The type-species of the genus *Girvanella* (calcareous algae). *Palaeontology,* v. 1, p. 22—28, pls. 5, 6, fig. 1, 1957.

Zalessky, M. D., Premières observations micros-copiques sur le schiste bitumineux du Volgien in-férieur: *Annls Soc. Géol. N.,* v. 51, p. 65—104, pls. 2—6, figs. 1, 2, 1926.

Zobell, C. E., *Marine microbiology: a monograph on hydrobacteriology.* Waltham, Mass.: Chronica Botanica, 240 p., 1946.

Zobell, C. E., Bacteria, p. 693—698. *In* H. S. Ladd (Ed.), *Paleoecology* (v. 2 of *Treatise on marine ecology and paleoecology*) Mem. Geol. Soc. Am., 67, v. 2, 1957.

THE RHODOPHYTA

It is not generally appreciated how little is actually known about most marine or freshwater algae and the Rhodophyta is probably the worst known of any algal group.

Peter S. Dixon, 1973

The Rhodophyta include some of the most primitive algae and some of the most specialized, some of the former long having been misidentified as green algae and some of the latter having been misidentified as animals (foraminifera, corals, sponges). In pigmentation and lack of any flagellated members or life stage, they resemble blue-green algae and may represent the most ancient eucaryotes (see Chapter 1), but only the calcified representatives are common as fossils, those with unmineralized thallus being relatively rarely reported. Others undoubtedly have been assigned to the blue-green, green, or brown algae, in the absence of definitive characteristics.

THE LIVING AND FOSSIL RHODOPHYTA

The Organism

A few Porphyridiales are unicellular, but most other red algae have a comparatively large multicellular filamentous, crustose, or foliaceous thallus. The larger taxa may appear morphologically complex, but all are constructed of aggregates of filaments, so that the structure is pseudoparenchymatous. A single cell of the Florideophycidae may increase in size from origin to maturity by as much as 35,000 times, and many of the cells are multinucleate. An alternation of morphologically unlike successive generations, some of which were once thought to be distinct taxa, is characteristic of the red algae.

Cytoplasm

The cytoplasm of young cells is dense and occupies most of the cell interior. In older cells that are greatly enlarged, vacuoles may occupy a major proportion of the cell interior, and the cytoplasm, nuclei, plastids, and other organelles and storage products are peripherally located.

Nucleus

The simpler red algae, the Bangiophycidae, are invariably uninucleate. The approximately 3-μm-diameter nucleus is centrally located in young cells, although it may lie at one side of the older vacuolated cells. In contrast, the Florideophycidae may have uninucleate or multinucleate apical cells, and the number of nuclei may increase with cell age. Apical cells of some *Griffithsia* may contain from 12 to 500 nuclei, and the very large oldest cells may have as many as 4000 nuclei. A single nucleus of the Florideophycidae may be central, but in multinucleate cells all nuclei lie near the cell periphery.

Nuclei that are peripheral in position may be somewhat flattened and of irregular outline.

Pigments, Thylakoids, and Plastids

One or several plastids may be present. A single axially located stellate plastid with large central pyrenoid is typical of many Bangiophycidae (see Figure 2.1), although plastids may be peripherally located in others. More complex Florideophycidae commonly have several plastids scattered throughout the cytoplasm, although the innermost cells of compact thalli may lack any plastids.

The thylakoids are enclosed in a double membrane as in other algae, although this is not connected to the endoplasmic reticulum as in others. Furthermore, the thylakoids are single as in the blue-green algae, rather than being grouped in twos or threes to form lamellae, as in other eucaryotic algae (Gibbs, 1970). All thylakoids may be nearly parallel in orientation, or the outermost one or two may form a circle just within the plastid membrane.

Phycobilisomes, the regularly aligned 35-nm-diameter granules containing the red algal accessory pigments phycocyanin and phycoerythrin, are located on the outer surfaces of the thylakoids as in the blue-green algae. Areas of electron-dense DNA fibrils are present between the thylakoids.

Like all oxygen-producing photosynthesizers, red algae have the pigment chlorophyll *a,* and a few taxa also have chlorophyll *d* (which may be only a slightly altered form of chlorophyll *a*). Among the accessory pigments, β-carotene is the most important carotenoid, and lutein the most ubiquitous of the xanthophylls. Other pigments include α-carotene, zeaxanthin, taraxanthin, violaxanthin, and neoxanthin. The characteristic colors of the red algae are due largely to the red or blue water-soluble biliproteins, phycoerythrin and phycocyanin, located in the phycobilisomes, and found only in the red algae, cryptophytes, and blue-greens. Variations of these pigments were earlier thought to be systematically restricted, and termed R-phycocyanin and R-phycoerythrin in the red algae, C-phycocyanin and C-phycoerythrin in the cyanophytes, and B-phycoerythrin in the bangiophycid red algae. Possibly any of the three groups may contain one, two, or all of the phycobilin variations, however, the blue-greens generally having only one of the three, although varying in different species. The relative amounts of the particular pigments may produce the various red, brown, or bluish thalli of the different rhodophytes.

Although the characteristic pigments allow the red algae to utilize the reduced light at the greater depths of the deep subtidal regions, some species also occur in the higher intertidal areas (*Porphyra* and *Bangia*). Other red algae occur beneath the ice at high latitudes and in areas where there is little or no light for long periods, suggesting that they may survive heterotrophically. Laboratory tests also indicate that some species (*Corallina, Gelidium*) can absorb organic material from the water.

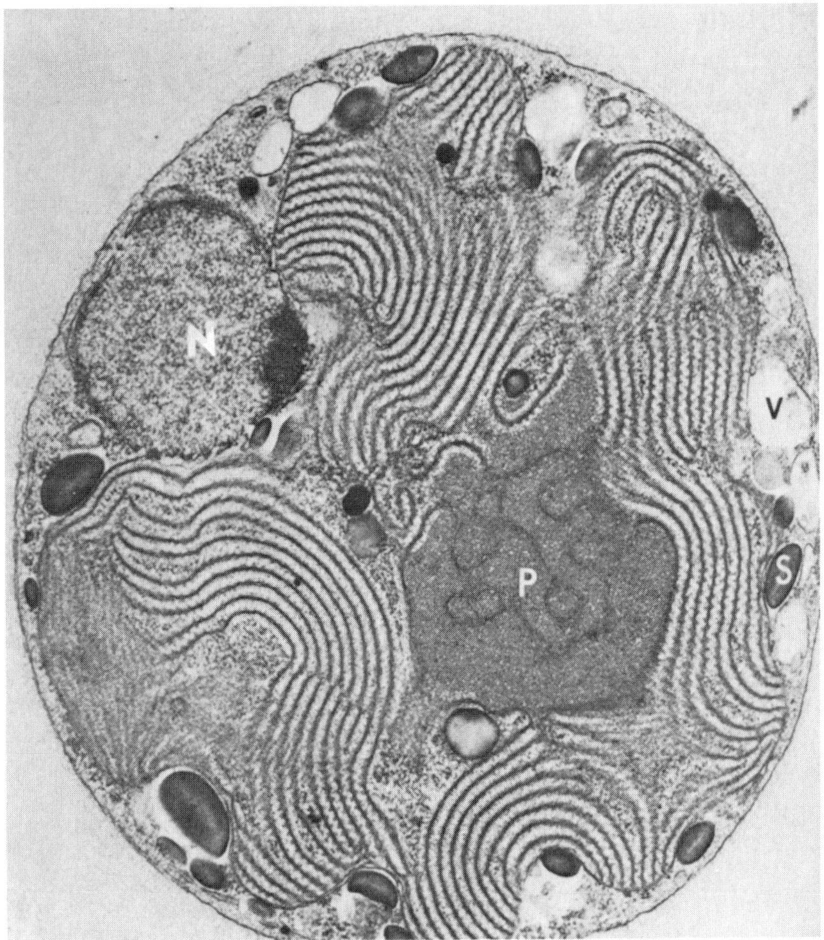

Figure 2.1
Porphyridium aerugineum Geitler, TEM of sectioned specimen grown in liquid medium. Cell surrounded by irregular cell membrane and diffuse fibrous sheath; nucleus (N) excentric in position, with large dark-colored nucleolus facing adjacent plastid; vacuoles (V) and starch grains (S) near cell periphery, pyrenoid (P) at center of plastid; plastid lamellae with small phycobiliprotein granules (phycobilisomes) on surface; ×23,500. From Gantt et al., 1968.

Storage Products

Floridean starch, the storage product of the red algae, occurs as grains lying free in the cytoplasm, commonly in the vicinity of the pyrenoids or the nucleus. An additional compound, trehalose, is found in freshwater red algae and in blue-greens and some fungi. Other storage compounds also have been reported in red algae.

Iodine, Bromine, and Chlorine

Species representing most of the orders of red algae have been shown to synthesize organic compounds with large amounts of iodine, bromine, and chlorine (Fenical, 1975). The halides may provide osmotic balance with the seawater, and at least some are sufficiently toxic to deter grazing on the algae.

The Wall and Sheath

Cells of the simplest red algae, such as the unicellular *Porphyridium* (see Figure 2.2) have a single external membrane surrounded by a thick mucilaginous sheath. More complex bangiophycid taxa such as *Porphyra* have a distinct cell wall composed of randomly arranged xylan microfibrils embedded in mucilage, and an outer proteinaceous cuticle of granular mannan.

In contrast to the bangiophycid xylan and mannan, cellulose is present in some florideophycid rhodophytes, about 25 percent of the wall of the ceramialean *Griffithsia* consisting of randomly arranged cellulose microfibrils. The coralline algal wall is noncellulosic, consisting instead of pectinlike material (Baas-Becking and Galliher, 1931).

Water-soluble mucilages comprise a large proportion of the wall, such as the agar obtained from the living *Ahnfeltia*, *Gelidium*, *Gracilaria*, and *Pterocladia*. This polysaccharide is used as a culture medium in the laboratory and for gelling or thickening a wide variety of processed foods. The similar gel compound carrageenan, obtained from *Chondrus* and *Gigartina*, is used for thickening cosmetics, drugs, and instant puddings and sauces.

Calcification

The mucilage component of the wall may be calcified, particularly in representatives of the Corallinaceae, but also in various other red algae. The calcium carbonate is deposited as calcite in the Corallinaceae (see Figure 2.3) and as aragonite in the calcified Nemaliales, as well as in the Peyssoneliaceae (Cryptonemiales). Both magnesium and strontium may be associated with the calcium carbonate, the proportion of magnesium being greatest in calcitic taxa, where it may comprise up to 30 percent of the carbonate (Chave, 1954); in contrast, the strontium is greater in aragonite-depositing taxa. Brucite, $Mg(OH)_2$, has been reported to occur with the magnesian calcite in *Goniolithon* (Schmalz, 1965; Weber and Kaufman, 1965). This composition results in an unusually high solubility of the *Goniolithon* skeletal material. The

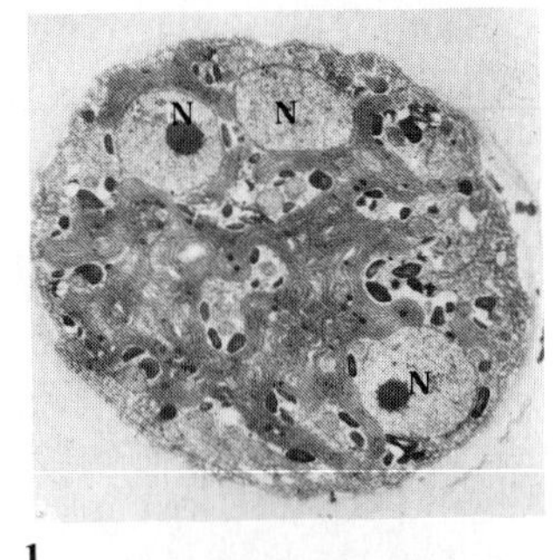

1

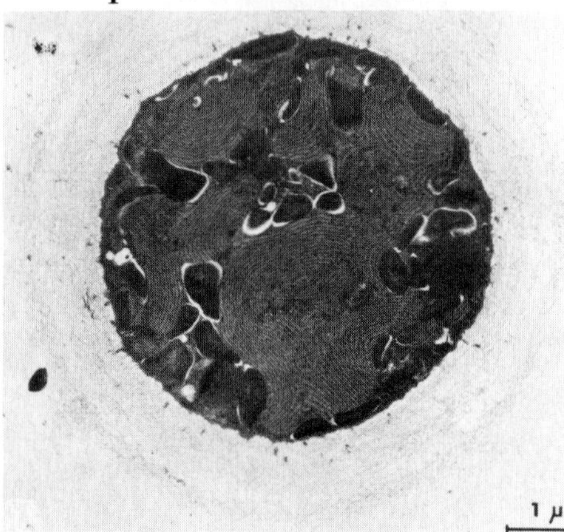

2

Figure 2.2
Porphyridium aerugineum. TEM of sectioned specimens from culture. **1.** Cell grown on agar, sectioned in process of division, with three of four nuclei (N) visible as lighter in color, those at upper left and lower right with dark-colored central nucleoli; cell surface irregular, cytoplasm filled with lamellar plastid, with dark ovoid starch grains and granular appearing mitochondria between lamellar folds; ×4250, from Gantt et al., 1968. **2.** Deeply stained cell to show thick-layered surrounding capsule, irregular cell surface beneath capsule, and lamellar plastid filling most of the cell; ×21,000, bar = 1 μm, from Ramus, 1972.

percentage of magnesium present in the calcite of the red algae appears to be related to the rate of growth, the slowly growing denser layers having a higher magnesium content than does the more rapidly growing part of the skeleton. Water temperature does not seem to have an appreciable effect on composition (Moberly, 1968).

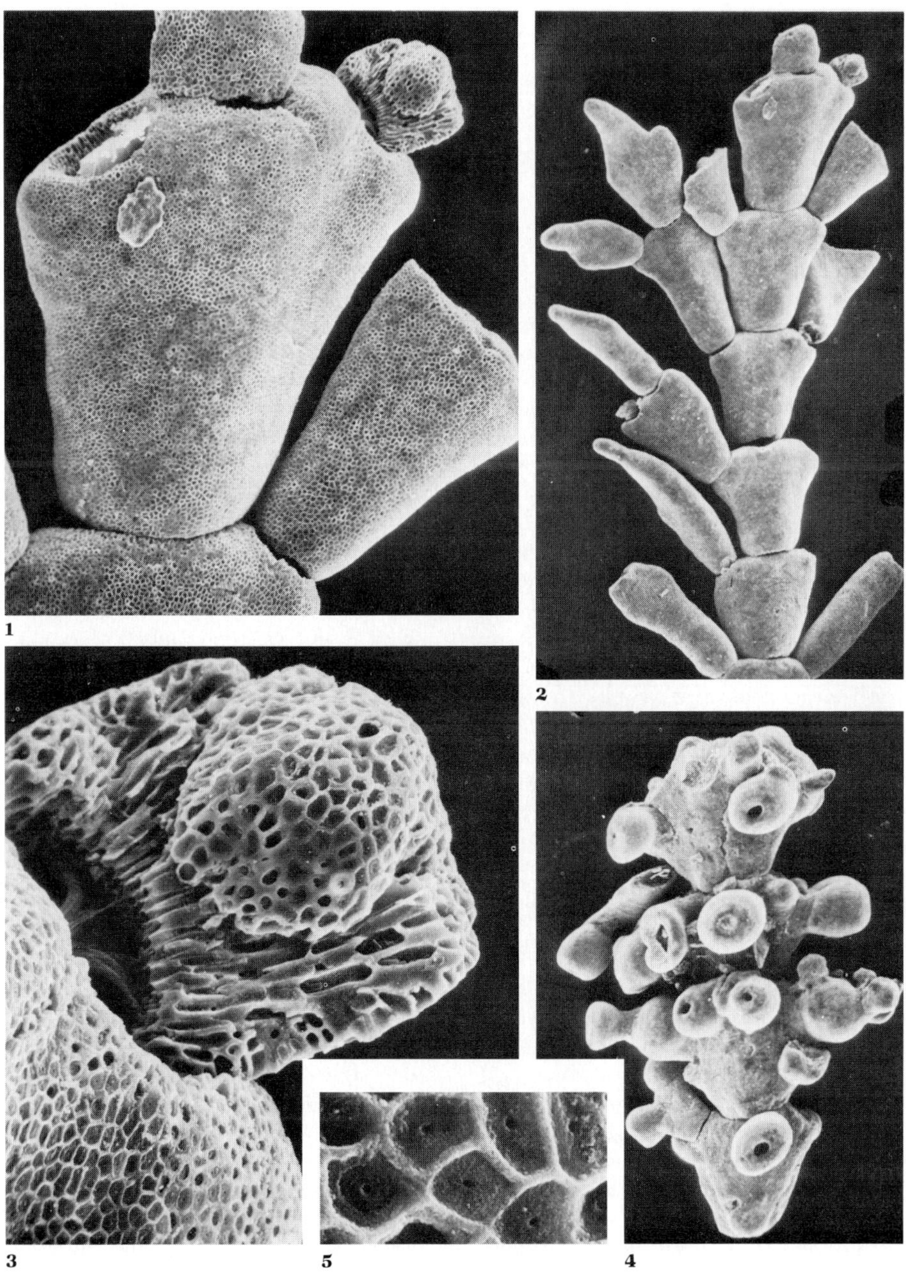

Figure 2.3

Corallina officinalis Linnaeus var. *chilensis* (Decaisne) Kützing, Cabrillo Beach, southern California, in SEM. **1,2.** Tip of vegetative thallus showing genicula (joints) and intergenicula (segments); 1, ×80; 2, ×24. **3.** Enlargement of broken part of same specimen showing closely parallel filaments, ×320. **4,5.** *C.* cf. *officinalis* var. *chilensis* in SEM. **4.** Exterior showing many conceptacles, probably carposporangial, ×24. **5.** Enlargement of part of surface showing inner wall of cells with porelike pit connection where outer wall has been destroyed, ×1600.

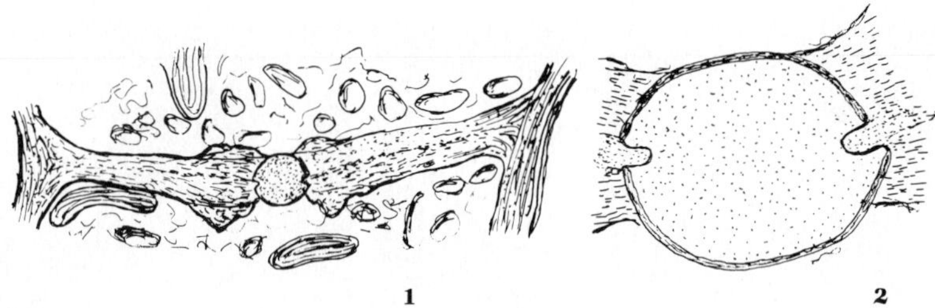

Figure 2.4
Porphyra leucostica Thuret. Pit connections in conchocelis stage;
sketches based on TEM of sectioned cells in Lee, 1971. **1.** Location of pit
connection in center of septum separating adjacent cells of a filament,
with septal thickening adjacent to the pit, ×4550. **2.** Enlargement of pit
connection with an ovoid plug and plug caps separating it from adjacent
cell contents of fertile cell row, ×25,500.

The aragonite of *Liagora* (Nemaliales) is deposited in the intercellular spaces as randomly arranged needlelike crystals 3 to 6 μm long and 0.5 to 0.7 μm wide, tapering to a flat tip, and similar in form to those of the green alga *Halimeda* (Borowitzka et al., 1974). The more massive calcite deposits in the corallines, such as *Lithophyllum* and *Corallina,* lie within the cell walls so that the cells are completely embedded, with the C-axes of the calcite crystals oriented perpendicular to the cell wall (Baas-Becking and Galliher, 1931).

Pit Connections

A characteristic feature of the Florideophycidae is the presence of **pit connections** in the walls separating successive cells of a filament. Although originally thought to be perforations in the wall occupied by intercellular cytoplasmic strands, electron microscopy has shown that true connections do not exist between cells. Instead the primary pit connections consist of an opening in the wall filled by a biconvex membrane-bound plug containing material of two different densities. The lenticular to discoidal plug generally is between 0.3 and 2.5 μm in diameter, but may be up to 40 μm in *Bonnemaisonia.* It consists of proteinaceous granular material of a composition different from that of both cytoplasm and wall. Both ends of the plug are covered by an electron-dense plug cap. Unlike the wall of the coralline algae, the plug is not calcified, and the pit connections appear as perforations in the skeletal remains (see Figure 2.3, part 5).

Primary pit connections are formed as new cross walls arise by annular ingrowth from the existing side walls at cell division, but leave a central region unclosed. The pore plug is formed within this temporary opening in the wall, while the plasmalemma remains continuous from cell to cell, external to the limiting membrane of the plug caps. Pit connections are found in the Bangiophycidae only in the filamentous conchocelis phase of *Bangia* and *Porphyra* (see Figure 2.4). Some pit connections in *Polysiphonia* are intracellular, flaring out into one of the adjacent cells; these **transfer connections** may serve as pathways for nutritive transport into specialized regions of the plant (Wetherbee, 1979).

Secondary pit connections also may be formed between cells of adjacent filaments, particularly in *Polysiphonia* (Rhodomelaceae) and the Corallinaceae. In the latter, dissolution of a part of the walls in contact allows intercellular communication, but a plug later is laid down in the opening, as in the true pit connections. Secondary pit connections in the Rhodomelaceae are more complex in development, involving nuclear division and transfer of the adjacent cells.

Specialized Cells

Many Florideophycidae have specialized secretory cells that contain little cytoplasm and no plastids or pigments, but have large vacuoles containing highly refractive contents. In the Bonnemaisoniaceae they are associated with high concentrations of iodine in the thallus, but in other taxa they may be related to the production of mucilage.

Iridescent cells, as seen in reflected light, are common in the Ceramiales and Rhodymeniales. The iridescence is from small spherical paired bodies or more elongate single ones, but their possible function is unknown.

External cells of many Florideophycidae have hairs that first appear as lateral protuberances with a large central vacuole. Lengthening rapidly, the **hair cell** or **trichocyte** becomes cylindrical, with only a small amount of cytoplasm, and when fully lengthened lacks a nucleus. The hair cells may be seasonally present, in which case they are most abundant in the spring and early summer.

Small terminal **cap cells** on the upright filaments of Corallinaceae form prior to the growth of the filament. As the cap cells never divide again, filament development results from the division of the second cell to arise from the cells of the prostrate system, as a type of intercalary cell division. The cap cells may serve to protect the dividing cell beneath and also are related to the deposition of calcium carbonate.

Growth and Reproduction

Red algae range from the extremely simple unicells such as *Porphyridium,* which may be similar to the earliest eucaryotes, to some of the most elaborate of the algae, with complex thalli and two or three distinct phases of the life cycle. The life history is better known for the Florideophycidae than for the Bangiophycidae, in which the true nature of certain reproductive bodies may still be questioned.

Morphologic Phases of the Bangiophycidae

Unicellular Rhodophyta. The Porphyridiaceae include unicellular red algae, such as *Porphyridium* and *Rhodella,* that range from the open ocean plankton to living in shore sands or as epiphytes to a freshwater habitat or subaerial on soil or damp walls. Only asexual reproduction is known to occur in these unicellular algae, generally by simple median fission. In dividing, the cell constricts to appear dumbbell-like, and gradually splits in two (see Figure 2.5). When grown in liquid culture, all cells of *Porphyridium purpureum* (Bory) Drew & Ross are spherical, but when grown on agar, the lower surface of the cell is flattened at the contact with the substrate. During rapid growth, cells commonly are from 5 to 7.5 μm in diameter, but in old cultures with a low rate of division the cells might attain as much as 60 μm diameter. These large cells occasionally divide into more than two daughter cells, variously producing 4, 8, or 16 cells. On agar, the larger cells also occasionally produce small buds or cells from the periphery, with anywhere from 1 to 15 such buds being held in the cell mucilage (Sommerfeld and Nichols, 1970). In the plankton, or in liquid cultures, the cells have very little mucilage cover, but on soil, as an epiphyte, or grown on agar, the bright red cell masses may be surrounded by large quantities of mucilage.

The similarly unicellular *Rhodella maculata* Evans has a thick mucilage cover and readily sticks to the bottom or sides of a culture vessel. Grown on agar, it also produces large cells up to 30 μm in diameter that form small peripheral buds, as does *Porphyridium* (Paasche and Throndsen, 1970).

Gliding Movement in the Porphyridiaceae. No true motile cells occur in the red algae, as no flagellated stage is present. The unicellular *Porphyridium purpureum* has a gliding movement; cells may be slightly elongated in the direction of movement but are not deformed as by amoeboid movement. A tapered mucilaginous stalk may occur at one end of the cell and the posterior part of moving cells con-

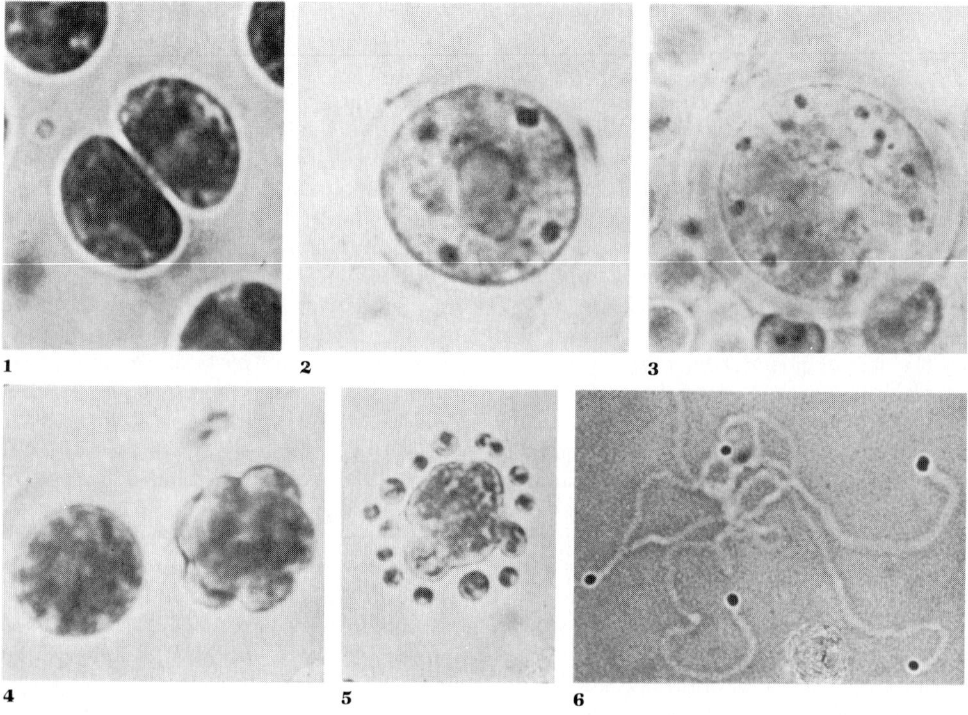

Figure 2.5

Porphyridium purpureum in culture. **1.** Binary fission, producing two daughter cells, ×3300. **2.** Quadrinucleate large cell, ×3000. **3.** Multinucleate large cell, ×1670. **4,5.** Cells forming more than two daughter cells, that of part 5 with peripheral buds held in cell mucilage; 4, ×1650; 5, ×1200. **6.** Showing trails left in agar surface by moving cells under overhead illumination, ×250. All as *P. cruentum,* from Sommerfeld and Nichols, 1970.

tains large vesicular areas filled with sheathlike fibrillar material (Lin et al., 1975). Release of the sheathlike material at the rear of the cell may aid the gliding movement. On agar surfaces, the moving cells left a distinct trail as a shallow groove. Secretory activity apparently solubilized the agar beneath the moving cell.

Multicellular Bangiophycidae. All red algae other than the Porphyridiaceae are multicellular, and show some organization of cell orientation. Some Porphyridiales have simple uniserial filaments, others are multiseriate, and variations may occur within a single species. The Bangiales may be differentiated into a prostrate basal portion or disk and erect fronds, or have only erect fronds arising directly from the germination of a spore. The thallus of *Compsopogon* is uniserial at first, but later segmentation produces a surface cover of small cells.

Both *Porphyra* and *Bangia* have two major growth phases (see Figure 2.6): a leafy thallus, and a filamentous phase formerly described as a separate genus *Conchocelis*. Some **conchocelis** forms are free-living, but others are perforating in shells. Each phase arises initially from spores produced by the other phase, but also may be repeated by spores of another type. The nature of the spore produced by either phase is affected by the temperature and amount of light available, hence the two phases vary in relative abundance over the year. At 9°C, the *Bangia* phases produced both **mono-**

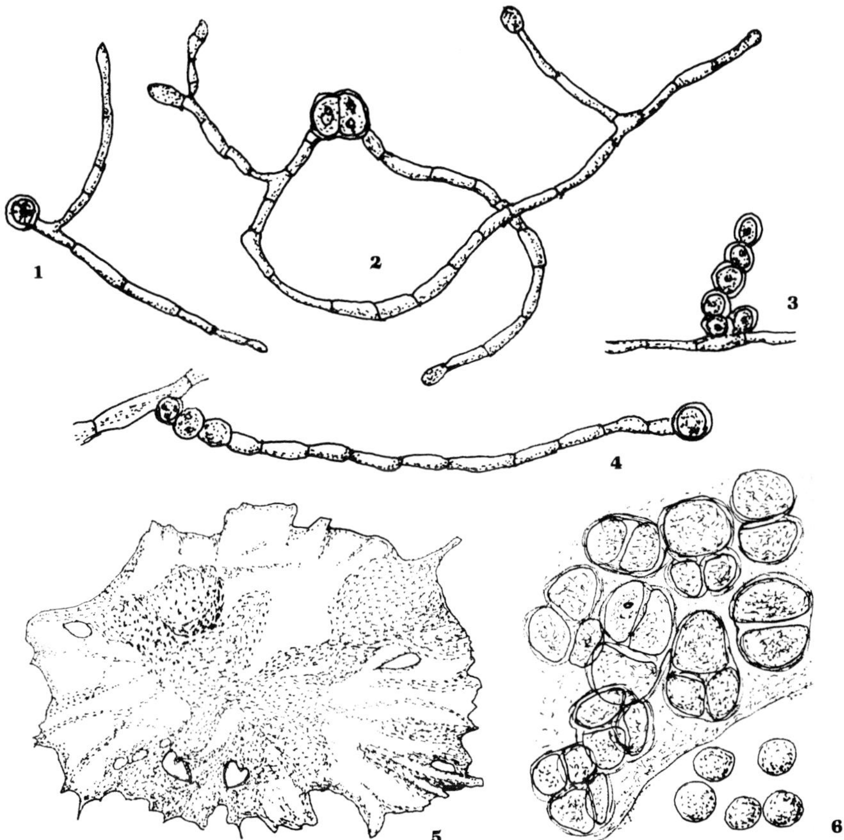

Figure 2.6

Porphyra umbilicalis (Linnaeus) J. Agardh. Life history. **1.** Germinating spore in five-month-old culture, producing branching conchocelis filament. **2.** Conchocelis stage in five-month-old culture, showing swollen apical cells. **3.** Branch of conchocelis filament with cells rounding off to produce detached conchospores that upon germination will produce the typical foliose *Porphyra* vegetative stage. **4.** Year-old culture, with sporelike cells or conchospores just following branching, and with terminal monospores that may produce new conchocelis stage directly. 1—4, ×370, redrawn from Conway, 1964b. **5,6.** Foliose *Porphyra* stage. **5.** Entire specimen, ×0.25, redrawn from photo in Conway, 1964b. **6.** Edge of thallus with formation of many large type 3α-spores that germinate into the conchocelis filaments, freed spores at lower right; ×500, from photo in Conway, 1964a.

spores, which grew into additional *Bangia* filaments, and **carpospores**, which were favored by longer days and germinated to produce the conchocelis phase. At 22°C, only monospores were produced, repeating the *Bangia* phase (Sommerfeld and Nichols, 1973). Similarly, *Porphyra* spores produce a conchocelis phase under long-day conditions, whereas with less than 12 hours of light the leafy *Porphyra* stage is repeated.

Discovery that the conchocelis stage alternates with a leafy stage also showed the relationship between the Bangiophycidae and the Florideophycidae to be closer than had been recognized previously. The conchocelis stage resembles many Florideophycidae in its entirely filamentous thallus, apical as well as intercalary growth, the commonly ribbonlike **parietal plastids** (in contrast to the centrally located stellate one of the *Porphyra* stage), pit connections between cells, a central vacuole, and a peripheral thylakoid in the plastids (Cole and Conway, 1975). This degree of similarity suggests that although the two groups may be regarded as distinct subclasses (as is done herein), they are not sufficiently different as to be regarded as separate classes.

The nature of the spores, presence of gametes, or even any sexual reproduction in the Bangiophycidae are all uncertain. Some reported instances of cell fusion later were shown to represent cell division, and supposed fertilization canals or tubes were proven merely to be filamentous bacteria and fungi infecting the red alga. Certain colorless spores of *Porphyra* and *Bangia* have been erroneously regarded as spermatia, and some reports of meiosis are regarded as mitotic divisions by others, who found the same number of chromosomes in all phases of the life history. Although not yet satisfactorily demonstrated, sexual reproduction probably occurs in the more advanced Bangiophycidae. It may not be present in the unicellular Porphyridiaceae, although whether this is a retained primitive character or a secondary reduction is also uncertain.

Spore Production. Even the nature of the spores is poorly defined in the Bangiophycidae, as the spores are not always formed in characteristic sporangia like those of the Florideophycidae. Three types of spores occur (Drew, 1956).

Type 1 monospores are produced in differentiated sporangia, as in *Porphyropsis, Rhodochaete,* and *Erythrotrichia.* Some evidence suggests that these may be male gametes.

Type 2 monospores are produced by undifferentiated cells; they involve the release of separate vegetative cells or the cell contents as in *Asterocytis, Bangia, Erythrotrichia, Neevea,* and *Porphyra.* These spores appear to be asexual.

Type 3 spores are produced by successive divisions of a mother cell, and may be of two different types in *Bangia* and *Porphyra.* In one, divisions result in 4 to 8 spores, or rarely more. The mother cell may be a zygote, the spores themselves may be gametes, or they may be asexual carpospores. The second form of type 3 spore results from many more divisions of the mother cell, producing up to 128 spores that have lost their pigments in the process. These more numerous bodies have been regarded as spermatia or male gametes, but with very little evidence. Instead, the few large type 3 spores are better referred to as large spores or α-spores, and the others as small spores or β-spores (Conway, 1964a), avoiding any implication as to their nature or function.

Morphologic Phases of the Florideophycidae

Although the unicellular red algae apparently have very simple life histories, the more advanced rhodophytes may have a sequence of two or three distinct phases that may be morphologically similar or wholly unlike, and variously produce gametes or different types of spores. The differing morphology of the different stages in the life history has made necessary studies of cultures, in order to relate them. For example, *Bonnemaisonia asparagoides* (Woodward) C. Agardh alternates successively between haploid **gametophyte**, diploid **carposporophyte**, and diploid **tetrasporophyte** phases, the latter being a minute prostrate plant that originally was described as a distinct species *Hymenoclonium serpens* (Crouan) Batters; the tetrasporophytes of *Bonnemaisonia hamifera* Hariot and *B. nootkana* (Esper) Silva are filamentous forms known previously as *Traillliella intricata* Batters (Chihara, 1965).

Thalli of the Florideophycidae are formed by the aggregation of filaments, each filament produced by the successive divisions of an apical cell. As the new cell walls develop, each forms a pit connection.

Two major morphologic types of Florideophycidae are the encrusting thallus and the erect foliose thallus. The encrusting thallus is generally circular or irregularly lobed, with the basal layer of radiating filaments being produced by germination of a spore. All cells of this layer then produce simple or sparsely branched upright filaments, held together by mucilage or tightly compacted. The horizontal filaments of *Peyssonelia* are calcified, and the upright filaments of some species may similarly be lime impregnated. The coralline algae have both prostrate crustose thalli and erect foliose thalli. The crusts are organized into an inner **hypothallium**, composed of the lowermost filaments that are subparallel to the substrate and commonly lack plastids in the cells; a median **perithallium**, composed of the filaments that branch from the hypothallial filaments and arch upwards perpendicular to the substrate; and the outer or superficial **epithallium**, each of whose cells is the termination of a perithallial filament (see Figure 2.7).

The erect foliose thallus produces a greater variety of form and organization. It generally arises from a prostrate system of filaments, although this system may be much reduced. The erect filaments grow by apical cell division, and most species show some branching. In more complex forms, the principal upright filaments (of unlimited growth) may be surrounded by lateral filaments (of limited growth), the aggregations of filaments forming an axis. The greatest complexity is attained by forms with differentiation, both of the filaments within an axis and of the axes themselves. Where a single principal axial filament is present in each axis it is termed uniaxial; where many filaments are present in the axis, it is termed multiaxial (as in *Nemalion*).

The erect branches or fronds of the coralline algae may have calcified segments termed **intergenicula** alternating with the uncalcified **nodes** or **genicula** (see Figure 2.8). A core of long-celled filaments with plastid-lacking cells forms the axis and is termed the **medulla;** it is surrounded by short-celled filaments of the **cortex.** The filaments of the cortex also arch upwards to terminate in the epithallium. Thus in these branches the medulla is homologous to the hypothallium of the crustose thallus, and the cortex comparable to the perithallium (Johansen, 1976). Although the distinction between medulla and cortex is useful in description, the difference apparently is not fundamental, for tissue that originated as cortex may become medulla with additional growth of the thallus (Dixon, 1973).

Although the coralline algae are commonly separated into the encrusting forms and the erect jointed or articulated ones (as in the classification here followed), other divisions have been based on variations in the reproductive cycles. Cabioch (1972) divided the coralline algae into groups on the basis of spore germination characteristics, and Chihara (1974) recognized two distinct groups on the basis of reproductive cycles, spore size, and spore germination, each of which included both encrusting and articulated taxa. The group including *Lithophyllum* and *Amphiroa* also was characterized by having secondary pit connections between adjacent filaments, whereas the group that included *Lithothamnium* and *Corallina* had lateral cell fusion instead. Use of these varied criteria may produce a more realistic classification when data are available for all corallines, but as most of the definitive characters of the reproductive cycle and spore germination are not recognizable in fossils, the form of thallus is here retained as of taxonomic importance.

The Gametophyte. The special male and female gametangia commonly occur on separate plants (see Figure 2.9). Many antheridial branches may be present on male plants, each **antheridium** producing a nonmotile spermatium, which upon release is carried by the water to the female plant. Gametangia-bearing plants are gametophytes.

The female oogonium, termed a **carpogonium,** has an elongate tip, the **trichogyne,** which commonly is mucilage-covered and receives the spermatium. The carpogonium may be produced on a special branch lacking plastids, and may have a specialized **auxiliary cell** into which the zygote nucleus passes after fer-

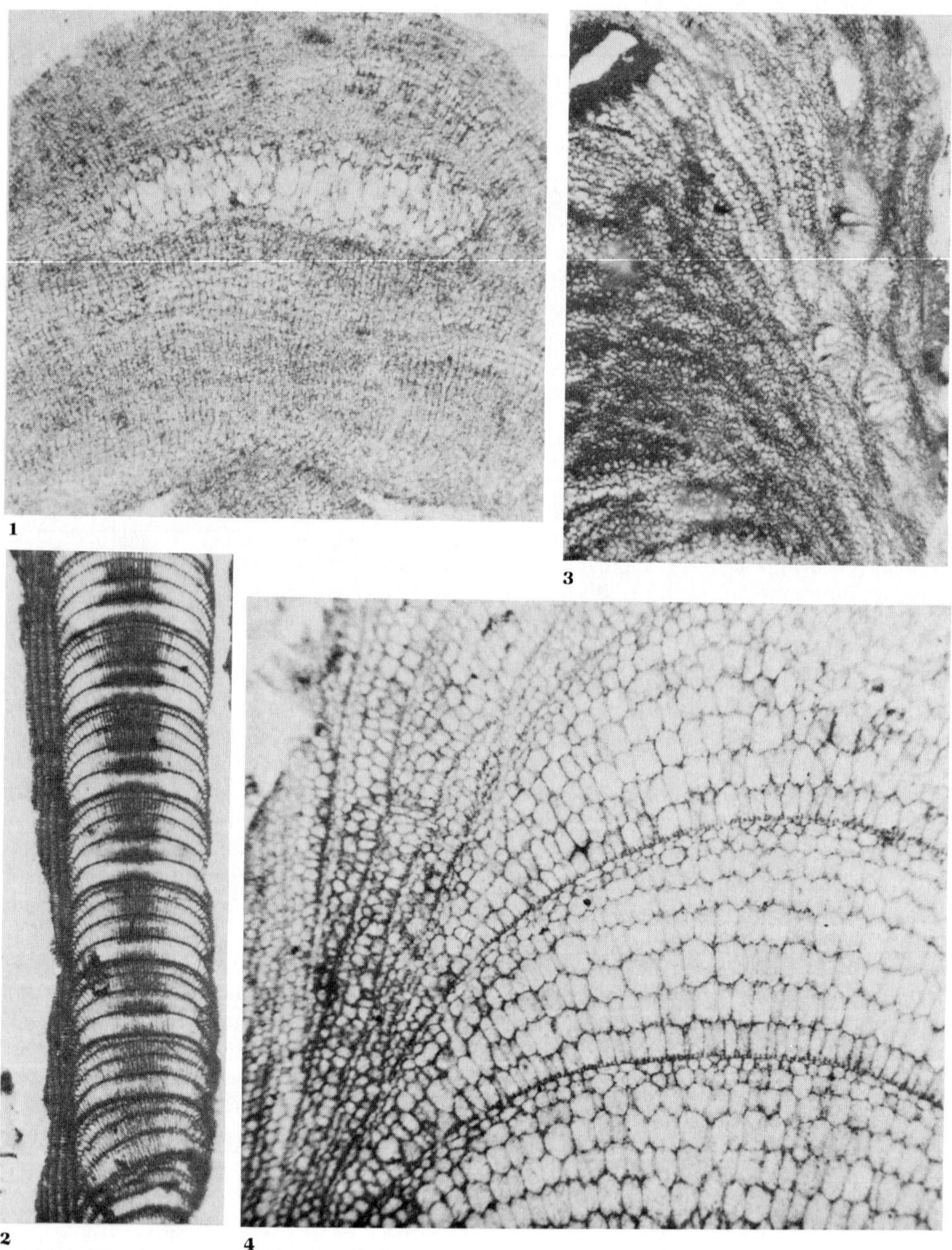

Figure 2.7

Figure 2.7 *(facing page)*
Coralline algae. **1.** *Lithothamnium* cf. *L. bofillii* Lemoine, Eocene, Ishigaki-shima, Ryukyu
Islands, section showing hypothallus, perithallus, and sporangia-filled conceptacle;
×100, from Johnson, 1964a. **2.** *Amphiroa prefragilissima* Lemoine, Miocene, Guam, section of specimen with much of perithallus worn off, but some remaining at left, and
with well-developed medullary hypothallus; ×40. **3,4.** *Aethesolithon grandis* Johnson,
Miocene, Guam, ×50. **3.** Sectioned margin of encrusted form with sporangia-filled conceptacles. **4.** Section showing medullary hypothallus grading into marginal perithallus at
left. 2−4, from Johnson, 1964b.

Figure 2.8 *(overleaf)*
Fossil and living articulated corallines. **1,2.** *Corallina* sp., Will Rogers State Beach, southern California, in SEM. **1.** Segment of branched and articulated thallus, ×30. **2.** Enlargement of part of center of broadest segment, showing individual cells, pit connections in
some of the intercellular septa appearing as holes in the calcified thallus from which the
noncalcified plugs have disintegrated; ×2000. **3.** *C.* cf. *C. vancouveriensis* Yendo, Cabrillo
Beach, southern California, SEM showing terminal conceptacles, probably tetrasporangial
in nature; ×30. **4.** *Jania guamensis* Johnson, Miocene, Guam, section of terminal segment, ×40, from Johnson, 1964b. **5.** *Amphiroa rigida* Lamouroux, South Africa. Section
showing uncalcified geniculum in center and calcified intergenicular segments with parts
of two conceptacles visible at left, and with young secondary branch as yet lacking genicular tissue; ×110, from Johansen, 1969b. **6.** *Corallina* cf. *C. cossmannii* Lemoine, Eocene,
Ishigaki-shima, Ryukyu Islands. Section through two intergenicula and the connecting
geniculum, showing successive growth zones; ×40, from Johnson, 1964a.

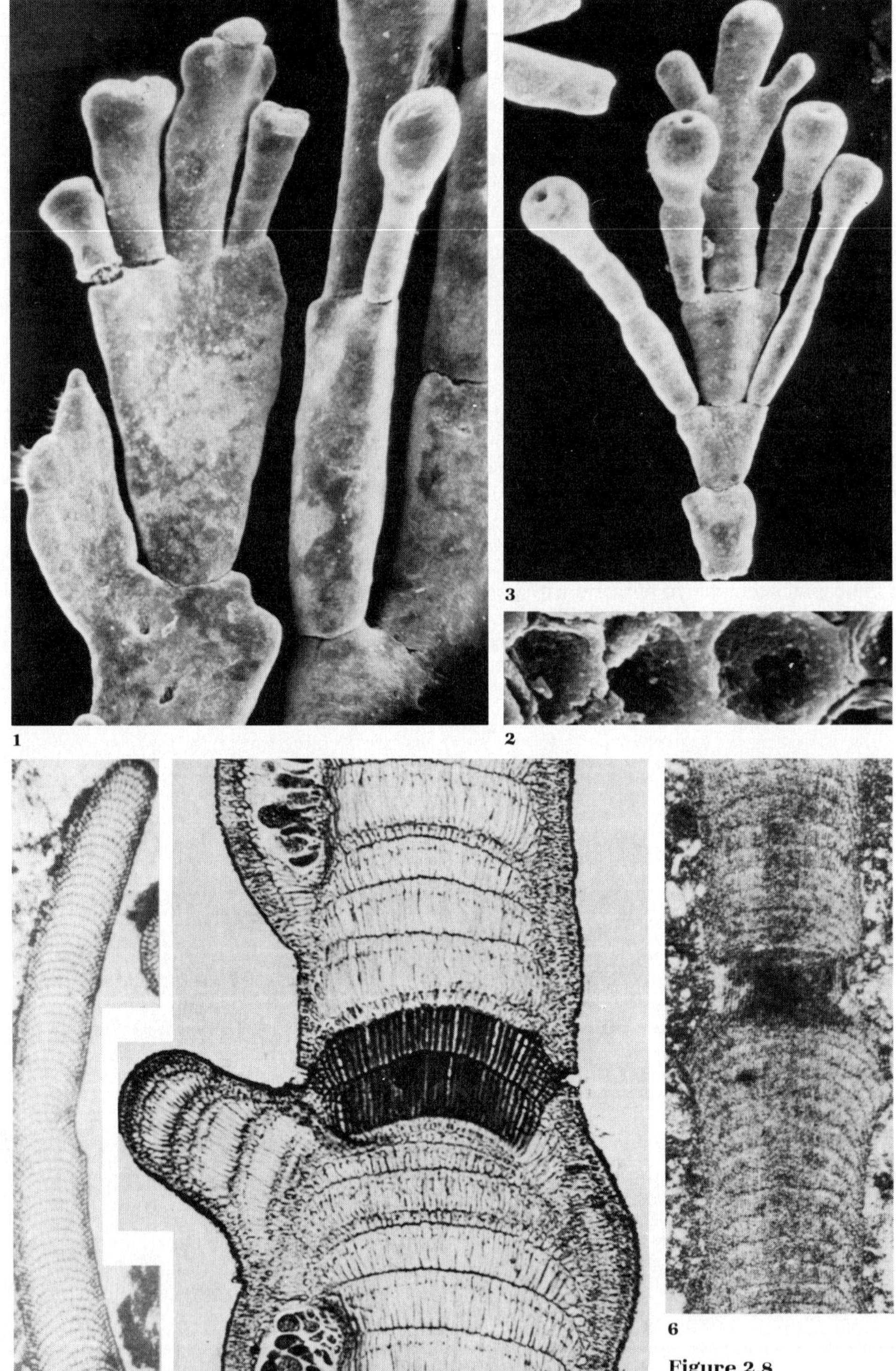

Figure 2.8

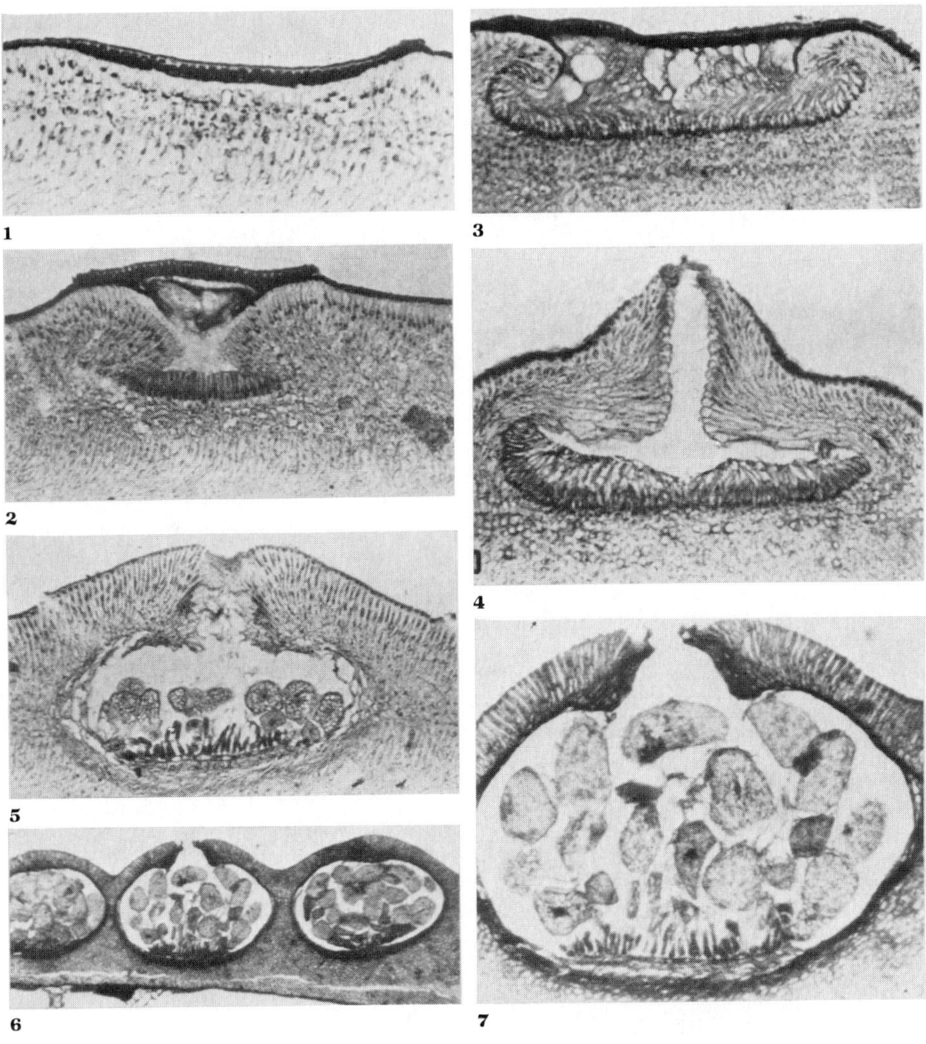

Figure 2.9

Bossiella californica (Decaisne) Silva subsp. *schmittii* (Manza) Johansen, California. Development of conceptacles. **1,2.** Young and mature stages of development of female conceptacles; **1**, ×160; **2**, ×128. **3,4.** Development of spermatangial conceptacles, ×144. **5 – 7.** Development of carposporangial conceptacles. **5.** Young stage, ×128. **6.** Mature stage, ×48. **7.** Enlargement, ×144. All from Johansen, 1973.

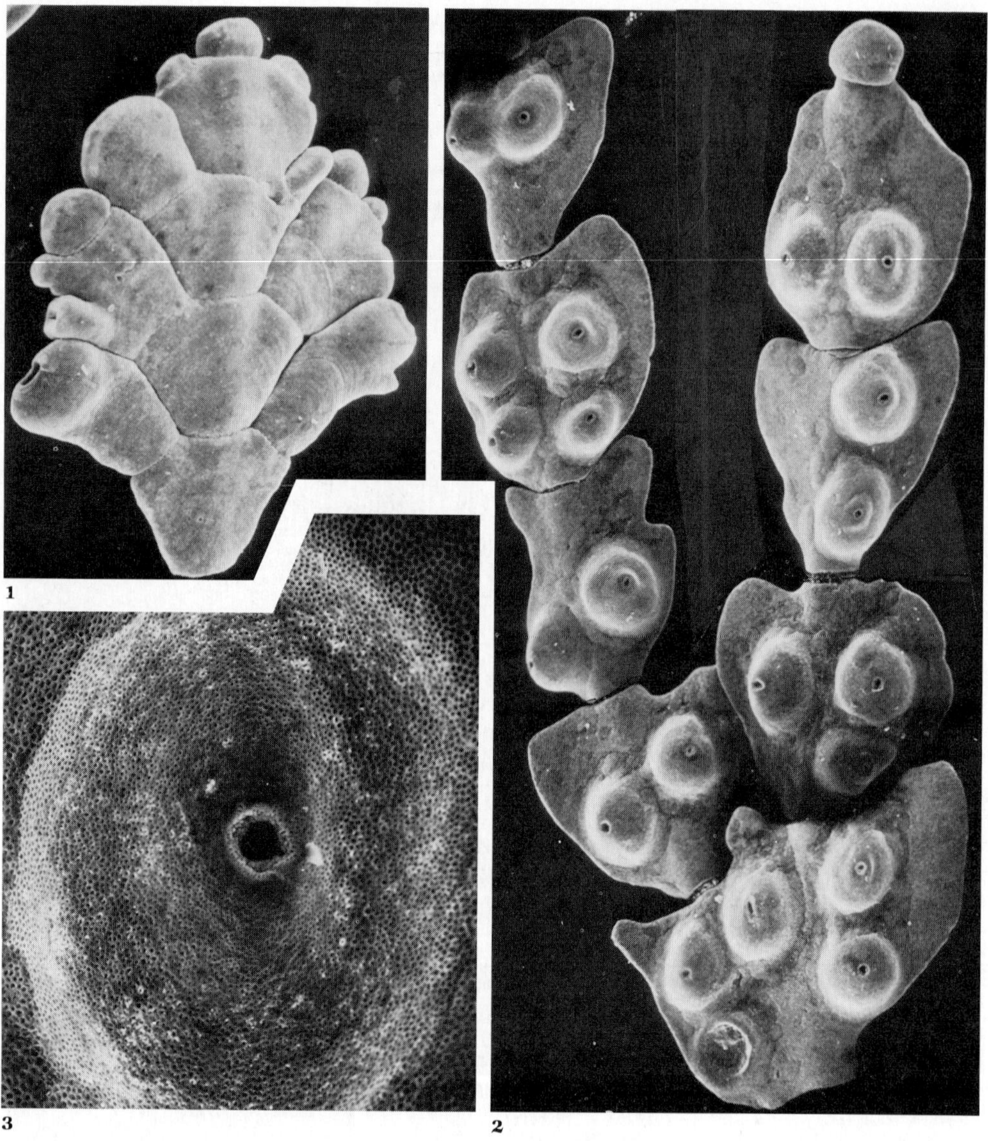

Figure 2.10

1. *Bossiella plumosa* (Manza) Silva, Cabrillo Beach, California. Tip of a branch showing broad intergenicula, SEM, ×24. **2, 3.** *B. chiloensis* (Decaisne) Johansen, Will Rogers State Beach, California. **2.** Part of two branches, showing numerous conceptacles on face of intergenicula; SEM, ×24. **3.** Enlargement of conceptacle at upper right of part 2, showing the many cells opening at the surface and the opening in the swollen conceptacle through which contents escaped; SEM, ×160.

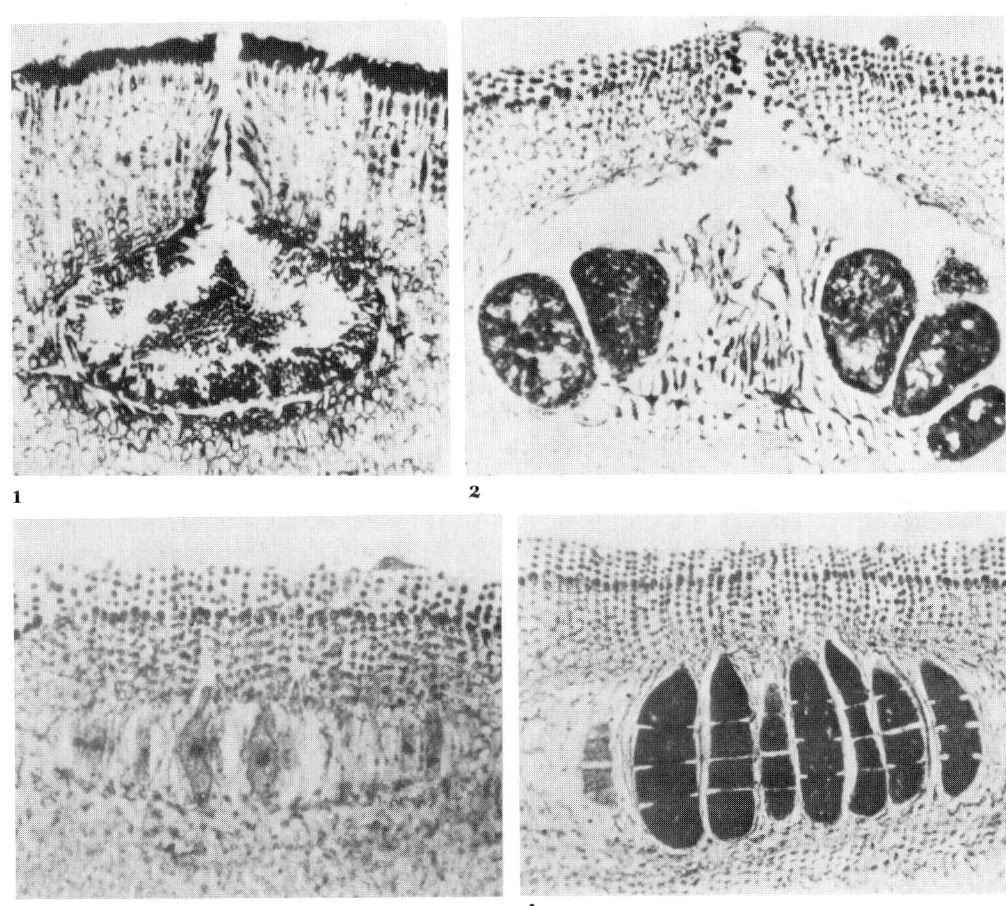

Figure 2.11
Conceptacles of crustose coralline algae. **1.** *Neopolyporolithon reclinatum* (Foslie) Adey & Johansen, California. Section of spermatangial conceptacle, ×200. **2–4.** *Clathromorphum parcum* (Setchell & Foslie) Adey. **2.** Carposporangial conceptacle, ×160. **3,4.** Tetrasporangial conceptacles. **3.** Early stage, pre-meiosis, ×160. **4.** Mature conceptacle with dark-stained zonate tetraspores, ×150. All from Adey and Johansen, 1972.

tilization. The presence or absence of an auxiliary cell and its origin separate the orders of Florideophycidae.

The Carposporophyte. After fertilization, the zygote remains on the female gametophyte. When it germinates, the diploid carposporophyte generation remains as a parasite on the female plant. It produces **carposporangia,** each of which commonly produces a single carpospore, but in some species four carpospores may be produced. The liberated carpospores germinate to produce diploid sporophyte plants. The intercalation of a diploid carposporophyte generation in the life history allows many sporophyte plants to develop from a single zygote.

Tetrasporophytes. The sporophyte resulting from the carpospore germination most commonly produces **tetrasporangia,** irregularly scattered over the surface, located in pits or **sori,** or in **conceptacles** in the Corallinaceae (see Figures 2.10 and 2.11). The arrangement of

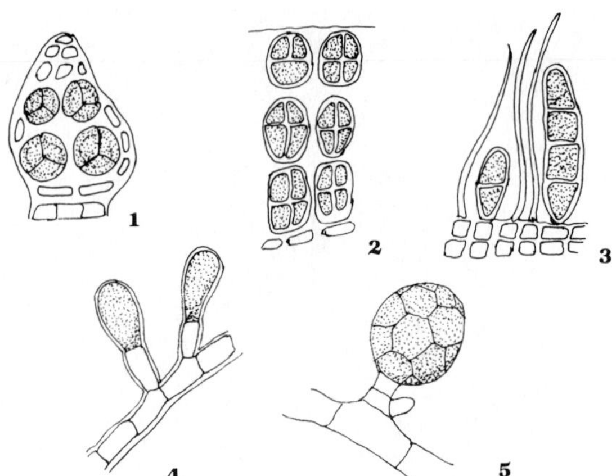

Figure 2.12

Types of sporangia in red algae. **1 – 3.** Tetrasporangia with tetraspores. **1.** Tetrahedral tetraspores, as in *Heterosiphonia*. **2.** Cruciate tetraspores, as in *Petroglossum*. **3.** Zonate tetraspores, as in *Hildenbrandia*. **4.** Monosporangia, with single spore, as in *Acrochaetium*. **5.** Polysporangia, with many spores, as in *Pleonosporium*.

the spores in these sporangia may be cruciate, zonate, or tetrahedral (see Figure 2.12). Although tetrasporangia are most frequent, some red algae produce only one spore in **monosporangia**, or two in **bisporangia**, and a few produce more than four in **polysporangia**. *Corallina* may have as many as 40 tetrasporangia within a single conceptacle (Johansen, 1970).

Both meiotic divisions apparently occur within the tetrasporangial mother cell nuclear envelope in *Polysiphonia* and others, producing the four daughter nuclei simultaneously (Scott and Thomas, 1975). The resulting spores are haploid.

The spores may be forcibly freed from the sporangia as a result of swelling of the surrounding mucilage. The spores themselves also may have a mucilage sheath that occupies as much as 88 percent of the total volume of the mucilage-covered spore. The mucilage retards sinking of the spore and thus aids in dispersal; when the spore has settled to the bottom, the mucilage cover provides sufficient drag to counteract water movement until cementing material is secreted by the spore (Boney, 1975).

Life Histories. The presence of two or three separate generations, or phases, and whether these are morphologically alike or not result in at least four types of life history in the Florideophycidae (Dixon, 1973).

1. Morphologically unlike gametophytes and tetrasporophytes; tetrasporophyte develops from zygote on the female gametophyte; meiosis at time of tetraspore production. *Liagora* (Nemaliales).

2. Morphologically unlike gametophytes and carposporophytes, the latter developing on the gametophyte; meiosis occurs during development of carposporophyte. *Lemanea* (Nemaliales).

3. Three phases present; gametophyte, carposporophyte, and tetrasporophyte, all morphologically dissimilar; carposporophyte develops on the gametophyte; meiosis at time of tetraspore formation. *Bonnemaisonia* (Nemaliales).

4. Three phases present; gametophyte, carposporophyte, and tetrasporophyte (see Figure 2.13), with first and last morphologically similar, carposporophyte developing on the gametophyte, meiosis occuring with formation of tetraspores. *Polysiphonia* (Ceramiales).

In cultures, some species of red algae occasionally develop both spermatangia and polysporangia on the male plants, and some tetrasporophytes produce both tetraspores and spermatangia (West and Norris, 1966). Whether these variants are present in nature is not certain, although some may be monoecious and others dioecious.

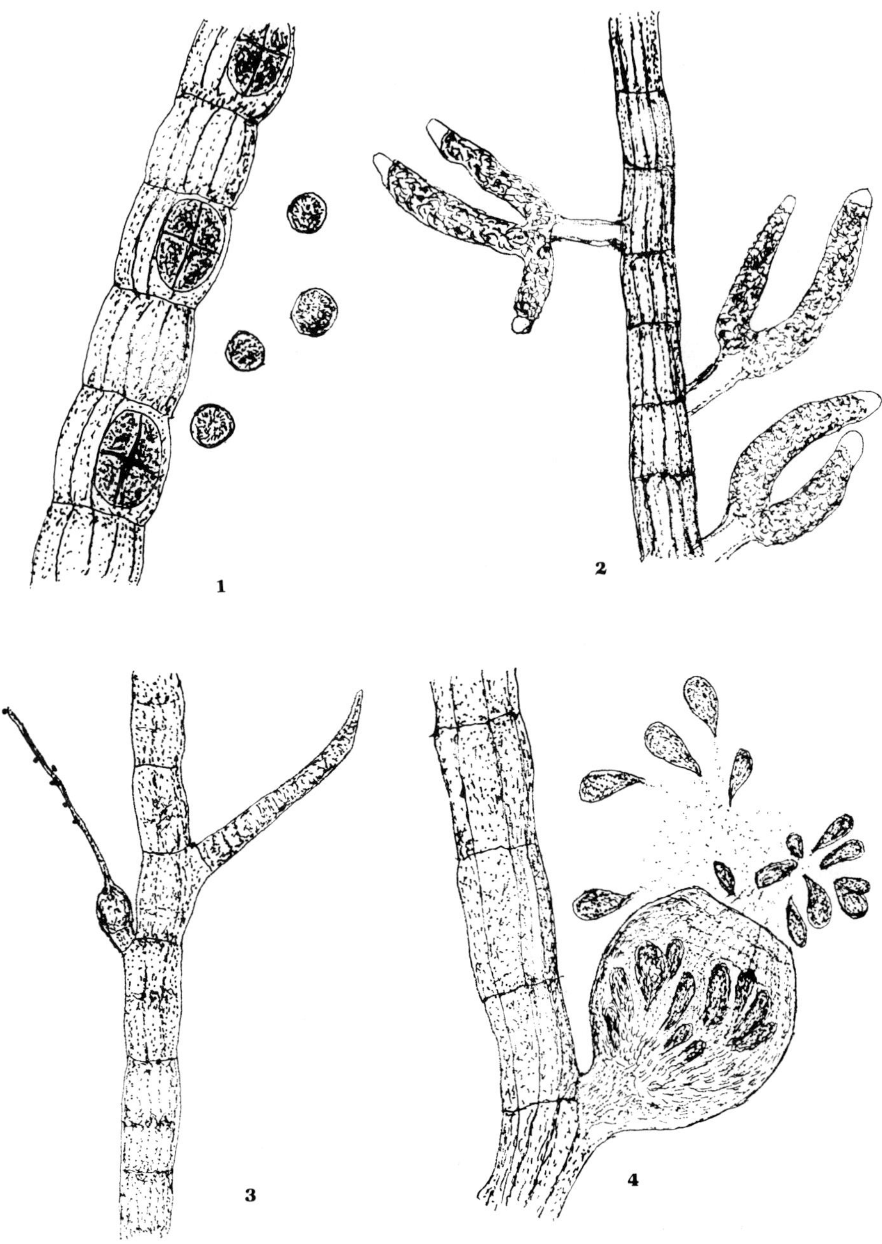

Figure 2.13
Polysiphonia boldii Wayne & Edwards, stages in the life history. **1.** Tetrasporophyte, showing tetrasporangia containing tetraspores, with released tetraspores of varied size; ×100. **2.** Male plant cultured from tetraspore, bearing antheridia; ×100. **3.** Female plant cultured from tetraspore, with carpogonium at left terminating in very long trichogyne to which spermatia are adhering; ×130. **4.** Carposporophyte formed by zygote; released carpospores germinate into tetrasporophyte; ×150. Sketches based on photos in Edwards, 1968.

Ecology and Distribution

Most red algae are marine, living in the rocky intertidal or sublittoral areas from the tropics to the poles. Approximately 800 genera and nearly 5100 marine species are known. In contrast, only some 36 genera and less than 200 species are found in fresh water (Bourrelly, 1970; Dixon, 1973).

Some unicellular species such as *Porphyridium purpureum* (also known as *P. cruentum*) grow in either marine or fresh water, at a wide range of salinity and pH. Rapid growth in culture occurred at salinities ranging from 4.5 to 35‰ and at a pH ranging from 5.2 to 8.3 (Jones et al., 1963). It grows only in the light with the use of inorganic sources of carbon. Growth was fair at salinities as high as 87.5‰, down to as low as 1 or 2‰, with the best growth between 35 and 52.5‰ (Sommerfeld and Nichols, 1970).

The unicellular *Rhodella* requires an external source of vitamin B_{12}, as do many other red algae, including *Bangia*, the conchocelis stage of *Porphyra*, *Asterocytis*, *Erythrotrichia*, *Polysiphonia*, and *Nemalion* (Turner, 1970). In nature, most red algae are contaminated with bacteria, which may provide the algae with the necessary vitamins.

The pigment complement of the red algae can utilize the blue light that is predominant at greater depths, hence a depth zonation was suggested to occur, with green algae in shallowest water and the reds at depth. Examination of the occurrence in detail shows a much more complex zonation. In the Caribbean, Taylor (1961) reported 759 algal species, of which 203 were green, 103 brown, and 453 red algae. In these very clear waters, many of the species have a wide depth range, but at 50 m depth, 17.7 percent of the green algae persisted, in contrast to only 9 percent of the brown and 3.5 percent of the red algae.

As already discussed, one stage of the life history of the bangiophycid *Bangia* and *Porphyra* is a free-living filamentous or foliose plant, and another stage is a perforating form in various shells, previously known as *Conchocelis*. Similarly, many Florideophycidae have unlike stages in their life history, requiring culturing in the laboratory to determine relationships. The sporophyte and gametophyte stages, as well as the conchocelis and foliose ones, have different ecological requirements, so that one stage may reproduce itself successfully without the alternation of the other stage in different geographic areas or at different seasons of the year.

The crustose coralline algae are important reef frame builders and sediment producers in the tropics, where they have been most thoroughly studied. Some taxa such as *Lithothamnium* are known from the tropics to the poles, although best developed in the colder waters of temperate and arctic seas (Littler, 1972). They may be of greatest importance in the high latitudes, where they are the main calcifiers (Adey and Macintyre, 1973). The crustose corallines have a wide depth range, from the intertidal region to the lower limit of the photic zone, forming thin crusts at depths of 400 m. In many Pacific reefs, a pronounced wave-resistant ridge formed by the crustose species was long termed the "*Lithothamnium* ridge," but many taxa are involved in these reefs. Many greatly exceed *Lithothamnium* in abundance, particularly *Neogoniolithon*, *Porolithon*, and *Lithoporella*, with less common *Mesophyllum* and *Archaeolithothamnium* (Wray, 1977b).

A sequence of algal growth has been described for the development of algal ridges (Adey and Vassar, 1975). The corals *Acropora* and *Millepora* first formed a pavement in the high energy waters less than 1 to 2 m deep. If occasionally calmer periods occur and grazers crop the fleshy algae, *Neogoniolithon* and *Porolithon* become more important. The association builds up toward sea level at a rate of 5 to 6 mm per year, and as sea level is reached, *Lithophyllum* replaces the other taxa, and rapidly builds to as much as 30 cm above sea level. An extensive ridge is formed in a region of fairly rapid sea level rise.

In spite of their importance in reef building, the calcareous algae of the reefs and atolls are slow growing. Their productivity is similar to that of other benthic algae, being of less importance than that of the algal symbionts of the associated corals (Marsh, 1970).

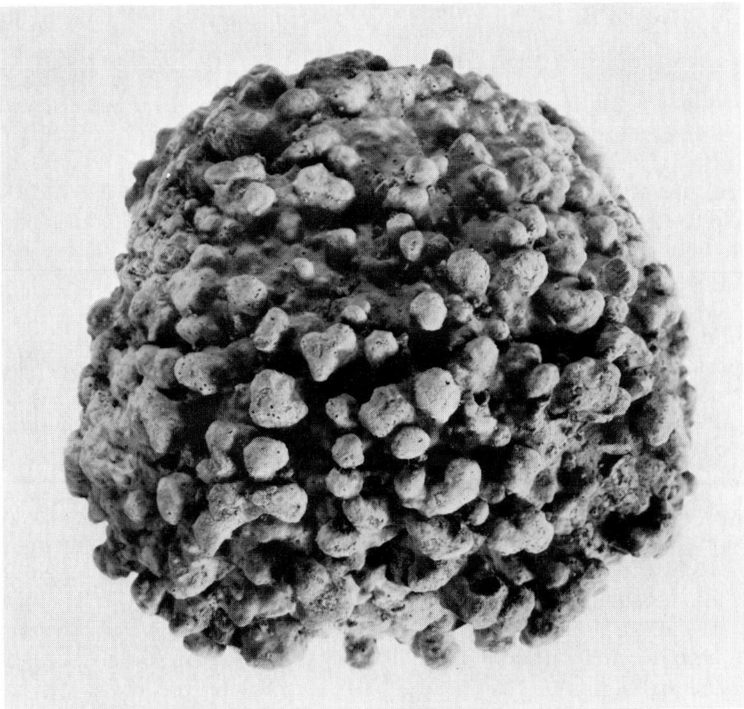

Figure 2.14
Rhodolith from Vema Seamount in Atlantic Ocean, lat. 31°38′ S, long. 8°19′ E; calcareous nodule formed by concentric layers of encrusting coralline algae, foraminifers, and other attached organisms; ×1.6.

Rhodoliths

Spherical algal balls, consisting of concentric layers of crustose coralline algae and associated foraminifers, were variously termed oncolites (but this term should be restricted to the blue-greens; see Chapter 1), **rhodolites** (Bosellini and Ginsburg, 1971), **rhodoids**, or **rhodoliths** (Barnes et al., 1970), the latter term being preferrable, as rhodolite is a mineral term. Such algal balls have been reported from the coast of Sptizbergen in the Arctic Sea to the slopes of Pacific atolls. They are found at depths of 50 to 150 m on various offshore oceanic banks and continental margins (see Figure 2.14). Although

they can grow actively to depths of at least 60 to 70 m, most of the deeper ones are relict structures. Growth is slow, probably averaging about 1 mm of thickness per 5 to 10 years, the rhodoliths of up to 30 cm diameter probably having an age of 500 to 800 years or more (Adey and Macintyre, 1973). Radiocarbon dating of rhodoliths from the Agulhas Bank off South Africa gave an age of 13,760 ± 120 years for specimens from 115 m depth and 12,990 ± 100 years for those from 120 m depth (Siesser, 1972).

Older fossil rhodoliths have been reported widely in the Eocene, Oligocene, and Miocene of southern Europe and the Mediterranean region.

GEOLOGIC OCCURRENCE

As previously mentioned, the best known red algal fossils are those of the lime-depositing Cryptonemiales; other groups have a more limited record. Because of the absence of distinguishing morphological features, simple unicellular fossils with mucilage sheath have most commonly been referred to the blue-green algae, although at least some probably were in fact eucaryotes, and perhaps red algae. *Eosphaera tyleri* Barghoorn from the 1.9-billion-year-old Gunflint Formation (see Figure 2.15) was originally regarded as of unknown affinity or a blue-green alga; however, the presence of apparent buds held in its mucilage sheath is unlike any known cyanophyte. Instead, it is quite similar to the unicellular *Porphyridium purpureum* (compare Figure 2.5), which may divide by simple fission into 2 daughter cells, by multiple fission into 4, 8, or 16 cells, or may produce buds at the periphery of a multinucleate cell (Sommerfeld and Nichols, 1970). The sheath is thin in planktonic cells, but thick and firm in cells grown on agar in the laboratory. Tappan (1976) suggested that cells of *Eosphaera* with tenuous sheath and no tubercles were such planktonic cells, whereas those with budlike spheres surrounding the larger cell probably were trapped in the algal mats of the Precambrian stromatolites; like the modern individuals similarly subject to desiccation, they retained a thick mucilage sheath that firmly held the newly formed buds.

The presence of lateral buds on specimens of the Gunflint *Huroniospora microreticulata* Barghoorn is reminiscent of another modern unicellular red alga, *Rhodella maculata,* described from shore sands, but also known as a truly planktonic species in the Oslo Fjord (Paasche and Throndsen, 1970) and as a terrestrial form. *Rhodella* has a broad temperature and salinity range of tolerance, as would be expected in a shore-living form. The wide occurrence of the Precambrian *Huroniospora* is compatible with a similar range in habitat (Tappan, 1976).

As indicated in Chapter 1, other unicells of the Precambrian also have been regarded as possible eucaryotes. Although generally these have been compared to green algae, a red algal affinity like the modern Bangiophycidae seems more probable (Allsopp, 1969; Phillips and Carr, 1977). In fact, if the principal requirement for preservation of the algal morphology in unmineralized fossils is the possession of an extracellular mucilage sheath like that of the blue-green algae, the unicellular red algae may be the only unicellular eucaryotes to have been preserved in the Precambrian sediments (Phillips and Carr, 1977). The Bitter Springs *Glenobotrydion* may in fact be related to the unicellular Porphyridiales, rather than to the green algae as suggested by Schopf and Blacic (1971), or to blue-greens as suggested by others.

The tetrad cells in the Bitter Springs Formation (Schopf and Blacic, 1971), described as *Eotetrahedrion princeps,* originally were regarded as of probable green or red algal nature (see Figure 2.16). Others have regarded these as fortuitously aligned cells, and have placed them with the blue-green algae (Knoll and Barghoorn, 1975; Phillips and Carr, 1977). That they might belong to the red algae was regarded as unlikely, as no evidence was found to indicate the presence of megascopic thalli of the Ceramiales or Rhodymeniales (Schopf and Blacic, 1971). However, these may have been restricted to facies other than stromatolitic ones, where only rare spores would be trapped. Additional tetrahedral tetrads or cells with triradiate markings found in other Precambrian deposits (Volkova, 1976), including the 1.5-billion-year-old Amelia Dolomite of Northern Territory, Australia (J. H. Oehler et al., 1976), were regarded as green algae; they were compared to the tetrahedral tetrads reported in a coccoid colorless green alga found on damp walls, *Mycacanthacoccus cellaris* Hansgirg. On the other hand, sheath-enclosed tetraspores are common in the red algae, tetrasporangia being the most widely distributed form of sporangia in the Florideophycidae (Dixon, 1973). Meiosis occurs during the development of the tetraspores. As many lines of evidence now suggest that the red algae are more primitive than the green algae, these Precambrian tetrahedral tetrads as well as certain unicells would appear best assigned to the present division.

Few other records of red algal spore fossils

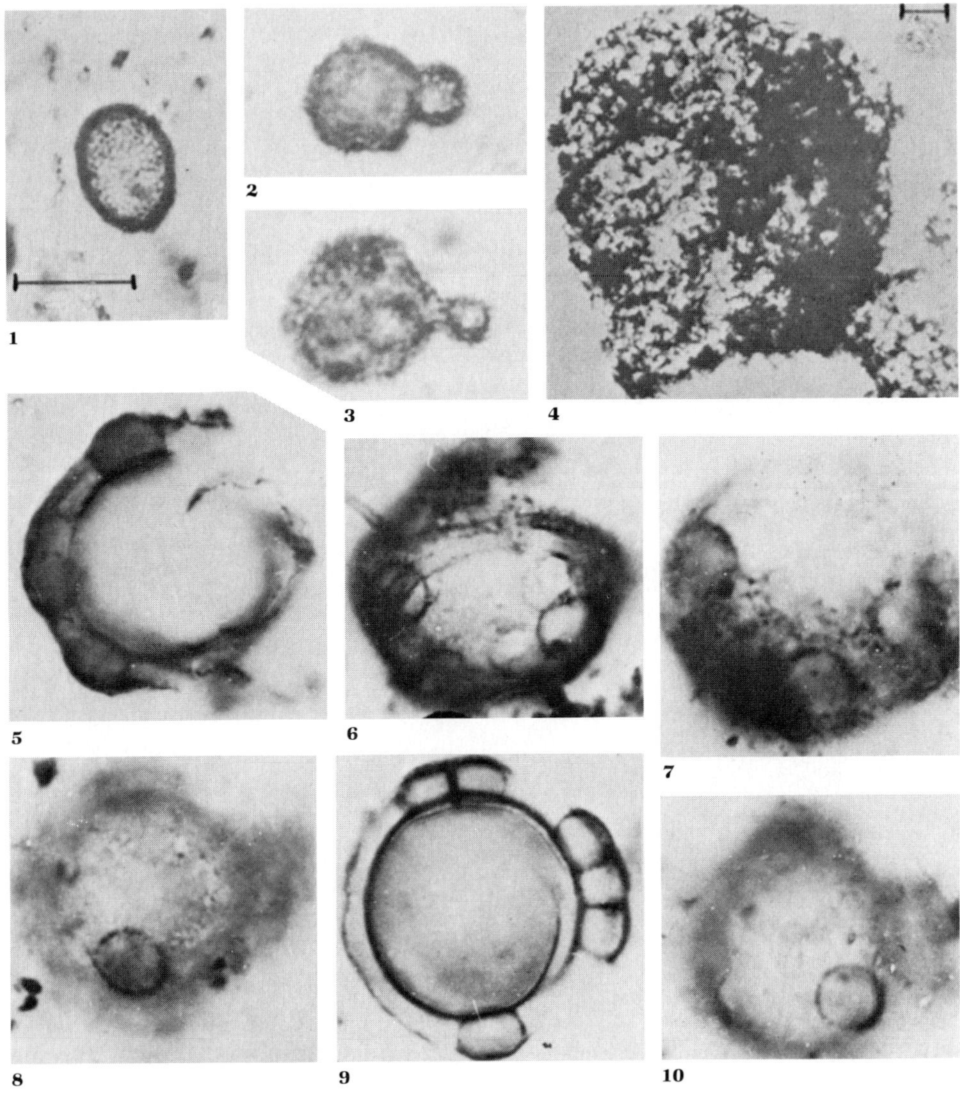

Figure 2.15
Probable Precambrian Porphyridiaceae. **1 – 4.** *Huroniospora microreticulata,* Gunflint Formation, Ontario, with irregularly reticulate surface (compare *Porphyridium aerugineum* in Figure 2.2). **1.** Single cell, ×1600, bar = 10 μm; from Barghoorn and Tyler, 1965. **2,3.** Showing budding, approximately ×2500 (magnification not given), from Darby, 1974, *Bull. Geol. Soc. Am.,* v. 85, p. 1595 – 1596, reproduced by permission. **4.** TEM showing apparent bud at lower left, ×6000, bar = 1 μm; from Cloud and Hagen, 1965. **5 – 10.** *Eosphaera tyleri,* Gunflint Formation, Ontario; from Barghoorn and Tyler, 1965, *Science,* v. 147, p. 563 – 577. Copyright 1965, American Association for the Advancement of Science, reproduced by permission. **5,6.** Same specimen at two levels of focus, showing apparent buds on surface held in mucilage sheath, ×1500 (compare *Porphyridium purpureum* in Figure 2.5). **7.** Another specimen, ×1650. **8 – 10.** Three levels of focus of additional specimen, ×1500.

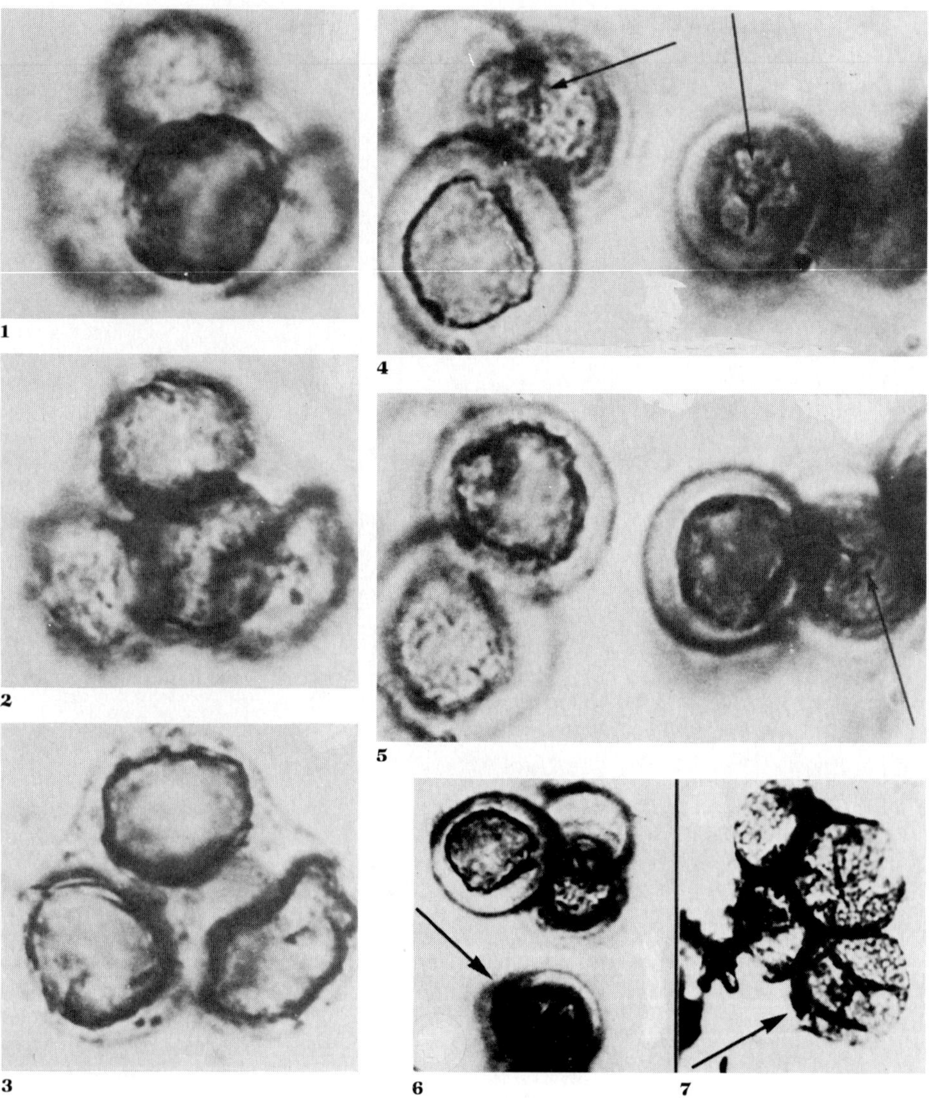

Figure 2.16

Eotetrahedrion princeps Schopf & Blacic, Bitter Springs Formation, Precambrian, central Australia. **1–3.** Cell tetrad, photographed at various levels of focus. **4,5.** Isolated cells showing triradiate mark (arrow) indicating tetrad origin, at two levels of focus. **1–5**, ×2000, from Schopf and Blacic, 1971. **6,7.** Additional specimens with triradiate scar (arrow), ×1200, previously unpublished photo from J. W. Schopf.

are known, although *Ovocalcisporites multidissaeptus* Fay & Diersche (see Figure 2.17) was reported to be calcified segmented spores like the (germinating) carpospores of *Bonnemaisonia* (Fay and Diersche, 1975). However, these fossils are somewhat large (150 to 200 μm diameter) in comparison to the usual 40- to 80-μm-diameter spores of the living *Bonnemaisonia*.

Megascopic algae in the Precambrian (Walter et al., 1976) such as *Proterotainia montana* Walter et al., *P. neihartensis* (Walcott) Walter et al.,

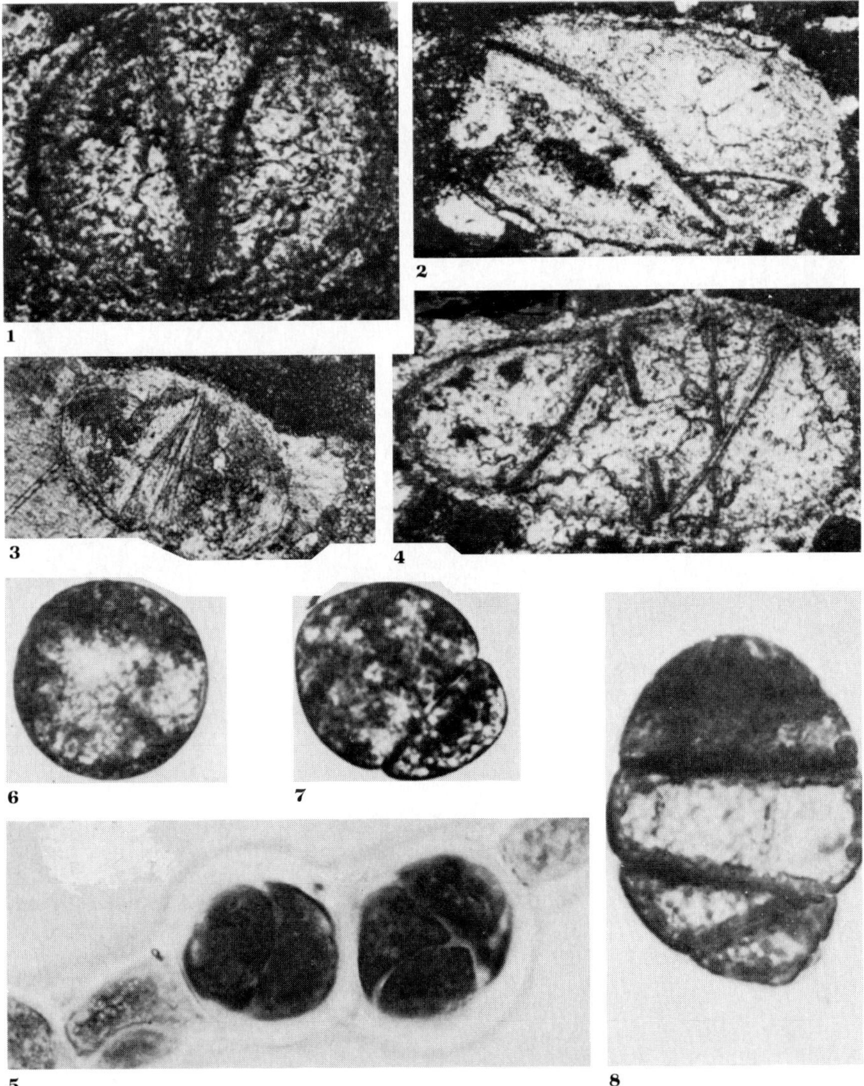

Figure 2.17

1—4. *Ovocalcisporites multidissaeptus.* Sectioned specimens, Upper Jurassic.
1,2,4, from Germany; 3, from Austria; 1, ×270; 2, ×140; 3, ×130; 4, ×155; from
Fay and Diersche, 1975. **5—8.** *Bonnemaisonia hamifera* spores for comparison
with the fossil spores of *Ovocalcisporites;* from Chen et al., 1969. **5.** Two mature
tetrasporangia, each with tetrad of cruciately arranged tetraspores, with third
sporangium beginning to differentiate at left, ×300. **6.** Mature tetraspore after release from sporangium, ×375. **7.** First cell division of tetraspore after two days,
×450. **8.** Subsequent divisions of germinating spore after two to three days,
×850.

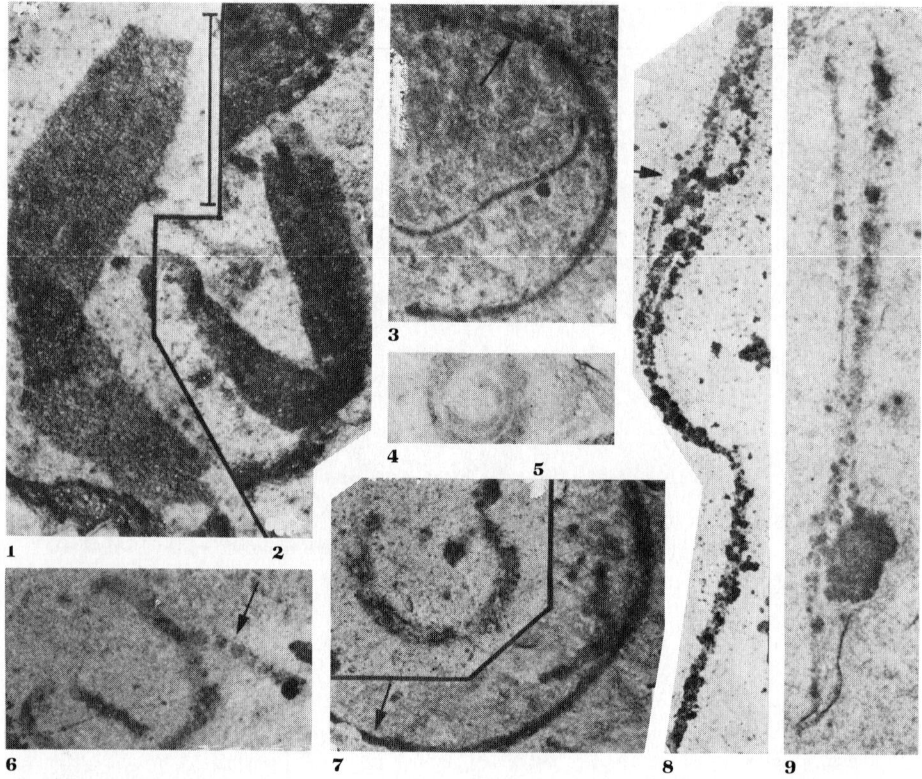

Figure 2.18
Megascopic Precambrian algae, 1.3-billion-year-old Greyson Shale, Belt Supergroup, Montana. **1,2.** *Lanceoforma striata,* foliose thalli, ×25. **3 — 7.** *Grypania spiralis,* showing coiled strands. **3, 6, 7.** Show segmentation at arrows. **4.** Tightly coiled specimen. **3,5 — 7,** ×4; **4,** ×1.8. **8, 9.** *Proterotainia neihartensis.* **8.** Showing ribbonlike thallus split into two strands (arrow), ×1.1. **9.** Two crossed strands, ×5.8. All from Walter et al., 1976.

Lanceoforma striata Walter et al., and *Grypania spiralis* (Walcott) Walter et al. (see Figure 2.18) may be related to such living red algae as *Nemalion* and *Lemanea* (Nemaliales), *Cryptonemia* (Cryptonemiales), or *Porphyrella* (Bangiales). Closer relationships cannot be determined in the absence of morphological detail in the fossils. Possibly similar forms were the source of the already mentioned tetrahedral tetrads, the macroscopic thallus living in a different environment than that of the blue-green algal mats that formed the stromatolites. Perhaps acid macerations or thin sections of the strata associated with the megascopic thalli also would produce such tetrads, their rarity in the stromatolitic cherts being due to the rare entrapment of spores in the mats at some distance from their normal habitat.

Early Paleozoic records of the Rhodophyta are scarce, but include the perforating conchocelis stage of the Bangiaceae from the Cambrian on (Meijer, 1969), the Cambrian Epiphytaceae and some Solenoporaceae (both Cryptonemiales), and *Waputikia* (Rhodomelaceae) and the Ordovician *Delesserites* (Delesseriaceae), the last two representing the Ceramiales.

The nodular encrusting Solenoporaceae were important beginning with the Ordovician, being associated with tetracorals and other reef

Figure 2.19
Ungdarellaceae. Many of these have been described as foraminifers or sponges as well as being referred to the red algae. 1 — 3. *Stacheia marginulinoides* Brady, Mississippian, England. ×64. **1,2.** Two views of one specimen. **3.** Sectioned specimen. **4.** *Stacheoides polytrematoides* (Brady) Cummings, specimen attached to a crinoid stem, Mississippian, Scotland, ×22. **5 — 8.** *Fourstonella fusiformis* (Brady) Cummings, Mississippian, England. **5.** Sectioned specimen. **6,7.** Opposite views of paratype. **8.** Lectotype. 5,8, ×65; 6,7, ×44. All from Loeblich and Tappan, 1964, from the *Treatise on Invertebrate Paleontology,* courtesy of the Geological Society of America and University of Kansas.

formers in the Paleozoic and Mesozoic, but declined as the Corallinaceae became important, to disappear entirely in the early Tertiary. Primitive or ancestral coralline algae appeared in the middle and later Paleozoic, and the modern types of Corallinaceae in the Mesozoic, becoming progressively more diverse. Jointed corallines appeared in the mid-Mesozoic. The calcareous red algae thus were the important reef-building algae in the Mesozoic and Cenozoic, occupying the dominant position held by the blue-green algae of the Precambrian and blue-green and green algae of the Paleozoic (Wray, 1977a).

Within the Nemaliales, the Gymnocodiaceae appeared in the Permian and persisted through the early Cretaceous. One other late Paleozoic family, the Ungdarellaceae, is of questionable affinity. Many taxa placed therein were originally described as foraminifers (see Figures 2.19 and 2.20), others as algae. *Komia* Korde has been regarded as a stromatoporoid (Wilson et al., 1963), others were transferred to the red algae (Petryk and Mamet, 1972), and these and additional taxa were regarded as Ischyrospongia by Termier et al. (1977). The latter is a class of sponges that includes the recent sclerosponges and fossil stromatoporoids and Pharetronida, as well as the Aoujgaliida and Moravamminida, the two latter orders including taxa previously

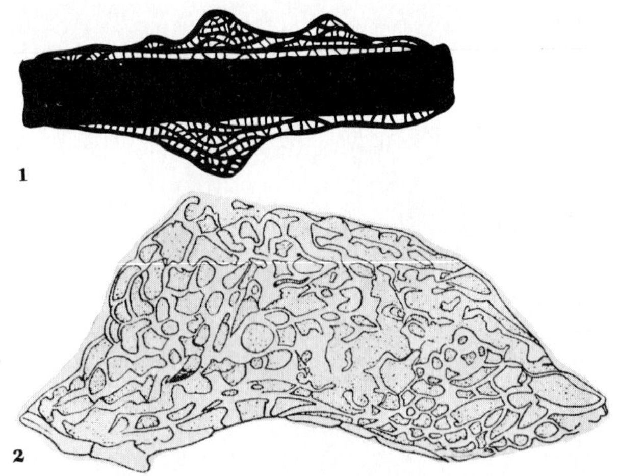

Figure 2.20

Ungdarellaceae. **1.** *Stacheoides polytrematoides,* diagrammatic section
showing internal structure, ×18, from
Cummings, 1955b. **2.** *Aoujgalia variabilis*
G. Termier & H. Termier, Mississippian, Morocco, sectioned specimen,
×40, after Termier and Termier, 1950.
Both from Loeblich and Tappan, 1964,
Treatise on Invertebrate Paleontology,
courtesy of the Geological Society of
America and University of Kansas.

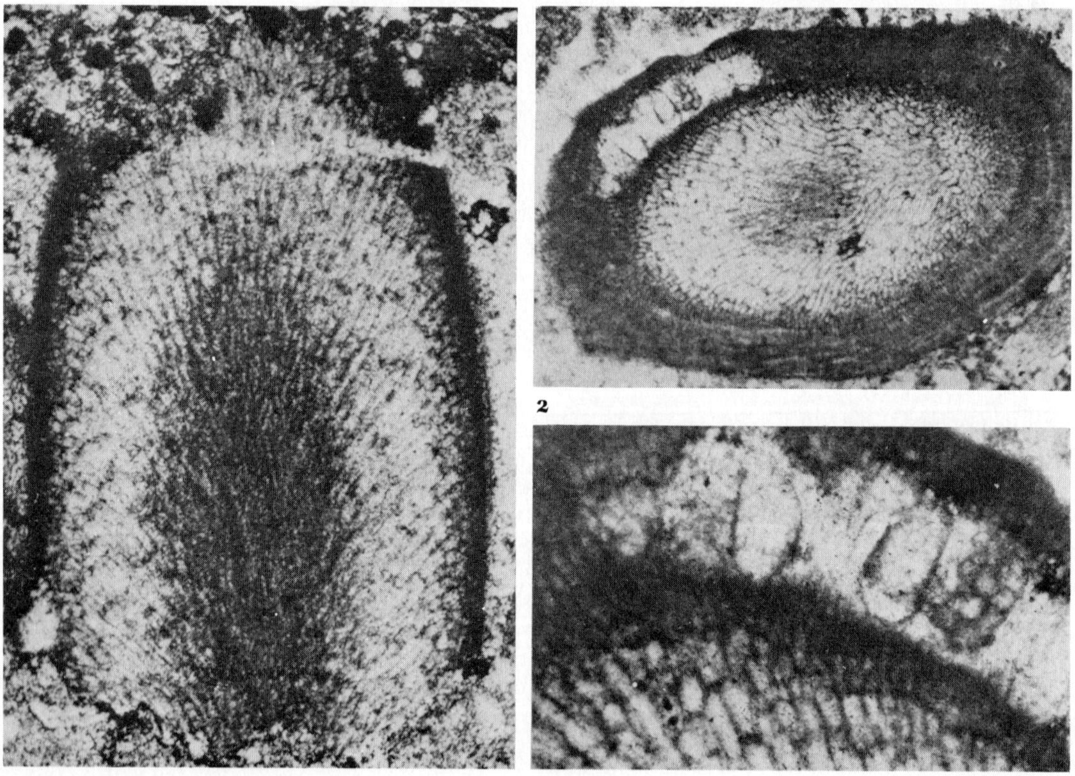

Figure 2.21

Agardhiellopsis cretacea Lemoine, Lower Cretaceous. **1.** Longitudinal secion, axial medullary region
dark, intermediate part lighter in color, with dark cortex, Albian, Spain, ×85. **2.** Oblique section, the
outer perithallus visible at upper left of figure, with conceptacle and sporangia, Albian, France, ×50.
3. Enlargement of outer part, showing sporangia in conceptacle, ×150. All from Lemoine, 1966.

assigned to the foraminifers. Although the problem of their relationships is still unresolved, the Ungdarellaceae are herein retained in the red algae, in view of the nature of their growth and the absence of spicules that would indicate a sponge affinity.

Mesozoic red algae are dominated by the cryptonemialean Corallinaceae, Solenoporaceae, and Peyssoneliaceae, with some Gymnocodiaceae (Nemaliales) and rare Gigartinales (see Figure 2.21). During the Cretaceous, the Corallinaceae replaced most of the Solenoporaceae (Poignant, 1977). Coralline algae are abundant in much of the Tertiary, particularly in the lower latitudes (see Figures 2.22 and 2.23), and a few noncalcified florideophycid fos-

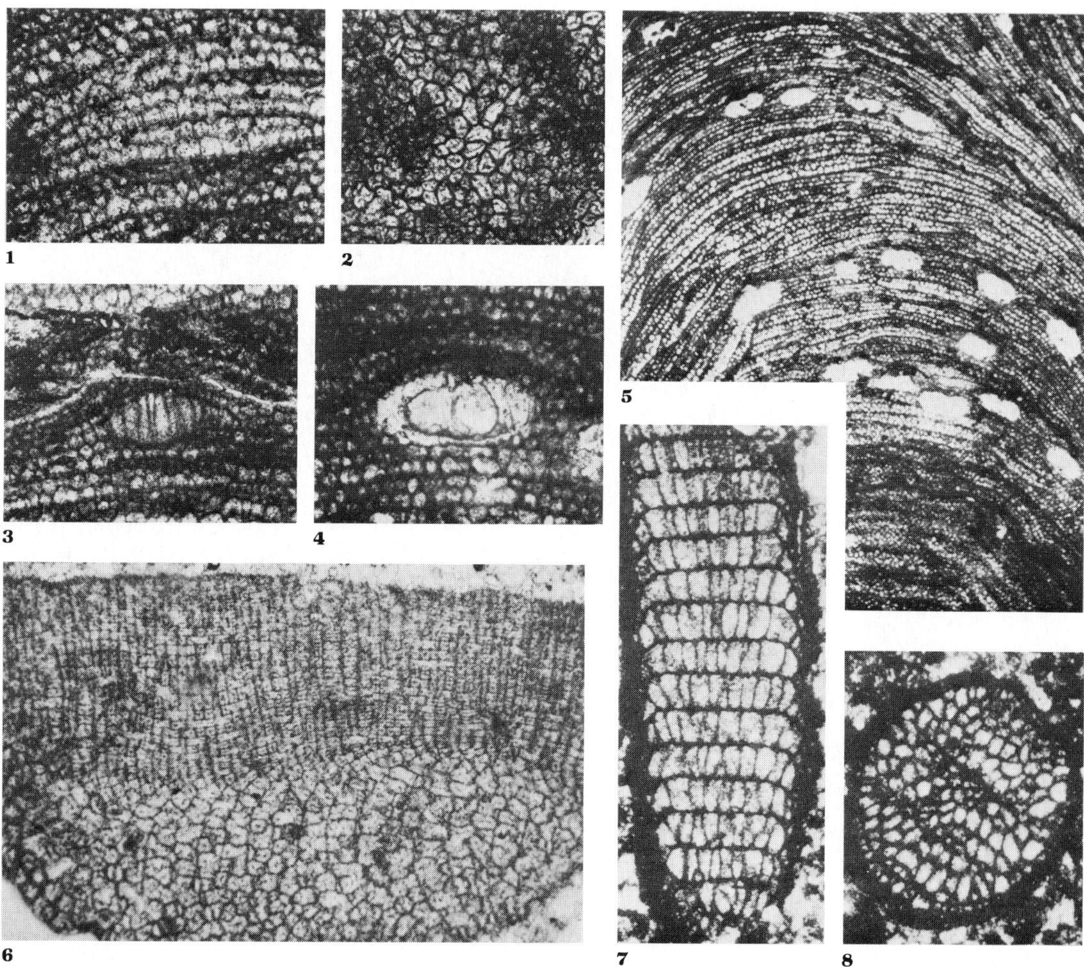

Figure 2.22

1 − 5. *Hydrolithon iranicum* (Elliott) Massieux, Miocene, Iran, from Elliott, 1970. **1.** Vertical section of perithallus. **2.** Horizontal section of perithallus. **3.** Developing conceptacle, with early differentiation of fertile and sterile areas. **4.** Conceptacle with mature sporangia. 1 − 4, ×90. **5.** Vertical section of part of branch with numerous successive layers and conceptacles, ×19. **6.** *Lithophyllum quadrangulum* Lemoine var. *welschii* Lemoine, Miocene, Guam. Section showing distinct hypothallus and perithallus, ×100; from Johnson, 1964b. **7,8.** *Subterraniphyllum thomasii* Elliott, Oligocene, Iran, ×50, from Elliott, 1957. **7.** Vertical section. **8.** Transverse section.

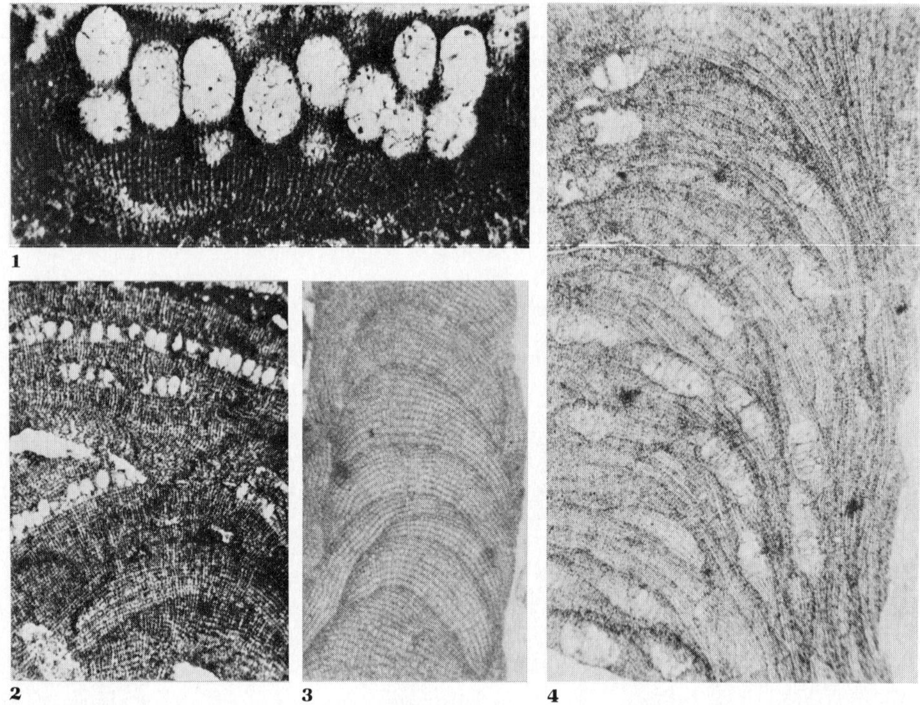

Figure 2.23

Cenozoic Melobesioideae. **1.** *Archaeolithothamnium megasporum* Johnson, Pleistocene, Saipan, section showing thin crust of perithallic tissue with large sporangia, ×80. **2.** *Archaeolithothamnium* cf. *A. lugeoni* Pfender, Miocene, Saipan, section showing hypothallus at base, perithallus, and rows of sporangia, ×32. 1,2, from Johnson, 1957. **3.** *Mesophyllum guamense* Johnson, Miocene, Guam, section of branch with strong lenticular growth zones, ×40. **4.** *M. ishigakiense* Johnson, Eocene, Ryukyu Islands, slightly oblique section of branch, with many irregular growth zones and conceptacles filled with sporangia; ×40. 3,4, from Johnson, 1964a.

sils also are known, in particular those associated with the Miocene diatomites (Parker and Dawson, 1965).

EVOLUTION

Many lines of evidence suggest that red algae are the most primitive eucaryotic algae. The pigments and single thylakoids of the plastids both are like the procaryotic blue-greens, showing very little evolutionary change in this respect, regardless of the method by which the eucaryotic cell evolved. Other algae have different pigments and thylakoids in stacks of twos or threes. The food reserves also are similar to those of blue-greens, both the rhodophyte floridean starch and the blue-green cyanophycean starch being α-1:4-linked polyglucans. Other features not associated with the plastids (hence not dependent on either theory of eucaryote origin) that place the red algae as primitive include the low ribosomal RNA molecular weight, gliding motility, and the pit connections that are paralleled by cell interconnections in blue-green algae, as well as the absence of cilia or flagella at any stage of the life history. Unique in the red algae among any major group of eucaryotes, the nonflagellar character probably is primitive rather than due to a secondary reduction, as the nuclei lack true centrioles (Dixon, 1973). As mitotic microtubules have been reported in the red algae, the microtubule system probably is more primitive than the flagellum-centriole one (Pickett-Heaps, 1975, p. 566). Other eucaryotes were derived from a red algal line that evolved the 9+2 flagella from aggregates of microtubules (Cavalier-Smith, 1975), followed by loss of phycobilin pigments, and evolution of others. Microtubules also evolved into other organelles of motility, including the haptonemata of the Haptophyta and the axopodia of the Radiolaria. In contrast, the unicellular Porphyridiaceae and male gametes and spores of other red algae have a gliding movement reminiscent of that of the blue-green algae and some bacteria (Allsopp, 1969).

The molecular weight of the heavy component of ribosomal RNA of the red algae *Porphyridium aerugineum* and *Griffithsia pacifica* is intermediate between that of procaryotes and other eucaryotes. The procaryote molecular weight of rRNA heavy subunits is about 0.56 to 1.07 m, that of the red algae is 1.21 m, that of dinoflagellates is 1.23 m, green algae and fungi range from 1.28 to 1.3 m, and angiosperms from 1.27 to 1.31 m. The increase in molecular weight is correlative with higher phylogenetic position; the intermediate weight of the red algae, lower than other eucaryotes but higher than the procaryotes, thus confirms their placement as the most primitive of the eucaryotes, and the dinoflagellates as next most primitive, although both are still much advanced over the procaryotes (Howland and Ramus, 1971; Gressel et al., 1975).

Possibly the absence of sexual reproduction in the Porphyridiaceae is a retained primitive character, and the sexual form of reproduction may have first appeared in the more advanced Bangiophycidae. Reliable evidence is not available as to whether the first appearance of sexual reproduction was simultaneous with or much delayed after the appearance of mitosis. Theo-

retical arguments have been given for both possibilities.

Although commonly regarded as the most primitive of eucaryotes, the red algae have been regarded by some as secondarily simplified, evolving from the Cryptophyta by loss of a flagellated stage (Lee, 1972; McQuade, 1977). Except for the absence of flagella, the red alga *Rhodosorus marinus* Geitler was regarded as most similar to the Cryptophyceae (Lee, 1974) in having a lobate plastid with starch-enclosed central pyrenoid. On the basis of pyrenoid characters, evolution was suggested to have taken three main directions, one to the Bangiales, a second line in the Florideophycidae, and the third through the Goniotrichales (the Goniotrichaceae is here included with the Porphyridiales). In each of the proposed lineages, morphological complexity increases, and the pyrenoid becomes buried within the plastid but loses its starch shell. On this basis, *Porphyridium* was regarded as more advanced than *Rhodosorus* or *Rhodella* and was suggested to have evolved from a spore of *Porphyra* or *Bangia* that retained the unicellular condition.

The presence of pit connections between successive cells of a filament or series is characteristic of the Florideophycidae and was thought to distinguish this group from the Bangiophycidae, which did not have pit connections. When pit connections were found in the filamentous conchocelis stage of *Porphyra* but not the foliose stage, Lee (1971) suggested that the absence was due to a later loss from an ancestral form with pit connections in all phases of the life history. The Bangiaceae were regarded as having too complex a life history to have been ancestral to all the pit-connection-possessing Florideophycidae. However, more recent studies have shown the conchocelis stage to have many of the other characters of the Florideophycidae, including peripheral plastid lamellae, bandlike peripheral plastids rather than central stellate ones, a central vacuole, and apical growth, as already noted, indicating a fairly close relationship of these two major groups of red algae.

McQuade (1977) has suggested that the primitive eucaryote was an amoeba, asexual, and lacking flagella. From this evolved the amoeboflagellates with paired whiplash flagella, and later the cryptomonads with paired tinsel flagella. The chrysomonads with one whiplash and one tinsel flagellum then arose, sexuality first arising in this group with heterokont flagella. Hence all algae and protozoa with sexual reproduction were derived from the chrysomonads. However, McQuade regarded both the amoeboid organisms, chrysophytes, and dinoflagellates as lacking sexual reproduction, although it has now been amply demonstrated in *Dinobryon* (Chrysophyta), *Ochrosphaera* (Haptophyta), the thecamoebians (conjugation and gametic), and dinoflagellates; not all flagellate protozoa and chlorophytes have paired whiplash flagella as McQuade stated; both euglenids and prasinophytes are heterokont, as are gametes of *Boderia* (an allogromiid foraminifer). More probably the amoeboid organisms are advanced or degenerate forms (like the parasitic dinoflagellates), and others such as radiolarians will undoubtedly be proven to have a sexual cycle like that of dinoflagellates if the difficulties in culturing them can be overcome.

Their many primitive characters suggest that the red algae, termed the Mesoprotista by Dougherty and Allen (1960), were the first eucaryotes. Thus if certain Precambrian unicells have correctly been assigned to eucaryotes, such as those of the 1.9-billion-year-old Gunflint Formation, or the Bitter Springs Formation (900 million years old) of Australia, they are more probably red algae like the Bangiophycidae (Allsopp, 1969; Tappan, 1976) than green algae as previously suggested (Schopf, 1968; Kaźmierczak, 1979).

Although some modern Bangiophycidae are more highly evolved than others, this is to be expected in a group (Porphyridiales) that probably has existed for 2 billion years. Apparently the Bangiales and simpler Florideophycidae also have a very long history, if *Lanceoforma* and *Proterotainia* are correctly assigned to these groups. The conchocelis type of perforating alga is known from the Cambrian, indicating the early development of the alternating foliose and filamentous generations in the life history of the red algae.

CLASSIFICATION

The classification given in Table 2.1 is based on Papenfuss (1955), Dixon (1973), and some later additions, such as by Johansen (1969a, 1976), emphasis being placed on those forms with a fossil record.

Table 2.1
Classification of the Division Rhodophyta.

Division Rhodophyta   Wettstein 1901

Class Rhodophyceae   Ruprecht 1851

I. Subclass Bangiophycidae de Toni 1897, orth. mut. L. M. Newton 1953
Simple morphology, including unicells and cells without cytoplasmic connections; plastids may be stellate with central pyrenoid; reproduction by fission, or by simple nonflagellated spores termed gonidiospores; sexual reproduction incompletely known. Marine and freshwater; Precambrian to Holocene.

A. Order Porphyridiales Kylin 1937
(incl. order Goniotrichales Kylin 1956)

Unicellular, solitary or in irregular masses, or with simple or branching filaments; mucilage abundant. Marine, fresh and brackish water, and subaerial. Precambrian to Holocene.

1. Family Porphyridiaceae Kylin 1937
Precambrian, Holocene.
Chroothece Hansgirg 1884; *Clonophycus* J. Oehler 1977 (Precambrian); ?*Cyanidium* Geitler 1933; *Eosphaera* Barghoorn 1965 (Precambrian); *Flintiella* Ott in Bourrelly 1970; ?*Glenobotrydion* Schopf 1968 (Precambrian); ?*Huroniospora* Barghoorn 1965 (Precambrian); *Porphyridium* Nägeli 1849; *Rhodella* Evans 1970; *Rhodosorus* Geitler 1930; *Rhodospora* Geitler 1927; *Vanhoeffenia* Wille 1924.

---

2. Family Goniotrichaceae (Rosenvinge) Skuja 1939 ?Precambrian, Holocene.
Asterocytis (Hansgirg) Gobi ex Schmitz 1896; *Bangiopsis* Schmitz in Engler & Prantl 1896; *Chroodactylon* Hansgirg 1885; *Goniotrichopsis* Smith in Smith & Hollenberg 1943; *Goniotrichum* Kützing 1843; *?Grypania* Walter, Oehler & Oehler 1976 (Precambrian).
3. Family Phragmonemataceae Skuja 1939
Cyanoderma Weber van Bosse 1887; *Kyliniella* Skuja 1926; *Neevea* Batters 1900; *Phragmonema* Zopf 1882.

B. Order Bangiales Engler 1892
Multicellular, filamentous, foliate or discoid; intercalary growth; central stellate plastids with pyrenoid, or peripheral without pyrenoid; reported sexual reproduction not substantiated. Marine and freshwater. Precambrian, Cambrian to Holocene.

1. Family Bangiaceae (Gray) Nägeli 1847
?Precambrian, Cambrian to Holocene.
Bangia Lyngbye 1819; *?Lanceoforma* Walter, Oehler & Oehler 1976 (Precambrian); *Porphyra* C. Agardh 1824 (syn.: *Conchocelis* Batters 1892; perforating stage, Cambrian to Holocene); *Porphyrella* Smith & Hollenberg 1943.
2. Family Boldiaceae Herndon 1964
Boldia Herndon 1964.
3. Family Erythropeltidaceae Skuja 1939
Erythrocladia Rosenvinge 1909; *Erythrotrichia* Areschoug 1850; *Porphyropsis* Rosenvinge 1909; *Smithora* Hollenberg 1959 (epiphytic).

C. Order Compsopogonales Skuja 1939
Thallus with cortex distinct from axis; uniseriate; branching filaments; monospores.

1. Family Compsopogonaceae Schmitz 1896
Compsopogon Montagne 1846.

D. Order Rhodochaetales Skuja 1939
Thallus without distinct cortex; growth terminal. Marine only.
Rhodochaete Thuret in Bornet 1892.

II. Subclass Florideophycidae (Lamouroux 1813) Engler 1892.
Multicellular filaments or thalli; cells with cytoplasmic connections; cell division rarely intercalary; complete sexual reproduction, female gamete in carpogonium, male gamete as one spermatium per spermatangium; zygote forms carposporophyte, carpospores then giving rise to tetraspores and then to sexual generation. Aragonite deposited by *Liagora* and *Galaxaura*, calcite by Corallinaceae. Marine and freshwater. ?Precambrian, Cambrian to Holocene.

A. Order Nemaliales Schmitz 1892 orth. mut. Christensen 1967
Axial filament or mass of filaments; single carpogonia; tetrasporangia cruciate if present, rarely tetrahedral. Marine and freshwater. ?Precambrian, Permian to Holocene.

1. Family Acrochaetiaceae Fritsch 1944
Acrochaetium Nägeli 1861; *Audouinella* Bory 1823.
2. Family Batrachospermaceae (C. Agardh) Dumortier 1829
Batrachospermum Roth 1797.
3. Family Lemaneaceae (S. F. Gray) Harvey orth. mut. Rabenhorst 1868
Miocene to Holocene.
Lemanea Bory 1808 (Miocene to Holocene).
4. Family Helminthocladiaceae (J. Agardh) Harvey orth. mut. Hauck 1883
?Precambrian, Holocene.
Helminthocladia J. Agardh 1852; *Helminthora* J. Agardh 1852; *Liagora* Lamouroux 1812; *Nemalion* Targioni-Tozetti 1818; *?Proterotainia* Walter, Oehler & Oehler 1976 (Precambrian).
5. Family Chaetangiaceae Kützing orth. mut. Hauck 1883
Permian to Holocene.
Galaxaura Lamouroux 1816; *Hapalophloea* Pia 1935 (Permian); *Pilodea* Pia 1935 (Permian); *Scinaia* Bivona 1822.
6. Family Gymnocodiaceae Elliott 1955
Permian to Lower Cretaceous.
Diversocallis Dragastan 1969 (Lower Cretaceous); *Getaeia* Dragastan 1973 (Lower Cretaceous); *Gymnocodium* Pia 1927 (Permian); *Permocalculus* Elliott 1955 (Permian to Lower Cretaceous).
7. Family Naccariaceae Kylin 1928
Atractophora Crouan & Crouan 1848; *Naccaria* Endlicher 1836.
8. Family Bonnemaisoniaceae Schmitz in Engler 1892
?Jurassic, Holocene.
Asparagopsis Montagne 1841; *Bonnemaisonia* C. Agardh 1822; *?Ovocalcisporites* Fay & Diersche 1975 (Jurassic).

(*continued*)

9. Family Gelidiaceae Kützing orth. mut. Harvey 1853
Gelidium Lamouroux 1813; *Pterocladia* J. Agardh 1852.

B. Order Cryptonemiales Schmitz in Engler 1892

Multicellular, axial filament or multiaxial; auxiliary cells produced prior to fertilization; some with heavy lime encrustation (*Lithothamnium*); marine, rare in fresh water (*Hildenbrandia*). Precambrian to Holocene.

1. Family Dumontiaceae Bory orth. mut. Schmitz 1889
Miocene to Holocene.
Dumontia Lamouroux 1813; *Paleopikea* Parker & Dawson 1965 (Miocene).
2. Family Weeksiaceae Abbott 1968
Weeksia Setchell 1901.
3. Family Gloiosiphoniaceae Schmitz in Engler 1892
Miocene to Holocene.
Gloiosiphonia Carmichael ex Berkeley 1833; *Paleosiphonia* Parker & Dawson 1965 (Miocene).
4. Family Endocladiaceae (J. Agardh) Kylin 1928
Endocladia J. Agardh 1841.
5. Family Rhizophyllidaceae Montagne orth. mut. Schmitz 1889
6. Family Solenoporaceae Pia 1927
Differ from Corallinaceae in the much larger cells and undifferentiated tissue. Precambrian to Miocene.
Bija Vologdin 1932 (Cambrian); *Lithocaulon* Meneghini 1857 (Cretaceous); *Metasolenopora* Yabe 1912 (Jurassic); *Neosolenopora* Mastrorilli 1955 (Miocene); *Parachaetetes* Denninger 1906 (Ordovician to Paleocene; syn.: *Petrophyton* Yabe 1912); *Pseudochaetetes* Haug 1883 (Mississippian to Cretaceous); *Pseudolithothamnium* Pfender 1936; *Solenomeris* Douville 1925 (Paleocene); *Solenopora* Dybowsky 1878 (Precambrian, Ordovician); *Uraimella* Chuvashov 1973 (Devonian).
7. Family Peyssoneliaceae Denizot 1968.
Syn.: Squamariaceae (J. Agardh) Hauck 1883
Aragonite-depositing. Lower Cretaceous to Holocene.
Ethelia Weber van Bosse 1913 (Lower Cretaceous to Holocene); *Metapeyssonnelia* Bou-douresque, Coppejans & Marcot 1976; *Peyssonelia* Decaisne 1841 (Paleocene, Holocene).

8. Family Hildenbrandiaceae (Trevisan) Rabenhorst 1868
Hildenbrandia Nardo 1834.

9. Family Corallinaceae (Lamouroux) Harvey 1849
Devonian to Holocene.
Following Adey and Johansen (1972), Adey and Macintyre (1973); and Johansen (1976), the first three subfamilies include crustose corallines, the others are the jointed corallines; other proposed classifications include crustose and jointed taxa in the same subfamilies (Cabioch, 1971, 1972; Chihara, 1974). The Upper Devonian *Keega* Wray 1967, formerly regarded as a primitive coralline alga, is now regarded as a stromatoporoid.
    a. Subfamily Lithophylloideae Setchell 1943
Lower Cretaceous to Holocene.
Ezo Adey, Masaki & Akioka 1974; *Lithophyllum* Philippi 1837 (Lower Cretaceous to Holocene); *Subterraniphyllum* Elliott 1957 (Oligocene); *Tenarea* Bory 1832.
    b. Subfamily Mastophoroideae (Svedelius) Setchell 1943
Miocene to Holocene.
Choreonema Schmitz 1889; *Fosliella* Howe 1920; *Goniolithon* Foslie 1898; *Heteroderma* Foslie 1909; *Hydrolithon* Foslie 1909 (Miocene, Holocene; syn.: *Pseudaethesolithon* Elliott 1970); *Lithoporella* Foslie 1909 (Upper Jurassic to Holocene); *Mastophora* Decaisne 1842; *Metamastophora* Setchell & Mason 1943; *Neogoniolithon* Setchell & Mason 1943; *Porolithon* Foslie 1909; *Pseudolithophyllum* Lemoine 1913.
    c. Subfamily Melobesioideae (J. Areschoug) Mason 1953
Devonian to Holocene.
Distichoplax was originally described as an alga, then transferred to the pterobranchs (similar to *Rhabdopleura*) by Lemoine (1960); discovery of specimens with conceptacles (Varma, 1962) resulted in its replacement in the red algae.
Aethesolithon Johnson 1964 (Miocene, Pliocene); *Archaeolithophyllum* Johnson 1956 (Mississippian to Permian); *Archaeolithothamnium* Rothpletz 1891 (Upper Jurassic to

Holocene); *Bicorium* Maslov 1956 (Devonian); *Clathromorphum* Foslie 1898; *Cuneiphycus* Johnson 1960 (Pennsylvanian); *Dermatolithon* Foslie 1898 (Miocene to Holocene); *Distichoplax* Pia 1934 (Upper Cretaceous to Eocene); *Katavella* Chuvashov 1965 (Devonian); *Kvaleya* Adey & Sperapani 1971; *Kymalithon* Lemoine & Emberger 1967 (Lower Cretaceous); *Leptophytum* Adey 1966; *Lithothamnium* Philippi 1837 (Jurassic to Holocene); *Lysvaella* Chuvashov 1971 (Permian); *Melobesia* Lamouroux 1816 (Oligocene to Holocene); *Mesolithon* Maslov 1955 (Paleocene); *Mesophyllum* Lemoine 1928 (Cretaceous to Miocene); *Neopolyporolithon* Adey & Johansen 1972; *Palaeophyllum* Maslov 1950 (Lower Cretaceous to Upper Cretaceous); *Palaeothamnium* Conti 1945 (Miocene); *Phymatolithon* Foslie 1898; *Polyporolithon* Mason 1953; *Pomotophyllum* Conti 1943 (Miocene); *Sclerothamnium* Airoldi 1936 (Triassic); *Sporolithon* Heydrich 1897; *Tharama* Wray 1967 (Devonian).

    d. Subfamily Metagoniolithoideae Johansen 1969

Metagoniolithon Weber van Bosse 1904.

    e. Subfamily Amphiroideae Johansen 1969
    Cretaceous to Holocene.

Amphiroa Lamouroux 1812 (Upper Cretaceous to Holocene); *Lithothrix* Gray 1867.

    f. Subfamily Corallinoideae (Areschoug) Foslie 1908
    Jurassic to Holocene.

Alatocladia (Yendo) Johansen 1969; *Archamphiroa* Steinmann 1930 (Jurassic); *Arthrocardia* Decaisne 1842 (Upper Cretaceous to Holocene); *Bossiella* Silva 1957 (syn.: *Bossea* Manza 1937, non Reichenbach 1841); *Calliarthron* Manza 1937; *Cheilosporum* Areschoug 1852; *Chiharaea* Johansen 1966; *Corallina* Linnaeus 1758 (Lower Cretaceous to Holocene); *Haliptylon* (Decaisne) Johansen 1971; *Jania* Lamouroux 1812 (Upper Cretaceous to Holocene); *Marginisporum* (Yendo) Ganesan 1967; *Serraticardia* (Yendo) Silva 1957; *Yamadaea* Segawa 1955.

    10. Family Cryptonemiaceae (J. Agardh) Harvey 1849

Cryptonemia J. Agardh 1842; *Grateloupia* C. Agardh 1822; *Halymenia* C. Agardh 1817.

    11. Family Kallymeniaceae (J. Agardh) Funk 1927

Callophyllis Kützing 1843; *Kallymenia* J. Agardh 1842.

    12. Family Choreocolacaceae Sturch 1926
Harveyella Schmitz & Reinke in Reinke 1889.

    13. Family Pustulariaceae Vologdin 1962
Precambrian.

Pustularia Vologdin 1955 (Precambrian).

    14. Family Epiphytaceae Korde 1959
Precambrian to Devonian.

Batinevia Korde 1966 (Cambrian); *Epiphyton* Bornemann 1886 (Precambrian to Devonian).

C. Order Gigartinales Schmitz in Engler 1892
    Single axial filament or multiaxial; intercalary cell becomes the auxiliary cell. Marine, one freshwater genus. Devonian to Holocene.

    1. Family Cruoriaceae (J. Agardh) Kylin 1928
Cretaceous to Holocene.

Cruoria Fries 1836; *Cruoriopsis* Dufour 1864; *Ethelia* Weber van Bosse 1913 (Cretaceous to Holocene).

    2. Family Calosiphoniaceae Kylin 1932
Calosiphonia Crouan & Crouan 1852; *Schmitzia* Silva 1959 (syn. *Bertholdia* Schmitz 1897, non Lagerheim 1889).

    3. Family Gymnophlaeaceae Kützing 1843
Schizymenia J. Agardh 1851.

    4. Family Chondriellaceae Levring 1941

    5. Family Sebdeniaceae Kylin 1932
Sebdenia Berthold 1882.

    6. Family Rissoëllaceae (J. Agardh) Kylin 1932

    7. Family Hypneaceae J. Agardh 1852

    8. Family Gracilariaceae (Nägeli) Kylin 1930
Gracilaria Greville 1830.

    9. Family Plocamiaceae Kützing orth. mut. Kylin 1930
Plocamium Lamouroux 1813.

    10. Family Sphaerococcaceae Dumortier orth. mut. Cohn 1872
    Oligocene to Holocene
Sphaerococcus Stackhouse 1797 (incl. *Sphaerococcides* Schimper 1869; Oliogcene to Holocene).

    11. Family Furcellariaceae Greville orth. mut. Kylin 1932
    ?Jurassic, Holocene.
Furcellaria Lamouroux 1813; ?*Nipponophycus* Yabe and Toyama 1928 (Jurassic).

    12. Family Stictosporaceae Kylin 1932
    13. Family Sarcodiaceae Kylin 1932
    14. Family Mychodeaceae (Schmitz & Hauptfleisch) Kylin 1932

(continued)

15. Family Dicranemaceae (Schmitz & Hauptfleisch) Kylin 1932

16. Family Solieriaceae (Harvey) Hauck 1885
Lower Cretaceous, Holocene.
Agardhiella Schmitz 1889; *Agardhiellopsis* Lemoine 1966 (Lower Cretaceous); *Solieria* J. Agardh 1842.

17. Family Acrotylaceae Schmitz in Engler 1892

18. Family Rhabdoniaceae Kylin 1932
Catenella Greville 1830.

19. Family Ungdarellaceae Maslov 1956
Syn.: Aoujgaliaceae Termier et al., 1975
Komia was regarded as a stromatoporoid (Wilson et al., 1963) but redescribed as a red alga by Mamet and Roux 1977; others here referred have been placed with the foraminifers, transferred to the red algae (Mamet and Roux, 1977; Petryk and Mamet 1972), or regarded as Ischyrospongia (Termier et al., 1977).
Amorfia Rácz 1964 (Pennsylvanian); *Aoujgalia* Termier & Termier 1950 (Mississippian); *Epistacheoides* Petryk & Mamet 1972 (Mississippian); *Fourstonella* Cummings 1955 (Mississippian to Permian); *Komia* Korde 1951 (Mississippian to Permian); *Mametella* Brenckel 1977 (Mississippian); *Parastacheia* Mamet & Roux 1977 (Mississippian); *Petschoria* Korde 1941 (Pennsylvanian); *Pseudokomia* Rácz 1964 (Pennsylvania); *Pseudostacheoides* Petryk & Mamet 1972 (Mississippian); *Stacheia* Brady 1876 (Mississippian); *Stacheoides* Cummings 1955 (Mississippian); *Ungdarella* Maslov in Korde 1951 (Mississippian to Permian); *Uralites* Chuvashov 1973 (Devonian).

20. Family Rhodophyllidaceae Schmitz in Engler 1892
Cystoclonium Kützing 1843; *Rhodophyllis* Kützing 1847.

21. Family Polyideaceae Kylin 1944
Polyides C. Agardh 1822.

22. Family Phyllophoraceae Nägeli 1847
Ahnfeltia Fries 1836; *Gymnogongrus* Martius 1828; *Phyllophora* Greville 1830; *Stenogramme* Harvey in Hooker & Arnott 1840; *Stenogrammites* Kretschetovitsch 1936 (Permian to Jurassic).

23. Family Gigartinaceae Bory orth. mut. Cohn 1880
Miocene to Holocene.
Chondrides Schimper 1869 (Miocene); *Chondrus* Stackhouse 1797; *Gigartina* Stackhouse 1809; *Petrocelis* J. Agardh 1851.

D. Order Rhodymeniales Schmitz in Engler 1892
Thallus multiaxial, in some forms with hollow interior. Small auxiliary cells arise from a support cell before fertilization. Marine, ?Precambrian, Triassic to Holocene.

1. Family Rhodymeniaceae Harvey 1849
?Precambrian, Triassic to Holocene.
Botryocladia Kylin 1931; ?*Lanceoforma* Walter, Oehler & Oehler 1976 (Precambrian); *Lomentarites* Fliche 1905 (Triassic); ?*Rhodomenites* Miquel 1851 (Tertiary); *Rhodymenia* Greville 1830.

2. Family Champiaceae Kützing orth. mut. Bliding 1928
Champia Desvaux 1808; *Lomentaria* Lyngbye 1819.

E. Order Ceramiales Oltmanns 1904
Basically uniaxial in construction, carpogonia with auxiliary cells produced after fertilization from a support cell. Marine, rare in fresh water.

1. Family Ceramiaceae (S. F. Gray) Harvey orth. mut. Rabenhorst 1847
Miocene to Holocene.
Antithamnion Nägeli 1847; *Bornetia* Thuret 1855; *Callithamnion* Lyngbye 1819; *Ceramium* Roth 1797; *Griffithsia* C. Agardh 1817; *Paleothamnion* Parker & Dawson 1965 (Miocene); *Plumaria* Schmitz 1889.

2. Family Rhodomelaceae (J. Agardh) Harvey 1849
Cambrian to Holocene.
Bostrychia Montagne 1838; *Chondria* C. Agardh 1817; *Donezella* Maslov 1929 (Mississippian to Pennsylvanian); *Phenacocladus* Cockerell 1926 (Eocene); *Polysiphonia* Greville 1824; *Polysiphonides* Schimper 1869 (Miocene); *Rhodomela* C. Agardh 1822; ?*Waputikia* Walcott 1919 (Cambrian).

3. Family Delesseriaceae Bory orth. mut. Nägeli 1847
?Ordovician, Cretaceous to Holocene.
Caloglossa J. Agardh 1876; *Delesseria* Lamouroux 1813; ?*Delesserites* Sternberg 1833 (?Ordovician, Upper Cretaceous to Miocene).

4. Family Dasyaceae Kützing orth. mut. Rosenberg 1933
Dasya C. Agardh 1824; *Heterosiphonia* Montagne 1842.

REFERENCES

Adey, W. H., and P. J. Adey, Studies on the biosystematics and ecology of the epilithic crustose Corallinaceae of the British Isles. *Br. Phycol. J.,* v. 8, p. 343−407, figs. 1−71, 1973.

Adey, W. H., and H. W. Johansen, Morphology and taxonomy of Corallinaceae with special reference to *Clathromorphum, Mesophyllum,* and *Neopolyporolithon* gen. nov. (Rhodophyceae, Cryptonemiales). *Phycologia,* v. 11, p. 159−180, figs. 1−69, 1972.

Adey, W. H., and I. G. Macintyre, Crustose coralline algae: a re-evaluation in the geological sciences. *Bull. Geol. Soc. Am.,* v. 84, p. 883−904, 31 figs., 1973.

Adey, W. H., and J. M. Vassar, Colonization, succession and growth rates of tropical crustose coralline algae (Rhodophyta, Cryptonemiales). *Phycologia,* v. 14, p. 55−69, figs. 1−14, 1975.

Adolphe, J. P., Reproductive cycle of an oncolithic red alga (Nemalionales) of the lacustrine Miocene of Euböa (Greece), p. 199−201, fig. 1. *In* Eric Flügel, *Fossil algae, recent results and developments.* Berlin, Heidelberg, New York: Springer-Verlag, 1977.

Airoldi, M., Sull' appartenenza della "Pelagosite" ad un nuovo genere di alga calcarea. *Atti Accad. Naz. Lincei Rc., Classe Sci. Fis., Mat., Nat.,* ser. 6, v. 24, p. 18−23, figs. 1−2, 1 pl., 1936.

Allsopp, A., Phylogenetic relationships of the Procaryota and the origin of the eucaryotic cell. *New Phytol.,* v. 68, p. 591−612, pl. 1, fig. 1, 1969.

Baas-Becking, L. G. M., and E. W. Galliher, Wall structure and mineralization in coralline algae. *J. Phys. Chem.,* v. 35, p. 467−479, figs. 1−7, 1931.

Barghoorn, E. S., and S. A. Tyler, Microorganisms from the Gunflint Chert. *Science,* v. 147, p. 563−577, figs. 1−10, 1965.

Barnes, J., D. J. Bellamy, D. J. Jones, B. A. Whitton, E. A. Drew, and J. N. Lythgoe, Sublittoral reef phenomena of Aldabra. *Nature,* v. 225, p. 268−269, figs. 1−3, 1970.

Boney, A. D., Mucilage sheaths of spores of red algae. *J. Mar. Biol. Ass. U.K.,* v. 55, p. 511−518, 1 fig., 1975.

Borowitzka, M. A., A. W. D. Larkum, and C. E. Nockolds, A scanning electron microscope study of the structure and organization of the calcium carbonate deposits of algae. *Phycologia,* v. 13, p. 195−203, figs. 1−18, 1974.

Bosellini, Alfonso, and R. N. Ginsburg, Form and internal structure of Recent algal nodules (Rhodolites) from Bermuda. *J. Geol.,* v. 79, p. 669−682, figs. 1−13, 1971.

Bourrelly, P., *Les algues d'eau douce Initiation à la systématique,* Tome III, *Les algues bleues et rouges, Les Eugléniens, Peridiniens et Cryptomonadines.* Paris: Ed. N. Boubee & Cie, 512 p., 134 pls., figs., 1970.

Brenckle, Paul, *Mametella,* a new genus of calcareous red algae (?) of Mississippian age in North America. *J. Paleont.,* v. 51, p. 250−255, pl. 1, figs. 1−3, 1977.

Cabioch, Jacqueline, Essai d'une nouvelle classification des Corallinacées actuelles. *C. R. Hebd. Séanc. Acad. Sci., Paris.,* v. 272, Sér. D, p. 1616−1619, 1971.

Cabioch, Jacqueline, Étude sur les Corallinacées. *Cah. Biol. Mar.,* v. 12, p. 121−186, pls. 1, 2, figs. 1−16, 1972.

Cavalier-Smith, T., The origin of nuclei and of eucaryotic cells. *Nature,* v. 256, p. 463−468, figs. 1−5, 1975.

Chave, K. E., Aspects of the biogeochemistry of magnesium: 1. Calcareous marine organisms. *J. Geol.,* v. 62, p. 266−283, figs. 1−16, 1954.

Chen, L. C. M., T. Edelstein, and J. McLachlan, *Bonnemaisonia hamifera* Hariot in nature and in culture. *J. Phycol.,* v. 5, p. 211−220, figs. 1−31. 1969.

Chihara, Mitsuo, Life cycle of the Bonnemaisoniaceous algae in Japan (1). *Sci. Rep. Tokyo Kyoiku Daig.,* Sec. B, v. 10 (153), p. 121−154, pls. 1−6, figs. 1−19, 1961.

Chihara, Mitsuo, Life cycle of the Bonnemaisoniaceous algae in Japan (2). *Sci. Rep. Tokyo Kyoiku Daig.,* Sec. B, v. 11 (161), p. 127−153, pls. 7−11, figs. 1−34, 1962.

Chihara, Mitsuo, Germination of the carpospores of *Bonnemaisonia nootkana,* with special reference to the life cycle. *Phycologia,* v. 5, p. 71−79, figs. 1, 2, 1965.

Chihara, Mitsuo, The significance of reproductive and spore germination characteristics to the systematics of the Corallinaceae: nonarticulated coralline algae. *J. Phycol.,* v. 10, p. 266−274, figs. 1−6, 1974.

Chuvashov, B. I., *Katavella*-novyy rod iskopaemykh bagryanok [*Katavella*-a new genus of fossil red algae]. *Paleont. Zh.,* no. 2, p. 144−146, pl. 12, 1965.

Chuvashov, B. I., Novye Devonskie vodorosli Urala [New Devonian algae of the Urals]. *Trudy Inst. Geol. Geokhim., Ural. Nauch. Tsentr., Akad. Nauk SSSR,* vyp. 99, Sb. po Voprosam Stratig., no. 18, p. 28−47, pls. 1−5, figs. 1, 2, 1973.

Cloud, P. E., and H. Hagen, Electron microscopy of the Gunflint microflora: preliminary results. *Proc. Natn. Acad. Sci. U.S.A.,* v. 54, p. 1−8, 24 figs., 1965.

Cockerell, T. D. A., An alga from the Eocene of Colorado. *Torreya,* v. 26, p. 111−112, 1 fig., 1926.

Cole, Kathleen, and Elsie Conway, Phenetic implications of structural features of the perennating phase in the life history of *Porphyra* and *Bangia* (Bangiophyceae, Rhodophyta). *Phycologia,* v. 14, p. 239−245, figs. 1−13, 1975.

Conti, Sergio, Contributo allo studio delle Corallinacee del terziario italiano. II. Corallinacee del Miocene Ligure-Piemontese. *Palaeontogr. Ital.,* v. 41, p. 37−61, pls. 5−8, figs. 1−3, 1943.

Conway, E., Autecological studies of the genus *Porphyra*: I. The species found in Britain. *Br. Phycol. Bull.*, v. 2, p. 342−348, pls. 1, 2, 1964a.

Conway, E., Autecological studies of the genus *Porphyra*: II. *Porphyra umbilicalis* (L.) J. Ag. *Br. Phycol. Bull.*, v. 2, p. 349−363, figs. 1−15, 1964b.

Cummings, R. H., New genera of foraminifera from the British Lower Carboniferous. *J. Wash. Acad. Sci.*, v. 45, p. 1−8, figs. 1−5, 1955a.

Cummings, R. H., *Stacheoides*, a new foraminiferal genus from the British Upper Paleozoic. *J. Wash. Acad. Sci.*, v. 45, p. 342−346, figs. 1−8, 1955b.

Darby, D. G., Reproductive modes of *Huroniospora microreticulata* from cherts of the Precambrian Gunflint Iron-Formation. *Bull. Geol. Soc. Am.*, v. 85, p. 1595−1596, 1 fig., 1974.

Dawson, E. Y., Marine red algae of Pacific Mexico— Part I. Bangiales to Corallinaceae Subf. Corallinoideae. *Allan Hancock Pacif. Exped.*, v. 17, p. 1−238, pls. 1−33, 1953.

Dawson, E. Y., Marine red algae of Pacific Mexico— Part II. Cryptonemiales (cont.). *Allan Hancock Pacif. Exped.*, v. 17, p. 241−409, pls. 34−44, 1954.

Dawson, E. Y., Marine red algae of Pacific Mexico— Part 7. Ceramiales: Ceramiaceae, Delesseriaceae. *Allan Hancock Pacif. Exped.*, v. 26, p. 1−207, pls. 1−50, 1962.

Dawson, E. Y., *Marine botany, An introduction*. New York, Chicago, San Francisco, Toronto, London: Holt, Rinehart and Winston, Inc., xii + 371 p., 1966.

Deninger, Karl, Einige neue Tabulaten und Hydrozoen aus mesozoischen Ablagerungen. *Neues Jb. Miner. Geol. Paläont.*, Jg. 1906, v. 1, 61−70, pls. 5−7, 1906.

Dixon, P. S., The Rhodophyta: some aspects of their biology. II. *Oceanogr. Mar. Biol. A. Rev.*, v. 8, p. 307−352, 1970.

Dixon, P. S., *Biology of the Rhodophyta*. Univ. Rev. Bot., v. 4, 285 p., 31 figs., 1973.

Dougherty, E. C., and M. B. Allen, Is pigmentation a clue to protistan phylogeny?, p. 129−144, figs. 1, 2. *In* M. B. Allen, *Comparative biochemistry of photoreactive systems*. New York: Academic Press, 1960.

Dragastan, Ovidiu, Un nouveau genre d'algue dans le Crétacé inférieur du bassin septentrional de Babadag-Dobrogea (Roumanie). *C. R. Séanc. Soc. Phys. Hist. Nat. Genève*, n.s., v. 7, p. 82−87, figs. 1−3, 1973.

Drew, K. M., Reproduction in the Bangiophycidae. *Bot. Rev.*, v. 22, p. 553−611, pls. 1−3, 1956.

Edwards, Peter, The life history of *Polysiphonia denudata* (Dillwyn) Kützing in culture. *J. Phycol.*, v. 4, p. 35−37, figs. 1−6, 1968.

Edwards, Peter, The life history of *Callithamnion byssoides* in culture. *J. Phycol.*, v. 5, p. 266−268, figs. 1−9, 1969.

Elliott, G. F., The Permian calcareous alga *Gymnocodium*. *Micropaleontology*, v. 1, p. 83−90, pls. 1−3, 1955.

Elliott, G. F., *Subterraniphyllum*, a new Tertiary calcareous alga. *Palaeontology*, v. 1, p. 73−75, pl. 13, 1957.

Elliott, G. F., The sexual organization of Cretaceous *Permocalculus* (Calcareous algae). *Palaeontology*, v. 4, p. 82−84, pl. 11, figs. 1, 2, 1961.

Elliott, G. F., Tertiary solenoporacean algae and the reproductive structures of the Solenoporaceae. *Palaeontology*, v. 7, p. 695−702, pls. 104−108, 1965.

Elliott, G. F., *Pseudaethesolithon*, a calcareous alga from the Fars (Persian Miocene). *Geologica Romana*, v. 19, p. 31−46, pls. 1−4, figs. 1, 2, 1970.

Emberger, Louis, *Les plantes fossiles dans leurs rapports avec les végétaux vivants*, 2nd ed. Paris: Masson & Cie., 758 p., 743 figs., 1968.

Evans, L. V., Electron microscopical observations on a new red algal unicell, *Rhodella maculata* gen. nov., sp. nov., *Br. Phycol. J.*, v. 5, p. 1−13, figs. 1−19, 1970.

Fay, Mario, and Volker Diersche, First evidence of red algal spores in Upper Jurassic carbonate sediments of southern Germany and the Austro-Bavarian Alps. *Neues Jb. Geol. Paläont., Mh.* 1975 (10), p. 577−640, 9 figs., 1975.

Fenical, William, Halogenation in the Rhodophyta, a review. *J. Phycol.*, v. 11, p. 245−259, figs. 1−9, 1975.

Flügel, P., Solenoporaceen (Algen) aus den Zamblach-Schichten (Rhät) d. Fischerwiese bei Alt-Aussee, Steiermark. *Neues Jb. Geol. Paläont., Mh.* 1960 (8), p. 339−354, figs. 1−3, 1960.

Fritsch, F. E., *The structure and reproduction of the algae*, v. 2, Foreword, Phaeophyceae, Rhodophyceae, Myxophyceae. Cambridge: University Press, xiv + 939 p., 336 figs., 1952.

Gantt, Elisabeth, M. R. Edwards, and S. F. Conti, Ultrastructure of *Porphyridium aerugineum* a blue-green colored rhodophytan. *J. Phycol.*, v. 4, p. 65−71, figs. 1−6, 1968.

Gibbs, S. P., The comparative ultrastructure of the algal chloroplast. *Ann. N.Y. Acad. Sci.*, v. 175, p. 454−473, figs. 1−3, 1970.

Gressel, J., T. Berman, and N. Cohen, Dinoflagellate ribosomal RNA; an evolutionary relic? *J. Molec. Evol.*, v. 5, p. 307−313, figs. 1, 2, 1975.

Hommersand, M. H., and R. B. Searles, Bibliography on Rhodophyta, p. 760−767. *In* J. R. Rosowski and B. C. Parker, *Selected Papers in Phycology*. Lincoln: University of Nebraska Department of Botany, 1971.

Howland, G. P., and J. Ramus, Analysis of blue-green and red algal ribosomal-RNA's by gel electrophoresis. *Arch. Mikrobiol.*, v. 76, p. 292−298, fig. 1, 1971.

Johansen, H. W., Morphology and systematics of

coralline algae with special reference to *Calliarthron*. *Univ. Calif. Publs. Bot.*, v. 49, vii + 78 p., 19 pls., 1969a.

Johansen, H. W., Patterns of genicular development in *Amphiroa* (Corallinaceae). *J. Phycol.*, v. 5, p. 118−123, figs. 1−19, 1969b.

Johansen, H. W., The diagnostic value of reproductive organs in some genera of articulated coralline red algae. *Br. Phycol. J.*, v. 5, p. 79−86, figs. 1−13, 1970.

Johansen, H. W., Changes and additions to the articulated coralline flora of California. *Phycologia*, v. 10, p. 241−249, figs. 1−22, 1971a.

Johansen, H. W., *Bossiella*, a genus of articulated corallines (Rhodophyceae, Cryptonemiales) in the eastern Pacific. *Phycologia*, v. 10, p. 381−396, figs. 1−32, 1971b.

Johansen, H. W., Ontogeny of sexual conceptacles in a species of *Bossiella* (Corallinaceae). *J. Phycol.*, v. 9, p. 141−148, figs. 1−37, 1973.

Johansen, H. W., Articulated coralline algae. *Oceanogr. Mar. Biol. A. Rev.*, v. 12, p. 77−127, figs. 1−15, 1974.

Johansen, H. W., Current status of generic concepts in coralline algae (Rhodophyta). *Phycologia*, v. 15, p. 221−244, figs. 1−90, 1976.

Johnson, J. H., Geologic importance of calcareous algae with annotated bibliography. *Colo. Sch. Mines Q.*, v. 38 (1), p. 1−102, figs. 1−23, 1943.

Johnson, J. H., An introduction to the study of rock building algae and algal limestones. *Colo. Sch. Mines Q.*, v. 49 (2), p. 1−117, pls. 1−62, 1954.

Johnson, J. H. *Archaeolithophyllum*, a new genus of Paleozoic coralline algae. *J. Paleont.*, v. 30, p. 53−55, pl. 14, 1956a.

Johnson, J. H., Ancestry of the coralline algae. *J. Paleont.*, v. 30, p. 563−567, pl. 68, 1956b.

Johnson, J. H., Calcareous algae. *Prof. Pap. U.S. Geol. Surv.*, 280−E, p. 209−246, pls. 36−60, 1957.

Johnson, J. H., The algal genus *Archaeolithothamnium* and its fossil representatives. *J. Paleont.*, v. 37, p. 175−211, pls. 25−30, 1 fig., 1963.

Johnson, J. H., Eocene algae from Ishigaki-shima Ryūkyū-rettō. *Prof. Pap. U.S. Geol. Surv.*, 399−C, p. C1−C13, pls. 1−7, 1964a.

Johnson, J. H., Fossil and Recent calcareous algae from Guam. *Prof. Pap. U.S. Geol. Surv.*, 403−G, p. G1−G40, pls. 1−15, 1964b.

Johnson, J. H., Paleocene calcareous red algae from northern Iraq. *Micropaleontology*, v. 10, p. 207−216, pls. 1−3, 1964c.

Johnson, J. H., Miocene coralline algae from northern Iraq. *Micropaleontology*, v. 10, p. 477−485, pls. 1−3, 1964d.

Johnson, J. H., A review of the Cambrian algae. *Colo. Sch. Mines Q.*, v. 61 (1), xiv + 162 p., 65 pls., 1966.

Johnson, J. H., A review of the Lower Cretaceous algae. *Colo. Sch. Mines Prof. Contr.*, no. 6, xvi + 180 p., 68 pls., 1969.

Johnson, J. H., and B. J. Ferris, Tertiary coralline algae from the Dutch East Indies. *J. Paleont.*, v. 23, p. 193−198, pls. 37−39, 1949.

Jones, R. F., H. L. Speer, and W. Kury, Studies on the growth of the red alga *Porphyridium cruentum*. *Physiologia Pl.*, v. 16, p. 636−643, figs. 1−6, 1963.

Kaźmierczak, J., The eukaryotic nature of *Eosphaera*-like ferriferous structures from the Precambrian Gunflint Iron Formation, Canada: a comparative study. *Precambrian Res.*, v. 9, p. 1−22, figs. 1−5, 1979.

Knoll, A. H., and E. S. Barghoorn, Precambrian eukaryotic organisms: a reassessment of the evidence. *Science*, v. 190, p. 52−54, fig. 1, 1975.

Korde, K. B., *Vodorosli Kembriya yugo-vostoka Sibirskoy Platformy* [*Cambrian algae of the southeastern Siberian Platform*]. Trudy Paleont. Inst., vyp. 89, p. 1−147, pls. 1−28, figs. 1−30, 1961.

Korde, K. B., Novye materialy k sistematike i evolyutsii krasnykh vodorosley Rannego Paleozoya [New material on the systematics and evolution of red algae of the Early Paleozoic]. *Dokl. Akad. Nauk SSSR*, v. 166, p. 1440−1442, fig. 1, 1966.

Kylin, Harald, *Die Gattungen der Rhodophyceen*. Lund: CWK Gleerups Förlag, xv + 683 p., 458 figs., 1956.

Lee, R. E., The pit connections of some lower red algae: ultrastructure and phylogenetic significance. *Br. Phycol. J.*, v. 6, p. 29−38, figs. 1−23, 1971.

Lee, R. E., Origin of plastids and the phylogeny of the algae. *Nature*, v. 237, p. 44−46, figs. 1, 2, 1972.

Lee, R. E. Chloroplast structure and starch grain production as phylogenetic indicators in the lower Rhodophyceae. *Br. Phycol. J.*, v. 9, p. 291−295, fig. 1, 1974.

Lemoine, Mme. Paul, Contribution à l'étude des corallinacées fossiles: *Bull. Soc. Géol. Fr.*, ser. 4, v. 17, p. 233−283, figs. 1−23, 1917.

Lemoine, Mme. Paul, sur l'attribution du *Distichoplax biserialis* (Dietrich) Pia aux pterobranches (*Rhabdopleura*). *C. R. hebd. Séanc. Acad. Sci., Paris*, v. 246, p. 2145−2148, 1958.

Lemoine, Marie, Comparison de *Distichoplax biserialis* et des *Rhabdopleura* fossiles et actuels. *Revue Micropaléont.*, v. 3, p. 95−102, pls. 1, 2, 1960.

Lemoine, Marie, Un nouveau genre d'algue du Crétacé Inférieur: *Agardhiellopsis* nov. gen. *Revue Micropaléont.*, v. 8, p. 203−210, 1 pl., 1 fig., 1966.

Lin, Hsiu-ping, M. R. Sommerfeld, and J. R. Swafford, Light and electron microscope observations on motile cells of *Porphyridium purpureum* (Rhodophyta). *J. Phycol.*, v. 11, p. 452−457, figs. 1−7, 1975.

Littler, M. M., The crustose Corallinaceae. *Oceanogr. Mar. Biol. A. Rev.*, v. 10, p. 311–347, 1972.

Loeblich, A. R., Jr., and Helen Tappan, *Treatise on invertebrate paleontology, Pt. C, Protista 2, Sarcodina chiefly "Thecamoebians" and Foraminiferida.* Lawrence, Kans.: Geol. Soc. Am. and Univ. Kansas Press, v. 1, xxxi + 511 p., 399 figs., v. 2, 390 p., 254 figs., 1964.

McQuade, A. B., Origins of the nucleate organisms. *Q. Rev. Biol.*, v. 52, p. 249–262, fig. 1, 1977.

Mamet, B., and A. Roux, Algues rouges Dévoniennes et Carbonifères de la Téthys Occidentale. 4me Partie. *Revue Micropaléont.*, v. 19, p. 215–266, pls. 1–9, figs. 1–13, 1977.

Mamet, B., and B. Rudloff, Algues Carbonifères de la partie septentrionale de l'Amérique du Nord. *Revue Micropaléont.*, v. 15, p. 75–114, pls. 1–10, figs. 1–3, 1972.

Manza, A. V., The genera of the articulated corallines. *Proc. Natn. Acad. Sci. U.S.A.*, v. 23, p. 44–48, 1937.

Manza, A. V., A revision of the genera of articulated corallines. *Philipp. J. Sci.*, v. 71, p. 239–316, pls. 1–20, 1940.

Marsh, J. A., Jr., Primary productivity of reef-building calcareous red algae. *Ecology*, v. 51, p. 255–263, figs. 1–6, 1970.

Mastrorilli, V. I., Nuovo contributo allo studio delle corallinacee del Oligocene Ligure-Piemontese: I reperti della tavoletta ponzone. *Atti Inst. Geol. Univ. Genova*, v. 5, p. 153–406, pls. 1–42, figs. 1–35, 1968.

Meijer, J. J. de, Fossil non-calcareous algae from insoluble residues of algal limestones. *Leid. Geol. Meded.*, v. 44, p. 235–263, pls. A–L, 1969.

Moberly, Ralph, Jr., Composition of magnesium calcites of algae and pelecypods by electron microprobe analysis. *Sedimentology*, v. 11, p. 61–82, figs. 1–11, 1968.

Oehler, D. Z., Pyrenoid-like structures in Late Precambrian algae from the Bitter Springs Formation of Australia. *J. Paleont.*, v. 51, p. 885–901, pls. 1–4, 1977.

Oehler, J. H., D. Z. Oehler, and M. D. Muir, On the significance of tetrahedral tetrads of Precambrian algal cells. *Origins Life*, v. 7, p. 259–267, figs. 1–4, 1976.

Paasche, E., and J. Throndsen, *Rhodella maculata* Evans (Rhodophyceae, Porphyridiales) isolated from the plankton of the Oslo Fjord. *Nytt Mag. Bot.*, v. 17, p. 209–212, fig. 1, 1970.

Papenfuss, G. F., Classification of the algae, p. 115–224. *In* E. L. Kessel, *A century of progress in the natural sciences 1853–1953.* San Francisco: Calif. Acad. Sci., 1955.

Parker, B. C., and E. Y. Dawson, Non-calcareous marine algae from California Miocene deposits. *Nova Hedwigia*, v. 10, p. 273–295, pls. 76–96, 1965.

Perkins, R. D., and C. I. Tsentas, Microbial infestation of carbonate substrates planted on the St. Croix Shelf, West Indies. *Bull. Geol. Soc. Am.*, v. 87, p. 1615–1628, 46 figs., 1976.

Petryk, A. A., and B. L. Mamet, Lower Carboniferous algal microflora, southwestern Alberta. *Can. J. Earth Sci.*, v. 9, p. 767–802, pls. 1–10, figs. 1–10, 1972.

Phillips, D. O., and N. G. Carr, Nucleic acid analysis and the endosymbiotic hypothesis. *Taxon*, v. 26, p. 3–42, figs. 1–16, 1977.

Pia, J., Die wichtigsten Kalkalgen des Jungpaläozoikums und ihre geologische Bedeutung. *C. R. Congrès Avanc. Étud. Stratigr. Carb., 2d Heerlen 1935*, v. 2, p. 765–856, 1937.

Pickett-Heaps, J. D., *Green algae, structure, reproduction and evolution in selected genera.* Sunderland, Mass.: Publ. Sinauer Associates Inc., vii + 606p., 1975.

Poignant, A. F., The Mesozoic red algae: A general survey, p. 177–189, 1 fig. *In* Eric Flügel, *Fossil algae, recent results and developments.* Berlin, Heidelberg, New York: Springer-Verlag, 1977.

Rácz, L., Carboniferous calcareous algae and their association in the San Emiliano and Lois-Ciguera Formations (Prov. León, NW Spain). *Leid. Geol. Meded.*, v. 31, p. 1–112, pls. 1–13, figs. 1–26, 1964.

Ramus, J., The production of extracellular polysaccharide by the unicellular red alga *Porphyridium aerugineum. J. Phycol.*, v. 8, p. 97–111, 1972.

Rich, Mark, Upper Mississippian (Carboniferous) calcareous algae from northeastern Alabama, south-central Tennessee, and northwestern Georgia. *J. Paleont.*, v. 48, p. 360–374, pls. 1–5, fig. 1, 1974.

Schmalz, R. F., Brucite in carbonate secreted by the red alga *Goniolithon* sp. *Science*, v. 149, p. 993–996, figs. 1, 2, 1965.

Schopf, J. W., Microflora of the Bitter Springs Formation, Late Precambrian, central Australia. *J. Paleont.*, v. 42, p. 651–688, pls. 77–86, figs. 1–6, 1968.

Schopf, J. W., and J. M. Blacic, New microorganisms from the Bitter Springs Formation (Late Precambrian) of the north-central Amadeus Basin, Australia. *J. Paleont.*, v. 45, p. 925–960, pls. 105–113, 1971.

Scott, J. L., and J. P. Thomas, Electron microscope observations of telophase II in the Florideophyceae. *J. Phycol.*, v. 11, p. 474–476, figs. 1, 2, 1975.

Segonzac, Geneviève, Réflexions sur le genre *Palaeothamnium* Conti (Algue Mélobésiée). *Bull. Bur. Rech. Geol. Min. Paris*, ser. 2, sec. 4, no. 4, p. 1–9, 4 pls., 1969.

Segonzac, Geneviève, and Juliette Villatte, Longévité du genre *Solenopora* Dyb. *C. R. Hebd. Séanc. Acad. Sci., Paris*, v. 249, p. 1785–1787, 1959.

Siesser, W. G., Relict algal nodules (Rhodolites) from the South African Continental Shelf. *J. Geol.*, v. 80, p. 611–616, figs. 1–3, 1972.

Sommerfeld, M. R., and H. W. Nichols, Comparative studies in the genus *Porphyridium* Naeg. *J. Phycol.,* v. 6, p. 67–78, figs. 1–33, 1970.

Sommerfeld, M. R., and H. W. Nichols, The life cycle of *Bangia fuscopurpurea* in culture. I. Effects of temperature and photoperiod on the morphology and reproduction of the *Bangia* phase. *J. Phycol.,* v. 9, p. 205–210, figs. 1–6, 1973.

Tappan, Helen, Possible eucaryotic algae (Bangiophycidae) among early Proterozoic microfossils. *Bull. Geol. Soc. Am.,* v. 87, p. 633–639, figs. 1–21, 1976.

Taylor, W. R., Distribution in depth of marine algae in the Caribbean and adjacent seas. *Recent Adv. Bot.,* v. 1, p. 193–197, 1961.

Termier, Geneviève, and Henri Termier, *Paléontologie Marocaine,* Tome II, *Invertébrés de l'Ère Primaire,* Fasc. 1 Foraminifères, Spongiares et Coelentérés. Paris: Hermann and Cie., 218 p., 51 pls., 1950.

Termier, H., G. Termier, and D. Vachard, On Moravamminida and Aoujgaliida (Porifera, Ischyrospongia) Upper Paleozoic "Pseudo Algae," p. 215–219, figs. 1, 2. *In* Eric Flügel, *Fossil algae, recent results and developments.* Berlin, Heidelberg, New York: Springer-Verlag, 1977.

Turner, M. F., A note on the nutrition of *Rhodella. Br. Phycol. J.,* v. 5, p. 15–18, 1970.

Varma, C. P., New observations on the index fossil alga *Distichoplax* Pia from Laki (Lower Eocene) Beds of the Punjab Salt Range. *Palaeobotanist,* v. 9, p. 26–31, pl. 1, fig. 1, 1962.

Volkova, N. A., O nakhodke Dokembriyskikh spor s tetradnym rubtsom [On some Precambrian spores with tetrad scar], p. 14–18, pl. 2. In *Paleontologiya Morskaya Geologiya, Mezhdunar. Geol. Kongr. XXV Sess.,* Akad. Nauk SSSR, Minist. Geol. SSSR. Moscow: Nauka, 1976.

Vologdin, A. G., *Drevneyshie vodorosli SSSR [Ancient algae of the USSR].* Moscow: Akad. Nauk SSSR, 656 p., 46 pls., 1962.

Walter, M. R., J. H. Oehler, and D. Z. Oehler, Megascopic algae 1300 million years old from the Belt Supergroup, Montana: a reinterpretation of Walcott's *Helminthoidichnites. J. Paleont.,* v. 50, p. 872–881, pls. 1, 2, 1976.

Weber, J. N., and J. W. Kaufman, Brucite in the calcareous alga *Goniolithon. Science,* v. 149, p. 996–997, fig. 1, 1965.

West, J. A., and R. E. Norris, Unusual phenomena in the life histories of Florideae in culture. *J. Phycol.,* v. 2, p. 54–57, figs. 1–6, 1966.

Wetherbee, R., "Transfer connections": specialized pathways for nutrient translocation in a red alga? *Science,* v. 204, p. 858–859, figs. 1–3, 1979.

Wilson, E. C., R. H. Waines, and A. H. Cooper, A new species of *Komia* Korde and the systematic position of the genus. *Palaeontology,* v. 6, p. 246–253, pls. 34, 35, 1963.

Wray, J. L., Algae in reefs through time. *Proc. N. Am. Paleont. Conv. Chicago 1969,* Pt. J, p. 1358–1373, figs. 1–20, 1971.

Wray, J. L., Late Paleozoic calcareous red algae, p. 167–176, figs. 1, 2. *In* Erik Flügel, *Fossil algae, recent results and developments.* Berlin, Heidelberg, New York: Springer-Verlag, 1977a.

Wray, J. L., Calcareous algae. *Dev. Palaeont. Stratigr.,* v. 4, 185 p., 170 figs., 1977b.

ACRITARCHA OR HYSTRICHOPHYTA

Precambrian and Paleozoic algal distribution, and the per-cent of total present production that is of marine origin, lead to the ... suggestion that too great an importance has been ascribed to terrestrial floras in the total oxygen balance. Through much of the geologic past, oceans and shallow epicontinental seas covered more than their present 71% of the global surface, allowing an even greater proportional area for phytoplankton production. ... Every major invertebrate animal phylum had arisen and diversified, and even the chordates were represented by graptolites, agnathids and placoderms before the first appearance of vascular plants. All were totally dependent upon the algal groups during this early period of evolution.

H. Tappan, 1968

*O*f the microfossils that are preserved as or-ganic material rather than mineralized, some have unquestioned affinity to bacteria, blue-green or green algae, dinoflagellates, fungi, or higher plants. Other globular to angular, smooth-walled to spiny structures, ranging from 5 to 500 μm in diameter, are of uncertain affinity; they occur in sediments of all ages, al-though characteristically early Paleozoic. For a time they were termed "hystrichospheres" and included with the genus *Hystrichosphaera* (= *Spiniferites*). However, *Hystrichosphaera* it-self, and many others once included with it, are now known to be cysts of dinoflagellates (Evitt, 1961, 1963). With their assignment to the di-noflagellates, a new group **Acritarcha** (from the Greek, meaning "of uncertain origin") was proposed for the unallocated residue (Evitt,

1963, p. 300), some of which have since been transferred to the Prasinophycean green algae.

Acritarchs are "Small microfossils of un-known and probably varied biological affinities consisting of a central cavity enclosed by a wall of single or multiple layers and of chiefly or-ganic composition; symmetry, shape, struc-ture, and ornamentation varied; central cavity closed or communicating with the exterior by varied means, for example: pores, a slitlike or irregular rupture, a circular opening (the pylome). ... The name chosen implies no affin-ity with any other organisms...[but] the ac-ritarchs as defined here are a polyphyletic assemblage. ... Some acritarchs are probably dinoflagellates ... . Surely most of them are not dinoflagellates, but, in aggregate, represent a variety of life stages (e.g., eggs, cysts, mature tests) of assorted one-celled and higher organisms, both plants and animals. ... I recommend that the acritarchs be treated under the Botanical Code" (Evitt, 1963, p. 300).

Following the removal to dinoflagellates or to prasinophytes, based on various lines of evidence, of many of the taxa originally included as acritarchs, the residue now is generally regarded to represent only the encysted stage of some form of phytoplankton, whether one of the extant algal divisions or a group now extinct.

ACRITARCH MORPHOLOGY

The Vesicle

The main body of the acritarch is variously termed a **vesicle**, central body, test, or shell. Connotations of the last two terms might imply a living chamber in contrast to a cyst, and the term *central body* might be confused with the inner body present in some dinoflagellates and perhaps some acritarchs, whereas *vesicle* is more strictly a morphological term and is preferred here. All vesicles are hollow.

Size

The acritarch vesicle has a fairly broad size range, from about 5 to 500 μm in diameter, but most commonly between 50 and 100 μm. Some Precambrian specimens are only 1 to 2 μm in diameter. Individual species rarely show much size variation, although Eisenack (1963a) noted that the size variation of some species either required ontogenetic enlargement of an originally smaller vesicle, and thus involved the unlikely resorption and redeposition of wall material of the resting cyst; or else reflected cyst formation by free-living cells that varied in size in the vegetative stage, as for example in modern diatoms. Some living prasinophycean green algae, to which the fossil *Leiosphaeridia* and *Tasmanites* are related, apparently do show size increase of the cyst or phycoma stage, but no evidence is available as to a similar habit of the acritarchs.

Symmetry

The general shape of the vesicle ranges from spherical, elliptical, or ovoid to discoid, arcuate, crescentic, fusiform, cylindrical, or flasklike. The periphery may be rounded or angular, and a variety of surface sculpture and processes may be regularly distributed or localized in occurrence to result in a symmetrical appearance.

Surface ornamentation and processes may result in radial symmetry in all planes as in *Baltisphaeridium* (see Figure 3.1) or symmetry may be limited to planes passing vertically through an axis of bipolarity and the plane perpendicular to the axis, resulting in similar poles, as in *Orthosphaeridium* and *Acanthodiacrodium*. Those with symmetry in all planes passing vertically through the axis but not in the plane perpendicular to the axis have dissimilar poles, as in *Ooidium*, *Aremoricanium*, and *Domasia*. Symmetry may be basically radial in all planes, except for the polarity imposed by an excystment opening, as in *Axisphaeridium* and *Rhopaliophora*. Variations in the processes of others may result in a polarity, as in *Unellium* or *Scuticabolus*. Other taxa, such as *Veryhachium*, have bilateral symmetry.

Clusters or "Colonies"

In defining the Acritarcha as organic microplankton of uncertain origin, certain groups were specifically excluded, such as Chitinozoa, spores, "microforaminifera" represented by organic chamber linings, and specimens consisting of intimately connected clusters of cells, such as *Gloeocapsomorpha*, *Pediastrum*, and *Botryococcus* (Downie et al., 1963), genera now regarded as Chlorophyceae. Such probable Prasinophyceae as *Leiosphaeridia* and *Tasmanites* also may occur in aggregates (Combaz, 1967). Other acritarchs predominantly occur as isolated cells, yet occasionally are found in clusters, and some definitely multicellular forms also have been included with the acritarchs, such as the Precambrian *Nevidia* and *Hymenophacoïdes*, the Ordovician *Pictonicopila* and *Symplassosphaeridium*, and the Silurian *Synsphaeridium* (see Figure 3.2).

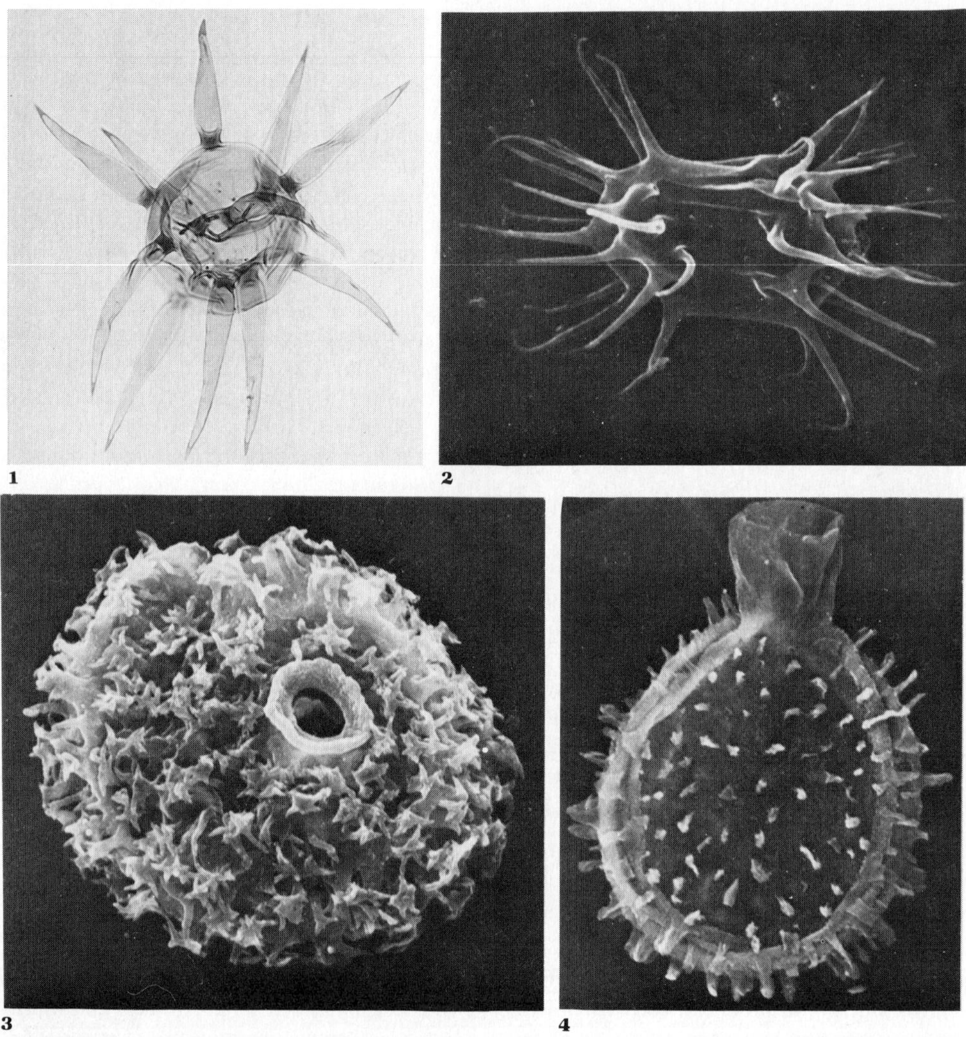

Figure 3.1

Symmetry in the acritarchs. **1.** *Baltisphaeridium perclarum* Loeblich & Tappan, Upper Ordovician, Oklahoma. Spherical vesicle, elongate hollow tapering processes in all directions; note inner plugs at junction of processes with vesicle; ×307. **2.** *Actinotodissus longitaleosus* Loeblich & Tappan, Middle Ordovician, Oklahoma, with bipolar symmetry, processes at each end; SEM ×1200, from Loeblich and Tappan, 1978. **3.** *Rhopaliophora impexa* Tappan & Loeblich, Upper Ordovician, Indiana; many short complex processes, radial symmetry modified by the presence of a cyclopyle with flaring rim; SEM, ×1040, from Tappan and Loeblich, 1971b. **4.** *Aremoricanium decoratum* Loeblich & MacAdam, Middle Ordovician, Oklahoma, flasklike, with elongate neck, short, flanged processes; SEM, ×640, from Loeblich and MacAdam, 1971.

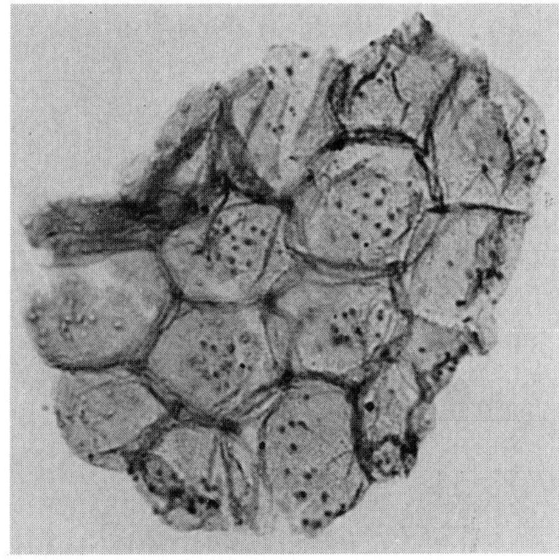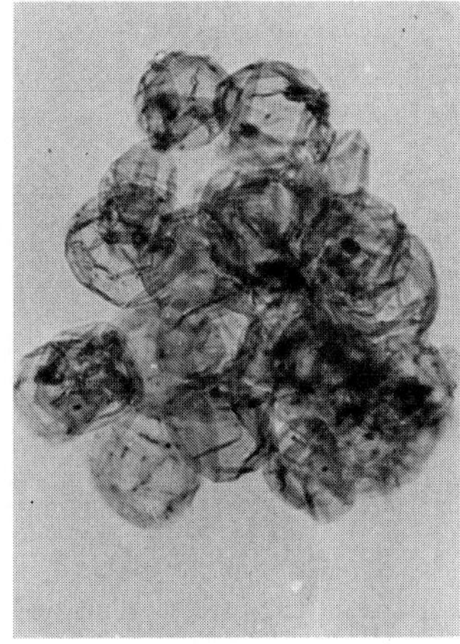

Figure 3.2
Acritarch aggregates. **1.** *Synsphaeridium* cf. *gotlandicum* Eisenack, ×960.
2. Cluster of leiospheres, or *Synsphaeridium,* ×518. Both Middle
Ordovician, Oklahoma.

Unispecific clusters of normally isolated cells have been reported from rocks of various ages, in thin sections and flakes of rock, and in acid macerations. All individuals of a single cluster generally are not only of a single species and identical morphology, but of identical size as well. Thus a species may have a broad size range within a given sample, but clusters of that species will include all large or all smaller individuals; separate clusters may include individuals of a different size, but no single cluster will have a mixture of morphologic types or different size categories (Cramer and Díez de Cramer, 1972a, p. 111).

Frequent clusters of individuals occur in the Lower Cambrian rocks of Esthonia (Volkova, 1968), representing *Micrhystridium, Baltisphaeridium,* and *Leiomarginata.* Clusters of species of *Priscogalea* and *Cymatiogalea* were present in the Lower Ordovician (Tremadocian) of England (Rasul, 1974), and aggregates of *Cymbosphaeridium pilar* (Cramer) Lister occurred in the Silurian San Pedro Formation of Spain (Cramer, 1966b, pl. 2, figs. 7, 8), with groups of from 2 to over 100 individuals in sheetlike masses of from 3 to 8 individuals thick. Other aggregates of acritarchs were present in the latter formation, as were similar clusters of chitinozoans.

Similar clusters of *Micrhystridium* and of *Veryhachium* in the Triassic of western Canada were compared to the colonial Chlorococcalean algae (Jansonius, 1962).

From Middle Jurassic cherts of France, Valensi (1953) described such a grouping of 13 individuals of *Micrhystridium* cf. *arachnoides* Valensi around four larger (20μm) spheres, and regarded the smaller specimens (all between 7 and 8 μm in diameter) as possible reproductive spores. This may instead be an accidental association.

A Lower Ordovician (Tremadocian) shale from Morocco provided some 20 to 25 individuals of *Impluviculus miloni* (Deunff) Loeblich & Tappan, all of nearly identical size and grouped in an organic mass, the central individuals of the mass being difficult to discern clearly (Deunff, 1968b). The apparent colonial organization was suggested to resemble similar assemblages among the green algae.

These clusters thus have been reported through much of the geologic range of the acritarchs, although they generally represent species of *Baltisphaeridium, Micrhystridium,*

Veryhachium, or the leiospheres. As some aggregates were reported in rock thin sections, they were believed to represent the preservation of the living assemblage. Cramer and Díez de Cramer (1972a, p. 111) regarded the acritarchs as living in colonylike agglomerations with individuals intertwined, although not physically interconnected; hence they were not true colonies. The rarity of such reports was suggested to be a result of the usual rigorous processing techniques, as they state that most samples would yield such clusters with sufficiently gentle processing techniques.

Although some platelike agglomerations may be original, some of the reported aggregates may be due to flocculation during processing, and some may represent a postmortem clumping on the sea floor at burial.

The Wall

Wall Structure

The vesicle wall may be constructed of one or two layers. If two are present, the inner is termed the **endophragm,** and the outer layer the **periphragm.** These terms are preferable to *endoderm* and *ectoderm,* which have also been used but should be avoided as they have quite different meanings in biology. Some acritarchs are reported to have a third layer or **mesophragm.**

The wall layers may differ in thickness, one being thicker and more rigid, and the other thinner and filmy. Both the number of layers and their relative thickness and rigidity are diagnostic for various species.

The endophragm is thicker in such genera as *Actipilion, Ascostomocystis,* and *Fimbriaglomerella* (see Figure 3.3). **Processes,** flanges, and other vesicle sculpture may involve only the outer wall layer.

Wall Ultrastructure

Studies of the wall ultrastructure tentatively suggest a broad separation of the acritarchs into two groups, one showing a resemblance to the Prasinophyceae and the other more similar to the dinoflagellates (Jux, 1971, 1975). In the former group are *Baltisphaeridium, Goniosphaeridium, Peteinosphaeridium,* and *Veryhachium,* all characterized by a single-layered wall penetrated by two size classes of winding radial ducts, whose spacings, size, and abundance result variously in a dense (*Baltisphaeridium*) to a spongy appearance (*Goniosphaeridium*). *Peteinosphaeridium* also appears to have a lamellar wall construction. The vesicle wall thickness ranges from 1.0 to 1.5 μm in *Baltisphaeridium longispinosum* (Eisenack) Eisenack and *Goniosphaeridium balticum* (Eisenack) Eisenack to 2 μm in *Peteinosphaeridium trifurcatum* (Eisenack) Staplin et al. and 3.5 μm in *B. multipilosum* (Eisenack) Eisenack. Processes have a thinner wall than the vesicle; the process wall is 0.1 to 1.5 μm thick in *B. pachyacanthum* Eisenack, 0.15 to 0.2 μm in *B. multipilosum,* and about 0.5 μm in *B. longispinosum.* The smaller ducts or pores are about 0.01 to 0.03 μm in diameter and 0.10 to 0.15 μm apart in *Peteinosphaeridium* and *B. multipilosum,* and as close as 0.05 to 0.10 μm apart in *G. balticum.* The larger ducts range from 0.15 to 0.2 μm in diameter in *B. multipilosum* and *G. balticum,* are randomly distributed, and vary from nearly straight and radial to slightly curved, winding, or sinuous. Sections of the processes of *Peteinosphaeridium trifurcatum* suggested that these arise from a single-layered porous wall. Both the single-layered wall and the radial ducts of two sizes in these acritarchs are reminiscent of similar features of the wall of *Tasmanites* and the living *Pachysphaera* of the Prasinophyceae. *Peteinosphaeridium majorfurcatum* Kjellström, which has a double-layered wall (Martin and Kjellström, 1973), belongs instead to *Ordovicidium.*

In contrast to *Baltisphaeridium,* the wall of *Acanthodiacrodium divisum* (Deunff) Deflandre & Deflandre-Rigaud is very thin, 0.3 to 0.5 μm in thickness, but dense in structure (Jux, 1971). At high magnification, it appears porous, with cavities of 0.5 to 2.0 μm in diameter in the wall of the mid-region. Vesicle and processes were suggested to have a double-layered construction, with the two layers separated by a pore zone, although this is not very distinct. The ducts and cavities are more variable in form and

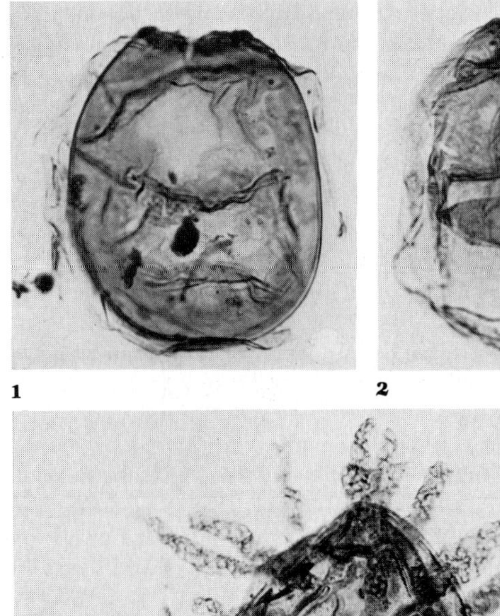

Figure 3.3
Wall layers of different thickness.
1,2. *Ascostomocystis hydria* Drugg &
Loeblich, lower Eocene, Alabama,
showing thicker darker-colored endophragm forming the inner body,
and thinner outer periphragm;
×480. **3.** *Actipilion druggii* Loeblich,
Upper Ordovician, Oklahoma, with
processes formed solely of the
periphragm, and the main vesicle
having both the thicker endophragm
and the periphragm; ×518.

size than are those of the single-layered porous wall of *Baltisphaeridium*. The wall construction of *Acanthodiacrodium* was said to be closer to that of such dinoflagellates as *Spiniferites* than to that of *Tasmanites* and other Prasinophyceae, or such acritarchs as *Baltisphaeridium* and *Veryhachium*.

While the results to date appear promising, studies of wall ultrastructure must be extended to many more of the 300± acritarch genera of the early Paleozoic than the half dozen yet studied before the wall characters can be successfully applied to the determination of relationships of this enigmatic group or to their classification.

Wall Composition

The chemical composition of the acritarch wall is not known in detail. It was at first regarded to be an alteration product of cellulose, because of the similar nature of the fossil dinoflagellates, when they were regarded as fossilized cellulosic thecae. The distinctness of this organic material from cellulose was emphasized by Eisenack (1963a), although he regarded the acritarchs and hystrichospheres (dinoflagellates) as a monophyletic group. The wall was described as a macromolecular substance, perhaps a polyterpene or a condensate of unsaturated fatty acids (Eisenack, 1963a), similar in composition to leaf cutin or the material (sporopollenin) of the pollen and spore coverings, which are similarly resistant to preparation methods.

When the acritarch vesicle is ashed, a small residue of silica (SiO_2) always remains (Eisenack, 1938b). Eisenack (1963a) suggested that this might have been postmortally adsorbed, hence result from preservation rather than reflecting the original composition. However, silica is deposited by many kinds of phytoplankton, leaving open the possibility of its original presence here as well.

Timofeev (1956a) chemically analyzed the organic residue from the Russian Lower Ordovician (Tremadocian) *Obolus* beds and reported a composition similar to fossil sporonins, hence plantlike in origin, with 71.88 percent carbon, 7.84 percent hydrogen, 2.16 percent nitrogen, and 18.12 percent combined oxygen and sulfur.

In spite of the absence of detailed information as to wall composition, variations in the wall are observable between different taxa, and even in separate parts of an individual specimen, as between processes and vesicle. These differences affect color and transparency, rigidity, resistance to oxidation and preparation methods, and varying susceptibility to dyes. Some acritarchs are colored only with great difficulty by the usual organic stains, whereas others color readily. Central body and processes also may stain differentially, as well as varying in original color and transparency.

Preservation

The very resistant organic compounds of the acritarch wall have been preserved under favorable conditions from rocks as old as the Precambrian, yet these fossils are susceptible to destruction by heat or pressure. Usually obtained for study by the acid maceration process, some are too altered for isolation by this method, but nevertheless may be recognized in thin sections and thus used stratigraphically even when the rock is badly fractured. Many such assemblages have been described from the lower Paleozoic (e.g., Burmann, 1968, 1970; Deunff, 1965; Deflandre and Ters, 1966; Downie and Ford, 1966; LeCorre and Deunff, 1969).

As they are planktonic, acritarchs are not strongly facies controlled, and may be obtained from silts, marls, or limestones. In general, the organic microplankton is well preserved in marls, and that in shales is likely to be compressed, although perhaps more successfully photographed as the necessary depth of focus is less. Specimens retain their three-dimensional appearance in limestones, but are poorly preserved in dolomites and recrystallized limestones. In dark shales of widely varying geographic occurrence and geologic age, the fossil specimens commonly have pyrite crystals in their interior (Deunff, 1959). Although the pyrite can be removed by the chemical treatment of the samples, the impression of the pyrite crystals may give a spurious ornamentation to the surface of the acritarch vesicle (see Figure 3.4). Similarly, Downie (1957, p. 416) suggested that in coccolith limestones, "a strong ornament had been imprinted on most of the plankton preserved in organic matter due to compaction among the coccoliths."

Surface Sculpture

The wall surface may be smooth, wrinkled, granulate, hispid, striate, tuberculate, reticulate, punctate, or perforate. The sculpture consists of excrescences on the ectophragm alone, and is optically continuous with it, hence does not modify the vesicle cavity. Insufficiently described for many species, either due to use of too low magnification for observation, poor preservation, or faulty preparation methods that obscured the details, wall surface sculpture nevertheless is highly diagnostic for those species in which it is known. The sculpture may change from vesicle to processes or vary on different areas of the vesicle. Flanges or crests may be equatorial, form "skirts" at each pole, or cover the surface in a reticulate pattern.

Terms commonly used for wall sculpture are those originally proposed for spore exine sculpture (Faegri and Iversen, 1950; Kremp, 1965), but some terms may be variously defined. Cramer and Díez de Cramer (1968) regard a smooth surface as the normal or basic condi-

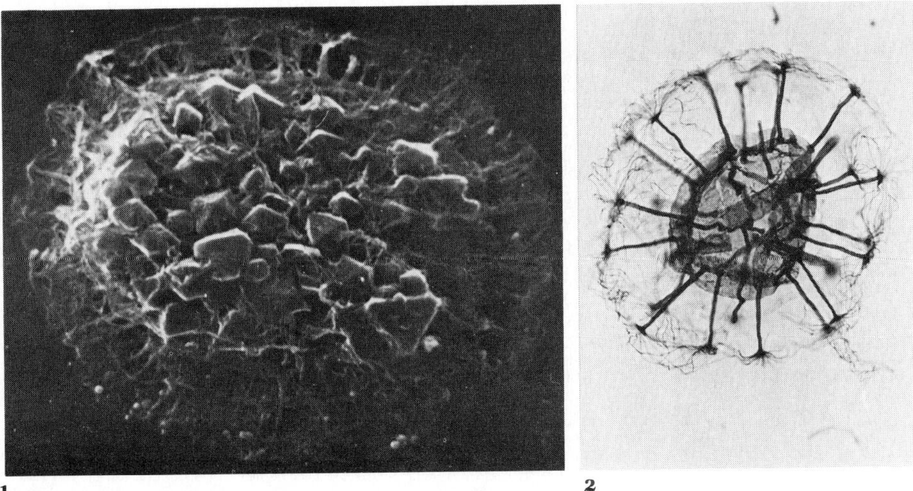

Figure 3.4
Tunisphaeridium concentricum Deunff & Evitt, Middle Silurian, New York. **1.** Spherical vesicle with numerous interconnected processes; surface of vesicle strongly imprinted by the pyrite crystals that had filled the interior; SEM, ×1120. **2.** Showing normally smooth surface of vesicle, terminally multifurcate, and interconnected elongate processes; ×414.

tion, with modifications including microscabrate, scabrate, microrugulate, rugulate, striate, spinate, echinate, mammilate, and verrucate sculpture, or a surface cover of hairlike or threadlike processes. Schopf (in Tschudy and Scott, 1969) and Evitt (in Tschudy and Scott, 1969) also referred to papillate, apiculate, spinulate (or spinose), psilate, and granulate sculpture. These terms generally refer to the appearance with light microscopy, but acritarch surfaces may have sculptural features of much lesser magnitude, requiring electron microscopy for their recognition or interpretation. Tappan and Loeblich (1971b) proposed a somewhat simplified terminology for acritarch sculpture as visible in the light microscope, using the prefix *micro-* to indicate features of less than one micron diameter or breadth that require electron microscopy for resolution. They include (see Figure 3.5):

1. Laevigate. Smooth and unornamented even at high magnification, as in *Ammonidium waldronensis* (Tappan & Loeblich) Wicander.

2. Granulate. Tiny granules, irregularly scattered or regularly aligned in rows to appear "pseudocostate." A granulate surface with sculpture of ultramicroscopic size, or microgranulate, is found in *Comptaluta pirellum* Tappan & Loeblich.

3. Punctate. Scattered shallow pits. Micropunctate sculpture has pits less than one micron in diameter (see Figure 3.6).

4. Foveolate. Closely arranged pits, greater in diameter than inter-pit areas; microfoveolate sculpture appears spongy in electron micrographs.

5. Corrugate. With irregular ridges.

6. Fossulate. Closely spaced, elevated granules, with irregular intervening grooves, resulting in a negative or incised reticulum.

7. Echinate. Surface with small thornlike spinules, tapering rapidly from the broader base, as in *Anomaloplaisium lumariacuspis* Tappan & Loeblich or *Disparifusa hystricosa* (see Figure 3.7).

8. Baculate. Modified echinate sculpture, with cylindrical spinules of equal diameter

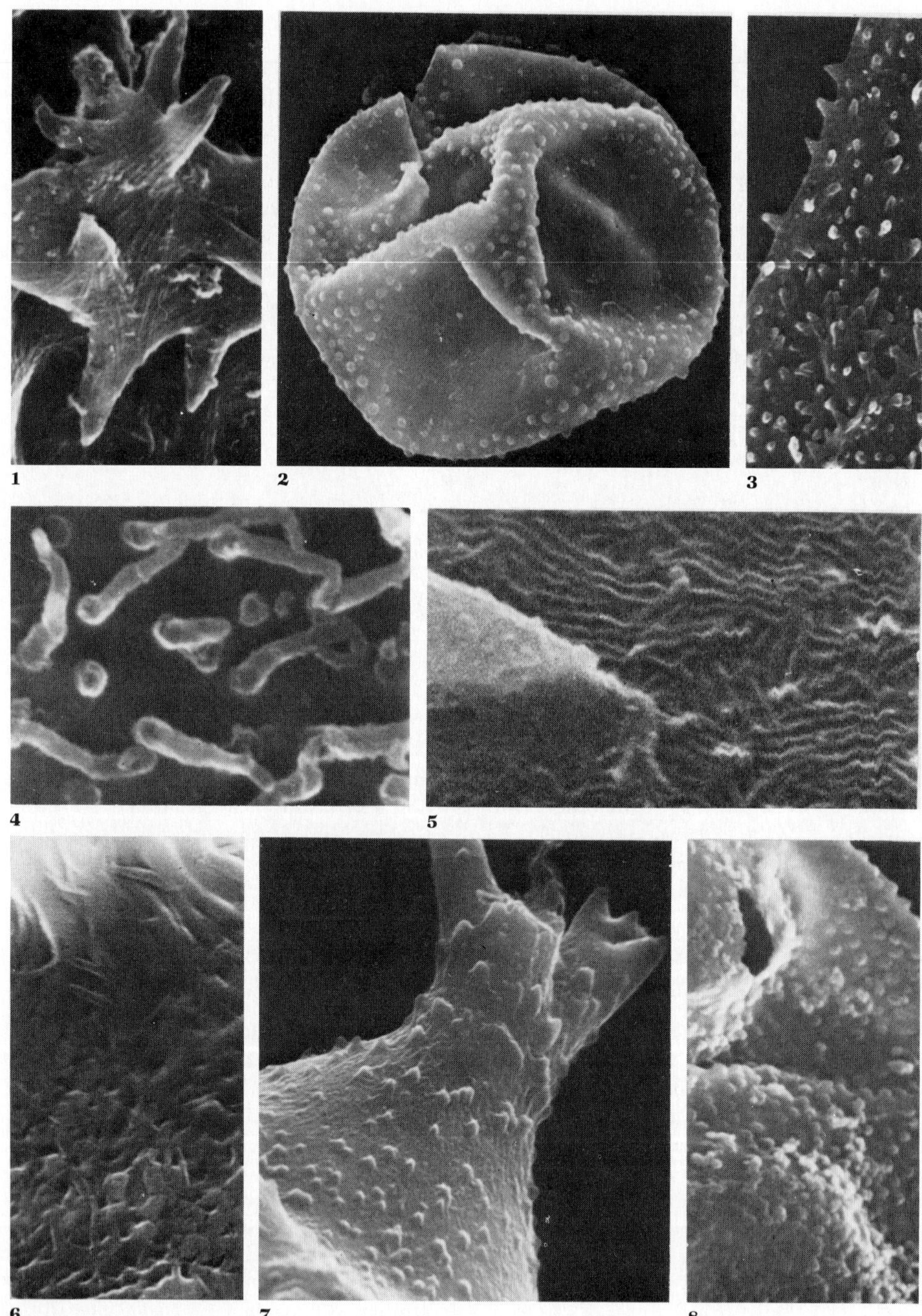

Figure 3.5

---

throughout their length and terminating bluntly.

9. Gemmate. Modified granulate sculpture with small spherical projections that are constricted at the attachment and then enlarge. Sculpture ranges from gemmate to microgemmate in *Psenotopus condrocheus.*

10. Rugulate. Finely and irregularly wrinkled surface; *Actipilion druggii* is microrugulate.

11. Costate. Parallel, narrow, elongate ridges, separated by wider inter-ridge areas. Costae or microcostae may be irregularly aligned, discontinuous and producing a vermicular appearance, very regularly aligned to simulate a pinnate venation in *Rhopaliophora foliatilis* Tappan & Loeblich, or anastomosing in a cross-hatched fashion in *Dicommopalla macadamii* Loeblich.

12. Cristate. Low, simple, flangelike crests that may anastomose to form a reticulum as in *Melikeriopalla* (a prasinophyte) or to produce an irregular labyrinthine pattern.

In addition to the standard types indicated above, various compound types of sculpture may occur, for example microgranulate-punctate, with both scattered pits and granules; or microcorrugate-foveolate, with irregular ridges and deep intervening pits. Different sculptural features may be found in particular areas of the vesicle, as in *Ooidium rossicum* Timofeev, which is coarsely granulate at one pole and for about two-fifths the height of the vesicle, then is costate, with discontinous ridges extending nearly to the opposite pole, but dying out just below the first of the terminal trabecular processes.

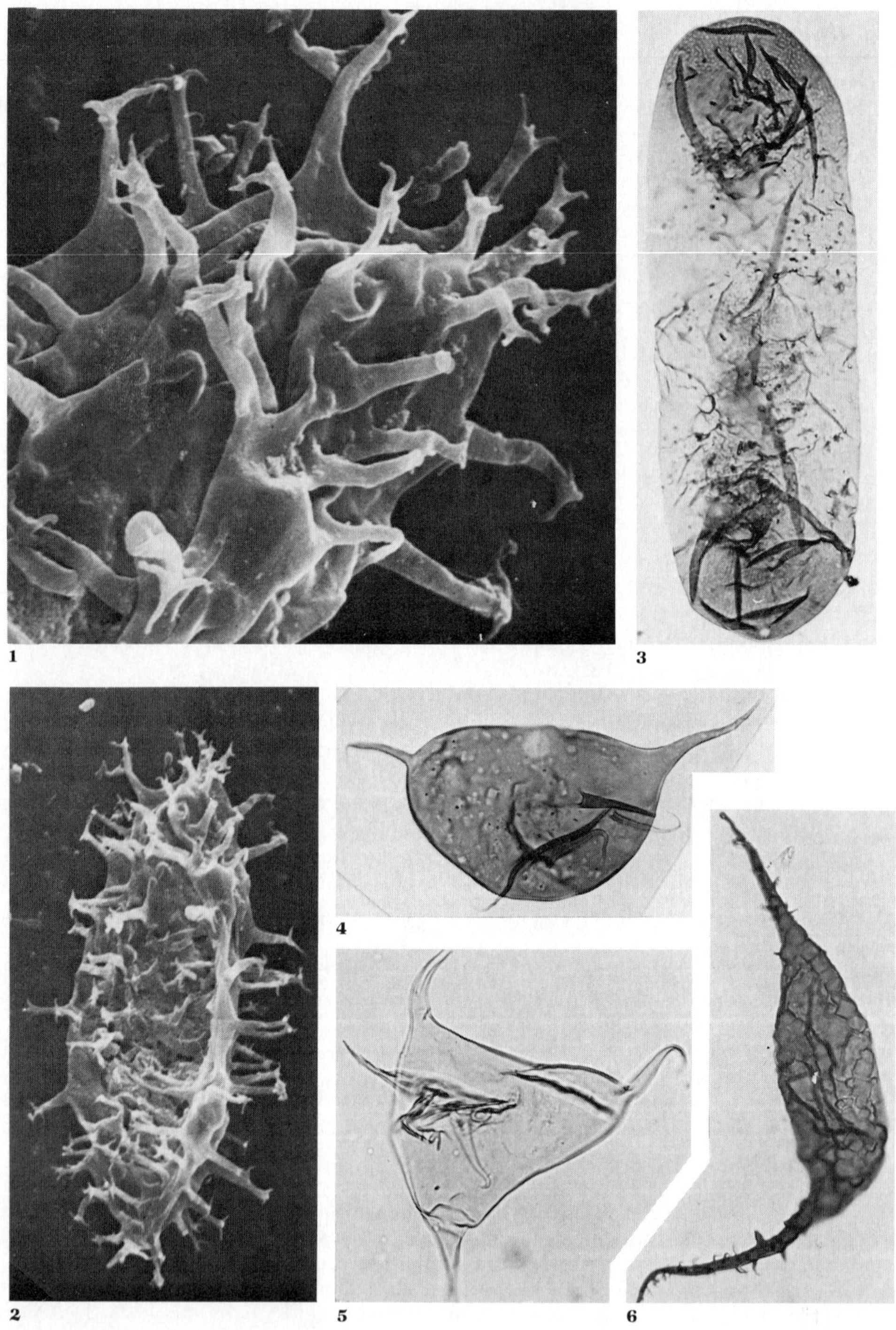

Figure 3.6

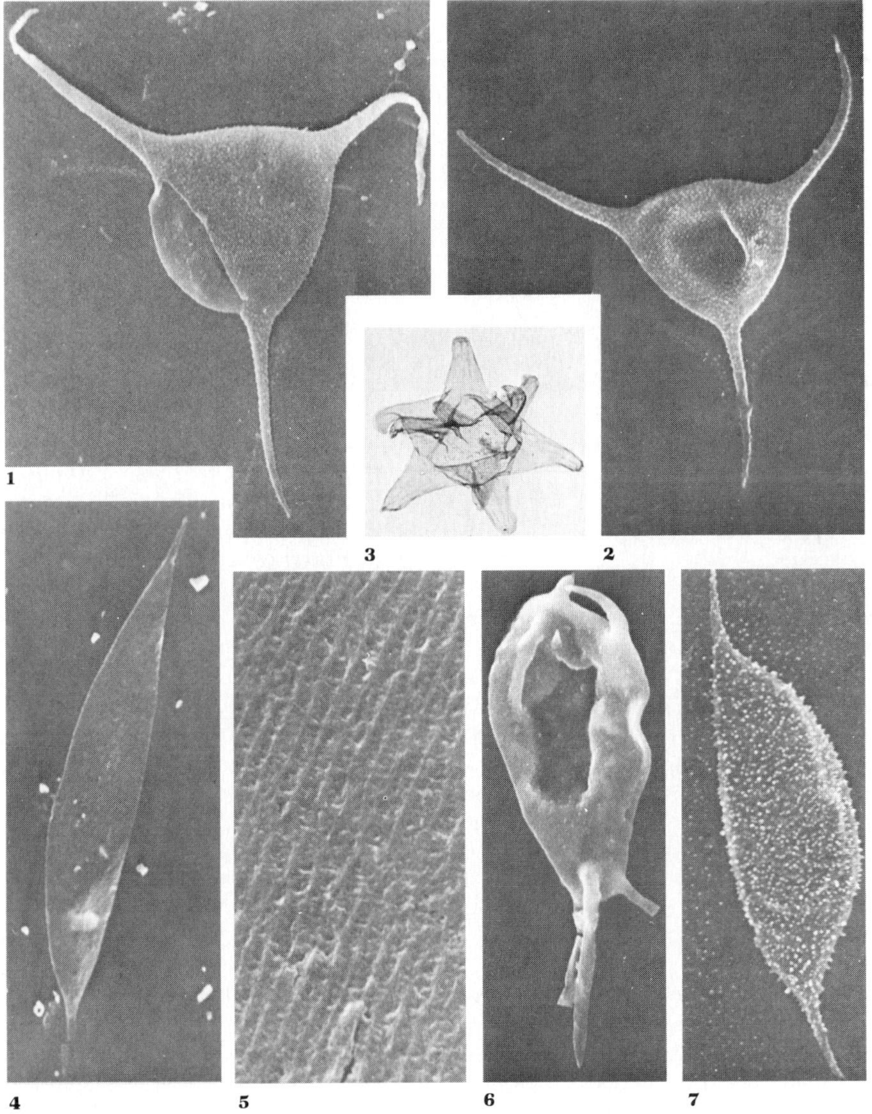

Figure 3.7
Middle Ordovician acritarchs, Oklahoma. **1,2.** *Veryhachium irroratum* Loeblich &
Tappan, SEM, ×800. **1.** With open epityche (compare Figure 3.19 part 4).
3. *Rhiptosocherma improcera* (Loeblich) Loeblich & Tappan, ×307, from
Loeblich, 1970b. **4,5.** *Cleithronetrum cancellatum* Loeblich & Tappan. **4.** Entire
vesicle, SEM, ×320. **5.** Portion of surface enlarged to show longitudinal ridges
and cross bars, SEM, ×2400. **6.** *Comptaluta pirellum,* SEM, ×600, from Tappan
and Loeblich, 1971b. **7.** *Disparifusa hystricosa,* entire vesicle (compare Figure
3.5 part 3), SEM, ×720, from Loeblich, 1970b.

Processes

One or more processes or appendages may project from the vesicle surface in random distribution, be restricted to the equatorial plane, or be present only at one or both poles of an elongate or ellipsoid vesicle. Processes range in length from short prickles to considerably more than the vesicle diameter, and in quantity from few to countless numbers. That this number may vary somewhat within a species was regarded by Eisenack (1963b) as precluding a dinoflagellate affinity. An individual specimen may have all processes alike or may have processes of two or more types, even showing considerable variety. The process nature generally is highly characteristic of the species, however.

Solid Processes

The simplest in appearance, solid processes are generally of small diameter, commonly appearing threadlike or hairlike as in *Elektoriskos* (see Figure 3.8), but may be subconical and blunt tipped, or even extremely elongate with multifurcate and anastomosing tips, as in *Tunisphaeridium*. They appear to arise from the outer wall layer if more than one layer is present.

Trabeculae

Thin, cylindrical, and apparently solid projections, or **trabeculae**, repeatedly branch and anastomose to form complex arches, as at one pole of *Ooidium* (see Figure 3.9). Originally thought to be tubercles, they differ from the processes of other acritarchs, and their true character was determined only with the SEM (Loeblich, 1970b, figs. 26, 27).

Tubular Processes

Some processes are hollow for all or part of their length, the cavity of the process interconnecting with that of the vesicle (see Figure 3.10). Such processes may be sharply separated from the main vesicle body, as in *Actinotodissus*, or may merge insensibly into the main body, as in many *Veryhachium*, *Leiofusa*, and *Micrhystridium*. All processes are closed distally, communicating only with the vesicle interior.

Processes may be open and tubular in part, but closed proximally by a plug or a solid or perforated diaphragm that prevents intercommunication with the vesicle interior (see Figure 3.11), as in *Orthosphaeridium* or *Ozotobrachion*. Some processes appear to be partially secondarily filled, as reported by Eisenack (1963a) for *Goniosphaeridium polygonale* (Eisenack) Eisenack and *Multiplicisphaeridium digitatum* (Eisenack) Eisenack. The solid processes were suggested to represent a more advanced or mature ontogenetic stage, and the hollow ones to represent immature cysts (Staplin et al., 1965). However, although the extent of the filling in processes may vary within individuals of a species, or even in different processes of a single specimen, no species are known to have both hollow and solid processes, suggesting that this feature is a diagnostic one for the particular taxon.

Some acritarchs have thick-walled vesicles and thinner-walled processes that appear to arise solely from the periphragm, or outer wall layer

---

Figure 3.8

Solid processes. **1,2.** *Ecthymapalla echinata* Loeblich & Wicander, Lower Devonian, Oklahoma, from Loeblich and Wicander, 1976. **1.** Equatorial focus, showing thick wall and solid processes, ×1300. **2.** SEM, showing subconical blunt-tipped processes, with proximal ridges; also note granulate-rugulate surface of vesicle; portion of excystment opening visible at top; ×6000. **3,4.** *Gorgonisphaeridium plerispinosum* Wicander, Upper Devonian, Indiana. **3.** Showing elongate tapering solid processes, ×1300. **4.** SEM, showing simple conical spinules, ×10,000. **5.** *Elektoriskos aurora* Loeblich, Middle Silurian, New York, showing elongate solid but flexible processes, ×960.

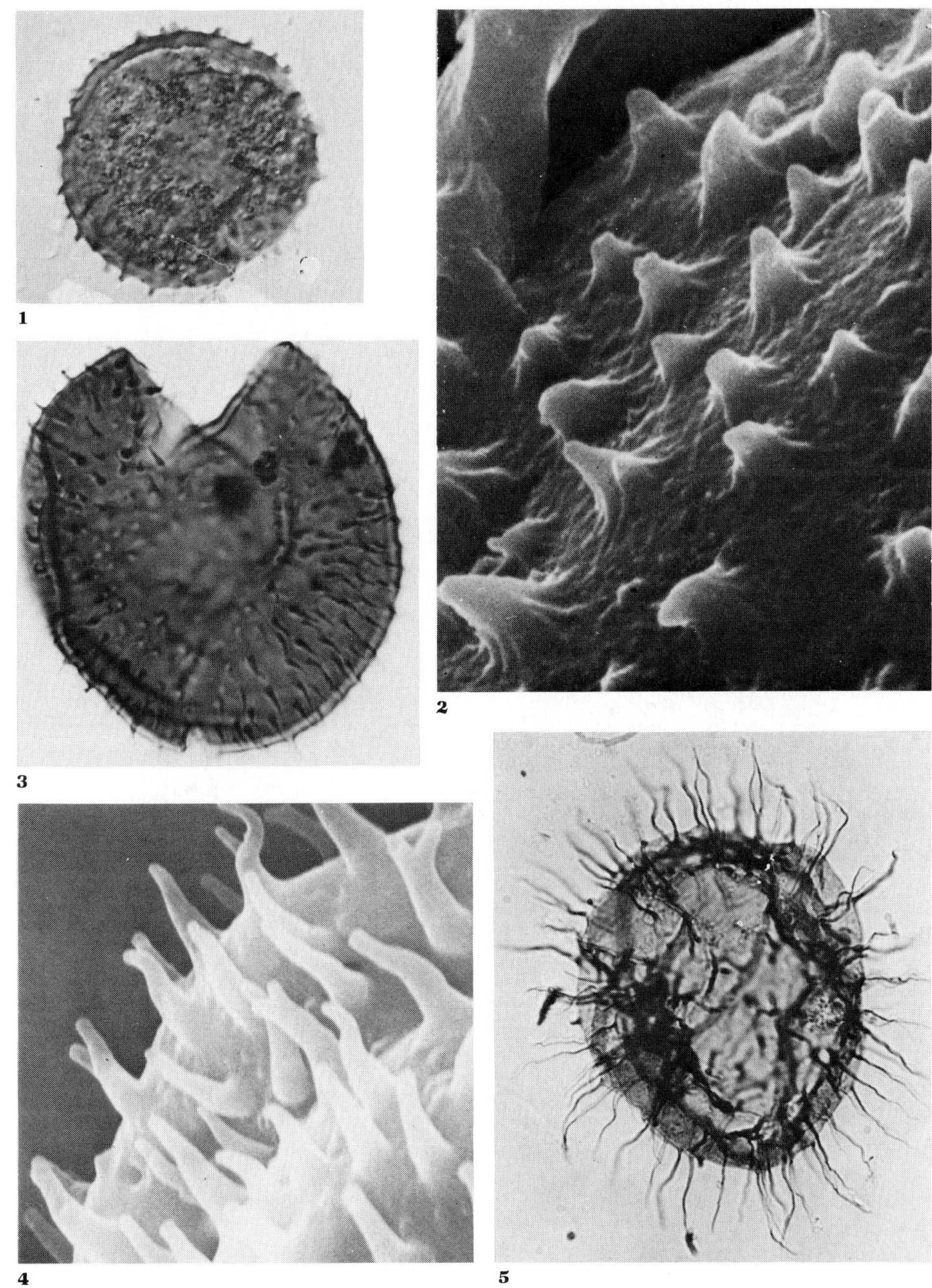

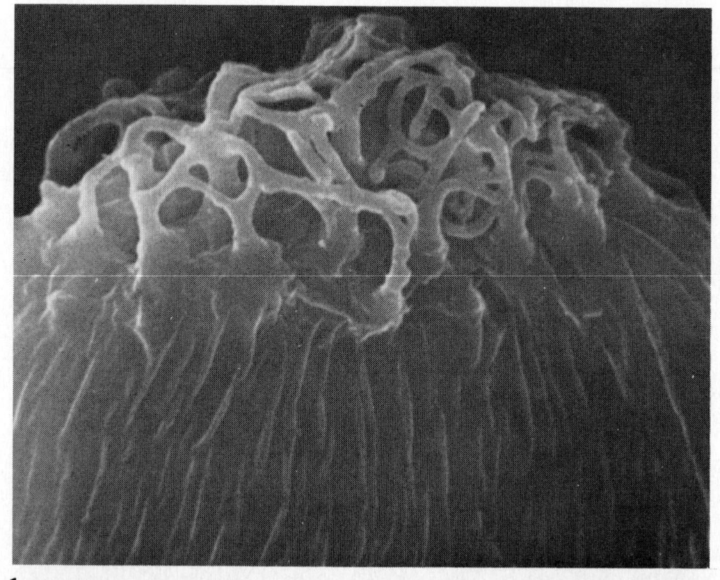

1

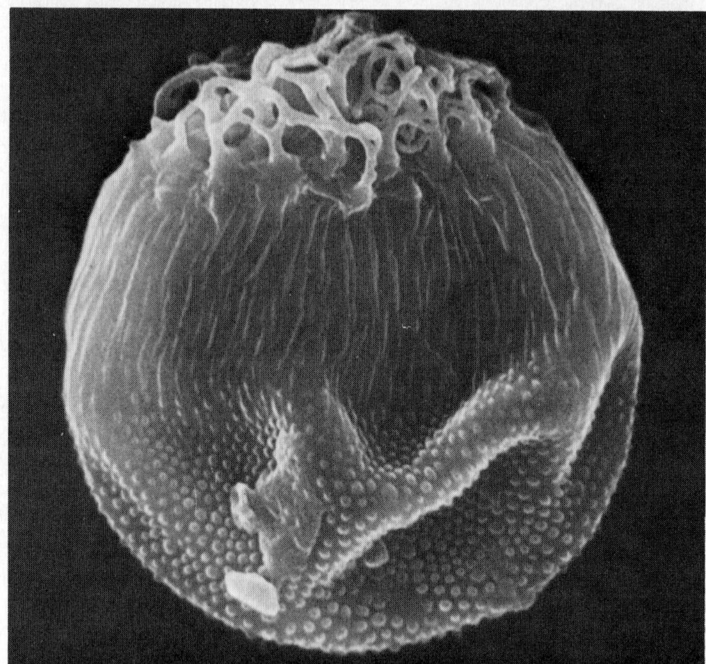

2

Figure 3.9

Anastomosing trabeculae. 1,2. *Ooidium rossicum,* Lower Ordovician (Tremadocian), USSR. **1.** Detail of trabeculae, SEM, ×4000. **2.** Entire vesicle, showing prominent granulate base merging into longitudinal discontinuous ridges in the upper half, with numerous solid, anastomosing trabeculae at top; SEM, ×2400. Both from Loeblich, 1970b.

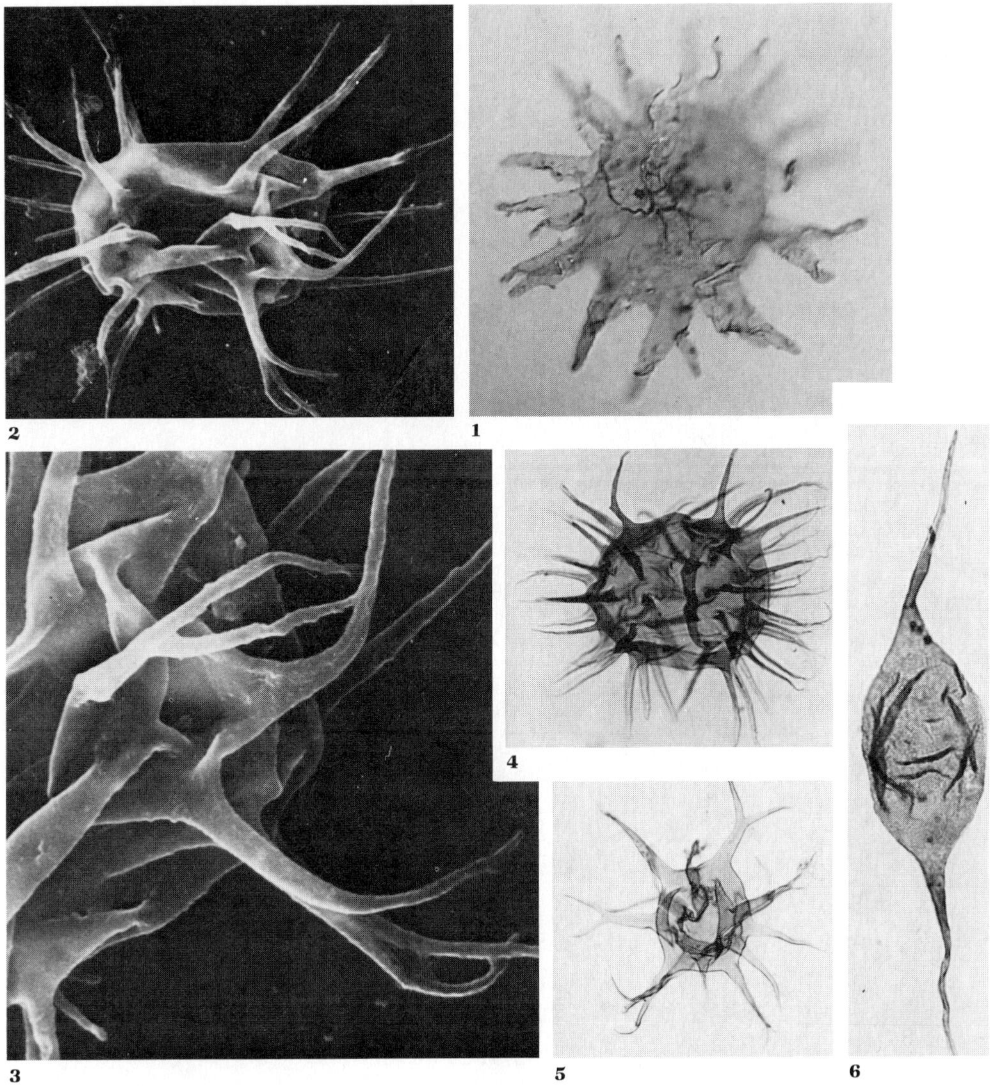

Figure 3.10

Open tubular processes, connecting with vesicle cavity. **1.** *Sol radians* Cramer, Upper Devonian, Ohio, processes radiating in one plane, ×1040. **2—4.** *Actinotodissus longitaleosus*, Middle Ordovician, Oklahoma. **2,3.** SEM showing elongate vesicle, with simple processes arising at each pole, rarely bifurcating; 2, ×1600; 3, ×4800. **4.** Showing hollow processes interconnecting with vesicle cavity, ×768. **5.** *Multiplicisphaeridium* cf. *bifurcatum* Staplin, Jansonius & Pocock, Upper Ordovician, Oklahoma, with simple and bifurcating processes, ×414. **6.** *Leiofusa granulacutis*, Middle Silurian, New York, showing long flexible polar processes that merge proximally into the vesicle cavity, ×768, from Loeblich, 1970b.

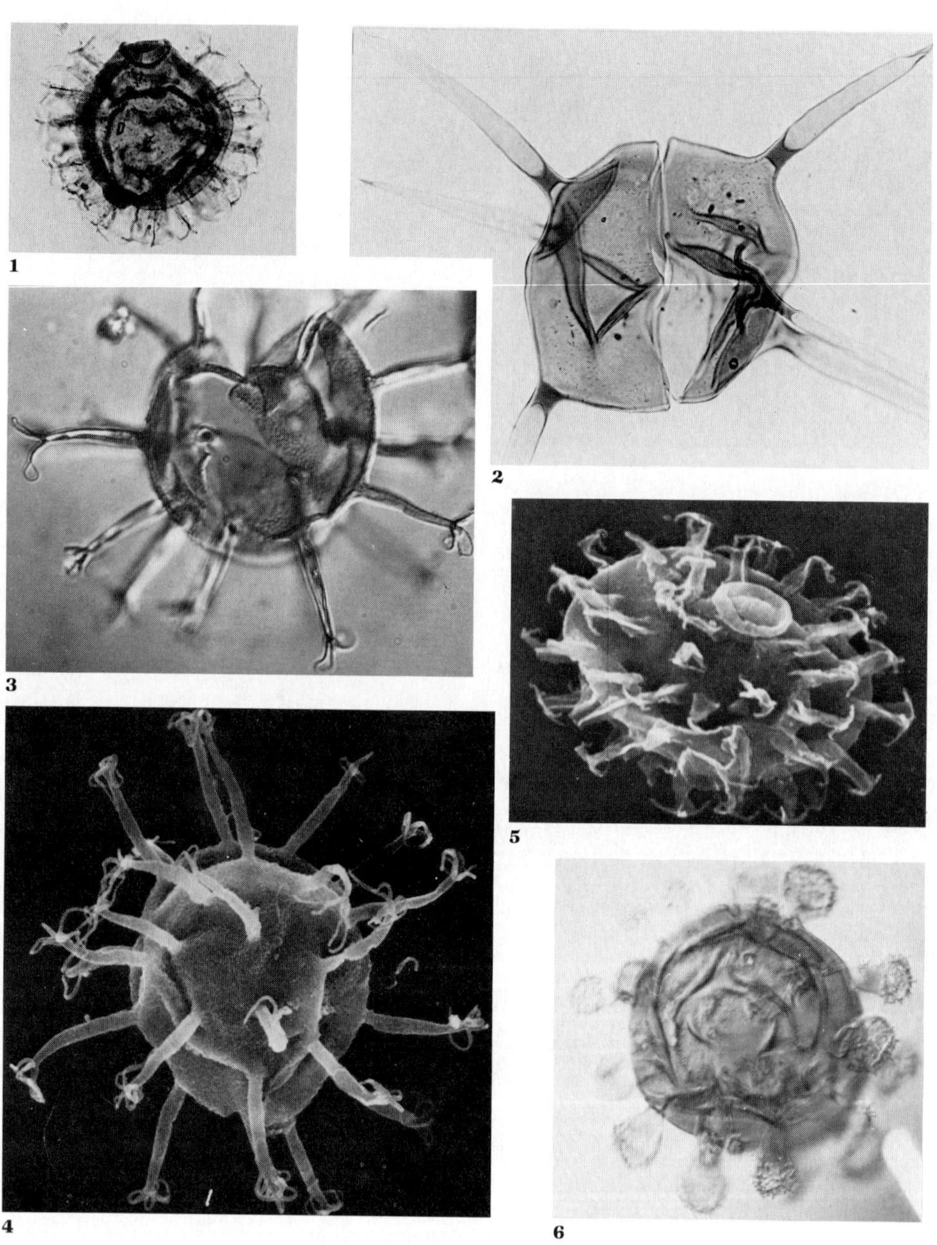

Figure 3.11

Processes that do not communicate with vesicle interior. **1,5.** *Polyancistrodorus columbariferus* Loeblich & Tappan, Middle Ordovician, Oklahoma. **1.** Showing thick-walled vesicle with thin-walled triradiate processes formed solely of the periphragm, terminally furcate and recurved at the ends, cyclopyle containing operculum visible at top, and pseudopylome present at lower left; ×414, from Loeblich and Tappan, 1969. **5.** SEM, showing appearance of the processes and cyclopyle with operculum, ×800. **2.** *Orthosphaeridium insculptum* Loeblich, Upper Ordovician, Oklahoma, showing plugs that close the processes proximally, near the juncture with the vesicle; specimen shows median rupture that is the characteristic excystment mode; ×414. **3,4.** *Gracilisphaeridium encantador* (Cramer) Eisenack & Cramer, Lower Silurian, Gotland. **3.** Scabrate vesicle surface; elongate processes with terminal furcations that rejoin the base to form distinct loops; excystment mechanism is a simple rupture; ×1040. **4.** SEM, ×1600. **6.** *Visbysphaera dilatispinosa* (Downie) Lister, Lower Silurian, Gotland. Thick-walled vesicle, and thinner-walled short inflated clublike processes that do not connect with vesicle interior, crowned with tiny prickles and formed only of periphragm; ×414.

(see Figure 3.12), whereas the inner layer is smooth and continuous beneath the proximal end of the process. Occasionally a process may be detached, and the inner layer of the wall then is seen to continue without interruption across the opening in the periphragm resulting from loss of the process.

Processes range from short to elongate threadlike or bristlelike structures; they may be conical, tubular, highly inflated and saclike, digitate, petaloid, or flanged, and variously ornamented (see Figures 3.13 and 3.14). Some are complexly buttressed against the vesicle wall. The distal end of a simple process may taper,

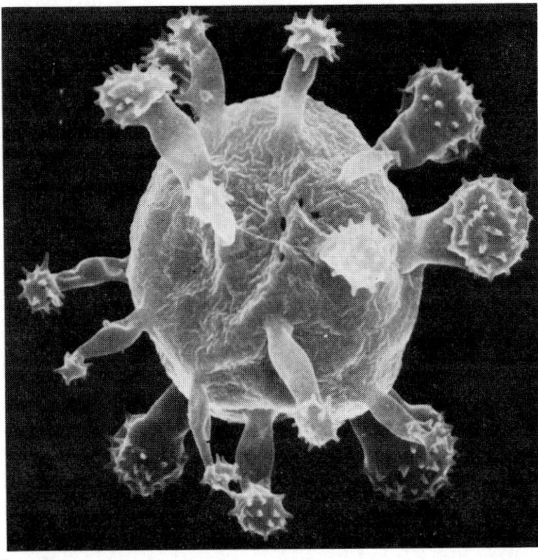

1

2

Figure 3.12

1. *Visbysphaera dilatispinosa.* Lower Silurian, Gotland, thick-walled vesicle with rugose surface and terminally inflated, thinner-walled, clublike processes; SEM, ×1600. **2.** *Psenotopus chondrocheus,* Middle Silurian, Indiana, SEM, ×800, from Tappan and Loeblich, 1971b.

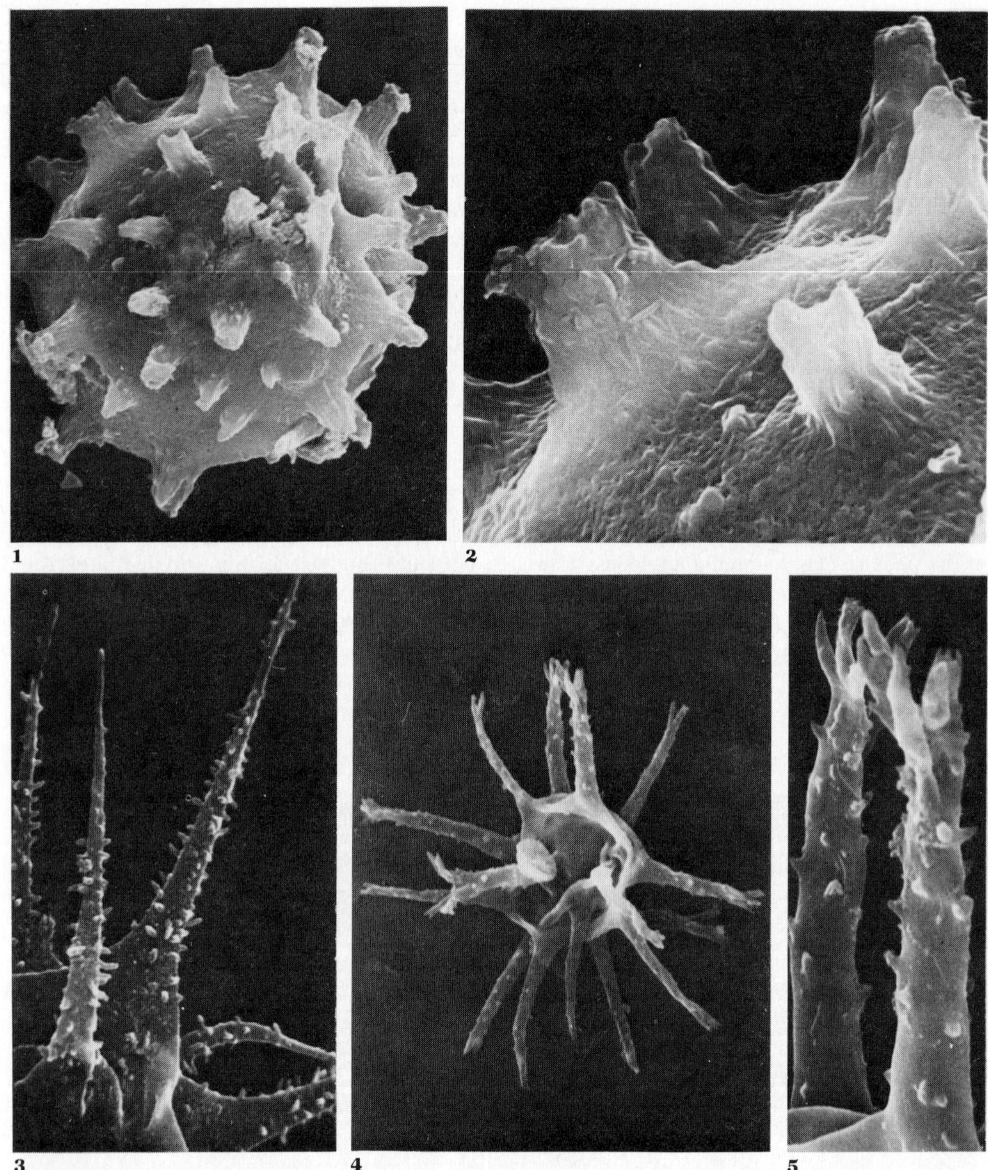

Figure 3.13
Ornamented hollow processes, in SEM. **1,2.** *Tylotopalla caelamenicutis*, Middle Silurian, New York. **1.** Entire vesicle, showing short, thick, blunt processes, ×1200. **2.** Part of vesicle, showing punctate surface, ridges extending along process length, and blunt to knobby process termination, ×3600 (see also Figure 3.5 part 6). **3.** *Polygonium* sp., Middle Ordovician, Oklahoma, showing numerous tiny spinules on the narrow elongate processes, ×1920. **4,5.** *Diexallophasis* cf. *denticulata* (Stockmans & Willière) Loeblich, Middle Silurian, Indiana. **4.** Entire vesicle, with elongate, terminally multifurcate processes covered with tiny spinules, ×560. **5.** Detail of two processes, ×2080.

whereas a complex one may bifurcate, or branch repeatedly to form an intricate network whose individual components have been separately designated (Cramer and Díez de Cramer, 1968). The basal part or **trunk** of a complexly branched process attaches to the main body of the vesicle; it may be cylindrical, taper or enlarge distally. The first subdivisions from the trunk are **pinnae**, the subdivisions or branches from the pinnae are termed first order **pinnulae**, and when present subsequent subdivisions are second (and third, etc.) order pinnulae (see Figure 3.15). Branching may be regular or irregular; processes may bifurcate or may be palmate, with two to five or more pinnae arising from the process trunk. In addition to the method of branching and the number of pinnae and pinnulae, other important features include the basal angle of the trunk (with the vesicle) and the angle of pinnulae and pinnae from their base, the form of the trunk (e.g., conical, cylindrical, clavate), and the relative size of the various components. In *Tunisphaeridium*, the palmately furcate processes anastomose and interconnect with the similar furcations of adjacent processes to produce a highly complex network (see Figure 3.16).

In correspondence with features of some chorate dinoflagellate cysts in which the position and number of processes reflect the position and number of thecal plates, Lister (1970b, p. 46) proposed a process formula for use in acritarch description. The polar axis and **apex** were determined by the position of the excystment opening. From the midline toward the apex, the "epicyst" processes were then divided into pre-equatorial (P), subapical (S), and apical (A), and on the "hypocyst" from the midline toward the posterior, they were termed equatorial (E), post-equatorial (Po), and antapical (An).

The process formula was then an indication of the number of each type, $A^1 S^5 P^{4-5} E^0 Po^{3-4} An^1$, for example, indicating one apical, five subapical, four to five pre-equatorial, no equatorial, three to four post-equatorial, and one antapical process. An attempt was made to follow such a procedure in the description of *Cymbosphaeridium bikidium* Lister. This formula implies the presence of a tabulate theca, such as is known to occur only in the dinoflagellates, but most acritarchs show no indication of such regularity of process position; a process formula thus appears to be of very limited usefulness in this group.

The taxonomic importance of the appendage character also has been controversial. All "spiny spheres," that is, all with processes, originally were placed in *Hystrichosphaera*. After typical hystrichospheres were shown to be dinoflagellates, those lacking evidence of such affinity were placed in *Baltisphaeridium*. Somewhat earlier, a lower size limit of 20 μm had been given for *Baltisphaeridium*, and the species less than 20 μm in diameter were placed in *Micrhystridium*. Although the type species of the latter genus was of Cretaceous age, many Paleozoic species were also placed in this genus. Frequency diagrams of Tremadocian acritarch size (Downie, 1958) were suggested to indicate a more natural size division at about 17 μm, as peaks occurred at about 10 μm and 25 μm. Staplin (1961) questioned the validity of any generic limit based on size alone, and separated from both genera those species with branching processes, as *Multiplicisphaeridium*. Downie and Sarjeant (1963) regarded both the spine terminations and the solid or hollow nature of the processes as highly variable even within a species, and did not accept the generic emendation of *Baltisphaeridium* and *Micrhys-*

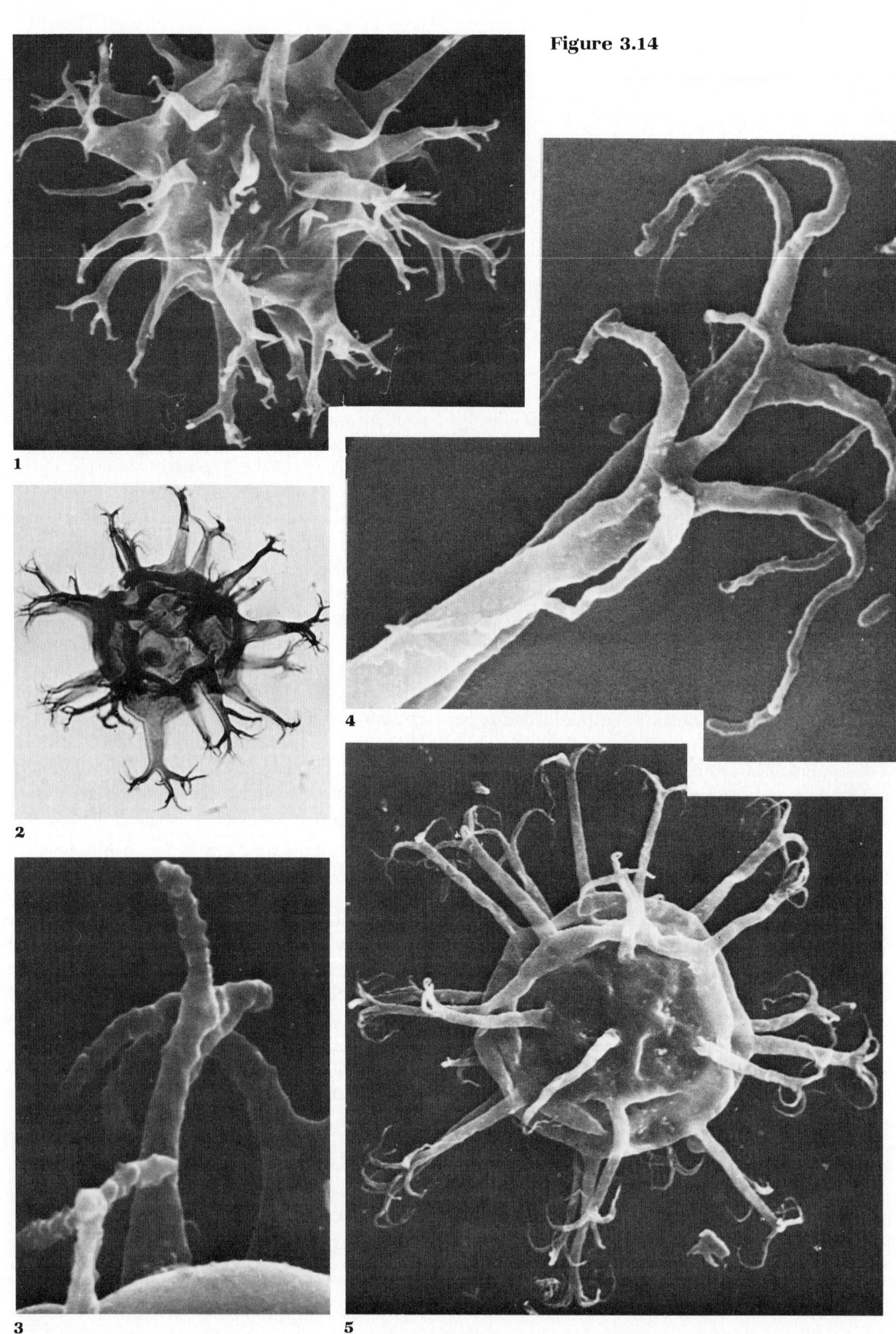

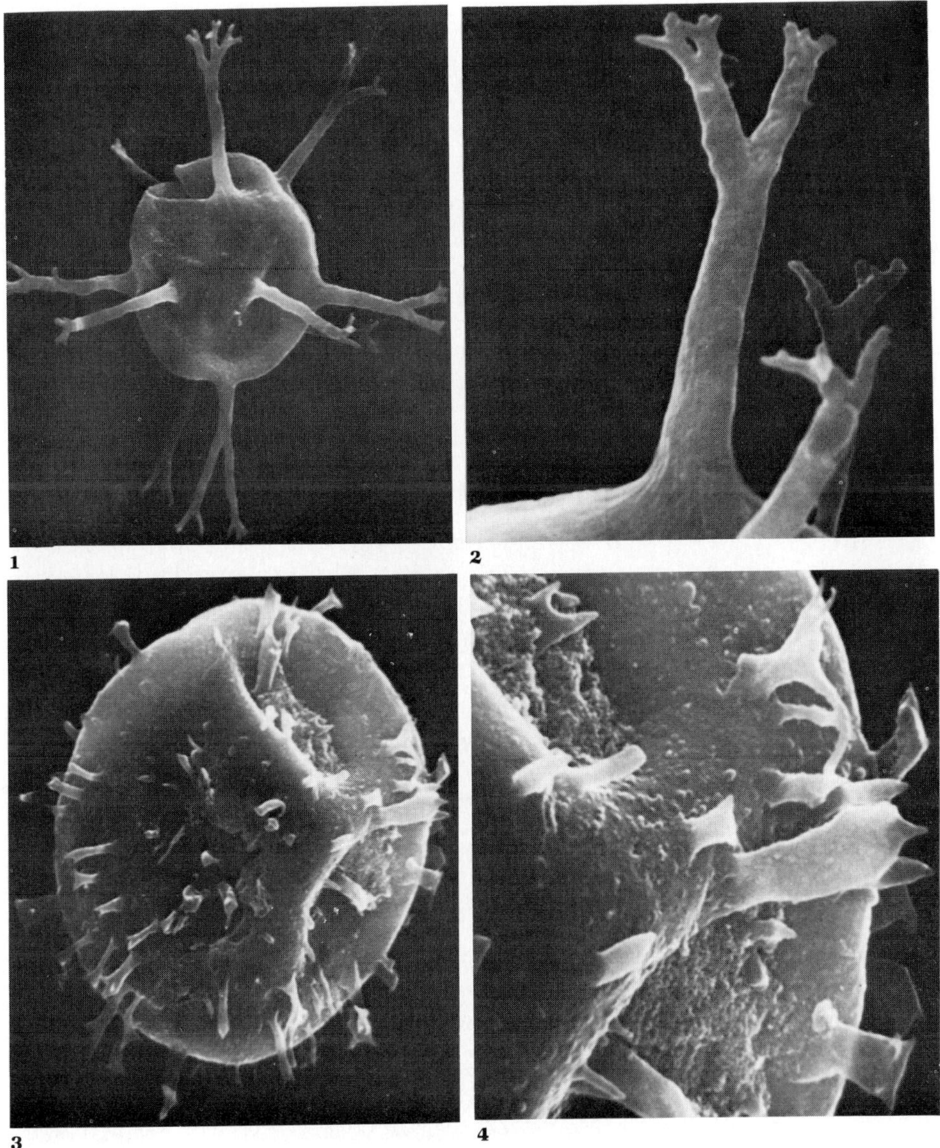

Figure 3.15

Middle Silurian acritarchs, New York, in SEM. **1,2.** *Multiplicisphaeridium fisheri* (Cramer) Lister. **1.** Entire vesicle, excystment split at top, and processes furcate to the second and third order, ×800. **2.** Enlargement of process, ×2880. **3,4.** *Visbysphaera erratica* (Eisenack) Lister. **3.** Entire vesicle, extremely variable processes flaring, tapering, cylindrical, to multifurcate, ×1280. **4.** Enlargement showing granulate vesicle and variable processes, ×3200.

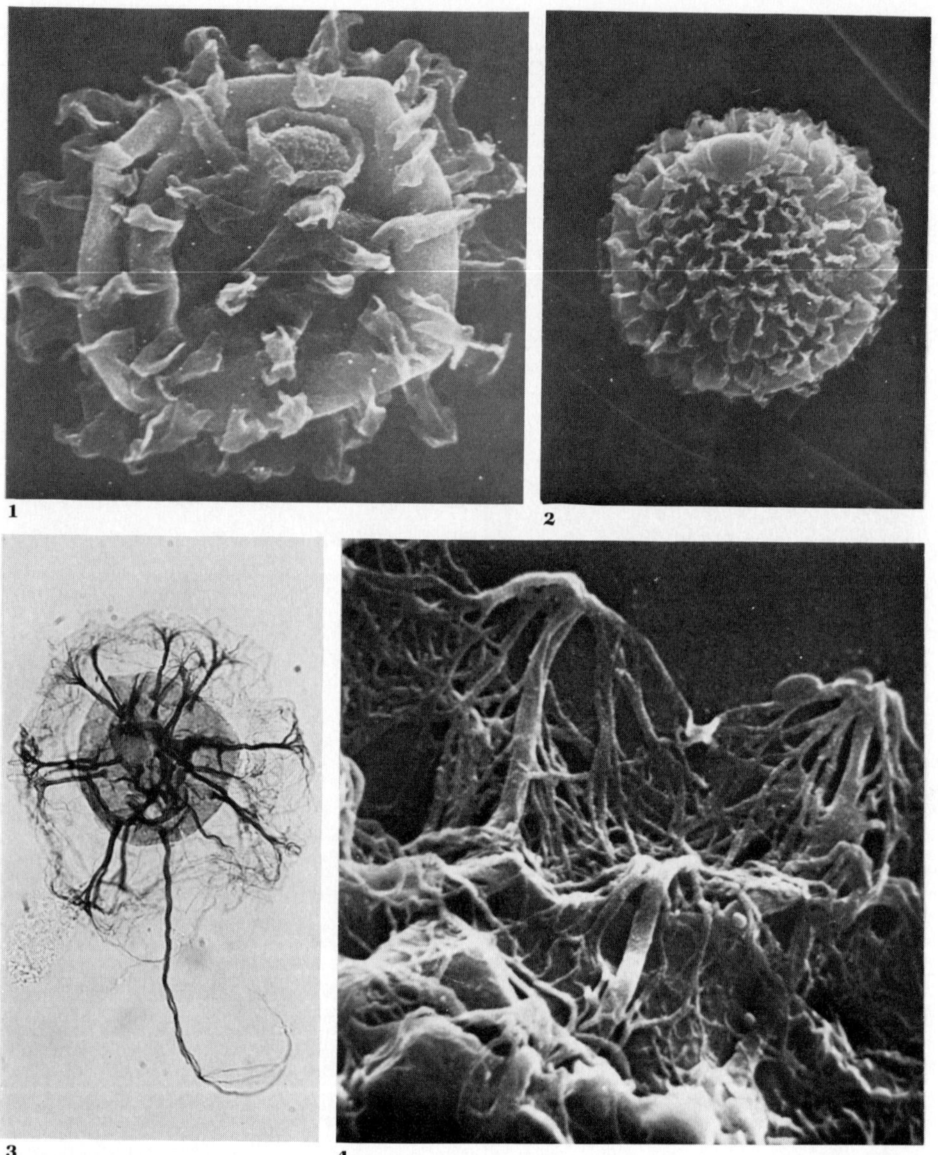

Figure 3.16

Buttressed and anastomosing processes. **1.** *Peteinosphaeridium* cf. *velatum* Kjellström, Middle Ordovician, Oklahoma, showing flangelike ribs on the processes, large pylome with neck, and granulate operculum still in place, SEM, ×1200. **2.** *Asketopalla formosula*, Middle Ordovician, Oklahoma, showing the short triradiate processes; cyclopylome with operculum visible in upper part of vesicle, SEM, ×800. **3.** *Tunisphaeridium caudatum* Deunff & Evitt, Middle Silurian, New York, showing equal-length solid processes that are palmately furcate and anastomose distally to form a netlike structure around the vesicle; one much longer process may extend free of the network for over half its length; ×768. **4.** *T. concentricum*, Middle Silurian, New York, SEM, ×5600, showing nature of the branching and anastomosing processes; detail of same specimen as shown in Figure 3.4 part 1.

tridium to include only species with simple process terminations. Later (Downie, 1973), *Multiplicisphaeridium* was recognized as distinct, although included with the *Micrhystridium-Veryhachium* group.

Filamentous Processes

Processes also may be extremely fine and elongate (see Figure 3.17). *Scuticabolus* (= *Ophiobolus* O. Wetzel 1933, non Reiss 1854) was described from Baltic Cretaceous cherts as a remarkably preserved ovoid cell 18 to 36 μm in length on which one or more elongate "flagella," about 5 to 10 times the length of the body and up to a length of 200 μm, were retained and preserved intact. The specimens may represent more than one species, although all have been referred to *S. lapidaris* (O. Wetzel) Loeblich III. "Flagella" were said to be apically attached, and one specimen was thought to represent a cell in process of division, as "flagella" were present at each end (Deflandre, 1936). *Dimastigobolus,* from the Upper Cretaceous (Senonian) of the Paris Basin, also was thought to be a soft-bodied flagellate (Deflandre, 1936), possessing two inequal flagella and a tegument with vermicular ornamentation.

Alberti (1961) later obtained similar specimens by acid maceration, disproving the flagellar nature of the elongate processes; he included them with the dinoflagellates and hystrichospherids. Similar well-preserved specimens were found in the Valanginian and Turonian of Germany (Alberti, 1961), up to the Maastrichtian, and in the Albian or Cenomanian to uppermost Cretaceous in France and North America (Canada, New Jersey, Alabama, Kansas, Texas) (Evitt, 1968). Preservation of true flagella would be truly remarkable in even one occurrence, as these organelles are seldom retained in preserved living material. In a restudy of the genus, Evitt (1968) found that one to seven processes commonly are grouped at one pole, but also may occur in a nonpolar position; they taper rapidly from the maximum diameter of about 0.25 to 0.75 μm near the attachment, then taper more slowly. The maximum length is uncertain, as they commonly are intricately twisted or tangled in a ball-like mass. Evitt concluded that the specimens represented egg cases of a planktonic organism, the filamentous processes perhaps aiding flotation by attachment to other plankton. In spite of his original definition of the acritarchs as including both animals and plants of unknown affinity, although treated under the Botanical Code, Evitt regarded this organism as an animal and thus retained for it the generic name *Ophiobolus* and family Ophiobolidae. However, neither in size, morphology, nor composition are egg cases of any known animals comparable to this organism, whereas similar cysts do occur in various modern and fossil algae. Because of the resistant nature of both wall and filaments, which remain intact after palynological preparation, an animal origin seems as impossible for these as Eisenack (1963a) argued for other acritarchs and "hystrichospheres." Elongate, flagellalike processes occur in other acritarchs, such as *Domasia, Pseudomasia, Geron,* and *Deunffia,* where the flagellumlike nature was noted by Downie (1960) in proposing the genus. Many specimens of *Scuticabolus* also show a slit or series of radiating slits at the pole opposite to the process attachment, which are strongly suggestive of the apical excystment openings of dinoflagellates and acritarchs. Because it fits the original definition of the Acritarcha quoted at the beginning of this chapter, and is certainly to be included among the algae and treated under the Botanical Code, the homonymous name *Ophiobolus* must be replaced by *Scuticabolus.*

Tubular Vellum

Some Silurian and Devonian acritarchs are characterized by a tubular **vellum,** formed of the outer wall layer attached near the midline of the spherical vesicle, which may flare outward toward one or both poles and appear skirtlike (see Figure 3.18). Riblike thickened areas may add strength to the vellum of *Chuttecloska athyrma* Loeblich & Wicander, which extends only in one direction; the two hemispheres are each covered with a flaring skirt in *Riculasphaera fissa* Loeblich & Drugg, and the tubular vellum is attached to the vesicle of *Carminella*

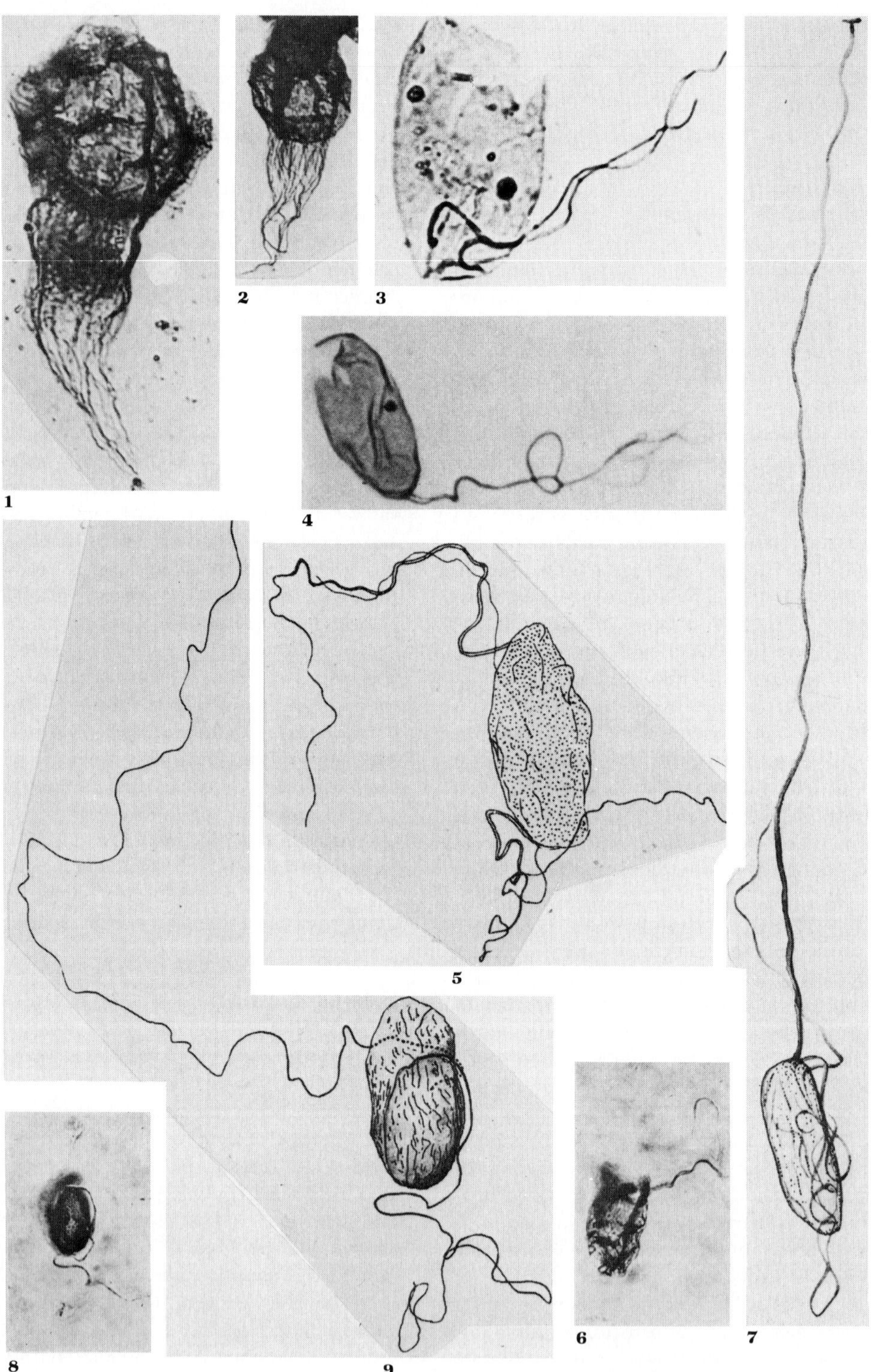

maplewoodensis F. H. Cramer by many processes of varied length, the longer processes also projecting from the vellum at its margins.

Excystment Openings

Early acritarch studies ascribed little importance to the presence or absence of an opening that was reported to occur in the vesicle of some species. Some Precambrian and Cambrian species were described from Russia as having a trilete opening such as characterizes spores of higher plants, but later workers generally could not confirm the presence of such openings except in rare smooth forms that probably represent red algal tetrad spores. Nevertheless, escape of the cell from the acritarch vesicle or cyst may be effected in various ways (Staplin et al., 1965; Eisenack, 1968a; Loeblich and Drugg, 1968; Loeblich and Tappan, 1969; Loeblich, 1970b; Lister, 1970b; Vavrdová, 1976). Some algae reproduce by both sexual and asexual means during different generations or stages in the life cycle of a single species; if both generations involved cyst formation, theoretically the two might have distinct means of protoplast escape, but no evidence is available from living species in support of this suggestion.

The opening produced in excystment is the only asymmetrical feature imposed on the basically radial symmetry of some acritarchs; in correspondence with those having a preformed aperture, the position of any excystment opening is regarded as defining the apical end of the vesicle. Some have maintained an even closer agreement with the terminology of the dinoflagellates, regarding this terminal or apical opening as defining the anterior end and the lateral openings as lying astride the dorsal midline (Lister, 1970b). However, Cramer (1970c) described *Hemibaltisphaeridium* as having processes and pylome at the posterior pole. Lister also regarded the half of the vesicle that contained the opening as the epicyst, and that opposite as the hypocyst (corresponding to the epitract and hypotract respectively of the dinoflagellates). Although some similarities exist (composition of the wall, cyst character, possession of excystment openings, varied surface ornamentation, processes, and flanges), most early Paleozoic acritarchs show no convincing evidence of dinoflagellate affinity, although many have been transferred to the Prasinophyceae. Other acritarchs may represent different or even extinct algal divisions, hence a more neutral terminology appears preferable for those that cannot be confidently assigned as yet.

Terminal or Lateral Rupture

The simplest and most primitive method of excystment involves simple **rupture** of the vesicle (Loeblich and Tappan, 1969) at a weakened area that is not visible as such prior to excystment. The split is only partial, the vesicle gaping open but remaining in one piece (e.g., in *Micrhystridium, Unellium,* and *Baltisphaeridium;* see Figure 3.19). Care must be taken to distinguish such openings from mere accidental breaks in damaged specimens, cracked by the process of sample preparation or damaged during fossilization (Eisenack, 1974). Accidental openings of this type may be recognized by the lack of conformity in different individuals of a species. In contrast, identical rupture of a large number of

---

Figure 3.17
Filamentous processes. **1,2.** *Geron guerillerus* Cramer, Middle Silurian, Spain. 1, ×860; 2, ×540; from Cramer, 1966a. **3 – 7.** *Scuticabolus lapidaris* (O. Wetzel) Loeblich III. **3.** Upper Cretaceous, Canada, ×1000; reproduced by permission of the National Research Council of Canada, from the *Canadian Journal of Earth Sciences,* v. 12, p. 1157 – 1174, from J. H. Wall and Singh, 1975. **4.** Danian, Paleocene, Jutland, ×850, from Morgenroth, 1968. **5,6.** Upper Cretaceous, France. **7.** Upper Cretaceous, Germany, showing less than one-third total length of filament. 5, ×1000; 6,7, ×900; from Deflandre, 1936. **8,9.** *Dimastigobolus longifilum* Deflandre, Upper Cretaceous, France. 8, ×450; 9, ×1000; from Deflandre, 1936.

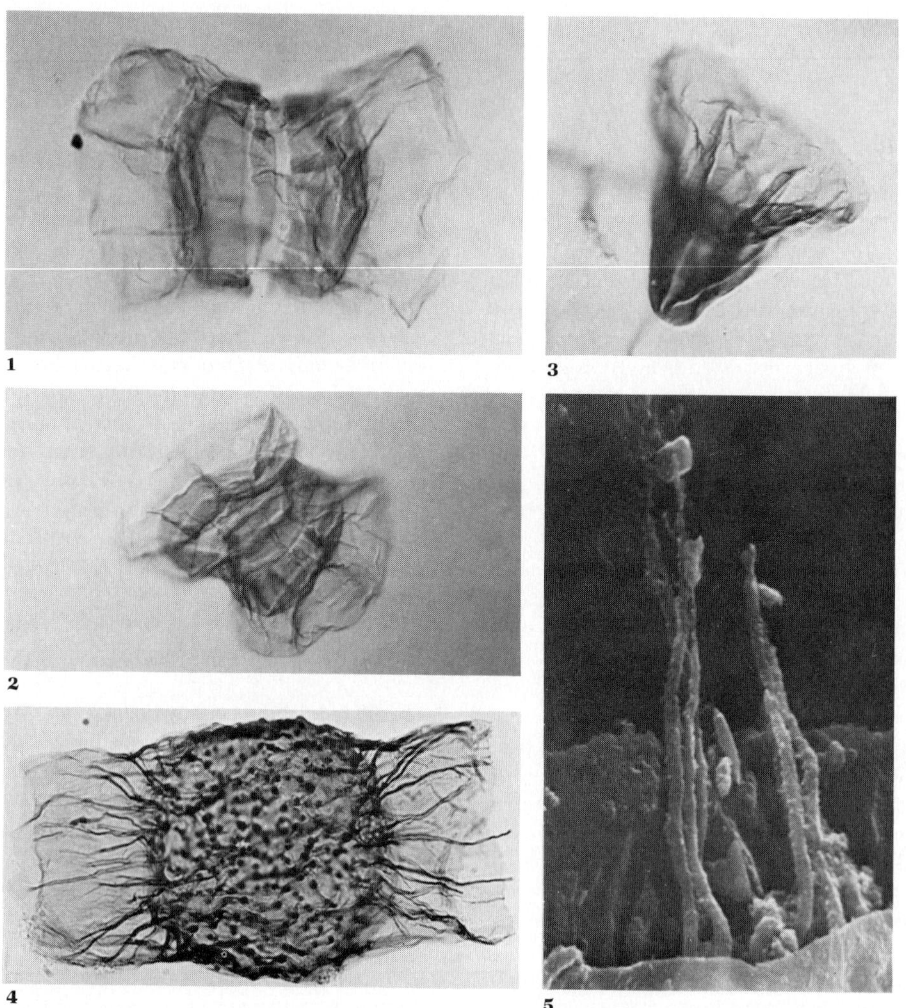

Figure 3.18

Vellum in acritarchs. **1,2.** *Riculasphaera fissa,* Lower Devonian, Oklahoma. **1.** Globular vesicle, with median split as method of excystment, the tubular, skirtlike vellum present on each half, ×1040. **2.** Entire specimen, ×600. **3.** *Chuttecloska athyrma,* Lower Devonian, Oklahoma; pyriform vesicle, with flaring vellum extending in only one direction, and strengthened by riblike thickenings, ×1040, from Loeblich and Wicander, 1976. **4,5.** *Carminella maplewoodensis,* Middle Silurian, New York. **4.** Entire vesicle, to which tubular enveloping vellum is attached by means of many short processes; processes longer toward the poles, and may extend beyond the edges of the vellum, ×768. **5.** SEM, showing elongate solid, finely granulate, narrow processes at edge of vellum, some entirely free, but with fragments of the vellum still attached at the distal ends, ×4800.

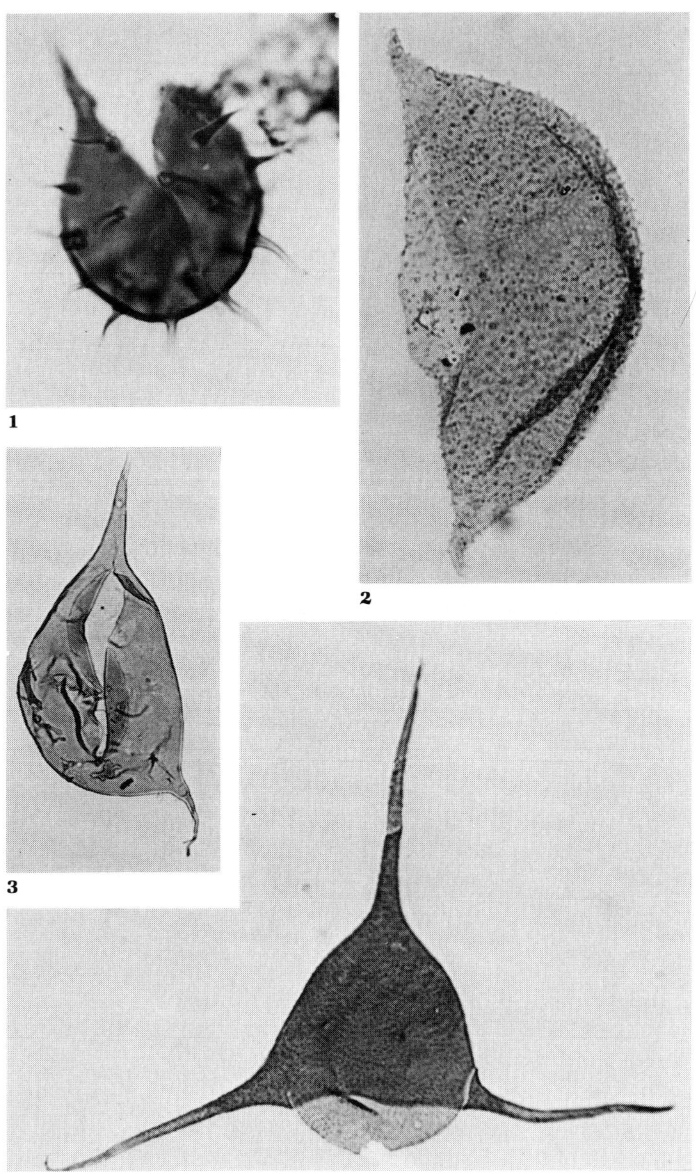

Figure 3.19
Excystment mechanisms; terminal or lateral rupture, epityche.
1. *Unellium ampullium* Wicander, Upper Devonian, Indiana, showing simple splitting of vesicle adjacent to the elongate process, ×1300, from Wicander and Loeblich, 1977. **2.** *Disparifusa hystricosa*, Middle Ordovician, Oklahoma, showing elongate slit at left, on the shorter, straighter margin, ×1300. **3.** *Sylvanidium paucibrachium*, Upper Ordovician, Oklahoma, with elongate slit on one of the lateral faces, ×384, from Loeblich, 1970b. **4.** *Veryhachium irroratum*, Middle Ordovician, Oklahoma, showing flaplike epityche and granulate surface, ×960.

well-preserved specimens probably represents their natural opening mechanism.

Certain elongate and fusiform acritarchs may have a lateral split or furrow that probably was for excystment (*Arbusculidium, Disparifusa, Petalosporites, Fusilites, Sylvanidium*); it is commonly longitudinal and less frequently transverse in orientation. Other fusiform or ovate taxa show other ways of excystment. *Scuticabolus* may have one or more slitlike ruptures at the end (apical?) opposite the filament attachments.

Extensive ruptures in the Tremadocian *Acanthodiacrodium ubui* F. Martin were suggested to indicate a possible tabulation, comparable to the development of a dinoflagellate archeopyle (Lister, 1970a), but this also has been interpreted as an accidental breakage resulting from compression in burial. Approximately half the observed specimens showed such an opening, although this seemed to differ in different individuals. If tabulation was reflected, a series of four equatorial plates, one antapical, and one or two apical plates may have been present, but the evidence is as yet inconclusive.

Epityche. A modification of the simple rupture, the **epityche** (Loeblich and Tappan, 1969), is a broad flap produced by an arched fissure, which folds outward and down like the flap of an envelope (see Figure 3.19, part 4). This slitlike rather than rounded excystment opening was illustrated on Ordovician and Silurian *Veryhachium* (Loeblich and Tappan, 1969; Loeblich, 1970b). It similarly characterizes the genus in Devonian, Permian, Triassic, and Cretaceous strata. The epityche differs from both the rupture and the simple pylome in its characteristic position and form, and differs from the latter generally in showing no indication of its future position prior to the onset of the excystment process, although some specimens show a slightly thinner line at this position that may indicate incipient excystment.

Occasional specimens have a small median tonguelike projection from the epityche margin, as in *Veryhachium irroratum* (Loeblich and Tappan, 1969, pl. 3, fig. 1; pl. 4, fig. 2). Termed the ventral tongue (Lister, 1970b), it is reflected in the opposite margin by a corresponding ventral notch.

Equatorial Rupture or Median Split

Certain acritarchs with bipolar symmetry separate into two discrete halves upon excystment (see Figure 3.20), a somewhat more complex method of excystment that involves a complete median or equatorial split of the vesicle. Entire specimens thus represent individuals that were fossilized without excysting. This method is characteristic of the Ordovician *Orthosphaeridium*, the Devonian *Riculasphaera* and *Fimbriaglomerella*, and the Permian *Circulisporites* and *Peltacystia*. Previously described as an equatorial rupture or dehiscence (Balme and Segroves, 1966), median splitting (Loeblich and Tappan, 1969), or a cingular cryptosuture (Lister, 1970b), the designation as a **median rupture** or split is descriptive without implying taxonomic relationships. The term *cingular* implies (without justification) the presence of the transverse flagellar groove or cingulum of the dinoflagellates, yet no dinoflagellate is known to have a cyst comparable to any of the above-mentioned acritarchs. Similarly, the term *dehiscence* is used in pollen and spore terminology, and might also have an unwarranted connotation.

Cryptopylome

Loss of a distinct portion of the former wall of the acritarch vesicle during excystment may leave a rounded to subangular opening, the **cryptopylome** of *Cymbosphaeridium pilar*. As in the simple rupture and the epityche, the position or nature of the cryptopylome is not apparent prior to excystment, hence all three features were together regarded as a "cryptosuture" (Lister, 1970b). The latter term may imply similarity to formation of the dinoflagellate archeopyle, and as sufficient evidence of plates and sutures, or of dinoflagellate affinity, is not available for taxa retained as acritarchs, a term that refers to the opening itself rather than to an implied origin is preferred.

Figure 3.20

Excystment by median rupture. **1,2.** *Peltacystia venosa* Balme & Segroves. **1.** Slightly oblique polar view, showing characteristic ornamentation of nodes and ridges. **2.** Inner view of one of the two halves after the median separation, showing the equatorial rows of nodes toward top of figure. **3,4.** *P. calvitium* Balme & Segroves. **3.** Polar view of entire vesicle, showing equatorial and median nodular ridges, apical boss, and reticulate surface. **4.** Equatorial view of specimen showing only partially complete median rupture. All from the Upper Permian, Western Australia, SEM, ×1600.

Pylome or Cyclopyle

The most obvious excystment structure is the **pylome**, a relatively large circular or ovate opening termed a **cyclopyle** by Eisenack (1969), although it also may be subpolygonal. The margin of the opening may be slightly thickened, and the opening lateral in position as in *Cyclopsiella* or *Leiovalia* (see Figure 3.21), or terminal as in *Saharidia* or *Ascostomocystis*. More commonly the border is marked by a low collar as in *Peteinosphaeridium* and *Priscogalea*, a tubular neck as in *Aremoricanium,* or flaring rim as in *Polyancistrodorus.* The opening is subquadrangular in *Impluviculus.* Unlike the previously mentioned excystment methods, the pylome is preformed. Prior to excystment, the opening is closed by a pluglike **operculum** which may be preserved still in place, and which is reminiscent of the superficially similar feature in the much smaller siliceous chrysophyte cysts. At excystment, the operculum may be extruded and lost in *Asketopalla* or *Polyancistrodorus* (see Figure 3.22), remain attached at one side of the opening as in *Operculites* and some *Cyclopsiella,* or collapse and be preserved within the vesicle body. All openings in acritarchs were originally regarded as pylomes, but Sarjeant (1967, p. 203) restricted the term to the circular openings, and it is here further restricted to exclude the cryptopylome.

"Pylomes" also were reported (Eisenack, 1968a) in *Tasmanites* and *Leiosphaeridia,* genera now included with the Prasinophycean green algae rather than with the Acritarcha. As their openings were regarded as homologous with those of the acritarchs, Eisenack suggested that the openings formed only during the last stage of ontogenetic development. Even the true pylome (Überhöhtes Pylom) was said to be constructed during the escape of the protoplast (Eisenack, 1963a, p. 124), and thus was not present in specimens that might have been fossilized prior to excystment. *Baltisphaeridium* was said to have a pylome only rarely, although breakage of a process at the contact with the vesicle might simulate a small round pylome (as also noted by Downie, 1958, p. 344; Eisenack, 1968a). Although still reported to possess a pylome (Eisenack, 1969), *Baltisphaeridium* probably excysts only by means of a terminal rupture. Some individuals that were illustrated as having a pylome instead are damaged specimens from which one of the long tubular processes was torn. Acritarchs with a true pylome or cryptopylome are now placed elsewhere, for example, *"Aremoricanium" cylindrosum* (Eisenack) Eisenack, Cramer & Diez, or *Cymbosphaeridium pilar.* These are quite distinct in morphology from Prasinophycean cysts. As specimens that were fossilized prior to excystment have the operculum still in place, the original formation of the pylome during the construction of the cyst appears indisputable. A similar argument holds for those rare specimens with twinned or double pylomes. These have been reported in *"Hystrichosphaeridium diploporum* Eisenack" illustrated by Eisenack, 1951, pl. 2, fig. 6), which is neither a *Hystrichosphaeridium* nor a *Baltisphaeridium, "Peteinosphaeridium breviradiatum* (Eisenack) Eisenack" (illustrated by Eisenack, 1968a, pl. 1, fig. 9), *Dicommopalla macadamii* (Loeblich, 1970a, figs. 3, 4) and *Asketopalla formosula* (Loeblich and Tappan, 1971a, pl. 103, figs. 1, 3).

Figure 3.21
Pylomes. **1,2.** *Dicommopalla macadamii,* Upper Ordovician, Indiana. **1.** Showing two pylomes, ×960, from Loeblich, 1970a. **2.** SEM, ×1500. **3.** *Axisphaeridium* sp., Upper Ordovician, Indiana, showing pylome at periphery at top, SEM, ×1000. **4.** *Cyclopsiella vieta* Drugg & Loeblich, Oligocene, Mississippi, pylome with operculum in place, ×480, from Drugg and Loeblich, 1967. **5,6.** *Priscogalea* sp., Lower Ordovician (Tremadocian), USSR. Two levels of focus, showing very large pylome with slightly raised and crenulated border, ×960, from Loeblich, 1970b. **7,8.** *Asketopalla formosula,* Middle Ordovician, Oklahoma, at two levels of focus, ×518. **7.** Showing pylome about one-third distance above the lower margin. **8.** Focused on opposite side, showing pylome with operculum at upper center on periphery.

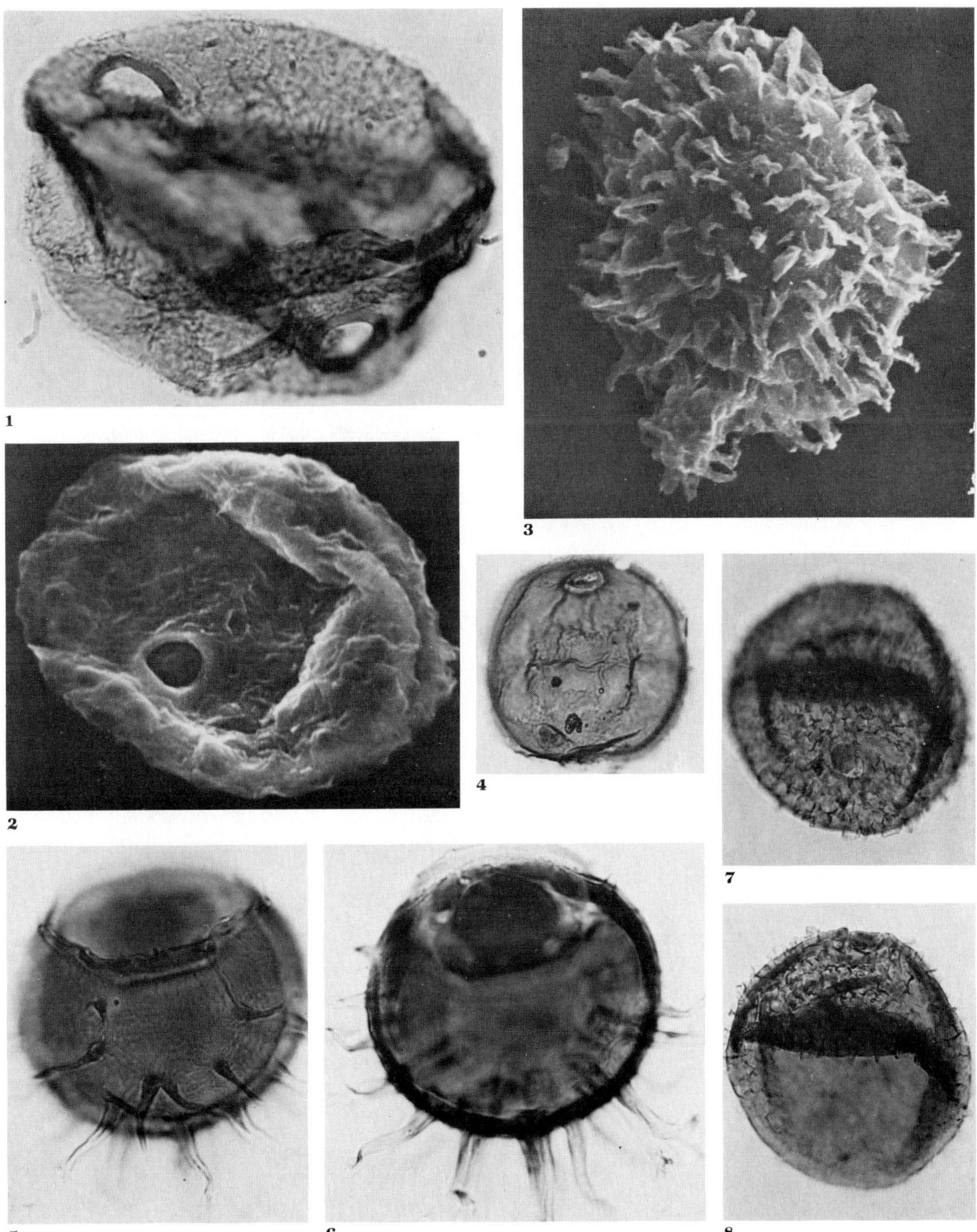

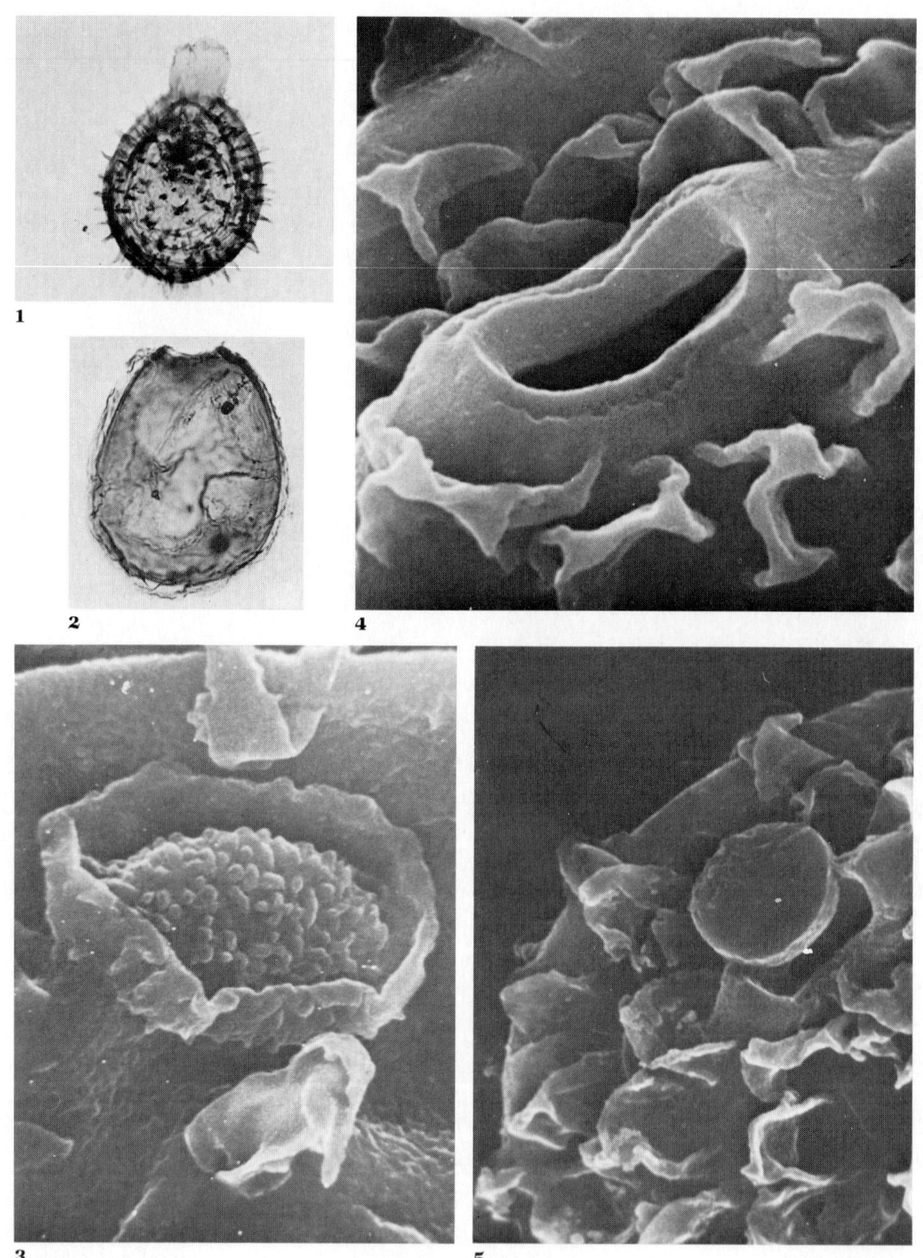

Pseudopylome

A second structure may be present at the aboral
or antapical end, opposite to the true pylome,
in *Polyancistrodorus*. This **pseudopylome** resembles a nonfunctional pylome and may represent an abortive attempt to construct an
opening (see Figure 3.11, part 1). The pseudopylome is short, narrow, and cylindrical,
and its margin is thick walled; it has no true
opening or operculum.

SYSTEMATIC POSITION
OF THE ACRITARCHS

Students of the acritarchs generally are aligned
in two groups, those regarding the group as
highly polyphyletic, including both plants and
animals classified in artificial morphological
"subgroups"; and those believing them to be of
undoubted algal origin, either representing
closely related organisms (Eisenack, 1963a, and
later), for which the term Hystrichophyta was
proposed (Mädler, 1963, 1967), or various algal
kingdoms and classes, possibly including the
Chrysophyceae, Dinophyceae, Chlorophyceae,
and Phaeophyceae, and perhaps additional extinct major taxa (Fisher, 1953; Volkova, 1965;
Staplin et al., 1965; Downie, 1967; Tappan, 1968;
Loeblich, 1970b; Tappan and Loeblich, 1971a).

The earliest report of these fossil organisms
was that of M. C. White (1862), who illustrated
specimens from the Ordovician (Black River

limestone), Silurian, and Devonian (Corniferous
limestone) of western New York, regarding
them as diatoms, sponge gemmules, or
sporangia of the desmid *Xanthidium* following a
similar consideration of fossils in Cretaceous
flint chips (which later were proven to be dinoflagellate cysts). Other spiny spheres such as
the *Sphaerosomatites* of Rothpletz (1880) variously were regarded as radiolaria or as
silicoflagellates but may also have been acritarchs; their original figures and illustrations
are insufficient for determination, and the
types were destroyed during the war.

Eisenack (1931, 1934, 1938a,b) described
many species from Ordovician and Silurian erratic pebbles in the Baltic glacial deposits, at
first referring them to *Bion* or *Ovum*, as possibly representing eggs of planktonic organisms.
From modern seas, Lohmann (1904) had described a great variety of floating eggs of
copepods, fish, and other organisms to which
the fossil specimens bore a superficial resemblance. Eisenack (1931) discussed previous interpretations as to the relationships of the
Paleozoic forms, including the suggestion that
they might represent eggs of the extinct graptolites or trilobites, bryozoan statoblasts, or desmids and considered them to be cysts, observing that their chemical nature was quite distinct
from that of animals, as well as different from
desmids.

Additional studies of European Paleozoic taxa
(O. Wetzel, 1932; Deflandre, 1942, 1946; Deunff,
1951, 1954a–c, 1955a,b) generally regarded
them as "hystrichospheres," resembling the
Mesozoic forms that are now considered to be

dinoflagellates. Other species were illustrated and described in America as desmids and other freshwater taxa (Laird, 1935; Baschnagel, 1942) or as hystrichospheres (Deunff, 1957).

Specimens described as having a surficial triradiate mark from the early Paleozoic of Russia (Naumova, 1949; Timofeev, 1956b, 1959) and the late Precambrian (Timofeev, 1958, 1960a), were regarded as trilete spores of vascular plants (Bryophyta or Pteridophyta), or as primitive Psilophytales (Timofeev, 1960c). Others stated that specimens with trilete marks were absent in rocks older than Silurian, regarding those figured by the Russian workers as resulting from misinterpretation of crushed and poorly preserved specimens (Volkova, 1965; Ważyńska, 1967). Tiny triradial marks on sporelike bodies in the upper Precambrian Bitter Springs Formation of Australia (Schopf, 1970) were regarded as indicating red algal affinity (Ceramiales). Volkova (1976) found similar specimens that had separated, bearing small triradiate tetrad scars, and considered them of algal origin.

As most of the tiny spherical or polygonal specimens of the lower Paleozoic lack such triradiate marks, Eisenack (1938b, 1951) became convinced of their close relationship, regarding them as phytoplankton on the basis of vesicle composition (he made chemical analyses of these in 1931), their many morphologic similarities, and the presence of excystment openings or pylomes, some with operculum in place, and others with operculum missing or having opened like a trapdoor. Such features eliminated any possibility of a relationship to the spores of higher plants. Fisher (1953, p. 15) regarded the lower Paleozoic "hystrichospheres" as spores of marine algae and suggested that their brown color might indicate Phaeophycean affinity. However, all fossil organic matter is yellow to brown in color unless coalified, whereas the color of modern algae is due to the pigments in the intracellular plastids and not to the appearance of the cell wall.

The resemblance of the fossil "hystrichospheres" to dinoflagellates was noted by O. Wetzel (1933, p. 77) and Deflandre (1947) as many new assemblages were described. Evitt (1961, 1963) then removed the typical hystricho-spherids to the dinoflagellates and proposed the Group Acritarcha for the non-dinoflagellate residue. Those removed as dinoflagellates had such distinctive features as a recognizable indication of tabulation, an archeopyle (excystment opening) produced by the removal of definite and constant plate equivalents, a median girdle, and so forth, although Evitt noted that other acritarchs also might prove to be dinoflagellates. Sarjeant (1970a, p. 246) remarked about the relatively late appearance of fossil dinoflagellate cysts in spite of the very primitive character of the organisms. Because the "decline of the acritarchs coincides with the increase of dinoflagellate cysts," he suggested that perhaps "the majority of acritarchs are dinoflagellate cysts." Lister (1970b, p. 38) stated that "the parallels shown by the excystment behavior of acritarchs and dinoflagellate cysts suggest that many acritarchs are either cysts of dinoflagellates or that they are the cysts of an extinct group of organisms related to the dinoflagellates and which exhibit a mechanism of excystment closely allied to that found in the cysts of dinoflagellates." He regarded the acritarchs as cysts of naked dinoflagellates, the differences between the hystrichospheres and acritarchs being the difference between armored and unarmored dinoflagellates. This hypothesis does not explain the scarcity of modern acritarchs in spite of the abundance of living unarmored dinoflagellates, and has not been substantiated in cultures.

In contrast to Sarjeant's statement of the coincident inverse relationship between acritarch and dinoflagellate diversification, many workers have noted the major early Paleozoic diversification and middle Paleozoic decline of the acritarchs (see Figure 3.23). Deunff (1961a, p. 45) observed that the Cambrian to Silurian was the most favorable of all geologic time for the development of the fossil microplankton, and Eisenack (1963b) noted that acritarch abundance continued into the Devonian. Many workers have noted the abrupt reduction in acritarch diversity from the Devonian into the Carboniferous (Eisenack, 1954a; Downie, 1967; Tappan, 1968, 1970; Tappan and Loeblich, 1971a, 1972, 1973a,b; Wicander, 1975), and their very low diversity in Permian and Triassic rocks

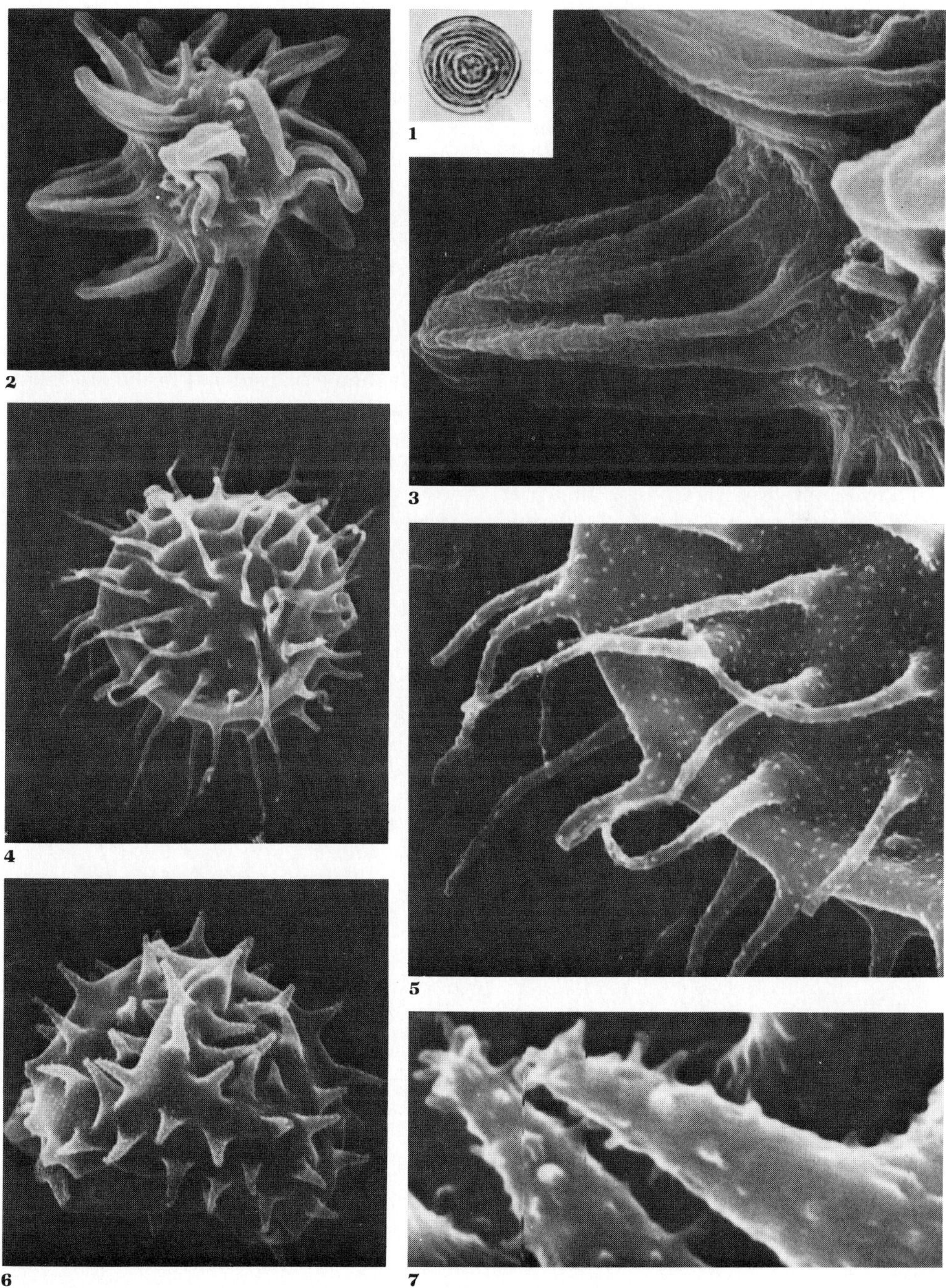

Figure 3.23

Permian acritarchs, Western Australia. **1.** *Circulisporites parvus* de Jersey, polar view of discoidal vesicle bearing concentric or spiralling ridges and furrows, ×600, from Segroves, 1967. **2,3.** *Mehlisphaeridium fibratum* Segroves, in SEM. **2.** Entire vesicle, with prominent ridges on processes, ×800. **3.** Process showing rugose surface, ×2400. **4,5.** *Micrhystridium* sp. in SEM, with many flexible processes. 4, ×1600, 5, ×4800. **6, 7.** *Micrhystridium* sp. in SEM with short conical processes. 6, ×1600. 7, Detail of spinulate processes, ×4800.

(see Figure 3.24). Dinoflagellate diversification in the Jurassic followed the acritarch decline after the lapse of some 170 million years, and no group of organisms is known to have undergone a second major radiation following near extinction during a long intervening period. Instead, each group that becomes extinct is replaced by a quite distinct or only distantly related group that diversified after the disappearance of its former competitor. It thus seems unlikely that the majority of early Paleozoic acritarchs are in fact peridinean dinoflagellates, although a few Paleozoic dinoflagellates are known.

Other groups (tasmanitids and leiospheres) that formerly were treated with the hystrichospheres have been shown to be related to the modern Prasinophycean green algae and are discussed with that group (Chapter 10). Recent studies (Jux, 1971) of their wall ultrastructure show that such Paleozoic acritarchs as *Baltisphaeridium, Peteinosphaeridium,* and *Goniosphaeridium* have densely spaced radially arranged canals in the wall, resembling the wall structure of the tasmanites but differing from that of the fossil dinoflagellates. Jux suggested that many acritarchs thus might be more closely related to the Prasinophyceae than to the Pyrrhophyta. Although this supports their separation from dinoflagellates, the nature of the wall structure is not as yet known for enough acritarchs to indicate whether this is characteristic of all, or of only a few, which as a result should be removed to the Prasinophyta. Acritarchs represent an encysted stage of some form of phytoplankton, but this alone is insufficient to indicate a relationship, for the cysts as well as the vegetative stages may be morphologically similar in various algal divisions. Modern algae may produce cysts as a regular part of the

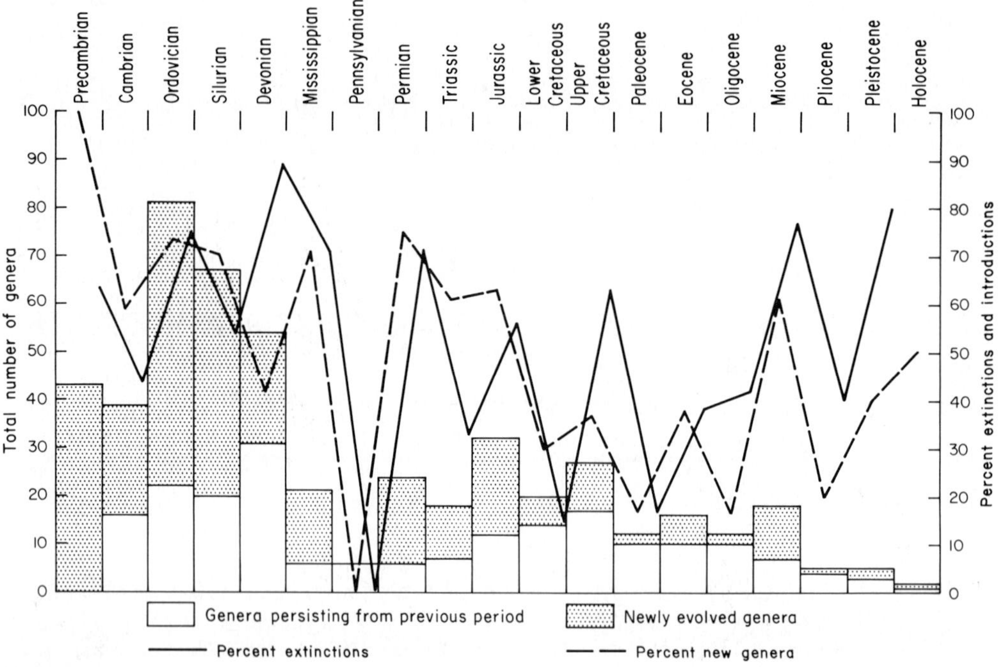

Figure 3.24
Generic diversity of acritarchs and Prasinophyceae through geologic time, showing number of known genera per period divided into those persisting from the preceding period and those newly evolved, and percentage of the generic assemblage that first appeared or that became extinct within each period. From Tappan and Loeblich, 1972.

reproductive cycle, commonly as resting zygote cysts after sexual reproduction; possibly others produce resting cysts as a response to unfavorable conditions. Although both the vegetative cell and cyst of diatoms and some coccolithophorids may be mineralized and preservable as fossils, naked vegetative cells of other organisms may produce highly resistant cysts. Eisenack (1969) considered the acritarchs as phytoplankton algae, but added that whether the fossils were of vegetative individuals or resting cysts was irrelevant for their classification, as they were entire, self-supporting organisms in either case, and thus should be described as true genera, species, and families under the botanical classification.

Cysts resembling one or another of the acritarchs are known in the Chlorophyta, Chrysophyta, and Xanthophyta, as well as the Pyrrhophyta. A simple flask with operculum or plugged neck opening is known in the green algae; for example, *Coelastrum* of the Chlorococcales, *Pleurangium* of the Chaetophorales, and *Acetabularia* of the Dasycladales, the latter being calcified. Similar, lightly silicified Chrysophycean cysts are known for *Chrysapsis, Thallochrysis,* and others. Cysts that separate medially to form two similar halves are known in *Chlorobotrys* and *Pseudotetraëdron* of the Xanthophyceae, and in *Coelastrum* and *Desmatractum* of the Chlorophyceae. A great variety of spiny, smooth, and reticulated cysts is now known for the dinoflagellates, including a few that are calcified as well as the more common organic-walled ones. Still greater variety may be recognized as information accumulates concerning the encysted stages of living species. Other smooth, spiny, or otherwise ornamented cysts or zygospores are known in the Chlorophyceae (for example, in *Sphaeroplea* and the desmids), the Prasinophyceae (*Halosphaera, Pterosperma*), and calcified in the Coccolithophorales.

Although the chemical nature of some cysts is known, such as the calcified ones of *Acetabularia* (Chlorophyta) and the Coccolithophorales (Haptophyta), or the silicified ones of the diatoms, Chrysophyta, and Xanthophyta, for many other algae this has not yet been determined. Too often, the cyst has not even been described for modern taxa, thus hindering comparisons for the fossil taxa. The presently available information concerning the fossil record suggests that no modern representatives exist of such early Paleozoic acritarchs as the diacrodians, or those with prominent preformed pylome (*Asketopalla, Axisphaeridium*). Perhaps relic species may eventually be found living that are related to such acritarchs as *Micrhystridium* or *Veryhachium,* and the pigmentation and other characters of the cell will then supply conclusive evidence as to the affinity of these.

OCCURRENCE

Geologic Distribution

Acritarchs are extremely abundant in rocks of early Paleozoic age, increasing rapidly in diversity and abundance. Many species of late Precambrian age also have been described from Western Europe, USSR, North America, Africa, and Australia; these are relatively simple forms, spherical, with smooth, pebbly, or reticulate surfaces, many probably being referrable to the Prasinophyceae. In the latest Precambrian and Early Cambrian, leiospheres dominated the assemblage, although a few small forms, generally 10 to 15 μm in diameter, have short solid processes. Rare in the latest Precambrian, these spiny forms increased to as much as 30 percent of the assemblage (Volkova, 1972) in the pretrilobite Lower Cambrian rocks. In the lowest trilobite-bearing Lower Cambrian strata, the acanthomorphitic (spiny, or process-bearing) acritarchs are much more important, somewhat larger in size, and some (*Baltisphaeridium*) have hollow, tubular processes. Upper Cambrian assemblages are better known, and Cambrian or Lowest Ordovician (Tremadocian) acritarchs have been described from the USSR (see Figure 3.25), England, Ireland, France, Germany, Sweden, the Baltic region, North Africa, and Canada. Eight faunal zones, characterized by the acritarch assemblages, have been proposed for the Cambrian through Early Ordovician Tremadocian in

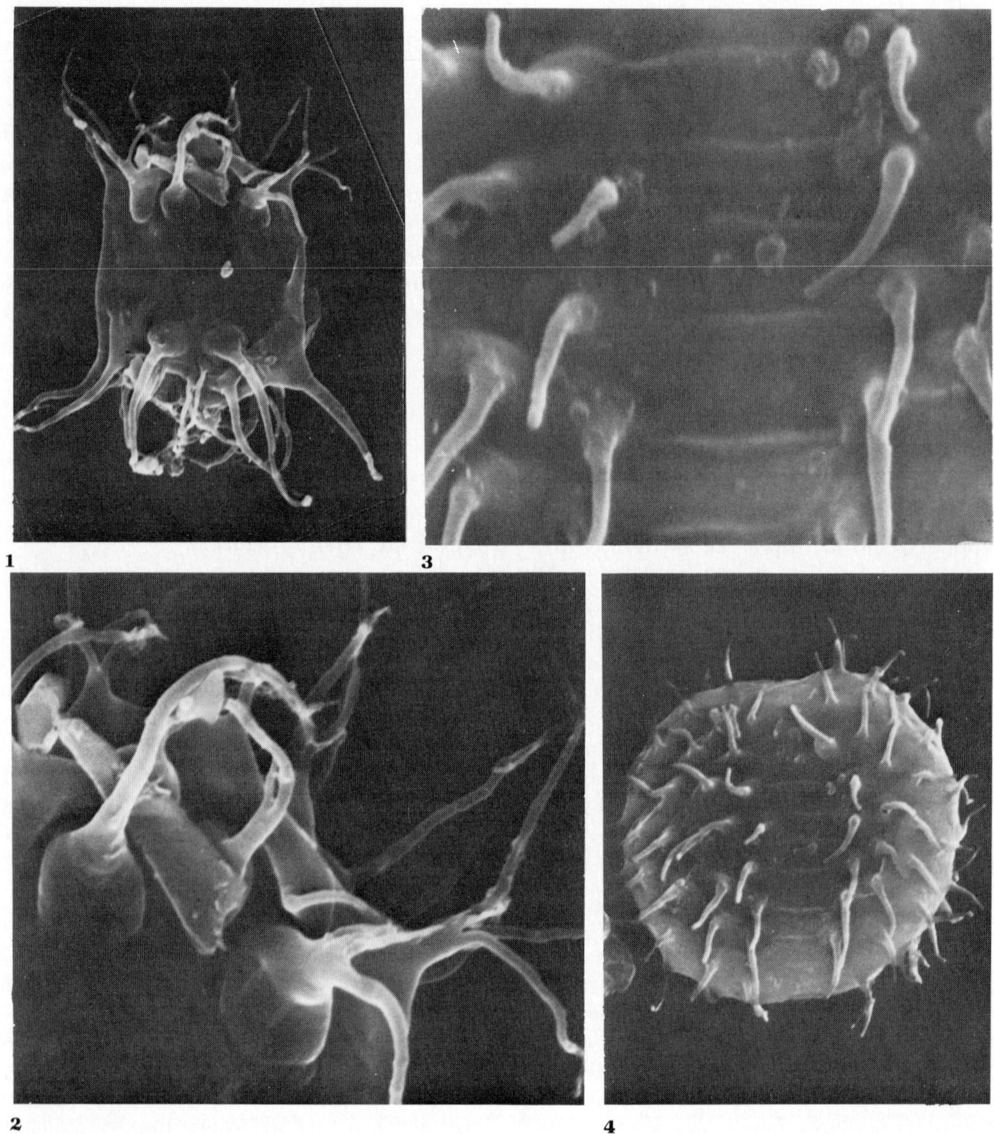

Figure 3.25
Lower Ordovician (Tremadocian) acritarchs from the USSR, in SEM. **1,2.** *Acanthodiacrodium selectum* Timofeev. **1.** Entire vesicle, ×1600. **2.** Enlargement of processes, ×4800. **3,4.** *A. numerosum* Timofeev. **3.** Enlargement of median region showing longitudinal ridges and short tapering processes, ×4800. **4.** Entire vesicle, ×1600.

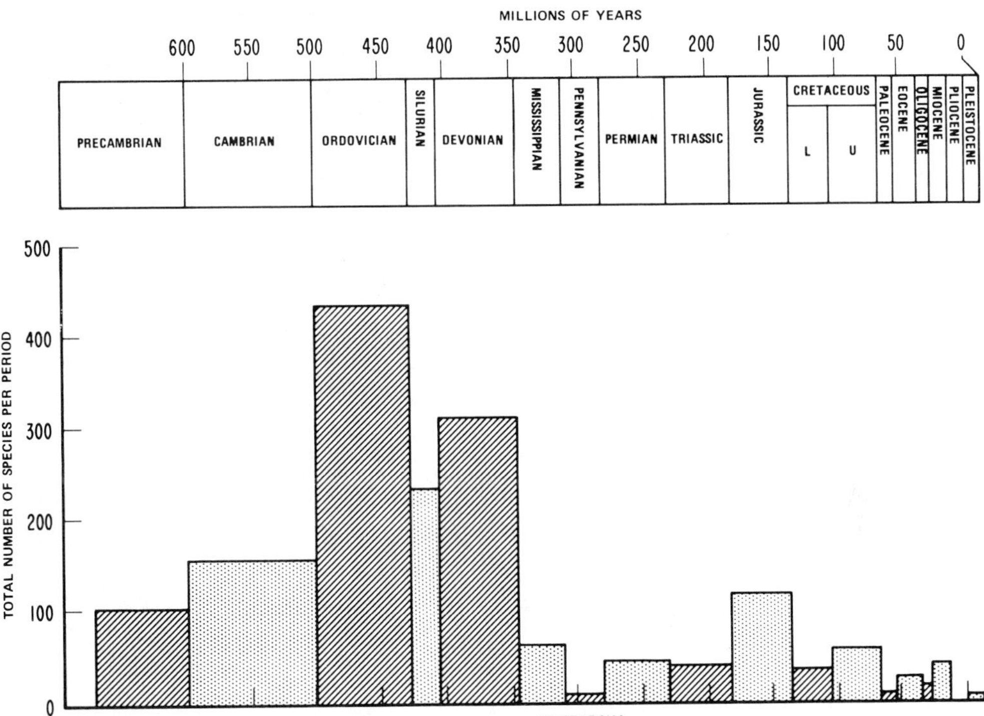

Figure 3.26
Acritarch species diversity totalled by geologic periods, with horizontal scale in millions of years. Note that Ordovician total diversity is about double that of the Silurian, although its duration was about three times as great. The marked increase from Cambrian to Ordovician is apparent, as is the even more sudden post-Devonian decline. From Tappan and Loeblich, 1973b.

France and Belgium (Vanguestaine, 1974), during which the acritarchs increased rapidly in abundance and diversity. The amount of organic matter reaches 4 percent of the total rock in the English Tremadocian (Downie, 1958), where acritarchs were estimated to occur in numbers of 100,000 per gram of rock. In French Ordovician rocks, the median acritarch number was determined in thin sections as about 500,000 per cubic centimeter of sediment (Deunff, 1959, p. 11). In Brittany, abundance ranged from 320,000 (about 50 percent *Cymatiogalea*) up to about 900,000 per cubic centimeter with three coequal genera, *Veryhachium, Micrhystridium,* and the prasinophyte *Cymatiosphaera,* comprising most of the assemblage (Henry, 1969).

Acritarch diversification was steady throughout the Ordovician, from the distinctive assemblages of the Early Ordovician (Tremadocian) through the Middle and Late Ordovician (see Figure 3.26). Ordovician acritarchs include many species with large cyclopyle and short, complexly formed processes, such as *Asketopalla* and *Polyancistrodorus* (see Figure 3.27); others had variously ornamented fusiform and elongate vesicles, or spherical or polygonal ones with few to many processes (see Figure 3.28). Middle Ordovician species are diverse and abundant in Oklahoma, and Upper

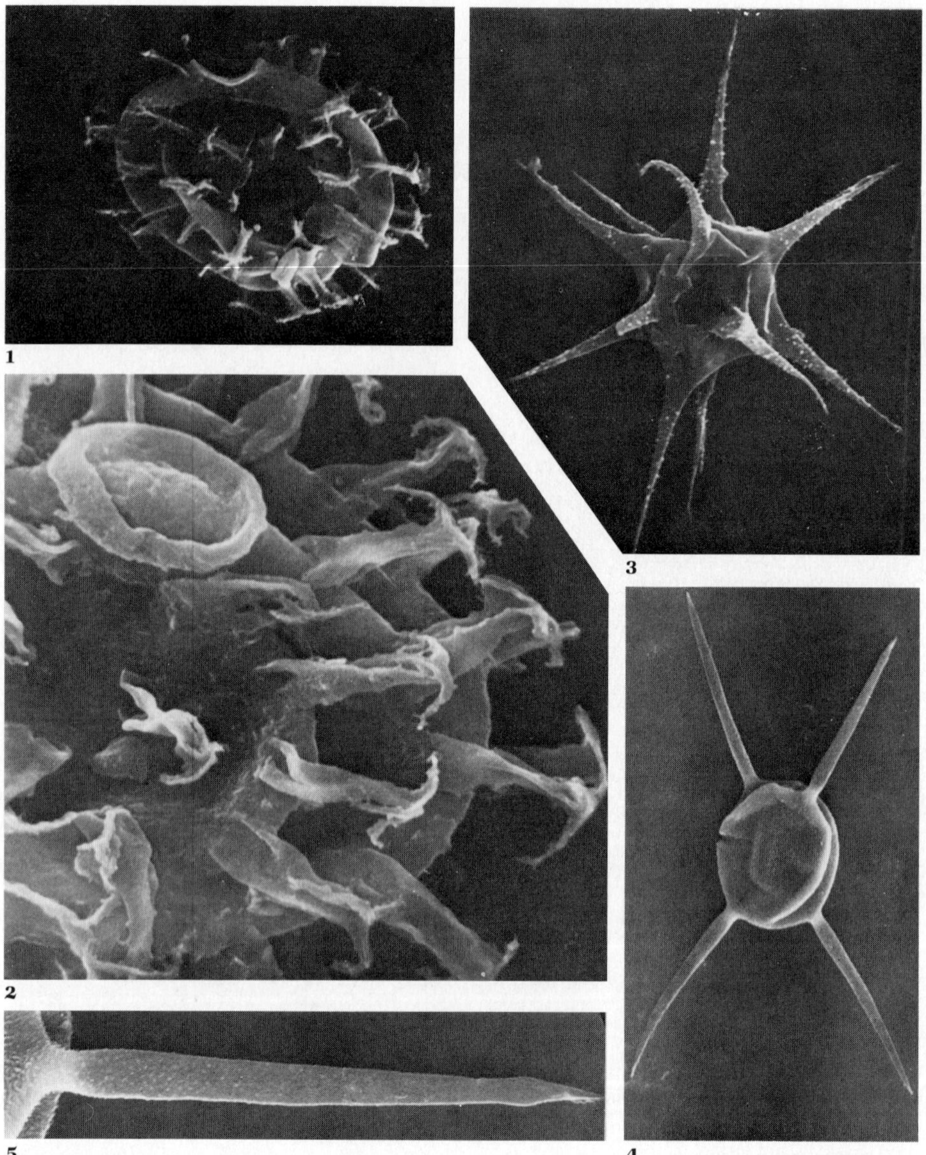

Figure 3.27

Middle Ordovician acritarchs, Oklahoma, in SEM. **1,2.** *Polyancistrodorus columbariferus.* **1.** Entire vesicle, showing open cyclopyle at top and projecting flanged processes, ×800. **2.** Enlargement of upper part of specimen with operculum in place (same specimen as in Figure 3.11 part 5), ×2080. **3.** *Polygonium* sp., entire vesicle, showing polygonal vesicle and tapering elongate spinulate processes (same specimen as enlargement in Figure 3.13 part 3), ×640. **4,5.** *Orthosphaeridium vibrissiferum.* **4.** Entire vesicle, showing incipient median split for excystment, ×280; from Loeblich and Tappan, 1971b. **5.** Enlargement of one process, showing basal constriction at position of internal plug and spinulate vesicle surface grading into granulose processes that become laevigate distally, ×880 (see also Figure 3.5 part 4).

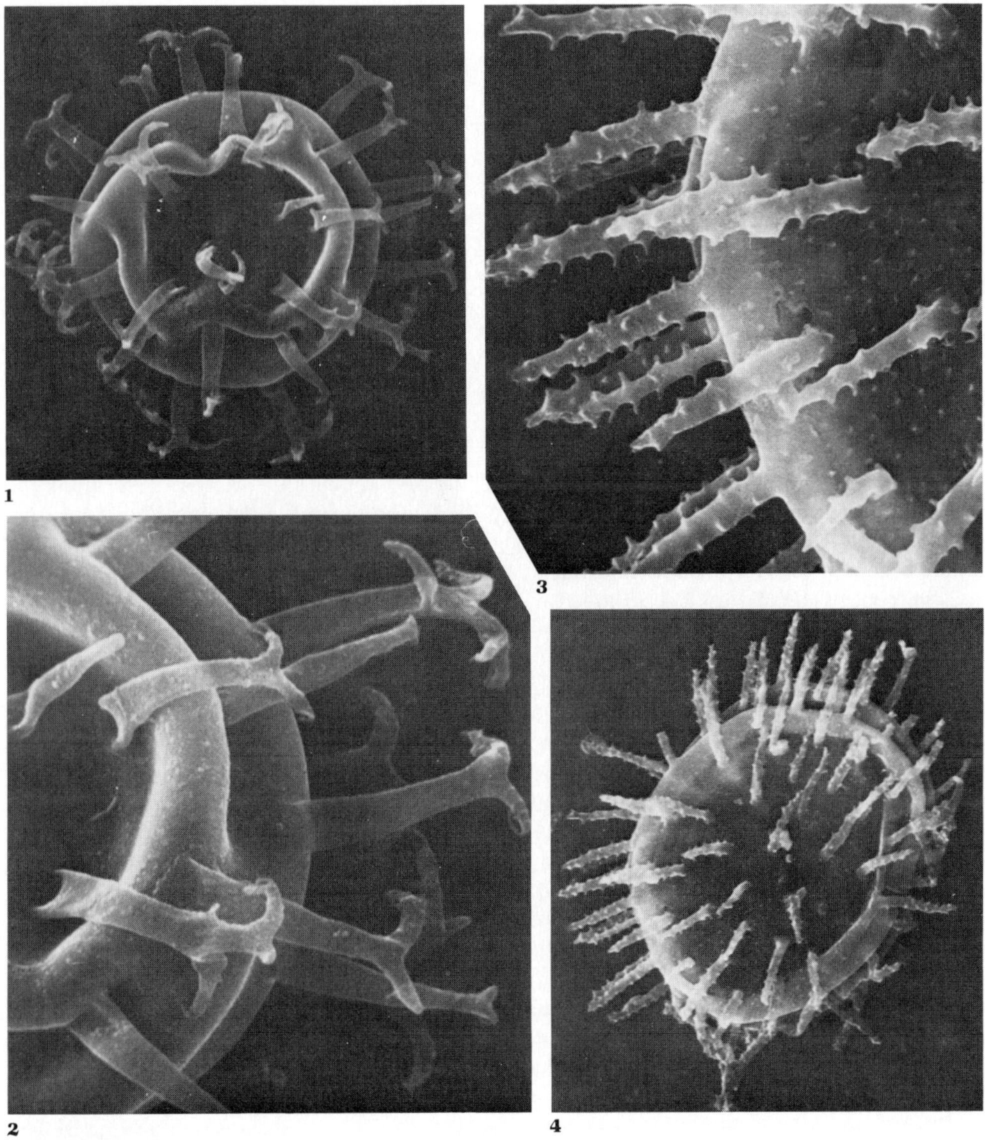

Figure 3.28
Middle Ordovician acritarchs, Oklahoma, in SEM. **1,2.** *Ordovicidium elegantulum.* **1.** Entire vesicle, ×680. **2.** Enlarged area showing multifurcate processes, ×1760. **3,4.** *Baltisphaeridium accinctum* Loeblich & Tappan, from Loeblich and Tappan, 1978. **3.** Portion enlarged to show granulate vesicle surface, and numerous short, tapering, highly spinulate processes, ×2400. **4.** Entire vesicle, ×800.

Ordovician ones in Oklahoma and Indiana. Silurian species similarly are excellently preserved through the mid-continent region and in western New York (see Figure 3.29). Acritarchs may have attained their acme in the Silurian (Silurian stratigraphic occurrences have been tabulated by Cramer, 1971a), but early Paleozoic assemblages are still not completely documented. In general, the Silurian species are somewhat larger and of more varied morphology than those of the Ordovician; many have long and complex processes and ornate wall surfaces. In the Silurian Wenlock Shale of England, acritarchs occur in numbers of 1000 to 10,000 individuals per cubic centimeter of rock (Downie, 1959), or about 5000 individuals per gram (Downie, 1963). Slightly less than some measurements for the Ordovician, this may be affected by dilution with other organic matter, as spores gradually become more important, comprising up to 4 percent of the assemblage in the Wenlock. In addition, the prasinophycean leiospheres remained an important part of the total microplankton (about one-third), and range from 20 to 90 percent of the assemblage in different samples.

Acritarchs were still highly varied in the Early Devonian (see Figures 3.30, 3.31, and 3.32), but diversity began to fluctuate and decline, leading to a marked drop in Late Devonian time. Locally a single species of *"Baltisphaeridium"* might comprise 90 percent of the assemblage (Deunff, 1965). In the Upper Devonian Chagrin Shale of Ohio, the 38 species of acritarchs comprised 5 to 17 percent of the organic matter, leiospheres from 17 to 55 percent, and land plant spores from 34 to 76 percent (Tappan et al., 1971, 1972). The acritarchs and leiospheres decrease in relative abundance upward in the section, as plant spores become more abundant. This is not wholly an environmental effect, however, for acritarch diversity becomes slightly greater as the abundance drops (see Figure 3.33). On a worldwide basis, the latest Devonian is characterized by a major decrease in acritarch diversity; the decline began near the end of the Frasnian and continued through the last of the Devonian and Early Mississippian. Late Paleozoic acritarchs are extremely rare, the total diversity of the Carboniferous being less than 10 percent that of the Devonian. A few specialized forms appeared in the Permian, particularly globular and discoidal species that excysted by median splitting (*Circulisporites, Peltacystia*), some simple ones persisted (*Micrhystridium, Veryhachium*), and a few other process-bearing forms appeared for the first time (*Mehlisphaeridium*); although individual species might locally be abundant, the assemblage was sparse.

Microplankton in general increased sharply in numbers and variety in the Jurassic, but the organic-walled phytoplankton was dominated by recognizable dinoflagellates after Early Jurassic and post-Paleozoic acritarchs are rare and unimportant. Prasinophycean leiospheres and tasmanitids continued to the present, but the large spinose and process-bearing acritarchs of the Paleozoic had disappeared. The few acanthomorphic types persisting in the Cenozoic, such as *"Micrhystridium,"* are small, and none have the characteristic large pylome of the early Paleozoic taxa. Although Upper Jurassic rocks of England were reported to contain about 3 percent of combustible organic matter (Downie, 1957), about 65 percent was provided by the dinoflagellates and the remainder by pollen and spores. The total also was less than the amount of organic matter present in Lower Ordovician rocks, which had been due solely to the acritarchs and prasinophyceans.

Figure 3.29
Middle Silurian acritarchs, New York. **1,2.** *Baiomeniscus camarus* Loeblich. **1.** Arcuate vesicle, with longitudinal ridges, ×768. **2.** SEM of surface, showing prominent grana on the ridges, ×8000.
3. *Pulvinosphaeridium parvum* Loeblich, ×768. **4,5.** *Lophosphaeridium* sp., in SEM. **4.** Entire vesicle, showing gemmate projections, ×1600. **5.** Portion of vesicle enlarged, surface microrugulate between gemmae, ×4800. **6,7.** *Tylotopalla caelamenicutis*, ×768. **6.** Focused on surface to show ridges. **7.** Short conical processes. **8,9.** *T. digitifera* Loeblich. Terminally furcate processes, ×768. Figures 1–3, 6–9, from Loeblich, 1970b.

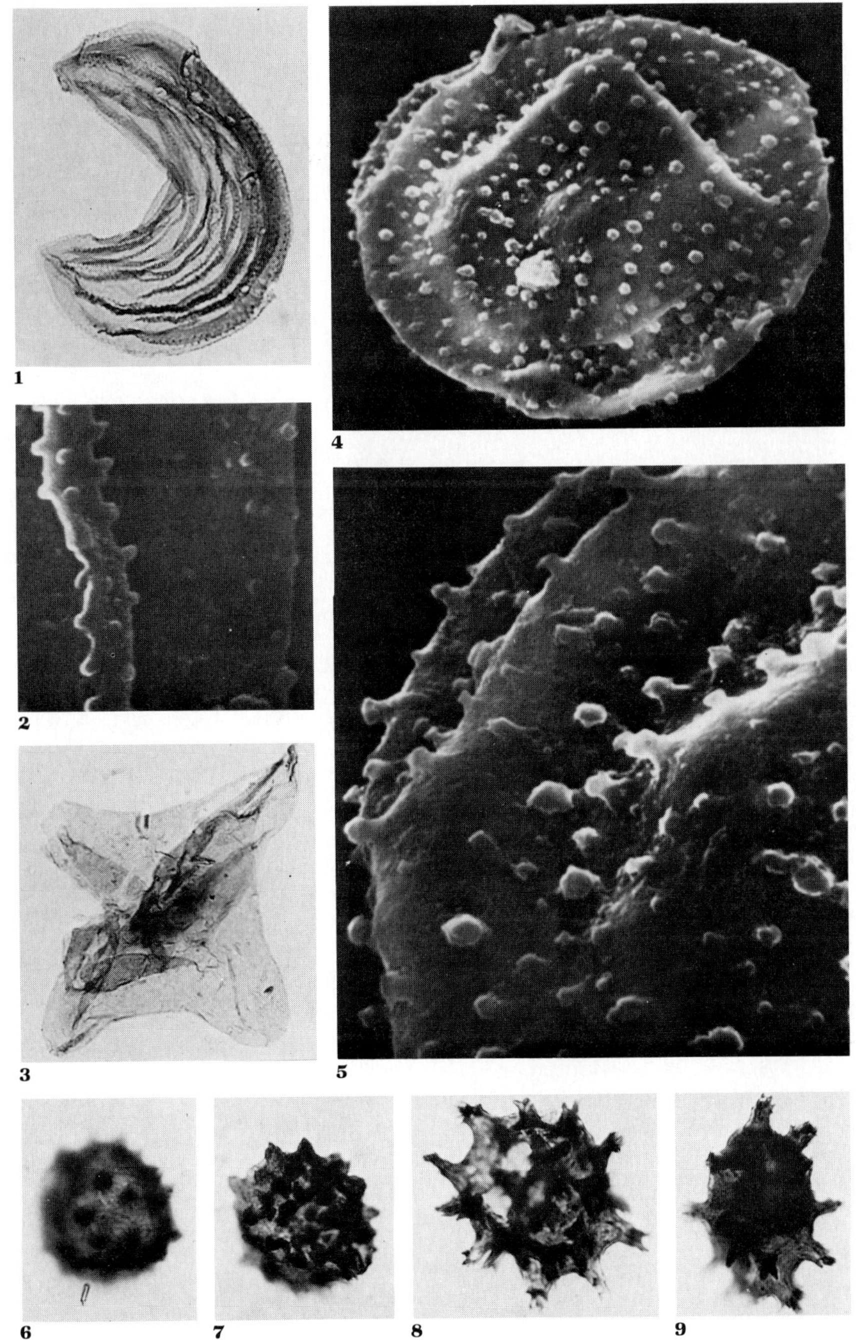

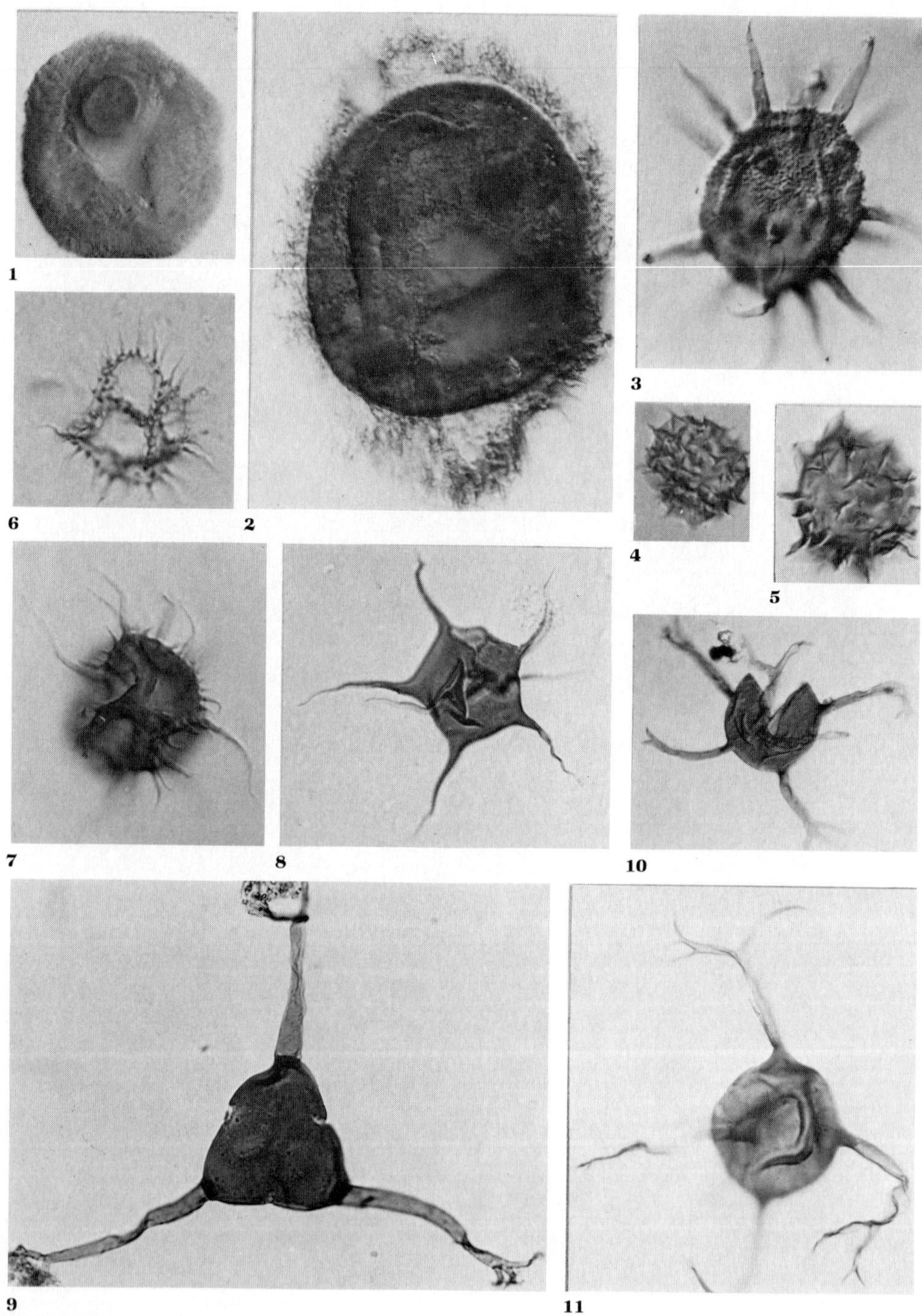

192

Figure 3.30
Lower Devonian acritarchs, Oklahoma. **1.** *Nanocyclopia aspratilis* Loeblich & Wicander, with operculum in place in cyclopyle, ×600. **2.** *Dasypilula compacta* Loeblich & Wicander, thick-walled vesicle with fibrous mat of thin, solid processes, ×1040. **3.** *Ectypolopus elimatus* Loeblich & Wicander, proximally plugged processes and granulate surface, ×1040. **4,5.** *Induoglobus latipenniscus* Loeblich & Wicander, triradiate processes bearing thin flanges, ×1040. **6,7.** *Jubatasphaera dispariluma* Loeblich & Wicander, ×1040. **6.** Partially degraded specimen, with preserved process-bearing ridges. **7.** Specimen with proximally plugged hollow processes aligned on ridges, and smooth intervening area. **8.** *Perissolagonella amsdenii* Loeblich & Wicander, pentagonal vesicle with long flexible process at each angle, and one from face, small excystment opening at top of photo ×600. **9.** *Ozotobrachion dicros* Loeblich & Drugg, showing thick-walled vesicle, and hollow, terminally furcate processes that do not connect with vesicle cavity, ×768. **10,11.** *Oppilatala vulgaris* Loeblich & Wicander. **10.** Proximally plugged terminally furcate processes, and excystment opening by simple rupture, ×600. **11.** Globular vesicle, and processes that furcate to the third order, ×1040. Figures 1–8, 10, 11, from Loeblich and Wicander, 1976.

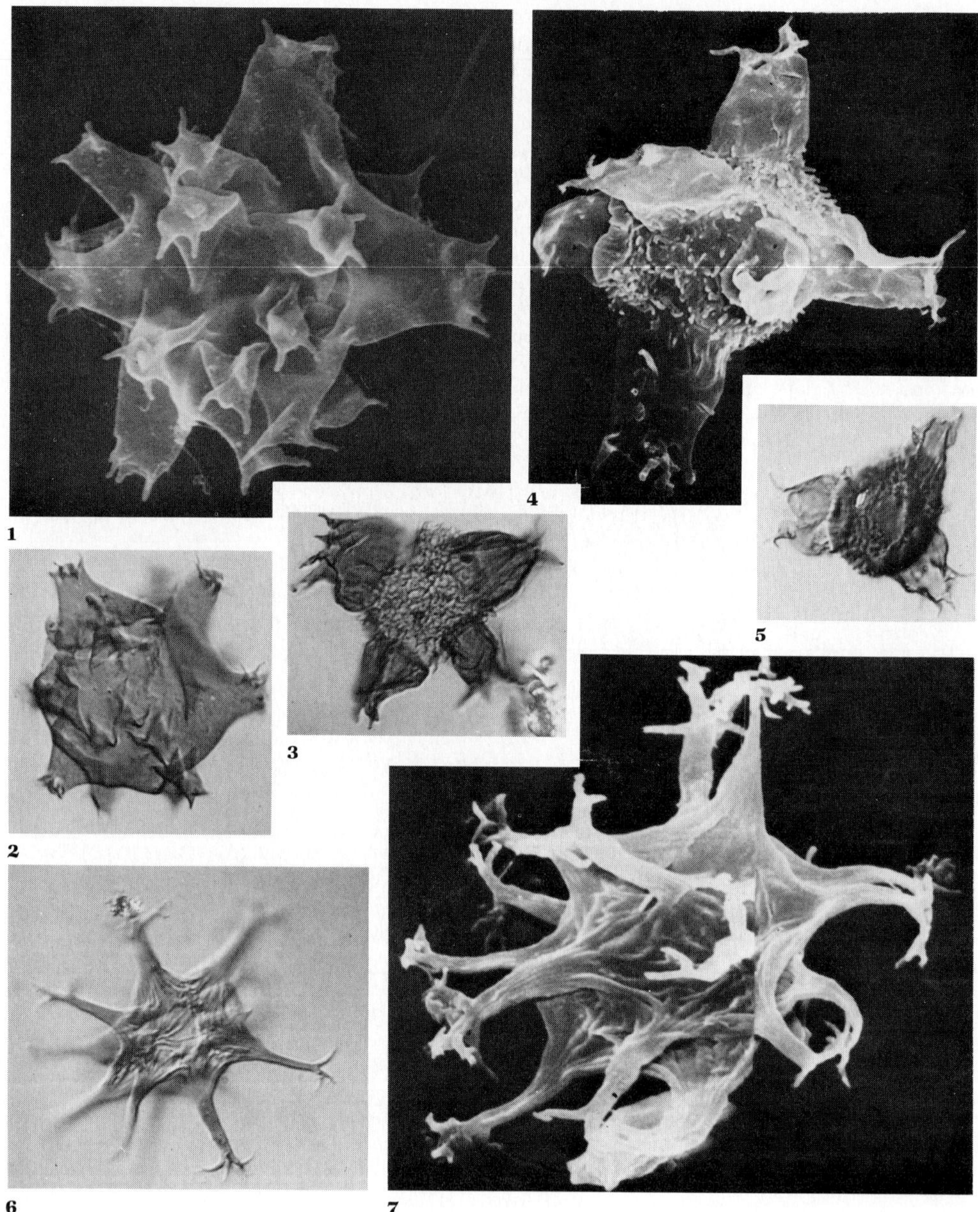

Figure 3.31

Lower Devonian acritarchs, Oklahoma. **1,2.** *Thysanoprobolus polykion* Loeblich & Tappan. **1.** SEM, broad, conical, terminally multifurcate processes, ×1600. **2.** Intercommunication of processes and vesicle cavity, ×768. **3—5.** *Demorhethuim lappaceum* Loeblich & Wicander. **3,5.** Thick-walled spinulate vesicle and thin-walled tubular processes with multifurcate terminations, ×1040. **4.** SEM, ×1840. **6,7.** *Actinotophasis complurilata* Loeblich & Wicander. **6.** Phase contrast, ×414. **7.** SEM, parallel ridges on processes join on the vesicle, with furcations to fourth order on processes, ×3200. 3,4, from Loeblich and Wicander, 1974; 5, from Loeblich and Wicander, 1976.

194

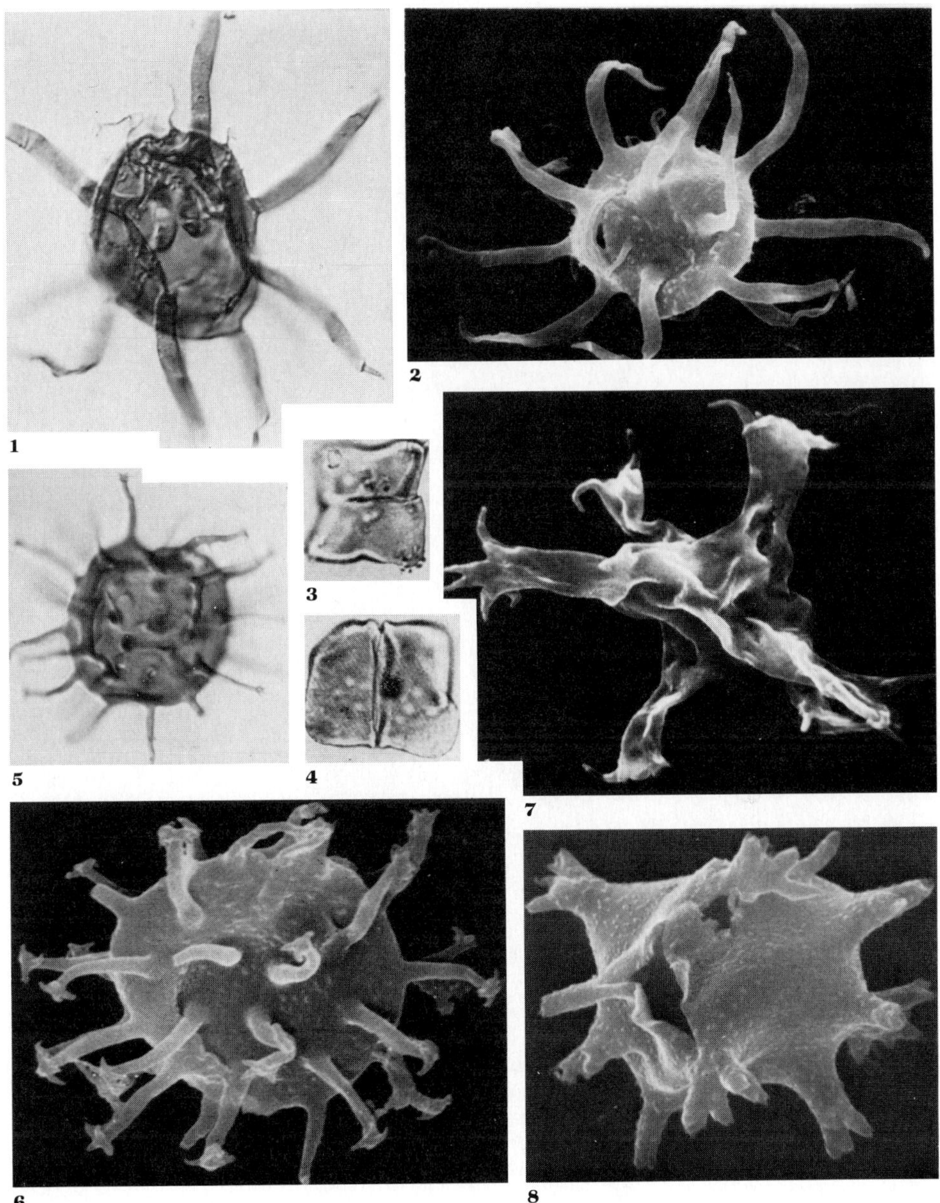

Figure 3.32

Lower Devonian acritarchs. **1,2.** *Hetermorphidion echinopillum* Loeblich & Wicander, Oklahoma. **1.** Proximally plugged elongate, straplike processes, and tiny flexible processes that open into vesicle cavity, ×1040. **2.** SEM, ×1600. **3,4.** *Schizocystia pilosa* Jardiné, Combaz, Magloire, Peniguel & Vachey, Algeria; subquadrate vesicle excysting by median split; ×640, from Jardiné et al., 1972. **5,6.** *Ammonidium uncinum* Loeblich & Wicander, Oklahoma, showing flexible processes with quadrifurcate tips. 5, ×1040; 6, SEM, ×3200. **7.** *Ozotobrachion pulvinus* Loeblich & Wicander, Oklahoma, SEM, ×1200. **8.** *Evittia cymosa*, Oklahoma, vesicle with granulate wall and multifurcate short processes (same specimen as in Figure 3.5 part 7), SEM, ×800, from Loeblich, 1970b.

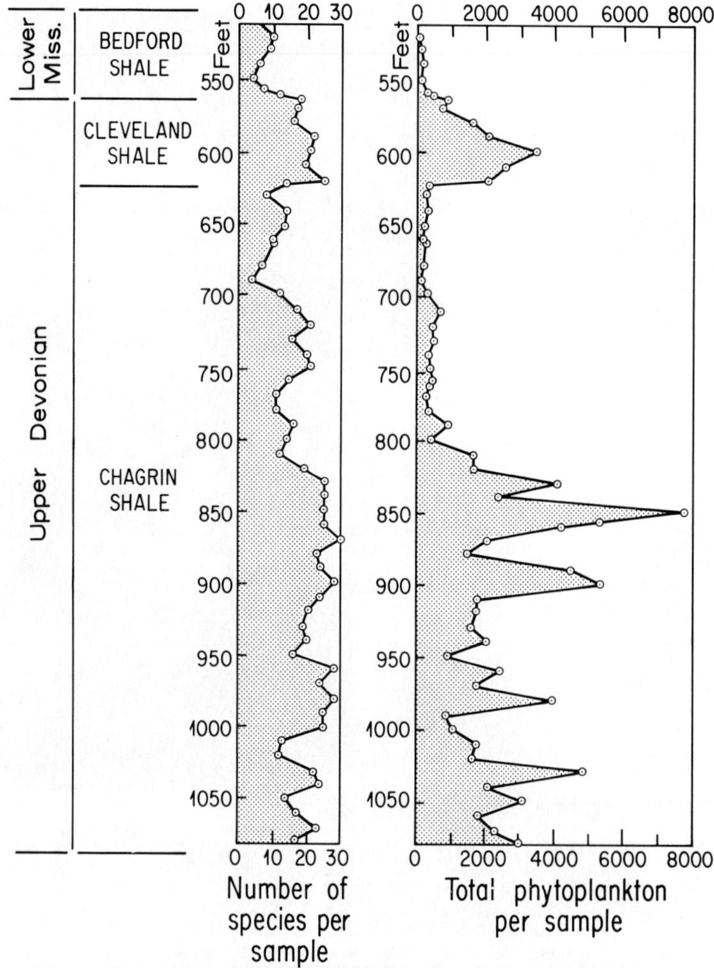

Figure 3.33
Fluctuating diversity and abundance of acritarchs and prasinophycean algae at the Devonian-Mississippian boundary. Total number of specimens of phytoplankton fluctuates, but is relatively high until about 65 meters (200 feet) below the top of the Chagrin Shale, then is low in the upper Chagrin. Both diversity and abundance increase temporarily in the Cleveland Shale before dropping to very low levels within the Mississippian. Redrawn from Wicander, 1975.

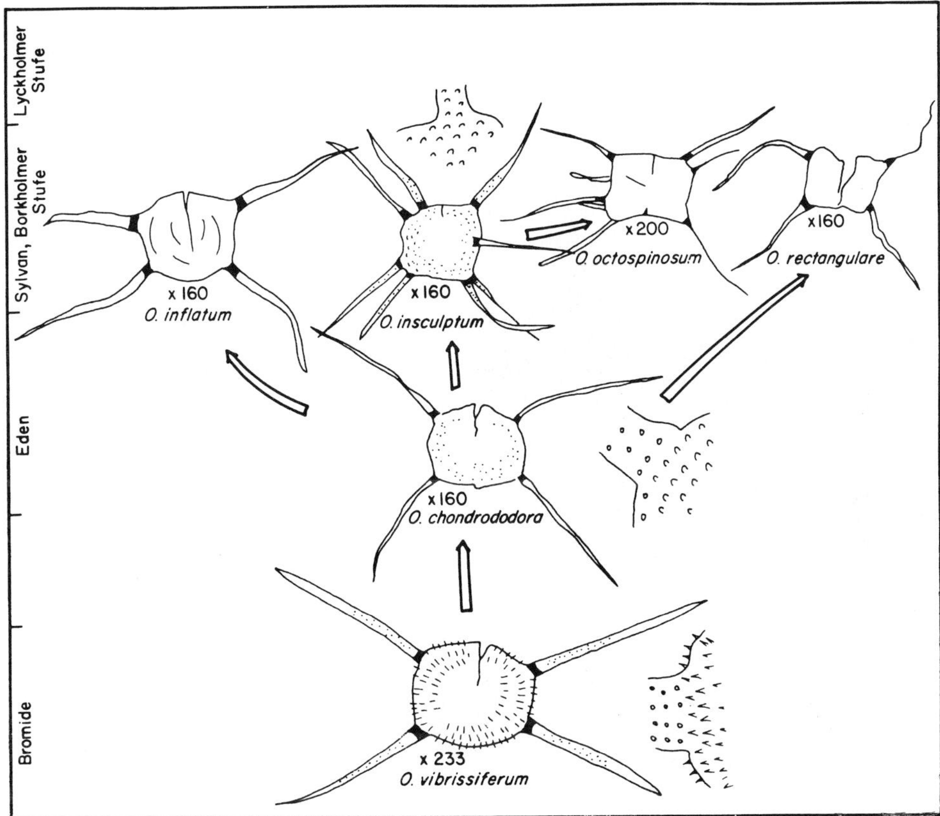

Figure 3.34
Suggested phylogenetic trends in *Orthosphaeridium* in the Middle and Late Ordovician. Composite scale at left indicates relative position of strata from which the different species were described. From Loeblich and Tappan, 1971b.

Evolutionary Trends

Other than the previously mentioned general trends such as increase in acritarch size and increase in length and complexity of their processes from the Precambrian through Devonian, little evidence as to possible evolutionary trends within a lineage or within a genus has been presented. Gradations between genera such as *Micrhystridium* and *Veryhachium* have been suggested, but these are not regarded as evolutionary patterns, as these genera are extremely long-ranged. Species of *Ortho-sphaeridium* in the Middle and Late Ordovician were suggested to show an evolutionary trend toward an increased number of processes and decreased vesicle inflation and ornamentation (see Figure 3.34). In contrast, Middle Ordovician species of *Aremoricanium* are generally smooth surfaced and have only short processes, whereas those of the Late Ordovician have few, narrow, elongate tubular processes (Loeblich and MacAdam, 1971; see Figure 3.35). These trends in the Ordovician, if substantiated, were not long persistent, however, for both genera have only a restricted geologic occurrence.

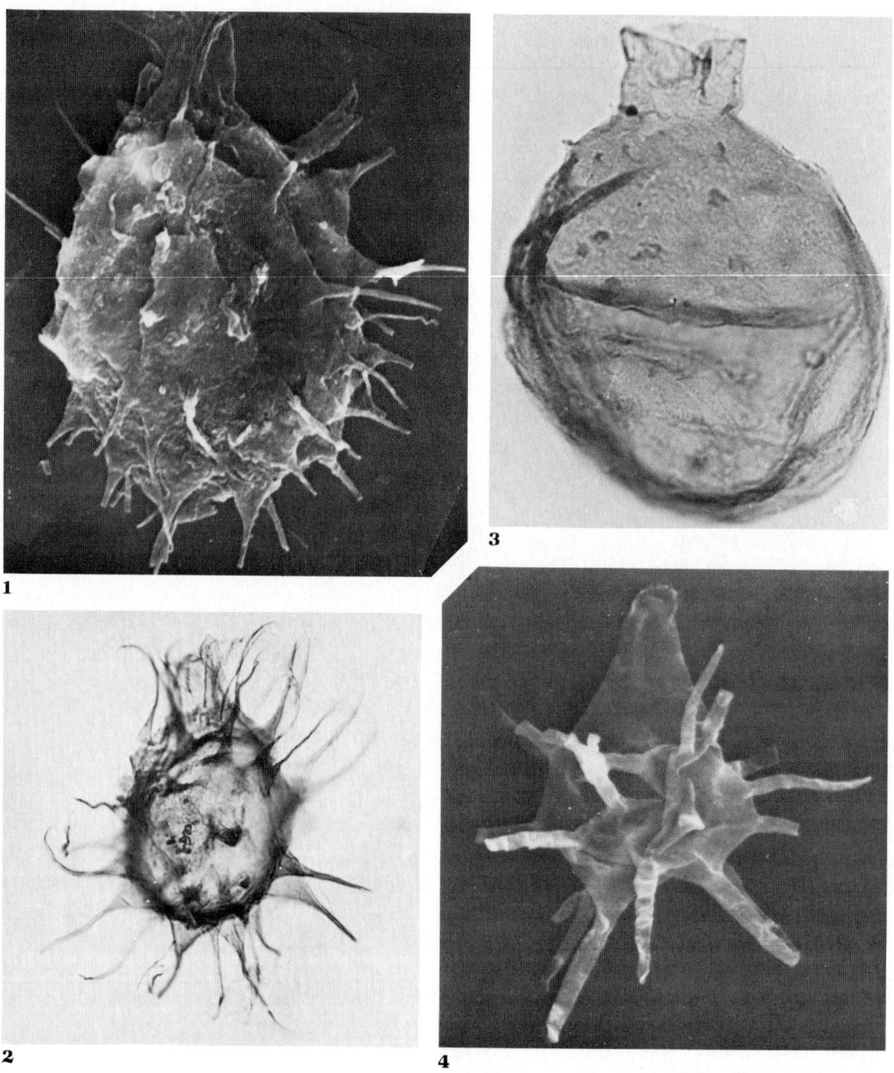

Figure 3.35

Evolutionary trends in *Aremoricanium*. **1,2.** *Aremoricanium rigaudi* Deunff, Middle Ordovician, Oklahoma. Short, narrow spines and elongate cylindrical neck. 1, SEM, ×1280; 2, ×768. **3.** *A. simplex* Loeblich & MacAdam, Middle Ordovician, Oklahoma. Flask-shaped vesicle, tubular neck and pylome, granulate surface, and no processes other than short blunt spinules; ×768. **4.** *A. squarrosum* Loeblich & MacAdam, Upper Ordovician, Indiana. Tapering neck and smaller pylome, and elongate strap-like processes, SEM, ×480. All from Loeblich and MacAdam, 1971.

Contamination and Reworking

As is true of most smaller microfossils, contamination in the laboratory, mixing in the field, and reworking in the sediments are continual problems hampering their study. Contamination by airborne pollen and spores can be minimized by care in sample preparation, but reworking must be recognized on other evidence. An assemblage described by Davey (1970) from the English Cenomanian contained many acritarchs as well as dinoflagellates, including 10 species of *Micrhystridium* and 9 of *Veryhachium,* 4 *Cymatiosphaera,* a *Leiofusa,* and a *Pterospermella.* Most of the acritarchs were quite rare, and some were very poorly preserved and strongly resembled various early Paleozoic taxa. *Micrhystridium* cf. *variabile* Valensi, *M. minutispinum* Wall, *Veryhachium reductum* (Deunff) de Jekhowsky, and *V.* sp. A, all were stated to be possible examples of reworking.

Because the indigenous fossils occur only in the cementing material between grains of sand or clay, Cramer (1970) suggested that very mild processing without complete dissolution of siliceous material would limit the amount of contamination from reworked fragments. In practice, this is not an effective solution, however, as incomplete processing results in too high a quantity of extraneous matter obscuring specimens; commonly the cementing material is more resistant to preparation techniques than are the small grains present.

Paleoecology

Because acritarch studies are relatively young, taxonomic work has had a history of only about 40 years, and the greatest volume has appeared within the last decade. Detailed studies of morphology and ultrastructure largely are limited to the 1970s. Hence, although the quantity of basic information is expanding rapidly, it still is insufficient for definitive statements concerning the ecologic requirements of the group.

The general worldwide distribution of the acritarchs, their presence in a wide variety of rock types, and the biota with which they occur all indicate a planktonic habit; the previously mentioned chemical composition of the wall indicates their plant rather than animal nature. By analogy with modern ecosystems, their small size and great abundance also suggest that they must have been phytoplankton and thus primary producers of organic matter rather than zooplankton consumers. Modern phytoplankton assemblages not only vary geographically, but also seasonally. Seasonal variations of course would be obliterated in the geologic record, but such present-day fluctuations suggest ecologic controls that may have been equally important for the acritarchs.

The primary requirement for phytoplankton is the availability of light, hence it is restricted to the photic zone during the actively growing stages. Acritarchs must have lived only in the uppermost waters at high latitudes but could have lived somewhat deeper in the water column in clear waters at lower latitudes, but such effects are difficult to separate from those of temperature or depth. Other important physical parameters that may have been species-specific were temperature, salinity, availability of nutrients, and turbidity. Biologic factors would have included the limitation of the standing crop by grazing zooplankton and suspension feeders; perhaps the amount of infaunal bioturbation of the bottom sediment would have controlled their destruction or preservation as fossils. Mutual inhibition by other phytoplankton or benthic algae, by competition for nutrients, or by the production of inhibiting external metabolites also may have affected the nature and abundance of the acritarch assemblages. Most of these ecologic parameters have been suggested as controlling acritarch distribution and abundance at various times and places.

Temperature

Latitudinal control of paleotemperatures was regarded by Cramer (1969a) as most important in the distribution of Silurian acritarch assemblages. He regarded as unimportant variations in salinity, light, turbidity, or nutrient supply; as all assemblages were present on the relatively shallow shelf seas, the effects of depth and distance from shore also were discounted. Plotting the distribution of the known occurrences of acritarchs on a palinspastic base map reconstructing continental positions in the Silurian, various temperature-controlled contemporaneous assemblages were described as paralleling paleolatitudes (Cramer, 1970a, 1971a,b; Cramer and Díez de Cramer, 1972a). An impoverished *Neoveryhachium carminae* association (later termed the Netromorphitae facies), characterized also by *Leiofusa, Dactylofusa,* and *Eupoikilofusa,* was present in Brazil, southern Libya, and Nigeria, nearest the Gondwana Pole, from paleolatitude 70° to 40° S. The maximum acritarch diversity during the Silurian was between paleolatitudes 30° and 50°S; it decreased toward higher and lower paleolatitudes, thus suggesting an optimum temperature less than that of the paleoequatorial region (Cramer and Díez de Cramer, 1972a).

The typical *Neoveryhachium carminae* association represented paleolatitudes 40° to 20° S, occurring in Spain, North Africa, Saudi Arabia, and in Georgia and Florida. At still lower paleolatitudes the warmer water *Domasia elongata* Downie association (with rare *N. carminae*) occupied from about 15° to 35° south paleolatitude. It is found in England, Wales, Nova Scotia, Belgium, France, Germany, and Saudi Arabia, and in Pennsylvania, Indiana, Ohio, Illinois, Kentucky, and Missouri. Later (Cramer, 1971a), the *Deunffia* facies was separated from this; it occurs in Ohio, Indiana, Kentucky, New York, and in England, between paleolatitudes 15° and 20° S (see Figure 3.36). The equatorial facies lay north of paleolatitude 20° S, and was later termed the *Pulvinosphaeridium-Estiastra* facies. Containing *Multiplicisphaeridium corallinum* (Eisenack) Eisenack and *Baltisphaeridium digitatum* (Eisenack) Eisenack, it occurs in the present Baltic region and Anticosti Island. Still later, the *Gloeocapsomorpha prisca* Zalessky facies was defined (Cramer and Díez de Cramer, 1972a) on the basis of a record of this chlorophycean in the Canadian Arctic and one in Yunnan, China. These Silurian assemblages were said to be time-transgressive, the warmer associations (*Pulvinosphaeridium, Deunffia,* and *Domasia*) gradually overlapping the *Neoveryhachium* facies over a distance of about 200 km, corresponding with the northward drift of eastern North America and western Europe during the Middle Silurian. This shift in facies was said to indicate a continental plate movement in a N 45° W to N 65° W direction, at a rate of between 2 and 3 cm annually (Cramer, 1971b).

Although the suggested facies distribution is compatible with the Gondwana reconstruction upon which they were plotted, the described boundaries overlap and are not sharply delineated, hence do not provide positive evidence as to the global reconstruction. For example, the facies marker *N. carminae* not only occurs in the association named for it, but also in the Netromorphitae, *Deunffia,* and *Domasia* assemblages. The *Pulvinosphaeridium-Estiastra* facies reported from Gotland and Anticosti Island is said to be dominated by *Baltisphaeridium* species that also occur in the *Deunffia* and *Domasia* associations; *Domasia* also is found in the *Pulvinosphaeridium* assemblage. Similarly, paleolatitudinal boundaries are drawn for all facies, although only three assemblages have been reported in both North America and Europe, and of these, the *N. carminae* association was only reported in Spain, and that with *Deunffia* only in England. The equatorial assemblage is reported only in two localities in the world, scarcely sufficient for the latitudinal boundaries postulated. In fact, adequate locality and stratigraphic data are not available for most of the records, and as continuous stratigraphic sequences are available for less than a half-dozen localities, they are insufficient as yet for worldwide reconstructions and interpretations.

Proximity to Shore

Three morphologic classes of acritarchs were recognized as paleoenvironmental indicators for the Middle and Late Ordovician (Jacobson, 1979). A leiospherid class represents nearshore shallow water, a peteinosphaerid-*Dicommopalla* class is

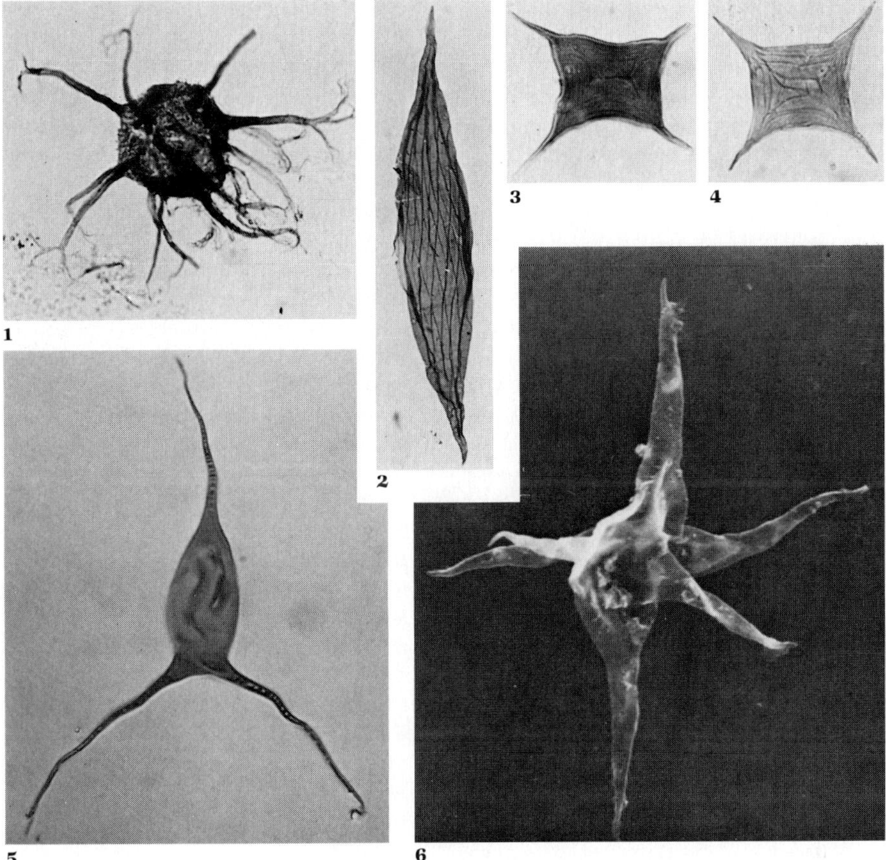

Figure 3.36
Silurian acritarchs. **1.** *Piliferosphaera setosa* Loeblich, Middle Silurian, New York, ×768. **2.** *Leiofusa rhikne* Loeblich, Middle Silurian, New York, ×307. **3,4.** *Neoveryhachium carminae* (Cramer) Cramer, Middle Silurian, New York, ×768. **5.** *Domasia trispinosa* Downie, Lower Silurian, Gotland, ×768. **6.** *Estiastra stellata* Loeblich, Middle Silurian, New York. Elongate pointed processes, with rare granules on the surface, SEM, ×720. 1,3,4, from Loeblich, 1970b.

characteristic of a shoal habitat, and a baltisphaerid-veryhachid-*Polygonium* class occurred in the open sea area.

The Silurian assemblages defined by Cramer and outlined in the preceding section were regarded as solely reflecting temperature belts but correspond equally well to distance from land. The Gondwana pole centered on a land area, and the "high latitude" *Neoveryhachium carminae* facies closely parallels the distribution of platform mudstones, as these are indicated by Berry and Boucot (1967). Cramer (1971b) also suggested that *N. carminae* may have favored a highly turbid, near-coast environment, as abundance is unusually high in shaly portions of the tidal flat sequence in Spain and Tunisia. Other species are more common in calcareous sediments. The Netromorphitae association, which Cramer (1971b) indicates as locally overwhelmed by prasinophycean tasmanitids and leiospheres, occurs closest to land and includes brackish deposits, a facies common to the tasmanitids.

In a preliminary study of Silurian acritarch distribution in Pennsylvania, the open marine facies contained acritarchs, none were found in the fluvial Tuscarora-Shawangunk, and poorly preserved rare individuals were present in the barrier bar and tidal flat lagoonal facies, suggesting partial current control of distribution (Smith and Saunders, 1970). Middle Silurian calcareous shale and argillaceous micrite of the Rochester Formation of southern Ontario contains abundant acritarchs (Thusu, 1972). The dominant spiny spheres (acanthomorphid species) were regarded as characteristic of inshore basinal environments; the somewhat fewer polygonomorphids (*Veryhachium*) and netromorphids (elongate to fusiform taxa) are more characteristic of an open-sea environment. The Rochester Formation was regarded as having been deposited in a fairly shallow epicontinental sea, under calm subtidal inner neritic conditions with good circulation, supporting a great variety of invertebrate benthos.

Environmental control of assemblages in Late Devonian reefs was indicated by Staplin (1961). Near the reef, only smooth leiospheres were present, whereas numerous acritarchs with long thin processes were found at distances greater than a mile from the reef; beyond four miles thick-processed species and saccate and polygonal ones also were present. The progressive increase in abundance with distance from the reef was suggested to reflect an optimum environment in quiet deeper water. Selective destruction by wave action, diagenesis, or scavengers was discounted as Chitinozoans and plant spores were not similarly restricted. In the Late Devonian of Ohio (see Figure 3.37), fluctuations in abundance and diversity appeared to be less influenced by distance from shore, and all of the types coexisted that were restricted in occurrence in the reefal areas of the Canadian section (Wicander, 1975). In this local area, the acritarch diversity dropped sharply in the Late Devonian and Early Mississippian, as had earlier been indicated for the Devonian worldwide (Tappan, 1968; Tappan and Loeblich, 1973b).

The changes in relative abundance of *Micrhystridium* and dinoflagellates in the English Cenomanian (Davey, 1970, p. 385) also were ascribed to the effect of shore proximity, although *Micrhystridium* here was said to prefer a nearshore and perhaps shallow water environment. A similar inshore preference was indicated for *Micrhystridium* and *"Baltisphaeridium"* in the English Early Jurassic (Wall, 1965), whereas *Veryhachium, Domasia,* and similar forms favored an open-sea area. Single-species populations appeared to indicate inshore conditions, whereas highly diverse heterogeneous assemblages accumulated offshore, closely corresponding to modern phytoplankton assemblages. Those species with very reduced processes appeared tolerant of turbulent conditions, and those with long delicate processes were restricted to areas of quiet deposition. During relative deepening of the basin, populations were uniform in composition and less varied; during transgressive phases and under open-sea conditions, the assemblage was more varied, both in species and in their morphology. Diversity again decreased during regression of the seas and the deposition of coarser sediments.

Depth of Water

Although plankton is not controlled by the nature of the bottom, a depth zonation within the water column may occur. Inverse abundance maxima shown by prasinophycean leiospheres and tasmanitids from those of the acritarchs in the Polish Ordovician and Silurian (Górka, 1969) and the English Silurian (Downie, 1963) were suggested to result from different environmental preferences such as depth restrictions. Similarly, assemblages that were thought to have been controlled by distance from shore may be depth correlated (Gray and Boucot, 1972). Thus Silurian shallow water assemblages in Ohio contained a single morphological type (one or few species), and deep water assemblages contained a variety of growth forms, for the resting stages could accumulate at the bottom from the

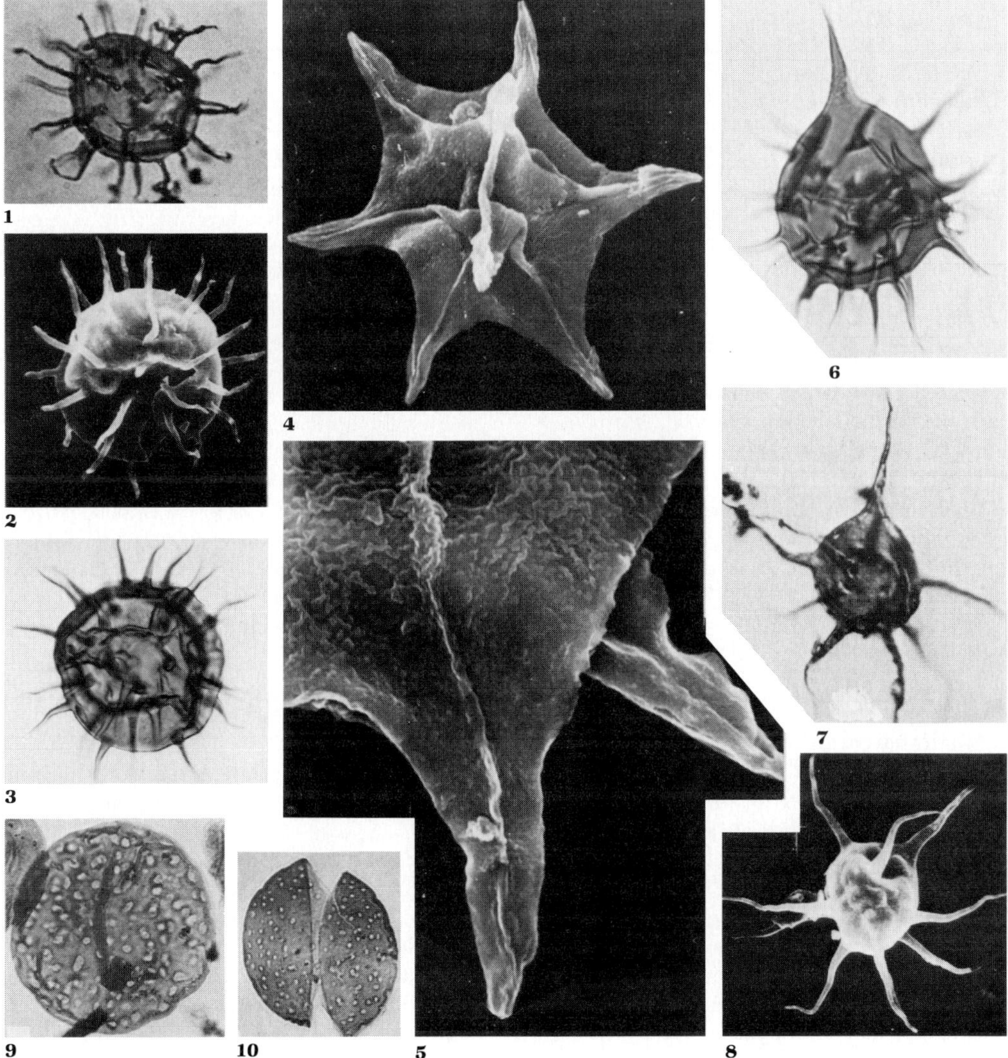

Figure 3.37

Upper Devonian (1−8) and Mississippian (9,10) acritarchs. **1,2.** *Solisphaeridium apodasmion* (Wicander) Wicander & Loeblich, Ohio, from Wicander, 1974. **1.** ×1040. **2.** SEM, ×800. **3.** *Gorgonisphaeridium ohioense* (Winslow) Wicander, Indiana, ×414. **4,5.** *Stellinium micropolygonale* (Stockmans & Willière) Playford, Oklahoma. **4.** Polyhedral vesicle and median ridge on each process. SEM, ×800. **5.** Enlargement of another specimen to show rugulate surface, SEM, ×2400. **6.** *Unellium ampullium*, Indiana, one large polar process and many narrower ones, ×768. **7,8.** *Guttatisphaeridium pandum* Wicander, Ohio, from Wicander, 1974. 7, ×600; 8, SEM, ×600. **9,10.** *Lacunalites sphaericus* Hemer & Nygreen, Mississippian, Saudi Arabia, ×400, from Hemer and Nygreen, 1967. **9.** Vesicle. **10.** Showing median splitting for excystment.

entire sequence of vertically (depth) stratified assemblages. Gray and Boucot suggested that the Sphaeromorphitae inhabited surface waters, hence were ubiquitous in distribution and alone were represented near shore, whereas the acanthomorphitic acritarchs were restricted to deeper waters, occurring in more diverse assemblages.

Salinity

As far as known, all Precambrian and early Paleozoic acritarchs are marine in occurrence (Smith and Saunders, 1970; Vidal, 1976), and later freshwater occurrences are doubtful, particularly after the dinoflagellates and Prasinophyceae are eliminated. Some Quaternary nonmarine taxa have been described, but many have been later reassigned to the dinoflagellates.

Diversity decreases rapidly toward shore in the marine environment, the more complex taxa being typically offshore in occurrence. Rock lithology similarly suggests this relationship, as acritarch diversity is greatest in fine-grained silty shales and siltstones and in rocks of considerable carbonate content, providing the enclosing rocks have not been subjected to oxidation, either at the sea floor or during diagenesis. Varied abundances in certain strata, independent of facies, have been suggested to result from suddenly appearing algal blooms, even in rocks as old as the Precambrian (Vidal, 1976). Some species also showed an inverse relationship of the population size and the lithologic grain size.

An Upper Permian assemblage found in coals and silts associated with coals in Western Australia contains *Peltacystia, Tetraporina, Schizosporis,* and *Circulisporites* together with the green alga *Botryococcus,* and is regarded (Balme and Segroves, 1966) as representing continental, fresh or brackish water conditions.

Species described as *Baltisphaeridium* from Flandrian (Holocene) peats in Southwest Australia (Churchill and Sarjeant, 1963; Harland and Sarjeant, 1970) later were described as a distinct genus, *Creberlumectum.* Associated taxa included the green alga *Pediastrum,* as well as some dinoflagellate cysts, spores, and pollen; *Creberlumectum* probably also is referrable to the dinoflagellates.

Various levels of Pleistocene peats, representing late Glacial to post-Glacial time, also contained *Micrhystridium* and *Leiosphaeridia* in Staffordshire, England (Sarjeant and Strachan, 1968). Elsewhere, however, acritarchs are rare or absent from freshwater deposits, or have been regarded as reworked from older marine deposits. In general they provide a good indication of marine conditions.

In contrast to the brackish limitation, the Middle Triassic assemblage of Switzerland was seemingly limited by excess salinity (Brosius and Bitterli, 1961). Species of *Veryhachium, Micrhystridium,* and *?Domasia* were abundant in limestones of both the lower and upper Muschelkalk, but were absent from the anhydrite and dolomite of the middle Muschelkalk, where the high salinity was regarded as having been unfavorable for the microplankton.

Bioturbation of Sediments

In studying the microplankton of the Cenomanian of England, Davey (1970) noted that the chalks at Speeton in Yorkshire and in the lower part of the section at Hunstanton, Norfolk, contained thick-walled algal remains but no dinoflagellates, acritarchs, or **miospores.** The abundant macrofauna included brachiopods, lamellibranchs, and echinoids, suggesting that the organic microplankton and spores had been depleted by the mud-feeding invertebrates so that none were incorporated in the sediments. Elsewhere, rocks of the same age had an excellent microflora.

However, as previously noted, the cysts that comprise the organic-walled microplankton are highly resistant to chemical treatment and probably would survive passage through the invertebrate digestive systems. More probably, diagenetic oxidation of the sediments at the sea floor resulted in destruction of the thinner-walled organic remains; similar diagenetic alteration commonly has affected even the calcareous microfossils (nannoplankton, foraminifers) in many of the Cretaceous chalks.

CLASSIFICATION

Acritarcha were defined as cysts of unknown organisms and probably polyphyletic, hence their classification was regarded as artificial (Evitt, 1963; Deflandre, 1947; Downie et al., 1963). Genera and species are defined under the Botanical Code (Downie et al., 1961), although their assignment to a particular division remains uncertain. The nature of the taxa recognized by the various workers demonstrates the same dichotomy of philosophy that characterizes much of the study of microfossils. Thus some workers regard the acritarchs as probably representing a closely interrelated group of taxa that can be defined and recognized on the basis of their fossilized cysts. At the opposite extreme, these fossils are regarded merely as stratigraphic tools, and emphasis is placed on simplicity of differentiation regardless of its natural basis; any morphologic characters that do not remain recognizable even during incipient metamorphism or kerogenization are disregarded. The same characters that are assigned major importance by some workers, including presence or absence of an inner body, a single or double wall, proximally open or plugged processes, and the nature of the excystment opening, if present, may be regarded by others as facultative characters of only a particular stage in the life history of the individual and hence taxonomically unreliable.

The current very rapid proliferation of acritarch taxa is in large part due to the very short time over which they have been studied, with a total of less than a dozen genera having been proposed as long as 20 years ago. The large number of taxa described during the last two decades has already made the acritarchs highly useful biostratigraphically, and as stratigraphic ranges become still better determined, they will increase in utility. The attempt to recognize "natural" species and eventually higher taxonomic categories will not impede their value.

The dichotomy in philosophy of acritarch taxonomy is also evident in their classification. Because of their uncertain relationships, the group was subdivided into a number of non-Linnaean "subgroups" (Downie et al., 1963).

Neither groups nor subgroups are recognized under the Botanical or Zoological Codes. Names of subgroups originally were morphologically derived, rather than based on the name of a typical included genus (Downie et al., 1963), to avoid the necessity of their reconstitution if the true relationships of the typical genus were later discovered, leading to its removal. Of the original 13 subgroups, some have since been transferred wholly or in part to the green algae (Prasinophyceae) or dinoflagellates, but following the philosophy of Downie, Evitt, and Sarjeant (1963) as to the desirability of non-Linnaean suprageneric categories for the Acritarcha, additional groups and subgroups have been proposed (Brito, 1967a, 1969; Segroves, 1967; Sinha, 1968; Timofeev, 1969; Pocock, 1972; Naumova and Umnova in Umnova, 1975; German in Timofeev et al., 1976).

Some additional groups and subgroups also have been based on the names of included genera (Staplin et al., 1965; Cramer and Díez de Cramer, 1972a; Vavrodová, 1973; Naumova and Umnova in Umnova, 1975), and some true families have been proposed. However, if acritarch genera are regarded as "form genera," they would not be assignable to families under the Botanical Code. All subgroups, groups, and additional categories proposed for the acritarchs are listed in the following section, although some now are referrable to Prasinophyta or Chlorophyta.

Even farther outside the Linnaean system, G. and M. Deflandre (1964b) redefined the subgroups as "parafamilies," regarding them as wholly artificial categories, although this has generally not been followed.

At the opposite extreme, all hystrichospherids and acritarchs are regarded as representing a homogeneous assemblage (Eisenack, 1963a), or even a separate algal class, the Hystrichophyta (Mädler, 1963, 1967). Genera were then assigned to true families, which carry priority under the biological codes of nomenclature (Eisenack, 1938b, 1954a; 1963a; Timofeev, 1957, 1966; Mädler, 1963).

Although the commonly used subgroups were defined to simplify the problem of their unknown relationships and to avoid questions of priority, thus allowing greater stability, in fact the opposite result has been obtained. Lacking rules of priority, various authors have proposed subgroup names, and others use Linnaean families, with botanical endings (-aceae) or zoological ones (-idae).

The subgroups were based on gross morphology, such as vesicle form, presence or absence of an inner body, and presence and nature of the processes. Most workers agree that this form classification separates some intergrading taxa (e.g., *Micrhystridium* and *Veryhachium*) and that it obscures the close relationships of others, as for example the "Leiofusidae" (Eisenack, 1938b, as discussed by Combaz et al., 1967), or the Ooidaceae of Timofeev (1957).

Downie (1973) utilized a different system, with 10 groups identified by characteristic genera. Of these, his *Pachysphaera, Tasmanites,* and *Pterospermopsis* groups are here considered as prasinophyceans. The *Baltisphaeridium* group includes those of *Baltisphaeridium* type and *Priscogalea* type (many taxa with large preformed pylome were included in each of these types, whereas typical *Baltisphaeridium* has none), *Orthosphaeridium* type (that excysts by a median split), and *Goniosphaeridium* type. Taxa with no large processes, some with processes opening into the body cavity, and some with plugged processes were thus included in this group, which Downie suggested might be related to the tasmanitids (Prasinophyceae). The *Navifusa* group includes nonspinose elongate forms that may have cyclopyles. The Sphaeromorph acritarchs are relatively simple, lack processes, have a rather featureless structure, and may be prasinophyceans. *Leiosphaeridia* is a typical example, although numerous other similar genera have been described from the Precambrian, where they are dominant, and from the lower Paleozoic; they are rare after the Devonian. The *Micrhystridium-Veryhachium* group has large processes communicating with the vesicle and excysts by epityche or "cryptosuture." Some resemble certain living dinoflagellate cysts, and this group eventually may be referred there. This is an extremely important group in the Ordovician and Silurian, and locally may be common in some Mesozoic strata. Downie's *Leiofusa* Group (dominant in the Silurian and

Devonian) is characterized by an elongate body and excystment through lateral slit or epityche, and includes those with bipolar symmetry as well as others with processes only at one pole. The Acanthodiacrodian (= Diacrodian) group has bipolar ornamentation, is reported to have a double-layered wall like the dinoflagellates (Jux, 1971), and some may show indications of tabulation similar to that of dinoflagellates (Lister, 1970a); it is most characteristic of the basal Ordovician (Tremadocian). The last of Downie's groups, the *Visbysphaera* group, tends to have an inner body, and opens by cryptosuture; it includes the *Visbysphaera* type (with inner cyst), *Tyligmasoma* type, and *Diexallophasis* type (complex processes, but rarely with inner body).

Recognizing the problems inherent in an artificial system of groups and subgroups, and to avoid the implication that the acritarchs might all be related, some authors currently arrange taxa alphabetically, and avoid any suprageneric categories. Nearly 400 genera have been proposed to date for the nondinoflagellate acritarchs (including about 10 percent that are now placed with the Prasinophyceae), making an alphabetical arrangement increasingly cumbersome, as it separates some closely related or intergrading taxa and places highly unlike forms in close proximity. However, too little information is yet available for the development of a comprehensive Linnaean classification for the acritarchs. Some taxa are still unrecognizable, being based on inadequate original descriptions and figures, and as yet have not been restudied to provide details of the wall characters, nature of the processes, surface sculpture, and mechanism of excystment. Greater emphasis on such characters will result in improved stratigraphic utility, and may allow a better assessment of their relationships and their eventual assignment to definite families and higher taxa.

Suprageneric Categories Proposed for the Acritarcha

The various groups, subgroups, and families are indicated in chronological order, including the original or a modified definition, synonyms, and the present allocation of some that were later assigned elsewhere. Although listed for the sake of reference, as many are found in the current literature on this group, the non-Linnaean system is not recommended, and insufficient data are available as yet for a complete family classification.

1. Subgroup Acanthomorphitae Downie, Evitt & Sarjeant 1963

(= Subgroup Baltisphaeritae Staplin, Jansonius & Pocock 1965; part of Group Sphaerohystrichomorphida Timofeev 1966; Family Baltisphaeridiaceae Eisenack 1969)

Spherical or ellipsoidal vesicle, no inner body, no surface crests; simple or branching processes, solid or hollow; simple pylome, or no observed opening. Originally included both forms with preformed pylome and those that excyst by splitting.

2. Subgroup Polygonomorphitae Downie, Evitt & Sarjeant 1963

(= part of Group Sphaerohystrichomorphida Timofeev 1966)

Polygonal vesicle, no inner body, no crests; processes isolated or basally fused, few in number, rarely branching.

3. Subgroup Sphaeromorphitae Downie, Evitt & Sarjeant 1963

(= Family Leiosphaeridae Eisenack 1954; Family Leiosphaeridaceae Timofeev 1956; Subfamilies Protoleiosphaerideae and Leiosphaerideae Timofeev 1959; Family Leiosphaeridiaceae Mädler 1963; Subgroup Monosphaeritae Timofeev 1966; Group Sphaeromorphida and Subgroup Monosphaeritae Timofeev 1966; Subgroup Leiosphaeridae Maithy 1969)

Spherical to ellipsoidal vesicle, no inner body, surface smooth, punctate, or granular; no processes; may have pylome or open by splitting. Prasinophyta.

4. Subgroup Netromorphitae Downie, Evitt & Sarjeant 1963

(= Family Leiofusidae Eisenack 1938; Group Fusomorphida Timofeev 1966; Family Leiofusidaceae Eisenack 1969; Group Scaphomorphida Timofeev 1973a)

Elongate to fusiform vesicle, no inner body, smooth to granular surface; distally closed spines may be present at one or both poles; rare pylomes in *Leiofusa,* but most with no observed opening.

5. Subgroup Diacromorphitae Downie, Evitt & Sarjeant 1963

(= Family Diacrodiaceae Timofeev 1959; Family Diornatosphaeridae Downie 1958; Subfamilies Homodiacrodeae and Heterodiacrodeae Timofeev 1959; Family Trachydiacrodidae Deflandre & Deflandre-Rigaud 1961; Group Ellipsoidomorphida and Subgroup Homodiacromorphitae Timofeev 1966)

Spherical to ellipsoidal vesicle, no inner body; equatorial region with smooth or wrinkled surface, poles ornamented; no observed opening.

6. Subgroup Herkomorphitae Downie, Evitt & Sarjeant 1963

(= Cymatiosphaeraceae Mädler 1963; part of Group Sphaerohystrichomorphida Timofeev 1966; Group Reticulosphaeromorphitae Sinha 1969)

Spherical to ellipsoidal or subpolygonal vesicle, no inner body; surface with crests subdividing surface into polygonal fields; no median girdle; no observed opening. Prasinophyta.

7. Subgroup Pteromorphitae Downie, Evitt & Sarjeant 1963

(= Family Pterospermopsidae Eisenack 1954; Family Pterosphaeridiaceae Mädler 1963; Family Pterospermellaceae Eisenack 1972)

Spherical, ellipsoidal to polygonal vesicle, compressed; no inner body; equatorial flange may be supported by processes or radial folds; no other processes or crests; slitlike opening, probably preformed. Prasinophyta.

8. Subgroup Prismatomorphitae Downie, Evitt & Sarjeant 1963

(= Family Polyedryxidae Deunff 1961; Group Edromorphida Timofeev 1973a)

Prismatic to polygonal vesicle, sharp margin may be produced into a flange; surface smooth, granular, or reticulate; no observed openings. Prasinophyta.

9. Subgroup Oomorphitae Downie, Evitt & Sarjeant 1963

(= Family Ooidaceae Timofeev 1957; Group Ooidomorphida Timofeev 1965)

Spherical to elliptical vesicle, no inner body, surface smooth, striate, or granular, cluster of intertwining trabeculae, solid or hollow processes at one pole; no observed opening.

10. Subgroup Stephanomorphitae Downie, Evitt & Sarjeant 1963

Spherical or ovoid vesicle, no inner body,

flange at one pole; tubular processes, opening not observed. Dinoflagellates.

11. Subgroup Disphaeromorphitae Downie, Evitt & Sarjeant 1963

Spherical to ovoid cyst, with inner body, no crests or processes; simple opening, or none observable. Dinoflagellates.

12. Subgroup Dinetromorphitae Downie, Evitt & Sarjeant 1963

(= Family Diplotestidae Cookson & Eisenack 1960)

Fusiform to elongate body, may be curved; with inner body; no crests, rarely with processes; pylome or no observed opening. Dinoflagellates.

13. Subgroup Platymorphitae Downie, Evitt & Sarjeant, 1963

Circular, oval, or triangular outline, flattened; inner body of similar form; no crests or processes; surface smooth or granular; no observed opening. Probably dinoflagellates.

14. Group Tasmanititae Staplin, Jansonius & Pocock 1965

(= Family Tasmanaceae Sommer 1956)

Laminated wall with radial pores. Prasinophyta.

15. Subgroup Polysphaeritae Timofeev 1966 (Subgroup of Sphaeromorphida)

Clusters of spherical or subspherical individuals. ?Chlorophyceae.

16. Group Versimorphida Timofeev 1966

Pillow-shaped and wedge- or tongue-shaped forms. Questionable organisms.

17. Family Scuticabolaceae Loeblich III 1967 (=Ophiobolidae Deflandre, 1952)

Ovoid, with up to seven much elongated delicate processes from one pole; may open with a series of radial slits at opposite pole.

18. Subgroup Schizomorphitae Segroves 1967

Spherical, ellipsoidal, discoidal, or polygonal, variously sculptured; complete or incomplete equatorial rupture. Some are Prasinophyta.

19. Subgroup Scutellomorphitae Brito 1967

Rounded, flattened, discoid, undulating periphery resulting from numerous thick, blunt projections. Prasinophyta.

20. Subgroup Retractomorphitae Brito 1969

Triangular to polygonal, flattened, thick-walled vesicle, thin-walled processes that do not connect with vesicle cavity.

21. Subgroup Megasphaeromorphida Timofeev 1969 (of Group Sphaeromorphida)

Exceptionally large spherical forms that otherwise are referrable to Monosphaeritae.

22. Group Carminellidae Cramer & Díez de Cramer 1972

Spherical central body surrounded by open-ended outer wall (periderm) to which it is attached by radially arranged supporting filaments.

23. Group Geronidae Cramer & Díez de Cramer 1972

Oval to ellipsoidal body, surrounded by loose outer membrane that at one pole extends to form one or more elongate processes that may be surrounded by a cylindrical ectodermal skirt.

24. Subgroup Coryphomorphitae Vavrdová 1973

Spherical to polyhedral, sculpture concentrated at corners (angles).

25. Subgroup Porata Naumova & Umnova in Umnova 1975

Spherical to ovoid, with double wall and diverse surface sculpture, characterized by large, permanently present pylome, with smooth or corrugated margin and thickened encircling rim.

26. Group Mutabilimorpha Hermann [German] in Timofeev et al., 1976

Body undergoes major alterations in form during growth.

Acritarch Genera

In other chapters of this volume, many if not all described genera have been placed in a classification, with the generic citation including author and date. Because no such allocation is made here, for reasons discussed in the preceding section, validly described acritarch genera and their authors and date are merely listed here alphabetically for reference purposes. Lack of type material, and incomplete descriptions and inadequate illustrations in some cases, do not allow either positive recognition or placement in synonymy for all taxa. Thus, their presence in this list may not necessarily indicate acceptance as distinct genera.

Acanthodiacrodium Timofeev 1958 (syn.: *Dior-*

natosphaera Downie 1958; *Lophodiacrodium* Timofeev 1958; *Lophorytidodiacrodium* Timofeev 1958)

Acanthosphaera Naumova 1968

Aciculites Piel 1971

Acriora Wicander 1974

Acrum Fombella 1977

Actinotodissus Loeblich & Tappan 1978

Actinotophasis Loeblich & Wicander 1976

Actipilion Loeblich 1970

Adara Fombella 1977

Adarve Fombella 1977

Adornofusa Loeblich & Tappan 1978

Alisum Górka 1965

Alliumella Fanderflit in N. I. Umnova & Fanderflit 1971

Amazonites Sommer & Costa 1972

Ammonidium Lister 1970 (syn.: *Caiacorymbifer* Tappan & Loeblich 1971)

Amosopollis Cookson & Balme 1962

Angularia Samoylovich ex Norris & Sarjeant 1965

Angulisphaerina Lopukhin 1966

Anguloplanina Rudavskaya 1973

Annulatopsophosphaera Pykhova 1969

Anomaloplaisium Tappan & Loeblich 1971

Anthatractus Deunff 1961

Antruejadina Cramer, Díez, Rodrigues & Fombella 1976

Aranidium Yankauskas 1975

Arbusculidium Deunff 1968

Archaeodiscina Naumova 1960

Archaeofavosina Naumova 1960

Archaeomassulina Naumova 1968

Archaeoperisaccus Naumova 1960

Archaeopertusina Naumova 1960

Archaeopsophosphaera Naumova 1968

Archaeosacculina Naumova in Naumova & Pavlovskiy 1961

Aremoricanium Deunff 1955

Arkonia Burmann 1970

Arkonites Legault 1973

Ascostomocystis Drugg & Loeblich 1967

Asketopalla Loeblich & Tappan 1969

Attritasporites Combaz 1968

Aureosphaera Cramer, Díez, Rodrigues & Fombella 1976

Aureotesta Vavrdová 1972

Axisphaeridium Eisenack 1967

Bacisphaeridium Eisenack 1962

Baiomeniscus Loeblich 1970

Balmeella Pant & Mehra 1963

Baltisphaeridium Eisenack 1958

Bambuites Sommer 1971

Barakella Cramer & Díez 1977

Barathrisphaeridium Wicander 1974

Bavlinella Shepeleva 1962

Brazilea Tiwari & Navale 1967

Brochopsophosphaera Shepeleva & Timofeev 1963

Buedingiisphaeridium Schaarschmidt 1963

Candelasphaeridium Deunff 1978

Cantabrica Cramer, Díez, Rodrigues & Fombella 1976

Carminella Cramer 1968

Celtiberium Fombella 1977

Cephalia Sah & Kar 1974

Cepillum Cramer 1964

Chabiosphaera Drábek 1972

Cheleutochroa Loeblich & Tappan 1978

Chomotriletes Naumova 1939

Chuttecloska Loeblich & Wicander 1976

Circulisporites de Jersey 1962

Cleithronetrum Loeblich & Tappan 1978

Comasphaeridium Staplin, Jansonius & Pocock 1965

Comptaluta Tappan & Loeblich 1971

Concentrites Venkatachala, Bhandari, Chaube & Rawat 1974

Congoites Tiwari & Navale 1967

Conradidium Stockmans & Willière 1969

Constrictosphaerina Lopukhin 1966

Cornplanktona Sah & Kar 1974

Cornutosphaera Sin in Sin & Liu 1973

Coryphidium Vavrdová 1972 (syn.: *Octogonium* F. Martin 1974)

Costatilobus Playford 1977

Costatosphaerina Lopukhin 1966

Crassiangulina Jardiné, Combaz, Magloire, Peniguel & Vachey 1972

Crassisphaeridium Wicander 1974

Cryptosphaera Sah & Kar 1974

Cryptostromatium Moreau-Benoit 1974

Cuatrifolia Cramer, Díez, Rodrigues & Fombella 1976

Cyclopsiella Drugg & Loeblich 1967

Cymatiogalea Deunff 1961 (syn.: *Archaeohystrichosphaeridium* Timofeev ex Loeblich & Tappan 1976)

Cymatiomorpha Elsik & Dilcher 1974

Cymbosphaeridium Lister 1970

Cystidiopsis Nagy 1965

Dactylofusa Brito & Santos 1965

Daillydium Stockmans & Willière 1969

Dasydiacrodium Timofeev ex Deflandre &
Deflandre-Rigaud 1961

Dasypilula Loeblich & Wicander 1976

Dateriocradus Tappan & Loeblich 1971

Demorhethium Loeblich & Wicander 1974

Deunffia Downie 1960

Deusilites Hemer & Nygreen 1967

Diaphorochroa Wicander 1974

Dicommopalla Loeblich 1970

Dicrodiacrodium Burmann 1968

Dictyosphaera Sin & Liu 1973

Diexallophasis Loeblich 1970

Dilatisphaera Lister 1970

Dimastigobolus Deflandre 1936

Disectispora Tiwari & Navale 1967

Disparifusa Loeblich 1970

Divietipellis Wicander 1974

Domasia Downie 1960

Domasiella Eisenack 1969

Dominopolia Kir'yanov 1974

Ecmelostoiba Wicander 1974

Ecthymabrachion Wicander 1974

Ecthymapalla Loeblich & Wicander 1976

Ectypolopus Loeblich & Wicander 1976

Eisenackidium Cramer & Díez 1968 (syn.:
Crameria Lister 1970)

Elektoriskos Loeblich 1970

Eliasum Fombella 1977

Ellipsaletes Cramer 1966

Eomicrhystridium Deflandre 1968

Ephelopalla Wicander 1974

Ericanthea Cramer & Díez 1977

Estiastra Eisenack 1959

Ethmosphaeridium Timofeev 1973 (syn.: *Dictyosphaeridium* Timofeev 1969, non W. Wetzel, 1952)

Eupoikilofusa Cramer 1971 (syn.: *Pteroidea*
Sheshegova 1975)

Evittia Brito 1967

Excultibrachium Loeblich & Tappan 1978

Exilisphaeridium Wicander 1974

Exochoderma Wicander 1974

Favososphaera Burmann 1972

Favososphaeridium Timofeev 1966

Filisphaeridium Staplin, Jansonius & Pocock
1965

Fimbriaglomerella Loeblich & Drugg 1968

Florisphaeridium Lister 1970

Foveofusa Lele & Chandra 1972

Frankea Burmann 1970

Fueloepia Nagy 1965

Fusilites Hemer & Nygreen 1967

Geron Cramer 1969

Glottimorpha Timofeev 1966

Gochtia Eisenack 1968

Goniolopadion Playford 1977

Goniosphaeridium Eisenack 1969

Gorgonisphaeridium Staplin, Jansonius &
Pocock 1965

Goryniella Timofeev 1973

Gracilisphaeridium Eisenack & Cramer in
Eisenack, Cramer & Díez 1973

Granomarginata Naumova 1960

Grebespora Jansonius 1962

Guttatisphaeridium Wicander 1974

Gyalorhethium Loeblich & Tappan 1978

Gyratosphaerina Lopukhin 1966

Hapsidopalla Playford 1977

Helosphaeridium Lister 1970

Hemibaltisphaeridium Cramer 1971

Hemideunffia Thusu 1973 (syn.: *Downiea* Pöthe
de Baldis 1975, non Hussain in Caro 1973)

Hemisphaerium Hemer & Nygreen 1967

Hetermorphidion Loeblich & Wicander 1976

Hispanaediscus Cramer 1966

Holothuriadeigma Loeblich 1970

Iberosphaeridium Cramer, Díez, Rodrigues &
Fombella 1976

Impluviculus Loeblich & Tappan 1969

Induoglobus Loeblich & Wicander 1976

Induropilarius Deflandre & Ters in Ters &
Deflandre 1966

Jubatasphaera Loeblich & Wicander 1976

Lacunalites Hemer & Nygreen 1967

Lacunopsophosphaera Pykhova 1969

Lanveocia Deunff 1978

Laterilites Hemer & Nygreen 1967

Lecaniella Cookson & Eisenack 1962

Leioaletes Staplin 1960

Leiodiscina Naumova 1968

Leiofusa Eisenack 1938

Leioligotriletes Timofeev 1960

Leiomarginata Naumova 1960

Leiopsophosphaera Naumova 1968

Leiovalia Eisenack 1965

Leoniella Cramer 1964

Lophominuscula Naumova 1960

Lophosphaeridium Timofeev ex Downie 1963

Lunulidia Eisenack 1958

Lusata Naumova 1968

Lusatia Burmann 1970

Macroptycha Timofeev in Timofeev, German & Mikhaylova 1976

Mamillata Naumova 1968

Margosphaera Nagy 1965

Marrocanium Cramer, Kanes, Díez & Christopher 1974

Mecsekia Hajós 1966

Megasacculina Naumova 1960

Mehlisphaeridium Segroves 1967

Membranolimbus Malyavkina ex Norris & Sarjeant 1965

Metaleiofusa Wall 1965

Micrhystridium Deflandre 1937

Microconcentrica Naumova 1960

Moyeria Thusu 1973

Multiplicisphaeridium Staplin 1961

Naevisphaeridium Wicander 1974

Nanocyclopia Loeblich & Wicander 1976

Navifusa Combaz, Lange & Pansart 1967

Neoveryhachium Cramer 1971

Nevidia Vavrdová 1966

Nothooidium Loeblich & Tappan 1976

Nucellohystrichosphaera Timofeev & German in Timofeev, German & Mikhaylova 1976

Nucellosphaeridium Timofeev 1966

Ocridosphaeridium Timofeev 1973

Octaplata Sah & Kar 1974

Octoedryxium Rudavskaya 1973 (syn.: *Quadratimorpha* Sin & Liu 1973)

Odontomorpha Hajós 1966

Oligopogonium Loeblich & Tappan 1976

Onondagella Cramer 1966

Ooidium Timofeev 1957 (syn.: *Zonooidium* Timofeev 1957)

Operculites Newman 1965

Oppilatala Loeblich & Wicander 1976

Ordovicidium Tappan & Loeblich 1971

Orthosphaeridium Eisenack 1968 (syn.: *Baltisphaera* Burmann 1970)

Ozotobrachion Loeblich & Drugg 1968

Pachysphaeridium Burmann 1970

Palacanthus Wicander 1974

Palaeobion O. Wetzel 1961

Palaeocryptidium Deflandre 1955

Palamphimorphium Deflandre & Ters in Ters & Deflandre 1966

Palanaea Sah & Kar 1974

Paleopedicystus Staplin 1961

Papinochium Deflandre & Ters in Ters & Deflandre 1966

Pavlovskya Naumova in Naumova & Pavlovskiy 1961

Peltacystia Balme & Segroves 1966

Percultisphaera Lister 1970

Perissolagonella Loeblich & Wicander 1976

Petalosporites Agasie 1969

Peteinosphaeridium Staplin, Jansonius & Pocock 1965

Pheoclosterium Tappan & Loeblich 1971

Picostella Cramer, Allam, Kanes & Díez 1974

Pilasporites Balme & Hennelly 1956

Piliferosphaera Loeblich 1970

Pirea Vavrdová 1972

Podoliella Timofeev 1973

Podolina German in Timofeev, German & Mikhaylova 1976

Poikilofusa Staplin, Jansonius & Pocock 1965

Polyancistrodorus Loeblich & Tappan 1969

Polydeunffia Cramer 1971

Polygonium Vavrdová 1966

Polyplanifer Cramer 1964

Portalites Hemer & Nygreen 1967

Priscogalea Deunff 1961

Priscotheca Deunff 1961

Prismatocystis Habib 1969

Protoarchaeosacculina Naumova 1960

Prototrematosphaeridium Timofeev 1962

Psenotopus Tappan & Loeblich 1971

Pseudomasia De Coninck 1968

Psilosphaera Sah & Kar 1974

Pulvinomorpha Timofeev 1966

Pulvinosphaeridium Eisenack 1954

Pustulisphaeridium Wicander 1974

Puteoscortum Wicander & Loeblich 1977

Pyramidosporites Segroves 1967

Quadraditum Cramer 1964

Quadrisporites Hennelly 1958

Quadruilobus Tappan & Loeblich 1971

Radialetes Playford 1962

Retisphaeridium Staplin, Jansonius & Pocock 1965

Revinotesta Vanguestaine 1974

Rhachosoarium Tappan & Loeblich 1971

Rhiptosocherma Loeblich & Tappan 1978

Rhopaliophora Tappan & Loeblich 1971

Riculasphaera Loeblich & Drugg 1968

Rifeniata Naumova 1968

Rugocystis Venkatachala, Bhandari, Chaube & Rawat 1974

Rugulasphaeridium Bharadwaj, Tiwari & Venkatachala 1971

Rugulidium Cramer, Kanes, Díez & Christopher 1974

Saharidia Combaz 1968

Saturnus Cramer 1966

Scaphita Timofeev in Timofeev, German & Mikhaylova 1976

Scaphospinosa Timofeev in Timofeev, German & Mikhaylova 1976

Schepelevia Aristova 1972

Schismatosphaeridium Staplin, Jansonius & Pocock 1965

Schizocystia Cookson & Eisenack 1962

Schizodiacrodium Burmann 1968 (syn.: *Adorfia* Burmann 1970)

Scuticabolus Loeblich III 1967 (syn.: *Ophiobolus* O. Wetzel 1933, non Riess 1854)

Senispora Bandyopadhyay 1969

Sigmopollis Hedlund 1965

Singraulipollenites Sinha 1969

Sol Cramer 1964 (syn.: *Solita* Cramer 1966)

Solatisphaera Wicander & Loeblich 1977

Solisphaeridium Staplin, Jansonius & Pocock 1965

Sphaeroporalites Hemer & Nygreen 1967

Spheripollenites Couper 1958

Spumiosa Naumova 1968

Spumosina Naumova 1968

Spurimoyeria Wicander & Loeblich 1977

Stelliferidium Deunff, Górka & Rauscher 1974

Stellinium Jardiné, Combaz, Magloire, Peniguel & Vachey 1972

Stenozonoligotriletes Timofeev 1960

Stictosphaeridium Timofeev 1966

Striatotheca Burmann 1970

Stylodiniopsis Eisenack 1954

Sulcatosphaeridium N. I. Umnova in N. I. Umnova & Yakovlev 1970

Sylvanidium Loeblich 1970

Symplassosphaeridium Timofeev 1960

Synsphaeridium Eisenack 1965

Tectitheca Burmann 1968

Temporina Sah & Kar 1974

Tetraedrixium Timofeev 1973

Tetraletes Cramer 1966

Tetraporina Naumova 1939

Tetrapterites Sullivan & Hibbert 1964

Thysanoprobolus Loeblich & Tappan 1970

Tophoporata Pykhova 1973

Tornacia Stockmans & Willière 1966

Toroformis Lopukhin 1972

Trachyhystrichosphaera Timofeev & German in Timofeev, German & Mikhaylova 1976

Trachyoligotriletes Timofeev 1960

Trachyrytidodiacrodium Timofeev 1959

Trachyzonodiacrodium Timofeev 1959

Trematoligotriletum Timofeev 1959

Trematophora Eisenack 1965

Trematosphaeridium Timofeev 1959

Triangumorpha Sin & Liu 1973

Trichosphaeridium Timofeev 1966

Trunculumarium Loeblich & Tappan 1976

Tubulospina Davey 1970

Tunisphaeridium Deunff & Evitt 1968

Tyligmasoma Playford 1977 (syn.: *Triangulina* Cramer 1964, non Malyavkina 1949)

Tylotopalla Loeblich 1970

Umbellasphaeridium Jardiné, Combaz, Magloire, Peniguel & Vachey 1972

Uncinasphaera Wicander 1974

Unellium Rauscher 1969

Vallenia Pedersen in Bondesen, Pedersen & Jørgensen 1967

Vavososphaeridium Timofeev 1959

Vavrodovella Loeblich & Tappan 1976 (syn.: *Tetradinium* Vavrdová 1973, non Klebs 1912; *Tetraniveum* Vavrdová 1976)

Veryhachium Deunff ex Downie 1959 (syn.: *Wilsonastrum* Jansonius 1962; *Simsangia* Baksi 1962; *Hystrichotriangulatum* Andreeva 1966; *Dorsennidium* Wicander 1974)

Vesiculosphaerina Lopukhin 1966

Villosacapsula Loeblich & Tappan 1976

Virgatasporites Combaz 1968

Visbysphaera Lister 1970

Vogtlandia Burmann 1970

Vulcanisphaera Deunff 1961

Winwaloeusia Deunff 1977

REFERENCES

Alberti, Gerhard, Zur Kenntnis mesozoischer und alttertiärer Dinoflagellaten und Hystrichosphaerideen von Nord- und Mitteldeutschland sowie einigen anderen europäischen Gebieten. *Palaeontographica Abt. A,* v. 116, p. 1–58, pls. 1–12, 1961.

Andreeva, E. M., Rastitel'nye mikrofossilii neyasnogo sistematicheskogo polozheniya [Plant microfossils of unestablished systematic position], p. 114—135 (Tom 3, pls. 1—29). *In* I. M. Pokrovskaya, *Paleopalinologiya*, Tom 1, *Metodika paleopalinologischeskikh issledovaniy i morfologiya nekotorykh iskopaemykh spor, pyl'tsy i drugikh rastitel'nykh mikrofossiliy*, Trudȳ Vses. Nauchno-issled. Geol. Inst., vyp. 141, 1966.

Bachmann, Alfred, and M. E. Schmid, Mikrofossilien aus dem österreichischen Silur. *Verh. Geol. Bundesanst., Wien*, 1964 no. 1, p. 53—64, pls. 1—6, 1964.

Balme, B. E., and K. L. Segroves, *Peltacystia* gen. nov.; a microfossil of uncertain affinities from the Permian of Western Australia. *J. Proc. R. Soc. West. Aust.*, v. 49, p. 26—31, figs. 1—4, 1966.

Baschnagel, R. A., Some microfossils from the Onondaga Chert of central New York. *Bull. Buffalo Soc. Nat. Sci.*, v. 17 (3), p. 1—8, pl. 1, 1942.

Berry, W. B. N., and A. J. Boucot, Continental stability—a Silurian point of view. *J. Geophys. Res.*, v. 72, p. 2254—2256, 1967.

Bharadwaj, D. C., R. S. Tiwari, and B. S. Venkatachala, An upper Devonian mioflora from New Albany Shale, Kentucky, U.S.A. *Palaeobotanist*, v. 19, p. 29—40, pls. 1—5, 1971.

Bouché, P. M., Revue bibliographique des Hystrichosphères (Acritarches) du Paléozoique et du Permo-Trias. *Revue Inst. Fr. Pétrole*, v. 19, p. 3—31, 1964.

Brito, I. M., Novos microfósseis Devonianos do Maranhão. *Publçāo Avuls. Univ. Bahia Esc. Geol.*, no. 2, p. 1—4, 1 pl., 1965.

Brito, I. M., Novo subgrupo de Acritarcha do Devoniano do Maranhão. *Anais Acad. Bras. Cienc.*, v. 39, p. 163—166, pls. 1—3, figs. 1, 2, 1967a.

Brito, I. M., Contribuição ao conhecimento dos microfósseis Devonianos de Pernambuco II—Acritarcha Pteromorphitae. *Anais Acad. Bras. Cienc.*, v. 39, p. 285—287, pl. 1, 1967b.

Brito, I. M., Os Acritarcha. *Notas prelim. Estud. Div. Geol. Miner. Bras.*, no. 138, p. 1—21, pls. 1, 2, 1967c.

Brito, I. M., Silurian and Devonian Acritarcha from Maranhão Basin, Brazil. *Micropaleontology*, v. 13, p. 473—482, pls. 1, 2, 1967d.

Brito, I. M., Contribuição ao conhecimento dos microfósseis Silurianos e Devonianos da Bacía do Maranhão. III—O gênero *Veryhachium* (Acritarcha). *Bolm Geol. Inst. Geosciénc. Univ. Fed. Rio de J.*, no. 2, p. 11—17, 2 pls., 1968.

Brito, I. M., Un nouveau sous-groupe d'Acritarche. *Bolm Geol. Inst. Geosciénc. Univ. Fed. Rio de J.*, no. 4, p. 11—13, figs. 1—4, 1969.

Brito, I. M., Contribuição ao conhecimento dos microfósseis Silurianos e Devonianos da Bacía do Maranhão. V—Acritarcha Herkomorphitae e Pris-matomorphitae. *Anais Acad. Bras. Cienc.*, v. 43 (Supl.), p. 201—208, figs. 1—10, 1971.

Brito, I. M., and Aymar da Silva Santos, Contribuição ao conhecimento dos microfósseis Silurianos e Devonianos da Bacía do Maranhão. Parte 1. Os Netromorphitae (Leiofusidae). *Notas Prelim. Estud. Div. Miner. Bras.*, no. 129, p. 1—23, pls. 1, 2, 1 fig., 1965.

Brosius, M., and P. Bitterli, Middle Triassic hystrichosphaerids from salt-wells Riburg-15 and -17, Switzerland. *Bull. Verein. Schweiz. Petrol.-Geol. u.-Ing.*, v. 28 (74), p. 33—49, 2 pls., 8 figs., 1961.

Burmann, Gusti, Diacrodien aus dem unteren Ordovizium. *Paläont. Abh., Abt. B*, v. 2, p. 639—652, pls. 1—7, 1968.

Burmann, Gusti, Weitere organische Mikrofossilien aus dem unteren Ordovizium. *Paläont. Abh., Abt. B*, v. 3, p. 289—332, pls. 2—19, 1970.

Christopher, R. A., Morphology and taxonomic status of *Pseudoschizaea* Thiergart and Frantz ex R. Potonié emend. *Micropaleontology*, v. 22, p. 143—150, pl. 1, fig. 1, 1976.

Churchill, D. M., and W. A. S. Sarjeant, Freshwater microplankton from Flandrian (Holocene) peats of southwestern Australia. *Grana Palynol.*, v. 3 (3) 1962, p. 29—53, pls. 1, 2, figs. 1—17, 1963.

Cloud, Preston, and Alta Germs, New Pre-Paleozoic nannofossils from the Stoer Formation (Torridonian), northwest Scotland. *Bull. Geol. Soc. Am.*, v. 82, p. 3469—3474, figs. 1—14, 1971.

Combaz, André, Un microbios du trémadocien dans un sondage d'Hassi-Messaoud. *Act. Soc. Linn. Bordeaux*, v. 104, sér. B, no. 29, p. 1—26, pls. 1—4, figs. 1—5, 1968.

Combaz, André, F. W. Lange, and J. Pansart, Les "Leiofusidae" Eisenack, 1938. *Rev. Palaeobot. Palynol.*, v. 1, p. 291—307, pls. 1, 2, figs. 1—3, 1967.

Combaz, André, and Gérard Peniguel, Étude palynostratigraphique de l'Ordovicien dans quelques sondages du Bassin de Canning (Australie Occidentale). *Bull. Centre Rech. Pau, SNPA*, v. 6, p. 121—167, pls. 1—4, figs. 1—3, 1972.

Cookson, I. C., Additional microplankton from Australian late Mesozoic and Tertiary sediments. *Aust. J. Mar. Freshwat. Res.*, v. 7, p. 183—191, 2 pls., 1956.

Cookson, I. C., and Alfred Eisenack, Microplankton from Australian and New Guinea upper Mesozoic sediments. *Proc. R. Soc. Vict.*, v. 70, p. 19—79, pls. 1—12, figs. 1—20, 1958.

Cookson, I. C., and Alfred Eisenack, Some Cretaceous and Tertiary microfossils from Western Australia. *Proc. R. Soc. Vict.*, v. 75, p. 269—273, pl. 37, 1962a.

Cookson, I. C., and Alfred Eisenack, Additional microplankton from Australian Cretaceous sediments. *Micropaleontology*, v. 8, p. 485—507, pls. 1—7, figs. 1, 2, 1962b.

Cookson, I. C., and Alfred Eisenack, Mikroplankton aus Australischen mesozoischen und tertiären Sedimenten. *Palaeontographica Abt. B*, v. 148, p. 44–93, pls. 20–29, figs. 1–3, 1974.

Cramer, F. H., Microplankton from three Palaeozoic formations in the Province of León, NW-Spain. *Leid. Geol. Meded.*, v. 30, p. 253–361, pls. 1–24, figs. 1–56, 1964.

Cramer, F. H., Some acritarchs from the San Pedro Formation (Gedinnien) of the Cantabric Mountains in Spain. *Bull. Soc. Belge Géol. Paléont. Hydrol.*, v. 73 (1964), p. 33–38, pls. 1, 2, 1965.

Cramer, F. H., Palynology of Silurian and Devonian rocks in northwest Spain. *Boln Inst. Géol. Min. Esp.*, v. 77, p. 225–286, pls. 1–4, figs. 1–7, 1966a.

Cramer, F. H., Additional morphographic information on some characteristic acritarchs of the San Pedro and Furada Formations (Silurian Devonian Boundary) in León and Asturias, Spain. *Notas Comun. Inst. Geol. Min. Esp.*, no. 83, p. 27–48, pls. 1–5, 1966b.

Cramer, F. H., Palynomorphs from the Siluro-Devonian boundary in NW Spain. *Notas Comun. Inst. Geol. Min. Esp.*, no. 85, p. 71–82, pls. 1–3, 1966c.

Cramer, F. H., Hoegispheres and other microfossils incertae sedis of the San Pedro Formation (Siluro Devonian boundary) near Valporquero, León, NW Spain. *Notas Comun. Inst. Geol. Min. Esp.*, no. 86, p. 75–94, pls. 1, 2, figs. 1, 2, 1966d.

Cramer, F. H., Palynologic microfossils of the Middle Silurian Maplewood Shale in New York. *Revue Micropaléont.*, v. 11, p. 61–70, 1 pl., 1 fig., 1968a.

Cramer, F. H., Silurian palynologic microfossils and paleolatitudes. *Neues Jb. Geol. Paläont. Mh.* 1968 (10), p. 591–597, 1 fig., 1968b.

Cramer, F. H., Possible implications for Silurian paleogeography from phytoplancton assemblages of the Rose Hill and Tuscarora Formations of Pennsylvania. *J. Paleont.*, v. 43, p. 485–491, pl. 70, 2 figs., 1969a.

Cramer, F. H., *Geron*, an acritarch genus from the Silurian. *Bull. Soc. Belge Géol. Paléont. Hydrol.*, v. 76, p. 217–225, pl. 1, figs. 1–5, 1969b.

Cramer, F. H., Middle Silurian continental movement estimated from phytoplankton-facies transgression. *Earth Planet. Sci. Letters*, v. 10, p. 87–93, 3 figs., 1970a.

Cramer, F. H., Acritarchs and chitinozoans from the Silurian Ross Brook Formation, Nova Scotia. *J. Geol.*, v. 78, p. 745–749, fig. 1, 1970b.

Cramer, F. H., Distribution of selected Silurian acritarchs. *Revta Esp. Micropaleont., Numero Extraord. 1970 [1971]*, p. 1–202, pls. 1–23, 1971a.

Cramer, F. H., Implications from middle Paleozoic palynofacies transgressions for the rate of crustal movement, especially during the Wenlockian. *Anais Acad. Bras. Cienc.*, v. 43 (Supl.), p. 51–66, pls. 1–4, figs. 1–6, 1971b.

Cramer, F. H., Benssyah Allam, W. H. Kanes, and M. C. R. Díez, Upper Arenigian to lower Llanvirnian acritarchs from the subsurface of the Tadla Basin in Morocco. *Palaeontographica Abt. B*, v. 145, p. 182–190, pls. 27, 28, 1 fig., 1974.

Cramer, F. H., and M. del C. Díez de Cramer, Consideraciones taxonómicas sobre las acritarcas del Silúrico Medio y Superior del Norte de España. Las acritarcas acantomorfíticas. *Boln. Geol. Min., Esp.*, v. 79, p. 541–574, pls. 1–21, figs. 1–9, 1968.

Cramer, F. H., and Maria C. Díez de Cramer, Acritarchs from the Lower Silurian Neahga Formation, Niagara Peninsula, North America. *Can. J. Earth Sci.*, v. 7, p. 1077–1085, figs. 1, 2, 1970.

Cramer, F. H., and M. C. Díez de Cramer, North American Silurian palynofacies and their spatial arrangement: Acritarchs. *Palaeontographica Abt. B*, v. 138, p. 107–180, pls. 31–36, figs. 1–8, 1972a.

Cramer, F. H., and M. C. Díez de Cramer, Acritarchs from the upper middle Cambrian Oville Formation of León, northwestern Spain. *Revta Esp. Micropaleont., Numero Extraord. XXX Aniver. E. N. Adaro*, p. 39–50, pls. 1, 2, 1972b.

Cramer, F. H., and M. C. R. Díez, Early Paleozoic palynomorph provinces and paleoclimate, p. 177–188, figs. 1–4. *In* C. A. Ross, *Paleogeographic provinces and provinciality*. Spec. Publs. Soc. Econ. Paleont. Miner., Tulsa, no. 21, 1974.

Cramer, F. H., and M. C. R. Díez, Acritarchs from the La Vid Shales (Emsian to lower Couvinian) at Colle, León, Spain. *Palaeontographica Abt. B*, v. 158, p. 72–103, pls. 1–7, 1976a.

Cramer, F. H., and M. C. R. Díez, Seven new late Arenigian species of the Acritarch genus *Coryphidium* Vavrdová, 1972. *Paläont. Z.*, v. 50, p. 201–208, pl. 23, figs. 1, 2, 1976b.

Cramer, F. H., and M. C. R. Díez, Late Arenigian (Ordovician) acritarchs from Cis-Saharan Morocco. *Micropaleontology*, v. 23, p. 339–360, pls. 1–6, figs. 1–3, 1977.

Cramer, F. H., M. D. C. R. Díez, R. M. Rodriguez, and M. A. Fombella, Acritarcos de la formacion San Pedro (Silurico Superior) de Torrestio, Provincia de Leon, España. *Revta Esp. Micropaleont.*, v. 8, p. 439–452, pls. 1, 2, figs. 1, 2, 1976.

Cramer, F. H., W. H. Kanes, M. C. D. Díez, and R. A. Christopher, Early Ordovician acritarchs from the Tadla Basin of Morocco. *Palaeontographica Abt. B*, v. 146, p. 57–64, pls. 25, 26, 1974.

Davey, R. J., Non calcareous microplankton from the Cenomanian of England, northern France and North America, Part II. *Bull. Br. Mus. Nat. Hist., Geology*, v. 18, p. 333–397, pls. 1–10, figs. 1–9, 1970.

DeConinck, Jan, Dinophyceae et Acritarcha de l'Yprésien du Sondage de Kallo. *Mém. Inst. R. Sci. Nat. Belg.,* no. 161 (1968), p. 1—67, pls. 1—17, figs. 1, 2, 1969.

Deflandre, Georges, Microfossiles des silex crétacés. I. Généralités. Flagellés. *Annls. Paléont.,* v. 25, p. 151—191, pls. 1—10, 1936.

Deflandre, Georges, Sur les hystrichosphères des calcaires siluriens de la Montagne Noire. *C. R. Hebd. Séanc. Acad. Sci., Paris,* v. 215, p. 475—476, 16 figs., 1942.

Deflandre, Georges, Microfossiles des calcaires siluriens de la Montagne Noire. *Annls. Paléont.,* v. 31 (1945), p. 39—75, pls. 1—3, 41 figs., 1946.

Deflandre, Georges, Le problème des hystrichosphères. *Bull. Inst. Océanogr. Monaco,* v. 44 (918), p. 1—23, figs. 1—5, 1947.

Deflandre, Georges, Systématique des hystrichosphaeridés: sur l'acception du genre *Cymatiosphaera* O. Wetzel. *C. R. Somm. Séanc. Soc. Géol. Fr.,* no. 12, p. 257—258, 1954.

Deflandre, Georges, Sur l'existence, dès le Précambrien, d'Acritarches du type Acanthomorphitae: *Eomicrhystridium* nov. gen. Typification du genre *Palaeocryptidium* Defl. 1955. *C. R. Hebd. Séanc. Acad. Sci., Paris,* v. 266, Sér. D, p. 2385—2389, pl. 1, figs. 1—3, 1968.

Deflandre, Georges, and M. Deflandre-Rigaud, Nomenclature et systematique des hystrichosphères (sens. lat.). Observations et rectifications. *Revue Micropaléont.,* v. 4, p. 190—196, 11 figs., 1962.

Deflandre, Georges, and M. Deflandre, Acritarches I Polygonomorphitae-Netromorphitae pro parte, Appendice: genres *Deflandrastrum* Combaz et *Wilsonastrum* Jansonius. *Fichier Micropaléont. Général Sér. 12,* Arch. Orig. Centr. Document., Centre Nat. Rech. Sci., no. 392, p. i—x, fiches 1948—2175, 1964a.

Deflandre, Georges, and M. Deflandre, Notes sur les acritarches. *Revue Micropaléont.,* v. 7, p. 111—114, 1964b.

Deflandre, Georges, and M. Deflandre, Acritarches II, Acanthomorphitae 1, Genre *Micrhystridium* Deflandre sens. lat. *Fichier Micropaléont. Général Sér. 13,* Arch. Orig. Centre Document., Centre Nat. Rech. Sci., no. 402, p. i—v, fiches 2176—2521, 1965a.

Deflandre, Georges, and M. Deflandre, Remarques critiques sur le genre *Micrhystridium* Defl. *Revue Micropaléont.,* v. 8, p. 85—89, 1965b.

Deflandre, Georges, and J. C. Foucher, *Diacrocanthidium* nov. gen., Diacrodien présumé du Crétacé pourvu d'un archéopyle. Affinités péridiniennes des Diacrodiens? *Cah. Micropaléont.,* sér. 1 (5), Arch. Orig. Centre Document. C.N.R.S., no. 439, p. 1—5, pls. 1, 2, 1967.

Deflandre, Georges, and W. A. S. Sarjeant, Nouvel examen de quelques holotypes de dinoflagellés fossiles et d'acritarches. *Cah. Micropaléont.,* sér. 2 (1), Arch. Orig. Centre Document., C.N.R.S., no. 466, p. 1—10, 1 pl., 1970.

Deflandre, Georges, and Mireille Ters, Sur la présence d'acritarches ordoviciennes dans les schistes subardoisiers de la région de la Mothe-Achard (Vendée). Extension du Silurien (grès armoricain et schistes d'Angers) en Vendée littorale. *C. R. Hebd. Séanc. Acad. Sci., Paris,* v. 262, sér. D, p. 237—240, 1 pl., 1966.

Deunff, Jean, Sur la présence des microorganismes (Hystrichosphères) dans les schistes ordoviciens du Finistère. *C. R. Hebd. Séanc. Acad. Sci., Paris,* v. 233, p. 321—323, figs. 1—6, 1951.

Deunff, Jean, Microorganismes planctoniques (Hystrichosphères) dans le Dévonien du Massif Armoricain. *C. R. Somm. Séanc. Soc. Géol. Fr.,* no. 11, p. 239—242, 8 figs., 1954a.

Deunff, Jean, *Veryhachium* genre nouveau d'Hystrichosphère du Primaire. *C. R. Somm. Séanc. Soc. Géol. Fr.,* no. 13, p. 305—307, 1954b.

Deunff, Jean, Sur un microplancton du Dévonien du Canada recelant des types nouveaux d'Hystrichosphères. *C. R. Hebd. Séanc. Acad. Sci., Paris,* v. 239, p. 1064—1066, 17 figs., 1954c.

Deunff, Jean, Un microplancton fossile Dévonien à Hystrichosphères du continent Nord-Americain. *Bull. Microsc. Appl.,* sér. 2, v. 5, p. 138—150, pls. 1—4, 1955a.

Deunff, Jean, *Aremoricanium,* genre nouveau d'Hystrichosphères du Silurien breton. *C. R. Somm. Séanc. Soc. Géol. Fr.,* no. 11, p. 227—229, figs. 1—3, 1955b.

Deunff, Jean, Microorganismes nouveaux (Hystrichosphères) du Dévonien de l'Amérique du Nord. *Bull. Soc. Géol. Minér. Bretagne,* n. sér., v. 2, p. 5—14, 2 pls., 1957.

Deunff, Jean, Microorganismes planctoniques du Primaire armoricain I. Ordovicien du Veryhac'h (Presqu'île de Crozon). *Bull. Soc. Géol. Minér. Bretagne,* n. sér., fasc. 2 (1958), p. 1—41, pls. 1—12, 1959.

Deunff, Jean, Formations récifales et Hystrichosphères. *Bull. Soc. Géol. Fr.,* sér. 7, v. 1, 409—410, 1960.

Deunff, Jean, Un microplancton à Hystrichosphères dans le Trémadoc du Sahara. *Revue Micropaléont.,* v. 4, p. 37—52, pls. 1—3, 1961a.

Deunff, Jean, Quelques précisions concernant les Hystrichosphères du Dévonien du Canada. *C. R. Somm. Séanc. Soc. Géol. Fr.,* no. 8, p. 216—218, 1961b.

Deunff, Jean, Le genre *Duvernaysphaera* Staplin. *Grana Palynol.,* v. 5, p. 210—215, figs. 1—7, 1964a.

Deunff, Jean, Systématique du microplancton fossile à acritarches. Révision de deux genres de l'Ordovicien inférieur. *Revue Micropaléont.,* v. 7, p. 119—124, pl. 1, 1964b.

Deunff, Jean, Acritarches du Dévonien Supérieur de

la Presqu'île de Crozon. *C. R. Somm. Séanc. Soc. Géol. Fr.*, fasc. 5, p. 162−164, figs. 1−8, 1965.

Deunff, Jean, *Recherches sur les microplanctons du Dévonien (Acritarches et Dinophyceae).* Thesis, Univ. Rennes, p. 1−168, pls. 1−14, I−XII, 1966a.

Deunff, Jean, Acritarches du Dévonien de Tunisie. *C. R. Somm. Séanc. Soc. Géol. Fr.*, fasc. 1, p. 22−24, figs. 1−8, 1966b.

Deunff, Jean, Présence d'Acritarches dans une série dévonienne du Lac Huron (Canada). *C. R. Somm. Séanc. Soc. Géol. Fr.*, fasc. 6, p. 258−260, figs. 1−19, 1967.

Deunff, Jean, *Arbusculidium*, genre nouveau d'Acritarche du Trémadocien marocain. *C. R. Somm. Séanc. Soc. Géol. Fr.*, fasc. 3, p. 101−102, figs. 1−14, 1968a.

Deunff, Jean, Sur une forme nouvelle d'acritarche possédant une ouverture polaire (*Veryhachium miloni* n. sp.) et sur la présence d'une colonie de *Veryhachium* dans le Trémadocien marocain. *C. R. Hebd. Séanc. Acad. Sci., Paris*, v. 267, p. 46−49, 1 pl., 1968b.

Deunff, Jean, Le genre *Polyedryxium*, p. 7−49, pls. 1−8. In *Commission Internationale de Microflore du Paléozoique (C.I.M.P.), Microfossiles organiques du Paléozoique*, Fasc. 3, *Les Acritarches.* Paris: Centre National Recherche Scientifique, 1971.

Deunff, Jean, *Lanveocia* et *Candelasphaeridium*, genres nouveaux d'acritarches des Grès de Landévennec (Gédinnien inférieur de la Rade de Brest-Finistère). *Geobios, Lyon*, v. 11, p. 113−117, 1 pl., 1978.

Deunff, Jean, and W. R. Evitt, *Tunisphaeridium*, a new acritarch genus from the Silurian and Devonian. *Stanford Univ. Publs. Geol. Sci.*, v. 12 (1), p. 1−13, pls. 1, 2, figs. 1−11, 1968.

Deunff, Jean, Hanna Górka, and Raymond Rauscher, Observations nouvelles et précisions sur les acritarches à large ouverture polaire de Paléozoique Inférieur. *Geobios, Lyon*, v. 7, p. 5−18, pls. 1−7, 1974.

Deunff, Jean, J. P. Lefort, and F. Paris, Le microplancton Ludlovien des formations immergées des Minquiers (Manche) et sa place dans la distribution du Paléoplancton Silurien. *Bull. Soc. Géol. Minér. Bretagne*, sér. C, v. 3 (1), p. 9−28, pls. 1−4, 1971.

Döring, Harry, Planktonartige Fossilien des Jura/Kreide-Grenz-bereichs der Bohrungen Werle (Mecklenburg). *Geologie, Beih.*, v. 10 (Beih. 32), p. 110−120, pls. 16, 17, 1 fig., 1961.

Downie, Charles, Microplankton from the Kimeridge Clay. *Q. Jl. Geol. Soc. Lond.*, v. 112, p. 413−434, pl. 20, figs. 1−6, 1957.

Downie, Charles, An assemblage of microplankton from the Shineton Shales (Tremadocian). *Proc. Yorks. Geol. Soc.*, v. 31, p. 331−350, pls. 16, 17, 1958.

Downie, Charles, 'Hystrichospheres' (acritarchs) and Wenlock Shale of England. *Palaeontology*, v. 2, p. 56−71, pls. 10−12, 1959.

Downie, Charles, *Deunffia* and *Domasia*, new genera of hystrichospheres. *Micropaleontology*, v. 6, p. 197−203, 1 pl., 5 figs., 1960.

Downie, Charles, 'Hystrichospheres' (acritarchs) and spores of the Wenlock Shales (Silurian) of Wenlock, England. *Palaeontology*, v. 6, p. 625−652, pls. 91, 92, figs. 1−4, 1963.

Downie, Charles, The geological history of the microplankton. *Rev. Palaeobot. Palynol.*, v. 1, p. 269−281, 3 figs. 1967.

Downie, Charles, Observations on the nature of the acritarchs. *Palaeontology*, v. 16, p. 239−259, pls. 24−27, 1973.

Downie, Charles, W. R. Evitt, and W. A. S. Sarjeant, Dinoflagellates, hystrichospheres and the classification of the acritarchs. *Stanford Univ. Publs. Geol. Sci.*, v. 7 (3), p. 3−16, 1963.

Downie, Charles, and T. D. Ford, Microfossils from the Manx Slate series. *Proc. Yorks. Geol. Soc.*, v. 35, p. 307−322, pls. 17−18, figs. 1, 2, 1966.

Downie, Charles, and W. A. S. Sarjeant, On the interpretation and status of some hystrichosphere genera. *Palaeontology*, v. 6, p. 83−96, 1963.

Downie, Charles, and W. A. S. Sarjeant, Bibliography and index of fossil dinoflagellates and acritarchs. *Mem. Geol. Soc. Am.*, 94 (1964), p. 1−180, 1965.

Downie, Charles, and W. A. S. Sarjeant, Group Acritarcha Evitt 1963, p. 207−209. *In* W. B. Harland et al., *The fossil record.* London: Geol. Soc. London, 1967.

Downie, Charles, G. L. Williams, and W. A. S. Sarjeant, Classification of fossil microplankton. *Nature*, v. 192, p. 471, 1961.

Drugg, W. S., Palynology of the upper Moreno Formation (Late Cretaceous-Paleocene) Escarpado Canyon, California. *Palaeontographica Abt. B*, v. 120, p. 1−71, pls. 1−9, 5 figs., 1967.

Drugg, W. S., and A. R. Loeblich, Jr., Some Eocene and Oligocene phytoplankton from the Gulf Coast, U.S.A. *Tulane Stud. Geol. Paleont.*, v. 5, p. 181−194, pls. 1−3, figs. 1−8, 1967.

Eisenack, Alfred, Neue Mikrofossilien des baltischen Silurs. I. *Paläont. Z.*, v. 13, p. 74−118, pls. 1−5, figs. 1−5, 1931.

Eisenack, Alfred, Neue Mikrofossilien des baltischen Silurs III. und neue Mikrofossilien des böhmischen Silurs I. *Paläont. Z.*, v. 16, p. 52−76, pls. 4, 5, figs. 1−35, 1934.

Eisenack, Alfred, Neue Mikrofossilien des baltischen Silurs IV. *Paläont. Z.*, v. 19 (1937), p. 217−243, pls. 15, 16, figs. 1−22, 1938a.

Eisenack, Alfred, Hystrichosphaerideen und verwandte Formen im baltischen Silur. *Z. Geschiebeforsch. Flachldgeol.*, v. 14, p. 1−30, pls. 1−4, 1938b.

Eisenack, Alfred, Chitinozoen und Hystrichosphaerideen im Ordovicium des Rheinischen Schiefergebirges. *Senckenbergiana,* v. 21, p. 135–152, pls. A, B, figs. 1–20, 1939.

Eisenack, Alfred, Über Hystrichosphaerideen und andere kleinformen aus baltischem Silur und Kambrium. *Senckenbergiana,* v. 32, p. 187–204, pls. 1–4, figs. 1–6, 1951.

Eisenack, Alfred, Mikrofossilien aus Phosphoriten des samländischen Unteroligozäns und über die Einheitlichkeit der Hystrichosphaerideen. *Palaeontographica Abt. A,* v. 105, p. 49–95, pls. 7–12, figs. 1–8, 1954a.

Eisenack, Alfred, Hystrichosphären aus dem baltischen Gotlandium. *Senckenbergiana,* v. 34, p. 205–211, 1 pl., 7 figs., 1954b.

Eisenack, Alfred, Chitinozoen, Hystrichosphären und andere Mikrofossilien aus dem *Beyrichia*-Kalk. *Senckenberg. Leth.,* v. 36, p. 157–188, pls. 1–5, figs. 1–13, 1955.

Eisenack, Alfred, Mikrofossilien in organischer Substanz aus dem Lias Schwabens (Süddeutschland). *Neues Jb. Geol. Pälaont. Abh.,* v. 105, p. 239–249, pls. 19, 20, figs. 1, 2, 1957.

Eisenack, Alfred, Mikroplankton aus dem norddeutschen Apt nebst einigen Bemerkungen über fossile Dinoflagellaten. *Neues Jb. Geol. Pälaont. Abh.,* v. 106, p. 383–422, pls. 21–27, figs. 1–10, 1958a.

Eisenack, Alfred, Mikrofossilien aus dem Ordovizium des Baltikums. I. Markasitschicht, *Dictyonema*-Schiefer, Glaukonitsand, Glaukonitkalk. *Senckenberg. Leth.,* v. 39, p. 389–405, pls. 1, 2, 4 figs., 1958b.

Eisenack, Alfred, Neotypen baltischer Silur-Hystrichosphären und neue Arten. *Palaeontographica Abt. A,* v. 112, p. 193–211, pls. 15–17, figs. 1–11, 1959.

Eisenack, Alfred, Hystrichosphären als Nahrung ordovizischer Foraminiferen. *Neues Jb. Geol. Paläont. Mh.* 1961, p. 15–19, figs. 1–4, 1961.

Eisenack, Alfred, Mitteilungen über Leiosphären und über das Pylome bei Hystrichosphären. *Neues Jb. Geol. Paläont. Abh.,* v. 114, p. 58–80, pls. 2–4, figs. 1, 2, 1962a.

Eisenack, Alfred, Einige Bemerkungen zu neueren Arbeiten über Hystrichosphären. *Neues Jb. Geol. Paläont. Mh. 1962,* p. 92–101, 1962b.

Eisenack, Alfred, Mikrofossilien aus dem Ordovizium des Baltikums. 2. Vaginatenkalk bis Lyckholmer Stufe. *Senckenberg. Leth.,* v. 43, p. 349–366, pl. 44, figs. 1–7, 1962c.

Eisenack, Alfred, Neue problematische Mikrofossilien. *Neues Jb. Geol. Paläont. Abh.,* v. 114, p. 135–141, pl. 5, 1 fig., 1962d.

Eisenack, Alfred, Hystrichosphären. *Biol. Rev.,* v. 38, p. 107–139, pls. 1, 2, 1963a.

Eisenack, Alfred, Mitteilungen zur Biologie der Hystrichosphären und über neue Arten. *Neues Jb. Geol. Paläont. Abh.,* v. 118, p. 207–216, pls. 19, 20, 1 fig., 1963b.

Eisenack, Alfred, Mikrofossilien aus dem Silur Gotlands, Hystrichosphären, Problematika. *Neues Jb. Geol. Paläont. Abh.,* v. 122, p. 257–274, pls. 21–24, 2 figs., 1965a.

Eisenack, Alfred, Die Mikrofauna der Ostseekalk l. Chitinozoen, Hystrichosphären. *Neues Jb. Geol. Paläont Abh.,* v. 123, p. 115–148, pls. 9–13, figs. 1, 2, 1965b.

Eisenack, Alfred, *Axisphaeridium* n. g., eine axialsymmetrische Hystrichosphäre aus dem baltischen Ordovizium. *Neues Jb. Geol. Paläont. Mh.,* 1967, p. 398–400, fig. 1, 1967.

Eisenack, Alfred, Über die Fortpflanzung paläozoischer Hystrichosphären. *Neues Jb. Geol. Paläont. Abh.,* v. 131, p. 1–22, pls. 1–3, fig. 1, 1968a.

Eisenack, Alfred, Mikrofossilien eines Geschiebes der Borkholmer Stufe, baltisches Ordovizium, F$_2$. *Mitt. Geol. StInst. Hamb.,* Heft 37, p. 81–94, pls. 23–25, figs. 1–13, 1968b.

Eisenack, Alfred, Zur Systematik einiger paläozoischer Hystrichosphären (Acritarcha) des baltischen Gebietes. *Neues Jb. Geol. Paläont. Abh.,* v. 133, p. 245–266, 1969.

Eisenack, Alfred, Mikrofossilien aus dem Silur Estlands und der Insel Ösel. *Geol. För. Stockh. Förh.,* v. 92, p. 302–322, figs. 1–7, 1970.

Eisenack, Alfred, Die Mikrofauna der Ostseekalke (Ordovizium), 3. Graptolithen, Melanoskleriten, Spongien, Radiolarien, Problematika nebst 2 Nachträgen über Foraminiferen und Phytoplankton. *Neues Jb. Geol. Paläont. Abh.,* v. 137, p. 337–357, figs. 1–62, 1971.

Eisenack, Alfred, Beiträge zur Acritarchen-Forschung. Contribution to acritarch research. *Neues Jb. Geol. Paläont. Abh.,* v. 147, p. 269–293, figs. 1–50, 1974.

Eisenack, Alfred, F. H. Cramer, and M. D. C. R. Díez Rodriguez, *Katalog der fossilen Dinoflagellaten, Hystrichosphären und verwandten Mikrofossilien.* Band III *Acritarcha, 1. Teil.* Stuttgart: E. Schweizerbart'sche Verlagsbuchhandlung, 1104 p., 4 pls., 452 figs., 1973.

Eisenack, Alfred, F. H. Cramer, and M. C. R. Díez, *Katalog der fossilen Dinoflagellaten, Hystrichosphären und verwandten Mikrofossilien.* Band IV *Acritarcha, 2. Teil.* Stuttgart: E. Schweizerbart'sche Verlagsbuchhandlung, 863 p., 317 figs., 1976.

Evitt, W. R., Observations on the morphology of fossil dinoflagellates. *Micropaleontology,* v. 7, p. 385–420, pls. 1–9, 1961.

Evitt, W. R., A discussion and proposals concerning fossil dinoflagellates, hystrichospheres, and acritarchs, I, II. *Proc. Natn. Acad. Sci. U.S.A.,* v. 49, p. 158–164, 298–302, 1963.

Evitt, W. R., The Cretaceous microfossil *Ophiobolus lapidaris* O. Wetzel and its flagellum-like filaments. *Stanford Univ. Publs. Geol. Sci.,* v. 12 (3), p. 1—9, pl. 1, figs. 1—6, 1968.

Faegri, K., and J. Iversen, *Text-book of modern pollen analysis.* Copenhagen: Munksgaard, 168 p., 17 figs., 1950.

Fisher, D. W., A microflora in the Maplewood and Neahga Shales. *Bull. Buffalo Soc. Nat. Sci.,* v. 21, p. 13—18, pl. 7, 1953.

Fombella Blanco, M. A., Acritarcos de edad Cambrico Medio-Inferior de la Provincia de Leon, España. *Revta Esp. Micropaleont.,* v. 9, p. 115—124, pl. 1, fig. 1, 1977.

Fridrikhsone, A. I., Akritarkhi *Baltisphaeridium* i gistrikhosfery (?) iz kembriyskikh otlozheniy Latvii [The acritarchs *Baltisphaeridium* and hystrichospheres (?) from Cambrian sediments of Latvia]. *Paleontologiya i Stratigrafiya Pribaltiki i Belorussii,* Sbornik 3, Uprav. Geol. Sov. Minist. Litovskoy SSSR, Litovskiy Nauchno-Issledov. Geologorazved. Inst., Vil'nyus, p. 5—22, pls. 1—3, 1971.

Gardiner, P. R. R., and M. Vanguestaine, Cambrian and Orodovician microfossils from south-east Ireland and their implications. *Bull. Geol. Surv. Ireland,* v. 1, p. 163—210, pls. 1, 2, figs. 1—9, 1971.

Gitmez, G. U., Dinoflagellate cysts and acritarchs from the basal Kimmeridgian (Upper Jurassic) of England, Scotland, and France. *Bull. Br. Mus. Nat. Hist., Geology,* v. 18, p. 233—331, pls. 1—14, figs. 1—34, 1970.

Górka, Hanna, Quelques nouveaux acritarches des silexites du Trémadocien supérieur de la région de Kielce (Montagne de Ste-Croix, Pologne). *Cah. Micropaléont.,* sér. 1 (6), p. 1—7, pls. 1, 2, Arch. Orig. Centre Docum. C.N.R.S. no. 441, 1967.

Górka, Hanna, Microorganismes de l'Ordovicien de Pologne. *Palaeont. Pol.,* no. 22, 102 p., 31 pls., 44 figs., 1969.

Górka, Hanna, Les acritarches de concrétions calcaires du Famennien Supérieur de Łagów (Monts de Sainte Croix, Pologne). Acritarcha z konkrecji wapiennych Famenu górnego Łagowa (Góry Świętokrzyskie, Polska). *Acta palaeont. Pol.,* v. 19, p. 225—250, pls. 10—18, 2 figs., 1974.

Gray, Jane, and A. J. Boucot, Palynological evidence bearing on the Ordovician Silurian paraconformity in Ohio. *Bull. Geol. Soc. Am.,* v. 83, p. 1299—1314, 5 figs., 1972.

Harland, Rex, and W. A. S. Sarjeant, Fossil freshwater microplankton (Dinoflagellates and acritarchs) from Flandrian (Holocene) sediments of Victoria and Western Australia. *Proc. R. Soc. Vict.,* v. 83, p. 211—234, pls. 21, 22, figs. 1—9, 1970.

Hemer, D. O., and P. W. Nygreen, Algae, acritarchs and other microfossils incertae sedis from the Lower Carboniferous of Saudi Arabia. *Micropaleontology,* v. 13, p. 183—194, pls. 1—3, 1 fig., 1967.

Henry, J. L., Sur la conservation d'un rassemblement d'Acritarches (micro-organismes incertae sedis) sous forme de colonie dans l'Ordovicien du Finistère. *C. R. Hebd. Séanc. Acad. Sci. Paris,* v. 258, p. 1001—1003, figs. 1—5, 1964.

Henry, J. L., Quelques acritarches (micro-organismes incertae sedis) de l'Ordovicien de Bretagne. *C. R. Somm. Séanc. Soc. Géol. Fr.,* fasc. 7, p. 265—267, figs. a—c, 1966.

Henry, J. L., Micro-organismes incertae sedis (acritarches et chitinozoaires) de l'Ordovicien de la Presqu'île de Crozon (Finistère), Gisements de Mort-Anglaise et de Kerglintin. *Bull. Soc. Géol. Minér. Bretagne* (1968), p. 59—100, pls. 1—13, 1969.

Jacobson, S. R., Acritarchs as paleoenvironmental indicators in Middle and Upper Ordovician rocks from Kentucky, Ohio and New York. *J. Paleont.,* v. 53, p. 1197—1212, 12 figs., 1979.

Jansonius, Jan, Palynology of Permian and Triassic sediments, Peace River area, Western Canada. *Palaeontographica Abt. B,* v. 110, p. 35—98, pls. 11—16, figs. 1—3, 1962.

Jardiné, Serge, Microplancton (Acritarches) et limites stratigraphiques du Silurien terminal au Dévonien supérieur. *C. R. VII Congr. Int. Stratig. Géol. Carbonifère, Krefeld 23—28 Aug. 1971,* v. 1, p. 313—323, figs. 1, 2, 1972.

Jardiné, Serge, A. Combaz, L. Magloire, G. Peniguel, and G. Vachey, Acritarches du Silurien terminal et du Dévonien du Sahara Algérien. *C. R. VII Congr. Int. Stratig. Géol. Carbonifère, Krefeld 23—28 Aug. 1971,* v. 1, p. 295—311, pls. 1—3, 1972.

Jardiné, Serge, and L. Yapaudjian, Lithostratigraphie et palynologie de Dévonien-Gothlandien gréseux du bassin de Polignac (Sahara). *Revue Inst. Fr. Pétrole,* v. 23, p. 439—469, pls. 1—6, figs. 1—8, 1968.

Jekhowsky, B. de, Sur quelques hystrichosphères Permo-Triasiques d'Europe et d'Afrique. *Revue Micropaléont.,* v. 3, p. 207—212, pls. 1, 2, 1961.

Jux, Ulrich, Über den Feinbau der Wandungen einiger paläozoischer Baltisphaeridiaceen. *Palaeontographica Abt. B.,* v. 136, p. 115—128, pls. 41—44, figs. 1—7, 1971.

Jux, Ulrich, Phytoplankton aus dem mittleren Oberdevon (Nehden-Stufe) des sudwestlichen Bergischen Landes (Rheinisches Schiefergebirge). *Palaeontographica Abt. B,* v. 149, p. 113—138, pls. 1—7, figs. 1—3, 1975.

Kjellström, Göran, Ordovician microplankton (Baltisphaerids) from the Grötlingbo Borehole No. 1 in Gotland, Sweden. *Årsb. Sver. Geol. Unders.,* no. 65 (1), Ser. C, nr. 655, p. 1—75, pls. 1—4, 1971a.

Kjellström, Göran, Middle Ordovician microplankton from the Grötlingbo Borehole No. 1 in Gotland, Sweden. *Årsb. Sver. Geol. Unders.,* no. 65 (15), Ser. C, nr. 669, p. 1—35, figs. 1—22, 1971b.

Kjellström, Göran, Lower Viruan (Middle Ordovician) microplankton from the Ekön Borehole No. 1 in Östergötland, Sweden. *Årsb. Sver. Geol. Unders.,* 70(6), Ser. C, nr. 724, p. 1—44, figs. 1—34, 1976.

Klement, K. W., Dinoflagellaten und Hystrichosphaerideen aus dem unteren und mittleren Malm Südwestdeutschlands. *Palaeontographica Abt. A,* v. 114, p. 1—104, pls. 1—10, figs. 1—37, 1960.

Konzalová, Magda, Acritarchs from the Bohemian Precambrian (Upper Proterozoic) and Lower-Middle Cambrian. *Rev. Palaeobot. Palynol.,* v. 18, p. 41—56, 1974.

Kremp, G. O. W., *Morphologic encyclopedia of palynology, an international collection of definitions and illustrations of spores and pollen.* Program in Geochronology, Contr. 100. Tucson: Univ. Arizona Press, xiii + 186 p., 38 pls., 1965.

Laird, H. C., The nature and origin of chert in the Lockport and Onondaga Formations of Ontario. *Trans. R. Can. Inst.,* v. 20, p. 231—304, pls. 16—20, figs. 1—11B, 1935.

LeCorre, Claude, and Jean Deunff, Sur la présence d'acritarches au sommet des schistes de l'Ordovicien moyen de Sud de Rennes. *Bull Soc. Géol. Minér. Bretagne,* Sér. C, v. 1, p. 45—48, 1 pl., 1 fig., 1969.

Legault, J. A., Chitinozoa and Acritarcha of the Hamilton Formation (Middle Devonian), southwestern Ontario. *Bull. Geol. Surv. Can.,* 221, p. 1—103, pls. 1—13, figs. 1—7, 1973.

Lewis, H. P., The microfossils of the Upper Carodocian [sic] phosphate deposits of Montgomeryshire, North Wales. *Ann. Mag. Nat. Hist.,* ser. 2, v. 5, p. 1—7, 27—39, pls. 1—4, 1940.

Lister, T. R., The method of opening, orientation and morphology of the Tremadocian acritarch, *Acanthodiacrodium ubui* Martin. *Proc. Yorks. Geol. Soc.,* v. 38, p. 47—55, pl. 5, 1970a.

Lister, T. R., A monograph of the acritarchs and Chitinozoa from the Wenlock and Ludlow Series of the Ludlow and Millichope areas, Shropshire. *Palaeontogr. Soc.* [*Monogr.*] , v. 124(1), p. 1—100, pls. 1—13, 1970b.

Lister, T. R., and D. W. Holliday, Phytoplankton (acritarchs) from a small Ordovician inlier in Teesdale (County Durham), England. *Proc. Yorks. Geol. Soc.,* v. 37, p. 449—460, pl. 19, 1970.

Loeblich, A. R., Jr., *Dicommopalla,* a new acritarch genus from the Dillsboro Formation (Upper Ordovician) of Indiana, U.S.A. *Phycologia,* v. 9, p. 39—43, figs. 1—12, 1970a.

Loeblich, A. R., Jr., Morphology, ultrastructure and distribution of Paleozoic acritarchs. *Proc. N. Am. Paleont. Conv.,* Chicago, Pt. G., p. 705—788, figs. 1—38, 1970b.

Loeblich, A. R., Jr., and W. S. Drugg, New acritarchs from the Early Devonian (Late Gedinnian) Haragan Formation of Oklahoma, U.S.A. *Tulane Stud. Geol. Paleont.,* v. 6, p. 129—137, pls. 1—4, 1968.

Loeblich, A. R., Jr., and R. B. MacAdam, North American species of the Ordovician acritarch genus *Aremoricanium. Palaeontographica Abt. B,* v. 135, p. 41—47, pls. 14—19, 1971.

Loeblich, A. R., Jr., and Helen Tappan, Acritarch excystment and surface ultrastructure with descriptions of some Ordovician taxa. *Revta Esp. Micropaleont.,* v. 1, p. 45—57, pls. 1—4, fig. 1, 1969.

Loeblich, A. R., Jr., and Helen Tappan, *Thysanoprobolus,* a new acritarch genus from the Early Devonian (Late Gedinnian) Haragan Formation of Oklahoma, U.S.A. *Proc. Biol. Soc. Wash.,* v. 83, p. 261—266, figs. 1—12, 1970.

Loeblich, A. R., Jr., and Helen Tappan, New observations of the ultrastructure of *Asketopalla,* an Ordovician acritarch. *J. Paleont.,* v. 45, p. 899—901, pls. 103, 104, 1971a.

Loeblich, A. R., Jr., and Helen Tappan, Two new *Orthosphaeridium* (Acritarcha) from the Middle and Upper Ordovician. *Trans. Am. Microsc. Soc.,* v. 90, p. 182—188, figs. 1—12, 1971b.

Loeblich, A. R., Jr., and Helen Tappan, Some new and revised organic-walled phytoplankton microfossil genera. *J. Paleont.,* v. 50, p. 301—308, 1976.

Loeblich, A. R., Jr., and Helen Tappan, Some Middle and Late Ordovician microphytoplankton from central North America. *J. Paleont.,* v. 52, p. 1233—1287, pls. 1—16, 2 figs., 1978.

Loeblich, A. R., Jr., and E. R. Wicander, New Early Devonian (Late Gedinnian) microphytoplankton: *Demorhethium lappaceum* n. g., n. sp., from the Bois d'Arc Formation of Oklahoma, U.S.A. *Neues Jb. Geol. Paläont. Mh.* 1974, p. 707—711, figs. 1—6, 1974.

Loeblich, A. R., Jr., and E. R. Wicander, Organic-walled microplankton from the Lower Devonian Late Gedinnian Haragan and Bois d'Arc Formations of Oklahoma, U.S.A., Part I. *Palaeontographica Abt. B,* v. 159, p. 1—39, pls. 1—12, 1976.

Loeblich, A. R. III, Nomenclatural notes in the Pyrrhophyta, Xanthophyta and Euglenophyta. *Taxon,* v. 16, p. 68—69, 1967.

Lohmann, Hans, Eier und sogennannte Cysten der Plankton-Expedition. *Ergebn. Plankton-Exped. Humboldt-Stiftung,* Bd. IV. N, p. 1—61, pls. 1—7, 1904.

Mädler, K. A., Die figurierten organischen Bestandteile der Posidonienschiefer. *Beih. Geol. Jb.,* v. 58, p. 287—406, pls. 15—30, figs. 51—54, 1963.

Mädler, K. A., Hystrichophyta and acritarchs. *Rev. Palaeobot. Palynol.,* v. 5, p. 285—290, 1967.

Martin, Francine, Les Acritarches du Sondage de la Brasserie Lust à Kortrijk (Courtrai) (Silurien belge). *Bull. Soc. Belge Géol. Paléont. Hydrol.,* v. 74 (1965), p. 354—402, pl. 1, figs. 1—35, 1966a.

Martin, Francine, Les Acritarches de Sart-Bernard (Ordovicien belge.) *Bull. Soc. Belge Géol. Paléont. Hydrol.,* v. 74 (1965), p. 423−444, figs. 1−17, 1966b.

Martin, Francine, Les Acritarches du parc de Neuville-sous-Huy (Silurien belge). *Bull. Soc. Belge Géol. Paléont. Hydrol.,* v. 75 (1966), p. 306−335, pl. 1, figs. 1−10, 1967.

Martin, Francine, Les Acritarches de l'Ordovicien et du Silurien Belges, Détermination et valeur stratigraphique. *Mém. Inst. R. Sci. Nat. Belg.,* 160 (1968), p. 1−175, pls. 1−8, figs. 1−87, I−XL, 1969.

Martin, Francine, Les Acritarches de l'Ordovicien Inférieur de la Montagne Noire (Herault, France). *Bull. Inst. R. Sci. Nat. Belg., Sci. Terre,* v. 48, p. 1−61, pls. 1−11, 1972.

Martin, Francine, Ordovicien Supérieur et Silurien Inférieur a Deerlijk (Belgique). Palynofacies et microfacies.*Mém. Inst. R. Sci. Nat. Belg.,* 174, p. 1−71, pls. 1−8, 1974.

Martin, Francine, and Göran Kjellström, Ultrastructural study of some Ordovician acritarchs from Gotland, Sweden. *Neues Jb. Geol. Paläont. Mh.* 1973, p. 44−54, figs. 1−17, 1973.

Martinez Macchiavello, J. C., Quelques acritarches d'un échantillon du Dévonien Inférieur (Cordobés) de Blanquillo, Département de Durazno, Uruguay. *Revue Micropaléont.,* v. 11, p. 77−84, pl. 1, figs. 1−13, 1968.

Morgenroth, Peter, Zur Kenntnis der Dinoflagellaten und Hystrichosphaeridien des Danien. *Geol. Jb.,* v. 86, p. 553−578, pls. 41−48, text-figs. 1−4, 1968.

Nagy, Esther, A Mecsek hegység Miocén retegeinek palynológiai vizsgálata. Palynological elaborations the Miocene layers of the Mecsek Mountains. *Magy. Allami Földt. Intéz. Évk.,* v. 52, p. 233−650, pls. 1−56, figs. 1−62, 1969.

Naumova, S. N., Spory nizhnego Kembriya. *Izv. Akad. Nauk SSSR, Ser. Geol.,* no. 4, p. 49−56, 1949.

Norris, Geoffrey, Triassic and Jurassic miospores and acritarchs from the Beacon and Ferrar Groups, Victoria Land, Antarctica. *N. Z. Jl. Geol. Geophys.,* v. 8, p. 236−277, figs. 1−75, 1965.

Norris, Geoffrey, and W. A. S. Sarjeant, A descriptive index of genera of fossil Dinophyceae and Acritarcha. *Palaeont. Bull., Wellington* [N.Z. Geol. Surv.], v. 40, p. 1−72, 1965.

Pflug, H. D., Einige Reste Niederer Pflanzen aus dem Algonkium. *Palaeontographica Abt. B,* v. 117, p. 59−74, pls. 25−29, figs. 1, 2, 1966.

Pflug, H. D., Structured organic remains from the Fig Tree Series (Precambrian) of the Barberton Mountain Land (South Africa). *Rev. Palaeobot. Palynol.,* v. 5, p. 9−29, pls. 1−4, fig. 1, 1967.

Pichova, N. G. [Pykhova], Microfossils of Lower Cambrian and Precambrian deposits in eastern Siberia.*Rev. Palaeobot. Palynol.,* v. 5, p. 31−38, pls. 1−3, 1967.

Playford, Geoffrey, Lower to Middle Devonian acritarchs of the Moose River Basin, Ontario. *Bull. Geol. Surv. Can.,* 279, p. 1−87, pls. 1−20, figs. 1−16, 1977.

Pocock, S. A. J., Palynology of the Jurassic sediments of western Canada. *Palaeontographica Abt. B,* v. 137, p. 85−153, pls. 22−29, figs. 1−45, 1972.

Pöthe de Baldis, E. D., Microplancton adicional del Silurico Superior de Santiago del Estero, Republica Argentina. *Ameghiniana,* v. 11, p. 313−327, pls. 1−4, 1974.

Pöthe de Baldis, E. D., Microplancton del Wenlockiano de la Precordillera Argentina. *Revta. Esp. Micropaleont.,* v. 7, p. 489−505, pls. 1−5, 1975a.

Pöthe de Baldis, E. D., Microplancton de la Formacion Los Espejos Provincia de San Juan, Republica Argentina. *Revta Esp. Micropaleont.,* v. 7, p. 507−518, pls. 1−3, 1975b.

Pykhova, N. G., Akritarkhi dokembriya yuzhnogo Urala, Sibiri, Vostochno-Evropeyskoy Platformy i ikh znachenie dlya stratigrafii [Acritarchs of Precambrian sections of the southern Urals, Siberia and eastern European platform and their stratigraphic significance], p. 15−17, pls. 1, 2. In *Mikrofossilii Drevneyshikh Otlozheniy,* Akad. Nauk SSSR, Sibirsk. Otdel, Inst. Geol. Geofiz., Moscow: "Nauka," 1973a.

Pykhova, N. G., Akritarkhi Verkhnemotskogo gorizonta Irkutskogo Amfiteatra [Acritarchs from the Upper Motskian Horizon of the Irkutsk Amphitheatre]. *Izv. Akad. Nauk SSSR, Ser. Geol.,* no. 6, p. 127−132, 1 pl., 1973b.

Rasul, S. M., The Lower Palaeozoic acritarchs *Priscogalea* and *Cymatiogalea. Palaeontology,* v. 17, p. 41−63, pls. 3−7, figs. 1−5, 1974.

Rasul, S. M., and Charles Downie, The stratigraphic distribution of Tremadoc acritarchs in the Shineton Shales succession, Shropshire, England. *Rev. Palaeobot. Palynol.,* v. 18, p. 1−9, 1974.

Rauscher, Raymond, Présence d'une forme nouvelle d'Acritarches dans le Dévonien de Normandie. *C. R. Hebd. Séanc. Acad. Sci., Paris,* v. 268, sér. D, p. 34−36, pl. 1, 1969.

Rauscher, Raymond, *Recherches micropaléontologiques et stratigraphiques dans l'Ordovicien et le Silurien en France. Etude des Acritarches, des Chitinozoaires et des Spores.* Sci. Géol. Mém. 38, Inst. Géol., Univ. Louis Pasteur, Strasbourg, 209 p., 12 pls., 1973.

Rauscher, Raymond, Les Acritarches de l'Ordovicien en France. *Rev. Palaeobot. Palynol.,* v. 18, p. 83−97, pls. 1, 2, figs. 1, 2, 1974.

Roblot, M. M., Nouveaux Acritarches du Précambrien normand. Leur étude à la microsonde electronique. *C. R. Hebd. Séanc. Acad. Sci., Paris,* v. 264, sér. D, p. 1263−1265, pl. 1, 1967.

Rossignol, Martine, Hystrichosphères du Quaternaire

en Méditerranée orientale, dans les sédiments Pléistocènes et les boues marines actuelles. *Revue Micropaléont.*, v. 7, p. 83−99, pls. 1−3, figs. A−H, 1964.

Rothpletz, A., Radiolarien, Diatomaceen und Sphärosomatiten im silurischen Kieselschiefer von Langenstriegis in Sachsen. *Z. Dt. Geol. Ges.*, v. 32, p. 446−467, pl. 21, 1880.

Rudavskaya, V. A., Akritarkhi pogranichnykh otlozheniy Rifeya i Kembriya yuga vostochnoy Sibiri [Acritarchs from the Riphean-Cambrian boundary deposits in southeast Siberia], p. 17−21, pls. 1, 2. In *Mikrofossilii drevneyshikh otlozheniy*, Trudy *III Mezhdun. Palinol. Konfer., Novosibirsk, 1971*, Akad. Nauk SSSR, Sibirsk. Otdel., Inst. Geol. Geofiz., Moscow: "Nauka." 1973.

Sannemann, D., Hystrichosphaerideen aus dem Gotlandium und Mittel-Devon des Frankenwaldes und ihr Feinbau. *Senckenberg. Leth.*, v. 36, p. 321−346, 6 pls., 19 figs., 1955.

Sarjeant, W. A. S., Microplankton from the Cornbrash of Yorkshire. *Geol. Mag.*, v. 96, p. 329−346, pl. 13, figs. 1−8, 1959.

Sarjeant, W. A. S., New Hystrichospheres from the Upper Jurassic of Dorset. *Geol. Mag.*, v. 97, p. 137−144, pl. 6, figs. 1−4, 1960a.

Sarjeant, W. A. S., Microplankton from the Corallian rocks of Yorkshire. *Proc. Yorks. Geol. Soc.*, v. 32, p. 389−408, pls. 12−14, 1960b.

Sarjeant, W. A. S., Taxonomic notes on hystrichospheres and acritarchs. *J. Paleont.*, v. 38, p. 173−177, 1964.

Sarjeant, W. A. S., Observations on the acritarch genus *Micrhystridium* (Deflandre). *Revue Micropaléont.*, v. 9, p. 201−208, pl. 1, fig. 1, 1967.

Sarjeant, W. A. S., Microplankton from the Upper Callovian and Lower Oxfordian of Normandy. *Revue Micropaléont.*, v. 10, p. 221−242, pls. 1−3, figs. 1−5, 1968.

Sarjeant, W. A. S., Xanthidia, palinospheres and 'Hystrix.' A review of the study of fossil unicellular microplankton with organic cell walls. *Microscopy: J. Quekett Microsc. Club*, v. 31, p. 221−253, figs. 1−12, 1970a.

Sarjeant, W. A. S., Acritarchs and tasmanitids from the Chhidru Formation, uppermost Permian of West Pakistan, p. 277−304, pls. 1, 2, figs. 1−10. *In* B. Kummel and C. Teichert, *Stratigraphic boundary problems: Permian and Triassic of West Pakistan.* Univ. Kansas Dept. Geol., Spec. Publ. 4, 1970b.

Sarjeant, W. A. S., Acritarchs and tasmanitids from the Mianwali and Tredian Formations (Triassic) of the Salt and Surghar Ranges, West Pakistan, p. 35−73, pls. 1−4, figs. 1−5. *In* A. Logan and L. V. Hills, *The Permian and Triassic Systems and their mutual boundary*, Calgary: Can. Soc. Petrol. Geol., Mem. 2, 1973.

Sarjeant, W. A. S., and Isles Strachan, Freshwater acritarchs in Pleistocene peats from Staffordshire, England. *Grana Palynol.*, v. 8, p. 204−209, figs. 1−3, 1968.

Schaarschmidt, F., Sporen und Hystrichosphaerideen aus dem Zechstein von Büdingen in der Wetterau. *Palaeontographica Abt. B*, v. 113, p. 38−91, pls. 11−20, 1963.

Schön, M., Hystrichosphaerideen aus dem Mittleren Buntsandstein von Thüringen. *Mber. Dt. Akad. Wiss. Berl.*, v. 9, p. 527−535, 1 pl., 2 figs., 1967.

Schopf, J. W., Precambrian micro-organisms and evolutionary events prior to the origin of vascular plants. *Biol. Rev.*, v. 45, p. 319−352, pls. 1−6, figs. 1−7, 1970.

Schultz, E., Sporae dispersae aus der Trias von Thüringen: *Abh. Zent. Geol. Inst.*, v. 1 (1965), p. 257−287, pls. 20−23, 1 fig., 1966.

Segroves, K. L., Cutinized microfossils of probable nonvascular origin from the Permian of Western Australia. *Micropaleontology*, v. 13, p. 289−305, pls. 1−3, figs. 1, 2, 1967.

Shepeleva, E. D., Rastitel'nye ? ostatki neizvestnoy sistematicheskoy prinadlezhnosti iz otlozheniy Bavlinskoy Serii Volgo-Ural'skoy neftenosnoy provintsii [Plant remains of unknown taxonomic origin from deposits of the Bavlin Series of the Volgo-Uralian oil province]. *Dokl. Akad. Nauk SSSR*, v. 142, p. 456−457, fig. 1, 1962.

Sheshegova, L. I., Akritarkhi Paleozoya, p. 9−35, pls. 5−8. In *Vodorosli Paleozoya i Mezozoya Sibiri, III Mezhdun. Palinol. Konfer. SSSR, Novosibirsk 1971*, Akad. Nauk SSSR, Sibirsk Otdel., Inst. Geol. Geofiz., Moscow: "Nauka," 1971a.

Sheshegova, L. I., O nekotorykh Siluriyskikh akritarkhakh Podolii [On some Silurian acritarchs from Podolia], p. 36−49, pls. 9−13. In *Vodorosli Paleozoya i Mezozoya Sibiri, III Mezhdun. Palinol. Konfer. SSSR, Novosibirsk 1971*, Akad. Nauk SSSR, Sibirsk. Otdel., Inst. Geol. Geofiz., Moscow: "Nauka," 1971b.

Sheshegova, L. I., Verkhnedevonskie akritarkhi Anzherskogo rayona Kuzbassa [Upper Devonian acritarchs of the Angerian region of Kuzbass], p. 50−60, pls. 14−18. In *Vodorosli Paleozoya i Mezozoya Sibiri, III Mezhdun. Palinol. Konfer. SSSR, Novosibirsk 1971*, Akad. Nauk SSSR, Sibirsk. Otdel., Inst. Geol. Geofiz., Moscow: "Nauka," 1971c.

Sheshegova, L. I., Akritarkhi Silura i nizov Devona Podolia [Silurian and lower Devonian acritarchs of Podolia]. *Trudy Inst. Geol. Geofiz. Sib. Otd.*, vyp. 81, p. 36−69, pls. 18−24, 1974.

Sheshegova, L. I., Fitoplankton Silura Tuvy (razrez "Elegest") [Phytoplankton of the Silurian of Tuva (the "Alegest" section)]. *Trudy Inst. Geol. Geofiz. Sib. Otd.*, vyp. 224, p. 1−99, pls. 1−51, 1975.

Sinha, Veena, Some "acritarchs" and other microfossils from Barakar Stage of Lower Gondwanas, India. *Palaeobotanist,* v. 17, p. 326—331, 1 pl., 1 fig., 1968.

Smith, N. D., and R. S. Saunders, Paleoenvironments and their control of acritarch distribution: Silurian of east-central Pennsylvania. *J. Sedim. Petrol.,* v. 40, p. 324—333, figs. 1—6, 1970.

Spode, F., A new record of hystrichospheres from the Mansfield Marine Band, Westphalian. *Proc. Yorks. Geol. Soc.,* v. 34, p. 357—370, pl. 38, figs. 1, 2, 1964.

Staplin, F. L., Reef-controlled distribution of Devonian microplankton in Alberta. *Palaeontology,* v. 4, p. 392—424, pls. 48—51, 9 figs., 1961.

Staplin, F. L., Jan Jansonius, and S. A. J. Pocock, Evaluation of some acritarchous hystrichosphere genera. *Neues Jb. Geol. Paläont. Abh.,* v. 123, p. 167—201, pls. 18—20, 13 figs., 1965.

Staplin, F. L., S. A. J. Pocock, and Jan Jansonius, Applications of hystrichospherids in Western Canada. *Can. Oil. Gas Inds.,* v. 13, p. 107 (Abstr.), 1960.

Stockmans, François, and Yvonne Willière, Hystrichosphères du Dévonien belge (Sondage de l'Asile d'aliénés à Tournai). *Senckenberg. Leth.,* v. 41, p. 1—11, pls. 1, 2, 1960.

Stockmans, François, and Yvonne Willière, Description de trois hystrichosphères trouvés dans les schistes de l'horizon marin de petit buisson. *Publs. Cent. Natn. Geol. Houill.,* no. 5, p. 29—30, figs. a—c, 1962a.

Stockmans, François, and Yvonne Willière, Hystrichosphères du Dévonien belge (sondage de l'Asile d'aliénés à Tournai). *Bull. Soc. Belge Géol. Paléont. Hydrol.,* v. 71, p. 41—77, 2 pls., 34 figs., 1962b.

Stockmans, François, and Yvonne Willière, Les hystrichosphères ou mieux les acritarches du Silurien belge, Sondage de la Brasserie Lust à Courtrai (Kortrijk.). *Bull. Soc. Belge Géol. Paléont. Hydrol.,* v. 71 (1962), p. 450—481, pls. 1—3, figs. 1—36, 1963.

Stockmans, François, and Yvonne Willière, Les acritarches du Dinantien du Sondage de Vieux Leuze à Leuze (Hainaut, Belgique). *Bull. Soc. Belge Géol. Paléont. Hydrol.,* v. 75 (1966), p. 233—242, 1 pl., 4 figs., 1967.

Stockmans, François, and Yvonne Willière, Acritarches du Famennien Inférieur. *Mém. Acad. R. Belg. Cl. Sci. 8°,* 2ᵉ ser., v. 38(6), p. 1—62, pls. 1—4, 1969.

Stockmans, François, and Yvonne Willière, Acritarches de la "Tranchée de Senzeille" (Frasnien supérieur et Famennien inférieur). *Mém. Acad. R. Belg. Cl. Sci. 8°,* 2ᵉ ser., v. 41(5), p. 1—79, pls. 1—4, 1974.

Tappan, Helen, Primary production, isotopes, extinctions and the atmosphere. *Palaeogeogr. Palaeoclimat. Palaeoecol.,* v. 4, p. 187—210, 1 fig., 1968.

Tappan, Helen, Phytoplankton abundance and Late Paleozoic extinctions: a reply. *Palaeogeogr. Palaeoclimat. Palaeoecol.,* v. 8, p. 56—66, 1970.

Tappan, Helen, Microplankton, ecological succession, and evolution. *Proc. N. Am. Paleont. Conv. Chicago, Sept. 1969,* pt. H, p. 1058—1103, figs. 1—8, 1971.

Tappan, Helen, W. C. Cornell, and E. R. Wicander, Quantification of fossil phytoplankton fluctuations. *Abstr. Res., 15th Ann. Rept. Res., Am. Chem. Soc., Petrol. Res. Fund,* p. 195—196 (Abstr.), 1971.

Tappan, Helen, W. C. Cornell, and E. R. Wicander, Quantification of fossil phytoplankton fluctuations. *Abstr. Res., 16th Ann. Rept. Res., Am. Chem. Soc., Petrol. Res. Fund,* p. 64—65 (Abstr.), 1972.

Tappan, Helen, and A. R. Loeblich, Jr., Geobiologic implications of fossil phytoplankton evolution and time-space distribution, p. 247—340, 1 pl., 12 figs. *In* R. Kosanke and A. T. Cross, Symposium on palynology of the Late Cretaceous and Early Tertiary. *Spec. Pap. Geol. Soc. Am.,* no. 127, 1971a.

Tappan, Helen, and A. R. Loeblich, Jr., Surface sculpture of the wall in Lower Paleozoic acritarchs. *Micropaleontology,* v. 17, p. 385—410, pls. 1—11, 1971b.

Tappan, Helen, and A. R. Loeblich, Jr., Fluctuating rates of protistan evolution, diversification and extinction. *24th Int. Geol. Congr., Sec. 7 Paleont.,* p. 205—213, figs. 1—6, 1972.

Tappan, Helen, and A. R. Loeblich, Jr., Smaller protistan evidence and explanation of the Permian-Triassic crisis, p. 465—480. *In* A. Logan and L. V. Hills, *The Permian and Triassic Systems and their mutual boundary.* Calgary: Can. Soc. Petrol. Geol., Mem. 2, 1973a.

Tappan, Helen, and A. R. Loeblich, Jr., Evolution of the oceanic plankton. *Earth-Sci. Rev.,* v. 9, p. 207—240, figs. 1—9, 1973b.

Taugourdeau, Philippe, Sur un curieux microfossile incertae sedis du Frasnien du Boulonnais *Frasnacritetrus* nov. gen. (Acritarche). *Cah. Micropaléont.,* ser. 1(10), p. 1—4, 1 pl., 1968.

Thusu, Bindra, Depositional environments of the Rochester Formation (Middle Silurian) in southern Ontario. *J. Sedim. Petrol.,* v. 42, p. 930—934, figs. 1, 2, 1972.

Thusu, Bindra, Acritarches provenant de l'Ilion Shale (Wenlockien), Utica, New York. *Revue Micropaléont.,* v. 16, p. 137—146, pls. 1, 2, 1973a.

Thusu, Bindra, Acritarchs of the Middle Silurian Rochester Formation of southern Ontario. *Palaeontology,* v. 16, p. 799—826, pls. 104—106, 1973b.

Timofeev, B. V., Early Paleozoic flora of the Baltic provinces and main stages of its development. *XX Int. Geol. Congr., Resumenes Trab. Present.,* p. 187—188, 1956a.

Timofeev, B. V., Hystrichosphaeridae Kembriya [Cambrian Hystrichosphaeridae]. *Dokl. Akad. Nauk SSSR,* v. 106, p. 130—132, fig. 1, 1956b.

Timofeev, B. V., O novoy gruppe iskopaemykh spor [On a new group of fossil spores]. *Ezheg. Vses. Paleont. Obshch.*, vyp. 16, p. 280−284, 1 pl., 1957.

Timofeev, B. V., Über das Alter sächsischer Grauwacken. *Geologie*, v. 7, p. 826−835, 3 pls., 1958.

Timofeev, B. V., *Drevneyshaya flora Pribaltiki i ee stratigraficheskoe znachenie* [*The ancient flora of the Prebaltic and its stratigraphic importance*]. Trudy Vses. Neft. Nauchno-Issled. Geol.-Razv. Inst., vyp. 129, 319 p., 25 pls., 4 figs., 1959.

Timofeev, B. V., Spory Dokembriya [Precambrian Spores]. *Mezhdunar. Geol. Kongr. XXI Sess., Dokl. Sov. Geol.*, Prob. IX, Stratigrafiya i korrelyatsiya Dokembriya, p. 138−147, 1 pl., 1960a.

Timofeev, B. V., Sur la caractéristique micropaléontologique de la formation de Visingsö. *Geol. För. Stockh. Förh.*, v. 82, p. 28−42, pl. 2, 1960b.

Timofeev, B. V., Spory i fitoplankton Proterozoya i rannego Paleozoya Evrazii [Spores and phytoplankton of the Proterozoic and early Paleozoic of Eurasia]. *Mezhdunar. Geol. Kongr. XXI Sess., Dokl. Sov. Geol.*, Prob. VI, p. 177−188, pls. 1, 2, 1960c.

Timofeev, B. V., O sistematike drevneyshikh fitoplanktonnykh organizmov i dispernykh spor [On the systematics of ancient phytoplankton organisms and dispersed spores], p. 227−229. In *Sistematika i metody izucheniya iskopaemykh pyl'tsy i spor*, Akad. Nauk SSSR, Sibir. Otdel., Inst. Geol. Geofiz., 1964.

Timofeev, B. V., *Mikropaleofitologicheskoe issledovanie drevnikh svit* [*Micropaleophytological investigations of ancient strata*]. Akad. Nauk SSSR, Lab. Geol. Dokembriya, Moscow. 147 p., 89 pls., 1966.

Timofeev, B. V., *Sferomorfidy Proterozoya* [*Proterozoic Sphaeromorphida*]. Akad. Nauk SSSR, Inst. Geol. Geokhronol. Dokembriya, Leningrad: "Nauka," p. 1−145, pls. 1−39, 1969.

Timofeev, B. B., Mikrofitofossilii Proterozoya i rannego Paleozoya [Proterozoic and Early Paleozoic microfossils], p. 7−12, pl. 1. In *Mikrofossilii drevneyshikh otlozheniy*, Trudy *III Mezhdunar. Palinol. Konf., Novosibirsk 1971*, Akad. Nauk SSSR, Sibirsk. Otdel., Inst. Geol. Geofiz. Moscow: "Nauka," 1973a.

Timofeev, B. V., *Mikrofitofossilii Dokembriya Ukrainy* [*Microphytofossils of the Precambrian in the Ukraine*]. Akad. Nauk SSSR, Inst. Geol. Geokhronol. Dokembriya, Leningrad: "Nauka," p. 1−58, pls. 1−40, 1973b.

Timofeev, B. V., T. N. German, and N. S. Mikhaylova, *Mikrofitofossilii Dokembriya, Kembriya, i Ordovika* [*Microphytofossils of the Precambrian, Cambrian, and Ordovician*]. Akad. Nauk SSSR, Inst. Geol. Geokhronol. Dokembriya, Leningrad: "Nauka," p. 1−106, pls. 1−48, 1976.

Tschudy, R. H., and R. A. Scott, *Aspects of palynology.* New York, London, Sydney, Toronto: Wiley-Interscience, vii + 510 p., 1969.

Umnova, N. I., *Akritarkhi Ordovika i Silura Moskovskoy sineklizy i Pribaltiki* [*Acritarchs of the Ordovician and Silurian of the Moscow region and Prebaltic*]. Minist. Geol. RSFSR, Territor. Geol. Upravlenie Tsentral'. Rayonov Mosk. Kompl. Geol. Gidrogeol. Eksped. Moscow: "Nedra," 167 p., 20 pls., 1975.

Umnova, N. I., and E. K. Fanderflit, Kompleksy akritarkh Kembriyskikh i nizhneordovikskikh otlozheniy zapada i severo-zapada Russkoy Platformy [Acritarch complex of the Cambrian and Lower Ordovician deposits of western and northwest Russian Platform], p. 45−73, pls. 1, 2. In *Palinologicheskie issledovaniya v Belorussii i drugikh rayonakh SSSR*, Uprav. Geol. Sovete Minist. BSSR, Beloruss. Nauchno-Issledov. Geol. Inst. (BelNIGRI). Minsk: "Nauka i Tekhnika," 1971.

Umnova, N. I., and B. A. Yakovlev, Akritarkhi Yaroslavskoy Serii severnoy chasti Russkoy Platformy. *Materialy po Geologii i Poleznym Iskopaemym tsentral'nykh rayonov evropeyskoy chasti SSSR*, vyp. 6, Minist. Geol. RSFSR, Territ. Geol. Uprav. Tsentral. Rayonov Nauchno-Tekhn. Gornoe Obshchestvo, Moscow, p. 92−112, pls. 1, 2, figs. 1, 2, 1970.

Valensi, Lionel, Sur quelques microorganismes planctoniques des silex du Jurassique moyen du Poitou et de Normandie. *Bull. Soc. Géol. Fr.*, ser. 5, v. 18 (1948), p. 537−550, figs. 1−6, 1949.

Valensi, Lionel, Microfossiles des silex du Jurassique Moyen. Remarques pétrographiques. *Mém. Soc. Géol. Fr.*, n. sér., v. 32(4), mém. 68, p. 1−100, pls. 1−16, 1953.

Vanguestaine, Michel, Découverte d'acritarches dans le Revinien supérieur du massif de Stavelot. *Annls. Soc. Géol. Belg.*, v. 90 (1966−1967), p. 585−600, pls. 1−3, 1967.

Vanguestaine, Michel, Les acritarches du Sondage de Grand Halleux (Note préliminaire). *Annls. Soc. Géol. Belg.*, v. 91, p. 361−375, 1 pl., 1968.

Vanguestaine, Michel, New acritarchs from the Upper Cambrian of Belgium, p. 28−30, 1 pl. In *Mikrofossilii Drevneyshikh otlozheniy*, Trudy *III Mezhdunar. Palinol. Konfer. Novosibirsk 1971*, Akad. Nauk SSSR Sibirsk. Otdel., Inst. Geol. Geofiz. Moscow: "Nauka," 1973.

Vanguestaine, Michel, Espèces zonales d'acritarches du Cambro-Trémadocien de Belgique et de l'Ardenne Française. *Revue Palaeobot. Palynol.*, v. 18, p. 63−82, pls. 1, 2, figs. 1−5, 1974.

Varma, C. P., Do dinoflagellates and hystrichosphaerids occur in fresh water sediments? *Grana Palynol.*, v. 5, p. 124−128, 1964.

Vavrdová, Milada, Ordovician acritarchs from central Bohemia. *Vest. Ústřed. Úst. Geol.*, v. 40, p. 351−357, pls. 1−4, figs. 1−7, 1965.

Vavrdová, Milada, Mikroorganismy (Acritarcha) z Proterozoických buližníku čech. *Čas. Národ. Mus.*, v. 135, p. 93−96, pls. 1, 2, figs. 1−3, 1966a.

Vavrdová, Milada, Palaeozoic microplankton from central Bohemia. *Čas. Miner. Geol.*, v. 11, p. 409 – 414, pls. 1 – 3, figs. 1 – 4, 1966b.

Vavrdová, Milada, Acritarchs from Klabava Shales (Arenig). *Vest. Ústřed. Úst. Geol.*, v. 47, p. 79 – 86, pls. 1, 2, figs. 1 – 4, 1972.

Vavrdová, Milada, New acritarchs from Bohemian Arenig (Ordovician). *Vest. Ústřed. Úst. Geol.*, v. 48, p. 285 – 289, pls. 1, 2, figs. 1 – 4, 1973.

Vavrdová, Milada, Excystment mechanism of Early Paleozoic acritarchs. *Čas. Miner. Geol.*, v. 21, p. 55 – 64, pls. 1 – 4, figs. 1 – 7, 1976.

Vidal, Gonzalo. Late Precambrian microfossils from the basal sandstone unit of the Visingsö Beds, South Sweden. *Geol. Palaeont.*, v. 8, p. 1 – 14, pl. 1, figs. 1 – 4, 1974.

Vidal, Gonzalo, Late Precambrian microfossils from Visingsö Beds in southern Sweden. *Fossils Strata*, no. 9, p. 1 – 57, figs. 1 – 23, 1976.

Volkova, N. A., Spory dokembriya Pridnestrov'ya [Precambrian spores from the Dniester Region]. *Dokl. Akad. Nauk SSSR*, v. 142, p. 893 – 895, figs. 1 – 23, 1962.

Volkova, N. A., O prirode i klassifikatsii mikrofossiliy rastitel'nogo proiskhozhdeniya iz Dokembriya i nizhnego Paleozoya [Nature and classification of microfossils of plant origin from Precambrian and Lower Paleozoic]. *Paleont. Zh.*, 1965(1), p. 13 – 25, 1965.

Volkova, N. A., Akritarkhi Dokembriyskikh i Nizhnekembriyskikh otlozheniy Estonii [Acritarchs of the Precambrian and Lower Cambrian deposits of Esthonia]. *Trudy Inst. Geol. Nauk, Mosk.*, vyp. 188, p. 8 – 36, pls. 1 – 12, 1 fig., 1968.

Volkova, N. A., Akritarkhi severo-zapada Russkoy Platformy [Acritarchs of the northwestern Russian Platform]. *Trudy Inst. Geol. Nauk, Mosk.*, vyp. 206, p. 224 – 236, pls. 46 – 51, 1969.

Volkova, N. A., Lower Cambrian "Hystrichosphaerids." *J. Palynol.*, v. 7, p. 26 – 29, 1972.

Volkova, N. A., Akritarkhi iz pogranichnykh sloev nizhnego-srednego Kembriya zapadnoy Latvii [Acritarchs from Lower-Middle Cambrian subsurface sediments of western Latvia], p. 194 – 198, pls. 27, 28. In *Biostratigrafiya i paleontologiya nizhnego Kembriya Evropy i severnoy Azii*, Akad. Nauk SSSR, Geol. Inst., Sibirsk. Otdel. Inst. Geol. Geofiz. Moscow: "Nauka," 1974.

Volkova, N. A., O nakhodke Dokembriyskikh spor s tetradnym rubtsom [On some Precambrian spores with tetrad scar], p. 14 – 18, pl. 2. In *Paleontologiya Morskaya Geologiya, Mezhdunar. Geol. Kongr., XXV Sess.*, Akad. Nauk SSSR, Minist. Geol. SSSR, "Nauka," 1976.

Wall, David, Microplankton, pollen and spores from the Lower Jurassic in Britain. *Micropaleontology*, v. 11, p. 151 – 190, pls. 1 – 9, 1965.

Wall, David, and Charles Downie, Permian hystrichospheres from Britain. *Palaeontology*, v. 5, p. 770 – 784, pls. 112 – 114, 1963.

Wall, J. H., and Chaitanya Singh, A Late Cretaceous microfossil assemblage from the Buffalo Head Hills, north-central Alberta. *Can. J. Earth Sci.*, v. 12, p. 1157 – 1174, pls. 1 – 5, figs. 1 – 3, 1975.

Ważyńska, Hanna, Wstępne badania mikroflorystyczne osadów sinianu i kambru z obszaru Białowieży [Preliminary microfloristic examinations of the Sinian and Cambrian deposits from the Białowieża area]. *Kwart. Geol.*, v. 11, p. 10 – 19, pls. 1 – 4, fig. 1, 1967.

Wetzel, Otto, Die Typen der baltischen Geschiebefeuersteine beurteilt nach ihrem Gehalt an Mikrofossilien. *Z. Geschiebeforsch. Flachldgeol.*, v. 8, p. 129 – 146, pls. 1 – 3, figs. 1, 2, 1932.

Wetzel, Otto, Die in organischer Substanz erhaltenen Mikrofossilien des baltischen Kreidefeuersteins mit einem Sedimentpetrographischen und stratigraphischen anhang. *Palaeontographica Abt. A*, v. 78, p. 47 – 156, pls. 3 – 7, 1933.

White, M. C., Discovery of microscopic organisms in the siliceous nodules of the Paleozoic rocks of New York. *Am. J. Sci.*, ser. 2, v. 33, p. 385 – 386, 27 figs., 1862.

Wicander, E. R., Upper Devonian-Lower Mississippian acritarchs and prasinophycean algae from Ohio, U.S.A. *Palaeontographica Abt. B*, v. 148, p. 9 – 43, pls. 5 – 19, 1974.

Wicander, E. R., Fluctuations in a Late Devonian – Early Mississippian phytoplankton flora of Ohio, U.S.A. *Palaeogeogr. Palaeoclimat. Palaeoecol.*, v. 17, p. 89 – 108, figs. 1 – 9, 1975.

Wicander, E. R., and A. R. Loeblich, Jr., Organic-walled microphytoplankton and its stratigraphic significance from the Upper Devonian Antrim Shale, Indiana, U.S.A. *Palaeontographica Abt. B*, v. 160, p. 129 – 165, pls. 1 – 11, 3 figs., 1977.

Yankauskas, T. V., and G. K. Vaytekunene, Akritarkhi iz Silura Pribaltika. *Paleont. Zh.*, 1972 (2), p. 113 – 121, pl. 17, figs. 1 – 3, 1972.

DINOFLAGELLATES

In the Glass of Sea Water I send with this are some of the Animalcules which cause the Sparkling Light in Sea Water: they may be seen by holding the Phial up against the Light, resembling very small Bladders or Air Bubbles, and are in all Places of it from Top to Bottom, but mostly towards the Top, where they assemble when the Water has stood still some Time, unless they have been killed by keeping them too long in the Phial.

Placing one of these Animalcules before a good Microscope; an exceeding minute Worm may be discovered, hanging with its Tail fixed to an opake Spot in a kind of Bladder, which it has certainly a Power of contracting or distending, and thereby of being suspended at the Surface, or at any Depth it pleases in the including Water.

H. Baker, 1753

*T*his description of a *Noctiluca* by Baker (1753) was the first reference to the individual organism of this group, whose periodic blooms result in luminescent seas or the toxic red tides of many coasts. The reddish waters result from the characteristic pigments, particularly peridinin, whose coloration also supplied the name for the Division Pyrrhophyta (the Greek *pyrrhos* means "flame colored" or "tawny"). The coastal phenomenon known as red water or red tides caused by the annual blooms of the tiny "flame-colored plants" in the late spring or summer was even mentioned in the Old Testament (Exodus 7:20−21) and in the writings of the ancient Greeks. The other common name for this group, the Dinoflagellata, is from the Greek *dinos,* meaning "whirling," and refers to the rotational progression of these flagellates in movement; the common generic termination *-dinium* is from the same stem (it is not to be confused with the Greek *deinos,* which has the meaning "fearful" or "terrible," as in dinosaurs).

Although tiny, dinoflagellates may occur in astronomical numbers, hence not only may cause the luminescent seas and red waters that long have been a source of interest, but as primary producers in the marine environment they are surpassed in importance only by the diatoms. The individual cells of these organisms are equally astonishing in their variety: they are unique in combining characteristics of both the procaryotes and eucaryotes (they were even placed in a separate kingdom, the Mesocaryota, on the basis of these features), and of both plants and animals, some being photosynthetic, whereas others are grazing or predacious, parasitic or endo- or ectosymbionts. The major classifications, monographs, and texts concerned with either protozoa or algae always include the dinoflagellates, regarding them as animals (Chatton, 1952; Grell, 1973; Honigberg et al., 1964; Kiselev, 1950; Kofoid and Skogsberg, 1928; Kofoid and Swezy, 1921; Lebour, 1925; and many other texts on protozoa), or as plants (Fott, 1971; Kiselev, 1951; Lemmermann, 1910; Papenfuss, 1955; Schiller, 1931—1937; Schilling, 1913; and other volumes on algae).

The above-mentioned luminescent *Noctiluca* is large (up to 1500 to 2000 μm across) and nonphotosynthetic, with a taillike "prod" rather than flagella (see Figure 4.1). The more typical dinoflagellate (see Figure 4.2) is free-living, photosynthetic, biflagellate, and much smaller,

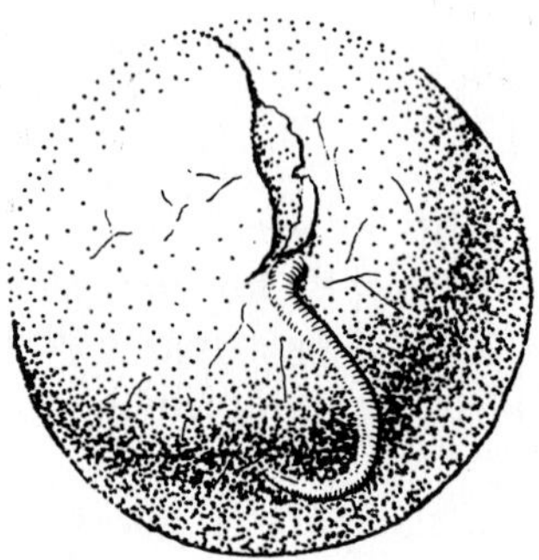

Figure 4.1
Noctiluca miliaris Suriray. Vegetative cell, with elongate prod or tentacle, approximately ×65. Redrawn from Kofoid, 1920.

Figure 4.2
Gymnodinioid or naked dinoflagellate, diagrammatic. Left of figure shows cell exterior, right is optical section to show interior. Bar = 10 μm.

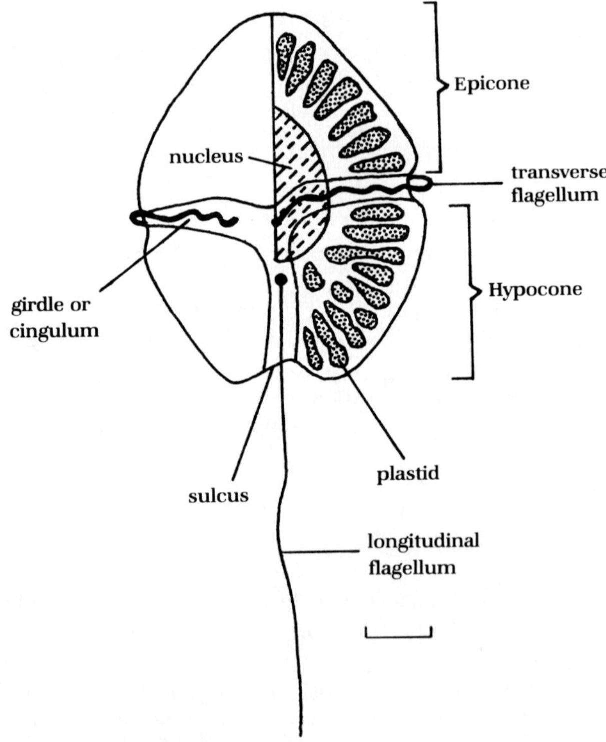

generally between 30 and 60 μm in length. An oblique transverse constriction or groove, the **girdle** or **cingulum**, divides the cell into an anterior and posterior half, and a **longitudinal groove** or **sulcus** on the ventral side connects the ends of the girdle and generally extends somewhat to the posterior. The two flagella lie within these grooves, arising laterally from separate pores within the girdle and sulcus. The flagellum from the anterior pore near the proximal end of the girdle circles to the organism's left within the girdle. The second flagellum emerges from the posterior pore within the sulcus and trails behind the cell. Because of their small size and transparency the upper and lower surfaces of the cell as seen in the light microscope have at times been confused, and individual cells incorrectly illustrated with the girdle obliquity opposite to its true nature and the transverse flagellum circling to the right. A considerable degree of torsion of the cell in *Cochlodinium* produces a sulcus that also is wound about the cell rather than being short and straight. A few highly modified species lack flagella, and instead have a tentacle or prod (as in *Noctiluca*), or a suckerlike stalk (in parasitic species). Only the motile zoospores of such highly modified forms indicate their dinoflagellate relationship.

Some dinoflagellates have only a thin pellicle covering the cell and are known as naked or nonthecate dinoflagellates; the armored dinoflagellates have a cellulosic theca constructed of a series of plates of distinctive number and arrangement (see Figure 4.3). A few have a bipartite theca with apically inserted flagella (see Figure 4.4).

Equally as distinctive as the outer cell morphology is the unusually large dinoflagellate nucleus. This **dinokaryon** or **mesokaryon** nucleus has continuously condensed and visible chromosomes that appear as beadlike strands. Nuclear division similarly is unusual, lacking spindle formation. Much recent information suggests that the usual vegetative cell is haploid, whereas the diploid stage is of relatively short duration in the life cycle.

The biflagellate motile stage may alternate with an encysted stage that may have a highly resistant imperforate covering. Once thought to result only from unfavorable conditions, evidence now indicates this to be a **hypnozygote** stage in sexual reproduction, meiosis occurring just prior to or at the time of excystment. The resting cyst may closely resemble the motile stage, even to the presence of girdle, sulcus, and pseudotabulation, or may be quite dissimilar, consisting of a globular or elongate main body from which extend elongate processes. Excystment occurs through an opening, the **archeopyle**, that reflects the position and number of particular thecal plates of the motile cell.

The first dinoflagellates to be formally described (both by O. F. Müller, 1773) were *Bursaria hirundinella* and *Vorticella cincta*, now referred to *Ceratium* and *Peridinium*, respectively. More detailed descriptions were given by Ehrenberg in the 1830s, but perhaps because of presumed analogy with the numerous ciliates he had described, dinoflagellates were regarded as possessing both a flagellum and a transverse row of cilia. His illustrations (1836a, pl. 2, figs. 1−5) of various species correctly showed the longitudinal flagellum, but the undulating transverse flagellum partially emerging from the girdle was interpreted as a row of cilia (see Figure 4.5). Claparède and Lachmann (1859, p. 393) followed this interpretation of the anterior border of the cingulum as ciliated (this is now oriented as posterior), and regarded the dinoflagellates as the "infusoires cilio-flagellés," intermediate between the ciliates and flagellates. The Protozoa then included the Ciliata, Suctoria, Cilioflagellata, and Flagellata. Others (e.g., Bergh, 1881; Stein, 1883) figured various dinoflagellates with ciliated girdles, until Klebs (1883, p. 351) correctly recognized the biflagellate nature of the "cilioflagellata." Illustrating both the straight longitudinal flagellum and the undulating transverse flagellum, Klebs stated that no cilia were present, hence this group was regarded as quite distinct from the Protozoan ciliates and flagellates and placed with the "Thallophytes" or algae.

Ehrenberg (1836b) described the first known fossil dinoflagellates from chert fragments; these were long regarded as preserved thecae, although evidence gradually accumulated to indicate that only the resistant cysts are fossilized. Development of such resting cysts in living

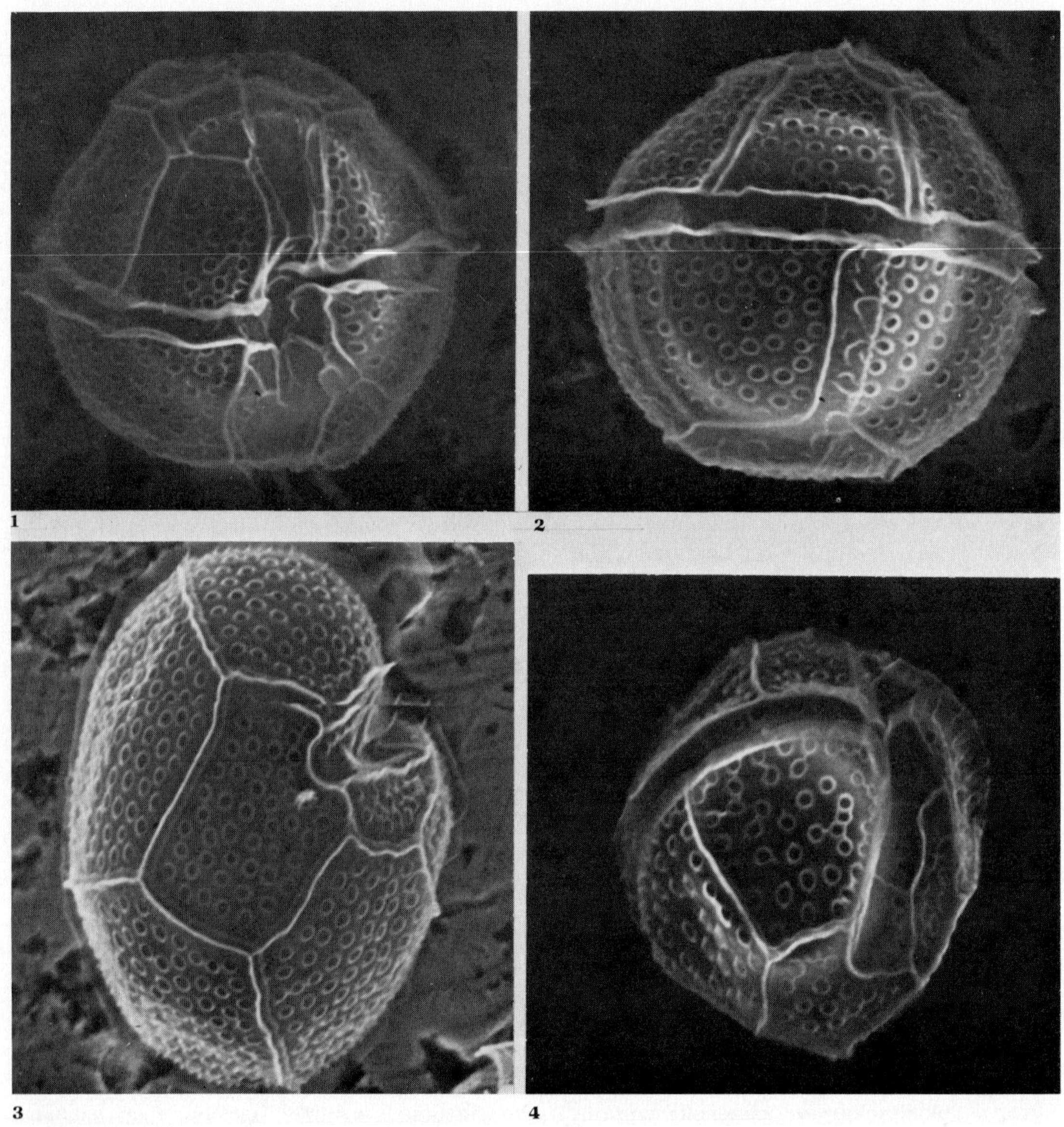

Figure 4.3

Gonyaulax polyedra Stein. SEM of theca of glutaraldehyde-fixed and freeze-dried cell. **1.** Ventral view, showing descending cingulum as it spirals to cell's left, and sulcus connecting ends of cingulum and extending toward antapex. **2.** Dorsal view, showing intercalary bands at the sutures. **3.** Antapical view, showing hypotheca, with base of sulcus on right. **4.** Right side of smaller cell, viewed obliquely from base, sulcus at right, ×2000. Other photos, ×1500. Plate borders marked by ridges, and plates sculptured with bordered pits that are not open pores. From Loeblich III.

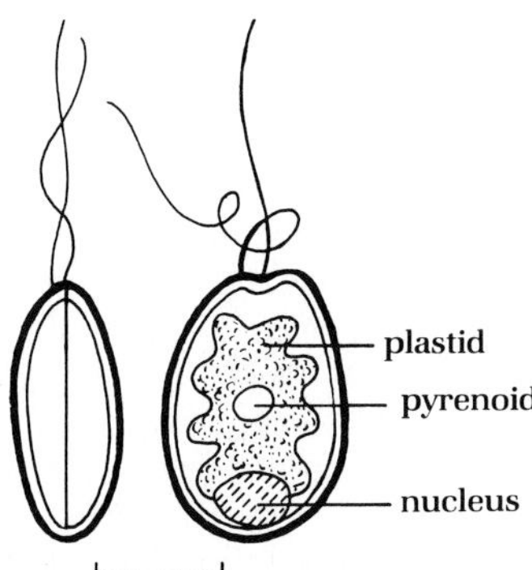

Figure 4.4
Prorocentrum. Left, edge view; and right, valve view of this bivalved dinoflagellate, with apical flagella, large nucleus, plastid, and pyrenoid. Bar = 20 μm.

species was first observed by Carter (1858) who described the *Peridinium* in red tides along the Bombay coast. The encystment process in living dinoflagellates was also described by Claparède and Lachmann (1859). Many other dinoflagellate genera and species were described and beautifully illustrated by Stein (1883), who reversed the earlier orientation to one with a posteriorly trailing longitudinal flagellum and observed the oblique nature of the cell division during binary fission. As the previous term Cilioflagellata had been shown no longer to be appropriate, Bütschli (1885) proposed the new name Dinoflagellata.

THE CELL

The greatest variation in size is found in the nonthecate dinoflagellates; although the average is about 100 μm in length, adult motile cells range between 7 and 212 μm, and some parasitic species attain a length of as much as 700 μm. The smallest species yet described (see Figure 4.6) is *Gymnodinium simplex* (Lohmann) Kofoid & Swezy, which has been reported as small as

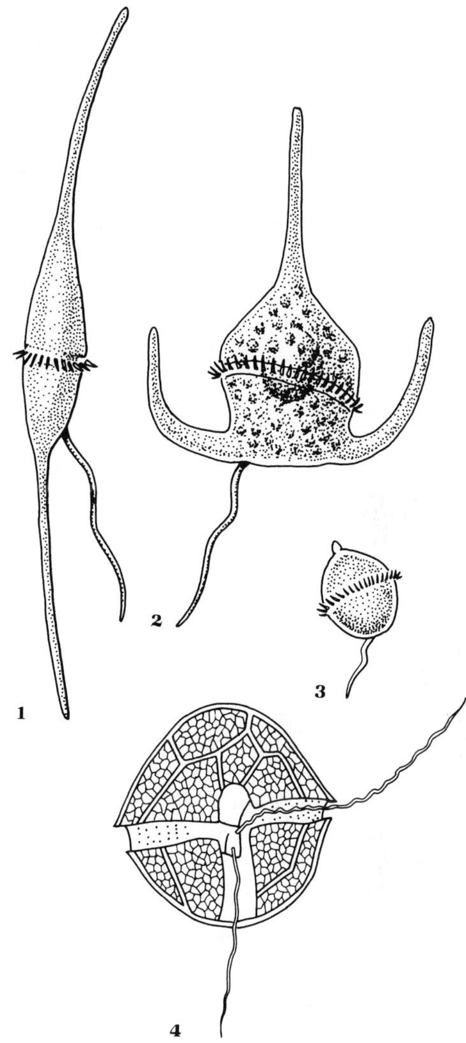

Figure 4.5
Early interpretations of dinoflagellate morphology. **1 – 3.** After Ehrenberg (1836a), showing cingulum occupied by supposed row of cilia. **1.** *Ceratium fusus* (Ehrenberg) Dujardin, length 200 μm. **2.** *C. tripos* (O. F. Müller) Nitzsch, length 175 μm. **3.** *Scrippsiella trochoidea* (Stein) Loeblich III, length 36 μm. **4.** From Klebs (1883), *Peridinium bipes* Stein, correctly showing transverse flagellum rather than cilia in cingulum, length 80 μm.

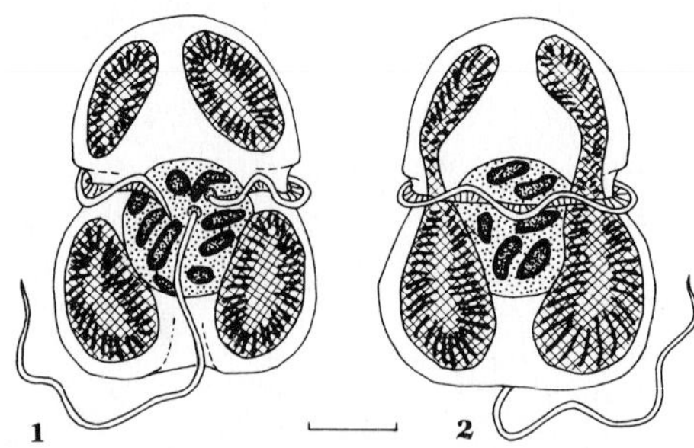

Figure 4.6
Gymnodinium simplex. **1.** Ventral view. **2.** Dorsal view of this smallest living dinoflagellate species, showing central nucleus and peripheral plastids. Bar = 2 μm. With permission, from Dodge, 1974a, *J. Mar. Biol. Ass. U.K.,* v. 54, p. 171−177, published by Cambridge University Press; slightly modified.

1.7 μm in length (Lebour, 1925), although these small rounded individuals lacked the usual grooves and flagella.

The transverse girdle of the usual nonthecate dinoflagellate separates anterior and posterior portions of the cell, termed **epicone** and **hypocone**, respectively. A longitudinal furrow or sulcus is present on the ventral side, extending to the posterior from the girdle. The two flagella lie within these depressions. **Armored** dinoflagellates have a cellulosic **theca** that similarly is divided into **epitheca** and **hypotheca.** The armored species commonly are between 30 and 60 μm long, with *Ceratium hirundinella* (O. F. Müller) Bergh up to 400 μm (see Figure 4.7) and others as little as 15 μm long. A single species may show a variation of 30 percent or more in size (Kofoid and Swezy, 1921).

Figure 4.7
Some freshwater dinoflagellates. **1,2.** *Ceratium hirundinella.* **1.** Ventral view of motile cell, from lake at Fullerton, California, ×300, from Loeblich Jr. and Loeblich III, 1966. **2.** Ventral view of cyst, Lake Zurich, Switzerland, ×440, from Wall and Evitt, 1975. **3−5.** *Peridinium cinctum,* ventral view; slightly to left in ventral view; and dorsal view, from Fullerton, California, ×300.

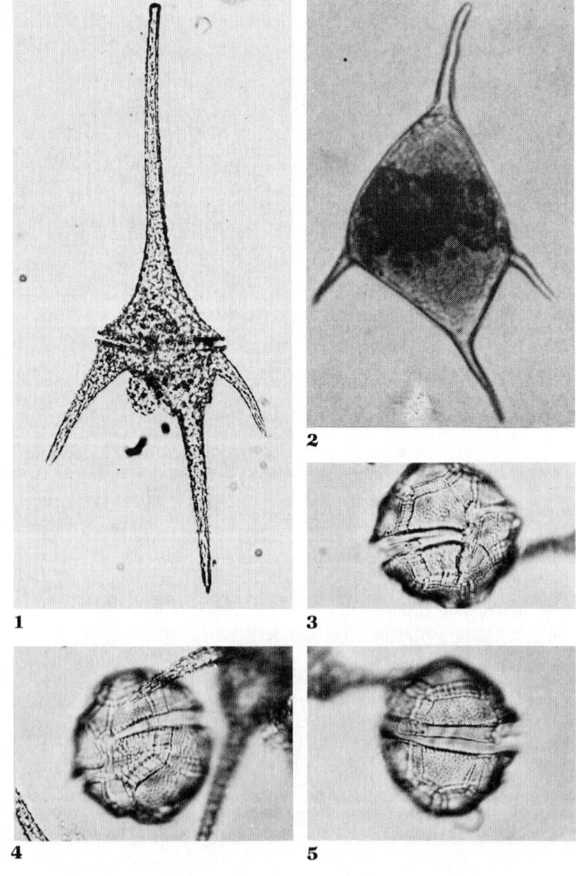

The Protoplast

The two unlike flagella, nucleus, varicolored plastids, and certain other organelles are discussed separately, as some provide strong indications concerning the evolutionary relationships of the dinoflagellates. Following the earliest study of dinoflagellate cells by electron microscopy of thin sections (Grell and Wolfarth-Bottermann, 1957), a wealth of data accumulated by such studies of the wall, flagella, plastids, nuclei, chromosomes, and other organelles. Dinoflagellate ultrastructure (reviewed by Dodge, 1971b, 1973) has been useful in morphologic studies, in understanding symbiotic relationships, and in indicating probable phylogenetic relationships.

Cytoplasm

Photosynthetic species may have a single large plastid or many small ones that appear green, yellow, or brown, or intermediate shades of these colors. Both these and the non-photosynthetic species also may have highly colored cytoplasm. The background cytoplasm is commonly a soft pearly gray, but with an overriding coloration resulting variously from pigmented oil drops, organelles, or the yellowish to greenish or reddish plastids. The ocellus or eyespot of some species may be deep red to almost black. Extra-plastid pigments of the photosynthetic species may be widely dispersed or limited to oil drops or other localized regions of the cell. As a result of these various pigmented organelles or inclusions, the general appearance of *Noctiluca* is colorless to bluish green; *Gymnodinium* may appear violet to a deep red purple; *Erythropsidinium*, *Warnowia*, and some *Gymnodinium* may be rose colored; *Peridinium*, a bright coral red; and various *Gyrodinium* may be violet, red, purple, yellow-green, or greenish with numerous red spots.

The cytoplasm also may contain vacuoles, pyrenoids, an osmo-regulatory organelle (pusule), ejectile organelles (nematocysts and trichocysts), an eyespot or ocellus, mitochondria, Golgi bodies, starch grains, and oil droplets, and a few even have internal skeletal spicules.

Nucleus

The dinoflagellate nucleus is unique among eucaryotes, other than the Euglenids, in that the rounded to rodlike beaded chromosomes of nearly uniform size never lose the metaphase-anaphase condensation of the chromatin and can be readily fixed and stained throughout interphase (see Figure 4.8). During mitosis (nuclear division) there is no centromere or spindle formation. As such features of the dinoflagellate nucleus (dinokaryon) are distinct from those of the typical eucaryote nucleus, Dodge (1965a) termed it mesocaryotic and placed the dinoflagellates in the new Kingdom Mesocaryota, intermediate in position between the Procaryota and Eucaryota. Radiolaria and Heliozoa were suggested as possible other members of the new kingdom.

The dinoflagellate nucleus is relatively large in proportion of cell volume, although various species show a broad range in nuclear size, as well as in the size, shape, and number of their chromosomes.

A simple spherical or subspherical nucleus occurs in *Katodinium rotundatum* (Lohmann) Loeblich III, *Oxyrrhis marina* Dujardin, *Cachonina niei* Loeblich III, and *Scrippsiella trochoidea;* the nucleus of *Amphidinium herdmanii* Kofoid & Swezy and *Prorocentrum micans* Ehrenberg is flattened, and triangular or crescentic in plan; and that of *Gonyaulax* spp. and certain freshwater *Peridinium* is large and U-shaped (Dodge, 1966).

A **perinuclear capsule** encloses the nuclear region, separating it from the remaining cytoplasm, although as in other plants perforations in the capsule generally allow some cytoplasmic communication. The normally perforate capsule is not found in a few species; pores are lacking in *Noctiluca* (Zingmark, 1970a), but their function is served by minor vesicles. Pores are restricted to vesicular structures in the capsule of *Gymnodinium fuscum* (Ehrenberg) Stein (Dodge and Crawford, 1969a). The perinuclear capsule of the Warnowiaceae has two layers constructed of flattened lamellae and is differentiated only at a particular stage of growth, rather than being a permanent feature (Greuet, 1972b). These features clearly distinguish the dinoflagellate capsule from that of the

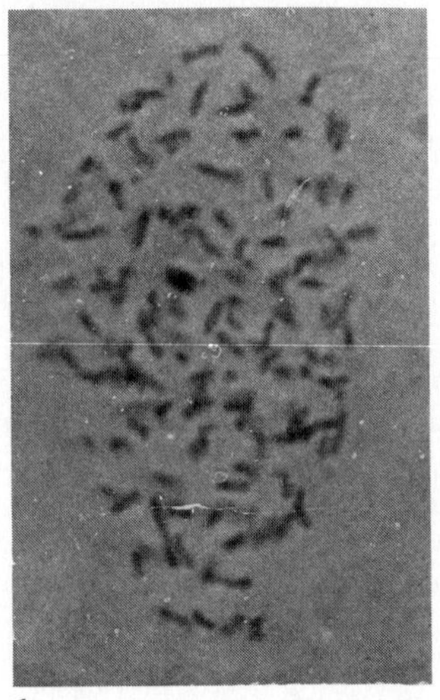

Figure 4.8
1. Squashed acetocarmine-stained nucleus of *Crypthecodinium cohnii* (Seligo) Chatton in Grassé, showing 108 chromosomes. **2.** Drawing of same. Both from Allen et al., 1975.

radiolaria, with which it has been compared. Ultrastructural study also has proven that the perinuclear capsule is merely a sporadic development of endoplasmic reticulum (Mignot, 1970), and the radiolarian capsule has since been compared to the dinoflagellate theca instead (Hollande et al., 1970).

Most dinoflagellates have nucleoli within the nucleus, although none is clearly recognizable in some (e.g., *Prorocentrum*). Some freshwater *Peridinium,* and the marine *Gessnerium tamarense* (Lebour) Loeblich III & L. A. Loeblich and *G. catenellum* (Whedon & Kofoid) Loeblich III & L. A. Loeblich have a large nucleolus-like central body inside the arms of the U-shaped nucleus that is not surrounded by nuclear material, but divides at the time the chromosomes divide in mitosis (Dodge, 1964b, 1966).

The chromosomes themselves are tiny spherical bodies, 0.5 to 1.0 μm in diameter in *Amphidinium klebsii* Kofoid & Swezy, *Katodinium rotundatum,* and *Prorocentrum triestinum* Schiller; they are rod-shaped and up to 7 μm long and 0.5 μm in diameter (Dodge, 1963b) in *Gessnerium tamarense, Cachonina niei, Scrippsiella trochoidea,* and *Prorocentrum minimum* (Pavillard) Schiller (as *P. mariaelebouriae* (Parke & Ballantine) Loeblich III, a synonym). The large chromosomes of *P. micans*

are about 15 μm long and 1 μm in diameter. Squashing the interphase nucleus allows separation and counting of the chromosomes, which vary widely in number in different taxa. Thus the parasitic *Syndinium turbo* Chatton has a mean of 5, *Prorocentrum balticum* (Lohmann) Loeblich III has 20, *Cachonina niei* has 112, the freshwater *Ceratium hirundinella* has 274, and there are as many as 325 in *Endodinium chattonii* Hovasse (see Table 4.1).

The many chromosomes also may differ in arrangement. Those of *Prorocentrum micans* show helical coiling as in higher plants, although they are quite different in their continuously condensed state and distinctive method of division (Dodge, 1964a). In contrast, chromosomes of other species may be aligned in parallel rows or otherwise variously arranged to fill the nucleus. Autoradiography of tritiated thymine of dividing cells of *P. micans* showed that all nuclei incorporated substantial amounts, indicating continuous replication of nuclear DNA throughout the cell cycle, as in the procaryotes, and suggestive of a very primitive eucaryote (Filfilan and Sigee, 1977).

The dinoflagellate nucleus is also unusual among eucaryotes in that cytochemical staining shows the chromosomes to consist solely of fibrillar DNA. Generally no RNA and no protein is present in the chromosomes, but protein and

Table 4.1

Mean number of chromosomes reported in some dinoflagellate nuclei.

	Chromosome number	Reference
Class Syndiniophyceae		
Order Syndiniales		
Family Syndiniaceae		
Syndinium turbo Chatton	5	(Loeblich Jr. and Loeblich III, 1966)
Class Dinophyceae		
Order Chytriodiniales		
Family Chytriodiniaceae		
Myxodinium pipiens J. & M. Cachon & Bouquaheux	40	(Cachon et al., 1970a)
Order Gymnodiniales		
Family Gymnodiniaceae		
Amphidinium carterae Hulburt	32	(Dodge and Crawford, 1968)
A. klebsii	32	(Dodge, 1963a)
Endodinium chattonii	325	(D. L. Taylor, 1971a)
Gymnodinium dodgei Sarma & Shyam	195	(Sarma and Shyam, 1974)
G. vitiligo Ballantine	44	(Dodge, 1963a)
G. zachariasii Lemmermann	64	(Entz, 1921)
Katodinium rotundatum	68	(Dodge, 1963b)
Oxyrrhis marina	55	(Dodge, 1963a)
Order Peridiniales		
Family Crypthecodiniaceae		
Crypthecodinium cohnii	99	(Kubai and Ris, 1969)
Family Ceratiaceae		
Ceratium hirundinella	274	(Entz, 1921)
C. tripos var. *subsalsum* Ostenfeld	200	(Borgert, 1910)
Family Gonyaulacaceae		
Gonyaulax polygramma Stein	100	(Entz, 1921)
Gessnerium tamarense	144	(Dodge, 1963a)
Family Peridiniaceae		
Cachonina niei	112	(Loeblich III, 1968)
Scrippsiella trochoidea	44	(Dodge, 1963a)
Order Prorocentrales		
Family Prorocentraceae		
Prorocentrum balticum	20	(Dodge, 1963a)
P. minimum	32	(Dodge, 1963a)
P. micans	68	(Dodge, 1963a)
P. pusillum (Schiller) Loeblich III	24	(Dodge, 1963a)
P. triestinum	24	(Dodge, 1963a)

lipid are present as a granular amorphous matrix of the nucleus, surrounding the chromosomes, much as in the bacterial nucleus (Dodge, 1964a). The DNA fibrils are about 25 Å in diameter (Kubai and Ris, 1969), whereas DNA and the combined protein together form fibrils of 100 to 250 Å diameter in the typical eucaryote. The presence of acid-soluble proteins has been demonstrated in log-phase (rapidly dividing) cells, but none or much fewer occur in stationary phase cells (Rizzo, 1973). Some RNA and protein is present in *Crypthecodinium cohnii* in much smaller amounts than the DNA. The RNA/DNA ratio is 0.32, and the protein/DNA ratio 1.19; the nuclei are about 5 μm in diameter and contain 6.9 picograms (6.9 $\times$ 10^{-12} grams) of DNA per nucleus. Nuclei of *Scrippsiella trochoidea* are about 15 μm in diameter and contain 34 pg DNA per nucleus; those of *Gymnodinium nelsonii* Martin range from 10 to 40 μm in diameter, although most are about 20 to 25 μm, and each contains about 143 pg DNA. As these amounts are from two to four orders of magnitude greater than the amounts of DNA per nucleus in other algae, such as the green *Chlorella* or *Euglena,* and much closer in magnitude to the amount in the diploid higher plants, the nuclei were suggested possibly to be polyploid or to contain a substantial amount of redundant DNA (Rizzo and Noodén, 1973). The proteins present also differ significantly from those of other eucaryotes (Loeblich III, 1976); the dinoflagellates resemble procaryotes in lacking histone.

Rae (1976) found dinoflagellate DNA to contain a considerable amount of the unusual base hydroxymethyluracil, previously reported only from bacteriophages that infect *Bacillus*. This base substantially raises the density and lowers the thermal stability of the DNA, resulting in a discrepancy between the guanine + cytosine content determinations obtained from buoyant density measurements and the temperature at which 50 percent of the DNA was denatured in a thermal gradient. This discrepancy was noted in varying amounts in all dinoflagellates studied, *Prorocentrum cassubicum* (Woloszynská) Dodge & Bibby, *Amphidinium carterae, Crypthecodinium cohnii, Peridinium triquetrum* (Ehrenberg) Lebour, and *Zooxanthella micro-*

adriatica (Freudenthal) Loeblich III & Sherley, but did not occur in rhodophyte, cryptophyte, chloromonadophyte, chrysophyte, diatom, chlorophyte, and euglenid algae examined. Although not necessarily primitive, this unusual feature of the DNA is characteristic only of the dinoflagellates.

The protein-synthesizing ribosomes also contain RNA, and this similarly is anomalous in the dinoflagellates. The molecular weight of the light subunits of ribosomal RNA is 0.7 $\times$ 10^6 and that of the heavy subunits is 1.23 $\times$ 10^6 in the freshwater *Peridinium cinctum* fa. *westii* (Lemmermann) Lefèvre. The light fraction is typically eucaryotic, but the molecular weight of the heavy fraction is much lower than in other eucaryotes, where cytoplasmic RNA ranges from 1.3 $\times$ 10^6 in protozoa, green plants, and fungi, and 1.4 $\times$ 10^6 in arthropods, to 1.5 $\times$ 10^6 in birds, and 1.75 $\times$ 10^6 in mammals. In contrast, that of procaryotes ranges from 1.05 to 1.1 $\times$ 10^6. A similar anomalously heavy subunit of ribosomal RNA occurs in some red algae, which also may represent a more primitive eucaryotic condition than that of the fungi or green plants (Gressel et al., 1975).

The amount of dinoflagellate whole cell DNA supports their eucaryotic nature however. Resting cells have only one-half as much DNA as do the logarithmically growing ones. The stationary phase of *C. cohnii* has 3.25 pg/cell of DNA whereas the log stage contains up to 15.6 pg/cell; stationary cells of *Gyrodinium resplendens* Hulburt have a minimum of 22pg/cell, and the log phase a maximum of 74 pg/cell; the maximum yet reported for a dinoflagellate is 200 pg/cell in *Gonyaulax polyedra*. This is much greater than in bacteria, *Escherichia coli* having 0.004 pg/cell, but is comparable to eucaryotes, the human haploid genome containing 2.7 pg of DNA (Allen et al., 1975).

Fibrils in the dinoflagellate chromosome are arranged in archlike whorls, as seen in electron microscopy (see Figure 4.9), similar to those of the bacterial nucleoid. However, the low level of chromosomal proteins may allow the similar folding pattern (Loeblich III, 1976) and thus, regardless of the closeness of the relationship, would result in the distinctive appearance of the dinoflagellate chromosomes.

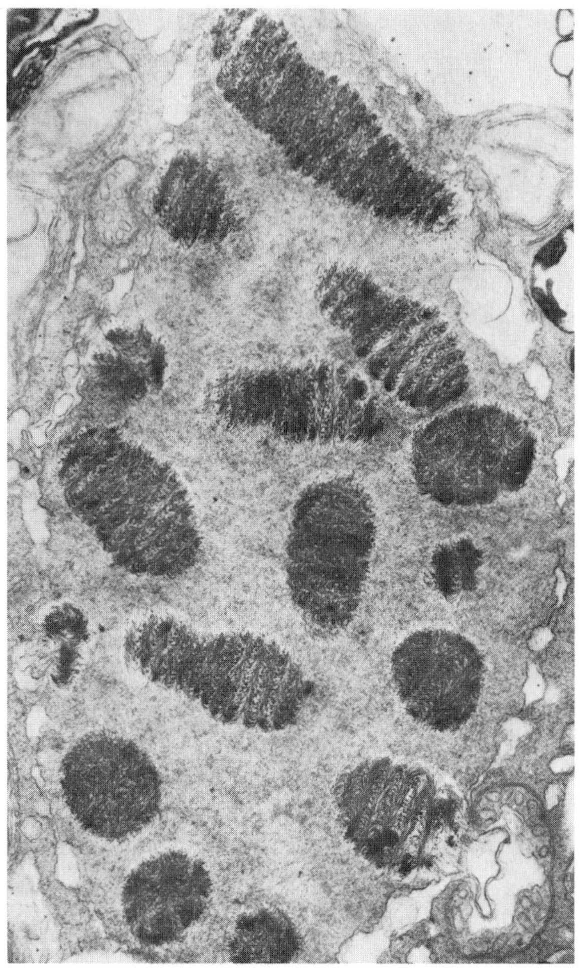

Figure 4.9
Nucleus of *Prorocentrum minimum,* TEM
of glutaraldehyde-fixed, sectioned cell,
showing large condensed chromosomes in
various orientations, with distinctive fibrillar swirl pattern, ×46,000; from Loeblich
III, 1976.

A great variety of evidence indicates that the
typical dinocaryotic nucleus with continuously
visible discrete chromosomes of fibrillar structure represents the haploid vegetative stage,
and that in these dinoflagellates only the zygote
is diploid, meiosis being post-zygotic. This evidence includes experiments involving mutants,
both spontaneous mutations and those induced
chemically or by ultraviolet irradiation, the frequency of the mutants and recombinants being
consistant only with a haploid organism
(Roberts et al., 1974b; Beam and Himes, 1974).
The large number of morphologically similar
chromosomes, from 99 to 110 in *Crypthecodinium cohnii,* might suggest polyploidy,

as previously indicated, with three or more sets
of homologous chromosomes. The large size,
permanently condensed state, and appearance
of the chromosomes in electron microscopy
also led to the suggestion of polyteny with mul-
tistranded diploid chromosomes, each 30-Å-
wide fibril in *Prorocentrum micans* consisting
of a single DNA double helix. The twisted fibrils
in relaxed state result in the beaded appearance
of the chromosome, but when stretched appear
as a series of figure 8s in electron microscopy
(Haapala and Soyer, 1973).

Mutational data appear to rule out both the
possibility of simple polyteny and polyploidy,
however, as well as the extreme polyploidy in

which all chromosomes are alike. Studies of whole cell DNA by means of renaturation kinetics allow a sharp distinction between procaryote and eucaryote genomes, the former having an extremely small amount of repeated sequence DNA, whereas eucaryotes have repeated sequences interspersed with unique DNA. The dinoflagellate *C. cohnii* showed eucaryote affinities in the renaturation kinetic experiments. Extreme polyploidy would show at least 100 copies of repeated sequences (in the 100 chromosomes), and the polytene model would indicate many repeats in each chromosome; but renaturation indicated a maximum of 1 to 3 repetitions in *C. cohnii,* hence the highly complex DNA, the nature of the repeated sequences, and the frequency of haploid mutation all rule out either a highly polytene or a polyploid nucleus and resemble typical eucaryote patterns (Allen et al., 1974, 1975). Visco-elastic molecular weight measurements of the chromosomal DNA indicated that most DNA molecules correspond to a single piece per chromosome to a maximum of six, but probably between one and three (Allen et al., 1974; Roberts et al., 1974b).

An additional unusual feature of the dinoflagellate nuclear organization is shown cytologically. The chromosomes are attached to the nuclear envelope, which remains intact through the division cycle and is thought to control the mitotic distribution of the chromosomes to the daughter cells. Chromosomes show no distinct alignment on a metaphase plate as in other eucaryotes. As dinoflagellate mitosis occurs without the typical eucaryotic spindle fibers and centromeres for segregation of daughter chromosomes, a different mechanism must be involved.

Three types of chromosome segregation have been reported among typical dinoflagellates (Tomas, 1974). Segregation is vertical in *Prorocentrum micans,* with the free ends of the chromatids sliding in the direction of the poles of division (Dodge, 1963b). In the nuclear division of *Crypthecodinium cohnii, Gymnodinium micrum* (Leadbeater & Dodge) Loeblich III, and *Gonyaulax polyedra,* cytoplasmic invaginations containing microtubules extend into the nucleus, and the chromatids slide along the nu-

clear envelope at these invaginations. A third type of nuclear division characterizes such *Gessnerium* as *G. tamarense* and *G. catenellum,* in which a central body is found near the diverging arms of the U-shaped nucleus (Dodge, 1964b; Tomas, 1974). Composed of ribosomelike granules and lacking DNA, the central body is regarded as an endosome, or persistant nucleolus, and seems to control the chromosome movement. As the nucleolus divides and separates, it segregates the chromosomes, but the nuclear envelope remains intact. The cells remain motile as cytokinesis or cell division begins, and the new flagella appear about midway in the process (Tomas, 1974).

Syndinium nuclear division is still different. Unlike most dinoflagellates, *Syndinium* has only a few (four to six) chromosomes, and these are V-shaped with the apex toward a centriole, giving a rosettelike appearance. The centrioles also are connected by a central spindle during division. This condition is intermediate between the simple attachment to a membrane (as in procaryotes and typical dinoflagellates) and the mitotic spindle of the typical eucaryote, as a microtubule organizing center is differentiated in the *Syndinium* nucleus at the attachment site and is linked to spindle poles by microtubules (Ris and Kubai, 1974). The low number of chromosomes in *Syndinium,* the V-shape of the chromosomes that have sufficient basic protein to be detectable histochemically, and the possession of centrioles associated with mitosis were regarded as sufficient basis for the segregation of these intracellular parasites that lack a theca and have no fossil record as the separate Order Syndiniales and Class Syndiniophyceae (Loeblich III, 1976).

Noctiluca does not have the usual dinokaryon nucleus, but has a nucleus devoid of evident chromosomes during interphase (Zingmark, 1970a). Its chromosomes resemble the prophase state of the usual eucaryotic dividing nucleus. Surrounded by a granular mass, the chromosomes themselves occupy only a small portion of the nuclear volume (Afzelius, 1962). Life history studies suggest that this "nocticaryotic" nucleus is diploid, and the previously reported "chromosomes" are in fact only differentiated nucleoli. True dinokaryotic nuclear structure is present only in the presumably haploid biflagellate gametes of *Noctiluca* (Zingmark, 1970a).

Thus, while resembling procaryotes in some primitive or reduced characters, many other aspects of the dinoflagellate nucleus are characteristic of the eucaryotes. The DNA sequence properties are like those of eucaryotes, although the scarcity of basic protein associated with the DNA and the dimensions of the chromatin fibrils are procaryotelike characters. Dinoflagellates may have diverged early from the eucaryote line, but whether or not they represent an evolutionary intermediate or a direct evolutionary line (hence mesocaryotic) is as yet uncertain (Allen et al., 1975).

In addition to a true dinokaryotic nucleus, *Peridinium foliaceum* (Stein) Biecheler and *Peridinium balticum* (Levander) Lemmermann have been reported to have a second and eucaryotic nucleus (Dodge, 1971a; Blanchard-Babillot, 1972). Detailed cultural and ultrastructural studies have shown the latter in fact to belong to an intracellular symbiont (Tomas and Cox, 1973), and only the dinocaryotic nucleus belongs to the dinoflagellate itself. This is mentioned later in the discussion of dinoflagellate symbiosis.

Plastids and Pigments

Most dinoflagellates are holophytic, possessing plastids with the characteristic photosynthetic pigments (see Figure 4.10) that provided the basis for the divisional name Pyrrhophyta or "flame plants." However, some lack plastids and are heterotrophic, although they may contain pigmented oil drops. Some nonphotosynthetic dinoflagellates harbor endosymbionts, but many do not. *Crypthecodinium cohnii* is nonphotosynthetic, but like some colorless Chlorophyceae, euglenids, and diatoms contains nonfunctional degenerate plastids or proplastids (Kubai and Ris, 1969). In contrast, *Oxyrrhis marina* lacks any remnant of either plastids or pyrenoids (Dodge and Crawford, 1971b).

Like many features of the dinoflagellates, their plastids are varied in character. The saucer-shaped plastids of the tiny *Aureodinium pigmentosum* Dodge are located peripherally within the cell; those of the larger species such as *Gessnerium tamarense* are lenticular and radial in position. Simple discoid plastids are found in *Ceratium furca* (Ehrenberg) Claparède & Lachmann, the tiny *Gymnodinium simplex*, which has radially aligned lamellae, *Gyrodinium dorsum* Kofoid & Swezy, and *Peridinium cinctum* (O. F. Müller) Ehrenberg (Dodge, 1974a, 1975a).

The typical discoid dinoflagellate plastid contains numerous parallel lamellae that extend for the length of the plastid, paralleling its long axis and commonly the cell surface as well (Dodge, 1974a). Each lamella consists of 2 or 3 firmly attached layers or thylakoids, but these appear single in *Pyrodinium bahamense* Plate and *Gonyaulax grindleyi* Reinecke (Gaudsmith and Dawes, 1972). Cells in old cultures may have as many as 30 thylakoids in a stack. Most frequent are 3-layered thylakoids, as in most algae other than the primitive-appearing red algae, with single thylakoids, and the Cryptophyceae, with double ones. Like the plastids, the thylakoids are circular in plan view, 0.15 to 3.6 μm across, and about 240 Å thick (Dodge, 1968, 1975a).

A second type of plastid consists of a peripheral chloroplast reticulum in which the lamellae parallel the cell boundary. In this type, as in the disc type, lamellae stop within the plastid, with no contact of the thylakoids and plastid envelope. Reticular plastids occur in *Amphidinium carterae, A. klebsii, Cachonina niei, Katodinium rotundatum, Prorocentrum balticum,* and *P. cassubicum*. Some species reportedly have both discoid and reticular plastids (Dodge, 1975a).

Gonyaulax polyedra demonstrates typical circadian rhythms in cell division, photosynthesis, and bioluminescence, as do many other dinoflagellates. Structural changes in the plastids also seem to be related to these rhythmic changes in photosynthetic activity (Herman and Sweeney, 1975). Within the radially arranged plastids, the thylakoid stacks nearer the cell interior are more widely separated during the day than are those toward the cell periphery, and the rhythmicity in spacing is maintained

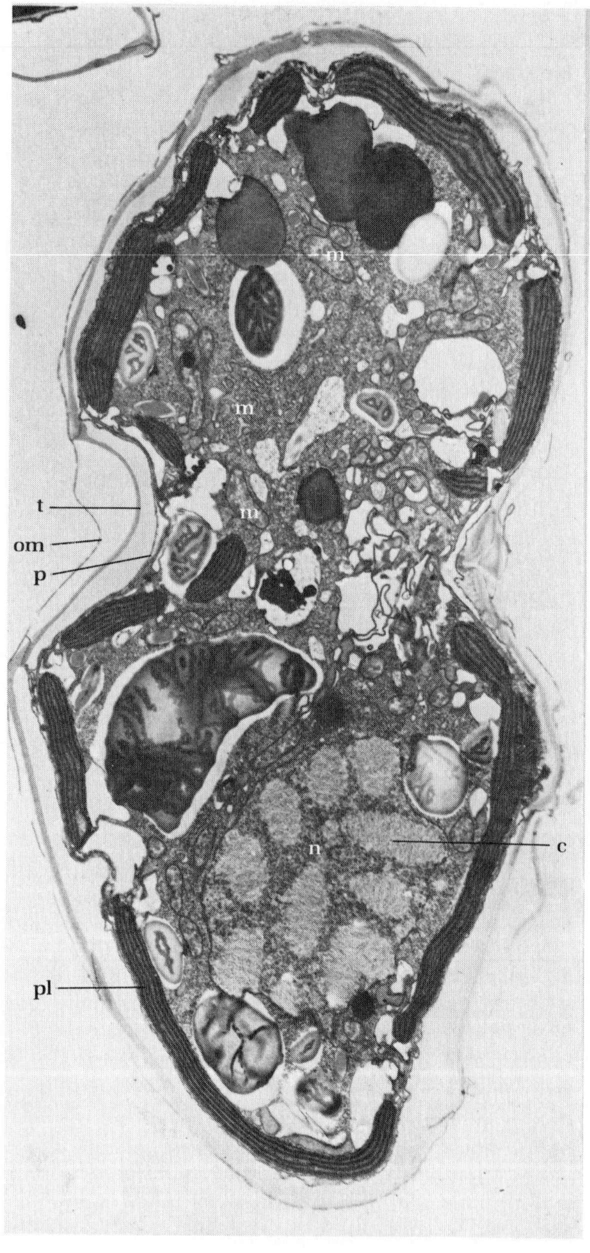

even in cells kept at constant light. The changes occur only within a given radius from the cell center, which contains a densely packed aggregation of cytoplasmic ribosomes. The ribosomes are postulated to control the arrangement of the cell organelles and the rhythmic changes in thylakoid spacing, which may allow greater photosynthetic activity.

Surrounding the thylakoids is the chloroplast matrix of protein, with occasional lipid globules and DNA areas that in some species appear nucleoluslike or form lenticular chloroplast nu-

Table 4.2

Pigments identified by thin-layer chromatography in selected dinoflagellates (Data from Jeffrey et al., 1976).

	β-carotene	chlorophyll a	chlorophyll c_1	chlorophyll c_2	diadinoxanthin	dinoxanthin	peridinin	neoperidinin	fucoxanthin	diatoxanthin	neofucoxanthin	γ-carotene
Prorocentraceae												
Prorocentrum cassubicum	X	X	X	X	X	X	X					
Gymnodiniaceae												
Amphidinium carterae	X	X		X	X	X	X	X				
Amphidinium klebsii	X	X		X	X	X	X					
Gymnodinium simplex	X	X		X	X	X	X	X				
Gymnodinium splendens Lebour	X	X		X	X	X	X					
Gyrodinium dorsum	X	X		X	X	X	X					
Peridiniaceae												
Cachonina niei	X	X		X	X	X	X	X				
Ensiculifera loeblichii Cox & Arnott	X	X		X	X	X	X					
Scrippsiella trochoidea	X	X		X	X	X	X					
Gonyaulacaceae												
Gonyaulax grindleyi	X	X		X	X	X	X					
Gonyaulax polyedra	X	X		X	X	X	X					
Phytodiniaceae												
Pyrocystis lunula (Schütt) Schütt	X	X		X	X	X	X	X				
Prorocentraceae												
Prorocentrum sp.	X	X	X	X					X		X	
Peridiniaceae												
Peridinium balticum	X	X	X	X	X				X	X	X	
Peridinium foliaceum	X	X	X	X	X				X	X	X	X

cleoids (as in *Scrippsiella trochoidea*). The entire plastid is surrounded by an envelope consisting of three membranes, each about 230 Å in thickness. This three-membrane envelope is similar to that of the Euglenophyceae, in contrast to the double membrane envelope of other algae.

Photosynthetic pigments present in the plastids include both chlorophylls and carotenoids (see Table 4.2), the common carotenoid/chlorophyll ratio being about 3:4 or 3:5 (Jeffrey et al., 1976). Both chlorophyll *a* and *c* occur (a combination elsewhere known only in the cryptomonads). Chlorophyll *a* generally dominates, although under some extreme conditions of high light intensity, low temperature, or nutrient deficiency chlorophyll *c* may be dominant. Phycobilin pigments (phycocyanin) have been reported (although not verified in the laboratory) to give a bluish tinge to such freshwater species as *Gymnodinium aeruginosum* Stein and *G. coeruleum* Dogiel (West, 1916).

Although somewhat less abundant than the chlorophylls, carotenoids are important in this group, producing their characteristic color. *Scrippsiella trochoidea*, with a cell diameter of between 23 and 28 μm, has 11 pg (11×10^{-12} g) of carotenoids per cell, as compared to about 40 pg of combined chlorophyll *a* and *c* (Eppley and

Sloan, 1966; Jeffrey et al., 1976). Of the carotenoids, between 50 and 70 percent may consist of the distinctive dinoflagellate xanthophyll peridinin (= sulcatoxanthin). Peridinin absorbs light over a series of wavelengths between the two peaks of maximum absorption of the two chlorophylls, thus greatly increasing photosynthetic efficiency for this group. Another orange xanthophyll, pyrrhoxanthin, is known only from the dinoflagellates, as are the carotenoids dinoxanthin, neoperidinin, neodiadinoxanthin, and neodinoxanthin (Loeblich III and Smith, 1968). The predominant carotene of photosynthetic species is β-carotene, and that of nonphotosynthetic species is γ-carotene; some dinoflagellates also may contain α-carotene.

A few dinoflagellates lack peridinin but have fucoxanthin as their major xanthophyll. These include *Gymnodinium micrum, Peridinium foliaceum,* and *Gymnodinium veneficum* Ballantine (reported by Mandelli, 1968). A relationship was postulated to the diatoms because of this pigment complex, which is typical of chrysophytes and phaeophytes. Later studies showed that some of the dinoflagellates with the chrysophytelike pigments also had two nulei, one being dinokaryon and one typically eucaryotic. These species, *P. foliaceum* and *P. balticum,* also had discoid plastids with four-membrane chloroplast envelopes and chrysophytelike true girdle lamellae radially arranged around the cell periphery, enclosing small DNA areas at the ends of the plastids; they possessed diatoxanthin and neofucoxanthin as well as fucoxanthin. These fucoxanthin-containing species also had a chlorophyll $a:c$ ratio of 4, in contrast to the usual $a:c$ ratio of 2 in other dinoflagellates (Jeffrey et al., 1976). The plastids containing fucoxanthin and the eucaryotic nucleus are enclosed within a membrane inside the dinoflagellate; this membrane-enclosed portion is now believed to represent an endosymbiont in a heterotrophic species that lacked plastids of its own (Tomas and Cox, 1973). The endosymbiont has variously been regarded as a chrysophyte (Tomas and Cox, 1973), haptophyte or bacillariophyte (Dodge, 1975a), or chloromonad (Fine and Loeblich III, 1976a). Other species with fucoxanthin

and no peridinin lack a second nucleus, and their plastids do not have girdle lamellae; these include *Prorocentrum* sp. (Jeffrey et al., 1976), *G. micrum, G. veneficum,* and *G. vitiligo.* Plastids of *G. micrum* are cup-shaped with lobate rims, and all of this group have both discoid and reticular plastids; thylakoids occur in twos and threes and appear branched and interconnected (Dodge, 1975a). Possibly some of these morphological features may be artifacts produced in sectioning, but the unusual nature of the plastids and pigments suggest that these too originated as symbionts, although of possibly different origin, as no eucaryotic nuclei are now present.

Oil Drops

Abundant red pigments characterize living *Peridinium wisconsinense* Eddy collected in August, the red color masking the usual yellow-brown pigment. Probably due to a carotenoid, the color is not associated with the plastids, as it escapes in small droplets from cells whose wall is broken during examination under the microscope. Eren (1969b) reported that red pigments concentrated in bean-shaped areas in the center of *Peridinium cinctum* fa. *westii* during encystment. The red bodies were divided among the daughter cells resulting from division at germination of the cyst. Because the red cells commonly were large, with broad striated intercalary growth bands, they have been regarded as characteristic of old cells preparing for encystment (Lefèvre, 1932a; Eren, 1969b; Nicholls, 1973). The large size also may indicate that the cells are zygotes. *Woloszynskia pascheri* (Suchlandt) von Stosch also is characterized by red pigments, and its thick-walled oval cysts may occur in such abundance as to form red snow and ice along the shores of Lake Davos, Switzerland (Suchlandt, 1916).

Pyrenoids

Pyrenoids are dense, refractive, proteinaceous bodies that commonly are associated with algal chloroplasts (see Figure 4.11), although a few dinoflagellates, such as the very tiny *Gym-*

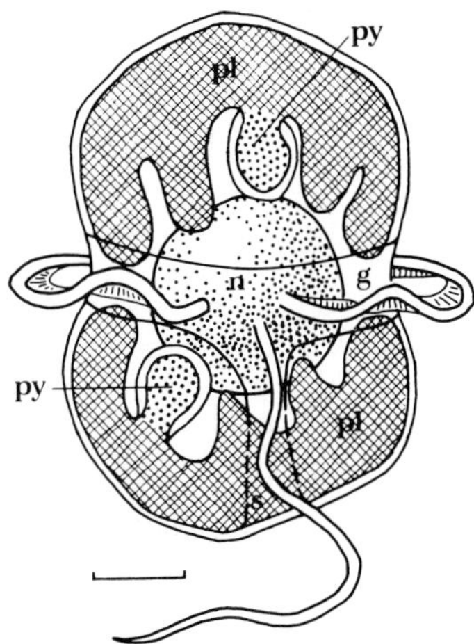

Figure 4.11
Gymnodinium pigmentosum. Ventral view; pl, plastids; py, single-stalked pyrenoid; g, girdle; s, sulcus; n, nucleus. Bar = 2 μm. Redrawn from Dodge, 1967.

nodinium simplex, lack pyrenoids (Dodge, 1974a). When present, the pyrenoids may be entirely enclosed within the plastid or protrude from its margin, but are always bounded by the plastid envelope.

The morphologically simplest pyrenoids are oval to fusiform, one per plastid, and as in *Gymnodinium micrum* are not pierced by the chloroplast lamellae. Other simple pyrenoids are larger; for example, those of *Prorocentrum micans* or *P. minimum,* with the interlamellar thylakoids running through the center.

Somewhat more complex pyrenoids (see Figure 4.12) may be stalked and range from the simple starch-sheathed stalked one on the inner side of the plastids of *Aureodinium pigmentosum* (Dodge, 1968) to the usual four simple-stalked ones suspended from the inner surface of the plastid of the symbiotic *Endodinium chattonii.* Multiple-stalked pyrenoids are present in *Cachonina niei* and *Scrippsiella trochoidea;* the single large pyrenoid of *Amphidinium carterae* also is suspended by multiple branches from the plastid, and traversed by a few of the two-thylakoid lamellae; several starch plates surround the pyrenoid. Most complex is the multiple-stalked pyrenoid of *Peridinium triquetrum,* whose outer margin is pierced by multiple tubules (Dodge and Crawford, 1971a; Dodge, 1973; D. L. Taylor, 1971a).

As different species within a genus may have different types of pyrenoid, or some have a pyrenoid and others lack any, these characters are useful only in species delimitation. Most of the pyrenoid types also occur in other algal groups as well (Dodge and Crawford, 1971a).

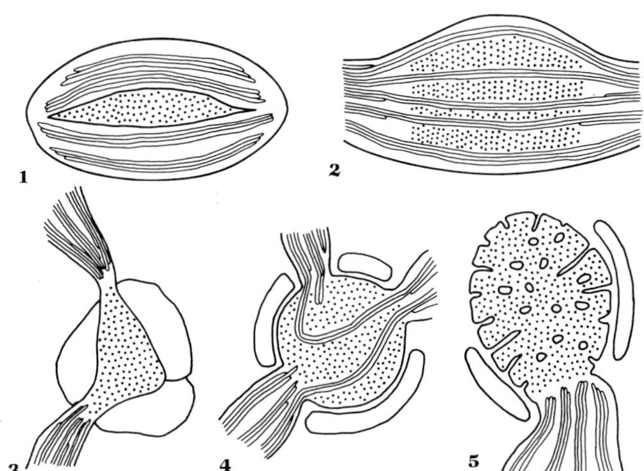

Figure 4.12
Pyrenoid types. **1.** Fusiform interlamellar pyrenoid within plastid. **2.** Compound pyrenoid. **3.** Multiple-stalked pyrenoid of *Cachonina niei.* **4.** Multiple-stalked pyrenoid traversed by thylakoids, as in *Amphidinium carterae.* **5.** Invaginated pyrenoid, pierced by tubules, in *Peridinium triquetrum.* Pyrenoids, stippled; plastids shown with lamellar thylakoids; 3 – 5 show starch plates surrounding pyrenoids. 3, from TEM by Loeblich III; others redrawn after Dodge and Crawford, 1971a.

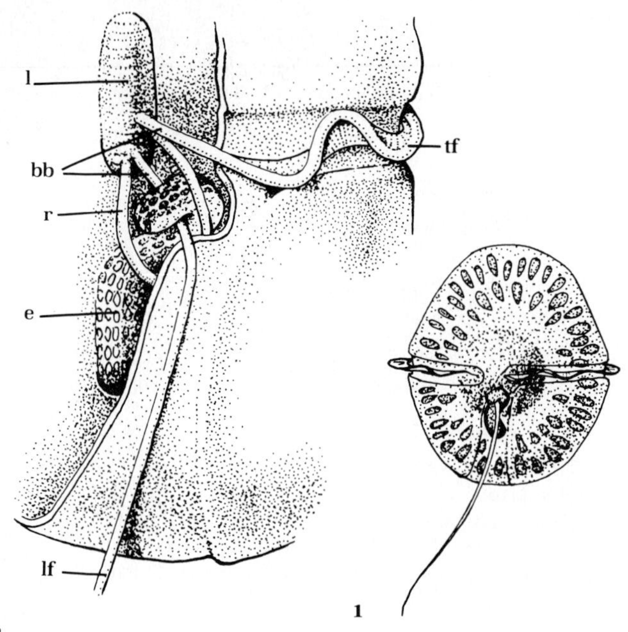

Figure 4.13
Peridinium foliaceum. **1.** Cell showing central nucleus and large hook-shaped eyespot near base of longitudinal flagellum. Modified from Lebour, 1925, *The dinoflagellates of northern seas,* Mar. Biol. Assoc. U.K., published by Cambridge University Press; used with permission. **2.** Enlargement of central region to show eyespot (e), lamellar body (l), basal body of flagella (bb), banded roots (r), transverse flagellum (tf) and longitudinal flagellum (lf). Based on Dodge and Crawford, 1969b.

Eyespots and Ocelli

Several dinoflagellate species have an **eyespot** or **stigma** consisting of a collection of osmiophilic red or orange granules usually located near the flagellar bases on the ventral surface. Like similar bodies in the Chlorophyta and Euglenophyta, the eyespot may aid in detecting a light source and thus be related to phototactic movement. However, eyespots are known in only some 5 percent of the dinoflagellates, including all freshwater species, yet phototaxis characterizes most dinoflagellates. Others have suggested that the eyespot may shade a photoreceptor, rather than itself acting as such (Dodge and Crawford, 1969b).

The simplest eyespot in the dinoflagellates is that of *Woloszynskia coronata* (Woloszynská) Thompson, which consists of an ovoid mass of the osmiophilic granules just behind the longitudinal groove. No enclosing membrane is known. The eyespot of *Peridinium cinctum* fa. *westii* is part of one of the chloroplasts adjacent to the sulcus and consists of a single layer of globules just beneath the plastid envelope

(Dodge, 1973). The eyespot of *Peridinium foliaceum* (see Figure 4.13), a flattened trapezoidal orange structure on the ventral side near the flagellar bases at the anterior end of the sulcus, appears to be a highly modified chloroplast. Surrounded by a triple membrane like that of the chloroplasts, the eyespot has no thylakoids associated with it. An anterior portion is bent in a hook over the basal part of the longitudinal flagellum. The eyespot itself consists of two layers of osmiophilic granules that range in size from 0.08 to 0.2 μm in diameter, the layers being separated by granular material. A distinctive lamellar body, 3 μm in length and constructed of a stack of up to 50 flattened discoid vesicles, is adjacent to the eyespot; its lamellae may show connections with the cell endoplasmic reticulum. Possibly light passing through the "hook" opening of the eyespot activates the lamellar body, if the eyespot serves as a shade (Dodge and Crawford, 1969b; Dodge, 1973).

An eyespot in *Peridinium limbatum* (Stokes) Lemmermann is large, bright-red, and persistent, being prominent in the anterior end of

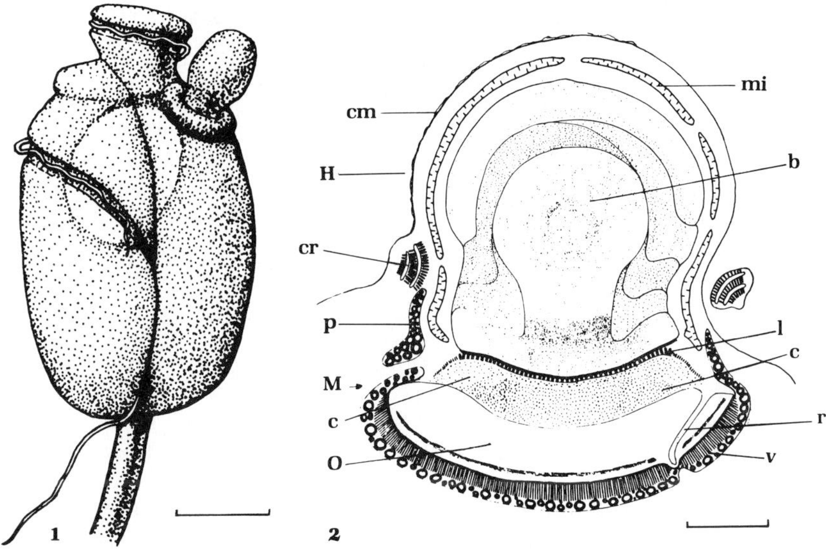

Figure 4.14

Erythropsidinium pavillardii. **1.** Ventral view of cell, showing proximal part of tentacle at posterior, girdle and transverse flagellum encircling cell one-and-a-half times, with large ocellus projecting at right, ovoid hyalosome encircled at base by ringlike melanosome. Bar = 20 μm. Redrawn from Greuet, 1968a. **2.** Diagrammatic section of ocellus, ventral side to left; H, hyalosome; M, melanosome; b, crystalline body; c, core; cm, cell membrane; cr, constricting ring; l, ladderlike pigmented layer; mi, mitochondria; o, ocellar chamber; p, pigmented ring; r, reticulum; v, vesicular layer. Bar = 5 μm. Modified from Greuet, 1965.

cyst cell contents obtained from freshwater bottom sediments. Remains of the eyespot could be discerned after acetolysis as small granular bodies (Evitt and Wall, 1968), leading to the suggestion that similar bodies in fossil dinoflagellates that had been interpreted as nuclei (Eisenack, 1954) or as endosymbionts (Eisenack, 1966a) might represent such eyespot remnants.

The more complicated **ocellus** occurs in some nonphotosynthetic Warnowiaceae, and as these are parasitic attached species, its function is uncertain. This complex photoreceptive organelle has a distinct lens, pigment mass, and sensory core, resembling animal light perception organs. First described in *Warnowia* (Pouchet, 1884, 1885; Schütt, 1895), it also occurs in *Erythropsidinium* (see Figure 4.14), and *Nematodinium.* Although correctly described

originally, later workers mistook the sessile dinoflagellate for the ciliate *Vorticella* and thought the ocellus to have been ingested from a medusoid. *Erythropsidinium* was correctly transferred to the dinoflagellates by Delage and Hérouard (1896).

Details of the large ocellus of *Erythropsidinium pavillardii* (Kofoid & Swezy) P. C. Silva, which attains as much as 40 μm in size, have been elucidated by electron microscopy (Greuet, 1965, 1968a). The complexity is convergent toward that of the metazoan eye, although it has been suggested to represent a much modified plastid. The ocellus has two parts: the protruding hyaline spherical to elongate refractive **hyalosome** between 7 and 22 μm in diameter that contains a stratified "crystalline lens" covered by a "cornea" containing large flattened mitochondria; and a photosensitive part of similar size or several times larger, consisting of a heavily pigmented cup or **melanosome** lying deeper in the cell and moving freely about the lens (Kofoid and Swezy, 1921; Greuet, 1965, 1968b). Usually the ocellus is left of the intercingular sulcus, just posterior to the girdle in the anterior half of the cell; it may project from the surface so that it is exposed to light as the organism moves forward.

Pusules

Saclike vacuoles or **pusules** that open through a slender canal into the flagellar pore are found in most dinoflagellates. Commonly two occur in a cell, one associated with each flagellar pore. Bütschli (1885) compared them to the contractive vacuoles of ciliates, but Kofoid and Swezy (1921) noted that they have an enclosing membrane, contain fluid but no particles, and thus might represent a fluid intake system. Others regard an excretory function as more probable (Cachon et al., 1970b). Although the function remains uncertain, pusules are regarded as permanent organelles, rather than transient like contractile vacuoles (Leadbeater and Dodge, 1966). They are found only in the dinoflagellates, occurring in some marine and all freshwater species, and in both free-living and parasitic ones. Species that lack pusules include the photosynthetic *Aureodinium pigmentosum* and *Gessnerium tamarense,* the photosynthetic endosymbionts *Zooxanthella microadriatica* and *Endodinium chattonii,* and the nonphotosynthetic heterotrophic *Oxyrrhis marina* and *Crypthecodinium cohnii* (Dodge, 1972), so that neither presence nor absence of pusules appears to indicate their function.

The pusule may consist only of tubules or may be associated with pusule vesicles. A simple tubular pusule that may have a collecting chamber is the only type found both in marine species such as *Gonyaulax polyedra* and *Ceratium furca* and in freshwater ones, where it is the most common type, occurring in such species as *Ceratium hirundinella, Peridinium cinctum,* and *Wolozynskia tylota* (Mapletoft et al.) Bibby & Dodge. Tubular pusules with in-

vaginating walls are known in marine species only, being present in *Scrippsiella trochoidea, Peridinium foliaceum, P. triquetrum, Protoperidinium depressum* (Bailey) Balech, and *Katodinium rotundatum.* A sack pusule with invaginated walls characterizes the marine Prorocentrales, such as *Prorocentrum balticum, P. minimum, P. micans,* and *P. pusillum* (Dodge, 1972).

Pusules associated with vesicles are found in some marine and some freshwater species. Numerous vesicles open directly into the flagellar canals of the marine *Amphidinium carterae, A. klebsii,* and *Gymnodinium nelsonii.* Other marine species have pusule vesicles opening into a collecting chamber that branches from the flagellar canal, as in *Amphidinium herdmanii* and *Gymnodinium micrum* (see Figure 4.15). A complex pusule with an internal collecting chamber surrounded by vesicles occurs in the freshwater *Gymnodinium fuscum,* and a very complex tubular pusule associated with vesicles may occupy up to one-eighth of the cell volume of the freshwater species *Woloszynskia coronata* (Dodge and Crawford, 1968, 1969a; Dodge, 1972).

Trichocysts and Nematocysts

Most dinoflagellates have **trichocysts,** fusiform ejectile organelles that discharge a threadlike filament when stimulated by contact, heat, or chemicals. The trichocysts lie at the cell periphery, perpendicular to the cell surface. Thecal plates of armored dinoflagellates have characteristically arranged **trichocyst pores** or perforations through which these organelles may be ejected (see Figure 4.16). The banded proteinaceous trichocyst develops within swollen vesicles surrounded by Golgi vesicles; the mature trichocyst then moves elsewhere in the cytoplasm. Consisting of a dense rhomboidal shaft and a fibrous neck with longitudinally striated lining, each trichocyst is enclosed in a wavy single membrane (Leadbeater and Dodge, 1966; Dodge, 1973). The ejected filaments are banded threads several micrometers in length and square or rhombic in section (see Figure 4.17).

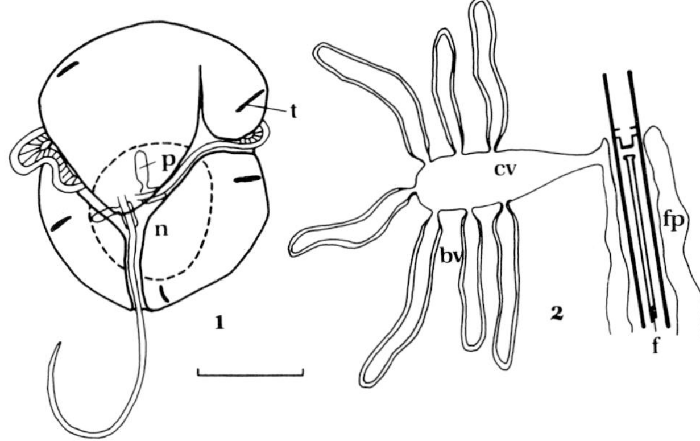

Figure 4.15
Gymnodinium micrum. **1.** Entire cell, to show location of pusules (p), trichocysts (t), and nucleus (n); bar = 5 μm. **2.** Pusule enlarged, approximately ×17,000. Long branch vesicles (bv) open through constricted connection into base of flasklike central vesicle (cv), which enters flagellar pore (fp) near base of flagellum (f). Modified from Leadbeater and Dodge, 1966.

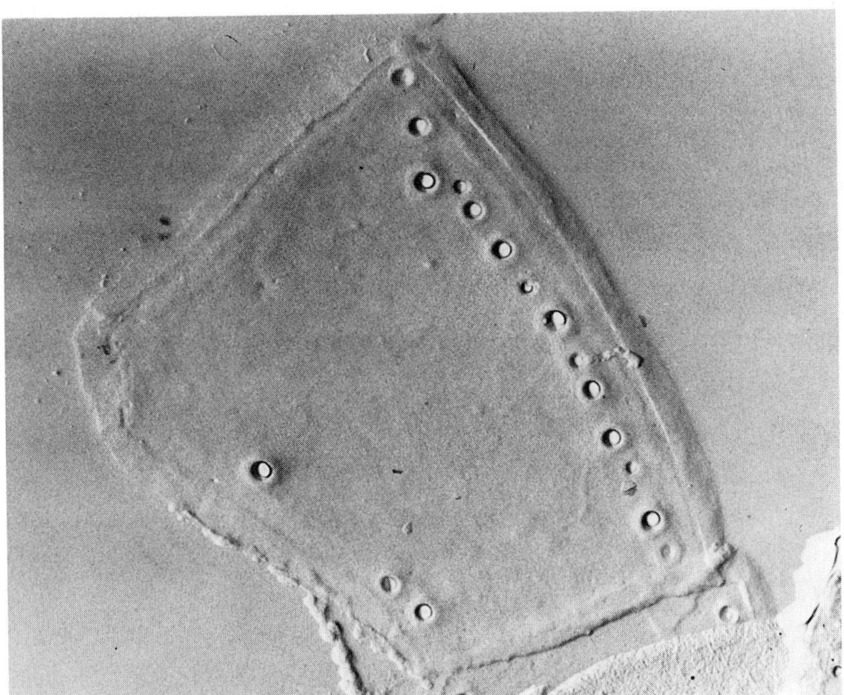

Figure 4.16
Cachonina niei. Carbon replica (TEM) of thecal plate, showing marginally arranged trichocyst pores, ×13,400. From Loeblich III, 1970.

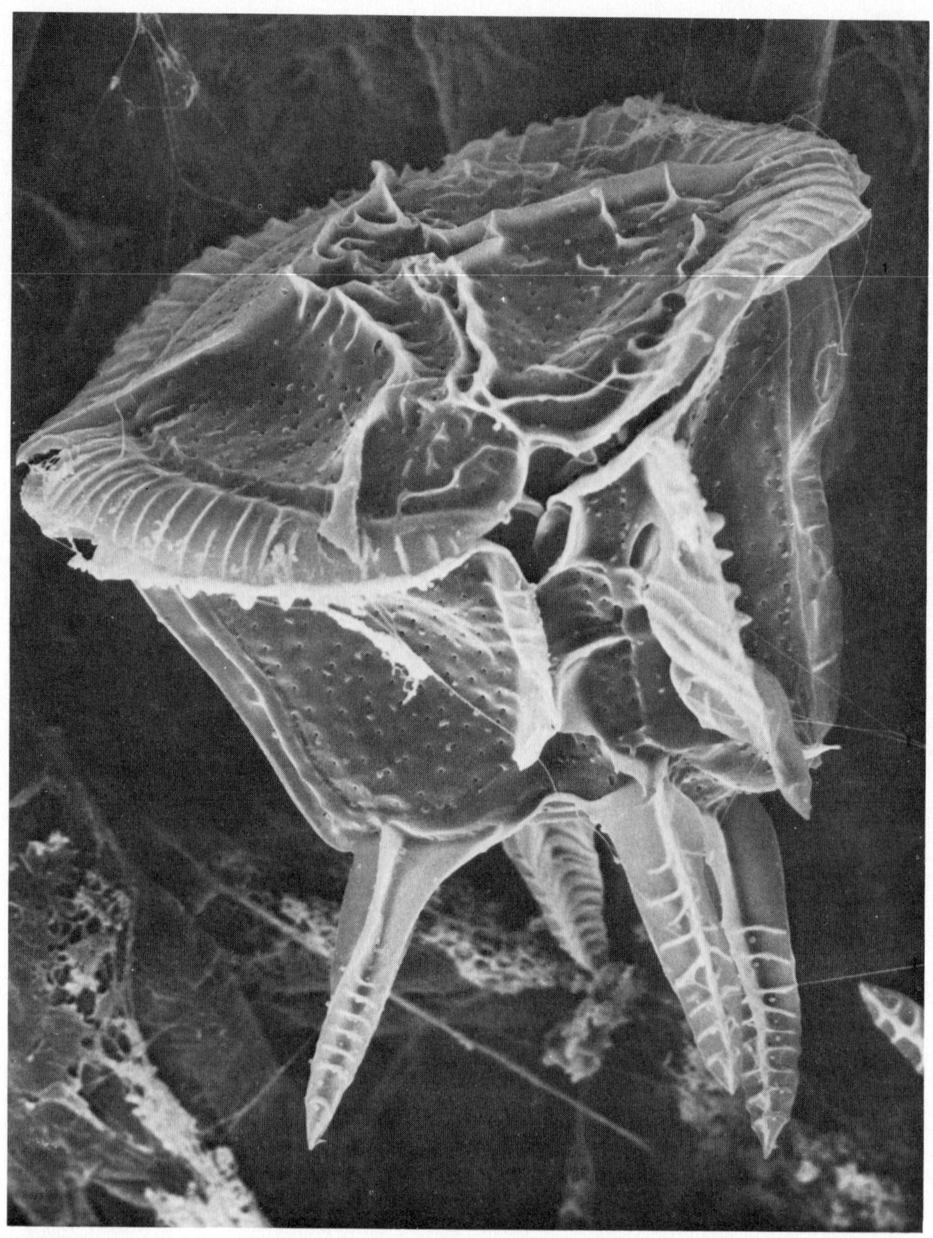

Figure 4.17
Ceratocorys horrida, SEM, ventral view, showing low epitheca, transversely ridged girdle and sulcal lists, and antapical spines, the numerous trichocyst pores, and ejected trichocyst threads on epitheca, at right of figure, and just to left of sulcus, ×1760. From Loeblich III.

Examination of *Gonyaulax polyedra, Gymnodinium nelsonii, Prorocentrum micans,* and *Scrippsiella trochoidea* showed that the charged and discharged trichocysts differ in structure; the change occurs almost instantaneously, probably in response to water uptake, which is thought to be the mechanism of discharge (Bouck and Sweeney, 1966). Dinoflagellate trichocysts differ in their banded character from similar structures in the Cryptophyceae, as the latter have tubular, nonbanded trichocysts.

Nematocysts or **cnidocysts** are found in a few dinoflagellates (see Figure 4.18). These structures are somewhat more complicated than trichocysts, but less so than the cnidocysts of the Cnidaria. The nematocyst of *Polykrikos,* 5 to 22 μm in length, has a posterior chamber (ampulla of Chatton, 1952; taeniocyst of Greuet, 1972a) that contains a spiral filament and a rigid external or anterior capsule or chamber with the trigger. This is closed externally by a valve and operculum, above which lies the **taeniocyst.** The tripartite structure of the taeniocyst consists of a main body, surmounted by a neck and a head that may be external to the cell (Greuet, 1972a). External stimulation of the nematocyst causes it to rupture explosively, ejecting the long filament (Kofoid and Swezy, 1921).

The rodlike nematocyst of *Nematodinium* is membrane-enclosed and has a central core that tapers to a point toward the cell exterior (Greuet, 1971).

Flagella

The flagella, like the nucleus, pigments, and plastids of the dinoflagellates, are highly distinctive. One of the two flagella generally is transverse in position, lying within the median girdle or cingulum, and the other is longitudinal, occupying the median sulcus and trailing behind. The two flagella arise from separate pores near the junction of the girdle and sulcus. The **transverse flagellum** arises from the anterior pore near the proximal end of the girdle, encircles the cell to the organism's left in a counterclockwise direction as viewed from the apex,

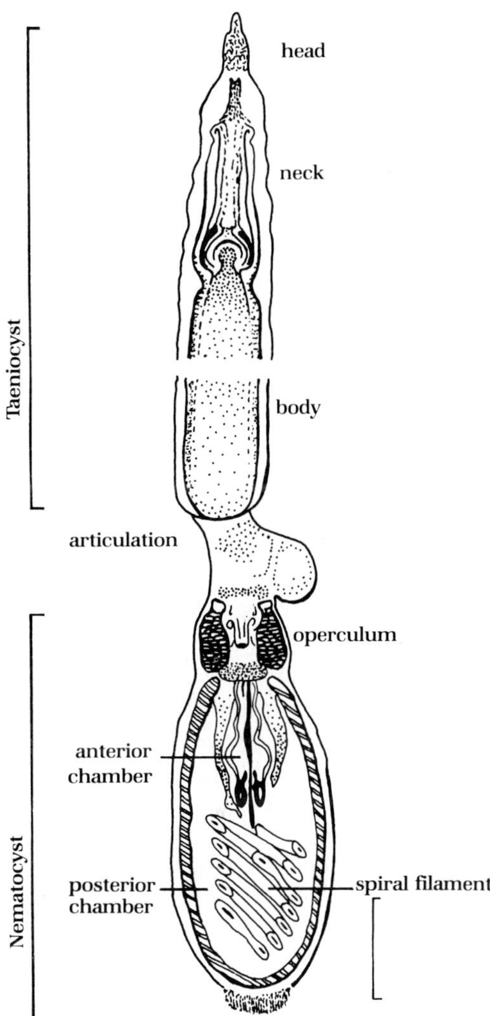

Figure 4.18
Trichocyst structure of *Polykrikos schwartzii* Bütschli. Inner part: Nematocyst contains spiral filament and is capped by operculum, just below articulation with elongate taeniocyst, which may be 10 – 12 μm in length. Bar = 2 μm. Redrawn from Greuet, 1972a.

and lies partly within the cingulum. It is about twice the length of the **longitudinal flagellum.** The latter emerges from the posterior pore near the anterior end of the sulcus, projects backward, and is more readily visible as it trails behind.

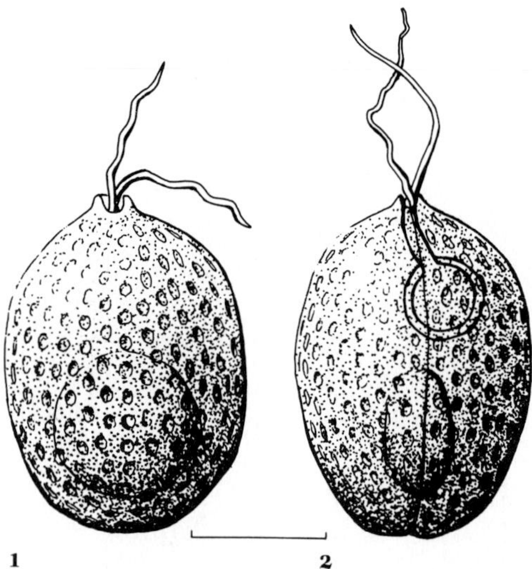

Figure 4.19

Prorocentrum marinum (Tsenkovskiy) Loeblich
III. **1.** Side view of single valve, showing terminal
flagella and large nucleus in cell posterior. **2.**
Edge view of two valves, showing large globular
pusule opening near flagellar bases. Bar = 20
μm. Redrawn from Schütt, 1895.

The Prorocentrales classically have been regarded as having terminal flagella (see Figure
4.19), but these are differentiated both structurally and functionally like those of the usual di-
nophycean, the longitudinal flagellum trailing
behind, and the undulating transverse flagellum
scribing a looplike path around the anterior end
of the cell. A reinterpretation of thecal structure
in the Prorocentrales (Loeblich III, 1976) regards
the two valves as homologous with the epitheca
and hypotheca; the flagellar insertion between
the two thus corresponds closely to the usual
dinoflagellate pattern. The two flagella also
were thought to arise from separate pores, but
SEM studies have shown that in *Prorocentrum
minimum*, *P. rhathymum* Loeblich III, Sherley,
& Schmidt, and *P. triestinum*, both flagella arise
from a single large pore. A nearby smaller
pore in the former two species probably serves
for release of the mucilage that both produce
copiously (Loeblich III et al., 1979).

As mentioned earlier, the transverse flagellum originally was interpreted as a row of cilia,
giving rise to the term Cilioflagellata for the
group. Some 50 years later, Klebs (1883, p. 351)
noted that the girdle was occupied by a ribbonlike flagellum rather than cilia. Still later the
ribbon was observed to have a solid fibril and an
undulating membrane (Schiller, 1931–1937, p.
3), and to be **stichonematic**, having numerous
fine **mastigonemes** along its margin (Deflandre, 1934a, pl. 3, figs. 3–5), in contrast to
the cylindrical **whiplash** or supposedly **acronematic** longitudinal flagellum; later studies
have demonstrated the presence of fine mastigonemes on the longitudinal flagellum also.
An additional variation has been observed in
Oxyrrhis marina, which has tightly imbricated
ellipsoidal scales, about 0.2 to 0.25 μm in
greatest diameter, arranged along the flagella
(Clarke and Pennick, 1972). Mastigonemes of
the transverse flagellum are inserted in bundles
composed of a group of long fibrils (2.6 μm
long) surrounded by somewhat shorter ones
and these surrounded by a shorter series (0.5
μm). Each bundle of mastigonemes is inserted
adjacent to a scale. Scales have not as yet been
reported in other dinoflagellates.

Electron microscopy has allowed further
elaboration of the flagellar inner structure (see
Figure 4.20). The normal flagellar base of the
transverse flagellum gives rise to the 9 + 2 fibrils of the sinuous axoneme. The flagellar
sheath is broadly expanded at one side of the
axoneme and encloses a somewhat shorter
and straight **striated strand** of fibrous protein
with a 660 Å periodicity banding. A sheathing
membrane encloses axoneme, striated strand,
and some intervening packing material; a single
row of long (2 μm) fine hairs is attached at one
side of the sheath (see Figure 4.21), except at the
distal tip, which includes neither mastigonemes
nor the striated strand. The tip may be lightly
attached near the flagellar bases, but is readily
detached upon fixation (Leadbeater and Dodge,
1967b). Because the longer axoneme and the
shorter striated strand are held together by
packing material within the sheathing membrane, a helical coiling of the axoneme about
the shorter striated strand was postulated
(Leadbeater and Dodge, 1966, 1967a). However,
SEM of fixed and freeze-dried specimens requires a slight modification of this model, as the

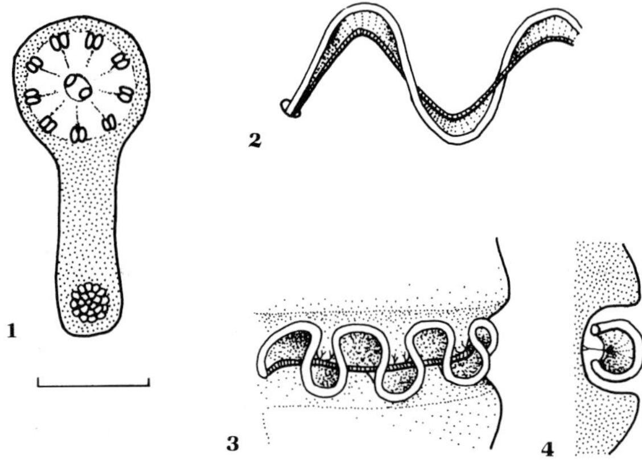

Figure 4.20

Flagellar structure. **1.** Diagrammatic section of transverse flagellum of *Gymnodinium micrum*, showing axoneme with 9+2 arrangement of fibrils connected to striated strand by band of packing material; bar = 0.25 μm; redrawn from Leadbeater and Dodge, 1967a. **2.** Supposed helical structure of transverse flagellum, based on Leadbeater and Dodge, 1967a. **3.** Revised interpretation, based on electron micrographs, with shorter striated strand held in girdle by anchoring threads, pulling axoneme into ruffled pattern. **4.** Cross section of periphery, showing axoneme arching peripherally to anchored striated strand. 3,4. redrawn from F. J. R. Taylor, 1975.

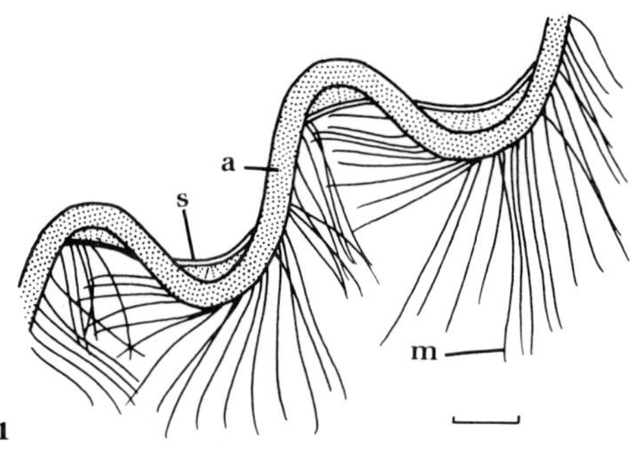

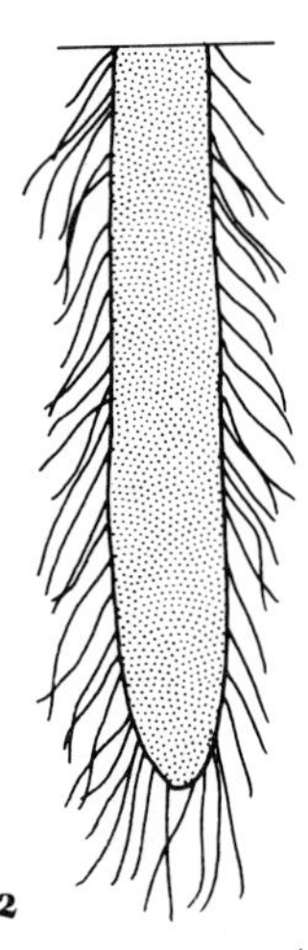

Figure 4.21

Flagellar mastigonemes. **1.** Part of transverse flagellum, showing axoneme (a, stippled) and striated strand (s), with single row of elongate mastigonemes (m). Redrawn, with permission, from TEM in Dodge, 1973, *The fine structure of algal cells*, copyright by Academic Press, Inc. (London) Ltd. **2.** Distal tip of longitudinal flagellum of *Gymnodinium micrum*, showing two rows of mastigonemes, based on TEM in Leadbeater and Dodge, 1966. Bar = 1 μm in both drawings.

axoneme is always external to the striated strand and does not spiral about it. Delicate threads link the margin of the striated strand to the membrane within the girdle, an anchorage that would not be possible if the axoneme was spiralled and helical waves passed along its length (F. J. R. Taylor, 1975). The structure, much like that of an Elizabethan ruffled collar, influences both the form and movement of the flagellum. A rapid undulatory motion of the transverse flagellum appears to produce a circular or elliptical helical wave (in *Ceratium*) traveling from base to tip, to the left of the cell on the ventral surface and to its right on the dorsal surface. The amplitude of the wave results in about half of the flagellum projecting above the girdle edges so that the undulating axoneme of the transverse flagellum is clearly visible when viewed from above or below (Jahn et al., 1963; Leadbeater and Dodge, 1966, 1967a). However, the periodic attachment of the flagellum would not agree with such movement. Instead the axonemal waves appear to follow a hemihelical course, and the undulation of the flagellum causes each segment to pass in an arc alternately above and below the striated strand (F. J. R. Taylor, 1975).

The longitudinal flagellum has two distinct regions. The proximal two-thirds has a normal 9 + 2 axoneme surrounded by organized packing material and an enclosing membrane. The distal third is narrower, about one-half the width of the proximal portion, and contains only the fibrils and tight-fitting sheath. Two rows of short (0.5 μm) fine hairs extend along the side of the longitudinal flagellum and are especially numerous distally (Leadbeater and Dodge, 1966, 1967b).

Viewed either dorsally or ventrally, the longitudinal flagellum shows a broad movement, but seen laterally it appears to be straight, indicating that its slow-beating motion describes a planar wave. The frequency of beating in *Ceratium* was determined to be about 30 per second. The undulatory wave from base to tip of the flagellum produces a strong posteriorly directed current that forms a three-dimensional vortex. In high-speed photography, the movement was shown not to have a sine wave form, but the bending waves of constant amplitude form a pattern composed of circular arcs and straight lines, such as is generated by a simple on/off control of the bending (Brokaw and Wright, 1963; Jahn et al., 1963).

Tentacle or Prod

An elongate projection from the cell may be present in somewhat modified dinoflagellates and may replace one or both of the flagella for movement or attachment. Not all such tentacles are homologous, however. The structure of striated fibers and microtubules in the **tentacle** of *Noctiluca* is an adaptation for movement, and only one flagellum is present. *Oxyrrhis marina* has a tentacle that lacks the power of movement and appears to be only a protrusion from the cell, perhaps representing a reduced hypocone (Dodge and Crawford, 1971b), as the species also is biflagellate. *Proterythropsis crassicaudata* Kofoid & Swezy has a continuation of the girdle on its **prod**, and the tentacle of the parasitic *Protoodinium* is modified to penetrate the host for ingestion of its cytoplasm.

Other dinoflagellates with a prod or tentacle include *Spatulodinium pseudonoctiluca* (Pouchet) J. Cachon & M. Cachon, *Pavillardia tentaculifera* Kofoid & Swezy, and *Erythropsidinium agile* (Hertwig) P. C. Silva.

Internal Skeleton

Siliceous internal stellate structures occur in the Actiniscaceae, such isolated skeletal spicules having been described long before the organism itself. *Actiniscus* commonly has two large arched pentaradial spicules capping the ends of a perinuclear membrane, oriented with their concave sides opposed. Additional smaller **pentasters** in the cell may represent different stages of growth.

Descriptions of the pentasters based on light microscopy have been incomplete; the structures were regarded as latticelike and suggested to indicate a relationship of this group to the radiolarians (Zimmermann, 1930). With SEM (see Figure 4.22), the inner surface of the stars is shown to form a solid plate, slightly or-

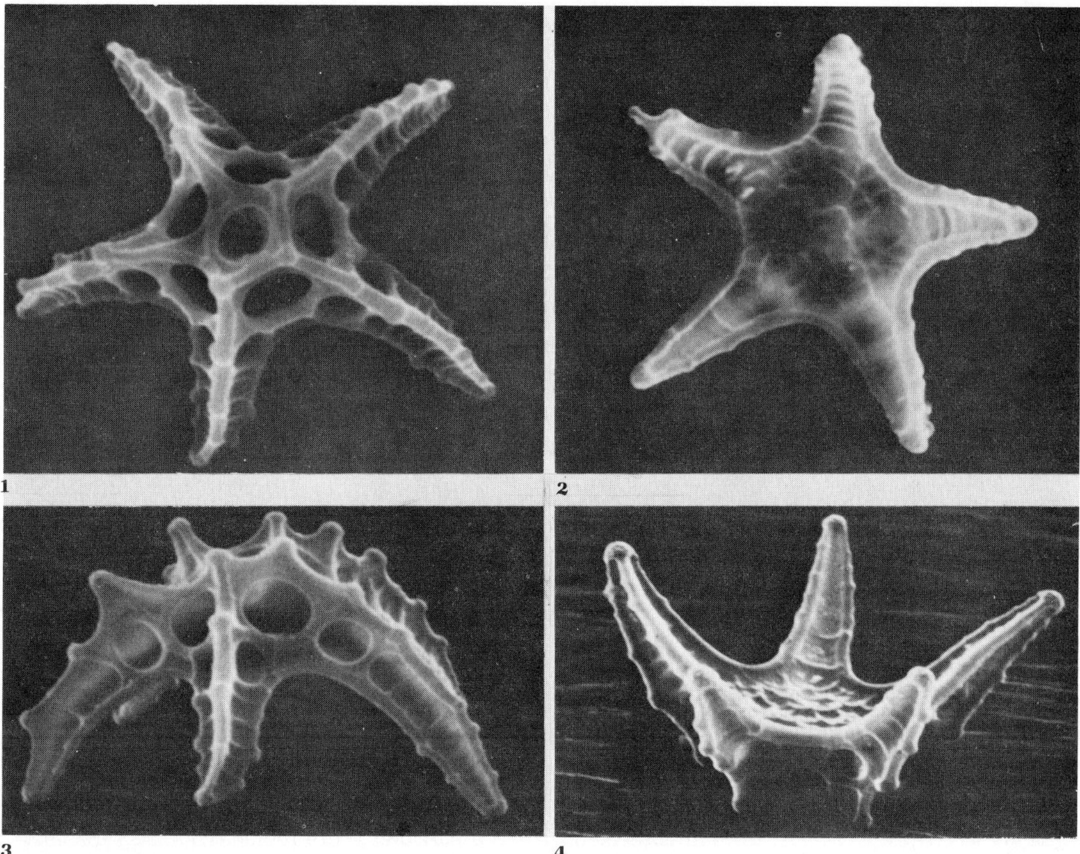

Figure 4.22

Actiniscus pentasterias (Ehrenberg) Ehrenberg. Gulf of California, off Guaymas, Mexico. SEM of pentasters. **1.** Convex surface, showing central ring, radiating ribs, and large alveolae. **2.** Basal view, showing continuous surface with scalloped ornamentation. **3.** Edge view. **4.** Oblique view of concave surface. All ×2000.

namented by a low network of ridges across the central area, and by low concentric arched ridges extending onto the rays to their tips. The external surface of the pentaster is strongly arched and perforated by large openings. From a ring surrounding the central opening, a row of progressively smaller openings extends along each arm on each side of its central ridge. Both the ring surrounding the central opening and the median ribs of each arm are ornamented at the exterior with strongly projecting knobs and transverse ridges. Thus in detail the pentaster has no structural similarity to the radiolarians,

and the living cell is typically a gymnodinioid dinoflagellate.

Actiniscus skeletal elements were originally described from the North Sea off Norway and as fossils from Greece (Ehrenberg, 1840). At first regarded as a subgenus of the silicoflagellate *Dictyocha, Actiniscus* was later elevated to separate generic status, although various species referred therein were in fact silicoflagellates and others were the calcareous nannoplankton discoasters. Species of *Actiniscus* were figured by Ehrenberg (1854) from the Miocene of Richmond, Hollis Cliffs, and Rappahannock,

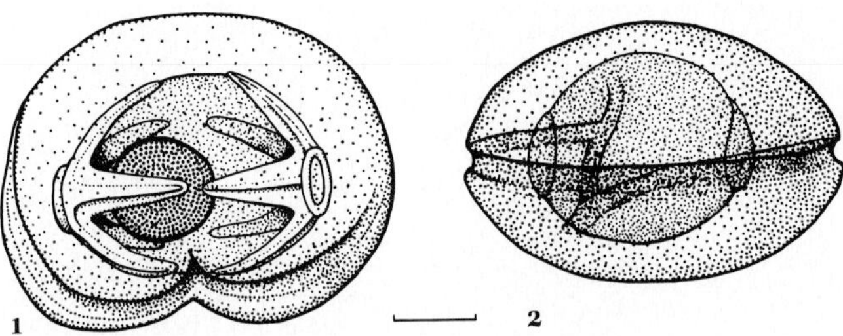

Figure 4.23

Actiniscus pentasterias. **1.** Oblique antapical view of cell, showing two pentasters enclosing central capsule and nucleus. **2.** Dorsal view, showing central capsule (pentasters and nucleus omitted). Bar = 10 μm. Redrawn from Zimmermann, 1930.

Virginia (pl. 18, figs. 59—62; pl. 33, figs. XV, 1, and XVII, 1), from Aegina and Zante, Greece, and Caltanisetta, Sicily (pl. 19, figs. 45—47; pl. 20, fig. 48; pl. 22, fig. 51), and from sea ice at Assistance Bay, North Pole (pl. 35A, XXIII, 1), and Nicobar Islands, Andaman Sea (pl. 36, fig. 36).

The nature of the living cell was first determined by Schütt (1891), who described from the Gulf of Naples a dinoflagellate possessing chromatophores and two to four inner siliceous starlike skeletal structures. Later reports of *Actiniscus* regard it as nonphotosynthetic (Steidinger and Williams, 1970) or either holophytic (containing greenish brown and yellow-brown plastids) or phagocytic (consuming other dinoflagellates, cryptomonads, and chrysophta) (Bursa, 1969). Although recognizing their identity, Schütt (1891) proposed the new but synonymous name *Gymnaster* and incorrectly transferred therein *Actiniscus pentasterias, A. sirius* (Ehrenberg) Ehrenberg, and *A. tetrasterias* Ehrenberg. Living *Actiniscus* now is known in the Mediterranean; Gulf of Mexico; Atlantic, Pacific, and Arctic Oceans; Hudson Bay, Hudson Strait, and freshwater Ogak Lake on Baffin Island; and Great Bear Lake, Northwest Territories, Canada. No ultrastructural studies have been made of the cytoplasmic contents, but the biflagellate cell is typically gymnodinioid (see Figure 4.23), with median girdle, and sulcus that extends both onto the epicone and to the posterior onto the dorsal surface. A strong noncellulosic pellicle is present, but no

theca; the cytoplasm contains many refractive bodies and numerous red fat globules. An organic membrane surrounding the nucleus of adult cells is capped at the two sides by the stellate siliceous spicules. Thin, smooth, and elastic, the membrane is insoluble in weak acetic acid, but susceptible to Javelle water (sodium hypochlorite) (Zimmermann, 1930; Bursa, 1969; Steidinger and Williams, 1970).

Zimmerman (1930) regarded the perinuclear capsule of *Actiniscus* as homologous to the central capsule of the radiolarians, and the siliceous skeleton and dinokaryon nucleus were regarded as supporting evidence of their close phylogenetic relationship. As the dinoflagellate "capsule" consists of endoplasmic reticulum, the homology is now discounted. Pentasters were suggested by Bursa (1969) to be relict skeletal elements derived from the more complex skeletal capsule of the radiolarians. However, the earliest fossil *Actiniscus* (late Oligocene or Miocene age) is fully developed, and the complex radial ultrastructure of the paired stellate bodies shows little similarity to a modified radiolarian skeleton. Nonsiliceous dinoflagellates occur from the Silurian onward, but none show any radiolarian affinities.

Fossil *Actiniscus* are widely distributed in the marine diatomites of Miocene age and later, but are never very abundant. In the western Mediterranean, living *Actiniscus* occurs in concentrations up to 15 cells per liter, is slightly more numerous near the surface in winter, but

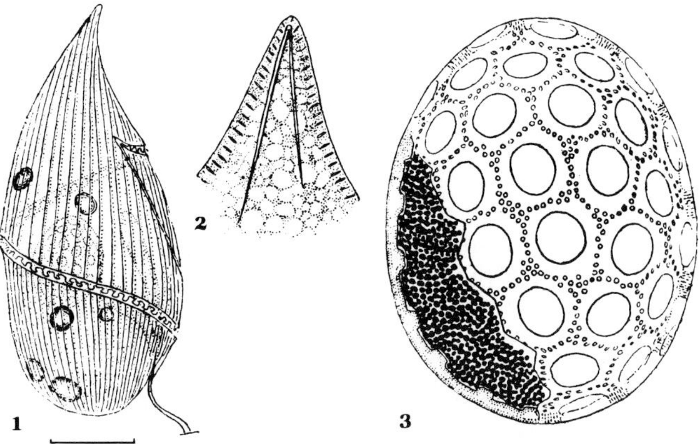

Figure 4.24
Plectodinium nucleovolvatum. **1.** Living cell from right side, with spiraling transverse flagellum and two longitudinal flagella (indicating a motile zygote); bar = 25 μm. **2.** Enlarged view of apex, showing V-shaped spicule in apical region and numerous trichocysts shown as dark fusiform bodies subjacent to and perpendicular to cell surface. **3.** Perinuclear capsule, enlarged, that closely surrounds the dinokaryon nucleus, as shown by optical section at lower left. From Biecheler, 1934a.

in summer occurs at depths of 50 to 100 m (Nival, 1969b). In the eastern Gulf of Mexico abundances ranged up to 100 cells per liter (Steidinger and Williams, 1970). Reports of *Actiniscus* in Canadian Arctic freshwater lakes, such as Great Bear Lake, Northwest Territories, and Ogak Lake, Baffin Island, may be relict occurrences; in at least the first locality the pentasters appear very weakly silicified (Bursa, 1969), and their identity is uncertain. Encysted cells also have been reported (Steidinger and Williams, 1970).

Additional endoskeletal dinoflagellates, as old as late Oligocene, include *Calicipedinium, Cinctactiniscus,* and *Foliactiniscus* (all of Dumitricǎ, 1973); *Carduifolia* Hovasse also has been transferred to the dinoflagellates, in agreement with the original comparisons to *Actiniscus,* although it was later included with the ebridians. All these endoskeletal taxa have a curved externally convex spicule, with central body and four to five radiating costate arms, and with progressively more prominent ornamentation toward the spicule center.

Although *Actiniscus* is much better known, a few other living dinoflagellates have been reported to have resistant internal skeletal structures, although some may not be mineralized. As they are nonthecate, the cells readily disintegrate when collected and are poorly known. Two such nonthecate flagellates, *Monaster* and *Amphitholus,* were named by Schütt (1895) and have not been since restudied in more detail. Both generic names are junior homonyms. The "skeleton" of the single specimen found of *Monaster* was said not to be siliceous. *Amphitholus elegans* Schütt also has an apparent honeycomblike skeleton that occupies most of the cell. Both are gymnodinioid cells, with distinct median girdle, and may represent cysts enclosed within a theca, as in the Lophodiniaceae.

The colorless dinoflagellate *Plectodinium nucleovolvatum* Biecheler (see Figure 4.24) is abundant in the Etang de Thau, on the southern coast of France; it resembles *Actiniscus* in having a distinct nuclear capsule and internal spicule, although the latter is quite distinct. The long, inverted, V-shaped, hollow rod or **acicule** is bent at an acute angle, has very thin arms that taper at their extremities, and lies at the anterior pole. The cell is pyriform, rounded posteriorly, 120 to 150 μm in length, and 50 to 60 μm in breadth, with girdle descending in a helical spire for about one complete turn. The narrow sulcus is reported always to contain two longitudinal flagella. The exterior of the cell is longitudinally striated, trichocysts being aligned with the striae, which are twice as numerous on the hypocone as on the epicone. The large central dinokaryon nucleus is surrounded by, but not in contact with, a thin nuclear membrane, as they are separated by a thick areolated envelope or capsule. Resistant to acetic acid, the capsule dissolved in Javelle water (sodium hypochlorite) somewhat more slowly than the cell cytoplasm and probably is proteinaceous in composition. In these features it is similar to the capsule of *Actiniscus.* The capsule surface is ornamented by hexagonal elements 2 to 3 μm in diameter that surround smaller circles; fine punctations occur along the median line of the hexagonal borders (Biecheler, 1934a).

A capsule structure that may be similar to that of *Plectodinium* is reported in *Gonyaulax pacifica* Kofoid. Specimens from the Bay of Bengal and Andaman Sea in the process of ecdysis were seen to have delicate rectangular, overlapping platelets surrounding the nucleus that were insoluble in glacial acetic acid (F. J. R. Taylor, 1969), but might be either organic or siliceous in nature. The platelets were also visible in thecate individuals that were encysting.

The biflagellate gymnodinioid *Dicroerisma* Taylor & Cattell from the Straits of Georgia, British Columbia, has a sharply defined girdle and contains a single inverted Y-shaped hollow, mineralized skeletal structure that extends from apex to antapex (see Figure 4.25). The three tips of the Y are bifurcated. No theca is present, but staining of the surface brings out an apparent reticular meshwork, termed the periplastic reticulum. The latter is reminiscent of the appearance of *Amphitholus,* or perhaps *Woloszynskia.* The composition of the spicule, termed an **axoskeleton,** is uncertain, although it is not soluble in acetic acid. The appearance is similar to that of some Ebriales, although *Di-*

croerisma was assigned to the "Gymnosclerotaceae," an invalid name for the Actiniscaceae. Like the Ebriales, it lacks plastids, but no evidence of phagotrophy has been observed. Possibly it ingests bacteria or particulate organic material. *Dicroerisma* dies not have the perinuclear capsule that is present in *Actiniscus* and *Plectodinium,* although the spicule was compared to the apical acicule of the latter genus. Like other skeleton-possessing dinoflagellates, it lacks a true theca, having only a thin external periplastic membrane (F. J. R. Taylor and Cattell, 1969).

The somewhat aberrant *Brachydinium* and *Asterodinium,* family Brachydiniaceae, have arms or horns like those of *Ceratium,* but these appear jointed and are movable. Oblong to rounded "chromatophores," very refringent bodies present in these arms, strongly resemble skeletal elements (Sournia, 1972a).

Pavillardinium kofoidii (Gaarder) Sournia from the San Diego area of California (see Figure 4.26) contains 15 to 20 perforate discoid skeletal inclusions (Kofoid, 1907b), mostly restricted to the epitheca. Similar perforated skeletal elements had been figured in a single gymnodinioid cell lacking chromatophores, which was obtained from the "house" or temporary casing of the filter-feeding pelagic tunicate *Oikopleura* (Lohmann, 1903, p. 65, pl. 1, fig. 9). Illustrated only with low magnification, they strongly resemble the fossil siliceous perforated discs described as the silicoflagellates *Rocella* and *Pseudorocella* from the Eocene and Miocene; *Rocella* (= *Stictodiscus*) is a centric diatom, *Pseudorocella* (= *Macrora*) a probable chrysophyte. No regular arrangement of the skeletal elements was observed, nor is their chemical composition known; whether the spicules are indigenous to the organism or are residues of some ingested prey is uncertain.

Achradina pulchra Lohmann is a colorless cell 23.5 μm in length, found in the interior of the tunicate *Oikopleura* and in the Atlantic Ocean from the surface to a depth of 30 to 50 m (see Figure 4.27). No flagella were observed; a skeleton that is insoluble in acetic acid may be siliceous. Additional species were described from the Gulf Stream, Brazil Current, equatorial and South Atlantic, and the western Mediterra-

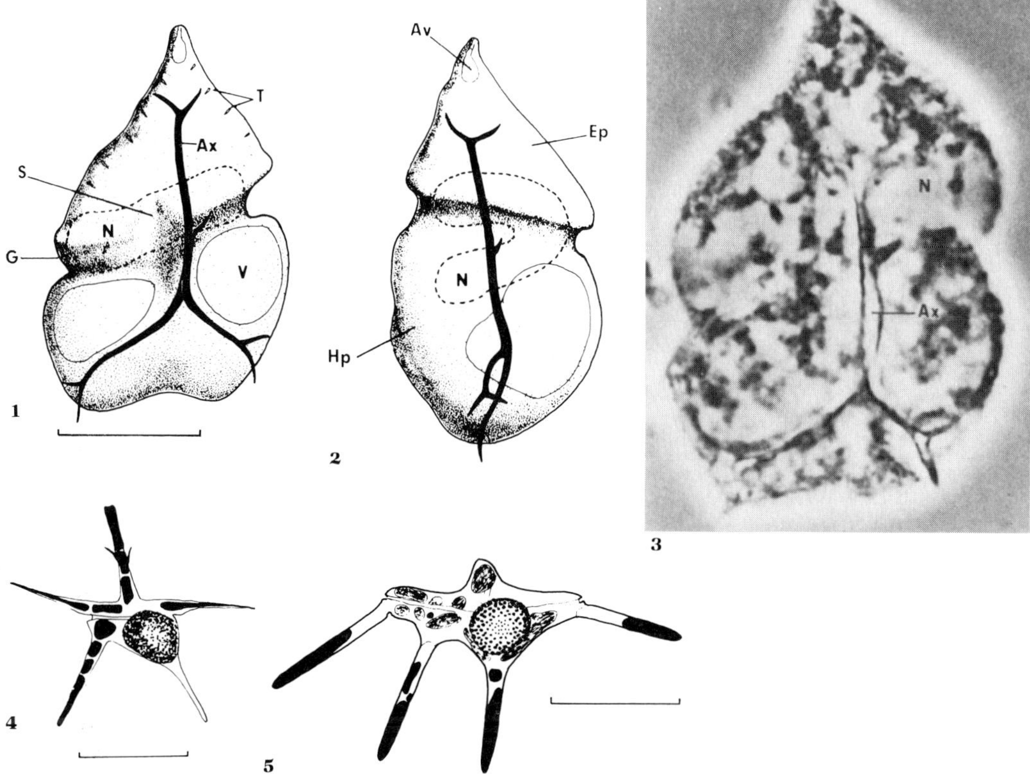

Figure 4.25
1–3. *Dicroerisma psilonereiella* F. J. R. Taylor & Cattell, showing epicone (Ep), hypocone (Hp), axoskeleton (Ax), apical vesicle (Av), vesicle (V), trichocysts (T), nucleus (N), girdle (G), and sulcus (S). From F. J. R. Taylor and Cattell, 1969. **1.** Ventral view; bar = 10 μm. **2.** Left side. **3.** Phase contrast, $\times$2560. **4.** *Asterodinium spinosum* Sournia, Indian Ocean, off Madagascar; bar = 25 μm. **5.** *Brachydinium capitatum* Taylor, Indian Ocean, off Madagascar; bar = 25 μm. Refringent bodies in the arms may be skeletal elements; 4,5, after Sournia, 1972a.

nean (Lohmann, 1920; Hentschel, 1933; Nival, 1969a).

The skeleton of *Achradina* is oblong, consisting of one or two ovoid to subspherical chambers (not three as erroneously interpreted by Lohmann) and commonly a projecting triangular apex. The larger chamber is posterior, lacks any opening to the exterior, but has an anterior pore connecting the chambers. The anterior chamber is subspherical, slightly smaller to nearly equal in size to the posterior one; a large pore opens laterally to the exterior and may occupy nearly an entire lateral surface of the chamber. In other specimens three such pores may occur, one of which is much smaller and difficult to discern. The surface of the skeleton has two to four longitudinal crests extending to the apex, where they may fuse to form a terminal spine. Along the surface of the chambers these crests are constructed of a series of open arches that are fused at their upper surfaces. The distinctly bilateral symmetry results in inequidistant crests. A single lateral pore in the anterior chamber may occupy the entire surface between adjacent crests. Elsewhere the intercrest surface is sculptured with a network in

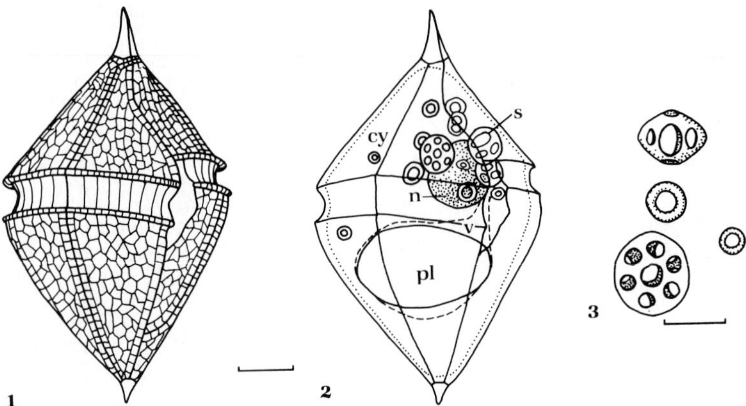

Figure 4.26
Pavillardinium kofoidii. **1.** Right side of cell; bar = 10 μm. **2.** Optical section, showing cytoplasm (cy), nucleus (n), plastid (pl) surrounded by vacuole (v), and skeletal inclusions (s). **3.** Skeletal elements enlarged; bar = 5 μm. Redrawn from Kofoid, 1907b.

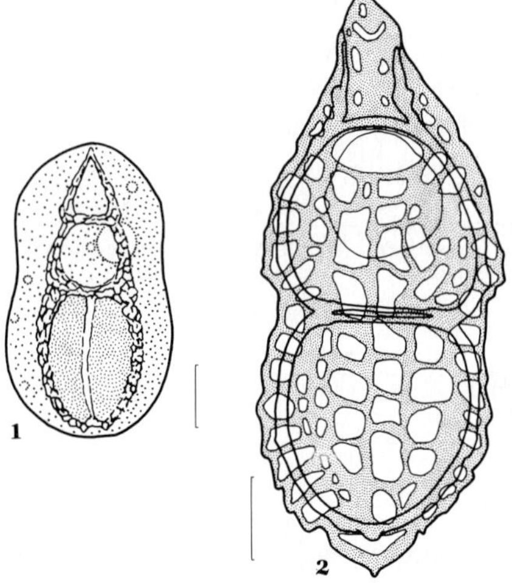

Figure 4.27
1. *Achradina pulchra* Lohmann, cell with internal skeleton, in optical section; bar = 5 μm; redrawn from Lohmann, 1903. **2.** *A. pulchra* forma *sulcata,* Mediterranean Sea; bar = 5 μm; redrawn from Nival, 1969a.

relief, but the inner surface of the chamber wall is smooth (Nival, 1969a). Many features of the skeleton of *Achradina* are reminiscent of such Ebriophyceae as *Hermesinum* and *Ammodochium,* which similarly have siliceous internal skeletons, rather than the paired spicules of *Actiniscus.*

The Amphiesma

The Wall

The armored dinoflagellates possess a cellulosic theca composed of plates overlying a pellicular layer. The unarmored, or naked, dinoflagellates lack the theca, but have a very fragile cover which appears to be homologous. Schütt (1895) noted the gradational character and proposed the term **amphiesma** for both types of cell coverings. Electron microscopy has substantiated the use of a single term, as corresponding membranes are found in both naked and thecate taxa (Loeblich III, 1970). From the exterior to the cytoplasmic membrane four separate layers were recognized: the outer membrane

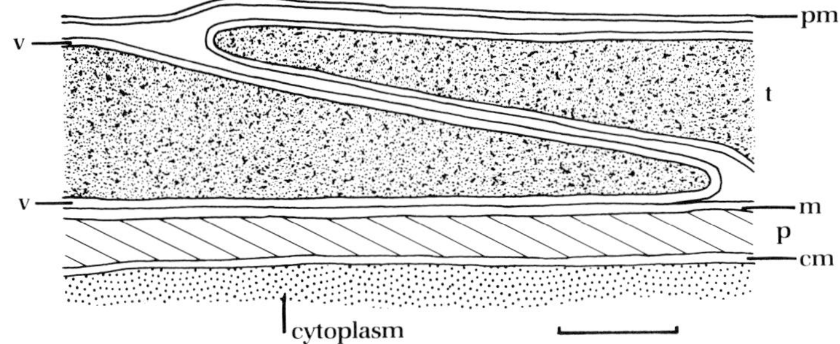

Figure 4.28
Diagrammatic section of wall of thecate dinoflagellate at a suture showing outer or plasma membrane (pm), thecal vesicle or vesicular layer (v), thecal plate (t), pellicular membrane (m), pellicle (p), and cytoplasmic membrane (cm). Bar = 0.1 μm. From Loeblich III.

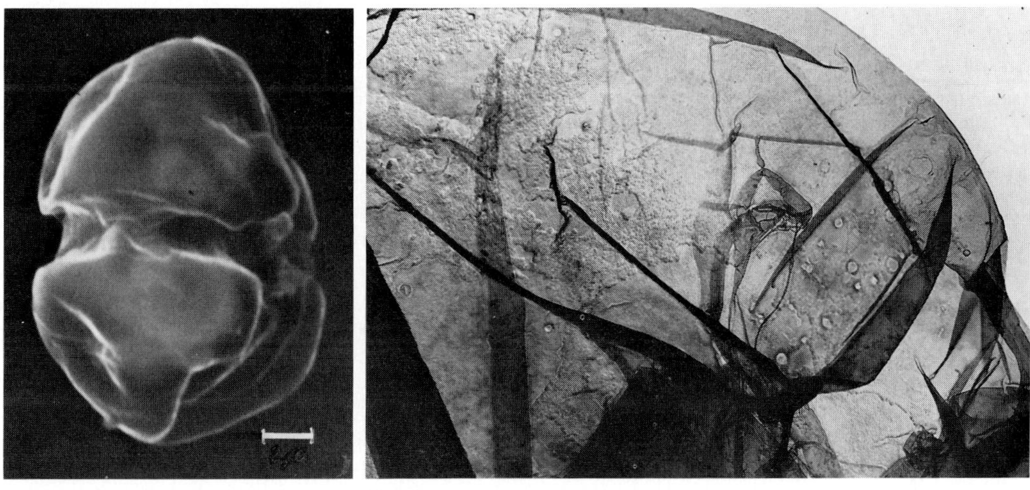

Figure 4.29
Cachonina niei. **1.** SEM of glutaraldehyde-fixed and freeze-dried cell, showing outer or plasma membrane that encloses theca and obscures tabulation; bar = 2 μm. **2.** Carbon replica of same, showing continuous membrane surface, ×14,640. From Loeblich III, 1970.

(also termed **plasma membrane**), **vesicular layer, thecal layer,** and **pellicular layer** (see Figure 4.28).

Outer or Plasma Membrane. The outer continuous unit membrane surrounds the cell, external to the theca of the armored species, where it obscures the plate pattern as seen in the SEM (see Figure 4.29). TEM of carbon replicas shows the presence of trichocyst pores in this membrane. This sole continuous membrane is perhaps best termed the plasma membrane (Dodge and Crawford, 1970b; Wetherbee, 1975a), but should not be termed the thecal membrane (as by Kalley and Bisalputra, 1971), as it might then be confused with the layer of thecal plates or the thin sheets of unarmored species that are homologous with plates.

Vesicular Layer. Beneath the outer or plasma membrane are flattened hexagonal vesicles that enclose the thecal plates of armored species.

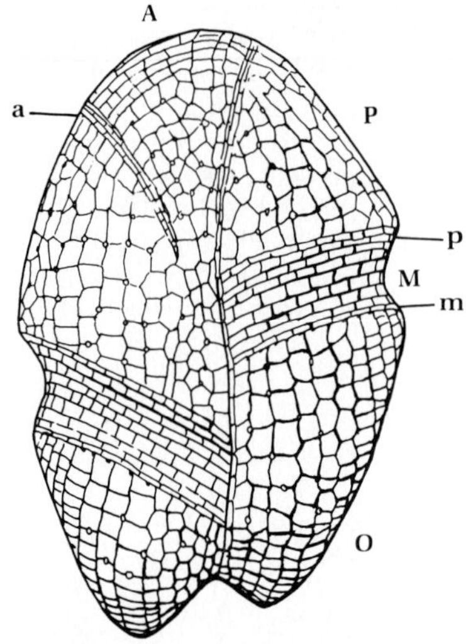

Figure 4.30
Gyrodinium pavillardii. Ventral view of silver-impregnated cell showing réseau argentophile that divides epicone surface into acromere (A) and prosomere (P), mesomere (M) of girdle region, and opisthomere (O) in hypocone, divided by the lines termed acrobase (a), prosobase (p), and mesobase (m). Redrawn from Biecheler, 1952.

Nonthecate species have similar vesicles, however, first described by Biecheler (1934b) as a silver-staining network (**réseau argentophile**) on *Gyrodinium pavillardii* Biecheler (see Figure 4.30). Impregnated with silver, a rectangular network of lines was revealed on the surface, with a generally longitudinal orientation, but forming a bricklike pattern adjacent to the girdle. Biecheler regarded these lines as **acrobase**, **prosobase**, and **mesobase**, dividing the surface into four major regions, termed the **acromere** and **prosomere** in the epicone, the **mesomere** in the girdle region, and the **opisthomere** in the hypocone. Other species were shown to have more numerous vesicles, *Amphidinium carteri* having approximately 150 polygonal units of about 1.1 μm diameter, as seen in sections, whereas *Woloszynskia tenuissima* (Lauterborn) Thompson has 250 such units (Crawford et al., 1971). Thin sheets within the vesicles of some unarmored species were interpreted as homologous with thecal plates (Dodge and Crawford, 1968, 1969a), as the plates of armored species are synthesized within such vesicles. Although prominent during cell division and shaping, the inner and outer membranes of the vesicles may be discontinuous and difficult to recognize in the mature theca and are identified separately as inner and outer plate membranes (Dodge and Crawford, 1970b; Wetherbee, 1975a; see Figure 4.31).

During ecdysis, the plasma membrane, thecal plates, and outer vesicular membrane are shed, leaving the inner alveolar or plate membrane to cover the cell (see Figure 4.32).

Thecal Layer. This layer consists of rigid units or plates of cellulose, whose arrangement and number (between 2 and over 100) are constant for a species and are utilized in classification. If not strongly indicated, the plate margins may be emphasized by staining, with silver, with 1 percent trypan blue, or with HI and iodine combined with chloral hydrate. Individual plates also may be separated for study by treatment with dilute Javelle waters, an aqueous solution of sodium hypochlorite (Graham, 1942; Stosch, 1969a).

Tai and Skogsberg (1934, p. 401) concluded from sections of *Dinophysis acuta* Ehrenberg that the theca did not grow and thicken either centripetally or centrifugally, but by intussusception; that is, by a constant addition and rearrangement of molecules throughout the thickness of the plates. This has not been substantiated, however, and at least in such forms as *Gonyaulax polyedra,* the growth of the theca occurs by additions at the plate margins, the amount added as such **intercalary bands** in the vicinity of the cingulum being sufficient to allow for the increase in diameter from the smallest to largest cells (Dürr and Netzel, 1974).

The thecal plates may be thick or thin, and variously sculptured, with ridges, spines, pits, reticulations, or flanges, and may be perforated by trichocyst pores. With the aid of electron microscopy many of these thecal surface features now are known in great detail (Fott and Ludvik, 1956; Loeblich III, 1970; Dodge, 1971b; Dürr and Netzel, 1974; Ricard, 1974; F. J. R. Taylor, 1971, 1973a).

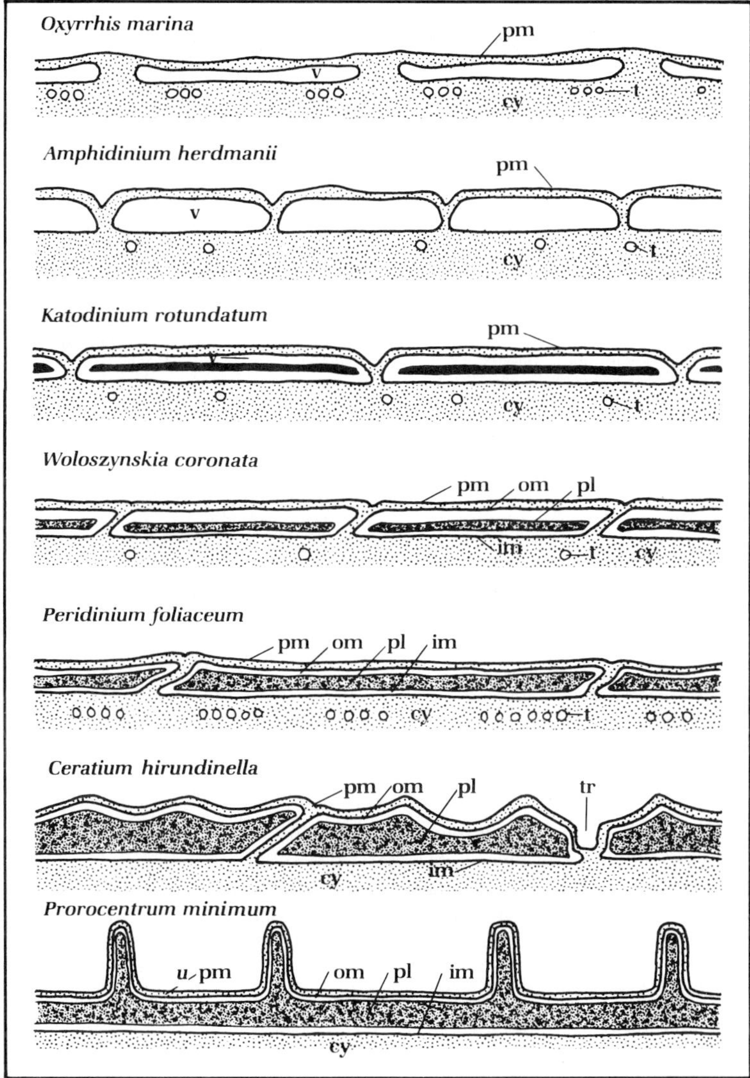

Figure 4.31
Structure of the dinoflagellate amphiesma. Exterior of cell at top of each section. From exterior, layers that may occur are the plasma membrane (pm), outer plate membrane (om), vesicle (v), plate of theca (pl), inner plate membrane (im), cytoplasm (cy). Microtubules (t) and location of trichocyst pore (tr) are also shown. Redrawn from Dodge and Crawford, 1970b.

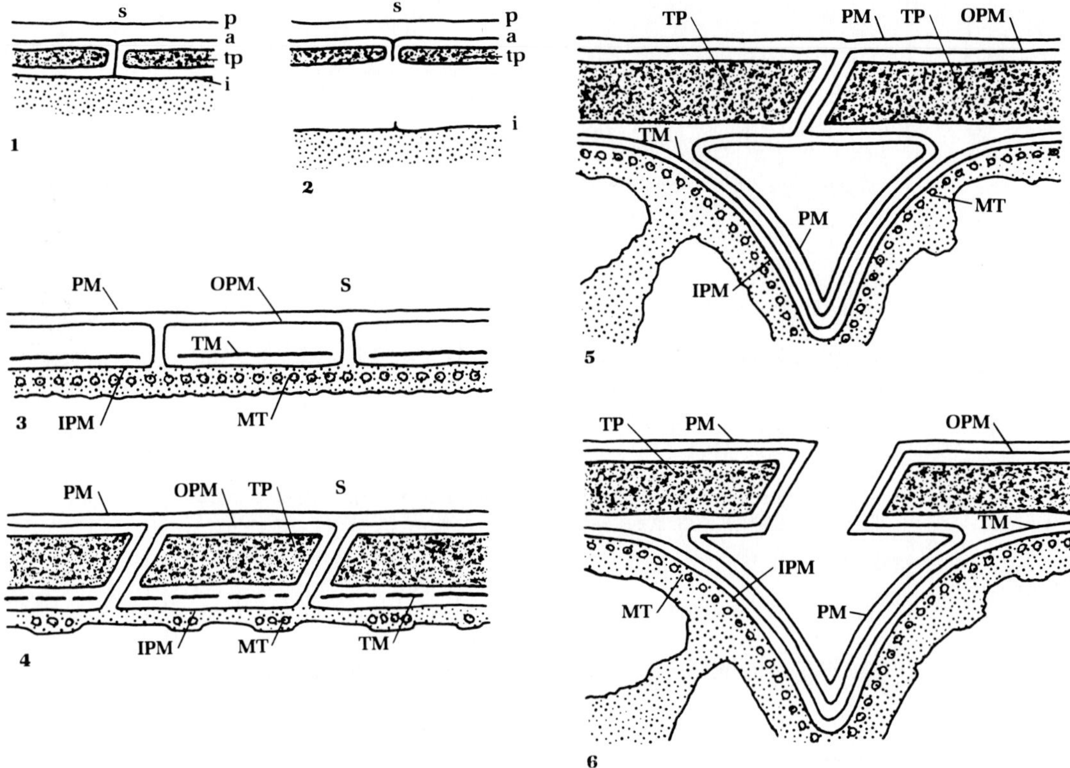

Figure 4.32

1,2. Structure of amphiesma, showing effect of ecdysis, redrawn from Dürr and Netzel, 1974.
1. Layers of amphiesma from exterior at suture (s), plasma membrane or plasmalemma (p), outer alveolar membrane (a, = outer plate membrane), cellulosic thecal plate (tp), inner alveolar membrane (i, = inner plate membrane), cytoplasm at bottom of drawing. **2.** Part of amphiesma shed in ecdysis (p, a, tp), and part remaining over the protoplast (i). **3 – 6.** Plate synthesis and thecal structure; PM, plasma membrane; outer (OPM) and inner (IPM) plate membranes fuse to form vesicles that contain thecal membrane (TM); after thecal plates (TP) form, thecal membrane is discontinuous; S, suture; dense layer of microtubules (MT) in early stage, later dispersed in small packets. **3.** Orientation of cell membranes before plate synthesis. **4.** Structure after plate synthesis. **5.** Cleavage system of membranes and relation to parent membranes before separation of plates in *Ceratium tripos.* **6.** Membrane systems after separation. 3,4, with permission, from Wetherbee, 1975a, *J. Ultrastruct. Res.,* v. 50, p. 58 – 64, copyright by Academic Press Inc. (London) Ltd.; 5,6, with permission, from Wetherbee, 1975b, *J. Ultrastruct. Res.,* v. 50, p. 65 – 76, copyright by Academic Press Inc. (London) Ltd.

Pellicular Layer. This innermost thinner and continuous layer of the amphiesma immediately overlies the cytoplasmic membrane in armored species, and also has been termed the **plasmalemma** (Kalley and Bisalputra, 1971). Pores may traverse this layer, but sutures do not. The pellicular layer is fibrous in structure, with fibers of 190 Å diameter, and is of highly resistant composition. This composition and the position beneath the thecal plates suggest (Loeblich III, 1970) that the pellicular layer may be analogous to the material of the dinoflagellate cyst. This layer was not observed in sections of *Ceratium tripos* (O. F. Müller) Nitzsch, a marine species that does not form cysts (Wetherbee, 1975a).

During cell division and plate synthesis, a continuous layer of microtubules may underlie the cell cover. In the mature theca, the microtubules are dispersed into smaller groups (Wetherbee, 1975a).

Chemical Composition. The firm tabulated wall of the Peridiniales has been variously reported to be cellulosic, chitinous, or siliceous, particularly by early workers. The latter two were eliminated as possibilities by Lankester (1885).

Using a method that freed the cell walls from the contents, the dry weight of the cell wall was found to consist of 15 percent ash(Loeblich III, 1970, p. 871). The organic material of the wall consists of

      84.4% glucan
       3.5% lipid
       2.3 to 3.9% protein
      <u> 9.8</u> to 8.2% unknown
     100.0%

Chemical tests indicate that the plates of the thecal layer are of cellulose covered by an acid labile material, or else the cellulose fibrils are so tightly interwoven that interaction with basic solvents is prevented. No chitin is present (Loeblich III, 1970).

Isolated walls of the freshwater *Peridinium cinctum* fa. *westii* do not dissolve in water, dilute acids or bases, cold strong phosphoric acid (72 percent), Schweitzer reagent (cuprammonium), or cadoxen 95 percent. The composition was said to be a polymer of D-glucose, but to differ from both cellulose and laminarin by its X-ray diffraction pattern. In addition, a small amount of nitrogen, phosphorus, and ash is present, and an inorganic fraction of silica (Nevo and Sharon, 1969).

In contrast to the cellulosic composition of the thecal plates, the pellicular layer may be an acidic polysaccharide, as it is highly resistant to cellulase and to strong acids and bases (Loeblich III, 1970).

Plate Patterns (Tabulation) of the Theca

Because of the difficulty in preserving cell contents, cytological studies of dinoflagellates lagged behind studies of the theca. Hence the genera and species are defined and classified largely on the basis of thecal features. Recognizing that the number and arrangement of plates in the dinoflagellate theca are constant within a taxon, and that these are taxonomically useful, separate nomenclatural systems were proposed for the plates of the three armored orders, the Peridiniales, Prorocentrales, and Dinophysiales.

Tabulation of the Peridiniales. The very regular arrangement of thecal plates was early recognized. Stein (1883) described the plate series bordering each side of the median girdle as pre- and post-basals and the terminal series as frontals and endplates (equivalent to apical and antapical series). The terminology now generally followed is that of Bütschli (1885), in which apical, preequatorial, postequatorial, and antapical plates are each numbered from the sulcus, counterclockwise around the cell as viewed from the apex. Later, Kofoid (1907a, p. 179) proposed a series of symbols to indicate the various plates (see Figure 4.33) and a formula to describe the tabulation. Although other variations have been proposed (Lefèvre, 1932a, see Figure 4.34; Graham, 1942), Kofoid's system has gained widest acceptance.

The portion of the theca anterior to the me-

Figure 4.33
Tabulation of the peridiniacean
Scrippsiella trochoidea, showing the
plate series indicated by various patterns as well as symbols of Kofoid system. **1.** Apical view of epitheca. **2.** Antapical view of hypotheca. **3.** Sulcal region. **4.** Ventral view of epitheca. **5.**
Dorsal view of hypotheca. **6.** Ventral
view of cell, showing epitheca, girdle,
sulcus, hypotheca, and nucleus (N).
Other drawings show apical plates
$(1-4')$, anterior intercalary plates
$(1-3a)$, precingular plates $(1-7'')$,
cingular plates $(1-6c)$, anterior sulcal
(Sa), left sulcal (Sl), right sulcal (Sr),
and posterior sulcal (Sp) plates, postcingular plates $(1-5''')$, and antapical
plates $(1-2'''')$. Bar = 10 μm. Redrawn from Balech, 1959.

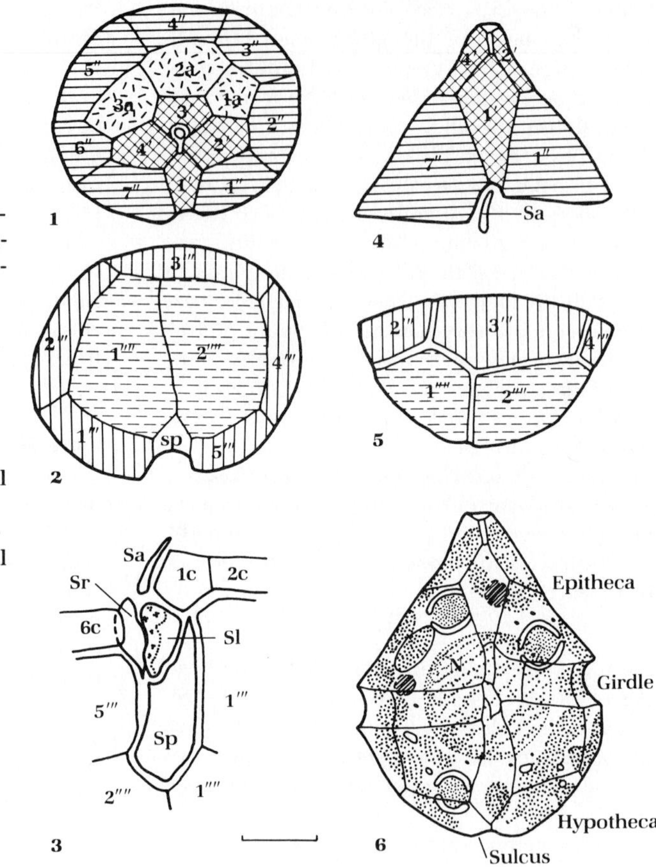

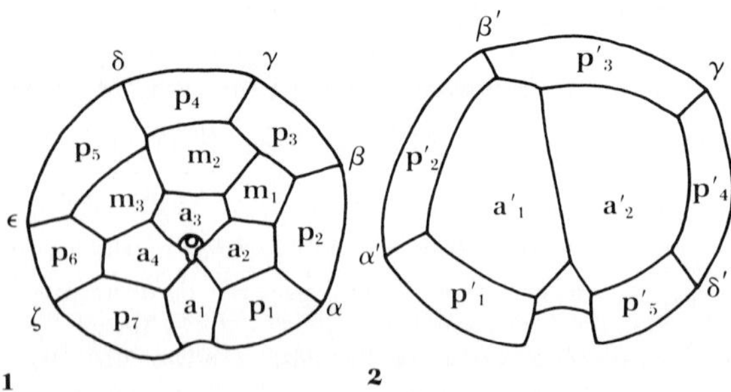

Figure 4.34
Tabulation of *Scrippsiella* in the
plate terminology of Lefèvre
(1932a). **1.** Apical view of
epitheca: a_1-a_4, apical plates;
m_1-m_3, median plates (equivalent to intercalary plates);
p_1-p_7, preequatorials (= precingulars); epithecal sutures
numbered α, β, γ … . **2.** Antapical view of hypotheca: $p'_1-p'_5$,
postequatorials (= postcingu-
lars); $a'_1-a'_2$, antapical plates;
hypothecal sutures indicated as
α', β', γ' … .

dian girdle is the epitheca; that to the posterior is the hypotheca. The girdle or cingulum, separating epitheca and hypotheca, may be symmetrically equatorial in position with the ends meeting ventrally, but asymmetry is more usual, one end of the girdle being somewhat displaced at the sulcus. If the right-hand side of the girdle lies posterior to the left-hand side, as is the rule in all nonthecate species and most thecate ones, the girdle is said to have a descending spiral, and to be left-handed.

Both epitheca and hypotheca are subdivided into latitudinal plate series, each named and designated by symbols. Separate plates within a series are numbered counterclockwise, as viewed from the anterior or apex, beginning at the ventral area. From the apex, the **apical** series (′) is anteriormost, followed by the more numerous and larger plates of the **precingular** (″) series, which lies adjacent to the median girdle with its **cingular** (c) series of plates. Posterior to the cingular series are the still larger but usually fewer plates of the **postcingular** series (‴), which is interrupted ventrally by the sulcus, and finally the **antapical** (′′′′) series of one or more plates.

Some genera (*Heterodinium*) have an equal number of plates in hypotheca and epitheca, some have more in the hypotheca (*Centrodinium*), and most have more epithecal plates. One to three additional plates, termed **anterior intercalaries** (a), may lie between the apical and precingular series on the dorsal side; a similar group between the postcingular and antapical series is designated as **posterior intercalary** (p) (see Figure 4.35). Although this system of plate nomenclature is generally followed, some prefer to use letter designations for the plates rather than the symbols, as apical platelet (pl), apical plate (ap), anterior intercalary (a), precingular (pr), girdle (g), postcingular (po), posterior intercalary (p), antapical (ant), and sulcal (s) (Graham, 1942).

Kofoid (1907a) did not study in detail the ventral area or sulcus with its very tiny plates, but Abé (1936) described these for *Protoperidinium* (see Figure 4.36) and Graham (1942) reported on the ventral area of other genera. The ventral

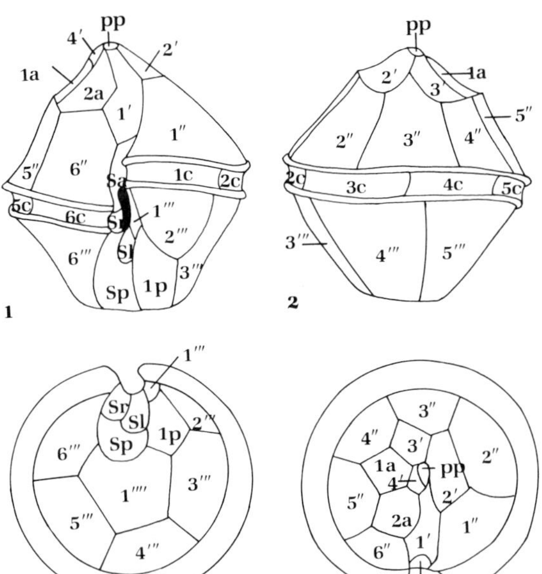

Figure 4.35
Tabulation of the gonyaulacacean *Gonyaulax polyedra*. Symbols as in Figure 4.33, with addition of pore plate (pp) and posterior intercalary (1p). Compare SEM in Figure 4.3. **1.** Ventral side. **2.** Dorsal side. **3.** Antapex. **4.** Apex. Redrawn from Kofoid, 1911.

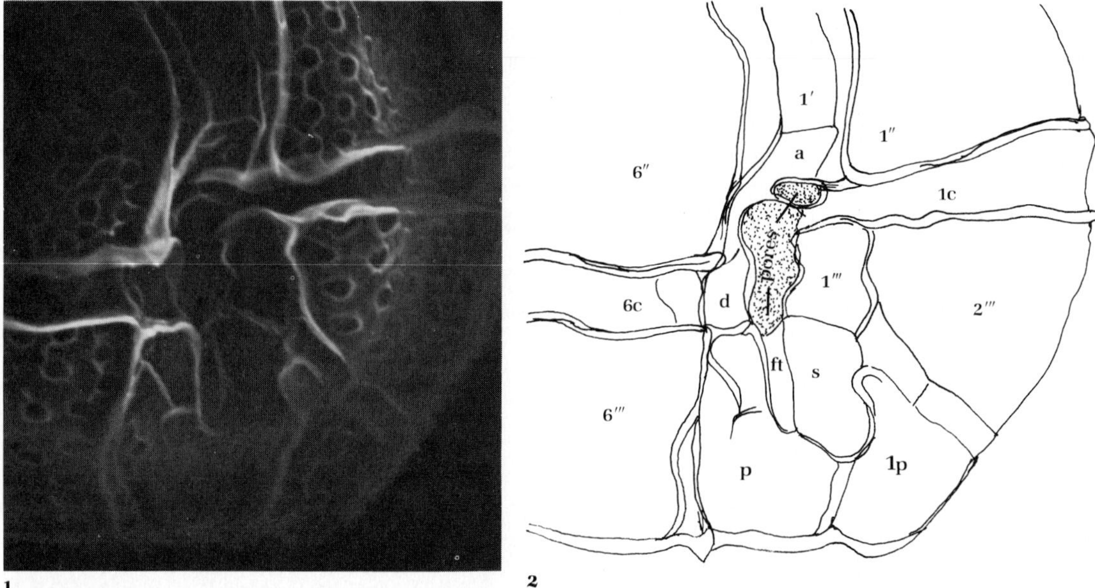

Figure 4.36
Ventral area of *Gonyaulax polyedra*. **1.** SEM, ×2625, from Loeblich III. **2.** Plate arrangement based on SEM of part 1, using terminology of Abé (1936); a, anterior plate of sulcus; d, dextral or right sulcal plate; s, sinistral or left sulcal plate; p, posterior sulcal plate; ft, flagellar trough, extending to posterior from flagellar pore. Compare with diagram of this species in Figure 4.35.

area commonly contains 5 to 7 plates, although there may be as few as four (*Cachonina*) or as many as 10 (*Ceratocorys*) (Balech, 1949; Loeblich III, 1970). The eight distinct types of **sulcal plates** recognized by Graham are an anterior sulcal plate; posterior, right, and left sulcal plates; right, left, and posterior accessory plates; and an intercalary plate. Except for the posterior sulcal and intercalary sulcal plates, all border on the flagellar pore. The anterior sulcal plate (as) lies at the anterior edge of the pore, against the proximal end of the girdle, and may extend into the epitheca (as in *Gonyaulax*). The posterior sulcal plate (ps) is always present, but is more varied than the preceding, flaring to a wide elliptical expanse on the ventral area of the hypotheca (*Gonyaulax*), being obscure and rectangular (*Ceratocorys*), or intricate, horseshoe-shaped, and nearly the size of the thecal plates (*Peridinium*). The left sulcal plate

(ls) always is present; it varies somewhat in position although invariably anterior to the posterior plate, bordering the pore at its left or posterior. The left accessory sulcal plate (la) is anterior to the flagellar pore to the left of the sulcus in *Acanthogonyaulax*. One to three plates may be present on the right of the ventral area between the anterior and posterior plates; a single one is the right sulcal plate (rs), as in most *Peridinium*. A right accessory sulcal plate (ra) may be anterior to the right sulcal plate and an intercalary sulcal plate (i) may be posterior to it. Either of the former two right plates or both may border the pore, but the intercalary sulcal plate never does. In addition to these usual plates, *Peridinium* and *Gonyaulax* may have a posterior accessory sulcal plate (pa) between the posterior plate and the posterior edge of the pore. The demarcation of the sulcus also varies widely. The ventral area of

Heteraulacus is flat, without a distinct boundary, and with little structure; *Peridinium* has a deep sulcal groove, so that plates cannot be discerned without dissection (Graham, 1942).

The very broad concave ventral area of *Ceratium* is specialized, including large, extremely hyaline thin plates and a small sulcus on the left. *Ceratium hirundinella* is said to lack a true sulcus, instead having a large opening (sulcal aperture) that leads into a large indentation or ventral chamber. The flagella are inserted in the upper right corner of this ventral chamber (Dodge and Crawford, 1970a). Although the ventral area has been regarded as containing a single plate, or two, four, or five plates, these were not regarded as part of the regular plate series. Restudying this ventral area, Wall and Evitt (1975) considered the larger plates to represent the sixth precingular, cingular, and postcingular plates, and only the smaller two or more plates to represent the sulcal series. *Ceratium* tabulation thus becomes 4', 0a, 6'', 5−6c, 2−?s, 6''', 1p, 1'''', hence is close to that of *Gonyaulax* type, such as the 4', 6'', 5s, 6''', 1p, 1'''' of *Gessnerium catenellum*.

Most Peridiniaceae have a plate formula of 4', 7'', 5''', 2''''; *Cachonina niei* has somewhat more plates, its formula being pp (pore plate), cp (canal plate), 5', 3a, 8'', 6c, as, rs, ls, ps, 5''', 2'''' (Loeblich III, 1968). The **pore plate** and canal plate were proposed to represent an additional preapical series (Loeblich III, 1970, p. 882), as similar pore plates have been reported in other genera (i.e., *Ceratium, Ceratocorys, Fragilidium, Gonyaulax, Heteraulacus, Heterodinium, Palaeophalacroma,* and *Pavillardinium*). Freshwater *Ceratium* have a tiny ringlike plate at the tip of the apical horn (Bourrelly and Couté, 1976). Formulae of these and other living dinoflagellates are summarized by Loeblich III (1970, p. 884−905).

The plate surfaces range from very simple to highly ornamented. The ornate plates of *Gonyaulax polyedra* have many small depressions (see Figure 4.37), each with one or more central pores that may have elevated margins. The plates also may have an elevated ridgelike border at the growing edge. Carbon replicas of the interior surface of the thecal plates similarly show the perforate depressions (Loeblich III, 1970, figs. 7−9). In contrast, the 35 plates of *Cachonina niei* have no ornamentation other than a few marginal trichocyst pores with very moderately raised borders.

The plate edges are bevelled at the contact with adjacent plates (see Figure 4.38); generally two edges of a plate overlap adjacent ones, and two lie beneath adjacent plate margins (Loeblich III, 1970, figs. 12−15). In *Gonyaulax polyedra*, plates of the pre- and post-cingular series have intercalary bands on an overlapping border on the apical or adapical border, respectively, and on the longitudinal margin closest to the ventral region. Plates 3'' and 4''' have such overlapping bands on both longitudinal margins, the ventral apical plate 1' has no growth band, and plates 6'' and 6''' have none on their apical and antapical borders, respectively. In ecdysis, plate 3'' readily separates like a door, hinged at the cingulum, as it is not overlapped on any margin, and plates 2', 1'', and 2'' separate from plates 1' and 3' to allow the protoplast to escape. Dürr and Netzel (1974) suggested an overlap formula, with superscript of accent marks indicating the direction of overlap. Thus '( ) indicates that the left margin of a plate overlaps that adjacent (left as viewed by the observer, not the organism's left), ( )' indicates right margin overlap, and '( )' indicates that both margins are overlapping. Thus *Gonyaulax polyedra* would be described as (1'), '(1''), '(2''), '(3'')', (4')', (5'')', (6'')' on the epitheca, and '(1'''), '(2'''), '(3'''), '(4''')', (5''')', and (6''')' on the hypotheca (see Figure 4.39).

Tabulation of the Prorocentrales. Two nearly equal major thecal plates characterize the Prorocentrales, these generally being regarded as right and left (Lebour, 1925, p. 12), hence the two flagella that emerge from between them are anterior in position. No distinct girdle and sulcus are present. In contrast, Bütschli (1885) and Loeblich III (1976) interpret the valves as equivalent to the epitheca and hypotheca of the Peridiniales, and the juncture between them as equivalent to the girdle. The flagella are dissimilar; one has hairs and a

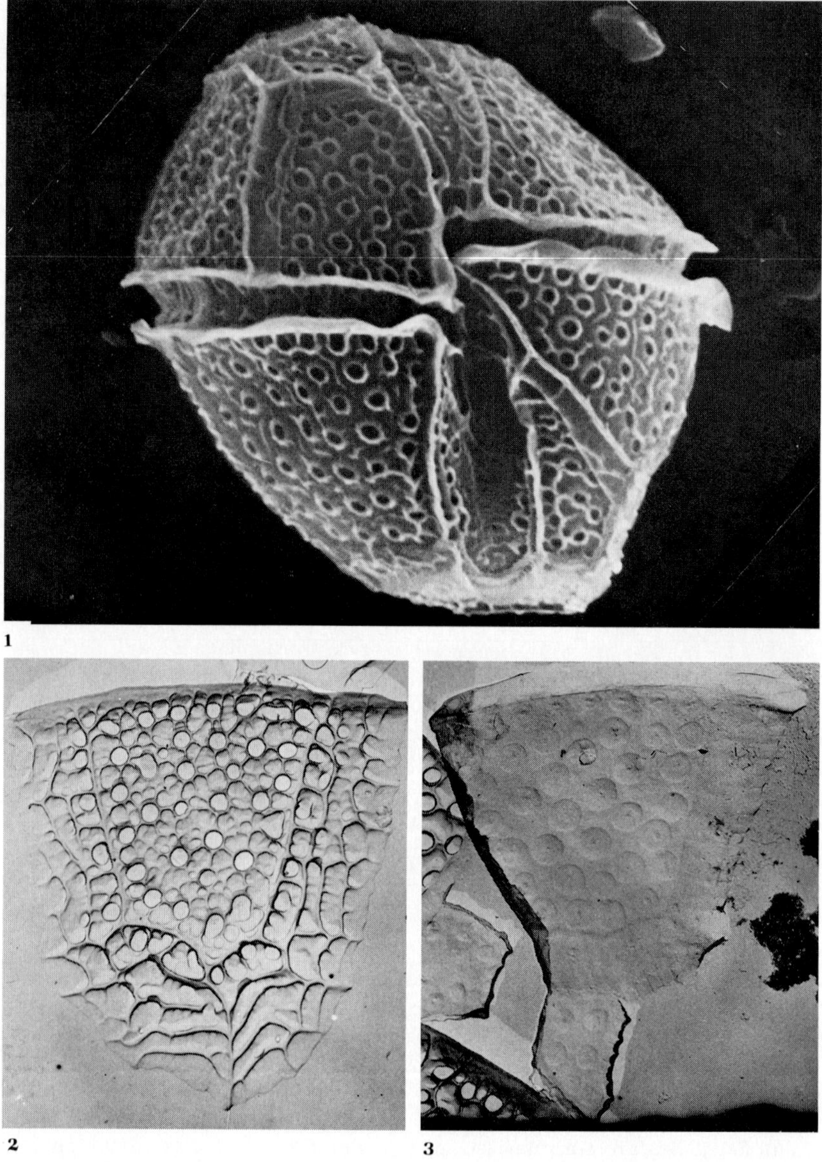

Figure 4.37

Gonyaulax polyedra. **1.** SEM of glutaraldehyde-fixed and freeze-dried cell, with ornamented thecal plates. **2.** Carbon replica of exterior of thecal plate, showing surface ridges and ringlike areas. **3.** Carbon replica of interior surface, showing continuous surface with tiny pores at bottom of the circular areas. All ×1900, from Loeblich III, 1970.

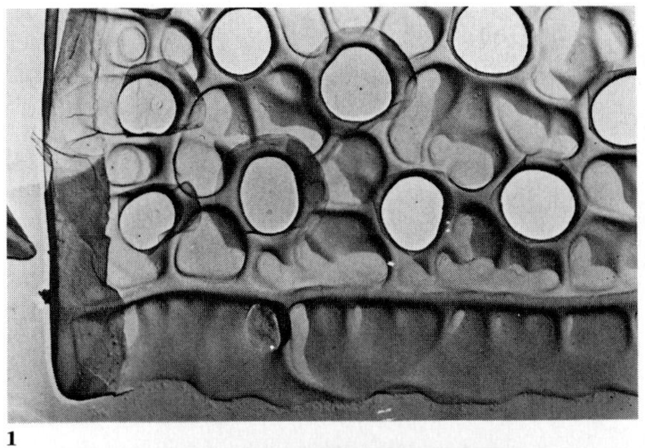

1

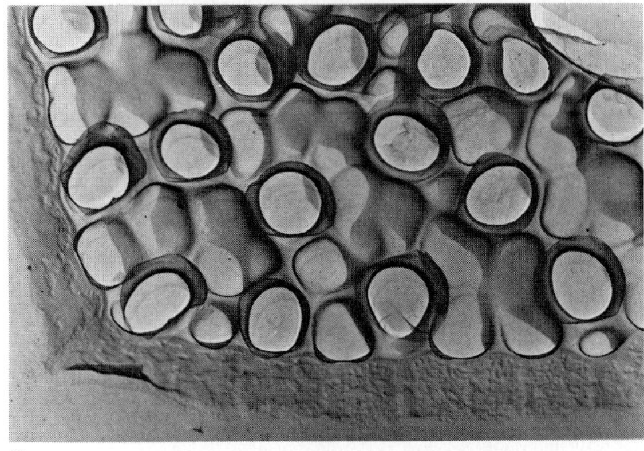

2

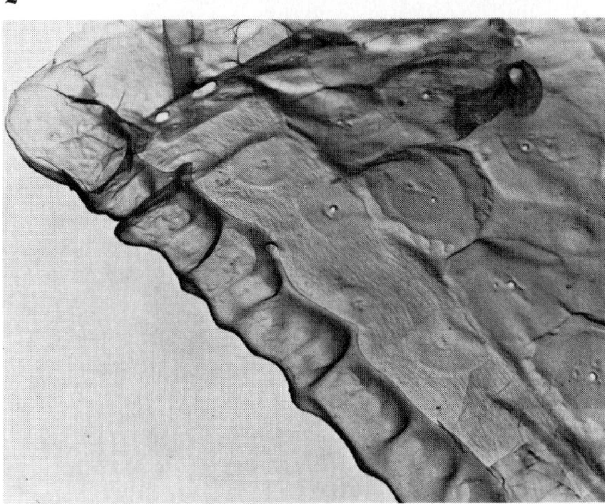

3

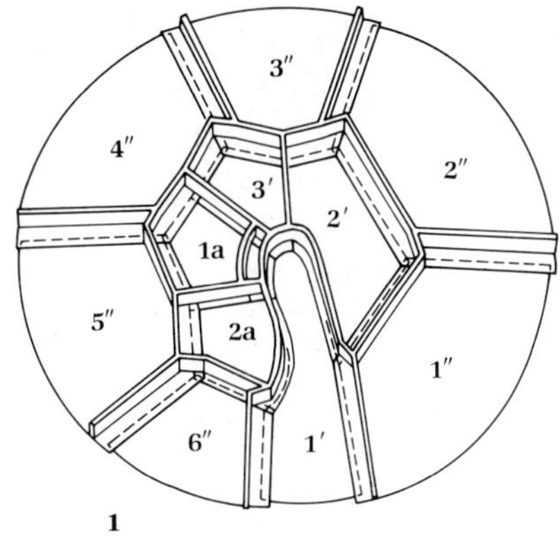

1

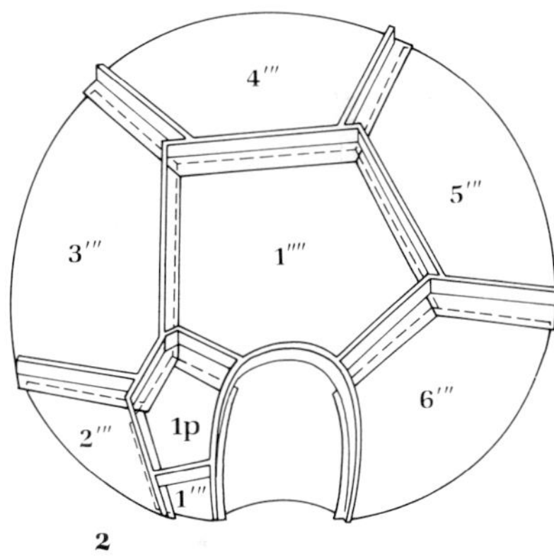

2

Figure 4.39
Gonyaulax polyedra tabulation, showing overlapping and underlapping (dashed lines) plate margins, with crests on the overlapping plates that parallel their margins. **1.** Epitheca. **2.** Hypotheca. Sulcus at bottom of each figure. Redrawn from Dürr and Netzel, 1974.

Figure 4.38
Gonyaulax polyedra. **1.** Carbon replica (TEM) of exterior surface of overlapping plate, showing bordering ridge. **2.** Same, for underlapping margin of plate. **3.** Carbon replica (TEM) of interior surface of an overlapping plate, showing continuous theca with very tiny trichocyst pores in centers of bordered pits. All ×7000, from Loeblich III, 1970.

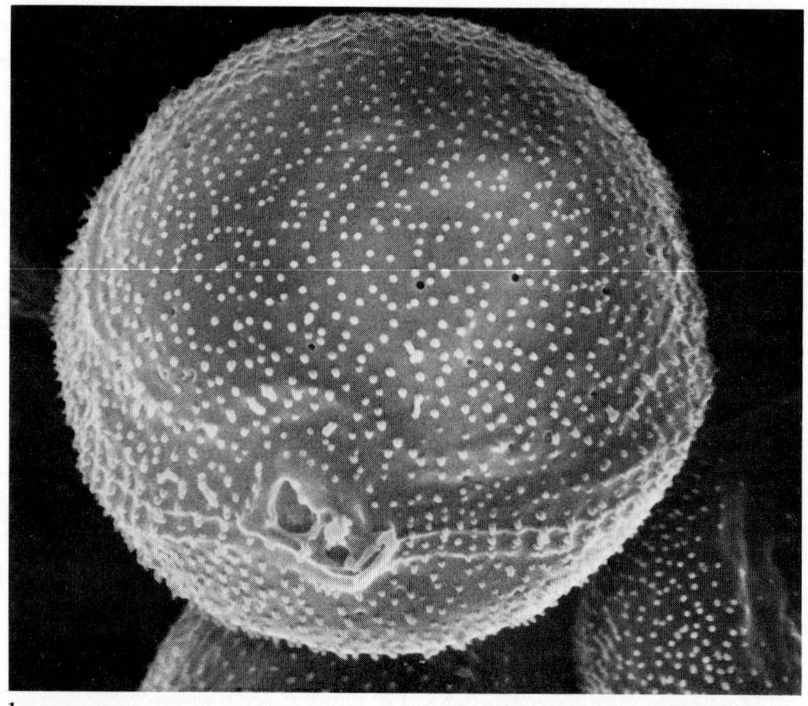

1

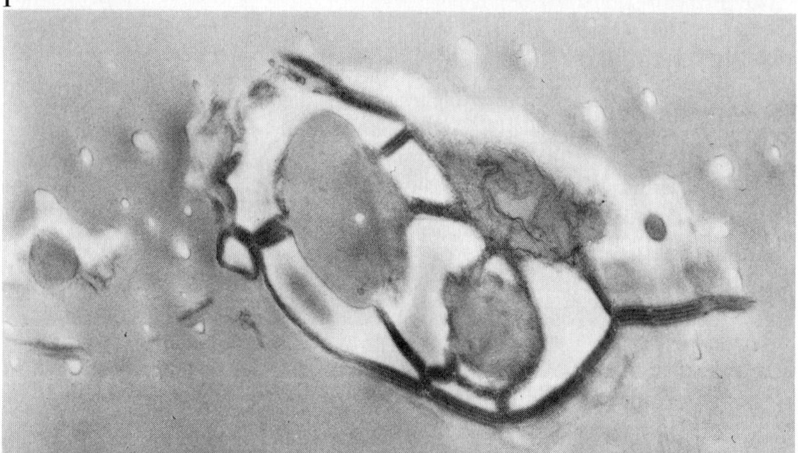

2

Figure 4.40

Prorocentrum minimum. **1.** SEM, oblique apical view, showing flagellar
pores at junction of two thecal plates and spines on thecal surface; ×7100.
2. TEM of section through apical region, showing small plates (white) around
two central pores that previously were regarded as separate openings for
the two flagella; both flagella apparently arise from the larger pore, and the
smaller one may be the opening for the pusule, serving for extrusion of
mucilage; ×14,100. Compare Figure 4.41. From Loeblich III, 1976.

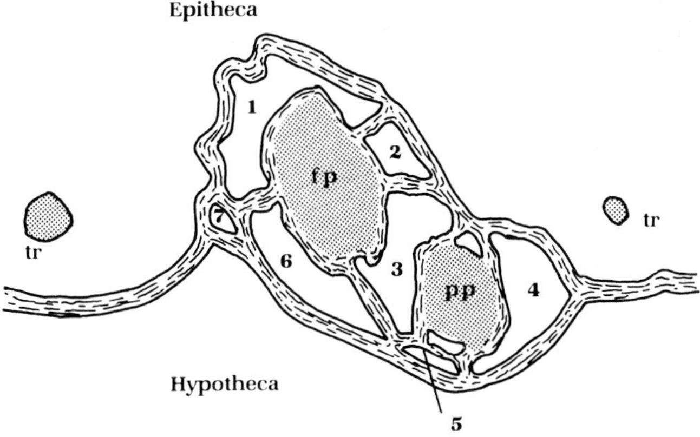

Figure 4.41
Diagram based on TEM shown in Figure 4.40 part 2, showing fp, large pore from which arise the flagella; and pp, smaller pore that may serve as opening of pusule. Pores are surrounded by smaller plates (1 − 7); other apparent plates at pore margins are extensions of major plates that project into the area. tr, trichocyst pores. Left valve is interpreted as epitheca, and right valve as hypotheca. Redrawn from Loeblich III, 1976.

striated band external to the axoneme but enclosed in the flagellar membrane, and thus corresponds to the transverse helical flagellum of the Peridiniales; the other lacks hairs and striated band and is straighter, corresponding to the whiplash longitudinal flagellum. The flagella emerge from a pore (see Figure 4.40) in an indentation in the margin of one of the valves (right valve of Lebour, 1925, p. 12, and Abé, 1967, p. 372, but left of Biecheler, 1952). The region of emergence was variously regarded as an opening between the valves, as containing a small "flagellar pore plate" (Abé, 1967), or a number of small plates (Biecheler, 1952, fig. III 3, p. 21), although neither their number nor their arrangement were determined in detail. A restudy of *Prorocentrum minimum* and *P. cassubicum* by electron microscopy (Loeblich III, 1976) showed that seven small plates are present, so arranged as to surround two separate pores (see Figure 4.41). The larger of the two pores lies closest to the left valve (of Biecheler) and contains the two flagella (Loeblich III et al., 1979). The left valve is regarded as equivalent to the epitheca, and the right valve to the hypotheca

(Loeblich III, 1976, p. 25). The helical flagellum coils around the valves.

Prorocentrum micans has numerous trichocyst pores at the base of small depressions in the large plates, and *P. minimum* is covered with both trichocyst pores and many small spines.

Tabulation of the Dinophysiales. A different arrangement of plates characterizes the Dinophysiales. The epitheca is generally much reduced, and during cell division the theca splits longitudinally into two halves, the division occurring along the **sagittal fission suture** separating a right and left valve. The reduced epitheca of *Amphisolenia* consists of only a small flat plate, and the hypotheca is much elongated. Generally 17 plates varying in shape and relative size are present in the Dinophysiales; as the tabulation scheme of the Peridiniales could not be adapted for these, a separate system was proposed (Tai and Skogsberg, 1934, p. 389 − 390; see Figure 4.42). Plates of the left valve are indicated in capital

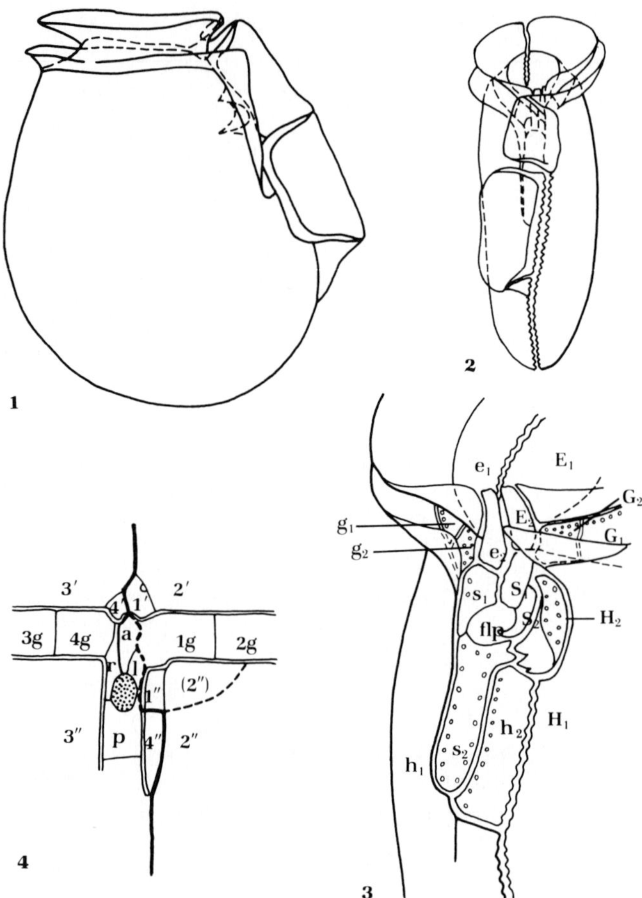

Figure 4.42
Dinophysis fortii Pavillard. **1.** Right side. **2.** Ventral view, 1,2, approximately ×600. **3.** Part of ventral view enlarged to show tabulation, left valve indicated in capitals; right in lowercase; subscript 1, dorsal member of pair; 2, ventral member; E,e, epithecal plates; G,g, girdle plates; H,h, hypothecal plates; S,s, sulcal plates (in sulcus, subscript 1 is anterior, 2 posterior of pair); flp, flagellar pore; redrawn from Tai and Skogsberg, 1934. **4.** Tabulation according to Abé, 1967; 1−4′, epithecal plates; 1−4g, cingular plates; 1−4″, hypothecal plates (2″ previously termed postcingular plate); a, r, l, p, respectively anterior, right, left, and posterior sulcal plates. Heavy line indicates fission suture for megacytic growth, separating 3′, 4′, a, l, p, 3″, 4″ on one side from 1′, 2′, 1g, 1″, 2″ on the other.

letters, those of the right valve with lowercase letters. Four series of plates were recognized; the epithecal group (E,e) consists of 4 epithecal plates (two pairs) and a pore plate (P) containing the apical pore. Balech (1967b, p. 101, 123) noted that right and left apical plates adjoin the pore plate of *Latifascia,* bringing its total to 19 plates. In the epithecal and other series, the dorsal pair of plates is numbered 1, and the ventral pair numbered 2 in the Tai and Skogsberg system. The very large dorsal epithecal plates comprise most of the epitheca and are bordered laterally by delicate flangelike lists. The girdle group (G,g) is the next series of plates, consisting of 4 cingular plates (two pairs) lying at the bottom of the transverse furrow; G_1 and g_1 are on the left and right, respectively, of the dorsal side, and G_2

and g_2 ventral in position. The hypothecal group of 4 plates (H,h) includes 2 very large and distinctly paired plates, and 2 small ventral ones that are not as clearly paired. Unlike the Peridiniales, the Dinophysiales have quite distinct sulcal plates, the sulcal group (S,s) comprising 4 plates that lie in the bottom of the longitudinal furrow. They also may be paired (S_1 and s_1 being anterior, and S_2 and s_2 posterior), although this is less obvious than for the epithecal or cingular plates.

In contrast to this system using capital and small letters, Balech (1967b, p. 123) proposed numbering the plates of each series from 1 to 4, beginning left of the sulcus and proceeding counterclockwise around the theca, as viewed from above. Thus epithecal plates E_1 and E_4 are

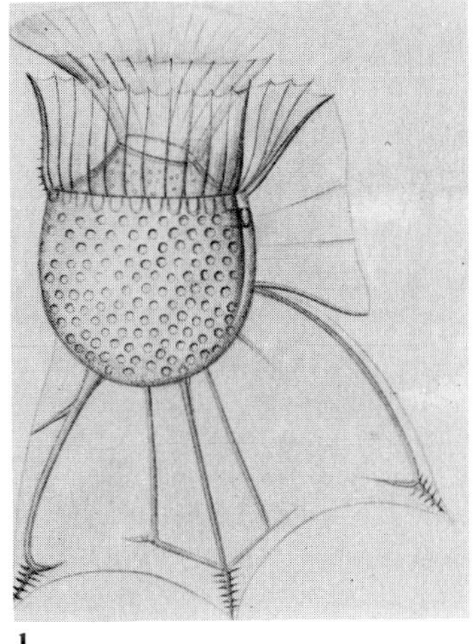
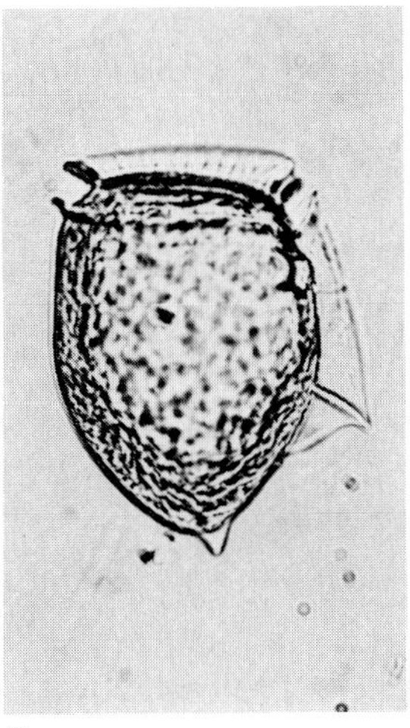

Figure 4.43
1. *Ornithocercus magnificus* Stein, right side of young individual, showing prominent lists, with supporting spines of the sulcal lists at right; approximately ×450, from Stein, 1883. **2.** *Dinophysis odiosa* (Pavillard) Tai & Skogsberg, right side; Pacific Ocean, off La Jolla, California; ×480, from Loeblich Jr. and Loeblich III, 1966.

to the left and right of the ventral side, and E_2 and E_3 are the main dorsal plates, on left and right sides, respectively. Of the four cingular plates (C_1 to C_4 from left to right) the dorsal two (C_2 and C_3) are much longer than the ventral two. The hypotheca consists of the large dorsal H_2 and H_3 in large part, as the two ventral hypothecal plates form only the narrow bases of the left sulcal list. Sulcal plates were termed Sp (the posterior main sulcal plate), Sd (the dextral or right sulcal), Ss (sinistral or left sulcal), and Sa (anterior sulcal).

As in the Peridiniales, the cellulose plates of the Dinophysiales theca form within vesicles, and the cell wall lies within the outermost cell membrane. As the cells divide by binary fission, each daughter cell inherits half of the parental theca and generates anew its missing plates and lists, the latter appearing to be extruded from the cell surface (F. J. R. Taylor, 1971).

Lists. The flangelike projections or **lists** (see Figure 4.43) arising from the margins of the girdle and sulcus, perpendicular to the thecal plate surface, consist of a single "membrane." These lists arise only at the plate edges, never within the plates. Primary lists are the anterior and posterior cingular ones bordering the girdle, and the right and left sulcal lists, these

commonly obscuring the grooves they border. The girdle lists are directed forward, and the anteriormost may be funnelshaped. The anterior cingular list is developed on the dorsal and ventral epithecal plates, and in the Dinophysiaceae the posterior cingular list generally arises only from the dorsal hypothecal plate. The right sulcal list is from the right dorsal hypothecal plate, whereas the left sulcal list, most complex of all, is from the left and right ventral hypothecal plates and generally also the right dorsal hypothecal plate. The longitudinal lists in *Dinophysis* are very broad and supported by spines. Secondary lists may form in addition to the primary lists, especially on the posterior and dorsal margins of the two dorsal hypothecal plates (Tai and Skogsberg, 1934, p. 405).

THE CYST

Dinoflagellates long have been known to produce thick-walled resting cysts as a stage in the life cycle, but little emphasis was given to the cyst structure or morphology. Thin-walled temporary cysts also are produced by some taxa, and individual cells that have shed the theca during ecdysis occasionally have been confused with cysts (as by Allman, 1855), although they have no resistant cover.

Although early studies of fossil dinoflagellates assumed that the cellulosic theca of the armored dinoflagellate was preserved, more and more evidence accumulated to indicate that only the resting cyst is fossilized (Evitt, 1961a). Studies of the dinoflagellate cyst then proceeded on two fronts, the biological studies relating to the life cycle and method of reproduction, and those of the paleontologist attempting to determine the relationships of the cyst to the respective motile cell in order to understand their morphologic differences and similarities. Toward this end, cysts of many living species were studied, either by finding cysts in process of formation and still retained within the parent theca or by germination in culture of cysts obtained from the plankton or the bottom sediments in order to identify the emerging motile cell. Motile stage and cyst are now known for about 40 to 50 species, including both marine and freshwater and thecate and nonthecate dinoflagellates.

The common resting cyst is a sexually produced diploid zygospore, generally produced in late summer or fall. Some have an organic wall that is remarkably resistant to acids and bases and is regarded as consisting of sporopollenin, although the exact composition is uncertain. Because of its remarkable resistance, such a cyst may be preserved in the sediments, fossilized, and eventually recovered by chemical maceration of the rocks, with no apparent change in composition. However, not all cysts are of sporopollenin; for example, living freshwater *Ceratium* has a cellulosic cyst which rapidly disintegrates, as does the cellulosic theca. A few dinoflagellates have a calcareous cyst, and some fossil ones of silica may represent a siliceous replacement of an originally organic or calcareous wall, or may be primarily siliceous. No similar living siliceous cysts are yet known.

Some cysts show marked similarities to the motile cell, but in other species the cysts, and hence the fossil representatives, differ in many respects (see Table 4.3). To date cysts have been described in detail for only a relatively small number of the living dinoflagellates, making difficult the determination of relationships of many of the fossil taxa that are known solely from such resting stages.

Background of Cyst Studies

Among the earliest records of dinoflagellate cysts is that of Carter (1858), who reported *Peridinium* red tides near Bombay. Cells were greenish during the early stage of the bloom, and later developed red globules in the interior and produced a cyst. The cyst might continue to float at the surface or sink to the bottom. Cell division within the germinating cyst produced from two to four new flagellated cells that were then liberated.

Undeniable ellipsoidal cysts with furcated spines were described (Stein, 1883) as *Cladopyxis brachiolata* Stein and compared to the fossil *Xanthidium ramosum* and *X. furcatum* of Ehrenberg, now known as *Spiniferites ramosus* (Ehrenberg) Mantell.

Stein also reported cysts from which new motile stages arose in *Gymnodinium vorticella* Stein, *Peridinium foliaceum,* and *P. triquetrum,* and illustrated (pl. 7, fig. 12) the cysts of *Heteraulacus polyedricus* (Pouchet) Drugg & Loeblich Jr., *Ceratium hirundinella* (pl. 14, figs. 9–11; see Figure 4.44), and *C. tetraceros* Schrank. Bütschli (1885) reported cysts in the latter two *Ceratium* species, that of *C. tetraceros* being illustrated with original figures prepared by Lieberkühn (pl. 53, figs. 7c, 7d), showing three to four daughter cells forming within the cyst. Bütschli also reported cysts in *Peridinium cinctum.* Schütt (1887, 1895) described cyst formation and later escape of the cyst from the theca for *Dissodium lenticulum* (Bergh) Loeblich III, *Goniodoma acuminatum, Gonyaulax spinifera* (Claparède & Lachmann) Diesing (Schütt, 1895, p. 7, pl. 9, fig. 34_{12}), and *Heteraulacus polyedricus* (pl. 8, figs. 30_4 and 30_5). Encystment of *Protoperidinium ovatum* Pouchet also was described by Schütt (1896, p. 14–15). In describing the formation of the cyst within the theca, Delage and Hérouard (1896, p. 379) noted that the cyst horns are shorter and

Table 4.3
Comparison of thecate cell and resting cyst of living dinoflagellates (modified from Evitt and Davidson, 1964; Evitt, 1970; Wall, 1971a).

	Thecate vegetative cell	Resting cyst
Morphology and structure	Theca of separate plates.	Cyst a single unit.
	Single-layered thecal plates.	One or more wall layers.
	Many sutures between plates.	Excystment suture only.
	Characteristic tabulation.	No true tabulation.
	Particular number and arrangement of plates.	Sculpture or processes may indicate pseudotabulation (paratabulation).
	Complex ultrastructure.	Simple ultrastructure.
	Flagellar pores in sulcal area.	No flagellar pores.
	No large aperture.	Excystment aperture (archeopyle) and operculum, of characteristic shape and position.
	Functional cingulum and sulcus, contain flagella.	Furrows nonfunctional.
	Plates may have trichocyst pores.	Wall not perforated.
Composition	Cellulose, destroyed by acids; plates separate after treatment with sodium hypochlorite.	Acid-resistant cutin (sporopollenin); few calcified; rare modern ones cellulosic (not preservable).
Behavior	Motile biflagellate cell.	Nonmotile resting cyst.
	Active metabolism (most photosynthetic; with plastids).	Metabolism reduced (contains stored starch, oil).
	May change in shape and size.	No change in shape or size.
	Cell divides by binary or multiple fission.	Meiotic cell division.
	May shed and replace wall in ecdysis.	No ecdysis.

more rounded than those of the cell, that fine ridges on the thick-walled cyst correspond to the sutures of the formerly enclosing theca, and that some cysts have spines or horns that are not present on the cell. The resistance of the cyst to chemical agents was suggested to indicate a composition of cellulose impregnated with silica. Mangin (1907, p. 1057) studied the cysts of *Ceratium cornutum* (Ehrenberg) Claparède & Lachmann and reported the thick wall to consist of an unusual association of cellulose, pectic compounds, and callose. Although no acetolysis-resistant compounds were reported, Mangin's study did not exclude the possible presence of additional components. The cellulosic composition of living *Ceratium* cysts may explain the absence of Cenozoic fossil cysts of this group (Wall and

Evitt, 1975); similar-appearing Cretaceous fossils are composed of sporopollenin, suggesting an evolutionary change in this lineage.

Cysts now regarded as of dinoflagellate affinity (*Nematosphaeropsis*) were described by Ostenfeld (1903, p. 578) as *Pterosperma labyrinthus* Ostenfeld, and later referred to *Pterococcus* (Lohmann, 1904). Paulsen (1908, p. 11) illustrated thick-walled cysts of marine dinoflagellates, some of which lacked evidence of a girdle, and compared these to the cysts of freshwater species previously reported by Stein (1883, pl. 13). Jollos (1910, p. 182) described cysts in *Gymnodinium fucorum* (= *Crypthecodinium cohnii*) whose contents later produced four daughter cells, as well as cysts of *Ceratium tripos, C. fusus,* and *C. furca;* West (1909) and Lemmermann (1910, p. 584) reported cysts in *C.*

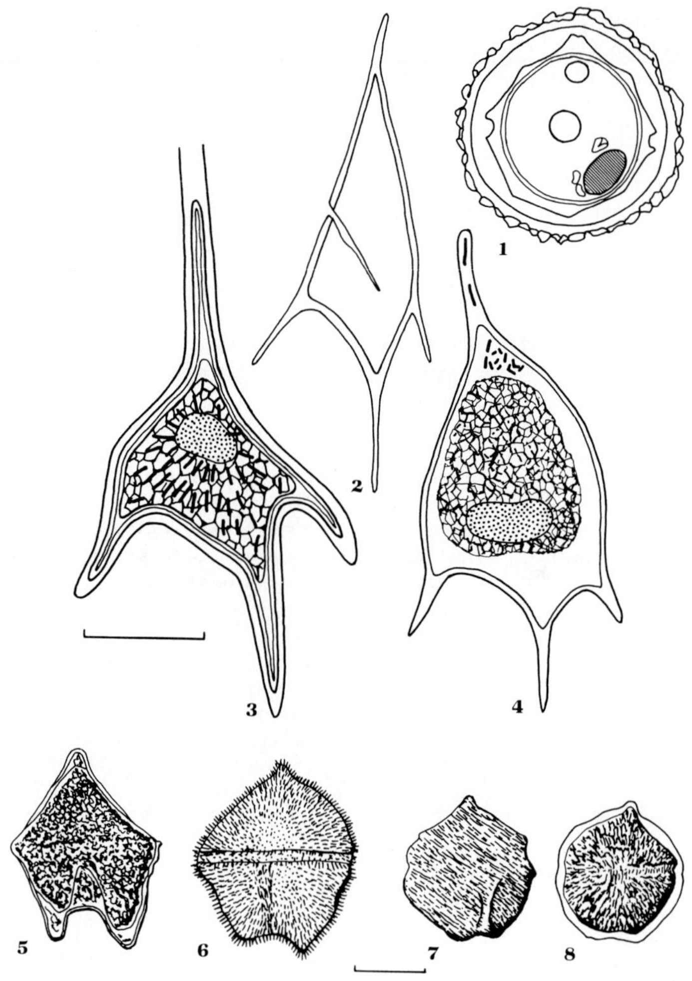

hirundinella and various *Peridinium,* and Meunier (1910) illustrated cysts forming within the vegetative cell of *Protoperidinium granii* (Ostenfeld) Balech (pl. 1bis, figs. 35 – 36), *Gonyaulax grindleyi* (pl. 1bis, fig. 37 as *Protoceratium splendens* and as *P. reticulatum* in fig. 38), and "*Glenodinium danicum* Paulsen" (pl. 2, figs. 40 – 42); cysts of *Peridinium nivale* (Meunier) Schiller were reported in yellow snow in the ice of the Kara Sea.

West (1909; 1916, p. 69 – 70) described the seasonal production of resting cysts in freshwater species of *Peridiniopsis, Peridinium,* and *Ceratium.* The entire populations encysted dur-

ing the winter in response to lowered temperatures, and their germination in the spring followed in sequence. Both encystment and germination were illustrated. Other records of cysts were not always recognized as such. Lindemann (1919, p. 219, text-figs. 6 – 9, pl. 1, fig. 1) described *Kolkwitziella salebrosa* Lindemann as a rare peridinean with thick-walled, light to dark brown theca of somewhat rounded form, only faintly impressed girdle and sulcus, and no apparent tabulation. Only empty "thecae" were observed, without cell contents. Later, Lindemann (1928) proposed the new family Kolkwitziellaceae and new "class" Kolkwitziellales for this form in which the theca was not divided into platelets. Paulsen (1949, p. 6) regarded the higher categories as invalid, being based on inadequately known taxa, and suggested that Lindemann's specimens might represent young individuals. However, the heavy wall, dark color, faint girdle, and lack of tabulation strongly suggest that the freshwater *Kolkwitziella* is based on a cyst of some other freshwater species, probably one of the associated species of *Peridinium* such as *P. elpatiewskyi* (Ostenfeld) Lemmermann or *Peridiniopsis cunningtoni* Lemmermann.

Varying degrees of detail were reported for cysts of *Ceratium hirundinella* (Meunier, 1919, pl. 21, figs. 7–8), cysts enclosed in the theca of *Ceratium cornutum* (pl. 21, fig. 12), and isolated cysts of *Protoperidinium conicum* (Gran) Balech (pl. 21, fig. 24). Cysts were illustrated within the theca of *C. hirundinella* (Entz, 1925) and *Peridiniopsis borgei* (by Entz, 1926), and the encystment process was described for *Ceratium* by Huber and Nipkow (1922, 1923) and by Hauge (1958), and for *Peridinium* by Lindemann (1919) and Diwald (1938). Cysts were described by Thompson (1947) for various freshwater *Ceratium*, *Peridinium*, and *Woloszynskia*, as were the various stages in their development, and the excystment in the laboratory of cysts obtained in the field also was described. E. Nordli (1951) described spinose cysts of *Gonyaulax polyedra*, both free and within the parent theca, and compared them to the *Xanthidium* of Ehrenberg; he also illustrated (p. 52, fig. 4) thick-walled cysts of *Protoperidinium granii*. Erdtman (1954) described hystrichospheridlike cysts in marine *Gonyaulax* and *Peridinium* (= *Protoperidinium*) and Cridland (1958) described the formation of spiny, sulfuric acid–resistant cysts by *Gymnodinium hippocastanum* Cridland.

Dinoflagellate cysts and "hystrichospherids" were increasingly reported in modern sediments, as well as in older strata: J. Muller (1959) found "hystrichospherids" in concentrations of more than 1000 per gram of sediment along the west coast and to the northwest of Trinidad; Traverse and Ginsburg (1966) found 1800 cysts per gram west of Andros Island in the Bahamas; and as many as 17,000 cysts per gram were reported in sediments from the southern part of the Gulf of California (Cross et al., 1966), the abundance increasing away from shore, perhaps as a result of lessened detrital dilution. Cysts may occur both in shallow-water littoral deposits and those of the abyssal deep sea, and in clays, calcareous muds, and fine silts (Wall, 1970). Erdtman (1954) recorded acid-resistant hystrichospherids in the plankton of the Firth of Gullmarn, which were identified by Braarud as cysts of dinoflagellates, including *Gonyaulax polyedra*. Cysts have now been reported in waters and modern sediments from many regions, particularly from the Atlantic (Wall, 1965; Wall and Dale, 1966, 1968a, 1969; Wall et al., 1967; D. B. Williams, 1971a,b; Reid, 1972, 1974), Caribbean (Wall, 1967), eastern Mediterranean (Rossignol, 1961, 1964), Pacific (Evitt and Davidson, 1964), and Red Sea (Wall and Warren, 1969).

Cyst distribution in the sediment seems to correlate with surface water conditions. The absolute frequency of cysts per unit weight of sediment seems related to factors that control total productivity, the seasonal stability of the surface waters, and the rate of sedimentation. The relative frequency of the cysts is closely correlated with the distribution of surface currents and water masses; for example, they occur in low frequencies in tropical regions such as the central North Atlantic. Nevertheless, they are found in all sediment types, ranging from the red clays to calcareous oozes (D. B. Williams, 1971b).

Other less well-known factors concerning fossil cyst distribution and abundance include the relationship between cyst abundance and the distribution and original abundance of the living population, the rate and total quantity of cyst production by the plankton, the effect of current transportation and settling rates, dilution by detrital material, predation by benthic organisms, and the nature of diagenesis.

Nature of the Cyst

Evidence from cultures indicates that most resting cysts are sexually produced diploid zygospores (Stosch, 1964, 1965, 1967, 1972, 1973), sexual reproduction having been demonstrated in various marine and freshwater species. In cultures, encystment follows a period of maximum growth, and critical combinations of light intensity, light duration, nutrient levels, and temperature may be involved in triggering the onset of the reproductive and encystment process (discussed in more detail in the section on sexual reproduction, pp. 329–337).

Cysts of freshwater *Ceratium* were found to require a resting period before they could germinate, germination then generally following an increase in temperature. Cysts could remain viable for as much as six to seven years (Hauge, 1958), after which the cellulosic cyst decayed. Similar resting periods prior to germination are required by the marine *Gonyaulax digitalis* (Pouchet) Kofoid in the laboratory (Wall and Dale, 1966).

Cysts have been collected from the plankton and surface sediments and germinated in the laboratory for a variety of marine and freshwater species, proving the cyst-nature of the fossil dinoflagellates, as well as the generic and specific relationships of the respective motile and cyst stages. For example, the freshwater *Peridinium limbatum* has a double-walled or capsulate cyst much like the fossil *Deflandrea* (see Figure 4.45). A relationship of *Deflandrea* to living *Peridinium* had been suggested earlier on various lines of evidence (Manum, 1963; Eisenack and Fries, 1965; Evitt and Wall, 1968; Wall and Dale, 1968a), although cysts of *P. limbatum* had been misinterpreted as thecae of the motile cells, and only the inner capsule regarded as a cyst (Eisenack and Fries, 1965). Separate taxa had even been described for these. Other *Deflandrea*-like cysts collected from bottom sediments have been induced to excyst in the laboratory (Evitt and Wall, 1968). The cyst forms within the theca and is released as the theca splits into dorsal and ventral halves along certain plate sutures, the ventral section comprising 11 plates in addition to those of the sulcus, and the dorsal section including the remaining 17 plates. The cyst resembles the theca in outline, with larger anterior portion, blunt and asymmetrical apical horn, and two inequal chisel-shaped antapical horns. Small surface granules and spines outline the areas on the cyst that correspond to thecal plates. The subcircular (front view) inner capsule forms soon after completion of the outer cyst wall, and individuals from bottom sediments lacking this capsule appear not to be viable, as their contents invariably were partially decayed.

In addition to the distinctive resting cysts, whose morphology and relationships are discussed later, other nonmotile stages are known in cultures. Some species produce a nonmotile coccoid stage with a smooth, transparent outer membrane enclosing the ovoid biflagellate cell. The flagella lie in grooves left where the protoplast draws away from the surrounding membrane. The coccoid stages may occur at excystment, or be produced by ecdysis by a previously motile cell under unfavorable conditions.

A **hypnoid stage** consists of a simple oval cell with transparent outer membrane (periplast) surrounding a nonflagellate protoplast that also lacks grooves, although plastids and stigma are present. By amoeboid movement, the hypnoid cell successively abandons its periplast and develops a new one, thus leaving behind a train of empty periplasts. Both hypnoid and coccoid stages are temporary; they may lead to a new motile cell, but are quite distinct from the resting cyst in morphology, nature of the enclosing membrane, and duration of the stage (Wall and Dale, 1970).

Other species may produce temporary cysts in the process of vegetative cell division, such as by multiple fission (discussed in the section on vegetative reproduction, pp. 319–329).

Figure 4.45
1–4. *Kolkwitziella salebrosa,* described as a distinct genus, but probably representing a peridinioid cyst; from Lindemann, 1919. **1.** Ventral view, showing thick, nontabulated wall. **2–4.** Dorsal, ventral, and lateral views. **5–12.** *Peridinium limbatum,* Round Pond, Falmouth, Massachusetts. **5–8.** Ventral surface, optical section, right side, and apical view, latter with ventral side at bottom. **9.** Isolated endocyst. **10.** Opened cyst with operculum at left. **5–10.** ×400. Note transapical archeopyle in 7–10. **11.** SEM of theca. **12.** SEM of cyst, 11,12, ×1000. 5–8,11,12, from Evitt and Wall, 1968; 9,10, from Evitt, 1975.

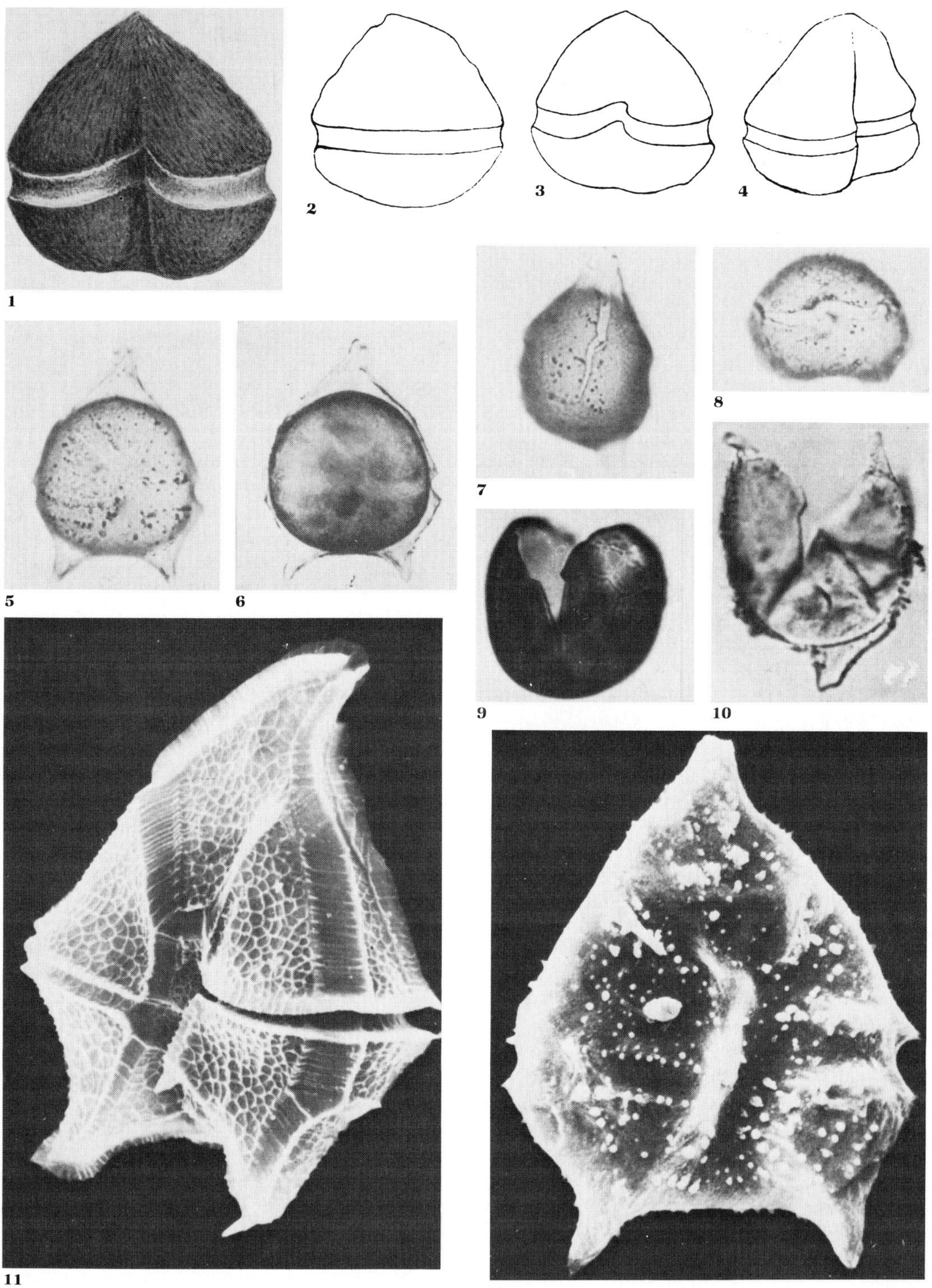

1
2
3
4
5
6
7
8
9
10
11
12

Cyst Wall Composition

Most resting cysts are composed of a highly resistant organic material, unlike the cellulose of the motile theca. Isolation of fossil specimens by acid maceration was noted by Eisenack (1931) to disprove a cellulosic composition for the early Paleozoic cysts (acritarchs), as well as the dinoflagellates and hystrichospheres of the Mesozoic and Cenozoic (Eisenack, 1935, 1936, 1938), and to indicate a distinctive composition (Eisenack, 1938, p. 184−185).

The resting cyst wall of *Woloszynskia apiculata* von Stosch was described (Stosch, 1973) as having a delicate outermost **perispore** (the theca of the motile zygote) lacking cellulose, which disintegrates as the zygote ages. Next within is a thin **exospore** of the cyst proper, which is resistant to strong sulfuric acid and expands by the pressure of the slime formed during germination. A thin **mesospore** within this exospore lacks cellulose or acidic polysaccharides and is destroyed by mild acid hydrolysis (this may be the source of the slime that fills and finally bursts the exospore during germination). The innermost thick stratified **endospore** shows a cellulose reaction when the reagent can penetrate, resists mild acid hydrolysis, and is the first layer to be dissolved at germination. Because only the exospore remains after the release of swarmers in such species, Stosch (1973) suggested that this exospore is the part preserved as fossil. Resting cysts of *Peridinium* produced in culture were resistant to various chemicals and became red-violet in a test for chitin (Pfiester, 1975). Other tests indicate that chitin is not present in most cysts, however.

Fossil dinoflagellates stain with various dyes, ranging from Safranin (red) to Sudan III (Eisenack, 1963a). The results of histological staining, examination in polarized light, and the index of refraction indicate (Deflandre, 1938, p. 162) that fossil dinoflagellates and hystrichospherids do not correspond exactly in composition to any of the fundamental or accessory constituents of the thecal wall of the living dinoflagellates. Deflandre (1938, p. 161−162) suggested that the anisotropy in polarized light of the cellulose of the living cells was lost in the fossils by a condensation to a higher molecular weight, with a conversion from the crystalline phase of cellulose to an amorphous one. Additional histochemical analyses (Pastiels, 1945) of lower Eocene hystrichospheres showed that they were neither siliceous nor of cellulose, hemicellulose, callose, chitin, cutin, or pectic material, although the possibility was recognized that some alteration could have occurred during fossilization. Later studies (Eisenack, 1963a; Evitt and Davidson, 1964) agreed in the absence of cellulose or chitin in fossil specimens, which withstood treatment in concentrated HCl, HNO_3 (even with the addition of $KClO_3$), potassium hydroxide, hot 50 percent HF, 5 percent sodium hypochlorite, potassium hypochlorite, concentrated ammonium hydroxide and boiling acetolysis mixture (acetic anhydride and concentrated H_2SO_4). Instead the cyst wall consists of sporopollenin, similar to the very resistant material of the exine of spores and pollen of the vascular plants. **Sporopollenin** is a high molecular weight polymer of C-H-O (primarily monocarboxylic or dicarboxylic fatty acids), for which the exact structural composition has not yet been resolved. Ultrastructural studies suggest that the cyst may be produced by the resistant pellicular layer that underlies the theca (Loeblich III, 1970).

In contrast to the highly resistant composition of some modern and most fossil cysts, the living *Peridinium ponticum* Wall & Dale (see Figure 4.46) has an organic-walled cyst covered with numerous small and delicate spines that do not withstand hot acid (HCl, HF). Only the spines are destroyed, and although they do not appear to consist of either silica or calcite, their composition differs from that of the remainder of the cyst wall (Wall et al., 1973).

Calcareous cysts were first known as fossils (Deflandre, 1947a, 1948; Gocht, 1959) ranging from the Jurassic to Miocene. They were believed to be originally calcareous, as they were isolated from clays in which calcareous nannoplankton and foraminifera were well preserved, as were the opaline siliceous sponge spicules (Deflandre, 1947a). Tabulation was like that of living *Peridinium* (Deflandre, 1947a, p. 1782). Similar calcareous cysts were found in marine muds (see Figure 4.47) and incubated

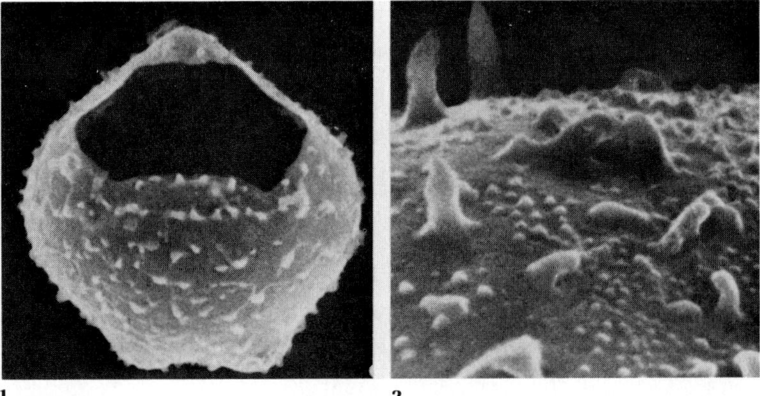

Figure 4.46
Peridinium ponticum Wall & Dale, cysts from Quaternary of the Black Sea, with small irregular spines that differ in composition from main cyst and do not withstand hot acid. **1.** Dorsal view of cyst with archeopyle, ×1000. **2.** Surface enlarged to show spinules, ×3800. Both from Wall et al., 1973.

(Wall and Dale, 1968c; Wall et al., 1970) to produce motile cells. They proved to be resting cysts of the peridiniaceans *Scrippsiella* "*sweeneyae*" (not *S. sweeneyae* Balech ex Loeblich III = *S. trochoidea,* fide Fine and Loeblich III, 1976b), *S. trochoidea,* and *Ensiculifera* cf. *mexicana* Balech. The fossil Calciodinellaceae are calcitic, as shown by X-ray diffraction, but with an inner acid-insoluble organic layer that may be comparable to the usual organic cyst.

Fütterer (1976, 1978) compared various species of *Thoracosphaera* to the dinoflagellate *Calciodinellum,* and regarded *Thoracosphaera* as a calcareous dinoflagellate cyst. The Miocene to Recent *T. albatrosiana* has a broad opening with zigzag margin, similar to the archeopyle of other dinoflagellates, but with no other indication of tabulation. Unquestionably this species is a dinoflagellate, as is *T. tuberosa* Kamptner. However, Cretaceous and Paleogene species lack the broad angular bordered opening, having instead only a small circular opening, unlike that of other dinoflagellates. Such species include the type species of *Thoracosphaera, T. heimii* (Lohmann) Kamptner, and the Cretaceous to Recent *T. saxea* Stradner, Paleocene to Oligocene *T. tesserula* Fütterer, Miocene *T. deflandrei* Kamptner, and Pliocene to Pleistocene

T. granifera Fütterer, and are not here regarded as dinoflagellates. Instead, they strongly resemble the Mesozoic Stomiosphaeraceae such as *Pithonella* (see Chapter 9), or the cysts of dasyclad algae such as *Acetabularia* (see Chapter 10), although the latter may be about three times as large. *Thoracosphaera albatrosiana* and *T. tuberosa* probably should be transferred to another genus within the dinoflagellates.

Siliceous cysts, such as *Peridinites* (see Figure 4.48), are not known to have modern counterparts; they occur in diatomaceous deposits and may be a secondary siliceous replacement. Deflandre (1933, p. 269) noted that breakage of the siliceous specimens did not follow the sutures that would probably have been lines of least resistance, hence suggested that some silica may have been deposited during fossilization. Others did break along the tabulation boundaries, and hollow specimens with no extraneous silica as well as some intact isolated plates were also observed with perfect preservation, suggesting an original siliceous wall. Deflandre noted that other dinoflagellates may secrete silica, hence these siliceous cysts also may be primarily formed.

In studying dinoflagellates preserved in flint erratics, Gocht (1970b) observed specimens in

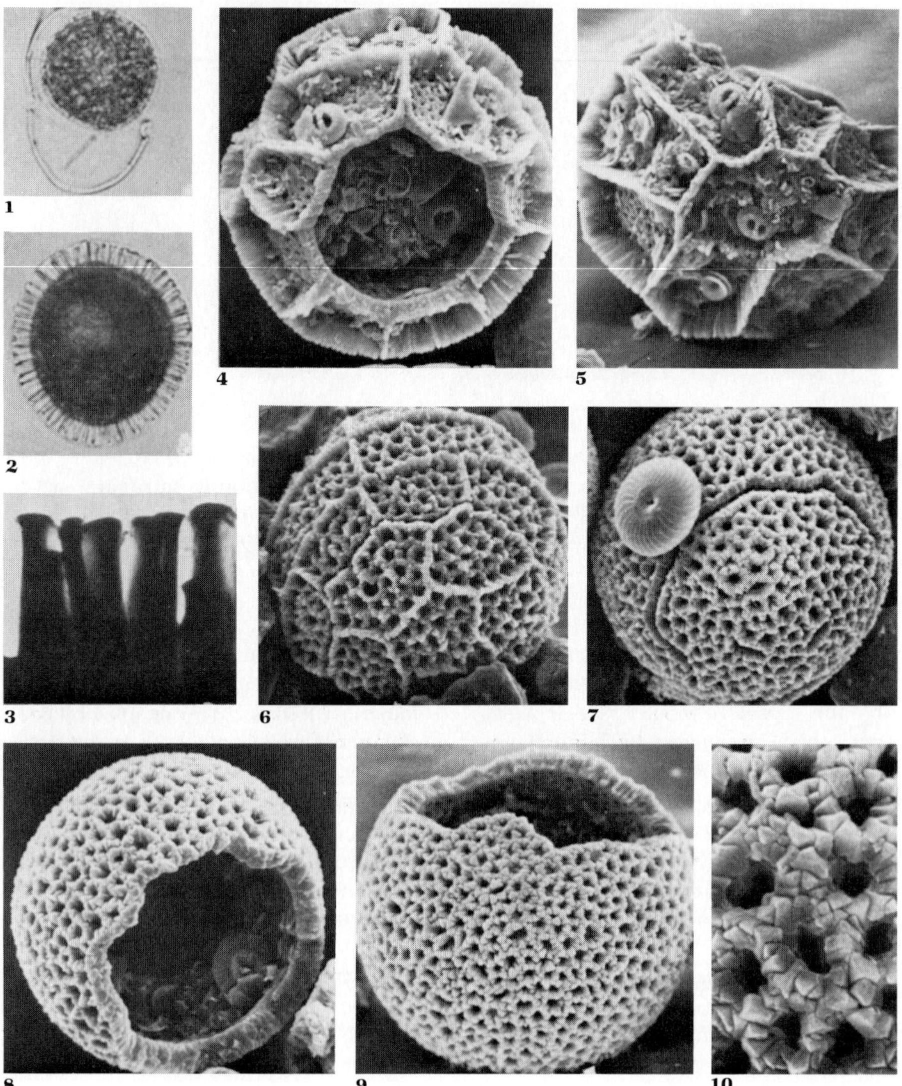

Figure 4.47

1–3. *Scrippsiella trochoidea* cysts, from Wall et al., 1970. **1.** Immature cyst being released from theca. **2.** Mature cyst with dense cell contents, and covered with numerous calcitic spines. 1,2, ×480. **3.** TEM of spines, ×6800. **4–6.** *Calciodinellum operosum* Deflandre. **4.** Equatorial Atlantic Ocean, oblique apical view, showing archeopyle, ventral area at left. **5.** Same, ventral side. 4,5. SEM, ×1040. **6.** Pleistocene, Atlantic Ocean, oblique apical dorsal view, with operculum in place, SEM, ×1600. **7–10.** *"Thoracosphaera" albatrosiana* Kamptner, probably also referrable to *Calciodinellum*, and not related to typical *Thoracosphaera* (see Chapter 9). **7.** Pleistocene, equatorial Atlantic Ocean, oblique apical view, showing archeopyle suture, but operculum in place. **8,9.** Recent, equatorial Atlantic, oblique view, showing large apical archeopyle, the angular border indicating paratabulation; and lateral ventral view, the projection at archeopyle margin probably corresponding to first apical plate (1′). 7–9, SEM, ×1600. **10.** Central Pacific, detail of distal surface, showing arrangement of crystallites that result in reticular surface, SEM, ×5280. 4–10, from Fütterer, 1976.

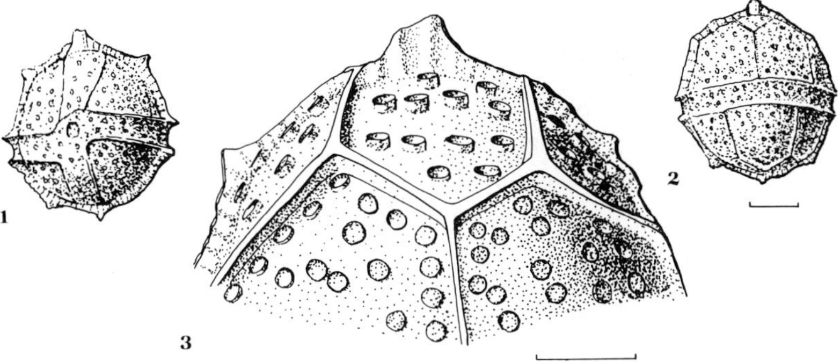

Figure 4.48

Peridinites oamaruense (Deflandre) Deflandre, upper Eocene, Oamaru, New Zealand. **1.** Ventral view. **2.** Dorsal view, bar = 10 μm. **3.** Upper part of epitheca, bar = 5 μm. Redrawn from Deflandre, 1933.

chips (as had Ehrenberg, Wetzel, and Deflandre earlier) but also obtained specimens by acid maceration. After dissolution in 40 percent HF, some specimens still had the original organic material preserved, but others of the same species were replaced by a brittle whitish mineral. If the original replacement had been of calcite, the sample preparation method would have converted these to fluorite, but the index of refraction of the pseudomorphs in polarized light was too high (1.557 – 1.558, instead of 1.434 of calcium fluoride). Recovery was not better at low concentrations of HF, as would be expected if the conversion of calcite was involved, hence the chemical nature of these fossils remains questionable.

Cyst Wall Structure

The cyst wall typically is two-layered, the layers either in contact throughout or separated by a more or less continuous space. Less common is a single-layered wall or one with three layers (see Figure 4.49). The single-layered wall is termed an **autophragm** (Evitt, 1969) and may result from fusion of the original two, or may be primitive.

When two layers are present, the outermost (outer wall of Evitt, 1967a) is termed the **periphragm** (Downie and Sarjeant, 1966). Generally, it alone carries the surface sculpture, spines, crests, and processes that provide evidence of tabulation. The inner layer (**endophragm**) may be in close contact with the outer one, even to being differentiated only by its denser nature. As seen in TEM of sectioned *Cordosphaeridium inodes* (Klumpp) Eisenack, the periphragm of the body and its processes is labyrinthic (Jux, 1968a), the many cavities elongated parallel to the surface resulting in its superficially fibrous appearance as seen with optical microscopy. Although it has a denser labyrinthic structure and fewer cavities, the inner layer is not sharply differentiated in this species. *Wetzeliella solida* (Gocht) Williams & Downie and some *Deflandrea* possess an additional layer, the **mesophragm** (Evitt, 1969), between the periphragm and endophragm. In addition, a thin and discontinuous filmy layer may connect the processes distally; this veillike reticulum or a coarser network of trabeculae may form an additional layer outside the periphragm termed the **ectophragm** (Downie and Sarjeant, 1966). *Pareodinia ceratophora* Deflandre may have a delicate outer mantle of unorganized flocculent organic matter that is destroyed by the slightest touch; termed the **calyptra** (Gocht, 1970a), it was compared to similar structures in *Netrelytron* (= *Kalyptea*) and *Paranetrelytron* (= *Pareodinia*). A similar agglutinating outer layer of the cyst wall of the living *Pyrophacus*

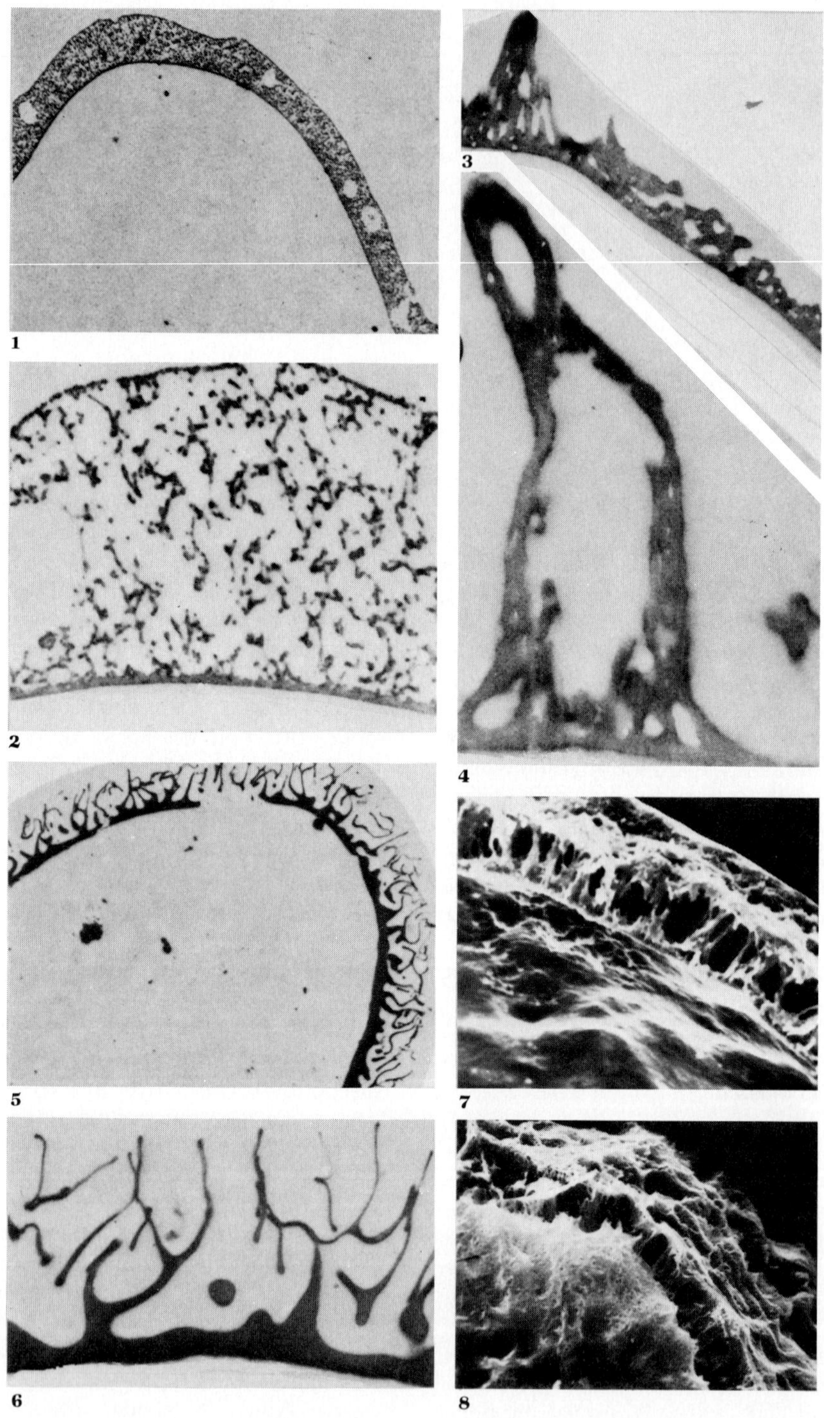

Figure 4.49
Structure of the wall of dinoflagellate cysts. **1,2.** *Apteodinium spiridoides* Benedek, Oligocene, Germany, from Benedek, 1972. **1.** TEM of section of part of wall, ×1400. **2.** Enlargement, ×9500. **3,4.** *Operculodinium centrocarpum* (cyst of *Gonyaulax grindleyi?*) Holocene, Grand Banks, Newfoundland, TEM of sectioned wall of the chorate cyst, ×9000, and enlargement, ×23,000. **5,6.** *Bitectatodinium tepikiense*, proximochorate cyst, possibly of *Gonyaulax spinifera* type; Holocene, from Grand Banks, Newfoundland; section of wall (TEM), ×2200, and enlargement, ×9000. 3–6, from Jux, 1976. **7.** *Eodinia pachytheca* Eisenack, Upper Jurassic, Germany, section of the wall in SEM, archeopyle border on cingulum, showing inner and outer layers, ×2000. **8.** *Aldorfia aldorfensis,* Jurassic, Germany, SEM of wall structure at archeopyle border on precingular wall, ×1600. 7,8, from Gocht, 1975.

---

horologium Stein has been compared to the calyptra of these fossil species (Wall and Dale, 1971).

In contrast to those species that show a gradational contact between closely appressed layers, other species have distinctly separated layers, the intervening space being termed the **pericoel.** The cavity within the endophragm which would be occupied by the resting cell contents is the **endocoel.** In such forms the endophragm forms a separate **capsule** or **inner body** within the main cyst cavity, as in some living *Peridinium* and the fossil *Deflandrea* where the capsule was long recognized as a cyst (Deflandre, 1936, p. 186) although the outer wall was believed to represent the fossilized theca. Still other species show a structure intermediate between these extremes. *Spiniferites bentori* (Rossignol) Sarjeant, the resting cyst of *Gonyaulax digitalis* (Pouchet) Kofoid, has a smooth and denser inner layer and a more spongy outer layer that alone forms the processes. The two layers separate at the base of the processes, leaving a large open chamber between the layers beneath each process (Jux, 1968b). *Polysphaeridium simplex* (White) Davey & Williams similarly has an outer matted spongy layer and inner denser layer, but in addition perforations or ducts pierce the wall and that of the processes (Jux, 1971). In *Aldorfia aldorfensis* (Gocht) Stover & Evitt, the inner layer (endophragm) may be connected to the outer (periphragm) by one or a cluster of fibrous connective struts (see Figure 4.50). The pericoel in this species consists only of the hollow areas between the struts; the periphragm dips slightly into the larger of the hollow areas, giving it an undulating surface.

The struts are most closely spaced at the margins of the archeopyle (Gocht, 1970a). In SEM, the wall surface of many fossil dinoflagellates shows tiny pores, tubercles, or papillae. Marked elevations, such as the crests of *Ctenidodinium* or *Gonyaulacysta,* are of much denser structure than the remainder of the surface and lack the surface ultra-ornamentation (Gocht, 1970a).

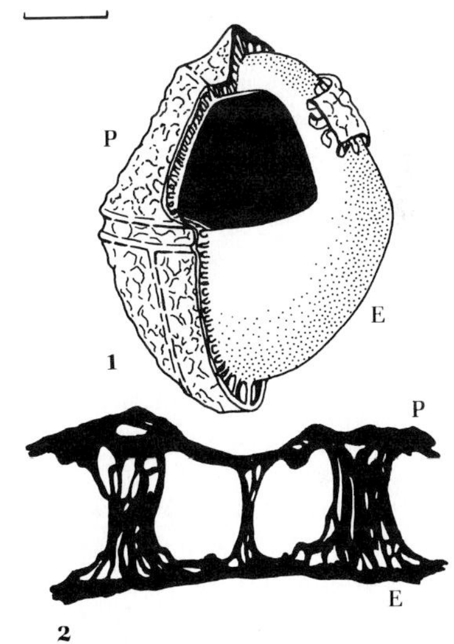

Figure 4.50
Aldorfia aldorfensis, Upper Jurassic, Germany. **1.** Cyst, showing outer periphragm (P) connected to endophragm (E) by clusters of fibrous connective struts; archeopyle opening shown in black; bar = 20 μm. **2.** Section of wall, reconstructed from electron micrographs. Redrawn from Gocht, 1970a.

Morphology of the Resting Cyst

Much of the terminology used in description of external cyst morphology, wall structure, and excystment mechanisms has been based on fossil material, but also is applicable to resting cysts of living species. The dinoflagellate cyst fossils variously have been termed a cyst, shell, **tract**, or test. They range from specimens with distinctly dinoflagellate morphology, including median girdle and indication of tabulation, to a nondescript globular form whose relationship is indicated solely by the excystment opening or by the nature and arrangement of surface sculpture and processes. The compact central portion of those with prominent projections or processes is termed the main body. The cyst outline as viewed from the dorsal or ventral side is termed the **ambitus.** In some cysts, an indication of the position of the thecal girdle or cingulum, the **paracingulum**, may allow recognition of the anterior **epitract** or **epicyst,** and the posterior **hypotract** or **hypocyst,** corresponding to the epitheca and hypotheca of the motile cell. As does the theca, the cyst may have major projections from the body, normally in the apical or antapical regions, less commonly postcingular in location (as in modern *Ceratium*). These reflect the position of the horns of the motile cell theca, although the cyst horns are shorter and blunter. Commonly the body cavity extends into the horns, but these may be solid terminally.

Major Cyst Types. Morphologically, cysts may be divided into three groups, **proximate, chorate,** and **cavate.** These were originally described as correlative with the amount of contraction of the protoplast from the thecal wall during encystment and the resultant gradation in the degree of similarity to thecal form and reflected tabulation (Downie and Sarjeant, 1966). More recently, thecae germinated from cysts of *Gonyaulax digitalis* were shown to have a cell diameter like that of the cyst lumen, and not that of cyst body and processes (Wall and Dale, 1968a, p. 291), arguing against the proposed origin of chorate cysts by contraction within the parent theca. A cyst having even

greater volume than a vegetative cell might result from the fusion of the relatively large gametes characteristic of some species or by continued growth of a motile zygote for a time. Thus correspondence of terminology to mode of cyst formation seems doubtful (Davey, 1969a, p. 2) but may be useful in a purely descriptive sense.

Proximate and chorate cysts originally were defined as having closely appressed wall layers, whereas the layers of cavate cysts separated to form a distinct central body. Corresponding with the wall terminology, a cyst with a single wall or autophragm was regarded as an **autocyst** (Evitt et al., 1977), the **autoblast** of Evitt (1969). A cyst with separated layers consists of a **pericyst** (Evitt et al., 1977; = the **periblast** of Evitt, 1969, or **pericorpus** of Norris and McAndrews, 1970), formed of the periphragm, and an inner **endocyst** (also termed **endoblast** or **endocorpus**) constructed of the endophragm. In the single known example with a mesophragm, this layer forms the **mesocyst** (or **mesoblast**).

To avoid usage of the complex terminology proposed for the various wall layers and the potential difficulty if additional layers with intervening cavities were to be discovered, Cox (1971) suggested use of only the terms *wall, body,* and *cavity,* numbered from the interior outward. Thus the inner body of the cavate cyst of *Deflandrea* would be body 1, bounded by wall 1 and enclosing cavity 1 (the endocoel or space enclosed by the inner capsule). The space (**pericoel**) between capsule and outer wall (**periphragm**) is cavity 2, and the outer wall or periphragm is wall 2. Three-walled cysts would consist from inside to exterior of cavity 1, wall 1 of body 1, cavity 2, wall 2 forming body 2, cavity 3, and wall 3 of body 3. Although the terminology is simplified, and the relative position conveyed by the number, difficulty arises with such genera as *Deflandrea,* which may have an inner body or lack one. Because in this genus the outer wall might be either wall 1 or wall 2, depending on the absence or presence of this inner capsule, the numbering system appears more confusing than useful.

1. *Proximate cysts.* The cyst wall appears to have formed adjacent to the thecal wall, with

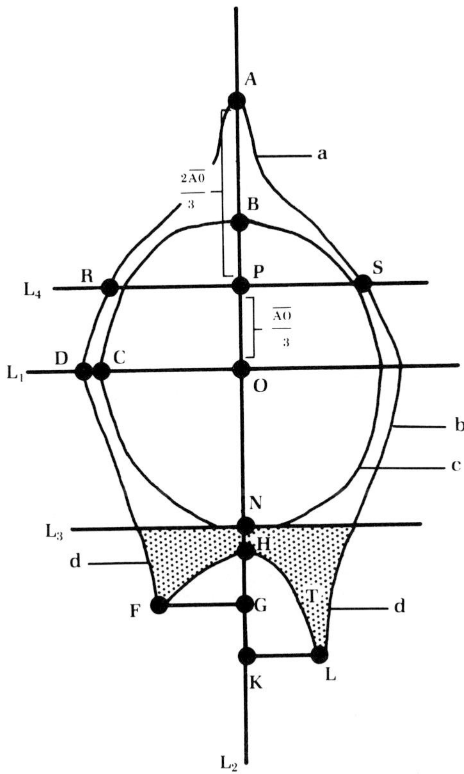

Figure 4.53
Basis for geometric formulae defined for cavate cyst *Deflandrea*, indicating proportions of apical horn (a), periphragm (b), central capsule (c), and antapical horns (d), by constructing median and tangential lines to locate points. Ratios of the different lengths and areas describe the species. Redrawn from Sellberg and Kjellström, 1974.

formulae appear to supply little information beyond that gained from the measurements alone (length and breadth of the cyst, inner body, and horns), although emphasizing the need for accurate and precise measurements.

Reflected Tabulation (Paratabulation or Pseudotabulation). Tabulation of the motile stage may be indicated by the location of ridges, flanges, processes, process-groups, and grooves on the cyst surface. However, the individual plate equivalents cannot be separated in the cyst as they can in the theca, the tabulation being indicated only by variations in wall thickness, or by the sculptural features that are arranged in polygonal fields or that may indicate the position of median girdle and sulcus. The plate equivalents of the cyst thus indicate a **pseudotabulation** (Balech, 1967a, p. 9) or **paratabulation** (Lentin and Williams, 1975a).

These plate equivalents are termed apical, intercalary, precingular, cingular, postcingular, and antapical, like the corresponding areas of the theca, and are described by plate formulae like those of the theca. Sutural structures mark the position of the sutures (**parasutures**) between the plate equivalents; intratabular features occupy the mid-area of the paraplates, and nontabular features show no apparent relationship to tabulation of the motile stage. The polygonal "fields," marked by crests on the surface of the hystrichospheres (see Figure 4.54), were at first regarded as not suturelike (Lejeune, 1937a, p. 247), as broken ones never

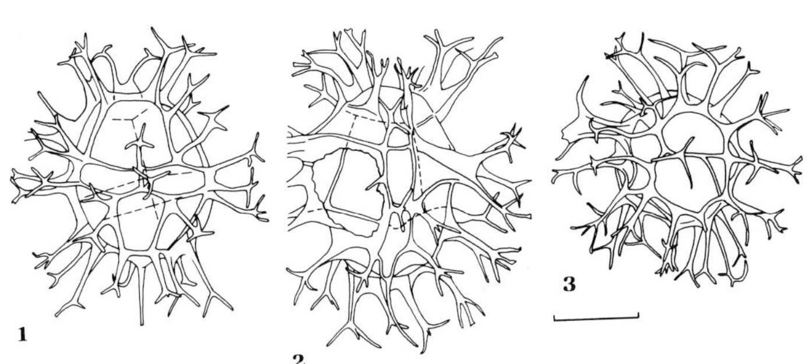

Figure 4.54
Spiniferites ramosus, three specimens from Upper Cretaceous, Limbourg Basin, Belgium. **1–3.** Dorsal, ventral, and antapical views. Dashed lines in 1 and 2 show cingulum continuing on opposite side and offset in sulcal region. Bar = 20 μm. Redrawn from Lejeune-Carpentier, 1937a.

separated at these so-called sutures in the manner of dinoflagellate thecae. When specimens later (Lejeune-Carpentier, 1940) were found with polygonal openings (= archeopyles), the areas were believed only to represent "lines of weakest resistance" where breakage was most likely to occur. Eisenack (1954, p. 51−52) even suggested that the "fields" reported by Lejeune-Carpentier were artifacts due to shrinking of the original silica gel as it hardened to flint around the specimens during fossilization.

The polygonal opening (archeopyle) left by excystment not only indicates cyst rather than thecal affinity of the fossils, but provides evidence for orientation and determination of the paratabulation. Deductions based on the fossils thus allowed Evitt (1961a) definitely to allocate the fossil hystrichospheres to the dinoflagellates, an identification later upheld by cultures of living dinoflagellates.

The reflected tabulation of the cyst agrees with the number and general location of thecal plates, but the correspondence is not exact. Some cyst sutures do not exactly underlie those of the theca; for example, the distance between the displaced ends of the cingulum of *Spiniferites ramosus* cysts is greater than that between the ends of the cingulum of the motile theca (Wall, 1970). On the other hand, in cavate cysts of the living *Peridinium limbatum*, which strongly resemble the common fossil *Deflandrea*, "the outer surface of the [inner] capsule is smooth, but the inner surface shows a fine wrinkling on which is superimposed a larger network of distinct grooves," whose pattern closely resembles that of the thecal sutures, in spite of the distance separating the inner capsule from the original theca. This pattern on the inner capsule must be determined solely by the protoplast itself, as it cannot be due to a mold-cast effect of the enveloping theca (Evitt, 1970, p. 41). Thecal ultrastructure of living dinoflagellates suggests that the cyst may be homologous to the acid-resistant pellicular layer that lies between the thecal membrane and cytoplasmic membrane, whose form may be genetically controlled and is not dependent on proximity to the theca (Loeblich III, 1970).

Pandasutural Zone. In some proximate cysts, the intercalary zone of the theca may also be reflected as a wide and commonly transversely striated band between the paraplates (Lejeune-Carpentier, 1938b, figs. 5−10; Gocht and Netzel, 1976, figs. 3a,b, 8a, 19−21). This zone is the **pandasutural zone** (Lentin and Williams, 1975a).

Horns, Processes, and Crests. Horns are relatively large projections from the cyst, derived from similar projections on the theca, and generally contain a relatively large expansion of the body cavity. Like the thecal horns, they are of restricted location and designated by the position of origin. Common locations are apical (single) and antapical (one or two); if two antapical horns are inequal in size, the larger is left-hand in position. Horns located in the median or cingular area may project from the right and left sides of the paracingulum. Other horns may be postcingular in position, commonly a single postcingular one occurring on the right side. Both right and left sides may have horns, the right being larger if they differ in size. An apical and two antapical horns occur in cysts of *Peridinium* and the fossil *Deflandrea*. These and cingular horns may occur in *Wetzeliella,* and an antapical and one or two postcingular horns may occur in cysts of modern *Ceratium* or the fossil *Odontochitina*.

Surface features constructed solely of the periphragm wall layer include ridgelike or flangelike linear crests and needlelike or tubular processes (see Figure 4.55). Crests may be solid or perforated, with smooth, serrate, or denticulate edge, and generally are sutural in position, occurring at the approximate position on the cyst that is occupied by the sutures separating plates of the theca. Use of terms such as *list, membrane,* or *septum* for such structures should be avoided, in view of their other distinct biological connotations.

Processes, the tubular, cylindrical, or ribbonlike outgrowths from the periphragm, may occur on all cyst types, proximate, chorate, or cavate. They may taper or flare distally, be straight, curved, or sinuous, a single one cen-

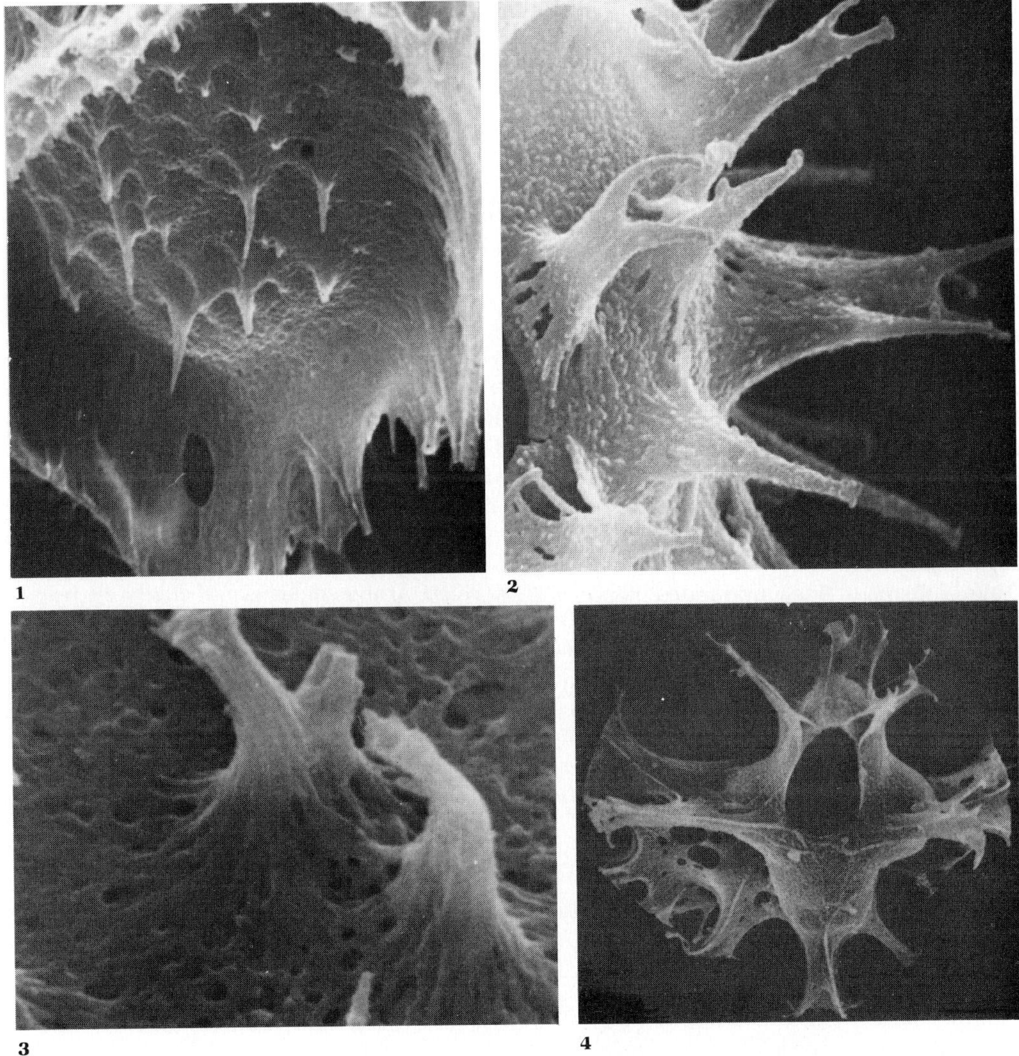

Figure 4.55

Surface crests and processes on dinoflagellate cysts. **1.** *Muratodinium fimbriatum* (Cookson & Eisenack) Drugg, lower Eocene, Alabama, SEM of plate equivalent 3''', showing portion of cingular border in upper left, and elevated fibrous crests at right and bottom of figure that mark the location of the reflected sutures; in addition, the paraplate surface is reticulate, with smaller spinelike processes arising at the angles of the reticulum, the surface of the remainder of the wall being finely granulate; ×2080. **2.** *Lingulasphaera spinula* Drugg, lower Eocene, Alabama, apical view, SEM, showing archeopyle border at lower left, granulate surface, and numerous tapering, tubular processes that may bifurcate and may be capitate at their ends; ×4640. **3.** *Fibrocysta lappacea* (Drugg) Stover & Evitt, lower Eocene, Alabama, SEM of epitractal surface on right side, showing short, intratabular fibrous, closed processes; ×8000. **4.** *Spiniferites cruciformis* Wall & Dale, Quaternary, Black Sea, SEM of entire cyst in dorsal view, showing tubular processes located on the plate equivalents, and some connected by broad flangelike membranes; ×880. 4, from Wall et al., 1973; others from Drugg, 1970.

tered on a plate equivalent, or numerous processes aligned either along the plate margins and sutural in position or randomly spaced, without apparent reference to paratabulation. Processes may be solid or hollow, terminally acuminate, rounded, capitate, or variously divided, from bifurcate to multifurcate. Adjacent processes may be connected distally by narrow solid trabeculae (*Cannosphaeropsis*) or by a filamentous layer or ectophragm (*Membranilarnacia*).

Cyst Wall Surface. The surface morphology of the dinoflagellate cysts ranges widely. In addition to the major appendages, ridges, and crests, the surface ultrastructure, as seen in scanning electron microscopy, is highly differentiated (Gocht, 1970a; Drugg, 1970). It may be smooth to granulate in *Scriniodinium luridum* (Deflandre) Klement, finely perforated by pores of 0.1 to 0.2 μm diameter in *Ctenidodinium continuum* Gocht, highly spongy in *Muratodinium fimbriatum* (see Figure 4.56), or have tiny capitate processes closely aligned in a reticulate pattern in *Batiacasphaera compta* Drugg. Surface corrosion occurring during fossilization, or during the preparation for study, may modify the apparent surface structure.

Flagellar Markings. During the controversy as to whether fossil dinoflagellates represented the motile theca or a resting cyst, specimens of *Deflandrea* that have apparent flagellar pores (Eisenack, 1954, p. 53, pl. 9, figs. 8, 9; 1966b, p. 208) were regarded as proof of the thecal nature. Apparent pores in the ventral area were found in both proximate and cavate genera, including species of *Gonyaulacysta* (Cookson and Eisenack, 1958, text-fig. 10, pl. 3, fig. 13); *Diacanthum* (Gocht, 1970a, pl. 27, fig. 1a,b; see Figure 4.57), *Ginginodinium* (Eisenack, 1961, p. 288; Cookson and Eisenack, 1965c, pl. 19, figs. 5, 6, text-fig. 3a), *Deflandrea* (Cookson and Eisenack, 1965a, pl. 11, fig. 11; 1965b, pl. 16, fig. 1, text-fig. 1; 1965c, pl. 18, figs. 1, 3, 4, text-fig. 1), and *Wetzeliella* (Gocht, 1967, pls. 13, 14; 1969, pl. 9, figs. 1, 2, text-fig. 5), and all were regarded as indications of the thecal stage. Such structures in the highly spinose *Wetzeliella* invariably lie below the midpoint on the ventral side and commonly slope obliquely toward the slightly smaller right antapical horn; they could not merely represent the imprint left by the theca as the cyst formed, as the spines would effectively protect the surface of *Wetzeliella* from contact with an enclosing theca (Gocht, 1967). However, when the ventral area was examined with the SEM, a very thin membranous layer generally was seen to cover the sulcal region, and although a paratabulation was evident in *Gonyaulacysta* and *Ctenidodinium*, no true perforation was present for the emergence of flagella (see Figure 4.58). Similarly, examination of the cavate *Deflandrea* and *Wetzeliella*, with flagellar markings evident in transmitted light, showed no indication of openings in SEM, proving that these also were merely structural or thickness variations in an unperforated surface (Gocht, 1970a),

---

Figure 4.56
Wall surfaces of fossil dinoflagellate cysts in SEM. **1.** *Caligodinium amiculum* Drugg, Paleocene, Alabama, side view of epicyst, showing partly opened archeopyle, single-layered wall, and smooth to finely punctate surface, ×3500. **2.** *Tectatodinium psilatum* Wall & Dale, Quaternary, Black Sea, showing micropunctate surface of smooth or psilate wall at archeopyle margin, ×5000. **3.** *Fibrocysta lappacea*, lower Eocene, Alabama, with fibrous network forming periphragm, ×10,000. **4.** *Batiacasphaera compta*, upper Eocene, Alabama, with surface reticulum formed of tiny rodlike elements, ×4500. **5.** *Lingulasphaera spinula*, lower Eocene, Alabama, showing microgranulate surface, ×10,000. **6.** *Muratodinium fimbriatum*, lower Eocene, Alabama, porous fibrous structure of the periphragm, ×8000. **7.** *Eodinia pachytheca*, Upper Jurassic, Germany, showing large circular bordered reticulations that resemble the bordered pits of some dinoflagellate thecae, ×2200. 2, from Wall et al., 1973; 7, from Gocht, 1975; others from Drugg, 1970.

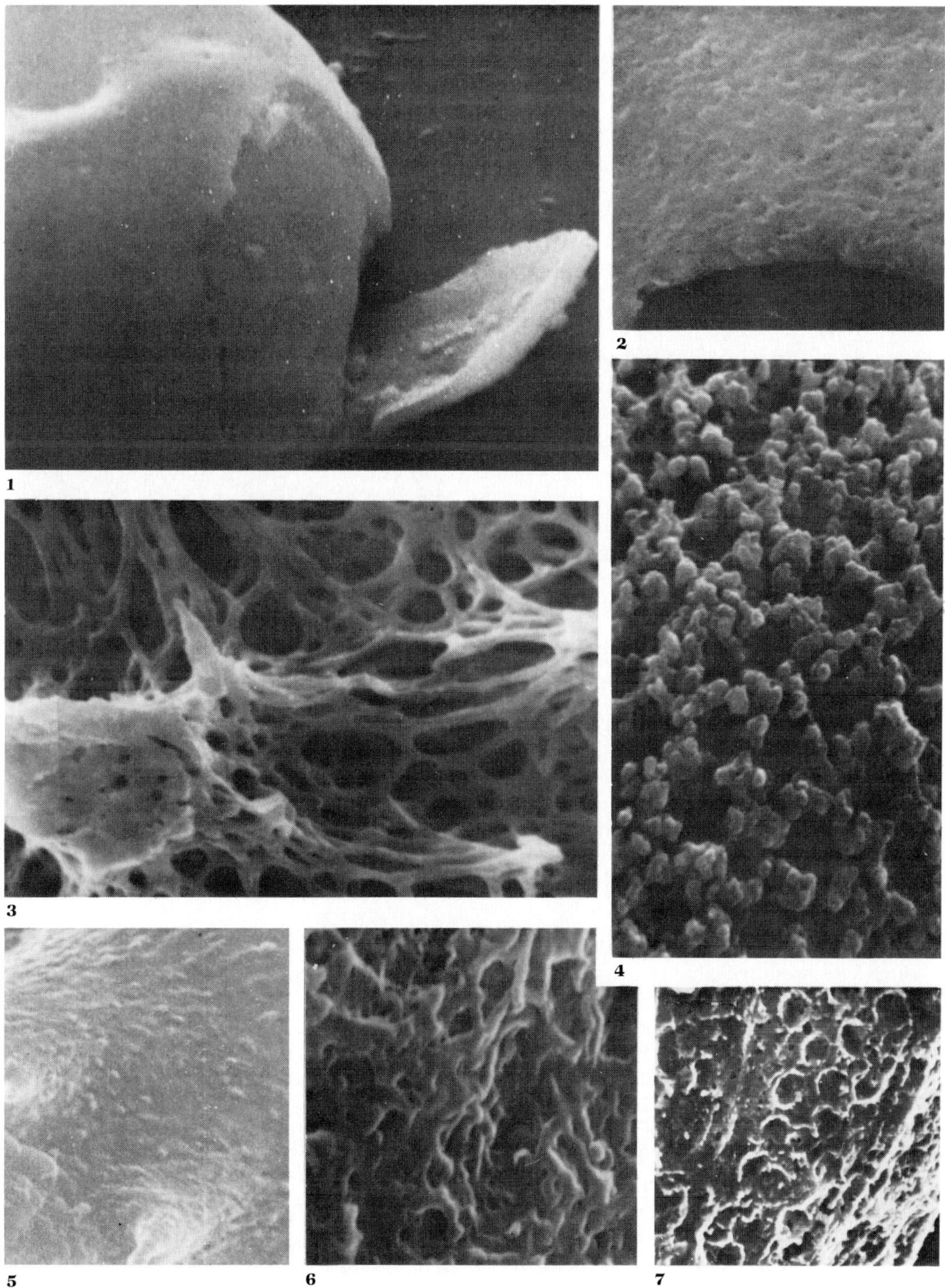

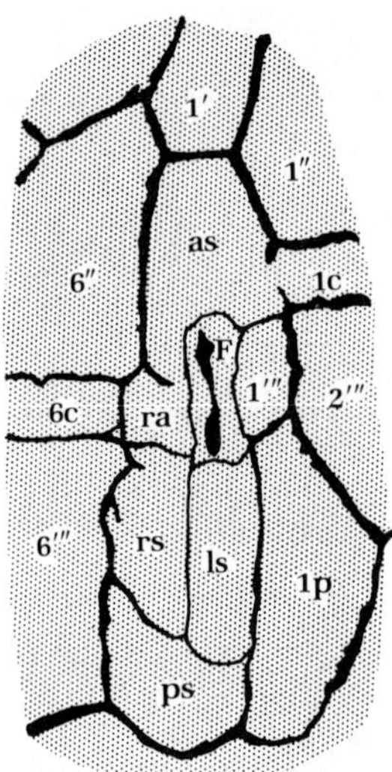

Figure 4.57
Diacanthum filapicatum (Gocht) Stover & Evitt, Middle Jurassic, Germany. Ventral area of proximate cyst, showing paratabulation and flagellar pore (F). Redrawn from Gocht, 1970a.

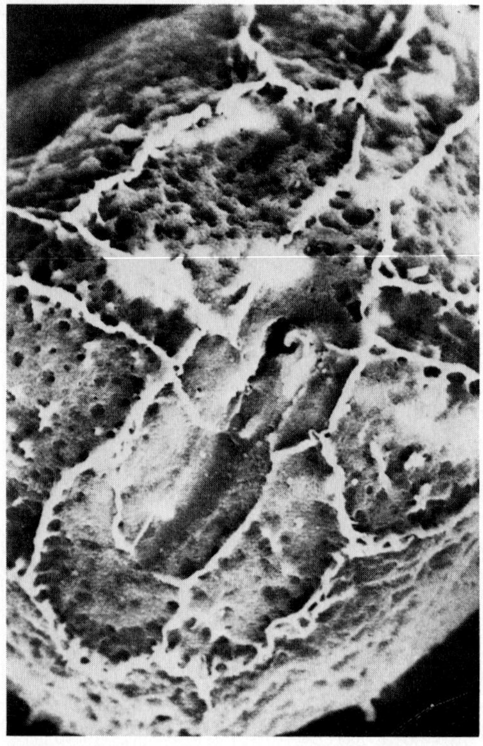

Figure 4.58
Diacanthum filapicatum, Middle Jurassic, Germany. Ventral area, in SEM, showing intersection of cingulum and sulcus, well-marked sulcal plate equivalents, and centrally located indication of flagellar pore, usually covered by thin membrane, ×1600. From Gocht, 1970a.

for which the nonfunctional term **flagellar markings** was proposed.

Fossil *Dinogymnium* from the Upper Cretaceous of New Jersey was reported to have flagellar pores in the ventral region, hence the fossils were regarded as motile cells rather than cysts (May, 1976). Like those of the Lower Cretaceous *Gonyaulacysta* of Germany (Gocht, 1970a), these flagellar markings are an indication of the position of the flagella in the motile cell, but the fossil specimens invariably are cysts. Even the specimens of *Dinogymnium* illustrated by May (1976, figs. le, 3a) show clearly the apical archeopyle or excystment opening. The Gymnodiniales, to which *Dinogymnium* has

been referred, are nonthecate, and the thin delicate cell membrane could not be fossilized. Modern *Gymnodinium* does produce resting cysts as a result of sexual reproduction, and not only may retain the flagella during early production of the thickened wall, but may eventually escape as motile cells from the cyst (see discussion under sexual reproduction, p. 331). As those living *Gymnodinium* resting cysts yet known are spiny spheres rather than gymnodinioid in appearance, the fossil *Dinogymnium* seems more probably a peridinialean dinoflagellate cyst related to the Lophodiniaceae, for which paratabulation is not apparent, although cingulum, sulcus, flagellar markings, and apical archeopyle clearly indicate its dinoflagellate nature (Tappan and Loeblich, 1977).

In explanation of the nonfunctional flagellar pores or markings in cysts, Gocht (1970a) noted that earlier workers had suggested that the cyst was produced in some Ceratia by a substance originating through a plasma duct in the longitudinal furrow, hence might have been functional in the encystment process. If the cyst is produced from the pellicular layer, as has been suggested (Loeblich III, 1970), such thinner areas might be residual from variations in this layer, regardless of the proximity of the thecal layer of the amphiesma. Some modern species retain the flagella of both gametes in a motile zygote stage for some time prior to developing the thickened wall of the resting cyst, hence the openings may remain, or the wall may be thinner at that location. Some cysts have a resistant layer external to a cellulosic one (the latter not preserved).

Inclusions Within the Cyst. Organic globules have been reported as cell inclusions within fossil specimens of *Amphidiadema*, *Cordosphaeridium*, *Deflandrea*, and *Dinogymnium*. These were early identified as "nuclei" by Eisenack (1954, p. 84), but Cookson (1956, p. 185) argued that the maceration process in preparing the fossil material with Schultze's solution would have destroyed any cytoplasmic remnants or organelles that might have escaped decay during fossilization. Later, Eisenack (1961, p. 316) suggested that they might represent symbionts or parasites, as too many were present in a single specimen to have been nuclei. Similar organic bodies were later reported in fossil pseudochitinous foraminifera in the Silurian, Cretaceous, Paleocene, and Eocene, in addition to occurring in the dinoflagellates. Evitt and Wall (1968, p. 9) reported that the eyespots were particularly resistant in living encysted dinoflagellates and suggested that these might be preserved in the fossils, but Eisenack (1969b, p. 103 – 104) noted their occurrence in all ages and as many as nine in a single specimen, again indicating that they could not be eyespots or any structured protoplast remnants.

Archeopyle and Operculum. Although the process of excystment had been noted by many of the earlier dinoflagellate workers (Stein, 1883; Bütschli, 1885; Schütt, 1887, 1895; Jollos, 1910; West, 1909, Meunier, 1919; Huber and Nipkow, 1922), emphasis was placed on the number and appearance of the emerging cells, and little attention focussed on the excystment opening. Only Huber and Nipkow (1922, figs. 4a – c) mentioned the form of the opening, which was described as a simple elongate slit in the wall of the *Ceratium hirundinella* cyst.

Because only the cyst is present in the fossil record, paleontologists have necessarily utilized its every recognizable feature and have observed the excystment opening to be important and constant in character in both living and fossil cysts. The opening, termed the **archeopyle** (Evitt, 1961a, p. 389; definition restricted, 1967a, p. 6), ranges in different taxa from a tiny opening to one equal in breadth to the maximum cyst diameter. Generally the archeopyle results from the release of one or more polygonal pieces of the cyst wall, which reflect the position of one or more paraplates, although they may be slightly smaller or larger than the plates of the theca. The part of the cyst wall that is removed, as one or more pieces, is termed the **operculum.** The operculum may detach completely from the cyst body, or may remain attached at one side. The archeopyle is preformed, as Rossignol (1964) noted that its position is indicated by a slightly more transparent line "of least resistance" a short distance from the border of the third precingular plate in *Spiniferites bentori*, shown by Wall and Dale (1970) to be the cyst form of *Gonyaulax digitalis*.

Even before the fossils were recognized as representing cysts, the presence of a characteristic opening had been noted. O. Wetzel (1933b, p. 61) reported "windows" in hystrichospherids, Lejeune-Carpentier (1937a, p. 256, pl. 1, fig. 4) noted large openings in Ehrenberg's types of *Spiniferites ramosus* (visible in Ehrenberg's figures, 1837b, pl. 1, fig. XV, 5), and Deflandre (1936, p. 170) observed that part of the "theca" is commonly missing from *Spongodinium delitiense* (Ehrenberg) Deflandre. The characteristic loss of paraplate 3″ on the dorsal side of the Cretaceous *Gonyaulacysta wetzelii*

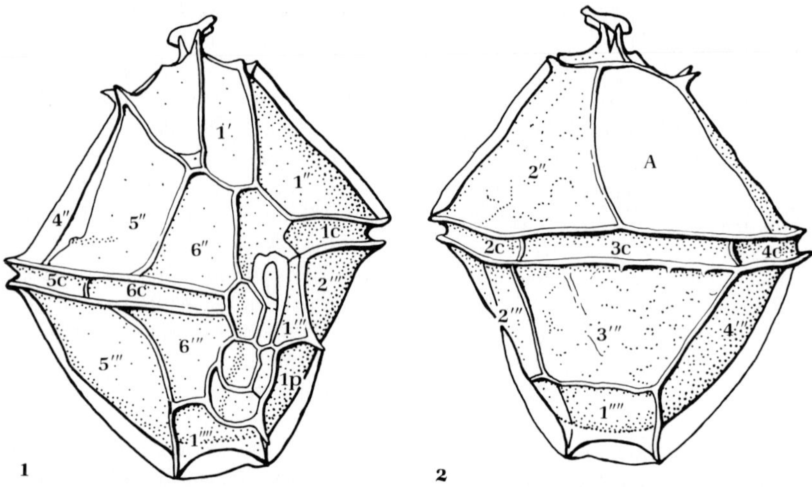

Figure 4.59
Gonyaulacysta wetzelii. Cretaceous glacial pebble from Germany. Proximate cyst. **1.** Ventral view. **2.** Dorsal view, showing tabulation and archeopyle (A), occupying position of plate 3″. Redrawn from Lejeune-Carpentier, 1939.

(Lejeune-Carpentier) Sarjeant (see Figure 4.59) was reported by Lejeune-Carpentier (1939), as were the polygonal openings in the hystrichospherids, which resembled the similar-shaped fields surrounding each tubelike process (Lejeune-Carpentier, 1940). Deflandre (1938) observed that other species of *Gonyaulacysta* similarly lacked the third precingular and suggested that this was related to the process of reproduction. The very regular form and size of the opening in *Operculodinium tiara* (Klumpp) Stover & Evitt was regarded by Klumpp (1953) as indicating a "preformed orifice." Eisenack (1954, p. 79–80) noted that the sharply angled openings with irregular outline in Tertiary specimens were constant in position, but might be present or absent in a species, hence indicated a relationship to a reproductive state or the change from an encysted to naked cell.

Gocht (1955, p. 86, 91) commented on the almost invariable presence of "slip-holes" in *Deflandrea, Wetzeliella,* and *Rhombodinium,* and later (Gocht, 1957) suggested that the common bisection of *Pseudoceratium* specimens was related to escape of the cell contents. Because the *Deflandrea* group lacked evident tabulation, it was originally placed with the Gymno-

diniaceae. Klement (1957, p. 410) in describing *Scriniodinium* observed that the constant form and position of the opening in the inner body provided proof of a latent tabulation, whereas neither tabulation nor an inner body had been reported in any *Gymnodinium.* The fossil taxa were regarded as having a typically peridinean form.

Eisenack (1958, p. 413) observed that the position of the opening in *Deflandrea* and other genera must be regarded as dorsal, thus allowing their orientation. On the basis of their possession of an archeopyle, *Spiniferites, Hystrichosphaeridium,* and other genera early were regarded as cysts (Evitt, 1961a), although other fossils were still regarded as fossilized motile stages (*Palaeoperidinium, Nannoceratopsis*) or a motile stage with enclosed cyst (*Deflandrea, Pseudoceratium*). Considering the peridinian-like fossils as preserved thecae, Eisenack (1958, p. 411) nevertheless noted that the openings were not homologous to the oblique separation of the theca of modern species in division; because of the regular form and position of the openings they were considered to be taxonomically significant in the fossils. Based on additional evidence from cultured living species, all fossil dinoflagellates are now believed to represent cysts, and the character and position of the archeopyle is an important morphological feature of these.

Some workers have also referred to openings or separation of the theca of the living vegetative cell as an archeopyle (Boltovskoy, 1973a; Ricard, 1974), but the dissolution of the boundary of the operculum by the resting zygote at the time of meiosis and germination, to which the term *archeopyle* is restricted (Evitt, 1967a), is considerably different from the process of thecal rupture or ecdysis that results from unfavorable conditions, or the process of vegetative binary fission, in which the theca of the mother cell is divided among the resultant progeny. Thus these features of the theca should not be confused with the archeopyle, an excystment opening in the resting cyst. Interestingly, however, the protoplast of *Cachonina niei* in ecdysis leaves the empty theca through the anterior intercalary region, similar to the anterior dorsal region in which excystment occurs in the peridinioid cysts (Loeblich III, 1968), perhaps due to thinner walls in this region of both theca and cyst, as well as the nature of thecal plate overlap. (See later discussions of ecdysis, p. 320, and thecal partition, p. 323, under vegetative reproduction, and of cyst formation, under sexual reproduction, p. 329).

Because the archeopyle is constant in position and its predetermined location recognizable in sectioned specimens as a very thin place in the wall, it is not an accidental rupture (Jux, 1968a). The wall structure of the operculum is like that of the cyst proper, although with less evident layering.

Twelve archeopyle types were defined (Evitt, 1967a) and appropriate symbols proposed to indicate the paraplates represented by archeopyle and operculum, the number of pieces, and whether the operculum was freed or remained attached (see Figure 4.60). Most archeopyles are restricted to the epicyst, and apical archeopyles are common, as in *Hystrichosphaeridium* and *Areoligera*. Intercalary archeopyles may consist of one, two, or three paraplates and are found in *Deflandrea*, *Wetzeliella*, and *Trithyrodinium*. Precingular archeopyles may have one (those consisting of

paraplate 3″ are most frequent), two, three, or six parts, and characterize such genera as *Lingulodinium*, *Operculodinium*, and *Spiniferites*. The archeopyle of cysts of the living *Gonyaulax polyedra* consists of three precingular paraplates (2″, 3″, 4″). Combination archeopyle types may include paraplates from more than one series; apical and intercalary series are represented in the archeopyle of *Ovoidinium;* apical, intercalary and precingular paraplates are represented in the archeopyle of *Palaeoperidinium pyrophorum* (Ehrenberg) Sarjeant. The opercula may be simple and consist of a single piece, or be compound and separate into two or more pieces at excystment. An additional combination archeopyle (not separately named by Evitt, 1967a) was later described as epithecal (Norris, 1965) and modified to epitractal (Downie and Sarjeant, 1966) or epicystal (Evitt et al., 1977), as it consists of the entire epicyst.

1. *Apical archeopyle.* Indicated by the symbol "A," the apical archeopyle is one of the most common types and generally includes all apical paraplates as a unit that remains attached ventrally; "tA" indicates that all apicals are involved. In addition, minor intercalary plates may be included, but if dorsal intercalaries are combined with the apical paraplate, the archeopyle is regarded as a combination type.

Cladopyxidium has a circular apical archeopyle bordered by a raised margin, the free operculum incorporating only the 2′ and 3′ apicals (McLean, 1972). A sulcal notch or reentrant in the margin may occur on the ventral side of the apical archeopyle and marks the cingular displacement of the first apical paraplate to interrupt the precingular series (Evitt, 1963). The sulcal tongue is a corresponding extension on the operculum (Evitt, 1967a).

2. *Intercalary archeopyles.* Represented by one or more middorsal paraplates of the intercalary series, this archeopyle may approach but does not completely extend to either the apex or cingulum. Whether representing one or more paraplates, the intercalary archeopyle (I) may have a simple operculum (indicated by a bar over the corresponding letter) or consist of two

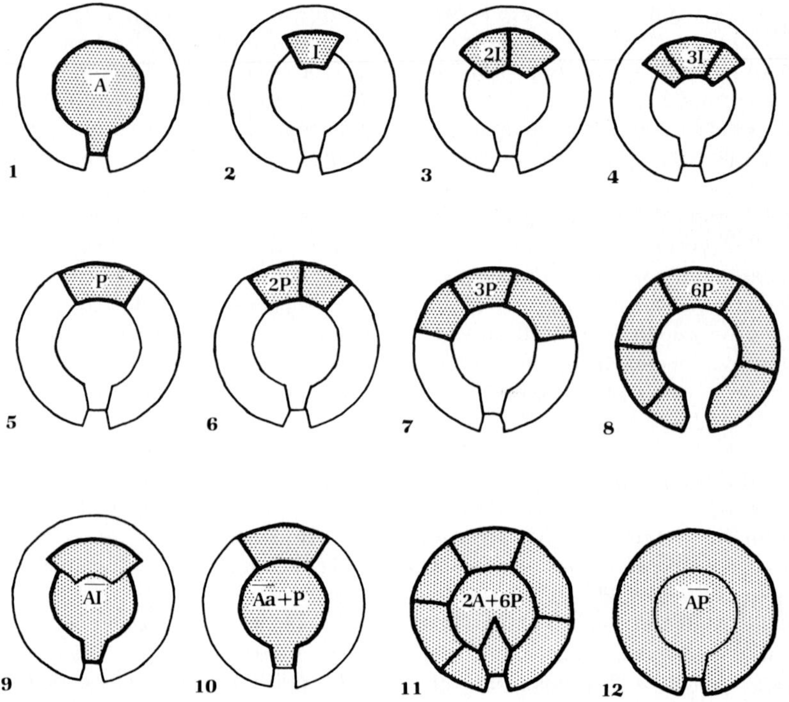

Figure 4.60

Diagrammatic representation of principal archeopyle types. Lines indicate plate series equivalents of epicyst; stippled area indicates operculum and archeopyle. Heavy lines indicate separate pieces of compound operculum. All views apical, with ventral area at bottom. **1.** Apical archeopyle, type $\overline{\mathrm{A}}$, with free operculum including all apical plate equivalents. **2–4.** Intercalary archeopyles. **2.** Type I, single intercalary. **3.** Type 2I, with two-piece compound operculum. **4.** Type 3I, with three-piece intercalary operculum. **5–8.** Precingular archeopyles, showing respectively one-(P), two-(2P), three-(3P), and six-piece (6P) opercula. **9–12.** Combination archeopyles. **9.** Type $\overline{\overline{\mathrm{AI}}}$, consisting of all apical and intercalary plate equivalents in a single piece. **10.** Type $\overline{\mathrm{A}}$a+P, with two-piece operculum, all apicals in one piece, remaining attached ventrally (indicated by "a"), and the precingular equivalents freed. **11,12.** Epicystal archeopyles, type 2A+6P losing all epicystal plates as eight separate pieces, and type $\overline{\mathrm{AP}}$ losing all as a single piece. Based on Evitt, 1967a.

or three separate pieces (indicated by a number preceding the letter symbol). The operculum may be free or attached (indicated by "a" following the symbol) at the edge closest to the cingulum (as in some *Deflandrea*) or that nearest the apex (as in some *Wetzeliella*). Type I, an intercalary archeopyle with simple operculum equivalent to the second anterior intercalary (2a), is another of the three most common archeopyle types known. Type 2I is an intercalary archeopyle with compound operculum representing two anterior intercalary paraplates. Type 3I is like the preceding, but with operculum representing three anterior intercalary paraplates.

3. *Precingular archeopyles.* These represent one or more precingular paraplates, with the posterior margin extending along the cingulum, although the anterior margin does not extend to the apex. The operculum is always free and is usually simple, corresponding to the third precingular (3″) as in *Aldorfia* and *Gonyaulacysta*

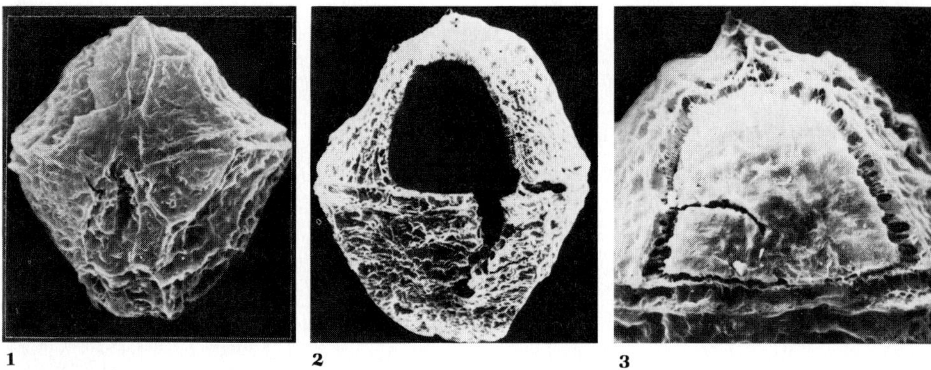

Figure 4.61
Archeopyle formation in *Aldorfia aldorfensis,* Middle Jurassic, Germany, SEM. **1.** Ventral view of cyst, ×550. **2.** Dorsal side, showing open archeopyle, ×500. **3.** Dorsal view of epicyst showing archeopyle border, with operculum in place just above reflected cingulum, ×750. From Gocht, 1970a.

(see Figure 4.61); last of the three most common archeopyle types, this is indicated as type $\overline{P}$. Type 2P has a compound operculum of two precingular paraplates (either 2″ and 3″ or 3″ and 4″ in all known examples). Type 3P is similar, representing three precingulars (2″, 3″, and 4″), and type 6P has a compound operculum representing all six precingular paraplates.

4. *Combination archeopyles.* These correspond to paraplates of more than one series. Type $\overline{AI}$ has a simple operculum representing all apical paraplates together with the *Peridinium*-like intercalary ones. In known examples, three middorsal intercalary paraplates are included. Type $\overline{Aa}$+P is a combination archeopyle with compound operculum in which one attached piece includes all apical paraplates, and a second piece represents only the third precingular (3″) and may be free or attached. *Rhaetogonyaulax* has a compound archeopyle consisting of an apical closing paraplate (equivalent to the canal plate or pore plate of *Cachonina*) and six apicals, and five additional pieces representing the anterior intercalary paraplates, hence type A+5I (Wiggins, 1973; see Figure 4.62).

The transapical archeopyle in cysts of the living *Peridinium limbatum,* and a similar opening in certain Lower Cretaceous dinoflagellate fossils, consists of a vertical division of the epicyst, involving the apical, intercalary, and precingular series of the dorsal side, along the ventral margins of 3″, 1a, 3′, 3a, and 5″. The fissure stops at the cingulum (Evitt and Wall, 1968). Accessory sutures in the cyst of Pleistocene specimens of *P. limbatum* may separate the three dorsal precingular fields to allow a flap to fold out (Norris and Hedlund, 1972).

5. *Epicystal archeopyle.* The entire epicyst becomes the operculum in those archeopyles variously termed epithecal (Norris, 1965), or better, epicystal or epitractal (Downie and Sarjeant, 1966), to distinguish between features of the cyst and those of the motile stage. This type is included with the combination archeopyles by Evitt (1967a). Some also separate as distinct a cingular archeopyle, in which separation occurs within the girdle equivalent, as in *Heteraulacacysta campanula* Drugg & Loeblich Jr., but this is here included as epicystal. The simplest epicystal form is type $\overline{AP}$, including all epicystal paraplates as a unit. Type 2A+6P is an epicystal archeopyle with compound operculum of eight pieces, one isolated apical paraplate, one piece representing a group of three apicals, and six separate pieces representing the precingular paraplates.

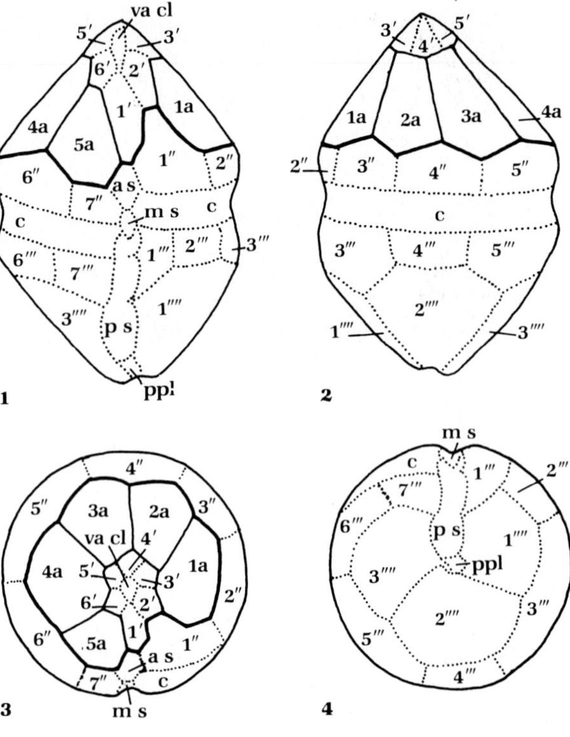

Figure 4.62
Tabulation of *Rhaetogonyaulax arctica,*
Triassic, Alaska. 1–4. Ventral, dorsal, apical, and antapical views; va cl, ventral apical closing plate; ppl, posterior ventral
plate. Redrawn from Wiggins, 1973.

6. *Antapical archeopyle.* Nearly all known archeopyles are restricted to the epicyst, but two instances of hypocystal archeopyles have been reported. *Tuberculodinium vancampoae* (Rossignol) Wall has an archeopyle on the posterior or hypocystal cyst surface, representing three to five antapicals (Wall and Dale, 1971). As the cyst proved to be only a stage in the life cycle of *Pyrophacus,* the species is now known as *P. vancampoae* (Rossignol) Wall & Dale. The motile stage has perhaps more thecal plates than does any other known living dinoflagellate (as many as 25 epithecal, 16 cingular, and 30 hypothecal plates).

An antapical archeopyle was also reported for *Inversidinium* (McLean, 1973b). Although no tabulation was evident, orientation was based on the general peridinioid outline. Both inner body and periphragm were ruptured at the antapical tip. Based solely on outline, the orientation is somewhat doubtful; if reversed, the archeopyle would instead be apical.

7. *Miscellaneous archeopyles.* Other reported archeopyles that fit none of the above categories occur in fossil dinoflagellates for which no indication of tabulation is preserved. *Ascodinium* has a nearly circular archeopyle, somewhat eccentrically apical in position, with ventrally attached operculum (Evitt, 1967a). The tiny archeopyle of *Dinogymnium* lies at the extreme apex and is dorsoventrally elongated, with the arcuate borders forming a lozenge-shaped operculum (Evitt, 1967b; Evitt et al., 1967) as in *Dinogymnium longicorne* (Vozzhennikova) Harland, *D. sibiricum* (Vozzhennikova) Lentin & Williams, or *D. nelsonense* (Cookson) Evitt, Clarke & Verdier.

If the archeopyle of a cavate cyst is constructed differently in the epicyst and endocyst, the epicyst archeopyle type designation is followed by a slash and the indication for the endocyst archeopyle, as I/3I. If the endocyst has no opening, this may be indicated as I/—.

Later modifications to this system have been

suggested; for example, indicating the exact paraplates by their symbol, as well as the number of these and the paraplate series involved (Lentin and Williams, 1975a). In another modification, parentheses were used instead of the superscript bar to indicate a single opercular piece, to facilitate typing or typesetting (Norris, 1978).

Systematic Relationships of Cysts and Thecae

Because all dinoflagellate fossils, other than the Actiniscaceae, are cysts rather than motile cells, their correspondence to the living cell is of major importance to paleontologists. Life cycles of only a few of the many living species have been studied in sufficient detail to indicate correspondence of cysts and motile stages (see Table 4.4), but even these indicate that cysts will be useful in more detailed discrimination of some modern genera. Living *Gonyaulax* may produce resting cysts of types referred to *Spiniferites, Tectatodinium, Nematosphaeropsis, Operculodinium,* and *Lingulodinium,* and many other fossil cyst-genera are characterized by *Gonyaulax*-like paratabulation (see Figure 4.63). Similarly, the living *Peridinium* is one of the larger genera, yet the many species previously included in that genus differ somewhat in tabulation and differ significantly in the nature of the cyst. Species recently transferred to *Scrippsiella* have calcareous cysts, and

Table 4.4
Motile cell and separately named equivalent resting cyst of some living species.

Motile cell	Encysted stage
Gonyaulax digitalis	*Spiniferites bentori*
G. digitalis	*S. nodosus* (Wall) Sarjeant
G. grindleyi (= *Protoceratium reticulatum*)	*Operculodinium centrocarpum* (Deflandre & Cookson) Wall
G. grindleyi	*?O. israelianum* (Rossignol) Wall
G. grindleyi	*O. psilatum* Wall
G. polyedra	*Lingulodinium machaerophorum* (Deflandre & Cookson) Wall
G. scrippsae Kofoid	*Spiniferites bulloideus* (Deflandre & Cookson) Sarjeant
G. scrippsae	*S. belerius* Reid
G. scrippsae	*S. elongatus* Reid
G. spinifera	*Bitectatodinium tepikiense* Wilson
G. spinifera	*Nematosphaeropsis labyrinthea* (Ostenfeld) Reid
G. spinifera	*Spiniferites mirabilis* (Rossignol) Sarjeant
G. spinifera	*S.* cf. *ramosus*
G. spinifera	*Tectatodinium pellitum* Wall
G. sp. indet.	*Spiniferites ramosus* v. *pachydermus* (Rossignol) Harland
Peridinium limbatum	*Peridinium limbatum* v. *minnesotense* Eisenack
Protoperidinium avellana (Meunier) Balech	*Chytroeisphaeridia cariacoensis* Wall
P. conicoides (Paulsen) Balech	*?C. simplicia* Wall
Pyrodinium bahamense	*Hemicystodinium zoharyi* (Rossignol) Wall
Pyrophacus vancampoae	*Tuberculodinium vancampoae*
Scrippsiella "sweeneyae"	*Calciodinellum* sp.
S. trochoidea	*Calciodinellum* sp.

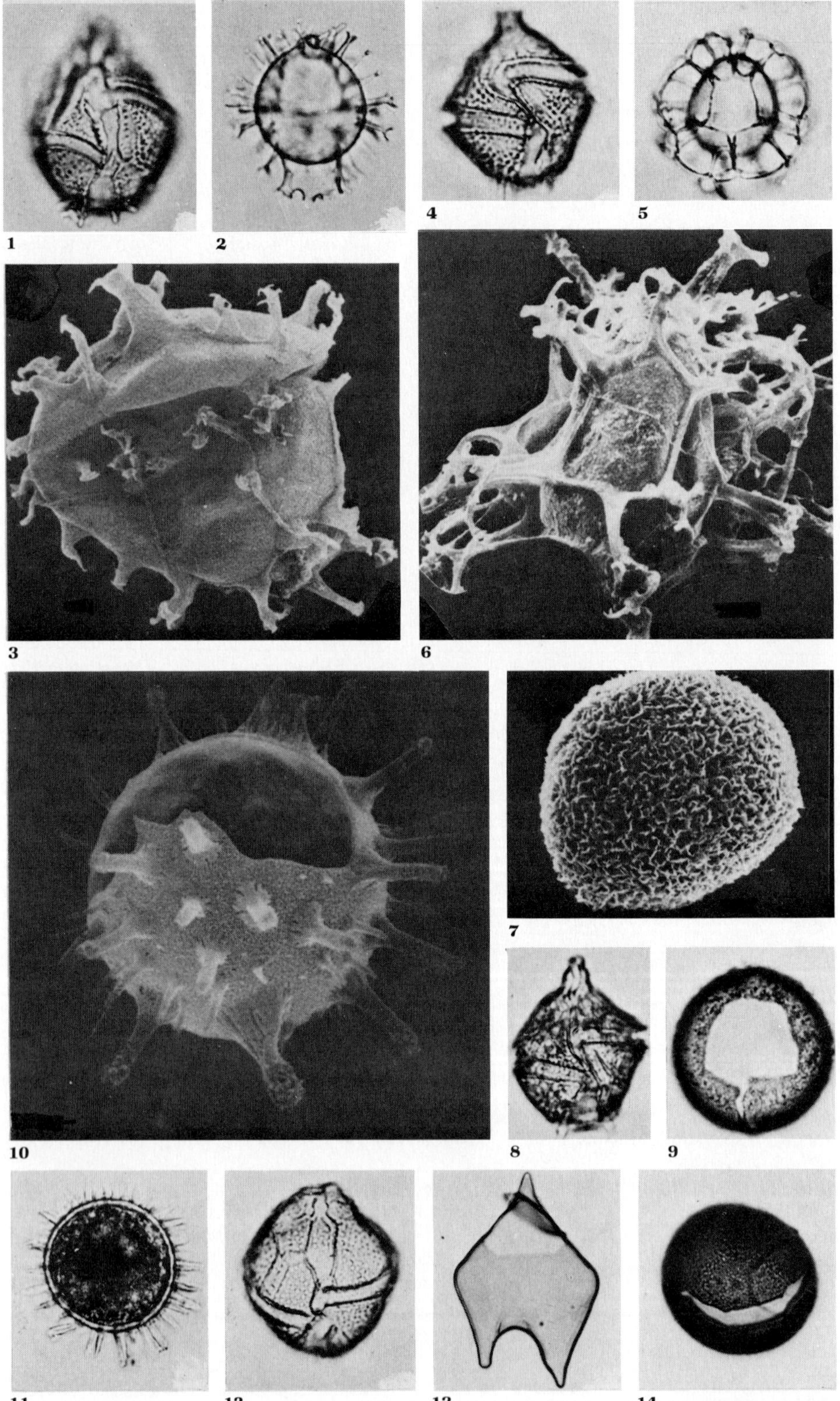

Figure 4.63 *(facing page)*
Motile cells and cysts. **1−9.** Motile cells of *Gonyaulax spinifera* type and cysts from which they germinate. **1,2.** Motile cell, ×446, and *Spiniferites mirabilis* cyst from which it germinated, ×340. **3.** SEM of *Spiniferites ramosus* cyst, ×1000. **4,5.** Motile cell, ×583, and *Nematosphaeropsis balcombiana* Deflandre & Cookson cyst from which it germinated, ×410. **6.** SEM of *N. balcombiana* cyst, ×1660. **7.** SEM of *Bitectatodinium tepikiense*, ×1000. **8,9.** Motile cell, ×500, and *Tectatodinium pellitum* cyst, ×410, from which it germinated. **10−12.** *Gonyaulax polyedra* and its cyst, referrable to *Lingulodinium machaerophorum*. **10.** SEM, ×1100. **11,12.** Cyst of *L. machaerophorum*, ×454, and *G. polyedra* motile cell that germinated from it, ×500. **13.** *Protoperidinium oceanicum* (Vanhöffen) Balech cyst, ×300. **14.** Cyst of *Diplopsalopsis orbicularis* (Paulsen) Lebour, ×400. 3,6,7, from Jux, 1976; 10, from Wall et al., 1973; others from Wall and Dale, 1967.

Scrippsiella, Cachonina, and *Ensiculifera* all have more girdle plates than the marine *Protoperidinium,* resembling the freshwater *Peridinium* in this regard. The type species of *Peridinium,* the freshwater *P. cinctum,* has the tabulation 4', 7'', 5c, ?s, 5''', 2'''', and *P. limbatum* has a similar tabulation with 2 apical platelets and 5 cingulars. Both these freshwater species have cysts with an inner capsule, similar to the fossil *Deflandrea,* whereas cysts of marine living *Protoperidinium* lack an inner capsule, excyst with a different type of archeopyle, and have 3 rather than 5 cingular paraplates. The thecal differences in tabulation that led to the recognition of *Protoperidinium*

for the marine species, such as *P. subinerme* (Paulsen) Loeblich III and *P. minutum* (Kofoid) Loeblich III, are further supported by the morphologic differences of their cysts.

Although these and other fossil cysts may have a paratabulation like that of some living genera, some fossils show no evidence of tabulation, or insufficient indication for recognition of the paratabulation formula. Some also have a paratabulation that has no known counterpart among living taxa. The majority of fossil cysts show a relationship with the Peridiniales, only *Nannoceratopsis* appearing to be related to the Dinophysiales (see Figure 4.64); no fossils show direct evidence of relationship to the Prorocen-

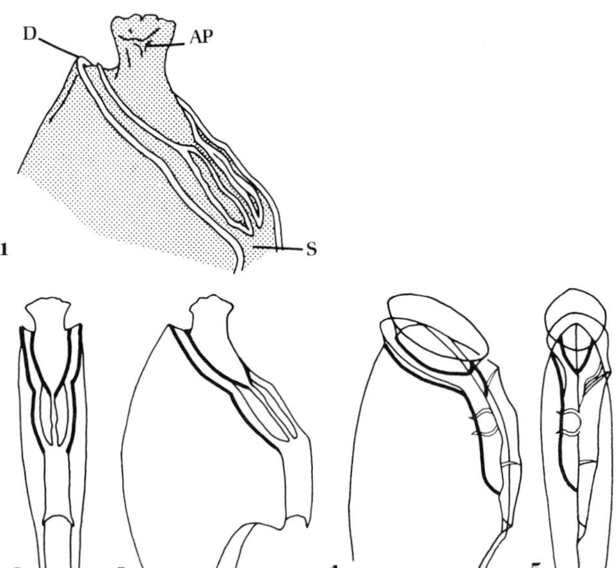

Figure 4.64
Comparison of *Nannoceratopsis* and living *Dinophysis*. **1.** Oblique view of anterior part of *Nannoceratopsis gracilis* Alberti, showing grooves, ridges; AP, apical prominence; D, dorsal angle; S, sagittal band. **2,3.** Ventral and right side of *Nannoceratopsis*. **4,5.** Similar views of *Dinophysis diegensis* Kofoid, for comparison of structures (heavy lines) and general form. Redrawn from Gocht, 1972, *Lethaia,* v. 5, p. 15−29, published by Universitetsforlaget, Oslo University.

trales, although perhaps the acritarchs represent this other major group of armored dinoflagellates (Loeblich III, 1976). Nonthecate dinoflagellates are poorly known as fossils, as very few data concerning resistant resting cysts are available for modern species. Such genera as *Dinogymnium* and perhaps *Cystodiniopsis* have been regarded as fossil Gymnodiniales, although they are quite unlike the cysts of living Gymnodinialeans, and *Dinogymnium* appears closer to the Lophodiniaceae of the Peridiniales (Tappan and Loeblich, 1977).

Most fossil dinoflagellate cysts that show some evidence of paratabulation on the surface, or by the nature of the archeopyle, have an affinity with one or another of three major living genera, *Peridinium* (and allies), *Gonyaulax*, and *Ceratium*, and a few are related to minor living genera. *Peridinium*-like tabulation is seen in the cavate cysts of the *Deflandrea* and *Wetzeliella* groups, and some living freshwater *Peridinium* species are known to have cavate cysts, with an inner capsule. *Gonyaulax*-like fossils include proximate cysts (*Rhynchodiniopsis*), chorate cysts (*Spiniferites*), and cavate cysts (*Scriniodinium*); *Pyrodinium*-like cysts include *Hemicystodinium* (like the cyst of living *Pyrodinium*), *Hystrichosphaeridium*, *Homotryblium*, and others. Cysts have not been described for living marine *Ceratium*, although cellulosic and nonresistant cysts have long been known for its freshwater species. Marine Cretaceous cyst genera such as *Odontochitina* and *Pseudoceratium* are *Ceratium*-like in general morphology, although they show no indication of paratabulation. Calcareous cysts like those of the fossil Calciodinellaceae are produced by the living *Scrippsiella*, including *S. trochoidea*, and by *Ensiculifera* cf. *mexicana*. Both motile stage and resting cyst are now known for various species of *Diplopsalis*, *Diplopsalopsis*, *Diplopeltopsis*, *Ensiculifera*, *Gonyaulax*, *Pyrophacus*, *Pyrodinium*, and many *Peridinium*, largely through the studies of Wall and Dale (see Figure 4.65). However, as no living counterpart has been recognized for many fossil cysts, a separate classification is used for those known only as cysts.

PHYSIOLOGY

Movement

Flagellar Action

Dinoflagellate movement results both from the action of the two dissimilar flagella, which may beat alone or simultaneously, and the asymmetrical form of the cell. The cell normally moves forward along a spiral path, simultaneously rotating about its longitudinal axis. Early confusion as to the cell anterior and posterior led Delage and Hérouard (1896, p. 377) to state that the longitudinal flagellum was held in front during movement, the cell generally rotating in a dextral sense, although it might also turn in the opposite direction. Their statement that the transverse flagellum determined the rotation was repeated by most later workers. Similarly, the longitudinal flagellum was regarded as producing the forward movement.

New techniques have allowed greater accuracy in determining the movement and separating the effects of the two flagella. Visual examination and high speed photography of cells in suspensions of 1.18-μm-diameter polystyrene spheres recorded the water currents produced

---

Figure 4.65
Motile cells and cysts. 1 —3. *Pyrophacus vancampoae*. 1. Hypothecal view of motile cell from Phosphorescent Bay, Puerto Rico, SEM, ×900, from Wall and Dale, 1971, 2,3. Cyst, originally described as *Tuberculodinium vancampoae*, side and apical views, ×450, from Wall, 1967. 4 — 7. *Gonyaulax grindleyi* (formerly *Protoceratium reticulatum*). 4. Theca, ×1000; from Stein, 1883. 5 — 7. SEM of cyst, originally described as *Operculodinium centrocarpum*, from Jux, 1976; 5, ×1000; 6, part of surface enlarged, ×2000; 7, appearance of processes, ×3300.

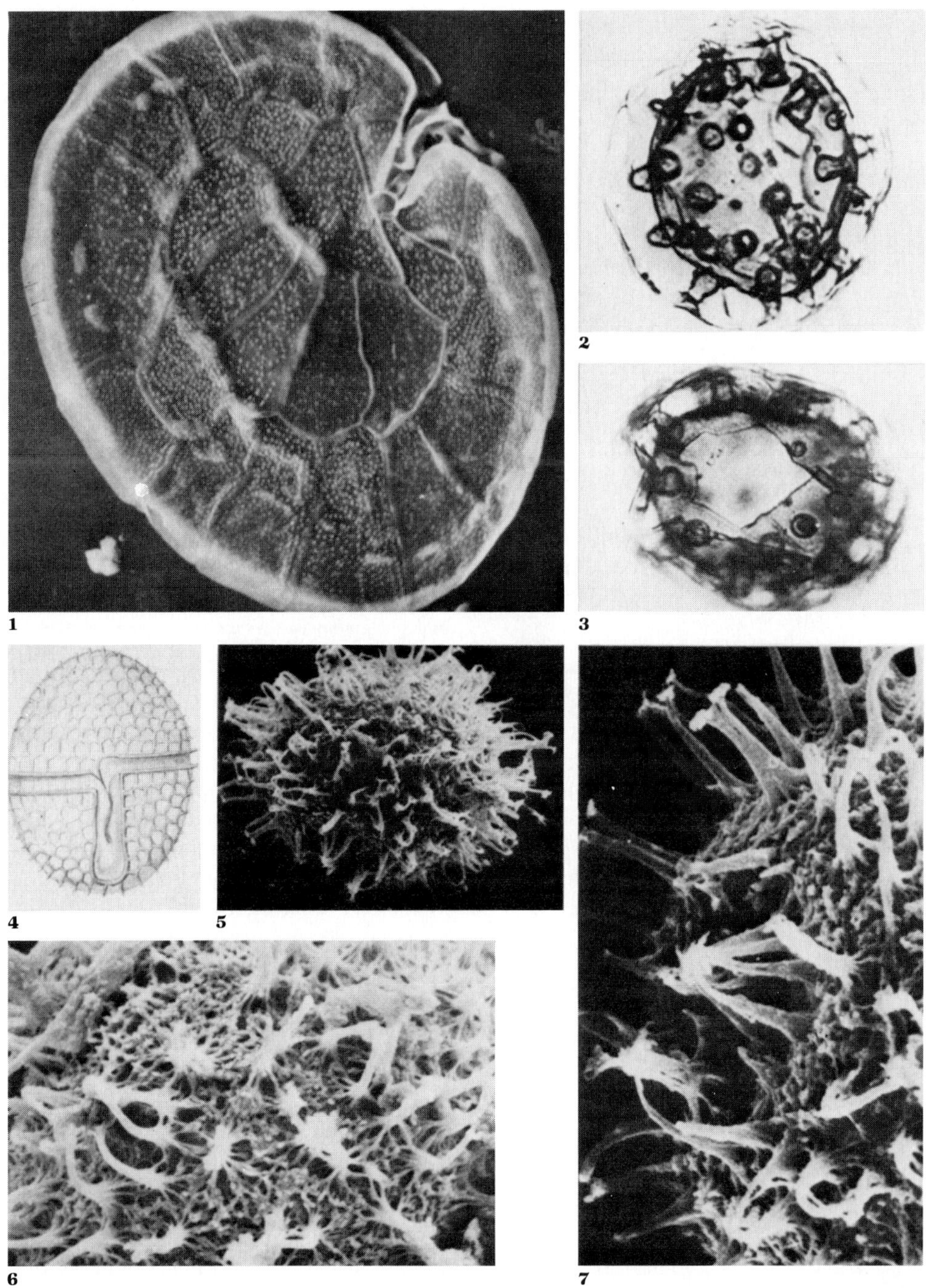

1
2
3
4
5
6
7

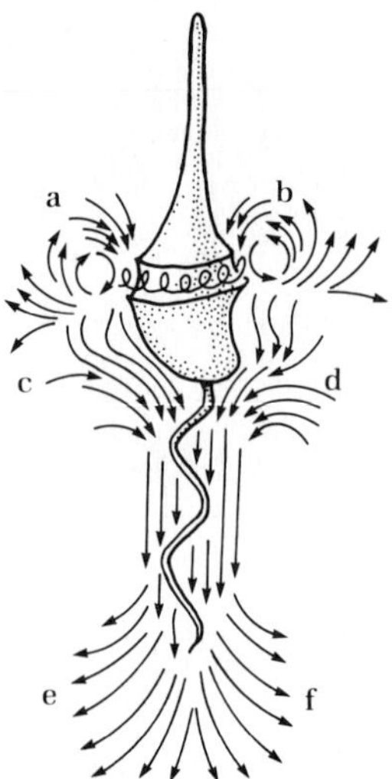

Figure 4.66

Flagellar movement in *Ceratium*. Helical waves along transverse flagellum push water to posterior. Longitudinal flagellum produces distally moving planar wave series that also pushes rearward, hence either may result in forward progress. Arrows indicate paths of tiny polystyrene spheres, moving to posterior at girdle, then out, returning forward at a,b. At c,d, spheres are driven backward by action of transverse flagellum, into current produced by longitudinal flagellum, then outward at e and f. Redrawn from Jahn and Votta, 1972; reproduced, with permission, from the *Annual Review of Fluid Mechanics*, v. 4; copyright 1972 by Annual Reviews Inc.

by flagellar motion (see Figure 4.66), demonstrating that the transverse flagellum supplies a forward component of movement. The current generated solely by the transverse flagellum extends around the organism and may even be the only means of forward propulsion. Cells may rotate either to the right or left, and some alternate from one rotational direction to the other while continuing to move forward. *Ceratium tripos* regularly rotates with ventral side to the right when the transverse flagellum is beating, but to the left when it does not. The flagellum apparently fits tightly within the girdle and protrudes only slightly, some of the force generated by its movement pushing against the body wall. The anterior flange of the girdle may

serve primarily as a pushing surface, whereas the posterior flange merely aids in confining the helical wave. Motion begins at the flagellar base and extends to the tip, the resultant force pushing toward the base and turning the ventral surface of the cell to the right, causing its predominantly dexiotropic rotation. The transverse flagellum functions principally in forward propulsion, whereas the planar wave of the longitudinal flagellum produces no direct rotation, only some forward movement (Jahn et al., 1963). As the transverse flagellum produces both forward and rotational movement, the trailing longitudinal flagellum may serve primarily as a rudder (Gittleson et al., 1974).

Cell Asymmetry

If movement of the longitudinal flagellum produces no rotational movement, and that of the transverse flagellum causes rotation to the right, as also is favored by much of the structural differentiation, the occasional leiotropic rotation must result in dinoflagellates, as in other flagellates, solely from the cell asymmetry.

Kofoid (1906a) stated that such asymmetrical morphological features as the descending right spiral of the girdle caused the predominantly dexiotropic rotational movement. The radius of the spiral path is influenced by the cell asymmetry, the greatest radius characterizing the most strongly asymmetrical cells.

Other features of the cell asymmetry appear to be adaptations for flotation. The distal deflection to the left of the tips of the antapical horns of *Triposolenia* orients the body in the position of greatest resistance to descent in response to gravity. This dinoflagellate commonly lives at a depth of about 200 meters, and sinking much below this level would remove it from the lighted zone. The lists about the flagellar pore tend to keep the axis upward, and fatty materials also accumulate in this end of the cell so that any flagellar action tends to move the cell upward (see Figure 4.67). When there is no active movement, the asymmetry of the antapical horns tends to turn the cell onto its broad right face and retard sinking. The cell turns over in this way after a descent of less than 10 cell

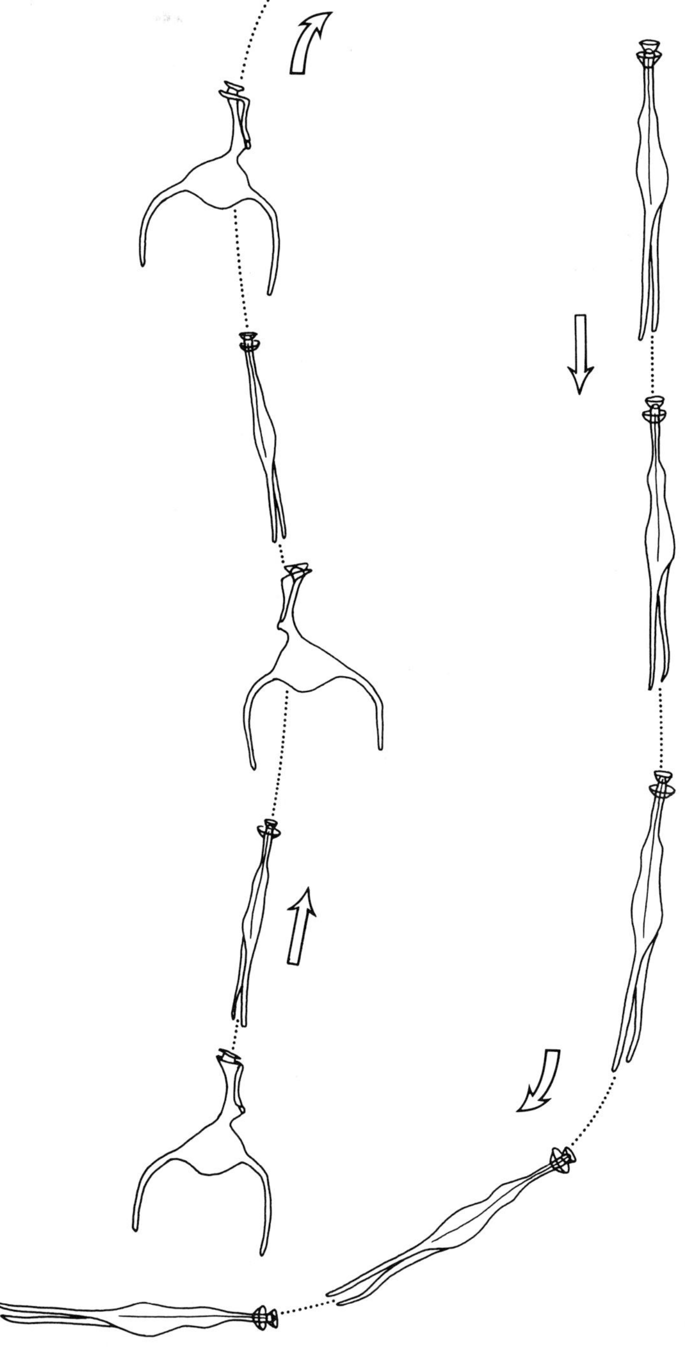

Figure 4.67
Diagram of locomotion in *Triposolenia*, showing forward movement (left), and orientation resulting from sinking of asymmetrical cell (right). Redrawn from Kofoid, 1906a.

lengths. The same effect results from the horns and lists in *Amphisolenia* and *Ceratium,* and from girdle asymmetry in *Protoperidinium depressum* (Kofoid, 1906a). *Protoperidinium mediterraneum* (Kofoid) Balech has two slender, equal, slightly curved antapical spines, that on the right side being most strongly curved, with tips inflexed ventrally and obliquely toward the median line. As in *Triposolenia,* the sinking organism is shifted from the vertical position, in which the least surface is exposed to displacement of the water, to lie on the dorsal face, which has the greatest form resistance. The result is a 30 percent increase in the surface exposed to the gravitational force (Kofoid, 1909b).

Although most dinoflagellates are observed to rotate during movement, a rotational movement due to anterior thrust is probably strongly resisted in *Ornithocercus* by the forward projecting collarlike girdle lists (F. J. R. Taylor, 1971). The inequal sizes of the right and left upper girdle lists would concentrate the flow of water over the sulcal area (as previously noted by Abé, 1967), and the keellike left sulcal list would inhibit rotation. Such an increased flow of water over the sulcus of the nonphotosynthetic *Dinophysis,* with its winglike lists, may be a highly useful specialization for feeding (F. J. R. Taylor, 1971).

Rate of Movement

Both forward movement and vertical migration in the water column may result from flagellar activity. *Katodinium rotundatum,* with a cell volume of about 350 μm^3, moves between 300 and 500 μm/sec, or at an average of 370 μm/sec, equivalent to about 30 cell lengths. It simultaneously has a rotation of about 5 to 10 gyres per second, although the path appears nearly straight. *Gyrodinium* sp., with cell size of 250 μm^3, moves from 200 to 245 μm/sec (average speed 230 μm/sec), with 2 to 3 gyres per second. *Peridinium gregarium* Lombard & Capon moved at a velocity of 245 μm/sec (Lombard and Capon, 1971). *Amphidinium carterae* in culture swims at speeds ranging from 14.1 to 123.3 μm/sec, with a mean speed of 75.1 μm/sec. Its movement was slowest along a straight path, and greatest along a helical path (Gittleson et

al., 1974). Rates of movement reported for other species include a mean velocity of 250 μm/sec (range 175 to 325 μm/sec) for *Gonyaulax polyedra,* and mean of 319 μm/sec (range 220 to 360 μm/sec) for *Gyrodinium* sp. (Hand et al., 1965). At such speeds, cells could migrate 10 to 15 m over a 12-hour day.

At the end of a red tide off Seal Beach, California, in November 1967, the main concentration of the *Ceratium furca* bloom migrated from a depth of about 2 m, down to 5 m by two hours past sunset; by four-and-a-half hours after sunset the concentration had dispersed between 5 and 16 m depth (Eppley et al., 1968). *Gonyaulax polyedra* and *Cachonina niei* migrate to the surface of an aquarium tank in the light, and downward during the dark period at a rate of 1 to 2 m per hour, a rate some 20 times as great as that resulting only from sinking. Diurnal migration allows the cells to obtain nitrogen from the deeper waters at night, as nitrogen would otherwise be depleted by the dense cell concentration in the upper water layers. The downward migration commences before dark, and the beginning of the next upward movement occurs before it is light. The migration in anticipation of light conditions suggests that the cell responds to gravity and an inner periodicity rather than to effects of the light and dark periods themselves (Eppley et al., 1968).

In addition to the flagellar movement and cell asymmetry that aid the cell in maintaining its position in the water column, buoyancy also may be controlled by the physiological state of the cells. The settling velocity of *Prorocentrum* and *Gymnodinium* is greatest at night from midnight to 1:00 A.M., and then decreases to the slowest settling rate from 8:00 to 11:00 A.M., when photosynthetic activity is strong. The rate of settling increased two to two-and-a-half times in aged cultures, or those deficient in phosphorus, indicating the strong effect of the physiological state of the cells upon their buoyancy (Akinina, 1969).

Phototaxis and Mechanism of Orientation

Movement that is a reaction to the stimulus of a light source is termed **phototaxis.** Much dinoflagellate movement appears to be related to the duration and intensity of light, from the normal rotation of the swimming cell, which equally exposes all plastids to the light, to the vertical diurnal migration in the water column or in interstitial waters of shore sands. The ocellus and eyespot of various species may serve as a photoreceptor, but others have no obvious pigmented organelle, although orientation to the light is important for photosynthesis.

Orientation of a cell to a light stimulus is thought to be accomplished by the differential illumination of a pair of light receptors located near the base of the flagella around a shading body. In experiments with cultures of *Gyrodinium dorsum,* a light stimulus signals the cell to cease the axial rotation, and then apparently to turn toward the most stimulated of the paired receptors until it is shaded; normal swimming then resumes. If two light sources are separated at an angle of 90°, the cell will turn to a pathway between the two, but if the sources are 180° apart, shading is not possible, and the cell will not reorient itself (Hand, 1970). A sudden localized light stimulus to *G. dorsum* causes the cell to cease forward movement for a few seconds, then to turn rapidly so that the anterior is directed toward the light. Forward movement then continues in the new direction. The change in direction is accomplished by the longitudinal flagellum, which in a single whiplike motion is brought to lie along the side of the cell facing the light stimulus, bending at the point where it projects from the sulcus, and remaining rigidly in this position for several seconds, depending upon the intensity of the light stimulus. It then begins to beat in a planar wave, which originates proximally. The cell is reoriented by this beat, and the flagellum then returns to its trailing position. The transverse flagellum appears not to beat during this process. Examination of cells photographed during the motionless period showed that nearly 90 percent stopped with the sulcus-girdle junction of the ventral side facing the light source, and 90 percent of these then accurately reoriented themselves to move toward the light source. Many dinoflagellates have eyespots associated with the sulcal region, and this region in *Gyrodinium* also appears to contain the light receptor, perhaps associated with one of the many plastids. As the cell rotates the receptors would be shaded by the plastids in the anterior portion. The sulcus of *Gyrodinium* is extended anteriorly from the girdle, leaving a channel through which light will enter if the cell deviates from a light-parallel path. The phototactic orientation thus results from positioning of the stopped cell with respect to the stimulus, followed by a definite flagellar response (Hand and Schmidt, 1975).

The mechanism of movement in dinoflagellates with flagella terminal rather than contained in a girdle and sulcus was indicated for *Prorocentrum minimum* by Parke and Ballantine (1957). During forward movement, the equivalent of the transverse flagellum projects forward to scribe a circular path over the cell apex, but the longitudinal flagellum is directed backward, where it vibrates rapidly (see Figure 4.68). To change direction, the longitudinal

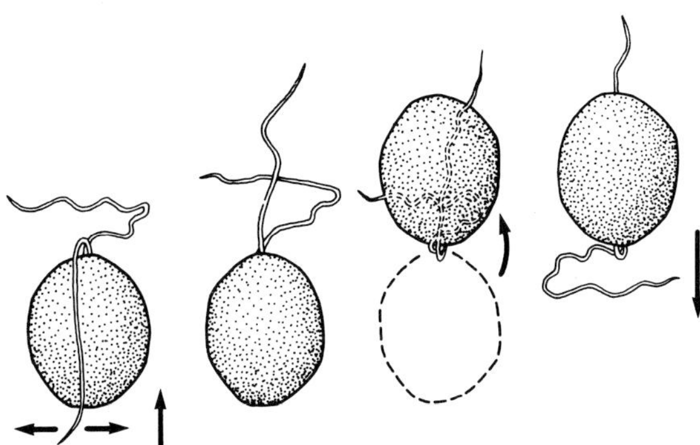

Figure 4.68
Movement in *Prorocentrum minimum.* Left to right: normal forward movement, with rapid vibration of backward-directed longitudinal flagellum; prior to change in direction, cell stops, and longitudinal flagellum sweeps forward; cell turns over onto the longitudinal flagellum; movement then proceeds in the opposite direction. Based on Parke and Ballantine, 1957, *J. Mar. Biol. Ass. U.K.,* v. 36, p. 643–650, published by Cambridge University Press, used with permission.

flagellum sweeps forward, and the cell slows its movement and stops. The body then turns to lie over the longitudinal flagellum to complete the change of direction.

Nutrition

Dinoflagellates vary greatly in their nutrition. Ultrastructural and biochemical evidence suggests that the dinoflagellates are a relatively primitive group, perhaps representing an offshoot prior to the development of the more typical eucaryotes. Their long evolutionary history, only partly represented by the known fossil record, has allowed a wide divergence in morphology, nutrition, and habitat.

Early difficulties involved in maintaining dinoflagellates in culture hindered the determination of the necessary requirements for their growth, as well as an understanding of their reproduction. Eventually, methods were developed for culturing dinoflagellates (Barker, 1935), and many later improvements in artificial and enriched seawater media have allowed the isolation and maintenance in culture collections of a variety of marine and freshwater species, including both photosynthetic and heterotrophic ones. Limiting factors in their growth are now relatively well understood (see summary in Loeblich III, 1967).

Autotrophs, Auxotrophs, and Myxotrophs

The typical dinoflagellate, most numerous in species, is the completely autotrophic biflagellate, photosynthetic cell, requiring no other preformed organic compounds. A common modification of this basic nutritional type is that of the auxotroph, which is photosynthetic but requires additional vitamins, such as B_{12}, thiamine, and biotin.

Some autotrophic dinoflagellates not only possess plastids, but also may ingest other organisms, hence are myxotrophic. The freshwater *Ceratium hirundinella* contains many chromoplasts, but its food vacuoles may contain remains of bacteria, bluegreen algae, and diatoms, all apparently undergoing digestion (Dodge and Crawford, 1970a). Ingestion is probably through a sulcal aperture that opens into a large ventral chamber.

Holozoic Dinoflagellates

Generally only the nonthecate dinoflagellates lack photosynthetic pigments and ingest their food in an animallike fashion. Such colorless dinoflagellates as *Oxyrrhis marina* lack any remnant of plastids or pyrenoids and do not have even the relic "proplastids" that are found in colorless euglenids, chlorophyceans, and diatoms to serve for starch storage. Some may live either on particulate food or saprobically (on decaying or dissolved organic matter). *Oxyrrhis marina* in culture utilized as food every one of 14 different microalgae offered, including examples of rhodophycean, chlorophycean, prasinophycean, chrysophycean, haptophycean, and bacillariophycean algae and yeast cells (Dodge and Crawford, 1971b, 1974).

Ingestion generally occurs in the region of the sulcus, through a pore or dispensable plate that allows extension of a pseudopodium, or by temporary separation of the plates to allow the prey to be engulfed.

In addition to the holozoic *Oxyrrhis marina, Polykrikos* has been observed with ingested *Ceratium* (Steidinger et al., 1967) and *Warnowia*, in addition to other particles (Kofoid and Swezy, 1921). After engulfing the prey, which may approach the size of the cell itself, the dinoflagellate rounds up, becoming shorter and broader to accommodate the increased mass, and may surround itself with an easily ruptured, thin-walled hyaline temporary digestive cyst. When digestion has been completed, the cell resumes its normal size and shape.

The tiny *Gymnodinium fungiforme* Anisimova attaches by its ventral face to the dorsal part of the girdle of much larger dinoflagellates, which it perforates in order to suck out some of the plastids from the prey. When filled, the *Gymnodinium* abandons the prey and rounds up during digestion; the prey appears unharmed and continues to swim. If a digestive cyst is produced by this species, it is extremely thin-walled (Biecheler, 1952). *Gyrodinium pavillardii* has apparently normal chromatophores, but also was observed to ingest the ciliate *Strombidium* after injecting it with numerous trichocysts. Strangely enough, the similarly predacious *Strombidium* did not eject its own trichocysts and was engulfed with little struggle, although it commonly preys upon dinoflagellates as well as other protists. The dinoflagellate opens in the region of the longitudinal sulcus to receive the prey, which may be 40 μm long and 25 μm in breadth, whereas the *Gyrodinium* itself is only about 50 μm long. Nevertheless, it has little difficulty in engulfing the almost equal-sized prey, the ensuing digestive process requiring some five to six hours. *Crypthecodinium* also has been observed to ingest prey.

A third method of predation is found in *Gyrodinium vorax* Biecheler, which develops a pseudopod that pierces the prey in order to obtain access to its protoplasm. *Peridinium gargantua* Biecheler combines these methods of feeding, as it first attaches by means of a pseudopod and then engulfs its prey (Biecheler, 1952).

Saprophytes

Some nonpigmented dinoflagellates are saprophytic, utilizing decaying organic matter. These include *Amphidinium pellucidum* Herdman and *Sinophysis ebriola* (Herdman) Balech, which live in the interstitial spaces in sands immediately overlying organic-rich black muds (Dragesco, 1965).

Parasites

Some dinoflagellates are external parasites on pteropods, copepods, or eggs of tunicates; internal intestinal parasites of siphonophores, annelids, or copepods; or coelomic parasites of copepods. Others may parasitize various protists, including the radiolarian and tintinnid protozoans and the algal prasinophyceans, diatoms, and even other dinoflagellates. Morphologically, the parasitic stage may be totally unlike the typical dinoflagellate, and some have in fact been assigned to other groups of organisms, including the rhizopod *Gromia* and a foraminifer (*Bargoniella* or *Salpicola* is in fact an *Oodinium*). The true relationships are indicated by the nature of the characteristically gymnodinioid motile stage.

Protoodinium, a parasite of jellyfish, retains many of the characters of a free-living form, including a tabulated theca with conspicuous girdle and sulcus, chloroplasts, and two flagella. It also has a funnel-shaped tentacle that penetrates the ectodermal cells of the scyphozoan and sucks the host cytoplasm, thus being a phagotrophic parasite rather than the more common osmotroph (Cachon and Cachon, 1971b). *Oodinium*, a parasite of appendicularian tunicates, annelids, and the acantharian protozoans, is osmotrophic. It has a peduncle composed of endoplasmic reticulum that extends from the cell posterior and attaches to the host cuticle by means of an adhesive terminal disk (Cachon and Cachon, 1971a). *Chytriodinium* is a similar parasite on the surface of floating copepod eggs, absorbing material from the egg by osmotrophy through a suckerlike stalk; feeding persists during sporogenesis by the dinoflagellate. *Myxodinium*, the related ectoparasite of the prasinophycean *Halosphaera*, appears as a gelatinous sphere at the surface until the algal cell is completely consumed (see Figure 4.69), after which the parasite undergoes division. *Myxodinium pipiens* has both micro- and macrospores, but although these may serve as gametes they also develop parthenogenetically and are eventually liberated by the bursting of the gelatinous wall (Cachon and Cachon, 1968b; Cachon et al., 1970a). *"Dissodinium" pseudocalani* Drebes (probably = *Sporodinium pseudocalani* Gönnert) is parasitic on copepod eggs, attaching by a suction organ until the egg contents have been consumed. It then forms a spherical cyst in which 8 to 32 ovate secondary cysts are produced by simultaneous division. From each of these cysts arise between 16 and 32 colorless gymnodinioid flagellate swarmers (Drebes, 1969).

Filodinium hovassei J. & M. Cachon, an ec-

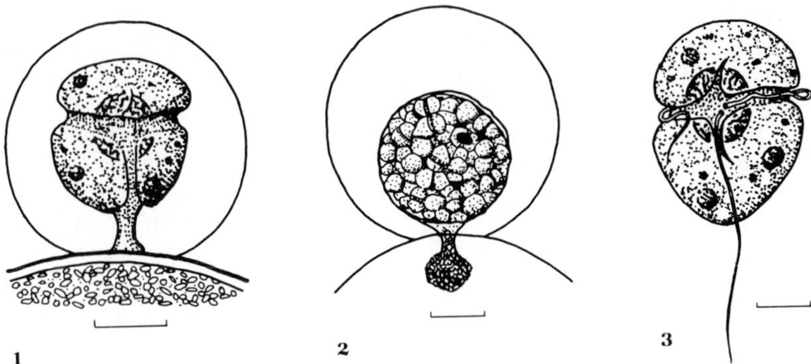

Figure 4.69

Myxodinium pipiens. A parasite on the prasinophyte *Halosphaera.* **1.** Juvenile settling on *Halosphaera* and penetrating surface with a suckerlike pseudopodium; bar = 10μm. **2.** After enlarging and consuming all host cell contents, new individual sporocysts develop within mucilage envelope; microspores and megaspores are released and may act as gametes; bar = 25 μm. **3.** Flagellated stage that reinfests new host cell; bar = 5 μm. Redrawn from Cachon et al., 1970a.

toparasite on the appendicularian tunicate *Oikopleura,* is *Gymnodinium*-like, but also has an extremely developed central pseudopodial system by which it attaches and crawls over the surface of the host. Young individuals have a thin conical posterior pseudopodium, and the larger adults have two pseudopodial networks that arise from the two ends of the sulcus. Its similarity to the naked *Gymnodinium* in contrast to the thecate *Protoodinium* and *Oodinium* suggests that not all parasitic dinoflagellates are as closely related as had been supposed (Cachon and Cachon, 1968a).

Erythropsidinium pavillardii attaches by a posterior tentacle that rhythmically elongates and retracts. A differentiated ingestion system, the stomopharyngian complex, consists of a "stomopod" that injects digestive juices into the prey and a canal that evacuates the material of the prey into an internal sac. A fibrillar system connects the tentacle, the two flagellar bases, and an anterior ocellus (Greuet, 1969). The cylindrical cell, 100 μm in length, has a girdle with a descending left spiral of one-and-a-half turns, the first whorl at the cell apex having a very small diameter. The girdle then spirals obliquely down the cell to a nearly longitudinal orientation on the ventral face. Early workers incorrectly oriented the cell with tentacle or prod at the anterior, but locomotion is with the prod posterior (Kofoid and Swezy, 1917).

Amoebophrya ceratii (Koeppen) Cachon is a parasitic dinoflagellate with both thecate and naked dinoflagellate hosts. *Gessnerium catenellum* chains from off Juan de Fuca Strait, Washington, included some exceptionally large cells, 70 to 80 μm in diameter, each of which contained a trophont or feeding stage of *Amoebophrya.* Nearly 40 percent of the individuals of this species were infested, although other species were not. The parasite forms a beehivelike structure, or cavity in the cell, and feeds on the nucleus of the host.

Atlanticellodinium tregouboffi Cachon & Cachon-Enjumet is an intracellular parasite of the Phaeodarian radiolarian *Planktonetta atlantica* Borgert. Very young specimens of the dinoflagellate, 5 to 6 μm in diameter, are found in the external protoplasm of the radiolarian rather than among ingested material in the inner part of the cell and probably actively infect their host. The roughly spherical young cells are uninucleate. With growth they become cylindrical and the nucleus divides repeatedly until the cell is 40 to 80 μm in length, with unequal regions corresponding to the smaller hyposome and larger episome of a free dinoflagellate. The parasite eventually produces numerous uninucleate, ovoid sporocytes, 12 to 15 μm in diameter, that are successively detached and released to infect a new host (Cachon and Cachon-Enjumet, 1965).

Other endocellular parasites include *Duboscquella tintinnicola* (Lohmann) Chatton on tintinnids, other *Duboscquella* on the dinoflagellate *Noctiluca* and various gymnodinians and peridinians, *Keppenodinium* on sphaerellarian radiolarians, *Amoebophrya* on tintinnids and radiolarians as well as dinoflagellates, *Paulsenella* on the diatom *Chaetoceros*, and *Trypanodinium* on copepod eggs. *Blastodinium* is an internal intestinal parasite of pelagic copepods and siphonophores, and *Syndinium* is a copepod coelomic parasite (Chatton, 1920; Cachon, 1964).

Host Symbionts

Still another mode of nutrition characterizes dinoflagellates that harbor photosynthetic symbionts within their cell. *Ornithocercus, Parahistioneis, Katodinium, Histioneis,* and *Citharistes* may contain blue-green algal symbionts, and the possible blue-green symbionts of *Amphisolenia,* formerly regarded as plastids, could fix both nitrogen and carbon for the host cell, thus being extremely valuable in such nitrogen-depleted regions as the Indian Ocean (F. J. R. Taylor, 1973b).

A strain of *Noctiluca miliaris* in New Guinea contains motile green flagellates in its vacuole, a single *Noctiluca* cell harboring from 6000 to 12,000 of the endosymbionts. During a month in culture with no additional food supply, the *Noctiluca* lived and divided normally in the light, the two daughter cells at each division dividing the symbionts. When kept in the dark, the flagellate symbiont died, as did the *Noctiluca* unless another green flagellate, *Dunaliella,* was supplied as food. The symbiont eventually was lost even from the cultures maintained in light (Sweeney, 1971).

Kofoidinium, a member of the Noctilucaceae, has a gymnodinioid early stage, and an eventually aberrant adult which normally ingests diatoms, silicoflagellates, and so on. However, *K. splendens* Cachon & Cachon may contain symbiotic xanthellae of yellowish green color whose systematic position is uncertain (Cachon and Cachon, 1967b). A host cell of 100 μm may contain two or three such symbionts, and larger cells may have up to a dozen. A few cells were found to lack xanthellae but to contain plastids that may have been retained from lysed xanthellae, although no other evidence could be found to support this.

The endosymbiont in other dinoflagellates has no separate cell wall, its organelles being separated from the dinoflagellate contents by a single membrane. *Peridinium foliaceum* and *P. balticum* were described as binucleate, with one characteristically mesocaryotic dinoflagellate nucleus, and one that was typically eucaryotic (Dodge, 1971a; Tomas et al., 1973). In culture, the mesocaryotic nucleus of *P. foliaceum* divided first, and the eucaryotic one later (Blanchard-Babillot, 1972), the resultant two to eight (commonly four) daughter cells obtaining one nucleus of each type. Mesocaryotic and eucaryotic nuclear division is synchronous in *P. balticum,* whose atypical plastids were unlike those of other dinoflagellates, and were thought to be closer to the Chrysophyta (exclusive of the Haptophyceae; Tomas and Cox, 1973), Bacillariophyceae, or Haptophyceae (Dodge, 1975a). Synchronous division of host and symbiont also occurs in other protists, such as in the thecamoebian *Paulinella* and its blue-green symbiont *Synecchococcus.* The membrane of the endosymbiont of *Peridinium foliaceum* encloses the eucaryotic nucleus and many plastids and mitochondria, and excludes the dinokaryon nucleus, starch granules, pusule, and trichocysts (see Figure 4.70). The many plastids, their three-thylakoid ultrastructure and four enclosing membranes, the presence of chlorophylls c_1 and c_2, as well as the absence of a cell wall, all suggest a chloromonad affinity for the endosymbiont (Fine and Loeblich III, 1976a) rather than a relationship to chrysophytes, haptophytes, cryptophytes, bacillariophytes, and phaeophytes, all of which had been previously suggested.

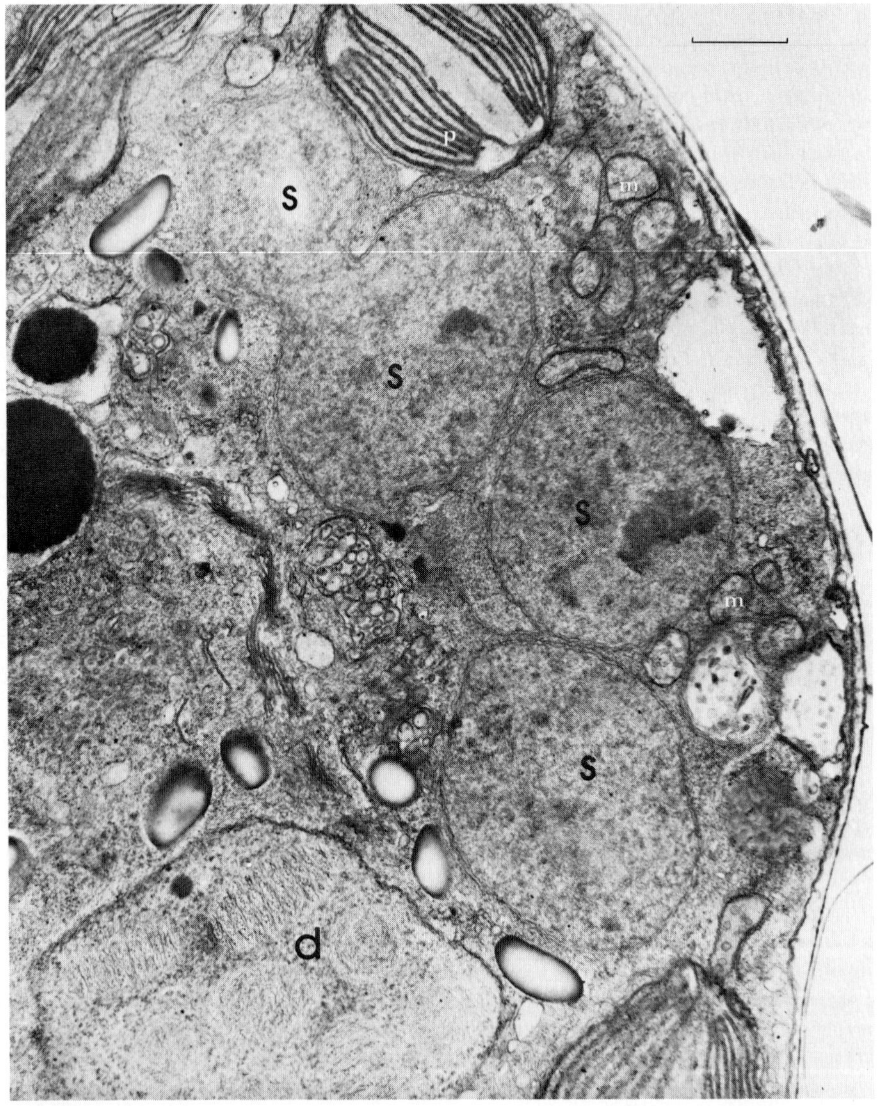

Figure 4.70
Peridinium foliaceum, containing dinokaryon (d) and eucaryotic nuclei, the four lobes of the latter (s) indicated in the section; eucaryotic nucleus, mitochondria (m), and plastids (p) all belong to the photosynthetic symbiont, of probable chloromonad affinity. TEM of sectioned glutaraldehyde-fixed cell, ×20,000. Bar = 1 μm. From Loeblich III, 1976.

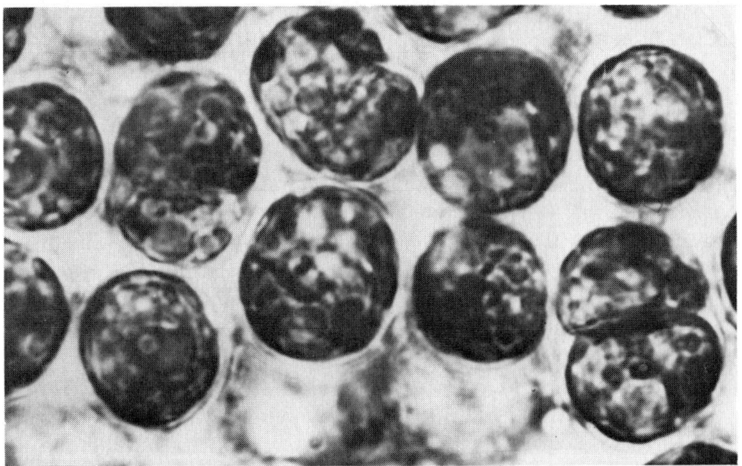

Figure 4.71
Zooxanthella microadriatica. Numerous cells, some in process of division, obtained as symbionts from tentacles of sea anemone *Anthopleura elegantissima,* San Juan Island, Washington, ×2100. From Loeblich III.

Endosymbionts

Symbiotic dinoflagellates, known as **zooxanthellae,** are the most widely occurring marine algal symbiont. Characteristic of nutrient-deficient waters, they are common in benthic and planktonic foraminifera, radiolaria, non-photosynthetic dinoflagellates such as *Noctiluca,* and in various coelenterate polyps. They similarly are common in benthic corals, anemones, and molluscs in nutrient-deficient tropical waters and reefs. In return for protection and the nitrogen, phosphorus, and CO_2 supplied by the waste products of the host (McLaughlin and Zahl, 1959), the symbiont provides oxygen and the photosynthetic products, chiefly in the form of glycerol (Fankboner, 1971), as a major or even sole nutrient source for the host.

Not all zooxanthellae are closely related. Earlier included with the parasitic dinoflagellates in the Blastodiniaceae, the photosynthetic symbionts have been separated as distinct species and genera placed in separate families and orders (see Figure 4.71). Some are reported to have a cellulosic wall or a thick amorphous layer regarded as homologous to the thecal layer, and five distinct membranes. The nature of the cell covering (Loeblich III, 1970, p. 899), presence of flagella in the symbiotic stage of some (D. L. Taylor, 1971b), vegetative reproduction by binary fission, and active photosynthetic metabolism as a symbiont are regarded as indications of a vegetative rather than encysted stage. Thus, although the motile cells of some symbionts do resemble *Gymnodinium* or *Amphidinium,* retention of the symbionts as a distinct group appears warranted on the basis of the distinct life cycle (Loeblich III, 1970; Hollande and Carré, 1974, p. 597), as is done for the Pyrocystaceae, Noctilucales, and Phytodiniales. The flagellated stage may be more distinctive than the symbiotic one, however, and life history studies may be necessary for identification. That *Symbiodinium* (correctly *Zooxanthella*) has a *Gymnodinium*-like motile stage (although now known to be thecate; Loeblich III and Sherley, 1979), and *Endodinium* an *Amphidinium*-like one, has even been regarded as evidence that the symbionts are not interrelated, only characterized by a similar habitat preference (D. L. Taylor, 1971a). *Zooxanthella microadriatica* is common in benthic hosts, including at least 13 different species of coelenterates and the large clam *Tridacna* (D. L. Taylor, 1969); it also is found in the pelagic Scyphozoan *Cassiopeia.* *Endodinium chattonii* occurs in such pelagic coelenterate hosts as *Velella* and *Porpita,* and other species are endosymbionts of the acoelomate *Amphiscoleps,* as well as the foraminifera and radiolarians (D. L. Taylor, 1973).

The symbionts are passed from parent to larval offspring by some hosts, but others require new infection by each generation. Corals lacking symbionts have been reinfected in the laboratory, the host readily accepting cells from symbiotic individuals of their own species, but rejecting those from other hosts. As motile zooxanthella cells entered the polyp gastrovascular cavity, they were engulfed by endoderm cells, and within two weeks were present in some species in numbers of 10,000 cells per millimeter of tentacle length (Muscatine et al., 1975). Others required a longer period for reestablishment of the symbiosis.

Symbionts may be readily isolated from their host and cultured separately, but the host organisms generally grow poorly without the symbiont unless other organic nutrients are supplied. Use of various tracers in coral-zooxanthella combinations has shown that 35 to 50 percent of the photosynthetic product excreted by the alga is passed on to the host, the dominant glycerol converted to lipids, and some glucose and alanine being converted by the coral to protein (Muscatine and Cernichiari, 1969). Similar transfer of nutrients has been documented for *Cassiopeia* (Balderston and Claus, 1969) and sea anemones (Muscatine and Hand, 1958). Experimental evidence from *Chlorohydra* indicates that algal photosynthesis always balances the respiration of both hydra and alga combined, even under conditions of reduced light. Similar results were obtained for the corals *Porites* and *Millepora* (Beyers, 1966).

Dinoflagellate symbionts occur in all tropical reef-building corals, their presence and photosynthetic activity affecting both the coral growth and its calcification. Deposition of calcium carbonate may be as much as nine times as rapid in the light, during photosynthesis by the symbiont, as in the dark (Goreau, 1959); if symbionts are lost, calcification not only is much slower, but also shows no variation from day to night. The nature of the interaction is uncertain; production of organic compounds by the symbiont may supply the coral with the energy necessary for calcification, removal of phosphates by the symbiont may enhance calcium deposition, or possibly the mere removal of carbon dioxide for photosynthesis by the symbiont may favor calcification (Pearse and Muscatine, 1971).

A drastic reduction in coral metabolism, as a result of sharply lowered salinity or available oxygen, unusually high temperatures, or starvation, generally results in the expulsion of the zooxanthellae, which degenerate as the host supply of nitrogen and phosphorus decreases. Extensive bleaching of shallow coral-reef communities in eastern Jamaica followed the severe rains during Hurricane Flora in October 1963. This loss of color resulted from the mass expulsion of their zooxanthellae by the hydrocoral *Millepora* and the hexacorallian Scleractinia, Zoanthidea, and Actiniaria in the shallower zones of the reef. In three days, the 550 mm of rain had resulted in a freshening of the surface waters to a depth of about 2.5 m, the lower osmotic pressure of the water causing physiological injury to the coelenterates, although mechanical destruction was minor. Different species reacted differently. The Gorgonaceans (octocorals) did not bleach but had a high mortality. Mildly bleached corals required six to eight weeks to regenerate the zooxanthellar populations from the few remaining. The polyps of the bleached individuals continued to expand and feed and many of the bleached colonies survived, although regeneration of the symbiont populations was slow, as the zooxanthellae apparently do not migrate from one part of the coral colony to another (Goreau, 1964).

The clam *Tridacna* contains abundant zooxanthellae in the tissue of the mantle edge, and the presence of occasional cells undergoing digestion in other parts of the clam led to the suggestion that they also provided a particulate food source. However, only old and degenerate cells are thus phagocytosed, and this seems to be a mechanism for eliminating inactive cells so that the host-symbiont balance is maintained. Coelenterates merely release such cells into the water, but the *Tridacna* utilizes them directly as food when they are no longer actively photosynthesizing (Fankboner, 1971).

Endodinium chattonii is a spherical cell of 12 to 14 μm diameter, occurring in the host *Velella velella*. The wall consists of four outer membranes in addition to the plasma membrane; a large dense yellow-brown peripherally located plastid has four pyrenoids suspended from it, each with two-thylakoid lamellae, as in typical free-living *Amphidinium*. The oval dinokaryon nucleus in the posterior end of the motile stage hypocone has between 250 and 400 chromosomes (D. L. Taylor, 1971a). Similarity of this motile stage to *Amphidinium* has been regarded as a basis for considering the symbiont congeneric and the symbiotic spherical stage as a cyst, although other evidence suggests separation of the symbionts as distinct genera of the Zooxanthellaceae.

The planktonic foraminifer *Globigerinoides sacculifer*, from the Gulf Stream near the Bahamas, contains symbiotic dinoflagellates that move into the test chambers in the evening, and in the morning move outward along the test spines to the most distal parts of the foraminiferan's ectoplasm. Three kinds of zooxanthellae were reported: a free-swimming *Gymnodinium*-like form found in the ectoplasm; some rare, large, spindle-shaped forms; and a common oval one with characteristic large mesokaryotic nucleus, peripheral plastids, and large stalked pyrenoids with prominent starch sheath. The ovate cells were regarded (Anderson and Bé, 1976) as most similar to the tiny marine *Aureodinium pigmentosum*, which has both a flagellate thecate stage and a nonmotile coccoid phase that lacks flagella and theca (Dodge, 1967). Hence, it is possible that all three types of zooxanthella in the planktonic foraminifer may represent only different phases of a single species.

The intracellular symbiotic dinoflagellates have a distinctive life cycle, with the nonmotile symbiotic stage being similar to an encysted state, whereas the motile stage is characteristically dinoflagellatelike (Freudenthal, 1962; Bibby and Dodge, 1972). Individual cells of *Zooxan-*

thella microadriatica in *Cassiopeia* are thin-walled but thecate, SEM indicating a tabulation of 1 pr (preapical), 5', 4a − 6a, 9 − 11'', approximately 20c (the cingular plates in two rows), 8 − 9s, 7 − 8''', 3'''' (Loeblich III and Sherley, 1979).

Numerous irregular to discoid plastids are peripherally located. In culture, cells divide frequently by binary fission. Some produce thick-walled cysts that germinate as flagellate zoospores, and others produce nonmotile phototrophic aplanospores. The flagellate stage is of short duration, the cells soon losing their flagella to become sessile. Occasionally, abundant spherical swarmers that had a faint girdle and trailing flagellum were produced. These swarmers may have been gametes, but as none were observed to fuse in the clonal cultures, possibly two separate clones may be required for pairing (Freudenthal, 1962).

A less formal symbiosis characterized another planktonic foraminiferan, *Hastigerina pelagica*, from the Florida Current in the Bahamas. This species has a gelatinous capsule surrounding the test and extending about one-half the length of the radial spines. From one to three different planktonic dinoflagellate species were found adhering to the capsule surface, in nearly constant numbers per unit area, on 80 percent of the 150 specimens collected and maintained in the laboratory. *Pyrocystis elegans* Pavillard was most frequent (as many as 79 individuals present on a single foraminiferan), and was observed to reproduce asexually every two to three days. The crescentic nonmotile parent cell divided to produce two motile swarmers that, upon release, quickly lost the flagella, swelled to the approximate size of the parent, and remained on the foraminiferal host. The flagellate stage could allow dispersal, but the frequency of the isolated dinoflagellate in the water was very low, suggesting that its presence on the foraminifer was highly concentrated. The nonmotile vegetative stage of the dinoflagellate could thereby maintain its position in the euphotic zone, while being protected from plankton grazers; it might obtain nutrients from the metabolic wastes or the capsule of the foraminifer, and the short-lived swarmer stage could be aided in dispersal. The *Hastigerina* apparently does not utilize the dinoflagellate as food, but may utilize its excreted photosynthetic product or oxygen. In addition to *Pyrocystis elegans,* the *Hastigerina* also harbored lesser numbers of *P. pseudonoctiluca* Thomson ex Murray, which was far more abundant in the water, and *P. fusiformis* Thomson ex Murray, rare both in the water and on the foraminiferan (Alldredge and Jones, 1973).

In contrast to the nonmotile intracellular symbionts, flagellated dinoflagellates also may be intercellular symbionts of marine acoelomate flatworms. The animals have a central core of digestive cells, a surrounding parenchyma in which the symbiotic algae form a conspicuous layer, a layer of muscle, and an outermost ciliated epithelium. The symbiont *Amphidinium klebsii* lies between cells of the host and shows constant orientation, with the cell apex pointed toward the central parenchyma and the broader hypocone with its large nucleus directed toward the exterior, perhaps to better orient the plastid with respect to the light. Both flagella are retained in the symbiotic state (D. L. Taylor, 1971b).

Bioluminescence

Bioluminescence is the production of cold light by an organism, a blue-green light resulting from the enzymatic reaction in which the enzyme luciferase oxidizes the substrate luciferin. All luminescent dinoflagellates are marine, but not all marine species are luminescent; some species even include both luminescent and nonluminescent strains. Nevertheless, dinoflagellates are responsible for most of the luminescence of surface waters of the sea, the intensity of the light being directly proportional to the quantity of dinoflagellates (see Figure 4.72). Apparently the number of photons emitted per cell is constant, as was shown (Seliger and McElroy, 1968) for *Pyrodinium bahamense,* Spectacular displays of bioluminescence, resulting from blooms of this species in concentrations of up to 10 million cells per liter, have given the name to various tropical bays, such as Fire Lake near Nassau in the Bahamas, and

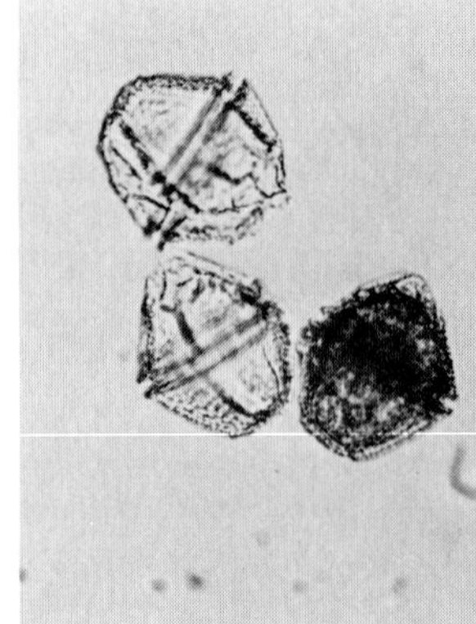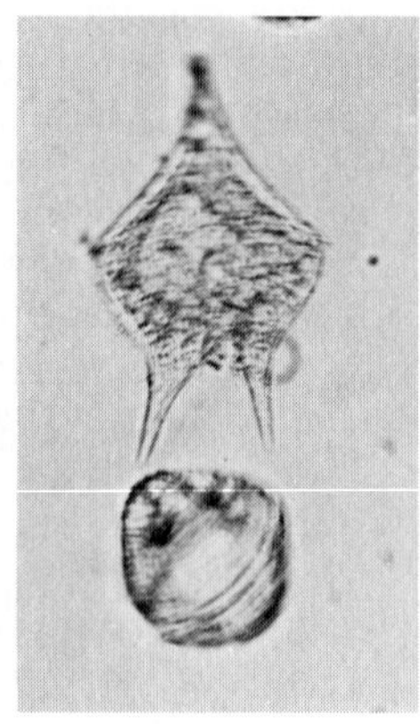

Figure 4.72
Some bioluminescent species of the Southern California coast. Left, *Gonyaulax polyedra,* ×340; upper right, *Protoperidinium oceanicum,* ×150.

Bahia Fosforescente in Puerto Rico. Similar effects have been reported from Falmouth Harbor, Oyster Bay, Jamaica (Seliger et al., 1970). In part these concentrations result from a marked density stratification of the water, a surface layer of reduced salinity keeping the phytoplankton concentrated at about 18 to 20 cm depth.

In other regions, *Noctiluca miliaris* is an important neritic luminescent species, as indicated in the quotation heading this chapter. A smaller (200 to 450 μm diameter instead of the usual 400 to 850 μm) strain of this species is present on the American west coast of California and Washington (Eckert and Findlay, 1962).

Pyrocystis is the characteristic bioluminescent dinoflagellate of the equatorial regions (noted by John Murray on the Challenger Expedition). *Pyrocystis* emits many more photons per amount of nitrogen in the cell (generally 7 to 50 times more) than does *Pyrodinium bahamense* or the California coastal species *Gonyaulax polyedra.* The two large species *Pyrocystis pseudonoctiluca* and *P. fusiformis* emit, respectively, 37 to 89 × 10^9 photons per cell and 23 to 62 × 10^9 photons per cell, both up to 1000 times more than does *Gonyaulax. Pyrocystis acuta* Kofoid emits 3 to 6 × 10^9 photons per cell, and *P. lunula* from 5 to 13 × 10^9 photons per cell (Swift et al., 1973).

Bioluminescent species include probably all *Gonyaulax*, such as *G. catenata* (Levander) Kofoid, *G. digitalis, G. hyalina* Ostenfeld & Schmidt, *G. polygramma, G. polyedra, G. sphaeroidea* Kofoid, and *G. spinifera,* but only some species of *Ceratium* and *Protoperidinium;* for example, *C. fusus* (some clones), and *P. brochii* (Kofoid & Swezy) Balech, *P. conicum, P. depressum, P. divergens* (Ehrenberg) Balech, *P. globulus* (Stein) Balech, *P. granii, P. leonis* (Pavillard) Balech, *P. oceanicum* (Vanhöffen) Balech, *P. pallidum* (Ostenfeld) Balech, *P. pentagonum* (Gran) Balech, *P. punctulatum* (Paulsen) Balech, *P. steinii,* and *P. subinerme* (Paulsen) Loeblich III. Other luminescent species include *Gessnerium monilatum* (Howell) Loeblich III, *G. catenellum, Gymnodinium flavum* Kofoid & Swezy, *Noctiluca miliaris* and *N. scintillans* (Macartney) Ehrenberg, *Polykrikos schwartzii, Pyrocystis lunula* and *P. pseudonoctiluca,* and *Pyrodinium bahamense* (summaries in Kelly, 1968; Sweeney, 1963; Tett, 1971).

Luminescence in the photosynthetic dinoflagellates is of two types, a flashing light and a dim glow. Flashing requires an external mechanical or electrical stimulus and the flash lasts about 0.1 second (Hastings et al., 1956). The dim glow occurs toward the end of the dark period, may last a few hours, and may be accompanied by flashing. After a period in the dark, an undisturbed sample of seawater containing luminescent dinoflagellates will spontaneously emit luminescence, or it may be stimulated mechanically. The number of flashes in 15 minutes per number of dinoflagellate cells in a liter of water ranges from 0.49 to 1.30 at the Isle of Cumbrae off Scotland, from 0.05 to 1.49 in Plymouth Sound, and from 0.04 to 1.08 at Millport, England. This variation from about 0.04 to 1.5 flashes per cell in 15 minutes shows a range of 50 times the lowest value to the highest (Tett, 1971).

In *Noctiluca,* apparently synchronous flashes arise from many microscopic organelle sources (microflashes) to produce the total luminescent flash (macroflash), which is propagated in the peripheral cytoplasm at a rate of 60 μm/msec. A cell of *Noctiluca* has about 2.5×10^5 cell inclusions of 0.5 to 1.5 μm diameter, of which about 5 percent emit light during a flash. The intensity is proportional to the particle size, and the total is about 10,000 microflashes per cell. About 5×10^5 photons are emitted per microflash, and 5×10^9 per macroflash. As the same proportion of cell inclusions also was stimulated by ultraviolet light to fluoresce blue-green, the same particles are thought to show both fluorescence and luminescence (Eckert, 1966).

Toxins

About 20 living dinoflagellates are known to produce toxins that poison various invertebrates (mollusks, annelids, sipunculids, holothurians, starfish, crabs) and cause mass mortality of fish. All highly toxic species are marine; they include species of *Amphidinium, Cochlodinium, Gessnerium, Gymnodinium, Noctiluca, Polykrikos, Prorocentrum,* and *Pyrodinium* (Russell, 1965; Worth et al., 1975). Toxic species may occur in large numbers during seasonal blooms known as red tides. Exceptionally poisonous blooms include those of *Gessnerium tamarense* (see Figure 4.73) on the Atlantic Coast in Eastern Canada, New England, and in northeastern England and Belgium (L. Loeblich and Loeblich III, 1975), *G. catenellum* on the Pacific Coast from northern California to Alaska, *G. monilatum* and *Gymnodinium breve* Davis in the Gulf of Mexico, and *Prorocentrum* sp. in Japan.

The dinoflagellate toxin that is concentrated in the shellfish is thermostable and survives cooking and canning (Ballantine and Abbott, 1957). The toxin of *Gessnerium catenellum,* termed saxitoxin, is a nitrogenous base, $C_{10}H_{15}N_7O_3 \cdot 2HCl$ (Wong et al., 1971), that interferes with the movement of sodium ions through excitable membranes. Highly water soluble, it is 100,000 times more active than cocaine, and 50 times more potent than curare. It accumulates in live mussels, clams, scallops, and oysters, may reach high levels in three to four days, and may be stored up to two years in the Alaskan butter clam (Steidinger, 1973). Symptoms of the paralytic shellfish poisoning resulting from human consumption of poisoned shellfish progress from peripheral

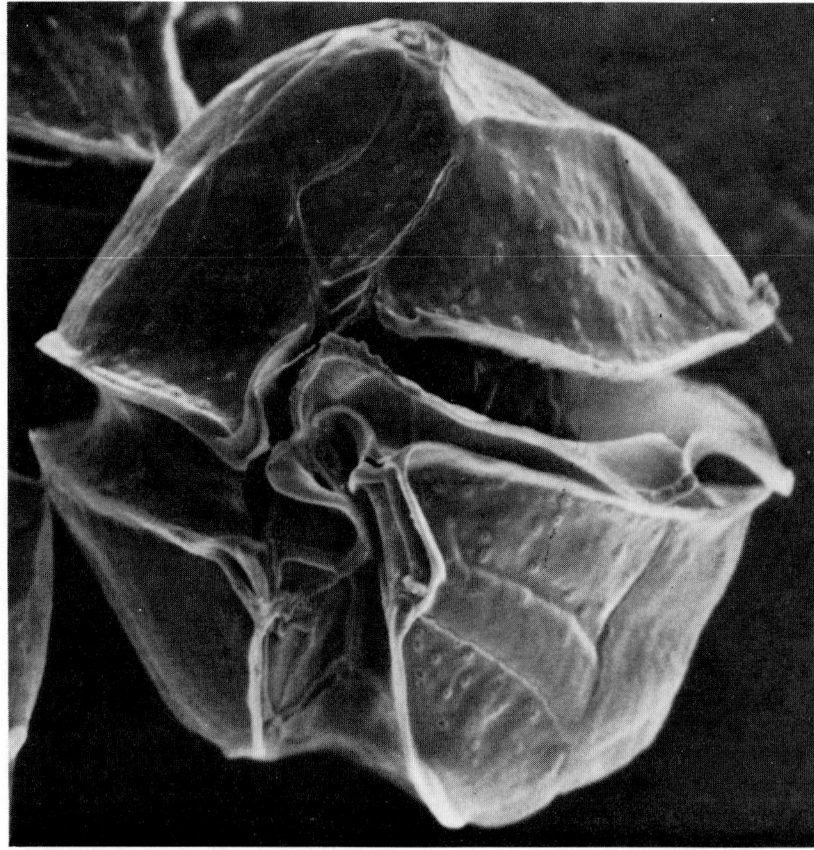

Figure 4.73
Gessnerium tamarense. Toxic dinoflagellate from the 1972 red tide at Glouces-
ter, Massachusetts. SEM of ventral side of glutaraldehyde-fixed and critical-
point-dried cell; note that both flagellar pores are visible in keyhole-shaped
upper part of sulcus at junction with cingulum. Drying has caused left side of
cell to separate between precingular plates and girdle. ×3800. From L. Loeblich and Loeblich III, 1975.

paralysis and numbness about the lips to loss of
strength in the muscles of the neck and extremities and death by respiratory failure. The
fatality rate is from 1 to 10 percent (Russell,
1965, p. 304). Some 250 human deaths and over
1600 instances of paralytic shellfish poisoning
have been reported worldwide, related to toxic
dinoflagellate blooms. Saxitoxinlike poisons also
are produced by *G. acatenellum* (Whedon &
Kofoid) Loeblich III & L. A. Loeblich from the
Pacific Ocean and *G. tamarense* on the Atlantic
coast, the latter having two separate toxins and
being even more virulent.

Toxin produced by *Gymnodinium breve,* the
Florida red tide species, is of low molecular
weight and alkali-labile, and quite unlike saxitoxin (Steidinger, 1973). It contains phosphorus
($C_{90}H_{162}O_{17}P$) and is soluble in lipid solvents and
lower alcohols. Three active substances, two
neurotoxins and one hemolytic fraction, all less

potent than saxitoxin, mostly affect those animals with a specialized central nervous system. Fish death has been attributed to respiratory failure; no human deaths have been reported, although tingling in extremities, ataxia, slowed pulse, and pupil dilation are among symptoms of poisoning from consumption of toxic bivalves. Toxic particles of lysed cells carried in the air in seaspray also may cause irritation to the upper respiratory system and eyes (Steidinger and Joyce, 1973).

Gessnerium monilatum was found to have its maximum toxicity during the declining phase of blooms rather than at the peak abundance, suggesting that the toxin is cell-bound and released by autolysis when the cells are dying. Fish death associated with *G. monilatum* blooms was shown to be correlated only with the toxin production and not related to any decrease in available oxygen (Aldrich et al., 1967).

Zooxanthellae of some gorgonian corals produce toxins that are unlike those of other dinoflagellates and may be protective for the coral host.

Reproduction

Vegetative Reproduction

Vegetative cell division in the Pyrrhophyta may be by simple fission in unarmored species, by shedding the theca (**ecdysis**) and dividing, or by a diagonal splitting of the theca itself, each daughter cell taking part of the parent theca. Vegetative division is the most frequent method of reproduction and once was thought to be the only method, as until recently sexual reproduction was not known to occur.

Under natural conditions, many photosynthetic dinoflagellates undergo division only at night, but at various periods. In culture, with controlled light-dark cycles, *Gymnodinium splendens* divided mostly at the beginning of a dark cycle, *Gonyaulax sphaeroidea* at the end of a dark cycle, *Gonyaulax polyedra* at the beginning of a light cycle, and *Prorocentrum micans* at the middle of the light cycle (Sweeney and Hastings, 1962).

Rates of Division.   Division rates are high for most microorganisms, but those of dinoflagellates average somewhat less than other phytoplankton, probably because of their commonly larger size. The highest rates are generally those for unarmored species such as *Amphidinium* and *Oxyrrhis*. The fastest reported reproductive rate is once every eight hours or less, routinely obtained in culture for *Crypthecodinium cohnii*, a heterotrophic, nonphotosynthetic species (Tuttle and Loeblich III, 1975, p. 7), suggesting that the photosynthetic process may in some way be rate-limiting for cell division in other species. *Peridinium triquetrum* and *Prorocentrum* may divide twice or more in 24 hours, and *Gonyaulax* and *Gymnodinium* average 1.75 divisions during that period. The larger *Ceratium* spp. have a division rate of 0.29 to 0.54 and *Dinophysis* averages 0.28 divisions in the duration of a day (see summary in Loeblich III, 1967).

Simple Binary or Multiple Fission.   The most common method of reproduction in the dinoflagellates is by simple binary fission of the vegetative cells, one cell splitting into two. In its simplest form, this can only take place in the unarmored species, such as *Amphidinium, Gymnodinium,* and *Oxyrrhis,* where the naked cell pinches apart while continuously synthesizing the amphiesma as the daughter cells separate. The new cells of most unarmored species separate immediately, but some remain attached for a time; newly divided daughter cells of *Gymnodinium pseudopalustre* Schiller remain connected as post-divisional pairs for at least 12 hours (Stosch, 1973). Some daughter cells may remain temporarily associated in chains, as in *Cochlodinium geminatum* (Schütt) Schütt, or even permanently associated in short colonies (*Polykrikos*), which generally have fewer nuclei than individuals.

Although binary fission is characteristic of most nonthecate dinoflagellates, others undergo multiple fission. *Woloszynskia apiculata* does not divide as motile cells, but forms a "division cyst" by rounding off and forming a thin cellulosic wall inside the usual reticulated, noncellulosic theca. The theca itself

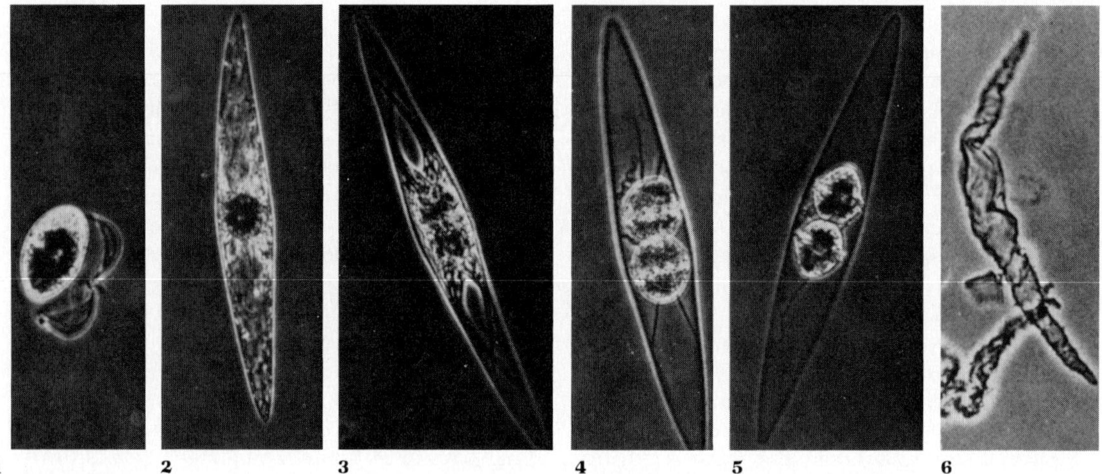

Figure 4.74

Pyrocystis acuta. Asexual reproduction. From left to right: protoplast leaving theca by ecdysis, although still temporarily attached by means of the flagella, ×320; protoplast greatly elongates, becoming fusiform, ×220; vegetative cell divides within the fusiform wall, ×260; daughter cells rounded and produce flagella, ×390; cell with two swarmers just prior to their release, ×220; abandoned fusiform cell wall, ×260. From Swift and Wall, 1972.

serves as a perispore, enveloping the cyst, and in turn is surrounded by mucilage. The division cyst lies on the substrate. One to three consecutive nuclear divisions are followed by cytoplasmic cleavage, producing two, four, or eight daughter cells, each developing a sulcus toward the outside of the original cell just beneath the wall. The new daughter cells are released by dissolution of the cellulosic wall and bursting of the perispore within some four to six hours after the cell originally becomes immobile and settles to the bottom. This vegetative division results in a great diversity of cell size (Stosch, 1973).

Amphidinium carterae also may produce more than two daughter cells by fission. A larger than average-sized cell, retaining a recognizable polarity during the process, divides longitudinally and at right angles into four, or occasionally six to eight, daughter cells that form a grapelike group before eventually separating (Vien, 1968).

Ecdysis. Dinoflagellates that have a cell wall may undergo cell division by shedding the wall and cell membranes, either before or after cell division. In cultures, globose cells of *Pyrocystis lunula* divide internally to produce 8, 16, or 32 crescentic cells (see Figures 4.74 and 4.75). After these are released, each crescentic cell produces one to two, rarely four, gymnodinioid flagellate swarmers, which are in turn released, lose the flagella, and enlarge rapidly to form a new crescentic cell. Within about 10 minutes volume increase is 22 times in *P. fusiformis* swarmer cells.

The motile cell of *Helgolandinium subglobosum* von Stosch sheds the theca and produces a thin-walled division cyst in which two daughter cells are formed (Stosch, 1969b). Similar ecdysis prior to cell division has been reported in other Peridiniales (*Crypthecodinium, Dissodium, Peridinium,* and *Protoperidinium*) and in some species of *Gonyaulax* and *Heteraulacus* (summarized by Loeblich III, 1970). In this method, none of the parental theca is transferred to the daughter cells and all membranes and thecal plates are newly formed.

Although ecdysis thus may occur prior to cell division and be related to the process of reproduction, it may also occur as a response to adverse conditions such as crowding (in culture or

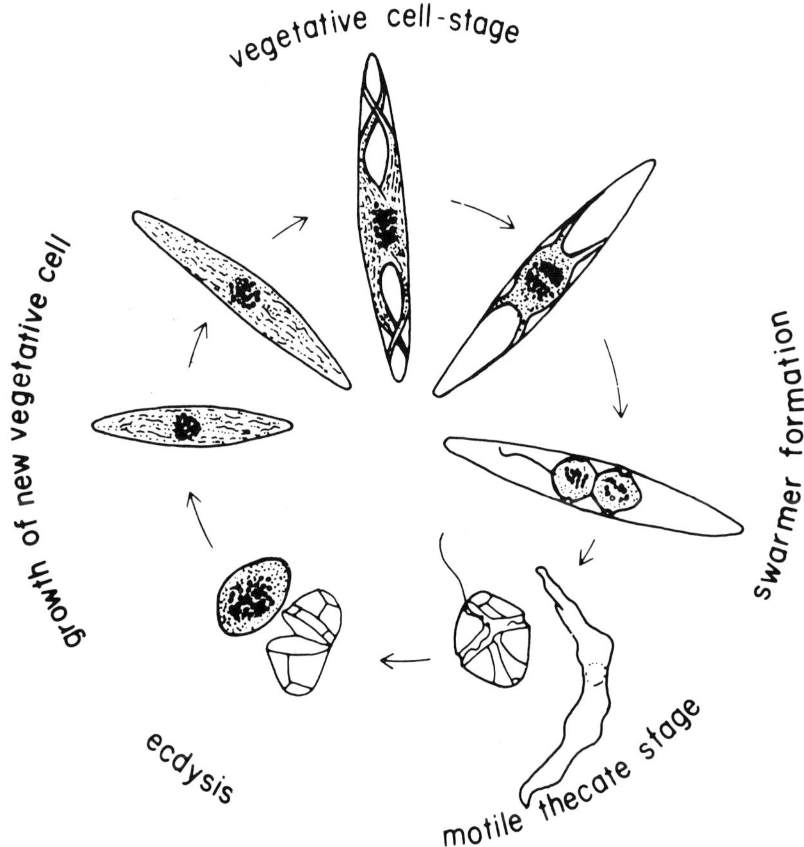

Figure 4.75
Life cycle of *Pyrocystis acuta*. Motile thecate stage undergoes ecdysis to produce fusiform vegetative stage, which later produces new thecate swarmers. From Swift and Wall, 1972.

during blooms in nature), abnormally high temperatures, unusually intense lighting, or abnormal salinity. Ecdysis may be induced in cultures by a change in the medium or by the presence of light during the usual dark phase. The shedding of the theca may be a reaction to concentration of metabolic products within the cell (Kofoid, 1908) or to other physiologic shocks. The resultant naked cell may regenerate a theca, within two or three days for *Gonyaulax polyedra*, or may undergo cell division by simple constriction of the protoplast, although the new daughter cells of this species usually divide the theca of the mother cell between them. The surface of the naked protoplast has ridges in the position of the former plate boundaries, indicating that the inner plate membrane, or innermost part of the vesicles, is not shed, but that the plasma membrane, plates, and outer plate membrane of the vesicle are lost in ecdysis (Dürr and Netzel, 1974).

The cell contents are released in ecdysis through the opening of specific plates. *Cachonina niei* commonly undergoes ecdysis in culture, releasing the cell through its anterior intercalary region so that entire empty epithecae are rarely observed (Loeblich III, 1968, 1969). Freshwater species of *Peridinium*, such as *P. bipes, P. volzii* Lemmermann, and *P. willei* Huitfeld-Kaas, left under unfavorable conditions before fixation, shed their 7 dorsal epithecal plates, allowing the protoplast to escape, and *P. gatunense* Nygaard sheds the entire epitheca (Boltovskoy, 1973b). *Gonyaulax*

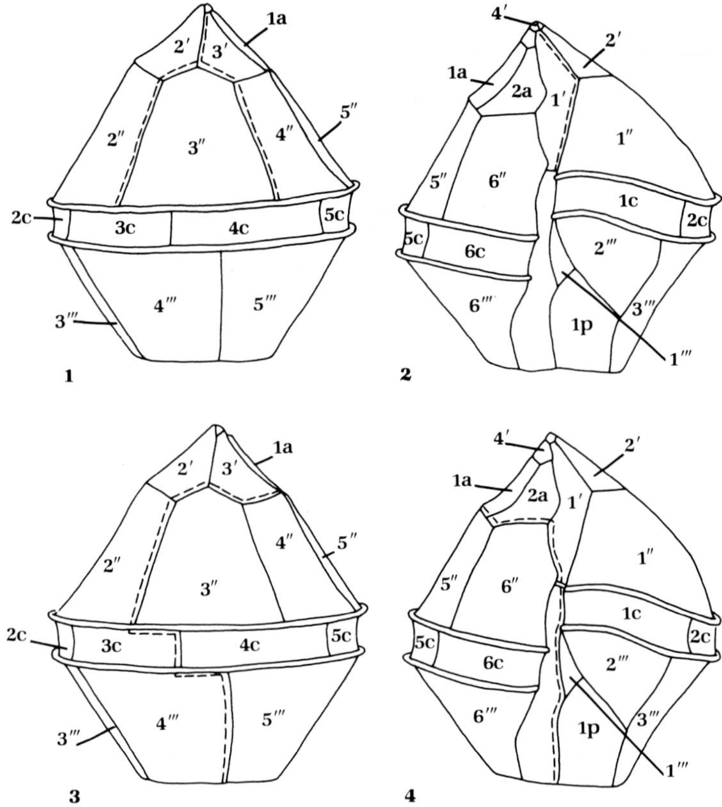

Figure 4.76

Gonyaulax polyedra. Fission lines in ecdysis and in cell division. **1,2.** Dorsal and ventral views respectively, the dashed line showing line of separation of the thecal plates during ecdysis. Plate 3″ opens first, as its margins overlap adjacent plates, but this is not an archeopyle (which is an opening in the resting cyst formed at germination). **3,4.** Dorsal and ventral views respectively of cell, showing line (dashed) of thecal separation at vegetative binary fission; the theca splits into anterosinistral and posterodextral parts. Separation in sulcus not observed, hence hypothetical. Redrawn from Dürr and Netzel, 1974.

polyedra first opens the dorsal precingular plate 3″ (see Figure 4.76), which overlaps all adjacent plates at the margins (see discussion of tabulation of the Peridiniales, pp. 261−265). The third precingular separates like a door hinged at the boundary with the girdle, followed by a separation of plates 2′, 1″, and 2″ from plates 1′ and 3′; the protoplast escapes with a jerk from the resulting fissure, with sufficient force at times to dislodge the entire left half of the epitheca (Dürr and Netzel, 1974).

The rounded cell resulting from ecdysis has been erroneously regarded as a cyst by some, but the method of development, absence of a resistant wall, and temporary nature of the nonthecate stage differ from the true process of encystment. Unlike the thick-walled resistant resting zygote, fossilization of a naked protoplast following ecdysis is most unlikely, and it should not be termed a cyst. Furthermore, the thecal opening that accompanies ecdysis is not correlative with the archeopyle of the resting

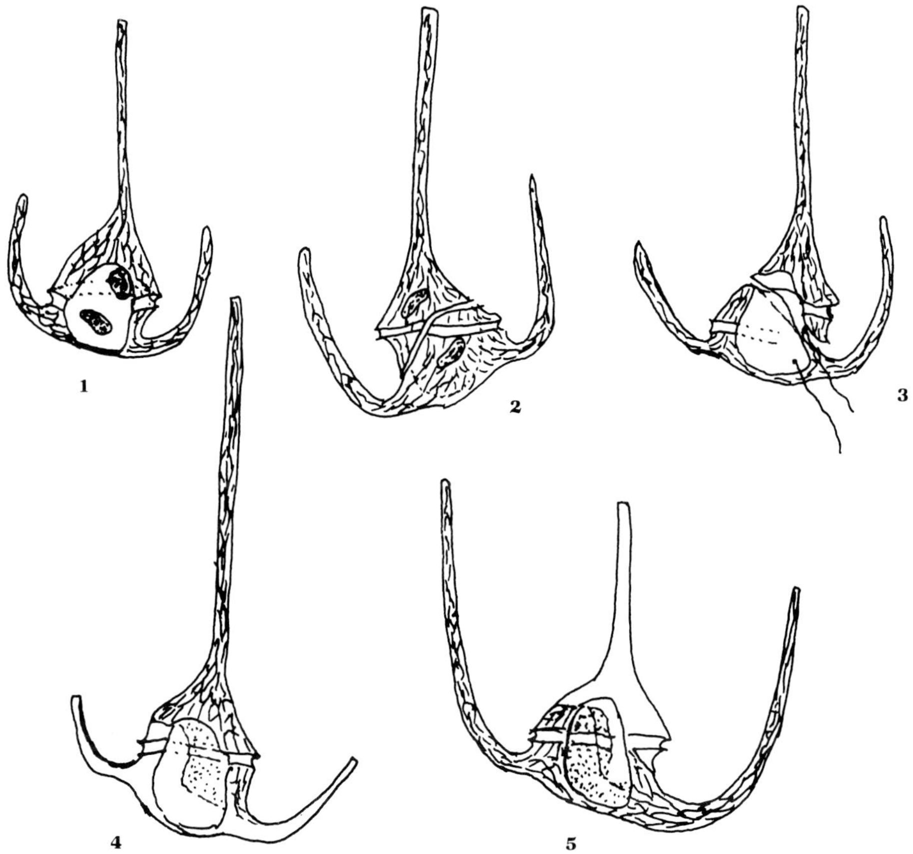

Figure 4.77
Thecal partition during cell division in *Ceratium tripos*. 1,3–5, ventral views; 2, dorsal view. **1.** Nuclear division completed. **2.** Diagonal suture beginning for separation of theca. **3.** Separation of cytoplasm, each daughter cell with separate flagellum. **4.** Daughter cell has inherited strongly ornamented anterosinistral moiety of mother cell theca and is regenerating posterior part. **5.** Daughter cell has inherited ornamented posterodextral half of mother cell theca and is regenerating anterior part. Flagella shown only in part 3. Redrawn from Bergh, 1887.

cyst that accompanies its germination and should not be so described as by Boltovskoy (1973a). As indicated above, the ecdysial opening of *Gonyaulax polyedra* consists of the hinged opening of plate 3'', with others merely separating. The cyst archeopyle of this living species represents paraplates 2'', 3'', and 4'', each forming separate free pieces and never remaining attached to the cyst after opening (Evitt, 1967a).

Thecal Partition. This third method of vegetative cell division (observed and illustrated for the Peridiniales by Bergh, 1881, 1887; Stein, 1883; Schütt, 1887; Pouchet, 1893; and many later workers) consists of a diagonal fissure splitting the theca of the mother cell along plate sutures into anterosinistral and posterodextral halves (see Figure 4.77). Vegetative reproduction involving partition of the parental theca also characterizes the Prorocentrales (*Prorocen-*

trum) and Dinophysiales (*Amphisolenia, Dinophysis, Ornithocercus,* and *Oxyphysis*), with division separating the two halves along a median line (Pouchet, 1893). Each of the two daughter cells retains one half of the parental theca and forms the missing part anew.

Recently divided cells may be recognized by the much thicker and more strongly ornamented inherited thecal part, which contrasts in appearance with the thinner, less ornamented newly synthesized part (Kofoid, 1911; Loeblich III, 1970). This newly produced portion of the theca of *Gonyaulax polygramma* was termed an archeopyle by Ricard (1974, p. 129), as it represented a part lost during reproduction, but (as in the previously mentioned ecdysis) this is an incorrect application of a term that was defined to indicate the opening in a resting cyst by which the protoplast escaped at germination or excystment, rather than an opening for any purpose of the theca itself.

Diagonal separation characterizes most Peridiniales in which ecdysis does not precede cell division (including *Cachonina, Ceratium, Ceratocorys, Pyrodinium, Spiraulaxina,* and some *Gonyaulax* and *Heteraulacus;* summary in Loeblich III, 1970). Because some species (of *Gonyaulax, Heteraulacus*) reproduce vegetatively by ecdysis and others by thecal partition, such genera might possibly require subdivision; however, at least one species, *Gonyaulax polyedra,* may reproduce either by thecal partition or by simple cell fission after ecdysis. Cell division with thecal partition occurs while the *G. polyedra* cell rotates and swims, each daughter cell taking half the parent theca. One takes the anterosinistral portion, including all apical plates, anterior intercalaries, plates 1″ and 2″, 1c to 3c, 1‴ to 4‴, and the posterior intercalary. The posterodextral daughter cell inherits plates 3″ to 6″, 4c to 6c, 5‴ to 6‴, and probably the antapical plate (Dürr and Netzel, 1974). Because in this species new cells rarely show a marked variation in degree of sculpture of the new and inherited portions of the theca, the new half may rapidly increase the wall thickness and ornamentation, although possibly the old portion may be shed after a new theca forms, or a completely new theca may be synthesized after the new cells undergo ecdysis (Dürr and Netzel, 1974).

During its rapid increase by vegetative division, the freshwater *Peridinium aciculiferum* Lemmermann produces temporary thin-walled gelatinous division cysts in which fission occurs (West, 1909). Cell division in *Pyrodinium bahamense* is most frequent just after sunrise, about 30 percent of the cells dividing within two hours; the entire process from a single cell to complete separation of the daughter cells occupies about 105 minutes (Buchanan, 1968).

The process of cell division was studied by electron microscopy of sectioned cells of *Ceratium tripos* at various stages of separation and thecal formation (Wetherbee, 1975b). During the last two hours of the dark period, nuclear division commences, with cytokinesis beginning later. Binary fission occurs along predetermined sutures, and each daughter cell retains half of the parental theca, beginning formation of the other half as the cytoplasm separates. Thecal development is well underway before cleavage is completed. The four membranes of the future cell covering are present around the developing cleavage furrow, which first differentiates just below the sutures in the region of separation. Separation occurs gradually as the cleavage furrow works its way around the cell circumference. Microtubules forming a layer just beneath the inner plate membrane are active in both cleavage and plate development. A plasma membrane is outermost in the developing furrow and eventually connects with the outermost plasma membrane that envelops the cell as the plates are pushed apart. Thus, the two membranes that bounded the adjacent thecal vesicles at the suture combine with the plasma membrane to become four just prior to the final split. Near the end of the process of cell division, only the outermost plasma membrane encloses both parts of the dividing cell while the new plates are produced beneath it. Sutures are already present in *Ceratium tripos* during the initial differentiation of the cell covering before any plates are formed. After separation of the two daughter cells, the sutures that were involved in the fission are again formed and active synthesis of the thecal plates commences. The microtubule layer, which had been conspicuous during cleavage and early development of cell shape, disperses into groups of one to four

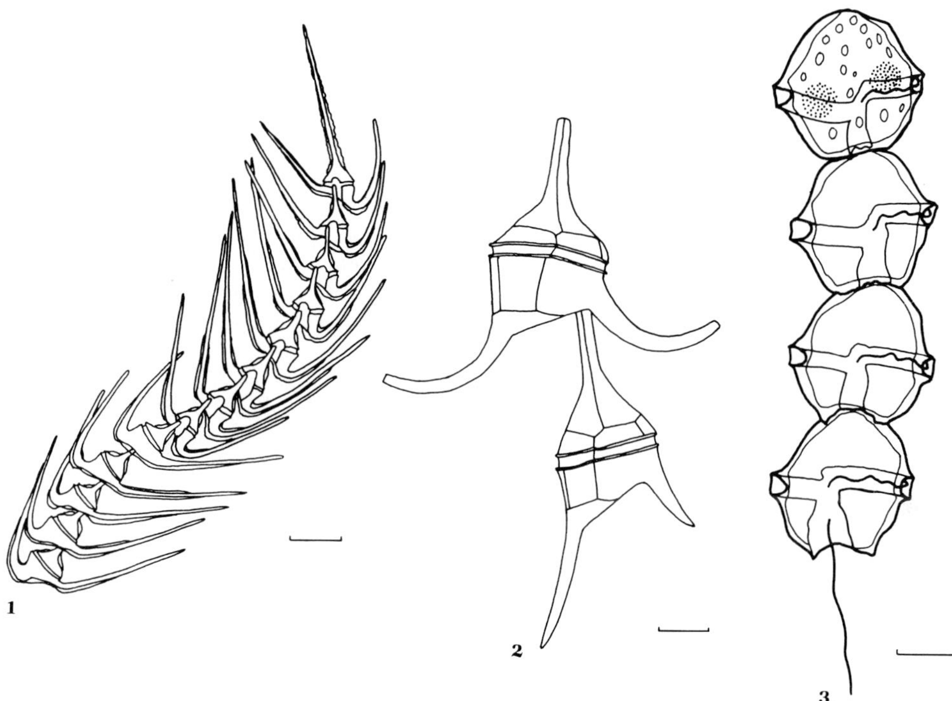

Figure 4.78
Chain formation. **1.** *Ceratium vultur* Cleve, chain twisted so that posteriormost five cells are seen from the dorsal side, the others from the ventral side; bar = 100 μm; redrawn from Kofoid, 1909a. **2.** Chain of two cells, anterior one representing *Ceratium massiliense*, and posterior, *C. macroceros;* bar = 20 μm; redrawn from Kofoid, 1909a. **3.** Chain of *Gessnerium fraterculum* (Balech) Loeblich III & L. A. Loeblich; bar = 20 μm; redrawn from Balech, 1964.

tubules in the maturing theca (Wetherbee, 1975a, 1975b).

Chain Formation. Commonly the daughter cells resulting from fission separate immediately (*Gonyaulax spinifera,* many *Ceratium, Ceratocorys,* most Dinophysiaceae), but the new cells of other species may remain attached temporarily. Attachment may be by a dorsal connection (*Dinophysis caudata* Saville-Kent, *Ornithocercus*), or more persistent chains may be produced by a cytoplasmic connection that is pulled out from the apical pore of the posterior cell and remains attached to the distal end of the girdle of the anterior one (*Gonyaulax catenata,* and many *Ceratium;* see Figure 4.78).

Chains of cells are frequent in *Ceratium pavillardii* Jörgensen, the end of the apical horn bearing a small platelet with a large central pore. In all members of a chain other than the anteriormost, the apical platelet is thickened to form a knobby process, which fits into a "socket" formed of the fourth girdle plate at the top of the sulcus of the next anterior individual. Isolated cells of *C. pavillardii* that are not in chains lack both the knobby apical horn and the socket at the end of the horn trough, and their apical horns are more elongated (Graham, 1942).

Chain-forming or **catenate** dinoflagellates such as *Ceratium* and *Glenodinium* were illustrated by Pouchet (1883, 1893), the separate cells of a chain involving as many as 20 to 30 very similar individuals. Some *Ceratium* produce chains during division at night and remain connected between 3:00 and 7:00 A.M., but separate as individual cells later in the day (Kofoid, 1909a). Motile pairs of thecate *Pyrodinium* *bahamense,* joined apex to antapex with ventral sides aligned, may result from incomplete separation following cell division (Buchanan, 1968).

Chains of cells with cytoplasmic continuity occur in *Gessnerium catenellum* (2 to 8 cells) (Tomas, 1974). The red-water-producing *Gymnodinium catenatum* Graham, which may occur in numbers of a million cells per liter in surface water of the Gulf of California, forms chains of as many as 29 cells (Graham, 1943), and the Florida East Coast red tide species *Gessnerium monilatum,* during the period of rapid increase in cultures, forms chains of as many as 32 cells (Aldrich et al., 1967).

An atypical chain was produced by *Gonyaulax series* Kofoid & Rigden, in which individuals had complete polar attachment of the cytoplasm. New individuals formed successively toward the distal end of the chain, the posterior cell having only a thin pellicle, with no visible sutures (Kofoid and Rigden, 1912).

Occasional chains have been reported to include individuals of more than one distinct species (Kofoid, 1909a; Peters, 1930, fig. 18). The posterior individual of one chain of four appeared to be a typical *Ceratium tripos,* and the three anterior ones "*C. californiense,*" now = *C. macroceros* (Ehrenberg) Cleve. The ancestral form was regarded as the *C. tripos,* the posterior portion of successive generations of cells being progressively modified toward "*C. californiense*" (Kofoid, 1909a). Another example involved two attached individuals, the anterior representing "*C. ostenfeldii,*" now = *C. massiliense* (Gourret) Jörgensen, and the posterior "*C. californiense.*" This was suggested to be an incomplete chain, the absent two end members representing the ancestral form, those remaining representing third generation "cousins." In both cases, the two species involved in a single chain represented different subgenera of *Ceratium.* Although Kofoid postulated that mutations were rapid in response to an environmental shock (temperature, salinity, nutrient supply, etc.) in an area of upwelling, it is probable that species of *Ceratium* include greater intraspecific variation than had been assumed.

Growth. Following cell division that produces two smaller daughter cells, and prior to the next division, the new cells must enlarge to adult size. The nonthecate *Crypthecodinium cohnii* in pure cultures shows a range of cell size from 4 to 45 μm for a volume ratio of 1:700 (Barker, 1935), although judging from later studies (Beam and Himes, 1974) the larger cells probably were motile zygotes. Even thecate dinoflagellates may show a large size increase, cells of *Gonyaulax* ranging from 30 to 70 μm for a twelvefold increase in size (Barker, 1935), and a necessary corresponding enlargement of the theca.

Wall formation in the thecate dinoflagellates is of two main types, the original formation of the plates within the large flattened vesicles formed by inner and outer plate membranes, and the later growth and thickening of the plates. Active wall thickening continues for some time after the cell shape is well defined and may result in banding as well as increased surface ornamentation, but in older cells the thickening is restricted to particular areas (Wetherbee, 1975a).

Cell volume increase in thecate dinoflagellates is accommodated by growth of the plates of the theca adjacent to the sutures. In the Dinophysiales, such growth involves formation of a **megacytic (intercalary) band** on each plate, at either side of the meridional suture separating right and left cell halves. In binary fission, separation takes place along this suture, whose undulating margin gives it a zigzag appearance. Unlike the original theca, the intercalary region of growth (megacytic zone) lacks perforations and stains a much deeper blue with trypan blue (Tai and Skogsberg, 1934, p. 403). Cells in this megacytic stage are so distinctive in appearance that some have been described as separate varieties. Following separation of the new daughter cells, the megacytic zone disappears as the new thecal half is formed (Pavillard, 1915). Resorption begins in the ventral area and terminates middorsally, the last area of attachment persisting temporarily as a dorsal megacytic bridge, even after the lists begin to form in *Ornithocercus,* but complete resorption of the megacytic zone generally precedes list development (F. J. R. Taylor, 1973a). Young cells rarely show any remnant of the intercalary growth.

Cell volume increase in the Peridiniales also is

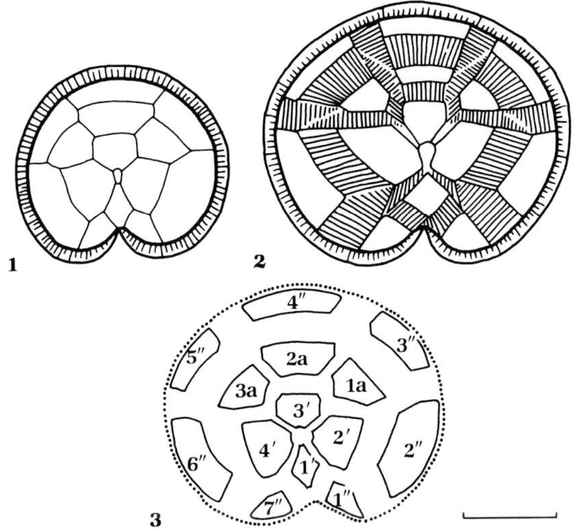

Figure 4.79
Protoperidinium subinerme. **1.** Apical view of epitheca of young specimen with simple sutures. **2.** Old cell with wide intercalary bands. **3.** Plates of small individual spread into position of those in old cell, showing that all cell enlargement is accommodated by the intercalary bands; bar = 30 μm. Redrawn from Peters, 1929.

accompanied by intercalary growth, with bands formed between adjacent plates. Small specimens of *Protoperidinium subinerme* (see Figure 4.79) have narrow intercalary bands and large individuals have broad bands, widest in the mid-region and narrower toward anterior and posterior poles; the plate area exclusive of the intercalary bands remains the same size as in the small cells (Peters, 1929, p. 90, 96). Loeblich III (1970) observed that the intercalary band is present on only one of the two plates in contact and may be on the underlying plate in *Ensiculifera,* or the overlying plate in *Gonyaulax polyedra.* Outer edges of the underlying plate are bevelled to fit beneath that adjacent, and the edges of the overlapping plate are thicker and more heavily sculptured, with ridge and pronounced margin (Loeblich III, 1970; Dürr and Netzel, 1974).

Electron microscopy of sectioned *Ceratium tripos* in process of division (cytokinesis) shows cytological activity in the area underlying the plates inherited from the parental thecal half, similar to that occurring in the region where new membranes and thecal plates are being formed. Possibly the inherited larger and thicker thecal plates are thus reduced in overall size and thickness to meet the volume limits of the new daughter cell in a process reversing the growth of intercalary bands (Wetherbee, 1975b).

Exuviation and Autotomy. After simple transverse fission involving thecal partition, the newly synthesized plates of the daughter cells develop the ornamentation, horns, ridges, and reticulations to match the inherited plates. However, some morphological features differ in degree within the same species at different seasons, including thecal plate thickness, length of horns, and rugosity. As half of the parental theca is inherited, it must be modified in order to maintain a symmetrical cell while undergoing seasonal variations. The process of **exuviation** (Kofoid, 1908) consists of shedding a few of the thicker old plates at a time, replacing these by thin and delicate new ones, until the entire theca is thus replaced. Kofoid suggested that such thecal modification might occur to agree with the new physical conditions when individual specimens are carried by currents into waters of different temperature. Similarly, the length of horns of various long-spined *Ceratium* may be modified by **autotomy** (Pouchet, 1893; Kofoid, 1908). A constriction first develops on the horn and gradually forms a section plane, cutting off the distal end with its included cytoplasm (see Figure 4.80). Shortening of horns is proportional, so that the cell continues to maintain the same proportions and symmetry after shortening. Specimens also can be observed in which originally short horns have been later

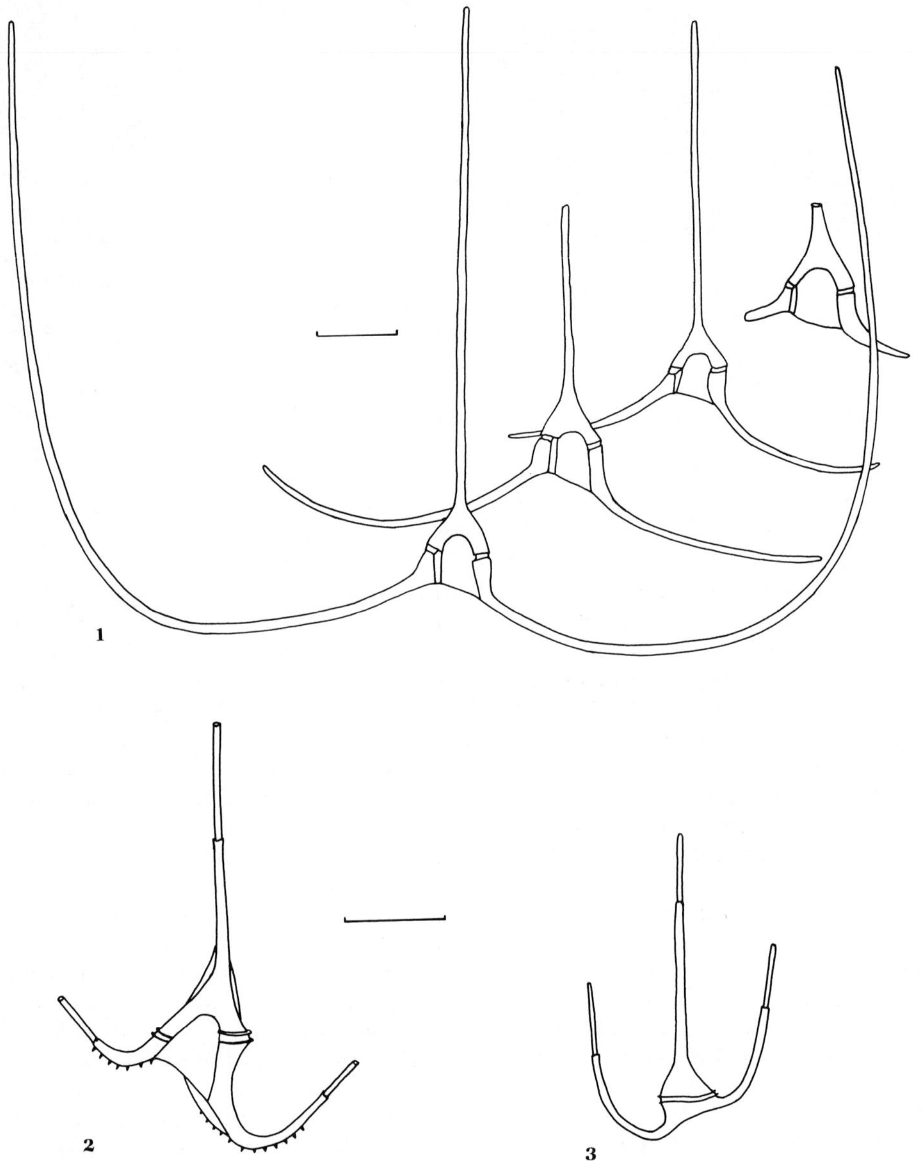

Figure 4.80
Seasonal or temperature-related modifications of thecal morphology. **1.** *Ceratium trichoceros* (Ehrenberg) Kofoid, showing series from left to right of progressive and proportionate reduction of antapical horns by autotomy from warm water long-horned form to colder water form; bar = 50 μm; redrawn from Kofoid, 1908. **2,3.** Ventral view of *C. massiliense* and dorsal view of *C. horridum* Gran, showing proportional regeneration of all three horns after autotomy, or seasonally as a modification to warm water form; bar = 100 μm; redrawn from Kofoid, 1908.

elongated, again maintaining their relative proportions, while adapting to changed temperature, salinity, and other density-related variables.

Sexual Reproduction

Various stages in sexual reproduction have been observed and described in the dinoflagellates, but many of these were incorrectly interpreted, and the occurrence of sexuality in these organisms was not generally accepted. Chatton (1934, p. 310) and Paulsen (1949, p. 19) regarded dinoflagellates as entirely asexual, hence incapable of producing hybrids. Much of the early difficulty in interpretation resulted from the common practice of studying dinoflagellates as fixed material from the plankton, because of the problems in culturing them. As any variations present could not be unquestionably placed in sequence, many early reports of cell fusion (Stein, 1883), production of microswarmers (Hensen, 1887), nuclear fusion (at times misidentified as involving cell division instead), cyst formation, and tetrad division at or following excystment were discounted as proof of sexuality. In addition, different types of cysts and the process of ecdysis also were confused with the stages in the sexual cycle.

Over the last few years, culture methods have been perfected, and both armored and unarmored species from marine and freshwater habitats alike have been followed through their life cycle with the use of cytological evidence, genetic recombination of experimentally induced mutant clones, and other techniques in recognition of dinoflagellate sexuality. Mutant clones of *Crypthecodinium cohnii* were chemically induced, the mutants being deficient in β- and γ-carotene pigments. Some albino clones lacked carotenes entirely; others were cream-colored and had only 10 percent of the normal amount of carotenes. The frequency of mutants was consistent with a usual haploid condition. In nitrogen and phosphorus-deficient media, pairs of motile cells fused and as many as 13 percent yellow recombinants resulted from the fusion (Tuttle and Loeblich, 1974a,b). Other mutants of *C. cohnii* were induced by ultraviolet irradiation, the mutants varying in cell surface features and motility. Fusion of swarmers resulted in a zygote that retained both longitudinal flagella. After isolation, the zygotes later produced four to eight recombinant daughter cells (Beam and Himes, 1974).

Various means of inducing the sexual cycle have been utilized with cultures, most of which occur in nature in the late summer and fall, or under conditions of crowding, as in red tides. Thus reduction of the light period to 10 hours, and decrease in temperature from 21° to 15°C, induced gamete formation in *Gymnodinium pseudopalustre*, and limitation of nitrogen and phosphorus resulted in gametogenesis in *Woloszynskia apiculata* (Stosch, 1973). Genetic recombination and gametic fusion of *Crypthecodinium cohnii* was enhanced in nitrogen and phosphorus-deficient media (Tuttle and Loeblich, 1974b), and nitrogen deficiency was used by Pfiester (1975) to induce sexual reproduction in *Peridinium cinctum* f. *ovoplanum* Lindemann. Resting cysts have been observed most frequently after the log phase of population growth and in overcrowded populations, both in the laboratory and in the late stages of natural blooms (Wall and Dale, 1968a, p. 268) that would result in nutrient depletion (Pfiester, 1975). All such cysts now appear to represent **hypnozygotes.**

As a result of the numerous studies of dinoflagellates in culture, the presence of a sexual reproductive phase now is well documented. As might be expected in such a large and varied group, the sexual cycle also includes numerous modifications. Most dinoflagellates have a haploid vegetative stage, but *Noctiluca* is diploid, with only short-lived haploid gametes. Many species have gametes that appear identical to the vegetative cell but are smaller; gametes of some are of two distinct sizes, others have isogametes, some are homothallous (individual gametes from a single clone may fuse), and others heterothallous, requiring two separate clones or mating types. Some invariably produce a resting zygote (hypnozygote cyst), others have a motile or **planozygote** stage before producing the hypnozygote, and others only a motile zygote and no resting cysts. Duration of the diploid state prior to meiosis also varies in different species.

Nuclear Cyclosis and Knot-Stage. The dinoflagellate nucleus is unusual in having condensed chromosomes throughout interphase; two other unusual features of the nucleus that have long been known are now recognized as an indication of the sexual cycle. These are the **nuclear cyclosis** first observed by Pouchet (1883) and the Knauel-stadium or **knot-stage,** originally interpreted as a unique form of mitosis (Borgert, 1910, pl. 1, fig. 3, pl. 2, fig. 1; 1912).

Nuclear cyclosis, in which the nuclear contents slowly rotate, was reported in exceptionally large cells of *Ceratium fusus* and *C. tripos* (Pouchet, 1883, 1885). A similar feature was described (Biecheler, 1952) in *Peridinium sociale* (Henneguy ex Labbé) Biecheler, *P. balticum,* *"Goniodoma" pseudogonyaulax* Biecheler, *Gyrodinium pavillardii,* and *Crypthecodinium cohnii.* In *P. sociale* it occurred only in nonmotile individuals with exceptionally large nucleus that had lost the flagella and settled to the bottom; cyclosis was followed by nuclear and cytoplasmic division, resulting in two daughter cells (Biecheler, 1952). Stosch (1972) followed in detail the fusion of gametes and development of the zygote of *Ceratium horridum.* The zygote grows for some days to become a large cell with greatly enlarged nucleus and then undergoes nuclear cyclosis, during which the chromosomal mass circulates about twice a minute inside the nuclear membrane, for a period of some hours. Chromosomal pairing occurs during the movement, and the first meiotic division results in two cells. After two to three days the second division results in four cells from the original zygote. Cyclosis thus is the beginning of the meiosis in the dinoflagellates. Varying with the species, the meiotic prophase and nuclear cyclosis may occur prior to the first division of the motile zygote; or in those species with resting cyst, meiosis may follow its germination, as in Biecheler's *Peridinium sociale.*

During the knot-stage (Knauel-stadium), the unusually large nucleus has an irregular mass of long, relatively straight condensed chromosomes that appear double. Borgert (1910) originally believed this stage to represent mitotic nuclear division, but Stosch (1964, 1972) showed

this also to be related to meiosis and to represent a late stage of cyclosis involving the first meiotic nuclear and cell division. Although cyclosis could only be observed in living material, the knot-stage was evident in preserved and fixed material as well. The process of cyclosis may aid in pairing the very numerous and elongate chromosomes, which, in the knot-stage, has taken place. Either feature of the nucleus in the dinoflagellates is an indication of meiotic division, and hence of sexual reproduction.

Isogamy. As the usual vegetative cell of the dinoflagellate is haploid, the nucleus requires no reduction division for gametogenesis, and, theoretically at least, simple conjugation is possible. Conjugation was reported in *Glenodinium pulvisculus* (Ehrenberg) Stein and *Peridinium triquetrum* (by Stein, 1883, as *Heterocapsa triquetra*), but in preserved material may have been confused with vegetative division. However, Zederbauer (1904) reported that conjugation in *Ceratium hirundinella,* in which cells attached by their ventral areas, with the anterior ends in opposite directions, resulted in production of a three-horned resting cyst. Entz (1924) observed that this coupling was accomplished by means of a tube between the two individuals, which first appeared as a papillate protrusion, those of two cells then joining. Such a tube is visible at the lower left of the cell of *Ceratium hirundinella* illustrated by Loeblich Jr. and Loeblich III (1966, pl. 1, fig. 3).

Pairs of cells, joined by the bridge, were common toward the end of the vegetative reproductive season in October, and particularly in late afternoon when cell division does not occur; hence the process was suggested by Entz to be sexual. He noted the occurrence of many cysts with the coupling cells. Coupled pairs of cells also are distinguishable from those in process of vegetative division in having an entire theca with normal horns on both cells, which are joined ventrally. Dividing cells divide the parent theca, hence each daughter cell has only a partial theca, and incompleted horns on the regenerated part; division is such that the cells have like orientation, with the ventral side of one cell abutting the dorsal side of the other (Entz, 1924).

Apparently in all cases, the cells acting as gametes and fusing are considerably smaller than the normal vegetative cells, although otherwise morphologically indistinguishable. Species that have identical gametes (isogamous) may be homothallous (no difference in mating types, so that any individuals of a clone may fuse) or heterothallous (with two distinct mating types); both marine and freshwater isogamous species are known, and the resultant zygote varies from a motile cell identical in appearance to the normal haploid vegetative cell to a resting cyst that may be considerably different in aspect.

Gametes are produced by *Gymnodinium pseudopalustre* under reduced temperature and light by frequent **depauperating division,** resulting in smaller cells with fewer plastids and less pigment. Whether more than one such division is required is uncertain, but the process is irreversible, and if fusion does not occur, the gametes die. The species is homothallous. Prior to gametic fusion, clusters of 2 to 10 gametes swarm together, swimming in a weaving motion as small "dancing groups" (Stosch, 1973). Pairs then fuse, the pairs being joined by a hyaline, somewhat globular projection that forms a bridge slightly below the respective intersections of cingulum and sulcus of the two cells. The actual fusion of the cytoplasm does not occur through this bridge, but through a fusion of one flange of the respective sulci. As fusion proceeds, both flagella of the separate individuals are retained, but one of the transverse flagella disappears by the time the fusing cells possess a single common cingulum and a common sulcus. The nuclei approach, but do not fuse as yet. The cell rounds off, secreting a very thin preliminary wall that inflates, leaving a hyaline space between it and the surface of the protoplast. Small granules appear on the protoplast surface and grow out as spines as the exospore wall layer is secreted. The thin preliminary wall then bursts and crumples to one side, the entire process being completed in about nine minutes. During the following 48 hours, the plastids lose their color and become inconspicuous, masses of red oil appear, and the nuclei fuse. After about four weeks in the dark, at lowered temperatures, the thick-walled, sulfuric acid-resistant, spiny cyst will germinate when returned to light and warmer temperature. A cingulum forms by median contraction of the protoplast before the spore bursts. The escaping flagellate swarmer still retains both longitudinal flagella (of the original gametes). In about 36 hours, the nucleus enlarges, and cyclosis accompanies the chromosome pairing of meiosis. Two successive meiotic divisions produce four haploid daughter cells, the two longitudinal flagella being divided and inherited by the cells of the first division (Stosch, 1973).

Amphidinium carterae gametes appear identical to usual vegetative cells; in culture the gametes approach in pairs, and the membranes disappear at the point of contact, allowing the cytoplasm to mix. A short time later the nucleus and cytoplasm is involved in cyclosis and the cell forms a large rounded nonmotile and naked zygote, with volume double that of the normal individuals. The zygote has a double membrane but does not form a resting cyst. At germination the cell enlarges and undergoes two meiotic divisions, the resultant cells remaining attached by the anterior ends for a time until the flagella appear. The full-sized cells than separate (Vien, 1967, 1968).

Peridinium cinctun f. *ovoplanum* thecate cells produce small naked gametes, their fusion producing a zygote that remains motile while enlarging. A very dark-colored warty cover is produced, and its wall thickens as the cell becomes a nonmotile hypnozygote. Germination after seven to 8 weeks releases a single large cell containing 4 nuclei; division of the cytoplasm results in two vegetative daughter cells, the remaining nuclei possibly aborting (Pfiester, 1975). Small thecate cells functioned as gametes of *Peridinium gatunense* in culture, the growth of the motile zygote and later development of nonmotile cyst being similar to these processes in *P. cinctum* (Pfiester, 1976).

Flagellate isogametes of the freshwater *Peridiniopsis lubiniensiforme* (Diwald) Bourrelly are produced within a mother cell by two successive divisions (see Figure 4.81). Somewhat smaller than the vegetative swarm spores that are similarly produced, they differ also in having the longitudinal furrow extending obliquely far onto the epicone. The species is heterothal-

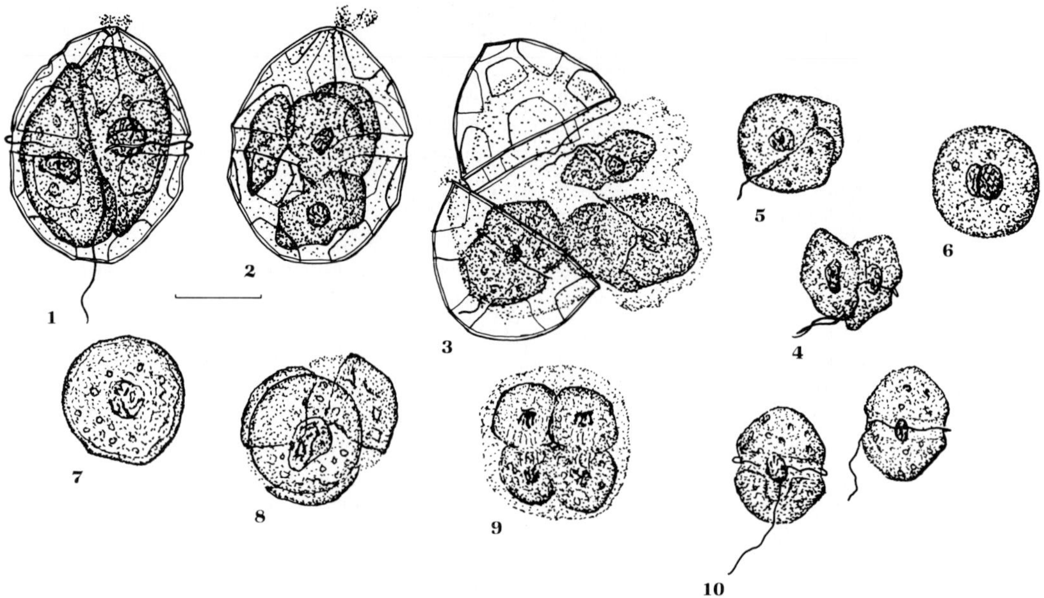

Figure 4.81

Peridiniopsis lubiniensiforme. Sexual reproduction and encystment in a freshwater dinoflagellate. Bar = 15 μm. **1,2.** Development of gametes within parent theca. **3.** Release of gametes. **4,5.** Gametes joining in pairs to form zygote. **6,7.** Resting zygote cyst. **8.** Germination, resulting in disruption of cyst. **9,10.** Production of four daughter cells (probable result of meiosis) that become new motile cells. Redrawn from Diwald, 1938.

lous, as two distinct mating types (+,−) are required as a stimulus for gamete formation (Diwald, 1938). The gametes fuse by their ventral sides and lose the flagella to form a spherical zygote enclosed in a simple brownish membrane. After a period of at least 10 days, germination produces four new cells by tetrad division and meiosis. Resting cysts appear readily in cultures if left 6 to 8 days without subculturing. The cyst forms within the theca by contraction of the protoplast and the production of a colorless, homogenous, nontabulate covering. These cysts required slow drying and a resting period of at least 10 days. Addition of fresh puddle water enriched with K_2HPO_4 and $Ca(NO_3)_2$ induced excystment (Diwald, 1938).

In sexual reproduction, *Woloszynskia apiculata* also is heterothallous and requires individuals from two different clones, hence indicating the existence of two mating types (see Figure 4.82). The gametes result from differ-

entiating divisions that produce smaller and less pigmented cells than normal vegetative ones, although no other distinct morphologic difference has been observed. Gametes swim quite rapidly, and many show negative phototaxis. When those from two different clonal cultures are mixed, pairing begins almost immediately, with the formation of large and more obvious "dancing groups" than in *Gymnodinium*. Pairs of cells then swim away, spinning around their common longitudinal axis. A protoplasmic bridge is formed between members of a fusing pair, as in *Gymnodinium*, but if pairs of *Woloszynskia* are separated, they may reproduce vegetatively again. Normally the cells coalesce, the respective longitudinal flagella beating in coordination; a common cingulum and sulcus is constructed; and nuclear fusion occurs. Zygotes remain motile, but the retention of the two longitudinal flagella allows recognition of the zygote as distinct from a haploid vegetative

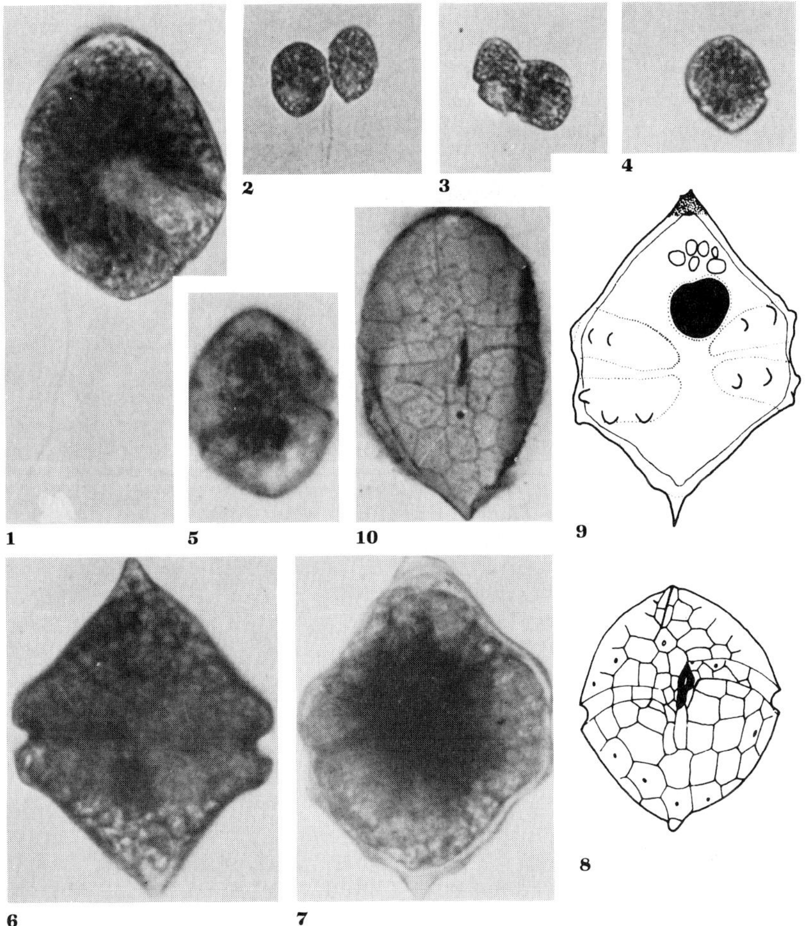

Figure 4.82

Woloszynskia apiculata, sexual reproduction and cyst formation. **1.** Motile cell, left side, sulcus with longitudinal flagellum to left, dark eyespot in lower left, ×1000. **2.** Gametes pairing with same posterior-anterior alignment. **3.** Fusion beginning eight minutes later. **4.** Thirty minutes later, pair has developed common cingulum. 2–4, ×500. **5.** Nine minutes later, young planozygote has two flagella and two eyespots. **6.** Old planozygote assuming form of resting stage. **7.** Ventral view of mature thick-walled hypnozygote cyst. **8.** Old cyst, exospore punctation shown at apex, eyespot in sulcus. **9,10.** Ventral views of thecae treated with laurylsulfate to show tabulation. 5–10, ×1000. All from Stosch, 1973.

cell. Later it is transformed into a rhomboid, tuberculate resting zygote cyst, the cell becoming conical at the apices, filled with starch and appearing dark in transmitted light. The cyst wall has a very thin, punctate, acid-resistant exospore, enclosing a thin mesospore that is widest at the apices, and a thick and stratified cellulosic endospore. The entire cyst is tightly enclosed in the outermost perispore, which is the envelope of the motile zygote. The first (and sometimes second) meiotic division occurs within the zygote; the resultant daughter cells are enclosed within the mucilage-filled exospore, as the endospore has disappeared. As the exospore finally ruptures, the swarmers are forcibly expelled. Commonly only two swarmers are released, the second meiotic division occurring within a later division cyst, followed rapidly by vegetative division (Stosch, 1973).

Hypnozygote resting cysts also have been reported in other *Woloszynskia*, although cysts of *W. tylota* are not resistant to acids, and those of the type of the genus, *W. reticulata* Thompson, are smooth and rounded. In addition to the tubercles, cysts of *W. pascheri* and *W. tylota* have a single horn, whereas *W. apiculata* has two. These differences in resting cyst features may provide a basis for dividing *Woloszynskia* into two subgenera (Stosch, 1973). The original description (Mapletoft et al., 1966) of *W. tylota*, describing "spore" formation, noted that the encysting cells became enlarged, and that many had two longitudinal flagella (also noted by Bibby and Dodge, 1972), remaining motile during thickening of the wall. Retention of two longitudinal flagella by an abnormally large cell allows its recognition as the motile zygote, distinct from the vegetative cells of *W. apiculata* and *Gymnodinium pseudopalustre*. These features of *W. tylota* also are an indication of the early zygote immediately after gametic fusion in the course of sexual reproduction. References to two posterior flagella in *Ceratium* (Claparède and Lachmann, 1859), *Plectodinium* (Biecheler, 1934a), *Spirodinium* (West, 1916), and *Thecadinium* (Dragesco, 1965; see Figure 4.83) also indicate motile zygotes. Such motile zygotes (planozygotes) may be indistinguishable from the normal haploid vegetative cell, other than by their larger size (Stosch, 1973; Bol-

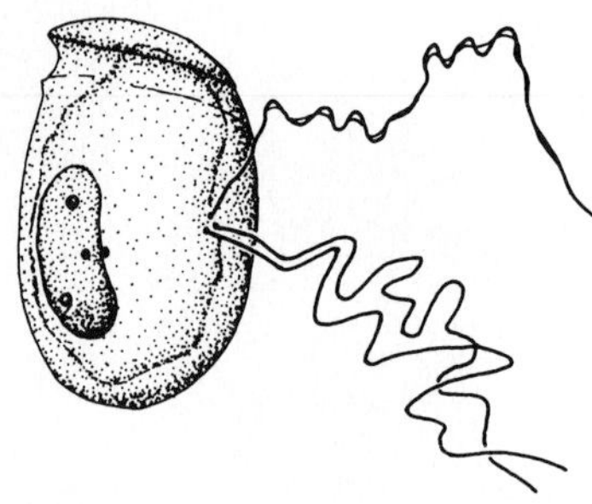

Figure 4.83
Thecadinium petasatum (Herdman) Kofoid & Skogsberg. Triflagellate cell, with crenulated transverse flagellum and two longitudinal flagella, hence probably a motile zygote (planozygote); approximately ×1000. Redrawn from Dragesco, 1965.

tovskoy, 1975), but may then produce resting cysts. The pre-meiotic zygotes also may emerge from the cysts as biflagellate swarmers, as in *Gymnodinium pseudopalustre* (Stosch, 1973), Similarly, the single motile cell released from the germinating cyst of *Gonyaulax polyedra* (fossil cysts of this type were described as *Lingulodinium machaerophorum*) undergoes two successive (hence probably meiotic) divisions, giving rise to four daughter cells (Wall and Dale, 1968a, p. 271, 280). In contrast, two motile cells are released from the cyst of *Dissodium lenticula* (Schütt, 1887; Wall and Dale, 1968a, p. 279), indicating that the first meiotic division apparently preceded germination as in *Woloszynskia* and *Helgolandinium*.

Oxyrrhis marina also has tiny isogametes that produce a similarly motile zygote. Two-stage meiosis produces four haploid swarmers, without prior development of a cyst (Stosch, 1964). Stosch also noted that the nuclear knot-stage in early meiosis had been erroneously illustrated for this species as stages in mitosis (Dodge, 1963b).

Heterogamy. In contrast to those dinoflagellates in which gametes are all identical, others have tiny microgametes and larger macrogametes. These have been observed in various species of *Ceratium*, although interpreted in a multitude of ways. Hensen (1887) first reported the abundant microswarmers and regarded them as minor variations or young individuals, whereas Kofoid (1909a) believed them to represent mutations, and Apstein (1911) described them as buds. Chains occasionally included alternating normal-sized cells and smaller ones with truncated horns (Peters, 1930). Only detailed study of cultures allowed the recognition of these small individuals as gametes (Stosch, 1964), produced by "impoverishing division." Transverse division of a parent cell produced a generation with either anterior or posterior (inherited portion) like the parent, but with the newly developed half being much reduced in size. If transferred to fresh culture medium, the next generation would revert to normal vegetative development. Otherwise, the next generation would consist of tiny forms 10 percent of normal size with reduced plastids, sparse pigments, thin-walled thecae, and condensed nuclei lacking nucleoli. *Ceratium horridum* may include microswarmers (male gametes) as 30 percent of the population during the peak production in September and October (see Figure 4.84). The female gamete appears identical to the usual vegetative cell, but has a more consolidated nucleus. Gametes fuse by their ventral sides, the microgamete gradually rounding off and being completely absorbed by the female. Its nucleus enters the female cell and occasionally may be observed surrounded by the dismantled and partially resorbed plates of its theca. Previous reports of asexual budding (Apstein, 1910; Silva, 1971) associated with great metabolic and reproductive activity actually represent the fusion of unlike gametes at their respective furrows (Stosch, 1964, 1972); studies based on preserved material could not determine the correct sequence of events. Similar reports of cannibalism (Bursa, 1963, fig. 60) of *"Gyrodinium" rubrum* (= *Gymnodinium rubrum* Kofoid & Swezy), in which a larger cell engulfs a much smaller individual at the girdle region, probably are instances of gametic fusion. In living marine *Ceratium*, the motile zygote remains unchanged in form, with a protracted growth phase preceding meiosis and de-

velopment of the four haplonts. In contrast, such freshwater species as *C. cornutum* produce an overwintering resting cyst, germination releasing a single swarmer that later undergoes meiosis (Stosch, 1972).

Resting cysts and details of excystment have long been known in *C. hirundinella* from lakes in Switzerland, Norway, Germany, and North America (Entz, 1921, 1925; Hauge, 1958; Huber and Nipkow, 1922; Peters, 1930; Stein, 1883; Thompson, 1947; Wall and Evitt, 1975). The relationships of these cysts to the sexual cycle was determined by Stosch (1964, 1972). Unlike many resting cysts, those of *Ceratium* are cellulosic in composition (Entz, 1925; Wall and Evitt, 1975), hence are not preservable as fossils, although they may be present in the surface sediments of lakes and may remain viable up to six-and-a-half years (Huber and Nipkow, 1922).

Diploid Vegetative Cell. The vegetative cell of *Noctiluca* (see Figure 4.85) is unusual in appearance, subspherical in outline, highly vacuolated, and nonpigmented, with a prominent tentacle or prod and an unusually large nucleus, which is diploid in contrast to the haploid vegetative nucleus of other dinoflagellates. Gametes of *Noctiluca*, gametic fusion, and the zygote development were illustrated by Gross (1934). When gametes are produced the tentacle and internal food vacuoles disappear, the cell becomes rounded, and the nuclear mass peripheral, a stage termed the gametocyte mother cell (Zingmark, 1970a, 1970b). The nucleus divides twice, resulting in a distinct tetrad of nuclei (the result of meiosis). Continuous synchronous division occurs at intervals of about 45 minutes, the nuclei remaining in the original four groups until they crowd the cell interior (usually the 256-nucleate stage, but large cells may produce up to 1024 nuclei). The gametocyte cell does not change in size throughout this process. Then uniflagellated gametes, 10 × 15 μm in diameter and with dinokaryotic nuclei, develop and are released. The flagellum is about 50 μm long, and beats with sinuous waves, so that the *Noctiluca* gamete moves in the manner of other dinoflagellates. Pairs conjugate (Gross, 1934), including those from a single mother

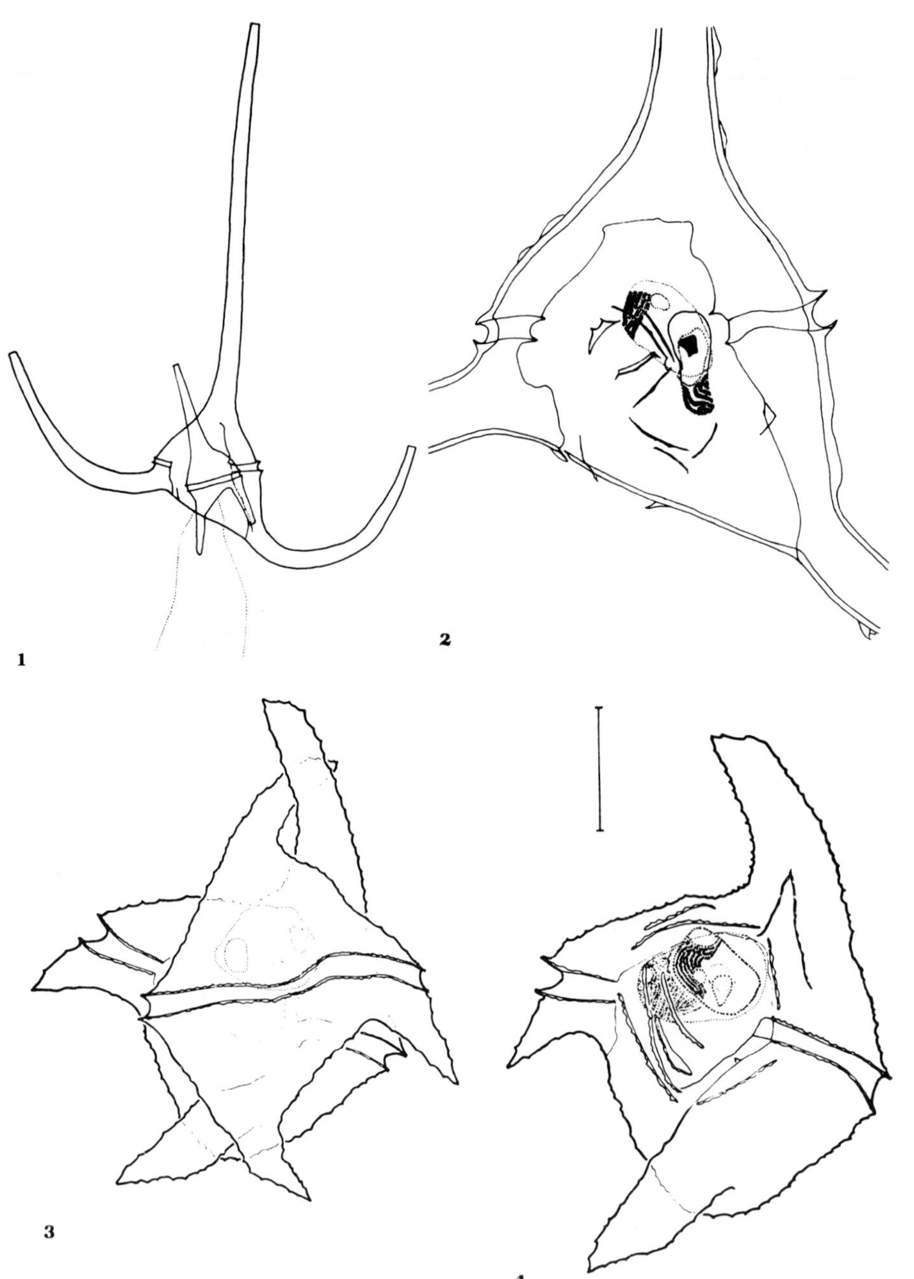

Figure 4.84
1. *Ceratium horridum.* Fusion of macrogamete and microgamete, at the respective sulcal areas, once interpreted as "budding" from preserved material. **2.** Same cells three hours later, showing microgamete cytoplasm absorbed, a few of its isolated thecal plates remaining visible in the ventral area of the macrogamete; nuclei not yet completely fused. 1,2, from Stosch, 1964. **3.** *Ceratium cornutum,* fusion of gametes by ventral sides, smaller male gamete in foreground. **4.** After six hours, only isolated plates of male gamete cell wall persist, nuclei not yet fused; bar = 20 μm. 3,4, from Stosch, 1972.

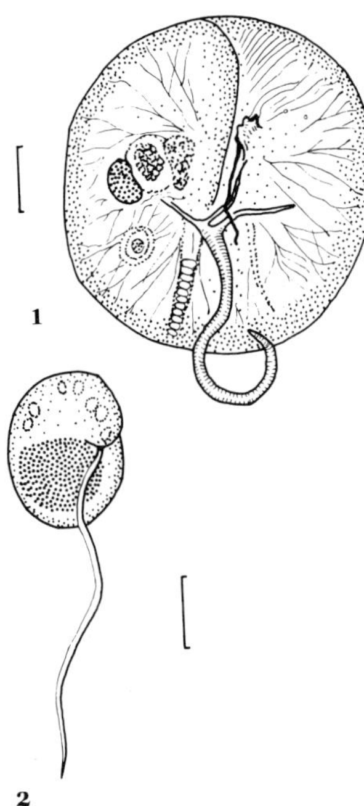

Figure 4.85
Noctiluca scintillans. **1.** Posteroventral view of mature diploid cell, showing sulcus as vertical line in top half, short longitudinal flagellum, and elongate tentacle; bar = 100 μm; redrawn from Kofoid, 1920. **2.** Single uniflagellated gamete of *Noctiluca miliaris,* with large posterior dinokaryon nucleus; bar = 5 μm; drawn from photo in Zingmark, 1970b.

gametocyte (homothallous). Both flagella are retained for a time, but eventually they cease to beat and are shortened, and the cell drops to the bottom. The two nuclei fuse to form the early zygote stage, the cell now being about 25 μm in diameter. Although few lived beyond this stage in cultures, one increased to about 200 μm in size, became vacuolate, and developed a tentacle like the usual vegetative *Noctiluca* (Zingmark, 1970b).

Summary. Although long questioned, sexual reproduction has now been demonstrated in many different dinoflagellates and involves many of the peculiar characters of the dinoflagellates that had already been described

but not understood: the resting cyst, microswarmers, double longitudinal flagella, conjugation, cannibalism, budding, and nuclear cyclosis and knot-stage. Difficulties in preserving living material and maintaining cultures through complete life cycles have hampered these studies, and many life cycles are still incompletely known. Some species have been studied in detail through the entire life cycle, others in the process of encystment, others from germination of cysts; sexuality has been described unequivocally for various species of *Amphidinium, Ceratium, Crypthecodinium, Gonyaulax, Gymnodinium, Gyrodinium, Helgolandinium, Heteraulacus, Noctiluca, Oxyrrhis, Peridiniopsis, Peridinium,* and *Woloszynskia,* and is strongly indicated for many others.

Gametes may be of nearly the same size as vegetative cells, or much smaller; they may be naked or thecate, isogamous or heterogamous. The resultant zygote may be motile or nonmotile, and may have a cellulosic cover, an acid-resistant sporopollenin one, a complex one of layers of different composition, or merely a double membrane, It appears probable that all resting cysts, such as comprise the fossil record of the dinoflagellates, are in fact hypnozygotes and the result of gametic fusion. Motile zygotes and those with only cellulosic or membranous cover are not fossilizable.

The sexual cycle is initiated at times of reduced temperature and light or nutrient depletion, hence commonly follows blooms in the late summer or fall. Some cysts require a resting period before germination and remain viable as long as 16 years (*Peridinium,* reported by Peters, 1930). Germination could take place in *Ceratium hirundinella* cysts at temperatures from 5°C to a maximum of 35°C, although optimum temperatures were between 23° and 26°C (Hauge, 1958). Marine *Gonyaulax spinifera* from near Bermuda germinated between 16° and 25°C from cysts referable to the fossil cyst species *Spiniferites ramosus* (Wall and Dale, 1970).

Both meiotic divisions may take place within the motile zygote or the hypnozygote. Only the first division may occur within the cyst, and the second in a later temporary cyst; or a single cell may be released from the cyst, both meiotic divisions occurring after germination.

ECOLOGY

Population Dynamics

Productivity

Dinoflagellates are among the more important primary producers of the present seas, the larger and thecate species being common in the warmer waters, and nonthecate species being more numerous in cooler waters. Temporary blooms may occur in any of the regions inhabited by various species, and the productivity at such periods may be extremely high. Because of the great variation in cell numbers and division rates through the year, as well as the patchiness of plankton distribution, total productivity is difficult to estimate with any degree of accuracy.

In the Mediterranean Sea, near the mouth of the Nile River, at Damietta, the phytoplankton count is generally about 66,000 cells per liter of seawater, dinoflagellates being among the most important components. As the seasonal flood discharge of the Nile results in nutrient enrichment, the plankton rapidly increases to about 1.24 million cells per liter four days after flooding and up to 2.4 million cells per liter by three weeks after flooding. In the rapid increase, the diatoms and ebriids dominate, while the dinoflagellates increase more slowly, probably because of their somewhat larger average size. At the height of the bloom, the dense photoplankton growth gives the sea an opaque, reddish-brown appearance, the scum clogging plankton nets in two to three minutes (Halim, 1960a).

The specialized *Achradina*, with inner and possibly siliceous spicule, is reported in the western Mediterranean in numbers up to 92 cells per liter, being most common at 13.5 to 14.0°C and salinity of 35.0 to 37.75%₀ and more frequent in the surface layers in winter than in summer or in depths below 100 m (Nival, 1969b). Along the coast of Guinea in the South Atlantic it occurs in numbers up to 108 cells per liter (Hentschel, 1933).

Over a three-year period, 198 dinoflagellate species were observed in a measurement of abundance at Villefranche-sur-Mer in the Mediterranean (Halim, 1960b). A spring growth phase is characterized by a massive increase in cell numbers during March or April to a May or June maximum of nearly 200 times the number present in the previous winter; the spring bloom is marked by high amplitude fluctuations of short duration. During the spring maximum, the water is stratified, the surface layer that contains the bloom being of lower density, higher temperature, and more alkaline pH (between 8.2 and 8.27), as well as being better lighted. Below a depth of 25 m, the abundance is similar to that of winter. Dominant species of the bloom are those of *Prorocentrum* and *Ceratium*, with *Protoperidinium* and *Gonyaulax* less frequent. As the nutrients, such as phosphorus, become depleted in the surface waters, the maximum is followed by a gradual decline in summer, particularly of the dominant species; by late July, *Gonyaulax* may be most frequent. The stationary phase of growth in summer is followed by a second but lesser maximum in the autumn; decreased production of dinoflagellates from the end of October until the spring bloom some five to six months later results in the winter minimum. The water has a minimum vertical stability during the winter, supporting only a much reduced population. Of the total 88 species identified, different species are dominant at different times; most abundant in winter are *Heterodinium*, *Oxytoxum*, and *Dinophysis*, while other species may be extremely rare. Not only are the local physical conditions important, such as availability of solar energy, temperature, and concentration of nutrient salts, but the productivity of a given species may depend largely on its own growth rate as compared to others.

The main zone of dinoflagellate photosynthesis in the sea is in the upper 25 m, as at Villefranche during blooms. However, many dinoflagellates are not adapted to high light intensity; these shade species generally live at greater depths. The light saturation point has been determined in the laboratory for various species of *Amphidinium*, *Prorocentrum*, *Gymnodinium*, and *Gyrodinium* as about 2500 to 3000 fc (footcandles). At higher intensities, of 8000 to 10,000 fc, comparable to noon sunlight, photosynthesis was reduced to only 20 to 30 percent of the maximum (Ryther, 1956).

Pyrocystis may have a population maximum near or below the bottom of the euphotic zone, between 100 and 200 m, regarded as the compensation depth for phytoplankton photosynthesis. In cultures with light controlled to simulate the amount available at such depths, the cells of *Pyrocystis fusiformis* continued to divide at the normal rate, but cell size decreased in proportion to the amount of light available. Nevertheless, this species is one of the most shade tolerant species observed in the laboratory (Swift and Meunier, 1976).

Many methods have been used to estimate local productivity, each with its own disadvantages and shortcomings (see reviews and additional references in Tappan and Loeblich, 1971, and in Steidinger and Williams, 1970, p. 8 – 12). Like other phytoplankton, dinoflagellates form an important part of the marine food chain. Examination of the amino acid composition of the plankton from a red tide off southern California (in which *Gonyaulax polyedra* was the dominant species) showed the protein to be of good nutritional value, and similar to the casein used in trial feeding of rats. These experiments suggest that dinoflagellates may have potential value as a human food source, and that plankton farming might be carried out in conjunction with obtaining fresh water and chemicals from seawater (Patton et al., 1967).

Phytoplankton may be of value to organisms other than grazing ones in their role as photosynthesizers. Zooxanthellae are known to provide organic compounds to their invertebrate hosts, the products of photosynthesis excreted by the dinoflagellate providing their main or even sole source of nutrition. Using radioactive carbon, Khmeleva (1967) found that the radiolarian-zooxanthella symbioses are from 1.5 to 3 times as productive in the Red Sea as the free-living phytoplankton. In such symbiosis the major product of photosynthesis is passed to the host, and the endosymbionts themselves have a reduced growth rate as compared to free-living cells (Smith et al., 1969). Free-living dinoflagellates also excrete organic molecules, however, both the toxic ones of some red tide organisms and a large proportion of the usual photosynthetic product of others. At various light intensities *Prorocentrum minimum* ex-

creted 2.4 percent of the ^{14}C assimilated at 281 fc, 7.7 percent of the ^{14}C assimilated at 2350 fc, and 9 percent of that assimilated at 11,250 fc. At the same high light intensity of 11,250 fc, *Gymnodinium nelsonii* excreted 18 percent of the assimilated ^{14}C. During the stationary phase of growth in cultures, both *G. nelsonii* and *Scrippsiella trochoidea* excreted more than 50 percent of the photoassimilated carbon (Hellebust, 1965; Loeblich III, 1967). This dissolved organic matter is then available to other organisms, adding greatly to the overall productivity.

Blooms and Red Tides

The characteristic reddish-brown pigments of the dinoflagellates result in the discoloration of the seas known as red tides, red water, or yellow water, when the cell numbers temporarily become very great, usually near shore. Generally the patches are small and last only a few days, but sometimes they extend along the coast for 160 kilometers or more, as in 1907 in southern California, when the red water spread from 0.8 to 5 km off the shore. In 1902 a red tide extended from Santa Barbara to San Diego and lasted for about two months (Kofoid and Swezy, 1921). The very abundant phytoplankton may result in mass fish mortality, particularly of bottom-feeding types, as the plankton clogs the gills of fish, but may have little effect on invertebrates. Other species, including several *Gonyaulax*, selectively affect invertebrates, whereas a third group, including *Gessnerium acatenellum* (see Figure 4.86), *G. catenellum*, and *G. tamarense*, kills few marine animals directly, but their toxins accumulate in the living filter-feeding bivalves and cause paralytic shellfish poisoning of those consuming the bivalves. Additionally, the death and decay of huge numbers of cells at a red tide climax results in high bacterial populations and oxygen depletion that may also be fatal to various marine organisms (Steidinger, 1973).

The size of the cells responsible for the bloom, as well as their numbers, will affect the degree of discoloration of the water. Only 20 cells/ml of *Noctiluca* will result in the same vis-

Figure 4.86
Gessnerium acatenellum. SEM of ventral side of cell from June 1965 red tide in Malaspina Inlet, B. C., Canada, showing prominent flanges bordering girdle and cingulum and paralleling other plate sutures, and numerous trichocyst pores. Toxins produced by this species accumulate in filter-feeding bivalves to cause paralytic shellfish poisoning. ×2025. From L. Loeblich and Loeblich III, 1975.

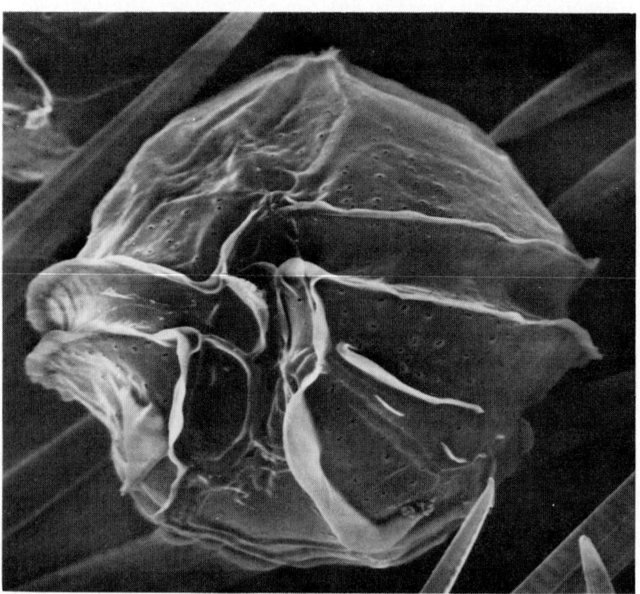

ible discoloration as 570 filaments of the blue-green alga *Trichodesmium,* or 6000 cells of *Peridinium triquetrum* (Hart, 1966). The ratio of surface area of these three species is about 3:2:1, *Noctiluca* having a diameter of from 650 to 2000 μm, and the *Peridinium* from 16 to 30 μm. Different species may bloom successively in the same area; for example *Prorocentrum* has a spring bloom and *Gonyaulax* an autumn one in La Jolla Bay, California, most blooms being monospecific at any given time. A few may include 2 or 3 dominant species in nearly equal numbers and other species of lesser frequency. A sample of one bloom containing 2,315,000 cells per liter contained 10 species with more than 5000 cells each, 4 of these numbering more than 150,000 cells, and 4 more numbering over 10,000 cells (Hart, 1966).

Red tides have been documented along the Florida coast since 1844 and may extend out from shore for 64 km along the entire west and much of the northwest coast of Florida, covering an area up to 36,257 km^2 and extending through 38 m in depth of the water column (Steidinger and Joyce, 1973). Low salinity generally prevents the red tides from entering es-tuarine regions. However, drought conditions in 1971 reduced this low salinity barrier sufficiently that the red tide affected Tampa Bay in Florida. Accumulation of cells was too rapid to result solely from reproduction, but apparently resulted in part from wind movement of surface populations and recruitment from large cyst populations. Floating dead fish removed from Tampa Bay for disposal totalled some 2367 tons in the 1971 bloom (Steidinger and Ingle, 1972).

Gymnodinium breve, the responsible organism in Florida red tides, is normally present in less than 1000 cells per liter and is harmless in such concentrations. Blooms of this species are a natural occurrence in the Gulf of Mexico each year, but only occasionally do severe outbreaks affect the coastal areas to the degree of the 1946−1947, 1953−1954, and 1957 red tides, resulting in major fish kills. The most extensive red tide conditions yet recorded on the Florida west coast lasted some 11 months in 1946−1947. Dead fish littered the beach; an estimated 100 pounds of fish were washed up per linear foot of beach between Cape Romano and Boca Grande Pass. In addition to the fish kills, human respiratory ailments resulted from the airborne dinoflagellate toxins (Steidinger and Williams, 1970). The July-August 1964 Florida red tides covered thousands of square kilometers of the northeast Gulf of Mexico, including Apalachee Bay, Florida. The discolored water and floating dead fish first were observed some 35 km from the coast, southwest of Steinhatchee River, and moved west-northwest to eventually dissipate in the open water. *G. breve* was present in numbers up to one million cells per liter at the surface, with 100,000 to 700,000 cells per liter at medium depths, and from 20,000 to 200,000 cells per liter at the bottom. Other dinoflagellates and diatoms were also abundant, including various species of *Gymnodinium, Peridiniopsis, Gyrodinium,* and *Gonyaulax,* together with large numbers of dinoflagellate cysts (from 3300 to 200,000 per liter). In one occurrence, *Pyrodinium bahamense* abundance resulted in discolored water over 800 m in length and 90 m wide, with 3 million cells per liter (Steidinger et al., 1966; Donnelly et al., 1966; Saunders and Wahlquist, 1966). The toxic occurrence of this species in New Guinea consisted of concentrated streaks containing 600,000 cells per liter, resulting in paralytic shellfish poisoning in 1972 (Worth et al., 1975).

Occasional *Gymnodinium breve* red tides occur on the east coast of Florida as well, probably arising in southwest Florida waters and being transported through the Florida Keys under particular current conditions (Murphy et al., 1975). Because red tides are associated with unusually heavy rains or increased runoff in Florida, an influx of nutrients was suggested as the main trigger. Laboratory experiments with *Gymnodinium breve,* the responsible organism in Florida blooms, indicated that its explosive growth was not related to an increase in organic matter, as this species did not show heterotrophic growth in the dark, but might be stimulated by B vitamins, trace metals, and chelators that were brought into the area by the increased runoff (Aldrich, 1962). Studies of conditions that might result in red tides indicated that phosphorus is present in the Gulf of Mexico in amounts equal or exceeding optimal requirements for a bloom, hence is probably neither a limiting nor a triggering factor. Salinity, pH, and temperature also are favorable, but insufficient evidence was available to indicate any part played by nitrogen. Extremely high iron concentrations in the water masses may have provided a stimulating growth factor (Donnelly et al., 1966). Steidinger (1975) suggested that *G. breve* blooms originate 18 to 74 km offshore in the late summer and fall. Patchy early occurrences were suggested to arise from germination of the benthic resting stages (cysts or hypnozygotes) that provide a seed population. Subsequent movement of the bloom inshore depends on the current and wind conditions. Although possibly the main cyst accumulations might be located offshore, the effects on the productivity of the Gulf of Mexico of any attempts at control might outweigh the benefits.

On a cruise in the western English Channel, the R.V. *Sarsia* found dense phytoplankton populations at the thermocline, the boundary between the nutrient-depleted surface waters and the cold, deeper nutrient-rich waters. Abundant *Gyrodinium aureolum* Hulburt gave chlorophyll *a* values of 100 mg/m^3 at the end of July; cell densities of up to 5 million cells per liter were recorded in the chlorophyll-rich layer of the thermocline in October (Denton, 1976, p. 14–16).

Hydrographic conditions on the Norwegian coast cause deep mixing of the waters. Resting spores or cysts of the dinoflagellates, which are abundant in the later stages of the blooms, settle to the bottom in the fall. Where depths exceed 50 m, convection currents do not return them to the surface during the winter, but in the spring those at shallowest depths near the coast are first returned to the photic zone (O. Nordli, 1951). Here, as elsewhere, dinoflagellate blooms are prevalent after the spring diatom bloom, and in late summer or fall. Discolored or "brown" water along the Norwegian coast in October and November 1966 resulted from the *Gyrodinium aureolum* bloom. In the outer Oslofjord the 26- to 40-μm-long cells numbered as much as 70 million per liter and resulted in mass mortality of sea trout. The inner polluted part of the fjord was dominated by *Prorocentrum micans,* in quantities of 20 million cells per liter, but perhaps because of the excessively strong upwelling caused by strong winds, no pollution reached the outer part of the fjord where blooms of *Gyrodinium* occurred (Braarud and Heimdal, 1970).

Red tides do not occur every year in California, but apparently result from the conjunction of various factors. Rainfall and runoff are limited and, unlike the Nile region of the Mediterranean, there appears to be no connection between stream influx of nutrients and red tide conditions. Even at the marine sewer outfalls, where nutrient influx is high, no blooms result (Lackey and Clendenning, 1965). Although the red tide organism *Gonyaulax polyedra* has high nutrient requirements, these are met by diurnal migration. The largest concentration of cells is at the surface in daylight, but moves to a depth of 10 to 15 m at night, where nitrate levels are high near shore as a result of upwelling. The species has a relatively low growth rate, with about 0.5 doublings per day at 20°C, under optimum light conditions (Eppley et al., 1969). The red water conditions are generally patchy, related to current patterns and upwelling conditions along the coast that result in concentration.

Although some blooms result largely from such concentration mechanisms as upwelling and wind-driven currents, with the cells recruited from cysts at the bottom, others apparently result from explosive growth rates. Small but dense concentrations of *Gymnodinium* sp. were observed in the uppermost 2 km of the 11 km length of the estuarine Duplin River at Sapelo Island, Georgia, in late February and early March 1955. This river receives no freshwater runoff. Cell concentrations during the bloom ranged from 1 million to 18 million cells per liter, the cells dividing every 7 to 12 hours just prior to the peak of the bloom. Over the entire bloom the upper 10 cm of water was estimated to contain a total of 2.08×10^{12} cells on March 4, 24.2×10^{12} on March 5, 11.3×10^{12} on March 6, and 1.76×10^{12} on March 7, for a total net production of 2.18 to 13.7 mg C per liter per day (Ragotzkie and Pomeroy, 1957).

The importance to the paleontologist of modern plankton blooms lies in both the position of dinoflagellates as primary producers for entire food chains and the relationship of cyst formation and abundance to the later stages of blooms, inasmuch as the fossil record is solely of these resting zygote cysts.

Habitat

Dinoflagellates are widely distributed in both marine and fresh waters. They occur in the plankton of freshwater lakes, rivers, bogs and ponds, brackish and salt water lakes, seas and oceans, as well as in snow and in the interstitial waters of the sandy beaches of lakes and seas.

Marine

As the vegetative stage of living dinoflagellates lacks a mineralized test or covering, this group is less readily preserved for study than some other phytoplankton. Nonthecate species disintegrate rapidly after collection, and many thecate species are so delicate, readily undergoing ecdysis, as to require study soon after collection. Smaller species pass through the usual plankton nets, and delicate ones are unlikely to withstand the centrifugation techniques for obtaining nannoplankton. Because of these factors, studies of living dinoflagellate distribution have lagged behind those of diatoms and coccolithophores. As noted by Steidinger and Williams (1970), fewer than 100 dinoflagellate species had been recorded in the Gulf of Mexico before 1967, but with the recognition of dinoflagellates as the red tide organisms in this region, interest in them increased greatly, so that over 400 species, varieties, and forms were recorded by 1970. A similar number of species occurs in the Caribbean Sea (Wood, 1968).

Physical Parameters. Some dinoflagellates are cosmopolitan in distribution, others more restricted by temperature, water depth, and nearness to land. Generally most species of *Ceratium, Dinophysis, Gonyaulax, Peridiniopsis, Prorocentrum,* and *Protoperidinium* (see Figure 4.87) are cosmopolitan, although some species of these genera may be restricted to tropical waters.

Figure 4.87
Protoperidinium steinii. A commonly neritic cosmopolitan
species. SEM of critical-point-dried cell, from red tide outbreak,
June 1965, in Malaspina Inlet, B.C., Canada; dorsal view, showing
prominent apical horn, crenulated girdle flanges, and reticulate
ornamentation of plates, which also have many tiny scattered
trichocyst pores; ×2900. From Allen et al., 1975.

The plankton was divided by Haeckel (1891) into **meroplanktonic** species, those producing benthic resting spores, and **holoplanktonic** species, lacking a benthic stage, and thus independent of the bottom. *Prorocentrum balticum* is a well-known holoplanktonic species of general occurrence in coastal and oceanic waters alike, in all oceans, and at all latitudes. Blooms of this species are reported from the Oslo Fjord, the North Atlantic, and the coast of Angola, Africa. Meroplanktonic species such as *Peridinium triquetrum* and *Scrippsiella trochoidea* have benthic resting stages, and thus may be absent from the plankton for at least part of the year (Braarud, 1962). Generally the meroplankton occurs in shallower coastal waters, as the germinating resting spores must be able to regain the photic zone. In some instances, meroplanktonic species are neritic, and holoplanktonic ones oceanic, but the terms are not entirely equivalent. The benthic resting stages of many meroplanktonic dinoflagellates seasonally may provide a seed population for reinvasion of the oceanic environment. The meroplankton has the greatest interest for the paleontologist, inasmuch as only the resting cysts are fossilized. The sole exception, *Actiniscus,* represented in the fossil record by its internal siliceous skeleton, lives in both coastal and oceanic areas of the eastern Gulf of Mexico

(Steidinger and Williams, 1970) as well as elsewhere, hence is cosmopolitan, rather than strictly neritic or oceanic.

Neritic dinoflagellate species typically are of generalized shape and lack conspicuous wings or horns. Simple, nonthecate *Gymnodinium*, *Gyrodinium,* and *Noctiluca* are accompanied by many armored species of *Protoperidinium*, *Gonyaulax*, *Prorocentrum*, *Zygabikodinium*, *Peridiniopsis, Dissodium, Dinophysis,* and some *Ceratium.*

Oceanic species commonly are ornate, with elongate horns, wings, lists, and other complex structures. These cell projections serve to increase surface area, both as an aid to flotation and an aid for nutrient uptake. Most nonthecate species are oceanic, especially the brightly colored species and those with ocelli. Oceanic species range from photosynthetic to holozoic or saprophytic in habit. Occasionally mixed with the neritic assemblage due to current distribution, the typically oceanic dinoflagellates include most *Ceratium,* many *Protoperidinium*, and the Dinophysiaceae with complex wings, lists, and spines. The elongate *Amphisolenia* and the ornate *Ornithocercus* similarly are oceanic (Lebour, 1925).

Salinity. Some dinoflagellates are euryhaline, tolerating a wide range of salinity variation, but others are more sensitive. *Scrippsiella trochoidea* has a salinity optimum of 15 to 20⁰/₀₀ (Braarud, 1961). *Amphidinium carterae* is euryhaline, although optimal growth occurs at a salinity of 25⁰/₀₀; growth ceased at the reduced salinity of between 2.5 and 10⁰/₀₀, but continued at the highest salinity tested, at 35⁰/₀₀ (McLachlan, 1961). The nonphotosynthetic supralittoral *Oxyrrhis marina* is subjected to wide salinity fluctuations in its tidepool habitat, and in culture thrives at salinities from 4⁰/₀₀ to 130⁰/₀₀, reaching its maximum division rate at the optimal salinity of 16⁰/₀₀, at a temperature of 22.5°C and pH between 8 and 10 (Droop, 1959). All marine species yet studied, except *Gymnodinium breve* (optimum of 27 to 37⁰/₀₀), have optimal growth in salinity well below that of the open sea (35⁰/₀₀), averaging an optimum of about 20⁰/₀₀, the lowest optimum being the 10⁰/₀₀ characteristic of *Prorocentrum balticum* (for references see Loeblich III, 1967).

As many peridiniaceans appear to prefer the reduced salinity near shore, and gonyaulacaceans prefer the more open marine environment, the ratio of gonyaulacaceans to peridiniaceans in cyst assemblages has been utilized as a salinity indicator (Harland, 1973). In deep sea cores in the Caribbean, this ratio is approximately 18 : 1 (Wall, 1967), and in the vicinity of Woods Hole, Massachusetts, the nearshore population is reported to have a ratio of 44 : 100. In keeping with this preference for reduced salinity, freshwater assemblages also are dominated by peridiniaceans.

Temperature. Many dinoflagellates have a wide temperature tolerance range, a requirement for such cosmopolitan species as *Prorocentrum balticum* and *P. micans. Ceratium* continues to grow at temperatures ranging from −1.0° to 29.5°C (E. Nordli, 1957), although most species have a temperature optimum between 18° and 25°C, warmer than normal ocean surface temperatures (Barker, 1935). The highest temperature tolerated in culture was 35°C by *Crypthecodinium cohnii.*

Generally correlative with temperature limitations, *Ceratium* assemblages characterize five distinct floras, some limited to the tropics and others to subpolar regions, with a cosmopolitan group of wider distribution. No distinctly temperate species are recognized, as temperate latitudes are populated by a mixture of tropical and cosmopolitan species (Graham, 1941). The five floras characterize the cold North Atlantic, warm Atlantic, cold North Pacific, warm Pacific, and southeast Pacific, the distribution being related to water mass characterisitcs (Graham and Bronikovsky, 1944). Although most species occur in both the Atlantic and Pacific, some 8 tropical species are restricted to the Pacific Ocean. As these tropical species are found only in waters warmer than 20°C, they could not move through the cold southern waters of the Cape Horn region. Among the 23 characteristic intolerant tropical ceratia are *C. breve* (Ostenfeld & Schmidt) Schröder, *C. digitatum* Schütt, *C. incisum* (Karsten) Jörgensen, and *C. trichoceros.* The group of 23 more tolerant tropical species includes *C. candelabrum* (Ehrenberg) Stein, *C. gravidium* Gourret, and *C. tripos.* Very tolerant tropical species number only 5, including *C. extensum* (Gourret) Cleve and *C. massiliense;* the 5 cosmopolitan species include *C. furca, C. fusus,* and *C. horridum,* and the 5 subpolar species include *C. arcticum* (Ehrenberg) Cleve and *C. macroceros.* About one-third of the 58 species of *Ceratium* studied by Graham were "shade species," being most abundant at a depth of about 100 m. Interestingly, actively dividing cells in chains, of *Ceratium vultur* and others, have been found to depths of 4389 m in the Eastern Mediterranean southwest of Turkey (Kimor and Wood, 1975).

A difficulty in utilizing the *Ceratium* species as environmental indicators results from their many temperature-related morphologic variations. Many supposed species thus are seen to be part of a continuous variation. In high latitudes and during the winter season, they are large and robust, with shortened horns and crests. In the tropics, and during the summer, the same species is represented by smaller, delicate forms, with very fine elongate horns. These intergradations are suggested to be only geographic or temperature-related phenotypes, although it is not certain that there is no genetic difference in addition (Sournia, 1966).

Nutrient Requirements. The living *Ceratocorys horrida* Stein is a tropical and warm temperate species, with distribution limited by the 19°C isotherm. If found in colder regions, it is an indication of tropical water incursions. Characteristically, warmer waters have low nutrient concentration, and this species is adapted to low levels of nitrates and phosphate (below 10 mg PO_4/m^3), occurring in many areas at levels of 5 mg PO_4/m^3. It is not particularly sensitive to changes in salinity. Other *Ceratocorys* and some *Gonyaulax* similarly are warm water species adapted to low nutrient levels (Graham, 1942).

Many of the dinoflagellates that are adapted to low nutrient concentrations (phosphate and nitrogen) not only occur in nutrient-poor waters, but their blooms characteristically follow those of the diatoms in the spring and fall in northern regions, as the nutrients are seasonally depleted (Graham, 1942). Generally, in tropical regions of low phosphate, dinoflagellate diversity may be high but represents low total numbers of individuals.

In addition to the requirements for phosphates and nitrates, various species, including the photosynthetic ones, need one or more vitamins, particularly vitamins B_{12} (cyanocobalamin), B_1 (thiamin hydrochloride), and H (biotin). Maintenance of bacteria-free cultures requires the addition of these vitamins to the media, and their availability in the oceans may affect the composition of the dinoflagellate assemblage and its abundance.

Distribution. *Cold Atlantic Waters.* Temperature tolerances result in a broadly latitudinal distribution of dinoflagellates, like that of other plankton. Lebour (1925) wrote a major monograph on those of northern seas, the distribution having since been determined in greater detail. A few subpolar species, which may occur in large numbers, characterize the cold North Atlantic flora north of lat. 40° N. Surface temperatures are less than 16°C, but phosphate levels are high, generally over 20 mg PO_4/m^3. Characteristic cold water species may be accompanied by rare cosmopolitan or tolerant tropical species (usually 1 or 2 up to as many as 10 species per station) that were carried northward by the west-wind drift (Graham and Bronikovsky, 1944).

South Atlantic dinoflagellate distribution has been documented by Böhm (1933) and Balech (1971b). In the South Atlantic from the Argentine coast to Antarctica, some 140 dinoflagellate species have been recorded, the greatest abundance being in the northernmost four stations, where various subtropical species were found. At many of the stations the dominant dinoflagellate is *Prorocentrum,* accompanied in some stations by *Ceratium.* Most dinoflagellates in these South Atlantic waters are shade species of deeper water, probably transported by deep currents (at depths of 200 to 400 m), and appear in shallower waters only in upwelling areas (Balech, 1971b). Antarctic phytoplankton species must tolerate a high and constant salinity, and extremely low temperatures, conditions to which only a few dinoflagellates are adapted (Smayda, 1958).

Cold Pacific Waters. In the cold waters of the North Pacific, temperatures are below 17°C, and phosphate levels about 25 mg/m^3, in places being as great as 100 mg PO$_4$/m^3. The phytoplankton includes a few subpolar and cosmopolitan species, but these are abundant (Graham and Bronikovsky, 1944). In the microbiota of San Diego Bay the over 100 species of dinoflagellates outnumber all other groups. Most abundant in the coastal stations, and fewest in the inner harbor, most of the latter were from interstitial waters (Lackey and Clendenning, 1965).

In the southwestern Pacific and Antarctic, dinoflagellates are relatively unimportant members of the plankton. Some *Protoperidinium* and *Dinophysis* occur, but many species described from southern waters are suggested to be only ecoforms of cosmopolitan species, with thicker walled theca and increased protuberances (Wood, 1964). In the eastern South Pacific, the surface temperatures range from 14°C to 27°C near Panama. The waters are always eutrophic, with high levels of phosphate, from 25 mg PO$_4$/m^3 to over 50 mg/m^3. Diversity is low, but the few cosmopolitan and tolerant tropical species present occur in large numbers (Graham and Bronikovsky, 1944).

Tropical Oceans. Phytoplankton living in tropical waters must tolerate very high temperatures and high salinities, such dinoflagellates as *Protoperidinium elegans* (Cleve) Balech being particularly successful in this respect (Smayda, 1958). South of lat. 40° N, in the warmer Atlantic, surface temperatures generally exceed 20°C, nutrient levels are low, and the plankton is sparse in numbers but of considerable diversity. The assemblage is distinctly tropical, but contains a few cosmopolitan species (Graham and Bronikovsky, 1944).

In the western equatorial Atlantic, some 218 species have been identified, including some surface-living species of *Ceratium, Gonyaulax* (including *G. hyalina, G. spinifera,* and *G. polygramma*), *Blepharocysta,* and *Palaeophalacroma.* In contrast, many other species were found only in deep tows and are regarded as shade species. These shade species are adapted to the low light levels of the deeper waters by increased surface area, very thin walls, and expanded cells with elongate processes. The elongate horns of shade species of *Ceratium* are crowded with plastids, as is the cell itself, in contrast to surface species of *Ceratium,* whose horns have only a few plastids. Other typically shade-living dinoflagellates of the equatorial region are *Gonyaulax pacifica, Heterodinium* spp., *Ornithocercus splendidus* Schütt, and *Triposolenia bicornis* Kofoid. When typical shade-dwelling forms are found in the upper waters, they then serve as biological indicators of upwelling, for example, various *Ceratium,* especially of the *C. vultur* group, *Dinophysis* spp., *Protoperidinium inflatum* (Okamura) Balech, and *Pyrocystis fusiformis* (Balech, 1971a). Shade species may occur at higher levels in the winter, and their distribution may not be simply a light reaction, but a response to nutrient levels, as most shade species have high nutrient requirements (nitrates and phosphates). As the nutrient availability is greater in deeper waters, particularly in the tropics where thermal stratification is continuous, these species only occur at depth or where upwelling brings nutrient-rich water to the surface (Graham and Bronikovsky, 1944).

The warmer Pacific areas have temperatures like those of the Atlantic, mostly above 20°C, but have more variable phosphate levels, ranging from less than 10 mg PO$_4$/m^3 in the central North Pacific up to 25 mg PO$_4$/m^3 to the north, and to the south reaching more than 25 mg PO$_4$/m^3. Many large species (30 or more) of *Ceratium* are present, although total abundance is low, most of the species being tolerant tropical ones (Graham and Bronikovsky, 1944). *Heteraulacus polyedricus* is another common tropical species that is not very temperature sensitive, as it persists when carried into cooler waters. Present in tropical waters of 20 to 30°C, it is carried into waters of 15 to 16°C off Japan and California and in the southeast Pacific, and in waters down to temperatures from 12 to 4°C in the Atlantic near the British Isles (Graham, 1942).

In the North Indian Ocean and Red Sea, 25 dinoflagellate genera and 186 species have been reported, including various species of *Prorocentrum, Pyrocystis, Gonyaulax, Protoperidinium, Ornithocercus,* and *Dinophysis* (Matzenauer, 1933).

Interstitial Flora

A rich microflora may be present in the interstitial waters in marine, brackish, and freshwater shore sands, as well as within the bottom muds. In addition to bacteria, protozoa, euglenids, diatoms, and coccolithophores, dinoflagellates are especially rich in species and may produce immense single-species concentrations. When abundant, their pigmented cells produce colored patches of brown, yellow, and green on the littoral sands at low tide.

An early study of the **interstitial flora** (Herdman, 1921, 1922, 1924a,b) of the Port Erin sands, Isle of Man, England, recorded 40-odd species, most of them rare and some of uncertain taxonomic rank. Relatively common species were of *Amphidinium, Noctiluca,* and *Polykrikos.* Their abundance and the resultant discoloration of the sands on the beach were characterized by a daily rhythm that seemingly is dependent on the stimuli of light and tides. The interstitial flora is exceptionally abundant and active at spring tides and scarcer at neap tides. Similar assemblages, including identical species, have been recorded at Woods Hole, Massachusetts (Herdman, 1924a,b; Lackey, 1961), Roscoff and Sieck in France (Balech, 1956; Dragesco, 1965), and in southern California (Lackey and Clendenning, 1965).

Concentrations of cells may be large, *Gymnodinium* being reported in numbers of 10,000 cells per cm^2, and totalling as much as 2 g of wet mass in a liter of sand. In similar concentrations at Sieck, Brittany, *Amphidinium herdmanii* populations covered some 200 m^2 of sand and were estimated to number about 30 billion individuals, representing some 500 g of wet dinoflagellate cells. Although the sand was completely reworked to a depth of tens of meters at each high tide, the *Amphidinium* population was again present at the next low tide, suggesting a very marked vertical migration for protection (Dragesco, 1965). *Thecadinium petasatum* in the same area was mainly restricted to deeper pools, where it occurred by millions, although it was much more restricted in occurrence in the sands (see Figure 4.88). The species has radiating and scale-like brown plastids, and when present in

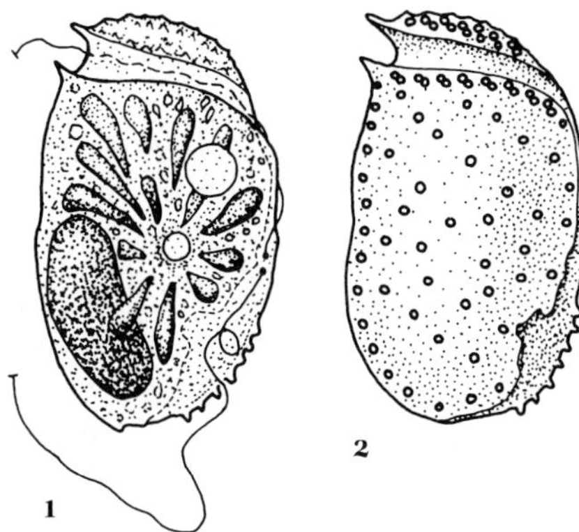

Figure 4.88
Thecadinium petasatum. **1.** Appearance of vegetative cell living in interstitial water of intertidal sand, showing flagella (shortened in figure), large nucleus at cell posterior, and radially arranged plastids. **2.** Empty theca, showing trichocyst pores. Slightly modified from Dragesco, 1965.

large numbers would form nearly black patches on the sands. Even where the dinoflagellates are less evident at the surface as patches, footprints produced by stamping on the wet sand are colored yellow by the many cells that sweep into the depressions as they fill with water (Dragesco, 1965).

Many of the dinoflagellates of the interstitial habitat are strongly compressed; some are photosynthetic, such as *Prorocentrum marinum, Polykrikos lebourae* Herdman, *Gymnodinium arenicolus* Dragesco, and *Thecadinium;* but others are colorless and saprophytic, being particularly numerous in the sands just above an organic black mud layer (*Amphidinium pellucidum* and *Sinophysis ebriola*). Perhaps those adapted to this environment are particularly euryhaline, as *Amphidinium pellucidum* has been reported in both brackish and marine sands, and *Gymnodinium arenicolus* occurs in marine sands at Roscoff, France, and in the freshwater Lake Leman (Dragesco, 1965).

An interstitial microflora obtained from cores of bottom muds at Woods Hole included more than 25 species of dinoflagellates. Some small saprozoic *Amphidinium, Gymnodinium, Hemi-*

dinium, and *Katodinium* are common in the anaerobic patches, where they are accompanied by sulfur bacteria, euglenids, and ciliates. Other *Amphidinium* and *Thecadinium* species are restricted to sandy aerobic patches of the bottom (Lackey, 1961).

Tidepools

Dinoflagellates may be abundant in tidepools as well as the open ocean. *Peridinium gregarium* forms dense populations at the bottom of small rocky tidepools along the southern California coast. It is benthic during heavy tidal action and at reduced temperatures, but at other times it becomes free of the substrate in the morning and collects in masses at the upper surface of the submerged rocks. These masses of cells are held together by a mucous matrix excreted from the antapical pores, while the oxygen from photosynthesis accumulates in the matrix and floats it to the pool surface. In the afternoon and evening the individual cells leave the mucus behind and return to the bottom. By means of this diurnal migration, they take advantage of the maximum available light, yet avoid copepods, the main predators of the tidepools, as these do not tolerate the high light intensity at the surface (Lombard and Capon, 1971).

Fresh Water

Although dinoflagellates are more diverse in the marine environment, many are found in fresh water, particularly in a somewhat acidic situation. Most freshwater species have a worldwide distribution; of 55 armored dinoflagellate species from the northern United States reported by Eddy (1930), 51 were previously known from Europe, 3 from Africa, 2 from Asia, and 7 from Australia. *Peridinium* is represented by 28 species and *Peridiniopsis* by 17 species.

The well-known *Ceratium hirundinella* is worldwide in distribution, in both brackish and fresh water. Other widely occurring genera include the thecate *Gonyaulax, Peridinium,* and *Peridiniopsis,* and nonthecate *Amphidinium*

and *Gymnodinium* (of which 23 species were reported in fresh water by Kofoid and Swezy, 1921); *Gyrodinium* and *Hemidinium* are represented mainly by small species of generalized type.

Early reports of freshwater dinoflagellates include those by Lemmermann (1910), Schilling (1913), Skuja (1956), and Thienemann (1950). More recently, Bourrelly (1970) has recorded 30 dinoflagellate genera and 230 species as present in fresh water.

Although less well known than their marine counterparts, massive blooms of dinoflagellates also occur in fresh water. A bloom of *Peridinium uberrimum* Allman resulted in brown waters in large ponds in June and July of 1854, the color of the water being so deep brown that a one-half-inch white disc became invisible at a depth of three to six inches (~7 to 15 cm). Many cells were in process of fission, and others formed rounded "cysts" (Allman, 1855), although whether these were true resting cysts or merely rounded cell contents after ecdysis is not certain.

Few cysts were observed in the plankton during heavy blooms of *Peridinium cinctum* forma *westii* in Lake Kinereth, northern Israel, but following the bloom, many hundreds of cysts were present per gram of sediment on the bottom. Empty cysts also were frequent, each showing a characteristic opening (the archeopyle) through which the germinating cell had emerged (Eren, 1969b).

A bloom of a species of *Gymnodinium* was observed to occur under the ice at Lake Baikal (Kozhova, 1959), and red snow and ice covering Lake Davos in the Swiss Alps resulted from massive accumulation of cysts of *Woloszynskia pascheri.* The encysted cells contained oil droplets and pigmented granules (haematochrome) of a bright red color. Motile cells are present at temperatures between 0 and 15°C, and the cells withstood an overnight temperature of −16°C. The thin-walled ovoid cysts became rigid when frozen, and, in fact, the "cysts" may represent only the frozen pellicle. Individual cells formed agglomerations or "colonies" up to 7 cm across, and winnowing of these agglomerations produced the strips of red snow. In culture, cells quickly become motile after thawing, and old cultures produce thick-walled cysts (Suchlandt, 1916). Similar red snow due to this species has been reported at Lake Simcoe, Ontario, Canada (Gerrath, 1974).

THE FOSSIL RECORD

Historical Background

Fossil dinoflagellates first were observed in thin sections of cherts, Ehrenberg (1836b, p. 114) reporting two species as *Peridinium pyrophorum* (now *Palaeoperidinium*) (see Figure 4.89) and *P. delitiense* (now *Spongodinium*). Four additional species that were referred to the desmid *Xanthidium* are typical "hystrichospherids," or chorate dinoflagellate cysts, *Xanthidium furcatum*, *X. aculeatum*, *X. pilosum*, and *X. ramosum*. These and other species that Ehrenberg later placed in *Xanthidium* have since been variously transferred to *Hystrichosphaeridium*, *Hystrichokolpoma*, *Operculodinium*, *Tenua*, and *Spiniferites*. The following year Ehrenberg (1837a) described *Peridinium priscum* and *Xanthidium delitiense* and recorded fossil *P. pyrophorum*, *X. furcatum*, *X. aculeatum*, and *X. hirsutum* (the last three are living desmid species that were originally described from Berlin). Illustrations were then given (Ehrenberg, 1837b) of *P. delitiense, P. pyrophorum, X. hirsutum, X. furcatum, X. ? ramosum, X. tubiferum,* and *X. bulbosum,* all from the Cretaceous cherts of Delitzsch, Germany. However, although Ehrenberg discovered and first reported the fossil dinoflagellates, he was not the first to publish illustrations of his material. Shortly after his discovery was announced to the Berlin Academy, some of his specimens, together with labelled sketches of the species, were sent by his associate, Baron von Humboldt (1837), to M. Arago of Paris to be shown to the Paris Academy and deposited in the Museum d'Histoire Naturelle at Paris. These illustrations were published in the Comptes Rendus of the Paris Academy by Turpin (1837), with a suggestion that they most probably represented fossil "eggs" of the bryozoan *Cristatella,* an assignment quickly and strongly re-

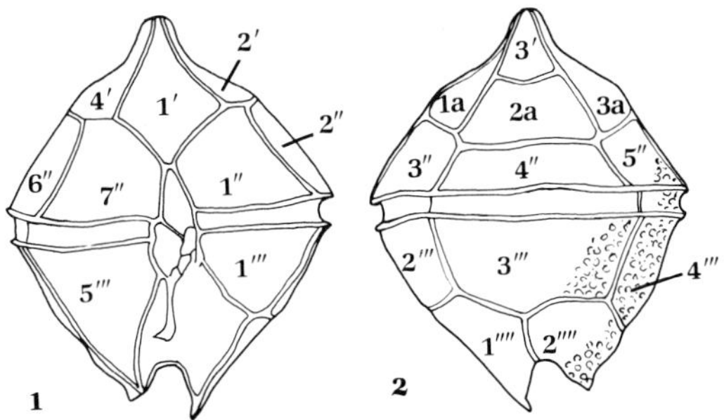

Figure 4.89
Palaeoperidinium pyrophorum. Ehrenberg's type specimen from Cretaceous flints of Delitzsch, Germany, showing **1**, ventral, and **2**, dorsal views of holotype. Redrawn from Lejeune-Carpentier, 1938b.

futed by Ehrenberg (1837b). In 1839 Ehrenberg reported the presence of *Peridinium pyrophorum* and two species of *Xanthidium* in the Upper Cretaceous chalk of Brighton, England, five species of *Xanthidium* from Gravesend, England, and two from Meudon, France. Some of Ehrenberg's original material has been redescribed since then (Lejeune-Carpentier, 1937a,b, 1938b; Sarjeant, 1967b).

Reporting on new species in the English flints, Reade (1839, p. 194) suggested a method of chipping the flints instead of grinding thin sections; nearly a century later this was redescribed by Deflandre (1935, p. 215−217) as providing better results than were obtainable by sectioning. Conrad (1941) used both methods, and Valensi (1953, p. 18−19) reported finding 320 microplankton specimens in 50 mm^3 of Jurassic flint nodules in chips, but found only some 60 specimens in a comparable volume in thin sections, suggesting that the flint chips more accurately reflected the true richness of the assemblage.

Examination of various layers of the flints has shown a marked difference in specimen abundance (W. Wetzel, 1965), ranging from 16 specimens per cubic centimeter (representing 4 species) to as many as 144 individuals representing 36 species in a similar volume. A flint specimen from Kiel had 800 microplankton specimens per cubic centimeter, representing 40 species (although including radiolaria and coccolithophorids as well as the dinoflagellates). In a reexamination of flint chips from Delitzsch about $\frac{1}{2}$ mm thick and 2 mm in diameter, originally studied by Ehrenberg, some 80 individual specimens were found, largely clumped at one side of the fragment (Lejeune-Carpentier, 1937b, p. 60–61, and some 500 specimens were reported in a 2-cm^2 area in thin sections (Lejeune-Carpentier, 1938b).

In other early studies, additional species were described from the English Cretaceous (White, 1842; Mantell, 1844), Mantell observing their composition to be organic rather than siliceous. Mantell's material also has been restudied (Sarjeant, 1967c) and the types refigured. Deane (1845) noted that the fossil "xanthidia" of the flints also were present in HCl residues that he obtained by dissolution of Cretaceous chalk from cliffs between Dover and Folkestone. Noting the openings (now known as archeopyles) in the specimens, he disclaimed any relationship of these fossils with the desmids such as the true *Xanthidium.* Wilkinson (1849) found additional specimens in the muds of the Thames River, but after this flurry of studies in the early 1800s, no further interest was shown until some 50 years later, when Merrill (1895) illustrated and described similar specimens from the Texas Lower Cretaceous, although identifying them as sponge spicules. Merrill's material also has been redescribed and assigned to various dinoflagellate species (Sarjeant, 1966a).

Reinsch (1905, p. 402, footnote) finally suggested the dinoflagellate nature of Ehrenberg's fossil *Xanthidium,* indicating that they were probably peridineans, but his comment was then forgotten. Nearly 40 years later, fossil dinoflagellates again were discovered, this time in the radiolarian earth of Springfield, Barbados, in material prepared for diatoms (Lefèvre, 1932b, 1933a–c, 1940). First seen in balsam mounts of radiolarians, specimens were later observed free of matrix. Most individuals of *Peridinites* were seen to be "incomplete," always lacking the same paraplates, due to disarticulation along part of the girdle and at the antapical sutures. Such openings were sug-

gested to have resulted from the liberation of a *Gymnodinium*-like stage in asexual reproduction. The thick siliceous wall, with its exaggerated ornamentation of ridges, striae, flagellar pores and perforations, was regarded as due to silica replacement of an originally cellulosic theca in the unusual environment. Deflandre (1933, 1940) described similar siliceous specimens from the diatomites of Oamaru, New Zealand, and the Paleocene of Kuznetsk, also regarding them as fossilized thecae, even after the organic-walled taxa had been recognized as cysts (Deflandre, 1965; Sarjeant and Downie, 1967).

During the 1930s and 1940s, many studies were made of the fossil dinoflagellates from the cherts in Germany (O. Wetzel, 1932, 1933a,b), France (Deflandre, 1935, 1936; Valensi, 1949, 1953), and Belgium (Conrad, 1941; Lejeune-Carpentier, 1937a and later publications). O. Wetzel (1935) also obtained specimens from the German upper Eocene clays, beginning the major expansion of studies based on palynological preparations of rocks of all types and from many ages.

Although some of the organic-walled specimens were described as dinoflagellates in the earliest studies, based on the median girdle or cingulum, indications of tabulations, and other morphologic features, the systematic position of the large group of hystrichospheres (including Ehrenberg's *Xanthidium* and Mantell's *Spiniferites*) was regarded as uncertain. New suggestions of their dinoflagellate nature, like that of Reinsch (1905), were based on their several features in common with modern cysts (Erdtman, 1954, p. 110). From bottom sediment samples collected in the fall, some were identified as the resting cysts of *Gonyaulax polyedra, Protoceratium reticulatum* (now *G. grindleyi*), and *Peridinium triquetrum.*

Additional evidence that the fossils were not thecae was provided by the unusual morphology of those even in very young sediments that did not resemble motile cells of any living taxa. Deflandre (1965, p. 392) observed that ridges present in the transverse groove or cingulum of *Phanerodinium* in Cretaceous cherts would have made it nonfunctional for a flagellum, as would the trabeculae covering the girdle of

Wetzelodinium. Even after some fossils were assigned to the cyst stage of the life cycle because of their similarity to cysts of living dinoflagellates, others were still thought possibly to be motile cells (Balech, 1967a; Deflandre, 1965; Sarjeant and Downie, 1967; Eisenack, 1966b; Gocht, 1967; May, 1976). Evitt (1961a) identified 39 fossil genera as motile cells, and 43 as cysts; the 30 remaining unallocated include some genera now placed with the Prasinophyceae and others in the Acritarcha.

Geologic History

The many primitive features of the dinoflagellates suggest that they must have had a long geologic history, yet they are best known in the Mesozoic and Cenozoic. Two possible explanations are obvious: They may be present but unrecognized, perhaps regarded as Acritarcha; or they may have had a long geologic history prior to developing the extremely resistant cysts that are preservable.

The first explanation is suggested by a possible relationship of the Prorocentrales to the acritarchs (Chapter 3), those phytoplankton cysts that represent the dominant members of the early Paleozoic microflora (Loeblich III, 1976, p. 23). Although cysts have not yet been demonstrated in the Prorocentrales, this stage may have been secondarily lost. This explanation suggests that the earliest dinoflagellates had a two-part theca, and that the distinctly tabulated peridinialean theca arose by segmentation from the bipartite one. The presence of empty vesicles in the naked dinoflagellate amphiesma in the position of the thecal plates of armored species suggests that the nonthecate dinoflagellates are a later offshoot rather than primitive.

If acritarchs are not dinoflagellates, the Silurian *Arpylorus* is the oldest fossil representative; any earlier dinoflagellate diversification may have preceded the development of a resistant-walled cyst. Only a few living dinoflagellates produce acid-resistant cysts, and these apparently all are among the Peridiniales. Some Gymnodinialean cysts are mildly acid-resistant, but do not withstand the maceration treatment used with fossil dinoflagellates. Cysts produced by some living Gymnodiniales, and by some Peridiniales as well (e.g., *Ceratium,* some *Woloszynskia*) are cellulosic and not acid-resistant or preservable as fossils. Others are largely cellulosic, but with an acid-resistant layer (*Woloszynskia,* possibly *Lophodinium*). Some mineralized cysts are known in the Peridiniaceae, and similar calcareous cysts are known as fossils. No information is available as to cysts in living Dinophysiales, although the fossil *Nannoceratopsis* is regarded as belonging to that group. The other 12 living orders have no fossil record, nor do most of the living families of the Peridiniales. Thus, the dinoflagellate fossil record may be more biased than that of such other major phytoplankton groups as the diatoms or coccolithophorids.

Over the past decade a wealth of data has accumulated from both fossils and living taxa, allowing the correlation of cyst and motile stage for an increasing variety of species and genera, and leaving little doubt as to the cyst nature of the fossil dinoflagellates (Eisenack, 1969a; Evitt, 1963; Sarjeant, 1967a, 1974; Tappan and Loeblich, 1977).

An understanding of the relationship of the fossils to the living cells has proceeded on two fronts, both by the theoretical assumptions that the processes, crests, and ridges of the fossilized cyst provide an indication of the tabulation of the parent motile cell (Evitt, 1963), and by the study of life cycles and development of cysts in culture (Stosch, 1973) or the study of cysts still within the theca of the parent cell and the incubation of cysts in the laboratory (Evitt and Davidson, 1964; Evitt and Wall, 1968; Wall, 1965; Wall and Dale, 1966 and later). Use of the SEM in conjunction with the light microscope also has led to a better understanding of the nature of the fossil cysts as well as the theca of the living species.

Like other fossil plankton, dinoflagellates have proven to be excellent index fossils, and some species have been recorded from equivalent strata in six continents. Summaries of the stratigraphic and geographic distribution of fossil dinoflagellates make their utilization much simpler (Drugg and Stover, 1975; Harker and Sarjeant, 1975; Millioud et al., 1975; Riley

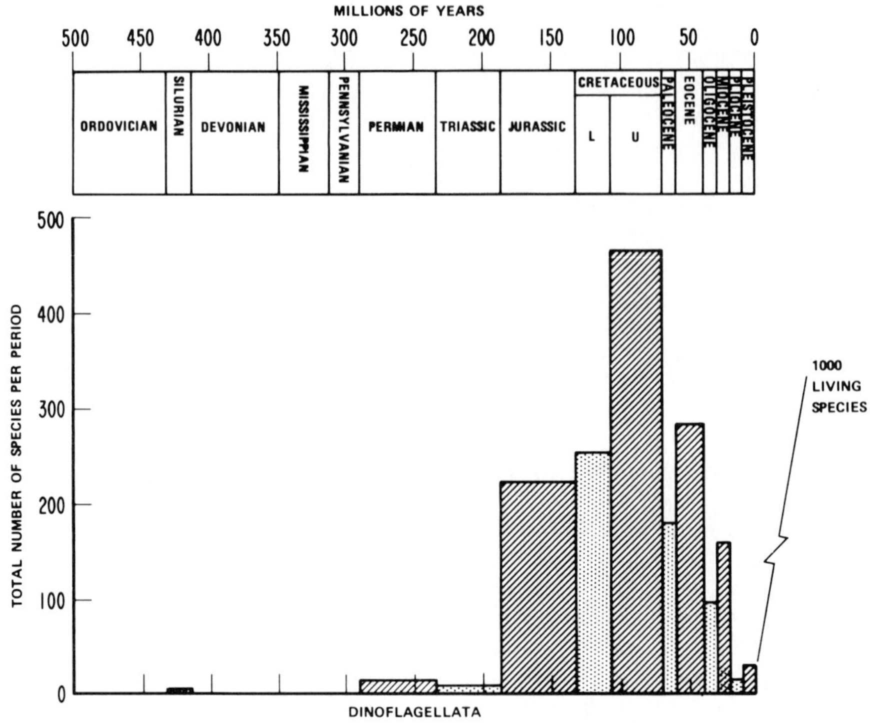

Figure 4.90
Dinoflagellate species diversity through geologic time. Total number of described species has increased since this chart was prepared, but the relative numbers for various periods have not changed greatly. Note rapid diversification during the Jurassic and Cretaceous periods, and post-Cretaceous continual decline in cyst-producing species. From Tappan and Loeblich, 1973.

and Sarjeant, 1972; Sarjeant, 1975a; Sarjeant and Downie, 1967). Some dinoflagellate lineages appear to have a long history, as recorded by their cysts, yet only a small number of the living genera and families are known as yet to have such resistant resting cysts, suggesting that in this group the fossils provide a relatively incomplete record of the true diversity (see Figure 4.90).

Paleozoic

The Silurian *Arpylorus antiquus* Calandra from Tunisia (see Figure 4.91) apparently has a precingular archeopyle and indications of tabula-tion. Its nature has been questioned (Eisenack, 1969a, p. 338) because of its rarity and the somewhat ambiguous appearance of the surface "tabulation" that may represent only surface folds. The specimens do appear dinoflagellatelike, however, and are generally accepted.

The gap of some 150 to 200 million years between this and the next fossil record, in spite of the intensive study of the intervening rocks, is also troublesome. Some reported dinoflagellates from the Permian (*Kofoidopsis* Tasch, *Nannoceratopsiella* Tasch) appear instead to be fragments of plant cuticle or mineral grains rather than phytoplankton, but *Palaeoperidinium* sp. (Jansonius, 1962) from the Permian of Canada appears to be an unquestionable dinoflagellate.

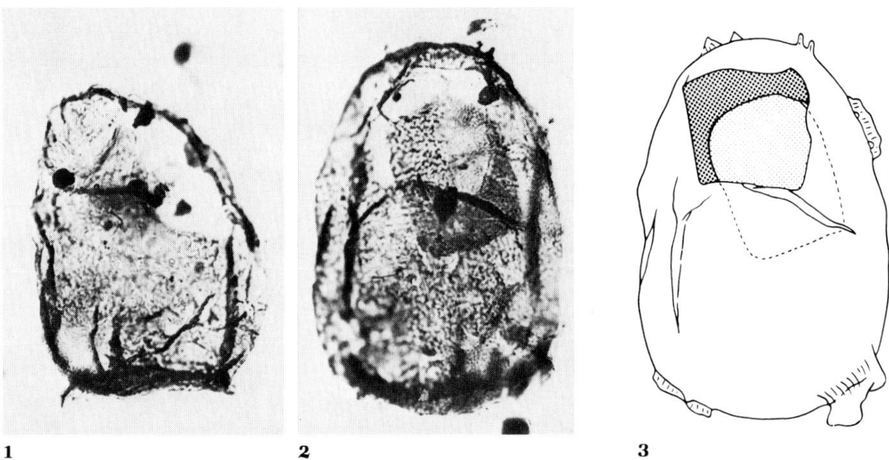

Figure 4.91

Arpylorus antiquus Calandra, Upper Silurian, North Africa. **1.** Left side. **2.** Dorsal view, with operculum slightly displaced. **3.** Diagram based on part 2. All ×320, from Evitt, 1967a.

Mesozoic

The Triassic dinoflagellate assemblage is sparse, although *Diconodinium* cf. *caulleryi* (Deflandre) Deflandre was reported from the Lower Triassic of western Canada (Jansonius, 1962), *Rhaetogonyaulax* (as *Shublikodinium*) from the Upper Triassic (Carnian) of northern Alaska (Wiggins, 1973), *Hebecysta, Heibergella, Noricysta,* and *Sverdrupiella* from the slightly younger Norian in Arctic Canada (Bujak and Fisher, 1976), and additional genera and species from the Middle and Upper Triassic of Israel (Horowitz, 1975) and from the uppermost Triassic (Rhaetic) of England and Austria (Sarjeant, 1963a; Morbey, 1975). Within the Triassic, both gonyaulacacean and peridiniacean dinoflagellates are known, the various cavate, proximate, chorate, and marginate cysts also showing a variety of archeopyle types. *Rhaetogonyaulax* has a characteristic gonyaulacacean paratabulation (1 apical closing plate, 6′, 5a, 7″, ?c, 5s, 7‴, 1ppl, 3⁗) and a combination apical-intercalary archeopyle (Wiggins, 1973). *Suessia swabiana* Morbey in contrast has an unusually large number of plates and additional plate series (Morbey,

1975), and probably indicates a different lineage (see Figure 4.92).

Assemblages of the Early Jurassic and early part of the Middle Jurassic also are sparse, consisting of small individuals of uniform and monotonous appearance (Gocht, 1964; Morgenroth, 1970), but with some interesting species. The early Liassic *Dapcodinium priscum* Evitt resembles *Suessia* in having a large number of plates, with paratabulation of 4′, 4a, 7″, 6c, 6‴, 2p, 1⁗, and an unusually broad ventral area (Evitt, 1961c). A still more interesting Jurassic genus, *Nannoceratopsis* Deflandre, has a morphology suggesting relationships with the modern Order Dinophysiales (Evitt, 1961b; not the Desmophyceae as indicated by Harland, 1972, p. 135). Widely distributed in Europe, North America, and Australia, the similarity of this genus to living *Dinophysis* and *Phalacroma* was demonstrated by Gocht (1972). However, no other fossil Dinophysiales are known to span the gap of 135 million years since the Late Jurassic. Modern Dinophysiales are not known to produce similar resting cysts, although none have yet been maintained in culture to determine the nature of their life cycle. The suggestion (Haskell, 1974) that *Nannoceratopsis* (see

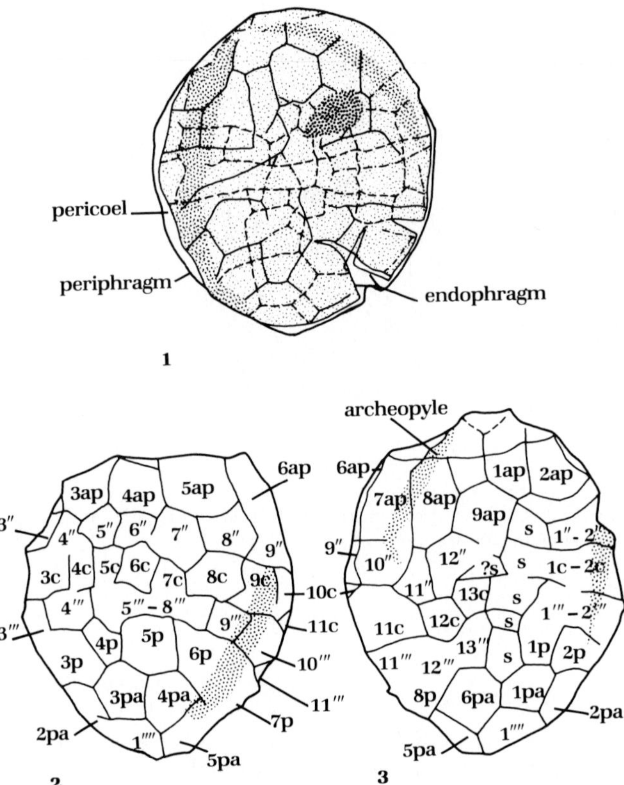

Figure 4.92
Tabulation of *Suessia swabiana* Morbey, Upper Triassic, Austria. **1.** Ventral side, stippled area is endophragm, with dorsal plates dashed, as seen in transparency; ap, postapical series; pa, preantapical series. **2.** Dorsal view. **3.** Ventral view, with archeopyle formed by loss of apical and anterior intercalary paraplates. Redrawn from Morbey, 1975.

Figure 4.93) is referrable to chamber linings of calcareous foraminiferans was refuted by Evitt (1974a) on the basis of dissimilar morphology, color, and size range. In addition, no Jurassic foraminifer with such specialized morphology is known, and the restricted geologic range of *Nannoceratopsis* would be even stranger with such an affinity.

Another interesting group appearing in the Cretaceous, the Pseudoceratiaceae, has a cavate cyst with elongate horns that strongly resembles cysts of living freshwater *Ceratium* and was probably ancestral to that genus. Although these Cretaceous representatives have resistant sporopollenin cysts, the living marine taxa seemingly produce none, and freshwater species have reverted to a cellulosic cyst. Hence a Cenozoic record is lacking for this important modern group.

Dinogymnium also is characteristically Cretaceous in distribution. Once thought to represent the Gymnodiniales, it probably is a peridinialean like the modern Lophodiniaceae, whose marine evolution also involved a loss of the acid-resistant cyst stage, restricting its fossil record.

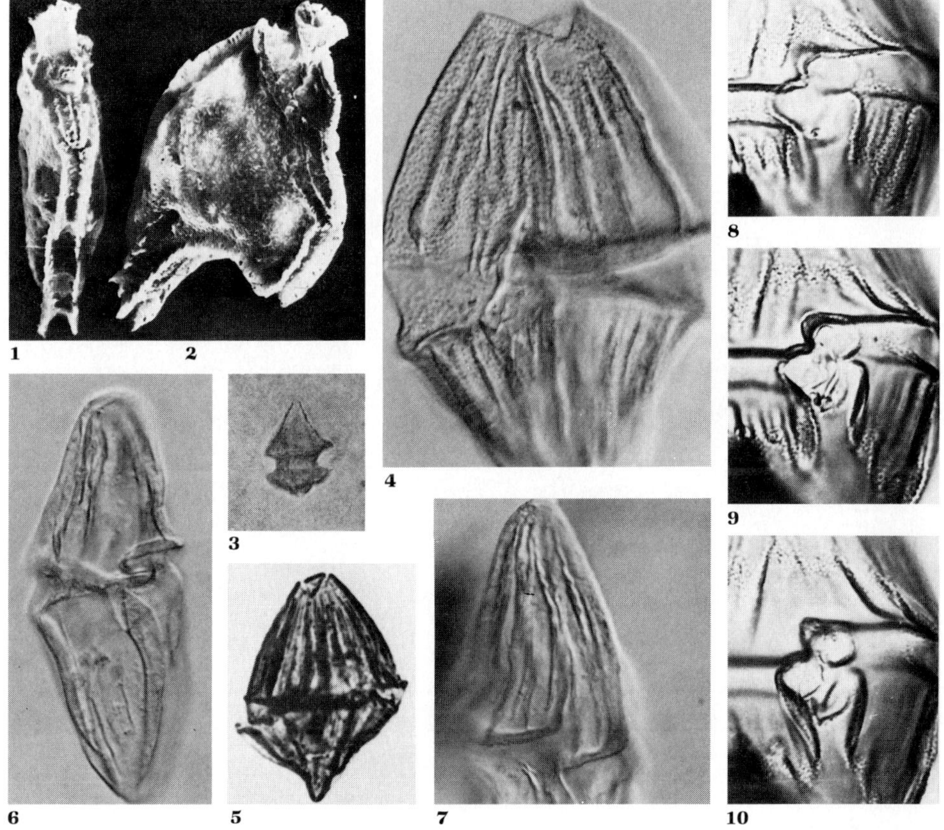

Figure 4.93

1,2. *Nannoceratopsis gracilis,* Lower Jurassic, Germany, SEM, from Gocht, 1972, *Lethaia,* v. 5, p. 15–29, published by Universitetsforlaget, Oslo University. **1.** Ventral view, ×695. **2.** Right side, ×760. **3.** *Dinogymnium strombomorphum* (Deflandre) Evitt et al., Upper Cretaceous, France, ×615, with permission, from Deflandre, 1943, *Bull. Soc. Géol. France,* sér. 5, v. 13, p. 499–509. **4,5.** *D. acuminatum* Evitt, Clarke & Verdier. **4.** Upper Cretaceous (Taylor), Texas, phase contrast, ×1040, from Tappan and Loeblich, 1977. **5.** Paleocene, probably reworked, ×320, from Evitt, 1973. **6,7.** *D. westralium* (Cookson and Eisenack) Evitt et al., Upper Cretaceous (Taylor), Texas, ×1040; 6, from Tappan and Loeblich, 1977. **8–10.** *D. acuminatum,* Upper Cretaceous, California, ventral view of girdle area at successively deeper focus levels, ×800, from Evitt et al., 1967.

Proximate and chorate cyst species were dominant during the Middle and Late Jurassic, although accompanied by a few membranate and trabeculate ones. Cavate cysts become significant in the Early Cretaceous, such genera as *Deflandrea* (see Figure 4.94), *Wetzeliella,* and allied forms remaining important through the Tertiary. Some biostratigraphic zonations have been proposed for the Mesozoic (Johnson and Hills, 1973; Pocock, 1976), although dinoflagellates are less well known than other microfossils for such uses.

During the Middle and Late Jurassic and through the Cretaceous, dinoflagellate diversity increased rapidly (see Figure 4.95), culminating in a described Cretaceous diversity equal to 75

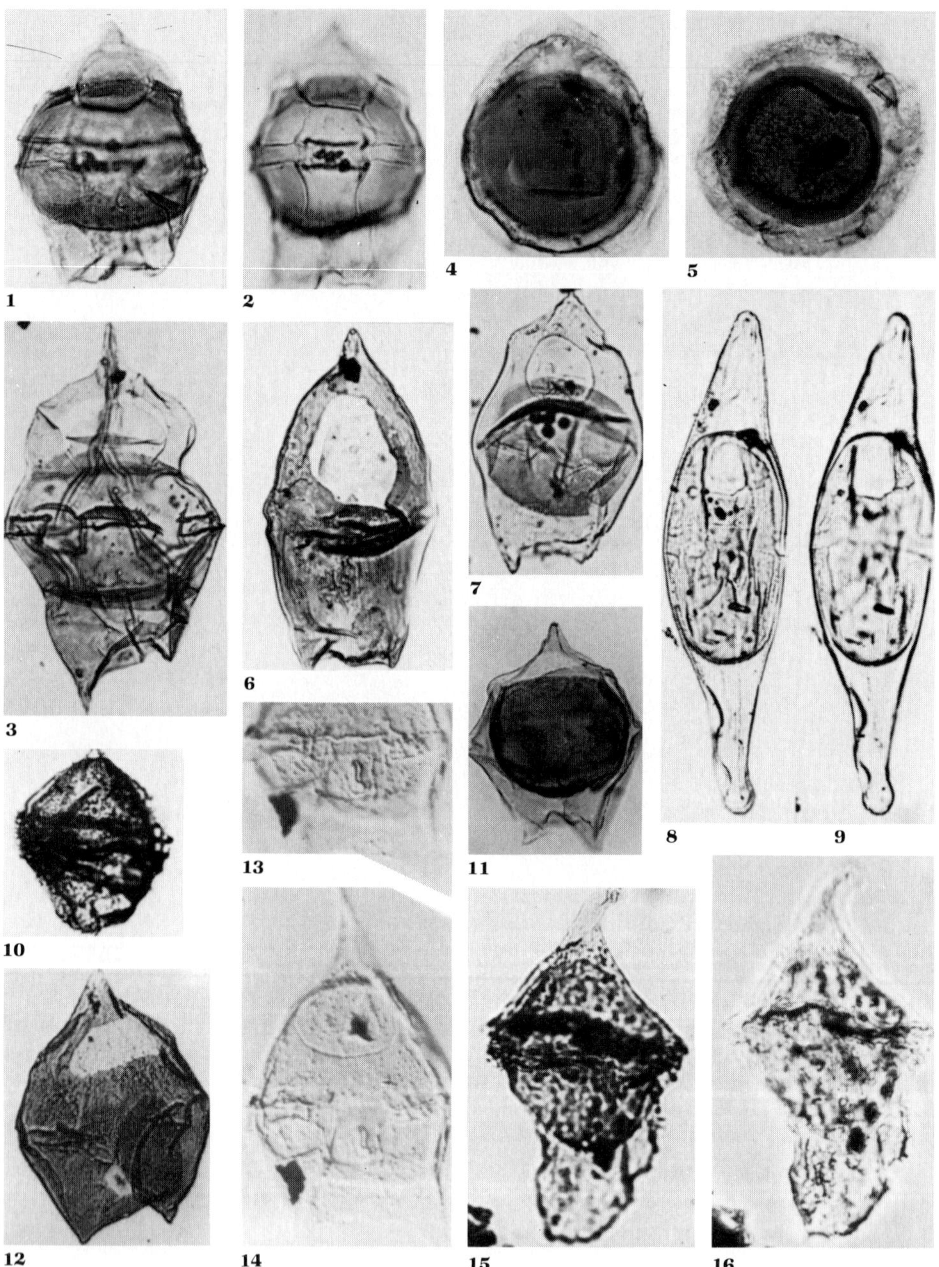

Figure 4.94 (*facing page*)

Deflandreaceae. **1,2.** *Chatangiella scheii* (Manum) Lentin & Williams, Cretaceous, Arctic Canada, ×400, from Manum, 1963. **3.** *C. ditissima* (McIntyre) Lentin & Williams, Upper Cretaceous, Arctic Canada, ×400, from McIntyre, 1975. **4,5.** *Catillopsis abdita* Drugg, lower Eocene, Alabama, dorsal and ventral views, ×390, from Drugg, 1970. **6,7.** Upper Cretaceous, Arctic Canada, from McIntyre, 1975. **6.** *Isabelidinium amphiatum* (McIntyre) Lentin & Williams, ×400. **7.** *I. microarmum* (McIntyre) Lentin & Williams, ×400. **8,9.** *Svalbardella cooksoniae* Manum, lower Tertiary, Spitzbergen, ×400, with permission, from Manum, 1960, *Nytt Mag. Bot.*, v. 8, p. 17–26, published by Universitetsforlaget, Oslo University. **10.** *Hebecysta brevicornuta* Bujak & Fisher, Upper Triassic, Arctic Canada, ×480, from Bujak and Fisher, 1976. **11,12.** *Deflandrea phosphoritica* Eisenack. **11.** Lower Eocene, Alabama, ×385. **12.** Specimen lacking inner body but with periphragm operculum inside cyst, Eocene London Clay, England, ×320, from Williams and Downie, 1966b. **13,14.** *D. medcalfii* Stover, Paleocene, Australia, phase contrast, focussed on sulcus and girdle; note flagellar markings in part 13; ×400, from Stover, 1974. **15,16.** *Sverdrupiella septentrionalis* Bujak & Fisher, Upper Triassic, Canada, focussed on ventral and dorsal surfaces, ×480, from Bujak and Fisher, 1976.

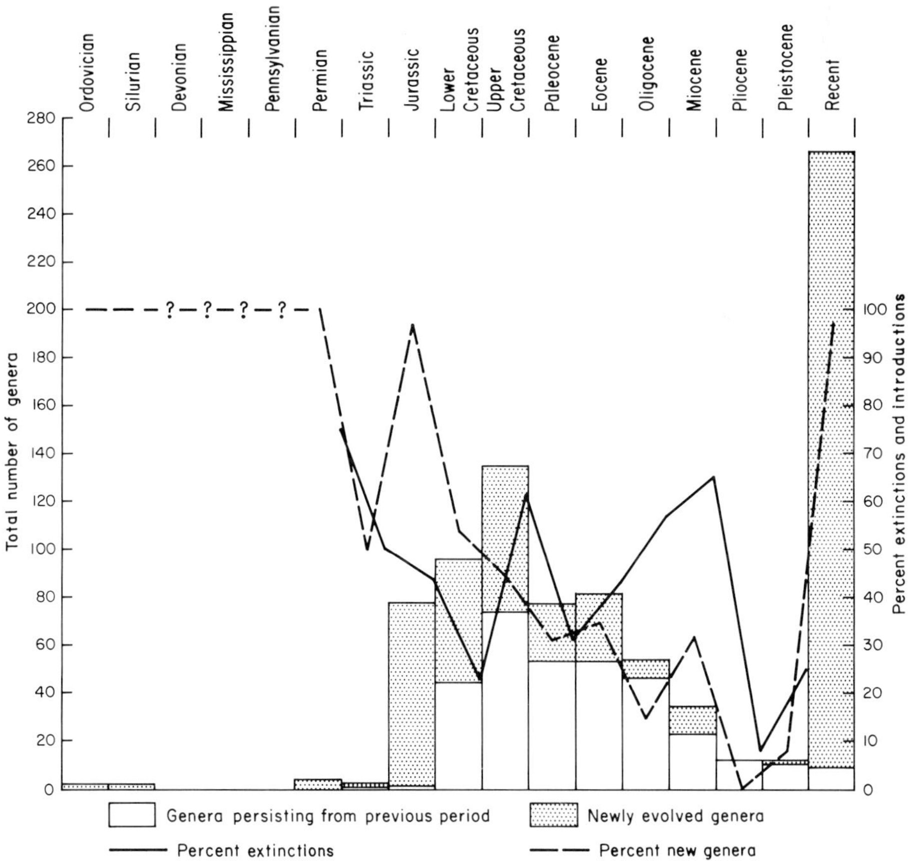

Figure 4.95

Generic diversity of dinoflagellates. Diversity increases rapidly in the Mesozoic, and declines steadily following the Cretaceous, except for a temporary and minor increase in the Eocene. From Tappan and Loeblich, 1972.

percent of the number now living (Tappan and Loeblich, 1973). As living dinoflagellates that produce fossilizable remains include only about 20 percent of the orders and 13 percent of the families, the total diversity in the Mesozoic was probably much greater, suggesting that dinoflagellates then had reached a climax and have declined through the Cenozoic. The increase in new generic taxa was greatest in the Jurassic, the rapid turnover then resulting in some 34 percent of these becoming extinct in the Jurassic; in the Early Cretaceous diversification continued and extinction rates declined to 23 percent. Like other plankton groups, the extinctions increased in the Late Cretaceous to 61 percent, leaving a limited Paleocene assemblage of opportunistic and cosmopolitan taxa.

The less widespread extinctions of dinoflagellates than of coccolithophores at the end of the Cretaceous may result from their lower nutrient requirements; modern dinoflagellates are most abundant in the nutrient-poor warmer waters, and their annual blooms commonly follow the spring diatom blooms as the nutrients in the water are depleted.

Cenozoic

Of the 43 families of dinoflagellate fossils listed herein (see Classification, pp. 386 − 430), only the monogeneric Heteraulacacystaceae, siliceous Peridinitaceae, and those with internal siliceous spicules (*Actiniscus;* see Figure 4.96) are wholly Cenozoic in occurrence. The latter probably are the only fossil representatives of the Gymnodiniales (they are not referrable to the Peridiniales as was done by Harland, 1972, p. 136).

In general, the early Tertiary assemblages include many Deflandreaceae (upon which a bio-

stratigraphic zonation was proposed for the Paleogene by Costa and Downie, 1976) and many taxa with complex processes, forming membranate, trabeculate (see Figure 4.97), and marginate cysts. These latter decrease rapidly in importance, however, and many genera and families disappear in the Oligocene or Miocene. The rate of generic extinctions increased through the Paleogene, from 31 percent in the Paleocene, 43 percent in the Eocene, 57 percent in the Oligocene, and 65 percent in the Miocene, leading to the much reduced diversity of the later Neogene (Tappan and Loeblich, 1972).

Biostratigraphic zonations have been proposed for the Cenozoic based on dinoflagellates (see Table 4.5; G. L. Williams, 1975; G. L. Williams and Bujak, 1977). The geologic ranges of other species are fairly well known, and evolutionary lineages have been proposed for some groups. Dinoflagellate associations also may be utilized as environmental indicators, *Deflandrea* being a typically near-shore taxon (D. B. Williams and Sarjeant, 1967). Other techniques include ratios of dinoflagellates to microforaminifera (the resistant organic lining of certain benthic foraminiferal tests that persists in acid macerations), or dinoflagellates to spores; in both ratios dinoflagellates are regarded as planktonic, and their higher percentages indicate a more offshore character of the environment (Burgess, 1971).

Associations of dinoflagellates from a Paleogene sequence in England were environmentally controlled (Downie et al., 1971), ranging from the estuarine habitat in the Woolwich beds (*Wetzeliella* and *Deflandrea*) to the open sea environment of the London Clay where *Areoligera* and *Cyclonephelium* are characteristic (see Figure 4.98), as are the associated *Hystrichosphaeridium,   Spiniferites,   Achomo-*

Figure 4.96
Miocene Actiniscaceae, in SEM. **1,2.** *Actiniscus pentasterias,* oblique views of convex and concave surfaces. **3,4.** *Actiniscus elegans* Ehrenberg, with much more elevated and rugose pentasters, convex surface and edge view. **1 − 4,** from Malaga Mudstone, Redondo Beach, California, ×1600. **5 − 8.** *Foliactiniscus folia* (Hovasse) Dumitrică, upper Miocene, Lompoc, California. **5,6.** Concave surface, ×2000, and oblique view of same, ×1600. **7,8.** Convex surface and edge view, ×2400; note absence of alveoli on convex surface such as occur in *Actiniscus.*

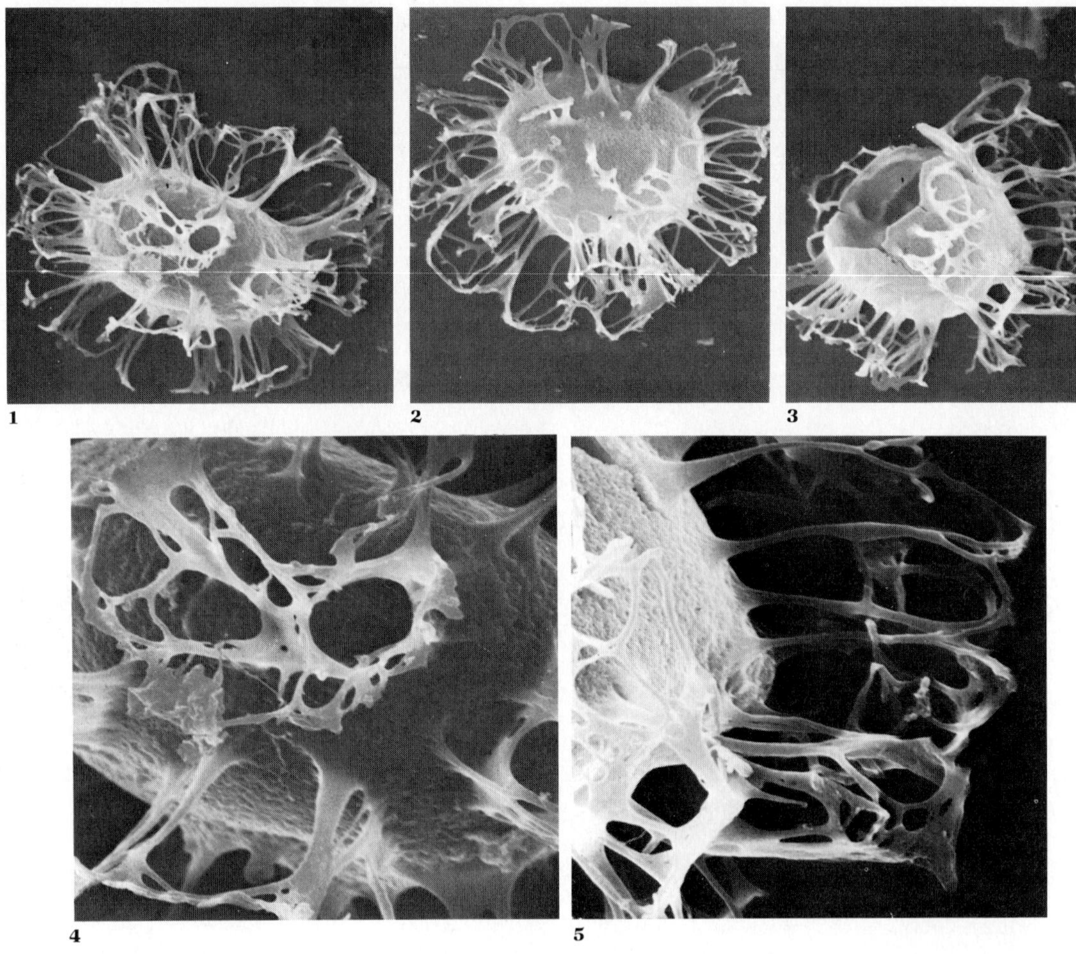

Figure 4.97
Adnatosphaeridium cf. *vittatum* Williams & Downie, upper Eocene, Mississippi, SEM. **1.** Antapical view. **2.** Lateral view. **3.** Oblique ventral view showing archeopyle. 1–3, ×400. **4.** Central region of antapex enlarged to show ringlike antapical process group, ×1300. **5.** Margin of oblique view enlarged to show trabeculae, ×1300.

Table 4.5
Biostratigraphic zonation based on dinoflagellate assemblages (from G. L. Williams, 1975, and G. L. Williams and Bujak, 1977). AZ = Assemblage Zone; PZ = Peak Zone; Z = Zone; SZ = Subzone.

Qt.	Pleistocene		*Spiniferites scabratus* (Wall) Sarjeant AZ	
Tertiary	Pliocene		*Hystrichosphaeridium choanophorum* Deflandre & Cookson AZ	
	Miocene	Late	*Cannosphaeropsis* sp. A. AZ	
		Middle	*Pentadinium laticinctum* Gerlach AZ	
		Early	*Apteodinium* sp. B. AZ	
	Oligocene	Late Middle	*Chiropteridium dispersum* Gocht AZ	
		Early	*Deflandrea heterophlycta* Deflandre & Cookson AZ	
	Eocene	Late	*Diphyes colligerum* (Deflandre & Cookson) Cookson AZ	
		Middle	*Adnatosphaeridium reticulense* (Pastiels) DeConinck PZ	
		Early	*Areoligera senonensis* Lejeune-Carpentier PZ	
	Paleocene	Late	*Deflandrea speciosa* Alberti AZ	
		Early	*Palaeoperidinium pyrophorum* AZ	
Cretaceous	Maastrichtian		*Dinogymnium euclaense* Cookson & Eisenack Z	
	Campanian		*Odontochitina operculata* (O. Wetzel) Deflandre & Cookson AZ	
	Santonian		*Hystrichosphaeridium truncigerum* Deflandre AZ	
	Coniacian		*Oligosphaeridium pulcherrimum* (Deflandre & Cookson) Davey & Williams AZ	
	Turonian		*Surculosphaeridium longifurcatum* PZ	
	Cenomanian		*Cleistosphaeridium polypes* (Cookson & Eisenack) Davey AZ	
	Albian		*Spinidinium vestitum* Brideaux—*Eucommiidites minor* Groot & Penny Z	*Rugubivesiculites rugosus* Pierce SZ
	Aptian		*Subtilisphaera perlucida* (Alberti) Jain & Millipied—*Hystrichosphaerina schindewolfii* Alberti Z	*Canningia attadalica* (Cookson & Eisenack) Stover & Evitt SZ

(continued)

Table 4.5 (*continued*)

	Barremian	*Aptea anaphrissa* (Sarjeant) Sarjeant & Stover Z
	Hauterivian	*Ctenidodinium elegantulum* Millioud Z
	Valanginian	*Phoberocysta neocomica* Z
	Berriasian	
Jurassic	Tithonian	*Ctenidodinium panneum* (Norris) Lentin & Williams Z
	Kimmeridgian	*Hystrichogonyaulax cladophora* (Deflandre) Stover & Evitt Z
	Oxfordian	*Gonyaulacysta jurassica* (Deflandre) Norris & Sarjeant Z
	?Callovian	*Valensiella vermiculata* Gocht Z
	Bathonian	

Figure 4.98 (*facing page*)

Areoligeraceae. **1.** *Cyclonephelium pastielsi* Deflandre & Cookson, upper Eocene, Mississippi, showing apical archeopyle and trabeculate processes, ×600. **2–4.** *Adnatosphaeridium vittatum* Williams & Downie. **2.** Upper Eocene, Mississippi, SEM showing apical archeopyle and trabeculate processes, ×480. **3,4.** Lower Eocene, England, ×400, from G. L. Williams and Downie, 1966b; 3, entire specimen; 4, with archeopyle. **5.** *A. reticulense* (Pastiels) DeConinck, Eocene, England, entire specimen, ×400, from G. L. Williams and Downie, 1966b. **6.** *Glaphrocysta ordinata* (Williams & Downie) Stover & Evitt, Eocene, England, dorsal view with apical archeopyle, ×265, from G. L. Williams and Downie, 1966b. **7.** *G. retiintexta* (Cookson) Stover & Evitt, Upper Cretaceous, Australia, showing operculum in place in apical archeopyle, ×385, from Cookson, 1965a. **8–10.** *Areoligera cassicula* Drugg, lower Eocene, Alabama, from Drugg, 1970. **8.** SEM, showing lacelike trabeculae, ×800. **9,10.** Light microscope, ×307.

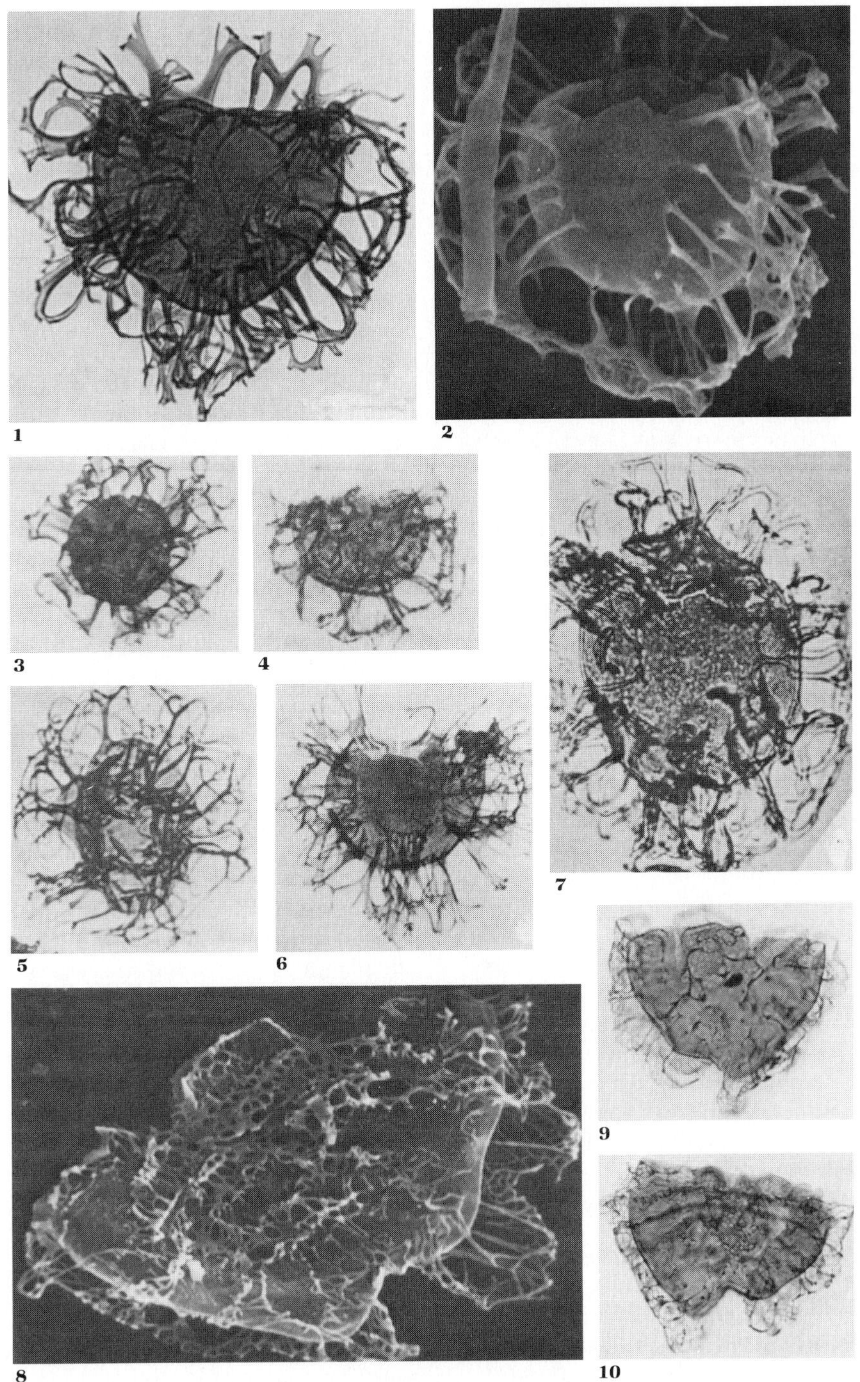

1
2
3
4
5
6
7
8
9
10

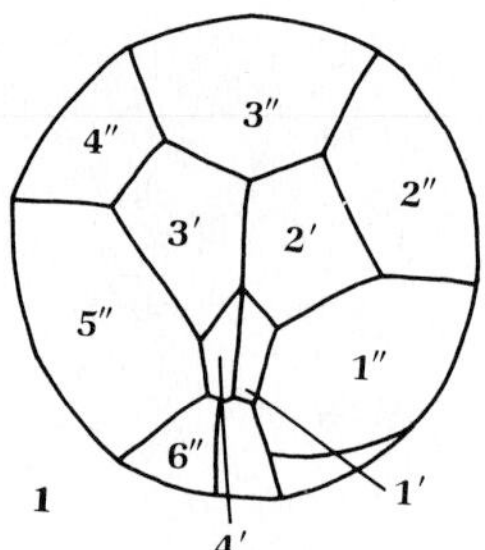
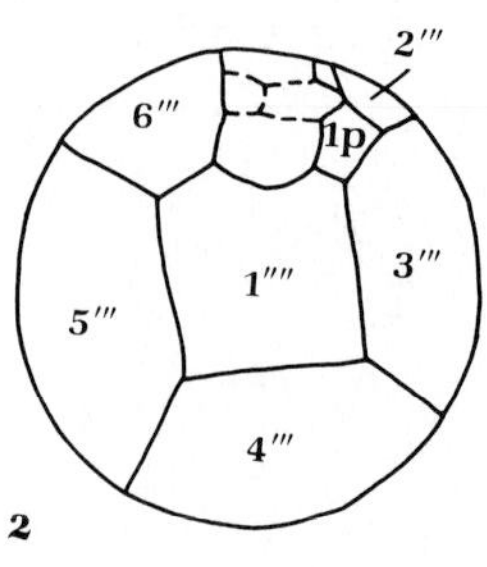
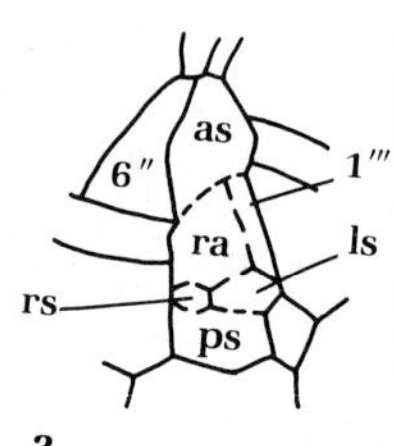

Figure 4.99
Tabulation of *Spiniferites bentori* cyst, believed to be resting stage of *Gonyaulax digitalis*, shown diagrammatically. **1–3.** Apical view, antapical view, sulcal area. Redrawn from Wall, 1967.

sphaera, and *Cordosphaeridium.* A different assemblage (containing *Comasphaeridium* as well as the acritarch *Micrhystridium*) characterized the initial and closing stages of a marine transgression.

Some Pleistocene studies have identified modern taxa, but relatively few cyst taxa have been related to a motile stage, hence cyst taxa are commonly utilized (Harland, 1974; Harland and Downie, 1969; Rossignol, 1961, 1964; Takahashi, 1971; Wall, 1970, 1971b; Wall and Dale, 1968b, 1968c, 1973, 1974; Wall et al., 1973; Wall and Warren, 1969). Distribution of dinoflagellate cysts in subrecent sediments suggests that they may not exactly reflect distribution of the motile cells. Subrecent cyst distribution has been studied in relation to season of the year, distance from shore, and climatic conditions (Rossignol, 1961, 1964; Downie and Singh, 1969; D. B. Williams, 1971a,b; Reid, 1972, 1974). In bottom sediments, dinoflagellate cysts are more abundant in December than in October. Furthermore, the distribution of the cyst *Spiniferites bentori* differs from that of its motile stage *Gonyaulax digitalis* (see Figure 4.99), the cyst distribution possibly being passively modified by movement of the water masses and currents.

Nonmarine Fossil Dinoflagellates

Cysts have been reported for many living freshwater dinoflagellates (Stein, 1883; Bütschli, 1885, and many later workers), but fossil dinoflagellates generally were regarded as strictly marine. The first reported nonmarine fossil dinoflagellate (Traverse, 1955; Evitt, 1974b) is *Saeptodinium hansonianum* (Traverse) Stover & Evitt, from the Oligocene of Vermont, probably related to such living species as *Peridinium limbatum.* Freshwater dinoflagellates also occur in the Eocene of Germany (Krutzsch, 1962), Miocene of Alaska (Engelhardt, 1976), and in southeastern Australia (Harris, 1973) in Paleocene, Eocene, and lower Miocene strata of fluvial and lacustrine origin, where the diversity is low, although the distinctive species may occur in considerable abundance. Other nonmarine dinoflagellate fossils largely are of Pleistocene or postglacial age (see Figure 4.100; Churchill and Sarjeant, 1962, 1963; Harland, 1971; Harland and Sarjeant, 1970; Norris and McAndrews, 1970). Churchill (1960) had originally regarded some as blue-green or green algal spores. The rarity of freshwater fossils is not unexpected, however, as living freshwater dinoflagellate genera and species are much less common than their marine relatives. Possibly dinoflagellates invaded the freshwater habitat contemporaneously with the diatoms in the earliest Cenozoic, as no Cretaceous nonmarine species have been recorded.

CLASSIFICATION

Phylogenetic Relationships

The main divisions of algae are classically divided on the basis of characteristic pigments, hence those with chlorophylls *a* and *c* (chromophytes) have been regarded as closely

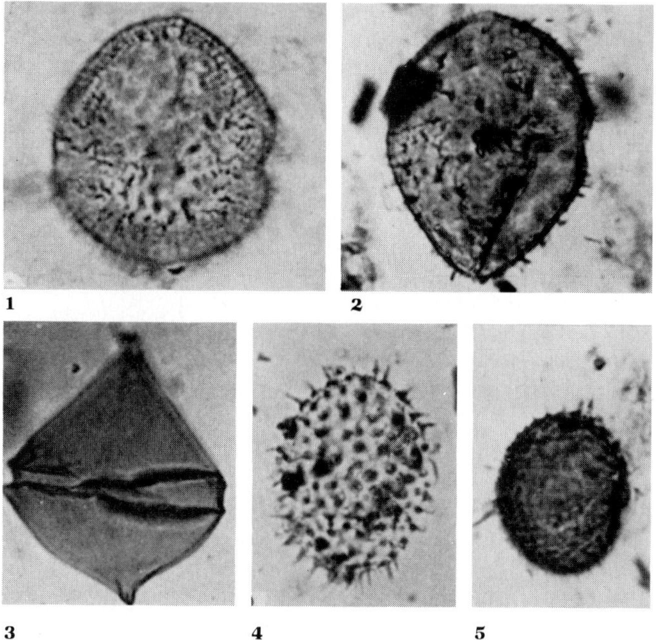

Figure 4.100
Freshwater fossil dinoflagellates, from Holocene peats of Australia. **1.** *Aquadulcum myalupensis* (Churchill & Sarjeant) Harland & Sarjeant, ventral view, ×750. **2.** *A. pikei* (Churchill & Sarjeant) Harland & Sarjeant, ventral view, ×750. **3.** *Muiradinium dorsispirale* (Churchill & Sarjeant) Harland & Sarjeant, dorsal view, ×750. 4. *Creberlumectum tinglewoodense* (Churchill & Sarjeant) Harland & Sarjeant, ×750. **5.** *C. echiniplax* (Churchill & Sarjeant) Harland & Sarjeant, ×750. All from Churchill and Sarjeant, 1963.

related, and distinct from those with chlorophylls *a* and *b* (chlorophytes). Dinoflagellates belong to the first group, and euglenids belong to the second, although similar to dinoflagellates in a variety of other features. The food reserve product in the dinoflagellates is starch, as in the chlorophytes and prasinophytes, yet their cell organization is much closer to that of the chrysophytes.

Prior to the description of their various ultrastructural details, Kofoid and Swezy (1921, p. 83–84) suggested that dinoflagellates were derived from the cryptomonads, the nonthecate dinoflagellates being primitive and the armored species more advanced, thus repeating their phylogeny in the ontogenetic development of the armored cell from the naked zoospore or half-thecate divisional stage. These relationships are not supported by data now available. The dinoflagellate mesocaryotic nucleus, with primitive and even procaryotic similarities and with condensed chromosomes of fibrillar structure, is distinct and more primitive in character than the typically eucaryotic cryptomonad nucleus; the plastids have three rather than two or four membranes as in all other algae except the euglenids; the thylakoids are in stacks of three as in most other algae, although the cryptomonads have two, and they are more numerous in some prasinophytes and chlorophytes; the two dissimilar flagella separate the dinoflagellates from the chlorophytes and prasinophytes whose similar flagella are smooth, or scaly and hairy, respectively, and from the euglenids with single emergent hairy flagellum. These flagellar features are more like those of the chloromonads, xanthophytes, chrysophytes, and phaeophytes in having a smooth posterior flagellum and hairy anterior one, although the unique dinoflagellate anterior (transverse) flagellum has a paraflagellar rod or striated strand that pulls it into a ruffle, the longitudinal flagellum also may have short mastigonemes, and the flagella of *Oxyrrhis* have both scales and mastigonemes.

The ultrastructural features of the flagella, chloroplast membranes, and thylakoids were suggested (Dodge, 1974b) to provide a better indication of relationships than the classically used pigment complex, in view of the endosymbiotic theory that plastids may have resulted from algal symbiosis. However, the recent expansion of this theory (Margulis, 1970) to regard not only the plastids as the result of endosymbiosis, but also the mitochondria, flagella, and even the nuclear division spindles, would preclude assigning these features any greater reliability in postulating relationships than pigments and plastid ultrastructure. A combination of both ultrastructural and biochemical characters thus provides the best evidence available from the living cells.

Primitive features of the dinoflagellate nucleus that are procaryotelike include the chromatin fibrillar diameter, low level of chromosomal basic proteins, the membrane attachment of the chromosomes, the swirl pattern evident in sectioned chromosomes, and the continuous replication of nuclear DNA (in *Prorocentrum*); the molecular weight of the RNA in dinoflagellates and red algae also is more like that of procaryotes than that of other algae. Dinoflagellate nuclear features that are typically eucaryotic are the repeated and highly complex DNA, the nature of DNA synthesis in the cell cycle, the presence of basic proteins, and the extranuclear microtubules reinterpreted as a spindle. The dinoflagellates thus are primitive eucaryotes, probably geologically old, having diverged before the evolution of eucaryotic chromatin, but after the evolution of repeated DNA. Variations in the pigment complex, especially of the carotenoids, suggest a two-stage capture of plastids from endosymbionts, most being procaryotic, but those with fucoxanthin perhaps resulting from capture of a eucaryotic plastid (as in *Peridinium balticum* and *P. foliaceum,* which still retain the endosymbiotic eucaryotic nucleus and plastids, but perhaps also *Gymnodinium micrum* and *G. veneficum*) (Loeblich III, 1970).

Obviously, all living taxa have a long evolutionary history following their divergence, but a combination of data suggests that the red algae, dinoflagellates, and prasinophytes diverged very early, the latter two groups respectively giving rise to the other "brown-pigmented" and "green-pigmented" algae and their derivatives.

The discovery of a distinct layer of vesicles in nonthecate species, homologous to the vesicles in which plates of the theca are synthesized, reduces the formerly apparently sharp distinction between the two major dinoflagellate groups, and indicates that thecate species appeared first, the nonthecate ones retaining a layer of vesicles although failing to synthesize the cellulosic plates. Interestingly, Gourret (1883) regarded *Gymnodinium* as derived from the thecate taxa, but his suggestion was ignored. Development of a layer of functionless vesicles, such as now are known to occur in the nonthecate dinoflagellates, with these later becoming a site for cellulose deposition, appears to be a less likely evolutionary sequence.

As previously noted, if the two valves of the Prorocentrales are reinterpreted as equivalent to epitheca and hypotheca, surrounding a flagellar region of seven plates (Loeblich III, 1976), and this is regarded as the primitive condition, as previously suggested by Kofoid and Swezy (1921, p. 83–84), it is possible that the acritarchs with terminal pylome may represent the oldest dinoflagellates. No cyst is known in living Prorocentrales, to either support or negate this hypothesis.

Within the thecate dinoflagellates, the Dinophysiales and Peridiniales evolved independently. Early workers regarded them as evolving separately from the Gymnodiniaies (Bütschli, 1885, p. 1016; Kofoid and Swezy, 1921, p. 84; Tai and Skogsberg, 1934, p. 389), but if the Prorocentrales are primitive, the Dinophysiales and Peridiniales may have diverged from this group and the Gymnodiniales also may be derived from thecate ancestors. The hypothetical primitive dinoflagellate has been regarded either as having a solid theca, or one of few plates as in the Prorocentrales, that later became subdivided; or as having many plates (as suggested by the many vesicles in the Gymnodiniales) that gradually fused to attain the present number. Graham (1942, p. 6) suggested that plate increase and decrease both occurred in separate areas of the theca and that the primitive state in the Peridiniales consisted of 8 hypothecal plates, 5 sulcal plates, and 6 girdle plates, but with greater epithecal variation, ranging from 10 to 16 plates, no two species being identical. He suggested that epithecal evolution thus was independent of the remainder of the theca.

A central capsule surrounding the nucleus of some dinoflagellates was suggested to indicate a relationship to the radiolarians (Zimmerman, 1930; Chatton, 1934, p. 310). *Actiniscus* has an elliptical capsule consisting of a thin membrane surrounding and somewhat larger than the nucleus. The two ends of the ellipse are capped by the siliceous stellate spicules characteristic of this dinoflagellate. Similarly, *Plectodinium nucleovolvatum* has a spherical, thick membrane that is closely appressed to the nucleus and has a hexagonal-patterned ornamentation. *Peridinium depressum* has a capsular membrane that is relatively thin and is concentric to the nuclear membrane, the intracapsular cytoplasm being vacuolated and enclosing many mitochondria (Hollande et al., 1962). The capsule of the dinoflagellates has been shown to consist of endoplasmic reticulum (see discussion of the nucleus on pp. 231–237), and that of the radiolarians now is regarded as external, perhaps corresponding to the dinoflagellate theca (Hollande et al., 1970). Thus, neither the superficial similarity of their capsules nor the presence of siliceous skeletons in the Actiniscaceae and radiolarians provide reliable evidence of close relationship; the mesocaryotic dinoflagellate nucleus, presence of flagella rather than pseudopodia, and the many other cytological differences appear more important. Furthermore, *Actiniscus* is not primitive, being known only from the late Oligocene to the present, some 25 or 30 million years, in contrast to the 440 million years of the earliest dinoflagellate *Arpylorus;* the even older radiolarians were well diversified by the Early Ordovician, 500 million years ago. *Actiniscus* is a gymnodinioid, nonthecate dinoflagellate, probably characterized by a secondarily reduced amphiesma, instead of a primitive one; the capsule and skeleton probably are morphologically convergent features rather than indicating a close relationship.

Based on the primitive features of the living cell, dinoflagellates probably diverged from the ancestral chromophytes in the late Precambrian (Loeblich Jr., 1974). The oldest certain fossil representatives of the Peridiniales are Silurian, the oldest Dinophysiales Jurassic, the Ebriales are first known in the Paleocene, and the Gymnodiniales in the Oligocene. Within the Peridiniales, the dominant fossil lineages are those of Gonyaulacacean type (perhaps including the Silurian *Arpylorus*, but certainly represented by the Late Triassic *Rhaetogonyaulax*, and a variety of Jurassic and younger taxa), and the Peridiniaceans (represented by the Triassic *Heibergella, Noricysta,* and *Sverdrupiella* and the Jurassic Calciodinellaceae, but rapidly diverging in the Early Cretaceous). *Ceratium*-like taxa appeared in the Cretaceous, but are lacking from Tertiary sediments, suggesting that the nature of the cysts of this group may have changed through geologic time.

While retaining certain of their primitive features, dinoflagellates have become highly differentiated in other respects. The ocellus or eyespot, trichocysts, and nematocysts are elaborate organelles that occur in some advanced types. The wide variety of nutritional methods and habitats also is consistent with a highly evolved group, as the dinoflagellates range from photosynthetic to saprophytic, host symbionts to holozoic, and from free-living to endosymbiotic and parasitic, in fresh, brackish, or marine waters.

No fossil evidence of endosymbiotic or parasitic dinoflagellates is known, although both the host symbionts and hosts to the parasitic forms are geologically old taxa that may early have evolved such relationships (Loeblich III, 1976, p. 21). Probably the primitive dinoflagellates were photosynthetic, and the parasitic and holozoic taxa are later derivations. Holozoic dinoflagellates have no plastids, no proplastids, and no pyrenoids, and have become completely "animallike" in nutrition.

Evolutionary relationships proposed for this group must depend more on comparisons with living taxa than generally is necessary for other groups with a fossil record, inasmuch as only the resting cyst is preserved and only a small percentage of even the living dinoflagellates have such a fossilizable stage.

Classification of Living Dinoflagellates

Unlike some groups of plant protists, the dinoflagellates are well indexed, so that they may be readily identified. All living species of dinoflagellates were monographed by Schiller (1931−1937); later new taxa and taxonomic revisions of marine taxa were indexed by Sournia (1973) and Sournia et al. (1975). Freshwater taxa were monographed by Schilling (1913), were included by Schiller (1931−1937), and recently were monographed by Bourrelly (1970). Genera of both Recent and fossil dinoflagellates have been indexed (Loeblich Jr. and Loeblich III,

1966, 1968, 1969, 1970a,b, 1971, 1974). An illustrated catalogue (Eisenack and Klement, 1964; Eisenack and Kjellström, 1971, and supplements) reproduces the original descriptions and figures of the fossil dinoflagellates; a bibliography and index of fossil genera and species was published by Downie and Sarjeant (1965), and original and translated descriptions of genera were compiled by Norris and Sarjeant (1965). The index to fossil genera and species has been updated by Lentin and Williams (1973, 1975a, 1977).

The classification of living dinoflagellates shown in Table 4.6 is based on amphiesmal morphology and mode of nutrition in the living cell (Schiller, 1931−1937; Chatton, 1952; see summary in Loeblich III, 1970, 1976). The classification summarized herein is modified from the latter by the recognition of the Ornithocercaceae (Dinophysiales) as a separate family, the transfer of the Pyrocystaceae from a separate Order Pyrocystales to the Peridiniales in view of the *Gonyaulax*-like tabulation of the motile stage of *Pyrocystis* (Swift and Wall, 1972), and the addition of the Order Arthrodiniales and Family Brachydiniaceae, both of Sournia (1972a). The Class Dinophyceae includes 15 orders and 47 families, the Class Ebriophyceae is covered in a separate chapter, and the wholly parasitic classes Ellobiophyceae and Syndiniophyceae, with 5 families, are only mentioned.

Table 4.6
Classification of living dinoflagellates, with representative genera.

Division Pyrrhophyta
I. Class Dinophyceae Fritsch 1929

Motile cells biflagellate, with unlike, heterodynamic flagella; large nucleus with many permanently condensed chromosomes, characterized by continuous DNA synthesis, and lacking histochemically detectable histones; chromatin fibrillar in structure, fibrils in arched swirls; chromosomes attached to nuclear membrane during division, and lack centrioles and spindle. Fossils are resting cysts (Dinophysiales, Peridiniales) or internal siliceous spicules of the Actiniscaceae (Gymnodiniales). Silurian to Holocene.

A. Order Desmocapsales Pascher 1914

Ovoid or compressed bivalved motile cells with apically inserted flagella, one encircling the apex in a plane perpendicular to the cell axis, and one flagellum directed forward; theca cellulosic; photosynthetic.
Marine, no fossils.

1. Family Adinomonadaceae Schiller 1928
Adinomonas Schiller 1928.
2. Family Desmocapsaceae Pascher 1914
Desmocapsa Pascher 1914.
3. Family Desmomastigaceae Loeblich III 1970
Desmomastix Pascher 1914.

B. Order Prorocentrales Lemmermann 1910

1. Family Prorocentraceae Stein 1883 (see Figure 4.101)

Theca of two large plates, equivalent to epitheca and hypotheca of Peridiniales, and seven smaller plates around the two pores from which flagella emerge; transverse flagellar-equivalent with axoneme, striated strand, and mastigonemes; other lacking strand and hairs; flagellar pore plates set in notch in epithecal equivalent (= left) valve; photosynthetic.

Marine interstitial sand, neritic, pelagic, rare in brackish and fresh water; no fossils.

Mesoporos Lillich 1937; *Prorocentrum* Ehrenberg 1834.

C. Order Dinophysiales Lindemann 1928 (see Figure 4.102)

Armored dinoflagellates, theca with two nearly equal valves joining at a sagittal suture; flanges or lists border the longitudinal and transverse grooves, epitheca commonly much reduced; tabulation includes five epithecal plates (seven in *Latifascia*), four girdle plates, four hypothecal plates, four sulcal plates; probable fossils in Jurassic.

(*continued*)

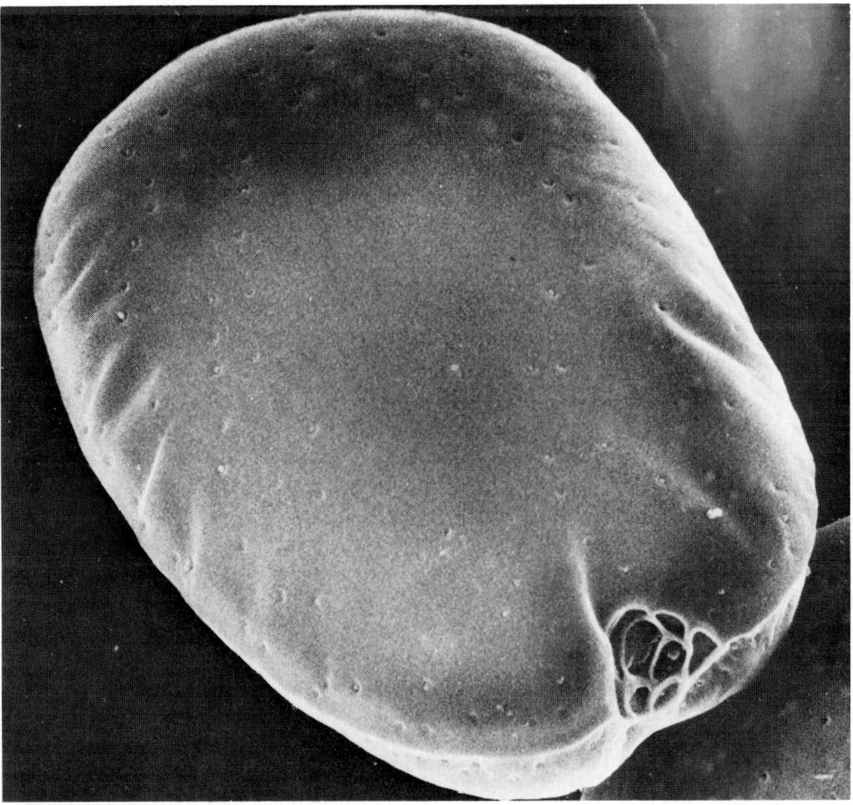

Figure 4.101
Prorocentrum cassubicum. SEM of glutaraldehyde-fixed and critical-point-dried cell, showing scattered trichocyst pores on the bivalved theca, with flagellar pore in indentation at apex (to lower right of figure) surrounded by seven smaller plates. ×8500. From Loeblich III, 1976.

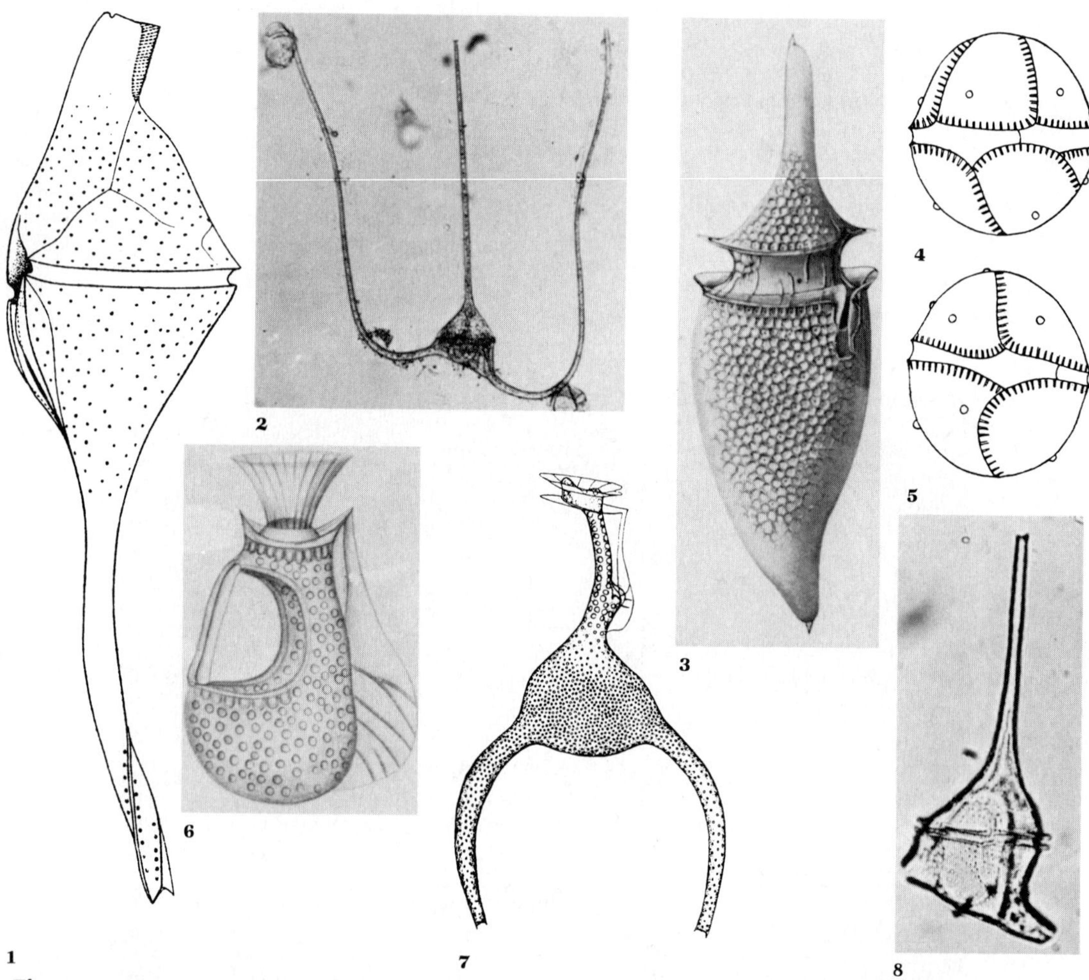

Figure 4.102
Dinophysiales, Archaeosphaerodiniopsaceae, Centrodiniaceae, and Ceratiaceae. **1.** *Centrodinium elongatum* Kofoid, ×450, from Kofoid, 1907c. **2.** *Ceratium contrarium* (Gourret) Pavillard, off La Jolla, California, ventral view, showing elongate horns, ×120, from Loeblich Jr. and Loeblich III, 1966. **3.** *Oxyphysis oxytoxoides* Kofoid, North Pacific Ocean, ×1167, from Kofoid, 1926. **4,5.** *Archaeosphaerodiniopsis verrucosa* Rampi, after Rampi, 1943. **6.** *Citharistes regius* Stein, Atlantic Ocean, from Stein, 1883. **7.** *Triposolenia truncata* Kofoid, Pacific Ocean, off San Diego, California, ×506, from Kofoid, 1906b. **8.** *Ceratium divaricatum* Kofoid, off La Jolla, California, empty theca, focussed through specimen to dorsal side, but also showing broad sulcal area; ×300, from Loeblich Jr. and Loeblich III, 1966.

1. Family Amphisoleniaceae Lindemann 1928

Cell much elongated, antapex attenuated or divided into two, three, or more horns; compressed laterally.

Marine, pelagic, warm water.

Amphisolenia Stein 1883; *Oxyphysis* Kofoid 1926; *Triposolenia* Kofoid 1906.

2. Family Dinophysiaceae (Bergh) Stein 1883

As described for the order. Marine, pelagic.

Dinophysis Ehrenberg 1839; *Latifascia* Loeblich Jr. & Loeblich III 1966 (syn.: *Heteroschisma* Kofoid & Skogsberg 1928, non Wachsmuth 1883; nec Simroth 1895); *Sinophysis* Nie & Wang 1944.

3. Family Ornithocercaceae Kofoid & Skogsberg 1928

Highly expanded girdle and sulcal lists, with radial or reticular thickenings or veins.

Marine, pelagic, warm water.

Citharistes Stein 1883; *Histioneis* Stein 1883; *Ornithocercus* Stein 1883; *Parahistioneis* Kofoid & Skogsberg 1928.

D. Order Peridiniales Haeckel 1894

Armored dinoflagellates, with characteristic tabulation of the cellulosic theca. Families listed alphabetically; fossils known (see separate classification, Table 4.7).

1. Family Archaeosphaerodiniopsaceae Deflandre in Grassé 1952

Subspherical, reduced tabulation of 5″, 4‴, each plate perforated by one or two pores; poorly known.

Marine, planktonic.

Archaeosphaerodiniopsis Rampi 1943.

2. Family Centrodiniaceae Schiller 1937

Flattened fusiform theca with horns at apex and antapex; girdle deeply incised, sulcus restricted to hypocone; tabulation 2′, 3a, 7″, 5c, ?s, 5‴, 1p, 1⁗; photosynthetic.

Marine, warm water.

Centrodinium Kofoid 1907.

3. Family Ceratiaceae Lindemann 1928

Theca triangular or fusiform in dorsal or ventral view; dorsally convex, with apical or antapical horns, apical horn open distally; tabulation 4′, 0a, 6″, 5–6c, 2–?s, 6‴, 1p, 1⁗; most photosynthetic, some also ingest particulate

food, especially those in fresh water; unlike gametes, zygote motile or forming resting cysts.

Marine and freshwater, planktonic.

Ceratium Schrank 1793.

4. Family Ceratocoryaceae Lindemann 1928

Epitheca reduced to flattened disc, hypotheca deep and cuplike, girdle lists prominent, hypotheca with 4 to 10 spines or horns or winglike expansions; tabulation 3′, 1a, 5″, 8c, 6‴, 1p, 1⁗; photosynthetic.

Marine, planktonic.

Ceratocorys Stein 1883.

5. Family Cladopyxidaceae Stein 1883 (see Figure 4.103)

Spheroidal; apical pore, simple or terminally bifid spine arising from each plate; tabulation 4′, 0a, 8″, 6‴, 2⁗; photosynthetic.

Marine, planktonic.

Cladopyxis Stein 1883; *Palaeophalacroma* Schiller 1928; *Sinodinium* Nie 1945.

6. Family Crypthecodiniaceae Biecheler 1938

Similar to *Gyrodinium*, but girdle not completely encircling cell; tabulation visible after silver impregnation as 4′, 3a, 5″, x, 6c, 5‴, 5s, 3⁗, the additional large plate (x) occupying region of missing girdle on both epitheca and hypotheca; nonphotosynthetic; sexual reproduction demonstrated; isogamous, resting zygote cyst.

Marine, common on *Fucus* and other algae.

Crypthecodinium Biecheler 1938.

7. Family Dinosphaeraceae Lindemann 1928

Tabulation 3′, 1a, 6″, 6c, 6s, 5‴, 1⁗, similar to *Gonyaulax*.

Freshwater.

Dinosphaera Kofoid & Michener 1912.

8. Family Glenodiniopsaceae Schiller 1935

Freshwater species with thin-walled theca, marine ones thicker-walled, tabulation 4′, 4a, 7–8″, ?c, 6s, 6‴, 2⁗; photosynthetic.

Marine, brackish, and freshwater.

Glenodiniopsis Wołoszyńska 1916; *Hemidinium* Stein 1878; *Sphaerodinium* Wołoszyńska 1916.

9. Family Gonyaulacaceae Lindemann 1928

Groove extending from sulcus to apex; tabulation pp, 3–4′, 0–3a, 6″, 6c, 5–10s, 5–6‴, 1p, 1⁗; includes many of the toxic dinoflagellates; photosynthetic; some bioluminescent (*Gonyaulax* and *Pyrodinium*); isogamous; resting cysts of *Pyrodinium* = *Hemicystodinium*.

(continued)

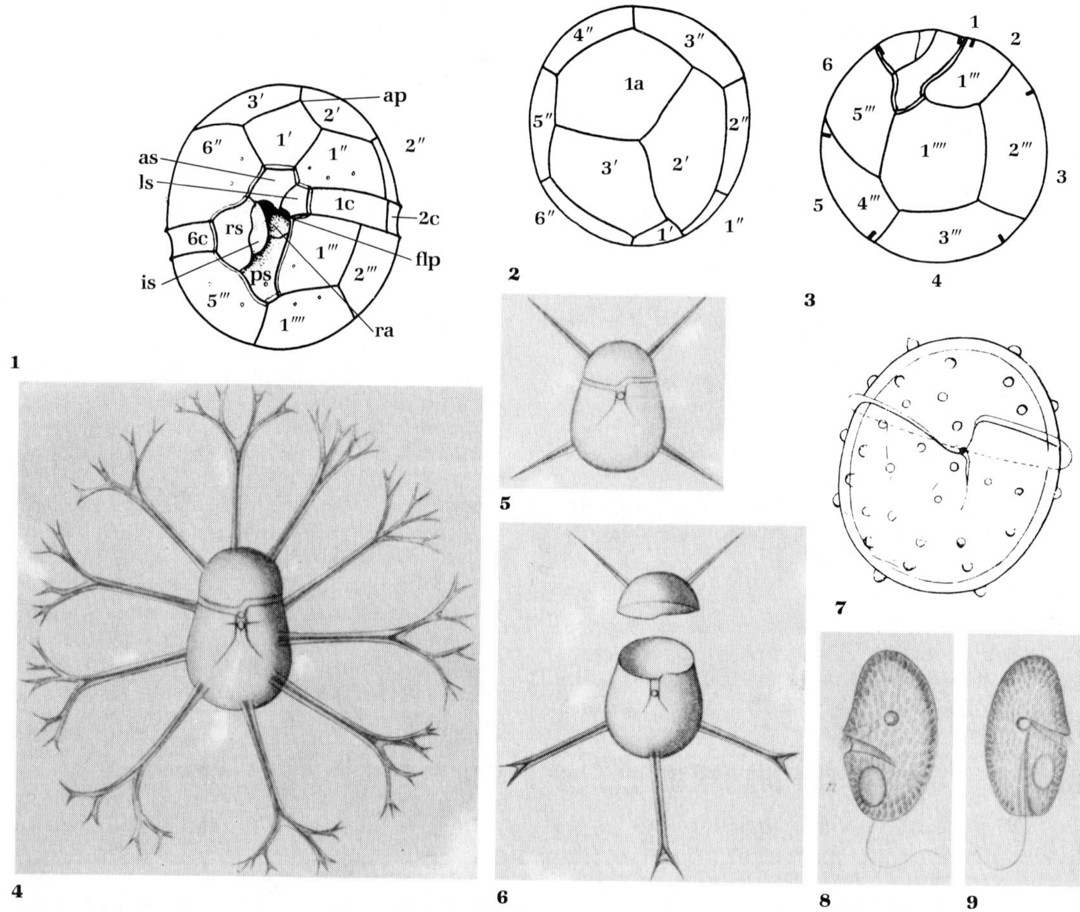

Figure 4.103

Cladopyxidaceae, Dinosphaeraceae, Glenodiniopsaceae. **1–3.** *Dinosphaera palustris* (Lemmermann) Kofoid & Michener. **1.** Ventral view, showing tabulation; ventral area with flagellar pore (flp), anterior sulcal (as), right sulcal (rs), left sulcal (ls), right accessory sulcal (ra), intercalary sulcal (is), and posterior sulcal (ps) plates. **2,3.** Apical and antapical views, the latter showing position of cingular plates by checks and numbers on periphery; ×1120, from Kofoid and Michener, 1912. **4–6.** *Cladopyxis brachiolata*, from Stein, 1883. **7.** *Palaeophalacroma verrucosa* Schiller, from the Adriatic, ×1440, from Schiller, 1928. **8,9.** Dorsal and ventral views of *Hemidinium nasutum* Stein, from Stein, 1883.

Marine; rare in fresh water.

Acanthogonyaulax Kofoid 1911; *Amphidoma* Stein 1883; *Gessnerium* Halim 1969; *Gonyaulax* Diesing 1866; *Pyrodinium* Plate 1906; *Spiraulaxina* Loeblich III 1970.

10. Family Heteraulacaceae Loeblich Jr. & Drugg 1968

Subspherical to polygonal, apical pore with operculum; short, broad sulcus; tabulation pp, 3′, 7″, 6c, 6s, 5‴, 1p, 1⁗; photosynthetic.

Marine.

Heteraulacus Diesing 1850 (syn.: *Goniodoma* Stein 1883, non Zeller 1849).

11. Family Heterodiniaceae Lindemann 1928 (see Figure 4.104)

Similar to *Peridinium,* sculptured theca; epitheca larger than hypotheca, two antapical horns; tabulation pp, 3′, 1a, ventral pore plate, 6″, 6c, ?s, 6‴, 3⁗; photosynthetic.

Marine, especially tropical.

Heterodinium Kofoid 1906.

(continued)

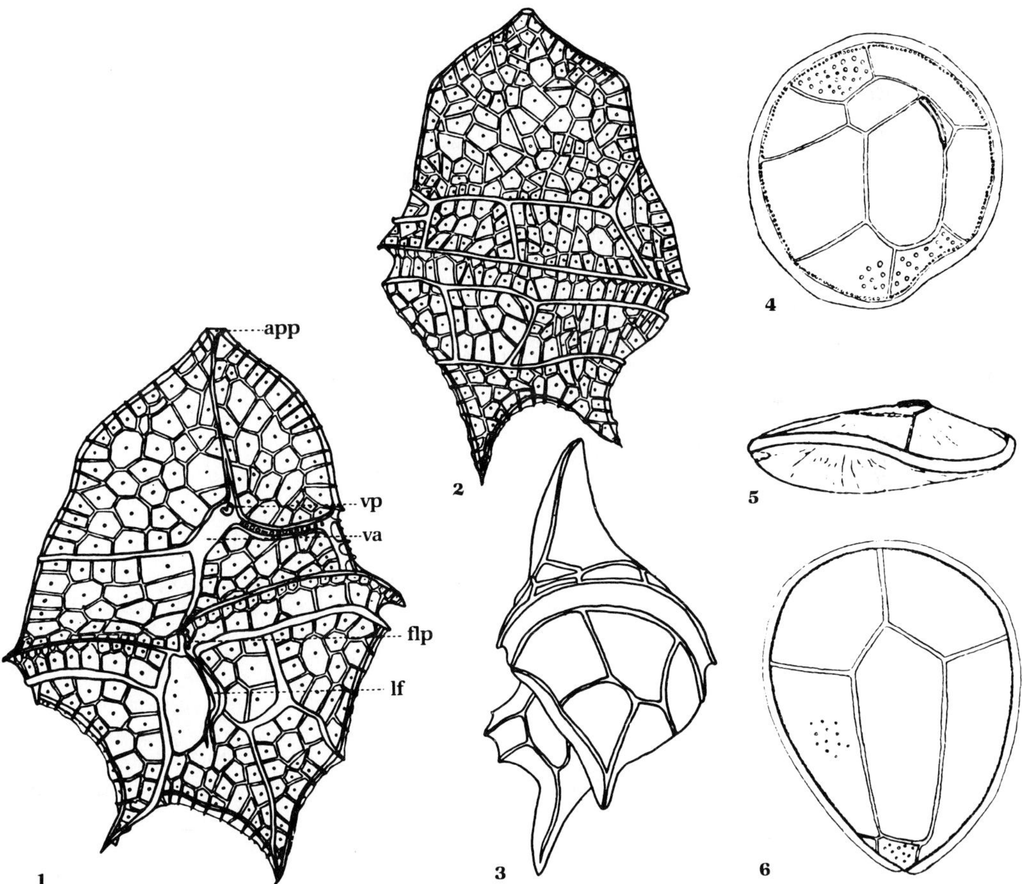

Figure 4.104

Heterodiniaceae, Ostreopsidaceae. **1–3.** *Heterodinium scrippsi* Kofoid, from Kofoid, 1906c. **1.** Ventral view, ×725, showing apical pore (app), ventral pit (vp), ventral area (va), flagellar pore (flp), and longitudinal furrow (lf). **2.** Dorsal view, ×420. **3.** Oblique left side, ×420. **4–6.** *Ostreopsis siamensis* Schmidt, apical, edge, and antapical views; approximately ×500, from Schmidt, 1902.

12. Family Lophodiniaceae Osorio-Tafall 1942 (see Figure 4.105)

Theca of many small very thin hexagonal plates that are perforated by fine to large pores; photosynthetic; isogamous sexual reproduction produces resting zygote cysts.

Freshwater, few marine.

Aureodinium Dodge 1967; *Katodinium* Fott 1957 (syn.: *Massartia* Conrad 1926, non Wildemann 1897); *Lophodinium* Lemmermann 1910; *Woloszynskia* Thompson 1951. (*Amphitholus* Schütt 1895 non Haeckel 1887, and *Monaster* Schütt 1895 non Etheridge 1892 need restudy but may represent cysts of marine Lophodiniaceae).

13. Family Ostreopsidaceae Lindemann 1928

Flattened theca, excentric apex, tabulation pp, 3′, 7″, 6c, 4s + t (a transitional plate), 5‴, 2⁗.

Marine, brackish water.

Ostreopsis Schmidt 1902 (syn.: *Coolia* Meunier 1919).

14. Family Oxytoxaceae Lindemann 1928 (see Figure 4.106)

Fusiform, hypotheca larger than epitheca; tabulation (*Oxytoxum*) 3′, 2a, 5−6″, 5c, 4−8s, 5‴, 1⁗.

Marine, warm water.

Adenoïdes Balech 1956; *Oxytoxum* Stein 1883; *Pavillardinium* de Toni 1936 (syn.: *Murrayella* Kofoid 1907, non Schmitz 1893).

15. Family Peridiniaceae Ehrenberg 1832

Spherical, ellipsoidal to polyhedral, sulcus limited to hypocone, or only slightly extended; tabulation commonly 4′, 2−3a, 7″, 3−5c, 4−6s, 5‴, 2⁗. Freshwater *Peridinium* have 5c; marine *Cachonina*, *Ensiculifera*, *Scrippsiella*, and *Protoperidinium* have 3c and a transitional plate, and tend toward heterotropy, whereas those with more cingular plates are photosynthetic; some also are characterized by containing endosymbionts; some bioluminescent; some marine species toxic; isogamous sexual reproduction, zygotes motile, or resting cyst; cysts proximate in some *Peridinium*, cavate in some marine *Protoperidinium*; *Scrippsiella* and *Ensiculifera* have proximate calcareous cysts. (Includes Glenodiniaceae.)

Marine and freshwater.

Cachonina Loeblich III 1968; *Diplopsalopsis* Meunier 1909; *Dissodium* Abé 1941 (syn.: *Diplopsalis* Bergh 1881, non Sclater 1866); *Ensiculifera* Balech 1967; *Glenodinium* Ehrenberg 1837; *Oblea* Balech 1964; *Peridiniopsis* Lem-

(*continued*)

Figure 4.105

Lophodiniaceae. **1−3.** *Woloszynskia apiculata,* showing numerous thin plates evident when prepared by laurylsulphate, ×1000, from Stosch, 1973. **1,2.** Ventral and apical views of epitheca. **3.** Antapical view of hypotheca. **4.** *Aureodinium pigmentosum,* theca detached from motile cell, TEM of gold-palladium-shadowed cell; poorly known, "theca" may be only empty vesicles of wall and taxon synonymous with *Gymnodinium,* ×6000, from Dodge, 1967. **5−9.** *Lophodinium dadayi* Osorio-Tafall, Veracruz, Mexico, all approximately ×500, from Osorio-Tafall, 1942. **5.** Antapical view. **6.** Ventral view. **7.** Optical section. **8.** Apical view. **9.** Dorsal view. Parts 6 and 9 show vertical ridges with large perforations, regarded as a skeletal supporting structure by Osorio-Tafall, but probably the resting cyst. Parts 5 and 8 show the covering of tiny plates, probably representing the theca within which the cyst forms (compare *Woloszynskia* and *Aureodinium*). Part 7 shows resting cyst within the theca of encysted cell. **10.** *Monaster rete* Schütt, showing what was described as an intracellular skeleton in the epitheca and a netlike structure in the hypotheca. Not seen since Schütt's description, it probably also represents a *Lophodinium*-like cyst within a thin-walled tabulated theca; ×560, from Schütt, 1895. **11−13.** *Amphitholus elegans* Schütt. **11.** Cell showing nucleus (N) and cytoplasm that surrounds the theca after treatment with caustic potash. **12.** Coloration of living cell. **13.** Originally described as a "skeleton," but not found again, this probably also represents a cyst. 11,12, ×400; 13, ×560; from Schütt, 1895. **14−17.** *Lophodinium polylophum* (Daday) Lemmerman, Laguna Estia Postillon, Paraguay, apical, antapical, ventral, and dorsal views of preserved material, probably representing a resting cyst, ×600, from Daday, 1905.

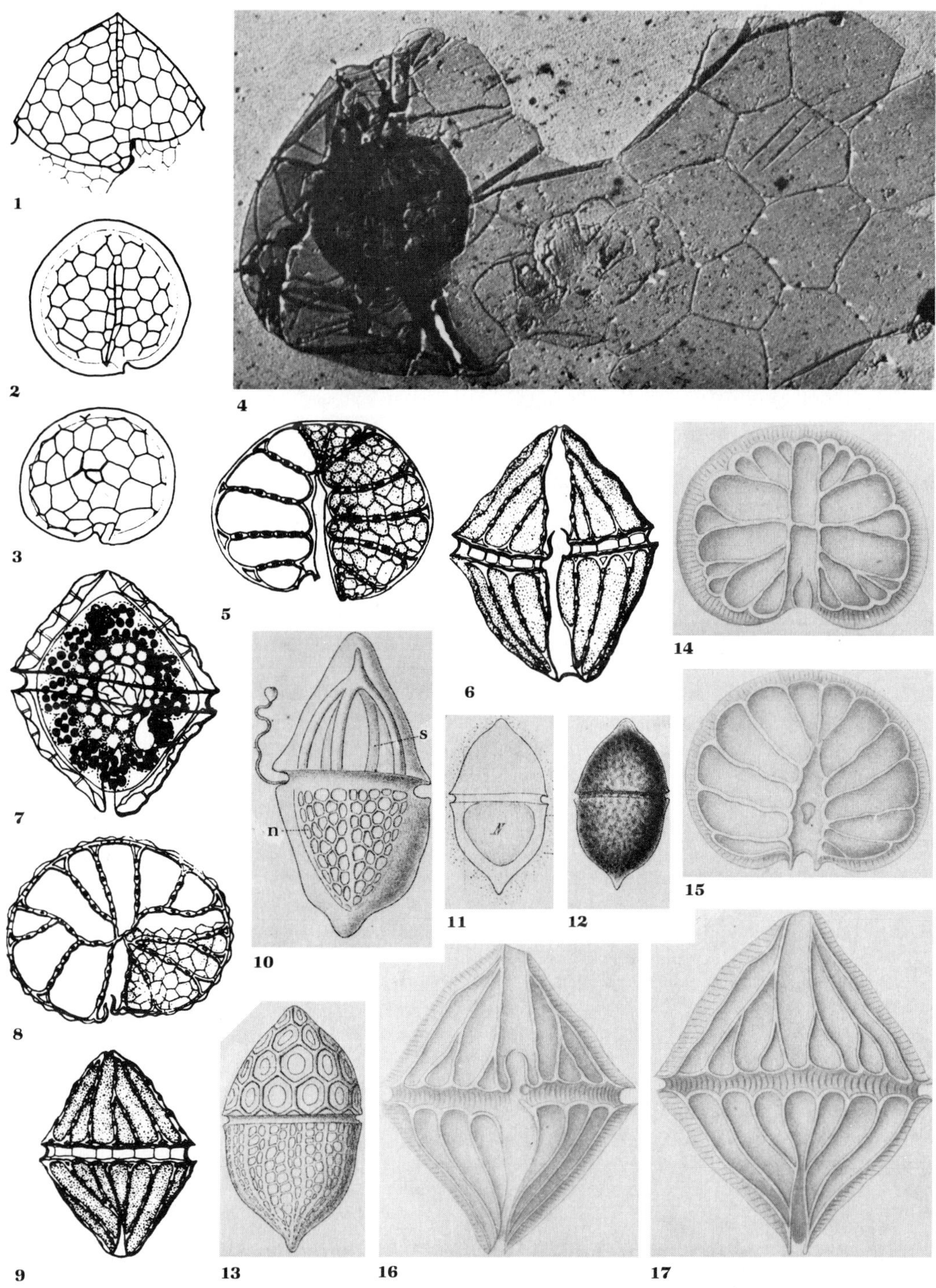

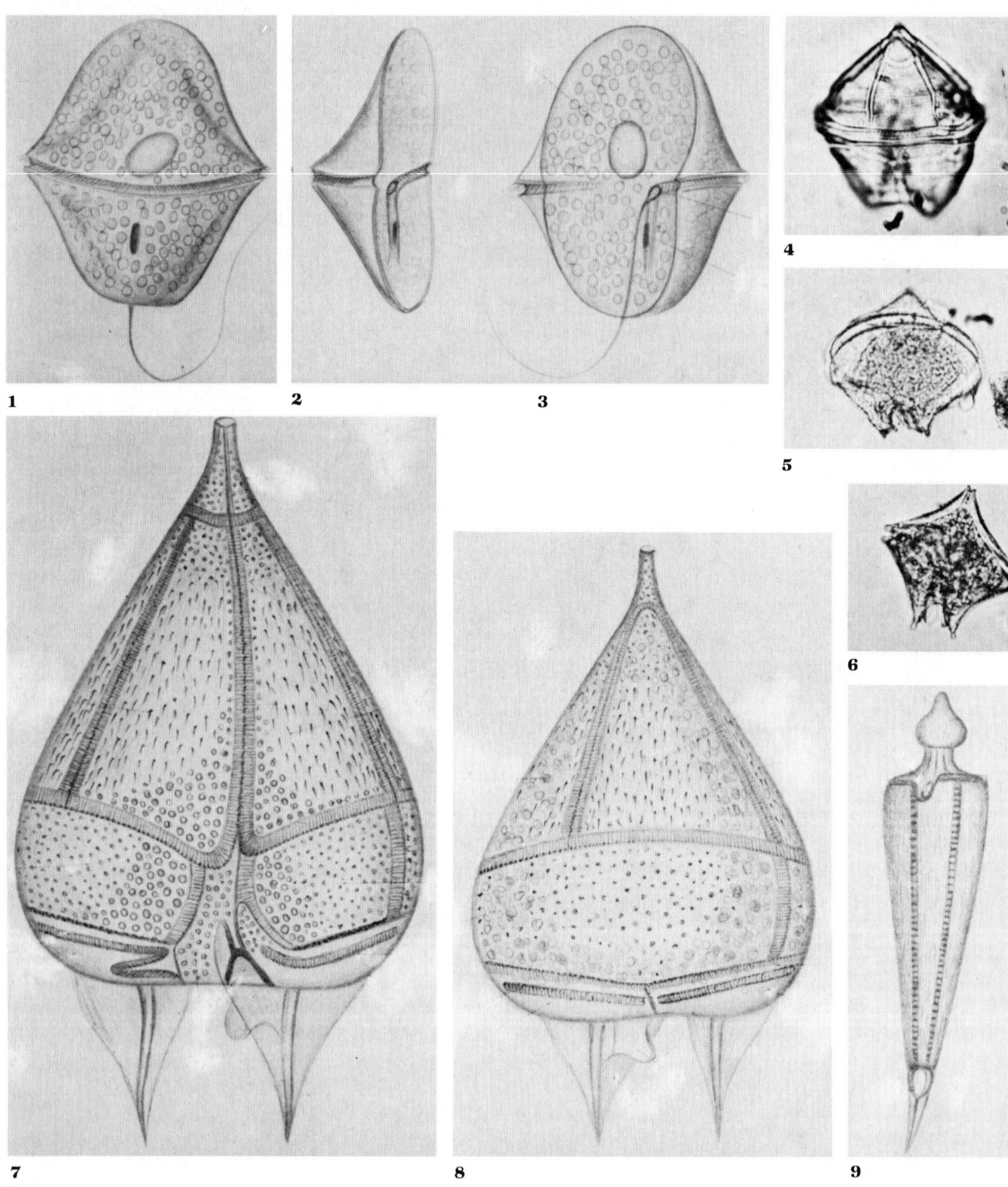

1

2

3

4

5

6

7

8

9

mermann 1904; *Peridinium* Ehrenberg 1830; *Protoperidinium* Bergh 1881; *Scrippsiella* Balech ex Loeblich III 1965; *Zygabikodinium* Loeblich Jr. & Loeblich III 1970.

16. Family Podolampaceae Lindemann 1928

Tabulation variously interpreted as lacking a girdle series, hence 3′, 1a, 5″, 4s, 3‴, 5⁗, or possessing a modified one, and 3′, 1a, 5″, 3c, 5s, 3‴, 3⁗, in *Podolampas;* epitheca larger than hypotheca, apex may be acuminate; photosynthetic.

Marine, especially tropical.

Blepharocysta Ehrenberg 1873 (syn.: *Lissodinium* Matzenauer 1933); *Podolampas* Stein 1883 (syn.: *Parrocelia* Gourret 1883).

17. Family Pyrocystaceae (Schütt) Apstein 1909

Short-lasting motile thecate stage alternates with larger fusiform to crescentic nonmotile vegetative stage ("cysts") that may have cellulosic wall; nonmotile stage produces two to eight motile cells that enlarge; gonyaulacacean tabulation of *Pyrocystis* 4′, 0a, 6″, 6c, 6‴, 1p, 1⁗; photosynthetic.

Marine, planktonic.

Dissodinium Klebs in Pascher 1916; *Pyrocystis* Murray in Thomson & Murray 1885.

18. Family Pyrophacaceae Lindemann 1928 (see Figure 4.107)

Numerous plates in theca, tabulation of

(continued)

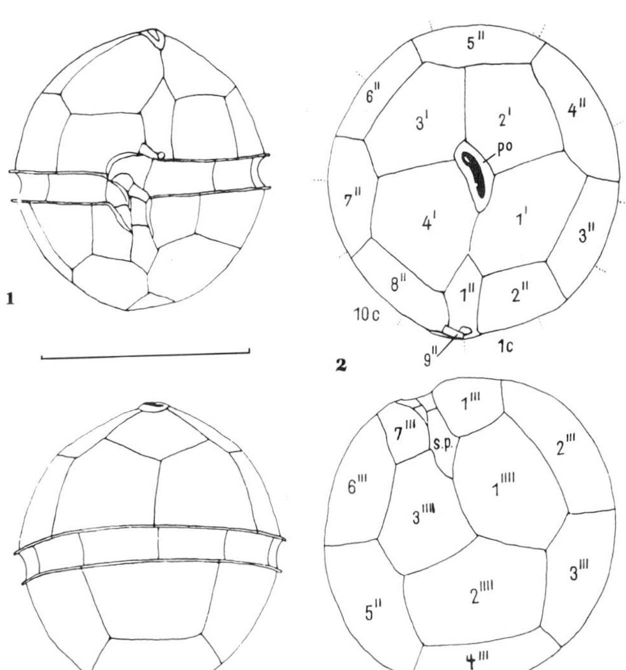

Figure 4.107

Pyrophacaceae. **1–4.** *Helgolandinium subglobosum,* ventral, apical, dorsal, and antapical views. Bar = 30 μm. From Stosch, 1969b.

Pyrophacus, 5 − 9′, 0 − 8a, 9 − 15 (16?)″, 12 − 13c, ?s, 9 − 16‴, 0 − 6p, 3 − 11‴″; may produce two to four daughter cells within theca; anisogamous sexual reproduction reported in *Helgolandinium; Pyrophacus* cysts (= *Tuberculodinium*) may have agglutinating gelatinous cover; photosynthetic.

Marine, planktonic.

Fragilidium Balech ex Loeblich III 1965; *Helgolandinium* von Stosch 1969; *Pyrophacus* Stein 1883.

19. Family Thecadiniaceae Balech 1956

Tabulation pp, 4 − 5′, 0 − 3a, 7 − 8″, 5 − 6c, 5 − 6s, 4 − 5‴, 2p, 1‴″; flattened laterally.

Marine, interstitial water in beach sands.

Roscoffia Balech 1956; *Thecadinium* Kofoid & Skogsberg 1928.

E. Order Arthrodiniales Sournia 1972

1. Family Brachydiniaceae Sournia 1972

?Armored, nontabulate, cingulum present, sulcus not observed; reduced body largely occupied by nucleus; radiating arms (horns) mobile, contain refringent elongate "plastids" that may be skeletal components.

Marine, planktonic, warm water.

Asterodinium Sournia 1972; *Brachydinium* F. J. R. Taylor 1963.

F. Order Gymnodiniales Lemmermann 1910

Amphiesma of free-living motile cells has no thecal plates, only a layer of vesicles.

1. Family Actiniscaceae Kützing 1844

Gymnodinioid motile cell, with one or more perinuclear stellate or **Y**-shaped siliceous spicules, some of unknown composition. Motile zygote in *Plectodinium.*

Marine; rarely lacustrine.

Actiniscus Ehrenberg 1840 (syn.: *Gymnaster* Schütt 1891); *Achradina* Lohmann 1903; *Dicroerisma* F. J. R. Taylor & Cattell 1969; *Plectodinium* Biecheler 1934. Also see fossil genera listed in Table 4.7.

2. Family Gymnodiniaceae (Bergh) Lankester 1885 (see Figure 4.108)

Membranous cell envelope, but no cellulosic plates; no tentacle, no ocellus, no internal skeleton; photosynthetic or with holozoic nutrition; isogamous sexual reproduction; zygote nonmotile in *Amphidinium, Gymnodinium,* motile in *Gyrodinium.* Some produce resting cyst zygotes.

Marine, fresh and brackish water.

Amphidinium Claparède & Lachmann 1859; *Bernardinium* Chodat 1923; *Cochlodinium* Schütt in Engler & Prantl 1896; *Gymnodinium* Stein 1878; *Gyrodinium* Kofoid & Swezy 1921 (syn.: *Spirodinium* Schütt in Engler & Prantl 1896, non Fiorentini 1890).

3. Family Polykrikaceae Kofoid & Swezy 1921

Pseudocolonial, as cytokinesis is incomplete and karyokinesis also may be incomplete; common pellicle with surface argentophilic network encloses the two, four, or eight "cells" of half (or fewer) as many nuclei; common sulcus extends from apex to antapex; each member has a separate left-descending girdle; each pseudocolony may have up to 32 flagella and is 100 to 140 μm long; may have nematocysts and trichocysts; photosynthetic and holozoic.

Marine.

Polykrikos Bütschli 1873; *Pheopolykrikos* Chatton 1933.

4. Family Pronoctilucaceae Lebour 1925 (see Figure 4.109)

Girdle and sulcus rudimentary; tentacle more or less developed; flagella apical or ventral; amphiesma with thin irregular vesicles beneath exterior punctate membrane; body and flagella of *Oxyrrhis* with imbricated scales, flagella also have mastigonemes; sexual reproduction produces motile zygote (*Oxyrrhis*); phagotrophic. (Includes Protodiniferidae Kofoid & Swezy 1921, Entomosigmaceae Chatton in Grassé 1952.)

Marine, neritic, warm water.

Entomosigma Schiller 1925; *Oxyrrhis* Dujardin 1841; *Pronoctiluca* Fabre-Domergue 1889 (syn.: *Protodinifer* Kofoid & Swezy 1921).

5. Family Warnowiaceae Lindemann 1928

Resemble Gymnodiniaceae but with ocellus in sulcal region; amphiesma has vesicular layer beneath surface argentophilic network; some have tentacle in addition to flagella; nematocysts, cnidocysts may occur; nonphotosynthetic.

(continued)

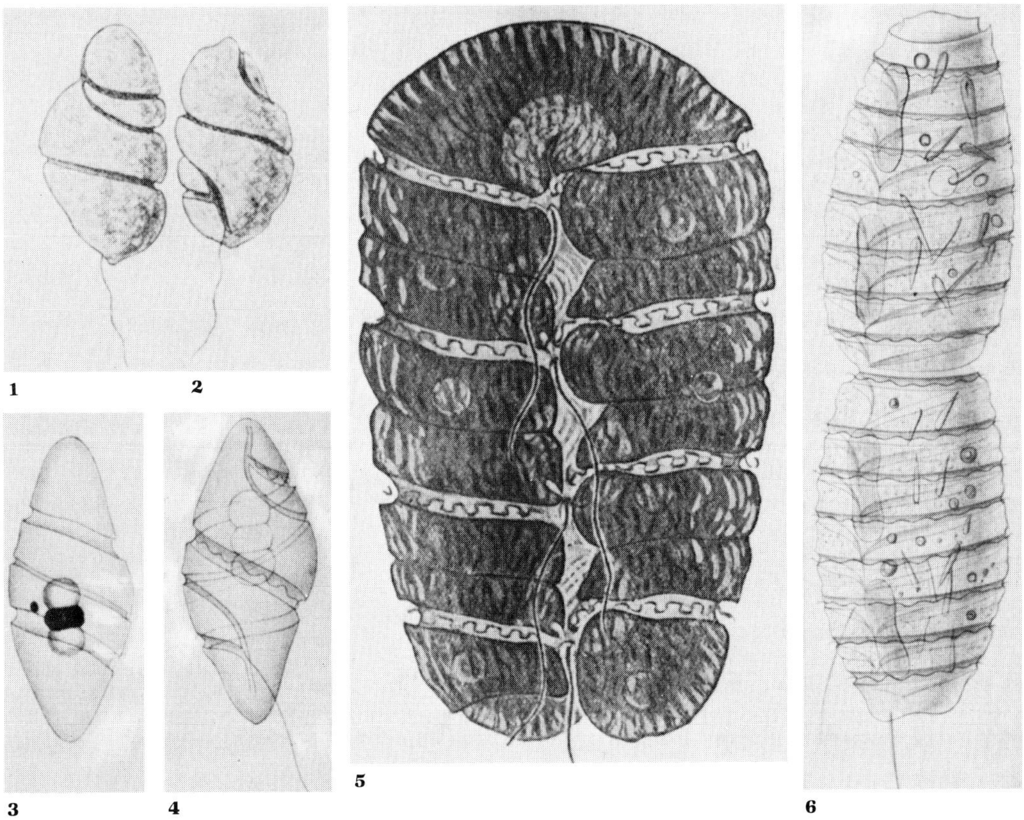

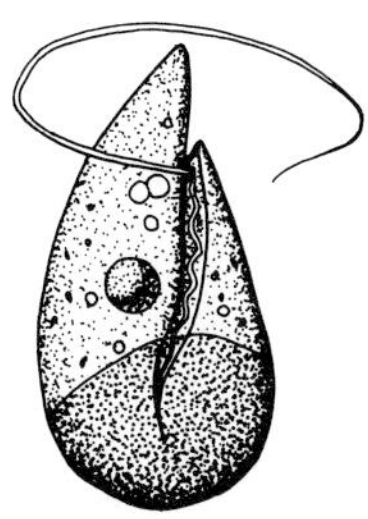

Figure 4.108

Gymnodiniaceae, Polykrikaceae, Warnowiaceae. **1,2.** *Cochlodinium strangulatum* Schütt, dorsal and ventral views, ×160, from Schütt, 1896. **3,4.** *Warnowia fusus* (Schütt) Lindemann, ×440, from Schütt, 1895. **5.** *Pheopolykrikos beauchampi* Chatton, ×900, from Chatton, 1933. **6.** *Polykrikos schwartzi* in process of division, dorsal view, four elongate nuclei at left of cell; ×375, from Bütschli, 1873.

Figure 4.109

Pronoctilucaceae. *Entomosigma peridinioides* Schiller. Ventral view, showing position of two flagella and plastid in posterior of cell. Bar = 5 μm. Redrawn from Schiller, 1925.

Marine, planktonic.

Erythropsidinium Silva 1960 (syn.: *Erythropsis* Hertwig 1884 non Lindley 1827); *Nematodinium* Kofoid & Swezy 1921; *Proterythropsis* Kofoid & Swezy in Kofoid 1920; *Warnowia* Lindemann in Engler & Prantl 1928 (syn.: *Pouchetia* Schütt 1895 non Richard 1830).

G. Order Noctilucales Haeckel 1894

Diploid vegetative cell with nocticaryotic nucleus; cell highly vacuolated, inflated, subspherical to reniform; protoplasmic strands between vacuoles extend from periphery to central mass around nucleus; vacuolar cell sap with osmotic pressure like seawater, but reduced specific gravity as chloride ions replace the divalent iron sulfide; lipids also abundant and aid flotation. Gametes haploid with dinokaryotic nucleus and resemble *Gymnodinium;* young zygote also gymnodinioid with distinct sulcal region; adult with girdle reduced to short trough at left of reduced sulcus, which consists of deep oral groove at midventral line; longitudinal flagellum present, but short; transverse flagellum represented by toothlike or ribbonlike process at left of sulcus. Tentacle or prod arises from posterior part of original sulcus, extends to cell posterior; consists of striated fibers and microtubules and is adapted to movement. Cell amphiesma consists of outermost membrane, vesicular layer, and plasma membrane, and is underlaid by microtubular layer. May have winglike expansions.

1. Family Noctilucaceae Kent 1881

Characters of the order; life cycle complex, some may reproduce asexually in any of the stages, which may resemble other genera.

Marine, neritic, cosmopolitan.

Abedinium Loeblich Jr. & Loeblich III 1966 (syn.: *Leptophyllus* J. Cachon & M. Cachon 1964, non Hope 1842, nec Quenstedt 1876); *Craspedotella* Kofoid 1905; *Cymbodinium* J. Cachon & M. Cachon 1967; *Kofoidinium* Pavillard 1928; *Leptodiscus* Hertwig 1877; *Noctiluca* Suriray in Lamarck 1816; *Pavillardia* Kofoid & Swezy in Kofoid 1920; *Pomatodinium* Cachon & Cachon-Enjumet 1966; *Scaphodinium* Margalef 1963; *Spatulodinium* J. Cachon & M. Cachon 1968.

H. Order Phytodiniales Loeblich III 1970

(Syn.: Dinococcales)

1. Family Phytodiniaceae Klebs 1912 (see Figure 4.110)

Nonmotile stage with stalk produced from apical pore, or coccoid; motile stage gymnodinioid; nucleus dinokaryon.

Freshwater.

Dinastridium Pascher 1927; *Tetradinium* Klebs 1912.

I. Order Zooxanthellales Loeblich III 1970

1. Family Zooxanthellaceae Hovasse & Tessier 1923

Endosymbionts; coccoid vegetative stage with multiple membranes like theca, nucleus dinokaryon; binary fission in vegetative stage; biflagellate gymnodinioid motile stage (swarmers) of short duration; photosynthetic.

Marine.

Endodinium Hovasse 1922; *Zooxanthella* Brandt 1881 (syn.: *Symbiodinium* Freudenthal 1962).

J. Order Blastodiniales Schiller 1935

Parasitic, generally nonthecate, but may have amphiesmal vesicles without plates (*Blastodinium* Chatton 1906) or may have a tabulated layer (*Oodinium* Chatton 1912, *Protoodinium* Hovasse 1935); motile swarm spores gymnodinioid or like *Gyrodinium.*

Marine, parasitic on protozoa or metazoans; no fossils.

  1. Family Apodiniaceae Chatton 1920
  2. Family Blastodiniaceae Cavers 1913
  3. Family Diplomorphaceae Cachon 1964
  4. Family Haplozooaceae Chatton 1920
  5. Family Oodiniaceae Chatton 1920
  6. Family Protoodiniaceae Cachon 1964
  7. Family Chytriodiniaceae Cachon & Cachon 1968

Ectoparasitic on protists or crustacean eggs, continuing to feed on host during development of reproductive spores.

Marine, no fossils.

Chytriodinium Chatton 1912; *Myxodinium* Cachon, Cachon & Bouquaheux 1968.

K. Order Dinamoebidiales Fott 1959

1. Family Dinamoebidiaceae Fott 1959

Fusiform nonmotile stage with cellular

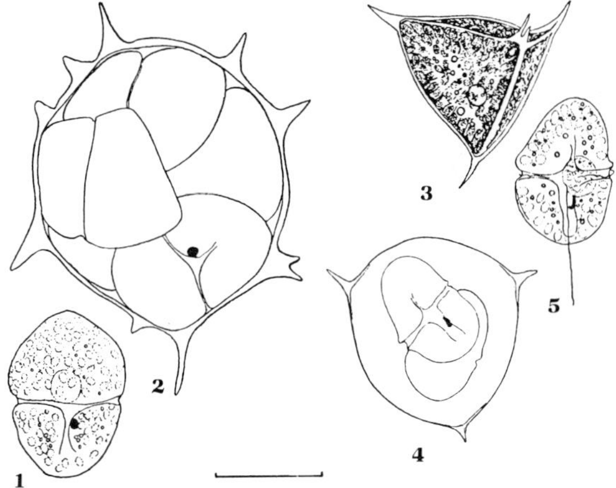

Figure 4.110
Phytodiniaceae. **1,2.** *Dinastridium sexangulare* Pascher. **1.** Motile swarmer. **2.** Four swarm cells developing within the angular vegetative swarm cell. **3–5.** *Tetradinium minus* Pascher. **3.** Vegetative cell. **4.** Two swarm cells developing within wall of vegetative cell. **5.** Motile swarmer. Bar = 20 μm. From Pascher, 1927.

membrane; motile stage amoeboid and gymnodinioid; one dinokaryon nucleus in all stages; nonphotosynthetic.

Marine, no fossils.

Dinamoebidium Pascher 1916 (syn.: *Dinamoeba* Pascher 1916 non Leidy 1875).

L. Order Dinotrichales Pascher 1914
Filamentous vegetative stage, with cell wall; motile stage nonthecate.

Marine, no fossils.

1. Family Dinocloniaceae Pascher 1927
Dinoclonium Pascher 1927.

2. Family Dinotrichaceae Pascher 1931
Dinothrix Pascher 1914.

M. Order Gloeodiniales Loeblich III 1970

1. Family Gloeodiniaceae Schiller 1937
Coccoid, external gelatinous layer; flagellate swarmers (like *Hemidinium*) with thin thecal plates; photosynthetic.

Freshwater, no fossils.

Gloeodinium Klebs 1912.

N. Order Protaspidales Loeblich III 1970

1. Family Protaspidaceae Skuja 1939 (Figure 4.111)

Membranous amphiesma, oval, flattened; longitudinal groove contains posterior flagellum, anterior flagellum directed forward; pseudopodia may project from groove; dinokaryon nucleus, nonphotosynthetic.

Freshwater, no fossils.

Protaspis Skuja 1939.

II. Class Ebriophyceae Loeblich III 1970

A. Order Ebriales Honigberg et al., 1964
See Chapter 5.

III. Class Ellobiophyceae Loeblich III 1970

A. Order Thalassomycetales Loeblich III 1970

1. Family Thalassomycetaceae Niezabitowski 1913 (see Figure 4.112)
Parasitic, cell wall lacking during parasitic stage.

No fossils.

Thalassomyces Niezabitowski 1913.

IV. Class Syndiniophyceae Loeblich III 1976

A. Order Syndiniales Loeblich III 1976
Differ from Dinophyceae in low number

(*continued*)

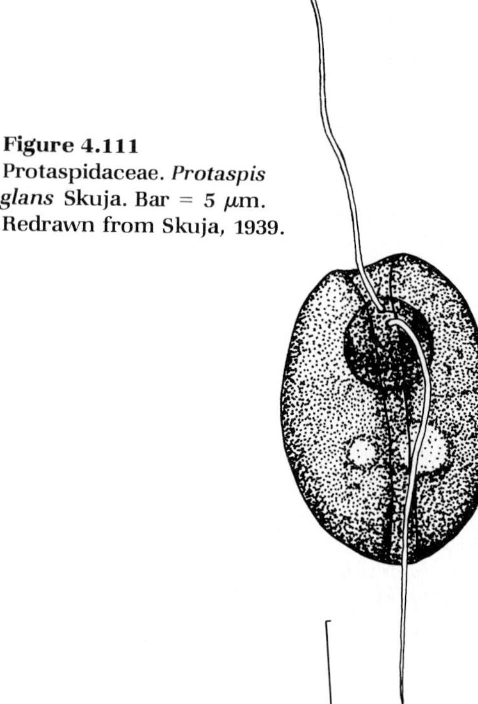

Figure 4.111
Protaspidaceae. *Protaspis glans* Skuja. Bar = 5 μm. Redrawn from Skuja, 1939.

(4 – 10) of V-shaped chromosomes containing histochemically detectable basic proteins; centrioles associated with mitosis; no thecal plates; nutrition by means of intracellular parasitism; isogamous sexual reproduction. Formerly Coccidiniales, but nature of *Coccidinium* Chatton & Biecheler 1934 is questionable, and may not belong here.

1. Family Amoebophryaceae Loeblich III 1970
No fossils.
Amoebophrya Koeppen 1894; *Atlanticellodinium* Cachon & Cachon-Enjumet 1965.
2. Family Dubosquellaceae Loeblich III 1970
No fossils.
Dubosquella Chatton 1920; *Keppenodinium* Loeblich Jr. & Loeblich III 1966 (syn.: *Hollandella* Cachon 1964 non Gill 1901).
3. Family Sphaeriparaceae Loeblich III 1970
No fossils.
Sphaeripara Poche 1911 (syn.: *Lohmannia* Neresheimer 1903, non Michael 1898; *Lohmannella* Neresheimer 1904, non Trouessart 1901; *Neresheimeria* Uebel 1912).
4. Family Syndiniaceae Chatton 1920
No fossils.
Syndinium Chatton 1910; *Trypanodinium* Chatton 1912.

Figure 4.112
Parasitic dinoflagellate, *Thalassomyces fagei,* growing on back of euphausid *Euphausia pacifica,* ×8, from R. Emerson.

Classification of Fossil and Living Cysts

Because fossil dinoflagellates at first were believed to be thecae, fossil species were assigned to modern families and even genera; only a few distinctive fossils were described as new genera or made the basis for description of new families (Deflandre, 1936, 1947a, 1952a,b; Eisenack, 1954, 1961, 1964; Evitt, 1963; Gocht, 1957; Neale and Sarjeant, 1962; Vozzhennikova, 1961, 1963, 1965, 1967, 1979; O. Wetzel, 1933a,b). Classifications (Deflandre, 1952a,b; Eisenack, 1961, 1964; Vozzhennikova, 1963, 1965, 1967, 1979) combined the families and genera based on living types with those erected for the fossil examples, whether the dinoflagellates were regarded as protozoans or as algae. Upon the recognition that all fossil dinoflagellates (other than the skeletal elements of the Actiniscaceae) represent cysts rather than the motile cell, a system of "cyst-families" was proposed (Sarjeant and Downie, 1966) based on cyst morphology, independent of the thecate stage. These "cyst-families" have no nomenclatural standing under the Botanical Code, and their effective date is that at which they were proposed as true families (Sarjeant and Downie, 1974; Sarjeant, 1974). The date for *Palaeoperidinium* and the family Palaeoperidiniaceae is discussed elsewhere (Loeblich Jr., 1968).

In view of the difficulty in relating fossil cysts to living species and genera, some specialists on the group use no family classification and arrange taxa alphabetically. Others (Harland, 1971, 1972) use a classification based on motile cells for Quaternary fossil taxa, and one based on cyst taxa for Tertiary and geologically older dinoflagellates. Reflected tabulation and other morphologic features may indicate a general relationship of some fossil cysts to the modern *Ceratium*, *Gonyaulax*, and *Peridinium*, hence Wall and Dale (1968a) proposed two major groups of "lineages" for classifying the fossils, combining these with the living taxa in a single classification. Their Gonyaulacacean lineages included those cysts (dominantly proximate or chorate) with *Gonyaulax*-like paratabulation (see Figure 4.113) and precingular or apical ar-

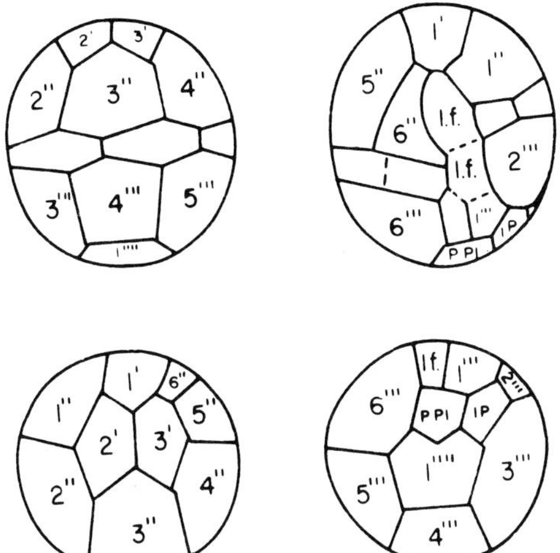

Figure 4.113
Tabulation of *Corrudinium incompositum* (Drugg) Stover & Evitt, Oligocene, Mississippi; lf, longitudinal furrow; from Drugg, 1970.

cheopyles, as well as the *Ceratium*-like ones that commonly have antapical and postcingular horns and apical archeopyles. The Peridiniacean lineages have such *Peridinium*-like features as a peridinioid shape and *Peridinium*-like tabulation (see Figure 4.114), and an archeopyle involving one or more of the large dorsal intercalary areas, alone or in combination with apical or precingular areas. The cavate *Deflandrea* and similar genera and proximate *Pareodinia* are peridiniacean, as are the calcareous cysts such as *Calciodinellum* (resembling the similar cysts of modern *Scrippsiella*) and the siliceous *Peridinites*.

In earlier work on this group, an alphabetical arrangement proved satisfactory, as the numbers of described taxa were not large. With the great increase in described taxa, however, a more systematic arrangement appears useful. Interestingly, of the 15 orders of the Class Dinophyceae and their 47 families, only 3 orders and 6 families of living dinoflagellates are known to have fossil representatives: the Dinophysiales (family Dinophysiaceae), Gym-

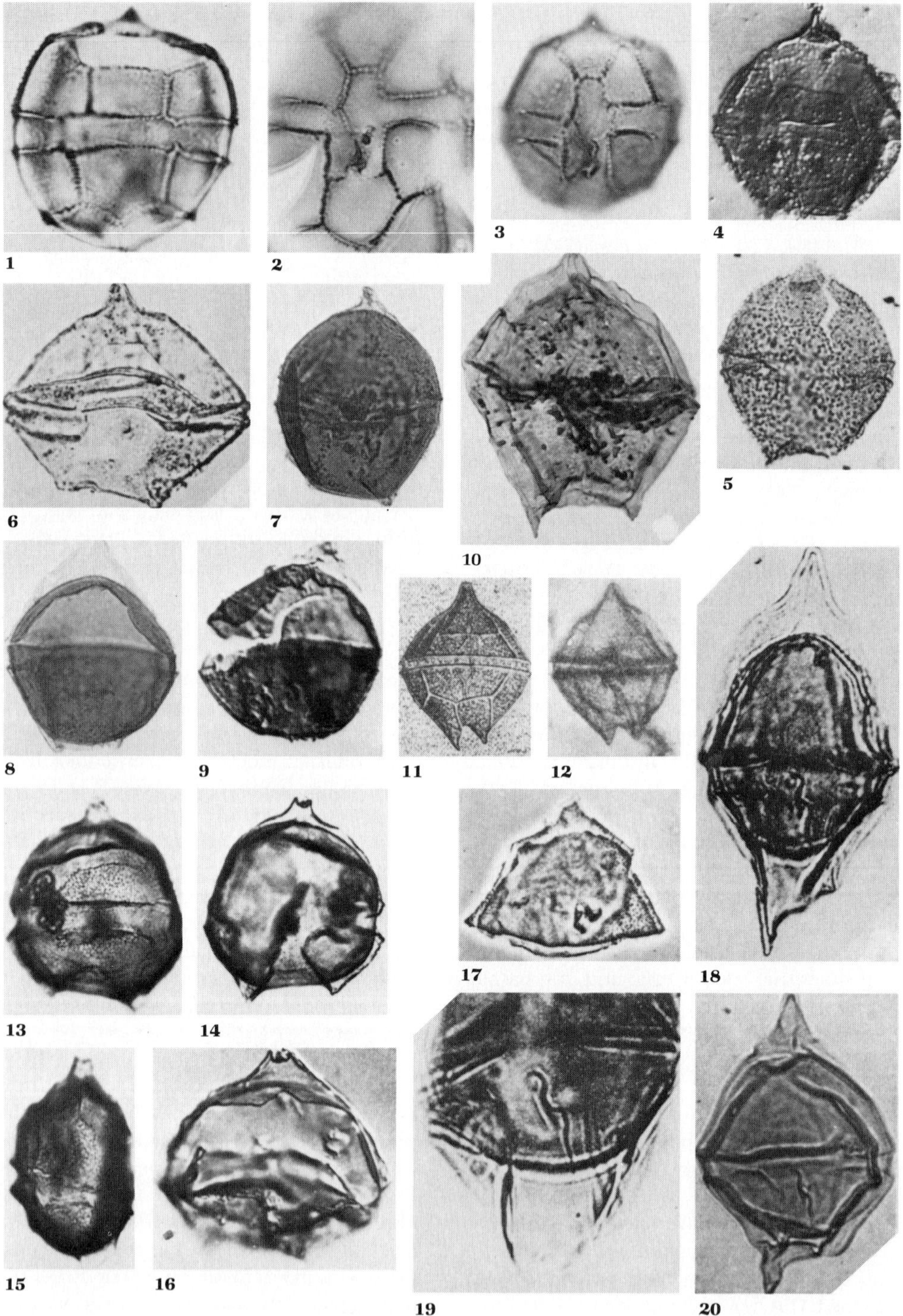

1 2 3 4 5 6 7 8 9 10 11 12 13 14 15 16 17 18 19 20

Figure 4.114
Phthanoperidiniaceae, Palaeoperidiniaceae. **1–3.** *Phthanoperidinium amoenum* Drugg & Loeblich, Oligocene, Mississippi, ×768, from Drugg and Loeblich, 1967. **1.** Dorsal view. **2,3.** Sulcal region and ventral view. **4,5.** *Ginginodinium ornatum* (Felix & Burbridge) Lentin & Williams, Upper Cretaceous, Canada, ×400, from McIntyre, 1975. **6.** *Maduradinium pentagonum* Cookson & Eisenack, Upper Cretaceous, Australia, ×400, from Cookson et al., 1970. **7–9.** *Saeptodinium eurypylum* (Manum & Cookson) Stover & Evitt, Cretaceous, Arctic Canada. **7,8.** Ventral and dorsal views, ×400, with permission, from Manum and Cookson, 1964, *Skr. Norske Vidensk-Akad. Mat.-Naturv. Kl.*, n.s., no. 17, published by Universitetsforlaget, Oslo University. **9.** Dorsal view, ×400, from Evitt, 1975. **10–12.** *Palaeoperidinium pyrophorum.* **10.** Paleocene, Alabama, ×307, from Drugg, 1970. **11,12.** Cretaceous, Germany; 11, from Ehrenberg, 1838b. 12, Photo of original type, ×240, from Lejeune-Carpentier, 1938b. **13–17.** *Saeptodinium hansonianum,* upper Oligocene, freshwater, Vermont, ×320, from Evitt, 1974b. **13,14.** Dorsal focus and ventral side. **15.** Right side. **16.** Apical region, showing archeopyle. **17.** Operculum, showing both ectophragm and endophragm layers. **18–20.** *Subtilisphaera trendalli* (Cookson & Eisenack) Lentin & Williams, Cretaceous, Australia. **18.** Ventral surface, ×800. **19.** Part of ventral surface showing S-shaped thickening or "flagellar marking," ×1120. 18,19, from Cookson et al., 1970. **20.** ×600, from Cookson et al., 1971.

---

nodiniales (family Actiniscaceae), and Peridiniales (Ceratiaceae, Gonyaulacaceae, Peridiniaceae, and Pyrophacaceae). *Palaeotetradinium* Deflandre 1936 has been referred to the Phytodiniales (= Dinococcales), but this allocation is doubtful. As more living taxa are studied in culture, their life cycles and relationships will be better understood, and many more may prove to have fossilizable resting cysts such as those already described. The variety of cysts already known, even in the small number of species yet described, suggests that this stage of the life cycle will be useful in subdividing some of the large modern genera (e.g., *Gonyaulax, Peridinium*) as well as in indicating their long evolutionary history.

Modern taxa are described and classified on the basis of data from the motile cell (morphology of the amphiesma and cell contents) and aspects of the life history, but only rarely including information about their cysts. Species with very similar motile stages produce a variety of encysted, parasitic, or symbiotic stages: At least eight genera based on cysts are regarded as referable to *Gonyaulax* motile cells (Reid, 1974), and four species based on cysts, representing three genera, have a *Gonyaulax spinifera* motile cell, suggesting that it needs subdivision. Other similar cysts may be produced by quite distinctive motile cells. Ideally, all stages should be described, but as our knowledge of most will remain incomplete, the taxa based on living cells cannot be used for fossil cysts below the ordinal level. The classification summarized in Table 4.7 is based on the cysts alone (the Actiniscaceae is based on internal spicules). This classification (with examples of representative genera) generally is similar to that of Sarjeant and Downie (1974), with some modifications (the inclusion of additional families and genera proposed since their article, recognition of different authorship and dates for some genera and families, synonymizing of some, and the recognition of some earlier family names).

In the brief family descriptions, an indication is given of the equivalent lineages of Wall and Dale (1968a), as well as the family taxa that are combined. Not all fossil genera are included, as many relationships are obscure, and even some of those genera included in the classification remain doubtful. Fossil genera mentioned in the text and not allocated in the table are listed following the classification, with an indication of authorship and geologic occurrence. Details of others may be located through the various indices.

Table 4.7
Classification of fossil dinoflagellates (cysts and spicules).

A. Order Gymnodiniales Lindemann 1928

1. Family Actiniscaceae Kützing 1884 (see Figure 4.115)
Motile living cell is gymnodinioid, and contains one or more Y-shaped or stellate siliceous spicules around the nuclear region; such stellate bodies are common in siliceous pelagic sediments.
Upper Oligocene to Recent.
Actiniscus Ehrenberg 1840; *Calicipedinium* Dumitrică 1973; *Carduifolia* Hovasse 1932; *Cinctactiniscus* Dumitrică 1973; *Foliactiniscus* Dumitrică 1973.

B. Order Dinophysiales Lindemann 1928

1. Family Nannoceratopsaceae Gocht 1970
Proximate cysts; subtriangular in lateral view; strongly flattened; minute epicyst, subapical paracingulum, broad and high hypocyst, parasulcus on narrow ventral surface, two prominent antapical horns separated by concave antapex, ventral horn may be reduced; archeopyle epicystal, operculum remaining attached.
Jurassic.
Nannoceratopsis Deflandre 1938.

C. Order Peridiniales Schütt 1896

1. Family Arpyloraceae Sarjeant 1978
Proximate peridiniacean cysts, ovoid to polygonal in outline, epicyst larger than hypocyst, tabulation uncertain but probably ?4′, 5−6a, 5−7″, 5−7c, 5−7‴, ?p, ?3‴″; type I or 2I archeopyle.
Silurian.
Arpylorus Calandra 1964.

2. Family Peridinitaceae Sarjeant & Downie 1967
Proximate siliceous peridiniacean cysts, ovoid; paratabulation 4′, 3a, 7″, 5‴, 2‴″; paracingulum and sulcus well marked; wall thick, surface with raised ornamentation; archeopyle postcingular. Includes family Lithoperidiniaceae Deflandre 1952.
Paleocene to Oligocene; Barbados, USSR, New Zealand.
Jusella Vozzhennikova 1963; *Peridinites* Lefèvre 1933 (syn.: *Lithoperidinium* Deflandre 1933).

3. Family Calciodinellaceae Deflandre 1947
Proximate calcareous peridiniacean cysts; paracingulum well marked or obscure; paratabulation 4′, 3a, 7″, 6c, 5‴, 0p, 2‴″; type 2−4A + 1−3I archeopyle. Similar cysts characterize living *Scrippsiella trochoidea* and *Ensiculifera* cf. *mexicana*.
Upper Jurassic to Quaternary; Europe, Africa, N. America; Atlantic Ocean, Caribbean, Mediterranean.
Alasphaera Keupp 1979; *Bicarinellum* Deflandre 1948; *Biechelerella* Deflandre 1948; *Calcicarpinum* Deflandre 1948; *Calcigonellum* Deflandre 1948; *Calciodinellum* Deflandre 1947; *Calciogranellum* Deflandre 1948; *Calcipterellum* Deflandre 1948; *Calcisphaerellum* Deflandre 1948.

4. Family Dinogymniaceae Sarjeant & Downie 1974
Proximate cysts; elongate, biconical to ovoid; paracingulum and parasulcus may be indicated; wall may be pierced by numerous mural canals; may have longitudinal ribs or folds; small type A archeopyle; possibly related to living Lophodiniaceae.
Upper Cretaceous; N. America, Europe, Africa, USSR, India, Australia.
Dinogymnium Evitt, Clarke & Verdier 1967.

5. Family Pareodiniaceae Gocht 1957 (see Figure 4.116)
Syn.: Netrelytraceae Sarjeant & Downie 1966.
Proximate ovoid to polygonal peridiniacean cysts, may have apical horn and rarely one or two antapical horns; may have enclosing kalyptra; paratabulation weak if present, 6′, 6a, 6″, 7‴, ?p, 3‴″; type I, 2I, or 3I archeopyle. Part of "pseudoceratioid lineage."
Upper Triassic to middle Miocene; N. America, S. America, Europe, USSR, Australasia.
Broomea Cookson & Eisenack 1958; *Caligodinium* Drugg 1970; *Glomodinium* Dodekova 1975; *Heibergella* Bujak & Fisher 1976; *Kalyptea* Cookson & Eisenack 1960 (syn.: *Komewuia* Cookson & Eisenack 1960; *Netrelytron* Sarjeant 1961); *Pareodinia* Deflandre 1947 (syn.: *Imbatodinium* Vozzhennikova 1967; *Paranetrelytron* Sarjeant 1966; *Pluriarvalium* Sarjeant 1962).

(continued)

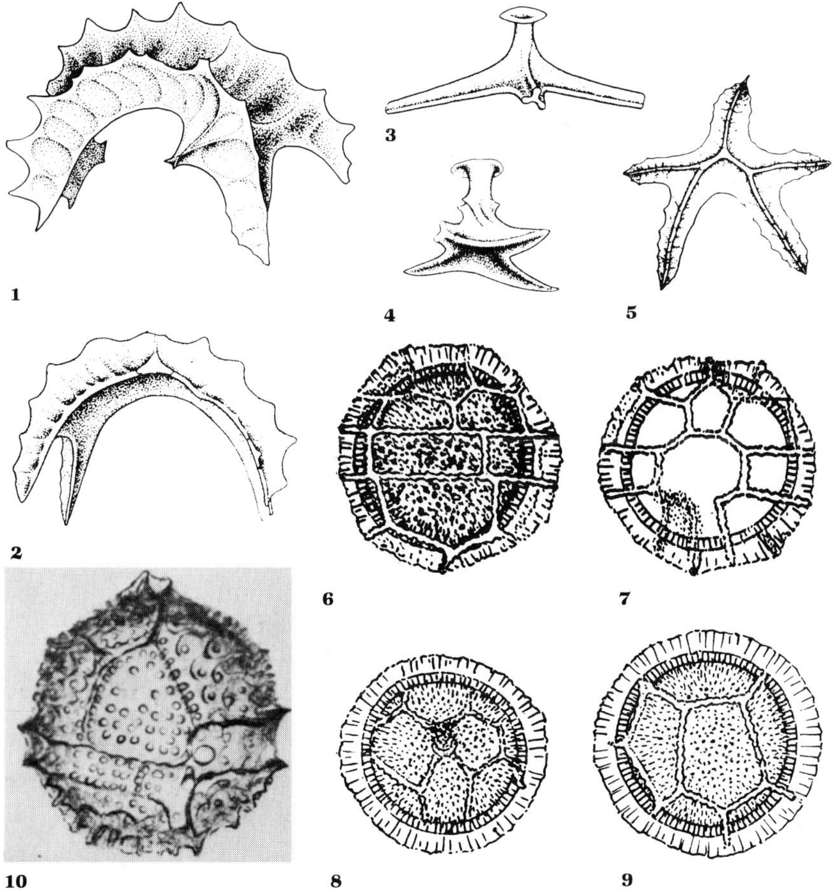

Figure 4.115

Actiniscaceae, Calciodinellaceae, Peridinitaceae. **1,2.** *Cinctactiniscus cinctus* (Hovasse) Dumitrică, upper Oligocene, southwest Pacific, ×950. **3,4.** *Calicipedinium quadripes* Dumitrică, middle Miocene, Romania, ×1450. **5.** *Foliactiniscus folia*, middle Miocene, southwest Pacific, ×950. 1–5, from Dumitrică, 1973. **6—9.** *Calciodinellum operosum*, Miocene, Oran, from Deflandre, 1948. **6,7.** Holotype, dorsal view and focus through specimen to ventral side. **8,9.** Paratype, in apical and antapical views. 6–8, ×1000; 9, ×1140. **10.** *Peridinites maculatum* (Vozzhennikova) Lentin & Williams, lower Oligocene, Kazakhstan, ×420, from Vozzhennikova, 1967.

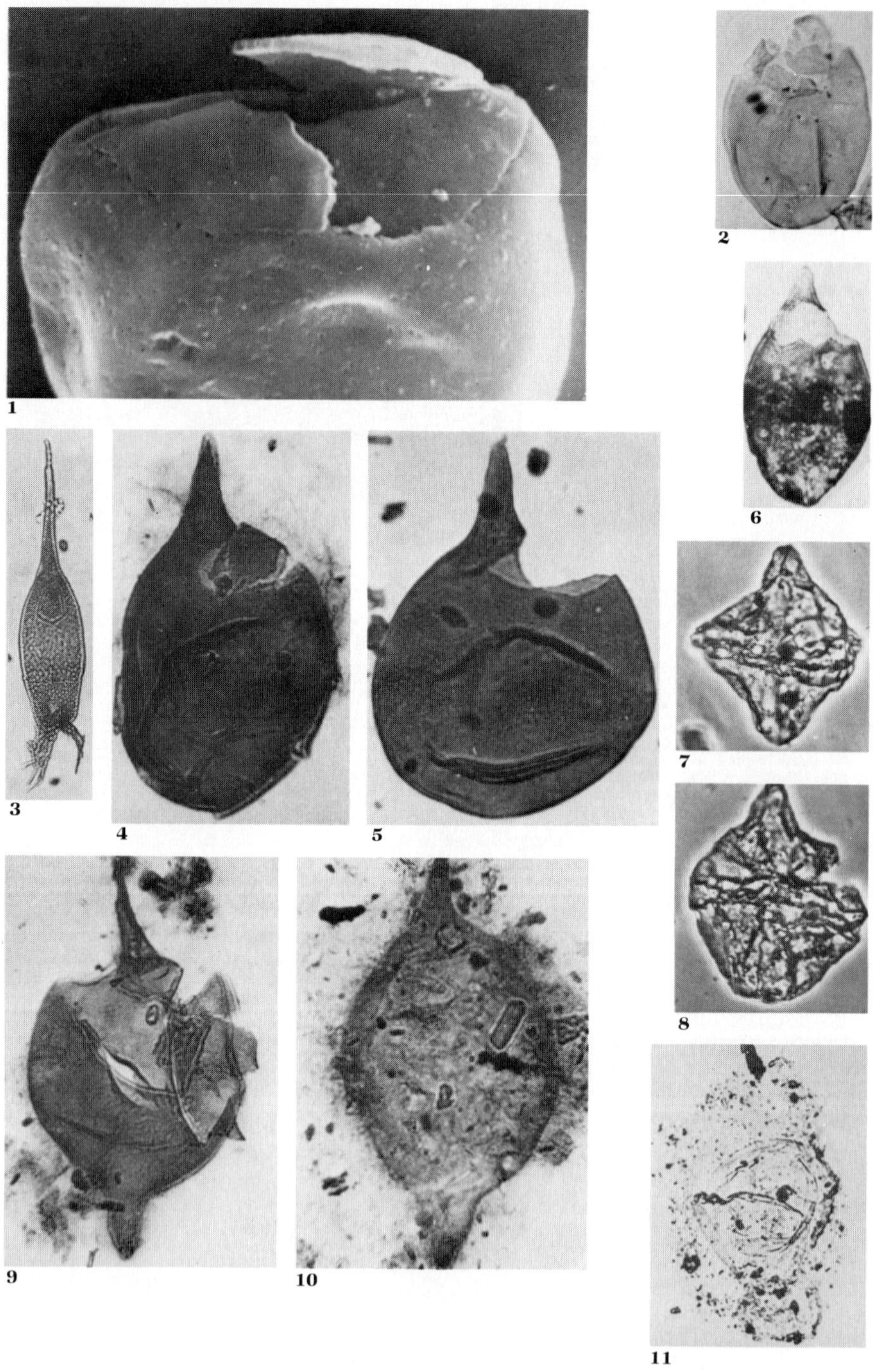

Figure 4.116
Pareodiniaceae. **1,2.** *Caligodinium amiculum,* Paleocene, Alabama, from Drugg, 1970. **1.** SEM, dorsal view, showing partly open archeopyle, ×1760. **2.** Entire specimen, ×307. **3.** *Broomea ramosa* Cookson & Eisenack, Jurassic, Western Australia, dorsal view, ×240, from Cookson and Eisenack, 1958. **4,5,11.** *Pareodinia ceratophora;* 4,5, Upper Jurassic, Alaska, ×670, from Wiggins, 1975. **4.** Left side, showing operculum and partly enclosing calyptra. **5.** Specimen with open archeopyle. **11.** Specimen originally described as *Paranetrelytron,* ×400, from Sarjeant, 1966b. **6.** *P. evitti* (Pocock) Wiggins, Jurassic, Canada, oblique dorsal view, ×480, from Johnson and Hills, 1973. **7,8.** *Heibergella asymmetrica* Bujak & Fisher, Upper Triassic (Norian), Arctic Canada, dorsal and left side views, ×480, from Bujak and Fisher, 1976. **9,10.** *Kalyptea stegasta* (Sarjeant) Wiggins, Jurassic, Alaska, specimen with archeopyle and specimen with well-developed calyptra, ×480, from Wiggins, 1975.

6. Family Phthanoperidiniaceae Drugg & Loeblich Jr. 1967

Proximate peridiniacean cysts, ovoid to polygonal, may have apical and antapical horns; paratabulation 4′, 2−3a, 7″, 5‴, 2⁗, type I archeopyle. Living *Protoperidinium subinerme* has cyst of *Omanodinium* type; *P. conicum* and *P. nudum* (Meunier) Balech produce *Multispinula* cysts, and *P. compressum* (Abé) Balech has a *Stelladinium* cyst form.

Upper Cretaceous to Recent; N. America, Europe, Australasia.

Leipokatium Bradford 1975; *Maduradinium* Cookson & Eisenack 1970; *Multispinula* Bradford 1975; *Omanodinium* Bradford 1975; *Phthanoperidinium* Drugg & Loeblich Jr. 1967; *Stelladinium* Bradford 1975.

7. Family Palaeoperidiniaceae Vozzhennikova ex Sarjeant 1967

Proximate peridiniacean cysts, ovoid to polygonal, may have apical and two antapical horns; paratabulation 1−4′, 1−3a, 7″, 5−?7c, ?5s, 5‴, 2⁗, type $\overline{A3I3P}$ archeopyle (epicystal), or transapical. Includes cysts of living freshwater *Peridinium.*

Lower Cretaceous to lower Pliocene; N. America, Europe, USSR.

Astrocysta Davey 1970; *Ginginodinium* Cookson & Eisenack 1960; *Morkallacysta* Harris 1973; *Palaeoperidinium* Deflandre ex Sarjeant 1967; *Pentagonum* Vozzhennikova 1963; *Saeptodinium* Harris 1973; *Subtilisphaera* Jain & Millipied 1973.

8. Family Hexagoniferaceae Sarjeant & Downie ex Vozzhennikova 1967 (see Figure 4.117)

Cavate peridiniacean cysts, ovoid to polygonal, may have apical and one or two antapical horns; weak paratabulation if any, type tA, I, or AI archeopyle. Part of "deflandreoid lineage." Includes living *Peridinium* species.

Upper Triassic to upper Oligocene, Recent; N. America, Europe, Africa, USSR, India, Australasia.

Ascodinium Cookson & Eisenack 1960; *Hexagonifera* Cookson & Eisenack 1961; *Inversidinium* McLean 1973; *Noricysta* Bujak & Fisher 1976; *Ovoidinium* Davey 1970 (syn.: *Craspedodinium* Cookson & Eisenack 1974; *Evittia* Pocock 1972, non Brito 1967; *Pocockia* Lentin & Williams 1973); *Parvocavatus* Gitmez 1970; *Polygonifera* Habib 1972; *Pyxidiella* Cookson & Eisenack 1958; *Senoniasphaera* Clarke & Verdier 1967.

9. Family Deflandreaceae Eisenack 1954 (see Figures 4.118 and 4.119)

Cavate peridiniacean cysts, ovoid to polygonal, may have apical horn, one or two antapical and one or two lateral horns; paratabulation weak, 4′, 3a, 7″, 5‴, 2⁗; type I or IP archeopyle with operculum of one or more pieces. "Deflandreoid lineage"; includes families Wetzeliellaceae Vozzhennikova 1961, Chatangiellaceae Vozzhennikova 1967, and Cooksoniellaceae Vozzhennikova 1967.

Upper Triassic to Middle Miocene (*Geiselodinium:* Eocene to Miocene, Freshwater); N. America, S. America, Europe, Africa, USSR, India, Australasia.

(*continued*)

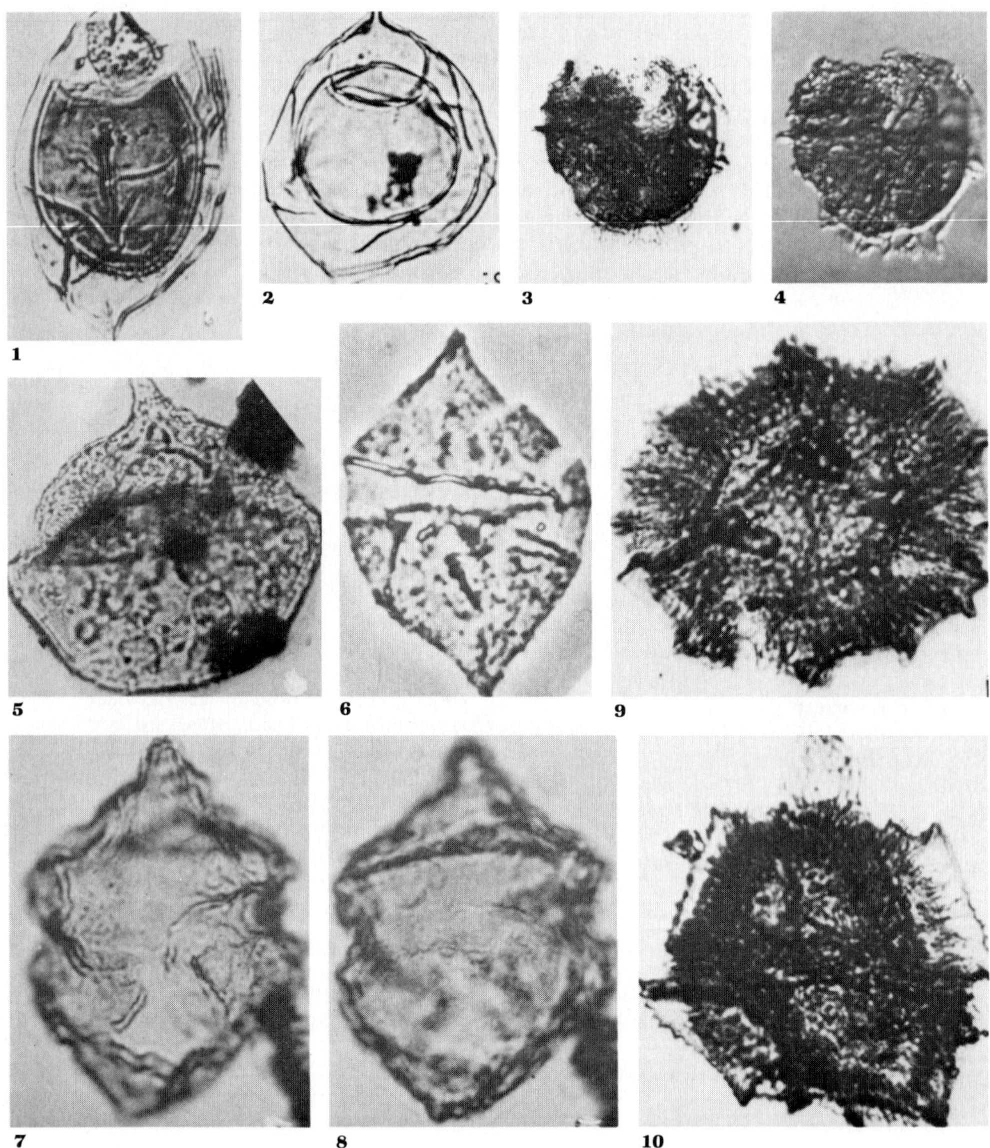

Figure 4.117

Hexagoniferaceae, Nelsoniellaceae, Shublikodiniaceae, Nelchinopsaceae. **1.** *Ascodinium ovale* Cookson & Eisenack, middle Cretaceous, Western Australia, ×700, from Cookson et al., 1970. **2.** *A. acrophorum* Cookson & Eisenack, middle Cretaceous, Australia, ×550, from Cookson and Eisenack, 1960a. **3,4.** *Noricysta fimbriata* Bujak & Fisher, Upper Triassic, Arctic Canada, archeopyle includes all of epitract except precingular plate equivalents, ×600, from Bujak and Fisher, 1976. **5.** *Nelsoniella tuberculata* Cookson & Eisenack, Tertiary, Argentina, ×460, from Pöthe de Baldis, 1966. **6.** *Rhaetogonyaulax rhaetica* (Sarjeant) Loeblich & Loeblich, Upper Triassic, Austria, phase contrast, ×875, from Harland et al., 1975. **7,8.** *R. arctica* (Wiggins) Stover & Evitt, Upper Triassic, Alaska, ventral and dorsal views, ×1000, from Wiggins, 1973. **9,10.** *Nelchinopsis kostromiensis* (Vozzhennikova) Wiggins, Lower Cretaceous, Alaska, apical and side views, ×1000, from Wiggins, 1972.

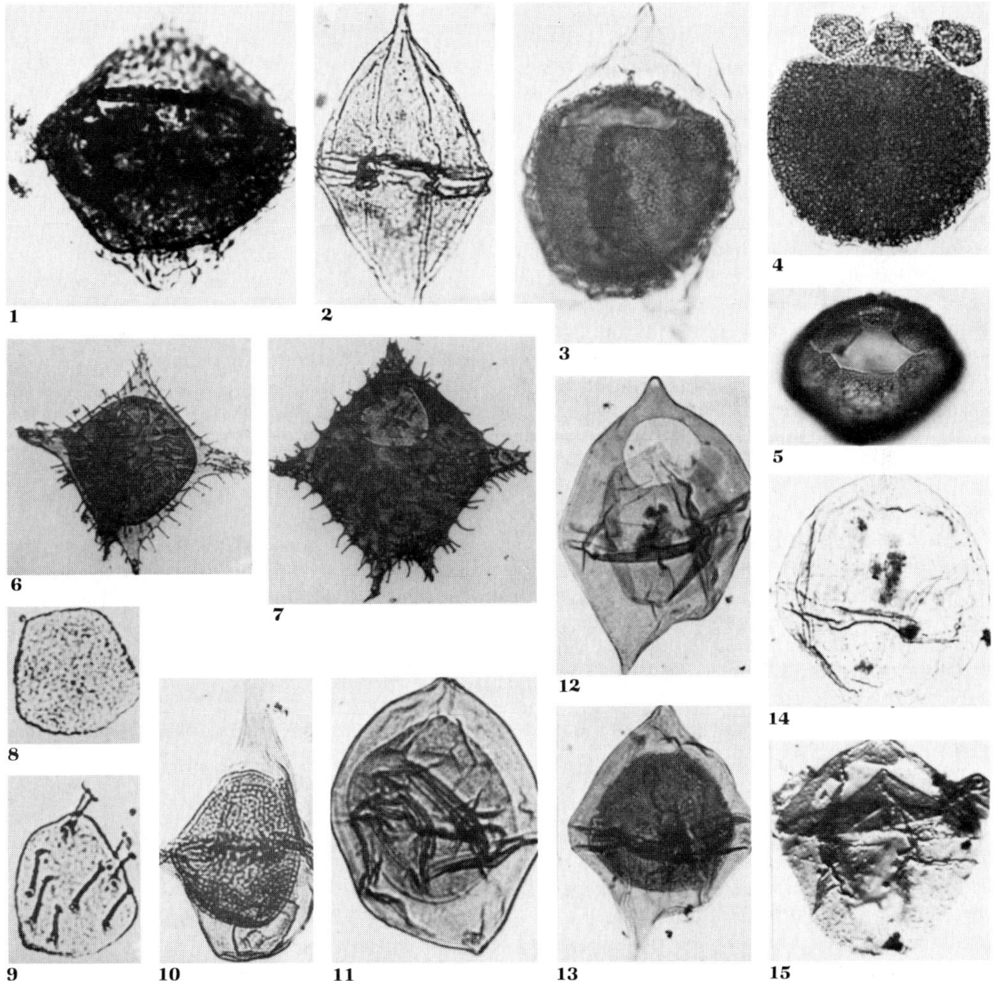

Figure 4.118

Deflandreaceae. **1.** *Uvatodinium marginatum* Vozzhennikova, Upper Cretaceous, Canada, ×400, from J. H. Wall and Singh, 1975. Reproduced by permission of the National Research Council of Canada, from the *Canadian Journal of Earth Sciences*, v. 12, p. 1157–1174, 1975. **2.** *Diconodinium multispinum* (Deflandre & Cookson) Eisenack & Cookson, Lower Cretaceous, Australia, ×400, from Eisenack and Cookson, 1960. **3–5.** *Trithyrodinium suspectum* (Manum & Cookson) Davey, Upper Cretaceous, Arctic Canada, ×400, with permission from Manum and Cookson, 1964, *Skr. Norske Vidensk-Akad. Mat.-Naturv. Kl.*, n.s., no. 17, published by Universitetsforlaget, Oslo University. **3.** Dorsal view. **4.** Partially detached opercular pieces. **5.** Apical view. **6–9.** *Wetzeliella articulata* Eisenack, Eocene London Clay, England, from G. L. Williams and Downie, 1966a. **6,7.** Ventral and dorsal sides, ×265. **8,9.** Opercula from endophragm and periphragm, respectively, ×560. **10.** *Dingodinium cerviculum* Cookson & Eisenack, Cretaceous, Australia, ×360, from Cookson and Eisenack, 1958. **11.** *Eurydinium ingrami* (Cookson & Eisenack) Stover & Evitt, Upper Cretaceous, Australia, ×440, from Cookson et al., 1971. **12,13.** *Isabelidinium acuminatum* (Cookson & Eisenack) Stover & Evitt, Upper Cretaceous, Canada, dorsal view and optical section, ×400, from McIntyre, 1975. **14,15.** *Laciniadinium orbiculatum* McIntyre, Upper Cretaceous, Canada, ×400, from McIntyre, 1975.

Alterbia Lentin & Williams 1976 (syn.: *Albertia* Vozzhennikova 1967, non Schimper 1837); *Andalusiella* Riegel 1974; *Bulbodinium* O. Wetzel 1960; *Catillopsis* Drugg 1970; *Chatangiella* Vozzhennikova 1967 (syn.: *Australiella* Vozzhennikova 1967; *Cooksoniella* Vozzhennikova 1967); *Deflandrea* Eisenack 1938 (syn.: *Ceratiopsis* Vozzhennikova 1963; *Cerodinium* Vozzhennikova 1963); *Diconodinium* Eisenack & Cookson 1960; *Dingodinium* Cookson & Eisenack 1958; *Dioxya* Cookson & Eisenack 1958 (syn.: *Vozzhennikovia* Lentin & Williams, 1975); *Evittodinium* Deflandre 1964; *Geiselodinium* Krutzsch 1962; *Hebecysta* Bujak & Fisher 1976; *Isabelidinium* Lentin & Williams 1977 (syn.: *Isabelia* Lentin & Williams 1975 non Barbosa-Rodrigues 1877); *Kisselevia* Vozzhennikova 1967; *Laciniadinium* McIntyre 1975; *Lejeunia* Gerlach 1961; *Luxadinium* Brideaux & McIntyre 1976; *Moesiodinium* Antonescu 1974; *Palaeocystodinium* Alberti 1961; *Phelodinium* Stover & Evitt 1978; *Pseudodeflandrea* Alberti 1959; *Rhombodinium* Gocht 1955; *Senegalinium* Jain & Millepied 1973; *Spinidinium* Cookson & Eisenack 1962; *Svalbardella* Manum 1960; *Sverdrupiella* Bujak & Fisher 1976; *Teneridinium* Krutzsch 1962; *Trithyrodinium* Drugg 1967; *Uvatodinium* Vozzhennikova 1963; *Wetzeliella* Eisenack 1938 (syn.: *Dracodinium* Gocht 1955); *Wilsonidium* Lentin & Williams 1975; *Xenikoon* Cookson & Eisenack 1960.

10. Family Nelsoniellaceae Eisenack 1961

Cavate peridiniacean cysts, ovoid to polygonal, may have apical horn; paratabulation 4', 2−3a, 7'', 5''', 2'''', but may be obscure and paracingulum may be weakly indicated; type P archeopyle. Part of "deflandreoid lineage."

Upper Cretaceous; Australasia.

Amphidiadema Cookson & Eisenack 1960;

Nelsoniella Cookson & Eisenack 1960; *Sumatradinium* Lentin & Williams 1975.

11. Family Shublikodiniaceae Wiggins 1973

Proximate gonyaulacacean cysts, ovoidal, subequal epicyst and hypocyst; paratabulation 6', 5a, 7'', 7''', 3'''''; type A+5I archeopyle.

Upper Triassic, marine; N. America, Europe. *Rhaetogonyaulax* Sarjeant 1966 (syn.: *Shublikodinium* Wiggins 1973).

12. Family Nelchinopsaceae Wiggins 1972

Proximate gonyaulacacean cysts; ovoidal, with apical horn; paratabulation 5', 6−7a, 6'', 5−6''', 1''''; type A+I archeopyle.

Lower Cretaceous, marine; N. America (Alaska), USSR.

Nelchinopsis Wiggins 1973.

13. Family Gonyaulacystaceae Sarjeant & Downie ex Górka 1970 (see Figures 4.120 and 4.121)

Proximate gonyaulacacean cysts; ovoid to polygonal; paratabulation 1−6', 0−5a, 6−8'', 5−6+c, 5−7''', 1−2p, 1−1pv, 0−7pc, 0−1''''; type 1P (rarely 2P) archeopyle. Posterior intercalary (p), posterior ventral sulcal (pv) and a possible additional posterior circle (pc) of paraplates may occur. "Gonyaulacoid lineage"; some cysts of this type in living *Gonyaulax*.

Middle Jurassic to Quaternary; Europe, N. America, Africa, USSR, India, Australasia.

Acanthaulax Sarjeant 1968; *Aldorfia* Stover & Evitt 1978; *Carpodinium* Cookson & Eisenack 1962; *Corrudinium* Stover & Evitt 1978; *Cribroperidinium* Neale & Sarjeant 1962; *Cryptarchaeodinium* Deflandre 1939; *Diacanthum* Habib 1972; *Druggidium* Habib 1973; *Egmontodinium* Gitmez & Sarjeant 1971; *Gonyaulacysta* Deflandre 1964; *Herendeenia* Wiggins

(*continued*)

Figure 4.119

Deflandreaceae. **1,2.** *Lejeunia kozlowskii* Górka, Upper Cretaceous, Canada, ×400, from McIntyre, 1975. **3.** *Palaeocystodinium incertum* (Deflandre) G. Deflandre & M. Deflandre, Upper Cretaceous, in chert, ×240, from W. Wetzel, 1971. **4.** *Spinidinium sagittulum* (Drugg) Lentin & Williams, lower Eocene, Alabama, ×307, from Drugg, 1970. **5,6.** *Rhombodinium glabrum* (Cookson) Vozzhennikova. **5.** London clay, Eocene, England, ×400, from G. L. Williams and Downie, 1966a. **6.** Eocene, Alabama, ×384. **7,8.** *Wilsonidium echinosuturatum* (G. J. Wilson) Lentin & Williams, middle Eocene, New Zealand, focussed on dorsal and ventral sides, ×320, from Wilson, 1967c. **9,10.** *Kisselevia coleothrypta* (Williams & Downie) Lentin & Williams. **9.** Eocene, New Zealand, ×320, from Wilson, 1967c. **10.** London Clay, Eocene, England, ×265, from G. L. Williams and Downie, 1966a.

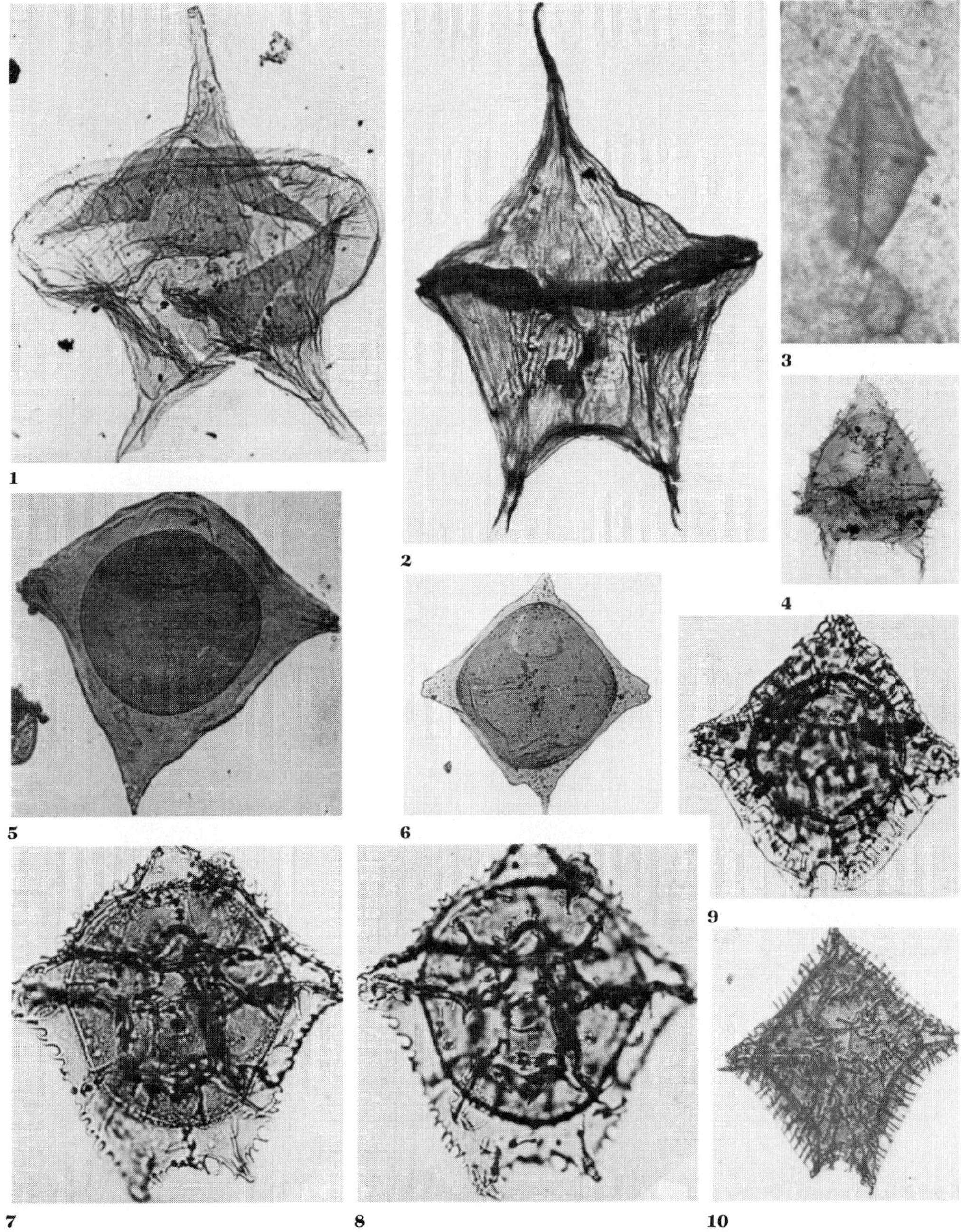

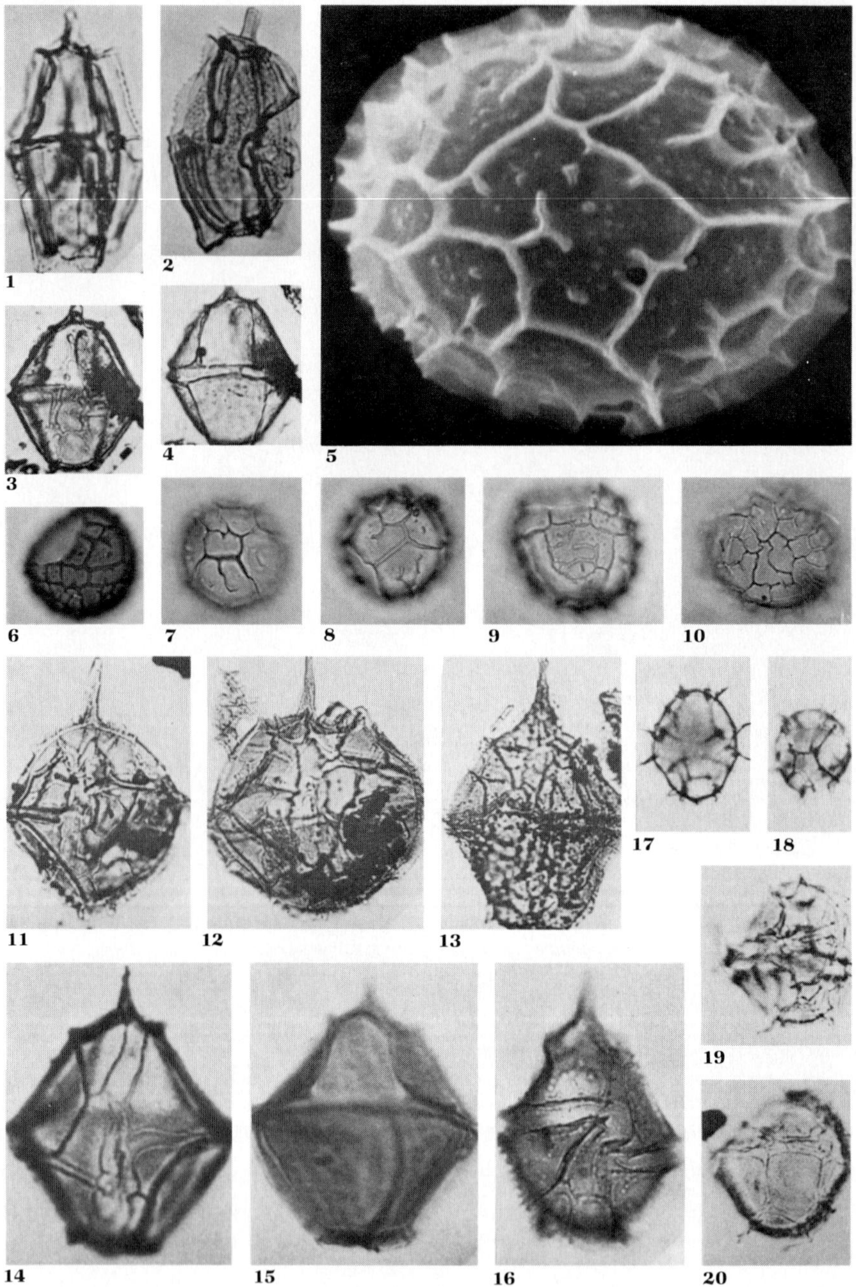

1969; *Hystrichogonyaulax* Sarjeant 1969; *Impagidinium* Stover & Evitt 1978; *Lanterna* Dodekova 1969; *Leptodinium* Klement 1960; *Muratodinium* Drugg 1970; *Occisucysta* Gitmez 1969; *Prionodinium* Leffingwell & Morgan 1977; *Rhynchodiniopsis* Deflandre 1935.

14. Family Apteodiniaceae Eisenack 1961 (see Figures 4.122 and 4.123)

Proximate gonyaulacacean cysts; ovoid to subpolygonal; paracingulum may be weakly indicated; paratabulation absent or indistinct; type 1P (rarely 2P) archeopyle. "Gonyaulacoid lineage" in part, including cysts referrable to living *Gonyaulax;* also " apteodinioid lineage."

Upper Jurassic to Quaternary; Europe, N. America, USSR, Australasia.

Apteodinium Eisenack 1958; *Archaeotectatum* Habib 1972; *Bitectatodinium* Wilson 1973; *Caledonidinium* Reid 1974; *Ellipsoidinium* Clarke & Verdier 1967; *Emslandia* Gerlach 1961; *Kenleyia* Cookson & Eisenack 1965; *Soaniella* Vozzhennikova 1967; *Spongodinium* Deflandre 1936; *Tectatodinium* Wall 1967; *Trichodinium* Eisenack & Cookson 1960; *Xenicodinium* Klement 1960.

15. Family Belodiniaceae Eisenack 1961 (see Figures 4.124, 4.125, 4.126, and 4.127)

Proximate gonyaulacacean cysts, ovoidal to polygonal, paratabulation $1-6'(?7)$, $0-6a$, $6-7''$, $5-6'''$, $1-2p$, $0-1pv$, $1''''$; type A archeopyle (one, two, or more opercular pieces), may have parasulcal notch. Includes "lithodinioid and eisenackioid lineages"; includes family Microdiniaceae Eisenack 1964.

Upper Triassic to upper Eocene; Europe, N. America, USSR, India, Australasia.

Belodinium Cookson & Eisenack 1960; *Cladopyxidium* McLean 1972; *Clathroctenocystis* Wiggins 1972; *Dapcodinium* Evitt 1961; *Eisenackia* Deflandre & Cookson 1955; *Fibradinium* Morgenroth 1968; *Glyphanodinium* Drugg 1964; *Lithodinia* Eisenack 1935 (syn.: *Meiourogonyaulax* Sarjeant 1966); *Maturodinium* Morgenroth 1970; *Microdinium* Cookson & Eisenack 1960; *Muiriella* Churchill & Sarjeant 1963; *Subtilidinium* Morgenroth 1968.

16. Family Fromeaceae Sarjeant & Downie 1966 ex 1974 (see Figure 4.128)

Proximate cysts; ovoid or with median constriction; paratabulation absent or indistinct; type A archeopyle, no parasulcal notch.

Middle Jurassic to upper Eocene; Europe, N. America, USSR, Australasia.

Fromea Cookson & Eisenack 1958; *Halophoridia* Cookson & Eisenack 1962; *Horologinella* Cookson & Eisenack 1962.

17. Family Canningiaceae Sarjeant & Downie 1966 ex Brideaux 1971

Proximate gonyaulacacean cysts, ovoid, paratabulation absent or indistinct; type A archeopyle, with parasulcal notch. Includes part of "lithodinioid and dictyopyxidioid lineages."

Upper Triassic to lower Oligocene; Europe, N. America, S. America, Africa, USSR, India, Australasia.

Batiacasphaera Drugg 1970; *Canningia* Cookson & Eisenack 1960; *Chytroeisphaeridia* Sarjeant 1962; *Ellipsoidictyum* Klement 1960 (syn.: *Dic-*

(continued)

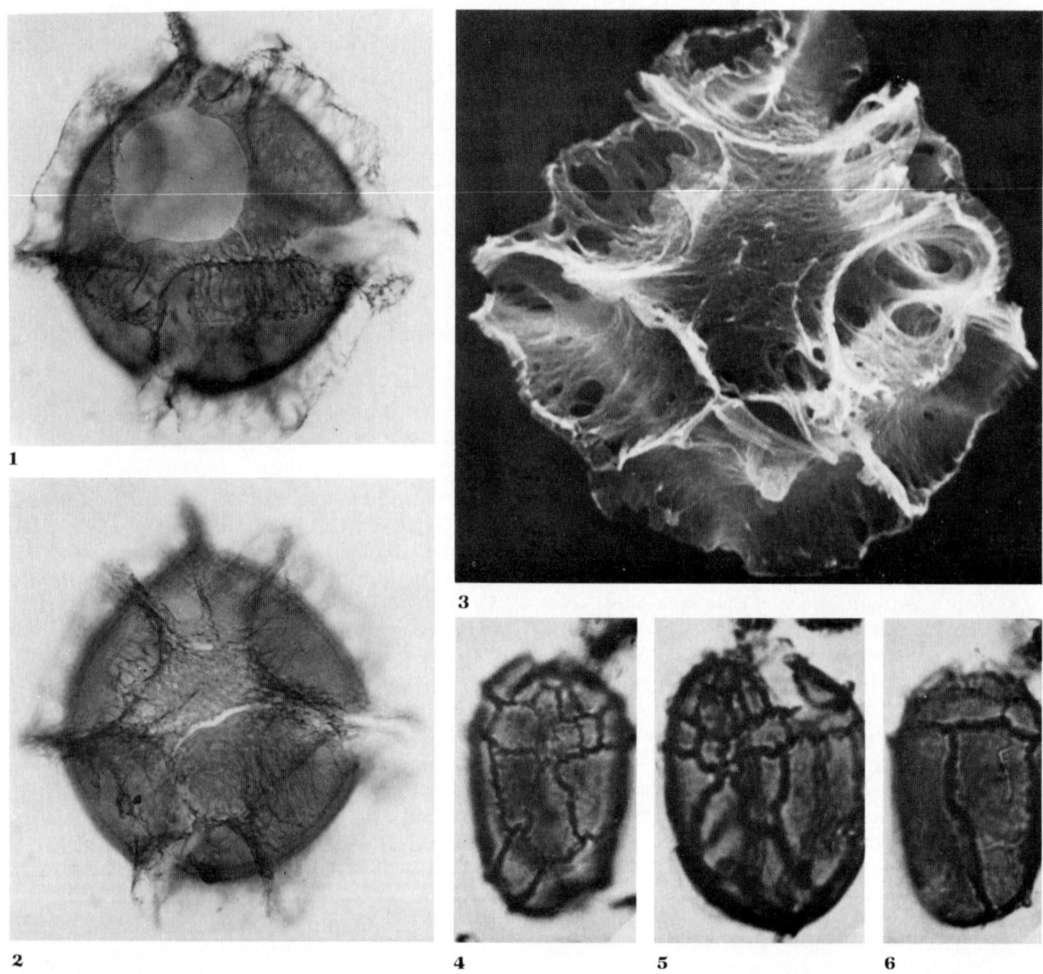

Figure 4.121

Gonyaulacystaceae. **1–3.** *Muratodinium fimbriatum,* lower Eocene, Alabama, from Drugg, 1970. **1,2.** Dorsal and ventral focus, ×384. **3.** SEM, ventral view, ×1000. **4–6.** *Druggidium apicopaucicum* Habib, Lower Cretaceous, Atlantic Ocean core, off North Carolina, ventral, left side, and dorsal views, ×850, from Habib, 1973.

Figure 4.122 (*facing page*)

Apteodiniaceae. **1,2.** *Trichodinium hirsutum* Cookson, Paleocene, Australia, dorsal and ventral surfaces, ×400, from Cookson, 1965b. **3.** *Xenicodinium lubricum* Morgenroth, Paleocene, Danian, Germany, dorsal view, ×1100, from Morgenroth, 1968. **4,5.** *Kenleyia lophophora* Cookson & Eisenack, Paleocene, Australia, dorsal and oblique right side views, ×650, from Cookson and Eisenack, 1967. **6–8.** *Tectatodinium psilatum,* Pleistocene, Black Sea, dorsal, ventral, and right side views of this variable species, ×545, from Wall et al., 1973. **9,10.** *Spongodinium delitiense,* Danian, Germany, ×370, from Morgenroth, 1968. **11–13.** *Apteodinium granulatum* Eisenack, Lower Cretaceous, Germany slightly oblique dorsal view, showing archeopyle; ventral side, with archeopyle showing by transparency; and right side view; ×400, from Eisenack, 1958.

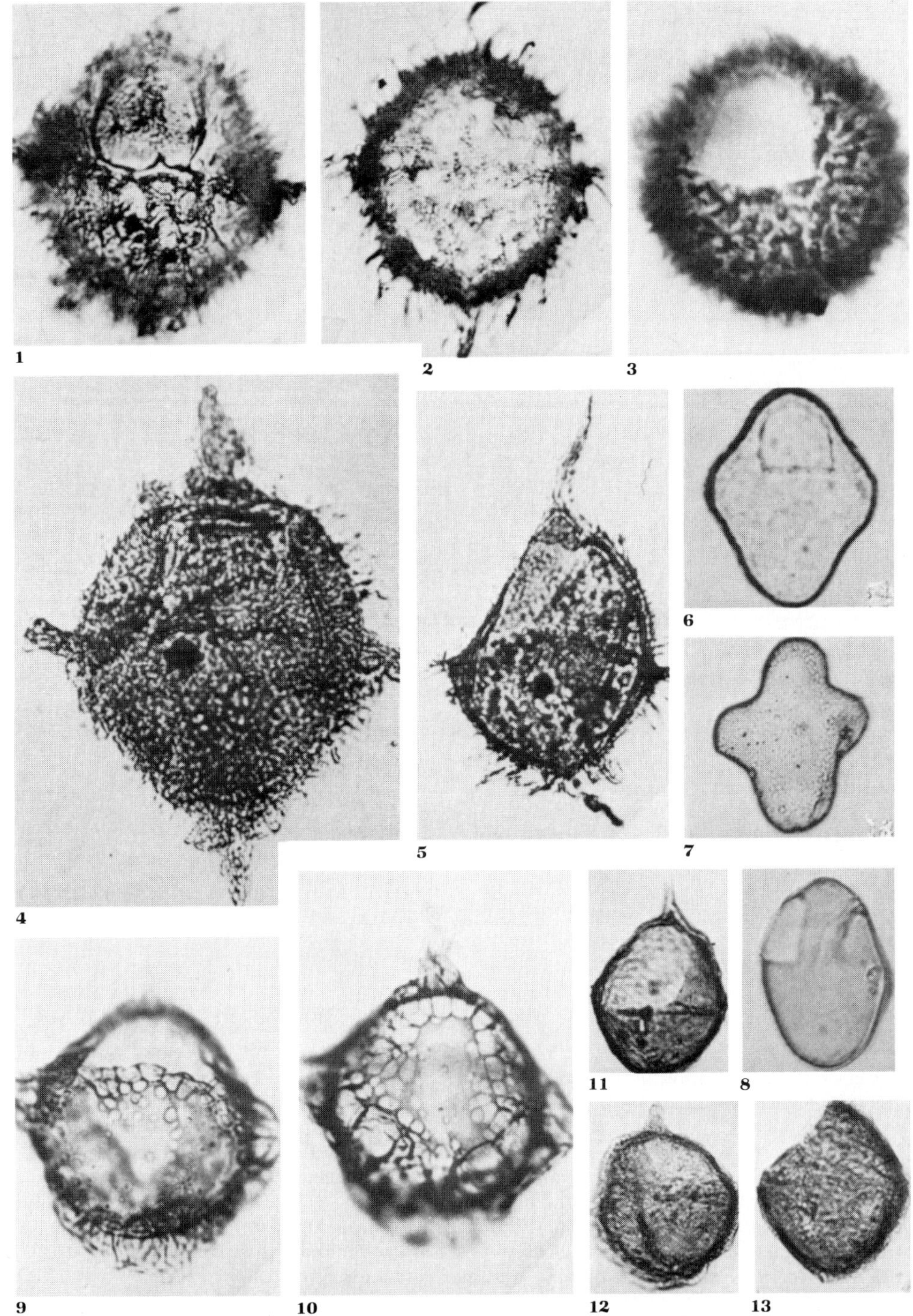

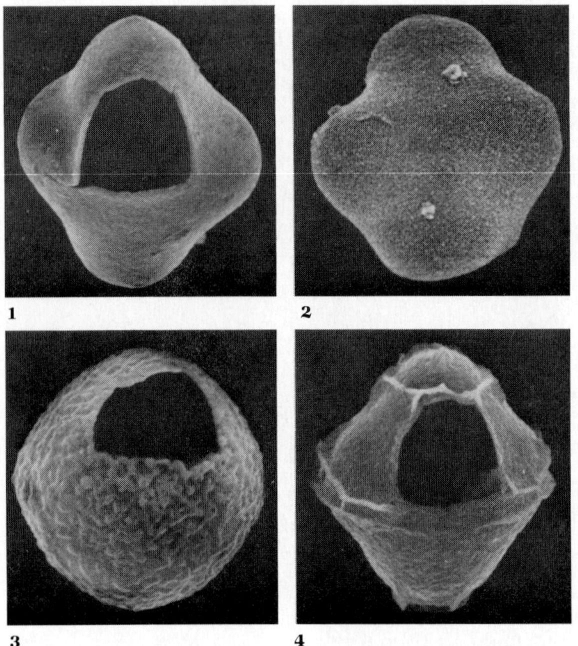

Figure 4.123
Tectatodinium psilatum, Pleistocene, Black Sea, SEM, showing variation. **1.** Dorsal view, ×810. **2.** Ventral view, ×735. **3.** Dorsal view, more rugose surface, ×772. **4.** Rare indication of tabulation, dorsal view, ×862. From Wall et al., 1973.

Figure 4.124 (*facing page*)
Belodiniaceae. **1–3.** *Fibradinium annetorpense* Morgenroth, Danian, Paleocene, Sweden, antapical pole, dorsal, and ventral views; 1,3, ×1400; 2, ×1350; from Morgenroth, 1968. **4.** *Belodinium dysculum* Cookson & Eisenack, Upper Jurassic, England, ×271, from Gitmez, 1970. **5–6.** *Eisenackia scrobiculata* Morgenroth, Danian, Germany, apical and dorsal views, ×1000, from Morgenroth, 1966a. **7–10.** *Alisocysta ornata* (Cookson & Eisenack) Stover & Evitt, Eocene, Australia, from Stover, 1975. **7.** Dorsal view, ×400. **8.** Dorsal view, focussed on ventral surface, ×500. **9.** Apical view, with archeopyle, ×400. **10.** Free operculum, ×500. **11–14.** *Microdinium ornatum* Cookson & Eisenack, Lower Cretaceous, Australia, from Cookson and Eisenack, 1960a. **11–12.** Ventral and dorsal views, ×800. **13.** Apical view, with archeopyle, ×700. **14.** Showing partially detached apical plate, ×800.

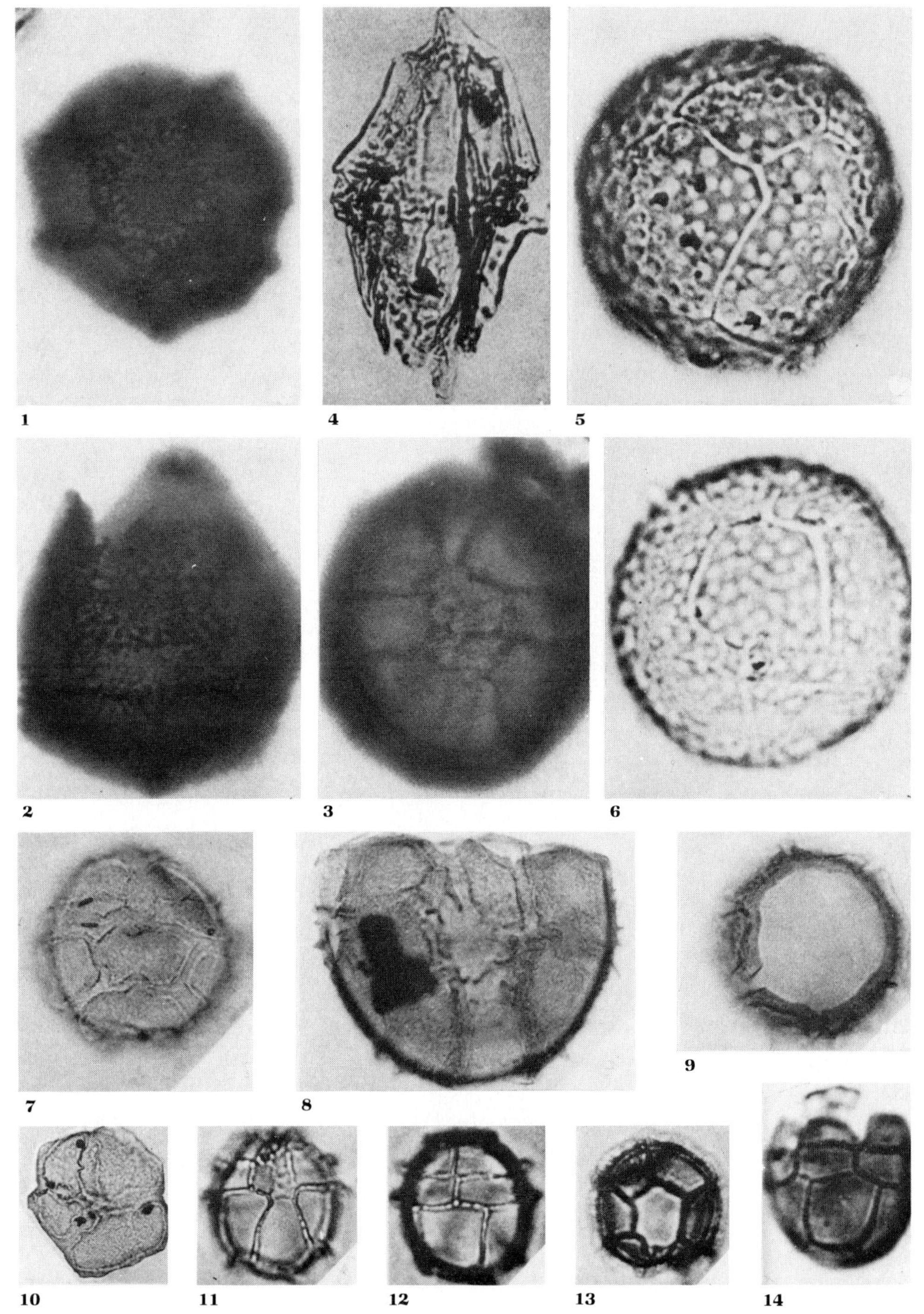

1
4
5
2
3
6
7
8
9
10
11
12
13
14

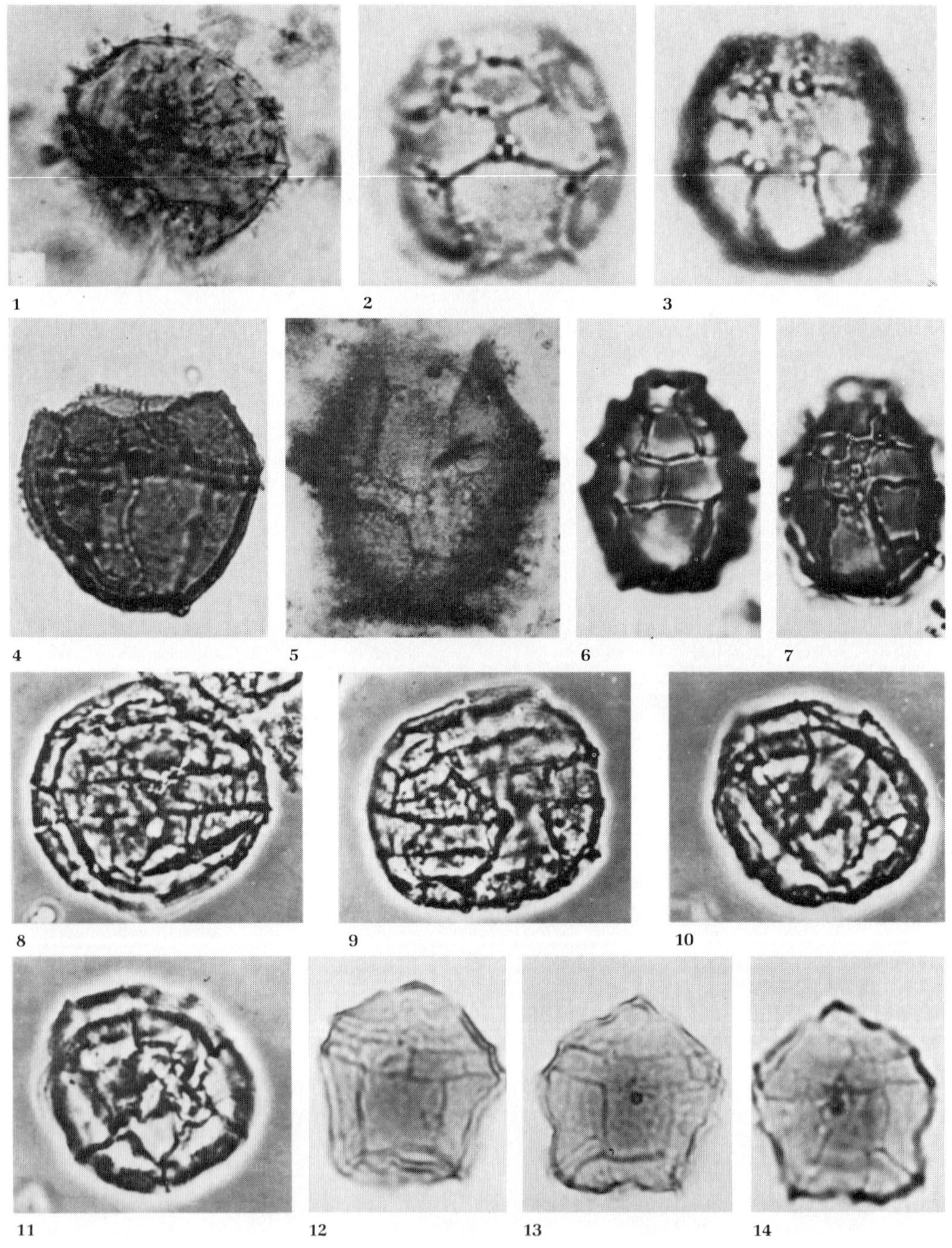

Figure 4.125 (*facing page*)

Belodiniaceae. **1.** *Muiriella plioplax* Churchill & Sarjeant, Holocene freshwater peat, Australia, ventral view, ×750, from Churchill and Sarjeant, 1963. **2,3.** *Subtilidinium minutum* Morgenroth, Danian, Paleocene, Jutland, dorsal and ventral views, ×2500, from Morgenroth, 1968. **4.** *Lithodinia bulloidea* (Cookson & Eisenack) Gocht, Lower Cretaceous, France, ×650, from Davey and Verdier, 1974. **5.** *L. valensii* (Sarjeant) Gocht, Upper Jurassic, France, ventral view, ×700, from Sarjeant, 1966c. **6,7.** *Cladopyxidium saeptum* (Morgenroth) Stover & Evitt, Paleocene, Virginia, dorsal and ventral views, ×945, from McLean, 1972. **8–11.** *Dapcodinium priscum* Evitt, Lower Jurassic, Denmark, ×1000, from Evitt, 1961c. **8,9.** Dorsal and ventral views. **10,11.** From apex, focussed on hypotheca and on epitheca. **12–14.** *Glyphanodinium facetum* Drugg, Danian, Paleocene, California, showing 12,13, dorsal, and 14, ventral views, ×1000, from Drugg, 1964.

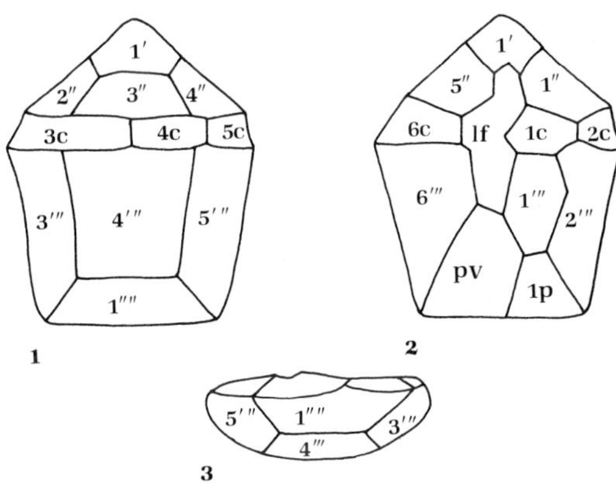

Figure 4.126

Tabulation of *Glyphanodinium facetum*, Paleocene, California **1.** Dorsal view. **2.** Ventral view; pv, posterior ventral plate; lf, longitudinal furrow. **3.** Antapical view. Redrawn from Drugg, 1964.

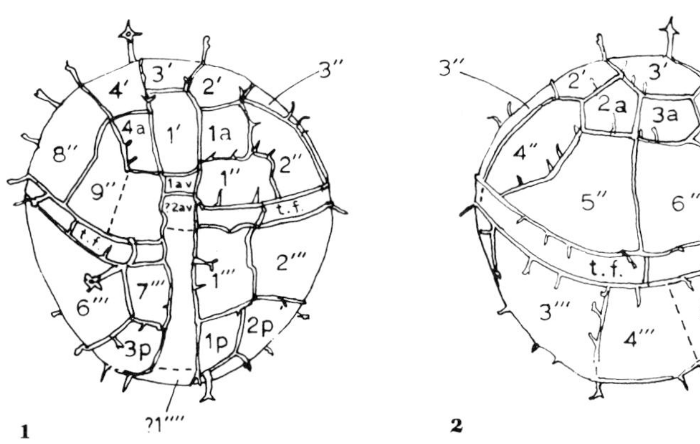

Figure 4.127.

Tabulation of *Muiriella plioplax*. **1.** Ventral view; av, anterior ventral plate. **2.** Dorsal view. Both ×700, from Churchill and Sarjeant, 1963.

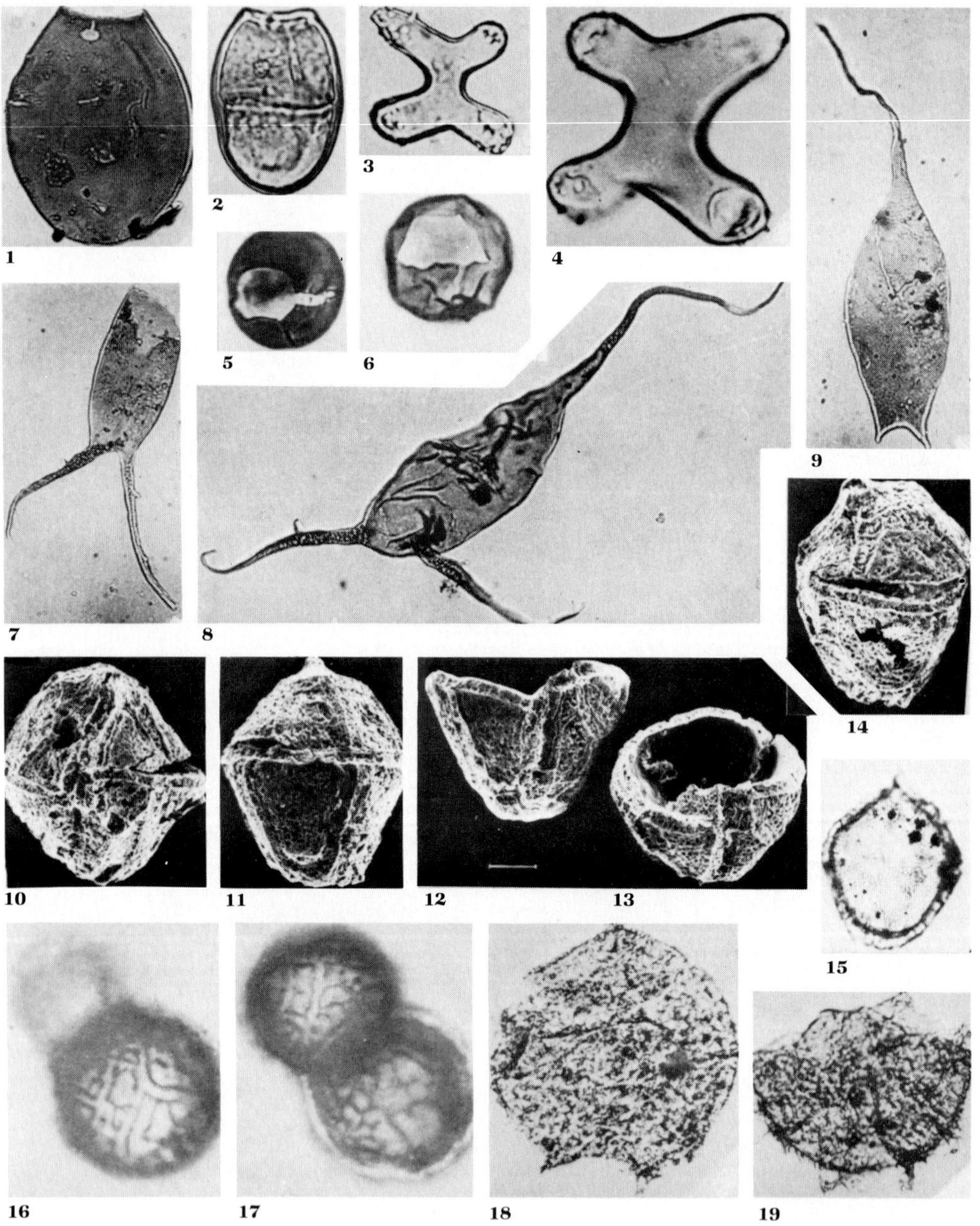

tyopyxis Cookson & Eisenack 1960, non Ehrenberg 1844; *Dictyopyxidia* Eisenack 1961); *Epelidosphaeridia* Davey 1969; *Histiocysta* Davey 1969; *Kallosphaeridium* De Coninck 1969; *Membranosphaera* Samoilovich ex Norris & Sarjeant 1965; *Necrobroomea* Wiggins 1975 (syn.: *Batioladinium* Brideaux 1975); *Rhombodella* Cookson & Eisenack 1962; *Tenua* Eisenack 1958.

18. Family Eodiniaceae Eisenack 1961

Proximate cysts, ovoid, no apparent paratabulation; paracingulum present, archeopyle epicystal.

Middle Jurassic; Europe, USSR.

Biorbifera Habib 1972; *Eodinia* Eisenack 1966.

19. Family Ctenidodiniaceae Sarjeant & Downie 1966 ex Gocht 1970 (see Figure 4.129)

Proximate gonyaulacacean cysts, ovoid to polygonal, elevated crests bordering paracingulum, paratabulation 3–5′, 0–3a, 6–7″, 6c, 5–7‴, 1p, 0–1pv, 1⁗, epicystal archeopyle. Part of "oodnadatioid and ctenidodinioid lineages;" includes family Dichadogonyaulacaceae Sarjeant & Downie 1974.

Lower Jurassic to Lower Cretaceous; N. America, Europe, Australasia.

(*continued*)

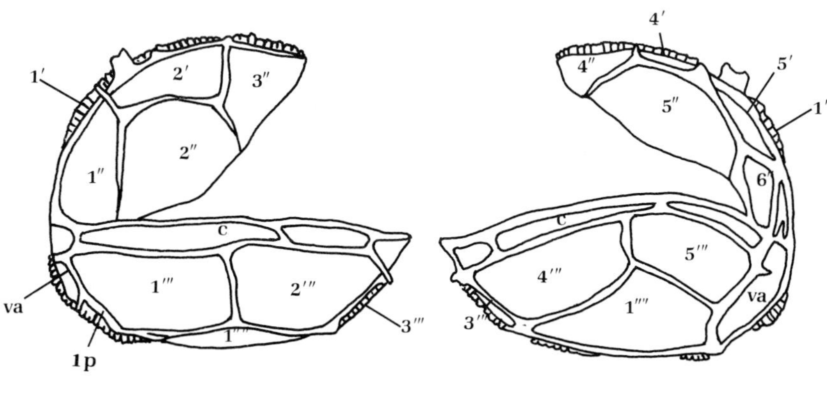

Figure 4.129
1,2. Tabulation of *Ctenidodinium schizoblatum* (Norris) Lentin & Williams, Upper Jurassic, England; left lateral and right lateral views, showing formation of epitractal archeopyle; va, ventral area. Redrawn from Norris, 1965.

Ctenidodinium Deflandre 1938 (syn: *Dichado-gonyaulax* Sarjeant 1966); *Energlynia* Sarjeant 1976; *Luehndea* Morgenroth 1970; *Manco-dinium* Morgenroth 1970; *Wanaea* Cookson & Eisenack 1958.

20. Family Heteraulacacystaceae Drugg & Loeblich Jr. 1967 (see Figure 4.130)

Ovoid to polygonal proximate gonyaulaca-cean cysts, paratabulation 3′, 7″, 5‴, 3‴′, epicystal archeopyle.

Lower Eocene to upper Oligocene; Europe, N. America.

Heteraulacacysta Drugg & Loeblich Jr. 1967.

21. Family Hystrichodiniaceae Deflandre 1936 (see Figures 4.131 and 4.132)

Proximochorate gonyaulacacean cysts, ovoid to polygonal, may have apical horn or pro-cesses, parasutures with ridges to crests, bear-ing spines or processes; paratabulation 3−6′, 0−1a, 6″, 6c, 5−6‴, 0−1p, 0−1pv, 1‴′; archeopyle type P. Includes parts of "gonyaulacoid and hystrichodinioid lineages"; cysts of this type in many living *Gonyaulax*, in-

cluding *G. spinifera, G. digitalis, G. scrippsae;* in-cludes Spiniferitaceae Sarjeant 1970, Hesler-toniaceae Vozzhennikova 1979.

Upper Jurassic to Quaternary; N. America, S. America, Europe, USSR, Africa, India, Australasia.

Achomosphaera Evitt 1963; *Danea* Morgenroth 1968; *Hafniasphaera* Hansen 1977; *Heliodinium* Alberti 1961; *Heslertonia* Sarjeant 1966; *Hys-trichodinium* Deflandre 1935; *Hystrichostrogy-lon* Agelopoulos 1964; *Nematosphaeropsis* De-flandre & Cookson 1955; *Ochetodinium* Damassa 1979; *Planinosphaeridium* Eisenack 1965 (syn.: *Ataxiodinium* Reid 1974); *Protoellip-soidinium* Davey & Verdier 1971; *Pterodinium* Eisenack 1958; *Spiniferites* Mantell 1850 (syn.: *Hystrichosphaera* O. Wetzel ex Deflandre 1937); *Turbiosphaera* Archangelsky 1968.

22. Family Xiphophoridiaceae Sarjeant & Downie 1966 ex 1974 (see Figure 4.133)

Proximochorate gonyaulacacean cysts; ovoid to polygonal, parasutures with spinose ridges;

(*continued*)

Figure 4.130
Tabulation of *Heteraulacacysta campanula* Drugg & Loeblich, middle Eocene, Alabama. **1−4.** Dorsal, ventral, apical, and antapical views; lf, sulcus or lon-gitudinal furrow. Development of cingular archeopyle suggested in Part 1. All ×455, redrawn from Drugg and Loeblich, 1967.

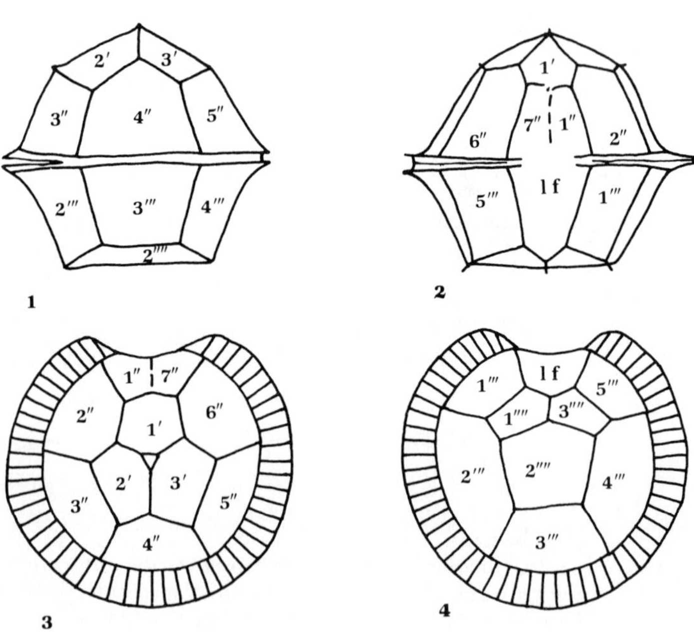

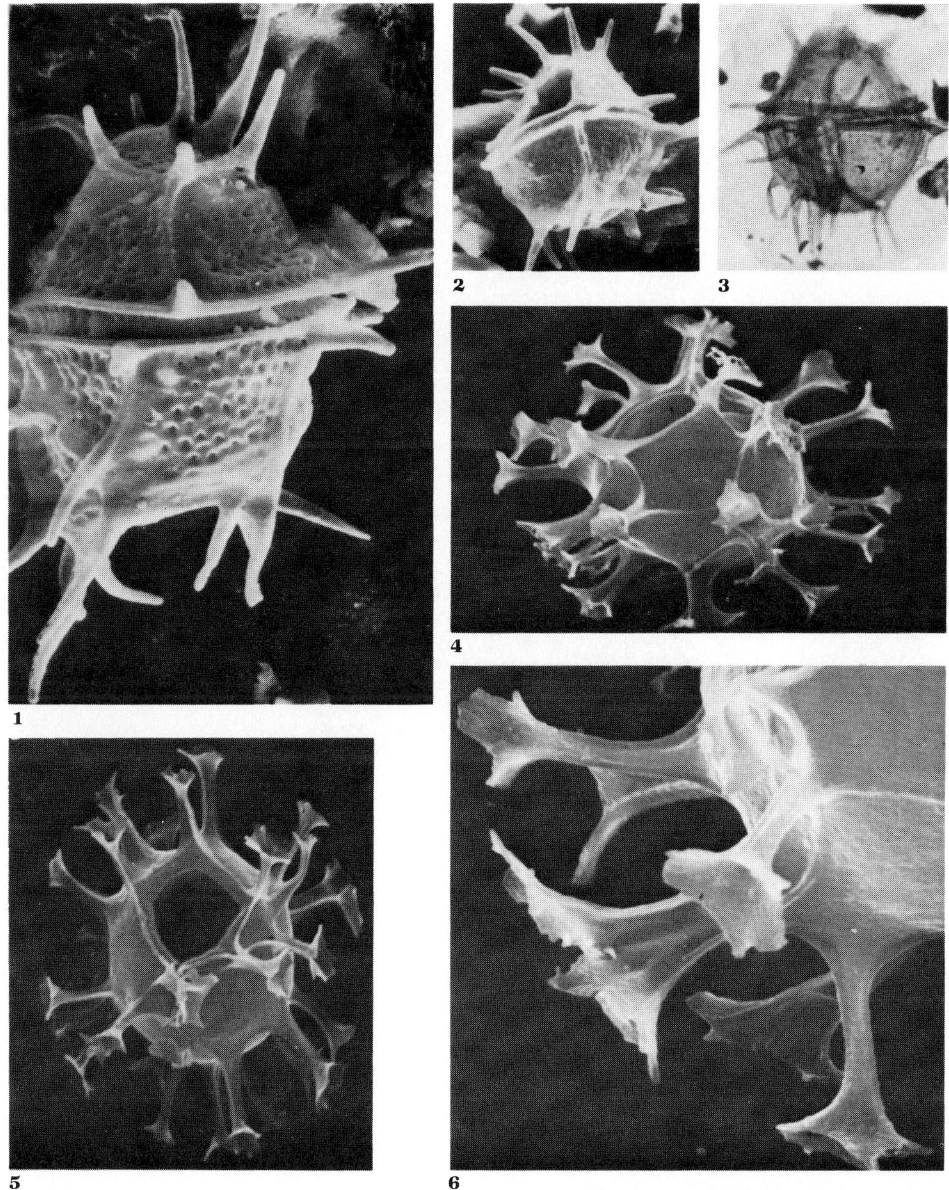

Figure 4.131

Hystrichodiniaceae. **1–3.** *Hystrichodinium pulchrum* Deflandre, Lower Cretaceous, Netherlands, from Jardiné and Raynaud, 1972. **1.** SEM, view of left side, ×800. **2.** SEM, right side, showing open archeopyle to upper left, ×400. **3.** Optical microscope view, ×400. **4–6.** *Spiniferites buccinus* (Davey & Williams) Sarjeant, upper Eocene, Mississippi, SEM. **4.** Hypotheca, showing archeopyle to upper left. **5.** Dorsal view (apex at upper left of photo), ×480. **6.** Processes enlarged, ×1600.

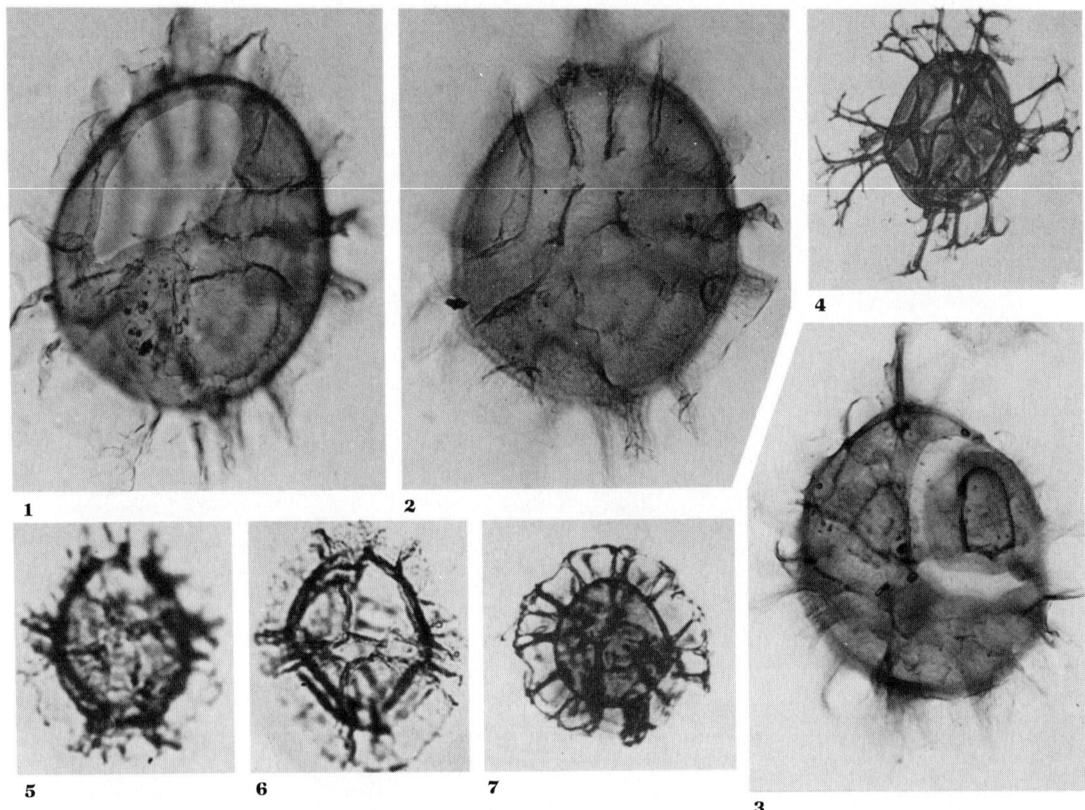

Figure 4.132

Hystrichodiniaceae. **1–3.** *Danea californica* (Drugg) Damassa, Paleocene, Alabama, ×384, from Drugg, 1970. **1,2.** Dorsal and ventral focus, with archeopyle. **3.** Oblique left side, showing operculum in place within archeopyle. **4.** *Achomosphaera ramulifera* (Deflandre) Evitt var. *perforata* Davey & Williams, Eocene, England, showing archeopyle, ×425, from Davey and Williams, 1966a. **5,6.** *Pterodinium aliferum* Eisenack, Lower Cretaceous, Germany, ventral side, and dorsal side showing archeopyle, ×500, from Eisenack, 1958. **7.** *Nematosphaeropsis balcombiana*, Quaternary, Caribbean Sea cores, ×375, from Wall, 1967. **8,9.** *Pentadinium circumsutum* (Morgenroth) Stover & Evitt, lower Oligocene, Germany, dorsal side with archeopyle and operculum, and ventral side, ×500, from Morgenroth, 1966b. **10.** *Achomosphaera ramulifera* (Deflandre) Evitt, Cretaceous, France, ×450, from Deflandre, 1937b. **11.** *A. alcicornu* (Eisenack) Davey & Williams, Eocene, England, ×425, from Davey and Williams, 1966a. **12,13.** *Nematosphaeropsis philippoti* (Deflandre) De Coninck, Upper Cretaceous, France, two levels of focus, ×540, from Foucher, 1972.

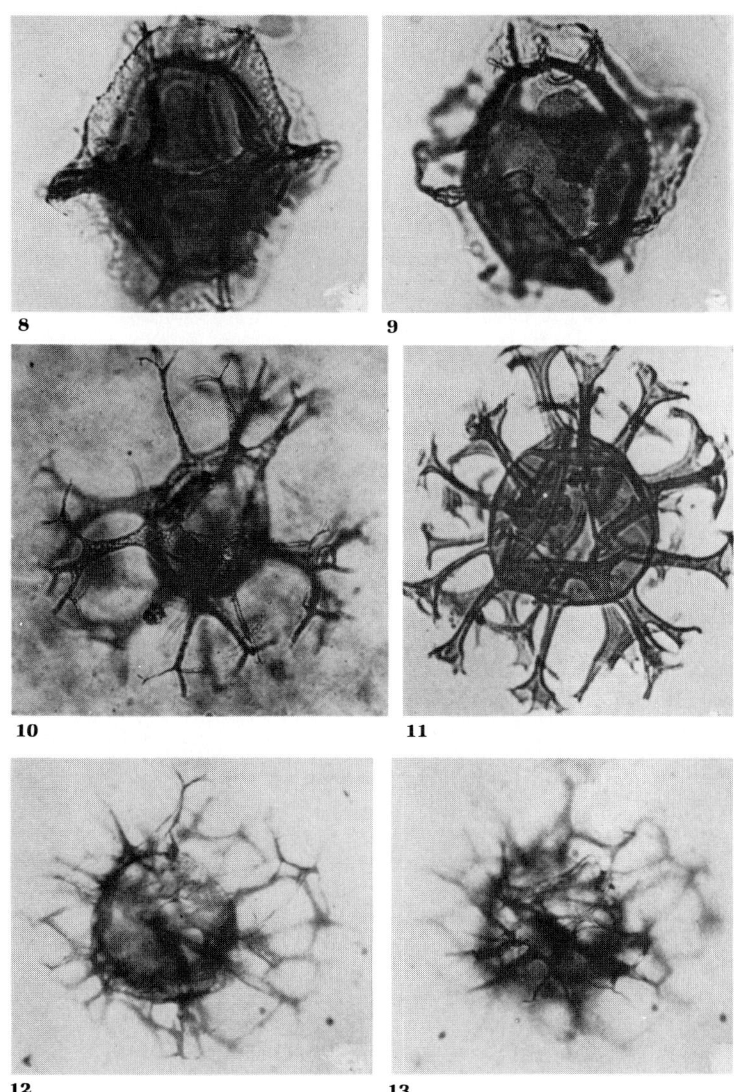

8 9

10 11

12 13

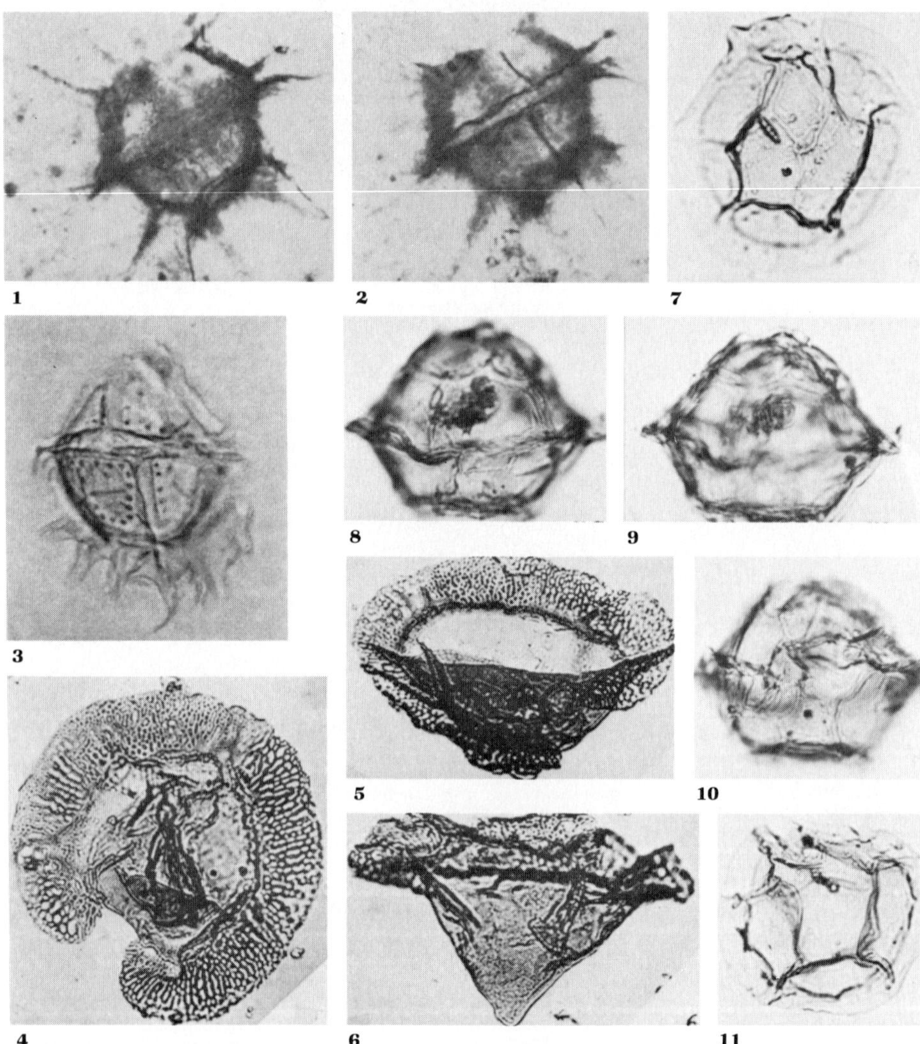

Figure 4.133

Ctenidodiniaceae, Heteraulacacystaceae, Xiphophoridiaceae. **1–3.** *Xiphophoridium alatum* (Cookson & Eisenack) Sarjeant, Upper Cretaceous. **1,2.** In Coniacian chert, France, ×540, from Foucher, 1972. **3.** From France, ×500, from Davey and Verdier, 1973. **4–6.** *Wanaea clathrata* Cookson & Eisenack, Upper Jurassic, Australia; 4, ×280; 5,6, ×300; from Cookson and Eisenack, 1958. **7–11.** *Heteraulacacysta campanula*, middle Eocene, Alabama, ×384, from Drugg and Loeblich, 1967. **7.** Apical view of epicyst. **8–10.** Focus on dorsal side, ambitus, and ventral side. **11.** View of hypocyst (antapical).

paracingulum with high bordering crests; type A archeopyle; paratabulation ?4', 1a, 6'', 0−1p, 6''', 1''''. Includes part of "lithodinioid lineage."

Lower Cretaceous to Upper Cretaceous; Europe, USSR, N. America, Australasia.
Xiphophoridium Sarjeant 1966.

23. Family Toolongiaceae Sarjeant & Downie 1966 ex 1974 (see Figure 4.134)

Proximochorate gonyaulacacean cysts, ovoid to polygonal, parasutures with elevated crests, paratabulation 4', 6'', 6c, 5''', 1p, 1s, 1'''', type tAtP (epicystal) archeopyle. Similar to modern *Pyrodinium*. Includes part of "oodnadattioid lineage."

Lower Cretaceous to upper Oligocene; Europe, USSR, Australasia.
Dinopterygium Deflandre 1935 (syn.: *Oodnadattia* Eisenack & Cookson 1960); *Toolongia* Cookson & Eisenack 1960.

24. Family Membranilarnaciaceae Eisenack 1963 (see Figure 4.135)

Chorate gonyaulacacean cysts, with enclosing thin-walled ovoid membrane supported by the processes, may have apical horn; paratabulation 1−4', 6'', 0−4c, 5''', 1p, 1''''; type tA archeopyle. Includes parts of "hystrichosphaeridioid and dictyopyxidioid lineages."

Lower Cretaceous to lower Oligocene; N. America, Europe, Australasia.
Chlamydophorella Cookson & Eisenack 1958; *Dorocysta* Davey 1970; *Eatonicysta* Stover & Evitt 1978; *Membranilarnacia* Eisenack 1963; *Omatia* Cookson & Eisenack 1958; *Samlandia* Eisenack 1954 (syn.: *Palmnickia* Eisenack 1954); *Tubidermodinium* Morgenroth 1966; *Valensiella* Eisenack 1963.

25. Family Scriniocassiaceae Sarjeant & Downie 1966 ex 1974 (see Figures 4.136 and 4.137)

Chorate gonyaulacacean cysts, with enclosing spheroidal membrane supported by spines or ridges; type 1P or 2P, or IP archeopyle. Includes parts of "gonyaulacoid and lithodinioid lineages."

Lower Jurassic to Quaternary; N. America, Europe, USSR.
Gardodinium Alberti 1961; *Scriniocassis* Gocht 1964; *Tuberculodinium* Wall 1967 (= living *Pyrophacus*).

26. Family Hystrichosphaeridiaceae Evitt 1963 (see Figures 4.138, 4.139, and 4.140)

Chorate gonyaulacacean cyst, ovoid to polygonal; processes centered on paraplates (intratabular), size commonly corresponding to tabular size; paratabulation 1−5', 0−1a, 6'', 0−6c, ?s, 5−6''', 0−1p, 0−1''''; type tA archeopyle. "Hystrichosphaeridioid lineage." Includes Hystrichokolpomaceae, Compositosphaeridiaceae and Tanyosphaeridiaceae of Vozzhennikova 1979.

Upper Triassic to Quaternary; N. America, S. America, Europe, USSR, India, Australasia.
Areosphaeridium Eaton 1970; *Compositosphaeridium* Dodekova 1974; *Conosphaeridium* Cookson & Eisenack 1969; *Hystrichokolpoma* Klumpp 1953; *Hystrichosphaeridium* Deflandre 1937; *Litosphaeridium* Davey & Williams 1966; *Oligosphaeridium* Davey & Williams 1966; *Perisseiasphaeridium* Davey & Williams 1966; *Surculosphaeridium* Davey, Downie, Sarjeant & Williams 1966; *Tanyosphaeridium* Davey & Williams 1966.

27. Family Exochosphaeridiaceae Sarjeant & Downie 1966 ex 1974 (see Figure 4.141)

Chorate, gonyaulacacean cyst, ovoid to subpolygonal; processes intratabular or peritabular; apical process larger than others; type P archeopyle, paratabulation obscure.

Lower Cretaceous to upper Oligocene; N. America, Europe, Africa, USSR, India, Australasia.
Cometodinium Deflandre & Courteville 1939; *Cyclapophysis* Benson 1976; *Exochosphaeridium* Davey, Downie, Sarjeant & Williams 1966; *Lanternosphaeridium* Morgenroth 1966.

28. Family Actinothecaceae Cookson & Eisenack 1970 (see Figure 4.142)

Chorate gonyaulacacean cyst, ovoid to polygonal; intratabular tubular processes, corresponding in size variation to that of paraplates; paratabulation 4', 6'', 6c, 0−5s, 6''', 1p, 1s, 1''''. Type A+$\overline{3A}$+6P (epicystal) archeopyle, operculum of one or more pieces. "Homotryblioid lineage"; includes Homotrybliaceae Sarjeant & Downie 1966 ex 1974.

Lower Cretaceous to Quaternary; N. America, Europe, Red Sea, Caribbean Sea, North Atlantic and east Pacific Oceans.

(continued)

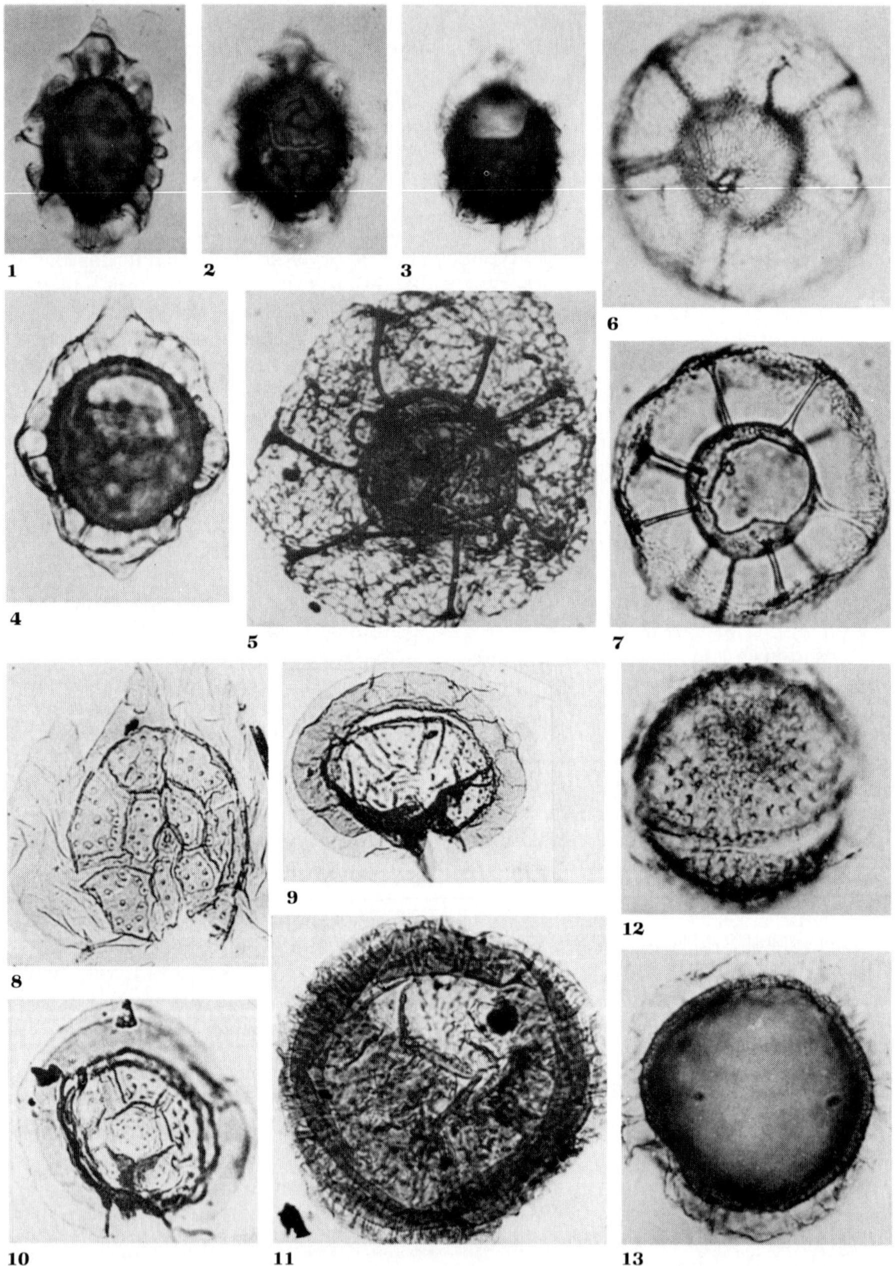

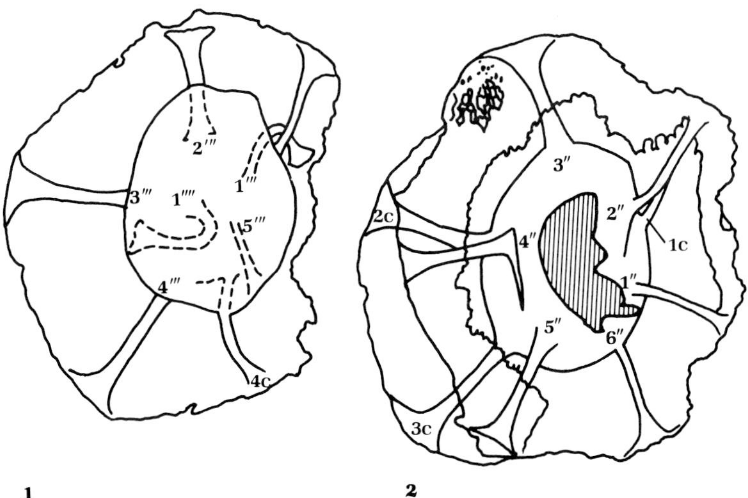

Figure 4.135
Tabulation of *Eatonicysta ursulae*. **1.** Antapical surface, seen through the cyst. **2.** Apical surface, archeopyle shaded, sulcal notch present between plate equivalents 1″ and 6″. Redrawn from Williams and Downie, 1966b.

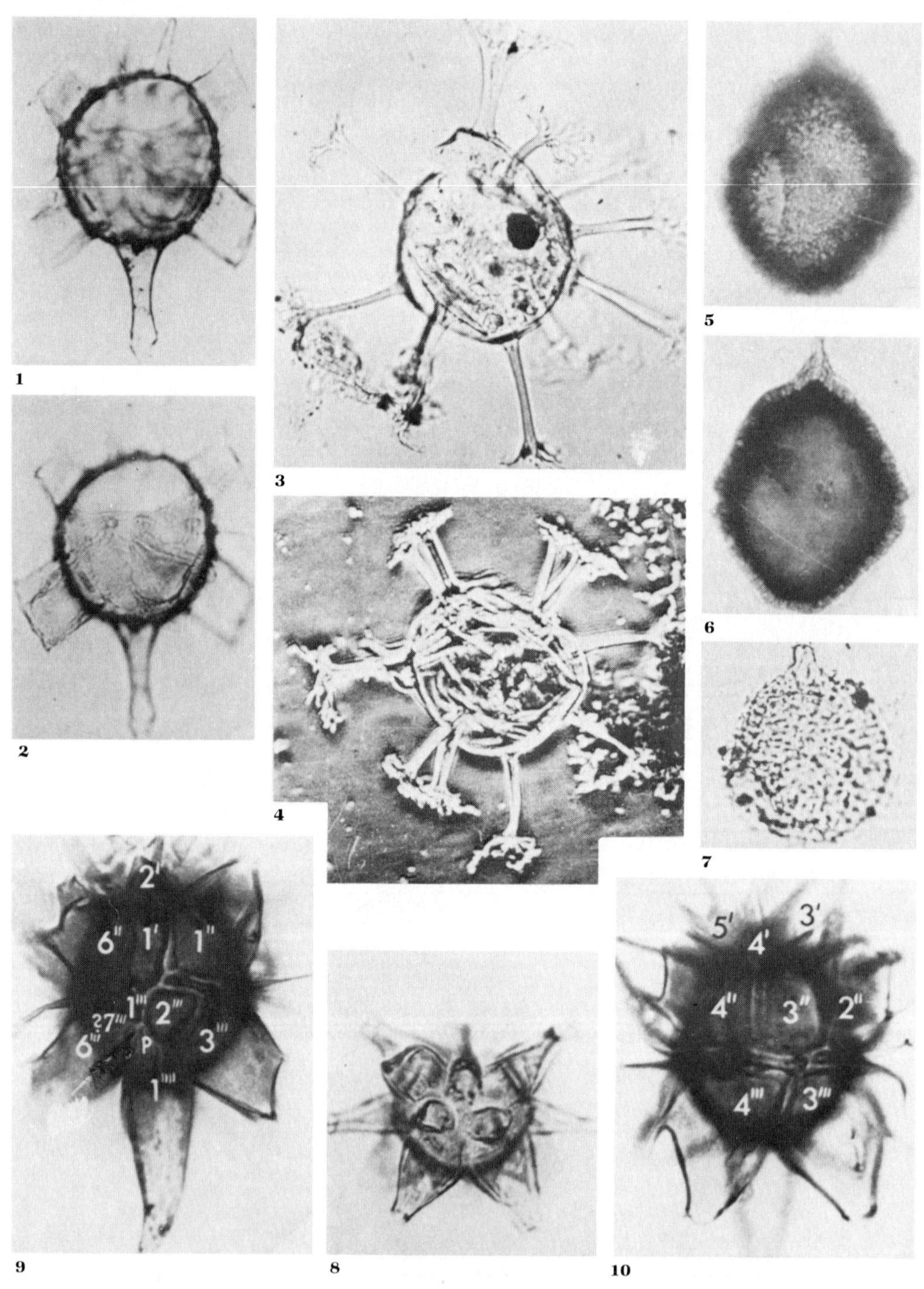
1
2
3
4
5
6
7
2'
6" 1' 1"
1‴ 2‴
7‴ 2
6‴ P 3‴
1‴
9
8
5' 3'
4'
4" 3" 2"
4‴ 3‴
10

Figure 4.136 (*facing page*)

Scriniocassiaceae, Hystrichosphaeridiaceae. **1,2.** *Achilleodinium biformoides* (Eisenack) Eaton, Eocene, Germany, optical section and dorsal side showing archeopyle, ×400, from Morgenroth, 1966a. **3,4.** *Areosphaeridium diktyoplokus* (Klumpp) Eaton, Eocene, Germany. **3.** Showing partly detached operculum, ×425, from Agelopoulos, 1967. **4.** ×500, from Klumpp, 1953. **5,6.** *Aldorfia deflandrei* (Clarke & Verdier) Stover & Evitt, Upper Cretaceous, France, ×720, from Foucher, 1972. **7.** *Gardodinium trabeculosum* (Gocht) Alberti, Lower Cretaceous, England, ×500, from Sarjeant, 1966b. **8.** *Hystrichokolpoma bulbosum* (Ehrenberg) Morgenroth, Paleocene, Germany, dorsal side, ×700, from Morgenroth, 1968. **9,10.** *H. cinctum* Klumpp, Paleocene, Germany, ventral and dorsal sides, ×600, from Morgenroth, 1968.

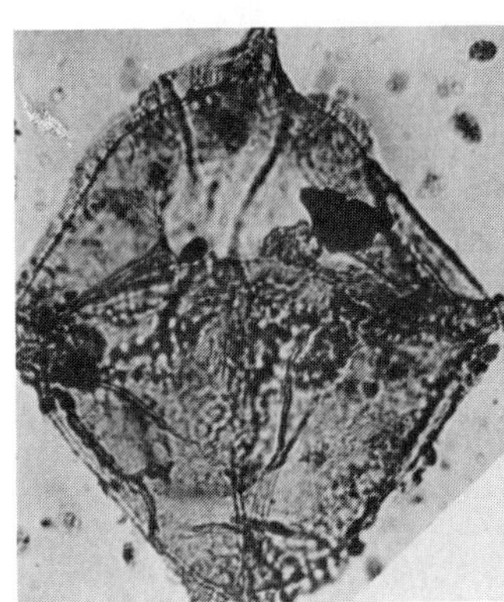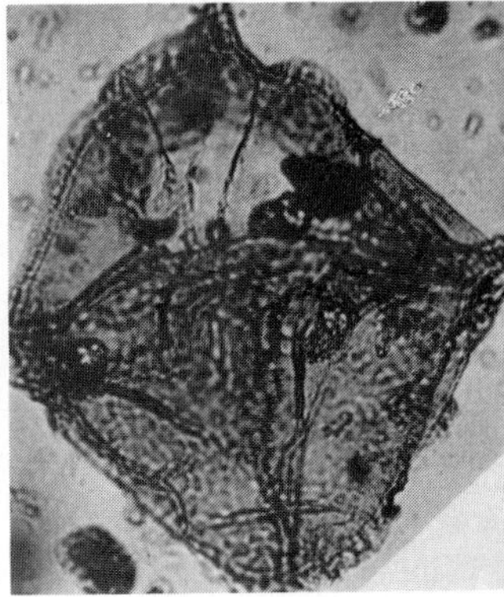

Figure 4.137

Scriniocassiaceae. **1,2.** *Scriniocassis dictyotus* (Cookson & Eisenack) Gocht, Jurassic, North Atlantic cores, ×252, from Habib, 1974. **1.** Dorsal focus. **2.** Ventral focus.

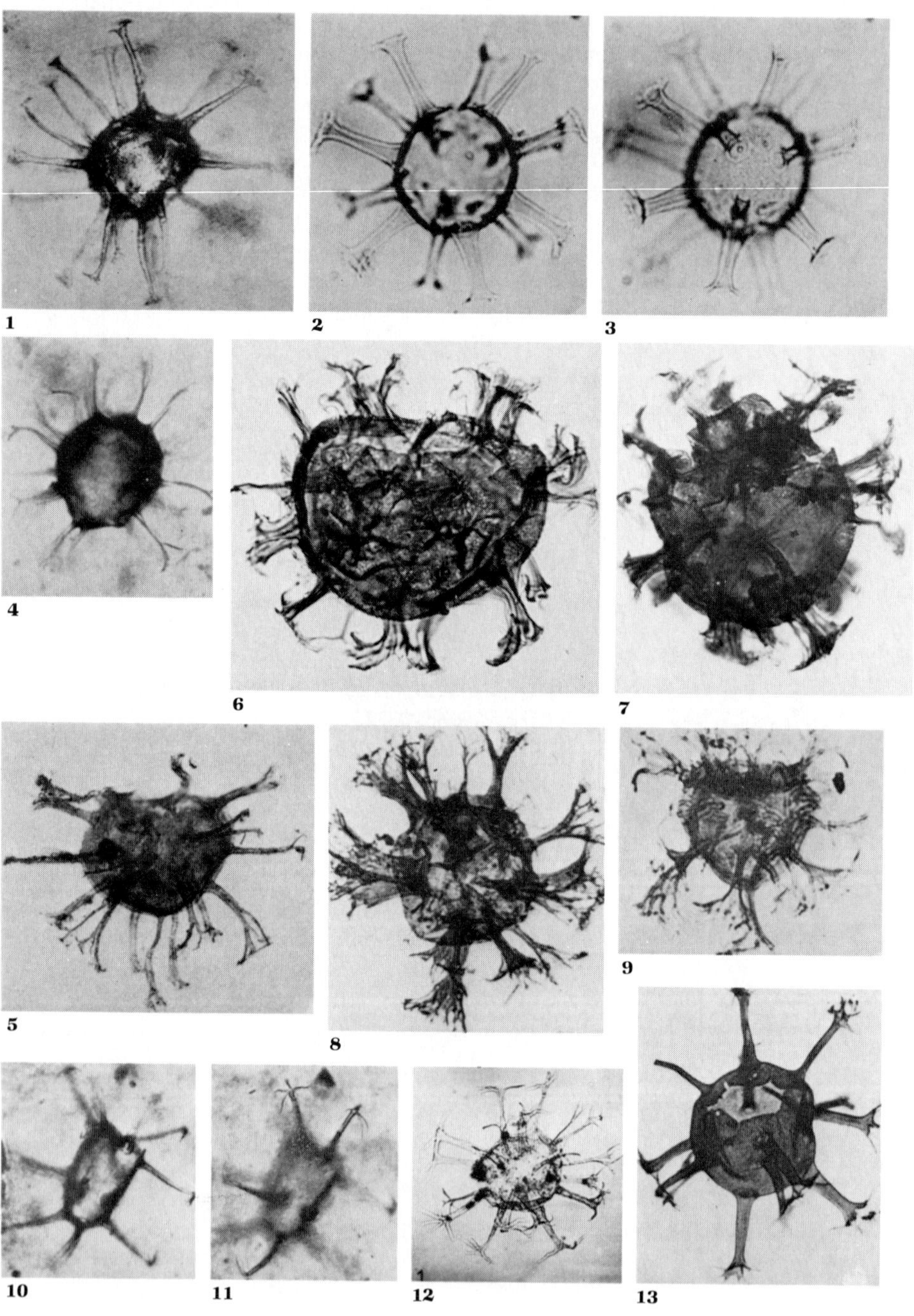

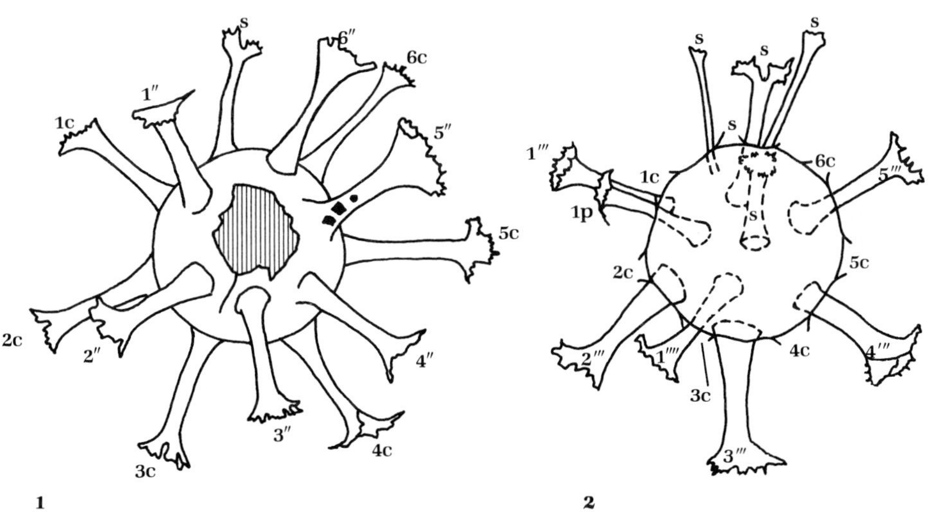

Figure 4.139
Tabulation of *Hystrichosphaeridium tubiferum*. **1.** Apical view, showing archeopyle, precingular and cingular processes, and a sulcal process. **2.** Antapex, viewed by transparency, showing post-cingular, sulcal, posterior intercalary, and antapical processes. Both ×1000, from Davey and Williams, 1966b.

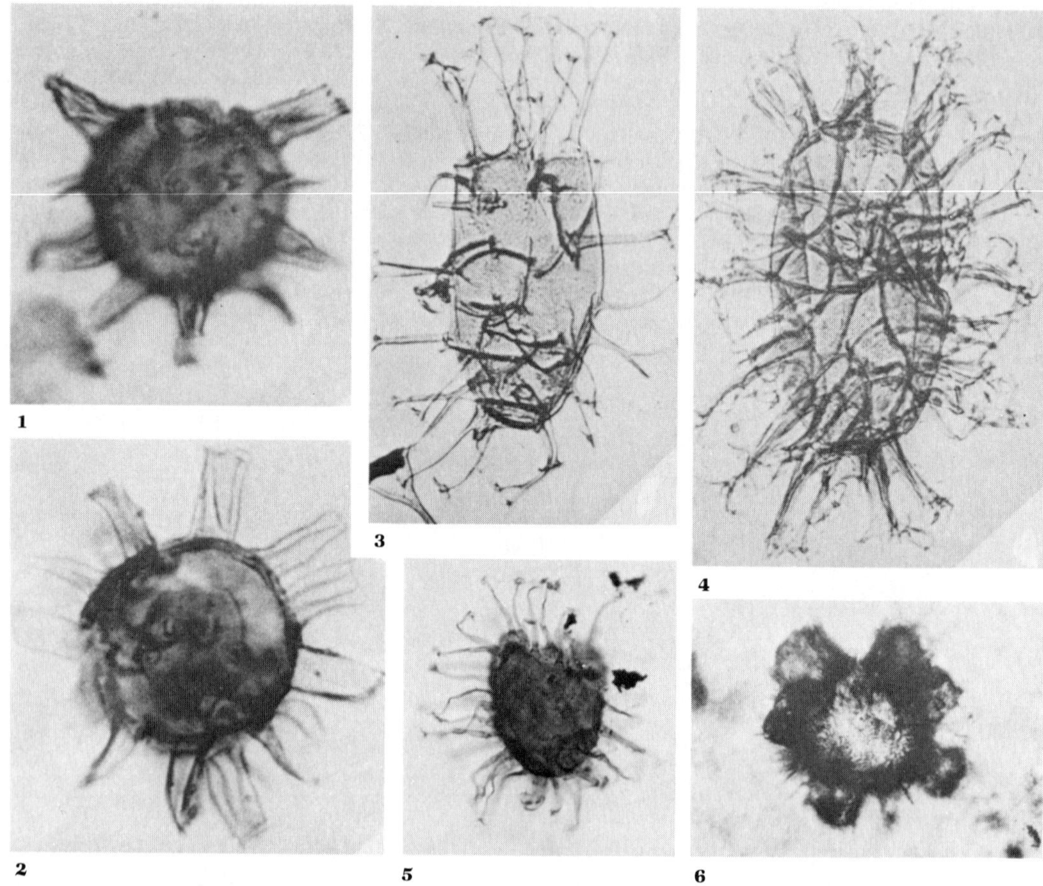

Figure 4.140

Hystrichosphaeridiaceae. **1,2.** *Litosphaeridium conispinum* Davey & Verdier, Lower Cretaceous, France, ×650, from Davey and Verdier, 1973. **1.** Dorsal view. **2.** Dorsal view with hexagonal operculum slightly displaced at center of figure, archeopyle opening at upper left. **3,4.** *Distatodinium ellipticum* (Cookson) Eaton, upper Eocene, Australia, ×670, from Cookson, 1965a. **3.** With archeopyle. **4.** Entire. **5.** *Tanyosphaeridium xanthiopyxides* (O. Wetzel) Stover & Evitt, Paleocene, Alabama, ×486, from Drugg, 1970. **6.** *Litosphaeridium flosculum* (Deflandre) Davey & Williams, Cretaceous, France, ×450, from Deflandre, 1937b.

Figure 4.141 (*facing page*)

Exochosphaeridiaceae. **1–3.** *Fibrocysta lappacea,* lower Eocene, Alabama, from Drugg, 1970. **1.** SEM of left side, with archeopyle at right, ×1000. **2,3.** Optical section and dorsal view, ×384. **4.** *Cometodinium obscurum* Deflandre & Courteville, Upper Cretaceous, France, ×700, from Deflandre and Courteville, 1939. **5,6.** *Coronifera striolata* (Deflandre) Stover & Evitt, Upper Cretaceous, France; 5, ×500, from Davey and Verdier, 1973; 6, ×540, from Foucher, 1972. **7,8.** *Fibrocysta radiata* (Morgenroth) Stover & Evitt, lower Eocene, Germany, dorsal and ventral sides, ×500, from Morgenroth, 1966a. **9.** *Exochosphaeridium pseudhystrichodinium* (Deflandre) Davey et al., Upper Cretaceous, France, ×540, from Foucher, 1972.

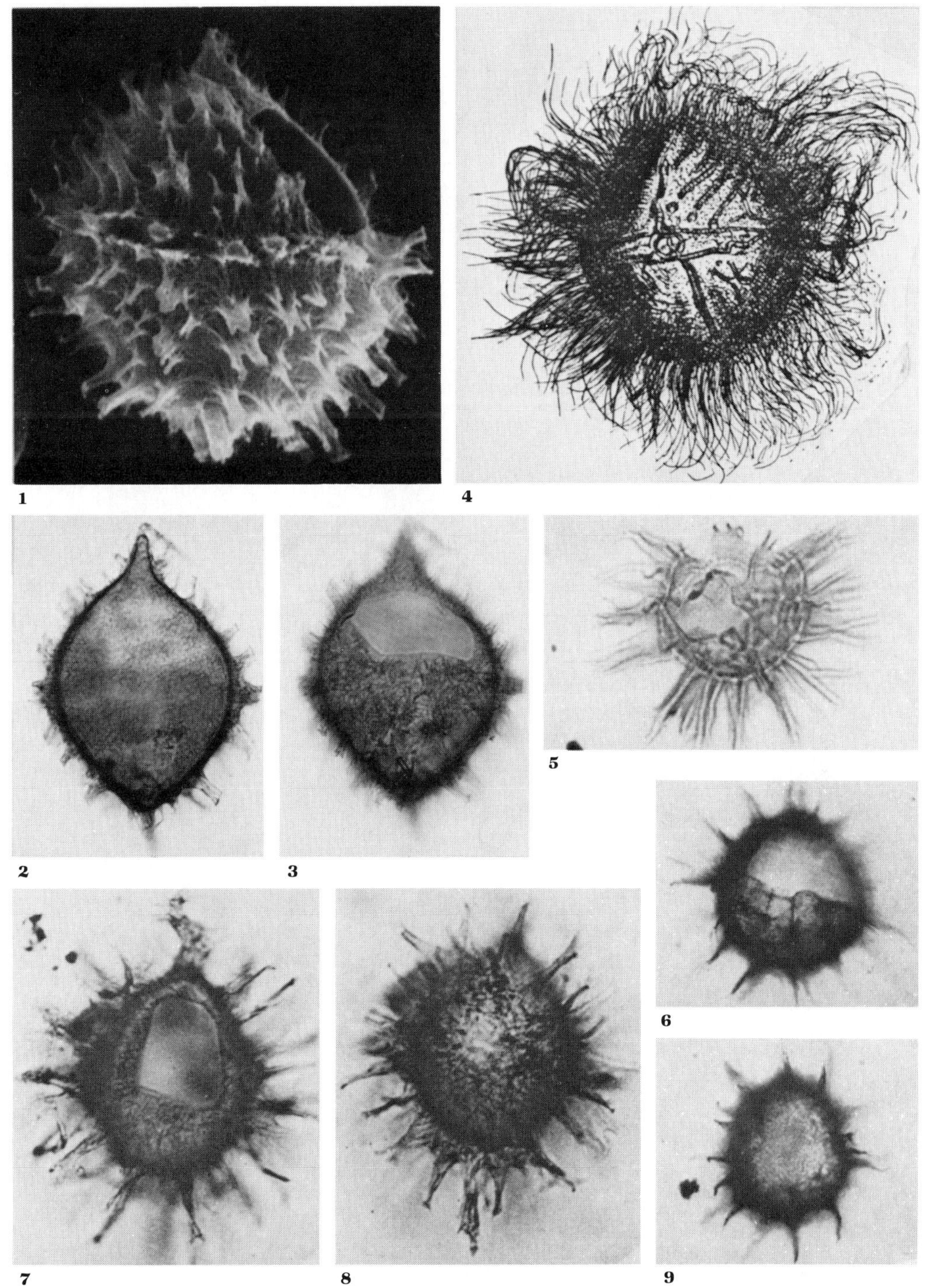

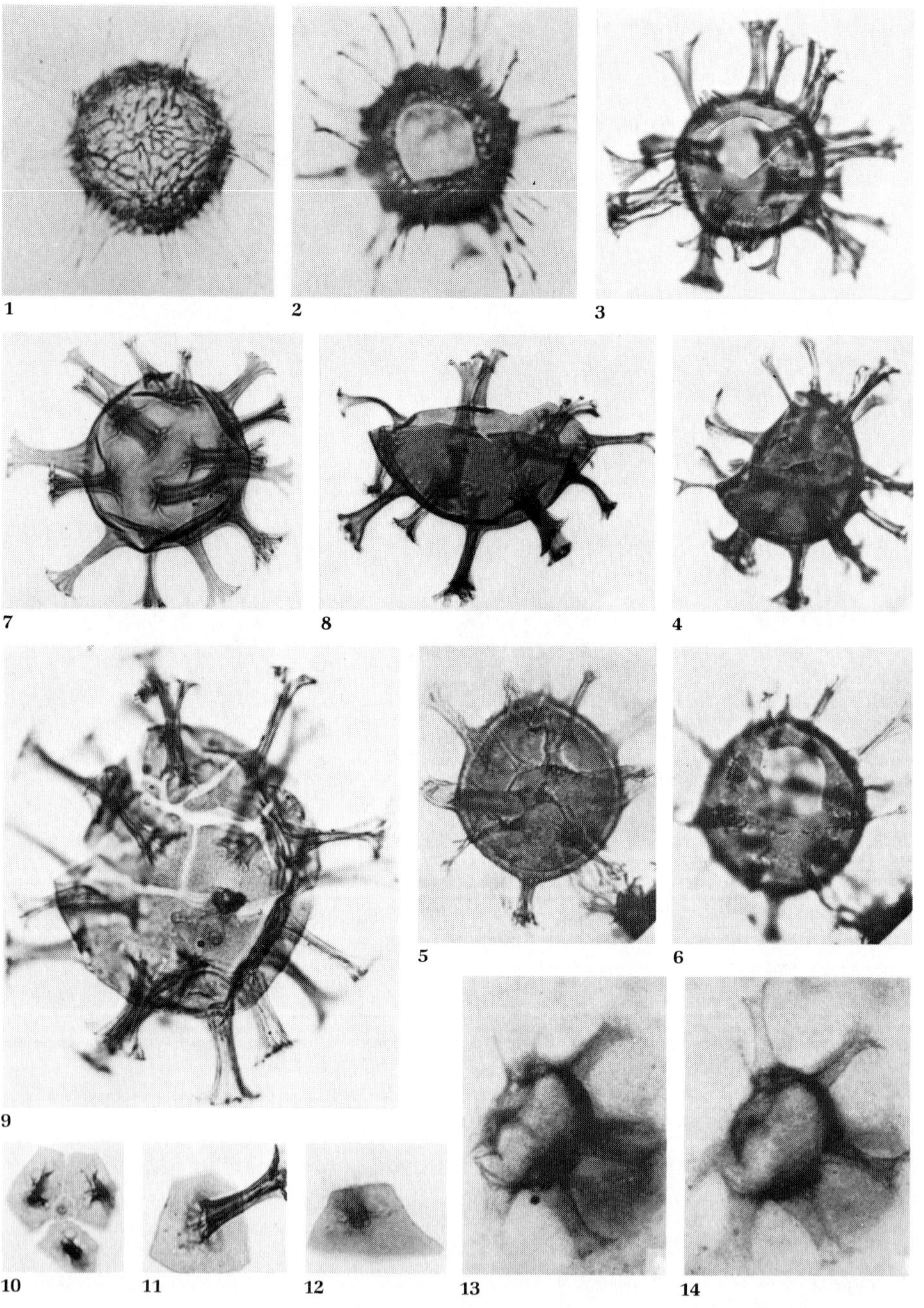

Figure 4.142
Homotrybliaceae, Cordosphaeridiaceae. **1,2.** *Cordosphaeridium funiculatum* Morgenroth, lower Eocene, Germany, ventral and dorsal surfaces, showing archeopyle, ×400, from Morgenroth, 1966a. **3– 6.** *Cordosphaeridium inodes.* **3.** Upper Eocene, Alabama, showing archeopyle, ×307. **4– 6.** Paleocene, Germany, from Morgenroth, 1968; 4, left side, ×272; 5,6, ventral and dorsal focus, ×296. **7– 12.** *Homotryblium plectilum* Drugg & Loeblich, Oligocene, Mississippi, from Drugg and Loeblich, 1967. **7.** Antapical view of hypocyst, ×385. **8.** Side view of hypocyst, ×385. **9.** Complete individual, with epicyst beginning to break up, ×768. **10.** Isolated plates 2 −4′ of operculum, ×458. **11.** Isolated plate 3″ of operculum. **12.** Isolated plate 1″, from inside. 11,12, ×538. **13,14.** *Callaiosphaeridium asymmetricum* (Deflandre & Courteville) Davey & Williams, Upper Cretaceous, France, ×360, from Foucher, 1972.

Actinotheca Cookson & Eisenack 1960; *Callaiosphaeridium* Davey & Williams 1966; *Eocladopyxis* Morgenroth 1966; *Hemicystodinium* Wall 1967 (= cyst of living *Pyrodinium bahamense*); *Homotryblium* Davey & Williams 1966.

29. Family Cordosphaeridiaceae Sarjeant & Downie 1974

Chorate gonyaulacacean cyst, ovoid to subpolygonal; intratabular tubular processes, size generally corresponding to that of paraplates; type P archeopyle. Part of "hystrichosphaeridioid lineage."

Lower Cretaceous to Upper Miocene; N. America, Europe, Africa, India.

Amphorosphaeridium Davey 1969; *Cordosphaeridium* Eisenack 1963; *Diversispina* Benson 1976.

30. Family Systematophoraceae Sarjeant & Downie 1974 (see Figures 4.143 and 4.144)

Chorate gonyaulacacean cyst, ovoid to subpolygonal; peritabular raised crests or lines of processes, simple to anastomosing, tubular or solid; paratabulation 1 − 5′, 0 − 1a, 6″, 0 − 6c, ?s, 5 − 6‴, 0 − 1p, 0 − 1″″; type tA archeopyle. Includes parts of "systematophoroid, eisenackioid, hystrichokolpomoid, and dictyopyxidioid lineages" and Coroniferaceae Vozzhennikova 1979.

Lower Cretaceous to middle Miocene; N. America, Europe, Africa, USSR, India, Australasia.

Aireiana Cookson & Eisenack 1965; *Amphorula* Dodekova 1969; *Coronifera* Cookson & Eisenack 1958; *Diphyes* Cookson 1965; *Duosphaeridium* Davey & Williams 1966; *Em-*

metrocysta Stover 1975; *Epiplosphaera* Klement 1960; *Hemiplacophora* Cookson & Eisenack 1955; *Heterosphaeridium* Cookson & Eisenack 1968; *Polystephanephorus* Sarjeant 1961; *Prolixosphaeridium* Davey, Downie, Sarjeant & Williams 1966; *Schematophora* Deflandre & Cookson 1955; *Systematophora* Klement 1960; *Taeniophora* Klement 1960.

31. Family Cleistosphaeridiaceae Sarjeant & Downie 1974 (see Figure 4.145)

Chorate gonyaulacacean cyst, ovoid to subpolygonal; nontabular processes, type tA archeopyle.

Upper Triassic to middle Miocene; N. America, S. America, Europe, Africa, India, Australasia.

Anthosphaeridium Cookson & Eisenack 1968; *Cleistosphaeridium* Davey, Downie, Sarjeant & Williams 1966; *Impletosphaeridium* Morgenroth 1966; *Lingulasphaera* Drugg 1970; *Polysphaeridium* Davey & Williams 1966.

32. Family Florentiniaceae Harker & Sarjeant 1975 (see Figure 4.146)

Chorate gonyaulacacean cyst, subspherical; one hollow intratabular process per paraplate, large tubular antapical process open distally; paratabulation 3 − 4′, 6″, 6c, 2 − 5s, 5‴, 1p, 1″″; type 1P or tA + 1P archeopyle. Includes Silicisphaeraceae Vozzhennikova 1979.

Lower Cretaceous to lower Paleocene; Europe, Australia, N. America.

Florentinia Davey & Verdier 1973; *Kleithriasphaeridium* Davey 1974; *Silicisphaera* Davey & Verdier 1976.

(*continued*)

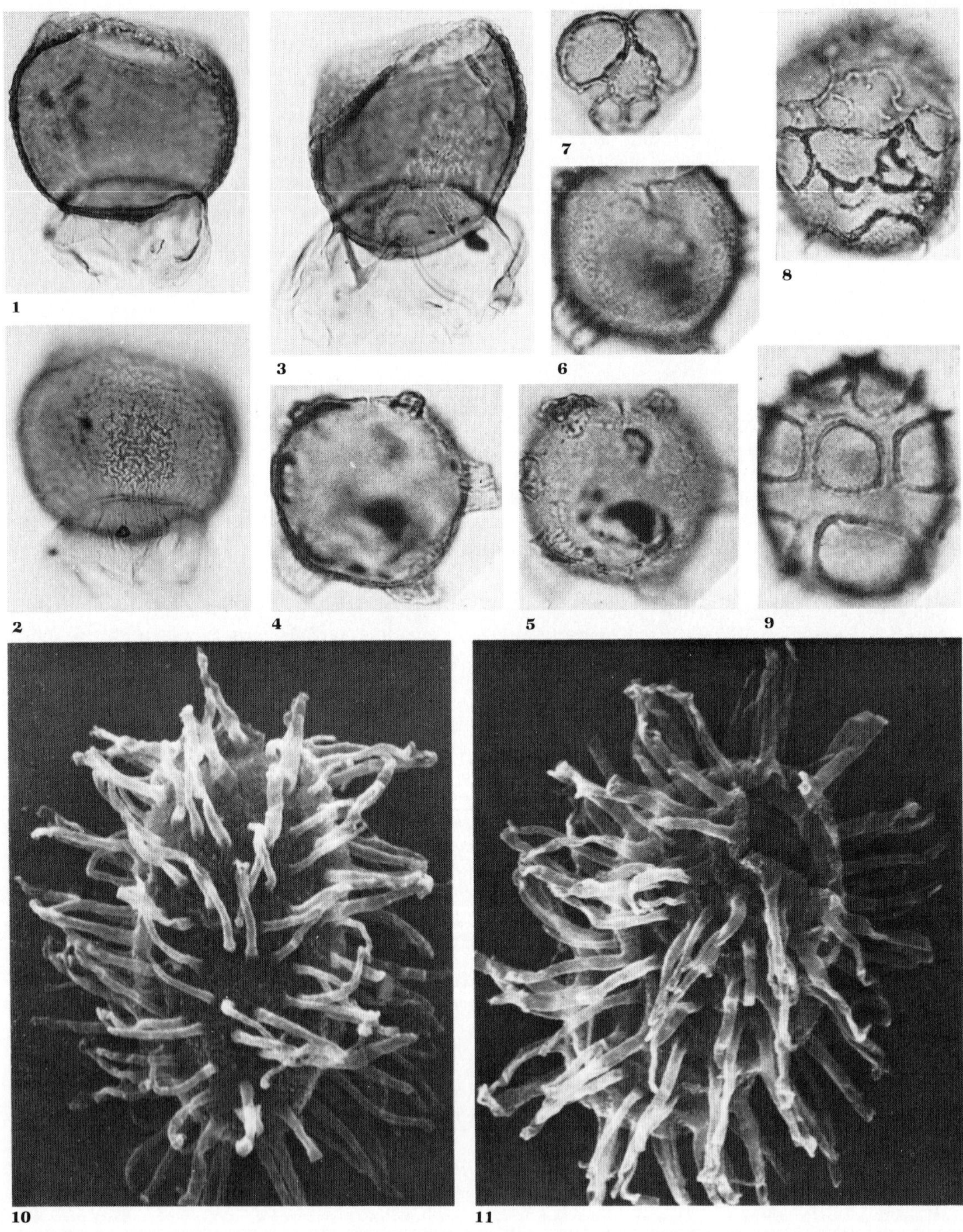

Figure 4.143 (*facing page*)
Systematophoraceae. **1–3.** *Duosphaeridium rugosum* Drugg, Paleocene, Alabama, ×486, from Drugg, 1970. **1,2.** Optical section and surface. **3.** Showing skirtlike periphragm. **4–6.** *Aireiana verrucosa* Cookson & Eisenack, Eocene, Australia, optical section and focus on opposite surfaces, ×500, from Stover, 1975. **7–9.** *Schematophora speciosa* Deflandre & Cookson, Eocene, Australia, ×625, from Stover, 1975. **7.** Isolated operculum. **8,9.** Ventral and dorsal focus, **10,11.** *Prolixosphaeridium mixtispinosum* (Klement) Davey et al., Upper Jurassic, western Atlantic. **10.** SEM of entire specimen, ×2000. **11.** Oblique view of specimen with apical archeopyle, ×1500.

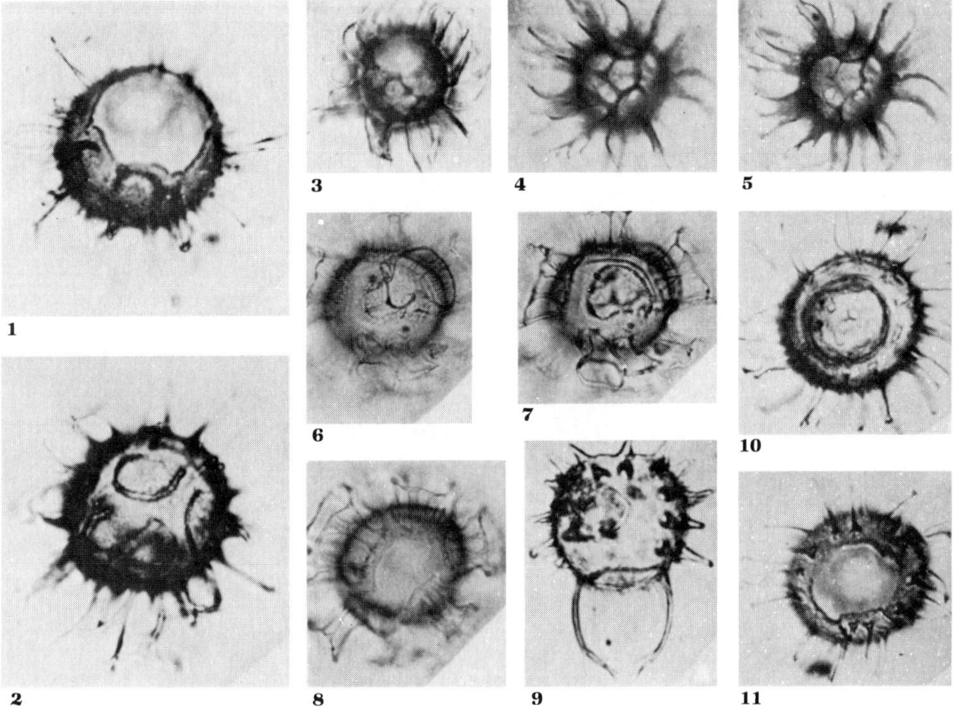

Figure 4.144
Systematophoraceae. **1, 2.** *Systematophora placacantha* (Deflandre & Cookson) Davey et al., Eocene, Germany, view of archeopyle and opposite surface, ×640, from Morgenroth, 1966a. **3–5.** *Coronifera oceanica* Cookson & Eisenack, Upper Cretaceous, France. **3.** Showing archeopyle, ×432, from Foucher, 1974. **4,5.** Two levels of focus, ×432, from Foucher, 1972. **6–8.** *Emmetrocysta urnaformis* (Cookson) Stover, Eocene, Australia, three levels of focus, from antapical side (6) to apical (8), ×400, from Stover, 1975. **9–11.** *Diphyes colligerum* (Deflandre & Cookson) Cookson, Eocene, Australia, from Cookson, 1965a. **9.** Ventral view of specimen with short stiff processes, ×536. **10,11.** Two levels of focus, dorsal side with archeopyle, ×385.

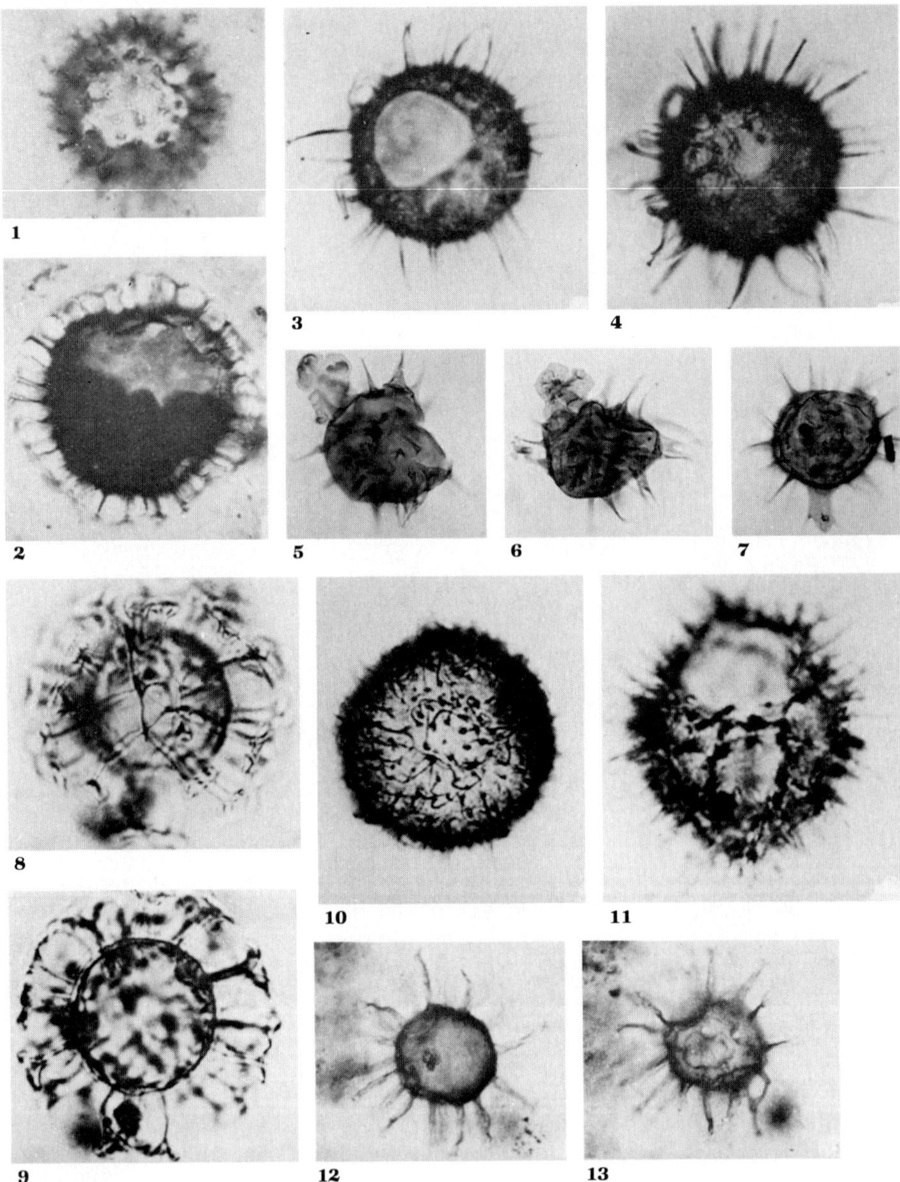

Figure 4.145
Cleistosphaeridiaceae. and Lingulodiniaceae. **1,2.** *Cleistosphaeridium huguoniotii* (Valensi) Davey, Upper Cretaceous, France, surface and optical section, ×864, from Foucher, 1974. **3,4.** *Operculodinium tiara*, Eocene, Germany, two focus levels, ×640, from Morgenroth, 1966a. **5—7.** *Lingulasphaera spinula*, lower Eocene, Alabama, ×390, from Drugg, 1970. **5,6.** Side views of specimens showing open archeopyle with attached opercula. **7.** Showing antapical horn. **8,9.** *Impleto-*

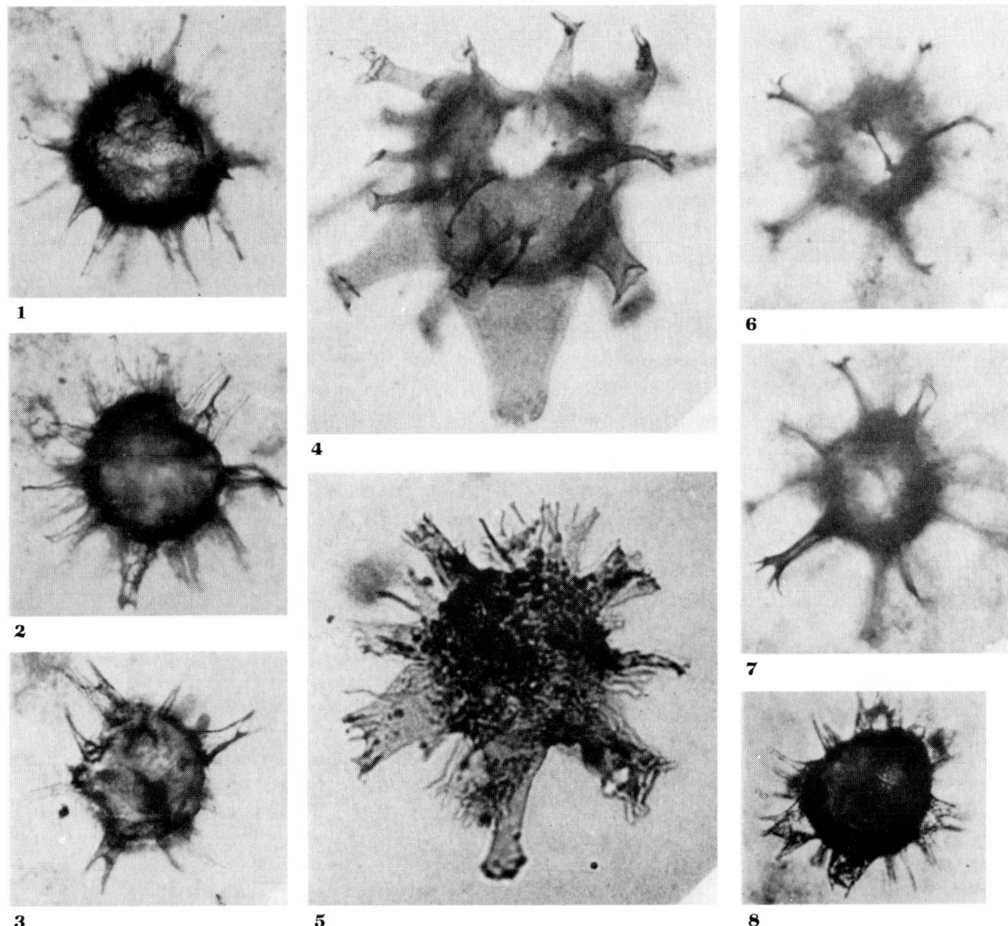

Figure 4.146

Florentiniaceae. **1,2.** *Florentinia clavigera* (Deflandre) Davey & Verdier, Upper Cretaceous, France, surface and optical section, ×540, from Foucher, 1972. **3.** *Silicisphaera tridactylites* (Valensi) Davey & Verdier, Upper Cretaceous, France, ×540, from Foucher, 1972. **4.** *Florentinia deanei* (Davey & Williams) Davey & Verdier, Lower Cretaceous, France, ventral view, with archeopyle visible through the specimen, ×500, from Davey and Verdier, 1973. **5.** *F. laciniata* Davey & Verdier, Lower Cretaceous, ×500, from Davey and Verdier, 1973. **6,7.** *Kleithriasphaeridium readei* (Davey & Williams) Davey & Verdier, Upper Cretaceous, France, two levels of focus, ×450, from Foucher, 1974. **8.** *Silicisphaera ferox* (Deflandre) Davey & Verdier, Cretaceous, France, ×450, from Deflandre, 1937b.

sphaeridium rugosum Morgenroth, lower Eocene, Germany, surface and optical section, ×1040, from Morgenroth, 1966a. **10.** *I. densicomatum* (Maier) Morgenroth, lower Eocene, Germany, ×400, from Morgenroth, 1966a. **11.** *Operculodinium erinaceum* (Morgenroth) Stover & Evitt, lower Eocene, Germany, showing archeopyle, ×800, from Morgenroth, 1966a. **12,13.** *Polysphaeridium* cf. *pumilum* Davey & Williams, Upper Cretaceous, France, two levels of focus, ×575, from Foucher, 1974.

33. Family Lingulodiniaceae Sarjeant & Downie 1974 (see Figure 4.147)

Chorate gonyaulacacean cysts, ovoid to subpolygonal; processes nontabulate; cysts similar to modern *Gonyaulax polyedra* and *G. grindleyi;* tabulation of motile stage 3−6′, 0−4a, 6″, 6c, 6‴, 1p, 1⁗; type 1P to 5P archeopyle. "Lingulodinioid lineage."

Lower Cretaceous to Quaternary; N. America, Europe, India, Australasia.

Lingulodinium Wall 1967 (syn.: *Trioperculodinium* Drugg 1970); *Operculodinium* Wall 1967. Living *Gonyaulax polyedra* has a *Lingulodinium* cyst, and *G. grindleyi* an *Operculodinium* one; whether all fossil lingulodiniacean taxa are referrable to this living genus is uncertain.

34. Family Areoligeraceae Evitt 1963 (see Figure 4.148)

Marginate to trabeculate gonyaulacacean cysts, ovoid to polygonal; processes may be interconnected distally by trabeculae or a membranous cover, intratabular, aligned in rows, or fused as crests; type tA archeopyle, with parasulcal notch offset to left. "Areoligeroid lineage"; includes Canninginopsaceae Vozzhennikova 1967 and Adnatosphaeridiaceae Sarjeant & Downie 1966 ex 1974.

Upper Triassic to upper Miocene; N. America, S. America, Europe, India, USSR, Africa, Australasia.

Adnatosphaeridium Williams & Downie 1966; *Areoligera* Lejeune-Carpentier 1938; *Canninginopsis* Cookson & Eisenack 1962; *Chiropteridium* Gocht 1960; *Comparodinium* Morbey 1975; *Cyclonephelium* Deflandre & Cookson 1955; *Glaphrocysta* Stover & Evitt 1978; *Renidinium* Morgenroth 1968; *Riculacysta* Stover 1977.

35. Family Cannosphaeropsitaceae Sarjeant & Downie 1966 ex 1974

Trabeculate, chorate gonyaulacacean cysts, ovoid to polygonal; processes parasutural in position; type P archeopyle. Part of "gonyaulacoid lineage."

Middle Jurassic to lower Oligocene; N. America, Europe, Australasia.

Cannosphaeropsis O. Wetzel 1932; *Hapsidaulax* Sarjeant 1975.

36. Family Pseudoceratiaceae Eisenack 1961 (see Figures 4.149 and 4.150)

Cavate gonyaulacacean cysts with ceratioid relationship, enclosing body oval to triangular, with apical and antapical horns and a lateral bulge or horn; paratabulation distinct or indistinct; type tA archeopyle. Pseudoceratioid lineage of the peridinacean group of Wall and Dale; but *Ceratium* is gonyaulacacean; includes Diplotestaceae Cookson & Eisenack 1960; Muderongiaceae Neale & Sarjeant 1962, and Odontochitiniaceae and Endoceratiaceae Vozzhennikova 1965.

Cretaceous; N. America, USSR, Europe, Africa, India, Australasia.

Aptea Eisenack 1958; *?Ceratocystidiopsis* Deflandre 1937; *?Diplofusa* Cookson & Eisenack 1960; *Endoceratium* Vozzhennikova 1965; *Odontochitina* Deflandre 1935; *Odontochitinopsis* Deflandre 1964; *Pseudoceratium* Gocht 1957; *Muderongia* Cookson & Eisenack 1958; *Phoberocysta* Millioud 1969; *Prismatocystis* Habib 1969; *Wallodinium* Loeblich Jr. & Loeblich III 1968 (syn.: *Diplotesta* Cookson & Eisenack 1960 non Brongniart 1874); *Xenascus* Cookson & Eisenack 1969.

37. Family Thalassiphoraceae Gocht 1968 ex Sarjeant & Downie 1974 (see Figure 4.151)

Cavate gonyaulacacean cyst, ovoid to

(*continued*)

Figure 4.147

Lingulodiniaceae. **1−3.** *Lingulodinium siculum* (Drugg) Wall & Dale, Oligocene, Mississippi, ×486, from Drugg, 1970. **1,2.** Median focus showing processes, and surface of dorsal side with operculum in place. **3.** Dorsal view, showing archeopyle left by loss of three-part operculum. **4,5.** *Operculodinium centrocarpum*, Pleistocene, Caribbean Sea cores, ×390, from Wall, 1967. **6−9.** *O. placitum* Drugg & Loeblich, from Drugg and Loeblich, 1967; 6−8, Oligocene, Mississippi. **6,7.** Ventral and dorsal focus, showing archeopyle, ×672. **8.** Lateral view showing surface, ×960. **9.** Upper Eocene, Alabama, lateral view, showing displaced operculum that has fallen within cyst, ×672. **10−12.** *Lingulodinium solarum* (Drugg) Wall & Dale, Oligocene, Mississippi, antapical view, optical section, and apical view showing upper part of the precingular archeopyle, ×384, from Drugg, 1970.

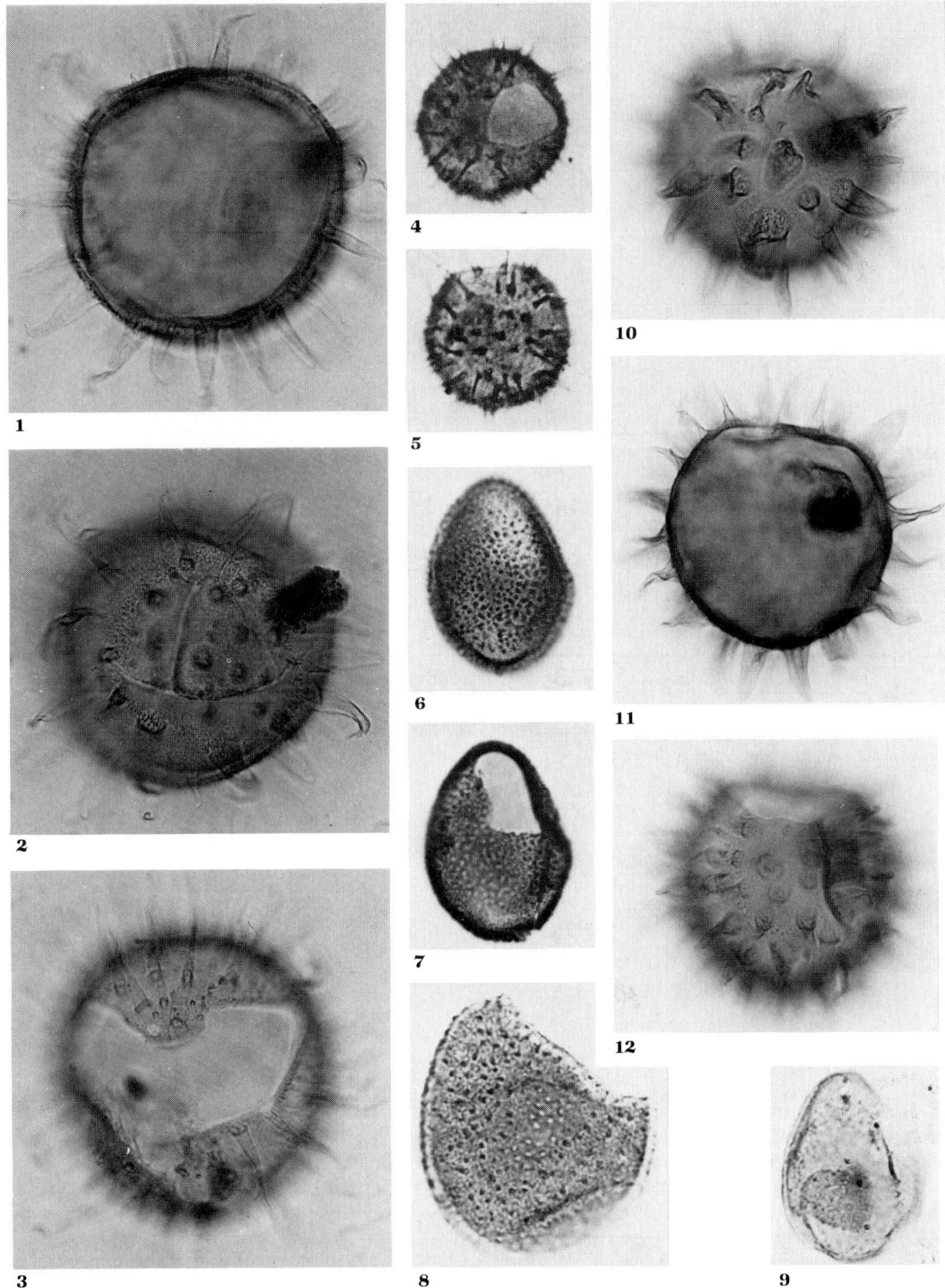

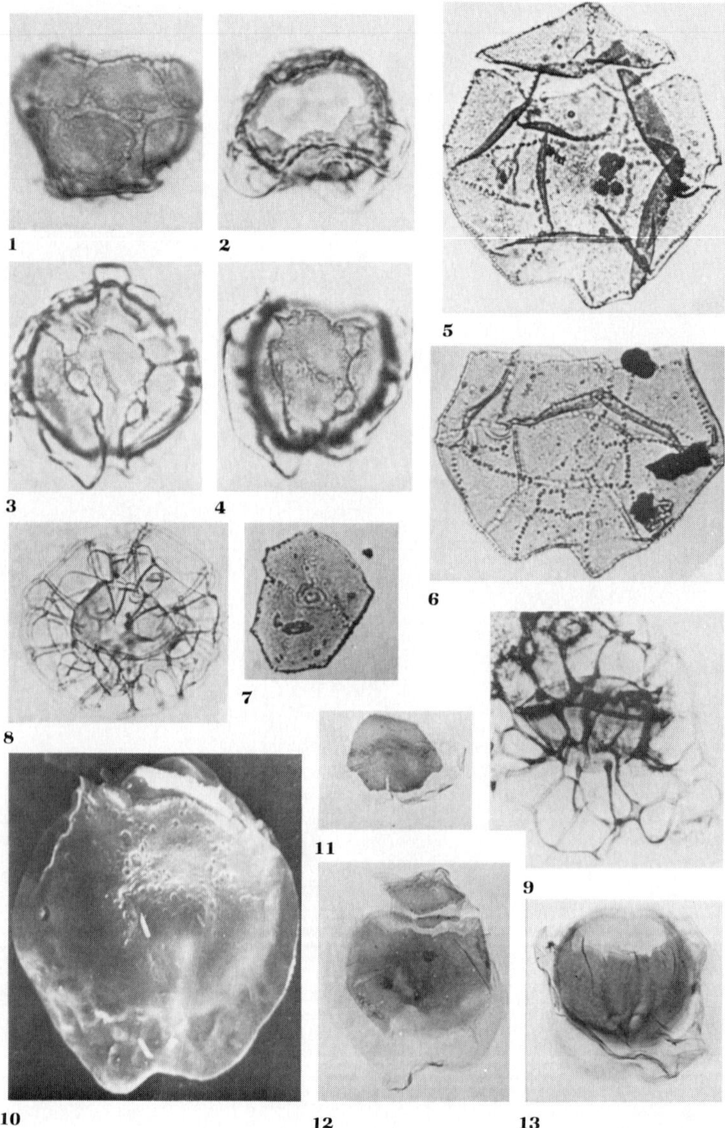

Figure 4.148

Areoligeraceae, Hystrichodiniaceae. **1—4.** *Renidinium membraniferum* Morgenroth, Paleocene, Germany, from Morgenroth, 1968. **1.** Dorsal side, ×390. **2,3.** Apical view with archeopyle, and side view showing operculum in place, ×375. **4.** Ventral view, ×325. **5—7.** *Canninginopsis denticulata* Cookson & Eisenack, Cretaceous, Australia, dorsal view with operculum in place and ventral view with apical archeopyle, ×341, and detached operculum, ×364, from Cookson and Eisenack, 1962. **8.** *Nematosphaeropsis pusulosa* (Morgenroth) Stover & Evitt, lower Oligocene, Germany, ×325, from Morgenroth, 1966b. **9.** *Adnatosphaeridium robustum* (Morgenroth) Eaton, lower Eocene, Germany, ×390, from Morgenroth, 1966a. **10—13.** *Senoniasphaera inornata* (Drugg) Stover & Evitt, Paleocene, Alabama, from Drugg, 1970. **10.** SEM, ×488. **11.** Isolated operculum, ×250. **12,13.** With operculum and with open archeopyle, ×250.

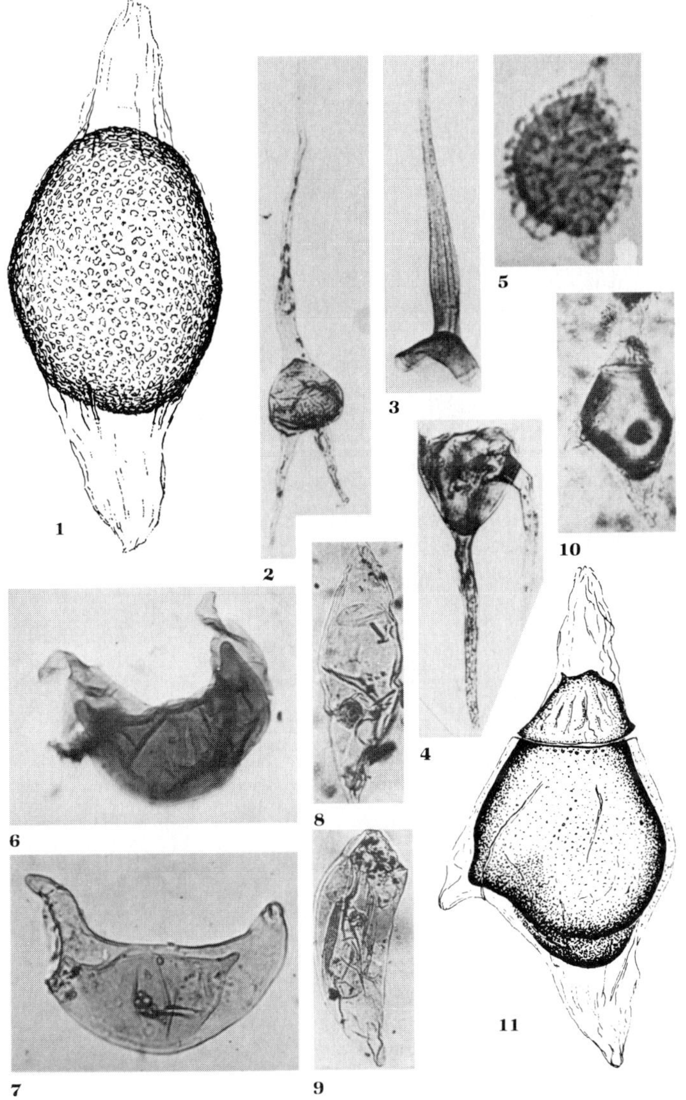

Figure 4.149

Pseudoceratiaceae. **1.** *Ceratocystidiopsis simplex* Deflandre, Upper Cretaceous, France, ×650, from Deflandre, 1937b. **2–4.** *Odontochitina costata* Alberti. **2.** Upper Cretaceous, England, entire specimen, showing elongate horns, ×166, from Davey, 1970. **3,4.** Cretaceous, Australia, isolated operculum (apex) and cyst with archeopyle, ×228, from Cookson et al., 1970. **5.** *Aptea polymorpha* Eisenack, Lower Cretaceous, Germany, ventral view, ×276, from Wall and Evitt, 1975. **6,7.** *Wallodinium bidigitatum* (Manum & Cookson) Lentin & Williams, with permission from Manum and Cookson, 1964, *Skr. Norske Vidensk-Akad. Mat.-Naturv. Kl.,* n.s., no. 17, published by Universitetsforlaget, Oslo University. **6.** Upper Cretaceous, Arctic Canada. **7.** Australia, ×422. **8,9.** *W. krutzschi* (Alberti) Habib, Lower Cretaceous, Germany, entire specimen, and one with apex lost as operculum, ×325, from Alberti, 1961. **10,11.** *Odontochitinopsis molesta* (Deflandre) Eisenack, Upper Cretaceous, France; 10, ×292; 11, ×650; from Deflandre, 1937b.

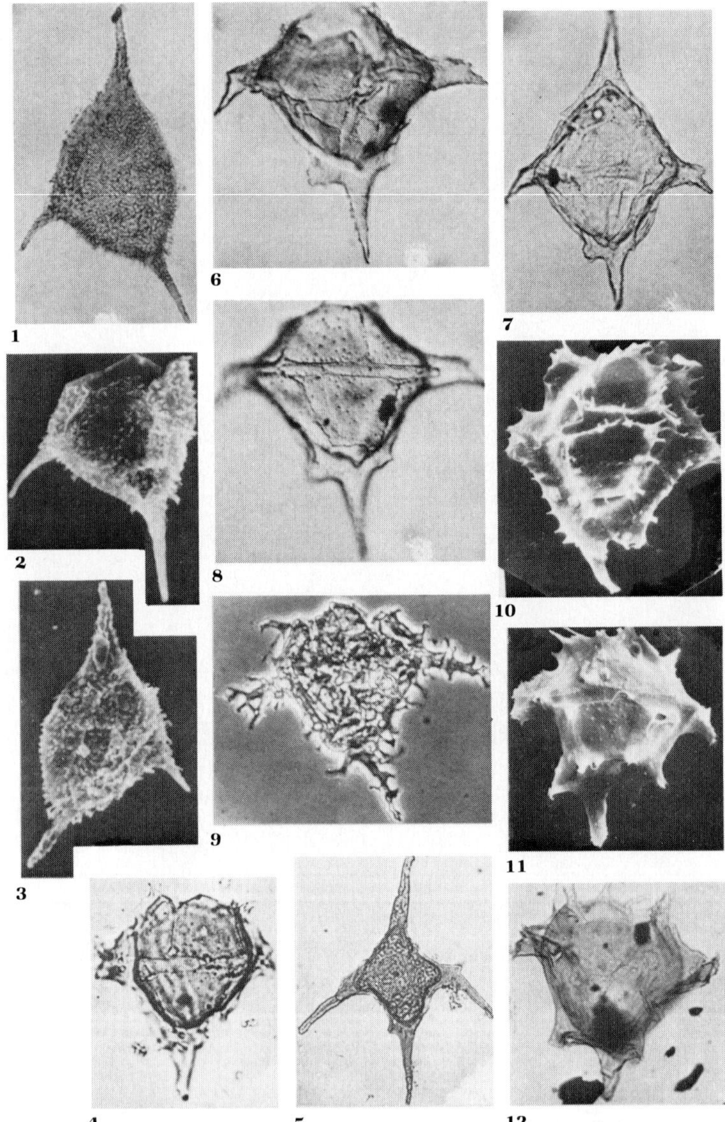

Figure 4.150
Pseudoceratiaceae of the Lower Cretaceous. **1–3.** *Pseudoceratium pelliferum* Gocht, England, from Wall and Evitt, 1975. **1.** Ventral view, ×270. **2.** SEM of ventral side, showing archeopyle, ×325. **3.** SEM, dorsal side, ×332. **4,5.** *Muderongia mcwhaei* Cookson & Eisenack, Australia, from Cookson and Eisenack, 1958. **4.** With archeopyle, ×374. **5.** Entire specimen, ×208. **6–8.** *M.* cf. *mcwhaei*, California, from Wall and Evitt, 1975. **6,8.** Ventral and dorsal focus, ×260. **7.** Ventral view, focussed on dorsal side, showing partly opened archeopyle suture, ×221. **9–12.** *Phoberocysta neocomica* (Gocht) Millioud, France. **9.** Phase contrast, showing archeopyle border, ×369, from Millioud, 1969. **10,11.** SEM, ×325. **12.** Optical microscope, ×325, from Jardiné and Raynaud, 1972.

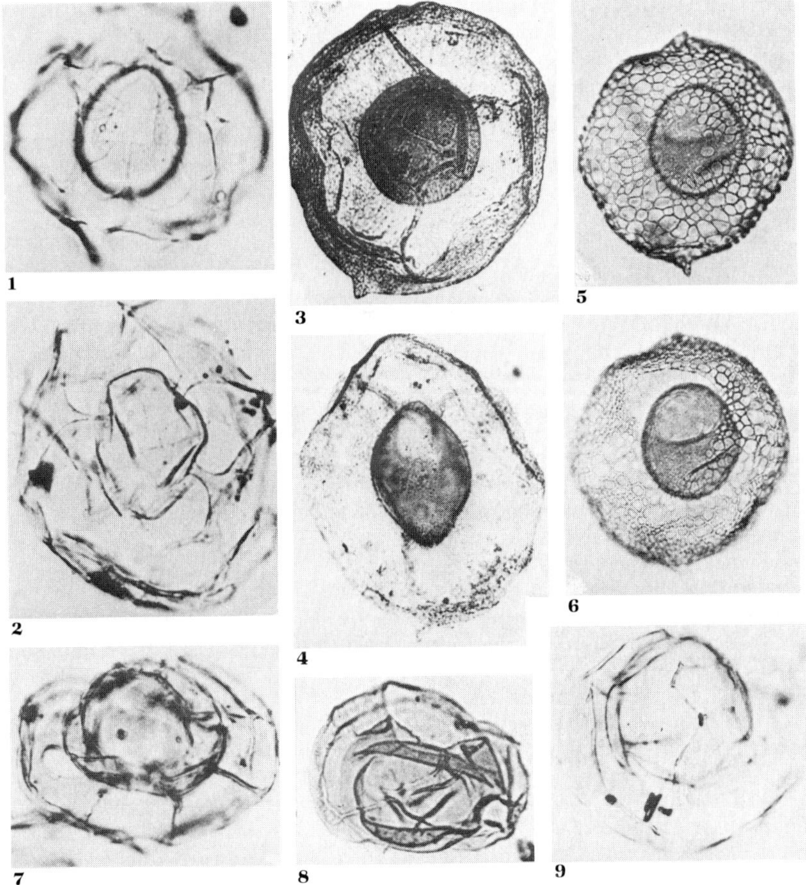

Figure 4.151

Thalassiphoraceae, Stephodiniaceae. **1,2.** *Thalassiphora dynamica* (Morgenroth) Stover & Evitt, lower Eocene, Germany, ×560, from Morgenroth, 1966a. **1.** Dorsal side, showing archeopyle in inner capsule. **2.** Showing connections of capsule and outer membrane. **3,4.** *T. pelagica* (Eisenack) Eisenack & Gocht. **3.** Eocene, England, showing operculum in place, with bordering archeopyle suture, ×192, from G. L. Williams and Downie, 1966b. **4.** Lower Eocene, Germany, showing open archeopyle, ×200, from Morgenroth, 1966a. **5,6.** *T. reticulata* Morgenroth, lower Oligocene, Germany, ventral and dorsal focus, latter with archeopyle, ×176, from Morgenroth, 1966b. **7–9.** *Stephodinium coronatum* Deflandre. **7,9.** Middle Cretaceous, England, apical view with archeopyle in outer membrane, and antapical view showing antapical plate equivalent, both ×400, from Davey, 1970. **8.** Lower Cretaceous, France, specimen with precingular archeopyle at top of figure, ×400, from Davey and Verdier, 1973.

polyhedral; pericyst may have apical or antapical horn or prominence; paratabulation 1′, 1a, 5″, 4‴, 1‴′ when indicated, paracingulum may be elevated; type IP archeopyle (loss of third precingular paraplate indicates gonyaulacacean relationship).

Lower Cretaceous to Quaternary; N. America, S. America, Europe, Australasia.

Thalassiphora Eisenack & Gocht 1960 (syn.: *Erikania* Morgenroth 1966).

38. Family Stephodiniaceae Eisenack 1964

Cavate gonyaulacacean cyst, endocyst oval to polygonal; pericyst equatorially inflated, surrounding endocyst excentrically, and may be supported by fibrous structures; type P archeopyle.

Lower Cretaceous to upper Eocene; N. America, Europe, Australasia.

Stephodinium Deflandre 1936.

39. Family Endoscriniaceae Vozzhennikova 1965 (see Figures 4.152, 4.153, and 4.154.

Cavate gonyaulacacean cysts, ovoid to polygonal; may have apical or antapical horns or protuberances; periphragm may be spinose or reticulate; crests or linear ornamentation indicating paratabulation, if any, of 4′, 0−1a, 6″, 6c, 5−6‴, 0−1p, 0−1‴′; type P (rarely 2P) archeopyle. Part of "gonyaulacoid lineage."

Middle Jurassic to upper Miocene; Recent; N. America, S. America, Europe, USSR, Australasia.

Hystrichosphaeropsis Deflandre 1935; *Palaeohystrichophora* Deflandre 1935; *Pentadinium* Gerlach 1961; *Psaligonyaulax* Sarjeant 1966; *Rottnestia* Cookson & Eisenack 1961; *Scriniodinium* Klement 1957 (syn.: *Endoscrinium* Klement 1960); *Smolenskiella* Vozzhennikova 1967; *Triblastula* O. Wetzel 1933; *Tubotuberella* Vozzhennikova 1967.

40. Family Lecaniellaceae Cookson & Eisenack 1970

Discoid, no evident paratabulation, epicystal archeopyle. Marine, Upper Cretaceous, Tertiary.

Eyrea Cookson & Eisenack in Cookson, Eisenack & Ingram 1971; *Lecaniella* Cookson & Eisenack 1962; *Paralecaniella* Cookson & Eisenack 1970.

41. Family Stephanelytraceae Stover, Sarjeant & Drugg 1977 (see Figure 4.155)

Subspherical to ellipsoidal, proximochorate; two distinctly separated wall layers, ectocoel with parasutural or nontabular processes that are tubiform and expanded distally; paratabulation ?1′, 5″, x−6c, 5‴, 1‴′, 2s; type A archeopyle, corona antapical.

Middle Jurassic to Upper Jurassic; Europe, North America.

Stephanelytron Sarjeant 1961.

Dinoflagellate fossils of uncertain relationships, most of which are in the text or illustrated but not assigned to a family, include *Aquadulcum* Harland & Sarjeant 1970, freshwater, Holocene; *Codoniella* Cookson & Eisenack 1961 (syn.: *Codonia* Cookson & Eisenack 1960, non Huebner 1823), Upper Cretaceous; *Creberlumectum* Harland & Sarjeant 1970, freshwater, Holocene; *Diacrocanthidium* Deflandre & Foucher 1967, Upper Cretaceous; *Eyachia* Gocht 1979, Lower to Middle Jurassic; *Muiradinium* Harland & Sarjeant 1970, freshwater, Holocene; *Palaeostomocystis* Deflandre 1937, Jurassic to Paleocene (Danian); *Palaeotetradinium* Deflandre 1936, Upper Cretaceous; *Raphidodinium* Deflandre 1936, Upper Cretaceous; *Sirmiodinium* Alberti 1961, Lower Cretaceous; *Suessia* Morbey 1975, Upper Triassic.

Figure 4.152

Endoscriniaceae. **1−3.** *Triblastula utinensis* O. Wetzel, Upper Cretaceous, New Jersey, ventral surface, optical section, and oblique view of dorsal surface with archeopyle, ×500, from Evitt, 1961a. **4,5.** *Psaligonyaulax deflandrei* Sarjeant, Upper Cretaceous, England, ventral and dorsal focus, from ventral side, ×500, from Sarjeant, 1966c. **6,7.** *Scriniodinium galeritum* (Deflandre) Klement, Upper Jurassic, Germany, ventral and dorsal views, ×580, from Klement, 1960. **8−10.** *Rottnestia borussica* (Eisenack) Cookson & Eisenack. **8,9.** Lower Eocene, Germany, dorsal side, ×800, and oblique left side showing archeopyle, ×700, from Morgenroth, 1966a. **10.** Upper Oligocene, Germany, ×525, from Benedek, 1972.

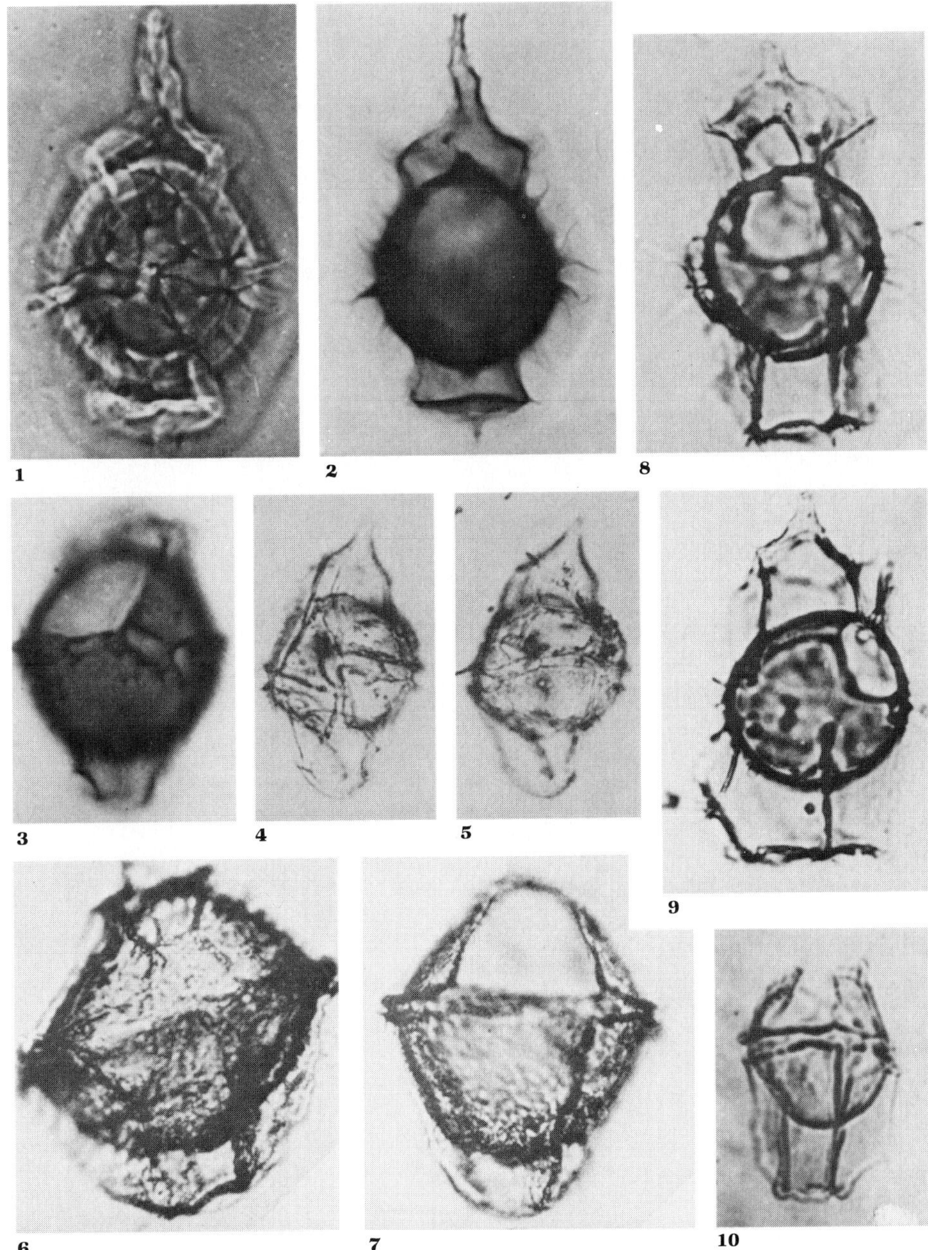

1

2

8

3

4

5

9

6

7

10

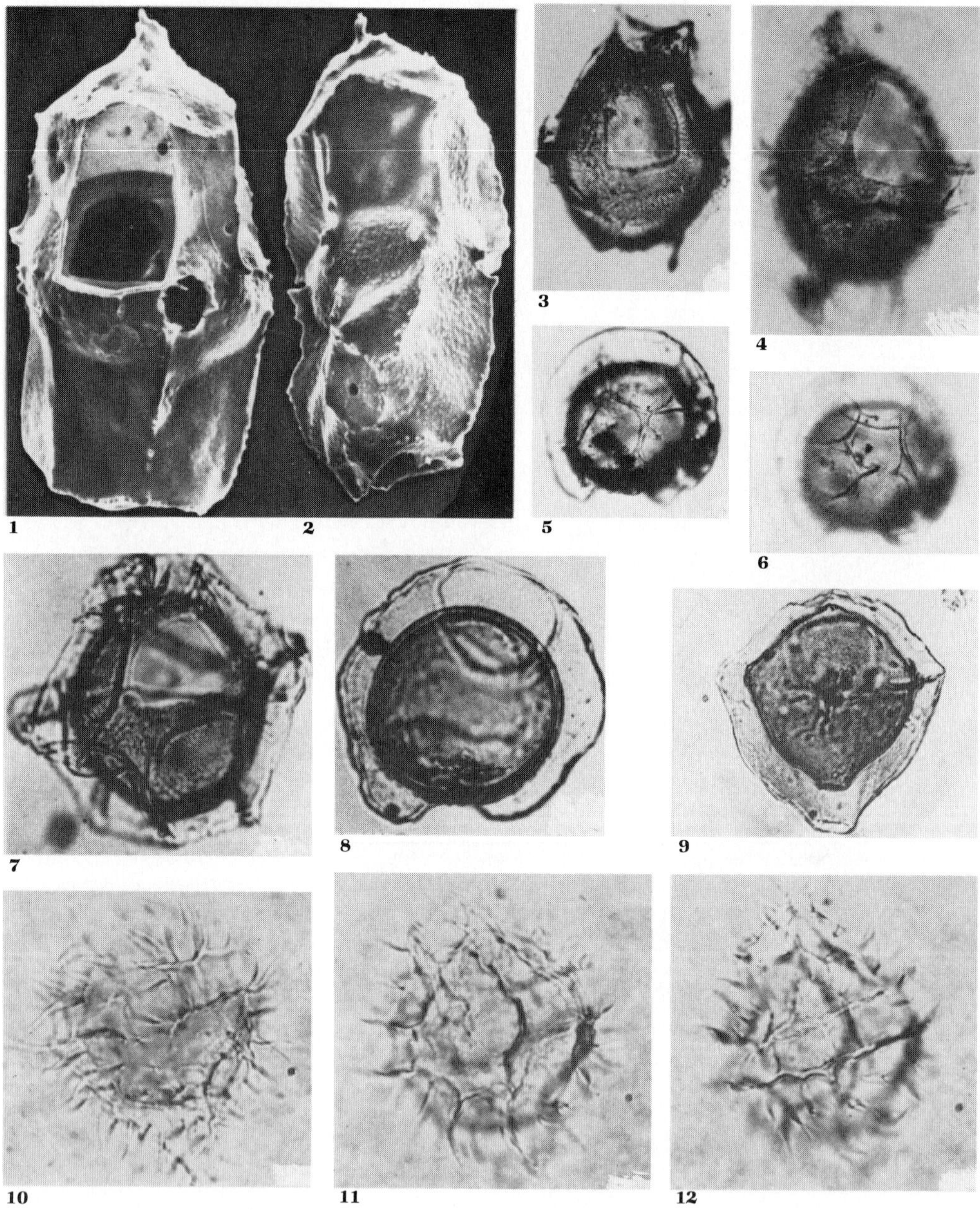

Figure 4.153 (*facing page*)
Endoscriniaceae. **1,2.** *Hystrichosphaeropsis quasicribrata* (O. Wetzel) Gocht, Upper Cretaceous, Germany, SEM, dorsal side, ×1000, and left ventral view, ×800, from Gocht, 1976. **3,4.** *H. complanata* Eisenack, upper Eocene to lower Oligocene, from Eisenack, 1965. **3.** Dorsal view, containing operculum in unopened archeopyle, ×400. **4.** Left side, showing open archeopyle, ×450. **5,6.** *Pentadinium laticinctum* Gerlach, middle Oligocene, Germany, ventral and dorsal views, ×275, from Gerlach, 1961. **7.** *P. taeniagerum* Gerlach, Oligocene, Germany, dorsal view, with archeopyle, ×860, from Benedek, 1972. **8.** *P. laticinctum* subsp. *imaginatum* Benedek, upper Oligocene, Germany, ×360, from Benedek, 1972. **9.** *Scriniodinium attadalense* (Cookson & Eisenack) Downie & Sarjeant, Lower Cretaceous, Australia, ×400, from Cookson and Eisenack, 1958. **10–12.** *Palaeohystrichophora infusorioides* Deflandre, Upper Cretaceous, France, three levels of focus, ×1080, from Foucher, 1972.

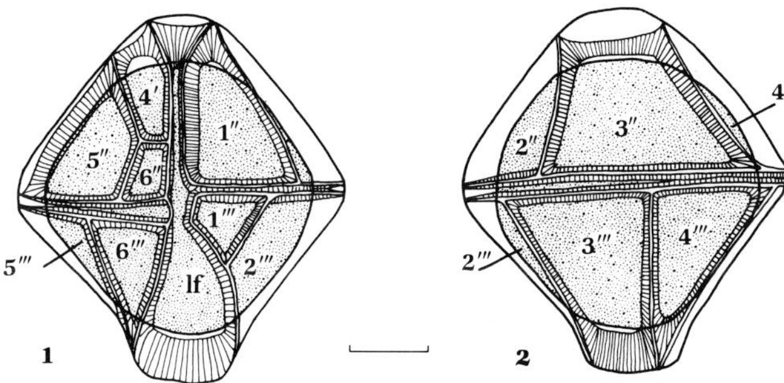

Figure 4.154
Tabulation of *Scriniodinium subvallare* Sarjeant, ventral (left) and dorsal sides; lf, longitudinal furrow. Bar = 20 μm. Redrawn from Sarjeant, 1962a.

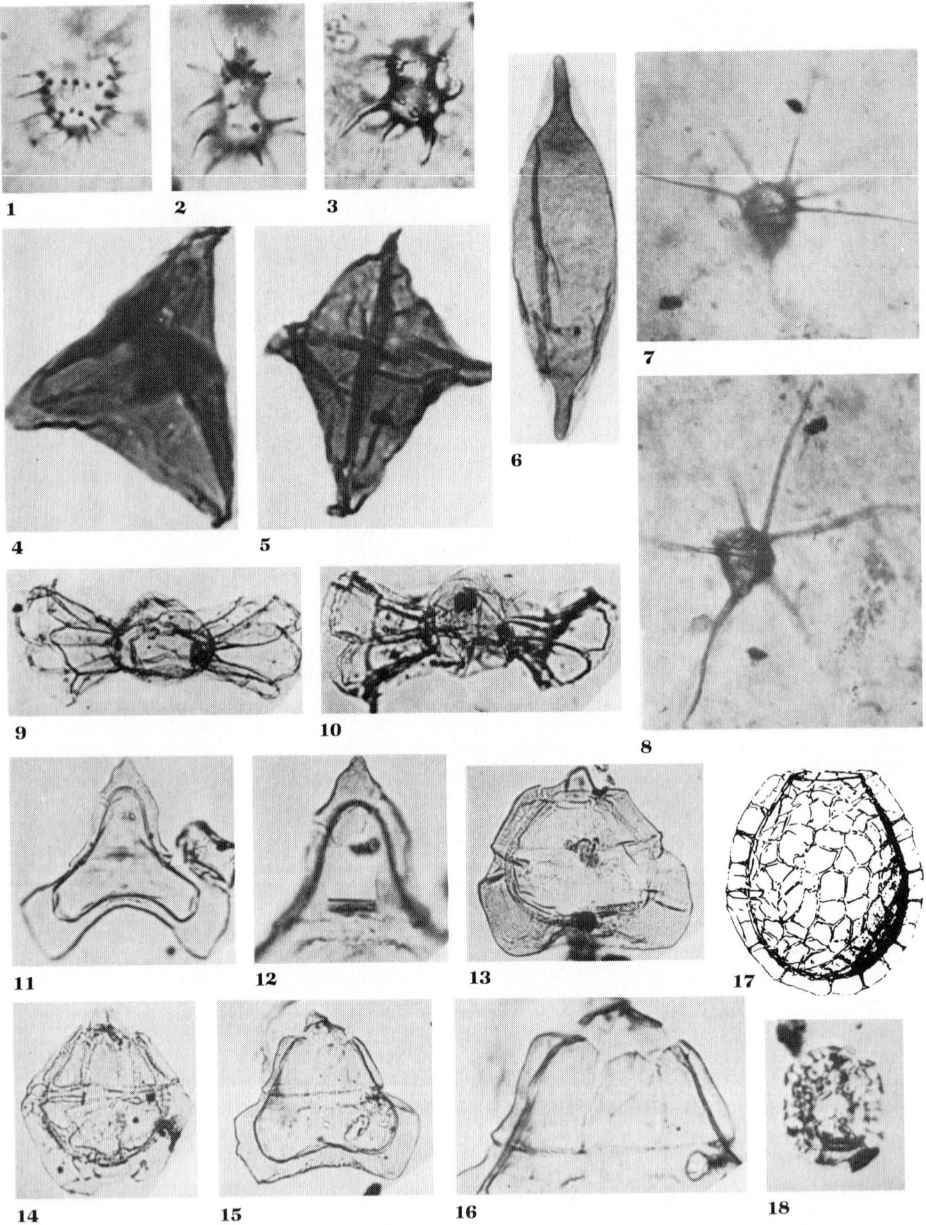

Figure 4.155

1 – 3. *Diacrocanthidium echinulatum* (Deflandre) Loeblich Jr. & Loeblich III, Upper Cretaceous, France, ×1080, from Foucher, 1974. **1.** With apical archeopyle. **4,5.** *Palaeotetradinium silicorum* Deflandre, Upper Cretaceous, Germany, ×1360, from Morgenroth, 1968. **4.** With inner body. **6.** *Diplofusa gearlensis* Cookson & Eisenack, Cretaceous, Australia, ×432, from Cookson and Eisenack, 1960a. **7,8.** *Raphidodinium fucatum* Deflandre, Cretaceous, France, ×360, from Deflandre, 1936. **9,10.** *Codoniella campanulata* (Cookson & Eisenack) Downie & Sarjeant, Upper Cretaceous, Australia, ×320, from Cookson and Eisenack, 1960a. **11 – 16.** *Sirmiodinium grossi* Alberti, Upper Jurassic to Lower Cretaceous, California, from Warren, 1973; 11 – 13, ventral views; 14 – 16, dorsal views. **12.** Enlargement of apex in part 11, showing precingular opercular piece folded back over girdle area, apical segment slightly displaced, ×455. **16.** Enlargement of apex in part 15, showing slightly displaced precingular and apical opercular pieces, ×585. 11,13 – 15, ×288. **17.** *Palaeostomocystis reticulata* Deflandre, Cretaceous, France, showing apical archeopyle, ×1720, from Deflandre, 1937b. **18.** *Stephanelytron redcliffense* Sarjeant, Jurassic, England, with skirtlike corona at base, ×400, from Sarjeant, 1961a.

REFERENCES

Abé, T. H., Notes on the protozoan fauna of Mutsu Bay. III Subgenus *Protoperidinium* Genus *Peridinium Sci. Rep. Tôhoku. Univ., Ser. 4 Biol.,* v. 11, p. 19 – 48, figs. 1 – 50, 1936.

Abé, T. H., The armoured Dinoflagellata: II Prorocentridae and Dinophysidae (A.). *Publs. Seto Mar. Biol. Lab.* 14, p. 369 – 389, 1967.

Afzelius, B. A., The nucleus of *Noctiluca scintillans*. Aspects of nucleocytoplasmic exchange and the formation of nuclear membrane. *J. Cell Biol.,* v. 19, p. 229 – 238, figs. 1 – 6, 1962.

Agelopoulos, Johann, *Hystrichosphären, Dinoflagellaten und Foraminiferen aus dem eozänen Kieselton von Heiligenhafen, Holstein.* Dissn. Math. -Naturw. Fak. Eberhard-Karls-Univ. Tübingen, p. 1 – 74, 1967.

Akinina, D. K., Otnositel'naya skorost' osedaniya dinoflagellat v svyazi so skorost'yu ikh deleniya [Relative velocity of settling of Dinoflagellata as dependent on their division rates]. *Okeanologiya,* v. 9, no. 2, p. 301 – 305, figs. 1 – 4, 1969.

Alberti, Gerhard, Über *Pseudodeflandrea* n. gen. (Dinoflag.) aus dem Mittel-Oligozän von Deutschland. *Mitt. Geol. StInst. Hamb.,* v. 28, p. 91 – 92, 1 fig., 1959a.

Alberti, Gerhard, Zur Kenntnis der Gattung *Deflandrea* Eisenack (Dinoflag.) in der Kreide und im Alttertiär Nord-und Mitteldeutschlands. *Mitt. Geol. StInst. Hamb.,* v. 28, p. 93 – 105, pls. 8, 9, 1 fig., 1959b.

Alberti, Gerhard, Zur Kenntnis mesozoischer und alttertiärer Dinoflagellaten und Hystrichosphaerideen von Nord-und Mitteldeutschland sowie einigen anderen europäischen Gebieten. *Palaeontographica Abt. A,* v. 116, p. 1 – 58, 12 pls., 1961.

Aldrich, D. V., Photoautotrophy in *Gymnodinium breve* Davis. *Science,* v. 137, p. 988 – 990, 1962.

Aldrich, D. V., S. M. Ray, and W. B. Wilson, *Gonyaulax monilata:* population growth and development of toxicity in cultures. *J. Protozool.,* v. 14, p. 636 – 639, figs. 1 – 3, 1967.

Alldredge, A. L., and B. M. Jones, *Hastigerina pelagica:* foraminiferal habitat for planktonic dinoflagellates. *Mar. Biol.,* v. 22, p. 131 – 135, figs. 1, 2, 1973.

Allen, J. R., T. M. Roberts, R. C. Tuttle, L. C. Klotz, and A. R. Loeblich III, Dinoflagellate DNA organization and chromosome structure. *J. Protozool.,* v. 21, p. 426 (Abstr.), 1974.

Allen, J. R., T. M. Roberts, A. R. Loeblich III, and L. C. Klotz, Characterization of the DNA from the dinoflagellate *Crypthecodinium cohnii* and implications for nuclear organization. *Cell,* v. 6, p. 161 – 169, 1975.

Allman, G. J., Observations on *Aphanizomenon flosaquae,* and a species of Peridinea. *Q. Jl. Microsc. Sci.,* v. 3, p. 21 – 25, 1855.

Anderson, O. R., and A. W. H. Bé, The ultrastructure of a planktonic foraminifer, *Globigerinoides sacculifer* (Brady), and its symbiotic dinoflagellates. *J. Foramin. Res.,* v. 6, p. 1 – 21, pls. 1 – 9, 1976.

Antonescu, Emanuel, Un nouveaux genre de dinoflagellé dans le Jurassique Moyen de Roumanie. *Rev. Micropaléont.,* v. 17, p. 61 – 65, pl. 1, 1975.

Apstein, C., Knospung bei *Ceratium tripos* var. *subsalsa. Int. Revue Ges. Hydrobiol. Hydrogr.,* v. 3, p. 34 – 36, figs. 1 – 8, 1910.

Apstein, C., Biologische Studie über *Ceratium tripos* var. *subsalsa* Ostf. *Wiss. Meeresunters., Abt. Kiel,* v. 12, p. 137−162, 10 figs., 1911.

Archangelsky, Sergio, Sobre el paleomicroplancton del Terciario inferior de Rio Turbio, Prov. de Santa Cruz. *Ameghiniana,* v. 5, p. 406−416, 2 pls., 1968.

Baker, Henry, Of luminous water insects, p. 399−403, pl. 15. *In* Henry Baker, *Employment for the microscope.* London: R. Dodsley, 1753.

Balderston, W. L., and George Claus, A study of the symbiotic relationship between *Symbiodinium microadriaticum* Freudenthal, a zooxanthella and the upside down jellyfish, *Cassiopea* sp. *Nova Hedwigia,* v. 17, p. 373−383, pls. 95, 96, 1969.

Balech, Enrique, Estudio de *"Ceratocorys horrida"* Stein var. *"extensa"* Pavillard. *Physis, B. Aires,* v. 20, p. 165−173, 1 pl., 1949.

Balech, Enrique, Étude des Dinoflagellés du sable de Roscoff. *Revue Algol.,* n.s., v. 2, p. 29−52, figs. 1−66, 1956.

Balech, Enrique, Two new genera of dinoflagellates from California. *Biol Bull. Mar. Biol. Lab., Woods Hole,* v. 116, p. 195−203, figs. 1−3, 1959.

Balech, Enrique, El plancton de Mar del Plata durante el periódo 1961−1962 (Buenos Aires, Argentina). *Boln. Inst. Biol. Mar., Mar del Plata,* no. 4, p. 1−49, pls. 1−5, 1964.

Balech, Enrique, Observaciónes sobre dinoflagelados fosiles. *Comun. Mus. Argent. Cienc. Nat. Bernardino Rivadavia, Paleont.,* v. 1, no. 2, p. 5−20, figs. 1−6, 1967a.

Balech, Enrique, Dinoflagelados nuevos o interesantes del Golfo de Mexico y Caribe. *Revta Mus. Argent. Cienc. Nat. Bernardino Rivadavia Inst. Nac. Invest. Cienc. Nat., Hidrobiol.,* v. 2, no. 3, p. 77−126, pls. 1−9, 1967b.

Balech, Enrique, Microplancton del Atlantico Ecuatorial Oeste (Equalant I). *Publnes. Serv. Hidrogr. Nav. Repub. Argent.,* H 654, 103 p., 12 pls., 1971a.

Balech, Enrique, Microplancton de la Campaña Oceanografica Productividad III. *Revta Mus. Argent. Cienc. Nat. Bernardino Rivadavia Inst. Nac. Invest. Nat., Hidrobiol.,* v. 3, no. 1, p. 1−202, 39 pls., 1971b.

Balech, Enrique, El genero *"Protoperidinium"* Bergh, 1881 (*"Peridinium"* Ehrenberg, 1831, partim). *Revta Mus. Argent. Cienc. Nat. Bernardino Rivadavia Inst. Nac. Invest. Nat., Hidrobiol.,* v. 4, no. 1, p. 1−79, 1974.

Ballantine, Dorothy, and B. C. Abbott, Toxic marine flagellates: their occurrence and physiological effects on animals. *J. Gen. Microbiol.,* v. 16, p. 274−281, figs. 1−3, 1957.

Barker, H. A., The culture and physiology of the marine dinoflagellates. *Arch. Mikrobiol.,* v. 6, p. 157−181, figs. 1−11, 1935.

Baumeister, Willy, Dinophyceen aus perennierenden Gewässern des Schwingrasenmoores bei Burgberg, sowie aus Alpsee, Freibergsee und dem Moorweiher in Oberstdorf (Allgäu). *Arch. Protistenk.,* v. 106, p. 535−552, figs. 1−10, 1963.

Beam, C. A., and Marion Himes, Evidence for sexual fusion and recombination in the dinoflagellate *Crypthecodinium* (*Gyrodinium*) *cohnii. Nature,* v. 250, p. 435−436, figs. 1, 2, 1974.

Benedek, P. N., Phytoplanktonten aus dem Mittel- und Oberoligozän von Tönisberg (Niederrheingebiet). *Palaeontographica Abt. B,* v. 137, p. 1−71, pls. 1−16, 28 figs., 1972.

Benson, D. G., Jr., Dinoflagellate taxonomy and biostratigraphy at the Cretaceous-Tertiary boundary, Round Bay, Maryland. *Tulane Stud. Geol. Paleont.,* v. 12, p. 169−233, pls. 1−15, figs. 1−4, 1976.

Bergh, R. S., Der Organismus der Cilioflagellaten. Eine phylogenetische Studie. *Morph. Jb.,* v. 7, p. 177−288, pls. 12−16, 1881.

Bergh, R. S., Ueber den Theilungsvorgang bei den Dinoflagellaten. *Zool. Jb.,* v. 2, p. 73−86, pl. 5, 1887.

Beyers, R. J., Metabolic similarities between symbiotic coelenterates and aquatic ecosystems. *Arch. Hydrobiol.,* v. 62, p. 273−284, 1966.

Bibby, B. T., and J. D. Dodge, The encystment of a freshwater dinoflagellate: a light and electron microscopical study. *Br. Phycol. J.,* v. 7, p. 85−100, figs. 1−14, 1972.

Bibby, B. T., and J. D. Dodge, The fine structure of the chloroplast nucleoid in *Scrippsiella sweeneyae* (Dinophyceae). *J. Ultrastruct. Res.,* v. 48, p. 153−161, figs. 1−9, 1974.

Biecheler, Berthe, Sur un dinoflagellé à capsule périnucléaire, *Plectodinium,* n. gen. *nucleovolvatum,* n. sp. et sur les relations des péridiniens avec les radiolaires. *C. R. Hebd. Séanc. Acad. Sci., Paris,* v. 198, p. 404−406, 3 figs., 1934a.

Biecheler, Berthe, Sur le réseau argentophile et la morphologie de quelques péridiniens nus. *C. R. Séanc. Soc. Biol.,* v. 115, p. 1039−1042, 1 fig., 1934b.

Biecheler, Berthe, Recherches sur les Péridiniens. *Suppl. Bull. Biol. Fr. Belg.* 36, vi + 149 p., 1952.

Blanchard-Babillot, Claude, Recherches sur le mode de division du péridinien *Glenodinium foliaceum. C. R. Hebd. Séanc. Acad. Sci., Paris,* v. 275, sér. D, p. 2635−2638, pls. 1−4, 1972.

Böhm, Anton, Zur Verbreitung einiger Dinoflagellaten im Südatlantik. *Bot. Arch.,* v. 35, p. 397−407, figs. 1−11, 1933.

Boltovskoy, Andrés, Formacion del arqueopilo en tecas di dinoflagelados. *Revta. Esp. Micropaleont.,* v. 5, p. 81−98, pls. 1−3, 1973a.

Boltovskoy, Andrés, *Peridinium gatunense* Nygaard. Estructura y estereoultraestructura tecal (Dinoflagellida). *Physis, Sec. B, B. Aires,* v. 32 (85), p. 331—344, 1973b.

Boltovskoy, Andrés, Estructura y estereoultraestructura tecal de dinoflagellados. II. *Peridinium cinctum* (Müller) Ehrenberg. *Physis, Sec. B, B. Aires,* v. 34 (89), p. 73—84, figs. 1—38, 1975.

Borgert, Adolf, Kern- und Zellteilung bei marinen *Ceratium*-Arten. *Arch. Protistenk.,* v. 20, p. 1—46, pls. 1—3, 1910.

Borgert, Adolf, Eine neue Form der Mitose bei Protozoen. *Verh. Int. Congr. Zool. VIII, Graz,* p. 408—418, 1912.

Bouck, G. B., and B. M. Sweeney, The fine structure and ontogeny of trichocysts in marine dinoflagellates. *Protoplasma,* v. 61, p. 205—223, figs. 1—16, 1966.

Bourrelly, Pierre, Notes sur les péridiniens d'eau douce. *Protistologica,* v. 4, p. 5—14, pls. 1, 2, 1968.

Bourrelly, Pierre, *Les algues d'eau douce. Initiation à la systématique.* Tome III: *Les algues bleues et rouges. Les Eugléniens, Peridiniens et Cryptomonadines.* Paris: Ed. N. Boubée and Cie, 512 p., 1970.

Bourrelly, Pierre, and Alain Couté, Observations en microscopie électronique à balayage des *Ceratium* d'eau douce (Dinophycées). *Phycologia,* v. 15, p. 329—338, 1976.

Braarud, Trygve, Morphological observations on marine dinoflagellate cultures (*Porella perforata, Goniaulax tamarensis, Protoceratium reticulatum*). *Avh. Norske VidenskAkad. Oslo, Math.-Naturv. Kl.,* 1944, no. 11, p. 1—18, pls. 1—4, 1945.

Braarud, Trygve, Taxonomical studies of marine dinoflagellates. *Nytt Mag. Naturvid.,* v. 88, p. 43—48, 1951.

Braarud, Trygve, A red water organism from Walvis Bay (*Gymnodinium galatheanum* n. sp.). *Galathea Rep.,* v. 1, p. 137—138, figs. 1, 2, 1957.

Braarud, Trygve, Cultivation of marine organisms as a means of understanding environmental influences on populations, p. 271—298, figs. 1—8. *In* M. Sears, *Oceanography.* Washington: Amer. Assoc. Adv. Sci., Publ. 67, 1961.

Braarud, Trygve, Species distribution in marine phytoplankton. *J. Oceanogr. Soc. Japan, 20th Anniv. Vol.,* p. 628—649, figs. 1—6, 1962.

Braarud, Trygve, and B. R. Heimdal, Brown water on the Norwegian coast in autumn 1966. *Nytt Mag. Bot.,* v. 17, p. 91—97, figs. 1—4, 1970.

Bradford, M. R., New dinoflagellate cyst genera from the Recent sediments of the Persian Gulf. *Can. J. Bot.,* v. 53, p. 3064—3074, figs. 1—28, 1975.

Brideaux, W. W., Recurrent species groupings in fossil microplankton assemblages. *Palaeogeogr. Palaeoclimat. Palaeoecol.,* v. 9, p. 101—122, pls. 1—4, figs. 1—7, 1971a.

Brideaux, W. W., Palynology of the Lower Colorado Group central Alberta, Canada I. Introductory remarks. Geology, and microplankton studies. *Palaeontographica Abt. B,* v. 135, p. 53—114, pls. 21—30, text figs. 1—10, 1971b.

Brideaux, W. W., Taxonomic note: redefinition of the genus *Broomea* and its relationship to *Batioladinium* gen. nov. (Cretaceous). *Can. J. Bot.,* v. 53, p. 1239—1243, figs. 1—3, 1975.

Brideaux, W. W., and D. J. McIntyre, *Lunatadinium dissolutum* gen. et sp. nov., a dinoflagellate cyst from Lower Cretaceous rocks, Yukon Territory and Northern District of MacKenzie. *Bull. Can. Petrol. Geol.,* v. 21, p. 395—402, pl. 1, fig. 1, 1973.

Brideaux, W. W., and D. J. McIntyre, Miospores and microplankton from Aptian-Albian rocks along Horton River, District of Mackenzie. *Bull. Can. Geol. Surv.,* 252 (1975), 85 p., 14 pls., 1976.

Brokaw, C. J., and Leigh Wright, Bending waves of the posterior flagellum of *Ceratium. Science,* v. 142, p. 1169—1170, figs. 1, 2, 1963.

Brosius, Marita, Plankton aus dem nordhessischen Kasseler Meeressand (Oberoligozän). *Z. Dt. Geol. Ges.,* v. 114, p. 32—56, pls. 1—8, figs. 1, 2, 1963.

Buchanan, R. J., Studies at Oyster Bay in Jamaica, West Indies. IV. Observations on the morphology and asexual cycle of *Pyrodinium bahamense* Plate. *J. Phycol.,* v. 4, p. 272—277, pls. 1, 2, fig. 1, 1968.

Bujak, J. P., and M. J. Fisher, Dinoflagellate cysts from the Upper Triassic of Arctic Canada. *Micropaleontology,* v. 22, p. 44—70, pls. 1—9, 1976.

Burgess, J. D., Palynological interpretation of Frontier environments in central Wyoming. *Geosci. Man,* v. 3, p. 69—82, pl. 1, figs. 1—13, 1971.

Bursa, A. S., Some morphogenetic factors in taxonomy of dinoflagellates. *Grana Palynol.,* v. 3 (1962), p. 54—66, figs. 1—68, 1963.

Bursa, A. S., *Actiniscus canadensis* n. sp., *A. pentasterias* Ehrenberg v. *arcticus* n. var., *Pseudoactiniscus apentasterias* n. gen., n. sp. Marine relicts in Canadian Arctic lakes. *J. Protozool.,* v. 16, p. 411—418, figs. 1—22, 1969.

Bursa, A. S., Morphogenesis and taxonomy of fossil and contemporary Dinophyta secreting discoasters, p. 129—143, pls. 1, 2, figs. 1—22. *In* A. Farinacci, *Proc. II Planktonic Conference, Roma 1970,* v. 1. Rome: Edizioni Tecnoscienza. 1971.

Bütschli, Otto, Einiges über Infusorien. *Arch. Mikrosk. Anat.,* v. 9, p. 657—678, pl. 25, 26, 1873.

Bütschli, Otto, Abtheilung II: Mastigophora, p. 865—1088. In *Dr. H. G. Bronn's Klassen und Ordnungen des Thier-Reichs, Wissenschaftlich dargestellt in Wort und Bild,* v. 1, *Protozoa.* Leipzig: C. F. Winter'sche Verlagshandlung, 1885.

Cachon, Jean, Contribution à l'étude des Péridiniens parasites. Cytologie, Cycles Évolutifs. *Annls. Sci. Nat.,* v. 6, ser. 12, p. 1—158, pls. 1—36, 1964.

Cachon, Jean, and Monique Cachon-Enjumet, *Atlanticellodinium tregouboffi* nov. gen. nov. sp., Péridinien Blastuloidae Neresheimer, parasite de *Planktonetta atlantica* Borgert, Phaeodarié Atlanticellidae. Cytologie, cycle biologique, évolution nucléaire au cours de la sporogenèse. *Archs. Zool. Exp. Gen.,* v. 105, p. 369—379, pl. 1, figs. 1—5, 1965.

Cachon, Jean, and Monique Cachon, *Cymbodinium elegans* nov. gen. nov. sp., péridinien Noctilucidae Saville-Kent. *Protistologica,* v. 3, p. 313—318, pl. 1, figs. 1, 2, 1967a.

Cachon, Jean, and Monique Cachon, Contribution à l'étude des Noctilucidae Saville-Kent. I. Les Kofoidininae Cachon J. et M. Évolution morphologique et systématique. *Protistologica,* v. 3, p. 427—444, pls. 1—5, figs. 1—9, 1967b.

Cachon, Jean, and Monique Cachon, *Filodinium hovassei* nov. gen. nov. sp., péridinien phorétique d'Appendiculaires. *Protistologica,* v. 4, p. 15—18, 1 fig., 1968a.

Cachon, Jean, and Monique Cachon, Cytologie et cycle évolutif des *Chytriodinium* (Chatton). *Protistologica,* v. 4, p. 249—262, pls. 1, 2, figs. 1—6, 1968b.

Cachon, Jean, and Monique Cachon, Contribution à l'étude des Noctilucidae Saville-Kent. Évolution morphologique, cytologie, systématique. II. Les Leptodiscinae Cachon J. et M. *Protistologica,* v. 5, p. 11—33, figs. 1—10, 1969.

Cachon, Jean, and Monique Cachon, Ultrastructures du genre *Oodinium* Chatton. Différenciations cellulaires en rapport avec la vie parasitaire. *Protistologica,* v. 7, p. 153—169, 14 figs., 1971a.

Cachon, Jean, and Monique Cachon, *Protoodinium chattoni* Hovasse manifestations ultrastructurales des rapports entre le Péridinien et la Méduse-hôte: fixation, phagocytose. *Arch. Protistenk.,* v. 113, p. 293—305, figs. 1—10, 1971b.

Cachon, Jean, Monique Cachon, and F. Bouquaheux, *Myxodinium pipiens* gen. nov., sp. nov., péridinien parasite d'*Halosphaera. Phycologia,* v. 8 (1969), p. 157—164, figs. 1—5, 1970a.

Cachon, Jean, Monique Cachon, and Claude Greuet, Le système pusulaire de quelques péridiniens libres ou parasites. *Protistologica,* v. 6, p. 467—476, 7 figs., 1970b.

Calandra, François, Sur un présumé Dinoflagellé, *Arpylorus* nov. gen. du Gothlandien de Tunisie. *C. R. Hebd. Séanc. Acad. Sci., Paris,* v. 258, p. 4112—4114, figs. 1—3, 1964.

Caro, Yves, Contribution à la connaissance des dinoflagelles du Paleocene-Eocene Inférieur des Pyrenees Espagnoles. *Revta Esp. Micropaleont.,* v. 5, p. 329—372, pls. 1—5, figs. 1—4, 1973.

Carter, H. J., Note on the red colouring matter of the sea round the shores of the island of Bombay. *Ann. Mag. Nat. Hist.,* ser. 3, v. 1, p. 258—262, 1858.

Chatton, Édouard, Les péridiniens parasites. Morphologie, reproduction, ethologie. *Archs. Zool. Exp. Gen.,* v. 59, p. 1—475, pls. 1—18, 1920.

Chatton, Édouard, *Pheopolykrikos beauchampi* nov. gen., nov. sp. dinoflagellé polydinide autotrophe, dans l'Etang de Thau. *Bull. Soc. Zool. Fr.,* v. 58, p. 251—255, 1 fig., 1933.

Chatton, Édouard, L'origine péridinienne des Radiolaires et l'interprétation parasitaire de l'anisosporogénèse. *C. R. Hebd. Séanc. Acad. Sci., Paris,* v. 198, p. 309—312, 1934.

Chatton, Édouard, Classe des dinoflagelles ou Péridiniens, p. 309—390. *In* P. P. Grassé, *Traité de Zoologie, Anatomie, Systématique, Biologie,* v. 1, pt. 1, Protozoaires, Généralités, Flagellés, Paris: Masson et Cie, 1952.

Chatton, Édouard, and Berthe Biecheler, Documents nouveaux relatifs aux Coccidinides (Dinoflagellés parasites). La sexualité du *Coccidinium Mesnili* n. sp. *C. R. Hebd. Séanc. Acad. Sci., Paris,* v. 203, p. 573—576, 1 fig., 1936.

Churchill, D. M., Living and fossil unicellular algae and aplanospores. *Nature,* v. 186, p. 493—494, fig. 1, 1960.

Churchill, D. M., and W. A. S. Sarjeant, Fossil dinoflagellates and hystrichospheres in Australian freshwater deposits. *Nature,* v. 194, p. 1094, fig. 1, 1962.

Churchill, D. M., and W. A. S. Sarjeant, Freshwater microplankton from Flandrian (Holocene) peats of southwestern Australia. *Grana Palynol.,* v. 3 (1962), p. 29—53, fig. 1—35, 1963.

Claparède, Édouard, and Johannes Lachmann, Etudes sur les Infusoires et les Rhizopodes. *Mém. Inst. Natn. Génev.,* v. 6, p. 392—484, pls. 14—24, 1859.

Clarke, K. J., and N. C. Pennick, Flagellar scales in *Oxyrrhis marina* Dujardin. *Br. Phycol. J.,* v. 7, p. 357—360, figs. 1—8, 1972.

Clarke, K. J., and N. C. Pennick, The occurrence of body scales in *Oxyrrhis marina* Dujardin. *Br. Phycol. J.,* v. 11, p. 345—348, figs. 1—10, 1976.

Clarke, R. F. A., R. J. Davey, W. A. S. Sarjeant, and J. P. Verdier, A note on the nomenclature of some Upper Cretaceous and Eocene dinoflagellate taxa. *Taxon,* v. 17, p. 181—183, 1968.

Clarke, R. F. A., and J. P. Verdier, An investigation of microplankton assemblages from the Chalk of the Isle of Wight, England. *Verh. K. Ned. Akad. Wet., Afd. Natuurk.,* ser. 1, v. 24, no. 3, p. 1—96, pls. 1—17, 1967.

Conrad, Walter, Quelques microfossiles des silex cretacés. *Bull. Mus. R. Hist. Nat. Belg.,* v. 17, no. 36, p. 1—10, pl. 1, figs. 1—3, 1941.

Cookson, I. C., Records of the occurrence of *Botryococcus braunii, Pediastrum* and the Hystrichosphaerideae in Cainozoic deposits of Australia. *Mem. Natn. Mus., Melb.,* no. 18, p. 107–123, pls. 1, 2, 1953.

Cookson, I. C., Additional microplankton from Australian Late Mesozoic and Tertiary sediments. *Aust. J. Mar. Freshwat. Res.,* v. 7, p. 183–191, pls. 1, 2, 1956.

Cookson, I. C., Cretaceous and Tertiary microplankton from south-eastern Australia. *Proc. R. Soc. Vict.,* v. 78, p. 85–93, pls. 9–11, 1965a.

Cookson, I. C., Microplankton from the Paleocene Pebble Point Formation, south-western Victoria. *Proc. R. Soc. Vict.,* v. 78, p. 137–141, pls. 24, 25, 1965b.

Cookson, I. C., and Alfred Eisenack, Microplankton from Australian and New Guinea Upper Mesozoic sediments. *Proc. R. Soc. Vict.,* v. 70, p. 19–79, pls. 1–12, figs. 1–20, 1958.

Cookson, I. C., and Alfred Eisenack, Microplankton from Australian Cretaceous sediments. *Micropaleontology,* v. 6, p. 1–18, pls. 1–3, 1960a.

Cookson, I. C., and Alfred Eisenack, Upper Mesozoic microplankton from Australia and New Guinea. *Palaeontology,* v. 2, p. 243–261, pls. 37–39, 1960b.

Cookson, I. C., and Alfred Eisenack, Tertiary microplankton from the Rottnest Island Bore, Western Australia. *J. Proc. R. Soc. West. Aust.,* v. 44, p. 39–47, pls. 1, 2, fig. 1, 1961a.

Cookson, I. C., and Alfred Eisenack, Upper Cretaceous microplankton from the Belfast No. 4 Bore, South-western Victoria. *Proc. R. Soc. Vict.,* v. 74, p. 69–76, pls. 11, 12, 1 fig., 1961b.

Cookson, I. C., and Alfred Eisenack, Additional microplankton from Australian Cretaceous sediments. *Micropaleontology,* v. 8, p. 485–507, pls. 1–7, figs. 1, 2, 1962.

Cookson, I. C., and Alfred Eisenack, Microplankton from the Browns Creek Clays, S. W. Victoria. *Proc. R. Soc. Vict.,* v. 79, p. 119–131, pls. 11–15, 1965a.

Cookson, I. C., and Alfred Eisenack, Microplankton from the Dartmoor Formation, S. W. Victoria. *Proc. R. Soc. Vict.,* v. 79, p. 133–137, pls. 16, 17, 1965b.

Cookson, I. C., and Alfred Eisenack, Microplankton from the Paleocene Pebble Point Formation, south-western Victoria Part 2. *Proc. R. Soc. Vict.,* v. 79, p. 139–146, pls. 18, 19, figs. 1–3, 1965c.

Cookson, I. C., and Alfred Eisenack, Some Early Tertiary microplankton and pollen grains from a deposit near Strahan, western Tasmania. *Proc. R. Soc. Vict.,* v. 80, p. 131–140, pls. 17–21, 1967.

Cookson, I. C., and Alfred Eisenack, Microplankton from two samples from Gingin Brook No. 4 Borehole, Western Australia. *J. Proc. R. Soc. West. Aust.,* v. 51, p. 110–122, figs. 1–6, 1968.

Cookson, I. C., and Alfred Eisenack, Some microplankton from two bores at Balcatta, Western Australia.

J. Proc. R. Soc. West. Aust., v. 52, p. 3–8, figs. 1, 2, 1969.

Cookson, I. C., and Alfred Eisenack, Mikroplankton aus Australischen Mesozoischen und Tertiären Sedimenten. *Palaeontographica Abt. B,* v. 148, p. 44–93, pl. 20–29, figs. 1–3, 1974.

Cookson, I. C., Alfred Eisenack, and B. S. Ingram, Cretaceous microplankton from the Eucla Basin, Western Australia. *Proc. R. Soc. Vict.,* v. 83, p. 137–157, pls. 10–14, 1970.

Cookson, I. C., Alfred Eisenack, and B. S. Ingram, Cretaceous microplankton from Eyre No. 1 Bore Core 20, Western Australia. *Proc. R. Soc. Vict.,* v. 84, p. 217–226, pls. 7–11, 1971.

Cookson, I. C., and N. F. Hughes, Microplankton from the Cambridge Greensand (mid-Cretaceous). *Palaeontology,* v. 7, p. 37–59, pls. 5–11, 1964.

Corradini, D., Non-calcareous microplankton from the Upper Cretaceous of the northern Apennines. *Boll. Soc. Paleont. Ital.,* v. 11 (1972), p. 119–197, pls. 19–39, figs. 1–9, 1973.

Costa, L. I., and Charles Downie, The distribution of the dinoflagellate *Wetzeliella* in the Palaeogene of north-western Europe. *Palaeontology,* v. 19, p. 591–614, pl. 92, 1976.

Cox, R. L., Dinoflagellate cyst structures: walls, cavities, and bodies. *Palaeontology,* v. 14, p. 22–33, pl. 7, figs. 1–4, 1971.

Crawford, R. M., J. D. Dodge, and C. M. Happey, The dinoflagellate genus *Woloszynskia.* I. Fine structure and ecology of *W. tenuissimum* from Abbot's Pool, Somerset. *Nova Hedwigia,* v. 19, p. 825–840, figs. 1–14, 1971.

Cridland, A. A., A note on two freshwater dinoflagellates. *New Phytol.,* v. 57, p. 285–287, 1958.

Cross, A. T., G. C. Thompson, and J. B. Zaitzeff, Source and distribution of palynomorphs in bottom sediments, southern part of Gulf of California. *Mar. Geol.,* v. 4, p. 467–524, figs. 1–14, 1966.

Daday, Eugen von, Untersuchungen über die Süsswasser-Mikrofauna Paraguays. *Zoologica, Stuttgart* 18 (44), 374 p., 23 pls., 1905.

Dale, Barrie, New observations on *Peridinium faeroense* Paulsen (1905), and classification of small orthoperidinioid dinoflagellates. *Br. Phycol. J.,* v. 12, p. 241–253, figs. 1–35, 1977.

Damassa, S. P., Eocene dinoflagellates from the Coastal Belt of the Franciscan Complex, northern California. *J. Paleont.,* v. 53, p. 815–840, 8 pls., 8 figs., 1979.

Davey, R. J., Some dinoflagellate cysts from the Upper Cretaceous of northern Natal, South Africa. *Palaeont. Afr.,* v. 12, p. 1–23, pls. 1–4, 1969a.

Davey, R. J., The evolution of certain Upper Cretaceous hystrichospheres from South Africa. *Palaeont. Afr.,* v. 12, p. 25–51, pls. 1–4, figs. 1–3, 1969b.

Davey, R. J., Non-calcareous microplankton from the Cenomanian of England, northern France and North America. Part I. *Bull. Br. Mus. Nat. Hist., Geology*, v. 17, p. 103−180, 11 pls., 16 figs., 1969c.

Davey, R. J., Non-calcareous microplankton from the Cenomanian of England, northern France and North America. Part II. *Bull. Br. Mus. Nat. Hist., Geology*, v. 18, p. 333−397, 10 pls., 9 figs., 1970.

Davey, R. J., Charles Downie, W. A. S. Sarjeant, and G. L. Williams, Studies on Mesozoic and Cainozoic dinoflagellate cysts. *Bull. Br. Mus. Nat. Hist., Geology*, Suppl. 3, p. 1−248, pls. 1−26, figs. 1−64, 1966.

Davey, R. J., Charles Downie, W. A. S. Sarjeant, and G. L. Williams, Appendix to "Studies on Mesozoic and Cainozoic dinoflagellate cysts." *Bull. Br. Mus. Nat. Hist., Geology*, Appendix to Suppl. 3, p. 1−24, 1969.

Davey, R. J., and J. P. Verdier, An investigation of microplankton assemblages from the Albian of the Paris Basin. *Verh. K. Ned. Akad. Wet., Afd. Natuurk.*, ser. 1, v. 26(2), 58 p., 7 pls., 1971.

Davey, R. J., and J. P. Verdier, An investigation of microplankton assemblages from latest Albian (Vraconian) sediments. *Revta. Esp. Micropaleont.*, v. 5, p. 173−212, pls. 1−5, figs. 1−12, 1973.

Davey, R. J., and J. P. Verdier, Dinoflagellate cysts from the Aptian type sections at Gargas and La Bédoule, France. *Palaeontology*, v. 17, p. 623−653, pls. 91−93, 1974.

Davey, R. J., and J. P. Verdier, A review of certain non-tabulate Cretaceous hystrichospherid dinocysts. *Rev. Palaeobot. Palynol.*, v. 22, p. 307−335, pls. 1−4, 1976.

Davey, R. J., and G. L. Williams, The genera *Hystrichosphaera* and *Achomosphaera*. *Bull. Br. Mus. Nat. Hist., Geology*, Suppl. 3, p. 28−52, pls. 1−5, 1966a.

Davey, R. J., and G. L. Williams, The genus *Hystrichosphaeridium* and its allies. *Bull. Br. Mus. Nat. Hist., Geology*, Suppl. 3, p. 53−106, pls. 6−12, 1966b.

Deane, Henry, On the existence of fossil *Xanthidia* in the Chalk. *Ann. Mag. Nat. Hist.*, v. 16, p. 64−65, 1845.

Deane, Henry, On the occurrence of fossil *Xanthidia* and Polythalamia in Chalk. *Trans. Microsc. Soc. Lond.*, v. 2, p. 77−79, pl. 9, 1849.

DeConinck, Jan, Dinophyceae et Acritarcha de l'Yprésien du Sondage de Kallo. *Mém. Inst. R. Sci. Nat. Belg.*, no. 161, 67 p., 17 pls., 2 figs., 1968.

Deflandre, Georges, Note préliminaire sur un péridinien fossile *Lithoperidinium oamaruense* n.g. n.sp. *Bull. Soc. Zool. Fr.*, v. 58, p. 265−273, figs. 1−7, 1933.

Deflandre, Georges, Sur la structure des flagelles. *Annls. Protist.*, v. 4, p. 31−53, pls. 1−5, 1934a.

Deflandre, Georges, Existence, sur les flagelles, de filaments latéraux ou terminaux (mastigonèmes). *C. R. Hebd. Séanc. Acad. Sci., Paris*, v. 198, p. 497−499, 4 figs., 1934b.

Deflandre, Georges, Considérations biologiques sur les micro-organismes d'origine planctonique conservés dans les silex de la craie. *Bull. Biol. Fr. Belg.*, v. 69, p. 213−244, pls. 5−9, figs. 1−11, 1935.

Deflandre, Georges, Microfossiles des silex Crétacés. Première Partie. Généralités. Flagellés. *Annls. Paléont.*, v. 25, p. 151−191, pls. 1−10, 1936.

Deflandre, Georges, *Phanerodinium*, genre nouveau de dinoflagellé fossile des silex. *Bull. Soc. Fr. Microsc.*, v. 6, p. 109−115, 5 figs., 1937a.

Deflandre, Georges, Microfossiles des silex Crétacés. Deuxième partie. Flagellés *incertae sedis*, Hystrichosphaeridés, Sarcodinés, Organismes divers. *Annls. Paléont.*, v. 26, p. 51−103, pls. 8−15, 1937b.

Deflandre, Georges, Microplancton des mers jurassiques conservé dans les marnes de Villers-Sur-Mer (Calvados), étude liminaire et considérations générales. *Trav. Stn. Zool. Wimereux*, v. 13, p. 147−200, pls. 5−11, 1938.

Deflandre, Georges, Sur les Dinoflagellés des schistes bitumineux d'Orbagnoux (Jura). *Bull. Soc. Fr. Microsc.*, v. 8, p. 141−145, pl. 6, 1939.

Deflandre, Georges, Sur un nouveau Péridinien fossile, à thèque originellement siliceuse. *C. R. Hebd. Séanc. Acad. Sci., Paris*, v. 211, p. 265−268, figs. 1−4, 1940.

Deflandre, Georges, Le microplancton Kiméridgien d'Orbagnoux et l'origine des huiles sulfurées naturelles. *Mém. Acad. Sci. Inst. Fr.*, v. 65, ser. 2, no. 5, p. 1−32, pls. 1−7, 1941.

Deflandre, Georges, Sur quelques nouveaux dinoflagelles des silex Crétacés. *Bull. Soc. Géol. Fr.*, sér. 5, v. 13, p. 499−509, pl. 17, 1943.

Deflandre, Georges, *Calciodinellum* nov. gen., premier représentant d'une famille nouvelle de dinoflagellés fossiles à thèque calcaire. *C. R. Hebd. Séanc. Acad. Sci., Paris*, v. 224, p. 1781−1782, figs. 1−6, 1947a.

Deflandre, Georges, Le problème des Hystrichosphères. *Bull. Inst. Océanogr. Monaco*, no. 918, p. 1−23, figs. 1−5, 1947b.

Deflandre, Georges, Sur quelques microorganismes planctoniques des silex jurassiques. *Bull. Inst. Océanogr. Monaco*, no. 921, p. 1−12, figs. 1−23, 1947c.

Deflandre, Georges, Les Calciodinellidés Dinoflagellés fossiles à thèque calcaire. *Botaniste*, v. 34, p. 191−219, figs. 1−37, 1948.

Deflandre, Georges, Classe des dinoflagellés, p. 116−124. *In* J. Piveteau, *Traité de Paléontologie*, v. 1, *Les Stades inférieurs d'organisation du Règne Animal*. Paris: Masson et Cie., 1952a.

Deflandre, Georges, Dinoflagellés fossiles, p. 391−406. *In* P. P. Grassé, *Traité de Zoologie, Anatomie, Systématique, Biologie*, v. 1, pt. 1, Protozoaires, Généralités, Flagellés. Paris: Masson et Cie., 1952b.

Deflandre, Georges, Remarques sur la classification des Dinoflagellés fossiles, à propos d'*Evittodinium* nouveaux genre crétacé de la famille des Deflandreaceae. *C. R. Hebd. Séanc. Acad. Sci., Paris,* v. 258, p. 5027 − 5030, figs. 1 − 9, 1964a.

Deflandre, Georges, Quelques observations sur la systématique et la nomenclature des dinoflagellés fossiles. *Lab. Micropaléont. Ecole Pratique Hautes Etudes, Inst. Paléontol. Museum, Paris,* 8 p., 13 figs., 1964b.

Deflandre, Georges, État actuel de nos connaissances sur l'ancienneté des dinoflagellés. *Archs Zool. Exp. Gen.,* v. 105, p. 381 − 394, 1 pl., 1965.

Deflandre, Georges, *Stenopyxinium* n. g. *grassei* n. sp., nouveau type de kyste de dinoflagellé fossile d'un silex crétacé. *Protistologica,* v. 3, p. 423 − 426, figs. 1 − 6, 1967.

Deflandre, Georges, and I. C. Cookson, Fossil microplankton from Australian late Mesozoic and Tertiary sediments. *Aust. J. Mar. Freshwat. Res.,* v. 6, p. 242 − 313, pls. 1 − 9, figs. 1 − 58, 1955.

Deflandre, Georges, and H. Courteville, Note préliminaire sur les microfossiles des silex crétacés du Cambrésis. *Bull. Soc. Fr. Microsc.,* v. 8, p. 95 − 106, pls. 2 − 4, 1939.

Deflandre, Georges, and J. C. Foucher, *Diacrocanthidium* nov. gen., Diacrodien présumé du Crétacé, pourvu d'un archéopyle. Affinités péridiniennes des Diacrodiens? *Cah. Micropaléont.,* sér. 1, no. 5, p. 1 − 5, pls. 1, 2, Arch. Orig. Centre Document. C. N. R. S. No. 439, 1967.

Deflandre, Georges, and W. A. S. Sarjeant, Nouvel examen de quelques holotypes de dinoflagellés fossiles et d'acritarches. *Cah. Micropaléont.,* sér. 2, no. 1, p. 1 − 10, 1 pl., Arch. Orig. Centre Document. C. N. R. S. No. 466, 1970.

Delage, Yves, and Edgard Hérouard, *Traité de zoologie concrète.* Paris: C. Reinwald, v. 1, xxx + 584 p., 870 figs., 1896.

Denton, E. J., Report of the Council for 1975 − 6. *Mar. Biol. Ass. U.K.,* p. 1 − 50, 1976.

Diwald, Karl, Die ungeschlechtliche und geschlechtliche Fortpflanzung von *Glenodinium lubiniensiforme* spec. nov. *Flora, Jena,* v. 132, p. 174 − 192, figs. 1 − 4, 1938.

Diwald, Karl, Ein Beitrag zur Variabilität und Systematik der Gattung *Peridinium* (nach dem auf den österreichischitalienischen Terminfahrten der "Najade" in den Jahren 1911 − 1914 gesammelten Material). *Arch. Protistenk.,* v. 93, p. 121 − 184, figs. 1 − 23, 1939.

Dodekova, L. D., Les dinoflagellés et acritarches de l'Oxfordien-Kimeridgien de la Bulgarie du Nord-Est. *God. Sof. Univ., Geol.,* v. 60, p. 9 − 30, pls. 1 − 3, figs. 1, 2, 1967.

Dodekova, L. D., Dinoflagellés et acritarches du Tithonique aux environs de Pleven, Bulgarie Central du Nord. *Izv. Geol. Inst., Sof., Ser. Paleont.,* v. 18, p. 13 − 24, pls. 1 − 5, figs. A − H, 1969.

Dodekova, L. D., Dinoflagelati i Akritarkhi ot Titona v tsentralna severna Bulgariya. *Izv. Geol. Inst., Sof., Ser. Paleont.,* v. 20, p. 5 − 22, pls. 1 − 5, 1971.

Dodge, J. D., Chromosome numbers in some marine dinoflagellates. *Botanica Mar.,* v. 5, p. 121 − 127, 1963a.

Dodge, J. D., The nucleus and nuclear division in the Dinophyceae. *Arch. Protistenk.,* v. 106, p. 442 − 452, pls. 37 − 39, fig. 1, 1963b.

Dodge, J. D., Chromosome structure in the Dinophyceae II. Cytochemical studies. *Arch. Mikrobiol.,* v. 48, p. 66 − 80, figs. 1 − 20, 1964a.

Dodge, J. D., Nuclear division in the dinoflagellate *Gonyaulax tamarensis. J. Gen. Microbiol.,* v. 36, p. 269 − 276, pls. 1 − 4, 1964b.

Dodge, J. D., Chromosome structure in the dinoflagellates and the problem of the mesocaryotic cell. *Progress in Protozoology, Int. Conf. on Protozoology, no. 2, London 1965,* Abstracts Int. Congress Ser. 91, Abst. 339, p. 264 − 265, Excerpta Medica Foundation, 1965a.

Dodge, J. D., Thecal-fine-structure in the dinoflagellate genera *Prorocentrum* and *Exuviaella. J. Mar. Biol. Ass. U.K.,* v. 45, p. 607 − 614, pls. 1 − 4, figs. 1 − 6, 1965b.

Dodge, J. D., The Dinophyceae, p. 96 − 115, fig. 3.1 − 3.10. *In* M. B. E. Godward, *The chromosomes of the algae.* London: Edward Arnold Ltd., 1966.

Dodge, J. D., Fine structure of the dinoflagellate *Aureodinium pigmentosum* gen. et sp. nov. *Br. Phycol. Bull.,* v. 3, p. 327 − 336, figs. 1 − 7, 1967.

Dodge, J. D., The fine structure of chloroplasts and pyrenoids in some marine dinoflagellates. *J. Cell Sci.,* v. 3, p. 41 − 48, figs. 1 − 13, 1968.

Dodge, J. D., A dinoflagellate with both a mesocaryotic and a eucaryotic nucleus. I. Fine structure of the nuclei. *Protoplasma,* v. 73, p. 145 − 157, figs. 1 − 13, 1971a.

Dodge, J. D., Fine structure of the Pyrrhophyta. *Bot. Rev.,* v. 37, p. 481 − 508, figs. 1 − 9, 1971b.

Dodge, J. D., The ultrastructure of the dinoflagellate pusule: a unique osmo-regulatory organelle. *Protoplasma,* v. 75, p. 285 − 302, figs. 1 − 15, 1972.

Dodge, J. D., *The fine structure of algal cells.* London and New York: Academic Press, 261 p., figs. 1.1 − 14.3, 1973.

Dodge. J. D., A redescription of the dinoflagellate *Gymnodinium simplex* with the aid of electron microscopy. *J. Mar. Biol. Ass. U.K.,* v. 54, p. 171 − 177, pls. 1 − 4, fig. 1, 1974a.

Dodge, J. D., Fine structure and phylogeny in the algae. *Sci. Prog., Oxf.,* v. 61, p. 257 − 274, figs. 1 − 16, 1974b.

Dodge, J. D., A survey of chloroplast ultrastructure in

the Dinophyceae. *Phycologia,* v. 14, p. 253–265, figs. 1–11, 1975a.

Dodge, J. D., The Prorocentrales (Dinophyceae) II. Revision of the taxonomy within the genus *Prorocentrum. Bot. J. Linn. Soc., Lond.,* v. 71, p. 103–125, 1975b.

Dodge, J. D. and B. T. Bibby, The Prorocentrales (Dinophyceae) I. A comparative account of fine structure in the genera *Prorocentrum* and *Exuviaella. Bot. J. Linn. Soc., Lond.,* v. 67, p. 175–187, pls. 1–7, 1973.

Dodge, J. D., and R. M. Crawford, Fine structure of the dinoflagellate *Amphidinium carteri* Hulbert [sic]. *Protistologica,* v. 4, p. 231–242, pls. 1–6, figs. 1–4, 1968.

Dodge, J. D., and R. M. Crawford, The fine structure of *Gymnodinium fuscum* (Dinophyceae). *New Phytol.,* v. 68, p. 613–618, figs. 1–12, 1969a.

Dodge, J. D., and R. M. Crawford, Observations on the fine structure of the eyespot and associated organelles in the dinoflagellate *Glenodinium foliaceum. J. Cell Sci.,* v. 5, p. 479–493, figs. 1–13, 1969b.

Dodge, J. D., and R. M. Crawford, The morphology and fine structure of *Ceratium hirundinella* (Dinophyceae). *J. Phycol.,* v. 6, p. 137–149, figs. 1–26, 1970a.

Dodge, J. D., and R. M. Crawford, A survey of thecal fine structure in the Dinophyceae. *Bot. J. Linn. Soc., Lond.,* v. 63, p. 53–67, pls. 1–7, fig. 1, 1970b.

Dodge, J. D., and R. M. Crawford, A fine-structural survey of dinoflagellate pyrenoids and food-reserves. *Bot. J. Linn. Soc., Lond.,* v. 64, p. 105–115, pls. 1–7, 1971a.

Dodge, J. D., and R. M. Crawford, Fine structure of the dinoflagellate *Oxyrrhis marina* I. The general structure of the cell. *Protistologica,* v. 7, p. 295–304, figs. 1–18, 1971b.

Dodge, J. D., and R. M. Crawford, Fine structure of the dinoflagellate *Oxyrrhis marina* II. The flagellar system. *Protistologica,* v. 7, p. 399–409, figs. 1–22, 1971c.

Dodge, J. D., and R. M. Crawford, Fine structure of the dinoflagellate *Oxyrrhis marina* III-phagotrophy. *Protistologica,* v. 10, p. 239–244, figs. 1–8, 1974.

Donnelly, P. V., J. Vuille, M. C. Jayaswal, R. A. Overstreet, J. Williams, M. A. Burklew, and R. M. Ingle, A study of contributory chemical parameters to red tide in Apalachee Bay. *Fla. Board Conserv. Mar. Lab. Prof. Pap.,* ser. 8, p. 43–84, 1966.

Downie, Charles, Microplankton from the Kimeridge Clay. *Q. Jl. Geol. Soc. Lond.,* v. 112, p. 413–434, pl. 20, figs. 1–6, 1957.

Downie, Charles, The geological history of the microplankton. *Rev. Palaeobot. Palynol.,* v. 1, p. 269–281, figs. 1–3, 1967.

Downie, Charles, W. R. Evitt, and W. A. S. Sarjeant, Dinoflagellates, hystrichospheres, and the classification of the acritarchs. *Stanford Univ. Publs. Geol. Sci.,* v. 7(3), p. 1–16, 1963.

Downie, Charles, M. A. Hussain, and G. L. Williams, Dinoflagellate cyst and acritarch associations in the Paleogene of south-east England. *Geosci. Man,* v. 3, p. 29–35, pls. 1, 2, figs. 1–6, 1971.

Downie, Charles, and W. A. S. Sarjeant, Bibliography and index of fossil dinoflagellates and acritarchs. *Mem. Geol. Soc. Am.* 94 (1964), p. 1–180, 1965.

Downie, Charles, and W. A. S. Sarjeant, The morphology, terminology and classification of fossil dinoflagellate cysts. *Bull. Br. Mus. Nat. Hist., Geology,* Suppl., 3, p. 10–17, figs. 1–4, 1966.

Downie, Charles, and Gurdip Singh, Dinoflagellate cysts from estuarine and raised beach deposits at Woodgrange, Co. Down, N. Ireland. *Grana Palynol.,* v. 9, p. 124–132, figs. 1–12, 1969.

Dragesco, Jean, Étude cytologique de quelques flagellés mésopsammiques. *Cah. Biol. Mar.,* v. 6, p. 83–115, pls. 1–2, figs. 1–17, 1965.

Drebes, Gerhard, *Dissodinium pseudocalani* sp. nov., ein parasitischer Dinoflagellat auf Copepodeneiern. *Helgoländer Wiss. Meeresunters.,* v. 19, p. 58–67, figs. 1–4, 1969.

Droop, M. R., A note on some physical conditions for cultivating *Oxyrrhis marina. J. Mar. Biol. Ass. U. K.,* v. 38, p. 599–604, figs. 1–4, 1959.

Drugg, W. S., *Glyphanodinium,* a new dinoflagellate genus from the Paleocene of California. *Proc. Biol. Soc. Wash.,* v. 77, p. 237–240, figs. 1–6, 1964.

Drugg, W. S., Palynology of the Upper Moreno Formation (Late Cretaceous-Paleocene) Escarpado Canyon, California. *Palaeontographica Abt. B,* v. 120, p. 1–71, pls. 1–9, figs. 1–5, 1967.

Drugg, W. S., Some new genera, species and combinations of phytoplankton from the Lower Tertiary of the Gulf Coast, U.S.A. *Proc. N. Am. Paleont. Conv., Chicago,* Part G, p. 809–843, figs. 1–19, 1970.

Drugg, W. S., and A. R. Loeblich, Jr., Some Eocene and Oligocene phytoplankton from the Gulf Coast, U.S.A. *Tulane Stud. Geol. Paleont.,* v. 5, p. 181–194, pls. 1–3, figs. 1–8, 1967.

Drugg, W. S., and L. E. Stover, Stratigraphic range charts, selected Cenozoic dinoflagellates. *Am. Ass. Stratigr. Palynol., Contr. Ser.,* no. 4, p. 73–76, pls. 1–7, 1975.

Dumitrică, Paulian, Cenozoic endoskeletal dinoflagellates in southwestern Pacific sediments cored during Leg 21 of the DSDP. *Init. Repts. Deep Sea Drill. Proj.,* v. 21, p. 819–835, pls. 1–5, figs. 1, 2, 1973.

Dürr, Georg, and Harald Netzel, The fine structure of the cell surface in *Gonyaulax polyedra* (Dinoflagellata). *Cell Tissue Res.,* v. 150, p. 21–41, figs. 1–10, 1974.

Eckert, Roger, Subcellular sources of luminescence in *Noctiluca*. *Science*, v. 151, p. 349–352, 3 figs., 1966.

Eckert, Roger, and Margaret Findlay, Two physiological varieties of *Noctiluca miliaris*. *Biol. Bull. Mar. Biol. Lab., Woods Hole*, v. 123, p. 494–495, 1962.

Eddy, Samuel, The fresh-water armored or thecate dinoflagellates. *Trans. Am. Microsc. Soc.*, v. 49, p. 277–321, pls. 28–35, 1930.

Ehrenberg, C. G., Das Leuchten des Meers. Neue Beobachtungen nebst Übersicht der Hauptmomente der geschichtlichen Entwicklung dieses merkwürdigen Phänomens. *Abh. K. Akad. Wiss. Berlin*, 1834, p. 411–575, pls. 1, 2, 1836a.

Ehrenberg, C. G., Mittheilungen über die in den Feuersteinen bei Delitzsch vorkommenden mikroskopischen Algen und Bryozoen als Begleiter der fossilen Infusorien. *Verh. Preuss. Akad. Wiss., Berlin*, p. 114–115, 1836b.

Ehrenberg, C. G., Die neuesten Fortschritte in der Erkenntniss der Infusorien als Felsmassen. *Amtlicher Ber. Versammlung Deutsch. Naturforsch. u. Ärzte zu Jena im Sept. 1836*, p. 69–76, 1837a.

Ehrenberg, C. G., Über das Massenverhältniss der jetzt lebenden Kiesel-Infusorien und über ein neues Infusorien-Conglomerat als Polirschiefer von Jastraba in Ungarn. *Abh. Preuss. Akad. Wiss.*, 1836 (1838), p. 109–135, pls. 1–2 [separate 1837], 1837b.

Ehrenberg, C. G., [Feuersteinen der Gegend von Delitzsch bei Leipzig]. *Mitt. Verh. Naturforschender Freunde zu Berlin, Viertes Quartal*, 1836, p. 46–48, 1837c.

Ehrenberg, C. G., Über das Massenverhältniss der jetzt lebenden Kiesel-Infusorien als Meeres-Absatz in Nord-Amerika und eine Vergleichung derselben mit den organischen Kreide-Gebilden in Europa und Africa. *Ber. Verh. K. Preuss. Akad. Wiss. Berlin*, p. 57–97, 1838.

Ehrenberg, C. G., Über die Bildung der Kreidefelsen und des Kreidemergels durch unsichtbare Organismen. *Abh. Preuss. Akad. Wiss.*, 1838 (1840), p. 59–147, pls. 1–4 [separate 1839], 1839.

Ehrenberg, C. G., Über noch jetzt zahlreich lebende Thierarten der Kreidebildung und den Organismus der Polythalamien. *Abh. K. Akad. Wiss. Berlin*, 1839, p. 81–174, pls. 1–4, 1840.

Ehrenberg, C. G., *Mikrogeologie, das Erden und Felsen schaffende Wirken des unsichtbaren kleinen selbständigen Lebens auf der Erde.* Leipzig: Leopold Voss, 1–374 p., Namenregister 31 p., index to Atlas 88 p., pls. 1–40, 1854.

Eisenack, Alfred, Neue Mikrofossilien des baltischen Silurs I. *Paläont. Z.*, v. 13, p. 74–118, pls. 1–5, figs. 1–5, 1931.

Eisenack, Alfred, Mikrofossilien aus Doggergeschieben Ostpreussens. *Z. Geschiebeforsch.*

Flachldgeol., v. 11, p. 167–184, pls. 4, 5, figs. 1–9, 1935.

Eisenack, Alfred, Dinoflagellaten aus dem Jura. *Annls. Protist.*, v. 5, p. 59–63, pl. 4, figs. 1–5, 1936.

Eisenack, Alfred, Die Phosphoritknollen der Bernsteinformation als Überlieferer tertiären Planktons. *Schr. Phys.-Ökon. Ges. Königsb.*, v. 70, p. 181–188, figs. 1–6, 1938.

Eisenack, Alfred, Mikrofossilien aus Phosphoriten des Samlandischen Unteroligozäns und über die Einheitlichkeit der Hystrichosphaerideen. *Palaeontographica Abt. A*, v. 105, p. 49–95, pls. 7–12, figs. 1–8, 1954.

Eisenack, Alfred, Mikroplankton aus dem norddeutschen Apt nebst einigen Bemerkungen über fossile Dinoflagellaten. *Neues Jb. Geol. Paläont. Abh.*, v. 106, p. 383–422, pls. 21–27, 1958.

Eisenack, Alfred, Einige Erörterungen über fossile Dinoflagellaten nebst Übersicht über die zur Zeit bekannten Gattungen. *Neues Jb. Geol. Paläont. Abh.*, v. 112, p. 281–324, pls. 33–37, figs. 1–8, 1961.

Eisenack, Alfred, Hystrichosphären. *Biol. Rev.*, v. 38, p. 107–139, pls. 1, 2, figs. 1–4, 1963a.

Eisenack, Alfred, Zur *Membranilarnax*-Frage. *Neues Jb. Geol. Paläont. Mh.*, 1963, p. 98–103, 1963b.

Eisenack, Alfred, Sind die Hystrichosphären Zysten von Dinoflagellaten? *Neues Jb. Geol. Paläont. Mh.*, 1963, p. 225–231, 1963c.

Eisenack, Alfred, *Cordosphaeridium* n.g. ex *Hystrichosphaeridium*, Hystrichosphaeridea. *Neues Jb. Geol. Paläont. Abh.*, v. 118, p. 260–265, pl. 29, 1963d.

Eisenack, Alfred, Erörterungen über einige Gattungen fossiler Dinoflagellaten und über die Einordnung der Gattungen in das System. *Neues Jb. Geol. Paläont. Mh.*, 1964, p. 321–336, figs. 1–3, 1964.

Eisenack, Alfred, Über einige Mikrofossilien des samländischen und norddeutschen Tertiärs. *Neues Jb. Geol. Paläont. Abh.*, v. 123, p. 149–159, pls. 14, 15, figs. 1–3, 1965.

Eisenack, Alfred, Symbionten in fossilen Protisten. *Paläont. Z.*, v. 40, p. 103–107, pl. 8, 1966a.

Eisenack, Alfred, Über einige Probleme bei fossilen Dinoflagellaten. *Arch. Protistenk.*, v. 109, p. 207–222, pl. 60, figs. 1, 2, 1966b.

Eisenack, Alfred, *Katalog der fossilen Dinoflagellaten, Hystrichosphären und verwandten Mikrofossilien*, v. 1, *Dinoflagellaten* 1. Erganzungslieferung. Stuttgart: E. Schweizerbart'sche Verlagsbuchhandlung, p. 889–895, 1967.

Eisenack, Alfred, Bemerkungen zur Systematik der fossilen Dinoflagellaten. *Neues Jb. Geol. Paläont. Mh.*, 1969, p. 337–343, 1969a.

Eisenack, Alfred, Kritische Bemerkungen und Richtigstellungen im Gebiet der fossilen Dinoflagellaten und Acritarchen. *Neues Jb. Geol. Paläont. Abh.*, v. 134, p. 101–116, pls. 18, 19, figs. 1–3, 1969b.

Eisenack, Alfred, and I. C. Cookson, Microplankton from Australian Lower Cretaceous sediments. *Proc. R. Soc. Vict.*, n.s., v. 72, p. 1–11, pls. 1–3, 1960.

Eisenack, Alfred, and Magnus Fries, *Peridinium limbatum* (Stokes) verglichen mit der Tertiären *Deflandrea phosphoritica* Eisenack. *Geol. För. Stockh. Förh.*, v. 87, p. 239–248, figs. 1, 2, 1965.

Eisenack, Alfred, and Hans Gocht, Neue Namen für einige Hystrichosphären der Bernsteinformation Ostpreussens. *Neues Jb. Geol. Paläont. Mh.*, 1960, p. 511–518, figs. 1–4, 1960.

Eisenack, Alfred, and Göran Kjellström, *Katalog der fossilen Dinoflagellaten, Hystrichosphären und verwandten Mikrofossilien*, Band II, *Dinoflagellaten*. Stuttgart: E. Schweizerbart'sche Verlagsbuchhandlung, 1130 p., 6 pls., 1971.

Eisenack, Alfred, and K. W. Klement, *Katalog der fossilen Dinoflagellaten, Hystrichosphären und verwandten Mikrofossilien*, Bd. 1, *Dinoflagellaten*. Stuttgart: E. Schweizerbart'sche Verlagsbuchhandlung, 888 p., 9 pls., 1964.

Engelhardt, D. W. *Geiselodinium tyonekensis* sp. nov., a dinoflagellate cyst from the nonmarine Tertiary (Miocene) of Alaska. *Geosci. Man*, v. 15, p. 121–124, 1 pl., 1976.

Entz, Géza, Über die mitotische Teilung von *Ceratium hirundinella*. *Arch. Protistenk.*, v. 43, p. 415–430, pls. 13, 14, figs. 1–10, 1921.

Entz, Géza, On chain formation in *Ceratium hirundinella*. *Biologica Hung.*, v. 1(3), p. 1–4, pl. 2, 1924.

Entz, Géza, Über Cysten und Encystierung der Süsswasser-Ceratien. *Arch. Protistenk.*, v. 51, p. 131–183, 50 figs., 1925.

Entz, Géza, Beiträge zur Kenntniss der Peridineen. I. Zur Morphologie und Biologie von *Peridinium Borgei* Lemmermann. *Arch. Protistenk.*, v. 56, p. 397–446, 1 pl., 1926.

Eppley, R. W., O. Holm-Hansen, and J. D. H. Strickland, Some observations on the vertical migration of dinoflagellates. *J. Phycol.*, v. 4, p. 333–340, figs. 1–9, 1968.

Eppley, R. W., J. N. Rogers, and J. J. McCarthy, Half-saturation constants for uptake of nitrate and ammonium by marine phytoplankton. *Limnol. Oceanogr.*, v. 14, p. 912–920, 1969.

Eppley, R. W., and P. R. Sloan, Growth rates of marine phytoplankton: correlation with light absorption by cell chlorophyll *a*. *Physiologia Pl.*, v. 19, p. 47–59, figs. 1–6, 1966.

Erdtman, G., On pollen grains and dinoflagellate cysts in the Firth of Gullmarn, SW. Sweden. *Bot. Notiser*, v. 2, p. 103–111, 4 figs., 1954.

Eren, Jacob, Cyst formation in *Peridinium cinctum* fa. *westii. J. Protozool.*, v. 16 Suppl., p. 35 (Abstr.), 1969a.

Eren, Jacob, Studies on development cycle of *Peridinium cinctum* f. *westii. Verh. Int. Verein. Limnol.*, v. 17, p. 1013–1016, pl. 7, 1969b.

Evitt, W. R., Observations on the morphology of fossil dinoflagellates. *Micropaleontology*, v. 7, p. 385–420, pls. 1–9, figs. 1–8, 1961a.

Evitt, W. R., The dinoflagellate *Nannoceratopsis* Deflandre: morphology, affinities and infraspecific variability. *Micropaleontology*, v. 7, p. 305–316, pls. 1, 2, 1961b.

Evitt, W. R., *Dapcodinium priscum* n. gen., n. sp., a dinoflagellate from the Lower Lias of Denmark. *J. Paleont.*, v. 35, p. 996–1002, pl. 119, 20 figs., 1961c.

Evitt, W. R., Dinoflagellate synonyms: *Nannoceratopsis deflandrei* Evitt junior to *N.? gracilis* Alberti. *J. Paleont.*, v. 36, p. 1129–1130, 1962.

Evitt, W. R., A discussion and proposal concerning fossil dinoflagellates, hystrichospheres, and acritarchs. *Proc. Natn. Acad. Sci. U.S.A.*, v. 49, p. 158–164, 298–302, figs. 1–4, 1963.

Evitt, W. R., Dinoflagellate studies II. The archeopyle. *Stanford Univ. Publs. Geol. Sci.*, v. 10(3), p. 1–83, pls. 1–11, figs. 1–50, 1967a.

Evitt, W. R., Progress in the study of fossil *Gymnodinium* (Dinophyceae). *Rev. Palaeobot. Palynol.*, v. 2, p. 355–363, 1 pl., 1 fig., 1967b.

Evitt, W. R., Dinoflagellates and other organisms in palynological preparations, p. 439–479. *In* R. H. Tschudy and R. A. Scott, *Aspects of Palynology*. New York: Wiley-Interscience, 1969.

Evitt, W. R., Dinoflagellates—a selective review. *Geosci. Man*, v. 1, p. 29–45, 1970.

Evitt, W. R., Maestrichtian *Aquilapollenites* in Texas, Maryland, and New Jersey. *Geosci. Man*, v. 7, p. 31–38, 2 pls., 1973.

Evitt, W. R., Foraminifera affinities. . . . Comment. *Geology*, v. 2, p. 582–583, 1974a.

Evitt, W. R., Restudy of an Oliogocene freshwater dinoflagellate from Vermont. *Geosci. Man*, v. 9, p. 1–6, 1 pl., 1974b.

Evitt, W. R., The archeopyle in Cretaceous *Palaeoperidinium eurypylum* (Manum and Cookson) comb. nov., and similar dinoflagellates. *Geosci. Man*, v. 11, p. 77–86, 1 pl., 1975.

Evitt, W. R., R. F. A. Clarke, and J. P. Verdier, Dinoflagellate studies. III. *Dinogymnium acuminatum* n. gen., n. sp. (Maastrichtian) and other fossils formerly referable to *Gymnodinium* Stein. *Stanford Univ. Publs. Geol. Sci.*, v. 10(4), p. 1–27, pls. 1–3, figs. 1–22, 1967.

Evitt, W. R., and S. E. Davidson, Dinoflagellate studies I. Dinoflagellate cysts and thecae. *Stanford Univ. Publs. Geol. Sci.*, v. 10(1), p. 1–12, pl. 1, figs. 1, 2, 1964.

Evitt, W. R., J. K. Lentin, M. E. Millioud, L. E. Stover, and G. L. Williams, Dinoflagellate cyst terminology. *Geol Surv. Pap. Can.*, 76−24, p. 1−11, figs. 1−6, 1977.

Evitt, W. R., and David Wall, Dinoflagellate studies IV. Theca and cyst of Recent freshwater *Peridinium limbatum* (Stokes) Lemmermann. *Stanford Univ. Publs. Geol. Sci.*, v. 12(2), p. 1−15, pls. 1−4, 1968.

Fankboner, P. V., Intracellular digestion of symbiotic Zooxanthellae by host amoebocytes in giant clams (Bivalvia: Tridacnidae), with a note on the nutritional role of the hypertrophied siphonal epidermis. *Biol. Bull. Mar. Biol. Lab., Woods Hole*, v. 141, p. 222−254, figs. 1−8, 1971.

Filfilan, S. A., and Sigee, D. C., The replication of nuclear DNA in the dinoflagellate *Prorocentrum micans*. *J. Protozool.*, v. 24 (4), p. 46A (Abstr.), 1977.

Fine, K. E., and A. R. Loeblich III, Endosymbiosis in the marine dinoflagellate *Kryptoperidinium foliaceum*. *J. Protozool.*, v. 23(2) Suppl., p. 8A (Abstr.), 1976a.

Fine, K. E., and A. R. Loeblich III, Similarity of the dinoflagellates *Peridinium trochoideum, P. faeroënse* and *Scrippsiella sweeneyae* as determined by chromosome numbers, cell division studies and scanning electron microscopy. *Proc. Biol. Soc. Wash.*, v. 89, p. 275−288, figs. 1−21, 1976b.

Fott, Bohuslav, *Algenkunde.* Jena: Gustav Fischer Verlag, 581 p., 303 figs., 1971.

Fott, Bohuslav, and L. Ludvik, Über den submikroskopischen Bau des Panzers von *Ceratium hirundinella. Preslia,* v. 28, p. 278−280, 1956.

Foucher, J. C., Étude micropaléontologique des silex coniaciens du puits 19 de Lens-Liévin (Pas-de-Calais). *Bull. Mus. Natn. Hist. Nat., Paris,* sér. 3, no. 21, Sci. Terre 5, 1971, p. 77−129, pls. 1−14, 1972.

Foucher, J. C., Microfossiles des silex du Turonien Supérieur du Ruyaulcourt (Pas-de-Calais). *Annls. Paléont., Invertébrés,* v. 60(2), p. 113−164, pls. 1−11, 1974.

Freudenthal, H. D., *Symbiodinium* gen. nov. and *Symbiodinium microadriaticum* sp. nov., a zooxanthella: taxonomy, life cycle, and morphology. *J. Protozool.,* v. 9, p. 45−52, figs. 1−18, 1962.

Fütterer, Dieter, Kalkige Dinoflagellaten ("Calciodinelloideae") und die systematische Stellung der Thoracosphaeroideae. *Neues Jb. Geol. Paläont. Abh.,* v. 151, p. 119−141, figs. 1−25, 1976.

Fütterer, Dieter, Distribution of calcareous dinoflagellates in Cenozoic sediments of Site 366, Eastern North Atlantic. *Init. Repts. Deep Sea Drill. Proj.,* v. 41, p. 709−737, pls. 1−9, 1978.

Gaarder, K. R., Dinoflagellatae from the "Michael Sars" North Atlantic deep-sea expedition, 1910. *Rep. Scient. Results Michael Sars N. Atlant. Deep Sea Exped.,* v. 2(3), p. 1−62, 77 figs., 1954.

Gaudsmith, J. T., and C. J. Dawes, The ultrastructure of several dinoflagellates with emphasis on *Gonyaulax polyedra* Stein and *Gonyaulax monilata* Davis. *Phycologia,* v. 11, p. 123−132, figs. 1−20, 1972.

Gerlach, Ellen, Mikrofossilien aus dem Oligozän und Miozän Nordwestdeutschlands, unter besonder Berücksichtigung der Hystrichosphaeren und Dinoflagellaten. *Neues Jb. Geol. Paläont. Abh.,* v. 112, p. 143−228, pls. 25−29, figs. 1−23, 1961.

Gerrath, J. F., Two rare flagellates new to North America. *Br. Phycol. J.,* v. 9, p. 219 (Abstr.), 1974.

Gitmez, G. U., Dinoflagellate cysts and acritarchs from the basal Kimmeridgian (Upper Jurassic) of England, Scotland and France. *Bull. Br. Mus. Nat. Hist., Geology,* v. 18, p. 231−331, pls. 1−14, figs. 1−34, 1970.

Gitmez, G. U., and W. A. S. Sarjeant, Dinoflagellate cysts and acritarchs from the Kimmeridgian (Upper Jurassic) of England, Scotland and France. *Bull. Br. Mus. Nat. Hist., Geology,* v. 21, p. 171−257, pls. 1−17, figs. 1−27, 1972.

Gittleson, S. M., S. K. Hotchkiss, and F. G. Valencia, Locomotion in the marine dinoflagellate *Amphidinium carterae* (Hulburt). *Trans. Am. Microsc. Soc.,* v. 93, p. 101−105, figs. 1, 2, 1974.

Gocht, Hans, Hystrichosphaerideen und andere Kleinlebewesen aus Oligozänablagerungen Nord- und Mitteldeutschlands. *Geologie,* v. 1, p. 301−320, pls. 1, 2, 1952.

Gocht, Hans, *Rhombodinium* und *Dracodinium,* zwei neue Dinoflagellaten-Gattungen aus dem norddeutschen Tertiär. *Neues Jb. Geol. Paläont. Mh.,* 1955, p. 84−92, figs. 1−5, 1955.

Gocht, Hans, Mikroplankton aus dem nordwestdeutschen Neokom (Teil I). *Paläont. Z.,* v. 31, p. 163−185, pls. 18−20, figs. 1−16, 1957.

Gocht, Hans, Mikroplankton aus dem nordwestdeutschen Neokom. *Paläont. Z.,* v. 33, p. 50−89, pls. 3−8, 1959.

Gocht, Hans, Die Gattung *Chiropteridium* n. gen. (Hystrichosphaeridea) im deutschen Oligozän. *Paläont. Z.,* v. 34, p. 221−232, pls. 17, 18, figs. 1−28, 1960.

Gocht, Hans, Planktonische kleinformen aus dem Lias-/Dogger-Grenzbereich Nord- und Süddeutschlands. *Neues Jb. Geol. Paläont. Abh.,* v. 119, p. 113−133, pls. 15−17, figs. 1−57, 1964.

Gocht, Hans, Geisselansatzstellen bei *Wetzeliella* (Dinoflagellata, Deflandreaceae). *Neues Jb. Geol. Paläont. Abh.,* v. 128, p. 195−200, pls. 13, 14, 1967.

Gocht, Hans, Zur Morphologie und Ontogenie von *Thalassiphora* (Dinoflagellata). *Palaeontographica Abt. A,* v. 129, p. 149−156, pls. 25−27, figs. 1, 2, 1968.

Gocht, Hans, Formengemeinschaften alttertiären Mikroplanktons aus Bohrproben des Erdölfeldes

Meckelfeld bei Hamburg. *Palaeontographica Abt. B,* v. 126, p. 1−100, pls. 1−11, figs. 1−49, 1969.

Gocht, Hans, Dinoflagellaten-Zysten aus dem Bathonium des Erdölfeldes Aldorf (NW-Deutschland). *Palaeontographica Abt. B,* v. 129, p. 125−165, pls. 26−35, figs. 1−30, 1970a.

Gocht, Hans, Dinoflagellaten-Zysten aus einem Geschiebefeurstein und ihr Erhaltungszustand. *Neues Jb. Geol. Paläont. Mh.,* 1970, p. 129−140, figs. 1−5, 1970b.

Gocht, Hans, Zur Morphologie der Dinoflagellaten-Gattung *Nannoceratopsis* Deflandre. *Lethaia,* v. 5, p. 15−29, figs. 1−6, 1972.

Gocht, Hans, Neuuntersuchung von *Eodinia pachytheca* Eisenack 1936 (Dinoflagellata, Oberjura). *Neues Jb. Geol. Paläont. Abh.,* v. 148, p. 12−32, figs. 1−23, 1975.

Gocht, Hans, *Hystrichosphaeropsis quasicribrata* (O. Wetzel) ein Dinoflagellat aus dem Maastricht Nordeuropas. *Neues Jb. Geol. Paläont. Mh.,* 1976, p. 321−336, figs. 1−18, 1976.

Gocht, Hans, and Harald Netzel, Rasterelektronenmikroskopische Untersuchungen am Panzer von *Peridinium* (Dinoflagellata). Scanning electron microscope studies of the theca of *Peridinium* (Dinoflagellata). *Arch. Protistenk.,* v. 116, p. 381−410, pls. 43−52, figs. 1−9, 1974.

Gocht, Hans, and Harald Netzel, Reliefstrukturen des Kreide-Dinoflagellaten *Palaeoperidinium pyrophorum* (Ehr.) im Vergleich mit Panzer-Merkmalen rezenter *Peridinium*-Arten. *Neues Jb. Geol. Paläont. Abh.,* v. 152, p. 380−413, figs. 1−31, 1976.

Gönnert, R. *Sporodinium pseudocalani* n.g., n. sp., ein parasit aus Copepodeneiern. *Z. ParasitKde.,* v. 9, p. 140−143, figs. 1−7, 1936.

Goreau, T. F., The physiology of skeleton formation in corals under different conditions. *Biol. Bull. Mar. Biol. Lab., Woods Hole,* v. 116, p. 59−75, 1959.

Goreau, T. F., Mass expulsion of Zooxanthellae from Jamaican reef communities after Hurricane Flora. *Science,* v. 145, p. 383−386, figs. 1, 2, 1964.

Górka, Hanna, Coccolithophoridés, dinoflagellés, hystrichosphaeridés et microfossiles incertae sedis du Crétacé supérieur de Pologne. *Acta Palaeont. Pol.,* v. 8, p. 3−90, pls. 1−11, text-pls. 1−7, 1963.

Górka, Hanna, Les microfossiles du Jurassique Supérieur de Magnuszew (Pologne). *Acta Palaeont. Pol.,* v. 10, p. 291−334, pls. 1−5, 1965.

Gourret, Paul, Sur les péridiniens du Golfe de Marseille. *Annls. Mus. Hist. Nat. Marseille, Zool.,* v. 1, mém. 8, p. 5−114, pls. 1−4, 1883.

Graham, H. W., An oceanographic consideration of the dinoflagellate genus *Ceratium. Ecol. Monogr.,* v. 11, p. 99−116, figs. 1−14, 1941.

Graham, H. W., *Studies in the morphology, taxonomy and ecology of the Peridiniales.* Publs. Carnegie Instn, no. 542, v + 129 p., 1 pl., figs. 1−67, 1942.

Graham, H. W., *Gymnodinium catenatum,* a new dinoflagellate from the Gulf of California. *Trans. Am. Microsc. Soc.,* v. 42, p. 259−261, figs. 1, 2, 1943.

Graham, H. W., and N. Bronikovsky, *The genus Ceratium in the Pacific and North Atlantic Oceans.* Publs. Carnegie Instn. no. 565, vii + 209 p., figs. 1−27, charts 1−54, 1944.

Grell, K. G., *Protozoology.* New York, Heidelberg, Berlin: Springer-Verlag, 554 p., 437 figs., 1973.

Grell, K. G., and K. E. Wohlfarth-Bottermann, Licht- und elektronenmikroskopische Untersuchungen an den Dinoflagellaten *Amphidinium elegans* n. sp. *Z. Zellforsch. Mikrosk. Anat.,* v. 47, p. 7−17, figs. 1−6, 1957.

Gressel, J., T. Berman, and N. Cohen, Dinoflagellate ribosomal RNA; An evolutionary relic? *J. Molec. Evol.,* v. 5, p. 307−313, figs. 1, 2, 1975.

Greuet, Claude, Structure fine de l'ocelle d'*Erythropsis pavillardi* Hertwig, Péridinien Warnowiidae Lindemann. *Cr. R. Hebd. Séanc. Acad. Sci., Paris,* v. 261, p. 1904−1907, 1 pl., 1965.

Greuet, Claude, Organisation ultrastructurale de l'ocelle de deux péridiniens Warnowiidae, *Erythropsis pavillardi* Kofoid et Swezy et *Warnowia pulchra* Schiller. *Protistologica,* v. 4, p. 209−230, pls. 1−6, figs. 1−6, 1968a.

Greuet, Claude, *Leucopsis cylindrica,* nov. gen., nov. sp., péridinien Warnowiidae Lindemann. Considérations phylogénétiques sur les Warnowiidae. *Protistologica,* v. 4, p. 419−422, fig. 1, 1968b.

Greuet, Claude, Étude morphologique et ultrastructurale du trophonte d'*Erythropsis pavillardi* Kofoid et Swezy. *Protistologica,* v. 5, p. 481−503, pls. 1−5, figs. 1−7, 1969.

Greuet, Claude, Étude ultrastructurale et évolution des cnidocystes de *Nematodinium,* Péridinien Warnowiidae Lindemann. *Protistologica,* v. 7, p. 345−355, figs. 1−16, 1971.

Greuet, Claude, La nature trichocystaire du cnidoplaste dans le complexe cnidoplaste nématocyste de *Polykrikos schwartzi* Bütschli. *C. R. Hebd. Séanc. Acad. Sci., Paris,* sér. D, v. 275, p. 1239−1242, 2 pls., 2 figs., 1972a.

Greuet, Claude, Intervention de lamelles annelées dans la formation de couches squelettiques au niveau de la capsule périnucléaire de péridiniens Warnowiidae. *Protistologica,* v. 8, p. 155−168, figs. 1−14, 1972b.

Greuet, Claude, Les critères de détermination chez les péridiniens Warnowiidae Lindemann. *Protistologica,* v. 8 (1972), p. 461−469, figs. 1−12, 1973.

Gross, Fabius, Zur Biologie und Entwicklungsgeschichte von *Noctiluca miliaris. Arch. Protistenk.,* v. 83, p. 178−196, pl. 6, figs. 1−6, 1934.

Haapala, O. K., and M. O. Soyer, Structure of dinoflagellate chromosomes. *Nature, New Biol.,* v. 244, p. 195–197, figs. 1, 2, 1973.

Habib, Daniel, Middle Cretaceous palynomorph assemblages from clays near the Horizon Beta deep-sea outcrop. *Micropaleontology,* v. 16, p. 345–379, pls. 1–10, 1970.

Habib, Daniel, Dinoflagellate stratigraphy Leg 11, Deep Sea Drilling Project. *Init. Repts. Deep Sea Drill. Proj.,* v. 11, p. 367–425, pls. 1–22, 1972.

Habib, Daniel, Taxonomy, morphology, and suggested phylogeny of the dinoflagellate genus *Druggidium. Geosci. Man,* v. 7, p. 47–55, pls. 1–3, figs. 1–4, 1973.

Habib, Daniel, Morphogenetic relationship between *Scriniodinium campanula* Gocht and *Scriniodinium dictyotum* Cookson and Eisenack. *Geosci. Man,* v. 9, p. 45–51, pls. 1, 2, 1974.

Haeckel, Ernst, Plankton-Studien. *Jena. Z. Naturw.,* v. 25 (N. Folge 18), p. 232–336, 1891.

Halim, Youssef, Observations on the Nile bloom of phytoplankton in the Mediterranean. *J. Cons. Perm. Int. Explor. Mer,* v. 26, p. 57–67, 1960a.

Halim, Youssef, Étude quantitative et qualitative du cycle écologique des dinoflagellés dans les eaux de Villefranche-Sur-Mer. *Annls. Inst. Océanogr., Monaco,* n. sér., v. 38(2), p. 123–232, pls. 1–5, 1960b.

Halim, Youssef, Dinoflagellates of the South-East Caribbean Sea (East Venezuela). *Int. Revue Ges. Hydrobiol. Hydrogr.,* v. 52, p. 701–755, pls. 1–10, 1967.

Hand, W. G. Phototactic orientation by the marine dinoflagellate *Gyrodinium dorsum* Kofoid. I. A mechanism model. *J. Exp. Zool.,* v. 174, p. 33–38, 1970.

Hand, W. G., P. A. Collard, and Demorest Davenport, The effects of temperature and salinity change on swimming rate in the dinoflagellates *Gonyaulax* and *Gyrodinium. Biol. Bull. Mar. Biol. Lab., Woods Hole,* v. 128, p. 90–101, figs. 1–6, 1965.

Hand, W. G., and J. A. Schmidt, Phototactic orientation by the marine dinoflagellate *Gyrodinium dorsum* Kofoid. II Flagellar activity and overall response mechanism. *J. Protozool.,* v. 22, p. 494–498, figs. 1–11, 1975.

Hansen, J. M., Dinoflagellate stratigraphy and echinoid distribution in Upper Maastrichtian and Danian deposits from Denmark. *Bull. Geol. Soc. Denmark,* v. 26, p. 1–26, figs. 1–22, 1977.

Harker, S. D., and W. A. S. Sarjeant, The stratigraphical distribution of organic-walled dinoflagellate cysts in the Cretaceous and Tertiary. *Revue Palaeobot. Palynol.,* v. 20, p. 217–315, 1975.

Harland, Rex, Fossil dinoflagellate cysts from Lake Gnotuk, Victoria, Australia. *Proc. R. Soc. Vict.,* v. 84, p. 245–254, pls. 15–16, figs. 1–19, 1971.

Harland, Rex, A summary review of the morphology and classification of the fossil Peridiniales (Dinoflagellates) with respect to their modern representatives. *Geophytology,* v. 1, p. 135–150, figs. 1–4, 1972.

Harland, Rex, Dinoflagellate cysts and acritarchs from the Bearpaw Formation (Upper Campanian) of southern Alberta, Canada. *Palaeontology,* v. 16, p. 665–706, pls. 84–88, 1973.

Harland, Rex, Quaternary (Flandrian?) dinoflagellate cysts from the Grand Banks, off Newfoundland, Canada. *Revue Palaeobot. Palynol.,* v. 16 (1973), p. 229–242, 4 pls., 1 fig., 1974.

Harland, Rex, and Charles Downie, The dinoflagellates of the interglacial deposits at Kirmington, Lincolnshire. *Proc. Yorks. Geol. Soc.,* v. 37, p. 231–237, pl. 7, 1969.

Harland, Rex, S. J. Morbey, and W. A. S. Sarjeant, A revision of the Triassic to Lowest Jurassic dinoflagellate *Rhaetogonyaulax. Palaeontology,* v. 18, p. 847–864, pls. 100–104, 1975.

Harland, Rex, and W. A. S. Sarjeant, Fossil freshwater microplankton (dinoflagellates and acritarchs) from Flandrian (Holocene) sediments of Victoria and western Australia. *Proc. R. Soc. Vict.,* v. 83, p. 211–234, pls. 21, 22, figs. 1–9, 1970.

Harris, W. K., Tertiary non-marine dinoflagellate cyst assemblages from Australia. *Spec. Publs. Geol. Soc. Aust.,* no. 4, p. 159–166, pls. 1, 2, 1973.

Hart, T. J., Some observations on the relative abundance of marine phytoplankton populations in nature, p. 375–393. *In* H. Barnes, *Some contemporary studies in marine science.* London: George Allen and Unwin Ltd., 1966.

Haskell, T. R., Foraminiferal affinities exhibited by the dinoflagellate *Nannoceratopsis* Deflandre. *Geology,* v. 2, p. 579–582, figs. 1–5, 1974.

Hastings, J. W., M. Wiesner, J. Owens, and B. M. Sweeney, The rhythm of luminescence in a marine dinoflagellate. *Anat. Rec.,* v. 125, p. 611, 1956.

Hauge, H. V., On the freshwater species of *Ceratium. Nytt Mag. Bot.,* v. 6, p. 97–119, figs. 1–9, 1958.

Hellebust, J. A., Excretion of some organic compounds by marine phytoplankton. *Limnol. Oceanogr.,* v. 10, p. 192–206, 1965.

Hensen, Victor, Über die Bestimmung des Planktons oder des im Meere treibenden Materials an Pflanzen und Tieren. 5. *Ber. Commn. Wiss. Unters. Dt.-Meere.,* 12–16 Jg., p. 1–108, 1887.

Hentschel, E., Allgemeine Biologie des Südatlantischen Ozeans. *Wiss. Ergebn. Dt. Atlant. Exped. 'Meteor',* v. 9, p. 1–343, 1933.

Herdman, E. C., Notes on dinoflagellates and other organisms causing discolouration of the sand at Port Erin. *Proc. Trans. Lpool. Biol. Soc.,* v. 35, p. 59–63, 1921.

Herdman, E. C., Notes on dinoflagellates and other organisms causing discolouration of the sand at Port

Erin II. *Proc. Trans. Lpool. Biol. Soc.,* v. 36, p. 15—30, 1922.

Herdman, E. C., Notes on dinoflagellates and other organisms causing discolouration of the sand at Port Erin III. *Proc. Trans. Lpool. Biol. Soc.,* v. 38, p. 58—63, 1924a.

Herdman, E. C., Notes on dinoflagellates and other organisms causing discolouration of the sand at Port Erin IV. *Proc. Trans. Lpool. Biol. Soc.,* v. 38, p. 75—84, 1924b.

Herman, E. M., and B. M. Sweeney, Circadian rhythm of chloroplast ultrastructure in *Gonyaulax polyedra,* concentric organization around a cluster of ribosomes. *J. Ultrastruct. Res.,* v. 50, p. 347—354, figs. 1—7, 1975.

Hollande, André, Étude comparée de la mitose syndinienne et de celle des péridiniens libres et des hypermastigines infrastructure et cycle évolutif des syndinides parasites de radiolaires. *Protistologica,* v. 10, p. 413—451, figs. 1—23, 1975.

Hollande, André, Jean Cachon, and Monique Cachon-Enjumet, Mise en évidence, par la microscopie électronique, d'une capsule central chez divers Péridiniens. Considérations sur les affinités entre Dinoflagellés et Radiolaires. *C. R. Hebd. Séanc. Acad. Sci., Paris,* v. 254, p. 2069—2071, pls. 1, 2, 1962.

Hollande, André, Jean Cachon, and Monique Cachon, La signification de la membrane capsulaire des radiolaires et ses rapports avec le plasmalemme et les membranes du réticulum endoplasmique. Affinités entre radiolaires, heliozoaires et péridiniens. *Protistologica,* v. 6, p. 311—318, figs. 1—6, 1970.

Hollande, André, and D. Carré, Les xanthelles des radiolaires Sphaerocollides, des Acanthaires et de *Velella velella:* infrastructure-cytochemie-taxonomie. *Protistologica,* v. 10, p. 573—601, 1974.

Honigberg, B. M., W. Balamuth, E. C. Bovee, J. O. Corliss; M. Gojdics, R. P. Hall, R. R. Kudo, N. D. Levine, A. R. Loeblich Jr., J. Weiser, and D. H. Wenrich, A revised classification of the Phylum Protozoa. *J. Protozool.,* v. 11, p. 7—20, 1964.

Horowitz, Aharon, Kystes de dinoflagellés du Trias du Sud d'Israël. *Revue Micropaléont.,* v. 18, p. 23—27, pl. 1, 1975.

Huber, Gottfr., and Fr. Nipkow, Experimentelle Untersuchungen über die Entwicklung von *Ceratium hirundinella* O.F.M. *Z. Bot.,* v. 14, p. 337—371, 12 figs., 1922.

Huber, Gottfr., and Fr. Nipkow, Experimentelle Untersuchungen über Entwicklung und Formbildung von *Ceratium hirundinella. Flora, Jena,* v. 116, p. 114—215, 20 figs., 1923.

Huber-Pestalozzi, G., Das Phytoplankton des Susswassers. Teil 3, Cryptophyceae, Chloromonadophyceae, Dinophyceae. *Binnengewässer,* v. 16, p. 1—322, 1950.

Humboldt, Alexander von, *Amphicora sabella*–Infusoires fossiles (Extrait d'une lettre de M. de Humboldt à M. Arago). *C. R. Hebd. Séanc. Acad. Sci., Paris,* v. 4, p. 26—27, 1837.

Jahn, T. L., W. M. Harmon, and M. Landon, Mechanisms of locomotion in flagellates: I. *Ceratium. J. Protozool.,* v. 10, p. 358—363, figs. 1—11, 1963.

Jahn, T. L., and J. J. Votta, Locomotion of Protozoa. *Ann. Rev. Fluid Mech.,* v. 4, p. 93—116, figs. 1—20, 1972.

Jansonius, Jan, Palynology of Permian and Triassic sediments, Peace River Area, western Canada. *Palaeontographica Abt. B,* v. 110, p. 35—98, pls. 11—16, figs. 1—3, 1962.

Jardiné, Serge, and J. F. Reynaud, Utilisation du microscope électronique à balayage en palynologie. *Revue Inst. Fr. Pétrole,* v. 27, p. 201—216, pls. 1—3, 1972.

Jeffrey, S. W., M. Sielicki, and F. T. Haxo, Chloroplast pigment patterns in dinoflagellates. *J. Phycol.,* v. 11 (1975), p. 374—384, figs. 1, 2, 1976.

Johannes, R. E., J. Alberts, C. D'Elia, R. A. Kinzie, L. R. Pomeroy, W. Sottile, W. Wiebe, J. A. Marsh, Jr., P. Helfrich, J. Maragos, J. Meyer, S. Smith, D. Crabtree, A. Roth, L. R. McCloskey, S. Betzer, N. Marshall, M. E. Q. Pilson, G. Telek, R. I. Clutter, W. D. DuPaul, K. L. Webb, and J. M. Wells, Jr., The metabolism of some coral reef communities, a team study of nutrient and energy flux at Eniwetok. *Bioscience,* v. 22, p. 541—543, 1972.

Johnson, C. D., and L. V. Hills, Microplankton zones of the Savik Formation (Jurassic), Axel Heiberg and Ellesmere Islands, District of Franklin. *Bull. Can. Petrol. Geol.,* v. 21, p. 178—218, pls. 1—3, figs. 1—12, 1973.

Jollos, Victor, Dinoflagellatenstudien. *Arch. Protistenk.,* v. 19, p. 178—206, pls. 7—10, 1910.

Jörgensen, Eugen, Die Ceratien. Eine kurze Monographie der Gattung *Ceratium* Schrank. *Int. Revue Ges. Hydrobiol. Hydrogr.,* v. 4 (Biol. Suppl. Heft 1), p. 1—124, pls. 1—10, 184 figs., 1911.

Jux, Ulrich, Über den Feinbau der Wandung bei *Cordosphaeridium inodes* (Klumpp, 1953). *Palaeontographica Abt. B.,* v. 122, p. 48—54, pls. 13, 14, 1 fig., 1968a.

Jux, Ulrich, Über den Feinbau der Wandung bei *Hystrichosphaera bentori* Rossignol 1961. *Palaeontographica Abt. B,* v. 123, p. 147—152, pl. 31, fig. 1, 1968b.

Jux, Ulrich, Über den Feinbau der Wandungen einiger Tertiärer Dinophyceen-Zysten und Acritarcha, *Hystrichosphaeridium, Impletosphaeridium, Lingulodinium. Palaeontographica Abt. B,* v. 132, p. 165—174, pls. 25—28, figs. 1—3, 1971.

Jux, Ulrich, Über den Feinbau der Wandungen bei *Operculodinium centrocarpum* (Deflandre & Cookson) Wall 1967 und *Bitectatodinium tepikiense* Wilson

1973. *Palaeontographica Abt. B,* v. 155, p. 149–156, pls. 1–5, 1976.

Kalley, J. P., and T. Bisalputra, *Peridinium trochoideum:* the fine structure of the thecal plates and associated membranes. *J. Ultrastruct. Res.,* v. 37, p. 521–531, 12 figs., 1971.

Kelly, M. G., The occurrence of dinoflagellate luminescence at Woods Hole. *Biol. Bull. Mar. Biol. Lab., Woods Hole,* v. 135, p. 279–295, figs. 1–7, 1968.

Khmeleva, N. N., Rol' radiolyariy pri otsenke pervichnoy produktsii v krasnom more i adenskom zalive [The role of Radiolaria in evaluation of the primary production in the Red Sea and Gulf of Aden]. *Dokl. Akad. Nauk SSSR,* v. 172, p. 1430–1433, 1967.

Kimor, B., and E. J. F. Wood, A plankton study in the eastern Mediterranean Sea. *Mar. Biol.,* v. 29, p. 321–333, figs. 1–5, 1975.

Kiselev, I. A., Pantsyrnye zhgutikonostsy (Dinoflagellata) morey i presnykh vod SSSR. Opredeliteli po faune SSSR [Armored flagellates (Dinoflagellates) of the marine and fresh water of the USSR. Determination of the fauna of the USSR]. *Izd. Zool. Inst., Akad. Nauk SSSR,* v. 33, p. 1–279, Moskva and Leningrad, 1950.

Kiselev, I. A., Kriptomonadovye i peridinei evropeyskogo severa SSSR (Pyrrophyta) [Cryptomonads and peridinians of European northern USSR (Pyrrhophyta)]. *Trudy Bot. Inst. V. L. Komarova Akad. Nauk SSSR,* ser. 2, vyp. 7, p. 13–164, pls. 1–17, 1951.

Kjellström, Göran, Archaeopyle formation in the genus *Lejeunia* Gerlach, 1961 amend. *Geol. För. Stockh. Förh.,* v. 94, p. 467–469, figs. 1, 2, 1972.

Kjellström, Göran, Maastrichtian microplankton from the Höllviken Borehole No. 1 in Scania, southern Sweden. *Sver. Geol. Unders.,* ser. C, no. 688, Årsbok 67, no. 8, p. 1–59, figs. 1–48, 1973.

Klebs, Georg, Ueber die Organisation einiger Flagellaten-Gruppen und ihre Beziehungen zu Algen und Infusorien. *Unters. Bot. Inst. Tübingen,* v. 1, p. 233–361, pls. 2, 3, 1883.

Klement, K. W., Revision der Gattungszugehörigkeit einiger in die Gattung *Gymnodinium* Stein eingestufter Arten jurassicher Dinoflagellaten. *Neues Jb. Geol. Paläont. Mh.,* 1957, p. 408–410, 1 fig., 1957.

Klement, K. W., Dinoflagellaten und Hystrichosphaerideen aus dem unteren und mittleren Malm Südwestdeutschlands. *Palaeontographica Abt. A,* v. 114, p. 1–104, pls. 1–10, figs. 1–37, 1960.

Klement, K. W., Armored dinoflagellates of the Gulf of California. *Bull. Scripps Instn. Oceanogr.,* v. 8, p. 347–372, pls. 1–3, 1964.

Klumpp, Barbara, Beitrag zur Kenntniss der Mikrofossilien des mittleren und oberen Eozän. *Palaeontographica Abt. A,* v. 103, p. 377–406, pls. 16–20, figs. 1–5, 1953.

Kofoid, C. A., On the significance of the asymmetry in *Triposolenia. Univ. Calif. Publs. Zool.,* v. 3, p. 127–133, figs. A. B, 1906a.

Kofoid, C. A. Dinoflagellata of the San Diego region. II. On *Triposolenia,* a new genus of the Dinophysidae. *Univ. Calif. Publs. Zool.,* v. 3, p. 93–116, pls. 15–17, 1906b.

Kofoid, C. A. Dinoflagellata of the San Diego region. I. On *Heterodinium,* a new genus of the Peridinidae. *Univ. Calif. Publs. Zool.,* v. 2, p. 341–368, pls. 17–19, 1906c.

Kofoid, C. A., The plates of *Ceratium* with a note on the unity of the genus. *Zool. Anz.,* v. 32, p. 177–183, figs. 1–8, 1907a.

Kofoid, C. A., Dinoflagellata of the San Diego region III. Descriptions of new species. *Univ. Calif. Publs. Zool.,* v. 3, p. 299–340, pls. 22–33, 1907b.

Kofoid, C. A., New species of dinoflagellates. *Bull. Mus. Comp. Zool. Harv.,* v. 50, p. 163–207, pls. 1–18, 1907c.

Kofoid, C. A., Exuviation, autotomy and regeneration in *Ceratium. Univ. Calif. Publs. Zool.,* v. 4, p. 345–386, figs. 1–33, 1908.

Kofoid, C. A., Mutations in *Ceratium. Bull. Mus. Comp. Zool. Harv.,* v. 52, p. 213–257, pls. 1–4, figs. A–E, 1909a.

Kofoid, C. A., On *Peridinium steini* Jörgensen, with a note on the nomenclature of the Peridinidae. *Arch. Protistenk.,* v. 16, p. 25–47, 1 pl., 1909b.

Kofoid, C. A. Dinoflagellata of the San Diego region IV. The genus *Gonyaulax,* with notes on its skeletal morphology and a discussion of its generic and specific characters. *Univ. Calif. Publs. Zool.,* v. 8, p. 187–286, pls. 9–17, 1911.

Kofoid, C. A., Significance of certain forms of asymmetry of dinoflagellates. *VII Int. Congr. Zool. Boston 1907,* p. 928–931, 1912.

Kofoid, C. A. A new morphological interpretation of the structure of *Noctiluca,* and its bearing on the status of the Cystoflagellata (Haeckel). *Univ. Calif. Publs. Zool.,* v. 19, p. 317–334, pl. 18, 1920.

Kofoid, C. A., On *Oxyphysis oxytoxoides* gen. nov., sp. nov. a dinophysoid dinoflagellate convergent toward the peridinioid type. *Univ. Calif. Publs. Zool.,* v. 28, p. 203–216, pl. 18, 1926.

Kofoid, C. A., and J. R. Michener, On the structure and relationships of *Dinosphaera palustris* (Lemm.). *Univ. Calif. Publs. Zool.,* v. 11, p. 21–28, figs. 1–8, 1912.

Kofoid, C. A., and E. J. Rigden, A peculiar form of schizogony in *Gonyaulax. Bull. Mus. Comp. Zool. Harv.,* v. 54, p. 335–348, pls. 1, 2, 1912.

Kofoid, C. A., and Tage Skogsberg, The Dinoflagellata: the Dinophysoidae. *Mem. Mus. Comp. Zool. Harv.,* v. 51, p. 1–766, pls. 1–31, 103 figs., 1928.

Kofoid, C. A., and Olive Swezy, On the orientation of

Erythropsis. Univ. Calif. Publs. Zool., v. 18, p. 89−102, figs. 1−12, 1917.

Kofoid, C. A., and Olive Swezy, *The free-living unarmored Dinoflagellata.* Mem. Univ. Calif., no. 5, vii + 538 p., pls. 1−12, 1921.

Kozhova, O. M., O podlednom "tsvetenii" ozera Baykal [On the water bloom under the ice in Lake Baikal]. *Bot. Zh. SSSR,* v. 44, p. 1001−1004, 1959.

Krutzsch, Wilfried, Die Mikroflora der Geiseltalbraunkohle. Teil III Süsswasserdinoflagellaten aus subaquatisch gebildeten Blätterkohlenlagen des mittleren Geiseltales. *Hallesches Jb. Mitteld. Erdgesch.,* v. 4, p. 40−45, pls. 10, 11, fig. 1, 1962.

Kubai, D. F., and Hans Ris, Division in the dinoflagellate *Gyrodinium cohnii* (Schiller). A new type of nuclear reproduction. *J. Cell Biol.,* v. 40, p. 508−528, 1969.

Lackey, J. B., Bottom sampling and environmental niches. *Limnol. Oceanogr.,* v. 6, p. 271−279, fig. 1, 1961.

Lackey, J. B., and K. A. Clendenning, Ecology of the microbiota of San Diego Bay, California. *Trans. S. Diego Soc. Nat. Hist.,* v. 14, p. 9−40, figs. 1−3, 1965.

Lankester, E. R., Protozoa, p. 830−866, In *The Encyclopaedia Britannica,* Ed. 9, v. 19, 1885.

Leadbeater, B. S. C., and J. D. Dodge, The fine structure of *Woloszynskia micra* sp. nov., a new marine dinoflagellate. *Br. Phycol. Bull.,* v. 3, p. 1−17, pls. 1−6, 1966.

Leadbeater, B. S. C., and J. D. Dodge, Fine structure of the dinoflagellate transverse flagellum. *Nature,* v. 213, p. 421−422, figs. 1−5, 1967a.

Leadbeater, B. S. C., and J. D. Dodge, An electron microscope study of dinoflagellate flagella. *J. Gen. Microbiol.,* v. 46, p. 305−314, pls. 1−4, figs. 1−5, 1967b.

Lebour, M. V., *The dinoflagellates of northern seas.* Plymouth: Mar. Biol. Assoc. U.K., vi + 250 p., 35 pls., 1925.

Lefèvre, Marcel, *Monographie des espèces d'eau douce du genre* Peridinium. Archs. Bot. Mém., v. 2, Mém. 5, p. 1−210, pls. 1−6, figs. 1−915, 1932a.

Lefèvre, Marcel, Sur la présence de Péridiniens dans un dépôt fossile des Barbades. *C. R. Hebd. Séanc. Acad. Sci., Paris,* v. 194, p. 2315−2316, 1932b.

Lefèvre, Marcel, Sur la structure de la thèque chez les *Peridinites. C. R. Hebd. Séanc. Acad. Sci., Paris,* v. 197, p. 81−83, 1933a.

Lefèvre, Marcel, Recherches sur les Péridiniens fossiles des Barbades. *Bull. Mus. Natn. Hist. Nat., Paris,* sér. 2, v. 5, p. 415−418, 1933b.

Lefèvre, Marcel, Les *Peridinites* des Barbades. *Revue Mycol.,* v. 6, p. 215−229, figs. 1−30, 1933c.

Lefèvre, Marcel, Sur la nature de la thèque originelle des *Peridinites. C. R. Hebd. Séanc. Acad. Sci., Paris,* v. 211, p. 599−601, 1940.

Lejeune, Maria, L'étude microscopique des silex. Un fossile anciennement connu et pourtant méconnu: *Hystrichosphaera ramosa* Ehrbg. (Deuxième Note). *Annls. Soc. Géol. Belg.,* v. 60, Bull. 7, p. 239−260, 2 pls., 1937a.

Lejeune-Carpentier, Maria, L'étude microscopique des silex. Une intéressante préparation d'Ehrenberg. (Quatrième Note). *Annls. Soc. Géol. Belg.,* v. 61, Bull. 2, p. 59−71, 2 pls., 1937b.

Lejeune-Carpentier, Maria, L'étude microscopique des silex. *Areoligera:* nouveau genre d'Hystrichosphaeridée (Sixième Note). *Annls. Soc. Géol. Belg.,* v. 62, p. 163−174, 7 figs., 1938a.

Lejeune-Carpentier, Maria, *Peridinium pyrophorum* Ehrenberg. *Bull. Mus. R. Hist. Nat. Belg.,* v. 14 (44), p. 1−13, 10 figs., 1938b.

Lejeune-Carpentier, Maria, L'étude microscopique des silex. Un nouveau Péridinien crétacique: *Gonyaulax wetzeli* (Septième Note). *Annls. Soc. Géol. Belg.,* v. 62, p. 525−529, 2 figs., 1939.

Lejeune-Carpentier, Maria, L'étude microscopique des silex. Systématique et morphologie des "Tubifères" (Huitième Note). *Annls. Soc. Géol. Belg.,* v. 63, p. 216−237, figs. 1−14, 1940.

Lejeune-Carpentier, Maria, L'étude microscopique des silex. Sur *Hystrichosphaeridium hirsutum* (Ehrenberg) et quelques formes voisines (Neuvième Note). *Annls. Soc. Géol. Belg.,* v. 64, Bull. 3, p. 71−92, figs. 1−9, 1941.

Lejeune-Carpentier, Maria, L'étude microscopique des silex. Une Hystrichosphaeridée à classer parmi les Péridiniens (Onzième Note). *Annls. Soc. Géol. Belg.,* v. 57, p. 22−28, 6 figs., 1943.

Lejeune-Carpentier, Maria, L'étude microscopique des silex. Espèces nouvelles ou douteuses de *Gonyaulax* (Douzième Note). *Annls. Soc. Géol. Belg.,* v. 69, Bull. 4, p. 187−197, 6 figs., 1946.

Lejeune-Carpentier, Maria, L'étude microscopique des silex. *Gymnodinium* et *Phanerodinium* (Dinoflagellates) de Belgique (Treizième Note). *Annls. Soc. Géol. Belg.,* v. 74, p. 307−313, figs. 1−8, 1951.

Lemmermann, Ernst, III Klasse Peridiniales, p. 563−682. *In* Ernst Lemmermann, *Kryptogamenflora der Mark Brandenburg und angrenzender Gebiete,* III Algen I (Schizophyceen, Flagellaten, Peridineen). Leipzig: Gebrüder Borntraeger, 1910.

Lentin, J. K., and G. L. Williams, Fossil dinoflagellates: index to genera and species. *Geol. Surv. Pap. Can.,* 73−42, p. 1−176, 1973.

Lentin, J. K., and G. L. Williams, *A monograph of fossil peridinioid dinoflagellate cysts.* Bedford Inst. Oceanogr., Dartmouth, Nova Scotia, Rept. Ser. BI-R-75-16, vi + 237 p., 21 pls., 1975a.

Lentin, J. K., and G. L. Williams, Fossil dinoflagellates: index to genera and species. Supplement 1. *Can. J. Bot.,* v. 53, p. 2147−2157, 1975b.

Lentin, J. K., and G. L. Williams, *Fossil dinoflagellates: index to genera and species, 1977 edition.* Bedford Inst. Oceanogr., Dartmouth, Nova Scotia, Rept. Ser. BI-R-77-8, v + 209 p., 1977.

Lindemann, Erich, Untersuchungen über Susswasserperidinien und ihre Variationsformen. *Arch. Protistenk.,* v. 39, p. 209−262, pl. 17, figs. 1−144, 1919.

Lindemann, Erich, Abteilung Peridineae (Dinoflagellatae), p. 3−104, figs. 1−92. *In* A. Engler and K. Prantl, *Die Natürlichen Pflanzenfamilien nebst ihren Gattungen und wichtigeren Arten insbesondere den Nutzpflanzen,* Bd. 2. Leipzig: Wilhelm Engelmann, 1928.

Loeblich, A. R., Jr., *Palaeoperidinium* Deflandre, 1934, or *Palaeoperidinium* Deflandre ex Sarjeant, 1967 (Pyrrhophyta). *J. Paleont.,* v. 42, p. 583, 1968.

Loeblich, A. R., Jr., Protistan phylogeny as indicated by the fossil record. *Taxon,* v. 23, p. 277−290, 2 figs., 1974.

Loeblich, A. R., Jr., and W. S. Drugg, Heteraulaceae nom. nov. *Taxon,* v. 17, p. 90, 1968a.

Loeblich, A. R., Jr., and W. S. Drugg, Heteraulacaceae (Pyrrhophyta), a correction. *J. Paleont.,* v. 42, p. 1486, 1968b.

Loeblich, A. R., Jr., and A. R. Loeblich III, Index to the genera, subgenera, and sections of the Pyrrhophyta *Stud. Trop. Oceanogr. Miami,* no. 3, x + 94 p., 1 pl., 1 fig., 1966.

Loeblich, A. R., Jr., and A. R. Loeblich III, Index to the genera, subgenera, and sections of the Pyrrhophyta, II. *J. Paleont.,* v. 42, p. 210−213, 1968.

Loeblich, A. R., Jr., and A. R. Loeblich III, Index to the genera, subgenera, and sections of the Pyrrhophyta, III. *J. Paleont.,* v. 43, p. 193−198, 1969.

Loeblich, A. R., Jr., and A. R. Loeblich III, Index to the genera, subgenera, and sections of the Pyrrhophyta, IV. *J. Paleont.,* v. 44, p. 536−543, 1970a.

Loeblich, A. R., Jr., and A. R. Loeblich III, Index to the genera, subgenera, and sections of the Pyrrhophyta [V]. *Phycologia,* v. 9, p. 199−203, 1970b.

Loeblich, A. R., Jr., and A. R. Loeblich III, Index to the genera, subgenera, and sections of the Pyrrhophyta, VI. *Phycologia,* v. 10, p. 309−314, 1971.

Loeblich, A. R., Jr., and A. R. Loeblich III, Index to the genera, subgenera, and sections of the Pyrrhophyta, VII. *Phycologia,* v. 13, p. 57−61, 1974.

Loeblich, A. R. III, Aspects of the physiology and biochemistry of the Pyrrhophyta. *Phykos, Iyengar Meml. Vol.,* v. 5 (1966), p. 216−255, 1967.

Loeblich, A. R. III, A new marine dinoflagellate genus, *Cachonina,* in axenic culture from the Salton Sea, California, with remarks on the genus *Peridinium. Proc. Biol. Soc. Wash.,* v. 81, p. 91−96, figs. 1−7, 1968.

Loeblich, A. R. III, Thecal ultrastructure and composition of modern dinoflagellates. *J. Paleont.,* v. 43, p. 892 (Abstr.), 1969.

Loeblich, A. R. III, The Amphiesma or dinoflagellate cell covering. *Proc. N. Am. Paleont. Conv., Chicago,* Pt. G, p. 867−929, figs. 1−26, 1970.

Loeblich, A. R. III, Dinoflagellate evolution: speculation and evidence. *J. Protozool.,* v. 23, p. 13−28, figs. 1−9, 1976.

Loeblich, A. R. III, and M. F. Hedberg, Desmokont mitosis: involvement of several microtubule containing channels. *Botanica Mar.,* v. 19, p. 255−257, figs. 1−3, 1976.

Loeblich, A. R. III, L. C. Klotz, T. M. Roberts, R. C. Tuttle, and J. R. Allen, The dinoflagellate nucleus. *J. Phycol.,* v. 10 Suppl., p. 14−15 (Abstr.), 1974.

Loeblich, A. R. III, and J. L. Sherley, Observations on the theca of the motile phase of free-living and symbiotic isolates of *Zooxanthella microadriatica* (Freudenthal) comb. nov. *J. Mar. Biol. Ass. U.K.,* v. 59, p. 195−205, pls. 1−4, 1979.

Loeblich, A. R. III, J. L. Sherley, and R. J. Schmidt, The correct position of flagellar insertion in *Prorocentrum* and description of *Prorocentrum rhathymum* sp. nov. (Pyrrhophyta). *J. Plankton Res.,* v. 1, p. 113−120, figs. 1−17, 1979.

Loeblich, A. R. III, and V. E. Smith, Chloroplast pigments of the marine dinoflagellate *Gyrodinium resplendens. Lipids,* v. 3, p. 57−65, figs. 1−6, 1968.

Loeblich, L. A., and A. R. Loeblich III, The organism causing New England red tides: *Gonyaulax excavata,* p. 207−224, figs. 1−19, *In* V. R. Lo Cicero, *Proc. first international conference on toxic dinoflagellate blooms.* Wakefield: Massachusetts Science and Technology Foundation, 1975.

Lohmann, Hans, Neue Untersuchungen über den Reichthum des Meeres an Plankton und über die Brauchbarkeit der verschiedenen Fangmethoden. Zugleich auch ein Beitrag zur Kenntniss des Mittelmeerauftriebs. *Wiss. Meeresunters. Abt. Kiel,* n. folge, v. 7, p. 1−86, pls. 1−4, 1903.

Lohmann, Hans, Eier und sogenannte Cysten der Plankton-Expedition. *Ergebn. Plankton-Exped. Humboldt-Stiftung,* Bd. IV N, 61 p., 7 pls., 1904.

Lohmann, Hans, Untersuchungen zur Feststellung des vollständigen Gehaltes des Meeres an Plankton. *Wiss. Meeresunters. Abt. Kiel,* n. folge, v. 10, p. 129−370, pls. 1−17, 1908.

Lohmann, Hans, Die Bevölkerung des Ozeans mit Plankton nach den Ergebnissen der Zentrifugenfänge während der Ausreise der "Deutschland" 1911. Zugleich ein Beitrag zur Biologie des Atlantischen Ozeans. *Arch. Biontol.,* v. 4(3), p. 1−617, 16 pls., 113 figs., 1920.

Lombard, E. H., and B. Capon, Observations on the tidepool ecology and behavior of *Peridinium gregarium. J. Phycol.,* v. 7, p. 188−194, figs. 1, 2, 1971.

McIntyre, D. J., Morphologic changes in *Deflandrea* from a Campanian section, District of Mackenzie, N.W.T., Canada. *Geosci. Man,* v. 11, p. 61—76, pls. 1—4, 1975.

McLachlan, Jack, The effect of salinity on growth and chlorophyll content in representative classes of unicellular marine algae. *Can. J. Microbiol.,* v. 7, p. 399—406, figs. 1—3, 1961.

McLaughlin, J. J. A., and P. A. Zahl, Axenic zooxanthellae from various invertebrate hosts. *Ann. N.Y. Acad. Sci.,* v. 77, p. 55—72, 1959.

McLean, D. M., *Cladopyxidium septatum,* n. gen., n. sp., possible Tertiary ancestor of the modern dinoflagellate *Cladopyxis hemibrachiata* Balech, 1964. *J. Paleont.,* v. 46, p. 861—863, 1 pl., 1972.

McLean, D. M., Emendation and transfer of *Eisenackia* (Pyrrhophyta) from the Microdiniaceae to the Gonyaulacaceae. *Geol. För. Stockh. Förh.,* v. 95, p. 261—265, fig. 1, 1973a.

McLean, D. M., A problematical dinoflagellate from the Tertiary of Virginia and Maryland. *Palaeontology,* v. 16, p. 729—732, pl. 90, 1973b.

Maier, Dorothea, Planktonuntersuchungen in tertiären und quartären marinen sedimenten. Ein Beitrag zur Systematik, Stratigraphie und Ökologie der Coccolithophorideen, Dinoflagellaten und Hystrichosphaerideen vom Oligozän bis zum Pleistozän. *Neues Jb. Geol. Paläont. Abh.,* v. 107, p. 278—340, pls. 27—33, figs. 1—4, 1959.

Malloy, R. O., An Upper Cretaceous dinoflagellate cyst lineage from Gabon, West Africa. *Geosci. Man,* v. 4, p. 57—65, pl. 1, figs. 1—5, 1972.

Mandelli, E. F., Carotenoid pigments of the dinoflagellate *Glenodinium foliaceum* Stein. *J. Phycol.,* v. 4, p. 347—348, 1968.

Mangin, L. A., Observations sur la constitution de la membrane des Péridiniens. *C. R. Hebd. Séanc. Acad. Sci., Paris,* v. 144, p. 1055—1057, 1907.

Mangin, L. A., Modifications de la Cuirasse chez quelques Péridiniens. *Int. Revue Ges. Hydrobiol. Hydrogr.,* v. 4, p. 44—54, pls. 7, 8, 1911.

Mangin, L. A., Sur la flore planctonique de la Rade de Saint-Vast-la-Hougue 1908—1912. *Nouv. Archs Mus. Hist. Nat., Paris,* sér. 5, v. 5, p. 147—241, figs. 1—16, 1913.

Mantell, G. A., *The medals of creation: or, first lessons in geology and the study of organic remains.* London: H. G. Bohn, v. 1, p. 239—242, 1844.

Manum, Svein, Some dinoflagellates and hystrichosphaerids from the Lower Tertiary of Spitsbergen. *Nytt Mag. Bot.,* v. 8, p. 17—26, 1 pl., 1960.

Manum, Svein, Some new species of *Deflandrea* and their probable affinity with *Peridinium*. *Årbok Norsk Polarinst.,* 1962, p. 55—67, 3 pls., 4 figs., 1963.

Manum, Svein, and I. C. Cookson, Cretaceous microplankton in a sample from Graham Island, Arctic Canada, collected during the second "Fram"-expedition (1898—1902) with notes on microplankton from the Hassel Formation, Ellef Ringnes Island. *Skr. Norske Vidensk-Akad. Mat.-Naturv. Kl.,* n.s., no. 17, 36 p., 7 pls., 1964.

Mapletoft, H., M. Montgomery, J. Waters, and P. Wells, A new freshwater dinoflagellate. *New Phytol.,* v. 65, p. 54—58, fig. 1, 1966.

Margulis, L., *Origin of eucaryotic cells.* New Haven: Yale Univ. Press, 349 p., 1970.

Matzenauer, Lothar, Die Dinoflagellaten des Indischen Ozeans (mit Ausnahme der Gattung *Ceratium*). *Bot. Arch.,* v. 35, p. 437—510, figs. 1—77, 1933.

May, F. E., Dinoflagellates: fossil motile-stage tests from the Upper Cretaceous of the northern New Jersey Coastal Plain. *Science,* v. 193, p. 1128—1130, figs. 1—3, 1976.

Merrill, J. A., Fossil sponges of the flint nodules in the Lower Cretaceous of Texas. *Bull. Mus. Comp. Zool. Harv.,* v. 28, p. 1—26, 1 pl., 1895.

Metzner, P., Bewegungsstudien an Peridineen. *Z. Bot.,* v. 22, p. 225—265, figs. 1—12, 1929.

Meunier, Alphonse, *Microplankton des Mers de Barents et de Kara: Duc d'Orléans Campagne Arctique de 1907,* pt. 1. Bruxelles: C. Bulens, 355 p., 37 pls., 1910.

Meunier, Alphonse, Microplankton de la Mer Flamande. Partie 3. Les Peridiniens. *Mém. Mus. R. Hist. Nat. Belg.,* v. 8(1), p. 1—116, pls. 15—21, 1919.

Michael, Erhard, Mikroplankton und Sporomorphe aus dem nordwestdeutschen Barrême. *Mitt. Geol. Inst. Techn. Hochsch. Hannover,* Heft 2, p. 22—48, pls. 1—5, fig. 1—3, 1964.

Mignot, J. P., Remarques sur le développement du reticulum endoplasmique et du système vacuolaire chez les Gymnodiniens. *Protistologica,* v. 6, p. 267—281, 1970.

Millioud, M. E., Dinoflagellates and acritarchs from some western European Lower Cretaceous type localities, p. 420—434, pls. 1—3. *In* P. Bronnimann and H. H. Renz, *Proc. First Int. Conf. Planktonic Microfossils, Geneva, 1967,* v. 2. Leiden: E. J. Brill, 1969.

Millioud, M. E., G. L. Williams, and J. K. Lentin, Stratigraphic range charts, selected Cretaceous dinoflagellates. *Am. Ass. Stratigr. Palynol., Contr. Ser.,* no. 4, p. 65—71, pls. 1—12, 1975.

Morbey, S. J., The palynostratigraphy of the Rhaetian Stage, Upper Triassic in the Kendelbachgraben, Austria. *Palaeontographica Abt. B,* v. 152, p. 1—75, pls. 1—19, figs. 1—37, 1975.

Morgenroth, Peter, Mikrofossilien und Konkretionen des Nordwesteuropäischen Untereozäns. *Palaeontographica Abt. B,* v. 119, p. 1—53, pls. 1—11, 1966a.

Morgenroth, Peter, Neue in organischer Substanz erhaltene Mikrofossilien des Oligozäns. *Neues Jb. Geol. Paläont. Abh.*, v. 127, p. 1–12, pls. 1, 2, 1966b.

Morgenroth, Peter, Zur Kenntnis der Dinoflagellaten und Hystrichosphaeridien des Danien. *Geol. Jb.*, v. 86, p. 553–578, pls. 41–48, text-figs. 1–4, 1968.

Morgenroth, Peter, Dinoflagellate cysts from the Lias Delta of Lühnde/Germany. *Neues Jb. Geol. Paläont. Abh.*, v. 136, p. 345–359, pls. 9–13, 1 fig., 1970.

Muller, Jan, Palynology of Recent Orinoco Delta and shelf sediments, Reports of the Orinoco Shelf Expedition; v. 5. *Micropaleontology*, v. 5, p. 1–32, pl. 1, figs. 1–23, 1959.

Müller, O. F., *Vermium terrestrium et fluviatilum seu animalium infusoriorum, helminthicorum et testaceorum, non marinorum secincta historia.* Heineck et Faber typis Martini Hallager, Havniae et Lipsiae 1(1) [32] + 135 p., 1773.

Murphy, E. B., K. A. Steidinger, B. S. Roberts, J. Williams, and J. W. Jolley, Jr., An explanation for the Florida East Coast *Gymnodinium breve* red tide of November 1972. *Limnol. Oceanogr.*, v. 20, p. 481–486, 1975.

Muscatine, Leonard, and Else Cernichiari, Assimilation of photosynthetic products of zooxanthellae by a reef coral. *Biol. Bull. Mar. Biol. Lab., Woods Hole,* v. 137, p. 506–523, 1969.

Muscatine, Leonard, and Cadet Hand, Direct evidence for the transfer of materials from symbiotic algae to the tissues of a coelenterate. *Proc. Natn. Acad. Sci. U.S.A.,* v. 44, p. 1259–1263, figs. 1–3, 1958.

Muscatine, Leonard, R. R. Pool, and R. K. Trench, Symbiosis of algae and invertebrates: aspects of the symbiont surface and the host-symbiont interface. *Trans. Am. Microsc. Soc.,* v. 94, p. 450–469, figs. 1–17, 1975.

Nagy, Esther, Investigations into the Neogenic microplankton of Hungary. *Palaeobotanist,* v. 15, p. 38–46, pls. 1, 2, figs. 1–3, 1966.

Nagy, Esther, Palynological elaborations in the Miocene layers of the Mecsek Mountains. *Magy. Allami Földt. Intéz Évk.,* v. 52, p. 233–649, pls. 1–56, 62 figs., 1969.

Neale, J. W., and W. A. S. Sarjeant, Microplankton from the Speeton Clay of Yorkshire. *Geol. Mag.,* v. 99, p. 439–458, pls. 19, 20, figs. 1–9, 1962.

Nevo, Zvi, and Nathan Sharon, The cell wall of *Peridinium westii,* a non cellulosic glucan. *Biochim. Biophys. Acta,* v. 173, p. 161–175, figs. 1–7, 1969.

Nicholls, K. H., Observations on red colored cells of *Peridinium wisconsinense* Eddy from Buckhorn Lake, Ontario. *Trans. Am. Microsc. Soc.,* v. 92, p. 526–528, fig. 1, 1973.

Nival, Paul, Nouvelles observations sur *Achradina pulchra* Lohmann, dinoflagellé, Gymnosclerotidae (= Amphilotales) en Méditerranée. *Protistologica,* v. 5, p. 125–136, pls. 1, 2, figs. 1–5, 1969a.

Nival, Paul, Données écologiques sur quelques protozoaires planctoniques rares en Méditerranée. *Protistologica,* v. 5, p. 215–225, figs. 1–5, 1969b.

Nordli, Erling, Resting spores in *Goniaulax polyedra* Stein. *Nytt Mag. Naturvid.,* v. 88, p. 207–212, 3 figs., 1951.

Nordli, Erling, Experimental studies on the ecology of Ceratia. *Oikos,* v. 8, p. 200–265, 2 pls., 22 figs., 1957.

Nordli, Ottar, Dinoflagellates from Lofoten. *Nytt. Mag. Naturvid.,* v. 88, p. 49–55, figs. 1–7, 1951.

Norris, Geoffrey, Archeopyle structures in Upper Jurassic dinoflagellates from southern England. *N. Z. Jl. Geol. Geophys.,* v. 8, p. 792–806, figs. 1–17, 1965.

Norris, G., Phylogeny and a revised supra-generic classification for Triassic-Quaternary organic-walled dinoflagellate cysts (Pyrrhophyta). Part I. Cyst terminology and assessment of previous classifications. *Neues Jb. Geol. Paläont. Abh.,* v. 155, p. 300–317, 2 figs., 1978.

Norris, Geoffrey, and R. W. Hedlund, Transapical sutures in dinoflagellate cysts. *Geosci. Man,* v. 4, p. 49–56, pls. 1–3, figs. 1, 2, 1972.

Norris Geoffrey, and J. H. McAndrews, Dinoflagellate cysts from post-glacial lake muds, Minnesota (U.S.A.). *Rev. Palaeobot. Palynol.,* v. 10, p. 131–156, pls. 1–3, 5 figs., 1970.

Norris, Geoffrey, and W. A. S. Sarjeant, A descriptive index of genera of fossil Dinophyceae and Acritarcha. *Paleont. Bull. Geol. Surv. N.Z.,* 40, p. 1–72, 1965.

Norvick, M. S., and D. Burger, Palynology of the Cenomanian of Bathurst Island, Northern Territory, Australia. *Bull. Bur. Miner. Resour. Geol. Geophys. Aust.,* v. 151, p. 1–169, pls. 1–34, figs. 1–28, 1976.

Orr, W. N., and Steven Conley, Siliceous dinoflagellates in the northeast Pacific rim. *Micropaleontology,* v. 22, p. 92–99, pls. 1, 2, 1976.

Osorio Tafall, B. F., Estudios sobre el plancton de Mexico. 1. El género *Lophodinium* Lemm. (Dinophyceae Peridiniales). *Ciencia, Méx.,* v. 3, p. 114–119, 1 fig., 1942.

Ostenfeld, C. H., Phytoplankton from the sea around the Faeröes, p. 558–611, figs. 119–144. In *Botany of the Faeröes based upon Danish investigations.* Part 2. Copenhagen: Det Nordiske Forlag, 1903.

Papenfuss, G. F., Classification of the Algae, p. 115–224. *In* E. L. Kessel, *A century of progress in the natural sciences, 1853–1953.* San Francisco: Calif. Acad. Sci., 1955.

Parke, D. L., Une nouvelle espèce de dinoflagellé siliceux: *Actiniscus. Revue Micropaléont.,* v. 17, p. 81–85, pl. 1, 1975.

Parke, Mary, and Dorothy Ballantine, A new marine

dinoflagellate: *Exuviaella mariae-lebouriae* n. sp. *J. Mar. Biol. Ass. U.K.,* v. 36, p. 643—650, figs. 1—18, 1957.

Pascher, Adolf, Die braune Algenreihe aus der Verwandtschaft der Dinoflagellaten (Dinophyceen). *Arch. Protistenk.,* v. 58, p. 1—58, figs. 1—38, 1927.

Pastiels, André, Étude histochemique des coques d'Hystrichosphères. *Bull. Mus. R. Hist. Nat. Belg.,* v. 21(17), p. 1—20, 1945.

Pastiels, André, Contribution a l'étude des microfossiles de l'Éocène Beige. *Mém. Mus. R. Hist. Nat. Belg.,* no. 109, p. 1—77, pls. 1—6, figs. 1—4, 1948.

Patton, Stuart, P. T. Chandler, E. B. Kalan, A. R. Loeblich III, G. Fuller, and A. A. Benson, Food value of red tide (*Gonyaulax polyedra*). *Science,* v. 158, p. 789—790, 1967.

Paulsen, Ove, Nordisches Plankton. XVIII Peridiniales, p. 1—124, figs. 1—155. *In* K. Brandt and C. Apstein, *Nordisches Plankton, Botanischer Teil.* Kiel and Leipzig: Lipsius and Tischer, 1908.

Paulsen, Ove, Observations on dinoflagellates. *Biol. Skr.,* v. 6(4), p. 1—67, figs. 1—30, 1949.

Pavillard, Jules, Accroissement et scissiparité chez les Péridiniens. *C. R. Hebd. Séanc. Acad. Sci., Paris,* v. 160, p. 372—375, 1 fig., 1915.

Pearse, V. B., and Leonard Muscatine, Role of symbiotic algae (Zooxanthellae) in coral calcification. *Biol. Bull. Mar. Biol. Lab., Woods Hole,* v. 141, p. 350—363, figs. 1—7, 1971.

Peters, Nicolaus, Beiträge zur Planktonbevölkerung der Weddellsee nach den Ergebnissen der Deutschen Antarktischen Expedition 1911—1912. Beitrag III. Die Peridineen bevölkerung der Weddellsee mit besonderer Berüchsichtigung der Wachstums- und Variationsformen. *Int. Revue Ges. Hydrobiol. Hydrogr.,* v. 21(1928) p. 17—146, figs. 1—33, 1929.

Peters, Nicolaus, Peridinea, p. 13—84, 174 figs. *In* G. Grimpe and E. Wagler, *Tierwelt der Nord- und Ostsee,* v. 1, IId$_2$. Leipzig: Akad. Verlags., Becker and Erler Kom.-Ges., 1930.

Pfiester, L. A., Sexual reproduction of *Peridinium cinctum* f. *ovoplanum* (Dinophyceae). *J. Phycol.,* v. 11, p. 259—265, figs. 1—3, 1975.

Pfiester, L. A., Sexual reproduction of *Peridinium gatunense* Nygaard. *J. Phycol.,* v. 12 (Suppl.), p. 16—17 (Abstr.), 1976.

Pocock, S. A. J., Palynology of the Jurassic sediments of western Canada, Part 2 Marine species. *Palaeontographica Abt. B,* v. 137, p. 85—153, pls. 22—29, figs. 1—45, 1972.

Pocock, S. A. J., A preliminary zonation of the uppermost Jurassic and lower part of the Cretaceous, Canadian Arctic, and possible correlation in the western Canada Basin. *Geosci. Man,* v. 15, p. 101—114, 2 pls., 1976.

Pothe de Baldis, E. D., Microplancton del Terciario de Tierra del Fuego. *Ameghiniana,* v. 4, p. 219—228, pls. 1, 2, 1966.

Pouchet, Georges, Contribution à l'histoire des cilioflagellés. *J. Anat. Physiol.,* v. 19, p. 399—455, pls. 19—22, 1883.

Pouchet, Georges, D'un oeil véritable chez les Protozoaires. *C. R. Séanc. Soc. Biol.,* sér. 8, v. 1 (v. 36), p. 593—596, 1884.

Pouchet, Georges, Nouvelle contribution à l'histoire des Péridiniens marins. *J. Anat. Physiol.,* v. 21, p. 28—88, pls. 2—4, 1885.

Pouchet, Georges, Histoire naturelle, Chapter X. *In* Voyage de "la Manche" à l'Ile Jan Mayen et au Spitzberg Jouillet-Août, 1892. *Nouv. Archs. Missions Scient. Litt.,* v. 5, p. 154—217, pl. 22, 1893.

Prager, J. C., Fusion of the family Glenodiniaceae into the Peridiniaceae, with notes on *Glenodinium foliaceum* Stein. *J. Protozool.,* v. 10, p. 195—204, figs. 1—3, 1963.

Rae, P. M. M., Hydroxymethyluracil in eukaryote DNA: a natural feature of the Pyrrhophyta (Dinoflagellates). *Science,* v. 194, p. 1062—1064, 1976.

Ragotzkie, R. A., and L. R. Pomeroy, Life history of a dinoflagellate bloom. *Limnol. Oceanogr.,* v. 2, p. 62—69, 2 figs., 1957.

Rampi, Leopoldo, Su qualche altra Peridinea nuova o rara delle acque di Sanremo. *Atti Soc. Ital. Sci. Nat.,* v. 82(2), p. 151—157, figs. 1—9, 1943.

Reade, J. B., On some new organic remains in the flint of the Chalk. *Ann. Mag. Nat. Hist.,* v. 2, p. 191—198, 2 pls., 1839.

Reid, P. C., Dinoflagellate cyst distribution around the British Isles. *J. Mar. Biol. Ass. U.K.,* v. 52, p. 939—944, figs. 1—3, 1972.

Reid, P. C., Gonyaulacacean dinoflagellate cysts from the British Isles. *Nova Hedwigia,* v. 25, p. 579—637, pls. 1—4, figs. 1—15, 1974.

Reinsch, P. F., Die Palinosphärien, ein mikroskopischer vegetabiler Organismus in der Mucronatenkreide. *Zentbl. Miner. Geol. Paläont.,* Jg. 1905, p. 402—407, figs. 1, 2, 1905.

Ricard, M., Quelques dinoflagellés planctoniques marins de Tahiti étudiés en microscopie à balayage. *Protistologica,* v. 10, p. 125—135, pls. 1—5, 1974.

Riegel, Walter, New forms of organic-walled microplankton from an Upper Cretaceous assemblage in southern Spain. *Revta. Esp. Micropaleont.,* v. 6, p. 347—366, pls. 1—3, 1974.

Riley, L. A., and W. A. S. Sarjeant, Survey of the stratigraphical distribution of dinoflagellates, acritarchs and tasmanitids in the Jurassic. *Geophytology,* v. 2, p. 1—40, 1972.

Ris, Hans, and D. F. Kubai, An unusual mitotic mechanism in the parasitic protozoan *Syndinium* sp. *J. Cell Biol.*, v. 60, p. 702–720, figs. 1–33, 1974.

Rizzo, P. J., Isolation and partial characterization of dinoflagellate nuclei and chromatin. *Diss. Abst. Int.*, v. 33, p. 5175B (Abstr.), 1973.

Rizzo, P. J., and L. D. Noodén, Isolation and chemical composition of dinoflagellate nuclei. *J. Protozool.*, v. 20, p. 666–672, figs. 1–6, 1973.

Roberts, T. M., R. C. Tuttle, J. R. Allen, A. R. Loeblich III, and L. C. Klotz, New genetic and physicochemical data on structure of dinoflagellate chromosomes. *Nature*, v. 248, p. 446–447, 1 fig., 1974a.

Roberts, T. M., L. C. Klotz, A. R. Loeblich III, and J. R. Allen, Measurement of molecular weight of dinoflagellate chromosomal DNA. *J. Phycol.*, v. 10 Suppl., p. 15 (Abstr.), 1974b.

Rossignol, Martine, Analyse pollinique de sédiments marins Quaternaires en Israel. 1: Sediments Recents. *Pollen Spores*, v. 3, p. 303–324, pls. 1, 2, figs. 1–17, 1961.

Rossignol, Martine, Aperçus sur le développement des Hystrichosphères. *Bull. Mus. Natn. Hist. Nat., Paris*, sér. 2, v. 35 (2), p. 207–212, pls. 1, 2, figs. 1–17, A–H, 1963.

Rossignol, Martine, Hystrichosphères du Quaternaire en Méditerranée orientale, dans les sédiments Pléistocènes et les boues marines actuelles. *Revue Micropaléont.*, v. 7, p. 83–99, pls. 1–3, figs. A–H, 1964.

Russell, F. E., Marine toxins and venomous and poisonous marine animals. *Adv. Mar. Biol.*, v. 3, p. 255–384, figs. 1–20, 1965.

Ryther, J. H., Photosynthesis in the ocean as a function of light intensity. *Limnol. Oceanogr.*, v. 1, p. 61–70, figs. 1–7, 1956.

Sarjeant, W. A. S., Microplankton from the Cornbrash of Yorkshire. *Geol. Mag.*, v. 96, p. 329–346, pl. 13, figs. 1–8, 1959.

Sarjeant, W. A. S., New hystrichospheres from the Upper Jurassic of Dorset. *Geol. Mag.*, v. 97, p. 137–144, pl. 6, figs. 1–4, 1960.

Sarjeant, W. A. S., Microplankton from the Kellaways Rock and Oxford Clay of Yorkshire. *Palaeontology*, v. 4, p. 90–118, pls. 13–15, figs. 1–15, 1961a.

Sarjeant, W. A. S., *Systematophora* Klement and *Polystephanosphaera* Sarjeant. *J. Paleont.*, v. 35, p. 1094–1096, 1961b.

Sarjeant, W. A. S., Upper Jurassic microplankton from Dorset, England. *Micropaleontology*, v. 8, p. 255–268, pls. 1, 2, 1962a.

Sarjeant, W. A. S., Microplankton from the Ampthill Clay of Melton, South Yorkshire. *Palaeontology*, v. 5, p. 478–497, pls. 69, 70, figs. 1–13, 1962b.

Sarjeant, W. A. S., Fossil dinoflagellates from Upper Triassic sediments. *Nature*, v. 199, p. 353–354, figs. 1–3, 1963a.

Sarjeant, W. A. S., *Favilarnax*, a new genus of Mesozoic hystrichospheres. *J. Paleont.*, v. 37, p. 719–721, 1963b.

Sarjeant, W. A. S., Taxonomic notes on hystrichospheres and acritarchs. *J. Paleont.*, v. 38, p. 173–177, 1964.

Sarjeant, W. A. S., The supposed "sponge spicules" of Merrill, 1895, from the Lower Cretaceous (Albian) of Texas. *Breviora*, no. 242, p. 1–15, pl. 1, fig. A, 1966a.

Sarjeant, W. A. S., Further dinoflagellate cysts from Speeton Clay. *Bull. Br. Mus. Nat. Hist., Geology*, Suppl. 3, p. 199–214, pls. 21–23, 1966b.

Sarjeant, W. A. S., Dinoflagellate cysts with *Gonyaulax*-type tabulation. *Bull. Br. Mus. Nat. Hist., Geology*, Suppl. 3, p. 107–156, pls. 13–16, figs. 23–40, 1966c.

Sarjeant, W. A. S., The stratigraphical distribution of fossil dinoflagellates. *Revue Palaeobot. Palynol.*, v. 1, p. 323–343, 1967a.

Sarjeant, W. A. S., The genus *Palaeoperidinium* Deflandre 1934 (Dinophyceae). *Grana Palynol.*, v. 7, p. 241–258, figs. 1–6, 1967b.

Sarjeant, W. A. S., The rediscovery of a lost species of dinoflagellate cyst: *Hystrichosphaera* (ex: *Spiniferites*) *reginaldi* (Mantell, 1844) comb. nov. *Microscopy: J. Quekett Microsc. Club*, v. 30, p. 241–250, 1 pl., 1967c.

Sarjeant, W. A. S., Microplankton from the Upper Callovian and Lower Oxfordian of Normandy. *Revue Micropaléont.*, v. 10, p. 221–242, pls. 1–3, figs. 1–5, 1968.

Sarjeant, W. A. S., Xanthidia, palinospheres and 'Hystrix'. *Microscopy: J. Quekett Microsc. Club*, v. 31, p. 221–253, figs. 1–12, 1970a.

Sarjeant, W. A. S., The genus *Spiniferites* Mantell, 1850 (Dinophyceae). *Grana*, v. 10, p. 74–78, 1970b.

Sarjeant, W. A. S., Dinoflagellate cysts and acritarchs from the Upper Vardekløft Formation (Jurassic) of Jameson Land, East Greenland. *Meddr. Grønland*, v. 195 (4), p. 1–64, pls. 1–7, figs. 1–9, 1972.

Sarjeant, W. A. S., *Fossil and living dinoflagellates*. London and New York: Academic Press, 182 p., 45 figs. 1974.

Sarjeant, W. A. S., Stratigraphic range charts: Triassic and Jurassic dinoflagellates. *Am. Ass. Stratig. Palynol., Contr. Ser.*, No. 4, p. 51–63, pls. 1–3, 1975a.

Sarjeant, W. A. S., *Hapsidaulax*, new genus of dinoflagellate cysts from the Jurassic (Bathonian) of the Isle of Skye. *Scott. J. Geol.*, v. 11, p. 143–149, pls. 1, 2, figs. 1–3, 1975b.

Sarjeant, W. A. S., Jurassic dinoflagellate cysts with

epitractal archeopyles. *Grana,* v. 14 (1974), p. 49—56, pls. 1—3, 1975c.

Sarjeant, W. A. S., *Energlynia,* new genus of dinoflagellate cysts from the Great Oolite Limestone (Middle Jurassic: Bathonian) of Lincolnshire, England. *Neues Jb. Geol. Paläont. Mh.,* 1976, p. 163—173, 16 figs., 1976a.

Sarjeant, W. A. S., Dinoflagellate cysts and acritarchs from the Great Oolite Limestone (Jurassic: Bathonian) of Lincolnshire, England. *Geobios, Lyon,* no. 9, p. 5—46, 6 pls., 6 figs., 1976b.

Sarjeant, W. A. S., *Arpylorus antiquus* Calandra, emend., a dinoflagellate cyst from the Upper Silurian. *Palynology,* v. 2, p. 167—179, pls. 1—4, 1978.

Sarjeant, W. A. S., and Charles Downie, The classification of dinoflagellate cysts above generic level. *Grana Palynol.,* v. 6, p. 503—527, 1966.

Sarjeant, W. A. S., and Charles Downie, Class Dinophyceae, p. 195—207, figs. 2.3A—2.6B. *In* W. B. Harland et al., *The fossil record.* London: Geol. Soc. London, 1967.

Sarjeant, W. A. S., and Charles Downie, The classification of dinoflagellate cysts above generic level: a discussion and revisions. *Birbal Sahni Inst. Palaeobot.,* Spec. Publ. 3, p. 9—32, 1974.

Sarma, Y. S. R. K., and R. Shyam, On the morphology, reproduction and cytology of two freshwater dinoflagellates from India. *Br. Phycol. J.,* v. 9, p. 21—29, 1974.

Saunders, R. P., and C. L. Wahlquist, Diatoms and their relationship to *Gymnodinium breve* Davis. *Fla. Board Conserv., Mar. Lab. Prof. Pap. Ser.,* 8, p. 16—33. 1966.

Schiller, Josef, Über die Besiedlung europäischer Meere mit Cryptomonaden und über einen Flagellaten peridinieenähnlicher Organisation (*Entomosigma peridinioides*). *Öst. Bot. Z.,* v. 74, p. 194—198, figs. a—d, 1925.

Schiller, Josef, Die planktischen Vegetation des adriatischen Meeres. C. Dinoflagellata. 1 Teil. Adiniferida, Dinophysidaceae. Systematischer Teil. *Arch. Protistenk.,* v. 61, p. 45—99, pl. 3, 1928.

Schiller, Josef, Dinoflagellatae (Peridineae) in monographischer Behandlung, p. 1—617, figs. 1—631. In *Dr. L. Rabenhorst's Kryptogamen-Flora von Deutschland, Österreich und der Schweiz,* Bd. 10, Abt. 3, Teil 1—4. Leipzig: Akad. Verlags., 1931—1937.

Schilling, A. J., Dinoflagellatae (Peridineae). *In* A. Pascher, *Die Süsswasser-Flora Deutschlands, Österreichs und der Schweiz,* v. 3. Jena: Gustav Fischer, iv + 66 p., 69 figs., 1913.

Schmidt, Johs, Flora of Koh Chang. Contributions to the knowledge of the vegetation in the Gulf of Siam. Peridiniales. *Bot. Tidsskr.,* v. 24, p. 212—221, figs. 1—8, 1902.

Schütt, Franz, Ueber die Sporenbildung mariner Peridineen. *Ber. Dt. Bot. Ges.,* v. 5, p. 364—374, pl. 18, 1887.

Schütt, Franz, Sulla formazione scheletrica intracellulare di un dinoflagellato. *Neptunia,* v. 1, p. 407—426, pl. 3, 1891.

Schütt, Franz, Die Peridineen der Plankton-Expedition. *Ergebn. Plankton-Exped. Humboldt-Stiftung,* Bd. 4, M. a A, p. 1—170, pls. 1—27, 1895.

Schütt, Franz, Peridiniales, p. 1—30. *In* A. Engler and K. Prantl, *Die natürlichen Pflanzenfamilien, nebst ihren Gattungen und wichtigeren Arten, insbesondere den Nutzpflanzen, unter Mitwirkung Zahlreicher hervorragender Fachgelehrten.* I Teil 1. Abt. b. Gymnodiniaceae, Prorocentraceae, Peridiniaceae, Bacillariaceae. Leipzig: W. Engelmann, 1896.

Seliger, H. H., J. H. Carpenter, M. Loftus, and W. D. McElroy, Mechanisms for the accumulation of high concentrations of dinoflagellates in a bioluminescent bay. *Limnol. Oceanogr.,* v. 15, p. 234—245, figs. 1—9, 1970.

Seliger, H. H., and W. D. McElroy, Studies at Oyster Bay in Jamaica, West Indies. I. Intensity patterns of bioluminescence in a natural environment. *J. Mar. Res.,* v. 26, p. 245—255, 1968.

Sellberg, Björn, and Göran Kjellström, Geometric characterization of cavate dinoflagellate cysts of the genus *Deflandrea. Neues Jb. Geol. Paläont. Mh.,* 1974, p. 315—320, 2 figs., 1974.

Sellberg, Björn, and Göran Kjellström, Geometric characterization of dinoflagellate cysts: an addendum. *Neues Jb. Geol. Paläont. Mh.,* 1975, p. 51—53, figs. 1, 2, 1975.

Silva, E. S., The 'small form' in the life cycle of dinoflagellates and its cytological interpretation. *Proc. II Planktonic Conf. Roma 1970,* p. 1157—1167, pls. 1—3, 1971.

Skuja, Henrich, Beitrag zur Algenflora Lettlands II. *Acta Horti Bot. Univ. Latv.,* v. 11—12 (1—3), p. 41—169, pls. 1—11, 1939.

Skuja, Henrich, Taxonomische und biologische Studien über das Phytoplankton Schwedischer Binnengewässer. *Nova Acta R. Soc. Scient. Upsal. IV,* v. 16(3), p. 1—404, 1956.

Smayda, T. J., Biogeographical studies of marine phytoplankton. *Oikos,* v. 9, p. 158—191, figs. 1—7, 1958.

Smith, D. L., Leonard Muscatine, and D. Lewin, Carbohydrate movement from autotrophs to heterotrophs in parasitic and mutualistic symbiosis. *Biol. Rev.,* v. 44, p. 17—90, 1969.

Sournia, Alain, Sur la variabilité infraspécifique du genre *Ceratium* (Péridinien planctonique) en milieu marin. *C. R. Hebd. Séanc. Acad. Sci., Paris,* v. 263, p. 1980—1983, 1966.

Sournia, Alain, Contribution à la connaissance des Péridiniens microplanctoniques du Canal de Mozambique. *Bull. Mus. Natn. Hist. Nat., Paris*, sér. 2, v. 39, p. 417–438, pl. 1, figs. 1–8, 1967a.

Sournia, Alain, Le genre *Ceratium* (Péridinien planctonique) dans le Canal de Mozambique. Contribution à une révision mondiale. *Vie Milieu, Sér. A, Biol. Mar.*, v. 18, p. 375–500, pls. 1–3, 1967b.

Sournia, Alain, Une période de poussées phytoplanctoniques près de Nosy-Bé (Madagascar) en 1971. I. Espèces rares ou nouvelles du phytoplancton. *Cah. Off. Rech. Scient. Tech. Outre-Mer (ORSTOM), Sér. Oceanogr.*, v. 10, p. 151–159, figs. 1–27, 1972a.

Sournia, Alain, Quatre nouveaux dinoflagellés du plancton marin. *Phycologia*, v. 11, p. 71–74, figs. 1–8, 1972b.

Sournia, Alain, Catalogue des espèces et taxons infraspécifiques de dinoflagellés marins actuels publiés depuis la revision de J. Schiller, I. Dinoflagellés libres. *Nova Hedwigia, Beihefte*, v. 48, x + 92 p., 1973.

Sournia, Alain, Jean Cachon, and Monique Cachon, Catalogue des espèces et taxons infraspécifiques de dinoflagellés marins actuels publiés depuis la revision de J. Schiller. II Dinoflagellés parasites ou symbiotiques. *Arch. Protistenk.*, v. 117, p. 1–19, 1975.

Soyer, M.-O., Étude cytologique ultrastructurale d'un dinoflagellé libre, *Noctiluca miliaris* Suriray: Trichocystes et inclusions paracristallines. *Vie Milieu, Sér. A, Biol. Mar.*, v. 19, p. 305–314, pls. 1–3, 1 fig., 1968.

Stanley, E. A., Upper Cretaceous and Paleocene plant microfossils and Paleocene dinoflagellates and hystrichosphaerids from northwestern South Dakota. *Bull. Am. Paleont.*, v. 49, p. 177–384, pls. 19–49, figs. 1–3, 1965.

Steidinger, K. A., Phytoplankton ecology: a conceptual review based on Eastern Gulf of Mexico research. *CRC Critical Rev. Microbiol.*, v. 3, issue 1, p. 49–68. Cleveland: Chemical Rubber Co., 1973.

Steidinger, K. A., Implications of dinoflagellate life cycles on initiation of *Gymnodinium breve* red tides. *Envir. Letters*, v. 9, p. 129–139, 1975.

Steidinger, K. A., J. T. Davis, and Jean Williams, Observations of *Gymnodinium breve* Davis and other dinoflagellates. *Fla. Board Conserv. Mar. Lab., Prof. Pap. Ser.*, no. 8, p. 8–15, 1966.

Steidinger, K. A., J. T. Davis, and Jean Williams, A key to the marine dinoflagellate genera of the west coast of Florida. *Fla. Board Conserv. Mar. Lab., Tech. Ser.*, no. 52, p. 1–45, 1967.

Steidinger, K. A., and R. M. Ingle, Observations on the 1971 summer red tide in Tampa Bay, Florida. *Envir. Letters*, v. 3, p. 271–278, 1972.

Steidinger, K. A., and E. A. Joyce, Jr., Florida red tides. *Fla. Dept. Nat. Resources, Educ. Ser.*, no. 17, p. 1–26, 1973.

Steidinger, K. A., and Jean Williams, *Dinoflagellates: Memoirs Hourglass Cruises*, v. 2. St. Petersburg: Marine Research Lab., Fla. Dept. Nat. Resources, p. 1–251, 45 pls., 8 figs., 1970.

Stein, F. R. von, *Der Organismus der Infusionsthiere nach eigenen Forschungen in systematischer Reihenfolge bearbeitet*. Abt. III, Hälfte II, Die Naturgeschichte der arthrodelen Flagellaten. Leipzig: Wilhelm Engelmann, 30 p., 25 pls., 1883.

Stone, J. F., Palynology of the Almond Formation (Upper Cretaceous), Rock Springs Uplift, Wyoming. *Bull. Am. Paleont.*, v. 64 (278), p. 1–135, pls. 1–20, 1973.

Stosch, H. A. von, Zum Problem der sexuellen Fortpflanzung in der Peridineengattung *Ceratium*. *Helgoländer Wiss. Meeresunters.*, v. 10, p. 140–152, figs. 1–7, 1964.

Stosch, H. A. von, Sexualität bei *Ceratium cornutum* (Dinophyta). *Naturwissenschaften*, v. 52, p. 112–113, 1 fig., 1965.

Stosch, H. A. von, Dinophyta, p. 626–636, figs. 18–26. *In* W. Ruhland, *Handbuch der Pflanzenphysiologie*, v. 18. Berlin, Heidelberg, New York: Springer Verlag, 1967.

Stosch, H. A. von, Dinoflagellaten aus der Nordsee I. Über *Cachonina niei* Loeblich (1968), *Gonyaulax grindleyi* Reinecke (1967) und eine methode zur Darstellung von Peridineenpanzern. *Helgoländer Wiss. Meeresunters.*, v. 19, p. 558–568, figs. 1–18, 1969a.

Stosch, H. A. von, Dinoflagellaten aus der Nordsee II. *Helgolandinium subglobosum* gen. et spec. nov. *Helgoländer Wiss. Meeresunters.*, v. 19, p. 569–577, figs. 1–6, 1969b.

Stosch, H. A. von, La signification cytologique de la 'cyclose nucléaire' dans le cycle de vie des dinoflagelles. *Mém. Soc. Bot. Fr.*, 1972, p. 201–212, figs. 1–8, 1972.

Stosch, H. A. von, Observations on vegetative reproduction and sexual life cycles of two freshwater dinoflagellates, *Gymnodinium pseudopalustre* Schiller and *Woloszynskia apiculata* sp. nov. *Br. Phycol. J.*, v. 8, p. 105–134, figs. 1–87, 1973.

Stover, L. E., Palaeocene and Eocene species of *Deflandrea* (Dinophyceae) in Victorian coastal and offshore basins, Australia. *Spec. Publs. Geol. Soc. Aust.*, v. 4, p. 167–188, pls. 1–5, (1973), 1974.

Stover, L. E., Observations on some Australian Eocene dinoflagellates. *Geosci. Man*, v. 11, p. 35–45, pls. 1–3, 1975.

Stover, L. E., Oligocene and early Miocene dinoflagellates from Atlantic Corehole 5/5B, Blake Plateau. *Contr. Ser. Am. Ass. Stratigr. Palynol.*, no. 5A, p. 66–89, pls. 1–3, figs. 1, 2, 1977.

Stover, L. E., and W. R. Evitt, Analyses of Pre-Pleistocene organic-walled dinoflagellates. *Stanford Univ. Publs. Geol. Sci.,* v. 15, p. i—iii, 1—298, 1978.

Stover, L. E., W. A. S. Sarjeant, and W. S. Drugg, The Jurassic dinoflagellate genus *Stephanelytron:* emendation and discussion. *Micropaleontology,* v. 23, p. 330—338, pl. 1, figs. 1—3, 1977.

Strain, H. H., Pigments, p. 243—262. *In* G. M. Smith, *Manual of phycology.* Waltham: Chronica Botanica, 1951.

Suchlandt, Otto, Dinoflagellaten als Erreger von rotem Schnee. *Ber. Dt. Bot. Ges.,* v. 34, p. 242—246, pl. 3, fig. 1, 1916.

Sweeney, B. M., Bioluminescent dinoflagellates. *Biol. Bull. Mar. Biol. Lab., Woods Hole,* v. 125, p. 177—181, 1963.

Sweeney, B. M., Laboratory studies of a green *Noctiluca* from New Guinea. *J. Phycol.,* v. 7, p. 53—58, figs. 1—4, 1971.

Sweeney, B. M., and J. W. Hastings, Rhythms, p. 687—700. *In* R. A. Lewin, *Physiology and biochemistry of algae.* New York, London: Academic Press, 1962.

Swift, Elijah, W. H. Biggley, and H. H. Seliger, Species of oceanic dinoflagellates in the genera *Dissodinium* and *Pyrocystis:* Interclonal and interspecific comparisons of the color and photon yield of bioluminescence. *J. Phycol.,* v. 9, p. 420—426, figs. 1, 2, 1973.

Swift, Elijah, and E. G. Durbin, Similarities in the asexual reproduction of the oceanic dinoflagellates *Pyrocystis fusiformis, Pyrocystis lunula,* and *Pyrocystis noctiluca. J. Phycol.,* v. 7, p. 89—96, figs. 1—33, 1971.

Swift, Elijah, and Valerie Meunier, Effects of light intensity on division rate, stimulable bioluminescence and cell size of the oceanic dinoflagellates *Dissodinium lunula, Pyrocystis fusiformis* and *P. noctiluca. J. Phycol.,* v. 12, p. 14—22, figs. 1—5, 1976.

Swift, Elijah, and C. C. Remsen, The cell wall of *Pyrocystis* spp. (Dinococcales). *J. Phycol.,* v. 6, p. 78—86, figs. 1—13, 1970.

Swift, Elijah, and David Wall, Asexual reproduction through a thecate stage in *Pyrocystis acuta* Kofoid, 1907 (Dinophyceae). *Phycologia,* v. 11, p. 57—65, figs. 1—19, 1972.

Tai, Li-sun, and Tage Skogsberg, Studies on the Dinophysoidae, marine armored dinoflagellates of Monterey Bay, California. *Arch. Protistenk.,* v. 82, p. 380—482, pls. 11—12, figs. 1—14, 1934.

Takahashi, K., Microfossils from the Pleistocene sediments of the Areaka Sea area, West Kyushu. *Trans. Proc. Palaeont. Soc. Japan,* no. 81, p. 11—26, pls. 2—5, 1971.

Tappan, Helen, and A. R. Loeblich, Jr., Geobiologic implications of fossil phytoplankton evolution and time-space distribution, p. 247—340, 1 pl., 12 figs. *In* R. Kosanke and A. T. Cross, Symposium on palynol-

ogy of the Late Cretaceous and Early Tertiary, *Geol. Soc. Am., Spec. Pap.,* 127, 1971.

Tappan, Helen, and A. R. Loeblich, Jr., Fluctuating rates of protistan evolution, diversification and extinction. *XXIV Int. Geol. Congr., Sec. 7 Paleont.,* p. 205—213, 6 figs., 1972.

Tappan, Helen, and A. R. Loeblich, Jr., Evolution of the oceanic plankton. *Earth-Sci. Rev.,* v. 9, p. 207—240, 9 figs., 1973.

Tappan, Helen, and A. R. Loeblich, Jr., Peridinialean cyst affinity, rather than gymnodinialean motile stage, of the Late Cretaceous dinoflagellate *Dinogymnium. Trans. Am. Microsc. Soc.,* v. 96, p. 497—505, figs. 1—21, 1977.

Tasch, Paul, Hystrichosphaerids and dinoflagellates from the Permian of Kansas. *Micropaleontology,* v. 9, p. 332—336, pl. 1, 1963.

Tasch, Paul, Ken McClure, and Orrin Oftedahl, Biostratigraphy and taxonomy of a hystrichosphere-dinoflagellate assemblage from the Cretaceous of Kansas. *Micropaleontology,* v. 10, p. 189—206, pls. 1—3, 1964.

Taylor, D. L., Identity of zooxanthellae isolated from some Pacific Tridacnidae. *J. Phycol.,* v. 5, p. 336—340, figs. 1—4, 1969.

Taylor, D. L., Ultrastructure of the 'zooxanthella' *Endodinium chattonii* in situ. *J. Mar. Biol. Ass. U.K.,* v. 51, p. 227—234, pls. 1—3, 1971a.

Taylor, D. L., On the symbiosis between *Amphidinium klebsii* [Dinophyceae] and *Amphiscolops langerhansi* [Turbellaria: Acoela]. *J. Mar. Biol. Ass. U.K.,* v. 51, p. 301—313, pls. 1—5, figs. 1—4, 1971b.

Taylor, D. L., The cellular interactions of algal-invertebrate symbiosis. *Adv. Mar. Biol.,* v. 11, p. 1—56, figs. 1—8, 1973.

Taylor, F. J. R., Parasitism of the toxin-producing dinoflagellate *Gonyaulax catenella* by the endoparasitic dinoflagellate *Amoebophrya ceratii. J. Fish. Res. Bd. Can.,* v. 25, p. 2241—2245, figs. 1—10, 1968.

Taylor, F. J. R., Peri-nuclear structural elements formed in the dinoflagellate *Gonyaulax pacifica* Kofoid. *Protistologica,* v. 5, p. 165—167, pl. 1, 1969.

Taylor, F. J. R., Scanning electron microscopy of the theca of the dinoflagellate genus *Ornithocercus. J. Phycol.,* v. 7, p. 249—258, figs. 1—33, 1971.

Taylor, F. J. R., Unpublished observations on the thecate stage of the dinoflagellate genus *Pyrocystis* by the late C. A. Kofoid and Josephine Michener. *Phycologia,* v. 11, p. 47—55, figs. 1—10, 1972.

Taylor, F. J. R., Topography of cell division in the structurally complex dinoflagellate genus *Ornithocercus. J. Phycol.,* v. 9, p. 1—10, pls. 1—4, figs. A-D, 1973a.

Taylor, F. J. R., General features of dinoflagellate material collected by the *"Anton Bruun"* during the International Indian Ocean Expedition, p. 155—169, figs. 1—6. *In* B. Zeitzschel, *Ecological studies,* v. 3, *Analysis and synthesis.* Berlin, New York: Springer Verlag, 1973b.

Taylor, F. J. R., Non-helical transverse flagella in dinoflagellates. *Phycologia,* v. 14, p. 45—47, figs. 1—8, 1975.

Taylor, F. J. R., and S. A. Cattell, *Discroerisma* [sic] *psilonereiella* gen. et sp. n., a new dinoflagellate from British Columbia coastal waters. *Protistologica,* v. 5, p. 169—172, pl. 1, 1969.

Tett, P. B., The relation between dinoflagellates and the bioluminescence of sea water. *J. Mar. Biol. Ass. U.K.* v. 51, p. 183—206, 2 figs., 1971.

Thienemann, August, *Die Binnengewässer Einzeldarstellungen aus der Limnologie und ihren Nachbargebieten, Das Phytoplankton des Süsswassers. Systematik und Biologie.* 3 Teil Cryptophyceen, Chloromonadinen, Peridineen. Stuttgart: E. Schweizerbart'sche Verlagsbuchhandlung, v. 16 (3), ix + 310 p., 69 pls., 1950.

Thompson, R. H., Fresh-water dinoflagellates of Maryland. *Contr. Chesapeake Biol. Lab.,* Publ. 67, p. 3—24, pls. 1—4, 1947.

Throndsen, Jahn, Motility in some marine nanoplankton flagellates. *Norw. J. Zool.,* v. 21, p. 193—200, figs. 1—5, 1973.

Tomas, R. N., Cell division in *Gonyaulax catenella,* a marine catenate dinoflagellate. *J. Protozool.,* v. 21, p. 316—321, figs. 1—20, 1974.

Tomas, R. N., and E. R. Cox, Observations on the symbiosis of *Peridinium balticum* and its intracellular alga. I. Ultrastructure. *J. Phycol.,* v. 9, p. 304—323, figs. 1—45, 1973.

Tomas, R. N., E. R. Cox, and K. A. Steidinger, *Peridinium balticum* (Levander) Lemmermann, an unusual dinoflagellate with a mesocaryotic and an eucaryotic nucleus. *J. Phycol.,* v. 9, p. 91—98, figs. 1—15, 1973.

Traverse, Alfred, Pollen analysis of the Brandon Lignite of Vermont. *Rep. U. S. Bur. Mines,* no. 5151, p. 1—107, figs. 1—13, 1955.

Traverse, Alfred, and R. N. Ginsburg, Palynology of the surface sediments of Great Bahama Bank, as related to water movement and sedimentation. *Mar. Geol.,* v. 4, p. 417—459, figs. 1—13, 1966.

Turpin, C. R., Analyse ou étude microscopique des différents corps organisés et autres corps de nature diverse qui peuvent, accidentellement, se trouver enveloppés dans la pâte translucide des silex. Première partie. *C. R. Hebd. Séanc. Acad. Sci., Paris,* v. 4, p. 304—314, figs. A—C', 1837.

Tuttle, R. C., and A. R. Loeblich III, The discovery of genetic recombination in the dinoflagellate *Cryp-*

thecodinium cohnii. J. Phycol., v. 10 Suppl., p. 16 (Abstr.), 1974a.

Tuttle, R. C., and A. R. Loeblich III, Genetic recombination in the dinoflagellate *Crypthecodinium cohnii. Science,* v. 185, p. 1061—1062, 1974b.

Tuttle, R. C., and A. R. Loeblich III, An optimal growth medium for the dinoflagellate *Crypthecodinium cohnii. Phycologia,* v. 14, p. 1—8, figs. 1—10, 1975.

Valensi, Lionel, Sur quelques microorganismes planctoniques des silex du Jurassique moyen du Poitou et de Normandie: *Bull. Soc. Géol. Fr.,* sér. 5, v. 18, p. 537—550, figs. 1—6, 1949.

Valensi, Lionel, Microfossiles des silex du Jurassique Moyen, Remarques pétrographiques. *Mém. Soc. Géol. Fr.,* v. 32(4), mém. 68, p. 1—100, pls. 1—16 (26—41), 1953.

Valensi, Lionel, Sur quelques microorganismes des silex Crétacés du Magdalénien de Saint-Amand (Cher). *Bull. Soc. Géol. Fr.,* sér. 6, v. 5, p. 35—40, figs. 1, 2, 1955.

Valensi, Lionel, Etude micropaléontologique des silex du magdalénien de Saint-Amand (Cher). *Bull. Soc. Préhist. Fr.,* v. 52, p. 584—596, pls. 1—5, 1956.

Vien, Cao, Sur l'existance de phénomènes sexuels chez un Péridinien libre, l'*Amphidinium carteri. C. R. Hebd. Séanc. Acad. Sci., Paris,* v. 264, p. 1006—1008, pls. 1, 2, 1967.

Vien, Cao, Sur la germination du zygote et sur un mode particulier de multiplication végétative chez le Péridinien libre *Amphidinium carteri. C. R. Hebd. Séanc. Acad. Sci., Paris,* v. 267, sér. D, p. 701—703, pls. 1, 2, 1968.

Vozzhennikova, T. F., K voprosu o sistematike iskopaemykh Peridiney [A contribution to the problem of taxonomy of fossil Peridinae]. *Dokl. Akad. Nauk SSSR,* v. 139, p. 1461—1462, 1961.

Vozzhennikova, T. F., Tip Pyrrhophyta. Pirrofitovye vodorosli [Division Pyrrhophyta. Pyrrhophyte algae], p. 171—186, figs. 1—20. *In* Yu. A. Orlov, *Osnovy Paleontologii,* v. 14, *Vodorosli, Mokhoobraznye, Psilofitovye, Plaunovidnye, Chlenistostebel'nye, Paporotniki.* Moscow: Akad. Nauk SSSR, 1963.

Vozzhennikova, T. F., *Vvedenie v izuchenie iskopaemykh peridineevykh vodorosley* [Introduction to the study of fossil peridinean algae]. Akad. Nauk SSSR, Sibirskoe Otdel., Inst. Geol. Geofiz., 156 p., 51 figs. Moscow: Nauka, 1965.

Vozzhennikova, T. F., *Iskopaemye peridinei Yurskikh, Melovykh i Paleogenovykh otlozheniy SSSR* [Fossil peridineans of Jurassic, Cretaceous and Paleogene deposits of USSR]. Akad. Nauk SSSR, Sibirskoe Otdel., Inst. Geol. Geofiz., 347 p., 121 pls., 17 figs. Moscow: Nauka, 1967.

Vozzhennikova, T. F., Dinotsisty i ikh stratigraficheskoe znachenie [Dinocysts and their stratigraphic importance]. *Akad. Nauk SSSR, Sibirskoe otdel.,*

Trudy Inst. Geol. Geofiz., vyp. 422, 222 p., 27 pls., 19 figs., 1979.

Wall, David, Modern hystrichospheres and dinoflagellate cysts from the Woods Hole region. *Grana Palynol.,* v. 6, p. 297—314, figs. 1—29, 1965.

Wall, David, Fossil microplankton in deep-sea cores from the Caribbean Sea. *Palaeontology,* v. 10, p. 95—123, pls. 14—16, figs. 1—8, 1967.

Wall, David, Quaternary dinoflagellate micropaleontology 1959—1969. *Proc. N. Am. Paleont. Conv., Chicago, 1969,* Part G, p. 844—866, 1970.

Wall, David, Biological problems concerning fossilizable dinoflagellates. *Geosci. Man,* v. 3, p. 1—15, pls. 1, 2, figs. 1, 2, 1971a.

Wall David, The lateral and vertical distribution of dinoflagellates in Quaternary sediments, p. 399—405. *In* B. M. Funnel and W. R. Riedel, *The micropaleontology of oceans.* Cambridge: University Press, 1971b.

Wall, David, and Barrie Dale, "Living fossils" in western Atlantic plankton. *Nature,* v. 211, p. 1025—1026, figs. 1—6, 1966.

Wall, David, and Barrie Dale, The resting cysts of modern marine dinoflagellates and the palaeontological significance. *Revue Palaeobot. Palynol.,* v. 2, p. 349—354, pl. 1, 1967.

Wall, David, and Barrie Dale, Modern dinoflagellate cysts and evolution of the Peridiniales. *Micropaleontology,* v. 14, p. 265—304, pls. 1—4, figs. 1—7, 1968a.

Wall, David, and Barrie Dale, Early Pleistocene dinoflagellates from the Royal Society Borehole at Ludham, Norfolk. *New Phytol.,* v. 67, p. 315—326, pl. 1, fig. 1, 1968b.

Wall, David, and Barrie Dale, Quaternary calcareous fossil dinoflagellates (Calciodinellidae) and their natural affinities. *J. Paleont.,* v. 42, p. 1395—1408, pl. 172, figs. 1—3, 1968c.

Wall, David, and Barrie Dale, The "hystrichosphaerid" resting spore of the dinoflagellate *Pyrodinium bahamense* Plate, 1906. *J. Phycol.,* v. 5, p. 140—149, figs. 1—34, 1969.

Wall, David, and Barrie Dale, Living hystrichosphaerid dinoflagellate spores from Bermuda and Puerto Rico. *Micropaleontology,* v. 16, p. 47—58, pl. 1, figs. 1—22, 1970.

Wall, David, and Barrie Dale, A reconsideration of living and fossil *Pyrophacus* Stein, 1883 (Dinophyceae). *J. Phycol.,* v. 7, p. 221—235, figs. 1—40, 1971.

Wall, David, and Barrie Dale, Paleosalinity relationships of dinoflagellates in the Late Quaternary of the Black Sea—a summary. *Geosci. Man,* v. 7, p. 95—102, figs. 1—5, 1973.

Wall, David, and Barrie Dale, Dinoflagellates in Late Quaternary of the Black Sea. *Micropaleontology,* v. 19, 364—380, figs. 1—5. *In* E. T. Degens and D. A. Ross, The Black Sea: geology, chemistry and biology. *Mem. Am. Ass. Petrol. Geol.,* no. 20, 1974.

Wall, David, Barrie Dale, and Kenichi Harada, Descriptions of new fossil dinoflagellates from the Late Quaternary of the Black Sea. *Micropaleontology,* v. 19, p. 18—31, pls. 1—3, fig. 1, 1973.

Wall, David, Barrie Dale, G. P. Lohmann, and W. K. Smith, The environmental and climatic distribution of dinoflagellate cysts in modern marine sediments from regions in the North and South Atlantic Oceans and adjacent seas. *Mar. Micropaleont.,* v. 2, p. 121—200, figs. 1—32, 1977.

Wall, David, and W. R. Evitt, A comparison of the modern genus *Ceratium* Schrank, 1793, with certain Cretaceous marine dinoflagellates. *Micropaleontology,* v. 21, p. 14—44, pls. 1—3, figs. 1—11, 1975.

Wall, David, R. R. L. Guillard, and Barrie Dale, Marine dinoflagellate cultures from resting spores. *Phycologia,* v. 6, p. 83—86, figs. 1—4, 1967.

Wall, David, R. R. L. Guillard, Barrie Dale, E. J. Swift, and Norimitsu Watabe, Calcitic resting cysts in *Peridinium trochoideum* (Stein) Lemmermann, an autotrophic marine dinoflagellate. *Phycologia,* v. 9, p. 151—156, figs. 1—10, 1970.

Wall, David, and J. S. Warren, Dinoflagellates in Red Sea piston cores, p. 317—328, figs. 1—4. *In* E. T. Degens and D. A. Ross, *Hot brines and Recent heavy metal deposits in the Red Sea.* New York: Springer-Verlag, 1969.

Wall, J. H., and Chaitanya Singh, A Late Cretaceous microfossil assemblage from the Buffalo Head Hills, north-central Alberta. *Can. J. Earth Sci.,* v. 12, p. 1157—1174, pls. 1—5, figs. 1—3, 1975.

Warren, J. S., Form and variation of the dinoflagellate *Sirmiodinium grossi* Alberti, from the Upper Jurassic and Lower Cretaceous of California. *J. Paleont.,* v. 47, p. 101—114, pls. 1—3, 1973.

Weiler, Helmut, Über einen Fund von Dinoflagellaten, Coccolithophoriden und Hystrichosphaerideen im Tertiär des Rheintales. *Neues Jb. Geol. Paläont. Abh.,* v. 104, p. 129—147, pls. 11—13, figs. 1—18, 1956.

West, G. S., A biological investigation of the Peridineae of Sutton Park, Warwickshire. *New Phytol.,* v. 8, p. 181—196, figs. 20—26, 1909.

West, G. S., *Algae,* vol. I, *Myxophyceae, Peridinieae, Bacillarieae, Chlorophyceae, together with a brief summary of the occurrence and distribution of freshwater algae.* Cambridge: Univeristy Press, viii + 475 p., 271 figs., 1916.

Wetherbee, Richard, The fine structure of *Ceratium tripos,* a marine armored dinoflagellate I. The cell covering (theca). *J. Ultrastruct. Res.,* v. 50, p. 58—64, figs. 1—11, 1975a.

Wetherbee, Richard, The fine structure of *Ceratium tripos,* a marine armored dinoflagellate II. Cytokinesis and development of the characteristic cell shape. *J. Ultrastruct. Res.,* v. 50, p. 65—76, figs. 1—22, 1975b.

Wetherbee, Richard, The fine structure of *Ceratium*

tripos, a marine armored dinoflagellate III. Thecal plate formation. *J. Ultrastruct. Res.*, v. 50, p. 77−87, figs. 1−15, 1975c.

Wetzel, Otto, Die typen der baltischen Geschiebefeuersteine, beurteilt nach ihrem Gehalt an Mikrofossilien. *Z. Geschiebeforsch. Flachldgeol.*, v. 8, p. 129−146, pls. 1−3, 2 figs., 1932.

Wetzel, Otto, Die in organischer Substanz erhaltenen Mikrofossilien des baltischen Kreide-Feuersteins. Mit einem sedimentpetrographischen und stratigraphischen Anhang. *Palaeontographica Abt. A*, v. 77, p. 141−188, figs. 1−10, 1933a.

Wetzel, Otto, Die in organischer Substanz erhaltenen Mikrofossilien des baltischen Kreide-Feuersteins. Mit einem sedimentpetrographischen und stratigraphischen Anhang (Schluss). *Palaeontographica Abt. A*, v. 78, p. 1−110, pls. 1−7, figs. 11−15, 1933b.

Wetzel, Otto, Die Mikropalaeontologie des Heiligenhafener Kieseltones (Ober-Eozän). *Jber. Niedersächs. Geol. Ver.*, v. 27, p. 41−75, pls. 8−10, 1935.

Wetzel, Otto, Eine neue Dinoflagellaten-Gruppe aus dem baltischen Geschiebefeuerstein. *Schr. Naturw. Ver. Schlesw.-Holst.*, v. 31, p. 81−86, pl. 1, 1960.

Wetzel, Otto, New microfossils from the Baltic Cretaceous flintstones. *Micropaleontology*, v. 7, p. 337−350, pls. 1−3, 1961.

Wetzel, Walter, Beitrag zur Kenntnis des danzeitlichen Meeresplanktons. *Geol. Jb.*, v. 66, p. 391−419, 1 pl., figs. 1−35, 1952.

Wetzel, Walter, Versuch einer quantitativen Erfassung des Oberkreide-Planktons. *Paläont. Z.*, v. 39, p. 240−248, figs. 1−4, 1965.

Wetzel, Walter, Organogene Inhalte der Flintsteinkerne oberkretazischer Seeigel aus dem baltisch-norddeutschen Raum. *Meyniana*, v. 21, p. 87−94, pls. 1, 2, 1971.

White, H. H., On fossil Xanthidia. *Microsc. J.*, v. 2, p. 35−40, pl. 4, 1842.

Wiggins, V. D., Two Lower Cretaceous dinoflagellate species from Alaska. *Micropaleontology*, v. 15, p. 145−150, pls. 1, 2, figs. 1−3, 1969.

Wiggins, V. D., Two new Lower Cretaceous dinoflagellate genera from southern Alaska (U.S.A.). *Rev. Palaeobot. Palynol.*, v. 14, p. 297−308, pls. 1−3, figs. 1−3, 1972.

Wiggins, V. D., Upper Triassic dinoflagellates from arctic Alaska. *Micropaleontology*, v. 19, p. 1−17, pls. 1−5, 1973.

Wiggins, V. D., The dinoflagellate family Pareodiniaceae: a discussion. *Geosci. Man*, v. 11, p. 95−115, 5 pls., 1975.

Wilkinson, S. J., Observations on *Xanthidium*, both fossil and Recent. *Trans. Microsc. Soc. Lond.*, v. 2, p. 89−92, pl. 13, 1849.

Williams, D. B., The distribution of marine dinoflagellates in relation to physical and chemical conditions, p. 91−96. *In* B. M. Funnel and W. R. Riedel, *The micropalaeontology of oceans.* Cambridge: University Press, 1971a.

Williams, D. B., The occurence of dinoflagellates in marine sediments, p. 231−243, figs. 14.1−14.11. *In* B. M. Funnel and W. R. Riedel, *The micropalaeontology of oceans.* Cambridge: University Press. 1971b.

Williams, D. B., and W. A. S. Sarjeant, Organic-walled microfossils as depth and shoreline indicators. *Mar. Geol.*, v. 5, p. 389−412, figs. 1−16, 1967.

Williams, G. L., Dinoflagellate and spore stratigraphy of the Mesozoic-Cenozoic, offshore Eastern Canada. *Geol. Surv. Pap. Can.*, 74−30, p. 107−161, 1975.

Williams, G. L., and J. P. Bujak, Cenozoic palynostratigraphy of offshore eastern Canada. *Contr. Ser. Am. Ass. Stratigr. Palynol.*, no. 5A, p. 14−47, pls. 1−5, figs. 1−9, 1977.

Williams, G. L., and Charles Downie, *Wetzeliella* from the London Clay. *Bull. Br. Mus. Nat. Hist., Geology*, Suppl. 3, p. 182−198, pls. 18−20, 1966a.

Williams, G. L., and Charles Downie, Further dinoflagellate cysts from the London Clay. *Bull. Br. Mus. Nat. Hist., Geology*, Suppl. 3, p. 215−235, pls. 24−26, 1966b.

Williams, G. L., W. A. S. Sarjeant, and E. J. Kidson, A glossary of the terminology applied to dinoflagellate amphiesmae and cysts and acritarchs. *Am. Ass. Stratig. Palynol., Contr. Ser.*, no. 2, 222 p., 14 pls., 1973.

Wilson, G. J., Some new species of Lower Tertiary dinoflagellates from the McMurdo Sound, Antarctica. *N. Z. Jl. Bot.*, v. 5, p. 57−83, figs. 1−42, 1967a.

Wilson, G. J., Microplankton from the Garden Cove Formation, Campbell Island. *N. Z. Jl. Bot.*, v. 5, p. 223−240, figs. 1−29, 1967b.

Wilson, G. J., Some species of *Wetzeliella* Eisenack (Dinophyceae) from New Zealand Eocene and Paleocene strata. *N.Z. Jl. Bot.*, v. 5, p. 469−497, figs. 1−34, 1967c.

Wilson, G. J., Palynology of the middle Pleistocene Te Piki Bed, Cape Runaway, New Zealand. *N.Z. Jl. Geol. Geophys.*, v. 16, p. 345−354, figs. 1, 2, 1973.

Wong, J. L., R. Oesterlin, and H. Rapoport, The structure of saxitoxin. *J. Am. Chem. Soc.*, v. 93, p. 7344−7345, 1971.

Wood, E. J. F., Studies in microbial ecology of the Australasian region. I. Relation of oceanic species of diatoms and dinoflagellates to hydrology. *Nova Hedwigia*, v. 8, p. 5−20, pls. 3−5, 1964.

Wood, E. J. F., *Dinoflagellates of the Caribbean Sea and adjacent areas.* Coral Gables: Univ. Miami Press, 143 p., figs. 1−404, A−F, 1968.

Worth, G. K., J. L. MacLean, and M. J. Pierce, Paralytic shellfish poisoning in Papua, New Guinea, 1972. *Pacif. Sci.*, v. 29, p. 1−5, 1975.

Zaitzeff, J. B., and A. T. Cross, The use of dinoflagellates and acritarchs for zonation and correlation of the Navarro Group (Maestrichtian) of Texas, p. 341—377, pls. 1—6, figs. 1—11. *In* R. M. Kosanke and A. T. Cross, Symposium on palynology of Late Cretaceous and Early Tertiary. *Geol. Soc. Am., Spec. Pap.,* 127, 1971.

Zederbauer, E., Geschlechtliche und ungeschlechtliche Fortpflanzung von *Ceratium hirundinella. Ber. Dt. Bot. Ges.,* v. 22, p. 1—8, pl. 1, 1904.

Zimmermann, Walter, Neue und wenig bekannte Kleinalgen von Neapel I—V. *Z. Bot.,* v. 23, p. 421—442, 1 pl., 11 figs., 1930.

Zingmark, R. G., Ultrastructural studies on two kinds of mesocaryotic dinoflagellate nuclei. *Am. J. Bot.,* v. 57, p. 586—592, figs. 1—10, 1970a.

Zingmark, R. G., Sexual reproduction in the dinoflagellate *Noctiluca miliaris* Suriray. *J. Phycol.,* v. 6, p. 122—126, figs. 1—18, 1970b.

EBRIDIANS

...on conviendra que les Ebriidées constituent un groupe fertile en problèmes imprévues, ce qui n'est certainement pas un des moindres attraits de leur étude.

Georges Deflandre, 1936

Marine planktonic, silica-secreting flagellates, the ebridians are relatively rare both in modern seas and as fossils. Perhaps, therefore, it is not surprising that this class of flagellates is not even mentioned in the majority of monographs and texts in the English language on general zoology, botany, and paleontology, or even those more specialized ones on protozoology or phycology. It is better covered in various publications in French, German, and Spanish, and probably most completely discussed in the French zoological treatise (Deflandre, 1952).

The first species to be described was the fossil organism *Dicladia clathrata* Ehrenberg 1844, although it was not until 1854 that a specimen of this species was illustrated from the Miocene of Virginia. Some resting spores of diatoms were included with the ebriid, hence the genus has been used erroneously in both groups. Prior to the publication of Ehrenberg's drawing of *Dicladia* in *Mikrogeologie* (1854), another ebridian was illustrated as *Dictyocha* by Bailey (1844, pl., fig. 20) from Petersburg, Virginia or Piscataway, Maryland. Although details of ornamentation were omitted in these early illustrations, Bailey's *Dictyocha* is a species of *Ebriopsis*, probably *E. antiqua* (Schulz) Hovasse. The first report of a living ebriid was that of Schumann (1867), who described *Dictyocha tripartita* as a silicoflagellate, because of its similar siliceous internal skeleton; as *Ebria tripartita* (Schumann) Lemmermann, it is now the type species of *Ebria*.

In early publications the ebridians were included with diatoms or silicoflagellates, and as these groups have plastids and are photosynthetic, Lemmermann (1901a) placed both ebriids and silicoflagellates in the plant kingdom, although in separate orders, the Siphonotestales (silicoflagellates) characteristically with a skeleton of tubular hollow rods, and the Stereotestales including those with solid siliceous rods (ebridians). The two orders are now placed in distinct divisions, the silicoflagellates in the Chrysophyta, and the ebridians in the Pyrrhophyta, on the basis of such cytological features as the nuclear character and type and number of flagella. In addition to the skeletal differences noted by Lemmermann, the silicoflagellates have a single flagellum, and the ebridians have two dissimilar flagella; the silicoflagellates are photosynthetic, the ebridians holozoic, with a dinokaryon nucleus. On the basis of skeletal composition, ebridians also have been regarded as related to the monopylarian (perhaps larval) radiolarians (Hovasse, 1931a,b), although they have flagella rather than pseudopodia. The Ebriophycean cell lacks an external theca or vesicular wall layer such as characterizes the Dinophyceae (although erroneously stated to be thecate by Haq, 1978); it superficially resembles the Desmocapsales and Prorocentrales in having apically inserted flagella.

Comparatively little study has been made of living ebridians, whereas greater emphasis has been placed on the fossils, particularly in France during the 1930s and early 1950s, and more recently with a resurgence of interest coinciding with the Deep Sea Drilling Program.

The rare living species, representing only *Ebria*, *Hermesinella*, and *Hermesinum*, are the sole relics of a once diverse group of the mid-Tertiary.

THE LIVING ORGANISM

The Protoplast

The ovate unicellular body, about 10 μm in length, consists of colorless to slightly yellowish or rose-colored cytoplasm, which is not differentiated into ectoplasm and endoplasm. The

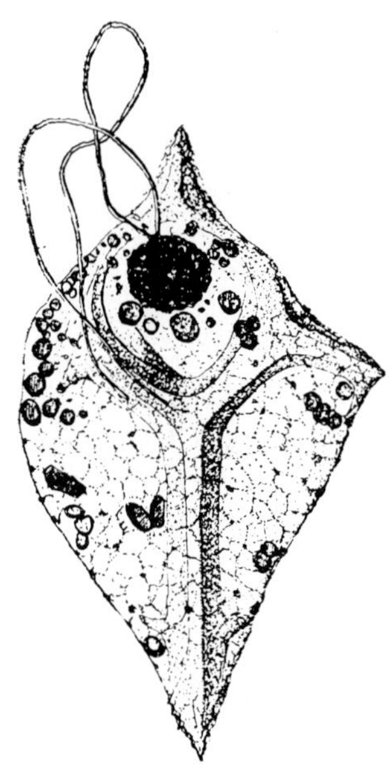

Figure 5.1
Hermesinum adriaticum Zacharias. From fixed living material, showing subapically inserted flagella attaching near the large nucleus; large spicule visible within cell. From Hovasse, 1934.

cytoplasm contains numerous tiny refringent fat granules, and completely encloses the siliceous skeleton. The large nucleus is of classic dinokaryon type in *Hermesinum*, whereas that of *Ebria* also has a relatively large nucleolus (Hovasse, 1932e). The two dissimilar flagella are subapical in position and inserted near the nucleus (see Figure 5.1). They may be long and thin as in *Ebria*, or very fine and short as in *Hermesinum*. No pseudopodia have been observed. Ebridians may further resemble the dinoflagellates in having flagella beating in dissimilar ways, although at least one flagellum beats toward the front. When disturbed, one of the flagella may be shed. The cell swims rather rapidly in a helical path, the irregular movement providing the source of the name *Ebria* (from the Latin *ebrius*, "drunken").

Lacking plastids, the ebriids ingest their food in an animallike manner; diatoms reportedly are a major food source, having been often observed in the cytoplasm. Symbiotic zooxanthellae also have been reported in *Ebria tripartita*. However, the yellow-green material filling the cells of *Hermesinum adriaticum* was regarded as the remnants of ingested flagellates (Hargraves and Miller, 1974).

The Skeleton

The internal skeleton consists of a solid framework of narrow (3 to 4 μm) silica rods, with branches radiating at angles of 120° apart when viewed from above (see Figure 5.2). The skeleton thus has a basic tetraxial (in *Hermesinum*) or triradial (in *Ebria*) type of symmetry. The system of terminology for the skeleton proposed by Deflandre (1934) was based on that of sponge spicules. The initial four-rayed or tridentlike spicule is a **triaene** (see Figure 5.3) and the triradial type (in which one ray is atrophied) is a **triode** (see Figure 5.4). The individual branches are **actines,** the lower or axial one of the triaene spicule is the **rhabde.** The branches at the extremities of the actines are **clades;** rather than being free terminally as in the sponges, they may connect adjacent actines. The clades are also subdivided according to location; those from the ends of the upper actines that are directed toward the anterior or nuclear pole are termed **proclades. Opisthoclades** leave the same point, but are directed backwards, and they even may rejoin the distal extremity of the triaene rhabde (see Figure 5.5). Anterior and posterior **synclades** may connect the distal ends of the branches from adjacent proclades and opisthoclades. Additional bars (**mesoclades**) may connect the divergent branches of single proclades or opisthoclades.

The various skeletal parts may leave openings between; the **upper windows** (fenêtres supérieurs) lie between proclade branches; the **lower windows** (fenêtres inférieurs) occur between opisthoclades; **median windows** (fenêtres medianes) may lie between the mesoclades and actines. The ebridian skeleton is never truly symmetrical, although it may have a roughly axial symmetry, as suggested by the initial spicule. Commonly the symmetry is somewhat bilateral, with a displacement of the center of the triaene to one side. The side to which it is displaced is commonly somewhat convex, and is regarded as ventral. There may be a further displacement of the axis to the right of the skeleton.

Teratological double skeletons may occur in ebridians as in the silicoflagellates, probably as an accidental result of cementation of the daughter skeleton to that of the parent during the reproductive process (see Figure 5.6). However, the species *Hermesinum geminum* Dumitrică & Perch-Nielsen apparently is represented entirely by double specimens (see Figure 5.7) (Perch-Nielsen, 1975). Fossil double specimens are common in *Ammodochium* and *Pseudammodochium* (see Figure 5.8), in which the two skeletons are joined by siliceous bridges, or have a joint collar at the nuclear pole. Although these features have been suggested to indicate conjugation, binucleate specimens, or the zygote resulting from cell and nuclear fusion, the presence of the two spicules in all stages of development of *Hermesinum geminum* indicate instead an incomplete separation during asexual reproduction.

Skeletal Ornamentation

Determination with the light microscope of the surface ornamentation of the skeleton of the ebridians was hampered by their very small size, hence it was known only in a general way to vary somewhat in rugosity. Use of the carbon-replica technique and the TEM (Jerković, 1968) gave the first detailed information about the surface of an ebriid skeleton, that of *Hermesinum schulzii* Hovasse, although the open network of the skeleton prevents good replication for TEM study. The external surface of the skeleton is reticulate, and the inner surfaces of the same rods, as well as the initial spicule, are smooth. More recently, various studies with the SEM (see Figure 5.9) have aided resolution of the ornamentation, which may be extremely complex, reticulate, nodular, or punctate, as well as the general structure of the skeleton (Hargraves and Miller, 1974; Ling and McPherson, 1974; Perch-Nielsen, 1975; Martini, 1976).

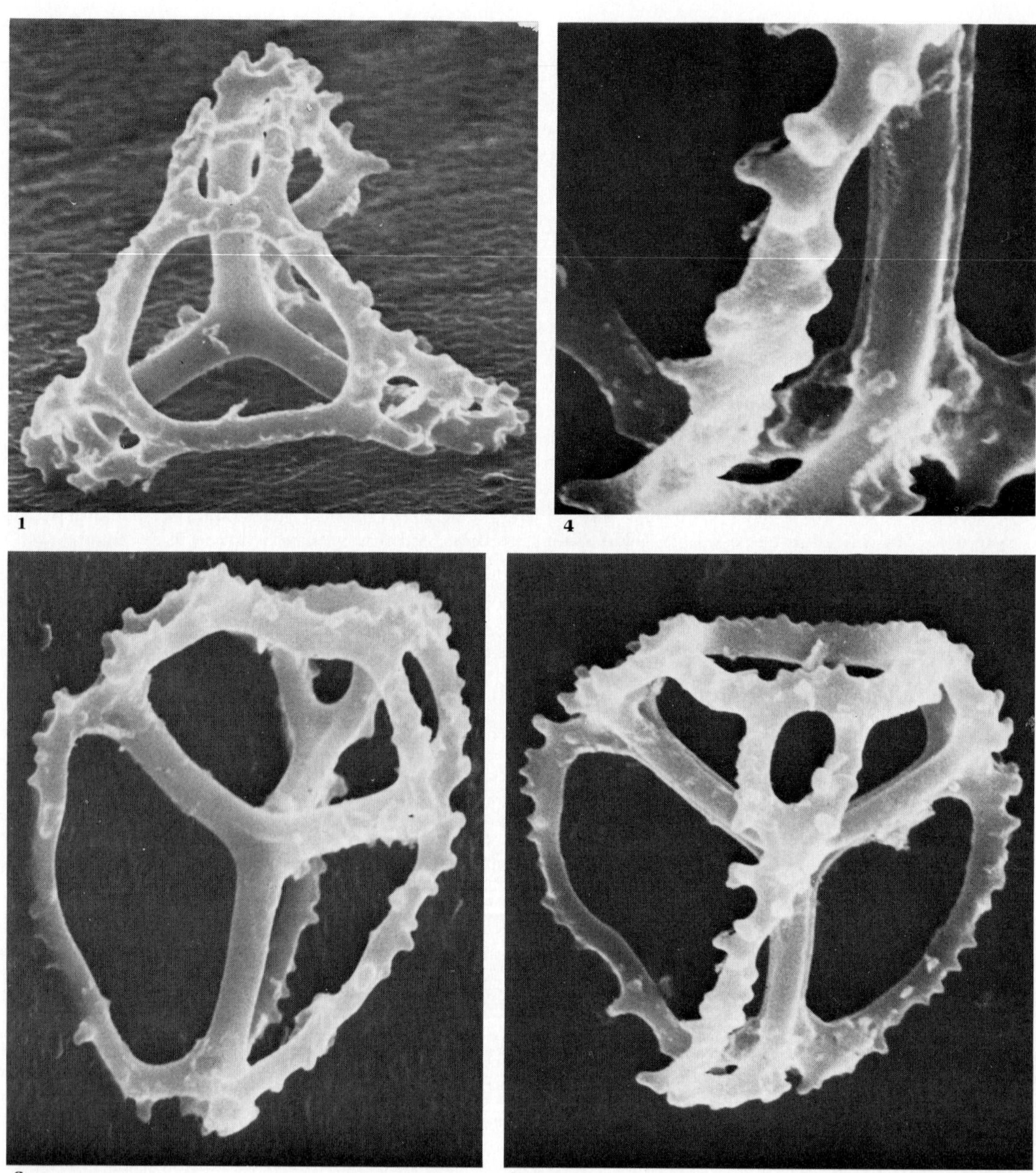

Figure 5.2

Hermesinum praesidis Deflandre, Miocene, Lompoc, California, SEM. **1.** Apical view, showing smooth surface of initial triaene, visible through the apical or upper window. **2.** View between two opisthoclades showing triaene with elongate rhabde and three divergent actines. **3.** Side view facing an actine with bifurcating proclade and opisthoclade; note asymmetry, especially near base. 1—3, ×2500. **4.** Enlargement of part of part 3, showing surface ornamentation, ×6000.

466

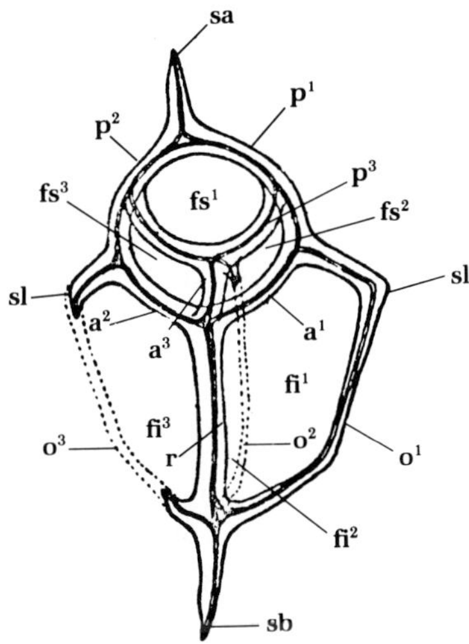

Figure 5.3
Hermesinum skeletal terminology. r, axial shaft or rhabde; a¹, a², a³, the three actines of the triaene; p¹, p², p³, anterior clades or proclades; o¹, o², o³, posterior clades or opisthoclades; fs¹, fs², fs³, upper windows, or fenêtres supérieurs; fi¹, fi², fi³, lower windows, or fenêtres inférieures; sa, apical spine; sb, basal spine; sl, lateral spine. Modified from Frenguelli, 1938.

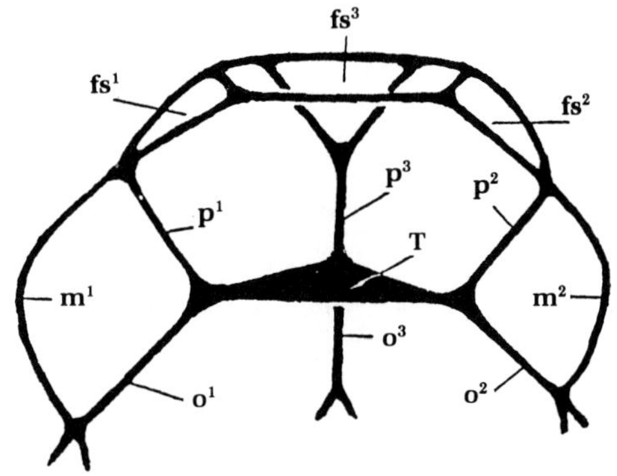

Figure 5.4
Ditripodium fenestratum Deflandre, diagrammatic to show terminology. T, triode, the initial triangular spicule; p¹, p², p³, proclades; m¹, m², mesoclades; o¹, o², o³, opisthoclades, directed toward base; fs¹, fs², fs³, upper windows. From Deflandre, 1951.

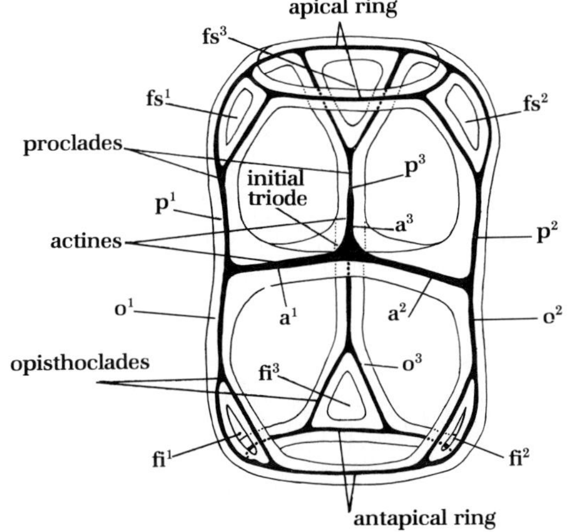

Figure 5.5
Ammodochium ampulla Deflandre. Skeletal diagram. Actines (a¹, a², a³) from initial triode; proclades (p) and opisthoclades (o) are directed anteriorly and posteriorly respectively from ends of the actines, and in turn branch to surround the upper (fs) and lower (fi) windows; apical and antapical rings are formed by synclades connecting the branched proclades and opisthoclades, respectively. Redrawn from Deflandre, 1934.

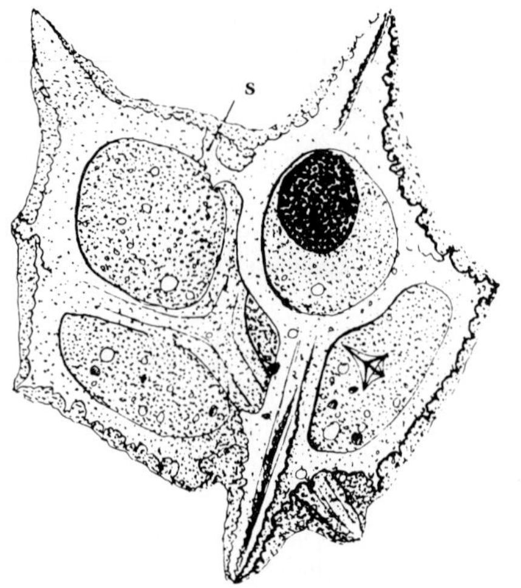

Figure 5.6
Hermesinum adriaticum. Living cell with double skeleton; s, suture between the two skeletons; large nucleus shown in center of right skeleton, and early development of third skeleton visible at lower right; ×2500, from Hovasse, 1932e.

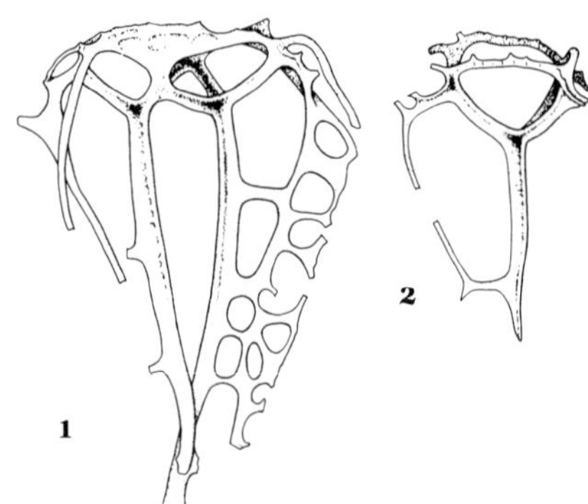

Figure 5.7
Hermesinum geminum, Oligocene, Norwegian Sea. **1.** Adult, showing two vertical rhabdes, ×460. **2.** Young individual, ×645. From Perch-Nielsen, 1975.

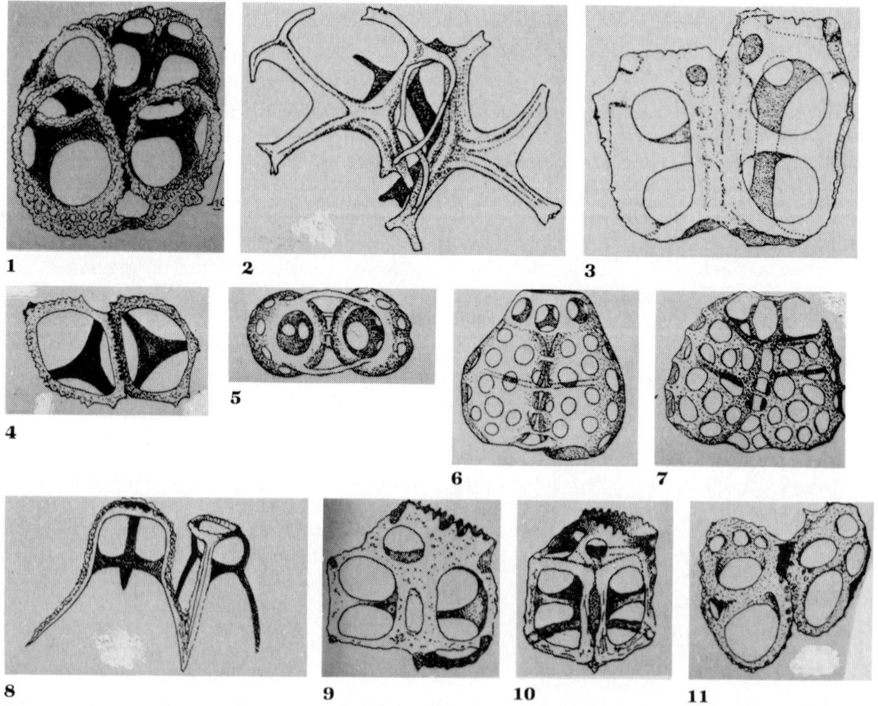

Hypersilicification may produce such other skeletal variations as highly developed spines, supplementary crests, and reduced openings between the bars. Hypersilicification commonly (although not invariably) precedes development of a lorica.

The Lorica

A **lorica**, or inflated, chamberlike structure, may be formed at the nuclear pole surmounting the nuclear ring, and may also have a reticulate ornamentation. The lorica is always developed on a double or triple skeleton in *Ammodochium,* but only on a single skeleton in other forms.

In early development the loricate or *Podamphora* stage has a basal opening, but later this is closed, so that the only remaining opening is at the top. As this opening is large and the lorica occasionally perforated, the lorica is not believed to represent a cyst. The loricate stage is best known in Paleocene and Eocene specimens of *Ebriopsis, Hovassebria,* and *Podamphora,* but is also found in Miocene *Ebriopsis* and *Podamphora,* and more rarely in *Pseudammodochium.* An exception to this general development of the loricate stage is found in *Micromarsupium,* which produces a lorica around the entire skeleton. The process commences as hypersilicification of the skeleton, and finally completely encloses it, leaving an opening only at the posterior end, instead of at the top as in the *Podamphora* type of lorica. This suggests an inversion of polarity during the ontogeny of the individual.

Although the lorica is not now regarded as representing a cyst, another modification found in *Ammodochium* and *Parathranium* (see Figure 5.10) may represent an encysted stage. This "cyst" develops within the normal skeleton, and has only a very tiny pore as an opening, suggesting a restriction of the cytoplasm around the nucleus within the more easily closed capsule. Such cysts are rare, but have been observed in specimens from the Miocene of California and Japan.

Figure 5.8 (*facing page*)
Double skeletons in fossil ebridians. **1.** *Hermesinum schulzii* Hovasse, quadruple skeleton, ×700, from Hovasse, 1932d. **2.** *Ditripodium elephantinum* Hovasse, Miocene, Hungary, ×1000, from Deflandre, 1951. **3.** *Ammodochium danicum* Deflandre, Paleocene, Jutland, ×1000, from Deflandre, 1951. **4.** *A. prismaticum* Hovasse var. *paradoxum* Hovasse, upper Eocene, New Zealand, apical view of double skeleton, ×1000, from Hovasse, 1932d. **5.** *Pseudammodochium sphericum* Hovasse, upper Eocene, New Zealand, apical view of double skeleton, ×1000, from Deflandre, 1951. **6,7.** *P. dictyoides* Hovasse, upper Eocene, New Zealand, lateral view, ×1000; 6, from Deflandre, 1951; 7, from Hovasse, 1932d. **8.** *Parathranium tenuipes* Hovasse, Miocene, California, ×500, from Hovasse, 1932b. **9,10.** *Ammodochium rectangulare* (Schulz) Deflandre, upper Eocene, New Zealand, ×1000, from Deflandre, 1951. **11.** *Craniopsis minor* Hovasse, upper Eocene, New Zealand, ×900, from Hovasse, 1932d.

Figure 5.9 (*overleaf*)
Surface ornamentation of the ebridian skeleton, in SEM. **1,2.** *Hermesinum* cf. *adriaticum,* living, Salton Sea, California. **1.** Upper part, showing nuclear window as large opening at top, two actines of the central triaene with smooth surface, and reticulate and scalloped surface of proclades, which bifurcate to form three centrally perforated triangular areas surrounding the nuclear ring, ×4500. **2.** Lateral view of apical part, showing spinules that arise at the angles of the reticulations, ×3000. **3.** *Ditripodium latum* Hovasse, Miocene, Lompoc, California, showing pitted surface, ×6000.

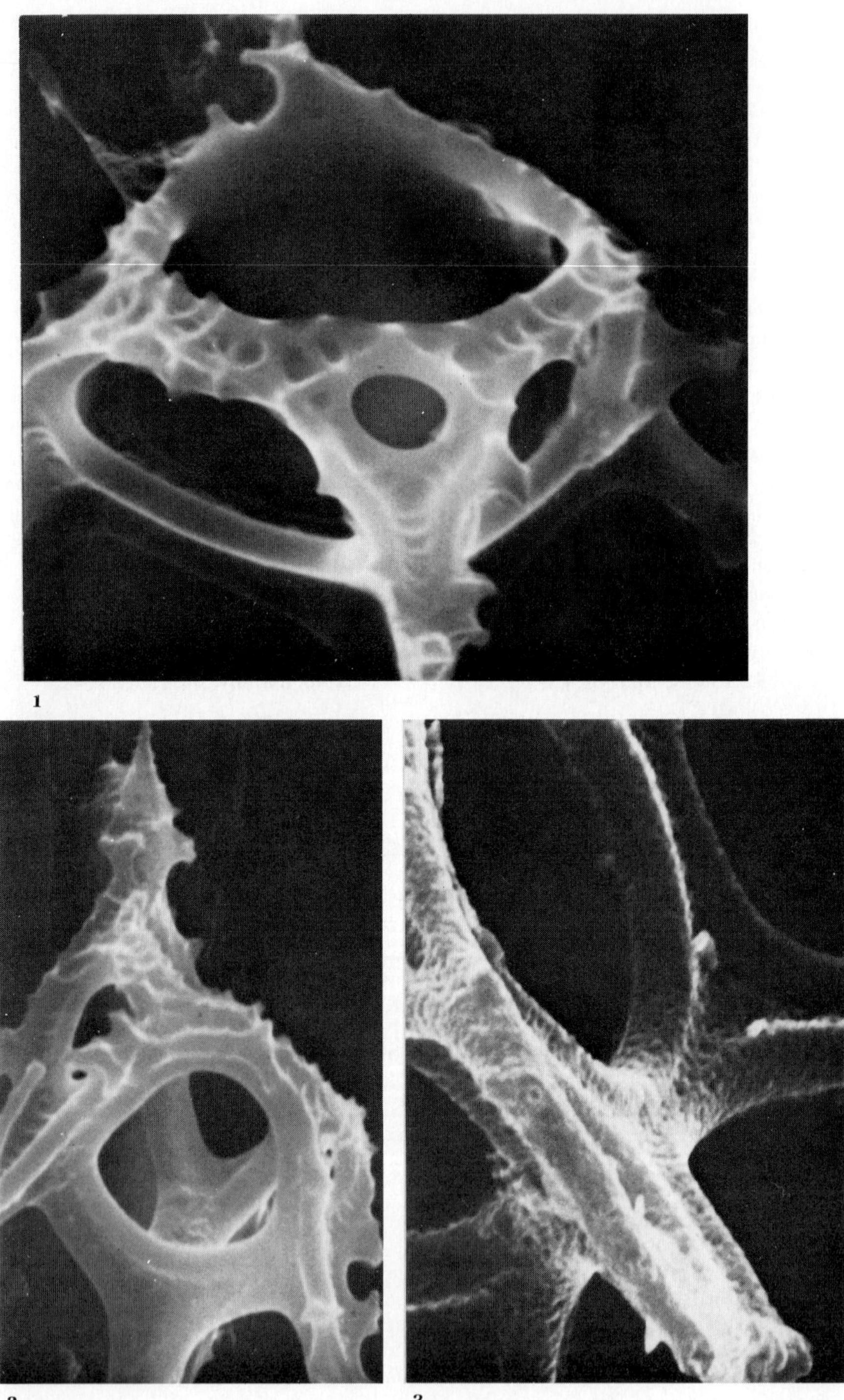

1

2

3

Figure 5.9

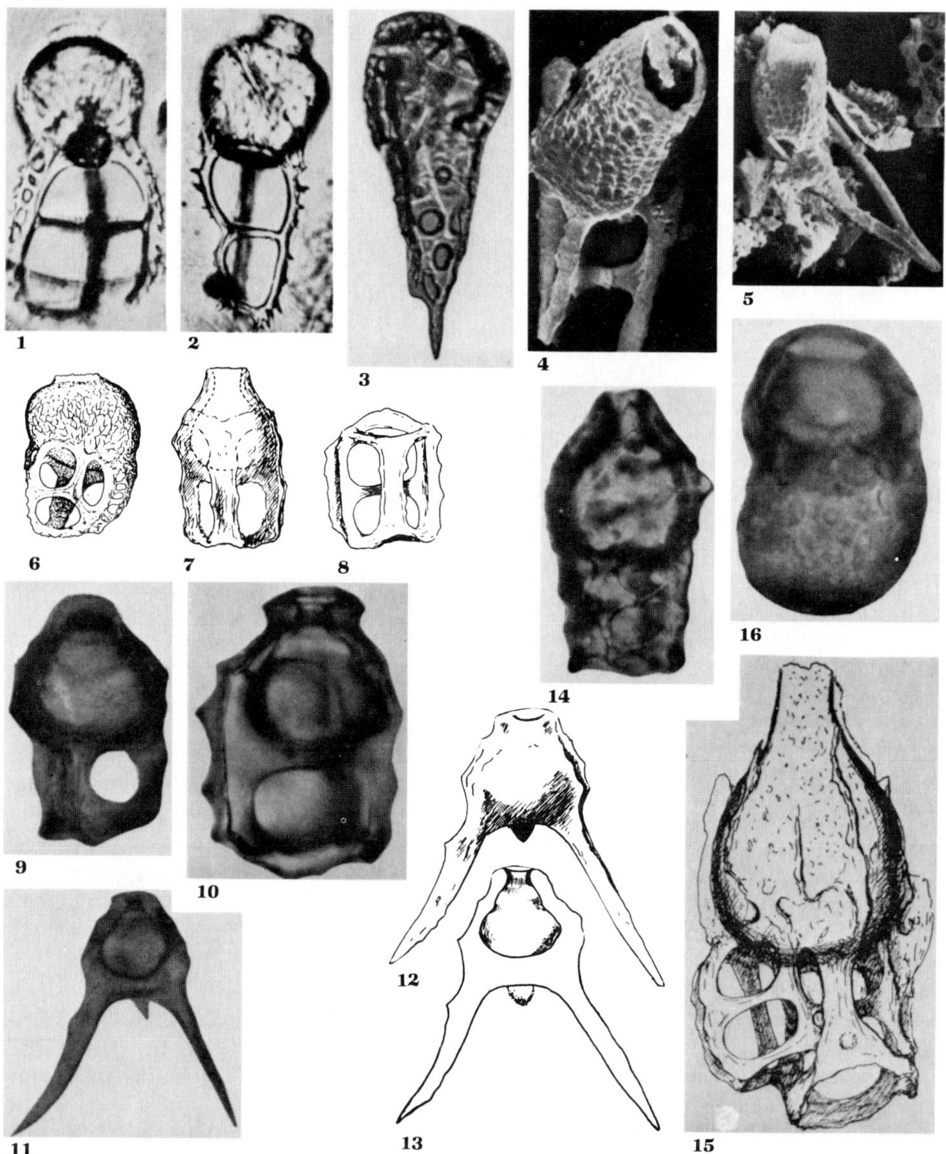

Figure 5.10

Lorica or cyst development. **1,2.** *Podamphora elgeri* Gemeinhardt, Miocene, Karand, Hungary, ×380 from Frenguelli, 1940. **3.** *Micromarsupium anceps* Deflandre var. *curticannum* Deflandre, upper Eocene, New Zealand, ×400; from Deflandre, 1951. **4,5.** *M. rostovense* Martini, Oligocene, Rostov, USSR, SEM; 4, ×1040; 5, ×520; from Martini, 1976. **6.** *Ebriopsis antiqua,* Miocene, Abashiri, Japan, ×410, from Deflandre, 1951. **7–10.** *Ammodochium rectangulare;* 8, skeleton without lorica; others with lorica or cyst. **7,8.** ×935, after Deflandre, 1933; **9,10.** Miocene, Japan, ×1360, after Deflandre, 1951. **11.** *Parathranium intermedium* Hovasse, Miocene, ×910, from Deflandre, 1951. **12,13.** *P. tenuipes,* Miocene, Setanaigori, Japan, ×935, from Deflandre, 1933. **12.** Exterior. **13.** Optical section. **14,15.** *Ammodochium ampulla.* **14.** ×585, from Deflandre 1933. **15.** Upper Eocene, Oamaru, New Zealand, ×935, from Deflandre, 1933. **16.** *Pseudammodochium robustum* Deflandre, Miocene, Abashiri, Japan, ×1360, from Deflandre, 1951.

REPRODUCTION

Little information is available concerning the ontogeny of living ebridians, and only vegetative reproduction by simple fission has been reported. Sexual reproduction is hypothesized on the basis of skeletal morphology in Tertiary forms, and it has been stated that the ebridians are the only unicellular organisms whose life cycle may be better known for Tertiary forms than for the living ones.

In asexual reproduction, the new daughter skeleton forms gradually within the parent cytoplasm, although at first not with any particular orientation (see Figure 5.11). Later the skeleton is oriented so that the apical pole of each is at the same end of the cell, nuclear division occurs, and the entire cell content is divided in two, each half taking one of the nuclei and one of the two skeletons (see Figure 5.12). Rarely the skeletal replication is accompanied by an apparently accidental cementation of the two skeletons, to form a double skeleton, and thus prevent the completion of the cell division process. Similar double skeletons have been observed in fossil material as old as the Paleocene.

The siliceous chamber or lorica surmounting the normal skeleton of some fossil specimens was thought at one time to be an allogromiid foraminifer that used the ebridian skeleton as foreign matter agglutinated to its test. However, modern allogromiids are rare in marine waters, are not silica-secreting, and the ebridian skeleton was invariably oriented with the lorica at the nuclear pole. It also has been postulated to represent a cyst or resting stage of the ebridian, or as a development related to sexual reproduction, because the lorica has been observed surmounting two or even three specimens of normal form, as well as on a single one. A cyst could be formed with either single or double skeletons, but if the lorica results from conjugation, there should always be more than one individual. A loricate stage has not yet been observed on any living material.

Although neither loricate skeletons nor true cysts are known in living species, fossil specimens that have been postulated to represent such stages have a long history, as the double skeletons with a common collar occur in many Paleocene rocks. In the later Paleocene of Kuznetsk are loricate forms of *Ebriopsis crenulata* Hovasse, and by the late Eocene of Oamaru, New Zealand, loricae are found on single skeletons of *Ebriopsis*, *Hovassebria*, and *Micromarsupium*, and on double skeletons of *Ammodochium*. In the Miocene, loricate stages are common in *Ebriopsis* and *Podamphora*, and rare in *Pseudammodochium*. Internal cysts are known in both *Ammodochium* and *Parathranium*.

OCCURRENCE

Ecology

All known ebridians, living and fossil, are marine in occurrence; they are present in the neritic zone on all coasts, in marine sloughs, and near river mouths, and even in inland saline seas, such as the Salton Sea of California. Most are found in cold or temperate waters, but locally they may be abundant in warm waters also.

Ebria has an optimum temperature of about 10°C, and is known in the North Sea, Baltic Sea, Barents Sea, the Mediterranean Sea, Sea of Marmara, Black Sea, North and South Atlantic (off Patagonia), Pacific (off British Columbia; Puget Sound, Washington; and San Diego, California), and in the Salton Sea of California. *Hermesinum* has a higher optimum temperature, about 20 to 25°C, and is living in the Adriatic Sea, Sea of Marmara, Black Sea, Atlantic Ocean off Rhode Island, Pacific Ocean off California, and in the inland Salton Sea of California. Both are known as fossils in the Atlantic, Pacific, and Mediterranean provinces.

Although rarely abundant, *Ebria tripartita* has been reported in the Baltic Sea, near Bornholm, in quantities of 500 cells per liter of seawater, and in a bloom in Long Island Sound in 1952 reached 49,000 cells per liter (Conover, 1956). Numerous *Ebria tripartita* with many double skeletons and many specimens in process of

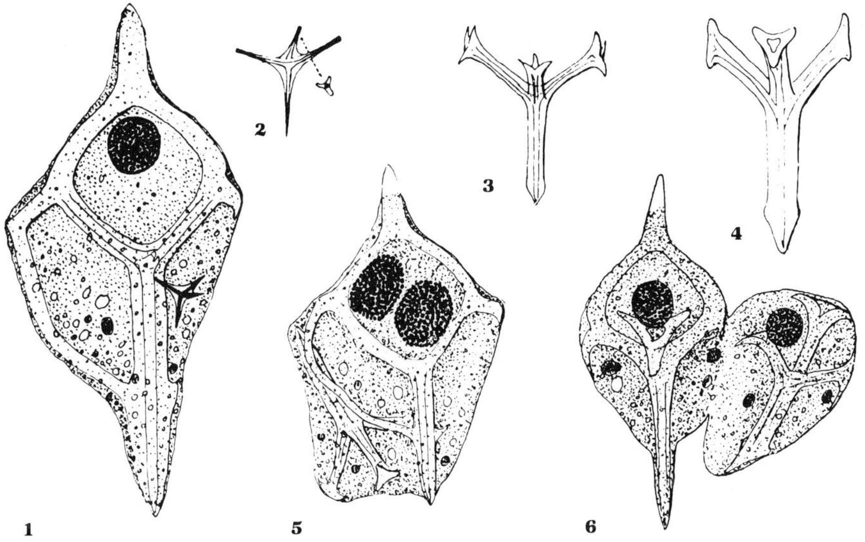

Figure 5.11
Cell division in *Hermesinum adriaticum*. **1.** Adult cell with early development of daughter skeleton as spicule at lower right. **2—4.** Successive growth stages of daughter skeleton. **5.** Nucleus divided, daughter skeleton at lower left, with orientation opposite that of parent. **6.** Cell division, daughter cell at right, with orientation like that of parent; flagella not shown. 1—4, ×1600; 5,6, ×1050; after Hovasse, 1932e.

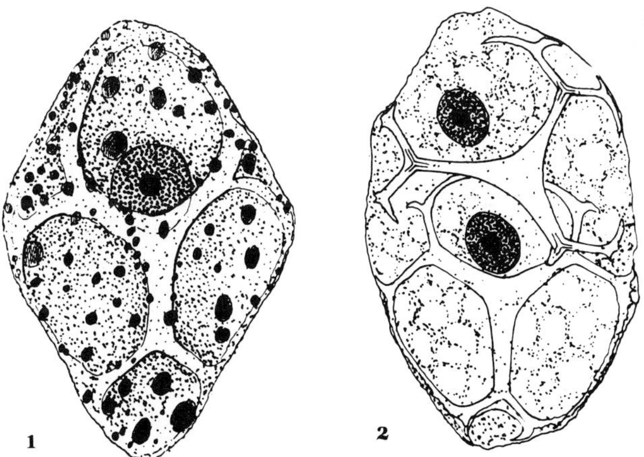

Figure 5.12
Ebria tripartita, living, from Bosporus, Turkey, on the Black Sea, ×2000. **1.** Showing skeleton, large nucleus with central nucleolus, and fatty inclusions. **2.** Daughter skeleton nearly complete, and nucleus divided in two, just prior to completion of the cell division. From Hovasse, 1932e.

division were observed off French Equatorial Africa in June 1955. During that year the water had been warm, over 25°C from January to April, with a salinity of less than 35°/₀₀. At the end of April the temperature dropped and salinity increased, and in June upwelling resulted in a temperature of 20°C combined with a salinity of 35°/₀₀, and an influx of abundant *Ebria* and *Hermesinum*. In an unusual occurrence in August 1956 at the mouth of the Nile River, ebridians comprised 43.3 percent of the total phytoplankton, with 15,000 cells per liter of *Hermesinum adriaticum* even surpassing the diatoms in number (Halim, 1960).

Hermesinum adriaticum was reported off Rhode Island in numbers up to 253,000 cells per liter in August 1970, although none were found at the same station the following year at that date. The species was generally most abundant between the end of July and mid-October, and no individual was observed after October 14 and prior to July 21 at any of the stations studied. A rapid increase to large numbers occurred only when the water temperature exceeded 25°C, and the decline coincided with the cooling trend in the fall. A heavy rainfall also caused a precipitous decline, as 98 percent of the population disappeared within 48 hours, hence the species appears sensitive to sudden freshening of the water (Hargraves and Miller, 1974). Such controlling conditions have been inferred from the distribution of the living species in nature, but information is lacking as to the relative importance of the various factors, as no ebridians have yet been maintained in cultures.

Hermesinum was reported in the Salton Sea (although identified as "a silicoflagellate, possibly a *Dictyocha*," by Carpelan, 1961) in numbers up to 450,000 per liter, the maximum number occurring in June, 1955. Originally fresh, the Salton Sea was formed by entrapment of water from the Colorado River, but by dissolution of deposits on the lake bed and accumulation of salts from runoff it has become more saline. Marine fish and invertebrates have been introduced from the Pacific Ocean, from the regions of San Diego, Los Angeles, and San Francisco, as well as from the Gulf of California, suggesting that the *Hermesinum* (see Figure 5.13) and *Ebria*, which also is present in the Salton Sea, are of Pacific origin.

The skeletons of these free-living planktonic organisms commonly are observed in the excrement of copepods that had ingested the cells, the accumulation of the skeletons on the sea floor perhaps facilitated by concentration in the coprolites. Most fossil species were described from marine diatomites, although they are there much less abundant than the associated silicoflagellates. The number of species is much less also, and like that of generic taxa, has decreased greatly since the Miocene to three described living species, *Ebria tripartita*, *Hermesinum adriaticum*, and *H. platense* Frenguelli. A living *Hermesinella* has been reported but not validly described.

Geologic Occurrence

A large percentage of the many fossil species have been described from poorly documented material, hence much remains to be done in obtaining a detailed record of their geologic and geographic distribution. There is no record of pre-Cenozoic ebridians, but eight genera occur in the Paleocene, representing three of the four families (see Figure 5.14) and attesting to a rapid early expansion and differentiation. An addi-

Figure 5.13
Hermesinum cf. *adriaticum*. Living, Salton Sea, California, collected by A. R. Loeblich III; in SEM. **1,2.** Same specimen in lateral views, showing inner triaene spicule, elongate rhabde, apical and basal spines, and one complete opisthoclade that attaches near the base of the rhabde, the other two ending in short recurved spines; note smooth triaene, ornate exterior parts of the proclades and opisthoclades, and apical or nuclear window visible just below apical spine in part 2. **3.** View from above, looking through nuclear window at the triradiate triaene, apical spine at lower left. **4.** Oblique posterior view, viewing rhabde and actines of the triaene from below, specimen oriented as in part 1. **5,6.** Views of another specimen, from side opposite that seen in parts 1 and 2. 1,3–6, ×1500; 2, ×2000.

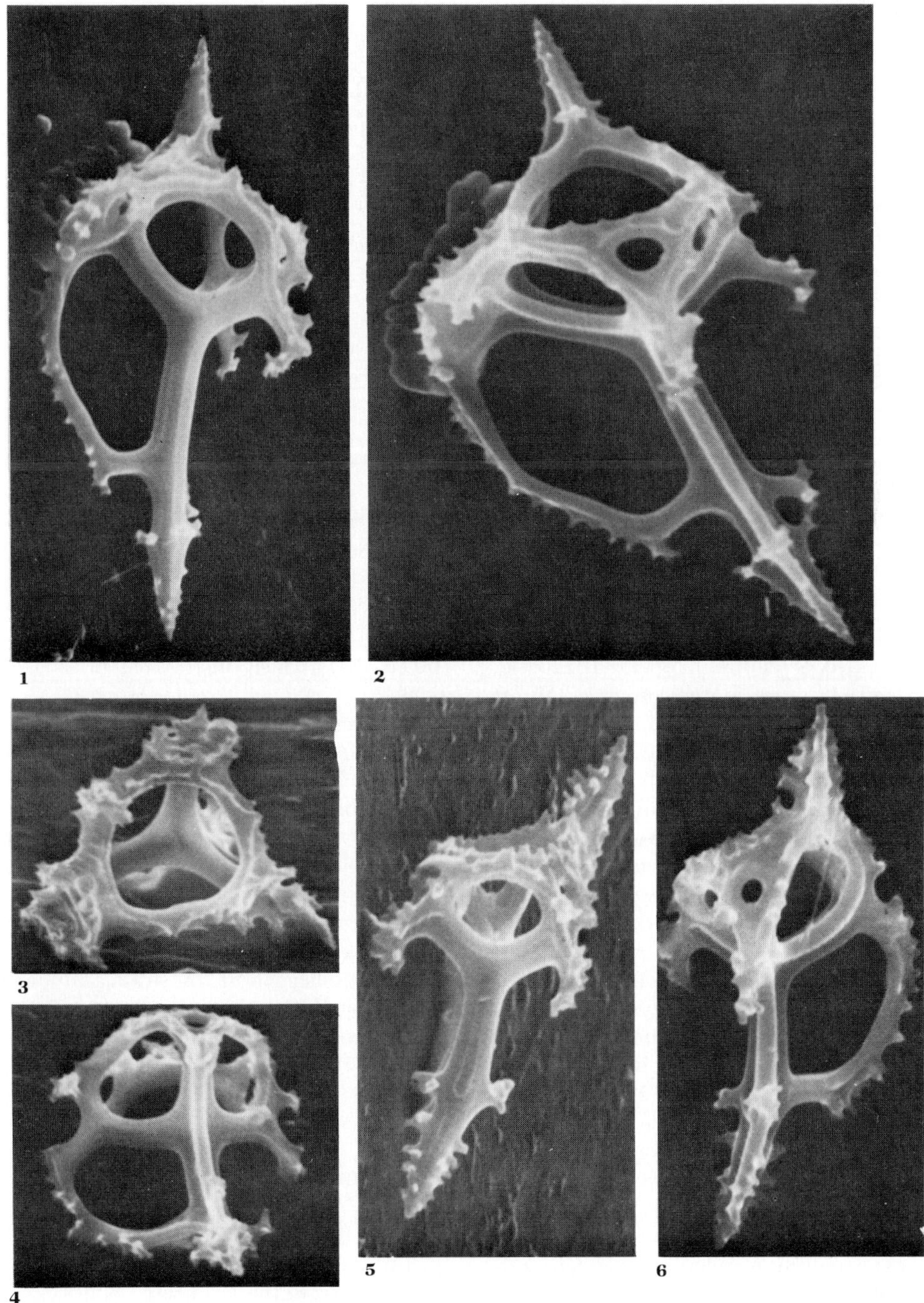

1
2
3
4
5
6

Figure 5.14 — Geologic range of ebridian genera. Shaded cells (marked X) show the range of each genus through the epochs.

Epoch	Falsebria	Spongebria	Ammodochium	Ebriopsis	Hovassebria	Podamphora	Pseudoammodochium	Hermesinum	Craniopsis	Ebrinula	Hermesinopsis	Micromarsupium	Parebriopsis	Polyebriopsis	Hermesinella	Dicladia	Ditripodium	Parathranium	Podamphoropsis	Thranium	Ebria
Holocene								X							X						X
Pleistocene								X							X						X
Pliocene								X							X						X
Miocene			X	X	X	X	X	X						X	X	X	X	X	X	X	X
Oligocene			X	X	X	X	X	X						X	X						
Eocene			X	X	X	X	X	X	X	X	X	X	X	X	X						
Paleocene	X	X	X	X	X	X	X														

Figure 5.14
Geologic range of ebridian genera, showing high diversity in the Eocene, Oligocene reduction, Miocene expansion, and subsequent decline.

tional seven genera first appeared in the Eocene, and six more in the Miocene. All of these genera died out by the close of Miocene time, with the exception of the living *Ebria*, *Hermesinella*, and *Hermesinum*.

Species diversity shows a similar pattern through the Cenozoic (see Figure 5.15). First appearing in the Paleocene, they underwent an early rapid turnover with the extinction of about 68 percent of the 25 Paleocene species. These extinctions were balanced by the introduction of new species in the Eocene (see Figure 5.16), but when some 77 percent of the Eocene taxa became extinct, almost no new species arose in the Oligocene in compensation, so that the flora became much less diverse. A greatly increased rate of evolution in the Miocene resulted in the most diverse assemblage of this group. A similar high rate of extinctions since the Miocene has not been balanced by new introductions, so that this group is at present merely a relic one (Tappan and Loeblich, 1972).

Ebria was particularly abundant in the North Pacific during the Tertiary, apparently migrating into the Atlantic at the end of the Tertiary, and occurring in the Baltic Quaternary deposits. It is widely distributed in the modern seas.

Because ebridians are less common than the associated radiolarians or diatoms, they are easily overlooked as fossils or in plankton collections from modern seas, and some of the smaller species may have been mistaken for radiolarians. The rapid evolution and expansion of the ebridians should result in their being excellent geologic indices, when detailed distributional data are available. Some have been used as zonal indices in the biostratigraphic studies, based largely on silicoflagellates, of material from the Deep Sea Drilling Project (see zonations in Figures 7.14 and 7.15).

Evolutionary Trends

Presumably the triaene type of skeleton is primitive, although both triaene and triode forms occur in the Paleocene. There is no pre-Tertiary record to suggest their ancestry or relationships, but the triode is thought to have arisen by a shortening of the lower axial branch of the triaene, the rhabde, until it disappears,

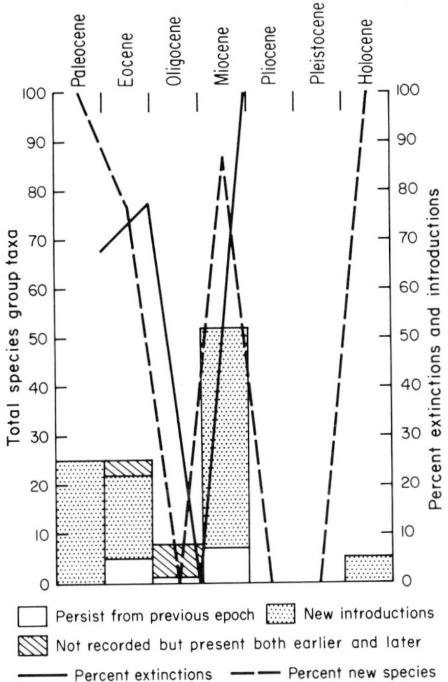

Figure 5.15
Species diversity in the Ebriales. Total species in each epoch are divided into those persisting from that preceding, and new introductions. Percentages of species becoming extinct and of those newly evolved are shown by solid and dashed lines, respectively. From Tappan and Loeblich, 1972.

coincident with an elongation of the opisthoclades.

Two general trends in skeletal shape have arisen. One of these is toward the development of a globular form, such as is found in many unrelated groups adapted to a planktonic existence. There is, in addition, a tendency in other forms such as *Hermesinum* to develop an elongated skeleton. Because *Hermesinum* moves relatively rapidly in its helical path, this shape may be an expression of streamlining for adapting to movement through the water.

CLASSIFICATION

Although this group is included in the protozoans by zoologists, generally as the Order Ebriida, Subclass Ebriideae (Deflandre, 1936, p. 65), Class Phytomastigophorea, Superclass Mastigophora, Subphylum Sarcomastigophora, it commonly is placed among the phytoplankton in the plant kingdom, together with the similar silica-secreting silicoflagellates. Thus the ebriids were regarded by botanists for nearly half a century as representing the Order Stereotestales, Subclass Silicoflagellatophycidae, Class Chrysophyceae, Phylum Chrysophyta (Lemmermann, 1901a; Papenfuss, 1955). However, their placement in the Chrysophyta with the silicoflagellates is not supported by flagellar number, nuclear morphology, skeletal structure, or mode of skeletal formation. Only the siliceous composition of the internal skeleton is found in common, but this also characterizes other chrysophytes, diatoms, some dinoflagellates, xanthophytes, radiolaria, heliozoans, and so forth.

A possible relationship with the radiolaria was suggested by Hovasse (1931b), considering the ebriids tentatively as larval radiolarians.

The ebridians are similar to other dinoflagellates in being biflagellate, with flagella differing in length as well as in pattern of motion. The dinokaryon nucleus is present in both; and the

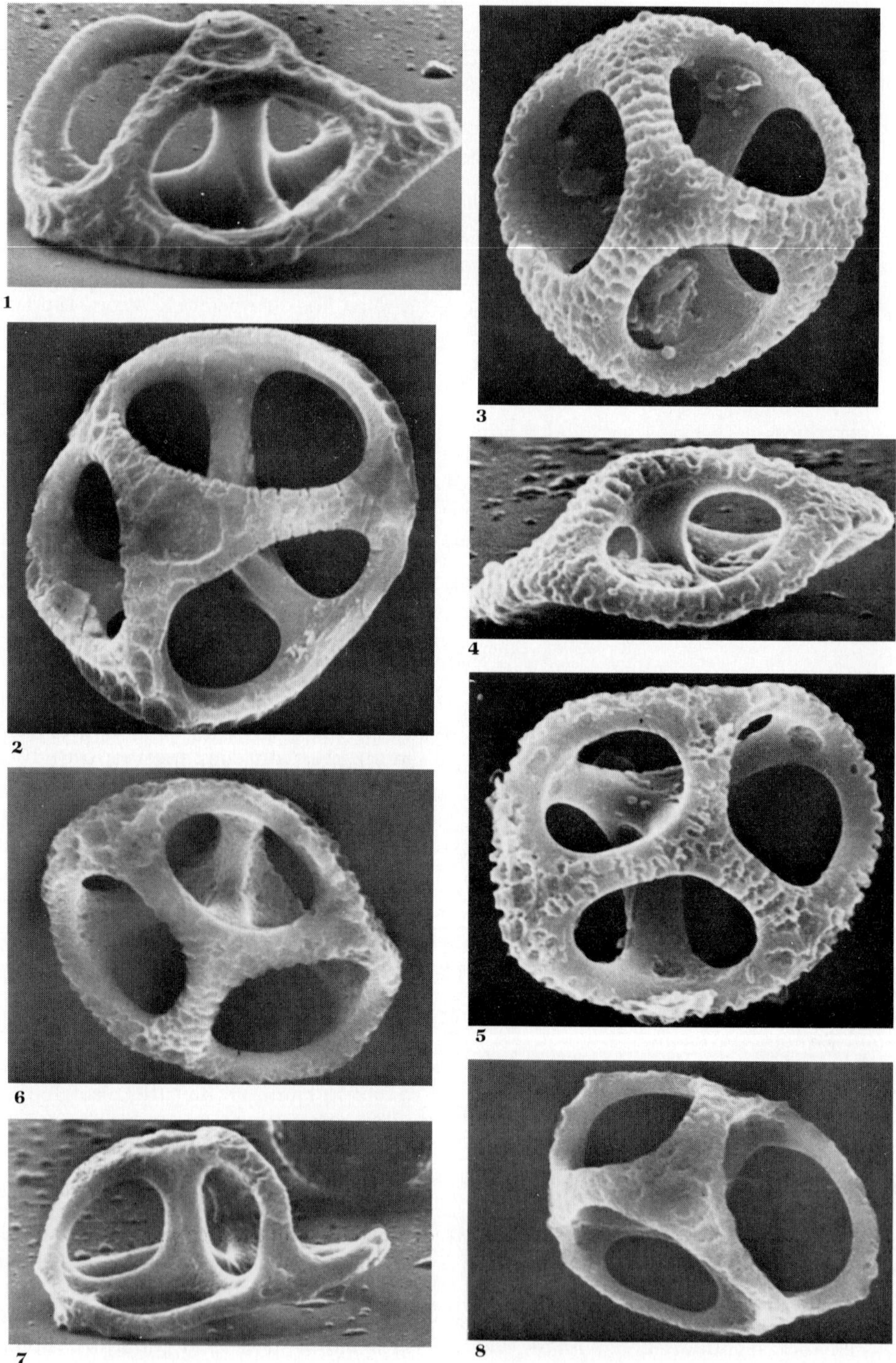

1
2
3
4
5
6
7
8

Figure 5.16
Ebriopsis crenulata, middle Eocene, Kellogg Shale, California, in SEM. **1,2.** Edge and upper view of same specimen, smooth-surfaced vertical rhabde visible in edge view. **3,4.** Top and oblique views of another specimen. **5,6.** Basal views. **7,8.** Edge and top of same specimen. 1, ×1760; 2−6,8, ×1600; 7, ×1680.

colorless, yellow, or rose-colored cytoplasm of the ebridian is reminiscent of the heterotrophic dinoflagellates. The heterotrophic nature of the ebridians is paralleled by the nonphotosynthetic, phagotrophic dinoflagellates, such as species of *Gymnodinium, Gyrodinium,* and *Polykrikos.* Both groups are present in the marine, neritic plankton, although dinoflagellates also include oceanic and freshwater species, as well as those that have developed a parasitic or symbiotic habit.

The skeleton of the ebridians resembles that of the dinoflagellate *Actiniscus* in being constructed of solid, rather than hollow, siliceous elements, and in having a reticulate surface. The formation of new skeletal elements during mitosis is similar in both groups, skeletons of the new daughter cells being initiated as miniatures of the adult, and subsequently enlarging to maturity.

The many common cytological features of the ebriids and dinoflagellates were suggested to indicate their relationship (Hovasse, 1934, 1943a), with the ebriids regarded as intermediate between dinoflagellates and radiolaria. *Ebria* and *Hermesinum* were compared to the silica-secreting dinoflagellate *Actiniscus,* and the three regarded as comprising a natural group (Hovasse, 1943a). As *Actiniscus* is a dinoflagellate, the ebriids were placed as a suborder Ebriida in the dinoflagellates (Hovasse, 1943a, p. 274), with a rank equivalent to the Adinida, Dinifera, and Cystoflagellata. Acknowledging the affinity of dinoflagellates and ebridians, Deflandre (1952) considered them to belong to distinct classes, although he did not propose a class name. Hajós (1968, p. 70) utilized the modified name Order Ebriidales, in the botanical system; and Loeblich III (1970) corrected the name to Order Ebriales, and formally proposed the Class Ebriophyceae within the Pyrrhophyta.

For this group with a relatively small number of genera and species, a proportionately large number of family group names have been proposed in both the zoological and botanical terminologies. Additional difficulties are involved in determining the correct name, as no name has priority outside of its own rank in botany, whereas family and subfamily names have coordinate status for priority in zoology, as do genera and subgenera. Some family group names have been used with no formal description or diagnosis, hence are invalid and carry no priority under either code of nomenclature.

A few taxonomic problems are present at the generic level as well. Designation of *D. clathrata* as the type species for *Dicladia* Ehrenberg was regarded (Loeblich III, 1965) as placing it as a prior synonym of *Parathranium* Hovasse 1932. However, the type of this species was not found in the Ehrenberg collections that have recently been restudied, nor was a similar form present in topotype material examined (Ling and McPherson, 1974). As a result, Ling and McPherson (1974) suggested that *D. clathrata* may in fact possess only two parallel opisthoclades, as originally described, rather than representing a *Parathranium*-like form with three opisthoclades, one having been broken in the specimen illustrated by Ehrenberg. The original drawing of *D. clathrata* shows straight rather than flaring opisthoclades, such as occur in all *Parathranium* from Ehrenberg's type samples. *Parathranium biclathratum* Hajós similarly has two opisthoclades. In view of this uncertain status, the genus *Parathranium* is recognized as distinct, and *Dicladia* is recognized questionably.

In the classification shown in Table 5.1, family synonyms are indicated as are generic ones, and strict priority is followed, according to the ICBN and ICZN.

Class Ebriophyceae Loeblich III 1970
Order Ebriales Honigberg et al., 1964
Unicellular, no resistant outer covering; two inequal and heterodynamic terminally inserted flagella, dinokaryon nucleus; nonphotosynthetic; internal siliceous skeleton composed of a basic three- or four-rayed spicule, the divergent and branching ends of the rays may rejoin terminally; loricate forms may occur; marine, neritic, planktonic; Paleocene to Holocene.

1. Family Hermesinaceae Hovasse 1943, nom. correct. herein (see Figures 5.17 to 5.19)
Syn.: Podamphoraceae Deflandre 1932 (nom. nud.); Hermesinidae Hovasse 1943; Craniopsidae Hovasse 1943; Ebriopsidae Deflandre 1950.

Skeleton with triaene initial spicule, or atrophied rhabde resulting in triode. Rhabde axial to slightly excentric in position; proclades simple or bifurcating, connected by synclades to form nuclear ring; opisthoclades simple or lacking. Double and multiple skeletons and loricate stage known in fossils.
Paleocene to Recent; marine, planktonic.
Craniopsis Frenguelli 1940; *Ebriopsis* Hovasse 1932 (syn.: *Haplohermesinum* Hovasse 1943); *Falsebria* Deflandre 1951; *Hermesinella* Deflandre 1934; *Hermesinopsis* Deflandre 1934; *Hermesinum* Zacharias 1906 (syn.: *Bosporella* Hovasse 1931); *Hovassebria* Deflandre 1936; *Micromarsupium* Deflandre 1934; *Parebriopsis* Hovasse 1932; *Podamphora* Gemeinhardt 1931 (syn.: *Parammodochium* Deflandre 1932; *Parebria* Hovasse 1932; *Podium* Hovasse 1932, non Fabricius 1805); *Podamphoropsis* Dumitrică 1972; *Spongebria* Deflandre 1950.

2. Family Ditripodiaceae Deflandre 1951, nom. correct. Hajós 1968 (see Figure 5.20)
Syn.: Subfamily Thraniinae Hovasse 1943; Tripodiidae Hovasse 1943 (part); Ditripodiidae Deflandre 1951.

Triode initial spicule; simple divergent proclades connected by synclades to form triangular or rounded nuclear ring; opisthoclades divergent, terminally free, pointed, or forked; mesoclades may occur. Double skeletons known; cyst known in *Parathranium*.
Miocene; marine, planktonic.
?Dicladia Ehrenberg 1844; *Ditripodium* Hovasse 1932; *Parathranium* Hovasse 1932; *Thranium* Hovasse 1932.

3. Family Ammodochiaceae Deflandre 1950, nom. correct. Hajós 1968 (see Figure 5.21)
Syn.: Subfamily Ammodochiinae Hovasse 1943; Tripodiidae Hovasse 1943 (part); Ammodochiidae Deflandre 1950.

Initial spicule triode, or atrophied or absent; proclades and opisthoclades parallel to divergent, simple or bifurcate, connected by synclades to form nuclear and antapical rings respectively; mesoclades may occur. Entire skeleton may be replaced by ovoid perforated shell. Double skeletons, loricate stage and cysts known.
Paleocene to Miocene; marine, planktonic.
Ammodochium Hovasse 1932 (syn.: *Tripodium* Hovasse 1932); *Pseudammodochium* Hovasse 1932.

4. Family Ebriaceae Lemmermann 1901 (see Figure 5.22)
Syn.: Ebriidae Poche 1913; tribe Ebrioidae Poche 1913; Ebriacidae Hovasse 1943.

(continued)

Figure 5.17
Hermesinaceae, SEM. **1—3.** *Hermesinum longispinosum* (Hovasse) Deflandre, Miocene, Lompoc, California. **1.** View toward mesoclades directly opposite the apical spine that form a triangle enclosing the small median window over actine a^2; complete opisthoclade o^1 is at left, and others are atrophied; elongate rhabde terminates in basal spine, apical spine arises from nuclear ring between actines (a^1 and a^3). **2.** Same specimen, facing rhabde between actines (a^2, at top, and a^1, foreground). **3.** Apical view of same, apical spine in foreground, three actine rays visible through nuclear window, basal part of triangle from actine a^2, which lacks an opisthoclade, is at top of figure. **4—6.** *Ebriopsis antiqua* (Schulz) Hovasse, Miocene (Luisian), Newport, California. **4.** Elongate form, showing axial anterior spine at lower right, and axial posterior spine at upper left. **5,6.** Two views of same individual; 5, with orientation as in part 4; 6, edge view, axial anterior spine at lower right. 1—5, ×1600; 6, ×1840.

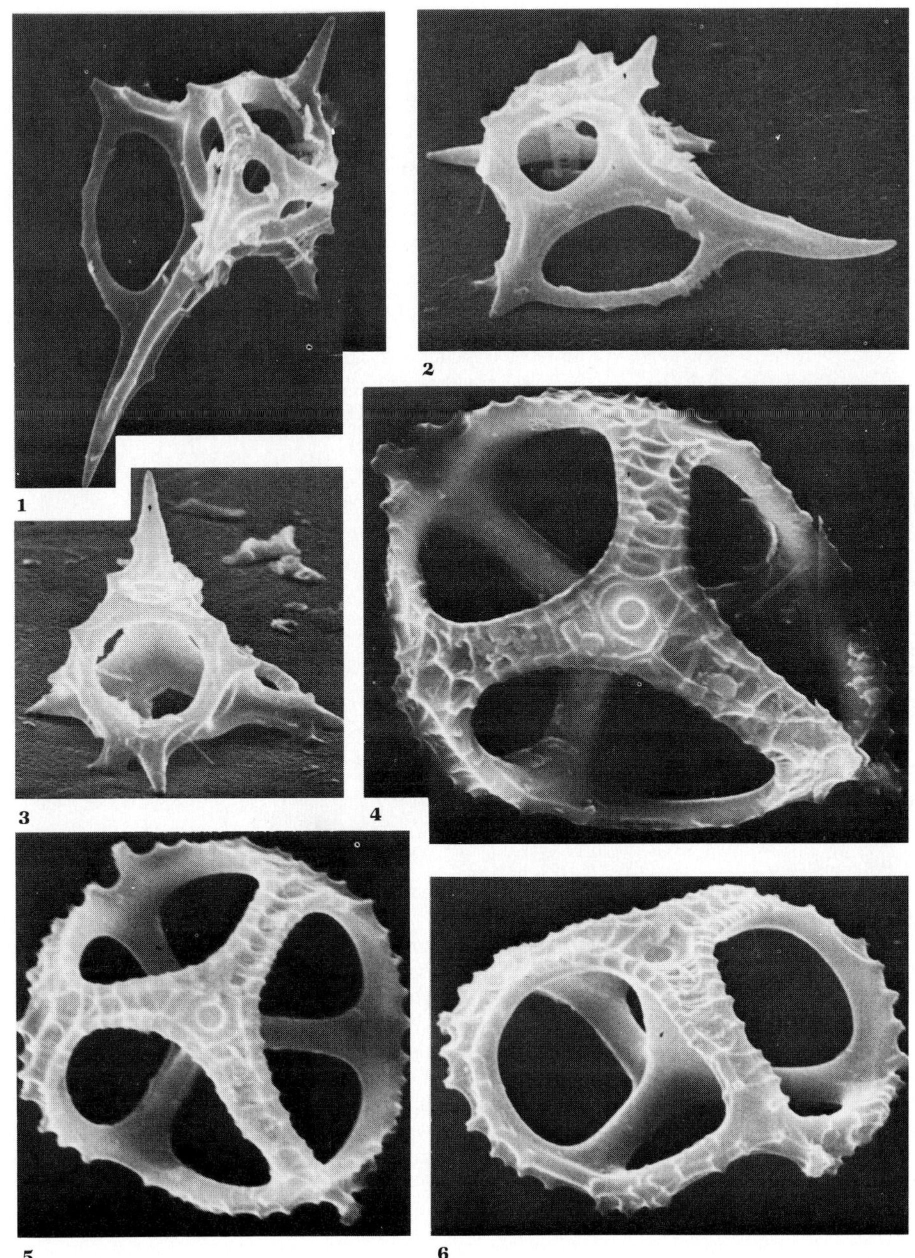

1

2

3

4

5

6

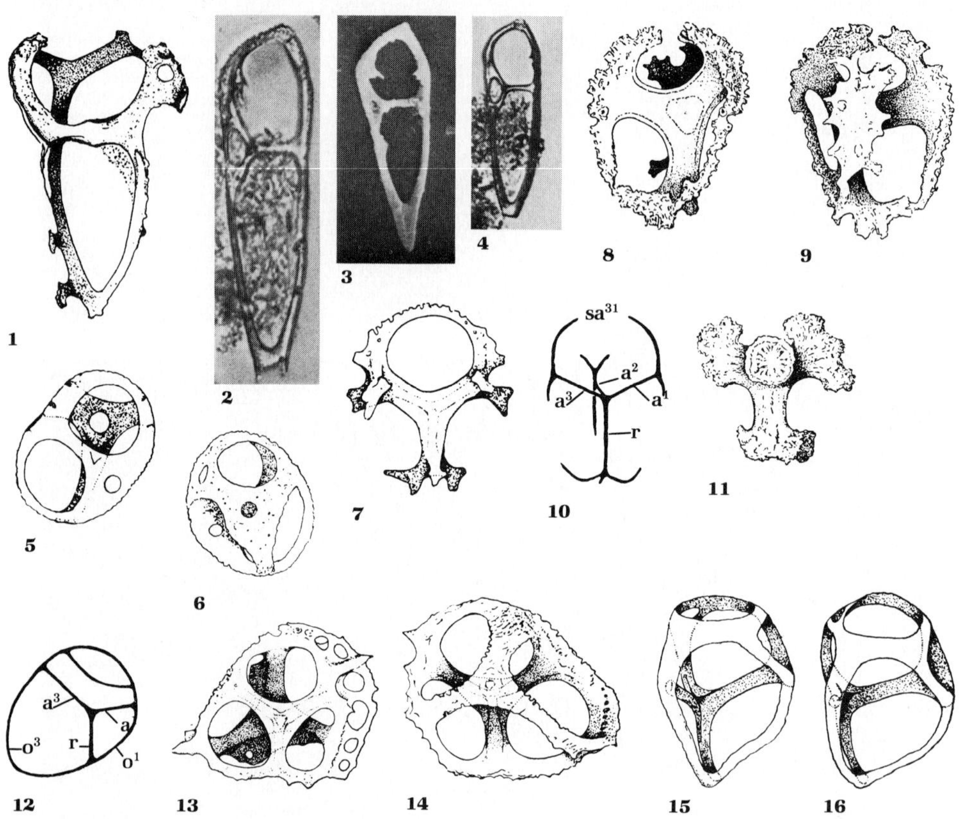

Figure 5.18

Hermesinaceae. **1.** *Micromarsupium anceps*, Eocene, Oamaru, New Zealand, ×745, from Deflandre, 1934. **2−4.** *M. rostovense*, Oligocene, USSR, from Martini, 1976; 2, ×625; 3, SEM, ×520; 4, ×400. **5,6.** *Hermesinopsis caulleryi* Deflandre, Eocene, Barbados, ×1120, from Deflandre, 1934. **5.** Frontal view. **6.** Oblique apical view. **7.** *Hovassebria brevispinosa* (Hovasse) Deflandre. Eocene, Oamaru, New Zealand, ventral face, ×800. **8,9.** *Falsebria ambigua* Deflandre, Paleocene, Jutland, ventral and dorsal views, ×800. **10.** *Falsebria* diagram, dorsal face, labels as in part 12, with sa^{31} indicating synclade between actines 1 and 3. **11.** *F. ambigua* f. *coralloidea* Deflandre, Paleocene, Jutland. Dorsal face, ×800. **12.** *Ebriopsis aplanata* Deflandre, diagram, ×800, with r, rhabde; a^3, a^1, actines; o^1, o^3, opisthoclades. **13,14.** *E. antiqua*, Miocene, Tukuro, Japan, ventral and dorsal faces, ×800. 7−14, from Deflandre, 1951. **15,16.** *Hermesinella transversa* Deflandre, Eocene, Oamaru, New Zealand, two views of same specimen, ×1120, from Deflandre, 1934.

Triode initial spicule; simple proclades perpendicular to triode, connected to rounded nuclear ring by anterior synclades; opisthoclades simple or bifurcating, connected by posterior synclades; opposite synclades may be connected by a rod.

Paleocene to Recent; marine, planktonic.

Ebria Borgert 1891 (syn.: *Ebriella* Deflandre 1934; *Proebria* Hovasse 1943); *Ebrinula* Deflandre 1950.

Some genera that were defined as ebridians are now placed elsewhere. *Carduifolia* Hovasse 1932 is included with the Actiniscaceae in the dinoflagellates; *Semantebria* Frenguelli 1941 and *Spyrebria* Frenguelli 1941 probably are radiolarians.

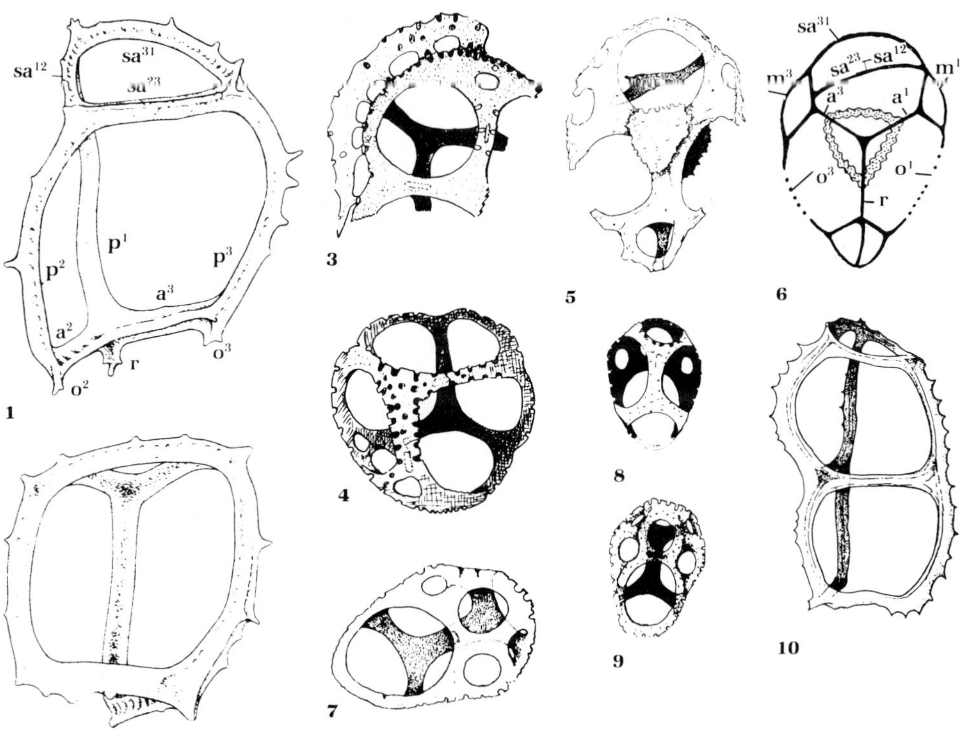

Figure 5.19

Hermesinaceae. **1,2.** *Podamphoropsis joidesi* Dumitrică, upper Miocene, Mediterranean, ×1080, with r, rhabde; o, opisthoclades; a, actines; p, proclades; sa, synclades; from Dumitrică, 1973. **3,4.** *Parebriopsis fallax* Hovasse, Eocene, Oamaru, New Zealand, ×800, from Hovasse, 1932d. **3.** Specimen with hypersilicified crest. **5,6.** *Spongebria marthae* Deflandre, Paleocene, Kuznetsk, USSR, ×800, from Deflandre, 1951; ventral view and diagram of same; symbols as above, with m, mesoclades. **7–9.** *Craniopsis octo* Hovasse ex Frenguelli, Eocene, Oamaru, New Zealand. **7.** ×1120, from Deflandre, 1934. **8,9.** Posterior and anterior faces, ×720, from Hovasse, 1932d. **10.** *Podamphora elgeri*, Miocene, Karand, Hungary, ×800, from Deflandre, 1951.

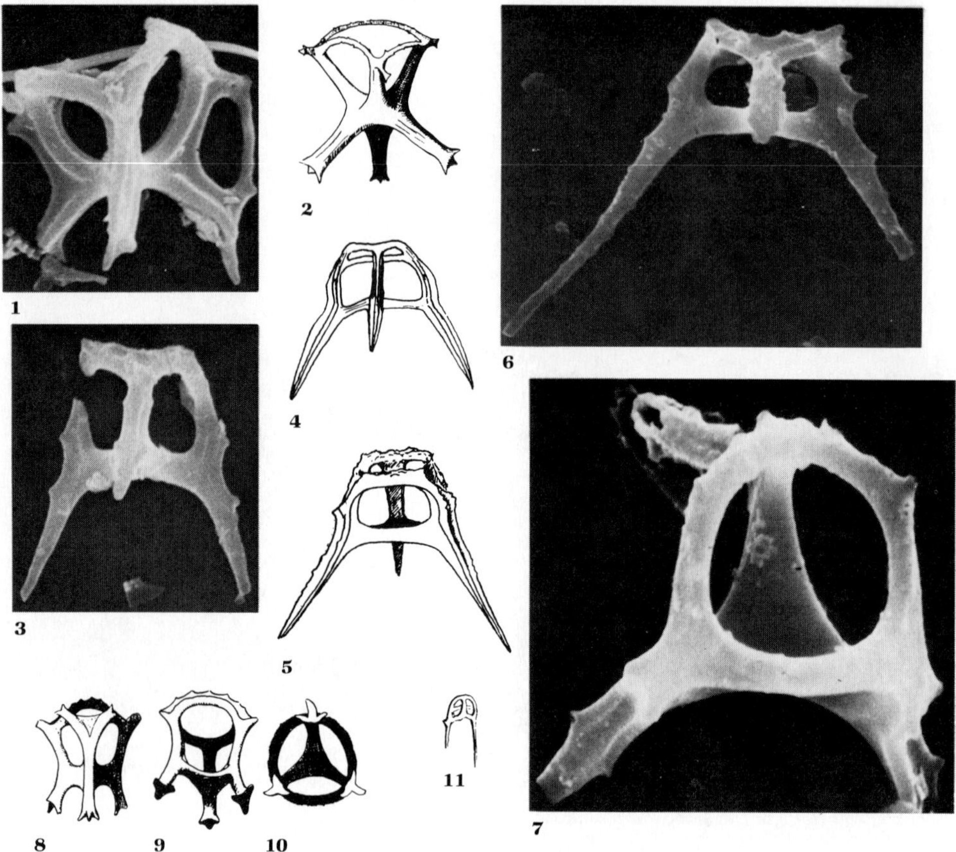

Figure 5.20

Ditripodiaceae from the Miocene. **1.** *Ditripodium latum,* Lompoc, California, SEM, ×1600. **2.** *D. elephantinum,* Sendai, Japan, ×440, from Hovasse, 1932b. **3.** *Parathranium quadrispinum* Deflandre, Lompoc, California, SEM, ×1600. **4—7.** *P. tenuipes.* **4,5.** Mejillones, Bolivia, ×1040, from Deflandre, 1932c. **6.** Side view, SEM, ×1600. **7.** Slightly oblique apical view, showing initial triode through the apical ring, SEM, ×3200. 6,7, Lompoc, California. **8—10.** *Thranium crassipes* Hovasse, California, side view, oblique basal view, and view from base showing central triode, ×560, from Hovasse, 1932b. **11.** *Dicladia clathrata,* Richmond, Virginia, ×240, from Ehrenberg, 1854.

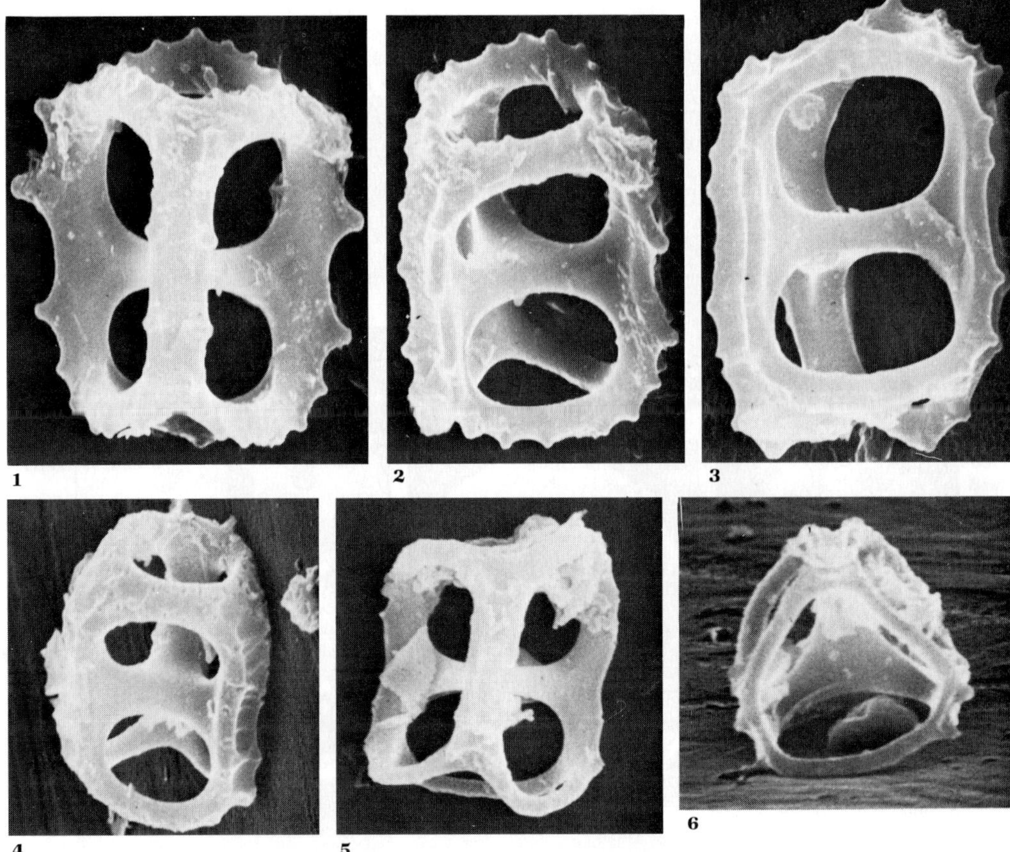

Figure 5.21

Ammodochiaceae. **1−3.** *Ammodochium rectangulare.* **1.** Lateral view, facing proclade and opisthoclade bar. **2.** Slightly oblique lateral view of same specimen, facing windows, and showing central triode. **3.** Lateral view of another specimen. **4−6.** *A. pyramidale* Hovasse. **4.** Oblique lateral view, showing apical opening, and triode seen through lateral windows. **5,6.** Side and basal views of same individual. All SEM, ×2000, from Miocene, Lompoc, California.

XANTHOPHYTA AND CHRYSOPHYTA

It is a safe prediction that under the discerning eyes of specialists other groups…will also contribute significantly to our understanding of paleolimnology and of paleoecology in general. Some of the ones that seem most promising at the moment are the chrysomonads, dinoflagellates, turbellarians, rotifers, oribatid mites, Notostraca, and Malacostraca. …

David G. Frey, 1964

Both the Xanthophyta and Chrysophyta (the former including the Xanthophyceae and Eustigmatophyceae, and the latter the Chrysophyceae, Bikosoecophyceae, and Craspedomonadophyceae in this chapter) are represented in the present marine environment, but their freshwater relatives are far better known. The Xanthophyta have a very sparse fossil record, although the silicified cysts and loricae are potentially preservable. Certain Chrysophyta are better represented; for example, the silicoflagellates, which are discussed in Chapter 7. Other Chrysophyta are more characteristic of fresh water, but have some very poorly known marine representatives. Part of the reason for this neglect lies in their small size of only a few microns in diameter. They are thus difficult to collect with the plankton nets, and because of their fragility are easily destroyed. Specimens must be studied immedi-

ately after collection, as they soon disintegrate, and many can neither be satisfactorily cultured nor be preserved for future study. They would thus seem to be unlikely candidates for a fossil record.

The chrysophytes are better known from their siliceous cysts, which are preserved in both marine and nonmarine deposits, particularly the diatomites. Modern species are largely known from fresh water, but marine fossils are better known than their freshwater counterparts, as micropaleontologic studies have concentrated on marine sediments. Unfortunately, much of the fossil chrysophyte record has poor stratigraphic documentation, but like other microfossils, they are being studied in connection with the DSDP, where ample stratigraphic documentation is provided. Chrysophyte cysts may eventually prove to be a useful adjunct to diatoms or radiolarians in some marine siliceous sediments, and they should be of even greater value in paleolimnology, where freshwater microfossil variety is more limited. Greater emphasis on the study of these freshwater fossils should be of considerable value for paleoecologic interpretation of the continental regions.

Similarly, results of a concentrated search for additional fossil Xanthophyceae might provide examples of fossils among these poorly known protists with silica-impregnated cell wall or their siliceous cysts, and allow a better interpretation of their past evolution and diversification. Information about the living cells thus may have potential value to the micropaleontologist in the event that some of these do appear in the fossil record.

The golden-brown (chrysophyte) and yellow-green (xanthophyte) flagellates are less well known than the many green algae or the dinoflagellates, and have variously been included with one group or another.

Pascher (1914) first combined the Chrysophyceae, Xanthophyceae, and Bacillariophyceae (diatoms) in a single division (Chrysophyta), because of their common pigments, storage products, membrane structure, fine structure of the flagella, and the similarity of their cysts in form, siliceous composition, and intracellular origin. Later Bourrelly (1957) added the brown algae (Phaeophyta) to these, terming the combined groups the Division Chromophyta. Diatoms, included by Pascher and Bourrelly with the chrysophytes, have chlorophyll *c* like that of the brown algae. None of the Chromophyta have chlorophyll *b* like that of green algae and the higher plants, but all have various accessory pigments, carotenes, and xanthophylls, that produce their characteristic golden-brown color.

Discovery of a distinctive organelle, the haptonema, in some of the former chrysophytes led to their separation as the Haptophyceae or Haptophyta (see Chapter 9); some regard the Chrysophyta, Xanthophyta, and Bacillariophyta as distinct divisions, and an additional group was separated from the Xanthophyceae as the Class Eustigmatophyceae of the Chromophyta (Hibberd and Leedale, 1970, 1971b), or even elevated to a separate division.

In the zoological classification all of the above groups as well as other flagellates are combined in a single class of phytoflagellates. As many of the phytoflagellates and the related coccoid or filamentous forms have had separate evolutionary paths since diverging, presumably in the Precambrian, the botanical system appears preferable; the Chrysophyta, Xanthophyta, Haptophyta, and Bacillariophyta are here regarded as distinct divisions, and the latter two are discussed in separate chapters.

DIVISION XANTHOPHYTA, CLASS XANTHOPHYCEAE

The Xanthophyceae or Heterokontae were early referred to the green algae, but were shown by Pascher (1914) to be more closely related to the Chrysophyceae and Bacillariophyceae, a conclusion that has been upheld by studies of the ultrastructure and pigments, following removal of the Eustigmatophyceae from the Xanthophyta and of the Haptophyceae from the Chrysophyta.

The Cell

The vegetative cell may be solitary, occur in coccoid or palmelloid masses, or form simple or branched septate or nonseptate filaments. It may be naked or have a firm cell wall, commonly consisting of two overlapping pieces (see Figure 6.1). A cell may have from 1 to 32 or more nuclei. Motile cells may be terminally biflagellate or amoeboid. However, the zoospores of the coccoid or filamentous species are very similar in appearance to the free-living flagellates.

Motile Cell or Zoospores

The zoospores (see Figure 6.2) of the coccoid *Ophiocytium* or the filamentous *Tribonema* resemble a "generalized Heterokont" in being

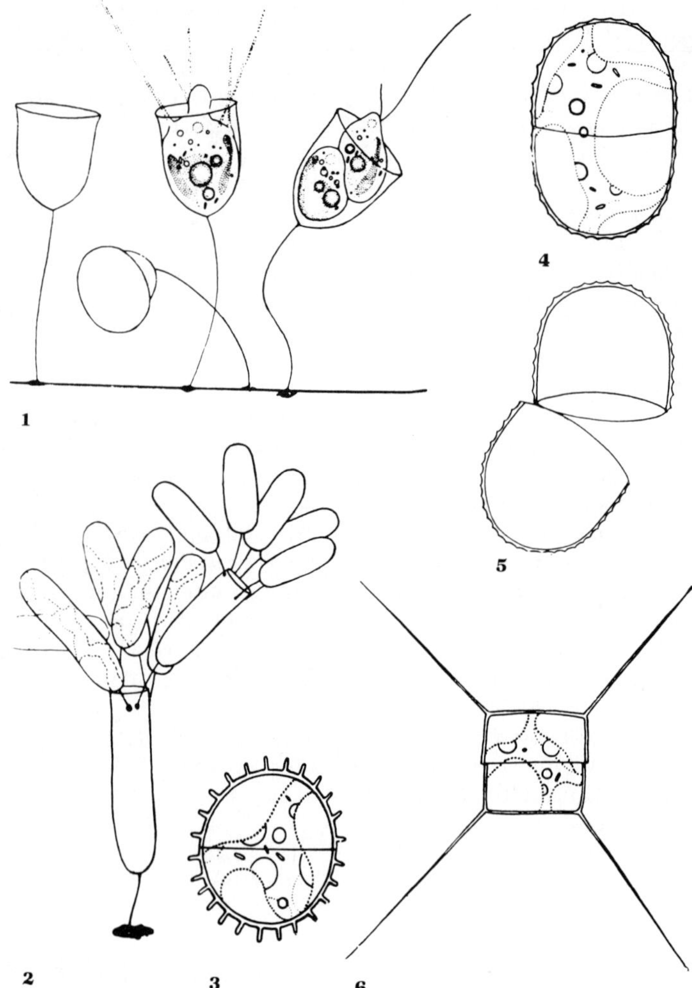

Figure 6.1

Xanthophyta. **1.** *Stipitococcus vas* Pascher. Empty loricae at left; center with pseudopod-bearing cell; right, cell division resulting in two zoospores with unlike flagella; ×1250. **2.** *Ophiocytium gracilipes* Rabenhorst, cylindrical cells in colonies, ×1000. **3.** *Akanthochloris bacillifera* Pascher, two-part cell wall with surface spines, may be silicified, ×2000. **4,5.** *Chlorallantus oblongus* Pascher, ×2000. **4.** Two-part cell wall that may be silicified, with reticulate sculpture. **5.** Empty cell wall after protoplast has escaped. **6.** *Pseudotetraëdron neglectum* Pascher, with two-part cell membrane bearing four much elongated needlelike projections, ×2000. All from Pascher, 1937–1939.

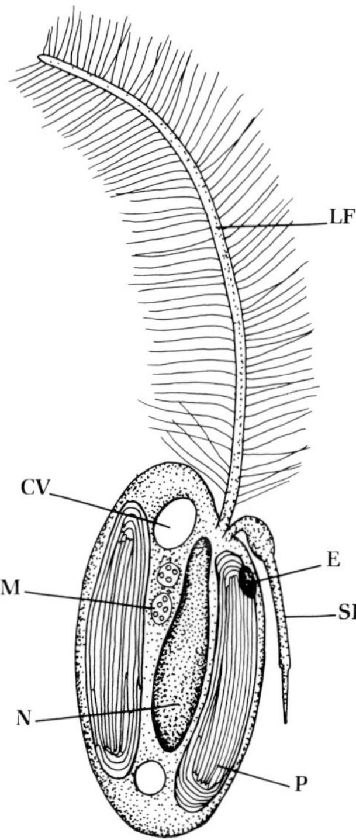

Figure 6.2
Ophiocytium majus Nägeli. Zoospore, showing LF, long mastigoneme-bearing flagellum; SF, short, smooth flagellum with basal swelling; N, nucleus; P, plastid with girdle lamellae; E, eyespot; M, mitochondrion; CV, contractile vacuole; ×750, based on electron micrographs and light microscopy of Hibberd and Leedale, 1971a.

bilaterally symmetrical, ovoid, from 5 to 8 μm in length, with truncated anterior end (Hibberd and Leedale, 1971a). A somewhat elongate or pyriform nucleus has a single nucleolus, and the anterior end of the nucleus lies very close to the flagellar basal bodies. The cell is biflagellate, one flagellum of a length equal that of the cell being directed forwards and bearing two rows of stiff hairs (mastigonemes). The other flagellum is shorter, from one-third to one-half the length of the anterior one, is smooth and hairless, but has a prominent swelling near its base.

Plastids and Pigments

The one to many discoid yellowish green plastids contain a high proportion of carotenoids that mask the green color of the chlorophyll and give the characteristic appearance. When heated in concentrated hydrochloric acid, the cells turn blue, whereas those of the Chlorophyta remain green or yellowish green.

The motile cells and zoospores of filamentous species typically have two plastids adjacent to the ventral and dorsal surfaces. **Girdle lamellae** are present in most Xanthophyceae, although lacking in *Bumilleria* (Massalski and Leedale, 1969). Pyrenoids are not prominent and may be lacking in some; when present, the pyrenoids are traversed by the three-thylakoid lamellae of the plastids. An eyespot is present as part of one of the plastids, and is located near the flagellar bases.

The Xanthophyceae have chlorophyll *a* and no chlorophyll *b*. Some also have small amounts of chlorophyll *c*; *Tribonema* has a ratio of 170:1 by weight of chlorophyll *a* to chlorophyll *c*. *Chloridella* has a slightly larger amount of chlorophyll *c*, whereas *Ophiocytium* has none (Whittle, 1976a,b). The major xanthophyll is violaxanthin. Other pigments include β-carotene, vaucheriaxanthin-ester, heteroxanthin, diatoxanthin, cryptoxanthin epoxide, and a neoxanthinlike pigment (Whittle and Casselton, 1975b).

Food reserves are stored as oil or leucosin, never as starch. Some colorless forms exist, and these and some pigmented ones may be saprophytic (ingesting dissolved material) or phagotrophic (ingesting particulate food).

Cell Wall

The cell wall is of cellulose in *Botrydium*, *Tribonema*, and *Vaucheria*, but is a noncellulosic polysaccharide in *Monodus*. In some, it may be impregnated with silica. Commonly the wall consists of two overlapping halves. The two halves form H-shaped sections in filamentous species, such as *Tribonema*, a single cell being enclosed by half of each of two adjacent H-pieces (see Figure 6.3). The cell wall of the

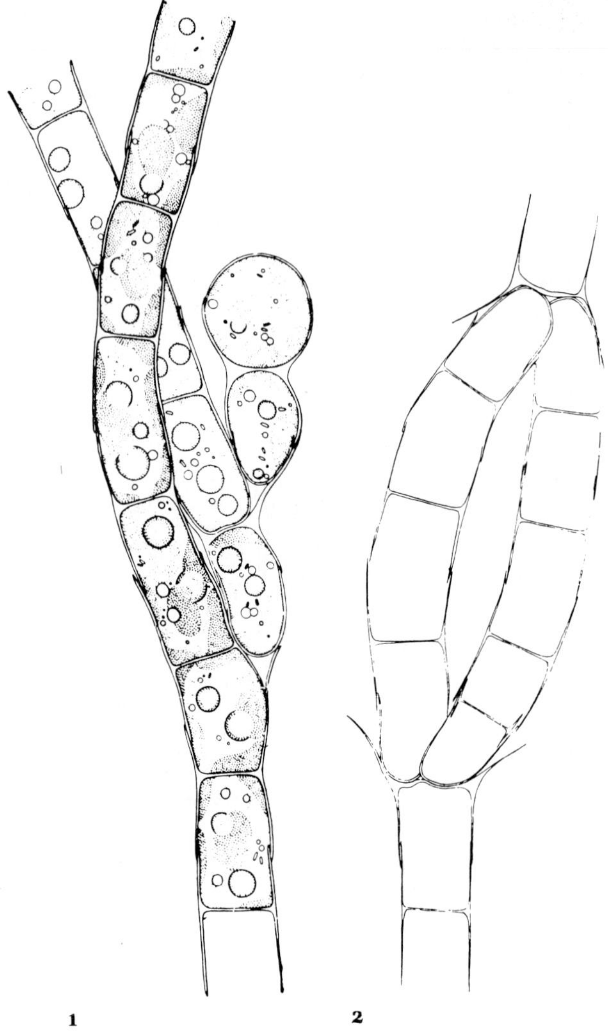

vegetative *Ophiocytium* similarly is bipartite, but consists of a tubular basal part that increases in length as the cell grows, being attached basally by a stalk to the substrate or to other cells, and topped by a rounded cap that remains of constant size. The rims of the tubular part and cap both taper and overlap for some distance, the edge of the cap lying external to the margin of the tubular part (Hibberd and Leedale, 1971a).

Reproduction

Multiplication may occur by simple cell division or by means of zoospores. The biflagellate zoospores may form singly or in quantity within a cell. **Aplanospores** or **statospores** (cysts) may also be formed. The entire cell may form an aplanospore, or the protoplast may divide to form a number of these. The aplanospores may germinate directly into a new vegeta-

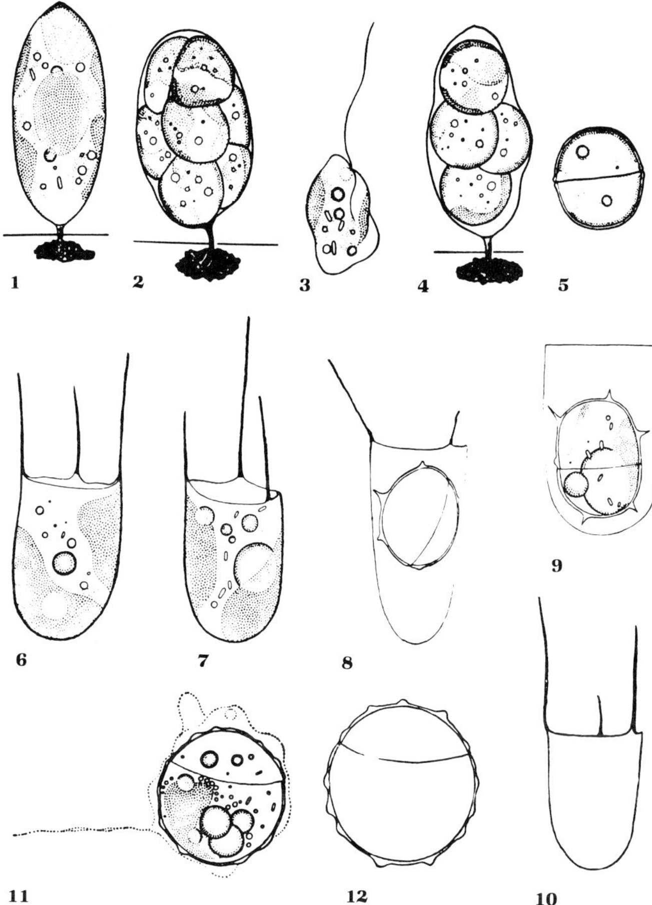

Figure 6.4
1—5. *Characiopsis gracilis* Pascher, ×1000. **1.** Attached vegetative cell. **2.** Formation of zoospores. **3.** Isolated zoospore. **4.** Formation of resting cysts. **5.** Single resting cyst. **6—10.** *Schilleriella anuraea* Pascher, ×2000. **6,7.** Vegetative cells. **8,9.** Formation of resting cyst. **10.** Empty cell wall. **11,12.** *Chloromeson agile* Pascher, cyst formation, approximately ×1500. **11.** Development of cyst within amoeboid cell that had lost flagella. **12.** Cyst, showing rough surface and inequally bipartite construction. All from Pascher, 1937—1939.

tive stage, or may first produce zoospores that grow into the next stage. The cysts or statospores, formed endogenously as in the Chrysophyta, consist of two overlapping halves of equal or inequal size, similar to the two-part walls of the vegetative cell (see Figure 6.4). At germination, two or four daughter protoplasts are liberated as amoeboid or flagellated zoospores. The two-part cysts of the Xanthophyceae thus resemble the two overlapping parts of a diatom frustule rather than the flasklike cysts of the Chrysophyceae. Although the composition is seldom reported, some cysts are at least lightly silicified.

Sexual reproduction is best known in such filamentous forms as *Vaucheria* and *Tribonema.* Identical gametes occur in *Tribonema;* the zoospores of *Botrydium* are facultative gametes that may fuse or develop asexually. Meiosis apparently takes place within the zygote, so that the haploid condition predominates through the life cycle. Sexual reproduction is more evident in *Vaucheria,* in view of the relatively large reproductive structures (see Figure 6.5). Protuberances arise from the siphonous tubes, into which the streaming cytoplasm carries many nuclei and plastids. In development of the oogonium, all but one of the nuclei migrate back into the siphonate portion, and a septum closes off the base. The antheridium is multinucleate when the septum forms to isolate it. The spermatozoids of *Vaucheria* are elongate, with two unlike flagella inserted laterally, the short front flagellum with

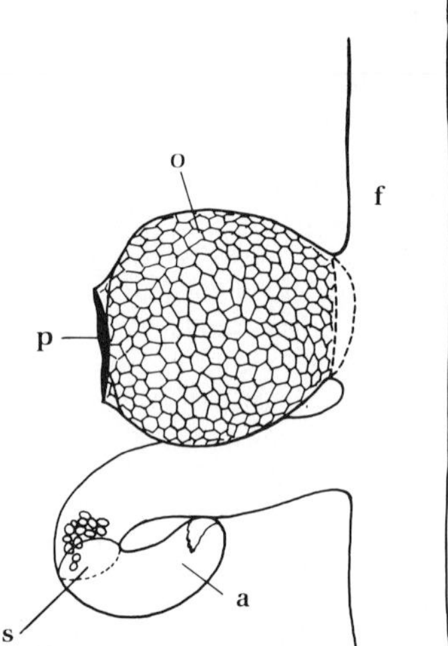

Figure 6.5
Vaucheria sp. f, nonseptate filament; o,
oogonium, with p, point of sperm entrance; s,
transverse septum formed to isolate antheridium; a, discharged antheridium; approximately ×500.

two rows of mastigonemes, and the posterior
flagellum elongated and smooth, greatly resembling the sperm cells of the brown alga
Fucus. In addition, they have similar microtubule structures near the anterior end and a
much elongated nucleus (Moestrup, 1970).

Occurrence

Most Xanthophyceae live in fresh water, and a
few are terrestrial in habit, living on tree trunks
or in the soil. They prefer an alkaline condition.
A few can live in the dark on organic compounds supplied in the laboratory (*Tribonema*),
but the xanthophytes are characteristically
photosynthetic. Marine species occur, but are
less well known.

Xanthophyceae have a very meager fossil record. *Vaucheria antiqua* Ludwig was described
from the Miocene Braunkohle of Germany, and
Vaucheria is reported to be important in the
deposition of the thick Pleistocene and

Holocene freshwater travertine, as in the
streams of the Arbuckle Mountains, Oklahoma
(Emig, 1917). In addition to the *Vaucheria*, various blue-green and green algae also cause the
deposition of the cavernous travertine, the lime
encasing the plant filaments persisting after the
organic matter decays and little evidence of the
plant structure remains.

Other fossils that have been referred to the
Xanthophyceae are correctly assigned elsewhere. Thus *Botryococcus*, important in the
formation of boghead coal and once regarded as
a xanthophyte, is now known to be a
chlorophyte. Similarly, Emberger (1968, p. 66)
placed the Leiosphaeridiaceae (*Leiosphaeridia,
Tasmanites*) and the modern *Pachysphaera* in
the Xanthophyceae, although these are correctly referred to the Prasinophyceae, of the
green algae.

Xanthophytes are less differentiated than
other algal divisions, such as the Chlorophyta or
Haptophyta, and include fewer flagellates. Perhaps because modern xanthophyte taxa are less
studied, possible fossil examples may not have
been recognized as such. Unfortunately, the
composition of the endogenously formed cysts
of most Xanthophyceae is not known. If
silicified, like the similarly formed cysts of the
Chrysophyceae, they should be equally as readily preserved in freshwater diatomites and recognized by the characteristic two-part structure. If merely organic, some of the thick-
walled ones might be preservable like those of
dinoflagellates, if of a similar sporopollenin
composition. The general lack of information
as to cyst morphology and composition of both
the marine and freshwater Xanthophytes that
would be useful in identification of possible fossils points up the necessity for additional life
history studies of many of the modern microalgae.

CLASS EUSTIGMATOPHYCEAE

The Eustigmatophyceae, recently proposed as a
class of the Chromophyta (Hibberd and Leedale,
1970, 1971b), are as yet imperfectly known.

Largely distinguished from the Xanthophyta on the basis of ultrastructural details, they also contain distinctive pigments (Whittle and Casselton, 1975a). A few species previously placed with green algae have similarly been transferred to the Eustigmatophyceae (Antia et al., 1975). Although here discussed with the Xanthophyceae (restricted sense), the Eustigmatophyceae were considered to be distinctly different, whereas the other Xanthophyceae from a natural grouping with the Chrysophyceae (exclusive of the Haptophyceae or Haptophyta), Bacillariophyceae, and Phaeophyceae, which have been included in the Chromophyta.

The Cell

Most Eustigmatophyceae are coccoid nonmotile cells (see Figure 6.6), but *Pseudocharaciopsis* is elongate, multinucleate, and has a short stalk and distinct attaching disc by which the cells may attach in clusters (Lee and Bold, 1973). A single elongate plastid has three-thylakoid lamellae, no girdle lamellae, and from one to four polygonal stalked pyrenoids projecting from the inner face of the plastid, surrounded by flat plates of storage material. The thylakoids do not pierce the pyrenoid. A large orange-red eyespot lies at the anterior end of the cell, independent of the plastid. The eyespot consists of pigmented droplets, which are not membrane-bound (Hibberd and Leedale, 1970). The pigmentation is distinct, as all contain violaxanthin as the major xanthophyll pigment, in contrast to the dominant diadinoxanthin and absence of violaxanthin in the true Xanthophyceae. Other abundant pigments present are chlorophyll *a*, β-carotene, and the xanthophyll vaucheriaxanthin-ester (Whittle and Casselton, 1975a).

The flask-shaped zoospores commonly have a single emergent flagellum, the anterior one, with two rows of stiff hairs (mastigonemes); a second basal body is present within the cell, and in a few species a second flagellum may be emergent, and is short and smooth (see Figure 6.7). A swollen area at the proximal end of the anterior flagellum near the eyespot is an expansion of the flagellar sheath.

To date, only a few species of some genera have been placed in the Eustigmatophyceae, although they are morphologically distinct as well as having different pigment complexes. A few produce cysts, and the thick cell wall of *Chlorobotrys* may be silicified, but as yet there is no report of fossil representatives.

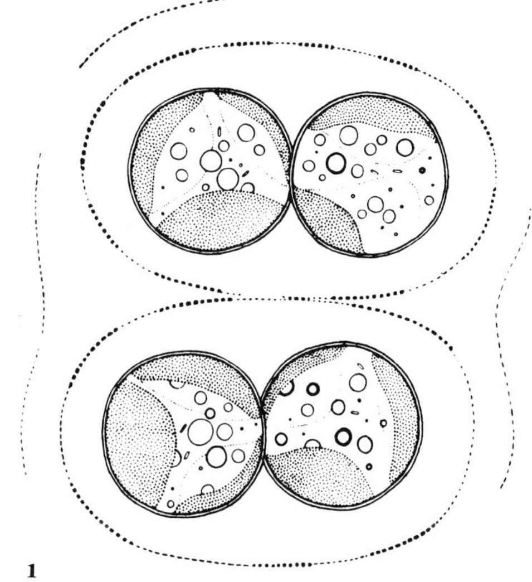

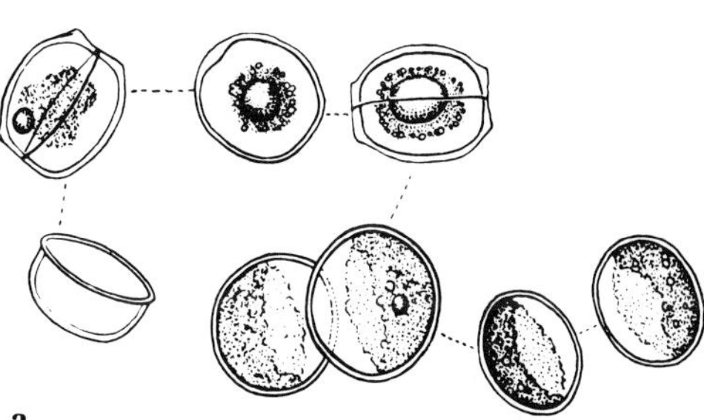

Figure 6.6
Eustigmatophyceae. *Chlorobotrys regularis* Bohlin. **1.** Four-celled colony, ×1000. **2.** Formation of bipartite cysts and young cells, ×500. Both from Pascher, 1937 — 1939; 2, after Bohlin.

Table 6.1
Classification of the Xanthophyta.

Division Xanthophyta Polyansky & Hollerbach 1951

I. Class Xanthophyceae Fritsch 1935

Usually biflagellate, with anterior flagellum bearing a double row of mastigonemes; posteriorly directed flagellum smooth, swollen proximally; two or more plastids, typically with girdle lamella and three-thylakoid lamellae that may enter the pyrenoids; pigments include chlorophyll *a* and *c* (in most) and various carotenes and xanthophylls; eyespot is part of plastid; may be coccoid, amoeboid, flagellate, siphonous, or palmelloid. Marine, freshwater, and soil-dwelling. Miocene to Recent.

A. Order Chloramoebales P. C. Silva 1962
Syn.: Heterochloridales Pascher 1912

Naked, solitary flagellated cell, with one or two inequal terminal flagella; also may have pseudopods; produces bivalved siliceous cysts. Marine, fresh and brackish water.
Chloromeson Pascher 1930; *Heterochloris* Pascher 1914.

B. Order Rhizochloridales Pascher 1925

Solitary amoeboid cell, partially enclosed in lorica; attached. Bipartite cysts in *Rhizochloris*. Freshwater and marine.
Rhizochloris Pascher 1918; *Stipitochloris* Pascher 1937; *Stipitococcus* W. West & G. S. West 1898.

C. Order Heterogloeales Pascher 1939
Syn.: Heterocapsales Pascher 1912

Solitary nonmotile coccoid cells in gelatinous aggregates, or palmelloid; cell wall may be silicified, especially in old cells; contractile vacuoles present; reproduction by nonmotile cells; produce thick-walled bivalved cysts. Marine and freshwater.
Pseudotetraëdron Pascher 1912.

D. Order Mischococcales Fritsch in West 1927
Syn.: Heterococcales Pascher 1912

Coccoid unicells, cell wall present, may be silicified in *Akanthochloris, Chlorallantus,* and *Tetragoniella;* cell may be attached by mucilaginous stalk; no contractile vacuoles; thick-walled cysts may be silicified. Marine and freshwater plankton.
Akanthochloris Pascher 1930; *Arachnochloris* Pascher 1930; *Characiopsis* Borzi 1895; *Chlorallantus* Pascher 1930; *Heterococcus* Chodat 1908; *Ophiocytium* Nägeli 1849; *Schilleriella* Pascher 1932; *Tetraëdriella* Pascher 1930.

E. Order Tribonematales Pascher 1939
Syn.: Heterotrichales Pascher 1912

Filamentous, simple or branched septate filaments, may be attached; may be symbiotic with iron bacteria. Freshwater, soil-dwelling, marine.
Tribonema Derbès & Solier 1856.

F. Order Vaucheriales Bohlin 1901

Vesicles or nonseptate branched siphonous filaments, multinucleate; may be calcium carbonate encrusted. Freshwater, soil-dwelling; Miocene to Recent.
Botrydium Wallroth 1815; *Vaucheria* de Candolle 1801.

II. Class Eustigmatophyceae Hibberd & Leedale 1970 (of Chromophyta)

Coccoid unicells, may be elongated and attached by short stalk (*Pseudocharaciopsis*). Zoospores have only one emergent flagellum, the anterior one, with two rows of stiff hairs (mastigonemes); also have a second flagellar basal body or less frequently a short emergent smooth second flagellum; large eyespot of pigmented droplets, not membrane enclosed, anterior in position, independent of plastid; plastid elongate, with three-thylakoid lamellae, no girdle lamella; polygonal stalked pyrenoid projects from inner face of plastid, not pierced by thylakoids; pigments include chlorophyll *a* and no auxiliary chlorophyll; dominant xanthophylls are vaucheriaxanthin-ester and violaxanthin; cell wall of polysaccharide (not cellulose), may be silicified in *Chlorobotrys*. Most are nonmarine, few marine.
Chlorobotrys Bohlin 1901; *Ellipsoidion* Pascher 1937; *Monallantus* Pascher 1937; *Monodus* Chodat 1913; *Pleurochloris* Pascher 1925; *Polyedriella* Pascher 1930; *Pseudocharaciopsis* Lee & Bold 1973; *Vischeria* Pascher 1938.

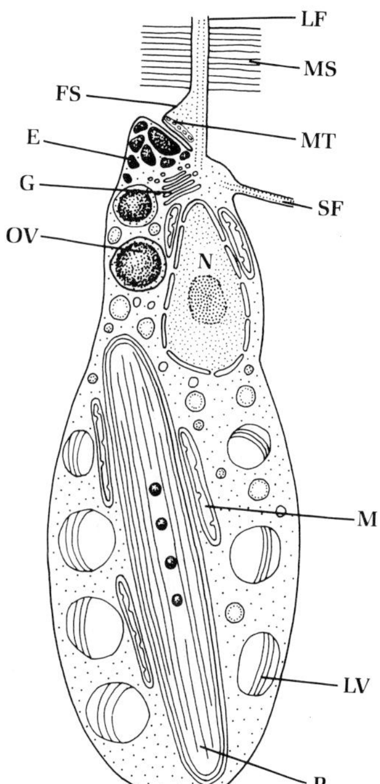

Figure 6.7
Eustigmatophyceae. *Pseudocharaciopsis texensis* Lee & Bold. Diagrammatic median section of zoospore, showing LF, long flagellum, with MS, mastigonemes, and FS, flagellar swelling; MT, microtubules; E, eyespot; SF, short flagellum; G, Golgi apparatus; OV, osmiophilic vesicles; N, nucleus; M, mitochondrion; LV, laminate vesicle; and P, plastid. Redrawn from Lee and Bold, 1973.

CLASSIFICATION OF THE XANTHOPHYTA

The abbreviated classification of the Xanthophyta shown in Table 6.1 is based on Papenfuss (1955), Bourrelly (1968), and Hibberd and Leedale (1970).

DIVISION CHRYSOPHYTA

The Chrysophyta, sometimes referred to as Chrysomonads, commonly have a uni- or biflagellate cell as a principal or temporary stage. Most are solitary, but filamentous, palmelloid, and coccoid colonial Chrysophyta are known.

The Cell

Cell Wall or Lorica

Most Chrysophyta are plantlike in having a cellulosic cell wall, an outer membrane that is not directly involved in the cleavage of a cell undergoing reproduction. The wall may be a simple covering, or may form a lorica that encloses the naked cell. The lorica may be cellulosic (*Dinobryon*), silicified (*Chrysococcus*), or calcareous (*Pseudokephyrion*), or consist of a basketlike structure of silica rods (*Pleurasiga*). The lorica of *Chrysolykos* is hyaline in appearance, but in electron microscopy is seen to be constructed of interwoven microfibrils of cellulose, each fibril from 100 to 150 Å in width. The fibrils form a cross-layered pattern with some longitudinally and others transversely aligned, with no matrix between them. Others have a scaly lorica (*Hyalobryon, Epipyxis*), the scales in electron microscopy also showing a structure of interwoven microfibrils (Kristiansen, 1969). The scales in *Syncrypta* (see Figure 6.8) may be present in layers as deep as twice the 4-μm diameter of the cell itself (Clarke and Pennick, 1975). The scales may be distinctive in form, discoid, sievelike, or crown- to goblet-shaped in *Paraphysomonas* (see Figure 6.9), but are so small as to be discernible only with the electron microscope. Other Chrysophyceae may have

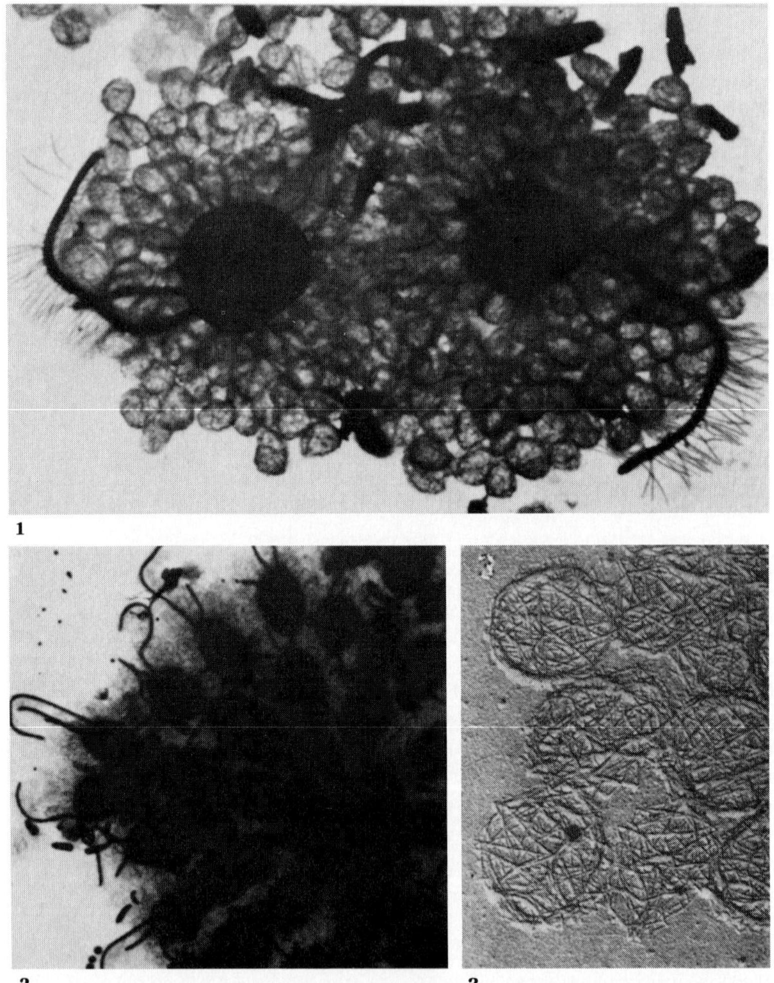

Figure 6.8
Syncrypta glomerifera Clarke & Pennick. Salt marsh pool, England.
1. TEM of solitary cells with scales; note mastigoneme-bearing long flagellum of each cell; ×6000. **2.** Part of colony showing many cells, ×1700. **3.** Shadowcast preparation of detached scales with open fibrous structure, ×18,900. From Clarke and Pennick, 1975.

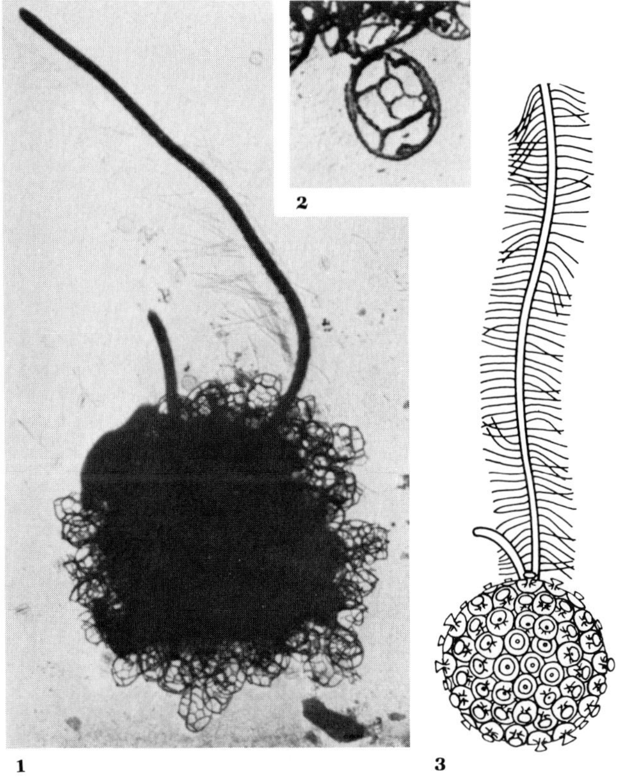

1

2

3

silicified scales, plates, and spines, as in *Synura*
or *Mallomonas* (see Figure 6.10), of similarly
very tiny size.

Not all Chrysophyceae have a cell wall; some
may be naked and amoeboid, with pseudopodia
as well as flagella, or palmelloid. Cell shape is
not as distinctive in those without a cell wall, as
the protoplast is readily deformed.

Cytology

A single nucleus is present, the nuclear envelope enclosing the yellow-green or golden-
brown bilobed plastid (see Figure 6.11). Three-
thylakoid lamellae are present, and girdle
lamellae occur in all chrysophytes except
Chrysamoeba. Various kinds of pyrenoids may
occur, most containing some of the thylakoid

lamellae. A simple eyespot is found in some.
Lipid globules are common in the cytoplasm,
and a large vacuole in the posterior part of the
cell contains the storage food reserves as
chrysolaminarin. Freshwater species have contractile vacuoles. Ejectile organelles termed
discobolocysts occur in *Ochromonas* (see Figure 6.12) and some others, forming a peripheral
layer of as many as 50 spherical alveoli in the
cell. Upon their explosive discharge, a discoid
projectile fastened at the end of a thin thread of
the same fibrillar-reticulate structure as the
disc is released (Hibberd, 1970).

Flagella

Motile cells, and zoospores of others, may be
biflagellate, with one pantonematic (hairy)
flagellum with two rows of mastigonemes, and

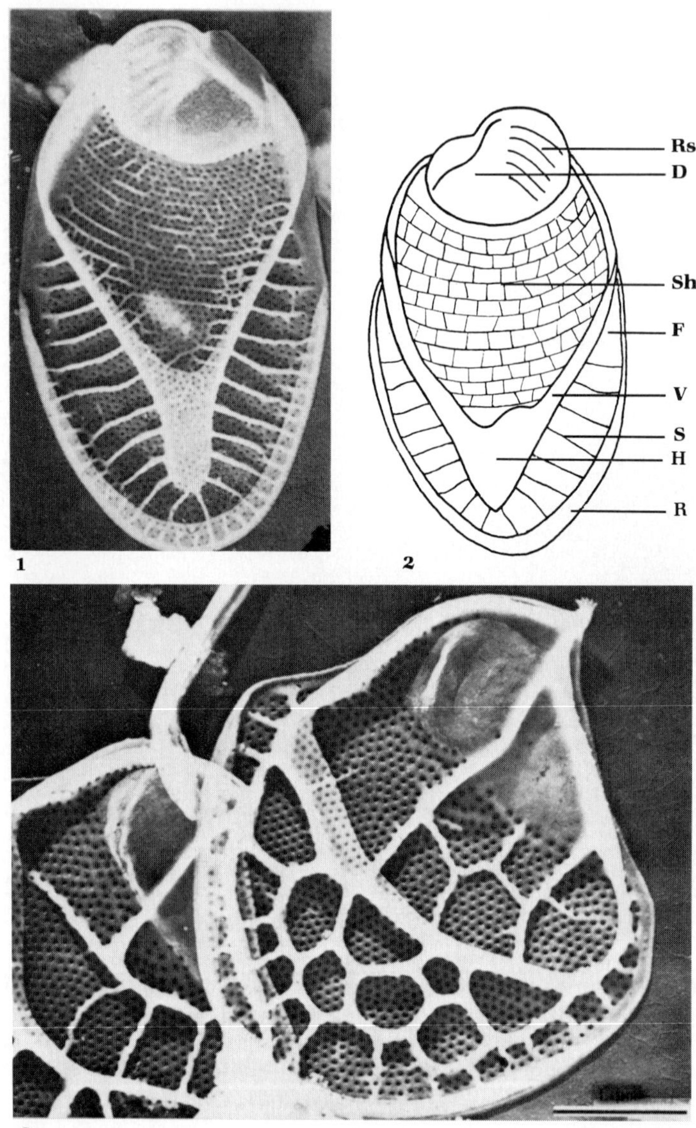

Figure 6.10

1,2. *Mallomonas leboimei* Bourrelly, fresh water, Denmark, single scale, TEM, ×7000. **2.** Diagram of scale, showing terminology, from distal to proximal end of scale; Rs, ribs; D, dome; Sh, shield; F, flange; V, V-rib; S, strut; H, hood; R, rim; from Fott, 1962. **3.** *M. harrisae* Takahashi, Japan, siliceous apical scale, TEM, negative print, showing coarse network on distal layer and fine perforations of inner layer, ×18,000 (bar = 1 μm), from Takahashi, 1975.

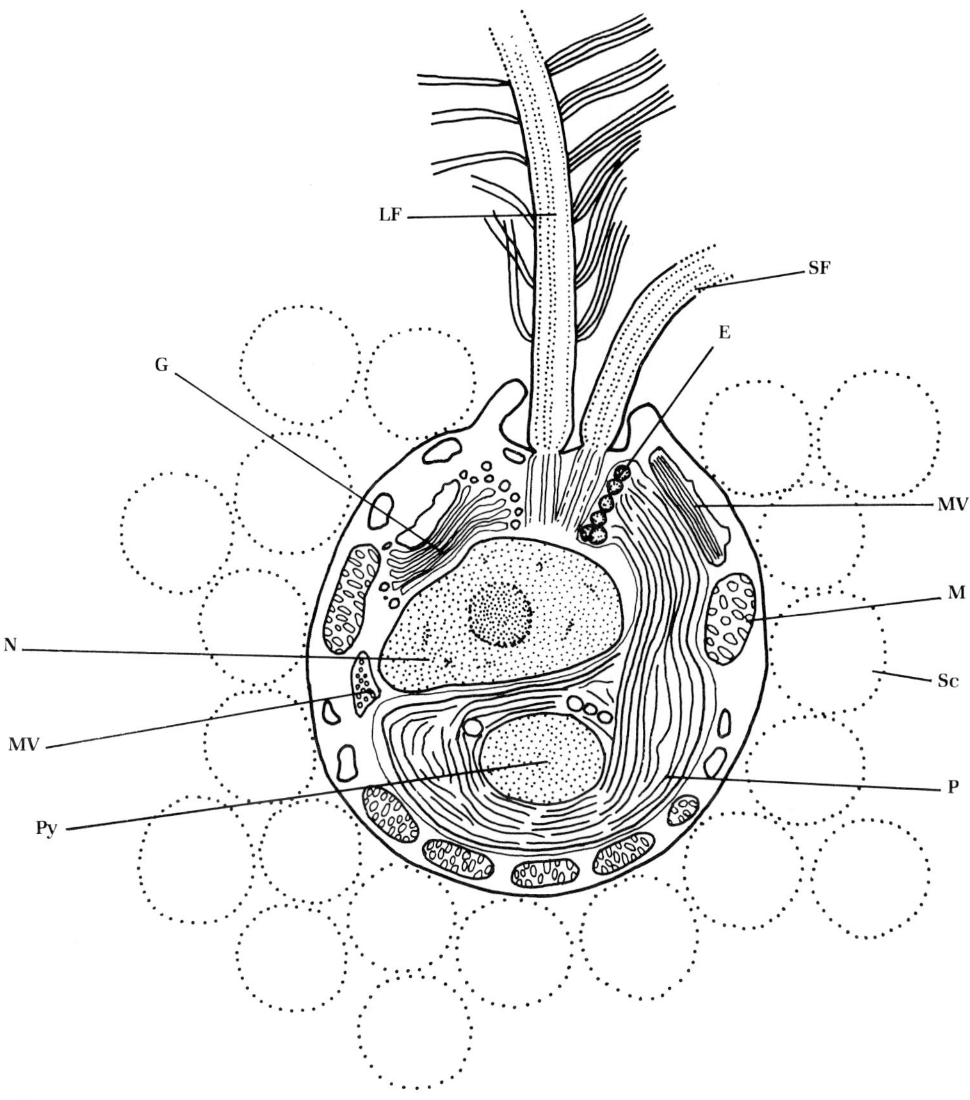

Figure 6.11

Syncrypta glomerifera, diagrammatic sectioned cell, showing LF, long flagellum with mastigonemes; SF, short smooth flagellum; E, eyespot; G, Golgi body; MV, mastigoneme vesicle; Sc, scale; M, mitochondrion; N, nucleus; P, plastid; Py, pyrenoid. Redrawn from Clarke and Pennick, 1975.

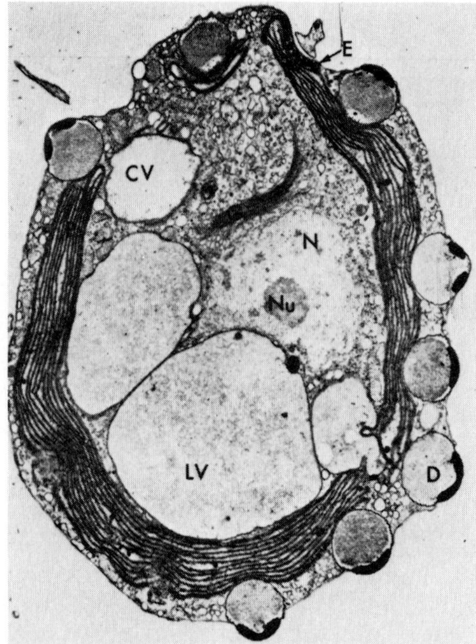

Figure 6.12
Ochromonas tuberculata Hibberd, fresh water,
England. TEM of sectioned cell, showing D, peripheral discobolocysts; E, eyespot; N, nucleus
with Nu, nucleolus; LV, leucosin vesicles; CV,
contractile vacuole; and peripheral plastid;
×5700, from Hibberd, 1970.

one acronematic (smooth). The flagella may be
of nearly equal length, but more commonly the
smooth one is shorter and may even be nonemergent, so that the cell appears uniflagellate.
The posterior flagellum of *Bikosoeca* may be
modified as an attachment stalk within the
lorica (see Figure 6.13). Flagellar scales are
present on the flagella of *Mallomonas, Sphaleromantis,* and *Synura.*

Choanoflagellate Collar

A ring of very fine tentacles or pseudopodia may
form a collar in the choanoflagellates such as
Campanoeca, Pleurasiga, and *Codonosiga,* and
may be retractible. The collar may serve to

catch food particles in a mucus, the particles
then being transported to a food vacuole.

Nutrition

Chrysophyta typically are photosynthetic, the
golden-brown plastids containing chlorophyll *a,*
fucoxanthin, chlorophyll *c* in some species,
β-carotene, violaxanthin, and diatoxanthin.
However, some colorless species occur that
may be otherwise identical to pigmented ones.
Thus *Ochromonas* (pigmented) and *Monas* (colorless) may appear identical in all other respects.

Pigmented species also may have the capability for heterotrophic nutrition. Under poor light
conditions, *Ochromonas* may ingest much of its
food, and its plastids become small and pale.
Whether such individuals would gradually become entirely colorless is not known, but some
reversion to a colorless state does occur. Under
such conditions the former producers would
thus become consumers.

Both the pigmented and colorless Chrysophyceae require for growth an exogenous
supply of one or more vitamins, such as
thiamine, the vitamin B_{12} requirements of
chrysophytes being like those of birds and
mammals. The vitamins are obtained in solution in coastal regions, having been washed
from the soil of nearby land areas where they
are produced in quantity by terrestrial bacteria
and blue-green and green algae. In the open
ocean these vitamins may be derived from extracellular metabolic secretions of bacteria or
of other algae, or regenerated by the decomposition of the cells of associated species.

Reproduction

Vegetative reproduction is characteristic, consisting of longitudinal fission in the flagellated
cell, after which the daughter cells separate.
When complete separation does not occur,
palmelloid, filamentous, or dendroid colonies
may result. These colonies may be of definite

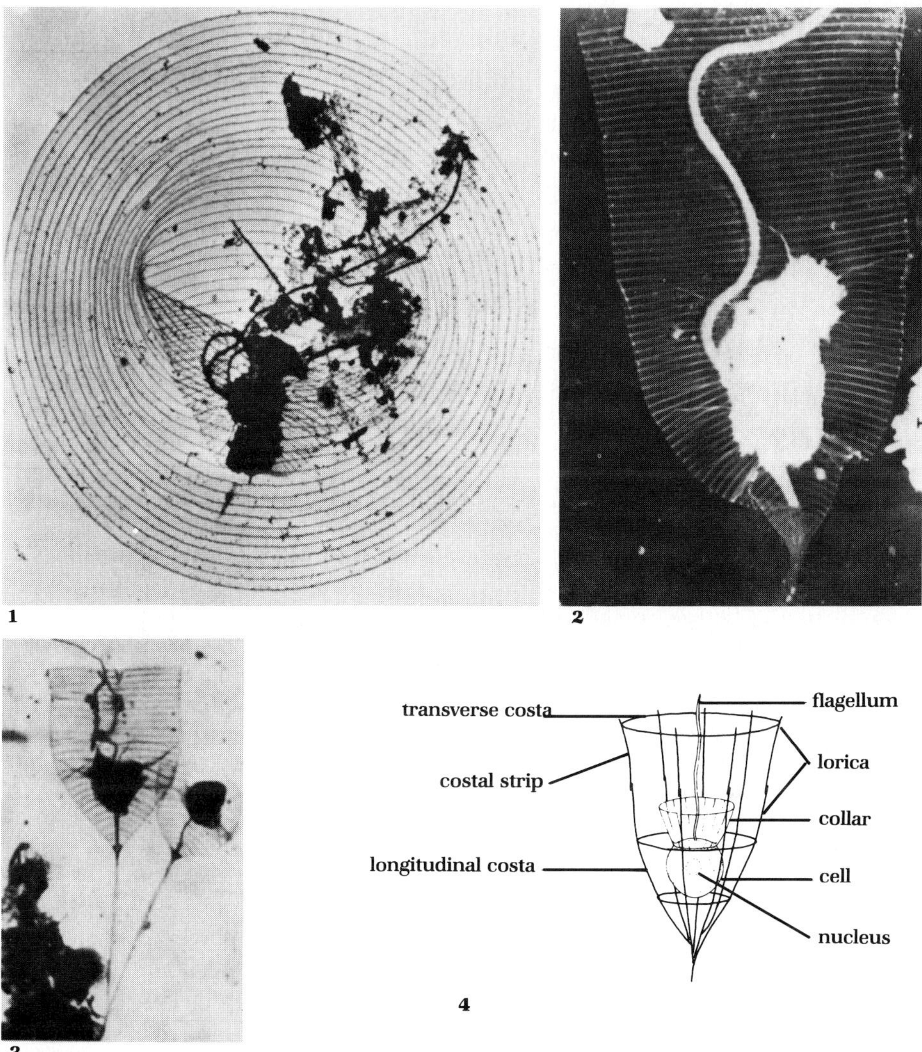

Figure 6.13

1–3. *Bikosoeca.* **1.** *B. crystallina* Skuja, fresh water, Denmark. TEM, showing spiral construction of lorica and flagellate cell in center, ×2000, from Kristiansen, 1972. **2.** *B. eurystoma* Hilliard, fresh water, Alaska, showing cell within lorica, ×3000. **3.** *B. epiphytica* Hilliard, fresh water, Alaska, showing two attached individuals, stained to show transverse banding of lorica, ×1500. 2,3, from Hilliard, 1971. **4.** Diagrammatic sketch of *Campanoeca dilatata* Throndsen, showing morphologic terms applicable to the choanoflagellates; from Throndsen, 1974.

form, but their individual cells remain completely independent, and may later separate and continue growth.

Sexual reproduction in *Dinobryon* consists of the fusion of two identical vegetative cells. In *Ochrosphaera*, haploid gametes fuse to form a diploid zygote; meiotic division results in four new haploid vegetative cells.

Cysts

An important feature common to the various Chrysophyceae is their formation of flasklike or bottle-shaped endogenous siliceous cysts that range in size from 6 to 10 μm, rarely as much as 20 μm (see Figure 6.14). They are basically of cellulosic or pectic composition, staining with the biological stains ruthenium red and methylene blue, but commonly are highly impregnated with silica. The exterior of the cyst may be smooth or sculptured, with fine elongate spines, short blunt or terminally inflated spines, nodes, ridges, or reticulations. The ornamentation may cover the entire surface or be localized at the oral or aboral end. The cyst opening may consist of a simple **pore**, have a flangelike collar, or be produced on a distinct neck of variable length. Some have a single arcuate projection (termed a **stomatocerque**) arising from one side of the opening, so that the cyst is asymmetrical. A small, organic **plug** or stopper is formed at a later stage in encystment, to seal the cyst opening. The plug contains little or no silica.

Because little is known of the life history of most Chrysophyceae, whether the cyst is an obligate part of the life cycle or formed only occasionally is uncertain. Cyst formation in *Kephyriopsis* (see Figure 6.15) and *Stenocalyx* occurs following sexual fusion of two cells (Fott, 1971, p. 58, 62). The cyst is commonly somewhat larger in size than the normal vegetative cell, and some binucleate cysts have been observed, whose nuclei subsequently fused.

In the process of encystment, the cell loses its flagella, rounds off, and becomes amoeboid. Globules of food reserves accumulate as oil or leucosin, and the cytoplasm becomes highly

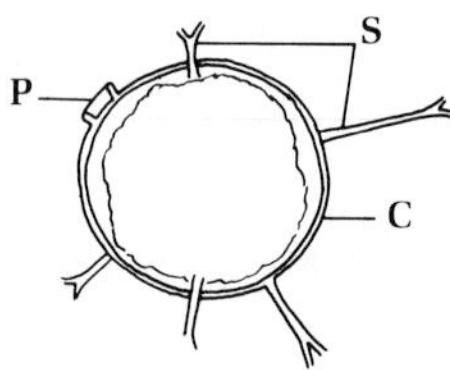

Figure 6.14
Chromulina sp.; C, cyst; S, silica spines; P, pore at end of neck for escape of flagellated cell at excystment; $\times$1500.

areolated. All discobolocysts that had formed a peripheral layer in *Ochromonas tuberculata* Hibberd were discharged prior to encystment, and contractile vacuoles were active. Coalescence of golgi vesicles results in a complete double-membraned sphere, the silica deposition vesicle. Silica is then deposited between the two membranes as small nodules, which coalesce to form the complete thin wall of the cyst. The collar and spines are added next by centrifugal deposition, the pore at the anterior end of the cyst resulting from a rupture of the membranes and absence of silica deposition. After the cyst is completed, a plug is formed in the pore, leaving some cytoplasm and the plasmalemma outside. The inner membrane of the silica deposition vesicle then becomes the functional plasmalemma (Hibberd, 1977).

The completed cyst encloses cytoplasm with many refringent granules that probably are fat globules, the yellow-gold to pale greenish plastids, one or two nuclei, and a large amount of food reserves and carotenoids.

At excystation, the unsilicified organic plug dissolves and one or several small motile cells resulting from fission of the original protoplast escape through the opening in the cyst.

Cysts are not accorded equal taxonomic importance in the various groups of algae. They are regarded as highly diagnostic in the green algal Zygnematales, and in some blue-green algae, but are accorded value only at the specific level in the modern Chrysophyta. Modern cysts have been reported almost solely from freshwater species, but have been known in these since 1865, when their relationship was first reported by Cienkowski. The cysts have not yet been de-

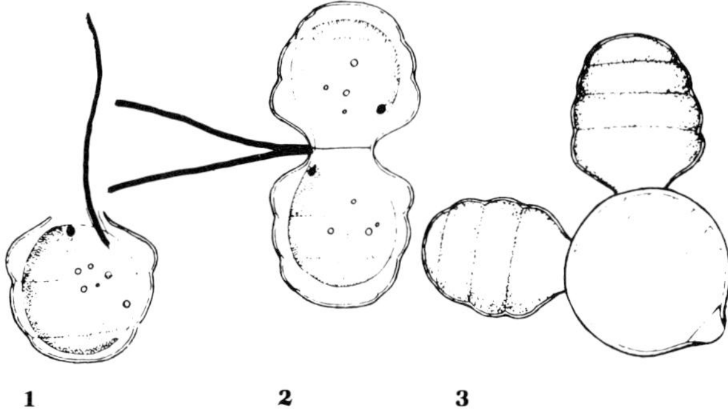

Figure 6.15
Kephyriopsis cincta Schiller. **1.** Single flagellated cell within lorica. **2.** Conjugation of two vegetative cells, functioning as gametes. **3.** Formation of cyst by the zygote. All ×2000, from Fott, 1971.

scribed for many species, but a few species and genera have been based only on the isolated cysts. Chodat (1922) erected a separate family, the Chrysostomataceae, for those nonmarine forms known only from the cysts (see Figure 6.16). They were thought to have only a short-lasting vegetative stage and a cyst stage of greater duration. A comparable family, the Archaeomonadaceae, was based on marine cysts. Separate family categories and genera for isolated cysts are convenient for paleontologists, for whom only this part is available as fossils, but the system is not used by biologists. The cyst genera are not comparable to modern chrysophyte genera, and may represent any of some 8 of the 12 presently recognized orders of Chrysophyceae. The surface sculpturing may differ markedly from species to species, within a genus or higher category, but neither genus, family, nor order can be determined from the cyst alone.

Conrad (1931) attempted to utilize the cysts in chrysophyte taxonomy. As he observed a great variety of cyst forms in various freshwater species of *Chromulina,* he proposed restricting the genus to include only those with smooth unornamented spherical cysts and a simple pore, a majority of the species previously referred to *Chromulina.* Species with cysts ornamented with somewhat spiralling ridges, both with and without a neck, were referred to the new genus *Spirochrysis.* The genus *Acanthochrysis* was proposed for species with spinose cysts, sometimes also possessing spiralling ridges, and with a neck. Due to the difficulty in determining the complete life cycle and thus relating the cysts to the vegetative

stages, this system has not been generally followed as yet. Many species have been described without mention of cysts, even though such other features as shape, size, and cell contents may be less definitive.

In studies of the cysts alone, the pore is the most important element, for it determines the anterior-posterior axis, and allows orientation. The chamber may have either a simple or double wall, a character used by Zanon (1947) as the basis for a separate section (Polymenidae) of the Chrysostomataceae, and for the definition of 13 new genera (*Parachrysostomum, Paraclericia, Paracarnegia, Paraoutesia, Paradeflandreia, Asterostomum, Corymbostomum, Anisostomum, Plagiostomum, Catastomum, Hemplectostomum, Archaeocorynbostomum,* and *Archaeocatastomum*), as opposed to the Section Monymenidae (genera *Carnegia, Outesia, Chrysostomum, Chrysastrella, Clericia, Deflandreia,* and *Trachelostomum*). This secondary layer in the wall is not generally regarded as of systematic importance in this group, as it is due to the later addition of silica, and is somewhat less refringent than the primary layer. Thickness of the secondary layer varies from one specimen to another in a single species, and may vary even on a single specimen, being thicker toward the pore region and thinner to absent in the posterior part. Andrieu (1936) observed that Hyrax, an excellent mounting medium for diatoms, has an index of refraction that obscures the median section when used for chrysophytes, and causes the appearance of parallel lines, suggesting many layers in the wall. The apparent double wall may in some instances be only an artifact of the mounting

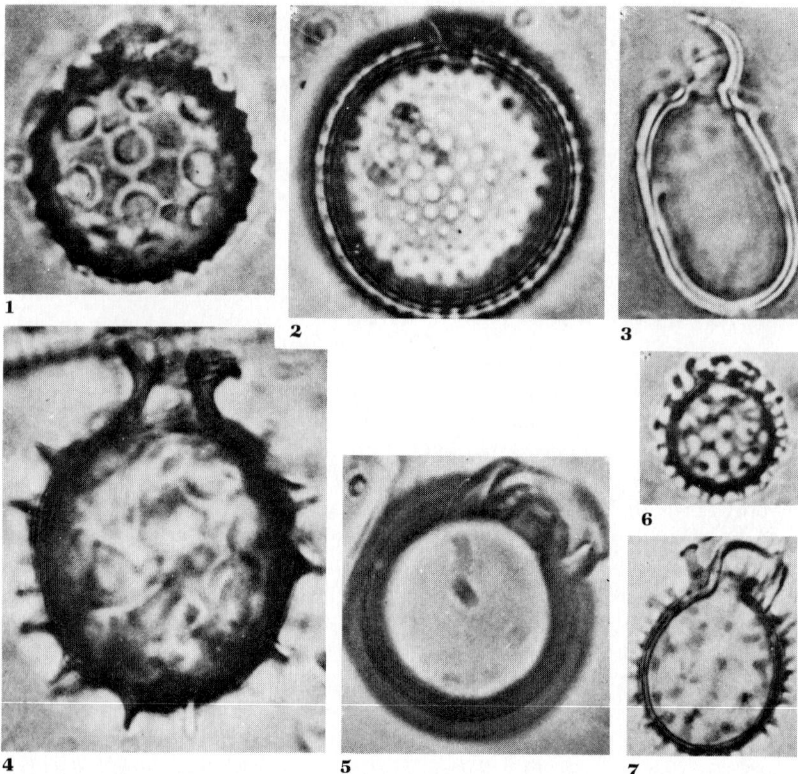

Figure 6.16
Chrysostomataceae, postglacial freshwater diatomite, Connecticut. **1.** *Clericia ansulata* Conrad. **2.** *C. pulcherrima* Conrad. **3.** *Carnegia operculata* Frenguelli. **4.** *Outesia yberiensis* Frenguelli var. *reticulatospinosa* Conrad. **5.** *O. robusta* Conrad. **6.** *Carnegia cristata* (Frenguelli) Deflandre var. *ornatissima* Conrad. **7.** *C. armata* (Frenguelli) Deflandre var. *uncinata* Frenguelli. 1,2,4,5, ×2300; 3,6,7, ×1650. All from Conrad, 1940.

medium. Canada balsam is better for use in this group, as the median optical section, needed for determination of important features such as the pore, is then clearer.

The use of electron microscopy has added greatly to an understanding of the surface morphology of the chrysophyte cysts (Cornell, 1970; Hajós, 1975; Perch-Nielsen, 1975; Stradner, 1971), but should be used in combination with transmitted light in order to see the details of the pore, wall thickness, and layering.

The pore may be bordered by a neck or col-larette, and may have a lateral projection or stomatocerque. The neck is a prolongation of the pore margin, either a slight upward extension of the thinner primary chamber wall or an extension of the secondary layer of silica. The collarette, an external secondary structure more or less concentric around the pore, may also arise from the original wall layer, or from the secondary layer covering the cyst exterior. When it arises from the original wall, the collarette may also have a secondary layer, and be double. The stomatocerque (a term proposed

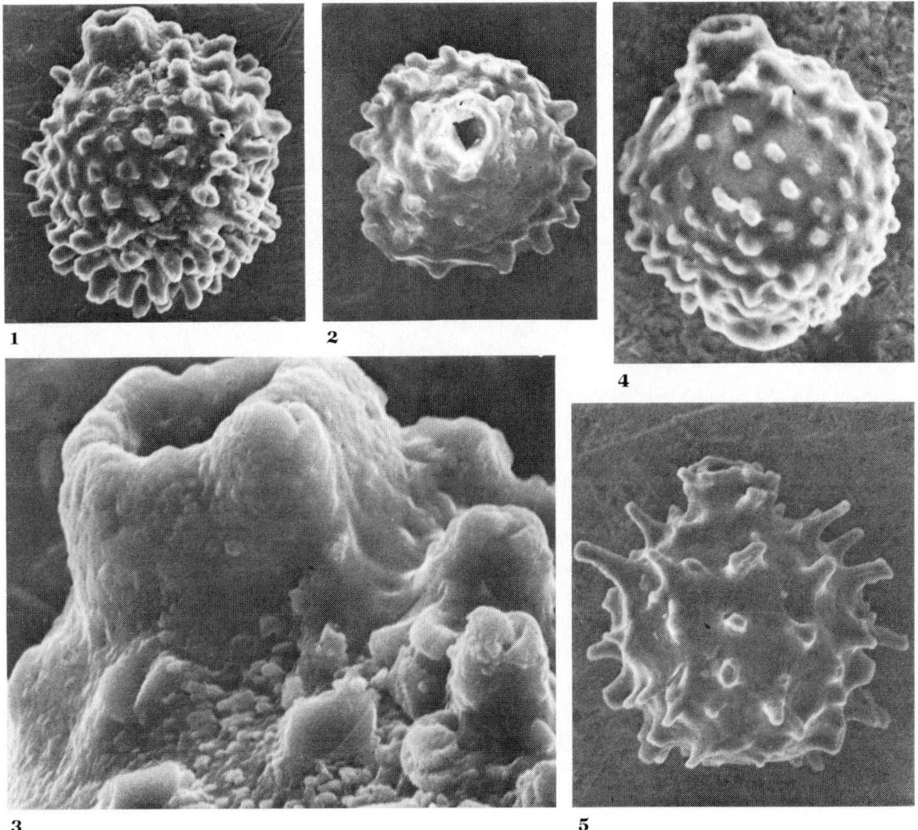

Figure 6.17

Bolboforma, a calcareous cyst, Oligocene, Mississippi. **1–3.** *B. spinosa* Daniels &
Spiegler. **1.** Side view showing thick radiating spines, granulate interspine surface,
and pore at end of projecting neck. **2.** Top view of same, showing partially preserved
membrane across opening. 1,2, ×320. **3.** Enlargement of upper part of part 1, ×1600.
4. *B. rotunda* Daniels & Spiegler, side view, ×360. **5.** *B.* sp., side view of specimen with
elongate spines, ×320.

by Andrieu, 1937) is an aliform, claviform, ros-
trate, or laminar projection from one side of
the pore. When present, it results in bilateral
symmetry of the cyst. It should be illustrated in
various views (frontal, lateral, and apical), although normally a side view is sufficient for
identification of other cysts.

In Oligocene and Miocene rocks, small calcareous cysts have been described that are
flasklike, formed of a single calcite crystal, have
a neck and terminal opening, but apparently
had a calcified membranelike partition closing
the opening within the neck, rather then the
pluglike stopper of most chrysophycean cysts.
Originally regarded as a probably protozoan
test (Daniels and Spiegler, 1974), the cyst nature
of *Bolboforma* (see Figure 6.17) was later indicated by the discovery of this partition (Rögl and
Hochuli, 1976). Some living chrysophytes are
reported to have calcite-encrusted loricas, as
well as siliceous cysts, hence the formation of
calcareous cysts is possible.

Occurrence

Habitat

Chrysophyceae are common in pure fresh acid waters of low mineral content, particularly in colder areas, or during the cooler seasons. Most are planktonic, although coccoid, filamentous, and palmelloid sessile forms also occur. Siliceous cysts of freshwater species may accumulate in large quantities in the bottom sediments, as many as 30,000 cysts being reported in a cubic centimeter of fresh sediment in Lake Galichskoye, Ukraine.

In addition to the strictly marine silicoflagellates, other Chrysophyceae (*Hydrurus, Chromulina, Meringosphaera*) occur in the salt marshes, and even the open ocean, but are less well known than the freshwater representatives because of the difficulty of obtaining living specimens. That the naked chrysomonads may be an important part of the marine plankton was indicated by Kolkwitz (1911), in records of the occurrence of plankton in the Mediterranean Sea near Monaco (see Table 6.2).

Lohmann (1912, p. 31) also reported abundant naked chrysomonads in tropical surface marine waters. The small cells, 6 to 9 μm in size, occurred in numbers of 10,000 to 33,000 cells per liter.

Geologic Record

As many modern Chrysophyceae occur in freshwater, fossil forms should prove useful in paleolimnology, particularly in view of the general scarcity of other useful freshwater microfossils.

The vegetative stage is seldom preserved, although Bradley (1946) reported the organic lorica of a chrysomonad flagellate, probably *Dinobryon* or *Cyrtophora*, from coprolites in the Eocene Bridger Formation of Wyoming. These and the associated desmids apparently were present in great abundance in the pond water drunk by a carnivore, whose digestive system did not attack the cellulosic plant material. Thus the microalgae were preserved in the silicified coprolites.

The lorica of some chrysophytes is mineralized and may somewhat resemble a cyst. The lorica of *Chrysococcus* or *Pseudokephyrion* may be ornate, but consists of a fine yellow-brown pectic membrane impregnated with calcite rather than silica, and is soluble in acetic acid. Other *Chrysococcus* have a siliceous lorica, and some *Pseudokephyrion* with a calcareous lorica also produce siliceous cysts. Synuraceae have a motile cell with a covering of siliceous scales instead of a lorica, and the lorica of the Acanthoecaceae (Craspedomonadophyceae) is constructed of siliceous costae or ribs (see Figure 6.18), as shown by

Table 6.2
Plankton cells per cubic centimeter of seawater (Kolkwitz, 1911).

	Near Monaco			Near Santa Margharita, 1 km from shore
	Surface	100 m	1200 m	Surface
Chrysomonads, naked, free-swimming	60	25	1	30
Coccolithophorids	2	1	—	—
Silicoflagellates	1	—	—	—
Dinoflagellate (*Gymnodinium*)	1	—	—	10
Diatoms	—	—	1 (dead) *Hemiaulus*	1 *Rhizosolenia*

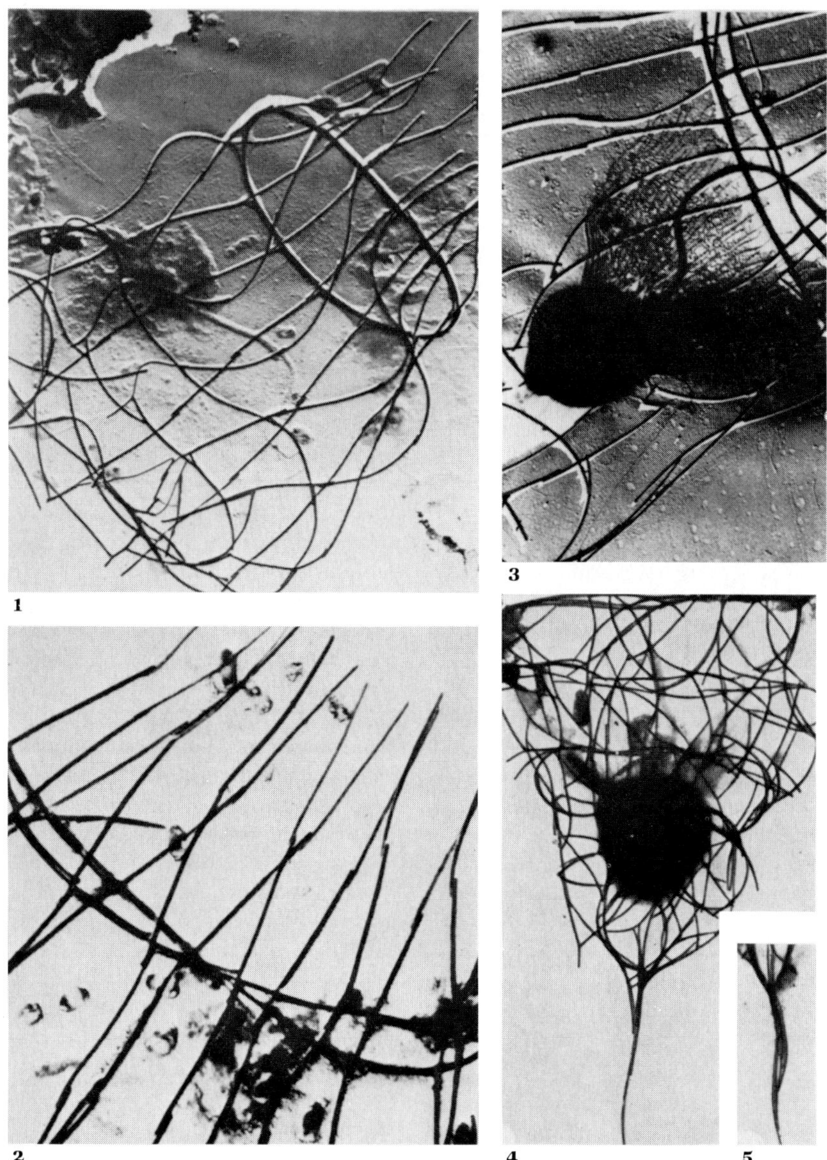

Figure 6.18

Choanoflagellates. **1–3.** *Diaphanoeca grandis* W. Ellis. **1.** Entire lorica. **2.** Portion at higher magnification to show overlapping costal segments. 1,2, TEM, magnification not given, from Gold et al., 1970. **3.** Central part of specimen, showing cell with collar and flagellum, and part of costal strips of lorica, TEM, ×5040, reproduced, with permission, from Throndsen, 1970, *Nytt. Mag. Bot.,* v. 17, p. 103–111, published by Universitetsforlaget, Oslo University. **4,5.** *Stephanoeca pedicellata* Leadbeater, shadowcast whole mount, in TEM, ×4000, from Leadbeater, 1972a. **4.** Whole cell showing cell within lorica. **5.** Posterior spine of a lorica, composed of six or more strips.

X-ray analysis (Leadbeater and Manton, 1974). This latter group is of worldwide distribution and constitutes one of the three most diverse groups in the modern marine nannoplankton. The lightly silicified scales, bristles, and loricae of various Chromulinales, Ochromonadales, and Craspedomonadales should be preservable under favorable conditions, like the chrysophyte cysts or silicoflagellate skeletons; yet perhaps because of their small size, no fossils of the motile cells have been identified as such.

Fossil siliceous discoid specimens described as *Pseudorocella* Deflandre 1938 were referred to the silicoflagellates, but were shown to be synonymous with *Macrora* Hanna 1932, originally described as a diatom (Bukry, 1977), but later as of uncertain position. Deflandre (1938b) had observed twin individuals (see Figure 6.19), which he regarded as in the process of reproduction, hence resembling the double skeletons of *Dictyocha*. However, the saucerlike discs fit within each other, the convex side of one against the concave side of the other, rather than with the two concave sides opposed in a mirror image orientation such as occurs in the silicoflagellates.

Similar tiny siliceous discs were described from a north German river (Gerloff, 1968) as a centric diatom, *Potamodiscus kalbei* Gerloff, believed related to *Stictodiscus*. No connecting bands were found. *Potamodiscus* discs later were observed (Gaarder et al., 1976) in other fresh water (Lake Erie) and marine coastal waters off South Africa, Japan, Spain, Norway, Sweden, Denmark, Finland, Arctic Canada, Oregon, California, Rhode Island, Mississippi,

and Texas. These discs also were found on complete cells in preserved material from Peru and Texas. Unlike diatoms, the cells are rounded, 14 to 20 μm in diameter, and covered with the imbricated siliceous scales. Individual scales, 3.8 to 4.9 μm in diameter, had two layers, both layers convex distally, perforated, and interconnected by struts. The preserved material lacked flagella, but other features suggest a relationship to *Mallomonas* or *Synura*. The strong similarity of *Macrora* and *Pseudorocella* to *Potamodiscus* indicates that all represent the isolated scales of synuraceans, and are unrelated to diatoms or silicoflagellates.

Other than the silicoflagellates (Chapter 7) the best fossil record of chrysophytes is that provided by the siliceous cysts, present in both marine and freshwater deposits. Fossils from fresh water are referred to the family Chrysostomataceae, and marine species are placed in the Archaeomonadaceae (Upper Cretaceous to Holocene). Genera present in the Cretaceous (see Figures 6.20 and 6.21) include *Acanthosphaeridium, Archaeomonas, Archaeomonadopsis, Archaeosphaeridium, Artisphaeridium, Litheusphaerella, Micrampulla,* and *Pararchaeomonas* (Cornell, 1970, 1972; Hajós, 1975; Perch-Nielsen, 1975; Rampi, 1940). Other genera occur in Cenozoic rocks, generally being common in siliceous rocks of the lower Eocene and middle Miocene, rare in Oligocene and lower Miocene, and rare from upper Miocene to the present (see Figure 6.22).

Ehrenberg (1854) illustrated chrysophyte cysts from freshwater diatomites in Berlin, from British Guiana, Australia, and Japan,

---

Figure 6.19
Synuraceae. **1—4.** *Potamodiscus kalbei.* **1—3.** Isolated scales in TEM, showing large perforations in distal layer and fine, regularly aligned pores of proximal layer; 1,2, ×12,000; 3, ×14,000. **4.** Entire cell in SEM, showing imbricated scales, ×4400. 1,3, Biloxi, Mississippi; 2,4, Galveston, Texas; all from Gaarder et al., 1976. **5—8.** *Macrora barbadensis* (Deflandre) Bukry, Eocene, Barbados, showing double-layered structure like that of *Potamodiscus;* 5,6, ×1735; 7,8, ×1010; from Deflandre, 1938b. **9.** *M. stella* (Azpeitia) Hanna, proximal view of specimen described as *Pseudorocella corona* Deflandre, upper Miocene, Hungary, ×1335, from Deflandre, 1947. **10,11.** *Mallomonas harrisae,* showing structure of the siliceous scales in TEM, Japan, from Takahashi, 1975. **10.** Scales and spines of a single cell, ×1920; bar = 5 μm. **11.** Body scales of one lorica, ×4800; bar = 2 μm. Compare Figure 6.10, part 3.

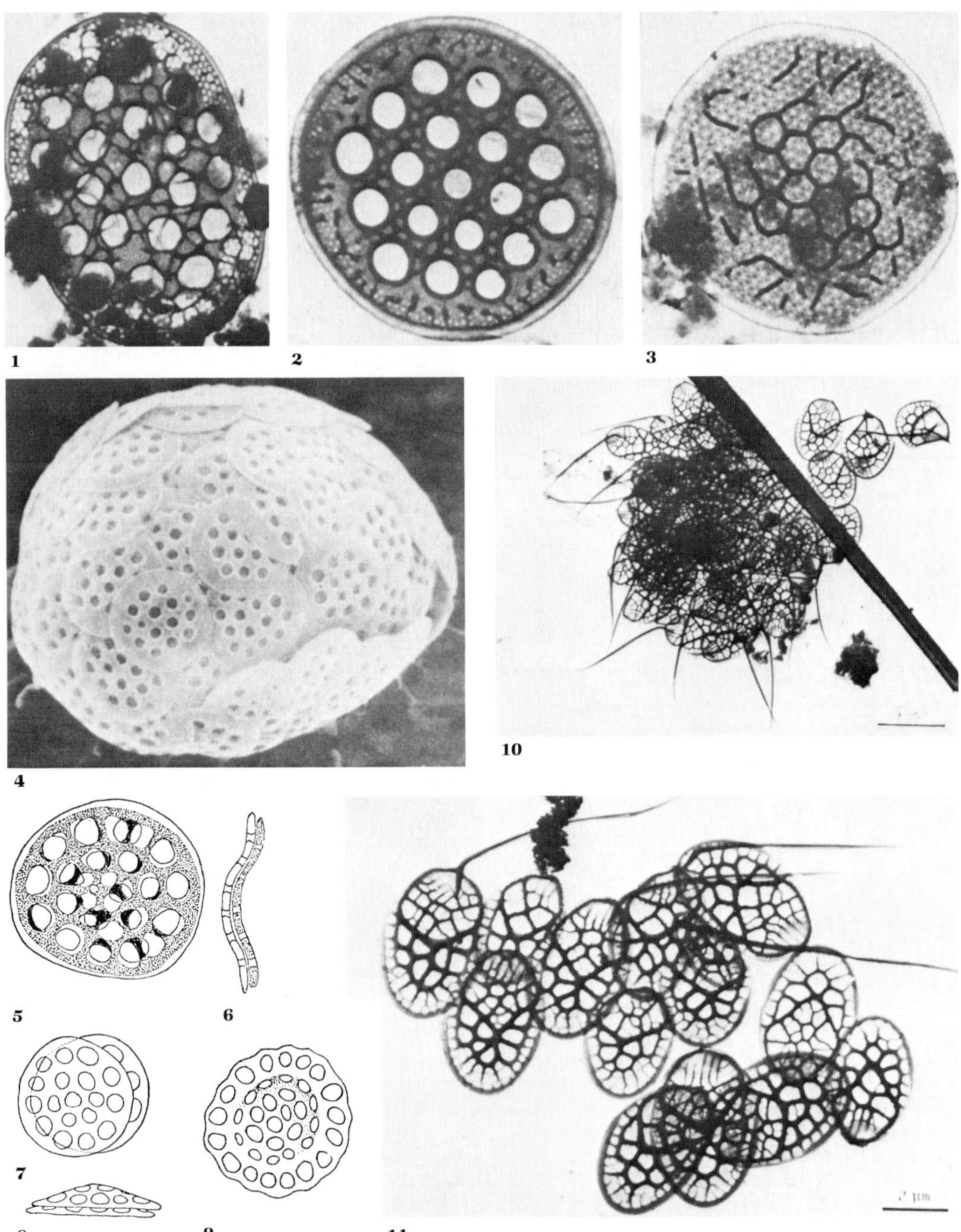

1
2
3
4
5
6
7
8
9
10
11

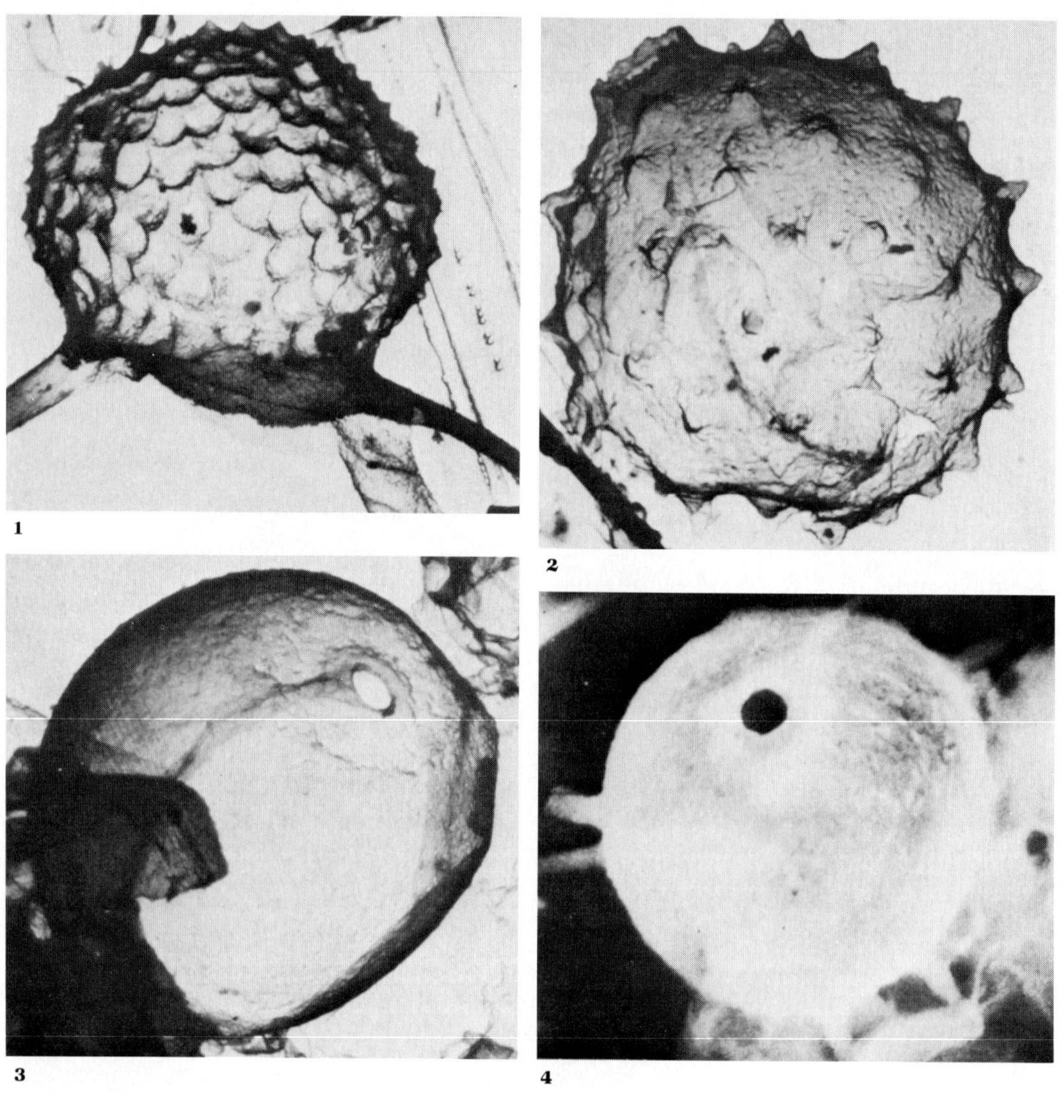

Figure 6.20
Cretaceous archaeomonads, South Pacific Ocean. **1.** *Acanthosphaeridium reticulatum* Hajós & Stradner, TEM, ×6000. **2.** *Archaeomonas chiarugii* Rampi, showing short neck and pore at upper left, TEM, ×12,600. **3.** *Archaeomonas cratera* Deflandre, with pore at upper right, TEM, ×15,000. 1−3, from Hajós, 1975. **4.** *A. paucispinosa* Deflandre, SEM, ×11,200, from Perch-Nielsen, 1975.

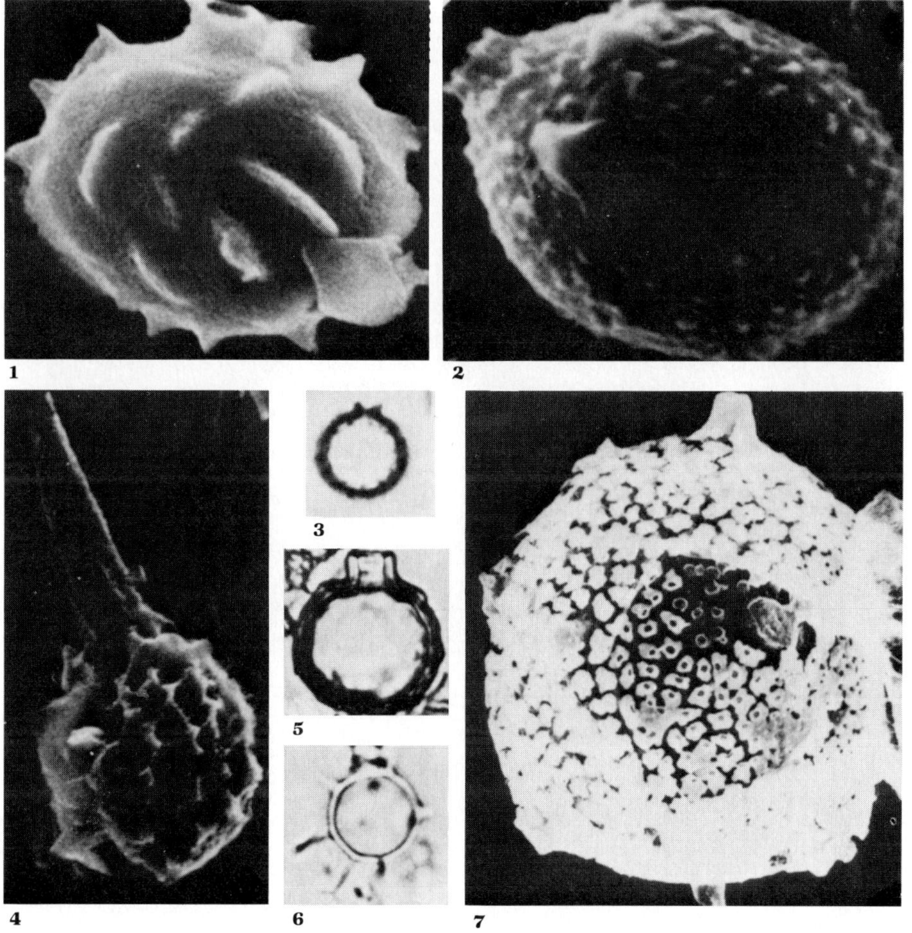

Figure 6.21

Cretaceous archaeomonads. **1.** *Archaeomonas vermiculosa* Deflandre, California, SEM, ×770. **2.** *A. membranosa* Rampi, California, SEM, ×8600. **3.** *A. cretacea* Rampi, DSDP, southwest Pacific, ×2000. **4.** *Archaeomonadopsis elegante* Rampi, California, SEM, ×3000. **5.** *Pararchaeomonas* sp., DSDP, southwest Pacific, ×800. **6.** *Litheusphaerella cretacea* Perch-Nielsen, DSDP, southwest Pacific, ×2000. **7.** *Artisphaeridium fragile* Stradner, TEM, ×9000. 1,2,4, from Cornell, 1970; 3,5,6, from Perch-Nielsen, 1975; 7, from Hajós, 1975.

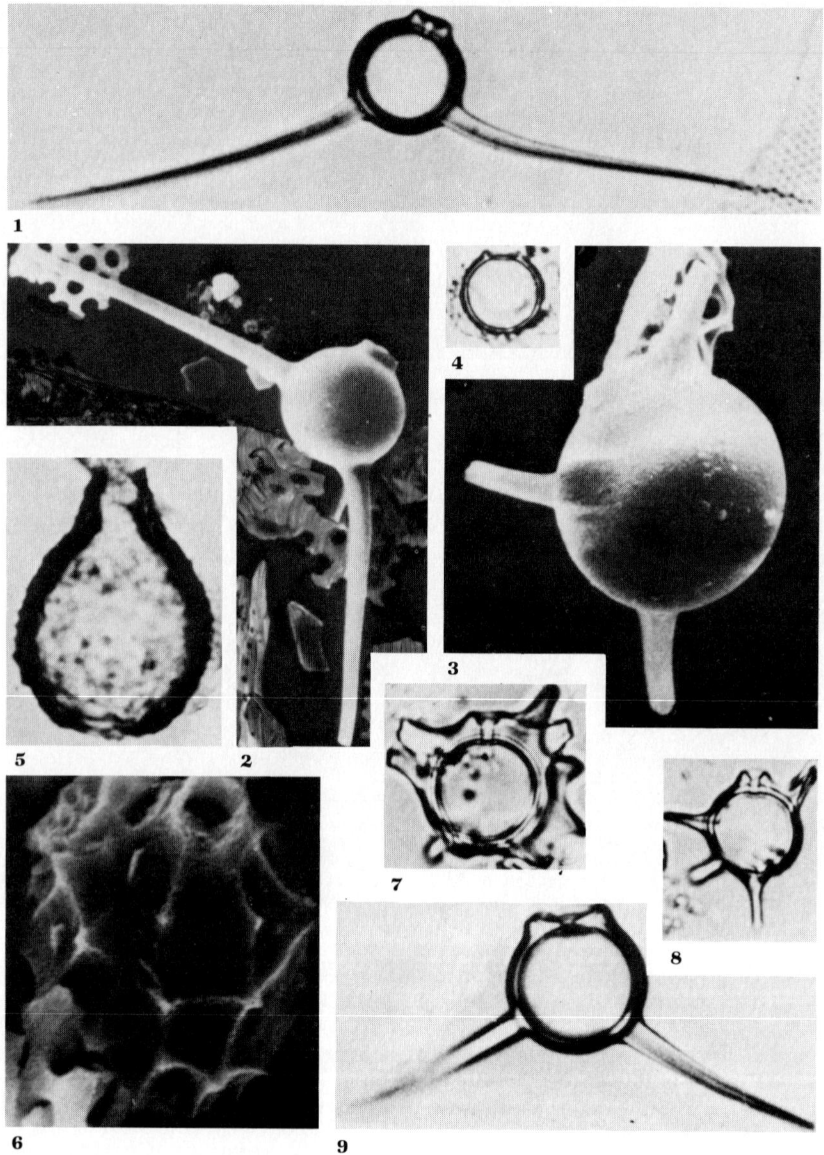

Figure 6.22

Cenozoic archaeomonads. **1–3.** *Archaeosphaeridium tasmaniae* Perch-Nielsen, DSDP, southwest Pacific; 1, ×800; 2,3, SEM; 2, ×1150; 3, ×2300.
4. *Archaeomonas cratera*, upper Eocene, southwest Pacific, ×800.
5. *Archaeomonadopsis* cf. *elegante*, Oligocene, southwest Pacific, ×800.
6. *Archaeomonas angulosa* Deflandre, Miocene, California, SEM, ×6400.
7,8. *Archaeosphaeridium dumitricae* Perch-Nielsen, upper Eocene, southwest Pacific, ×800. **9.** *A. australensis* Perch-Nielsen, Oligocene, southwest Pacific, ×800. 6, from Cornell, 1970; others from Perch-Nielsen, 1975.

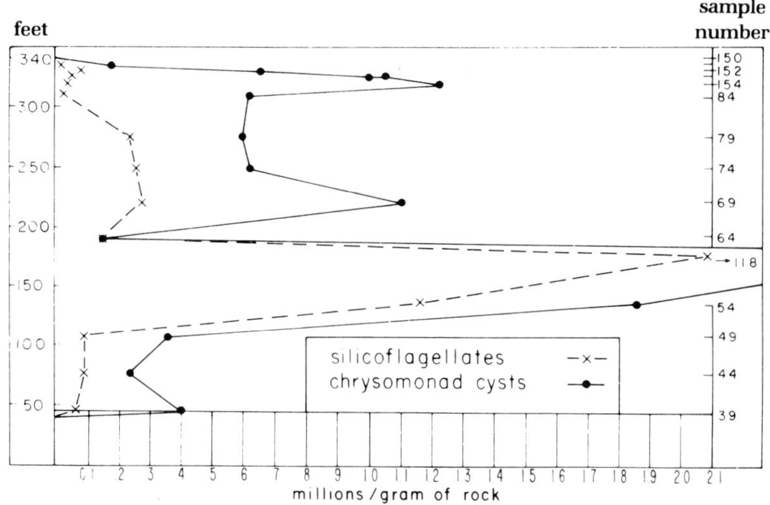

Figure 6.23
Abundance fluctuations of chrysomonad cysts and silicoflagellates in the Upper Cretaceous Marca Shale Member of the Moreno Formation, California. From Cornell, 1972.

referring them to the euglenid genera *Trachelomonas* and *Chaetotyphla*. However, the loricae of the euglenids, like some chrysophyte ones, are only superficially similar in form to the present cysts, and are organic in composition, and not silicified. Many fossil chrysophytes nevertheless were originally described as species of *Trachelomonas*, before Deflandre (1934) indicated their correct relationship.

Chrysophyte cysts are extremely abundant in some present and fossil peat bogs and nonmarine diatomites, ranging from 1 to 2 percent of the number of diatom cells in some diatomites, up to 25 percent. According to Conrad (1940), the marine archaeomonads may surpass the diatoms in abundance in Paleocene diatomites of Jutland, and in the upper Eocene of Austria locally form a nearly pure **archaeomonadite** (Tynan, 1971). In the Upper Cretaceous Marca Shale Member, Moreno Formation, California (see Figure 6.23), as many as 11.8 million cysts were found per gram of rock (Cornell, 1972).

Marine archaeomonad cysts have been reported from the Upper Cretaceous Moreno Formation of California; the Paleocene of Mors and Ile de Fuur, Jutland, Denmark, and of Kuznetsk, Volga River basin of USSR; the Eocene of Oamaru, New Zealand, and Springfield, Barbados; the Miocene of Santa Barbara, California, of Richmond, Virginia, of Poplein and elsewhere in Maryland, of Beach Haven, New Jersey, of Karand, Kekko, and Nyermegy (probably Nyiregyhaza), Hungary, and of Abashiri, Japan; the post-Quaternary of Santa Cruz, Argentina; and the Holocene sediments of the Caspian Sea, as well as from Cretaceous and Cenozoic rocks penetrated by the DSDP exploration.

Freshwater deposits containing Chrysostomataceae include those of the Miocene of Washington state, and of Bohemia; the Pliocene of Argentina, and of Santa Fiora and Monte Amiata, Tuscany, Italy; the Quaternary of Auvergne, France; and postglacial lakes from Connecticut.

In addition to contributing substantial amounts of opaline silica to the sediments, chrysophytes similarly provide a source of organic matter. The large amount of oil and other fatty food reserves in their cells may have been an important source of petroleum.

Classification

Subdivision of the Chrysophyta is classically based on general morphology, their solitary or colonial state, and their flagellate, coccoid, palmelloid, filamentous, or rhizoid nature (Pascher, 1913).

The former more inclusive Chrysophyta was subdivided to eliminate those taxa with similar flagella and a haptonema as the Haptophyta (Parke, 1961), hence the division is considerably reduced from that of Pascher (1914) or Bourrelly (1957, 1968). Matvienko (1960) modified the classification to recognize five classes, in part equivalent to earlier orders, based on the nature of the cell, whether amoeboid, flagellate, coccoid, filamentous, or palmelloid. The present classification is similar to that of Bourrelly and others, after eliminating the Haptophyta. The Class Craspedomonadophyceae, included here as by Bourrelly (1968) and Christensen (1962), is placed with the protozoans instead by some (see Parke and Leadbeater, 1977).

In view of the limted fossil record of the Chrysophyta, only an abbreviated classification is given for the living taxa (see Table 6.3), followed by the families based on the fossil and modern cysts (see Table 6.4).

Table 6.3
Classification of the Chrysophyta.

Division Chrysophyta Pascher 1914

I. Class Bikosoecophyceae A. Loeblich & L. Loeblich 1979

Family Bikosoecaceae Stein 1878

Loricate flagellates, hyaline lorica with spiral or annular elements and may be iron encrusted; cell attached to lorica by the smooth posterior flagellum that acts as a contractile stalk; anteriorly directed flagellum with mastigonemes; free or sessile, solitary or colonial, epiphytic or planktonic. Class status was suggested by Kristiansen (1972), but no name was then proposed. No Latin diagnosis has yet been given. Freshwater; Holocene.
Bikosoeca James-Clark 1868 (syn.: *Bicoeca* Stein 1878).

II. Class Chrysophyceae Fritsch in West 1927

A. Order Chromulinales Bourrelly 1957 (see Figures 6.24, 6.25)

Syn.: Chrysocapsales Pascher 1912

Solitary, palmelloid or rhizopodial; no definite cell wall; motile stage with single hairy flagellum; no protoplasmic collarette; with or without lorica, *Chrysococcus* and *Porochrysis* with silicified, calcified, or iron-impregnated lorica; may produce siliceous cysts. Freshwater and marine; may be psammophytic; Holocene.
Calycomonas Lohmann 1908 (syn.: *Codonomonas* van Goor 1925; *Codomonas* Lackey 1939); *Chromulina* Cienkowski 1870; *Chrysapsis* Pascher 1910; *Chrysococcus* Klebs 1892; *Chry-* *sopyxis* Stein 1878; *Hydrurus* C. Agardh 1824; *Kephyrion* Pascher 1911; *Pascherella* Conrad 1926; *Porochrysis* Pascher 1917; *Sphaleromantis* Pascher 1910.

B. Order Thallochrysidales Bourrelly 1957

Filamentous or thalloid; with definite cell wall, uniflagellate zoospores resemble *Chromulina;* may have siliceous endogenous cysts. Marine, fresh and brackish water; Holocene.
Thallochrysis Conrad 1914.

C. Order Chrysosphaerales Pascher 1914 (see Figure 6.26)

Coccoid, solitary, or colonial; definite cell wall; uniflagellate *Chromulina*-like zoospores; may have siliceous spines (*Aureosphaera, Meringosphaera*); cysts. Freshwater and marine (*Aureosphaera, Meringosphaera*); Holocene.
Aureosphaera Schiller 1916; *Chrysosphaera* Pascher 1914; *Meringosphaera* Lohmann 1903.

D. Order Chrysosaccales Bourrelly 1957

Palmelloid; no definite cell wall; no flagellate zoospores; may produce siliceous cysts. Freshwater; Holocene.
Chalkopyxis Pascher 1931.

E. Order Ochromonadales Pascher 1910

Colorless (*Monas*) or pigmented, no definite cell wall; biflagellate, with unlike flagella (rarely

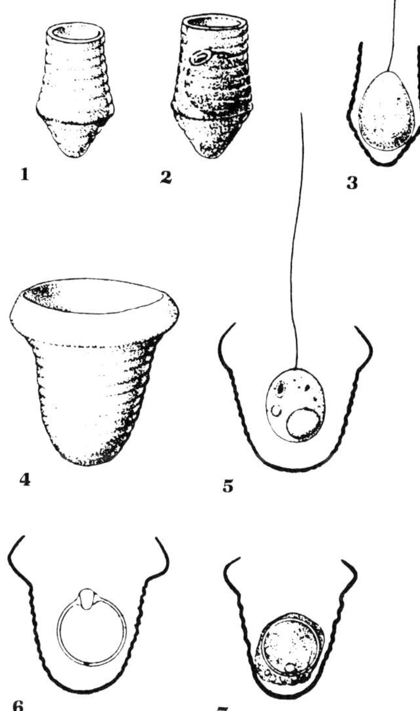

Figure 6.24
Chromulinales (Chrysococcaceae). 1 — 3.
Calycomonas vangoorii (Conrad) Lund, brackish
water, Belgium, ×2000. **1.** Empty lorica. **2.**
Lorica enclosing cyst. **3.** Sectioned lorica, enclosing flagellated cell. **4 — 7.** *C. pascheri* (van
Goor) Lund, brackish water, Belgium. **4.** Empty
lorica. **5 — 7.** Sectioned individuals, showing
flagellated cell, lorica with enclosed empty cyst,
and cell in process of encystment, with some
cytoplasm external to cyst; ×1550. All from
Conrad, 1938b.

with three or four flagella); solitary, filamentous, or palmelloid; naked or with cellulosic or
pectic lorica that may be iron- or manga-
nese-encrusted, or possibly calcareous in
Pseudokephyrion, or with siliceous scales;
siliceous cysts. Marine, fresh and brackish water. Eocene to Holocene.

1. Family Dinobryaceae (see Figure 6.27)
Chrysolykos Mack 1951; *Dinobryon* Ehrenberg
1834; *Pseudokephyrion* Pascher 1913; *Uroglena*
Ehrenberg 1834.

2. Family Ochromonadaceae (see Figure
6.28)
Boekelovia Nicolai & Bass-Becking 1935;
Chrysobotriella Strand 1928; *Entodesmis* Borzi
1892; *Monas* Ehrenberg 1838; *Ochromonas*
Wyssotzki 1887; *Olisthodiscus* N. Carter 1937;
Paraphysomonas De Saedeleer 1938; *Syncrypta*
Ehrenberg 1835.

3. Family Synuraceae (see Figure 6.29)
Chrysosphaerella Lauterborn 1896; *Conradiella*
Pascher 1925; *Macrora* Hanna 1932 (syn.:
Pseudorocella Deflandre 1938); *Mallomonas*
Perty 1851; *Microglena* Ehrenberg 1838;
Potamodiscus Gerloff 1968; *Synura* Ehrenberg
1834.

F. Order Chrysapionales Bourrelly 1957
Coccoid, nonmotile, solitary or colonial;
definite cell wall; biflagellate zoospores with unlike flagella; cysts unknown. Freshwater;
Holocene.
Chrysapion Pascher & Valkanov 1943; *Phaeogloea* Chodat 1932.

G. Order Phaeothamniales Bourrelly 1957
Simple or branching filaments; definite cell
wall; zoospores with two heterodynamic
flagella; siliceous cysts. Marine, fresh and brackish water; Holocene.
Nematochrysis Pascher 1914; *Phaeothamnion*
Lagerheim 1884.

H. Order Phaeoplacales Bourrelly 1957
Filamentous, parenchymatous, cell wall present; zoospores unknown. Freshwater and
marine; Holocene.
Stichochrysis Pringsheim 1955.

I. Order Rhizochrysidales Pascher 1925
Amoeboid, no flagellate stage; no definite cell
wall; naked or loricate. *Chrysothylakion, Heliak-
tis,* and *Stephanoporos* may have calcified and
iron-impregnated cover; solitary or colonial;
colorless or pigmented; some produce siliceous
cysts. Fresh and brackish water; rarely marine
(*Chrysothylakion*); Holocene.
Chrysamoeba Klebs 1892; *Chrysothylakion*
Pascher 1915; *Heliaktis* Pascher 1940; *Lagynion*
Pascher 1912; *Rhizochrysis* Pascher 1912;
Stephanoporos Conrad & Pascher 1940.

(continued)

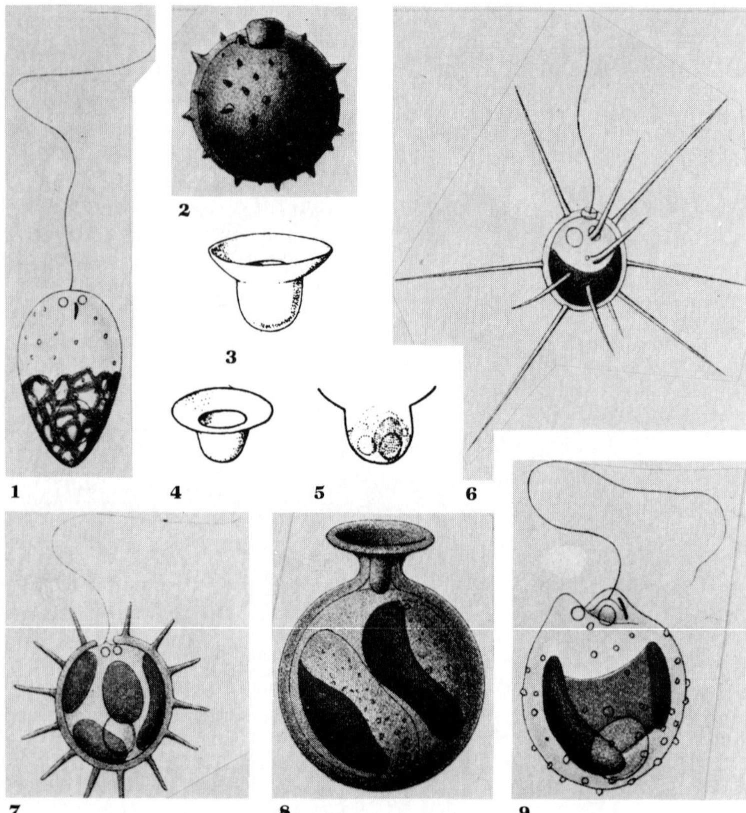

Figure 6.25
Chromulinales. **1,2.** *Chrysapsis yserensis* Conrad, brackish water,
×2000. **1.** Motile cell, with reticulumlike plastid in cell posterior, and
elongate flagellum. **2.** Cyst, with thick yellowish membrane. **3—5.**
Kephyrion petasatum Conrad, ×1750. **3,4.** Empty lorica. **5.** Section to
show cell (flagellum not shown). **6.** *Chrysococcus radians* Conrad,
flagellate cell in spherical spinose lorica that is strongly silicified, showing single plastid in posterior, ×2000. **7.** *C. dokidophorus* Pascher, with
four discoid plastids and two apical vacuoles; wall of lorica thick and
silicified; ×2000. **8,9.** *Chromulina pascheri* Hofeneder, ×2000. **8.** Cyst
containing two daughter cells, wall of cyst strongly silicified, plug closing pore. **9.** Motile cell, with single bandlike plastid. 3—5, after Conrad,
1938b; others from Conrad, 1926.

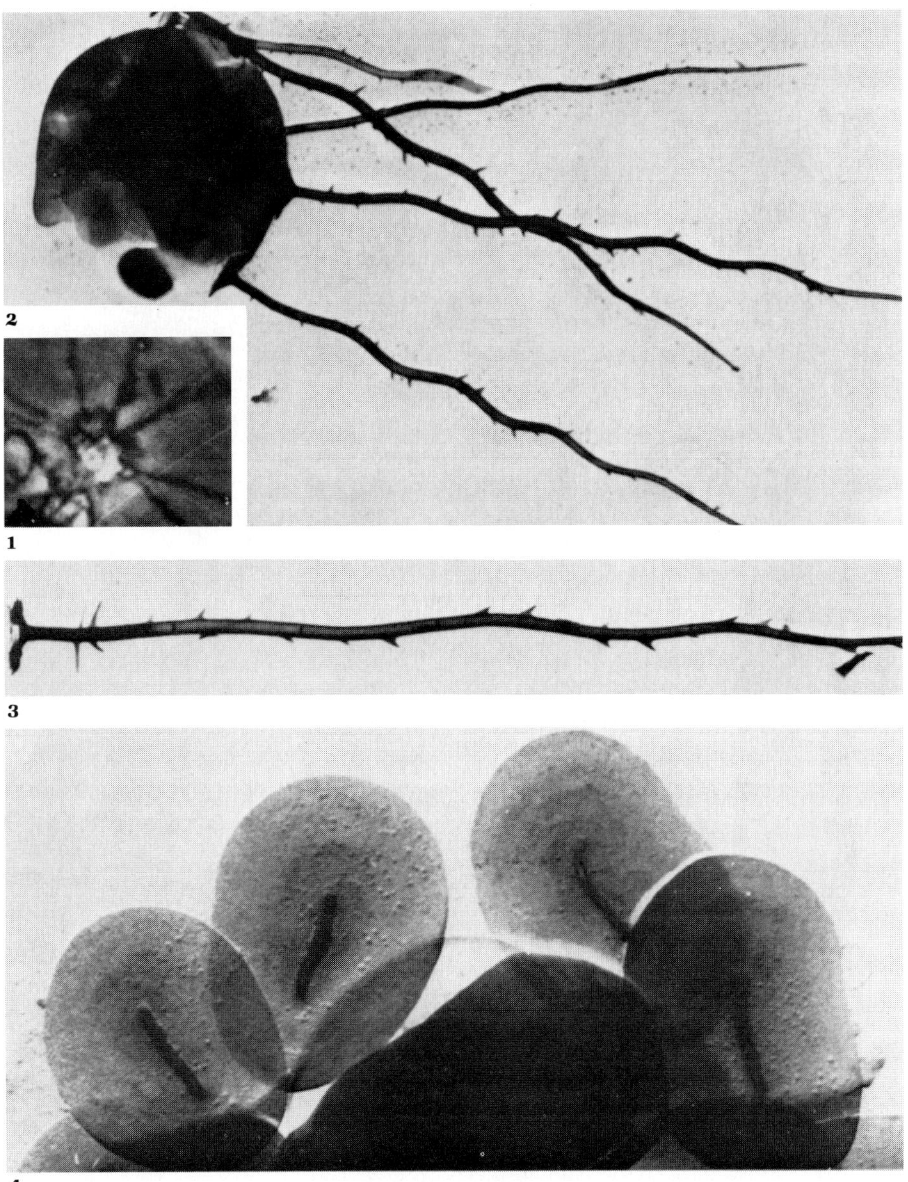

Figure 6.26

Meringosphaera mediterranea Lohmann, Mediterranean coast of Yugoslavia. **1.** Whole cell in phase contrast, ×1000. **2.** Shadowcast TEM, ×4000. **3.** Single spine, ×7000. **4.** Plate scales, ×20,000. All from Leadbeater, 1974, *J. Mar. Biol. Ass. U.K.*, v. 54, p. 179–196, published by Cambridge University Press; used with permission.

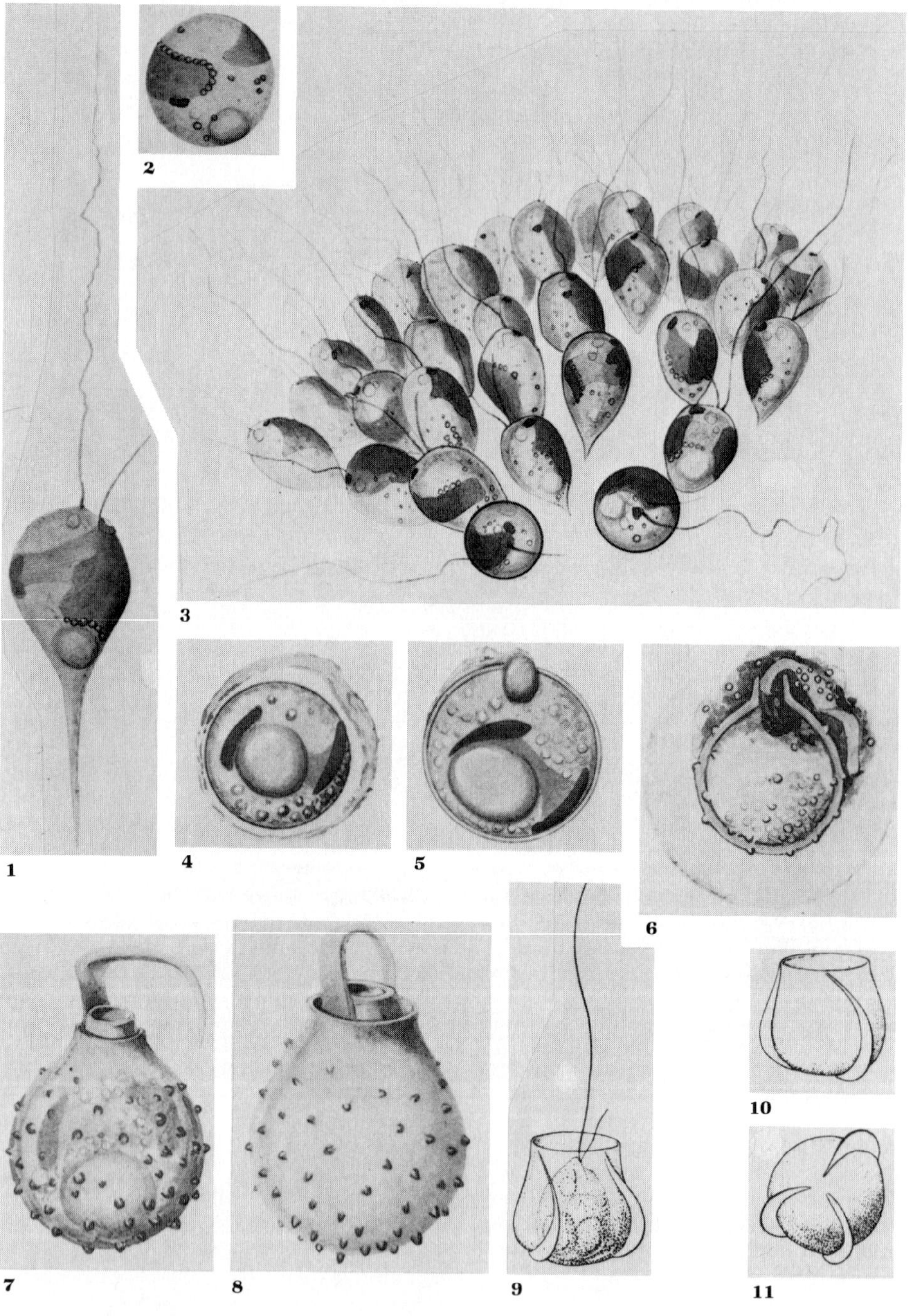

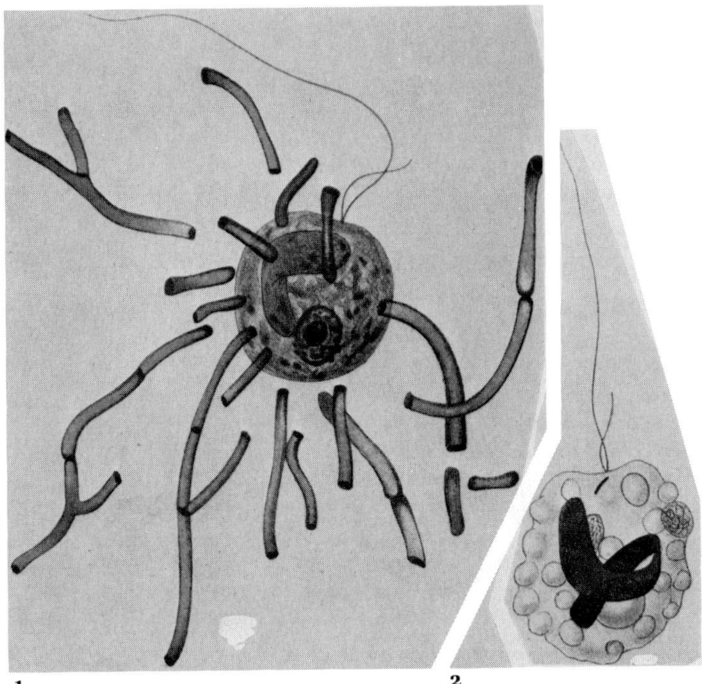

1   2

Figure 6.28
Ochromonadales: Ochromonadaceae. **1,2.** *Ochromonas crenata*
Klebs, flagellate cell with nucleus, single nucleolus, and bandlike
plastid; surface layer of discobolocysts appearing like vacuoles before discharge (part 2), and long bandlike structures after discharge
(part 1); ×2000, from Conrad, 1926.

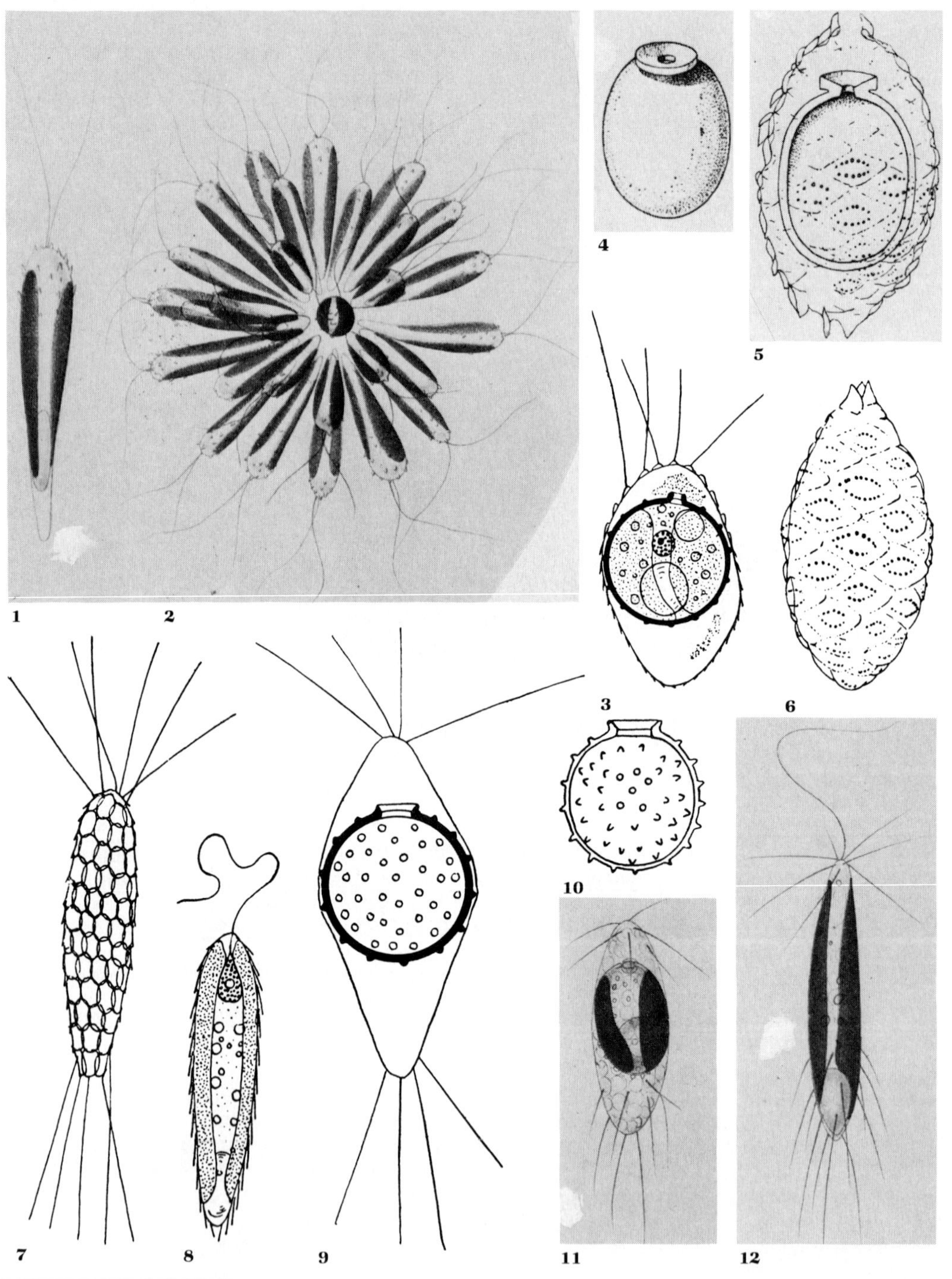

J. Order Sarcinochrysidales Gayral ex Gayral & Billard 1977

Motile or filamentous; unlike flagella laterally inserted on motile cells and zoospores. Marine and freshwater; Holocene.

Nematochrysopsis Chadefaud 1947; *Sarcinochrysis* Geitler 1930.

K. Order Stichogloeales Bourrelly 1957

Coccoid, solitary, or colonial; well-defined cell wall; no flagellate zoospores; cysts present. Marine and freshwater; Holocene.

Stichogloea Chodat 1897.

L. Order Vallacertales Glezer 1966 (see Chapter 7)

Solitary cell, photosynthetic, single flagellum, internal skeleton of hollow siliceous rods in basketlike structure; cysts unknown. Marine; Cretaceous to Recent.

III. Class Craspedomonadophyceae Hibberd 1976

Syn.: Class Craspedophyceae Chadefaud 1960; Subclass Craspedomonadophycidae Bourrelly 1965.

Order Craspedomonadales Stein 1878 (= Choanoflagellates)

Solitary or colonial, free or attached; single flagellum; collarette formed by pseudopodia or tentacles; most are nonphotosynthetic; lorica has silica fibers in basketlike network in Acanthoecaceae; cysts unknown. Commonly attach to surface film of water by extension of protoplast or lorica from end opposite flagellum. Marine, fresh and brackish water; Holocene, no fossils.

Although regarded as nonphotosynthetic relatives of the Chrysophyceae here as by Bourrelly (1968) and Christensen (1962), some ultrastructural features are in disagreement, the mitochondrial cristae being flat as in the green algae, metazoans, and metaphytes, rather than tubular as in the other Chromophyta and many Protozoa (Leadbeater and Manton, 1974).

1. Family Pedinellaceae Pascher 1910 (see Figure 6.30)

Naked cells with few tentacles and not a distinct collar around flagellum, each tentacle with a core of three microtubules; emergent flagellum with single row of hairs, "short" flagellum reduced to basal body only; some photosynthetic. Marine and freshwater. *Apedinella* Throndsen 1971; *Audomonas* Lackey 1942 (freshwater); *Cyrtophora* Pascher 1911 (freshwater); *Pedinella* Wyssotzki 1887; *Pseudopedinella* N. Carter 1937.

2. Family Codosigaceae Kent 1880

Cells naked or embedded in mucilage matrix; closely aggregated rhizopodia form distinct collar; rarely photosynthetic (*Stylochromonas*). Marine and freshwater.

Codonocladium Stein 1878; *Codosiga* James Clark 1867; Desmarella Kent 1878; *Dicraspedella* Ellis 1930; *Kentrosiga* J. Schiller 1953; *Monosiga* Kent 1881; *Proterospongia* Kent 1882 (syn. *Protospongia* Kent 1880, non Salter 1864); *Stylochromonas* Lackey 1940.

3. Family Salpingoecaceae Kent 1880

Cells with organic lorica, firm and smooth (chitin or cellulose?); nonphotosynthetic. Freshwater and marine.

Bicosta Leadbeater 1978; *Calliacantha* Leadbeater 1978; *Choanoeca* W. N. Ellis 1930; *Diploeca* Ellis 1930; *Pachysoeca* W. N. Ellis 1930; *Salpingoeca* James Clark 1867.

(continued)

Figure 6.29
Ochromonadales: Synuraceae. **1,2.** *Synura adamsii* S. M. Smith, brackish water, Belgium, ×1000, from Conrad, 1926. **1.** Single cell. **2.** Colony. **3.** *Mallomonas tonsurata* Teiling. Encysted cell, with globular cyst enclosed within scale-covered cell, ×1000, from Krieger, 1930. **4—6.** *M. lychenensis* Conrad, ×1250, from Conrad, 1938a. **4.** Cyst. **5.** Sectioned cell with cyst enclosed in cell. **6.** Exterior of scale-covered cell. **7—10.** *M. teilingii* Conrad, ×1000, from Krieger, 1930. **7.** Cell exterior, showing overlapping scales and elongate spines. **8.** Sectioned cell, showing peripheral plastids and terminal flagellum. **9.** Cyst within cell, causing median bulge. **10.** Isolated cyst. **11,12.** *M. litomesa* Stokes, ×1000, from Conrad, 1926. **11.** Elliptical cyst enclosed within cell and distending the median region; plastids visible in cyst, surface scales and spines of vegetative cell visible at exterior. **12.** Vegetative cell showing normal elongate narrow form and terminal flagellum.

4. Family Acanthoecaceae Norris 1965 (see Figure 6.31)

Cells with rhizopodial collar and single flagellum; lorica consisting of separate silica fibers or costal strips regularly arranged to form basketlike structure; one species of *Diaphanoeca* may be photosynthetic. All marine.

Acanthoeca W. N. Ellis 1930; *Acanthoecopsis* Norris 1965; *Campanoeca* Throndsen 1974; *Crinolina* Thomsen 1976; *Diaphonoeca* W. N. Ellis 1930; *Diplotheca* Valkanov 1970; *Parvicorbicula* Deflandre 1960; *Pleurasiga* Schiller 1925; *Polyoeca* Kent 1881; *Savillea* Loeblich III 1967 (syn.: *Ellisiella* Norris 1965, non Saccardo 1881, nec Batista 1956); *Sportelloeca* Norris 1965; *Stephanoeca* W. N. Ellis 1930.

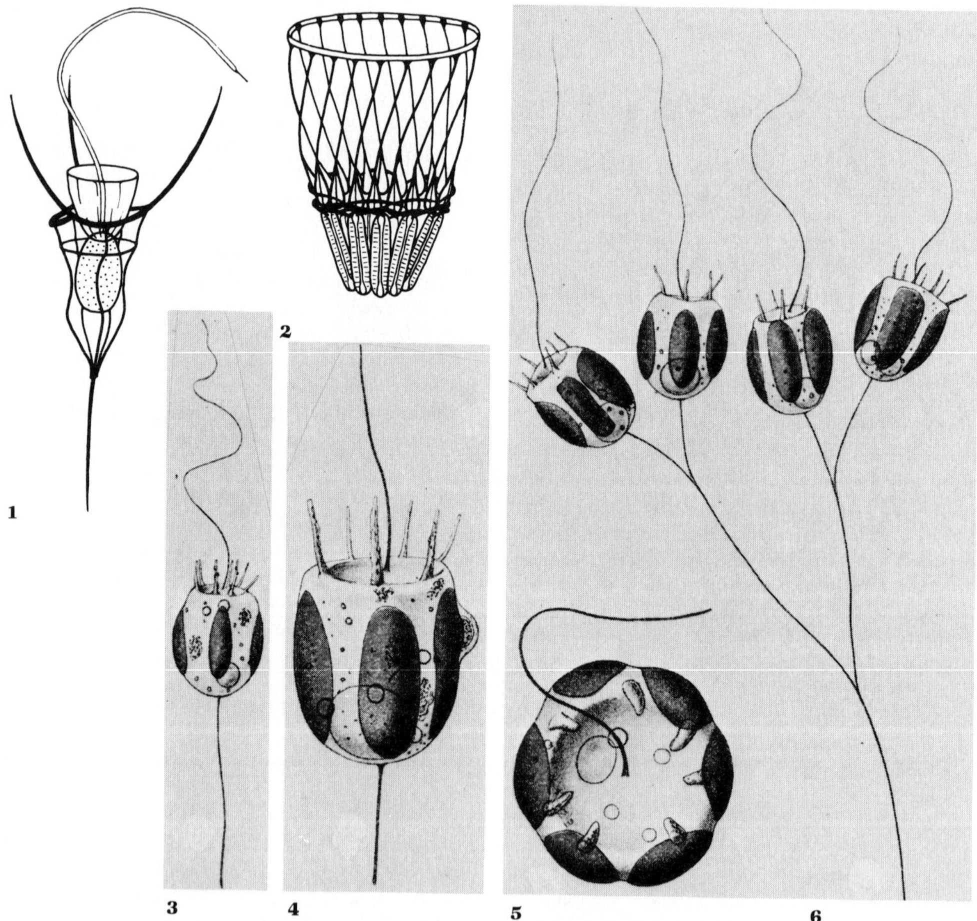

Figure 6.30

Pedinellaceae, Salpingoecaceae. **1.** *Calliacantha natans* (Grøntved) Leadbeater, flagellated cell in basketlike lorica. **2.** *Diploeca costata* Valkanov. 1,2, ×2100, from Leadbeater, 1972a. **3 – 6.** *Pedinella hexacostata* Wyssotski, showing ring of few pseudopodia, apical flagellum, large, ovate plastids, and attachment by peduncle, which may be used in locomotion by free-swimming cells; from Conrad, 1926. **5.** Apex. **6.** Colony of four individuals. 3,6, ×1000; 4, ×2000; 5, ×2500.

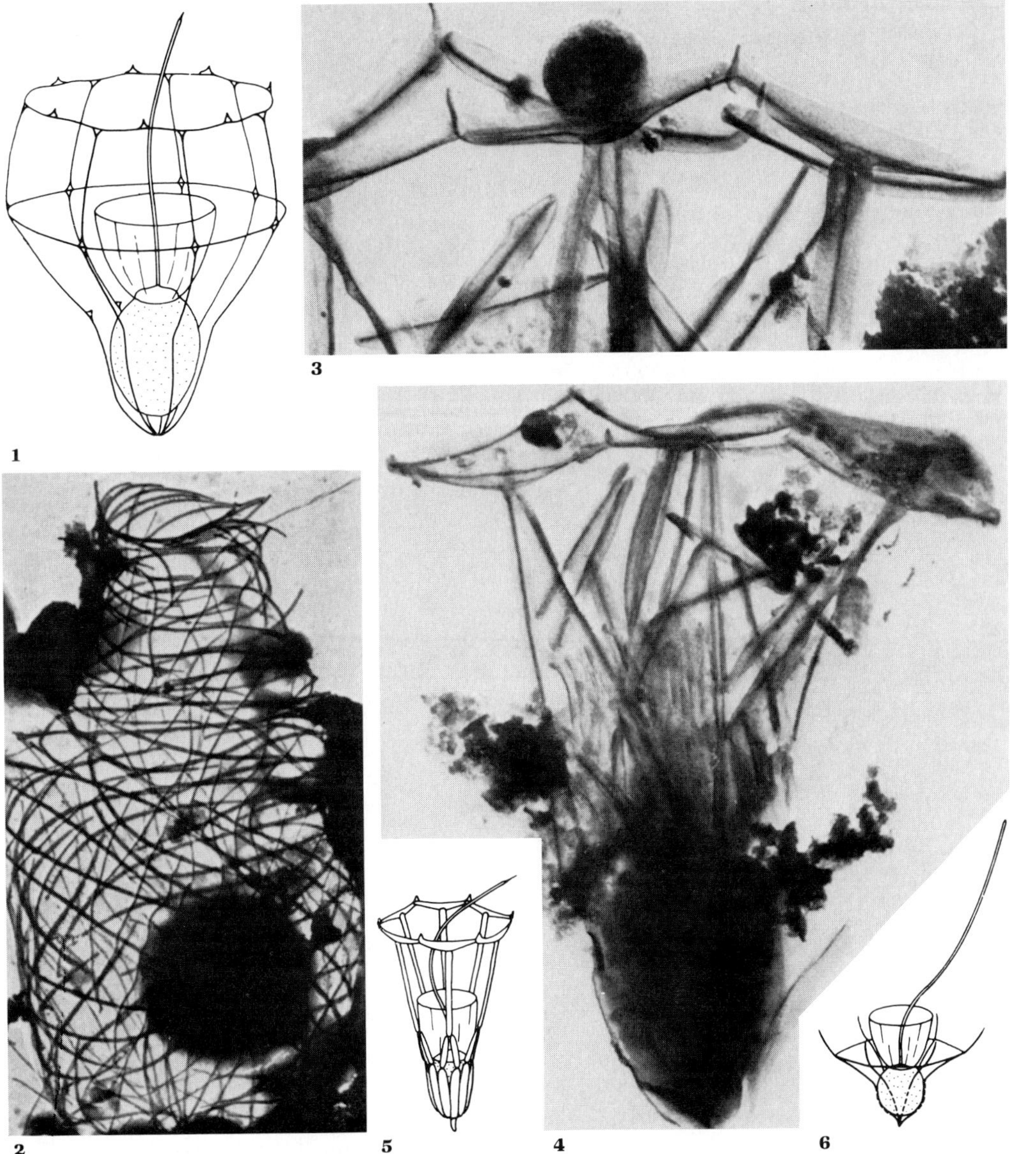

Figure 6.31

Acanthoecaceae. **1.** *Pleurasiga reynoldsii* Throndsen, coast of Yugoslavia, ×2500, from Leadbeater, 1973. **2.** *Savillea parva* (Ellis) Loeblich III, North Jutland, Denmark, shadowcast mount in TEM of cell with spirally wound costae, ×7500, from Leadbeater, 1972a. **3 – 5.** *Parvicorbicula campaniformis* Leadbeater, coast of Yugoslavia. **3.** Upper part of lorica showing costal strips joined in T-like formation, ×13,500. **4.** Shadowcast whole cell, TEM, ×11,000. **5.** ×2500. **6.** *P. spinifera* Leadbeater, ×2500. 3 – 6, from Leadbeater, 1973.

Table 6.4
Cyst genera and families of the Chrysophyceae.

1. Family Archaeomonadaceae Deflandre 1932 (see Figure 6.32)

Siliceous cysts, commonly 6 to 20 μm long, rarely 3 up to 35 μm, single pore at one pole, flush or produced. Marine; Cretaceous to Holocene.
Acanthosphaeridium Hajós & Stradner 1975; *Amphilithopyxis* Deflandre 1932; *Archaeomonadopsis* Deflandre 1938; *Archaeomonas* Deflandre 1932; *Archaeosphaeridium* Deflandre 1932; *Artisphaeridium* Stradner in Hajós, 1975; *Litharchaeocystis* Deflandre 1932; *Litheusphaerella* Deflandre 1932; *Lithuropyxis* Deflandre 1933; *Micrampulla* Hanna 1927; *Pararchaeomonas* Deflandre 1932.

2. Family Chrysostomataceae Chodat 1921 (see Figures 6.33 and 6.34)

Siliceous cysts similar to those of various Chrysophyceae, but of unknown relationship. Nonmarine; Tertiary to Holocene.
Carnegia Pantocsek 1912; *Chrysastrella* Chodat 1922; *Chrysostomum* Chodat 1922; *Clericia* Frenguelli 1925; *Deflandreia* Frenguelli 1938; *Outesia* Frenguelli 1925; *Trachelostomum* Frenguelli 1939.

3. Position uncertain

Calcareous, flasklike cysts, with wall and ornamentation comprising a single crystal of calcite, may have produced neck with terminal opening, closed by a thinner calcitic "membrane" or partition at a level below the lip of the neck; surface smooth or granulose, may have prominent blunt projections. Marine; Oligocene to Miocene.
Bolboforma von Daniels & Spiegler 1974.

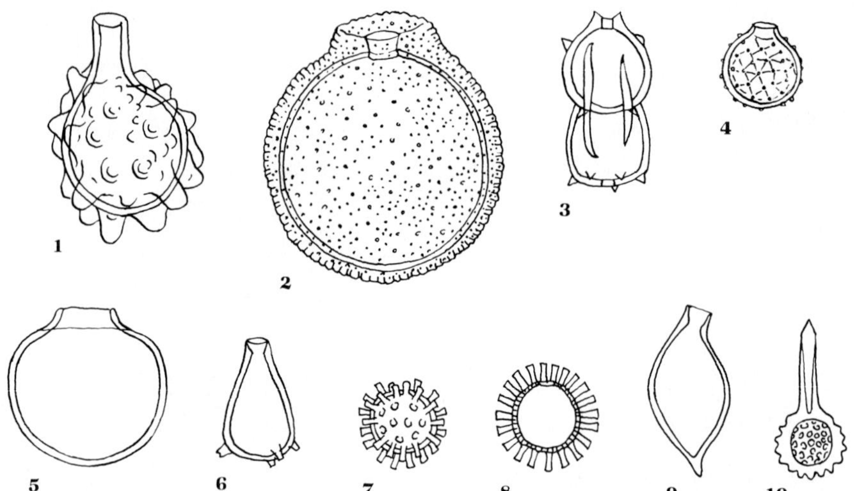

Figure 6.32
Archaeomonadaceae. **1.** *Archaeomonadopsis frenguellii* Rampi, Upper Cretaceous, California, ×1145. **2.** *Archaeosphaeridium dangeardianum* Deflandre, Miocene, Maryland, ×1240. **3.** *Amphilithopyxis aenigmatica* Deflandre, Paleocene, Jutland, ×1945. **4.** *Archaeomonas chiarugii* Rampi, Upper Cretaceous, California, ×1525. **5.** *Pararchaeomonas colligera* Deflandre, Paleocene, Jutland, ×1240. **6.** *Litharchaeocystis glabra* Rampi, Upper Eocene, New Zealand, ×935. **7,8.** *Litheusphaerella spectabilis* Deflandre, Paleocene, Jutland, ×1240. **7.** Exterior. **8.** Optical section. **9.** *Lithuropyxis barbadensis* Deflandre, Eocene, Barbados, ×1715. **10.** *Micrampulla parvula* Hanna, Upper Cretaceous, California, ×975. 1,4, redrawn from Rampi, 1940b; 2,3,5,7,8, redrawn from Deflandre, 1932c; 6, redrawn from Rampi, 1948; 9, redrawn from Deflandre, 1933.

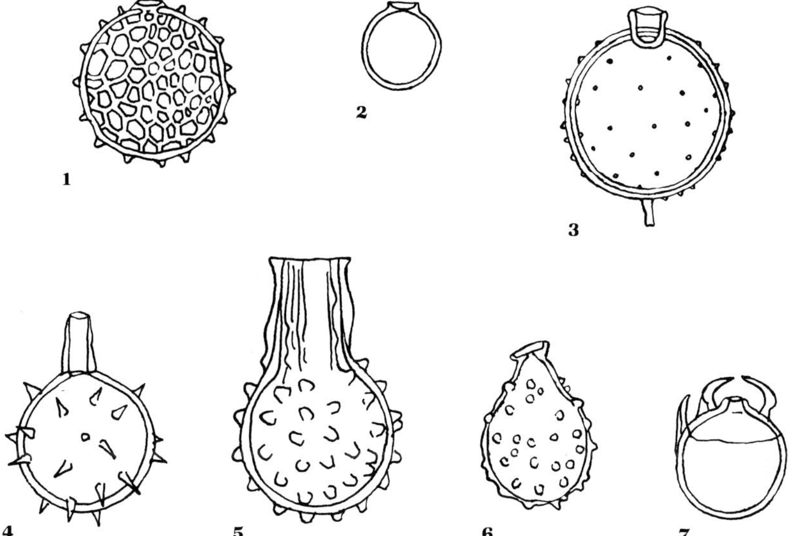

Figure 6.33

Chrysostomataceae. **1.** *Clericia elegantissima* Rampi, Isle of Roti, ×1400. **2.** *Chrysostomum simplex* Chodat, Argentina, ×1200. **3.** *Chrysastrella incerta* Rampi, Italy, ×1400. **4.** *Deflandreia longispina* Rampi, Italy, ×1400. **5.** *Trachelostomum rampii* Frenguelli, Argentina, ×2100. **6.** *Outesia perlata* Frenguelli, Argentina, ×1200. **7.** *Carnegia forcipata* Frenguelli, Isle of Roti, ×1400. All redrawn: 1,7, from Rampi, 1940a; 2,6, from Frenguelli, 1936; 3, 4, from Rampi, 1939; 5, from Frenguelli, 1939.

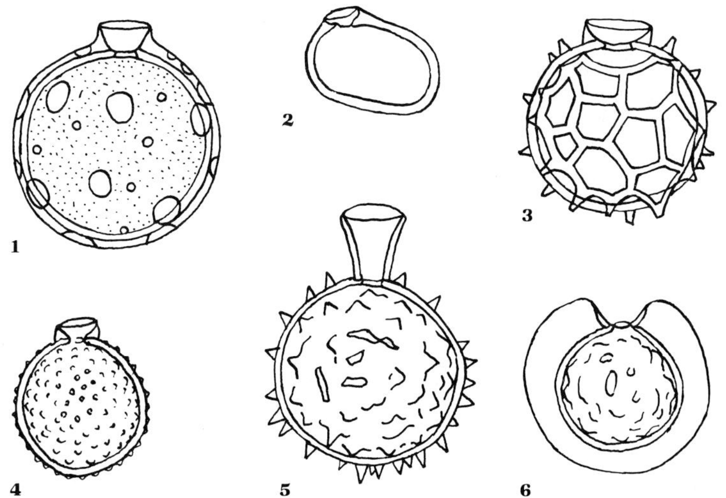

Figure 6.34

Variation in cyst ornamentation in *Clericia.* **1.** *C. zanoni* Rampi. **2.** *C. irregolare* Rampi. **3.** *C. retiforme* Rampi. **4.** *C. colligera* Rampi. **5.** *C. frenguellii* Rampi. **6.** *C. aureolata* Rampi. All from the Isle of Roti, ×1400, redrawn from Rampi, 1940a.

REFERENCES

Andrieu, Bernard, Note sur les Chrysostomatacées d'une tourbe de l'Ile Kerguelen. *Bull. Soc. Fr. Microsc.,* v. 5, p. 51—60, 28 text-figs., 1936.

Andrieu, Bernard, Les Chrysostomatacées d'Auvergne. I. Dépôt de Verneuge (Puy-de-Dôme). *Bull. Soc. Fr. Microsc.,* v. 6, p. 49—58, 17 text-figs., 1937.

Andrieu, Bernard, Les Chrysostomatacées d'Auvergne. II. Dépôt de Vassivière (Puy-de-Dôme). *Bull. Soc. Fr. Microsc.,* v. 7, p. 96—100, 12 text-figs., 1938.

Antia, N. J., T. Bisalputra, J. Y. Cheng, and J. P. Kalley, Pigment and cytological evidence for reclassification of *Nannochloris oculata* and *Monallantus salina* in the Eustigmatophyceae. *J. Phycol.,* v. 11, p. 339—343, figs. 1—6, 1975.

Belcher, J. H., Lorica construction in *Pseudokephyrion pseudospirale* Bourrelly. *Br. Phycol. J.,* v. 3, p. 495—499, figs. 1—3, 1968.

Belcher, J. H., The fine structure of the loricate colourless flagellate *Bicoeca planctonica* Kisselew. *Arch. Protistenk.,* v. 117, p. 78—84, figs. 1—12, 1975.

Bourrelly, Pierre, Recherches sur les Chrysophycées: morphologie, phylogénie, systématique. *Revue Algol., Mém. Hors-Ser.,* no. 1, p. 1—412, 11 pls., 1957.

Bourrelly, Pierre, Loricae and cysts in the Chrysophyceae. *Ann. N.Y. Acad. Sci.,* v. 108, p. 421—429, 10 figs., 1963.

Bourrelly, Pierre, La classification des Chrysophycées, ses problèmes. *Revue Algol.,* v. 8, p. 56—60, 1965.

Bourrelly, Pierre, *Les algues d'eau douce. Initiation à la systématique,* tome II, *Les algues jaunes et brunes, Chrysophycées, Phéophycées, Xanthophycées et Diatomées.* Paris: N. Boubée, 438 p., 114 pls., 1968.

Bradley, W. H., Coprolites from the Bridger Formation of Wyoming: their composition and microörganisms. *Am. J. Sci.,* v. 244, p. 215—239, 4 pls., 1946.

Bukry, David, Coccolith and silicoflagellate stratigraphy, South Atlantic Ocean, Deep Sea Drilling Project Leg 39. *Init. Repts. Deep Sea Drill. Proj.,* v. 39, p. 825—839, pls. 1—3, 1977.

Chodat, R., Matériaux pour l'histoire des algues de la Suisse. *Bull. Soc. Bot. Genève,* v. 13, p. 66—114, 20 figs., 1922.

Christensen, Tyge, *Systematisk botanik.* København: Munksgaard, 178 p., 75 figs., 1962.

Cienkowski, L., Beiträge zur Kenntniss der Monaden. *Arch. Mikrosk. Anat. EntwMech.,* v. 1, p. 203—232, pls. 12—14, 1865.

Clarke, K. J., and N. C. Pennick, *Syncrypta glomerifera* sp. nov., a marine member of the Chrysophyceae bearing a new form of scale. *Br. Phycol. J.,* v. 10, p. 363—370, figs. 1—15, 1975.

Clerici, E., Sulla diffusione di alcuni organismi microscopici delle roccie accompagnanti i tufi vulcanici romanici. *Atti Accad. Naz. Lincei Rc. Sed. Solen.,* ser. 6, v. 1, p. 733—736, 11 figs., 1925.

Colom, G., Arqueomonadineas, silicoflagelados, discoasteridos fosiles de España. *Ciencias,* v. 5, p. 343—356, 1940.

Conrad, W., Recherches sur les flagellates de nos eaux saumâtres. 2e Partie: Chrysomonadines. *Arch. Protistenk.,* v. 56, p. 167—231, pls. 7—9, figs. 1—28, 1926.

Conrad, W., Recherches sur les flagellates de Belgique. 1. Flagellates des etangs des "eaux douces" a Vieux-Héverlé-les-Louvain. *Mém. Mus. R. Hist. Nat. Belg.,* no. 47, 65 p., 1931.

Conrad, W., Revision du genre *Mallomonas* Perty (1851) incl. *Pseudo-Mallomonas* Chodat (1920). *Mém. Mus. R. Hist. Nat. Belg.,* no. 56, p. 3—82, 70 figs., 1933.

Conrad, W., Notes protistologiques, I. *Mallomonas lychenensis* n. sp. *Bull. Mus. R. Hist. Nat. Belg.,* v. 14(20), p. 1—4, figs. 1—8, 1938a.

Conrad, W., Notes protistologiques, III. Chrysomonadines intéressantes du nannoplankton saumâtre. *Bull. Mus. R. Hist. Nat. Belg.,* v. 14(29), p. 1—7, figs. 1—13, 1938b.

Conrad, W., Notes protistologiques, V. Observations sur *Uroglena soniaca* n. sp. et remarques sur le genre *Uroglena* Ehr. (incl. *Uroglenopsis* Lemm.). *Bull. Mus. R. Hist. Nat. Belg.,* v. 14(42), p. 1—27, pls. 1—4, figs. A—F, 1938c.

Conrad, W., Kystes de Chrysomonadines ou Chrysostomatacées? *Bull. Mus. R. Hist. Nat. Belg.,* v. 14, no. 46, p. 1—6, 17 figs., 1938d.

Conrad, W., Chrysomonadées fossiles des collections du Musée Royal d'Histoire naturelle de Belgique. *Bull. Mus. R. Hist. Nat. Belg.,* v. 16, no. 45, p. 1—15, 2 pls., 1940.

Cornell, W. C., The chrysomonad cyst-families Chrysostomataceae and Archaeomonadaceae: their status in paleontology. *Proc. N. Am. Paleont. Conv., Chicago 1969,* pt. G, p. 958—965, figs. 1—5, 1970.

Cornell, W. C., Late Cretaceous chrysomonad cysts. *Palaeogeogr. Palaeoclimat. Palaeoecol.,* v. 12, p. 33—47, figs. 1—7, 1972.

Cornell, W. C. *Archaeomonadopsis incerta* Rampi, 1940—a diatom fragment. *J. Paleont.,* v. 47, p. 1121—1122, figs. 1, 2, 1973.

Dales, R. P., On the pigments of the Chrysophyceae. *J. Mar. Biol. Ass. U.K.,* v. 39, p. 693—699, 1960.

Daniels, C. H. von, and D. Spiegler, *Bolboforma* n. gen. (Protozoa?)—eine neue stratigraphisch wichtige Gattung aus dem Oligozän/Miozän Nordwestdeutschlands. *Paläont. Z.,* v. 48, p. 57—76, pls. 7—10, figs. 1—3, 1974.

Deflandre, Georges, *Litharchaeocystis costata* nov. gen. nov. spec., Chrysophycée marine fossil. Remarques sur les Chrysostomatacées. *C. R. Hebd. Séanc. Acad. Sci., Paris,* v. 194, p. 1273–1275, 2 figs., 1932a.

Deflandre, Georges, Archaeomonadaceae, une famille nouvelle de protistes fossiles marins à loge siliceuse. *C. R. Hebd. Séanc. Acad. Sci., Paris,* v. 194, p. 1859–1861, 7 figs., 1932b.

Deflandre, Georges, Note sur les Archaeomonadacées. *Bull. Soc. Bot. Fr.,* v. 79, p. 346–355, 38 figs., 1932c.

Deflandre, Georges, Sur quelques protistes siliceux d'un sondage de la mer Caspienne. *Bull. Soc. Fr. Microsc.,* v. 1, p. 78–81, 1932d.

Deflandre, Georges, Second note sur les Archaeomonadacées. *Bull. Soc. Bot. Fr.,* v. 80, p. 79–90, 41 figs., 1933.

Deflandre, Georges, Sur l'abus de l'emploi, en paléontologie, du nom de genre *Trachelomonas* et sur la nature de quelques ex *"Trachelomonas"* siliceux (Chrysomonadines) tertiaires et quaternaires. *Annls. Protist.,* v. 4, p. 151–165, 10 figs., 1934.

Deflandre, Georges, Les flagellés fossiles: Aperçu biologique et paléontologique; role géologique. *Actual. Scient. Ind.,* no. 335, 98 p., 135 figs., 1936a.

Deflandre, Georges, Microfossiles des silex crétacés. I. Généralités. Flagellés. *Annls. Paléont.,* v. 25, p. 151–191, 10 pls., 1936b.

Deflandre, Georges, Troisième note sur les Archaeomonadacées. *Bull. Soc. Fr. Microsc.,* v. 7, p. 73–88, 43 figs., 1938a.

Deflandre, Georges, Sur deux microfossiles siliceux énigmatiques (Silicoflagellidées?). *Bull. Soc. Fr. Microsc.,* v. 7, p. 90–96, figs. 1–25, 1938b.

Deflandre, Georges, Sur les Archaeomonadacées. Action lithogénétique. Signification stratigraphique. *C. R. Somm. Séanc. Soc. Géol. Fr.,* p. 53–55, 1944.

Deflandre, Georges, *Phyllodictyocha* nov. gen., Silicoflagellidés et formes affines du Miocène de Hongrie. *Bull. Soc. Bot. Fr.,* v. 93, p. 335–337, figs. 1–5, 1947.

Deflandre, Georges, Chrysomonadines fossiles, p. 560–565. *In* P. P. Grassé, *Traité de Zoologie.* Paris: Masson et Cie., 1952.

Ehrenberg, C. G., *Mikrogeologie. Das Erden und Felsen schaffende Wirken des unsichtbaren kleinen selbständigen Lebens auf der Erde.* Leipzig: L. Voss, 374 p., 40 pls., 1854.

Emberger, Louis, *Les plantes fossiles dans leurs rapports avec les végétaux vivants,* 2nd ed. Paris: Masson et Cie., 758p., 743 figs., 1968.

Emig, W. H., The travertine deposits of the Arbuckle Mountains, Oklahoma, with reference to the plant agencies concerned in their formation. *Bull. Okla. Geol. Surv.,* 29, p. 1–76, figs. 1–4, 1917.

Esser, S. C., Surface ultrastructure of chrysomonad cysts. *J. Protozool.,* v. 22(3) (Program and Abstracts), p. 19A (Abstr.), 1975.

Firtion, F., Spongilles et Chrysostomatacées d'un lignite de la formation cinéritique du Lac Chambon (Puy-de-Dôme). *Bull. Soc. Géol. Fr.,* sér. 5, v. 14, p. 331–346, pls. 7, 8, 1944.

Firtion, F., Sur les Chrysostomatacées du bassin tertiaire de Menat (Puy-de-Dôme). *Bull. Soc. Géol. Fr.,* sér. 5, v. 15, p. 45–52, 1 pl., 1945.

Fott, Bohuslav, Taxonomy of *Mallomonas* based on electron micrographs of scales. *Preslia,* v. 34, p. 69–84, pls. 1–6, figs. 1–4, 1962.

Fott, Bohuslav, *Algenkunde,* 2nd ed. Jena: VEB Gustav Fischer Verlag, 581 p., 303 figs., 1971.

Frenguelli, Joaquín [Gioacchino], Sopra alcuni microrganismi a guscio siliceo. *Boll. Soc. Geol. Ital.,* v. 44, p. 1–8, 1 pl., 1925.

Frenguelli, Joaquín, Diatomee fossili delle conche saline del deserto cileno. *Boll. Soc. Geol. Ital.,* v. 47, p. 185–236, 1929a.

Frenguelli, Joaquín, *Trachelomonas* de los esteros de la region del Yberá en la provincia de Corrientes, Argentina. *Revta. Chil. Hist. Nat.,* v. 33, p. 563–568, pl. 32, 1929b.

Frenguelli, Joaquín, Analisis microscopico de una muestra de tripoli de Angostura (Provincia de Colchagua-Chile). *Revta. Chil. Hist. Nat.,* v. 35, p. 9–14, 1 pl., 1931.

Frenguelli, Joaquín [Gioacchino], Resti silicei di microrganismi dei travertini della Somalia. *Palaeontogr. Ital.,* v. 32, suppl. 1, p. 67–77, pl. 6 (1), 1933a.

Frenguelli, Joaquín [Gioacchino], Trachelomonadi del pliocene argentino. *Memorie Soc. Geol. Ital.,* v. 1, p. 1–44, 3 pls., 1933b.

Frenguelli, Joaquín, Einige Bemerkungen zu den Archaeomonadaceen. *Arch. Protistenk.,* v. 84, p. 232–241, 1 fig., 1935a.

Frenguelli, Joaquín, Traquelomónadas del Platense. *Notas Mus. La Plata,* v. 1 (Paleont. no. 2), p. 35–44, 32 figs., 1935b.

Frenguelli, Joaquín, Crisostomatáceas del Neuquen. *Notas Mus. La Plata,* v. 1 (Bot. no. 9), p. 247–275, 6 figs., 1936.

Frenguelli, Joaquín, *"Deflandreia"* nuevo género de Crisostomatáceas. *Notas Mus. La Plata,* v. 3 (Bot. no. 18), p. 47–54, 1938.

Frenguelli, Joaquín, Crisostomatáceas del Rio de la Plata. *Notas Mus. La Plata,* v. 4 (Bot. no. 25), p. 285–308, 1 pl., 7 figs., 1939.

Frenguelli, Joaquín, Nuevas especies argentinas del genero *Chrysastrella* (Chrysostomataceae). *Notas Mus. La Plata,* v. 10 (Bot. no. 49), p. 99–105, 2 figs., 1945.

Frey, D. G., Remains of animals in Quaternary lake and bog sediments and their interpretation. *Arch. Hydrobiol. Beih. Ergebn. Limnol.,* v. 2, p. 1–114, 2 pls., 1964.

Fritsch, F. E., *The structure and reproduction of the algae,* vol. I. Cambridge: University Press, xvii + 791 p., 245 figs., 1935.

Gaarder, K. R., G. A. Fryxell, and G. R. Hasle, *Potamodiscus kalbei* Gerloff: an organism with siliceous scales. *Arch. Protistenk.,* v. 118, p. 346−351, pls. 71−76, 1976.

Gayral, P., and C. Billard, Synopsis du nouvel ordre des Sarcinochrysidales (Chrysophyceae). *Taxon,* v. 26, p. 241−245, 1977.

Gerloff, J., Elektronenmikroskopische Untersuchungen an Diatomeenschalen. VI. *Potamodiscus kalbei* nov. gen. et nov. spec. *Willdenowia,* v. 4(3), p. 353−364, 1968.

Gold, Kenneth, R. M. Pfister, and V. R. Liguori, Axenic cultivation and electron microscopy of two species of Choanoflagellida. *J. Protozool.,* v. 17, p. 210−212, figs. 1−5, 1970.

Hajós, Márta, Mátraalja Miocén üledékeinek diatomái [Die diatomeen der Miozänen ablagerungen des Mátravorlandes]. *Geol. Hung., Ser. Paleont.,* fasc. 37, p. 1−401, pls. 1−68, 1968.

Hajós, Márta, Late Cretaceous Archaeomonadaceae, Diatomaceae and Silicoflagellatae from the South Pacific Ocean, Deep Sea Drilling Project, Leg 29, Site 275. *Init. Repts. Deep Sea Drill. Proj.,* v. 29, p. 913−1009, pls. 1−40, figs. 1−21, 1975.

Hajós, Márta, and István Pálflavy, Magyaregregy diatomás üledékeinek életföldtani vizsgálata [Examen biogeologique des dépôts à diatomées de Magyaregregy]. *Évi Jelent. Magy. K. Földt. Intéz.,* 1960, p. 89−119, pls. 1−3, figs. 1−5, 1963.

Hanna, G. D., Cretaceous diatoms from California. *Occ. Pap. Calif. Acad. Sci.,* 8, p. 5−39, pls. 1−5, 1927.

Hanna, G. D., The diatoms of Sharktooth Hill, Kern County, California. *Proc. Calif. Acad. Sci.,* ser. 4, v. 20 (6), p. 161−263, 17 pls., 1932.

Hibberd, D. J., Observations on the cytology and ultrastructure of *Ochromonas tuberculatus* sp. nov. (Chrysophyceae) with special reference to the discobolocysts. *Br. Phycol. J.,* v. 5, p. 119−143, figs. 1−45, 1970.

Hibberd, D. J., Ultrastructure of cyst formation in *Ochromonas tuberculata* (Chrysophyceae). *J. Phycol.,* v. 13, p. 309−320, figs. 1−38, 1977.

Hibberd, D. J., and G. F. Leedale, Eustigmatophyceae—a new algal class with unique organization of motile cell. *Nature,* v. 225, p. 758−760, figs. 1−3, 1970.

Hibberd, D. J., and G. F. Leedale, Cytology and ultrastructure of the Xanthophyceae. II. The zoospore and vegetative cell of coccoid forms, with special reference to *Ophiocytium majus* Naegeli. *Br. Phycol. J.,* v. 6, p. 1−23, figs. 1−27, 1971a.

Hibberd, D. J., and G. F. Leedale, A new algal class— the Eustigmatophyceae. *Taxon,* v. 20, p. 523−525, 1971b.

Hibberd, D. J., and G. F. Leedale, Observations on the cytology and ultrastructure of the new algal class, Eustigmatophyceae. *Ann. Bot.,* v. 36, p. 49−71, pls. 1−7, figs. 1, 2, 1972.

Hilliard, D. K., Notes on the occurrence and taxonomy of some planktonic chrysophytes in an Alaskan Lake, with comments on the genus *Bicoeca. Arch. Protistenk.,* v. 113, p. 98−122, pl. 10, figs. 1−7, 1971.

Jones, R. L., L. J. McKenzie, and A. H. Beavers, Opaline microfossils in some Michigan soils. *Ohio J. Sci.,* v. 64, p. 417−423, 1964.

Kolkwitz, R., Ueber das Kammerplankton des Süsswasser und der Meere. *Ber. Dt. Bot. Ges.,* v. 29, p. 386−402, 3 figs., 1911.

Krieger, W., Untersuchungen über Plankton-Chrysomonaden. Die Gattungen *Mallomonas* und *Dinobryon* in monographischer Bearbeitung. *Bot. Arch.,* v. 29, p. 257−329, figs. 1−63, 1930.

Kristiansen, Jørgen, Lorica structure in *Chrysolykos* (Chrysophyceae). *Bot. Tidsskr.,* v. 64, p. 162−168, figs. 1−4, 1969.

Kristiansen, Jørgen, Structure and occurrence of *Bicoeca crystallina,* with remarks on the taxonomic position of the Bicoecales. *Br. Phycol. J.,* v. 7, p. 1−12, figs. 1−6, 1972.

Leadbeater, B. S. C., Ultrastructural observations on some marine choanoflagellates from the coast of Denmark. *Br. Phycol. J.,* v. 7, p. 195−211, figs. 1−27, 1972a.

Leadbeater, B. S. C., *Paraphysomonas cylicophora* sp. nov., a marine species from the coast of Norway. *Norw. J. Bot.,* v. 19, p. 179−185, figs. 1−17, 1972b.

Leadbeater, B. S. C., External morphology of some marine choanoflagellates from the coast of Jugoslavia. *Arch. Protistenk.,* v. 115, p. 234−252, pls. 13−20, figs. 1−3, 1973.

Leadbeater, B. S. C., Ultrastructural observations on nanoplankton collected from the coast of Jugoslavia and the Bay of Algiers. *J. Mar. Biol. Ass. U.K.,* v. 54, p. 179−196, pls. 1−7, fig. 1, 1974.

Leadbeater, B. S. C., Renaming of *Salpingoeca* sensu Grøntved. *J. Mar. Biol. Ass. U.K.,* v. 58, p. 511−515, 1 fig., 1978.

Leadbeater, B. S. C., and Irene Manton, Preliminary observations on the chemistry and biology of the lorica in a collared flagellate (*Stephanoeca diplocostata* Ellis). *J. Mar. Biol. Ass. U.K.,* v. 54, p. 269−276, pls. 1, 2, 1974.

Lee, K. W., and H. C. Bold, *Pseudocharaciopsis texensis* gen. et sp. nov., a new member of the Eustigmatophyceae. *Br. Phycol. J.,* v. 8, p. 31−37, figs. 1−14, 1973.

Loeblich, A. R. III, Notes on the divisions Chlorophyta, Chrysophyta, Pyrrhophyta and Xanthophyta and the family Paramastigaceae. *Taxon,* v. 16, p. 230–236, 1967.

Lohmann, H., Untersuchungen über das Pflanzen– und Tierleben der Hochsee. Zugleich ein Bericht über die biologischen Arbeiten auf der Fahrt der "Deutschland" von Bremerhaven nach Buenos Aires in der Zeit vom 7. Mai bis 7. September 1911. *Veröff. Inst. Meeresk., Univ. Berl., N. F., A. Geogr.-naturwiss. Reihe,* hefte 1, p. 1–92, 2 pls., 14 figs., 1912.

Ludwig, R., Fossile Pflanzen aus der jüngsten Wetterauer Braunkohle. *Palaeontographica,* v. 5, 81 p., 1857.

Manton, Irene, J. Sutherland, and B. S. C. Leadbeater, Further observations on the fine structure of marine collared flagellates (Choanoflagellata) from arctic Canada and west Greenland: species of *Parvicorbicula* and *Pleurasiga. Can. J. Bot.,* v. 54, p. 1932–1955, figs. 1–71, 1976.

Massalski, Andrzej, and G. F. Leedale, Cytology and ultrastructure of the Xanthophyceae. I. Comparative morphology of the zoospores of *Bumilleria sicula* Borzi and *Tribonema vulgare* Pascher. *Br. Phycol. J.,* v. 4, p. 159–180, figs. 1–20, 1969.

Matvienko, O. M., Shlyakhi i napryamki evolyutsiynogo rozvitku zolotistikh vodorostey (Chrysophyta) [Ways and trends of evolutional development in Chrysophyta]. *Ukr. Bot. Zh.,* v. 17(4), p. 3–8, 1960.

Moestrup, Ø., On the fine structure of the spermatozoids of *Vaucheria sescuplicaria* and on the later stages in spermatogenesis. *J. Mar. Biol. Ass. U.K.,* v. 50, p. 513–523, pls. 1–6, 1970.

Norris, R. E., Neustonic marine Craspedomonadales (Choanoflagellates) from Washington and California. *J. Protozool.,* v. 12, p. 589–602, figs. 1–24, 1965.

Nygaard, Gunnar, Ancient and recent flora of diatoms and Chrysophyceae in Lake Gribsø. *Folia Limnol. Scand.,* v. 8, p. 32–94, pls. 1–12, 1956.

Pantocsek, József, A kopacseli andesittufa Kovamoszatai. *Bot. Közl.,* v. 12, p. 126–137, pls. 1, 2, 1913a.

Pantocsek, József, A lutillai ragpalában elöforduló Bacillariák vagy Kovamoszatok leirása. *Pozs. orvostermészettud. Egyes. Közl.,* n.f., v. 23 (v. 32), p. 19–35, pls. 1, 2, 1913b.

Papenfuss, G. F., Classification of the algae, p. 115–225. *In* E. L. Kessel, *A century of progress in the natural sciences.* San Francisco: Calif. Acad. Sci., 1955.

Parke, Mary, Some remarks concerning the Class Chrysophyceae. *Br. Phycol. Bull.,* v. 2, p. 47–57, pls. 1, 2, 1961.

Parke, Mary, and B. S. C. Leadbeater, Check-list of British marine Choanoflagellida—second revision. *J. Mar. Biol. Ass. U.K.,* v. 57, p. 1–6, fig. 1, 1977.

Pascher, A., *Die Süsswasser-Flora Deutschlands, Österreichs und der Schweiz.* Heft 2: Flagellatae II, Chrysomonadinae, Cryptomonadinae, Eugleninae, Chloromonadinae und Gefärbte Flagellaten unsicheren Stellung. Jena: G. Fischer, 192 p., 1913.

Pascher, A., Über Flagellaten und Algen. *Ber. Dt. Bot. Ges.,* v. 32, p. 136–160, 1914.

Pascher, A., Über die Verbreitung endogener bzw. endoplasmatisch gebildeter Sporen bei den Algen. *Beih. Bot. Zbl.,* v. 49, p. 293–308, 13 figs., 1932.

Pascher, A., Heterokonten, p. 1–1091. In *Dr. L. Rabenhorst's Kryptogamen-Flora von Deutschland, Österreich und der Schweiz,* 2nd ed., Bd. 11, Lief. 1–6. Leipzig: Akademische Verlagsgesellschaft, 1937–1939.

Pennick, N. C., and K. J. Clarke, *Paraphysomonas corbidifera* sp. nov., a marine, colourless, scale-bearing member of the Chrysophyceae. *Br. Phycol. J.,* v. 8, p. 147–151, figs. 1–10, 1973.

Perch-Nielsen, Katharina, Late Cretaceous to Pleistocene archaeomonads, ebridians, endoskeletal dinoflagellates, and other siliceous microfossils from the subantarctic southwest Pacific, DSDP, Leg 29. *Init. Repts. Deep Sea Drill. Proj.,* v. 29, p. 873–907, pls. 1–13, figs. 1–3, 1975.

Rampi, L., Note sur les Chrysostomatacées tertiaires de Santa Fiora. *Bull. Soc. Fr. Microsc.,* v. 6, p. 67–75, 54 figs., 1937a.

Rampi, L., Les Diatomées et les Chrysostomatacées d'une tourbe du Monte Amiata. *Bull. Soc. Fr. Microsc.,* v. 6, p. 129–136, 16 figs., 1937b.

Rampi, L., Note sur les Chrysostomatacées du dépôt de Crognuolo (Monte Amiata). *Bull. Soc. Fr. Microsc.,* v. 8, p. 15–20, 17 figs., 1939.

Rampi, L., Diatomee e Crisostomatacee dell'Isola di Rodi. *Nuovo G. Bot. Ital.,* v. 47, p. 572–578, 19 figs., 1940a.

Rampi, L., Archaeomonadacee del cretaceo Americano. *Atti Soc. Ital. Sci. Nat.,* v. 79, p. 60–67, 16 figs., 1940b.

Rampi, L., Su alcune Archaeomonadacee (Crisomonadine fossili marine) nuove od interessanti. *Atti Soc. Ital. Sci. Nat.,* v. 87, p. 185–188, 7 figs., 1948.

Rampi, L., Archaeomonadacées de la diatomite éocène de Kreyenhagen, California. *Cah. Micropaléont.,* ser. 1, v. 14, p. 1–12, pls. 1–3, 1969.

Rögl, Fred, and P. Hochuli, The occurrence of *Bolboforma,* a probable algal cyst in the Antarctic Miocene of DSDP Leg 35. *Init. Repts. Deep Sea Drill. Proj.,* v. 35, p. 713–719, pls. 1, 2, 1976.

Scherffel, A., Über die Cyste von *Monas. Arch. Protistenk.,* v. 48, p. 187–195, 1924.

Schussnig, B., *Ochrosphaera neapolitana,* nov. gen., nov. spec., eine neue Chrysomonade mit Kalkhülle. *Öst. Bot. Z.,* v. 79, p. 164–170, 4 figs., 1930.

Stradner, H., On the ultrastructure of Miocene Archaeomonadaceae (Phytoflagellates) from Limberg, Lower Austria. *In* A. Farinacci, *Proc. II Plankton Conf. Roma 1970,* v. 2, p. 1183−1199, pls. 1−6. Roma: Edizioni Tecnoscienza, 1971.

Takahashi, Eiji, Studies on genera *Mallomonas* and *Synura,* and other plankton in fresh water with the electron microscope IX. *Mallomonas harrisae* sp. nov. (Chrysophyceae). *Phycologia,* v. 14, p. 41−44, figs. 1−8, 1975.

Thomsen, H. A., Studies on marine choanoflagellates I. Silicified choanoflagellates of the Isefjord (Denmark). *Ophelia,* v. 12, p. 1−26, figs. 1−42, 1973.

Throndsen, J., Marine planktonic Acanthoecaceans (Craspedophyceae) from Arctic waters. *Nytt Mag. Bot.,* v. 17, p. 103−111, figs. 1−25, 1970.

Throndsen, J., Planktonic choanoflagellates from North Atlantic waters. *Sarsia,* v. 56, p. 95−122, figs. 1−47, 1974.

Tynan, E. J., The Archaeomonadaceae of the Calvert Formation (Miocene) of Maryland. *Micropaleontology,* v. 6, p. 33−39, 1 pl., 1960.

Tynan, E. J., Geologic occurrence of the archaeomonads. *In* A. Farinacci, *Proc. II Planktonic Conf. Roma, 1970,* v. 2, p. 1225−1230. Roma: Edizioni Tecnoscienza, 1971.

Van Landingham, S. L., Chrysophyta cysts from the Yakima Basalt (Miocene) in south-central Washington. *J. Paleont.,* v. 38, p. 729−739, pl. 120, 1964.

Whittle, S. J., The major chloroplast pigments of *Chlorobotrys regularis* (West) Bohlin (Eustigmatophyceae) and *Ophiocytium majus* Naegeli (Xanthophyceae). *Br. Phycol. J.,* v. 11, p. 111−114, 1976a.

Whittle, S. J., Chlorophyll *c* in the Xanthophyceae. *Br. Phycol. J.,* v. 11, p. 201 (Abstr.), 1976b.

Whittle, S. J., and P. J. Casselton, The chloroplast pigments of the algal classes Eustigmatophyceae and Xanthophyceae. I. Eustigmatophyceae. *Br. Phycol. J.,* v. 10, p. 179−191, fig. 1, 1975a.

Whittle, S. J., and P. J. Casselton, The chloroplast pigments of the algal classes Eustigmatophyceae and Xanthophyceae. II. Xanthophyceae. *Br. Phycol. J.,* v. 10, p. 192−204, 1975b.

Zanon, Vito, Saggio di sistematica della Crisostomatacee. *Acta Pontif. Acad. Sci.,* v. 11, p. 43−59, 1 pl., 1947.

Zanon, Vito, Crisostomatacee africaine. *Atti Accad. Naz. Lincei Memorie,* ser. 8, v. 3, sez. 3, fasc. 1, p. 1−34, pls. 1−4, 1950.

Chapter 7

SILICOFLAGELLATES

I have already pointed out the striking correspondence between the fossils of the infusorial stratum of Virginia with those of Oran in Africa. ... Believing that it will be of great interest to geologists both at home and abroad to trace out this correspondence of the fossils of regions so far distant, and of beds which are at present referred to different epochs, I have added to my plate ... a number of figures of siliceous bodies not before described, found in the infusorial stratum of Virginia. ... I suspect that these belong to the genus Dictyocha of Ehrenberg, several species of which occur at Oran, Caltanisetta, &c ... Should the true age of either the American or African deposits be determined by means of the fossil infusoria, it will be an additional instance of the importance of this branch of microscopic paleontology.

J. W. Bailey, 1842

The silicoflagellates are exclusively marine, but cosmopolitan, members of the siliceous microplankton, considered either as an order of flagellated protozoans or as an algal order.

Known from skeletal remains in fossil material and from modern oceanic deposits since 1837, they were variously regarded as "polygastrica," diatoms, and radiolarians, until Borgert (1890) noted the presence of a single flagellum in living *Distephanus* (see Figure 7.1). Realizing they could no longer be included in any of the above-mentioned groups, he proposed for them the order Silicoflagellata. Lemmermann (1901a) then divided the silicoflagellates into two orders: the Siphonotestales included uniflagellate forms (e.g., *Distephanus*) with skeletal framework of tubular hollow rods, and the Stereotestales included biflagellate forms (now the Ebriales) with solid rather than hollow skeletal framework. The latter order is now placed with the dinoflagellates rather than with true silicoflagellates in the Chrysophyta.

THE LIVING ORGANISM

The Protoplast

The individual cell, from 20 to 200 μm in diameter, consists of clear hyaline protoplasm surrounding a single relatively large, colorless, rounded to oval nucleus. The nucleus contains one or two nucleoli and is immediately surrounded by a denser layer of cytoplasm that

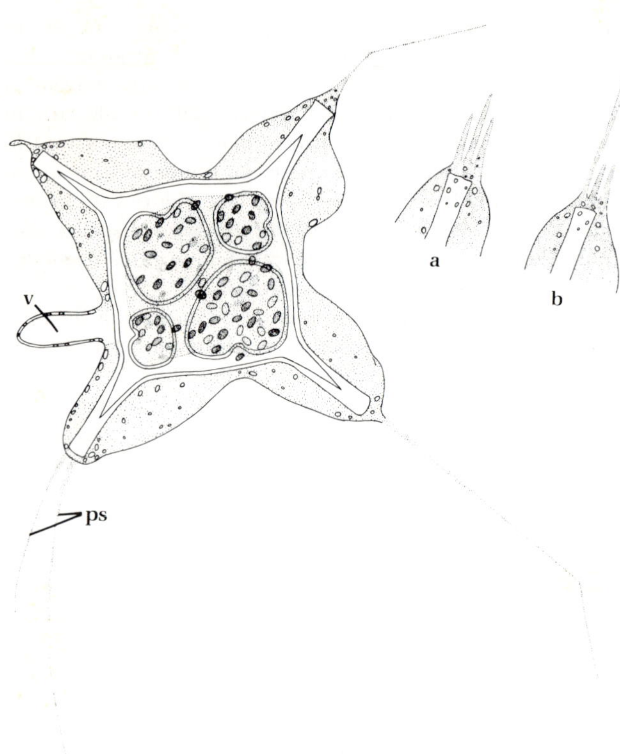

Figure 7.4
Dictyocha fibula, abapical
side, showing pseudopodia (ps) and
vacuole (v); a,b, showing pseudopodial alteration by amoeboid movement; redrawn from Marshall,
1934.

tion. A change of direction is attained by thrust-
ing the flagellum out stiffly in the new direction
(Van Valkenburg, 1971b).

Very fine mobile pseudopodia also may project from the outer protoplasmic layer and may
be visible on every spine as tiny threads (see
Figure 7.4). During skeletal secretion, the flagellum may disappear and the pseudopodia become more prominent, being clear (nongranular), rigid, and radial, extending along the
spines, and reaching a length from one to one-
and-a-half times the cell diameter. Observations
of the living organism generally were made on
recently collected material, as the silicoflagellates are delicate organisms, and only recently
have they been successfully maintained in culture (Van Valkenburg and Norris, 1970).

The Skeleton

The basketlike skeleton of the silicoflagellate
consists of opaline silica in the form of the
monomeric acid $Si(OH)_4$, with minor amounts
of other elements. Examination with the electron microprobe (Lipps, 1970, p. 969) showed
these other elements to consist of 0.05 to 1 percent Al, and 0.05 − 0.5 percent each of Cl, Ca,
Na, and Mg. No threads or veins of organic material are present within the silica skeleton (Van
Valkenburg, 1971a).

The skeleton is typically tubular in structure,
with a framework of hollow rods (see Figure
7.5), in contrast to the solid rods of the ebridian
skeleton. Whether the internal cavity is filled
with air or liquid during the life of the organism

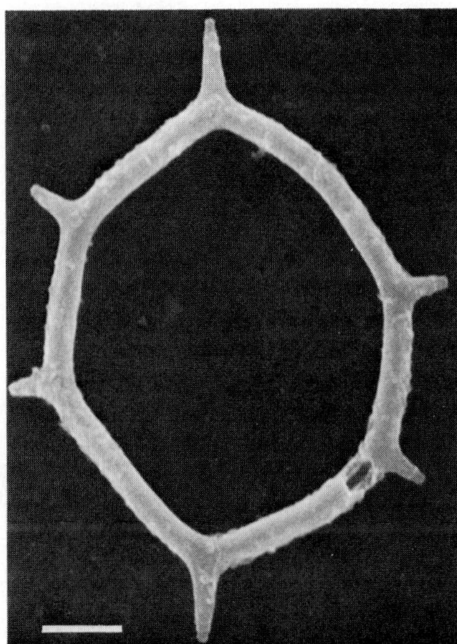

Figure 7.5
Mesocena hexagona Haeckel, Miocene, California; SEM, break in skeleton at lower right shows hollow cavity within the skeletal bars; ×1000, bar = 10 μm.

is as yet unknown, although no connection from the cavity to the exterior has been observed. In prepared slides it seems to be filled with air, but the air is replaced by prolonged immersion in toluene or xylol. Possibly the physical state of the opaline silica allows the penetration of liquids. Occasionally a delicate septum may cross the internal cavity; this was regarded as of systematic importance in the description of the genera *Septamesocena* Bachmann 1970 and *Bachmannocena* Locker 1974, but appears intermittantly in species of *Mesocena* and *Corbisema* and probably is not a constant feature (Dumitrică, 1973b).

Because the colorless and transparent skeleton has a refractive index of about 1.4, it is difficult to see in a water mount or in the usual type of permanent mount. The use of styrax (or storax; a resin obtained from the tree *Liquidambar*), whose refractive index is 1.65, or the synthetic resin hyrax, with a refractive index of 1.71, as with the diatoms, is much preferable for mounting and studying the silicoflagellate skeleton. For example, the surface of the skeleton may have a fine reticulate network, which is discernible with the light microscope only with such special preparation.

The usual skeleton is radially symmetrical, and like many other organisms adapted to a planktonic existence commonly has elongate processes or spines that project radially from the corners of the polygonal **basal ring** (see Figure 7.6). Supporting bars arise from the middle of each of the sides of the polygon, arching to meet centrally (*Dictyocha*) or attaching to support a smaller **apical ring** (*Distephanus*). In addition to the elongate spines at the angles of the basal ring, smaller spines may be present, and very tiny conical projections (microspines) may be scattered singly or occur in clusters on the basal ring and radial spines. Different terms have been proposed for the various parts of the silicoflagellate skeleton, including bars, beams, struts, rods, rings, dome, and plate for the larger features, and spine, tooth, pike, mammilon, and bulb for those of smaller scale; the openings between the bars are termed **windows** or **portals** (terminology given in varying detail by Deflandre, 1950a–c; Stradner, 1961; Glezer, 1966; Loeblich III et al., 1968; Jerković, 1969a; Ling, 1972; Bukry, 1976a).

Double skeletons occur seasonally in the plankton and are known in fossil materials; they represent a stage in division, the two skeletons lying with their basal rings adjacent, and in some instances held together by the spines on the basal ring, or even cemented together. Such double skeletons were known in preserved material and as fossils as old as Late Cretaceous before their true nature was recognized, hence they were variously thought to represent sexual reproduction and the conjugation of the adult cells, or to result from asexual division. Only the latter case has been substantiated by a study of living material.

Study of these double skeletons, and their formation, has shown that the two paired specimens are mirror images, with the same number of sides, spines, and rings (see Figure

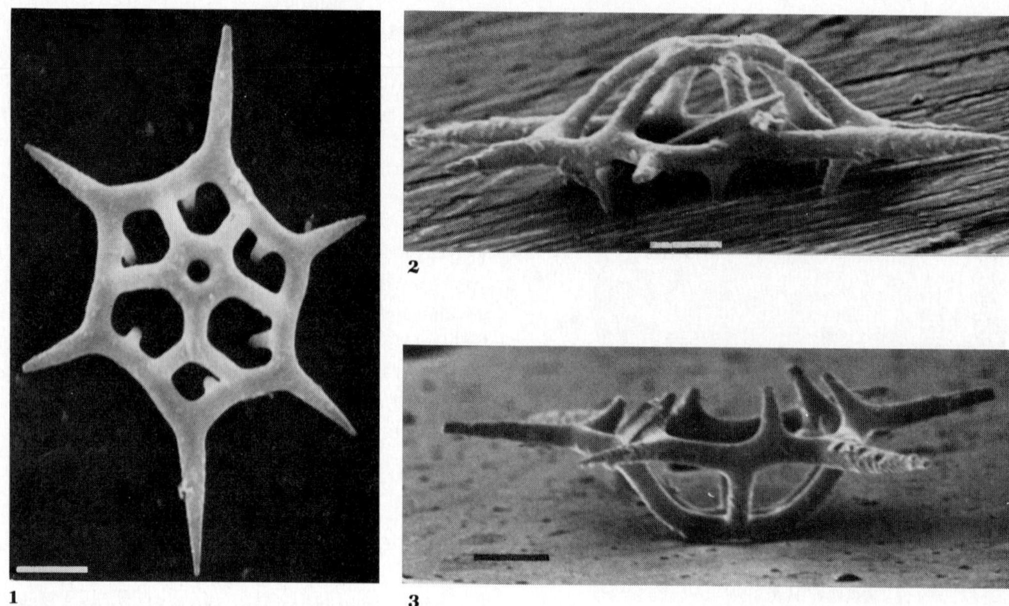

Figure 7.6

Distephanus speculum, Miocene, California. **1.** Apical view in SEM, showing apical and basal rings connected by lateral rods, leaving lateral windows between. Radial spines extend outward and basal accessory spines turn adapically from the basal ring. Apical ring surrounds the apical window. ×1000, bar = 10 μm. **2,3.** Lateral views in SEM of specimen resting on basal ring and basal accessory spines, and one resting on apical ring. Note rugose distal ends of basal spines. ×1250, bar = 8 μm.

7.7). The basal rings are most similar, as they are nearly in contact when formed, hence most strongly controlled by the morphology of the parent skeleton. More skeletal variations occur in the apical region, which is farther from the corresponding region of the parent cell and may not be completely developed at the onset of cell division. Occasional anomalous specimens show such modifications of the apical area, ranging from its subdivision to a cannopiloid form by the addition of supplementary bars, to the complete loss of the apical ring at the other extreme. Spines may be present or absent, may vary in length, and may be forked distally. Some species also have small supplementary spines, the entire surface may be rugose and hispid, and even the number of sides and spines may vary in a single species. Statistical studies have shown that in a living population about 70 to 76

percent of *Distephanus octonarius* (Ehrenberg) Haeckel have octagonal symmetry, but the remainder may have 6, 7, 9, or 10 sides. Similarly, although 95 to 98 percent of the population of *D. speculum* has a hexagonal symmetry, the remainder may have 5, 7, or 8 sides. The variations in the basal ring (e.g., a few individuals with one more or less angle then the normal form of the species) generally are due to slight variations in the parent skeleton that become intensified in later generations. A forked or split spine in one specimen may result in doubled spines in the daughter, and this may in turn result in formation of an extra side in its progeny. Aberrant skeletons were found in 1.78 percent of the *D. speculum* population in the Firth of Clyde (Boney, 1973).

All gradations may be observed between some of these variations, but others are obviously

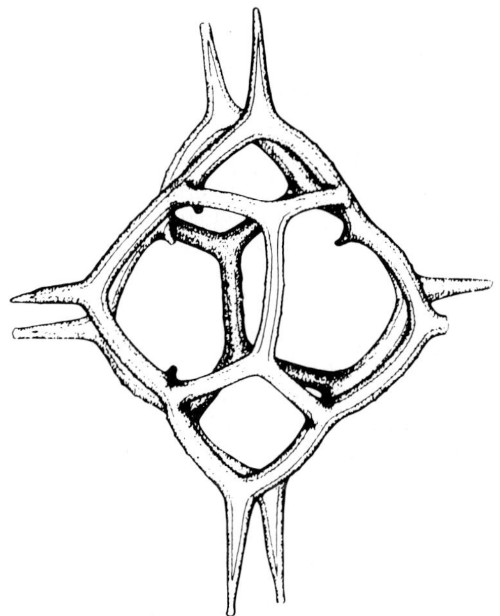

Figure 7.7
Double skeleton of *Dictyocha fibula* Ehrenberg, Miocene, Mediterranean Sea, ×835, from Dumitrică, 1973a.

mere abnormalities. Some variations are so regular and frequent that they are regarded as distinct taxa, although there has been little agreement as to the importance of different skeletal characters for specific or even generic discrimination. As a result of these morphologic differences, many varieties, subspecies, and formae have been proposed for the silicoflagellate species, resulting in numerous taxonomic problems (Loeblich III et al., 1968). Frenguelli (1935) first noted the great variation in a living species in nature, and cultures have also shown that a single clone of *Dictyocha fibula* Ehrenberg produced skeletons comparable to six taxa that were recognized by Gemeinhardt (1930), including *Distephanus speculum, D. speculum* var. *septenarius* (Ehrenberg) Jørgensen, *Dictyocha fibula, D. fibula* var. *pentagona* Schulz, "*D. hexacantha* Schulz" (= *Cannopilus*

hexacantha (Schulz) Deflandre), and "*Cannopilus calyptra* Haeckel" (Van Valkenburg and Norris, 1970). This degree of variation in clonal cultures leaves uncertain the validity of the skeleton for taxonomic differentiation, although perhaps aberrancies are more prevalent in cultures than in nature. As a similar degree of variation is not apparent in fossil taxa, Van Valkenburg (1971a) suggested that perhaps these asexually reproducing organisms are on the way to extinction and have developed a weakness in the timing mechanism that controls the predivision stage, resulting in the high degree of morphologic plasticity in living taxa. During their acme in the Miocene and earlier, species may have had more stability. Skeletal variations also may be seasonal, about 80 percent of the skeletons in the fall being thin and not completely silicified, whereas in January only about 5 percent are thin, and the remainder are much thicker. Cannopiloid forms are scarcer and dictyochoid ones more common in autumn, whereas the reverse is true in winter. Double skeletons with the two basal rings facing are abundant in the fall.

A more unusual variation has been reported in a few specimens, in which a portion of the hollow cavity of the skeleton is locally much swollen into a globular or elongate vesicle, which may have surficial irregularities or even a pore to the exterior. This may be due to a parasite, but as such forms have only been noted in fossil material, the cause cannot be definitely determined.

Silicification occurs sequentially in the silicoflagellates, rather than the entire skeleton being formed at once, and the number of spines and degree of elaboration of the apical apparatus in *Dictyocha* may depend on the time elapsed in skeletal deposition. Evidence from electron micrographs also suggests that any parts of the skeleton that come to touch during growth will fuse at the juncture, and that such junctures are of varied age in a single individual (Van Valkenburg, 1971a).

In addition to the small spines that may be present on the silicoflagellate skeleton, the surface of the skeleton may be smooth or ornamented with a network of crests and depressions (Deflandre, 1940c, 1950a; Bachmann and Keck, 1969; Jerković, 1969a,b, 1971; Lipps, 1970; Wornardt, 1971; Dumitrică, 1973b; McPherson and Ling, 1973; Martini and Müller, 1976). The ornamentation of the surface may consist of long crests with rare ridgelike knots; a netlike reticulation with ridges 0.30 μm across, either localized or covering the entire distal surface; a concentration of knotlike elevations, especially on the apical structure; or knotlike thickenings forming numerous transverse ridges across the bars of the basal ring (see Figure 7.8). Like the degree of silicification of the skeleton itself, the amount of ornamentation may be related to environmental factors or to the life span of the cell, as the surface complexity is most prominent in the adult. Difficult to see in the light microscope, these surface features are apparent with electron microscopy.

A few individuals also have globular expansions or thickenings of the skeleton, particularly at the corners of the basal ring. This feature was made the basis of the genus *Hannaites* Mandra (see Figure 7.9), but similar inflated structures have been found on both *Dictyocha* and *Mesocena* (although less regular in occurrence). Specimens identical in other respects to *Hannaites* occur both with and without these thickenings (Martini and Müller, 1976), those with the bulblike thickenings being common in the southern hemisphere, whereas specimens lacking this feature occur in the equatorial Pacific and Norwegian-Greenland Sea; *Hannaites* thus is a synonym of *Dictyocha*.

The importance to the paleontologist of a knowledge of the living organism and its ontogenetic development is well demonstrated by the silicoflagellates. Incompletely silicified daughter skeletons have been described as belonging to distinct genera and species, both from the modern seas and from the fossil record. Apparently, many mesocenoid forms described as separate species represent incompletely silicified skeletons, with only their basal ring developed. It has been suggested that probably only two or three fossil forms of *Mesocena* represent valid species, and the nearly 30 remaining taxa that have been proposed represent the mesocenoid, incomplete basal rings of various species of *Dictyocha* or *Distephanus*.

The skeleton of the silicoflagellate, like that of many other planktonic organisms, has elongate spines that may increase its resistance to sinking. The asymmetry also results in a preferred orientation in the water that similarly may minimize sinking. Thus silicoflagellates tend to sink with the broad surface perpendicular to the direction of descent, an orientation that not only slows the sinking rate, but also best exposes all plastids to the light (Lipps, 1970). The organisms are most common in regions of upwelling, hence little energy is required to maintain a position in the photic zone.

REPRODUCTION

Only vegetative reproduction, with simple division of the original protoplast and nucleus, is known in the silicoflagellates. Before the nuclear division takes place, the nucleus becomes somewhat enlarged, and the new daughter skeleton is secreted with its basal ring adjacent to that of the parent. The basal ring of the new

---

Figure 7.8

Skeletal surface ornamentation of silicoflagellates, in SEM. **1,2.** *Dictyocha aspera* (Lemmermann) Frenguelli, Miocene, California; **1**, ×1040; **2**, ×3200. **3.** *Deflandryocha cymbiformis* Jerković, ×3600. **4.** *Distephanus schauinslandii* Lemmermann, TEM, ×7200. **3,4**, Tertiary, Yugoslavia, from Jerković, 1971. **5,6.** *Vallacerta hortonii* Hanna, Upper Cretaceous, California, from McPherson and Ling, 1973. **5.** Apical view of plate, ×5440. **6.** Apical view, ×480. **7.** *Corbisema disymmetrica* (Dumitrică) Bukry, lower Paleocene, southwest Pacific, ×352, from Dumitrică, 1973b. **8.** *Corbisema geometrica* Hanna, Upper Cretaceous, California, ×400, from McPherson and Ling, 1973. **9.** *Dictyocha quadria* (Mandra) Martini & Müller, upper Eocene, one angle of skeleton, ×800. **10.** *D. fibula*, lower Oligocene, ×1920. **11.** *Mesocena diodon* Ehrenberg, ×2120. **12.** *M. elliptica* (Ehrenberg) Ehrenberg, ×815. **11,12**, upper Miocene. 9 − 12, Norwegian − Greenland Sea, from Martini and Müller, 1976.

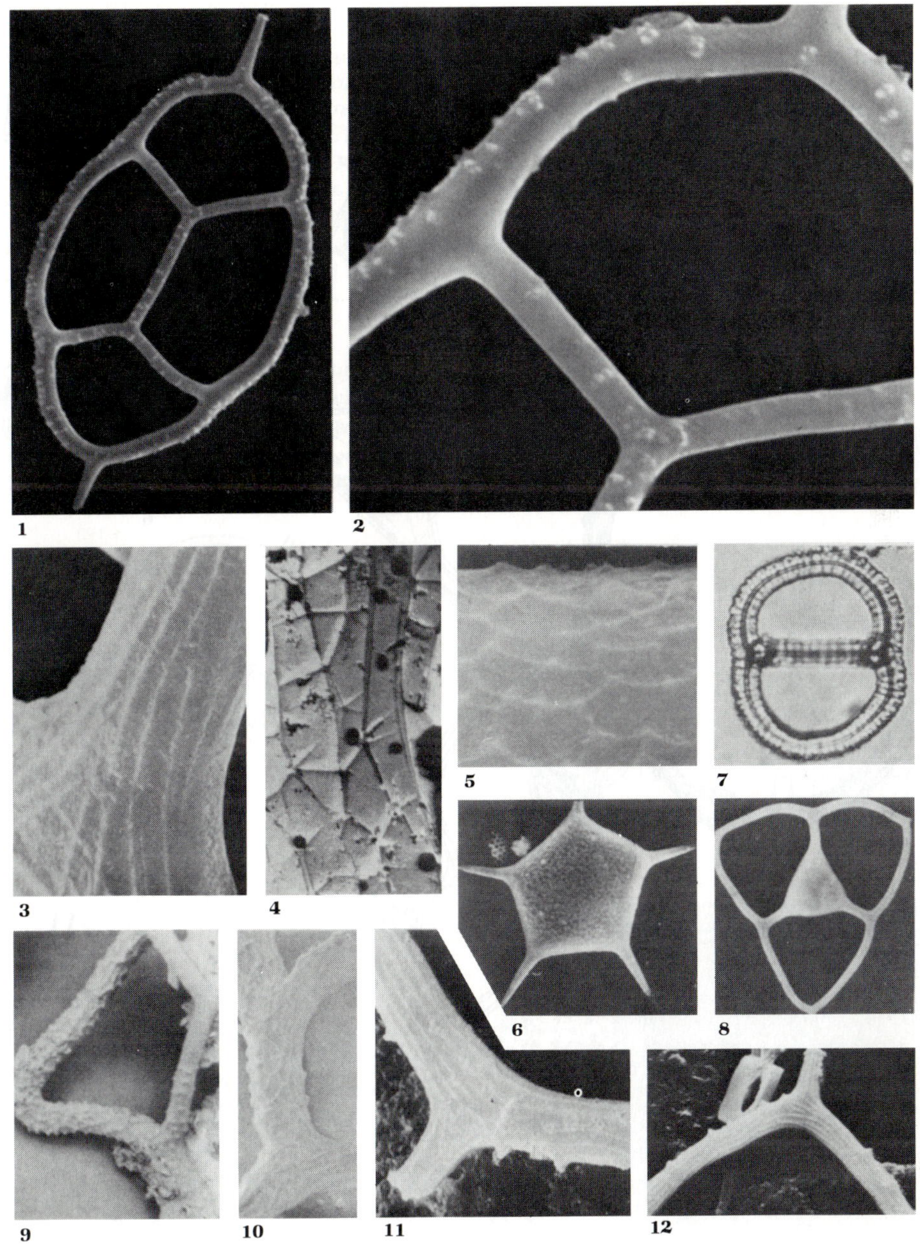

1 2 3 4 5 6 7 8 9 10 11 12

apparently seasonal, corresponding to features of water temperature and stability, as the number of double forms is much greater in autumn, when it may reach 44 percent, than during the remainder of the year (Boney, 1976).

In cultures, under controlled conditions, *Dictyocha* had an average generation time of 49 hours. Optimum growth occurred at 10°C and a salinity of 24°/₀₀ in enriched seawater (Van Valkenburg and Norris, 1970). Although *D. fibula* has been reported to occur in warm waters, with optimum of 18 to 20°C (Gemeinhardt, 1930), it disappears when temperatures exceed 15°C (Travers and Travers, 1968) and in cultures all cells died at 20°C. Reproduction was speeded, and cell counts were higher at 15° than at 10° (Van Valkenburg and Norris, 1970), but fewer cells survived; most shed the skeletons and settled to the bottom of the culture vessel, where they eventually disintegrated.

When cell density reached a certain level in cultures, the individuals suddenly died. Under unfavorable conditions, other cells shed the skeleton, produce a tough periplast, and lose their flagella, settling to the bottom and undergoing nuclear divisions to produce a multinucleate cell (coenocyte) with slow amoeboid movement. Such aberrant forms were produced at a maximum cell density at 10°C, and at any cell density at 15° or when maintained in continuous light. None of the coenocytes were observed to produce a new uninucleate swimming, skeleton-bearing form, hence whether this is a normal stage in the life cycle or an aberrancy resulting from culture conditions unfavorable to these delicate forms is uncertain (Van Valkenburg and Norris, 1970).

Formation of a resting stage or cyst has been reported in *Distephanus speculum* from British Columbia (Wailes, 1927). The double skeleton was said to separate into two halves; the protoplast then formed a resting spore enclosed in a colorless, gelatinous star-shaped investment. The single sketch of this cyst (see Figure 7.11) resembles the structure of *Actiniscus,* and perhaps may represent such a mistakenly identified form. No other report of such a development is known in the silicoflagellates.

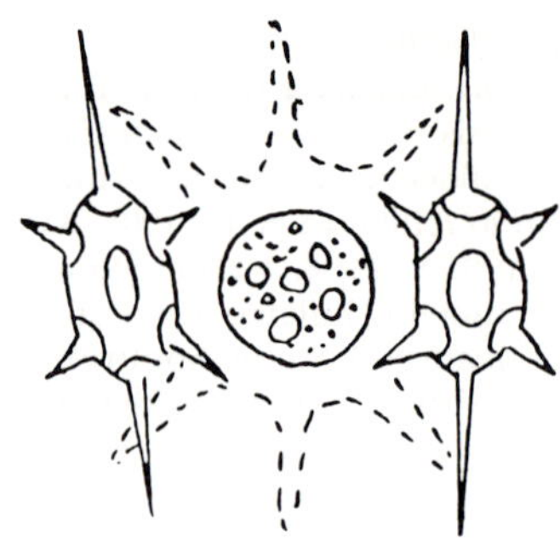

Figure 7.11
Formation of presumed "cyst" in *Distephanus speculum* var. *regularis* Lemmermann, from British Columbia; specimen reported to separate into two halves, and produce a resting spore enclosed in colorless, gelatinous stellate covering (dashed line); possibly a specimen of *Actiniscus* (dinoflagellate) was mistaken for the silicoflagellate; ×800, from Wailes, 1927.

OCCURRENCE

Ecology

Although seldom present in great abundance, silicoflagellates are widely distributed in the present oceans. All are marine, requiring a salinity of over 20°/₀₀. Because of their small size, not all can be obtained in the ordinary plankton net, and one of smaller mesh must be used. They are not restricted to the open ocean, as some may be found among the marine algae in coastal waters.

Physical Environment

The number of living species is small, probably less than half a dozen. Most abundant are *Dictyocha fibula, D. navicula* Ehrenberg, *Distephanus octonarius,* and *D. speculum.* Although silicoflagellates are known in all latitudes, individual species distribution seemingly is temperature controlled. The optimum temperature for the characteristically cold water species *D. speculum* is near 0°C, whereas *D.*

octonarius is abundant in temperatures between 12 and 20°C, diminishing rapidly in numbers below 12° and absent in waters from 2 to 5°C.

Dictyocha fibula occurs in warmer waters, although in the Aegean Sea maximum growth occurred at the coldest local temperatures (14 to 15°C) in March, and the species was rare from May through December, when temperatures were between 16 and 23°C (Ignatiades, 1970). Cell concentrations did not correlate with nutrient supply or salinity, but only with temperature, cell counts ranging from 20 to 600 cells per liter.

In the bay of Villefranche-sur-Mer, on the Mediterranean coast of France, *Dictyocha fibula* was present from January to March in concentrations of 4000 to 5000 cells per liter, was absent in summer, and rare in the fall. This occurrence also was related to the temperature and vertical circulation of the water, the species being absent during the warmer period and when the water was stratified with little vertical movement (Nival, 1965). In the Mediterranean Sea between Sardinia and Tunisia, cells numbered 1000 per liter at temperatures below 13.5°C (Coste et al., 1969), and during the period of maximum productivity in typical Atlantic water, the daily silicoflagellate carbon fixation was about 10 mg C per cubic meter.

As is true of diatoms, silica may be a limiting requirement of the silicoflagellates, and its availability in cold and upwelled waters may allow silicoflagellate development. Species diversity in the Cenozoic was suggested to correlate with fluctuations of temperature, intensity of circulation, and upwelling (Lipps, 1970). Thus when warmer temperatures prevailed at high latitudes, as in the Paleocene and Oligocene, upwelling was reduced and the silica-depositing plankton was of limited abundance and diversity. When high latitudes became colder and active circulation was restored, as in the Eocene and particularly during the Miocene, these organisms flourished. Although the cooling during the latest Cenozoic resulted in increased diatom abundance and diversification, silicoflagellates have not shown a similar increase, but have continued their post-Miocene decline.

Although less diverse, and generally less abundant at present than such other phytoplankton as diatoms, dinoflagellates, and coccolithophorids, silicoflagellates may temporarily be present in large numbers. Concentrations of about 50,400 cells per liter have been reported in the Oslo Fjord, and a peak of 109,000 per liter was reported in Long Island Sound in April 1952 (Conover, 1956). Much less information is available for these tiny forms than for the net plankton because of the increased difficulty of obtaining quantitative records. Silicoflagellates are an important source of food for certain organisms, and many have been obtained from copepod excrement and the guts of tunicates (*Salpa*) and echinoderm larvae. Many species were originally described from guano deposits, hence were the base of the food chain of the fish-eating birds. Specimens obtained from these various predators include many partly silicified daughter skeletons as well as double skeletons, representing individuals that were devoured while undergoing cell division.

Geologic Record

Abundance and Silica Deposition

Presently found in all seas, the silicoflagellates probably were also cosmopolitan in former times. Their more restricted occurrence in Mesozoic and Cenozoic rocks is in part due to selective preservation, as the opaline silica of which the skeleton is constructed is a relatively unstable mineral and dissolves in a basic (alkaline) medium.

Silicoflagellates occur in siliceous rocks, particularly in diatomites, their numbers increasing in the more strictly pelagic sediments. In some regions, silicoflagellates are sufficiently abundant as to comprise a large proportion of the rock, as much as 40 to 60 percent of the rock mass (Distanov and Glezer, 1974). Where they are more abundant than the diatoms, the rocks have been named as a separate rock type. First termed a silicoflagellithite (Filipescu,

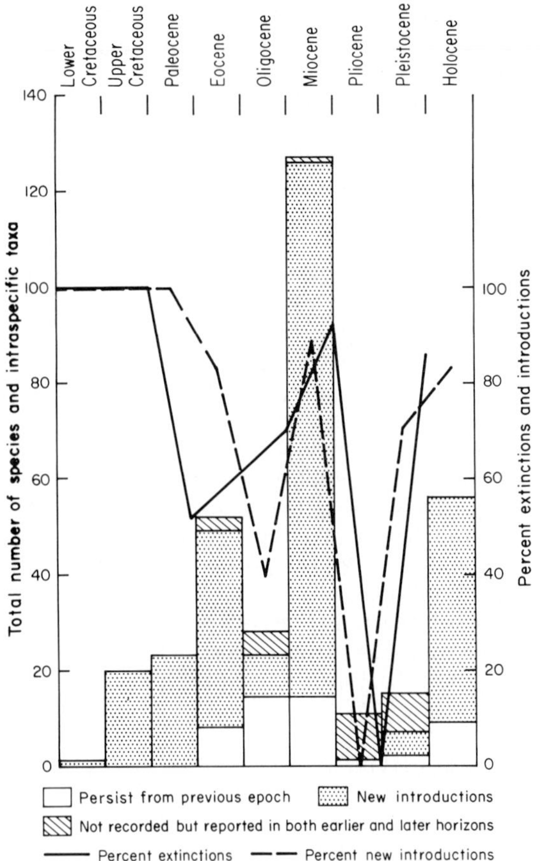

Figure 7.12
Species diversity of the silicoflagellates, showing appearance of new species exceeding extinctions, to result in a gradual increase in diversity through the Cretaceous, Paleocene, and Eocene, then an Oligocene decrease, and a major diversifiction in the Miocene, followed by an even more rapid decline. Recent studies of DSDP material have increased the total number of known species, but the time and pattern of major changes is not affected. From Tappan and Loeblich, 1972.

1943), or silicoflagellitite (Filipescu, 1968), in the Miocene of Rumania and for similar occurrences in Yugoslavia (Jerković, 1965) and the USSR (Proshkina-Lavrenko, 1959), the rock name later was modified to **silicoflagellite** (Distanov and Glezer, 1974).

Silicoflagellates first appeared in the middle Cretaceous and became more numerous in the Late Cretaceous, having been reported in California, Wyoming, Colorado, Germany, Denmark, Hungary, and Russia (see Figure 7.12). Like other microplankton they were low in diversity during the Paleocene, but became more abundant in the Eocene, when some 40 species are known in California, Barbados, Germany, Russia, and New Zealand. Oligocene diversity declined, as extinctions were not matched in

number by newly evolved taxa; the records are scarce, but include France, Austria, and Siberia. The silicoflagellate acme was attained in the Miocene; over 100 species and varieties have been described from Miocene rocks, the high diversity resulting from the evolution of many new species. Miocene silicoflagellate-bearing deposits occur in California, Maryland, New Jersey, and Virginia, Mexico, Chile, Japan, Algeria, Spain, France, Italy, Greece, Austria, Hungary, Rumania, Yugoslavia, and the USSR. Silicoflagellates are present in the Miocene, Pliocene, and Pleistocene in many areas of the Pacific and Antarctic Oceans, as has been shown in the DSDP (see following section on biostratigraphy). Elsewhere, they have been reported from the Pleistocene of Poland and from

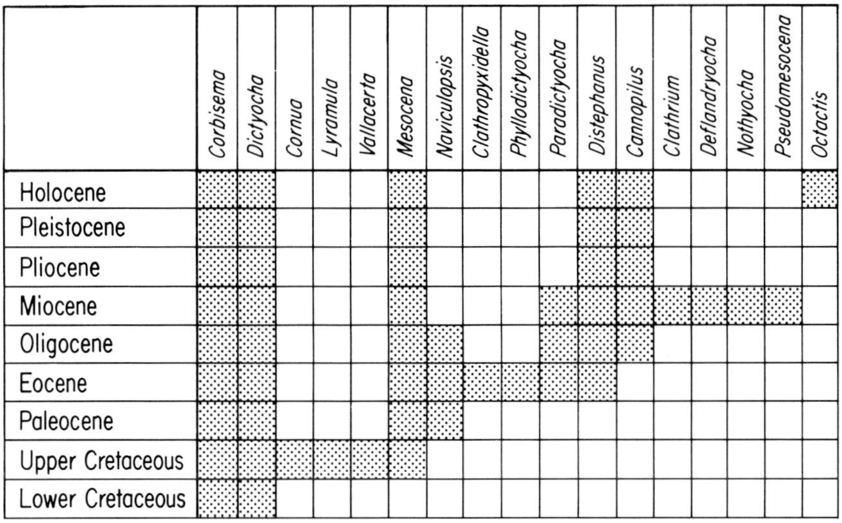

	Corbisema	Dictyocha	Cornua	Lyramula	Vallacerta	Mesocena	Naviculopsis	Clathropyxidella	Phyllodictyocha	Paradictyocha	Distephanus	Cannopilus	Clathrium	Deflandryocha	Nothyocha	Pseudomesocena	Octactis
Holocene	■	■				■					■	■					■
Pleistocene	■	■				■					■	■					
Pliocene	■	■				■					■	■					
Miocene	■	■				■				■	■	■	■	■	■	■	
Oligocene	■	■				■	■			■	■	■					
Eocene	■	■				■	■	■	■	■	■						
Paleocene	■	■				■	■										
Upper Cretaceous	■	■	■	■	■	■											
Lower Cretaceous	■	■															

Figure 7.13
Silicoflagellate generic ranges through geologic time.

guano deposits in Chile and southern Africa (Loeblich III et al., 1968; Tappan and Loeblich, 1971, 1973; see Figure 7.13).

Biostratigraphy

As noted in the quotation at the head of the present chapter, silicoflagellates were utilized for intercontinental correlation and age dating within five years of the first silicoflagellate description in the early part of the nineteenth century. Silicoflagellate-bearing rocks are relatively uncommon, however, and most records were from isolated exposures and discontinuous sections, commonly of doubtful stratigraphic position. In addition, much of the earlier literature is scattered in relatively inaccessible publications. Compilation of a comprehensive index of taxa and the reproduction of all original descriptions and illustrations to that date (Loeblich III et al., 1968) greatly facilitated the study of this group. The availability of continuous sections from ocean cores, through the auspices of the DSDP beginning in 1969, has made possible their renewed stratigraphic utility. Biostratigraphic zonations have been proposed for both high and low latitudes (see Figures 7.14 and 7.15) and have made possible the exact correlation of certain of the earlier described land exposures (Barbados, Denmark, New Zealand). Slightly different zonations have been utilized in the various oceans and at different latitudes, but their equivalence has been documented by reference to other planktonic organisms, particularly the calcareous nannoplankton and planktonic foraminifera.

Characteristic zone fossils for the Pleistocene in cold waters include *Dictyocha fibula* Ehrenberg var. *aculeata* Lemmermann (non *D. aculeata* Ehrenberg) = *D. epiodon* Ehrenberg; *Distephanus octangulatus* Wailes; *D. octogonus* (Ehrenberg) Dumitrică; *D. octonarius; D. speculum;* and *D. subarctios* Ling. Warm water Pleistocene species include *Mesocena elliptica* (Ehrenberg) Ehrenberg, and *M. quadrangula* Ehrenberg ex Haeckel, as well as *D. epiodon.*

Zonal fossils for the Pliocene in high latitudes include *Cannopilus hemisphaericus* (Ehrenberg) Haeckel; *Dictyocha aspera* var. *pygmaea* Ciesielski; *D. boliviensis* Frenguelli (a synonym of *Distephanus heptacanthus* (Ehrenberg) Dumitrică); *D. pseudofibula* (Schultz) Tsu-

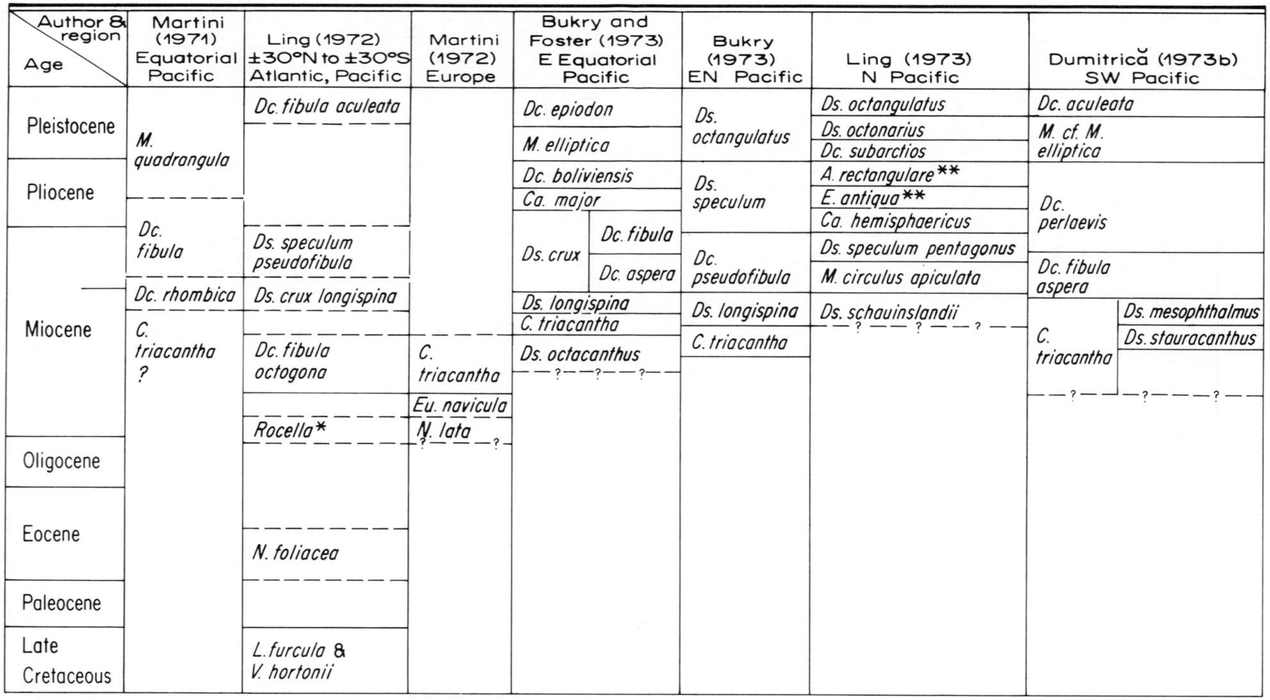

Author & region / Age	Martini (1971) Equatorial Pacific	Ling (1972) ±30°N to ±30°S Atlantic, Pacific	Martini (1972) Europe	Bukry and Foster (1973) E Equatorial Pacific		Bukry (1973) EN Pacific	Ling (1973) N Pacific	Dumitrică (1973b) SW Pacific	
Pleistocene	M. quadrangula	Dc. fibula aculeata		Dc. epiodon		Ds. octangulatus	Ds. octangulatus	Dc. aculeata	
				M. elliptica			Ds. octonarius	M. cf. M. elliptica	
							Dc. subarctios		
Pliocene	Dc. fibula			Dc. boliviensis		Ds. speculum	A. rectangulare**	Dc. perlaevis	
				Ca. major			E. antiqua**		
							Ca. hemisphaericus		
		Ds. speculum pseudofibula		Ds. crux	Dc. fibula	Dc. pseudofibula	Ds. speculum pentagonus	Dc. fibula aspera	
					Dc. aspera		M. circulus apiculata		
Miocene	Dc. rhombica	Ds. crux longispina		Ds. longispina		Ds. longispina	Ds. schauinslandii		Ds. mesophthalmus
	C. triacantha ?			C. triacantha		C. triacantha	? — ? — ?	C. triacantha	Ds. stauracanthus
		Dc. fibula octogona	C. triacantha	Ds. octacanthus					
				— ? — ? — ? —				— ? — ? —	
			Eu. navicula						
		Rocella*	N. lata ?						
Oligocene									
Eocene		N. foliacea							
Paleocene									
Late Cretaceous		L. furcula & V. hortonii							

Figure 7.14

Silicoflagellate biostratigraphy. Zonations proposed in various regions and oceans. Silicoflagellate genera abbreviated as *Ca, Cannopilus; C, Corbisema; Dc, Dictyocha; Ds, Distephanus; Eu, Eunaviculopsis; L, Lyramula; M, Mesocena; N, Naviculopsis; V, Vallacerta. Rocella* (*) is a diatom, and *Ammodochium* (*A*) and *Ebriopsis* (*E*) (both marked **) are ebridians that have been utilized in these zonations.

Age	Bukry (1974) nontropical	Martini (1974) Europe	Bukry and Foster (1974) deep sea	Perch-Nielsen (1975) SSW Pacific	Ciesielski (1975) Antarctic	Martini and Müller (1976) Norwegian Sea and Greenland Sea
Pleistocene	*Ds. octangulatus*			*Dc. aculeata*	*Ds. speculum A*	
	Ds. octonarius			*Ds. octogonus*		
Pliocene	*Ds. speculum*			*P. dumitricae*	*Ds. speculum B* / *Dc. boliviensis* / *Dc. aspera pygmaea* / *Dc. pseudofibula*	*Ds. speculum*
				— — — *M. diodon* — — —		*Dc. boliviensis*
Miocene	*Dc. pseudofibula*			*Ds. crux*	*M. diodon*	
	Ds. longispina		— —?— —?— —		*M. circulus*	*M. circulus*
	C. triacantha		*Ds. octacanthus*	*N. biapiculata*	*C. triacantha*	*C. triacantha*
	N. quadrata	*Eu. navicula*	*N. quadrata*	*N. robusta*	*Eu. navicula* / *N. regularis* / *N. robusta*	*Eu. navicula*
	Ds. speculum pentagonus	*N. lata*	*R. gemma**	*R. gemma**		*N. lata*
Oligocene	*N. biapiculata*	*N. biapiculata*	*Dc. deflandrei*	*C. spinosa*	*N. biapiculata*	*N. biapiculata*
	Dc. deflandrei → [*Dc. frenguellii* / *M. apiculata* / *N. trispinosa*]			*Dc. medusa*	*Dc. deflandrei*	
Eocene	*Dc. hexacantha*	*Dc. bimucronata*	*Dc. hexacantha*	*C. hastata, C. apiculata*		*Dc. quadria* / *C. bimucronata*
	N. constricta	*N. foliacea, N. minor, Dc. transitoria, Dc. naviculoidea, Dc. deflandrei*	*N. constricta*	*C. hexacantha*		*N. foliacea* / *N. minor* / *Dc. transitoria*
Paleocene	*C. hastata*		*C. hastata*			
Late Cretaceous	*L. furcula*		*L. furcula*	*L. furcula*		

Figure 7.15

Silicoflagellate biostratigraphy. Zonations proposed in various regions and seas. Generic abbreviations as for Figure 7.14, with the addition of *P, Paradictyocha. Dictyocha frenguellii* Subzone (Bukry, 1974) is now *D. fischeri* Bukry Subzone, as the zone marker is not = *D. frenguellii* Deflandre.

mura (syn.: *Distephanus speculum forma pseudofibula*); *Distephanus speculum;* and *Paradictyocha dumitricae* Perch-Nielsen. Warm water Pliocene zonal markers include *Cannopilus major* (Frenguelli) Bukry & Foster; *Dictyocha fibula;* and *D. perlaevis* Frenguelli.

Zone fossils for the Miocene in various regions are *Corbisema triacantha* (Ehrenberg) Hanna; *Dictyocha aspera* (also cited by some as *D. fibula* var. *aspera*); *D. pseudofibula; D. rhombica* (Schulz) Deflandre; *Distephanus crux* (Ehrenberg) Haeckel; *D. longispina* (Schulz) Bukry & Foster (syn.: *D. crux* f. *longispina*); *D. mesophthalmus* (Ehrenberg) Haeckel; *D. octacanthus* (Desikachary & Maheshwari) Bukry & Foster; *D. quinquangellus* Bukry & Foster (syn.: *D. speculum* var. *pentagonus* Lemmermann, 1901, non *D. pentagonus* Wailes 1939); *D. schauinslandii; D. stauracanthus* (Ehrenberg) Haeckel (syn.: *D. fibula* var. *octogona* Tsumura); *Eunaviculopsis navicula* (Ehrenberg) Ling; *Mesocena circulus* (Ehrenberg) Ehrenberg; *M. circulus* var. *apiculata* Lemmermann; *M. diodon; Naviculopsis biapiculata* (Lemmermann) Frenguelli; *N. lata* (Deflandre) Frenguelli; *N. quadrata* (Ehrenberg) Locker; *N. regularis* (Carnevale) Ling; and *N. robusta* Deflandre.

Oligocene silicoflagellate zonal markers include *Corbisema spinosa* Deflandre; *Dictyocha deflandrei* Frenguelli ex Glezer; *D. fischeri* (as *D. frenguellii*); *D. medusa* Haeckel; *Mesocena apiculata* (not the same as *M. circulus* var. *apiculata* Lemmermann); *Naviculopsis biapiculata;* and *N. trispinosa* (Schulz) Glezer.

Silicoflagellate zonal markers for the Eocene are *Corbisema apiculata* (Lemmermann) Hanna; *C. bimucronata* Deflandre (also as *Dictyocha bimucronata*); *C. hastata* (Lemmermann) Frenguelli; *C. hexacantha* (also as *Dictyocha hexacantha*); *Dictyocha deflandrei; D. "naviculoidea"; D. quadria; D. transitoria* Deflandre; *Naviculopsis constricta* (Schulz) Frenguelli; *N. foliacea* Deflandre; and *N. minor* (Schulz) Frenguelli.

Paleocene silicoflagellate zone species are *Corbisema hastata* and *C. hexacantha,* both present also in the Eocene; Cretaceous zone species are *Lyramula furcula* Hanna and *Vallacerta hortonii.*

Paleotemperature Interpretation

The known temperature preferences of living silicoflagellates have been applied to the study of paleotemperatures, particularly with respect to the abundance ratios of the genera *Dictyocha* and *Distephanus* (Mandra, 1969b; Mandra and Mandra, 1972; Jendrzejewski and Zarillo, 1972; Cornell, 1974b; Ciesielski and Weaver, 1974; Poelchau, 1976), and the modern surface-water temperatures corresponding to these ratios (see Table 7.1). Major changes in the ratios correspond to distinct oceanographic boundaries, and thus allow recognition of distinct water masses and their distribution through the later Cenozoic (Miocene to present). Thus, Pliocene temperatures in the Antarctic surface waters appear to have been as much as 10°C higher than at present (Ciesielski and Weaver, 1974).

Table 7.1
Surface water temperatures as indicated by the *Dictyocha/Distephanus* ratios.

Ratio	Surface temperature (°C) indicated	
	Mandra and Mandra, 1972	Ciesielski and Weaver, 1974
0	4°	1 − 1.5°
0.10 − 0.50	4 − 15°	2.5 − 5°
0.5 − 1.0	15 − 21°	5 − 7°
1.0 − 3.5	21 − 25°	7 − 10°
>3.5	25°	10 + °

Certain trends in silicoflagellate evolution have been observed from a study of the fossil representatives, but the same trends also may occur as skeletal variations in modern populations. One of these trends is toward the loss of the basal ring; the resultant *Cornua* type of skeleton is particularly characteristic in the Cretaceous. A second trend was noted in the Tertiary species, toward the loss of the apical apparatus and reduction of the skeleton to the basal ring only, as in *Mesocena*. A progressive subdivision of the apical region into numerous apical windows instead of a single opening, by the addition of supplementary bars, results in double, triple, or multiple apical windows, and eventually leads to the genus *Cannopilus* (see Figure 7.16), while an opposite trend toward simplification of the apical region results in loss of the apical window and its replacement

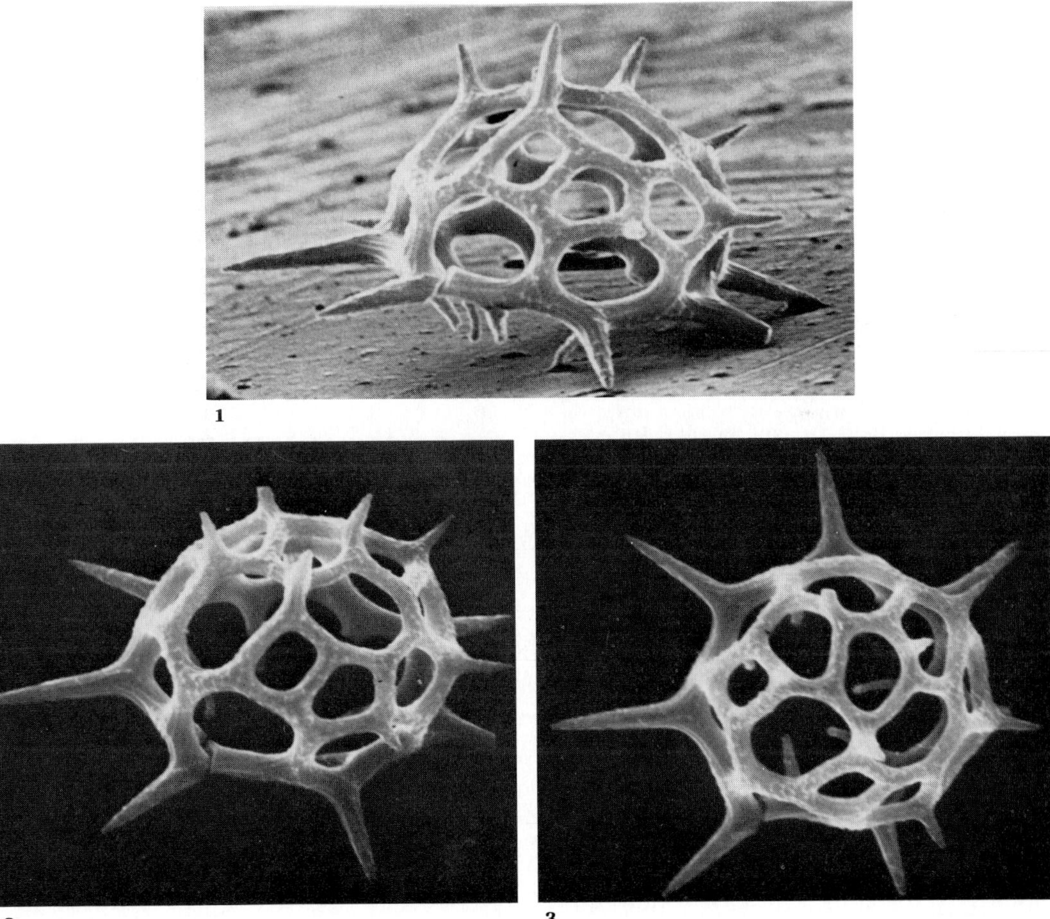

Figure 7.16
Cannopilus depressus (Ehrenberg) Locker, Miocene (Luisian), California. **1.** Side view. **2.** Oblique view. **3.** Apical view. All in SEM, showing numerous bars and windows, spines radiating from basal ring, vertical spines at junction of bars in apical region, and slightly oblique smaller spines on adapical side of basal ring that are inclined toward the center; surface of basal ring and spines with elongate striae; surface of apical structure reticulate; ×1100.

merely by an arched bar, as in the genus *Naviculopsis,* where the apical reduction is also accompanied by a simplification of the basal ring.

A different phylogenetic scheme suggests an increasing complexity of development, the simple U-shaped *Lyramula* of the Cretaceous giving rise to successively more complex forms, first to *Cornua,* and thence to *Dictyocha.* Gradual loss of one angle leads from the quadrate *Dictyocha* to the elongate *Naviculopsis* in the Eocene, and a different lineage showed an increase in number of sides and in size of the apical ring, leading from *Distephanus* to *Paradictyocha.* The triangular forms lasted from the Cretaceous until Miocene time and now appear to be extinct.

Continued search for additional species in the older rocks, particularly in Cretaceous strata, may aid in determining more clearly the trends of evolution in these tiny forms.

CLASSIFICATION

Although regarded only as of family status in many earlier publications and some recent treatises, the silicoflagellates are currently separated as a distinct order, whether regarded as protozoans or as algae. The Order Silicoflagellida, of protozoologists, is placed in the Class Phytomastigophorea, Superclass Mastigophora, Subphylum Sarcomastigophora, and Phylum Protozoa. Phycologists recognized the Order Dictyochales, but the Order Vallacertales Glezer 1966 has priority over the Order Dictyochales Christensen 1967 if the entire group is placed in a single order as is usual.

The silicoflagellates are plantlike in mode of nutrition, possessing numerous plastids, and are like the Chrysophyta in pigment composition and in possessing both a flagellum and pseudopodia. Many other chrysophytes secrete a siliceous cyst or lorica, although the silicoflagellates are alone in possessing the hollow tubular internal skeleton.

The classification within the order, as well as the distinction of genera and species of both living and fossil representatives, has been based solely on their skeletal characters, and at times skeletal variations occurring within a single population have been described as distinct taxa. Both ontogenetic studies and population statistics have been utilized occasionally to determine which characters are genetically controlled and thus useful for taxonomic purposes, and which were individual nonheritable variations.

Presently recognized families and genera are indicated in Table 7.2, although some specialists recognize more and others fewer than indicated here.

Table 7.2
Classification of the silicoflagellates.

Division Chrysophyta Pascher 1914	
Class Chrysophyceae (Pascher) Fritsch in West 1927 Order Vallacertales Glezer 1966 Syn.: Order Silicoflagellata Borgert 1890; Order Siphonotestales Lemmermann 1901; Order Dictyochales Christensen 1967	**1.** Family Dictyochaceae Wallich 1865 (see Figures 7.17 to 7.20) Siliceous skeleton, consisting of a basal ring, and an apical ring or area, or reduced to basal ring alone; skeletal framework typically tubular, but may have rare partitions, or may be *(continued)*

Figure 7.17
Dictyochaceae. **1,2.** *Dictyocha stapedia* Haeckel. **1.** Apical view of specimen with apiculate apical bar, SEM, ×1428. **2.** Apical view of specimen with almost smooth apical bar, SEM, ×1300. **3,4.** *D. fibula,* showing varying amount of apiculation. **3.** Apical view, ×1100. **4.** Adapical view of more elongate specimen, ×1300. **5.** *Distephanus crux,* apical view showing basal ring with subparallel ridges and reticulate surface of apical ring, ×2000. All from the Miocene of California.

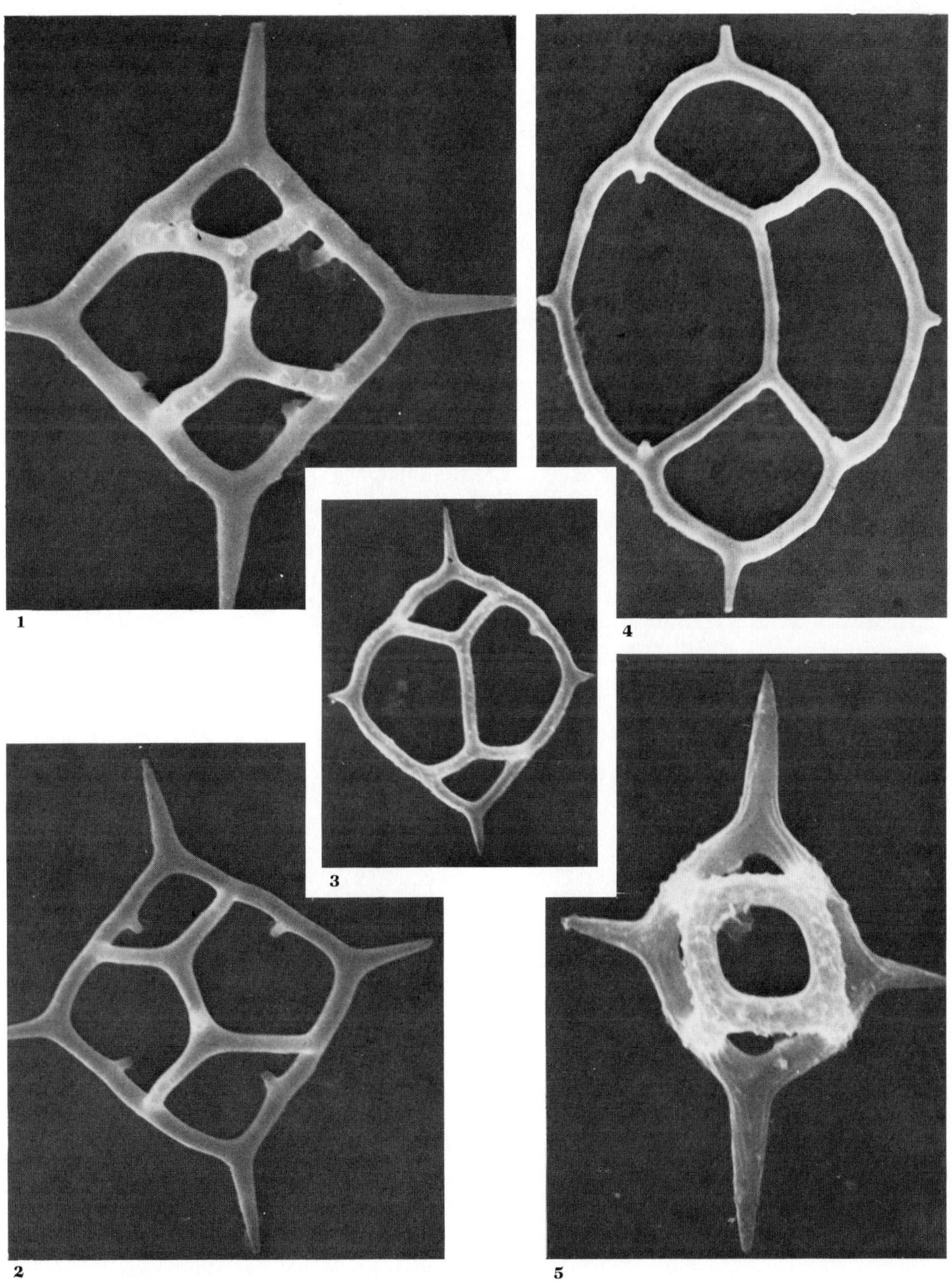

1
2
3
4
5

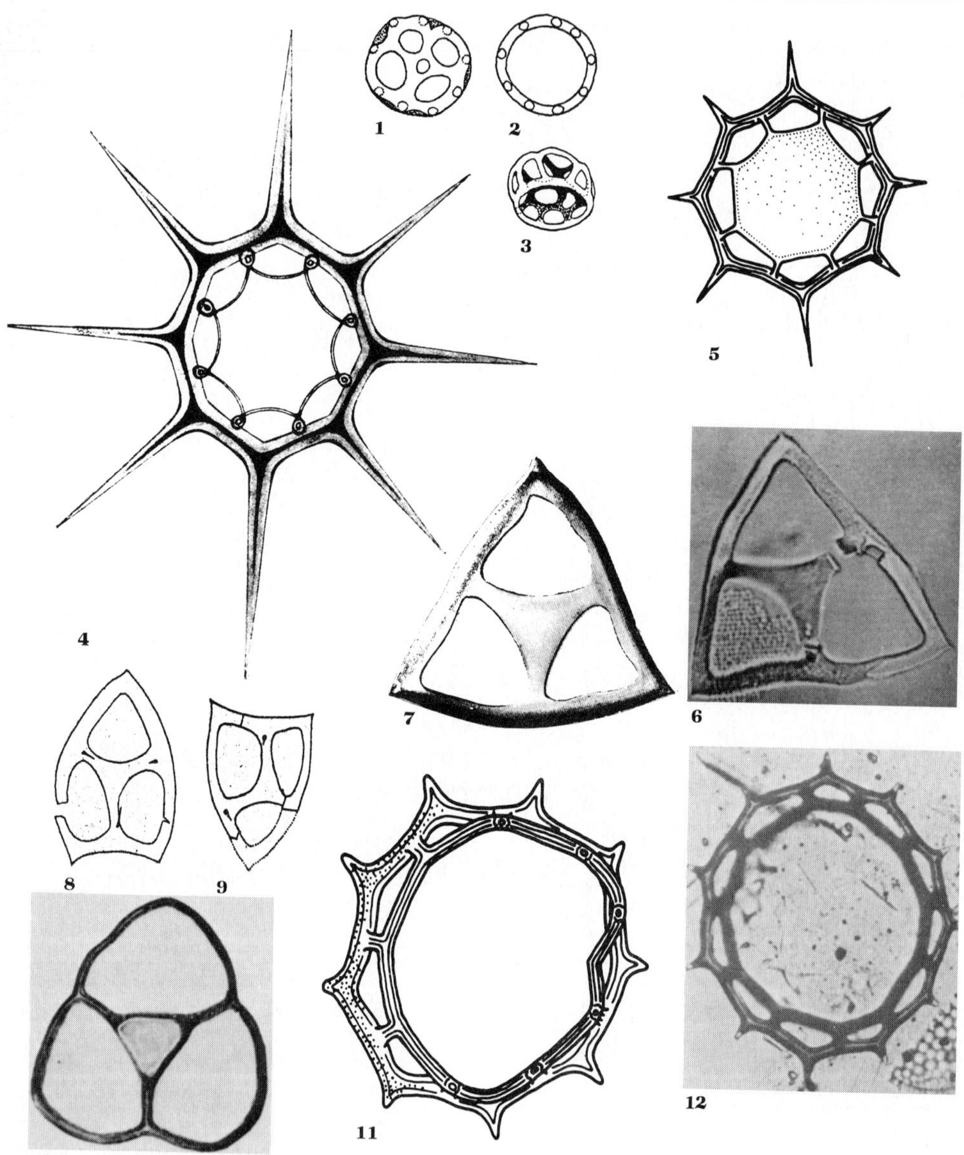

Figure 7.18

Dictyochaceae. **1–3.** *Clathropyxidella similis* Deflandre, apical view, basal ring, and oblique view, Eocene, Springfield, Barbados, West Indies, ×1736, from Deflandre, 1938. **4.** *Octactis pulchra* Schiller, Recent, Adriatic Sea, ×960, from Schiller, 1925. **5.** *Octactis ?bioctonaria* (Ehrenberg) Locker, showing apical membrane; lectotype, from fossil guano, Quaternary, Arica, Chile; ×800, from Locker, 1974. **6,7.** *Phyllodictyocha recta* (Schulz) Deflandre forma *lateradiata* Tsumura, upper Eocene, Springfield, Barbados, ×745, from Tsumura, 1963. **8,9.** *P. schulzii* Deflandre, ×500, from Deflandre, 1947; 8, upper Eocene, Springfield, Barbados; 9, upper Miocene, Hungary. **10.** *Corbisema geometrica*, Upper Cretaceous, California, ×420, from Hanna, 1928. **11,12.** *Paradictyocha polyactis* (Ehrenberg) Frenguelli. **11.** Holotype, Miocene from Caltanisetta, Sicily, ×800, illustrated by Locker, 1974. **12.** Middle Miocene, southwest Pacific, ×400, from Dumitrică, 1973b.

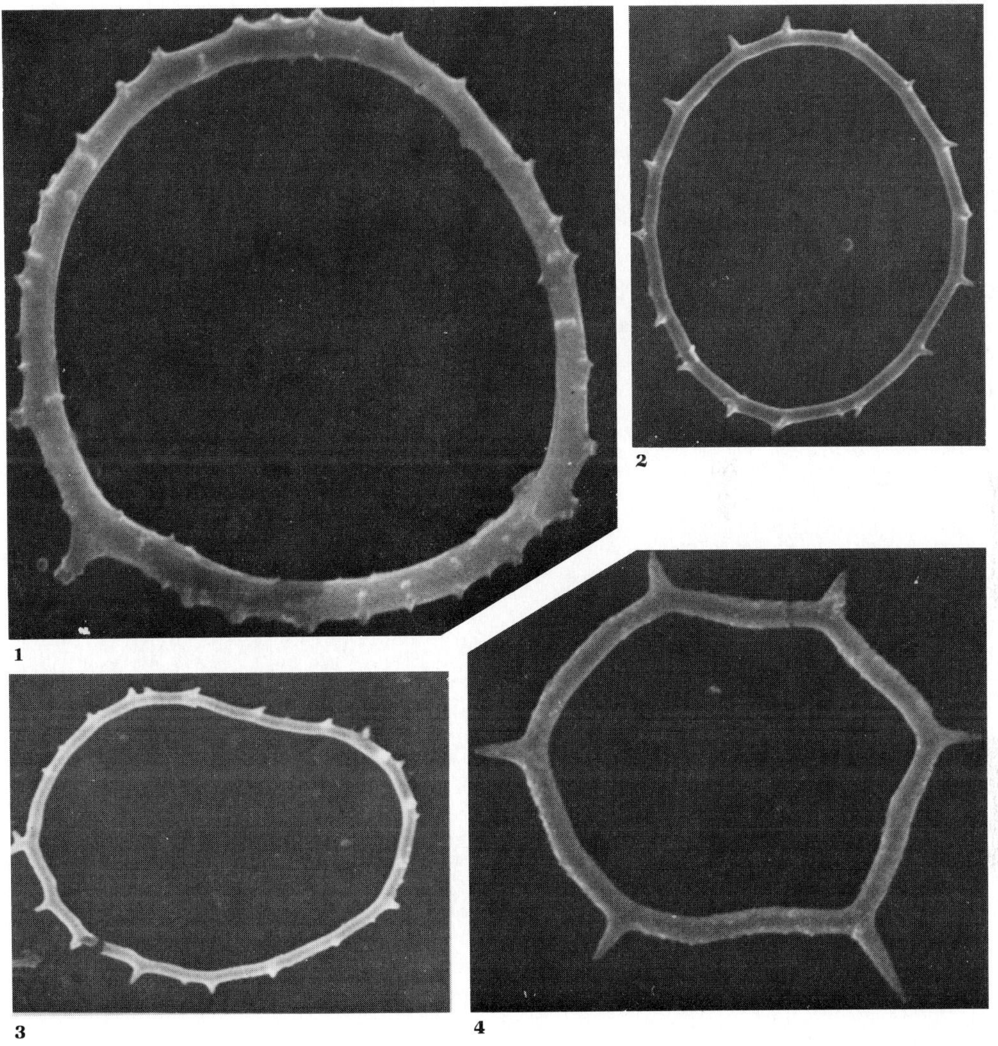

Figure 7.19
Dictyochaceae. **1.** *Mesocena circulus*, SEM, ×1690. **2,3.** *M. circulus* var. *apiculata*, SEM, ×562. **4.** *M. hexagona* Haeckel, SEM ×1000. All from the Miocene of California.

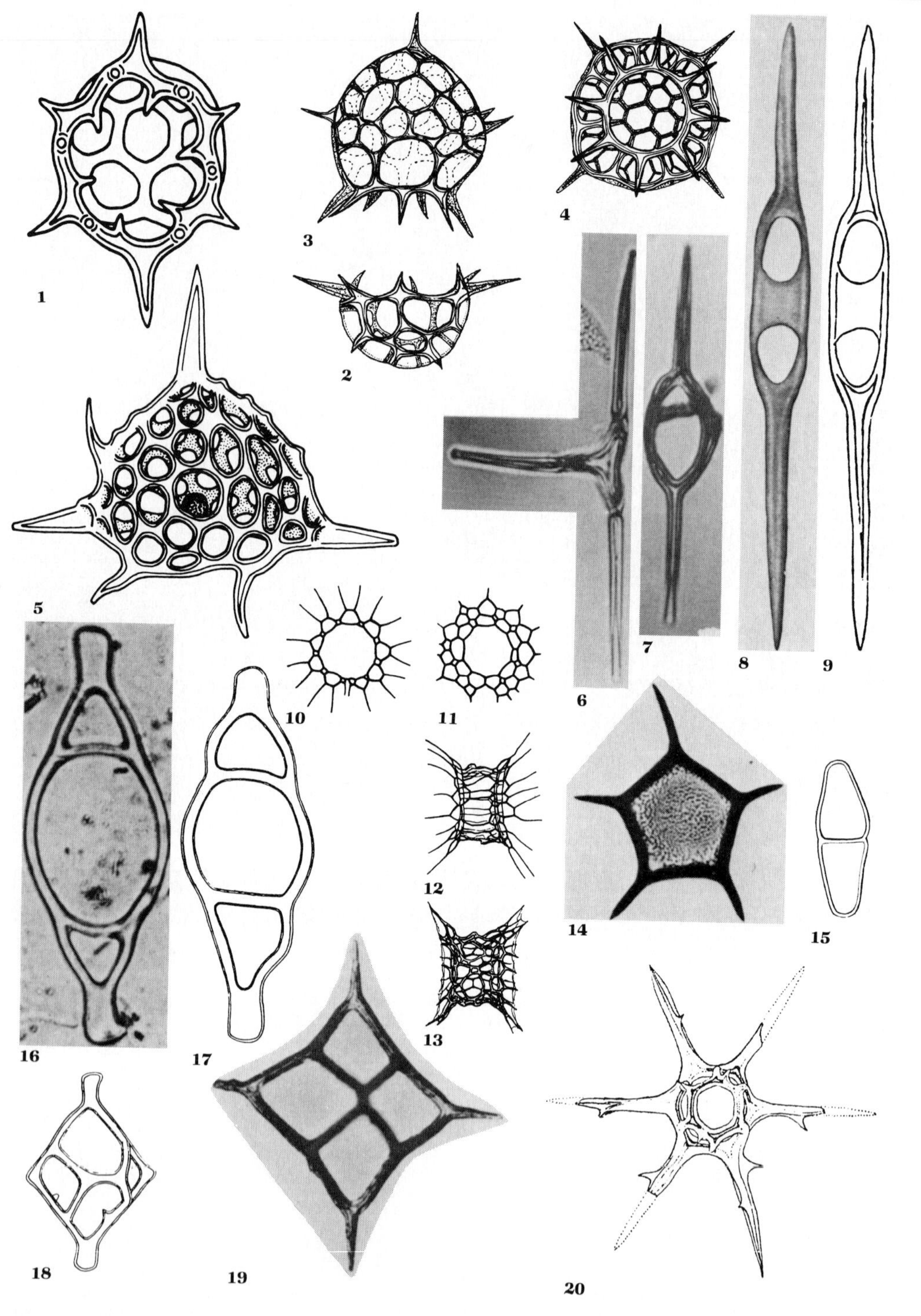

compressed or solid. Includes as synonyms the families Corbisemaceae Locker 1974 and Distephanaceae Locker 1974; as defined by Locker, these families each included genera upon which prior family names had been based, hence were superfluous.

Lower Cretaceous to Holocene; marine, planktonic.

Cannopilus Haeckel 1887; *Clathropyxidella* Deflandre 1938; *Corbisema* Hanna 1928; *Deflandryocha* Jerković 1963; *Dictyocha* Ehrenberg 1837 (syn.: *Hannaites* Mandra 1969); *Distephanus* Stohr 1880; *Eunaviculopsis* Ling 1977; *Halicalyptra* Ehrenberg 1847; *Mesocena* Ehrenberg 1843 (syn.: *Bachmannocena* Locker 1974; *Septamesocena* Bachmann 1970); *Naviculopsis* Frenguelli 1940; *Nothyocha* Deflandre 1949; *Octactis* Schiller 1925; *Paradictyocha* Frenguelli 1940; *Phyllodictyocha* Deflandre 1947; *Pseudomicromarsupium* Busen & Wise 1977.

2. Family Vallacertaceae Deflandre 1950

Siliceous skeleton hollow and tubular; reduced to elements of the apical apparatus, or with portions of the basal ring.

Upper Cretaceous; marine, planktonic.
Vallacerta Hanna 1928.

3. Family Cornuaceae Gemeinhardt 1930 (see Figure 7.21).

U- or Y-shaped vestiges of apical ring, or branching, with spines; may have vestiges of basal ring.

Upper Cretaceous; marine, planktonic.
Cornua Schulz 1928.

4. Family Lyramulaceae Tsumura 1963

Skeleton triradiate or branched, reduced to elements of the apical ring, or reduced to U-shaped vestiges.

Upper Cretaceous; marine, planktonic.
Lyramula Hanna 1928.

Genera that have been erroneously ascribed to the silicoflagellates include *Actiniscus* Ehrenberg 1843 (dinoflagellate), *Lutetianella* Filipescu 1943 and *Microdistephanus* Filipescu 1943 (both = *Actiniscus*); *Pseudomesocena* Hovasse 1932 (an incomplete radiolarian fragment); *Pseudorocella* Deflandre 1938; and *Rocella* Hanna 1930. *Pseudorocella* is a junior synonym of *Macrora* Hanna 1932, described as a diatom, but here placed in the Chrysophyta, Order Ochromonadales (see Chapter 6). The type species of *Rocella* is a synonym of *Stictodiscus gelidus* Mann, a diatom (see Chapter 8).

Figure 7.20

Dictyochaceae, Vallacertaceae. **1,2.** *Cannopilus hemisphaericus.* **1.** Lectotype, Miocene, Maryland, basal view, ×800 (showing possible relationship to *Distephanus* as suggested by Bukry, 1976b), from Locker, 1974. **2.** Miocene, Hungary, side view, ×600. **3,4.** *C. sphaericus* Gemeinhardt, Miocene, Hungary, side and basal views, ×600. 2−4, from Gemeinhardt, 1931a. **5.** *Halicalyptra virginica* Ehrenberg, holotype, Miocene, Virginia, ×800, from Locker, 1974 (type species of *Halicalyptra* as designated by Campbell, 1954; a different type species was given by Locker that would place this genus as a radiolarian, but is invalid until so designated by the ICZN). **6,7.** *Naviculopsis trispinosa* (Schulz) Glezer, southwest Pacific, ×545, from Perch-Nielsen, 1975; 6, lower Oligocene, edge view; 7, upper Eocene, basal view. **8,9.** *N. biapiculata* (Lemmermann) Frenguelli, upper Eocene, Oamaru, New Zealand, exterior surface and optical section, ×745, from Tsumura, 1963. **10−13.** *Clathrium reticulare* Frenguelli, Miocene, Chile, ×480; described as a silicoflagellate, but probably fragments of a diatom such as *Bacteriastrum;* from Frenguelli, 1938b. **10,11.** Showing increasing skeletal complexity. **12,13.** Reported as double skeletons, but lacking the symmetry and basal rings that characterize the silicoflagellates. **14.** *Vallacerta hortonii,* Upper Cretaceous, California, ×480, from Hanna, 1928. **15.** *Eunaviculopsis navicula,* upper Miocene, Greece, ×240, from Ehrenberg, 1854. **16,17.** *Deflandryocha cymbiformis,* Miocene, Yugoslavia; 16, ×800; 17, ×760; both from Jerković, 1965. **18.** *D. spathulata* Jerković, Miocene, Yugloslavia, ×600, from Jerković, 1965. **19.** *Dictyocha quadralta* Hanna, Upper Cretaceous, California, ×368, from Hanna, 1928. **20.** *Nothyocha insolita* Deflandre, Miocene, Sicily, ×610, from Deflandre, 1950c; described as a silicoflagellate, but may be a radiolarian fragment.

Clark, B. L., and A. S. Campbell, Radiolaria from the Kreyenhagen Formation near Los Banos, California. *Mem. Geol. Soc. Am.,* 10, p. 1—66, pls. 1—7, 1945.

Conover, S. A. M., Oceanography of Long Island Sound, 1952—1954 IV. Phytoplankton. *Bull. Bingham Oceanogr. Coll.,* v. 15, p. 62—112, 1956.

Cornell, W. C., Maastrichtian silicoflagellates of the Great Valley, California. *Geosci. Man,* v. 9, p. 37—44, 1 pl., figs. 1—5, 1974a.

Cornell, W. C., Silicoflagellates as paleoenvironmental indicators in the Modelo Formation (Miocene), California. *J. Paleont.,* v. 48, p. 1018—1029, pl. 1, figs. 1—5, 1974b.

Coste, Bernard, H.-J. Minas, and Pierre Nival, Distribution superficielle des Taux de production organique primaire et des silicoflagellés entre la Sardaigne et la Tunisie (Février 1968). *Tethys,* v. 1, p. 573—579, figs. 1—8, 1969.

Deflandre, Georges, Les silicoflagellés des terres fossiles à diatomées. *Bull. Soc. Fr. Microsc.,* v. 1, p. 10—20, text-figs. 1—60, 1932a.

Deflandre, Georges, Sur la systématique des silicoflagellés. *Bull. Soc. Bot. Fr.,* v. 79, p. 494—506, text-figs. 1—42, 1932b.

Deflandre, Georges, Formations énigmatiques du squelette chez quelques silicoflagellés. *Bull. Soc. Bot. Fr.,* v. 80, p. 809—814, pls. 9, 10, 1933.

Deflandre, Georges, Les flagellés fossiles: aperçu biologique et paléontologique; rôle géologique. *Actual. Scient. Ind.,* no. 335, 98 p., 135 text-figs. Paris: Hermann and Cie., 1936.

Deflandre, Georges, Sur deux microfossiles siliceux énigmatiques (Silicoflagellidées?). *Bull. Soc. Fr. Microsc.,* v. 7, p. 90—96, figs. 1—25, 1938.

Deflandre, Georges, Sur les affinités et la phylogenèse du genre *Vallacerta,* silicoflagellidée du Crétacé supérieur. *C. R. Hebd. Séanc. Acad. Sci., Paris,* v. 211, p. 445—448, text-figs. 1—8, 1940a.

Deflandre, Georges, L'origine phylogénétique des *Lyramula* et l'évolution des silicoflagellidées. *C. R. Hebd. Séanc. Acad. Sci., Paris,* v. 211, p. 508—510, text-figs. 1—12, 1940b.

Deflandre, Georges, Sur une structure réticulée méconnue du squelette des silicoflagellidées. *C. R. Hebd. Séanc. Acad. Sci., Paris,* v. 211, p. 597—599, text-figs. 1—8, 1940c.

Deflandre, Georges, Les notions de genre et de grade chez les silicoflagellidées et la phylogenèse des mutants naviculaires. *C. R. Hebd. Séanc. Acad. Sci., Paris,* v. 212, p. 100—102, text-figs. 1—24, 1941.

Deflandre, Georges, Remarques sur l'évolution des silicoflagellidées, à propos de deux espèces crétaciques nouvelles. *C. R. Hebd. Séanc. Acad. Sci., Paris,* v. 219, p. 463—465, text-figs. 1—9, 1944.

Deflandre, Georges, *Phyllodictyocha* nov. gen., silicoflagellidés et formes affines du Miocène de Hongrie. *Bull. Soc. Bot. Fr.,* v. 93, p. 335—337, text-figs. 1—5, 1947.

Deflandre, Georges, Extraits des procès-verbaux des séances, l'évolution des silicoflagellidés. *Microscopie,* v. 1, p. B21—B22, text-figs. 1—8, 1948.

Deflandre, Georges, Contribution a l'étude des silicoflagellidés actuels et fossiles. *Microscopie,* v. 2, p. 72—108, text-figs. 1—113, 1950a.

Deflandre, Georges, Contribution a l'étude des silicoflagellidés actuels et fossiles (suite). *Microscopie,* v. 2, p. 117—142, text-figs. 114—173, 1950b.

Deflandre, Georges, Contribution a l'étude des silicoflagellidés actuels et fossiles (suite et fin). *Microscopie,* v. 2, p. 191—210, text-figs. 174—243, 1950c.

Deflandre, Georges, Classe des Silicoflagellidés (Silicoflagellata Borgert, 1891), v. 1, p. 425—438, text-figs. 324—338. *In* Pierre-P. Grassé, *Traité de zoologie.* Paris: Masson et Cie., 1952.

Desikachary, T. V., and C. L. Maheshwari, Fossil silicoflagellates from Colebrook and Nancoori Islands. *J. Indian Bot. Soc.,* v. 35, p. 257—264, pl. 13, 1956.

Distanov, U. G., and Z. I. Glezer, Silikoflagellit-novyy tip opalovykh kremnistykh porod [Silicoflagellite—a new type of opaline siliceous rock]. *Litologiya i Poleznye Iskopaemye,* 1974(2), p. 112—114, 2 figs., 1974.

Dumitrică, Paulian, *Dictyocha bachmanni* n. sp. et considérations sur la lignée phylogénétique *Dictyocha crux—D. stauracantha—D. bachmanni. Cah. Micropaléont.,* sér. 1, no. 4, p. 1—6, pls. 1, 2, 1967.

Dumitrică, Paulian, Miocene and Quaternary silicoflagellates in sediments from the Mediterranean Sea. *Init. Repts. Deep Sea Drill. Proj.,* v. 13, p. 902—933, pls. 1—12, figs. 1—4, 1973a.

Dumitrică, Paulian, Paleocene, late Oligocene and post-Oligocene silicoflagellates in southwestern Pacific sediments cored on DSDP Leg 21. *Init. Repts. Deep Sea Drill. Proj.,* v. 21, p. 837—883, pls. 1—13, figs. 1—6, 1973b.

Ehrenberg, C. G., Über noch jetzt zahlreich lebende Thierarten der Kreidebildung und den Organismus der Polythalamien. *Abh. Preuss. Akad. Wiss. Berlin,* 1839, p. 81—174, pls. 1—4, 1840.

Ehrenberg, C. G., Verbreitung und Einfluss des mikroskopischen Lebens in Süd- und Nord-Amerika. *Abh. Preuss. Akad. Wiss. Berlin,* 1841, p. 291—446, pls. 1—4, 1843.

Ehrenberg, C. G., *Mikrogeologie, das Erden und Felsen schaffende Wirken des unsichtbaren kleinen selbstandigen Lebens auf der Erde.* Leipzig: Leopold Voss, p. 1—374, Namenregister 31 p., index to atlas 88 p., pls. 1—40, 1854.

Filipescu, M. G., Les dépôts à silicoflagellidées et à radiolaires du Miocène de la région Subcarpatique de

Roumanie. *Bull. Sect. Scient. Acad. Roum.,* v. 26, p. 261–270, pls. 1, 2, figs. 1–3, 1943.

Filipescu, M. G., Dépôts siliceaux organiques dans les Carpates et les Subcarpates; considérations sur les conditions de développement des organismes siliceux. *Revue Roum. Géol. Géophys. Géogr., sér. Géol.,* v. 12, p. 85–89, figs. 1–13, 1968.

Frenguelli, Joaquín, Sobre variaciones de un silicoflagelado *(Dictyocha fibula). An. Soc. Cient. Arg. Secc. S. Fé,* v. 5, p. 57–60, 1933.

Frenguelli, Joaquín, Variaciones de *Dictyocha fibula* en el Golfo de San Matías (Patagonia septentrional). *An. Mus. Argent. Cienc. Nat.,* v. 38, p. 265–281, pls. 1–14, 1 text-fig., 1935.

Frenguelli, Joaquín, Sílicoflagelados del Río de la Plata. *Notas Mus. La Plata,* v. 3 (*Zoologia* no. 14), p. 231–245, text-figs. 1–5, 1938a.

Frenguelli, Joaquín, *"Clathrium reticulare"* probable representante fosil de un nuevo tipo silicoflagelados. *Notas Mus. La Plata,* v. 3 (*Paleontologia* no. 12), p. 131–145, text-figs. 1–7, 1938b.

Frenguelli, Joaquín, Variaciones de *"Dictyocha speculum"* Ehr. en el Golfo de San Jorge (Patagonia). *Notas Mus. La Plata,* v. 3 (*Zoologia* no. 11), p. 117–123, text-figs. 1–3, 1938c.

Frenguelli, Joaquín, Consideraciones sobre los sílicoflagelados fósiles. *Revta Mus. La Plata,* v. 2 (*Paleontologia* no. 7), p. 37–112, pls. 1–4, text-figs. 1–38, 1940.

Frenguelli, Joaquín, Sílicoflagelados y radiolarios del Trípoli del valle de Til-Til (Chile). *Notas Mus. La Plata,* v. 6 (*Paleontologia* no. 28), p. 93–100, pls. 1–6, 1941.

Frenguelli, Joaquín, Sílicoflagelados del tripoli de Mejillones (Chile). *Physis, B. Aires,* v. 20, p. 272–284, 1951.

Gemeinhardt, Konrad, Silicoflagellatae, v. 10, p. 1–87, 69 text-figs. In *Dr. L. Rabenhorst's Kryptogamen-Flora von Deutschland, Österreich und der Schweiz.* Leipzig: Akademische Verlagsgesellschaft, 1930.

Gemeinhardt, Konrad, Organismenformen auf der Grenze zwischen Radiolarien und Flagellaten. *Ber. Dt. Bot. Ges.,* v. 49, p. 103–110, pl. 10, 1931a.

Gemeinhardt, Konrad, Die Silicoflagellaten der deutschen Südpolar-Expedition 1901–1903. *Dt. Südpol.-Exped.,* v. 20 (Zoologie v. 12), p. 221–258, pls. 42, 43, 1931b.

Gemeinhardt, Konrad, Die Silicoflagellaten des südatlantischen Ozeans. *Wiss. Ergebn. Dt. Atlant. Exped. "Meteor",* v. 12, p. 274–312, pls. 7, 8, text-figs. 116–125, tables 65–67, 1934.

Glezer, Z. I., Nekotorye novye dannye o semeystve Vallacertaceae Deflandre (Silicoflagellatae) [Some new data on the family Vallacertaceae Deflandre (Silicoflagellatae)]. *Inf. Sb. Vses. Geol. Inst.,* no. 10, p. 103–113, text-figs. 1–14, table 1, 1959.

Glezer, Z. I., Paleogenovye kremnevye zhgutikovye vodorosli (Silicoflagellatae) zapadnoy Sibiri [Paleogene siliceous flagellate algae (Silicoflagellatae) from western Siberia]. *Inf. Sb. Vses. Geol. Inst.,* no. 35, p. 125–136, pls. 2, 3, 1 table, 1960.

Glezer, Z. I., K voprosu o filogeneze kremnevykh zhgutikovykh vodorosley [On the phylogeny of siliceous flagellata]. *Paleont. Zh.,* no. 1, p. 146–156, text-figs. 1–6, 1962.

Glezer, Z. I., Klass Silicoflagellateae. Kremnevye zhgutikovye vodorosli ili silikoflagellaty [Class Silicoflagellateae. Siliceous flagellate algae or silicoflagellates], p. 161–170, figs. 1–19. *In* Yu. A. Orlov, *Osnovy Paleontologii, Vodorosli, Mokhoobraznye, Psilofitovye, Plaunovidnye, Chlenistostebel'nye, Paporotniki.* Akad. Nauk SSSR, 1963.

Glezer, Z. I., Novye kremnevye zhgutikovye vodorosli paleogena SSSR: Silicoflagellatae fossiles novae URSS [New siliceous flagellate algae of the Paleogene of the USSR], p. 46–58, pls. 1, 2. *In Novosti sistematiki nizshikh rasteniy.* Akad. Nauk SSSR, Bot. Inst. Komarova, Leningrad Otdel: "Nauka", 1964.

Glezer, Z. I., *Flora sporovykh rasteniy SSSR,* v. 7, *Kremnevye zhgutikovye vodorosli (Silikoflagellaty). [Flora of spore-plants of USSR, Siliceous flagellate algae (Silicoflagellates)].* Akad. Nauk SSSR, Bot. Inst. Komarova, Moscow, Leningrad: "Nauka," 330 p., 33 pls., 28 figs., 1966.

Glezer, Z. I., O nakhodke pozdneeotsenovykh diatomovykh, silikoflagellat i ebriidey na zapadnom sklone mugodzhar. De bacillariophytis, silicoflagellatis et ebriideis in aeocaeno superiore declivium occidentalium montium mugodzharensium inventis, p. 21–25, figs. 1–10. In *Novosti sistematiki nizshikh rasteniy.* Akad. Nauk SSSR, Bot. Inst. Komarova, Leningrad Otdel.: "Nauka." 1967.

Glezer, Z. I., Pozdneeotsenovye kompleksy diatomovykh, kremnevykh zhgutikovykh vodorosley i ebriidey yugo-zapadnoy chasti Turgayskogo progiba [Late Eocene complex of diatoms, siliceous flagellate algae and ebriids of the SW part of the Turgay Syncline]. *Trudy Vses. Nauchno-Issled. Geol. Inst. [Biostratigr. Sb.,* vyp. 4], n. ser., vyp. 130, p. 67–84, pls. 1–5, 1969.

Haeckel, Ernst, Report on the Radiolaria collected by *H.M.S. Challenger* during the years 1873–1876. *Rep. Scient. Res. H.M.S. Challenger,* v. 18, clxxxviii + 1803 p., 140 pls., 1 map, 1887.

Hajós, Márta, Mátraalja Miocén üledékeinek diatomái. Die diatomeen der Miozänen ablagerungen des Mátravorlandes. *Geologica Hung., Ser. Paleont.,* fasc. 37, p. 1–401, pls. 1–68, 1968.

Hajós, Márta, Late Cretaceous Archaeomonadaceae, Diatomaceae and Silicoflagellatae from the South Pacific Ocean, Deep Sea Drilling Project, Leg 29, Site 275. *Init. Repts. Deep Sea Drill. Proj.,* v. 29, p. 913–1009, pls. 1–40, figs. 1–21, 1975.

Hanna, G. D., Silicoflagellata from the Cretaceous of California. *J. Paleont.,* v. 1, p. 259–263, pl. 41, 1928.

Hanna, G. D., A new genus of Silicoflagellata from the Miocene of Lower California. *J. Paleont.,* v. 4, p. 415–416, pl. 40, 1930.

Hanna, G. D., Diatoms and silicoflagellates of the Kreyenhagen Shale. *Rep. Calif. St. Min. Bur.,* v. 27, p. 197–201, pls. 1–5, 1931.

Hovasse, Raymond, L'énantiomorphie des squelettes chez les silicoflagellés. *Bull. Soc. Zool. Fr.,* v. 57, p. 54–56, 1 fig., 1932a.

Hovasse, Raymond, Contribution à l'étude des silicoflagellés. Multiplication, variabilité, hérédité, affinités. *Bull. Biol. Fr. Belg.,* v. 66, p. 447–501, pls. 6, 7, text-figs. 1–11, 1932b.

Hovasse, Raymond, Nouvelles recherches sur les flagellés à squelette siliceux: ebriidés et silicoflagellés fossiles de la diatomite de Saint-Laurent-La-Vernède (Gard). *Bull. Biol. Fr. Belg.,* v. 77, p. 271–284, text-figs. 1–9, 1943a.

Hovasse, Raymond, Nouvelles recherches sur les flagellés à squelette siliceux: ebriidés et silicoflagellés fossiles de la diatomite de Saint-Laurent-La-Vernède (Gard). *Bull. Biol. Fr. Belg.,* v. 77, p. 285–294, text-figs. 10–16, 1943b.

Ichikawa, Wataru, Preliminary report on silicoflagellates from the Neogene Tertiary of the Hokuriku District, Japan. *Sci. Rep. Kanazawa Univ.,* v. 5, p. 31–37, pl. 1, text-figs. 1, 2, table 1, 1956.

Ichikawa, Wataru, Norio Fuji, and Alfred Bachmann, Fossil diatoms, pollen grains and spores, silicoflagellates and arachaeomonads [*sic*] in the Miocene Hojuji Diatomaceous Mudstone, Nota Peninsula, central Japan, *Sci. Rep. Kanazawa Univ.,* v. 9, p. 25–118, pls. 1–7, text-figs. 1–20, 1964.

Ignatiades, Lydia, The relationship of the seasonality of silicoflagellates to certain environmental factors. *Botanica Mar.,* v. 13, p. 44–46, fig. 1, 1970.

Jendrzejewski, J. P., and G. A. Zarillo, Late Pleistocene paleotemperature oscillations defined by silicoflagellate changes in a subantarctic deep-sea core. *Deep Sea Res.,* v. 19, p. 327–329, figs. 1, 2, 1972.

Jerković, Lazar, Sur un nouveau type de silicoflagellidé fossile, *Deflandryocha* nov. gen., à cornes radiales spatulées. *C. R. Hebd. Séanc. Acad. Sci., Paris,* v. 256, p. 2202–2204, text-figs. 1–10, 1963.

Jerković, Lazar, Sur quelques silicoflagellidés de Yougoslavie. *Revue Micropaléont.,* v. 8, p. 121–130, pls. 1, 2, 1965.

Jerković, Lazar, Les silicoflagellidés fossiles des environs de Zagreb, de Bosanska kostajnica et de Derventa (Yougoslavie). *Godišnjak. Biol. Inst. Saraj.,* v. 22, p. 21–127, pls. 1–12, 4 figs., 1969a.

Jerković, Lazar, Les nouvelles recherches de la superficie du squelette des silicoflagellides. *Godišnjak. Biol. Inst. Saraj.,* v. 22, p. 129–176, pls. 1–28, 1969b.

Jerković, Lazar, Tendances évolutives de l'ultrastructure superficielle du squelette des silicoflagellidés (Crétacé–Récent), p. 659–662, figs. 1–6. *In* A. Farinacci, *Proc. II Plankton. Conf. Roma 1970.* Roma: Edizioni Tecnoscienza, 1971.

Jousé, A. P., Algae Diatomaceae aetatis supernecretaceae ex arenis argillaceis systematis fluminis bolschoy aktay in declivitate orientali Ural borealis. *Akad. Nauk SSSR, Bot. Inst. Bot. Mater., Otdel Sporovykh Rastenii,* v. 6, p. 65–78, pls. 1, 2, text-figs. 1–3, 1949.

Jousé, A. P., Diatomeae et Silicoflagellatae aetatis cretae superne e montibus uralensibus septentrionalibus. *Akad. Nauk SSSR, Bot. Inst., Bot. Mater., Otdel Sporovykh Rastenii,* v. 7, p. 42–65, pls. 1–6, text-figs. 1, 2, 1951.

Jousé, A. P., Silicoflagellatae aetatis paleogenae. *Akad. Nauk SSSR, Bot. Inst., Bot. Mater., Otdel Sporovykh Rastenii,* v. 10, p. 77–81, text-figs. 1–12, 1955.

Klement, K. W., Occurrence of *Dictyocha* (Silicoflagellates) in the Upper Cretaceous of Wyoming and Colorado. *J. Paleont.,* v. 37, p. 268–270, text-figs. 1–3, 1963.

Lemmermann, Ernst, Beiträge zur Kenntniss der Planktonalgen. *Ber. Dt. Bot. Ges.,* v. 19, p. 85–95, pl. 4, 1901a.

Lemmermann, Ernst, Silicoflagellatae. Ergebnisse einer Reise nach dem Pacific. H. Schauinsland 1896/97. *Ber. Dt. Bot. Ges.,* v. 19, p. 247–271, pls. 10, 11, 1901b.

Ling, H. Y., Silicoflagellates from central North Pacific core sediments. *Bull. Am. Paleont.,* v. 58, p. 85–129, pls. 18–20, figs. 1–5, 1970.

Ling, H. Y., Silicoflagellates and ebridians from the Shinzan diatomaceous mudstone member of the Onnagawa Formation (Miocene), northeast Japan, p. 689–703, pls. 1, 2. *In* A. Farinacci, *Proc. II Plankton. Conf., Roma 1970.* Roma: Edizioni Tecnoscienza, 1971.

Ling, H. Y., Upper Cretaceous and Cenozoic silicoflagellates and ebridians. *Bull. Am. Paleont.,* v. 62, p. 135–229, pls. 23–32, figs. 1–7, 1972.

Ling, H. Y., Silicoflagellates and ebridians from Leg 19. *Init. Repts. Deep Sea Drill. Proj.,* v. 19, p. 751–775, pls. 1–3, figs. 1–4, 1973.

Ling, H. Y., Distribution and biostratigraphic significance of *Dictyocha subarctios* (silicoflagellate) in the North Pacific. *Trans. Proc. Palaeont. Soc. Japan,* n.s., no. 101, p. 264–270, pl. 29, figs. 1–4, 1976.

Ling, H. Y., Late Cenozoic silicoflagellates and ebridians from the Eastern North Pacific region. *Proc. 1st Int. Congr. Pacific Neogene Stratig., Tokyo, 1976,* p. 205–233, pls. 1–3, figs. 1–10, 1977.

Ling, H. Y., L. M. McPherson, and D. L. Clark, Late Cretaceous (Maestrichtian?) silicoflagellates from the

Alpha Cordillera of the Arctic Ocean. *Science,* v. 180, p. 1360—1361, figs. 1, 2, 1973.

Lipps, J. H., Ecology and evolution of silicoflagellates. *Proc. N. Am. Paleont. Conv. Chicago 1969,* Pt. G, p. 965—993, figs. 1—11, 1970.

Locker, Sigurd, Revision der Silicoflagellaten aus der mikrogeologischen Sammlung von C. G. Ehrenberg. *Eclog. Geol. Helv.,* v. 67, p. 631—646, pls. 1—4, 1974.

Locker, Sigurd, *Dictyocha varia* sp. n., eine miozäne Silicoflagellaten-Art mit kompliziertem Variationsmodus. *Z. Geol. Wiss. Berlin,* v. 3, p. 99—103, figs. 1—7, 1975.

Loeblich, A. R. III, L. A. Loeblich, Helen Tappan, and A. R. Loeblich, Jr., Annotated index of fossil and Recent silicoflagellates and ebridians, with descriptions and illustrations of validly proposed taxa. *Mem. Geol. Soc. Am.* 106, xi + 319 p., 53 pls., 20 figs., 1968.

McPherson, L. M., and H. Y. Ling, Surface microstructure of selected silicoflagellates. *Micropaleontology,* v. 19, p. 475—480, pls. 1, 2, 1973.

Mandra, Y. T., Fossil silicoflagellates from California, U.S.A. *Int. Geol. Congr., 21 Copenhagen,* pt. 6, Proc. Sec. 6, p. 77—89, 1960.

Mandra, Y. T., Silicoflagellates of the Buttle Diatomite. *SEPM-AAPG Pacific Section, Guidebook to geology of Salinas Valley and San Andreas Fault, Annual Spring Field Trip,* p. 99—103, text-figs. 1—19, 1963.

Mandra, Y. T., Silicoflagellates from the Cretaceous, Eocene, and Miocene of California, U.S.A. *Proc. Calif. Acad. Sci.,* ser. 4, v. 36, p. 231—277, 83 figs., 1968.

Mandra, Y. T., A new genus of Silicoflagellata from an Eocene South Atlantic deep-sea core (Protozoa: Mastigophora). *Occ. Pap. Calif. Acad. Sci.,* no. 77, p. 1—7, figs. 1—7, 1969a.

Mandra, Y. T., Silicoflagellates: a new tool for the study of Antarctic Tertiary climates. *Antarctic J. U.S.,* v. 4, p. 172—174, 4 figs., 1969b.

Mandra, Y. T., A. L. Brigger, and Highoohi Mandra, Preliminary report on a study of fossil silicoflagellates from Oamaru, New Zealand. *Occ. Pap. Calif. Acad. Sci.,* no. 107, p. 1—11, 1973a.

Mandra, Y. T., A. L. Brigger, and Highoohi Mandra, Chemical extraction techniques to free fossil silicoflagellates from marine sedimentary rocks. *Proc. Calif. Acad. Sci.,* ser. 4, v. 39, p. 273—284, figs. 1—3, 1973b.

Mandra, Y. T., and Highoohi Mandra, Upper Eocene silicoflagellates from New Zealand. *Antarctic J. U.S.,* v. 6, p. 177—178, figs. 1, 2, 1971.

Mandra, Y. T., and Highoohi Mandra, Paleoecology and taxonomy of silicoflagellates from an Upper Miocene diatomite near San Felipe, Baja California, Mexico. *Occ. Pap. Calif. Acad. Sci.,* no. 99, p. 1—35, figs. 1—48, 1972.

Marshall, S. M., The Silicoflagellata and Tintinnoinea. *Scient. Rep. Gt. Barrier Reef Exped.,* v. 4, p. 623—664, text-figs. 1—43, table 1, 1934.

Martin, G. C., Radiolaria, p. 447—459, pl. 130. In *Miocene.* Baltimore: Md. Geol. Surv., 1904.

Martini, Erlend, Neogene silicoflagellates from the equatorial Pacific. *Init. Repts. Deep Sea Drill. Proj.,* v. 7, p. 1695—1708, pl. 1, figs. 1—3, 1971.

Martini, Erlend, Silicoflagellate zones in the late Oligocene and early Miocene of Europe. *Senckenberg. Leth.,* v. 53, p. 119—122, figs. 1—4, 1972.

Martini, Erlend, Silicoflagellate zones in the Eocene and early Oligocene. *Senckenberg. Leth.,* v. 54, p. 527—532, fig. 1, 1974.

Martini, Erlend, Neogene and Quaternary silicoflagellates from the central Pacific Ocean (DSDP Leg 33). *Init. Repts. Deep Sea Drill. Proj.,* v. 33, p. 439—449, pls. 1, 2, figs. 1—3, 1976.

Martini, Erlend, and Carla Müller, Eocene to Pleistocene silicoflagellates from the Norwegian-Greenland Sea (DSDP Leg 38). *Init. Repts. Deep Sea Drill. Proj.,* v. 38, p. 857—895, pls. 1—12, figs. 1—3, 1976.

Mukhina, V. V., Species composition of the Late Paleocene diatoms and silicoflagellates in the Indian Ocean. *Micropaleontology,* v. 22, p. 151—158, pls. 1, 2, fig. 1, 1976.

Nival, Paul, Sur le cycle de *Dictyocha fibula* Ehrenberg dans les eaux de surface de la rade de Villefranche-sur-Mer. *Cah. Biol. Mar.,* v. 6, p. 67—82, figs. 1—6, 1965.

Perch-Nielsen, Katharina, Late Cretaceous to Pleistocene silicoflagellates from the southern Southwest Pacific, DSDP, Leg 29. *Init. Repts. Deep Sea Drill. Proj.,* v. 29, p. 677—721, pls. 1—15, figs. 1—3, 1975.

Perch-Nielsen, Katharina, New silicoflagellates and a silicoflagellate zonation in north European Palaeocene and Eocene diatomites. *Meddr. Dansk Geol. Foren. [Bull. Geol. Soc. Denmark],* v. 25, p. 27—40, figs. 1—35, 1976.

Poelchau, H. S., Distribution of Holocene silicoflagellates in North Pacific sediments. *Micropaleontology,* v. 22, p. 164—193, pls. 1—6, figs. 1—13, 1976.

Proshkina-Lavrenko, A. I., Sovremennye i iskopaemye silikoflagellaty i ebriidei Chernomorskogo basseyna. Silicoflagellatae nec non ebriideae nostrorum temporum et fossiles ponti euxini [Recent and fossil silicoflagellates and ebriidians of the Black Sea basin]. *Trudy Bot. Inst. Akad. Nauk SSSR, Sporovye Rasteniya,* vyp. 12, p. 142—175, pls. 1—4, text-figs. 1—8, 1959.

Rampi, Leopoldo, Ricerche sul fitoplancton del mar Ligure. 8. I silicoflagellati della acque di Sanremo. *Atti Soc. Ital. Sci. Nat.,* v. 87, p. 64—67, figs. 1—9, 1948.

Schiller, Josef, Die planktontischen Vegetation des adriatischen Meers: B. Chrysomonadina, Heterokontae, Cryptomonadina, Eugleninae, Volvocales. I. Systematischer Teil. *Arch. Protistenk.*, v. 53, p. 59—123, pls. 3—6, 30 figs., 1925.

Schulz, Paul, Beiträge zur Kenntnis fossiler und rezenter Silicoflagellaten. *Bot. Arch.*, v. 21, p. 225—292, 83 text-figs., 1928.

Stradner, Herbert, Über fossile Silicoflagelliden und die Möglichkeit ihrer Verwendung in der Erdölstratigraphie. *Erdöl Kohle Erdgas Petrochem.*, v. 14, p. 87—92, pls. 1—3, fig. 1, 1961.

Tappan, Helen, and A. R. Loeblich, Jr., Geobiologic implications of fossil phytoplankton evolution and time-space distribution, p. 247—340, 1 pl., 12 figs. *In* R. Kosanke and A. T. Cross, Symposium on palynology of the Late Cretaceous and Early Tertiary. *Spec. Pap. Geol. Soc. Am.*, no. 127, 1971.

Tappan, Helen, and A. R. Loeblich, Jr., Fluctuating rates of protistan evolution, diversification and extinction. *24th Int. Geol. Congr., Sec. 7, Paleont.*, p. 205—213, 6 figs., 1972.

Tappan, Helen, and A. R. Loeblich, Jr., Evolution of the oceanic plankton. *Earth-Sci. Rev.*, v. 9, p. 207—240, 9 figs., 1973.

Travers, A., and M. Travers, Les silicoflagellés du Golfe de Marseille. *Mar. Biol.*, v. 1, p. 285—288, fig. 1, 1968.

Tsumura, Kôhei, A systematic study of Silicoflagellatae. *J. Yokohama Munic. Univ.*, C-45, no. 146, p. 1—84, pls. 1—28, text-figs. 1—4, 1963.

Tynan, E. J., Silicoflagellates of the Calvert Formation (Miocene) of Maryland. *Micropaleontology*, v. 3, p. 127—136, pl. 1, text-figs. 1—3, 1957.

Van Valkenburg, S. D., Observations on the fine structure of *Dictyocha fibula* Ehrenberg. I. The skeleton. *J. Phycol.*, v. 7, p. 113—118, 10 figs., 1971a.

Van Valkenburg, S. D., Observations on the fine structure of *Dictyocha fibula* Ehrenberg. II. The protoplast. *J. Phycol.*, v. 7, p. 118—132, 24 figs., 1971b.

Van Valkenburg, S. D., and R. E. Norris, The growth and morphology of the silicoflagellate *Dictyocha fibula* Ehrenberg in culture. *J. Phycol.*, v. 6, p. 48—54, figs. 1—12, 1970.

Wailes, G. H., The harvest of the sea. *Mus. Notes, Vancouver*, v. 2, p. 15—27, 4 pls., 1927.

Wornardt, W. W., Jr., Eocene, Miocene and Pliocene marine diatoms and silicoflagellates studied with the scanning electron microscope, p. 1277—1300, pls. 1—16. *In* A. Farinacci, *Proc. II Plankton. Conf. Roma 1970.* Roma: Edizioni Tecnoscienza, 1971.

Zanon, D. V., Silicoflagellate fossili Italiane. *Atti Accad. Pontif. Nuovi Lincei*, v. 87, p. 40—82, 1 pl., text-figs. 1—3, 1934.

Zhuze, A. P., and V. V. Mukhina, Zona *Mesocena elliptica* Ehr. v Pleystotsenovykh osadkakh Tikhogo Okeana [The *Mesocena elliptica* Ehr. zone in Pleistocene sediments of the Pacific Ocean]. *Okeanologiya*, v. 13, p. 467—475, figs. 1—3, 1973.

DIATOMS

The diatoms have many advantages as microfossils, some of which are possessed by no other group of organisms: They occur in virtually all bodies of nontoxic water exposed to light, and easily recognized taxa are characteristic of many different ecologic environments between truly marine conditions and potable fresh water and at widely varying temperatures, salinities, pH, and chemical composition. Similar fossil forms are found in sediments that were deposited under such environmental conditions. ...

Kenneth E. Lohman, 1964

*D*iatoms are tiny algae that are extremely abundant in the present seas and bodies of fresh water. Because of the beauty and ornate sculpture of their siliceous shells, they long have been of great interest to microscopists. For many years emphasis was placed on detailed description of the shells in light microscopy, neglecting other aspects of their biology or geologic occurrence. Many species were described merely as fossils, with no stratigraphic and little if any geographic documentation, making any stratigraphic application impossible, even in areas of abundant diatomaceous rocks.

Major changes have occurred in the study of the diatoms over the past 20 years. First, the development of electron microscopy allowed a complete change of emphasis in morphologic studies. Although the use of direct TEM or carbon replicas made available much greater detail, the true depth relationships of many features, including the wall fine structure, was only understood with use of SEM, and much new terminology has been proposed on the basis of the new information.

New culturing techniques combined with the improved microscopical techniques also allowed much better resolution of diatom life history, the nature of spore formation, and sexual reproduction. Biochemical studies have concerned silica deposition, nutrition, enzyme electrophoresis, DNA composition, and the amino acid composition of the organic component of the frustule.

The third major impetus to diatom studies has come from the DSDP of the past ten years, which has provided continuous cored sections of stratigraphically dated material. Biostratigraphic zonations have been proposed on the basis of the diatom assemblages and correlated with those based on various other microfossils.

The diatoms are placed by some workers as a class Bacillariophyceae together with the Chrysophyceae and Xanthophyceae in the Division Chrysophyta. All three groups may form similar endoplasmic cysts, secrete silica at some stage, store oils rather than starch, and possess a bipartite cell wall at some stage. Their differences are emphasized by others who rank the diatoms as a separate division, Bacillariophyta. The siliceous bipartite cell wall or **frustule** is of major importance in diatom classification. Two broad groups have been based on the frustule, those with basically radial symmetry, the Centrales or **centric diatoms**, and those with basically bilateral symmetry, the Pennales or **pennate diatoms** (see Figure 8.1). A third group, the Mediales, is sometimes used for the araphid bipolar diatoms. Although these groups remain useful as general descriptive terms, some recent classifications use only orders or suborders based on a variety of characters, without attempting to fit all diatoms into the two major categories.

Probably 100,000 diatom species have been described, of which perhaps 12,000 to 15,000 are truly distinct. During the early period of their study, little was known of the variability within a species. Many minor differences were regarded as valid bases for new taxa, prior to the detailed study of variations that occur within the life cycle and in successive generations.

Diatom cell size ranges over four orders of magnitude, from as little as 0.75 μm in diameter in *Chaetoceros galvestonensis* Collier & Murphy (Collier and Murphy, 1962) or 1.8 μm in *Cylindropyxis profunda* Hendey (Hendey et al., 1969) up to 2 mm in diameter in *Ethmodiscus rex* (Wallich in Rattray) Hendey in Wiseman & Hendey (McHugh, 1954). Chains of colonial species may attain more than 7 cm in length.

Diatoms are widely distributed in all lighted water bodies, marine and freshwater. They probably are the most important living marine primary producers. They may be free-living or attached, benthic or planktonic, single or colonial, and a few are endosymbiotic.

THE DIATOM CELL

The Protoplast

The protoplast consists of a layer of cytoplasm that lines the inside of the boxlike siliceous skeleton or frustule, and extends into all cavities within the wall. This cytoplasmic layer surrounds a large central vacuole (see Figure 8.2); a strand or strands of cytoplasm may bridge the vacuole, connecting the center points of the two valves or halves of the box. Pennate diatoms may have a large vacuole at each side of the central protoplasmic strand in the mid-region, and their thickest cytoplasm at the poles.

Nucleus

The nucleus is diploid, and commonly rounded or oval in form, although in some pennates it may be reniform. It commonly ranges between 2 and 47 μm in diameter, and contains 1 to 10 darker-colored nucleoli. The nucleus of centric diatoms is usually polar in location; if central in position the cell probably is commencing division (Reimann, 1960). Mitotic division involves formation of a central spindle. Between 19 and 24 chromosomes occur in *Lithodesmium undulatum* Ehrenberg (Manton et al., 1970).

The amount of DNA per cell in centric species ranges from 0.4 pg in *Skeletonema costatum* (Greville) Cleve, to 6 pg in *Thalassiosira fluviatilis* Hustedt, up to 60 to 80 pg in *Coscinodiscus asteromphalus* Ehrenberg. It generally is less in pennates, 0.11 pg in *Navicula pelliculosa* (Brébisson) Hilse, and 0.18 pg in *Cylindrotheca fusiformis* Reimann & Lewin (Loeblich and Loeblich, 1979).

The mole percentage of guanine and cytosine (G+C) in the DNA of diatoms is generally between 37 and 58, being somewhat higher in the pennate species than in centric diatoms. That of most diatoms is less than 50 percent G+C, in contrast to other algal groups that tend to be above 50 percent G+C (Sober, 1970). Representative amounts are 39.1 in *Chaetoceros decipiens* Cleve, 40.8 in *Cyclotella cryptica* Reimann, Lewin & Guillard, 43.4 in *Melosira italica* (Ehrenberg) Kützing, 40.3 in *Thalassiosira nor-*

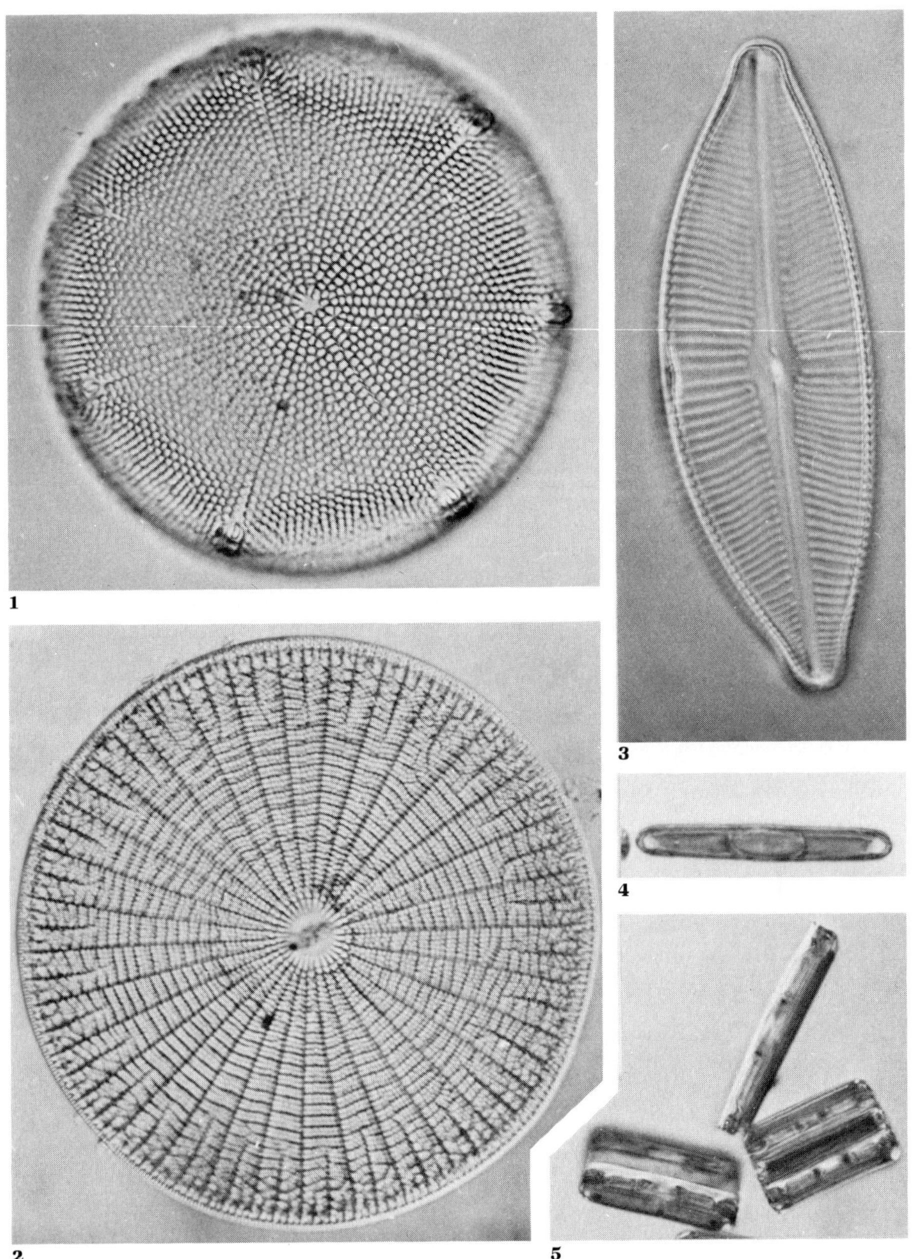

Figure 8.1

1,2. Centric diatoms, with radial symmetry, ×320. **1.** *Aulacodiscus margaritaceus* Ralfs in Pritchard. **2.** *Arachnoidiscus ornatus* Ehrenberg. **3 − 5.** Pennate diatoms, with bilateral symmetry. **3.** *Cymbella ehrenbergii* Kützing, ×750. **4,5.** *Grammatophora marina* (Lyngbye) Kützing; ×320. 4, valve view, and 5, girdle view of cells held together in short chains.

1

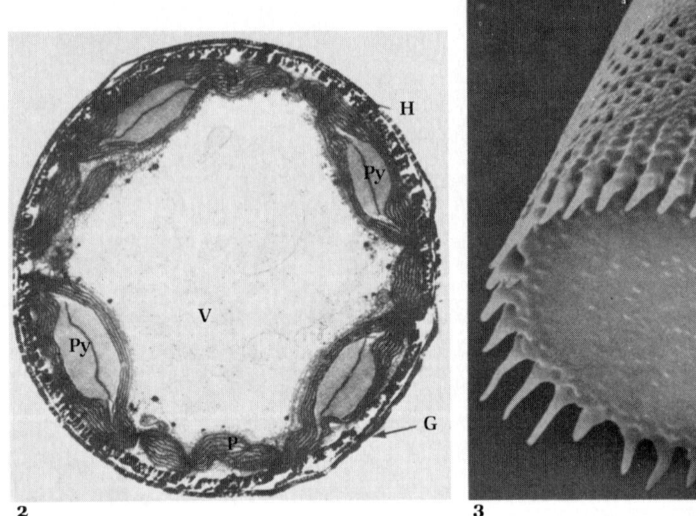

2

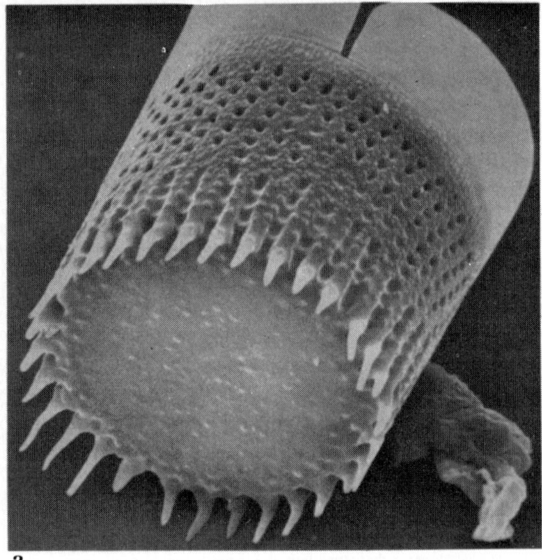

3

Figure 8.2

1,2. *Melosira varians* C. A. Agardh; H, hypovalve; E, epivalve; G, girdle bands; N, nucleus; No, nucleolus; P, peripheral plastids; Py, pyrenoids; V, vacuole; M, mitochondria; D, dictyosome; L, lipid drops; from Crawford, 1973a. **1.** Longitudinal section of cell in TEM, ×3040. **2.** Transverse section in TEM, ×2300. **3.** *M. granulata* Ralfs, Lake Rotoiti, New Zealand, showing linking spines at margin of valve, and incomplete girdle band at top of figure, SEM, ×2640, from V. Cassie.

denskioeldii Cleve, among the centrics, and 45.4 percent in *Cylindrotheca fusiformis,* 58 in *Navicula pelliculosa,* 46.9 in *Nitzschia angularis* W. Smith, and 36.9 in *Rhabdonema adriaticum* Kützing (Loeblich and Loeblich, 1979).

Plastids

The pigment-bearing plastids occur within the cytoplasm and differ in size, shape, and number, even among different species of a single genus. Those of the centrics are commonly rounded, irregular, or discoidal, and are numerous, whereas the pennate diatoms usually have only two plastids; these are laminate and extend along opposite sides of the protoplast. Some pennates have only a single, irregularly lobed plastid, that of *Gomphonema parvula* Kützing being H-shaped (P. A. Dawson, 1973). The plastids lie adjacent to the valve of free-living species, but next to the girdle band of colonial ones; more rarely they may occupy the center of the cell.

The plastids contain lamellae with three to nine thylakoids, most commonly three or four, and a girdle band of three thylakoids encircles the plastid. Each thylakoid is from 12 to 15 nm in thickness. The plastid DNA of *Amphipleura rutilans* (Trentepohl) Cleve occurs as a ring of fibrils at the rim of the plastid (Drum, 1969).

Diatoms characteristically are autotrophic, their plastids being the site of photosynthesis. Because of their abundance and wide distribution, they are important producers of organic matter, being sometimes termed "the grass of the sea." Nonphotosynthetic diatoms also are known, however, with a saprophytic mode of nutrition (*Nitzschia alba* Lewin & Lewin, *N. putrida* Benecke, and *"Synedra" hyalina* Prowazek). Morphologically, they differ from the colored diatoms in the absence of a lamellar plastid, but instead possess double-membrane-bounded proplastids 0.5 to 1.0 μm in diameter, with a homogeneous matrix. The proplastids appear like much reduced and nonfunctional plastids. Apparently, these apochlorotic diatoms evolved from a photosynthetic counterpart when they lost the ability to form the normal plastids or when plastids degenerated and lost their pigments (Lauritis et al., 1968).

Pigments

The pigments of the diatoms give the plastids a golden-brown color, ranging to various shades of yellow, olive-green, or brown. These pigments include chlorophyll a, c_1, and c_2 with the chlorophyll c comprising from 11 to 37 percent of the total chlorophyll), and accessory carotenoids and xanthophylls. The ratio of carotenoids to chlorophyll a in diatoms is between 0.26 and 2.11. The principal yellow pigment is β-carotene; additional ones are ϵ-carotene, very small amounts of diadinoxanthin and diatoxanthin, the photosynthetically active and abundant xanthophyll fucoxanthin, neofucoxanthin A and B, and an unknown additional xanthophyll (Healey et al., 1967; Hendey, 1964).

Pyrenoids

One or several ovoid, spherical, lenticular, or planoconvex pyrenoids occur within the plastids. The pyrenoids may be membrane-limited (Drum and Pankratz, 1964), and may be crossed by thylakoid bands with one to three discs. They do not have a starch sheath. Bright and glistening in appearance, they are particularly prominent in the pennate taxa, probably serving for reserve food storage. Oil droplets also occur in the cytoplasm or plastids of most diatoms. These may be localized on the cytoplasmic bridge in *Navicula cuspidata* Kützing, or may be scattered through the cell in *Pinnularia.* The oil globules first appear and are most numerous in the vicinity of the pyrenoids; usually somewhat larger than the pyrenoids, they have a dull, faintly bluish color. The usual storage product is a polysaccharide formerly regarded as leucosin, but now known as chrysolaminarin (Lewin and Guillard, 1963). However, the storage product of *Gomphonema parvula* is not chrysolaminarin but consists of a polyphosphate and lipid material instead (P. A. Dawson, 1973).

Various types of extracellular organic material are produced by diatoms, not all of similar composition. Filaments between cells that arise from the strutted processes of *Thalassiosira fluviatilis* and *Cyclotella cryptica* are of chitan (a chitinlike substance that following hydrolysis consists entirely of glucosamine); the capsule of *Navicula pelliculosa* is a polyuronide, and that of *Phaeodactylum tricornutum* Bohlin contains xylose, mannose, fucose, and galactose. The sheath of the tube-dwelling *Amphipleura rutilans* (Trentepohl) Cleve is a polymer of xylose, mannose, and traces of protein, and *Gomphonema olivaceum* (Lyngbye) Desmazières has a stalk of an acidic polysaccharide. Organic compounds also are excreted in solution, up to as much as 25 percent of the amount fixed by photosynthesis during exponential growth, and as much as 43 percent in the stationary phase (see section on photosynthesis and productivity pp. 614−615).

The Siliceous Frustule

Wall Composition

The wall consists of very pure amorphous silica (polymerized silicic acid), in part hydrated (opaline). The amount of silica (SiO_2) per cell varies in different species, from as much as 50 percent or more of total dry weight in heavily silicified species to less than 10 percent in some weakly silicified ones (Lewin et al., 1966). Aluminum has been reported in cleaned frustules, and Lewin (1962) stated that the frustules of marine planktonic species contain 1.5 percent Al_2O_3 or Fe_2O_3, with 96.5 percent SiO_2. *Biddulphia pulchella* Gray was examined spectrographically by Venkataraman et al. (1962). They reported aluminum, magnesium, iron, titanium, and sodium in addition to the silicon, and stated that the diatom wall contained crystalline silica, quartz, and traces of orthoclase or potassium feldspar (although the latter may have been a contaminant). Examination of both live and fossil diatoms by X-ray powder diffrac-

tion (Desikachary and Dweltz, 1961) indicated the presence of crystalline α-quartz, both in the marine *Isthmia nervosa* Kützing from California, and freshwater *Cyclotella*, *Gomphonema*, and *Fragilaria* from India. The amount of crystalline quartz was greater in fossil specimens from the Miocene of Lompoc, California, and the upper Eocene of Oamaru, New Zealand, indicating a recrystallization of the fossils.

The amount of water in the opaline silica also varies, from 9.6 percent in clean, dried frustules of the freshwater *Navicula pelliculosa* to only 1.9 percent in marine *Biddulphia* and *Coscinodiscus*. In comparison, siliceous sponge spicules contain from 5 to 13 percent water. Probably the amorphous silica has different states of hydration. Fossil specimens are less soluble in water than living ones, and have a specific gravity of 2.00 instead of the 2.07 of living diatom silica. The refractive index of 1.40 to 1.42 suggests a higher water content than does the specific gravity, but more data are needed as to specific gravity, refractive index, and degree of hydration of various marine and freshwater species, both living and fossil.

Electron microscopy suggests a spongelike substructure, indicating a silicic acid gel (Helmcke, 1954). The silica wall also adsorbs positively charged colloids as does a dried silica gel. This adsorption technique suggests for *Navicula pelliculosa* a specific surface area about five times that of the visible area, indicating a high degree of porosity. If spherical, the colloidal particles of the gel would be about 22 nm in average diameter. Particles about 6 nm in diameter were observed in partially silicified walls of *Navicula pelliculosa* (Lewin, 1962).

The cell wall also contains organic material, termed pectin as it stains with ruthenium red. The organic matrix contains amino acids and sugars and other unidentified proteinaceous and polysaccharide compounds. An amino acid known thus far only in diatom cell walls, 3, 4-dihydroxyproline, may determine the molecular structure of the organic matrix associated with silicification (Nakajima and Volcani, 1969). It has been identified in seven marine species of the diatom genera *Cyclotella*, *Cylindrotheca*, *Navicula*, *Nitzschia*, and *Phaeodactylum*. The amino acid composition of the organic matrix of the diatom frustule (mixed species) was determined by King (1977). As in other siliceous groups of organisms, the dominant amino acid is glycine, in contrast to higher amounts of aspartic acid in calcareous skeletal material. The high glycine content of the diatoms is nearly identical to that of the opal phytoliths of higher plants, in contrast to a slightly lower level of glycine in siliceous animal groups (radiolarians and sponges). Diatoms differed in having a sixfold higher proportion of threonine than is found in the opal phytoliths.

Electron microscopy of *Navicula pelliculosa* has shown the presence of an organic skin surrounding the silica shell. Removal of the silica shell with HF vapor leaves this skin intact, with a pattern of ribs, pores, and other ornamentation identical to that of the silica shell. The silica shell forms intracellularly, within a membrane-bounded vesicle, which probably becomes the skin of the completed shell (Reimann et al., 1966).

Silica Deposition

The diatom frustule is not solid, but is a tenuous or even laminate network of silica (see Figure 8.3). The area of the perforations may include 10 to 30 percent of the surface area of the valve in *Chaetoceros.* Such planktonic species have attained a maximum wall strength with minimum weight and volume of silica.

The loculate structure of the wall suggests that it forms as a mosaic, with simultaneous formation of the various parts instead of a surface deposition of silica. The process of wall formation has been studied in various species, with cultures in synchronous division. Such cell synchrony may be obtained by maintaining cultures in controlled alternate periods of light and dark, or by limiting the silicon in the medium until cell division stops. The cells divide simultaneously when silicon is added. Silicate uptake is thereby shown to utilize energy, being dependent upon growth and aerobic respiration.

In the division cycle of the diatom, the young cell first increases in size, then nuclear division and cytoplasmic division are followed by deposition of silica at the new separating membranes. Silica is taken from the medium only during the formation of the new valves following nuclear division (Coombs et al., 1967a). The silicic acid presumably is taken into the cytoplasm, where it may or may not be transformed or partially polymerized before being deposited in a three-layered organic membrane, termed the **silicalemma.** The silicalemma presumably is responsible for silica deposition, which first occurs at the center of the daughter cell interface following cytoplasmic division. In *Navicula pelliculosa* and others with central raphe, deposition begins near the raphe, but in *Nitzschia alba,* with marginal raphe, deposition begins in the center of the valve and appears last near the raphe. The silicon deposition vesicle is related to the mitotic apparatus, and the position controlled by the latter, rather than by the raphe (Lauritis et al., 1968; Coombs et al., 1968).

Separation of the daughter cells follows completion of the two new valves. Those species with numerous intercalary bands may continue to add these at later stages.

Shell deposition is fairly rapid. *Melosira varians* and *Stephanopyxis turris* (Greville & Arnott) Ralfs (see Figure 8.4) complete the shell formation process in 15 to 20 minutes, *Nitzschia linearis* W. Smith in 12 to 13 minutes (Reimann, 1960).

The silicalemma probably forms the basis of the organic skin of the mature wall, after secondary addition of organic matter. It protects the shell from solution during the life of the diatom; after the cell dies, decay of the membrane may result in leaching of the silicic acid.

Frustule Morphology

The diatom cell wall is a siliceous bipartite skeleton termed a frustule. One section (the **epitheca**) overlaps the other section (**hypotheca**) like the lid of a box (see Figure 8.5). Each half of the "box" also is constructed of two parts, the **valve,** whose major surface is a more or less flattened plate (the **valve face**) bending sharply at the edge to form a marginal flange (the **valve mantle**), and a hooplike **connecting band** attached at the edge of the flange or valve mantle. The connecting band of the older valve

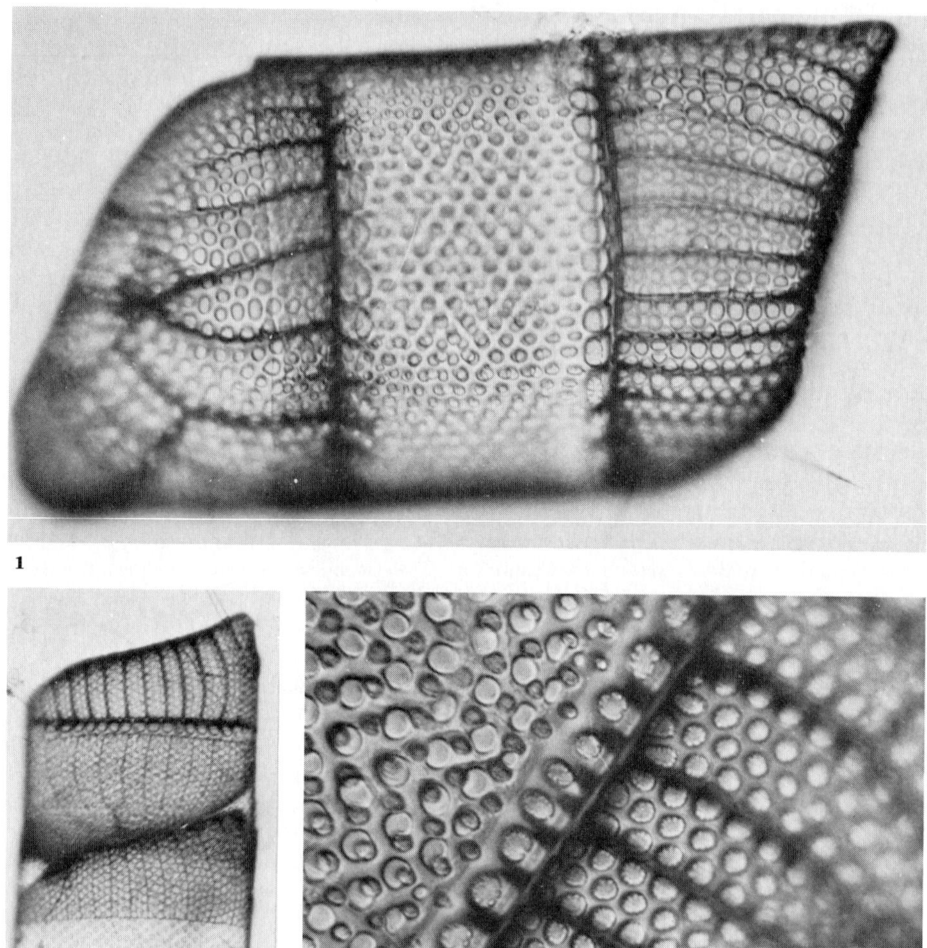

Figure 8.3

Isthmia nervosa. **1.** Cell showing coarsely perforate wall of the frustule, epivalve at right, hypovalve at left, ×320. **2.** Two cells that have recently divided, but still held within the girdle bands of the parent cell, ×120. **3.** Enlargement of part of part 1, at junction of valve and girdle band, showing complexly indented margin of perforations, ×750.

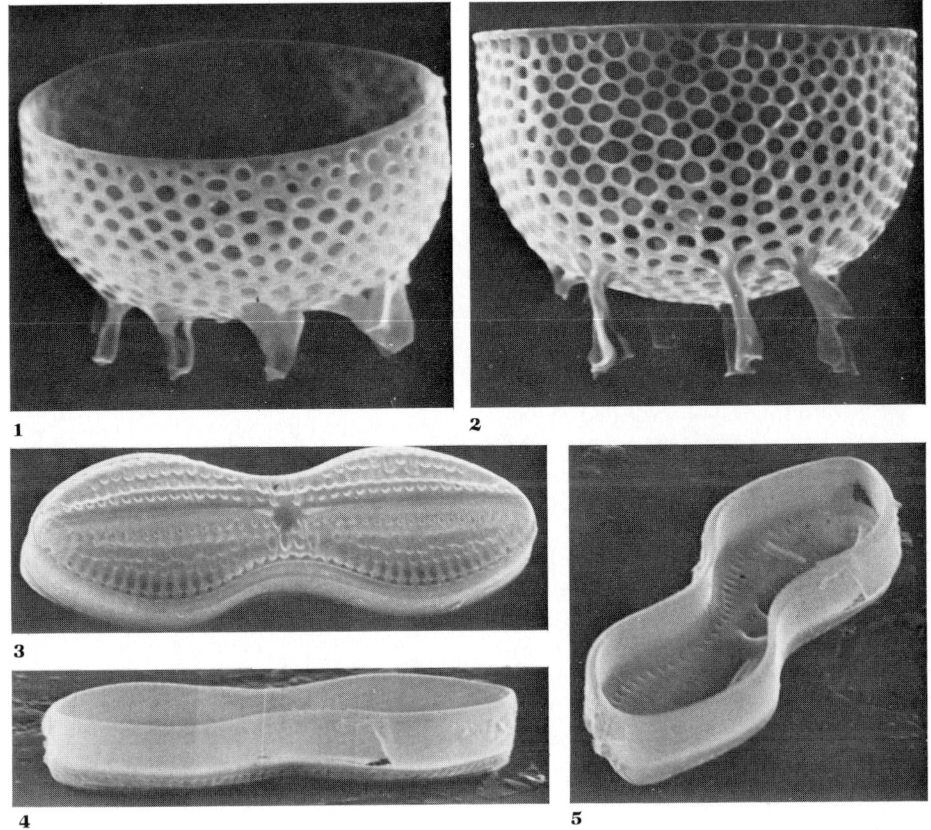

Figure 8.4

1,2. *Stephanopyxis turris,* Miocene, California, a species that is still living, single valves, SEM, ×1600. **3–5.** *Diploneis bombus* Ehrenberg, a pennate diatom, Miocene (Luisian), California, SEM. **3.** Slightly oblique valve view, ×880. **4.** Edge view with interior up, showing smooth mantle, ×720. **5.** Oblique view of interior, ×800.

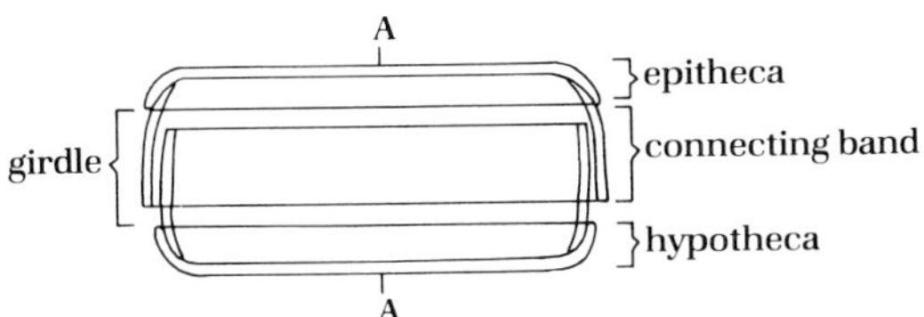

Figure 8.5
Frustule morphology. Diagrammatic girdle view, showing epitheca, hypotheca, and the two connecting bands that form the girdle. Pervalvar axis is that connecting the valve midpoints (AA).

or **epivalve** is termed an **epicingulum**, the two comprising the epitheca; the connecting band of the younger valve or **hypovalve**, the **hypocingulum**, is overlapped by the epicingulum. The joined cingula of a frustule comprise the **girdle** or **cincture** (cinctura).

Shape and Symmetry. Commonly the greatest dimension of the diatom is that of valve diameter or length, but in some the greatest dimension is seen in girdle view. This may result from an extremely deep valve mantle (some *Melosira*), a wide simple girdle (*Ethmodiscus*), a complex girdle of numerous imbricated plates, or hooplike intercalary bands between the valve and connecting bands. The intercalary bands may project inward as incomplete partitions or septa, parallel to the valves.

In girdle view, the diatom usually appears rectangular, but in valve view it may be circular, triangular, quadrate, elongate-ovate, wedge-shaped, or any of many other variations. Commonly the valve is thin, flattened, or slightly convex, but it may be strongly arched or even gablelike. Centric diatoms may have a radial symmetry about the **pervalvar axis** (connecting the midpoints of the two valves), or have fewer planes of symmetry (as in *Isthmia*). Pennate diatoms are more elongate; the lengthwise axis is termed **apical** or **sagittal**, and the **transapical** or **transverse axis** is perpendicular to both apical and pervalvar axes (see Figure 8.6). The transapical axis thus represents the breadth of the valve through the midpoint. Some pennate diatoms have only two planes of symmetry (apical and transapical), or may be symmetrical only about the transapical axis, and their valves curved parallel to the apical axis. The **valval plane** parallels the valves, representing the plane of division between them.

The Girdle. As noted above, the girdle or cincture consists of the part of the frustule between the valves, and consists of the epicingulum and hypocingulum (see Figure 8.7). The area of overlap of the epicingulum over the hypocingulum is the thecal junction. A single element of the girdle is termed a girdle band or

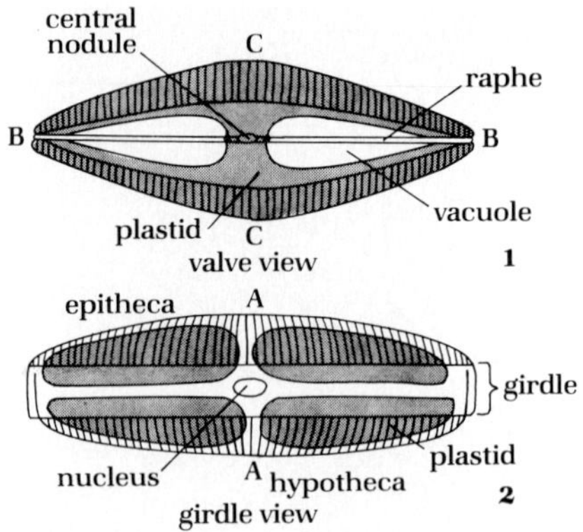

Figure 8.6
Frustule morphology of a pennate diatom; AA, pervalvar axis; BB, apical or sagittal axis; CC, transapical or transverse axis. **1.** Valve view, showing median slit or raphe at each end, separated at center of valve by central nodule; plastid and vacuole of cell interior also indicated. **2.** Girdle view, showing epitheca, girdle, and hypotheca, and position of nucleus and plastid.

segment, and the junction of any of these segments forms a suture. The girdle bands may be (1) complete rings or closed bands, (2) **incomplete open bands** that encircle more than half the girdle, (3) **half-bands** that extend around half the girdle or less, or (4) **segmented bands** with two or more segments around the girdle that are separated by gaps (Stosch, 1975).

Intercalary bands are the girdle bands proximal to the valve, and may differ in form or structure from distally located elements. The intercalary band (or two in the case of half-bands) that is adjacent to the valve also may be termed the **valvocopula**, and the next band or half-bands are termed **copulae.** More distal elements of the cingulum are termed connecting bands (these also have been termed **pleurae**), and are generally different in shape or ornamentation than the cingular bands or half-bands that adjoin the valvocopula. The bands, half-bands, or segments may have a short extension (the **ligula**) pointing in the advalvar direction and completely or partly filling a gap in the next adjacent band or half band by underlapping. A similar extension in an abvalvar

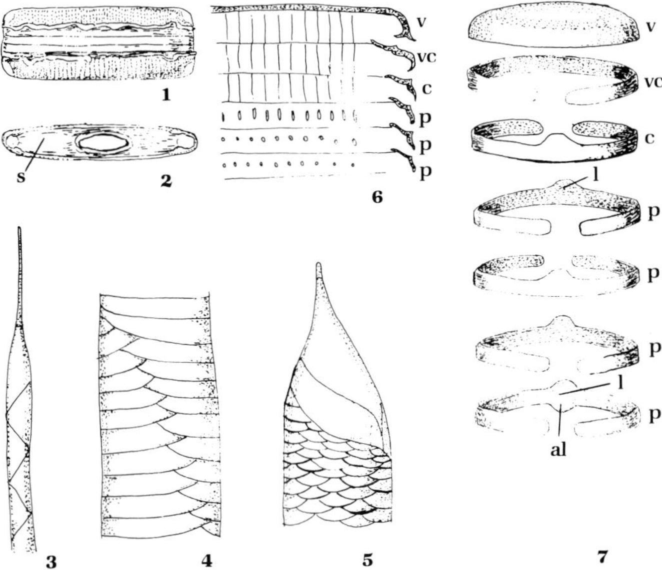

Figure 8.7

Girdle band modifications. **1,2.** *Grammatophora marina.* **1.** Girdle view showing valves separated by girdle bands. **2.** Valve view of girdle band, a complete band, with septum (s) closing all but a small central area. **3.** *Rhizosolenia hebetata* (Bailey) Gran f. *semispina* (Hensen) Gran, showing cuneate half-bands. **4.** *Dactyliosolen antarcticus* Castracane, portion showing half-bands, with junction of successive pairs slightly offset from those preceding, resulting in a twisted zigzag junction. **5.** *Rhizosolenia acuminata* (H. Peragallo) Pavillard, apical region, and part of the segmented girdle band area, with numerous segments in each band. **6,7.** Succession of incomplete bands, redrawn from Stosch, 1975. **6.** Cross section of valve (v), adjacent intercalary band or valvocopula (vc), next band or copula (c), and remaining connecting bands or pleurae (p). **7.** Same, expanded, showing nature of the incomplete bands, the gap being occupied by the ligula (l) of the next more distal band; some may have expansions at both margins, forming the ligula and antiligula (al).

direction, the **antiligula**, may close a similar gap in adjacent bands by overlapping (Stosch, 1975).

Raphe System. The **raphe** is a longitudinal slit in the valve of the pennate diatoms along the apical axis. It is commonly present on both valves, but in some (e.g., *Achnanthes, Cocconeis*) occurs only on one valve, the opposite valve having a smooth **hyaline area**, the **axial area** or **pseudoraphe**, along the apical axis (see Figure 8.8). Still others (the araphid diatoms) have only an axial area on both valves. A similar hyaline area at the poles is termed the **apical area**, and a median expansion the **central area**.

The raphe is interrupted at the valve midpoint, at a thickened central area or **central nodule.** A transapical expansion of the central nodule nearly to the valve margin is termed a **stauros.** Similar thickenings of the wall at each end of the raphe are termed **polar nodules.**

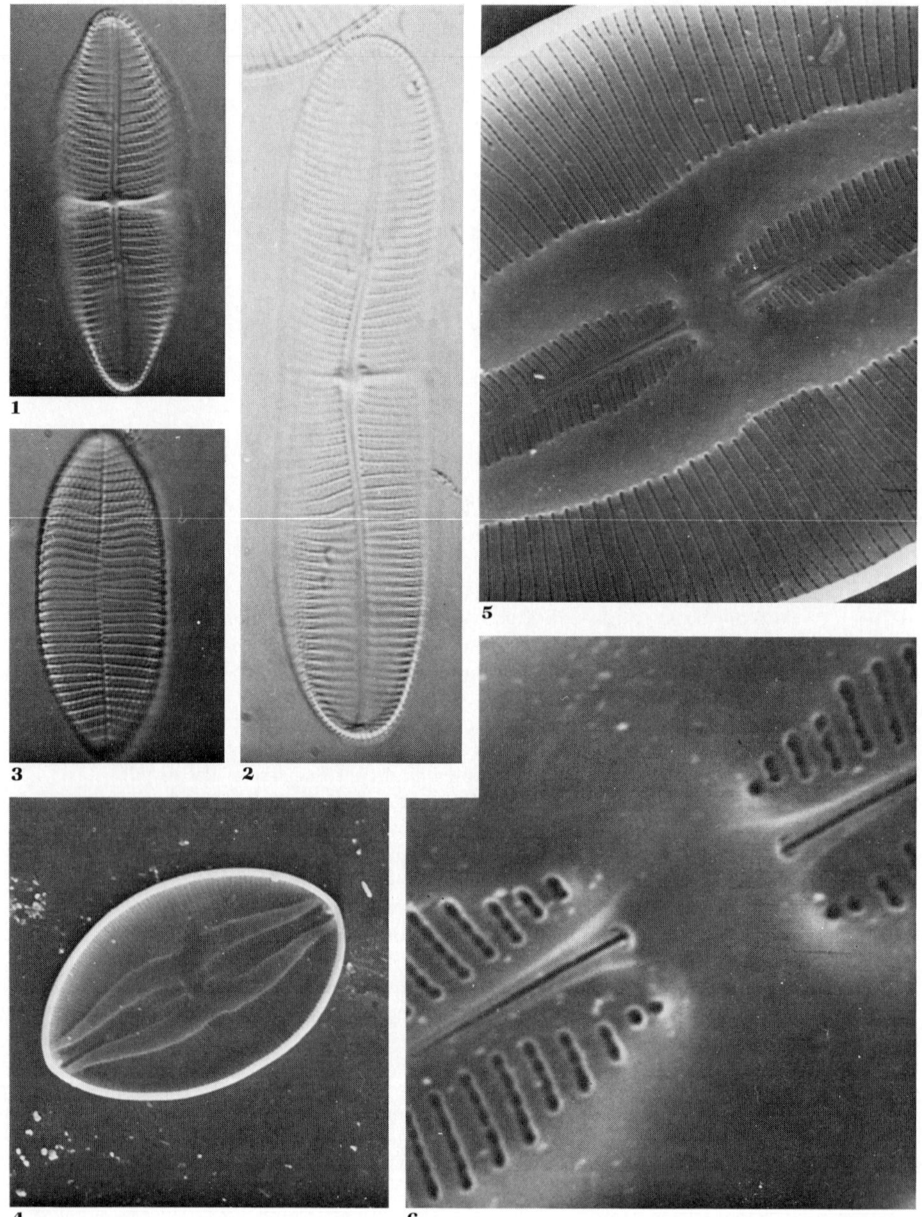

Figure 8.8

1–3. *Achnanthes longipes* Agardh, ×600. **1,2.** Raphe-bearing valves of small and large cells. **3.** Valve with median axial area or pseudoraphe only. **4–6.** *Navicula spectabilis* Gregory, Miocene, Valmonte diatomite, Southern California, SEM. **4.** Entire cell, ×520. **5.** Central area enlarged to show raphe slit, transverse rows of pores that appear as striae at low magnification, and H-shaped nonperforate hyaline area, ×1600. **6.** Enlargement of central area, showing central nodule, central pores at ends of raphe slits, hyaline area, and rows of pores, ×4000.

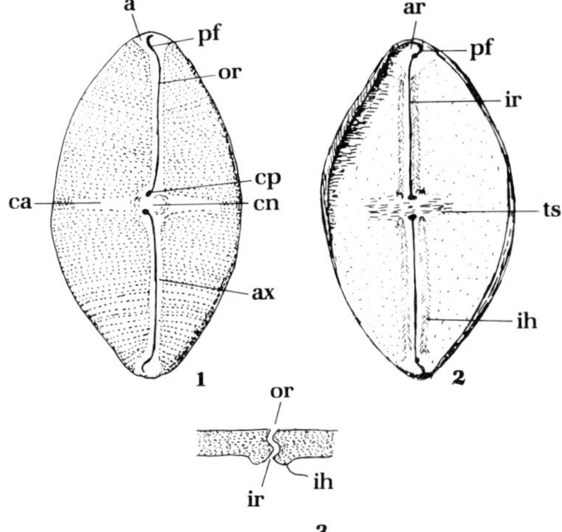

Figure 8.9
Raphe morphology, semidiagrammatic; or, outer raphe fissure; ir, inner raphe fissure; pf, polar fissure or groove; ar, apical ridge; ih, inner apical horn or ridge bordering raphe; ax, axial hyaline area; cn, central nodule; ca, central area; cp, central pore; ts, transapical stauros. **1.** Valve exterior. **2.** Valve interior. **3.** Much enlarged section through raphe between central area and apex.

Expansions of the raphe fissure at the central nodule form the **central pore,** and at each polar nodule a similar expansion forms the **polar pore.**

In cross section, the raphe is like a V on its side, the outer segment being termed the outer fissure and the inner part of the V the inner fissure (see Figure 8.9). A curved groovelike extension of the raphe, the central fissure, may extend onto the central nodule from the central pore, and similar grooves, the polar fissures, curve terminally around the polar nodules, both ends turning in the same direction on a single valve, but in the opposite direction on the opposing valve, resulting in a diagonal symmetry. The central and polar fissures, when present, are merely grooves in the valve, and do not penetrate it as does the raphe itself (Schrader, 1973b). A funnel-shaped expansion, the infundibulum, may be present on the inside of the valve at the terminal nodules.

Although the typical naviculoid raphe system consists on each valve of two grooves, one at each end of the valve from the central nodule to the polar pore, *Nanoneis* has a raphe only from the central nodule to one end of each valve, with a hyaline axial area or pseudoraphe at the other end of the valve. The raphe of one valve is at the end of the cell that is occupied by an axial area on the other valve (Norris, 1973). The monoraphid *Cocconeis* and *Achnanthes* have a fully developed raphe only on one valve. The central pores are not well developed and no central fissures occur, although the polar pore and fissures are present. This is regarded as a more primitive raphe system than that of the biraphid diatoms (Schrader, 1973b).

The raphe of some diatoms is carried on a raised keel or winglike extension on one or both margins of the valve. The raphe-bearing keel may be central in position (*Bacillaria*), excentric and near the valve margin (*Eunotia*), diagonally aligned (*Nitzschia*), spirally encircle the cell (*Cylindrotheca*), or be canallike on winglike projections (alae) on both margins of the valve (*Surirella*). The silica shell of *Cylindrotheca fusiformis* Reimann & Lewin is enclosed in organic material, but when carefully removed, the weakly silicified valve is shown to consist of four parallel strips (see Figure 8.10), the inner two adjacent to the raphe fissure being connected by silica arches, the **fibulae,** which cover the raphe (Reimann et al., 1965).

The **canal raphe** system consists of a raphe with a tubular canal on its inner margin, separated by fibulae from the remainder of the cell interior, and is found in representatives of the Epithemiaceae, Nitzschiaceae, and Surirel-

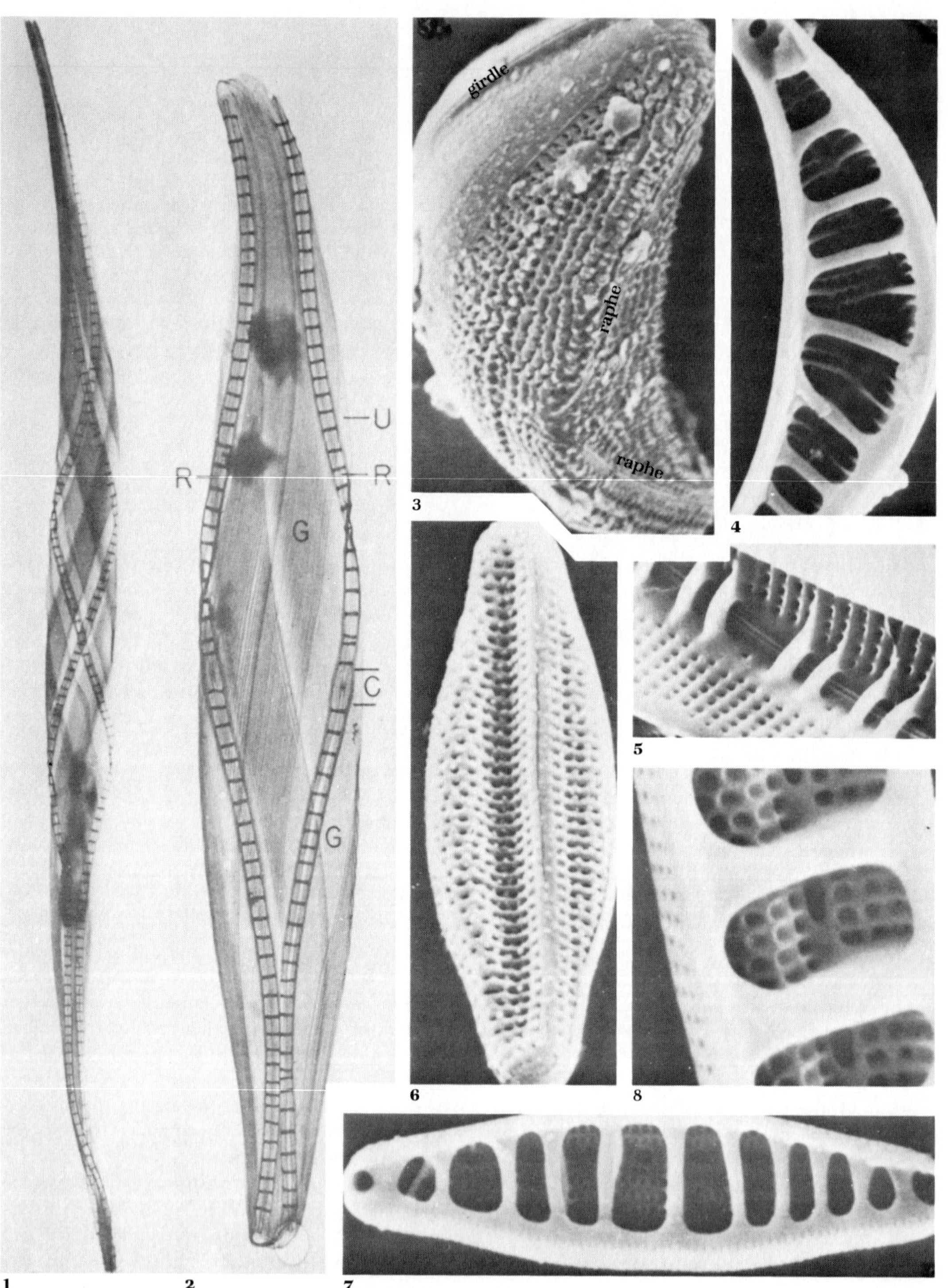
girdle
raphe
raphe
U
R
R
G
C
G
1
2
3
4
5
6
7
8

Figure 8.10
Raphe systems. **1,2.** *Cylindrotheca fusiformis,* TEM of cell wall, showing spiralling raphe crossed by many fibulae, ×4000. **2.** Same, showing frustule constructed of two parts, each with raphe (R), central nodule (C), and girdle (G) of many girdle bands, with unsilicified part (U) between raphe and girdle, ×7500; from Reimann et al., 1965. **3,4.** *Epithemia cistula* (Ehrenberg) Ralfs var. *lunaris* Grunow, SEM, from Paddock and Sims, 1977. **3.** Oblique exterior view; valve girdle at upper left, raphe visible as narrow line crossing valve from upper right toward lower left, then turning sharply toward lower right, ×2000. **4.** Valve interior, showing septa, ×2200. **5.** *Hantzschia virgata* (Roper) Grunow var. *intermedia* (Grunow) Round, SEM, part of interior of valve, showing raphe slit bridged by arched fibulae, but with no canal beneath raphe, ×4300, from Mann, 1977. **6–8.** *Denticula tenuis* Kützing, SEM, from Paddock and Sims, 1977. **6.** Exterior of valve, ×5500. **7.** Inside of valve showing costae that divide cavity into chambers, ×3900. **8.** Portion of interior enlarged to show elongate portulae opening from base of chambers into the raphe canal, ×8000.

---

laceae (see Figure 8.11). The fibulae are any raphe-bridging siliceous structures (Mann, 1977) separated by interfibular spaces (regarded in light microscopy as keel punctae). Recent studies (Paddock and Sims, 1977) indicate that as similar canal raphes also occur in some other genera, not all have a common origin. The term is thus descriptive for those diatoms in which fibulae form a duct into which the raphe opens from the valve exterior. This duct communicates with the cell cavity through **portulae**, holes in the inner raphe canal wall. In some instances the portula may represent merely an interfibular space. The canal raphe may be elevated on a ridge or keel, whose walls enclose a cavity termed a gable. A special elongate passageway through the raised keel of *Surirella robusta* Ehrenberg is termed an **alar canal;** it opens into the raphe canal at one end and into the frustule cavity at its other end. In some *Surirella,* an opening may be present in the keel between adjacent alar canals and is termed a fenestra (Paddock and Sims, 1977).

Fine Structure of the Wall.   Even the standard light microscope allowed recognition of the complexity of the diatom wall, indicating that siliceous membranes partially closed some of the perforations. Later, resolution was somewhat increased by the use of ultraviolet illumination. More recently, electron microscope studies have aided immeasurably in determination of wall fine structure. Most diatom genera have by now been examined in TEM (e.g. by Helmcke and Krieger, 1952a–c; Desikachary, 1954, 1957; Okuno, 1958, 1959, 1962; Cassie and Bertaud, 1960). Because TEM tends to flatten depth, stereo pairs were used in determining relief and wall structure. More recently SEM has greatly facilitated studies of diatom shell structure (e.g., Miller, 1969; Crawford, 1973b; Fryxell and Hasle, 1973; Gasse, 1970, 1971; Hasle, 1972a,b, 1973a–d; Jerković, 1972; Heimdal, 1973; Mann, 1977; Paddock and Sims, 1977; Ross and Sims, 1971, 1972, 1973; Schrader, 1973b; Sydow and Christenhuss, 1972; and many others).

Early electron microscope studies led to the description of many variations in wall structure (Kolbe, 1948; Desikachary, 1957; Hendey, 1959, 1971) represented by two basic types: a laminar wall of a single layer of silica, and a locular one with separate inner and outer layers connected by vertical partitions to form the chambers between. Study of many more taxa, particularly by SEM, has shown that a variety of wall structures is produced by different types of wall perforations (Ross and Sims, 1972).

Surface markings on the frustule result from the contrast of thinner and thicker parts of the shell as well as various types of perforations. These features generally are radial or concentric about a central point in the centric diatoms, or the symmetry may be modified by structures at the margins; in pennate diatoms the ornamentation is related to a main central axis.

Areolae, Pseudoloculi, and Alveoli.   As presently defined (Anonymous, 1975), the basal

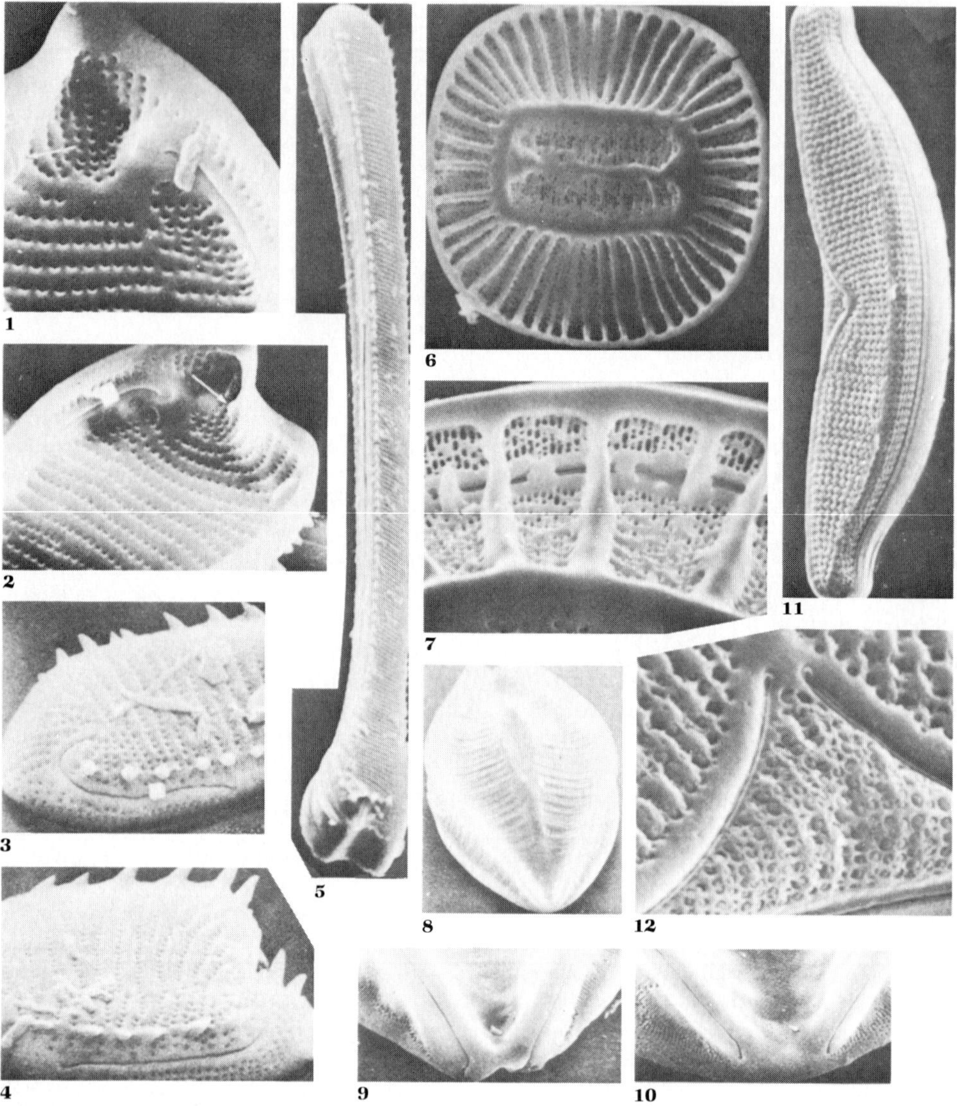

Figure 8.11

Raphe modifications in SEM. 1—5. *Actinella punctata* Lewis, from Hasle, 1973d. **1,2.** Internal views of broad end, showing short raphe in valve mantle ending at terminal nodule, apical pore field, and labiate process (arrow); **1**, ×7600; **2**, ×4960. **3,4.** Exterior of narrow and broad valve poles, showing short raphe slits, ×4320. **5.** Exterior of entire valve, ×1680. **6,7.** *Campylodiscus clypeus* Ehrenberg, from Paddock and Sims, 1977. **6.** Interior of entire valve, ×385. **7.** Enlargement of part of interior, showing prominent costae, raphe opening into tubular canal that is crossed by costae, and shorter fibulae, ×1360. **8—10.** *Surirella elegans* Ehrenberg, from Schrader, 1973b. **8.** Entire frustule, ×400. **9.** Central area showing canal raphe. **10.** Apical area showing ends of raphe canals. **9,10**, ×2000. **11,12.** *Epithemia turgida* Ehrenberg, from Paddock and Sims, 1977. **11.** Entire frustule, showing biarcuate raphe, at ventral margin for most of length, arching toward central nodule at valve midpoint, central part of raphe in groove, ×800. **12.** Enlargement of central area with central nodule and raphe ends at dilated central pores, ×3600.

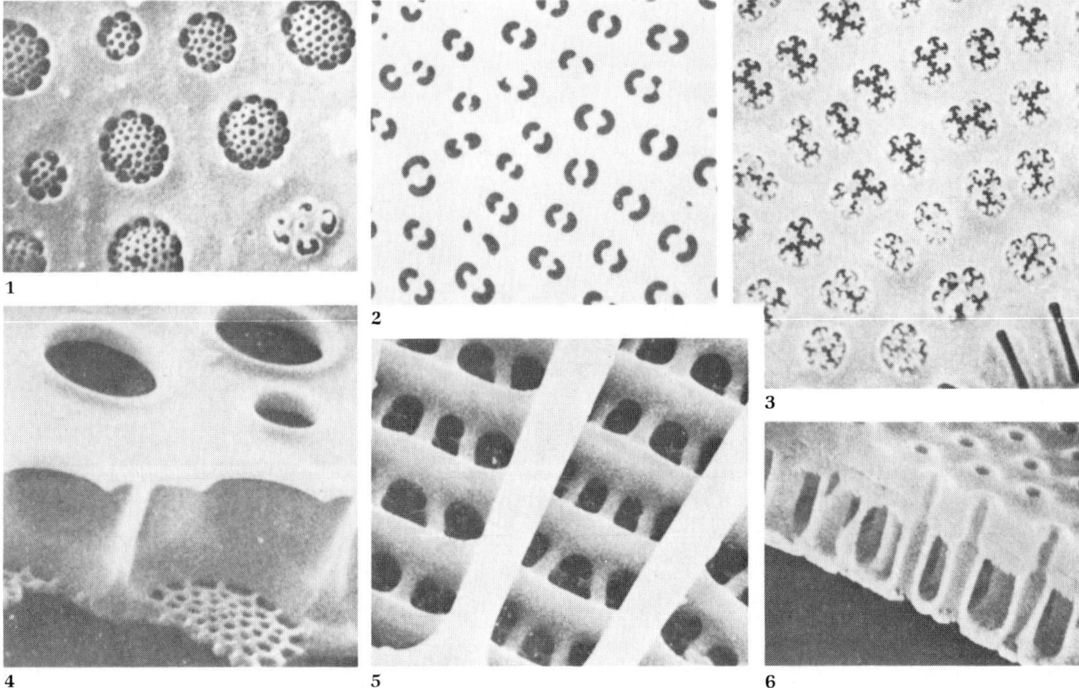

Figure 8.12

Wall perforations, SEM. **1.** *Triceratium antediluvianum,* Miocene, Spain, surface showing poroids with cribra, and strutted process at lower right, ×5000. **2.** *Triceratium shadboltianum* Greville, part of outer surface, showing poroids with rotae, ×5000. **3,5.** *Arachnoidiscus ehrenbergii* Bailey, Miocene, California, ×2500. **3.** Detail of outer surface near center, showing elongated pores at lower right, and poroids with complex volae projecting in from the margins. **5.** Detail of inner surface, showing poroids, strong radial costae, and lower circular ones. **4.** *Coscinodiscus oculus-iridis* showing broken margin, inner surface above, with large foramina at inner surface, and cribra at outer surface, ×10,000. **6.** *Actinocyclus octonarius* Ehrenberg, edge of broken valve, inner surface above, showing loculi with velae at outer surface, and foramina at inner surface, ×5000. All from Ross and Sims, 1972.

siliceous layer of the wall is perforated regularly by subcircular, elliptical, or hexagonal chamberlike structures, the **areolae,** that are normally covered by a very thin layer or **velum** that may be slit or perforated (see Figure 8.12). Four main types of areolae may be recognized: (1) the poroid areola (areola poroides), which is not markedly constricted at the valve surface, as in *Triceratium antediluvianum* (Ehrenberg) Grunow; (2) the loculate areola (areola loculata), occluded by a velum at one surface of the valve, and markedly constricted in diameter at the other surface to a small opening termed the foramen (also known as a cover-pore), as in

Coscinodiscus oculus-iridis Ehrenberg; (3) the pseudoloculus, a chamber formed by distal expansion of anastomosing or reticulate costae on the outside of the basal siliceous layer, as in *Melosira moniliformis* (O. F. Müller) C. Agardh (see Figure 8.13); and (4) the alveolus, an elongated chamber or channellike structure extending from the axial or central part of the valve to its margin, with perforated outer layer and large opening to the cell interior, as in *Pinnularia* and *Cyclotella.*

The valve surface may appear striate due to linear alignments of alveoli, hence forming alveolate striae; or punctate striae may be formed

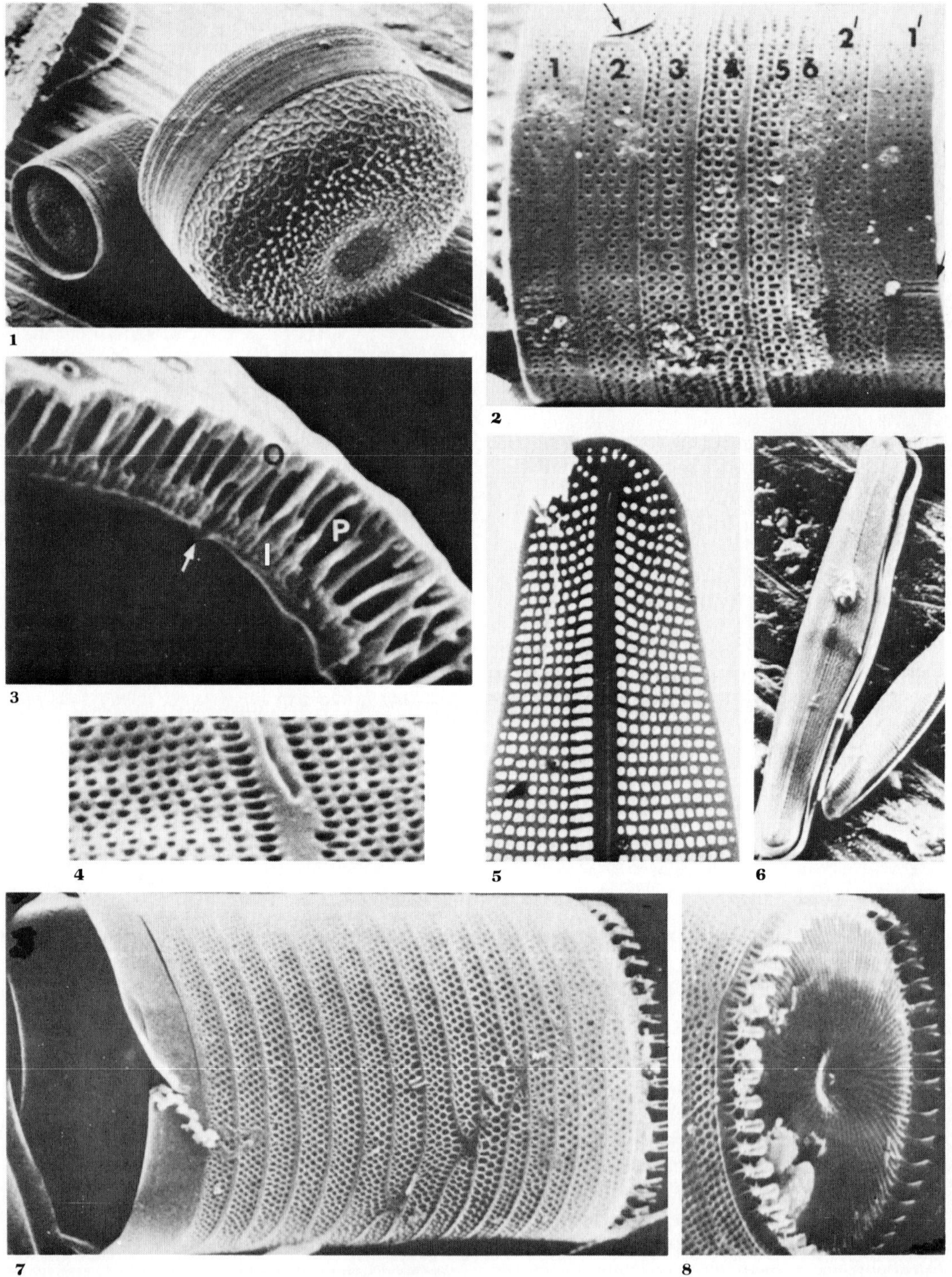
1
2
3
4
5
6
7
8
O
P
I
1 2 3 4 5 6 2' 1'

Figure 8.13
1−3. *Melosira moniliformis,* SEM, from Crawford, 1977. **1.** Frustules of large and small cells, ×800. **2.** Complete girdle, bands 1−6 of epicingulum, and bands 1′ and 2′ of hypocingulum; other hypocingular bands overlapped by epicingulum, with arrows indicating open ends of bands, ×2400. **3.** Fracture of valve mantle through pseudoloculi (P), a labiate process (arrow), and outer (O) and inner (I) wall layers, ×9000. **4,5.** *Berkeleya hyalina* (Round & Brooks) Cox. **4.** SEM of central end of raphe, showing central hyaline area and striae composed of aligned pores, ×16,000. **5.** TEM of valve apex, showing aligned pores in striae and row of enlarged pores at one side of raphe, ×5000. **6.** *B. scopulorum* (Brébisson) Cox, SEM of entire frustules, ×1200. 4−6, from Cox, 1975. **7,8.** *Detonula pumila* (Castracane) Schütt, SEM, from Hasle, 1973c. **7.** Lateral view, showing numerous girdle bands, and marginal circlet of strutted processes on the valve periphery at right, ×3200. **8.** Oblique valve view, showing strutted processes at valve edge, and radial costae on valve face leading to central strutted process, ×3300.

---

by similar rows of elements of uncertain character and termed puncta (but which may later be recognized as areolae, poroids, or loculi). Punctae are among the smallest structures, commonly from 0.5 to 1 μm in diameter, but as little as 0.037 μm in *Thalassiosira gravida* Cleve (Hendey et al., 1954). The punctae may be scattered, or occur in rows (striae), and occasionally were reported erroneously as beads when studied in the light microscope. The fine and regular punctate ornamentation of *Amphipleura pellucida* Kützing allowed its use as a classical test object for determining the resolving power of microscope lenses. The apparent fine striations of this species are seen under high magnification to be rows of very tiny perforations that in turn are seen in the electron microscope to contain a fine porous plate at each inner extremity, with perforations only 6 to 8 nm in diameter (Stoermer and Pankratz, 1964).

The velum is a thin perforated silica layer across an areola; it may (1) be perforated by regularly arranged pores to form a **cribrum**, as in *Triceratium antediluvianum;* (2) consist of a bar or many radial bars across an alveola, the **rota**, either with or without a widened area at the center, as in *T. shadboltianum;* or (3) consist of many separate elements projecting inward from the areolar margin, the **vola**, as in *Arachnoidiscus ehrenbergii.*

Three other types of openings or cavities may occur in the wall: (1) **passage pores** may connect adjacent loculate areolae; (2) a **bullula**, or bubblelike void, may occur between well-separated areolae and lack an opening to either surface of the basal siliceous layer, as in *Aulacodiscus reticulatus* Pantocsek; and (3) a continuous space or **hypocaust** may occur within the basal siliceous layer between well separated thin-walled areolae.

Elevations, Horns, and Folds. Other modifications of the diatom frustule include thickened areas of the wall or elevations above the general surface (see Figure 8.14). An **elevation** is a raised area of a structure similar to the remainder of the valve and not projecting laterally beyond the valve margin, as in *Biddulphia biddulphiana* (J. E. Smith) Boyer. An exceptionally long and narrow elevation is a **horn**, as found in *Hemiaulus polymorphus* Grunow. Similar elongated outgrowths of the valve projecting beyond the valve margin are the hollow **setae** (or **awns**) of *Chaetoceros* and *Bacteriastrum delicatulum* Cleve. The end cells of a colony may have terminal setae that differ in orientation from those of other cells. The valve wall also may be indented as a fold (*Hemiaulus*) or more deeply as a sulcus, as in *Melosira ambigua* (Grunow) O. Müller.

Hyaline Fields. Areas where the basal siliceous layer is not perforated by areolae or punctae are termed **hyaline fields.** These fields may be radiate or stellate as in the hyaline rays of *Asteromphalus.* Pennate diatoms may have an axial area, consisting of a hyaline field along the apical axis that is imperforate (previously termed a pseudoraphe in araphid diatoms, but

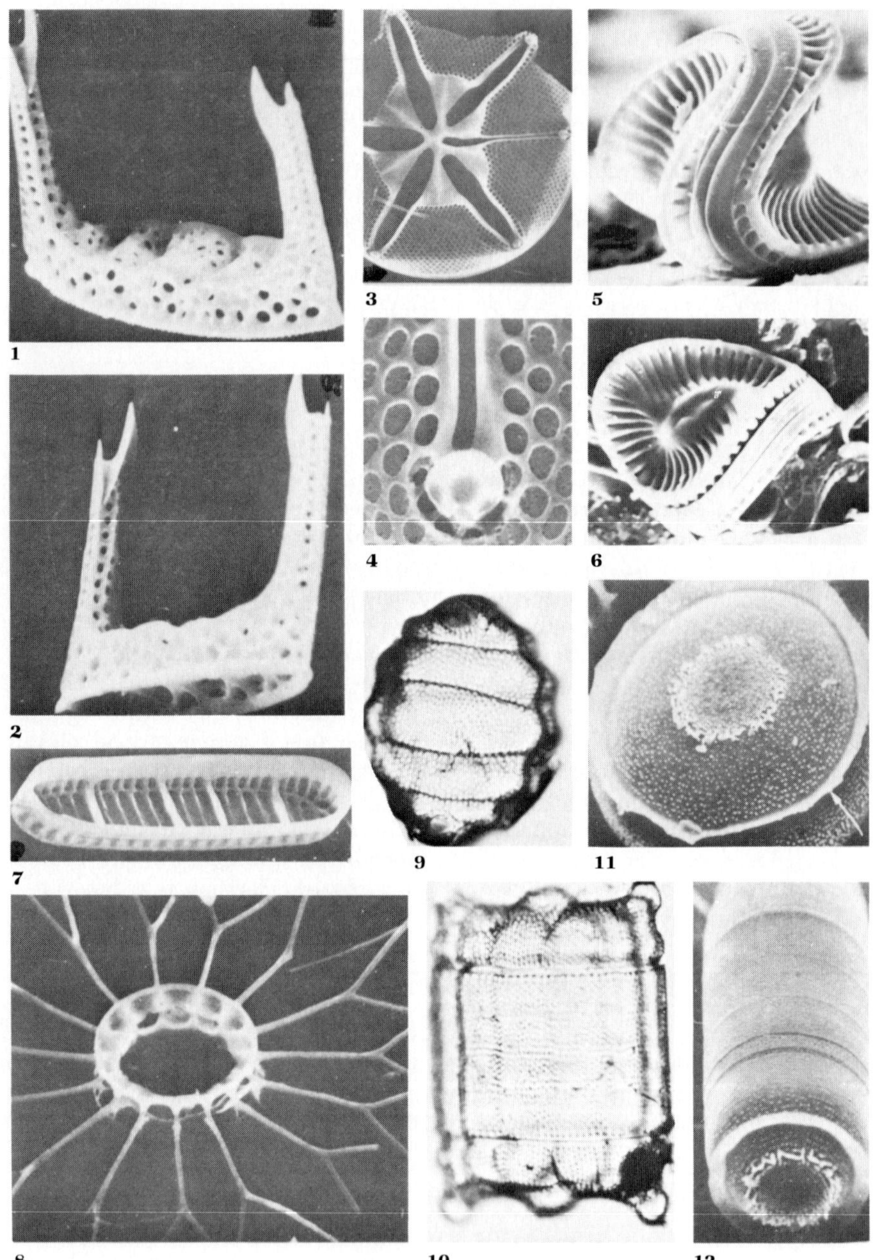

1
2
3
5
4
6
7
9
11
8
10
12

Figure 8.14
Morphologic features. **1,2.** *Hemiaulus polymorphus,* SEM, oblique views from above and below, show-ing linking spines on elevations at each end, with indented transverse folds visible from above, and internal costae seen from below; **1,** ×712; **2,** ×740; both from Ross et al., 1977. **3,4.** *Asteromphalus* sp. in SEM, from Hasle, 1972a. **3.** Interior of valve, showing stellate hyaline area, ×640. **4.** Enlargement of narrow hyaline ray showing marginal labiate process at the end, ×5520. **5,6.** *Plagiodiscus costatus* Jurilj, SEM, from Jerković, 1972. **5.** Girdle view, ×260. **6.** Oblique view, with raphe-bearing keel or alae at edge of valve, ×220. **7.** *Denticula seminae* Simonsen & Kanaya, inside view, SEM, showing pseudosepta, secondary pseudosepta, keel punctae, and raphe, ×2080, from Hasle, 1972b. **8.** *Bac-teriastrum delicatulum,* SEM, showing adjacent valves with bifurcating setae, ×2000, from Hargraves, 1976. **9,10.** *Biddulphia pulchella,* valve view and girdle view, showing transverse folds and knobby elevations, ×600. **11,12.** *Melosira nummuloides* (Dillwyn) C. Agardh, SEM. **11.** Valve view, showing marginal carina (arrow), and circlet of linking spines forming a corona near the center, ×610; from Crawford, 1973b. **12.** Double cell, with epivalves of both cells visible, separated by many girdle bands that conceal the hypovalves, ×1600, from Crawford, 1973c.

---

similar areas occur in both these and raphe-bearing taxa). A central hyaline area is an ex-panded or otherwise separate portion in the midregion of the axial area in the pennate diatoms. A central area that extends transapi-cally across the valve is termed a **fascia.** Some *Navicula* have a lateral area, an apical extension of the expanded hyaline central area that sepa-rates the striae into distinct areas on each side of the center, as in *Navicula spectabilis* (see Fig-ure 8.8 part 4).

Keels, Costae, Craticula, and Collars. Solid, elongate thickenings of the valve may be pres-ent in many diatoms. The thickening may ap-pear as a keel (also termed alae in some), the summit of the raphe-bearing ridge in taxa with a sharply angled valve, such as *Surirella elegans.* A marginal ridge, continuous or interrupted, lies between the valve face and valve mantle in *Eunotogramma weissei* Ehrenberg. A similar ringlike series of irregular projections or **link-ing spines** on the valve of *Melosira* is termed a **corona** (Crawford, 1973b). Costae are elongated solid linear thickenings, and may be external or internal in position. The internal costae of *Eunotogramma weissei* are also termed **pseudosepta** (see Figure 8.15). Both primary and secondary pseudosepta may occur in *Den-ticula seminae* and *D. lauta.* **Craticula** are structures consisting of thick siliceous bars within the valve of some pennate diatoms, such as *Navicula cuspidata* Kützing. Another special

costa is the **collar** that may be elevated on the outer side of the valve (also termed a carina by Crawford, 1973b) as in *Melosira nummuloides.*

Internal Ducts and Stigma. A specialized internal duct system is present in *Entogonia.* Each valve has three ducts that underlie and follow the border of the inner and outer trian-gles, probably attached to this border through-out their length. Near the valve angles, the ducts bend upward toward the valve surface, where they open into a pair of orifices. Rarely the ducts fuse below the surface and open into a single orifice. Because of their delicate nature, the ducts may not be preserved, but their pres-ence may be indicated by the paired orifices (Holmes and Brigger, 1977).

Other special features of diatom frustules in-clude the **stigma** of some pennate diatoms, a perforation of the basal siliceous layer near the central nodule; the stigma has an unoccluded external opening and may be elevated, and has a spongy or cracked appearance internally. *Cym-bella cistula* (Hemprich) Kirchner (see Figure 8.16) is an example.

Ocelli, Pseudocelli, and Pseudonodules. Api-cal and marginal fields may appear distinct from the remainder of the valve. Examples of these are (1) the **pseudocellus,** a field of areolae of smaller size than those of the remainder of the valve, as in *Trigonium* and *Biddulphia;* (2) the **ocellus,** a siliceous plate with thickened

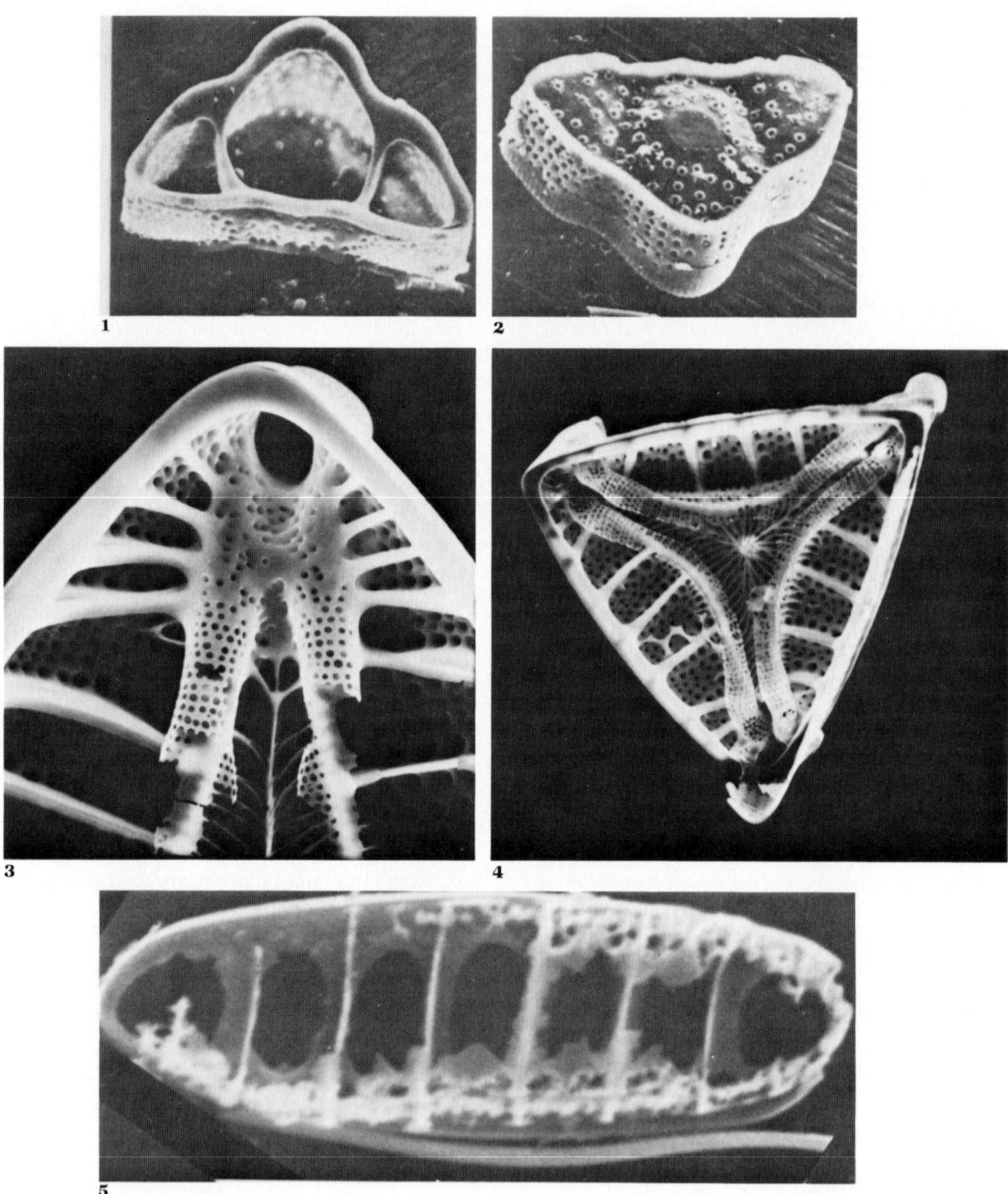

Figure 8.15

Internal features of diatoms, SEM. **1,2.** *Eunotogramma weissei,* Eocene, USSR, ×525, from Ross and Sims, 1972. **1.** Interior of valve, showing inner costae. **2.** Oblique exterior, showing marginal ridge between valve face and mantle. **3,4.** *Entogonia* sp., interior, showing pseudosepta, and internal perforated ducts that connect adjacent valve corners and open into orifice on exterior; 3, ×1500; 4, ×560; both from Holmes and Brigger, 1977. **5.** *Denticula lauta* Bailey, Miocene, Luisian, Southern California, showing pseudosepta, ×4500.

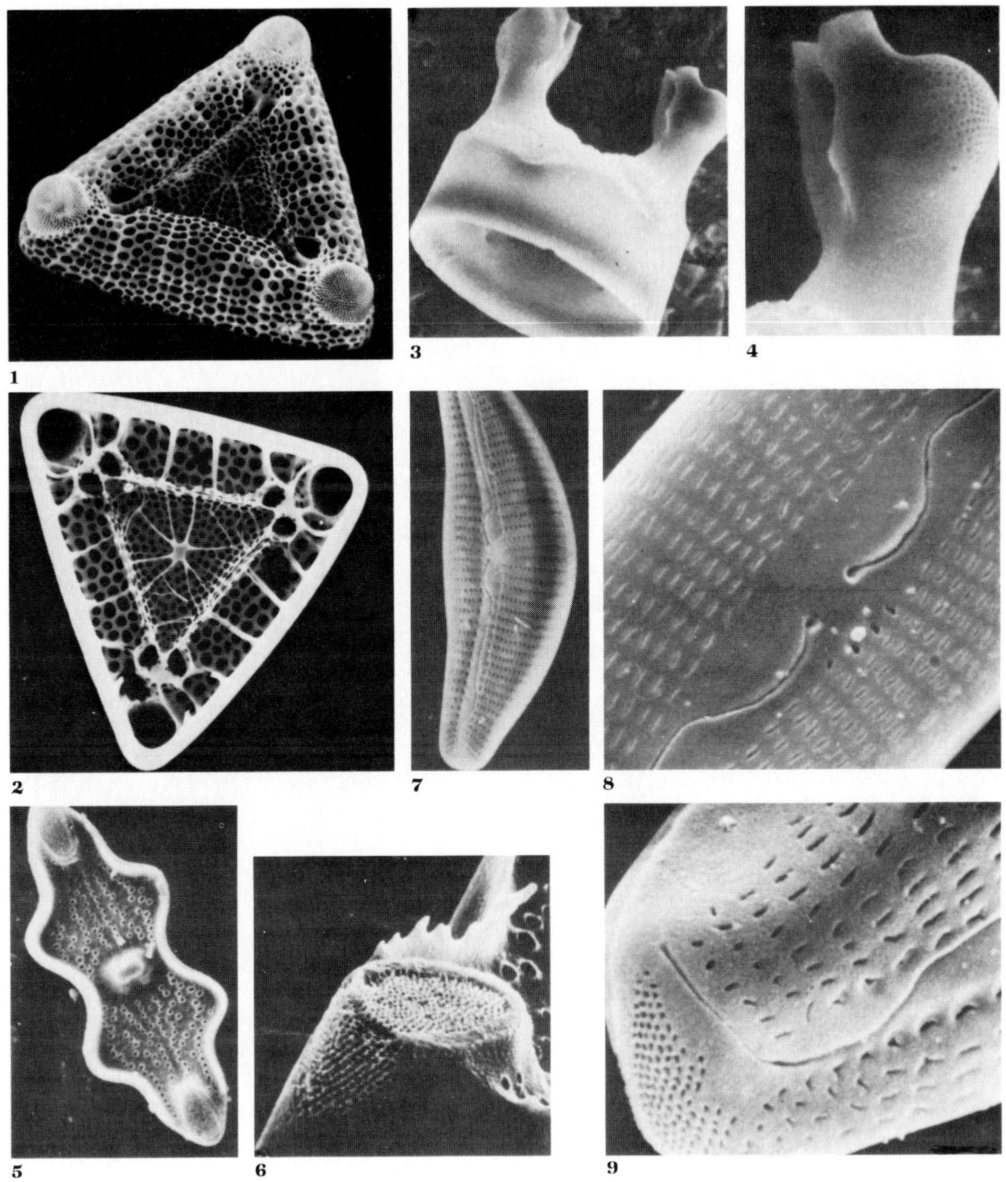

Figure 8.16

Stigma, pseudocelli, and ocelli in SEM. **1,2.** *Entogonia davyana* Greville var. *orbicularis* Holmes & Brigger, from Holmes and Brigger, 1977. **1.** Outside of valve, showing punctate elevations or pseudocelli at corners, ×256. **2.** Inside view, with pair of orifices at each angle of inner triangle, and pseudosepta dividing the outer triangle, ×270. **3,4.** *Hemiaulus haitensis* R. Ross, from Ross et al., 1977. **3.** Valve exterior from side, showing prominent elevations that bear linking spines, ×545. **4.** Enlargement of one bilobed elevation showing pores of pseudocellus at right, ×1470. **5,6.** *Plagiogramma nankoorense* Grunow, Miocene, California, from Ross and Sims, 1973. **5.** Entire valve from exterior, ×600. **6.** Enlarged apex of valve showing ocelluslike structure, ×2400. **7–9.** *Cymbella cistula*, a freshwater epiphyte on *Chara*, in SEM, from Hufford and Collins, 1972. **7.** Entire valve, ×1200. **8.** Central area, showing raphe ends and central pores, as well as four large isolated punctae forming the stigma, ×2960. **9.** End of valve, showing sharply angled raphe end, rows of elongate punctae down to the raphe, and many fine punctae at tip of valve, ×8000.

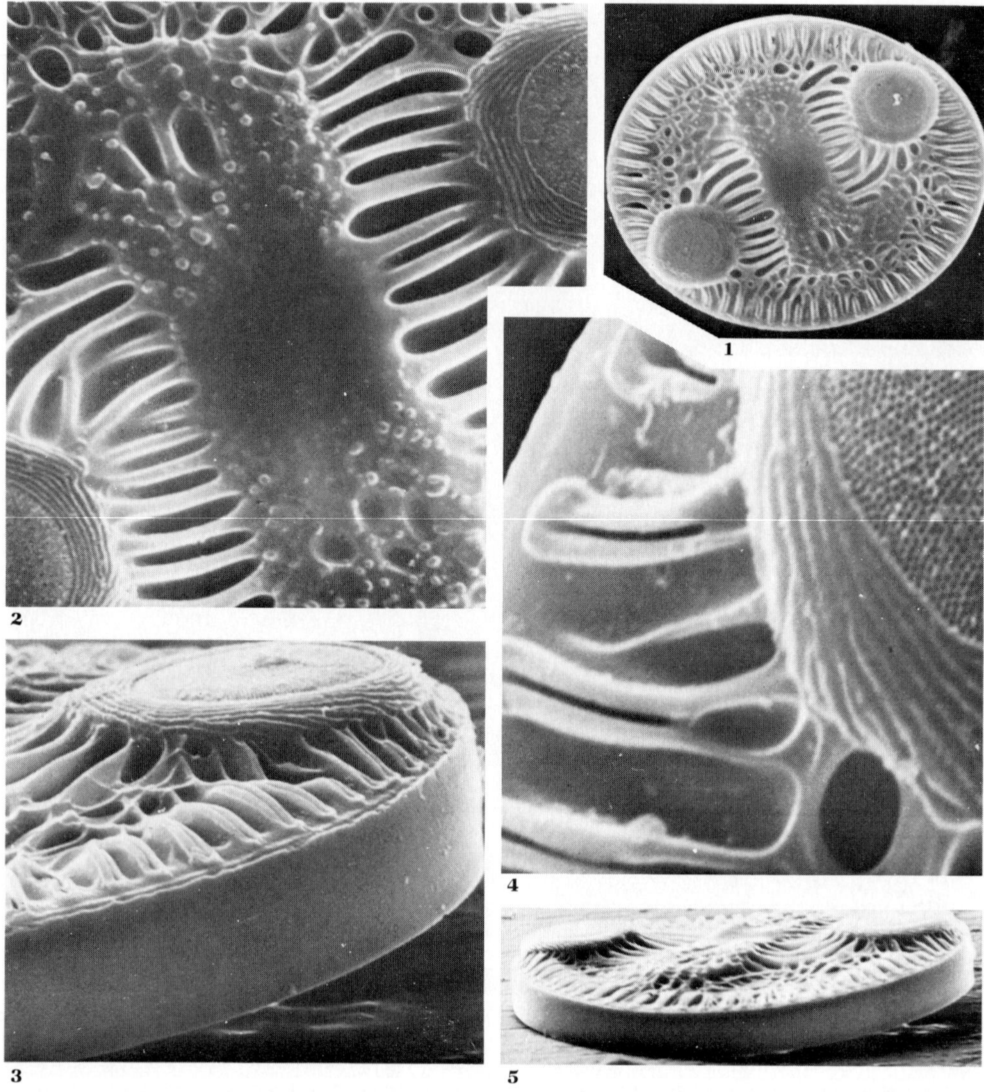

Figure 8.17

Auliscus caelatus J. W. Bailey, Miocene, Malaga Mudstone, Southern California, SEM. **1.** Exterior view of valve showing large elevated ocelli, central hyaline area, and complex ornamentation, ×480. **2.** Enlargement of central hyaline area, bordering area of elongate perforations leading toward the ocelli, and granules at extremities of the hyaline area, ×1600. **3.** Part of valve in edge view, showing smooth valve mantle, elevated ocellus with porelli, and radial slits with liplike borders at valve edge, ×1600. **4.** Vertical view of part of perforated ocellus at upper right, and radial marginal slits, ×4000. **5.** Entire valve, as in part 3, ×480.

structureless rim pierced by closely spaced holes or porelli (singular, porellus); in pennate diatoms such as *Synedra,* the porelli are arranged in longitudinal rows, whereas in the centrics such as *Auliscus,* they are radial, concentric, or irregular in arrangement (see Figure 8.17); (3) a **pseudonodule** is a marginal to submarginal indented structure, only one per valve (see Figure 8.18). An areolate pseudonodule (pseudonodulus areolatus; Simonsen, 1975) is an aggregation of somewhat smaller areolae in a limited depressed area near the valve margin, as in *Actinocyclus normanii* (Gregory) Hustedt. An areolate-operculate pseudonodule consists of a depression in the valve plane covered by a siliceous membrane (operculum) with a central multiple opening (the pseudonodulus multiplex of Ross and Sims, 1972), as in *Hemidiscus cuneiformis* Wallich. In the operculate pseudonodule (pseudonodulus operculatus) the areolae beneath the operculum fuse and disappear, leaving only the operculum (the pseudonodulus simplex of Ross and Sims, 1972), as in *Actinocyclus octonarius.* A luminate pseudonodule (pseudonodulus luminatus) consists of a simple circular opening in the siliceous valve membrane with neither areolae nor operculum; it is known only in *Roperia tesselata* (Roper) Grunow.

Processes.   A variety of processes with imperforate walls occur in the centric diatoms and are of considerable systematic importance (see Figure 8.19). The **labiate process** (**rimoportula** of Ross and Sims, 1972) is a pore or tube through the valve that may project from the surface, but has an unconstricted outer opening; it is restricted to form a straight or arcuate slit at the interior of the valve, where it commonly is bordered with thickened margins and may project somewhat (*Actinoptychus senarius* (Ehrenberg) Ehrenberg). A special form of labiate process is the **periplekton** of *Rutilaria,* whose distal extensions clasp similar structures of adjacent frustules of a colony.

An **occluded process,** the hollow tube being distally constricted, is known in *Thalassiosira angstii* (Gran) Makarova. A **strutted process** (previously termed an apiculus, the fultoportula of Ross and Sims, 1972, or strutted tubulus of Hasle, 1972a) consists of a tube through the valve, with internal arched supports and surrounded by satellite pores or chambers. One or more strutted processes may be central or marginal in position in the Thalassiosiraceae (see Figure 8.20). Mucilage threads may be extruded through the strutted process in *Skeletonema potamos* (Weber) Hasle.

Spines of varying size, closed or solid projections from the surface, may be termed spinules, granules (small rounded projections), or spines. **Linking spines** that connect frustules of adjacent cells in chains occur in *Hemiaulus capitatus* Greville (see Figure 8.21).

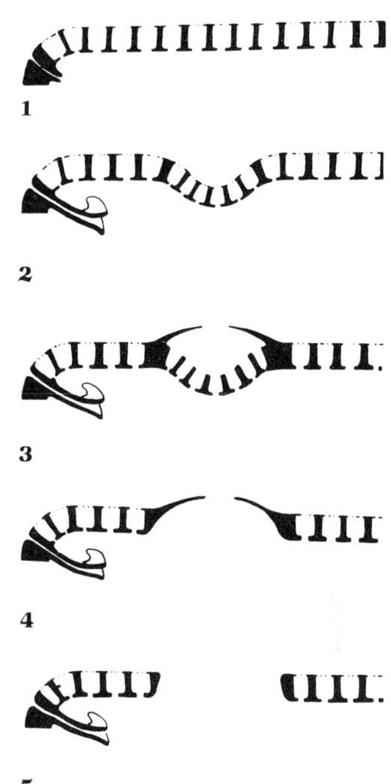

Figure 8.18
Diagrammatic cross sections of pseudonodules.
1. Possible ancestral type, as in *Coscinodiscus.*
2. Areolate pseudonodule. **3.** Areolate-operculate pseudonodule. **4.** Operculate pseudonodule. **5.** Luminate pseudonodule, perhaps derived from areolate or operculate type. All after Simonsen, 1975.

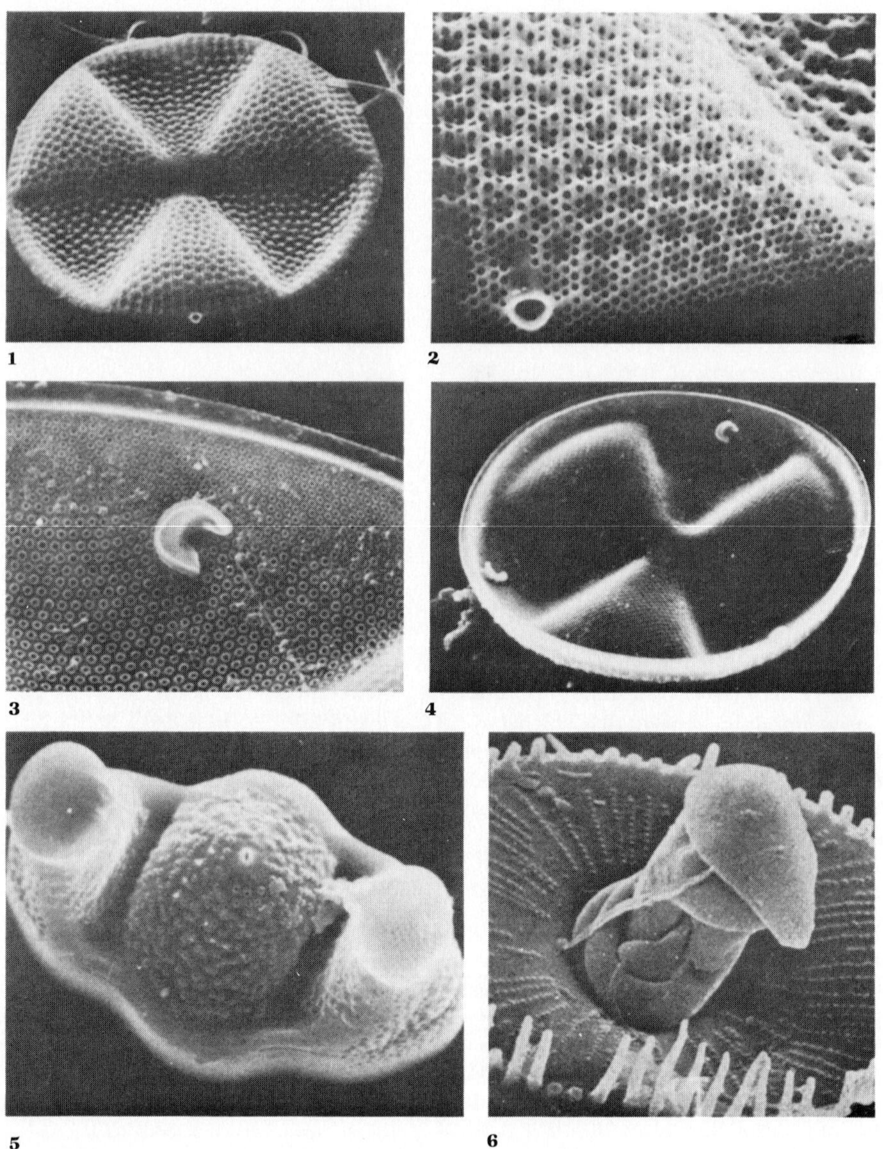

Figure 8.19

Labiate processes. **1 – 4.** *Actinoptychus* sp. cf. *senarius,* SEM, from Fryxell and Hasle, 1973. **1.** Exterior, with labiate process at center front, ×800. **2.** Same, enlarged, ×3050. **3.** Interior of valve, showing projecting curved form of labiate process, ×3850. **4.** Same, ×1150. **5.** *Hemiaulus includens* Grunow, valve view, with labiate process at center of median elevation, ×1750, from Ross et al., 1977. **6.** *Rutilaria radiata* Grove & Sturt, Eocene, Oamaru, New Zealand, central part of valve with periplekton, a modified labiate process, and other labiate processes on periphery, ×1500; from Ross and Sims, 1972.

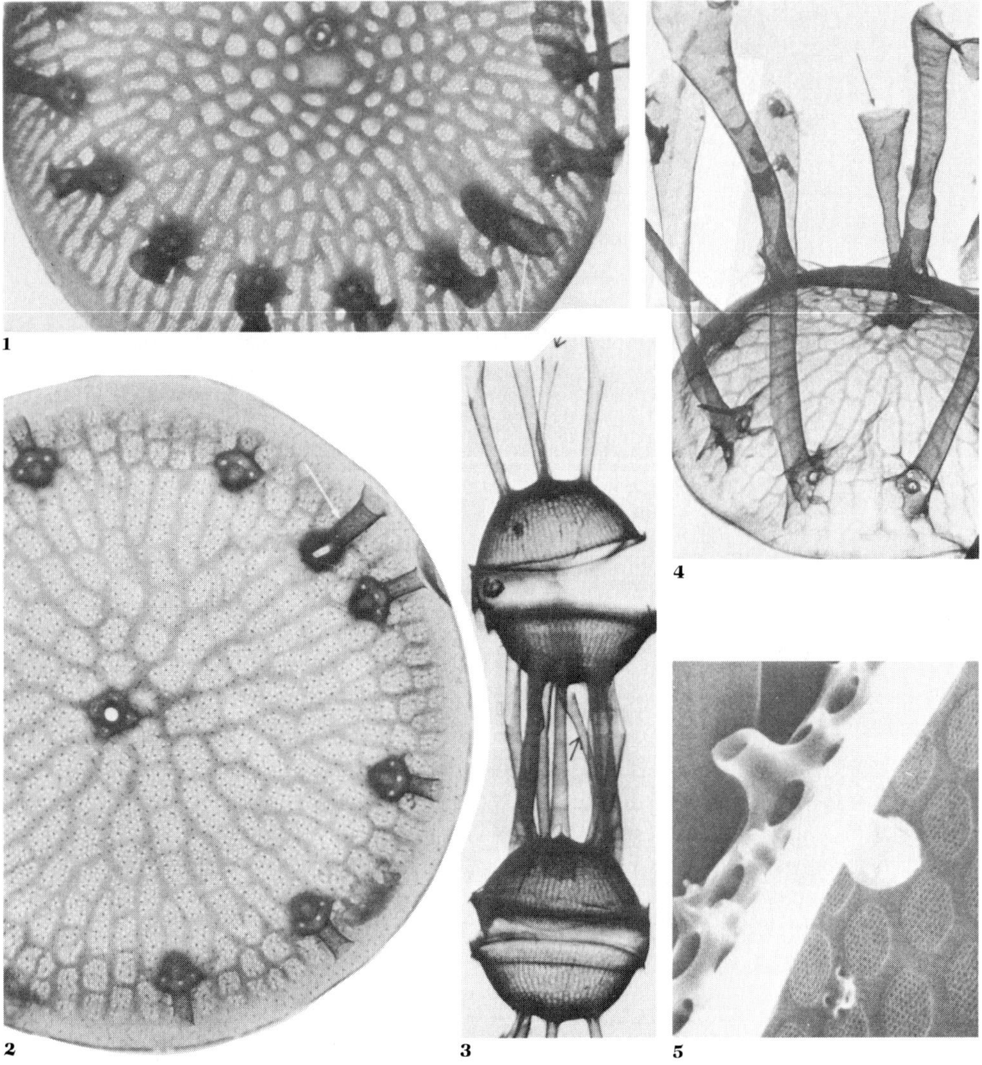

Figure 8.20

Labiate and strutted processes. **1.** *Detonula confervacea* (Cleve) Gran, part of valve in TEM, showing one central and a marginal ring of strutted processes, one labiate process (arrow), and many areolae with sieve membranes, ×14,400. **2.** *Thalassiosira* sp., with one central and one marginal ring of strutted processes, and a marginal labiate process (arrow), TEM, ×16,000. 1,2, from Hasle, 1972a. **3.** *Skeletonema potamos*, TEM, showing two cells, connected by means of strutted processes, and a labiate process (arrow), ×5600, from Hasle and Evensen, 1976. **4.** *S. costatum*, TEM, part of valve with marginal strutted processes and pores, and a central labiate process (arrow), ×14,400. **5.** *Thalassiosira* sp., valve margin with labiate process and strutted processes; note areolae with inner sieve membranes; SEM, ×12,000. 4,5, from Hasle, 1972a.

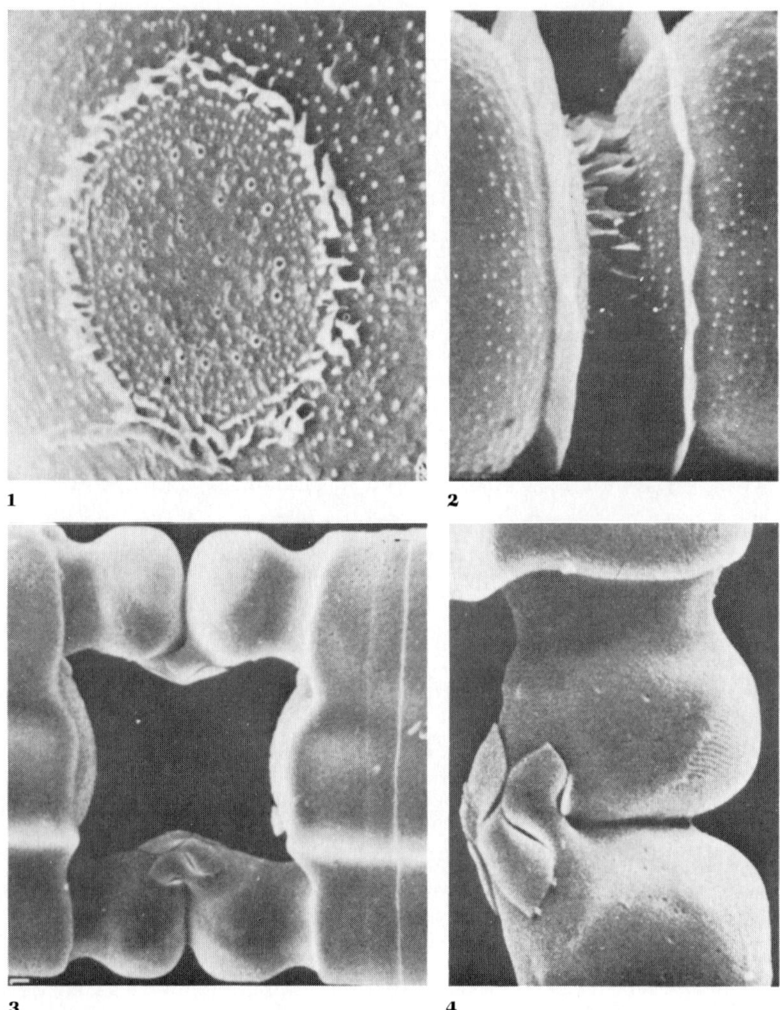

Figure 8.21

Linking spines, SEM. **1,2**, *Melosira nummuloides*, from Crawford, 1973b.
1. Valve view, showing ring of irregular spines forming the corona, and surrounding a ring of strutted processes, ×7000. **2.** Girdle view of interlocking coronas of linking spines of two cells, and marginal flangelike carina on each cell, ×3600. **3,4.** *Hemiaulus capitatus*, Eocene, Oamaru, New Zealand. **3.** Two valves linked by spines on the elevations, ×670, from Ross et al., 1977. **4.** Enlargement of one elevation and linking spines, ×1500, from Ross and Sims, 1972.

PHYSIOLOGY

Vegetative Reproduction

Two modes of reproduction occur in diatoms, the most usual being simple binary fission. The cell division process has been studied in the laboratory in synchronous cultures. Cell synchrony is induced in nature by the alternation of night and day, hence all cells of a filament commonly divide simultaneously (Lewin et al., 1966). In cell division, the protoplast first increases in size, the nucleus divides mitotically, and other organelles are divided or duplicated. If a single plastid is present it splits longitudinally, but if more than one is present, they are divided equally and do not increase in number until cell division is completed. Pyrenoids also increase in number by division, never arising *de novo*. After division or segregation of the various cell organelles, the cytoplasm divides transverse to the longitudinal axis (in the valvar plane).

Following cytoplasmic division, two new silica valves form in the center of the old cell, one for each new half of the dividing cell. As growth continues, the original two valves are pushed apart, and each daughter cell receives one old and one new valve. The newly formed valves lie inside those inherited from the parent, hence are hypovalves; those received from the mother cell are the epivalves of the new individuals. One daughter cell is the same size as the parent, having inherited the former epivalve. That whose new epivalve was the hypovalve of the parent is then smaller than the parent; thus the average size of successive generations is progressively smaller (Macdonald-Pfitzer Hypothesis; see Figure 8.22 and Table 8.1).

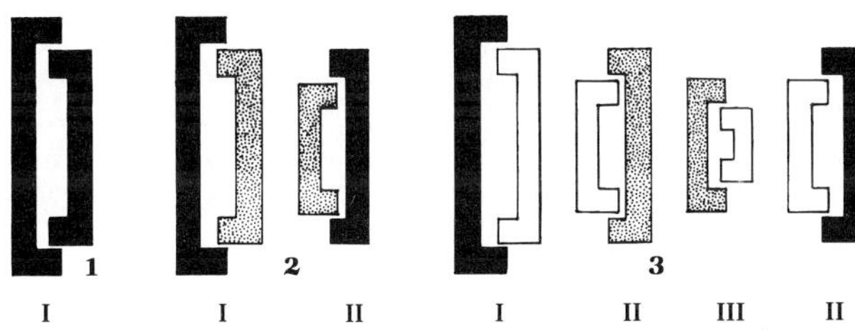

Figure 8.22
Diagram to illustrate decrease in size distribution in successive generations, according to the Macdonald-Pfitzer hypothesis. 1–3. Cells present at time of three successive generations; original parent cell valves shown black; next generation stippled; third generation white. I, size of original parent cell; II, second largest; III, third largest.

Table 8.1
Progressive decrease in mean size of diatom population with successive cell divisions, according to Macdonald-Pfitzer Hypothesis.

	a	> b	> c	> d	> e	> f	> g	Total cells
Original cell	1							1
First division	1	1						2
Second division	1	2	1					4
Third division	1	3	3	1				8
Fourth division	1	4	6	4	1			16
Fifth division	1	5	10	10	5	1		32
Sixth division	1	6	15	20	15	6	1	64

The cell dimensions and size ratios also change with repeated divisions. In pennate diatoms, the transapical and pervalvar axes decrease less rapidly than the apical axis in the smaller daughter cells, so that the cells appear more stumpy. The ornamentation is not merely reduced in size but becomes more closely spaced, and may even be differently oriented.

A few species do not show such reduction in size (e.g., *Navicula pelliculosa*). The cell wall may be weakly silicified and the girdle sufficiently plastic to allow resumption of the original size after cell division is completed (*Eunotia*), or the two valves may be identical in size, and overlap restricted to the connecting bands of the girdle.

During cell division, each daughter cell inherits one of the parental valves, and possibly one of the old girdle bands as well. However, *Stephanodiscus* produces two new girdle bands for each of the daughter cells while still enclosed within the girdle bands of the parent (Round, 1971). Examination of other diatoms during division is necessary to determine whether or not this is usual or atypical for diatoms.

Cylindrotheca fusiformis, Navicula pelliculosa, and most other diatoms only produce a silica shell following nuclear division and cytokinesis, but the requirement for completed cytokinesis prior to shell formation is not universal. *Cyclotella cryptica* apparently forms the silica shells as a function of cytoplasm volume (Badour, 1968). In culture, when cell division was inhibited by a deficiency of calcium or magnesium ions, the cells continued to elongate and nuclear division resulted in binucleate cells. The silica content of each cell also increased to an amount sufficient for the two valves that would normally have been produced. As cell division did not occur, two lateral silica shells formed in the region of the connecting bands, perpendicular to the normal valve orientation. These lateral shells had the usual valve fine structure, although the shape was slightly modified to fit the cell curvature. Nuclear division had preceded the lateral shell deposition, however, as in other diatoms.

Although cell division of other protists commonly takes place at night or in the early morning, that of *Skeletonema costatum, Ditylum brightwellii* (West) Grunow, and other diatoms maintained in light-dark cycles occurred during the light period (Jørgensen, 1966; Eppley et al., 1967).

Some diatoms may divide more than once per day. The highest rate recorded for a benthic diatom is 3.2 divisions per day, whereas planktonic species may divide up to 8 times daily. Their more rapid rate of cell division is correlative with a high photosynthetic rate per unit of biomass (Lewin and Guillard, 1963).

Sexual Reproduction and Auxospore Formation

The progressive decrease in cell size resulting from vegetative reproduction has a lower viable limit. Maximum size is then regained during the process of sexual reproduction by the formation of special rejuvenescent cells, the **auxospores**, thus avoiding the extinction of the clone. The few species with a more elastic girdle band (e.g., *Eunotia*), which do not decrease in size with successive divisions, do not produce auxospores (Geitler, 1935) and apparently have no sexual reproduction.

When first observed, auxosporulation was thought to be solely a vegetative process, but laboratory cultures and cytological studies have shown it to be an integral part of sexual reproduction in diatoms. First recognized in the pennate diatoms (Geitler, 1932), it was not thought to occur in centric species until oogamous reproduction first was demonstrated in *Melosira* by Stosch (1950). The process of sexual reproduction in diatoms thus has a dual advantage: genetic recombination and the restoration of cell size.

Diatoms are diploid organisms, with meiosis occurring at the time of gamete formation, and haploidy restricted to the gametes. Except for *Rhabdonema adriaticum,* all diatoms apparently are monoecious, a single clone being capable of sexual reproduction.

Production of auxospores is limited to cells of medium size, about 30 to 40 percent of the maximum, and does not occur in large cells or in

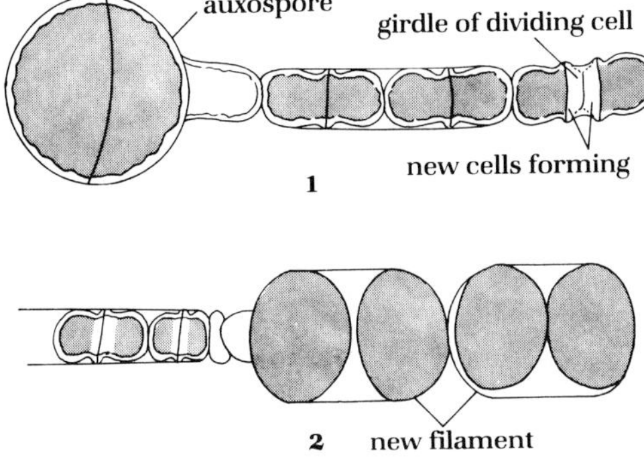

Figure 8.23
Auxospore of *Melosira*. **1.** Colony showing vegetative division at right, and auxospore at left of chain. **2.** Production of new large-sized filament at right from auxospore, ×1200.

those below a certain minimum (see Figure 8.23). Gametes are produced by cells less than 20 to 35 μm in diameter in *Coscinodiscus concinnus* W. Smith, whereas the average cell size is 27 to 48 μm. Male gametes of *Isthmia nervosa* were produced by cells between 81 and 122 μm in diameter, and female gametes in cells of 114 to 132 μm. In centric diatoms, the sex of the gametes is also a function of size, the onset of sexual reproduction and sex determination perhaps being related to the ratio of nuclear to cytoplasmic volume. The medium-sized cells may continue to reproduce vegetatively for a time, but under the influence of some environmental stimulus may instead produce gametes, a zygote and auxospore. In the laboratory, auxosporulation may be triggered by transfer to a weaker medium, by manganese deficiency, increased salinity or sodium ion, increased temperature or light intensity (although one species reacts to decreased light), or to a shortened photoperiod. In nature, it may be caused by seasonal modifications of light or temperature, or nutrient depletion. The degree of change necessary to result in auxospore formation is progressively less as the cell diameter of the population decreases, smaller cells responding more readily to the environmental stimuli (Gross, 1937).

As is true of other protists, the variety of modifications of the sexual process seems endless (Geitler, 1935; Gross, 1937; Stosch, 1950; Ramsfjell, 1959a; Dawson, 1966; Manton and von Stosch, 1966; Heath and Darley, 1972; Drebes, 1972, 1977).

Sexual Reproduction in the Centric Diatoms

Reproduction in the centric diatoms is oogamous, the female gamete or egg being nonmotile, and the male gametes being uniflagellate. The various modifications of the development of gametes, fertilization, and auxospore formation known to date have been summarized by Drebes (1977).

Male Gamete Formation. Successive cell divisions produce small diploid spermatogonia within the parent frustule. Two meiotic divisions then result in the production of four uniflagellate male gametes. A male gamete of *Coscinodiscus concinnus* is 4 to 8 μm in length, exclusive of the 32- to 42-μm-long flagellum (Holmes, 1967). The single flagellum has two opposed rows of hairs or mastigonemes along its length, thus resembling the flagella of the Chrysophyta (Stosch, 1958c). However, in cross section, the flagellum is unusual in lacking the central pair of microtubules in the axoneme that characterize the usual eucaryote 9+2 flagellum (Manton and Stosch, 1966; Heath and Darley, 1972), having only the ring of nine paired microtubules.

Vegetative cells of *Cyclotella* between 5 and 7 μm in diameter function directly as spermatogonia, each producing four uniflagellated male gametes (Schultz and Trainor, 1970). In *Melosira* and *Bacteriastrum* a sequence of mitotic divisions, termed depauperizing mitoses, produces 2, 4, 8, 16, or 32 spermatogonia, hence from 8 to 128 male gametes. *Bacteriastrum hyalinum* Lauder produces up to 16 naked spermatogonia that enlarge, escape from the mother cell, and undergo two meiotic nuclear and cytoplasmic divisions. After the first meiotic division, each daughter cell produces two flagella, which are divided between the two cells resulting from each of the second meiotic divisions, so that the final male gametes are uninucleate (Drebes, 1972). Probably the same two-stage nuclear and cell division with flagellar production interspersed is the explanation of the reports of 32 biflagellate microgametes in *Rhizosolenia hebetata* f. *semispina* (Hensen) Gran (Ramsfjell, 1959a), and of both uniflagellate and biflagellate gametes in cultures of *Cyclotella meneghiniana* Kützing (Schultz and Trainor, 1968). Some centric diatoms such as *Stephanopyxis turris* have thin silica shells covering the spermatogonia. *Guinardia flaccida* (Castracane) Peragallo has nuclear depauperizing mitoses without cytokinesis, resulting in multinucleate spermatogonangia.

Development of the gametes by the spermatogonia may occur in two different ways. In the first of these, as in *Bacteriastrum hyalinum*, the first meiotic nuclear division results in a binucleate cell that produces two flagella. The flagellar bases separate and migrate to the poles during subsequent division, one being retained by each of the resulting cells at the completion of the second meiotic division. The result of this method is the production of four small colorless uniflagellate gametes and a mass of residual cytoplasm and assorted organelles, including the plastids. It occurs in *Actinocyclus, Actinoptychus, Biddulphia, Melosira,* and *Stephanopyxis* (Heath and Darley, 1972; Drebes, 1977).

More common than the above is the method in which both meiotic nuclear divisions in the spermatogonia are accompanied by cytokinesis, and four uniflagellate male cells are produced, leaving no residual cytoplasm.

Female Gamete. Unlike the male gametes, the **oogonia** develop directly from vegetative cells, without an intervening depauperizing mitosis. The oogonia are the mother cells of the auxospores. In production of the oogonia, the vegetative cell elongates, the nucleus enlarges, and the plastids increase in number. The subsequent meiotic divisions and the fate of the resultant nuclei may occur in three different ways in the centric diatoms (Drebes, 1977).

In possibly the most primitive method, the first meiotic division is followed by cytokinesis, producing two egg cells. The second meiotic division is not accompanied by cell division, but instead one of the resulting nuclei degenerates. Each of the two egg cells thus contains a haploid nucleus and one pycnotic or nonfunctional nucleus, as in some *Biddulphia* and in *Lithodesmium.*

A modification of this method results in a single egg and a nonfunctional polar body. Following the first meiotic division, an unequal cell division takes place. The larger segment then undergoes a second meiotic division, followed by degeneration of one of the resulting nuclei. The single functional egg thus has a single nucleus and a pycnotic one, and the polar body remains nonfunctional. This method is known only in *Biddulphia rhombus* (Ehrenberg) W. Smith and *Cerataulus smithii* Ralfs ex Pritchard.

The third method produces a single egg, as each meiotic division is followed by degeneration of one nucleus, and no cytoplasmic division takes place. The final egg contains a large haploid nucleus and two pycnotic nuclei of different sizes. This is apparently the most advanced method, and occurs in most centric diatoms, such as *Stephanopyxis turris* and *Isthmia nervosa.*

Fertilization and Auxospore Formation. As the egg cell commonly is retained within the shell of the mother cell, sperm penetration is effected in various ways. *Melosira varians* has a narrow annular slit at the edge of the hypovalve and free margin of the epipleura; a fertilization pore 5 to 6 μm in diameter develops in one of the connecting bands in *Isthmia nervosa;* swelling of the protoplast of *Stephanopyxis turris* results in a small temporary opening; male gametes penetrate through a bristle opening in the girdle region of *Chaetoceros;* and the two eggs of *Biddulphia* sp. separate the oogonial valves, leaving a gap between them. The oogonial shells are shed entirely in *Bellerochea* and *Lithodesmium,* leaving the cells naked, and the weakly silicified membrane of *Biddulphia rhombus* may be penetrated by enzymatic action of the gametes (Drebes, 1977).

Following fertilization, the auxospore swells to some multiple of the diameter or apical length of the mother cell. The auxospore commonly differs in wall structure from the vegetative cell, and deposits an outer silicified series of rings, a **pre-perizonium**, similar to the perizonium of pennate diatoms, within which the new initial cell forms (see Figure 8.24).

Auxospores may be classified according to their relationship to the parent cell after formation (Anonymous, 1975). A free auxospore, as in *Lithodesmium undulatum,* has no direct connection to the parent cell. A terminal auxospore occurs at the end of the parent cell, as in *Rhizosolenia alata* Brightwell. When the auxospore is produced from the girdle of the parent cell, with its pervalvar axis transverse to the pervalvar axis of the parent cell, as in *R. shrubsolei* Cleve or *Bacteriastrum hyalinum,* it is termed a lateral auxospore. Intercalary auxospores are those with the parental epitheca and hypotheca still attached, while the cingula have broken away during enlargement of the auxospore, as in *Melosira varians* and the Coscinodiscaceae. A semi-intercalary auxospore may have one side still attached to a valve of the parent cell, and the other side either free or connected to a second auxospore, as in *Odontella regia* (Schultze) Simonsen and some *Biddulphia.*

The initial cell formed by the auxospore commonly differs in shape and structure from the normal vegetative cells, and some have even been described as separate taxa. Formation of each of the new valves is preceded by nuclear mitotic division, the extra nuclei becoming pycnotic. Later division of the initial cell produces the new series of vegetative cells, or chains of these.

Autogamy. An autogamous production of auxospores is reported in some centric diatoms, although the same species also may produce flagellated male cells. Possibly the male gametes are nonfunctional, or the autogamous development of auxospores may be facultative, occurring in the absence of fertilization. *Actinoptychus undulatus* Bailey in culture showed no nuclear fusion of gametes; instead, some 75 percent of the eggs have an incompleted

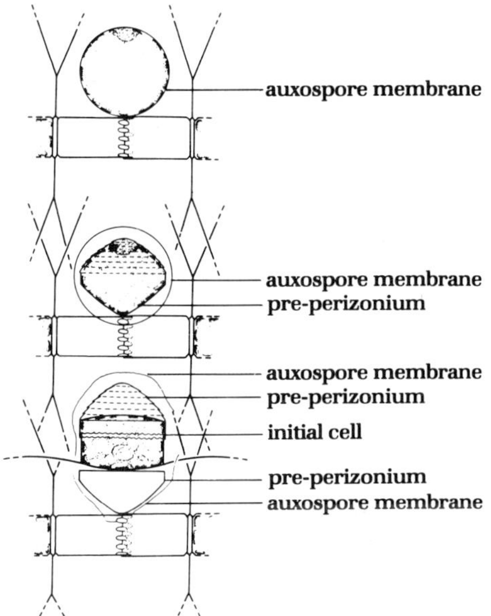

Figure 8.24
Bacteriastrum hyalinum, development of auxospore. Top, a lateral auxospore. Center, pre-perizonium secreted within the auxospore membrane. Below, initial cell produced within auxospore, the hypotheca with bristles and the epitheca lacking these; the pre-perizonium has split in two as the initial cell forms, and the auxospore membrane has begun to disintegrate. From Drebes, 1972.

meiosis, and parthenogenetic auxospore formation. The remaining 25 percent that did complete meiosis were not fertilized and died (Drebes, 1977).

Sexual Reproduction in the Pennate Diatoms

As in the centric diatoms, the pennate vegetative cell is diploid, meiosis occurring in the late stage of gametogenesis, and auxospores forming from the zygote. Unlike the centric oogamy, however, gametes of the pennates are morphologically alike and nonflagellate. Gametes are brought together by pairing of mother cells that act as gametangia. As in centric diatoms, various modifications of the basic process occur in the pennate diatoms, grouped in five different types (Drebes, 1977).

A transitional form of reproduction occurs in the araphid Diatomaceae. Vegetative bandlike colonies of *Rhabdonema adriaticum* differentiate male and female gametangia, the former (spermatogonia) arising from two or more depauperizing mitoses, as in the centrics. The end result is the production of two globular nonflagellate naked microgametes per spermatogonium, each with a haploid nucleus and a degenerating pycnotic one. The female equivalent arises by elongation of a vegetative cell, the first meiotic division producing an egg cell and a nonfunctional polar body, the second producing one halploid nucleus and a pycnotic one. Contact of the spermatogonia and oogonia is passive due to water movement, but upon contact, they become attached by means of mucous pads. By amoeboid movement the released spermatia approach the fertilizable area of the oogonium, the male nucleus alone enters the egg, and the zygote inflates to become an auxospore. A **perizonium** is formed within the zygote membrane.

The second and more usual type of pennate reproduction has morphologically alike male and female cells. The gametangial mother cells associate in pairs and become surrounded with mucilage. Both cells produce gametes like those of the egg cells of the centrics and araphid diatoms, each resulting in two similar nonmotile gametes that contain a normal haploid nucleus and a pycnotic one. Both gametes from a single cell may move, or one may be active and one remain passive within the mother cell, those from different mother cells fusing to form auxospores. As the auxospore enlarges, the outer membrane ruptures and a silicified perizonium is formed. The initial cell forms within this, each valve being produced in conjunction with a mitotic division. This type has been observed in *Amphora, Anomoeoneis, Cymbella, Navicula, Nitzschia,* and *Synedra.*

The third type of sexual reproduction in pennates is probably derived from the second type. Cytokinesis does not follow the first meiotic division, and only one functional gamete is produced by a mother cell. Paired mother cells thus produce a single auxospore. This type is known in some *Cocconeis, Eunotia, Navicula, Nitzschia,* and *Surirella.*

Type four involves automixis, in which auxospores arise from a single cell, either by pedogamy, in which a mother cell produces two gametes that then fuse (some *Amphora* and *Cymbella*), or by autogamy, in which the mother cell undergoes no cytokinesis, both nuclei remaining after the first meiotic division, the second division resulting in two haploid nuclei and two that abort. The two haploid nuclei fuse autogamously (*Cymbella, Denticula*).

The fifth type involves apomixis, the parthenogenetic formation of auxospores from a single diploid cell. Commonly a pseudopairing of mother cells occurs (*Cocconeis*), and a single mitotic division replaces the two meiotic divisions.

No positive evidence of purely vegetative development of auxospores is known in the pennate diatoms.

Statospores or Resting Cysts

The vegetative cell of some freshwater species may survive desiccation for months or even years. Other species have an incipient dimorphism, with a normal planktonic form alternating seasonally with a thicker-walled one that is less well adapted to a planktonic habit, such as the winter form of *Rhizosolenia hebetata* (Bailey) Gran, corresponding to a resting stage.

Following a period of active vegetative reproduction, neritic planktonic centric diatoms may produce thicker-walled or siliceous resting spores or cysts, with stored fats. This siliceous **statospore** has two overlapping valves as does the frustule, but does not have a girdle. The statospore valves are unlike, although both are convex; commonly the epivalve is spinose and the hypovalve smooth. The statospore thus differs from the frustule both in form and ornamentation. Those of *Chaetoceros* have even been described as distinct genera (*Goniothecium, Periptera, Omphalotheca,* and *Xanthiopyxis*). Resting spores of *Bacteriastrum* and *Leptocylindrus* have little resemblance to the vegetative cells (see Figure 8.25), those of *Rhizosolenia* may show similarities, and those of *Detonula* and *Stephanopyxis* have several features in common with the vegetative cells (Hargraves, 1976). Morphologically distinct resting cells occur also in *Cerataulina, Fragilaria, Melosira, Porosira,* and *Thalassiosira.*

Resting spores may be classified according to their relationship to the parent vegetative cell (Anonymous, 1975). An exogenous resting spore is isolated from the parent frustule when completed, as in *Detonula confervacea,* being neither partly nor wholly still enclosed. A semiendogenous resting spore remains with one valve still enclosed in the parent cell and the other valve free, as in *Stephanopyxis turris.* An endogenous resting spore remains entirely enclosed in the parent cell after it is completely formed, as in *Chaetoceros compressus* Lauder.

Chains of *Bacteriastrum hyalinum* cells with resting spores are common in the plankton. In producing the resting spores, the vegetative cell elongates and the cell contents accumulate at one end. The nucleus divides, a primary valve of the resting spore forms around the end of the protoplast, and one of the daughter nuclei degenerates. Following another mitotic division, the secondary spore valve is formed, the formation of every valve being preceded by mitosis. Both valves of the spore are arched, the primary one being spinose, the secondary one smooth. When the spore germinates, the spore valves are shed, and the normal bristled vegetative valves with 12 to 46 marginal bristles are formed, successive vegetative divisions producing a new chain (Drebes, 1972).

Commonly only one resting spore forms per cell, but two occur in cells of *Rhizosolenia setigera* Brightwell. Colonial forms such as *Chaetoceros* and *Hemiaulus* may produce cysts simultaneously in all cells of a chain, and *C. pseudocurvisetum* Mangin formed statospores in all cells of a chain and in lateral auxospores as well (LeBlanc, 1934). The diameter of the spores ranged from 9.8 μm in the normal cells to 27 μm in the auxospores, suggesting that diameter is no more diagnostic for cyst species determination than for the frustules.

The formation of resting spores by *Ditylum brightwellii* in culture (Gross, 1937, 1940a,b) oc-

curred in response to crowding, low temperature (below 12°C), and reduced light (see Figure 8.26). Statospores formed only when vegetative growth had resulted in a cell density of about 1000 per milliliter. After this density was attained, resting spores repeatedly formed at night and germinated during the day, for a period of one to two weeks. Meanwhile, other cells continued vegetative division. Finally the crowded resting spores developed a stronger siliceous membrane and persisted. Neither reduced temperature nor lessened light alone resulted in resting spore formation, but these factors triggered the process in densely crowded cultures, particularly when nutrient concentration was reduced. Transfer to new culture medium prevented spore formation. Seemingly, the important factor is the maintenance of an equilibrium between the cell sap and the medium. As the cell plasma membrane of *D. brightwellii* retracts, the protoplast shrinks into a small rounded spore only $\frac{1}{3}$ to $\frac{1}{20}$ the volume of the original cell, and containing only the solid components of the cell (nucleus, plastids, protoplasm), without the usual quantity of cell sap. Spore formation followed any conditions that resulted in decreased cell sap. Normal cell volume was maintained only by a number of coexistent conditions: (1) Maintenance of the sodium/calcium equilibrium controlled by the proportions of sodium and calcium chlorides; otherwise cell sap escaped and the plasma membrane retracted. (2) Maintenance of proper pH, as below 6.0 resulted in maximum plasmolysis. (3) Optimum metabolic activity, related to light and temperature, since maintenance of cell turgor expends energy. When the rate of photosynthesis is reduced, stored food rapidly is depleted until the available energy is no longer sufficient to control the equilibrium between cell water content and the external environment.

After resting spores formed, restoration of the proper conditions (e.g., transfer to new medium) results in a reversal of the process, with increase in cell water content and growth of the cell.

Germination of resting spores that remain within the parent frustule begins with the extension of fine processes of the cytoplasm into

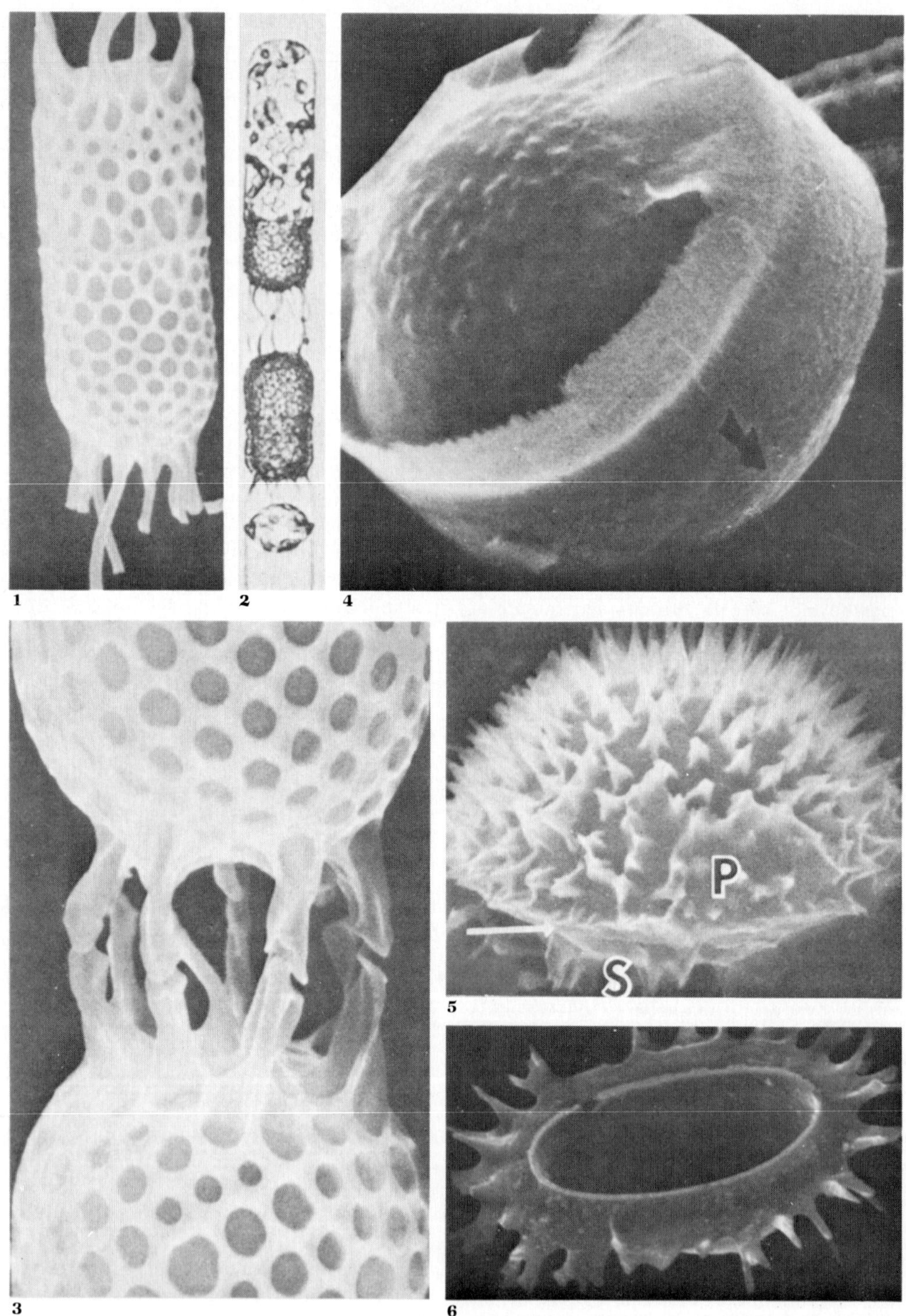

1 2 4

3 5 6

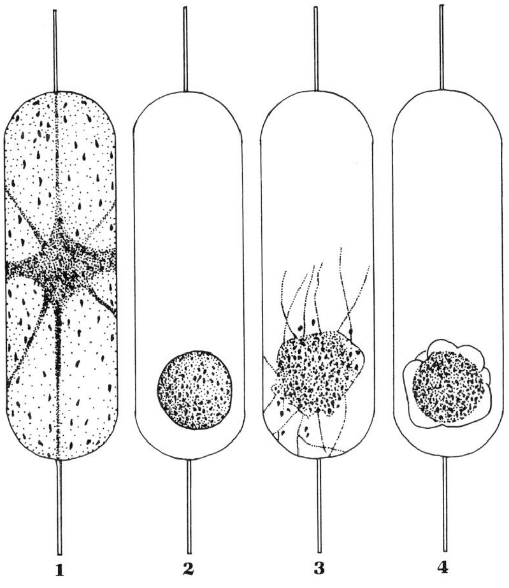

Figure 8.26
Resting spore formation in *Ditylum brightwellii.* **1.** Normal vegetative cell. **2.** Same cell after transfer into 3.5 percent NaCl solution. **3.** After three hours. **4.** After 24 hours. All ×280, redrawn from Gross, 1940b, *J. Mar. Biol. Ass. U.K.,* v. 24, p. 381—415, published by Cambridge University Press; used with permission.

the old frustule. The spore elongates and expands until the cell regains its normal appearance. If the resting spore had been isolated from the parent frustule, the extensions project in an amoeboid fashion, and the cytoplasm greatly expands. It then rounds off somewhat like an auxospore, and forms a new cell with about twice the diameter of the original parent, obtaining size re-establishment without the production of auxospores. This method may be common in nature, for auxospore formation is only annual or biannual in some species (Gross, 1940a). It may be induced in some species in culture by nutritional controls, but not in all.

Ditylum brightwellii maintains its size in North Sea populations largely by means of such vegetative enlargement (Stosch, 1965).

Intraspecific Variation

Size Range

The normal method of vegetative division results in a wide range of size among the frustules of a single species, their ornamentation also being affected by this size change. However, this regular decrease in size with successive generations is not found in some genera, such as *Skeletonema* and *Melosira*. In Vigo Bay, Spain, size distribution was not continuous in *Melosira sulcata* (Ehrenberg) Kützing, but showed peaks at definite intervals, suggesting a polymorphic population (Margalef, 1969). The diameter of cells along chains also showed no regular decrease. Cell size appeared to be related to temperature, salinity, and nutrient supply, large cells being more abundant at lower temperatures and at the lowered salinity of bays and estuaries. In lake sediment varves two species of *Melosira* and a *Stephanodiscus* showed simultaneous size changes, suggesting a climatic effect rather than synchronous auxosporulation. In various lakes, size changes appeared correlative, showing a decreasing diameter over a period of time followed by an increased size within the last 10 cm of bottom sediments, suggesting a correlation to temperature cycles (Margalef, 1969).

Size relationship to temperature also is not universal. Garstang (1937) found no correlation of size of the oceanic *Rhizosolenia styliformis* Brightwell to temperature. In contrast, size changes in *Coscinodiscus nodulifer* A. Schmidt in the equatorial Pacific were related to temperature, the largest specimens being present in areas of increased vertical circulation or upwelling (Burckle and McLaughlin, 1977).

Size and shape of the diatom cells probably are related to requirements for flotation; for nutrient absorption, particularly such modifications as would enable the cells to rise or sink or rotate in the water; and for resistance to predation. Smaller cells dominate in **eutrophic** waters (high nutrient levels) and large cells are dominant in **oligotrophic** waters, suggesting the importance of predation (or grazing) in this selection (Guillard and Kilham, 1977). Competition for resources is best in cells of less than 20 μm diameter, hence the smaller diatoms should be more abundant in nutrient-poor water if this were the limiting factor. However, in this environment predation also could be a major problem, and the cells tend to form chains, stellate colonies, or masses, or to increase the cell surface area or volume by means of projections and elongate spines. The adaptations for flotation and nutrient absorption follow those for limiting predation in importance, but may be morphologically similar.

Size does not necessarily indicate the contribution of the species to the total biomass, as the smaller cells may be much more numerous, and may have a much faster growth rate. In Florida coastal waters of the Gulf of Mexico, size and abundance of 58 species of diatoms were compared. The species with the largest total cell surface area, *Rhizosolenia alata,* was seventeenth in abundance. The most abundant, *Skeletonema costatum,* was fourth in cell surface area, about 177 μm^2, and averaged about 180,000 cells per liter. One of the rarer species, *Hemidiscus hardmannianus* (Greville) Mann, was 45th in abundance, with some 239 cells per liter, but had a surface area of 88,376 μm^2, hence contributed to the biomass an amount very similar to that of the most abundant species (Saunders and Glenn, 1969).

Life History Variations

Certain bipartite silicified resting cells were described as distinct genera and species before their relationship to centric diatoms such as *Chaetoceros* was recognized. Some cyst genera are still used for fossil representatives whose relationships are obscure.

Other morphological variations also occur. For example, cultures started from auxospores of *Ditylum brightwellii* contained mostly prismatic cells, although other cells had round, elliptical, or tetrangular valves such as had been previously described as separate formae (Gross, 1937). Furthermore, the species was observed to form secondary valves within the original cell, without cell division. Contraction of the protoplast toward the epivalve was followed by production of a new hypovalve, some specimens forming as many as eight successive hypovalves by successive contractions of the protoplast, culminating in production of a resting spore. Similar cells with secondary hypovalves in nature had been reported as probably due to incomplete cell division.

An unusual dimorphic diatom, *Phaeodactylum tricornutum,* has two distinct phenotypes that can undergo unlimited vegetative reproduction. One is oval, the other fusiform or even triradiate. As the various cell types have a simi-lar DNA content, they do not represent alternating haploid and diploid generations (Darley, 1968), such as occur in other protists. After endospore formation, oval cells were produced. The fusiform cell has no silica shell, only an organic wall that is divided into independent halves (Lewin et al., 1958). In section, the fusiform cell wall was determined by electron microscopy to be three-layered. The inner wall layer is thin and electron-opaque, the middle layer somewhat thicker and less opaque, and the outer wall is opaque and has a corrugated surface with numerous ridges in parallel sets (Reimann and Volcani, 1968). The oval form of the species has a single thin siliceous raphe-bearing valve (resembling that of *Cymbella*). The presence of this valve, the gliding cell movement, longitudinal vegetative division, auxospore formation, yellow-brown plastids, and storage of chrysolaminarin and oil allow its placement with the diatoms (Lewin, 1958), although the species was previously placed with the Chrysophyceae.

The initial cell of *Bacteriastrum hyalinum* resulting from lateral auxospore formation is identical to the solitary "species" *B. solitarium* Mangin. The latter was the only supposedly noncolonial species of *Bacteriastrum,* but was demonstrated in culture to be a stage prior to new chain formation of *B. hyalinum* (Drebes, 1967).

Lauritis et al. (1967) reported, from electron microscope observations of dividing cells, that *Nitzschia alba* clones gave rise to daughter cells referrable to both *Nitzschia* and *Hantzschia,* two genera previously believed to be characterized by different raphe positions. If this dimorphism is characteristic of all included species, the two genera are identical.

Holmes and Reimann (1966) noted three morphologic phases in the life cycle of *Coscinodiscus concinnus,* one of which had been previously regarded as a distinct species. The small cells just before auxospore formation were not identifiable as to species; valves of *C. concinnus* were three to five times larger after auxospore formation. As size decrease accompanied division, the valves developed a central rosette in valve view, and the girdle view changed from a drumlike outline to a trapezoi-dal one resembling *C. granii* Gough (Holmes, 1967). With further size decrease the central rosette also disappeared.

Other Morphologic Variations

Even stranger occurrences are the reports of diatoms in which the two valves are dissimilar, and otherwise referrable to distinct species and genera. Such interspecific variants have been reported in *Coscinodiscus, Actinocyclus,* and *Schimperiella.* Wood (1959) reported a specimen in which one valve was referrable to *Coscinodiscus lineatus* Ehrenberg, the other to *Asteromphalus roperianus* Ralfs ex Pritchard. Many individuals of both "species" occurred in the plankton sample, but only two specimens had valves representing both species. A similar occurrence of a frustule with dissimilar valves was reported by Morland (1887) from the upper Eocene of Oamaru, New Zealand. One valve represented *Porodiscus interruptus* Grove & Sturt, and the other *Craspedoporus elegans* Grove & Sturt. These may be merely due to accidental association of the isolated valves, however.

Variations in the morphology of the silica shell have been observed in clonal cultures, although diatom taxonomy is based almost exclusively on the frustule. Thus to test validity of frustule morphology the genetic constitution of various clones of the closely related centric species *Thalassiosira pseudonana* Hasle & Heimdal and *T. fluviatilis* was examined by an analysis of enzymes by electrophoresis (Murphy and Guillard, 1976). All clones of *T. fluviatilis* were distinctly similar and clearly separated from all *T. pseudonana,* although the various clones were obtained from different continents. However, a study of clones of *Skeletonema* showed a greater variability.

Habitat

Diatoms may be solitary or colonial, endosymbiotic, benthic, epiphytic or epizooic, or planktonic. Typically aquatic, in marine, brackish, and fresh water, they also may occur in muds and soils.

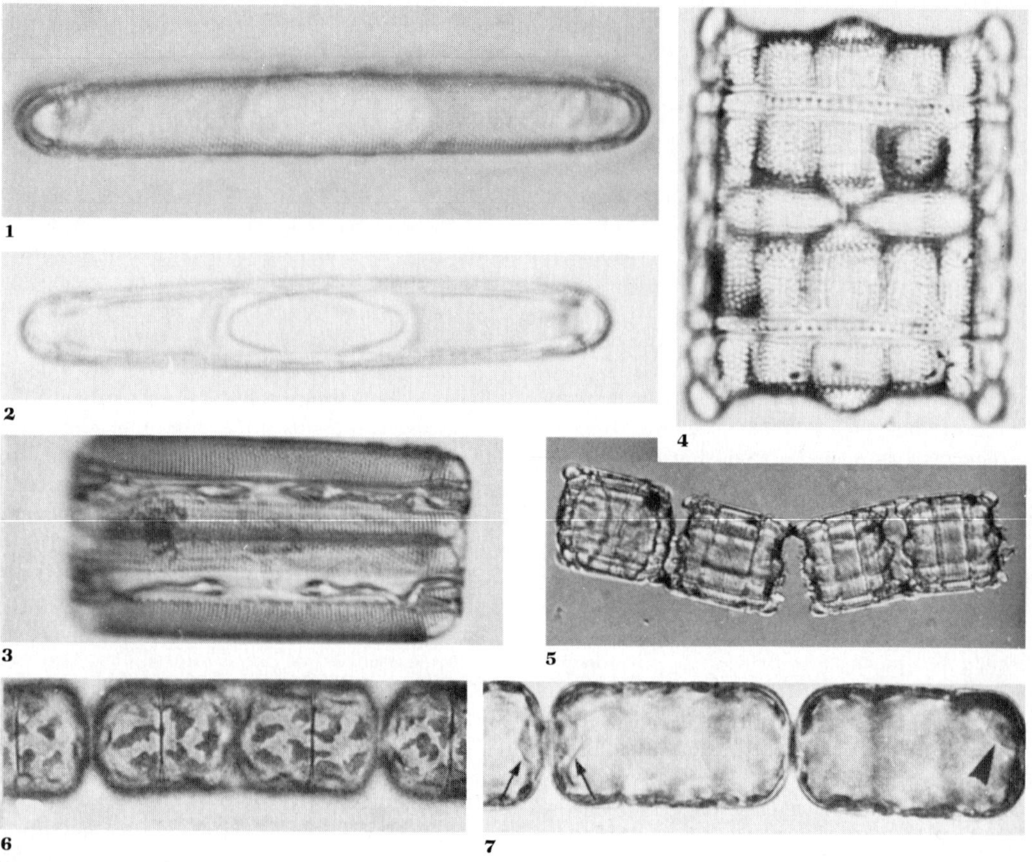

Figure 8.27

Colony formation. **1–3.** *Grammatophora marina,* ×750. **1.** Exterior. **2.** Focus on inner girdle band with wide septum from apices nearly to center of valve. **3.** Girdle view of short colony of tightly appressed smaller cells. **4,5.** *Biddulphia pulchella.* 4. Two cells just after separation, still held together by parent cell girdle, ×320. **5.** Part of colony, showing attachment by mucus pads at corners of cell, resulting in zigzag colonies, ×120. **6,7.** *Melosira moniliformis,* chains of cells held together by mucilage in central part of valve, ×550, from Crawford, 1977. **6.** Focus on surface to show outline of plastids, and girdle of cells. **7.** Midfocus of cells about to divide, showing mucilage connection between cells; arrows at left indicate nuclei in normal interphase position, arrowhead at right indicates nucleus migrating toward girdle region prior to mitosis.

All diatoms are capable of free and individual existence, but some may adhere in contiguous groups or colonies, whose form depends upon both species habit and cell shape (see Figure 8.27).

Substrate Attachment and Colonies

Many diatoms are attached to each other in colonies, attached to submerged rocks, or to various plants (epiphytes) or animals (epizooic species). In controlled tests, the epiphytes were found to be selective as to the seaweed on which they live; for example, *Gomphonema* was present in greatest numbers on *Utricularia,* and *Eunotia* predominated on *Naias* (Prowse, 1959). Among diatoms that attach to animals, *Cocconeis ceticola* Nelson covers the blue whale so abundantly that the latter has been termed "sulfur-bottom" (Bennett, 1920). *Sameioneis* is a stalked diatom that attaches to copepods in the eastern Pacific, South Atlantic, and Adriatic (Russell and Norris, 1971).

Epiphytic diatoms that are selective as to their preferred host reflect the distribution and abundance of the latter. Variations in the occurrence of the host plant will then affect not only the epiphytes, but also the fish or other zooplankton feeding upon them in turn, hence such distributions are of economic importance to fisheries.

Benthic diatoms commonly attach by means of small pads or cushions of secreted mucilage, or mucilage stalks. Mucilage cushions may occur only at the corners of the frustule and result in a zigzag colony in *Tabellaria* or a stellate one in *Asterionella.* Others join valve to valve in ribbonlike colonies, with the basal or oldest cell anchoring the ribbon to the substrate by its pad of mucus, as in *Melosira, Fragilaria,* or *Synedra.* Electron microscopy of sectioned *Biddulphia levis* Ehrenberg showed the mucus pad that connects adjacent cells by their horns to consist of many fine fibrils interspersed among still smaller granules (Heath and Darley, 1972). Mucilage stalks that attach to a substrate may be simple as in *Achnanthes* or *Gomphonema* or branching as in *Licmophora.*

A cell of *Amphipleura rutilans* secretes a sheath, and as the cell divides each newly formed cell adds to the sheath until it becomes a communal polysaccharide tube 20 to 50 μm in diameter in which the individual cells may move up and down (Drum, 1969).

Some planktonic species secrete large quantities of mucilage in which many cells may be embedded at random. Others are joined valve to valve by a single thread or by a network of mucus threads (*Thalassiosira*), or linked girdle to girdle, as in *Coenobiodiscus* (see Figure 8.28), the cells being separated within a compartmentalized organic matrix. All component cells of the *Coenobiodiscus* colony divide simultaneously to produce a new colony of the same size as the original, with a new organic matrix joining the cells (Loeblich et al., 1968).

Cells of *Bacillaria* colonies are attached raphe to raphe and glide against each other for the length of this attachment; the surface of contact thus varies, and the colony may form a short packet of cells or be greatly extended.

Linking spines may connect adjacent cells of a colony of *Hemiaulus,* or the elongate spines or setae may interlock, as in *Chaetoceros* or *Bacteriastrum.*

Planktonic Species and Flotational Adaptations

Most marine planktonic diatoms are centrics, as are some freshwater planktonic species of *Melosira, Cyclotella,* and *Rhizosolenia.* The dominantly freshwater pennates include a few planktonic representatives (*Tabellaria, Asterionella, Synedra*) although they are characteristically benthic.

Planktonic diatoms have numerous modifications that have been regarded as aids to flotation. Flotation is dependent in part upon water viscosity, density, and temperature, hence to compensate for seasonal changes of these physical conditions, specimens of a single species in summer may have thinner walls and more attenuated spines and setae than do winter individuals of the same species (Patrick, 1948).

Gran (1912) suggested that flotational adaptations were of four types: (1) those resulting in a ribbonlike shape, either a thin-walled flattened

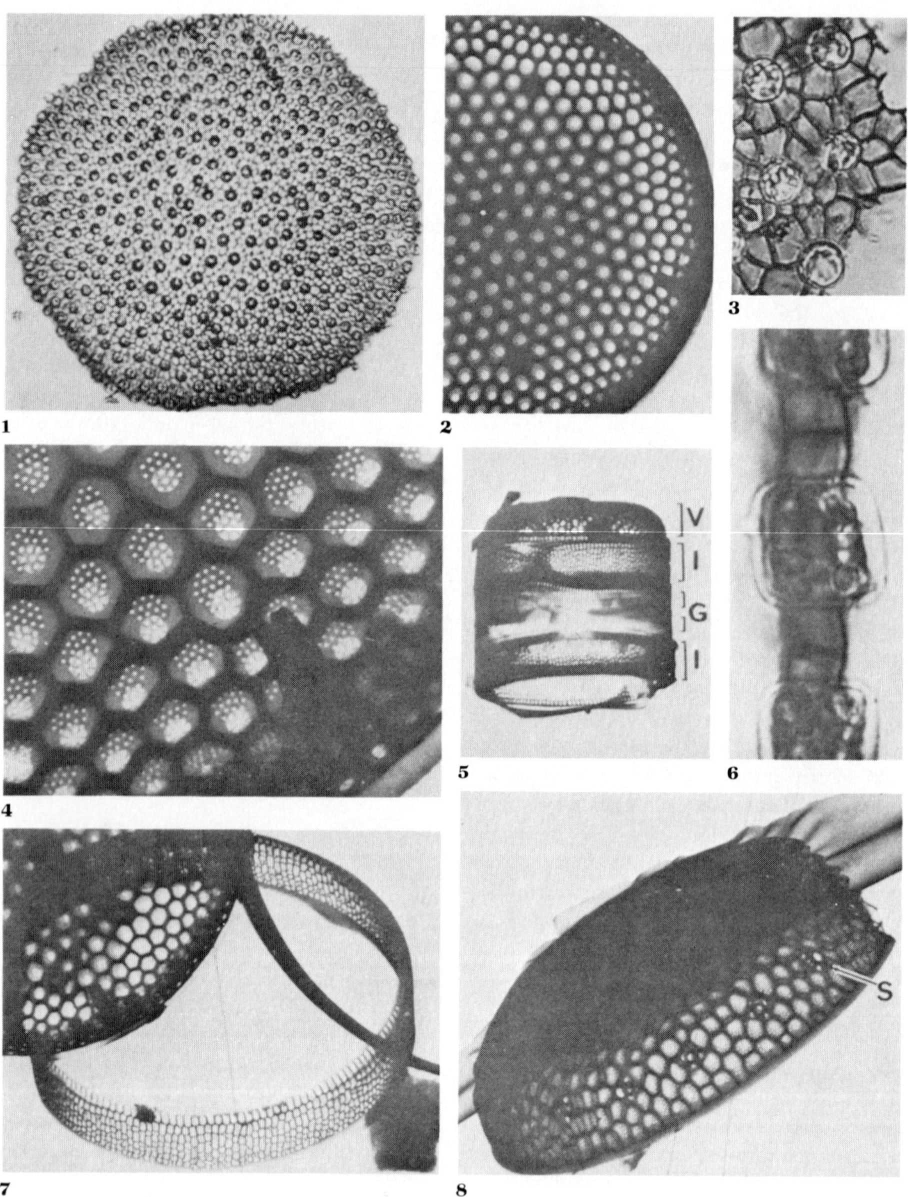

Figure 8.28

Coenobiodiscus muriformis Loeblich III, Wight & Darley. **1.** Colony, ×152. **2.** TEM of valve interior showing areolae, central strutted process (pore in center), and labiate processes (one visible at right), ×8000. **3.** Part of broken colony, showing cells joined by organic matrix, ×695. **4.** As in part 2, enlarged to show inner sieve membrane and outer open areolae, labiate process visible at lower right, TEM, ×34,080. **5.** Girdle view of frustule, showing convex valve (V), girdle bands (G), and intercalary bands (I), TEM, ×2560. **6.** Edge view of part of colony, showing cells and chambered organic matrix, ×1840. **7.** Isolated intercalary band, TEM, ×8960. **8.** Oblique edge view of valve, showing marginal strutted processes (S) that characterize the Thalassiosiraceae, TEM, ×10,400. All from Loeblich et al., 1968.

608

cell of large surface area, or ribbonlike colonies (*Fragilaria*); (2) the hair type, with cells much elongated in one dimension, or united into long narrow colonies (*Rhizosolenia*); (3) the branching type, with long spines that increase surface area, or setae used in chain formation (*Chaetoceros*); (4) the bladder type, consisting of a large cell with a thin film of cytoplasm around a large central vacuole containing cell sap with a specific gravity like that of seawater or lighter (*Coscinodiscus*).

Gross and Zeuthen (1948) noted that the specific gravity increases when cell sap is extruded during the formation of resting spores, hence the specific gravity of the cell sap is lighter than that of seawater. Resting spores, with or without the shell, sink in seawater, regardless of their shape, spines, colonial habit, or other modifications, whereas the otherwise similar vegetative cells float. Apparently, only reduced specific gravity is effective in maintaining the cell in the plankton. Braarud (1954) suggested that the much lighter cell sap of *Coscinodiscus centralis* Ehrenberg possibly results from a prevalence of divalent ions as compared to seawater. Beklemishev et al. (1961) made de-

terminations of six ions (Cl^-, Na^+, K^+, Ca^{2+}, Mg^{2+}, and SO_4^{2-}) in the cell sap of *Ethmodiscus rex*, noting that Na^+, K^+, Ca^{2+}, and SO_4^{2-} were reduced in amount as compared to seawater, and Mg^{2+} was absent from the cell sap. Amounts of these ions are low in fresh water, hence other means of reducing specific gravity are more important; walls are more lightly silicified, and oil droplets are common as inclusions in the cytoplasm.

Based on settling tube experiments, rates of sinking were determined for various marine species (see Table 8.2). This rate varies according to the condition of the population, being greatest in senescent cultures. Vigorously growing populations were more buoyant (Smayda and Boleyn, 1965, 1966a,b).

Although most planktonic diatoms are colonial, colony formation is probably not an adaptation for flotation, inasmuch as single and small cells have a greater surface to volume ratio than do colonies (Beklemishev, 1959, 1961). Colonies with cells in contact actually reduce their surface area, theoretically lessening their ability for flotation. Although specific gravity thus plays a more important role in flotation,

Table 8.2
Rate of sinking in the water column.

	Observed sinking rate (meters/day)		Calculated days required to sink 100 meters	
	min.	max.	at min. rate	at max. rate
Thalassiosira cf. *nana* Lohmann	0.05	5.40	2000	18.5
T. rotula Meunier	0.18	44.93	555	2.2
Nitzschia seriata Cleve	0.18	2.96	555	34.0
Skeletonema costatum	0.13	7.39	770	14.0
Rhizosolenia setigera	0.05	22.71	2000	5.0
Bacteriastrum hyalinum	0.22	7.49	455	13.5
Chaetoceros lauderi Ralfs	0.21	6.24	475	16.0

smaller cells may be preferentially used as food by grazing zooplankton, hence colony formation may be of selective value (especially for cells between 10 and 100 μm diameter) in limiting their susceptibility to grazers. The largest cells are solitary or show a tendency toward a loss of coloniality. The development of colonies by planktonic species may also be an adaptation for maintaining cells in close proximity for an increased likelihood of successful sexual reproduction (Smayda, 1970).

Various morphologic modifications of planktonic diatoms also may be important as mechanisms to cause twisting and vertical movement in the water column. Phytoplankton requires a means of sinking and rising and rotating as well as floating, in order to obtain the maximum availability of nutrients and light (Smayda, 1970).

Movement

Nature of Movement.   Passive current-related movement is characteristic of all planktonic species, but only the pennate diatoms show individual movement. Many pennate diatoms are benthic, living on the surface of sands and muds in an environment in which the frequent deposition of sediments makes necessary an ability to migrate back to the surface for light and air. Diatoms living in the interstitial waters of sands near the tide line may show a mass movement downward just prior to the incoming tide, returning later to the surface.

Movement is never in a continuous straight line, but is alternately backwards and forwards, in the general direction of the long axis of the cell, but in a series of curves or zigzags. The movement may be steady or in a series of jerks. Isolated cells may move freely, those of colonies move in relation to each other, or an entire colony may move together as one. The most rapid motion is found in species with true raphe, especially of the canal raphe type. Movement of *Amphora montana* Krasske is between 2 and 5 μm per second; that of *Navicula radiosa* Kützing is as much as 14 to 20 μm per second (Jarosch, 1962). The average speed of different species ranges from 0.2 to 25 μm per second, with the

maximum rate about twice the average (Harper, 1977).

Motile diatoms that are symmetrical about their transverse axis move in either direction with equal frequency, but asymmetrical species (*Surirella, Gomphonema*) move more frequently in the direction of the narrower end. Cells move vertically against a substrate as readily as horizontally, without the aid of any buoyancy factor (Drum and Hopkins, 1966).

Although motile diatoms generally move as individual cells, *Bacillaria* shows a concerted rhythmic colonial movement, due to a semi-coordinated movement of the individual cells. Gliding begins with the terminal cell moving over the adjacent one; the latter then moves forward in the same way, the entire colony following in turn until it forms an almost straight line of cells, joined tip to tip (Drum and Pankratz, 1965). The colony then closes up and may repeat the movement in the same direction, or in the opposite one. *Bacillaria* has a central raphe, with a siliceous hook protruding from one side of the raphe slit in each valve. These hooks may interlock as the cells slide against each other. While the colony is in the plankton, the cells use each other as a substrate for movement for better nutrient availability.

Mechanism of Movement.   The close relation of movement and the presence of a raphe is substantiated by the trimorphic species *Phaeodactylum tricornutum*, which variously appears as fusiform, triradiate, or ovate cells. The first two morphologic variants have no rigid frustule (Hendey, 1954) and are nonmotile. Only the oval phase was observed to move out over an agar plate, as do raphe-bearing diatoms (Lewin et al., 1958); this oval phenotype is seen in electron micrographs to possess a silica valve on one side, with a well-developed raphe.

Contact of the raphe with the substrate is necessary for diatom movement. The membrane that envelops all other parts of the diatom frustule does not cover the raphe (Reimann et al., 1965), allowing the raphe contact.

Although movement is related to the raphe, suggestions as to the exact mechanism have included pseudopodial action, hypothetical rows of cilia, cytoplasmic streaming, the expulsion of gas molecules such as oxygen, the presence of osmotic currents, differential surface tension, capillary pressure, and mucilage secretion. Electron microscopy has eliminated some of these possibilities, such as pseudopodia or cilia, and as the raphe may or may not contain protoplasm, cytoplasmic streaming is apparently not involved. The two raphes on a side may act in concert, both moving material in the same direction, and resulting in locomotion. If they act in opposite directions, adhering debris may move along the raphe slits toward the central nodule.

Liebisch (1929) suggested that currents of water along the raphe fissure were initiated by cytoplasmic contractions and that the cytoplasmic membrane was actually continuous beneath the raphe fissure. Jarosch (1962) suggested that oscillating protein fibrils are attached to the cell wall, those inside the cell causing protoplasmic streaming, with submicroscopic transverse waves in the surface fibrils causing the gliding of the cell. Waves visible in the microscope were thought to represent the superposition of such submicroscopic waves. However, electron microscopy shows no such protein filaments to be present in the raphe (Harper, 1977).

Locomotion and adhesion to a substrate are closely related. As secretion of a mucilagelike material is necessary for adhesion, and adhesion is necessary for locomotion, this secretion has been regarded as the main factor in diatom locomotion. Diatom movement occurs only when the diatom is adhering to a substrate, even if the substrate consists only of the secreted trail itself (Drum and Hopkins, 1966). The trails are not normally visible, but can be demonstrated to occur either by staining techniques or by electron microscopy. The secreted material is present along the entire length of an active raphe system, its active extrusion having been previously erroneously regarded as streaming cytoplasm or water currents.

Fibrillar elements were found to project through the raphe fissure of *Nitzschia alba* (Lauritis et al., 1968). Similar fibrils are present in the mucilage of the gliding unicellular red algae (see Chapter 2), hence might be involved in locomotion in diatoms. The source of the secreted polysaccharide fibrils was suggested to be the crystalloid bodies in the cytoplasm near the raphe slits (Hopkins and Drum, 1966). Such fibrils are seen in electron microscopy adjacent to the raphe system, at the cell apices, or throughout the cytoplasm. The fibrils were suggested to expand and spiral as they pass through the raphe slit, and push the diatom along where they adhere to the substrate. An alternative suggestion was that fibrillar bundles are contractile and pump the mucilage from the raphe; this would cause the mucilage to be extruded at right angles to the raphe and thus could not cause the movement that is parallel to the long axis of the diatom (Harper, 1977).

The adhesive and tractive forces exerted by diatoms were measured in the laboratory by their resistance to pressure from a glass wool fibre (Harper and Harper, 1967). Maximum adhesion, requiring the greatest force to detach a specimen from its substrate, was found in specimens living attached to beach sands. This force ranges from a minimum of 100 to a maximum of 2900 millidynes in *Diploneis ovalis* (Hilse) Cleve (one millidyne being approximately equal to the weight of 1.02 micrograms). *Amphora ovalis* (Kützing) Kützing had a minimum of 230 and maximum of 480 millidyne adhesive force when moving, and 140 and 630, respectively, when stationary. Others tested showed less adhesiveness, *Nitzschia linearis* showing between 0.36 and 6.6 millidynes when moving, and 0.1 to 12.0 when stationary. Traction forces ranged from 48 millidynes in *A. ovalis* to a mean of 1.1 in *N. linearis*. As movement results from the trail secretion, speed of movement is limited by the secretion rate, and not by water resistance. The average trail diameter is about 25 nm. Movement at 10 μm per second for six hours would entail a polysaccharide excretion of $\frac{1}{48}$ of the diatom's volume, greater speeds requiring even more. During locomotion, this thin strand of mucopolysaccharide is secreted continuously from a pore into the raphe, and swells in contact with the water to become fairly solid, although flexible and sticky. It is held against the entire length of the raphe either by osmosis lowering the pressure within

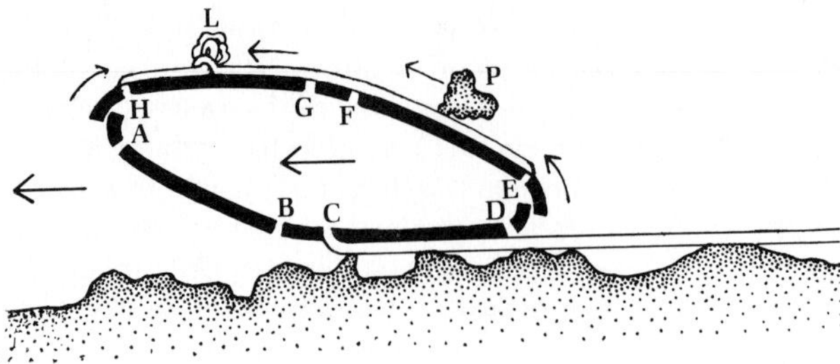

Figure 8.29
Diatom trail secretion and movement. Diagrammatic section of pennate diatom in the apical plane, frustule shown as heavy black line. A to H, pores; raphe slits along AB, CD, EF, GH; P, particle carried along by secretion as it moves along the raphe; L, lump of trail substance collecting at junction of moving secreted substance. Redrawn from Harper and Harper, 1967.

the slit, or by an interfacial tension between the liquid within the raphe and that of the surrounding water. When the strand of mucilage meets a rigid surface it adheres, and the pressure of the lengthening strand as it is secreted pushes the diatom along on its raphe. Water resistance or a convex valve may cause the diatom to be tilted slightly backward. Such a mechanism does not require the presence of protoplasm or water in the raphe; the trail alone must be kept against the raphe as it moves along. Small particles of organic matter or detritus may be moved up and down the dorsal raphe system in either longitudinal direction, with no obvious relation of either direction or velocity to the diatom movement (see Figure 8.29). Trail secretion through more than one pore simultaneously also explains the movement of particles along its upper surface while the diatom moves along the substrate (Drum and Hopkins, 1966; Harper and Harper, 1967).

Propulsion by the mucilage is difficult to explain solely by secretion, as the mucilage is extruded from minute pores at the ends of the raphe, which would have considerable frictional resistance to a viscous slime. But as the trail must come from the cell, it may polymerize outside the constriction, or pass through it in an aligned crystalline form. The force may result from the swelling against the side of the pores. Gordon and Drum (1970) regard the raphe as a parallel plate capillary containing fluid. The fluid reacts at the trailing end and turns into a form that doesn't wet the raphe walls, and is left behind as a trail. Additional fluid is drawn from the anterior end of the raphe by capillary pressure, and fibrils in the raphe fluid that are oriented perpendicular to the diatom shell may swell by reaction with seawater. However, there is no morphological evidence of excretion vesicles penetrating the plasmalemma, as would be expected in this theory (Duke and Reimann, 1977). Furthermore, the capillary mechanism does not explain the measured strength of some diatoms, or how one can climb over the raphe of another without a leakage of raphe fluids (Harper, 1977). If the curvature of the interface of raphe fluid and seawater is sharper at the trailing end than at the leading end of the cell, pressure differences could drive the motion. This would happen if liquid was sucked back into the cell at the trailing end of the raphe, conserving the raphe fluid. Such conservation would be of value to diatoms buried in sediment to a depth of several centimeters, which would need to move continuously for several hours to regain the surface. Particle transport along the raphe of a stationary *Navicula* cell results in a lump at the central nodule, when the trails from one or both raphe ends are moving toward the center. This movement of particles may be aided by the polar fissures that extend as continuation grooves or "quasi raphes" (Harper, 1977) beyond the polar nodules, although not penetrating the cell wall. These grooves may allow the raphe fluid to flow in under the mucilage thread, and thus be conserved.

Symbiosis and Commensalism

True symbiosis is not common in the diatoms. Reports of symbiotic coccolithophorids and *Coscinodiscus* in reality represent the chance agglutination of the coccospheres to the mucilage secretions of the diatoms (Gaarder and Hasle, 1962). Various species of *Rhizosolenia, Chaetoceros, Bacteriastrum,* and *Hemiaulus* in waters of limited nutrients have an endosymbiont, *Richelia intracellularis* Schmidt, a heterocystous, nitrogen-fixing blue-green alga. The diatom *Licmophora hyalina* (Kützing) Grunow occurs as an endosymbiont of the flatworm *Convoluta roscoffensis* Graff (Ax and Apelt, 1965), and *Navicula* and *Cocconeis* occur in the intercellular substance of receptacles of the brown alga *Ascophyllum* (Hasle, 1968a).

Diatoms may be epizooic on organisms ranging from the whale on which *Cocconeis* occurs to the ciliate protozoan *Tintinnus,* to which various species of *Chaetoceros* and *Hemiaulus* may be attached. The mutualistic association with the tintinnids gives motility to the otherwise passively floating diatom, aiding it in obtaining nutrients from a wider area than would otherwise be possible (Tappan and Loeblich, 1973).

Parasitism and Grazers

The fungus *Olpidium* may parasitize diatoms (Meunier, 1910, 1915), the dinoflagellate *Paulsenella chaetoceratis* (Paulsen) Chatton occurs on the bristles of *Chaetoceros borealis* Bailey and *C. decipiens,* and various flagellates may occur as epiphytes on *Cerataulina, Chaetoceros, Dactyliosolen, Nitzschia,* and *Skeletonema* (Cupp, 1943).

More important in controlling diatom abundance than parasites or many of the possibly limiting physical conditions are the grazers and, in particular, the timing of heaviest grazing. If heavy grazing occurs in the early stages of a bloom, total phytoplankton growth will be greatly limited. If grazing is somewhat delayed, the microflora may become well established. In colder waters, diatom blooms may occur before the water temperature is sufficiently high for extensive bacterial growth. Similarly it usually precedes the reproduction of the zooplankton or other grazing organisms.

Nutrition

Heterotrophy

Although diatoms, like most plants, are typically photosynthetic, some are facultative heterotrophs and can utilize preformed organic compounds for growth in the dark, or in the light in the absence of CO_2. A few lack plastids and are obligate heterotrophs.

Lewin (1953) tested 42 bacteria-free cultures of pennate diatoms for heterotrophy. Thirteen (*Navicula* and *Nitzschia* species) could grow with glucose as a sole carbon source. Of 60 different compounds tested on *Navicula pelliculosa,* only glucose, glycerol, and fructose were used. The glucose could be used in the dark, but the other two compounds were effective only in the light in the absence of CO_2.

A later examination of 44 cultures of diatoms reported that 16 grew in the dark with glucose, 8 with glucose or lactate, 1 with glucose or acetate, 1 with glucose, acetate, or lactate, 2 with lactate only, and 16 had no growth in the dark with any of these substances (Lewin and Lewin, 1960). Those which showed no growth in the dark represented species of the genera *Achnanthes, Amphipleura, Amphiprora, Amphora, Navicula, Nitzschia, Stauroneis,* and *Synedra.* Other species of *Amphora, Navicula,* and *Nitzschia* and the centric diatom *Cyclotella* did grow in the dark with the above carbon sources.

All known colorless diatoms, obligate heterotrophs, are marine. These rapidly growing forms live on decaying seaweeds and other organic debris along the shore. Three species, *Nitzschia putrida, N. leucosigma* Benecke, and *N. alba,* were found in culture to use lactate or succinate as a carbon source, and some could also use glucose or glutamate. All require exogenous sources of the vitamins thiamine and cobalamin (Lewin and Lewin, 1967).

A suitable carbon source thus allows growth of many diatoms without photosynthesis (about two-thirds of those tested in the laboratory). Free sugars such as glucose, sucrose, and fructose have been reported in deep sea sediments, and littoral muds are probably even richer in these (Lewin and Lewin, 1960). Heterotrophy may be important for species living in deeper water, or in high latitudes during the winter, as well as for temporary survival after sinking below the lighted zone, before upwelling permits their return.

Photosynthesis and Productivity

About 98 percent of the CO_2 of the earth's oceans and atmosphere is dissolved in the oceans. Equilibrium of dissolved CO_2 with the atmosphere is dependent on the rate of turnover of deeper waters. As the solubility of CO_2 in seawater is greatest at lower temperatures, more is dissolved in polar areas and in the deeper waters. Marked diurnal fluctuations in the CO_2 equilibrium result from photosynthetic activity. Where the rate of photosynthesis is high, as during diatom blooms, the CO_2 may be temporarily depleted from surface waters during the day, resulting in an elevation of the pH. There is more dissolved CO_2 below the photic zone, due to respiration and oxidation of organic matter and lessened utilization of the CO_2 by photosynthesis.

The rate of diffusion of carbon dioxide from the atmosphere becomes limiting to growth at relatively low population densities in laboratory cultures, where maximum growth requires aeration by stirring or by bubbling air into the medium. Dense cultures may require CO_2-enriched air to maintain an optimal growth rate. Similarly, quantities of CO_2 may limit photosynthetic potentialities in the ocean under conditions of high light intensity in lower latitudes.

The abundance of diatoms in colder water may be in part related to the increased quantity of dissolved CO_2 within the photic zone and the resultant lower pH. As these factors also favor preservation rather than dissolution of the silica, they may result in the accumulation of diatomaceous oozes.

In lower latitudes, maximum diatom abundance is restricted to regions of upwelling, which brings to the surface the colder waters with various dissolved nutrients and higher amounts of CO_2. Diatom abundance during periods or in regions of volcanic activity may be related to an increased supply of CO_2 as much as to the increased silica.

Because of their abundance, wide distribution, and rapid rate of reproduction, diatom photosynthesis is estimated to contribute 20 to 25 percent of the world's net primary production; some 2×10^{13} kilograms of organic carbon being annually fixed by the various phytoplankton species (Werner, 1977). The rapid cell division is also correlated with a high rate of photosynthesis per unit of biomass. The doubling rate for various diatoms may be as little as 9.6 hours for *Asterionella formosa* Hassall, 10.0 hours for *Phaeodactylum tricornutum* and 13.1 hours for *Skeletonema costatum* (Fogg, 1965). The latter species has even been reported to divide as frequently as eight times daily (quoted by Lewin and Guillard, 1963).

In addition to the organic carbon that becomes a part of the cell, diatoms also excrete large amounts of organic compounds in solution, from 3.6 to 25 percent of the carbon assimilated during the exponential phase of growth, and about four times as much during the stationary phase in *Thalassiosira fluviatilis*, up to a maximum of 43 percent. This material is excreted largely as glycolic acid (as much as 38.5 percent), with some glycerol and proline. *Stephanodiscus hantzschii* Grunow excretes glycollate and smaller amounts of polysaccharides and amino acids (Loeblich and Loeblich, 1979). These compounds may be utilized by associated bacteria, but some also are inhibiting to bacteria (Darley, 1977).

As diatoms are the major component of the present marine food cycle, they are of considerable interest to biologists in relation to fisheries and water pollution, and numerous estimates and measurements of their productivity have been made in both marine and fresh water. Obtaining accurate and meaningful measurements of production is difficult both in nature and in the laboratory. The concentration of diatom cells ranges from about 1 per liter in the oceanic "deserts" to about 10^6 per milliliter in some polluted estuaries (Guillard and Kilham, 1977). Numerical estimates are based on four sampling methods. The larger species may be obtained with the usual nylon plankton nets. These have openings of about 35 μm, and vertical hauls integrate the population of the entire water column, hence obscure the maximum abundances. A settling technique, using a 300-ml sample in a settling chamber and observing it with an inverted microscope, obtains more of the smaller specimens, but the smaller volume will miss species that occur in numbers of less than 10 per liter. A membrane filter is used commonly by Soviet expeditions; this method concentrates the cells from a one-liter volume onto a 3-cm-diameter filter for examination, but smaller species may be lost. Continuous centrifuges used on oceanographic ships collect samples from 10 to 100 m^3. Estimates of abundance based on the centrifuge are generally about double those based on the membrane filter method.

Distribution is commonly very patchy, and the most efficient producers are the smallest ones that commonly are lost in sampling. Fogg (1965) states that over half of all photosynthesis is by algae that pass through a 35-μm-mesh net. These may not represent a high percentage of the total mass, but their high surface/volume ratio results in greater efficiency. The total cell surface area is more closely related to photosynthetic rate than either cell number or cell volume. A large proportion of freshwater diatoms is within this smaller size range, in addition to many of those in the marine environment. Collier and Murphy (1962) reported diatoms in the Gulf of Mexico that were in the size range of bacteria. *Chaetoceros galvestonensis* Collier & Murphy measures 1.5 μm on the apical axis, and about 3 μm on the pervalvar axis; other species were as small as 0.75 μm in diameter. These tiny species were well developed, with nuclei, plastids, pyrenoid bodies, and oil globules, and grew rapidly in unialgal cultures. They probably are an impor-

tant food source for filter-feeders, as well as for planktonic larval stages of various organisms.

Ecologic Factors

Light

As autotrophic growth requires light, it is restricted to the upper layers of the water; various algal groups are adapted to differing kinds and amounts of light, however.

As noted earlier, diatom photosynthetic pigments include chlorophyll *a* and *c*, and accessory carotenes and xanthophylls. Fucoxanthin has been shown in the laboratory to be equally as efficient as the chlorophylls in diatom photosynthesis (Tanada, 1951). Chlorophyll *c* is not only equally efficient in normal light, but absorbs blue light better than do either chlorophylls *a* or *b* (the latter is important in green algae and other green plants but does not occur in diatoms). Hence its chlorophyll *c* allows the diatom to utilize light wavelengths that penetrate deepest in the sea. Due to local physical conditions in the Antarctic convergence, the available light consists mainly of blue-green radiation (Mandelli, 1967). Furthermore, ^{14}C determinations showed a high diatom photosynthetic rate per unit of chlorophyll *a*, suggesting that these light conditions restricted chlorophyll *a* synthesis and function, whereas chlorophyll *c* and fucoxanthin became relatively more effective. The photosynthetically active carotenoids also are important in habitats where the chlorophylls are less efficient.

The rate of diatom photosynthesis increases with enhanced light to a maximum at 10 kilolux, and then declines, as too much light becomes retardant. The amount of 50 to 75 kilolux, about 75 percent of full daylight, reduces photosynthesis to only 5 to 10 percent of the maximum rate. This rate is independent of temperature at low light intensities, but at light saturation becomes higher at higher temperatures. Ultraviolet light also damages diatom cells more rapidly than those of the green algae (Lewin and

Guillard, 1963), giving an additional selective value to their greater depth capabilities.

Spring diatom blooms may occur in deeper water in response to the increase in available light, rather than the nutrient enrichment of the surface layers or increased temperature. In winter the mixed layer is deep and the phytoplankton largely is present below the compensation depth where respiration equals oxygen production. Below this level light is deficient, and organic matter is lost by grazing, sinking, and respiration. When the mixed layer becomes shallower than the critical depth and light is available, the standing stock increases. Winter growth is thus limited by available light from the poles through the temperate regions. As polar production is light-limited to three months of the year, there is one summer bloom. Temperate latitudes commonly have both spring and fall blooms, related to nutrient cycling. In the tropics and subtropics, light may penetrate below the mixed layer continuously, and the main production occurs in the winter (Anderson and Banse, 1963).

Temperature

Most diatoms have a wide temperature tolerance, hence their latitudinal distribution is not as closely related to temperature itself as to various temperature-related or other kinds of seasonal changes. Such indirect, yet limiting, seasonal effects include the changed solubility of gases (CO_2, oxygen) and salts, and corresponding changes in pH, solubility of silica, overturn and cycling of nutrients, changes in salinity due to melting ice, and amount and duration of light in the higher latitudes. Most of these are discussed separately.

Because of the temperature effect on water viscosity, warm water and subtropical floras commonly have more ornamented frustules, thinner walls, elongate spines, and other aids to flotation. In colder water, the elongate naviculoid diatoms are relatively more important.

Diatoms may occur in Antarctic water of $-2°C$, or at the other extreme in hot springs. The ability of eurythermal species to thrive at various temperatures is related to the amount of nutrients available. Boreal diatoms may grow actively at $20°C$ in polluted waters, whereas their growth is negligible at that temperature in nonpolluted water (Smayda, 1958). Other diatoms may have lesser temperature tolerances, *Thalassiosira antarctica* Comber being restricted to temperatures between $-1.77°$ and $3.49°C$. The obligately coldwater Antarctic species *Fragilariopsis sublinearis* (van Heurck) Heiden & Kolbe had different temperature responses in its photosynthesis and its respiration, resulting in a requirement for low temperature (Bunt et al., 1966).

Achnanthes exigua Grun occurs in hot springs in Yellowstone Park and elsewhere in Montana and Washington. At temperatures of 30 to $40°C$, the diatoms may be the dominant eucaryotes. Laboratory studies of cultures showed no growth of this species below $15°C$, an optimum growth with cell division twice a day at $40°C$, and $44°C$ the maximum temperature at which growth continued (Fairchild and Sheridan, 1974).

Salinity

Diatoms occur in widely different salinities, ranging from fresh to brackish to marine water. *Melosira nummuloides* occurs in water of chlorinity from 2.1 to $19.9°/_{oo}$, although most marine species are somewhat less tolerant (Wood, 1965). Salinity effects are most noticeable in inshore areas, where the variations may determine the relative abundance of species, although not their actual distribution. Many species are in fact quite tolerant of salinity variation (euryhaline), the total salinity being less important than the quantity of particular ions.

Hydrogen Ion

Because of the close relationship of the dissolved CO_2 and the pH, diatom maxima result in an increased pH. Rapid cell multiplication in cultures is similarly accompanied by a depletion of CO_2 and rise of pH. The influence of diatoms upon the pH is shown in the greater variability of the pH in the upper 150 meters of the ocean waters, within the zone of photosynthesis. During their exponential growth, diatom cultures did not maintain an equilibrium between the CO_2 in the solution and in the atmosphere until the nitrogen source was depleted. When growth was slowed, equilibrium was again attained, and the pH dropped (Grant et al., 1967). Very few species live in water of pH less than 3.5 (but these few species may be abundant); the greatest diversity occurs at about a neutral pH, and the flora is restricted above pH 8.

Silica is more soluble under relatively alkaline conditions and the shells are corroded at the higher pH. Kolbe (1957) suggested that a physiochemical selection occurs in the ocean as the more delicate diatoms are selectively dissolved. Sea floor assemblages are poor in thin-walled species, and rich in the coarse and thick-walled species, although the latter actually may be rare in the plankton. The effect of pH on silicon solubility is discussed in more detail in a later section.

Silicon

Silicon is present in the frustule of the diatoms, and diatoms have an absolute requirement for silicon if cell division is to occur. The silicon is used in the form of orthosilicic acid, $Si(OH)_4$ (Lewin, 1962).

Both marine and fresh waters are undersaturated in silica. The solubility of amorphous silica in seawater at 0° to 25°C is between 0.05 and 0.15‰, whereas the maximum amount in solution in seawater is 0.0001 to 0.01‰, and in river waters is 0.02 to 0.03‰. Silicon is never absent from seawater, but a much lowered concentration in surface waters may result from heavy diatom production. In fact, diatoms so efficiently extract silica from the dilute solutions in nature that they may lower the concentration to as little as 0.00006 to 0.00008‰. At this stage, solution of quartz may result, particularly of finely dispersed particles. In silicon-deficient cultures, diatom activity will result in solution of silica from the glass container (Lisitsyn, 1967a). *Nitzschia* has been reported to decompose kaolin, utilizing the silicon from the crystal structure of the clay mineral, and liberating hydrated alumina (Vinogradov and Boichenko, 1942).

About 0.16 percent of the silica in solution in the ocean is fixed annually by the plankton in its skeletons (Lisitsyn and Bogdanov, 1968). Much of the skeletal silicon is recycled, as weakly silicified diatom frustules rapidly dissolve after the cell's death and the decomposition of the protecting organic skin. Accumulation of large quantities of diatom frustules as diatomaceous oozes at the sea floor thus requires special conditions.

The amount of silica in the diatom frustule may vary according to the depth of the water in which it lived and the stage of the growth cycle as well as the rate of reproduction. Rapidly dividing cells commonly have thinner walls. As the cell number increases in an actively growing culture, the amount of silicon per cell necessarily decreases with decrease in cell size. Cells of a species that does not decrease in size, *Navicula pelliculosa*, grown in a silicon-deficient medium, also absorbed additional silica when transferred to a silicon-rich medium, suggesting that the actual amount per cell may also vary. As this was not due to change in cell shape or length, only valve thickness could vary. The silicon-starved cells were arrested in the early stages of new wall formation in the division cycle, so that silicification was incomplete. Lewin (1957) reported that *N. pelliculosa* in low silicon concentration had an average valve thickness of 20 nm, and in high concentrations had a thickness of 60 nm. Other freshwater pennate diatoms have a frustule wall thickness of 80 to 150 nm, in part in proportion to cell size. If other nutrients are present in adequate amounts in the medium, the final yield of cells in a culture is proportional to the amount of silicon present. In controlled cultures of *Navicula pelliculosa*, growth and reproduction were maintained until silicon was depleted, and no further division occurred. When silicon was again added, walls formed within four hours. A sharp but temporary increase in respiration was followed by a drop in the amount of a coenzyme, and then by silica deposition, indicating the use of energy in the silicon uptake

(Coombs et al., 1967b). When the silicon-deficient cells were resupplied with silicon in the dark, and in a nitrogen-free medium, growth could not occur, but the diatoms nevertheless continued to take up enough silicon to have formed new valves. They merely added it to the existing valves instead. Similarly, if some nutrient other than silicon is limiting, so that division does not occur, the excess silicon taken up by the cell will be added to the existing wall.

Silicon deficiencies in cultures of five centric diatoms slowed or stopped their division, and resulted in an increase in chlorophyll *a* and an increased number of plastids and mitochondria per cell, until the cytoplasm was crowded. Shell morphology also was abnormal (Holmes, 1966). With severe silicon starvation, *Cylindrotheca fusiformis* produced shorter and wider cells, with a volume six times that of normal cells; plastids were enlarged and nuclear structure was changed (Lewin et al., 1966).

Silica is necessary both for the deposition of the cell wall and for net DNA synthesis and nuclear division in diatoms (Darley, 1970). Silicon-starved cells of *Cylindrotheca* enlarged but failed to divide. Normally the new cell wall forms only after completion of the cytoplasmic division, with the two daughter protoplasts still inside the parent frustules, separated only by a plasma membrane. After deposition of the silica shell, the cells separate. Silica starvation at the extremes that would prevent nuclear and cytoplasmic division probably is rare in nature, but may be a factor in limiting diatom blooms (Tappan and Loeblich, 1973).

No other element can be substituted for silicon by the diatoms. Although germanium, also in group IV of the periodic table of elements, is similar in many characteristics to silicon, it is inhibitory to diatoms at relatively low concentrations in cultures; it does not attain even this concentration in nature, however. The nature of the inhibitory effect of germanium is uncertain, but increased concentration of silicon lessens its severity (Lewin, 1966b).

Nitrogen

Diatoms can use nitrate, nitrite, and ammonia as a source of nitrogen, but not organically bound nitrogen. In cultures, *Cylindrotheca closterium* (Ehrenberg) Reimann & Lewin var. *californica* (Mereschkowsky) Reimann & Lewin could use ammonia, urea, or nitrate, and if all were available, preferentially used these sources in the order listed (Grant et al., 1967).

Some diatoms prefer high nitrogen concentrations (even pollution), and others prefer a low nitrogen/phosphorus ratio. The normal ratio in the sea is 15:1, whereas that of the plankton is 16:1. This would suggest that nitrogen is limiting, as it would be depleted sooner (Anderson and Banse, 1963). However, it may be regenerated more rapidly, and the supply may be continually increased through the action of nitrogen-fixing blue-green algae.

Holmes (1966) observed the effects of nitrogen deficiency in cultures of five species of marine centric diatoms. Deficiency resulted in a reduction of chlorophyll *a* per cell, and a decrease in the number of plastids, as these failed to divide. Each dividing cell received half those of the parent, and gradually fewer and fewer occurred per cell. Mitochondria also were reduced in number. *Skeletonema* has only one plastid, but this became too weakly pigmented to be recognized in the light microscope. Nitrogen-deficient cultures of diatoms, as well as of green algae and xanthophytes, accumulated large quantities of fats (Fogg, 1965).

Phosphorus

The required amount of phosphate for diatom growth has been determined with laboratory cultures. The temperate neritic planktonic species *Asterionella japonica* Gran is limited by phosphate concentrations less than 0.25 μg-atoms/liter, and *Phaeodactylum tricornutum* requires 0.55 μg-atoms/liter. The oceanic tropical species *Chaetoceros gracilis* Schütt requires even less, about 0.22 μg-atoms/liter. As mean surface concentrations in the northeast tropical Pacific range from 0.30 to 0.76 μg-atoms/liter, phosphate does not appear to be limiting in the Pacific (Thomas and Dodson, 1968).

Diatom cells in cultures cannot divide if the available phosphorus falls below a minimal value of about 6×10^{-14} g/cell for some species. If an excess is available, the cells may store up to 30 times that amount, using it when the external supply becomes reduced (Lewin and Guillard, 1963). Phosphorus deficiency caused the amount of chlorophyll *a* to decrease, the number of plastids to decrease, and the mitochondria to become darker in color, although they did not decrease in number as occurred with nitrogen deficiency (Holmes, 1966). Cultures of *Skeletonema* included many cells of large diameter, as auxospore formation was stimulated by the phosphorus deficiency.

High phosphorus levels also may be limiting. In fresh water they stimulate growth of blue-green and green algae, at the expense of diatoms and chrysophytes (Lund, 1965).

Sulfur

Sulfate in seawater of 35.00%/₀₀ salinity, at 20°C, is present at a concentration of 2.76 g/liter. Sulfate concentrations of 0.012 to 0.024 g/liter were sufficient to support growth of *Cylindrotheca fusiformis, Phaeodactylum tricornutum, Cyclotella cryptica,* and the nonphotosynthetic *Nitzschia alba.* The sulfate requirements were thus only 1 to 2 percent of the quantity in seawater (Lewin and Busby, 1967).

The freshwater *Navicula pelliculosa* may require sulfur in the process of silicification. When cells were washed in distilled water, they could not assimilate silicic acid, but the addition of divalent sulfur compounds restored this ability. The nonphotosynthetic *Nitzschia alba* did not show the same response to the addition of sulfur, but was restored by the addition of amino acids. Sulfhydryl groups are apparently important in the silica uptake (Lewin et al., 1966).

Boron

Twelve marine pennate diatoms, four marine centrics, and eight freshwater species were found by Lewin (1966a) to require boron, as cell division did not occur when this element was absent. The freshwater species have a lower requirement than the marine species, perhaps because of an enhancement of the deficiency by other salts present in the marine environment. The requirement was noted both for the photo-synthetic species and the colorless *Nitzschia.* Boron similarly was required by *Cylindrotheca fusiformis* grown autotrophically (in the light) and in the dark with preformed organic material.

Seawater contains 4.6 mg/liter of boron, and fresh water has 0.013 mg/liter; some red, green, and brown algae are known to concentrate boron, but this has not been reported for diatoms.

Barium

Rhizosolenia and *Chaetoceros* may have a high barium content. *Chaetoceros curvisetus* Cleve contains 4000 parts of barium per million of dry ash, and *Rhizosolenia calcar-avis* Schultze has between 20,000 and 30,000 parts per million. These delicately silicified forms are seldom well preserved, and the solution of the silica wall may result in a barium enrichment of the sediment. Barium is abundant in pelagic oozes, but as calcareous shells in this environment (foraminifera or coccolithophorids) contain very little barium, the deposition is probably not as barium carbonate (Brongersma-Sanders, 1967), but may be in part due to diatoms.

Manganese

Stosch (1942) noted that a manganese deficiency prevented the formation of new diatom valves. The cytoplasm detached from the frustule and the protoplast emerged. All smaller cells were naked, and larger ones became swollen spherical plasmodia with hundreds of nuclei. After manganese was added, shells were formed, but these were abnormal in shape until after auxospore formation.

External Metabolites

Diatoms and many other algae produce growth-inhibiting or stimulating substances of selective effect. In the laboratory, production of these substances is most notable in older cultures. Diatoms were inhibited by such substances produced by the green algae *Chlorella* and *Scenedesmus,* but *Scenedesmus* was stimulated by the secretions of *Chlorella* and of the diatom *Nitzschia*. Lake studies by Jørgensen (1957) indicated that maxima of planktonic and epiphytic diatoms never occurred together, suggesting that the metabolites of the plankton were inhibiting to the epiphytic flora. Where the plankton algae were abundant throughout the year (both diatoms and green algae), the epiphytic flora was small.

External metabolites also are bactericidal. Droop and Elson (1966) noted that healthy pelagic diatoms were largely free of bacteria during exponential increase. During the later stages of slower population growth or decline, bacterial activity increased, particularly affecting the smaller cells before auxospore rejuvenation. Cell extracts of 14 species of marine phytoplankton were tested for antibacterial activity (Duff et al., 1966). Diatoms included were *Cyclotella*, *Skeletonema*, and *Phaeodactylum*. The last two were both among the 6 species that showed strong antibacterial effects on the 9 marine bacterial species. Green and blue-green algae tested showed no antibacterial effects.

Vitamins

Like many other phytoplankton, some diatoms require an exogenous source of vitamin B_{12}. Over half the species in culture have such requirements. Of 43 cultures of pennates and 1 culture of centric diatoms examined, 24 needed no added vitamins, 6 needed thiamine, 11 needed cobalamin, and 3 needed both. *Cylindrotheca fusiformis* requires the pyramidine moiety of the thiamine, but is unable to grow with the thiazole moiety alone (Lewin, 1965).

Most culture studies of algae are based on bacteria-free cultures. Because bacterial growth in a water sample is proportional to the available surface area (of the containing vessel), their growth results in laboratory conditions quite different from those in nature. However, some diatoms will not grow in bacteria-free culture (*Asterionella japonica*), and others grow much better in enriched seawater with bacteria than in the absence of bacteria (e.g., *Skeletonema costatum*). This may be because of the bacterial production of vitamin B_{12} or other organic molecules necessary for growth (Fogg, 1965).

DISTRIBUTION

To understand the distribution of diatoms or other phytoplankton, records of the actual distribution must be obtained and correlated to laboratory culture studies to determine which factors control the occurrence and relative abundance of the organisms. In cultures, limiting factors may be the exhaustion of some nutrient (N, P, Si), rate of supply of oxygen or CO_2 (particularly in relatively stagnant water), alteration of the pH by the selective use of certain constituents of the medium, reduction of light by self-shading (probably more noticeable in laboratory cultures than in nature, because abundance differs by several orders of magnitude), the accumulation of auto-inhibiting substances (which may reach a critical level very suddenly), and grazing. All are difficult to estimate with any certainty. Nutrient quantities can be determined exactly in the laboratory, but in nature a low amount in the water may merely mean a balance between supply and consumption, and turnover may in fact be quite rapid. Furthermore, different species are controlled by different factors, their varying tolerances resulting in the well-known successions of dominant species through the year. Considerably more data are needed for a variety of species, including both the extremes they can tolerate and their optimal conditions.

Present Occurrence

Marine Habitat

Marine diatoms may be planktonic or benthic; the latter are littoral because they must live within the photic zone. Most regions of the oceans report between 100 and 200 species, up to nearly 300. Some 400 species have been reported in the Indonesian seas at various times, and records indicate that 350 species occur in the Antarctic, although some of the latter records probably are synonymous (Guillard and Kilham, 1977).

Plankton. The oceanic plankton is nearly always holoplanktonic, completely independent of the sea bottom and without a sedentary or resting spore stage, as in *Chaetoceros atlanticus* Cleve, *Thalassiothrix* sp., and *Planktoniella sol* (Wallich) Schütt. However, some spore producers do occur in oceanic waters (*Thalassiosira gravida*), so that the terms *oceanic* and *holoplanktonic* are not completely synonymous.

Some neritic plankton may be similarly holoplanktonic, although living close to the coastline; perhaps the distribution is related to nutrient supply (*Rhizosolenia fragilissima* Bergon). Other neritic plankton is meroplanktonic, and a resting spore stage occurs in the life cycle (*Thalassiosira nordenskioeldi, Chaetoceros laciniosus* Schütt, *C. debilis* Cleve). Other species are indifferent to the coastline proximity, have fewer requirements and greater tolerances, and may exist in both coastal and oceanic regions; for example, the meroplanktonic *Thalassiosira decipiens* (Grunow) Jørgensen and the holoplanktonic *Nitzschia delicatissima* Cleve (Smayda, 1958; Hendey, 1964).

Although distribution is patchy, phytoplankton may be present in great abundance over large elliptical areas of the ocean surface. These patches range from about a meter across to a total of 48 to 64 km wide and 193 to 290 km long, with an estimated mean size about 16 by 64 km. They probably are a response to upwelling caused by wind-induced water movements. Long narrow streaks of abundant cells, commonly about a meter wide, may contain about 10 million diatoms per liter of water, and 50 million flagellate cells per liter. Bainbridge (1957) quoted reports of 600,000 diatoms per liter in Kiel Bay and 2600 specimens per liter of *Biddulphia sinensis* Greville in the North Sea. One record from the Norwegian Balsfjord at a depth of 10 meters included 5400 ciliates, 26,000 dinoflagellates, 278,000 coccolithophorids, and 549,500 diatoms in a liter of seawater, a total of

nearly 859,000 cells. Kolbe (1927) reported 12 million diatom cells per liter in Grays Harbor, Washington, and Guillard and Ryther (1962) reported blooms of *Thalassiosira pseudonana* in numbers of 10 to 200 million cells per liter.

In a study of the diatoms of the Gulf of California, one quarter of the catches contained over 100,000 cells/liter, and 10 had over 1,000,000 cells/liter (Cupp and Allen, 1938). The greatest number observed was a catch containing 5,058,800 cells/liter, of which 4,945,000 were *Chaetoceros radicans* Schütt. Of the total 69 species, representing 32 genera, 41 were typically neritic species, and only 15 oceanic.

Although these figures are an indication of their extreme abundance, exact numerical abundance is less important than the particular species concerned. Diatom cell volume ranges from 51 μm^3 in *Nitzschia delicatissima* to 11,740,000 μm^3 in *Rhizosolenia acuminata* (Peragallo) Gran. Thus one *Rhizosolenia* cell equals in volume 230,000 cells of *Nitzschia*. Cell volume may also be affected by the size of the central vacuole, some species having a very thin cytoplasmic lining of the wall and a very large vacuole (Smayda, 1965).

Diatom abundance generally is greatest in coastal regions where nutrients are supplied from the land and from decaying organic matter, in regions of upwelling, or where melting ice adds to turbulence and water overturn, renewing the supply of such nutrients as nitrogen, phosphorus, silica, and CO_2.

Benthos. The benthos has a more restricted habitat than the plankton. Benthic species may be attached by mucus pads or stalks, or may occur as a film coating the substrate. Some live in the interstitial spaces in sand or mud flats. Nonmotile diatoms that are attached to sand grains, the **epipsammon**, are commonly less than 20 μm in length and may remain in the cracks or depressions of the grains. Most are araphid Diatomaceae or monoraphid Achnanthaceae (McIntire and Moore, 1977). Somewhat larger **epipelic** diatoms, which live on the silt grains but are not firmly attached, generally represent motile pennates.

Benthic species may require very little light

(*Auliscus, Hyalodiscus*) although some require intense light levels. Many are epiphytic on benthic algae and seagrasses. The turtle grass *Thalassia testudinum* König provides a substrate for 42 pennate diatom species representing 20 genera in the Biscayne Bay of Florida (Reyes-Vasquez, 1970). These diatoms provide a food source for numerous small crustaceans, gastropods, and fish.

Hendey (1964) subdivided the littoral benthos into three occurrence zones, sublittoral, littoral, and supralittoral. The sublittoral zone ranges from immediately offshore up to the level of extreme low spring tides. Frondose colonies and valve-to-valve chains and ribbons are common, attach to any firm substrate, and usually occur in water of 10 m depth or less (*Grammatophora, Rhabdonema, Achnanthes, Navicula, Synedra*).

Some normally benthic species may temporarily break free, due to the action of waves, currents, and so forth, and may then float for a time (being termed tychoplanktonic) before becoming reattached to the substrate. Most such tychoplanktonic species probably do not reproduce during their floating existence, although *Biddulphia aurita* (Lyngbye) Brébisson & Godey may reproduce in either the neritic plankton or the littoral zone.

The littoral zone lies between the extreme low water and extreme high water spring tides, covering a range of perhaps 5 m on a cliff coastline to as much as 4 km on a gently sloping beach. In an area such as the latter, separate lower, middle, and upper littoral zones may be delimited, although the first two commonly merge. On a soft mud or sand surface the diatoms may form a yellow-brown film, and their secreted mucus may act as a cementing medium. Here they may migrate vertically in response to the light, moisture, or tidal cycle, their vertical movement generally being on the order of a few millimeters. Such communities are short lived because of the instability of the environment. The diatoms disappear from the surface during flood tide, and reappear after low tide exposure. Nutrient supply is abundant in this environment, phosphates and nitrates being released by the rapidly decaying organic matter exposed on the beach.

The upper littoral area has a firmer substrate, and a more permanent assemblage containing up to 40 species. In clean sand, just after the tide recedes, small brown monospecific colonies may occur, with a diameter of 1 to 2 cm each. Some of these macroscopic colonies of *Navicula, Nitzschia,* and *Amphipleura,* occurring on the sand surface, have even been described as distinct genera. The diatom flora of the interstitial shore sands may be so abundant that at Port Erin, Isle of Man, England, "their photosynthetic activity was distinctly audible as a gentle sizzling to an observer standing upright upon the discolored area, while the sand was frothy with bubbles of gas, presumably oxygen, given off by them" (Herdman, 1924).

The supralittoral zone lies just above extreme high water and includes salt marshes, streams, pools, and similar microhabitats characterized by varying salinity. It has an abundant and varied flora.

The littoral benthos may include different species at different seasons, as does the plankton, but the distribution is always very patchy. Sampling methods used for plankton studies are not applicable to studies of the littoral flora.

Geographic Distribution in the Seas

Marine diatom occurrence is controlled by the interactions of the many varied factors discussed above, and does not necessarily correlate closely with abundance in the sediments, which will be discussed separately. However, some general characteristics of polar, tropical, and cosmopolitan diatoms have been suggested (Smayda, 1958).

Antarctic species must tolerate a high and fairly constant salinity and extremely low temperatures, requirements met by very few phytoplankton species other than diatoms (species of *Thalassiosira, Chaetoceros, Eucampia, Fragilaria,* and *Synedra*).

The Arctic has conditions similar to those of the Antarctic, but less uniform, because the effects of the warm Gulf Stream and Japan Current are carried farther toward the polar region. There are, as a result, very few typically

Arctic species, two being *Chaetoceros furcellatus* Bailey, and *Thalassiosira hyalina* (Grunow) Gran.

Species that inhabit tropical areas must tolerate very high temperatures, and secondarily high salinity. The very few typically tropical diatoms include *Chaetoceros laevis* Leuduger-Fortmorel and *Rhizosolenia robusta* Norman. *Planktoniella sol* is present in all tropical current systems, but is eurythermal and occurs also in temperate regions. It withstands temperatures ranging from 1.9° to 30.5°C, although the lower part of this temperature range may be due to transport from the area of its reproduction. Coccolithophores are more characteristic of tropical and subtropical waters than are diatoms, although the latter also are present. In the tropical and subtropical Atlantic, Wood (1968) reported *Hemiaulus, Rhizosolenia,* and *Coscinodiscus* as consistently present in sufficient numbers as to provide a bloom when conditions became favorable.

Cosmopolitan phytoplankton must have a wide temperature tolerance in order to live in both the colder waters and the equatorial regions of all oceans. It is restricted to the upper photic zone, in contrast to the zooplankton that can migrate vertically to seek preferred temperature levels. Among the more important cosmopolitan diatoms are *Chaetoceros compressus, Leptocylindrus danicus,* and *Skeletonema costatum. Thalassionema nitzschioides* Grunow, with circumglobal distribution, occurs at salinities of as much as 34.34%/$_{ooo}$ to as low as 3.6%/$_{oo}$ in the Barents Sea. It is equally eurythermal, living at temperatures of 22.2° to 30.7°C in tropical regions, and in the Oslofjord, where the mean temperature is 8.5°C. Possibly clones of less broad temperature tolerance occur within this distribution.

Less tolerant species may serve as indicators of various water masses that may have been introduced into an area, whether the cells remain viable or occur only as dead individuals. *Planktoniella sol* is an indicator species for the Gulf Stream where it enters the Norwegian Sea. As these currents pass through other regions, their physical characteristics become rapidly modified, and few species can maintain themselves under such changes. Other species may indicate polluted waters or dilution by fresh water (Smayda, 1958).

No particular latitude is most productive. Diatom abundance is seasonal in both the polar and tropical areas. In general, diversity is greatest in the tropical and subtropical plankton, but diatoms are relatively rare, and their occurrence in a given sample is largely chance. Diatoms are dominant in polar regions, where the diversity is less but occurrence of a given species more reliable.

Many comprehensive studies have been made of the diatom floras of various seas (including Castracane, 1886; Hustedt, 1927 – 1966; Schmidt, 1874 – 1959; and Schütt, 1896). Many studies have covered parts of the Atlantic Ocean (Karsten, 1905b; Kolbe, 1955; Lebour, 1930; Maynard, 1976), the northeast Atlantic (Hendey, 1964; Heurck, 1896; Peragallo and Peragallo, 1897 – 1907), the southeast Atlantic (Hendey, 1958), and the southwest Atlantic (Frenguelli, 1928). The seas of the Arctic regions were studied by Gemeinhardt (1935) and Meunier (1910, 1913, 1915); the Antarctic by Frenguelli and Orlando (1958), Hendey (1937), and Karsten (1905a); the Mediterranean by Pavillard (1925); and the Black Sea by Proshkina-Lavrenko (1963b). Many studies have been made in the Pacific (Kozlova and Mukhina, 1967; Venrick, 1971; Zhuze et al., 1967), the eastern Pacific (Cupp, 1943; Cupp and Allen, 1938; and Gran and Angst, 1931), and the western Pacific (Hustedt, 1937a, 1938a, 1939; Mann, 1925). Diatoms were described from the Java Sea by Allen and Cupp (1935), and from the Indian Ocean by Karsten (1907) and Kozlova (1971).

In the Antarctic, more than half of the common taxa are species of *Chaetoceros, Coscinodiscus, Nitzschia, Thalassiosira,* and *Rhizosolenia* (Guillard and Kilham, 1977). Three latitudinal zones could be recognized, the Subantarctic zone having a high diversity, including *Chaetoceros neglectus* Karsten and *Nitzschia barkleyi* Hustedt. The North Antarctic diatom flora includes *Chaetoceros dichaeta* Ehrenberg, *C. neglectus, Nitzschia pseudonana* (Hasle) Hasle, and other *Nitzschia* species. The South Antarctic assemblage includes *N. curta* (van Heurck) Hasle and *N. cylindrus* (Grunow) Hasle, in addition to certain endemic species of the ice pack. Ice floras include species of *Amphiprora,* *Fragilaria, Gomphonema, Navicula, Nitzschia, Pinnularia, Pleurosigma,* and *Synedra.*

In the Pacific Ocean, seven diatom assemblages that differed from the living assemblages because of selective dissolution of the frustules were recognized in bottom sediments (Zhuze et al., 1967). Long cored sections indicated that the relative abundances and the geographic distribution of the thanatocoenoses shifted in position through time. From north to south the thanatocoenoses were arctoboreal, north boreal, subtropical, eurytropical, equatorial, subantarctic, and antarctic. The arcotoboreal assemblage occurs in the northern margins of the Pacific, Bering Sea, and Sea of Okhotsk, and includes *Thalassiosira nordenskioeldii, T. gravida, Bacterosira fragilis* Gran, *Chaetoceros furcellatus,* and *Biddulphia aurita.* The north boreal assemblage is represented largely by oceanic plankton, although in some areas, such as the Kamchatka and Kurile regions, and off North America, some arctoboreal species also may be present. Typical species are *Thalassiosira excentrica* (Ehrenberg) Cleve, *Coscinodiscus curvatulus* Grunow, *C. marginatus* Ehrenberg, *Actinocyclus divisus* (Grunow) Hustedt, *Rhizosolenia hebetata, Thalassiothrix longissima* Cleve & Grunow, and *Denticula seminae.* Characteristic species of the subtropical association are *Thalassiosira decipiens, T. lineata* Jousé, *Coscinodiscus radiatus* Ehrenberg, *Thalassionema nitzschioides, Pseudoeunotia doliolus* (Wallich) Grunow, *Roperia tesselata, Nitzschia bicapitata* Cleve, *N. interrupta* Heiden, and *N. sicula* (Castracane) Hustedt. The eurytropical assemblage includes *Coscinodiscus crenulatus* Grunow, *C. nodulifer* A. Schmidt, *Hemidiscus cuneiformis, Thalassiosira oestrupii* (Ostenfeld) Proshkina-Lavrenko, and *Rhizosolenia bergonii* Peragallo. The well-preserved *C. nodulifer* and *Nitzschia marina* Grunow, Cleve & Grunow dominate the assemblage in the sediments in many localities, although locally *Ethmodiscus rex* comprises nearly 100 percent of the deposit. Species characteristic of the equatorial assemblage are *Asteromphalus imbricatus* Wallich, *Coscinodiscus africanus* Janisch, *Triceratium cinnamomeum* Greville, and *Asterolampra marylandica* Ehrenberg. Diatoms are rare in the sediments from lat. 5° to 45° S in the Pacific. The subantarctic assemblage is circumpolar in occurrence, and consists of almost 100 percent endemic species, including *Fragilariopsis antarctica* (Castracane) Hustedt, *Coscinodiscus lentiginosus* Janisch, *Thalassiothrix antarctica* Cleve & Grunow, and *Schimperiella antarctica* Karsten. The last assemblage characterizes the Antarctic, and includes *Fragilariopsis curta* Hustedt, *F. cylindrus* (Grunow) Helmcke & Krieger, *Eucampia balaustium* Castracane, *Thalassiosira gracilis* (Karsten) Hustedt, *Charcotia actinochilus* (Ehrenberg) Hustedt, *C. oculoides* Karsten, *C. symbolophorus* Grunow, and *Biddulphia weissflogii* Janisch.

The diatom thanatocoenoses in the Pacific show the influence of the major oceanic currents and the effect of the equatorial upwelling. In some regions, only 0.06 to 6 percent of the plankton frustules survive to reach the sediments, whereas in the Antarctic probably most of the crop survives dissolution. Kozlova and Mukhina (1967) recognized eight slightly different regions in the Pacific, a subarctic one, boreal, north tropical, and equatorial, a very poor south tropical one, intermediate (containing a mixture of species from the adjacent subantarctic and subtropical regions), subantarctic, and antarctic. Diversity varied from 13 species in the subarctic to 54 in the intermediate and 41 in the subantarctic zones.

Venrick (1971) studied seasonal changes in recurrent groups of species in the plankton of the North Pacific. In summer, three associations of neritic species occurred, four subarctic associations, two in the transitional area, and one in the central Pacific. The neritic assemblage was less rich in winter, and only one assemblage of neritic diatoms was recognized, three of subarctic, one transitional, and three in the central Pacific. The recurrent groups included both some species that were present all year and others that were seasonal.

The diatom assemblages of the North Atlantic have been studied by many of the oceanographic expeditions, but the equatorial Atlantic is much poorer in diatoms than the same area of the Pacific, averaging only about 50 cells per liter. However, the area of upwelling off the African coast is extremely rich, with more than 100,000 cells per liter. The distribution of diatoms in the surface sediments of the Atlantic accurately reflects their relative abundance in the overlying water, and shows no evidence of lateral drift during settling (Maynard, 1976). Abundance is correlated with the pattern of primary productivity and the quantities of dissolved phosphates and silica. As such water mass characteristics can be inferred from the distribution of the diatoms in the sediments, the latter is useful in reconstruction of paleooceanographic circulation patterns. Six assemblages were recognized by Maynard (1976), an Antarctic—South Atlantic assemblage dominated by *Fragilariopsis kerguelensis* (O'Meara) Hustedt and *Coscinodiscus lentiginosus;* a gyre margin assemblage at the major North and South Atlantic gyres containing *Pseudoeunotia doliolus;* a warm water equatorial assemblage with *Coscinodiscus nodulifer;* a North Atlantic subpolar assemblage at the north edge of the North Atlantic Drift containing *Thalassiosira gravida;* a subtropical assemblage in the central part of the North Atlantic Drift; and a subpolar assemblage characterized by *Thalassiothrix longissima, Rhizosolenia styliformis, R. bergonii,* and *Thalassiosira decipiens.*

Hendey (1937) recognized four diatom distributional zones in the South Atlantic that varied seasonally in width and position. The Antarctic Zone extends from the ice shelf to the Antarctic Convergence, the Subantarctic Zone extends from the Antarctic Convergence to the Subtropical Convergence, the Subtropical Zone extends from the Subtropical Convergence to the Tropical Zone, and the Tropical Zone lies near the equator. Kolbe (1955) noted that diatoms were scarce in the equatorial Atlantic as compared to the Pacific. Some of their more abundant species were common to sediments of the tropical regions of the two oceans (*Coscinodiscus nodulifer, Nitzschia marina, Hemidiscus cuneiformis*). A few were common to the tropical regions of both oceans and to the North Atlantic (*Coscinodiscus lineatus, Thalassiosira excentrica, Thalassionema* spp., *Thalassiothrix* spp.).

Although dinoflagellates generally dominate the plankton of the Indian Ocean, Kozlova (1971) recognized five diatom plankton biocoenoses there, representing tropical, subtropical, temperate, subantarctic, and antarctic zones.

The Marine Succession

Because plankton species are each distinctive in their tolerances to and preferences for various physical characteristics of the environment, the seasonal succession of dominant phytoplankton species is strongly affected by local conditions, including temperature, light, progressive nutrient depletion, and the accumulation of stimulatory or inhibiting secretions. Margalef (1967) suggested that during the succession, average cell size and species diversity generally increase, together with the proportion of motile species, whereas a decrease occurs in the rate of production per biomass and rate of cell division.

The first stage begins with the enrichment of the euphotic zone by upwelling, mixing, or increased runoff, and is characterized by small species, with high surface to volume cell ratio (1 $\mu m^2/1\ \mu m^3$). Diatoms include *Chaetoceros* and *Skeletonema.* Cell division occurs about once per day, cell density is about 100 to 1000 cells/ml, chlorophyll *a* content is high, and excretion of soluble and mucilaginous material is heavy.

In the second stage of succession, the number of planktonic diatoms increases, particularly large and medium-sized *Chaetoceros.* Cells are of medium size, colonies abundant, and the surface to volume ratio $\mu m^2/\mu m^3$ is 0.2 to 0.5. Species diversity is high, cell density less (20 to 100 cells/ml), and turnover is on the order of a few days.

The third stage is a continuation of the preceding, occurring somewhat later in the year. The *Chaetoceros* of stage two have formed resting cysts. The few diatoms present occurred in low numbers during the preceding stage, but are more tolerant of low nutrient conditions, and persist after the other species disappear. These include large cylindrical species such as *Bacteriastrum, Nitzschia, Corethron,* and *Rhizosolenia.* The amount of chlorophyll *a* is low.

The fourth stage occurs only in an oceanic situation or under poor nutritional conditions. Surface nutrients are depleted, and the waters have a stratified character. Diatoms almost completely disappear, except for a few associated with blue-green algae or ciliates. Dinoflagellates and coccolithophorids are relatively more important, the surface/volume ratio is less, cell division may occur only weekly, cell density is about 10 cells/ml, and chlorophyll *a* is still more reduced. Less pigmented forms are more common, and many may be heterotrophic.

Freshwater Habitat

The freshwater habitat includes both planktonic and benthic species in running or standing water, soil, and even an aerial environment. Their distribution has been summarized by Patrick and Reimer (1966) and Patrick (1977).

The freshwater plankton is mostly represented by the floating stage of a normally benthic population. Most are small species (nannoplankton), although a few are larger colonial forms. Characteristic centric freshwater planktonic species represent *Melosira*, *Cyclotella*, and *Rhizosolenia*. Much more abundant are the pennates, including *Asterionella*, *Fragilaria*, *Surirella*, *Synedra*, and *Tabellaria* (see Figure 8.30). Commonly the spring bloom consists of pennates and the fall one of centric diatoms.

Epiphytes attach to rocks or to rooted or floating vegetation by means of a mucilage stalk or cushion. Freshwater epiphytes include *Achnanthes*, *Cocconeis*, *Cymbella*, *Epithemia*, *Navicula*, *Rhoicosphenia*, and *Synedra*. Attached species commonly characterize running water, their shape varying with the nature of the current. *Desmogonium* is more elongate in rapidly flowing water than in standing water.

Freshwater benthic species attached to silt, sand, or a hard substrate in lakes, ponds, and slow-flowing streams are usually motile raphe-bearing pennates such as *Cymatopleura*, *Frustulia*, *Gyrosigma*, *Navicula*, *Neidium*, *Nitzschia*, *Pinnularia*, *Stauroneis*, and *Surirella*, as well as the araphid *Fragilaria*.

Eutrophic water, rich in dissolved nutrients or polluted, has a flora distinct from that of oligotrophic waters of low nutrient level. The trophic level of the lake is indicated by the ratio of centric to pennate diatoms; the higher the ratio, the more eutrophic the water. The dystrophic water of bogs and swamps has high quantities of humus, but is low in dissolved nutrients and has low pH and oxygen levels. These dystrophic regions also are characterized by a specialized flora, particularly *Eunotia* and *Pinnularia*. Muddy water, with considerable suspended load, generally has a poor microflora.

The soil habitat is characterized by raphe-bearing and motile pennate diatoms such as *Hantzschia*, *Navicula*, and *Pinnularia*. All are small and largely restricted to the top centimeter of the soil. If the soil dries out, they migrate to a greater depth. The diatom flora of the dry lakes of the western United States may be a modified soil assemblage (VanLandingham, 1966b). Aerial diatoms may live in mosses, on tree trunks, damp stones, leaves, and the spray zone of lakes and rivers. This rigorous environment contains mostly small and specialized species. A dry rock habitat may contain *Eunotia*, *Melosira*, and *Navicula;* the species may with-

---

Figure 8.30
Some freshwater diatoms. **1,2.** *Nitzschia denticula* Grunow, fresh water, Scotland, SEM, from Paddock and Sims, 1977. **1.** Exterior of frustule, showing keel that bears the raphe, ×2000. **2.** Interior of part of valve, showing fibulae that cross the raphe canal, and poroid (arrow) in wall of raphe canal, ×4100. **3 – 5.** *Tabellaria flocculosa* (Roth) Kützing var. *flocculosa*, Minnesota lakes, TEM, from Koppen, 1975. **3.** One apex of valve, showing transverse rows of pores, narrow longitudinal hyaline area or pseudoraphe, and row of short spines at valve margin, ×5000. **4.** Isolated complete intercalary band showing septum (s), and rudimentary septum (rs), ×5000. **5.** Entire valve as in part 3, showing median inflation and polar areas of fine pores, ×4760.

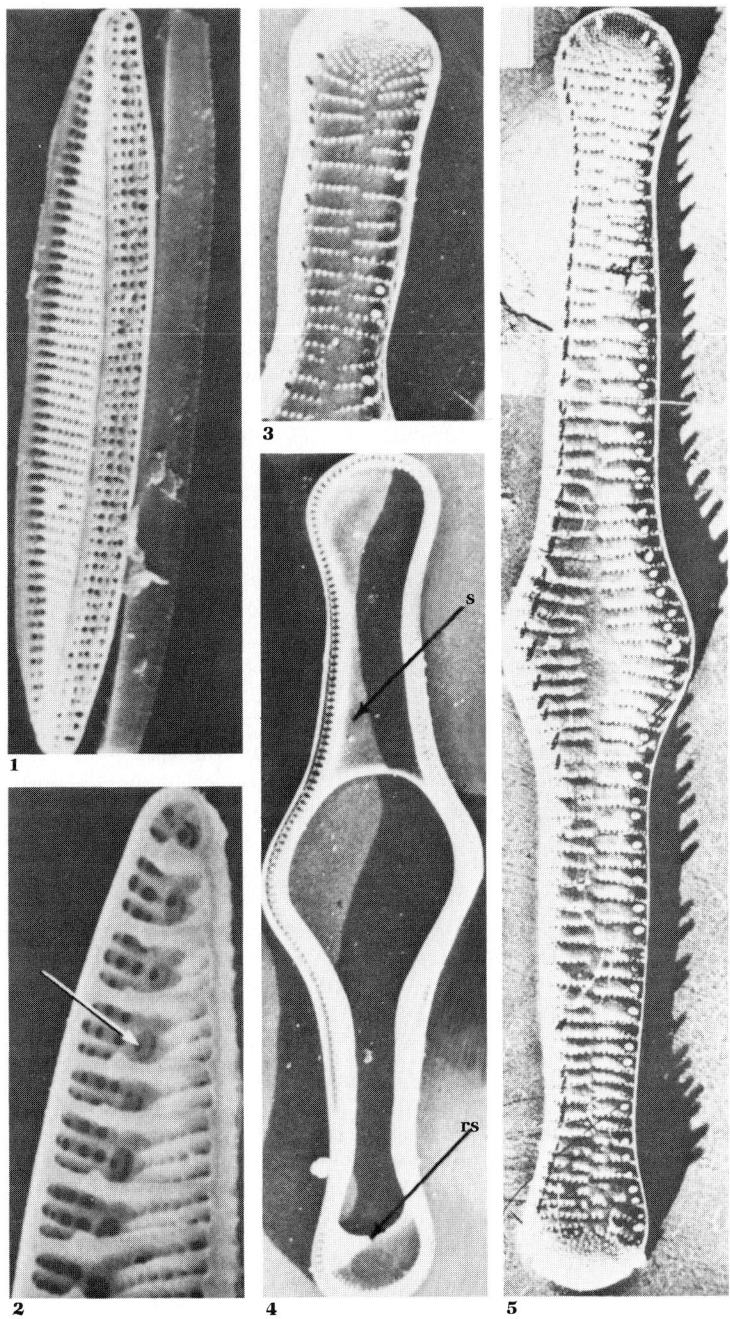

1
2
3
4
5
s
rs

stand desiccation by storing large amounts of oil, or may build inner siliceous plates. More diverse diatoms inhabit the wet rocks and moss, including species that may live either in or out of water, especially those of *Pinnularia, Melosira, Navicula,* and some *Cymbella, Gomphonema, Synedra, Achnanthes,* and *Epithemia* (Patrick, 1977).

Diatomaceous Oozes and Diatomites

Diatomaceous earth or diatomite is a very pure organically formed sediment, consisting of the accumulated siliceous frustules of diatoms. Because the rock is composed of numerous tiny, hollow frustules with their abundantly perforated walls, diatomite is very porous and light in weight. The opaline silica of which it is composed has a specific gravity of 2.1 to 2.2, but the porosity is so great that a dry block of diatomite has a specific gravity of only 0.45, and will even float on water. The great porosity, light weight, absorbent nature, filtering and abrasive qualities, and resistance to acids (other than hydrofluoric) combine to make it highly useful for insulation, structural materials, abrasives, filtering aids, and ceramic glazes, among other applications.

Distribution of Diatomaceous Sediments

Siliceous sediments are widespread in the present seas, covering about 40 percent of the sea floor (Riedel, 1959). Diatoms are the present major contributor to the siliceous oozes, supplying over 70 percent of the total, and followed in importance by radiolarians and sponges. Silica has no "compensation" depth, as does calcium carbonate, below which solution occurs, but the distribution of siliceous sediments is nevertheless more restricted than the presence of the organisms in the surface waters would suggest. Silica presently accumulates in three great belts, a southern siliceous belt around the Antarctic continent, a northern siliceous belt on the northern margins of the Pacific Ocean, and an equatorial belt in the Pacific and Indian Oceans (Lisitsyn, 1967a). Freshwater diatomites also occur, but are more limited in extent.

Covering a width of 900 to 2000 km, between lat. 45° and 77° S, the southern oceanic siliceous belt comprises about 75 percent of the present siliceous accumulations (see Figures 8.31 and 8.32). The sediments may consist of 72 percent amorphous silica. Counts of 100 to 440 million diatoms per gram of dry marine sediment have been made (Lisitsyn, 1967b). As nonmarine species may be smaller, freshwater lake beds may contain up to one billion diatom valves per gram of diatomite (Zhuze, 1968a).

The northern siliceous belt is largely restricted to the Pacific semienclosed marginal seas, such as the Okhotsk and Bering Seas, the Sea of Japan, and the Gulf of California. Diatom-rich sediments occur in the deeper areas, although in part this may result from lessened influx of detrital sediments.

The equatorial siliceous belt is prominent in the Pacific and Indian Oceans, but almost unrecognizable in the Atlantic. Most accumulations in this belt are at great depths, commonly below 4600 to 4900 m in the oceanic trenches where all calcareous material is gone, to a depth of 9394 m in the Kurile-Kamchatka Trench. In places, the equatorial belt contains sediments that are almost monospecific, with great numbers of the very large discoid diatom *Ethmodiscus rex* (see Figure 8.33), which was illustrated in the Challenger Report from 10,505 m depth. Such sediments, in which 80 percent of the frustules and fragments are of *Ethmodiscus,* have been termed *Ethmodiscus* silts for this species, which is commonly between 0.7 and 1.4 mm in diameter and may attain 2 mm (McHugh, 1954).

Although *Ethmodiscus* oozes are common in the tropical western Pacific between lat. 5° and 20° N and long. 130° to 160° E, occurring mostly below 5000 m depth, as at the bottom of the Philippine and Marianas trenches, *Ethmodiscus* is rare in the plankton, being present in numbers of less than one cell per m^3 in the upper 100 m (Belyaeva, 1968; Jousé et al., 1971). Similar *Ethmodiscus* oozes occur in the eastern Pacific (McHugh, 1954; Belyaeva, 1972), where they comprise 4 to 6 m of sediment; they occur

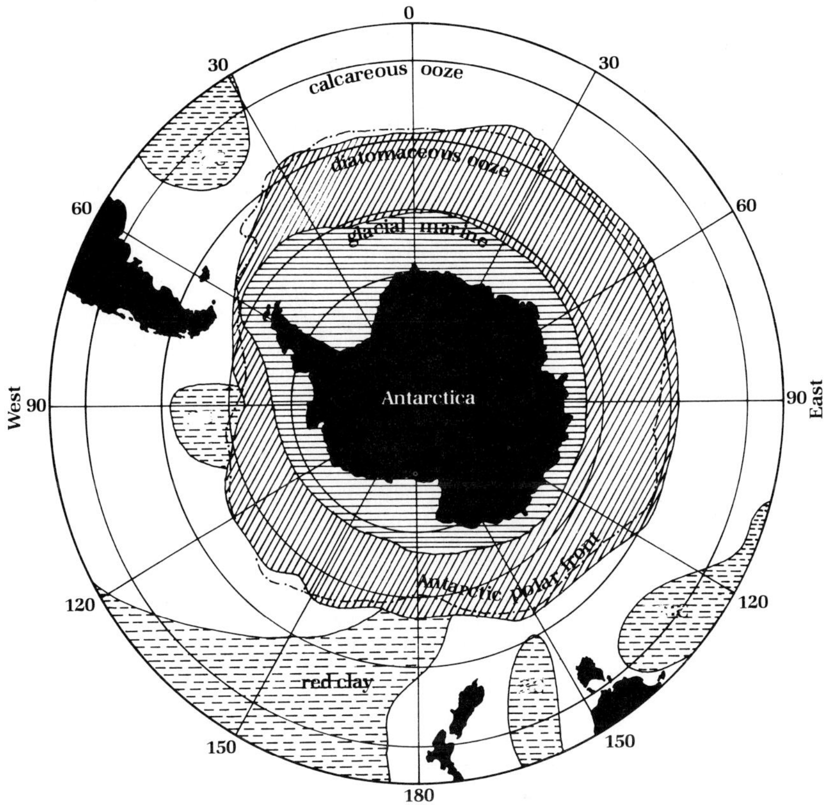

Figure 8.31
Distribution of sediment types around Antarctica, showing circumpolar belt of diatomaceous ooze extending to the Antarctic polar front, where it gives way to calcareous oozes. Redrawn and reproduced with permission from Hays, 1967, *Prog. Oceanogr.*, v. 4, p. 117–131, copyright 1967, Pergamon Press, Ltd.

at 12 levels in the Indian Ocean from the equator to lat. 20° S (Kozlova, 1971; Schrader, 1974). *Ethmodiscus rex* also has been observed in the plankton of the Sargasso Sea in the Atlantic (Swift, 1973), and in bottom sediments in the eastern equatorial Atlantic, particularly those representing the Pleistocene glacial episodes (Gardner and Burckle, 1975). Their abundant accumulation in the sediment, as opposed to their rarity in the plankton, was variously regarded as indicating a temporary high productivity (Wiseman and Hendey, 1953; Gardner and Burckle, 1975); related to low concentrations of dissolved phosphate (Belyaeva, 1970) or areas of upwelling (McHugh, 1954; Belyaeva, 1972); indi-

cating a concentration of scarce frustules by ocean currents (Mann, 1907); an accumulation from eroded fossil material at the sea floor (Riedel, 1954), inasmuch as the Challenger stations were reported to be of Miocene age on the basis of associated radiolarians; due to more rapid sinking of the larger frustules (Smayda, 1970) and hence less solution (Belyaeva, 1968; Schrader, 1974); or a combination of these factors (Schrader, 1974). Gardner and Burckle (1975) discounted selective solution in the Atlantic, as foraminifera in the same samples showed little solution effect. However, accumulation of siliceous deposits requires acidic conditions, high dissolved CO_2 and low pH, whereas

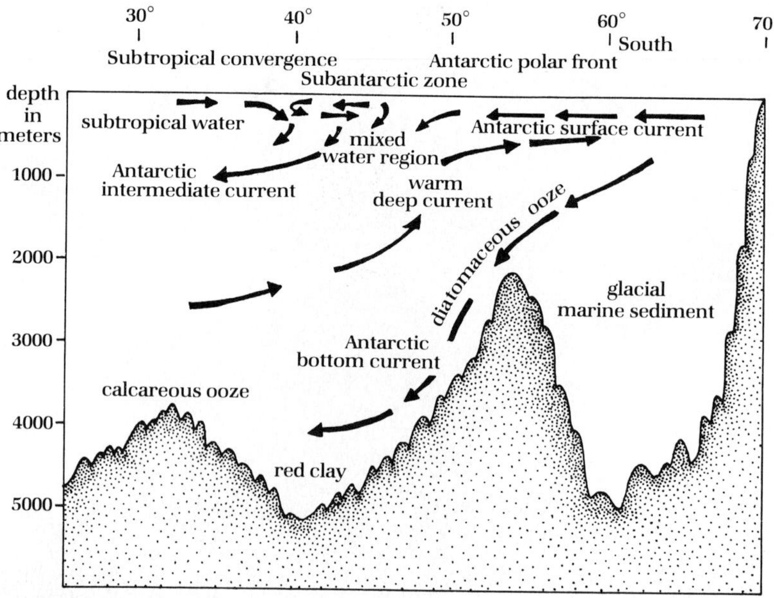

Figure 8.32
Vertical circulation in the Antarctic and its relationship to the distribution of diatomaceous ooze; vertical exaggeration 500 times. From various sources.

silica dissolves under relatively alkaline conditions that would favor the preservation of calcium carbonate (Kolbe, 1957; Tappan and Loeblich, 1971). The lightly silicified diatoms in the surface sediments could be selectively dissolved, leaving only the morphologically simple, large, heavily silicified frustules.

A siliceous sample from the central Pacific (lat. 2°01′ N, long. 172°32′ W) at a depth of 5328 m was determined to contain 9.7 million diatoms, 0.7 million silicoflagellates, and 8.1 million radiolarians per gram of sediment (Lisitsyn, 1967b). Because of the greater average size of the latter (except for *Ethmodiscus*), the radiolarians may supply the greatest part of the total weight of silica in the tropics, but are not as abundant in the siliceous zone of higher latitudes.

Between these major siliceous belts, amorphous silica is relatively rare in the sediments. Almost no modern siliceous sediments are forming between lat. 20° and 45° S or between lat. 20° and 40° N, and none occur in the Atlantic or Mediterranean.

Not only modern sediments have such a patchy distribution, for siliceous sediments of earlier ages are equally limited in occurrence. Their distribution is not quite like that of modern sediments, for extensive accumulations are present in the Mediterranean region and in both southern Europe and North Africa. The diatomites from Oran in Africa were among the earliest from which diatoms were described (Ehrenberg, 1838, 1840), and Algerian commercial production is second only to that of the United States. South Africa also has important diatomites. Others occur throughout Europe (Italy, Spain, Germany, Austria, Hungary, Norway, Denmark, and the USSR), South America, North America (Canada and both the Atlantic and Pacific coasts of the United States), Australia (New South Wales), New Zealand, Japan, and Siberia (Calvert, 1930; Deflandre, 1961). Perhaps the purest and thickest occurrences are those

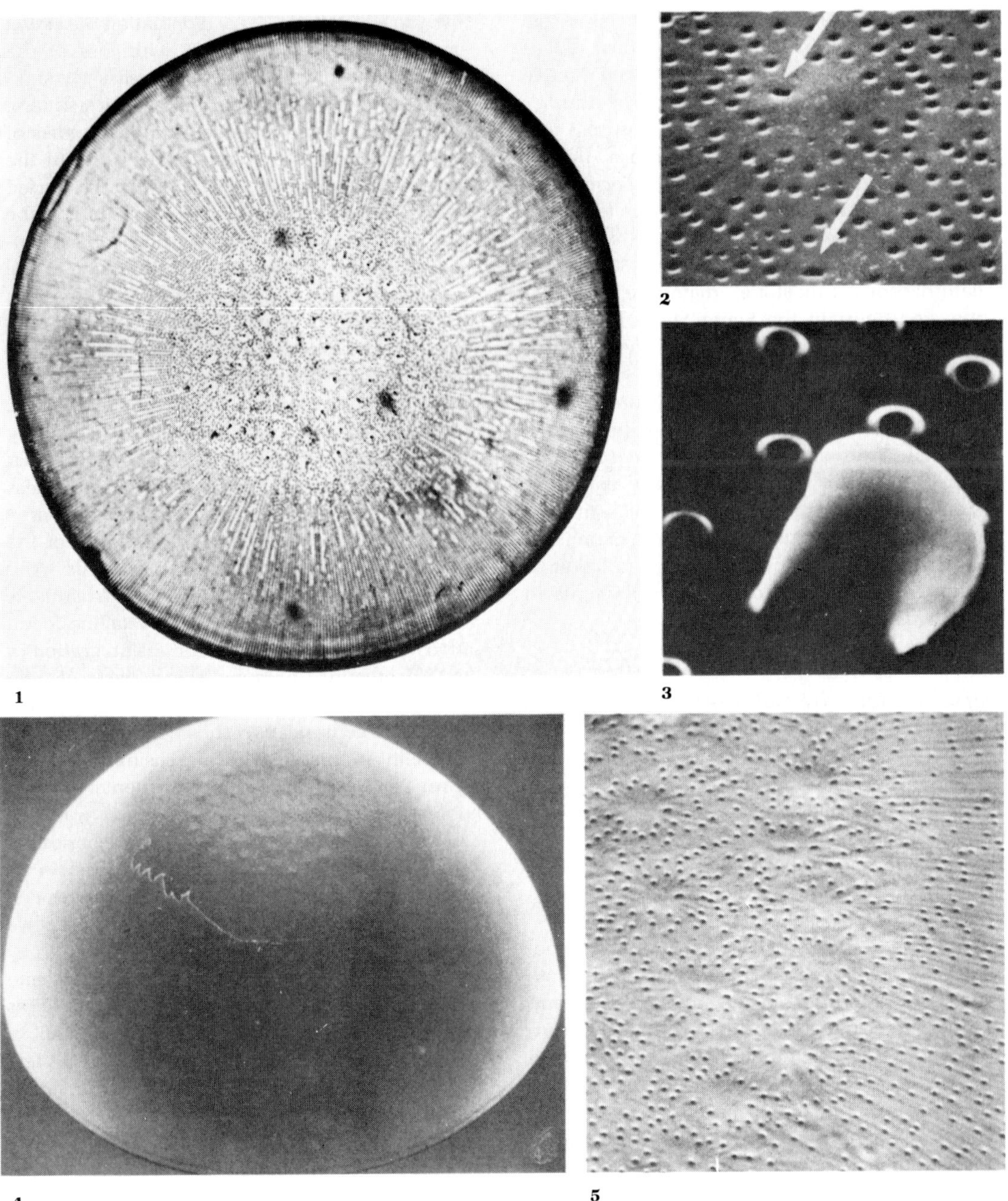

Figure 8.33
Ethmodiscus rex, the large diatom that forms the *Ethmodiscus* oozes on the sea floor. **1.** Valve view, from western Pacific, ×150, reproduced with permission from Wiseman and Hendey, 1953, *Deep Sea Research,* v. 1, p. 47−59, copyright 1953, Pergamon Press, Ltd. **2−5.** From Sargasso Sea, Atlantic Ocean, SEM, from Swift, 1973. **2.** Center of valve exterior, showing slits of labiate processes (arrows), ×2040. **3.** Interior of same, with labiate process, ×10,000. **4.** Entire valve, showing flattened central area, ×144. **5.** Part of central area of valve, showing at right the rows of punctae converging toward center with its irregularly arranged punctae and open areas.

occurs in a lake with silicon concentration less than 0.25 mg/liter.

Diagenesis and Preservation of Diatom Silica

Both the selective effect of the various environmental factors on the original microflora of an area and selective preservation on the sea floor (because of differences in silica wall thickness and morphology and hence susceptibility to dissolution in the water column) influence the original occurrence of the siliceous sediments. Diagenesis further affects both the assemblage that may occur and its preservation. Siliceous deposits generally are more susceptible to diagenesis than are calcareous ones, even over a few thousand years (King, 1977). The Monterey Formation of California contains mostly soft diatomaceous muds in the upper part, more indurated ones lower in the section, porcelanite, and finally chert in the lowest part of the sequence. Early studies of the rocks led Davis (1918) to regard submarine siliceous springs as a source of silica gels that later hardened into cherts, for some tuffaceous beds were present in the lower part of the Monterey. He did not regard the cherts as altered diatomites. Because both marine and fresh waters are undersaturated in silica, volcanic activity had long been related indirectly to diatomite accumulation. Whitney (1867) suggested that volcanic ash was a source of the silica utilized in the formation of freshwater diatomites. A similar source of the silica for the Monterey diatomites was suggested by Bramlette (1946), although he regarded the diatomites as being later altered secondarily. Dissolution of the finest siliceous particles and redeposition of the silica were related to the overlying load and metamorphism, which converted the diatomites to porcelanous shale and eventually to dense cherty rocks. Pre-Cretaceous siliceous fossils are unknown in an uncemented matrix, and are largely restricted to sediments of Tertiary or Quaternary age. Absence of diatoms in the younger rocks is due to solution preventing their accumulation, or to their later destruction by post-depositional leaching. Lohman (1960) suggested

that the Late Cretaceous Moreno diatomites of California were probably never exposed to a pH greater than 4.

Laboratory confirmation of the conversion of diatomite to porcelanite and chert was presented by Ernst and Calvert (1969). Experiments were conducted at approximately 2000 bars confining pressure, and at temperatures of 300°, 400°, and 500°C. The percentage of material recrystallized was found to be constant per unit time, but the rate was dependent upon the presence of an aqueous fluid. Extrapolation of the time required for complete conversion suggested that the Monterey porcelanite should recrystallize completely to chert in about 180 million years, at temperatures less than 20°C. The rate would be greater at higher temperatures, only 4 to 5 million years being required at 50°C. Similarly, circulating alkaline fluids would greatly increase the conversion rate. Much of the chertification of the Monterey has been along bedding planes and fractures where alkaline aqueous fluids were present. Deeper portions of the formation were subjected to slightly greater temperatures.

Because of the different effects on calcite and silica of the acidic or alkaline waters, diatomaceous oozes and diatomites are characteristic of colder water, low oxygen, high CO_2, and low pH. Such marine deposits are abundant along the east coast of North America, on both sides of the Pacific Ocean, and in the Mediterranean (North Africa, Italy), closely paralleling the mobile belts of volcanic activity and high coastal relief. Calcareous deposits, in contrast, characterize tropic and subtropic areas and gently sloping coastlines as in the Gulf Coast and Caribbean areas, or in France.

Geologic Occurrence

Contamination is a major problem in study of diatoms. Their small size and light weight allow air transport and may lead to contamination during sample preparation; specimens may occur in tap water, and vertical transport within the sediments by groundwater has even been suggested. Diatoms have been described from windblown dust, which may indicate an important means of transportation and colonization for freshwater and littoral marine species as well as a source of contamination of samples. Sampling of the air above the soil at a beach (Saar, 1967) indicates the extent of air transport. Samples were taken from resin-coated slides exposed to the air, perpendicular to air direction, for 30 to 45 minutes just before low tide. The wet beach was 40 m wide, and the dry beach extended 70 m to the base of a dune. At a location 10 m from the wet beach, one benthic species and 222 planktonic diatoms were caught, of which 217 were living. At the upper end of the dry beach 291 specimens of marine plankton diatoms were caught, with 261 living, 12 specimens of marine benthos, and 2 valves of freshwater species. Although some of the spray flora represented delicate species that would not be preserved, the coarser valves might be picked up from the sand and transported widely.

Marine Record

Diatoms have a good geologic record from the middle Cretaceous through the Cenozoic. Possible Jurassic diatoms from the German Lias were described by Rothpletz (1896) as *Pyxidicula bollensis* and *P. liasica*. The small caplike valves, between 6 and 14 μm in diameter, somewhat resemble resting cysts of centric diatoms, but probably are schizosphaerellids, referrable to the calcareous nannoplankton (Deflandre, 1959). All older records (for example, those cited as Paleozoic) have been shown to be due to contamination or erroneous age assignment (Pia, 1927, 1931; Frenguelli, 1932).

Aulacodiscus, Coscinodiscus, Ditylum, Endictya, and *Melosira* have been reported from Lower Cretaceous (Barremian-Aptian) black shales in the Polish Carpathians (Geroch, 1978). Although the specimens are pyritized, some frustules occur in short chains, and their surface perforations, girdle bands, and even processes are preserved.

Eleven middle Cretaceous (Albian) species were described from the Gault phosphorites of Hannover, Germany (Forti and Schulz, 1932).

Genera recorded were *Craspedodiscus, Dactyliosolen, Ditylum, Gladius, Isthmia, Ktenodiscus, Pyrgodiscus, Radiodiscus,* and *Stephanopyxis*. Müller (1912) described a supposed centric diatom, *Actinoclava,* from the Turonian of Westfalia, and Strel'nikova (1968) recorded *Coscinodiscus* and *Pyxilla* from the Santonian, (1965) 11 species from the Campanian of the eastern Urals, and (1974) 132 species from western Siberia. Schulz (1935) reported 55 species from the Campanian of Danzig, representing 17 genera. The Upper Cretaceous Marca Shale Member of the Moreno Formation of California (Late Maastrichtian) has yielded an excellent flora of some 130 species, belonging to 45 genera (Hanna, 1927a, 1934; Long et al., 1946; Barker and Meakin, 1947, 1949), as have the Upper Cretaceous of the USSR (Zhuze, 1949, 1951a; Strel'nikova, 1965, 1968, 1974) and deep sea cores in the South Pacific (Hajós, 1975). Dominant genera in diversity of species in the Cretaceous are *Triceratium, Stephanopyxis,* and *Hemiaulus* (see Figure 8.35); all Cretaceous diatoms are marine centric species, except for the rare araphid forms such as *Sceptroneis* (Hajós, 1975). Some 70 genera and nearly 300 Cretaceous species are known, but few zonations have been proposed. Nine biostratigraphic zones have been based on diatoms in the Late Campanian and Maastrichtian of the South Pacific (see Table 8.3).

Tertiary diatomites occur widely, particularly in California, New Jersey, Maryland and Virginia, Mexico, Peru and Chile, New Zealand, Japan, the USSR, Italy, Greece, Hungary, Poland, Rumania, Austria, and Yugoslavia. Particularly characteristic Paleogene genera include *Aulacodiscus, Biddulphia, Brightwellia, Hemiaulus, Kittonia, Melosira, Riedelia, Stephanopyxis, Syringidium,* and *Triceratium*.

Paleocene diatoms are rather poorly known, but some 120 species were reported from the USSR (Zhuze, 1951b; Glezer, 1974). Eocene diatomaceous deposits are much more widespread, in California (Hanna, 1927b, 1931; Kanaya, 1957); Sweden (Cleve-Euler, 1941); the USSR (Glezer, 1967); Oamaru, New Zealand (originally thought to be younger in age but now regarded as late Eocene; Grove and Sturt, 1886, 1887; Schrader, 1969a,b); Barbados (Hanna and

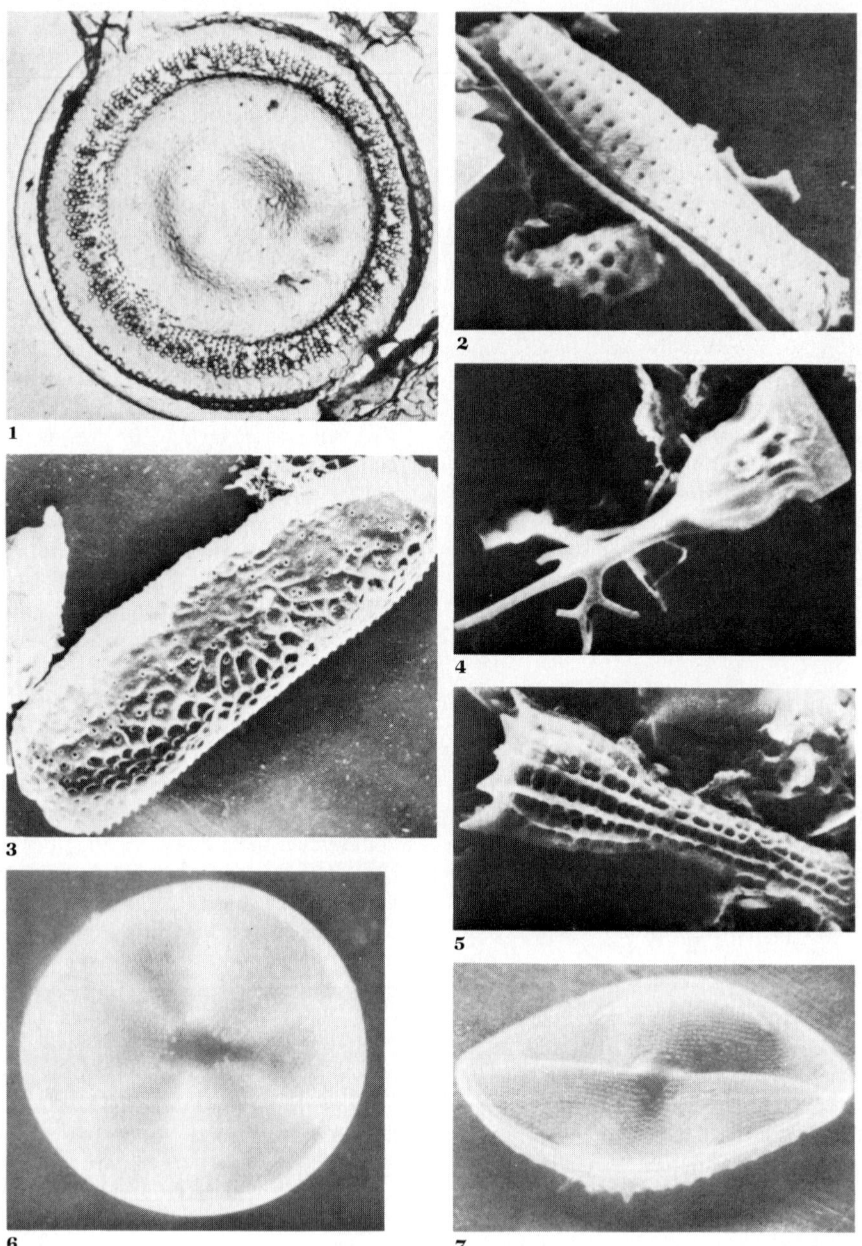

Figure 8.35
Cretaceous diatoms from the South Pacific. **1.** *Pseudopodosira westii* (Smith) Sheshukova & Glezer, TEM, ×3200. **2.** *Sceptroneis grunowii* Anissimova, SEM, ×2990. **3.** *Xanthiopyxis grantii* Hanna, TEM, ×3300. **4.** *Pterotheca crucifera* Hanna, SEM, ×2080. **5.** *P. cretacea* Hajós & Stradner, SEM, ×2080. **6,7.** *Actinoptychus packi* Hanna, valve view and oblique view, SEM, ×1200. All from Hajós, 1975.

Table 8.3
Upper Cretaceous (Late Campanian and Maastrichtian) diatom zones of Hajós (1975).

Pseudopyxilla jouseae Hajós Zone
Acanthodiscus antarcticus Hajós Zone
Cerataulus-Odontotropis Zone
Kentrodiscus armatus Hajós Zone
Biddulphia sparsepunctata Hajós Zone
Anaulus subantarcticus Hajós Zone
Epithelion russicum Pantocsek Zone
Chasea ornata Hajós & Stradner Zone
Horodiscus rugosus Hajós Zone

Brigger, 1964); and deep sea cores from the South Atlantic (Ross, 1976; Hanna et al., 1976), southwest Pacific (Hajós, 1976), and the Indian Ocean (Holmes and Brigger, 1977). Far fewer Oligocene floras have been described, although some are known from the USSR and from southwestern Pacific cores (Hajós, 1976). Other Paleogene diatoms have been described from the Norwegian Sea (Schrader and Fenner, 1976), Falkland area (Gombos, 1977), and USSR (Glezer and Sheshukova-Poretskaya, 1967; Zhuze and Sheshukova-Poretskaya, 1971). The Miocene was the period of greatest expansion of the diatoms, during which they reached their maximum development. About 400 species have been described from the California Monterey alone (see Figures 8.36 and 8.37). Typical Neogene genera are *Actinocyclus, Asterolampra, Asteromphalus, Chaetoceros, Coscinodiscus, Cosmiodiscus, Denticula, Hemidiscus, Rhizosolenia, Thalassionema, Thalassiosira,* and *Thalassiothrix.* Many Miocene centric genera disappeared by the end of the Miocene, but new ones appeared, although some areas of diatomite deposition in the Miocene, particularly in the Mediterranean region and the margins of the Atlantic, no longer have active diatomite accumulation.

Many articles have described Neogene diatoms, from California (Hanna, 1932a; Lohman, 1938; Wornardt, 1967; Barron, 1975a,b, 1976a,b), Oregon (Orr et al., 1971), Maryland (see Figure 8.38; Boyer, 1904; Lohman, 1948; Andrews, 1975, 1976, 1977), Mexico (Hanna and Grant, 1926), Trinidad (Lohman, 1974), Argentina (Frenguelli, 1933), Japan (Kanaya, 1959; Ichikawa, 1964; Koizumi, 1968; Komura, 1975, 1976), Java (Reinhold, 1937), Hungary (Hajós, 1968), and the USSR (Sheshukova-Poretskaya and Glezer, 1962; Makarova and Kozyrenko, 1966; Sheshukova-Poretzkaya, 1967; Cheremisinova, 1968, 1971; Moiseeva, 1971), and from the Pleistocene of Italy (Bonadonna, 1963, 1964) and Sweden (Miller, 1964). Diatoms described from the Cenozoic of the deep sea cores have provided the basis for detailed biostratigraphical zonations (see Table 8.4); such assemblages have been described from the North Pacific (Donahue, 1970; Koizumi, 1973), northwest Pacific (Koizumi, 1975a,b), northeast Pacific (Schrader, 1973a,c), equatorial Pacific (Burckle, 1972), South Pacific (Schrader, 1976), Antarctic

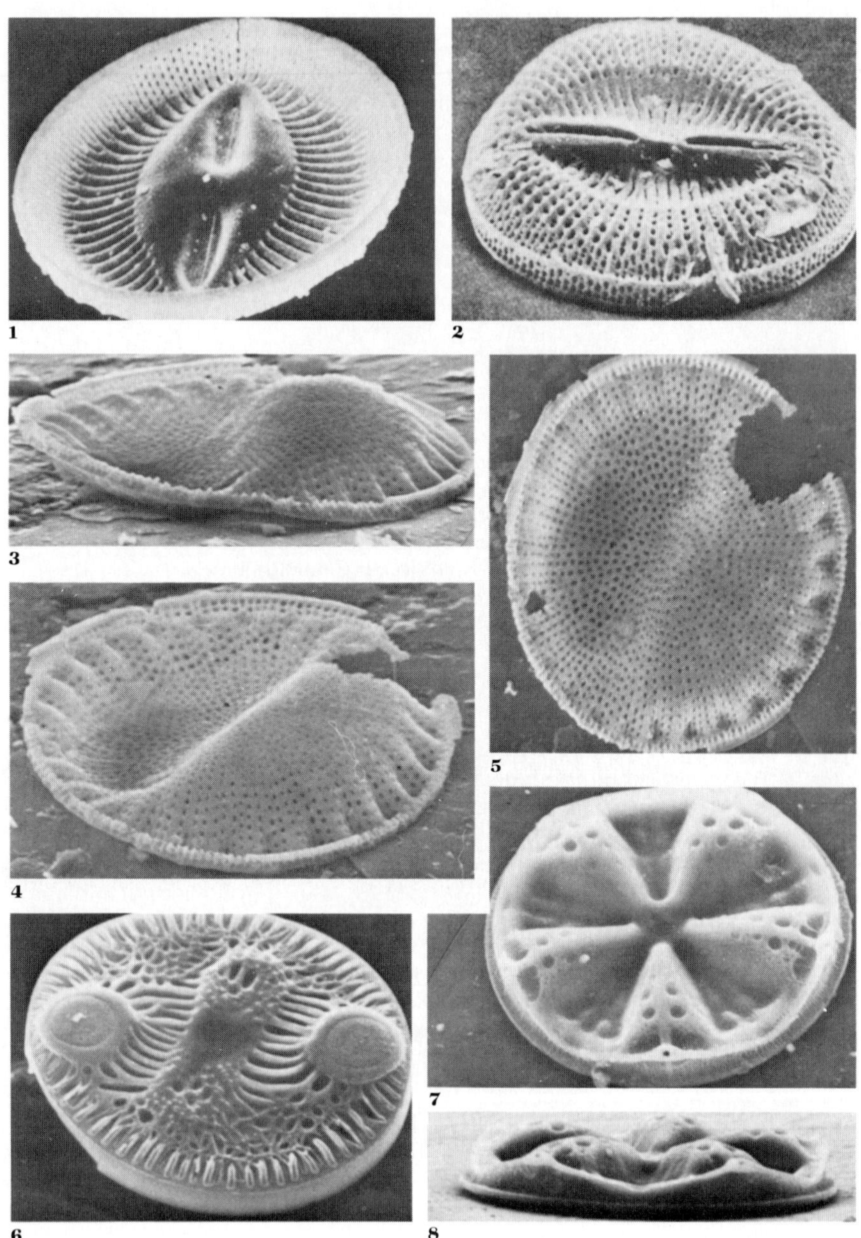

(McCollum, 1975; Gombos, 1977), Indian Ocean (Schrader, 1974), and Norwegian Sea (Schrader and Fenner, 1976).

Some general works on Cenozoic diatoms include those by Ehrenberg (1854), Schmidt (1874–1959), Proshkina-Lavrenko (1949a,b, 1950, 1974), and Lohman (1961).

The living marine plankton largely consists of centric diatoms, but the benthos is dominated by the pennate taxa. Living marine assemblages have been described (among many others) from near Philadelphia (Boyer, 1916), the west coast of North America (Cupp and Allen, 1938; Cupp, 1943; Gran and Angst, 1931), Great Britain (Heurck, 1896); Belgium (Meunier, 1913, 1915), the northern seas (Lebour, 1930), the southern seas (Hendey, 1937; Frenguelli, 1928; Frenguelli and Orlando, 1958), the Pacific (Jousé et al., 1971), the Philippines (Mann, 1925), the Java Sea (Allen and Cupp, 1935), and the Indian Ocean (Kozlova, 1971), as well as the Black Sea (Proshkina-Lavrenko, 1963b), Caspian Sea (Proshkina-Lavrenko and Makarova, 1968), and Sea of Azov (Proshkina-Lavrenko, 1963a).

Freshwater Record

In general, most freshwater diatoms are pennates. This includes all benthic species, and all but four genera that occur in the freshwater plankton.

Paleocene pennates have been reported to occur in freshwater deposits (Proshkina-Lavrenko, 1960), but none are known prior to the late Eocene in North America (Lohman and Andrews, 1968). Freshwater limestone blocks in the Wagon Bed Formation of Wyoming (upper Eocene) contain a diatom flora of nine genera and 34 species. Both centric and pennate taxa occur, including species of the centric *Anaulus* and *Melosira,* the araphid pennate *Ambistria* and *Fragilaria,* and the typically biraphid *Anomoeoneis, Navicula, Pinnularia, Stauroneis* (all Naviculaceae), and *Nitzschia* (Nitzschiaceae), although previously biraphid diatoms were thought to appear first in the lower Miocene (Schulz, 1935).

Oligocene freshwater diatoms have been described from France (Ehrlich, 1969), Oligocene

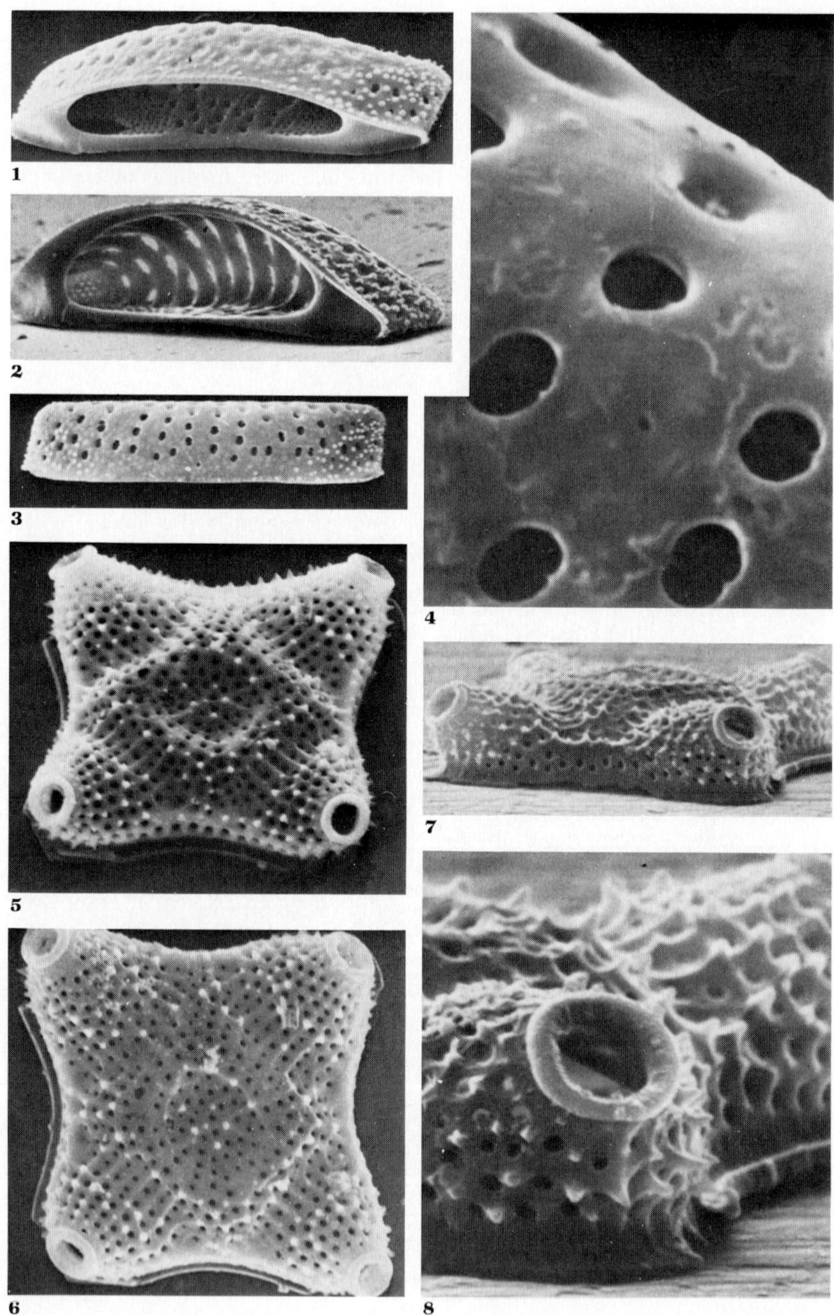

Figure 8.37

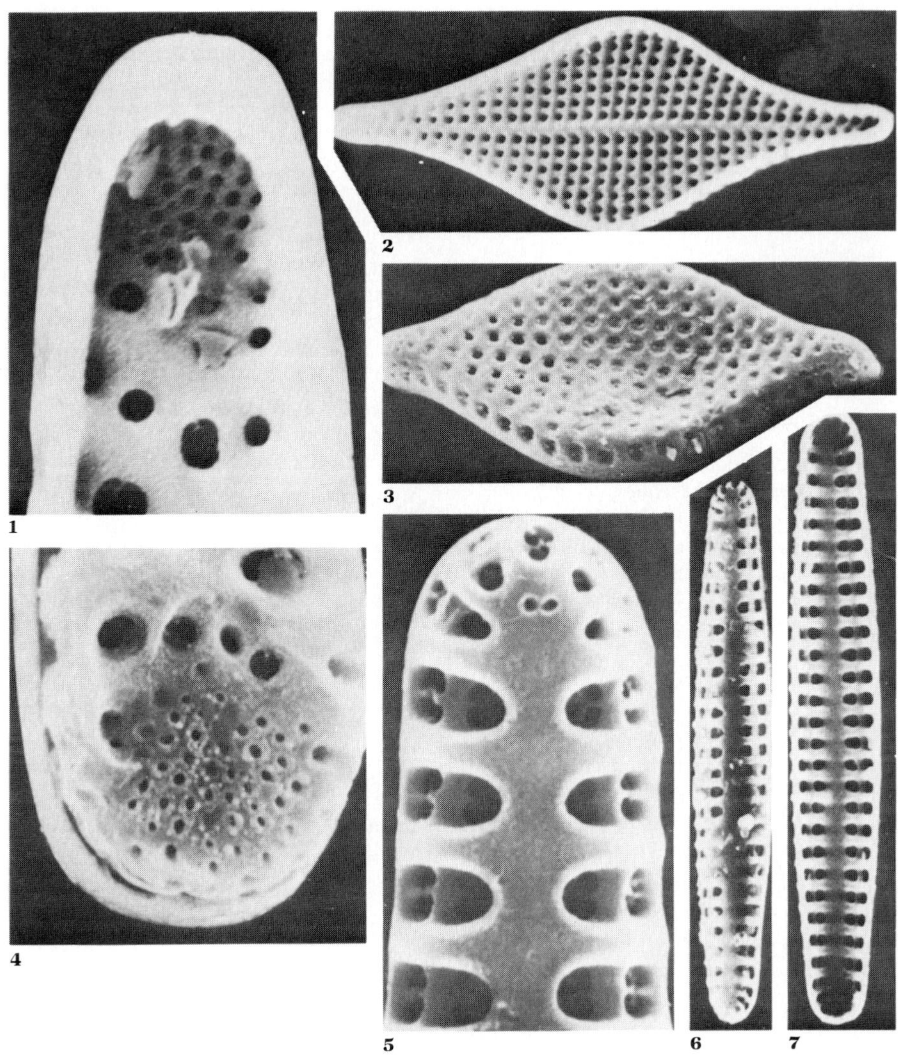

Figure 8.38

Miocene diatoms, Maryland, SEM. **1 — 4.** *Rhaphoneis margaritata* Andrews, from Andrews, 1975. **1,2.** Interior and exterior of tip of valve, ×5905. **3.** Inside of entire valve, ×950. **4.** Outside of valve, ×1330. **5 — 7.** *Delphineis angustata* (Pantocsek) Andrews, from Andrews, 1977. **5.** Exterior of tip, ×6000. **6.** Valve exterior, ×1320. **7.** Interior, ×1440.

Table 8.4

Cenozoic diatom biostratigraphy, largely from data of the Deep Sea Drilling Program.
RZ, Range Zone; PRZ, Partial Range Zone; CRZ, Concurrent Range Zone; Z, Zone

Region	Tropical Indian Ocean	North Pacific		Eastern Equatorial Pacific	Tropical Pacific	Antarctic
Author / **Age**	Schrader, 1974	Schrader, 1973a	Koizumi, 1973	Burckle, 1972	Zhuze, 1974	McCollum, 1975
Pleistocene	1	I	*Denticula seminae* PRZ			*Coscinodiscus lentiginosus* PRZ
	2	II	*Rhizosolenia curvirostris* Jousé PRZ	*Pseudoeunotia doliolus* RZ	*Pseudoeunotia doliolus* Z	*Coscinodiscus elliptiporus* Donahue – *Actinocyclus ingens* Rattray CRZ
	3, 4	III	*Actinocyclus oculatus* Jousé PRZ			
	5	IV	*Thalassiosira zabelinae* Jousé PRZ			*Rhizosolenia barboi* Brun – *Nitzschia kerguelensis* (O'Meara) Hasle PRZ
Pliocene	6	V		*Rhizosolenia praebergonii* Mukhina PRZ	*R. praebergonii* Z	
		VI				*Coscinodiscus kolbei* Jousé – *Rhizosolenia barboi* RZ
	7	VII	*Denticula seminae* – *D. kamtschatica* Zabelina CRZ		*Thalassiosira convexa* Mukhina Z	*Coscinodiscus insignis* Jousé PRZ
		VIII				
	8	IX				
	9–11	X		*Nitzschia jouseae* Burckle PRZ		*Nitzschia interfrigidaria* McCollum PRZ
	12				*Nitzschia jouseae* Z	*Nitzschia praeinterfrigidaria* McCollum PRZ
	13	XI		*Thalassiosira convexa* PRZ		
	14					
Upper Miocene	15	XII	*Denticula hustedtii* Simonsen Kanaya PRZ	*Nitzschia miocenica* Burckle PRZ	*N. miocenica* Z	*D. hustedtii* PRZ
	16	XIII				
	17				*Thalassiosira praeconvexa* Burckle Z	
	18	XIV		*Nitzschia porteri* Ehrenberg PRZ		
	19	XV	*Denticula lauta* PRZ		*N. porteri* Z	*D. hustedtii* – *D. lauta* PRZ
	20	to		*Coscinodiscus yabei* Kanaya PRZ		
	21	XVIII			*C. yabei* Z	
Middle Miocene		XIX		?		*D. lauta* – *D. antarctica* McCollum PRZ
					Coscinodiscus marginatus var. Z	
		XX				*D. antarctica* – *Coscinodiscus lewisianus* Greville Z
		XXI				
		XXII				
		to				
		XXIV				

California	Western Atlantic	Norwegian Sea
Schrader, 1976	Barron, 1975a (upper Miocene to Pliocene); Abbott, 1978 (middle Miocene)	Schrader and Fenner, 1976
		Thalassiosira oestrupii PRZ
		Rhizosolenia barboi PRZ
		Thalassiosira kryophila (Grunow) Jørgensen PRZ
	Thalassiosira hyalinopsis Barron AZ	
	Nitzschia reinholdii Kanaya & Koizumi CRZ	
		Coscinodiscus marginatus PRZ
	Rhaphoneis amphiceros v. *elongata* Peragallo PRZ	
	Thalassiosira antiqua (Grunow) Cleve-Euler CRZ	
Hemidiscus karstenii Jousé PRZ	*Nitzschia fossilis* (Frenguelli) Kanaya PRZ	*Denticula hustedtii* PRZ
C. yabei RZ	*Coscinodiscus kurzii* Grunow CRZ	*Cymatosira biharensis* Pantocsek PRZ
Denticula dimorpha Schrader PRZ	*Actinoptychus gruendleri* Schmidt AZ	*Goniothecium tenue* Brun PRZ
	?	
D. antarctica PRZ		*Rhizosolenia miocenica* Schrader PRZ
D. nicobarica Grunow PRZ		*Thalassiosira gravida* v. *fossilis* Jousé PRZ
C. lewisianus PRZ		*Actinocyclus ingens* Rattray PRZ
D. lauta PRZ		*Nitzschia* sp. RZ

(continued)

Region		Tropical Pacific	Antarctic
Author / Age		Zhuze, 1974	McCollum, 1975
Middle Miocene (*continued*)		XXV ?	*D. antarctica-* *D. nicobarica* Grunow PRZ
Lower Miocene		*Bogorovia veniamini* Jousé Z	*Coscinodiscus* sp. Z
Upper Oligocene		*Coscinodiscus vigilans* A. Schmidt Z *Craspedodiscus coscinodiscus* Ehrenberg Z	*Pyrgupyxis prolongata* (Brun) Hendey PRZ
Middle Oligocene		*Cestodiscus muhinae* Jousé Z	
Lower Oligocene		*Coscinodiscus excavatus* Castracane v. *quadriocellatus* Grunow Z	
Eocene		*Hemiaulus polycystinorum* Ehrenberg Z *Triceratium barbadense* Greville Z	
Paleocene			

California	Western Atlantic	Norwegian Sea
Gombos, 1977	Abbott, 1978	Schrader and Fenner, 1976
Nitzschia pusilla Schrader PRZ	*Coscinodiscus plicatus* PRZ	*Sceptroneis caducea* Ehrenberg PRZ
Thalassiosira spumellaroides Schrader PRZ	*Delphineis penelliptica* Andrews — *C. plicatus* CRZ	*Coscinodiscus plicatus* PRZ
T. spinosa Schrader PRZ	*D. penelliptica* PRZ	
Raphidodiscus marylandicus PRZ	*D. ovata* Andrews — *D. penelliptica* CRZ	*Denticula hyalina* Schrader PRZ
Nitzschia maleinterpretaria Schrader PRZ		
	D. ovata PRZ	*Rhizosolenia bulbosa* Schrader PRZ
		Thalassiosira fraga Schrader PRZ
B. veniamini PRZ		*Nitzschia maleinterpretaria* PRZ
	Actinoptychus heliopelta Grunow CRZ	*Coscinodiscus vigilans* PRZ
		Rhizosolenia norwegica Schrader PRZ
		Synedra jouseana Sheshukova-Poretskaya PRZ
		Pseudodimerogramma elegans Schrader PRZ
Pyrgupyxis prolongata PRZ		*Coscinodiscus praenitidus* Fenner PRZ
		Thalassiosira irregulata Schrader PRZ
Melosira architecturalis Brun PRZ		*Pseudodimerogramma filiformis* Schrader & Fenner PRZ
Hemiaulus incisus Hajós PRZ		*Sceptroneis pupa* Schrader & Fenner PRZ
Pyrgupyxis eocenica Hendey —		*Coscinodiscus oblongus* Greville RZ
Pterotheca aculeifera Grunow CRZ		*Triceratium barbadense* Greville PRZ
Hemiaulus inaequilaterus Gombos PRZ		
Sceptroneis sp. A PRZ		
Odontotropis klavensii Debes ex Hustedt RZ		

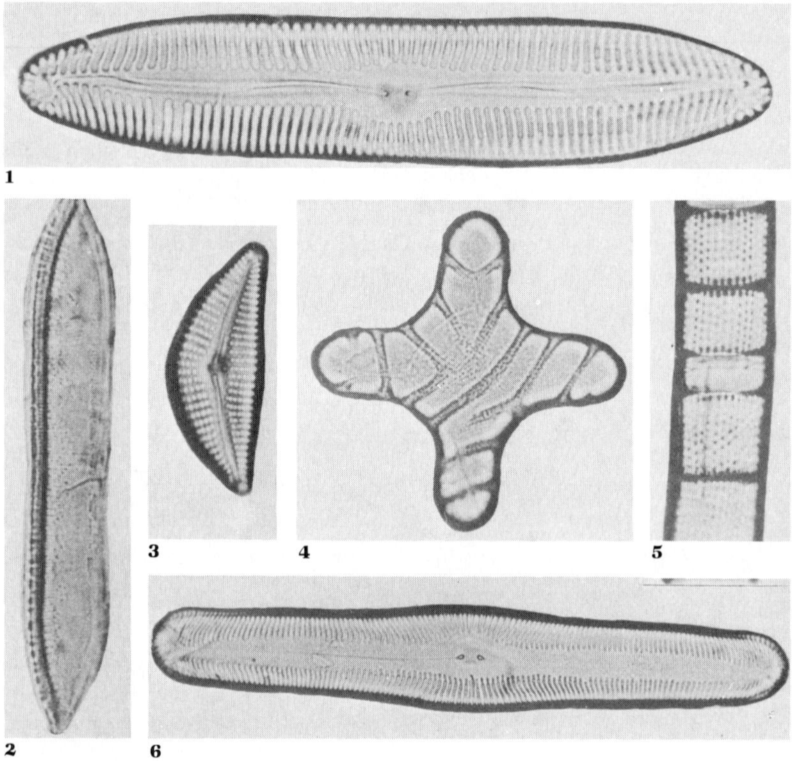

Figure 8.39

Late Miocene freshwater diatoms, Nebraska. **1.** *Pinnularia viridis* (Nitzsch)
Ehrenberg. **2.** *Nitzschia calva* Andrews. **3.** *Cymbella tumida* (Brébisson) van
Heurck. **4.** *Tetracyclus cruciformis* Andrews. **5.** *Melosira granulata*. **6.** *Pin-
nularia torta* (Mann) Patrick. 1–5, ×960; 6, ×475. All from Andrews, 1970.

and Miocene ones from the USSR (Rubina,
1976), and other Neogene freshwater diatoms
from the USSR (Golovenkina, 1967).

The middle Oligocene Florissant Lake Beds
contain 30 diatom species (Lohman, 1960), and
the Miocene is represented by many freshwater
diatomites in Europe and North America, in
Nebraska (see Figure 8.39), California, Nevada,
Oregon, and Washington (Andrews, 1970;
Okuno, 1958, 1959; VanLandingham, 1964,
1967a). The Pit River Basin of northern California contains about 350 m of relatively pure
freshwater diatomite, which probably accumulated at a rate of about one centimeter annually.
In places it consists almost exclusively of the
single species *Melosira granulata* (Ehrenberg)
Ralfs (Hanna and Grant, 1931). Miocene freshwater diatomites also occur in Patagonia (Frenguelli, 1933). The Great Basin region of the
United States contains freshwater diatomites of
Miocene, Pliocene, and Pleistocene age
(Lohman, 1961), with a total of more than 350
species and varieties. Nonmarine diatomites of
Pliocene and Pleistocene age are present in Cali-

fornia (Hanna and Grant, 1929; Lohman, 1938, 1964; Wornardt, 1964), others have been described from the Pliocene of Kansas (Hanna, 1932b), the Pleistocene of Oklahoma (Bond, 1968) and Wisconsin (Andrews, 1966), the Pleistocene of Lake Bonneville, Utah (Patrick, 1936), Lake Lahontan, Nevada (Hanna and Grant, 1931), lake beds in New Mexico (Lohman, 1935), Chile (Frenguelli, 1929; Dingman and Lohman, 1963), in the vicinity of Mexico City, Mexico (Bradbury, 1970, 1971), and lake beds of undetermined age in New Zealand (Barber, 1961). Living freshwater diatoms of Great Britain were monographed by West and Fritsch (1927), those of central Europe by Hustedt (1927–1966, 1930), those of the United States by Patrick and Reimer (1966), and all living freshwater diatoms were covered in a monograph by Bourrelly (1968).

Freshwater diatoms are useful for paleolimnological studies. Diatom succession in a South Dakota lake deposit (Haworth, 1972) could be divided into five stages or zones, correlated with characteristics of the water and of the surrounding vegetation. The oldest part of the core (820 to 680 cm depth) contained mostly benthic or littoral species that prefer alkaline conditions, and the assemblage was dominated by *Fragilaria brevistriata* Grunow, with other *Fragilaria* sp. and *Mastogloia* spp. The lowest part of the core had some acid-loving species. This Zone I is generally associated with a spruce forest, grading upward to a deciduous woodland habitat. Zone II (680 to 580 cm) contained fewer diatoms, and a low number of planktonic species. *Fragilaria* and *Mastogloia* remained dominant, and acid-loving species had disappeared. This zone is transitional to Zone III (580 to 420 cm), which represents an eutrophic or nutrient-rich environment. *Melosira granulata* and *Fragilaria crotonensis* Kitton were common; total diatom numbers were greater, as was the number of planktonic species; more brackish water taxa indicated a higher mineral content of the water related to increased evaporation, and the surrounding area was one of prairie conditions and increased soil erosion. Zone IV (420 to 330 cm) again is transitional, with woodlands developing around the lake margins. The diatom flora was abundant, but had fewer planktonic species. *Gomphonema*, *Cocconeis*, and *Melosira* were dominant. The final Zone V (330 to 160 cm) had an increased number of species that inhabit the surf zone of lakes. The water was probably shallower because of sedimentation. The dominant species was *Cocconeis placentula* Ehrenberg var. *lineata* (Ehrenberg) van Heurck.

A paleolimnological study of lakes of Minnesota and South Dakota (Bradbury, 1975) related changes in the diatom assemblage to human settlement in the area. Here also the fossil assemblages indicated an increase in productivity due to nutrient enrichment following soil erosion caused by the land clearance, later accentuated by waste disposal. The human disturbance of the area resulted in lower diatom diversity, dominance of a few species, a shift to spring-blooming plankton diatoms, with bluegreen algae dominating in the summer, and a decrease in benthic and epiphytic taxa. The original pre-human lakes had highly distinctive diatom floras, according to their local conditions, but following the human disturbance all the nutrient enriched lakes had a similar floral content, most such eutrophic lakes being dominated by the diatom *Stephanodiscus hantzschii*.

Surprisingly, freshwater diatoms also have been found in abundance in some deep sea cores (Kolbe, 1957, Rigby and Burckle, 1958). The occurrence was variously attributed to aeolian transport, ocean currents, subsiding land areas, and turbidity currents. The latter would require transport of 900 km from the African coast, and uphill 1000 m to the top of submarine hills.

Marine and halophilic species have been found in mountain streams and in the Munster Valley, perhaps transported there by aquatic migratory birds (Werner, 1960). Empty frustules of marine and brackish diatoms occur in the water and sediments of the southern Amazon delta, as far as 350 km from the mouth, but the water is fresh. Living phytoplankton does not occur in the river because of the high turbidity, but freshwater desmids brought in by tributary rivers are mixed in the Amazon with the marine diatom frustules that were transported upstream during the tidal cycle (Gessner and Simonsen, 1967).

Paleotemperatures

Within the Neogene, paleotemperature analyses have been based on the diatom assemblages, by use of the formula

$$Td = \frac{Xw(100)}{Xw + Xc}$$

where Xw is the total number of specimens of still extant warm water taxa, Xc is the total number of specimens of still extant cold water species, and Td is the diatom temperature value (Donahue, 1970). In the North Pacific Pleistocene, cold water species determined by Kanaya and Koizumi (1966) were *Denticula seminae*, *Rhizosolenia hebetata* forma *hiemalis* Gran,

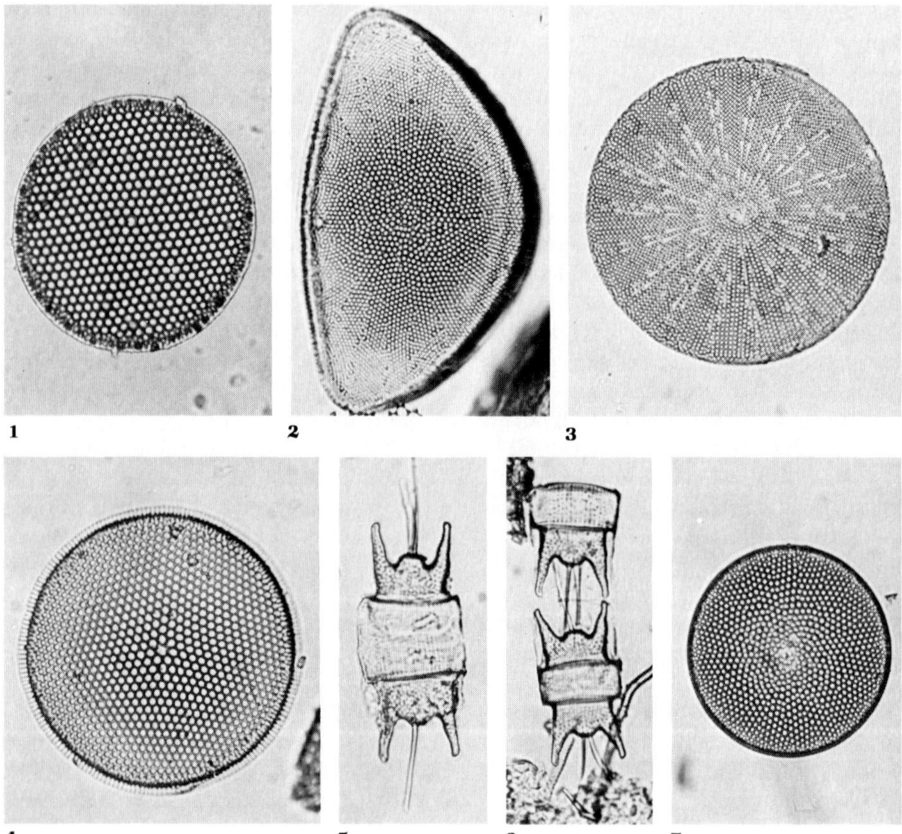

Figure 8.40
Diatoms characterizing temperature-controlled assemblages in the late Miocene and early Pliocene of California as interpreted from their present distribution in the Pacific Ocean. 1,2, tropical species; 3, subtropical species; 4–7, north boreal species. **1.** *Coscinodiscus lineatus* var. *leptopus* Grunow in van Heurck, Monterey Formation (Mohnian). **2.** *Hemidiscus cuneiformis*, Sisquoc Formation. **3.** *Actinocyclus ehrenbergii* Ralfs in Pritchard, Sisquoc Formation. **4.** *Coscinodiscus excentricus* Ehrenberg, Sisquoc Formation. **5,6.** *Biddulphia aurita*, Monterey Formation. **6,** Two cells attached in chain. **7.** *Coscinodiscus curvatulus*, Sisquoc Formation. All ×512.

Coscinodiscus marginatus, Actinocyclus ochotensis Jousé, *A. curvatulus* Janisch, *A. divisus, Thalassiosira gravida,* and *T.* spp. Warm water species were *Thalassiosira oestrupii, Nitzschia marina, Coscinodiscus nodulifer, C. radiatus, C. lineatus, Hemidiscus cuneiformis,* and *Pseudoeunotia doliolus.* The diatom numbers based on these taxa in cores in the North Pacific showed marked fluctuations, reflecting alternate cooling and warming of the surface water and the resultant modification of the diatom assemblages (Donahue, 1970).

Paleotemperature curves were obtained for the late Miocene and early Pliocene of California at Lompoc (Barron, 1973) by determining the number of specimens per 200 count that were referable to the tropical, subtropical, and north boreal assemblages of the present Pacific as described by Jousé et al. (1971). North boreal and subtropical species dominate the assemblages (see Figure 8.40) and fluctuate in relation to each other in abundance (see Figure 8.41). The fluctuations were then calibrated by comparing the present diatom distribution with winter isotherms in the Pacific (see Figure 8.42). These data suggested a colder temperature in the Miocene than at Lompoc today, but warmer than the present during the latest Miocene and early Pliocene. The warming trend may have been general in the latest Miocene, may merely represent fluctuations in the cold water California Current, or may reflect local conditions related to coastal configuration or local currents (Barron, 1973). Labeyrie (1974) suggested using the $^{18}O - ^{16}O$ ratios in diatom silica to determine surface water temperatures for the Pleistocene. Oxygen isotope ratios in the $CaCO_3$ of planktonic foraminifers had been used for this purpose, but as planktonic foraminifers live at various depths in the water column and would thus form their calcium carbonate at varied temperatures, diatoms might prove more reliable, since they are restricted to life in the photic zone.

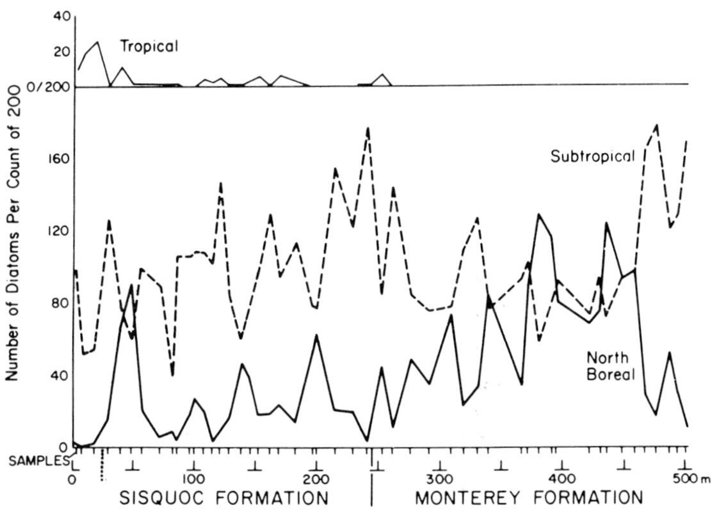

Figure 8.41
Relative numbers of diatoms characteristic of different temperature-controlled assemblages in the late Miocene and early Pliocene of Lompoc, California. From Barron, 1973.

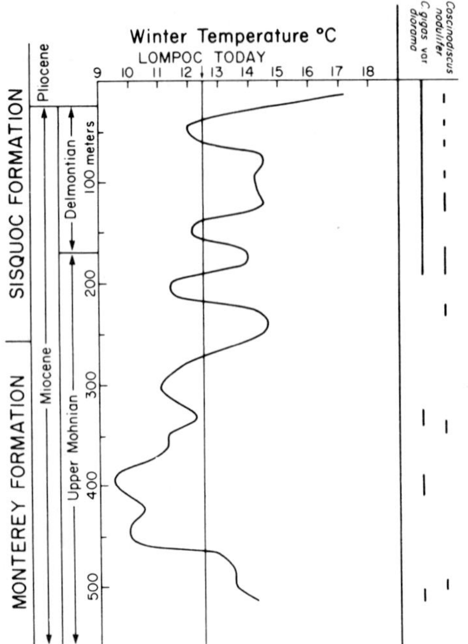

Figure 8.42
Paleotemperature curve for the late Miocene to early Pliocene in the area of Lompoc, California, based on diatom assemblages. Winter temperatures are seasonal minima. Vertical line indicates present temperature in this region. At right are indicated the local occurences of two rare warm water species, *Coscinodiscus nodulifer* and *C. gigas* Ehrenberg var. *diorama* (Schmidt) Grunow. Slightly modified from Barron, 1973.

EVOLUTION OF THE DIATOMS

The abundance of diatoms in the present seas, their importance in the deposition of siliceous sediments, and evidence that diatomaceous sediments are gradually converted to chert with time suggest that the known history of the diatoms is far from complete. Similarly, the disparity between the living assemblage and that present in the sediments has made suspect even the geologic distribution of the fossils of the Cretaceous and Cenozoic.

That the fossil record of any group of organisms is incomplete is obvious from the limited part that is fossilized of any organism, the effects of diagenesis, metamorphism, weathering, and removal of material, the limited environments recorded, the small percentage of the sediments available for study from outcrop, drilling, or ocean coring, and finally the minute chance of their being collected or studied. Nevertheless, the ornate beauty of the diatom frustules has led to an unsurpassed concentration of interest in them. In spite of the lack of geologic detail presented by some of the studies, much information has been accumulated, in particular by the DSDP. The picture that emerges is not only consistent within itself, but closely agrees with evidence of relationships obtained by comparative studies of living species.

The gradual increase in the diatoms from the middle Cretaceous suggests that the relative importance of siliceous organisms has varied through time. Paleozoic cherts undoubtedly are the remains of radiolarians, or locally of sponges, rather than having an origin as diatomaceous silts. Unindurated sediments might not allow preservation of the delicate siliceous microfossils, but both radiolarians and sponge spicules are well preserved throughout the Paleozoic sequence in limestones and in concretions; similar preservation of diatoms would be expected if they also had been present.

Diatoms thus probably first appeared in the late Mesozoic oceans. As yet, there are no undoubted pre-Mesozoic records of Chrysophyta (the closest relatives of the diatoms). If any existed, they may not have had mineralized coverings. Both diatoms and chrysomonads appeared in the Cretaceous and showed a rapid evolutionary divergence. Both have modern freshwater representatives, although their Mesozoic ancestors were entirely marine. Among the diatoms, only the centrics were well represented, but every suborder of centric diatoms was present in the Cretaceous, and araphid pennates appeared late in the period.

The earliest raphe-bearing pennates appeared in the Paleocene (*Mastogloia, Navicula*). Pennate diatoms also invaded the freshwater habitat early in the Cenozoic.

The relationship of pennate movement to mucilage secretion by pores in the raphe system suggests that the labiate processes or

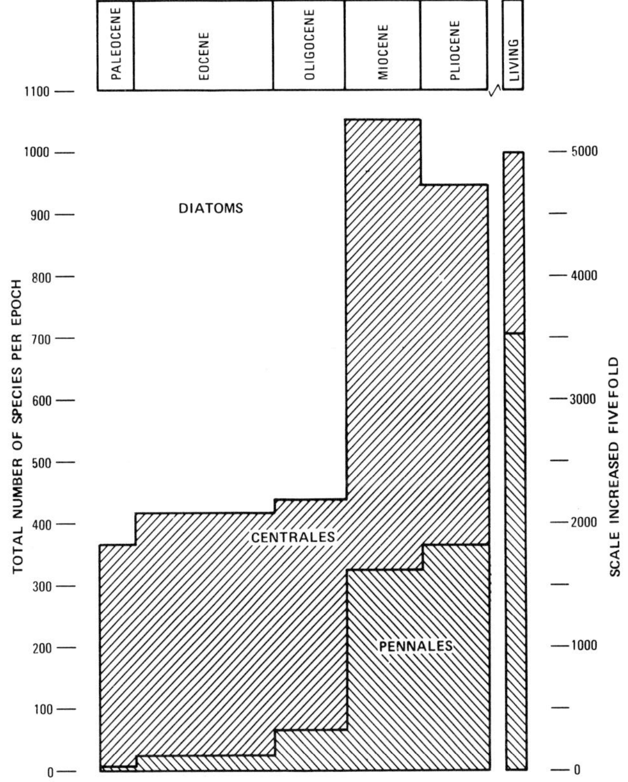

Figure 8.43
Species diversity of diatoms in the Tertiary, showing dominance of centric species in the Paleogene, with importance of the pennate species increasing rapidly through the Neogene. Scale at left side for Tertiary species, that at right (increased fivefold) for living species. Centric species are more diverse today than previously, but are much surpassed by the pennate taxa. From Tappan and Loeblich, 1973.

mucilage pores of the centric diatoms may have been modified to become the pennate apical mucus pores, terminal raphe pores, and even the entire raphe system (Hasle, 1973d). Neither monoraphid nor biraphid diatoms have such labiate processes. Within the centric diatoms, the pennate ancestry was suggested to lie in the Eupodiscaceae (Ross and Sims, 1973). The pennate *Plagiogramma* has ocelluslike structures at the poles, resembling those of *Auliscus* or *Triceratium,* as well as areolae that are similar to those of the centric diatoms.

Another link between the centric diatoms and araphids is indicated by the reproductive characters of the living *Rhabdonema adriaticum.* The oogamy of this species, with a large immotile "egg" cell and smaller motile male cells, resembles the reproductive system of the centric diatoms, although the sperm are not flagellated. Valve morphology is like that of the pennates (elongate), although no raphe is present (Stosch, 1958a). Another araphidean, *Grammatophora marina,* is still more pennatelike in reproduction, as it has similar gametes in both "sexes."

The monoraphid genera, such as *Cocconeis* and *Achnanthes,* are thought to be more primitive than the typically biraphid taxa (Schrader, 1973b). Central pores are not well developed and no central fissures are present, but both polar pores and polar fissures are present. Although this concept of a gradual progression in raphe complexity is appealing, the monoraphids appear to occur first in the late Eocene, whereas biraphid taxa are present in the middle Eocene. However, as these earliest Cenozoic floras are still quite imperfectly known, it is probably too early to determine if the monoraphids are primitive or secondarily reduced.

The Miocene was the time of greatest expansion of centric diatoms (see Figure 8.43). Although continuing to diversify, they have gradually been surpassed by the pennates (Tappan and Loeblich, 1973). Only a few centric species have invaded fresh water, and these are restricted to the plankton.

The rapid Tertiary expansion of the pennate diatoms was compared by Small (1950) to the stages of growth of a culture, with a lag or stationary phase of little growth, explosive growth, climax, and decline. The very small number of Paleocene species (three known to Small) increased to the present 3526 living species by a regular geometric progression. New species arose, and prior ones persisted, with only a relatively small number becoming extinct. Production of a new form (perhaps by minor changes) left the parent species unchanged, resulting in a constant doubling of taxa, rather than a replacement of older taxa by the new. Pennate diatoms are regarded as still within their explosive growth stage, whereas the older centric diatoms as a group reached this stage in the Miocene, and are now at a stationary climax, or in a gradual decline.

The large number of synonyms and varied species concepts make an accurate count of species living at each period somewhat difficult to assess, but the general pattern of diatom diversification appears similar to that described by Small.

CLASSIFICATION

The diatoms may be regarded as only the Class Bacillariophyceae in the algal division Chrysophyta, because of similar pigments, oil rather than starch storage, bipartite cell wall at some stage, and their silicified nature. The nature of the flagella of the male gamete also suggests a relationship to the silica-secreting Chrysophyta (Stosch, 1958c; Takahashi, 1964), and distinction from the previously associated calcite-depositing Haptophyta (such as the coccolithophores). Some silica-depositing chrysomonads (*Mallomonas* and *Synura*) have scales with a fine structure resembling that of the diatom frustule: complex perforated plates bearing spines, ribs, and bristles.

However, the many distinctive features of the diatoms also have been regarded as sufficient evidence for their status as a separate division, the Bacillariophyta, as is followed herein.

The wide interest in diatoms since their first description has resulted in the preparation of many monographic studies, with beautiful and extremely accurate illustrations. Unfortunately the geographic and particularly geologic data for the material was greatly abbreviated or omitted altogether, and the geologic distribution for many taxa has yet to be determined. Nevertheless, identification of material is greatly facilitated by monographs such as those of Smith (1853, 1856), Ehrenberg (1854), Schütt (1896), Heurck (1896), Pia (1927), Proshkina-Lavrenko (1949a,b, 1950), Schmidt (1874—1959), Boyer (1927a,b), West and Fritsch (1927), Hustedt (1927—1966, 1930), and Bourrelly (1968), and the indices of Mills (1933—1935) and Van-Landingham (1967b, 1968, 1969, 1971, 1975, 1978a,b, 1979).

Colonial diatoms were early believed to be similar to filamentous algae, whereas solitary ones were regarded as animals. Nitzsch (1817) separated the diatoms into two groups, animal and plant, on the basis of movement (pennate forms) or lack of movement (centric diatoms and some pennates).

Schütt (1896) separated the diatoms into two main groups on the basis of surface structures, the Centricae and Pennatae; others utilized for subdivision the presence of raphe, pseudoraphe, or neither; various features of the chromatophores; or the type of auxospores. Hendey (1937, 1964) recognized a single order and 10 suborders based on various shell features, and Silva (1962) recognized two classes (equivalent to Schütt's Centricae and Pennatae) and 10 orders (equivalent to Hendey's suborders, with names based on included genera, rather than descriptive). Others (Hustedt, 1930; Papenfuss, 1955; Ross, 1967) utilized Schütt's two groups as orders (renamed) with various included families.

As in other protistan groups whose taxonomy is based largely on the skeletal remains, the validity of taxa so based is occasionally questioned. To test the usefulness of shell characters in diatoms the genetic constitution of various clones of two supposedly closely related diatom species was studied by electrophoresis to determine the amino acid composition of the enzymes (Murphy and Guillard, 1976). Four clones of *Thalassiosira fluviatilis* from New York, Rhode Island, and Portugal, were compared to clones of *T. pseudonana* obtained from New York, Massachusetts, and Virginia, the Gulf Stream, Sargasso Sea, Virgin Islands, Germany, and Australia. All *T. fluviatilis* formed a clearly distinct and clearly separated group from those of *T. pseudonana,* although obtained from different continents, and isolated up to 10 years apart. *T. pseudonana* could be separated into 4 estuarine clones and one reefal one that were distinct from the four oceanic ones. A clone from the continental slope region was intermediate between the neritic and oceanic ones.

Within the Division Bacillariophyta, the classification followed in Table 8.5 is close to that of Hendey (1964), with some changes and modifications as suggested by Simonsen (1972, 1975), Hasle (1973a), Ross and Sims (1973), Round (1973), and Hendey (1974). As in other chapters, not all genera are listed, but most of those mentioned in the text are placed in the classification, together with various others.

Table 8.5
Classification of the diatoms.

Division Bacillariophyta
Engler & Gild 1924

I. Class Centrobacillariophyceae Silva 1962

Unicellular or colonial, valves with radially arranged perforations, no raphe or pseudoraphe (hyaline axial area); reproduction oogamous, uniflagellate male gamete; resting spores. Dominantly marine, few in fresh water. Lower Cretaceous to Holocene.

A. Order Eupodiscales Bessey 1907

Discoid or cylindrical cells, valves circular, flat or convex, rarely hemispherical; orientation of valve structure related to central pole; spines common, horns or knobs small or absent.

1. Family Melosiraceae Kützing 1844

Syn.: Corethronaceae Lebour 1930; Leptocylindraceae Lebour 1930

Cells subspherical or cylindrical, valve face flattened, mantle perpendicular to face; wall loculate with porous membranes at both surfaces; small strutted processes localized or scattered, no labiate processes; solid spines at margin of valve interlock with adjacent cells; several incomplete girdle bands may be present; resting spores. Marine and freshwater. Lower Cretaceous to Holocene.

Corethron Castracane 1886; *Dactyliosolen* Castracane 1886; *Druridgea* Donkin 1861; *Endictya* Ehrenberg 1845; *Groentvedia* Hendey 1964; *Hyalodiscus* Ehrenberg 1845; *Hydrosirella* Hustedt 1952; *Leptocylindrus* Cleve 1889; *Melosira* C.

Agardh 1824; *Muelleriopsis* Hendey 1972 (syn.: *Muelleriella* van Heurck 1896 non Hepp in Müller 1896); *Paralia* Heiberg 1863; *Podosira* Ehrenberg 1840; *Pyxidicula* Ehrenberg 1833; *Stephanopyxis* Ehrenberg 1844.

2. Family Thalassiosiraceae Lebour 1930

Syn.: Skeletonemaceae Lebour 1930

Cells circular in valve view with radial areolar pattern, wall poroid or loculate, with sieve membranes at interior; valves commonly with one labiate process; cells may be interconnected by a central or marginal circle of strutted processes or by organic material extruded through these; spines and tubular processes prominent; resting spores. Upper Cretaceous to Holocene.

Bacterosira Gran 1900; *Coenobiodiscus* Loeblich, Wight & Darley 1968; *Cyclotella* Kützing 1834; *Detonula* Schütt 1893 (syn.: *Schroderella* Pavillard 1913); *Ethmodiscus* Castracane 1886; *Lauderia* Cleve 1873; *Minidiscus* Hasle 1973; *Planktoniella* Schütt 1893; *Porosira* Jorgensen 1950; *Skeletonema* Greville 1865 (syn.: *Microsiphona* Weber 1907); *Stephanodiscus* Ehrenberg 1845; *Thalassiosira* Cleve 1873 (syn.: *Cylindropyxis* Hendey 1964; *Micropodiscus* Grunow in van Heurck 1883).

3. Family Coscinodiscaceae Kützing 1844

Cells discoid or as short cylinders, solitary; valves with loculate wall structure, with foramen on inner surface of valve wall and sieve plate at exterior surface; ring of labiate pro-

(continued)

cesses; two to many girdle bands, of incomplete porose rings. Marine and freshwater. Lower Cretaceous to Holocene.

Annellus Tempère 1915; *Brightwellia* Ralfs 1861; *Charcotia* Peragallo 1921; *Coscinodiscus* Ehrenberg 1838; *Coscinosira* Gran 1900; *Cosmiodiscus* Greville 1866; *Craspedodiscus* Ehrenberg 1845; *Cymatodiscus* Hendey 1958; *Cymatotheca* Hendey 1958; *Gossleriella* Schütt 1893; *Gutwinskiella* de Toni 1894; *Heterodictyon* Greville 1863; *Horodiscus* Hanna 1927; *Kozloviella* Jousé 1973; *Liradiscus* Greville 1865; *Meretrosulus* Hanna 1927; *Phacodiscus* Meunier 1910; *Pomphodiscus* Barker & Meakin 1947; *Porodiscus* Greville 1865; *Pseudopodosira* Jousé 1949; *Schimperiella* Karsten 1907; *Schuettia* de Toni 1894; *Stoschia* Janisch 1899; *Trochosira* Kitton 1870.

4. Family Asterolampraceae H. L. Smith 1872

Valves divided into sectors by hyaline rays that terminate in labiate processes; central area hyaline, fine punctae in interray areas. ?Eocene; Miocene to Holocene.

Asterolampra Ehrenberg 1845; *Asteromphalus* Ehrenberg 1844; *Bergonia* Tempère 1891; *Rylandsia* Greville & Ralfs 1861.

5. Family Actinodiscaceae Kützing 1844

Syn.: Heliopeltaceae H. L. Smith 1872

Discoidal or flattened valves, may be divided into radial segments that alternately are elevated or depressed; four or more prominent labiate processes near margin. Lower Cretaceous to Holocene.

Actinodiscus Greville 1863; *Actinoptychus* Ehrenberg 1839; *Aulacodiscus* Ehrenberg 1844; *Benetorus* Hanna 1927; *Brunia* Tempère 1891; *Centroporus* Pantocsek 1889; *Cladogramma* Ehrenberg 1844; *Compositus* Vekshina 1960; *Glorioptychus* Hanna 1927; *Lepidodiscus* Witt 1886; *Polymyxus* L. W. Bailey 1861; *Pseudostictodiscus* Grunow in Schmidt 1882; *Sturtiella* Simonsen & Schrader 1973; *Truania* Pantocsek 1886; *Tryblioptychus* Hendey 1958.

6. Family Stictodiscaceae Schütt 1896

Circular disklike valves, solitary, with prominent radial ribs, no horns or prominent spines. Upper Cretaceous to Holocene.

Arachnoidiscus Deane ex Pritchard 1852; *Stictodiscus* Greville 1861, *Rocella* Hanna 1930.

Rocella has been placed with the silicoflagellates, but is related to *Stictodiscus*, a diatom.

7. Family Hemidiscaceae Hendey 1937

Cells solitary, free, cuneate in pervalvar view; valves semicircular, surface punctate, more densely areolate or otherwise distinct ornamentation on margin, ornamentation interrupted centrally by hyaline area or zone of areolae; pseudonodulus present, marginal labiate processes do not protrude externally, but are expanded internally. Miocene to Holocene.

Actinocyclus Ehrenberg 1838; *Hemidiscus* Wallich 1860 (syn.: *Palmeria* Greville 1865); *Roperia* Grunow 1881.

8. Family Eupodiscaceae Kützing 1844

Syn.: Auliscaceae Hendey 1937; Rutilariaceae Pantocsek ex de Toni 1894

Cells short, cylindrical, valves flat or convex; may have marginal processes and usually two or three large ocelli in submarginal position. Upper Cretaceous to Holocene.

Auliscus Ehrenberg 1844 (syn.: *Fenestrella* Greville 1863); *Baxteriopsis* Karsten 1928; *Cerataulus* Ehrenberg 1844; *Craspedoporus* Greville 1863; *Eupodiscus* Bailey 1851; *Hendeya* Long, Fuge & Smith 1946; *Huttonia* Grove & Sturt 1887; *Kisseleviella* Sheshukova-Poretskaya 1962; *Kittonia* Grove & Sturt 1887; *Odontella* Agardh 1832; *Pseudocerataulus* Pantocsek 1889; *Pseudorutilaria* Grove & Sturt 1887; *Rattrayella* de Toni 1889; *Rutilaria* Greville 1863 (syn.: *Syndetocystis* Ralfs 1866; *Syndetoneis* Grunow 1888); *Stictocyclus* Mann 1925; *Triceratium* Ehrenberg 1840 (syn.: *Amphipentas* Ehrenberg 1830; *Amphitetras* Ehrenberg 1840); *Zygoceros* Ehrenberg 1839.

B. Order Rhizosoleniales Silva 1962

Cells elongate, cylindrical, or subcylindrical, may occur in chains; valves circular or subcircular, domed or conical, may be apiculate or have circlet of spines; may have numerous intercalary bands. Upper Cretaceous to Holocene.

1. Family Pyxillaceae Schütt 1896

Syn.: Thaumatodiscaceae Cleve 1885

Elongate cylindrical valves, with central horn much extended, symmetry radial. Upper Cretaceous to Holocene.

Gladius Schulz 1935; *Gyrodiscus* Witt 1885; *Mastogonia* Ehrenberg 1844; *Pterotheca* Grunow 1883 (may be based on spore); *Pyrgupyxis* Hendey 1969; *Pyxilla* Greville 1865.

2. Family Rhizosoleniaceae Petit 1889

Cylindrical cell, usually with many intercalary bands of varying type; single central or ex-

centric long or short labiate process; resting spores. Marine, rarely freshwater, planktonic. Lower Cretaceous to Holocene.
Ditylum Bailey 1862; *Guinardia* H. Peragallo 1892; *Lithodesmium* Ehrenberg 1840; *Rhizosolenia* Brightwell 1858.

Note: *Kentrodiscus* Pantocsek 1889 and *Pseudopyxilla* Forti 1909 may be spores of the Rhizosoleniaceae.

3. Family Chaetoceraceae H. L. Smith 1872
Syn.: Bacteriastraceae Lebour 1930
Cylindrical cells with much elongate central or lateral bristles or setae that may unite adjacent cells as chains. Resting spores common. Marine and freshwater plankton. Spores known from Upper Cretaceous (Coniacian) to Holocene; frustules known from lower Eocene to Holocene.
Attheya T. West 1860; *Bacteriastrum* Shadbolt 1854; *Chaetoceros* Ehrenberg 1844.

Note: *Periptera* and *Xanthiopyxis* Ehrenberg 1844 probably are resting spores.

C. Order Biddulphiales Krieger 1954
Valves angular, with processes, thickness exceeds diameter; ornamentation related to angles. Marine and freshwater; neritic, benthic, epiphytic, and planktonic. Lower Cretaceous (Albian) to Holocene.

1. Family Biddulphiaceae Kützing 1844
Cells oval to polygonal in valve view; united in chains by mucus pads formed by processes or spines; valves with two or more poles; have ocellus or pseudocellus; resting spores. Marine and freshwater; epiphytic, benthic, and planktonic. Lower Cretaceous (Albian) to Holocene.
Anaulus Ehrenberg 1844; *Bellerochea* van Heurck 1885; *Biddulphia* S. F. Gray 1821; *Climacodium* Grunow 1868; *Entogonia* Greville 1863; *Eunotiopsis* Grunow 1883 (syn.: *Leudugeria* Tempère 1893); *Eunotogramma* J. F. Weisse 1854; *Hydrosera* Wallich 1858; *Isthmia* Agardh 1832; *Neostreptotheca* von Stosch 1977; *Odontotropis* Grunow 1884; *Ploiaria* Pantocsek 1889; *Porpeia* Bailey in Pritchard 1861; *Sphynctolethus* Hanna 1927; *Streptotheca* Shrubsole 1890; *Terpsinoe* Ehrenberg 1841; *Trigonium* Cleve 1868; *Xystotheca* Hanna 1932.

2. Family Hemiaulaceae Heiberg 1863
Valves with two, three, or four poles, each angle with prominent horn, commonly horns are tipped with linking spines that attach to adjacent cells in chain; resting spores. Marine and freshwater plankton. Spores from Upper Cretaceous (Coniacian) to Holocene, frustules from lower Eocene to Holocene.
Cerataulina H. Peragallo 1892; *Eucampia* Ehrenberg 1839; *Goniothecium* Ehrenberg 1844; *Hemiaulus* Ehrenberg 1845; *Monobrachia* Schrader 1976; *Riedelia* Jousé & Sheshukova-Poretskaya 1971; *Trinacria* Heiberg 1863.

The following probably are spores of centric diatoms, but are of uncertain position: *Acanthodiscus* Pantocsek 1892; *Chasea* Hanna 1934; *Haynaldia* Pantocsek 1889; *Hercotheca* Ehrenberg 1844; *Keratophora* Pantocsek 1889; *Pantocsekia* Grunow 1886; *Poretzkia* Jousé 1949; *Pseudoaulacodiscus* Vekshina 1960; *Pyrgodiscus* Kitton 1883; *Stephanogonia* Ehrenberg 1844; *Syndendrium* Ehrenberg 1845; *Syringidium* Ehrenberg 1845.

II. Class Pennatibacillariophyceae Silva 1962
Cells solitary or colonial, valves tend to be elongate, with pores in bilateral arrangement; with raphe or central longitudinal hyaline area (pseudoraphe); sexual reproduction isogamous, with amoeboid gametes; dominant in fresh water, but also present in marine waters. Upper Cretaceous (Maastrichtian) to Holocene, dominantly Cenozoic.

A. Order Fragilariales (G. S. West) Silva 1962
Araphid diatoms, valves linear, lanceolate to spathulate, straight to arcuate; commonly with intercalary bands and septa; lack raphe, but may have median hyaline area or pseudoraphe, commonly with transapical striae or ribs. Marine and freshwater; benthic, epiphytic, planktonic. Upper Cretaceous (Maastrichtian); Paleocene to Holocene.

1. Family Diatomaceae Dumortier 1823
Cells oval to elongated, rectangular in girdle view; may have median pseudoraphe; commonly form ribbonlike colonies or zigzag chains. Upper Cretaceous (*Rhaphoneis* and *Sceptroneis* present from Maastrichtian); Paleocene (*Opephora* and *Synedra*); Eocene to Holocene.
Ambistria Lohman & Andrews 1968; *Amphicampa* Ehrenberg 1854; *Auriculopsis* Hendey

(*continued*)

1964; *Asterionella* Hassall 1855 (syn.: *Peronia* de Brébisson & Arnott ex Kitton 1868); *Bogorovia* Jousé 1973; *Campylosira* Grunow 1882; *Campylostylus* Shadbolt 1849; *Catenula* Mereschkowsky 1902; *Catillus* Hendey 1977; *Centronella* Voight 1902; *Ceratoneis* Ehrenberg 1840; *Clavicula* Pantocsek 1886; *Climacosphenia* Ehrenberg 1843; *Cyclophora* Castracane 1878; *Cymatosira* Grunow 1862; *Delphineis* Andrews 1977; *Diatoma* Bory 1824; *Dimeregramma* Ralfs in Pritchard 1861; *Drepanotheca* Schrader 1969; *Entopyla* Ehrenberg 1841; *Fragilaria* Lyngbye 1819; *Fragilariella* Hendey 1958; *Gephyria* Arnott 1860; *Glyphodesmis* Greville 1862; *Grammatophora* Ehrenberg 1840; *Grunowiella* van Heurck 1892; *Hannaea* Patrick & Reimer 1966 (incl. *Ceratoneis* auct. non Ehrenberg 1840); *Helminthopsis* van Heurck 1892; *Ikebea* Komura 1975; *Jousea* Glezer 1967; *Lamella* J. Brun 1894; *Leptoscaphos* Schrader 1969; *Licmophora* Agardh 1827; *Licmosphenia* Mereschkowsky 1902; *Meridion* Agardh 1824; *Opephora* P. Petit 1888; *Plagiogramma* Greville 1859; *Podocystis* Kützing 1844; *Praeepithemia* Jousé 1952; *Pseudodimerogramma* Schrader 1976; *Pseudosynedra* Leuduger-Fortmorel 1892; *Rhabdonema* Kützing 1844; *Rhaphoneis* Ehrenberg 1844; *Sawamuraia* Komura 1976; *Sceptroneis* Ehrenberg 1844 (syn.: *Incisoria* Hajós 1975); *Striatella* Agardh 1832; *Subsilicea* von Stosch & Reimann 1970; *Synedra* Ehrenberg 1830; *Synedrosphenia* Peragallo 1900; *Tabellaria* Ehrenberg 1840; *Tetracyclus* Ralfs 1843; *Thalassionema* Grunow ex Hustedt 1932; *Thalassiothrix* Cleve & Grunow 1880; *Trachysphenia* P. Petit 1877; *Tubularia* J. Brun 1894.

2. Family Protoraphidaceae Simonsen 1970
Araphid diatoms with sigmoid pseudoraphe; cells attached by gelatinous stalk. Holocene. *Protoraphis* Simonsen 1970; *Pseudohimantidium* Hustedt & Krasske 1941.

B. Order Eunotiales Silva 1962
Valves arcuate, sublunate, hemispherical, clavate; rudimentary raphe on one or both valves; no central nodule. Freshwater, epiphytic. Upper Eocene to Holocene.

1. Family Eunotiaceae Kützing 1844
Actinella Lewis 1864; *Desmogonium* Ehrenberg 1848; *Eunotia* Ehrenberg 1837.

C. Order Achnanthales (G. S. West) Silva 1962
Valves ovate, lanceolate, and dissimilar, with true raphe on one valve, and pseudoraphe on other valve; cells may attach in chains. Freshwater and marine, epiphytes; upper Eocene to Holocene.

1. Family Achnanthaceae Kützing 1844
Achnanthes Bory 1822; *Anorthoneis* Grunow 1867; *Campyloneis* Grunow 1862; *Cocconeis* Ehrenberg 1838; *Peroniopsis* Hustedt 1952; *Pseudoperonia* Manguin 1964; *Rhoicosphenia* Grunow 1860.

D. Order Naviculales Bessey 1907
Valves lanceolate to ovate, sigmoid, arcuate, or sublunate; flat or twisted; true raphe on both valves, consisting of either a simple inner and outer fissure, or the more complex canal raphe elevated on keel; may have polar and central nodules. Marine and freshwater; benthic, epiphytic, and planktonic. Paleocene to Holocene.

1. Family Naviculaceae Kützing 1844
Valves usually bilaterally symmetrical, rarely sigmoid; with straight raphe, consisting of two axially oriented slits on each valve from central to polar nodules; cell rectangular in girdle view. Paleocene (*Mastogloia* and *Navicula* reported in USSR, according to Glezer, 1974); Eocene to Holocene.
Amphipleura Kützing 1844; *Amphiprora* Ehrenberg 1843; *Anomoeoneis* Pfitzer 1871; *Berkeleya* Greville 1827; *Brebissonia* Grunow 1860; *Caloneis* Cleve 1891; *Cistula* Cleve 1894; *Cymatoneis* Cleve 1894; *Diatomella* Greville 1855; *Dictyoneis* Cleve 1890; *Didymosphenia* M. Schmidt 1899; *Dimidiata* Hajós 1973; *Diploneis* Ehrenberg 1844; *Donkinia* Ralfs 1860; *Frickia* Heiden 1874; *Frustulia* Rabenhorst 1853; *Gomphocymbella* O. Müller 1905; *Gyrosigma* Hassall 1845; *Lophotheca* Schrader 1969; *Mastogloia* Thwaites in W. Smith 1856; *Nanoneis* Norris 1973; *Navicula* Bory 1824; *Neidium* Pfitzer 1871; *Oestrupia* Heiden 1906; *Pinnularia* Ehrenberg 1840; *Pleurosigma* W. Smith 1852; *Progonoia* Schrader 1969; *Pseudamphiprora* Cleve 1881; *Raphidodiscus* H. L. Smith 1887; *Scoliopleura* Grunow 1860; *Scoliotropis* Cleve 1894; *Stauroneis* Ehrenberg 1843; *Stenoneis* Cleve 1894; *Trachyneis* Cleve 1894; *Tropidoneis* Cleve 1891; *Toxonidea* Donkin 1858; *Vanheurckia* de Brébisson 1868.

2. Family Auriculaceae Hendey 1964

Valves symmetrical on the transapical axis, asymmetrical on the apical axis, auricular or reniform in outline, with dorsal margin elevated as a ridge that bears the raphe. Marine, planktonic. Holocene.

Auricula Castracane 1873.

3. Family Cymbellaceae Kützing 1844

Syn.: Phaeodactylaceae Lewin 1958

Cells usually solitary, free, valves lunate, arcuate, or triradiate, symmetrical on the transapical axis, asymmetrical on the apical axis; valves with distinct raphe, polar and central nodules; surface striate, commonly with rows of punctae. Marine and freshwater, benthic, planktonic. Upper Eocene to Holocene.

Amphora Ehrenberg 1840; *Cymbella* Agardh 1830; *Okedonia* Eulenstein in van Heurck 1884; *Phaeodactylum* Bohlin 1897; *Sameioneis* Russell & Norris 1971 (syn.: *Homophora* Jurilj 1957, non J. G. Agardh 1892).

4. Family Gomphonemaceae Kützing 1844

Cells slightly cuneate in girdle view, sessile, attached by mucus stipes; valves clavate, symmetrical on apical axis, asymmetrical on transapical axis, poles different; distinct raphe, polar and central nodules. Freshwater, benthic. Miocene to Holocene.

Gomphoneis Cleve 1894; *Gomphonema* Ehrenberg 1831.

5. Family Epithemiaceae Grunow 1860

Cells free or epiphytic, solitary; valves arcuate or lunate, isopolar, symmetrical on transapical axis; asymmetrical on apical axis; surface with transverse costae separating rows of punctae; canal raphe at margin of valve. Marine and freshwater, epiphytic and benthic. Lower Miocene to Holocene.

Denticula Kützing 1844; *Epithemia* de Brébisson 1838; *Katahiraia* Komura 1976; *Rhopalodia* O. Müller 1895; *Yoshidaia* Komura 1976.

6. Family Nitzschiaceae Grunow 1860

Cells usually solitary, or may occur in packet-like colonies, join tip to tip, or be held in mucilage; cells elongate or slightly sigmoid; rhomboid in transverse section; canal raphe marginal in position; poles alike, punctae along keel, no polar nodules. Marine and freshwater, benthic. Upper Eocene to Holocene.

Bacillaria Gmelin 1788; *Cylindrotheca* Rabenhorst 1859; *Cymbellonitzschia* Hustedt 1924; *Gomphonitzschia* Grunow 1867; *Hantzschia* Grunow 1877; *Nitzschia* Hassall 1844 (syn.: *Fragilariopsis* Hustedt 1913); *Pseudoeunotia* Grunow 1865; *Rouxia* Brun & Héribaud 1893.

E. Order Surirellales (West) Silva 1962

Valves ovate, lanceolate, obovate, or subrectangular; flat or twisted; no nodules; pseudoraphe in polar axis; peripheral canal raphe elevated on winglike projection at valve margin. Marine and freshwater; benthic, epiphytic. Upper Oligocene to Holocene.

1. Family Surirellaceae Kützing 1844

Campylodiscus Ehrenberg 1840; *Cymatopleura* W. Smith 1851; *Hydrosilicon* Brun 1891; *Stenopterobia* de Brébisson 1867; *Surirella* Turpin 1828.

REFERENCES

Abbott, W. H., Temporal and spatial distribution of Pleistocene diatoms from the southeast Indian Ocean. *Nova Hedwigia*, v. 25, p. 291−347, pls. 1−11, figs. 1−8, 1974.

Abbott, W. H., Correlation and zonation of Miocene strata along the Atlantic margin of North America using diatoms and silicoflagellates. *Mar. Micropaleontol.*, v. 3, p. 15−34, pls. 1, 2, figs. 1−4, 1978.

Allen, W. E., and E. E. Cupp, Plankton diatoms of the Java Sea. *Ann. Jard. Bot. Buitenz.*, v. 44, p. 101−174, figs. 1−127, 1935.

Anderson, G. C., and Karl Banse, Hydrography and phytoplankton production, p. 61−90. *In* M. S. Doty, *Proceedings of the conference on primary productivity measurement, marine and freshwater, held at University of Hawaii, August 21−September 6, 1961*, 1963.

Andrews, G. W., Late Pleistocene diatoms from the Trempealeau Valley, Wisconsin. *Prof. Pap. U.S. Geol. Surv.*, 523−A, p. A1−A27, pls. 1−3, figs. 1−4, 1966.

Andrews, G. W., Late Miocene nonmarine diatoms from the Kilgore area, Cherry County, Nebraska. *Prof. Pap. U.S. Geol. Surv.*, 683−A, p. A1−A24, pls. 1−3, figs. 1−3, 1970.

Andrews, G. W., Systematic position and stratigraphic signifance of the marine Miocene diatom *Raphidodis-*

cus marylandicus Christian. *Beih. Nova Hedwigia*, v. 45, p. 231−250, pls. 1−5, 1973.

Andrews, G. W., Taxonomy and stratigraphic occurrence of the marine diatom genus *Rhaphoneis. Beih. Nova Hedwigia*, v. 53, p. 193−227, pls. 1−5, fig. 1, 1975.

Andrews, G. W., Miocene marine diatoms from the Choptank Formation, Calvert County, Maryland. *Prof. Pap. U.S. Geol. Surv.*, 910, p. 1−24, pls. 1−7, figs. 1, 2, 1976.

Andrews, G. W., Morphology and stratigraphic significance of *Delphineis*, a new marine diatom genus. *Beih. Nova Hedwigia*, v. 54, p. 243−260, pls. 1−4, 1977.

Anonymous, Proposals for a standardization of diatom terminology and diagnoses. *Beih. Nova Hedwigia*, v. 53, p. 323−354, figs. 1−48, 1975.

Ax, Peter, and Gieselbert Apelt, Die "Zooxanthellen" von *Convoluta convoluta* (Turbellaria Acoela) entstehen aus Diatomeen. Erster Nachweis einer Endosymbiose zwischen Tieren und Kieselalgen. *Naturwissenschaften*, v. 15, p. 444−446, figs. 1−7, 1965.

Badour, S. S., Experimental separation of cell division and silica shell formation in *Cyclotella cryptica. Arch. Mikrobiol.*, v. 62, p. 17−33, figs. 1−10, 1968.

Bailey, J. W., A sketch of the Infusoria, of the family Bacillaria, with some account of the most interesting species which have been found in a recent or fossil state in the United States. *Am. J. Sci.*, v. 41, p. 284−305, pl. 3, 1841; Part 2, v. 42, p. 88−105, pl. 2, 1842; Part 3, v. 43, p. 321−332, pl. 5, 1842.

Bailey, J. W., Account of some new infusorial forms discovered in the fossil Infusoria from Petersburg, Va., and Piscataway, Md. *Am. J. Sci.*, v. 46, p. 137−141, 1843.

Bainbridge, R., The size, shape and density of marine phytoplankton concentrations. *Biol. Rev.*, v. 32, p. 91−115, 1957.

Barber, H. G., The fossil freshwater diatoms of the Ongaroto Valley deposit, North Island, New Zealand. *J. Quekett Micr. Cl.*, v. 28 (ser. 4, v. 5), p. 387−391, pls. 1, 2, 1961.

Barber, H. G., The collection and preparation of Recent diatoms. *J. Quekett Micr. Cl.*, v. 29, p. 21−25, 1962.

Barker, I. W., and S. H. Meakin, New diatoms from the Moreno Shale. *J. Quekett Micr. Cl.*, ser. 4, v. 2 (1946), p. 143−144, pl. 22, 1947.

Barker, I. W., and S. H. Meakin, New and rare diatoms. *J. Quekett Micr. Cl.*, ser. 4, v. 2, p. 301−303, pls. 37, 38, 1949.

Barron, J. A., Late Miocene−Early Pliocene paleotemperatures for California from marine diatom evidence. *Palaeogeogr. Palaeoclimat. Palaeoecol.*, v. 14, p. 277−291, pls. 1−3, figs. 1−4, 1973.

Barron, J. A., Marine diatom biostratigraphy of the Upper Miocene-Lower Pliocene strata of southern California. *J. Paleont.*, v. 49, p. 619−632, figs. 1−8, 1975a.

Barron, J. A., Late Miocene−Early Pliocene marine diatoms from southern California. *Palaeontographica Abt. B*, v. 151, p. 97−170, pls. 1−15, figs. 1−6, 1975b.

Barron, J. A., Marine diatom and silicoflagellate biostratigraphy of the type Delmontian Stage and the type *Bolivina obliqua* Zone, California. *J. Res. U.S. Geol. Surv.*, v. 4, p. 339−351, figs. 1−11, 1976a.

Barron, J. A., Middle Miocene−Lower Pliocene marine diatom and silicoflagellate correlations in the California area. *Neogene Symposium, Soc. Econ. Paleont. Mineral. Ann. Meeting, San Francisco*, p. 117−124, figs. 1−7, 1976b.

Barron, J. A., Revised Miocene and Pliocene diatom biostratigraphy of Upper Newport Bay, Newport Beach, California. *Mar. Micropaleont.*, v. 1, p. 27−63, pls. 1−3, figs. 1−7, 1976c.

Beklemishev, C. W., Sur la colonialité des diatomées planctoniques. *Int. Revue Ges. Hydrobiol. Hydrogr.*, v. 44, p. 11−26, pls. 1, 2, figs. 1, 2, 1959.

Beklemishev, K. V., O znachenii kolonial'nosti dlya planktonnykh diatomey [On the role played by the colonies of plankton diatoms]. *Trudy Inst. Okeanol.*, v. 51, p. 16−30, figs. 1−3, 1961.

Beklemishev, K. V., M. N. Petrikova, and G. I. Semina, O prichine plavuchesti planktonnykh diatomovykh [On the cause of buoyancy of plankton diatoms]. *Trudy Inst. Okeanol.*, v. 51, p. 33−36, 1961.

Belyaeva, T. V., Raspredelenie i chislennost' diatomey roda *Ethmodiscus* Castr. v planktone i v osadkakh Tikhogo Okeana [Distribution and numbers of diatoms of the genus *Ethmodiscus* Castr. in the plankton and bottom sediments of the Pacific Ocean] *Okeanologiya*, v. 8, p. 102−110, figs. 1−3, 1968.

Belyaeva, T. V., Chislennost' *Ethmodiscus* v planktone Tikhogo Okeana [Abundance of the diatom *Ethmodiscus* in plankton of the Pacific Ocean]. *Okeanologiya*, v. 10, p. 847−850, 1 fig., 1970.

Belyaeva, T. V., Raspredelenie krupnykh form diatomovykh vodorosley v yugo-vostochnoy chasti Tikhogo Okeana [Distribution of large diatom algae in the southeastern part of the Pacific Ocean]. *Okeanologiya*, v. 12, p. 475−484, figs. 1−5, 1972.

Bennett, A. G., On the occurrence of diatoms on the skin of whales (with an appendix by E. W. Nelson). *Proc. R. Soc.*, ser. B, v. 91, p. 352−357, figs. 1, 2, 1920.

Bonadonna, F. P., Studi sul Pleistocene del Lazio; il bacino diatomitico di Prima Porta. *Geol. Romana*, v. 2, p. 179−193, pls. 1, 2, figs. 1−7, 1963.

Bonadonna, F. P., Studi sul Pleistocene del Lazio II. Il bacino diatomitico di Cornazzano (Bracciano, Roma). *Geol. Romana*, v. 3, p. 383−404, 18 figs., 1964.

Bond, T. A., A postglacial diatom assemblage from Caddo County, Oklahoma. *Micropaleontology*, v. 14, p. 484–488, pl. 1, 1968.

Bourrelly, Pierre, *Les algues d'eau douce. Initiation à la systématique*, tome II, *Les algues jaunes et brunes, chrysophycées, phéophycées, xanthophycées et diatomées*. Paris: N. Boubée, 438 p., 114 pls., 1968.

Boyer, C. S., Thallophyta—Diatomacea, p. 487–507, pls. 134, 135. In *Miocene*. Maryland Geol. Survey. Baltimore: Johns Hopkins Press, 1904.

Boyer, C. S., *The Diatomaceae of Philadelphia and vicinity*. Philadelphia: J. B. Lippincott Co., 143 p., 40 pls. 1916.

Boyer, C. S., Synopsis of North American Diatomaceae. Part I. Coscinodiscatae, Rhizoselenatae, Biddulphiatae, Fragilariatae. *Proc. Acad. Nat. Sci. Philad.*, v. 78 (Supplement), p. 1–228, 1927a.

Boyer, C. S., Synopsis of North American Diatomaceae. Part II. Naviculatae, Surirellatae. *Proc. Acad. Nat. Sci. Philad.*, v. 79 (supplement), p. 229–583, 1927b.

Braarud, Trygve, Ecology of marine phytoplankton. *VIII Int. Bot. Congr., Rapp. Comm.*, sec. 17, p. 178–185, 1954.

Bradbury, J. P., Diatoms from the Pleistocene sediments of Lake Texcoco, Mexico. *Revue Géogr. Phys. Géol. Dyn.*, ser. 2, v. 12, p. 161–168, figs. 1, 2, 1970.

Bradbury, J. P., Paleolimnology of Lake Texcoco, Mexico. Evidence from diatoms. *Limnol. Oceanogr.*, v. 16, p. 180–200, figs. 1–6, 1971.

Bradbury, J. P., Diatom stratigraphy and human settlement in Minnesota. *Spec. Pap. Geol. Soc. Am.* 171, p. 1–74, pls. 1–6, figs. 1–17, 1975.

Bramlette, M. N., The Monterey Formation of California and the origin of its siliceous rocks. *Prof. Pap. U.S. Geol. Surv.*, 212, 57 p., 19 pls., 1 map, 1946.

Brigger, A. L., A mounting medium for diatoms. *J. Quekett Micr. Cl.*, ser. 4, v. 5, p. 275–277, 1960.

Brongersma-Sanders, Margaretha, Barium in pelagic sediments and in diatoms. *Proc. K. Ned. Akad. Wet.*, ser. B, v. 70, p. 93–99, 1967.

Brun, J., *Diatomées des Alpes et du Jura et de la région suisse et française des environs de Genève*. Genève: H. Georg, 146 p., 9 pls., 1880.

Bunt, J. S., Olga van H. Owens, and G. Hoch, Exploratory studies on the physiology and ecology of a psychrophilic marine diatom. *J. Phycol.*, v. 2, p. 96–100, 4 figs., 1966.

Burckle, L. H., Late Cenozoic planktonic diatom zones from the eastern equatorial Pacific. *Beih. Nova Hedwigia*, v. 39, p. 217–246, pls. 1–3, figs. 1–4, 1972.

Burckle, L. H., and R. B. McLaughlin, Size changes in the marine diatom *Coscinodiscus nodulifer* A. Schmidt in the equatorial Pacific. *Micropaleontology*, v. 23, p. 216–222, figs. 1–3, 1977.

Busby, W. F., and Joyce Lewin, Silicate uptake and silica shell formation by synchronously dividing cells of the diatom *Navicula pelliculosa* (Bréb.) Hilse. *J. Phycol.*, v. 3, p. 127–131, 1967.

Calvert, Robert, *Diatomaceous earth*. Am. Chem. Soc. Monogr. Ser. New York: Chemical Catalog Co., 251 p., 69 figs. 1930.

Calvert, S. E., Factors affecting distribution of laminated diatomaceous sediments in Gulf of California, p. 311–330, 14 figs. *In* T. H. van Andel and G. G. Shor, Jr., *Marine geology of the Gulf of California*. Tulsa: Mem. Am. Ass. Petrol. Geol. 3, 1964.

Calvert, S. E., Accumulation of diatomaceous silica in the sediments of the Gulf of California. *Bull. Geol. Soc. Am.*, v. 77, p. 569–596, 16 text-figs., 1966a.

Calvert, S. E., Origin of diatom-rich varved sediments from the Gulf of California. *J. Geol.*, v. 74, p. 546–565, 8 text-figs., 1966b.

Cassie, Vivienne, Diatoms, dinoflagellates and hydrology in the Hauraki Gulf, 1964–1965. *N.Z. Jl. Sci.*, v. 9, p. 569–585, 1966.

Cassie, Vivienne, and W. S. Bertaud, Electron microscope studies of New Zealand marine plankton diatoms. *Jl. R. Microsc. Soc.*, ser. 3, v. 79, p. 89–94, pls. 30–32, fig. 1, 1960.

Castracane, A. F., Report on the Diatomaceae collected by *H.M.S. Challenger* during the years 1873–1876. *Rep. Scient. Results H.M.S. Challenger, Botany*, v. 2, pt. 4, 178 p., 30 pls., 1886.

Cheremisinova, E. A., Novye dannye o diatomeyakh neogenovykh otlozheniy Pribaykal'ya [New data on the Neogene diatoms of the Pre-Baikal region]. *Iskopaemye Diatomovye Vodorosli SSSR, Akad. Nauk SSSR, Sibirskoe Otdel., Inst. Geol. Geofiz.*, Moskva: "Nauka," p. 71–74, pls. 3–6, 1968.

Cheremisinova, E. A., Novye i interesnye vidy diatomovykh vodorosley iz Neogena Pribaykal'ya [Species novae et curiosae Bacillariophytorum e Neogeno regionis Baicalensis]. *Nov. Sist. Nizsh. Rast., Akad. Nauk SSSR, Bot. Inst. Komarova* 1971, p. 52–66, pls. 1–4, 1971.

Cholnoky, B. J., Beiträge zur Kenntnis der Ökologie der Diatomeen in dem Swartkops-Bache nahe Port Elizabeth (Südost-Kaapland). *Hydrobiologia*, v. 16, p. 229–287, figs. 1–53, 1960.

Cleve, P. T., Synopsis of the naviculoid diatoms. *K. Svenska VetenskAkad. Handl.*, n.s., v. 26, no. 2, p. 1–193, pls. 1–5; v. 27, no. 3, p. 1–219, pls. 1–4; 1894–1896.

Cleve-Euler, A., Alttertiäre Diatomeen und Silicoflagellaten im inneren Schwedens. *Palaeontographica Abt. A*, v. 92, p. 165–208, pls. 16–25, text-figs. 1–7, 1941.

Cleveland, G. B., Rapid method of sampling diatomaceous earth. *Spec. Rep. Short Contr. Calif. Geol., Calif. Div. Mines Geol.*, 100, p. 67–68, fig. 1, 1969.

Collier, A., and A. Murphy, Very small diatoms: preliminary notes and description of *Chaetoceros galvestonensis. Science,* v. 136, p. 780−781, 1 fig., 1962.

Coombs, J., P. J. Halicki, O. Holm-Hansen, and B. E. Volcani, Studies on the biochemistry and fine structure of silica shell formation in diatoms. Changes in concentration of nucleoside triphosphates during synchronized division of *Cylindrotheca fusiformis* Reimann & Lewin. *Expl. Cell Res.,* v. 47, p. 302−314, figs. 1−5, 1967a.

Coombs, J., P. J. Halicki, O. Holm-Hansen, and B. E. Volcani, Studies on biochemistry and fine structure of silica shell formation in diatoms. II. Changes in concentration of nucleoside triphosphates in silicon-starvation synchrony of *Navicula pelliculosa* (Bréb.) Hilse. *Expl. Cell Res.,* v. 47, p. 315−328, figs. 1−7, 1967b.

Coombs, J., J. A. Lauritis, W. M. Darley, and B. E. Volcani, Studies on the biochemistry and fine structure of silica shell formation in diatoms. V. Effects of colchicine on wall formation in *Navicula pelliculosa* (Bréb.) Hilse. *Z. PflPhysiol.,* v. 59, p. 124−152, 18 figs., 1968.

Coombs, J., and B. E. Volcani, Studies on the biochemistry and fine structure of silica shell formation in diatoms. Silicon-induced metabolic transients in *Navicula pelliculosa* (Bréb.) Hilse. *Planta,* v. 80, p. 264−279, figs. 1−7, 1968.

Cox, E. J., Further studies on the genus *Berkeleya* Grev. *Br. Phycol. J.,* v. 10, p. 205−217, figs. 1−35, 1975.

Crawford, R. M., The protoplasmic ultrastructure of the vegetative cell of *Melosira varians* C. A. Agardh. *J. Phycol.,* v. 9, p. 50−61, figs. 1−28, 1973a.

Crawford, R. M., The structure and formation of the siliceous wall of the diatom *Melosira nummuloides* (Dillw.) Ag. *Beih. Nova Hedwigia,* v. 45, p. 131−141, pls. 1−4, 1 fig., 1973b.

Crawford, R. M., The organic component of the cell wall of the marine diatom *Melosira nummuloides* (Dillw.) C. Ag. *Br. Phycol. J.,* v. 8, p. 257−266, figs. 1−15, 1973c.

Crawford, R. M., The taxonomy and classification of the diatom genus *Melosira* C. Ag. II. *M. moniliformis* (Müll.) C. Ag. *Phycologia,* v. 16, p. 277−285, figs. 1−23, 1977.

Cupp, E. E., Marine plankton diatoms of the West Coast of North America. *Bull. Scripps Instn. Oceanogr. Tech. Ser.,* v. 5, no. 1, p. 1−238, pls. 1−5, text-figs. 1−168, 1943.

Cupp, E. E., and W. E. Allen, Plankton diatoms of the Gulf of California obtained by Allan Hancock Pacific Expedition of 1937. *Allan Hancock Pacif. Exped.,* v. 3(5), p. 61−99, pls. 4−15, 1938.

Darley, W. M., Deoxyribonucleic acid content of the three cell types of *Phaeodactylum tricornutum* Bohlin. *J. Phycol.,* v. 4, p. 219−220, 1968.

Darley, W. M., Silicon and the division cycle of the diatoms *Navicula pelliculosa* and *Cylindrotheca fusiformis. Proc. N. Am. Paleont. Conv., Chicago 1969,* G, p. 994−1009, 1970.

Darley, W. M., Biochemical composition, p. 198−223. *In* Dietrich Werner, *The biology of diatoms.* Bot. Monogr., v. 13. Oxford, London, Edinburgh, Melbourne: Blackwell Scientific Publs., 1977.

Davis, E. F., The radiolarian cherts of the Franciscan Group. *Univ. Calif. Publs. Geol. Sci.,* v. 11, p. 235−432, 1918.

Dawson, E. Yale, *Marine botany, an introduction.* New York: Holt, Reinhart and Winston, Inc., xii + 371 p., 1966.

Dawson, P. A., Observations on the structure of some forms of *Gomphonema parvulum* Kützing II. The internal organization. *J. Phycol.,* v. 9, p. 165−175, figs. 1−30, 1973.

Deflandre, Georges, Sur les nannofossiles calcaires et leur systématique. *Revue Micropaléont.,* v. 2, p. 127−152, pls. 1−4, 1959.

Deflandre, Georges, *La vie créatrice de roches,* 6th ed. Paris: Presses Universitaires de France, 128 p., 25 figs. 1961.

Desikachary, T. V., Electron microscope study of diatom wall structure. III. *Isthmia nervosa* Kütz. *Am. J. Bot.,* v. 41, p. 616−619, figs. 1−11, 1954.

Desikachary, T. V., Electron microscope studies on diatoms. *Jl. R. Microsc. Soc.,* ser. 3, v. 76 (1956), p. 9−36, pls. 1−4, 1957.

Desikachary, T. V., and Kanwar Bahadur, Electron microscope study of diatom wall structure V. *Trans. Am. Microsc. Soc.,* v. 73, p. 274−277, 1 pl., 1954.

Desikachary, T. V., and N. E. Dweltz, The chemical composition of the diatom frustule. *Proc. Indian Acad. Sci.,* v. 53B, p. 157−165, pls. 9−11, 1961.

Dingman, Robert, and K. E. Lohman, Late Pleistocene diatoms from the Arica area, Chile. *Prof. Pap. U.S. Geol. Surv.,* 475-C, p. C69−C72, fig. 78. 1, 1963.

Donahue, J. G., Diatoms as indicators of Pleistocene climatic fluctuations in the Pacific sector of the Southern Ocean. *Progr. Oceanogr.,* v. 4, p. 133−140, 1 pl., 3 figs., 1967.

Donahue, J. G., Pleistocene diatoms as climate indicators in North Pacific sediments. *Mem. Geol. Soc. Am.,* 126, p. 121−138, pls. 1−3, figs. 1−6, 1970.

Drebes, Gerhard, *Bacteriastrum solitarium* Mangin, a stage in the life history of the centric diatom *Bacteriastrum hyalinum. Mar. Biol.,* v. 1, p. 40−42, figs. 1−4, 1967.

Drebes, Gerhard, The life history of the centric diatom *Bacteriastrum hyalinum* Lauder. *Beih. Nova Hedwigia,* v. 39, p. 95−110, figs. 1−7, 1972.

Drebes, Gerhard, Sexuality, p. 250−283, 10 figs. *In* Dietrich Werner, *The biology of diatoms.* Bot.

Monogr., v. 13. Oxford, London, Edinburgh, Melbourne: Blackwell Scientific Publs., 1977.

Drew, G., and W. Nultsch, Spezielle Bewegungsmechanismen von Einzellern (Bakterien, Algen). *Encyl. Plant Physiol.*, v. 17, p. 876—919, figs. 1—24, 1962.

Droop, M. R., and K. G. R. Elson, Are pelagic diatoms free from bacteria? *Nature*, v. 211, p. 1096—1097, 1 fig., 1966.

Drum, R. W., Light and electron microscope observations on the tube-dwelling diatom *Amphipleura rutilans* (Trentepohl) Cleve. *J. Phycol.*, v. 5, p. 21—26, figs. 1—4, 1969.

Drum, R. W., and J. T. Hopkins, Diatom locomotion: an explanation. *Protoplasma*, v. 62, p. 1—33, figs. 1—14, 1966.

Drum, R. W., and H. S. Pankratz, Pyrenoids, raphes, and other fine structure in diatoms. *Am. J. Bot.*, v. 51, p. 405—418, figs. 1—23, 1964.

Drum, R. W., and H. S. Pankratz, Locomotion and raphe structure of the diatom *Bacillaria*. *Nova Hedwigia*, v. 10, p. 315—317, pls. 97—100, 1965.

Drum, R. W., H. S. Pankratz, and E. F. Stoermer, Electron microscopy of diatom cells, 25 p., pls. 514—613. *In* J. G. Helmcke and W. Krieger, *Diatomeenschalen in Elektronenmikroskopischen Bild,* Teil VI. Lehre: J. Cramer, 1966.

Duff, D. C. B., D. E. Bruce, and N. J. Antia, The antibacterial activity of marine planktonic algae. *Can. J. Microbiol.,* v. 12, p. 877—884, 1 fig., 1966.

Duke, E. L., and B. E. F. Reimann, The ultrastructure of the diatom cell, p. 65—109, 2 figs. *In* Dietrich Werner, *The biology of diatoms.* Bot. Monogr., v. 13. Oxford, London, Edinburgh, Melbourne: Blackwell Scientific Publs., 1977.

Ehrenberg, C. G., *Die Infusionsthierchen als volkommene Organismen.* Leipzig: Leopold Voss, xviii + 547 p., 64 pls., 1838.

Ehrenberg, C. G., Über noch jetzt zahlreich lebende Thierarten der Kreidebildung und den Organismus der Polythalamien. *Abh. Preuss. Akad. Wiss. Berlin,* 1839, p. 81—174, pls. 1—4, 1840.

Ehrenberg, C. G., Verbreitung und Einfluss des mikroskopischen Lebens in Sud- und Nord-Amerika. *Abh. Preuss. Akad. Wiss. Berlin,* 1841, p. 291—446, pls. 1—4, 1843.

Ehrenberg, C. G., Mittheilung über zwei neue Lager von Gebirgsmassen aus Infusorien als Meeres-Absatz in Nord-Amerika und eine Vergleichung derselben mit den organischen Kreide-Gebilden in Europa und Afrika. *Ber. Preuss. Akad. Wiss. Berlin,* p. 57—97, 1844a.

Ehrenberg, C. G., Resultate seiner Untersuchungen der ihm von der Südpolreise des Capitaine Ross, so wie von den Herren Schayer und Darwin zugekommenen Materialien über das Verhalten des Kleinsten Lebens in den Oceanen und den grössten bisher zugänglichen Tiefen des Weltmeers vor. *Ber. Preuss. Akad. Wiss. Berlin,* p. 182—207, 1844b.

Ehrenberg, C. G., Untersuchungen über die kleinsten Lebensformen im Quellenlande des Euphrats und Araxes, so wie über eine an neuen Formen sehr reiche marine Tripelbildung von den Bermuda-Inseln vor. *Ber. Preuss. Akad. Wiss. Berlin,* p. 253—275, 1 pl., 1844c.

Ehrenberg, C. G., Neue Untersuchungen über das kleinste Leben als geologisches Moment. *Ber. Preuss. Akad. Wiss. Berlin,* p. 53—87, 1845.

Ehrenberg, C. G., *Mikrogeologie. Das Erden und Felsen schaffende Wirken des unsichtbar kleinen selbständigen Lebens auf der Erde.* Leipzig: Leopold Voss, 374 + 31 p. [index to names] + 88 p. [index to atlas], 40 pls., 1854.

Ehrenberg, C. G., Mikrogeologische Studien über das kleinste Leben der Meeres-Tiefgründe aller Zonen und dessen geologischen Einfluss. *Abh. Preuss. Akad. Wiss. Berlin,* 1872, p. 131—397, pls. 1—12, tables 1—3, 1873.

Ehrlich, Aline, Révision du gisement à diatomées Oligocène du Puy-de-Mur (Puy-de-Dôme). *Bull. Soc. Géol. Fr.,* ser. 7, v. 10 (1968), p. 510—515, pl. 29a, 1969.

Eppley, R. W., R. W. Holmes, and Eystein Paasche, Periodicity in cell division and physiological behavior in *Ditylum brightwellii,* a marine planktonic diatom, during growth in light-dark cycles. *Arch. Mikrobiol.,* v. 56, p. 305—323, 7 text-figs., 1967.

Ernst, W. G., and S. E. Calvert, An experimental study of the recrystallization of porcelanite and its bearing on the origin of some bedded cherts. *Am. J. Sci.,* v. 267-A, p. 114—133, pl. 1, figs. 1—8, 1969.

Fairchild, Eugene, and R. P. Sheridan, A physiological investigation of the hot spring diatom, *Achnanthes exigua* Grün. *J. Phycol.,* v. 10, p. 1—4, figs. 1—3, 1974.

Fenner, Juliane, H.-J. Schrader, and Harry Wienigk, Diatom phytoplankton studies in the southern Pacific Ocean, composition and correlation to the Antarctic Convergence and its paleoecological significance. *Init. Repts. Deep Sea Drill. Proj.,* v. 35, p. 757—813, pls. 1—14, figs. 1—15, 1976.

Fogg, G. E., *Algal cultures and phytoplankton ecology.* Madison and Milwaukee: Univ. Wisconsin Press, xiii + 126 p., 31 figs., 1965.

Forti, A., and P. Schulz, Erste Mitteilung über Diatomeen aus dem hannoverschen Gault. *Beih. Bot. Zbl.,* v. 50, abt. 2, p. 241—246, pl. 3, figs. 1—6, 1932.

Frenguelli, Joaquín, Diatomeas del océano Atlántico frente a Mar del Plata (República Argentina). *An. Mus. Nac. Hist. Nat. B. Aires,* v. 34, p. 497—572, pls. 1—21, 1928.

Frenguelli, Gioacchino, Diatomee fossili delle Conche Saline del Deserto Cileno-Boliviano. *Boll. Soc. Geol. Ital.,* v. 47, p. 185—236, pls. 10—14, 1929.

Frenguelli, Gioacchino, A proposito della diatomee del Paleozoico. *Boll. Soc. Geol. Ital.*, v. 51, p. 101–114, 1932.

Frenguelli, Gioacchino, Tecamebiani e diatomee nel Miocene del Neuquén (Patagonia settentrionale). *Boll. Soc. Geol. Ital.*, v. 52, p. 33–43, pl. 5, 2 text-figs., 1933.

Frenguelli, Joaquín, Diatomeas y silicoflagelados recogidas en Tierra Adélia durante las Expediciones Polares Francesas de Paul-Emile Victor (1950–1952). *Revue Algol.*, n. sér., v. 5, p. 3–48, 1960.

Frenguelli, Joaquín, and H. A. Orlando, Diatomeas y silicoflagélados del sector Antártico Sudamericano. *Inst. Antártico Argentino*, Publ. 5, p. 1–191, pls. 1–17, figs. 1–19, 1958.

Fritsch, F. E., *The structure and reproduction of the algae*, v. 1. Cambridge: University Press, xvii + 791 p., 245 figs., 1935.

Fryxell, G. A., and G. R. Hasle, Coscinodiscineae: some consistent patterns in diatom morphology. *Beih. Nova Hedwigia*, v. 45, p. 69–84, pls. 1–12, 1973.

Gaarder, K. R., and G. R. Hasle, On the assumed symbiosis between diatoms and coccolithophorids in *Brenneckella. Nytt Mag. Bot.*, v. 9, p. 145–149, pls. 1, 2, 1962.

Gardner, J. V., and L. H. Burckle, Upper Pleistocene *Ethmodiscus rex* oozes from the eastern equatorial Atlantic. *Micropaleontology*, v. 21, p. 236–242, figs. 1, 2, 1975.

Garstang, Walter, On the size-changes of diatoms and their oceanographic significance. *J. Mar. Biol. Ass. U.K.*, v. 22, p. 83–96, figs. 1–9, 1937.

Gasse, Françoise, Ultrastructure et organisation coloniale de la diatomée *Fragilaria construens* (Ehr.) Grun. révélée par le microscope électronique à balayage. *C. R. Hebd. Séanc. Acad. Sci., Paris*, v. 271, sér. D, p. 1975–1977, 1 pl., 2 figs., 1970.

Gasse, Françoise, Organisation ultrastructurale des colonies de quelques diatomées pennées araphidées, révélée par le microscope électronique à balayage. *C. R. Hebd. Séanc. Acad. Sci., Paris*, v. 272, sér. D, p. 3169–3172, pls. 1–3, 1971.

Geissler, U., and J. Gerloff, Elektronenmikroskopische Beiträge zur Phylogenie der Diatomeenrhaphe. *Nova Hedwigia*, v. 6, p. 339–352, pls. 97–103, 1963.

Geissler, U., J. Gerloff, J. G. Helmcke, W. Krieger, and B. Reimann, *In* J. G. Helmcke and W. Krieger, *Diatomeenschalen im Elektronenmikroskopischen Bild*, Teil III, 44 p., pls. 201–300, text-figs. 1–14, 1961; Teil IV, 51 p., pls. 301–413, 1963.

Geitler, Lothar, Der Formwechsel der pennaten Diatomeen (Kieselalgen). *Arch. Protistenk.*, v. 78, p. 1–226, 125 figs., 1932.

Geitler, Lothar, Reproduction and life history in diatoms. *Bot. Rev.*, v. 1, p. 149–161, 1935.

Geitler, Lothar, Die sexuelle Fortpflanzung der pennaten Diatomeen. *Biol. Rev.*, v. 32, p. 261–295, 1957.

Gemeinhardt, Konrad, Beiträge zur Kenntnis der Diatomeen, I, II, III. *Ber. Dt. Bot. Ges.*, v. 44, p. 517–532, pls. 12, 13, 1926.

Gemeinhardt, Konrad, Beiträge zur Kenntnis der Diatomeen, IV. *Ber. Dt. Bot. Ges.*, v. 45, p. 570–578, pl. 15, 1927.

Gemeinhardt, Konrad, Beiträge zur Kenntnis der Diatomeen, V. *Ber. Dt. Bot. Ges.*, v. 46, p. 285–290, pl. 7, 1928.

Gemeinhardt, Konrad, Diatomeen von der Westküste Norwegens. *Ber. Dt. Bot. Ges.*, v. 53, p. 42–142, pls. 3–16, text-figs. 1, 2, 1935.

Gerloff, Johannes, Elektronenmikroskopische Untersuchungen an Diatomeenschalen II-V. *Ber. Dt. Bot. Ges.*, v. 72, p. 75–83, pl. 6, 1959.

Geroch, Stanisław, Lower Cretaceous diatoms in the Polish Carpathians. *Roczn. Pol. Tow. Geol.*, v. 48, p. 283–295, pls. 1–5, 2 figs., 1978.

Gessner, Fritz, and Reimer Simonsen, Marine diatoms in the Amazon? *Limnol. Oceanogr.*, v. 12, p. 709–711, 1 fig., 1967.

Glezer, Z. I., O nakhodke pozdneeotsenovykh diatomovykh, silikoflagellat i ebriidey na zapadnom sklone Mugodzhar [De bacillariophytis, silicoflagellatis et ebriideis in Aeocaeno superiore declivium occidentalium montium Mugodzharensium inventis]. *Akad. Nauk SSSR, Novosti Sistematiki Nizhikh Rasteniy, Otdel. Ottisk*, p. 21–25, pl. 1, 1967.

Glezer, Z. I., Paleotsenovye diatomey [Paleocene diatoms], p. 109–115, pls. 13–17. *In* Z. I. Glezer et al., *Diatomovye vodorosli SSSR (iskopaemye i sovremennye)*, v. 1. Leningrad: Akad. Nauk SSSR, Bot. Inst. Komarova and "Nauka," 1974.

Glezer, Z. I., and V. S. Sheshukova-Poretskaya, O barbadosskikh i novozelandskikh tretichnykh vidakh v pozdneeotsenovoy diatomovoy flore Ukrainy [De speciebus tertiariis Barbadossicis ac Novozeelandicis in Aeocaeno Superiore Ucrainae]. *Nov. Sist. Nizsh. Rast., Akad Nauk SSSR, Bot. Inst. Komarova*, 1967, p. 25–38, pls. 1–3, 1967.

Golovenkina, N. I., Interesnye i redkie diatomovye vodorosli iz neogenovykh kontinental'nykh otlozheniy Sisianskogo rayona Armyanskoy SSR [Bacillariophyta aquae dulcis curiosa et rara e Neogeno Armeniae]. *Nov. Sist. Nizsh. Rast., Akad. Nauk SSSR, Bot. Inst. Komarova*, 1967, p. 38–46, pls. 1, 2, text-fig., 1967.

Gombos, A. M., Jr., Fossil diatoms from Leg 7, Deep Sea Drilling Project. *Micropaleontology*, v. 21, p. 306–333, pls. 1–8, 1975.

Gombos, A. M., Jr., Paleogene and Neogene diatoms from the Falkland Plateau and Malvinas outer basin: leg 36 Deep Sea Drilling Project. *Init. Repts. Deep Sea*

Drill. Proj., v. 36, p. 575—687, pls. 1—42, fig. 1—6, 1977.

Gordon, Richard, and R. W. Drum, A capillarity mechanism for diatom gliding locomotion. *Proc. Natn. Acad. Sci. U.S.A.,* v. 67, p. 338—344, 1970.

Gran, H. H., Pelagic plant life, p. 307—386. *In* J. Murray and J. Hjort, *Depths of the ocean.* London: Macmillan, 1912.

Gran, H. H., and E. C. Angst, Plankton diatoms of Puget Sound. *Publs. Puget Sound Mar. Biol. Stn.,* v. 7, p. 417—516, figs. 1—95, 1931.

Grant, B. R., J. Madgwick, and G. Dal Pont, Growth of *Cylindrotheca closterium* var. *californica* (Mereschk.) Reimann & Lewin on nitrate, ammonia and urea. *Aust. J. Mar. Freshwat. Res.,* v. 18, p. 129—135, figs. 1—3, 1967.

Gross, F., The life history of some marine plankton diatoms. *Phil. Trans. R. Soc.,* ser. B, v. 228, p. 1—47, pls. 1—4, 1937.

Gross, F., The development of isolated resting spores into auxospores in *Ditylum brightwelli* (West). *J. Mar. Biol. Ass. U.K.,* v. 24, p. 375—380, 2 text-figs., 1940a.

Gross, F., The osmotic relations of the plankton diatom *Ditylum brightwelli* (West). *J. Mar. Biol. Ass. U.K.,* v. 24, p. 381—415, 22 text-figs., 1940b.

Gross, F., and E. Zeuthen, The buoyancy of plankton diatoms: a problem of cell physiology. *Proc. R. Soc.,* ser. B, v. 135, p. 382—389, 1 fig., 1948.

Grove, E., and G. Sturt, On a fossil marine diatomaceous deposit from Oamaru, Otago, New Zealand. Part I. *J. Quekett Micr. Cl.,* ser. 2, v. 2, p. 321—330, pls. 18, 19, 1886; Part II, v. 3, p. 7—12, pls. 2, 3, 1887; Part III, v. 3, p. 63—78, pls. 5, 6, 1887; Appendix, v. 3, p. 131—148, pls. 10—14, 1887.

Grunow, A., Some critical remarks by Herr. A. Grunow on the Oamaru diatom papers of Messrs. Grove and Sturt, translated by G. C. Karop with annotations by E. Grove. *J. Quekett Micr. Cl.,* ser. 2, v. 3, p. 387—391, 1888.

Guillard, R. R. L., and Peter Kilham, The ecology of marine planktonic diatoms, p. 372—469, figs. 1—3. *In* Dietrich Werner, *The biology of diatoms.* Bot. Monogr., v. 13. Oxford, London, Edinburgh, Melbourne: Blackwell Scientific Publs., 1977.

Guillard, R. R. L., and J. H. Ryther, Studies of marine planktonic diatoms. I. *Cyclotella nana* Hustedt, and *Detonula confervacea* (Cleve) Gran. *Can. J. Microbiol.,* v. 8, p. 229—239, pls. 1, 2, figs. 1—4, 1962.

Guinard, M., On the disintegration of diatomaceous deposits. *J. Quekett Micr. Cl.,* ser. 2, v. 3, p. 188—189, 1887.

Hajós, Márta, Mátraalja Miocén Uledékeinek Diatomái. Die Diatomeen der miozänen Ablagerungen des Mátravorlandes. *Geologica Hung., Ser. Palaeont.,* fasc. 37, p. 1—402, pls. 1—63, text-figs. 1—27, 1968.

Hajós, Marta, The Mediterranean diatoms. *Init. Repts. Deep Sea Drill. Proj.,* v. 13, pt. 2, p. 944—969, pls. 1—9, 1973.

Hajós, Marta, Late Cretaceous Archaeomonadaceae, Diatomaceae, and Silicoflagellatae from the South Pacific Ocean, Deep Sea Drilling Project, Leg 29, Site 275. *Init. Repts. Deep Sea Drill. Proj.,* v. 29, p. 913—1009, pls. 1—40, figs. 1—21, 1975.

Hajós, Marta, Upper Eocene and Lower Oligocene Diatomaceae, Archaeomonadaceae, and Silicoflagellatae in southwestern Pacific sediments, DSDP Leg 29. *Init. Repts. Deep Sea Drill. Proj.,* v. 35, p. 817—883, pls. 1—25, figs. 1—4, 1976.

Hajós, Marta, and István Pálfalvy, Magyaregregy diatomás üledékeinek életföldtani vizsgálata [Examen biogéologique des dépôtes à diatomées de Magyaregregy]. *Evi Jelent. Magy. K. Földt. Intéz., Az. 1960. Évről,* p. 89—119, pls. 1—3, text-figs. 1—5, 1963.

Hanna, G. D., Cretaceous diatoms from California. *Occ. Pap. Calif. Acad. Sci.,* 8, p. 5—39, pls. 1—5, 1927a.

Hanna, G. D., The lowest known Tertiary diatoms in California. *J. Paleont.,* v. 1, p. 103—127, pls. 17—21, 1927b.

Hanna, G. D., Fossil diatoms dredged from Bering Sea. *Trans. S. Diego Soc. Nat. Hist.,* v. 5, p. 287—296, pl. 34, 1929.

Hanna, G. D., A review of the genus *Rouxia. J. Paleont.,* v. 4, p. 179—188, pl. 14, 1930a.

Hanna, G. D., The growth of *Omphalotheca. J. Paleont.,* v. 4, p. 192, pl. 14, 1930b.

Hanna, G. D., A new genus of Silicoflagellata from the Miocene of Lower California. *J. Paleont.,* v. 4, p. 415—416, pl. 40, figs. 8—18, 1930c.

Hanna, G. D., Diatoms and silicoflagellates of the Kreyenhagen Shale. *Rep. Calif. St. Min. Bur.* [Mining in Calif.], v. 27, p. 187—201, pls. A—E, 1931.

Hanna, G. D., The diatoms of Sharktooth Hill, Kern County, California. *Proc. Calif. Acad. Sci.,* ser. 4, v. 20, p. 161—263, pls. 2—18, 1932a.

Hanna, G. D., Pliocene diatoms of Wallace County, Kansas. *Bull. Univ. Kans.,* v. 33, no. 10 [Contr. Paleont. Kans. 2, Univ. Kans. Sci. Bull., v. 20, no. 21], p. 369—394, pls. 31—34, 1932b.

Hanna, G. D., Additional notes on diatoms from the Cretaceous of California. *J. Paleont.,* v. 8, p. 352—355, pl. 48, 1934.

Hanna, G. D., Index to Atlas der Diatomaceen-Kunde von Adolf Schmidt, continued by Martin Schmidt, Friedrich Fricke, Heinrich Heiden, Otto Müller, Friedrich Hustedt, Plates 1—480 (except 421—432). Hefte 1—120, 1874—1959. *Bibliotheca Phycologica,* v. 10, 208 p., 1969.

Hanna, G. D., Fossil diatoms from the Pribilof Islands, Bering Sea, Alaska. *Proc. Calif. Acad. Sci.,* ser. 4, v. 37, p. 167—234, figs. 1—105, 1970.

Hanna, G. D., and A. L. Brigger, Some fossil diatoms from Barbados. *Occ. Pap. Calif. Acad. Sci.,* 45, 27 p., 5 pls., 1964.

Hanna, G. D. and W. M. Grant, Miocene marine diatoms from Maria Madre Island, Mexico. *Proc. Calif. Acad. Sci.,* ser. 4, v. 15, p. 115–193, pls. 11–21, 1 text-fig., 1926.

Hanna, G. D., and W. M. Grant, Brackish-water Pliocene diatoms from the Etchegoin Formation of central California. *J. Paleont.,* v. 3, p. 87–100, pls. 11–14, 1929.

Hanna, G. D., and W. M. Grant, Diatoms of Pyramid Lake, Nevada. *Trans. Am. Microsc. Soc.,* v. 50, p. 281–297, pls. 25–27, 1931.

Hanna, G. D., N. I. Hendey, and A. L. Brigger, Some Eocene diatoms from South Atlantic cores, Part 1. New and rare species of *Arachnoidiscus. Occ. Pap. Calif. Acad. Sci.,* 123, p. 1–19, pls. 1–3, fig., 1, 1976.

Hargraves, P. E., Studies on marine plankton diatoms. II. Resting spore morphology. *J. Phycol.,* v. 12, p. 118–128, figs. 1–42, 1976.

Harper, M. A., Movements, p. 224–249, 6 figs. *In* Dietrich Werner, *The biology of diatoms.* Bot. Monogr., v. 13. Oxford, London, Edinburgh, Melbourne: Blackwell Scientific Publs., 1977.

Harper, M. A., and J. F. Harper, Measurements of diatom adhesion and their relationship with movement. *Br. Phycol. Bull.,* v. 3, p. 195–207, figs. 1–5, 1967.

Hart, T. J., Notes on practical methods for the study of marine diatoms. *J. Mar. Biol. Ass. U.K.,* v. 36, p. 593–597, 1957.

Hart, T. J., Speciation in marine phytoplankton. *Publs. Syst. Ass.,* 5, p. 145–155, 1963.

Hasle, G. R., *Navicula endophytica* sp. nov., a pennate diatom with an unusual mode of existence. *Br. Phycol. Bull.,* v. 3, p. 475–480, figs. 1–9, 1968a.

Hasle, G. R., The valve processes of the centric diatom genus *Thalassiosira. Nytt Mag. Bot.,* v. 15, p. 193–201, figs. 1–16, 1968b.

Hasle, G. R., Two types of valve processes in centric diatoms. *Beih. Nova Hedwigia,* v. 39, p. 55–78, figs. 1–45, 1972a.

Hasle, G. R., *Fragilariopsis* Hustedt as a section of the genus *Nitzschia* Hassall. *Beih. Nova Hedwigia,* v. 39, p. 111–118, pl. 1, 1972b.

Hasle, G. R., Thalassiosiraceae, a new diatom family. *Norw. J. Bot.,* v. 20, p. 67–69, 1973a.

Hasle, G. R., Morphology and taxonomy of *Skeletonema costatum* (Bacillariophyceae). *Norw. J. Bot.,* v. 20, p. 109–137, figs. 1–112, 1973b.

Hasle, G. R., Some marine plankton genera of the diatom family Thalassiosiraceae. *Beih. Nova Hedwigia,* v. 45, p. 1–49, pls. 1–18, figs. 1–4, 1973c.

Hasle, G. R., The "mucilage pore" of pennate diatoms. *Beih. Nova Hedwigia,* v. 45, p. 167–186, pls. 1–8, 1973d.

Hasle, G. R., and D. L. Evensen, Brackish water and freshwater species of the diatom genus *Skeletonema.* II. *Skeletonema potamos* comb. nov. *J. Phycol.,* v. 12, p. 73–82, figs. 1–31, 1976.

Hasle, G. R., and B. R. Heimdal, Some species of the centric diatom genus *Thalassiosira* studied in the light and electron microscopes. *Beih. Nova Hedwigia,* v. 31, p. 543–581, pls. 1–15, 1970.

Hasle, G. R., and B. R. E. de Mendiola, The fine structure of some *Thalassionema* and *Thalassiothrix* species. *Phycologia,* v. 6, p. 107–125, figs. 1–53b, 1967.

Haworth, E. Y., Diatom succession in a core from Pickerel Lake, northeastern South Dakota. *Bull. Geol. Soc. Am.,* v. 83, p. 157–172, 11 figs., 1972.

Hays, J. D., Quaternary sediments of the Antarctic Ocean. *Prog. Oceanogr.,* v. 4, p. 117–131, figs. 1–8, 1967.

Healey, F. P., J. Coombs, and B. E. Volcani, Changes in pigment content of the diatom *Navicula pelliculosa* (Bréb.) Hilse in silicon-starvation synchrony. *Arch. Mikrobiol.,* v. 59, p. 131–142, figs. 1–7, 1967.

Heath, I. B., and W. M. Darley, Observations on the ultrastructure of the male gametes of *Biddulphia levis* Ehr. *J. Phycol.,* v. 8, p. 51–59, figs. 1–22, 1972.

Heimdal, B. R., The fine structure of the frustules of *Melosira nummuloides* and *M. arctica* (Bacillariophyceae). *Norw. J. Bot.,* v. 20, p. 139–149, figs. 1–30, 1973.

Helmcke, J. G., Feinbau von Diatomeenschalen in Einzeldarstellungen. *Z. Wiss. Mikrosk.,* v. 60, p. 85–100, text-figs. 1–13, 1951.

Helmcke, J. G., Die Feinstruktur der Kieselsäure und ihre physiologische Bedeutung in Diatomeenschalen. *Naturwissenschaften,* v. 11, p. 254–255, 1954.

Helmcke, J. G., and W. Krieger, Feinbau von Diatomeenschalen in Einzeldarstellungen. *Z. Wiss. Mikrosk.,* v. 60, p. 197–202, figs. 1–11, 1951.

Helmcke, J. G., and W. Krieger, Neue Erkenntnisse über dem Schalenbau von Diatomeen. *Naturwissenschaften,* v. 39, p. 146–149, figs. 1–17, 1952a.

Helmcke, J. G., and W. Krieger, Elektronenmikroskopische Untersuchungen über den Kammerbau der Diatomeenmembran. *Ber. Dt. Bot. Ges.,* v. 64, Schlussheft, p. (27–29), pl. 1. 1952b.

Helmcke, J. G., and W. Krieger, Feinbau der Kieselschalen der Diatomee *Cyclotella comta* (Ehrbg) Kütz. *Ber. Dt. Bot. Ges.,* v. 65, p. 70–72, pl. 2, 1 fig., 1952c.

Helmcke, J. G., and W. Krieger, *Diatomeenschalen im Elektronenmikroskopischen Bild.* Teil I, 2. unveränderte Auflage. Weinheim: J. Cramer, 20 p., 102 pls., 1962; Teil II, 2. unveränderte Auflage, 24 p., pls. 103−200, 1962.

Hendey, N. I., A preliminary note on the distribution of marine diatoms during the Tertiary period. *J. Bot., Lond.,* v. 71, p. 111−118, 1933.

Hendey, N. I., The plankton diatoms of the Southern Seas. *'Discovery' Rep.,* v. 16, p. 151−364, pls. 6−13, 1937.

Hendey, N. I., Littoral diatoms of Chichester Harbour with special reference to fouling. *Jl. R. Microsc. Soc.,* ser. 3, v. 71, p. 1−81, pls. 1−18, 1951.

Hendey, N. I., Note on the Plymouth *'Nitzschia'* culture. *J. Mar. Biol. Ass. U.K.,* v. 33, p. 335−339, pl. 1, 1954.

Hendey, N. I., Marine diatoms from some West African ports. *Jl. R. Microsc. Soc.,* ser. 3, v. 77 (1957), p. 28−85, pls. 1−6, 1958.

Hendey, N. I., The structure of the diatom cell wall as revealed by the electron microscope. *J. Quekett Micr. Cl.,* ser. 4, v. 5 (v. 28), p. 147−175, pls. 1, 2, 1959.

Hendey, N. I., *An introductory account of the smaller algae of British coastal waters.* Part V. Bacillariophyceae (Diatoms). London: Fishery Invest., ser. 4, xxii + 317 p., 45 pls., 9 text-figs., 1964.

Hendey, N. I., Electron microscope studies and the classification of diatoms, p. 625−631, fig. 1. *In* B. M. Funnell and W. R. Riedel, *The micropalaeontology of oceans.* Cambridge: University Press, 1971.

Hendey, N. I., A revised check-list of British marine diatoms. *J. Mar. Biol. Ass. U.K.,* v. 54, p. 277−300, 1974.

Hendey, N. I., D. H. Cushing, and G. W. Ripley, Electron microscope studies of diatoms. *Jl. R. Microsc. Soc.,* v. 74, p. 22−34, 1954.

Hendey, N. I., H. A. von Stosch, and K. Kowallik, The fine structure of the frustules of *Cylindropyxis profunda* Hendey (1964), Coscinodiscaceae, a very small centric marine diatom. *Nova Hedwigia,* v. 18, p. 175−180, pl. 1, 1969.

Herdman, E. C., Notes on dinoflagellates and other organisms causing discolouration of the sand at Port Erin III. *Proc. Trans. Lpool. Biol. Soc.,* v. 38, p. 58−63, 1924.

Heurck, Henri van, *A treatise on the Diatomaceae containing introductory remarks on the structure, life history, collection, cultivation and preparation of diatoms, and a description and figure typical of every known genus, as well as a description and figure of every species found in the North Sea and countries bordering it, including Great Britain, Belgium, & c.* Translated by Wynne E. Baxter. London: William Wesley & Son, 558 p., 35 pls., 291 text-figs., 1896.

Holmes, R. W., Light microscope observations on cytological manifestations of nitrate, phosphate and silicate deficiency in four marine centric diatoms. *J. Phycol.,* v. 2, p. 136−140, figs. 1−10, 1966.

Holmes, R. W., Auxospore formation in two marine clones of the diatom genus *Coscinodiscus. Am. J. Bot.,* v. 54, p. 163−168, figs. 1−6, 1967.

Holmes, R. W., and A. L. Brigger, The morphology and occurrence of *Entogonia* from an Eocene locality in the Indian Ocean. *Beih. Nova Hedwigia,* v. 54, p. 215−241, figs. 1−5, 1977.

Holmes, R. W., and B. E. F. Reimann, Variation in valve morphology during the life cycle of the marine diatom *Coscinodiscus concinnus. Phycologia,* v. 5, p. 233−244, figs. 1−15, 1966.

Hopkins, J. T., and R. W. Drum, Diatom motility: an explanation and a problem. *Br. Phycol. Bull.,* v. 3, p. 63−67, figs. 1−3, 1966.

Hufford, T. L., and G. B. Collins, Some morphological variations in the diatom *Cymbella cistula. J. Phycol.,* v. 8, p. 192−195, figs. 1−24, 1972.

Humphrey, G. F., and D. V. Subba Rao, Photosynthetic rate of the marine diatom *Cylindrotheca closterium. Aust. J. Mar. Freshwat. Res.,* v. 18, p. 123−127, figs. 1, 2, 1967.

Hustedt, Friedrich, Untersuchungen über den Bau der Diatomeen. I. Raphe und Gallertporen der Eunotioideae. *Ber. Dt. Bot. Ges.,* v. 44, p. 142−150, pl. 2, 1926a.

Hustedt, Friedrich, Untersuchungen über den Bau der Diatomeen. II-III. *Ber. Dt. Bot. Ges.,* v. 44, p. 394−402, pl. 5, 1926b.

Hustedt, Friedrich, Fossile Bacillariaceen aus dem Loa-Becken in der Atacama-Wüste, Chile. *Arch. Hydrobiol.,* v. 18, p. 224−251, pls. 7−9, 1 text-fig., 1927.

Hustedt, Friedrich, Die Kieselalgen. In *Dr. L. Rabenhorst's Kryptogamen-Flora von Deutschland, Österreich und der Schweiz,* v. 7. Leipzig: Akademischeverlags, 1. Teil, p. 1−920, text-figs. 1−542, 1927−1930; 2. Teil, p. 1−845, text-figs. 543−1179, 1931−1959; 3. Teil, p. 1−816, text-figs. 1180−1788, 1961−1966.

Hustedt, Friedrich, Untersuchungen über den Bau der Diatomeen. IV-VI. *Ber. Dt. Bot. Ges.,* v. 46, p. 148−164, pl. 3, 1928.

Hustedt, Friedrich, Untersuchungen über den Bau der Diatomeen. VII, VIII. *Ber. Dt. Bot. Ges.,* v. 47, p. 101−110, pl. 3, 1929a.

Hustedt, Friedrich, Untersuchungen über den Bau der Diatomeen. IX. *Ber. Dt. Bot. Ges.,* v. 47, Supplement (Generalversammlungs-Heft), p. (59)-(69), text-figs. 1−5, 1929b.

Hustedt, Friedrich, Bacillariophyta (Diatomeae). *In* A. Pascher, *Die Süsswasser-flora Mitteleuropas,* Part 2, no. 10. Jena: G. Fischer, 466 p., 1930.

Hustedt, Friedrich, Untersuchungen über den Bau

der Diatomeen, X und XI. *Ber. Dt. Bot. Ges.*, v. 53, p. 3—41, pls. 1, 2, 1935a.

Hustedt, Friedrich, Untersuchungen über den Bau der Diatomeen. XII. *Ber. Dt. Bot. Ges.*, v. 53, p. 246—264, pl. 20, 1935b.

Hustedt, Friedrich, Systematische und ökologische Untersuchungen über die Diatomeen-Flora von Java, Bali und Sumatra. *Arch. Hydrobiol. Suppl.-Bd.* 15, Tropische Binnengewässer, heft 1, p. 131—177, pls. 9—12; heft 2, p. 187—295, pls. 13—20; heft 3, p. 393—506, pls. 21—28; heft 4, p. 638—790, pls. 36—43, 16 text-figs.; 1937a.

Hustedt, Friedrich, Zur Systematik der Diatomeen. I. *Ber. Dt. Bot. Ges.*, v. 55, p. 185—193, 1937b.

Hustedt, Friedrich, Zur Systematik der Diatomeen. II-III. *Ber. Dt. Bot. Ges.*, v. 55, p. 465—472, 1937c.

Hustedt, Friedrich, Systematische und ökologische Untersuchungen über die Diatomeen-Flora von Java, Bali und Sumatra. *Arch. Hydrobiol., Suppl.-Bd.* 16, Tropische Binnengewasser, heft 1, p. 1—155, 1938a; heft 2, p. 274—394, 1939.

Hustedt, Friedrich, Diatomeen aus den Pyrenäen. *Ber. Dt. Bot. Ges.*, v. 56, p. 543—572, pl. 25, text-figs. 1—21, 1938b.

Hustedt, Friedrich, Neue und wenig bekannte Diatomeen. III. Phylogenetische Variationen bei den rhaphidioiden Diatomeen. *Ber. Dt. Bot. Ges.*, v. 65, p. 133—144, pl. 5, 1952.

Ichikawa, Wataru, Fossil diatoms, p. 29—59, pls. 1—7, text-fig. 3. *In* Ichikawa, Wataru, Norio Fuji, and Alfred Bachmann, Fossil diatoms, pollen grains and spores, silicoflagellates and archaeomonads in the Miocene Hojuji diatomaceous mudstone, Noto Peninsula, Central Japan. *Sci. Rep. Kanazawa Univ.*, v. 9, 1964.

Jarosch, R., Gliding, p. 573—581, figs. 1—4. *In* R. A. Lewin, *Physiology and biochemistry of algae.* New York and London: Academic Press, 1962.

Jerković, Lazar, L'ultrastructure des frustules de quelques espèces endemiques des diatomées de la Yougoslavie. *Arch. Hydrobiol. Suppl.*, 41, Algol. Stud. 6, p. 1—10, pls. 1—6, 1972.

Jørgensen, Erik G., Diatom periodicity and silicon assimilation. Experimental and ecological investigations. *Dansk Bot. Ark.*, v. 18, p. 1—54, figs. 1—16, 1957.

Jørgensen, E. G., Photosynthetic activity during the life cycle of synchronous *Skeletonema* cells. *Physiologia Pl.*, v. 19, p. 789—799, 5 figs., 1966.

Jousé, A. P., O. G. Kozlova, and V. V. Muhina, Distribution of diatoms in the surface layer of sediment from the Pacific Ocean, p. 263—269, 2 figs. *In* B. M. Funnell and W. R. Riedel, *The micropalaeontology of oceans.* Cambridge: University Press, 1971.

Kanaya, Taro, Eocene diatom assemblages from the Kellogg and "Sidney" Shales, Mt. Diablo area, California. *Sci. Rep. Tôhoku Univ.*, ser. 2, Geol., v. 28, p. 27—124, pls. 3—8, 1957.

Kanaya, Taro, Miocene diatom assemblages from the Onnagawa Formation and their distribution in the correlative formations in northeast Japan. *Sci. Rep. Tôhoku Univ.*, ser. 2, Geol., v. 30, p. 1—130, pls. 1—11, text-figs. 1, 2, 1959.

Kanaya, Taro, and Itaru Koizumi, Interpretation of diatom thanatocoenoses from the North Pacific applied to a study of core V20-130 (Studies of a deep-sea core V20-130. Pt. 4). *Sci. Rep. Tôhoku Univ.*, ser. 2, Geol., v. 37, p. 89—130, 6 figs., 1966.

Karsten, G., Phytoplankton des Antarktischen Meeres nach dem Material der deutschen Tiefsee-Expedition 1898—1899. *Wiss. Ergebn. Dt. Tiefsee-Exped. 'Valdivia',* v. 2, pt. 2, p. 1—136, pls. 1—9, 1905a.

Karsten, G., Phytoplankton des Atlantischen Oceans nach dem Material der deutschen Tiefsee-Expedition 1898—1899. *Wiss. Ergebn. Dt. Tiefsee-Exped. 'Valdivia',* v. 2, pt. 2, p. 139—219, pls. 20—34, 1905b.

Karsten, G., Das Indische Phytoplankton nach dem Material der deutschen Tiefsee-Expedition 1898—1899. *Wiss. Ergebn. Dt. Tiefsee-Exped. 'Valdivia',* v. 2, pt. 2, p. 223—548, pls. 35—54, 1907.

King, Kenneth, Jr., Amino acid survey of Recent calcareous and siliceous deep-sea microfossils. *Micropaleontology*, v. 23, p. 180—193, figs. 1—5, 1977.

Köhler, A., Über die Feinstruktur von *Navicula (Pinnularia) nobilis* Ehbg. *Z. Bot.*, v. 22, p. 442—454, pl. 2, 1 fig., 1929.

Koizumi, Itaru, Tertiary diatom flora of Oga Peninsula, Akita Prefecture, northeast Japan. *Sci. Rep. Tôhoku Univ.*, ser. 2, Geol., v. 40, p. 171—240, pls. 32—35, 15 figs., 1968.

Koizumi, Itaru, The Late Cenozoic diatoms of Sites 183—193, Leg 19 Deep Sea Drilling Project. *Init. Repts. Deep Sea Drill. Proj.*, v. 19, p. 805—855, pls. 1—8, figs. 1—13, 1973.

Koizumi, Itaru, Neogene diatoms from the western margin of the Pacific Ocean, Leg 31, Deep Sea Drilling Project. *Init. Repts. Deep Sea Drill. Proj.*, v. 31, p. 779—819, pls. 1—5, figs. 1—9, 1975a.

Koizumi, Itaru, Neogene diatoms from the northwestern Pacific Ocean Deep Sea Drilling Project. *Init. Repts. Deep Sea Drill. Proj.*, v. 32, p. 865—889, pls. 1—4, 1975b.

Kolbe, R. W., Zur Ökologie, Morphologie und Systematik der Brackwasser-Diatomeen. Die Kieselalgen des Sperenberger Salzgebiets. *Pflanzenforschung*, v. 7, v + 146 p., 3 pls., 10 text-figs., 1937.

Kolbe, R. W., Elektronenmikroskopische Untersuchungen von Diatomeenmembranen. *Ark. Bot.*, v. 33A, no. 17, p. 1—21, pls. 1—10, text-figs. 1—5, 1948.

Kolbe, R. W., Diatoms from equatorial Atlantic cores. *Rep. Swed. Deep Sea Exped.*, v. 7, *Sediment Cores from the North Atlantic Ocean*, No. 3, p. 149—184, pls. 1, 2, text-figs. 1—6, 1955.

Kolbe, R. W., Freshwater diatoms from Atlantic deep-sea sediments. *Science,* v. 126, p. 1053 – 1056, figs. 1, 2, 1957.

Kolbe, R. W., and E. Tiegs, Zur mesohaloben Diatomeen-Flora des Werragebietes. *Ber. Dt. Bot. Ges.,* v. 47, p. 408 – 420, pl. 7, text-figs. 1, 2, 1929.

Komura, Seiichi, *Ikebea,* eine neue Gattung der pennaten Bacillareacean aus dem Neogen Japans. *Trans. Proc. Palaeont. Soc. Japan,* n.s., no. 99, p. 133 – 142, pl. 12, 1975.

Komura, Seiichi, *Sawamuraia, Katahiraia* and *Yoshidaia,* drei neue Diatomgattungen aus dem Neogen Japans. *Trans. Proc. Palaeont. Soc. Japan,* n.s., no. 103, p. 379 – 397, pls. 40, 41, figs. 1 – 14, 1976.

Koppen, J. D., A morphological and taxonomic consideration of *Tabellaria* (Bacillariophyceae) from the north central United States. *J. Phycol.,* v. 11, p. 236 – 244, figs. 1 – 45, 1975.

Korshikov, A. A., On the origin of the diatoms. *Beih. Bot. Zbl.,* 1 Abt., v. 46, p. 460 – 469, 1 fig., 1930.

Kozlova, O. G., *Diatomovye vodorosli Indiyskogo i Tikhookeanskogo Sektorov Antarktiki [Diatom algae of the Indian and Pacific Ocean sectors of the Antarctic].* Akad. Nauk SSSR, Inst. Okeanol. Moscow: "Nauka," 168 p., 6 pls., 72 figs., 1964.

Kozlova, O. G., The main features of diatom and silicoflagellate distribution in the Indian Ocean, p. 271 – 275, 2 figs. *In* B. M. Funnell and W. R. Riedel, *The micropalaeontology of oceans.* Cambridge: University Press. 1971.

Kozlova, O. G., and V. V. Mukhina, Diatoms and silicoflagellates in suspension and floor sediments of the Pacific Ocean. *Int. Geol. Rev.,* v. 9, p. 1322 – 1342, 13 figs., 1967.

Krauskopf, K. B., The geochemistry of silica in sedimentary environments. *Spec. Publs. Soc. Econ. Paleont. Miner., Tulsa,* 7, p. 4 – 19, 1959.

Kützing, Friedrich Traugott, *Die kieselschaligen Bacillarien oder Diatomeen.* Nordhausen, p. 1 – 152, pls. 1 – 30, 1844.

Labeyrie, Laurent, New approach to surface seawater palaeotemperatures using $^{18}O/^{16}O$ ratios in silica in diatom frustules. *Nature,* v. 248, p. 40 – 42, figs. 1, 2, 1974.

Lauritis, J. A., J. Coombs, and B. E. Volcani, Studies on the biochemistry and fine structure of silica shell formation in diatoms IV. Fine structure of the apochlorotic diatom *Nitzschia alba* Lewin and Lewin. *Arch. Mikrobiol.,* v. 62, p. 1 – 16, figs. 1 – 9, 1968.

Lauritis, J. A., B. B. Hemmingsen, and B. E. Volcani, Propagation of *Hantzschia* sp. Grunow daughter cells by *Nitzschia alba* Lewin and Lewin. *J. Phycol.,* v. 3, p. 236 – 237, figs. 1 – 3, 1967.

LeBlanc, R., Sur la reproduction du *Chaetoceros pseudocurvisetum* Mangin. *C. R. Hebd. Séanc. Acad. Sci., Paris,* v. 198, p. 601 – 603, figs. 1 – 4, 1934.

Lebour, M. V., *The planktonic diatoms of Northern Seas.* Ray Soc. Publs., v. 116, 244 p., 4 pls., 181 figs., 1930.

Lewin, J. C., Heterotrophy in diatoms. *J. Gen. Microbiol.,* v. 9, p. 305 – 313, pl. 1, 1953.

Lewin, J. C., Silicon metabolism in diatoms. IV. Growth and frustule formation in *Navicula pelliculosa. Can. J. Microbiol.,* v. 3, p. 427 – 433, figs. 1 – 7, 1957.

Lewin, J. C., The taxonomic position of *Phaeodactylum tricornutum. Can. J. Microbiol.,* v. 18, p. 427 – 432, 1 pl., 1958.

Lewin, J. C., The dissolution of silica from diatom walls. *Geochim. Cosmochim. Acta,* v. 21, p. 182 – 198, 1961.

Lewin, J. C., Silicification, p. 445 – 455, 1 pl. *In* R. A. Lewin, *Physiology and biochemistry of algae.* New York, London: Academic Press, 1962.

Lewin, J. C., The thiamine requirement of a marine diatom. *Phycologia,* v. 4, p. 141 – 144, fig. 1, 1965.

Lewin, J. C., Boron as a growth requirement for diatoms. *J. Phycol.,* v. 2, p. 160 – 163, 1966a.

Lewin, J. C., Silicon metabolism in diatoms. V. Germanium dioxide, a specific inhibitor of diatom growth. *Phycologia,* v. 6, p. 1 – 12, 4 figs., 1966b.

Lewin, J. C., and W. F. Busby, The sulfate requirements of some unicellular marine algae. *Phycologia,* v. 6, p. 211 – 217, 1 fig., 1967.

Lewin, J. C., and Ching-hong Chen, Silicon metabolism in diatoms. VI. Silicic acid uptake by a colorless marine diatom, *Nitzschia alba* Lewin and Lewin. *J. Phycol.,* v. 4, p. 161 – 166, figs. 1 – 10, 1968.

Lewin, J. C., and R. R. L. Guillard, Diatoms. *A. Rev. Microbiol.,* v. 17, p. 373 – 414, 1963.

Lewin, J. C., and R. A. Lewin, Auxotrophy and heterotrophy in marine littoral diatoms. *Can. J. Microbiol.,* v. 6, p. 127 – 134, pls. 1 – 3, 1960.

Lewin, J. C., and R. A. Lewin, Culture and nutrition of some apochlorotic diatoms of the genus *Nitzchia. J. Gen. Microbiol.,* v. 46, p. 361 – 367, 1 pl., 1967.

Lewin, J. C., R. A. Lewin, and D. E. Philpott, Observations on *Phaeodactylum tricornutum. J. Gen. Microbiol.,* v. 18, p. 418 – 426, pls. 1, 2, 1958.

Lewin, J. C., B. E. Reimann, W. F. Busby, and B. E. Volvani, Silica shell formation in synchronously dividing diatoms, p. 169 – 188, figs. 1 – 14. *In* I. L. Cameron and G. M. Padilla, *Cell synchrony; studies in biosynthetic regulation.* New York: Academic Press, 1966.

Liebisch, Werner, Experimentelle und kritische Untersuchungen über die Pektinmembran der Diatomeen unter besonderer Berücksichtigung der Auxosporenbildung und der Kratikularzustände. *Z. Bot.,* v. 22, p. 1 – 65, pl. 1, figs. 1 – 14, 1929.

Lisitsyn, A. P., Basic relationships in distribution of modern siliceous sediments and their connection with climatic zonation. *Int. Geol. Rev.,* v. 9, p. 631—652, 13 figs., 1967a.

Lisitsyn, A. P., Basic relationships in distribution of modern siliceous sediments and their connection with climatic zonation. *Int. Geol. Rev.,* v. 9, p. 842—865, 7 figs., 1967b.

Lisitsyn, A. P., Basic relationships in distribution of modern siliceous sediments and their connection with climatic zonation. *Int. Geol. Rev.,* v. 9, p. 1114—1130, text-figs. 38—45, 1967c.

Lisitsyn, A. P., and Yu. A. Bogdanov, Vsveshennyy amorfnyy kremnezem v vodakh Tikhogo Okeana [Suspended amorphous silica in the waters of the Pacific Ocean], p. 5—41, figs. 1—13. *In* A. P. Lisitsyn, *Rezul'taty issledovaniy po Mezhdunarodnym Geofizicheskim proektam, Okeanologischeskie Issledovaniya 18.* Moscow: Akad. Nauk SSSR, Mezhduvedomstvennyy Geofiz Komitet, "Nauka." 1968.

Loeblich, A. R. III, and L. A. Loeblich, Division Bacillariophyta, p. 425—450, 1 fig. *In* A. I. Laskin and H. A. Lechevalier, *Handbook of Microbiology,* 2nd ed., v. 2, *Fungi, algae, protozoa, and viruses.* West Palm Beach, Florida: CRC Press, Inc. (1978), 1979.

Loeblich, A. R. III, W. W. Wight, and W. M. Darley, A unique colonial marine centric diatom, *Coenobiodiscus muriformis* gen. et sp. nov. *J. Phycol.,* v. 4, p. 23—29, figs. 1—17, 1968.

Lohman, K. E., Diatoms from Quaternary lake beds near Clovis, New Mexico. *J. Paleont.,* v. 9, p. 455—459, 1935.

Lohman, K. E., Pliocene diatoms from the Kettleman Hills, California. *Prof. Pap. U.S. Geol. Surv.,* 189-C, p. 81—102, pls. 20—23, 1938.

Lohman, K. E., Geology and biology of North Atlantic deep-sea cores between Newfoundland and Ireland, Part 3. Diatomaceae. *Prof. Pap. U.S. Geol. Surv.,* 196-B, p. xx + 55—86, 1941.

Lohman, K. E., Middle Miocene diatoms from the Hammond Well. *Bull. Md. Dep. Geol. Mines,* 2, p. 151—187, pls. 5—11, 1948.

Lohman, K. E., The ubiquitous diatom: a brief survey of the present state of knowledge. *Am. J. Sci.,* v. 258-A, p. 180—191, 1960.

Lohman, K. E., Geologic ranges of Cenozoic nonmarine diatoms. *Prof. Pap. U.S. Geol. Surv.,* 424-D, p. D234—D236, fig. 373.1, 1961.

Lohman, K. E., Stratigraphic and paleoecologic significance of the Mesozoic and Cenozoic diatoms of California and Nevada. *Spec. Publs. Soc. Econ. Paleont. Miner., Tulsa,* 11, p. 58—64, 1964.

Lohman, K. E., A procedure for the microscopical study of diatomaceous sediments. *Beih. Nova Hedwigia,* v. 39, p. 267—283, figs. 1—13, 1972.

Lohman, K. E., Lower Middle Miocene marine diatoms from Trinidad. *Verh. Naturf. Ges. Basel,* v. 84, p. 326—360, pls. 1—6, 1974.

Lohman, K. E., and G. W. Andrews, Late Eocene nonmarine diatoms from the Beaver Divide area, Fremont County, Wyoming. *Prof. Pap. U.S. Geol. Surv.,* 593-E, p. E1—E26, pls. 1, 2, figs. 1—7. 1968.

Long, J. A., D. P. Fuge, and J. Smith, Diatoms of the Moreno Shale. *J. Paleont.,* v. 20, p. 89—118, pls. 13—19, figs. 1, 2, 1946.

Lund, J. W. G., The ecology of the freshwater phytoplankton. *Biol. Rev.,* v. 40, p. 231—293, 1965.

Lupikina, E. G., Novye i interesnye diatomovye vodorosli iz Ermanovskikh otlozheniy Zapadnoy Kamchatki. Diatomeae novae et curiosae e stratis Ermanicis partis Kamczatkae occidentalis. *Novosti Sistematiki Nizshikh Rasteniy 1965, Akad. Nauk SSSR, Bot. Inst. Komarova,* p. 15—22, pls. 1—4, figs. 1—5, 1965.

McCollum, D. W., Diatom stratigraphy of the Southern Ocean. *Init. Repts. Deep Sea Drill. Proj.,* v. 28, p. 515—571, pls. 1—16, figs. 1—12, 1975.

McHugh, J. L., Distribution and abundance of the diatom *Ethmodiscus rex* off the west coast of North America. *Deep Sea Res.,* v. 1, p. 216—222, figs. 1—3, 1954.

McIntire, C. D., and W. W. Moore, Marine littoral diatoms: ecological considerations, p. 333—371, 2 figs. *In* Dietrich Werner, *The biology of diatoms.* Bot. Monogr., v. 13. Oxford, London, Edinburgh, Melbourne: Blackwell Scientific Publs., 1977.

Magne-Simon, M. F., Note sur la processus de l'auxosporulation chez une diatomée marine, *Grammatophora marina* (Lyngb.) Kütz. *C. R. Hebd. Séanc. Acad. Sci., Paris,* v. 251, p. 3040—3042, text-figs. 1—8, 1960.

Magne-Simon, M. F., L'auxosporulation chez une Tabellariacée marine, *Grammatophora marina* (Lyngb.) Kütz. (Diatomée). *Cah. Biol. Mar.,* v. 3, p. 79—89, pls. 1—6, 1962.

Makarova, I. V., and T. F. Kozyrenko, *Diatomovye vodorosli iz morskikh Miotsenovykh otlozheniy yuga Evropeyskoy chasti SSSR [Diatom algae from Miocene marine deposits in the south European part of the USSR].* Bot. Inst. Komarova, Akad. Nauk SSSR, 69 p., 11 pls., 7 figs., 1966.

Mandelli, E. F., Enhanced photosynthetic assimilation ratios in Antarctic Polar Front (convergence) diatoms. *Limnol. Oceanogr.,* v. 12, p. 484—491, figs. 1—7, 1967.

Mann, Albert, Report on the diatoms of the Albatross voyages in the Pacific Ocean, 1888—1904: *Contr. U.S. Nat. Herb.,* v. 10(5), p. 221—442, pls. 44—54, 1907.

Mann, Albert, Marine diatoms of the Philippine Islands. *Bull. U.S. Natn. Mus.,* 100, v. 6, pt. 1, p. 1—182, pls. 1—39, 1925.

Mann, D. G., The diatom genus *Hantzschia* Grunow—an appraisal. *Beih. Nova Hedwigia,* v. 54, p. 323 – 354, pls. 1 – 11, 1977.

Manton, Irene, K. Kowallik, and H. A. von Stosch, Observations on the fine structure and development of the spindle at mitosis and meiosis in a marine centric diatom (*Lithodesmium undulatum*). IV. The second meiotic division and conclusion. *J. Cell Sci.,* v. 7, p. 407 – 443, figs. 1 – 34, 1970.

Manton, Irene, and H. A. von Stosch, Observations on the fine structure of the male gemete of the marine centric diatom *Lithodesmium undulatum. Jl. R. Microsc. Soc.,* v. 85, p. 119 – 134, figs. 1 – 13, 1966.

Margalef, Ramón, The food web in the pelagic environment. *Wiss. Meeresunters. Helgoländer,* v. 15, p. 548 – 559, 1967.

Margalef, Ramón, Size of centric diatoms as an ecologic indicator. *Mitt. Int. Verein. Theor. Angew. Limnol.,* v. 17, p. 202 – 210, figs. 1 – 4, 1969.

Maynard, N. G., Relationship between diatoms in surface sediments of the Atlantic Ocean and the biological and physical oceanography of overlying waters. *Paleobiology,* v. 2, p. 99 – 121, figs. 1 – 13, 1976.

Meunier, Alphonse, *Microplankton des Mers de Barents et de Kara.* Brussels: Duc d'Orléans Campagne Arctique de 1907, 2 vols., text 355 p., atlas 36 pls., 1910.

Meunier, Alphonse, Microplankton de la Mer Flamande. 1. Partie, Le genre *"Chaetoceros"* Ehr. *Mém. Mus. R. Hist. Nat. Belg.,* v. 7, fasc. 2, p. 1 – 58, pls. 1 – 7, 1913.

Meunier, Alphonse, Microplankton de la Mer Flamande. 2. Partie, Les Diatomacées (Suite) (Le genre *Chaetoceros* excepté). *Mém. Mus. R. Hist. Nat. Belg.,* v. 7, fasc. 3, p. 1 – 118, pls. 8 – 14, 1915.

Mikkelsen, Naja, Silica dissolution and overgrowth of fossil diatoms. *Micropaleontology,* v. 23, p. 223 – 226, figs. 1, 2, 1977.

Miller, Urve, Diatom floras in the Quaternary of the Göta River Valley (western Sweden). *Sver. Geol. Unders. Afh.,* ser. Ca, no. 14, 14:0, p. 1 – 67, pls. 1 – 8, text-figs. 1 – 20, 1964.

Miller, Urve, Fossil diatoms under the scanning electron microscope. A preliminary report. *Årsb. Sver. Geol. Unders.,* v. 63, no. 5, Ser. C, no. 642, p. 1 – 65, pls. 1 – 25, 1969.

Mills, F. W., *An index to the genera and species of the Diatomaceae and their synonyms 1816 – 1932.* London: Weldon & Wesley Ltd., pts. 1 – 21, p. 1 – 1726, 1933 – 1935.

Moiseeva, A. I., Atlas neogenovykh diatomovykh vodorosley Primorskogo Kraya [Atlas of Neogene diatoms of the Primorya Territory]. *Trudy Vses. Nauchno-issled. Geol. Inst.,* n.s., v. 171, p. 1 – 151, pls. 1 – 21, 1971.

Morland, H., Do *Porodiscus interruptus* and *Craspedoporus elegans* belong both to one form? *J. Quekett Micr. Cl.,* ser. 2, v. 3, p. 167 – 168, 1887.

Mukhina, V. V., Siliceous organisms in suspension and in the surface layer of Indian Ocean bottom sediments (from data obtained on the 36th cruise of the Research Vessel 'Vityaz'). *Oceanology,* v. 6, p. 660 – 667, 5 figs., 1966.

Müller, Otto, Diatomeenrest aus den Turonschichten der Kreide. *Ber. Dt. Bot. Ges.,* v. 29 (1911), p. 661 – 668, pl. 26, 1912.

Murphy, L. S., and R. R. L. Guillard, Biochemical taxonomy of marine phytoplankton by electrophoresis of enzymes. I. The centric diatoms *Thalassiosira pseudonana* and *T. fluviatilis. J. Phycol.,* v. 12, p. 9 – 13, figs. 1 – 7, 1976.

Nakajima, Tadashi, and B. E. Volcani, 3,4-Dihydroxyproline: a new amino acid in diatom cell walls. *Science,* v. 164, p. 1400 – 1401, figs. 1 – 3, 1969.

Nikolaev, V. A., Novye vidy diatomovykh vodorosley iz Tretichnykh otlozheniy Dal'nego Vostoka. Species diatomearum novae e sedimentis tertiariis orientis extremi. *Novosti Sistematiki Nizshikh Rasteniy 1965, Akad. Nauk SSSR, Bot. Inst. Komarova,* p. 13 – 15, figs. 1 – 5, 1965.

Nitzsch, C. L., *Beitrag zur Infusorienkunde oder Naturbeschreibung der Zerkarien und Bazillarien.* Halle: J. C. Hendel, viii + 128 p., 6 pls., 1817.

Norris, R. E., A new planktonic diatom, *Nanoneis hasleae* gen. et sp. nov. *Norw. J. Bot.,* v. 20, p. 321 – 325, figs. 1 – 6, 1973.

Okuno, Haruo, Electron-microscopic fine structure of fossil diatoms. V. Observation on some diatoms found in the "Celatoms". *Trans. Proc. Palaeont. Soc. Japan,* n.s., no. 31, p. 237 – 242, pls. 34, 35, 1958.

Okuno, Haruo, Electron-microscopic fine structure of fossil diatoms. 6. Stereoscopic observation. *Trans. Proc. Palaeont. Soc. Japan,* n.s., v. 36, p. 185 – 191, pls. 21, 22, 1959.

Okuno, Haruo, Electron-microscopical study on fine structures of diatom frustules XIX. *Bot. Mag., Tokyo,* v. 75, p. 119 – 126, pl. 1, text-figs. 1 – 4, 1962.

Okuno, Haruo, Fossil diatoms, 48 p., pls. 414 – 513 + 1 unnumbered pl. In J. G. Helmcke and W. Krieger, *Diatomeenschalen im Elektronenmikroskopischen Bild,* Teil V. Weinheim: J. Cramer, 1964.

Orr, W. N., Judi Ehlen, and J. B. Zaitzeff, A Late Tertiary diatom flora from Oregon. *Proc. Calif. Acad. Sci.,* ser. 4, v. 37, p. 489 – 500, figs. 1 – 30, 1971.

Paddock, T. B. B., and P. A. Sims, A preliminary survey of the raphe structure of some advanced groups of diatoms (Epithemiaceae-Surirellaceae). *Beih. Nova Hedwigia,* v. 54, p. 291 – 322, pls. 1 – 10, figs. 1 – 5, 1977.

Papenfuss, G. F., Classification of the algae, p. 115—224. *In* E. L. Kessel, *A century of progress in the natural sciences 1853—1953*. San Francisco: Calif. Acad. Sci., 1955.

Parsons, T. R., Infection of a marine diatom by *Lagenidium* sp. *Can. J. Bot.*, v. 40, p. 523, 1 fig., 1962.

Patrick, Ruth, Some diatoms of Great Salt Lake. *Bull. Torrey Bot. Club*, v. 63, p. 157—166, pl. 6, fig. 1, 1936.

Patrick, Ruth, Factors effecting the distribution of diatoms. *Bot. Rev.*, v. 14, p. 473—524, 1948.

Patrick, Ruth, Sexual reproduction in diatoms, p. 82—99, pl. 1. *In* D. H. Wenrich, *Sex in microorganisms*. Amer. Assoc. Adv. Sci., 1954.

Patrick, Ruth, New subgenera and two new species of the genus *Navicula* (Bacillariophyceae). *Notul. Nat.*, no. 324, p. 1—11, figs. 1, 2, 1959.

Patrick, Ruth, Ecology of freshwater diatoms and diatom communities, p. 284—332, 6 figs. *In* Dietrich Werner, *The biology of diatoms*. Bot. Monogr., v. 13. Oxford, London, Edinburgh, Melbourne: Blackwell Scientific Publs., 1977.

Patrick, Ruth, and C. W. Reimer, *The diatoms of the United States exclusive of Alaska and Hawaii*, v. 1, *Fragilariaceae, Eunotiaceae, Achnanthaceae, Naviculaceae*. Monogr. Acad. Nat. Sci. Philad., no. 13, 688 p., 64 pls., 1966.

Pavillard, J., Bacillariales. *Rep. Dan. Oceanogr. Exped. Mediterr.*, v. 2, Biol., no. J4, p. 1—72, figs. 1—116, 1925.

Peragallo, H., and M. Peragallo, *Diatomées marines de France et des districts maritimes voisins*. Grez-sur-Loing (S.-et M.). M. J. Tempère, texte, v. 1 (Tableaux synoptique & systematique), xii + 491 p., Atlas, v. 2, pls. 1—24, 1897; 25—48, 1898; 49—72, 1899; 73—80, 1900; 81—96, 1901; 97—110, 112—113, 1902; 111, 114—119, 1908; 120—124, 1907; 125—131, 1904; 132—135, 1905; 136—137, 1907.

Pia, J., Klasse Diatomeae (Bacillariophyta), p. 44—55, figs. 29—32. *In* Max Hirmer, *Handbuch der Paläobotanik*, Bd. 1, *Thallophyta, Bryophyta, Pteridophyta*. München & Berlin: R. Oldenbourg, 1927.

Pia, J., Vorliassische Diatomeen? *Neues Jb. Miner. Geol. Paläont. Ref.*, Abt. 3, no. 1, p. 107—131, 1931.

Pritchard, Andrew, *A history of Infusoria, including the Desmidiaceae and Diatomaceae British and foreign*. London: Whittaker, 4th ed., xii + 968 p., 40 pls., 1861.

Proshkina-Lavrenko, A. I., Ed., *Diatomovyy analiz, Knigi 1. Obshchaya i paleobotanicheskaya kharakteristika diatomovykh vodorosley [Diatom analysis. Book 1. Generalities and paleobotanical characteristics of diatom algae]*. Vses. Nauchno-Issledov. Geol. Inst. Minist. Geol., Bot. Inst. Komarova, Akad. Nauk SSSR, p. 1—239, pls. 1—11, 1949a.

Proshkina-Lavrenko, A. I., Ed., *Diatomovyy analiz, Knigi 2. Opredelitel' iskopaemykh i sovremennykh diatomovykh vodorosley. Poryadki Centrales i Mediales [Diatom analysis. Book 2. Diagnoses of fossil and recent diatom algae. Orders Centrales and Mediales]*. Vses. Nauchno-Issledov. Geol. Inst. Minist. Geol., Bot. Inst. Komarova, Akad. Nauk SSSR, p. 1—238, pls. 1—101, 1949b.

Proshkina-Lavrenko, A. I., Ed., *Diatomovyy analiz, Knigi 3. Opredelitel' iskopaemykh i sovremennykh diatomovykh vodorosley, poryadok Pennales [Diatom analysis. Book 3. Diagnoses of fossil and recent diatom algae. Order Pennales]*. Vses. Nauchno-Issledov. Geol. Inst. Minist. Geol., Bot. Inst. Komarova, Akad. Nauk SSSR, p. 1—398, pls. 1—117, 1950.

Proshkina-Lavrenko, A. I., K evolyutsii diatomovykh vodorosley [On the evolution of diatoms]. *Byull. Mosk. Obshch. Ispȳt. Prir., Otd. Biol.*, v. 65, no. 5, p. 52—62, 1960.

Proshkina-Lavrenko, A. I., *Diatomovye vodorosli planktona Azovskogo Morya [Diatom algae of the plankton of the Azov Sea]*. Moscow-Leningrad: Akad. Nauk SSSR, Bot. Inst. Komarova, 190 p., 9 pls., 1963a.

Proshkina-Lavrenko, A. I., *Diatomovye vodorosli bentosa Chernogo Morya [Benthic diatoms of the Black Sea]*. Moscow-Leningrad: Akad. Nauk SSSR, Bot. Inst., 243 p., 16 pls., 1963b.

Proshkina-Lavrenko, A. I., and I. V. Makarova, *Vodorosli planktona Kaspiyskogo Morya [Plankton algae of the Caspian Sea.]*. Leningrad: Akad. Nauk SSSR, "Nauka," 295 p., 10 pls., 78 figs., 1968.

Prowse, G. A., Relationship between epiphytic algal species and their macrophytic hosts. *Nature*, v. 183, p. 1204—1205, 1959.

Ramsfjell, E., Dimorphism and the simultaneous occurrence of auxospores and microspores in the diatom *Rhizosolenia hebetata* f. *semispina* (Hensen) Gran. *Nytt Mag. Bot.*, v. 7, p. 169—173, pl. 1, 1959a.

Ramsfjell, E., The occurrence of bi-flagellate swarmers within cells of the pennate diatom *Amphiprora hyperborea* (Grunow) Gran. *Nytt Mag. Bot.*, v. 7, p. 179—180, pl. 2, 1959b.

Reimann, B. E. F., Bildung, Bau und Zusammenhang der Bacillariophyceenschalen (electronenmikroskopische Untersuchungen). *Nova Hedwigia*, v. 2, p. 349—373, pls. 64—71, 1960.

Reimann, B. E. F., Deposition of silica inside a diatom cell. *Expl. Cell. Res.*, v. 34, p. 605—608, figs. 1—4, 1964.

Reimann, B. E. F., and J. C. Lewin, The diatom genus *Cylindrotheca* Rabenhorst (with a reconstruction of *Nitzschia closterium*). *Jl. R. Microsc. Soc.*, v. 83, p. 283—296, pls. 121—130, 1964.

Reimann, B. E. F., J. C. Lewin, and B. E. Volcani, Studies on the biochemistry and fine structure of silica in diatoms. I. The structure of the cell wall of *Cylin-*

drotheca fusiformis Reimann and Lewin. *J. Cell Biol.,* v. 24, p. 39—55, figs. 1—25, 1965.

Reimann, B. E. F., J. C. Lewin, and B. E. Volcani, Studies on the biochemistry and fine structure of silica shell formation in diatoms. II. The structure of the cell wall of *Navicula pelliculosa* (Bréb.) Hilse. *J. Phycol.,* v. 2, p. 74—84, figs. 1—21, 1966.

Reimann, B. E. F., and B. E. Volcani, Studies on the biochemistry and fine structure of silica shell formation in diatoms. III. The structure of the cell wall of *Phaeodactylum tricornutum* Bohlin. *J. Ultrastruct. Res.,* v. 21, p. 182—193, figs. 1—20, 1968.

Reimer, C. W., Diatoms and their physico-chemical environment. *Bull. Publ. Hlth. Serv.,* 999 WP 25—1965, p. 19—28, 1962.

Reinhold, Th., Fossil diatoms of the Neogene of Java and their zonal distribution. *Verh. Geol.-Mijnb. Genoot. Ned. Kolon., Geol. Ser.,* v. 12, p. 43—132, pls. 1—21, 1937.

Reyes-Vasquez, Gregorio, Studies on the diatom flora living on *Thalassia testudinum* König in Biscayne Bay, Florida. *Bull. Mar. Sci. Gulf Caribb.,* v. 20, p. 105—134, figs. 1—3, 1970.

Riedel, W. R., The age of the sediment collected at Challenger (1875) Station 225 and the distribution of *Ethmodiscus rex* (Rattray). *Deep Sea Res.,* v. 1, p. 170—175, pl. 1, 1954.

Riedel, W. R., Siliceous organic remains in pelagic sediments. *Spec. Publs. Soc. Econ. Paleont. Miner., Tulsa,* 7, p. 80—91, 1959.

Rigby, J. K., and L. H. Burckle, Turbidity currents and displaced fresh-water diatoms. *Science,* v. 127, p. 1504, 1958. Comment by R. W. Kolbe, p. 1504—1505.

Ross, Robert, Class Bacillariophyceae (Diatoms), p. 193—195. *In* W. B. Harland et al., *The Fossil record.* London: Geol. Soc. London, 1967.

Ross, Robert, The current state of diatom taxonomy at the species level, with special reference to some species of *Navicula* Sect. Lyratae. *Beih. Nova Hedwigia,* v. 39, p. 1—36, figs. 1—6, 1972.

Ross, Robert, Some Eocene diatoms from South Atlantic cores. Part II. *Rutilaria* Greville. *Occ. Pap. Calif. Acad. Sci.,* 123, p. 21—27, pl. 1, fig. 1, 1976.

Ross, Robert, and P. A. Sims, Studies of *Aulacodiscus* with the scanning electron microscope. *Beih. Nova Hedwigia,* v. 31, p. 49—88, pls. 1—13, 1970.

Ross, Robert, and P. A. Sims, Generic limits in the Biddulphiaceae as indicated by the scanning electron microscope, p. 155—177, pls. 1—5. *In* V. H. Heywood, *Scanning electron microscopy, systematic and evolutionary applications.* Systematics Assoc. Spec. Vol. 4. London, New York: Academic Press, 1971.

Ross, Robert, and P. A. Sims, The fine structure of the frustule in centric diatoms: a suggested terminology. *Br. Phycol. J.,* v. 7, p. 139—163, figs. 1—35, 1972.

Ross, Robert, and P. A. Sims, Observations on family and generic limits in the Centrales. *Beih. Nova Hedwigia,* v. 45, p. 97—121, pls. 1—7, 1973.

Ross, Robert, P. A. Sims, and G. R. Hasle, Observations on some species of the Hemiauloideae. *Beih. Nova Hedwigia,* v. 54, p. 179—213, pls. 1—9, figs. 1—3, 1977.

Rothpletz, A., Ueber die Flysch-Fucoiden und einige andere fossile Algen, sowie über liasische Diatomeen führende Hornschwämme. *Z. Dt. Geol. Ges.,* v. 48, p. 854—914, pls. 22—24, 1896.

Round, F. E., Observations on girdle bands during cell division in the diatom *Stephanodiscus. Br. Phycol. J.,* v. 6, p. 135—143, figs. 1—10, 1971.

Round, F. E., On the diatom genera *Stephanopyxis* Ehr. and *Skeletonema* Grev. and their classification in a revised system of the Centrales. *Botanica Mar.,* v. 16, p. 148—154, pls. 1—9, 1973.

Rubina, N. V., Novye i redkie vidy rodov *Coscinodiscus* Ehr. i *Aulacodiscus* Ehr. iz kontinental'nykh verkhneoligotsenovykh i nizhnemiotsenovykh otlozheniy zapadno-Sibirskoy nizmennosti [Species Coscinodisci Ehr. et Aulacodisci Ehr. e sedimentis continentalibus Oligocaeni superioris et Miocaeni inferioris depressionis Sibireae occidentalis]. *Nov. Sist. Nizsh Rast., 1976, Akad. Nauk SSSR, Bot. Inst. Komarova,* p. 54—59, pls. 1—3, 1976.

Russell, D. J., and R. E. Norris, Ecology and taxonomy of an epizoic diatom. *Pacif. Sci.,* v. 25, p. 357—367, 5 figs., 1971.

Saar, A. du, Results and problems in diatom investigation at the geological survey in Haarlem. *Geologie Mijnb.,* Jg. 46, p. 105—108, 1967.

Saunders, R. P., and D. A. Glenn, *Diatoms: memoirs of the Hourglass Cruises,* v. 1, pt. III. Fla. Dept. Nat. Resources, Marine Research Lab., St. Petersburg, Fla., p. 1—119, pls. 1—16, 1969.

Schmidt, A., *Atlas der Diatomaceen-Kunde.* Fortgesetzt von A. Schmidt, Fr. Fricke, O. Müller, H. Heiden und Fr. Hustedt. Leipzig: O. R. Reisland, 472 pls., 1874—1959.

Schrader, H. J., Die pennaten Diatomeen aus dem Obereozän von Oamaru, Neuseeland. *Beih. Nova Hedwigia,* v. 28, 124 p., 39 pls., 1969a.

Schrader, H. J., Neue Gattungen und Arten fossiler Diatomeen. Diagnosen der neuen Gattungen und Arten zu der Arbeit: die pennaten Diatomeen aus dem Obereozän von Oamaru, Neuseeland. *Nova Hedwigia,* v. 17, p. 427—432, 1969b.

Schrader, H. J., Cenozoic diatoms from the northeast Pacific, Leg 18. *Init. Repts. Deep Sea Drill. Proj.,* v. 18, p. 673—797, pls. 1—26, figs. 1—36, 1973a.

Schrader, H. J., Types of raphe structures in the diatoms. *Beih. Nova Hedwigia,* v. 45, p. 195—217, pls. 1—12, figs. 1—3, 1973b.

Schrader, H. J., Stratigraphic distribution of marine species of the diatom *Denticula* in Neogene North Pacific sediments. *Micropaleontology,* v. 19, p. 417−430, pl. 1, figs. 1−4, 1973c.

Schrader, H. J., Cenozoic marine planktonic diatom stratigraphy of the tropical Indian Ocean. *Init. Repts. Deep Sea Drill. Proj.,* v. 24, p. 887−967, pls. 1−22, figs. 1−12, 1974.

Schrader, H. J., Cenzoic planktonic diatom biostratigraphy of the southern Pacific Ocean. *Init. Repts. Deep Sea Drill. Proj.,* v. 35, p. 605−671, pls. 1−15, figs. 1−14, 1976.

Schrader, H. J., and Juliane Fenner, Norwegian Sea Cenozoic diatom biostratigraphy and taxonomy. *Init. Repts. Deep Sea Drill. Proj.,* v. 38, p. 921−1099, pls. 1−45, figs. 1−42, 1976.

Schultz, M. E., and F. R. Trainor, Production of male gametes and auxospores in the centric diatoms. *Cyclotella meneghiniana* and *C. cryptica. J. Phycol.,* v. 4, p. 85−88, figs. 1−5, 1968.

Schultz, M. E., and F. R. Trainor, Production of male gametes and auxospores in a polymorphic clone of the centric diatom *Cyclotella. Can. J. Bot.,* v. 48, p. 947−951, figs. 1−6, 1970.

Schulz, P., Diatomeen aus senonen Schwammgesteinen der Danziger Bucht. Zugleich ein Beitrag zur Entwicklungsgeschichte der Diatomeen. *Bot. Arch.,* v. 37, p. 383−413, 2 text-pls., text-figs. 1−7, 1935.

Schütt, F., Bacillariales, p. 31−150, text-figs. 44−282. *In* A. Engler and K. Prantl, *Die natürlichen Pflanzenfamilien,* Teil I, Abt. 1 b. Leipzig: Wilhelm Engelmann, 1896.

Sheshukova-Poretskaya, V. S., Klass Pennatae. Pennatnye, ili peristye [Class Pennatae. Pennates or pinnates], p. 122−151, figs. 123−200. *In* Yu. A. Orlov, *Osnovy Paleontologii, Vodorosli, Mokhoobraznye, Psilofitovye, Plaunovidnye, Chlenistostebel'nye, Paporotniki.* Moscow: Akad. Nauk SSSR, 1963.

Sheshukova-Poretskaya, V. S., *Neogenovye morskie diatomovye vodorosli Sakhalina i Kamchatki [Neogene marine diatom algae of Sakhalin and Kamchatka].* Leningrad Ordena Lenina Gosudarst. Univ., p. 1−432, pls. 1−50, 1967.

Sheshukova-Poretskaya, V. S., and Z. I. Glezer, Diatomovye vodorosli, silikoflagellaty i ebriidei iz Maykopskikh otlozheniy s R. Shibik (Krasnodarskiy kray) [Diatom algae, silicoflagellates and ebriids from Maykopian deposits of the Shibik region (Krasnodar territory)]. *Uchenye Zapiski, Leningrad Gosudarst. Univ.,* no. 313, Ser. Biol. Nauk, vyp. 49, p. 171−202, pls. 1−5, 1962.

Shropshire, R. F., Preparation of diatomaceous earth. *Trans. Am. Microsc. Soc.,* v. 50, p. 48−49, 1931.

Silva, P. C., Classification of algae, p. 827−837. *In* R. A. Lewin, *Physiology and biochemistry of algae,* Appendix A. New York: Academic Press, 1962.

Simonsen, Reimer, Ideas for a more natural system of the centric diatoms. *Beih. Nova Hedwigia,* v. 39, p. 37−54, fig. 1, 1972.

Simonsen, Reimer, On the pseudonodulus of the centric diatoms, or Hemidiscaceae reconsidered. *Beih. Nova Hedwigia,* v. 53, p. 83−94, pls. 1−3, fig. 1, 1975.

Simonsen, Reimer, and Taro Kanaya, Notes on the marine species of the diatom genus *Denticula* Kütz. *Int. Revue Ges. Hydrobiol. Hydrogr.,* v. 46, p. 498−513, pl. 1, fig. 1, index maps 1, 2, 1961.

Small, James, Quantitative evolution XVI. Increase of species-number in diatoms. *Ann. Bot.,* n.s., v. 14, p. 91−113, 4 figs., 1950.

Smayda, T. J., Biogeographical studies of marine phytoplankton. *Oikos,* v. 9, p. 158−191, 7 figs., 1958.

Smayda, T. J., Occurrence of unusual bodies in a marine pennate diatom. *Nature,* v. 196, p. 191, 1962.

Smayda, T. J., A quantitative analysis of the phytoplankton of the Gulf of Panama II. On the relationship between C^{14} assimilation and the diatom standing crop. *Bull. Inter.-Am. Trop. Tuna Commn.,* v. 9, p. 467−529, 12 figs., 1965.

Smayda, T. J., The suspension and sinking of phytoplankton in the sea. *Oceanogr. Mar. Biol. A. Rev.,* v. 8, p. 353−414, figs. 1−7, 1970.

Smayda, T. J., The growth of *Skeletonema costatum* during a winter-spring bloom in Narragansett Bay, Rhode Island. *Norw. J. Bot.,* v. 20, p. 219−247, figs. 1−8, 1973.

Smayda, T. J., and B. J. Boleyn, Experimental observations on the flotation of marine diatoms. I. *Thalassiosira* cf. *nana, Thalassiosira rotula* and *Nitzschia seriata. Limnol. Oceanogr.,* v. 10, 499−509, 4 figs., 1965.

Smayda, T. J., and B. J. Boleyn, Experimental observations on the flotation of marine diatoms. II. *Skeletonema costatum* and *Rhizosolenia setigera. Limnol. Oceanogr.,* v. 11, p. 18−34, 7 figs., 1966a.

Smayda, T. J., and B. J. Boleyn, Experimental observations on the flotation of marine diatoms. III. *Bacteriastrum hyalinum* and *Chaetoceros lauderi. Limnol. Oceanogr.,* v. 11, p. 35−43, 2 figs., 1966b.

Smith, William, *A synopsis of the British Diatomaceae; with remarks on their structure, functions and distribution; and instructions for collecting and preserving specimens.* London: John van Voorst, v. 1, xxxiii + 89 p., 31 pls., 1853; v. 2, xxx + 107 p., pls. 32−42, A−E, 1856.

Sober, H. A., *Handbook of biochemistry. Selected data for molecular biology.* 2nd ed., Sect. H. Cleveland: Chemical Rubber Company, p. H1−H65, 1970.

Stoermer, E. F., and H. S. Pankratz, Fine structure of the diatom *Amphipleura pellucida.* I. Wall structure. *Am. J. Bot.,* v. 51, p. 986−990, 8 figs., 1964.

Stosch, H. A. von, Form und Formwechsel der Diatomee *Achnanthes longipes* in Abhängigkeit von der Ernährung. Mit besonderen Berücksichtigung der Spurenstoffe. *Ber. Dt. Bot. Ges.,* v. 60, p. 2−16, 1942.

Stosch, H. A. von, Öogamy in a centric diatom. *Nature,* v. 165, p. 531−532, 1950.

Stosch, H. A. von, Entwicklungsgeschichtliche Untersuchungen an zentrischen Diatomeen. II. Geschlechtszellenreifung, Befruchtung und Auxosporenbildung einiger grundbewohnender Biddulphiaceen der Nordsee. *Arch. Mikrobiol.,* v. 23, p. 327−365, 55 text-figs., 1956.

Stosch, H. A. von, Kann die oogame Araphidee *Rhabdonema adriaticum* als Bindeglied zwischen den beiden grossen Diatomeen-gruppen angesehen werden?. *Ber. Dt. Bot. Ges.,* v. 71, p. 241−249, figs. 1−3, 1958a.

Stosch, H. A. von, Entwicklungsgeschichtliche Untersuchungen an zentrischen Diatomeen III. Die Spermatogenese von *Melosira moniliformis* Agardh. *Arch. Mikrobiol.,* v. 31, p. 274−282, text-figs. 1−5, 1958b.

Stosch, H. A. von, Einige Bemerkungen zur Phylogenie der Diatomeen. *Naturwissenschaften,* v. 45, p. 141, fig. 1, 1958c.

Stosch, H. A. von, Manipulierung der Zellgrösse von Diatomeen im Experiment. *Phycologia,* v. 5, p. 21−44, figs. 1−12, 1965.

Stosch, H. A. von, An amended terminology of the diatom girdle. *Beih. Nova Hedwigia,* v. 53, p. 1−28, pls. 1−7, figs. 1−18, 1975.

Stosch, H. A. von, and Gerhard Drebes, Entwicklungsgeschichtliche Untersuchungen an zentrischen Diatomeen. IV. Die Planktondiatomee *Stephanopyxis turris*−ihr Behandlung und Entwicklungsgeschichte. *Helgoländer Wiss. Meeresunters.,* v. 11, p. 209−257, figs. 1−18, 1964.

Stosch, H. A. von, and B. E. F. Reimann, *Subsilicea fragilarioides* gen. et spec. nov., eine Diatomee (Fragilariaceae) mit vorwiegend organischer Membran. *Beih. Nova Hedwigia,* v. 31, p. 1−36, pls. 1−8, 1970.

Stosch, H. A. von, G. Theil, and K. V. Kowallik, Entwicklungsgeschichtliche Untersuchungen an zentrischen Diatomeen. V. Bau und Lebenszyklus von *Chaetoceros didymum,* mit Beobachtungen über einige andere Arten der Gattung. *Helgoländer Wiss. Meeresunters.,* v. 25, p. 384−445, 26 figs., 1973.

Strel'nikova, N. I., Redkie i novye vidy pozdnemelovykh diatomovykh vodorosley vostochnogo sklona polyarnogo Urala. De diatomeis cretae superioris raris et novis declivis orientalis montium Uralensium Polarium. *Nov. Sist. Nizsh. Rast. 1965, Akad. Nauk SSSR, Bot. Inst. Komarova,* p. 29−37, pls. 1−6, 1965.

Strel'nikova, N. I., Pozdnemelovye diatomovye vodorosli [Late Cretaceous diatom algae]. *Iskopaemye Diatomovye Vodorosli SSSR, Akad. Nauk SSSR, Sibirskoe Otdel. Inst. Geol. Geofiz.* Moskva: "Nauka," p. 17−21, fig. 1, 1968.

Strel'nikova, N. I., Novye vidy diatomovykh vodorosley iz pozdnemelovykh otlozheniy vostochnogo sklona polyarnogo i pripolyarnogo Urala [Species novae Bacillariophytorum e sedimentis Cretae Posterioris in declivitate Orientali partis polaris ac praepolaris montium Uralensium]. *Nov. Sist. Nizsh. Rast. 1971, Akad. Nauk SSSR, Bot. Inst. Komarova,* p. 41−51, pls. 1−3, 1971.

Strel'nikova, N. I., *Diatomei pozdnego Mela* (*Zapadnaya Sibir'*) *[Diatoms of the Late Cretaceous (western Sibiria)].* Moscow: Akad. Nauk SSSR, Nauchnyy Sovet po Probleme, "Nauka," 201 p., 57 pls., 1974.

Swift, Elijah 5th, Cleaning diatom frustules with ultraviolet radiation and peroxide. *Phycologia,* v. 6, p. 161−163, 1 fig., 1967.

Swift, Elijah 5th, The marine diatom *Ethmodiscus rex:* its morphology and occurrence in the plankton of the Sargasso Sea. *J. Phycol.,* v. 9, p. 456−460, figs. 1−13, 1973.

Sydow, Burkard von, and Robert Christenhuss, Rasterelektronenmikroskopische Untersuchungen der Hohlräume in der Schalenwand einiger centrischer Kieselalgen. *Arch. Protistenk.,* v. 114, p. 256−271, pls. 23−28, 1 fig., 1972.

Takahashi, Eiji, Studies on genera *Mallomonas, Synura* and other plankton in freshwater with the electron microscope (V). On the similarity of the fine structure between scale of *Mallomonas* and frustule of diatom. *Bull. Yamagata Univ. Agric. Sci.,* v. 4, p. 137−145, 2 pls., 1 fig., 1964.

Tanada, T., The photosynthetic efficiency of carotenoid pigments in *Navicula minima. Am. J. Bot.,* v. 38, p. 276−283, figs. 1−5, 1951.

Tappan, Helen, and A. R. Loeblich, Jr., Geobiologic implications of fossil phytoplankton evolution and time-space distribution, p. 247−340, 1 pl., 12 figs. *In* R. Kosanke, and A. T. Cross, Symposium on palynology of the Late Cretaceous and Early Tertiary. *Geol. Soc. Am. Spec. Pap.,* 127, 1971.

Tappan, Helen, and A. R. Loeblich, Jr., Evolution of the oceanic plankton. *Earth-Sci. Rev.,* v. 9, p. 207−240, 9 figs., 1973.

Tempère, J. (with collaboration of others), *Le diatomiste, journal spécial s'occupant exclusivement des diatomées et de tout ce qui s'y rattache.* Paris: J. Tempère, v. 1, p. 1−180, pls. 1−24, 1890−1893; v. 2, p. 188, 197−260, pls. 1−24, 1893−1896 [in v. 2, p. 189−196 and pl. 18 omitted in error].

Thomas, W. H., and A. N. Dodson, Effects of phosphate concentration on cell division rates and yield of a tropical oceanic diatom. *Biol. Bull. Mar. Biol. Lab., Woods Hole,* v. 134, p. 199−208, figs. 1−4, 1968.

Tsumura, Kôhei, Additional confirmation of some rare or curious diatoms, Japanese and foreign. *J. Yokohama Munic. Univ.*, Ser. C-51, no. 168, p. 1—24, pls. 1—8, 1967.

VanLandingham, S. L., Miocene non-marine diatoms from the Yakima region, South Central Washington. *Nova Hedwigia*, v. 14, p. 1—80, pls. 1—56, 1964.

VanLandingham, S. L., Index to the genera in Atlas der Diatomaceen-Kunde 1875—1959. *Trans. Am. Microsc. Soc.*, v. 85, p. 295—297, 1966a.

VanLandingham, S. L., Diatoms from Dry Lakes in Nye and Esmeralda Counties, Nevada, USA. *Nova Hedwigia*, v. 11, p. 221—241, pls. 27—30, 1966b.

VanLandingham, S. L., Paleoecology and microfloristics of Miocene diatomites from the Otis Basin-Juntura region of Harney and Malheur Counties, Oregon. *Beih. Nova Hedwigia*, v. 26, p. 1—77, pls. 1—25, 1967a.

VanLandingham, S. L., *Catalogue of the fossil and Recent genera and species of diatoms and their synonyms, a revision of F. W. Mills' "An index to the genera and species of the Diatomaceae and their synonyms."* Part 1. *Acanthoceras* through *Bacillaria*. Weinheim: J. Cramer, xi + 493 p., 1967b.

VanLandingham, S. L., *Catalogue of the fossil and Recent genera and species of diatoms and their synonyms.* Part II. *Bacteriastrum* through *Coscinodiscus*. Lehre: J. Cramer, p. 494—1086, 1968.

VanLandingham, S. L., *Catalogue of the fossil and Recent genera and species of diatoms and their synonyms.* Part III. *Coscinophaena* through *Fibula*. Lehre: J. Cramer, p. 1087—1756, 1969.

VanLandingham, S. L., *Catalogue of the fossil and Recent genera and species of diatoms and their synonyms.* Part IV. *Fragilaria* through *Naunema*. Lehre: J. Cramer, p. 1757—2385, 1971.

VanLandingham, S. L., *Catalogue of the fossil and Recent genera and species of diatoms and their synonyms.* Part V. *Navicula*. Vaduz: J. Cramer, p. 2386—2963, 1975.

VanLandingham, S. L., *Catalogue of the fossil and Recent genera and species of diatoms and their synonyms.* Part VI. *Neidium* through *Rhoicosigma*. Vaduz: J. Cramer, p. 2964—3605, 1978a.

VanLandingham, S. L., *Catalogue of the fossil and Recent genera and species of diatoms and their synonyms.* Part VII. *Rhoicosphenia* through *Zygoceros*. Vaduz: J. Cramer, p. 3606—4241, 1978b.

VanLandingham, S. L., *Catalogue of the fossil and Recent genera and species of diatoms and their synonyms.* Part VIII. Supplementary taxa (through 1964), supplementary references, synonymy addendum, corrections, additions. Vaduz: J. Cramer, p. 4242—4654, 1979.

Venkataraman, G. S., S. C. Mehta, S. C. Das, and K. V. Natarajan, Fine structure of diatom valves IV. *Nova Hedwigia*, v. 4, p. 127—129, pls. 49—50, 1962.

Venrick, E. L., Recurrent groups of diatom species in the North Pacific. *Ecology*, v. 52, p. 614—625, figs. 1—5, 1971.

Vinogradov, A. P., and E. A. Boichenko, Razlozhenie kaolina diatomovymi vodoroslyami [Decomposition of kaolin by diatoms]. *Dokl. Akad. Nauk SSSR*, v. 37, p. 135—138, 1942.

Werner, Dietrich, Introduction with a note on taxonomy, p. 1—17. *In* Dietrich Werner, *The biology of diatoms*. Bot. Monogr., v. 13. Oxford, London, Edinburgh, Melbourne: Blackwell Scientific Publs., 1977.

Werner, R. G., Diatomées marines vivant en eau douce continentale. *C. R. Hebd. Séanc. Acad. Sci., Paris*, v. 251, p. 413—415, 1960.

West, G. S., and F. E. Fritsch, *A treatise on the British freshwater algae, in which are included all the pigmented Protophyta hitherto found in British freshwaters*. Cambridge: University Press, 534 p., 207 figs., 1927.

Whitney, J. D., On the freshwater infusorial deposits of the Pacific Coast and their connection with the volcanic rocks. *Proc. Calif. Acad. Sci.*, v. 3, p. 319—324, 1867.

Wiseman, J. D. H., and N. I. Hendey, The significance and diatom content of a deep-sea floor sample from the neighbourhood of the greatest oceanic depth. *Deep Sea Res.*, v. 1, p. 47—59, pls. 1, 2, figs. 1, 2, 1953.

Wood, E. J. F., An unusual diatom from the Antarctic. *Nature*, v. 184, p. 1962—1963, 1959.

Wood, E. J. F., *Marine microbial ecology*. London: Chapman and Hall Ltd.; New York: Reinhold Publishing Corporation; 243 p., 14 pls., 18 figs., 1965.

Wood, E. J. F., Studies of phytoplankton ecology in tropical and subtropical environments of the Atlantic Ocean. Part 3. Phytoplankton communities in the Providence Channels and the Tongue of the Ocean. *Bull. Mar. Sci.*, v. 18, p. 481—543, 7 text-figs., 1968.

Wornardt, W. W. Jr., Pleistocene diatoms from Mono and Panamint Lake Basins California. *Occ. Pap. Calif. Acad. Sci.*, 46, 27 p., 2 pls., 2 figs., 1964.

Wornardt, W. W. Jr., Miocene and Pliocene marine diatoms from California. *Occ. Pap. Calif. Acad. Sci.*, 63, 108 p., 217 figs., 1967.

Wornardt, W. W. Jr., Stratigraphic distribution of diatom genera in marine sediments in western North America. *Palaeogeogr. Palaeoclimat. Palaeoecol.*, v. 12, p. 49—74, pls. 1—9, 1972.

Zhuze, A. P., Novye diatomovye i kremnevye zhgutikovye vodorosli verkhnemelovogo vozrasta iz glinistykh peskov basseyna R. B. Aktay (Vostochnyysklon severnogo Urala). Algae diatomaceae aetatis supernecretaceae ex arenis argillaceis systematis fluminis Bolschoy Aktay in declivitate orientali Ural Borealis. *Bot. Materialy, Otdel. Spor. Rasteniy, Bot. Inst., Akad. Nauk SSSR*, v. 6, p. 65—78, pls. 1, 2, figs. 1—3, 1949.

Zhuze, A. P., Diatomovye i kremnevye zhgutikovye vodorosli Verkhnemelovogo vozrasta iz severnogo Urala. Diatomeae et silicoflagellatae aetatis cretae superne e montibus uralensibus septentrionalibus. *Bot. Materialy, Otdel. Spor. Rasteniy, Bot. Inst., Akad. Nauk SSSR,* v. 7, p. 42−65, pls. 1−6, figs. 1, 2, 1951a.

Zhuze, A. P., Diatomovye Paleotsenovogo vozrasta severnogo Urala. Diatomeae aetatis Palaeocaeni uralh septentrionalis. *Bot. Materialy, Otdel. Spor. Rasteniy, Bot. Inst., Akad. Nauk SSSR,* v. 7, p. 24−42, pls. 1−4, figs. 1, 2, 1951b.

Zhuze, A. P., Diatomovye v poverkhnostnom sloe osadkov Beringova Morya [Diatoms in the upper sediment layers of the Bering Sea]. *Trudȳ Inst. Okeanol.,* v. 32, p. 171−205, text-figs. 1−15, 1959.

Zhuze, A. P., Tip Bacillariophyta. Diatomovye vodorosli [Division Bacillariophyta. Diatom algae], p. 55−121, text-figs. 1−122. *In* Yu. A. Orlov, *Osnovy Paleontologii, Vodorosli, Mokhoobraznye, Psilofitovye, Plaunovidnye, Chlenistostebel'nye, Paporotniki.* Moscow: Akad. Nauk SSSR, 1963.

Zhuze, A. P., Drevnie diatomei i diatomovye porody Tikhookeanskogo basseyna [Ancient diatoms and diatomaceous rocks of the Pacific Ocean basin]. *Litologiya i Poleznye Iskopaemye,* 1968(1), p. 16−32, figs. 1−4, 1968a.

Zhuze, A. P., Novye vidy diatomovykh vodorosley v donnykh osadkakh Tikhogo Okeana i Okhotskogo Morya [Species novae Bacillariophytorum in sedimentis fundis Oceani Pacifici et Maris Ochotensis inventae]. *Nov. Sist. Nizsh. Rast., 1968, Akad. Nauk SSSR, Bot. Inst. Komarova,* p. 12−21, pls. 1−3, 1968b.

Zhuze, A. P., Novye i interesnye vidy i formy diatomey iz donnykh osadkov Tikhogo Okeana [Species formaeque nova et curiosae Bacillariophytorum in sedimentis fundi Oceani Pacifici]. *Nov. Sist. Nizsh. Rast., 1971, Akad. Nauk SSSR, Bot. Inst. Komarova,* p. 12−18, figs. 1−4, 1971.

Zhuze, A. P., Oligotsen-Miotsenovye biostratigraficheskie zony diatomey tropicheskoy oblasti Tikhogo Okeana [Oligocene-Miocene biostratigraphic diatom zones of the tropical part of the Pacific Ocean], p. 34−48, figs. 1−5. *In* A. P. Zhuze, *Mikropaleontologiya Okeanov i Morey.* Moscow: Akad. Nauk SSSR, Okeanograficheskaya Komissiya, "Nauka," 1974.

Zhuze, A. P., O. G. Kozlova, and V. V. Mukhina, Vidovoy sostav i zonal'noe raspredelenie diatomey v poverkhnostnom sloe osadkov Tikhogo Okeana [Species composition and zonal distribution of diatoms in the surface layer of Pacific Ocean sediments]. *Dokl. Akad. Nauk SSSR,* v. 172, p. 1183−1186, figs. 1, 2, 1967.

Zhuze, A. P., and V. S. Sheshukova-Poretskaya, Novyy rod *Riedelia* Jousé et Sheshuk. (Bacillariophyta) [Genus novum *Riedelia* Jousé et Sheshuk. (Bacillariophyta)]. *Nov. Sist. Nizsh. Rast., 1971, Akad. Nauk SSSR, Bot. Inst. Komarova,* p. 19−25, pls. 1, 2, 1971.

HAPTOPHYTA, COCCOLITHOPHORES, AND OTHER CALCAREOUS NANNOPLANKTON

The extinction at the end of the Mesozoic seems…to be most obviously and strikingly reflected by the planktonic life (plant and animal) of the oceans and by the larger forms dependent on plankton … . The data on calcareous nannoplankton and planktonic foraminifera are now adequate to indicate this world-wide extinction of most of the distinctive taxa, and to show that the extinction of these large populations was so abrupt that the stratal record of transition still remains obscure. … The hiatus … may involve many thousands of years but probably much less than a million; comparable changes in the fossil record normally require some millions of years. … It required several million years also for the meager assemblages of the nannoplankton and planktonic foraminifera surviving into the earliest Cenozoic to develop diversification comparable to that found in the late Mesozoic.

M. N. Bramlette, 1965

The calcareous nannoplankton includes the calcite-secreting haptophytes, coccolithophores, and certain fossil forms commonly associated with them, discoasters, nannoconids, schizosphaerellids, and pithonellids, which may or may not be related. The term **nannoplankton** was proposed by Lohmann (1909) for the very small forms that passed through the phytoplankton nets (60 μm or less)

and were collected by centrifuging seawater and catching the residue on filter paper. A more restrictive definition considers as nannoplankton only forms of less than 2 μm, and regards as **ultramicroplankton** those between 2 and 20 μm. Under the latter definition, most coccolithophores would not be included in the nannoplankton; for although isolated coccoliths, the separate platelike or rodlike calcareous bodies produced as a cell cover by the coccolithophores, may range from 1.3 μm to 50 μm in size, the complete cell is from about 5 μm to as much as 100 μm in diameter.

THE LIVING ORGANISM

The motile haptophyte cell is globular, ovate, or elongate, and bears two fine, similar flagella, of a length about 1.5 to 2.5 times the cell body length (see Figure 9.1). A third appendage, the **haptonema**, lies between the flagella, may be straight or coiled, and has a characteristic inner structure; in some haptophytes, the haptonema may be reduced to a mere vestige, or be nonemergent. The motile cell alternates during the life history with a nonmotile stage, which may be a spherical pelagic cyst, a packet of cells, or a branching filament.

Either the motile or nonmotile stage, or both, may be covered by organic scales, visible only with electron microscopy; or by calcified plates, the **coccoliths;** or by both scales and coccoliths.

The cell of *Emiliania huxleyi* is bounded by two double membranes, the outer one being discontinuous and perhaps corresponding to an organic scale cover underlying the coccoliths of other species. An external gelatinous layer, bearing the tiny calcareous plates or coccoliths (see Figure 9.2) either externally or embedded within it, may itself become calcified, and thoroughly cement the coccoliths to the surface. All coccoliths on a single coccosphere may be alike, but some species have more than one type on the cell, perhaps with one kind localized near the flagellar pole or as a median girdle, and with a different type covering the remainder of the cell.

Nucleus

A single spherical to variably formed nucleus lies in the posterior part of the cell. The nucleus has a perforated double membrane, a single dense nucleolus, and about 40 spheroidal to

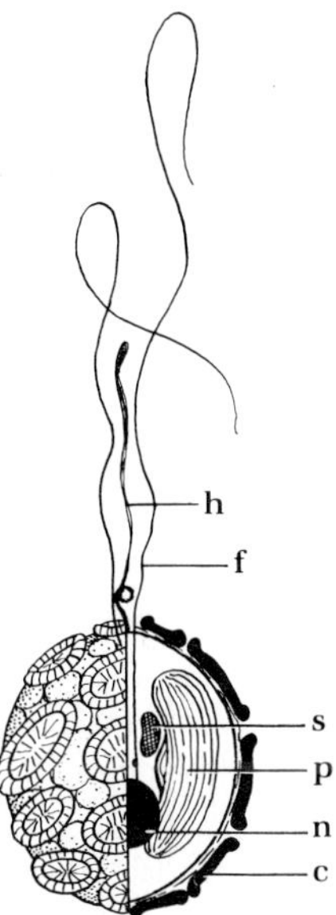

Figure 9.1
Generalized diagram of a coccolithophorid motile cell, showing exterior at left, and section at right. Cell has a covering of scales and coccoliths (c), two flagella (f), and a haptonema (h), and contains a nucleus (n), plastids (p) with pyrenoid bulge at the inner margin, and storage body (s) of leucosin.

oval chromosomes in *Cricosphaera roscoffensis* (P. Dangeard) Gayral & Fresnel. The motile cell of *C. carterae* (Braarud & Fagerland) Braarud has 42 ± 2 chromosomes, hence is regarded as diploid, whereas the haploid benthic stage has only 21 ± 1 (Klaveness, 1972).

Cricosphaera elongata (Droop) Braarud contains 5 μg DNA per 10^6 cells, and *Pavlova lutheri* (Droop) Green has only 0.1 μg DNA per 10^6 cells. The mole percent of guanine and cytosine (G+C) in the DNA of various haptophytes is between 58 and 65, hence somewhat higher than in diatoms. Representative amounts for the Isochrysidales are 64.8 percent G+C in *Emiliania huxleyi,* 59.2 percent in *Cricosphaera elongata,* 60.2 percent in *C. carterae,* and 61 percent in *Isochrysis galbana* Parke; and for the Prymnesiales, 58 mole percent G+C in *Prym-*

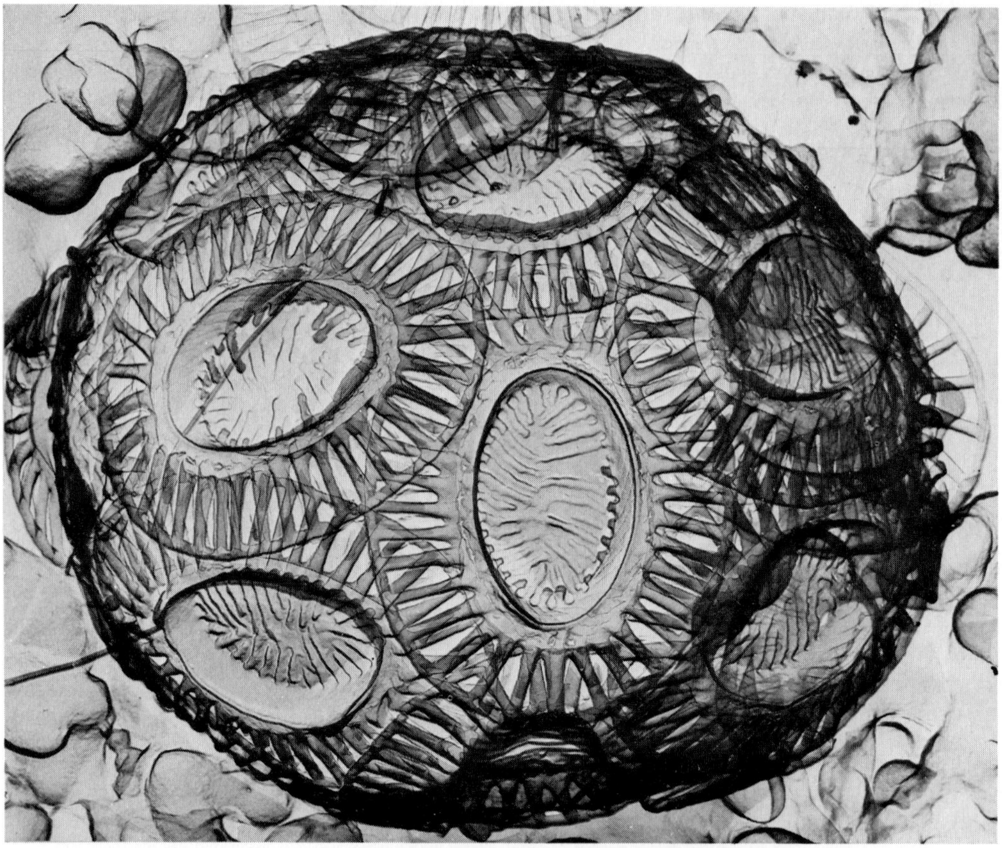

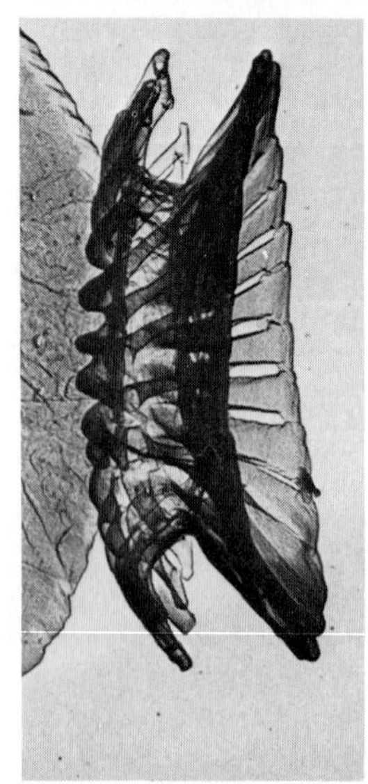

2

3

nesium parvum Carter (Loeblich and Loeblich, 1979).

Plastids, Pigments, and Pyrenoids

The two (or rarely four) large, clear, yellow-brown plastids have three-thylakoid lamellae. Unlike the Chrysophyta, with which the Haptophyta had previously been placed, the plastids have no peripheral girdle lamella, but are surrounded by a sheath of endoplasmic reticulum that may be continuous with the nuclear envelope. Plastid DNA is scattered through the matrix, and not localized. A single cup-shaped plastid is present in a cell of *Hymenomonas coronata* Mills, whereas *Chrysochromulina kappa* Parke & Manton has two to four saucer-shaped plastids in the motile phase, each with a pyrenoid on the inner face, and lobate to stellate plastids in the nonmotile phase (Parke et al., 1955).

The characteristic pigments are chlorophyll *a* and chlorophyllides c_1 and c_2, the latter two in a total amount approximately equal to that of chlorophyll *a* in *Emiliania huxleyi*, whereas *Cricosphaera carterae* has about five times as much chlorophyll *a*. Additional pigments include α- and β-carotene, diatoxanthin, diadinoxanthin, fucoxanthin, and neofucoxanthin A and B, as well as two unidentified xanthophylls in *E. huxleyi*. Large amounts of the deep orange fucoxanthin mask the other pigments, resulting in the golden brown color.

At its inner face, each plastid has a pyrenoid whose proteinaceous core is ensheathed by starch platelets. Thylakoids of the plastids may penetrate the pyrenoid of some taxa.

Other Organelles and Cytoplasmic Inclusions

Long and narrow mitochondria occur in the peripheral cytoplasm, just inside the region of the plastids. The Golgi body of the Haptophyta is very large and characteristic (see Figure 9.3). It is not adjacent to the nucleus as is that of the Chrysophyta, but is associated with one or more cisternae of endoplasmic reticulum. The cisternae are bunched together beneath the basal bodies of flagella and haptonema, then flare out in a fanlike form. Organic scales are produced within the Golgi body, progressively from the side adjacent to the endoplasmic reticulum to the opposite face (Hibberd, 1976). In many sections of *Cricosphaera carterae*, all elongated vesicles of the Golgi body contained scales, suggesting that scale formation is an active and continuous process (Leadbeater, 1970). In addition, the Golgi body is the site of formation of most of the mineralized coccoliths (as discussed on pp. 688−692).

Pavlova gyrans Butcher has a small **eyespot** at the anterior end of the plastid, consisting of about 15 osmiophilic pigmented droplets (Green and Manton, 1970). Similar pigmented granules occur in *Platychrysis*. Eyespots are unknown in other haptophytes, although present in most Chrysophyta.

The cytoplasm also may contain fat globules and red- or brownish-pigmented granules. Commonly a globule of chrysolaminarin, a whitish insoluble food reserve, is present in the central or posterior part of the cell. It is similar to the storage product of the brown algae in composition. *Pavlova mesolychnon* van der Veer has two types of inclusions. One occurs in granules in the membrane, and by X-ray pow-

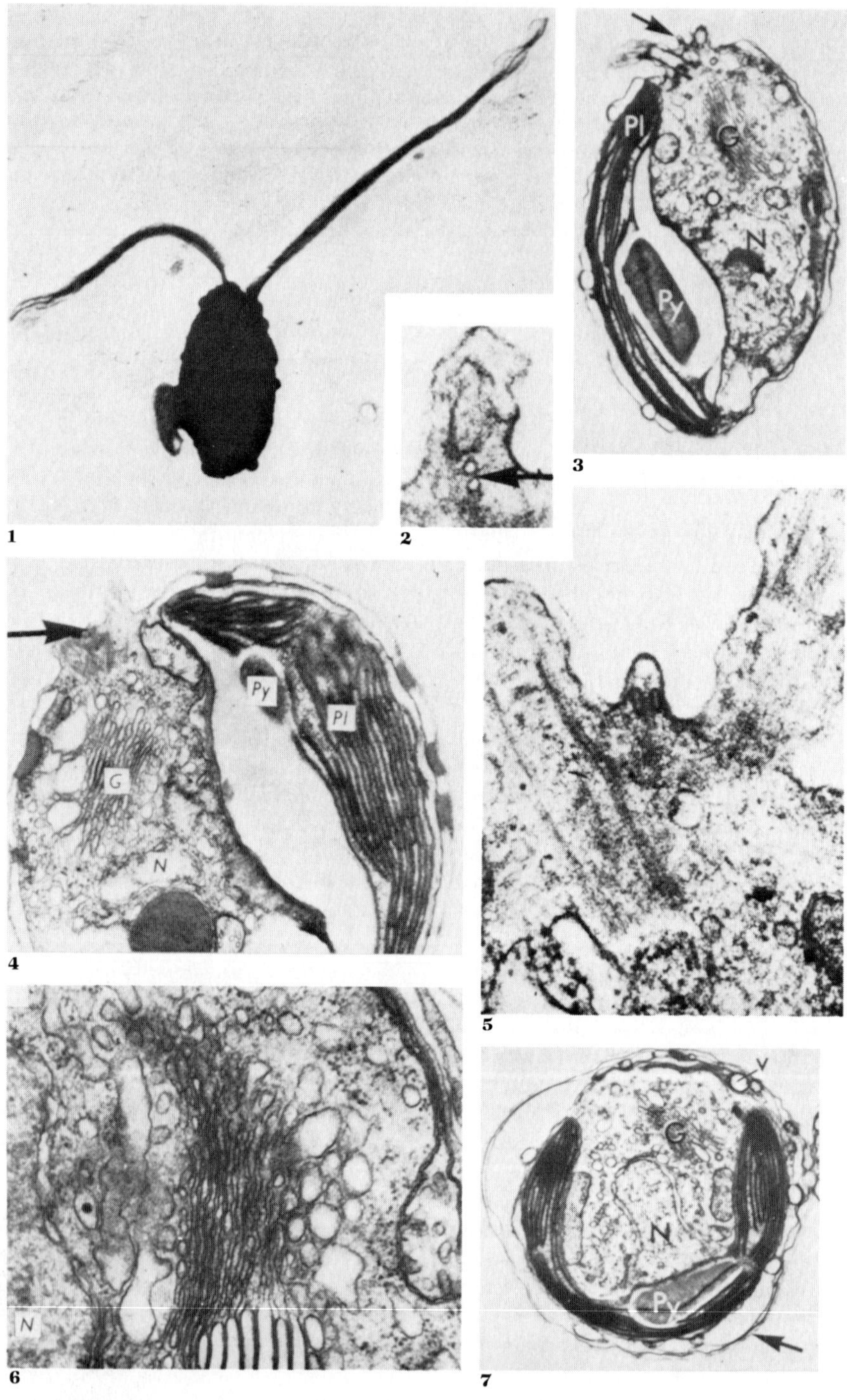

1
2
3
Pl
G
N
Py
4
Py
Pl
G
N
5
6
N
7
V
G
N
Py

Figure 9.3

Chrysotila lamellosa zooids; N, nucleus; Pl, plastid; Py, pyrenoid; G, Golgi body; V, peripheral vesicle. **1.** Shadowcast whole mount, showing elongate cell and flagella, ×7000. **2.** Section, showing microtubules (arrow) of the reduced haptonemal base, ×80,000. **3.** Longitudinal section, showing haptonema (arrow) between flagellar bases, ×9560. **4.** Longitudinal section, with haptonema (arrow), ×15,000. **5.** Flagellar pole, with reduced haptonema in center, and flagella at each side, ×60,000. **6.** Longitudinal section of Golgi body with dilated cisternae, ×35,000. **7.** Oblique transverse section of zooid, layer of body scales visible at arrow, ×9560. 3,7, reproduced with permission from Green and Parke, 1974, *J. Mar. Biol. Ass. U.K.*, v. 54, p. 539 – 550, published by Cambridge University Press. Other figures reproduced with permission from Green and Parke, 1975, *J. Mar. Biol. Ass. U.K.*, v. 55, p. 109 – 121, published by Cambridge University Press.

---

der diffraction appears to have the same composition as the paramylon of the Euglenophyta, and to be a glucose polymer. The other granules are small, globular, and strongly birefringent, but of unknown composition (Kreger and van der Veer, 1970).

A large contractile vacuole, located near the base of the flagella in the anterior region of freshwater species, serves as an excretory organelle.

Flagella and Haptonema

Motile cells of the Haptophyta typically have two smooth, similar flagella, that in *Chrysochromulina kappa* may be 1.5 to 2.5 times the length of the cell itself. A third appendage, the haptonema (Parke et al., 1955), arises between the two flagella and is the characteristic organelle from which the division derives its name. Once regarded as a third flagellum, the haptonema differs in structure, having three concentric membranes enclosing a ring of six to seven microtubules in the outer part of the haptonema, increasing to eight or nine microtubules within the cell. Unlike the flagellar microtubules, these are not paired, and no central pair of microtubules is present. In addition, the haptonema lacks a basal body like that of the flagella. Some haptophytes have a reduced haptonema, that of *Pavlova gyrans* having only four microtubules, which decrease to three, two, and eventually only one at the tip. Still more reduced is the haptonema of *Chrysotila*

lamellosa Anand, which contains only three microtubules (Parke et al., 1955; Green and Manton, 1970). The haptonema of *Ochrosphaera neapolitana* Schussnig consists only of a group of five microtubules associated with the flagellar bases, and *Imantonia* has a small haptonemal process that lacks any microtubules.

Hymenomonas pringsheimii Parke & Green and the motile stage of *Coccolithus pelagicus* (Wallich) Schiller have a short and thin haptonema, with a length in the former species of about one-half to three-fourths the cell diameter when extended, and about 2 to 10 μm in *Coccolithus.* The haptonema with a club-shaped tip and a basal swelling lies between the two flagella; short and usually coiled once or twice, it can be seen only when the motile cell is naked, before development of the coccolith cover.

Cell Movement and Haptonemal Function

The haptonema of *Chrysochromulina kappa* when fully extended may be many times the length of the flagella, and its adhesive tip may serve for temporary attachment to the substrate. During cell movement, the flagella commonly are directed backwards and the haptonema is tightly coiled, but at other times, the cell may move more slowly with the flagella directed forward and the haptonema stiffly and fully extended. The haptonemal coiling is rapid, only occupying about $\frac{1}{80}$ second. All species rotate and gyrate about the line of movement, the gyration being especially noticeable in larger species (Leadbeater, 1971). Movement is rapid, considering the size of the organism; *Calyptrosphaera* and *Syracosphaera* reportedly cover an average of 5 to 8 mm per minute at 18°C or a maximum of about 10 to 20 m per day. This is probably less than the rate of current movement, hence they are regarded as relatively passive plankton.

In addition to the motile stage, many haptophytes have a nonmotile stage that lacks flagella, although the haptonema may be retained. The haploid benthic filamentous stage of *Cricosphaera* and *Hymenomonas* may produce two types of motile cells, one with a haptonema and one lacking this structure (Parke, 1971). Whether one of these represents gametes and the other zoospores is uncertain.

Cell Covering

Scales

The cell surface of motile stages is commonly covered by a layer of scales. Scales of *Chrysochromulina brevifilum* Parke & Manton are ellipsoid, rimmed, about 0.7 μm in diameter, and have a central spine supported by four buttresslike ridges (Parke et al., 1955). *C. camella* Leadbeater & Manton and *C. adriatica* Leadbeater have an underlayer of discoidal scales and an outer layer of cup-shaped ones (see Figure 9.4), and *C. kappa* has oval to polygonal scales 0.3 × 0.4 to 0.5 × 0.8 μm in size, scales at the flagellar pole bearing a central spine like those of *C. brevifilum*.

Motile cells of *Isochrysis, Chrysotila, Sarcinochrysis, Ochrosphaera,* and *Apistonema* have a scale cover of round rimless scales (Parke, 1971); *Cricosphaera carterae* has two size classes of scales, smaller circular ones of about 0.7 μm diameter and larger elliptical ones from 0.7 × 1.1 up to 0.85 × 1.3 μm across that may be mineralized as calcareous coccoliths (Franke and Brown, 1971).

All scales are transparent to the electron beam and are visible only with electron microscopy. Commonly the scales have concentric ridges on the distal surface and radiating ridges

on the proximal surface, some may have a crenulate margin where the ridges extend to the edge. The scales are composed of cellulosic glycoprotein; the cellulosic component of *Hymenomonas pringsheimii* consists of fibrils 1.0 to 2.5 nm in diameter, forms a complex network, and has the chemical reaction of cellulose or celluloselike glucans (Brown et al., 1969). The scales of *Cricosphaera carterae* are constructed of a highly ordered network of radial and concentric fibrils similar to those of *H. pringsheimii* (Franke and Brown, 1971).

The scales are produced internally, within the Golgi body, and then extruded to the surface. Some haptophytes have only a scale cover; others may have a covering of calcareous coccoliths that are similarly produced within the cell or deposited on the organic base provided by the scales. An organic "skin" may surround both the entire cell and the coccolith cover.

Coccoliths

The typical coccolithophore cell is covered by an external layer of small regular calcareous plates termed coccoliths. In many living cells the cell membrane is flexible and the coccoliths loosely arranged over the surface, so that when examined under the microscope the cell can be flattened by pressure on a cover slip, but can recover its original shape undamaged.

Flexibility of the cell of *Cricosphaera carterae* (see Figure 9.5) is obtained by local movement of the coccoliths, suggesting that the adjacent coccoliths may be hooked together at their rims, abruptly detach, and hook together again in a new position (Outka and Williams, 1971). In some species calcite may also be deposited in the cell membrane, which becomes hard and brittle and holds the coccoliths firmly cemented (*Coccolithus* spp., *Syracosphaera* spp.). Such cemented coccospheres have been known for well over a century both from modern oceanic deposits and as fossils.

Organic Matrix of the Coccolith. The calcium carbonate of the coccoliths is deposited within an organic matrix, as was first noted by

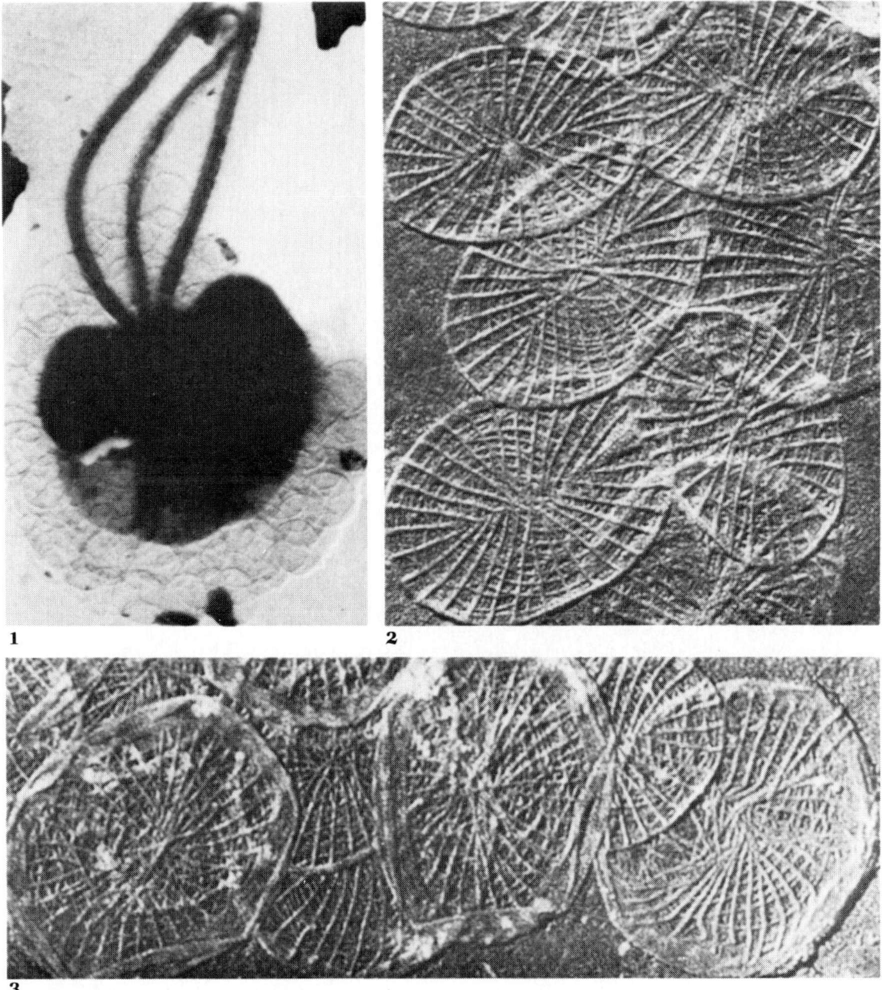

Figure 9.4

Chrysochromulina adriatica. **1.** Entire cell, TEM of shadowcast whole mount, showing terminally coiled haptonema inserted between the two flagella and outer covering of scales, ×8000. **2.** Under layer of scales. **3.** Outer layer of scales. 2,3, ×20,000. All reproduced with permission from Leadbeater, 1974, *J. Mar. Biol. Ass. U.K.,* v. 54, p. 179–196, published by Cambridge University Press.

Dixon (1900). This mode of calcification is characteristic of animal cells, and is known among protists only in the foraminifera and coccolithophores. The X-ray diffraction pattern obtained from coccoliths is identical to that of inorganic calcite except for a darker central region resulting from the included organic matter. After all calcium carbonate is removed by treatment with acid, the remaining organic matrix, largely a highly insoluble protein closely linked to a sulfated polysaccharide, retains the original form of the coccolith. After demineralization this organic matrix can reconcentrate and hold calcium ions. Separation (by dialysis)

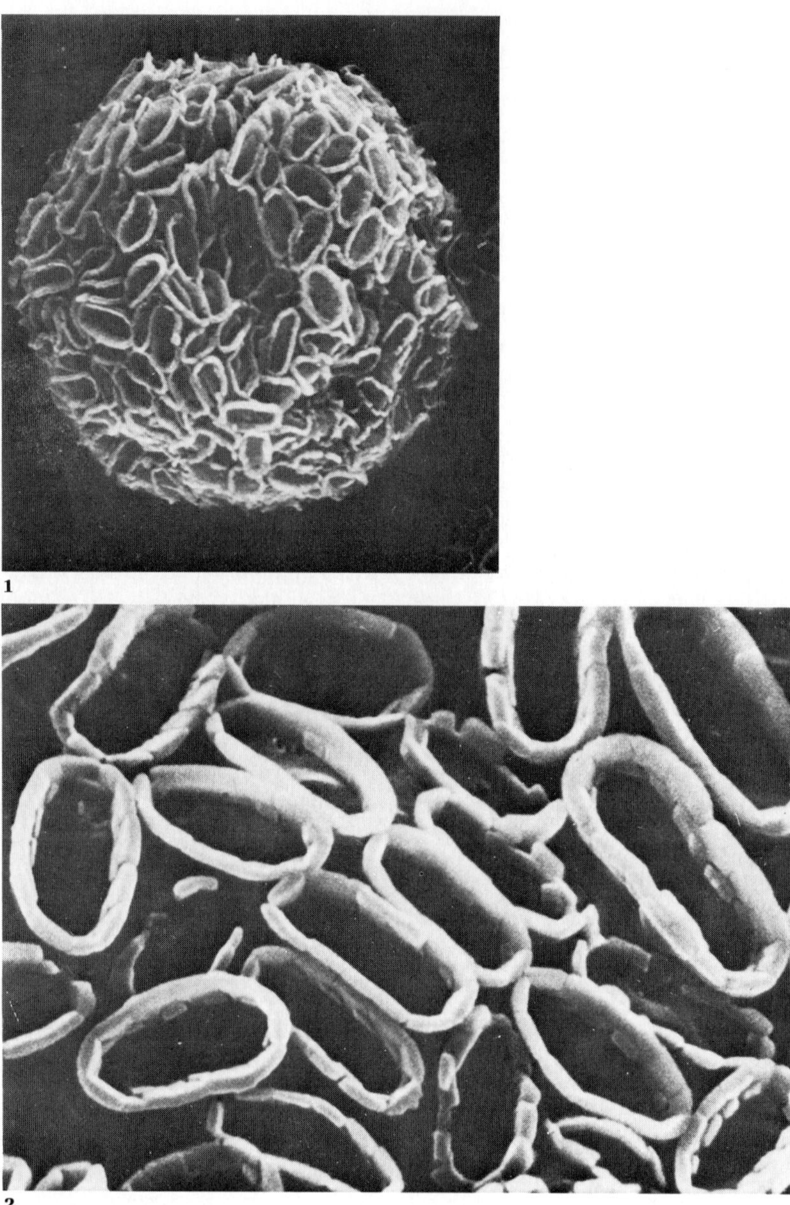

Figure 9.5
Cricosphaera carterae. **1.** SEM of cells of clone obtained from North Pacific
Gyre, ×6400. **2.** Enlargement of cricoliths showing alternating inner and
outer rim elements, ×22,500. From A. R. Loeblich III.

of the materials in solution in the decalcification liquor yielded material also capable of recombination with calcium. Analysis of the coccolith extracts in HCl, H_3PO_4, and HI indicated the presence of 16 amino acids (cysteic acid, methionine sulphone, aspartic acid, threonine, serine, proline, glutamic acid, glycine, alanine, valine, isoleucine, leucine, phenylalanine, lysine, hystidine, and arginine), traces of methionine sulphoxide, methionine, taurine, and ornithine, and extracts in H_3PO_4 and HI (not HCl) showed small amounts of tyrosine and cysteine (Isenberg et al., 1965).

Relationship of Coccolith Formation and Photosynthesis. Calcium deposition in various organisms has been shown to be related to the photosynthesis of the organism or of associated symbionts. Such a relationship is found in the coccolithophores, as in cultures coccolith production does not occur or is greatly reduced in the dark. By increasing the amount of CO_2 in the medium in which they were grown, cells of *Emiliania huxleyi* could be decalcified in life. The hydrogen ion concentration increased from pH 8 to pH 5.9 to 5.3, and the coccoliths disappeared. However, intracellular coccoliths in process of formation were not removed. After transfer to a normal medium, no coccoliths were produced if cells were kept in the dark, and after being kept in the dark they required one half hour of light before coccolith deposition was renewed. The maximum rate of coccolith production occurred in the light, the cells producing from five to ten coccoliths within 6 hours, and a complete coccolith cover in about 30 hours. Maximum coccolith production continued for about 1 hour after the commencement of a dark period as well. An individual coccolith formed in about an hour, and one or two could be formed at a time. Rate of formation was traced by the use of C^{14} in the growth medium. Radioactivity of the cells was measured after their growth in the C^{14} medium and again after removing the coccoliths from cells dried on a slide, by exposure for 20 minutes to fuming HCl. Loss of radioactivity indicated the C^{14} uptake of the coccoliths, as only the organic carbon of the cell remained after

solution of the $CaCO_3$. After the complete coccolith cover was replaced, additional coccoliths formed and were shed into the medium (Paasche, 1963, 1964, 1966a; Wilbur and Watabe, 1963). A single cell of *Emiliania huxleyi* is about 2.5 μm in radius, and averages about 15 coccoliths per cell. A single coccolith contains 5.5×10^{-13} g of calcium, and the cell has an organic carbon content of 0.1 g/cm^3. Thus the normal ratio of coccolith carbon to organic carbon in this species is 0.4.

Experiments with the uptake of C^{14} by coccolithophores, under varied light intensities and wavelengths, indicated that coccolith formation is catalyzed by two separate photochemical reactions, one being related to the plastid pigments, and the other to a pigment absorbing specifically in the blue part of the spectrum, or perhaps a combination of chlorophylls *a* and *c* and the carotenoids (Paasche, 1966b).

Two stages of the life cycle of *Coccolithus pelagicus* are characterized by different types of coccoliths; the larger, nonmotile form has a few heavy coccoliths (heterococcoliths), and uses several times as much carbon in the form of the coccolith calcium carbonate, as it does in photosynthesis. In contrast, the motile form has many delicate coccoliths on organic scales (holococcoliths), and coccolith production uses less than 2 percent of the amount of carbon used in photosynthesis (Paasche, 1969).

The constant ratio between coccolithogenesis and photosynthesis at varied light intensities suggests that these independent processes might be linked by a common route of carbon supply. The principal carbonate source used in coccolith formation is the bicarbonate ion; with the simultaneous uptake of this ion and calcium ions, $CaCO_3$ is deposited and the remaining CO_2 utilized in photosynthesis. Calcium deposition appears to serve as a quasi-detoxifying mechanism for excessive intracellular carbonate. However, studies of the light-dependent coccolith formation by *E. huxleyi* suggest that it is not related to carbon assimilation in photosynthesis, but may involve photoactivation of an enzyme, as coccolith deposition probably involves energy utilization (Paasche, 1966a).

In culture, coccoliths continued to form within *Emiliania huxleyi* cells at calcium concentrations so low as to result in dissolution of the coccoliths at the cell surface, although coccolith production is retarded by calcium concentrations about one-half that of normal seawater. Thus, coccolith calcification is only indirectly dependent on the factors controlling the nonbiological deposition of calcium carbonate (Paasche, 1964).

Cytology of Coccolith Formation. Dixon (1900) found that about 80 percent of the complete coccospheres of *Coccolithus pelagicus* contain an internal oval colorless body, within which lies a single coccolith, commonly complete and perfect, but in some specimens consisting only of a simple oval ring or collar representing the early stage of coccolith calcification. A few specimens contained one complete internal coccolith and another in the early stage of formation. The coccolith calcification was shown to begin with a narrow ring, which deepened into a collar, then formed flanges at the extremities to become the two valves or shields of the coccolith (placolith). A transverse bar was last to be formed, shortly before extrusion of the coccolith to the outside to become a part of the cell cover. Newly forming coccoliths were noted to be as large or larger than any already at the exterior, suggesting that the young cell first formed coccoliths of somewhat smaller size than those secreted by the mature cell. Later observers noted single intracellular coccoliths between the plastids; the internal position of these newly forming coccoliths was demonstrated by adding Ca^{45} to the cultures and testing autoradiographically. Information obtained from microtome sections made by Dixon was expanded by ultramicrotome studies and electron micrographs of *Emiliania huxleyi* coccospheres (Wilbur and Watabe, 1963). Coccolith production occurs within a homogeneous nongranular material located central to a "reticular body," and between it and the nucleus. This "reticular body" appears to control the formation of an organic matrix that first assumes the shape of the coccolith, although somewhat larger in size than the completed coccolith. After development of the organic matrix, calcification begins at many centers within it and near the base of the future coccolith, progres-

sing upward and peripherally to form the upper elements, and progressing peripherally at the base to form the basal elements. In the early stages, a thin membrane connects the centers of calcification in the matrix material and marks the location of the central extensions of the base. At the periphery, the membrane appears to divide and the two laminae pass above and below the center of calcification, as if the calcification resulted in spreading of the membrane. The completed coccoliths then are pushed outward to surround the cell. As additional coccoliths form, as many as three concentric layers envelop the cell, with the interlocked coccoliths held firmly.

Emiliania huxleyi is atypical in that it does not produce scales beneath the coccoliths. Like other species with large coccoliths, it produces a single coccolith at a time in special Golgi cisternae that are independent of the Golgi body itself, although probably formed by it. Matrix material of the coccolith is formed within the endoplasmic reticulum (probably the "reticular body" of Wilbur and Watabe, 1963), at the same time as mineralization occurs. As no organic scale is present, coccolith shape is directly controlled by the coccolith vesicle membrane (Klaveness, 1973a, 1976). The coccolith of *E. huxleyi* is constructed of many units, each consisting of an upper and lower element. Electron diffraction analysis suggests that each unit is a single crystal in the crystallographic sense, having the form of a hollow crystal or a mosaic of three individual crystals, with the C-axis parallel to the direction of the elongation of the element (see Figure 9.6; Watabe, 1967).

Other species, with smaller coccoliths, such as *Hymenomonas pringsheimii,* produce these directly within the Golgi apparatus itself. Lying at the cell periphery near the flagellar bases, distended **cisternae** of the Golgi body are oriented toward the cell surface. Many coccoliths may be produced simultaneously in these cisternae, coccoliths at different stages of maturity being present in various parts of the cell. From 41 to 82 Golgi generations may be required to synthesize the wall of an actively growing cell of *Hymenomonas pringsheimii,* one coccolith being produced every two minutes on an average (Brown, 1969). The Golgi apparatus has direct control of the coccolith

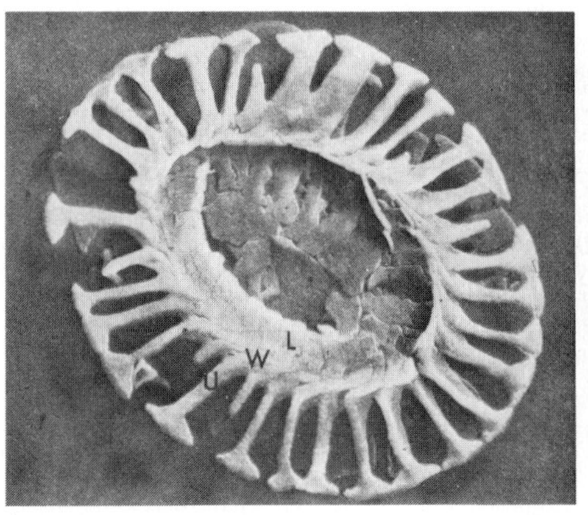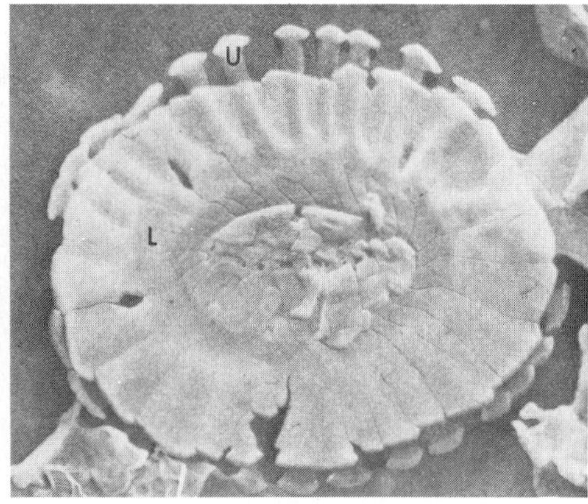

Figure 9.6
Emiliania huxleyi. Structure of a coccolith in TEM of carbon replica. **1.** Distal view showing continuous units formed by the radially arranged upper elements (U) connected to the lower unit (L) through the wall (W) of the central cylinder, ×20,200. **2.** Proximal view showing lower shield of the placolith formed by the lower elements (L), with portions of the outer shield formed by the upper elements (U) visible at the periphery, ×21,150. From Watabe, 1967.

production, forming electron-dense granules termed **coccolithosomes** that fuse to form the granular organic matrix in *Hymenomonas pringsheimii* and *Cricosphaera carterae* (see Figure 9.7). Independently produced, highly structured coccolith bases (or scales) first appear within the Golgi vesicle, and rims are built up on the bases by material from the coccolithosomes. The calcification of the rim sheath and the later extracellular emplacement of the coccolith and underlying scale follows (Outka and Williams, 1971; Dorigan and Wilbur, 1973; Klaveness, 1973a, 1976). In culture, *C. carterae* cells have three to four layers of scales and one layer of coccoliths. Small oval scales are located near the haptonema; round scales of medium size with radially ornamented lower surface and concentric whorls on the upper surface are evenly distributed over the cell exterior. Large oval scales also are present, one beneath each of the approximately 200 coccoliths (Paddock, 1968; Outka and Williams, 1971). A large number of vacuoles, containing scales, may lie at the cell anterior, perhaps being involved in the extrusion of the coccoliths.

The close interrelationship of the coccolithosomes, scale production, and coccolithogenesis is indicated by experiments showing that inhibitors of protein synthesis also inhibited calcium uptake, coccolith formation, and cell division in *C. carterae* (Dorigan and Wilbur, 1973).

A third type of coccolith formation occurs in a smaller number of species or during particular parts of the life cycle. The motile stage of *Coccolithus pelagicus,* long known by the separate name *Crystallolithus hyalinus,* does not produce coccoliths within the cell; coccoliths apparently are formed outside the cell membrane, but beneath the outer "skin." Organic scales originate within the Golgi body, several forming within a vesicle before they are moved to the cell exterior. As in *Cricosphaera carterae,* the scales are of two sizes, transparent, and rimless, with concentric markings on the outer face and radial ones on the inner face (see Figure 9.8). After the scales move outward, minute calcite rhombohedra are deposited at the cell exterior on the appropriate place on each scale, although not covering it completely. These coccoliths with little modification of the calcite crystal form are termed **holococcoliths**, and appear to be characteristic of the motile stages of various taxa that may have the more usual internally formed **heterococcoliths** during a

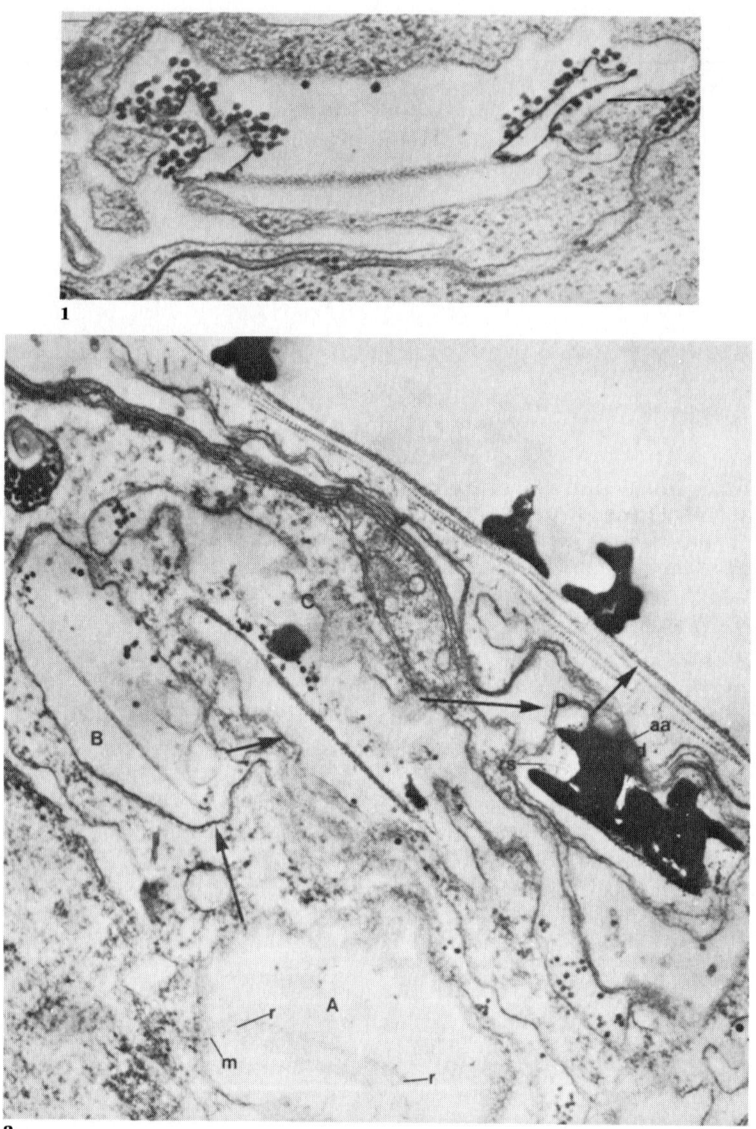

Figure 9.7

Cricosphaera carterae. **1.** Development of the coccolith in vacuole, with coccolithosomes densely clustered in region of rim matrix, arrow indicating coccolithosomes in peripheral cisterna; TEM of sectioned decalcified specimen, ×45,000. **2.** Section showing various stages of coccolith formation; vacuole at A contains part of base and specialized rim area (r) connected to membrane (m); vacuole at B shows base with coccolithosomes; obliquely cut vacuole, C, contains coccolithosomes and matrix, and shows beginning calcification; vacuole at D contains mature coccolith (black), residual coccolithosomes (rs), partially decalcified area (d), and anvil-shaped element of the coccolith (aa); arrows show direction of movement of developing coccoliths within the vaculoles, to the final position at cell exterior; ×39,500. From Outka and Williams, 1971.

690

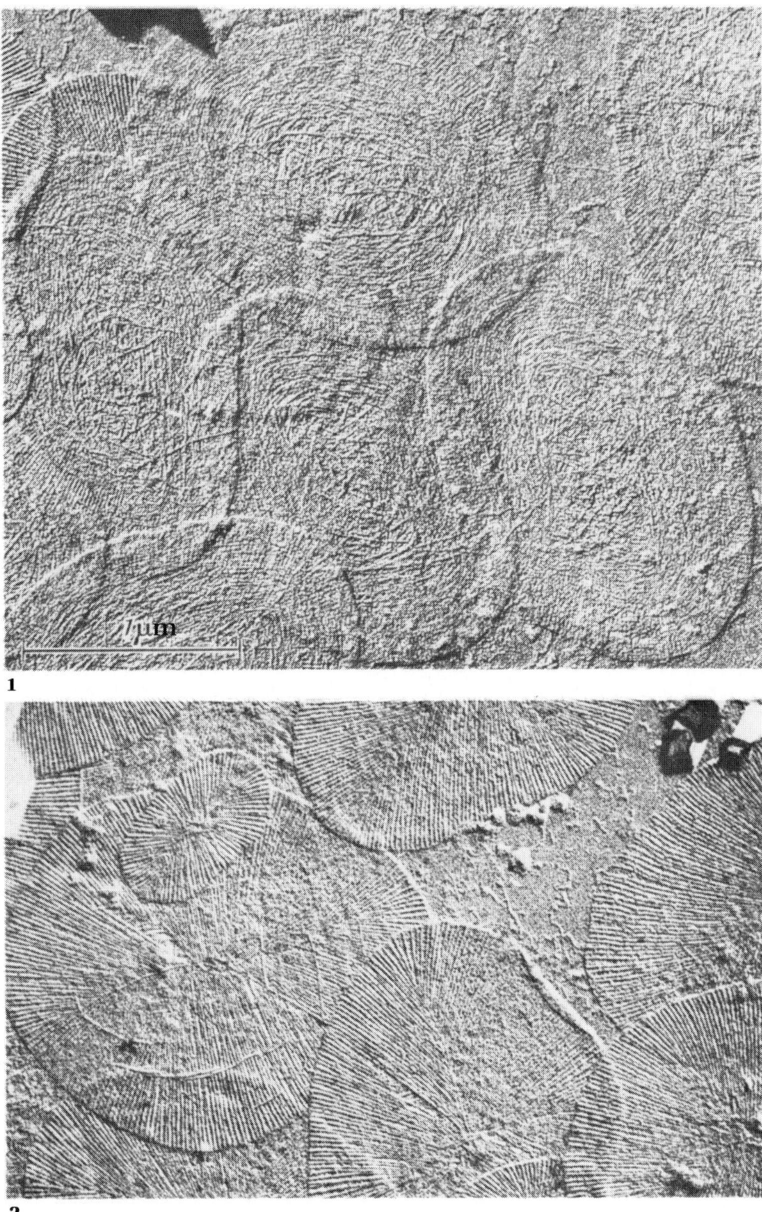

Figure 9.8

Organic scales of motile stage of *Coccolithus pelagicus* (previously known as *Crystallolithus hyalinus* Gaarder & Markali), in TEM. **1.** Distal surface showing concentric markings. **2.** Proximal surface with radial ridges. Both ×30,000, from Manton and Leedale, 1963.

nonmotile phase, or may have only a scale-covered benthic phase. On a cell, not all of the tightly imbricated scales develop holococcoliths, and the ability to construct ordered coccoliths may be lost in old cultures, so that there are merely scattered crystals. These coccoliths are not truly external, as a layer of "skin" covers them; just as more than one layer of internally formed coccoliths may accumulate at the outside of the cell, there may be several successive layers of scales, crystalloliths, and skin at the cell exterior in this stage. Crystals only occur in the space between the skin and next inner layer of organic scales (Parke and Adams, 1960; Manton and Leedale, 1963).

Although the holococcoliths are believed to be deposited externally, both *Calyptrosphaera sphaeroidea* Schiller and the motile *Crystallolithus* phase of *Coccolithus pelagicus* have a system of peripheral vesicles in the cell that contain granular stainable material very similar in appearance to the coccolithosomes of *Hymenomonas pringsheimii* and *Cricosphaera carterae* (Klaveness, 1973b).

Coccolith Mineralogy. The calcium carbonate in the coccoliths is in the crystalline form of calcite. Although first proven by X-ray diffraction patterns, the crystal form is recognizable in the electron microscope, and the crystal angles measurable. Early chemical tests had suggested that the tiny coccoliths and discoasters were of aragonite (Tan Sin Hok, 1926), but chemical tests are unreliable with such tiny

particles and X-ray determinations showed that in nature all known coccoliths are formed only of calcite, never of aragonite or vaterite, the polymorphs of $CaCO_3$. However, in a controlled culture in a nitrogen-deficient medium, all three polymorphs were formed by *Emiliania huxleyi*. In optimal salinity, temperature had no effect on the mineral form, but in the nitrogen-deficient medium the percentage of vaterite was slightly greater at 18°C than at 13°C, and decreased to 0 percent at 35°C. Aragonite had a reverse behavior, increasing in percentage at the higher temperatures.

Similar aberrant mineralization characterized *Cricosphaera carterae* cells that had been decalcified, then grown in a magnesium-deficient medium; about 60 percent of the coccoliths were calcite as usual, but up to 40 percent were aragonite. If magnesium was entirely absent from the medium, all coccoliths were of calcite, however (Blackwelder et al., 1976).

Calcium carbonate has an exceptionally strong power of double refraction, and the calcite crystals are so arranged that a negative uniaxial pseudofigure is seen in polarized light, each individual coccolith showing a well-defined black cross. This feature was first noted by Sorby (1861) and illustrated by Rothpletz (1896). Kamptner (1954) observed that each type of coccolith had a distinctive interference pattern, and the extinction position of the calcite elements and curvature of the extinction lines are important for species recognition (see Figure 9.9). They may also be used to indicate whether the distal or proximal side of an iso-

---

Figure 9.9
Texas Cretaceous coccoliths viewed in transmitted light, cross-polarized light, and SEM. **1—5.** *Manivitella pemmatoidea* (Deflandre) Thierstein, Del Rio Clay. **1—3.** Low focus, high focus, and cross-polarized light, ×3600. **4,5.** Distal and proximal sides in SEM, ×3300. **6—10.** *Reinhardtites anthophorus* (Deflandre) Perch-Nielsen. **6.** High focus. **7.** Rotated 45°. **8.** In cross-polarized light. **9.** Same, rotated 40°. **6—9,** Denton Marl, ×3600. **10.** SEM of specimen from Kiamichi Formation, proximal view, ×7000. **11—15.** *Broinsonia dentata* Bukry. **11.** Low focus. **12.** Same, rotated 30°. **13.** In cross-polarized light. **14.** Same, rotated 30°. **11—14,** Grayson Formation, ×3600. **15.** SEM in distal view, Denton Formation, ×7000. **16—20.** *Helicolithus trabeculatus* (Górka) Verbeek. **16,17.** Distal views in SEM, from Kiamichi Formation and Pawpaw Formation, ×5000 and ×6670, respectively. **18.** Normal light. **19.** Polarized light. **20.** Same, rotated 45°. **18—20,** Grayson Formation, ×3600. **21—25.** *Cribrosphaerella ehrenbergii* (Arkhangelskiy) Deflandre. **21.** Normal light. **22.** Polarized light in high focus. **23.** Same, in low focus. **21—23,** Pawpaw Formation, ×3600. **24.** Proximal view in SEM, Weno Formation. **25.** Distal view in SEM, Main Street Limestone. **24,25,** ×8000. All from Hill, 1976.

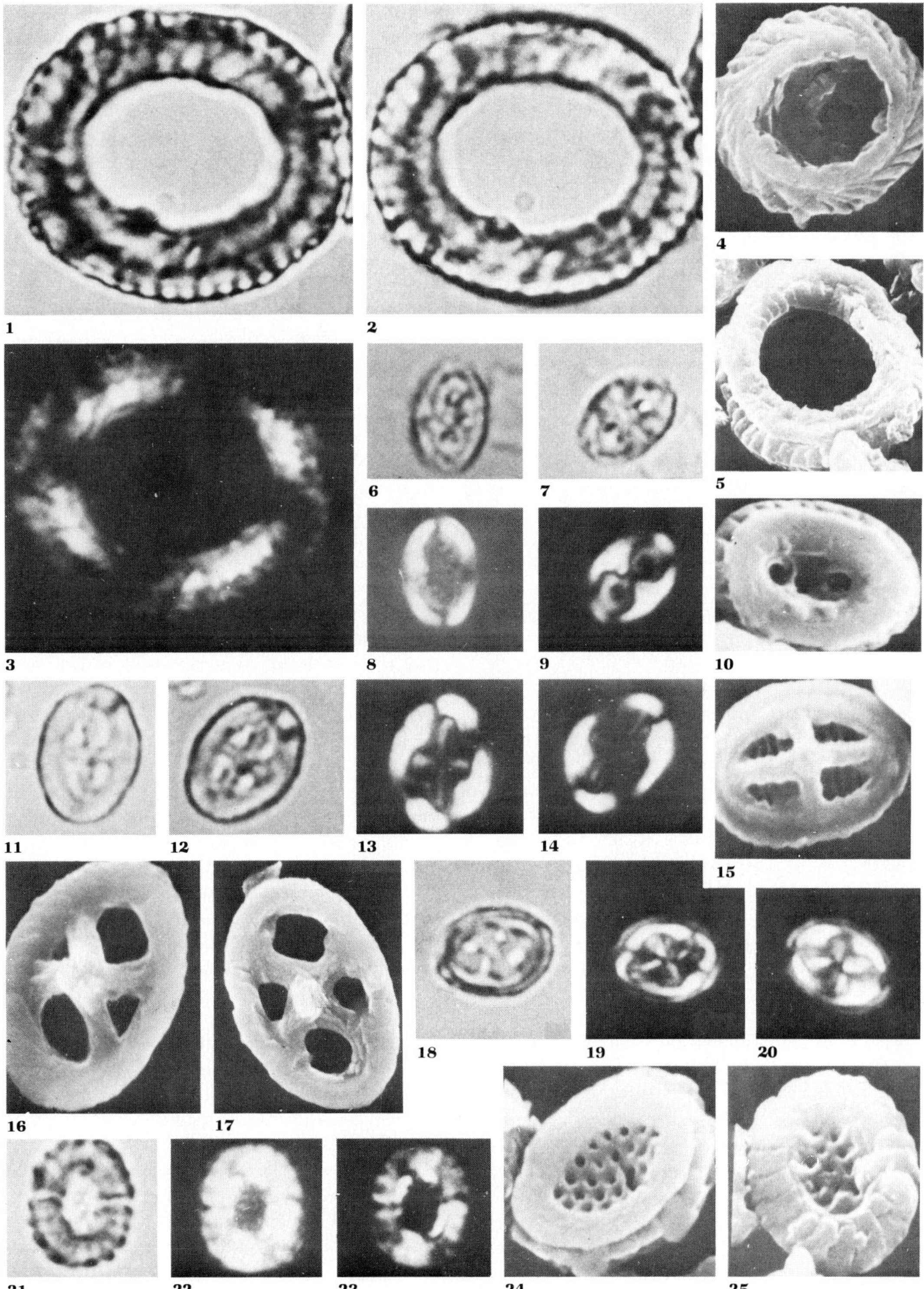

lated coccolith is under study. For example, a sinistral curvature of the crossed extinction lines indicates the distal side of *Discolithina,* but in other groups the same side may have dextral curvature. Care must be taken not to focus too far down, to avoid an erroneous appearance. Many coccoliths are more conspicuous and show more easily recognizable features under polarized light than in normal light.

In polarized light, a coccolith like that of the Cenozoic *Coccolithus* shows a simple interference figure. Although the coccolith is constructed of two shields or plates, connected centrally, the distal shield is not involved in the interference figure, as the optic axes of its crystals are at right angles to the plane of the shield.

Mesozoic *Ellipsagelosphaera* have a more complicated optical figure, as the crystals of the distal shield also show strong birefringence, and the black crosses of the two shields are superposed.

When the lines of the interference figure curve to the left, as traced out from the center in distal view of the coccolith, it is termed **laevogyre;** when these lines curve to the right, the figure is termed **dextrogyre.**

Coccolith Morphology and Crystal Structure. The varying degree of organic control of the crystal growth in coccoliths is the basis for subdivision into holococcoliths and heterococcoliths. In the former, the crystal form is unmodified and the crystals retain their geometrical faces and angles. The hexagonal or rhombohedral crystals merely show an ordered arrangement over the surface of the organic scales upon which they formed. In contrast, the internally formed heterococcoliths are elaborately constructed, their crystal faces and angles are suppressed, and the delicate form of the coccolith is completely controlled by the organism.

Both holococcoliths and heterococcoliths are constructed of numerous tiny crystals, commonly with a rotary arrangement, although rarely exactly radially symmetrical. These are termed **heliolithae.** Coccoliths consisting of a single or very few crystals may have a radial symmetry, as do discoasters, and are termed **ortholithae.** The appearance of the coccolith under crossed nicols of a petrographic microscope indicates whether the microstructure is of heliolithid or ortholithid character.

Inorganic calcite may have rhombohedral, prismatic, tabular, or scalenohedral crystal form. Most coccoliths are of rhombohedral calcite crystals in radial arrangement (including all with fibrous structure), a few have hexagonal prisms, and some have a combination of the two forms in a single coccolith. Discoasters have tabular crystals, and the scalenohedral crystal form occurs only in the Stephanolithiaceae. The two dominant crystal forms may be distinguished in polarized light, for the optic axis is parallel with the length of hexagonal prisms, and oblique to the rhombohedra.

Holococcoliths. Some holococcoliths are constructed of simple uncrowded rhombohedra scattered over the surface of the organic scale but not completely covering it, as in the motile crystallolithus phase of *Coccolithus pelagicus.* One crystal face of each rhombohedron rests against the scale. The platelike or saucerlike coccolith is bordered by a marginal row of crystals, one rhombohedron wide and two rhombohedra high (see Figure 9.10). Such a coccolith is termed a **crystallolith** (with reference to *Crystallolithus,* the motile phase of *Coccolithus*), or a disciform holococcolith.

Other holococcoliths may be constructed of hexagonal crystals. Those of *Sphaerocalyptra papillifera* have the form of a proximally open cap, and are termed **calyptroliths.** The upper part is formed of an assemblage of hexagonal prisms with basal pinacoid, arranged in a perfect lattice with the C-axis of the crystals perpendicular to the cell surface. The openings left are perfect hexagons, each bordered by six adjacent crystals (see Figure 9.11). All prism edges are parallel with the optic axis, the prismatic fibers giving a straight extinction between crossed nicols. The coccolith rim consists of partially disordered crystals of rhombohedral form.

Holococcoliths formed largely of hexagonal prisms, although some also have rhombohedra, include the calyptroliths of *Calyptrosphaera oblonga* Lohmann, and the **zygoliths** (each

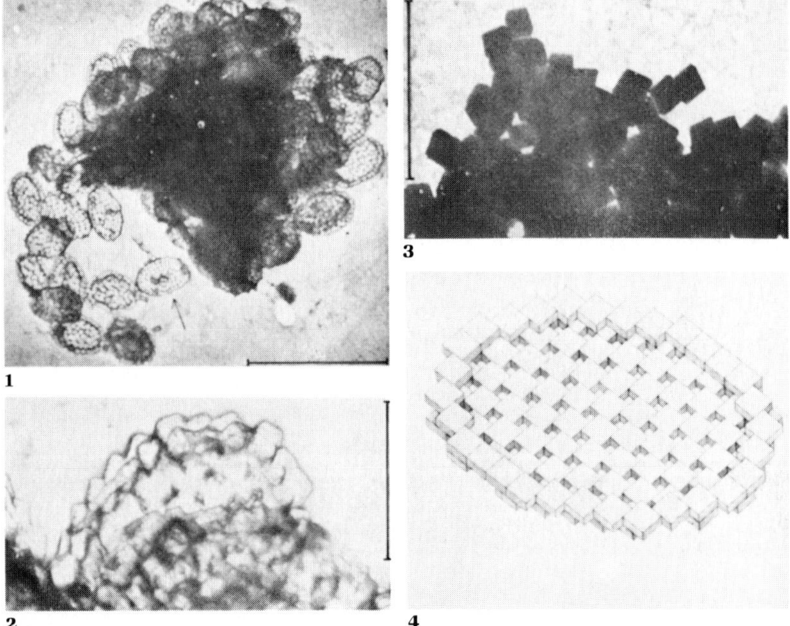

Figure 9.10

Crystallolith structure of motile *Coccolithus pelagicus* previously known as *Crystallolithus hyalinus*. **1.** Decalcified disintegrated cell, showing detached coccoliths, TEM, ×4000, bar = 5 μm. **2.** Part of same, enlarged to show peripheral row of rhombohedra, TEM, ×21,000, bar = 1 μm. **3.** Part of coccolith that is not decalcified, showing individual rhombohedra, TEM, ×24,000, bar = 1 μm. **4.** Model of the holococcolith structure, showing rhombohedra (compare coccolith marked by arrow in part 1). Reproduced with permission from Gaarder and Markali, 1956, *Nytt Mag. Bot.*, v. 5, p. 1−5, published by Universitetsforlaget, Oslo University.

an elliptical ring with slightly or strongly arched crossbar) of *Zygosphaera divergens* Halldal & Markali, *Homozygosphaera triarcha* Halldal & Markali, *H. tholifera* (Kamptner) Halldal & Markali, *H. wettsteinii* (Kamptner) Halldal & Markali, and *Periphyllophora mirabilis*. Crystal growth in holococcoliths is confined to directions parallel with the underlying scale surface, except at the margin and center. Once the outer rim is formed, later growth is only inwards, no growth occurring outward beyond the scale margin.

The small size of the crystals and their loosely packed arrangement allow the holococcoliths to disintegrate rapidly, and far fewer occur in the bottom sediments or are found as fossils than their proportion on living cells would indicate. Nevertheless, a few are known from sediments as old as Early Cretaceous (see Figure 9.12).

Heterococcoliths. Heterococcoliths are more complex than holococcoliths in structure, the crystals continuing to grow laterally, so that the adjoining rhombohedra overlap at their margins and the crystals become somewhat modified; the characteristic crystal edges and faces may still be recognizable in the electron mi-

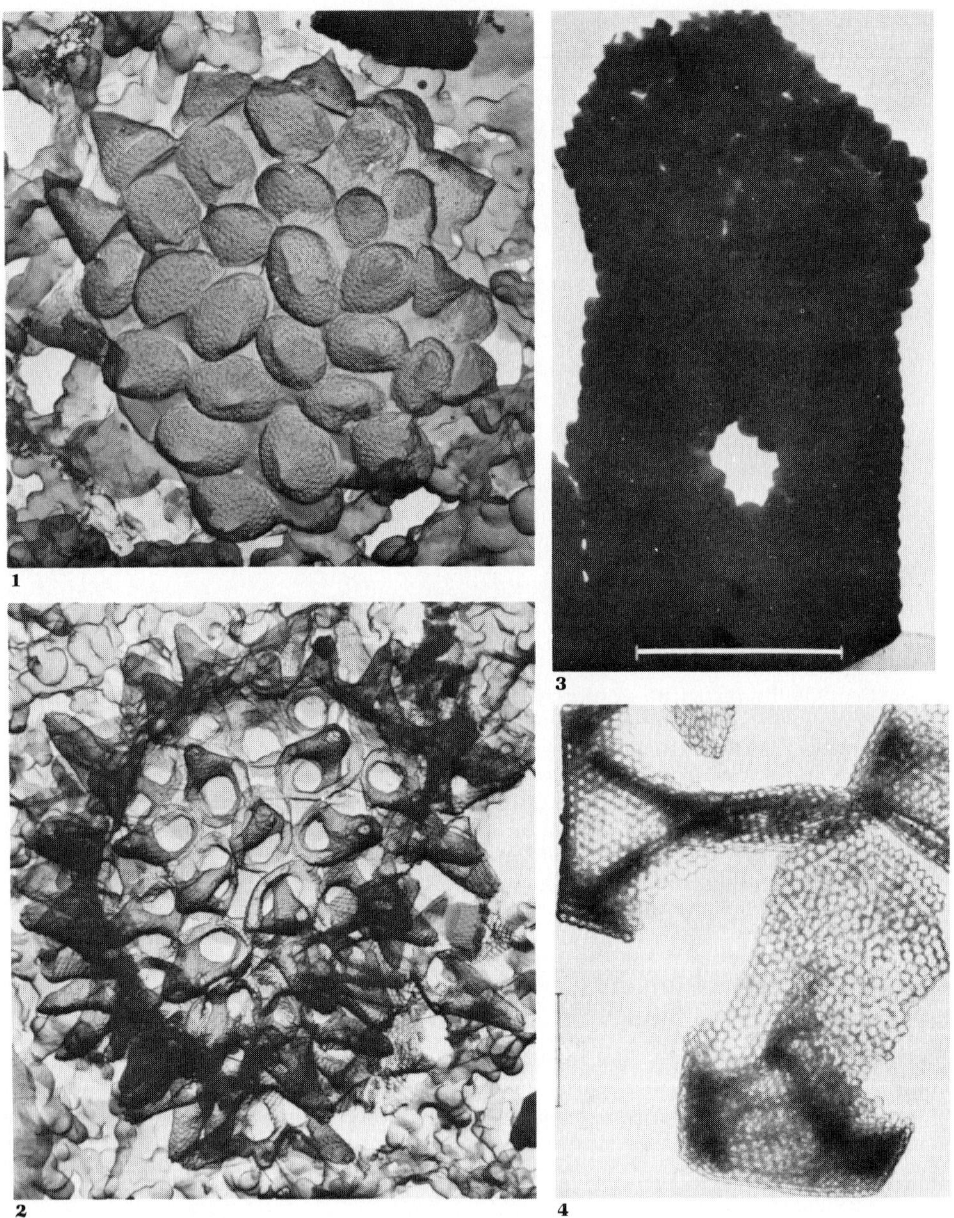

Figure 9.11

Holococcoliths. **1.** *Sphaerocalyptra* cf. *papillifera* (Halldal) Halldal, from plankton at 34°26′ N, 75°44′ W, TEM, ×8055. **2.** *Homozygosphaera triarcha* Halldal & Markali, from plankton at 34°27′ N, 76°06′ W, TEM, ×6248. 1,2, from A. McIntyre. **3,4.** *Periphyllophora mirabilis* (Schiller) Kamptner, Gulf Stream. Reproduced with permission from Halldal and Markali, *Avh. Norske VidenskAkad. Oslo*, 1955 (1), p. 1−30, published by Universitetsforlaget, Oslo University. **3.** TEM of isolated coccolith, a high zygoform holococcolith, ×28,000, bar = 1 μm. **4.** Decalcified coccoliths, showing closely packed hexagonal prisms, TEM, ×15,000 bar = 1 μm.

croscope. The coccolith of *Ahmuellerella phaseolus* (Black & Barnes) Reinhardt is a **discolith**, resembling a crystallolith but with more crowded crystals; a raised rim of small rhombs surrounds a disk formed by rhombohedra that are expanded as plates and crowded toward the center, where they pile up in a short mound or spirally twisted spike. Crystal structure here also suggests that growth began at the circumference and spread inward (see Figure 9.13).

The most complex coccoliths are formed internally, the crystal outlines are suppressed, and the shapes much more delicately molded. Dixon (1900) early had determined from sectioned living material that **placoliths**, which have two shields or plates connected by a cylindrical column, and thus resemble a collar button (see Figure 9.14), begin calcification as a narrow ring, which deepens into a collar, and then spreads laterally at the extremities to form the two shields. Study of the crystal structure has substantiated this (Black, 1963), crystal growth having begun near the top of the column. In the outer plate or shield a ring of rhombohedra surrounds the broad central pore, the crystals tapering somewhat toward the pore and expanding outward. The crystal form is preserved near the center of the plate, suggesting that growth probably began in that region and spread outward. The placolith column consists of rhombohedral crystals packed like barrel staves, or sometimes aligned obliquely, resulting in a spiral structure.

Other coccolith forms have been recognized in light microscopy, and their structure elucidated on the basis of electron microscopy.

In contrast to the proximally open calyptroliths, basketlike coccoliths with their opening distal in orientation on the cell are termed **lopadoliths**, as in *Scyphosphaera amphora* and *S. apsteinii* (see Figure 9.15). With a much shallower rim, similar to a discolith, but with many central perforations, the coccolith becomes a **cribrilith**, as in *Pontosphaera discopora* or *P. syracusana*. Another modification results in a **caneolith**, a discolithlike form with lath-filled central area, and petaloid upper and lower rims. **Cyrtoliths** are basketlike coccoliths with a projecting central structure, which may be very elongate and spinelike as in the specialized coccoliths at one pole of *Acanthoica*, or broad and stubby with a spike of fibrous crystals, as in the rhabdoliths of *Rhabdosphaera clavigera* and *R. stylifera*. The spines of *Discosphaera tubifera* are distally flared and trumpet shaped as a salpingiform cyrtolith (see Figure 9.16).

Some living species have two kinds of coccoliths on a single individual, one being a modification of the other in most cases. Thus the rhabdoliths of *Rhabdosphaera stylifera* are nearly identical with *Ahmuellerella phaseolus* except for the presence of the elongate spine. Similarly, *Syracosphaera pirus* and *S. pulchra* have simple caneoliths over most of the pyriform cell, but modified coccoliths surrounding the flagellar pole have a central spine

---

Figure 9.12 (*overleaf*)
Fossil holococcoliths. 1−3. Oligocene, Red Bluff Clay, Alabama, TEM, ×9600. 1,2. *Orthozygus aureus* (Stradner) Bramlette & Wilcoxon, side and proximal views. 3. *Lanternithus minutus* Stradner, distal view. 4−6. *Scampanella cornuta* Forchheimer & Stradner, Upper Cretaceous (Coniacian), Austria. 4. Side view, TEM, ×4080. 5,6. Oblique proximal view and side view, SEM, ×3300. 7,8. *Naninfula deflandrei* Perch-Nielsen, upper Eocene, Denmark, oblique and lateral views, TEM, ×8800. 9. *Peritrachelina joidesa* Bukry & Bramlette, upper Eocene, Shubuta Member of the Yazoo Formation, Mississippi, from side showing peripheral rim, TEM, ×8000. 10,11. *Holodiscolithus macroporus* (Deflandre) Roth, Oligocene Red Bluff Clay, Alabama, TEM, ×9600. 12,13. *Zygrhablithus bijugatus* (Deflandre) Deflandre. 12. Oblique distal view of basal disc, Red Bluff Clay, Alabama, TEM, ×9600. 13. Side view of complete specimen, from offshore sediments, western Atlantic, TEM, ×8000. 14. *Daktylethra punctulata* Gartner, middle Eocene, Lisbon Formation, Alabama, side view, TEM, ×7200. 1−3, 9−13, from Gartner and Bukry, 1969; 4−6, from Forchheimer and Stradner, 1973; 7,8, from Perch-Nielsen, 1968b.

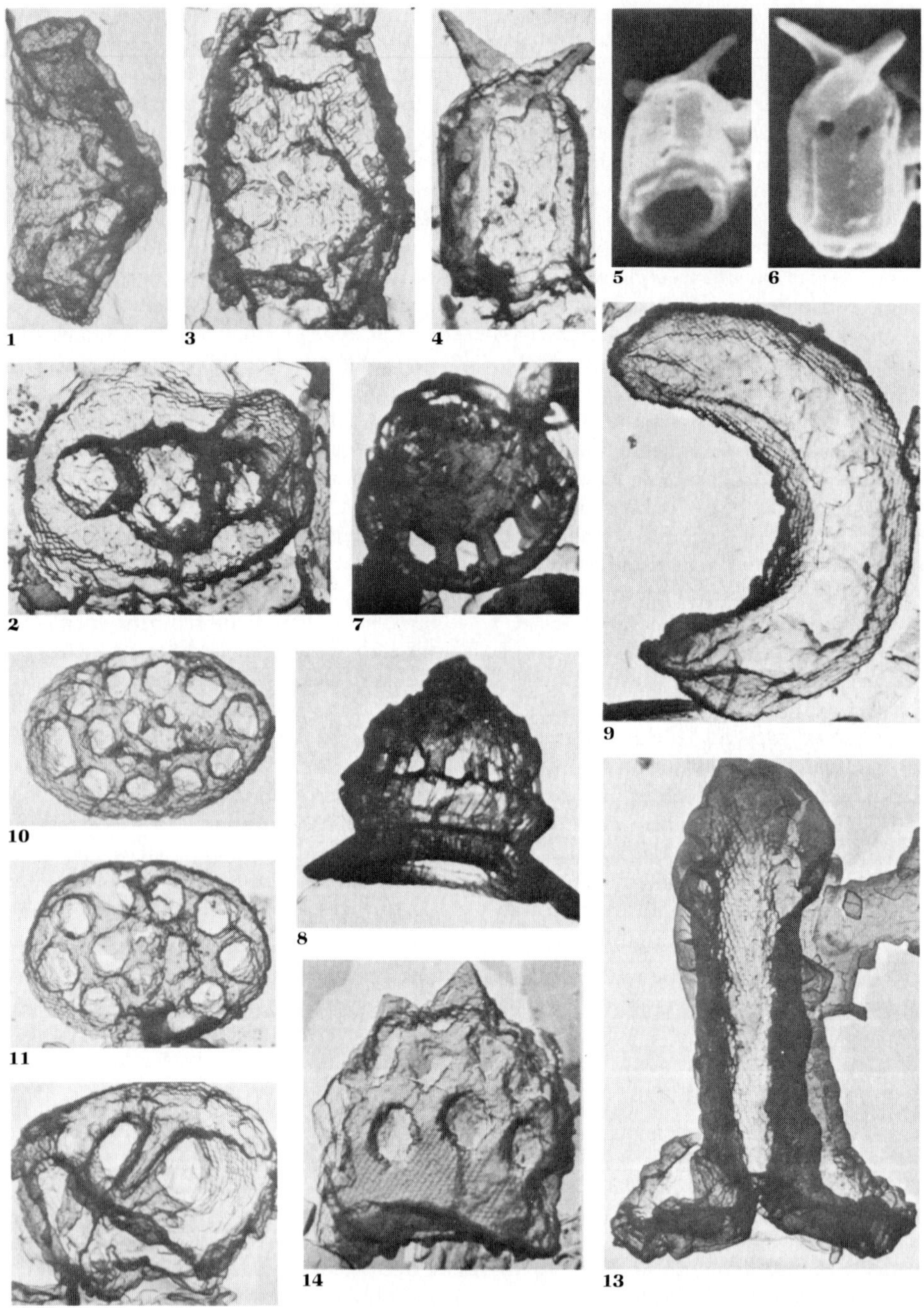

Figure 9.12

Figure 9.13

1. Coccosphere of *Rhabdosphaera stylifera* Lohmann, showing rhabdoliths and some coccoliths without spines, from plankton at 34°52′ N, 75°25′ W, TEM, ×5328. Smaller specimen is *Emiliania huxleyi*. Reproduced with permission from McIntyre and Bé, 1967, *Deep Sea Research*, v. 14, p. 561–597, copyright 1967, Pergamon Press, Ltd. **2.** *Rhabdosphaera clavigera* Murray & Blackman, isolated coccolith, SEM, ×6100, reproduced with permission from Black, 1965, *Endeavor*, v. 24, p. 131–137, copyright 1965, Pergamon Press, Ltd. **3,4.** *Ahmuellerella phaseolus* (Black & Barnes) Reinhardt, bottom sediments, South Atlantic, TEM, ×10,000, from Black and Barnes, 1961. Note similarity of latter specimens to plates at the center of the coccosphere of *R. stylifera* (part 1) that lack the long spines.

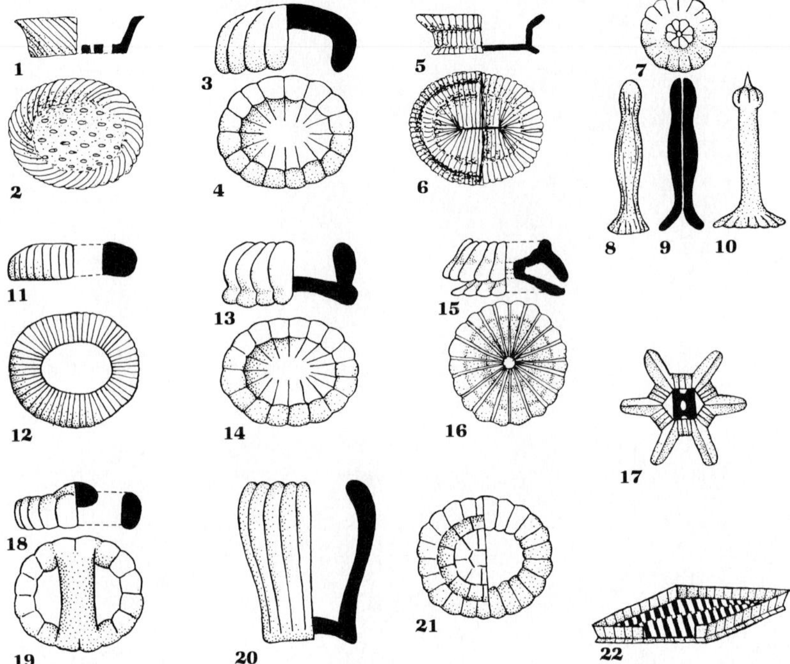

Figure 9.14

Shape of coccoliths, shown diagrammatically in plan and edge or profile views; edge views show exterior at left and section at right. **1,2.** Cribrilith, edge and distal views. **3,4.** Calyptrolith, edge and proximal views. **5,6.** Caneolith, edge and plan views, the latter showing proximal view at left and distal view at right. **7–10.** Rhabdolith. **7.** Plan view. **8,10.** Side views. **9.** Section. **11,12.** Cricolith, edge and distal views. **13,14.** Discolith, edge and distal views. **15,16.** Placolith, edge and proximal views. **17.** Stephanolith, in plan view. **18,19.** Zygolith, edge and distal views. **20,21.** Lopadolith, side and plan views, the latter showing proximal view at left and distal view at right. **22.** Scapholith, oblique view.

Figure 9.15

1. *Pontosphaera syracusana* Lohmann, ×7000. **2.** *P. discopora* Schiller, ×11,188. 1,2, from sediment core, 31°21′ S, 36°49′ W. **3.** *Scyphosphaera amphora* Deflandre, from 610 cm in core from 21°12′ N, 45°21′ W, ×5936. 4. *S. apsteinii* Lohmann, from sediment core at 17°12′ S, 16°13′ W, ×5052. All TEM of carbon replicas, from A. McIntyre.

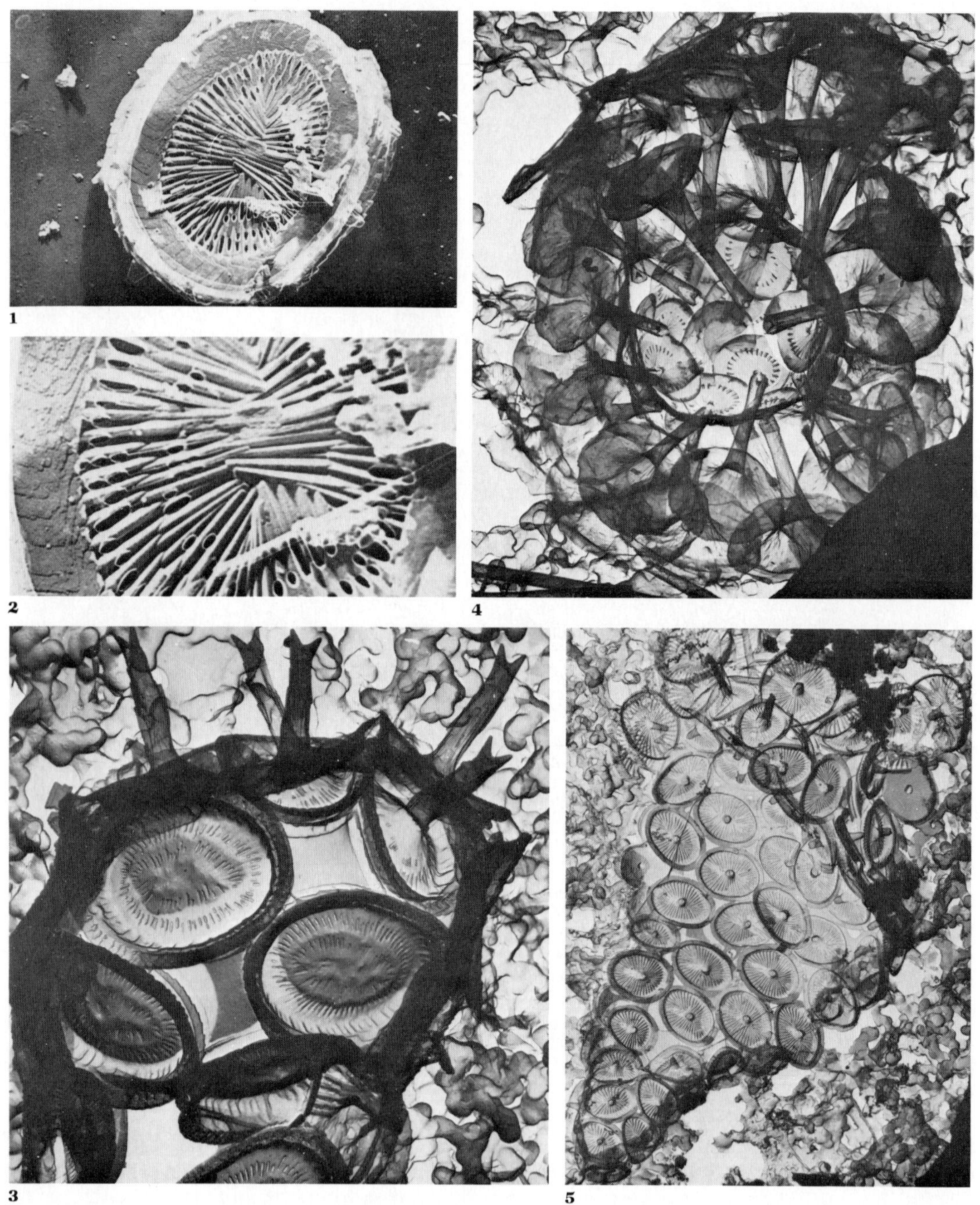

1
2
4
3
5

like that of a rhabdolith, terminally bifurcate in *S. pulchra.*

Another modification of the basal discolith form is shown in the **helicolith,** such as found in *Helicosphaera carteri;* the marginal flange begins as a series of short radial parallel elements that increase in height around the periphery of the oval disk, and overlap the early part of the flange in a spiral way. An elongate, diamond-shaped coccolith with a central area of parallel laths and a shallow rim is termed a **scapholith.** It characterizes the living *Anoplosolenia brasiliensis* (see Figure 9.17) and *Calciosolenia;* fossil isolated scapholiths are referred to *Scapholithus.*

Somewhat simpler morphology characterizes the **cyclolith,** a simple circular ring of elements, and the **cricolith,** which is elliptical in form as in *Cricosphaera carterae.*

Many Mesozoic coccoliths have the basic structure of a **loxolith ring,** having an outer wall of many narrow, steeply inclined calcite laths. These resemble barrel staves that have been sharply twisted (see Figure 9.18). Such a loxolith ring commonly has a single layer, but a few may have a second inner layer with the laths at a different angle. The inclination of the laths is the same when the coccolith is viewed from either side and resembles the threading of a left-handed screw, hence is described as a left-handed **imbrication.** This structure occurs in the simple unmodified cricolithlike ring of *Loxolithus;* in *Zygolithites* a transverse bridge spans the loxolith ring, whereas a cross connects the opposite sides of the loxolith ring of *Staurolithites* and *Chiastozygus* (Black, 1972a).

The separate shields of a placolith may be constructed of radially arranged wedgelike crystals or petaloid elements, lying adjacent to each other, in a nonimbricated, **jointive** arrangement. Others have overlapping segments, in an imbricate pattern such as the left-handed or counterclockwise imbrication of the distal shield of *Cyclagelosphaera,* which resembles the loxolith structure in this respect. More complex coccoliths may have separate cycles of clockwise and counterclockwise imbricated elements. Still other types may change the direction of element inclination across a convex shield, as in *Calcidiscus leptoporus* (see Figure 9.19).

An additional feature in the arrangement of elements of a coccolith is the **precession,** defined as the deviation from a radial direction

1

2

3

4

5

Figure 9.17

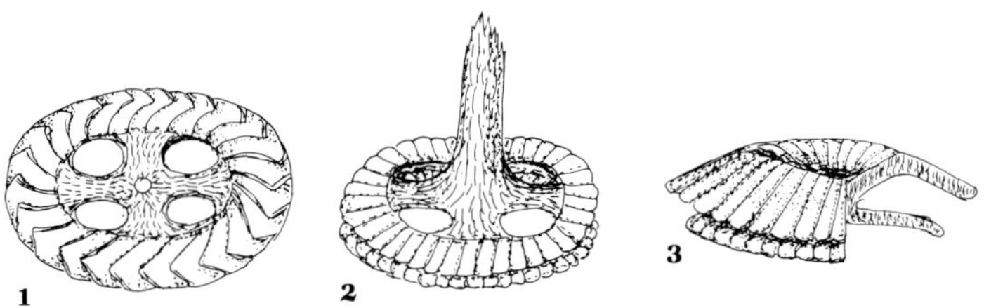

Figure 9.18

Coccolith structure, diagrammatic. **1.** Loxolithid structure, plan view, with outer wall of steeply inclined segments with sinistral (left-handed) imbrication. **2.** Podorhabdid structure, oblique view, with outer rim of two cycles of petaloid or jointive elements. **3.** Coccolithid structure, oblique side view, shown in section at right, with three series of elements forming proximal and distal shields, with a connecting tube or pillar.

Figure 9.19

1–4. *Calcidiscus leptoporus* (Murray & Blackman) Loeblich & Tappan, SEM. **1.** Entire coccosphere, Pleistocene, North Atlantic, ×3550. **2.** Single coccolith in proximal view, showing smaller proximal shield and larger diameter of distal shield, Pliocene, Algeria, ×6100. **3.** Distal shield in proximal view, from which the smaller proximal shield has broken away, Miocene, Algeria, ×4800. **4.** Distal shield in distal view, showing imbricated elements, Miocene, North Atlantic, approximately ×4900. 1,4, from Perch-Nielsen, 1972; 2,3, from Clocchiatti, 1971a. **5.** *Blackites amplus* Roth & Hay, Oligocene, Denmark, various cycles of elements showing different degrees of imbrication, and peripheral cycle of jointive, nonimbricative elements, TEM, ×10,500, reproduced with permission from Black, 1965, *Endeavor*, v. 24, p. 131–137, copyright 1965, Pergamon Press, Ltd. **6.** *Coronocyclus prionon* (Deflandre & Fert) Stradner, Miocene, North Atlantic, SEM, ×9000, from Perch-Nielsen, 1972.

undergone by the sutures between elements. If this suture deviates to the right in crossing the shield from the center to the circumference, as observed from the distal side, the precession is clockwise. In contrast to the imbrication, the precession appears to reverse in direction when viewed from the proximal side of the disk.

Early TEM of carbon replicas occasionally were reversed in printing to obtain dark shadows, or the carbon replica might be reversed on the grid, giving a false appearance to the direction of precession and imbrication. This difficulty is not found in SEM.

A distinctive coccolith of the Jurassic and Cretaceous is the **stephanolith,** a hollow form with polygonal or cylindrical wall, containing an open framework of radiating rods, which may extend as spines outside the ring. Each spine is a single scalenohedral crystal with optic axis perpendicular to the central axis of the coccolith. Another specialized form, the **sphenolith,** has a prismatic base of radial elements surmounted by a cone. In polarized light, a black cross is visible in longitudinal view, rather than in apical view as in other coccoliths.

Ortholithae. Coccoliths composed of one or very few crystals have a less varied morphology (see Figure 9.20). The **asteroliths** of *Discoaster* are stars or rosettes with a concave face, each ray formed of a single crystal, with vertical C-axis perpendicular to the plane of the disk. Individual crystals show a bilateral symmetry, suppressing the trigonal symmetry of the inorganic crystals. In crossed nicols, with the asterolith flat against the glass, they appear dark, hence represent a single crystal or a group of crystals with parallel C-axes. When viewed with SEM each ray is seen to be an independent crystal. The simplest structure is that of *Discoaster obtusus* Gartner, consisting of six rhombohedra (see Figures 9.21 and 9.22), whereas the rhombohedra of *D. adamanteus* Bramlette & Wilcoxon are modified to produce a stellate outline (Black, 1972b).

Braarudosphaera has **pentaliths** with five pieces, each consisting of a single crystal, oriented 72° apart. The cleavage plane of the crystals is in the plane of the pentalith, and perpendicular to the principal axis, to which

the optic axes are oblique. The calcite is internally laminated parallel to the cleavage, as can be seen in slightly eroded specimens in electron microscopy. Pentaliths may appear to be radially symmetrical, but most show slight variation in the dimensions of the separate crystals, giving a rotary symmetry like that of most other coccoliths. Even if no variation is visible in the symmetry, the extinction pattern in polarized light indicates the basic rotary structure (Black, 1972b).

Ortholithid structure also characterizes the **prismatoliths** or poroliths of *Thoracosphaera,* which are constructed of polygonal prisms that may be solid or axially perforated. No culture studies have as yet been made of forms with ortholithid structure, in order to determine how these coccoliths are formed, but it is interesting that all lack openings, and the tightly constructed cells of *Braarudosphaera* and *Thoracosphaera* seem to have characteristics of a cyst rather than of a vegetative cell (see Figure 9.23). Apparently some *Thoracosphaera* species are calcareous dinoflagellate cysts, but, as discussed later, typical *Thoracosphaera* is distinct. Discoasters also may be isolated cyst elements.

Ceratoliths are large horseshoe-shaped forms, known first as fossils, but later found as a single ceratolith per cell in a living coccolithophore (Norris, 1965, 1971). Each ceratolith acts as a single crystal of calcite, but in different genera the optic axis is distinctly oriented, so that some show maximum birefringence (*Ceratolithus*) and others show little or none (*Amaurolithus*). The C-axis of *C. cristatus* is in the plane of the ceratolith, and perpendicular to its long axis (Gartner and Bukry, 1975).

Micrococcoliths and Macrococcoliths. Coccoliths of two different sizes have been described on cells of *Umbellosphaera tenuis* (Kamptner) Paasche and *U. irregularis* Paasche (Markali and Paasche, 1955; McIntyre and Bé, 1967; Okada and Honjo, 1970). As they are identical in morphology, **micrococcoliths** may be those formed in the early stages of development of the cell, whereas **macrococcoliths** were formed after the cell attained its full size. Other species may have separate layers of distinct coccolith forms, the outer layer commonly hav-

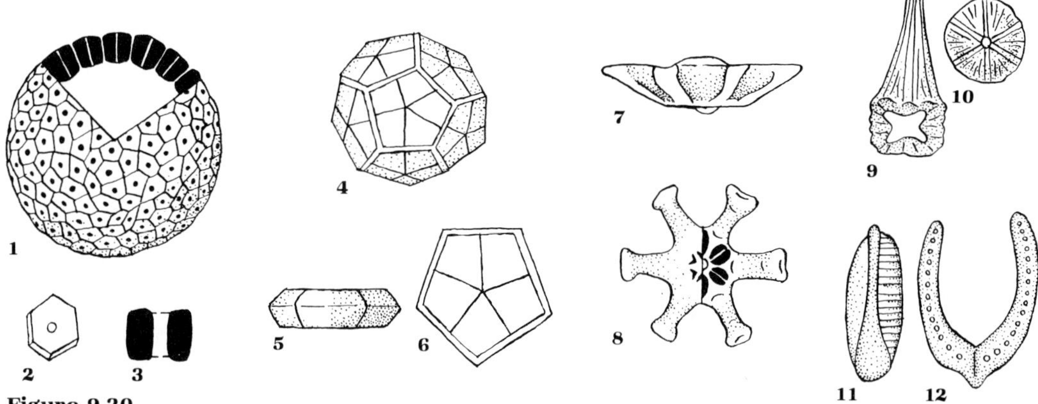

Figure 9.20

Ortholithae, diagrammatic. **1—3.** Prismatoliths; entire cell, with one segment shown in section; plan view and section of an isolated prismatolith. **4—6.** Pentaliths; entire cell, edge and plan views of isolated pentalith. **7,8.** Asterolith in edge and plan views, the latter showing appearance of opposite faces at left and right. **9,10.** Sphenolith, side and top views. **11,12.** Ceratolith, edge and side views.

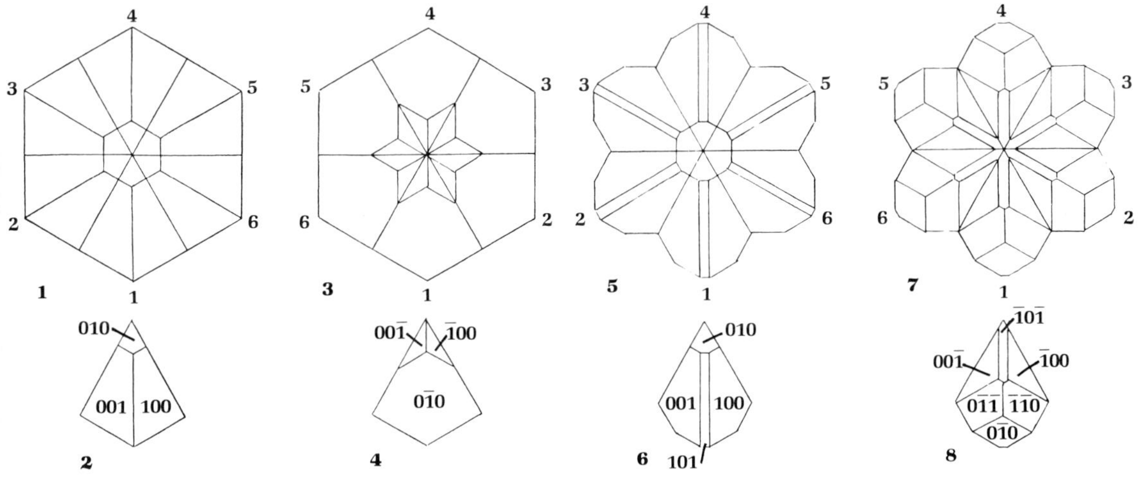

Figure 9.21

Crystallographic structure of discoasters, shown in orthographic projection on the principal plane, the basal pinacoid (111). **1—4.** *Discoaster obtusus* Gartner; each of the six arms is part of a rhombohedron, and has a crystallographic plane of symmetry along its length. **1,2.** From the steeply conical F surface, so called because a ring of crystal faces surrounds the center, with the face (010) toward the center; entire asterolith and a single arm. **3,4.** E surface, generally flatter, with crystal edges radiating away from the center; entire asterolith and a single arm. **5—8.** *D. adamanteus* Bramlette & Wilcoxon; the arms are parts of crystals combining the primtive rhombohedron, r = (100), and the obtuse rhombohedron, e = (101), with unequal development of the faces. **5,6.** F surface, entire asterolith and single arm. **7,8.** E surface, entire asterolith and single arm. From Black, 1972b.

Figure 9.22

Crystallographic structure of discoasters (compare Figure 9.21). **1,2.** *Discoaster obtusus*. **1.** Lower Miocene, Jamaica, E surface, (1̄00) and (001̄) form an arrowhead pair of faces pointing to the center, and (01̄0) is larger, sloping outwards, ×9000. **2.** Lower Miocene, Pacific Ocean, F surface, face (010) developed as ring of tiny faces at center, ×7500. **3–5.** *D. adamanteus*. **3.** Lower Miocene, Pacific Ocean, E surface, (1̄00) and (001̄) present as very small faces pointing to center, (1̄10) and (01̄1̄) large, ×4500. **4.** Middle Miocene, Trinidad, F surface, three of the arms united at the center to form a single crystal, ×7500. **5.** Miocene, North Atlantic, F surface, ×4500. 1–4, reversed prints of carbon replicas in TEM, from Black, 1972b; 5, SEM from Perch-Nielsen, 1972.

Figure 9.23 (*facing page*)

Ortholithoid coccoliths. **1.** *Braarudosphaera bigelowii* (Gran & Braarud) Deflandre, Cretaceous, New Jersey; a pentalith, TEM of carbon replica, ×9718. **2,3.** *Ceratolithus cristatus* Kamptner, TEM of carbon replicas. **2.** From sediment core, at 13°15′ N, 40°40′ W, ×5873. **3.** From 630 cm depth in core, at 21°12′ N, 45°21′ W, ×6903. 1–3, from A. McIntyre. **4–7.** *Thoracosphaera heimii* (Lohmann) Kamptner, SEM, northeast Atlantic, from Fütterer, 1976. **4.** Surface enlarged, showing sharp-edged crystallites, Holocene, ×6600. **5.** Entire specimen with distinct prismatoliths, Pleistocene, ×3400. **6.** Part of surface of collapsed Holocene specimen, showing outer surface at left, and inner surface at right, ×6600. **7.** Pleistocene specimen, with slight calcite overgrowth rounding the crystallites, ×3600. **8.** *T.* cf. *deflandrei* Kamptner, showing prismatoliths, upper Eocene, North Pacific, ×2900, from Haq and Lipps, 1971.

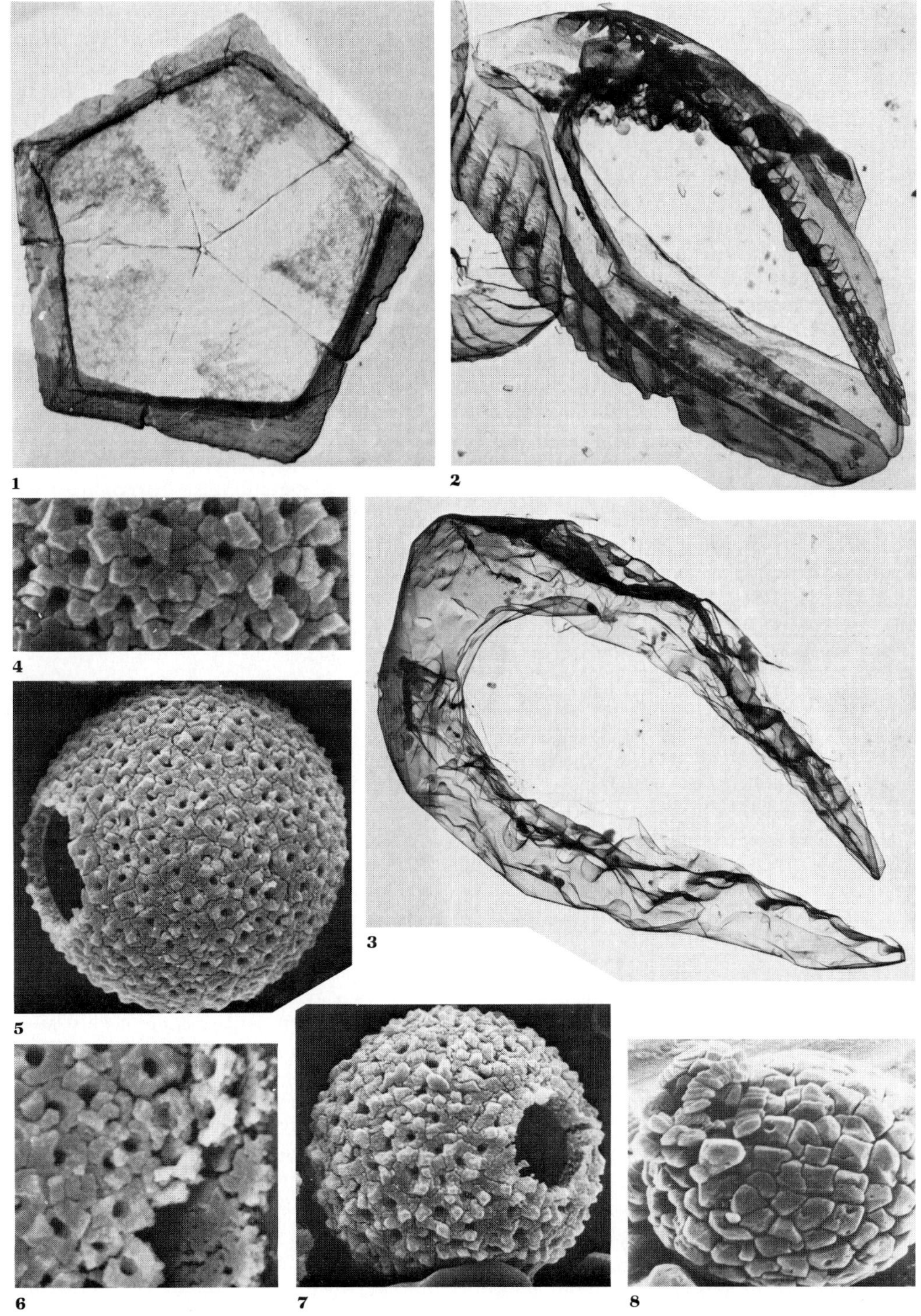

1
2
4
3
5
6
7
8

ing more prominent projections or other ornamentation.

Malformed Coccoliths. Many neritic species in the Red Sea, Yellow Sea, Timor Sea, and East and South China Seas were found to have irregularly developed coccoliths (Okada and Honjo, 1975), more than 90 percent of the cells having the malformed coccoliths in most stations in the South China Sea. Coccoliths of *Gephyrocapsa oceanica* Kamptner had an irregular central bar or lacked this structure completely, lacked a floor in the central opening, and the elements of the disk were narrow and barely flush, rather than imbricated as in normal coccoliths. Some coccospheres had a cover of randomly oriented elements, and only a few if any normal coccoliths. Similarly aberrant coccoliths of *Emiliania huxleyi* cells had twisted elements of distal shield and grid, coccoliths of *Syracosphaera pulchroides* Halldal & Markali had angular margins with adherent grid elements, and *Michaelsarsia elegans* Gran had irregular coccoliths with angular outlines and distorted grids. Malformed *E. huxleyi* also had been reported in the subarctic Pacific (Okada and Honjo, 1973).

No correlation of the malformations was observed with temperature, salinity, or dissolved oxygen, ammonium, and phosphorus, but some relationship was suggested with nitrite levels less than 0.1 μg-atoms per liter, although this was not conclusive (Okada and Honjo, 1975).

REPRODUCTION AND THE LIFE CYCLE

The maximum reported division rate for coccolithophores in laboratory culture is about 2.25 divisions per day in *Cricosphaera elongata*. Other species have a generation time ranging from 20 to 36 hours (*C. carterae*), and from 20 to 60 hours in *Emiliania huxleyi*. These times were determined under varying experimental conditions, hence they may not accurately reflect the normal generation time in nature, although the shortest time probably indicates the maximum possible rate of division for these species.

Many haptophytes show an alternation of two quite distinct stages in their life history (a dimorphic life cycle). Commonly one is coccolith-covered, but this may be either the motile or nonmotile part of the cycle, and some species have both generations coccolith-covered, even possessing different types of coccoliths. The nonmotile stage may be a floating sphere, or it may be filamentous and attached, or consist of accumulated masses of cells. The greatly varied appearance and even different habitat of the separate stages in the life cycle has led to their description as distinct living species and genera, and adds to the complications in recognizing fossil species.

The complete life cycle and type of cell present in the various stages are known for only a few haptophytes, yet many variations are represented. Eventually relationships may be better determined on the basis of life cycles than from cell or coccolith morphology alone, for a single species may possess distinct types of cells, variously covered with scales or different coccolith forms in different phases of its development.

Coccolith-Covered Motile and Nonmotile Pelagic Stages

The best known form of *Coccolithus pelagicus* is a nonmotile ovoid cell up to 25 to 40 μm in diameter and bearing 25 to 35 heavy simple ringlike coccoliths (placoliths), each between 4.5 and 13 μm in size (see Figure 9.24). Two types of organic scales also cover the cell. A few large thick oval scales, about 7.5 $\times$ 5 μm in size, form the outermost layer, and each is attached to the proximal surface of one of the coccoliths. Beneath this layer of coccolith-covered scales are several layers of many smaller thinner scales, ranging from circular to oval in plan, and approximately 2 μm in diameter, adjacent to the cell surface. Both scales and coccoliths are produced in the cisternae of the Golgi apparatus that is located near the flagellar bases, which persist in the nonmotile stage, although not bearing flagella. A single coccolith is formed at one time within the cell (Manton and Leedale, 1969).

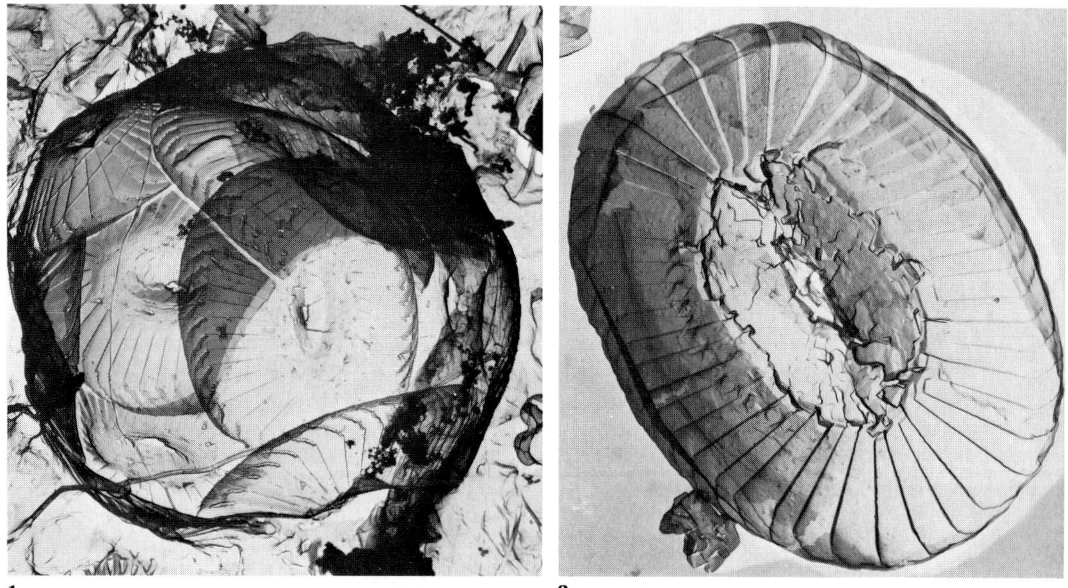

Figure 9.24
Coccolithus pelagicus, nonmotile stage, North Atlantic, TEM of carbon replicas. **1.** Entire coccosphere from plankton, at 61°31′ N, 36°03′ W, ×4770. Reproduced with permission from McIntyre and Bé, 1967, *Deep Sea Research,* v. 14, p. 561−597, copyright 1967, Pergamon Press, Ltd. **2.** Distal view of single coccolith, from sediment core at 55°49′ N, 52°27′ W, ×8325, from A. McIntyre.

This nonmotile stage of *C. pelagicus* can reproduce itself by simple fission, but eventually divides doubly (perhaps involving meiosis) into four spheroidal daughter cells, each from 10 to 22.5 μm in diameter and with 8 to 17 placoliths. Later one of the coccoliths is discarded, and a motile biflagellate cell escapes from the resultant opening in the coccosphere. Once described as a distinct taxon, *Crystallolithus hyalinus* (see Figure 9.25), the motile stage has thin, transparent, oval to elliptical plate scales embedded in the cell surface. One face of the scale has radiating ridges, and the other a network of concentric ridges. Calcite crystals are deposited on the external surface of the scales as a distinct type of coccolith (crystallolith). The motile biflagellate cell has a coiling haptonema, and swims rather slowly with a slow rotation. Food particles are ingested and digested at the nonflagellar pole, and the waste gathered into a vacuole that is periodically pushed out, taking with it part of the external scale layer and some of the coccoliths. Reproduction by binary fission produces two cells of unequal size, or a double fission may result in three or four daughter cells. As the naked daughter cells escape, the empty coccosphere settles to the bottom. The new cells gradually produce a scale cover and later the coccoliths. This motile stage also may reproduce itself for many generations. In cultures, the cells may settle to the bottom of the flask after five to eight weeks, losing their flagella, but retaining the haptonema for attachment. Larger cells arise (whether solely by vegetative growth or after fusion is not known) and develop large reserve food vesicles (of leucosin) that may occupy three-quarters of the cell volume. This leucosin reserve is utilized and disappears as the cell size increases. Other large vesicles then appear within the cell, in which are developed the few, very heavy and characteristic heterococcoliths of the nonmotile stage of *Coccolithus pelagicus.* In this species, both motile and nonmotile stages bear

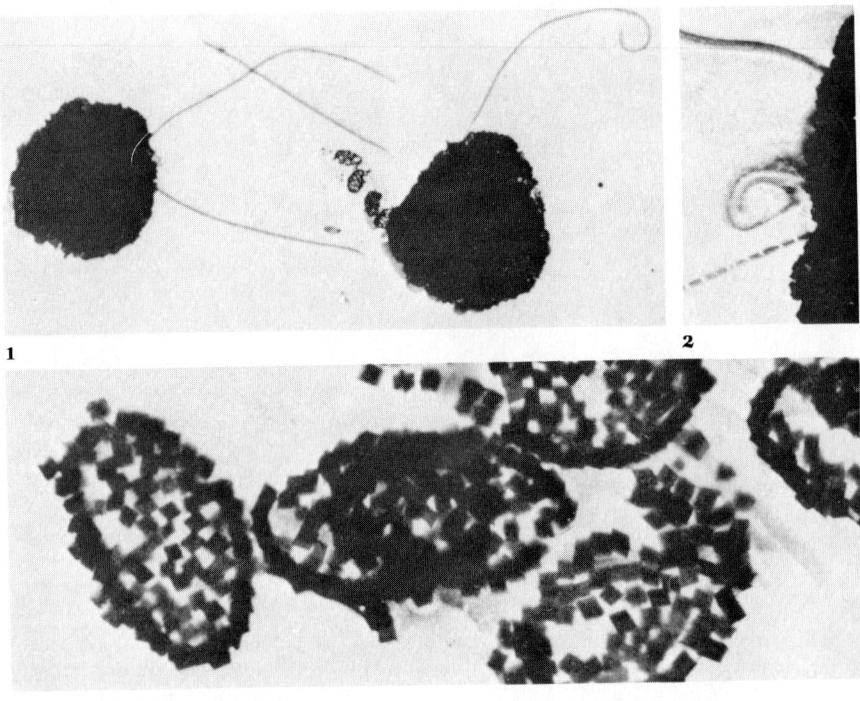

Figure 9.25

Coccolithus pelagicus motile stage (previously known as *Crystallolithus hyalinus*), TEM of shadowcast whole mounts. **1.** Two cells with crystallolith-covered bodies and two flagella, ×1400. **2.** Apex of cell, showing the coiled haptonema between the flagellar bases, ×8000. **3.** Crystalloliths, showing rhombohedra arranged on the plate scales, ×23,000. All reproduced, with permission, from Parke and Adams, 1960, *J. Mar. Biol. Ass. U.K.*, v. 39, p. 263−274, published by Cambridge University Press.

coccoliths, the motile one with crystalloliths of holococcolith form (the individual calcite crystals being easily recognizable), and the nonmotile form with heterococcoliths. Both stages are common in the plankton (Parke and Adams, 1960).

Coccolith-Covered Nonmotile Stage; Naked or Scale-Covered Motile Stage

Unlike *Coccolithus pelagicus,* only the nonmotile stage of *Emiliania huxleyi* is coccolith-covered, and it does not have the organic scales.

This coccolith-covered cell may divide as such, the coccoliths being divided among the daughter cells. At other times a naked swarmer may escape from the coccolith-cover as an amoeboid cell that rounds off and divides to produce other similar naked cells, the daughter cells gradually obtaining a coccolith cover. At other times, the naked cell will produce a scale cover, develop flagella, and then become motile, reproducing by fission to give rise to the nonflagellate coccolith-bearing form. The biflagellate variant and the nonmotile coccolith-covered stage may alternate in nature as part of the life cycle; the naked nonmotile cells, which contain the usual coccolith-forming structures

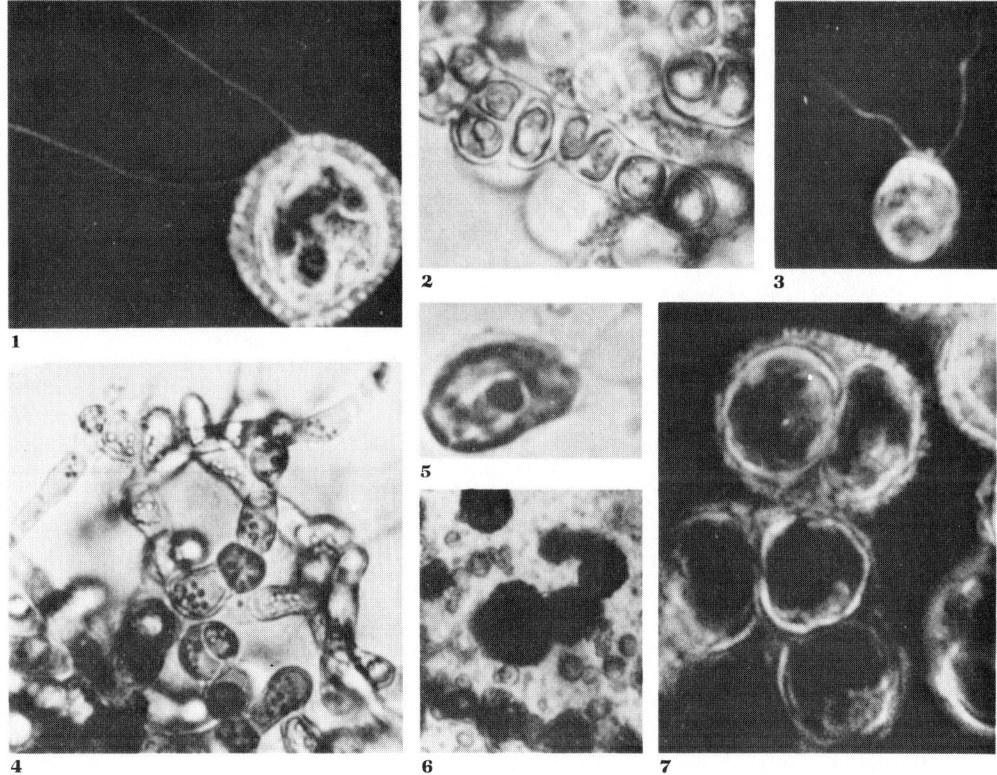

Figure 9.26
Coccoid and filamentous nonmotile stages in coccolithophores. *1,2. Hymenomonas pringsheimii.*
1. Motile cell, ×1650. **2.** Filamentous stage, ×800. **3,4.** *Cricosphaera carterae.* **3.** Motile cell
(compare Figure 9.5), ×1650. **4.** Filamentous *Apistonema*-like stage, ×375. **5—7.** *Ochrosphaera
neapolitana.* **5.** Motile cell, ×2300. **6.** Cyst stage, the cysts appearing as dark round masses, ×375.
7. Coccoid stage dividing, some with coccolith cover, ×1650. All from Parke, 1961a.

inside, but have no coccoliths or scales, may result from a loss of the ability to calcify due to laboratory culture conditions (Klaveness and Paasche, 1971). The naked clone and the coccolith-covered one have similar cell volume, protein content, and DNA content, hence these are not alternate phases of the sexual cycle. When placed in a nitrate-deficient medium, the naked cells produce a coccolith cover, suggesting that these atypical cells revert to the wild type under a physical stimulus (Paasche and Klaveness, 1970). Coccolith formation by the nonmotile stage is rapid, the entire cover with more than one layer of coccoliths being produced in a few hours. In the natural habitat the outermost coccoliths probably are detached by the mechanical effect of water turbulence, but they remain attached to the cell in laboratory cultures (Braarud, 1963).

The scale-covered biflagellate stage of *Ochrosphaera verrucosa* Schussnig is motile for only a short time, and alternates with the more common nonmotile coccoid, coccolith-bearing stage (Schwarz, 1932; Parke, 1971). The similarly nonmotile coccolith-bearing stage of *O. neapolitana* may divide by simple fission, or may produce scale-covered flagellate swarmers that again give rise to the coccolith-covered nonmotile stage (see Figure 9.26). A **pseudocyst,** produced by fusion of the coccoliths of this nonmotile cell (Gayral and Fresnel-Morange, 1971), strongly resembles *Thoracosphaera* (described only from empty cells).

Both *Ochrosphaera neapolitana* and *O. verrucosa* may have coccoliths of two types, one resembling the tremaliths of *Hymenomonas,* and one much more elevated (Gayral and Fresnel-Morange, 1971, 1976). *Ochrosphaera* has also been suggested to be a stage in the life cycle of *Cricosphaera carterae* (Lefort, 1971, 1975), but strains producing the *Ochrosphaera* phase in culture are regarded as not true *C. carterae* (Parke and Green in Parke and Dixon, 1976), whose coccoliths are in the form of cricoliths.

In addition to the scale- and coccolith-covered cells, *Ochrosphaera* may have a nonmotile stage consisting of packets of mucilage-enclosed cells, with extracellular calcareous deposits that are somewhat similar in appearance to the fossil genus *Tetralithus,* although structurally distinct (Parke, 1971).

Coccolith-Covered Motile Stage; Benthic Filamentous Nonmotile Stage

Some freshwater and near-shore marine coccolithophores have a coccolith-covered motile stage, but the naked nonmotile stage is quite dissimilar in appearance and may be a filament or packet of cells. The filamentous stages originally were described as distinct genera, *Apistonema, Chrysonema, Chrysotila, Chrysosphaera, Gloeothamnion, Nematochrysis, Ochrosphaera, Pleurochrysis,* and *Thallochrysis.* Motile stages that give rise to these filaments have variously been referred to *Hymenomonas, Cricosphaera, Syracosphaera,* and *Ochrosphaera.*

Most frequently studied is the species *Cricosphaera carterae.* Chromosome counts indicate that the motile stage of this species is diploid and the filamentous benthic stage is haploid (Rayns, 1962). The complex life cycle of this species includes both motile and nonmotile stages with coccolith cover, a filamentous stage, and naked motile cells. The typical motile cell has a cover of two sizes of unmineralized scales and an outer layer of cricolith-shaped coc-

coliths. As many as six of these relatively small coccoliths may be present in various stages of formation within a cell at the same time (Manton and Leedale, 1969).

The motile cell may reproduce vegetatively, at a rate in cultures of 0.85 to 1.0 divisions per day, the daughter cells dividing the coccolith cover. If the medium is diluted, the motile cells may instead produce four naked flagellate cells, 10 to 12 μm in diameter, each with a small bulbous haptonema inserted between the two 15-μm-long flagella. If the culture medium is renewed, these cells produce a coccolith cover, but if not, the flagella are shed, they settle to the substrate, attach, and germinate into *Apistonema*-like filaments. The cell divides twice, the daughter cells remaining as a tetrad in a single plane, or some of the daughter cells may slide past the others and continue dividing as a filament. Cultures maintained in a gelatinous medium also produce a filamentous stage that appears as a small dark green mass of spherical or elongate cells, about 20 μm long, with a wall of closely packed, imbricated scales that superficially resembles a mucilage sheath. The layer of scales may serve for protection against dehydration in the inshore or estuarine marine or brackish pools they inhabit (Paddock, 1968; Leadbeater, 1970; Outka and Williams, 1971; Lefort, 1975). Benthic stages were found to survive exposure to air at 0°C, freezing of seawater for three to four days, and exposure to air of high humidity at temperatures of 35 to 37.5°C, as well as a salinity range of 3 to 23.6%₀₀ (Boney and Burrows, 1966).

If the filamentous form is transferred from low to high light intensity, motile swarmers will appear within 24 hours. These rounded or oblong biflagellate cells with cuplike plastid have no haptonema, and eventually settle, attaching by the posterior end to the substrate to form a new filamentous stage. At times a second type of motile cell appears. Pyriform in outline, it has two long equal flagella and a short haptonema and swims more actively. It eventually produces the typical *Cricosphaera* coccoliths (Leadbeater, 1970). As these are diploid, fusion must occur at some stage, but has not as yet been observed.

Coccolith-Covered Motile Stage; Cysts and Palmella Stage

The motile cell of *Oolithotus fragilis* (Lohmann) Martini & Müller has a rigid spherical cover of from 7 to 20 coccoliths, but under unfavorable conditions, as when phosphate is limited, the cell content becomes dark and opaque, and a spherical, granulose, or spiny calcareous cyst may be produced. The cyst has a wall of closely packed calcareous prisms and surrounds the normal coccolith-bearing cell. From 4 to 64 orange-colored spores develop within the cyst and are liberated upon its decay or destruction. These spores give rise to a young flagellate cell, which may divide many times and form a thick mass of cells. A coastal form regarded as a phase of *O. fragilis* has an ovoid body, with coccoliths held loosely at the surface. Repeated cell divisions have been reported to produce chains and masses of cells, termed a palmella, that may contain thousands of individual cells. This has not been confirmed in culture studies, however.

Scale-Covered Motile Stage; Coccoid or Palmelloid Nonmotile Stage

Not all haptophytes possess coccoliths at some stage of the life history. Scale-bearing motile cells are produced by Isochrysidaceae (Isochrysidales) and by the Prymnesiales, and scaleless ones occur in the Pavlovales. Among the Isochrysidaceae, the motile scaly biflagellate cell has a reduced haptonema or lacks one entirely. The scaleless motile cell of the Pavlovales has two unequal flagella, those of *Exanthemachrysis* being 5.5 and 12 μm long, respectively, and a straight haptonema, about 2 μm long.

The motile cell eventually settles to the bottom and loses the flagella. *Chrysotila* (see Figure 9.27) then produces filaments, cubes, or packets of coccoid cells, and *Dicrateria, Imantonia, Isochrysis,* and *Exanthemachrysis* produce large masses of cells as a palmella stage. Benthic *Isochrysis* and *Chrysotila* cells are surrounded by a thick mucilaginous matrix in which are deposited small calcareous rods and cruciform structures identical to the fossil genus *Tetralithus. Chrysotila* also may produce calcareous structures like the fossil *Marthasterites* (Parke, 1971; Green and Parke, 1975). Eight to 64 daughter cells are released as swarmers by each nonmotile cell. *Dicrateria* and *Isochrysis* had previously been reported to produce siliceous cysts like those of the Chrysophyta, but these later were shown not to belong to the same organisms (Hibberd, 1976), and Haptophyta are not known to produce siliceous cysts.

Among the Prymnesiales, the scale-bearing motile phase of *Chrysochromulina kappa* produces two to four daughter cells by fission. The nonmotile phase that alternates with the motile one consists of naked amoeboid cells, lacking scales. Successive fissions of the amoeboid cells result in four cells with external walls that probably give rise to the motile phase again (Parke et al., 1955). *Phaeocystis pouchetii* (Hariot) Lagerheim has alternate free-living motile cells and a gelatinous colonial stage. Three different types of biflagellate, haptonema-bearing swarmers are produced: the normal zoospores that form new gelatinous colonies, some macrozoospores that leave the parent colony to form new ones, and microzoospores that represent an independent free-living phase and may reproduce by fission.

ECOLOGY AND DISTRIBUTION

Physical Parameters

Depth of Water

Coccolithophores occur in surface waters and in all depths down to 1000 m. They have been reported as deep as 4000 m in the Atlantic. The greatest abundance may be somewhat below the surface, commonly at a depth of about 50 m in the tropics. Certain species of wide geographic distribution may occur near the surface in the northern and southern limits of

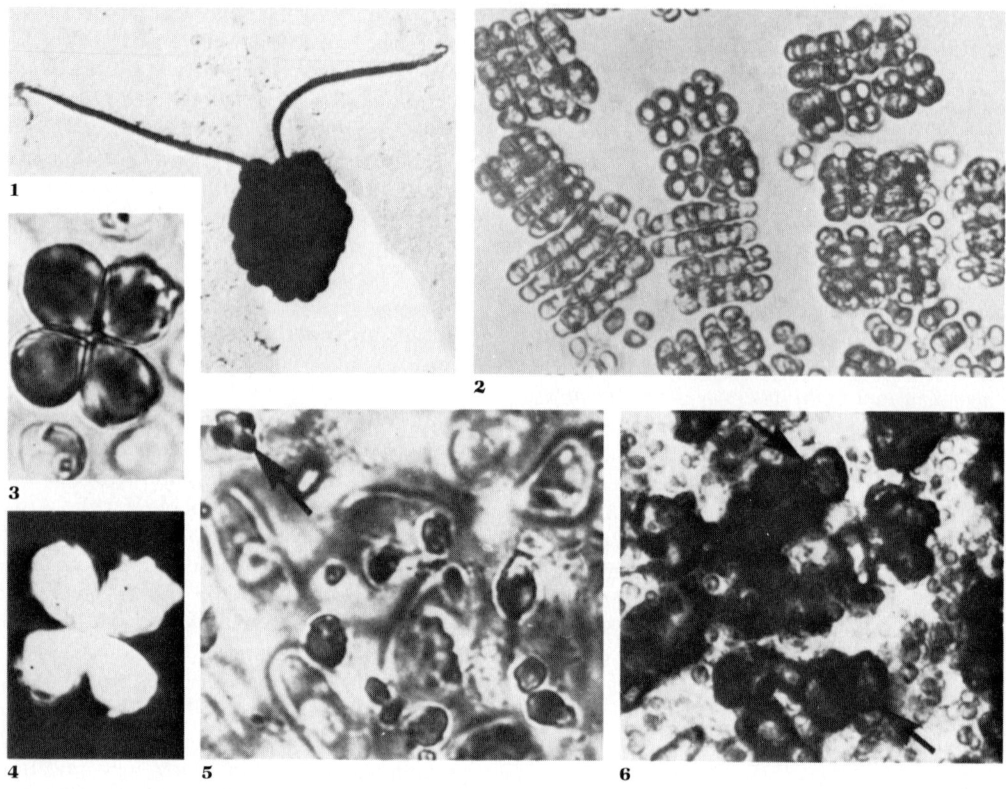

Figure 9.27

1,5. *Chrysotila stipitata* Anand. **1.** Shadowcast flagellate cell, ×7000. **5.** Mucilage stalks of benthic filamentous stage, with development of calcareous rods and *Tetralithus*-forms (arrow), ×1500. **2—4,6.** *C. lamellosa.* **2.** Blocks of cells in coccoid phase, ×500. **3,4.** *Tetralithus* elements in normal and cross-polarized light, ×1500. **6.** Masses of calcareous bodies, those at arrows having dumbbell shape, ×400. 1,3—6, reproduced with permission from Green and Parke, 1975, *J. Mar. Biol. Ass. U.K.,* v. 55, p. 109—121, published by Cambridge University Press; 2, reproduced with permission from Green and Parke, 1974, *J. Mar. Biol. Ass. U.K.,* v. 54, p. 539—550, published by Cambridge University Press.

their occurrence, as does *Emiliania huxleyi,* but are most abundant at about 50 m depth in the tropics. Some species may occur over widely varying depths in the same area, as does *Oolithotus fragilis,* but other species are depth zoned, their vertical distribution remaining constant regardless of the season. Although light at a given depth fluctuates from summer to winter, the influence of light, salinity, and nutrient availability may be more important in the depth zonation than temperature effect (see Table 9.1).

A study of the vertical and latitudinal distribution of coccolithophores in photic waters (0—200 m depth) of the mid-Pacific along the 155° W meridian (Honjo and Okada, 1974) recognized 90 species. Six surface water latitudinal zones had previously been recognized, from north to south, as the Subarctic Zone, Transitional Zone, North Central Zone, North Equatorial Zone, South Equatorial Zone, and South Central Zone. In the Subarctic, coccospheres were abundant from the surface to 30 m, sharply fewer below the thermocline at

Table 9.1

Comparison of coccolithophore cell number per liter, salinity, and temperature at Utsira, in the southern part of the west coast of Norway, 1945 (from Braarud et al., 1958).

Depth	April 23	May 22	June 6	June 18	June 29	August 18	September 3
Number of coccolithophore cells per liter							
1 meter	7000	36,500	632,500	575,000	—	—	273,500 (incl. 267,500 E. huxleyi)
10 meters	—	144,500	1,082,500	—	55,000	—	333,000 (incl. 325,000 E. huxleyi)
25 meters	—	14,500	200,000	—	—	62,500	—
Salinity ‰							
1 meter	32.50	30.19	28.84	31.40	—	—	32.19
10 meters	32.50	30.94	29.07	—	32.10	—	33.47
25 meters	—	32.78	32.22	—	—	32.89	—
Temperature °C							
1 meter	5.88	—	10.75	9.77	—	—	12.95
10 meters	5.83	—	10.67	—	11.10	—	11.87
25 meters	—	—	8.92	—	—	13.98	—

from 30 to 50 m, and declined to none at 100 m depth. In this region the total water column contained only about five species. In the Transitional Zone, cell numbers increased with depth to about an order of magnitude greater at 50 to 100 m depth than at the surface. Surface species diversity at lower latitudes ranged from 10 to 15 species to a maximum of 24 species at about 30° north latitude, where the total water column diversity was over 50 species. Total diversity then decreased again to about 26 species through the water column at the equator, and increased again in the southern hemisphere South Central Zone. The Central Zone contained fewer individuals in the surface water, but nearly equal abundance at all depths observed. Cell density was highest in the Equatorial Zone, with the maximum number occur-

ring at about 50 m depth. In the Subarctic Zone, *Emiliania huxleyi* was dominant at all depths, whereas *Umbellosphaera irregularis* (see Figure 9.28) was most abundant in the surface waters of the Central Zone. *Gephyrocapsa oceanica* and *Calcidiscus leptoporus* were found in the upper and mid-depth waters of the Equatorial Zone. *Umbilicosphaera hulburtiana* Gaarder, *Discosphaera tubifera*, and *Rhabdosphaera clavigera* were most common in surface waters, the former in the Equatorial Zone, and *D. tubifera* in the Transitional Zone. In contrast, *Umbellosphaera tenuis* was rare in surface waters, and common from 50 to 100 m in the Central Zone. *Oolithotus fragilis* increased in abundance down to the mid-depths of the photic zone in the Equatorial Zone. *Thorosphaera flabellata* Halldal & Markali and *Florisphaera profunda*

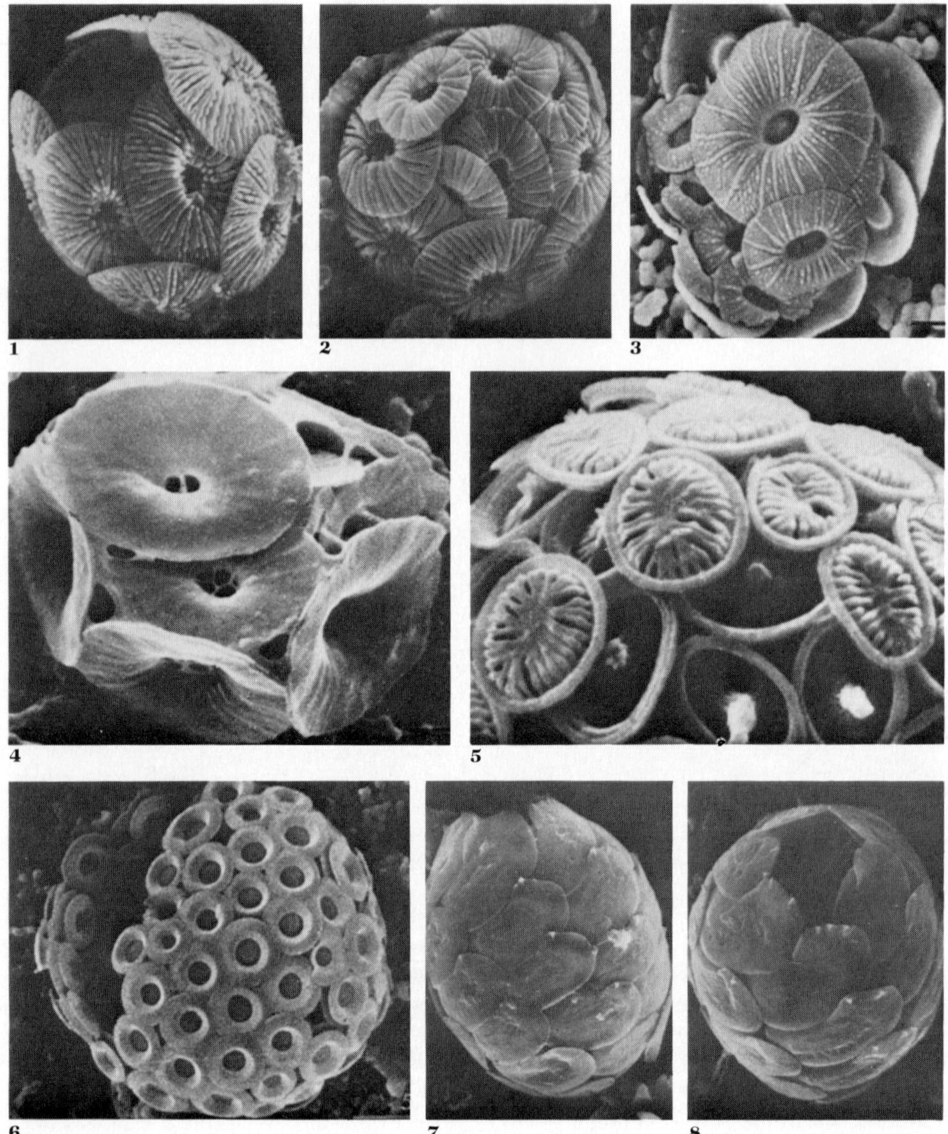

Figure 9.28

1–3. *Umbellosphaera tenuis* (Kamptner) Paasche. **1,2.** Tyrrhenian Sea, showing opening at one pole (part 1), ×3600, from Borsetti and Cati, 1972. **3.** From southwest Pacific, showing coccoliths of two sizes, ×3735. **4.** *U. irregularis* Paasche, southwest Pacific, with trumpetlike coccoliths, ×2935. **5.** *Syracosphaera nodosa*, portion of coccosphere from Southwest Pacific, showing different coccolith morphology of separate layers, ×8800. 3–5, from Okada and Honjo, 1970. **6.** *Umbilicosphaera sibogae* (Weber van Bosse) Gaarder, southwest Pacific, coccosphere, ×1760, from Honjo. **7,8.** *Helicosphaera hyalina* Gaarder, Tyrrhenian Sea, coccospheres, from side, ×2025, and oblique view showing terminal opening, ×2240; from Borsetti and Cati, 1972. All photos SEM.

Okada & Honjo were found only at depths below 100 m, increasing to a maximum between 150 and 200 m.

The differences in assemblages at various depths result in a bottom sediment assemblage that differs from that of the surface water flora, in addition to modifications due to dissolution and diagenesis. In the South Pacific, latitudinal zones characterized by bottom sediment assemblages (Burns, 1973a) were a Tropical Zone dominated by *Umbellosphaera irregularis*, together with *Rhabdosphaera stylifera* and *R. clavigera*, *Discosphaera tubifera*, and *Ceratolithus cristatus*. A Subtropical Zone contained mostly warm water species, such as *Umbellosphaera tenuis* and *D. tubifera*, with some cold water species present in the southern part. A Subtropical Convergence Zone is characterized by a drastic change to a cold water assemblage, and the Subantarctic Zone is a cold water flora, dominated by *Calcidiscus leptoporus*.

Temperature

Coccolithophores, together with dinoflagellates, are the characteristic phytoplankton of the marine tropical regions, but also occur in lesser numbers in cooler oceanic waters where diatoms and naked flagellates are numerically more important. Some coccolithophores are highly cosmopolitan and may also be dominant in numbers.

About 300 living species have been described, the richest floras being those of the Atlantic, where they have been described by many oceanographic expeditions. Many of the early published reports were of the species *Coccolithus pelagicus*, *Emiliania huxleyi*, and *Calcidiscus leptoporus*, and included at least 10 distinct species under these names. Their small size has hindered the correct identification of living cells, and modern records commonly are based on identification by means of electron microscopy.

Over 70 coccolithophore species are living in the Atlantic, and characterize five temperature-controlled assemblages (see Figure 9.29; McIntyre and Bé, 1967). Only a few of the species have a good fossil record, however. A Tropical assemblage contains *Umbellosphaera irregularis*, *U. tenuis*, *Cyclolithella annulus* (Cohen) McIntyre & Bé, *Oolithotus fragilis*, *Discosphaera tubifera*, *Rhabdosphaera stylifera*, *Helicosphaera carteri*, *Gephyrocapsa oceanica*, *Emiliania huxleyi*, and *Calcidiscus leptoporus*. The Subtropical assemblage includes some of the same species, *U. tenuis*, *R. stylifera*, *D. tubifera*, *C. annulus*, *G. oceanica*, *H. carteri*, *C. leptoporus*, *O. fragilis*, *E. huxleyi*, and the additional species *Umbilicosphaera sibogae*, previously known as *U. mirabilis* Lohmann. In the Transitional Zone, dominant species are *E. huxleyi*, *C. leptoporus*, *Gephyrocapsa ericsonii* McIntyre & Bé, *G. oceanica*, *R. stylifera*, *U. tenuis*, and *Coccolithus pelagicus*. A Subarctic assemblage includes the cosmopolitan *C. pelagicus*, *E. huxleyi* and *C. leptoporus*, and the Subantarctic assemblage is limited to *E. huxleyi* and *C. leptoporus*.

The most abundant living species is *Emiliania huxleyi*, a cosmopolitan species reported in numbers of 89,000 cells per liter of water in the Gulf of Panama at lat. 8° N and a temperature of 23 to 26°C (Smayda, 1966), and 200,000 cells per liter in the subantarctic Pacific at lat. 62° S and a temperature of 2 to 3°C (Hasle, 1960). It has been reported in numbers of 75 million per liter in the Oslo Fjord, and 115 million cells per liter on the Norwegian west coast at 11°C and 33‰ salinity (Berge, 1962). In culture, *E. huxleyi* grew at temperatures ranging from 7 to 27°C, with optimum growth at 20 to 23°C. It grows at a rate one-fourth as rapid at lower temperatures (10°C), hence is present in greater numbers at the higher temperatures. Although 7°C was the lowest temperature at which growth occurred in the laboratory, the species has been observed in nature at temperatures of 2 to 3°C. The upper limit observed in nature is about the same as the maximum of 27°C for growth of *E. huxleyi* in the laboratory (Paasche, 1968b). The species reproduces at all seasons in temperate and warm seas, but the populations in the northernmost and southernmost locations may be replenished yearly from the warmer waters. *Hymenomonas* sp. in culture grew only at temperatures between 12° and 22°C, but when the micronutrients in the medium were doubled in

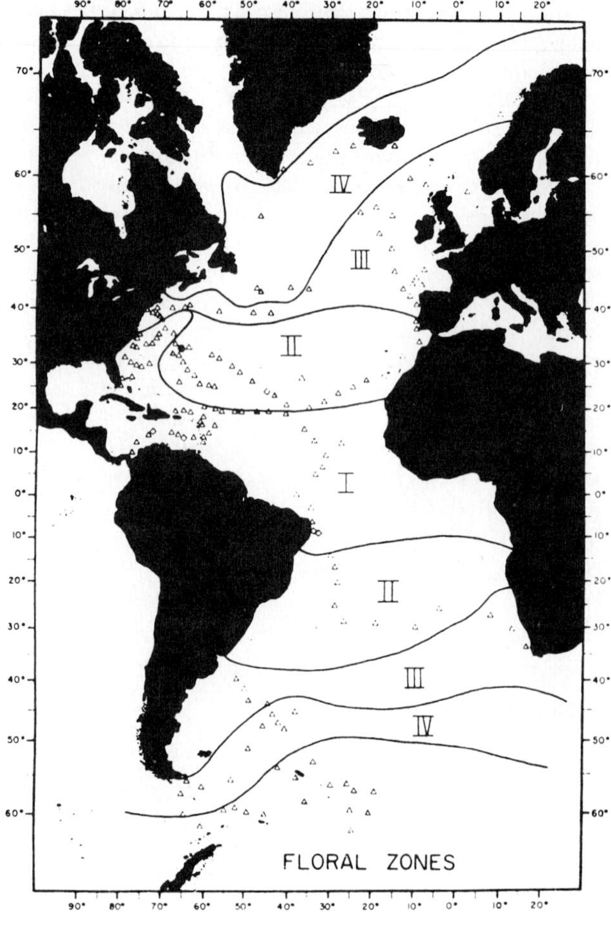

Figure 9.29
Coccolithophorid floral zones in the
Atlantic Ocean. I, Tropical; II, Subtropical; III, Transitional; IV, Subarctic
and Subantarctic. Sample locations
shown as triangles. Reproduced,
with permission, from McIntyre and
Bé, 1967, *Deep Sea Research,* v. 14, p.
561 – 597, copyright 1967, Pergamon
Press, Ltd.

amount, growth also occurred at temperatures as high as 25°C.

The number of coccoliths produced by a cell of *E. huxleyi* representing a given biomass varies little with the differences in temperature conditions, but is more affected by the light available (Paasche, 1968b). However, the dimensions of the coccoliths were found to differ significantly at different temperatures (Watabe and Wilbur, 1966). The width and length of the coccolith and the length of the elements of the upper disk decreased at temperatures above 18°C in culture, the maximum width of elements being obtained at a temperature of 24°C. Coccoliths were of normal form at 18°C, but abnormal mor-

phology increased at temperatures above or below 18°C. These temperature modifications also are evident in nature, distinct warm water and cold water ecophenotypes being recognizable at the extremes of the temperature range of this cosmopolitan species (McIntyre et al., 1970). The cold water variant has fewer rim elements, ranging from 23 to 33 with a mean of 29, and a solid proximal shield, with solid cover over the central opening. The warm water form has a mean of 35 elements in each rim, both rims being spokelike, and has a delicate grille over the proximal side of the central opening (McIntyre and Bé, 1967). The cold water phenotype occurs throughout the range of the species, but is present in very small numbers at temperatures above 14°C, where the warm water form dominates.

Coccolithophores with nonmotile benthic stages may survive temperature and salinity extremes in the benthic form. Such benthic filamentous phases of *Cricosphaera* in the laboratory survive temperatures as high as 35 to 40°C for an hour or more, and may survive freezing up to four days (Boney and Burrows, 1966).

Other modern species are restricted in temperature tolerance. *Coccolithus pelagicus* is a cold water species with a distribution restricted to the North Atlantic and North Pacific, at temperatures between 6 and 14°C, but with highest concentrations between 9 and 12°C (McIntyre et al., 1970). However, it is present in the bottom sediments of both hemispheres, as young as postglacial, 12,000 years BP. Perhaps moderate postglacial warming in the Southern Hemisphere caused its disappearance in that region. Another characteristically cold water species, *Pontosphaera borealis* Ostenfeld, described from the North Atlantic, has been suggested to be synonymous with *C. pelagicus*.

At the opposite temperature extreme, equatorial species include *Umbellosphaera irregularis* and species of *Rhabdosphaera* and *Discosphaera* that are restricted to temperatures of 18°C or warmer. The tropical and subtropical species *Oolithotus fragilis* has been reported in numbers of 30 million per liter of seawater in the Atlantic Ocean near Senegal.

In Norwegian fjords, four species comprised 90 percent of the coccolithophore flora; *Calciopappus caudatus* Gaarder & Ramsfjell (see Figure 9.30) was present in maximum abundance in May and June, while the waters were still cold, hence probably survived over winter in the fjords (Schei, 1975). The other three dominant taxa bloomed later in the summer, and probably were carried in from the Norwegian coastal current. Of these, *E. huxleyi* reached 95,000 cells per liter in September, *Anthosphaera robusta* (Lohmann) Kamptner reached a maximum of 60,440 cells per liter, and *Ophiaster hydroideus* (Lohmann) Lohmann attained a maximum of 17,000 cells per liter.

The quantity of cells present during a bloom is not necessarily indicative of the relative biomass, however, as large species in lesser numbers may represent a greater volume than some of the very abundant smaller species. Generally a phytoplankton bloom is followed by an increase in the number of plankton feeders, especially copepods and various invertebrate larvae, whose grazing rapidly reduces their numbers.

Salinity

Although characteristically oceanic, some coccolithophores occur in coastal waters, some prefer brackish water, and a few occur in fresh water. *Cricosphaera elongata* is found in supralittoral pools influenced by wave splash, in which salinity varies from 0.4 to over 25‰. *C. roscoffensis* has been found as a psammophyte, living in interstitial water in the sands of the intertidal area at Roscoff on the northwest coast of France (August, 1956), in quantities such that the sand was covered with a brownish coat. The film consisted almost exclusively of the coccolithophore, which was observed to swim rapidly in the collecting bottle.

In the Dead Sea, *Coccolithus pelagicus* has been found at salinities as high as 250‰. In cultures, growth of *E. huxleyi* is good at salinities above 20‰ but irregular below that, it does not grow in the laboratory at salinities less than 16‰ or more than 45‰ (Paasche, 1968b). Large populations have been recorded in the Oslo Fjord at salinities as low as 18‰, however. Studies of cores in the Black Sea show the first appearance of *E. huxleyi* at about 3000 years before the present, indicating the increasing salinity in these waters (Bukry, 1974c).

As was true of temperature, the nonmotile phase of *Cricosphaera* also may tolerate a wide range of salinity. This nonmotile phase withstands salinities up to 236‰ (normal oceanic salinity is 35‰), whereas the motile phase tolerates a range down to 4 to 8‰, up to 45 to 50‰, and a maximum of 90‰ (reviewed in Paasche, 1968b).

Coccolithophores also are found in fresh water, bogs, lakes, or rivers. About nine freshwater species are known of the genera *Hymen-*

722

Figure 9.30 (*facing page*)
1,2. *Calciopappus caudatus,* Norwegian Sea, TEM, reproduced with permission from Gaarder et al., *Avh. Norske VidenskAkad. Oslo,* 1954 (1), p. 1–9, published by Universitetsforlaget, Oslo University. **1.** Entire cell, ×5720. **2.** Decalcified coccoliths, showing structure, ×3000. **3,4.** *Ophiaster hydroideus.* **3.** Southwest Pacific, with two distinct types of coccoliths, TEM of carbon replica, ×3600, from Honjo. **4.** From 35°05′ S, 08°22′ E, South Atlantic, TEM of carbon replica showing cell at upper right, and ring of distinct polar coccoliths at lower left, ×8568, from McIntyre.

omonas, Pontosphaera, Acanthoica, and *Anacanthoica,* the most widely reported being *Hymenomonas roseola* Stein (see Figure 9.31). Probably more than one species has been reported under this name, as was true of the marine coccolithophores.

Calcium, Magnesium, Strontium, and Iron

Calcium carbonate is readily available in seawater, which is supersaturated even at the surface. Where the supersaturation is greatest, especially in low latitudes, coccolithophores comprise a greater percentage of the total phytoplankton. Cultures can be maintained in a calcium-deficient medium, but although growth is otherwise normal, no coccoliths are produced. When transferred to normal seawater concentration the coccolith cover rapidly develops. In a low calcium medium, barium and magnesium are not substituted for the calcium. About 0.1 to 0.15 percent $MgCO_3$ was found in coccoliths of *Cricosphaera carterae.* When grown in cultures of known calcium concentration, in the absence of strontium, coccoliths were formed of pure $CaCO_3$, but when strontium was present, the ratio of Sr/Ca concentration in the coccoliths to Sr/Ca in the medium was approximately 0.02, showing an apparent discrimination against strontium.

In cultures, maximal populations of *Emiliania huxleyi* were obtained in a medium with low iron concentration.

Light

Light is necessary for photosynthesis, hence is an important factor for most species that normally live in the photic zone. Deeper water forms apparently are less influenced by this

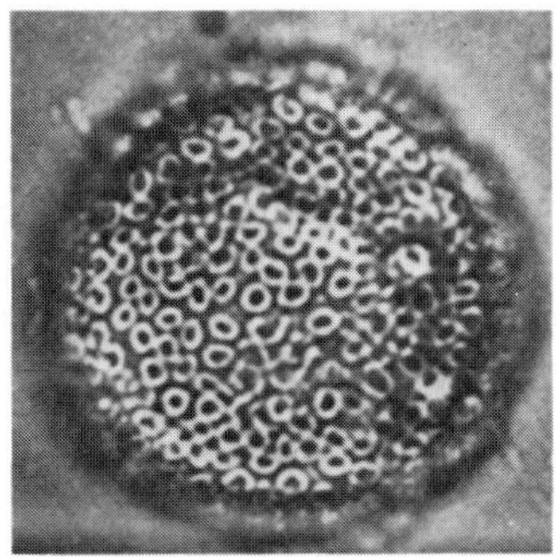

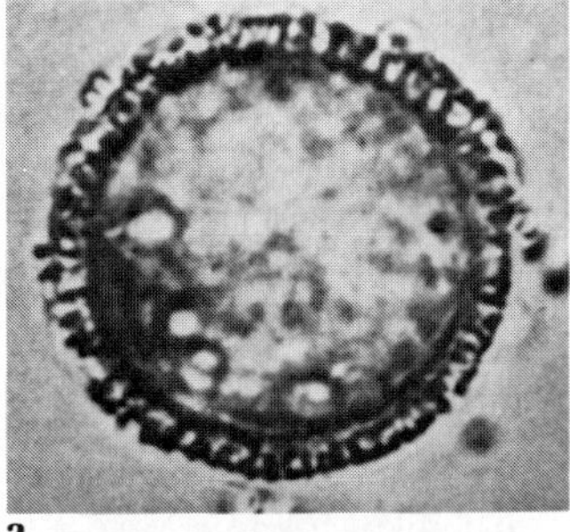

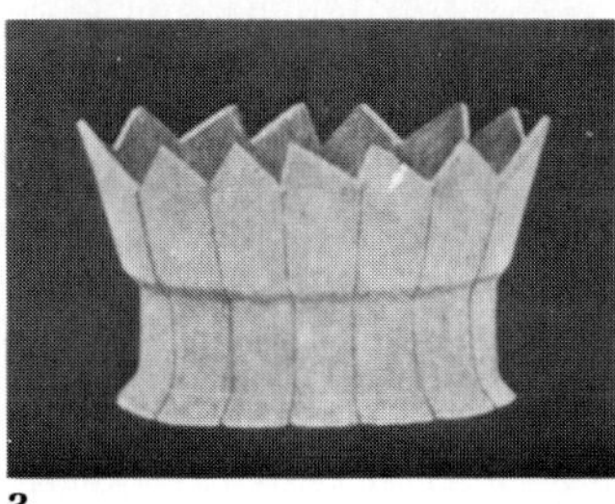

Figure 9.31
Hymenomonas roseola, a freshwater coccolithophorid. **1.** Exterior of cell. **2.** Optical section. **2,3,** ×1520. **3.** Model of a coccolith, based on electron micrographs, ×30,000. Reproduced with permission from Braarud, 1954a, *Nytt Mag. Bot.,* v. 3, p. 1–4, published by Universitetsforlaget, Oslo University.

factor. In cultures, most species ceased growth in the total absence of light, but could withstand large fluctuations in the amount of light. *Emiliania huxleyi* has a high degree of tolerance to light, as well as to fluctuations in temperature and salinity. It utilized light at an optimum between 10,000 and 25,000 lux (one lux = 0.0929 footcandles), but grew well from 1600 lux up to 50,000 lux. Light saturation for photosynthetic CO_2 fixation was at 35,000 lux for both *E. huxleyi* and *Hymenomonas* sp.

The high level of light in low latitudes has an inhibiting effect on some plants, reducing the amount of photosynthesis occurring in the uppermost water layers. It has been suggested that coccolithophores are favored in the warmer regions by the reflecting action of their coccolith cover, which reduces the amount of sunlight reaching the plastids within the cell. In culture a naked clone of *E. huxleyi* grew 15 percent slower at light saturation than did a coccolith-covered one. The naked clone had slightly smaller plastids, photosynthesized at a slower rate, and cell growth and division were slower than in the coccolith-covered one (Paasche and Klaveness, 1970). This lower photosynthetic rate of naked cells apparently is due to reduced plastids, and suggests that the coccolith cover does not act as a light-protective screen. Similarly, decalcification of coccolith-bearing cells should have resulted in reduced efficiency if light screening was a factor, but they acted exactly as did the coccolith-covered cells. In studies of varied light periods, a greater amount of coccolith carbonate was produced by *E. huxleyi* per cell volume in a 10-hour day than in a 16-hour lighted period or in continuous light (Paasche, 1968a). Coccolith production decreased with increase in light intensity or with photoperiod duration, again in contrast to expectations if the coccolith cover served for a light screen. When cells produced large numbers of coccoliths, many were shed into the medium.

Oxygen

Some species have a considerable tolerance to lack of oxygen, down to 20 percent of saturation, but *Emiliania huxleyi* and *Oolithotus fragilis* may encyst if conditions become too unfavorable. Other species need 70 percent or more of oxygen saturation for abundant occurrence and cannot tolerate saturation below 40 percent.

Nitrogen

In place of the nitrate normally used, some species in culture can utilize a wide variety of organic compounds. The presence of these organic compounds stimulated growth of some species. One strain of *Emiliania huxleyi* could utilize uric acid as a source of nitrogen, and interestingly, this species is commonly found in polluted waters. In nitrogen-deficient cultures, various polymorphs of calcite were formed in the coccoliths. One non-coccolith-bearing strain of *E. huxleyi* was led to produce coccoliths in a nitrogen-deficient medium, although it produced none in a medium of normal concentration.

Vitamins

Although photosynthetic, the Haptophyta, like many Chrysophyta, require an exogenous source of one or more vitamins, particularly vitamin B_1 (thiamine) and vitamin B_{12} (cobalamin). Most haptophytes and chrysophytes appear to respond best to the pyrimidine half of the thiamine, in contrast to the dinoflagellate *Oxyrrhis marina*, which requires the thiazole half (Droop, 1958). Vitamin B_{12} is synthesized by some bacteria, and probably by the blue-green, brown, and red algae. As these vitamins are produced in quantity by soil bacteria, they may be washed from the soils in solution and as a result are abundant in coastal waters. In the open ocean, the vitamins may be derived from extracellular production or cellular decomposition of the local bacteria.

All species investigated require a source of either thiamine or cobalamin or both. *Emiliania huxleyi* does not require vitamin B_{12} and is therefore abundant at times and in places where this may be lacking, and other phytoplankton thereby restricted. The vitamin requirements of the haptophytes and chrysophytes are among their animallike characteristics, and their cobalamin requirements are very similar to those of birds and mammals (Droop, 1954; Ford and Hutner, 1955; Lewin, 1961). Such vitamin or other nutrient requirements of the coccolithophores have been postulated to be a limiting factor leading to widespread extinctions of these pelagic organisms at the end of the Cretaceous period (Bramlette, 1965).

Heterotrophy

Although photosynthetic, many haptophytes ingest solid particles (bacteria, small algae) at the end of the cell opposite that of the flagella (Parke et al., 1955). However, not all haptophytes have this capability, neither *Prymnesium parvum* nor *Chrysochromulina parva* Lackey being phagotrophic (Leadbeater, 1971).

The motile stage of *Coccolithus pelagicus* has been seen to ingest bacteria, graphite, and plant cells up to about 5 μm in diameter, from which it could obtain both the required vitamins and the necessary organic carbon. The heterotrophic capability of this species may explain its northern distribution, as a varied nutritional possibility may allow it to survive better the long dark season in the Arctic.

Species cultured in the laboratory to date are apparently only limited heterotrophs, as they utilize some soluble organic matter and produce some by means of photosynthesis. Neither deepwater species nor any completely heterotrophic ones have yet been successfully cultured in the laboratory. When placed in the dark they cease growth. Only one species of *Hymenomonas* was found to grow at a very limited rate in darkness. Various species in culture that are known to ingest some organic matter are also partially dependent on a light source for photosynthesis.

Bernard (1963, 1967) reported *Oolithotus fragilis* in the Mediterranean down to 700 m depth, and to occur as large palmelloid aggregates of up to 900,000 cells that sank rapidly, carrying organic matter to the deeper parts of the sea. These cells at depth were also thought to be heterotrophic. However, Fournier (1968) observed that all particulate matter present in these waters, including inorganic particles, non-

viable cell remnants, and species of dinoflagellates, diatoms, and other coccolithophores was only about one-third the volume suggested for *O. fragilis* alone, hence regarded the previous estimates of its importance as far too great. Whether or not *O. fragilis* is heterotrophic is still uncertain.

Certain diatoms (*Brenneckella* and *Coscinodiscus*) have been observed with living coccolithophores attached near their girdle. These were variously thought to be symbiotic, or to be in an association favoring only one or the other of the two organisms. However, the coccolithophores were found to be merely caught in the large amount of mucilaginous material secreted by the diatoms in warmer waters, where both the diatoms and coccolithophores were common components of the plankton. No symbiosis was involved (Gaarder and Hasle, 1962).

FOSSIL CALCAREOUS NANNOPLANKTON

Background Studies

Isolated fossil coccoliths in Cretaceous rocks were first observed and illustrated by Ehrenberg (1836b), but were thought to be of inorganic origin. Sorby (1861) reported coccospheres from the English chalk and illustrated coccoliths and rhabdoliths, and Dawson (1874) reported coccoliths and rhabdoliths from the Cretaceous of Manitoba. Huxley (in Dayman, 1858) reported coccoliths from dredgings in the Atlantic Ocean, and Wallich (1861) proposed the term "coccosphere" for the coccolith-covered globules he observed in the sediments. Because of their abundance in planktonic ooze containing also the tests of pelagic foraminifera, Wallich at first believed the coccospheres to represent a foraminiferal growth stage. Others regarded the isolated coccolith as a separate organism, supposing the coccospheres to be a stage in their reproduction. When the coccospheres were found in the surface waters of the ocean, they were finally recognized as being a distinct pelagic organism (Wallich,

1865). The yellow-green plastids were observed by Murray and Blackman (1898) and the nucleus was described by Ostenfeld (1900). Dixon (1900) noted the organic pellicle enclosing the coccosphere with its coccoliths, the presence of an organic base within the coccolith calcite, and various stages in the development of new coccoliths in the interior of the coccosphere before their extrusion to become a part of the covering sphere. Lohmann (1902) first observed the flagella and placed the coccolithophores with the chrysomonad algae. Many oceanographic expeditions then resulted in a better understanding of the living species, and eventually laboratory studies provided information as to life histories, ultrastructure, coccolithogenesis, and growth requirements.

After the very early reports of fossil coccoliths, they received little attention until the early studies in the East Indies (Tan Sin Hok, 1926, 1927b, 1931), France (L. Dangeard, 1932; Deflandre, 1934, 1939), Algeria (Deflandre, 1942a), Spain (Colom, 1940), and central Europe (Kamptner, 1931, 1938, 1948; Klumpp, 1953). After their stratigraphic value was demonstrated by Bramlette and Reidel (1954), fossil coccoliths and discoasters were described in a host of studies from many areas of the world. Biostratigraphic zonations, at first based on land outcrop sections, were elaborated and refined on the basis of oceanic sections obtained by the DSDP over the past decade.

Problems in Studying Fossil Coccoliths

Many of the difficulties inherent in the study of fossil coccoliths are related to their small size. Both the normal light microscope and the petrographic microscope, for determination of their characteristic interference effects in polarized light, should be used in their study, and phase contrast equipment is also helpful. Concentration of the coccoliths is possible by treatment of the sediment with hydrofluoric acid to eliminate clay particles, but although the coccoliths are converted to fluorite and can then be studied in normal light, the original appearance in polarized light is destroyed. The coccoliths may be studied by mounting in a viscous medium such as silicone oil, in order to determine their orientation and view the specimen in both edge and plan views. Even with use of all these methods the smallest forms, from 2 to 4 μm in diameter, can only be studied in the electron microscope, and are thus less useful for routine stratigraphic work.

Much greater detail as to the morphology and crystal structure of coccoliths may be determined by study with the electron microscope, and many have been so studied since the first electron micrographs were published of the decalcified organic matrix of coccoliths (Braarud et al., 1952) and the calcite coccoliths themselves (Deflandre, 1952a). As calcite is opaque to the electron beam, carbon replicas (first used by Deflandre and Durrieu, 1957) proved superior to direct TEM. Techniques for preparing carbon replicas were summarized by Haq (1966, 1968) and Gartner (1968). Unfortunately specimens so studied could not always be related to the species described under the light microscope, and preparation of a carbon replica destroyed the coccolith. The difficulties of obtaining satisfactory carbon replicas also apparently favor the smallest specimens, which are the least useful in the light microscope. It was not always possible to determine whether illustrated carbon replicas were of the proximal or distal side of the coccolith, whether they were the normal aspect of the coccolith or a mirror image, or even whether the coccolith consisted of a single shield (discolith) or of two (placolith). Stereo pairs of the electron micrographs helped to add depth to the illustrations, and eventually techniques were devised by which the same specimens could be viewed with polarized light and by electron microscopy (Perch-Nielsen, 1967a; Smith, 1975a). SEM solved other difficulties as to orientation and added depth to the figures.

The small size of the calcareous nannoplankton, comparable to that of fine silt or clay particles, allows for their reworking and redeposition in sediments, and such derived material sometimes has been regarded of younger age than its true origin, and even nonmarine. Care must be used in interpreting the geologic record of the calcareous nannoplankton in areas where erosion of older strata may allow reworking of an earlier assemblage, such as in the middle and upper Tertiary of the Alpine region, or in the Pyrenees where Cretaceous, Paleocene, and lower Eocene specimens have been reworked into the upper Eocene strata. Coccoliths and discoasters dredged from the ocean bottom, or fossil material brought up in drilling, has occasionally been mistaken for living material when observed in plankton tows. However, recognition of the fossil specimens and knowledge of their provenance might allow interpretation of ancient drainage patterns, as in the Adriatic region (Cohen, 1965).

Coccoliths are usually preserved as the original calcite, with the crystalline structure intact. Under certain conditions, some may show recrystallization, etching of the surface, or may even be partially or completely dissolved. Other species in the same samples may show overgrowths of secondary calcite, modifying their appearance. Rarely, coccoliths may become silicified, in the form of opal or chalcedony, occurring in the powdery filling of hollow flints. They are most easily studied from the softer marls or clays, where their recovery is limited only by the degree of lithification and recrystallization that has occurred, but they are also recognizable on fractured surfaces, in thin sections of rocks, and in cherts.

Geologic Occurrence

Coccoliths and other calcareous nannoplankton species are generally abundant in rocks of Jurassic age to the present, and are rare in Triassic rocks. Most are known from isolated coccoliths or elements, although entire coccospheres may be preserved, generally representing particularly well-cemented forms, or possibly calcareous cysts. Because some living species are known to possess more than one form of coccolith on a coccosphere, or at different stages of the life cycle, it has been suggested that fossil isolated coccoliths might not represent valid species. However, in coccospheres that possess more than one kind of coccolith, one is most commonly a modification of the other, with a longer central spike, or some

similar modification. Even if completely different coccoliths had occurred on a fossil cell, one or two synonyms might be described for the fossil taxa, but this would not reduce their importance as biostratigraphic markers.

The presence of calcareous nannoplankton in large numbers in a rock implies a marine origin, deposition in the open sea, and a temperate or warm climate.

Paleozoic Calcareous Nannoplankton

Although the calcareous nannoplankton is typically post-Paleozoic in occurrence, Paleozoic coccoliths have been reported as well. Some of these appeared so close to Mesozoic or Cenozoic species as to be most certainly contaminants in the laboratory, for example those reported from the Pennsylvanian near Tulsa, Oklahoma. Other supposed Paleozoic coccoliths appear unrelated; for example, specimens in the Silurian and Devonian of North Africa (Deflandre, 1970) that are not calcareous, and have a concentric structure, rather than the radial structure characteristic of coccoliths.

A probable true coccolith was described from carbon replicas of the cleavage surface of shale in the Pennsylvanian of Missouri (Gartner and Gentile, 1973). The specimen consisted of lathlike imbricated elements in an ovoid form that strongly resembled a coccolith, although unlike any described younger species. No collar or central tube was discernible, nor was it possible to determine whether one or two shields were represented. Only a single occurrence is known for *Paleococcolithus missouriensis* (see Figure 9.32), but this simple disk could have given rise to the placolith structure by addition of a tube and second shield, or to the rhabdolith form by addition of a stem; tighter imbrication of the radial elements would produce a typical discolith (Gartner and Gentile, 1973).

Carbon replicas were described of Permian coccoliths from eastern Turkey (Pirini Radrizzani, 1971), identified as probable *Cricolithus*, and possibly *Discolithus*, *Lithastrinus*, or *Scapholithus*.

These rare records suggest that search of shale surfaces of the likely facies in other late

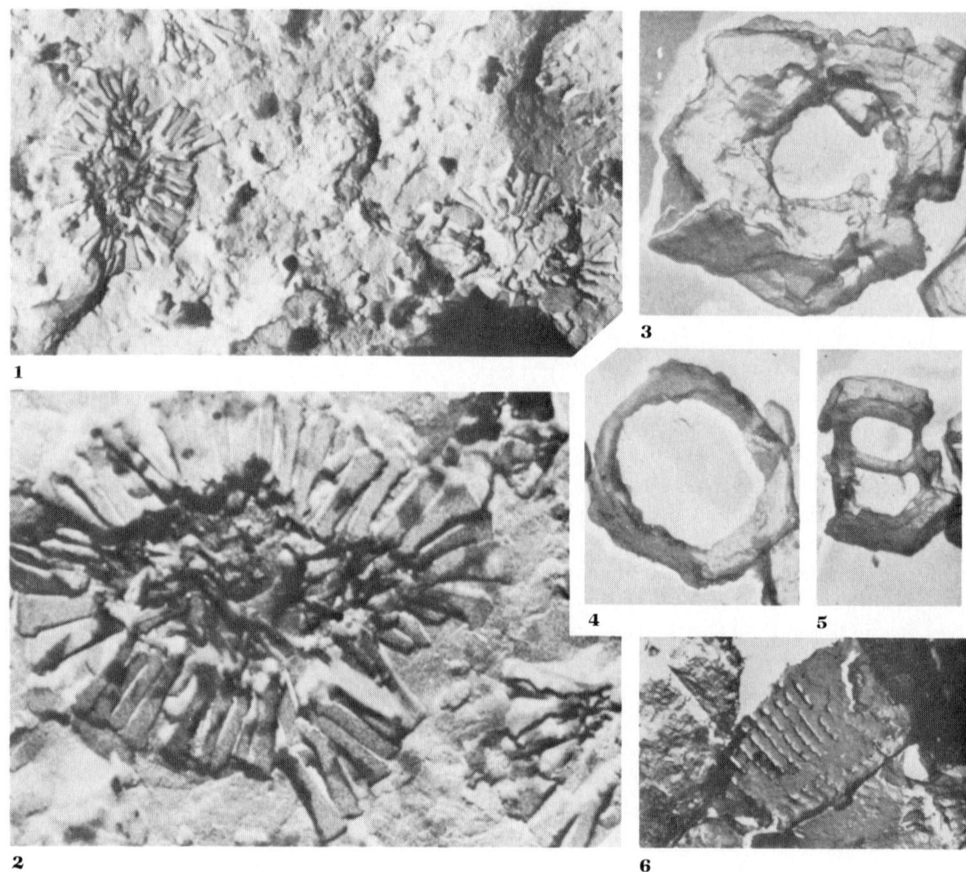

Figure 9.32

Paleozoic coccolithophorids. **1,2.** *Paleococcolithus missouriensis* Gartner & Gentile, Pennsylvanian, Tecumseh Shale, Missouri, from Gartner and Gentile, 1973. **1.** Parts of three specimens on the slab, SEM, ×3625. **2.** Enlargement, ×10,000. **3 – 6.** TEM of carbon replicas, Permian of Eastern Turkey, from Pirini Radrizzani, 1971. **3.** *?Lithastrinus*, ×6500. **4.** *Cricolithus* sp., ×8800. **5.** *?Discolithus*, ×9100. **6.** *?Scapholithus*, ×4550.

Paleozoic rocks might add further information as to the early evolution of the Haptophyta.

Mesozoic Calcareous Nannoplankton

Triassic records are nearly as rare as those of the Paleozoic, but some coccoliths and other calcareous nannoplankton are found in the Middle and Upper Triassic of the central Mediterranean region (Di Nocera and Scandone, 1977). These represent mostly *Nan-noconus* and rare Zygodiscaceae in the Middle Triassic, accompanied in the Upper Triassic by rare *Schizosphaerella* and the Podorhabdaceae, forming a nannofossil limestone. As the specimens were partly destroyed by recrystallization, they were not described as new taxa.

Nannoconids. *Nannoconus* is particularly abundant in Jurassic and Lower Cretaceous strata (see Figure 9.33), and some species occur in Upper Cretaceous deposits. It is commonly associated with radiolaria, tintinnids, and smooth ammonites in the Tethyan deposits where it was first observed, and it may be present in such quantities as to form a nearly pure nannoconid limestone. Nannoconids have been reported from the Alps, Apennines, Rumania, the Balearic Islands, North Africa, Cuba, Guatemala, Mexico, and Texas. Individual species have short geologic ranges and have been used in zonation of the Late Jurassic and Cretaceous (see Figure 9.34).

The individual **nannoconid** may be globular, conical, cylindrical, or pyriform in shape and may have a median constriction. A specimen is generally from 4 to 9 μm in diameter, and up to 12 μm in length. A central canal or cavity may be present, that of globular forms being enlarged centrally. The wall is constructed of tiny wedges or plates of calcite oriented perpendicular to the surface, and arranged in an apparent sinistral spiral, with the inclination about 30° from the horizontal.

The nannoconids have variously been suggested to be related to foraminifera, to represent gyrogonites of charophytes, or to be coccolithophores. The barrellike form was compared to the lopadoliths of true coccolithophores, although the latter are thin walled, and not the nearly solid cylinders of *Nannoconus*. Even more suggestive was the discovery of rosettelike groups of *Nannoconus* in thin sections of Neocomian limestone of Mexico (Trejo, 1960; Colom, 1965).

Because some specimens were observed to have a transverse constriction, Aubry (1974) suggested that the globular individuals resulted from transverse fission of the elongate cylindrical multilocular tests, hence were protozoans and unrelated to the coccolithophores. Division of a heavy calcareous test by dividing cells appears most unlikely, however. Instead, the multilocular tests may represent distinct species, or perhaps aberrant, malformed coccoliths, such as occur in modern coccolithophores under certain nutrient limitations. The abundance of nannoconids in some of the nannoconid limestones suggests that nutrient limitation was a distinct probability, particularly in their bathyal environment. Nannoconids generally are associated with a minimum of land-derived detritus, hence a similar lack of nutrient influx is probable. Blooms of modern phytoplankton rapidly deplete the nutrients available in the surface waters except where they are replenished near shore.

Schizospheres. These tiny globular calcareous nannofossils are constructed of two hemispherical to cap-shaped valves that may have a flared rim (see Figure 9.35). In general outline and their bipartite nature they resemble diatom resting spores, and probably were so regarded when first observed (reported Jurassic *Pyxidicula* are in fact schizospheres, as was observed by Deflandre, 1959). They differ from diatoms in being calcareous, in their distinctive ultrastructure, and in the different geologic occurrence. Their original calcareous nature is indicated by their identical present composition, regardless of the nature of the enclosing sediments. In microstructure the wall consists

---

Figure 9.33 (*overleaf*)
Nannoconids. **1.** *Nannoconus steinmannii* Kamptner, section, showing 3 rosettes (coccospheres?), Lower Cretaceous, (Hauterivian), Tamaulipas, Mexico, ×520, from Trejo. **2 — 6.** *N. abundans* Stradner & Grün, Lower Cretaceous (Barremian), Germany. **2.** Oblique view of concave distal side. **3.** Plan view of distal side. **4.** Proximal axial view, showing fluted margin of pillar. 2 — 4, SEM, ×9600. **5.** Oblique proximal view, showing imbrication of calcite crystals, ×8000. **6.** Axial view of specimen with open canal, ×8400. 5,6, TEM. 2 — 6, from Stradner and Grün, 1973. **7,8.** *N. truittii* Brönnimann, Lower Cretaceous, Denton Marl, Texas, side view in normal light and in crossed nicols. **9 — 12.** *N. elongatus* Brönnimann. **9,12.** Denton Marl, Texas. **10,11.** Fort Worth Limestone, Texas, in normal light and in crossed nicols. **13,14.** *N. multicadus* Deflandre & Deflandre-Rigaud, Pawpaw Shale, Texas. 7 — 14, ×2640, from Hill, 1976.

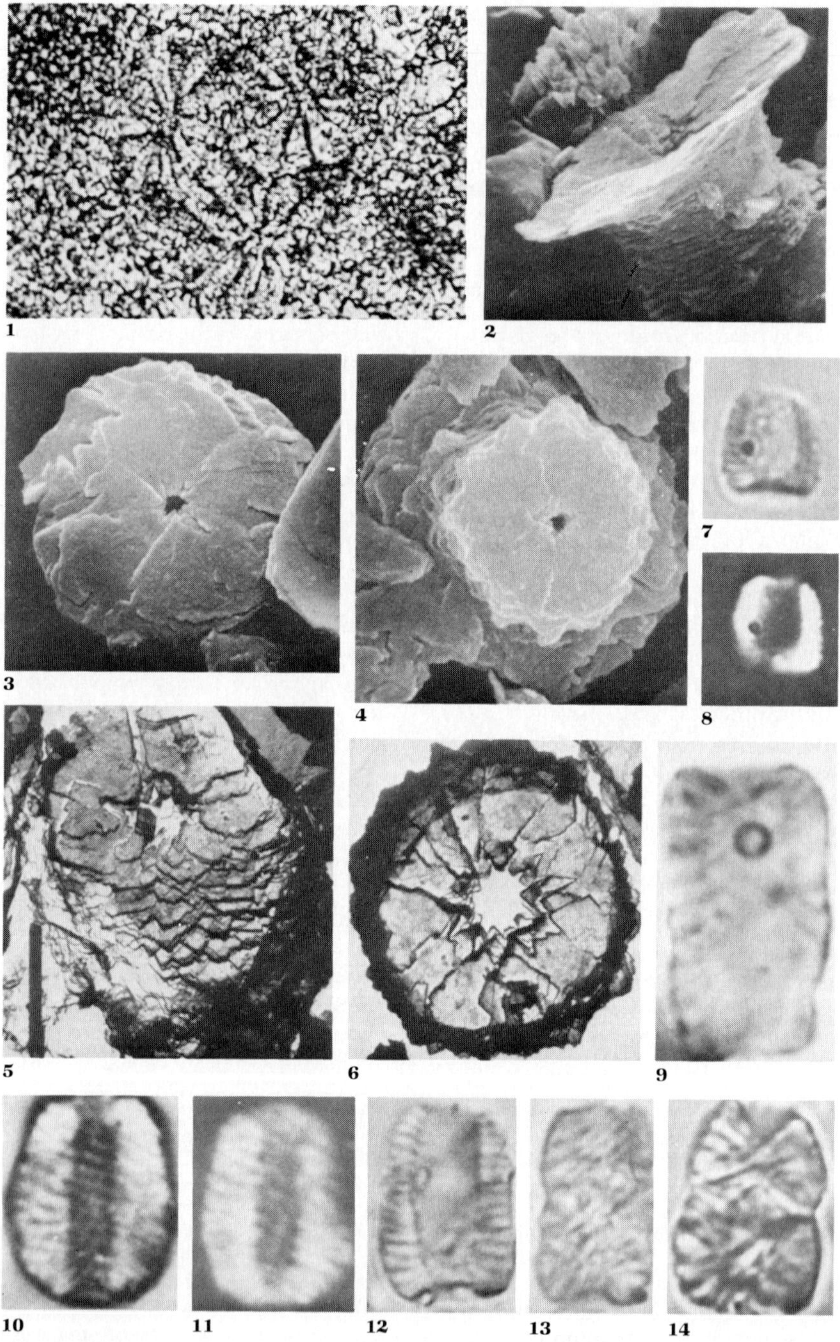

Figure 9.33

730

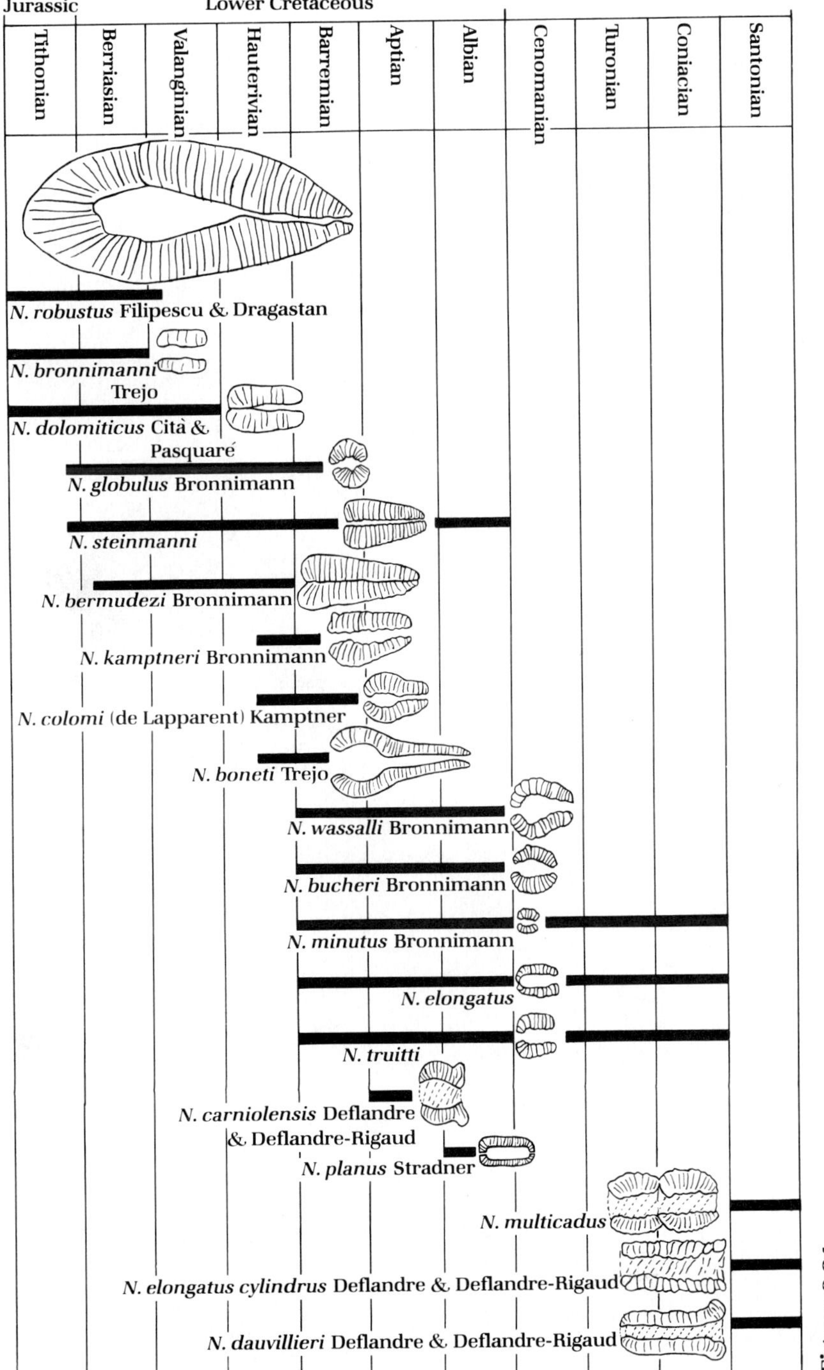

Figure 9.34
Geologic occurrence of some species of *Nannoconus.*

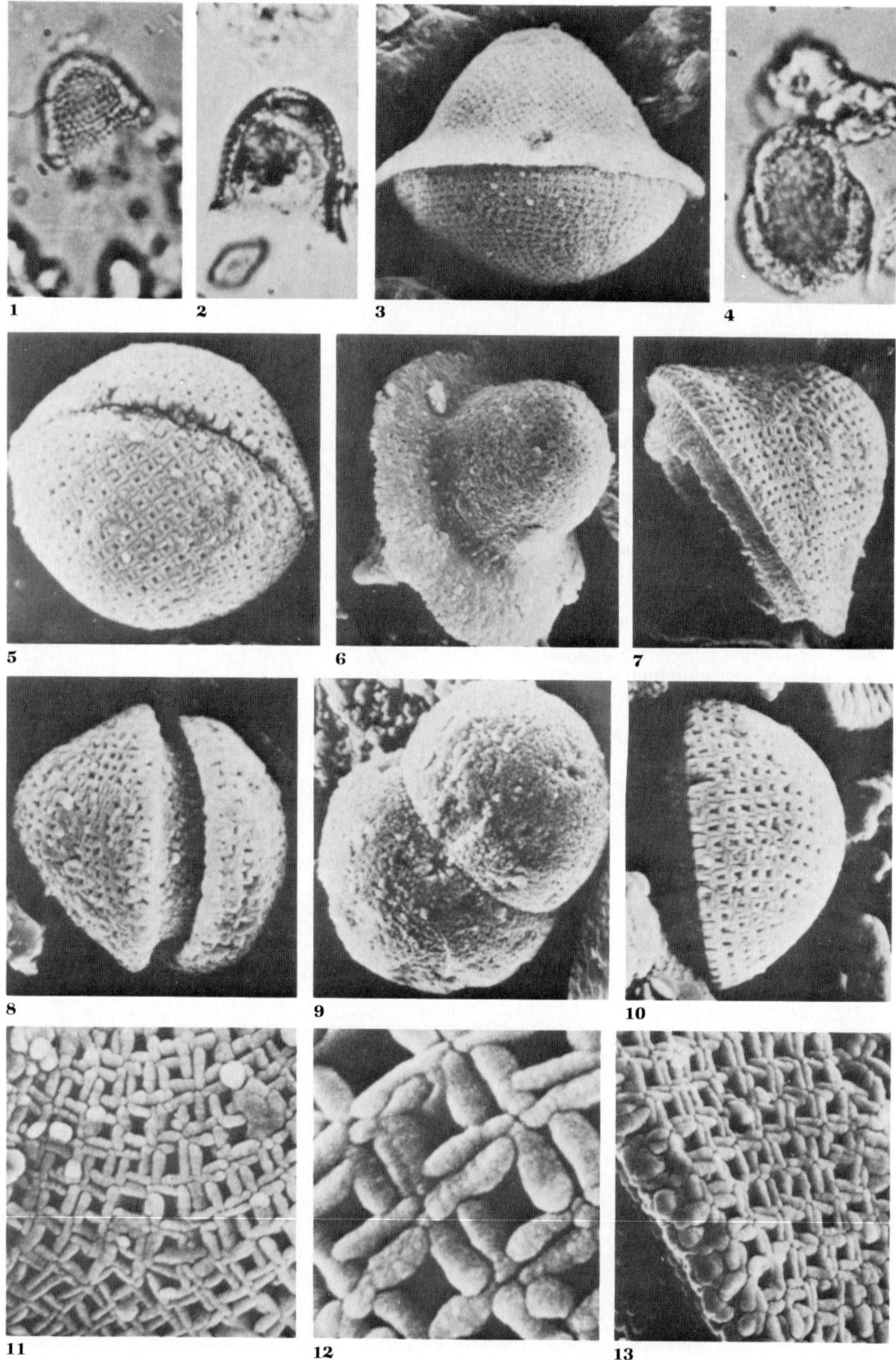

1
2
3
4
5
6
7
8
9
10
11
12
13

Figure 9.35 (*facing page*)
Schizospheres, Upper Jurassic, Oxfordian, France. **1,2,4.** Free valves, and one with valves attached by their rims, in transmitted light, ×1125. **3.** Specimen like that originally described as *Schizosphaerella punctulata* Deflandre & Dangeard, with outer valve having a flared rim, ×1800. **5.** ×2700. **6.** ×1890. **7.** Single valve, ×2025. **8.** ×2700. **9.** ×2250. **10.** Single valve, ×2700. **11,12.** Part of surface enlarged to show structure, ×6750 and ×20,250. **13.** Part of surface of specimen in part 7, ×6750. 3,5−13, SEM. All from Aubry and Depeche, 1974.

---

of elongate crystals, sets of four joining in a crosslike pattern, with adjacent sets of four overlapping so that each crystal lies against one of an adjacent cross; these crystal groups outline intervening square alveoli (see Figure 9.36; Aubry and Depeche, 1974), which in the light microscope appear as a cancellate surface. The crystals probably extend within the sphere as platelike structures to form partitions around the alveoli, but the internal structure is not certain. Originally described from the Jurassic, associated with nannoconids, coccoliths, tintinnids, and radiolaria, schizospheres also occur rarely in the Upper Triassic (Di Nocera and Scandone, 1977). The bipartite nature is unlike anything known in coccolithophores, but might represent an encysted or nonmotile phase, similar to the braarudospheres.

Stomiosphaeraceae. Commonly associated with the nannoconids and other calcareous nannoplankton are the similarly calcareous fossils variously referred to as Stomiosphaeridae, Cadosinidae, Pithonellidae, or Calcisphaerulidae (in zoological classification). Most were originally described from thin sections of the nannoplankton limestones of Jurassic and Cretaceous age in central Europe, and variously regarded as foraminifers, motile plankton flagellates or ciliates, or planktonic spores of benthic algae. Some genera included with these later have been regarded as calpionellids or early tintinnid ciliates, but others are undoubtedly encysted stages. Globular to ovoid, from 30 to 70 μm in diameter, and up to 200 μm in length (although commonly 55 to 60 μm), they are calcitic, the sectioned specimens of *Pithonella* and *Stomiosphaera* showing a black extinction cross between crossed nicols. The wall is layered (Nowak, 1968), the separate

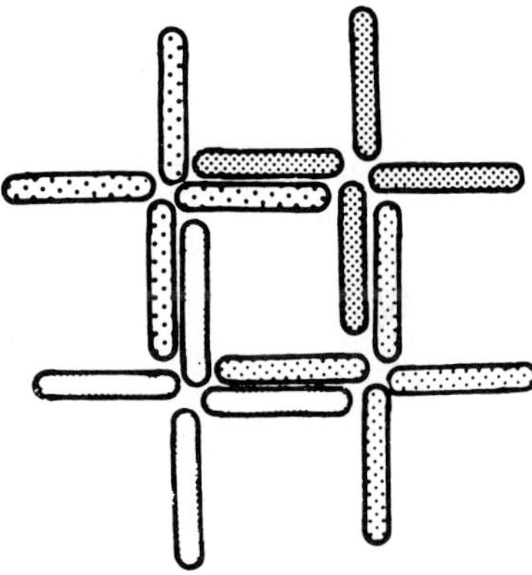

Figure 9.36
Diagram to show microstructure of the schizospheres, crystal rods forming crosses (each group of four shown in different pattern) that overlap to produce alveoli. From Aubry and Depeche, 1974.

layers having a spherolitic (*Stomiosphaera*) or radial fibrous structure (*Colomisphaera, Hemistomiosphaera*), or a combination of both (*Parastomiosphaera*). *Pithonella ovalis* (Kaufmann) Lorenz has a low magnesium calcite wall with an inner layer of anvil-shaped crystals that are aligned with the axis perpendicular to the surface, and an outer layer of elongate crystals. The layers probably were separated originally by an organic layer, now left as a cavity (Bein and Reiss, 1976).

The calcite composition of *Pithonella* was determined by X-ray analysis (Bolli, 1974), after an original aragonitic composition had been suggested (Banner, 1972). As is true of the schizospheres, however, all known specimens of the Stomiosphaeraceae are calcitic, regardless of the nature of the enclosing sediments, thus

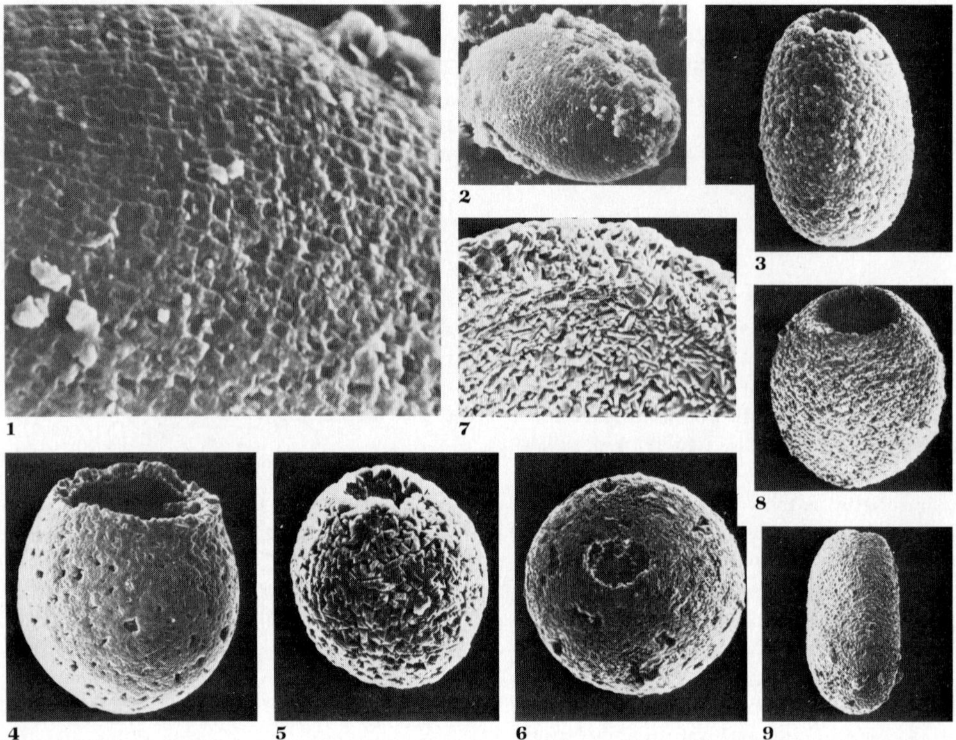

Figure 9.37

1,2. *Pithonella ovalis* (Kaufman) Lorenz, Cretaceous (Cenomanian), southern England, from Banner, 1972. **1.** Enlarged portion of surface, ×2208. **2.** Entire specimen, ×472. **3.** *P. cooki* Bolli, Upper Cretaceous (Coniacian-Santonian), ×400. **4.** *P. francadecimae* Bolli, ×520. **5.** *P. edgari* Bolli, ×600. 4,5, Lower Cretaceous (Albian). **6,7.** *P. carteri* Bolli, Upper Jurassic (Tithonian). **6.** Top, ×640. **7.** Cross section showing wall structure, ×2000. **8.** *P. quiltyi* Bolli, ×360. **9.** *P. veeversi* Bolli, ×336. 8,9, Lower Cretaceous (Albian). 3—9, Indian Ocean, from Bolli, 1974.

supporting its primary nature. Free specimens obtained from softer sediments allowed details of the surface to be observed (Bonet and Trejo, 1958; Banner, 1972; Bolli, 1974; Andri and Aubry, 1974). The surface is covered with longitudinal ridges connected to adjacent ones by weaker transverse ridges, resulting in a cancellate or reticulate appearance (see Figure 9.37).

An opening may be present at one pole, commonly filled with an operculumlike lid. This opening is not always present, hence is thought to have formed only at the latest stage of development of *Pithonella* and other taxa.

Pithonella occurs in outer shelf sediments, in relatively narrow belts between the shelf platform and the continental slope (Bein and Reiss, 1976), hence may not be a typical representative of the oceanic plankton. A benthic habitat was suggested by Adams et al. (1967), who related it to the unilocular foraminifera, although noting the absence of an aperture, and the imperforate wall. Bein and Reiss (1976) suggested that *Pithonella* instead represents free-floating parts of benthic algae. *Pithonella* very strongly resembles the calcareous spores produced by the dasyclad alga *Acetabularia* (see

Chapter 10), but the restricted geologic range (Jurassic to Paleocene) of the pithonellidlike forms is then unexplained. *Acetabularia* itself is known only since the Oligocene, although other genera of the Acetabulariaceae occur in the Jurassic and later periods. *Pithonella* also was compared to *Thoracosphaera,* and suggested to be a calcareous dinoflagellate cyst (Keupp, 1978) (see discussion of Thoracosphaeraceae on pages 747–750).

Coccoliths. The earliest abundant and well-preserved coccolith, in the uppermost Triassic and Lower Jurassic, is *Crucirhabdus primulus* Prins ex Rood, Hay & Barnard, a form that has distinct proximal and distal rims surrounding a shield bearing a central spine. Evolutionary lines from this species were proposed by Prins (1969) as giving rise to the Ahmuellerellaceae, Crepidolithaceae, and Podorhabdaceae. Many other coccolith types seem to appear suddenly, with little evidence as to their ancestry. Possibly the earlier haptophytes were scale bearing, and only later did various lineages commence the deposition of calcite. The wide variety of scales of living *Chrysochromulina* adds support to this theory.

Characteristic genera of the Lower Jurassic include *Alvearium, Axopodorhabdus, Crepidolithus,* and *Parhabdolithus* (see Figures 9.38 and 9.39), and in the later Jurassic the wheellike *Corollithion* and *Stephanolithion* appeared. Species that are zonal markers for the Lower Jurassic or whose appearance forms a stratigraphically important biohorizon include, in order of decreasing age, *Annulithus arkellii* Rood, Hay & Barnard, *Crucirhabdus primulus, Parhabdolithus marthae* Deflandre, *P. liasicus* Deflandre, *Palaeopontosphaera dubia* Noël, *Crepidolithus crassus* (Deflandre) Noël, *Axopodorhabdus cylindratus* (Noël) Wind & Wise, and *Discorhabdus tubus* Noël. Middle Jurassic indices include *Stephanolithion speciosum* Deflandre, *Diazomatolithus lehmanii* Noël, *S. speciosum* var. *octum* Rood & Barnard, *S. hexum* Rood & Barnard, *S. bigotii* Deflandre, *Polypodorhabdus escaigii* Noël (although Thierstein, 1976, regards this species as first appearing in the Upper Jurassic),

Podorhabdus rahla Noël, and *Discorhabdus jungii* Noël. Upper Jurassic marker species are *Diadozygus dorsetense* Rood, Hay & Barnard, *Corollithion geometricum* (Górka) Manivit, *Vekshinella stradneri* Rood, Hay & Barnard, *Ellipsagelosphaera communis,* and *Parhabdolithus embergeri* (Noël) Stradner (see Figure 9.40).

Although diversity was limited in the Jurassic, coccolithophores were present in large numbers in the Jurassic seas, and their coccoliths and entire coccospheres formed many layers and complete beds of limestone, particularly in the Jurassic of the English Dorset Coast, the North Sea, and the German Solnhofen Limestone.

Major Jurassic publications include those of Noël (1957, 1959, 1965b), Rood et al. (1971), Wilcoxon (1972), and Worsley (1971a).

Some Jurassic species continued into the Lower Cretaceous, but many new species accompanied them. The extensive pelagic limestones of the Lower Cretaceous of southern Europe and the marls and chalks of the American Gulf Coast and Mexico contain abundant coccoliths, commonly associated with nannoconids, and calpionellid tintinnids. Most Lower Cretaceous assemblages contain smaller individual coccoliths than those of the Upper Cretaceous and Tertiary, hence they received little study at first. Electron microscopy was particularly useful in developing the biostratigraphic zonation for the Lower Cretaceous.

Coccolith assemblages also first showed a latitudinal restriction in occurrence during the Lower Cretaceous, and other species showed a correlation with marginal and epicontinental seas, being absent from typically oceanic sequences. Among the Lower Cretaceous taxa that are restricted to tropical and subtropical paleolatitudes are *Polycostella beckmannii* Thierstein, *Diadorhombus rectus* Worsley, *Calcicalathina oblongata* (Worsley) Thierstein (see Figure 9.41), *Rucinolithus wisei* Thierstein, *R. irregularis* Thierstein, and *Lithraphidites bollii* (Thierstein) Thierstein. Many others are more cosmopolitan, and a single species, *Cribrosphaerella primitiva* Thierstein, seems to be restricted to the high paleolatitudes, in the boreal and austral regions (Thierstein, 1976).

Additional Lower Cretaceous species of

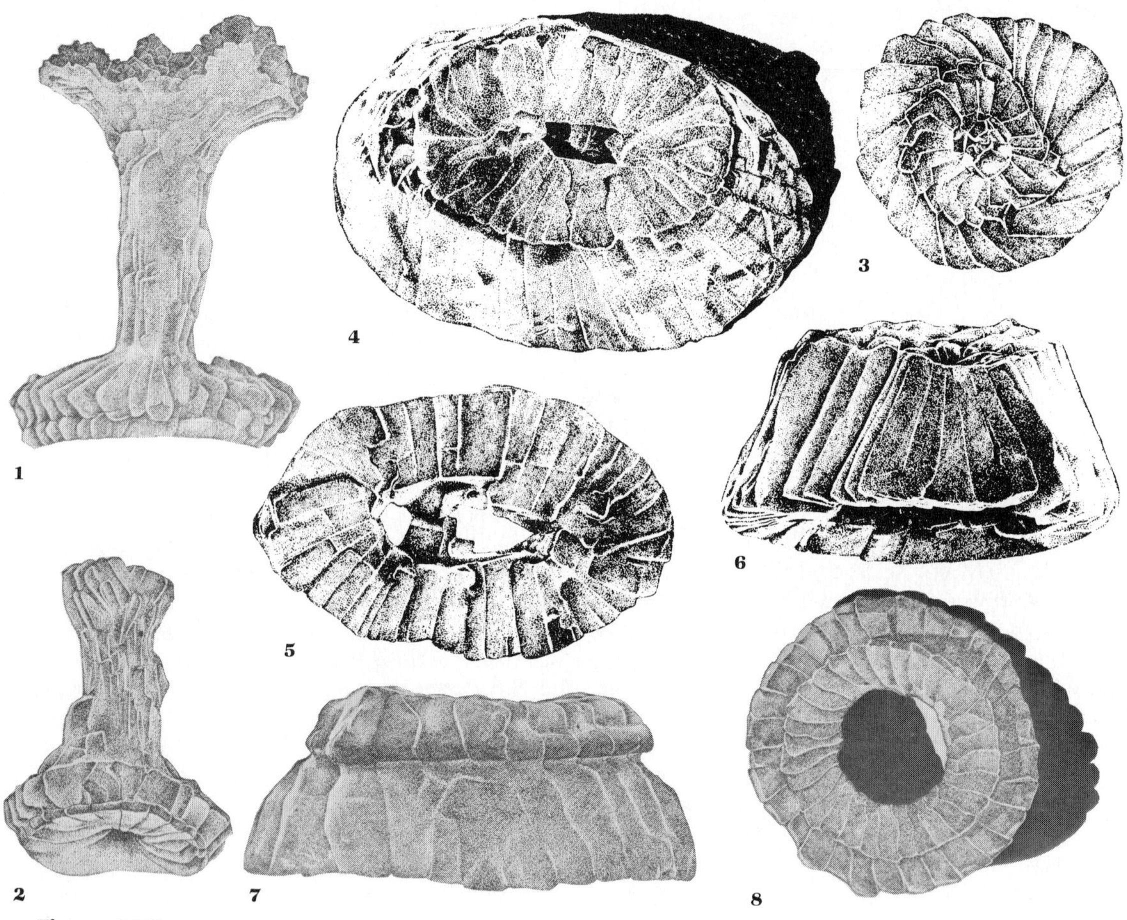

Figure 9.38

Jurassic coccolithophorids. **1,2.** *Discorhabdus patulus* (Deflandre) Noël, Upper Jurassic (Oxfordian), France, edge views, ×14,300 and ×16,000, respectively. **3.** *Cyclagelosphaera margereli* Noël, Upper Jurassic (Portlandian), Algeria, ×16,000. **4—6.** *Crepidolithus crassus* (Deflandre) Noël, Lower Jurassic, France, ×16,000. **4.** Proximal view. **5.** Distal view. **6.** Edge view. 4,6, Pliensbachian; 5, Toarcian. **7,8.** *Diazomatolithus lehmani* Noël, Upper Jurassic (Oxfordian), France, edge view, ×8000, and distal view, ×12,000. All based on TEM, from Noël, 1965b.

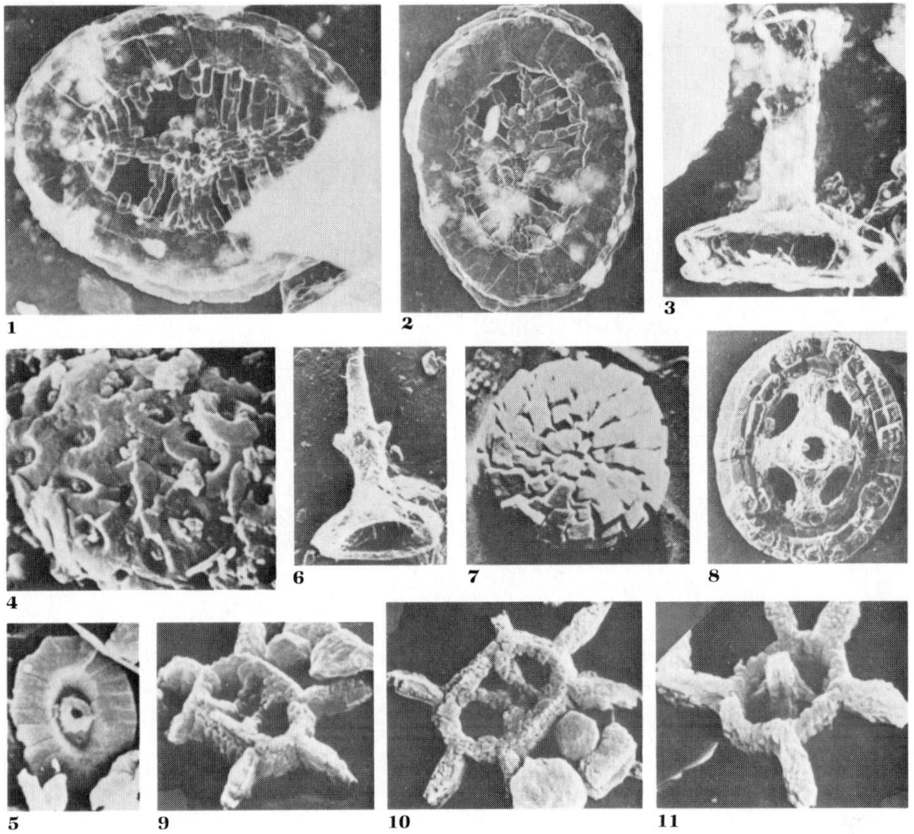

Figure 9.39

Lower and Middle Jurassic zonal markers. **1,2.** *Polypodorhabdus escaigi,* distal and proximal views, ×12,800. **3.** *Discorhabdus tubus,* side view. 1−3, Oxfordian, France, TEM, ×12,800, from Noël, 1965b. **4,5.** *Palaeopontosphaera dubia,* SEM, ×5330, from Noël, 1972. **4.** Entire coccosphere, Kimmeridgian, France. **5.** Isolated coccolith in distal view, Toarcian, France. **6.** *Podorhabdus rahla,* Oxfordian, England, oblique side view, TEM, ×2160, from Rood et al., 1971. **7.** *Alvearium dorsetense* Black, Liassic, England, ×6400. Reproduced with permission from Black, 1965, *Endeavor,* v. 24, p. 131−137, copyright, 1965, Pergamon Press, Ltd. **8.** *Axopodorhabdus cylindratus,* Oxfordian, England, distal view, TEM, ×2880, from Black, 1968. **9−11.** *Stephanolithion bigotii,* Oxfordian, France, SEM, ×5840, from Noël, 1972. **9,10.** Proximal face, in oblique and plan views. **11.** Distal face.

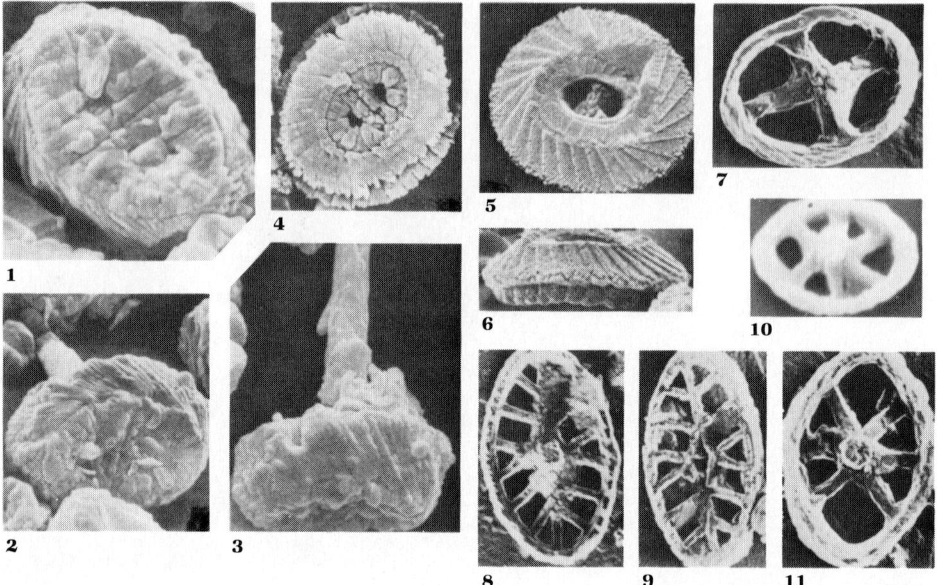

Figure 9.40

Late Jurassic zonal markers. **1—3.** *Parhabdolithus embergeri,* Lower Cretaceous (Berriasian), Spain, SEM, from Grün and Allemann, 1975. **1.** Proximal view, ×6455. **2.** Oblique view of proximal side. **3.** Lateral view. 2,3, ×4320. **4—6.** *Ellipsagelosphaera communis* (Reinhardt) Perch-Nielsen, Oxfordian, France, proximal face, distal face, and edge view, ×4000, from Noël, 1972. **7.** *Vekshinella stradneri,* Oxfordian, England, distal side, ×12,480. **8,9.** *Diadozygus dorsetense,* Oxfordian, England, distal view, ×15,200, and proximal view, ×14,400. **10,11.** *Corollithion geometricum.* **10.** Specimen from Lower Cretaceous, Kiamichi Formation, Texas, distal view, ×6400, from Hill, 1976. **11.** Oxfordian, England, distal view, ×10,000. 7—9,11, TEM, from Rood et al., 1971; others, SEM.

Figure 9.41 (*facing page*)

Characteristic Cretaceous species. **1—6.** *Watznaueria barnesae* (Black) Perch-Nielsen. **1.** Coniacian, Indian Ocean, coccosphere, SEM, ×4800, from Thierstein, 1974. **2.** Turonian, England, proximal view of isolated coccolith, TEM of carbon replica, ×8000, from Black, 1965; **3,4.** Del Rio Clay, Texas, distal view of coccolith, ×5280, and entire coccosphere, ×2640. **5,6.** Kiamichi Formation, Texas, entire coccosphere, ×2640, and proximal view of coccolith, ×3200. 3—6, SEM, from Hill, 1976. **7,8.** *Calcicalathina oblongata,* Lower Cretaceous, western Atlantic, SEM, oblique view of proximal side, ×7200, and plan view of distal side, ×1800. **9,16—19.** *Rucinolithus wisei,* Lower Cretaceous, western Atlantic. **9.** Transmitted light, ×2560. **16.** SEM of distal side, ×5200. **17.** SEM of proximal side, ×4000. **18,19.** Cross-polarized light and phase contrast of same specimen shown in part 9, ×2560. **10—12.** *Lithraphidites bollii,* Lower Cretaceous, western Atlantic. **10.** SEM, ×2400. **11,12.** Same specimen in SEM, ×2800, and in phase contrast, ×2560. 7—12, 16—19, from Thierstein, 1971. **13,14.** *Ellipsagelosphaera britannica,* Aptian, Spain. **13.** Coccosphere, SEM, ×6455, from Grün and Allemann, 1975. **14.** Kimmeridgian, England, proximal side, TEM of carbon replica, ×6400, from Black, 1965. **15.** *Lithraphidites carniolensis,* Del Rio Clay, Texas, spindle-shaped nannofossil, SEM, ×2400, from Hill, 1976. **20,21.** *Cruciellipsis cuvillieri* (Manivit) Thierstein, Lower Cretaceous, Spain, SEM, from Grün and Allemann, 1975. **20.** Distal view, ×4320. **21.** Proximal view, ×6455. 2,14, reproduced with permission from Black, 1965, *Endeavor,* v. 24, p. 131—137, copyright 1965, Pergamon Press, Ltd.

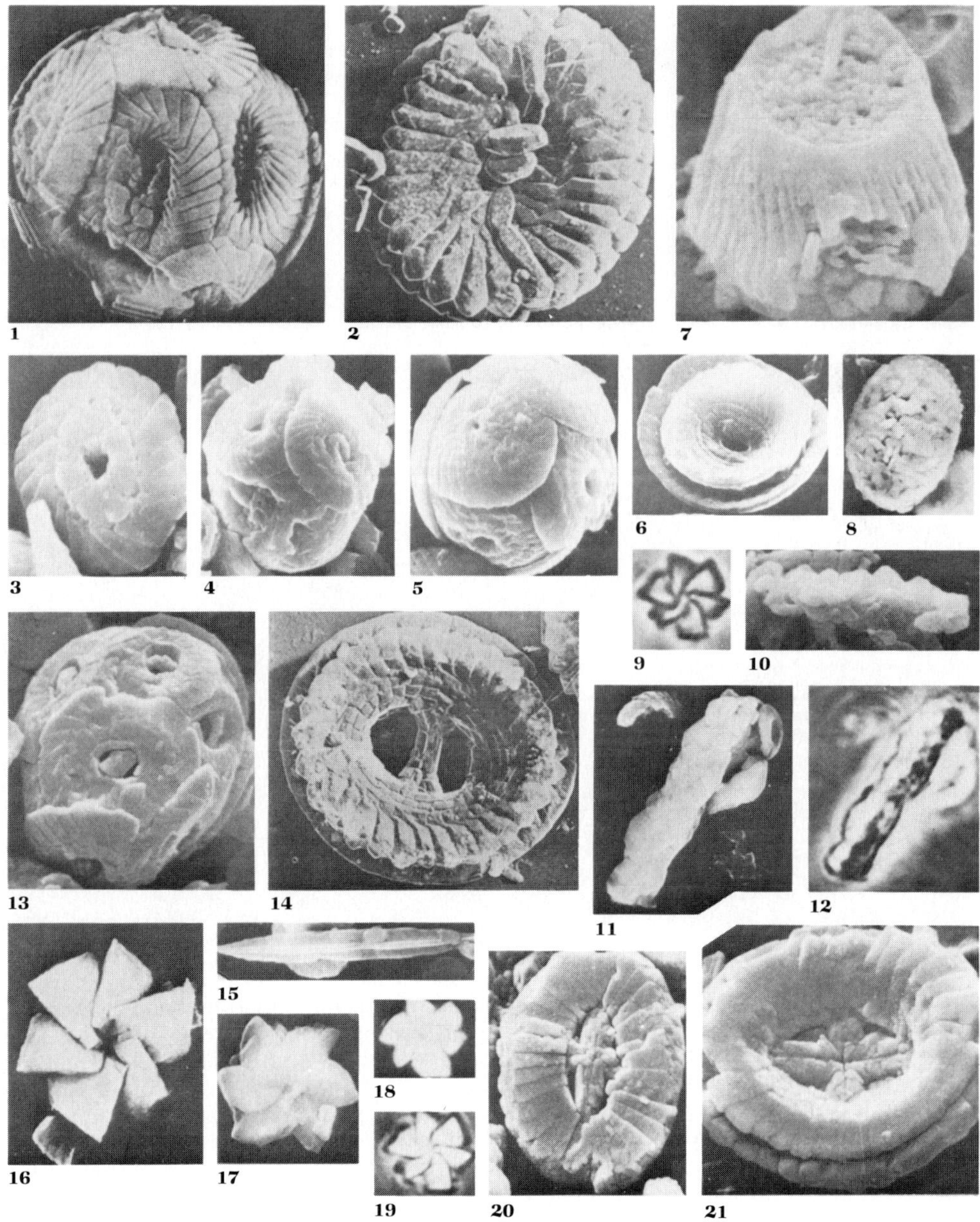

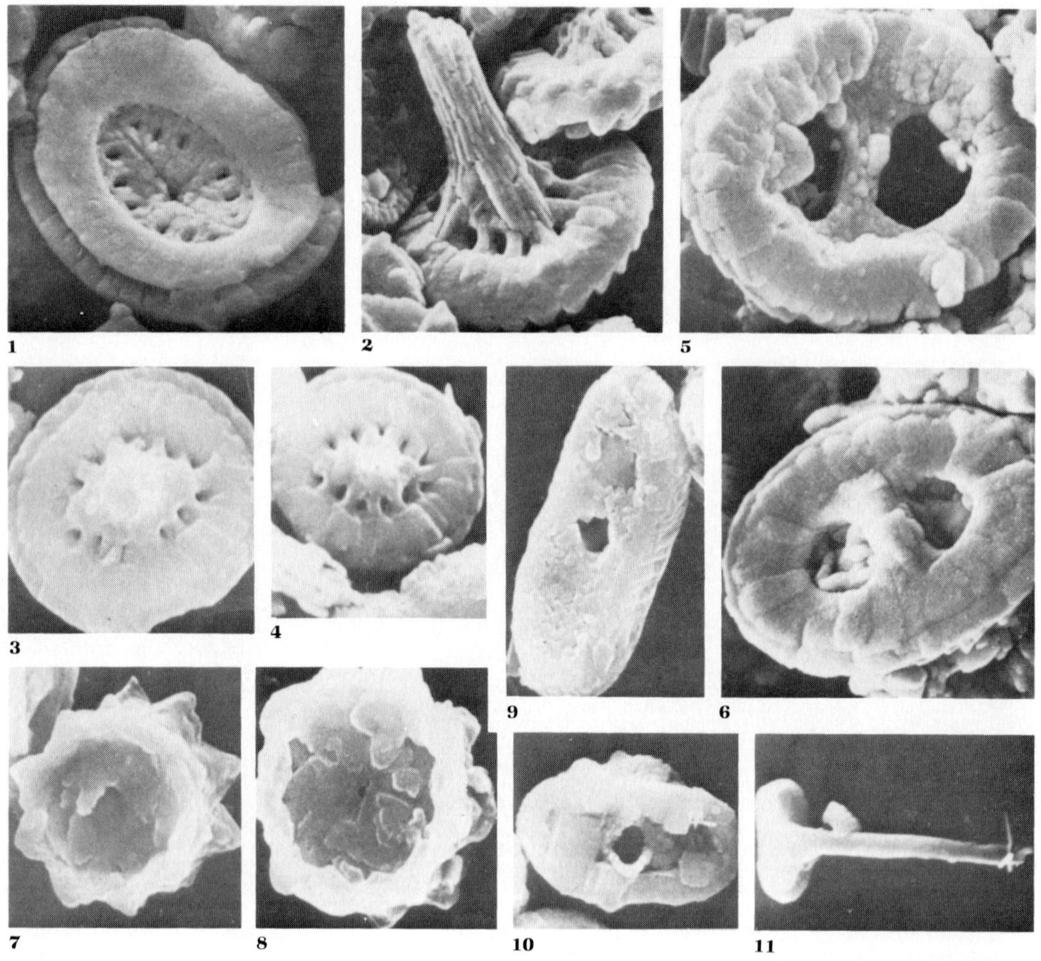

Figure 9.42

Cretaceous species, all SEM. **1—4.** *Retecapsa crenulata* (Bramlette & Martini) Grün. **1.** Aptian, Netherlands, proximal view, ×9700. **2.** Cenomanian, England, oblique distal view, ×8070. **3,4.** Distal views, from Main Street Limestone, ×5670, and Weno Formation, ×4170. **5,6.** *Speetonia colligata* Black, Valanginian, Spain, proximal and distal views, ×8070. **7,8.** *Eprolithus floralis* (Stradner ex Bukry) Stover, proximal views, from Main Street Limestone, ×4330, and Duck Creek Formation, Texas, ×6000. **9—11.** *Rhagodiscus angustus.* **9.** Duck Creek Formation, Texas, proximal view, ×5670. **10.** Del Rio Clay, distal view, ×3670. **11.** Main Street Limestone, oblique side view, ×3330. 1,2,5,6, from Grün and Allemann, 1975; others from Hill, 1976.

stratigraphic importance, in order of decreasing age, include the Berriasian *Lithraphidites carniolensis* Deflandre and *Cretarhabdus angustiforatus* (Black) Bukry, the Valanginian *Ellipsochiastus quadriserratus* Worsley and *Tubodiscus jurapelagicus* (Worsley) Roth, the Hauterivian *Watznaueria diaphanae* Worsley, the Barremian and Aptian *Micrantholithus obtusus* Stradner, *Watznaueria oblonga* Bukry, *Tetralithus malticus* Worsley, *Chiastozygus litterarius* (Górka) Manivit, *Rucinolithus irregularis, Eprolithus floralis,* and *Rhagodiscus angustus* (Stradner) Reinhardt (see Figure 9.42). Albian biohorizon and zonal markers include *Deflandrius columnatus* Stover, *Podorhabdus albianus* Black, *Eiffellithus turriseiffeli* (Deflandre) Reinhardt, *Broinsonia gammation* M. Hill, and *Staurolithites matalosus* (Stover) Čepek & Hay (see Figure 9.43).

Lower Cretaceous assemblages have been described by Black (1971b, 1971c, 1972a, 1973, 1975), Grün and Allemann (1975), Hill (1976), Manivit (1971), Manivit et al. (1977), Perch-Nielsen (1977), Thierstein (1973), and Verbeek (1977), among others.

The worldwide chalky deposits of the Upper Cretaceous, to which the period owes its name, are largely composed of coccoliths and coccolith debris. The maximum diveristy, as well as abundance, of the calcareous nannoplankton was attained in the Late Cretaceous, many of the species lasting for about 5 to 10 million years before disappearing, hence spanning one or two geologic stages. Use of first and last occurrences as biohorizons and combinations of species in zones allowed a Cretaceous zonation based on the coccoliths to about a 3-million-year resolution. Diagenetic effects add to the problem of studying the coccoliths in chalky sediments, in which they generally are less well preserved than in the clays. As a result various taxa have been described that proved to be fragmentary, partially destroyed by dissolution, or affected by overgrowths of secondary calcite. Selective dissolution effects also may modify the character of the assemblage remaining.

Among the characteristic Upper Cretaceous coccoliths, *Deflandrius* is present from the Albian through the Maastrichtian. Some taxa are more restricted stratigraphically, but widely distributed geographically, hence are excellent zonal markers. Others, such as *Kamptnerius magnificus* Deflandre, appear in the Turonian at the higher paleolatitudes, only reaching the paleoequatorial regions in the Campanian and Maastrichtian (see Figure 9.44) (Thierstein, 1976). However, the Turonian specimens have a flange of constant width, gradually evolving into a protruding, highly asymmetrical flange in the later part of the range.

Cenomanian and Turonian coccolith mar-

Figure 9.43 (*overleaf*)
Lower and Middle Cretaceous species. **1,2.** *Eiffellithus turriseiffeli.* **1.** Del Rio Clay, Texas, oblique distal view, SEM, ×3200. **2.** Grayson Formation, Texas, side view, ×2400. **3,4.** *Chiastozygus litterarius,* distal views, SEM. **3.** Goodland Formation, Texas, ×4000. **4.** Kiamichi Formation, Texas, ×4265. 1–4, from Hill, 1976. **5,6.** *Biscutum ellipticum* (Górka) Grün. **5.** Coccosphere, Barremian, Spain, SEM, ×6455. **6.** Berriasian, Spain, distal view of coccolith, SEM, ×17,225. **7–9.** *Bidiscus ignotus* (Górka) Hoffmann. **7.** Proximal view, SEM, ×9680. **8.** Coccosphere, SEM, ×5200. **9.** Distal view, SEM, ×12,880. 8, Hauterivian, Spain; 7,9, Berriasian, Spain, 5–9, from Grün and Allemann, 1975. **10,11.** *Podorhabdus albianus,* SEM, from Hill, 1976. **10.** Del Rio Clay, Texas, distal view, ×4000. **11.** Grayson Formation, oblique side view, ×3465. **12.** *Staurolithites matalosus,* Lower Cretaceous, western Atlantic, proximal side, SEM, ×10,400, from Thierstein, 1973. **13–15.** *Deflandrius cretaceus* (Arkhangelskiy) Bramlette & Martini, TEM of carbon replicas, from Bukry, 1969. **13,14.** Taylor Marl, Texas, side view, ×5135, and distal view, ×5495. **15.** Austin Chalk, Texas, proximal view, ×3880. **16.** *Lithraphidites acutus* Verbeek & Manivit, Cenomanian, Indian Ocean, SEM, ×4800, from Thierstein, 1974. **17,18.** *Micula staurophora,* Albian, Sweden, SEM, top view, ×3200, and side view, ×3520, from Forchheimer, 1972.

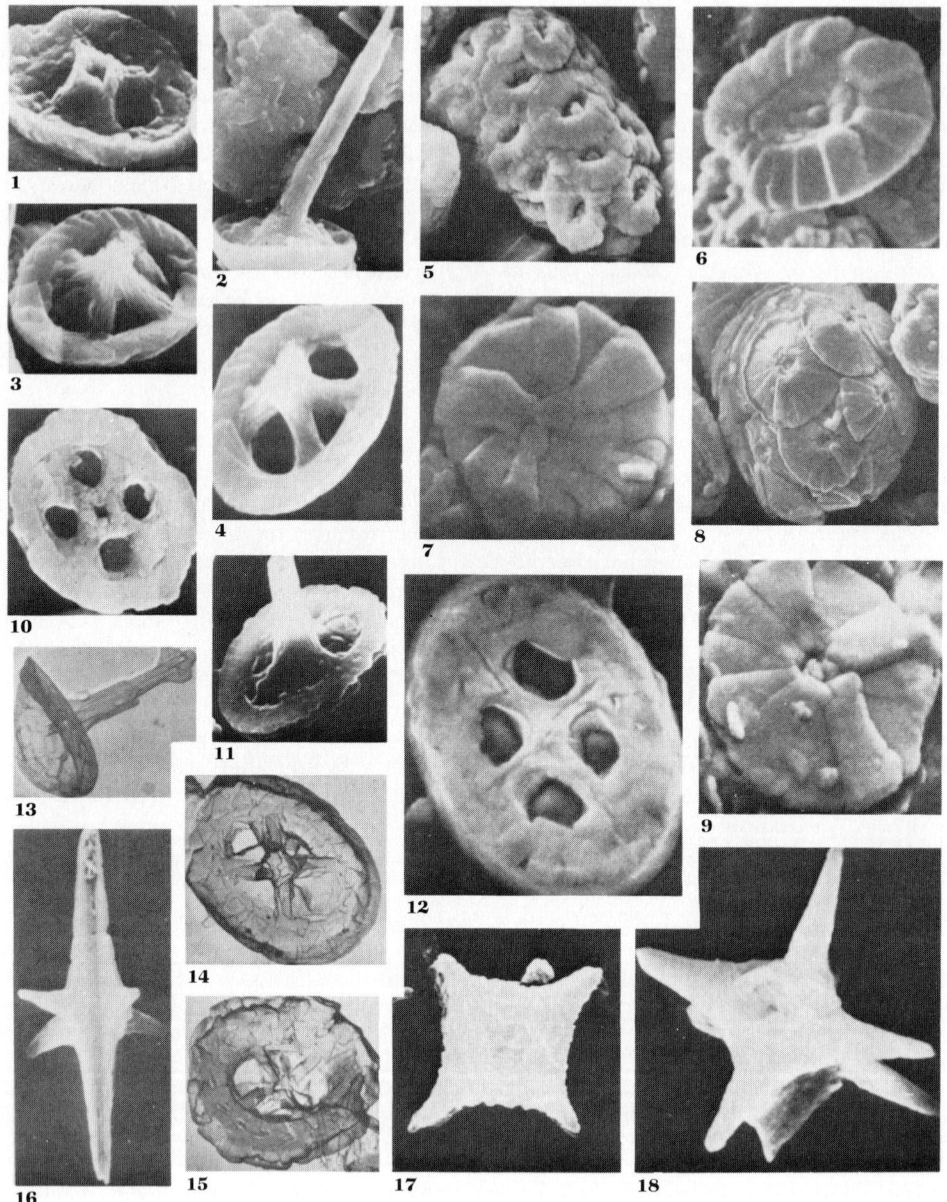

Figure 9.43

Figure 9.44

1—4. *Kamptnerius magnificus.* **1,2.** Upper Cretaceous (Coniacian) Indian Ocean, SEM, proximal side, ×3600, and distal side, ×3680, the latter showing secondary calcite overgrowth; from Thierstein, 1974. **3,4.** Prairie Bluff Formation, Alabama, TEM, distal and proximal views of narrow-winged form, ×2430, from Black, 1968. **5—7.** *Eiffellithus augustus* Bukry; 5,7, Taylor Marl, Texas; 6, Santonian, France; TEM, from Bukry, 1969. **5.** Distal view, ×3615. **6.** Proximal view, ×2715. **7.** Distal view, ×3075. **8.** *Lithraphidites quadratus,* Maastrichtian, TEM, ×6400, from Reinhardt, 1970b. **9—12.** *Cribrocorona gallica* (Stradner) Perch-Nielsen, Maastrichtian, Madagascar, from Perch-Nielsen, 1973. **9.** Proximal view, ×4800. **10.** Distal view, ×6400. **11.** Side view of broken specimen, ×7600. **12.** Oblique distal view, ×8400.

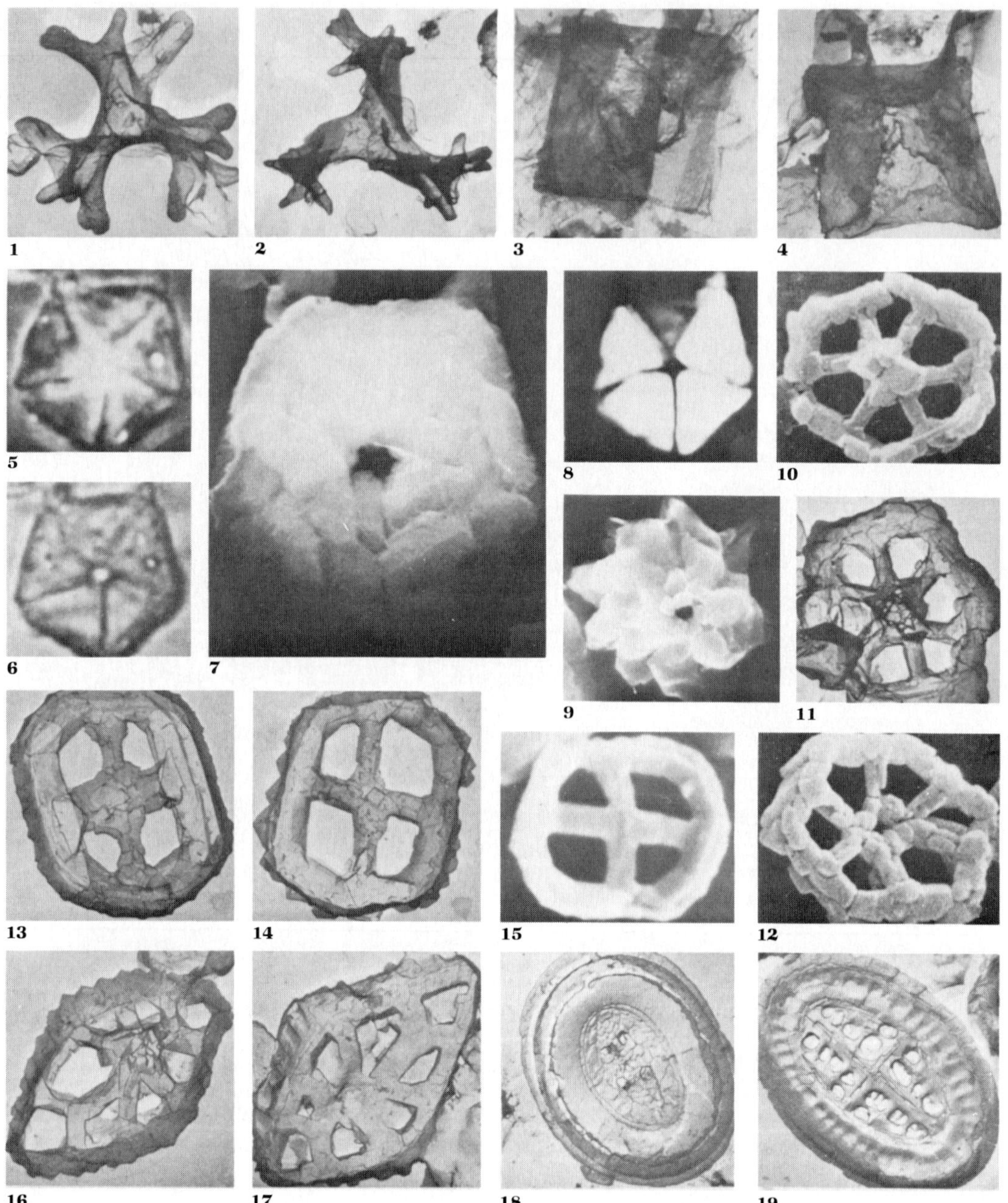

744

Cretaceous species. **1,2.** *Marthasterites furcatus* (Deflandre) Deflandre, Taylor Marl, Texas, TEM; 1, ×5320; 2, ×3840. **3,4.** *Micula decussata* Vekshina, Taylor Marl, Texas, TEM; 3, ×6130; 4, ×7280. 1−4, from Bukry, 1969. **5−8.** *Micrantholithus hoschulzi* (Reinhardt) Thierstein, Lower Cretaceous, western Atlantic, from Thierstein, 1971. **5,6,8.** Viewed in phase contrast, transmitted light, and cross-polarized light, respectively, all ×3200. **7.** SEM, ×7000. **9.** *Hayesites albiensis* Manivit, Goodland Formation, Texas, distal view, SEM, ×8000, from Hill, 1976. **10−12.** *Corollithion exiguum.* **10,12.** Maastrichtian, Madagascar, SEM, distal side, ×10,600, and proximal side, ×9800, from Perch-Nielsen, 1973. **11.** Austin Chalk, Texas, distal view, TEM, ×6650, from Bukry, 1969. **13−15.** *C. signum.* **13.** Distal view, ×8470. **14.** Proximal view, ×8960. 13,14, from Taylor Marl, Texas, TEM, from Bukry, 1969. **15.** Del Rio Clay, Texas, SEM, distal view, ×5000, from Hill, 1976. **16,17.** *Rhombolithion rhombicum* (Stradner & Adamiker) Black. **16.** Distal view, ×9450. **17.** Proximal view, ×9940. **18,19.** *Broinsonia parca.* **18.** Proximal view, ×3395. **19.** Distal view, ×4900. 16−19, Taylor Marl, Texas, TEM, from Bukry, 1969.

---

kers include *Podorhabdus albianus, Lithraphidites acutus, Chiastozygus irregularis* Čepek, *Corollithion signum* Stradner, *C. exiguum* Stradner, *Gartnerago obliquum* (Stradner) Noël, and *Micula staurophora* (Gardet) Stradner (see Figures 9.45 and 9.46). Zonal markers for the Coniacian, Santonian, and Campanian include *Tetralithus pyramidus* Gardet, *Marthasterites furcatus, Calculites obscurus* (Deflandre) Prins & Sissingh, *Broinsonia parca* (Stradner) Bukry, *Arkhangelskiella ethmopora* Bukry, *Eiffellithus eximius* (Stover) Perch-Nielsen, *Ceratolithoides aculeus* (Stradner) Prins & Sissingh, and *Tetralithus trifidus* (Stradner) Bukry. Maastrichtian biohorizon and zonal markers are *Chiastozygus initialis* (Górka) Risatti, *Lithraphidites quadratus* Bramlette & Martini, *Nephrolithus frequens* Górka, and *Micula murus* (Martini) Bukry.

The calcareous nannoplankton suffered a major crisis at the end of the Cretaceous, as documented by Bramlette and Martini (1964). Of some 35 Maastrichtian species, all but one became extinct. Only a few species appeared in the earliest Paleocene Danian, leaving the assemblage less diverse than at any time since the Triassic.

Important publications concerning the Upper Cretaceous coccoliths are those by Black (1964), Black and Barnes (1959), Bramlette and Martini (1964), Bukry (1969), Caratini (1963), Čepek and Hay (1970), Forchheimer (1968, 1970, 1972, 1974), Gartner (1968), Górka (1957, 1963), Noël (1971), Perch-Nielsen (1968a, 1972, 1973), Pienaar (1968), Reinhardt (1964, 1966, 1969, 1970a, 1970b, 1971), Reinhardt and Górka (1967), Risatti (1973), Shamray (1963), Shumenko (1971, 1976), Stradner (1962b), Thierstein (1976), Vekshina (1959), and Verbeek (1977).

The biostratigraphic zonation for the Mesozoic (see Table 9.2) was based in part on outcrop sections, and in part on data from the DSDP, with later taxonomic modifications.

Cenozoic Calcareous Nannoplankton

Following the major extinction of Cretaceous taxa, new taxa arose in the Tertiary, many of much larger size than those of the Cretaceous. In addition, new and unusual types appeared, such as *Sphenolithus, Triquetrorhabdulus, Thoracosphaera, Ceratolithus,* and *Discoaster.* A single type, *Braarudosphaera,* appears to have persisted from the Cretaceous to the present with very little change.

Braarudosphaeraceae. The living species *Braarudosphaera bigelowii,* whose distinctive pentalith cover has been mentioned earlier, has

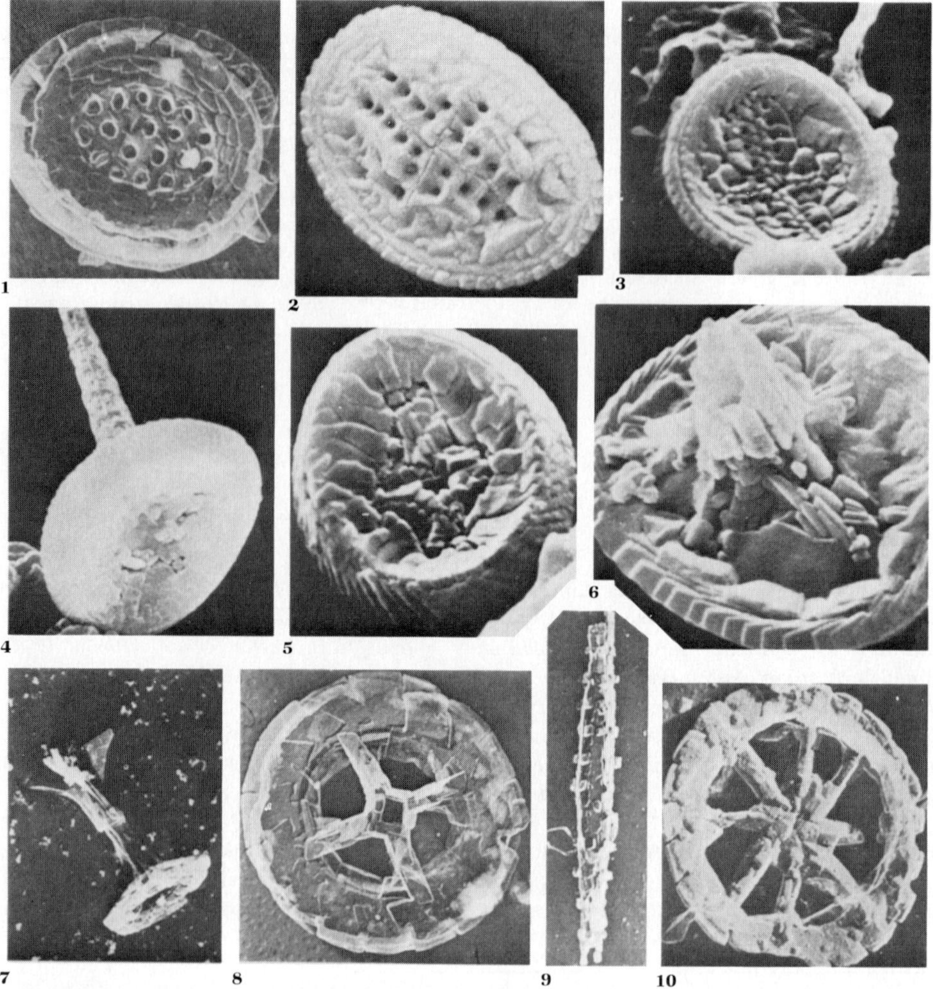

Figure 9.46
Cretaceous species. **1.** *Cribrosphaerella murrayi* (Arkhangelskiy) Vekshina, Upper Cretaceous, England, TEM, ×9750, from Black, 1965. **2,3.** *Gartnerago obliquum*, Aptian, Sweden, SEM. **2.** Distal view, ×6825. **3.** Oblique proximal view, ×5775. **4—6.** *Eiffellithus eximius*, SEM; 4,6, Hauterivian; 5, Aptian, Sweden. **4.** Oblique side view, ×3120. **5.** Oblique proximal view, ×6300. **6.** Oblique distal view, ×6900. 2—6, from Forchheimer, 1972. **7,8.** *Deflandrius columnatus*, Upper Cretaceous, England. **7.** Oblique side view, showing well-developed stem with finlike expansions, TEM, ×3710. **8.** Distal view of specimen with broken hollow stem, supported by central cross, and with 16 modified rhombohedra in outer rim, TEM, ×7000. **9.** *Microrhabdulus belgicus* Hay & Towe, Upper Cretaceous, England, with small calcite rhombs on surface, TEM, ×4200. **10.** *Cylindralithus laffittei* (Noël) Black, Lower Cretaceous, England, TEM, ×10,500. 1, 7—10, reproduced with permission from Black, 1965, *Endeavor*, v. 24, p. 131—137, copyright, 1965, Pergamon Press, Ltd.

seemingly persisted without change since the Paleocene, and earlier species were present in the Cretaceous, from at least the Aptian. *B. discula* Bramlette & Riedel is among the few species persisting through the Late Cretaceous crisis, having existed from the middle Albian (Lower Cretaceous) to the Eocene (Hill, 1976). Cenozoic and living *Braarudosphaera* and the related *Micrantholithus* are generally restricted to near-shore sediments and absent from typically oceanic deposits of the same age (with the exception of *B. rosa* Levin & Joerger, which forms an extensive Oligocene marine chalk in the South Atlantic). This ecologic restriction is less evident in the Cretaceous representatives (Hill, 1976). Although observed living in the modern seas, *Braarudosphaera bigelowii* has not yet been cultured in the laboratory, nor has the cell itself been described in detail, other than as possessing two plastids, each with a pyrenoid on its inner face, surrounding numerous lipid globules (Lefort, 1972). As the pentalith cover of *B. bigelowii* appears to leave no opening for flagella, the cell probably represents a nonmotile stage or cyst. In contrast, *B. magnei* Lefort has numerous circular overlapping pentaliths surrounding the cell and has two flagella, although no haptonema has been observed (Lefort, 1972).

Thoracosphaeraceae. *Thoracosphaera* was described originally from the present seas, ranging from the Adriatic to the South Pacific. Fossil species later referred to *Thoracosphaera* ranged from Pleistocene to Paleocene, with rare reports from the latest Upper Cretaceous. This spherical calcareous form, with terminal rounded opening that may be filled by an operculum, has a wall that consists of numerous irregularly shaped, interlocking units—angular plates that may be centrally perforated, or small irregular crystals of random orientation. Because of the operculum that may or may not be present, the individual specimens have been regarded as possible cysts, much like *Braarudosphaera*. The pseudocyst stage of *Ochrosphaera neapolitana* consists of large cells with indistinct coccoliths that coalesce to form an irregular calcareous cover (Gayral and Fresnel-Morange, 1971). *Thoracosphaera* may be a similar form.

Fütterer (1976) showed that *Thoracosphaera albatrosiana* Kamptner has an angular bordered opening strongly resembling a dinoflagellate archeopyle, and placed the genus in the peridinioid dinoflagellates, with the Calciodinellaceae. The wall of *T. albatrosiana* has a network of crystals bounding tiny irregular fields, very much like the surface of a single field of *Calciodinellum operosum,* between the sutural flanges (see Chapter 4). In contrast, *T. pelagica* Kamptner (correctly *T. heimii* (Lohmann) Kamptner), the type species of *Thoracosphaera,* has a smaller and circular opening (illustrated earlier in Figure 9.23), unlike the angular bordered one of the dinoflagellate archeopyles. The wall is constructed of many platelets, with central "pore," although these pores do not appear always to penetrate the entire wall thickness. No dinoflagellate cyst has a wall of separate platelets (although the theca does), but the thecal sutures instead are indicated by thickened areas or flanges in the cyst, as in *Calciodinellum operosum.*

Thoracosphaera albatrosiana and *T. tuberosa* Kamptner appear to be Calciodinellacean dinoflagellate cysts, incorrectly attributed to *Thoracosphaera,* but the type species, *T. heimii,* and various other fossil species do not appear to have dinoflagellate affinities, and are here retained in the Haptophyta, probably related to such coccolithophores as *Braarudosphaera* or *Ochrosphaera. Pithonella* was compared to *Thoracosphaera,* and also regarded as a probable calcareous dinoflagellate cyst (Keupp, 1978), but like typical *Thoracosphaera* has no distinctive features to support this assignment, hence is here retained with the calcareous nannoplankton.

The Maastrichtian genus *Centosphaera* has one or two peripheral flanges, resembling *Calciogranellum,* but is said to have an aperture between the two rims, suggesting that it probably also is unrelated to the dinoflagellates, hence is here placed with *Thoracosphaera.*

Rhabdothorax regalis (Gaarder) Gaarder has a calcareous covering with a thick fragile continuous crust, although individual elements can be resolved, each consisting of a plate bearing a

Table 9.2
Mesozoic biostratigraphic zonation based on calcareous nannoplankton. Z, Zone; SZ, Subzone. Compiled from Bukry, 1974a; Bukry and Bramlette, 1970; Čepek and Hay, 1970; Hill, 1976; Manivit, 1971; Manivit et al., 1977; Smith, 1975b; Thierstein, 1971, 1973; and Verbeek, 1977. Names of taxa have been modified in some instances to agree with later studies.

System	Stage	MY			
Upper Cretaceous	Maastrichtian	63	*Nephrolithus frequens* Z	*Micula murus* Z	
		67	*Lithraphidites quadratus* Z		
		69	*Chiastozygus initialis* Z	*Quadrum trifidum* Z	
	Campanian	75	*Ceratolithoides aculeus* Z	*Broinsonia parca* Z	*Quadrum gothicum* Z
			Broinsonia parca Z		*C. aculeus* Z
			Kamptnerius magnificus Z	*Eiffellithus eximius* Z	*B. parca* Z
			Arkhangelskiella ethmopora Z		
	Santonian	80	*Marthasterites furcatus* Z	*Gartnerago obliquum* Z	*Zygodiscus spiralis* Z
					Rucinolithus hayii Z
				M. furcatus Z	*Micula concava* Z
					Broinsonia lacunosa Z
	Coniacian	84	*Tetralithus pyramidus* Z	*Micula staurophora* Z	*M. furcatus* Z
	Turonian	86	*Corollithion exiguum* Z	*C. exiguum* Z	*E. eximius* Z
			C. signum Z		
		91	*Chiastozygus irregularis* Z		*Quadrum gartneri* Z
	Cenomanian		*Podorhabdus albianus* Z	*Lithraphidites acutus* Z	*Gartnerago obliquum* SZ
					Cruciellipsis chiastus SZ
Lower Cretaceous	Albian	95	*Staurolithites matalosus* Z	*Eiffellithus turriseiffelii* Z	*Prediscosphaera spinosa* SZ
					Hayesites albiensis SZ
				Podorhabdus albianus Z	
				100	
				Deflandrius columnatus Z	*Broinsonia gammation* SZ
				105	
	Aptian	106		*Rhagodiscus angustus* Z	
				109	
					Chiastozygus litterarius Z

Series	Stage	Age (Ma)	Zone			
Lower Cretaceous (cont.)	Barremian	112 – 118		*Tetralithus malticus* Z	*Micrantholithus hoschulzi* Z	*Watznaueria oblonga* Z
Lower Cretaceous (cont.)	Hauterivian	118 – 124 (121)	*Watznaueria diaphanae* Z	*Cruciellipsis cuvillieri* Z	*Lithraphidites bollii* Z	
Lower Cretaceous (cont.)	Valanginian	124 – 127	*Ellipsochiastus quadriserratus* Z	*Tubodiscus jura-pelagicus* Z	*Calcicalathina oblongata* Z	
Lower Cretaceous (cont.)	Valanginian	127 – 130	*Diadorhombus rectus* Z	*Ellipsagelosphaera britannica* Z	*Cretarhabdus angustiforatus* Z	
Lower Cretaceous (cont.)	Berriasian	131 – 136	*Nannoconus colomii* Z	139		
Upper Jurassic	Tithonian Purbeckian	136 – 146	*Parhabdolithus embergeri* Z			
Upper Jurassic	Kimmeridgian	146 – 151	*Ellipsagelosphaera communis* Z *Vekshinella stradneri* Z			
Upper Jurassic	Oxfordian	151 – 157	*Corollithion geometricum* Z *Diadozygus dorsetense* Z			
Middle Jurassic	Callovian	157 – 162	*Discorhabdus jungii* Z *Podorhabdus rahla* Z *Polypodorhabdus escaigi* Z *Stephanolithion bigotii* Z *Stephanolithion hexum* Z			
Middle Jurassic	Bathonian	162 –	*Stephanolithion speciosum* var. *octum* Z *Diazomatolithus lehmanii* Z *Stephanolithion speciosum* Z			
Middle Jurassic	Bajocian		*Discorhabdus tubus* Z			
Lower Jurassic	Toarcian		*Axopodorhabdus cylindratus* Z			
Lower Jurassic	Pliensbachian		*Crepidolithus crassus* Z *Palaeopontosphaera dubia* Z			
Lower Jurassic	Sinemurian		*Parhabdolithus liasicus* Z *P. marthae* Z *Crucirhabdus primulus* Z			
Lower Jurassic	Hettangian		*Annulithus arkellii* Z			

Figure 9.47
Rhabdothorax regalis, off New Zealand. **1,2.** Entire cell in light microscope and SEM, ×910. **3,4.** Enlargements of part of part 2, showing thick crust and irregular outline of separate calcareous elements, ×4770. Reproduced with permission from Gaarder and Heimdal, 1973, *Norw. J. Bot.,* v. 20, p. 89—97, published by Universitetsforlaget, Oslo University.

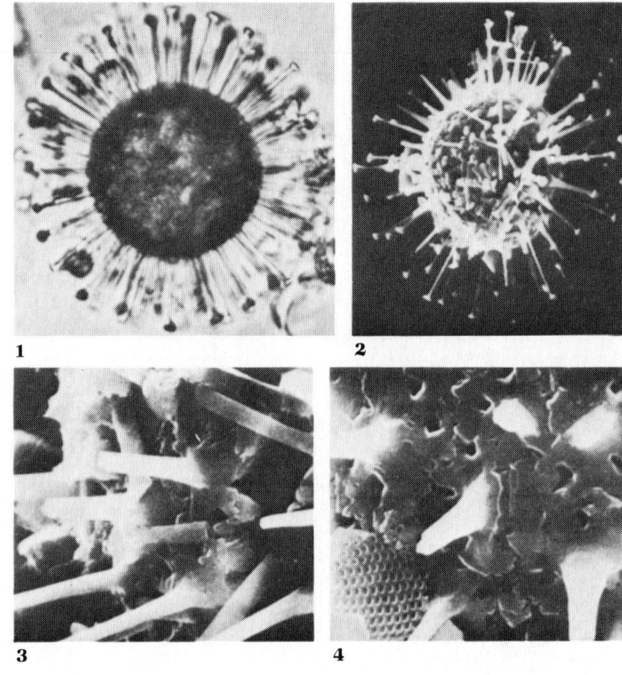

long slender flat-topped spine that is brittle and easily broken (see Figure 9.47). In polarized light, the optic axes appear parallel within a single element or crystallite. This structure was compared to that of cysts of the dinoflagellate *Scrippsiella trochoidea* (see Chapter 4) by Gaarder and Heimdal (1973), who suggested that *Rhabdothorax* also might be a dinoflagellate. For definite reallocation, culture of the cells from the cyst of *Rhabdothorax* is necessary, however. The cyst of *Scrippsiella trochoidea* consists of an acid-insoluble organic membrane, with the calcitic spines irregularly scattered over its surface, beneath an outer membrane. The membrane remains after removal of the spines in acid. The thick continuous crust of *Rhabdothorax* is not present in *Scrippsiella,* nor has an organic membrane been reported in *Rhabdothorax,* hence reassignment of this genus also remains uncertain in the absence of culture studies.

Ceratolithaceae. The distinctive asymmetrical horseshoe-shaped coccolith *Ceratolithus* was originally described from isolated fossil ceratoliths. The fossils were known for some 15 years before a living cell bearing ceratoliths was found. The ceratolith is ortholithid, consisting of a single crystallographic unit, and is highly birefringent in polarized light. It is brightest with the horns at about 45° to the direction of polarization, and goes to extinction with the horns about parallel to that direction (Gartner and Bukry, 1975). *Amaurolithus* has its optic (C) axis approximately perpendicular to the preferred orientation, with the ceratolith plate oriented perpendicular to the direction of light.

Living cells of *Ceratolithus* (Norris, 1965) were covered by a layer of simple circular scales or a coccosphere of ringlike coccoliths (Norris, 1971). These rings consisted of 35 to 40 elements, triangular in section, joined by alternate tiny elements, in a tenon-and-mortise arrangement. Within each cell was a large ceratolith, or in some cells, more than one. The ceratolith horns extend beyond the cell itself (see Figure 9.48). No flagella were observed.

Discoasters. The discoasters are one of the most useful groups of nannoplankton for Tertiary stratigraphy. Resembling tiny stars or rosettes and from 10 to 35 μm in diameter, they were at first thought to be inorganically formed, then later compared to holothurian spicules, although they are much too small and far too abundant for this possibility to be considered seriously. Because of their worldwide distribution, small size, huge numbers, and occurrence in deepwater facies, they are generally agreed to be the remains of a planktonic organism and may be isolated coccolithlike bodies, of either a motile cell or a cyst. They show no characteristic interference figures in polarized light, as do coccoliths, hence Tan Sin Hok (1926) suggested that they might be aragonitic. However, the asteroliths are shown by X-ray determination to consist of a single unit of calcite (ortholithid structure). The absence of birefringence in crossed nicols is because the C-axis of the calcite crystal is perpendicular to the radiate plane of the discoaster, although the latter may be somewhat curved or saucerlike rather than a true plane. Some species have a tendency to develop excess calcification or overgrowths so that accurate specific determination becomes difficult.

Discoasters first appeared in the late Paleocene about 60 million years ago (see Figure 9.49) and were particularly abundant in the Eocene, somewhat less varied in the Oligocene, and again highly differentiated in the Miocene, although many of the Miocene and later forms are very small. Many of the Paleocene and Eocene species are compact, multiradiate rosettes, whereas upper Eocene to middle Miocene species have fewer, relatively broad rays; typically stellate forms with relatively large central area or knob also became important (see Figure 9.50). In the late Tertiary, very narrow-rayed species predominate, some with three, four, or five rays as well as the more common six rays (see Figure 9.51).

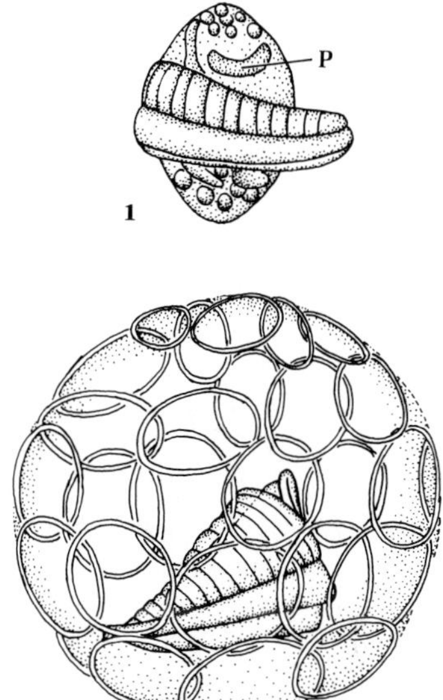

Figure 9.48
Ceratolithus cristatus living cells. **1.** Vegetative cell, showing plastid (p) and large ceratolith that projects beyond cell periphery. **2.** Empty sphere of ring-shaped coccoliths, enclosing a ceratolith. Both ×1000, redrawn from Norris, 1965.

The discoasters became scarcer in the later Tertiary, and only a single species, *Discoaster brouweri* Tan Sin Hok, continues to the end of the Pliocene. Although long regarded as indicating the Pliocene-Pleistocene boundary, the extinction of *D. brouweri* (see Figure 9.52) also has been suggested to lie within the lower Pleistocene at the Nebraskan glacial-Aftonian interglacial boundary (McIntyre et al., 1967), within the Olduvai normal event in the Matuyama reversed epoch of magnetic stratigraphy, at about 1.8 million years BP (Gartner, 1973) or somewhat above the top of the Olduvai event, at 1.6 million years (Hays and Berggren, 1971).

No certain records of living discoasters are known. *Discoaster planktonicus* Lecal is not a true discoaster, as it merely has surface markings on the spherical cell and no calcite asteroliths. It appears very close to the prasinophyte *Pterosperma ornatum* Schiller, which has a similar irregularly stellate surface pattern of crests and fields. Gaarder and Heimdal (1973) suggest that it might be synonymous with *Rhabdothorax regalis* from which the elongate spines were broken, but the absence of calcification makes this unlikely.

Discoaster brouweri, D. aster Bramlette & Riedel, and *D. challengeri* Bramlette & Riedel were reported in suspension in the Pacific Ocean, but no cells were observed, and the specimens were not corroded (Bogdanov and Ushakova, 1966). However, these fossil species are characteristic of Oligocene, Miocene, and Pliocene strata, and as noted by Martini (1971, discussion of Bursa, 1971), the material was found in suspension in the station immediately following coring of Pliocene bottom sediments.

Cells containing discoasters were reported to be living under the surface of ice at Devon Island, Canadian Arctic (Bursa, 1965). *Discoasteromonas calciferus* Bursa contained discoasters representing various fossil species and was observed to ingest pennate diatoms, green algal cells, euglenids, and dinoflagellates. The species was regarded as a dinoflagellate (Bursa, 1971), who stated that the discoasters might be decalcified. If the cells lack calcified discoasters, *Discoasteromonas* may be a prasinophyte like the *"Discoaster planktonicus"* of Lecal, or if a dinoflagellate, the discoasters probably were ingested from nearby fossil material that had been disturbed. If climatic deterioration caused the extinction of the discoasters elsewhere, the occurrence of living relics only as ice floras, rather than in the tropics, seems unlikely. No specimens of the *Discoasteromonas* discoasters themselves have yet been examined with the petrographic microscope to determine if they are calcitic, or with electron microscopy to make the identification certain. Thus the weight of evidence points to the extinction of the discoasters at the beginning of the Pleistocene. Occasionally corroded specimens may be found reworked into younger sediments, however.

Coccoliths. Holococcoliths are generally rare in the sediments, as the tiny hexagonal or rhombohedral crystals commonly are disaggregated and the crystallites of less than 0.1 μm diameter then are not recognizable. Among those that have been recorded as fossils are the Eocene and Oligocene species *Clathrolithus spinosus* Martini, *Discolithina macropora* (Deflandre) Levin & Joerger, *Orthozygus aureus* (Stradner) Bramlette & Wilcoxon, *Lanternithus minutus* Stradner, *Peritrachelina joidesa* Bukry & Bramlette, *Zygrhablithus bijugatus* (Deflandre) Deflandre, and *Daktylethra punctulata* Gartner (Stradner and Adamiker, 1966; Gartner and Bukry, 1969).

Other coccoliths in general show a reduction in size through the Cenozoic. Very large coccoliths up to 10 to 25 μm in diameter are present only in the early Tertiary. Most Miocene and Pliocene species are from 6 to 12 μm in diameter, and Pleistocene ones generally from 1 to 3 μm (Bukry, 1971c). Possibly this is related to the climatic cooling trend, as calcium carbonate is more readily available in warmer waters.

Zonation of the Cenozoic on the basis of the calcareous nannoplankton is the result of many contributions, variously based on outcrop sec-

---

Figure 9.49
Paleogene coccoliths and discoasters. **1,2.** *Heliolithus kleinpellii,* Paleocene, Bay of Biscay, from Perch-Nielsen, 1971c. **1.** Proximal side, ×4800. **2.** Oblique distal view, ×6080. **3—5.** *Fasciculithus tympaniformis,* Paleocene, southwest Pacific, ×4560, from Edwards and Perch-Nielsen, 1975. **3.** Distal view. **4,5.** Oblique proximal views. **6.** *Heliolithus riedeli* Bramlette & Sullivan, Lodo Formation, Paleocene, California, ×6000, from Perch-Nielsen, 1971c. **7,8.** *Discoaster multiradiatus* Bramlette & Riedel, Paleocene, southwest Pacific, a many-rayed species shown in proximal and distal views (note heavy overgrowth), ×4560, from Edwards and Perch-Nielsen, 1975. **9.** *Marthasterites tribrachiatus* (Bramlette & Riedel) Deflandre, lower Eocene, Denmark, ×1465, from Black, 1968. **10,11.** *Discoasteroides kuepperi,* Eocene, North Atlantic, proximal view, ×5065, and side view, ×5440 from Perch-Nielsen, 1972. **12,13.** *Discoaster binodosus* Martini, Eocene, North Atlantic oblique proximal view, ×5040, and proximal side, ×5440, from Perch-Nielsen, 1972. **14.** *D. gemmeus* Stradner, Eocene, northwest Pacific, ×8720, from Bukry, 1971b.

Figure 9.50
Tertiary coccoliths and discoasters. **1—4.** *Markalius inversus* (Deflandre) Bramlette & Martini. **1,2.** Eocene, North Atlantic, coccosphere, ×3360, and proximal view of coccolith, ×4800, TEM, from Perch-Nielsen, 1972. **3,4.** Eocene, Denmark, distal view, ×7040, and proximal view, ×7680, TEM, from Perch-Nielsen, 1971d. **5.** *Discoaster elegans* Bramlette & Sullivan, a multiradiate species, Paleocene, North Atlantic, ×2880, SEM, from Perch-Nielsen, 1972. **6.** *D. sublodoensis* Bramlette & Sullivan, Eocene, central Pacific, showing low stem, SEM, ×1920, from Haq, 1969. **7,8.** *D. saipanensis* (also referred to *Gyrodiscoaster*), Miocene, Algeria, oblique views, ×5440 and ×4560, SEM, from Clocchiatti, 1971a. **9.** *D. lodoensis* Bramlette & Riedel, Eocene, ×2000, TEM, from Black, 1965. **10.** *D. barbadiensis* Tan (a *Heliodiscoaster*), Eocene, ×4800. 9,10, TEM, reproduced with permission from Black, 1965, *Endeavor*, v. 24, p. 131—137, copyright 1965, Pergamon Press, Ltd. **11.** *D. tani* Bramlette & Riedel subsp. *nodifer* Bramlette & Riedel, Eocene, central Pacific, SEM, ×4000, from Haq, 1969.

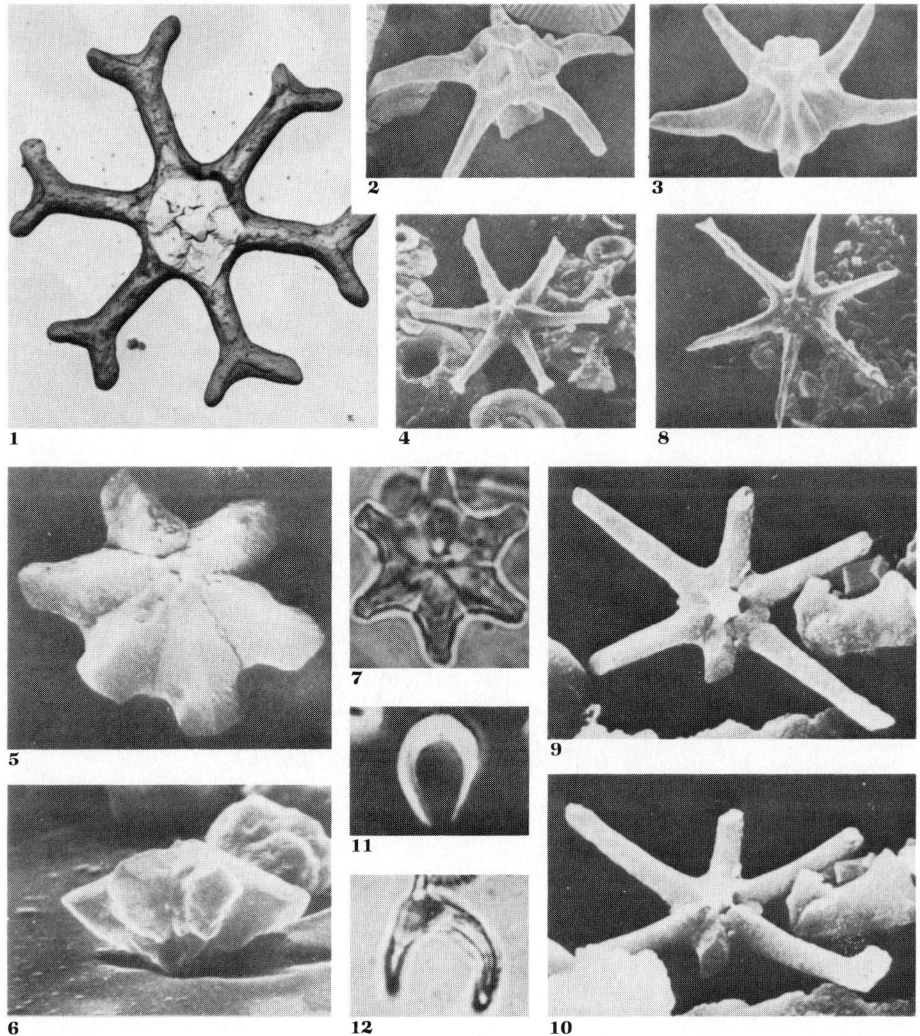

Figure 9.51

Neogene discoasters and ceratoliths. **1.** *Discoaster challengeri*, Miocene, at 1120 cm in core, 13°15′ N, 40°40′ W, Atlantic Ocean, TEM, ×6000, from A. McIntyre. **2,3.** *D. quinqueramus* Gartner, Miocene, North Atlantic, oblique views of opposite faces, SEM, ×3096 and ×5544, from Perch-Nielsen, 1972. **4.** *D. exilis* Martini & Bramlette, Miocene, western Indian Ocean, proximal view, SEM, ×2000, from Müller, 1974a. **5,6.** *D. druggii* Bramlette & Wilcoxon, Miocene, Algeria, distal face at different angles, SEM, ×6720 and ×5520, from Clocchiatti, 1971a. **7.** *D. kugleri* Martini & Bramlette, Miocene, Mediterranean Sea, ×1495, from Stradner, 1973. **8—10.** *D. calcaris* Gartner. **8.** Miocene, western Indian Ocean, distal view, SEM, ×2000, from Müller, 1974a. **9,10.** Miocene, Algeria, SEM, ×3920, from Clocchiatti, 1971a. **11,12.** *Amaurolithus tricorniculatus* (Gartner) Gartner & Bukry, lower Pliocene, Mediterranean Sea, in negative phase contrast and in normal light, ×1495, from Stradner, 1973.

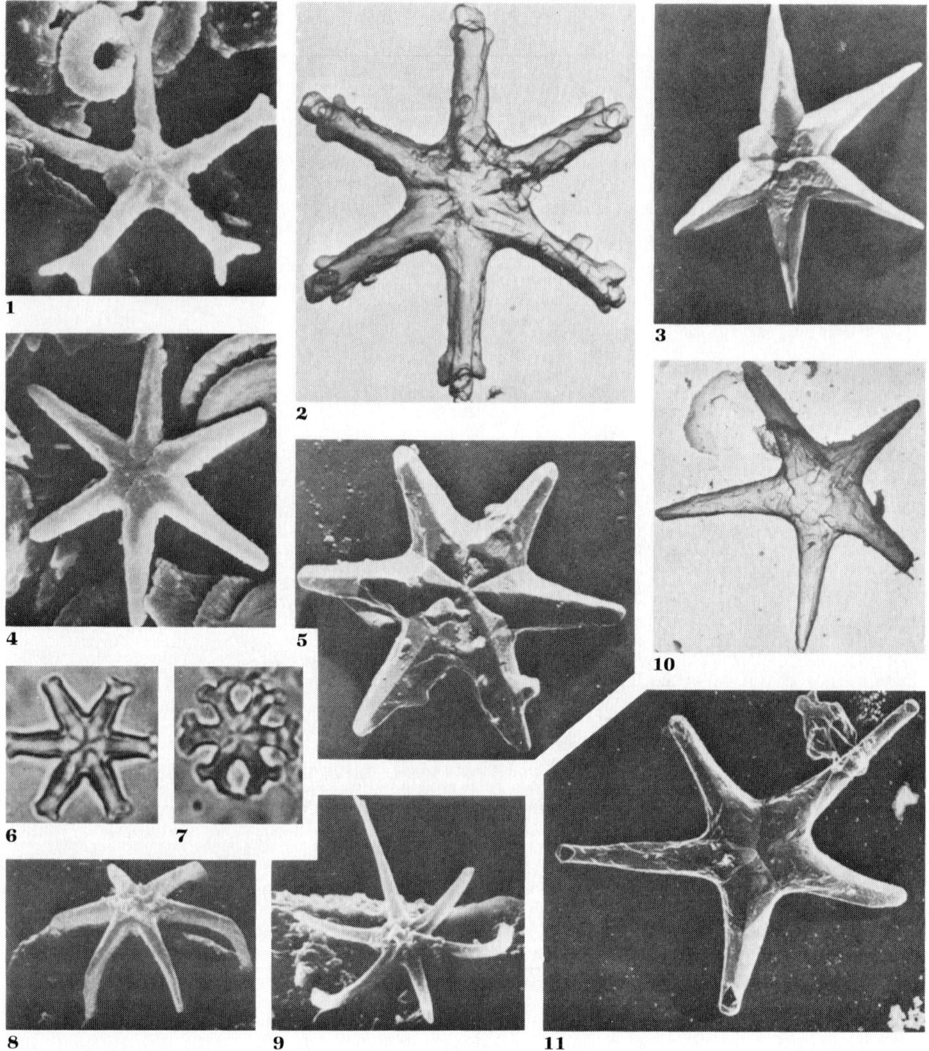

Figure 9.52

Neogene discoasters. **1.** *Discoaster pentaradiatus* Tan Sin Hok, lower Pliocene, Mediterranean Sea, proximal view, SEM, ×7200, from Stradner, 1973. **2.** *D. surculus* Martini & Bramlette, Pliocene, Pacific Ocean, ×4480, TEM, from Bukry, 1971b. **3—5,8,9.** *D. brouweri.* **3,5.** Pliocene, Pacific Ocean, five-rayed form, with sixth ray undeveloped, and normal six-rayed form, convex surface, TEM, ×4000 and ×8000, from Black, 1972b. **4.** Lower Pliocene, Mediterranean Sea, SEM, ×6935, from Stradner, 1973. **8,9.** Pliocene, western Indian Ocean, distal and proximal views, SEM, ×2000, from Müller, 1974a. **6,7.** Miocene, Mediterranean Sea, normal light, ×1495, from Stradner, 1973. **6.** *D. dilatus* Hay. **7.** *D. divaricatus* Hay. **10.** *D.* cf. *asymmetricus* Gartner, lower Pliocene, Mediterranean Sea, proximal view, TEM, ×4400, from Stradner, 1973. **11.** *D. asymmetricus,* Pliocene, Pacific Ocean, TEM, ×7200, from Black, 1972b.

tions or on the material cored in the ocean basins. Compilation of a standard sequence was made for the Paleogene, Zones NP 1–25, by Martini (1970), and for the Neogene, NN 1–21, by Martini and Worsley (1970). Some later modifications have been proposed, and these 46 biostratigraphic zones and their subzones correlated to the radiometric scale, so that the nannoplankton zonation of the Cenozoic provides a resolution to about 1 to 1.5 million years.

Lower Paleocene zone indices (see Figure 9.53) include *Conococcolithus panis* Edwards, *Markalius inversus* (Deflandre) Bramlette & Martini, *M. astroporus* (Stradner) Mohler & Hay, *Cruciplacolithus tenuis* (Stradner) Hay & Mohler, *Chiasmolithus danicus* (Brotzen) Hay & Mohler, *Ellipsolithus macellus* (Bramlette & Sullivan) Sullivan, and *Calcidiscus robustus* (Bramlette & Sullivan) Loeblich & Tappan.

Middle Paleocene species include (see Figure 9.54) *Fasciculithus tympaniformis* Hay & Mohler, *Heliolithus kleinpellii* Sullivan, *Discoaster gemmeus*, and *D. mohleri* Bukry & Percival; and upper Paleocene taxa are *Heliolithus riedeli, Discoaster nobilis* Martini, *D. multiradiatus, Chiasmolithus bidens* (Bramlette & Sullivan) Hay & Mohler, and *Campylosphaera eodela* Bukry & Percival.

Lower Eocene zonal markers are *Marthasterites contortus* (Stradner) Deflandre, *M. tribrachiatus, Discoaster binodosus, D. diastypus* Bramlette & Sullivan, and *D. lodoensis*. The middle Eocene is characterized by *D. sublodoensis, D. tani* Bramlette & Riedel, *D. tani* subsp. *nodifer* Bramlette & Riedel,

Sphenolithus radians Deflandre, *Nannotetrina alata* (Martini) Haq & Lohmann, *N. quadrata* (Bramlette & Sullivan) Bukry, *Discoasteroides kuepperi* (Stradner) Bramlette & Sullivan, *Rhabdosphaera inflata* Bramlette & Sullivan, *Discoaster mirus* Deflandre, *D. strictus* Stradner, *Chiasmolithus gigas* (Bramlette & Sullivan) Radomski, and *Birkelundia staurion* (Bramlette & Sullivan) Perch-Nielsen.

Upper Eocene zone fossils are *Discoaster saipanensis* Bramlette & Riedel, *D. bifax* Bukry, *Sphenolithus furcatolithoides* Locker, *Reticulofenestra umbilica* (Levin) Martini & Ritzkowski, *Chiasmolithus solitus* (Bramlette & Sullivan) Locker, *C. oamaruensis* (Deflandre) Hay, Mohler & Wade, *Pemma papillatum* Martini, *Bramletteius serraculoides* Gartner, *Chiasmolithus grandis* (Bramlette & Sullivan) Radomski, *Helicosphaera compacta* Bramlette & Wilcoxon, *Hayella situliformis* Gartner, *Discoaster tani* subsp. *tani, Cyclicargolithus reticulatus* (Gartner & Smith) Bukry, *Isthmolithus recurvus* Deflandre, *Sphenolithus pseudoradians* Bramlette & Wilcoxon, and *Discoaster barbadiensis* Tan (see Figure 9.55).

Species characteristic of the Oligocene are *Ericsonia subdisticha* (Roth & Hay) Roth, *Helicosphaera reticulata* Bramlette & Wilcoxon, *Calcidiscus formosus* (Kamptner) Loeblich & Tappan, *C. margaritae* (Roth & Hay) Loeblich & Tappan, *Reticulofenestra laevis* Roth & Hay, *Sphenolithus predistentus* Bramlette & Wilcoxon, *S. distentus* (Martini) Bramlette & Wilcoxon, *S. ciperoensis* Bramlette & Wilcoxon, *Dictyococcites abisectus* (C. Müller) Bukry & Per-

---

Figure 9.53 (*overleaf*)
Paleogene coccoliths. **1.** *Goniolithus fluckigeri* Deflandre, Eocene, Denmark, TEM, ×5600, from Perch-Nielsen, 1971d. **2,3.** *Micrantholithus altus* Bybell & Gartner, middle Eocene, Alabama, plan and side views, TEM, ×4000, from Bybell and Gartner, 1972. **4,5.** *Chiasmolithus gigas*, Eocene, Blake Plateau, phase contrast and cross-polarized light, ×2000, from Gartner, 1970. **6.** *Transversopontis exilis* (Bramlette & Sullivan) Perch-Nielsen, Eocene, Denmark, oblique proximal view, SEM, ×3680, from Perch-Nielsen, 1971d. **7,8.** *Ellipsolithus macellus*, Paleocene, North Atlantic, ×6640 and ×4400, SEM, from Perch-Nielsen, 1972. **9.** *Koczyia fimbriata* (Bramlette & Sullivan) Perch-Nielsen, Eocene, Denmark, oblique proximal view, SEM, ×3360, from Perch-Nielsen, 1971d. **10.** *Cruciplacolithus tenuis*, Paleocene, Alabama, distal view, TEM, ×8000, from Gartner, 1970. **11–14.** *Chiasmolithus danicus*. **11.** Paleocene, southwest Pacific, distal view, with some overgrowth in center, SEM, ×4560, from Edwards and Perch-Nielsen, 1975. **12–14.** Paleocene (Danian), Denmark; 12,13, polarized light and transmitted light, ×2000; 14, TEM, proximal side, ×8000; from Gartner, 1970.

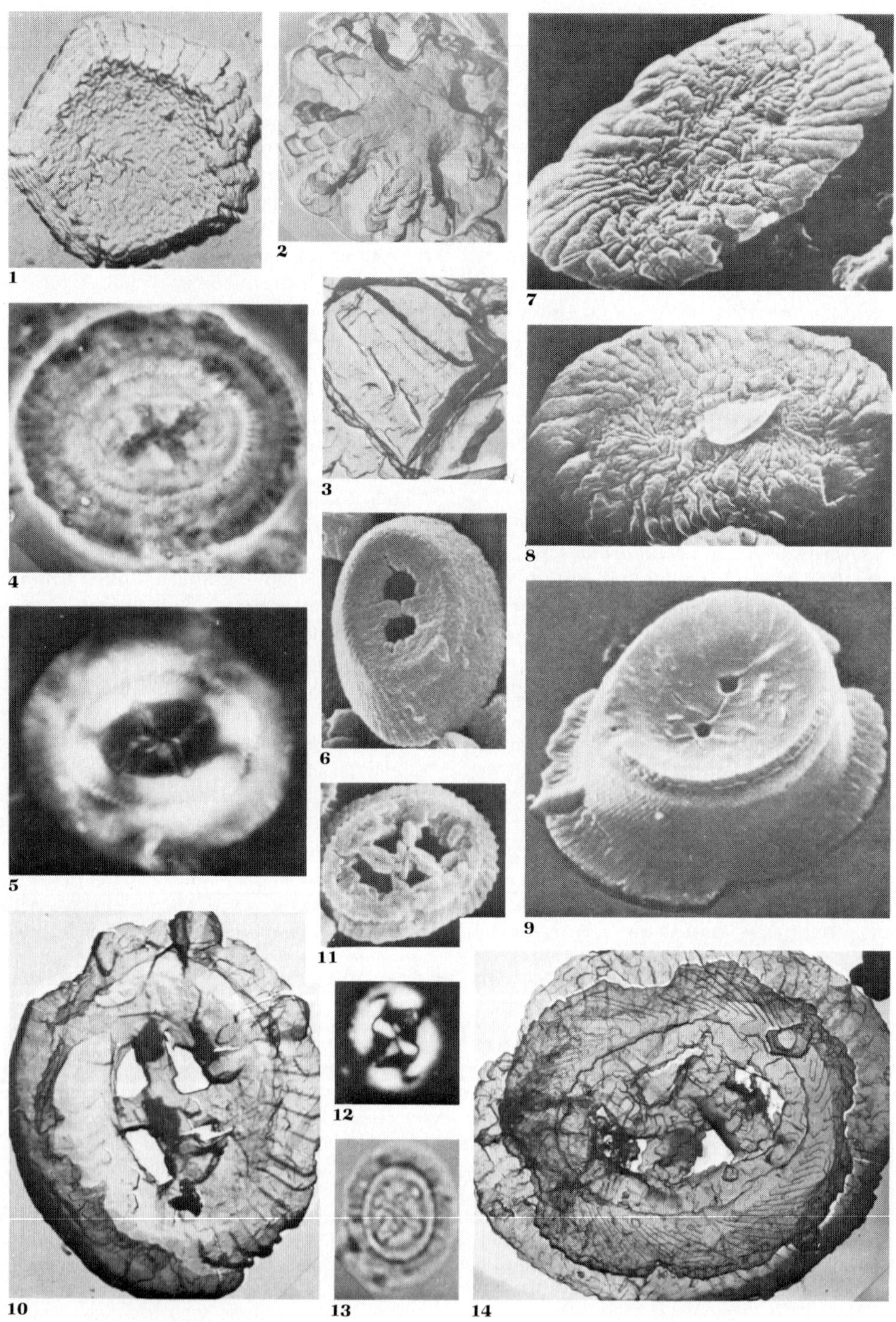

Figure 9.53

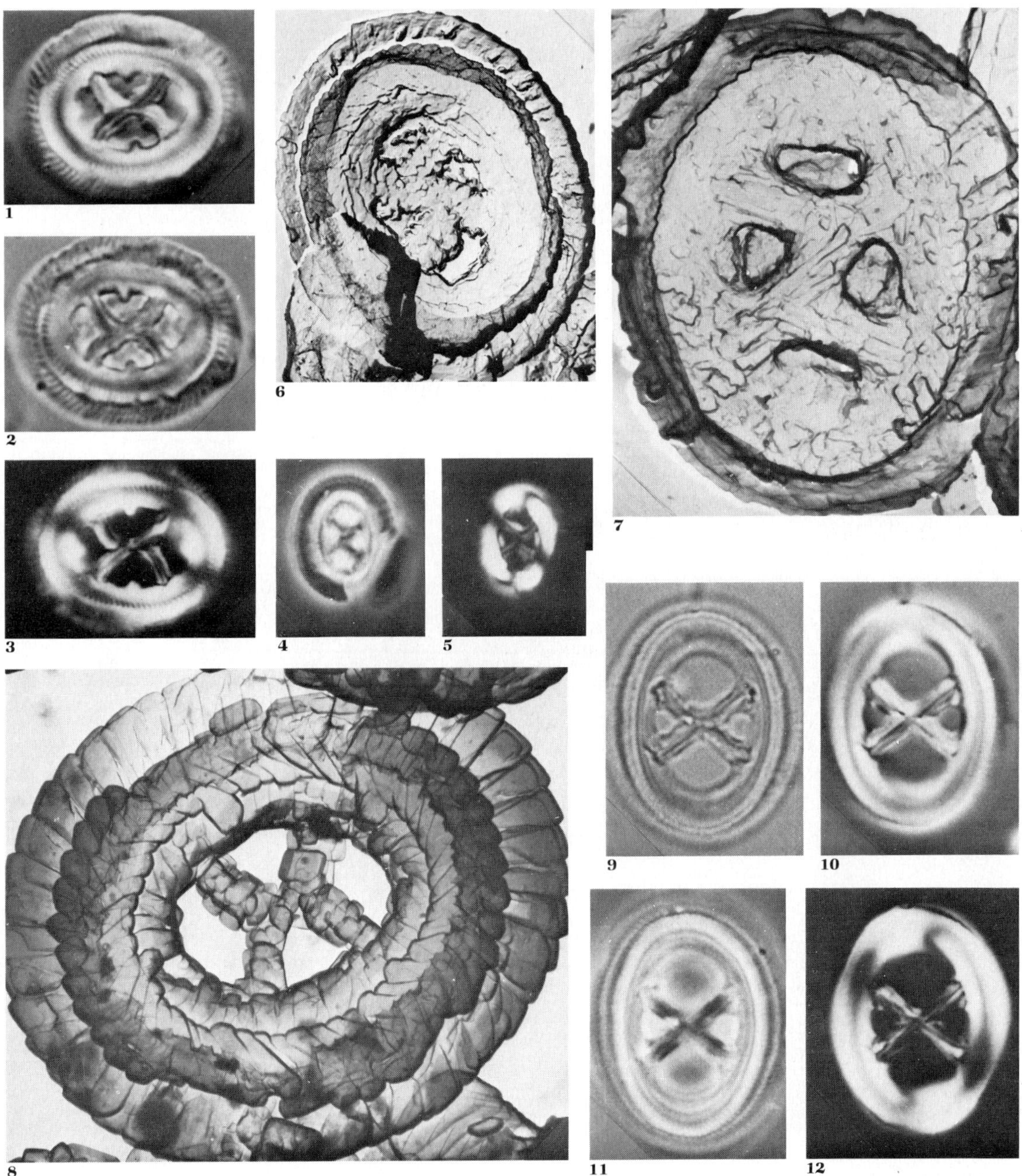

Figure 9.54

Paleogene *Chiasmolithus*. **1–3.** *C. grandis,* lower Eocene, in phase contrast, normal light, and cross-polarized light, ×1875. **4–7.** *C. bidens,* Paleocene, California. **4,5.** Phase contrast and polarized light, ×1875. **6,7.** TEM of proximal and distal sides, ×7500. **8.** *C. solitus,* Eocene, TEM of proximal side, ×7500. **9–12.** *C. oamaruensis,* upper Eocene, Crimea, transmitted light, interference contrast, phase contrast, and polarized light, ×1875. All from Gartner, 1970.

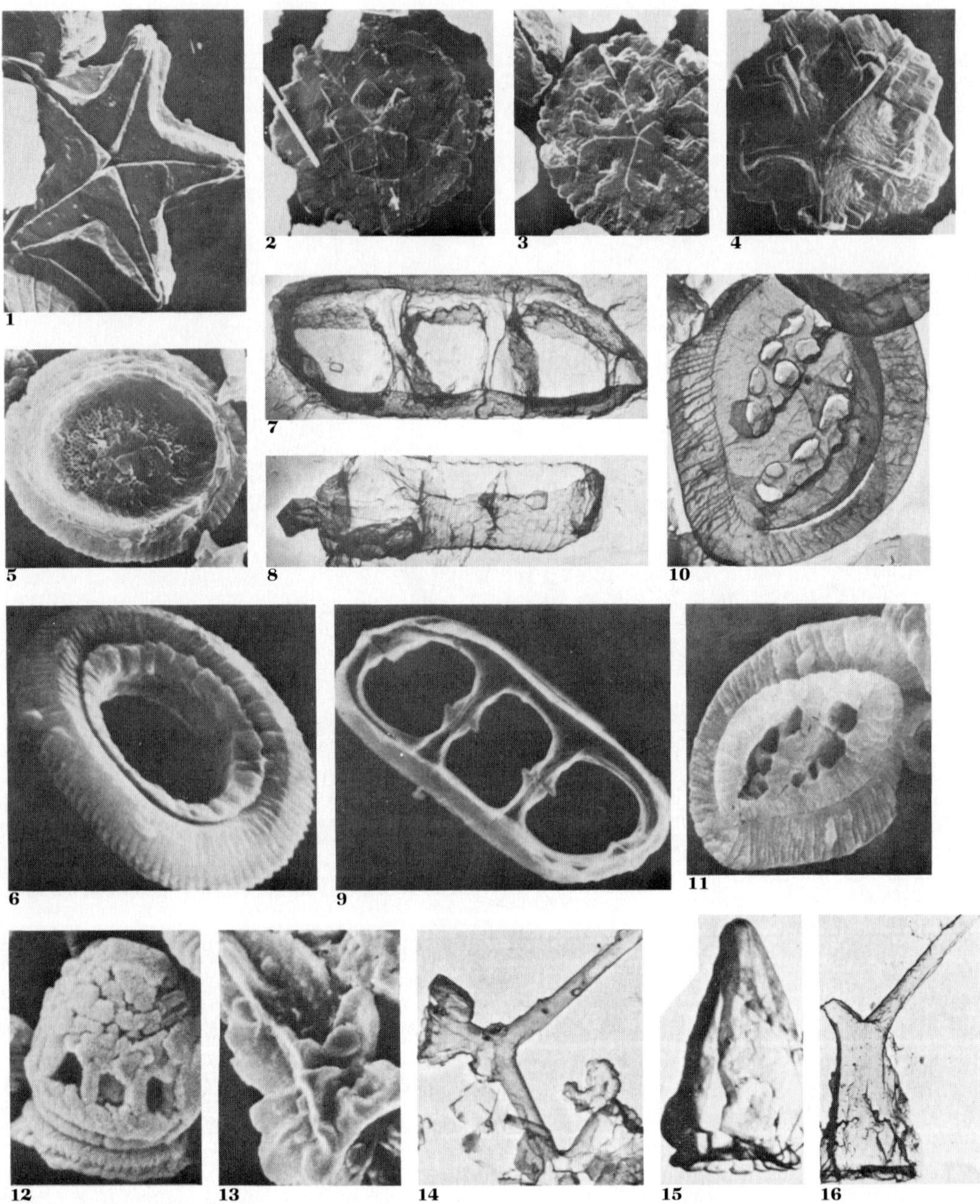

Figure 9.55
Paleogene coccoliths. **1.** *Micrantholithus pinguis* Bramlette & Sullivan, Paleocene, Lodo Formation, California, ×4800. **2 — 4.** *Pemma rotundum* Klumpp, middle Eocene, England. **2.** Nonpunctate surface, the central group of rhombohedra with (0$\bar{1}$0) crystal face visible, ×2400. **3,4.** Punctate surface, slightly corroded, ×2400, and deeply corroded, showing the lamination, ×3600. 1 — 4, TEM, from Black, 1972b. **5,6.** *Reticulofenestra umbilica*, distal views, SEM. **5.** Eocene, North Atlantic, ×3040, from Perch-Nielsen, 1972. **6.** Lower Oligocene, Denmark, ×9200, from Mikkelsen, 1975. **7 — 9.** *Isthmolithus recurvus*. **7,8.** Upper Eocene, New Zealand, TEM, plan and edge views, slightly oblique, ×12,000, from Stradner and Edwards, 1968. **9.** Lower Oligocene, Denmark, SEM, ×7200, from Mikkelsen, 1975. **10,11.** *Helicosphaera reticulata*, upper Eocene, Mississippi. **10.** TEM, ×4000, from Gartner, 1971. **11.** SEM, proximal view, ×3200, from Haq, 1973a. **12.** *Naninfula deflandrei* Perch-Nielsen, Eocene, Denmark, oblique distal view, SEM, ×7600, from Perch-Nielsen, 1971d. **13.** *Sphenolithus pseudoradians*, Oligocene, western Indian Ocean, SEM, ×4000, from Müller, 1974a. **14.** *S. distentus*, Oligocene, western Atlantic, TEM, ×2800. **15.** *S. predistentus*, Oligocene, western Atlantic, TEM, ×5600. 14,15, from Bramlette and Wilcoxon, 1967. **16,** *S. ciperoensis*, Oligocene, Trinidad, TEM, ×660, from Haq, 1971c.

cival, *D. bisectus* (Hay, Mohler & Wade) Bukry & Percival, and *Cyclicargolithus floridanus* (Roth & Hay) Bukry (see Table 9.3).

Within the Neogene, lower Miocene zonal markers are *Triquetrorhabdulus carinatus* Martini, *Discoaster druggii, D. deflandrei* Bramlette & Riedel, *Sphenolithus belemnos* Bramlette & Wilcoxon, *Helicosphaera ampliaperta* Bramlette & Wilcoxon, and *Sphenolithus heteromorphus* Deflandre. Middle Miocene species are *Discoaster exilis, D. kugleri, Coccolithus miopelagicus* Bukry, *Helicosphaera carteri, Catinaster calyculus* Martini & Bramlette, and *C. coalitus* Martini & Bramlette. Upper Miocene species are *Discoaster hamatus* Martini & Bramlette, *D. calcaris, D. bellus* Bukry & Percival, *D. neohamatus* Bukry & Bramlette, *D. neorectus* Bukry, *D. quinqueramus, D. berggrenii* Bukry, *Triquetrorhabdulus rugosus* Bramlette & Wilcoxon, *Amaurolithus primus* (Bukry & Percival) Gartner & Bukry, and *A. tricorniculatus* (Gartner) Gartner & Bukry (see Figure 9.56).

Pliocene zonal markers are *Ceratolithus rugosus* Bukry & Bramlette, *C. acutus* Gartner & Bukry, *Amaurolithus amplificus* (Bukry & Percival) Gartner & Bukry, *Discoaster asymmetricus, Reticulofenestra pseudoumbilica* (Gartner) Gartner, *Sphenolithus neoabies* Bukry & Bramlette, *Discoaster surculus, D. tamalis* Kamptner, *D. pentaradiatus,* and *D. brouweri.* Pleistocene species include *Calcidiscus macintyrei* (Bukry & Bramlette) Loeblich & Tappan, *Helicosphaera sellii* (Bukry & Bramlette) Jafar & Martini, *Cre-*

nalithus doronicoides (Black & Barnes) Roth, *Emiliania annula* (Cohen) Bukry, *E. ovata* Bukry, *Gephyrocapsa caribbeanica* Boudreaux & Hay, *G. oceanica, Ceratolithus cristatus,* and *Emiliania huxleyi* (see Figure 9.57 and Table 9.4).

Coccolith Oozes and Chalks

Sorby (1861) first observed that coccoliths were a very important constituent of deep sea deposits, after a comparison of those in the English Chalk with the coccoliths of modern deep sea oozes. The Solnhofen Limestone contains about 500,000 coccoliths per cubic millimeter, representing about 50,000 individual coccospheres (Flügel and Franz, 1967), and up to 1,850,000 coccoliths per cubic millimeter occur in the Upper Jurassic calpionellid limestones of southern France. From 500,000 to 6 million coccoliths were present per cubic millimeter in the shaly chalks of the Cretaceous Greenhorn Limestone of Kansas (Hattin and Darko, 1971), whereas for other more oceanic deposits with less detrital dilution estimates have ranged from 800,000 (Gümbel, 1873) to 69 million (Tan Sin Hok, 1927a) coccoliths and up to 5.5 million discoasters per cubic millimeter.

The appearance of calcareous planktonic microorganisms in the Mesozoic (see Figure 9.58) is said to have changed the location of the major carbonate deposition. Paleozoic lime-

Table 9.3

Paleogene biostratigraphic zonation based on calcareous nannoplankton. NP, Standard Paleogene nannoplankton zones; Z, Zone; SZ, Subzone.

Epoch			MY		Martini, 1976	Gartner, 1971, 1977; Roth, 1973, 1974		Bukry, 1971a, 1973b, 1974b, 1975; Perch-Nielsen, 1971c	
Oligocene	Upper		— 24.0	NP 25	Sphenolithus ciperoensis Z	Dictyococcites abisectus Z S. ciperoensis Z		S. ciperoensis Z	Dictyococcites bisectus SZ Cyclicargolithus floridanus SZ
	Middle		— 26.0	NP 24	Sphenolithus distentus Z	S. distentus Z	Reticulofenestra laevis Z	S. distentus Z	— 26.5
			— 32.0	NP 23	Sphenolithus predistentus Z			S. predistentus Z	
	Lower		— 34.0	NP 22	Helicosphaera reticulata Z	H. reticulata Z	Calcidiscus margaritae Z	H. reticulata Z	Calcidiscus formosus Z
			— 36.5	NP 21	Ericsonia subdisticha Z	E. subdisticha Z			— 37.0 E. subdisticha Z.
Eocene	Upper		— 37.5	NP 20	Sphenolithus pseudoradians Z	Discoaster barbadiensis Z		Discoaster barbadiensis Z	Cyclicargolithus reticulatus Z
			— 41.0	NP 19	Isthmolithus recurvus Z	I. recurvus Z			Discoaster tani tani Z
			— 42.0	NP 18	Chiasmolithus oamaruensis Z	Hayella situliformis Z Helicosphaera compacta Z Chiasmolithus grandis Z Bramletteius serraculoides Z Pemma papillatum Z			
			— 42.5	NP 17	Discoaster saipanensis Z	Reticulofenestra umbilica Z		R. umbilica Z	D. saipanensis Z
						Sphenolithus furcatolithoides Z		Nannotetrina quadrata Z	D. bifax Z — 45.0 Birkelundia staurion Z
			— 45.5						

Epoch	Sub-epoch	Age (Ma)	NP zone	Zone (col. 1)	Zone (col. 2)	Zone (col. 3)	Zone (col. 4)
Eocene	Middle	45.5	NP 16	*Discoaster tani nodifer* Z	*Discoaster tani* Z / *S. radians* Z	*Nannotetrina quadrata* Z (*cont.*)	*C. gigas* Z / *D. mirus* Z | *D. strictus* Z
		47.3	NP 15	*Nannotetrina alata* Z	*N. quadrata* Z		48.0 — *Rhabdosphaera inflata* Z
		48.5	NP 14	*Discoaster sublodoensis* Z	*D. sublodoensis* Z	*D. sublodoensis* Z	*Discoasteroides kuepperi* Z
		49.2	NP 13	*Discoaster lodoensis* Z	*D. lodoensis* Z	*D. lodoensis* Z	49.5
	Lower	50.0	NP 12	*Marthasterites tribrachiatus* Z	*M. tribrachiatus* Z	*M. tribrachiatus* Z	
		51.5	NP 11	*Discoaster binodosus* Z	*D. binodosus* Z	*Discoaster diastypus* Z	*D. binodosus* Z
		52.5	NP 10	*Marthasterites contortus* Z	*M. contortus* Z		*M. contortus* Z
Paleocene	Upper	53.0	NP 9	*Discoaster multiradiatus* Z	*D. multiradiatus* Z	*D. multiradiatus* Z	53.5 — *Campylosphaera eodela* Z / *Chiasmolithus bidens* Z
		55.0	NP 8	*Heliolithus riedeli* Z	*H. riedeli* Z	*Discoaster nobilis* Z	55.5
	Middle	56.0	NP 7	*Discoaster gemmeus* Z	*D. gemmeus* Z	*Discoaster mohleri* Z	
		57.0	NP 6	*Heliolithus kleinpellii* Z	*H. kleinpellii* Z	*H. kleinpellii* Z	
		58.0	NP 5	*Fasciculithus tympaniformis* Z	*F. tympaniformis* Z	*F. tympaniformis* Z	
	Lower	59.5	NP 4	*Ellipsolithus macellus* Z	*Calcidiscus robustus* Z	*E. macellus* Z	
		61.0	NP 3	*Chiasmolithus danicus* Z		*C. danicus* Z	
		63.5	NP 2	*Cruciplacolithus tenuis* Z	*C. tenuis* Z	*C. tenuis* Z	
		64.0	NP 1	*Markalius astroporus* Z	*M. astroporus* Z / *Conococcolithus panis* Z	*Biantholithus sparsus* Z	

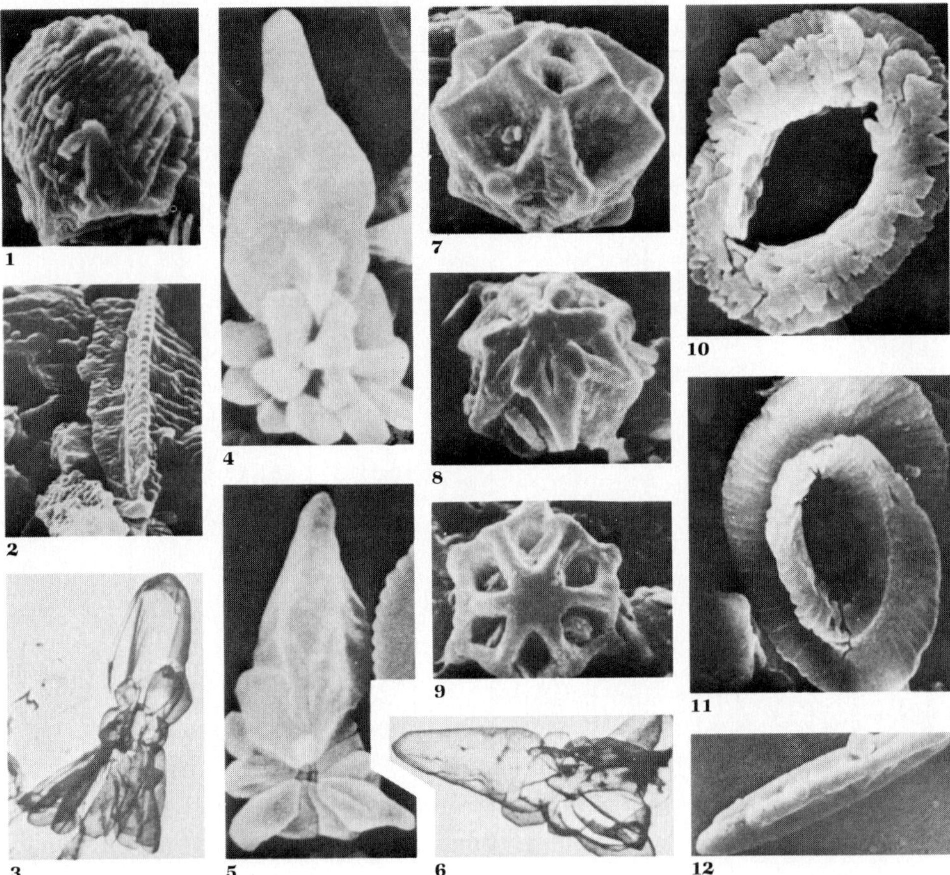

Figure 9.56

Neogene coccoliths. **1.** *Calciopilleus obscurus* Müller, Miocene, western Indian Ocean, SEM, ×4000, from Müller, 1974a. **2.** *Triquetrorhabdulus rugosus*, Miocene, North Atlantic, SEM, ×3255, from Perch-Nielsen, 1972. **3,6.** *Sphenolithus belemnos*, lower Miocene, equatorial Pacific, TEM, ×6400, from Haq and Lipps, 1971. **4,5.** *S. heteromorphus*, lower Miocene, southwest Pacific, with varied amount of overgrowth, SEM, ×9120, from Edwards and Perch-Nielsen, 1975. **7—9.** *Catinaster coalitus*, Miocene, western Indian Ocean, side, proximal, and distal views, SEM, ×4000, from Müller, 1974a. **10,11.** *Helicosphaera ampliaperta*, lower Miocene, Trinidad, SEM, distal view, ×4800, and proximal view, ×4000, from Haq, 1973a. **12.** *Triquetrorhabdulus carinatus*, Oligocene-Miocene, North Atlantic, SEM, ×3785, from Perch-Nielsen, 1972.

Figure 9.57 *(facing page)*

Neogene coccoliths. **1—3.** *Reticulofenestra pseudoumbilica*. **1.** Lower Pliocene, Mediterranean Sea, SEM, coccosphere, ×6500, from Stradner, 1973. **2,3.** Pliocene, northwest Pacific, coccoliths in TEM, ×8600, from Bukry, 1971b. **4.** *Calcidiscus macintyrei*, Pliocene, North Atlantic, distal view, SEM, ×4930, from Perch-Nielsen, 1972. **5.** *Emiliania annula*, upper Pliocene, Mediterranean Sea, complete coccosphere, SEM, ×11,000, from Stradner, 1973. **6.** *Ceratolithus rugosus*, Pliocene, western Indian Ocean, SEM, ×5000, from Müller, 1974a. **7.** *Emiliania huxleyi*, distal view of coccolith of warm water form, TEM, ×16,000, reproduced with permission from Black, 1965, *Endeavor*, v. 24, p. 131—137, copyright, 1965, Pergamon Press, Ltd. **8,9.** *Nannocorbis challengeri* Müller, Miocene, western Indian Ocean, SEM, ×5000, from Müller, 1974a. **10.** *Gephyrocapsa oceanica*, sediments, South Atlantic, distal view of coccolith, TEM, ×10,000, from Black and Barnes, 1961.

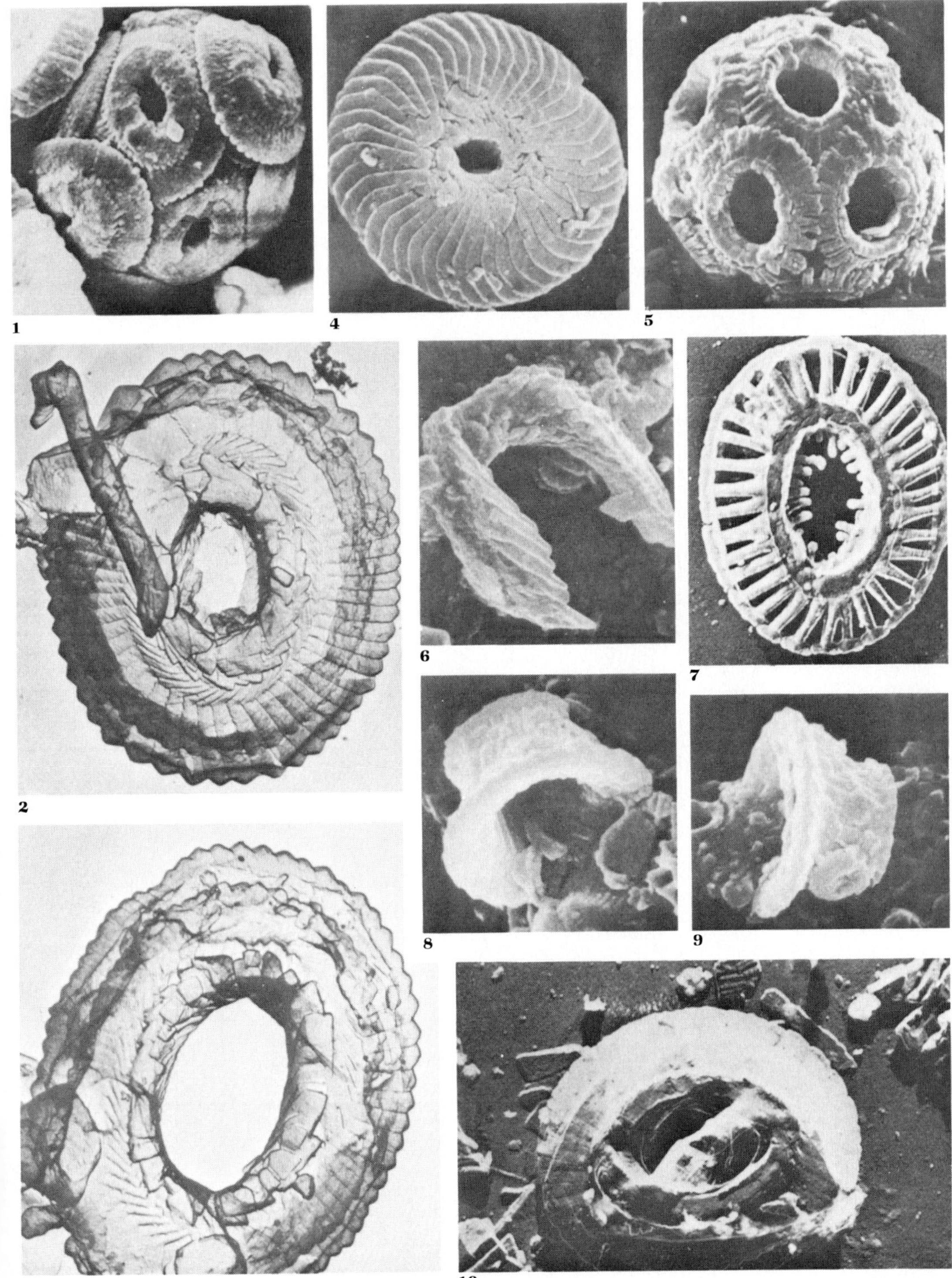

1
2
3
4
5
6
7
8
9
10

Table 9.4

Neogene biostratigraphic zonation based on calcareous nannoplankton. NN, Standard Neogene nannoplankton zones; Z, Zone; SZ, Subzone.

Epoch		MY	NN	Martini, 1976	Gartner, 1971, 1977; Roth, 1973, 1974	Bukry, 1971a, 1973b, 1974b, 1975; Perch-Nielsen, 1971c	
Pleistocene			NN 21	*Emiliania huxleyi* Z	*E. huxleyi* Z	*E. huxleyi* Z	
		— 0.2	NN 20	*Gephyrocapsa oceanica* Z	*G. oceanica* Z	*G. oceanica* Z	*Ceratolithus cristatus* SZ
		— 0.6	NN 19	*Emiliania annula* Z	*G. caribbeanica* Z; *E. annula* Z; *Helicosphaera sellii* Z	*Crenalithus doronicoides* Z	*Emiliania ovata* SZ — 0.9; *G. caribbeanica* SZ; *E. annula* SZ
Pliocene	Upper	— 1.8	NN 18	*Discoaster brouweri* Z	*Calcidiscus macintyrei* Z; *D. brouweri* Z	*D. brouweri* Z	*C. macintyrei* SZ — 2.0
		— 2.5	NN 17	*D. pentaradiatus* Z	*D. pentaradiatus* Z		*D. pentaradiatus* SZ; *D. surculus* SZ
		— 2.7	NN 16	*D. surculus* Z	*D. tamalis* Z		*D. tamalis* SZ
	Lower	— 3.5	NN 15	*Reticulofenestra pseudoumbilica* Z	*R. pseudoumbilica* Z	*R. pseudoumbilica* Z	*D. asymmetricus* SZ; *Sphenolithus neoabies* SZ
		— 3.8	NN 14	*Discoaster asymmetricus* Z			
		— 4.0	NN 13	*Ceratolithus rugosus* Z	*C. rugosus* Z	*A. tricorniculatus* Z	*C. rugosus* SZ
Miocene		— 4.6	NN 12	*Amaurolithus tricorniculatus* Z	*Amaurolithus amplificus* Z; *A. tricorniculatus* Z		*A. amplificus* SZ | *C. acutus* SZ; *Triquetrorhabdulus rugosus* SZ — 5.6
		— 5.0					

Miocene	Age (Ma)	NN	Zone	Zone	Zone	Subzone	Age (Ma)
Upper	5.0						
Upper		NN 11	Discoaster quinqueramus Z	A. primus Z	D. quinqueramus Z	A. primus SZ	5.6
Upper		NN 11		Discoaster berggrenii Z		D. berggrenii SZ	6.6
Upper		NN 11		Discoaster neohamatus Z	D. neohamatus Z	D. neorectus SZ	7.0
Upper							7.5
Upper	9.5	NN 10	Discoaster calcaris Z	D. bellus Z		D. bellus SZ	
Middle	11.0	NN 9	Discoaster hamatus Z	D. hamatus Z	D. hamatus Z	Catinaster calyculus SZ	
Middle		NN 9				Helicosphaera carteri SZ	
Middle	12.0	NN 8	Catinaster coalitus Z	C. coalitus Z	C. coalitus Z		
Middle	12.2	NN 7	Discoaster kugleri Z	D. kugleri Z	D. exilis Z	D. kugleri SZ	
Middle	13.0	NN 6	Discoaster exilis Z	D. exilis Z		Coccolithus miopelagicus SZ	
Lower	14.0	NN 5	Sphenolithus heteromorphus Z	S. heteromorphus Z	S. heteromorphus Z		
Lower	17.0	NN 4	Helicosphaera ampliaperta Z	H. ampliaperta Z	H. ampliaperta Z		
Lower	18.5	NN 3	Sphenolithus belemnos Z	S. belemnos Z	S. belemnos Z		
Lower	19.0	NN 2	Discoaster druggii Z	D. druggii Z	Triquetrorhabdulus carinatus Z	D. druggii SZ	
Lower	20.5					D. deflandrei SZ	
Lower		NN 1	T. carinatus Z	T. carinatus Z		D. abisectus SZ	23.0
Lower	24.0						

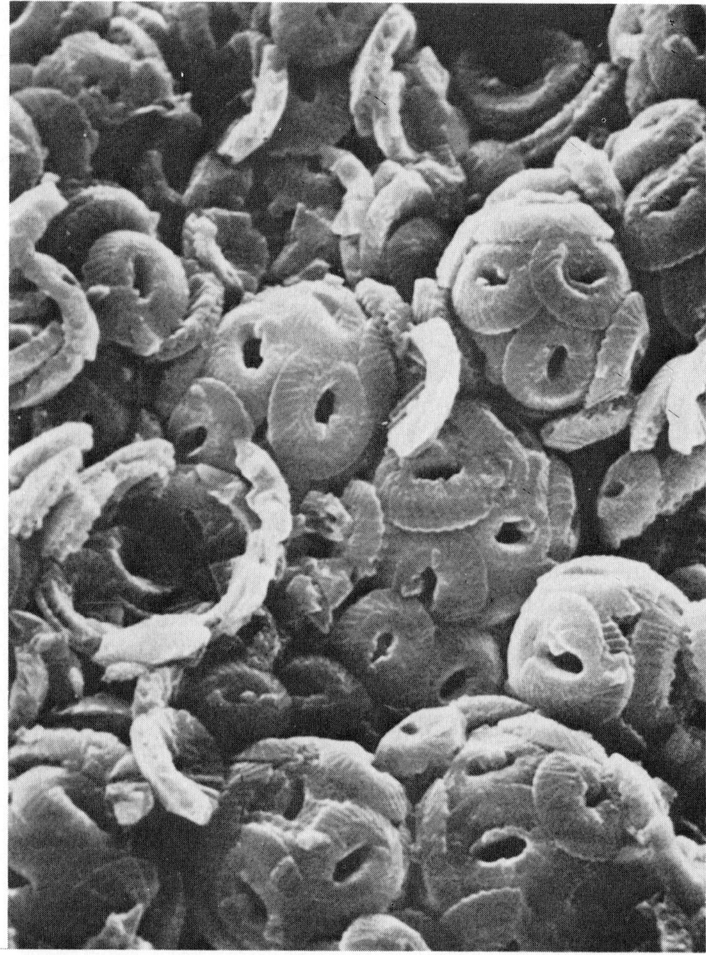

Figure 9.58
Surface of Upper Jurassic limestone, England, showing cocco-
spheres and isolated coccoliths of *Ellipsagelosphaera communis,*
SEM, ×3000, from D. Noël.

stones accumulated in shallow epicontinental seas where the calcium could later be recycled, but since the middle Mesozoic oceanic basins have been the major site of calcium deposition as planktonic oozes. However, the apparent scarcity of modern shallow water carbonates may in part be due to the masking effect of clastic material, because of the relatively greater topographic relief of the present continents.

Bramlette (1958) found that coccoliths and other calcareous nannoplankton supplied 20 percent of the total calcium carbonate of a Cretaceous chalk from Texas, 43 percent of a middle Eocene coccolith ooze from the deep Pacific, 58 percent of an Oligocene ooze, and 26 percent of a Miocene ooze. Probably they were more important in Cretaceous sedimentation than the above percentage suggests, as the sample used was deposited in a relatively shallow epicontinental sea.

Post-Tertiary oceanic sediments have a smaller percentage of coccoliths and greater percentage of pelagic foraminiferal remains. At present little calcareous material accumulates below 4500 m in the Pacific, and slightly deeper in the Atlantic, but there is evidence that the Tertiary oceans were warmer than at present, the sea bottom being about 12°C. As this would lower the solubility limits of calcium deposition to about 6700 m, calcareous oozes could have accumulated at greater depths in the Tertiary oceans than at present.

Because of their small size, coccoliths would be expected to sink very slowly in the water column and to show progressively greater dissolution as they pass through it. Instead, the assemblage on the bottom closely mirrors that of the surface waters, indicating little current transport during settling, and coccoliths obtained in sediment traps at various depths show relatively little effect of dissolution. The coccolithophores are grazed by zooplankton such as copepods, and the coccoliths and entire coccospheres are aggregated into fecal pellets. In addition to sinking more rapidly, the fecal pellets have an organic covering that prevents coccolith calcite dissolution during the settling process. Eventual biodegradation of the pellicle covering the pellets allows the coccoliths to spill out on the sea floor. A single pellet may carry 100,000 coccoliths (about 1 μg of $CaCO_3$) to the sea floor at a depth of 5000 m within a month, and the coccoliths are meanwhile protected from dissolution. In the equatorial Pacific 92 percent of the coccoliths produced in the euphotic layer may reach the bottom in this way (Honjo, 1976), and both dissolution and lateral transport are prevented.

The Upper Cretaceous Greenhorn Limestone, Carlile Shale, and Niobrara Chalk of Kansas and the Cretaceous "speckled shales" of Canada have a markedly speckled appearance. The white specks are ellipsoidal calcareous pellets about 0.12 mm in diameter, and consist of 80 to 90 percent coccoliths and coccolith debris, whereas the surrounding sediments are not coccolith-rich. These pellets probably were fecal pellets of the Cretaceous zooplankton (Hattin, 1975).

Other sediments contain a smaller amount of coccoliths than might be expected, and those present may be corroded or show secondary overgrowths. More delicate species may be selectively removed, and in indurated sediments, many of the smaller coccoliths in particular seem to have been selectively removed. Various alterations due to partial dissolution and secondary overgrowths may have been regarded as distinct taxa; for example, many species of *Broinsonia* (Burns, 1975). Solution may destroy the central area of the coccoliths, making *Emiliania huxleyi* and *Gephyrocapsa oceanica* coccoliths appear alike. Cold water variants of *E. huxleyi* and *Umbilicosphaera sibogae* are more solid and dense than the warm water variants. As the delicate, open-structured variations are less resistant, selective dissolution may suggest colder paleotemperatures than actually were present (Berger, 1973).

Experiments with solution rates have been conducted with Pliocene (Adelseck et al., 1973) and Cretaceous (Hill, 1975; Thierstein, 1976) assemblages. The assemblage changed in character and the number of species decreased with progressive dissolution. In general, placoliths were most resistant, as were some cyrtoliths, whereas caneoliths and scapholiths rapidly disappeared, and the holococcoliths were most fragile of all. Overgrowths occurred mostly on large coccoliths and on the discoasters in the Tertiary sediments (A. and R. McIntyre, 1971). Five categories of etching and five of overgrowths were recognized as various stages of relative solution and preservation (Bukry, 1973b).

Paleoecology

Coccolithophores at present are latitudinally distributed, and include both typically oceanic and typically neritic representatives. Various means of analysis have used the fossil coccoliths as paleoecologic indicators.

The abundant coccoliths of the Jurassic Solnhofen Limestone were suggested to indicate coccolith blooms (Buisonjé, 1972a,b). The laminated sediments consist of alternating limestones and thin marly intercalations in laminae averaging 2 cm in thickness. The laminae consist of coccolith fragments, particularly of *Watznaueria barnesae* and *Ellipsagelosphaera lucasi* Noël, with lesser amounts of additional species. Possibly an upwelling condition bringing nutrients to the surface caused the blooms, like modern red tides, benthic invertebrates being inhibited by the lack of oxygen. As organic matter settled and decayed, the bottom became anoxic and the laminated sediments were preserved.

Similarly abundant coccoliths in bands in the English Kimmeridge consist largely of *Ellipsagelosphaera britannica.* These Jurassic blooms were suggested to have been the source of the North Sea petroleum accumulations (Gallois, 1976).

Latitudinally related assemblages have been recognized for the Cretaceous (Thierstein, 1976), Paleocene through Oligocene (Haq and Lohmann, 1976; Haq et al., 1977; see Table 9.5), and Pleistocene (Gartner, 1972). Assemblages of the earliest Paleocene contain only cosmopolitan species, however. The later Paleogene floras were temperature-controlled, and shifted laterally through time in the Atlantic Ocean, generally in agreement with the temperature fluctuations as determined from oxygen isotope analysis. In the Eocene sediments, the ratio of discoasters to *Chiasmolithus* also is used as a temperature indication (Bukry, 1973a). Lower Eocene sediments at lat. 14° S contained 76 percent warm water *Discoaster* and 24 percent *Chiasmolithus,* whereas the higher latitude (37° S) had 46 percent *Discoaster* and 54 percent *Chiasmolithus.* During the middle Eocene, at lat. 19° N, the assemblage included 98 percent *Discoaster* and 2 percent *Chiasmolithus,* with the high latitude (54° N) sediments containing 33 percent *Discoaster* and 67 percent *Chiasmolithus.*

During the Pleistocene, periods of maximum temperature were indicated by the presence of *Ceratolithus cristatus* and *Umbellosphaera tenuis,* whereas temperature minima in the Caribbean region were characterized by *Coccolithus pelagicus, Discoaster perplexus* Bramlette & Riedel, *Discolithina japonica* Takayami, *Discosphaera tubifera, Gephyrocapsa caribbeanica, Oolithotus antillarum* (Cohen) Reinhardt, and *Umbellosphaera irregularis. Gephyrocapsa oceanica* exceeds the other species in abundance through the Pleistocene, and is represented by a cold water and a warm water morphotype (Gartner, 1972).

The oxygen isotopic composition of the coccolith calcite varies systematically in the Pleistocene sediments like that of the foraminiferal calcite. In sediments from the eastern Pacific this fractionation appears compatible with precipitation in equilibrium with the ocean surface waters, but in the Caribbean the coccolith calcite appears excessively enriched in ^{18}O, and may indicate some biological control of the composition (Anderson and Cole, 1975).

Although many coccolithophores are typically oceanic, others occur near shore. The Braarudosphaeraceae (*Braarudosphaera, Micrantholithus, Pemma, Pentaster*) as well as *Clathrolithus, Daktylethra, Discolithina, Lanternithus, Peritrachelina,* some *Rhabdosphaera, Scyphosphaera, Transversopontis,* and *Zygrhablithus* generally are not present in open ocean deposits (Bukry et al., 1971; Bybell and Gartner, 1972). Modern *Braarudosphaera* also occurs in coastal bays, such as the Bay of Fundy (Gran and Braarud, 1935). These near-shore species can withstand relatively high or low salinities, whereas the oceanic taxa can tolerate only limited variations in salinity.

In contrast to the usual near-shore habitat of *Braarudosphaera, B. rosa* Levin & Joerger was found to form a nearly monospecific chalk over a broad region of the South Atlantic. The Oligocene *Braarudosphaera* chalk is only 0.5 cm thick at DSDP Leg 3, Site 19, but at Site 22 it consists of several layers. The main bed, about 80 cm thick, consists entirely of *B. rosa,* whereas the other thinner beds contain from 50 to 70 percent *B. rosa.* Multiple laminations of the chalk are found at Site 20; and at the easternmost occurrence, Site 22 on the Rio Grande Rise, the bed is sufficiently lithified to be an acoustical reflector. This *Braarudosphaera* chalk thus extends as a recognizable layer for some 2800 km (Saito and Percival, 1970). It apparently formed at bathyal depths, and may record a long-lasting "bloom" of several hundred or thousand years (Saito and Percival, 1970), may reflect the warming of a cooler Antarctic current system aiding calcification in association with the high productivity (Wise and Hsü, 1971), or may possibly be due to the salinity tolerance of the species. Living *B. bigelowii* is present in the Black Sea at 17 to 18°/$_{00}$ salinity,

Table 9.5
Paleogene latitudinal zonation of coccolithophores. Data from Haq and Lohmann (1976).

	High latitude	Mid-latitude	Low latitude
Oligo-cene	*Coccolithus pelagicus* Assemblage *Dictyococcites bisectus* Assemblage	*Cyclicargolithus flori-danus – Dictyococcites hesslandii* (Haq) Haq & Lohmann Assemblage	Sphenolith-*Discoaster* Assemblage
Eocene	*Toweius craticulus* Hay & Mohler – *Coccolithus pelagicus* Assemblage *C. pelagicus – Calcidiscus formosus – C.* aff. *gammation* (Bramlette & Sullivan) Loeblich & Tappan Assemblage	*Discoaster* Assemblage *Reticulofenestra umbilica – Dictyococcites bisectus – C. pelagicus* Assemblage *Cribrocentrum reticulatum* (Gartner & Smith) Perch-Nielsen Assemblage	Reticulofenestrid Assemblage *Calcidiscus formosus –* Sphenolith Assemblage
Upper to Middle Paleo-cene	*Prinsius martinii* (Perch-Nielsen) Haq Assemblage *P. bisulcus* (Stradner) Hay & Mohler Assemblage	*Ericsonia subpertusa* Hay & Mohler Assemblage *Coccolithus pelagicus* Assemblage	*Toweius craticulus – C. pelagicus* Assemblage *Discoaster – Calcidiscus formosus* Assemblage *T. craticulus – Ericsonia subpertusa – Discoaster* Assemblage Fasciculith-*Discoaster* Assemblage
Lower Paleo-cene	Thoracosphaerid – *Markalius astroporus* Assemblage	Braarudosphaerid Assemblage	

but not in the Red Sea, where the salinity is 37 to 41‰ (Bukry, 1974c). Heavy rainfall or melting Antarctic ice locally may have reduced the South Atlantic surface water salinity, or excess evaporation may have increased the salinity level, while the poor oceanic circulation in the region failed to restore normal salinity to the photic zone (Perch-Nielsen, 1977), resulting in conditions tolerated by *B. rosa* but not by other taxa.

Elsewhere, *Braarudosphaera* appears to be indicative of near-shore sediments, whereas *Thoracosphaera* may be dominant in the offshore sediments.

EVOLUTION OF THE CALCAREOUS NANNOPLANKTON

From a few rare occurrences in the Paleozoic and Triassic, the calcareous nannoplankton increased rapidly both in diversity and abundance in the Jurassic and Cretaceous (see Figure 9.59), when their remains accumulated as extensive coccolith limestones and chalks. Maximum species diversity characterized the Late Cretaceous (Tappan and Loeblich, 1973), and was followed by the widespread extinctions simultaneously occurring in Europe, North Africa,

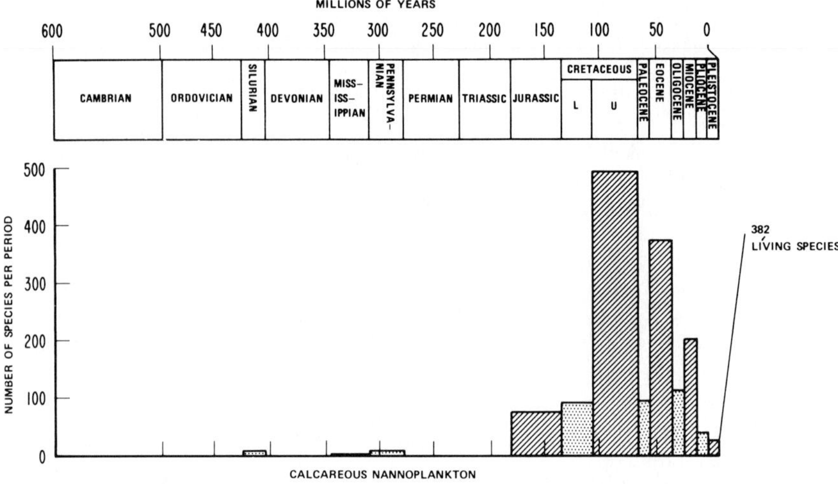

Figure 9.59

Described species diversity of calcareous nannoplankton, by geologic periods. Some Paleozoic records are doubtful. Note rapid Jurassic and Cretaceous diversification, marked drop to low numbers in the Paleocene, and general post-Cretaceous decline. Recent publications of the DSDP have added more species to these totals, particularly in the Jurassic and Cretaceous, but have not changed the general pattern. From Tappan and Loeblich, 1973.

North America, and elsewhere in the late Maastrichtian. Of some 40 Maastrichtian species, only 4 continued into the Danian (Bramlette and Martini, 1964). In addition, at least two-thirds of the Mesozoic genera disappeared, and only a few new genera and species evolved in the earliest Paleocene (Danian) to replace those that had disappeared, leaving an extremely reduced nannoflora (Tappan and Loeblich, 1972). Diversification occurred in the late Paleocene and Eocene, paralleling the similar diversification and extinction pattern of planktonic foraminifera, dinoflagellates, silicoflagellates, and diatoms. However, the calcareous nannoplankton reduction in the Oligocene was not followed by a major Miocene expansion like those of the other planktonic microorganisms. Instead, the post-Eocene decline continued sporadically through the remainder of the Cenozoic (see Figure 9.60). The present living coccolithophore flora is less diverse than that of the Cretaceous, even with the inclusion of holococcoliths and other poorly calcified taxa

such as the Halopappaceae and Calciosoleniaceae. Similar diversity fluctuations characterize both the Cenozoic coccoliths and the discoasters (see Figures 9.61 and 9.62; Haq, 1971d), being carried to an extreme in the latter group, which is now extinct.

The Late Cretaceous extinction of most coccolithophores was suggested to have resulted from a lack of nutrients, the warm equable seas of much of the Cretaceous resulting in poor circulation and the burial of much organic matter, while the nearly base-leveled continents provided little renewed influx (Bramlette, 1965). The lessened primary productivity as extinctions occurred might have affected the CO_2/O_2 balance in the seas, resulting in widespread submarine solution of calcium carbonate and the production of hardgrounds at the Cretaceous-Tertiary boundary (Tappan, 1968), thus accentuating the decrease in calcium carbonate production by the small remaining coccolithophore populations. Paraconformities occur widely at this stratigraphic position, not

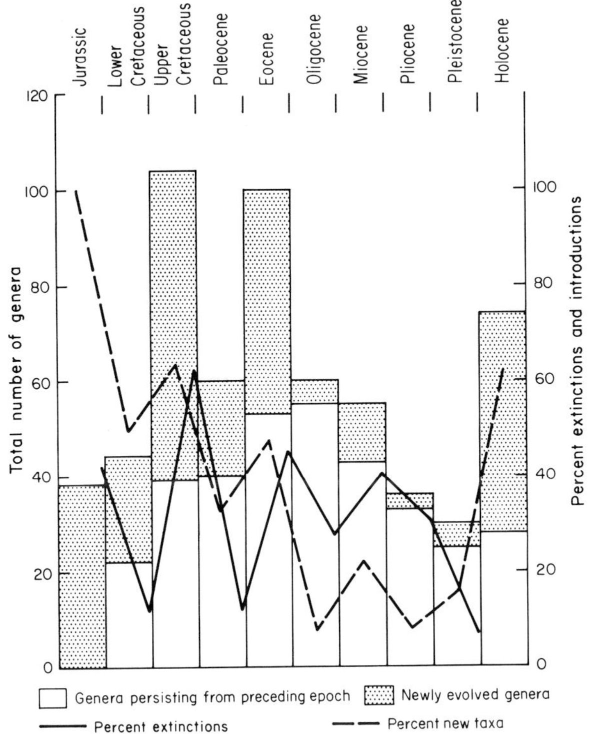

Figure 9.60
Generic diversity and pattern of new introductions and extinctions of calcareous nannoplankton through geologic time. Coincidence of peaks in percentages of new introductions and extinctions indicates rapid biologic turnover; low points show stable periods of relatively little change. From Tappan and Loeblich, 1972.

only in outcrops exposed on land, but even more prominently in the ocean basins, as shown by the DSDP (Worsley, 1971b). This widespread unconformity, the hardgrounds, and abundant glauconite and phosphates all indicate extensive submarine solution, and an elevation of the carbonate compensation depth (CCD) to near the surface of the oceans in the latest Cretaceous. The magnitude of the Cretaceous-Tertiary hiatus is greatest in the deepest parts of the oceans, and the CCD migrated vertically more than 1.5 km in the Atlantic Ocean during the Tertiary (Worsley, 1971b). Extensive limestone and chalk deposition in the Jurassic and Cretaceous may have resulted in a decrease in atmospheric CO_2, leading to climatic deterioration; polar cooling would increase the thermal gradients in the ocean and increase the CO_2 solubility in the higher latitudes and in deeper water. The CCD would appear first in

the deeper basins and migrate progressively to shallower levels. The most continuous record thus is found on the shallower portion of the continental shelves, and as noted by Worsley (1971b), the sparse nannoflora that survived the Cretaceous extinctions included such characteristically near-shore taxa as *Braarudosphaera*. As nannoplankton carbonate production and photosynthesis decreased, CO_2 could eventually accumulate in the atmosphere to previous levels, with a moderation of climatic conditions and the downward migration of the CCD. Only when nutrients and carbonate were again supplied in quantity from the rejuvenated continents could the sparse Paleocene nannoflora diversify (Tappan, 1968).

A change in the depth of the CCD, as suggested by Worsley (1971b), was discounted by Romein (1977), for although a reduction in pH may result in cultures of naked cells, these are

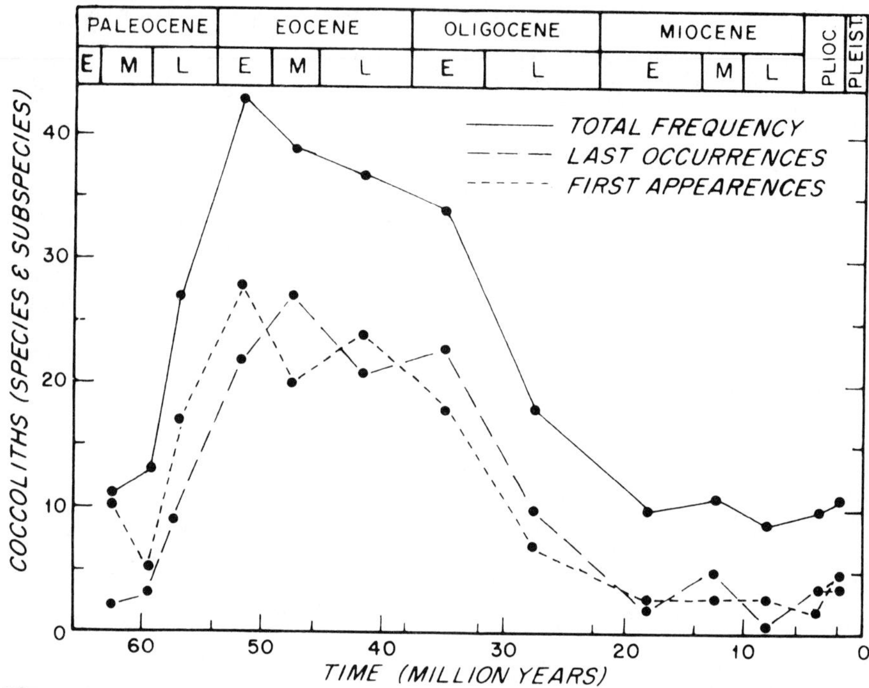

Figure 9.61
Diversity of coccoliths in the Cenozoic, shown by subdivisions of the epochs. As in Figures 9.59 and 9.60, the late Cretaceous extinctions are indicated by the low diversity of the early Paleocene, which consists mostly of newly appearing taxa; the maximum Eocene diversity is shown here largely to have been attained by the early Eocene, and a general decline to have followed the early Eocene maximum, with particularly rapid reduction characterizing the Oligocene. From Haq, 1971d.

viable, and a later increase in pH will allow them to produce their coccolith covers. Instead, a late Cretaceous cooling trend was regarded as the cause of the nannoplankton extinctions. Distinct biogeographic provinces were present by late Maastrichtian; continued cooling at the very end of the Cretaceous resulted in equatorial migration of the high latitude species and extinction of the low latitude warm water species. As the climate became warmer in the earliest Paleocene (Danian), the surviving species again migrated poleward and new species evolved in the lower latitudes (Romein, 1977).

Although not identifying its exact nature, a drastic physical-chemical event was suggested by Percival and Fischer (1977) to have caused the extinctions. In Spain, stresses beginning some 10,000 years before the end of the Cretaceous resulted in generally reduced, pyrite-containing sediments of the *Micula murus* Zone, and first affected the planktonic foraminifera. In addition, about 26 species of calcareous nannoplankton became extinct (these were termed the "vanishing species"). By the *Markalius astroporus* Zone, abundance was greatly reduced, and 6 disaster species ("persistent species") dominated. These included *Braarudosphaera, Thoracosphaera,* and other species that appear cystlike and characteristically occur in shallow water under brackish conditions or other extremes. Only these few species survived into the

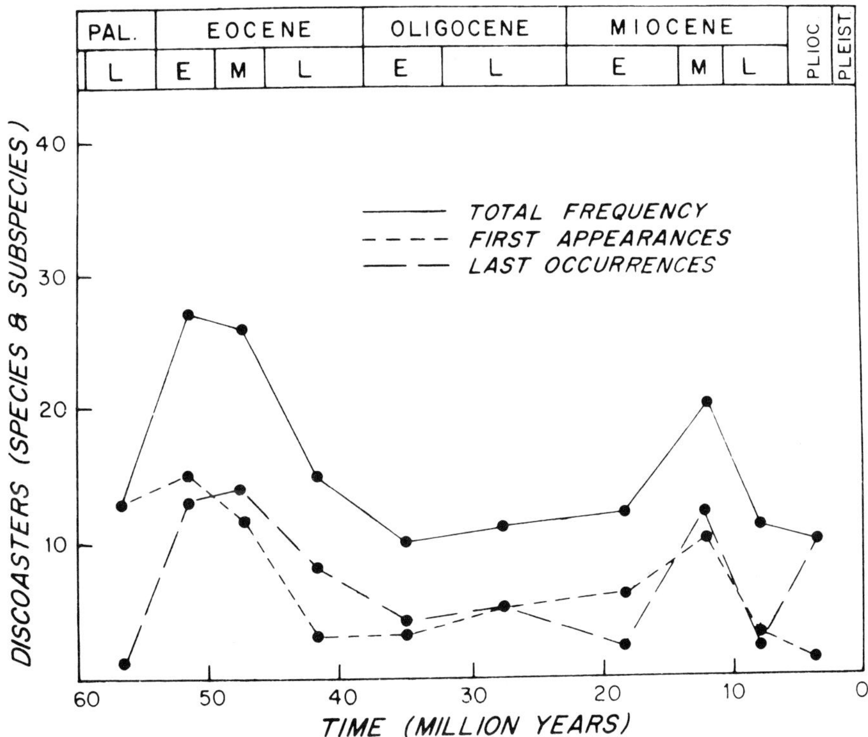

Figure 9.62
Diversity of discoasters in the Cenozoic, shown by subdivisions of the epochs.
Discoaster species diversity, like that of the coccoliths shown in Figure 9.61,
reached its maximum in the early Eocene, and after a later Eocene and Oligocene
decline, increased somewhat in the middle Miocene, before rapidly declining to the
Pliocene extinction. From Haq, 1971d.

Danian, and replacement was slow, as the unfavorable conditions persisted for about 1 million years. Within the Paleocene, some 9 new species appeared in the *Cruciplacolithus tenuis* Zone (the "incoming species"), and signalled the end of the crisis conditions, although diversity and abundance remained low (Percival and Fischer, 1977).

Within the Tertiary, the alternating diversification and decline of calcareous nannoplankton appear to be temperature related, the greatest diversity occurring in the warmer intervals, and less diverse assemblages consisting of cosmopolitan taxa in the cooler periods. These evolutionary patterns are reflected in the biostratigraphic zonation, as the zones of the cooler Oligocene are of approximately 3 million years duration, while those of warmer epochs average only 1.7 million years (Haq, 1971d). Species that tolerate only a limited range of temperature may spread widely in equable seas, but as climates become more continental or extreme their sharply reduced range results in extinction (Haq, 1973b).

The evolution, diversification, and extinction of the calcareous nannoplankton, like that of other phytoplankton, thus is closely related to the conditions of oceanic circulation, and ultimately to the distribution and elevation of the land masses through time. In turn, variations in the abundance and diversity of phytoplankton, the base of the aquatic food chains, had a marked effect on the evolutionary patterns, survival, or extinction of the contemporaneous animal life (Tappan and Loeblich, 1971).

Few evolutionary trends have been described for the Haptophyta as yet, although lineages have been suggested for species of *Chiasmolithus* (Gartner, 1970), *Helicosphaera* (Haq, 1973a), and *Ceratolithus* (Gartner and Bukry, 1975), among others.

CLASSIFICATION

The classification of the Haptophyta is less stabilized than that of most other algal divisions, in part because of the quite different morphology of separate phases of the life cycle, and in part because both the fossil record and many of the living taxa are based on the morphology of the coccoliths, with little information as to the nature of the living cell. Use of coccolith morphology also has varied, from the early gross morphological classification as discoliths, cricoliths, placoliths, and so forth, to the arrangement of the crystals in the separate coccoliths (heliolithid or ortholithid structure), degree of suppression of the inorganic crystal form (holococcoliths vs. heterococcoliths), and more recently the basic growth pattern of the individual coccoliths, such as the development from a basic loxolith rim, a coccolithid rim, or a podorhabdid one.

Living haptophytes include many that develop only a scale cover and lack coccoliths; some have a well-developed haptonema, but in others this is reduced to only a few microtubules, or may be lacking entirely. The rapid increase in knowledge of coccolith ultrastructure has led to rapid changes in the basis for classification of fossil taxa as well. As the large number of described taxa (indexed by Loeblich and Tappan, 1966, 1968, 1969, 1970a,b, 1972, 1973, continuation by Heck, 1979) make an alphabetical arrangement too cumbersome for utility, various modifications of the classification have been proposed in recent years (Black, 1968, 1971a−c, 1972a, 1973, 1975; Boudreaux and Hay, 1969; Bukry, 1969; Deflandre, 1963; Forchheimer, 1972; Gartner, 1968; Grün in Grün and Allemann, 1975; Haq, 1967; Hay, 1977; Hay and Mohler, 1967; Hoffmann, 1970; Jafar, 1975a; Lipps, 1969; Noël, 1965b, 1972; Norris, 1965; Reinhardt, 1965, 1967b; Rood et al., 1971, 1973; and Thierstein, 1973).

No name has yet been formally proposed for this division. First regarded as the separate Class Haptophyceae within the Chromophyta (Christensen, 1962), on the basis of distinctive features of the living cell, it was redefined as the Class Prymnesiophyceae by Hibberd (1976), following the procedure of typification of the higher taxa. However, the Class Coccolithophyceae Rothmaler 1951, based on *Coccolithus*, has priority. Other class level names had been based on *Coccolithophora* (an illegitimate synonym of *Coccolithus*). Similarly, the Division Haptophyta has been used (e.g., Hibberd, 1972; Parke and Green in Parke and Dixon, 1976) without formal diagnosis, as has the typified name Prymnesiophyta (Hibberd, 1976).

The classification given in Table 9.6 is a combination of those based on fossil taxa with that of modern haptophytes (Green, 1976; Parke and Green in Parke and Dixon, 1976; and Hibberd, 1976).

Table 9.6
Classification of the Haptophyta and other calcareous nannoplankton.

Division Haptophyta Hibberd 1972 Syn.: Prymnesiophyta Hibberd 1976 Class Coccolithophyceae Rothmaler 1951 Syn.: Coccolithophorales Lemmermann 1903; Coccolithophoridaceae Fritsch 1935; Coccolithophorides Stradner 1958; Coccolitho-	phoriinae Vekshina 1959; Haptophyceae Christensen 1962; Coccolithophorea Pitelka 1963; Coccolithophorineae Wood 1965; Prymnesiophyceae Hibberd 1976. Motile cell with two (usually) similar flagella and commonly with a haptonema, although the

latter may be rudimentary or even absent in some; cell covered with organic scales, and may also have cover of calcareous plates or coccoliths; motile stage alternates with nonmotile planktonic stage, or benthic filamentous or palmelloid stage. Marine, and few in fresh water.

A. Order Isochrysidales Pascher 1910

Syn.: Hymenomonadales Fritsch in West 1927

Cell membrane present or absent; motile cells with two similar isodynamic and acronematic flagella; haptonema absent or rudimentary; cells solitary or in groups, may have benthic filamentous stage alternating with motile one. Marine and freshwater.

1. Family Gephyrocapsaceae Black 1971

Haptophytes naked, with body scales or placolith cover, each coccolith consisting of two shields of nonimbricate radial elements around large central opening. Motile cells lack haptonema, although *Imantonia* has haptonemal process without microtubules. Cretaceous to Holocene.

Crenalithus Roth 1973; *Dicrateria* Parke 1949; *Emiliania* Hay & Mohler 1967 (syn.: invalid *Ellipsoplacolithus* Kamptner 1963; invalid *Pseudoemiliania* Gartner 1969); *Gephyrocapsa* Kamptner 1943; *Imantonia* Reynolds 1974; *Reticulofenestra* Hay, Mohler & Wade, 1966 (syn.: *Apertapetra* Hay, Mohler & Wade 1966; *Dictyococcites* Black 1967; *Stradnerius* Haq 1968); *Tremalithus* Kamptner ex Deflandre 1952.

2. Family Isochrysidaceae Parke 1949

Syn.: Marthasteritaceae Hay 1977

Motile cell with rudimentary haptonema containing microtubules; alternates with benthic filamentous stage; *Tetralithus*-like and *Marthasterites*-like calcareous elements formed externally. Upper Cretaceous to Eocene; Holocene.

Chrysotila Anand 1937 (syn.: *Ruttnera* Geitler 1942); *Isochrysis* Parke 1949; *Marthasterites* Deflandre 1959; *Tetralithus* Gardet 1955.

3. Family Ochrosphaeraceae Schussnig 1930

No emergent haptonema on motile cell, but group of five microtubules associated with flagellar bases; motile cell may have *Cricosphaera*-like coccoliths; some produce extracellular calcareous structures resembling *Tetralithus,* but not like true *Tetralithus.*

Ochrosphaera Schussnig 1930.

4. Family Thoracosphaeraceae Schiller 1930

Cell with thick wall constructed of closely packed prismatic elements or prismatoliths that may have central perforation (apparently not completely through wall); a small opening (for flagella or excystment?) may be present at one pole. Marine. Upper Cretaceous to Holocene.

Some species referred to *Thoracosphaera,* including *T. albatrossiana* Kamptner and *T. tuberosa* Kamptner, apparently are dinoflagellate cysts, referrable to the Calciodinellaceae, and resemble cysts of the living *Scrippsiella.* The type and other species of *Thoracosphaera* appear to resemble a pseudocyst of *Ochrosphaera.*

Brachiolithus Noël ex Loeblich & Tappan 1963; *Centosphaera* Wind & Wise 1977; *Thoracosphaera* Kamptner 1927.

5. Family Hymenomonadaceae Senn 1900

Motile cells with visible although rudimentary haptonema with microtubules; may have benthic filamentous stage alternating with motile one; coccoliths are simple rings (cricoliths) or rings subtending rods or central spikes at the scale margins. Marine, fresh and brackish water. Holocene.

Cricosphaera is regarded as a synonym of *Hymenomonas* by some, but was reinstated as distinct (Gayral & Fresnel-Morange, 1976) on the basis of coccolith morphology.

Apistonema Pascher 1925; *Calciarcus* Manton, Sutherland & Oates 1977; *Chrysonema* Anand 1937; *Coccochrysis* Conrad 1926; *Cricosphaera* Braarud 1960; *Heimiella* Lohmann 1913; *Hymenomonas* Stein 1878 (syn.: *Pleurochrysis* Pringsheim 1955); *Wigwamma* Manton, Sutherland & Oates 1977.

B. Order Zygosphaerales Hay 1977

Motile cell with holococcoliths constructed of tiny hexagonal or rhombohedral crystals. Some may be an alternate stage of other taxa (such as *Coccolithus*).

1. Family Calyptrosphaeraceae Boudreaux & Hay 1969

Syn.: Zygosphaeraceae Braarud & Gaarder 1961, nom. nud; Zygosphaeraceae Black 1971

Motile cell with zygoform or calyptroform holococcoliths, of very small calcite crystals in the form of rhombohedra or short hexagonal prisms. Rare as fossils, but known from Lower Cretaceous (Hauterivian) to Holocene.

(continued)

Acuturris Wind & Wise 1977; *Balaniger* Thomsen & Oates 1978; *Calyptrosphaera* Lohmann 1902 (syn.: *Calyptrolithus* Kamptner ex Deflandre 1952); *Clathrolithus* Deflandre 1954; *Corisphaera* Kamptner 1936; *Cyclocalyptra* Kamptner 1958; *Daktylethra* Gartner 1969; *Discoturbella* Roth 1970; *Helladosphaera* Kamptner 1936 (syn.: *Anthosphaera* Kamptner 1936); *Holodiscolithus* Roth 1970; *Homozygosphaera* Deflandre 1952; *Lacrymasphaera* Lecal 1965; *Lanternithus* Stradner 1962; *Lucianorhabdus* Deflandre 1959 (syn.: *Eurhabdus* Reinhardt 1965); *Naninfula* Perch-Nielsen 1968; *Okkolithus* Wind & Wise 1977; *Orastrum* Wind & Wise 1977; *Ottavianus* Risatti 1973; *Periphyllophora* Kamptner 1936; *Peritrachelina* Deflandre 1952; *Petalosphaera* Lohmann 1913; *Phanulithus* Wind & Wise 1977; *Pharus* Wind & Wise 1977; *Polycladolithus* Deflandre 1954; *Pyxolithus* Deflandre 1954; *Russellia* Risatti 1973; *Scampanella* Forchheimer & Stradner 1973; *Semihololithus* Perch-Nielsen 1971; *Sphaerocalyptra* Deflandre 1952; *Trochoaster* Klumpp 1953; *Turrispira* Manton, Sutherland & Oakes 1976; *Zygosphaera* Kamptner 1936 (syn.: *Orthozygus* Bramlette & Wilcoxon 1967); *Zygrhablithus* Deflandre 1959 (syn.: *Orthorhabdus* Bramlette & Wilcoxon 1967).

C. Order Discoasterales Hay 1977

Tubular, U-shaped, rosette, or stellate nannofossils.

1. Family Discoasteraceae Tan Sin Hok 1927 orth. mut. Vekshina 1959

Syn.: Discoasteridae Tan Sin Hok 1927

Asteroliths as tiny stars or rosettes 10 to 35 μm across, no crystal outline visible, but crystal orientation can be determined optically, and always has C-axis of calcite parallel to axis of symmetry, hence in flat specimens viewed with crossed nicols, specimens remain dark. Marine. Paleocene to Pliocene.

Agalmatoaster Klumpp 1953; *Catinaster* Martini & Bramlette 1963; *Clavodiscoaster* Prins 1971; *Discoaster* Tan Sin Hok 1927; *D. (Gyrodiscoaster)* Stradner 1961; *Discoasteroides* Bramlette & Sullivan 1961; *Hayaster* Bukry 1973; *Heliodiscoaster* Tan Sin Hok 1927; *Hemidiscoaster* Tan Sin Hok 1927; *Radiodiscoaster* Prins 1971; *Tribrachiatus* Shamray 1963; *Truncodiscoaster* Prins 1971; *Turbodiscoaster* Prins 1971.

2. Family Sphenolithaceae Deflandre 1952 orth. mut. Vekshina 1959

Syn.: Sphenolithidae Deflandre 1952

Discoid or platelike elements with conical projections; black cross visible in longitudinal position under crossed nicols, rather than in apical position as in typical coccoliths. Marine. Jurassic to Miocene.

Assipetra Roth 1973; *Calculites* Prins & Sissingh 1977; *Ceratolithina* Martini 1967; *Ceratolithoides* Bramlette & Martini 1964; *Furcatolithus* Martini 1965; *Micula* Vekshina 1959 (syn.: *Nannotetraster* Martini & Stradner 1960); *Quadrum* Prins & Perch-Nielsen 1977; *Rhomboaster* Bramlette & Sullivan 1961; *Sphenolithus* Deflandre 1952 (syn.: *Nannoturbella* Brönnimann & Stradner 1960); *Trochastrites* Stradner 1961.

3. Family Ceratolithaceae Norris 1965

Nonflagellate cell with single large horseshoe-shaped coccolith, ends projecting beyond cell; four ribbonlike plastids; cell enclosed in sphere of elliptical ringlike organic scales 6 to 9 μm in diameter, with ringlike calcareous coccoliths at the scale margins, the ring consisting of a single row of crystallites that alternate with smaller ones inserted in indentations in the larger crystallites; scaly sphere may enclose more than one cell, possibly as a reproductive stage. Marine. Miocene to Holocene.

Amaurolithus Gartner & Bukry 1975; *Angulolithina* Bukry 1973; *Ceratolithus* Kamptner 1950.

4. Family Heliolithaceae Hay & Mohler 1967

Coccoliths with two abutting shields, but lacking a connecting tube, one shield with subradial sutures producing a strong interference pattern between crossed polarizers; other shield smaller, with steeply inclined sutures and strongly imbricate segments. Paleocene.

Heliolithus Bramlette & Sullivan 1961.

5. Family Braarudosphaeraceae Deflandre 1947

Cell enclosed in cover of 12 pentaliths fit together in a regular pentagonal dodecahedron; each pentalith consists of five calcite crystals (rarely four or six), arranged radially in a single plane and oriented so that cleavage direction of each crystal is in same plane; pentalith sectors may leave small opening at their junction, and calcite of each segment may be interlaminated with thin sheets of organic matter. The usual absence of an opening in the sphere suggests that these may represent cysts, but none have as yet been maintained in culture. Marine. Lower Cretaceous to Holocene.

Astrionis Wilcoxon 1972; *Biantholithus* Bramlette & Martini 1964; *Braarudosphaera* Deflandre 1947 (syn.: *Eodiscoaster* Martini 1961); *Hexalithus* Gardet 1955; *Hexangulolithus* Bukry 1969; *Micrantholithus* Deflandre 1954; *Octolithites* Caratini 1963; *Pemma* Klumpp 1953; *Pentaster* Bybell & Gartner 1972; *Quinquerhabdus* Bukry & Bramlette 1971; *Vermiculithina* Bukry & Percival 1971.

6. Family Goniolithaceae Deflandre 1957
Syn.: Calyculaceae Noël 1972
Coccoliths pentagonal, with granular center and radially constructed border; 12 of the coccoliths fit together around the cell as a dodecahedron as in *Braarudosphaera,* although with distinct structure. Eocene.
Calyculus Noël 1972; *Carinolithus* Prins 1974; *Goniolithus* Deflandre 1957.

7. Family Lithostromatiaceae Deflandre 1959 orth. mut. herein.

Syn.: Lithostromationidae Deflandre 1959; Lithostromationaceae Haq 1967
Coccoliths similar to asteroliths but with both faces alike; triangular in form, 12 to 16 μm across, surface pitted. Marine. Eocene to Pliocene.
Lithostromation Deflandre 1942; *Martiniaster* Loeblich & Tappan 1963 (syn.: *Coronaster* Martini 1961, non Perrier 1884); *Trochoaster* Klumpp 1933.

8. Family Fasciculithaceae Hay & Mohler 1967
Subcylindrical nannofossils with wedge-shaped segments radiating from center; cylinder concave at one end, pointed or flat at other; surface may be ornamented; between crossed polarizers, side view shows interference figure as dark line in cylinder axis, but with incomplete extinction at edges; interference cross shows in side view. Marine. Paleocene.
Bomolithus Roth 1973; *Fasciculithus* Bramlette & Sullivan 1961.

9. Family Nannoconaceae Deflandre 1959 orth. mut. Noël 1965

Syn.: Nannoconidae Deflandre 1959; Nannoconidaceae Shumenko 1971
Isolated elements resembling spirally constructed lopadoliths; thick wall of wedges or plates of calcite oriented perpendicular to surface and surrounding an axial cavity or canal; opening at one or both poles. Marine, oceanic. Middle Triassic to Upper Cretaceous.
Conusphaera Trejo 1969 (syn.: *Cretaturbella* Roth 1971); *Nannoconus* Kamptner 1951; *Polycostella* Thierstein 1972.

D. Order Eiffellithales Rood, Hay & Barnard 1971
Coccoliths with loxolith or eiffellithid marginal rim of two closely superposed cycles of elements that may appear single except at the inner margin as viewed from the inner surface; central structure of bar, cross, or radial bars may support a central spine.

1. Family Eiffellithaceae Reinhardt 1965
Elliptical coccoliths with loxolith rim and partial or complete floor of a small number of large plates; large central spine may or may not have buttressed support. Marine. Cretaceous.
Eiffellithus Reinhardt 1965 (syn.: *Clinorhabdus* Stover 1966).
2. Family Rhagodiscaceae Hay 1977
Syn.: Actinozygaceae Rood, Hay & Barnard 1971; Alfordiaceae Black 1977
Coccoliths with loxolith rim and complete floor of small, equal-sized granules. Cretaceous.
Reinhardtites Perch-Nielsen 1968; *Rhagodiscus* Reinhardt 1967 (syn.: illeg., superfluous *Actinozygus* Gartner 1968; *Alfordia* Black 1975; *Rhabdolithina* Reinhardt 1967; *Viminites* Black 1975).
3. Family Ahmuellerellaceae Reinhardt 1965
Elliptical coccoliths with marginal wall like that of loxolith ring and central structure of a cross aligned with the axes of the ellipse. Jurassic and Cretaceous.
Ahmuellerella Reinhardt 1964; *Crucirhabdus* Prins ex Rood, Hay & Barnard 1973; *Helicolithus* Noël 1970; *Placozygus* Hoffmann 1970; *Staurolithites* Caratini 1960 (syn.: *Zygostephanos* Hoffmann 1970); *Vekshinella* Loeblich & Tappan 1963 (syn.: *Ephippium* Vekshina 1959, non Bolten 1798; *Vagalapilla* Bukry 1969; *Staurorhabdus* Noël 1972).

(continued)

4. Family Zygodiscaceae Hay & Mohler 1967 Syn.: invalid Zygolithaceae Noël 1965; Chiastozygaceae Rood, Hay & Barnard 1973

Elliptical coccoliths with outer wall of narrow steeply inclined laths or tabulae, like twisted barrel staves, forming a loxolith ring (as in *Pontosphaera,* but lacking a floor, and with fewer and more robust staves); center may be spanned by **I**-, **X**-, or **H**-shaped bar, symmetrical about the short axis of the ellipse; rim is dextrogyre in distal view between crossed nicols. Marine. Jurassic to Eocene.

Amphizygus Bukry 1969; *Barringtonella* Black 1973; *Bukrylithus* Black 1971; *Calcicalathina* Thierstein 1971; *Chiastozygus* Gartner 1968; *Chiphragmalithus* Bramlette & Sullivan 1961; *Cyclolithella* Loeblich & Tappan 1963 (syn.: *Cyclolithus* Kamptner ex Deflandre 1952, non Koenig 1825); *Glaukolithus* Reinhardt 1964; *Haslingfieldia* Black 1973; *Heteromarginatus* Bukry 1969; *Isthmolithus* Deflandre 1954; *Loxolithus* Noël 1965; *Misceomarginatus* Wind & Wise 1977; *Mitrolithus* Deflandre & Fert 1954 (syn.: *Alvearium* Black 1967); *Monomarginatus* Wind & Wise 1977; *Nannotetrina* Achuthan & Stradner 1967; *Neochiastozygus* Perch-Nielsen 1971; *Neccoccolithes* Sujkowski 1931 (syn.: *Zygolithus* Kamptner ex Matthes 1956; *Indumentalithus* Vekshina 1959; *Heliorthus* Brönnimann & Stradner 1960); *Tegumentum* Thierstein 1972; *Tranolithus* Stover 1966; *Tubirhabdus* Prins ex Rood, Hay & Barnard 1973; *Zeugrhabdotus* Reinhardt 1965; *Zygodiscus* Bramlette & Sullivan 1961; *Zygolithites* Black 1972.

5. Family Pontosphaeraceae Lemmermann 1908

Coccoliths of shallow lopadoliths, which may bear a central spine, or with simple or perforate floor. Freshwater (rare) and marine. Permian? Paleocene to Holocene.

Algirosphaera Schlauder 1945; *Crassapontosphaera* Boudreaux & Hay 1969; *Discolithina* Loeblich & Tappan 1963 (syn.: *Discolithus* Huxley 1868, non Fortis 1802); *Koczyia* Boudreaux & Hay 1969; *Lophodolithus* Deflandre 1954; *Pontolithina* Black 1968; *Pontosphaera* Lohmann 1902; *Transversopontis* Hay, Mohler & Wade 1966.

6. Family Scyphosphaeraceae Jafar 1975

Spherical cell covered with low elliptical lopadoliths, those of equatorial region exceptionally deep and enlarged, whether of similar or different shape. Marine. Eocene to Holocene.

Calciopilleus C. Müller 1974; *Lohmannosphaera* Schiller 1913; *Nannocorbis* C. Müller 1974; *Scyphosphaera* Lohmann 1902.

7. Family Stephanolithiaceae Black 1968 Syn.: Stephanolithionaceae Bukry 1969

Hollow coccoliths with cylindrical or polygonal wall consisting of elements that are not markedly imbricate, and with open framework of rods within the wall that may be radially or otherwise arranged; radial spines may project from the marginal ring or wall, each spine a single scalenohedral crystal with the optic axis at right angles to the central axis of the coccolith. Marine. Jurassic to Upper Cretaceous.

Angulofenestrellithus Bukry 1969; *Corollithion* Stradner 1961; *Diadorhombus* Worsley 1971; *Nodosella* Prins ex Rood, Hay & Barnard 1973; *Pontilithus* Gartner 1968; *Rhombolithion* Black 1973; *Rotelapillus* Noël 1972; *Stephanolithion* Deflandre 1939; *Stradnerlithus* Black 1971; (syn.: *Diadozygus* Rood, Hay & Barnard 1971); *Truncatoscaphus* Rood, Hay & Barnard 1971.

8. Family Polycyclolithaceae Forchheimer 1972

Syn.: Lithastrinaceae Thierstein 1973 (May); Eprolithaceae Black 1973 (Nov.)

Cylindrical coccoliths with one or more cycles of six imbricated elements or radial sectors; radial cross-polarized extinction lines parallel to nicols. Marine. Permian? Upper Jurassic to Eocene.

Bukryaster Prins 1971; *Cylindralithus* Bramlette & Martini 1964; *Eprolithus* Stover 1966; *Hayesites* Manivit 1971; *Lapideacassis* Black 1971; *Lithastrinus* Stradner 1962; *Polycostella* Thierstein 1971; *Polycyclolithus* Forchheimer 1968; *Radiolithus* Stover 1966; *Rhombogyrus* Black 1973; *Rucinolithus* Stover 1966.

9. Family Crepidolithaceae Black 1972

Thick elliptical coccoliths, consisting of a ring of calcite sectors with little or no imbrication, proximal side may have separately constructed floor and distal side may have massive spine. Marine. Jurassic to Cretaceous.

Carinolithus Prins 1974; *Crepidolithus* Noël 1965; *Mitosia* Worsley 1971; *Parhabdolithus* Deflandre 1952.

10. Family Helicosphaeraceae Black 1971

Ovoid cell with apical opening left by the cover of coccoliths, which have the form of asymmetrical helicoid helicoliths, with large elliptical central shield surrounded by spiralling flange constructed of radial elements. Marine.

Lower Miocene to Holocene.

Helicosphaera Kamptner 1954 (syn.: *Helicopontosphaera* Hay & Mohler 1967).

11. Family Calciosoleniaceae Kamptner 1937

Cylindrical or fusiform cell, enveloped by scapholiths with external surface an elongated parallelogram, interior occupied by grid of transverse bars; cell may have needlelike spines at one or both ends. Marine. Permian? Lower Cretaceous (Albian) to Holocene.

Acanthosolenia Bernard 1939; *Anoplosolenia* Deflandre 1952; *Calciopappus* Gaarder & Ramsfjell 1954; *Calciosolenia* Gran 1912; *Dictyolithus* Górka 1957; *Navisolenia* Lecal 1965; *Scapholithus* Deflandre 1954.

12. Family Microrhabdulaceae Deflandre 1963 orth. mut. Reinhardt 1966

Syn.: Microrhabdulidae Deflandre 1963

Rodlike, cylindrical to spindle-shaped nannofossils. Cretaceous.

Lithraphidites Deflandre 1963; *Microrhabdulinus* Deflandre 1963; *Microrhabduloidus* Deflandre 1963; *Microrhabdulus* Deflandre 1959; *Pseudolithraphidites* Keupp 1976.

13. Family Triquetrorhabdulaceae Lipps 1969

Spindle-shaped rods with triradial cross section and concave sides. Eocene to Miocene.

Pseudotriquetrorhabdulus Wise & Constans 1976; *Triquetrorhabdulus* Martini 1965.

E. Order Podorhabdales Rood, Hay & Barnard 1971 orth. mut. herein.

Syn.: Podorhabdinales Rood, Hay & Barnard 1971

Coccoliths with two or more cycles of petaloid rim elements (podorhabdid rim), only slightly if at all imbricate. Marine. Triassic? Jurassic to Cretaceous.

1. Family Podorhabdaceae Noël 1965

Syn.: Cretarhabdaceae Thierstein 1973; Retecapsaceae Grün 1975

Elliptical shield with narrow rim constructed of two to three layers of jointive petaloid or subquadrate elements; in polarized light this part shows a black cross with straight arms; petaloid rim surrounds a wide central area, spanned by various types of structures composed of many fine crystallites, with a solid or hollow spine or stalk commonly arising from the center. Upper Triassic? Jurassic to Upper Cretaceous.

a. Subfamily Podorhabdoideae Reinhardt 1967

Syn.: Ethmorhabdoideae Rood, Hay & Barnard 1971

Coccoliths with tubular spine. Jurassic to Cretaceous.

Axopodorhabdus Wind & Wise 1977; *Boletuvelum* Wind & Wise 1977; *Cleistorhabdus* Black 1972; *Dodekapodorhabdus* Perch-Nielsen 1968; *Ethmorhabdus* Noël 1965; *Hemipodorhabdus* Black 1971; *Hexapodorhabdus* Noël 1965; *Nephrolithus* Górka 1957; *Octocyclus* Black 1972; *Octopodorhabdus* Noël 1965; *Perissocyclus* Black 1971; *Podorhabdus* Noël 1965; *Teichorhabdus* Wind & Wise 1977; *Tetrapodorhabdus* Black 1971.

b. Subfamily Retecapsoideae Black 1972

Syn.: Helenoideae Grün 1975

Elliptical coccoliths with distal shield having two cycles of non- or only slightly imbricated elements; proximal shield with one cycle of inclined, non-, or only slightly imbricated elements; center of shield unperforated or with solid spine. Triassic? Jurassic to Cretaceous.

Allemannites Grün 1975; *Bipodorhabdus* Noël 1970; *Cretadiscus* Gartner 1968; *Cretarhabdella* Black 1971; *Cretarhabdus* Bramlette & Martini 1964 (syn.: *Stradneria* Reinhardt 1965; *Heterorhabdus* Noël 1970; *Helenea* Worsley 1971); *Cruciellipsis* Thierstein 1972; *Flabellites* Thierstein 1973; *Gephyrorhabdus* M. Hill 1976; *Grantarhabdus* Black 1971; *Microstaurus* Black 1971; *Miravetesina* Grün 1975; *Polypodorhabdus* Noël 1965; *Retecapsa* Black 1971.

c. Subfamily Cribrosphaerelloideae Tappan nom. subst.*

Syn.: Cribrosphaeroideae Hoffmann 1970

Central area with grid, but no process. Cretaceous.

Cribrosphaerella Deflandre 1952 (syn.: *Cribrosphaera* Arkhangelskiy 1912 non Popofsky 1906; *Favocentrum* Black 1964).

2. Family Deflandriaceae Black 1968

Syn.: Prediscosphaeraceae Rood, Hay & Barnard 1971

Elliptical to circular coccoliths with basal ring of one or more layers, each consisting of 16

---

* Subfamily Cribrosphaerelloideae, nom. subst. herein pro Cribrosphaeroideae Hoffmann 1970, as the latter is based on the homonym *Cribrosphaera*.

(*continued*)

primary cuneate granules and spanned by a cross of four buttresses; buttresses unite to support a central stem that consists of two bundles of elongate crystal laths placed end to end, with connecting node about one-half distance up the stem. Cretaceous.

Deflandrius was regarded as a synonym of *Prediscosphaera* Vekshina 1959 by Gartner (1968), but the original description of *Prediscosphaera* was based on an illegitimate type species, *P. decorata* Vekshina including in its synonymy a valid prior species, *Coccolithophora cretacea* Arkhangelskiy. Thus, the reinstatement of *Deflandrius* by Black (1973) is followed here.

Chrysochromulina has scales with a pattern like that of *Deflandrius,* and may represent the same lineage, the noncalcified scale-bearing form persisting after extinction of the calcifying one.
Deflandrius Bramlette & Martini 1964.

3. Family Percivaliaceae Black 1973
Coccoliths with broad marginal zone or proximal surface of elongate crystals arranged in concentric rings. Upper Cretaceous.
Percivalia Bukry 1969.

4. Family Sollasitaceae Black 1971
Elliptical coccoliths of two petaloid shields with jointive (nonimbricated) rays, enclosing a wide central area that is occupied by a grid or granular floor; no spine or central knob. Upper Jurassic to Upper Cretaceous, ?Oligocene.
Gaarderella Black 1973; *Heteromarginatus* Bukry 1961; *Sollasites* Black 1967 (syn.: *Costacentrum* Bukry 1969)

5. Family Speetoniaceae Black 1971
Elliptical coccoliths with two shields, distal shield with peripheral auxiliary ring and large central opening spanned by bridge, but lacking a spine. Lower Cretaceous.
Speetonia Black 1971.

6. Family Biscutaceae Black 1971
Syn.: Discorhabdaceae Noël 1972
Imperforate coccoliths of two closely appressed shields, each shield constructed of radial, nonimbricate petaloid elements. Marine. ?Jurassic, middle and Upper Cretaceous.
Bennocyclus Zweili 1974; *Bidiscus* Bukry 1969; *Biscutum* Black 1959; *Discorhabdus* Noël 1965; *Noellithina* Grün & Zweili 1974; *Palaeopontosphaera* Noël 1965; *Seribiscutum* Filewicz, Wind & Wise 1977; *Similicoronilithus* Bukry 1969.

7. Family Arkhangelskiellaceae Bukry 1969
Elliptical coccolith with podorhabdid rim of three or four tiers of rim elements, two to three tiers being visible from the proximal side and only two tiers visible on the distal side. Middle to Upper Cretaceous.

    a. Subfamily Arkhangelskielloideae Gartner 1968
Elliptical coccoliths with marginal rim of two to four tiers, without differentiation into separate shields, enclosing large central area that is divided into four quadrants by an axial cross or axial furrows; stem rare or absent. Cretaceous.
Acaenolithus Black 1973; *Arkhangelskiella* Vekshina 1959; *Aspidolithus* Noël 1969; *Broinsonia* Bukry 1969; *Cribricatillus* Black 1973; *Crucicribrum* Black 1973; *Laffittius* Noël 1969; *Prolatipatella* Gartner 1968.

    b. Subfamily Kamptnerioideae Bukry 1969
Large elliptical coccoliths with complex rim structure; one of three to four multielement rim tiers is very broad; central plate of many elements and perforate or closed but always convex distally; subaxial to axial bar structures prominent on proximal surface and insignificant on distal face. Upper Cretaceous.
Gartnerago Bukry 1969; *Kamptnerius* Deflandre 1959.

F. Order Coccolithales Schwarz 1932 orth. mut. Rood, Hay & Barnard 1971
    Syn.: Coccosphaerales Haeckel 1894; Coccolithophorales Schiller 1926; Coccolithinales Schwarz 1932; Coccolithophorida Martini 1958
Motile cell with two heterodynamic flagella and a haptonema; may have alternating nonmotile pelagic stage or benthic filamentous or palmelloid one; cell covered with layer of organic scales and calcareous coccoliths, coccoliths may all be similar, or more than one kind may be present on a cell or on the different stages of the life cycle, such as an alternation of a motile stage with holococcoliths and a nonmotile one with heterococcoliths; heterococcoliths of three cycles of elements, one forming a central tube or girdle (coccolithid rim), and the other two extending peripherally from it as proximal and distal shields. Marine, and brackish water. Permian to Holocene.

    1. Family Ellipsagelosphaeraceae Noël 1965
    Syn.: Watznaueriaceae Rood, Hay & Barnard 1971; Lotharingiaceae Noël 1972

Similar to Coccolithaceae, but with both shields strongly birefringent in polarized light, resulting in a complex interference pattern due to the two superposed black crosses (one from each shield); commonly with corona or ring of quadrate or keystone-shaped elements on distal surface separating outer zone of petaloid elements from the central area; coccospheres commonly preserved with interlocking coccoliths and possibly a calcified outer membrane; may represent an encysted or nonmotile stage. Jurassic to Cretaceous.

Caterella Black 1971; *Coptolithus* Black 1973; *Cyclagelosphaera* Noël 1965; *Diazomatolithus* Noël 1965; *Ellipsagelosphaera* Noël 1965; *Esgia* Worsley 1971; *Lotharingius* Noël 1972 (syn.: *Striatomarginis* Prins ex Rood, Hay & Barnard 1973); *Manivitella* Thierstein 1971; *Margolatus* Forchheimer 1972; *Sphenoradiatus* Worsley 1971; *Tretosestrum* Wilcoxon 1972; *Tubodiscus* Thierstein 1973; *Watznaueria* Reinhardt 1964 (syn.: *Colvillea* Black 1964, non Bojer ex Hooker 1834; *Actinosphaera* Noël 1965; *Calolithus* Noël 1965; *Maslovella* Tappan & Loeblich 1966).

2. Family Coccolithaceae Poche 1913 orth. mut. Kamptner 1928

Syn.: Coccolithophoraceae Lemmermann in Brandt & Apstein 1908; Coccolithidae Poche 1913

Haptophyceae with placolith cover on nonmotile resting stage, each shield of radiating petaloid elements around a tubular or solid central pillar, elements of distal shield strongly imbricated, those of proximal shield without imbrication; distal shield faintly illuminated or showing no effect in polarized light from either distal or proximal view; proximal shield strongly birefringent; no spine. Permian to Holocene.

a. Subfamily Coccolithoideae Kamptner 1928

Elliptical placoliths; interference figure of proximal shield laevogyre in distal view. Upper Cretaceous to Holocene.

Birkelundia Perch-Nielsen 1971; *Chiasmolithus* Hay, Mohler & Wade 1966; *Coccolithus* Schwarz 1894 (syn.: *Cyatholithus* Huxley 1868, suppressed ICZN; *Coccosphaera* Wallich 1877, non Perty 1862; *Coccolithophora* Lohmann 1902; *Crystallolithus* Gaarder & Markali 1956, the holococcolith-bearing motile stage of the non-

motile *Coccolithus*); *Cruciplacolithus* Hay & Mohler 1967 (syn.: *Crucilithus* Stradner 1968); *Hornibrookina* Edwards 1973; *Oolithotus* Reinhardt 1968; *Rhabdolekiskus* Hill 1976.

b. Subfamily Tergestielloideae Reinhardt 1966

Syn.: Cyclococcolithoideae Hay & Mohler in Boudreaux & Hay 1969

Circular placoliths, with interference figure of proximal shield radial to subradial. Permian to Holocene.

Calcidiscus Kamptner 1950 (syn.: illeg. *Cyclococcolithus* Kamptner 1958; *Tiarolithus* Kamptner 1958; illeg. *Cycloplacolithus* Kamptner 1963; illeg. *Cycloplacolithella* Haq 1968; illeg. *Cyclococcolithina* Wilcoxon 1970); *Conococcolithus* Hay & Mohler 1967; *Corannulus* Stradner 1962 (syn.: *Diademopetra* Hay, Mohler & Wade 1966); *Coronocyclus* Hay, Mohler & Wade 1966; *Cribrocentrum* Perch-Nielsen 1972; *Cricolithus* Kamptner 1958; *Cyclicargolithus* Bukry 1971; *Heyneckia* Lohmann 1913; *Ilselithina* Stradner 1966 (syn.: *Hayella* Roth 1969); *Markalius* Bramlette & Martini 1964; *Neosphaera* Lecal-Schlauder 1950; *Pedinocyclus* Bukry & Bramlette 1971; *Striatococcolithus* Bukry 1971; *Tergestiella* Kamptner 1941; *Umbilicosphaera* Lohmann 1902 (syn.: *Craspedolithus* Kamptner 1963).

G. Order Syracosphaerales Hay 1977

Cell with loose cover of delicate coccoliths that may not touch adjacent ones; coccoliths are caneoliths, cribriliths, cyrtoliths, or lopadoliths; tube formed by marginal cycle of elements, with distal and proximal rims of petaloid elements, central area closed by laths. Freshwater (rare), marine. Paleocene to Holocene.

1. Family Syracosphaeraceae Lemmerman 1908

Coccoliths as caneoliths and cribriliths. Marine. Oligocene to Holocene.

Alisphaera Heimdal 1973; *Anthosphaera* Kamptner 1936; *Caneosphaera* Gaarder 1977; *Cepekiella* Roth 1970; *Coronosphaera* Gaarder 1977; *Syracolithus* Kamptner 1941; *Syracosphaera* Lohmann 1902 (syn.: *Eusyracosphaera* Kamptner 1941; invalid *Syracorhabdus* Lecal & Bernheim 1960).

(continued)

2. Family Prinsiaceae Hay & Mohler 1967

Oval, subcircular, or circular placoliths, interference figure between crossed polarizers is dextrogyre in distal view; separate distal and proximal segment cycles, imbricated; central area with grille, bridge, or plug. Marine. Paleocene to Oligocene.

Bramletteius Gartner 1969; *Campylosphaera* Kamptner 1963; *Ellipsolithus* Sullivan 1964; *Ericsonia* Black 1964; *Prinsius* Hay & Mohler 1967; *Repagulum* Forchheimer 1972; *Toweius* Hay & Mohler 1967.

3. Family Acanthoicaceae Hay 1977 orth. mut. herein

Syn.: Acanthoiceae Hay 1977

Coccoliths in the form of simple or styliform cyrtoliths. Freshwater and marine. Holocene.

Acanthoica Lohmann 1903; *Anacanthoica* Deflandre 1952.

4. Family Halopappaceae Kamptner 1928

Syn.: Deutschlandiaceae Kamptner 1928

Calcified homogeneous cell cover, or with calcified plate scales of relatively few quadrangular elements or simple discoliths; coccoliths variously modified as elongate elements around apical opening, may aid flotation. Marine. Holocene.

Calcioconus Schiller 1916; *Deutschlandia* Lohmann 1912 (syn.: *Florisphaera* Okada & Honjo 1973); *Halopappus* Lohmann 1912; *Michaelsarsia* Gran 1912; *Ophiaster* Gran 1912 (syn.: *Bernardosphaera* Lecal 1951); *Pappomonas* Manton & Oakes 1975; *Papposphaera* Tangen 1970; *Thalassopappus* Kamptner 1941; *Thorosphaera* Ostenfeld 1910.

5. Family Rhabdosphaeraceae Ostenfeld 1900 orth. mut. Lemmermann 1908

Syn.: Family Rhabdosphaerales Ostenfeld 1900

Spherical cell with coccoliths in the form of discoliths bearing a central spine that consists of a solid bundle of calcite fibers, basal plate with closely packed elements in the floor, surrounded by a rim of larger elements. Eocene to Holocene.

Aspidorhabdus Hay & Towe 1962; *Blackites* Hay & Towe 1962; *Discosphaera* Haeckel 1894; *Lecalia* Loeblich & Tappan 1963 (syn.: *Rhabdocyclus* Lecal 1960 non Lang & Smith 1939); *Naninfula* Perch-Nielsen 1968; *Palusphaera* Lecal 1965; *Petasus* Perch-Nielsen 1971; *Pyrocyclus* Hay & Towe 1962; *Rhabdolithus* Kamptner ex Deflandre 1952; *Rhabdosphaera* Haeckel 1894

(syn.: *Rhabdolithes* Schmidt 1870, suppressed ICZN; *Clavosphaera* Lecal-Schlauder 1951; *Aspidorhabdus* Hay & Towe 1962); ?*Rhabdothorax* Kamptner 1973; *Umbellosphaera* Paasche 1955.

H. Order Prymnesiales (Fritsch) Papenfuss 1955

Flagellate monad or colonial (*Corymbellus*), or amoeboid with flagellate zoospores; no distinct cell membrane, two acronematic equal flagella and a distinct haptonema; surface of cell covered with organic scales. Marine, fresh, and brackish water. Holocene.

1. Family Prymnesiaceae Conrad 1926

Syn.: Platychrysidaceae Carter 1937

Motile cell with two smooth flagella and haptonema; body scales of two types present in separate layers on the cell surface; nonmotile stage alternates. Marine and freshwater (*Chrysochromulina*). Holocene.

Chrysochromulina Lackey 1939 (syn.: *Chrysocampanula* Fournier 1971); *Corymbellus* Green 1976; *Platychrysis* Geitler 1930; *Prymnesium* J. Massart 1920.

2. Family Phaeocystaceae Lagerheim 1896

Colonial vegetative stage, with cells scattered in spherical to elongate mucilaginous bladders; rounded motile cell with markedly heterodynamic flagella and short, noncoiling haptonema. Holocene.

Phaeocystis Lagerheim 1893.

I. Order Pavlovales Green 1976

Motile cells or palmelloid; nonmotile stage produces motile swarmers; motile cells with two unequal flagella and haptonema; shorter flagellum may be vestigial. Marine, fresh and brackish water. Holocene.

1. Family Pavlovaceae Green 1976

Cell with flagella and haptonema inserted ventrally or laterally, anterior flagellum with fine hairs and may also contain small dense bodies; posterior flagellum only with fine hairs; haptonema with distal reduction of number of microtubules; eyespot in *Pavlova*. Marine, fresh and brackish water.

Corcontochrysis Kalina 1970; *Diacronema* Prauser 1958; *Exanthemachrysis* Lepailleur 1970; *Pavlova* Butcher 1952.

Position uncertain.

1. Family Schizosphaerellaceae Deflandre 1959

Subspherical calcareous tests, 12 to 30 μm in diameter, constructed of two overlapping or interlocking hemispheres that may be disarticulated. Larger valve may have double rim, the opposite valve fitting into the groove between the rims; surface punctate. Upper Triassic to Jurassic.

Schizosphaerella Deflandre & Dangeard 1938 (syn.: *Nannopatina* Stradner 1961).

    2. Family Stomiosphaeraceae Wanner 1941

    Syn.: Pithonellidae Keller 1946; Calcisphaerulidae Bonet 1956

    Globular to ovoid specimens, rarely elongate; 30 to 70 μm in diameter, and 55 to 60 μm long, rarely up to 200 μm in length; wall nonperforate, of calcite by X-ray determination, sectioned specimens of *Pithonella* and *Stomiosphaera* showing an extinction cross between crossed nicols; wall originally about 2 μm in thickness, later additions may result in two or three or more layers, to a maximum of 10 μm thick; fibrous crystals in radial arrangement at extremities, may be oblique in equatorial region; surface with longitudinal ridges and weaker transverse ridges resulting in a cancellate or reticulate appearance; may have aperture at one end, apparently secondarily formed. Jurassic to Paleocene.

    Bonetocardiella Dufour 1968 (syn.: *Conejoconus* Knauer 1970) and *Cadosina* Wanner 1941 may be tintinnids, although sometimes included herein. *Pithonella* is regarded as distinct by some, as a spore of a dasyclad green algae, or as a dinoflagellate cyst.

Andriella Bolli 1974; *Cadosinopsis* Scheibner 1967; *Calcisphaerula* Bonet 1956; *Carpistomiosphaera* Nowak 1968; *Colomisphaera* Nowak 1968; *Hemisphaera* Nowak 1968; *Palinosphaera* Reinsch 1905; *Parastomiosphaera* Nowak 1968; *Pithonella* Lorenz 1902 (syn.: *Cadosinella* Vogler 1941; *Pleurozonaria* O. Wetzel 1933); *Stomiosphaera* Wanner 1941.

REFERENCES

Achuthan, M. V., and Herbert Stradner, Calcareous nannoplankton from the Wemmelian stratotype, p. 1−13, pls. 1−5, figs. 1, 2. *In* P. Brönnimann and H. H. Renz, *Proc. First Int. Conf. Planktonic Microfossils, Geneva, 1967,* v. 1. Leiden: E. J. Brill, 1969.

Adams, T. D., M. Khalili, and A. K. Said, Stratigraphic significance of some oligosteginid assemblages from Lurestan Province, northwest Iran. *Micropaleontology,* v. 13, p. 55−67, pl. 1, figs. 1−4, 1967.

Adelseck, C. G., Jr., G. W. Geehan, and P. H. Roth, Experimental evidence for the selective dissolution and overgrowth of calcareous nannofossils during diagenesis. *Bull. Geol. Soc. Am.,* v. 84, p. 2755−2762, 3 figs., 1973.

Anderson, T. F., and S. A. Cole, The stable isotope geochemistry of marine coccoliths: a preliminary comparison with planktonic foraminifera. *J. Foramin. Res.,* v. 5, p. 188−192, figs. 1, 2, 1975.

Andri, E., Mise au point et donées nouvelles sur la famille des Calcisphaerulidae Bonet 1956: les genres *Bonetocardiella, Pithonella, Calcisphaerula* et *"Stomiosphaera." Revue Micropaléont.,* v. 15, p. 12−34, pls. 1−3, figs. 1−12, 1972.

Andri, E., and M. P. Aubry, Recherches sur la microstructure des tests de *Pithonella ovalis* (Kaufmann) et *Pithonella perlonga* Andri. *Revue Micropaléont.,* v. 16 (1973), p. 159−167, pls. 1−3, 1974.

Aubry, M. P., Remarques sur la systématique des *Nannoconus* de la Craie. *Cah. Micropaléont.,* 1974(4), p. 3−22, pls. 1−9, 1974.

Aubry, M. P., and F. Depeche, Recherches sur les Schizosphères I. Les Schizosphères de Villers-sur-Mer variation morphologique ultrastructure et modifications diagénétiques. *Cah. Micropaléont.,* 1974(1), p. 3−15, pls. 1−6, 1974.

Báldi-Beke, Mária, Magyarországi miocén Coccolithophoridák rétegtani jelentösege [Die stratigraphische Bedeutung miozäner Coccolithophoren aus Ungarn]. *Földt. Közl.,* v. 90, p. 213−223, pl. 14, 1960.

Báldi-Beke, Mária, Coccolithophorida vizsgálatok a Mecseki miocénben [Untersuchungen an Coccolithophoriden aus dem Miozän des Mecsek-Gebirges]. *Évi Jelent. Magy. K. Földt. Intéz.,* 1961, no. 1, p. 161−173, 2 pls., 1964a.

Báldi-Beke, Mária, A *Nannoconus* nemzetség földtani szerepe [Geological importance of the genus *Nannoconus*]. *Évi Jelent. Magy. K. Földt. Intéz.,* 1961, no. 2, p. 169−181, 1 pl., 2 text-figs., 1964b.

Báldi-Beke, Mária, Alsó-Kréta képzödményeink Coccolithophorida Faunája [Lower Cretaceous Coccolithophorid fauna from Hungary]. *Évi Jelent. Magy. K. Földt. Intéz.,* 1962, p. 131−144, 1 pl., 1964c.

Banner, F. T., *Pithonella ovalis* from the early Cenomanian of England. *Micropaleontology,* v. 18, p. 278−284, pls. 1, 2, fig. 1, 1972.

Barnard, Tom, and W. W. Hay, On Jurassic coccoliths: a tentative zonation of the Jurassic of southern England and north France. *Eclog. Geol. Helv.,* v. 67, p. 563−585, pls. 1−6, figs. 1, 2, 1974.

Barrier, Janine, Nannofossiles calcaires des marnes de l'Aptien inférieur type: Bédoulien de Cassis-La Bédoule (Bouches-du-Rhône). *Bull. Mus. Natn. Hist. Nat., Paris,* ser. 3, no. 437, Sci. Terre 59, p. 1−68, pls. 1−15, 1977.

Bartolini, Carlo, Coccoliths from sediments of the western Mediterranean. *Micropaleontology,* v. 16, p. 129−154, pls. 1−8, 9 figs., 1970.

Bein, A., and Z. Reiss, Cretaceous *Pithonella* from Israel. *Micropaleontology,* v. 22, p. 83−91, pls. 1−3, fig. 1, 1976.

Berge, G., Discoloration of the sea due to *Coccolithus huxleyi* bloom. *Sarsia,* v. 6, p. 27−40, 1 pl., 5 text-figs., 1962.

Berger, W. H., Deep-sea carbonates: evidence for a coccolith lysocline. *Deep Sea Res.,* v. 20, p. 917−921, figs. 1, 2, 1973.

Bernard, Francis, Coccolithophorides nouveau ou peu connus observés à Monaco en 1938. *Archs. Zool. Exp. Gén.,* v. 81 (Notes et Revue), p. 33−44, 2 figs., 1939.

Bernard, Francis, Essai sur les facteurs de répartition des Flagellés calcaires. *Annls. Inst. Océanogr., Monaco,* ser. 2, v. 21, p. 29−112, 37 figs., 1942.

Bernard, Francis, Recherches sur le cycle du *Coccolithus fragilis* Lohm., flagellé dominant des mers chaudes. *J. Cons. Perm. Int. Explor. Mer,* v. 15, p. 177−188, 2 figs., 1948a.

Bernard, Francis, Rôle de flagellés dans la transparence marine en Méditerranée. *Bull. Soc. Hist. Nat. Afr. N.,* v. 38, p. 74−79, 1948b.

Bernard, Francis, Rôle des flagellés calcaires dans la fertilité et la sédimentation en mer profonde. *Deep Sea Res.,* v. 1, p. 34−46, 3 figs., 1953.

Bernard, Francis, Density of flagellates and Myxophyceae in the heterotrophie layers related to environment, p. 215−228. *In* C. H. Oppenheimer, *Symposium on marine microbiology.* Springfield, Illinois: C. C. Thomas Publ., 1963.

Bernard, Francis, Le nannoplancton en zone aphotique des mers chaudes. *Pelagos, Bull. Inst. Oceanogr., Alger,* v. 2, p. 5−32, 3 figs., 1965.

Bernard, Francis, Research on phytoplankton and pelagic protozoa in the Mediterranean Sea from 1953 to 1966. *Oceanogr. Mar. Biol. Ann. Rev.,* v. 5, p. 205−229, 1967.

Bignot, G., and L. Lezaud, Contribution a l'étude des *Pithonella* de la craie Parisienne. *Revue Micropaléont.,* v. 7, p. 138−152, pls. 1−3, figs. 1−5, 1964.

Black, Maurice, Fossil coccospheres from a Tertiary outcrop on the continental slope. *Geol. Mag.,* v. 99, p. 123−127, pls. 8, 9, 1962a.

Black, Maurice, [Stratigraphical distribution of coccoliths]. *Proc. Geol. Soc.,* 1961/62, p. 133−136, 1962b.

Black, Maurice, The fine structure of the mineral parts of the Coccolithophoridae. *Proc. Linn. Soc. Lond.,* v. 174, p. 41−46, 1 pl., 2 text-figs., 1963.

Black, Maurice, Cretaceous and Tertiary coccoliths from Atlantic seamounts. *Palaeontology,* v. 7, p. 306−316, pls. 50−53, 1964.

Black, Maurice, Coccoliths. *Endeavour,* v. 24(93), p. 131−137, figs. 1−25, 1965.

Black, Maurice, Taxonomic problems in the study of coccoliths. *Palaeontology,* v. 11, p. 793−813, pls. 143−154, 1968.

Black, Maurice, The systematics of coccoliths in relation to the paleontological record, p. 611−624, 4 pls. *In* B. M. Funnell and W. R. Riedel, *The micropalaeontology of oceans.* Cambridge: University Press, 1971a.

Black, Maurice, Problematical microfossils from the Gault Clay. *Geol. Mag.,* v. 108, p. 325−327, pls. 1−3, 1971b.

Black, Maurice, Coccoliths of the Speeton Clay and Sutterby Marl. *Proc. Yorks. Geol. Soc.,* v. 38, p. 381−424, pls. 30−34, 1971c.

Black, Maurice, British Lower Cretaceous coccoliths. I. Gault Clay. *Palaeontogr. Soc. [Monogr.],* v. 126(534), p. 1−48, pls. 1−16, figs. 1−38, 1972a.

Black, Maurice, Crystal development in Discoasteraceae and Braarudosphaeraceae (planktonic algae). *Palaeontology,* v. 15, p. 476−489, pls. 87−96, figs. 1, 2, 1972b.

Black, Maurice, British Lower Cretaceous coccoliths. I. Gault Clay. Part 2. *Palaeontogr. Soc. [Monogr.],* v. 127(537), p. iii−iv, 49−112, pls. 17−33, figs. 39−51, 1973.

Black, Maurice, British Lower Cretaceous coccoliths. I. Gault Clay. Part 3. *Palaeontogr. Soc. [Monogr.],* v. 129(543), p. 113−142, pl. 34, 1975.

Black, Maurice, and Barbara Barnes, The structure of coccoliths from the English Chalk. *Geol. Mag.,* v. 96, p. 321−328, pls. 8−12, 1959.

Black, Maurice, and Barbara Barnes, Coccoliths and discoasters from the floor of the South Atlantic Ocean. *Jl. R. Microsc. Soc.,* ser. 3, v. 80, p. 137−147, pls. 19−26, 1961.

Blackwelder, P. L., R. E. Weiss, and K. M. Wilbur, Effects of calcium, strontium, and magnesium on the coccolithophorid *Cricosphaera (Hymenomonas) carterae.* I. Cell division. *Mar. Biol.,* v. 34, p. 11−16, figs. 1−3, 1976.

Bogdanov, Yu. A., and M. G. Ushakova, Kokkolity gruppy *Discoaster* Tan Sin Hok v vodnoy vsvesi Tikhogo Okeana [Coccoliths of the *Discoaster* Tan Sin Hok group in suspensions from the Pacific Ocean]. *Dokl. Akad. Nauk SSSR,* v. 171, p. 465−467, figs. 1−4, 1966.

Bolli, H. M., Jurassic and Cretaceous Calcisphaerulidae from DSDP Leg 27, Eastern Indian Ocean. *Init. Repts. Deep Sea Drill. Proj.*, v. 27, p. 843−907, pls. 1−24, 1974.

Bóna, J., Coccolithophorida-vizsgálatok a mecseki neogén rétegekben [Coccolithophoriden-Untersuchungen in der neogenen Schichtenfolge des Mecsekgebirges]. *Földt. Közl.*, v. 94, p. 121−131, pls. 13−15, 3 text-figs., 1964.

Bonet, F., Zonification microfaunistica de las Calizas Cretacicas del este de Mexico. *Bol. Asoc. Méx. Geól. Petrol.*, v. 8(7−8), p. 3−102, pls. 1−35, 1956.

Bonet, F., and M. Trejo, Nuevos datos sobre la familia Calcisphaerulidae (Protozoa). *An. Esc. Nac. Cienc. Biol., Méx.*, v. 9, p. 43−48, pls. 1, 2, 1958.

Boney, A. D., and Angela Burrows, Experimental studies on the benthic phases of haptophyceae I. Effects of some experimental conditions on the release of coccolithophorids. *J. Mar. Biol. Ass. U.K.*, v. 46, p. 295−319, pl. 1, figs. 1−9, 1966.

Borsetti, A. M., and Franco Cati, Il nannoplancton calcareo vivente nel Tirreno Centro-Meridionale. *G. Geol.*, ser. 2, v. 38 (1970), p. 395−414, pls. 39−57, 1972.

Borsetti, A. M., and Franco Cati, Il nannoplancton calcareo vivente nel Tirreno Centro-Meridionale Parte II. *G. Geol.*, ser. 2A, v. 40, p. 209−240, pls. 12−18, 1976.

Bouché, P. M., Nannofossiles calcaires du Lutétien du bassin de Paris. *Revue Micropaléont.*, v. 5, p. 75−103, 4 pls., 32 text-figs., figs. A−C, 1962.

Boudreaux, J. E., and W. W. Hay, Calcareous nannoplankton and biostratigraphy of the Late Pliocene-Pleistocene-Recent sediments in the Submarex cores. *Revta. Esp. Micropaleont.*, v. 1, p. 249−292, pls. 1−10, 1 fig., 1969.

Braarud, Trygve, Coccolith morphology and taxonomic position of *Hymenomonas roseola* Stein and *Syracosphaera carterae* Braarud & Fagerland. *Nytt Mag. Bot.*, v. 3, p. 1−4, 2 pls., 1 text-fig., 1954a.

Braarud, Trygve, The ecology of marine phytoplankton. *Rapp. et Commun. VIIIᵉ Int. Bot. Congr.*, v. 17, p. 178−185, 1954b.

Braarud, Trygve, Species distribution in marine phytoplankton. *J. Oceanogr. Soc. Japan*, 20th anniversary vol., p. 628−649, 6 figs., 1962.

Braarud, Trygve, Reproduction in the marine coccolithophorid *Coccolithus huxleyi* in culture. *Pubbl. Staz. Zool. Napoli*, v. 33, p. 110−116, 2 figs., 1963.

Braarud, Trygve, G. Deflandre, P. Halldal, and E. Kamptner, Terminology, nomenclature, and systematics of the Coccolithophoridae. *Micropaleontology*, v. 1, p. 157−159, 1955.

Braarud, Trygve, and E. Fagerland, A coccolithophoride in laboratory culture, *Syracosphaera carterae* n.sp. *Avh. Norske VidenskAkad. Oslo*, 1946, no. 2, 10 p., 1 pl., 1946.

Braarud, Trygve, K. R. Gaarder, J. Markali, and E. Nordli, Coccolithophorids studied in the electron microscope. Observations on *Coccolithus huxleyi* and *Syracosphaera carterae. Nytt Mag. Bot.*, v. 1, p. 129−134, 2 pls., 4 text-figs., 1952.

Braarud, Trygve, K. R. Gaarder, and O. Nordli, Seasonal changes in the phytoplankton at various points off the Norwegian west coast. *FiskDir. Skr., ser. Havundersøkelser*, v. 12, no. 3, 77 p., 12 figs., 1958.

Braarud, Trygve, and E. Nordli, Coccoliths of *Coccolithus huxleyi* seen in an electron microscope. *Nature*, v. 170, p. 361−362, 2 figs., 1952.

Bramlette, M. N., *Discoaster* and some related microfossils. *Prof. Pap. U.S. Geol. Surv.*, 280 E−J, p. 247−253, pl. 61, 1957.

Bramlette, M. N., Significance of coccolithophorids in calcium-carbonate deposition. *Bull. Geol. Soc. Am.*, v. 69, p. 121−126, 1958.

Bramlette, M. N., Pelagic sediments, p. 345−366. *In* M. Sears (ed.), *Oceanography*. Washington, D.C.: Publs. Am. Ass. Advmt. Sci., 67, 1961.

Bramlette, M. N., Massive extinctions in biota at the end of Mesozoic time. *Science*, v. 148, p. 1696−1699, 1 fig., 1965.

Bramlette, M. N., and E. Martini, The great change in calcareous nannoplankton fossils between the Maestrichtian and Danian. *Micropaleontology*, v. 10, p. 291−322, 7 pls., 1 text-fig., 1964.

Bramlette, M. N., and W. R. Riedel, Stratigraphic value of discoasters and some other microfossils related to Recent coccolithophores. *J. Paleont.*, v. 28, p. 385−403, pls. 38, 39, 3 text-figs., 1954.

Bramlette, M. N., and F. R. Sullivan, Coccolithophorids and related nannoplankton of the early Tertiary in California. *Micropaleontology*, v. 7, p. 129−188, 14 pls., 1 text-fig., 1961.

Bramlette, M. N., and J. A. Wilcoxon, Middle Tertiary calcareous nannoplankton of the Cipero section, Trinidad, W. I. *Tulane Stud. Geol. Paleont.*, v. 5, p. 93−131, pls. 1−10, 1967.

Brönnimann, Paul, Microfossils incertae sedis from the Upper Jurassic and Lower Cretaceous of Cuba. *Micropaleontology*, v. 1, p. 28−51, 2 pls., 10 text-figs., 1955.

Brown, R. M., Jr., Observations on the relationship of the Golgi apparatus to wall formation in the marine chrysophycean alga, *Pleurochrysis scherffelii* Pringsheim. *J. Cell. Biol.*, v. 41, p. 109−123, figs. 1−14, 1969.

Brown, R. M., W. W. Franke, H. Kleinig, H. Folk, and P. Sitle, Cellulosic wall component produced by the golgi apparatus of *Pleurochrysis scherffelii. Science*, v. 166, p. 894−896, 1969.

Buisonjé, P. H. de, Recurrent red tides, a possible origin of the Solnhofen Limestone. I. *Proc. K. Ned. Akad. Wet.*, ser. B, v. 75, p. 152−166, 2 figs., 1972a.

Buisonjé, P. H. de, Recurrent red tides, a possible origin of the Solnhofen Limestone. II. *Proc. K. Ned. Akad. Wet.*, ser. B, v. 75, p. 167–177, 1972b.

Bukry, David, Upper Cretaceous coccoliths from Texas and Europe. *Paleont. Contr. Univ. Kans.*, Art. 51 (Protista 2), p. 1–79, pls. 1–50, fig. 1, 1969.

Bukry, David, Cenozoic calcareous nannofossils from the Pacific Ocean. *Trans. S. Diego Soc. Nat. Hist.*, v. 16, p. 303–327, pls. 1–7, 1971a.

Bukry, David, Coccolith stratigraphy Leg 6, Deep Sea Drilling Project. *Init. Repts. Deep Sea Drill. Proj.*, v. 6, p. 965–1004, pls. 1–8, 1971b.

Bukry, David, *Discoaster* evolutionary trends. *Micropaleontology*, v. 17, p. 43–52, pls. 1–3, 1971c.

Bukry, David, Coccolith and silicoflagellate stratigraphy, Tasman Sea and southwestern Pacific Ocean, Deep Sea Drilling Project Leg 21. *Init. Repts. Deep Sea Drill. Proj.*, v. 21, p. 885–893, pl. 1, figs. 1–5, 1973a.

Bukry, David, Coccolith stratigraphy, eastern equatorial Pacific, Leg 16 Deep Sea Drilling Project. *Init. Repts. Deep Sea Drill. Proj.*, v. 16, p. 653–711, pls. 1–5, figs. 1–4, 1973b.

Bukry, David, Low-latitude coccolith biostratigraphic zonation. *Init. Repts. Deep Sea Drill. Proj.*, v. 15, p. 685–703, figs. 1, 2, 1973c.

Bukry, David, Coccolith stratigraphy, offshore Western Australia, Deep Sea Drilling Project Leg 27. *Init. Repts. Deep Sea Drill. Proj.*, v. 27, p. 623–630, figs. 1–7, 1974a.

Bukry, David, Phytoplankton stratigraphy, offshore East Africa, Deep Sea Drilling Project Leg 25. *Init. Repts. Deep Sea Drill. Proj.*, v. 25, p. 635–646, figs. 1–4, 1974b.

Bukry, David, Coccoliths as paleosalinity indicators—evidence from Black Sea. *Mem. Am. Ass. Petrol. Geol.*, 20, p. 353–363, figs. 1–6, 1974c.

Bukry, David, Coccolith and silicoflagellate stratigraphy, northwestern Pacific Ocean, Deep Sea Drilling Project Leg 32. *Init. Repts. Deep Sea Drill. Proj.*, v. 32, p. 677–701, pls. 1–4, figs. 1–5, 1975.

Bukry, David, E. E. Brabb, and J. G. Vedder, Correlation of Tertiary nannoplankton assemblages from the Coast and Peninsular Ranges of California. *Spec. Publ. Boln. Geol. Minist. Minas Venez.*, 7, p. 1461–1483, pl. 1, fig. 1, 1977.

Bukry, David, and M. N. Bramlette, Coccolith age determinations Leg 3, Deep Sea Drilling Project. *Init. Repts. Deep Sea Drill. Proj.*, v. 3, p. 589–611, 1970.

Bukry, David, R. G. Douglas, S. A. Kling, and V. Krasheninnikov, Planktonic microfossil biostratigraphy of the northwestern Pacific Ocean. *Init. Repts. Deep Sea Drill. Proj.*, v. 6, p. 1253–1300, pls. 1–4, figs. 1–9, 1971.

Bukry, David, and S. F. Percival, Jr., New Tertiary calcareous nannofossils. *Tulane Stud. Geol. Paleont.*, v. 8, p. 123–146, pls. 1–7, 1971.

Burns, D. A., The latitudinal distribution and significance of calcareous nannofossils in the bottom sediments of the south-west Pacific Ocean (Lat. 15–55° S) around New Zealand, p. 221–228, figs. 1–6. *In* R. Fraser (Compiler), *Oceanography of the South Pacific 1972.* Wellington: New Zealand National Commission for UNESCO, 1973a.

Burns, D. A., Structural analysis of flanged coccoliths in sediments from the south west Pacific Ocean. *Revta. Esp. Micropaleont.*, v. 5, p. 147–160, pls. 1, 2, fig. 1, 1973b.

Burns, D. A., Changes in nannofossil morphology by diagenetic processes in Cretaceous chalk deposits. *N.Z. Jl. Geol. Geophys.*, v. 18, p. 469–476, figs. 1–26, 1975.

Bursa, A. S., *Discoasteromonas calciferus* n. sp., an Arctic relict secreting *Discoaster* Tak [sic] Sin Hok 1927. *Grana Palynol.*, v. 6, p. 148–165, pls. 1–4, figs. 1–22, 1965.

Bursa, A. S., Morphogenesis and taxonomy of fossil and contemporary Dinophyta secreting discoasters, p. 129–143, pls. 1, 2, 22 figs. *In* A. Farinacci, *Proc. II Planktonic Conf. Roma 1970*, v. 1. Rome: Edizioni Tecnoscienza, 1971.

Bybell, L. M., Middle Eocene calcareous nannofossils at Little Stave Creek, Alabama. *Tulane Stud. Geol. Paleont.*, v. 11, p. 179–252, pls. 1–24, 1975.

Bybell, L. M., and Stefan Gartner, Provincialism among mid-Eocene calcareous nannofossils. *Micropaleontology*, v. 18, p. 319–336, pls. 1–5, 1 fig., 1972.

Bystrická, Hedviga, Die Unter-Eozänen Coccolithophoridae (Flagellata) des Myjavaer Palaeogens. *Geol. Sb., Bratisl.*, v. 14, p. 269–281, pls. 1–4, 1963.

Bystrická, Hedviga, Les Coccolithophoridés (Flagellés) de l'Éocène supérieur de la Slovaquie. *Geol. Sb., Bratisl.*, v. 15, p. 203–225, pls. 5–8, 1964.

Bystrická, Hedviga, Der stratigraphische Wert von Discoasteriden im Palaeogen der Slowakei. *Geol. Sb., Bratisl.*, v. 16, p. 7–10, 1965.

Bystrická, Hedviga, Calcareous nannoplankton of the *Marthasterites tribrachiatus* Zone from the Lower Eocene sediments of Slovakia. *Geol. Sb., Bratisl.*, v. 27, p. 273–297, pls. 1–11, 1976.

Caratini, Claude, Sur la découverte de Nannoconidés dans le Cénomanien et le Turonien du bassin de Paris. *C. R. Somm. Séanc. Soc. Géol. Fr.*, fasc. 5, p. 106–107, 6 figs., 1960.

Caratini, Claude, Contribution à l'étude des coccolithes du Cénomanien supérieur et du Turonien de la région de Rouen. Thèse, Faculté des Sciences, Université d'Alger. *Publ. Lab. Géol. Appliquée*, 1960. v. 12, 61 p., 5 pls., 2 + 7 text-figs., 1963.

Carter, Nellie, New or interesting algae from brackish water. *Arch. Protistenk.*, v. 90, p. 1–68, pls. 1–8, 3 text-figs., 1937.

Čepek, Pavel, and W. W. Hay, Zonation of the Upper Cretaceous using calcareous nannoplankton. *Paläont. Abh., Abt. B, Paläobot.,* v. 3, p. 333−340, pls. 20, 21, 1970.

Christensen, T., *Botanik, Bind 2. Systematisk botanik. Nr. 2. Alger.* København: Munksgaard, 178 p., 75 figs., 1962.

Clocchiatti, Micheline, Contribution a l'étude du nannoplancton calcaire du Neogene d'Afrique du Nord. *Mém. Mus. Natn. Hist. Nat., Paris,* n. sér., sér. C, *Sci. Terre,* v. 23, Fasc. Unique, p. 1−135, pls. 1−40, figs. 1−21, 1971a.

Clocchiatti, Micheline, Remarques sur quelques rhabdolithes de Méditerranée. *Cah. Micropaléont.,* sér. 2, no. 9, p. 1−8, pls. 1−3, 1971b.

Clocchiatti, Micheline, Sur l'existence de coccosphères portant des coccolithes de *Gephyrocapsa oceanica* et de *Emiliania huxleyi* (Coccolithophoridés). *C. R. Hebd. Séanc. Acad. Sci., Paris,* sér. D, v. 273, p. 318−321, 2 figs., 1971c.

Cohen, C. L. D., Coccoliths and discoasters from Adriatic bottom sediments. *Leid. Geol. Meded.,* v. 35, p. 1−44, pls. 1−25, (1966) separately published 1965.

Cohen, C. L. D., and Peter Reinhardt, Coccolithophorids from the Pleistocene Caribbean deepsea core CP-28. *Neues Jb. Geol. Paläont. Abh.,* v. 131, p. 289−304, pls. 19−21, 8 figs., 1968.

Colom, Guillermo, Arqueomonadíneas, Silicoflagelados, Discoastéridos fósiles de España. *Ciencias,* v. 5, p. 343−356, 11 figs., 1940.

Colom, Guillermo, Los sedimentos burdigalienses de las Baleares. *Estudios Geol. Inst. Invest. Geol. Lucas Mallada,* no. 3, p. 21−112, 16 pls., 1946.

Colom, Guillermo, Jurassic-Cretaceous pelagic sediments of the western Mediterranean zone and the Atlantic area. *Micropaleontology,* v. 1, p. 109−124, pls. 1−5, text-figs. 1−4, 1955.

Colom, Guillermo, Essais sur la biologie, la distribution géographique et stratigraphique des Tintinnoídiens fossiles. *Eclog. Geol. Helv.,* v. 58, p. 319−334, pls. 1−3, text-figs. 1−3, 1965.

Conrad, W., Sur les Coccolithophoracées d'eau douce. *Arch. Protistenk.,* v. 63, p. 58−66, 9 figs., 1928.

Dangeard, Louis, Les craies et les calcaires à coccolithes de la Limagne. *Bull. Soc. Géol. Fr.,* sér. 5, v. 2, p. 67−82, pl. 6, 2 text-figs., 1932.

Dawson, G. M., Note on the occurrence of Foraminifera, coccoliths, &c., in the Cretaceous rocks of Manitoba. *Canadian Nat.,* ser. 2, v. 7, p. 252−257, 2 figs., 1874.

Dayman, Joseph, *Deep-sea soundings in the North Atlantic Ocean between Ireland and Newfoundland, made in H.M.S. Cyclops.* London: Eyre and Spottiswoode, 73 p., 4 pls., 1858.

Deflandre, Georges, Les Discoastéridés, microfossiles calcaires incertae sedis. *Bull. Soc. Fr. Microsc.,* v. 3, p. 59−67, text-figs. 1−31, 1934.

Deflandre, Georges, Les Flagellés. Aperçu biologique et paléontologique. Rôle géologique. *Actual. Scient. Ind.,* no. 335, 98 p., 135 text-figs. Paris: Hermann and Cie., 1936.

Deflandre, Georges, Les Stéphanolithes, représentants d'un type nouveau de coccolithes du Jurassique supérieur. *C. R. Hebd. Séanc. Acad. Sci., Paris,* v. 208, p. 1331−1333, text-figs. 1−14, 1939.

Deflandre, Georges, Coccolithophoridées fossiles d'Oranie. Genres *Scyphosphaera* Lohmann et *Thorosphaera* Ostenfeld. *Bull. Soc. Hist. Nat. Toulouse,* v. 77, p. 125−137, 36 figs., 1942a.

Deflandre, Georges, Possibilités morphogénétiques comparées du calcaire et de la silice à propos d'un nouveau type de microfossile calcaire de structure complexe, *Lithostromation perdurum* n.g. n.sp. *C. R. Hebd. Séanc. Acad. Sci., Paris,* v. 214, p. 917−919, 9 figs., 1942b.

Deflandre, Georges, *Braarudosphaera* nov. gen., type d'une famille nouvelle de Coccolithophoridés actuels à éléments composites. *C. R. Hebd. Séanc. Acad. Sci., Paris,* v. 225, p. 439−441, 5 figs., 1947.

Deflandre, Georges, Observations sur les Coccolithophoridés, à propos d'un nouveau type de Braarudosphaeridé, *Micrantholithus,* à éléments clastiques. *C. R. Hebd. Séanc. Acad. Sci., Paris,* v. 231, p. 1156−1158, 11 figs., 1950.

Deflandre, Georges, Classe des Coccolithophoridés, v. 1, p. 439−470, text-figs. 339−364bis. *In* Pierre-P. Grassé, *Traité de Zoologie.* Paris: Masson et Cie., 1952a.

Deflandre, Georges, Classe des Coccolithophoridés, v. 1, p. 107−115. *In* J. Piveteau, *Traité de Paléontologie.* Paris: Masson et Cie., 1952b.

Deflandre, Georges, Hétérogénéité intrinsèque et pluralité des éléments dans les coccolithes actuels et fossiles. *C. R. Hebd. Séanc. Acad. Sci., Paris,* v. 237, p. 1785−1787, 7 figs., 1953.

Deflandre, Georges, Sur les nannofossiles calcaires et leur systématique. *Revue Micropaléont.,* v. 2, p. 127−152, pls. 1−4, 1959.

Deflandre, Georges, Sur les Microrhabdulidés, famille nouvelle de nannofossiles calcaires. *C. R. Hebd. Séanc. Acad. Sci., Paris,* v. 256, p. 3484−3486, figs. 1−25, 1963.

Deflandre, Georges, Présence de nannofossiles calcaires (coccolithes et incertae sedis) dans le Silurodévonien d'Afrique du Nord. *C. R. Hebd. Séanc. Acad. Sci., Paris,* v. 270, p. 2916−2921, pls. 1−4, 1970.

Deflandre, Georges, and Marthe Deflandre-Rigaud, Présence de Nannoconidés dans le Crétacé superieur du Bassin parisien. *Revue Micropaléont.,* v. 2, p. 175−180, 1 pl., 1960.

Deflandre, Georges, and L. Durrieu, Application de la technique d'empreintes de carbone à la systématique des Coccolithophorides fossiles. *C. R. Hebd. Séanc. Acad. Sci., Paris,* v. 244, p. 2948–2951, 2 figs., 1957.

Deflandre, Georges, and Charles Fert, Sur la structure fine de quelques coccolithes fossiles observées au microscope électronique. Signification morphogénetique et application à la systématique: *C. R. Hebd. Séanc. Acad. Sci., Paris,* v. 234, p. 2100–2102, 8 figs. [corrections, p. 2500], 1952.

Deflandre, Georges, and Charles Fert, Étude des Coccolithophorides des vases actuelles au microscope électronique. Orientation des particules élémentaires de calcaire en rapport avec les notions d'Heliolithae et d'Ortholithae. *C. R. Hebd. Séanc. Acad. Sci., Paris,* v. 236, p. 328–330, 9 figs., 1953a.

Deflandre, Georges, and Charles Fert, Application du microscope électronique à l'étude des Coccolithophoridés. Technique et résultats liminaires. *Bull. Soc. Hist. Nat. Toulouse,* v. 88, p. 301–313, 4 pls., 1953b.

Deflandre, Georges, and Charles Fert, Observations sur les Coccolithophoridés actuels et fossiles en microscopie ordinaire et électronique. *Ann. Paléont.,* v. 40, p. 115–176, 15 pls., 127 text-figs., 1954.

Di Nocera, Silvio, and Paolo Scandone, Triassic nannoplankton limestones of deep basin origin in the central Mediterranean region. *Palaeogeogr. Palaeoclimat. Palaeoecol.,* v. 21, p. 101–111, pl. 1, figs. 1–4, 1977.

Dixon, H. H., On the structure of coccospheres and the origin of coccoliths. *Proc. R. Soc.,* v. 66, p. 305–315, pl. 3, 1900.

Dodge, J. D., *The fine structure of algal cells.* London, New York: Academic Press, 261 p., 1973.

Dorigan, J. L., and K. M. Wilbur, Calcification and its inhibition in coccolithophorids. *J. Phycol.,* v. 9, p. 450–456, figs. 1–5, 1973.

Dragesco, Jean, Étude cytologique de quelques Flagellés mésopsammiques. *Cah. Biol. Mar.,* v. 6, p. 83–115, pls. 1, 2, text-figs. 1–17, 1965.

Droop, M. R., Cobalamin requirement in Chrysophyceae. *Nature,* v. 174, p. 520–521, 1954.

Droop, M. R., Requirement for thiamine among some marine and supralittoral Protista. *J. Mar. Biol. Ass. U.K.,* v. 37, p. 323–329, 1 fig., 1958.

Dufour, Thierry, Quelques remarques sur les organismes *incertae sedis* de la famille des Calcisphaerulidae Bonet (1956). *C. R. Hebd. Séanc. Acad. Sci., Paris,* v. 266, ser. D, p. 1947–1949, 1 pl., 1968.

Edwards, A. R., A calcareous nannoplankton zonation of the New Zealand Paleogene, p. 381–419, 7 figs. *In* A. Farinacci, *Proc. II Planktonic Conf. Roma 1970,* v. 1, Rome: Edizioni Tecnoscienza, 1971.

Edwards, A. R., Key species of New Zealand calcareous nannofossils. *N. Z. Jl. Geol. Geophys.,* v. 16, p. 68–89, 88 figs., 1973a.

Edwards, A. R., Calcareous nannofossils from the southwest Pacific, Deep Sea Drilling Project, Leg 21. *Init. Repts. Deep Sea Drill. Proj.,* v. 21, p. 641–691, pls. 1–15, figs. 1–3, 1973b.

Edwards, A. R., and Katharina Perch-Nielsen, Calcareous nannofossils from the southern southwest Pacific, Deep Sea Drilling Project, Leg 29. *Init. Repts. Deep Sea Drill. Proj.,* v. 29, p. 469–539, pls. 1–21, fig. 1, 1975.

Ehrenberg, C. G., Bemerkungen über feste mikroskopische, anorganische Formen in den erdigen und derben Mineralien. *Ber. Dt. Akad. Wiss. Berl.,* p. 84–85, 1836a.

Ehrenberg, C. G., Ueber mikroskopische neue Charaktere der erdigen und derben Mineralien. *Annln. Phys.,* v. 39, p. 101–106, pl. 1, 1836b.

Ehrenberg, C. G., *Mikrogeologie. Das Erden und Felsen schaffende Wirken des unsichtbar kleinen selbstandigen Lebens auf der Erde.* Leipzig: Voss, 374 + 31 [index to names] +88 [index to atlas] p., 40 pls., 1854.

Ericson, D. B., Maurice Ewing and Goesta Wollin, Pliocene-Pleistocene boundary in deep-sea sediments. *Science,* v. 139, p. 727–737, 14 figs., 1963.

Farinacci, A., Round table on calcareous nannoplankton Roma, September 23–28, 1970, p. 1343–1369, 17 figs. *In* A. Farinacci, *Proc. II Planktonic Conf. Roma 1970,* v. 2. Rome: Edizoni Tecnoscienza, 1971.

Filipescu, M. G., and O. Dragastan, Asupra prezenţei unor depozite cu *Nannoconus* în sedimentele Jurasico-Cretacice din R. P. Romînă. *Studii Cerc. Geol.,* v. 8, p. 185–193, 1 pl., 1 text-fig., 1963.

Flügel, Erik, and H. E. Franz, Elektronenmikroskopischer Nachweis von Coccolithen im Solnhofener Plattenkalk (Ober-Jura). *Neues Jb. Geol. Paläont. Abh.,* v. 127, p. 245–263, pls. 24–26, 1 fig., 1967.

Forchheimer, Sylvia, Die Coccolithen des Gault-Cenoman, Cenoman und Turon in der Bohrung Höllviken I, Südwest-Schweden. *Årsb. Sver. Geol. Unders.,* 62, no. 6, ser. C, no. 635, p. 1–84, pls. 1–9, 23 figs., 1968.

Forchheimer, Sylvia, Scanning electron microscope studies of some Cenomanian coccospheres and coccoliths from Bornholm (Denmark) and Köpingsberg (Sweden). *Årsb. Sver. Geol. Unders.,* 64, no. 4, ser. C, no. 647, p. 1–43, 44 figs., 1970.

Forchheimer, Sylvia, Scanning electron microscope studies of Cretaceous coccoliths from the Köpingsberg Borehole No. 1, SE Sweden. *Årsb. Sver. Geol. Unders.,* 65, no. 14, ser. C, no. 668, p. 1–141, pls. 1–27, figs. 1–12, 1972.

Forchheimer, Sylvia, The stratigraphical distribution of Cretaceous coccoliths. *Årsb. Sver. Geol. Unders.*, 68, no. 3, ser. C, no. 696, p. 1–32, 1974.

Forchheimer, Sylvia, and Herbert Stradner, *Scampanella*, eine neue Gattung kretazischer Nannofossilien. *Verh. Geol. Bundesanst., Wien*, Jg. 1973, p. 285–289, pl. 1, 1973.

Ford, J. E., and S. H. Hutner, Role of vitamin B_{12} in the metabolism of microorganisms. *Vitam. Horm.*, v. 13, p. 101–136, 1 fig., 1955.

Foucher, Jean-Claude, and Josette Taugourdeau-Lantz, A propos de la suppression du genre *Pleurozonaria* O. Wetzel, 1933. *C. R. Hebd. Séanc. Acad. Sci., Paris*, v. 280, sér. D, p. 1677–1680, pls. 1, 2, 1975.

Fournier, R. O., Observations of particulate organic carbon in the Mediterranean Sea and their relevance to the deep-living coccolithophorid *Cyclococcolithus fragilis. Limnol. Oceanogr.*, v. 13, p. 693–697, 1968.

Franke, W. W., and R. M. Brown, Jr., Scale formation in Chrysophycean algae, III. Negatively stained scales of the coccolithophorid *Hymenomonas. Arch. Mikrobiol.*, v. 77, p. 12–19, figs. 1–6, 1971.

Fütterer, D., Kalkige Dinoflagellaten ("Calciodinelloideae") und die systematische Stellung der Thoracosphaeroideae. *Neues Jb. Geol. Paläont. Abh.*, v. 151, p. 119–141, figs. 1–25, 1976.

Gaarder, K. R., Coccolithineae, Silicoflagellatae, Pterospermataceae and other forms from the "Michael Sars" North Atlantic Deep-Sea Expedition 1910. *Rep. Scient. Results Michael Sars N. Atlant. Deep Sea Exped.*, v. 2, 20 p., 21 figs., 1954.

Gaarder, K. R., Electron microscope studies on holococcolithophorids. *Nytt Mag. Bot.*, v. 10, p. 35–51, 12 pls., 2 text-figs., 1962.

Gaarder, K. R., Comments on the distribution of coccolithophorids in the oceans, p. 97–103. *In* B. M. Funnell and W. R. Riedel, *The micropalaeontology of oceans.* Cambridge: University Press, 1971.

Gaarder, K. R., and G. R. Hasle, On the assumed symbiosis between diatoms and coccolithophorids in *Brenneckella. Nytt Mag. Bot.*, v. 9, p. 145–149, pls. 1, 2, 1962.

Gaarder, K. R., and G. R. Hasle, Coccolithophorids of the Gulf of Mexico. *Bull. Mar. Sci. Gulf Caribb.*, v. 21, p. 519–544, 12 figs., 1971.

Gaarder, K. R., and B. R. Heimdal, Light and scanning electron microscope observations on *Rhabdothorax regale* (Gaarder) Gaarder nov. comb. *Norw. J. Bot.*, v. 20, p. 89–97, figs. 1–19, 1973.

Gaarder, K. R., and B. R. Heimdal, A revision of the genus *Syracosphaera* Lohmann (Coccolithineae). *"Meteor" Forsch.-Ergebn.*, Reihe D., no. 24, p. 54–71, pls. 1–8, 1977.

Gaarder, K. R., and J. Markali, On the coccolithophorid *Crystallolithus hyalinus* n. gen., n.sp. *Nytt Mag. Bot.*, v. 5, p. 1–5, 1 pl., 1 text-fig. ["1957," separate 1956].

Gaarder, K. R., J. Markali, and E. Ramsfjell, Further observations on the coccolithophorid *Calciopappas caudatus. Avh. Norske VidenskAkad. Oslo*, 1954, no. 1, 9 p., 4 pls., 2 text-figs., 1954.

Gallois, R. W., Coccolith blooms in the Kimmeridge Clay and origin of North Sea oil. *Nature*, v. 259, p. 473–475, figs. 1, 2, 1976.

Gardet, Monique, Contribution à l'étude des coccolithes des terrains Neogènes de l'Algérie. *Publs. Serv. Carte Géol. Algér.*, ser. 2, Bull. 5, p. 477–550, 11 pls., 1 text-fig., 1955.

Gartner, Stefan, Jr., Coccoliths and related calcareous nannofossils from Upper Cretaceous deposits of Texas and Arkansas. *Paleont. Contr. Univ. Kans.*, Art. 48 (Protista Art. 1), p. 1–56, pls. 1–28, figs. 1–5, 1968.

Gartner, Stefan, Jr., Correlation of Neogene planktonic foraminifera and calcareous nannofossil zones. *Trans. Gulf-Cst Ass. Geol. Socs.*, v. 19, p. 585–599, pls. 1, 2, figs. 1–7, 1969.

Gartner, Stefan, Jr., Phylogenetic lineages in the Lower Tertiary coccolith genus *Chiasmolithus. Proc. N. Am. Paleont. Conv. Chicago 1969*, Part G, p. 930–957, figs. 1–17, 1970.

Gartner, Stefan, Jr., Calcareous nannofossils from the JOIDES Blake Plateau Cores and revision of Paleogene nannofossil zonation. *Tulane Stud. Geol.*, v. 8, p. 101–121, pls. 1–5, figs. 1–5, 1971.

Gartner, Stefan, Jr., Late Pleistocene calcareous nannofossils in the Caribbean and their interoceanic correlation. *Palaeogeogr. Palaeoclimat. Palaeoecol.*, v. 12, p. 169–191, pls. 1, 2, figs. 1–7, 1972.

Gartner, Stefan, Jr., Absolute chronology of the Late Neogene calcareous nannofossil succession in the equatorial Pacific. *Bull. Geol. Soc. Am.*, v. 84, p. 2021–2033, figs. 1–6, 1973.

Gartner, Stefan, Jr., Calcareous nannofossil biostratigraphy and revised zonation of the Pleistocene. *Mar. Micropaleont.*, v. 2, p. 1–25, pls. 1–4, figs. 1–5, 1977.

Gartner, Stefan, Jr., and David Bukry, Tertiary holococcoliths. *J. Paleont.*, v. 43, p. 1213–1221, pls. 139–142, 1969.

Gartner, Stefan, Jr., and David Bukry, Morphology and phylogeny of the coccolithophycean family Ceratolithaceae. *J. Res. U.S. Geol. Surv.*, v. 3, p. 451–465, figs. 1–8, 1975.

Gartner, Stefan, Jr., and Richard Gentile, Problematic Pennsylvanian coccoliths from Missouri. *Micropaleontology*, v. 18 (1972), p. 401–404, pl. 1, 1973.

Gayral, P., and C. Billard, Chrysophycées et Haptophycées des côtes françaises: mise au point systématique et nouvelles observations sur *Ruttnera chadefaudii* Bourrelly et Magne (Haptophycées). *Bull. Soc. Phycol. Fr.*, no. 22, p. 135−149, 20 figs., 1977.

Gayral, P., and J. Fresnel-Morange, Results preliminaires sur la structure et la biologie de la Coccolithacée *Ochrosphaera neapolitana* Schussnig. *C. R. Hebd. Séanc. Acad. Sci., Paris,* sér. D, v. 273, p. 1683−1686, pls. 1, 2, 1971.

Gayral, P., and J. Fresnel-Morange, Nouvelles observations sur deux Coccolithophoracées marines: *Cricosphaera roscoffensis* (P. Dangeard) comb. nov. et *Hymenomonas globosa* (F. Magne) comb. nov. *Phycologia*, v. 15, p. 339−355, 38 figs., 1976.

Gayral, P., C. Haas, and H. Lepailleur, Alternance morphologique de génerations et alternance de phases chez les Chrysophycées. *Mém. Soc. Bot. Fr.*, 1972, p. 215−230, pls. 1−3, 1972.

Górka, Hanna, Coccolithophoridae z górnego mastrychtu Polski środkowej [Les Coccolithophoridés du Maestrichtien supérieur de Pologne]. *Acta Paleont. Pol.,* v. 2, p. 235−284, 5 pls., 1957.

Górka, Hanna, Coccolithophoridés, Dinoflagelles, Hystrichosphaeridés et microfossiles incertae sedis du Crétacé supérieur de Pologne. *Acta Palaeont. Pol.,* v. 8, p. 3−90, 11 pls., 8 text-pls., 2 figs. on suppl., 1963.

Gran, H. H., and Trygve Braarud, A qualitative study of the phytoplankton in the Bay of Fundy and the Gulf of Maine (including observations on hydrography, chemistry and turbidity). *J. Biol. Bd. Can.*, v. 1, p. 279−467, 69 figs., 1935.

Green, J. C., Notes on the flagellar apparatus and taxonomy of *Pavlova mesolychnon* van der Veer, and on the status of *Pavlova* Butcher and related genera within the Haptophyceae. *J. Mar. Biol. Ass. U.K.*, v. 56, p. 595−602, pls. 1, 2, 1976.

Green, J. C., and Irene Manton, Studies in the fine structure and taxonomy of flagellates in the genus *Pavlova* I. A revision of *Pavlova gyrans*, the type species. *J. Mar. Biol. Ass. U.K.*, v. 50, p. 1113−1130, pls. 1−6, fig. 1, 1970.

Green, J. C., and Mary Parke, A reinvestigation by light and electron microscopy of *Ruttnera spectabilis* Geitler (Haptophyceae), with special reference to the fine structure of the zoids. *J. Mar. Biol. Ass. U.K.*, v. 54, p. 539−550, pls. 1−4, figs. 1, 2, 1974.

Green, J. C., and Mary Parke, New observations upon members of the genus *Chrysotila* Anand, with remarks upon their relationships within the Haptophyceae. *J. Mar. Biol. Ass. U.K.*, v. 55, p. 109−121, pls. 1−3, 1975.

Green, J. C., and R. N. Pienaar, The taxonomy of the order Isochrysidales (Prymnesiophyceae) with special reference to the genera *Isochrysis* Parke, *Dicrateria* Parke and *Imantonia* Reynolds. *J. Mar. Biol. Ass. U.K.*, v. 57, p. 7−17, pls. 1−6, 1977.

Grün, Walter, and Franz Allemann, The Lower Cretaceous of Caravaca (Spain); Berriasian calcareous nannoplankton of the Miravetes section, Subbetic Zone, Prov. of Murcia. *Eclog. Geol. Helv.,* v. 68, p. 147−211, pls. 1−10, figs. 1−35, 1975.

Grün, Walter, Ben Prins, and Fred Zweili, Coccolithophoriden aus dem Lias Epsilon von Holzmaden (Deutschland). Coccolithophorids from the Lias Epsilon, Holzmaden (S-Germany). *Neues Jb. Geol. Paläont. Abh.,* v. 147, p. 294−328, figs. 1−22, 1974.

Gümbel, C. W., Coccolithen in Eocänmergel. *Neues Jb. Miner. Geol. Paläont.,* p. 299−304, 1873.

Halldal, P., Comparative observations on coccolithophorids in light and electron microscopes and their taxonomical significance. *Rapp. et Commun. VIIIe Int. Bot. Congr.*, v. 17, p. 122−124, 1954.

Halldal, P., and J. Markali, Morphology and microstructure of coccoliths studied in the electron microscope. Observations on *Anthosphaera robusta* and *Calyptrosphaera papillifera*. *Nytt Mag. Bot.*, v. 2, p. 117−119, 2 pls., 1954a.

Halldal, P., and J. Markali, Observations on coccoliths of *Syracosphaera mediterranea* Lohm., *S. pulchra* Lohm., and *S. molischi* Schill. in the electron microscope. *J. Cons. Perm. Int. Explor. Mer,* v. 19, p. 329−336, 6 figs., 1954b.

Halldal, P., and J. Markali, Electron microscope studies on coccolithophorids from the Norwegian Sea, the Gulf Stream and the Mediterranean. *Avh. Norske VidenskAkad. Oslo,* 1955, no. 1, 30 p., 27 pls., 1955.

Haq, Bilal ul, Electron microscope studies on some Upper Eocene calcareous nannoplankton from Syria. *Stockh. Contr. Geol.,* v. 15, p. 23−37, pls. 1−6, 1 fig., 1966.

Haq, Bilal ul, Calcareous nannoplankton from the Lower Eocene of the Zinda Pir, District Dera Ghazi Khan, West Pakistan. *Geol. Bull. Panjab Univ.,* no. 6, p. 55−83, pls. 1−8, 1967.

Haq, Bilal ul, Studies on Upper Eocene calcareous nannoplankton from NW Germany. *Stockh. Contr. Geol.,* v. 18, p. 13−74, pls. 1−11, 3 figs., 1968.

Haq, Bilal ul, The structure of Eocene coccoliths and discoasters from a Tertiary deep-sea core in the central Pacific. *Stockh. Contr. Geol.,* v. 21, p. 1−19, pls. 1−5, 4 figs., 1969.

Haq, Bilal ul, Paleogene calcareous nannoflora Part I: the Paleocene of west-central Persia and the Upper Paleocene-Eocene of West Pakistan. *Stockh. Contr. Geol.,* v. 25, p. 1−56, pls. 1−14, 2 figs., 1971a.

Haq, Bilal ul, Paleogene calcareous nannoflora Part II: Oligocene of western Germany. *Stockh. Contr. Geol.,* v. 25, p. 57−97, pls. 1−18, fig. 1, 1971b.

Haq, Bilal ul, Paleogene calcareous nannoflora Part III: Oligocene of Syria. *Stockh. Contr. Geol.,* v. 25, p. 99−127, pls. 1−25, 1 fig., 1971c.

Haq, Bilal ul, Paleogene calcareous nannoflora Part IV: Paleogene nannoplankton biostratigraphy and evolutionary rates in Cenozoic calcareous nanno-plankton. *Stockh. Contr. Geol.,* v. 25, p. 129−158, figs. 1−15, 1971d.

Haq, Bilal ul, Evolutionary trends in the Cenozoic coc-colithophore genus *Helicopontosphaera. Micro-paleontology,* v. 19, p. 32−52, pls. 1−7, figs. 1, 2, 1973a.

Haq, Bilal ul, Transgressions, climatic change and the diversity of calcareous nannoplankton. *Mar. Geol.,* v. 15, p. M25−M30, figs. 1−4, 1973b.

Haq, Bilal ul, and J. H. Lipps, Calcareous nanno-plankton. *Init. Repts. Deep Sea Drill Proj.,* v. 8, p. 777−789, pls. 1−6, 1 fig., 1971.

Haq, Bilal ul, and G. P. Lohmann, Early Cenozoic cal-careous nannoplankton biogeography of the Atlantic Ocean. *Mar. Micropaleont.,* v. 1, p. 119−194, pls. 1−14, figs. 1−19, 1976.

Haq, Bilal ul, Isabella Premoli-Silva, and G. P. Lohmann, Calcareous plankton paleobiogeographic evidence for major climatic fluctuations in the early Cenozoic Atlantic Ocean. *J. Geophys. Res.,* v. 82, p. 3861−3876, figs. 1−12, 1977.

Hasle, G. R., Plankton coccolithophorids from the Subantarctic and Equatorial Pacific. *Nytt Mag. Bot.,* v. 8, p. 77−88, pls. 1−3, 1960.

Hattin, D. E., Petrology and origin of fecal pellets in Upper Cretaceous strata of Kansas and Saskatche-wan. *J. Sedim. Petrol.,* v. 45, p. 686−696, figs. 1−19, 1975.

Hattin, D. E., and D. A. Darko, Technique for deter-mining coccolith abundance in shaly chalk of Greenhorn Limestone (Upper Cretaceous) of Kansas. *Bull. Kans. Geol. Surv.,* 202, pt. 2, p. 1−11, 5 figs., 1971.

Hay, W. W., Utilization stratigraphique des Discoas-téridés pour la zonation du Paléocène et de l'Éocène inférieur. *In* Colloque sur le Paléogène (Bordeaux, Septembre 1962).*Mém. Bur. Rech. Geol. et Min.* [Paris], no. 28, C.r. 2, p. 885−889, 1964.

Hay, W. W., Calcareous nannofossils, p. 1055−1200, pls. 1−31. *In* A. T. S. Ramsay, *Oceanic Micropalaeon-tology,* v. 2. London, New York, San Francisco: Aca-demic Press, 1977.

Hay, W. W., and H. P. Mohler, Calcareous nanno-plankton from early Tertiary rocks at Pont Labau, France, and Paleocene−Early Eocene correlations. *J. Paleont.,* v. 41, p. 1505−1541, pls. 196−206, 5 figs., 1967.

Hay, W. W., H. P. Mohler, P. H. Roth, R. R. Schmidt, and J. E. Boudreaux, Calcareous nannoplankton zo-nation of the Cenozoic of the Gulf Coast and Caribbean-Antillean area, and trans-oceanic correla-tion. *Trans. Gulf-Cst. Ass. Geol. Socs.,* v. 17, p. 428−480, pls. 1−13, 1967.

Hay, W. W., H. P. Mohler, and M. E. Wade, Calcareous nannofossils from Nal'chik (northwest Caucasus). *Ec-log. Geol. Helv.,* v. 59, p. 379−399, pls. 1−13, 1966.

Hay, W. W., and K. M. Towe, Electron-microscope studies of *Braarudosphaera bigelowi* and some related coccolithophorids. *Science,* v. 137, p. 426−428, 6 figs., 1962a.

Hay, W. W., and K. M. Towe, Electronmicroscopic ex-amination of some coccoliths from Donzacq (France). *Eclog. Geol. Helv.,* v. 55, p. 497−517, 10 pls. 2 text-figs, 1962b.

Hay, W. W., and K. M. Towe, *Microrhabdulus belgicus,* a new species of nannofossil. *Micropaleontology,* v. 9, p. 95−96, 1 pl., 1963.

Hays, J. D., and W. A. Berggren, Quaternary bound-aries and correlations, p. 669−691, 9 figs. *In* B. M. Funnell and W. R. Riedel, *The micropalaeontology of oceans.* Cambridge: University Press, 1971.

Heck, S. E. van, Bibliography and taxa of calcareous nannoplankton. *INA Newsletter* (Proc. Int. Nanno-plankton Ass.), v. 1, no. 1, p. AB1−AB5, A1−A12, B1−B27, 1979.

Hibberd, D. J., The ultrastructure and taxonomy of the Chrysophyceae and Prymnesiophyceae (Hap-tophyceae): a survey with some new observations on the ultrastructure of the Chrysophyceae. *Bot. J. Linn. Soc., Lond.,* v. 72, p. 55−80, pls. 1−4, figs. 1, 2, 1976.

Hill, M. E. III, Selective dissolution of Mid-Cretaceous (Cenomanian) calcareous nannofossils. *Micropaleon-tology,* v. 21, p. 227−235, pls. 1, 2, fig. 1, 1975.

Hill, M. E. III, Lower Cretaceous calcareous nannofos-sils from Texas and Oklahoma. *Palaeontographica Abt. B,* v. 156, p. 103−179, pls. 1−15, figs. 1−5, 1976.

Hofeneder, Heinrich, Die Coccolithophoriden. *Mikro-kosmos,* v. 19, p. 25−31, 52−54, 28 figs., 1925.

Hoffmann, Norbert, Coccolithineen aus der weissen Schreibkreide (Unter-Maastricht) von Jasmund auf Rügen. *Geologie,* v. 19, p. 846−879, pls. 1−7, 4 figs., 1970.

Honjo, Susumu, Coccoliths: production, transporta-tion and sedimentation. *Mar. Micropaleont.,* v. 1, p. 65−79, pls. 1, 2, figs. 1, 2, 1976.

Honjo, Susumu, and Hisatake Okada, Community structure of coccolithophores in the photic layer of the mid-Pacific.*Micropaleontology,* v. 20, p. 209−230, figs. 1−17, 1974.

Huxley, T. H., On some organisms living at great depths in the North Atlantic Ocean. *Q. Jl. Microsc. Sci.,* ser. 2, v. 8, p. 203−212, pl. 4, 1868.

Huxley, T. H., [*Bathybius,* coccoliths and cocco-spheres. Discussion on Captain S. Osborn's paper "On the geography of the bed of the Atlantic and Indian oceans and the Mediterranean Sea"]. *Proc. R. Geogr. Soc.,* v. 15, p. 37−39, 1871.

Isenberg, H. D., and L. S. Lavine, Comparative biology of mineral deposition. *In* J. P. Adams (ed.), *Current practice in orthopaedic surgery.* St. Louis, Mo.: C. V. Mosby Co., p. 177–219, figs. 94–126, 1964.

Isenberg, H. D., L. S. Lavine, C. Mandell, and H. Weissfellner, Qualitative chemical composition of the calcifying organic matrix obtained from cell-free coccoliths. *Nature,* v. 206, p. 1153–1154, 1965.

Isenberg, H. D., L. S. Lavine, M. L. Moss, D. Kupferstein, and P. E. Lear, Calcification in a marine coccolithophorid. *Ann. N.Y. Acad. Sci.,* v. 109, p. 49–64, 13 figs., 1963.

Isenberg, H. D., L. S. Lavine, M. L. Moss, M. H. Shamos, and H. Weissfellner, Calcium^{45} turnover in a mineralizing coccolithophorid protozoan. *J. Protozool.,* v. 11, p. 531–534, 3 figs., 1964.

Isenberg, H. D., L. S. Lavine, and H. Weissfellner, The suppression of mineralization in a coccolithophorid by an inhibitor of carbonic anhydrase. *J. Protozool.,* v. 10, p. 477–479, 1 fig., 1963.

Isenberg, H. D., L. S. Lavine, H. Weissfellner, and A. Spotnitz, The influence of age and heterotrophic nutrition on calcium deposition in a marine coccolithophorid protozoon. *Trans. N.Y. Acad. Sci.,* ser. 2, v. 27, p. 530–544, 9 figs., 2 tables, 1965.

Jafar, S. A., Calcareous nannoplankton from the Miocene of Rotti, Indonesia. *Verh. K. Ned. Akad. Wet., Afd. Natuurk., Eerste Reeks,* v. 28, p. 1–99, pls. 1–15, figs. 1, 2, 1975a.

Jafar, S. A., Some comments on the calcareous nannoplankton genus *Scyphosphaera* and the neotypes of *Scyphosphaera* species from Rotti, Indonesia. *Senckenberg. Leth.,* v. 56, p. 365–379, pls. 1–3, 1975b.

Jafar, S. A., and Erlend Martini, On the validity of the calcareous nannoplankton genus *Helicosphaera. Senckenberg. Leth.,* v. 56, p. 381–397, pl. 1, figs. 1, 2, 1975.

Jeffrey, S. W., and M. B. Allen, Pigments, growth and photosynthesis in cultures of two chrysomonads, *Coccolithus huxleyi* and a *Hymenomonas* sp. *J. Gen. Microbiol.,* v. 36, p. 277–288, 7 figs., 1964.

Jírovec, O., K. Wenig, B. Fott, E. Bartoš, J. Weiser, and R. Šrámek-Hušek, *Protozoologie.* Praha: Naklad. Česk. Akad. Věd., 643 p., 289 figs. in text and pls., 1953.

Kamptner, Erwin, Beitrag zur Kenntnis adriatischer Coccolithophoriden. *Arch. Protistenk.,* v. 58, p. 173–184, 6 figs., 1927.

Kamptner, Erwin, Über eine Coccolithophoride aus der "Alten Donau" bei Wien, nebst einigen systematischen Bermerkungen. *Arch. Protistenk.,* v. 61, p. 38–44, 2 figs., 1928a.

Kamptner, Erwin, Über das System und die Phylogenie der Kalkflagellaten. *Arch. Protistenk.,* v. 64, p. 19–43, 1928b.

Kamptner, Erwin, Die Kalkflagellaten des Süsswassers und ihre Beziehungen zu jenen des Brackwassers und des Meeres. *Int. Revue Ges. Hydrobiol. Hydrogr.,* v. 24, p. 147–163, 7 figs., 1930.

Kamptner, Erwin, *Nannoconus steinmanni* nov. gen., nov. spec., ein merkwürdiges gesteinsbildendes Mikrofossil aus dem jüngeren Mesozoikum der Alpen. *Paläont. Z.,* v. 13, p. 288–298, 3 figs., 1931.

Kamptner, Erwin, Über Dauersporen bei marinen Coccolithineen. *Sber. Akad. Wiss. Wien, Math.-Naturw. Kl.,* Abt. I, v. 146, p. 67–76, 1 pl., 2 text-figs., 1937a.

Kamptner, Erwin, Neue und bermerkenswerte Coccolithineen aus dem Mittelmeer. *Arch. Protistenk.,* v. 89, p. 279–316, pls. 14–17, 1937b.

Kamptner, Erwin, Einige Bermerkungen über *Nannoconus. Paläont. Z.,* v. 20, p. 249–257, 5 figs., 1938.

Kamptner, Erwin, Die Coccolithineen der Südwestküste von Istrien. *Annln. Naturh. Mus. Wien,* v. 51, p. 54–149, 15 pls., 1941.

Kamptner, Erwin, Coccolithen aus dem Torton des Inneralpinen Wiener Beckens. *Sber. Öst. Akad. Wiss., Math.-Naturw. Kl.,* Abt. I, v. 157, p. 1–16, 2 pls., 1948.

Kamptner, Erwin, Über den submikroskopischen Aufbau der Coccolithen. *Anz. Öst. Akad. Wiss., Math.-Naturw. Kl.,* v. 87, p. 152–158, 1950.

Kamptner, Erwin, Das mikroskopische Studium des Skelettes der Coccolithineen (Kalkflagellaten). Übersicht der Methoden und Ergebnisse. I. Die Gestalt des Gehäuses und seiner Bauelemente. *Mikroskopie,* v. 7, p. 232–244, figs. 1–16. II. Der Feinbau der Coccolithen. *Mikroskopie,* v. 7, p. 375–386, figs. 17–27, 1952.

Kamptner, Erwin, Untersuchungen über den Feinbau der Coccolithen. *Arch. Protistenk.,* v. 100, p. 1–90, 50 figs., 1954.

Kamptner, Erwin, Fossile Coccolithineen-Skelettreste aus Insulinde. Eine mikropaläontologische Untersuchung. *Verh. K. Ned. Akad. Wet., Afd. Natuurk.,* ser. 2, v. 50, no. 2, 105 p., 9 pls., 1955.

Kamptner, Erwin, Morphologische Betrachtungen über Skelettelemente der Coccolithineen. *Öst. Bot. Z.,* v. 103, p. 142–163, 18 figs., 1956a.

Kamptner, Erwin, Das Kalkskelett von *Coccolithus Huxleyi* (Lohm.) Kpt. und *Gephyrocapsa oceanica* Kpt. (Coccolithineae). *Arch. Protistenk.,* v. 101, p. 171–202, pl. 16, 1956b.

Kamptner, Erwin, Betrachtungen zur Systematik der Kalkflagellaten, nebst Versuch einer neuen Gruppierung der Chrysomonadales. *Arch. Protistenk.,* v. 103, p. 54–116, 1958.

Kamptner, Erwin, Coccolithineen-Skelettreste aus Tiefseeablagerungen des Pazifischen Ozeans. *Annln. Naturh. Mus. Wien,* v. 66, p. 139–204, 9 pls., 39 text-figs., 1963.

Keupp, Helmut, Kalkiges Nannoplankton aus den Solnhofener Schichten (Unter-Tithon, südliche Frankenalb). Calcareous nannoplankton from the Solnhofen Limestones (L. Tithonian, Bavaria). *Neues Jb. Geol. Paläont. Mh.,* Heft 6, p. 361–381, 31 figs., 1976.

Keupp, Helmut, Calcisphaeren des Untertithon der Südlichen Frankenalb und die systematische Stellung von *Pithonella* Lorenz 1901. *Neues Jb. Geol. Palaont. Mh.,* Heft 2, p. 87–98, 14 figs., 1978.

Klaveness, Dag, *Coccolithus huxleyi* (Lohm.) Kamptn. II The flagellate cell, aberrant cell types, vegetative propagation and life cycles. *Br. Phycol. J.,* v. 7, p. 309–318, figs. 1–17, 1972.

Klaveness, Dag, *Coccolithus huxleyi* (Lohmann) Kamptner. I. Morphological investigations in the vegetative cell and the process of coccolith formation. *Protistologica,* v. 8 (1972), p. 335–346, figs. 1–17, 1973a.

Klaveness, Dag, The microanatomy of *Calyptrosphaera sphaeroidea,* with supplementary observations on the motile stage of *Coccolithus pelagicus. Norw. J. Bot.,* v. 20, p. 151–162, figs. 1–21, 1973b.

Klaveness, Dag, *Emiliania huxleyi* (Lohmann) Hay & Mohler. III. Mineral deposition and the origin of the matrix during coccolith formation. *Protistologica,* v. 12, p. 217–224, figs. 1–12, 1976.

Klaveness, Dag, and Eyestein Paasche, Two different *Coccolithus huxleyi* cell types incapable of coccolith formation. *Arch. Mikrobiol.,* v. 75, p. 382–385, 2 figs., 1971.

Klumpp, Barbara, Beitrag zur Kenntnis der Mikrofossilien des Mittleren und Oberen Eozän. *Palaeontographica Abt. A,* v. 103, p. 337–406, pls. 16–20, 5 textfigs., 1953.

Knauer, József, *Calcisphaerula, Pithonella* és *Stomiosphaera* a bakonyi középsökrétából [*Calcisphaerula, Pithonella* and *Stomiosphaera* from Middle Cretaceous beds of the Bakony Mountains]. *Földt. Közl.,* v. 100, p. 88–90, pl. 1, fig. 1, 1970.

Kreger, D. R., and J. van der Veer, Paramylon in a chrysophyte. *Acta Bot. Neerl.,* v. 19, p. 401–402, 1970.

Lavine, L. S., H. D. Isenberg, and M. L. Moss, Intracellular calcification in a coccolithophorid. *Nature,* v. 196, p. 78, 1962.

Leadbeater, B. S. C., Preliminary observations on differences of scale morphology at various stages in the life cycle of "*Apistonema-Syracosphaera*" sensu von Stosch. *Br. Phycol. J.,* v. 5, p. 57–69, figs. 1–27, 1970.

Leadbeater, B. S. C., Observations by means of ciné photography on the behavior of the haptonema in plankton flagellates of the class Haptophyceae. *J. Mar. Biol. Ass. U.K.,* v. 51, p. 207–217, pls. 1–4, figs. 1–4, 1971.

Leadbeater, B. S. C., Ultrastructural observations on nannoplankton collected from the coast of Jugoslavia and the Bay of Algiers. *J. Mar. Biol. Ass. U.K.,* v. 54, p. 179–196, pls. 1–7, fig. 1, 1974.

Lebour, M. V., *Coccolithophora pelagica* (Wallich) from the Channel. *J. Mar. Biol. Ass. U.K.,* v. 13, p. 271–275, 1 fig., 1923.

Lecal [Lecal-Schlauder], Juliette, Sur un *Coccolithus* n.sp. épiphyte d'une Bacillariale, *Coscinodiscus* n.sp. *Bull. Soc. Hist. Nat. Afr. N.,* v. 39, p. 15–21, 6 figs., 1949.

Lecal [Lecal-Schlauder], Juliette, Notes préliminaires sur les Coccolithophorides d'Afrique du Nord. *Bull. Soc. Hist. Nat. Afr. N.,* v. 40, p. 160–167, pl. 6, 6 textfigs., ["1949"], 1950.

Lecal [Lecal-Schlauder], Juliette, Recherches morphologiques et biologiques sur les Coccolithophorides nord-africains. *Annls. Inst. Océanogr., Monaco,* ser. 2, v. 26, p. 255–362, pls. 9–13, 47 text-figs., 1951.

Lecal [Lecal-Schlauder], Juliette, Sur une Protiste pélagique rattachable aux Discoastéridés. *Archs Zool. Exp. Gén.,* v. 89 (Notes et Revue), p. 51–55, 2 figs., 1952.

Lecal [Lecal-Schlauder], Juliette, Un nouvel *Hymenomonas: H. prenanti* n.sp. [Coccolithophoridés]. *Ann. Limnol.,* v. 1, p. 155–162, figs. 1–5, 1965.

Lecal, Juliette, and A. Bernheim, Microstructure du squelette de quelques Coccolithophorides. *Bull. Soc. Hist. Nat. Afr. N.,* v. 51, p. 273–297, 22 pls., 1960.

Lecal, Juliette, and A. Bernheim, Structure et biologie de quelques Coccolithophoridés après observations au microscope électronique. *Bull. Soc. Hist. Nat. Toulouse,* v. 99, p. 450–458, 4 pls., 1964.

Lefort, Françoise, Sur l'appartenance à une seule et méme espèce de deux coccolithophoracées, *Cricosphaera carterae* et *Ochrosphaera verrucosa. C. R. Hebd. Séanc. Acad. Sci., Paris,* v. 272, sér. D, p. 2540–2543, pl. 1, figs. 1–5, 1971.

Lefort, Françoise, Quelques caractères morphologiques de deux espèces actuelles de *Braarudosphaera* (Chrysophycées, Coccolithophoracées). *Botaniste,* v. 55, p. 81–93, 1972.

Lefort, Françoise, Étude du quelques Coccolithophoracées marines rapportées aux genres *Hymenomonas* et *Ochrosphaera. Cah. Biol. Mar.,* v. 16, p. 213–229, pls. 1–4, fig. 1, 1975.

Lemmermann, E., Flagellatae, Chlorophyceae, Coccosphaerales und Silicoflagellatae, p. 1–40, text-figs. 1–135. *In* K. Brandt and C. Apstein, *Nordisches Plankton, Botanischer Teil.* Kiel and Leipzig: Lipsius and Tischer, 1908.

Lepailleur, Henri, Sur un nouveau genre de Chrysophycées: *Exanthemachrysis* nov. gen. (*E. Gayralii* nov. sp.). *C. R. Hebd. Séanc. Acad. Sci., Paris,* v. 270, sér. D, p. 928–931, 4 figs., 1970.

Lewin, R. A., Phytoflagellates and algae, p. 401–417. *In* W. Ruhland (ed.), *Handbuch der Pflanzen-*

physiologie, Encyclopedia of Plant Physiology. Berlin: Springer-Verlag, 1961.

Lewin, R. A., and T. J. Chow, La enpreno de strontio en kokolitoforoj. *Pl. Cell. Physiol. Tokyo,* v. 2, p. 203–208, 4 figs., 1961.

Lipps, J. H., *Triquetrorhabdulus* and similar calcareous nannoplankton. *J. Paleont.,* v. 43, p. 1029–1032, pl. 126, 1969.

Locker, Sigurd, Neue stratigraphisch wichtige Coccolithophoriden (Flagellata) aus dem norddeutschen Alttertiär. *Mber. Dt. Akad. Wiss. Berl.,* v. 9, p. 758–768, pls. 1, 2, 1967.

Locker, Sigurd, Coccolithineen aus dem Paläogen Mitteleuropas. *Paläont. Abh., Abt. B, Paläobot.,* v. 3, p. 735–853, pls. 1–17, figs. 1, 2, 1972.

Loeblich, A. R., Jr., and Helen Tappan, Type fixation and validation of certain calcareous nannoplankton genera. *Proc. Biol. Soc. Wash.,* v. 76, p. 191–196, 1963.

Loeblich, A. R., Jr., and Helen Tappan, Annotated index and bibliography of the calcareous nannoplankton. *Phycologia,* v. 5, p. 81–216, 1966.

Loeblich, A. R., Jr., and Helen Tappan, Annotated index and bibliography of the calcareous nannoplankton II. *J. Paleont.,* v. 42, p. 584–598, 1968.

Loeblich, A. R., Jr., and Helen Tappan, Annotated index and bibliography of the calcareous nannoplankton III. *J. Paleont.,* v. 43, p. 568–588, 1969.

Loeblich, A. R., Jr., and Helen Tappan, Annotated index and bibliography of the calcareous nannoplankton IV. *J. Paleont.,* v. 44, p. 558–574, 1970a.

Loeblich, A. R., Jr., and Helen Tappan, Annotated index and bibliography of the calcareous nannoplankton V. *Phycologia,* v. 9, p. 157–174, 1970b.

Loeblich, A. R., Jr., and Helen Tappan, Annotated index and bibliography of the calcareous nannoplankton VI. *Phycologia,* v. 10, p. 309–339, 1972.

Loeblich, A. R., Jr., and Helen Tappan, Annotated index and bibliography of the calcareous nannoplankton VII. J. Paleont., v. 47, p. 714–759, 1973.

Loeblich, A. R. III, and L. A. Loeblich, Division Haptophyta, p. 451–467, 1 fig. *In* A. I. Laskin and H. A. Lechevalier, *Handbook of Microbiology,* 2nd ed., v. 2, *Fungi, Algae, Protozoa, and Viruses.* West Palm Beach, Florida: CRC Press, Inc., (1978) 1979.

Lohmann, H., Die Coccolithophoridae, eine Monographie der Coccolithen bildenden Flagellaten, zugleich ein Beitrag zur Kenntnis des Mittelmeerauftriebs. *Arch. Protistenk.,* v. 1, p. 89–165, pls. 4–6, 1902.

Lohmann, H., Die Gehäuse und Gallertblasen der Appendicularien und ihre Bedeutung fur die Erforschung des Lebens im Meer. *Verh. Dt. Zool. Ges.,* v. 19, p. 200–239, 6 figs., 1909.

Lohmann, H., Untersuchungen über das Pflanzen- und Tierleben der Hochsee. Zugleich ein Bericht über die biologischen Arbeiten auf der Fahrt der *"Deutschland"* von Bremerhaven nach Buenos Aires in der Zeit vom 7. Mai bis 7. September 1911. *Veröff. Inst. Meeresk., Univ. Berl., N.F., Geogr.-Naturw.,* Reihe 1, viii + 92 p., 2 pls., 14 text-figs., 1912.

Lohmann, H., Beiträge zur Charakterisierung des Tier-und Pflanzenlebens in den von der *"Deutschland"* wahrend ihrer Fahrt nach Buenos Ayres durchfahrenen Gebieten des Atlantischen Ozeans. II Teil. (Schluss). *Int. Revue Ges. Hydrobiol. Hydrogr.,* v. 5, p. 343–372, figs. 7–17, 1913a.

Lohmann, H., Über Coccolithophoriden. *Verh. Dt. Zool. Ges.,* v. 23, p. 143–164, 19 figs., 1913b.

Lohmann, H., Die Bevolkerung des Ozeans mit Plankton nach den Ergebnissen der Zentrifugenfänge während der Ausreise der *"Deutschland"* 1911. *Arch. Biontol.,* v. 4, no. 3, 617 p., 16 pls., 113 text-figs., 1919.

McClung, C. E., Microscopic organisms of the Upper Cretaceous. *Bull. Kans. Univ. Geol. Surv.,* v. 4, p. 413–429, pl. 85, 1898.

McIntyre, Andrew, and A. W. H. Bé, Modern Coccolithophoridae of the Atlantic Ocean I. Placoliths and cyrtoliths. *Deep Sea Res.,* v. 14, p. 561–597, pls. 1–12, figs. 1–17, 1967.

McIntyre, Andrew, A. W. H. Bé, and Resa Preikstas, Coccoliths and the Pliocene-Pleistocene boundary. *Prog. Oceanogr.,* v. 4, p. 3–35, pls. 1–6, figs. 1–5, 1967.

McIntyre, Andrew, A. W. H. Bé, and M. B. Roche, Modern Pacific Coccolithophorida: a paleontological thermometer. *Trans. N.Y. Acad. Sci.,* ser. II, v. 32, p. 720–731, 9 figs., 1970.

McIntyre, Andrew, and R. McIntyre, Coccolith concentrations and differential solution in oceanic sediments, p. 253–261, pls. 16.1–16.2, figs. 16.1–16.5. *In* B. M. Funnell and W. R. Riedel, *The micropalaeontology of oceans.* Cambridge: University Press, 1971.

McIntyre, Andrew, W. F. Ruddiman, and R. Jantzen, Southward penetrations of the North Atlantic Polar Front: faunal and floral evidence of large-scale surface water mass movements over the last 225,000 years. *Deep Sea Res.,* v. 19, p. 61–77, 8 figs., 1972.

Magne, Francis, Sur la cytologie d'une Chrysophycée marine vivant en aquarium, *Ochrosphaera neopolitana* Schussnig. *Revue Gén. Bot.,* v. 59, p. 231–240, 1 fig., 1952.

Maier, Dorothea, III. Zur Flora und Fauna der marinen Ablagerungen. Coccolithophorideen aus dem nierderrheinischen Tertiär. *Fortschr. Geol. Reinld. Westf.,* v. 1, p. 179–183, 1 fig., 1958.

Maier, Dorothea, Planktonuntersuchungen in tertiären und quartären marinen Sedimenten. *Neues Jb. Geol. Paläont. Abh.,* v. 107, p. 278–340, pls. 27–33, 4 text-figs., 1959.

Malyshek, V. T., Kokkolity: porodoobrazuyushchie foraminiferovoy svity Severnogo Kavkaza [Coccoliths:

rock builders of foraminiferal beds of the northern Caucasus]. *Dokl. Akad. Nauk SSSR,* v. 59, p. 315−316, 1 fig., 1948.

Manivit, Hélène, Contribution à l'étude des coccolithes de l'Éocène. *Publs. Serv. Carte Géol. Algér.,* ser. 2, Bull. 25, p. 331−382, 10 pls., ["1959"], 1961.

Manivit, Hélène, *Nannofossiles calcaires du Crétacé Français (Aptien-Maestrichtien); Essai de biozonation appuyée sur les stratotypes.* Thèse Doctorate d'Etat, Fac. Sci. d'Orsay, Inst. Geol., p. 1−167, pls. 1−32, 4 figs., 1971.

Manivit, H., K. Perch-Nielsen, B. Prins, and J. W. Verbeek, Mid Cretaceous calcareous nannofossil biostratigraphy. *Proc. K. Ned. Akad. Wet.,* ser. B, v. 80, p. 169−181, 1 pl., 1 fig., 1977.

Manton, Irene, and G. F. Leedale, Observations on the micro-anatomy of *Crystallolithus hyalinus* Gaarder and Markali. *Arch. Mikrobiol.,* v. 47, p. 115−136, 34 figs., 1963.

Manton, Irene, and G. F. Leedale, Observations on the microanatomy of *Coccolithus pelagicus* and *Cricosphaera carterae,* with special reference to the origin and nature of coccoliths and scales. *J. Mar. Biol. Ass. U.K.,* v. 49, p. 1−16, pls. 1−14, 1969.

Manton, Irene, and K. Oates, Fine structural observations on *Papposphaera* Tangen from the southern hemisphere and on *Pappomonas* gen nov. from South Africa and Greenland. *Br. Phycol. J.,* v. 10, p. 93−109, figs. 1−22, 1975.

Manton, Irene, and Joan Sutherland, Further observations on the genus *Pappomonas* Manton et Oates with special reference to *P. virgulosa* sp. nov. from West Greenland. *Br. Phycol. J.,* v. 10, p. 377−385, figs. 1−12, 1975.

Manton, Irene, Joan Sutherland, and K. Oates, Arctic coccolithophorids: two species of *Turrisphaera* gen. nov. from West Greenland, Alaska, and the Northwest Passage. *Proc. R. Soc.,* B, v. 194, p. 179−194, pls. 1−4, figs. 1, 2, 1976.

Manton, Irene, Joan Sutherland, and K. Oates, Arctic coccolithophorids: *Wigwamma arctica* gen. et sp. nov. from Greenland and arctic Canada, *W. annulifera* sp. nov. from South Africa and S. Alaska and *Calciarcus alaskensis* gen. et sp. nov. from S. Alaska. *Proc. R. Soc.,* B, v. 197, p. 145−168, pls. 1−8, 2 figs., 1977.

Maresch, Otto, Die Erforschung von Nannofossilien mittels des Elektronenmikroskopes in der Erdölindustrie. *Erdoel-Erdgas Z.,* v. 82, p. 377−384, pls. 1−4, 1966.

Markali, Joar, and Eyestein Paasche, On two species of *Umbellosphaera,* a new marine coccolithophorid genus. *Nytt Mag. Bot.,* v. 4, p. 95−100, 6 pls., 1955.

Martini, Erlend, Discoasteriden und verwandte Formen im NW-deutschen Eozän (Coccolithophorida). 1. Taxionomische Untersuchungen. *Senckenberg. Leth.,* v. 39, p. 353−388, 6 pls., 1958.

Martini, Erlend, Discoasteriden und verwandte Formen im NW-deutschen Eozän (Coccolithophorida). 2. Stratigraphische Auswertung. *Senckenberg. Leth.,* v. 40, p. 137−157, 6 figs., 1959a.

Martini, Erlend, *Pemma angulatum* und *Micrantholithus basquensis,* zwei neue Coccolithophoriden-Arten aus dem Eozän. *Senckenberg. Leth.,* v. 40, p. 415−421, 1 pl., 1959b.

Martini, Erlend, Braarudosphaeriden, Discoasteriden und verwandte Formen aus dem Rupelton des Mainzer Beckens. *Notizbl. Hess. Landesamt. Bodenforsch. Wiesbaden,* v. 88, p. 65−87, pls. 8−11, 1960.

Martini, Erlend, Nannoplankton aus dem Tertiär und der obersten Kreide von SW-Frankreich. *Senckenberg. Leth.,* v. 42, p. 1−32, 5 pls., 3 text-figs., 1961a.

Martini, Erlend, Der stratigraphische Wert der Lithostromationidae. *Erdöl-Z. Bohr-U. Fördertech.,* v. 77, p. 100−103, 2 figs., 1961b.

Martini, Erlend, Ein vollständiges Gehäuse von *Goniolithus fluckigeri* Deflandre. *Neues Jb. Geol. Paläont. Abh.,* v. 119, p. 19−21, pl. 6, 1964a.

Martini, Erlend, Die Coccolithophoriden der Dan-Scholle von Katharinenhof (Fehmarn). *Neues Jb. Geol. Paläont. Abh,* v. 121, p. 47−54, pls. 6, 7, 1964b.

Martini, Erlend, *Ceratolithina hamata* n. g., n. sp., aus dem Alb von N-Deutschland (Nannoplankton, incertae sedis). *Neues Jb. Geol. Paläont. Abh.,* v. 128, p. 294−298, pl. 30, 1967.

Martini, Erlend, Calcareous nannoplankton from the type Langhian. *G. Geol.,* Ser. 2, v. 35, p. 163−172, figs. 1−4, 1968.

Martini, Erlend, Standard Palaeogene calcareous nannoplankton zonation. *Nature,* v. 226, p. 560−561, 1 fig., 1970.

Martini, Erlend, Standard Tertiary and Quaternary calcareous nannoplankton zonation, p. 739−785, pls. 1−4. *In* A. Farinacci, *Proc. II Planktonic Conf. Roma 1970,* v. 2. Rome: Edizioni Tecnoscienza, 1971.

Martini, Erlend, Cretaceous to Recent calcareous nannoplankton from the central Pacific Ocean (DSDP Leg 33). *Init. Repts. Deep Sea Drill. Proj.,* v. 33, p. 383−423, pls. 1−13, fig. 1, 1976.

Martini, Erlend, and M. N. Bramlette, Calcareous nannoplankton from the experimental Mohole drilling. *J. Paleont.,* v. 37, p. 845−856, pls. 102−105, 2 text-figs., 1963.

Martini, Erlend, and T. Worsley, Standard Neogene calcareous nannoplankton zonation. *Nature,* v. 225, p. 289−290, 1970.

Mikkelsen, Naja, Marine Lower Oligocene sediments in Denmark as indicated by coccoliths in the Viborg Formation. *Meddr. Dansk Geol. Foren.,* v. 24, p. 83−86, figs. 1−6, 1975.

Mills, J. T., *Hymenomonas coronata* sp. nov., a new coccolithophorid from the Texas coast. *J. Phycol.,* v. 11, p. 149−154, 15 figs., 1975.

Mjaaland, G., Some laboratory experiments on the coccolithophorid *Coccolithus huxleyi*. *Oikos*, v. 7, p. 251—255, 4 figs., 1956.

Müller, Carla, Kalkiges Nannoplankton aus Tiefseekernen des Ionischen Meers. *"Meteor" Forsch.-Ergebn.*, Reihe C, no. 10, p. 75—95, pls. 1—6, figs. 1, 2, 1972.

Müller, Carla, Calcareous nannoplankton, Leg 25 (Western Indian Ocean). *Init. Repts. Deep Sea Drill. Proj.*, v. 25, p. 579—633, pls. 1—19, fig. 1, 1974a.

Müller, Carla, Nannoplankton aus dem Mittel-Miozän von Walbersdorf (Burgenland). *Senckenberg. Leth.*, v. 55, p. 389—405, pls. 1—4, 1974b.

Murray, George, and V. H. Blackman, On the nature of the coccospheres and rhabdospheres. *Phil. Trans. R. Soc.*, v. 190B. p. 427—441, pls. 15, 16, 1898.

Nagy, Istvan, A *Stomiosphaera* és a *Cadosina* nemzetség rétegtani szerepe a mecseki felsőjúrában [Sur le rôle stratigraphique des genres *Stomiosphaera* et *Cadosina* dans le Jurassique supérieur de la Montagne Mecsek]. *Földt. Közl.*, v. 96, p. 86—104, pl. 5, 3 figs., 1966.

Noël, Denise, Coccolithes des terrains jurassiques de l'Algérie. *Publs. Serv. Carte Géol. Algér.*, ser. 2, Bull. 8, p. 303—345, 8 pls. ["1956"], 1957.

Noël, Denise, Étude de coccolithes du Jurassique et du Crétacé inférieur. *Publs. Serv. Carte Géol. Algér.*, ser. 2, Bull. 20, p. 155—196, 11 pls., ["1958"], 1959.

Noël, Denise, Revision du genre *Discoaster* Tan Sin Hok, 1927. *Bull. Soc. Hist. Nat. Afr. N.*, v. 51, p. 201—229, 3 pls., 1960.

Noël, Denise, Sur la présence de Coccolithophoridés dans des terrains primaires. *C. R. Hebd. Séanc. Acad. Sci., Paris*, v. 252, p. 3625—3627, 11 figs., 1961.

Noël, Denise, Note préliminaire sur des coccolithes jurassiques. *Cahiers Micropaléont.*, ser. 1, no. 1, Arch. Orig. Centre Document. C.N.R.S. No. 408, p. 1—12, figs. 1—60, 1965a.

Noël, Denise, *Sur les coccolithes du Jurassique Européen et d'Afrique du Nord. Essai de classification des coccolithes fossiles.* Paris: Centre National de la Recherche Scientifique, 209 p., 29 pls., 74 figs., 1965b.

Noël, Denise, Contribution à la révision des coccolithes secondaires: essai d'établissement d'une hiérarchie des caractères génériques, p. 879—898, pl. 1. *In* A. Farinacci, *Proc. II Planktonic Conf. Roma 1970*, v. 2. Rome: Edizioni Tecnoscienza, 1971.

Noël, Denise, Éléments de morphologie comparée: *Gartnerago* Bukry 1969 et *Kamptnerius* Deflandre 1959 (Coccolithes Crétacés). *Cah. Micropaléont.*, sér. 3, no. 4, p. 1—10, pls. 1—3, 1972.

Norris, R. E., Living cells of *Ceratolithus cristatus* (Coccolithophorineae). *Arch. Protistenk.*, v. 108, p. 19—24, pls. 11—13, 1965.

Norris, R. E., Extant calcareous nannoplankton from the Indian Ocean, p. 899—909, 16 figs. *In* A. Farinacci, *Proc. II Planktonic Conf. Roma 1970*, v. 2. Rome: Edizioni Tecnoscienza, 1971.

Nowak, Wiesław, Stomiosferidy warstw cieszyńskich (Kimeryd-Hoteryw) Polskiego Śląska Cieszyńskiego i ich znaczenie stratygraficzne [Stomiosphaerids of the Cieszyn Beds (Kimmeridgian-Hauterivian) in the Polish Cieszyn Silesia and their stratigraphic value]. *Roczn. Pol. Tow. Geol.*, v. 38, p. 275—327, pls. 25—31, 1968.

Nowak, Wiesław, *Stomiosphaerina* nov. gen. (incertae sedis) of the Upper Cretaceous in the Polish Flysch Carpathians. *Rocnz. Pol. Tow. Geol.*, v. 44, p. 51—63, pls. 1—4, figs. 1—5, 1974.

Okada, Hisatake, and Susumu Honjo, Coccolithophoridae distributed in Southwest Pacific. *Pacif. Geol.*, v. 2, p. 11—21, pls. 1—3, figs. 1—3, 1970.

Okada, Hisatake, and Susumu Honjo, The distribution of oceanic coccolithophorids in the Pacific. *Deep Sea Res.*, v. 20, p. 355—374, pls. 1, 2, 13 figs., 1973.

Okada, Hisatake, and Susumu Honjo, Distribution of coccolithophores in marginal seas along the Western Pacific Ocean and in the Red Sea. *Mar. Biol.*, v. 31, p. 271—285, figs. 1—10, 1975.

Okada, Hisatake, and Andrew McIntyre, Modern coccolithophores of the Pacific and North Atlantic Oceans. *Micropaleontology*, v. 23, p. 1—55, pls. 1—13, figs. 1, 2, 1977.

Olsson, R. K., and Iradj Youssefnia, Cretaceous Calcisphaerulidae from New Jersey. *J. Paleont.*, v. 53, p. 1085—1093, 2 pls., 2 figs., 1979.

Ostenfeld, C. H., Über *Coccosphaera* und einige neue Tintinniden im Plankton des nördlichen Atlantischen Oceans. *Zool. Anz.*, v. 22, p. 433—439, 2 figs., 1899.

Ostenfeld, C. H., Über *Coccosphaera*. *Zool. Anz.*, v. 23, p. 198—200, 1900.

Ostenfeld, C. H., *Thorosphaera*, eine neue Gattung der Coccolithophoriden. *Ber. Dt. Bot. Ges.*, v. 28, p. 397—400, 5 figs., 1910.

Outka, D. E., and D. C. Williams, Sequential coccolith morphogenesis in *Hymenomonas carterae*. *J. Protozool.*, v. 18, p. 285—297, 20 figs., 1971.

Paasche, Eyestein, Coccolith formation. *Nature*, v. 193, p. 1094—1095, 2 figs., 1962a.

Paasche, Eyestein, Erratum to "Coccolith formation". *Nature*, v. 194, p. 1024, 1962b.

Paasche, Eyestein, The adaptation of the carbon-14 method for the measurement of coccolith production in *Coccolithus huxleyi*. *Physiologia Pl.*, v. 16, p. 186—200, 3 figs., 1963.

Paasche, Eyestein, A tracer study of the inorganic carbon uptake during coccolith formation and photosynthesis in the coccolithophorid *Coccolithus huxleyi*. *Physiologia Pl.*, suppl. 3, 82 pp., 41 figs., 1964.

Paasche, Eyestein, The effect of 3 (p-chlorophenyl)-1, 1-dimethylurea (CMU) on photosynthesis and light-dependent coccolith formation in *Coccolithus huxleyi. Physiologia Pl.,* v. 18, p. 138−145, 1 text-fig., 1965.

Paasche, Eyestein, Adjustment to light and dark rates of coccolith formation. *Physiologia Pl.,* v. 19, p. 271−278, 2 figs., 1966a.

Paasche, Eyestein, Action spectrum of coccolith formation. *Physiologia Pl.,* v. 19, p. 770−779, 4 figs., 1966b.

Paasche, Eyestein, The effect of temperature, light intensity, and photoperiod on coccolith formation. *Limnol. Oceanogr.,* v. 13, p. 178−181, fig. 1, 1968a.

Paasche, Eyestein, Biology and physiology of coccolithophorids. *A. Rev. Microbiol.,* v. 22, p. 71−86, 1968b.

Paasche, Eyestein, Light-dependent coccolith formation in the two forms of *Coccolithus pelagicus* with remarks on the ^{14}C zero-thickness counting efficiency of coccolithophorids. *Arch. Mikrobiol.,* v. 67, p. 199−208, figs. 1, 2, 1969.

Paasche, Eyestein, and D. Klaveness, A physiological comparison of coccolith-forming and naked cells of *Coccolithus huxleyi. Arch Mikrobiol.,* v. 73, p. 143−152, 2 figs., 1970.

Paddock, T. B. B., A possible aid to survival of the marine coccolithophorid *Cricosphaera* and similar organisms. *Br. Phycol. Bull.,* v. 3, p. 519−523, figs. 1−5, 1968.

Paréjas, E., Sur quelques *Actiniscus* du Crétacé supérieur des Brasses (Préalpes médianes) et de l'Ile d'Elbe. *C. R. Séanc. Soc. Phys. Hist. Nat. Genève,* v. 51, p. 100−107, 32 figs., 1934.

Parke, Mary, Some remarks concerning the class Chrysophyceae. *Br. Phycol. Bull.,* v. 2, p. 47−55, 2 pls., 1961a.

Parke, Mary, [Abstr.] Stages in life-histories of coccolithophorids. *Ann. Rep. Challenger Soc.,* v. 3, no. 13, p. 30, 1961b.

Parke, Mary, The production of calcareous elements by benthic algae belonging to the class Haptophyceae (Chrysophyta), p. 929−937, pls. 1, 2. *In* A. Farinacci, *Proc. II Planktonic Conf. Roma 1970,* v. 2. Rome: Edizioni Tecnoscienza, 1971.

Parke, Mary, and Irene Adams, The motile (*Crystallolithus hyalinus* Gaarder & Markali) and non-motile phases in the life history of *Coccolithus pelagicus* (Wallich) Schiller. *J. Mar. Biol. Ass. U.K.,* v. 39, p. 263−274, 4 pls., 1960.

Parke, Mary and P. S. Dixon, Check-list of British marine algae—Third revision. *J. Mar. Biol. Ass. U.K.,* v. 56, p. 527−594, 1976.

Parke, Mary, Irene Manton, and B. Clarke, Studies on marine flagellates II. Three new species of *Chrysochromulina. J. Mar. Biol. Ass. U.K.,* v. 34, p. 579−609, pls. 1−9, figs. 1−72, 1955.

Perch-Nielsen, Katharina, Eine Präparationstechnik zur Untersuchung von Nannoplankton im Lichtmikroskop und im Elektronenmikroskop. *Meddr. Dansk Geol. Foren.,* v. 17, p. 129−130, 1 pl., 1967a.

Perch-Nielsen, Katharina, Nannofossilien aus dem Eozän von Dänemark. *Eclog. Geol. Helv.,* v. 60, p. 19−32, pls. 1−7, fig. 1, 1967b.

Perch-Nielsen, Katharina, Der Feinbau und die Klassifikation der Coccolithen aus dem Maastrichtien von Dänemark. *Biol. Skr.,* v. 16, p. 1−96, pls. 1−32, figs. 1−44, 1968a.

Perch-Nielsen, Katharina, *Naninfula,* genre nouveau de nannofossiles calcaires du Tertiaire danois. *C. R. Hebd. Séanc. Acad. Sci., Paris,* v. 267, p. 2298−2300, pls. 1, 2, 1968b.

Perch-Nielsen, Katharina, Durchsicht Tertiärer Coccolithen, p. 939−980, pls. 1, 2, figs. 1−22. *In* A. Farinacci, *Proc. II Planktonic Conf. Roma 1970,* v. 2. Roma: Edizioni Tecnoscienza, 1971a.

Perch-Nielsen, Katharina, Einige neue Coccolithen aus dem Paleozän der Bucht von Biskaya. *Meddr Dansk Geol. Foren.,* v. 20, p. 347−361, pls. 1−14, 1971b.

Perch-Nielsen, Katharina, Neue Coccolithen aus dem Paleozän von Dänemark, der Bucht von Biskaya und dem Eozän der Labrador See. *Meddr Dansk Geol. Foren.,* v. 21, p. 51−66, pls. 1−7, 1971c.

Perch-Nielsen, Katharina, Elektronenmikroskopische Untersuchungen an Coccolithen und verwandten Formen aus dem Eozän von Dänemark. *Biol. Skr.,* v. 18(3), p. 1−76, pls. 1−61, figs. 1, 2, 1971d.

Perch-Nielsen, Katharina, Remarks on Late Cretaceous to Pleistocene coccoliths from the North Atlantic. *Init. Repts. Deep Sea Drill. Proj.,* v. 12, p. 1003−1069, pls. 1−22, 1972.

Perch-Nielsen, Katharina, Neue Coccolithen aus dem Maastrichtien von Dänemark, Madagaskar und Ägypten. *Meddr Dansk Geol. Foren.,* v. 22, p. 306−333, pls. 1−10, 1973.

Perch-Nielsen, Katharina, Albian to Pleistocene calcareous nannofossils from the western South Atlantic, DSDP Leg 39. *Init. Repts. Deep Sea Drill. Proj.,* v. 39, p. 699−823, pls. 1−50, 1 fig., 1977.

Perch-Nielsen, Katharina, and H. E. Franz, *Lapideacassis* and *Scampanella,* calcareous nannofossils from the Paleocene at sites 354 and 356, DSDP Leg 39, southern Atlantic. *Init. Repts. Deep Sea Drill. Proj.,* v. 39, p. 849−862, pls. 1−6, figs. 1−3, 1977.

Percival, S. F., and A. G. Fischer, Changes in calcareous nannoplankton in the Cretaceous-Tertiary biotic crisis at Zumaya, Spain. *Evol. Theory,* v. 2, p. 1−35, figs. 1−14, 1977.

Pienaar, R. N., Upper Cretaceous coccolithophorids from Zululand, South Africa. *Palaeontology,* v. 11, p. 361−367, pls. 69−71, 1968.

Pirini Radrizzani, C., Coccoliths from Permian deposits of eastern Turkey, p. 993—1001, pls. 1—3, figs. 1, 2. *In* A. Farinacci, *Proc. II Planktonic Conf. Roma 1970*, v. 2. Rome: Edizioni Tecnoscienza, 1971.

Prins, Bernard, Evolution and stratigraphy of coccolothinids [sic] from the Lower and Middle Lias, p. 547—558, pls. 1—3, 1 table. *In* P. Brönnimann and H. H. Renz, *Proc. First Int. Conf. Planktonic Microfossils, Geneva, 1967*, v. 2. Leiden: E. J. Brill, 1969.

Prins, B., Speculations on relations, evolution, and stratigraphic distribution of discoasters, p. 1017—1037, pls. 1—8. *In* A. Farinacci, *Proc. II Planktonic Conf. Roma 1970*, v. 2. Rome: Edizioni Tecnoscienza, 1971.

Provasoli, Luigi, Growth factors in unicellular marine algae, p. 385—403. *In* A. A. Buzzati-Traverso (ed.), *Perspectives in marine biology.* Berkeley and Los Angeles: Univ. Calif. Press, 1960.

Rayns, D. G., Alternation of generations in a coccolithophorid, *Cricosphaera carterae* (Braarud & Fagerl.) Braarud. *J. Mar. Biol. Ass. U.K.,* v. 42, p. 481—484, 2 pls., 1962.

Reinhardt, Peter, Einige Kalkflagellaten-Gattungen (Coccolithophoriden, Coccolithineen) aus dem Mesozoikum Deutschlands. *Mber. Dt. Akad. Wiss. Berl.,* v. 6, p. 749—759, 2 pls., 8 text-figs., 1964.

Reinhardt, Peter, Neue Familien fur fossile Kalkflagellaten (Coccolithophoriden, Coccolithineen). *Mber. Dt. Akad. Wiss. Berl.,* v. 7, p. 30—40, pls. 1—3, 6 text-figs, 1965.

Reinhardt, Peter, Zur Taxionomie und Biostratigraphie der fossilen Nannoplanktons aus dem Malm, der Kreide und dem Alttertiär Mitteleuropas. *Freiberger ForschHft.,* C196, Paläont., p. 5—109, 23 pls., 29 figs., 1966.

Reinhardt, Peter, Zur Taxionomie und Biostratigraphie der Coccolithineen (Coccolithophoriden) aus dem Eozän Norddeutschlands. *Freiberger ForschHft.,* C213, Paläont., p. 201—241, pls. 1—7, 20 figs., 1967a.

Reinhardt, Peter, Fossile Coccolithen mit rhagoidem Zentralfeld (Fam. Ahmuellerellaceae, Subord. Coccolithineae). *Neues Jb. Geol. Paläont. Mh.,* p. 163—178, 12 figs., 1967b.

Reinhardt, Peter, Neue Coccolithen-Arten aus der Kreide. *Mber. Dt. Akad. Wiss. Berl.,* v. 11, p. 932—938, pl. 1, 1969.

Reinhardt, Peter, Synopsis der Gattungen und Arten der mesozoischen Coccolithen und anderer kalkiger Nannofossilien, Teil I. *Freiberger ForschHft.,* C260, Paläont., p. 5—32, pl. 1, 55 figs., 1970a.

Reinhardt, Peter, Synopsis der Gattungen und Arten der mesozoischen Coccolithen und anderer kalkiger Nannofossilien. Teil II. *Freiberger ForschHft.,* C265, Paläont., p. 41—111, pls. 1—8, 122 figs., 1970b.

Reinhardt, Peter, Synopsis der Gattungen und Arten der mesozoischen Coccolithen und anderer kalkiger Nannofossilien, Teil III. *Freiberger ForschHft.,* C267, Paläont., p. 19—41, pls. 1—3, 49 figs., 1971.

Reinhardt, Peter, *Coccolithen. Kalkiges Plankton seit Jahrmillionen.* Wittenberg Lutherstadt. Die Neue Brehm-Bücherei, 453, A. Ziemsen Verlag, 99 p., 188 figs., 1972.

Reinhardt, Peter, and Hanna Górka, Revision of some Upper Cretaceous coccoliths from Poland and Germany. *Neues Jb. Geol. Paläont. Abh.,* v. 129, p. 240—256, pls. 31—33, figs. 1—6, 1967.

Riedel, W. R., M. N. Bramlette, and F. L. Parker, "Pliocene-Pleistocene" boundary in deep-sea sediments. *Science,* v. 140, p. 1238—1240, 1 fig., 1963.

Risatti, J. B., Nannoplankton biostratigraphy of the Upper Bluffport Marl-Lower Prairie Bluff Chalk interval (Upper Cretaceous), in Mississippi, p. 8—57, pls. 1—10, figs. 1—4. *In* L. A. Smith and J. Hardenbol, *Proceedings of symposium on calcareous nannofossils.* Houston: Gulf Coast Section, Soc. Econ. Paleont. Mineral., 1973.

Rögl, Fred, Danian Calcisphaerulidae of DSDP Leg 35, Site 323, Southeast Pacific Ocean. *Init. Repts. Deep Sea Drill. Proj.,* v. 35, p. 701—711, pls. 1—4, 1976.

Romein, A. J. T., Calcareous nannofossils from the Cretaceous/Tertiary boundary interval in the Barranco del Gredero (Caravaca, Prov. Murcia, S.E. Spain). I. and II. *Proc. K. Ned. Akad. Wet.,* ser. B, v. 80, p. 256—279, pls. 1—3, figs. 1—3, 1977.

Rood, A. P., W. W. Hay, and T. Barnard, Electron microscope studies of Oxford Clay coccoliths. *Eclog. Geol. Helv.,* v. 64, p. 245—272, pls. 1—5, figs. 1—3, 1971.

Rood, A. P., W. W. Hay, and T. Barnard, Electron microscope studies of Lower and Middle Jurassic coccoliths. *Eclog. Geol. Helv.,* v. 66, p. 365—382, pls. 1—3, 1973.

Roth, P. H., Oligocene calcareous nannoplankton biostratigraphy. *Eclog. Geol. Helv.,* v. 63, p. 799—881, pls. 1—14, figs. 1—17, 1970.

Roth, P. H., Calcareous nannofossils-Leg 17, Deep Sea Drilling Project. *Init. Repts. Deep Sea Drill. Proj.,* v. 17, p. 695—795, pls. 1—27, figs. 1, 2, 1973.

Roth, P. H. Calcareous nannofossils from the northwestern Indian Ocean, Leg 24, Deep Sea Drilling Project. *Init. Repts. Deep Sea Drill. Proj.,* v. 24, p. 969—994, figs. 1—4, 1974.

Roth, P. H., P. Baumann, and V. Bertolino, Late Eocene-Oligocene calcareous nannoplankton from central and northern Italy, p. 1069—1097, 13 figs. *In* A. Farinacci, *Proc. II Planktonic Conf. Roma 1970,* v. 2. Rome: Edizioni Tecnoscienza, 1971.

Roth, P. H., and W. H. Berger, Distribution and dissolution of coccoliths in the South and Central Pacific. *Spec. Publs. Cushman Fdn.,* 13, p. 87—113, pls. 1—3, figs. 1—24, 1975.

Roth, P. H., H. E. Franz, and S. W. Wise, Jr., Morphological study of selected members of the genus

Sphenolithus Deflandre (incertae sedis, Tertiary), p. 1099—1119, pls. 1—6, figs. 1—5. *In* A. Farinacci, *Proc. II Planktonic Conf. Roma 1970,* v. 2. Rome: Edizioni Tecnoscienza, 1971.

Roth, P. H., and H. R. Thierstein, Calcareous nannoplankton: leg 14 of the Deep Sea Drilling Project. *Init. Repts. Deep Sea Drill. Proj.,* v. 14, p. 421—485, pls. 1—16, figs. 1, 2, 1972.

Rothpletz, A., Ueber die Flysch-Fucoiden und einige andere fossile Algen, sowie über liasische Diatomeen führende Hornschwämme. *Z. Dt. Geol. Ges.,* v. 48, p. 854—914, pls. 22—24, 3 text-figs., 1896.

Roukhiyaynen, M. I., Novyy vid roda *Ceratolithus* (Ceratolithaceae) iz tsentral'no-Amerikanskikh Morey [Species Ceratolithi (Ceratolithaceae) nova e maribus Americae Centralis]. *Nov. Sist. Nizsh. Rast., Akad. Nauk SSSR Bot. Inst. Komarova,* 1969, p. 3—7, figs. 1, 2, 1969.

Saito, Tsunemasa, and S. F. Percival, Mid-Atlantic ridge sequence, p. 444—445. *In* The shipboard scientific party, summary and conclusions. *Init. Repts. Deep Sea Drill. Proj.,* v. 3, p. 441—471, 1970.

Schei, Ben, Coccolithophorid distribution and ecology in coastal waters of north Norway. *Norw. J. Bot.,* v. 22, p. 217—225, figs. 1—8, 1975.

Scheibner, Ervin, *Cadosinopsis,* nouveau genre du Crétacé supérieur de la zone des Klippes (Carpathes occidentales). *Revue Micropaléont.,* v. 10, p. 42—47, pls. 1, 2, figs. 1—5, 1967.

Schiller, J., Bericht über Ergebnisse der Nannoplanktonuntersuchungen anlässlich der Kreuzungen S.M.S. *Najade* in der Adria. *Int. Revue Ges. Hydrobiol. Hydrogr., Biol. Suppl.* 6 (4. Hefte, Art. 5), 15 p., pl. 2, 1914.

Schiller, J., Coccolithineae, p. 89—267, 137 figs. + figs. A—F. In *Dr. L. Rabenhorst's Kryptogamen-Flora von Deutschland, Österreich und der Schweiz.* 10. Band, 2. Abt. Leipzig: Akademische Verlagsgesellschaft, 1930.

Schlauder, Juliette, *Recherches sur les Flagellés calcaires de la baie d'Alger.* Diplôme Faculté des Sciences, Université d'Alger. 51 p., 9 pls., 1945.

Schmid, O. [Schmidt], Über Coccolithen und Rhabdolithen. *Sber. Akad. Wiss. Wien, Math.-Naturw. Kl.,* v. 62, Abt. I, p. 669—682, 2 pls., 1870.

Schmid, O. [Schmidt], On coccoliths and rhabdoliths. *Ann. Mag. Nat. Hist.,* ser. 4, v. 10, p. 359—370, pls. 16, 17, 1872.

Schwarz, E. H. L., Coccoliths. *Ann. Mag. Nat. Hist.,* ser. 6, v. 14, p. 341—346, 27 figs., 1894.

Schwarz, Elisabeth, Beitrage zur Entwicklungsgeschichte der Protophyten. IX, Der Formwechsel von *Ochrosphaera neapolitana. Arch. Protistenk.* v. 77, p. 434—462, figs. 1—7, pl. 13, 1932.

Shamray, I. A., Nekotorye formy verkhnemelovykh i paleogenovykh kokkolitov i diskoasterov na yuge russkoy platformy [Certain forms of Upper Cretaceous and Paleogene coccoliths and discoasters from the southern Russian Platform]. *Izv. Vyssh. Ucheb. Zaved., Geol. I Razv.,* no. 4, p. 27—40, 2 pls., 1963.

Shamray, I. A., and E. P. Lazareva, Paleogenovye Coccolithophoridae i ikh stratigraficheskoe znachenie [Paleogene Coccolithophoridae and their stratigraphic significance]. *Dokl. Akad. Nauk SSSR,* v. 108, p. 711—714, 1 fig., 1956.

Shumenko, S. I., Primenenie metoda elektronnomikroskopicheskikh replik i izucheniyu verkhnemelovykh Coccolithophoridae [Application of the method of electron-microscope replicas to the investigation of Upper Cretaceous Coccolithophoridae]. *Dokl. Akad. Nauk SSSR,* v. 147, p. 471—473, 1 fig., 1962.

Shumenko, S. I., *Litologiya i porodoobrazuyushchie organizmy (Kokkolitoforidy) verkhnemelovykh otlozheniy vostoka ukrainy i oblasti Kurskoy Magnitnoy Anomalii [Lithology and rock forming organisms (Coccolithophorids) of the Upper Cretaceous in the eastern Ukraine and the area of the Kursk Magnetic Anomaly].* Kharkov: Khar'kovskogo Univ., 163 p., 23 pls., 19 figs., 1971.

Shumenko, S. I., Izvestkovyy nanoplankton iz otlozheniy na granitse Mela i Paleogena Kryma [Calcareous nannoplankton from Cretaceous-Paleogene boundary beds of the Crimea]. *Dokl. Akad. Nauk SSSR,* v. 209, p. 919—921, 1973.

Shumenko, S. I., *Izvestkovyy nanoplankton Mezozoya Evropeyskoy Chasti SSSR [Mesozoic calcareous nannoplankton of the European part of the USSR],* Moscow: Akad. Nauk SSSR, Paleont. Inst., "Nauka," 140 p., 30 pls., 11 figs., 1976.

Siesser, W. G., Calcareous nannofossils from the South African continental margin. *Joint Geol. Surv./ Univ. Cape Town Mar. Geol. Progm.,* Bull. 5, p. 1—135, pls. 1—5, 1975.

Siesser, W. G., Calcareous nannofossils in Pleistocene sediment cores from the South African continental slope. *Trans. R. Soc. S. Afr.,* v. 42, p. 107—147, 13 figs., 1976.

Smayda, T. J., A quantitative analysis of the phytoplankton of the Gulf of Panama. III General ecological considerations, and the phytoplankton dynamics at 8°45'N, 79°23'W from November 1954 to May 1957. *Bull. Inter-Am. Trop. Tuna Commn.,* v. 11, p. 355—612, 1966.

Smith, C. C., A new method for studying calcareous nannofossils using scanning and transmitted light optics. *Revta. Esp. Micropaleont., Num. Especial,* p. 43—48, 1 pl., 1975a.

Smith, C. C., Upper Cretaceous calcareous nannoplankton zonation and stage boundaries. *Trans. Gulf-Cst. Ass. Geol. Socs.,* v. 25, p. 263—278, 2 figs., 1975b.

Sorby, H. C., On the organic origin of the so-called "crystalloids" of the Chalk. *Ann. Mag. Nat. Hist.,* ser. 3, v. 8, p. 193—200, 5 figs., 1861.

Stosch, H. A. von, Ein morphologischer Phasenwechsel bei einer Coccolithophoride. *Naturwissenschaften,* v. 42, p. 423, 1955.

Stosch, H. A. von, Der Geisselapparat einer Coccolithophoride. *Naturwissenschaften,* v. 45, p. 140−141, 1958.

Stradner, Herbert, Die fossilen Discoasteriden Österreichs. 1. Teil. Die in den Bohrkernen der Tiefbohrung Korneuburg 1 enthaltenen Discoasteriden. *Erdöl-Z. Bohr-U. Fördertech.,* v. 74, p. 178−188, 38 figs., 1958.

Stradner, Herbert, First report on the discoasters of the Tertiary of Austria and their stratigraphic use. *Proc. Fifth Wld. Petrol. Congr.,* v. 1, p. 1081−1095, 30 figs., 1959a.

Stradner, Herbert, Die fossilen Discoasteriden Österreichs. II. Teil. *Erdöl-Z. Bohr-U. Fördertech.,* v. 75, p. 472−488, 77 figs., 1959b.

Stradner, Herbert, Vorkommen von Nannofossilien in Mesozoikum und Alttertiär. *Erdöl-Z. Bohr-U. Fördertech.,* v. 77, p. 77−89, 1961.

Stradner, Herbert, Über das fossile Nannoplankton des Eozän-Flysch von Istrien. *Verh. Geol. Bundesanst., Wien,* p. 176−186, 2 pls., 1962a.

Stradner, Herbert, Über neue und wenig bekannte Nannofossilien aus Kreide und Alttertiär. *Verh. Geol. Bundesanst., Wien,* p. 363−377, 3 pls., 1962b.

Stradner, Herbert, New contributions to Mesozoic stratigraphy by means of nannofossils. *Proc. Sixth Wld. Petrol. Congr.,* sect. 1, paper 4 [preprint]. 16 p., 6 pls., 1963.

Stradner, Herbert, The nannofossils of the Eocene Flysch in the Hagenback Valley (Northern Vienna Woods), Austria. *Roczn. Pol. Tow. Geol.,* v. 39, p. 403−432, pls. 80−89, 4 figs., 1969.

Stradner, Herbert, Catalogue of calcareous nannoplankton from sediments of Neogene age in the eastern North Atlantic and Mediterranean Sea. *Init. Repts. Deep Sea Drill. Proj.,* v. 13, p. 1137−1199, pls. 1−51, fig. 1, 1973.

Stradner, Herbert, and Dieter Adamiker, Nannofossilien aus Bohrkernen und ihre elektronen mikroskopische Bearbeitung. *Erdoel-Erdgas Z.,* 82 Jg., p. 330−341, pls. 1−3, figs. 1−16, 1966.

Stradner, Herbert, Dieter Adamiker, and Otto Maresch, Electron microscope studies on Albian calcareous nannoplankton from the Delft 2 and Leidschendam 1 Deepwells, Holland. *Verh. K. Ned. Akad. Wet., Afd. Natuurk, Eerste Reeks,* v. 24, no. 4, 107 p., 48 pls., 11 figs., 1968.

Stradner, Herbert, and A. R. Edwards, Electron microscopic studies on Upper Eocene coccoliths from the Oamaru Diatomite, New Zealand. *Jb. Geol. Bundesanst., Wien, Sonderband* 13, 66 p., 48 pls., 10 figs., 1968.

Stradner, Herbert, and Walter Grün, On *Nannoconus abundans* nov. spec. and on laminated calcite growth in Lower Cretaceous nannofossils. *Verh. Geol. Bundesanst., Wien,* p. 267−283, pls. 1−6, 1973.

Stradner, Herbert, and A. Papp, Tertiäre Discoasteriden aus Österreich und deren stratigraphischen Bedeutung mit Hinweisen auf Mexico, Rumänien und Italien. *Jb. Geol. Bundesanst., Wien,* Sonderband 7, 159 p., 42 pls., 1961.

Sujkowski, Z., Petrografja kredy Polski. Kreda z głębokiego wiercenia w Lublinie w porównaniu z kredą niektórych innych obszarów Polski [Etude pétrographique du Crétacé de Pologne. La série de Lublin et sa comparaison avec la craie blanche]. *Spraw. Pol. Inst. Geol.,* v. 6, p. [11] + 485−628, pls. 6−13, 4 text-figs., 1931.

Sullivan, F. R., Lower Tertiary nannoplankton from the California Coast Ranges. I. Paleocene. *Univ. Calif. Publs. Geol. Sci.,* v. 44, p. 163−227, 12 pls., 2 figs., 1964.

Sullivan, F. R., Lower Tertiary nannoplankton from the California Coast Ranges. II. Eocene. *Univ. Calif. Publs. Geol. Sci.,* v. 53, p. 1−74, 11 pls. 2 text-figs., 1965.

Tan Sin Hok, On a young-Tertiairy [sic] limestone of the Isle of Rotti with coccoliths, calci- and manganese-peroxide-spherulites. *Proc. Sect. Sc., K. Akad. Wet.,* v. 29, p. 1095−1105, 11 figs, 1926.

Tan Sin Hok, Over de samenstelling en het onstaan van krijt en mergelgesteenten van de Molukken. *Jaarb. Mijnw. Ned.-Indië,* v. 55, p. 1−165, pls. 1−16, 2 figs., 1927a.

Tan Sin Hok, Discoasteridae incertae sedis. *Proc. Sect. Sc., K. Akad. Wet.,* v. 30, p. 411−419, 14 figs., 1927b.

Tan Sin Hok, Discoasteridae, Coccolithinae and Radiolaria. *Leid. Geol Meded.,* v. 5, p. 92−114, 1931.

Tangen, Karl, *Papposphaera lepida,* gen. nov., n.sp., a new marine coccolithophorid from Norwegian coastal waters. *Norw. J. Bot.,* v. 19, p. 171−178, 14 figs., 1972.

Tappan, Helen, Primary production, isotopes, extinctions, and the atmosphere. *Palaeogeogr., Palaeoclimat., Palaeoecol.,* v. 4, p. 187−210, 1 fig., 1968.

Tappan, Helen, and A. R. Loeblich, Jr., Geobiologic implications of fossil phytoplankton evolution and time-space distribution, p. 247−340, 1 pl., 12 figs. *In* R. Kosanke and A. T. Cross, Symposium on palynology of the Late Cretaceous and Early Tertiary. *Spec. Pap. Geol. Soc. Am.,* no. 127, 1971.

Tappan, Helen, and A. R. Loeblich, Jr., Fluctuating rates of protistan evolution, diversification and extinction. *24th Int. Geol. Congr.,* Sec. 7 Paleont., p. 205−213, figs. 1−6, 1972.

Tappan, Helen, and A. R. Loeblich, Jr., Evolution of the oceanic plankton. *Earth-Sci. Rev.,* v. 9, p. 207−240, figs. 1−9, 1973.

Thierstein, H. R., Tentative Lower Cretaceous calcareous nannoplankton zonation. *Eclog. Geol. Helv.*, v. 64, p. 459—488, pls. 1—8, figs. 1—5, 1971.

Thierstein, H. R., Lower Cretaceous calcareous nannoplankton biostratigraphy. *Abh. Geol. Bundesanst., Wien*, v. 29, p. 1—52, pls. 1—6, figs. 1—25, 1973.

Thierstein, H. R., Calcareous nannoplankton-Leg 26, Deep Sea Drilling Project. *Init. Repts. Deep Sea Drill. Proj.*, v. 26, p. 619—667, pls. 1—12, figs. 1—4, 1974.

Thierstein, H. R., Mesozoic calcareous nannoplankton biostratigraphy of marine sediments. *Mar. Micropaleont.*, v. 1, p. 325—362, pls. 1—5, figs. 1—8, 1976.

Thomsen, H. A., and K. Oates, *Balaniger balticus* gen. et sp. nov. (Prymnesiophyceae) from Danish coastal waters. *J. Mar. Biol. Assoc. U.K.*, v. 58, p. 773—779, pls. 1—3, 2 figs., 1978.

Trejo, M., La familia Nannoconidae y su alcance estratigráfico en America (Protozoa, incertae sedis). *Bol. Asoc. Mex. Géol. Petrol.*, v. 12, p. 259—314, 3 pls., 1960.

Valkanov, A., Ueber die Entwicklung von *Hymenomonas coccolithophora* Conrad. *Revue Algol.*, n.s., v. 6, p. 220—226, pls. 17, 18, text-figs. A—E, 1962.

Vekshina, V. N., Kokkolitoforidy maastrikhtskikh otlozheniy Zapadno-Sibirskoy nizmennosti [Coccolithophoridae of the Maastrichtian deposits of the west Siberian lowland]. *Trudȳ Sib. Nauchno-Issled. Inst. Geol. Geofiz. Miner. Sȳr'*. [SNIIGGIMS], v. 2, p. 56—81, 2 pls., 1 text-fig., 1959.

Verbeek, J. W., Calcareous nannoplankton biostratigraphy of Middle and Upper Cretaceous deposits in Tunisia, southern Spain and France. *Utrecht Micropaleont. Bulls.*, v. 16, 157 p., 22 figs., 12 pls., 1977.

Villain, J. M., "Calcisphaerulidae" (incertae sedis) du Crétacé Supérieur du Limbourg (Pays-Bas), et d'autres regions. *Paläontographica Abt. A.*, v. 149, p. 193—242, pls. 1—9, figs. 1—76, 1975.

Vishnevskiy, A. V., and P. A. Menyaylenko, Kokkolitoforidy nizhnemelovykh (aptskikh) glin Bakhchisarayskogo rayona [Coccolithophorids from the Lower Cretaceous (Aptian) clay of the Bakhchisaraisky region]. *Izv. Vȳssh. Ucheb. Zaved. Geol. I Razved.*, no. 11, p. 47—53, 2 pls., 1963.

Voeltzkow, A., Über Coccolithen und Rhabdolithen nebst Bemerkungen über den Aufbau and die Entstehung der Aldabra-Inseln. *Abh. Senckenb. Naturforsch. Ges.*, v. 26, p. 465—537, 3 figs., 1902.

Wallich, G. C., Remarks on some novel phases of organic life, and on the boring powers of minute annelids, at great depths in the sea. *Ann. Mag. Nat. Hist.*, ser. 3, v. 8, p. 52—58, 4 + 3 figs., 1861.

Wallich, G. C., On the structure and affinites of the Polycystina. *Trans. Micrsc. Soc. London*, ser. 2, v. 13, p. 57—84, 1865.

Wallich, G. C., On the true nature of the so-called "*Bathybius*" and its alleged function in the nutrition of the Protozoa. *Ann. Mag. Nat. Hist.*, ser. 4, v. 16, p. 322—339, 1875.

Wallich, G. C., Observations on the coccosphere. *Ann. Mag. Nat. Hist.*, ser. 4, v. 19, p. 342—350, pl. 17, 1877.

Watabe, Norimitsu, Crystallographic analysis of the coccolith of *Coccolithus huxleyi*. *Calc. Tiss. Res.*, v. 1, p. 114—121, figs. 1—9, 1967.

Watabe, Norimitsu, and K. M. Wilbur, Effects of temperature on growth, calcification, and coccolith form in *Coccolithus huxleyi* (Coccolithineae). *Limnol. Oceanogr.*, v. 11, p. 567—575, 7 figs., 1966.

Wilbur, K. M., and N. Watabe, Experimental studies on calcification in molluscs and the alga *Coccolithus huxleyi*. *Ann. N.Y. Acad. Sci.*, v. 109, p. 82—112, 19 figs., 1963.

Wilcoxon, J. A., Upper Jurassic—Lower Cretaceous calcareous nannoplankton from the western North Atlantic Basin. *Init. Repts. Deep Sea Drill. Proj.*, v. 11, p. 427—457, pls. 1—12, fig. 1, 1972.

Wind, F. H., and S. W. Wise, Mesozoic holococcoliths. *Geology*, v. 6, p. 140—142, 2 figs., 1978.

Wise, S. W., Jr., and R. E. Constans, Mid Eocene planktonic correlations: northern Italy—Jamaica, W. I. *Trans. Gulf-Cst. Ass. Geol. Socs.*, v. 26, p. 144—155, pls. 1—4, 1976.

Wise, S. W., Jr., and K. J. Hsü, Genesis and lithification of a deep sea chalk. *Eclog. Geol. Helv.*, v. 64, p. 273—278, 3 figs., 1971.

Wise, S. W., Jr., and F. H. Wind, Mesozoic and Cenozoic calcareous nannofossils recovered by DSDP Leg 36 drilling on the Falkland Plateau, southwest Atlantic Sector of the Southern Ocean. *Init. Repts. Deep Sea Drill. Proj.*, v. 36, p. 269—491, pls. 1—89, figs. 1—3, 1977.

Woodward, A., and B. W. Thomas, The microscopical fauna of the Cretaceous in Minnesota, with additions from Nebraska and Illinois (Foraminifera, radiolaria, coccoliths, rhabdoliths), p. 23—54, pls. C—E. In *The Geology of Minnesota*, v. 3, pt. 1. Minneapolis: State Printers. Geol. and Nat. Hist. Surv. Minn., 1895.

Worsley, T. R., Calcareous nannofossil zonation of Upper Jurassic and Lower Cretaceous sediments from the western Atlantic, p. 1301—1321, pls. 1, 2. In A. Farinacci, *Proc. II Planktonic Conf. Roma 1970*, v. 2, Rome: Edizioni Tecnoscienza, 1971a.

Worsley, T. R., The terminal Cretaceous event. *Nature*, v. 230, p. 318—320, 2 figs., 1971b.

Wray, J. L., and C. H. Ellis, Discoaster extinction in neritic sediments, northern Gulf of Mexico. *Bull. Am. Ass. Petrol. Geol.*, v. 49, p. 98—99, 1 fig., 1965.

GREEN ALGAE:
Prasinophyta, Chlorophyta, and Euglenophyta

In no other class of the Algae do we find such a striking and diverse development of those types of plant-body that must be regarded as primitive and that probably illustrate something of the innumerable attempts at the production of a soma that characterized the earlier phases of the evolution of plants. This diversity has particular interest since the Chlorophyceae, in the pigmentation of their chromatophores and the course of photosynthesis, stand nearer to the main line of evolution of the higher plants than any of the other algal classes.

F. E. Fritsch, 1935

*T*he green algae, Chlorophyta and Prasinophyta, have a diverse and lengthy fossil record. Less closely related to these divisions are the similarly pigmented Euglenophyta, and the Charophyta (see Chapter 11). The green algae also are of evolutionary interest as the ancestors of the higher plants. All of these divisions of the algae were included by earlier workers as distinct classes of a single division, but have been separated on the basis of new information as to their ultrastructure, division mechanisms, and reproductive characteristics, as these features appear to have greater systematic importance than their joint possession of the photosynthetic pigments, chlorophylls *a* and *b*. Following the elevation of these former classes to divisional status, many new classes were proposed within the Chlorophyta, which continue to be an exceptionally diverse group, although no longer regarded as the most primitive among the eucaryotes.

THE PRASINOPHYTA

The distinctive characters of this group were first recognized to have major importance by Chadefaud (1950); studies of the cells with electron microscopy have verified and expanded this separation. As now recognized, the Prasinophyta are noncellulosic green algae with dominant or temporary scaly quadriflagellate (or biflagellate) stage, whose similarly scaly flagella arise from within an **apical pit** that is surrounded externally by four projections. Prasinophytes may be simple flagellated solitary cells, coccoid, or palmelloid, and some are symbiotic. The encysted stage of certain oceanic prasinophytes such as *Halosphaera* and *Pachysphaera* was known long before the motile cell was recognized, and similar acid-resistant cysts are abundant in the fossil record from the Precambrian to the present.

The simplest prasinophyte is a free-swimming quadriflagellate cell, such as *Pedinomonas, Pyramimonas,* or *Platymonas* (see Figure 10.1), all of which have similar morphological characteristics, although other distinctive features, such as the method of cell division, are markedly divergent even at this level.

Pedinomonas has been suggested to be closest to the ancestral green flagellate; like *Pyramimonas* it has a scaly cover as well as the characteristic **phragmoplast** cell division similar to that of the higher plants. During cell division, the nuclear envelope is dispersed in mitosis, and the spindles are open and persistent to form the phragmoplast. The daughter nuclei are separated throughout the division process, and the spindle microtubules are aligned perpendicular to the plane of division. From similar primitive prasinophytes, other green algae with phragmoplast arose (Loeblich, 1974), including the Zygnematales, the genus *Coleochaete,* and the Charophyta (the latter two also have scale-covered motile cells like the prasinophytes, in addition to the phragmoplast) (Pickett-Heaps, 1975, 1976). Another evolutionary line apparently branched from these simple prasinophytes, producing a **phycoplast** during cell division. Here the dividing cell retains the nuclear envelope, which is pierced by microtubules at the pole, forming a closed, fenestrated spindle. The phycoplast has a system of microtubules that lie in the plane of cell division, rather than perpendicular to it. Daughter nuclei are widely separated at first, then approach closely to flatten against each other while the phycoplast microtubules form the dividing wall between them. When this phycoplast structure was first reported in *Chlamydomonas,* it was thought to be primitive, prior to the discovery of the phragmoplast in other algae. It characterizes the Volvocales,

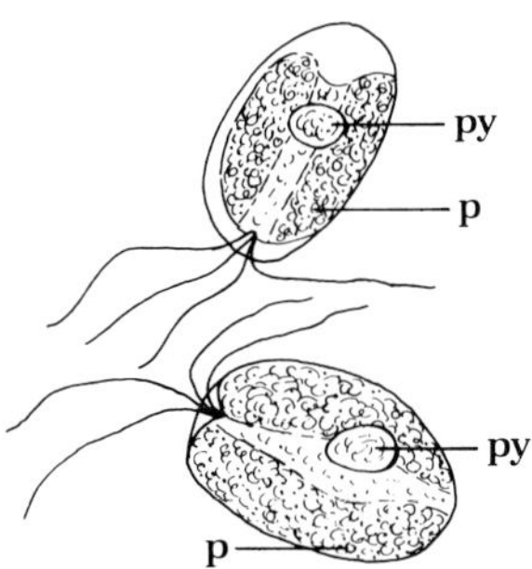

Figure 10.1
Platymonas sp., quadriflagellate motile cells, showing plastids (p) and pyrenoids (py), ×1900.

Chlorococcales, Oedogoniales, and Ulotrichales, and is found in its most primitive form in the prasinophyte *Platymonas,* which is probably similar to the ancestor of *Chlamydomonas* and other Chlorophyta (Stewart et al., 1974).

Based on the nature of their cell division, presence of scales, and absence of cellulose, as well as the distinctive ultrastructural features, the prasinophytes now are regarded as ancestral to all other green algae and to the higher plants, thus occupying an evolutionary position formerly ascribed to the Volvocales, or to the free-swimming *Chlamydomonas.* Some genera now included in the Prasinophyta were originally placed in various orders of the Chlorophyta and Cryptophyta; other prasinophytes probably will be recognized and transferred as details of their cytology become known. However, not all phycologists yet regard the differences as a basis for major taxonomic distinction, at least until a larger number of species have been similarly studied in detail.

The Cell

A typical prasinophyte such as *Pyramimonas grossii* (see Figure 10.2) is inversely pyramidal in form, the anterior end being widest and having four prominent lobes that may continue a short distance as longitudinal ridges. The lobes surround a shallow depression, from whose center arise the four similar flagella. The flagella do

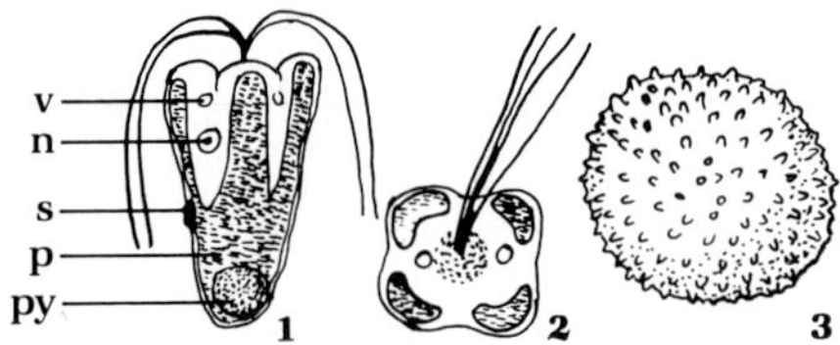

Figure 10.2

Pyramimonas grossii Parke. **1.** Quadriflagellate motile cell, showing plastid (p) with pyrenoid (py), eyespot or stigma (s), nucleus (n), and vacuoles (v). **2.** Anterior view. **3.** Cyst. All ×2500, redrawn with permission from Parke, 1949, *J. Mar. Biol. Ass. U.K.*, v. 28, p. 255–286, published by Cambridge University Press.

not taper, but end bluntly, and in electron microscopy are seen to be covered with fine hairs and scales. A bright yellowish green cup-shaped plastid is similarly four-lobed and occupies the base and entire cell periphery. A large starch-sheathed pyrenoid lies in the posterior part of the plastid. The nucleus is in the anterior part of the cell, and an eyespot or stigma at one side of the cell may be median to anterior in position (Parke, 1949). Electron microscopy shows the cell body to be covered with scales that may be of as many as three distinct types and form distinct layers (see Figure 10.3). The innermost layer consists of small, closely packed flat square scales, 0.04 to 0.05 μm in diameter; it is covered by a layer of larger, boxlike scales of about 0.25 μm width and with sides 0.15 μm high. This in turn is covered by an outermost layer of higher, crownlike scales (Swale, 1973; Norris and Pearson, 1975; Pennick et al., 1976). Flagellar inner scales are like the smaller body scales in size, but are commonly five- or six-sided. They are overlain by a layer of ovate rimless scales with median ridge that continues at one margin into a spine (see Figure 10.4). All scales are composed of a meshwork of regularly arranged fibrils, are formed in vesicles of the Golgi apparatus, and are stored in a scale reservoir lying near the flagellar root system, between the nucleus and a large vacuole, until they are transferred to the exterior.

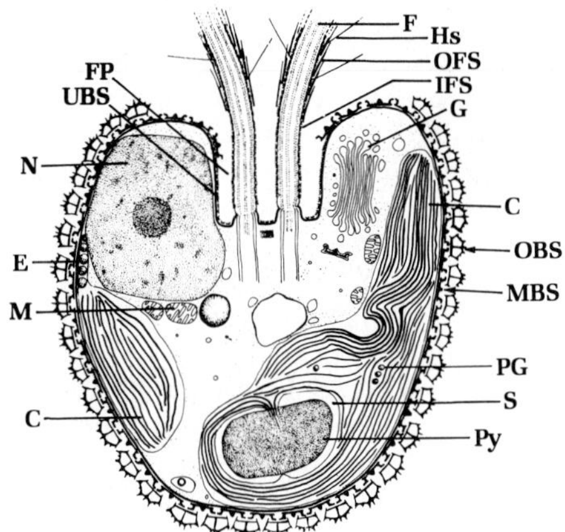

Figure 10.3

Pyramimonas obovata N. Carter, diagrammatic longitudinal section, showing flagellar pit (FP) containing flagella (F) with hairscales (Hs) and an outer (OFS) and inner (IFS) layer of flagellar scales; cell covered with outer (OBS), middle (MBS), and under (UBS) layers of scales, and contains plastid (C), pigment granules (PG), pyrenoid (Py), starch grains (S), mitochondria (M), eyespot (E), nucleus (N), and golgi apparatus (G). From Pennick et al., 1976.

Pyramimonas also may possess ejectile trichocysts, an organelle that is rarely found in green algae.

Reproduction

Only vegetative (asexual) reproduction is known in the Prasinophyta as yet. The flagellate cell of *Pyramimonas* divides longitudinally, each daughter cell inheriting one half of the divided plastid, as well as two of the flagella, and then regenerating the remainder.

Platymonas reproduces in a nonmotile stage. The old cells become heavily filled with starch and lose their four short flagella, becoming nonmotile. At cell division, one of the two daughter cells reverses position so that it is inverted; both are completely formed, including their flagella, prior to being released from the

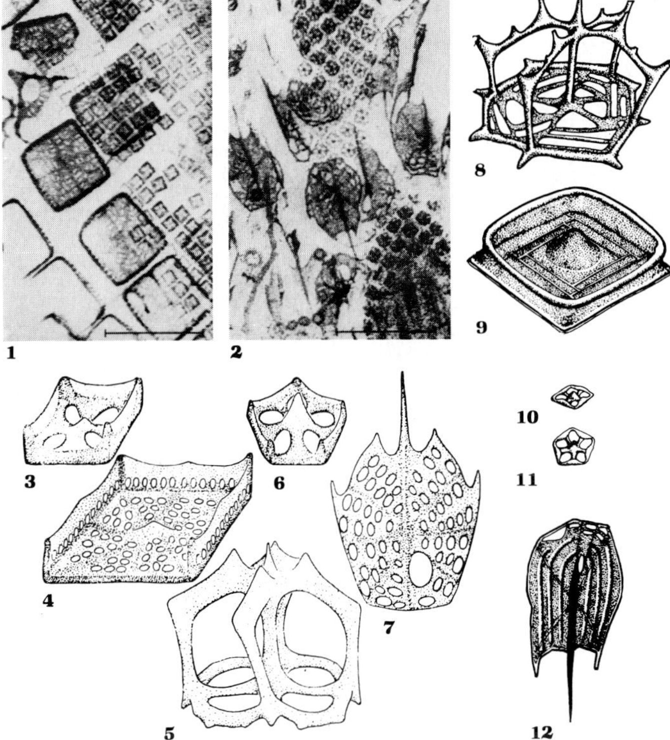

Figure 10.4
Prasinophyte scales. **1 — 7.** *Pyramimonas parkeae* Norris & Pearson, from Norris and Pearson, 1975. **1.** Body scales, showing the large rimmed scales that each overlie 16 of the smaller inner scales. **2.** Flagellar scales, showing large spine-bearing scales and spirally arranged smaller scales. 1,2, × 40,000. **3.** Small body scale. **4.** Rimmed body scale. **5.** Outer coronate body scale. **6.** Small flagellar scale. **7.** Large flagellar scale. **8 — 12.** *P. obovata,* scales, from Pennick et al., 1976. **8.** Outer layer body scale. **9.** Middle layer body scale. **10.** Inner layer body scale. **11.** Inner layer flagellar scale. **12.** Outer layer flagellar scale.

mother cell theca. The new theca is built from tiny stellate particles, manufactured within vesicles derived from Golgi bodies, that are homologous with the scales. These particles appear first in the vicinity of the pyrenoid and form last at the apical end, coalescing to form the theca. Flagella are developed prior to the completion of the wall in this area. *Platymonas* may occur in tidepools in such abundance as to form a greenish soup; the genus also occurs in fresh water.

Prasinocladus, one of the more advanced colonial dendroid forms, also grows on rocks or shells in tidepools. Asexual reproduction involves the liberation of quadriflagellate zoospores that resemble *Pyramimonas,* having blunt-tipped scaly flagella inserted in a depression at the anterior end. After swimming actively the zoospore settles anterior end down, loses the flagella, and secretes a new wall. It germinates to form an elongate tube, the zooid moving into the top of the tube. Division into additional cells with tubes may occur at any stage of growth. The sporelings from the zoospores may produce new flagellates within 13 days after settling to the substrate, but if conditions are unfavorable, the sporelings round off to form a thick-walled cyst. The latter germinates again upon the return of favorable conditions (Chihara, 1963).

Cyst or Phycoma

Some prasinophytes have a dominant quadriflagellate monad stage, and others a coccoid or palmelloid form, with motile zoospores. Others may have a long-lasting cyst or **phycoma** stage and a flagellate stage of lesser duration. The cysts may be smooth, spinose, or warty in appearance (Belcher, 1966), those of

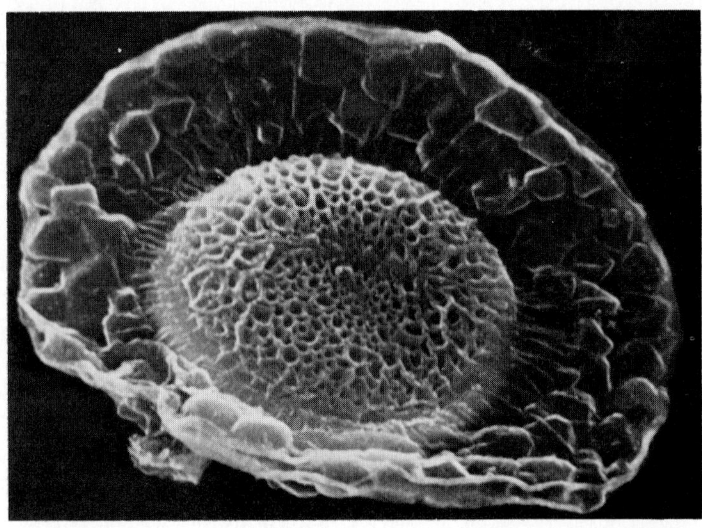

Pyramimonas ranging from 6 or 7 μm in diameter to as much as 24 μm in *P. amylifera* Conrad. The cystlike nonmotile stages of *Halosphaera*, *Pachysphaera*, and *Pterosperma* have long been known (Ostenfeld, 1910), but have been the object of more detailed study in recent years (Parke and Hartog-Adams, 1965; Parke, 1966; Parke et al., 1978), and have been related to the fossil *Tasmanites*, *Cymatiosphaera*, and the leiospheres (Boalch and Parke, 1971). The perforate, thick-walled phycoma of *Pachysphaera pelagica* Ostenfeld resembles *Tasmanites*; *Pterosperma moebii* (Jørgensen) Ostenfeld, with broad equatorial ala or flange, resembles the fossil *Pterospermella* (see Figure 10.5); and *Pterosperma marginatum* Gaarder has a reticulum of elevated crests resembling *Cymatiosphaera* (see Figure 10.6). Although Parke et al. (1978) regarded the fossil *Cymatiosphaera, Pterospermopsis,* and *Pterospermella* as congeneric with *Pterosperma*, they are here retained as distinct, as are the fossil dinoflagellates that similarly are distinct only in the encysted stage.

Development of the Cyst

Halosphaera (Parke and Hartog-Adams, 1965), *Pachysphaera* (Parke, 1966), and *Pterosperma* (Parke et al., 1978) have been shown to have two distinct phases in their life history, a motile quadriflagellate cell that can reproduce itself by simple fission, and a phycoma or cystlike phase that may develop directly from the motile one. A lunar periodicity was observed in the changeover from motile to phycoma stage in *Pterosperma* (Parke et al., 1978). Large numbers of young phycomata can be obtained from surface waters within 24 hours of the new or full moon, mature phycomata about four days before, and the empty cyst in abundance a few days before or after the new or full moon. As the motile cell and young phycomata are similar in different species, most are identifiable only as the empty phycoma.

After loss of flagella and early formation of the phycoma, it increases gradually in size to maturity, the protoplast then dividing to produce the motile stage. The cyst is at first about 10 μm in diameter, with a single nucleus, plastid, and pyrenoid. The plastid divides as the cyst enlarges; the pyrenoid enlarges and then also divides. The mature cyst may be 100 to 175 μm in diameter, having increased between 2 and 5 μm in diameter daily. The cytoplasm is crowded with small discoid plastids and globose starch-ensheathed pyrenoids. Repeated nuclear divisions and separation of cell contents produce many motile cells, each with a single nucleus, plastid, pyrenoid, and starch sheath. The inner wall swells (see Figure 10.7), the outer one

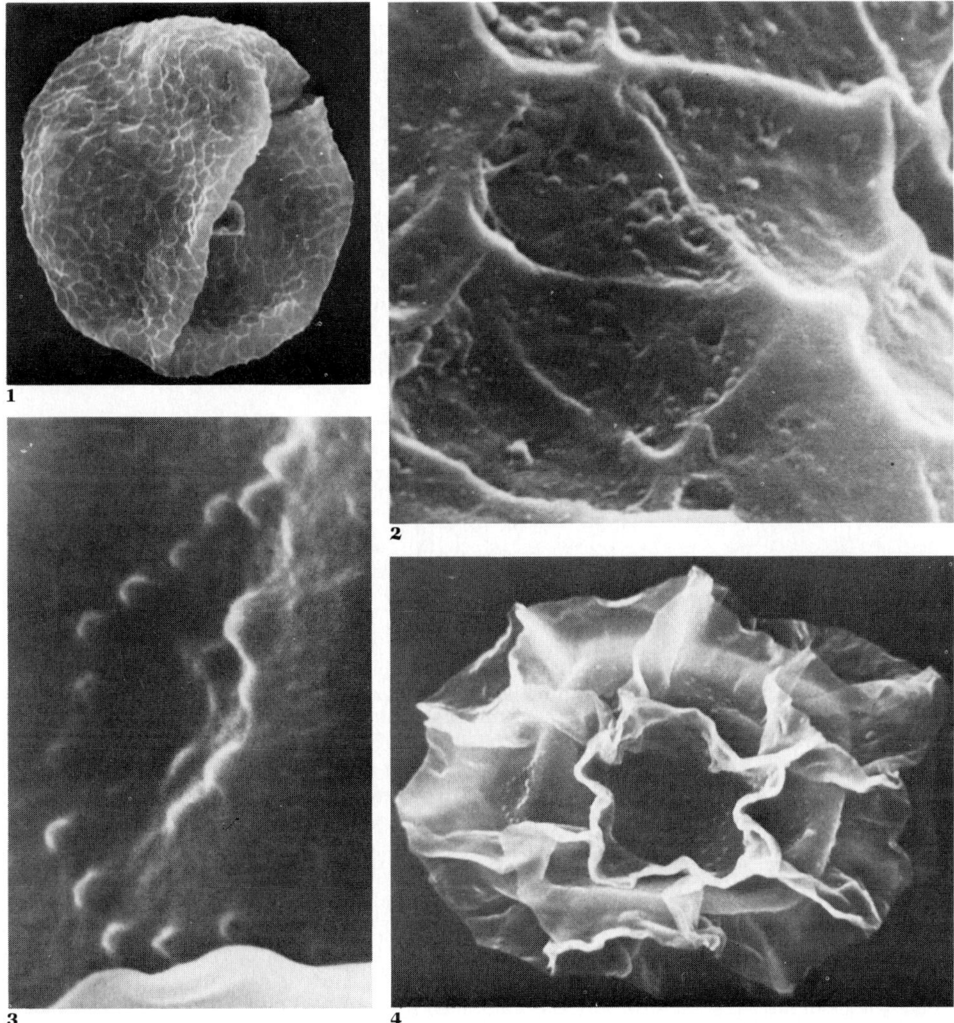

Figure 10.6

Cymatiosphaeraceae. **1,2.** *Melikeriopalla amydra* Tappan & Loeblich, Silurian of Indiana. **1.** Vesicle with split probably resulting from release of zoospores, SEM, ×800. **2.** Enlargement, showing grana in fields between crests of the reticulum, SEM, ×16,000. **3,4.** *Cymatiosphaera* sp., Lower Devonian, Oklahoma. **3.** Portion of a field showing rings of grana, SEM, ×14,400. **4.** Vesicle with extremely high filmy crests surrounding the sparsely granulate fields, SEM, ×1600.

splitting to liberate the inner wall and contents. Enlargement may continue; the contents then are released as quadriflagellate monads, with scale-covered body and flagella. The entire process from the formation of the cyst to release of new motile cells requires up to six hours in the laboratory.

Cyst Wall Structure and Composition

The pelagic cyst of *Pachysphaera* is spherical, with a double wall 8 to 10 μm thick pierced by radial canals that stop at the outer membrane. Although the perforations do not completely pierce the wall, they may allow an osmotic ex-

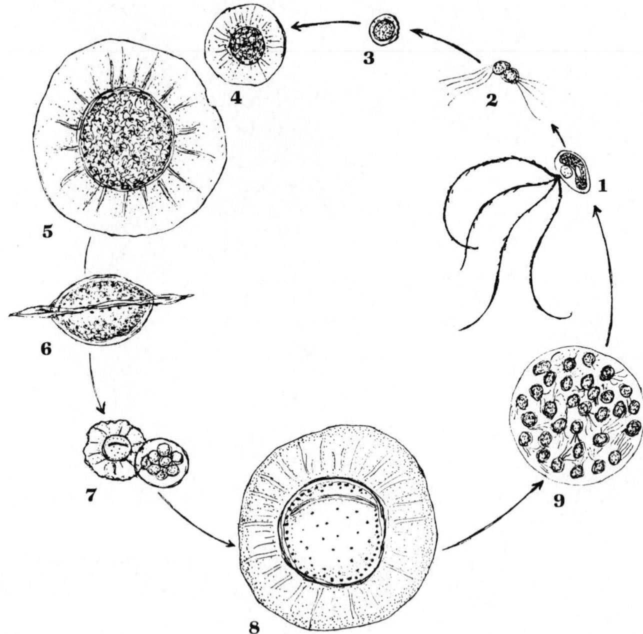

Figure 10.7
Life cycle of *Pterosperma*. **1.** Motile phase, a scale-covered cell with excentric nucleus, large saucer-shaped plastid lying against the convex side of the body, and starch-enclosed pyrenoid; four long excentrically inserted scale- and hair-covered flagella. Motile phase may reproduce independently and repeatedly by fission. **2.** Dividing motile cells. **3.** Beginning of phycoma phase as motile cell loses flagella, rounds off, and develops thickened wall and a flangelike ala. Phycoma increases rapidly in volume, plastid divides into many smaller ones that surround a central lipid globule; size increase continues to maturity, requiring from two weeks to three-and-a-half months in culture. **4,5.** Mature phycomata of varied size, with thickened inner wall, just prior to division of cell contents. **6.** Phycoma in edge view. **7.** Thickened inner wall of phycoma and contained cell material has escaped through the release suture, a slit in the outer phycoma wall; after escape, the former inner wall enlarges to double the previous diameter and the contents divide into two, four, or eight uninucleate masses that develop flagella. **8.** Empty phycoma, showing release suture (compare fossil prasinophytes of Figure 10.17 part 1 and Figure 10.5), peripheral ala, and ring of alaband pores, as well as other scattered pores. **9.** Phycoma inner wall, each of the early 8 quadriflagellate cells having divided twice to produce 32 motile cells; inner wall increases in diameter up to 16 times that prior to release from outer wall, accompanied by decrease in wall thickness and continued division of motile cells that finally burst the wall and escape. Based on various species, data from Parke et al., 1978. 1, ×1000; others, ×150.

change with the surrounding medium, and facilitate nutrient uptake (Combaz, 1967).

The inner part of the wall is pectic in nature, whereas the tough punctate outer wall probably consists of a complex "lipoid" substance. Radial elements in the wall are probably crystallites, perhaps consisting of fatty acid molecules, but may be of silica, although this has not been confirmed (Wall, 1962; Jux, 1969a). Various fossil taxa have distinct inner and outer walls, which may even represent a distinct inner body in *Maculatasporites* (see Figure 10.8).

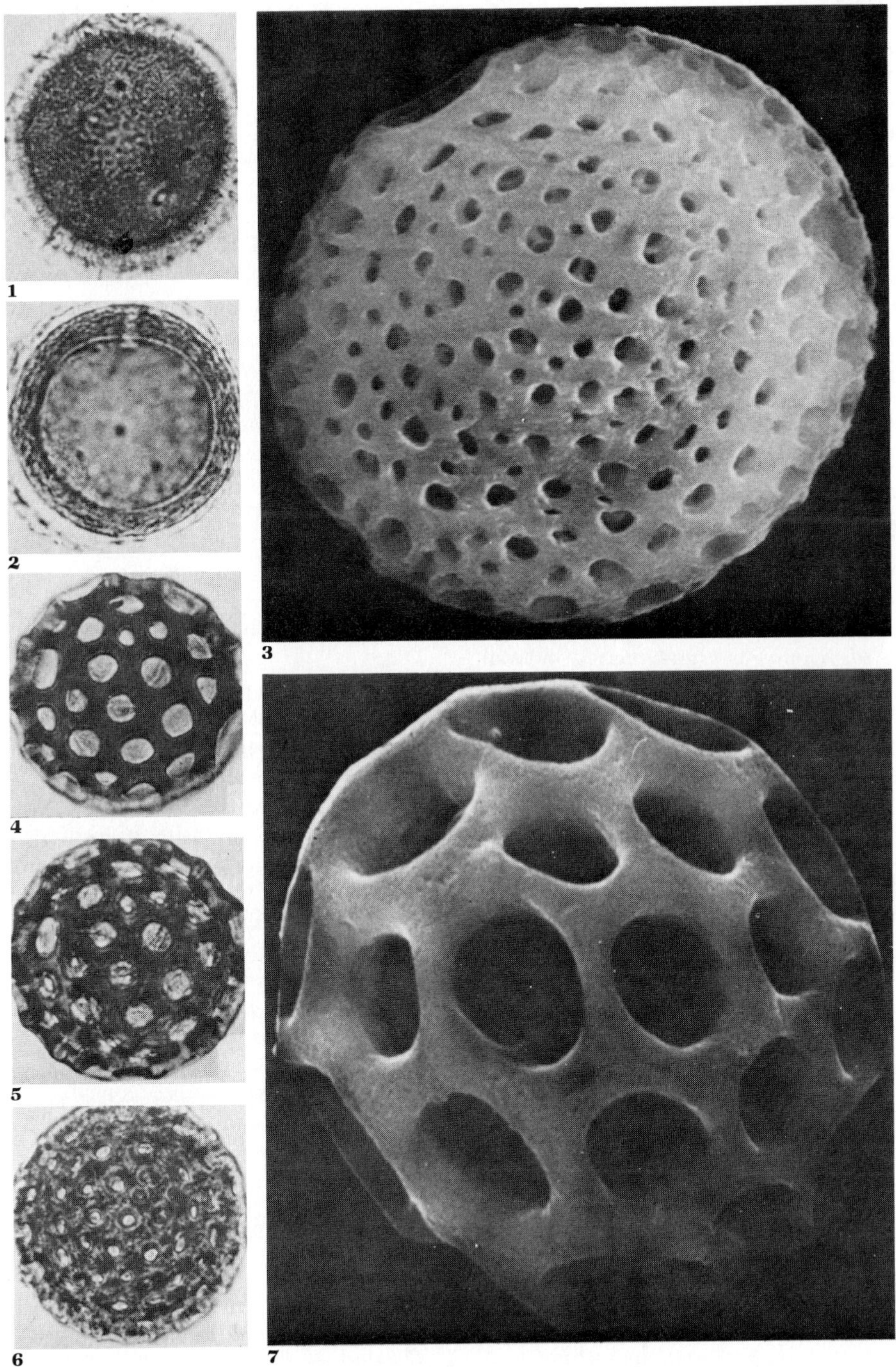

Figure 10.8

Permian Tasmanaceae, Australia. **1,2.** *Haplocystia pellucida* Segroves, surface view and optical section, ×750, from Segroves, 1967. **3.** *Maculatasporites amplus* Segroves, showing perforations of different sizes and large opening at upper left, SEM, ×2000. **4—7.** *M. minimus* Segroves. **4,5.** Surface and optical section of same specimen, showing very large pores of outer wall layer and somewhat collapsed inner body. **6.** Surface ×750, from Segroves, 1967. **7.** SEM, showing outer coarsely perforate wall, ×2500.

The cyst wall of *Halosphaera* is much thinner than that of *Pachysphaera*, seldom exceeding 1.5 to 5 μm, but like that of *Pachysphaera* also shows growth lamellae in electron micrographs (Jux, 1969b).

The wall chemistry of two fossil Prasinophyta, *Leiosphaeridia* and *Tasmanites*, was studied by infrared spectroscopy. Absorption peaks indicate the presence of COOH, CH_2, and CH_3. The relative abundance of the various groups suggests the presence of long aliphatic, saturated chains. The results were identical with specimens obtained from acid residues and with those from water dispersion and ultrasonic separation, indicating a high chemical stability. Even heating to 240°C did not noticeably affect the absorption peaks. However, ultrastructure studies showed a difference in the two organisms. The wall of *Leiosphaeridia* is completely homogeneous, with no observable ultrastructural features (see Figure 10.9). In contrast, *Tasmanites* is characterized by wall perforations, commonly with radially arranged pores of two distinct sizes; some also had a layered wall (Kjellström, 1968). Eisenack (1968) reported the presence of a pylome (a supposed excystment opening) in various species of *Leiosphaeridia* and *Tasmanites;* he did not regard their morphological similarity to be proof of a green algal affinity. A simple circular pylome was found in *Leiosphaeridia oelandica* Eisenack, a circular pylome with produced rim in *Tasmanites martinssoni* Eisenack, one with a thickened rim in *L. voighti* Eisenack, and a narrowed opening enlarging inward in *T. verrucosus* Eisenack.

As noted above, *Pachysphaera* may undergo repeated divisions of the protoplast, producing numerous motile cells that escape from the cyst. Suggestive of such a process are the smaller spheres (one to three) reported within the cyst of *Tasmanites medius* (Eisenack) Eisenack. These may have been a stage in the production of motile cells or may have represented a different type of reproduction. They do not appear to be accidental inclusions in at least some of the specimens described (Eisenack 1955, 1968). Specimens of *Campenia* from the German Jurassic were reported to have an elongate slit at one side of the globular, smooth-walled vesicle, within which could be seen smaller lanceolate "brood" bodies, pointed at each end. Similar lanceolate bodies that appeared to be rolled up were observed isolated in the same material and termed *Lancettopsis* (Mädler, 1963). They were regarded as possible "spores" that would develop into *Campenia*. However, as similar folded or collapsed individuals of many spherical species are found throughout the Paleozoic, these are probably neither spores nor a distinct taxon, but merely result from variations in the preservation of the normally globular specimens.

Occurrence

Present Distribution

Both *Halosphaera* and *Pachysphaera* occur in the marine plankton, at depths of 10 m or more. *Halosphaera viridis* Schmitz is widely distributed in the Mediterranean Sea, the North and South Atlantic, and particularly in the Gulf Stream, and is somewhat less abundant in the Pacific, Indian, and Antarctic Oceans. The Discovery Expedition reported it in surface waters of the Mediterranean; it is transported into the Atlantic by deeper currents, occurring in the tropical Atlantic only in the deeper waters below the photic zone (80 to 100 m deep). Specimens gradually ascend in the water column as they move northward with the currents (Ostenfeld, 1910; Yashnov, 1965). Monospecific blooms of *Halosphaera viridis* have been reported off Newfoundland (Ackman et al., 1970) and are thought to have resulted from an unusual enrichment of the coastal waters with nitrogen and phosphates from industrial wastes.

Other prasinophytes are common in tidepools, brackish and even fresh water, and *Platymonas* may be symbiotic with the marine worm *Convoluta*, although endosymbionts are less common in this division than in the Chlorophyta or Pyrrhophyta.

Figure 10.9
Leiosphaeridia spp. **1−3.** Middle Ordovician, Oklahoma. **1,2.** Views of
crushed and folded spherical cysts. **3.** Partially rolled up cyst, similar to
forms described as *Lancettopsis*. **4.** Upper Ordovician, Oklahoma, specimen split open to release contents (compare Figure 10.7 part 4). 1, ×518;
2−4, ×384.

Geologic Occurrence

As is true of most phytoplankton, the motile
flagellate cell is unlikely to be preserved unless
there are skeletal mineralized structures. Although the motile cell does possess tiny scales,
these are not calcified (as in the coccolithophores) or silicified (as in some chrysophytes).
The scales are constructed of fine fibrils,
but are not regarded as cellulosic; the wall of
Platymonas subcordiformis Hazen contains
amino acids and sugars but no glucose in

its breakdown product, and the possibility of cellulose was eliminated by paper chromatography, electron microscopy, X-ray diffraction, and staining techniques (Lewin, 1958). There is no indication of a possible resistance to chemical destruction and thus little likelihood of fossilization of the motile cell wall. However, the cyst stage of many prasinophytes is highly resistant, and both the thin-walled leiospheres and the thick-walled tasmanites, and those with surface ala and crests, appear closely related to these.

Although those fossils such as *Leiosphaeridia*, with thin-walled nonperforate cysts, resemble modern Halosphaeraceae (Wall, 1962), *Leiosphaeridia* lacks the lamellar wall structure that characterizes *Halosphaera* and *Pachysphaera* cysts (Jux, 1969b). Whether or not this is due to imperfect preservation is not certain.

Thin-walled leiospheres are found in the Precambrian, as well as throughout the Paleozoic. Some were originally described as spores of higher plants, but reported trilete marks were apparently an erroneous interpretation of folds due to compression, and later studies failed to confirm their presence. Leiospheres were even suggested to be trilobite eggs. The extremely abundant leiospheres may be so entangled in organic matter that some have been suggested to represent sac-enclosed spores of larger benthic algae, although their numbers and wide distribution indicate a planktonic rather than benthic habitat.

Thick-walled *Tasmanites*, ranging in diameter from 30 to 600 μm (see Figure 10.10), occur widely in the Paleozoic and locally in the Mesozoic. Geologically long-ranging (tasmanites and leiospheres both occur in the Precambrian), this group is abundant only under special conditions, but may then constitute 95 percent of the microfossil assemblage. The tasmanite black shales in North Africa in the Silurian (base of the Gothlandian) and Devonian (base of the Frasnian) were deposited in a transgressing sea with a low swampy coastline (Combaz, 1966). Associated microfossils include chitinozoans, acritarchs, and remains of graptolites. The strata also are unusually radioactive, perhaps an indication of unusual condi-

tions that restricted other types of organisms but allowed the proliferation of prasinophytes. Commonly occurring with conodonts and other marine fossils in Devonian and Mississippian black shales of Ohio, the abundance of *Tasmanites* is inversely proportional to that of landplant spores (Winslow, 1962).

In the Permian of Tasmania, "white coal" or "tasmanite," composed entirely of the tiny perforate-walled disclike *Tasmanites* (see Figure 10.11), accumulated in deposits several feet thick, extending laterally for several miles (Newton, 1875). A similar occurrence of tasmanite was reported in early Mesozoic rocks of the northern flank of the Brooks Range, Alaska. Beds from one to four feet thick consist almost entirely of the compressed *Tasmanites*, and are extremely organic-rich, yielding as much as 500 l of oil per tonne (150 gallons per ton; Tourtelot et al., 1968). As in other deposits of tasmanite, the discoid shape of the fossils results from compression, as the specimens undoubtedly were originally spherical.

The earliest Pterospermellaceae also appeared in the latest Precambrian but became much more important in the Paleozoic, particularly in the Devonian. Similar forms persist to the present time. Taxa with surface reticulum (Cymatiosphaeraceae) appear somewhat later, but are present at least by the Ordovician. Like the tasmanites and pterospermellaceans, they diversified greatly in the Devonian and later Paleozoic, but are rare in the Mesozoic and Cenozoic. The Pterosphaeridiaceae are less long ranging. As far as known, they appeared in the Permian and lasted only through the Jurassic. Possibly some Cymatiosphaeraceae that have been studied only at low magnification may prove to have the large perforations within the fields that characterize this family.

In general the fossil prasinophytes appear to be "disaster species," surviving the widespread extinctions that eliminated most acritarchs in the middle Paleozoic, and being most abundant in the absence of the other phytoplankton. When dinoflagellates and other algal groups diversified in the Mesozoic and Cenozoic, the prasinophytes became less important.

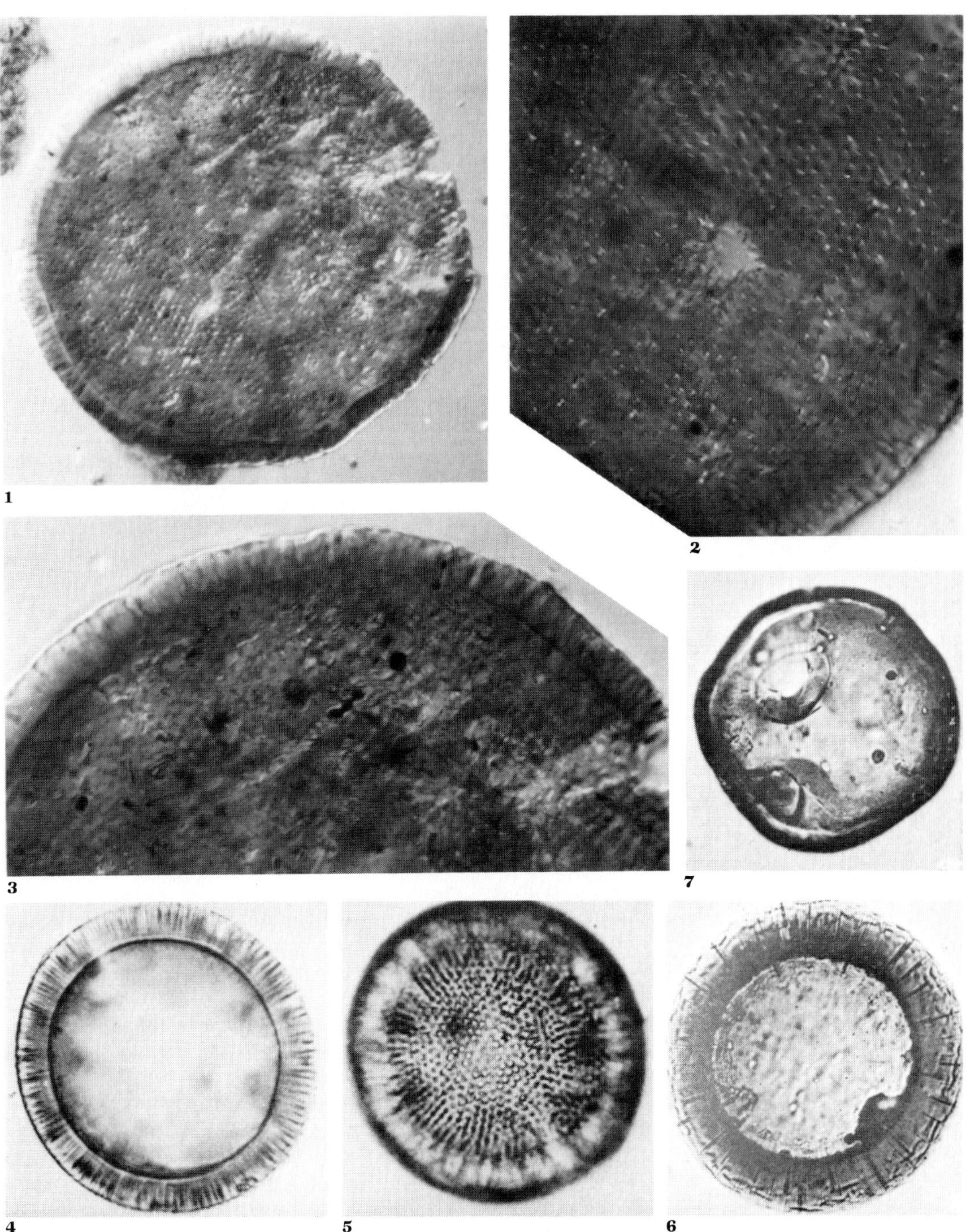

Figure 10.10

Tasmanites. **1—3.** *Tasmanites* sp., Upper Ordovician, Oklahoma. **1.** Showing thick wall in optical section, and punctate inner surface where pores open, ×518. **2,3.** Higher magnification focussed on surface and wall thickness, respectively, ×960. **4,5.** *T. globulus* (O. Wetzel) Morgenroth, lower Eocene, Germany, optical section and surface, ×600, from Morgenroth, 1966. **6,7.** *T. huronensis* (Dawson) Eisenack, Middle Silurian, Germany, optical section of very thick-walled specimen, ×175, and specimen with rounded pylome, ×280, from Eisenack, 1958.

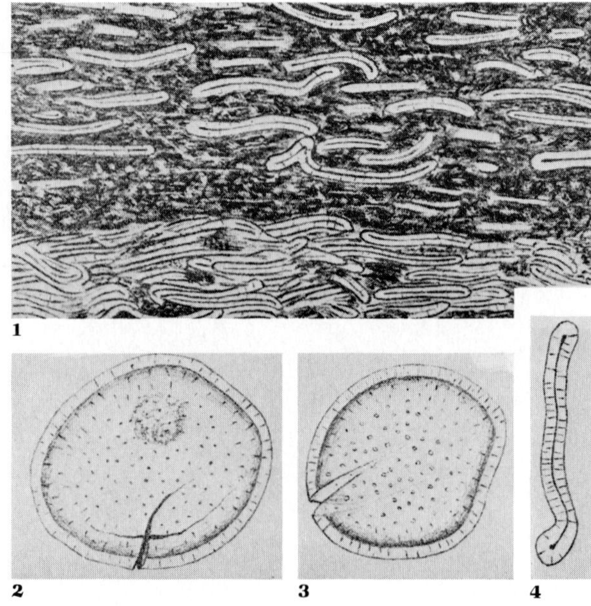

Figure 10.11
1. Section of tasmanite, cut perpendicular to bedding, showing flattened sections of *Tasmanites*, ×50. **2—4.** *Tasmanites punctatus* Newton, ×50. **2,3.** Showing punctations, thick wall, and rupture. **4.** Transverse section of compressed specimen. All from Newton, 1875.

Classification

As previously noted, separate recognition of the Prasinophyta is relatively recent, but the systematic importance assigned to their distinctive features varies among phycologists. The prasinophyte genera may be placed in families of the Chlorophyta, or in separate families and orders, class or division. Because of their position ancestral to other green algae, and their extremely long geologic history, the Prasinophyta are here regarded as deserving of divisional status.

From a fossil record at first suggested to include *Tasmanites,* and later other similar forms such as *Tytthodiscus, Noremia, Tapajonites,* and *Quisquilites* (see Figure 10.12), the scope has been gradually enlarged (Loeblich, 1974; Wicander, 1974), and additional genera are here regarded as best related to this group. Previously, some were included with the acritarchs as of unknown relationships, or as the orders Leiosphaeridiales, Tasmanales, and Hystrichosphaerales (= dinoflagellates) were placed in a Class "Hystrichophyta" (Mädler, 1963). In the present volume, fossils that appear related to the Prasinophyta are placed in two orders of modern taxa, but in distinct families and genera (see Table 10.1). Thus the family Leiosphaeridiaceae is in the Pyramimonadales, with the similar modern Halosphaeraceae; the families Tasmanaceae, Cymatiosphaeraceae, Pterosphaeridiaceae, and Pterospermellaceae are included in the order Pterospermatales with the modern Pterospermataceae, containing the similar *Pachysphaera* and *Pterosperma.* Although the fossils are thus regarded as distinct from the living genera and families, more information as to life cycles of additional living Prasinophyta may eventually allow some fossils (at least for the Cenozoic) to be related to the modern counterparts.

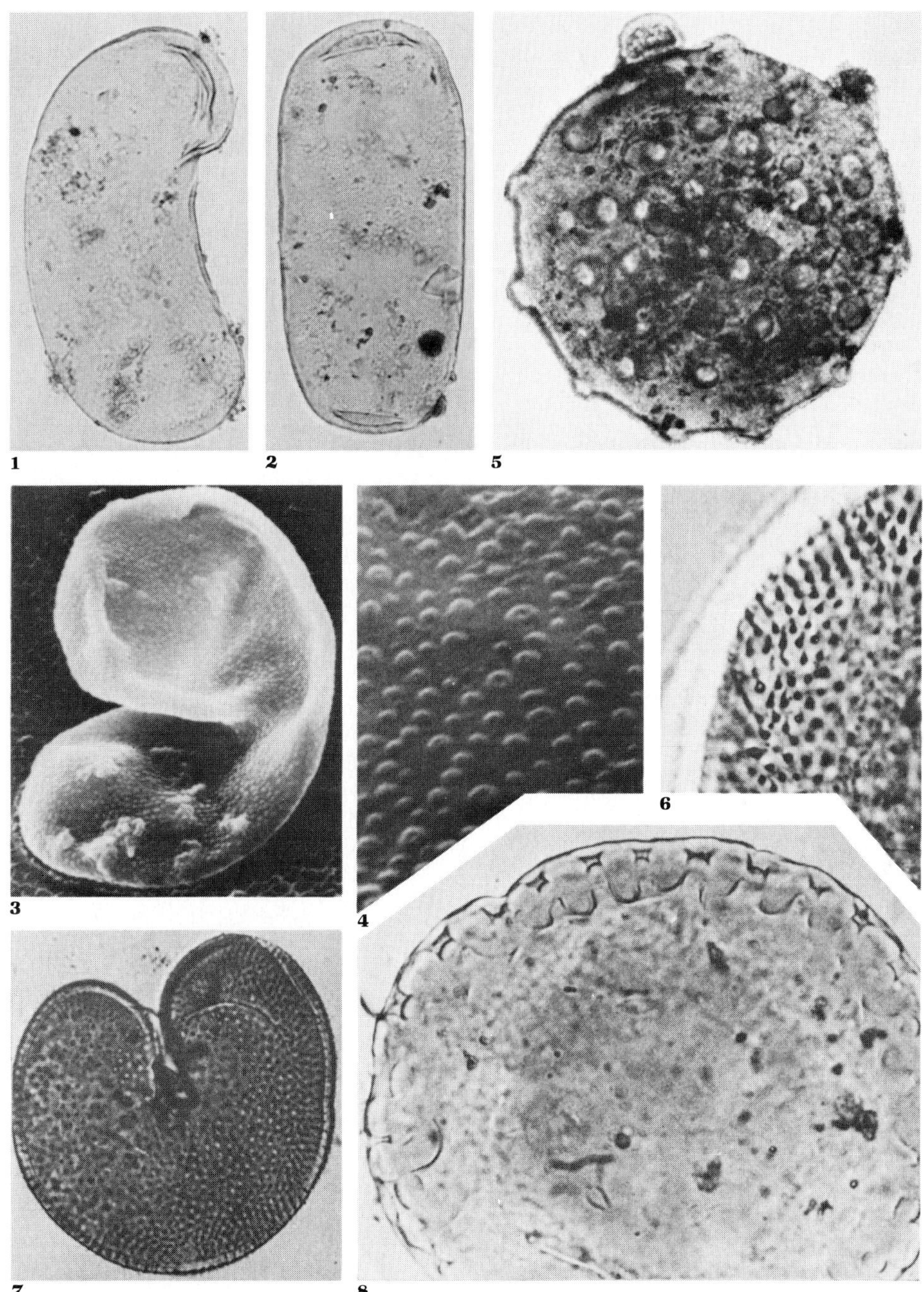

Figure 10.12

Devonian Tasmanaceae. **1—4.** *Quisquilites buckhornensis* Wilson & Urban, Upper Devonian, Oklahoma. **1,2.** ×916. **3.** SEM, ×1000. **4.** SEM, enlargement showing surface grana, ×3000. 3,4, from Wilson and Urban, 1971. **5.** *Tapajonites roxoi* Sommer & van Boekel, Devonian, Brazil, ×430, from Sommer and van Boekel, 1963. **6,7.** *Pseudolunulidia imperatrizensis* Brito & Santos, Middle Devonian, Brazil, from Brito, 1967c. **6.** Enlargement, showing spinulate surface, ×800. **7.** Entire specimen, ×320. **8.** *Maranhites brasiliensis* Brito, Middle to Upper Devonian, Brazil, ×500, from Brito, 1967c.

Table 10.1
Classification of the Prasinophyta.

Division Prasinophyta Round 1971

Motile stage with ultramicroscopic body scales, and 2 to 4 scaly flagella arising from apical pit that is surrounded by four apical projections; large axial plastid; may have colonial, coccoid, palmelloid, or branching stage; cysts or phycoma enlarge vegetatively. Planktonic and benthic; marine and freshwater; *Platymonas* may be symbiotic with the marine worm *Convoluta*. Precambrian to Holocene.

A. Order Pedinomonadales

1. Family Pedinomonadaceae Korshikov 1938
Holocene.
Pedinomonas Korshikov 1923.

B. Order Pyramimonadales Chadefaud 1950

1. Family Pyramimonadaceae
Holocene.
Asteromonas Artari 1913; *Chloraster* Ehrenberg 1848; *Pyramimonas* Schmarda 1850; *Stephanoptera* P. A. Dangeard 1910.
2. Family Halosphaeraceae
Holocene.
Halosphaera Schmitz 1878; *Hyalophysa* Cleve 1900.
3. Family Leiosphaeridiaceae Timofeev 1956 nom. corr. Mädler 1963
Syn: Family Leiosphaeraceae Eisenack 1954
Probably equivalent to living Halosphaeraceae. Some listed here may prove to be synonyms when thoroughly studied. Precambrian to Holocene.
Campenia Mädler 1963; *Chuaria* Walcott 1899 (syn.: *Fermoria* F. Chapman 1935; *Krishnania* Sahni & Shrivastava 1954; *Protobolella* F. Chapman 1935); *Granodiscus* Mädler 1963; *Halosphaeropsis* Mädler 1963; *Leiosphaeridia* Eisenack 1958 (syn: *Kildinella* Timofeev 1966; *Polyedrosphaeridium* Timofeev 1966; *Lancettopsis* Mädler 1963; *Protoleiosphaeridium* Timofeev 1960; *Synsphaeridium* Timofeev 1966 non Eisenack 1965; *Wendiella* Timofeev 1962); *Orygmatosphaeridium* Timofeev 1959; *Protosphaeridium* Timofeev 1966; *Pseudozonosphaeridium* Andreeva 1966; *Trachysphaeridium* Timofeev 1966 (syn.: *Menneria* Lopukhin 1971); *Zonaletes* Staplin 1960; *Zonosphaeridium* Timofeev 1969.

C. Order Prasinocladales

1. Family Prasinocladaceae
Holocene.
Platymonas G. S. West 1916; *Prasinochloris* Belcher 1966; *Prasinocladus* Kuckuck 1894; *Tetraselmis* Stein 1878.

D. Order Pterospermatales Schiller 1925

1. Family Nephroselmidaceae Pascher 1913
Holocene.
Bipedinomonas N. Carter 1937 (syn.: *Heteromastix* Korshikov 1923 non James-Clark 1965); *Micromonas* Manton & Parke 1960; *Nephroselmis* Stein 1878; *Pseudoscourfieldia* Manton 1975.
2. Family Pterospermataceae Lohmann 1904
Holocene.
Hexasterias Cleve 1900; *Pachysphaera* Ostenfeld 1899; *Pterosperma* Pouchet 1894; *"Sphaeropsis"* Meunier 1910 (non Saccardo 1880).
3. Family Tasmanitaceae Sommer 1956 nom. correct. herein (see Figures 10.13 and 10.14)
Syn.: Crassosphaeridae Simoncsics & Kedves 1961
Wall thick, perforated by canals that open as pores to the interior; some lacking wall perforations tentatively retained herein. Precambrian to Holocene.
Annulodiscus Hajós 1966; *Arabisphaera* Hemer & Nygreen 1967; *Cooksonella* Nagy 1965; *Crassosphaera* Cookson & Manum 1960; *Discoidella* Habib 1970; *Haplocystia* Segroves 1967; *Hungarodiscus* Krivan-Hutter 1962; *Inderites* Abramova & Marchenko 1964; *Krimodiscus* Aristova 1972; *Maculatasporites* Tiwari 1964; *Maranhites* Brito 1965; *Noremia* Kedves 1962; *Ocridoligotriletes* Timofeev 1958; *Pseudolunulidia* Brito & Santos 1965; *Pseudoschizaea* Thiergart & Frantz 1962 (syn.: *Concentricystes* Rossignol 1962); *Pseudotasmanites* Kir'yanov 1974; *Quisquilites* Wilson & Urban 1963; *Schizophacus* Pierce 1976; *Spongiocysta* Segroves 1967; *Tapajonites* Sommer & van Boekel 1963; *Tarsisphaeridium* Riegel 1974; *Tasmanites* Newton 1875 (syn.: *Perisaccus* Naumova 1939); *Trematoligotriletum* Timofeev 1959; *Tytthodiscus* Norem 1955.
4. Family Cymatiosphaeraceae Mädler 1963 (see Figure 10.15)

(continued)

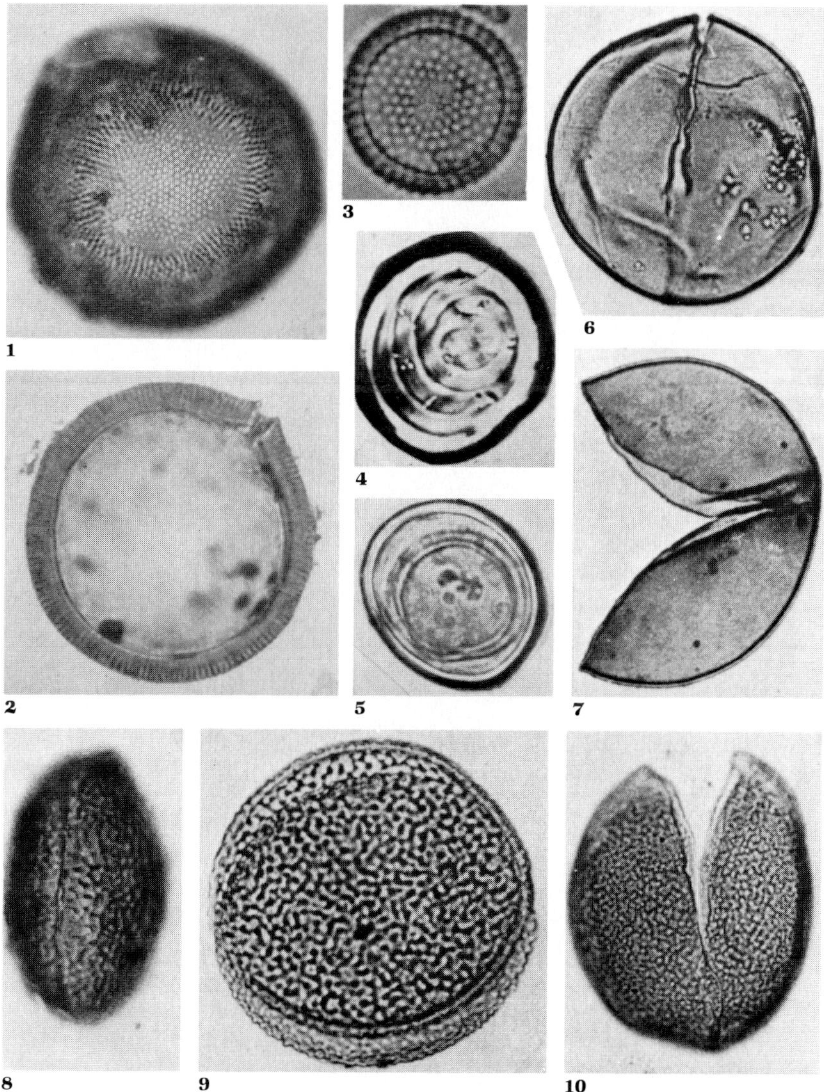

Figure 10.13

Mesozoic and Cenozoic Tasmanaceae. **1,2.** *Tytthodiscus suevicus* Eisenack, Lower Jurassic, Germany, from Eisenack, 1957. **1.** Surface view, ×400. **2.** Optical section, showing thick wall, ×330. **3.** *Discoidella hannae* Habib, middle Cretaceous, North Atlantic, ×850, from Habib, 1970. **4.** *Pseudoschizaea circula* (Wolff) Christopher. **5.** *P. rubina* Rossignol ex Christopher. 4,5, Pleistocene, Israel, ×800, from Rossignol, 1962. **6,7.** *Schizophacus spriggii* (Cookson & Dettmann) S. T. Pierce, equatorial views of specimens with incipient split and open split, ×400. **8−10.** *S. rugulatus* (Cookson & Dettmann) S. T. Pierce, ×400. **8,9.** Equatorial and polar views. **10.** Polar view of opened specimen. 6−10, Lower Cretaceous, Australia, from Cookson and Dettmann, 1959.

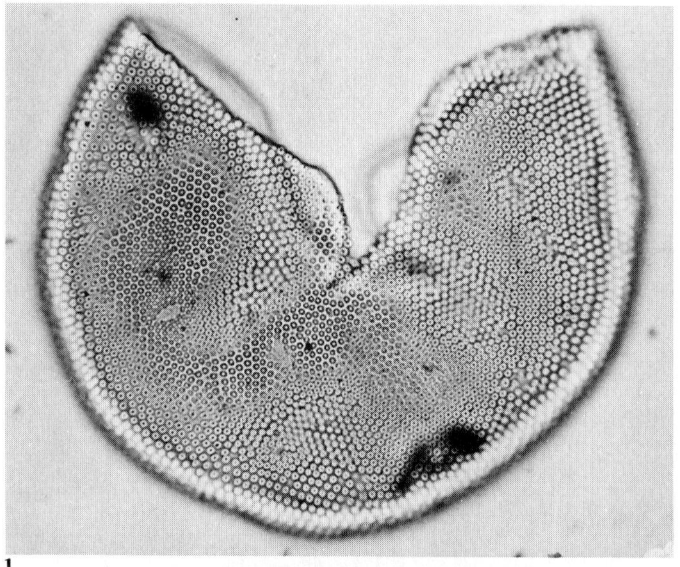

1

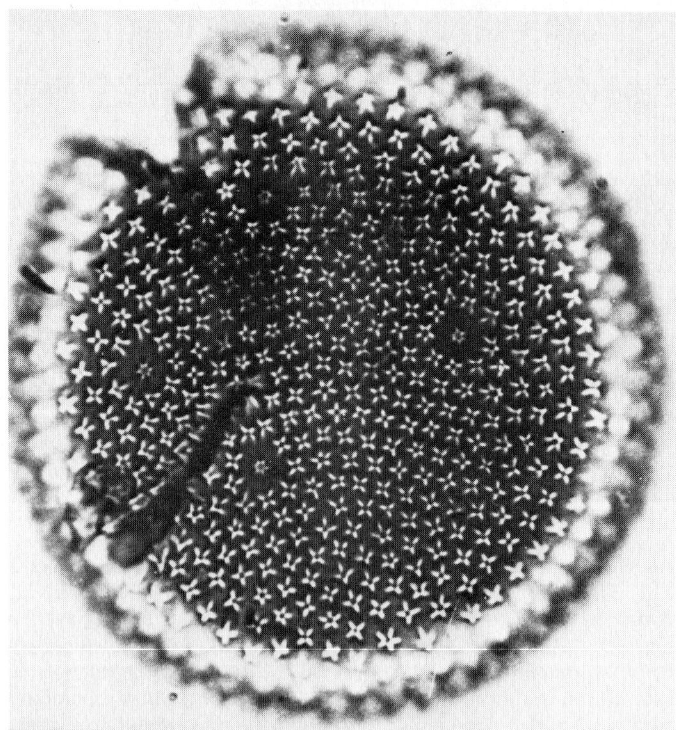

2

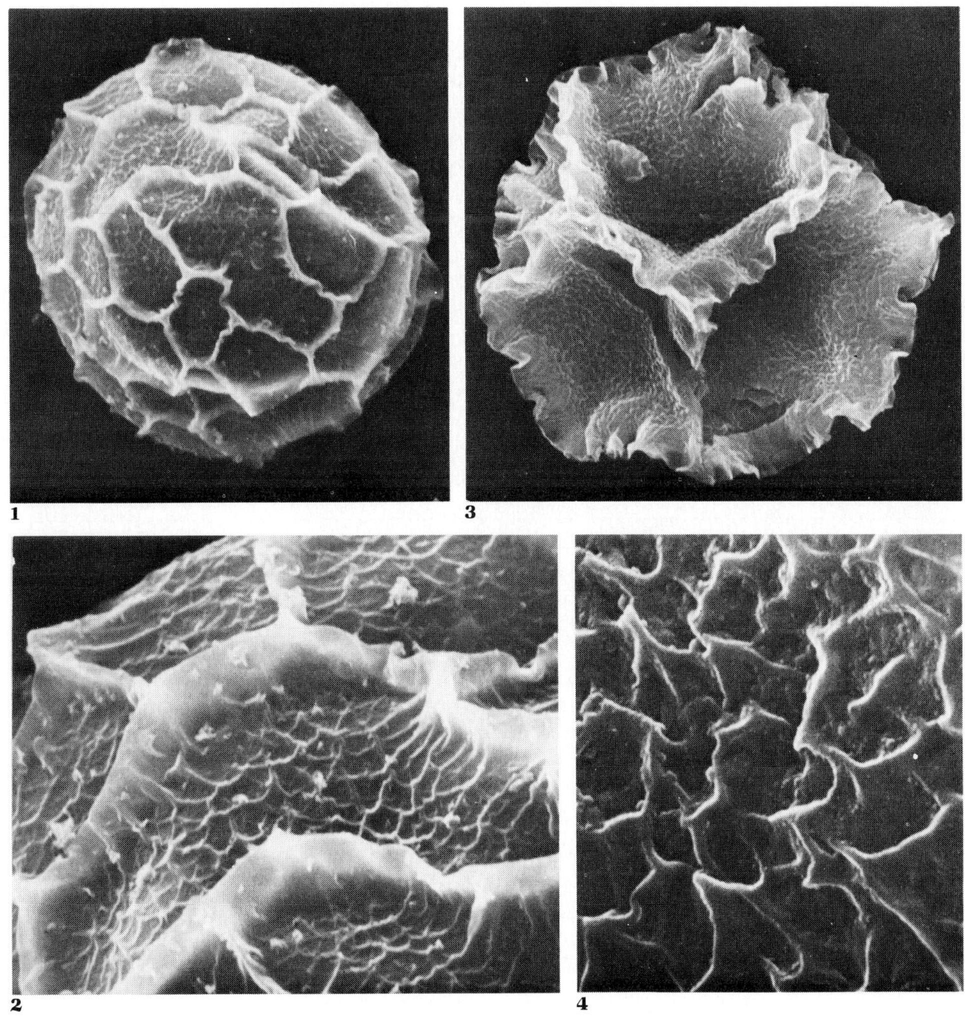

Figure 10.15
Cymatiosphaeraceae. 1,2. *Cymatiosphaera* sp., Lower Devonian, Oklahoma. 1. SEM, ×1200. 2. Part of surface enlarged to show second order reticulations on surface of fields between major crests, SEM, ×3600. 3,4. *Muraticavea enteichia* Wicander, Upper Devonian, Oklahoma. 3. Showing high, ruffled crests, SEM, ×800. 4. Surface of a field enlarged to show the second order reticulations, SEM, ×8000.

Surface divided into fields by a reticulum of intersecting crests or ridges. Ordovician to ?Holocene.

Apiroschadon Loeblich & Tappan 1976 (syn.: *Retialetes* Sah & Dutta 1966, non Staplin 1960); *Cymatiosphaera* O. Wetzel ex Deflandre 1954 (syn.: *Dictyotidium* Eisenack 1955); *Cymatiosphaeropsis* Mädler 1963; *Dasyhapsis* Loeblich & Wicander 1976; *Dictyochroa* Górka 1965; *Dictyosphaeridium* W. Wetzel 1952 (non Timofeev 1969); *Incurvatina* Naumova & N. I. Umnova in N. I. Umnova 1975; *Lophodictyotidium* Pocock 1972; *Melikeriopalla* Tappan & Loeblich 1971; *Muraticavea* Wicander 1974; *Ovnia* Cramer & Díez 1977; *Pterotosphaerula* Cramer 1966; *Ptilotoscolus* Loeblich & Wicander 1976; *Retialetes* Staplin 1960 (non Sah & Dutta 1966); *Romasphaera* Cookson & Eisenack 1974; *Schizosporis* Cookson & Dettmann 1959.

5. Family Pterosphaeridiaceae Mädler 1963 (see Figure 10.16)

Surface with fields, each bearing a large pore. Permian to Jurassic.

Pterosphaeridia Mädler 1963.

6. Family Pterospermellaceae Eisenack 1972 (see Figure 10.17)

Spherical to pillowlike vesicle with wide peripheral flange or ala. Precambrian to ?Holocene.

Duvernaysphaera Staplin 1961 (syn.: *Helios* Cramer 1964; *Veliferites* Brito 1967[*]); *Heliospermopsis* Nagy 1965; *Hidasia* Nagy 1965; *Polyedryxium* Deunff 1954; *Pterospermella* Eisenack 1972; *Pterospermopsimorpha* Timofeev 1966 (syn.: *Polynucella* Sin & Liu 1973); *Savitrinia* Nagy 1966; *Senzeillea* Stockmans & Willière 1969; *Staplinium* Jansonius 1962.

---

[*]As *Veliferites* Brito 1967 is regarded as a junior synonym of *Duvernaysphaera* Staplin 1961, the species *Veliferites tenuimarginatus* Brito 1967, *Micropaleontology*, v. 13, p. 477, pl. 1, figs. 4—8, is here transferred as *Duvernaysphaera tenuimarginata* (Brito) Tappan n. comb. See Figure 10.17 parts 5, 6.

Figure 10.16

Cymatiosphaeraceae, Pterosphaeridiaceae. **1.** *Dictyochroa ovalis* Górka, Upper Cretaceous, Poland, ×440, from Górka, 1965. **2—6.** *Schizosporis reticulatus* Cookson & Dettmann, Lower Cretaceous, Maryland, from Pierce, 1976. **2.** Polar view, ×240. **3.** Polar view, ×200. **4.** Enlargement of part of surface showing open pores in some fields, ×1280. **5.** Part of margin of part 2, showing vesicles and inner wall, ×1280. **6.** Sectioned specimen, showing thin-walled vesicles, and thick inner wall that is divided at top and bottom of figure at equatorial suture, ×480. **7,8.** *Pterosphaeridia* sp., Permian, Western Australia. **7.** Showing elevated crests enclosing fields with closely spaced grana and central pore elevated on a neck, SEM, ×2000. **8.** Enlargement of one field, SEM, ×8000.

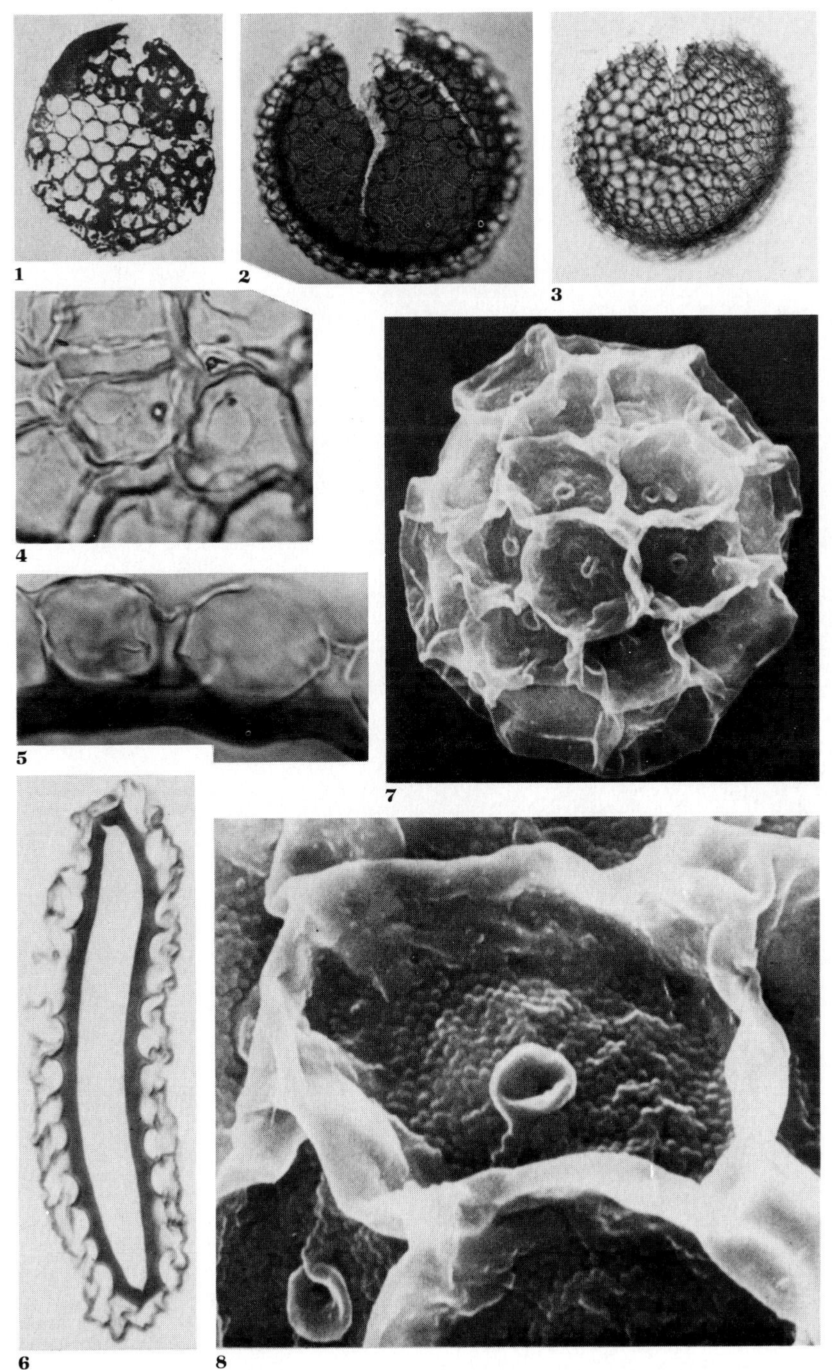

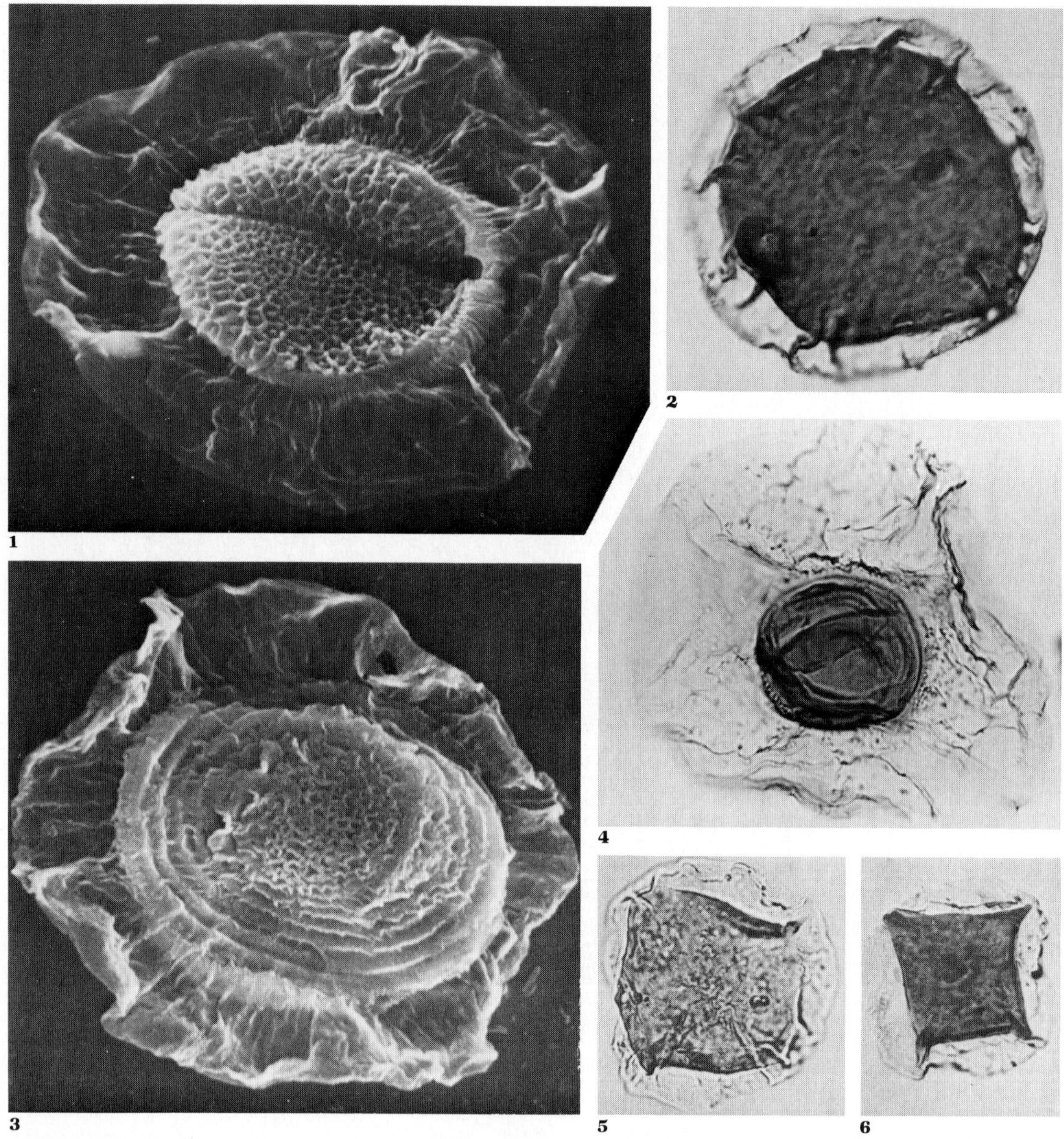

Figure 10.17

Pterospermellaceae. **1.** *Pterospermella reticulata*, Lower Devonian, Oklahoma, SEM, showing reticulate surface of vesicle, slitlike excystment opening, and wide equatorial flange, ×1000. **2.** *Duvernaysphaera* sp., Upper Devonian, Ohio, ×1300. **3.** *Pterospermella dichlidosis* Loeblich & Wicander, Lower Devonian, Oklahoma, SEM, showing concentric ridges, reticulate surface, and broad equatorial flange, ×1600. **4.** *Pterospermella* sp., Silurian, Indiana, showing extremely wide equatorial flange, ×1300. **5,6.** *Duvernaysphaera tenuimarginata*, Lower Devonian, Brazil; 5, ×800; 6, ×750; from Brito, 1967c.

THE CHLOROPHYTA

The Chlorophyta have the greatest diversity in morphology, habit, and habitat of any of the algal divisions. Although most lack any mineralized structures, some deposit calcium carbonate (for example, the dasyclads) and have an excellent fossil record from the Precambrian to the present. Even some organic-walled forms have been preserved, ranging from the coccoidal *Botryococcus,* with wide geologic and geographic distribution, to the spheroidal cells assigned to the Chlorophyta that are preserved in remarkable detail in Australian cherts of a billion years in age (Schopf, 1968).

Algal divisions are characterized by a marked parallelism but, in addition, have a close intradivisional relationship between their unicellular, filamentous, and even massive coenocytic forms. A wholly morphologic basis is thus not valid for separation of higher categories. As the Chlorophyta include both the typical protistan coccoid and flagellate cells and the more complex dendroid, palmelloid, and siphonate forms, the multicellular algae are also discussed here.

Like the blue-green algae, the green algae are widely distributed. They occur in marine, brackish, and fresh water, subaerially on rocks, in snow, and in soil. Some are single-celled and planktonic, others epiphytic or attached to animals, and others typically benthic and siphonous or multicellular. Some are endocellular symbionts of protozoa, coelenterates, and sponges, and others form the algal component of lichens.

The Organism

Vegetatively simple, the green algae nevertheless include a greater variety in their life histories and reproduction than does any other plant division.

Green algae are highly diversified morphologically. They include motile and nonmotile unicellular forms and motile and nonmotile colonial ones, simple or branched multicellular or nonpartitioned filaments, and massive colonies of irregularly arranged cells. Gametes of the sexual part of the life cycle are produced in modified vegetative cells, the latter particularly specialized in those forms with an oogamous cycle (with morphologically distinct motile and nonmotile gametes).

Pigments, Plastids, and Storage Products

The Chlorophyta are grass-green in the vegetative stage, although a few species contain additional pigments that may mask the green color. The plastids of the green algae vary greatly in form. Some are cup-shaped, occupying much of the cell; others are flat, twisted, or spiralling ribbons, small simple discs, or large and elaborately lobed plates. They generally occur in the peripheral cytoplasm but are axial in desmids. A single plastid or many may be present in the cell. The more elaborate plastids are found singly or two per cell; the small and discoid ones commonly are more numerous, up to several million in *Acetabularia acetabulum* (Linnaeus) Silva (Bonotto et al., 1976). Electron micrographs of the plastids show a bounding double membrane and thylakoid bands of two to six closely appressed pairs of lamellae. Motile cells may also have an eyespot at one side of the plastid.

The pigments are the chlorophylls *a* and *b,* xanthophylls (lutein, neoxanthin, violaxanthin, and zeaxanthin), and α- and β-carotenes that also characterize higher plants. In addition, the Codiales contain two xanthophylls (siphonein and siphonaxanthin) that are peculiar to them. A few "green" algae are nonphotosynthetic (*Polytomella*).

A pyrenoid associated with the plastid may be traversed by its lamellae. The pyrenoid may be enclosed in an envelope of starch plates, starch being the customary food reserve of Chlorophyta. In addition to the starch storage product, lipids also may be present. The amount of fat accumulation may be more closely related to environmental conditions than to an innate tendency of the organism, however. For example, starvation results in fat accumulation. In nitrogen-limited cultures of *Chlorella,* the fat content of the cells increased from 22 percent of the dry weight to 70 percent when growth had ceased 23 days later. The Chlorophyta contain a wide variety of sterols. Sterols do not occur in blue-green algae, although they are of general occurrence in other plants and animals. Interestingly, nearly all that have been reported in other than the red algae are found in the greens. For example, sitosterol and fucosterol are present in the Cladophorales as in the Charophyta, and various Chlorococcales contain ergosterol, chondrillasterol, and possibly zymosterol. Thus, the various green algae contain all sterols yet identified as occurring in the xanthophytes, diatoms, chrysophytes, and euglenids. Sitosterol is the most widely distributed sterol in the green algae, as well as in the higher plants (Miller, 1962).

Nucleus

Most green algae are uninucleate, although representatives of the Cladophorales, Siphonocladales, many Chlorococcales, and some Ulotricales have multinucleate cells; the nonseptate filaments of the Caulerpales, Codiales, and Derbesiales also are multinucleate. As in other eucaryotic organisms, the nucleus has a perforate nuclear membrane and one or more nucleoli. Embedded in colorless cytoplasm in the cell center, the nucleus commonly is surrounded by the plastids.

The nucleus of the dasyclad *Acetabularia acetabulum* is very large, up to 100 μm in diameter, and located at the base of the stalk (see Figure 10.18). Cells from which the basal part of the stalk has been removed, including the nucleus, are still able to differentiate the stalk, vegetative whorls, and even the reproductive cap, although no nuclei are present for the formation of gametes. The large nucleus of the normal cell divides repeatedly when forming the cysts of the cap, forming as many as 10,000 to 20,000 secondary nuclei of about 5 μm in diameter (Bonotto et al., 1969, 1976).

An experiment in which the base of the stalk and the included nucleus was removed from *A. acetabulum* (also known as *A. mediterranea*) allowed determination of the quantity of DNA present in the plastids, as no nuclear DNA could contaminate the experiment (Green and Bur-

ton, 1970). The plastids were found to contain large amounts of supercoiled DNA, the quantity within each plastid being within the range of the total DNA content of a bacterium, and similar to that of plastids of other green algae (*Chlorella, Chlamydomonas*) and of *Euglena*.

Flagella

Some of the unicellular and colonial green algae are motile, possessing flagella throughout their existence. Others have flagellate gametes or zoospores. The flagella are like those of all eucaryotic organisms in their ultrastructure, having two central fibrils within a ring of nine pairs of fibrils. All are surrounded by a sheath for most of their length. Because the fibrils project a short distance beyond the sheath, like a short whip from its handle, the smooth flagellum is termed whiplash. The commonly two or four flagella of like size and morphology are terminal; there may be three, four, five, or more flagella in some. When numerous, as in *Derbesia* and the Oedogoniales, the flagella may form a subterminal ring.

Cell Wall

The Chlorophyta commonly have a rigid cell wall, consisting of a firm cellulosic inner layer and an outer pectic layer that may be impregnated with calcium carbonate, as in the Dasycladales. True cellulose, like that of higher plants based on a glucose structure, is present in the Siphonocladales (*Dictyosphaeria, Siphonocladus*), a few Trentepohliales (*Trentepohlia*), and a few Cladophorales (*Cladophora, Chaetomorpha, Rhizoclonium*). The cellulose forms partially crystalline microfibrils of great length, embedded in cementing polysaccharides. Other green algae have mannans rather than true cellulose, as do the Codiales (*Codium*) and Dasycladales (*Acetabularia, Halicoryne, Dasycladus*). Xylans replace the cellulose in others, such as *Bryopsis* (Codiales), *Caulerpa, Halimeda,* and *Penicillus* (all Caulerpales). Hemicellulose components of the Chlorophyta include arabinose, galactose, glucose, mannose,

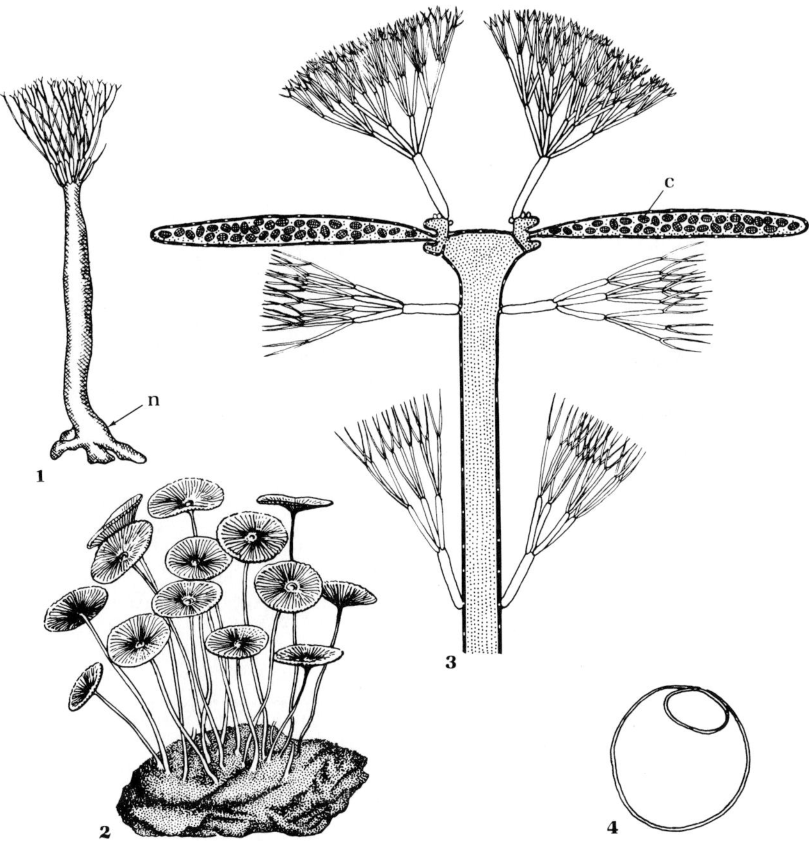

Figure 10.18

Acetabularia acetabulum. **1.** Young vegetative individual, showing position occupied by nucleus (n), ×18. **2.** Group of mature individuals, ×¾ natural size. **3.** Longitudinal section of mature individual, showing much subdivided branches of the whorls, and the cysts (c) within the rays of the disc, ×12.5. **4.** Cyst, ×140. 1−3, redrawn from Pia, 1926.

rhamnose, and xylose, their relative proportions differing in various species (Kreger, 1962; Preston, 1968). Even more strange is the presence in *Derbesia* of a xylan-based wall in the haplont (*Halicystis*) stage, whereas the diplont stage of this alga (*Derbesia* generation) has crystalline mannan in the wall and no cellulose or xylan.

The microstructure of the wall shows microfibrils that are commonly 100 to 200 Å in diameter, but in *Chlorella* and *Hydrodictyon* (Chlorococcales) are only 30 to 50 Å, and in *Va-lonia* (Siphonocladales) range from 100 to 350 Å (Kreger, 1962). The microfibrils of pure cellulose are straight, producing a network of distinct orientation. Microfibrils of xylan have a curved molecular linkage but a helical structure, resulting also in a straight fibril. In contrast, the mannan does not produce fibrils but has a granular structure, although the granules show an ordered array (Preston, 1968). In addition to these water-insoluble components, the cell wall may contain mucilaginous or pectic substances that are soluble in boiling water.

The Volvocaceae all lack a compact cell wall component, the Polyblepharidaceae entirely lack a cell wall, but the Chlamydomonadaceae do have a compact cell wall component. These differences may be of evolutionary significance. *Pediastrum* (Chlorococcales) has a cell wall framework of a network of submicroscopic chains of rings, the chains being within the size range of cellulose microfibrils, although their form differs greatly (Parker, 1964).

Oedogonium (Oedogoniales) has a thick wall with numerous lamellae of parallel microfibrils, successive lamellae oriented approximately at right angles. The microfibrils were indicated by chromatographic analysis, staining, and solubility to be of cellulose. A spongy mucilage encrusted the lamellae. No evidence of chitin was found, although it had been suggested to occur in *Oedogonium* (Parker, 1964).

Calcium Carbonate, Silica, and Cell Inclusions

The pectic substances in the cell wall may be associated with deposits of calcium carbonate in most desmids (Zygnematales) and Dasycladales and in some Siphonocladales, Cladophorales, Caulerpales, and Chlamydomonadales (*Phacotus*). The lime is commonly in the form of acicular crystals of aragonite, as shown by X-ray diffraction for *Acetabularia*. The aragonite is low in magnesium and strontium, containing 98.5 percent $CaCO_3$, 1.3 percent $SrCO_3$, and 0.2 percent $MgCO_3$ in *Acetabularia* (Marszalek, 1975), but up to 1 percent $MgCO_3$ and 2.3 percent $SrCO_3$ in some other green algae. In contrast, the calcitic red algae have only 0.25 to 0.37 percent $SrCO_3$ (Lewin, 1962).

Various others participate in the deposition of extracellular limy deposits in fresh water (as discussed in a later section on the rock-forming capabilities; see pages 831 – 833). These include representatives of the Chlorococcales, Volvocales, Cladophorales, Oedogoniales, and Zygnematales.

Silica is less common in the green algae than in the Chrysophyta, but small amounts are present in the wall of all green Hydrodictyaceae such as *Pediastrum* and *Hydrodictyon* and in *Tetraëdron* (Chlorococcaceae). The silica is isotropic, noncrystalline opal, with reportedly a small amount of α-quartz, although the latter may be due to contamination. In *Hydrodictyon* the silica occurs in irregular pieces up to 50 μm in maximum diameter, whereas in *Pediastrum* and *Tetraëdron* the extraction process may leave an intact siliceous cell wall (Millington and Gawlik, 1967; Parker, 1969).

In addition to the common presence of lime or of silica, gypsum crystals have been reported in the desmids *Closterium, Pleurotaenium,* and *Micrasterias,* and oxalate crystals in *Spirogyra* (A. Fischer, 1883).

Reproduction

Asexual Reproduction

Ordinary vegetative cell division is a common method of reproduction, particularly among the unicellular chlorophytes. Many species also may reproduce asexually by means of flagellated zoospores, which may form singly or in numbers within a cell. Some produce nonmotile spores (aplanospores). During simple division of flagellated cells, such as *Chlamydomonas,* the parent cell loses the flagella, the cell contents dividing into two or more daughter cells. These produce their own flagella, become motile, and escape by rupture of the mother cell wall. After growth to the size characteristic of the species adult, division may be repeated.

Some flagellated green algae are colonial (see Figure 10.19), *Volvox* having thousands of cells arranged at the periphery of a spherical colony. In the early stages all cells of the colony are alike, but as growth continues, the cells of the posterior half become larger. The enlarged cells, or **gonidia**, may divide to form daughter colonies, the remaining purely vegetative cells disintegrating as the new colonies are liberated. The number of cells in the colony is determined before liberation of the daughter colony (**coenobium**) from the parent one. No increase by cell division occurs later.

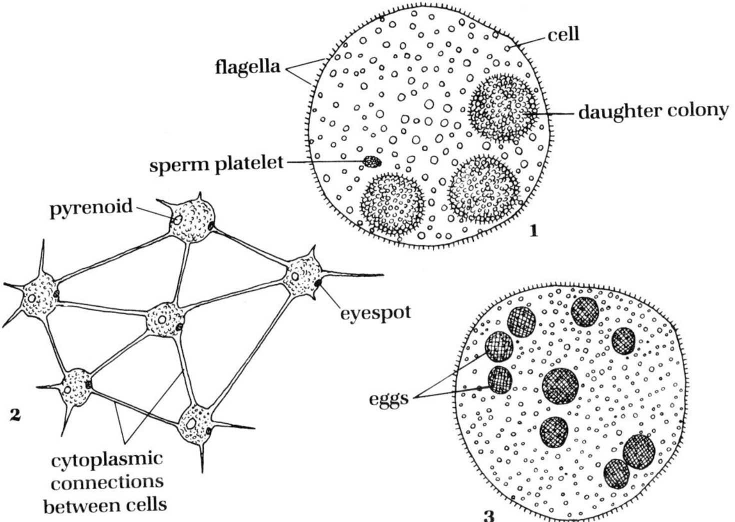

Figure 10.19
Volvox aureus Ehrenberg. **1.** Male colony (coenobium) of many individuals, with apex and flagella oriented outward, developing daughter colonies within, and developing sperm platelet. **2.** Enlargement of part of surface, showing individual cells with eyespot and pyrenoid, and cytoplasmic connections between cells. **3.** Female colony, similar to the male colony but containing egg cells. 1,3, ×100; 2, ×350.

Some nonmotile solitary cells also form nonmotile spores. *Chlorella,* a spherical cell with cuplike plastid, divides medially two or more times to form four to eight daughter protoplasts. These autospores are nonflagellated and resemble the mother cell. As they grow, they are freed by rupture of the parent cell wall.

Some nonmotile colonial green algae may produce flagellate zoospores (*Pediastrum*), whereas others (*Scenedesmus*) commonly lack motile cells but produce an autocolony within the adult cell, which is later freed. The filamentous *Ulothrix* produces flagellate zoospores, but the similarly filamentous *Spirogyra* has no asexual reproduction other than simple filament formation and fragmentation.

Sexual Reproduction

Sexual reproduction, involving the fusion of two like (see Figure 10.20) or dissimilar gametes, their nuclei, chromosomes, and genes, and later nuclear division in which the chromosome number is halved (meiosis), is known in most green algae. Nearly all green algae in fresh water are haploid, meiosis having occurred in the germinating zygote (e.g., *Volvox*), as is true of most higher plants. Many marine species are diploid, particularly the coenocytic Siphonocladales, Codiales, and Dasycladales; meiosis occurs when gametes are formed, as is more characteristic of the animal kingdom. Some representatives of the Volvocales, Ulotrichales, Cladophorales, and perhaps some Chlorococcales and Codiales show an alternation of haploid and diploid generations. These are termed **diplobiontic,** having two distinct, free-living stages, as opposed to the **haplobiontic** ones. The two stages of the diplobiontic life cycle may be morphologically alike (*Ulva*), or the haploid and diploid generations may so differ in appearance (*Derbesia*) as to have been originally described as distinct plants, the gametophyte stage of *Derbesia* long having been known as *Halicystis.*

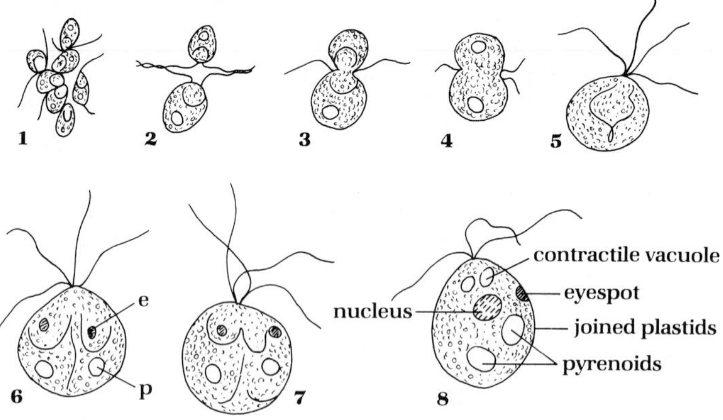

Figure 10.20

Chlamydomonas sp. Sexual reproduction with isogametes. **1.** Actively swimming cells beginning to pair (3:15 P.M.). **2.** Paired cells moving backwards and forwards (3:25 P.M.). **3,4.** Fusing gametes (3:27 P.M.). **5 — 7.** Zygote with double numbers of flagella, eyespots (e), pyrenoids (p), and plastids (3:29 P.M.). **8.** Zygote with fused nucleus and plastids but with more than the usual number of flagella, contractile vacuoles, and pyrenoids (3:45 P.M.). 1, 2, ×445; 3 — 5, ×526; 6 — 8, ×1000.

Gametes of both sexes may be produced by one individual or its asexual descendents (clone), or the sexes may be restricted to different individuals or clones. Some green algae produce gametes only in specially differentiated structures or gametangia. These may be differentiated as oögonia (producing a single non-motile egg, except in *Sphaeroplea*) and antheridia (producing motile male gametes).

Sexuality may be sporadic, rather than an obligate part of the cycle. Zoospores of *Tetracystis* and many other genera may function as isogamous (like) gametes and fuse to form zygotes, thus acting as facultative gametes. Alternately, the zoospores may be regarded as gametes that develop parthenogenetically if fertilization does not occur.

The zygote of most freshwater species is thick walled (see Figure 10.21) and serves as a resting stage; that of marine species is commonly thin walled and germinates soon after formation. The relatively few species that do

Figure 10.21

Volvox sp. **1.** Portion of female colony, showing small biflagellate vegetative or somatic cells separated by vertical intercellular membranes, and one large reproductive cell containing megagamete or egg; ×750. **2.** Top view of part of colony, showing individual cells separated by the angular membranes, which in some species do not extend for the entire cell length, the cells being connected by intercellular cytoplasmic bridges beneath the membranes; ×250. **3.** Section of colony after fertilization of the megagamete to form the thick-walled zygote, sectioned at left, and surface view at right; small vegetative cells present above the zygotes; ×500.

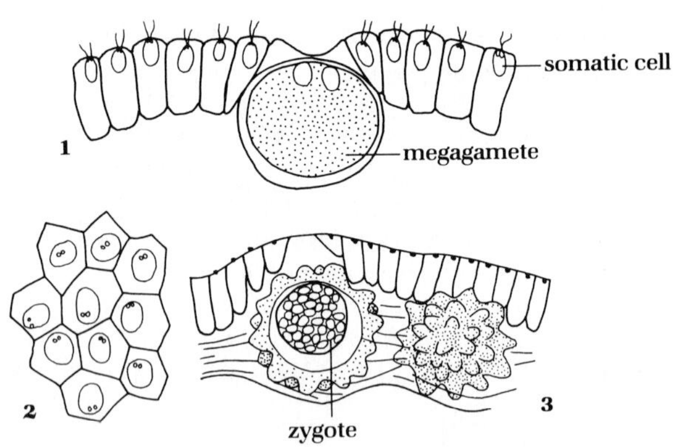

not hold to this generalization are believed to have recently entered their present habitat, having been derived from those typical of another.

Rock-Forming Chlorophyta

Many of the siphonate algae deposit aragonite at the exterior of the thallus. The abundance of certain of these, particularly the marine Caulerpales and Dasycladales, has resulted in the formation of algal limestones (Johnson, 1961).

Important lime-depositing green algae are the modern *Halimeda, Udotea, Penicillus,* and *Rhipocephalus* (all belonging to the family Udoteaceae, Order Caulerpales), as well as most species of the Dasycladales. The carbon isotope fractionation in photosynthesis results in ^{12}C-enriched organic matter, yet the aragonite deposited by the green algae is enriched in ^{13}C in comparison to the calcite of the lime-depositing red algae. The green algae contain a lower percentage of $MgCO_3$. The aragonite commonly occurs as needlelike crystals 2 to 9 μm long and about 0.5 μm in diameter (Lowenstam, 1955, p. 271), the needles being tangled in interlacing mats. Such deposits of aragonitic needlelike crystals were once regarded as inorganically precipitated, but much evidence now suggests that a major source of these aragonite needles, as in Florida and the Bahamas, is from disintegration of the thallus of such algae as *Penicillus, Rhipocephalus,* and *Halimeda* (Studer, 1963). In fact, up to 70 percent of the body weight of *Halimeda* in Hawaii has been determined to be aragonite (Moberly, 1962). *Acetabularia antillana* (Solms-Laubach) Egerod is extremely common in May and June in southern Florida, where it forms a continuous overgrowth on the bottom. The standing crop was estimated to produce 720 g of skeletal aragonite per square meter, and more than one "crop" probably is produced in a season (Marszalek, 1975).

The distribution of the calcareous algae is closely related to the degree of $CaCO_3$ saturation of the water. They are important contributors to reefs, chiefly between lat. 30° S and lat. 30° N. As the lime deposition of the green algae is largely external to the cells, it has been related to the process of CO_2 utilization in photosynthesis. Yet it is difficult to explain why only a few algae deposit the lime, and these commonly deposit it only in certain parts of the thallus that may not be related to the distribution of photosynthetic tissues (Lewin, 1962). In addition, larger green algae are most commonly lime-depositing, and in general the metabolism per gram decreases with increased size of the individual. The efficiency in primary productivity depends strongly on the thickness or massiveness of the thallus. For example, the noncalcareous *Cladophora* has a rate of production per gram of body weight that is 25 times that of the brown alga *Fucus* (Odum et al., 1958). If seawater is saturated with $CaCO_3$, then two grams of lime should be deposited for each gram of CO_2 reduced in photosynthesis to organic carbon, and by weight, the ratio of $CaCO_3$ to organic carbon should be about 8 : 1. Instead, this ratio is about 30 : 1 in *Halimeda*. Possibly the aragonite deposited as a result of the photosynthetic process may act as a nucleus for further precipitation by the organism (Lewin, 1962). Most calcareous algae occur in shallow water and are progressively less heavily calcified in deeper water (*Penicillus, Udotea, Rhipocephalus*). However, *Halimeda* is relatively successful in low light intensities to a depth of about 60 meters and may even be more heavily calcified in its deeper range (Goreau, 1963).

Because modern species of calcareous algae are largely restricted to the shallower portions of the warm seas, the fossil representatives are considered to have had a similar habitat. In general the calcareous green algae are rare in the earliest Paleozoic, expand in the Silurian, are rare in the Devonian, increase to a Mississippian climax, and then decline with a sharp drop at the end of the Permian. The Triassic is relatively poor in calcareous algae, except in the Tethyan Sea, where dasyclads were important rock builders. The Dasycladales first appeared in the Precambrian of Norway and the USSR but became important rock-builders in the Ordovician in various localities in northern Europe (Sweden, Norway, England, Esthonia, Gotland) and in Texas. Locally they were abundant in the Silurian. The dasyclads again became

Figure 10.22
Halimeda gracilis Harvey, living, Saipan, ×1, from Johnson, 1957.

important in the Permian (especially *Mizzia*) in Yugoslavia, Timor, Manchuria, and in the United States (western Texas and New Mexico). They attained their maximum in the Triassic and Jurassic across much of central Europe, developing then into most of the modern types with ramified branching. Although they are locally important in the Cretaceous in Mexico and Texas, they have since declined in importance. Many have been described from the Tertiary, for example in the Paris Basin, but these are generally smaller and less heavily calcified than earlier representatives.

Other siphonous green algae also were important in the Mississippian and Permian and then continued to expand to a maximum in the middle Cretaceous. Since then they have gradually declined to the present level. Although *Halimeda* is among the most important of the rock-forming algae in the Cenozoic (see Figure 10.22), green algae now in general are exceeded in limestone contribution by the red algae. *Halimeda* is at present abundant in the lagoon at Bikini Atoll and has been recognized in borings as deep as 580 m (1900 feet), comprising about 20 percent of the rock at a depth of about 12 m (40 feet) (Johnson, 1954a). *Halimeda* also was present in all the limestones of Saipan (Eocene, Miocene, Pleistocene) and comprised as much as 65 percent of the volume there (Johnson, 1957). Similarly it occurs in the Eocene of Austria, Eocene and Oligocene of France, and the Tertiary of Italy (L. and J. Morellet, 1922a).

The formation of **boghead coal** by *Botryococcus* (Order Chlorococcales) is mentioned in the discussion of that group. Remains of the organisms can be identified in the rock, the peat equivalent being termed **coorongite** and the coal stage **torbanite** or **kuckersite**

(Zalessky, 1917a). The vast quantities of organic matter in the oil shales could have been derived only from the lipid substances in the organisms, as other compounds would have decayed differently (Cane, 1967). Cellulose would convert to methane by bacterial action, proteins would be destroyed by putrefaction, and even the more stable lignins would decompose to humic matter. Most green algae are low in total fat content under normal conditions, although in nitrogen-limited cultures, *Chlorella* may produce lipids up to 75 percent of its dry weight. *Botryococcus*, in contrast, normally produces a large amount of oil. The later ageing, polymerization, and oxidation processes convert the oils to coorongite.

An unusual occurrence of abundant desmids near the River Irkutsk in Siberia resulted in the formation of a rock termed **closterite**, as the dominant component was a new species of the desmid *Closterium*. The fossil specimens were clear and yellowish, in contrast to the reddish matrix of the coal. Some specimens were reported to contain remnants of the plastids and perhaps even of the pyrenoids. Its age was uncertain but probably Tertiary (Zalessky, 1917b).

Green algae also participate in the formation of the thick Pleistocene and Holocene deposits of travertine in freshwater streams, as in the Arbuckle Mountains of Oklahoma (Emig, 1917). Various blue-green algae and *Vaucheria* (Xanthophyta) are more characteristic of this calcareous deposition, but some green algae also participate in the formation of cavernous travertine. Among these are *Trochiscia* (Chlorococcales), *Haematococcus* (Volvocales), *Cladophora* (Cladophorales), and *Oedogonium* (Oedogoniales). The cavernous travertine accumulates as the lime encases the plant filaments; the hollow calcareous tubules persist after decay of the organic matter. Two other varieties of travertine form under similar conditions. Banded travertine is somewhat less common but may form masses 30 to 90 cm in thickness. It accumulates at the rate of about a millimeter per month, largely due to the blue-green *Oscillatoria* and *Lyngbya*, but it may also involve the green *Haematococcus*. The pisolitic type of travertine consists of granules 1 to 8 mm

in diameter that gradually enlarge and fuse to produce a homogeneous deposit. It also largely results from blue-green activity but involves some *Trochiscia* and *Haematococcus*.

Life History and Distribution

Because of the great diversity characteristic of the green algae, the life history, ecology, and geologic record are discussed by orders. General discussion of the methods of reproduction, skeletal or cell wall material, and morphology is given above, but both the stages present in the individual life cycles and the morphology characteristic of the various stages show considerable variation.

The green algae, together with the diatoms, are the most important freshwater primary producers. In addition to the organic matter produced and stored within the cell, an appreciable quantity may be liberated into the water in solution, to be utilized by other organisms. The freshwater *Chlamydomonas* excreted up to 0.1 g of soluble polysaccharides per liter of culture medium over a two-week period, this dissolved material amounting to 20 to 25 percent of the total amount of organic matter produced (Lewin, 1956).

Rate of growth of green algae varies greatly according to the species. In general, the smaller cells have the most rapid doubling time or shortest life span. In culture *Pseudochlorella pyrenoidosa* (Zeitler) Lund varied in doubling time from 60.2 hours at 10°C to 2.6 hours at 39°C (Fogg, 1965). Recent data also indicate relatively rapid growth of some benthic green algae. *Udotea* was observed in situ by personnel of the Undersea Project Tektite 1 to have a life span of only three to four weeks from budding to senility (Clifton et al., 1970).

Green algae have the fewest growth requirements of any of the eucaryotic species, some growing on inorganic media alone, as do certain blue-greens. In contrast, 14 of 35 diatoms studied required an external vitamin source, as did all Euglenids, Cryptophyceae, Chrysophyceae, and Dinophyceae examined in the laboratory

(Provasoli, 1961). The nutritional self-sufficiency of the Chlorophyta is another indication of their relative simplicity.

Because water filters out various light wavelengths selectively, the separate photosynthetic pigments of algae are differently adapted to depth. Green algae use the red and blue parts of the spectrum, and as the red light is almost completely absorbed in about five meters, their occurrence depends on the depth of blue penetration. This is closely related to the amount of suspended matter in the water, which absorbs the shorter wavelengths. Thus, the benthic green algae occur in shallow water, particularly in the littoral and uppermost sublittoral zones, and in deeper water only when this is quite clear. Other algae, red and brown, may occur to greater depths (Levring, 1961).

This general depth distribution may be affected by other factors, however, for in the western Atlantic, Taylor (1961) found that 17.7 percent of the 203 green algal species occurred at depths of 5C m in very clear water, in contrast to only 9 percent of the 103 brown algal species, and 3.5 percent of the 453 red algae. The green algae at depths were *Caulerpa* and various Codiaceae with extensive rhizoidal systems for attachment. Seven of the *Caulerpa* species occurred below 100 m, as did ten of the Codiaceae. This depth range does indicate the great adaptability of the Chlorophyta.

Green algae of the temperate littoral zone include *Ulva* and *Enteromorpha;* if the water is somewhat brackish, *Caulerpa, Penicillus,* and *Acetabularia* may be present. In the subtropics this zone is occupied by *Caulerpa* and *Bryopsis;* the lower littoral zone includes *Boodlea.* In the supralittoral zone, *Prasiola* and *Ulothrix* may be present in high latitudes, and *Dunaliella* occurs in tidepools of varying salinity. In the sublittoral zone to lowest tide level, *Codium* and *Batophora* are present in the subtropics, and many calcareous species in the tropics, such as *Udotea, Halimeda,* and various dasyclads (Prescott, 1968).

Other green algae live in fresh water, soil, form part of the lichen associations, and even may live in salt pans or deserts. Correlative with this extensive geographic and ecologic distribution is their geologic history, beginning in the Precambrian. Fossil examples range from the rare preservation of organic material of vegetative cells and filaments to the calcareous skeletons of the dasyclads and Caulerpales.

Orders Chlamydomonadales, Volvocales, and Tetrasporales

The simplest of the Chlorophyceae are those with motile vegetative cells. They range from single cells (*Chlamydomonas*) to complex colonies (*Volvox*).

Chlamydomonas is a primitive-appearing biflagellate alga (see Figure 10.23), seldom over 25 μm in length, with cellulosic wall, a single large plastid, one or more pyrenoids, contractile vacuoles, a red pigment body or eyespot, and a single nucleus. Reproduction may be asexual, with the original haploid cell producing 4, 8, or 16 biflagellate haploid daughter cells, or sexual, with the production of 4 (or multiples of 4) gametes. These are similar in appearance to the asexual zoospores but somewhat smaller. They fuse in pairs, forming motile quadriflagellate zygotes. The flagella are then shed and the cell rounds up and becomes a heavy-walled diploid resting zygospore. This later germinates, nuclear meiosis resulting in new haploid zoospores. In cultures on agar, the entire vegetative cell may function as a gamete, the cells first clumping together, then pairing, losing their cellulosic cover, and fusing to produce zygotes.

The colonial Volvocales may be extremely simple, with the individuals merely connected by strands of cytoplasm, or the colony may be more complex, as in *Volvox* itself. The slightly ovoid *Volvox* colony consists of between 500 and 50,000 individuals arranged to form a hollow sphere. In early stages all cells of the colony appear similar. Later, certain cells in the posterior hemisphere enlarge and become somewhat depressed beneath the general surface. These cells (gonidia) alone produce the new colonies, the other vegetative cells disintegrating as the new daughter colonies mature, allowing them to escape. As the daughter colonies form within the parent one, the flagella of the new cells are directed inward. Later, by invagination of the lower end, the entire colony turns

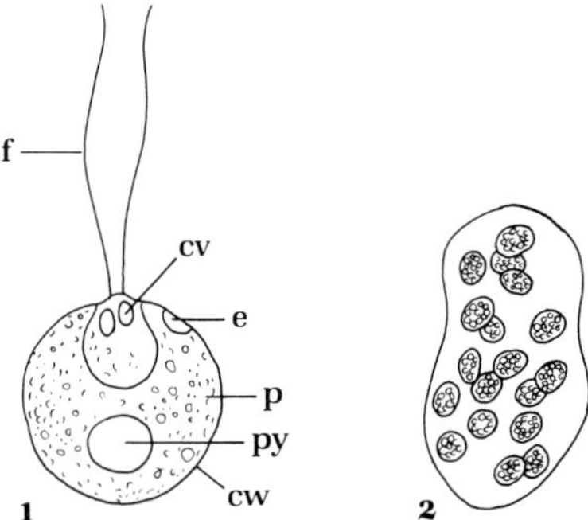

Figure 10.23
Chlamydomonas sp. **1.** Vegetative motile cell, with similar flagella (f), contractile vacuoles (cv), eyespot (e), plastid (p), pyrenoid (py), and definite cell wall (cw), ×1200. **2.** Palmella stage, with individual cells held in a mucilaginous matrix, ×350.

itself inside out so that the cells of the new colony then have their flagella directed outwards. The new colony escapes from the mucilaginous matrix of the mother colony sphere and lies free within the interior, where it continues to grow until breakdown of the mother colony allows its separation. Oogamous sexual reproduction also may occur, with some enlarged cells of the colony producing gametes. Male and female gametes are differentiated. The resultant zygote becomes thick-walled; at germination it produces a single juvenile colony.

Eovolvox silesiensis Kázmierczak (see Figure 10.24) was described from the Upper Devonian black, bituminous limestones in Poland (Kázmierczak, 1975, 1976a,b) and suggested to be related to the formation of the Paleozoic calcispheres. The Precambrian *Eosphaera* was also suggested to belong to the volvocalean algae. *Eovolvox* was stated to be similar in size of colony and individual cells to *Eudorina elegans* Ehrenberg. However, the Volvocalean colonies are merely held in a mucilage; those of *Eudorina* are evenly spaced around the colony, and commonly in tiers of cells. *Volvox* has a single layer of many cells. Although superficially similar in appearance, it appears unlikely that the delicate walls of the volvocalean cells would be preserved, hence the relationship to this group of *Eovolvox* is somewhat doubtful, and that of the calcispheres or *Eosphaera* even less probable.

The Chlamydomonadales and Volvocales are found mostly in fresh water, although *Dunaliella* occurs in marine neritic and estuarine environments, in fresh water, and in briny tidepools and salt flats. *Chlamydomonas nivalis* (Bauer) Wille is a snow-living alga that forms red blooms in snow at high elevations in summer in the Beartooth Mountains of Montana and Wyoming (Mosser et al., 1977). Other species of *Chlamydomonas* as well as *Carteria* and *Haematococcus* may be marine. The Chlamydomonadales are characteristically photosynthetic, but a few are colorless (e.g., *Polytoma*). *Chlamydomonas hedleyi* Lee et al. is an endosymbiont in the foraminifer *Archaias* (Lee et al., 1974).

Fossil representatives are rare due to the general instability of cellulosic coverings. However, *Chlamydomonas* has been reported in a coprolite from the upper Eocene Bridger Formation of southwestern Wyoming (Bradley, 1946).

Phacotus has the general appearance of *Chlamydomonas,* but the cell is surrounded by a

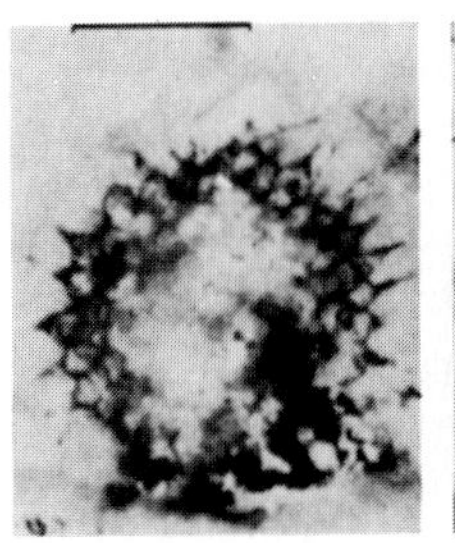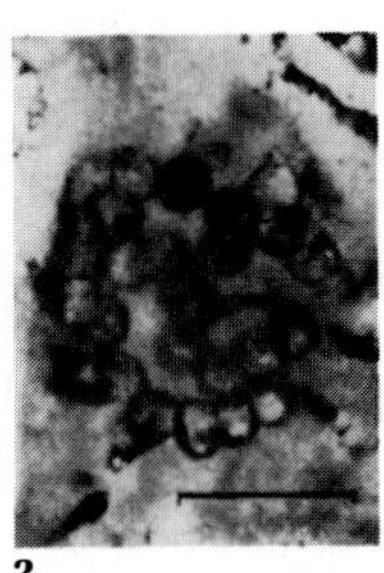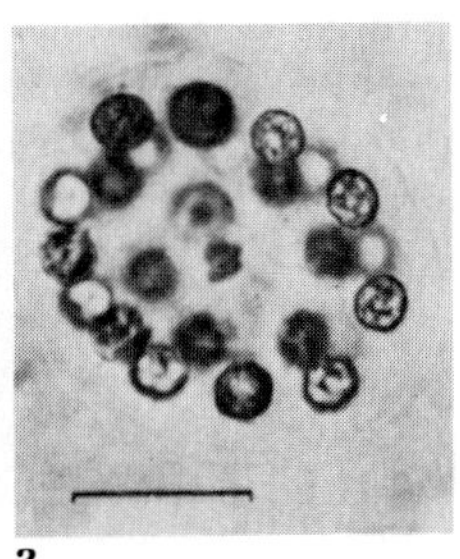

Figure 10.24
Volvocaceae. **1,2.** *Eovolvox silesiensis,* Upper Devonian, Poland, tangential section of colonies showing pyriform to ovoid cells. **3.** *Eudorina elegans,* a modern volvocalean from Poland, showing similar colony organization. Bar in each photo = 50 μm. From Kázmierczak, 1975.

thickened, rugose, and calcified membrane, divided longitudinally in two parts to form a bivalved shell. Division is between these valves during vegetative reproduction. The cells are commonly less than 22 μm in length but have been found as fossils even in rock-forming abundance. *Phacotus lenticularis* (Ehrenberg) Stein has been reported from the Oligocene and Miocene of Germany (Rutte, 1953) and the Pleistocene and postglacial deposits of Hungary, Sweden, Denmark, and Germany (Lagerheim, 1902). *Coccomonas* also has a calcareous membrane forming a thick, hard cover potentially capable of preservation. The lime may be in small plates or rods longitudinally arranged or may occur as fine granules (Bourrelly, 1966).

The family Lecaniellaceae includes Late Cretaceous and Tertiary forms with ovoid body that separates into two hemispheres, hence was regarded as probably close to the Phacotaceae (Cookson and Eisenack, 1970), but was later transferred to the dinoflagellates.

The Tetrasporales also have a meager fossil record, *Palmogloea* (as *Gloeocystis oxfordiensis* Lignier) being reported from the Jurassic (Oxfordian) of Caen, France. Small masses or tetrads of yellow spheres, 12 to 14 μm in diameter, seem to be spores, some apparently germinating at the time of fossilization. They were found in the tracheid cavities of fossil *Araucarioxylon* wood (Lignier, 1906). *Palmophylloites europaea* Strauss from the German Pliocene also was referred to the Tetrasporaceae. However, modern species are very difficult to place generically without culturing or to distinguish from the palmelloid stages of the Chlamydomonadaceae or some Ulotricales (Bourrelly, 1966).

Schizochlamys haywellensis Bradley was described from the Eocene Green River Formation of Wyoming. Living *Schizochlamys* is generally assigned to the Tetrasporaceae because of its gelatinous colony and the tuft of four to eight pseudocilia that arises from each cell. Frequent hemispherical fragments of old mother cell walls generally also are present in the colony. The presence of oil rather than starch as a storage product leaves some question as to whether the genus is correctly placed in the green algae. Placement of these fossil specimens in the Tetrasporales is thus still uncertain.

Order Chlorococcales

With motility restricted to the reproductive stages, the Chlorococcales are somewhat more advanced than the Volvocales and include both unicellular and colonial species. Most occur in fresh water, but several species of *Chlorella* are marine, and *Palmellococcus* (Oocystaceae) is characteristic of marine rocky pools (Wood, 1965). *Chlorochytrium* is a unicellular endophyte in various higher plants, such as mosses and grasses, but one species is endophytic in red algae in the temperate Pacific and may be so abundant as to give a distinct green color to the host plant. Because some modern Cladophorales have a stage in the life cycle that is morphologically identical to *Chlorochytrium*, this "genus" may eventually prove to be merely a stage in other species. *Chlorella* may be symbiotic in ciliates, sponges, or *Hydra;* the symbiotic green organism is also termed *Zoochlorella*, although it may be merely the common *Chlorella vulgaris* Beijerinck. The genus *Coccomyxa* is symbiotic in some echinoderms, and *Scenedesmus*, the only green colonial alga known to be a symbiont, may occur within the freshwater sponge *Carterius* (Fritsch, 1935).

Some of the Chlorococcales form the algal partner of lichens, including species of *Chlorella*, *Coccomyxa*, *Myrmecia*, *Pseudochlorella*, *Trebouxia*, and *Trochiscia* (Ahmadjian, 1967). Some may also live free, not lichenized, in soil, on the surface of soil, or on rocks and in fissures, even under such extreme conditions as in the Negev desert. They include *Chlorococcum*, *Bracteacoccus*, *Spongiochloris*, *Trebouxia*, *Trochiscia*, and *Protosiphon* (Friedmann et al., 1967).

In contrast to the symbionts, whose photosynthesis may support themselves and their host, many species may lose their green color in cultures that contain dissolved organic nutriment. Some Chlorophyta are saprophytic in nature, and *Characium* (Chlorococcaceae) and *Prototheca* (Chlorellaceae) are wholly non-photosynthetic. *Glaucocystis* (Oocystaceae) similarly lacks plastids but contains a blue-green algal symbiont, which has resulted in some confusion as to its systematic placement (Pascher, 1929).

Chlorococcaceae. An example of the Chlorococcalean life cycle is provided by soil-dwelling *Tetracystis* (Brown and Bold, 1964). Vegetative cells occur in groups of two, four, or eight to form globular colonies. In asexual reproduction, two to eight biflagellate zoospores develop from a mother cell. These are motile for a period, then lose the flagella, enlarge, and divide to form two to four nonmotile cells. The cells of the tetrad may enlarge and divide again to form a tissue of cells. The outer cell wall remains in place for a time but may split to allow separation of the cell clusters. Individual cells may also produce zoospores. In some species these zoospores function as isogamous gametes that unite to form a zygote. As mentioned earlier, the zoospores may thus be regarded either as facultative gametes or as gametes that develop parthenogenetically if fertilization does not occur. Some Chlorococcales produce thick-walled, rugose or spiny resting zygotes (as in *Chlorococcum*).

The genus *Tetraëdron* has commonly been included in the Oocystaceae, but recent studies show that it possesses biflagellate zoospores, as well as autospores, requiring its reassignment. Placed with the Hydrodictyaceae by Starr (1954), it was transferred to the Chlorococcaceae by Bourrelly (1966) in view of the absence of a coenobial habit. The Chlorococcaceae is represented in the Eocene by *Tetraëdron* (see Figure 10.25), occurring in the Green River Formation of Colorado (Bradley, 1929, 1931) and in the Miocene Mascall Formation of Oregon (Gray,

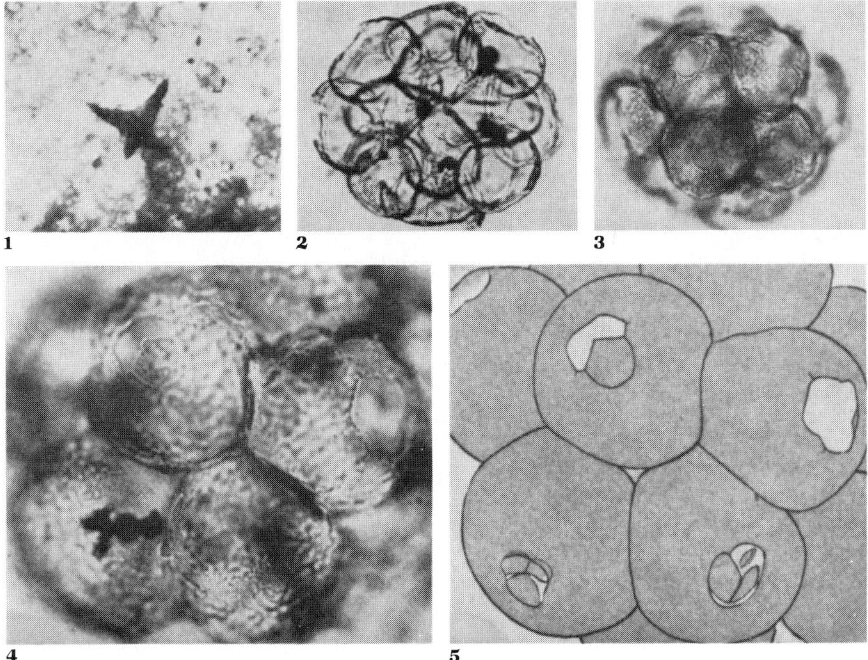

Figure 10.25

Chlorococcaceae. **1.** *Tetraëdron regulare* var. *torsum* (Turner) Brunnthaler, Eocene, Colorado, ×200, from Bradley, 1929. **2–5.** *Palambages morulosa* O. Wetzel, Upper Cretaceous, Germany, from Gocht and Wille, 1972. **2,3,** Two 16-celled colonies, optical section and surface showing openings, approximately ×400. **4,5.** Part of 16-celled colony, photo and diagram, showing all stages of opening, upper left cell with one platelet in place; upper right, all three lacking; lower left, all three in place; lower right, two platelets present, the third fallen within the cell; ×700.

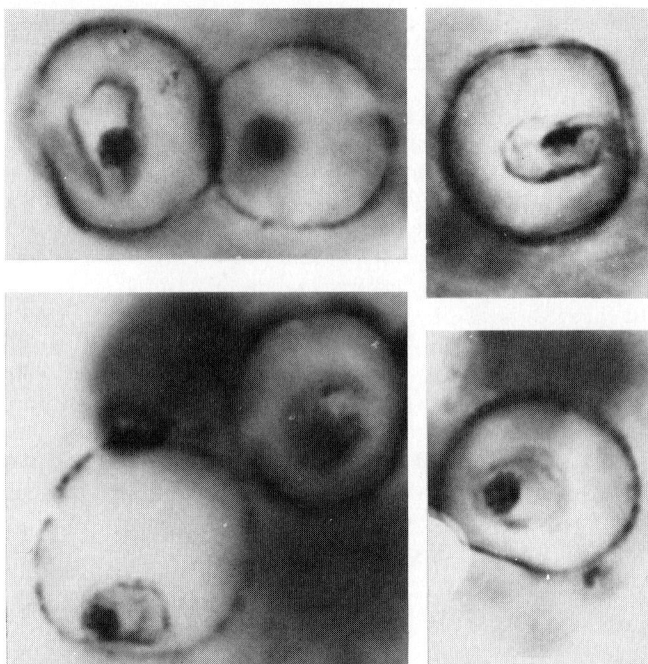

Figure 10.26
Chlorellaceae. *Caryosphaeroides pristina* J. W. Schopf, Bitter Springs Formation, Precambrian, central Australia, in thin section of chert, with outer and inner membranes preserved, together with dark inner body that may be an organellar remnant or degraded cytoplasm, but appears of constant size and appearance in many specimens; ×2190, from Schopf, 1968.

1960). Reports of this genus in the Devonian of New York are an erroneous identification of the acritarch *Veryhachium*, but the Cretaceous *Palambages* may belong here.

Churchill (1960) illustrated without naming various algal aplanospores from Holocene peats at Walpole, Western Australia and from the Eocene and Paleocene of the Perth Basin. These were compared to the green algal genera *Octogoniella* and *Desmatractum*, which Churchill retained in their classical position in the Oocystaceae. Bourrelly (1966) considers both genera as better placed with *Tetraëdron* in the Chlorococcaceae. Playford (1963) noted the resemblance of the Australian specimens to *Tetraporina* (see Chapter 3), originally regarded as angiosperm pollen, but which he stated to be algal in origin. Similar specimens occur in the Mississippian of the Moscow Basin, Canada, and Saudi Arabia, and in the Permian and Lower Triassic (with *Diporina* and *Triporina*) of Australia (Hemer and Nygreen, 1967; Balme, 1963; Segroves, 1967).

Chlorellaceae. *Chlorella*, the common soil alga, lacks zoospores but divides by fission to form four or eight nonmotile asexual reproductive daughter cells or autospores that resemble the parent cell.

Fossil Chlorellaceae include *Chlorellopsis* from the German Miocene and also are reported to occur in Jurassic lacustrine iron deposits of Normandy, France. The round to oval calcite-filled cells are 100 to 130 μm in diameter, occurring in a calcareous mass within the ferruginous oolitic Bajocian deposits (Dangeard, 1930). In addition to *Chlorellopsis, Caryosphaeroides* (see Figure 10.26) from the late Precambrian of Australia also is referred to this family. Somewhat similar forms described from the same area as green algae of uncertain assignment are *Glenobotrydion* and *Globophycus* (Schopf, 1968).

Hydrodictyaceae. Some of the Chlorococcales produce a coenobic colony from the original cell, *Pediastrum* producing flat, platelike colonies, and *Hydrodictyon* a cylindrical colony of individual tubular cells arranged in a honeycomblike net. Both have motile zoospores that are similar in appearance to isogamous gametes. *Pediastrum biradiatum* Meyen has a coenobium of 8 to 16 cells. During asexual reproduction each cell produces a number of motile zoospores that escape from the mother cell, swarm for about four minutes, and then over a period of eight to ten minutes the biflagellate oval zoospore develops into a flat round plate and finally a four-pronged adult cell. The chloroplast seemingly contains no pyrenoids until the adult form is attained, these then arising from a plastid (Moner and Chapman, 1960). *Pediastrum boryanum* (Turpin) Meneghini in cultures eventually lost the chlorophyll and produced resting cells. These were orange due to the carotenoid pigments in the cell fat inclusions. Although the wall was not thickened, dried resting cells remained viable for at least four years. The carotenoids were suggested to act as a filter for the photosynthetically active blue light and thus aid the cell to maintain its resting condition (Davis, 1962).

The wall of *Pediastrum* is double, with an inner continuous layer and an outer part of loose meshlike construction. The cell wall and meshlike layer are both highly resistant to NaOH, ethanol ether, and H_2SO_4 (Moner, 1955; Moner and Chapman, 1963). The inner layer was suggested to be cellulosic, but the composition was unproven. Resistance to palynologic treatment also suggests a compound other than cellulose. Later chemical studies indicate that the wall is relatively high in silica, probably concentrated in its outer part (Millington and Gawlik, 1967). Although silica would be removed in the maceration treatment, a resistant organic compound remains. After zoospores escape through a semicircular pore in the living cell wall, the empty cell remains almost intact.

If the Libyan Silurian *Deflandrastrum* (see Figure 10.27) is regarded as correctly assigned to the Hydrodictyaceae, that family also has a long geologic record. However, it is of marine occurrence, whereas modern representatives of the family are dominantly of freshwater habitat. Another modern genus, *Pediastrum*, also has been reported widely as fossil from the Lower Cretaceous (Albian) of Pakistan, the Upper Cretaceous of California (Evitt, 1963b), the Paleocene Fort Union Formation of Wyoming (Newman, 1965), the Eocene Green River Shale of Colorado and Wyoming (Bradley, 1964), the Eocene of Sumatra (Wilson and Hoffmeister, 1953), the Tertiary of Australia (Cookson, 1953), the Eocene Headon Beds of southern England (Borge and Erdtman, 1954), the Oligocene Florissant Formation of Colorado, the Miocene Mascall Formation of Oregon (Gray, 1960), and the Pleistocene and postglacial deposits of numerous areas (for example, the Scandinavian region, reported by Borge, 1895, and Lagerheim, 1902). Most of these occurrences are undoubtedly fluviatile or lacustrine, but possibly some of the early Tertiary and Cretaceous occurrences may be marine. Evitt (1963b) suggested that the resistant cells may have been transported into the marine sediments by streams, rather than living in the area in which they were fossilized.

Hydrodictyolites occurs in the Permo-Carboniferous, *Plaesiodictyon* in the Triassic of Luxemburg, and *Pediastrites* in the Late Jurassic of the USSR. In addition, the genus *Paleodictyon* Meneghini 1850 is regarded by some as a fossil record of an alga similar to the living *Hydrodictyon*. It has an extensive record, but some Ordovician, Pennsylvanian, Triassic, and later cited occurrences may be of different origin. Some are undoubtedly inorganic, others (perhaps all) regarded as complex feeding trails of worms or other invertebrates (Seilacher, 1954; Häntschel, 1962; Webby, 1969). Others are regarded as undoubted algal records, as in the Jurassic, Cretaceous, and early Tertiary of Japan and the Eocene, Oligocene, Miocene, and Pliocene of Italy and Austria (Koriba and Shigeru, 1939).

Coelastraceae. The single possible fossil record for this family is that of the specimen identified as *Coelastrum,* but which may instead be a *Pediastrum,* in the Eocene Green River Formation of Colorado (Bradley, 1929, 1931).

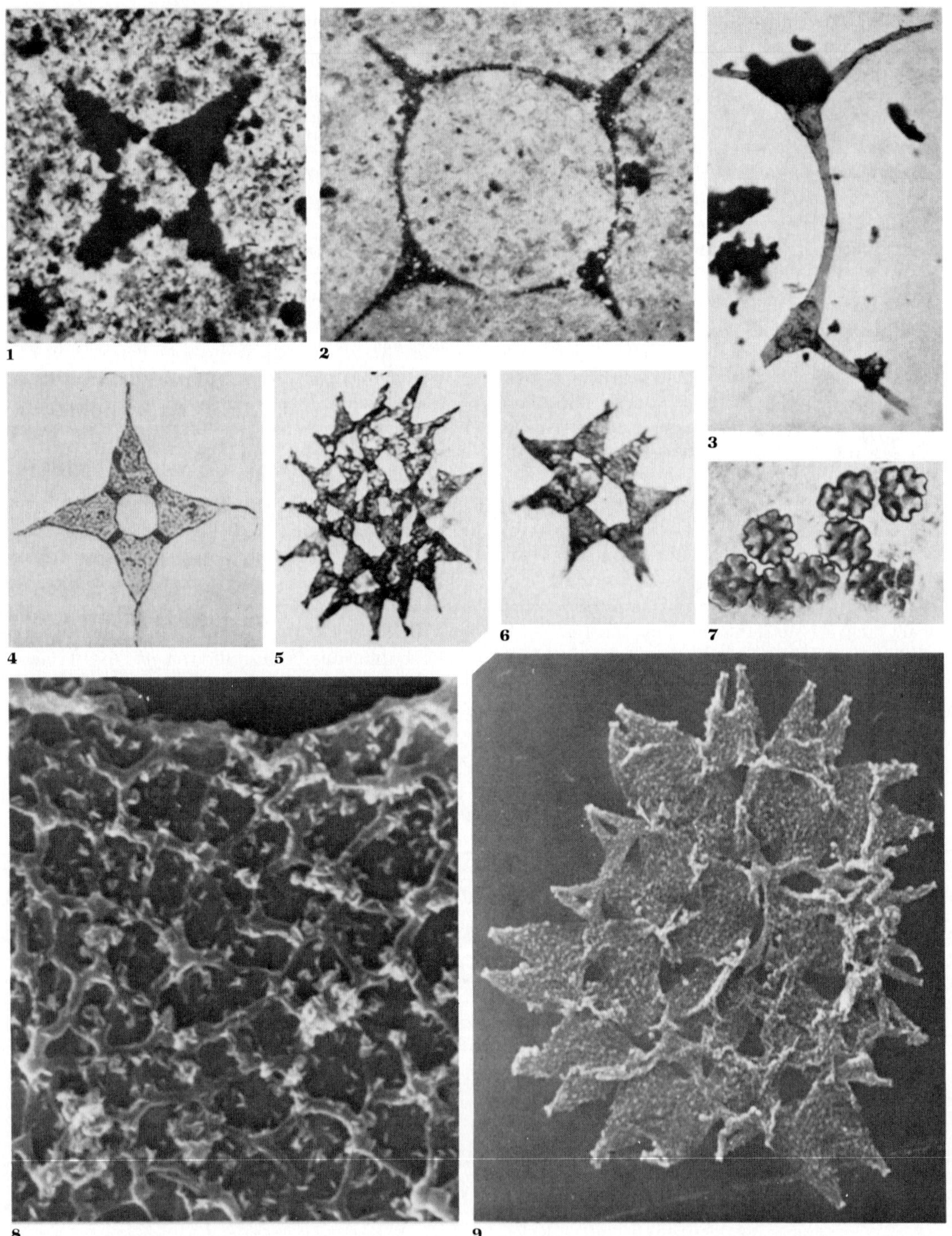

Botryococcaceae. Colonies of the lacustrine planktonic *Botryococcus* consist of large spherical aggregates of radially arranged cells, surrounded by a mucous envelope. Longitudinal cell division results in pairs of cells; the continuing process gradually produces large colonies. The individual pyriform cells lie within a cup-shaped cuticular thimble, whose distal end contains a thick fatty deposit covered by a cellulosic and pectic cap. The cells secrete an abundant red-colored oil that possibly has a hydrostatic function, aiding in flotation of the colony. Individual cells are about 5 to 15 μm in length, the colonies ranging from 10 to 100 μm in diameter. New colonies form by detachment of a few cells from an older colony.

Botryococcus was previously placed with the Xanthophyta but was shown to be a green alga (Blackburn in Blackburn and Temperley, 1936) because of the starch storage products and the presence of chlorophylls *a* and *b* (Belcher and Fogg, 1955), the latter not present in the xanthophytes. *Botryococcus* is resistant to a wide variety of conditions. For example, it withstands a range of pH 4.5 to pH 8 and may thus be important where conditions do not allow other algae to grow. Growth is relatively slow. During exponential growth the mean generation time is about one week, slowing to about five weeks at the stationary phase of growth. Perhaps diffusion into the cell is limited to the tip, as the lipoid-saturated cup encloses up to 90 percent of the cell. Nevertheless, *Botryococcus* may form such a dense growth over the surface of a lake that it exhausts the oxygen, resulting in heavy fish mortality (Belcher, 1968). Thick, oily gelatinous and highly decay-resistant algal de-

posits due to *Botryococcus* may accumulate to a thickness of 5 to 10 m on the lake bottoms. In time these become black and compact and contain large amounts of fats and paraffinlike material. Such deposits are termed boghead coal, the peat equivalent being termed Coorongite in the salt lakes and lagoons of Australia, Balkhashite in similar lakes in Russia (Thiessen, 1925), or Torbanite in the Mississippian of Scotland.

Botryococcus has an extensive geologic record (see Figure 10.28), although assigned erroneously to blue-green algae or xanthophytes at times, and reported under various names, such as *Gloeocapsomorpha*, *Elaeophyton*, *Epipolaia*, and *Reinschia*. It is known from the Precambrian of Bohemia (Konzalová, 1973), Gabon (Feys et al., 1967), and West Pakistan (Nagappa, 1957), and from all later periods from many localities in Europe, North America, India, and Australia (Bertrand, 1898; Bradley, 1924; Traverse, 1955; Gray, 1960; Kriván-Hutter, 1963; Chandra, 1964; Ramanujam, 1966; Konzalová and Mašek, 1972). The abundance of these organisms is such that about 300,000 cells are estimated to occur per cubic centimeter of rock in the Scottish oil shales, where the large, clear, lemon-yellow to orange-colored fossil thalli comprise about 11 percent of the rock volume.

Oocystaceae. The Oocystaceae are usually solitary, rarely form loose colonies, and do not form distinct coenobia. They reproduce solely by the formation of two to eight equally motile autospores. No vegetative division or zoospore formation is known. The modern *Eremo-*

Figure 10.28
Botryococcaceae. *Botryoccoccus* sp.,
Middle Ordovician, Oklahoma, SEM,
×1000.

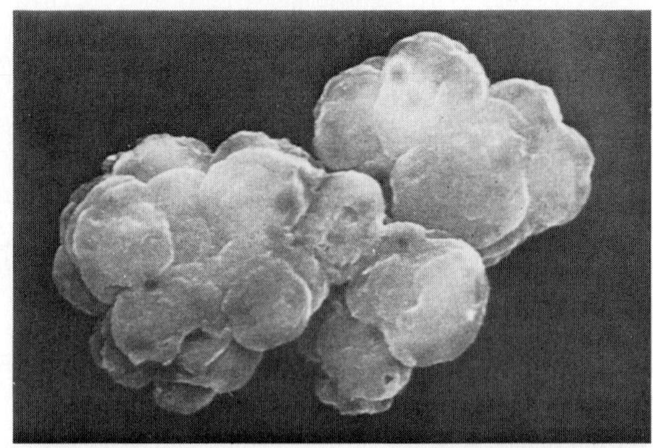

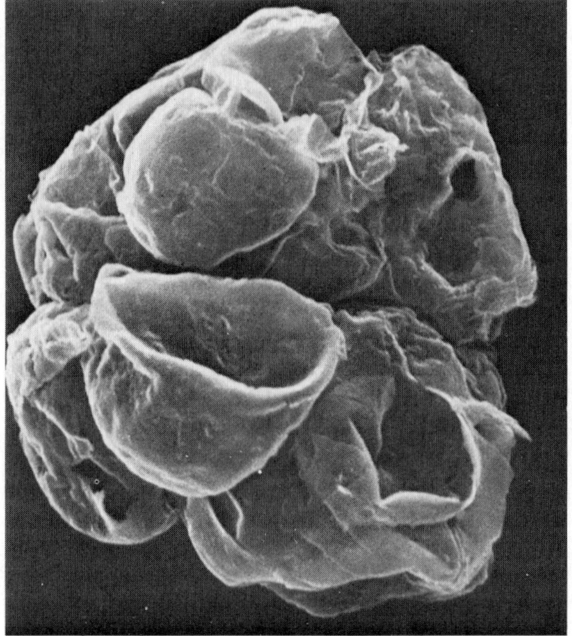

sphaera is relatively large, attaining about 200 μm in diameter, and growing abundantly on the bottom of ponds, swamps, and *Sphagnum* bogs.

Fossil representatives of the Oocystaceae include *Eremosphaera* in the Eocene Green River Formation of Colorado (Bradley, 1931), and *Lageniastrum*, parasitic on Carboniferous *Lepidodendron* spores.

Scenedesmaceae. Representatives of this family rarely produce zoospores. A four-celled colony (or with multiples of two) has laterally united cells. A new autocolony forms within each cell of the adult and is liberated by rupture of the parent wall. Genera previously placed here that proved to have zoospores, such as *Tetraëdron*, have been transferred elsewhere, hence the fossil record of this genus is no longer referrable here. However, Trainor (1963) reported *Scenedesmus* producing biflagellate zoospores in nitrogen-starved cultures, perhaps indicating a sexual phase.

Class Codiolophyceae

Phases of the life cycle that included a *Codiolum*-like stage were made the basis for a separate class (Kornmann, 1973). Included are the Orders Ulotrichales, Microsporales, Chaetophorales, Trentepohliales, Ulvales, and Prasiolales.

The common *Ulothrix* consists of unbranched filaments attached to the substrate in lakes, streams, and in the sea. All cells are identical, except that the basal one is modified for attachment. Quadriflagellate zoospores may be produced, and sexual reproduction involves biflagellate isogametes. Prior to germination of the zygote, meiosis occurs, four zoospores are produced, and each grows into a new filament.

Species of *Trentepohlia*, *Phycopeltis*, and *Cephaleuros* (all Trentepohliaceae), *Leptosira*, *Pseudopleurococcus*, and *Coccobotrys* (all Chaetophoraceae) may form the algal partner of lichens (Ahmadjian, 1967).

Ulotricaceae. Filaments in the Devonian Onondaga chert of New York, consisting of cells 15 to 20 μm in length, were referred to the modern Ulotricalean genus *Geminella* (Baschnagel, 1942), and *Hormosporites* was described from the German Devonian (Grüss, 1928). *Palaeogeminella* from the Upper Devonian of Texas (Fairchild et al., 1973) may be only an artifact, however (Schopf, 1974). An epiphytic alga associated with Upper Devonian *Archaeopteris* fossils and named *Courvoisiella ctenomorpha* (see Figure 10.29) consists of nonseptate, uniserial and biserial projecting tubular branches from a similar nonseptate horizontal part. Spherical structures at the side and tip of the projecting tubes were regarded as gametangia (Niklas, 1976).

Ulothrix also has been reported as fossil in the Tertiary of India (Mahadevan and Sarma, 1947), and from the Deccan Intertrappean cherts (Eocene), where Dwivedi (1959b) described cellular filaments, some of which contained a "nucleus-like" body in the cells. As Dwivedi stated, "some important characters could not be studied due to imperfect preservation," and the cell inclusions probably are not nuclear residues. A similar species of *Ulothrix*, also unnamed, was preserved as unbranched filaments in Eocene oil shale and lignite in India (Rao, 1957).

Sexual reproduction of *Ulothrix* is oogamous; the egg cell is in an enlarged oogonium that opens by a pore or fissure prior to fertilization. Sperm cells are produced in short, boxlike antheridia, the multiflagellate sperm morphologically resembling the zoospores. The zygote develops a double or multiple-layered wall and becomes a resting stage. It is liberated by disintegration of the oogonial wall and after meiosis produces four zoospores that in turn produce a new filament.

The genera *Eothrix* and *Globochaete* were described by Lombard (1937, 1945) from thin sections of Upper Jurassic limestones of the Alpine region and suggested to represent thalli of green algae. Although reported as such by later workers (Colom, 1948, 1955; Durand Delga, 1956), Verniory (1954) stated that they instead are random sections of the ossicles of the pelagic crinoid *Saccoma*. Serial sections were made to demonstrate this relationship, and each specimen was observed to be optically a single crystal of calcite, a feature characteristic of all echinoderms but not of the calcareous algae. Later Bronnimann (1955) described *Eothrix* and *Globochaete* from Cuba, as well as a new genus *Lombardia*, with three species, regarded also as possibly an echinoderm. Verniory (1956) regarded all Bronnimann's *Lombardia* species as synonymous with *Saccoma*, but Knauer (1966), in identifying the various *Saccoma* species that had been placed in *Eothrix* or *Lombardia*, nevertheless regarded *Lombardia arachnoidea* Bronnimann as sufficiently distinct from the others to be maintained as a distinct genus and species. All are currently regarded as crinoids and have no relationship with the green algae.

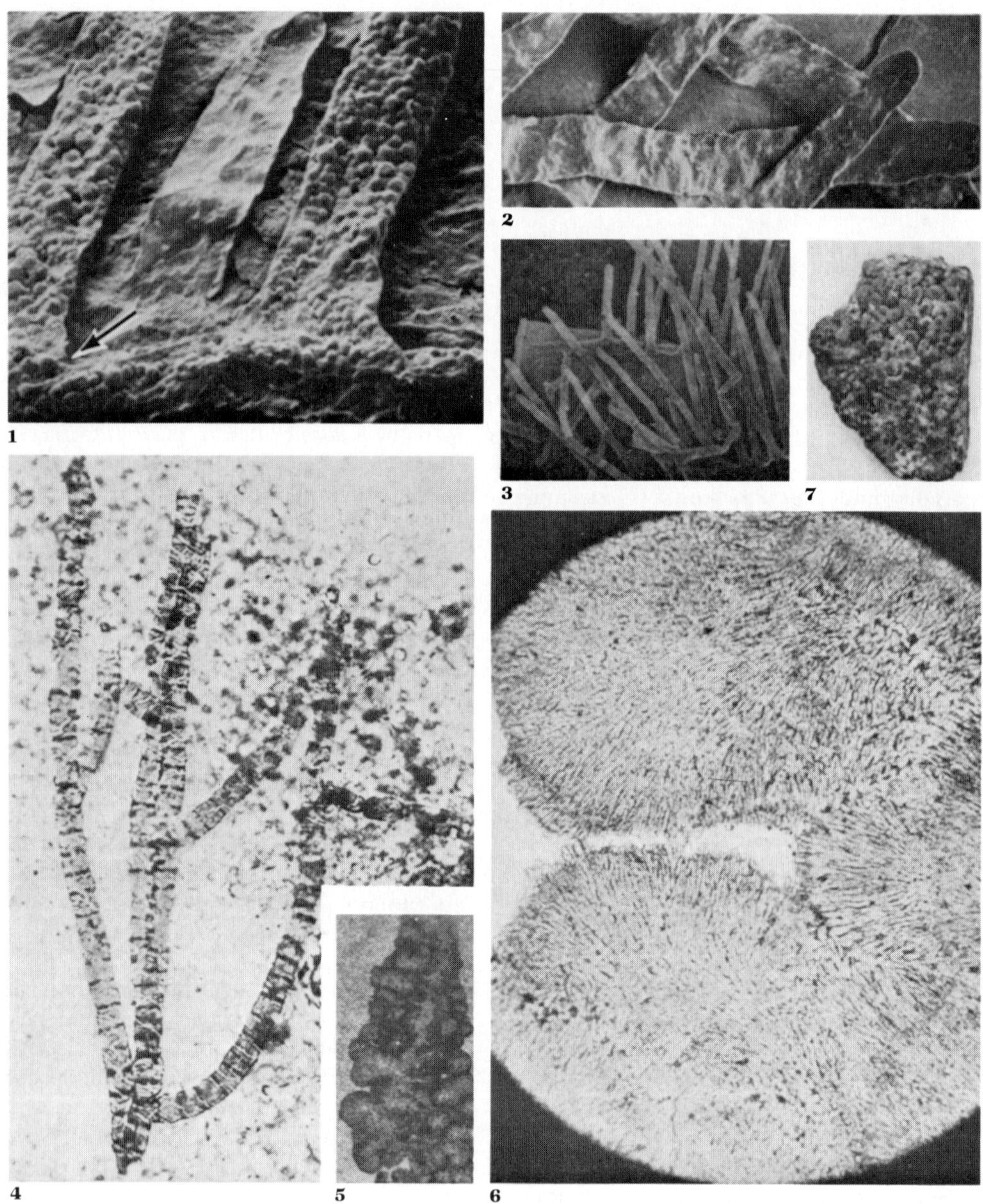

Figure 10.29

Ulotrichaceae, Chaetophoraceae. 1—3. *Courvoisiella ctenomorpha* Niklas, Upper Devonian, West Virginia, epiphyte associated with *Archaeopteris*, SEM, from Niklas, 1976. **1.** Part of thallus with procumbent (arrow) and projecting tubules, enclosing many pyrite crystals, ×1050. **2.** Intertwining tubules, ×1000. **3.** Parts of three thalli, ×500. *Stigeoclonium* sp. cf. *lubricum* (Dillwyn) Kützing, Eocene, Colorado, ×410, from Bradley, 1929. **5—7.** *Chlorotylites berryi* M. A. Howe, Paleocene, Alabama, from Howe, 1932. **5,6.** Sections of thallus; 5, ×2; 6, ×51. **7.** Exterior, ×1.

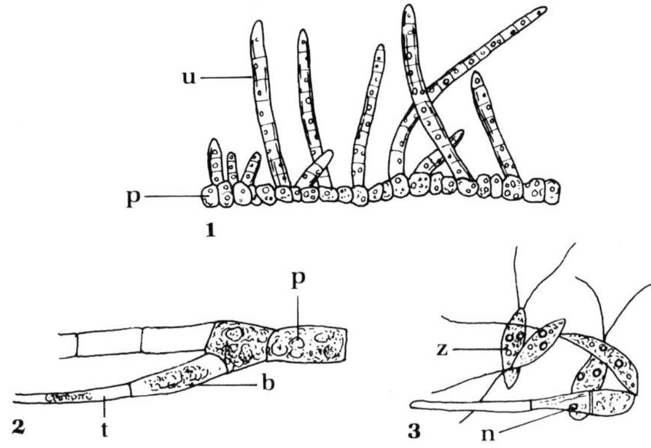

Figure 10.30
Stigeoclonium sp. **1.** Colony, showing upright filaments (u) arising from prostrate filament (p), ×325. **2.** New branch (b) with terminal cell (t), arising from prostrate filament (p), ×450. **3.** Biflagellate zoospores (z), and beginning development of new filament (n), ×350.

Microsporaceae. *Microspora* cf. *M. pachyderma* (Wille) Lagerheim was reported by Bradley (1929, 1931) from the Eocene Green River Formation of Colorado.

Chaetophoraceae. Fossils here referred include *Chlorotylites* in a Midway lignite (Paleocene) of Alabama. It consists of a silicified thallus with dichotomously branching filaments of uniseriate cells, possibly originally living on a decaying leaf as does modern *Chlorotylium* in freshwater streams (Howe, 1932). *Stigeoclonium* (see Figure 10.30) was reported from the Eocene Green River Formation of Colorado (Bradley, 1929, 1931); and the boring organisms *Chaetophorites tertiarius* Fliche from the Oligocene of France, and *C. gomontoides* Pratje from the Lower Jurassic, may belong to this family. *Chaetophorites* has also been reported from the Silurian (Pia, 1937), and the similarly boring *Foreliella*, originally described as perforating modern *Anodonta* shells, was reported to occur in lacustrine Tertiary *Unio* marls of the Aquitaine Basin of France (Caudwell, 1968). Accurate identification of these perforating algae is difficult, as various blue-green and red algae and fungi may produce similar excavations.

Trentepohliaceae. *Phycopeltis* is a modern epiphytic alga to which fossil specimens from the German Eocene and Oligocene have been referred. *Phycopeltis microthyrioides* Kirchheimer seemingly shows both gametangia and zoosporangia preserved on the discoid thalli. *Phycopeltis* found on Eocene fossil leaves in the Missisippi embayment region consists of a circular to linguiform thallus of a single layer of cells, arranged in radiating rows; some sporelike structures were also preserved. The epiphytes adhered to either surface of the leaves.

Ulvaceae. The Ulvales include the marsh plant *Ulva* (see Figure 10.31), which grows first in a filamentous form like *Ulothrix,* then the cells divide medially to produce a two-layered

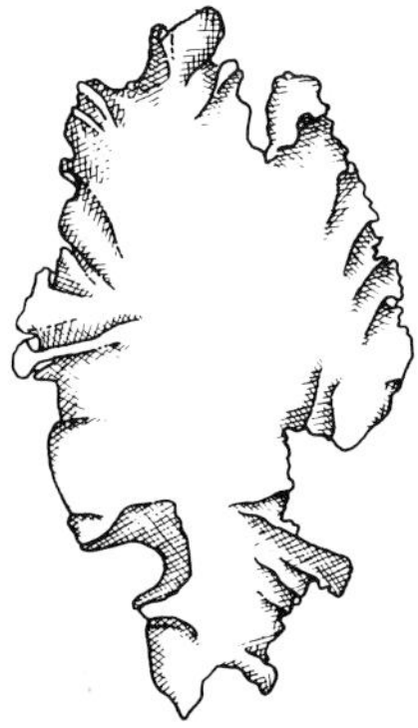

Figure 10.31
Ulva sp., the sea lettuce, a thallus consisting of two layers of cells, ×0.25.

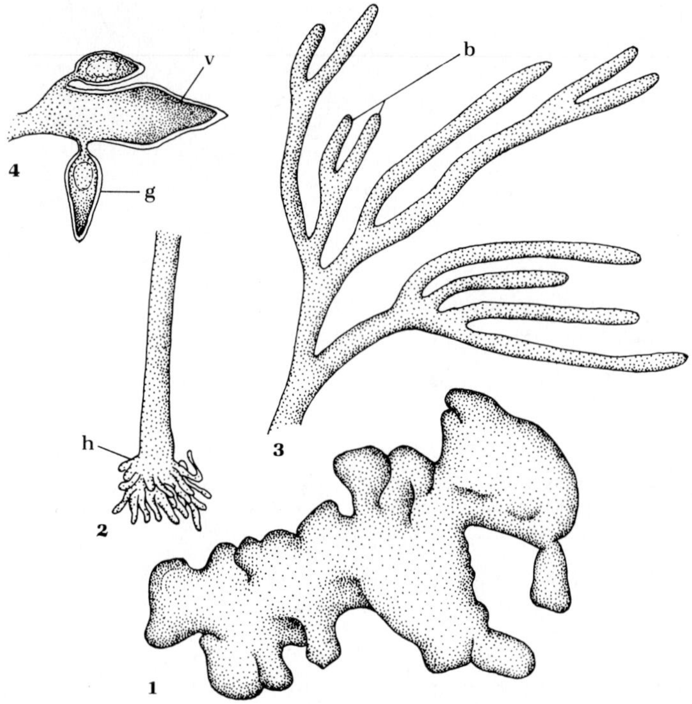

Figure 10.32
Codium fragile (Suriray) Hariot.
1. Encrusting form. **2.** Base of
erect form, showing holdfast (h).
3. Erect form with dichotomous
branching (b). 1−3, ×0.5. **4.** Vesicle (v) and gametangia (g), ×150.

thallus of uninucleate cells. Common in marine and brackish water, it is referred to as the sea lettuce. Both quadriflagellate zoospores and unlike biflagellate gametes are produced.

Orders Cladophorales, Codiales, Caulerpales, and Derbesiales

Cladophora (Cladophorales) has a complicated life cycle. Filaments consist of large multinucleate cells. The diploid plants of marine species that produce zoospores are morphologically similar to the haploid plants that produce biflagellate gametes. Meiosis thus is delayed for a generation. The wall of the Cladophorales contains cellulose in ribbonlike microfibrils, forming sheets of parallel fibrils with those of successive lamellae lying nearly at right angles. In composition the wall resembles that of higher plants, although more crystalline (Preston, 1961).

Cladophora has been reported as fossil in Tertiary lacustrine limes of the French Aquitaine Basin. Other fossil specimens re-ferred to *Cladophorites* have been reported in Tertiary freshwater deposits such as the Eocene (Bartonian) of France, the Oligocene (Chattian) near Rennes, and the Miocene of the Rhine Valley (Milon, 1932). They may occur as far back as the Jurassic and have also been reported in the Pleistocene of Australia.

Siphonous marine algae were long placed in the single order Siphonales, but this group has been subdivided by phycologists into three orders, the Codiales, Derbesiales, and Caulerpales. In contrast, paleontologists commonly have included these three and the Dasycladales in the Siphonales or in the Codiales. Although some assignments must be only tentative, an attempt is made here to place the fossil taxa in the more current classification. For example, the Precambrian siphonate *Palaeosiphonella* is here placed with the Cladophorales rather than the Siphonales. As presently restricted, the Codiales probably include no fossil representatives, most having been transferred to the Caulerpales.

Codium is a siphonous green alga that may be matlike or cylindrical and dichotomously

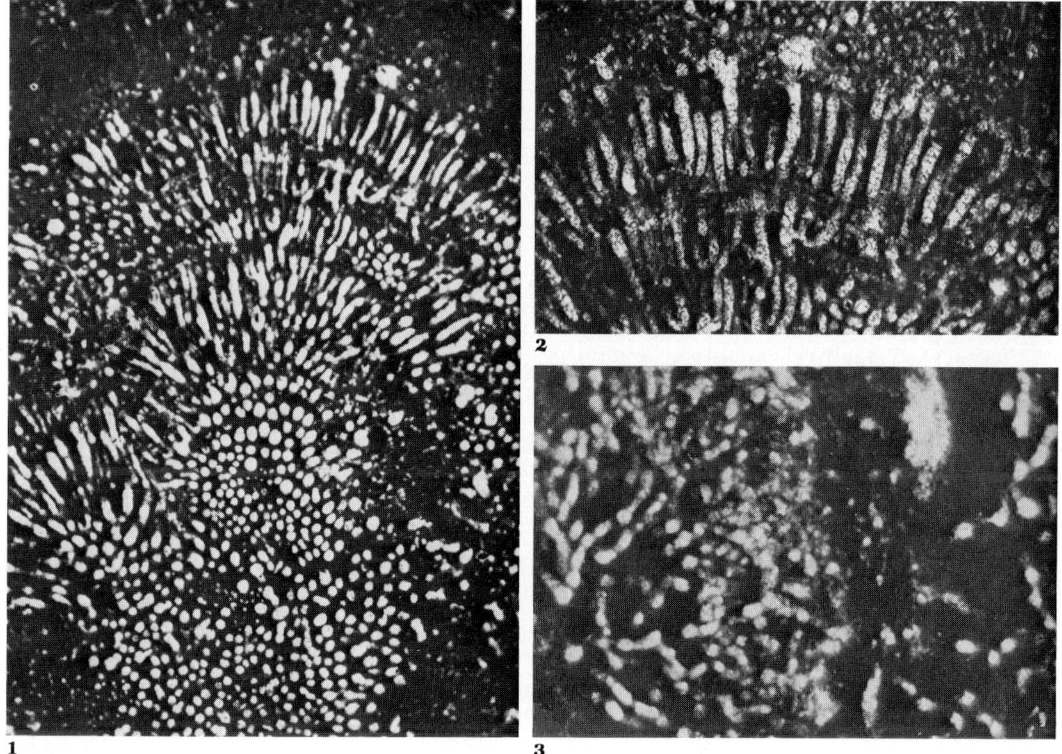

Figure 10.33
Garwoodiaceae. **1,2.** *Garwoodia gregaria* (Nicholson) Wood, Mississippian, Scotland, from Wood, 1941.
1. Showing variation in diameter of tubes, ×25. **2.** Portion enlarged, ×38. **3.** *Bevocastria conglobata*
Garwood, Mississippian, England, ×50, from Garwood, 1931.

branched. *C. fragile* (see Figure 10.32) may be 30
cm long, and *C. magnum* Dawson is believed to
be the largest living siphonous, multinucleate
alga, attaining up to 8 m length; most *Codium*
species are only 6 to 20 cm in height. The diploid thallus consists of numerous compactly interwoven filaments. Meiosis occurs during
formation of the biflagellate gametes. Unlike
most Codiales, *C. magnum* may occur in large
detached masses on mudflats, although it has a
stipelike base. Only one truly planktonic
species is known in this order, *Pseudobryopsis*
planktonica Cassie, which forms small thalli of
3 to 6 mm length and lacks rhizoids. It has been
observed in dense blooms in the waters of
Whale Bay, New Zealand (Cassie, 1969).

Among the siphonous green algae that previously were included in the Siphonales are the
Caulerpales. As here recognized, they include
the extinct Garwoodiaceae, the Udoteaceae,
Dichotomosiphonaceae, and the Caulerpaceae.
Most of the siphonous, lime-depositing fossils
here placed in the Udoteaceae were formerly
classed in the Siphonales or Codiales. Most
modern Codiales are noncalcareous, however,
whereas the living *Udotea* is commonly lime-
encrusted, with an elaborate development of
cortical layers. A separate order Dichotomo-
siphonales is recognized by some.

The Garwoodiaceae generally are crustose or
nodulose, with many tightly packed branching
tubular filaments (see Figure 10.33). *Cayeuxia*
and *Ortonella* are occasionally referred here or
to the Codiaceae but have been placed in the
blue-green algae by others (see discussion by
Hudson, 1970, p. 23), as has the entire family.

In the living *Udotea minima* Ernst, coarse creeping threads produce erect filaments that may be distinct or interwoven to form a stalk. Distally the filaments spread out like a brush or fan (see Figure 10.34), and laterals may arise from various positions. Fossil *Udotea palmetta* Fiore was described from the lower Eocene of Italy (Fiore, 1936).

Living *Halimeda* (see Figure 10.22) has a highly developed fanlike structure, its branched thallus consisting of calcified segments separated by uncalcified joints. The fossil *Ovulites* (see Figure 10.35) similarly consists of ovate to elongate segments that are linearly arranged and occasionally bifurcate. Other fossil Udoteaceae (see Figure 10.36) may be globular, ovate, flattened, or cylindrical and commonly are segmented. The central region has loosely packed erect tubular filaments that branch into smaller tubes that in turn curve outward to become nearly perpendicular to the outer surface at the dermal layer. The dermal layer consists of very finely divided branches. *Calcifolium* Schvetzov & Birina 1935 formerly was included in the Udoteaceae, but has been reassigned to the sponges (Perret and Vachard, 1975).

The Caulerpaceae have a less extensive fossil record, although the Ordovician *Dimorphosiphon* (see Figure 10.37) may belong here. *Cretosiphon caulerpoides* O. Wetzel, from the Cretaceous flints in Germany, also is regarded as very similar to modern *Caulerpa*.

The living *Caulerpa sertularioides* (Gmelin) Howe may form a heavy growth some 10 cm thick over a wide area of the Gulf of California (Dawson, 1966). The genus is typically tropical in distribution, occurring in relatively shallow quiet water. Unlike most marine algae, it is not always attached to a rocky substrate but possesses long creeping rhizomes up to a meter in length for attachment in sand or mud. Vegeta-

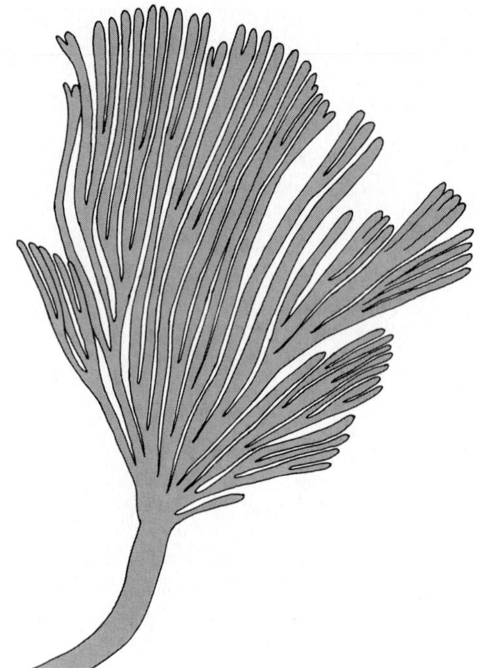

Figure 10.34
Udotea javensis (Montagne) A. & E. S. Gepp, ×25.

tive shoots arise from this rhizoidal base, attaining as much as 30 cm in height. They may consist largely of irregularly branched threads like those of the base and similar to some of the Codiales, but in other species possess flattened lateral branches that may be whorled, resembling *Chara*, consist of tight imbricated structures, or appear mosslike. Still other species have flattened leaflike laterals in two rows, presenting a bilateral symmetry. Some even have an external appearance suggestive of the vascu-

---

Figure 10.35
Udoteaceae. **1,2.** *Ovulites oehlerti* Munier-Chalmas. **1.** ×100. **2.** Broken margin showing perforations connecting through the calcareous wall, ×400. **3–5.** *O. elongatus* Lamarck, individual segments, with opening between segments visible in part 5; 3,5, ×50. **4.** Showing surface pores, ×400. **6–8.** *O. margaritula* Lamarck. **6.** ×150. **7.** Broken fragment, showing pores at inner side of wall and perforations through wall, ×400. **8.** Upper part of part 6, showing large openings for connection to next segment at a bifurcation, and wall pores at the exterior, ×200. All SEM, from middle Eocene (Lutetian), Montmirail, France.

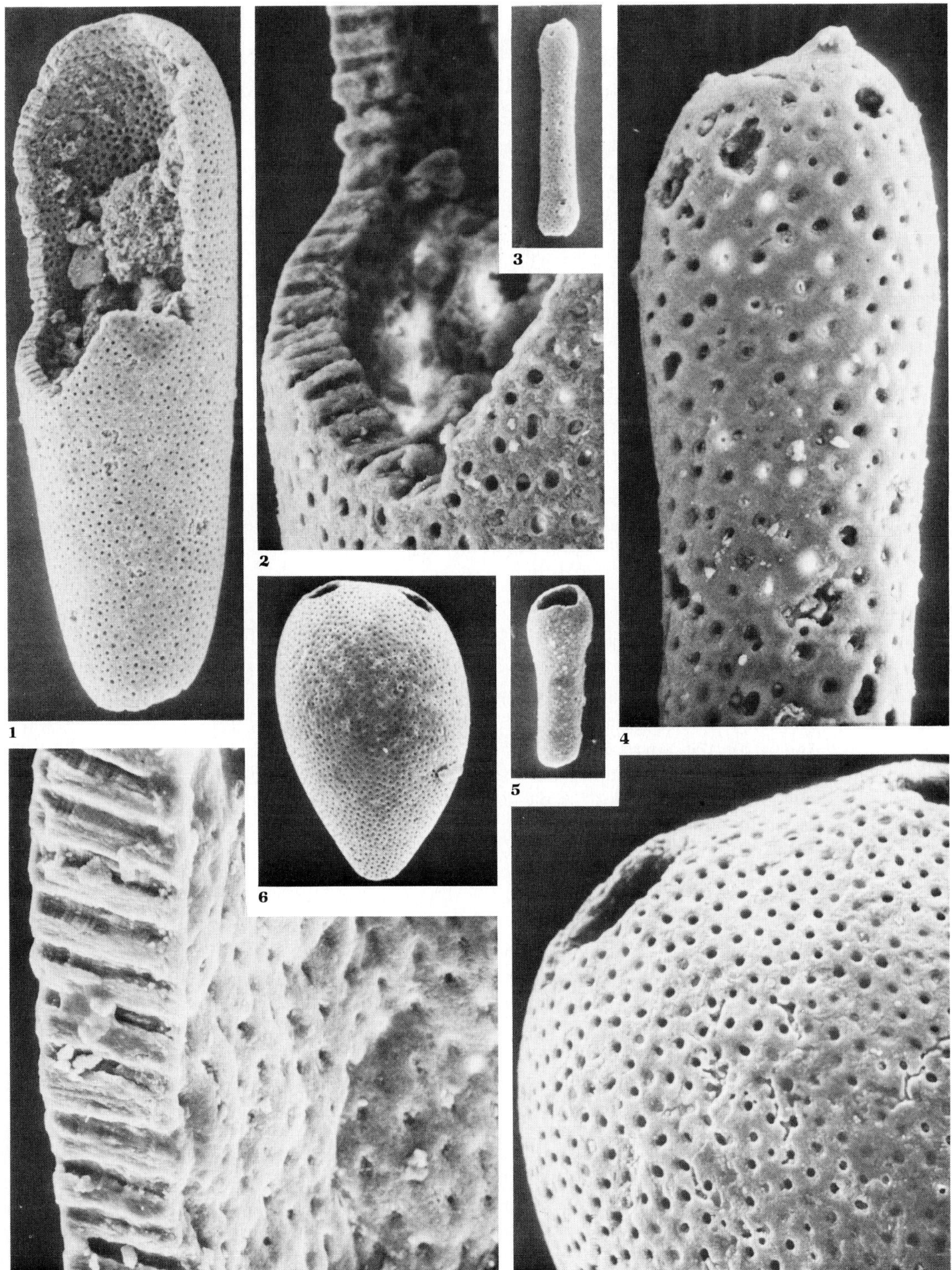

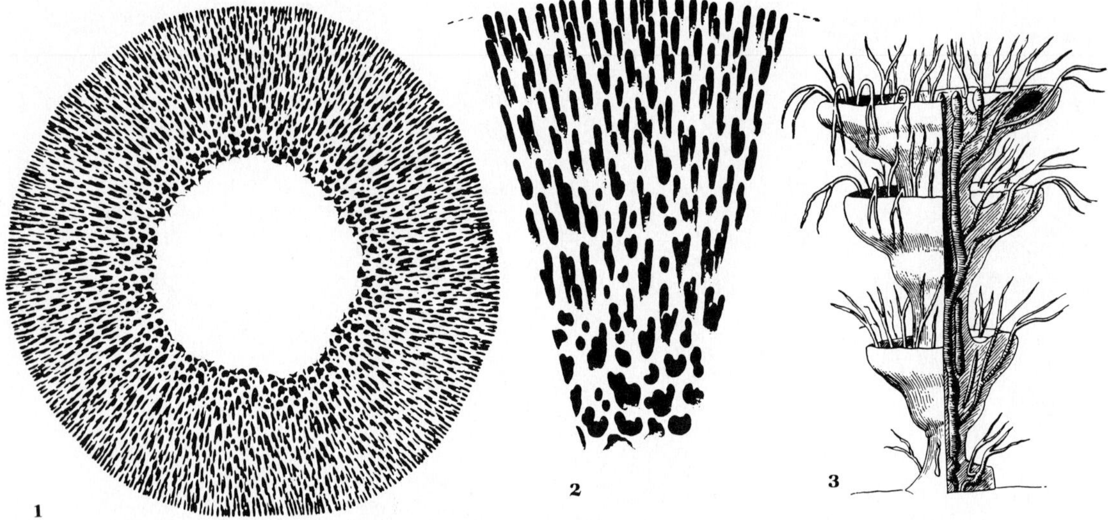

Figure 10.36

Udoteaceae. **1,2.** *Palaeocodium saharianum* Chiarugi, Mississippian, Libya, cross section, ×2, and portion enlarged, ×6, from Chiarugi, 1947. **3.** *Lancicula alta* Maslov, Lower Devonian, USSR, reconstruction, exterior at left, section at right; ×15, from Maslov, 1956.

lar plants. In all, the internal structure is siphonous, completely lacking septa (see Figure 10.38). A central vacuole is surrounded by an outer layer of cytoplasm containing the plastids and the many nuclei. Cylindrical skeletal strands traverse the central cavity, forming a complete network. Wall and skeletal strands show progressive thickening and present a stratified appearance in section. Both are composed largely of callose and pectin with no cellulose. The skeletal strands also are covered with cytoplasm. Reproduction may be by fragmentation of the thallus or by the liberation of biflagellate swarmers, of two sizes in some species and thus probably representing gametes, although no proof is as yet available.

The Order Derbesiales is characterized by a markedly dissimilar alternation of generations. The principal stage of *Derbesia* is filamentous and produces globose sporangia as short lateral branches cut off by an ingrowth of the wall. The zoospores have a whorl of flagella and settle to produce a morphologically different sexually reproducing generation that was for a long time referred to a separate genus *Halicystis*. It consists of an unbranched vesicular thallus with perennial rhizoidal system. This stage produces biflagellate gametes of two sizes, which upon fertilization produce the new *Derbesia* stage.

Pedobesia produces many elongate upright filaments from a discoidal thalluslike calcified attachment pad (MacRaild and Womersley, 1974). The attachment disk consists of two layers joined by many short cylindrical pillars, the upper layer being perforated by many elongated pores, which may be about three times as numerous as the inner pillars (see Figure 10.39). The platelike structure consists of aragonite, with a very minor amount of magnesium calcite (Feldmann et al., 1975). Multiflagellate zoospores are produced in terminal sporangia, and a *Halicystis*-like gametophytic stage has not been observed.

No fossil representatives are known for the Derbesiales, although the calcified attachment pads of *Pedobesia* could be preserved.

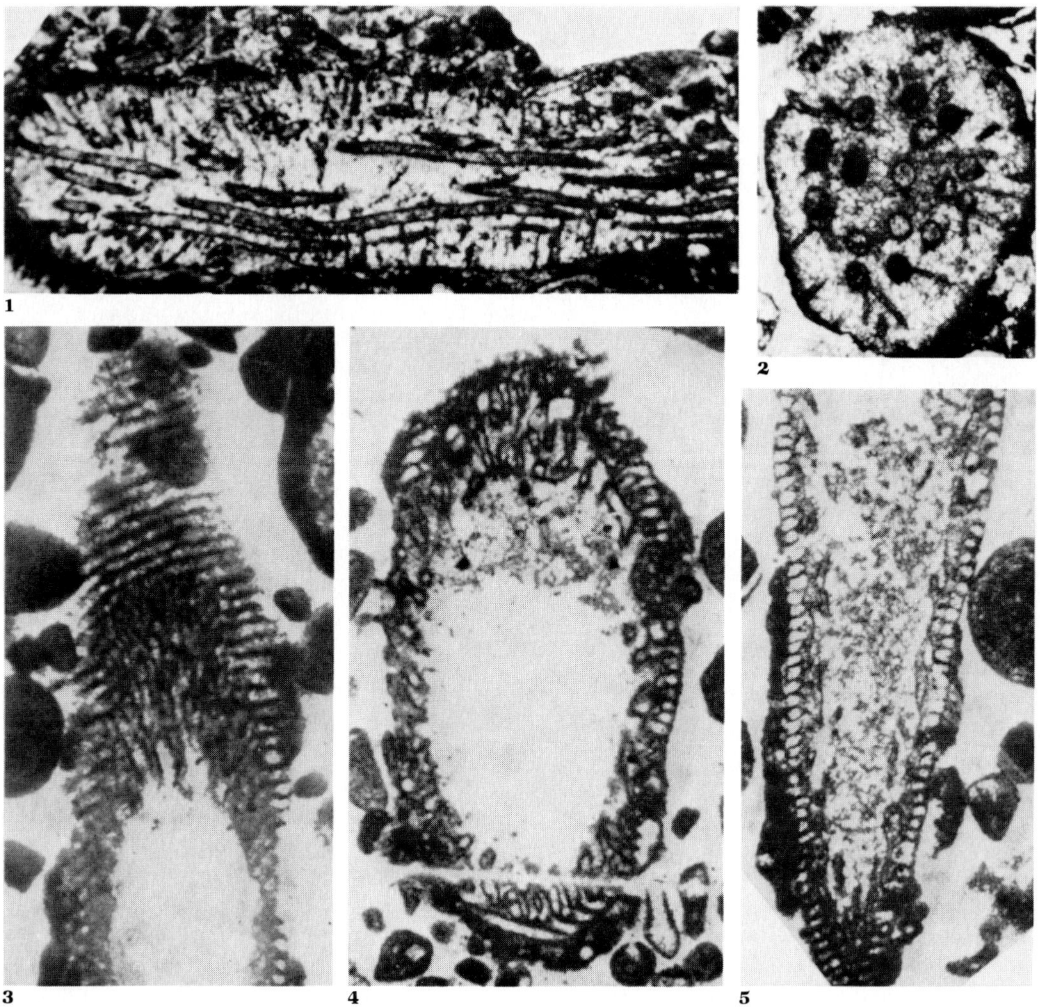

Figure 10.37

Udoteaceae. **1,2.** *Dimorphosiphon rectangularis* Høeg, Middle Ordovician, Norway, from Elliott, 1972. **1.** Longitudinal section, ×9. **2.** Transverse section, ×14. **3—5.** *Pseudocodium convolvens* Praturlon, Upper Jurassic, Italy, ×38, from Praturlon, 1964. **3.** Oblique section, showing relationship of inner tubules and outer spiral. **4.** Oblique transverse section. **5.** Oblique longitudinal section, showing that peripheral spiralled tube does not communicate with exterior.

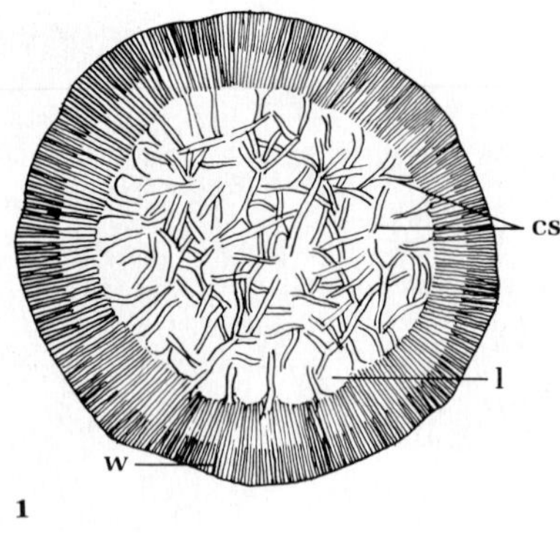

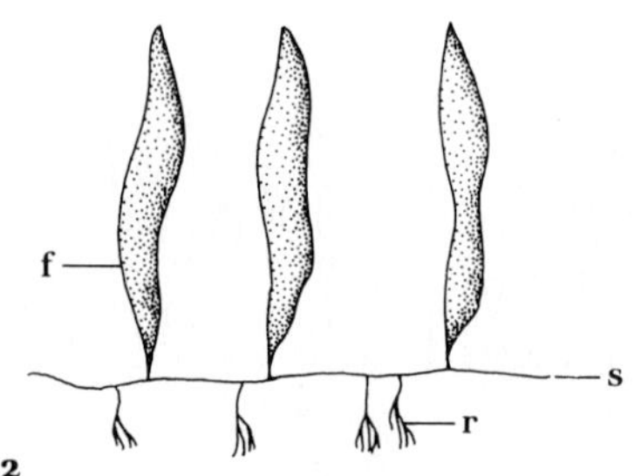

Figure 10.38
Caulerpa prolifera (Forsk.) Lamouroux. **1.** Cross section of stolon, approximately ×100, showing numerous wall layers (w), strengthening cytoplasmic strands (cs), and the lumen (l) separating these. **2.** Portion of thallus, showing stolon (s), attached to the substrate by the rhizoids (r), and giving rise to the erect foliose part (f), which is leaflike in appearance, but lacks septa, hence is functionally a single cell; ×0.25.

Order Siphonocladales

Like the following two orders (Receptaculitales and Dasycladales), the Siphonocladales are almost exclusively marine, but in contrast are poorly represented by fossils.

The Protovaloniaceae and Mongoloporellaceae are monotypic late Paleozoic families (see Figure 10.40); the Valoniaceae is represented in the Jurassic by *Verticillodesmis,* and the Siphonocladaceae by the Jurassic and Cretaceous *Pycnoporidium.*

Order Receptaculitales

The Paleozoic Receptaculitales has at times been placed either in the animal kingdom, with sponges or archaeocyathids (e.g., Laubenfels, 1955; Müller, 1968), corals, foraminifera, cystids, or tunicates, or in the plant kingdom, as an alga or even a fossilized cycad cone. A considerable body of evidence, in particular the mode of growth and enlargement and presence of gametangia (Nitecki, 1972b), indicates a relationship to the calcareous algae, and most closely to the Dasycladales (Kesling and Graham, 1962; Nitecki, 1969a,b). Some recognize merely a separate family within the Dasycladales (Nitecki, 1972b), but others recognize a separate Order Receptaculitida (in the animal kingdom; Müller, 1968) or Receptaculitales (plant kingdom; Rietschel, 1969, 1977).

Individuals range from 2 mm to 300 mm in diameter but are more commonly between 60 and 150 mm in diameter. They occur in shallow, quiet water deposits of Middle Ordovician to Late Devonian age (see Figures 10.41 and 10.42). Kesling and Graham (1962) interpreted the fossil *Ischadites* as a thallus with many branches radiating from an uncalcified central axis. The distal ends of the branches were said to be modified as gametangia, with gametocysts occurring between the inner and outer walls. In

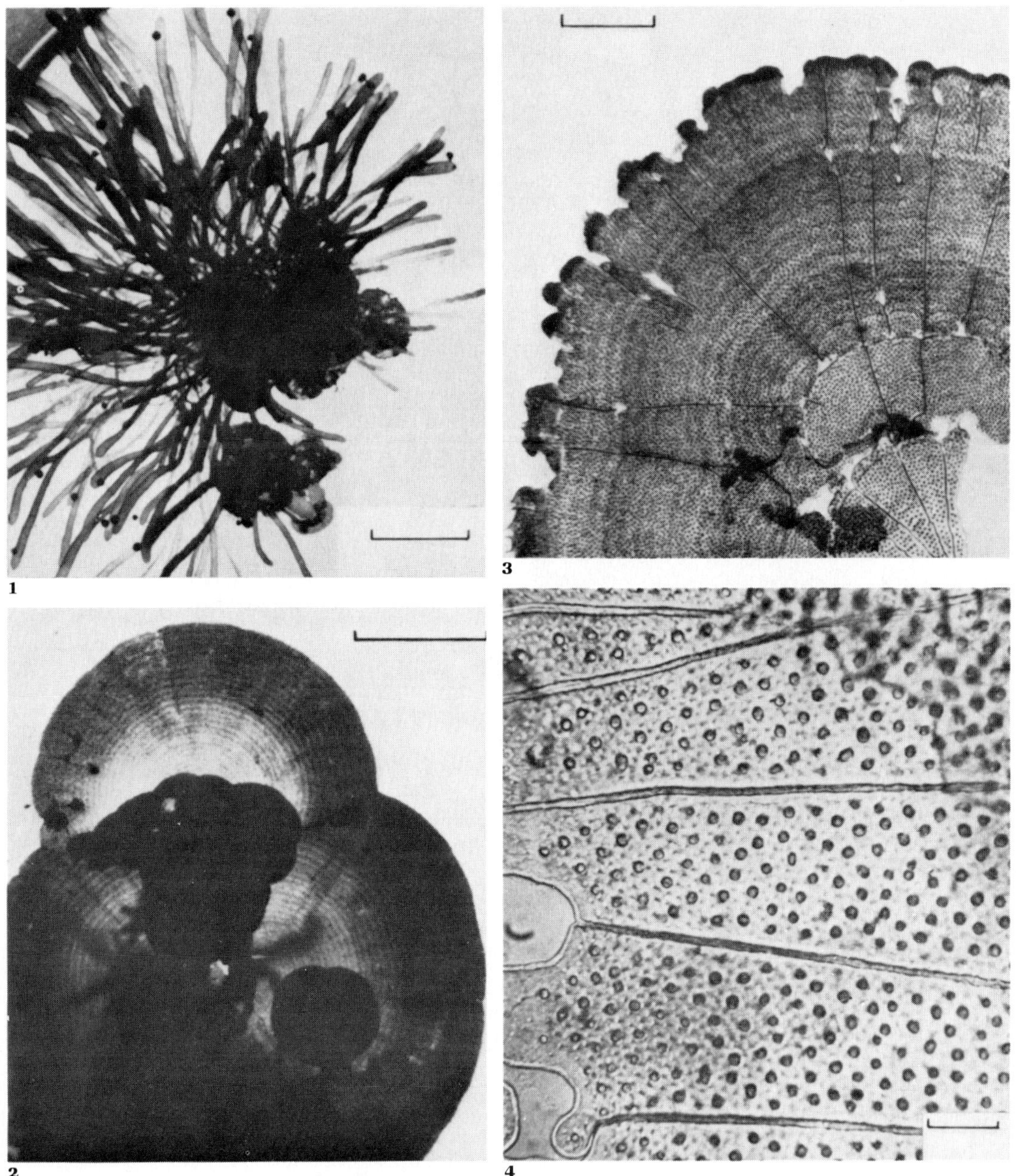

Figure 10.39
Pedobesia clavaeformis (J. Agardh) MacRaild & Womersley. **1.** Group of attachment discs, giving rise to many erect clavate branches, some with globular sporangia; bar = 10 mm. **2.** Disc in surface view, with secondary discs and mass of filaments; bar = 1 mm. **3.** Part of disc with three concentric rings of growth, radiating lines indicating separate growth sectors; bar = 150 μm. **4.** Enlargement, showing cell walls, pillars (appearing as distinct small circles), and pores (faint dots); bar = 20 μm. All from MacRaild and Womersley, 1974.

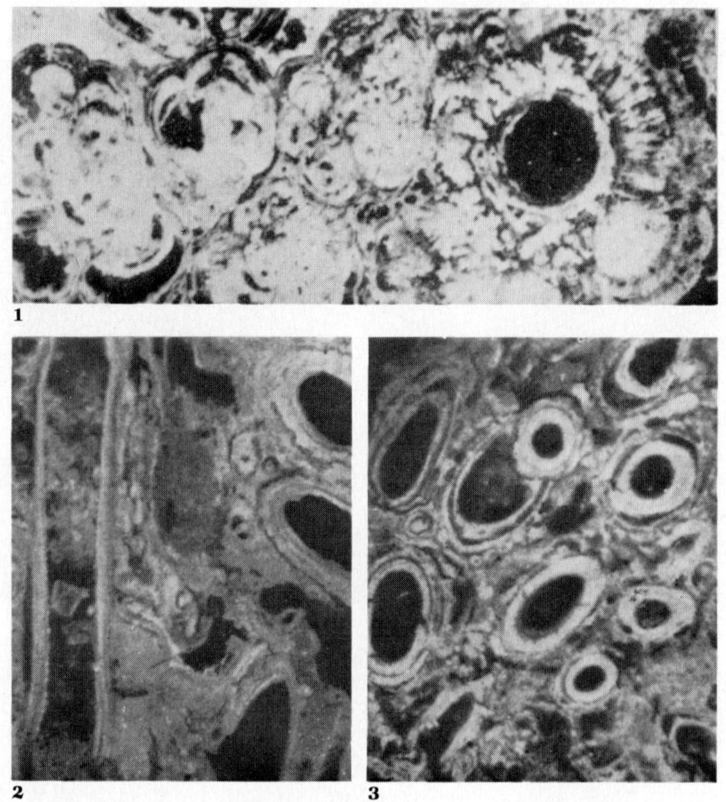

Figure 10.40

Protovaloniaceae, Mongoloporellaceae. **1.** *Protovalonia ichin-zaganense* Vologdin, ×7. **2,3.** *Mongoloporella rozovi* Vologdin, ×8. All from Permian, eastern Mongolia, from Vologdin, 1966.

Figure 10.41 (*facing page*)

Receptaculitaceae. **1.** *"Acanthochonia" barrandei* Hinde, Silurian, Czechoslovakia, ×1.6, from Rietschel, 1969. **2.** *Tettragonis orbis* Eichwald, Ordovician, Germany, ×0.8. **3,4.** *Receptaculites neptuni* Defrance, Middle Devonian, Germany. **3.** Radial section, oriented to cut the plates diagonally, showing meromes flaring at surface and interlocking, ×4. **4.** Portion of outer surface, showing intersecting meromes and plates, ×4. 2−4, from Rauff, 1892b. **5,6.** *Epimastopora malaysiana* Elliott, Middle Permian, Malaya, from Elliott, 1968a. **5.** Distal end enlarged, ×8.1. **6.** Entire specimen, ×2.8. **7.** *Ischadites murchisoni* Eichwald, Ordovician, Germany, ×0.8, from Rauff, 1892b.

1
2
3
4
5
6
7

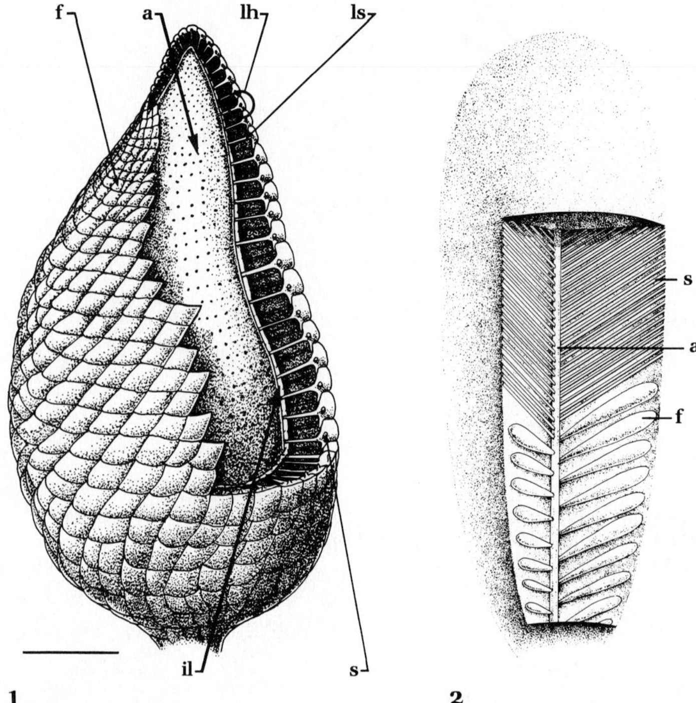

Figure 10.42

Receptaculitaceae. **1.** *Calathium egerodae* Nitecki, Silurian, Illinois; reconstruction, showing inner calcareous layer (il), stellate structure (s), facets (f), central axis (a), lateral head (lh), and lateral shaft (ls); bar = 1 cm; from Nitecki, 1972a. **2.** *Amphispongia oblonga* Salter, Silurian, Scotland. Reconstruction, showing axis (a), sterile laterals (s) at top, fertile ones (f) near base; length between 1 and 6 cm; from Nitecki, 1971.

contrast, Byrnes (1968) and Rietschel (1969) state that the supposed gametangia were merely sections of the radial skeletal elements, termed **cruciform rays** by Byrnes and **meromes** by Rietschel and Müller, following Rauff (1892b). Much more certain gametangia have now been described in *Ischadites hemisphericus* (Hall) Winchell & Schuchert (see Figure 10.43; Nitecki, 1972b).

The surface of the globular, pyriform, cylindrical, or platelike *Ischadites* is covered with regularly aligned rhombic or hexagonal plates spiralling from one pole to that opposite. One pole (probably the lower one) is completely closed by these plates, which may grade some-

what in size from the apices to the equatorial region; after they reach a given maximum, new plate series are intercalated. The central region of the upper side may have numerous tiny plates, but these may not be sufficiently calcified for preservation, leaving open the region once regarded as the "sponge" osculum. Underlying the plates and perpendicular to the surface lie the skeletal bodies termed meromes. Immediately beneath the surface cruciform rays project from the main shaft of the merome, those of adjacent meromes overlapping and interlocking to produce a firm skeleton; but weathering may produce varied surface features (see Figure 10.44).

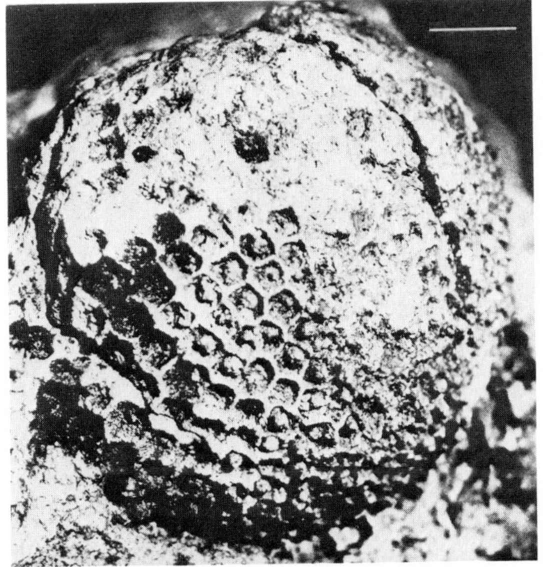

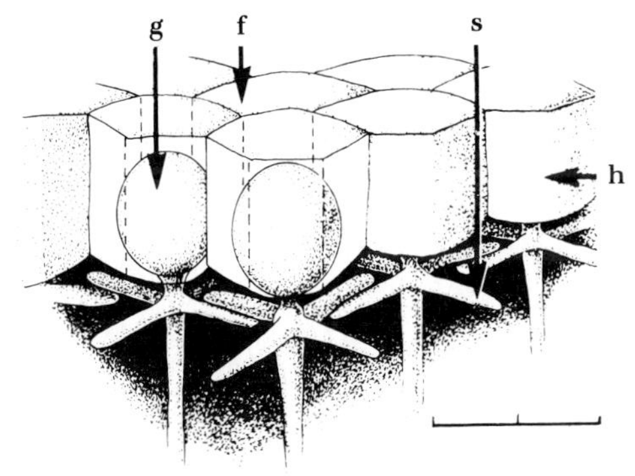

Figure 10.43
Ischadites hemisphericus. **1.** Silurian, Ohio, showing globular structures associated with many facets that are regarded as gametangia; bar = 5 mm. **2.** Diagram, showing gametangia (g), facet (f), stellate structure (s), and lateral head (h); bar = 2 mm. From Nitecki, 1972b.

The wall, consisting of the meromes and their capping plates, is most commonly of calcite but may show various degrees of recrystallization, hence may originally have been deposited as aragonite. In some regions specimens are secondarily silicified or even preserved as pyrite. The surrounding matrix may also contain isolated "branches" or meromes, suggested by Byrnes (1968) to indicate their loss with progressive growth. Rietschel (1969) suggested that the specimens may have been free-living in the soft mud without a true attachment (see Figure 10.45). Growth occurred from one pole, new plates being inserted as growth continued, and the individual plates enlarged by marginal growth. The central cavity of the fossil specimens was thought to contain the mother cell surrounded by the lateral branches (meromes) that gradually calcified internally rather than calcifying within a mucous membrane between the branches as in the dasyclads (see Figure 10.46).

The systematic position of the Cyclocriniteae is also uncertain. Placed by many with the Dasycladales, the Tribe Cyclocriniteae was transferred to the Receptaculitaceae by Nitecki (1972a). The calcified thallus is globose, with an apically inflated main axis from which the laterals radiate, to terminate in hexagonal facets (see Figure 10.47).

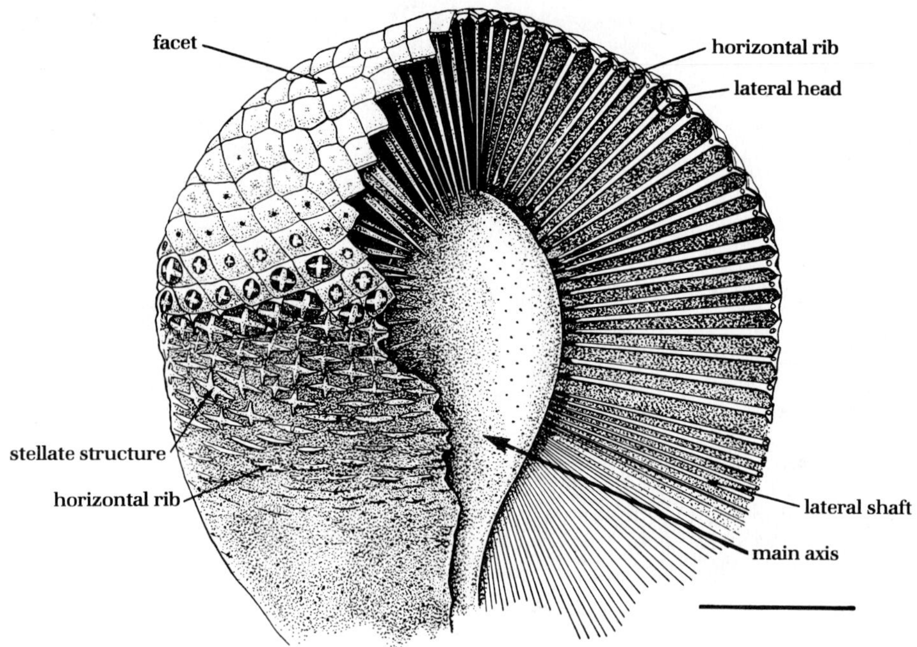

Figure 10.44
Ischadites sp., diagrammatic reconstruction to show the effect of differential weathering, with the upper part well preserved, and the lower part progressively removed to show only the cross section of the lateral shafts. Bar = 1 cm. From Nitecki, 1972a.

Figure 10.45
Diagram to show different interpretations of living position of *Ischadites*. **1.** Oldest area at base around the uncalcified stalk, and growing tip solid, as interpreted by Kesling and Graham (1962) and Byrnes (1968). **2.** Oldest meromes at the closed base, anchored in the soft substrate, and growing tip uncalcified, as suggested by Rietschel (1969).

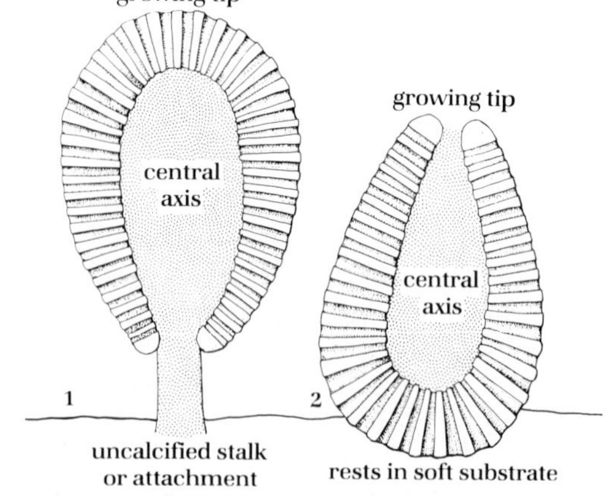

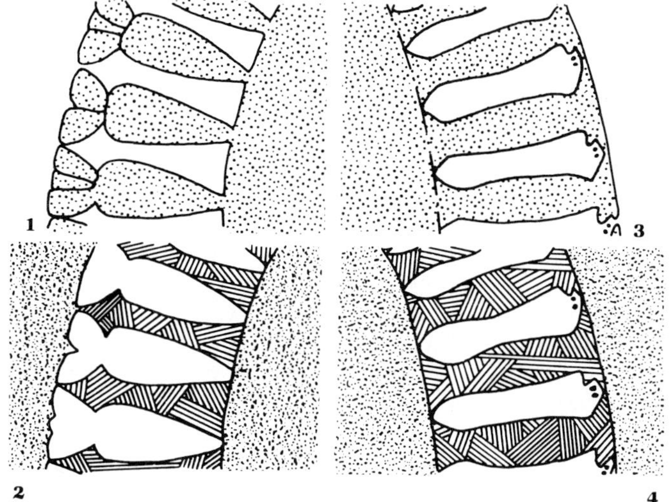

Figure 10.46
Comparison of calcification in the Dasycladales and Receptaculitales. **1,3.** Reconstruction of part of living specimen in section, with organic material of central axis and laterals shown in stippled pattern, and intervening spaces white. **2,4.** Same region in fossil material, with rock matrix shaded, spaces white, and calcification in region of laterals cross-striped. **1,2.** *Goniolina* sp., showing calcification occurring in the space between the laterals. **3,4.** *Receptaculites* sp., showing calcification of the laterals themselves. Diagram redrawn from Rietschel, 1969.

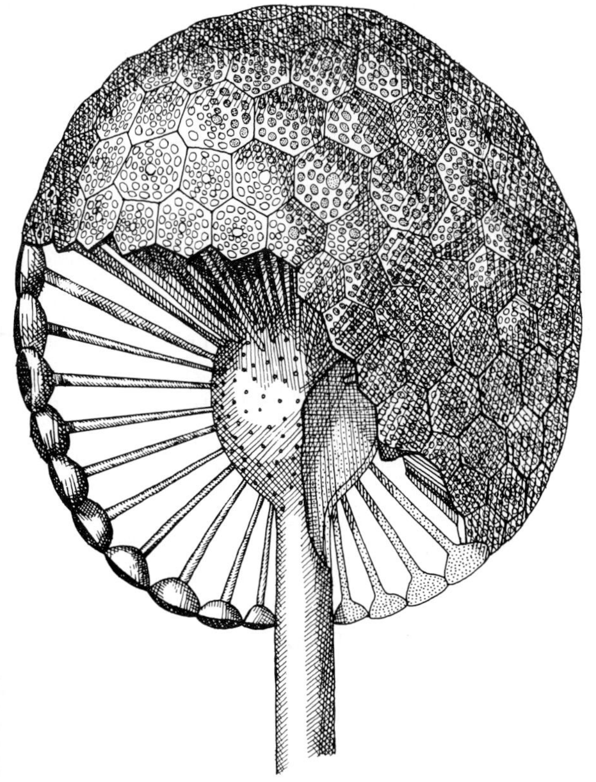

Figure 10.47
Cyclocrinus porosus Stolley, Ordovician, Germany. Reconstruction to show apically inflated central axis and radiating laterals that end distally in hexagonal plates, $\times 10$, redrawn from Pia, 1926.

Order Dasycladales

Only some eight genera of the Dasycladales are living in the present seas, although these tropical to subtropical lime-depositing algae have an extensive fossil record. Like the Receptaculitales, many were assigned to protozoans, sponges, coelenterates, or echinoderms until recognition of the reproductive structures resolved their position. Some recognize various tribes in a single family, but some also recognize more than one family.

The typical dasyclad thallus has an undivided erect axis, bearing one to many whorls of sterile lateral branches that may be simple or bifurcated to as much as the seventh order. The radially symmetrical thallus is attached to the substrate by means of an irregularly branching rhizoidal base. The numerous plastids commonly are rounded to elliptical in form. The vegetative thallus is uninucleate and diploid and has no separating cell walls, except in the formation of the specialized reproductive structures, when it also becomes multinucleate. Modified branches of the laterals produce operculate cysts or aplanospores that may serve as a resting stage upon liberation. Following meiosis, biflagellate isogametes develop within these cysts, still inside the specialized gametangia, and after escaping through the opening, may fuse in pairs to produce a diploid zygote that immediately settles to germinate into the new plant. Similar cysts may give rise to biflagellate zoospores that directly produce a new plant (Puiseux-Dao, 1962).

New branches are produced terminally beneath a pectic membrane or velum at the summit of the main axis (Valet, 1967). Calcification is characteristic in the dasyclads but may be localized in occurrence. Commonly it begins around the central axis and may also enclose the lateral branches to a varying degree. Some fossil species are known only from the calcified gametangia, and others from the entire calcified thallus in which closely packed radiating laterals form a nearly solid sphere or cylinder. Pores in fossil specimens represent the branches in life.

Modern species deposit lime in the form of aragonite, and fossils in the Tertiary at least are

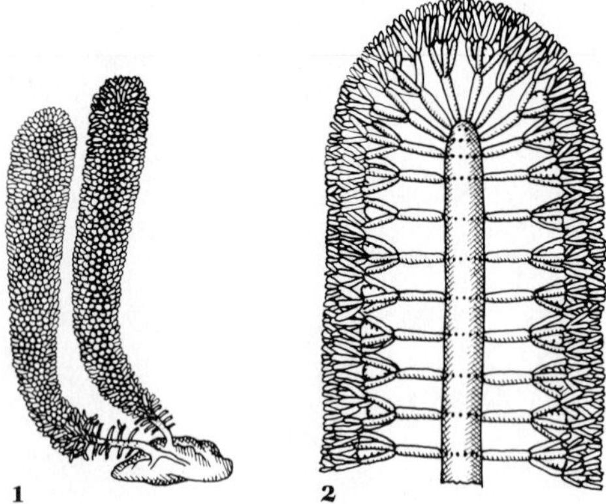

Figure 10.48
Dasycladus clavaeformis (Roth) Agardh. **1.** Living vegetative specimens, ×1. **2.** Section of upper part of spore-bearing specimen, ×6. Redrawn and modified from Pia, 1926, after Stolley.

preserved in this form. However, the abundant Paleozoic fossils referred to the Dasycladales are now calcitic, whether originally so or due to later alteration or replacement of the aragonite. Characters of taxonomic importance include the general form (spherical, cylindrical, tubular, simple, or branched), presence or absence of articulation, the character, number, and dimensions of lateral branches, and the position of the reproductive organs.

The morphology and development of these living dasyclads aid in an understanding of the fossil species. *Dasycladus* (see Figure 10.48), typical of the family, has an unbranched main axis up to 5 cm in length. Nonseptate rhizoids attach it to the rocky substrate, where the individuals may form dense growths in the quiet, shallow waters of the Mediterranean. Except at the base, the erect tubular axis is covered with densely packed whorls, each consisting of 10 to 15 narrow lateral branches. These laterals bifurcate to the second or third degree to produce successive orders of segments. All laterals of a whorl are of equal length, and a rather compact, hairy surface results. As in all dasyclads, numerous small plastids are present, each with a small pyrenoid.

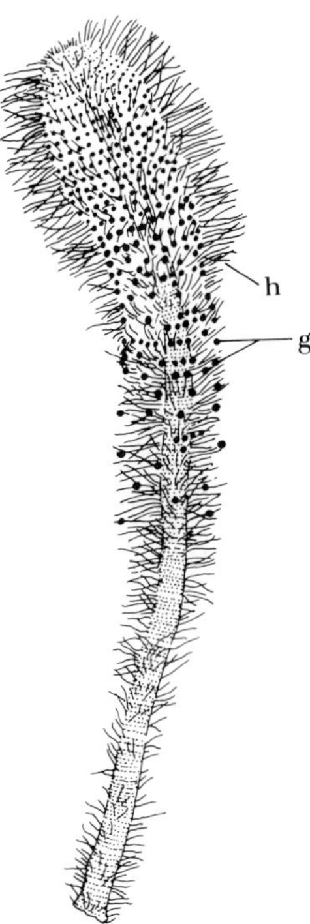

Figure 10.49
Neomeris sp., showing radiating hairs (h) and gametangia (g); approximately ×5.

The plant organization of *Dasycladus* is not highly differentiated, as a new apex can be regenerated if the first is removed; if the rhizoidal end is removed and the plant inverted, the former base will produce a new apex (Wulff, 1910).

Gametangia develop in the upper half of the plant. They appear to be terminal on the primary branches but really represent modified short second order branches that are swollen into spheres. The gametangia receive most of the contents of adjacent branches and thus become much deeper green in color before being cut off by a septum. They are surrounded by sterile second order branches.

The living *Batophora* is similar in structure to *Dasycladus* but has loosely placed whorls of branches and is not lime impregnated.

Neomeris is more complex and more heavily calcified (see Figure 10.49). The thallus of *N. an-*

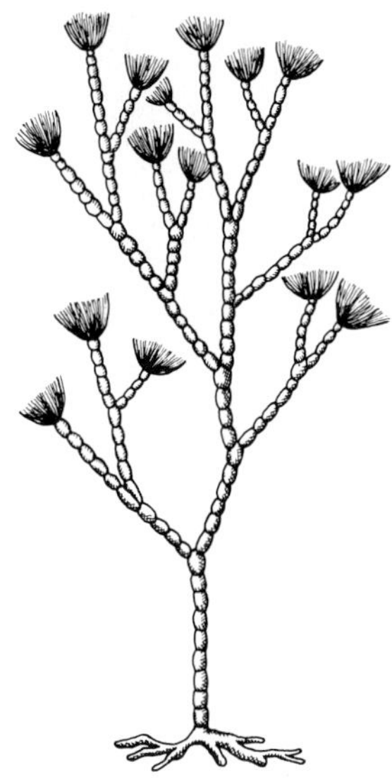

Figure 10.50
Cymopolia barbata (Linnaeus) Lamouroux, entire plant, ×0.7, redrawn from Pia, 1926.

nulata Dickie is an elongate calcified mass 1 to 2 cm in length, with an apical tuft of hairs. It is similarly covered with whorls of branched laterals. The outermost branch (third order) is hairlike, but is soon shed, the long delicate hairs being present only in the youngest (apical) part of the thallus. Each primary lateral produces only two second order branches. Their dilated tips press against adjacent ones so that they become six sided to produce a compact, faceted surface. Plastids lie near this outer surface. The wall of the main axis is also thickened except for a tiny pore opening around the laterals. Lime is deposited as a continuous layer on the inner surfaces of the facets, forming a cylinder perforated by tiny pores that represent the passageways for the branches. Primary branches and gametangia are similarly lime sheathed. The gametangia develop beneath the faceted surface and are homologous with the second order branches between which they lie. In some species the gametangia are interconnected by a common ringlike calcified mass.

The dichotomously branched thallus of *Cymopolia* (see Figure 10.50) has a series of cal-

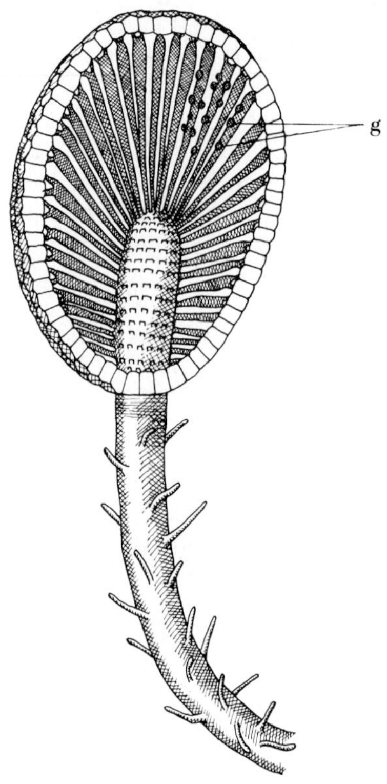

Figure 10.51
Bornetella capitata (Harvey) Agardh. Diverging radials arise from central axis and bifurcate just before they terminate in hexagonal plates. Gametangia (g) are borne in interior of globular structure, at the side of the laterals. ×9.

Figure 10.52
Halicoryne sp., showing whorled appearance of gametangia (g) that contain many cysts, ×5.

cified segments separated by narrow uncalcified joints. The second order branches of the calcified segments form a faceted surface reminiscent of *Neomeris*, although less regular. The space between the axis and the faceted surface at first contains mucilage. Later it becomes filled with a continuous cylinder of lime that is perforated by the canals due to the primary and second order branches, with spherical cavities denoting the position of the gametangia.

A thin cylinder of calcium carbonate, mixed with oxalate of lime, is deposited as an annular thickening around the side wall of the facets of *Bornetella* (see Figure 10.51). The primary branches occur in whorls, and gametangia develop along their sides in an alternating ar-

rangement. Each contains a number of cysts with membranous lids.

Among the living dasyclads, the Acetabulariaceae (see Figure 10.52) have the most specialized structure. The main axis of the Australian *Halicoryne* has alternate whorls of 8 sterile and 16 fertile laterals. The axis is slightly constricted at the position of the whorls of hairlike sterile laterals that branch to the third degree. Fertile laterals have a broad base and a podlike gametangium. Both axis and gametangia have a calcareous coating of varied thickness; the remainder of the thallus is uncalcified. The gametangia may contain a coherent mass of highly calcified cysts (Fritsch, 1935).

Acetabularia acetabulum (see Figure 10.53) may require two to three years of growth before developing fertile branches. In the first year the germinating zygote produces a threadlike upright branch and a rhizoidal holdfast. During the fall the upright dies, and only the base persists to produce a new upright cylinder in the spring. One or more whorls of sterile branches appear at the apex. Throughout most of the period of growth, the single large nucleus is situated in a lobe of the holdfast. At maturity,

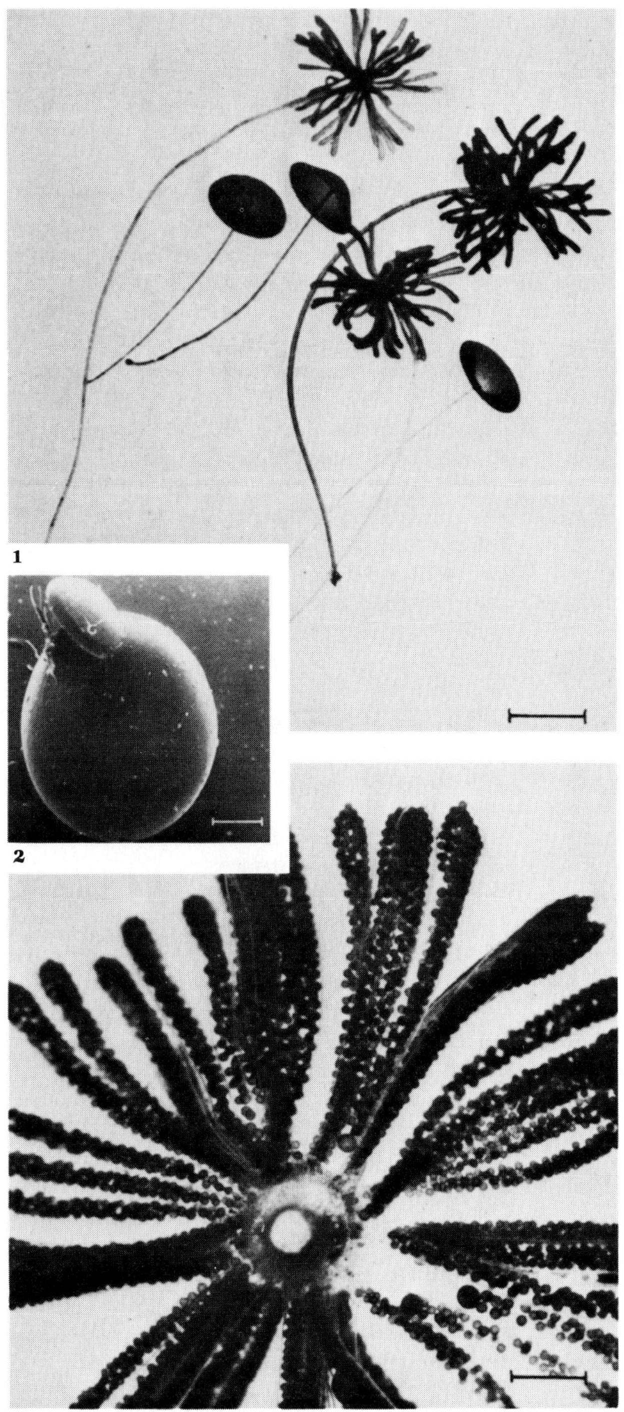

Figure 10.53
Acetabularia. **1.** Cells of *A. major* (larger) and *A. acetabulum* (smaller, with parasollike cap), ×1, bar = 10 mm. **2.** SEM of cyst, showing lid and opening, from which gametes have been released, ×385, bar = 13 μm. **3.** *A. major,* part of mature cap with rays filled with the spherical cysts, ×12, bar = 0.8 mm. All from Schweiger et al., 1974.

and after numerous mitotic divisions, the sterile laterals are detached, leaving a ring of small scars. A reproductive "umbrella" is formed by a whorl of gametangial rays, the elongate gametangia being completely joined at their margins. Both the outer wall of the upright and the gametangial umbrella are encrusted with lime carbonate and lime oxalate (Leitgeb, 1888). Daughter nuclei (also diploid) migrate into these rays and develop into many uninucleate operculate aplanospores or cysts. The cysts of *Acetabularia antillana* are from 140 to 185 μm in diameter, with a wall 10 to 25 μm thick, a pore at one end being closed by an operculum (Marszalek, 1975). Various living species have from 50 to 80 gametangial rays in the cap, each ray containing many cysts (up to 360 in *Acetabularia major* Martens), so that a cap may contain up to 20,000 separate cysts (Schweiger et al., 1974). The cysts of some species are uncalcified, but others are heavily calcified. They are released when the umbrella disintegrates in the fall, and remain dormant until spring. Meiosis then occurs, and about 5000 haploid biflagellate gametes escape from each cyst by detachment of the operculum. Pairing of the isogametes produces a quadriflagellate zygote that settles to produce a new plant. The only fossil representative of the genus known is *Acetabularia chiavonica* Squinabol from the Oligocene.

Acicularia is similar to *Acetabularia* but also deposits lime within the gametangial cavity. This results in a calcareous spicule, pointed at one end and rounded at the other, honeycombed with cavities that contain the cysts. Similar fossil spicules were recognized by Munier-Chalmas (1879) as related to such structures in modern *Acetabularia*. The living *Acetabularia schenckii* Moebius (formerly regarded as an extant species of *Acicularia*) later was found to corroborate his observations.

Fossil Dasyclads. The Family Dasycladaceae is a true relict family, with only 2 of the 19 or 20 tribes of the subfamily Dasycladoideae still extant; about 95 percent of the included genera also are extinct. No living genus has a large number of species, but dasyclads have been abundant and diverse at various periods of the geologic past. Most fossils are small cylindrical tubes pierced by many pores whose arrangement, form, and method of division allows reconstruction of the plant habit.

The Order Dasycladales includes five families as here recognized. The monotypic Timanellaceae is represented by the Precambrian *Timanella*, which may attain 10 cm in breadth and 20 cm in height. The Early Cambrian Tannuolaiaceae is monotypic. In the Seletonellaceae (see Figure 10.54), the Papillomembraneae occurs in the Precambrian of Norway. Described from rock thin sections as having a dasycladlike structure, *Papillomembrana* is about 0.4 to 0.5 mm in diameter, but is not calcified (Spjeldnaes, 1963). As other dasyclads are only partly calcified, including the living *Batophora*, this is not a contradiction for its assignment herein. Another Precambrian dasyclad, *Templuma sinica* Zhang, occurs in the 1200- to 1400-million-year-old Wumishan Formation of China, but its tribal or family assignment is uncertain.

The tribes Cambroporelleae, Amgaelleae, Seletonelleae, and Vologdinelleae are restricted to the Cambrian, the Rhabdoporelleae ranges from the Cambrian to Permian, and the Primicorallineae is found in the Ordovician and Silurian. The sparse laterals of *Primicorallina* (see Figure 10.55) branch to the third degree, as do those in the modern *Dasycladus*. The uncalcified *Inopinatella lawsoni* Elliott (see Figure 10.56), from the Silurian of England, and *Callithamniopsis fruticosa* (Hall) Whitfield, from the Ordovician of the United States, are preserved as flattened carbonaceous residues, but resemble early growth stages of such living dasyclads as *Neomeris* (Elliott, 1971). The earliest Rhabdoporelleae are characterized by a main axis with irregularly arranged unbranched laterals, as in *Rhabdoporella*. *Vermiporella* has a dichotomously branched main axis (see Figure 10.57).

The Tribe Verticilloporeae, formerly included in the dasyclads, has recently been referred to the Rhodophyta; some also consider the Dasyporelleae to belong elsewhere, although they are here included with the dasyclads.

By far the most important family is the

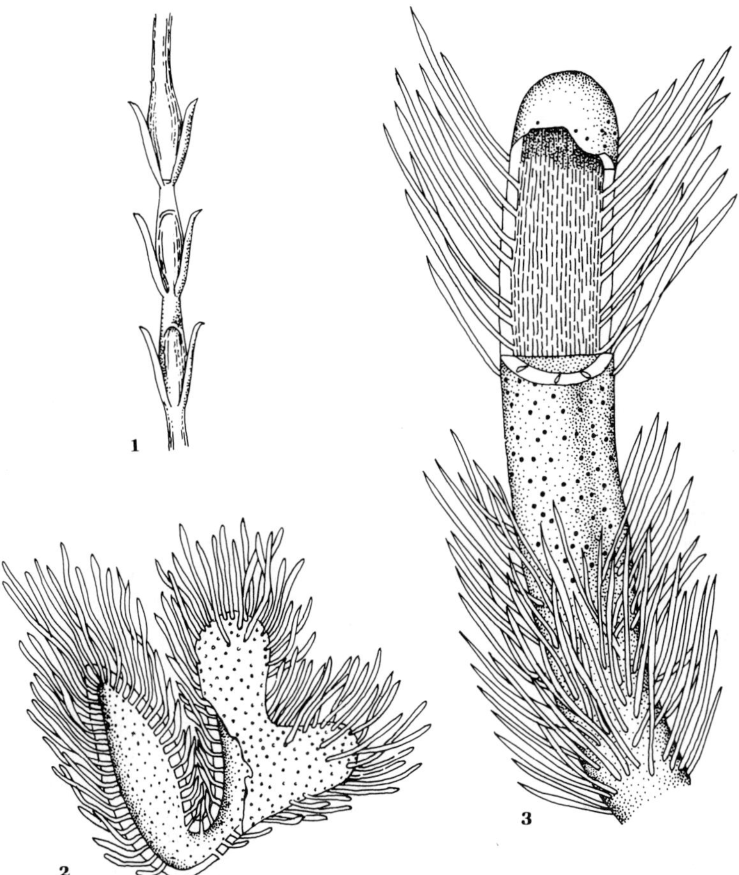

Figure 10.54
Seletonellaceae. **1.** *Vologdinella fragilis* Korde, Lower Cambrian, USSR, reconstruction, ×55. **2.** *Mejerella ramosa* Korde, Upper Cambrian, USSR, reconstruction, ×3.5. **3.** *Jacutiella aciculata* (Korde) Korde, Middle Cambrian, reconstruction, ×13.5. All redrawn from Korde in Korde et al., 1963.

Dasycladaceae, whose well over 125 genera are placed in three subfamilies and 20 tribes, ranging from the Cambrian to the Holocene. The thallus is a nonseptate tube or siphon, attached by means of rhizoids, and bearing many infertile lateral branches, which may be simple or branch repeatedly; many Mesozoic and Cenozoic taxa have separate fertile branches bearing sporangia or gametangia. In living dasyclads, the fertile branches may act as gametocysts, producing gametes directly, but in some taxa they are sporangia, producing motile spores that in turn produce gametes. As their nature is impossible to differentiate in the fossils, the fertile branches are variously termed sporangia or gametangia.

Of the Subfamily Dasycladoïdeae, the Palaeoporelleae and Scribroporelleae (see Figure 10.58) are found in the early Paleozoic, and the Palaeobereselleae and Bereselleae (see Figure 10.59) in the late Paleozoic. Some of these Paleozoic forms have been transferred to the

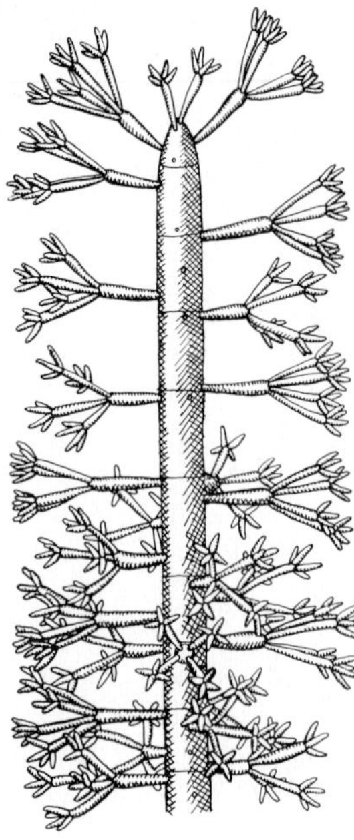

Figure 10.55
Primicorallina trentonensis Whitfield, Ordovician, New York, reconstruction, with part of laterals removed in upper part, ×7.5, redrawn from Pia, 1926, after Ruedemann.

Figure 10.56
1,2. *Nanopora anglica* Wood, Mississippian, England, ×230, from Wood, 1964. **1.** Oblique section, showing pores at base, and section through thallus at top. **2.** Transverse section. **3,4.** *Inopinatella lawsoni,* Silurian, England, from Elliott, 1971. **3.** Reconstruction of the uncalcified species showing appearance of the upper part of the plant as in life, ×5. **4.** Fossil specimen, showing main stem and primary branches, ×7.5.

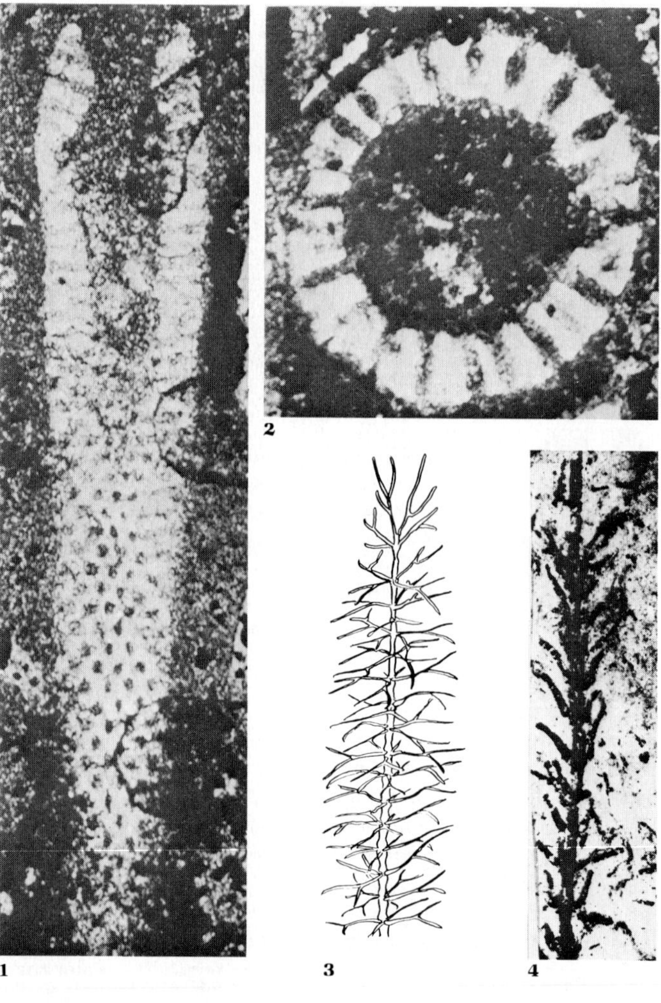

1   2   3   4

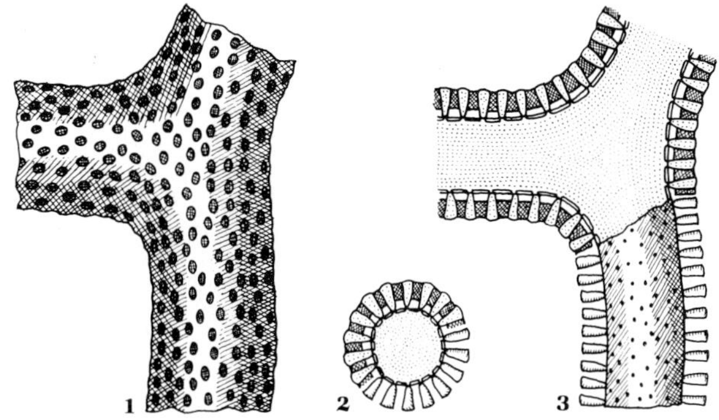

Figure 10.57
Vermiporella sp., reconstruction. **1.** Exterior. **2.** Cross section. **3.** Longitudinal section. In 2 and 3 the upper part of the drawing shows the calcified thallus, the calcareous part in cross hatching, and the organic matter stippled. Redrawn from Pia, 1920.

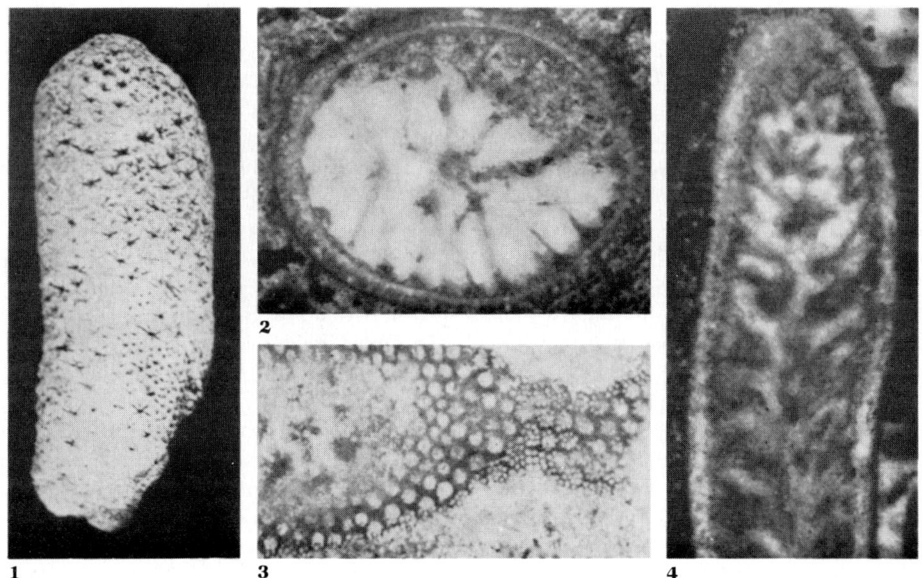

Figure 10.58
Scribroporella socialis Spriestersbach, Middle Devonian, Germany, from Rietschel, 1966. **1.** ×5. **2.** Transverse section, ×7. **3.** Longitudinal tangential section, base at right of figure showing the fine outer pores, and larger inner ones; central axis sectioned to left; ×10. **4.** Longitudinal section of small specimen, ×10.

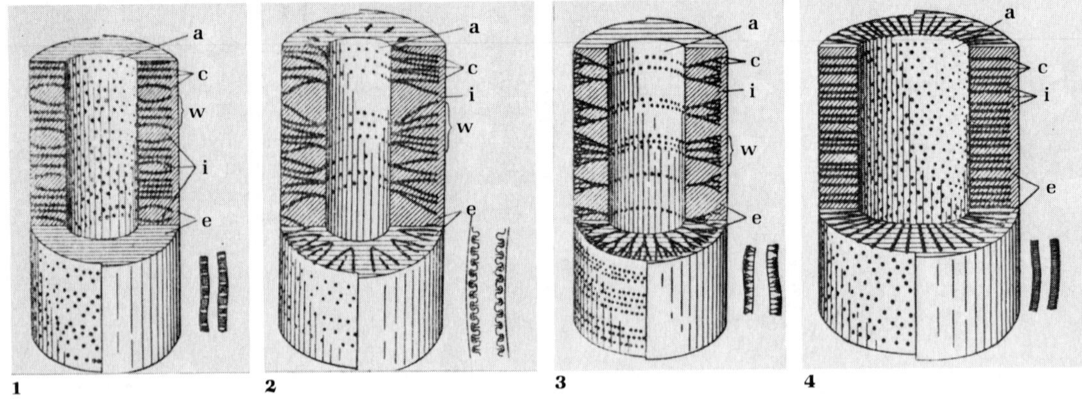

Figure 10.59

Bereselleae, reconstructions; from Pennsylvanian, USSR; a, central axis; c, canals; w, whorled laterals; i, intermediate zone; e, external hyaline region; small figures at right of reconstructions show appearance in thin section. **1.** *Beresella erecta* Maslov & Kulik. **2.** *Dvinella comata* Chvorova. **3.** *D.* (*Trinodella*) *bifurcata* Maslov & Kulik. **4.** *Samarella setosa* Maslov & Kulik. All from Maslov and Kulik, 1956.

sponge Class Ischyrospongia (Termier et al., 1977), including *Kamaenella* of the Palaeobereselleae and *Beresella, Dvinella, D.* (*Trinodella*), *Goksuella,* and *Uraloporella* of the Bereselleae, *Donezella,* and *Praedonezella.* They are here retained with the Chlorophyta in their classical position, pending additional study. Nothing is known of reproductive structures in these Paleozoic Dasycladaceae, the absence of fossil cysts or sporangia suggested to be because these were borne within the main axis, hence termed **endosporic** (Pia, 1920), although no fossil evidence for this occurs in genera prior to the Triassic (see Figure 10.60). On the contrary, perhaps they were lateral and homologous to the primary or secondary branching external to the main thallus and were shed rather than calcified and preserved (Emberger, 1968, p. 92). This type of development characterizes the Receptaculitales, as well as all known Dasycladaceae from the Cretaceous to the present. If the laterals functioned as sporangia or gametangia, as in *Triploporella* and most Mesozoic dasyclads, the structure is termed **cladosporic**, but if definite reproductive organs were present, they are **choristosporic**, as in most Cenozoic and all living dasyclads (e.g., *Batophora*). *Archaeobatophora* from the Upper Ordovician of Michigan is noncalcified like the modern *Batophora* and is regarded as a member of the same tribe (Nitecki, 1976).

The last of the tribes to appear in the Paleozoic was the Diploporeae, persisting from Mississippian through the Eocene. Diploporeae show varying degrees of branching complexity (see Figure 10.61), ranging from simple unbranched laterals in *Actinoporella,* to a series of segments, each with terminally inflated laterals of varied length that produce the compact spherical surface of each segment in *Mizzia,* and finally developing repeatedly branched laterals in *Diplopora.* In addition to complexity of the laterals themselves, variations in their arrangement also are of systematic importance. The term **aspondyl** (Pia, 1920) defines those with randomly placed laterals, as in *Gyroporella;* and the more complex form in which the laterals are distinctly arranged in whorls, as in *Eovelebitella* or *Salpingoporella,* is termed **euspondyl.** A modification of the euspondyl type, termed **metaspondyl,** has laterals spaced regularly in tufts, as in *Diplopora,* which has multibranched laterals. Many of the dasyclad algae are known from thin sections of limestone, and although interpretation of the randomly oriented sections may be difficult (see Figure 10.62), study of many different sections

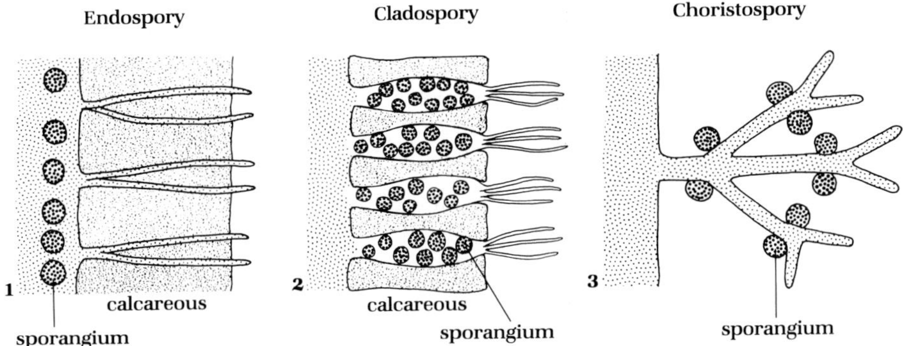

Figure 10.60

Proposed evolutionary changes in location of sporangia in dasyclads. **1.** *Diplopora tubispora* Ott, Triassic, showing endospore reproductive bodies with calcified walls, regarded as in the primitive location, the laterals partially enclosed in calcareous cover. **2.** *Triploporella* sp., Jurassic to Cretaceous, showing cladospore structure, the primary laterals crowded with sporangia and lime encrusted. **3.** *Batophora oerstedii* J. Agardh, living, noncalcified choristosporic species, with reproductive bodies grouped at junctions of primary, secondary, and tertiary laterals. In all drawings, central axis partly shown at left, and exterior with laterals at right. Area of calcification shown in parts 1 and 2; light stipple shows vegetative tissue; heavy stipple marks the reproductive bodies or sporangia. Modified from Elliott, 1972.

may allow the reconstruction of the appearance of the alga during life (see Figures 10.63 and 10.64).

The dasyclad thallus underwent a progressive increase in complexity during the Paleozoic. From a simple stem cell with densely arranged simple side-branches evolved those with large hollow thallus and external calcification. By the late Paleozoic, some dasyclads had whorled side branches or laterals, although nonwhorled types persisted. All post-Paleozoic species have a whorled arrangement.

During the Mesozoic, the dasyclads underwent a major diversification, although some of the newly evolved taxa were short-lived, such as the Kopetdagarieae (see Figure 10.65), the Triassic Teutloporelleae, and the Jurassic Petrasculeae, Stichoporelleae, Coniporeae (see Figure 10.66), and Linoporelleae. A few tribes that arose within the Mesozoic persisted into the Paleocene or Eocene, as did the Triploporelleae (see Figure 10.67), which ranges from the Jurassic to Eocene, the Thyrsoporelleae (see Figure 10.68), lasting from the Triassic to Eocene, and Dactyloporeae (see Figure 10.69), ranging

from Cretaceous to Eocene. Some of these, as well as others restricted to the Paleogene, such as the Ferganelleae, Morelletporeae, and Uterieae, were common and diversified in the European Tertiary, and particularly in the Paris Basin. The only living dasyclads in the subfamily are those of the tribe Dasycladeae, long regarded as lacking fossil representatives until the recent assignment there of the Paleocene-Eocene *Pagodaporella*.

The two remaining subfamilies of the Dasycladaceae both arose in the Mesozoic and have living representatives. Some are known only from the calcified sporangia, described as *Terquemella* and *Frederica* (see Figure 10.70), but which may be referrable at least in part to *Bornetella*. The Neomeridoïdeae are more heavily calcified, with the calcified branches arranged in whorls, so that separation between whorls may produce small discs (see Figures 10.71 and 10.72), as in *Larvaria*, a genus sometimes regarded as a synonym of *Neomeris*, but separated as distinct by others. Ranging from Cretaceous to the present, the Neomeridoïdeae have whorled laterals that are simple and un-

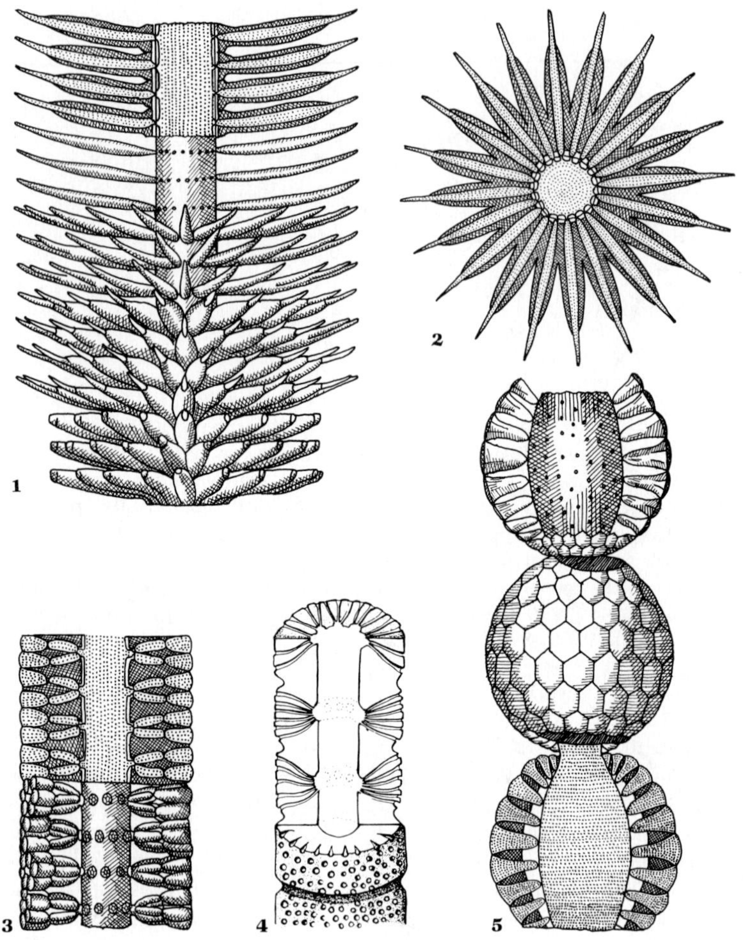

Figure 10.61

Reconstructions of Diploporeae. **1,2.** *Actinoporella podolica* Alth.
1. From top, a schematic longitudinal section, part of the decalcified
plant with laterals removed in front, then a section with all laterals,
and at the base the fully calcified plant. **2.** Transverse section
through one segment, partially decalcified. **3.** *Diplopora helvetica*
Pia, lower part showing plant without calcification; upper part as in
fossilized specimen, longitudinal section, with former organic
matter stippled, calcareous part crosshatched. **4.** *Eovelebitella
occitanica* Vachard, Mississippian, France, reconstruction, shown
from exterior at base, and in section at top of figure, ×40, from
Vachard, 1974. **5.** *Mizzia velebitana* Schubert, Permian, showing
decalcified segment at top, with front laterals removed (insertion of
laterals indicated by pores); central segment as appearing in life;
bottom segment as the calcified thallus in longitudinal section.
1–3,5, redrawn from Pia, 1920.

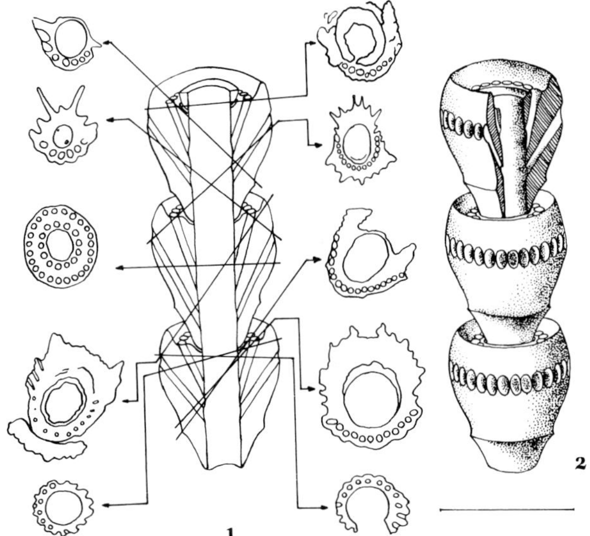

Figure 10.62
Salopekiella velebitana Milanović,
Permian, Yugoslavia. **1.** Longitudinal
section (center) and cross sections
resulting from randomly oriented
transverse and oblique sections. **2.** Reconstruction of thallus, with upper
segment showing internal structure.
Bar = 1 mm Redrawn from Milanović,
1965a.

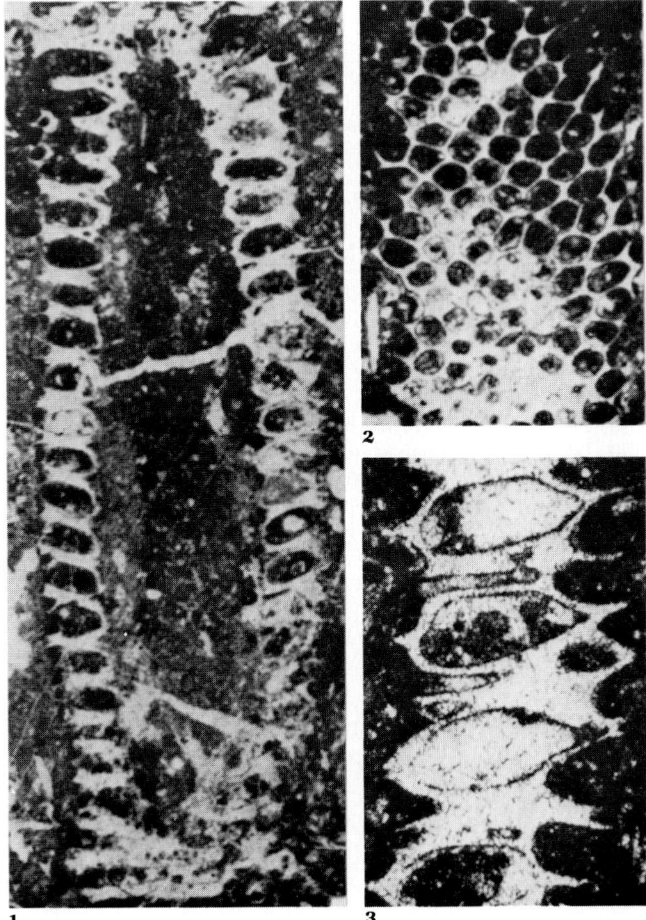

Figure 10.63
Diplopora annulatissima Pia,
Triassic, Yugoslavia. **1.** Slightly
oblique longitudinal section, ×8.9.
2. Tangential section, showing
honeycomblike appearance, ×21.3.
3. Longitudinal section through one
wall, ×31. From Sokač, 1968.

Figure 10.64

Diplopora annulatissima, reconstruction. **1.**
Portion of thallus, showing a, main axis; b, an-
nulation; c, terminally inflated lateral branch; d,
calcareous deposit; e, honeycomblike surface.
2. Single decalcified segment, showing arrangement and form of laterals. Compare Figure 10.63. From Sokač, 1968.

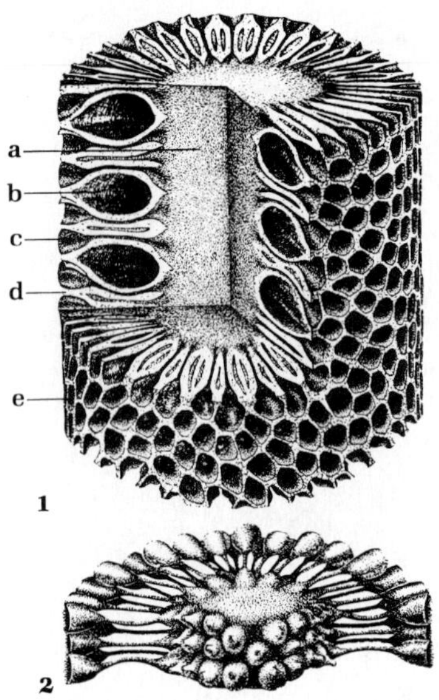

Figure 10.65 (*facing page*)

Kopetdagarieae, Coniporeae. **1,2.** *Sestrosphaera liasina* Pia, lower Jurassic,
Italy. **1.** Slightly oblique longitudinal section, ×6.4. **2.** Axial longitudinal sec-
tion, ×2.4. **3.** *Conodictyum striatum* Goldfuss, Jurassic, Germany, after
Goldfuss. **4,5.** *Coniporella clavaeformis* (d'Archiac) Fischer & Thierry,
Jurassic, France, basal and lateral views, ×1.6. **6,7.** *Eodascladus ogilviae*
Cros & Lemoine, Lower Jurassic, Italy. **6.** Oblique longitudinal section,
×4.8. **7.** Transverse section, ×24. **8,9.** *Fanesella dolomitica* Cros & Lemoine,
Lower Jurassic, Italy. **8.** Axial longitudinal section, ×1.2. **9.** Oblique section,
×8.8. 1,2,6−9, from Cros and Lemoine, 1966; 3−5, from Fischer and
Thierry, 1971.

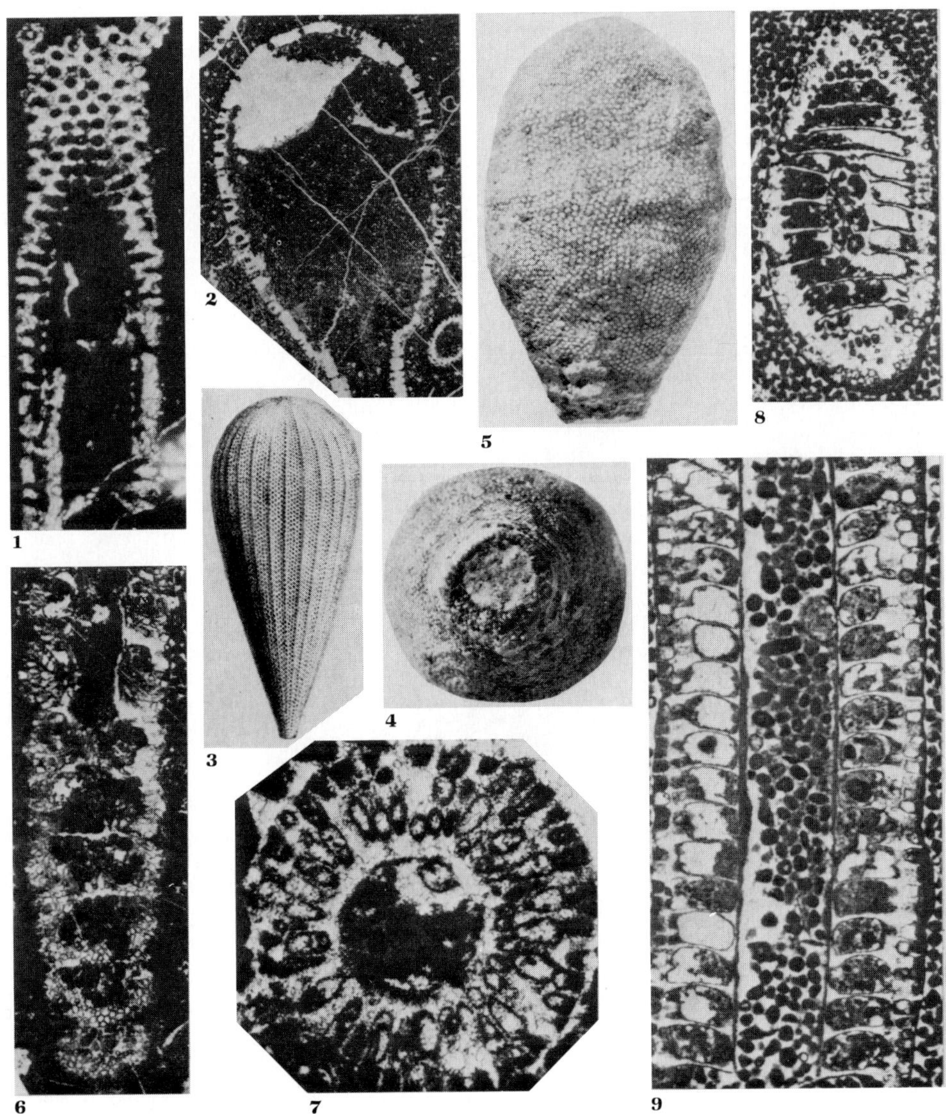

1
2
3
4
5
6
7
8
9

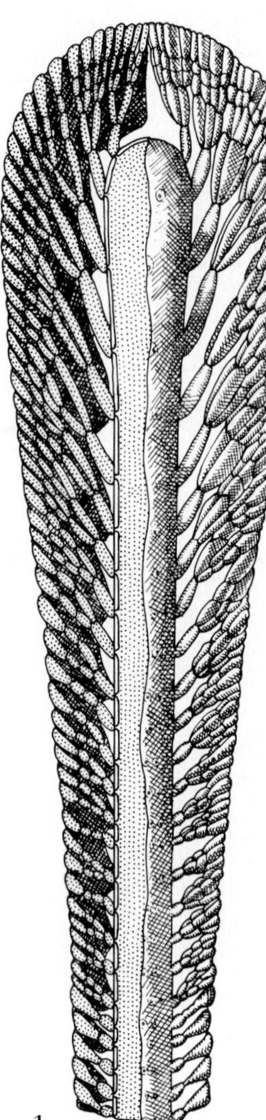

Figure 10.66
Coniporeae. *Palaeodasycladus mediterraneus* (Pia) Pia, Lower Jurassic, Italy. **1.** Reconstruction, showing longitudinal section with calcification between branches at left, and view of exterior, decalcified, and with some branches removed, at right. **2,3.** Transverse sections showing part with calcareous deposit, part decalcified. **2.** Many branched laterals near apex. **3.** Simple, unbranched laterals near base. All ×18, redrawn from Pia, 1920.

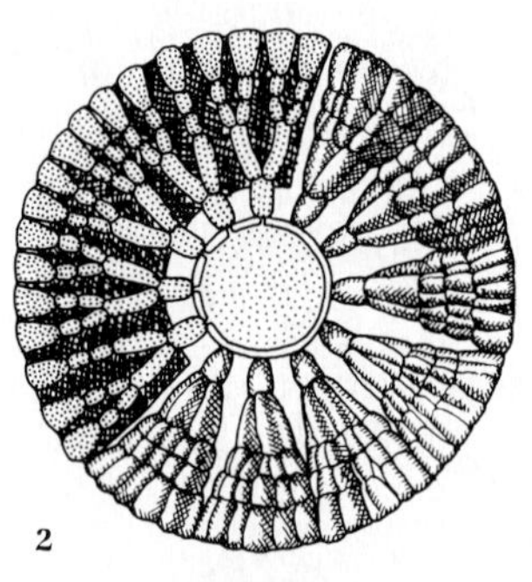

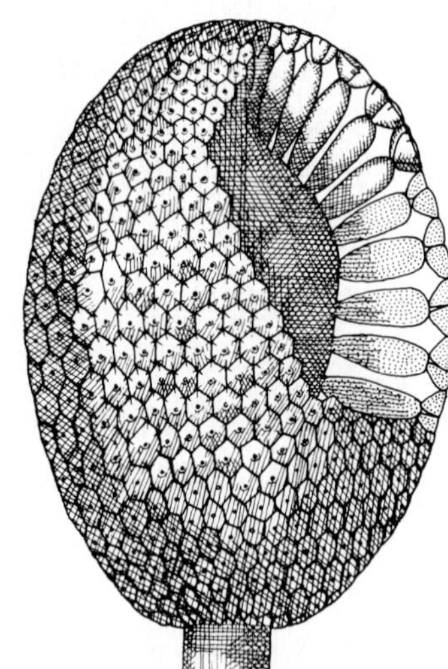

Figure 10.67
Triploporelleae. *Goniolina geometrica* (Roemer) Buvignier, Upper Jurassic, Germany. Reconstruction, approximately ×5. Redrawn from Pia, 1920.

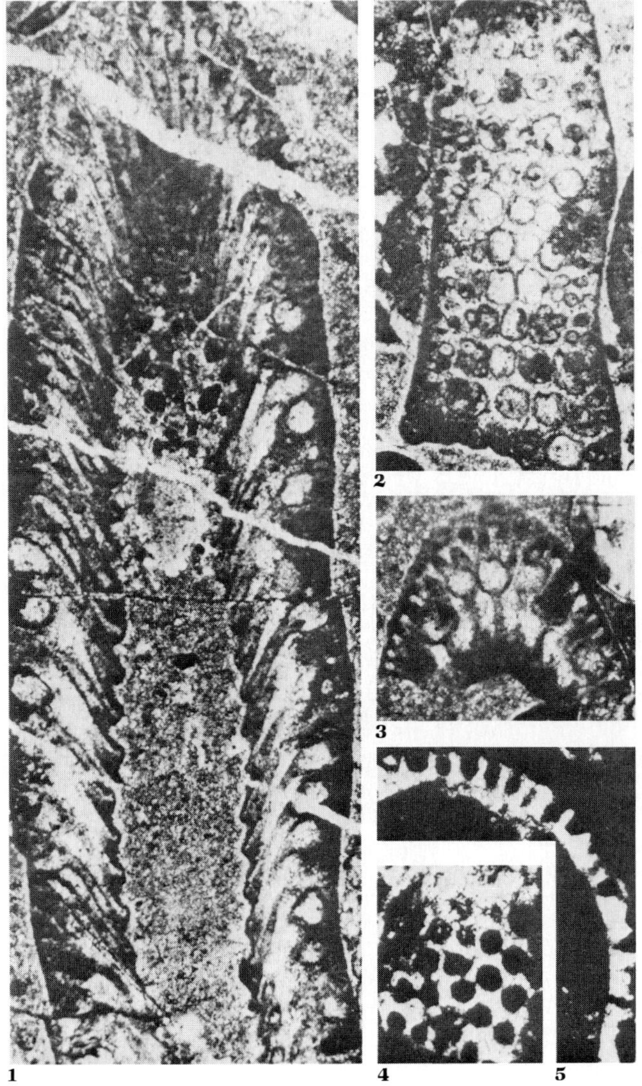

Figure 10.68
Thyrsoporelleae. **1 − 3.** *Suppiluliumaella polyreme* Elliott, Lower Cretaceous, Turkey. **1.** Longitudinal section. **2.** Outer tangential section, showing swellings and secondary branches. **3.** Oblique transverse section, showing form of branches. **4,5,** *Harlanjohnsonella annulata* Elliott, Upper Cretaceous, Yugoslavia. **4.** Tangential section. **5.** Transverse section. All ×24, from Elliott, 1968a.

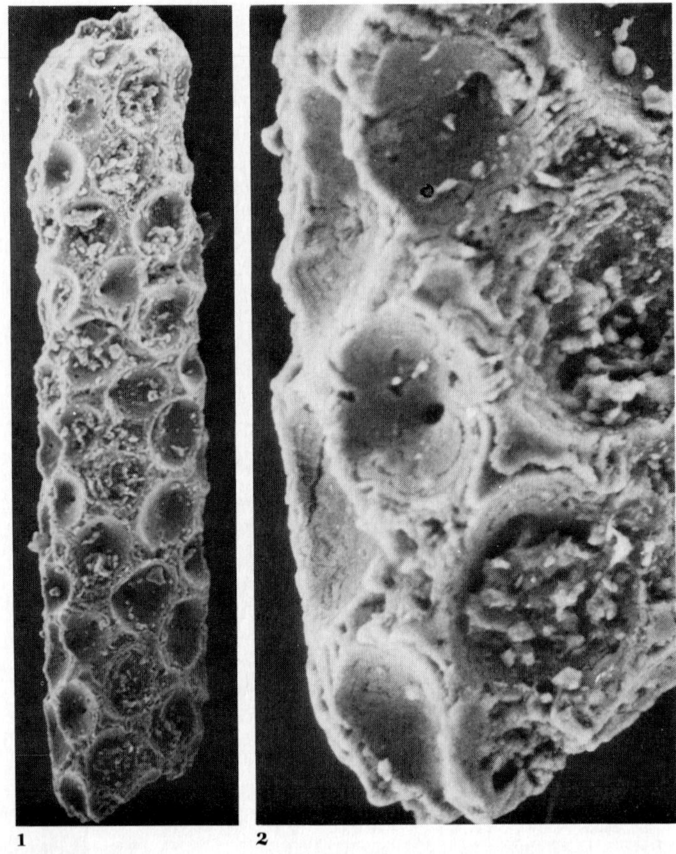

Figure 10.69
Dactyloporeae. **1,2.** *Dactylopora pauperata* Carpenter, Eocene, Lutetian, Damery, France, SEM. **1.** ×100. **2.** Enlargement, showing openings in pit center for cytoplasmic connection to the principal ramifications, ×400.

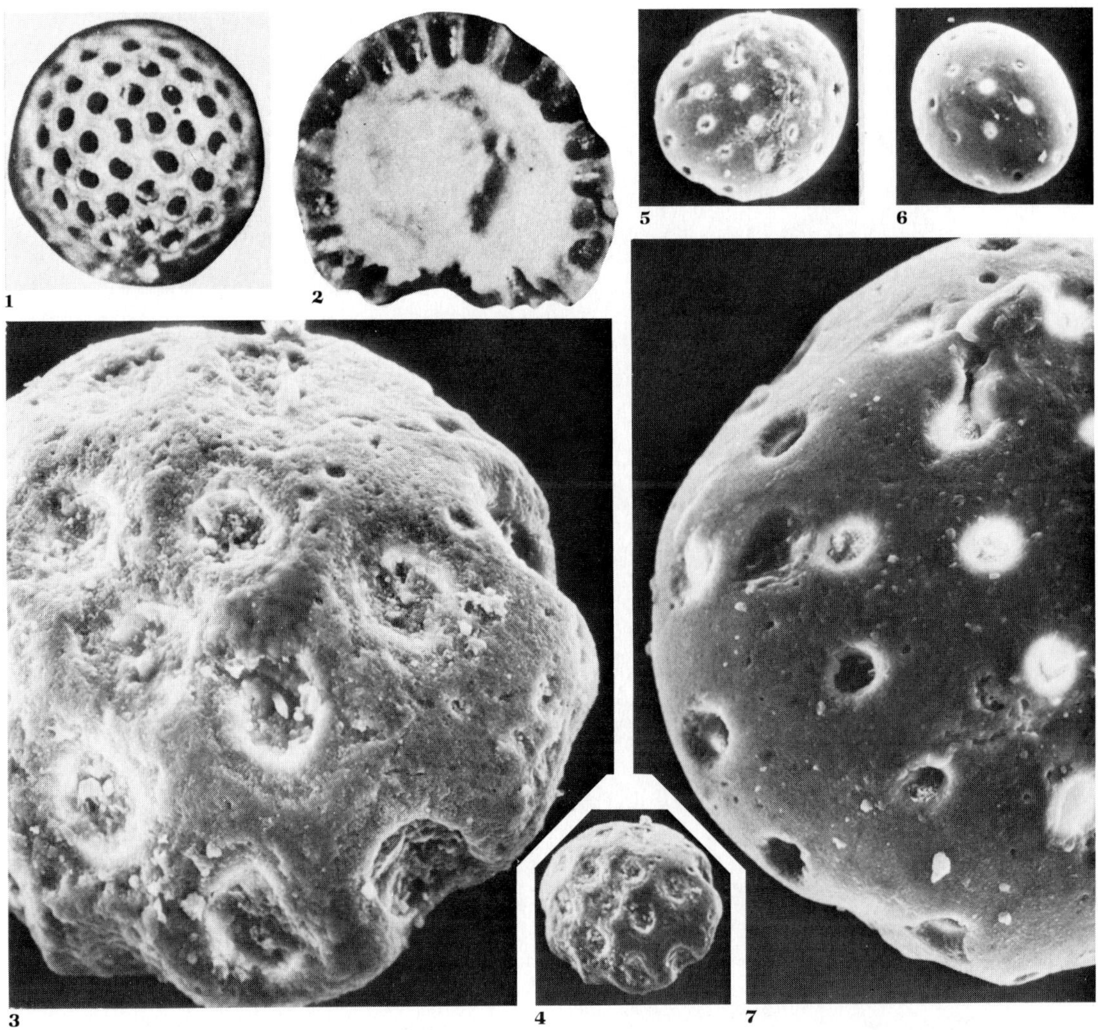

Figure 10.70

Bornetelloideae. **1,2.** *Frederica villiersi* Barta-Calmus, Eocene, Lutetian, France, ×70. Reproduced with permission, from Barta-Calmus, 1966, *Bull. Soc. Géol. Fr.*, sér. 7, v. 7, p. 906—910. **3,4.** *Terquemella parisiensis* Munier-Chalmas. **5—7.** *T. parvula* Morellet. 3—7, SEM; 3,7, ×400; 4—6, ×100. 5,7, Lutetian, Ezanville, France; 3,4,6, Lutetian, Montmirail, France.

Figure 10.71 (*overleaf*)

Neomeridoïdeae. *Larvaria limbata* Defrance, Eocene, Lutetian, SEM. Isolated rings that separate between whorls. **1—5.** Flattened upper surface, showing large axial cavity and calcified radiating branches. **6,7.** Convex lower surface, showing ring of pores that open into the fertile branches (gametangia). 1,4,5, from Montmirail, France; 2,3,6,7, from Grignon, France. 1,3,5,6, ×80; others, ×320.

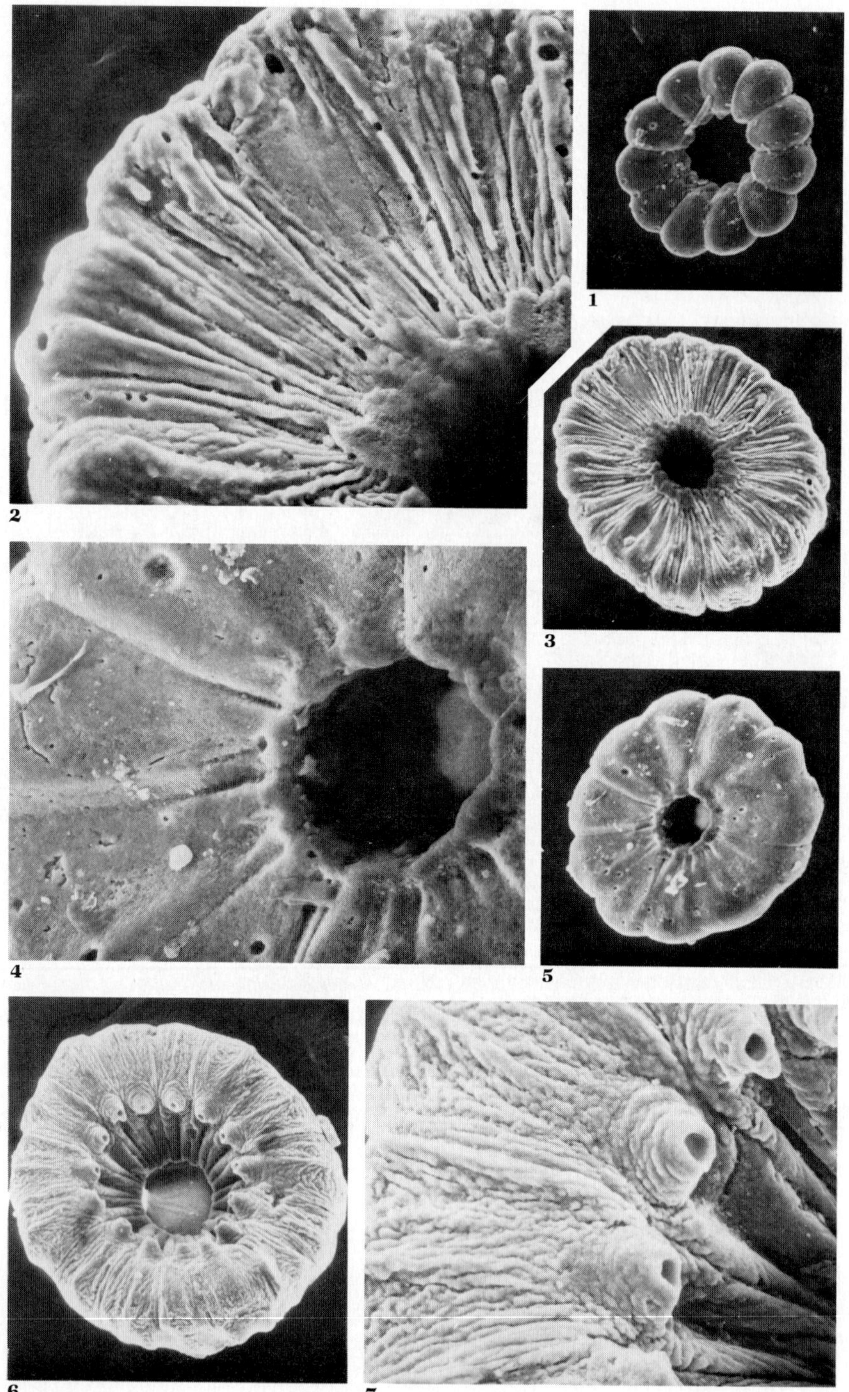

Figure 10.71

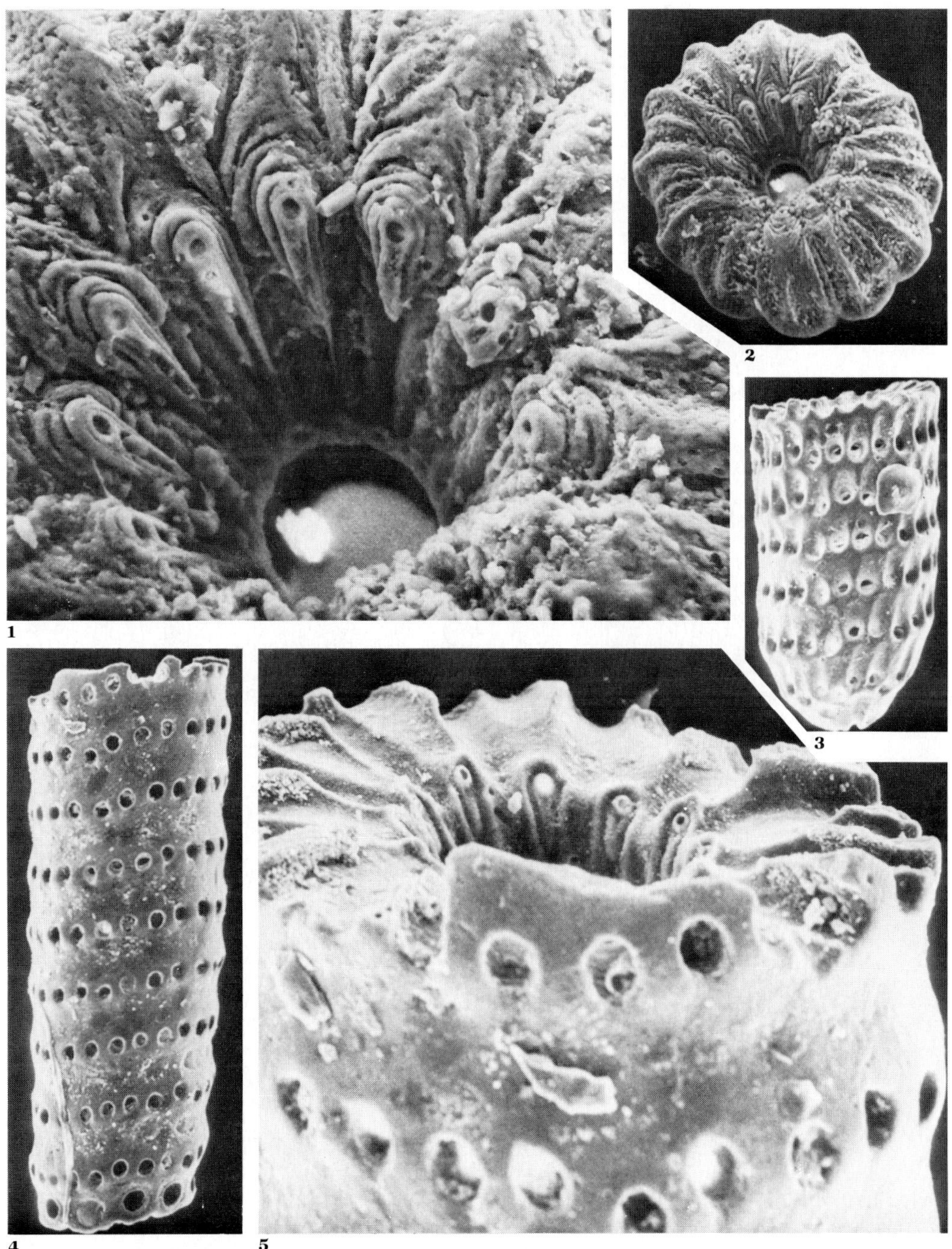

Figure 10.72

Neomeridoïdeae. **1,2.** *Larvaria limbata*, Eocene, Montmirail, France, SEM, lower surface of a segment showing pores into the infertile branches at the base of the central axial cavity, and the wider spaced pores opening into the fertile branches; ×400 and ×100, respectively. **3–5.** *L. encrinula* Defrance, Eocene, Lutetian, Damery, France, with numerous whorls, and lateral pores, SEM. **3,4.** ×50. **5.** Enlargement of part 4, showing break between whorls, and ring of pores at the edge of the central cavity, ×200.

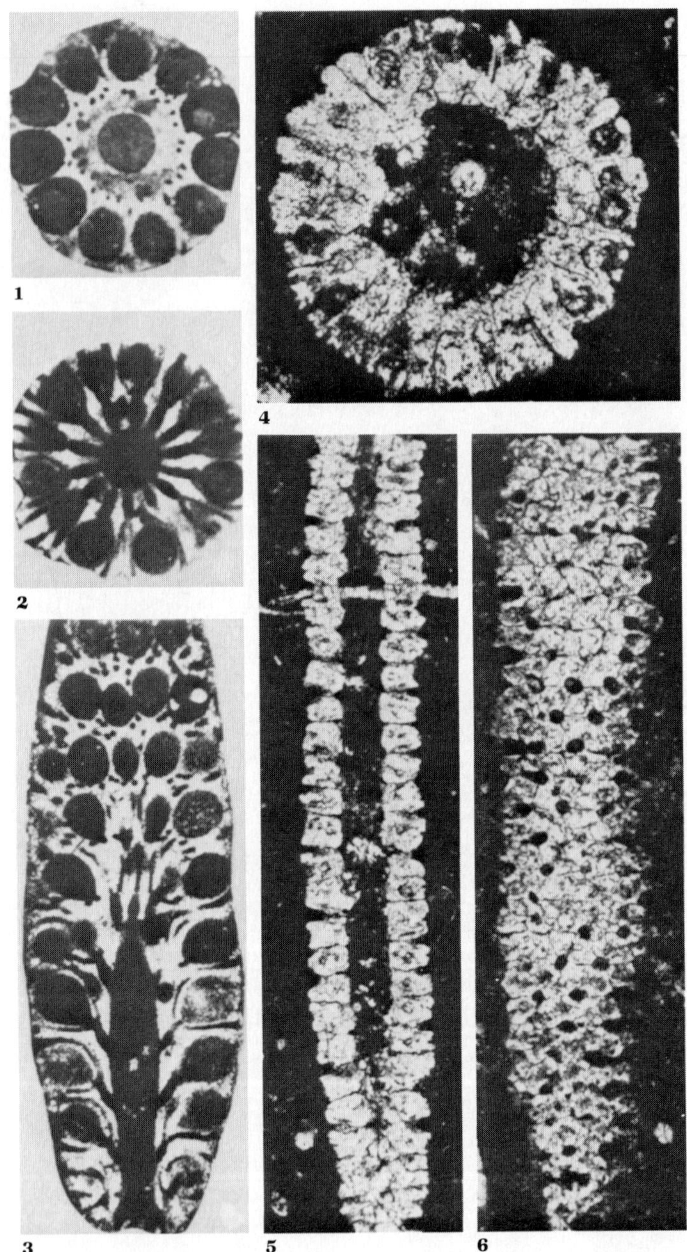

Figure 10.73

Neomeridoïdeae. **1 — 3.** *Cymopolia paktia* Kaever, middle Eocene, Afghanistan, ×30, from Kaever, 1969. **1,2.** Transverse sections, showing the axial cavity, from which radiate the infertile laterals, represented as pores or as the narrower tubes (black), and the fertile branches or gametangia, which are large spherical cavities. **3.** Longitudinal section, becoming slightly oblique at the top. **4 — 6.** *Heteroporella lepina* Praturlon, Upper Cretaceous, Italy, from Praturlon, 1967. **4.** Transverse section, mainly intersecting the whorl of fertile laterals, ×65. **5.** Longitudinal section, showing sterile branches as black pores, and fertile ones calcite-filled. **6.** Slightly oblique tangential section, showing sterile branches in black, and fine pores of the fertile branches at the lower part. 5,6, ×40.

branched associated with the inflated fertile branches (see Figures 10.73 and 10.74). The problematic genus *Campbelliella* was recently restudied and regarded as a dasyclad (Bernier, 1974), and is here tentatively placed with the Neomeridoïdeae. Known from thin sections of Upper Jurassic and possibly Lower Cretaceous limestones, where they appeared as isolated or nested U-shaped or V-shaped structures from 1 to 2.5 mm long, they were first described as "Organism C" (Favre and Richard, 1927), then described as a pteropod, *Vaginella striata* Carozzi, and later made the basis for the genera *Campbelliella*, *Daturellina*, and *Hadziella*, *Hadziina*, *Metacyclina*, and *Zetella* as aberrant tintinnids (Radoičić, 1959, 1963). They had also been regarded as pteropods, annelids, and the interlocking pallets from the posterior of a teredinid lamellibranch like the modern *Bankia* (Farinacci, 1963), and as algae *incertae sedis*. The oldest valid names for the organism are *Campbelliella* and *Daturellina*, of which the former was chosen by the reviser, Bernier (1974), who found specimens containing the sporangia as well as the sectioned thallus.

The family Acetabulariaceae ranges from the Jurassic to the present, and is characterized by a whorl of sporangia at the tip of the stalk or trunk, as in the common living *Acetabularia* discussed earlier.

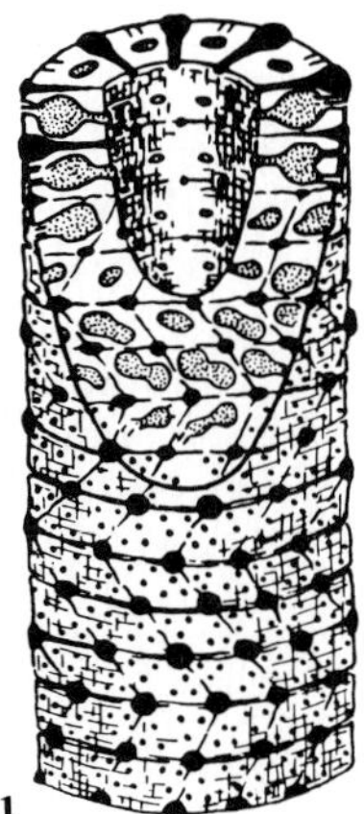

Figure 10.74
Neomeridoïdeae. **1.** *Heteroporella lepina*, reconstruction, showing black infertile branches and large pores, with the fertile sporangia or gametangia (stippled) possessing secondary branches that appear at the surface as tiny pores (compare sectioned specimens in Figure 10.73), approximately ×40, from Praturlon, 1967. **2.** *Turkmeniaria adducta* Maslov, Lower Cretaceous, USSR, reconstruction, with thallus indicated at lower right, calcified part in remainder, ×35, from Maslov, 1960b.

Distribution of Dasyclads. Living dasyclads are predominantly marine, and tropical to subtropical in distribution; the great majority of fossil taxa similarly lived in shallow warm marine waters, where they might occur in great numbers. A few freshwater representatives are known. *Diploporundus* is a noncalcified dasyclad occurring in the freshwater Lockatong Formation of Pennsylvania, of Late Triassic age (Bock, 1961). The monotypic genus *Goniolinites* is represented by the freshwater Quaternary *G. hemisphaerica* Maslov in Liskun & Maslov, and the living *Batophora* similarly occurs in lakes in New Mexico (Proctor, 1961).

Both the Dasycladales and the siphonous Caulerpales now occur in warm waters, yet their distribution in the Paleozoic ranged from northeastern Greenland (lat. 80° N) to Peru (lat. 15° S), where present climatic conditions are quite different. The most prominent floral change was in the Early Permian (Leonardian), at which time these siphonous green algae developed their present more restricted distribution. An abundance of these calcareous algae characterized the Tethyan Sea, whereas the probably cooler Zechstein Sea province has only noncalcareous fossil algae. The Late Permian dasyclad *Mizzia* occurs at lat. 60° N in British Columbia, at nearly the same latitude as the Zechstein province, however. Konishi (1960) noted this distribution and suggested that the Paleozoic climatic zonation was extremely complex. Less complexity is indicated, however, if the late Paleozoic fossil distribution is plotted on a paleomagnetic and plate tectonic based paleogeographic map.

Order Oedogoniales

Oedogonium (Oedogoniales) is an unbranched filamentous freshwater alga with complex, segmented, netlike plastids containing pyrenoids. Growth is localized in certain cells, hence intercalary in nature. During asexual reproduction a single multiflagellate zoospore with a subapical crown of about 120 flagella is formed by a vegetative cell. Although the biflagellate and quadriflagellate zoospores of other green algae are paralleled by free-living biflagellate and quadriflagellate monads (e.g., *Chlamydomonas, Carteria*), no known free-living green alga is similarly multiflagellate as the zoospores of *Derbesia* or *Oedogonium* (Bold, 1967).

Few fossils have been referred to the Oedogoniales other than *Paleooedogonium* (see Figure 10.75), with simple filaments of cylindrical cells interpreted as possessing both oogonia and antheridia. It was stated to have been a freshwater species that floated into the marine waters, where it was preserved in the Middle Devonian Onondaga cherts of Buffalo, New York. Other fossil representatives referred to this order are much younger geologically. *Oedogoniites palanensis* was described by Rao (1957) in a sequence of Eocene oil shales and lignites in India. Its unbranched filaments also consist of uniseriate cells, the basal one somewhat narrowed to serve for attachment. The elongate cells have no preserved cell contents, but circular holes were present in the outer wall of some cells.

Oedogonites was also proposed as a new generic name by Dwivedi (1959a) for another Eocene fossil specimen from India related to modern *Oedogonium.* As no named species were included in *Oedogonites* by Dwivedi, it lacks a type species, hence is both a homonym and invalid (ICBN Art. 37.1). This occurrence in the Deccan Intertrappean cherts is of interest in that three colonial mother cells appear in series in the filament, and dwarfed male cells are attached to the side. Remnants of a nucleus and chloroplast strands were also reported to be preserved in one cell, although preservation of such organelles appears improbable.

Order Zygnematales

Members of this exclusively freshwater order have markedly symmetrical cells. The very elaborate plastids are characteristically axile; that is, they are located in the central region of the cell. None of the Zygnematales have a motile reproductive stage. Asexual reproduction is by simple cell division, and sexual reproduction by the conjugation of amoeboid gametes, which are commonly merely the undivided cell protoplasts. This characteristic method of reproduction is indicated in the name Conjugales, by which this order has also been known.

Zygnemataceae. The unbranched filamentous Zygnemataceae (see Figure 10.76) are among the commonest algae in fresh water. These "pond scums" are especially abundant in stagnant water, where they form a bright green slimy or frothy mass on the surface; they may also occur near the margins of lakes and rarely even in flowing water. Growing by cell division and elongation, the filaments commonly have an external pectose mucilage sheath. Plastids of the Zygnemataceae and the Mesotaeniaceae may be of three types: a flat axile plate containing many pyrenoids; a pair of stellate plastids, each containing a single large central pyrenoid; or one or more parietal or marginal plastids in the form of spiral bands containing a series of pyrenoids. The spiralling of the coiled plastid is normally to the right in *Spirogyra* and to the left in *Genicularia.* The single nucleus lies in the center of the cell.

Asexual reproduction is accomplished by fragmentation of the filaments. Sexual reproduction is by conjugation (see Figure 10.77), which in *Spirogyra* begins when two filaments become closely aligned and held together by their mucilage. A short protrusion grows from each opposing cell of the two filaments, that from one filament meeting that from the neighboring filament. The projections gradually lengthen, pushing apart the filaments. Meanwhile the cells' contents lose much of their cell sap and become condensed. As the end walls of the two opposing cell protrusions dissolve, a

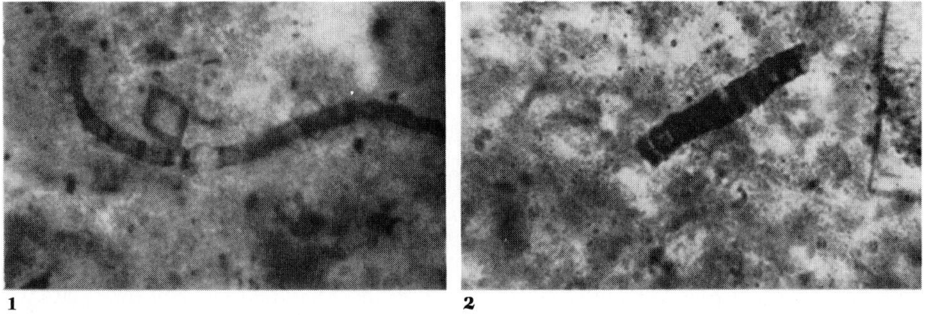

Figure 10.75
Oedogoniales. **1,2.** *Paleooedogonium micrum* Baschnagel, Middle Devonian, New York, in thin section of chert. **1.** Filament with oogonium. **2.** Filament with three antheridia. Both ×1200, from Baschnagel, 1966.

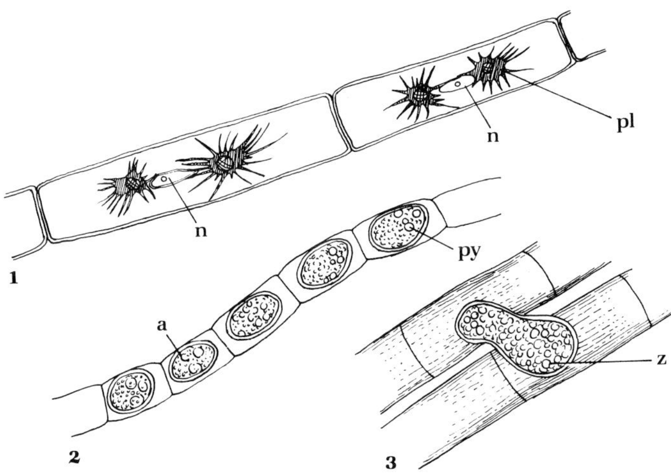

Figure 10.76
Zygnemataceae. *Zygnema* sp. **1.** Vegetative filament, showing n, nucleus; and pl, the two stellate plastids in each cell, each with a central pyrenoid. **2.** Filament contains a, azygospores or "akinetes," in which the pyrenoid (py) is discernible. **3.** Empty filaments, and a zygospore (z) that formed at their junction. All ×325.

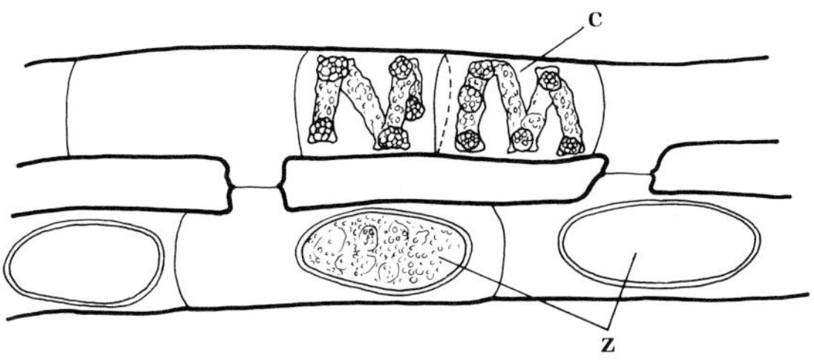

Figure 10.77
Spiroygyra sp. Two filaments that are joined by the conjugation tubes between adjacent cells to result in a ladderlike structure; cells (c) lacking an opposing partner remain vegetative, with their spiralling plastids; contents of cells connected by a conjugation tube have fused to form zygotes (z); ×325.

tubular connection is formed between the cells, through which the contents of one moves into the other, fusing to form a zygote. The sequence of connections between individual cells of the two filaments results in a ladderlike appearance. The zygotes then develop a thick, three-layered wall and lie dormant for a period. *Spirogyra* is thus said to show morphological isogamy, as the cells that fuse are identical, but to have physiological heterogamy, as one protoplast moves into the other cell and is thus regarded as the male gamete. Meiosis occurs prior to germination of the zygote to form a new filament, hence only the resting zygote is diploid. Occasionally a filament may be isolated and not sufficiently close to another for conjugation. Adjacent cells of a single filament may then fuse, their short projections looping across the intervening cell partition. Commonly conjugation occurs only between recently divided cells that are somewhat shorter than the average.

Fossil Zygnemataceae are rare, as no very resistant walls or skeletal elements are present in this group. However, an Eocene *Spirogyra*-like organism was reported, as *"Spirogyrites,"* from the Deccan Intertrappean beds of India (Shukla, 1950; Dwivedi, 1959a), with preserved spiral plastid. A similar specimen with preserved plastid was described as *Spirogyra wyomingia* Bradley from the Green River Formation lake beds of Wyoming, of Eocene age (Davis, 1916; Bradley, 1962, 1970). The opaque, reddish brown plastid appeared to be suspended in a colorless matrix, as if the cell were mummified with no distortion or shrinkage. In addition, zygospores referred to *Spirogyra* and *Mougeotia* were reported from soils in a ditch fill of a Bronze Age settlement in the Netherlands, and absent from subsoils at either side, indicating that they were the remains of organisms living while these ditches were in use (Van Geel, 1976).

The Desmids. The Mesotaeniaceae, or saccoderm desmids, are probably not closely related to the Desmidiaceae, or true (placoderm) desmids. Saccoderm desmids are found in pools and peat bogs, and a few are terrestrial. The single-layered cell wall is simple and structureless, is without pores, and consists of a single piece, in contrast to the true desmids. The cell is not medially constricted, and the single plastid in each cell is relatively simple. Following conjugation the zygote produces four individuals. None of the saccoderm desmids have been reported as fossil.

The placoderm desmids, Desmidiaceae, are mostly free-floating, solitary cells, although a few are united into filaments (*Micrasterias*), and some are attached. The cell may be cylindrical, curved, flattened, or discoidal, usually has a medial constriction or sinus, and a bridge or **isthmus** between the two distinct cell halves or **semicells** that are mirror images (see Figure 10.78). The large complex plastids are usually axile in position, appearing stellate, lobulate, spiral, or bandlike, and containing numerous pyrenoids; one (rarely two) is present in each semicell. *Closterium* also may produce 1 or more (up to 10) tiny crystals in the cytoplasm that are subsequently transferred to the cell vacuoles. These are said to be gypsum crystals that may act as statoliths, although this function has not been proven (Fischer, 1883; Kopetzky-Rechtperg, 1931). They may be merely excretory products.

The cell wall is constructed of two pieces, rarely more, is perforated by pores, and has an outer firm and thick cuticle layer of cellulose and pectic and iron compounds, and an inner structureless cellulosic layer. The wall also may contain traces of silica. The edges of the two halves are bevelled and fit closely together. They are firmly cemented during the life of the cell but separate during conjugation or may separate upon the death of the cell. The wall surface may be variously ornamented with granules, warts, and spines. A thin mucilaginous coating surrounds the cell, the pads of mucilage serving for adherence of attached species and to connect individual cells of those that form filamentous colonies. The wall pores through which mucilage is secreted may be evenly distributed over the surface, except at the isthmus, or may be grouped in relation to surface ornamentation. Desmid movement is also effected by mucilage secretion against a substrate, through pores at the ends of the cells. Because most desmids are planktonic in habit,

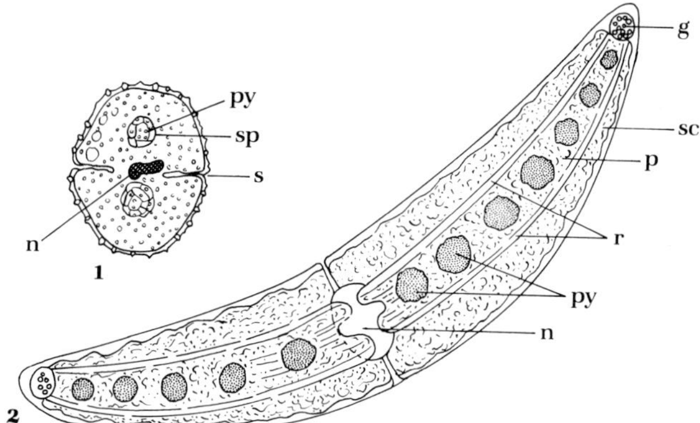

Figure 10.78
Desmidiaceae. 1. *Cosmarium* sp. **2.** *Closterium* sp. Both consist of two semicells (sc), joined at a sinus (s) at the midline, which also contains the nucleus (n); cells filled by the large plastids (p), which may have longitudinal ridges (r); plastids in each semicell contain one or more pyrenoids (py), which may be surrounded by starch plates (sp); gypsum crystals (g) are present in a terminal vacuole of each semicell of *Closterium;* ×650.

their varying shapes have been suggested as adaptations for flotation. *Micrasterias* is platelike, some *Closterium* are elongate and needlelike, *Staurastrum* has elongate processes, and *Xanthidium* and *Arthrodesmis* have spines. The mucilage envelope of the desmid cell also increases its surface area but does not add weight.

Vegetative reproduction is by simple fission, each daughter cell inheriting one of the parent semicell walls and constructing one anew. Cell division may occur daily in small species, but successive divisions are separated by several days in large ones. The wall first grows inward at the junction of the two semicells until they split apart. As the new semicell forms and enlarges to equal the original one, the single plastid from the prior semicell also enlarges and part of it extends through the isthmus from the old semicell into the new, which it gradually fills. The plastid then divides at the isthmus, so that a separate plastid lies in each semicell.

Sexual reproduction involves conjugation, as in other members of the order. Two cells come together, joined in a common mass of pectin mucilage. Filamentous species dissociate into separate cells prior to conjugation. The cells then open at the isthmus, liberating the protoplasts, which generally fuse outside the original walls. The heavy-walled resting zygote usually lies between the two cells; however, that of *Desmidium cylindricum* Greville forms within one of the original cells. Meiosis and germination

take place following a period of dormancy; the zygote produces only two new individuals, hence results in no increase in number over the parents. The desmid zygote, or zygospore, has a three-layered wall, the inner thin and colorless, the median one firm and brownish. The outer layer may have a smooth surface (*Closterium*) but is commonly ornamented with spines, warts, and furcate processes, superficially resembling the dinoflagellate cysts previously termed "hystrichospheres." The latter were in fact believed to be desmid zygospores when first observed.

As presently recognized, some desmid genera seem to intergrade so that species may be difficult to allocate. They may differ in side view, in end view, or in surface ornamentation.

Desmids may be abundant in freshwater ponds, slightly acidic (pH 5 to 6) *Sphagnum* bogs, and along the margins of clear oligotrophic lakes, low in dissolved nutriment. A single species, *Cosmarium salinum* Hansgirg, is found in brackish water. Most desmids prefer water with little dissolved calcium. However, *Oocardium* lives in weakly flowing limy water, where it may form extensive deposits of calcareous tuff (see Figure 10.79). This consists of hollow, elongate, and dichotomously forked tubes, encrusting the surface of mosses. Each tube widens toward the surface of the colony, is filled with mucilage, and encloses a single cell at the free end. The cells are between 19 and 25 μm long and up to 22 μm in breadth. The cal-

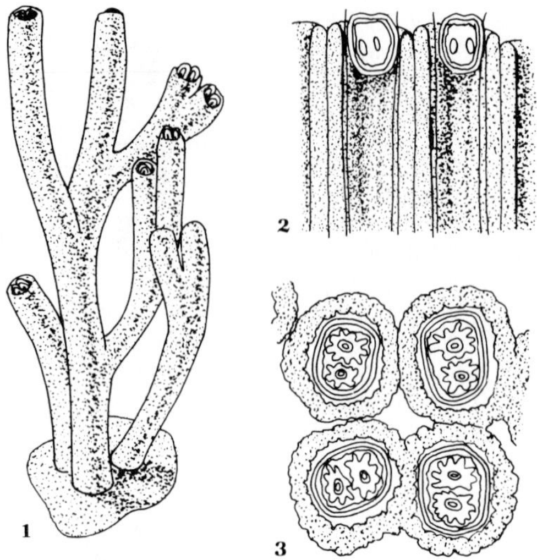

Figure 10.79

Oocardium stratum Naegeli. **1.** Calcareous tubes
(shaded), containing the *Cosmarium*-like cells at
their tips; redrawn from Wallner, 1933. **2,3.** Side
and top view of colony, showing thick-walled
calcareous tubes containing the small cells with
stellate plastids; 2, ×320; 3, ×523; redrawn from
Senn, 1899.

cium carbonate seemingly is deposited from
the water in relation to the assimilatory action
of the cell. As the tube grows longer, the cell
secretes additional mucilage beneath, so that it
maintains its position at the surface (Wallner,
1933).

Geographically, desmids are widely distributed, occurring in the freshwater deposits of
most continents. None have yet been reported
from Antarctica, although in northern latitudes
they may be acclimatized to withstand very cold
temperatures. Desmid-rich sediments have
been frozen and covered with snow for over 70
days with only a 20 percent reduction in the
number living. The cells are killed by rapid
freezing but not by gradually decreased temperature, allowing their survival in these cold
northerly regions (Duthie, 1964).

Their wide distribution in freshwater ponds
is commonly regarded as due to transportation
of zygospores on the feet or feathers of aquatic
birds. However, most species produce zygospores only rarely, and as ordinary vegetative individuals are sensitive to drying and to seawater,
dispersal would be slow. Proctor (1966) suggested that dispersal might more readily be effected by transportation within the digestive
tracts of birds. Viable cells were recovered from
the feces of ducks after at least an hour in the
digestive tract, although similar cells were rapidly killed by desiccation.

The nature of the cell composition makes
preservation as fossils unlikely, although a few
occurrences are known. Baschnagel (1942) reported *Arthrodesmis* and other species similar
to *Staurastrum* and *Cosmarium* from the Devonian of New York, but these are unrelated to
desmids, resembling instead various acritarchs
typical of the lower and middle Paleozoic that
do have well-preserved walls. *Paleoclosterium*
(see Figure 10.80) does appear closer to the desmids, however, and was described as containing the preserved remains of plastids (Baschnagel, 1966). A younger and less controversial
record is of *Stenixis cosmarioides* T. M. Harris
in the Upper Triassic, and undoubted desmids
are reported from Tertiary rocks. Abundant desmids found in mammalian coprolites in the
upper Eocene Bridger Formation of southwestern Wyoming (Bradley, 1946) included both con-
jugating cells and resting zygospores. Some
even appeared to have preserved cell contents
suggesting the remains of nuclei or pyrenoids.
The plastids seem to be preserved in another
specimen, but some were less well preserved,
having only a mineralized interior. Species
present included *Staurastrum enteroxenum*
Bradley and *Cosmarium* sp. cf. *pachydermum*
Lund. The presence of the fossil desmids suggests deposition in somewhat acid swampy
waters.

Baccinellula cosmarioides Weyland, described from the Pliocene Brown Coal of Italy
(Weyland, 1963), resembles *Cosmarium* but has
spherical semicells. Accompanying it were
many smaller dumbbell-shaped specimens that
may represent developmental stages of the
same species.

Interglacial deposits have produced some
fossil desmids, but many more are found in

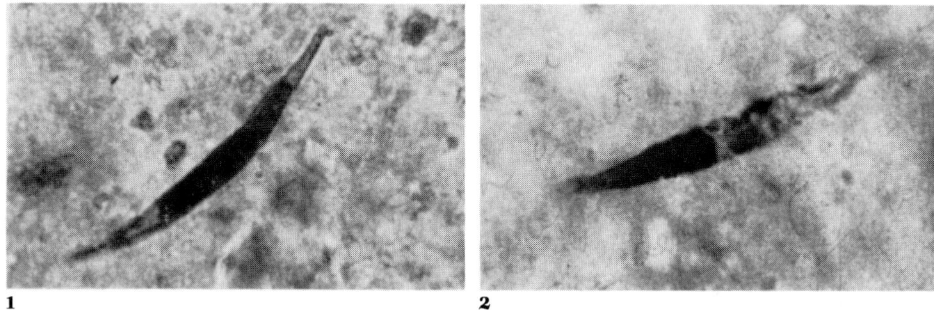

Figure 10.80
Paleoclosterium leptum Baschnagel, Middle Devonian, New York. **1.** Cell with terminal vacuoles and transverse striations. **2.** With reported plastid remains in left semicell. Both ×2500, from Baschnagel, 1966.

postglacial deposits (Borge, 1895, 1896). Lagerheim (1902) reported *Cosmarium, Staurastrum,* and *Euastrum,* and Messikommer (1938) observed some 18 genera as fossils from many localities in Germany, Poland, Sweden, Norway, Switzerland, England, Italy, and Finland.

Classification

Perhaps because of their great geologic age, the green algae are highly diversified in life history, physiology, morphology, and distribution. As a result, different classifications have been proposed with emphasis on one or another of these features. As is true generally of the algae, the nature of the pigments is of major importance. As a result, some still regard the Prasinophyta, Euglenophyta, and Charophyta as green algae, although most regard one or more as distinct from the main group of Chlorophyta. Classically the main form of the vegetative stage was used as a basis for subdivision. More recently, Chadefaud (1950) and Christensen (1962) utilized the nature of the flagellate stage for a basic dichotomy into the classes Prasinophyceae and Chlorophyceae, now regarded by some as divisions. Another system was based on the location of the plastids within the cell (Meyer, 1962); the Centroplastophyceae included five orders in which the plastids occupy the center of the cell (Centromonadales, Centropalmellales, Centrococcales, Centrothricales, and Centrostomatales); the Parietoplastophyceae included the Volvocales, Palmellales, Chlorococcales, Ulotrichales, Chaetophorales, Oedogoniales, Cladophorales, Ulvales, and Bryopsidales (which included the former Derbesiales, Dasycladales, Caulerpales, Codiales, and Siphonocladales); and a third class, Conjugatophyceae, was retained for the Conjugales or Zygnematales.

As noted earlier, the use of electron microscopy has brought new methods of separating various green algae, ranging from the method of cell division (phragmoplast or phycoplast) to the presence or absence of scales on body or flagella, and a variety of approaches to the life history. Unfortunately, some of the proposed new groups included no Latin diagnoses, hence according to the Botanical Code they are invalid. In addition, most categories were superfluous, as prior names of equivalent rank were already available; some proposed new family names were not even based on those of included genera. The classification followed herein is thus a combination of various current proposals, but will undoubtedly need modification as more information accumulates (see Table 10.2).

Table 10.2
Classification of the Chlorophyta.

Division Chlorophyta Pascher 1914

I. Class Chlorophyceae Kützing 1843
Motile stage with smooth (not scaly) flagella, and without apical pit.
Precambrian, Silurian to Holocene.

A. Order Chlamydomonadales Fritsch in West 1927
Dominantly motile unicells.

1. Family Dunaliellaceae Christensen 1967
Marine, fresh and brackish water; Holocene.
Dunaliella Teodoresco 1905; *Polytomella* Arago 1910.

2. Family Chlamydomonadaceae Stein 1878
Free-living in marine and fresh water; Eocene to Holocene.
Carteria Diesing 1866; *Chlamydomonas* Ehrenberg 1835 (Eocene, Holocene); *Polytoma* Ehrenberg 1833.

3. Family Phacotaceae (Bütschli) Oltmanns 1904
Freshwater; Oligocene to Holocene.
Coccomonas Stein 1878; *Phacotus* Perty 1852 (Oligocene to Holocene).

4. Family Haematococcaceae (Trevisan) Marchand orth. mut. Smith 1950
Marine, freshwater; Holocene.
Haematococcus Agardh 1828.

B. Order Volvocales Oltmanns 1904
Colonial, no distinct cell wall; ?Devonian, Holocene.

1. Family Volvocaceae Ehrenberg orth. mut. Cohn 1856
Freshwater, ?Devonian, Holocene.
?Eovolvox Kaźmierczak 1975 (Devonian); *Volvox* Linnaeus 1758.

C. Order Tetrasporales Lemmermann in Pascher 1915
Resemble Chlamydomonadales, but form palmelloid or dendroid colonies. In fresh water and soil; planktonic, epiphytic; Upper Jurassic to Holocene.

1. Family Palmellaceae (Endlicher) Kützing 1843 orth. mut. Nägeli 1847
Jurassic, Holocene.
Palmogloea Kützing 1843 (syn: *Gloeocystis* Nägeli 1849).

2. Family Tetrasporaceae (Nägeli) Klebs orth. mut. Wille in Warming 1884
Eocene to Holocene.
Actinochloris Korchikoff 1953; *Palmophylloites* Straus 1952 (Pliocene); *Schizochlamys* A. Braun ex Kützing 1849 (Eocene, Holocene); *Tetraspora* Link 1809.

D. Order Chlorococcales Marchand 1895 orth. mut. Pascher 1915
Unicellular or colonial; well-defined cell membrane; nonmotile vegetative stage, reproduce by zoospores or aplanospores; no contractile vacuoles. Mostly freshwater, some colorless saprophytes; few marine; some endosymbionts; some form algal partner of lichens; Precambrian to Holocene.

1. Family Chlorococcaceae Blackman & Tansley 1902
Upper Cretaceous to Holocene.
Bracteacoccus Tereg 1922; *Characium* Pascher 1929; *Chlorochytrium* Cohn 1872; *Chlorococcum* Meneghini 1842; *Desmatractum* West & West 1902 (Paleocene, Eocene, Holocene); *Myrmecia* Printz 1920; *Octogoniella* Pascher 1930 (Paleocene, Eocene, Holocene); *Palambages* O. Wetzel 1961 (Upper Cretaceous, Paleocene); *Spongiochloris* Starr 1955; *Tetracystis* Brown & Bold 1964; *Tetraëdron* Kützing 1845 (Syn.: *Closteridium* Reinsch 1888) (Eocene, Miocene, Holocene); *Trebouxia* de Puymaly 1924.

2. Family Chlorellaceae (N. Wille) Brunnthaler 1913
Precambrian to Holocene.
Chlorella Beijerinck 1890; *?Caryosphaeroides* Schopf 1968 (Precambrian); *Chlorellopsis* Reis 1923 (Jurassic, Miocene); *?Glenobotrydion* Schopf 1968 (Precambrian); *?Globophycus* Schopf 1968 (Precambrian); *Prototheca* Kruger 1894; *Pseudochlorella* Lund 1955; *Trochiscia* Kützing 1845.

3. Family Protosiphonaceae Blackman & Tansley 1902
Holocene.
Protosiphon Klebs 1896.

4. Family Hydrodictyaceae (S. F. Gray) Dumortier 1829
Silurian to Holocene.
Deflandrastrum Combaz 1962 (Silurian); *Hydrodictyolites* Elovski 1930 (Carboniferous, Per-

mian); *Hydrodictyon* Roth 1800; *Pediastrites* Zalessky 1926 (Jurassic); *Pediastrum* Meyen 1829 (Lower Cretaceous to Holocene; *Plaesiodictyon* W. Wille 1970 (Upper Triassic).

  5. Family Coelastraceae (West) N. Wille 1909 Eocene to Holocene.
Coelastrum Nägeli 1849 (Eocene; Holocene).
  6. Family Botryococcaceae N. Wille 1909 Precambrian to Holocene.
Botryococcus Kützing 1849 (Precambrian to Holocene).
  7. Family Oocystaceae Bohlin 1901 Carboniferous to Holocene.
Eremosphaera de Bary 1858 (Eocene, Holocene); *Glaucocystis* Itzigsohn in Rabenhorst 1868; *Lageniastrum* Renault 1894 (Carboniferous); *Oocystis* Nägeli in A. Braun 1855; *Palmellococcus* Chodat 1894.
  8. Family Scenedesmaceae Oltmanns 1904 Holocene.
Scenedesmus Meyen 1829.

E. Order Chlorosarcinales Groover & Bold 1969

  1. Family Chlorosarcinaceae Herndon 1958 Precambrian, Holocene
Chlorosarcina Gerneck 1907; *Latisphaera* Licari 1978 (Precambrian); *Maculosphaera* Licari 1978 (Precambrian).
  2. Family Coccomyxaceae (Chodat) G. M. Smith 1933
  Soil alga, Holocene.
Coccomyxa Schmidle 1901.

II. Class Codiolophyceae Kornmann 1973
  Heteromorphic life cycle, including a *Codiolum* stage, in which the germinating spore produces a basal disk, stalk, and club; marine and freshwater; Devonian to Holocene.

A. Order Ulotrichales Borzi 1895
  Simple, little differentiated, unbranched or branching filaments, or thallus or packet of cells; cell division by furrowing, no microtubules involved, hence no phycoplast or phragmoplast. Marine, freshwater, terrestrial; Devonian to Holocene.

  1. Family Ulotricaceae (Kützing) Borzi 1883
Courvoisiella Niklas 1976 (Upper Devonian); *Geminella* Turpin 1828 (Devonian, Holocene);

Hormosporites Grüss 1928 (Devonian); *Klebshormidium* Silva, Mattox & Blackwell 1972; *Ulothrix* Kützing 1836 (Eocene, Holocene).

B. Order Microsporales Bohlin 1901
  Cell division by furrowing, involving transverse microtubules (phycoplast). Freshwater, Eocene to Holocene.

  1. Family Microsporaceae Bohlin 1901
Microspora Thuret 1850 (Eocene, Holocene).

C. Order Chaetophorales West 1904
  Cell division by cell plate formation, phycoplast present.

  1. Family Chaetophoraceae (Greville) Harvey 1841
  Thallus of branching filaments of uniseriate cells; may be lime encrusted (*Chlorotylium*). Fresh water, Lower Jurassic to Holocene.
Chaetophora Schrank 1789; *Chaetophorites* Fliche 1886 (Jurassic, Oligocene); *Chlorotylites* M. A. Howe 1932 (Paleocene); *Chlorotylium* Kützing 1843; *Coccobotrys* Chodat 1913; *Foreliella* Chodat 1898 (Tertiary, Holocene); *Fritschiella* Iyengar 1932; *Leptosira* Borzi 1883; *Pseudopleurococcus* Snow 1899; *Stigeoclonium* Kützing 1843 (Eocene, Holocene).

D. Order Trentepohliales Chadefaud & Emberger 1960
  Terrestrial, epiphytic, to freshwater; may have wind-dispersed zoospores released from sporangia. Eocene to Holocene.

  1. Family Trentepohliaceae Hansgirg 1886
Cephaleuros Kunze 1827; *Phycopeltis* Millardet 1870 (Eocene, Oligocene, Holocene); *Trentepohlia* Martius 1817.

E. Order Ulvales Blackman & Tansley 1902
  Foliaceous or tubular thallus; isomorphic alternation of generations; cell division as in the Ulotrichales, with phycoplast. Marine and freshwater; Holocene.

  1. Family Ulvaceae Lamouroux 1813
Enteromorpha Link in Nees 1820; *Ulva* Linnaeus 1753.

(continued)

F. Order Prasiolales Fritsch in West 1927

Unbranched filament of flat, cylindrical cells, or thallus of two or more cells, stellate plastids; asexual reproduction by fragmentation; sexual reproduction oogamous.

Marine, freshwater, terrestrial; Holocene.

1. Family Prasiolaceae (Rabenhorst) Borzi 1895
Prasiola (Agardh) Meneghini 1838.

III. Class Bryopsidophyceae Round 1963

Siphonate green algae. Cambrian to Holocene.

A. Order Cladophorales West 1904

Simple or branched, free or anastomosing filaments, large bi- or multinucleate cells, elaborate plastids; some lime encrusted; alternation of generations is typical. Marine and freshwater; ?Precambrian, Jurassic to Holocene.

1. Family Cladophoraceae (Hassall) Cohn 1880
Chaetomorpha Kützing 1845; *Cladophora* Kützing 1843; *Cladophorites* Reis 1923 (Jurassic to Holocene); *?Palaeosiphonella* Licari 1978 (Precambrian); *Rhizoclonium* Kützing 1843.
2. Family Anadyomenaceae Kützing 1884
Holocene.
Anadyomene Lamouroux 1812.

B. Order Sphaeropleales Fritsch in West 1927
Freshwater; Holocene.

1. Family Sphaeropleaceae Kützing 1849
Sphaeroplea Agardh 1824.
2. Family Monostromataceae Kunieda ex Suneson 1947
Monostroma Thuret 1854.

C. Order Codiales Setchell 1929

Compact, nonseptate cellulosic thallus of intertwined coenocytic threads; hollow interior, no internal network of trabeculae; multinucleate, may have bi- or tripinnate fronds; free plastids discoid or fusiform, contain xanthophylls siphonein and siphonoxanthin; gametes in distinct gametocysts separated from vegetative plant. Mostly benthic, rarely floating; marine, tropical; Holocene.

1. Family Bryopsidaceae (Bory) De Toni 1888

Bryopsis Lamouroux 1809; *Pseudobryopsis* Berthold 1904.
2. Family Codiaceae (Trevisan) Zanardini 1843
Codium Stackhouse 1797.

D. Order Caulerpales Setchell 1929

Siphonous algae with internal network of trabeculae; no cellulose, some deposit aragonite; two kinds of plastids; gametogenic vesicles external on sporangiophores that are in communication with vegetative thallus. Marine and freshwater; Cambrian to Holocene.

1. Family Garwoodiaceae

Regarded as probably blue-green algae by some. Marine, tropical; Ordovician to Lower Cretaceous.
Bevocastria Garwood 1931; *Garwoodia* A. Wood 1941; *Hedstroemia* Rothpletz 1908; *Kitakamiania* Ishijima 1943; *Mitcheldeania* Wethered 1886; *Ortonella* Garwood 1914 (syn.: *Paragarwoodia* Poncet 1974); *Polymorphocodium* Derville 1930; *Stylocodium* Derville 1931.
2. Family Dichtomosiphonaceae Chadefaud ex Feldmann 1946
Freshwater; Holocene.
Dichotomosiphon Ernst 1902.
3. Family Udoteaceae (Endlicher) Agardh 1888
Marine, tropical; Cambrian to Holocene.
Abacella Maslov 1956; *Anchicodium* Johnson 1946; *Aphroditicodium* Elliott 1970; *Arabicodium* Elliott 1957; *Avrainvillea* Decaisne 1842; *Avrainvilleopsis* Forti 1926; *Bacinella* Radoičić 1959; *Boodleopsis* Gepp & Gepp 1911; *Boueina* Toula 1883; *Callipsygma* Agardh 1887; *Chlorodesmis* Bailey & Harvey 1858; *Cladocephalus* Howe 1905; *Clibeca* Poncet 1975; *Consinocodium* Endo 1961; *Dimorphosiphon* Höeg 1927; *Eugonophyllum* Konishi & Wray 1961; *Flabellaria* Lamouroux 1813; *Gemeridella* Borza & Misik 1975; *Geppella* Børgesen 1940; *Halimeda* Lamouroux 1812; *Hikorocodium* Endo 1951; *Ivanovia* Khvorova 1946; *Lancicula* Maslov 1956; *Litanaia* Maslov 1956; *Lithocodium* Elliott 1956; *Marinella* Pfender 1939; *Maslovina* Obrhel 1968; *Mizziella* Maslov 1956; *Neoanchicodium* Endo 1954; *Nuia* Maslov 1954 (syn.: *Bogutschanophycus* Korde 1954); *Orthriosiphon* Johnson & Konishi 1956; *Orthriosiphonoides*

Petryk & Mamet 1972; *Ovulites* Lamarck 1816; *Palaeocodium* Chiaruggi 1947; *Palaeoporella* Stolley 1893; *Paradella* Maslov 1956; *Penicilloides* Paul 1938; *Penicillus* Lamarck 1813; *Pseudochlorodesmis* Børgesen 1925; *Pseudocodium* Weber van Bosse 1895; *Rhipidodesmis* Gepp & Gepp 1911; *Rhipilia* Decaisne 1859; *Rhipiliopsis* Gepp & Gepp 1911; *Rhipocephalus* Kützing 1843; *Shecodium* Konishi 1954; *Shigaporella* Endo 1960; *Succodium* Konishi 1954; *Tauridium* Güvenç 1966; *Thaiporella* Endo 1965; *Tydemania* Weber van Bosse 1901; *Udotea* Lamouroux 1812 (Eocene, Holocene); *Uva* Maslov 1956; *Zaporella* Rácz 1964.

4. Family Caulerpaceae Greville 1830 orth. mut. de Toni 1889

Marine, tropical; Cretaceous to Holocene.
Caulerpa Lamouroux 1809; *Cretosiphon* O. Wetzel 1951 (Cretaceous).

E. Order Derbesiales Feldmann 1954

Alternation of siphonous spore-producing and saclike gamete-producing generations. *Pedobesia* has an aragonitic basal expansion with two platelike perforated layers connected by pillars. Marine, tropical; Holocene.

1. Family Derbesiaceae (Thuret) Kjellman 1883
Derbesia Solier 1847 (syn.: *Halicystis* Areschoug 1850); *Pedobesia* MacRaild & Womersley 1974.

F. Order Siphonocladales (Blackman & Tansley) Oltmanns 1904

Primary coenocytic thallus, divided into various multinucleate diploid parts to appear multicellular; attached; noncalcareous; chloroplasts reticulate; biflagellate swarmers formed by ordinary segments of thallus (segregative division). Marine; tropical and subtropical; Mississippian to Holocene.

1. Family Protovaloniaceae Vologdin 1966
Mississippian to Permian.
Protovalonia Vologdin 1966
2. Family Mongoloporellaceae Vologdin 1966
Permian.
Mongoloporella Vologdin 1966.
3. Family Valoniaceae Nägeli 1847
Upper Jurassic, Holocene.
Dictyosphaeria Decaisne 1842; *Valonia* C.

Agardh 1822; *Verticillodesmis* Dragastan & Misik 1975.
4. Family Siphonocladaceae Schmitz 1879
Upper Jurassic to Holocene.
Pycnoporidium Yabe and Toyama 1928 (syn.: *Solonoporidium* Pfender 1930); *Siphonocladus* Schmitz 1879.
5. Family Boodleaceae Børgesen 1925
Holocene.
Boodlea Murray & de Toni 1889.

G. Order Receptaculitales Sushkin 1962, nom. corr. A. H. Müller 1968

1. Family Receptaculitaceae Eichwald 1860
Globular to cylindrical or platelike thallus; internal cavity of fossils represents mother cell when living; lateral branches (meromes) grow at apex, are aligned in spiral or verticillate; outer end of meromes tetraradiate, surface of each merome covered with polygonal plate, meromes also calcified in later stages, but not area between laterals as in dasyclads; shallow water marine; benthic, tropical; ?Cambrian, Ordovician to Permian.

a. Tribe Amphispongieae Nitecki 1971
Upper Ordovician to Silurian.
Amphispongia Salter 1861; *Anomaloides* Ulrich 1878.

b. Tribe Cyclocrineae Pia 1920
Ordovician to Permian.
Apidium Stolley 1896; *Catena* Maslov 1956; *Coelosphaeridium* Roemer 1885; *Cyclocrinites* Eichwald 1840 (syn.: *Cerionites* Meek & Worthen 1868; *Cyclocrinus* Eichwald 1860; *Mastopora* Eichwald 1840; *Nidulites* Salter 1851; *Pasceolus* Billings 1857); *Epimastopora* Pia 1922; *Globuliferoporella* Chuvashov 1974; *Koninckopora* Lee 1912 (syn.: *Coeloceratioides* Derville 1931); *Koninckoporoides* Rich 1974; *Lepidolites* Ulrich 1889; *Pseudoepimastopora* Endo 1961; *Unjaella* Korde 1951.

c. Tribe Calathieae Nitecki 1969
?Cambrian, Ordovician to Silurian.
Calathella Rauff 1894; *Calathium* Billings 1865.

d. Tribe Receptaculiteae Nitecki 1969
Ordovician to Devonian.
Cerionites Meek & Worthen 1868; *Dictyocrinus* Hall 1859; *Ehlersospongia* Fagerstrom 1961; *Fisherites* Finney & Nitecki 1979; *Hexabactron* Campbell, Holloway & Smith 1974; *Ischadites*

(continued)

Murchison 1839 (syn.: *Acanthochonia* Hinde 1884); *I.* (*Neoischadites*) Byrnes 1968; *Palaeospongia* d'Orbigny 1849; *Receptaculites* Deshayes 1828; *Selenoides* Owen 1852; *Sphaerospongia* Pengelly 1861 (syn.: *Sphaeronites* Phillips 1841 non Hisinger 1828; *Polygonosphaerites* F. A. Roemer 1880); *Tettragonis* Eichwald 1842.

H. Order Dasycladales Pascher 1931

Vegetative plant suberect, with undivided axis surrounded by a dense aggregation of whorled branches, radially symmetrical; all segments with cytoplasmic continuity through pore in limiting wall, hence structurally unicellular and uninucleate, nucleus in holdfast; sexual plants multinucleate, specially differentiated reproductive organs; biflagellate gametes; plastids discrete, rounded or elliptical; wall of xylans or mannans, no cellulose; surface tends to be lime encrusted (aragonite), with calcification occurring in the mucus between branches; in fossils the aragonite may be converted to calcite. Precambrian to Holocene.

1. Family Timanellaceae Vologdin 1966
Precambrian, Sinian.
Timanella Vologdin in Vologdin & Kochetkov 1966.
2. Family Tannuolaiaceae Vologdin 1967
Lower Cambrian.
Tannuolaia Vologdin 1967.
3. Family Seletonellaceae Korde 1971
Precambrian to Permian.
    a. Tribe Papillomembraneae Korde 1971
Precambrian.
Papillomembrana Spjeldnaes 1963.
    b. Tribe Cambroporelleae Korde 1961
Lower Cambrian.
Cambroporella Korde 1950.
    c. Tribe Amgaelleae Korde 1957
Syn.: Sibirielleae Korde 1957
Middle Cambrian to Lower Devonian.
Amgaella Korde 1957; *Jacutiella* Korde 1964 (= *Sibiriella* Korde 1957 non Rudchenko 1955); *Thibia* Shuyskiy 1973.
    d. Tribe Seletonelleae Korde 1950
Upper Cambrian.
Mejerella Korde 1950; *Seletonella* Korde 1950.
    e. Tribe Vologdinelleae Korde 1957
Lower to Middle Cambrian.
Vologdinella Korde 1957; *Proaulopora* Vologdin 1962.

    f. Tribe Rhabdoporelleae Korde 1969
Cambrian to Permian.
Ajakmalajsoria Korde 1957; *Anatolipora* Konishi 1956; *Anthracoporella* Pia 1920; *Edelsteinia* Vologdin 1940; *Jansaella* Mamet & Roux 1975; *Kazakhstanelia* Korde 1957; *Pseudonanopora* Mamet & Roux 1975; *Rhabdoporella* Stolley 1893; *Unella* Poncet 1974; *Vermiporella* Stolley 1893 (syn.: *Uralella* Korde 1957).
    g. Tribe Primicorallineae Pia 1920
Ordovician to Silurian.
Callithamniopsis Whitfield 1894; *Inopinatella* Elliott 1971; *Primicorallina* Whitfield 1894.

    4. Family Dasycladaceae Kützing 1843
Cambrian to Holocene.
      a. Subfamily Dasycladoïdeae Valet 1969
      1) Tribe Dasyporelleae Pia 1920
Ordovician to Devonian.
Dasyporella Stolley 1893; *Intermurella* Elliott 1972; *Novantiella* Elliott 1972.
      2) Tribe Scribroporelleae Rietschel 1966
Originally described as a sponge, now placed with the algae. Middle Devonian.
Scribroporella Spriestersbach 1935.
      3) Tribe Palaeobereselleae Mamet & Roux 1975
Devonian to Mississippian.
Anthracoporellopsis Maslov 1956; *Exvotarisella* Elliott 1970; *Kamaena* Antropov 1967; *Kamaenella* Mamet & Roux 1975; *Palaeoberesella* Mamet & Roux 1975; *Parakamaena* Mamet & Roux 1975; *Pseudokamaena* Mamet in Petryk & Mamet 1972.
      4) Tribe Bereselleae Maslov & Kulik 1956
Mississippian to Pennsylvanian. Regarded as sponges by Termier et al., 1977.
Beresella Machaev 1937; *Donezella* Maslov 1929 (syn.: *Goksuella* Güvenç 1966); *Dvinella* Khvorova 1949; *Dvinella* (*Trinodella*) Maslov & Kulik 1956; *Exvotarisella* Elliott 1970; *Samarella* Maslov & Kulik 1956.
      5) Tribe Diploporeae Pia 1920
Mississippian to Eocene.
Aciculella Pia in Hirmer 1927; *Acroporella* Praturlon 1964; *Actinoporella* Guembel in Alth 1882; *Albertaporella* Johnson 1966; *Audrusoporella* Bystrický 1962; *Anisoporella* Botteron 1961; *Cabrieropora* Mamet & Roux 1975; *Carpathoporella* Dragastan 1967; *Clavaphysoporella* Endo 1958; *Clavaporella* Kochansky & Herak 1960; *Clypeina* Michelin 1845; *Connexia* Kochansky-Devidé 1970; *Cylin-*

droporella Johnson 1954; *Diplopora* Schafhaütl 1863; *Diploporundus* Bock 1961 (freshwater, Triassic); *Eovelebitella* Vachard 1974; *Griphoporella* Pia 1915; *Gyroporella* Gümbel 1872; *Herakella* Kochansky-Devidé 1970; *Holosporella* Pia 1930; *Likanella* Milanović 1965; *Macroporella* Pia 1912; *Mizzia* Schubert 1907; *Munieria* Deecke 1883; *Neogyroporella* Yabe and Toyama 1949; *Neomacroporella* Crescenti 1964; *Neomizzia* Lévy 1966; *Oligoporella* Pia 1912; *Pekiskopora* Mamet 1975; *Permoperplexella* Elliott 1968; *Physoporella* Steinmann 1903; *Pianella* Radoičić 1962; *Poikiloporella* Pia 1943; *Radoiciciella* Dragastan 1971; *Salopekiella* Milanović 1965; *Salpingoporella* Pia in Trauth 1927; *Selliporella* Sartoni & Crescenti 1962; *Thaumatoporella* Pia 1927; *Uragiella* Pia 1925; *Velebitella* Kochansky-Devidé 1964; *Verticilloporella* Raviv & Lorch 1970; *Windsoporella* Mamet & Rudloff 1972.

6) Tribe Stichoporelleae Fily & Rioult 1976

Middle Jurassic.

Stichoporella Pia in Hirmer 1927.

7) Tribe Kopetdagarieae Maslov 1965

Triassic to Lower Cretaceous.

Kopetdagaria Maslov 1960; *Sestrosphaera* Pia 1920.

8) Tribe Teutloporelleae Pia 1920

Triassic.

Teutloporella Pia 1912.

9) Tribe Petrasculeae (Pia 1920) Pia in Hirmer 1927

Jurassic.

Petrascula Guembel 1873.

10) Tribe Coniporeae Pia 1920.

Jurassic.

Archaeocladus Endo 1956; *Coniporella* Fischer & Thierry 1971; *Conodictyum* Goldfuss 1832 (syn.: *Conulina* Münster in Goldfuss 1832; *Conipora* de Blainville 1834); *Eodasycladus* Cros & Lemoine 1966; *Fanesella* Cros & Lemoine 1966; *Palaeodasycladus* Pia 1927 (syn.: *Palaeocladus* Pia 1920 non Ettingshausen 1885); *Pseudoclypeina* Radoičić 1970.

11) Tribe Linoporelleae Pia in Hirmer 1927

Upper Jurassic.

Linoporella Steinmann 1899; *Myrmekioporella* Pia 1925.

12) Tribe Triploporelleae Pia 1920

Jurassic to Eocene.

Broeckella L. & J. Morellet 1922; *Eogoniolina* Endo 1953; *Goniolina* d'Orbigny 1850; *Trinocladus* Rainieri 1922; *Triploporella* Steinmann 1880.

13) Tribe Ferganelleae Maslov 1955

Paleocene.

Ferganella Maslov 1955.

14) Tribe Morelletporeae Varma 1955

Paleocene.

Morelletpora Varma 1950.

15) Tribe Uterieae Pia in Hirmer 1927

Paleocene to Eocene.

Uteria Michelin 1845.

16) Tribe Thyrsoporelleae Pia 1927

Upper Triassic to Eocene.

Belzungia L. Morellet 1908; *Dobunniella* Elliott 1975; *Harlanjohnsonella* Elliott 1968; *Imperiella* Elliott & Süssli 1975; *Pentaporella* Senowbari-Daryan 1978; *Placklesia* Bilgütay 1968; *Suppiluliumaella* Elliott 1968; *Thyrsoporella* Gümbel 1871.

17) Tribe Dissocladelleae Elliott 1977

Upper Triassic to Paleocene.

Dissocladella Pia & Rao 1936.

18) Tribe Dactyloporeae Pia in Hirmer 1927

Upper Cretaceous to Eocene.

Dactylopora Lamarck 1816; *Digitella* L. & J. Morellet 1913; *Orioporella* Munier-Chalmas in L. & J. Morellet 1922; *Zittelina* L. & J. Morellet 1913.

19) Tribe Batophoreae Valet 1969

Ordovician to Holocene.

Archaeobatophora Nitecki 1976; *Batophora* J. Agardh 1855 (Marine and freshwater).

20) Tribe Dasycladeae Endlicher 1843

Paleocene to Holocene.

Chlorocladus Sonder 1871; *Dasycladus* C. Agardh 1827; *Pagodaporella* Elliott 1956.

b. Subfamily Bornetelloideae Valet 1969

Jurassic to Holocene.

Bornetella Munier-Chalmas 1877; *Catellaria* Maslov 1955; *Frederica* Barta-Calmus 1966; *Ollaria* Maslov 1955; *Terquemella* Munier-Chalmas 1877.

c. Subfamily Neomeridoideae Valet 1969

Upper Jurassic to Holocene.

Afghanopolia Kaever 1969; *Angioporella* Masse, M. A. Conrad & Radoičić 1973; *Campbelliella*

(*continued*)

Radoičić 1959 (syn.: *Daturellina* Radoičić 1959; *Hadziella* Radoičić 1962; *Hadziina* Radoičić 1969; *Metacyclina* Radoičić 1969; *Zetella* Radoičić 1969);*Cymopolia* Lamouroux 1816;*Heteroporella* Cros & Lemoine ex Praturlon 1967 (syn.: *Chinianella* Ott 1967); *Indopolia* Pia 1936; *Jodotella* L. & J. Morellet 1913; *Karreria* Munier-Chalmas 1877; *Larvaria* Defrance 1822; *Lemoinella* L. & J. Morellet 1913; *Meminella* L. & J. Morellet 1913; *Montiella* L. & J. Morellet 1922; *Morelletina* Maslov 1969; *Neomeris* Lamouroux 1816; *Parkerella* Munier-Chalmas 1877;*Pseudocymopolia* Elliott 1970; *Sakkionella* Segonzac 1970; *Turkmeniaria* Maslov 1960.

5. Family Acetabulariaceae Hauck 1884
Jurassic to Holocene.
Acetabularia Lamouroux 1816; *Acicularia* d'Archiac 1843; *Coscinoconus* Leupold 1935; *Halicoryne* Harvey 1860; *Polyphysa* Lamouroux 1816 (syn.: *chalmasia* Solms-Laubach 1895); *Rostroporella* Segonzac 1972.
Dasycladales, position uncertain
Anfractuosoporella Chuvashov 1974; *Carpenterella* Munier-Chalmas in L. & J. Morellet 1922; *Furcoporella* Pia 1918; *Goniolinites* Maslov in *Litopora* Johnson 1964;*Sphinctoporella* Mamet & Rudloff 1972;*Tersella* J. Morellet in Morellet & Ters 1952.

IV. Class Oedogoniophyceae Round 1971

A. Order Oedogoniales Blackman & Tansley ex West 1904

1. Family Oedogoniaceae (Thuret) de Bary 1900
Simple or branched filaments, with characteristic phycoplast division method; sexual reproduction oogamous, male gametes and zoospores with terminal crown of flagella; reticulate plastids. Fresh water; Devonian; Eocene-Holocene.
Oedogonium Link 1820; *Oedogoniites* Rao 1957 (Eocene); *Paleooedogonium* Baschnagel 1966 (Devonian).

V. Class Zygnemaphyceae Round 1971
?Devonian, ?Upper Triassic, Eocene to Holocene.

A. Order Coleochaetales Stewart, Mattox & Floyd 1973
Cell division with cell plate formation, phragmoplast present. Freshwater, Holocene.

1. Family Coleochaetaceae Naegeli 1860
Coleochaete de Brébisson 1844; ?*Stichococcus* Naegeli 1849.

B. Order Zygnematales Borge & Pascher 1931
Unicellular or filamentous colonies, elaborate plastids; binary cell fission with development of phragmoplast; sexual reproduction by conjugation; no motile stage. Freshwater.

1. Family Zygnemataceae (Meneghini) Kützing 1898
Eocene to Holocene.
Spirogyra Link 1820 (Eocene, Holocene); *Zygnema* C. Agardh 1817.
2. Family Mesotaeniaceae Oltmanns 1904
Holocene.
Cylindrocystis Meneghini 1838; *Mesotaenium* Naegeli 1849.
3. Family Desmidiaceae Kützing 1833
　　1) Tribe Cosmarieae Carter 1899
?Devonian, ?Upper Triassic, Tertiary to Holocene.
Arthrodesmis [Ehrenberg 1838] Ralfs 1848; *Baccinellula* Weyland 1963; *Cosmarium* [Corda 1834] Ralfs 1848; *Cosmocladium* de Brébisson 1856; *Desmidium* [Agardh 1825] Ralfs 1848, *Docidium* [de Brébisson 1844] Ralfs 1848; *Euastrum* [Ehrenberg 1832] Ralfs 1848; *Hyalotheca* [Ehrenberg 1840] Ralfs 1848; *Micrasterias* [Agardh 1827] Ralfs 1848; *Oocardium* Naegeli 1849; *Paleoclosterium* Baschnagel 1966; *Pleurotaenium* Naegeli 1849; *Sphaerozosma* [Corda 1825] Ralfs 1848; *Spondylosium* [de Brébisson 1844] Ralfs 1848; *Staurastrum* [Meyen 1829] Ralfs 1848; *Stenixis* Harris 1938; *Xanthidium* [Ehrenberg 1837] Ralfs 1848.
　　2) Tribe Penieae Lütkemüller 1902
Holocene.
Penium [de Brébisson 1844] Ralfs 1848.
　　3) Tribe Closterieae Lütkemüller 1902
Holocene.
Closterium [Nitzsch 1817] Ralfs 1848.

DIVISION EUGLENOPHYTA

From the earliest discussions of the euglenids, as Infusoria, they have been variously assigned to the animal and plant kingdoms. Colorless motile species with holozoic nutrition are regarded as animallike, whereas the plastids of some euglenids and the attached or palmelloid nature of others (*Colacium*) are plantlike. However, the close relationship of the colorless and the pigmented species is easily demonstrated. Pigmented forms may ingest dissolved or particulate food when light is insufficient for optimum photosynthesis. In cultures, they may continue saprophytic or phagotrophic nutrition after loss of the chloroplasts is induced by an antibiotic (Provasoli and Schatz, 1948; Pringsheim and Pringsheim, 1952). In fact, the presence or absence of chloroplasts is regarded as only of generic importance in the euglenids.

The Euglenid Cell

The cell is elongate, either circular or somewhat flattened in section, and has one or more emergent flagella (see Figure 10.81). The exterior may be soft and flexible (*Euglena*), or rigid, with a striate or spirally ridged surface (*Phacus*). A few genera are provided with a **lorica** outside the cell body (*Trachelomonas*). Open at the anterior end, to allow the flagella to protrude, the lorica consists of a firm gelatinous substance without cellulose. Although commonly colorless, the lorica may be yellow to brown due to included iron compounds.

The flagella arise from the base of a hollow depression or **reservoir**, formed by the invaginated anterior end. The flagella are featherlike (pantonematic), being provided throughout their length with a single row of tiny cilia. If only one flagellum is visible, another very short one is generally present, although it may not project from the reservoir. The single emergent flagellum is directed forward. Species with two flagella may have similar forward-directed ones, or unequal flagella, with one forward in position and the other trailing.

The nucleus is large, with permanently condensed chromosomes, similar to the dinokaryon nucleus of the Pyrrhophyta. No spindle or centromeres are present in the nuclear mitotic division. Euglenids have generally been regarded as lacking any sexual reproduction, but reports of unusually large numbers of chromosomes in some euglenids may indicate meiosis, and hence the presence of sexual reproduction, perhaps in relation to encystment (Leedale, 1967), as in the dinoflagellates.

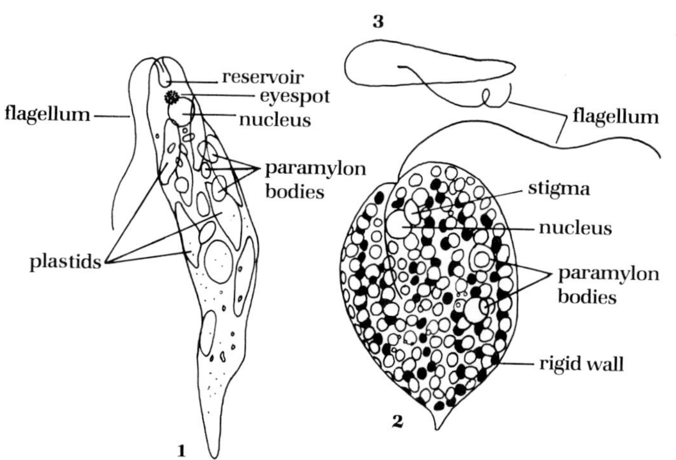

Figure 10.81
Euglenid cells. **1.** *Euglena mesnili* Deflandre & Dusi, ×1065. **2,3.** Side view and outline of top of *Phacus* sp., ×800.

Nutrition

Both pigmented and colorless species occur. The former have bright green chloroplasts, commonly several to a cell, that may be discoidal, band-shaped, or stellate. The pigments chlorophyll *a* and *b* of the euglenids represent their closest similarity to the green algae (Chlorophyta), as the two divisions are unlike in the organization of their cells. Accessory pigments of the euglenids are β-carotene and xanthophylls. Food is stored as paramylon, an insoluble polysaccharide compound. Like the Chlorophyta, the Euglenophyta that are colorless and lack plastids (e.g., *Astasia*) do have **proplastids** that serve for starch storage, whereas colorless dinoflagellates and chrysophytes have none (Dodge and Crawford, 1971).

Euglenophyta that lack plastids may ingest particulate food, hence are heterotrophic (e.g., *Peranema*), or utilize dissolved substances, as do the osmotrophic *Astasia, Rhabdomonas,* or *Trachelomonas.* Food is taken in at a softer part of the cell termed the **cytostome**, and some have a distinctive ingestion apparatus (*Peranema*). Phototrophic euglenids also require an external vitamin source, and some green euglenids are facultatively heterotrophic, capable of growth in the dark if supplied with suitable organic compounds. Many species have a contractile vacuole.

Reproduction

Reproduction is by longitudinal cell division, beginning at the anterior end. Division may occur while the cell is actively motile. Loricate types divide inside the lorica, one of the resultant cells escaping to form a new lorica, the other remaining in place.

Cysts or thick-walled resting stages, usually spherical or polygonal in shape, occur in many species. The cyst contents may later divide into many new cells. Whether this is related to possible sexual reproduction, as in dinoflagellates, is unknown.

Occurrence

As it is easily cultured, the freshwater *Euglena gracilis* Klebs has been intensively studied in the laboratory, yet its ecological niche in nature is still unknown. In general, the freshwater euglenids have been better studied than their marine counterparts. The latter have commonly been lumped with the naked chrysophytes and identified in phytoplankton surveys as "naked flagellates" or "monads." As a result, their importance in the oceans is generally underestimated, although in some areas these naked flagellates may surpass in abundance both the dinoflagellates and coccolithophorids. In Danish fjords they have been reported to occur in numbers of four million cells per liter of seawater.

Fossil Euglenids

Few naked flagellates are known as fossils, but rare occurrences have been reported from Eocene, Miocene, and Pliocene lignitic and bituminous shales. Thorough study of such deposits might provide a better understanding of the abundance and ecologic distribution of these soft-bodied organisms through geologic time.

Bradley (1929) described a fossil euglenid from the Eocene Green River Formation of Colorado, referring it to the modern genus *Phacus* (see Figure 10.82). The pyriform cell, about 46 μm in length, was preserved in the organic ooze of the Eocene lake bed, as were various blue-green and green algae, fungi, protozoa, insects, and spores, pollen, and plant cuticle.

Euglena sp. was reported from the Eocene lignites of India, associated with other green algae such as *Oedogoniites* and *Ulothrix* (Rao, 1957).

Many siliceous flasklike fossils were erroneously referred to *Trachelomonas,* until Deflandre (1934) showed them to represent cysts of chrysomonads. The euglenid genus *Trachelomonas* has a cellulosopectic test, which may be iron-impregnated, but is not resistant to oxidation. True *Trachelomonas* has

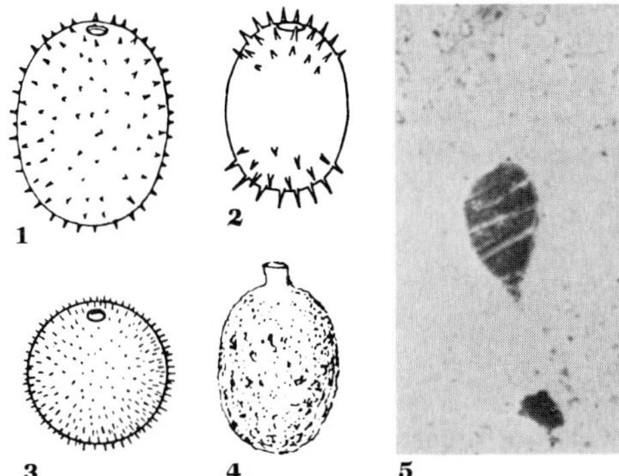

Figure 10.82

1−4. *Trachelomonas* loricae, ×600, from Conrad, 1938. **1.** *Trachelomonas hispida* (Perty) Stein, Bali. **2.** *T. raciborskii* Wołoszyńska, Java. **3.** *T. scabra* Playfair, Bali. **4.** *T. woycickii* Koczwara, Mentawei. **5.** *Phacus* cf. *caudata* Huber, Eocene, Colorado, ×410, from Bradley, 1929.

been found beautifully preserved in the Pliocene bituminous shales of Madagascar (Deflandre and Lenoble, 1948), and the specimens even referred to living species, but such records are extremely rare.

Classification

The early classification of the euglenids was based on their nutrition (Klebs, 1883); they were divided into green phototrophic euglenids, colorless, phagotrophic ones, and colorless saprophytic ones. However, the many pairs of closely related colorless and unpigmented euglenids suggest the artificiality of this system. Leedale (1967) proposed a classification based on flagellar characters and other cytologic features, which is followed herein (see Table 10.3, overleaf). Others (Asaul, 1970) maintain the more classical subdivisions with minor modification.

REFERENCES

Ackman, R. G., R. F. Addison, S. N. Hooper, and A. Prakash, *Halosphaera viridis:* fatty acid composition and taxonomical relationships. *J. Fish. Res. Bd. Can.,* v. 27, p. 251−255, 1970.

Ahmadjian, V., A guide to the algae occurring as lichen symbionts: isolation, culture, cultural physiology, and identification. *Phycologia,* v. 6, p. 127−160, 1967.

Antropov, I. A., Vodorosli Devona i nizhnego Karbona (Turne) tsentral'noy chasti vostoka Russkoy Platformy [Devonian and Lower Carboniferous (Tournaisian) algae of the central part of the eastern Russian Platform], p. 118−125, pls. 27, 28. In *Iskopaemye Vodorosli SSSR,* Akad. Nauk SSSR, Sibirskoe Otdel., Inst. Geol. Geofiz. Moscow: "Nauka," 1967.

Asaul, Z. I., Osnovni printsipi klasifikatsii evglenovikh vodorostey [Main principles of classification of euglenid algae]. *Ukr. Bot. Zh.,* v. 27, p. 545−556, 1970.

Azpeitia Moros, F., Datos para el estudio paleontológico del Flysch de la costa Cantábrica y de algunos otros puntos de España. *Boln. Inst. Geol. Min. Esp.,* v. 53, p. 1−65, pls. 1−19, 1933.

Bailey, G. P., Richard Rezak, and E. R. Cox, A revision of generic concepts of living members in the subfamily Acetabularieae (Dasycladaceae, Dasycladales) based on scanning electron microscopy. *Phycologia,* v. 15, p. 7−18, figs. 1−25, 1976.

Balme, B. E., Plant microfossils from the Lower Triassic of western Australia. *Palaeontology,* v. 6, p. 12−40, pls. 4−6, figs. 1−3, 1963.

Barta-Calmus, S., Algues Dasycladacées de Lutétien de Villiers-Saint-Frédéric (Yvelines). *Bull. Soc. Géol. Fr.,* sér. 7, v. 7 (1965), p. 906−910, pl. 39, figs. 1, 2, 1966.

Baschnagel, R. A., Some microfossils from the Onondaga Chert of central New York. *Bull. Buffalo Soc. Nat. Sci.,* v. 17, no. 3, p. 1−8, pl. 1, 1942.

Baschnagel, R. A., New fossil algae from the Middle Devonian of New York. *Trans. Am. Microsc. Soc.,* v. 85, p. 297−302, figs. 1−6, 1966.

Bassoullet, J. P., Paul Bernier, Raoul Deloffre, Patrick Genot, Michel Jaffrezo, A. F. Poignant, and Geneviève Segonzac, Réflexions sur la systématique des Dasycladales fossiles. Étude critique de la terminologie et importance relative des critères de classification. *Geobios,* v. 8, p. 259−290, figs. 1−6, 1975.

Belcher, J. H., *Prasinochloris sessilis* gen. et sp. nov., a coccoid member of the Prasinophyceae, with some remarks upon cyst formation in *Pyramimonas. Br. Phycol. Bull.,* v. 3, p. 43−51, fig. 1, 1966.

Belcher, J. H., Notes on the physiology of *Botryococcus braunii* Kützing. *Arch. Mikrobiol.,* v. 61, p. 335−346, figs. 1−6, 1968.

Table 10.3
Classification of the Euglenophyta.

Division Euglenophyta Pascher 1931
A. Order Eutreptiales Leedale 1967

Two similar but heterodynamic emergent flagella, one directed anteriorly, and one held laterally or to posterior during movement; green or colorless, but not phagotrophic; cell solitary, no cell envelope. Marine, fresh and brackish water; Holocene.
Distigma Ehrenberg 1838; *Distigmopsis* Hollande 1942; *Eutreptia* Perty 1852; *Eutreptiella* da Cunha 1913.

B. Order Euglenales Engler 1898
Green or colorless, phototrophic or osmotrophic, rarely phagotrophic; solitary, colonial, or palmelloid; may have eyespot and flagellar swelling; two flagella, one emergent and functional for locomotion, other short and not emergent from reservoir. May have cell envelope encrusted with manganese or iron. Freshwater and marine; Eocene to Holocene.
Astasia Dujardin 1841; *Colacium* Ehrenberg 1838; *Euglena* Ehrenberg 1838; *Phacus* Dujardin 1841; *Strombomonas* Deflandre 1930; *Trachelomonas* Ehrenberg 1833.

C. Order Rhabdomonadales Leedale 1967
Solitary, free-swimming, no envelope; colorless, osmotrophic but not phagotrophic; one emergent flagellum, mobile throughout length when swimming; no eyespot or flagellar swelling. Freshwater, Holocene.
Rhabdomonas Fresenius 1858.

D. Order Sphenomonadales Leedale 1967
Free-swimming, solitary, no envelope, cell rigid, with keels or grooves; one or two emergent flagella, straight anterior one not mobile throughout length; colorless, phagotrophic or osmotrophic, no eyespot or flagellar swelling. Freshwater and marine; Holocene.
Anisonema Dujardin 1841; *Petalomonas* Stein 1878; *Sphenomonas* Stein 1878.

E. Order Heteronematales Leedale 1967
Free-swimming, solitary, no envelope; no eyespot or flagellar swelling; colorless, phagotrophic, special ingestion organelle present; one or two emergent flagella, one stretched to anterior during swimming, movement of tip only resulting in gliding motion. Freshwater and marine; Holocene.
Heteronema Dujardin 1841; *Peranema* Dujardin 1841; *Urceolus* Mereschkowsky 1879.

F. Order Euglenamorphales Leedale 1967
Elongate, unflattened solitary cell, no envelope, nonrigid; green or colorless, not phagotrophic, endozoic, living in tadpole digestive tract; three or more emergent flagella of same length, size, and activity. Holocene.
Euglenamorpha Wenrich 1924.

Belcher, J. H., and G. E. Fogg, Biochemical evidence of the affinities of *Botryococcus*. *New Phytol.*, v. 54, p. 81—83, 1955.

Bernier, Paul, *Campbelliella striata* (Carozzi): algue Dasycladacée? une nouvelle interprétation de l'"Organisme C" Favre et Richard, 1927. *Geobios*, v. 7, p. 155—175, pls. 32—34, figs. 1—6, 1974.

Bertrand, C. E., Conclusions générales sur les charbons humiques et les charbons de purins. *C. R. Hebd. Séanc. Acad. Sci., Paris*, v. 127, p. 822—825, 1898.

Bertrand, Paul, Les Botryococcacées actuelles et fossiles et les conséquences de leur activité biologique. *C. R. Séanc. Soc. Biol.*, v. 96, p. 695—697, 1927.

Bilgütay, U., Some Triassic calcareous algae from Plackles (Hohe Wand, Lower Austria). *Verh. Geol. Bundesanst., Wien*, p. 65—79, pls. 1—3, figs. 1—6, 1968.

Blackburn, K. B., and B. N. Temperley, *Botryococcus* and the algal coals; Part 1, A reinvestigation of the alga *Botryococcus braunii* Kützing (by K. B. Blackburn); Part 2, The boghead controversy and the morphology of the boghead algae (by B. N. Temperley). *Trans. R. Soc. Edinb.*, v. 58, p. 841—868, 2 pls., 1936.

Boalch, G. T., and Mary Parke, The prasinophycean genera (Chlorophyta) possibly related to fossil genera, in particular the genus *Tasmanites*. *Proc. II Plankton Conf. Roma 1970*, v. 1, p. 99—105, pl. 1, 1971.

Bock, W., New fresh water algae of the eastern American Triassic. *Proc. Pa. Acad. Sci.,* v. 35, p. 77 — 81, figs. 1 — 3, 1961.

Bold, H. C., *Morphology of plants,* 2nd ed. New York: Harper and Row, xxix + 541 p., 1967.

Bonotto, S., B. Felluga, and J. Aksiyote, Quelques observations sur la morphologie des cystes d'*Acetabularia mediterranea. Protoplasma,* v. 67, p. 407 — 412, figs. 1 — 3, 1969.

Bonotto, S., P. Lurquin, and A. Mazza, Recent advances in research on the marine alga *Acetabularia. Adv. Mar. Biol.,* v. 14, p. 123 — 250, figs. 1 — 24, 1976.

Borge, O., Ueber subfossile Süsswasseralgen aus Gotland. *Bot. Zbl.,* v. 63, p. 56 — 58, 1895.

Borge, O., Nachtrag zur subfossilen Desmidiaceen-Flora Gotlands. *Bot. Notiser,* p. 111 — 113, figs. 1 — 10, 1896.

Borge, O., and G. Erdtman, On the occurrence of *Pediastrum* in Tertiary strata in the Isle of Wight. *Bot. Notiser,* p. 112 — 113, 1954.

Borza, Karol, and Milan Mišík, *Gemeridella minuta* n. gen., n. sp. aus der oberen Trias der Westkarpaten. *Geol. Sb. Bratisl.,* v. 26, p. 77 — 81, pl. 1, fig. 1, 1975.

Botteron, G., Étude géologique de la région du Mont d'Or (Préalpes romandes). *Eclog. Geol. Helv.,* v. 54, p. 29 — 106, pls. 1 — 12, figs. 1 — 16, 1961.

Bouroullec, J., and R. Deloffre, Les algues du Néocomien d'Aquitaine. *Bull. Centre Rech. Pau, SNPA,* v. 2, p. 213 — 261, 6 pls., 3 figs., 1968.

Bourrelly, P., *Les algues d'eau douce. Initiation à la systématique.* Tome 1, *Les algues vertes.* Paris: N. Boubée and Cie., 511 p., 117 pls., 1966.

Bradley, W. H., An oil shale and its microörganisms from the Fuson Formation of Wyoming. *Am. J. Sci.,* v. 208, p. 228 — 234, 4 figs., 1924.

Bradley, W. H., Freshwater algae from the Green River Formation of Colorado. *Bull. Torrey Bot. Club,* v. 56, p. 421 — 428, pls. 22, 23, 1929.

Bradley, W. H., Origin and microfossils of the oil shale of the Green River Formation of Colorado and Utah. *Prof. Pap. U.S. Geol. Surv.,* 168, p. 1 — 58, 28 pls., 3 text-figs., 1931.

Bradley, W. H., Coprolites from the Bridger Formation of Wyoming: their composition and microörganisms. *Am. J. Sci.,* v. 244, p. 215 — 239, 4 pls., 1946.

Bradley, W. H., Chloroplast in *Spirogyra* from the Green River Formation of Wyoming. *Am. J. Sci.,* v. 260, p. 455 — 459, 4 figs., 1962.

Bradley, W. H., Geology of Green River Formation and associated Eocene rocks in southwestern Wyoming and adjacent parts of Colorado and Utah. *Prof. Pap. U.S. Geol. Surv.,* 496-A, p. 1 — 86, figs. 1 — 20, 1964.

Bradley, W. H., Eocene algae and plant hairs from the Green River Formation of Wyoming. *Am. J. Bot.,* v. 57, p. 782 — 785, figs. 1 — 10, 1970.

Brito, I. M., Novos microfósseis Devonianos do Maranhão. *Publçao. Univ. Bahia Esc. Geol.,* 2, p. 1 — 4, 1 pl., 1965.

Brito, I. M., Novo subgrupo de Acritarcha do Devoniano do Maranhão. *Anais Acad. Bras. Cienc.,* v. 39, p. 163 — 166, pls. 1 — 3, figs. 1, 2, 1967a.

Brito, I. M., Contribuição ao conhecimento dos microfósseis Devonianos de Pernambuco II: Acritarcha Pteromorphitae. *Anais Acad. Bras. Cienc.,* v. 39, p. 285 — 287, pl. 1, 1967b.

Brito, I. M., Silurian and Devonian Acritarcha from Maranhão Basin, Brazil. *Micropaleontology,* v. 13, p. 473 — 482, pls. 1, 2, 1967c.

Brito, I. M., Contribuição ao conhecimento dos microfósseis Silurianos e Devonianos da Bacia do Maranhão IV: Os Tasmanaceae. *Bolm. Geol. Inst. Geociên. Univ. Fed. Rio de J.,* no. 3, p. 15 — 20, pl. 1, 1969.

Brito, I. M., Contribuição ao conhecimento dos microfósseis Silurianos e Devonianos da Bacia do Maranhão V: Acritarcha Herkomorphitae e Prismatomorphitae. *Anais Acad. Bras. Cienc.,* v. 43 (Supl.), p. 201 — 208, figs. 1 — 10, 1971.

Bronnimann, P., Microfossils incertae sedis from the Upper Jurassic and Lower Cretaceous of Cuba. *Micropaleontology,* v. 1, p. 28 — 51, pls. 1, 2, figs. 1 — 10, 1955.

Brown, R. M., Jr., and H. C. Bold, Phycological studies. V. Comparative studies of the algal genera *Tetracystis* and *Chlorococcum. Univ. Tex. Publs.,* 6417, p. 1 — 213, figs. 1 — 240, text-figs. 1 — 5, 1964.

Burgess, I. C., *Calcifolium* (Codiaceae) from the Upper Viséan of Scotland. *Palaeontology,* v. 8, p. 192 — 198, pls. 21, 22, figs. 1 — 3, 1965.

Byrnes, J. G., Notes on the nature and environmental significance of the Receptaculitaceae. *Lethaia,* v. 1, p. 368 — 381, figs. 1 — 5, 1968.

Bystrický, Jan, Nové Dasycladaceae triasu Slovenskéko Krasu [New Dasycladaceae of the Triassic in the Slovak karst]. *Geol. Sb. Bratisl.,* v. 13, p. 227 — 240, pls. 3, 4, 1962.

Bystrický, Jan, Die obertriadischen Dasycladazeen der Westkarpaten. *Geol. Sb. Bratisl.,* v. 18, p. 285 — 309, pls. 1 — 18, figs. 1 — 3, 1967.

Campbell, K. S. W., O. J. Holloway, and W. D. Smith, A new receptaculitid genus *Hexabactron* and the relationships of the Receptaculitaceae. *Palaeontographica Abt. A,* v. 146, p. 52 — 77, pls. 12 — 17, figs. 1 — 12, 1974.

Cane, R. F., The constitution and synthesis of oil shale. *Wld. Petrol. Congr. 7,* v. 3, p. 681 — 689, figs. 1 — 4, 1967.

Carozzi, A. V., L'Organisme "C" J. Favre (1927) est une *Vaginella* Portlandienne. *Archs. Sci. Phys. Nat.,* v. 7, p. 107 — 111, 1954.

Carozzi, A. V., Dasycladacées du Jurassique supérieur du bassin de Genève. *Eclog. Geol. Helv.*, v. 48, p. 31—67, pls. 5, 6, figs. 1—19, 1955.

Cassie, V., A free-floating *Pseudobryopsis* (Chlorophyceae) from New Zealand. *Phycologia*, v. 8, p. 71—76, figs. 1—10, 1969.

Caudwell, C., Restes végétaux mis en évidence dans les calcaires lacustres tertiaires d'Aquitaine. *Bull. Soc. Géol. Fr.*, sér. 7, v. 10, p. 618—621, pls. 30—32, 1968.

Chadefaud, M., Les cellules nageuses des algues dans l'embranchement des Chlorophycées. *C. R. Hebd. Séanc. Acad. Sci., Paris*, v. 231, p. 989—990, figs. 1—7, 1950.

Chadefaud, M., Les végétaux non vasculaires (Cryptogamie), Tome 1, 1016 p., 706 figs. *In* M. Chadefaud and L. Emberger, *Traité de botanique systématique.* Paris: Masson et Cie., 1960.

Chandra, A., A note on the occurrence of *Botryococcus* in the Miocene lignites of Kerala. *Curr. Sci.*, v. 33, p. 214—215, 7 figs., 1964.

Chapman, V. J., The Chlorophyta. *Oceanogr. Mar. Biol. Ann. Rev.*, v. 2, p. 193—228, 1964.

Chiarugi, Alberto, *Palaeocodium saharianum* n. gen., n. sp., nuova Codiacea Paleozoica del deserto Libico. *Palaeontogr. Ital.*, v. 41, n.s., v. 11 (1942—1943), p. 121—130, pl. 9, figs. 1, 2, 1947.

Chihara, M., The life history of *Prasinocladus ascus* as found in Japan, with special reference to the systematic position of the genus. *Phycologia*, v. 3, p. 19—28, figs. 1—3, 1963.

Christensen, T., *Botanik*, Bind 2. *Systematisk botanik.* Nr. 2. *Alger.* København: Munksgaard, 178 p., 75 figs., 1962.

Christopher, R. A., Morphology and taxonomic status of *Pseudoschizaea* Thiergart and Frantz ex R. Potonié emend. *Micropaleontology*, v. 22, p. 143—150, pl. 1, fig. 1, 1976.

Churchill, D. M., Living and fossil unicellular algae and aplanospores. *Nature*, v. 186, p. 493—494, 1 fig., 1960.

Clifton, H. E., C. V. W. Mahnken, J. C. van Derwalker, and R. A. Waller, Tektite 1, Man-in-the-sea project: marine science program. *Science*, v. 168, p. 659—663, figs. 1—3, 1970.

Colom, G., Sobre dos algas clorofíceas fósiles de las "Falsas Brechas" titónicas de los Alpides Españoles, La "*Globochaete alpina*" Lombard y "*Eothrix alpina*" Lombard. *Boln. Inst. Geol. Min. Esp.*, ser. 4, v. 61, p. 57—78, pls. 1—6, figs. 1—3, 1948.

Colom, G., Jurassic-Cretaceous pelagic sediments of the western Mediterranean zone and the Atlantic area. *Micropaleontology*, v. 1, p. 109—124, pls. 1—5, figs. 1—4, 1955.

Combaz, A., Sur un nouveau type de microplanctonte cénobial fossile du Gothlandien de Libye, *Deflandras-trum* nov. gen. *C. R. Hebd. Séanc. Acad. Sci., Paris*, v. 255, p. 1977—1979, figs. 1—7, 1962.

Combaz, A., Remarques sur les niveaux a Tasmanacées du Paléozoïque Saharien. *Palaeobotanist*, v. 15, p. 29—34, pls. 1, 2, 1966.

Combaz, A., Leiosphaeridaceae Eisenack, 1954, et Protoleiosphaeridae Timofeev, 1959—leurs affinités, leur rôles sédimentologique et géologique. *Rev. Palaeobot. Palynol.*, v. 1, p. 309—321, pls. 1—4, 1967.

Conrad, W., Flagellates des Iles de la Sonde (Euglénacéées). *Bull. Mus. R. Hist. Nat. Belg.*, v. 14 (8), p. 1—20, figs. 1—82, 1938.

Cookson, I. C., Records of the occurrence of *Botryococcus braunii, Pediastrum* and the Hystrichosphaeridae in Cainozoic deposits of Australia. *Mem. Natn. Mus. Vict.*, no. 18, p. 107—123, 2 pls., 1953.

Cookson, I. C., Cretaceous and Tertiary microplankton from south-eastern Australia. *Proc. R. Soc. Vict.*, v. 78, p. 85—93, pls. 9—11, 1965.

Cookson, I. C., and M. E. Dettmann, On *Schizosporis*, a new form genus from Australian Cretaceous deposits. *Micropaleontology*, v. 5, p. 213—216, 1 pl., 1959.

Cookson, I. C., and Alfred Eisenack, Die Familie der Lecaniellaceae n. fam.: Fossile Chlorophyta, Volvocales? *Neues Jb. Geol. Paläont. Mh.*, 1970, p. 321—325, figs. 1, 2, 1970.

Cookson, I. C., and Alfred Eisenack, Mikroplankton aus australischen mesozoischen und tertiären Sedimenten. *Palaeontographica Abt. B*, v. 148, p. 44—93, pls. 20—29, figs. 1—3, 1974.

Cookson, I. C., and S. Manum, On *Crassosphaera*, a new genus of microfossils from Mesozoic and Tertiary deposits. *Nytt Mag. Bot.*, v. 8, p. 5—9, pls. 1, 2, 1960.

Cramer, F. H., Microplankton from three Palaeozoic formations in the Province of León, NW-Spain. *Leid. Geol. Meded.*, v. 30, p. 253—361, pls. 1—24, figs. 1—56, 1964.

Crescenti, U., *Praerhapydionina murgiana* n. sp. (Foraminifero) e *Neomacroporella cretacica* n. gen., n. sp. (alga calcarea—Dasicladacea), nuovi microfossili del cretacico dell'Italia meridionale. *Boll. Soc. Geol. Ital.*, v. 83, fasc. 1, p. 5—15, pls. 1, 2, 1964.

Cros, P., and M. Lemoine, Dasycladacées nouvelles ou peu connues du Lias inférieur des Dolomites et de quelques autres régions Méditerranéennes. *Rev. Micropaléont.*, v. 9, p. 156—168, pls. 1, 2, text-figs. 1—10, 1966.

Currie, E. D., and W. N. Edwards, Dasycladaceous algae from the Girvan area. *Q. Jl. Geol. Soc. Lond.*, v. 98, p. 235—240, pl. 11, 1943.

Dangeard, L., Recifs et galets d'algues dans l'oölithe ferrugineuse de Normandie. *C. R. Hebd. Séanc. Acad. Sci., Paris*, v. 190, p. 66—68, 1930.

Dangeard, L., Les argiles noires éocènes de la Forét de la Londe (feuille géologique de Lisieux) contiennent des algues appartenant au genre *Botryococcus. C. R. Hebd. Séanc. Acad. Sci., Paris*, v. 201, p. 94–95, 1 fig., 1935.

Darden, W. H., Jr., Sexual differentiation in *Volvox aureus. J. Protozool.*, v. 13, p. 239–255, figs. 1–49, 1966.

Davis, C. A., On the fossil algae of the petroleum-yielding shales of the Green River Formation of Colorado and Utah. *Proc. Natn. Acad. Sci. U.S.A.*, v. 2, p. 114–119, 1916.

Davis, J. S., Resting cells of *Pediastrum. Am. J. Bot.*, v. 49, p. 478–481, figs. 1–7, 1962.

Dawson, E. Y., *Marine botany, an introduction.* New York: Holt, Rinehart and Winston, Inc., xii + 371 p., figs., 1966.

Deflandre, Georges, Monographie du genre *Trachelomonas* Ehr. *Revue Gén. Bot.*, v. 38, p. 358–380, 449–469, 518–528, 580–592, 646–658, 687–706, 1926.

Deflandre, Georges, Monographie du genre *Trachelomonas* Ehr. (suite et fin). *Revue Gén. Bot.*, v. 39, p. 73–98, pls. 15–29, 1927.

Deflandre, Georges, Sur l'abus de l'emploi, en paléontologie, du nom de genre *Trachelomonas* et sur la nature de quelques ex *"Trachelomonas"* siliceux (Chrysomonadines) tertiaires et quaternaires. *Annls. Protist.*, v. 4, p. 151–165, 10 figs., 1934.

Deflandre, Georges, Systématique des Hystrichosphaeridés: sur l'acception du genre *Cymatiosphaera* O. Wetzel. *C. R. Somm. Séanc. Soc. Géol. Fr.*, 1954, p. 257–258, 1954.

Deflandre, Georges, Sur la conservation de vestiges pyritisés de *Deflandrastrum* (Chlorophycées) et sur une curieuse forme nouvelle du Siluro-Dévonien d'Afrique du Nord. *C. R. Hebd. Séanc. Acad. Sci., Paris*, v. 265, p. 1776–1779, 1 pl., 1967.

Deflandre, Georges, and A. Lenoble, Sur la présence d'eugleniens fossiles du genre *Trachelomonas* Ehr. dans un schiste pliocène de Madagascar. *C. R. Hebd. Séanc. Acad. Sci., Paris*, v. 226, p. 509–511, 8 figs., 1948.

Deunff, Jean, Le genre *Duvernaysphaera* Staplin. *Grana Palynol.*, v. 5, p. 210–215, figs. 1–7, 1964.

Deunff, Jean, Le genre *Polyedryxium. Commission Internationale de Microflore du Paléozoique (C.I.M.P.), Microfossiles organiques du Paléozoique*, Fasc. 3, *Les Acritarches.* Edn. Cent. Natn. Rech. Scient., Paris, p. 7–49, pls. 1–8, 1971.

Dilcher, D. L., *Phycopeltis*—an alga epiphytic on Eocene leaves. *Am. J. Bot.*, v. 49, p. 699 [abstract], 1962.

Dodge, J. D., and R. M. Crawford, Fine structure of the dinoflagellate *Oxyrrhis marina* I. The general

structure of the cell. *Protistologica*, v. 7, p. 295–304, figs. 1–18, 1971.

Dragastan, O., Alge calcaroase in Jurasicul superior şi Cretacicul inferior din Munţii Apuseni [Algues calcaires du Jurassique Supérieur et du Crétacé Inférieur des Monts Apuseni]. *Studii Cerc. Geol.*, v. 12, p. 441–454, pls. 1–8, 1967.

Dragastan, O., Algues calcaires du Jurassique Supérieur et du Crétacé Inférieur de Roumanie. *Revue Micropaléont.*, v. 12, p. 53–62, pls. 1–3, 1969.

Dragastan, O., New algae in the Upper Jurassic and Lower Cretaceous in the Bicaz Valley East Carpathians (Romania). *Revta. Esp. Micropaleont.*, v. 3, p. 155–192, pls. 1–12, 1971.

Dragastan, O., and Milan Mišík, *Verticillodesmis clavaeformis* nov. gen. nov. sp. in the Upper Jurassic of Czorsztyn Series—Klippen Belt (Czechoslovakia). *Revta. Esp. Micropaleont.*, v. 7, p. 215–220, pl. 1, fig. 1, 1975.

Durand Delga, M., Repartition stratigraphique de certains microorganismes (*Globochaete, Eothrix, ...*) définis dans le Malm Mésogéen. *Bull. Serv. Carte Géol. Algér*, v. 8, p. 143–153, 5 pls., 5 figs., 1956.

Duthie, H. C., The survival of desmids in ice. *Br. Phycol. Bull.*, v. 2, p. 376–377, 1 fig., 1964.

Dwivedi, J. N., Fossil thallophytes from Mohgaon-Kalan locality, Chhindwara District, M.P. *Curr. Sci.*, v. 28, p. 285–286, figs. 1–4, 1959a.

Dwivedi, J. N., Occurrence of fossil alga, probably *Ulothrix*, in Tertiary beds of Mohgaon-Kalan locality (M.P.), India. *Curr. Sci.*, v. 28, p. 456–457, figs. 1, 2, 1959b.

Egerod, L. E., An analysis of the siphonous Chlorophycophyta with special reference to the Siphonocladales, Siphonales, and Dasycladales of Hawaii. *Univ. Calif. Publs. Bot.*, v. 25, p. 325–454, pls. 29–42, 23 figs., 1952.

Eisenack, A., Chitinozoen, Hystrichosphären und andere Mikrofossilien aus dem *Beyrichia* Kalk. *Senckenberg. Leth.*, v. 36, p. 157–188, pls. 1–5, 1955.

Eisenack, A., Probleme der Vermehrung und des Lebensraumes bei der Gattung *Leiosphaera* (Hystrichosphaeridea). *Neues Jb. Geol. Paläont. Abh.*, v. 102, p. 402–408, pl. 16, figs. 1, 2, 1956.

Eisenack, A., Mikrofossilien in organischer Substanz aus dem Lias Schwabens (Suddeutschland). *Neues Jb. Geol. Paläont. Abh.*, v. 105, p. 239–249, pls. 19, 20, 2 figs., 1957.

Eisenack, A., *Tasmanites* Newton 1875 und *Leiosphaeridia* n. gen. als Gattungen der Hystrichosphaeridea. *Palaeontographica Abt. A*, v. 110, p. 1–24, 12 pls., 9 figs., 1958.

Eisenack, A., Mitteilungen über Leiosphären und über das Pylom bei Hystrichosphären. *Neues Jb. Geol. Paläont. Abh.*, v. 114, p. 58–80, pls. 1–3, figs. 1, 2, 1962.

Eisenack, A., Über einige Arten der Gattung *Tasmanites* Newton, 1875. *Grana Palynol.*, v. 4, p. 203–216, 1 pl., 1963.

Eisenack, A., Über *Chuaria wimani* Brotzen. *Neues Jb. Geol. Paläont. Mh.*, p. 52–56, 2 figs., 1966.

Eisenack, A., Über die Fortpflanzung paläozoischer Hystrichosphären. *Neues Jb. Geol. Paläont. Abh.*, v. 131, p. 1–22, pls. 1–3, 1 fig., 1968.

Elliott, G. F., New calcareous algae from the Arabian Peninsula. *Micropaleontology*, v. 3, p. 227–230, 1 pl., 1957.

Elliott, G. F., The interrelationships of some Cretaceous Codiaceae (calcareous algae). *Palaeontology*, v. 8, p. 199–203, pls. 23, 24, 1965.

Elliott, G. F., Three new Tethyan Dasycladaceae (calcareous algae). *Palaeontology*, v. 11, p. 491–497, pls. 93–95, 1968a.

Elliott, G. F., Permian to Palaeocene calcareous algae (Dasycladaceae) of the Middle East. *Bull. Br. Mus. Nat. Hist., Geology*, suppl. 4, 111 p., 24 pls., 16 figs., 1968b.

Elliott, G. F., *Pseudocymopolia*, a Mesozoic Tethyan alga (Family Dasycladaceae). *Palaeontology*, v. 13, p. 323–326, pl. 60, 1970a.

Elliott, G. F., New and little-known Permian and Cretaceous Codiaceae (calcareous algae) from the Middle East. *Palaeontology*, v. 13, p. 327–333, pls. 61, 62, 1970b.

Elliott, G. F., Calcareous algae new to the British Carboniferous. *Palaeontology*, v. 13, p. 443–450, pls. 81–83, fig. 1, 1970c.

Elliott, G. F., A new fossil alga from the English Silurian. *Palaeontology*, v. 14, p. 637–641, pls. 120, 121, fig. 1, 1971.

Elliott, G. F., Lower Palaeozoic green algae from southern Scotland, and their evolutionary significance. *Bull. Br. Mus. Nat. Hist., Geology*, v. 22, p. 355–376, pls. 1–10, 1 fig., 1972.

Elliott, G. F., A consideration of the Tribe Thyrsoporelleae, dasyclad algae. *Palaeontology*, v. 20, p. 705–714, fig. 1, 1977.

Elliott, G. F., and Peter Süssli, *Imperiella* gen. nov., a new alga from the Ruteh Limestone, Upper Permian (Central Alborz Mountains, North Iran). *Eclog. Geol. Helv.*, v. 68, p. 449–455, pl. 1, figs. 1–3, 1975.

Emberger, Louis, *Les plantes fossiles dans leurs rapports avec les végétaux vivants*, 2nd ed. Paris: Masson & Cie., 758 p., 743 figs., 1968.

Emig, W. H., The travertine deposits of the Arbuckle Mountains, Oklahoma, with reference to the plant agencies concerned in their formation. *Bull. Okla. Geol. Surv.*, v. 29, p. 1–76, figs. 1–4, 1917.

Endo, R., Stratigraphical and paleontological studies of the later Paleozoic calcareous algae in Japan, XIII—a restudy of the genus *Physoporella. Trans. Proc.*

Palaeont. Soc. Japan, n.s., no. 31, p. 265–269, pl. 39, 1958.

Erdtman, G., New methods in pollen analysis. *Svensk Bot. Tidskr.*, v. 30, p. 154–164, figs. 1–21, 1936.

Evitt, W. R., A discussion and proposals concerning fossil dinoflagellates, hystrichospheres, and acritarchs, I and II. *Proc. Natn. Acad. Sci. U.S.A.*, v. 49, p. 158–164, 298–302, 1963a.

Evitt, W. R., Occurrence of freshwater alga *Pediastrum* in Cretaceous marine sediments. *Am. J. Sci.*, v. 261, p. 890–893, pl. 1, 1963b.

Fairchild, T. R., J. W. Schopf, and R. L. Folk, Filamentous algal microfossils from the Caballos Novaculite, Devonian of Texas. *J. Paleont.*, v. 47, p. 946–952, pls. 1, 2, 1973.

Farinacci, Anna, L'"Organismo C" Favre 1927 appartiene alle Teredinidae? *Geologica Romana*, v. 2, p. 151–176, pls. 1–5, figs. 1–6, 1963.

Favre, J., and A. Richard, Etude du Jurassique supérieur de Pierre-Chatel et de la Cluse de la Balme (Jura meridionale). *Mém. Soc. Paléont. Suisse*, v. 46, p. 1–39, 3 pls., 14 figs., 1927.

Feldmann, Jean, Jean-Paul Loreau, Louis Codomier, and Alain Couté, Morphologie et ultrastructure du squelette des thalles calcifiés de *Pedobesia* (ex *Derbesia*) *lamourouxii* (J. Ag.) comb. nov. *C. R. Hebd. Séanc. Acad. Sci., Paris*, v. 280, ser. D, p. 2641–2644, pls. 1–4, 1975.

Felix, C. J., Neogene *Tasmanites* and leiospheres from southern Louisiana, U.S.A. *Palaeontology*, v. 8, p. 16–26, pls. 5–8, 1965.

Feys, R., C. Greber, and M. Pascal, A propos de l'ancienneté de la flore continentale: découverte de "charbons" et de "phytomorphes" dans le Francevillien (Précambrien du Gabon). *Bull. Soc. Géol. Fr.*, sér. 7, v. 8 (1966), p. 638–641, pls. 15–16, 1967.

Fiore, M., Di un'alga fossile nuova per la "pesciara" di Bolca. *Rc. Accad. Sci. Fis. Mat., Napoli*, ser. 4, v. 5, p. 137–138, 1936.

Fischer, A., Ueber das Vorkommen von Gypskrystallen bei den Desmidieen. *Jb. Wiss. Bot.*, v. 14, p. 133–184, pls. 9, 10, 1883.

Fischer, J.-C., and Jacques Thierry, Révision de quelques Dasycladacées jurassiques et proposition d'un nouveau genre: *Coniporella. Bull. Mus. Natn. Hist. Nat., Paris*, ser. 3, no. 19, Sci. Terre 3, p. 25–34, figs. 1–7, 1971.

Fogg, G. E., *Algal cultures and phytoplankton ecology.* Madison: Univ. Wisconsin Press, xiii + 126 p., 4 pls., 31 figs., 1965.

Ford, T. D., and W. J. Breed, The problematical Precambrian fossil *Chuaria. Palaeontology*, v. 16, p. 535–550, pls. 61–63, 1973.

Frémy, P., and L. Dangeard, Observations sur le *Bot-*

ryococcus braunii Kützing actuel et fossile. *Annls. Paléont.,* v. 27, p. 115—136, 2 pls., 1938.

Friedmann, I., Y. Lipkin, and R. Ocampo-Paus, Desert algae of the Negev (Israel). *Phycologia,* v. 6, p. 185—200, figs. 1—17, 1967.

Fritsch, F. E., *The structure and reproduction of the algae,* vol. I. Cambridge: University Press, xvii + 791 p., 245 figs., 1935.

Garwood, E. J., The Tuedian beds of northern Cumberland and Roxburghshire east of the Liddel Water. *Q. Jl. Geol. Soc. Lond.,* v. 87, p. 97—156, pls. 7—16, 1931.

Génot, P., and A. F. Poignant, Contribution a l'étude, au microscope électronique à balayage, des dasycladacées de l'Éocène du Bassin de Paris. *Revue Micropaléont.,* v. 17, p. 66—74, pls. 1—3, figs. 1—3, 1975.

Gocht, Hans, and Wolfgang Wille, Untersuchungen an *Palambages morulosa* O. Wetzel (Chlorophyceae inc. sed.). *Neues Jb. Geol. Paläont. Mh.,* Jg. 1972, p. 146—161, figs. 1—24, 1972.

Goreau, T. F., Calcium carbonate deposition by coralline algae and corals in relation to their roles as reef builders. *Ann. N.Y. Acad. Sci.,* v. 109, p. 127—167, figs. 1—9, 1963.

Górka, Hanna, Les microfossiles du Jurassique Supérieur de Magnuszew (Pologne). *Acta Palaeont. Pol.,* v. 10, p. 291—334, pls. 1—5, 1965.

Gray, J., Fossil chlorophycean algae from the Miocene of Oregon. *J. Paleont.,* v. 34, p. 453—463, pl. 64, 2 figs., 1960.

Green, B. R., and Hugh Burton, *Acetabularia* chloroplast DNA: electron microscopic visualization. *Science,* v. 168, p. 981—982, figs. 1—3, 1970.

Grönblad, R., and J. Růžička, Zur Systematik der Desmidiaceen. *Bot. Notiser,* v. 112, p. 205—226, figs. 1—31, 1959.

Grüss, Johannes, Zur Biologie Devonischer Thallophyten. *Palaeobiologica,* v. 1, p. 487—517, pls. 39—41, 1928.

Güvenç, T., Représentants des Bereselleae (algues calcaires) dans le Carbonifère de Turquie et description d'un nouveau genre: *Goksuella* n.g. *Bull. Soc. Géol. Fr.,* sér. 7, v. 7 (1965), p. 843—850, pl. 32, 1966.

Gušić, I., The algal genera *Macroporella, Salpingoporella* and *Pianella* (Dasycladaceae). *Taxon,* v. 19, p. 257—261, 1970.

Habib, Daniel, Middle Cretaceous palynomorph assemblages from clays near the Horizon Beta deep-sea outcrop. *Micropaleontology,* v. 16, p. 345—379, pls. 1—10, 1970.

Häntzschel, W., Trace fossils and problematica, p. W177—W245, figs. 109—149. *In* R. C. Moore, *Treatise on invertebrate paleontology,* Pt. W. Lawrence, Kansas: Univ. Kans. Press and Geol. Soc. Am., 1962.

Hemer, D. O., and P. W. Nygreen, Algae, acritarchs and other microfossils incertae sedis from the Lower Carboniferous of Saudi Arabia. *Micropaleontology,* v. 13, p. 183—194, pls. 1—3, fig. 1, 1967.

Herak, M., Comparative study of some Triassic Dasycladaceae in Yugoslavia. *Geološki Vjesn.,* v. 18 (1964), p. 3—34, pls. 1—15, 1965.

Howe, M. A., *Chlorotylites,* a fossil green alga from Alabama. *Bull. Torrey Bot. Club,* v. 59, p. 219—220, pl. 15, 1932.

Hudson, J. D., Algal limestones with pseudomorphs after gypsum from the Middle Jurassic of Scotland. *Lethaia,* v. 3, p. 11—40, figs. 1—11, 1970.

Hurka, H., Über den anatomischen Bau und die systematische Stellung des paläozoischen Algengenus *Palaeoporella* Stolley. *Nova Hedwigia,* v. 15, p. 571—582, pl. 83, figs. 1—4, 1968.

Hurka, H., Umbildungstendenzen der Astformen in *Physoporella-Oligoporella*-Populationen (Dasycladaceen) aus dem Anis der Pragser Dolomiten (Italien). *Neues Jb. Geol. Paläont. Mh.,* p. 104—127, figs. 1—7, 1969.

Jaffrezo, Michel, Essai d'inventaire bibliographique des algues Dasycladacées du Jurassique et du Crétacé Inférieur. *Geobios,* no. 6, p. 71—99, 1973.

Johnson, J. H., Fossil calcareous algae from Bikini Atoll. *Prof. Pap. U.S. Geol. Surv.,* 260-M, p. 537—545, pls. 188—197, 1954a.

Johnson, J. H., Cretaceous Dasycladaceae from Gillespie County, Texas. *J. Paleont.,* v. 28, p. 787—790, pl. 93, 1954b.

Johnson, J. H., Calcareous algae. *In* Geology of Saipan Mariana Islands, Pt. 3. Paleontology. *Prof. Pap. U.S. Geol. Surv.,* 280-E, p. 209—246, pls. 36—60. 1957.

Johnson, J. H., *Limestone-building algae and algal limestones.* Boulder, Colo.: Johnson Publishing Co., 297 p., 139 pls., 1961.

Johnson, J. H., Lower Devonian algae and encrusting foraminifera from New South Wales. *J. Paleont.,* v. 38, p. 98—108, pls. 25—29, 1964.

Johnson, J. H., Lower Cretaceous algae from the Blake Escarpment, Atlantic Ocean, and from Israel. *Colo. Sch. Mines Prof. Contr.,* 5, vi + 46 p., 7 pls., 3 figs., 1968.

Johnson, J. H., and K. Konishi, Studies of Mississippian algae I. A review of Mississippian algae. *Colo. Sch. Mines Q.,* v. 51, p. 1—84, pls. 1—24, figs. 1—22, 1956a.

Johnson, J. H., and K. Konishi, Studies of Mississippian algae II. Mississippian algae from the western Canada Basin and Montana. *Colo. Sch. Mines Q.,* v. 51, p. 86—107, pls. 1—8, 1956b.

Jux, U., Über den Feinbau der Zystenwandung von *Pachysphaera marshalliae* Parke, 1966. *Palaeontographica Abt. B,* v. 125, p. 104—111, pls. 20—22, 1 textfig., 1969a.

Jux, U., Über den Feinbau der Zystenwandung von *Halosphaera* Schmitz, 1878. *Palaeontographica Abt. B.,* v. 128, p. 48−55, pls. 27−30, 1 fig., 1969b.

Kaever, M., Neue Dasycladaceen—*Afghanopolia fragilis* n. gen., n. sp. und *Cymopolia (Polytripa) paktia* n. sp.—aus dem Mittel-Eozän von Ost-Afghanistan. *Argumenta Palaeobot.,* v. 3, p. 15−42, pls. 9, 10, figs. 1−8, 1969.

Kamptner, E., Über das System und die Stammesgeschichte der Dasycladaceen (Siphoneae verticillatae). *Annln. Naturh. Mus. Wien,* v. 62, p. 95−122, 1958.

Kázmierczak, Józef, Colonial Volvocales (Chlorophyta) from the Upper Devonian of Poland and their palaeoenvironmental significance. *Acta Palaeont. Pol.,* v. 20, p. 73−85, pls. 17−20, 1 fig., 1975.

Kázmierczak, Józef, Devonian and modern relatives of the Precambrian *Eosphaera:* possible significance for the early eukaryotes. *Lethaia,* v. 9, p. 39−50, figs. 1−6, 1976a.

Kázmierczak, Józef, Volvocacean nature of some Paleozoic nonradiosphaerid calcispheres and parathuramminid "Foraminifera." *Acta Palaeont. Pol.,* v. 21, p. 245−258, pls. 19−22, fig. 1, 1976b.

Kedves, M., *Noremia,* a new microfossil genus from the Hungarian Eocene, and systematical and stratigraphical problems about the Crassosphaeridae. *Acta Miner. Petrogr., Szeged,* v. 15, p. 19−27, pls. 1, 2, 1962.

Kesling, R., and A. Graham, *Ischadites* is a dasycladacean alga. *J. Paleont.,* v. 36, p. 943−952, pls. 135, 136, 2 figs., 1962.

Khvorova, L. V., A new genus of Dasycladaceae from the Middle Carboniferous of the Moscow Basin. *Dokl. Akad. Nauk SSSR,* v. 65, p. 749−752, 3 figs., 1949.

Kjellström, G., Remarks on the chemistry and ultrastructure of the cell wall of some Paleozoic leiospheres. *Geol. För. Stockh. Förh.,* v. 90, p. 221−228, figs. 1−8, 1968.

Klebs, Georg, Ueber die Organisation einiger Flagellaten-Gruppen und ihre Beziehungen zu Algen und Infusorien. *Unters. Bot. Inst. Tübingen,* v. 1, p. 233−361, pls. 2, 3, 1883.

Knauer, J., A *Lombardia* kérdés [Sur le problème du genre *Lombardia*]. *Földt. Közl.,* v. 96, p. 195−199, 1966.

Kochansky, V., *Velebitella,* eine neue jungpaläozoische Diploporeengattung und ihre phylogenetischen Verhältnisse. *Geološki Vjesn.,* v. 17 (1963), p. 135−142, pls. 1−4, 1 fig., 1964.

Kochansky, V., and M. Herak, On the Carboniferous and Permian Dasycladaceae of Yugoslavia. *Geološki Vjesn.,* v. 13 (1959), p. 65−94, pls. 1−9, figs. 1−7, 1960.

Konishi, K., Studies of Mississippian algae III. *Anatolipora,* a new dasycladacean genus, and its algal associates from the Lower Carboniferous of Japan. *Colo. Sch. Mines Q.,* v. 51, p. 110−127, pls. 1−3, 1 fig., 1956.

Konishi, K., A prominent marine floral change during the Permo-Carboniferous. *Int. Geol. Congr. 21,* pt. 22, p. 36−38, 1960.

Konishi, K., Studies of Paleozoic Codiaceae and allied algae, Part I: Codiaceae (excluding systematic descriptions). *Sci. Rep. Kanazawa Univ.,* v. 7, p. 159−261, figs. 1−39, 1961.

Konishi, K., and R. C. Epis, Some early Cretaceous calcareous algae from Cochise County, Arizona. *Micropaleontology,* v. 8, p. 67−76, pl. 1, 1962.

Konzalová, Magda, Algal colony and rests of other microorganisms in the Bohemian Upper Proterozoic. *Věst. Ústřed. Úst. Geol.,* v. 48, p. 31−33, pls. 1, 2, 1 fig., 1973.

Konzalová, Magda, and Jan Mašek, Tufit se zbytky zelených řas rodu *Pila* ve středočeském karbonu (Westfal C) [Tuffite with remains of green algae in the Carboniferous of Central Bohemia (Westphalian C)]. *Věst. Ústřed. Úst. Geol.,* v. 47, p. 23−28, pls. 1, 2, figs. 1−3, 1972.

Kopetzky-Rechtperg, O., Die "Zersetzungskörperchen" der Desmidiaceenzelle. *Arch. Protistenk.,* v. 75, p. 270−283, pl. 17, 1931.

Korde, K. B., Vodorosli Kembriya yugo-vostoka Sibirskoy Platformy [Cambrian algae of the southeast Siberian Platform]. *Trudy Paleont. Inst.,* v. 89, p. 1−147, pls. 1−28, figs. 1−30, 1961.

Korde, K. B., Novoe rodovoe nazvanie *Jacutiella* Korde, nom. nov. [New generic name *Jacutiella* Korde, nom. nov.]. *Paleont. Zh.,* 1964, no. 2, p. 162, 1964.

Korde, K. B., K sistematike i evolyutsii vodorosley iz poryadka Dasycladales (Chlorophyta) [On the systematics and evolution of algae of the Order Dasycladales (Chlorophyta)]. *Byull. Mosk. Obshch. Ispyt. Prir.,* n.s., v. 76, Otd. Geol., v. 46(2), p. 134−135, 1971.

Korde, K. B., E. L. Kulik, V. P. Maslov, and T. A. Moskalenko, Tip Chlorophyta. Zelenye vodorosli [Division Chlorophyta. Green Algae], p. 198−223, figs. 1−60. *In* Yu. Orlov, *Osnovy paleontologii,* v. 14, *Vodorosli, mokhoobraznye, psilofitovye, plaunovidnye, chlenistostebel'nye, Paporotniki.* Moscow: Akad. Nauk SSSR, 1963.

Koriba, K., and M. Shigeru, On *Paleodictyon* and fossil *Hydrodictyon. Jubilee Publ. Commem. Prof. H. Yabe, M.I.A., Sixtieth Birthday,* v. 1, p. 55−68, pls. 4−5, figs. 1−7, 1939.

Kornmann, P., Codiolophyceae, a new class of Chlorophyta. *Helgoländer Wiss. Meeresunters.,* v. 25, p. 1−13, figs. 1−9, 1973.

Kreger, D. R., Cell walls, p. 315−335, figs. 1−3. *In* R. A. Lewin, *Physiology and biochemistry of algae.* New York: Academic Press, 1962.

Kriván Hutter, E., Microplankton from the Palaeogene of the Dorog Basin 2. *Annls. Univ. Scient. Bpest. Rolando Eőtvős, Geol.,* v. 6, p. 71−79, pls. 1−6, 1963.

Kühnel, J., *Udotea adnetensis* nov. spec., eine Alge des Rhät aus der Familie der Codiaceen. *Neues Jb. Miner. Geol. Paläont. BeilBd., Abt. B,* v. 69, p. 347–352, pls. 18, 19, 1932.

Kulik, E. L., Berezellidy Karbona Russkoy Platformy. *Paleont. Zh.,* 1964, no. 2, p. 99–114, pl. 8, figs. 1–3, 1964.

Lagerheim, G., Untersuchungen über fossile Algen, I, II. *Geol. För. Stockh. Förh.,* v. 24, p. 475–500, figs. 1–6, 1902.

Laporte, L. F., Codiacean algae and algal stromatolites of the Manlius Limestone (Devonian) of New York. *J. Paleont.,* v. 37, p. 643–647, 1 pl., 2 figs., 1963.

Laubenfels, M. W. de, Family Receptaculitidae Eichwald, 1860, p. E108–E110, fig. 89. *In* R. C. Moore, *Treatise on invertebrate paleontology,* Pt. E., Archaeocyatha and Porifera. Lawrence: Univ. Kansas Press and Geol. Soc. Am., 1955.

Lebouché, M. C., and M. Lemoine, Dasycladacées nouvelles du Lias calcaire (Lotharingien) du Languedoc Méditerranéen (St. Chinian, Boutenac). *Revue Micropaléont.,* v. 6, p. 89–101, pls. 1–3, figs. 1–7, 1963.

Lee, J. J., L. J. Crockett, J. Hagen, and R. J. Stone, The taxonomic identity and physiological ecology of *Chlamydomonas hedleyi* sp. nov., algal flagellate symbiont from the foraminifer *Archaias angulatus. Br. Phycol. J.,* v. 9, p. 407–422, figs. 1–9, 1974.

Leedale, G. F., *Euglenoid flagellates.* Englewood Cliffs: Prentice-Hall, Inc., xiii + 242 p., 176 figs., 1967.

Leitgeb, H., Die Incrustation der Membran von *Acetabularia. Sber. Akad. Wiss. Wien, Math.-Naturw. Kl.,* v. 96, p. 13–37, 1888.

Levring, T., Submarine illumination and vertical distribution of algal vegetation. *Recent Advances Bot.,* v. 1, p. 189–193, Univ. Toronto Press, 1961.

Lévy, J., *Neomizzia* (Dasycladacée) nouveau genre du Lias du Maroc. *Revue Micropaléont.,* v. 9, p. 37–39, pl. 1, 1966.

Lewin, J. C., Calcification, p. 457–465. *In* R. A. Lewin, *Physiology and biochemistry of algae.* New York: Academic Press, 1962.

Lewin, R. A., Extracellular polysaccharides of green algae. *Can. J. Microbiol.,* v. 2, p. 665–672, fig. 1, 1956.

Lewin, R. A., The cell walls of *Platymonas. J. Gen. Microbiol.,* v. 19, p. 87–90, pl. 1, 1958.

Lignier, O., Sur une algue oxfordienne (*Gloeocystis oxfordiensis* n. sp.). *Bull. Soc. Bot. Fr.,* v. 53, p. 527–530, fig. 1, 1906.

Liskum, I. G., and V. P. Maslov, Novaya izvestkovaya presnovodnaya mutovchataya sifoneya iz chetvertichnykh asadkov Altaya [New calcareous freshwater whorled siphonates from the Quaternary sediments of Altay]. *Dokl. Akad. Nauk SSSR,* v. 156, p. 1368–1370, 1 fig., 1964.

Loeblich, A. R., Jr., Protistan phylogeny as indicated by the fossil record. *Taxon,* v. 23, p. 277–290, 2 figs., 1974.

Loeblich, A. R., Jr., and Helen Tappan, Some new and revised organic-walled phytoplankton microfossil genera. *J. Paleont.,* v. 50, p. 301–308, 1976.

Loeblich, A. R., Jr., and E. R. Wicander, Organic-walled microplankton from the Lower Devonian Late Gedinnian Haragan and Bois d'Arc Formations of Oklahoma, U.S.A., Part I. *Palaeontographica Abt. B,* v. 159, p. 1–39, pls. 1–12, 1976.

Lombard, A., Microfossiles d'attribution incertaine du Jurassique supérieur alpin. *Eclog. Geol. Helv.,* v. 30, p. 320–331, pls. 19, 20, 1 fig., 1937.

Lombard, A., Attribution de microfossiles du Jurassique supérieur alpin à des Chlorophycées (Proto- et Pleurococcacées). *Eclog. Geol. Helv.,* v. 38, p. 163–173, figs. 1–3, 1945.

Lowenstam, H. A., Aragonite needles secreted by algae and some sedimentary implications. *J. Sedim. Petrol.,* v. 25, p. 270–272, 1955.

Lucas, G., and T. Pobeguin, Microstructure d'une algue calcaire: *Dactylopora* (Dasycladacée tertiaire). Remarques sur les organismes aragonitiques et sur leur fossilisation. *Revue Gén. Bot.,* v. 61, p. 325–336, pls. 5–7, figs. 1–4, 1954.

McBride, G. E., Cytokinesis and ultrastructure in *Fritschiella tuberosa* Iyengar. *Arch. Protistenk.,* v. 112, p. 365–375, pls. 42–50, 1970.

MacRaild, G. N., and H. B. S. Womersley, The morphology and reproduction of *Derbesia clavaeformis* (J. Agardh) De Toni (Chlorophyta). *Phycologia,* v. 13, p. 83–93, figs. 1–15, 1974.

Mädler, K. A., Die figurierten organischen Bestandteile der Posidonienschiefer. *Beih. Geol. Jb.,* v. 58, p. 287–406, pls. 15–30, figs. 51–54, 1963.

Mädler, K. A., Hystrichophyta and acritarchs. *Revue Palaeobot. Palynol.,* v. 5, p. 285–290, 1967.

Mahadevan, C., and S. R. Sarma, Deccan, Intertrappean Series (Vicarabad, Hyderabad State). *Palaeobot. India,* v. 6, p. 260, 1947.

Mamet, B. L., Sur deux Dasycladacées Carbonifères des Cordillères Nord-Américaines. *Revue Micropaléont.,* v. 17 (1974), p. 38–44, pls. 1–3, 1 fig., 1975.

Mamet, B., and A. Roux, Sur quelques algues tubulaires scalariformes de la Téthys Paléozoïque. *Revue Micropaléont.,* v. 17 (1974), p. 134–156, pls. 1–7, figs. 1–8, 1975a.

Mamet, B., and A. Roux, Dasycladacées Dévoniennes et Carbonifères de la Téthys Occidentale. *Revta. Esp. Micropaleont.,* v. 7, p. 245–295, pls. 1–13, figs. 1–5, 1975b.

Mamet, B., and A. Roux, Algues Dévoniennes et Carbonifères de la Téthys Occidentale. Troisième partie. *Revue Micropaléont.,* v. 18, p. 134–187, pls. 1–15, figs. 1–4, 1975c.

Mamet, B., and A. Roux, *Jansaella ridingi,* nouveau genre d'Algue? dans le Dévonien de l'Alberta. *Can. J. Earth Sci.,* v. 12, p. 1480–1484, figs. 1–7, 1975d.

Mamet, B., and B. Rudloff, Algues Carbonifères de la partie septentrionale de l'Amerique du Nord. *Revue Micropaléont.,* v. 15, p. 75–114, pls. 1–10, figs. 1–3, 1972.

Manton, Irene, Some phyletic implications of flagellar structure in plants. *Adv. Bot. Res.,* v. 2, p. 1–34, figs. 1–11, 1965.

Manton, I., and M. Parke, Observations on the fine structure of two species of *Platymonas* with special reference to flagellar scales and the mode of origin of the theca. *J. Mar. Biol. Ass. U.K.,* v. 45, p. 743–754, pls. 1–12, 1965.

Marszalek, D. S., Calcisphere ultrastructure and skeletal aragonite from the alga *Acetabularia antillana. J. Sedim. Petrol.,* v. 45, p. 266–271, figs. 1–5, 1975.

Maslov, V. P., O novykh formakh Tretichnykh vodorosley [On new forms of Tertiary algae]. *Dokl. Akad. Nauk SSSR,* v. 103, p. 145–148, 1 fig., 1955.

Maslov, V. P., Iskopaemye izvestkovye vodorosli SSSR [Fossil calcareous algae of the USSR]. *Trudy Inst. Geol. Nauk, Mosk.,* v. 160, p. 3–301, pls. 1–86, 1956.

Maslov, V. P., Atsikulyarii i ikh znachenie dlya stratigrafii SSSR (Aciculariae and their importance for stratigraphy in the USSR]. *Paleont. Zh.,* 1960, no. 3, p. 115–122, figs. 1–16, 1960a.

Maslov, V. P., Novye vodorosli Mela Kopet-Daga (Turkmeniya) [New algae from the Cretaceous of Kopet-Dag (Turkmenia)]. *Dokl. Akad. Nauk SSSR,* v. 134, p. 939–941, figs. 1–3, 1960b.

Maslov, V. P., Kopetdagarii—Novaya triba mutovchatykh sifoney (zelenye vodorosli) [Kopetdagarieae—new tribe of whorled siphonates (green algae)]. *Dokl. Akad. Nauk SSSR,* v. 164, p. 1154–1157, figs. 1–3, 1965.

Maslov, V. P., and E. L. Kulik, Novaya triba vodorosley (Bereselleae) iz Karbona SSSR [New algal tribe (Bereselleae) from the Carboniferous of the USSR]. *Dokl. Akad. Nauk SSSR,* v. 106, p. 126–129, 1 fig., 1956.

Maslov, V. P., and M. V. Yartseva, Mutovchasti sifoney (Dasycladiaceae) z vidkladiv paleotsenu Ukrayns'kogo shchita [Whorled siphonates (Dasycladiaceae) from Paleocene sediments of the Ukrainian Shield]. *Ukr. Bot. Zh.,* v. 26, no. 5, p. 31–38, 4 figs., 1969.

Masse, J.-P., M. A. Conrad, and Rajka Radoičić, *Angioporella fouryae* n. gen., n. sp., une algue calcaire (Dasycladaceae) du Barrémien du sud de la France. *Eclog. Geol. Helv.,* v. 66, p. 383–387, pl. 1, fig. 1, 1973.

Massieux, M., Présence d'*Ovulites* dans le "Calcaire Ypresien" des Corbières Septentrionales et discussion sur la nature de l'algue *Griphoporella arabica* Pfender. *Revue Micropaléont.,* v. 8, p. 240–248, pls. 1, 2, 1966.

Mathur, K., Occurrence of *Pediastrum* in Subathu Formation (Eocene) of Himachal Pradesh, India. *Sci. Cult.,* v. 29, p. 250, 1963.

Mathur, K., On the occurrence of *Botryococcus* in Subathu beds of Himachal Pradesh, India. *Sci. Cult.,* v. 30, p. 607–608, 1964.

Messikommer, E., Beitrag zur Kenntniss der fossilen und subfossilen Desmidiaceen. *Hedwigia,* v. 78, p. 107–201, pls. 2–10, 1 fig., 1938.

Meyer, K. I., Über das phylogenetische System der grünen Algen (Chlorophycophyta). *Preslia,* v. 34, p. 147–158, 1962.

Milanović, M., *Salopekiella,* novi rod familije Dasycladaceae iz permskih sedimenata Velebita [*Salopekiella,* a new genus of the family Dasycladaceae from the Permian sediments of the Velebit Range]. *Prirodosl. Istraž.,* Kn. 35, *Acta Geol.* 5, p. 373–380, pls. 1–3, figs. 1–4, 1965a.

Milanović, M., Zwei neue Gattungen der Familie Dasycladaceae aus dem Perm des Velebitgebirges. *Bull. Scient. Cons. Acads. RPF Yougosl.,* Sec. A, v. 10, p. 179–180, 2 figs., 1965b.

Miller, J. D. A., Fats and steroids, p. 357–370, 1 fig. *In* R. A. Lewin, *Physiology and biochemistry of algae.* New York: Academic Press, 1962.

Millington, W. F., and S. R. Gawlik, Silica in the wall of *Pediastrum. Nature,* v. 216, p. 68, 1967.

Milon, Y., Sur la présence de *Cladophorites* dans le calcaire bartonien de Saint-Pierre-de-Chevillé (Sarthe). *C. R. Somm. Séanc. Soc. Géol. Fr.,* p. 113–114, 1932.

Moberly, R., Jr., Calcium carbonate from noncoralline algae. *Bull. Am. Ass. Petrol. Geol.,* v. 46, p. 273, 1962.

Moner, J. G., Cell wall structure in *Pediastrum* as revealed by electron microscopy. *Am. J. Bot.,* v. 42, p. 802–806, figs. 1–6, 1955.

Moner, J. G., and G. B. Chapman, The development of adult cell form in *Pediastrum biradiatum* Meyen as revealed by the electron microscope. *J. Ultrastruct. Res.,* v. 4, p. 26–42, figs. 1–13, 1960.

Moner, J. G., and G. B. Chapman, Cell wall formation in *Pediastrum biradiatum* as revealed by the electron microscope. *Am. J. Bot.,* v. 50, p. 992–998, figs. 1–15, 1963.

Morellet, L., Deux algues siphonées verticillées du Thanetien du Boncourt (Oise). *Bull. Soc. Géol. Fr.,* sér. 4, v. 8, p. 96–99, 1908.

Morellet, L., and J. Morellet, Les Dasycladacées du Tertiaire parisien. *Mém. Soc. Géol. Fr. Paléont.,* v. 21 (mém. 47), p. 1–43, pls. 1–3, figs. 1–24, 1913.

Morellet, L., and J. Morellet, Les Dasycladacées Tertiaires de Bretagne et du Cotentin. *Bull. Soc. Géol. Fr.,* sér. 4, v. 17, p. 362–372, pl. 14, figs. 1, 2, 1917.

Morellet, L., and J. Morellet, Contribution à l'étude paléontologique du genre *Halimeda* Lamx. (Algue siphonée de la famille des Codiacées). *Bull. Soc. Géol. Fr.,* sér. 4, v. 22, p. 291–296, pls. 11, 12, figs. 1, 2, 1922a.

Morellet, L., and J. Morellet, Nouvelle contribution à l'étude des Dasycladacées Tertiaires. *Mém. Soc. Géol. Fr. Paléont.,* v. 25 (mém. 58), p. 1–35, pls. 1, 2, (9, 10), 1922b.

Morellet, L., and J. Morellet, *Tertiary siphoneous algae in the W. K. Parker collection with descriptions of some Eocene Siphoneae from England.* London: Brit. Mus. (Nat. Hist.), 55 p., 6 pls., 7 figs., 1939.

Morellet, L., and J. Morellet, Études sur les algues calcaires de l'Éocène du Cotentin. *Bull. Soc. Géol. Fr.,* sér. 5, v. 10, p. 201–206, figs. 1, 2, 1940.

Morgenroth, Peter, Mikrofossilien und Konkretionen des Nord-westeuropäischen Untereozäns. *Palaeontographica Abt. B,* v. 119, p. 1–53, pls. 1–11, 1966.

Mosser, J. L., A. G. Mosser, and T. D. Brock, Photosynthesis in the snow: the alga *Chlamydomonas nivalis* (Chlorophyceae). *J. Phycol.,* v. 13, p. 22–27, figs. 1–3, 1977.

Müller, A. H., Über *Receptaculites* (Miscellanea, Receptaculitida N.). *Freiberger ForschHft.,* C 221, Paläont., p. 1–13, pls. 1–4, figs. 1–8, 1968.

Munier-Chalmas, E., Observations sur les algues calcaires confondues avec les foraminifères et appartenant au groupe des Siphonées dichotomes. *Bull. Soc. Géol. Fr.,* sér. 3, v. 7, p. 661–670, 1879.

Nagappa, Y., Further occurrences of *Botryococcus* in western Pakistan. *Micropaleontology,* v. 3, p. 83, 1957.

Nagy, E., The microplankton occurring in the Neogene of the Mecsek Mountains. *Acta Bot. Hung.,* v. 11, p. 197–216, 6 [unnumbered] pls., 1965.

Newman, K. R., Significance of algae in Waltman Shale Member of Fort Union Formation (Paleocene), Wyoming. *Mountain Geologist,* v. 2, p. 79–84, 1965.

Newton, E. T., On tasmanite and Australian white coal. *Geol. Mag.,* v. 2, p. 337–342, pl. 10, 1875.

Niklas, K. J., Morphological and chemical examination of *Courvoisiella ctenomorpha* gen. and sp. nov., a siphonous alga from the Upper Devonian, West Virginia, U.S.A. *Rev. Palaeobot. Palynol.,* v. 21, p. 187–203, pls. 1, 2, figs. 1–3, 1976.

Nitecki, M. H., Redescription of *Ischadites koenigii* Murchison, 1839. *Fieldiana, Geol.,* v. 16, p. 341–359, figs. 1–15, 1969a.

Nitecki, M. H., Surficial pattern of Receptaculitids. *Fieldiana, Geol.,* v. 16, p. 361–376, figs. 1–11, 1969b.

Nitecki, M. H., North American cyclocrinitid algae. *Fieldiana, Geol.,* v. 21, xiv + 182 p., 53 figs., 1970.

Nitecki, M. H., Amphispongieae, a new tribe of Paleozoic dasycladaceous algae. *Fieldiana, Geol.,* v. 23(2), p. 11–22, figs. 1–3, 1971.

Nitecki, M. H., North American Silurian receptaculitid algae. *Fieldiana, Geol.,* v. 28, xii + 108 p., 45 figs., 1972a.

Nitecki, M. H., Gametangia of Silurian *Ischadites hemisphericus* (Receptaculitaceae, Dasycladales). *Phycologia,* v. 11, p. 1–4, figs. 1, 2, 1972b.

Nitecki, M. H., Ordovician Batophoreae (Dasycladales) from Michigan. *Fieldiana, Geol.,* v. 35(4), p. 29–40, figs. 1–5, 1976.

Norem, W. L., *Tytthodiscus,* a new microfossil genus from the California Tertiary. *J. Paleont.,* v. 29, p. 694–695, pl. 68, 1955.

Norris, R. E., and B. R. Pearson, Fine structure of *Pyramimonas parkeae,* sp. nov. (Chlorophyta, Prasinophyceae). *Arch. Protistenk.,* v. 117, p. 192–213, pls. 9–18, 1975.

Obrhel, J., *Maslovina meyenii* n.g. et sp.—neue Codiacea aus dem Silur Böhmens. *Vest. Ústř. Úst. Geol.,* v. 43, p. 367–370, pls. 1, 2, fig. 1, 1968.

Odum, E. P., E. J. Kuenzler, and Sister M. X. Blunt, Uptake of P^{32} and primary productivity in marine benthic algae. *Limnol. Oceanogr.,* v. 3, p. 340–345, figs. 1–3, 1958.

Osgood, R. G., Jr., and A. G. Fischer, Structure and preservation of *Mastopora pyriformis,* an Ordovician dasycladacean alga. *J. Paleont.,* v. 34, p. 896–902, pls. 117–118, 2 figs., 1960.

Ostenfeld, C. H., *Halosphaera* and Flagellata. *Bull. Trimest. Resumé Plankton. Cons. Perm. Int. Explor. Mer.,* 1, p. 20–38, pls. 3–5, 1910.

Ott, E., *Dissocladella cretica,* eine neue Kalkalge (Dasycladaceae) aus dem Mesozoikum der griechischen Inselwelt und ihre phylogenetischen Beziehungen. *Neues Jb. Geol. Paläont. Mh.,* p. 683–693, 7 figs., 1965.

Papenfuss, G. F., Classification of the algae, p. 115–225. *In* E. L. Kessel, *A century of progress in the natural sciences.* San Francisco: Calif. Acad. Sci., 1955.

Parke, Mary, Studies on marine flagellates. *J. Mar. Biol. Ass. U.K.,* v. 28, p. 255–286, pls. 1, 2, figs. 1–73, 1949.

Parke, Mary, The genus *Pachysphaera* (Prasinophyceae), p. 555–563, pls. 1, 2, figs. 1–6. *In* H. Barnes, *Some contemporary studies in marine science.* London: George Allen and Unwin Ltd., 1966.

Parke, Mary, G. T. Boalch, Rosemary Jowett, and D. S. Harbour, The genus *Pterosperma* (Prasinophyceae): species with a single equatorial ala. *J. Mar. Biol. Ass. U.K.,* v. 58, p. 239–276, pls. 1–14, figs. 1, 2, 1978.

Parke, Mary, and P. S. Dixon, A revised check-list of British marine algae. *J. Mar. Biol. Ass. U.K.,* v. 44, p. 499–542, 1964.

Parke, Mary, and P. S. Dixon, Check-list of British marine algae—third revision. *J. Mar. Biol. Ass. U.K.,* v. 56, p. 527–594, 1976.

Parke, Mary, and I. den Hartog-Adams, Three species of *Halosphaera*. *J. Mar. Biol. Ass. U.K.*, v. 45, p. 537 – 557, 1965.

Parker, B. C., The structure and chemical composition of cell walls of three chlorophycean algae. *Phycologia*, v. 4, p. 63 – 74, figs. 1 – 9, 1964.

Parker, B. C., Occurrence of silica in brown and green algae. *Can. J. Bot.*, v. 47, p. 537 – 540, pl. 1, 1969.

Pascher, A., Ueber einige Endosymbiosen von Blaualgen in Einzellern. *Jb. Wiss. Bot.*, v. 71, p. 386 – 462, 1929.

Pennick, N. C., K. J. Clarke, and J. P. Cann, Studies of the external morphology of *Pyramimonas* 2. *Pyramimonas obovata* N. Carter. *Arch. Protistenk.*, v. 118, p. 221 – 226, pls. 40 – 42, figs. 1, 2, 1976.

Perret, M.-F., and Daniel Vachard, Sur l'appartenance du genre *Calcifolium* (Alga auct.) aux Pharétrones (Porifera). *C. R. Hebd. Séanc. Acad. Sci., Paris*, sér. D, v. 280, p. 2649 – 2652, 1 fig., 1975.

Petryk, A. A., and B. L. Mamet, Lower Carboniferous algal microflora, southwestern Alberta. *Can. J. Earth Sci.*, v. 9, p. 767 – 802, pls. 1 – 10, figs. 1 – 10, 1972.

Pia, J., Die Siphoneae verticillatae vom Karbon bis zur Kreide. *Abh. Zool.-Bot. Ges. Wien*, v. 11, heft 2, p. 1 – 263, pls. 1 – 8, figs. 1 – 27, 1920.

Pia, J., Einige Dasycladaceen aus der Obertrias der Molukken, Geologische Onderzoekingen in den Oostelijken Oost-Indischen Archipel door H. A. Brouwer. *Jaarb. Mijnw. Ned.-Oost-Indië*, v. 52 (1923), 137 p., 1 pl., 1924.

Pia, J., *Pflanzen als Gesteinbildner.* Berlin: Gebrüder Borntraeger, 355 p., 1926.

Pia, J., 1 Abteilung: Thallophyta, p. 31 – 136, figs. 14 – 129. *In* M. Hirmer, *Handbuch der Paläobotanik,* Bd. 1, *Thallophyta-Bryophyta-Pteridophyta.* München and Berlin: R. Oldenbourg, 1927.

Pia, J., Upper Triassic fossils from the Burmo-Siamese frontier: a new Dasycladacea *Holosporella siamensis* nov. gen., nov. sp., with a description of the allied genus *Aciculella* Pia. *Rec. Geol. Surv. India*, v. 63, p. 177 – 181, pl. 4, fig. 1, 1930.

Pia, J., Calcareous green algae from Upper Cretaceous of Tripoli. *J. Paleont.*, v. 10, p. 3 – 13, 1936.

Pia, J., Die kalklösenden Thallophyten. *Arch. Hydrobiol.*, v. 31, p. 264 – 328, 341 – 398, 1937.

Pickett-Heaps, J. D., Scanning electron microscopy of some cultured desmids. *Trans. Am. Microsc. Soc.*, v. 93, p. 1 – 23, figs. 1 – 30, 1974.

Pickett-Heaps, J. D., *Green algae, structure, reproduction and evolution in selected genera.* Sunderland, Mass.: Publ. Sinauer Associates Inc., vii + 606 p., 1975.

Pickett-Heaps, J. D., Cell division in eucaryotic algae. *Bioscience*, v. 26, p. 445 – 450, figs. 1 – 5, 1976.

Pickett-Heaps, J. D., and H. S. Marchant, The phylogeny of the green algae: a new proposal. *Cytobios*, v. 6, p. 255 – 264, figs. 1 – 8, 1972.

Pierce, S. T., Morphology of *Schizosporis reticulatus* Cookson and Dettmann 1959. *Geosci. Man*, v. 15, p. 25 – 33, pls. 1, 2, figs. 1, 2, 1976.

Playford, G., Lower Carboniferous microfloras of Spitsbergen. Part two. *Palaeontology*, v. 5, (1962), p. 619 – 678, pls. 88 – 95, 1963.

Poncet, Jacques, Description de quelques algues calcaires éodévoniennes du Nord-Est du Massif Armoricain. *Bull. Soc. Géol. Fr.*, sér. 7, v. 16, p. 225 – 229, pls. 3 – 6, 1974.

Poncet, Jacques, *Clibeca devoniana* nov. gen., nov. sp. algue calcaire nouvelle de l'Éodevonien du NE. du Massif Armoricain (France). *Geobios*, v. 8, p. 119 – 123, pl. 11, fig. 1, 1975.

Pratje, O., Fossile kalkbohrende Algen (*Chaetophorites gomontoides*) in Liaskalken. *Zentbl. Miner. Geol. Paläont.*, p. 299 – 301, figs. 1 – 3, 1922.

Praturlon, A., Calcareous algae from Jurassic-Cretaceous limestone of Central Apennines (southern Latium-Abruzzi). *Geologica Romana*, v. 3, p. 171 – 202, 34 figs., 1964.

Praturlon, A., Algal assemblages from Lias to Paleocene in southern Latium-Abruzzi. A review. *Boll. Soc. Geol. Ital.*, v. 85, p. 167 – 194, 16 figs., 1966.

Praturlon, A., *Heteroporella lepina*, new dasyclad species from Upper Cenomanian-Lower Turonian of central Apennines. *Boll. Soc. Paleont. Ital.*, v. 5, (1966), p. 202 – 205, pls. 51, 52, 1 fig., 1967.

Prescott, G. W., Desmids. *Bot. Rev.*, v. 14, p. 644 – 676, 1948.

Prescott, G. W., *How to know the fresh-water algae: pictured key nature series.* Dubuque, Iowa: Wm. C. Brown Co., 272 p., 487 figs., 1964.

Prescott, G. W., *The algae: a review.* Boston: Houghton Mifflin Co., xi + 436 p., figs., 1968.

Preston, R. D., Wall structure in marine algae, p. 234 – 238. In *Recent advances in Botany: from lectures and symposia presented to the IX International Botanical Congress, Montreal, 1959*, v. 1. Toronto: Univ. Toronto Press, 1961.

Preston, R. D., Plants without cellulose. *Scient. Am.*, v. 218, p. 102 – 108, figs., 1968.

Pringsheim, E. G., and O. Pringsheim, Experimental eliminations of chromatophores and eye spot in *Euglena gracilis. New Phytol.*, v. 51, p. 65 – 76, 1952.

Proctor, V. W., *Batophora* from central New Mexico. *Phycologia*, v. 1, p. 160 – 163, 1961.

Proctor, V. W., Dispersal of desmids by waterbirds. *Phycologia*, v. 5, p. 227 – 232, 1966.

Provasoli, L., Growth factor requirements of algae and ecological implications, p. 247−249. In *Recent advances in botany: from lectures and symposia presented to the IX International Botanical Congress, Montreal, 1959*, v. 1. Toronto: Univ. Toronto Press, 1961.

Provasoli, L., and A. Schatz, Streptomycin-induced chlorophyll-less races of *Euglena. Proc. Soc. Exp. Biol. Med.*, v. 69, p. 279−282, 1948.

Puiseux-Dao, S., Recherches biologiques et physiologiques sur quelques Dasycladacées, en particulier *Batophora oerstedii* J. Ag. et l'*Acetabularia mediterranea* Lam. *Revue Gén. Bot.*, v. 69, p. 409−503, pls. 37−44, figs. 1−19, 1962.

Radoičić, Rajka, Krupne tintinine *Campbelliella* nov. gen. i *Daturellina* nov. gen.−preliminarna beleška [Large Tintinnina *Campbelliella* nov. gen. and *Daturellina* nov. gen.−preliminary notes]. *Vesn. Zav. Geol. Geofiz. Istraž. NR Srbije*, v. 17, p. 79−86, pls. 1, 2, 1959.

Radoičić, Rajka, *Hadziella zetae* gen. nov., spec. nov. aberantnikh Tintinina [*Hadziella zetae* gen. nov., spec. nov. aberrant Tintinnina]. *Zapisnici Srp. Geol. Društ.*, 1960−1961, p. 193−194, figs. 1, 2, 1963.

Radoičić, Rajka, Aberantna grana fosilnih Tintinina (Rodred Tintinnina) La branche aberrante des tintinnines fossiles (Sous-Ordre Tintinnina). *Palaeont. Jugosl.*, v. 9, p. 1−71, pls. 1−8, figs. 1−31, 1969.

Radoičić, Rajka, The new dasycladacean genus *Pseudoclypeina* (a preliminary report). *Bull. Scient. Cons. Acads. RPF Yougosl.*, sec. A, v. 15, p. 4−5, 1 fig., 1970.

Raineri, Rita, Alghe sifonee fossili della Libia. *Atti Soc. Ital. Sci. Nat.*, v. 61, p. 72−86, pl. 3, 1922.

Ralfs, J., *The British Desmidieae*. London: Reeve, Benham & Reeve, xxii + 226 p., 35 pls., 1848.

Ramanujam, C. G. K., Occurrence of *Botryococcus* in the Miocene lignite of South Arcot district, Madras. *Curr. Sci.*, v. 35, p. 367−368, 4 figs., 1966.

Rao, A. R., Algal remains from the Tertiary lignites of Palana (Eocene) Bikaner. *Curr. Sci.*, v. 26, p. 177−178, figs. 1−9, 1957.

Rauff, H., Kalkalgen und Receptaculiten. *Sber. Niederrhein. Ges. Nat.-U. Heilk.*, p. 74−90, 7 figs., 1892a.

Rauff, H., Untersuchungen über die Organisation und systematische Stellung der Receptaculitiden. *Abh. Bayer. Akad. Wiss., Math.-Phys. Kl.*, v. 17, p. 645−722, pls. 1−7, 12 figs., 1892b.

Raviv, Varda, and J. Lorch, *Verticilloporella*, a new Mesozoic genus of Dasycladaceae, with discussion on *Munieria* and *Actinoporella. Israel J. Bot.*, v. 19, p. 225−235, pls. 1, 2, figs. 1−3, 1970.

Reis, O. M., Kalkalgen und Seesinterkalke aus dem rheinpfälzischen Tertiär. *Geogn. Jh.*, v. 36, p. 103−130, 1923.

Rezak, R., Permian algae from Saudi Arabia. *J. Paleont.*, v. 33, p. 531−538, pls. 71, 72, 1 fig., 1959.

Rich, Mark, Upper Mississippian (Carboniferous) calcareous algae from northeastern Alabama, south-central Tennessee, and northwestern Georgia. *J. Paleont.*, v. 48, p. 360−374, pls. 1−5, fig. 1, 1974.

Rietschel, S., *Scribroporella*, eine Dasycladacee aus dem Mitteldevon des Bergischen Landes (Rheinisches Schiefergebirge). *Senckenberg. Leth.*, v. 47, p. 193−213, pls. 21−23, figs. 1−3, 1966.

Rietschel, S., Beiträge zur Kenntnis der Receptaculiten, 1. Die Receptaculiten. Eine Studie zur Morphologie, Organisation, Ökologie, und Überlieferung einer problematischen Fossil-Gruppe und die Deutung ihrer Stellung in System. *Senckenberg. Leth.*, v. 50, p. 465−517, pls. 1−4, figs. 1−14, 1969.

Rietschel, S., Beiträge zur Kenntnis der Receptaculiten, 2. Reconstruktionen als Hilfsmittel bei der Untersuchung von Receptaculiten (Receptaculitales, Thallophyta). *Senckenberg. Leth.*, v. 51, p. 429−447, pls. 1−3, figs. 1−7, 1970.

Rietschel, S., Receptaculitids are calcareous algae but no dasyclads, p. 212−214. *In* Eric Flügel, *Fossil algae, recent results and developments*. Berlin, Heidelberg, New York: Springer-Verlag, 1977.

Rossignol, Martine, Analyse pollinique de sédiments marines quaternaires en Israël. II. Sédiments Pleistocènes. *Pollen et Spores*, v. 4, p. 121−148, pls. 1, 2, 1962.

Round, F. E., The taxonomy of the Chlorophyta. *Br. Phycol. Bull.*, v. 2, p. 224−235, 1963.

Round, F. E., The taxonomy of the Chlorophyta. II. *Br. Phycol. J.*, v. 6, p. 235−264, 1971.

Rutte, E., Gesteinbildende Algen aus dem Eozän von Kleinkems an Isteiner Klotz in Baden. *Neues Jb. Geol. Paläont. Mh.*, p. 498−506, figs. 1−5, 1953.

Sartoni, S., and U. Crescenti, Ricerche biostratigrafiche nel Mesozoico dell'Apennino meridionale. *G. Geol.*, ser. 2, v. 29, p. 161−304, 42 pls., 1962.

Schopf, J. M., L. R. Wilson, and R. Bentall, An annotated synopsis of Paleozoic fossil spores and the definition of generic groups. *Rep. Invest. Ill. St. Geol. Surv.*, 91, p. 1−72, pls. 1−3, 1944.

Schopf, J. W., Microflora of the Bitter Springs Formation, late Precambrian, central Australia. *J. Paleont.*, v. 42, p. 651−688, pls. 77−86, figs. 1−6, 1968.

Schopf, J. W., Paleobiology of the Precambrian: the age of blue-green algae. *Evolut. Biol.*, v. 7, p. 1−43, figs. 1−10, 1974.

Schweiger, H. G., S. Berger, K. Kloppstech, K. Apel, and M. Schweiger, Some fine structural and biochemical features of *Acetabularia major* (Chlorophyta,

Dasycladaceae) grown in the laboratory. *Phycologia, v.* 13, p. 11–20, figs. 1–17, 1974.

Segonzac, Geneviève, Dasycladales nouvelles du Sparnacien des Pyrénées ariégeoises. *C. R. Hebd. Séanc. Acad. Sci., Paris, v.* 270, p. 1881–1884, 3 figs., 1970.

Segonzac, Geneviève, Algues calcaires du Sparnacien de Lavelanet (Ariège), Dasycladales, Caulerpale et Cryptonémiale. *Bull. Bur. Rech. Géol. Min. Paris,* 2nd Ser., sec. 4, no. 1, p. 5–19, pls. 1–4, 1971.

Segonzac, Geneviève, Une nouvelle Acétabulariacée tertiaire: *Rostroporella oviformis* n. g., n. sp. (Algue calcaire). *Bull. Soc. Géol. Fr.,* ser. 7, v. 13, p. 181–186, pls. 10–12 (1971) 1972.

Segroves, K. L., Cutinized microfossils of probable nonvascular origin from the Permian of western Australia. *Micropaleontology, v.* 13, p. 289–305, pls. 1–3, figs. 1, 2, 1967.

Seilacher, A., Die geologische Bedeutung fossiler Lebensspuren. *Z. dt. Geol. Ges., v.* 105, p. 214–227, pls. 7, 8, figs. 1–3, 1954.

Senn, G., Ueber einige coloniebildende einzellige Algen. *Bot. Ztg., v.* 57, p. 39–104, pls. 2, 3, figs. 1–39, 1899.

Shukla, V. B., *Spirogyrites* gen nov. *Palaeobot. India, v.* 7, p. 29, 1950.

Shuyskiy, V. P., *Izvestkovye rifoobrazuyushchie vodorosli nizhnego-Devona Urala [Calcareous reefforming algae of the Lower Devonian of the Urals].* Akad. Nauk SSSR, Ural. Nauch. Tsent. Inst. Geol. Geokh. Moscow: "Nauka," 155 p., 34 pls., 1973.

Sokač, B., A new genus of calcareous algae from the Middle Triassic of Velebit. *Geološki Vjesn., v.* 21 (1967), p. 207–212, pls. 1–4, fig. 1, 1968.

Sokač, B., and L. Nikler, Two new species of the genus *Petrascula* from the lower Lias of the Velebit Mountain. *Bull. Scient. Cons. Acads. RPF Yougosl.,* sec. A, v. 11, p. 7–8, 1 fig., 1966.

Solms-Laubach, H., Monograph of the Acetabularieae. *Trans. Linn. Soc. Lond.,* ser. 2, v. 5, p. 1–39, pls. 1–4, 1895.

Sommer, F. W., Os esporomorfos do Folhelho de Barreirinha. *Boln. Div. Geol. Miner., v.* 140, p. 1–49, 2 pls., 8 figs., 1953.

Sommer, F. W., South American Paleozoic sporomorphae without haptotypic structures. *Micropaleontology, v.* 2, p. 175–181, pls. 1, 2, 1956.

Sommer, F. W., and N. M. van Boekel, Some new Tasmanaceae from the Devonian of Pará. *Anais Acad. Bras. Cienc., v.* 35, p. 61–65, pls. 1–3, 1963.

Sommer, F. W., and N. M. van Boekel, Revisão dos Tasmanáceas Paleozóicas Brasileiras. *Anais Acad. Bras. Cienc., v.* 38, p. 53–64, figs. 1–13, 1966.

Spjeldnaes, N., A new fossil (*Papillomembrana* sp.) from the Upper Precambrian of Norway. *Nature, v.* 200, p. 63–64, figs. 1–3, 1963.

Spriestersbach, J., Beitrag zur Kenntnis der Fauna des rheinischen Devon. *Jb. Preuss. Geol. Landesant. BergAkad., v.* 55, p. 475–525, pls. 41–50, 3 figs., 1935.

Starr, R. C., Reproduction by zoospores in *Tetraëdron bitridens. Am. J. Bot., v.* 41, p. 17–20, figs. 1–16, 1954.

Stewart, K. D., K. R. Mattox, and C. D. Chandler, Mitosis and cytokinesis in *Platymonas subcordiformis,* a scaly green monad. *J. Phycol., v.* 10, p. 65–79, figs. 1–20, 1974.

Stewart, K. D., K. R. Mattox, and G. L. Floyd, Mitosis, cytokinesis, the distribution of plasmodesmata and other cytological characteristics in the Ulotrichales, Ulvales and Chaetophorales: phylogenetic and taxonomic considerations. *J. Phycol., v.* 9, p. 128–141, figs. 1–12, 1973.

Stockmans, F., and Y. Willière, Une couche à *Pachytheca* et à *Prototaxites* dans le Dévonien inférieur de la Belgique. *Bull. Mus. R. Hist. Nat. Belg., v.* 14, p. 1–5, 1938.

Stockmans, F., and Y. Willière, Hystrichosphères du Dévonien belge (Sondage de l'Asile d'aliénés à Tournai). *Senckenberg. Leth., v.* 41, p. 1–11, pls. 1, 2, 1960.

Stolley, E., Ueber silurische Siphoneen. *Neues Jb. Miner. Geol. Paläont.,* Jg. 1893, v. 2, p. 135–146, pls. 7, 8, 1893.

Strauss, A., Thallophyten, Kryptogamen und Gymnospermen aus dem Pliozän von Willershausen. *Ber. Dt. Bot. Ges., v.* 65, p. 74–79, pls. 3, 4, 1952.

Studer, H. P., Electron microscope study of aragonite crystals in marine sediments. *Bull. Am. Ass. Petrol. Geol., v.* 47, p. 371, 1963.

Swale, E. M. F., A third layer of body scales in *Pyramimonas tetrarhynchus* Schmarda. *Br. Phycol. J., v.* 8, p. 95–99, figs. 1–11, 1973.

Takahashi, K., and A. Yao. Plant microfossils from the Permian sandstone in the southern marginal area of the Tanba belt. *Trans. Proc. Palaeont. Soc. Japan,* n.s., no. 73, p. 41–48, pls. 4–6, figs. 1–4, 1969.

Tappan, Helen, and A. R. Loeblich, Jr., Surface sculpture of the wall in Lower Paleozoic acritarchs. *Micropaleontology, v.* 17, p. 385–410, pls. 1–11, 1971.

Taugourdeau, P., Le problème des *Leiosphaeridia:* un détail morphologique. *C. R. Somm. Séanc. Soc. Géol. Fr.,* p. 59, 1962.

Taugourdeau-Lantz, J., Premier aperçu sur les Tasmanacées françaises. *Bull. Soc. Géol. Fr.,* sér. 7, v. 10 (1968), p. 159–167, pls. 13, 14, 1969.

Taugourdeau-Lantz, J., Structure de quelques Tasmanacées en microscopie électronique. *C. R. Hebd.*

Séanc. Acad. Sci., Paris, v. 270, p. 682–685, 1 pl., 2 figs., 1970.

Taylor, W. R., Distribution in depth of marine algae in the Caribbean and adjacent seas, p. 193–197. In *Recent Advances in Botany: From lectures and symposia presented to the IX International Botanical Congress, Montreal, 1959,* v. 1. Toronto: Univ. Toronto Press, 1961.

Termier, H., G. Termier, and D. Vachard, On Moravamminida and Aoujgaliida (Porifera, Ischyrospongia). Upper Paleozoic "Pseudo Algae", p. 215–219, figs. 1, 2. In Eric Flügel, *Fossil algae, recent results and developments.* Berlin, Heidelberg, New York: Springer-Verlag, 1977.

Thiessen, R., Origin of the boghead coals. *Prof. Pap. U.S. Geol. Surv.,* 132 I, p. 121–135, pls. 27–40, 1925.

Timofeev, B. V., and T. N. German, Mitoz in rifeyskikh vodorosley [Mitosis in Riphean algae]. *Geol. Sb. Bratisl.,* v. 25, p. 167–172, pls. 1–4, 1974.

Tourtelot, H. A., J. R. Donnell, and I. L. Tailleur, Oil yield and chemical composition of shale from northern Alaska. *Wld. Petrol. Congr. 7, Proc.,* v. 3, p. 707–711, 1967.

Tourtelot, H. A., I. L. Tailleur, and J. R. Donnell, Tasmanite and associated organic-rich rocks, Brooks Range, northern Alaska. *Spec. Pap. Geol. Soc. Am.,* 101, p. 222–223, 1968 (Abst.).

Trainor, F. R., Zoospores in *Scenedesmus obliquus. Science,* v. 142, p. 1673–1674, figs. 1, 2, 1963.

Traverse, A., 1955, Occurrence of the oil-forming alga *Botryococcus* in lignites and other Tertiary sediments. *Micropaleontology,* v. 1, p. 343–350, pl. 1, figs. 1, 2, 1955.

Uteck, K., Über eine *Tasmanites*-Art aus dem mittleren Bundsandstein des Hildesheimer Waldes. *Neues Jb. Geol. Paläont., Mh.,* p. 90–91, 1 fig., 1962.

Vachard, Daniel, Sur les dayscladacées métaspondyles "vestibulaires," à propos d'un de leurs représentants viséens: *Eovelebitella occitanica* n. gen. n. sp. *C. R. Hebd Séanc. Acad. Sci., Paris,* Ser. D, v. 279, p. 1855–1858, figs. 1, 2, 1974.

Valet, Gabriel, Sur l'origine endogène des rameaux verticillés chez certaines Dasycladales. *C. R. Hebd. Séanc. Acad. Sci., Paris,* v. 265, p. 1175–1178, pl. 1, figs. 1–8, 1967.

Valet, Gabriel, Contribution à l'étude des Dasycladales, 1. Morphogenèse. *Nova Hedwigia,* v. 16, p. 21–82, pls. 4–26, 1968.

Valet, Gabriel, Contribution à l'étude des Dasycladales, 2. Cytologie et reproduction, 3. Révision systématique. *Nova Hedwigia,* v. 17, p. 551–644, pls. 133–157, 1969.

Valet, Gabriel, Revised systematics of Dasycladales. *Protoplasma,* v. 75, p. 488–489, (Abstr.), 1972.

Valet, Gabriel, and G. Segonzac, Les genres *Chalmasia* et *Halicoryne* (Algues Acétabulariacées). *Bull. Soc. Géol. Fr.,* sér. 7, v. 11, p. 124–127, pls. 2, 3, 1969.

Van Boekel, N. M. da Costa, Microfósseis Devonianos do Rio Tapajós, Pará, 1. Tasmanaceae. *Notas Prelim. Estud. Div. Geol. Miner. Bras.,* no. 145, p. 3–15, pls. 1, 2, figs. 1–6, 1968.

Van Geel, B., Fossil spores of Zygnemataceae in ditches of a prehistoric settlement in Hoogkarspel (The Netherlands). *Revue Palaeobot. Palynol.,* v. 22, p. 337–344, pl. 1, fig. 1, 1976.

Verniory, R., *Eothrix alpina* Lombard, Algue ou Crinoïde. *Archs. Sci.,* v. 7 (ann. 159), p. 327–330, 1 pl., 1954.

Verniory, R., La création du genre *Lombardia* Brönnimann est-elle justifiée? *Archs. Sci.,* v. 9 (ann. 161), p. 85–92, pls. 1–3, 1956.

Vidal, Gonzalo, Late Precambrian microfossils from the basal sandstone unit of the Visingsö Beds, South Sweden. *Geol. Palaeont.,* v. 8, p. 1–14, pl. 1, figs. 1–4, 1974.

Vidal, Gonzalo, Late Precambrian microfossils from Visingsö Beds in southern Sweden. *Fossils Strata,* no. 9, p. 1–57, figs. 1–23, 1976.

Vimal, K. P., Occurrence of *Botryococcus* in Eocene lignites of Cutch. *Curr. Sci.,* v. 22, p. 375–376, 1953.

Volkova, N. A., Akritarkhi dokembriyskikh i nizhnekembriyskikh otlozheniy Estonii [Acritarchs of the Precambrian and Lower Cambrian deposits of Esthonia], p. 8–36, pls. 1–12, figs. 1, 2. In N. A. Volkova, Z. A. Zhuravleva, V. E. Zabrodin, and B. Sh. Klinger, Problematiki Pogranichnykh Sloev Rifeya i Kembriya Russkoy Platformy, Urala i Kazakhstana. *Trudy Inst. Geol. Nauk, Mosk.,* v. 188, 1968.

Vologdin, A. G., *Drevneyshie vodorosli SSSR [Ancient algae of the USSR].* Moscow: Akad. Nauk SSSR, pt. 1, 448 p., 80 pls., 95 figs., pt. 2, p. 449–656, pls. 1–46, figs. 1–40, 1962.

Vologdin, A. G., Neskol'ko vidov verkhnepaleozoyskikh vodorosley vostochnoy Mongolii [Some species of Upper Paleozoic algae of eastern Mongolia], p. 31–47, pls. 1–6. In N. A. Marinov, *Materialy po geologii Mongol'skoy Narodnoy Respubliki.* Minist. Geol. SSSR, Nauchno-issledov. Labor. Geol. Zarubezhnykh Stran (NILZarubezh-geologiya) Minist. Geol. MNR, Akad. Nauk MNR. Moscow: "Nedra," 1966.

Vologdin, A. G., Ob ostatakakh sifoney iz Nizhnego Kembriya Khrebta Tannu-Ola (Tuva) [Concerning the remains of siphonate algae in the Lower Cambrian of the Tannu-Ola Mountains (Tuva)]. *Dokl. Akad. Nauk SSSR,* v. 174, p. 952–955, 2 figs., 1967.

Vologdin, A. G., and O. S. Kochetkov. Ob otkrytii ostatkov gigantiskikh sifoney v drevnikh sloyakh Timanskogo Kryazha [On the discovery of the remains of gigantic siphonate algae in ancient strata of the Timan Ridge]. *Dokl. Akad. Nauk SSSR*, v. 169, p. 672−675, 4 figs., 1966.

Wall, D., Evidence from recent plankton regarding the biological affinities of *Tasmanites* Newton 1875 and *Leiosphaeridia* Eisenack 1958. *Geol. Mag.*, v. 99, p. 353−362, pl. 17, 2 figs., 1962.

Wallner, J., *Oocardium stratum* Naeg. eine wichtige tuffbildende Alge Südbayerns. *Planta*, v. 20, p. 287−293, 3 figs., 1933.

Webby, B. D., Trace fossils (Pascichnia) from the Silurian of New South Wales, Australia. *Paläont. Z.*, v. 43, p. 81−94, pl. 10, 5 figs., 1969.

West, W., G. S. West, and N. Carter, *A monograph of the British Desmidiaceae*, v. 1−5. London: Ray Society, 1904−1923 [v. 1, 1904, Ray Soc. no. 82, 224 p., 32 pls. (West and West); v. 2, 1905, Ray Soc. no. 84, 204 p., pls. 33−64 (West and West); v. 3, 1908, Ray Soc. no. 88, 274 p., pls. 65−95 (West and West); v. 4, 1912, Ray Soc. no. 92, 194 p., pls. 96−128 (West and West); v. 5, 1923, Ray Soc. no. 108, 300 p., pls. 129−167 (Carter)].

Wetzel, O., Mikroskopische Reste von vermitlichen Algen im Feuerstein der Kreide Norddeutschlands. *Neues Jb. Geol. Paläont. Abh.*, v. 94, p. 101−110, pl. 13, 1951.

Weyland, H., Zwei neue Algen aus der Braunkohle von Baccinello (Toscana). *Palaeontographica Abt. B*, v. 113, p. 30−37, 1 pl., 1 fig., 1963.

Wicander, E. R., Upper Devonian-Lower Mississippian acritarchs and prasinophycean algae from Ohio, U.S.A. *Palaeontographica Abt. B*, v. 148, p. 9−43, pls. 5−19, 1974.

Wicander, E. R., Fluctuations in a Late Devonian-Early Mississippian phytoplankton flora of Ohio, U.S.A. *Palaeogeogr. Palaeoclimat. Palaeoecol.*, v. 17, p. 89−108, figs. 1−9, 1975.

Wille, W., *Plaesiodictyon mosellanum* n. g., n. sp., eine mehrzellige Grünalge aus dem Unteren Keuper von Luxemburg. *Neues Jb. Geol. Paläont. Mh.*, p. 283−310, figs. 1−15, 1970.

Wilson, L. R., and W. S. Hoffmeister, Four new species of fossil *Pediastrum*. *Am. J. Sci.*, v. 251, p. 753−760, pl. 1, 4 figs., 1953.

Wilson, L. R. and J. B. Urban, Electron microscope studies of the marine palynomorph *Quisquilites*. *Micropaleontology*, v. 17, p. 239−243, pls. 1, 2, 1971.

Winslow, M. R., Plant spores and other microfossils from Upper Devonian and Lower Mississippian rocks of Ohio. *Prof. Pap. U.S. Geol. Surv.*, 364, p. 1−93, pls. 1−22, figs. 1−12, 1962.

Wood, A., The Lower Carboniferous calcareous algae *Mitcheldeania* Wethered and *Garwoodia*, gen. nov. *Proc. Geol. Ass.*, v. 52, p. 216−226, pls. 13−15, 1941.

Wood, A., The algal nature of the genus *Koninckopora* Lee; its occurrence in Canada and western Europe. *Q. Jl. Geol. Soc. Lond.*, v. 98, p. 205−222, pls. 8−10, figs. 1−3, 1943.

Wood, A., A new dasycladacean alga, *Nanopora*, from the Lower Carboniferous of England and Kazakhstan. *Palaeontology*, v. 7, p. 181−185, pls. 31, 32, 1964.

Wood, E. J. F., *Marine microbial ecology*. New York: Reinhold Publ. Corp., 243 p., 14 pls., 18 figs., 1965.

Wulff, E., Ueber Heteromorphose bei *Dasycladus clavaeformis*. *Ber. Dt. Bot. Ges.*, v. 28, p. 264−268, 1910.

Yabe, H., and S. Toyama, On some rock-forming algae from the younger Mesozoic of Japan. *Sci. Rep. Tôhoku Univ.*, v. 12, p. 141−152, pls. 18−23, 1928.

Yashnov, V. A., *Halosphaera viridis* as an indicator of Mediterranean waters in the North Atlantic. *Oceanology*, v. 5, no. 5, p. 94−98, 1965. [Am. Geophy. Union translation of *Okeanologiya*, v. 5, no. 5, p. 884−890, 1965].

Zalessky, M. D., Sur le sapropélite marin de l'âge silurien formé par une algue cyanophycée. *Ezheg. Vser. Paleont. Obshch.*, v. 1, p. 25−42, pls. 2, 3, 1917a.

Zalessky, M. D., Sur quelques sapropélites fossiles. *Bull. Soc. Géol. Fr.*, sér. 4, v. 17, p. 373−379, 1917b.

Zalessky, M. D., Premières observations microscopiques sur le schiste bitumineux du Volgien inférieur. *Annls. Soc. Géol. N.*, v. 51, p. 65−104, pls. 2−6, 1926.

Zhang Pengyuan, A new fossil coenocytic Chlorophyceae from the Wumishan Formation, Jixian System in Jixian (County). *Acta Geol. Sinica*, v. 53 (2), p. 87−90, pl. 1, 1979.

CHAROPHYTES AND UMBELLINACEANS

A major problem is the change in floristic composition that took place between the first and the second great periods of the Charophyta —that of the Devonian and Lower Carboniferous on the one hand and the ages from the Upper Jurassic to the Oligocene on the other. How and when that change took place, how the morphological transition was manifested, at what time the dominant Paleozoic Trochiliscales and Sycidiales disappeared, and when the Charales of more or less Cenozoic character began to flourish, are all questions which for lack of adequate material cannot be answered at present.

Henning Horn af Rantzien, 1954b

The Charophyta, or stoneworts, are a relatively small group of freshwater algae that have comprised a significant part of the submerged vegetation of streams, ponds, and lakes, at least since the Devonian. Living Charophytes are widely distributed, being represented on nearly all continents. Although the number of genera and species has never been large at any single time, the wide geographic distribution and the limited geologic range of most taxa have allowed their use as stratigraphic indices in freshwater and brackish sediments. Of about 70 recognized living and fossil genera (exclusive of the Umbellinaceae), only 6 are living today, and these include a total of 81 living species (Wood and Imahori, 1965).

The Charophytes were regarded by Fritsch (1935) as a separate order of the green algae (Chlorophyta), perhaps representing a transition from the green algae to the bryophytes and higher plants. Others consider them sufficiently distinct from both other algae and higher plants to be placed in a separate division of the plant kingdom (Papenfuss, 1955).

The earliest known charophytes already were highly differentiated, with their reproductive organs protected in special gametangial structures rather than being exposed as in other algae (see Figure 11.1). Although indicating a higher level of organization than is characteristic of the Chlorophyta, these gametangial structures are quite unlike those of bryophytes.

Recent ultrastructural studies of cell division indicate that the charophytes diverged early from the main groups of Chlorophyta, as they are characterized by the phragmoplast type of

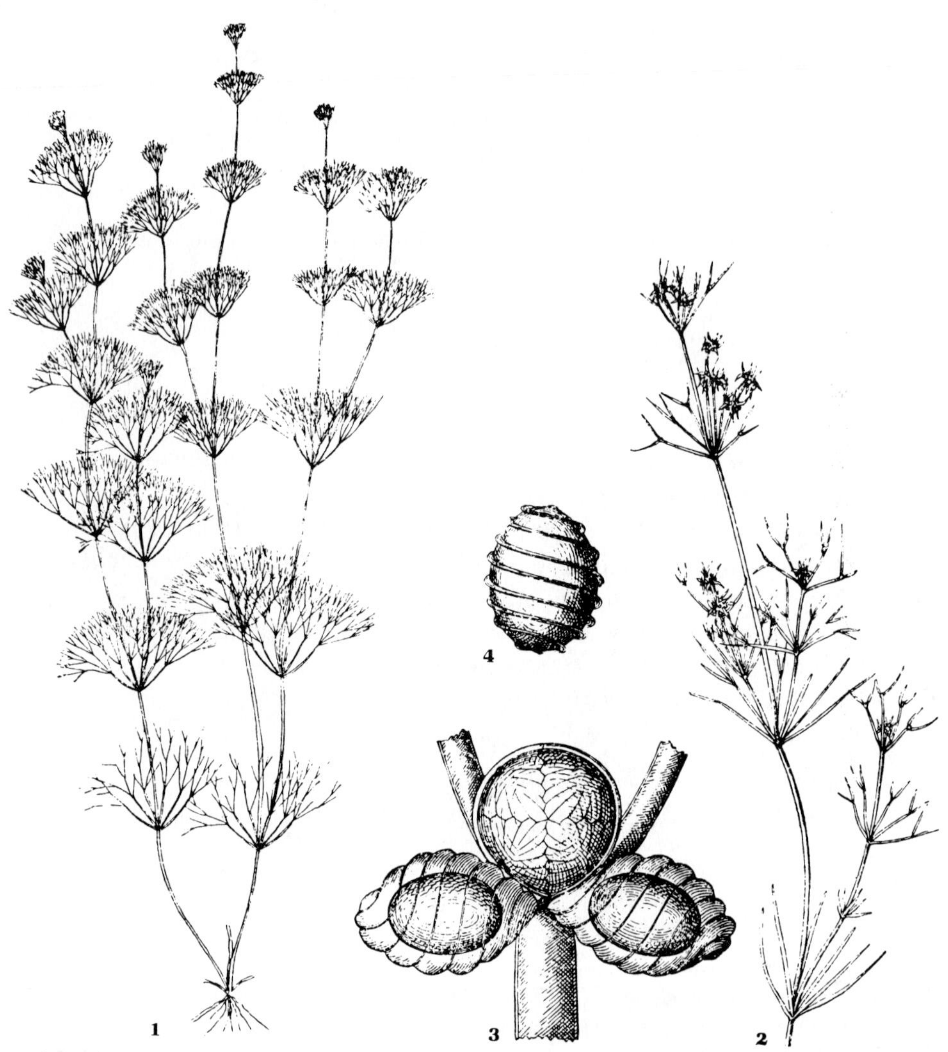

Figure 11.1
1. *Nitella gracilis* (Smith) Agardh, living, reduced. **2—4.** *Nitella flexilis* (Linnaeus) Agardh,
living. **2.** Part of plant, ×1.   **3.** Fertile branch, with spherical antheridium of shield plates,
and two oogonia, ×30. **4.** Limeshell, ×45. From Migula, 1897.

division (see Chapter 10), like that of the higher plants as well as the primitive Prasinophyta and the Zygnematales and Coleochaetales of the Chlorophyta. The latter group and the charophytes both have scale-bearing motile cells (gametes) like those of the prasinophytes, but not found in other green algae. Thus the charophytes may represent a distinct and early offshoot from the green algae that has persisted, although it did not result in the radiation that led to the higher plants.

THE LIVING ORGANISM

The charophyte thallus is an erect and intricately branched haploid plant ranging from 15 to 30 cm to as much as 2 m in height. It may be compact in growth or bushy, the size of the plant varying with depth, probably as a response to the available light.

The threadlike central axis or stalk, up to 4 mm in diameter, has regularly spaced whorls of laterals along its length. Unlike the superficially similar dasyclads that have whorled laterals, the charophyte plant consists of many distinct cells, whereas the vegetative dasyclad is a single cell, with one nucleus. In the charophytes, the axis and the laterals consist of very elongate biconvex **internodal cells** alternating with short, discoidal **nodal cells.** Internodal cells commonly are 2.5 to 5.0 cm long, and may be as much as 10 to 13 cm. Growth is from a dome-shaped apical cell that cuts off segments proximally. Each new segment divides transversely into a lower internodal cell, which continues to lengthen without further subdivision, and an upper biconcave nodal cell that divides vertically into several cells to give rise to the laterals (including cortical threads in *Chara* and *Lychnothamnus*) and to the sexual organs. Laterals of unlimited growth are termed **branches,** and those of limited growth are **branchlets.**

The plant is wholly submerged, being attached to a muddy or sandy substrate by rhizoids that also absorb nutrients (see Figure 11.2). Because charophytes are inhibited by an excess of phosphorus, they selectively occur on sandy bottoms where decaying organic matter is scarce.

The cell wall has an inner cellulosic layer and an outer gelatinous one. The wall of *Nitella* consists of a noncrystalline matrix of a pectic substance (hemicellulose, protein, etc.), reinforced with microfibrils of crystalline cellulose, whose orientation is roughly transverse to the cell axis. Some species, especially the living *Chara vulgaris* and *C. globularis* Thuillier, have a limy surface encrustation, but not all species deposit lime. When present, the lime commonly is calcite, although traces of aragonite occur in some brackish water forms, and an appreciable amount of magnesium may be present.

The peripheral region of the cells contains numerous small discoid plastids, similar to those of green algae although lacking pyrenoids. Food reserves are stored as starch. A single nucleus is centrally placed, the nodal cell nuclei being somewhat larger than those of internodal cells. In the elongated cells, cytoplasm continually streams up one side and down the other of a greatly enlarged central vacuole.

Gametangial structures are highly specialized, those in which the egg cell is formed being calcified and commonly preserved as fossils.

Charophytes are valuable water-clarifying agents and are among the most important food plants for waterfowl. They are also useful for cytological and physiological studies because of their freshwater habit, ease of cultivation, and extremely large cells.

Reproduction

Charophytes reproduce both vegetatively and sexually. Vegetative reproduction may occur by means of uni- to multicellular **bulbils** (see Figure 11.3) that arise from nodal cells; fragmentation of the mature plant may result in the separate fragments taking root to form new individuals, or secondary lateral shoots may arise from the rhizoids to produce new erect shoots.

In sexual reproduction, the plants commonly are monoecious, a single plant producing both types of gametes, although some species may

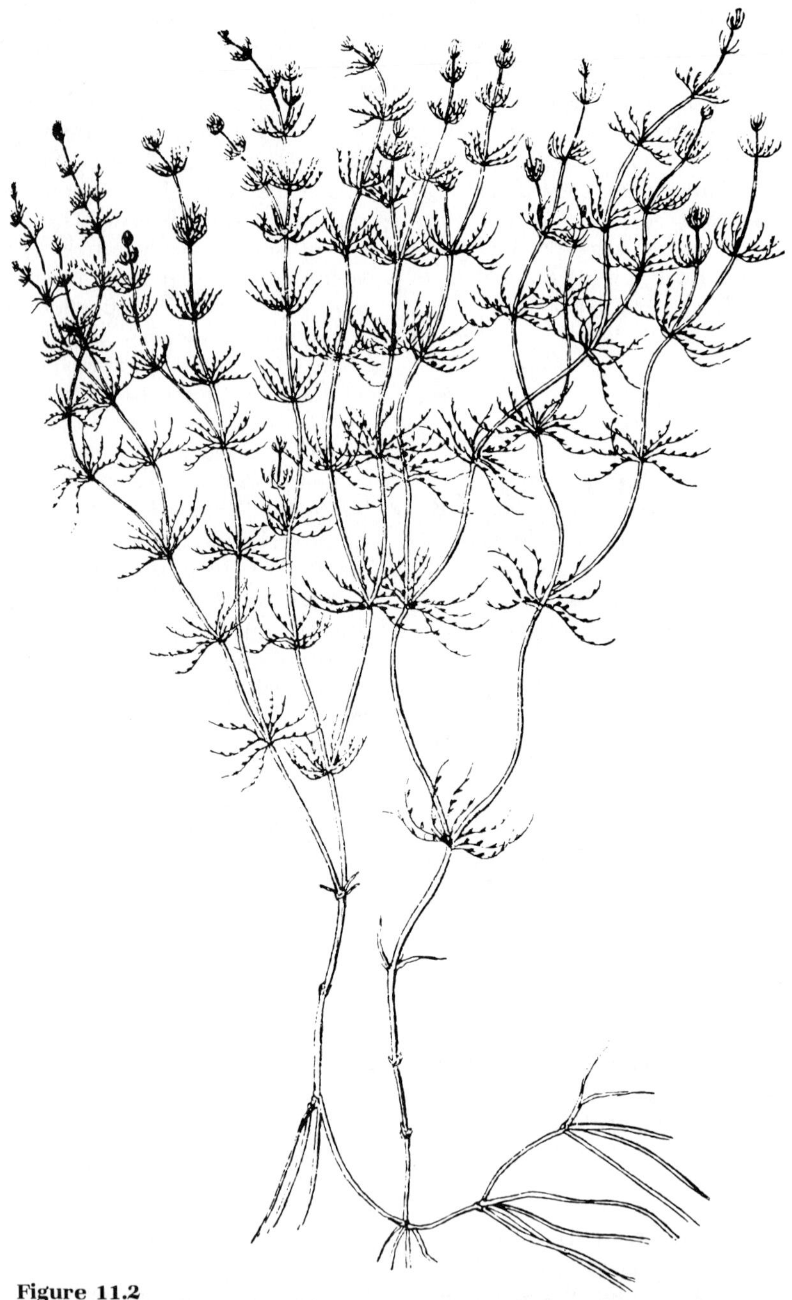

Figure 11.2
Chara vulgaris Vaillant ex Linnaeus, living. Vegetative plant showing elongate internodal cells and whorls arising from the very short nodal cells, and rhizoids at base for attachment, reduced. From Migula, 1897.

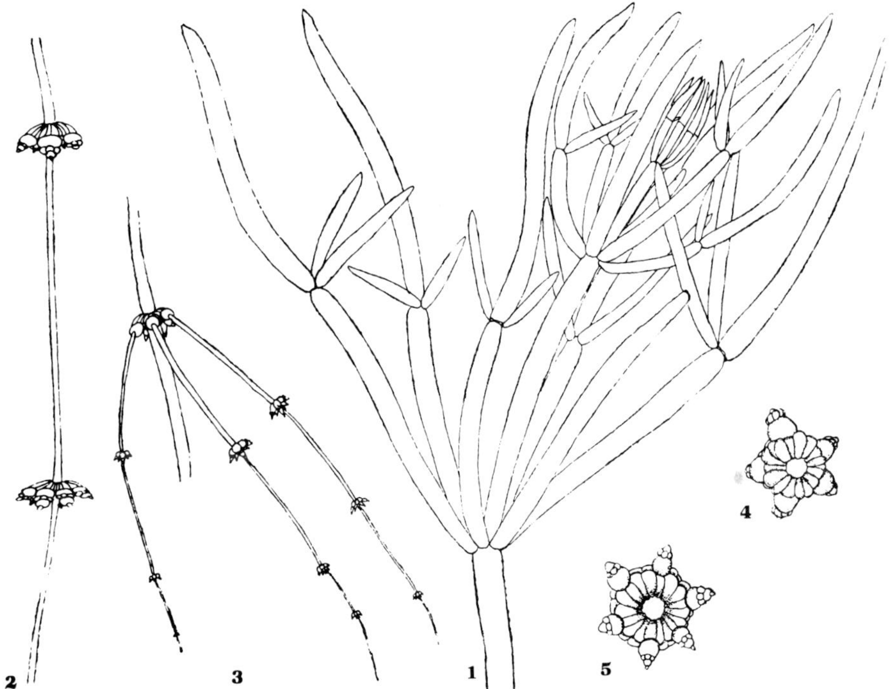

Figure 11.3
Tolypellopsis stelligera (Bauer) Migula, living. **1.** Vegetative branch, ×1. **2,3.** Showing bulbils arising at the nodal cells, ×4. **4,5.** Bulbils, from above and below, ×4. From Migula, 1897.

have dioecious populations. The gametangia arise on secondary vegetative shoots (branchlets) at the nodes, and are enclosed in a jacket of sterile cells (see Figure 11.4). Commonly the oogonia and antheridia occur close together on the plant. As the entire plant is haploid, no meiotic division immediately precedes gamete formation.

Experiments to determine breeding patterns in Charophyta showed that populations of the same species from different continents, or even widely separated areas on the same continent, generally were not interfertile (Proctor, 1975). The degree of morphologic variation included within a species by various workers thus may need modification, if members of a species must be interfertile.

The Antheridium

The male gametangium or **antheridium** is much more complex than that of any other algae. An outer series of eight sterile **shield cells** (rarely four) protects the inner fertile cells. The shield cells are nearly triangular plates with crenulate margins and radially grooved surfaces joined together to form a large spheroidal antheridium. In living species the anteridium is bright red or yellow in color, about 1 mm or less in diameter, and usually directed downward on the plant; the biflagellate and coiling male gametes form within it. The male gametangium of modern species is not calcified, hence normally is nonpreservable. Fossil antheridia have been reported in the Upper

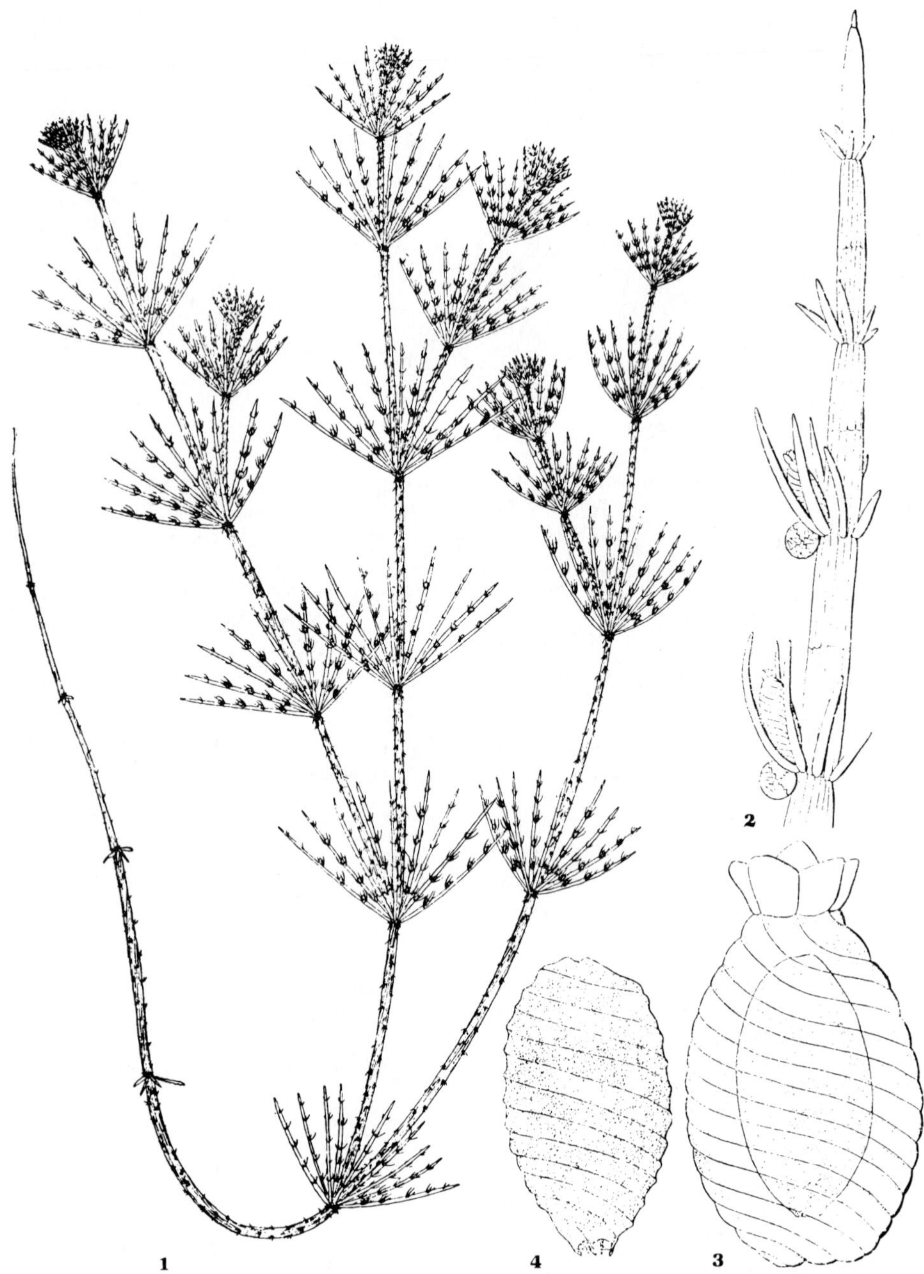

Figure 11.4
Chara intermedia A. Braun, living. **1.** Vegetative plant, ×1. **2.** Fertile branch, with sporophydia and antheridia, ×30. **3.** Oosporangium. **4.** Gyrogonite. 3,4, ×50. From Migula, 1897.

Jurassic of England and Lower Cretaceous of North America, but these are silicified or represent impressions or casts.

The Sporophydium

The female gametangium is termed the **sporophydium** (Horn af Rantzien, 1956a). Up to 1.5 mm in diameter, it is usually directed upward on the plant. It consists of three parts: (1) the **oogonium**, consisting of the egg cell and the sterile oogonial cells surrounding it; (2) the **sporostegium**, consisting of the nodal cell, **spiral enveloping cells**, and terminal **coronular cells;** and (3) the **pedicel cells.** The sporophydium itself is seldom preserved, but after fertilization of the egg cell, calcification commonly takes place within the spiral cells of the sporostegium. As the calcified structure contains the fertilized egg, it is properly termed the oosporangium.

The female gametangium of the charophytes is like that of other algae at first in being single celled and naked. Then the cell below the oogonium produces five bright green sausage-shaped cells that become much elongated and wind tightly around the egg cell. The cells wind in a sinistral manner in all members of the Charales, although the degree of torsion differs greatly among the various species, and the number of convolutions visible from the side varies. Commonly each cell turns from one to three times around the gametangium from the base to the apex.

The apex of the sporostegium has a coronula of five or ten cells, which are cut off from the spiralling cells by transverse septa. The subfamily Charoideae of the Characeae has five coronal cells, whereas the Nitelloideae has ten, in two superposed series of five each. At the opposite pole, a very short stalk cell or node cell serves to attach the sporostegium to the branchlet. Coronular cells of modern species are not calcified.

Within the sporostegium is produced the single egg cell, its nucleus being basal in position. In fertilization, the sperm cell penetrates through narrow slits at the apex and traverses the length of the ovum to fuse with its nucleus; the zygote nucleus then moves toward an apical position.

The Oosporangium and Gyrogonite or Lime Shell

After fertilization of the egg cell, the sporophydium becomes an **oosporangium.** The diploid nucleus of the oospore or zygote is enclosed in protoplasm containing starch grains and oil drops. The oospore is surrounded by a thickened yellow or brown cellulosic membrane and the calcified sterile enveloping cells of the sporophydium that form a **lime shell** or gyrogonite. The term **gyrogonite** is commonly used to designate the fossilized oosporangium, the name being derived from the form genus *Gyrogonites* Lamarck, a synonym of *Gyrogona.*

As the oospore matures, it presses into the furrows between the spiral cells, and spiral ridges develop on its surface. These spiral cells and ridges of gyrogonite and oospores are characteristic of living and most fossil charophytes. All Charales have sinistrally spiralling cells, commonly five in number (six or more in the fossil *Palaeochara* and *Eochara*), but the Devonian to Mississippian Trochiliscales have seven to ten spiral cells that spiral dextrally. The Devonian and Mississippian Sycidiales have 16 to 20 vertical units that may be transversely subdivided, but do not spiral.

Calcification. Calcification of the oosporangium takes place only in the living cells, but may begin at various stages in the oogonial development. Deposition begins on the concave inner wall of the spiral cells, against the oospore, and progresses until nearly the entire cell substance is replaced by laminated calcite (see Figure 11.5). The laminations consist of alternating dark and light lamellae, of colloidal organic matter (0.33 to 1 μm in thickness) and calcium carbonate (0.5 to 3.0 μm in thickness). If the calcification begins early, the spiral cells are nearly filled and the gyrogonite has broad spiral convolutions with narrow furrows, resembling the original oogonium. At maturity,

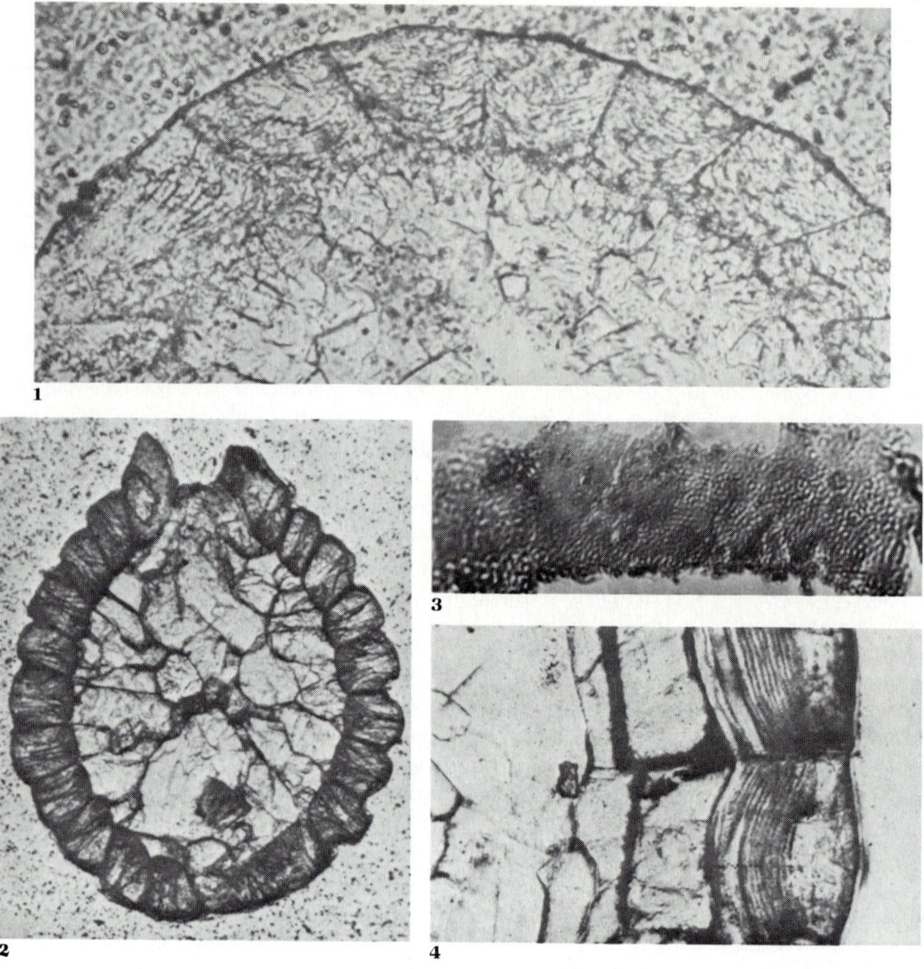

Figure 11.5
Wall structure of the gyrogonite. **1.** *Stellatochara maedleri* Horn af Rantzien, Middle Triassic, Sweden, thin section, showing cell walls and laminations of the calcite, ×400. **2,3.** *S. sellingii* Horn af Rantzien, Middle Triassic, Sweden. **2.** Entire gyrogonite, thin section, ×100. **3.** Fragment of oospore membrane, ×600. 1—3, from Horn af Rantzien, 1954b. **4.** *Nitellopsis (Tectochara) meriani* (L. & N. Grambast) Grambast & Soulié-Märsche, Pliocene, Austria, part of longitudinal section, showing part of two spirals with distinct lamination, ×200, from Horn af Rantzien, 1959a.

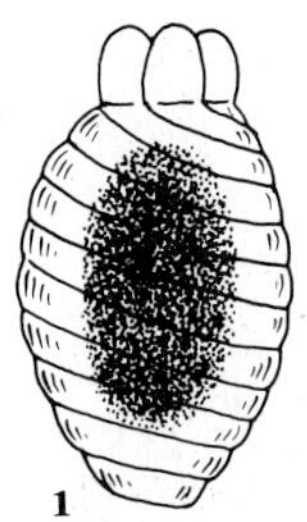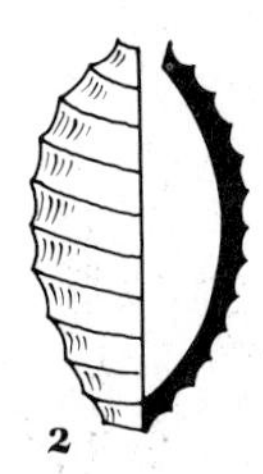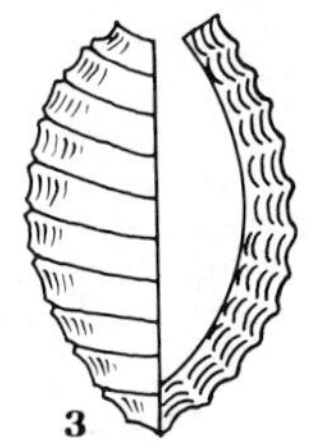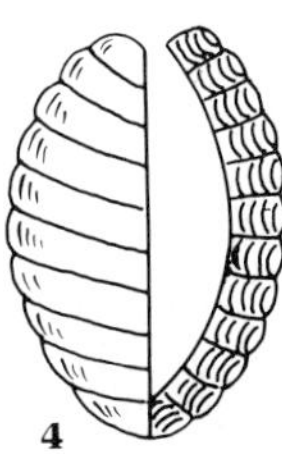

Figure 11.6
Development of the lime shell. **1.** Uncalcified sporophydium, showing spiralling tubular cells, with coronal cells at apex. **2–4.** Progressive calcification (cell material omitted), exterior at left, and cross section at right of each drawing. **2.** Thin calcareous deposit, forming ridges at the sutures (cell walls). **3.** Wall thickened by progressive laminar deposits. **4.** Mature gyrogonite with thick laminations, cell completely infilled, and depressions marking the sutures between cells. Varied surface characteristics may result from different degrees of gyrogonite calcification within a species.

the cell walls break down and adjacent limy deposits fuse to form a compact lime shell. If the calcification does not begin until late in development, only a thin calcite layer is formed, and each original cell will be represented by two sharp spiral ridges and a wide median furrow. Depending on the degree of calcification attained, the calcified surfaces of the cells may be concave, smooth, or convex, resulting in a variety of apparent surface characteristics and wall thicknesses (see Figure 11.6). The wall thickness of *Stomochara moreyi* (Peck) L. Grambast ranges from 14 to 70 μm, depending on the maturity of the gyrogonite (Peck and Eyer, 1963b).

Fossil and recent gyrogonites studied with the SEM had tiny pillars across the intercellular sutures (Feist-Castel, 1973), which were suggested to aid in the cohesion of the spiral cells (see Figure 11.7) during continued growth.

Only the spiral cells of the oosporangium are calcified in most charophytes, and not their coronal cells or stalk cell. However, the corona is apparently well preserved in some fossil forms such as *Karpinskya* (see Figure 11.8), and very rarely in other species. An assemblage from Pleistocene terraces of about 200 specimens of *Tectochara diluviana* (Mädler) L. Grambast, now

regarded as a synonym of *Nitellopsis obtusa* (Desvaux) J. Groves, included two specimens with calcified coronal cells. These consisted of five slightly twisted vertical calcareous projections, attached by their organic sheath (Maslov, 1963c). The swollen tips of the spiral cells of other species at times have been mistakenly reported as coronal cells.

One genus, *Tolypella*, differs in its method of calcification. Rather than an intracellular deposit, the lime is a surficial encrustation of the *Tolypella* gyrogonite, as well as of the vegetative part of the charophyte plant. Because this external calcification breaks down readily into a fine-grained calcareous mud when the plant dies, the vegetative part is seldom preserved. Similarly, fossil *Tolypella* gyrogonites are almost unknown.

Some modern genera, such as *Nitella*, have noncalcified oosporangia, hence are unlikely to be preserved.

Good preservation of the lime shell is favored by an alkaline matrix, whereas the organic material is better preserved in an acidic one. Thus well-preserved lime shells have poorly preserved organic remains, and the latter are best preserved when the lime shell is partially disintegrated. In a few rare occurrences, the vegeta-

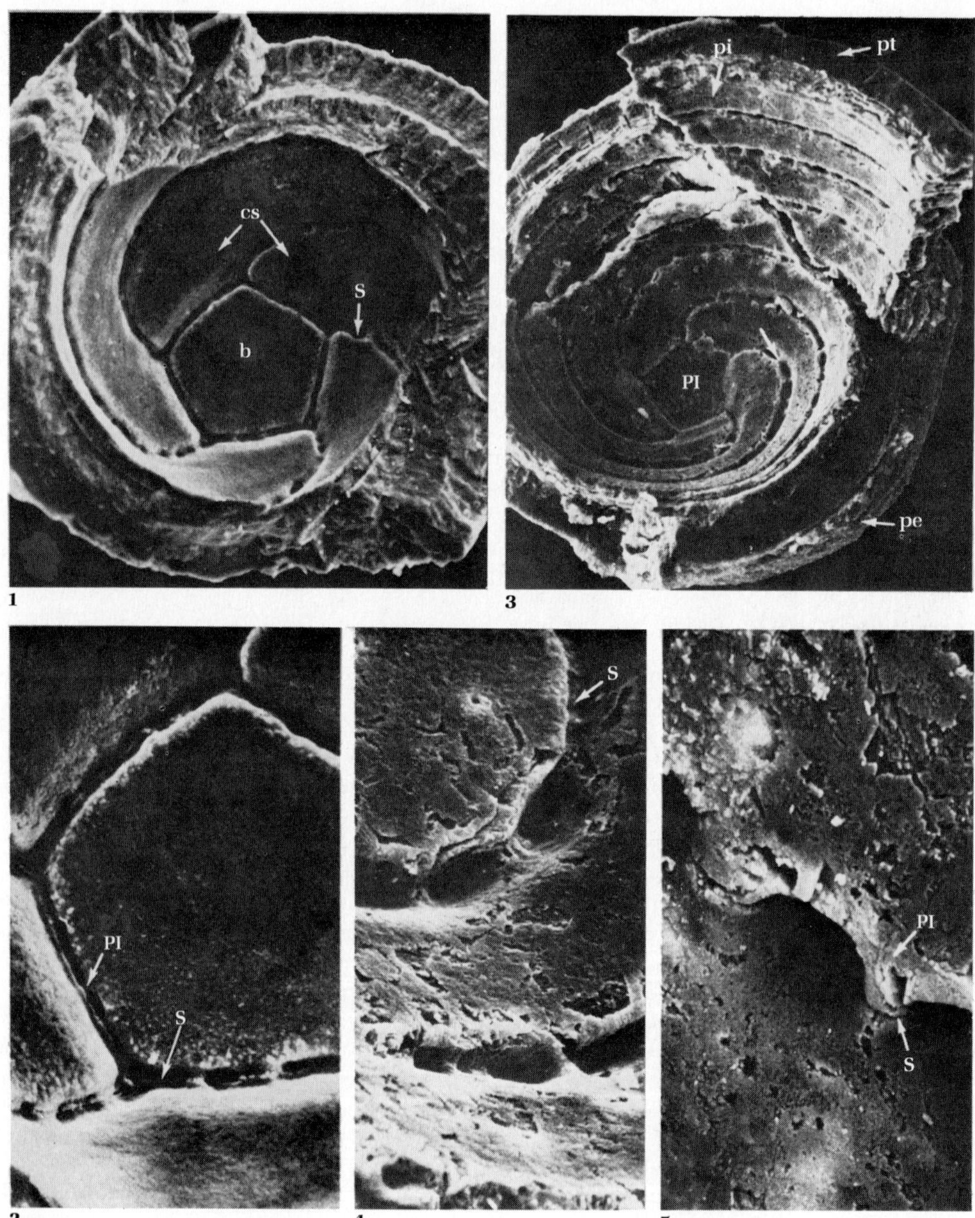

Figure 11.7

SEM of interior of basal part of gyrogonites showing b, basal plate; cs, spiral cell; S, suture; PI, pillars at sutures; pi, inner wall; pe, external wall; pt, transverse wall. **1,2.** *Lamprothamnium papulosum* (Wallroth) J. Groves, living; 1, ×160; 2, enlargement, ×405. **3−5.** *Harrisichara* sp., Paleogene, France; 3, ×135; 4, ×210; 5, ×625. All from Feist-Castel, 1973.

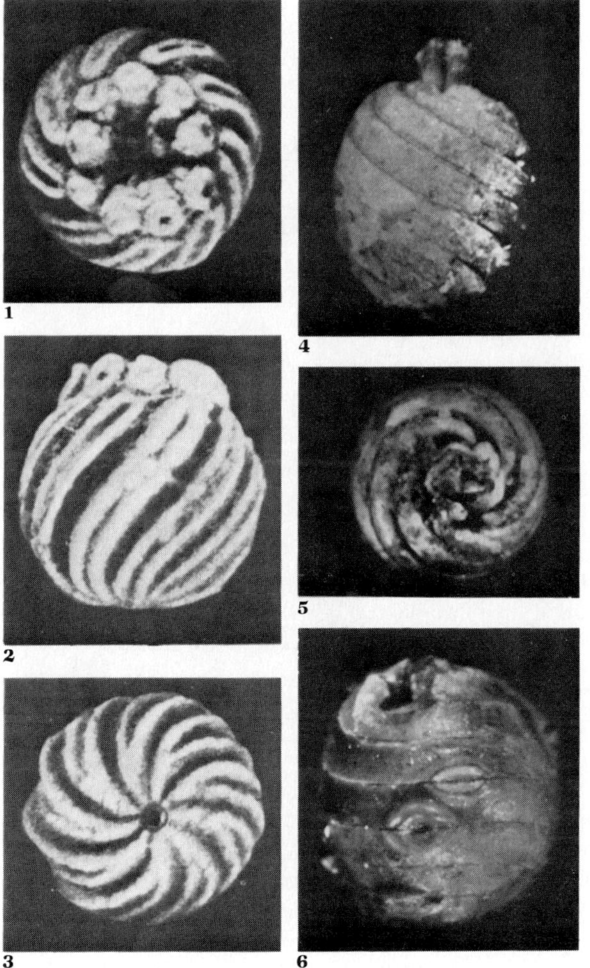

Figure 11.8
Preserved coronular cells. **1 – 3.** *Karpinskya oscolensis* (Samoylova) n. comb.*, as *Trochiliscus* (*Karpinskya*), Middle Devonian, Russian Platform, top, side, and basal views, showing nine calcified coronular cells, ×45, from Samoylova, 1961. **4 – 6.** *Nitellopsis obtusa* (as *Tectochara diluviana*) Quaternary, USSR, ×30, from V. P. Maslov. **4,5.** Side and top views of specimen with calcified coronular cells at apex. **6.** Side view of broken specimen with protrusions of unknown origin at sutures.

____

* *Karpinskya oscolensis* (Samoylova) Tappan, n. comb. herein for *Trochiliscus* (*Karpinskya*) *oscolensis* Samoylova, 1961, *Doklady Akad. Nauk SSSR*, v. 139, p. 206, fig. 1.

tive plant, gyrogonite with coronula and pedicel cell, and the oosporangial membrane are all silicified, and thus preserved.

Size and Shape of the Lime Shell. The oosporangium ranges in length from 275 to 300 μm in Oligocene *Tolypella,* to about 2 mm in early Tertiary *Chara.* Living species show all variations between these extremes. The general shape varies from spherical to cylindrical or even flasklike. Many gyrogonite features are characteristic of particular charophyte genera or species. Size is fairly constant within a modern species, as only the mature fruits are calcified; their size and shape, the width of the spiral cells (as measured in the equatorial region between adjacent sutures), the number of spirals visible in lateral view, and the equatorial angle of the spirals are all useful specific characters.

Wood (1959) tabulated the measurements of Recent charophyte species, including oogonium length, number of convolutions of the spiral cells, maximum oospore length, and so forth, for use in their identification and comparison to fossil forms.

A mathematical means of shape expression was suggested by Horn af Rantzien (1956a), based on the ratio of the length of the polar axis (PA) to the largest equatorial diameter (LED). This ratio is termed the isopolarity index (ISI). **Isopolar** forms are those having two like halves divided by their equator. **Anisopolar** forms have a greatest diameter other than at the equator, and an anisopolarity index (ANI) is based on the ratio between the distance from the apical pole to the largest diameter (AND) to the length of the polar axis (PA). Terminology of shape was then related to these indices (see Table 11.1).

The Basal Pole. Features of the basal pole may be diagnostic. A rounded to pentagonal **basal pore** was originally occupied by a nodal cell that was separated from the oospore by a membrane. External to this membrane the nodal cell may be calcified to form the **basal plug** or plate. The calcified basal plate was demonstrated in sectioned *Nitellopsis* (*Tectochara*) *merianii* by Maslov (1947), and shown to be an important morphologic feature because of its different form in various genera (Grambast, 1956c). The wall between the nodal cell and pedicel cell is not known in fossils, but outside the basal pore, which originally represented the wall, the lime shell may be depressed at the site of attachment of the pedicel cell. Because the basal plug develops later than the lime spirals, it may also appear in various stages of development and thickness. In diameter it ranges from 40 μm (*Chara minutissima*) up to 290 μm in *Nitellopsis* (*Tectochara*) *palaeohungarica* (Rasky) L. Grambast & Soulié-Märsche, and in thickness from a minimum of 20 μm in *N.* (*T.*) *merianii* up to 130 μm in *Gyrogona* sp. The thin basal plate of *N.* (*Tectochara*) has a concave upper face and slightly convex lower one, and the basal plate in *Chara* is a truncated pentagonal pyramid. Other morphological variations characterize other genera, and the presence of a basal plate characterizes the subfamily Charoideae. *Nitella* (subfamily Nitelloideae) does not have a single basal plate, but has three separate elements at the base of the spiral cells, as does fossil *Tolypella*.

A more complex basal structure, the **cage**, may be formed by a downward extension of the intercellular ridges (see Figure 11.9). The apical coronal cells of *Karpinskya* were interpreted as a basal cage, and the vertical orientation of the gyrogonite reversed by Conkin et al. (1974), but this reinterpretation is not widely accepted.

The Apex. The form and arrangement of the cells at the apex is also highly characteristic. The cells may be apically contracted, expanded, elevated or produced, oblique or vertical. They may meet tightly at the apex, or may leave an apical pore (see Figure 11.10). These apical structures are related to the means of dehiscence in germination, seven main types having been defined by Horn af Rantzien (1959a):

Table 11.1
Charophyta polarity indices.

Isopolar forms	ISI or $\dfrac{PA}{LED}$ in percent
peroblate	<50
oblate	50 − 75
suboblate	75 − 88
oblate spheroidal	88 − 100
prolate spheroidal	100 − 114
subprolate	114 − 133
prolate	133 − 200
perprolate	>200

Anisopolar forms	ANI or $\dfrac{AND}{PA}$ in percent
perovoidal	<15
ovoidal	15 − 29
subovoidal	29 − 43
ellipsoidal	43 − 57
subovoidal	57 − 71
obovoidal	71 − 85
perobovoidal	>85

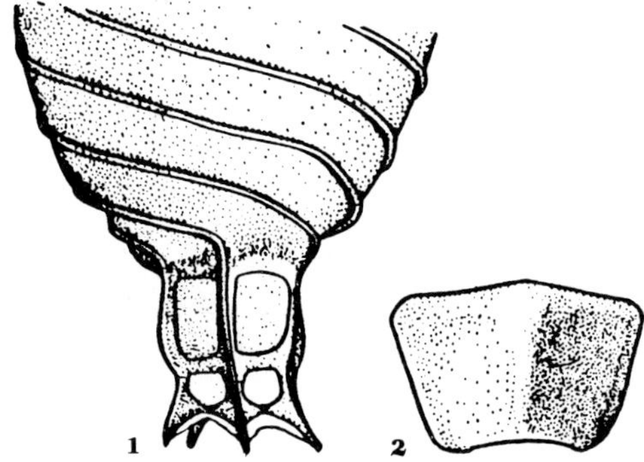

Figure 11.9

Chara hispida Linnaeus, living. **1.** Basal part of gyrogonite, in side view, showing the cage structure formed by three superposed cells; the larger upper cell produces the basal plate, the median thin nodal cell is not calcified, and the basal funnel is formed by the upper part of the bottom cell; ×100. **2.** Side view of the isolated basal plate, ×190. From Grambast, 1956c.

1. *Charoid.* In most modern species the calcareous spiral cells show decreased width and reduced thickness near the apex, forming a zone of dehiscence. No distinct apical depression occurs, but the apex may be weakly calcified; it is not developed as a rosette. *Chara.*

2. *Tectocharoid.* The zone of dehiscence is due to a sharp decrease in width and distinct reduction in thickness of the spirals. No distinct apical depression is present, but the center of the apex is strongly calcified, and developed as a rosette. *Nitellopsis (Tectochara), Nodosochara,* "*Grambastichara,*" some *Sphaerochara,* and possibly modern *Nitellopsis.*

3. *Brevicharoid.* The zone of dehiscence is due to a variable decrease in width, and reduced thickness of spirals in the distinctly depressed apical region. The apical center is strongly calcified, and is commonly developed as a rosette. "*Croftiella,*" *Peckichara, Grovesichara, Gyrogona.*

4. *Sphaerocharoid.* The zone of dehiscence is due to a variable, but usually slight, reduction in thickness of the spirals. No distinct depression occurs in the apical periphery. The spirals

Figure 11.10

Apical structures of charophyte gyrogonites, showing a, side view; b, top view with apical structure; c, basal view. **1.** Charoid, *Chara globularis,* living, ×35; redrawn from Maslov, 1963b. **2.** Tectocharoid, *Nitellopsis (Tectochara) meriani,* Miocene, USSR, ×17, redrawn from Maslov, 1966. **3.** Brevicharoid, *Gyrogona medicaginula* Lamarck, Oligocene, France, ×16, redrawn and used with permission, from Grambast and Paul, 1966, *Bull. Soc. Géol. Fr.,* sér. 7, v. 7, p. 239−247. **4.** Sphaerocharoid, *Maedleriella mangenoti* Grambast, Eocene, France, ×38. **5.** Aclistocharoid, *Rhabdochara stockmansi* Grambast, Lower Oligocene, France, ×30. 4,5, redrawn from Grambast, 1957.

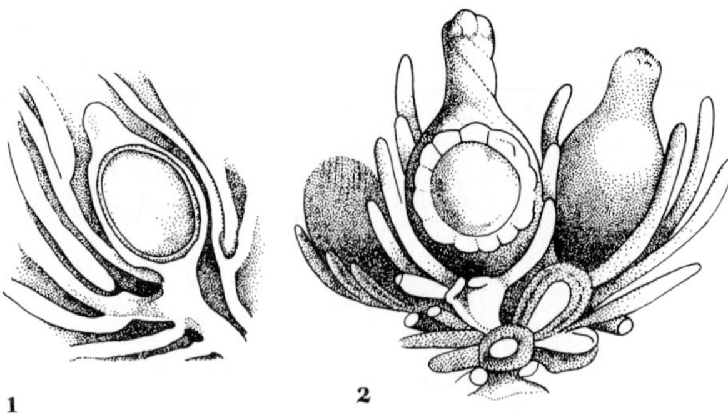

1

2

Figure 11.11
"Lagynophora liburnica Stache,"
Paleocene, Adriatic region, ×24, redrawn from Stache, 1889. **1.** Section.
2. Stache's reconstruction of apically
open, bottle-shaped gyrogonite enclosed in an aberrant utricle. Topotype
material has shown this to be *Microchara* with an external encrustation,
hence belonging to the Characeae
rather than the Clavatoraceae.

are of constant width, the apical center is moderately calcified, and is not developed as a
rosette. *Sphaerochara, Maedleriella.*

5. *Aclistocharoid.* The zone of dehiscence is
due to a variable (usually little or no) decrease in
spiral cell width, and a distinct depression in
the apical periphery. The apical center is
weakly calcified, and not developed as a rosette.
Aclistochara, Rhabdochara, Harrisichara.

6. *Raskyelloid.* The zone of dehiscence is due
to a distinct decrease in width and distinct reduction in thickness of the lime spirals, and by
their complete cessation in the apical
periphery. Five separate opercular cells, thinner than the lime spirals, form a rosette at the
apical center. *Raskyella.*

A seventh type was termed Lagynophoroid,
based on the forms described as *Lagynophora*
(see Figure 11.11), but these have been shown to
represent typical Characeae, such as *Microchara* (Bignot and Grambast, 1969), with external limy deposits.

The gyrogonites may also show surface
sculpture or ornamentation. This may be due
to the greater or lesser degree of calcification of
the lime spirals, resulting in central ridges or
depressions. Other types of ornamentation may
also be present; a nodose surface particularly
characterizes the early Tertiary species (pre-
Miocene), but is rare in Miocene and Pliocene
species and unknown in any living ones. This
ornamentation may be of specific but not
generic importance.

Oospore Membrane. The tough and resistant oospore membrane is of organic composition, although containing suberin and silicic
acid (Nordstedt, 1889), and may itself be preserved under favorable conditions, either
silicified or in the organic state. Attempts to isolate this organic membrane by solution of the
gyrogonite in weak (5 percent) acetic acid are
generally unsuccessful, as the membrane will
crumble. Mechanical removal of the lime shell
with needles may be possible, and it may then
be examined in a dry state, although it will collapse in liquids. Fragments of the membrane of
Tertiary charophytes examined in glycerine
jelly ranged from 3μm in thickness
(*Sphaerochara*) to 15 to 20 μm (*Gyrogona*). The
membrane may have two or three distinct
layers (Horn af Rantzien, 1959b), and also may
have a highly sculptured surface (Horn af Rantzien, 1956a), as shown in Table 11.2. If the
sculpture elements are indistinct, these types,
other than the first, may be prefixed by *semi-*.
To determine the type, the number of
sculptural elements along a 10-μm line is
counted. The sculpture may differ in various
parts of the gyrogonite, perhaps being tuberculate in the lower half and smooth in the upper
part.

Utricle. Members of the Clavatoraceae had an
outer covering or **utricle**, formed by vegetative
branches surrounding the gyrogonite and joining to form an envelope (see Figure 11.12). This

---

A. Simple Sculpture
   Smooth or psilate
   Granulate: maximum diameter of sculpture
      elements $< 2 \times$ smallest diameter; maxi-
      mum diameter < 1 μm.
   Tuberculate: maximum diameter of
      sculpture elements $< 2 \times$ smallest diam-
      eter, maximum diameter > 1 μm.
   Reticulate: maximum diameter of sculpture
      elements $> 2 \times$ smallest diameter, form-
      ing a network.
B. Combined Sculpture
   Granulate-reticulate: maximum diameter of
      sculpture elements < 1 μm, forming a
      network.
   Tuberculate-reticulate: maximum diameter
      of sculpture elements > 1 μm, forming a
      network.
   Granulate-tuberculate: sculpture elements
      of varying size, not forming a network.

---

adheres closely to the calcified gyrogonite, but
may be exfoliated in varying degree. In highly
developed species, the separate vegetative elements are no longer discernible, and the utricle
may have a highly ornamented surface with
ridges, pits, and other characteristic sculpture.
Within the calcified utricle, the gyrogonite may
also be calcified, although less so than is usual
in charophytes lacking the utricle. The
Oligocene *Anomalochara* (see Figure 11.13), described as utriculate, is based on poorly preserved material that leaves doubtful many morphological details, but is probably not a
charophyte (Grambast, 1964a). Its geologic occurrence is much younger than other
Clavatoraceae.

Germination. After the thickened oospore
membrane and the enclosing lime shell form
around the zygospore, the outer cell walls
decay and the gyrogonite falls to the bottom.
Here it may remain for weeks or longer before
germination. Nuclear reduction division
(meiosis) occurs prior to germination, so that
only the zygospore is diploid, the vegetative
plant being entirely haploid.

Growth of the developing zygospore within
may cause the rupture of the entire gyrogonite,
commonly near its equator. This primitive
method of germination is characteristic of
some Mesozoic and Tertiary charophytes. Others have various features of the apex that allow
easier germination through a weakened or less
calcified area or through an opening resulting
from the shedding of specialized opercular
cells.

The germinating zygospore gives rise to a
filamentous proembryo, from which develop
rhizoids and a short horizontally growing **protonemata** (see Figure 11.14). The new erect
plant arises as an axilary shoot from the protonemata.

Ecology

Charophytes are commonly abundant, and
occur nearly everywhere in freshwater pools
and lakes, the same species being found in
bodies of water of a wide size range (Langangen,
1974). They are present in running water only
where movement is slight. One species, *Nitella
terrestris* Iyengar, is adapted to live on moist
soil, where it forms small cushionlike plants in
south India. This only known terrestrial species
is not known to live submerged anywhere else
(Iyengar, 1960). Charophytes also are found in
brackish water of coastal bays, lagoons, and
fjords. They occur at present on all continents
except Antarctica, from sea level to an altitude
of 4764 m (15,630 feet) in the Himalaya Mountains, and from the tropics to the Arctic in Alaska
and Spitzbergen. They are more abundant in
temperate areas, however.

Depth

Although they grow submerged in the water,
charophytes may occur from near the shore to
depths of 18 m in the Finger Lakes of New York,

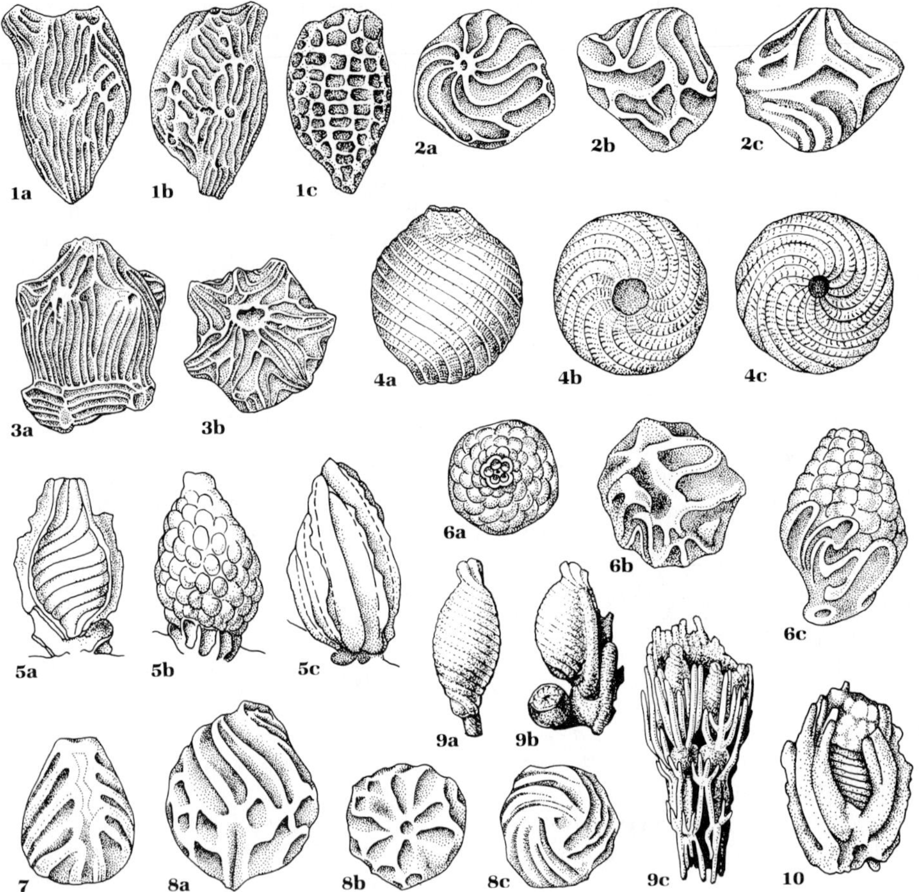

Figure 11.12

Clavatoraceae. **1.** *Ascidiella iberica* Grambast, Lower Cretaceous, Spain, opposite sides and inner layer, ×16, based on Grambast, 1966a. **2.** *Clypeator corrugatus* (Peck) Grambast, Lower Cretaceous, Idaho. 2a,b, top and side views, ×32, based on Peck, 1957; 2c, side view, ×46, redrawn from Maslov, 1963b. **3.** *Septorella brachycera* Grambast, Upper Cretaceous, France, side and top, ×28, redrawn from Grambast, 1964a. **4.** *Globator trochiliscoides* Grambast, Lower Cretaceous, Spain, side, top, and base, ×20, redrawn from Grambast, 1966b. **5.** *Clavator reidii* Groves, Upper Jurassic, England, longitudinal section, side view, and utricle, ×40, redrawn from Harris, 1939. **6.** *Nodosoclavator nodosus* (Peck) Maslov, Lower Cretaceous, South Dakota, top, base, and side views, ×27, based on Peck, 1957. **7.** *Flabellochara harrisii* (Peck) Grambast, Lower Cretaceous, Wyoming, side view, ×27, based on Peck, 1941. **8.** *Atopochara trivolvis* Peck, Lower Cretaceous, Oklahoma, side, base, and top, ×18, based on Peck, 1938. **9.** *Diectochara andica* Musacchio, Lower Cretaceous, Argentina; 9a, gyrogonite; 9b, gyrogonite and antheridium; 9a,b, ×27; 9c, vegetative branch with gyrogonites, ×14; all from Musacchio, 1971. **10.** *Echinochara spinosa* Peck, Upper Jurassic, Colorado, side view, ×27, based on Peck, 1957.

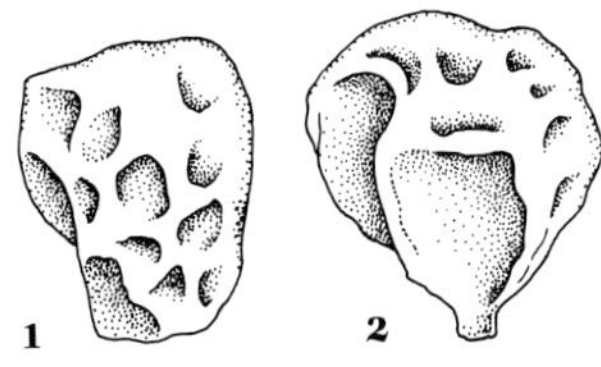

Figure 11.13
Anomalochara polymorpha Maslov, lower Oligocene, USSR, ×86, regarded as a clavatoracean, but now regarded as not a plant, redrawn from Maslov, 1961.

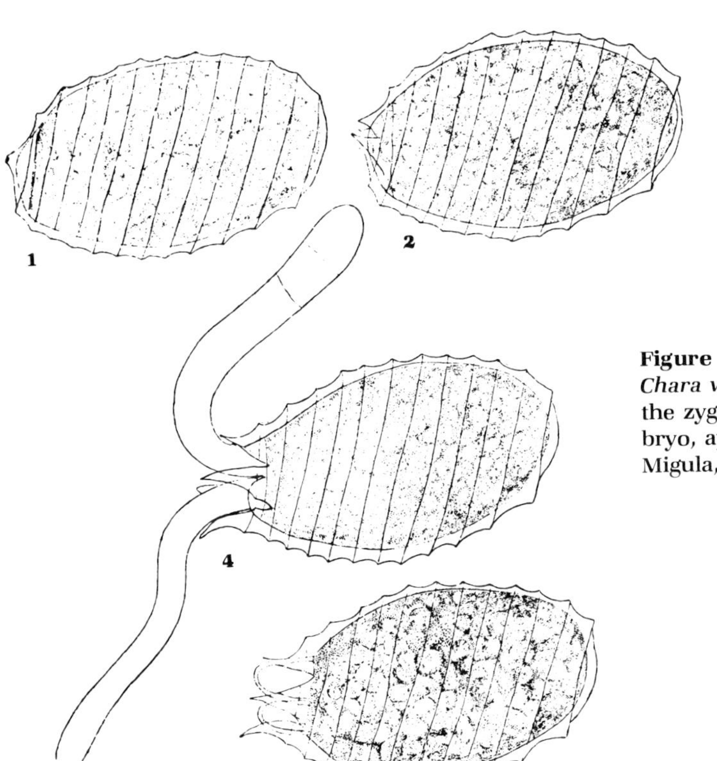

Figure 11.14
Chara vulgaris. Germination of the zygospore to form proembryo, apex at left, ×100. From Migula, 1897.

to nearly 30 m in the Konigsee in Germany, and commonly range to 30 m in some Japanese lakes. The depth to which they occur depends largely on the availability of light and the transparency of the water. The largest number are found in relatively shallow water, from about 1 to 8 m in depth. Different species characterize various depths, and a distinct succession of species of *Chara* and *Nitella* is interspersed with other types of vegetation, from the surface down to the greatest depths. Some are transitory elements of the succession in ponds, abundant for a few years and then replaced by other elements. Other species may persist for many years. These features seem also to characterize fossil floras; in correlative strata, one species may locally be more abundant than another. Commonly a particular environment has only one or very few species of charophytes at a single time.

Salinity

Although the charophytes may occur in fresh water to brackish water up to two-thirds the salinity of seawater, the individual species may be strongly controlled by salinity. Some modern species are restricted to brackish water; for example, *Chara canescens* Desvaux, *C. galioides* DeCandolle, *Tolypella hispanica* Norstedt ex T. F. Allen, and *Lamprothamnium papulosum*. The latter has been reported as characteristic of marginal brackish to marine waters in the Baltic Sea, English Channel, North Sea, Norway, western Mediterranean Sea, and the Atlantic coast. It is a common pioneer element of the coastal marshes. The brackish species are commonly found close to shore.

Oxygenation

Charophytes dominate the vegetation in oligotrophic lakes (little organic matter and well oxygenated) situated on a lime-rich substrate, to such an extent that such lakes are termed *Chara* lakes. They may also form dense belts in the *Lobelia*-lakes, similarly oligotrophic, but on lime-poor substrates. Although they can exist in poorly oxygenated water, and even on a substrate containing a considerable quantity of hydrogen sulfide, charophytes are sensitive to phosphorus and will not grow in polluted waters. They are abundant only in water containing less than 20 μg phosphorus per liter. In the clear water of oligotrophic lakes they form extensive subaqueous meadows; vascular plants such as the pondweeds offer more competition in eutrophic lakes (*Potamogeton* lakes) of higher productivity, although the same species of charophytes may occur as in the *Chara*-lakes or *Lobelia*-lakes (Langangen, 1974).

Lime

Chara lakes are commonly found in high-lime areas. The water is clear and phytoplankton sparse. Although some species may deposit lime, significant amounts are deposited only by a few species, such as *Chara vulgaris*. The presence of calcified oogonia is not related to the amount of lime present in the medium, as the intracellular deposition merely transforms organic calcium compounds in the cell sap (such as calcium succinate) into calcite. However, external calcification of the thallus is related to the calcium content of the water and to photosynthesis. The aquatic plants not only may use CO_2 in photosynthesis, but also H_2CO_3, HCO_3^-, and CO_3^{2-}. These may be combined with calcium ion and $Ca(OH)_2$ emitted. Reaction of the latter with $Ca(HCO_3)_2$ causes precipitation of $CaCO_3$ on the exterior of the plant in the form of tiny, irregularly placed crystals. In a calcium-rich environment, such external deposits may also appear at the exterior of the gyrogonites, even on the sporophydia of species that normally do not deposit lime (Nitelloideae).

Hydrogen Ion

Charophytes may occur in waters ranging from 5.2 to 9.8 pH, but are more characteristic of a pH near 7.0 or 8.0. Acidic waters are less favorable than alkaline ones.

Olsen (1944) related charophyte distribution to pH in Danish waters. Lakes of constantly highly acidic water and pH below 5.3 had no charophytes, nor did those with variable pH and acidity fluctuating between slightly acid and highly so. Prevalently acidic lakes with fluctuating acid-alkaline conditions, lime deficiency, and a minimum pH below 5.3, although pH fluctuations were rapid, contained species of *Nitella* (noncalcareous), such as *N. flexilis* and *N. translucens* (Persoon) Agardh (taxonomy modified to agree with Wood and Imahori, 1965). Prevalently neutrally alkaline lakes, but with fluctuating acid-alkaline waters, a minimum pH greater than 5.3, and a low lime concentration (down to 30 mg CaO per liter) but with some lime encrustation, included the above species of *Nitella*, and *N. gracilis* and "*Chara fragilis* Desvaux" = *C. globularis*. Lakes that were constantly alkaline, with little fluctuation in pH, and a lime content greater than 30 mg per liter of CaO, had the greatest number of species, including all the above (except *Nitella translucens*), in addition to *Chara canescens, C. hispida*

Figure 11.15
Tolypella nidifica (O. Müller) v. Leonhardi, living; a species of alkaline, high-calcium lakes; young plant, ×1. From Migula, 1897.

and varieties, *C. globularis* var. *aspera, C. tomentosa* Linnaeus, *C. vulgaris* and varieties, *Lamprothamnium papulosum, Nitella furcata* (Roxburgh ex Bruzelius) Agardh subsp. *mucronata, N. syncarpa* (Thuillier) Chevallier, *Nitellopsis obtusa, Tolypella intricata* (Trentepohl ex Roth) Leonhardi, *T. nidifica,* and *T. nidifica* var. *glomerata* (Desvaux in Loiseleur-Deslongchamps) R. D. Wood (see Figure 11.15).

FOSSIL CHAROPHYTES

Geologic Occurrence

Fossil charophytes have long been known, although some have been variously regarded as foraminifers, echinoderms, or molluscs. The genus *Gyrogona* was described by Lamarck (1804) among genera of foraminifers, which also were then regarded as molluscs. The only species then known to Lamarck was a fossil from the Paris Basin, hence in agreement with his system of indicating fossils by the generic termination *-ites,* the type species was described as *Gyrogonites medicaginula.* The generic diagnosis was given for *Gyrogona,* hence that name is correct, and the date of validation is 1804, not 1822 as sometimes stated. Lamarck was uncertain of their animal nature, and commented that the specimens had the form of a very small fruit of some species of lucerne (alfalfa). The earliest illustrations of the species appeared in 1807 (pl. 15, fig. 7a — c), and shortly thereafter the relationship with modern *Chara* was demonstrated.

The calcified gyrogonites commonly are associated with freshwater molluscs, ostracods, fish teeth, and wood fragments in nonmarine sediments, but they may also occur in brackish or near-shore sediments. Records of transportation by modern streams and currents suggest that fossil forms previously regarded as marine in habitat may have been similarly transported from a coastal swampy environment. Charophyte oogonia have been observed in modern sediments 80 km off the coast of Brittany, and co-occurrences with typical marine fossils have been reported from Jurassic, Lower Cretaceous, and Tertiary sediments. Possibly the Paleozoic Trochiliscales and Sycidiales similarly represent coastal brackish deposits, rather than truly marine ones.

In addition to the gyrogonites, a few calcified antheridia are known, although they are rare. Silicified stems, antheridia, and gyrogonites also have been recorded. In places the Morrison Formation may be almost a charophyte limestone, and silicified stems and other parts are known from the Jurassic Purbeck of England and the Lower Cretaceous of Idaho, Wyoming, and Texas.

Lime is deposited upon the vegetative portions of modern charophytes but is seldom preserved in this form, as the surficial calcification, unlike the laminated intercellular deposit of the gyrogonites, readily disintegrates into a fine-grained structureless lime mud. A fragment of a vegetative branchlet, preserved as an organic impression on a siltstone surface (Daily and Durham, 1966), was found to have an attached gyrogonite of *Nitellopsis* (*Tectochara*). Similar remains of vegetative branches have been described from scattered occurrences.

The value of charophytes for stratigraphy was at first obscured by too broad a species concept, allowing too wide a variability. Thus many species were referred to the genus *Chara,* although readily differentiated from the living genus. A similar problem is common to many microfossils. Use of closer limits for charophyte genera and species has shown that many had a short existence and are good index fossils (see Figures 11.16 and 11.17) for age determination of nonmarine beds at least to the period level. Characteristic Jurassic, Lower and Upper Cretaceous, and Tertiary assemblages have been differentiated. Locally, they have been useful for much closer stratigraphic delineation, at least to the level of stages in the near-shore Paleogene deposits of the Paris Basin (Grambast, 1962a).

Umbellinaceans

The umbellinaceans include small flasklike or globular specimens of Devonian and Mississippian age (see Figures 11.18 and 11.19), which are known extensively in Europe, Asia, Australia, and North America, but whose relationships are obscure. *Umbella* Maslov (in Bykova and Polenova, 1955) was described as a foraminifer, but as the name was a homonym it was renamed *Umbellina* (Loeblich and Tappan, 1961). However, when various species were shown to possess an operculum or lid, any relationship to the foraminifers was disproved (Maslov in Bykova and Polenova, 1955; Poyarkov, 1964, 1965, 1966; Mamet, 1970), and they were transferred to the charophytes (Miklukho-Maklay, 1961; Chuvashov, 1965; Poyarkov, 1965). Under the ICBN, *Umbellina* also has priority (March, 1961) in the plant kingdom over *Umbella* (Miklukho-Maklay, May — June, 1961), as well as being the correct name in the animal kingdom.*

---

* *Umbella lineata* Brazhnikova & Berchenko 1971, in O. I. Berchenko, Charophyta, *Atlas Fauny Turneyskikh Otlozheniy Donetskogo Basseyna* (*s opisaniem novykh vidov*), Akad. Nauk Ukrainskoy SSR, Inst. Geol. Nauk, p. 115, pl. 81, figs. 7 — 9, is herein transferred to *Umbellina lineata* (Brazhnikova & Berchenko) Tappan (see Figure 11.19 part 12).

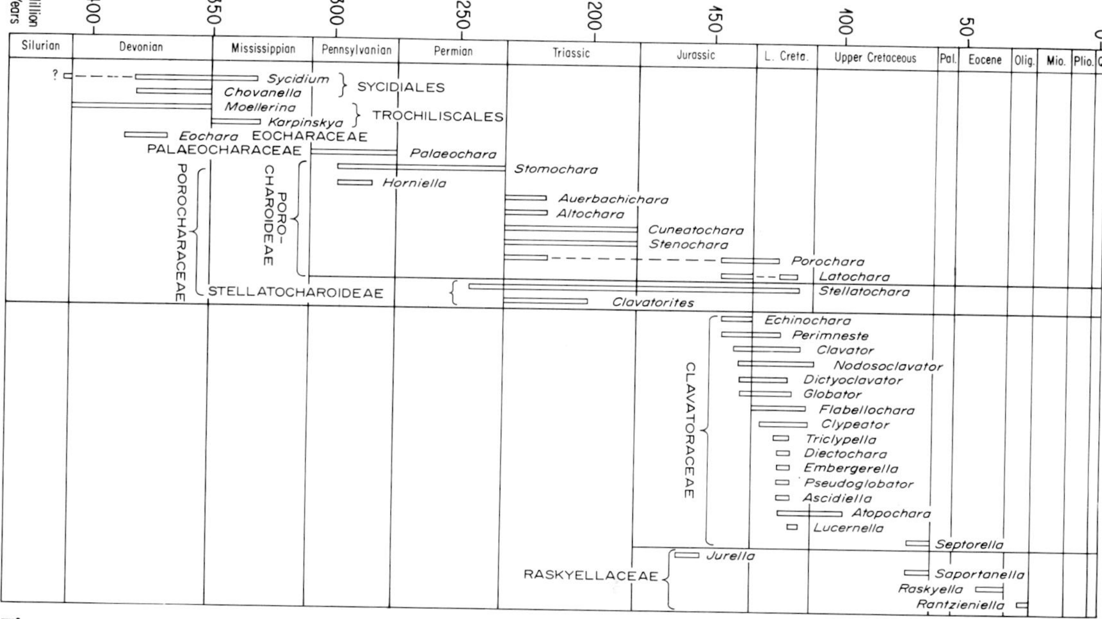

Figure 11.16
Geologic occurrence of charophyte genera, exclusive of the Characeae.

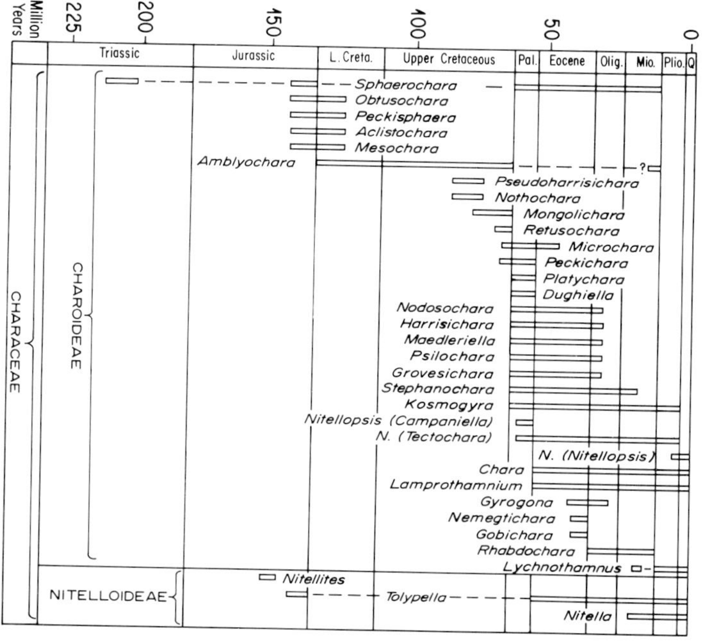

Figure 11.17
Geologic occurrence of genera
of the Characeae.

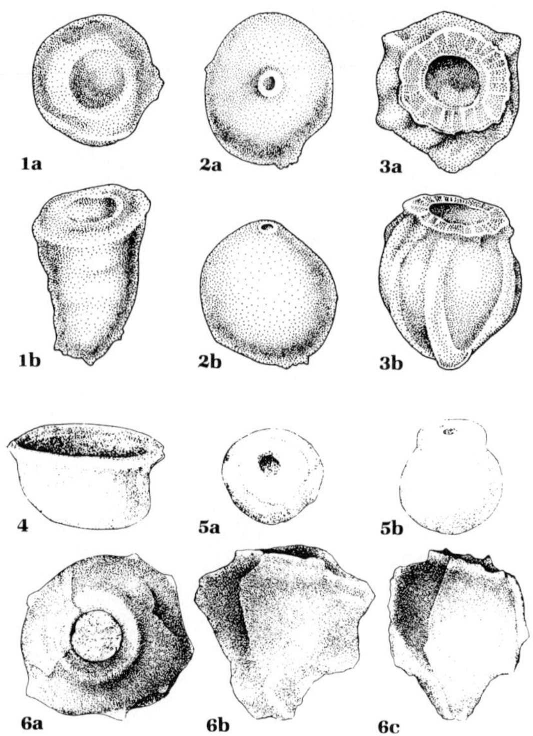

Figure 11.18
Umbellinaceae. All from the Upper Devonian, USSR; reconstructions; a, top view; b, side view. **1.** *Umbellina bella* (Maslov) Loeblich & Tappan, ×57. **2.** *Quasiumbella rotunda* (Bykova) Poyarkov, ×45. **3.** *Elenia famena* (Bykova) Poyarkov, ×40. **4.** *Planoumbella patella* (Bykova) Platonov, side view, ×50. **5.** *Protoumbella saccamminiformis* (Bykova) Mamet, ×50. **6.** *Eoumbella ollaria* (Bykova) Platonov; 6b,c, side views; ×50. All from Bykova and Polenova, 1955, in part redrawn.

Figure 11.19 (*facing page*)
Umbellinaceae. **1.** *Biumbella brazhnikovae* Mamet, Lower Mississippian, USSR, longitudinal sections, ×56, from Brazhnikova and Rostovtseva, 1966. **2.** *Protoumbella saccamminiformis*, Upper Devonian, USSR, transverse and longitudinal sections, ×80, from Bykova and Polenova, 1955. **3.** *Elenia aculeata* Berchenko, Lower Mississippian, USSR, longitudinal and transverse sections, ×56, from Berchenko, 1971. **4.** *Lagenumbella lageniformis* (Reytlinger) Mamet, Upper Devonian, USSR, longitudinal sections, ×80, from Reytlinger, 1966. **5.** *Archaelagena howchiniana* (Brady) Howchin, Mississippian, England, longitudinal section and exterior, ×40, from Brady, 1876. **6.** *Quasiumbelloides laevis* Berchenko, Lower Mississippian, USSR, longitudinal sections. 6a,b, ×72; 6c, ×56; from Berchenko, 1971. **7.** *Plavskina piriformis* Reytlinger, Upper Devonian, USSR, top and side views, ×48, from Reytlinger, 1960. **8.** *Planoumbella costata* Platonov, Upper Devonian, USSR, longitudinal sections, approximately ×56. **9.** *Spinumbella spinifera* Platonov, Upper Devonian, USSR, longitudinal and transverse sections, ×56. 8,9, from Platonov, 1974. **10.** *Pseudoumbella donbassica* Berchenko, Lower Mississippian, USSR, longitudinal and transverse sections, ×56, from Berchenko, 1971. **11.** *Eoumbella ollaria*, Upper Devonian, USSR, longitudinal and transverse sections, ×80, from Bykova and Polenova, 1955. **12.** *Umbellina lineata* (Brazhnikova & Berchenko) n. comb., Lower Mississippian, USSR, longitudinal section, ×56, from Berchenko, 1971. **13.** *Quasiumbella pseudorotunda* Brazhnikova & Berchenko, Lower Mississippian, USSR, longitudinal sections. 13a,c, ×64; 13b, ×56; 13a, from Brazhnikova and Rostovtseva, 1966; 13b,c, from Berchenko, 1971.

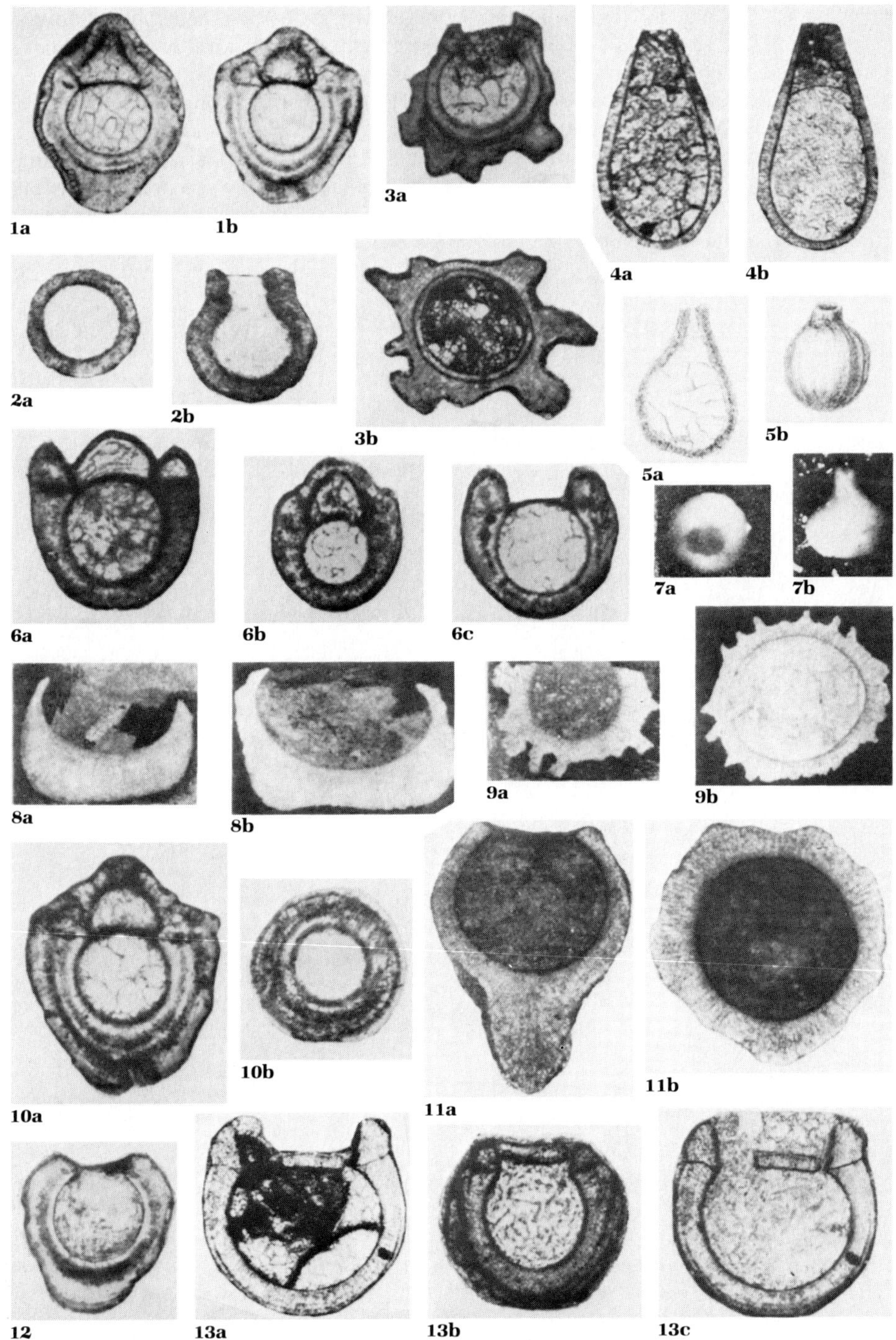

1a
1b
3a
4a
4b
2a
2b
3b
5a
5b
6a
6b
6c
7a
7b
8a
8b
9a
9b
10a
10b
11a
11b
12
13a
13b
13c

The umbellinaceans were considered either as primitive charophyte gyrogonia, somewhat similar to *Chovanella,* or perhaps as utricles, as most lack any vertical structures such as the vertical or spiralling cells of other charophyte gyrogonia (Poyarkov, 1964, 1965, 1966). Like the trochiliscaceans, the umbellinaceans are restricted to the littoral facies, probably in areas of decreased salinity, and are rare in strata containing foraminifers. Because of their general similarity in form, size and habitat, a charophyte affinity has been accepted by some, but many differences from other charophytes have been enumerated by others (Peck, 1974; Grambast, 1974). Umbellinaceans have openings at each pole; the apical one may be closed with a lid (see Figure 11.20), and a basal plate may be present at the opposite pole. Charophytes have no apical lid, and while gyrogonites may have a basal plate, the utricle does not. The charophyte gyrogonite also has a laminated wall structure, whereas the umbellinacean wall has a radial structure (see Figure 11.21). Allocation of the umbellinaceans to the

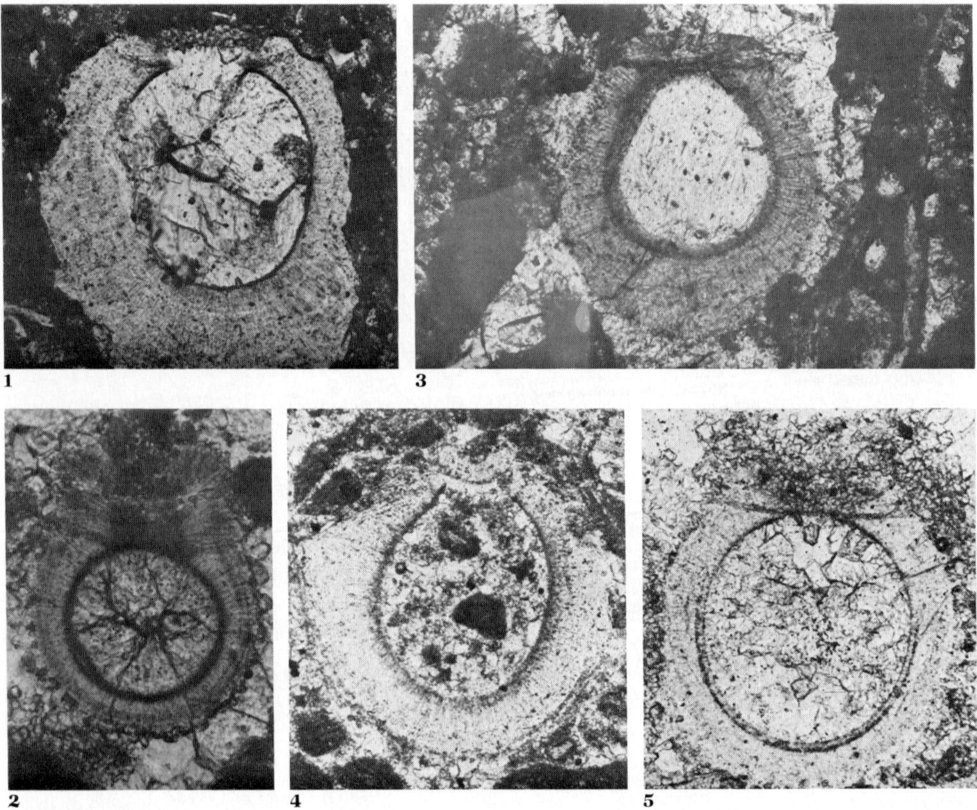

Figure 11.20
Umbellinaceae. **1.** *Umbellina* sp., section showing radial wall structure. **2.** *Protoumbella* sp. 1,2, Upper Devonian, Australia; reproduced by permission of the National Research Council of Canada, from Mamet, *Canadian Journal of Earth Sciences,* v. 7, p. 1164–1171, 1970. **3–5.** *Quasiumbella* spp., from B. L. Mamet. **3.** Upper Devonian, Australia, section showing thick radial wall, and apical opening with lid. **4.** Upper Devonian, Canada, showing apical lid. **5.** Lower Mississippian, Australia, showing apical cover. All ×97.

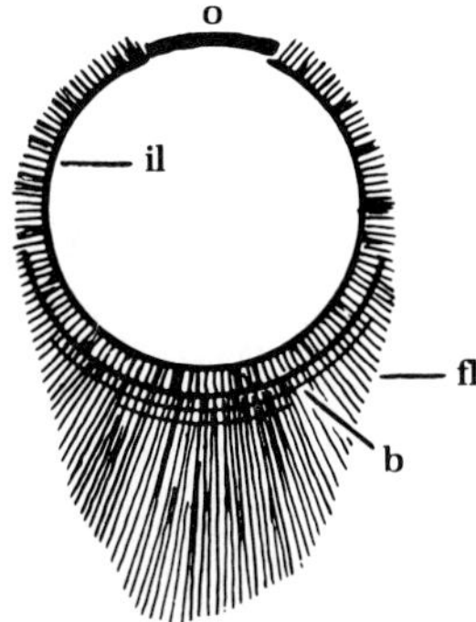

Figure 11.21
Umbellina, diagrammatic section, showing radial structure of the fibrous layer (fl) of the wall, the dark inner layer (il), buttresses (b), and operculum (o). Modified from Conil and Lys, 1964.

Charophyta remains questionable unless they represent an early offshoot that left no descendents.

The very tiny hollow calcareous spheres that have been reported from the Cambrian to the Permian as calcispheres also have been attributed to the charophytes by some, largely because two species of the Devonian *Moellerina* were originally described as *Calcisphaera lemoni* Knowlton and *Calcisphaera robusta* Williamson. Other calcispheres are not trochiliscids and show no differentiation of apical and basal regions as do charophytes. They may be of algal origin, although the exact affinities are unknown.

Paleozoic Charophytes

Post-Paleozoic charophytes have always been recognized as such, but the relationship of the Paleozoic Sycidiales and Trochiliscales was less certain. *Sycidium* was described by Sandberger (1849), and the trochiliscids by Pander (1856, p. 17), who used only the vernacular descriptive term "Trochilisken." Ehrenberg (1858) formally described the latter from the Upper Silurian or Lower Devonian of St. Petersburg, as a foraminiferan *Miliola* (*Holococcus?*) *panderi*, but also compared them to "*Chara*-seeds." He illustrated the species in 1862. These Paleozoic forms also were referred to lycopod spores, fish eggs, phyllopod eggs, echinoderms, and siphonalean algae, as well as foraminiferans. Peck (1934a) showed that the silicified trochiliscids were charophytes, and the preserved oospore membrane present on *Sycidium* also showed it to be a related form.

Gyrogonites of the Sycidiales lack the characteristic spiralling cells of the other orders, having vertically aligned cells that may be simple in *Chovanella* or subdivided by horizontal partitions in *Sycidium* (see Figure 11.22). Maslov (1961) suggested that the Paleozoic *Sycidium* and *Chovanella*, and possibly *Umbellina* and related forms, might be fossilized utricles rather than gyrogonites. Others regard *Sycidium* as a gyrogonite in which more numerous cells were necessary to enclose the oospore, because these did not spiral. Originally 9 cells were present, but these divided vertically to form about

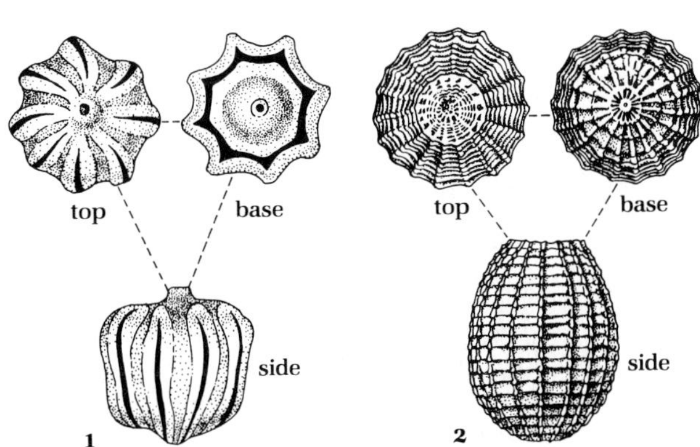

Figure 11.22
Sycidiales. **1.** *Chovanella kovalevi* Reytlinger & Yartseva, Upper Devonian, Russian Platform, ×43, redrawn from Reytlinger and Yartseva, 1958. **2.** *Sycidium melo* Sandberger, Devonian, USSR, ×21, redrawn from Karpinskiy, 1906.

Figure 11.23

1 – 3. *Sycidium* cf. *reticulatum* Sandberger. **1.** Longitudinal section, with pyrite replacement of organic material, ×43. **2.** Part of transverse section enlarged to show microstructure, ×133. **3.** Longitudinal section in reflected light, showing pyritized pore fillings, ×40. **4 – 6.** *S. volborthi* Karpinskiy subsp. *eifelicum* Langer, SEM. **4.** Base, viewed obliquely, ×67. **5.** Apical view of internal cast with unusually strong ribs, ×83. **6.** Lateral view, ×104. All from Middle Devonian, Germany, from Langer, 1976.

18, although *S. clathratum* Peck has only 16 cells. The numerous horizontal divisions of these vertical cells were compared to the formation of the coronula in modern charophytes, division merely beginning earlier on the sporostegium.

Unlike other charophytes, the wall of *Sycidium* is pierced by pores (see Figure 11.23), commonly pyrite-infilled in fossil specimens, hence *Sycidium* was assigned to a new Class Sycidiphyceae (Langer, 1976). *Kusjaella* also was referred to the Sycidiaceae (Chuvashov, 1973), but is known only from thin sections.

Gyrogonites of the Trochiliscales are closer in appearance to those of the Charales, but differ in having a larger number of cells that spiral dextrally rather than sinistrally (see Figure 11.24). Two genera are included, *Moellerina*, of which *Trochiliscus* is a junior synonym, and *Karpinskya* (Peck and Morales, 1966). The nature of *Moellerina* had earlier been uncertain, as the original illustrations were drawn directly on lithographic stone, and thus reversed in printing to show an apparent sinistral instead of dextral spiral. Also described as a foraminifer, *Weikkoella* was based on siliceous

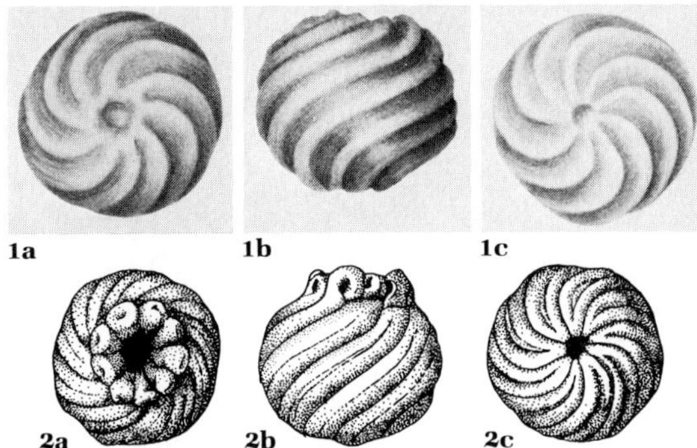

Figure 11.24
1. *Moellerina greenei* Ulrich, Middle Devonian, Indiana, apical, side, and basal views, ×27, from Peck and Morales, 1966. **2.** *Karpinskya laticostata* (Peck) Peck & Morales, Lower Mississippian, Missouri, ×18, redrawn from Peck, 1934a.

steinkerns of the oospore of *Moellerina,* hence is synonymous (Conkin et al., 1970). The vertical orientation of *Moellerina* was reversed and the coronal cells reinterpreted as a basal cage by Conkin et al. (1970, 1974), but the classical orientation is accepted herein.

Both the Trochiliscales and Sycidiales have short geologic ranges (Devonian to Lower Mississippian); a report of Silurian Sycidiales in Turkestan has not been confirmed as yet. Already well differentiated in the Early Devonian, the charophytes probably had a somewhat longer prior period of evolution and radiation, but no known early forms indicate the origin of the characteristic and complex gametangia of the charophytes. Not all modern charophytes deposit lime, and perhaps the early evolution was unrecorded as the ancestral forms had not developed the capability for calcification; this ability later arose in various branches at different times.

Possibly their early evolution was extremely rapid, however, in response to changing environmental conditions. Living charophytes are abundant only when the phosphorus level is below 20 μg per liter of water, other organic matter is limited, and the water is well oxygenated. The land plants that evolved in the Late Silurian and Devonian probably retained many nutrients on land that formerly could wash into the lakes and seas; the reduced supply of nutrients favored the rapid expansion of any group adapted to such conditions. Thus the Sycidiales and Trochiliscales may have appeared and evolved rapidly (Tappan and Loeblich, 1973).

When organic matter increased in the shallow waters, as is indicated by the extensive black shales of the Upper Devonian and Mississippian and the coal deposits of the Mississippian and Pennsylvanian, these primitive charophytes died out, and the charophytes in general remained rare in the late Paleozoic. Similarly, charophytes disappear from modern lakes when these become slightly more eutrophic and productive (*Potamogeton*-lakes).

Although the Trochiliscales and Sycidiales left no descendents, the Charales also first appeared in the Devonian. Gyrogonites of these rare early Charales, the Middle Devonian Eocharaceae, differed from more modern ones in having more numerous cells (greater than six) surrounding an apical pore, and from the Trochiliscales in the sinistral spiralling (see Figure 11.25). *Palaeochara,* known from a half-dozen specimens from the Nova Scotia Pennsylvanian, has six spiral cells, and was placed in a distinct family. However, rare specimens of various species of *Chara,* "*Grambastichara*", *Rhabdochara,* and *Tectochara* have been found with four or six spiral cells (Glukhovskaya, 1975), suggesting that *Palaeochara* also consists only of anomalous specimens and does not warrant separate family or generic status. The Porocharaceae also appeared in the Pennsylvanian, and have the five spiralling cells typical of other Charales.

The late Paleozoic thus has a very sparse record of charophytes, the previously abundant trochiliscaceans and sycidiaceans having disappeared completely, with only the rare poro-

Eocharaceae
Palaeocharaceae

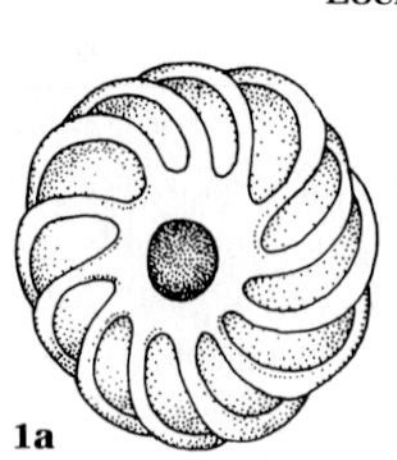
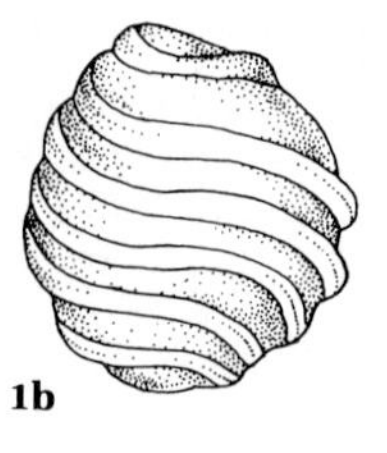

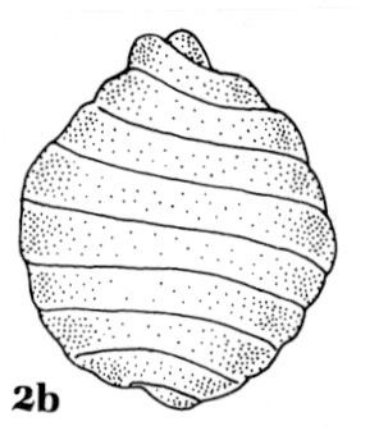

1a
1b
2a
2b

Porocharaceae

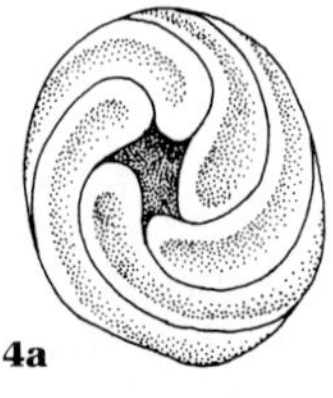
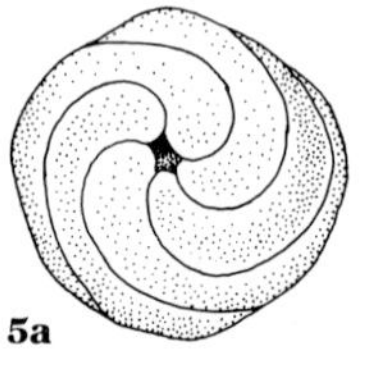

3a
4a
5a
6a

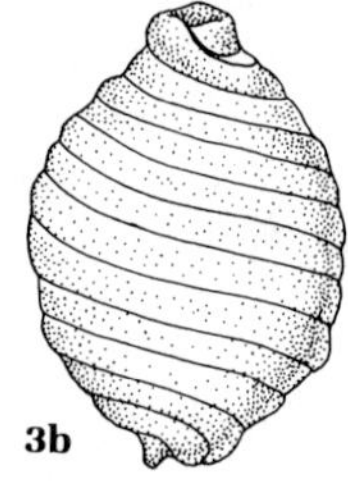
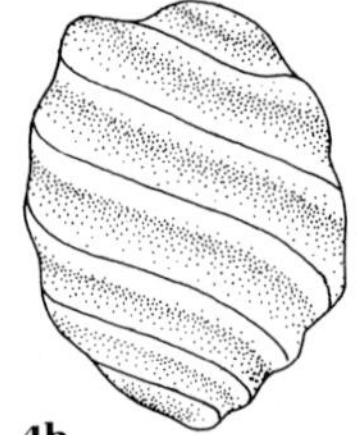
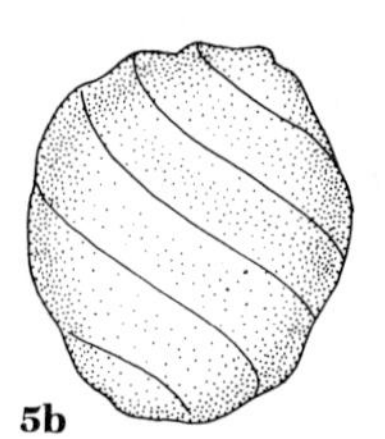

3b
4b
5b
6b

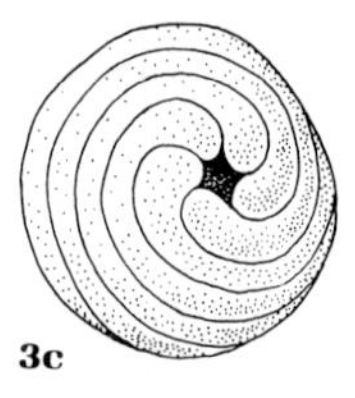
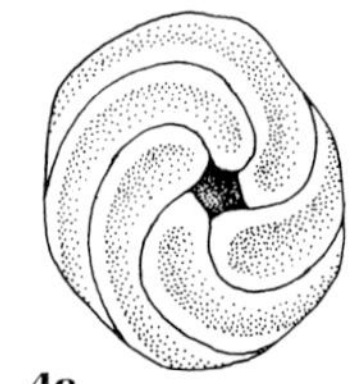
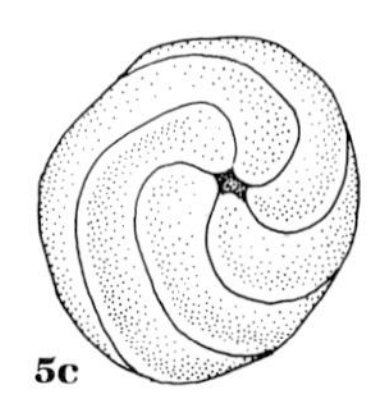
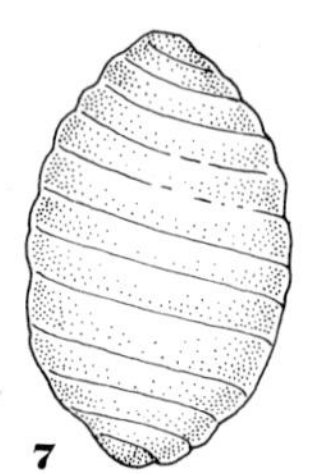

3c
4c
5c
7

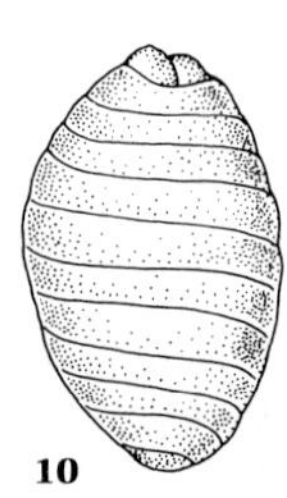

8a
8b
9a
9b
10

characeans providing the basis for the later Mesozoic expansion. Their rarity in the late Paleozoic obscures the record of the evolutionary changes in the group, as noted in the quotation at the head of this chapter.

Mesozoic Charophytes

Triassic charophytes have been described from the USSR (see Figure 11.26; Saydakovskiy, 1962, 1966b, 1968), Sweden (Horn af Rantzien, 1954b), Germany (Kozur and Reinhardt, 1969), and Denmark, and from the Upper Triassic of Arizona, Colorado (Peck and Eyer, 1963b), and Wyoming. All Triassic charophytes belong to the Porocharaceae, other than four species questionably assigned to *Sphaerochara* by Saydakovskiy (1966b, 1968).

Within the Porocharaceae, charophyte species diversity increased sharply in the Early Triassic over the number present in the Late Permian (see Figure 11.27), in marked contrast to the pattern in the phytoplankton groups of the same period, but in agreement with the greatly reduced level of organic matter in sediments of the Triassic, suggesting the presence of nutrient-deficient waters to which the charophytes are well adapted. Expansion of the charophytes continued in the Mesozoic, their maximum generic diversity occurring in the Late Jurassic and Early Cretaceous (see Figure 11.28).

Among the characteristic Mesozoic charophytes are the Clavatoraceae, which evolved in

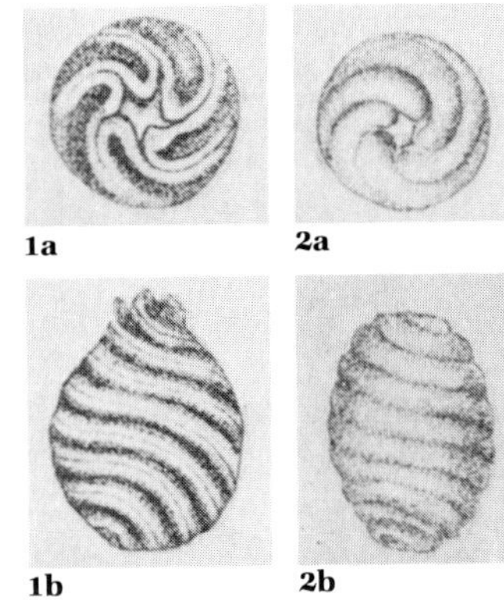

Figure 11.26
Lower Triassic charophytes, USSR. **1.** *Auerbachichara saidakovskyi* Kiselevskiy, apical and side views, ×57. **2.** *Altochara continua* Saydakovskiy, apical and side views, ×57. Both from Saydakovskiy, 1968.

the Jurassic from an ancestor similar to *Stellatochara* and expanded rapidly in the Early Cretaceous (see Figures 11.29 and 11.30). The Clavatoraceae have an external cover of vegetative elements around the gyrogonite, and an apical region produced as a neck that is topped by a pore. This utricle became progressively more specialized, culminating in the Cretaceous in the highly developed and ornate *Atopochara*, which originally was described from the United States, but later was reported from Europe, Asia, and North Africa. An evolutionary sequence from *Perimneste* to *Atopochara* was shown to involve a progressive

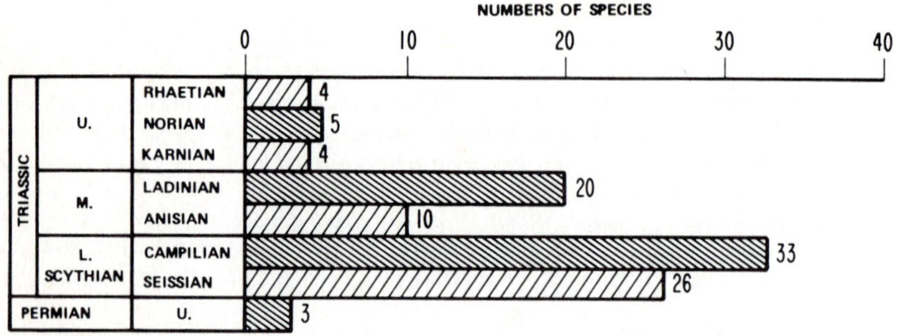

Figure 11.27

Species diversity of Charophyta at the Paleozoic-Mesozoic boundary. The Early and Middle Triassic diversification may be a reflection of low nutrient levels in the water, to which charophytes are adapted. As various groups of phytoplankton appeared and became abundant in the Late Triassic, suggesting an increased nutrient supply, the charophytes declined. From Tappan and Loeblich, 1973.

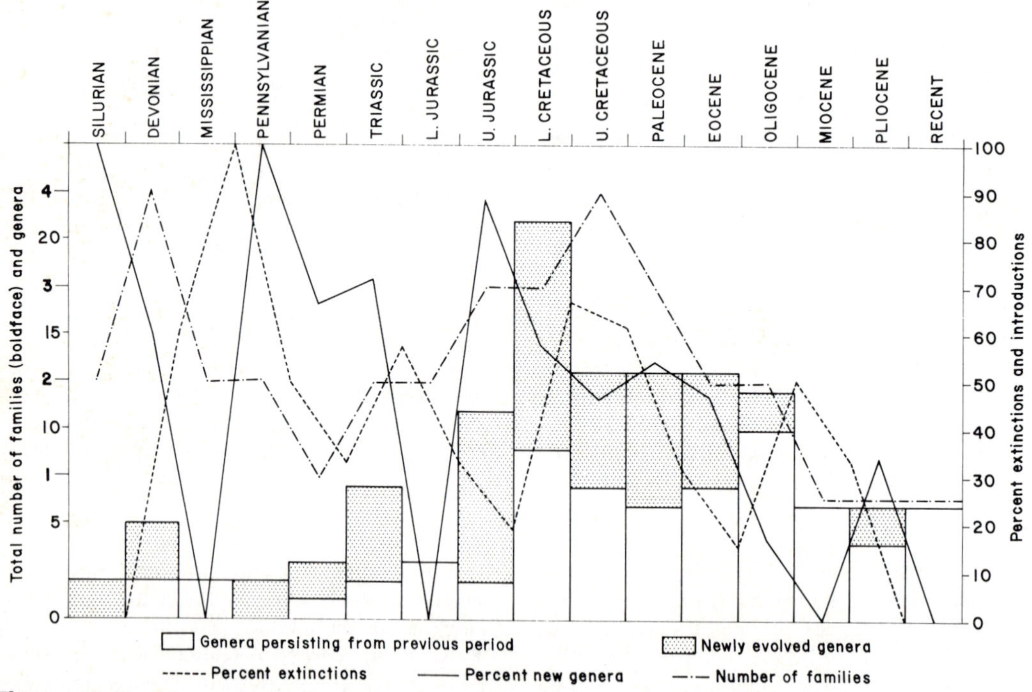

Figure 11.28

Generic and family diversity of charophytes. Generic diversity dropped sharply at the end of the Devonian and at the end of the Triassic, attained a maximum in the Lower Cretaceous, and showed major declines in the Upper Cretaceous and Oligocene. From Grambast, 1974.

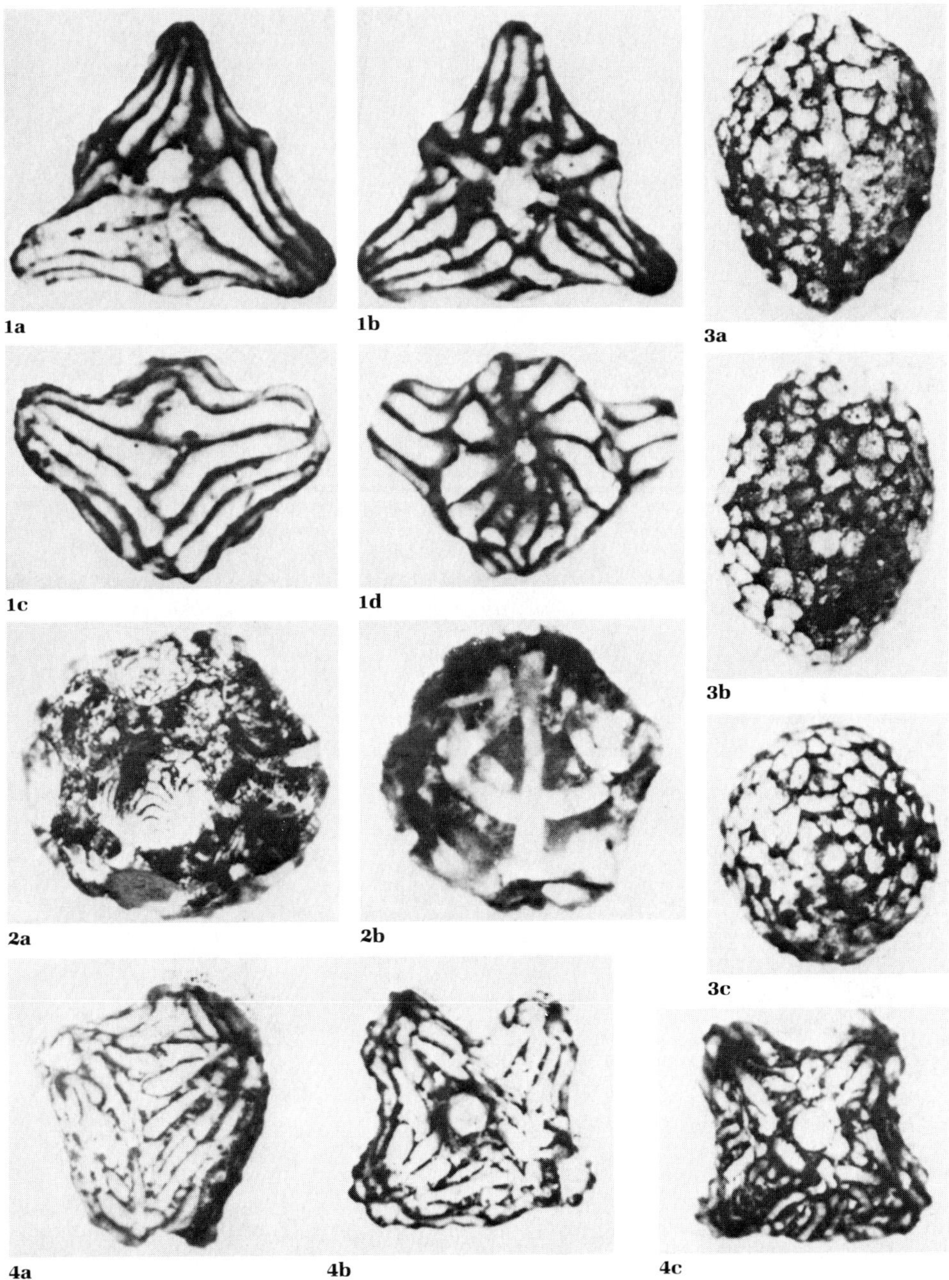

Figure 11.29

Clavatoraceae. **1.** *Triclypella calcitrapa* Grambast, Lower Cretaceous, Spain; 1a, apical view; 1b, base; 1c,d, side views, with 1d facing angle; ×56, from Grambast, 1969. **2.** *Perimneste horrida* Harris, Lower Cretaceous; 2a, side view of utricle; 2b, internal structure; ×32, from Grambast, 1967a, **3.** *Dictyoclavator fieri* (Donze) Grambast, Upper Jurassic, France; 3a,b, side views; 3c, apex; ×32, from Grambast, 1966a. **4.** *Embergerella cruciata* Grambast, Lower Cretaceous, Spain, side view, apex, and base of utricle, ×40, from Grambast, 1969.

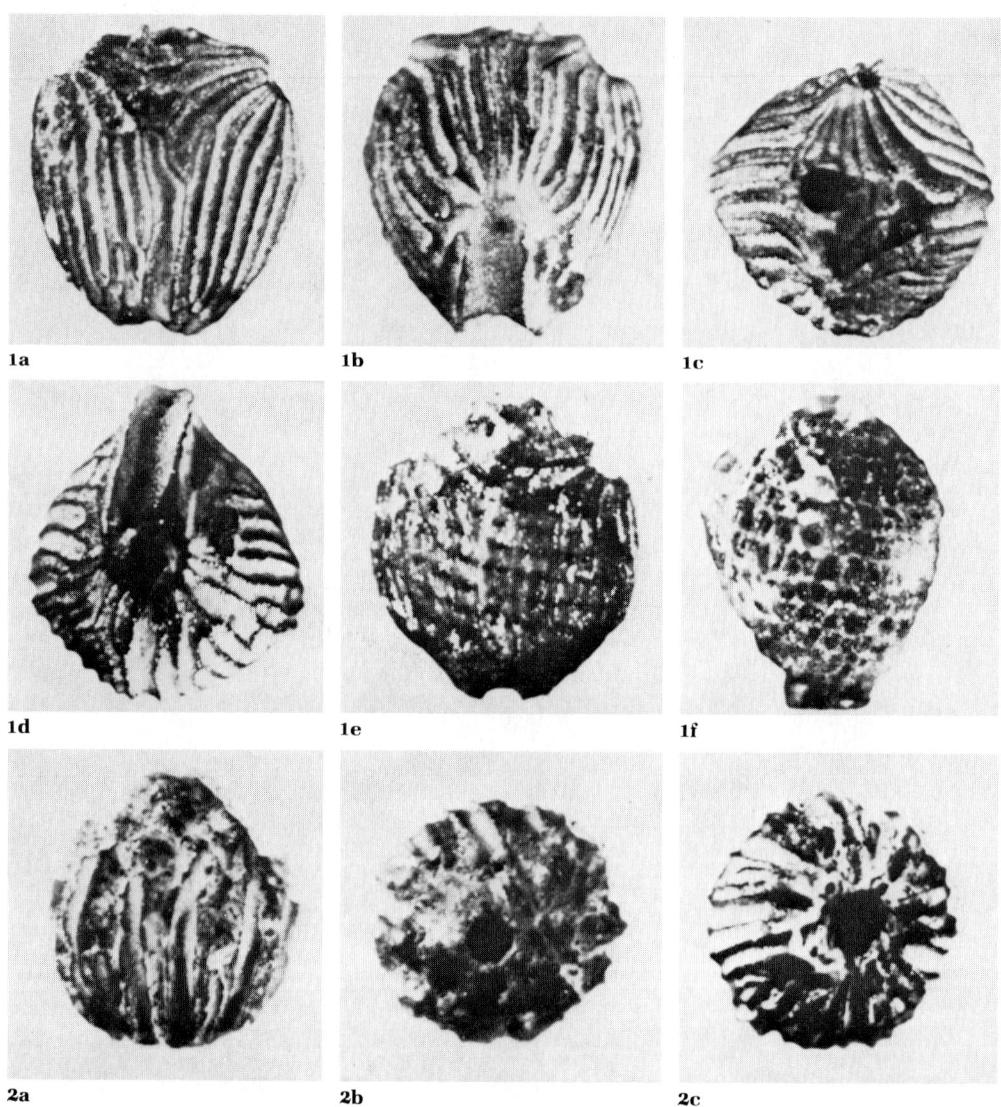

Figure 11.30

Clavatoraceae. **1.** *Lucernella ampullacea* Grambast & Lorch, Lower Cretaceous, Lebanon; 1a, dorsal face; 1b, ventral face; 1c, apex; 1d, base; 1e, dorsal face of different specimen showing gyrogonite beneath utricle; 1f, worn surface of specimen showing nodular layer of utricle; ×50, from Grambast and Lorch, 1968. **2.** *Pseudoglobator fourcadei* Grambast, Lower Cretaceous, Spain, side, apex, and base, ×40, from Grambast, 1969.

simplification of the utricle (see Figures 11.31 and 11.32), from the Berriasian *P. horrida* Harris with three similar ramules that ramify trichotomously and bear numerous antheridia, to the Barremian *P. ancora* Grambast, which had lost the uppermost forks and had only structureless antheridia; from this evolved the Aptian *Atopochara trivolvis* Peck, from which the basalmost cells and most antheridia had disappeared. Albian and Cenomanian *Atopochara* have no remaining antheridia (Grambast, 1974).

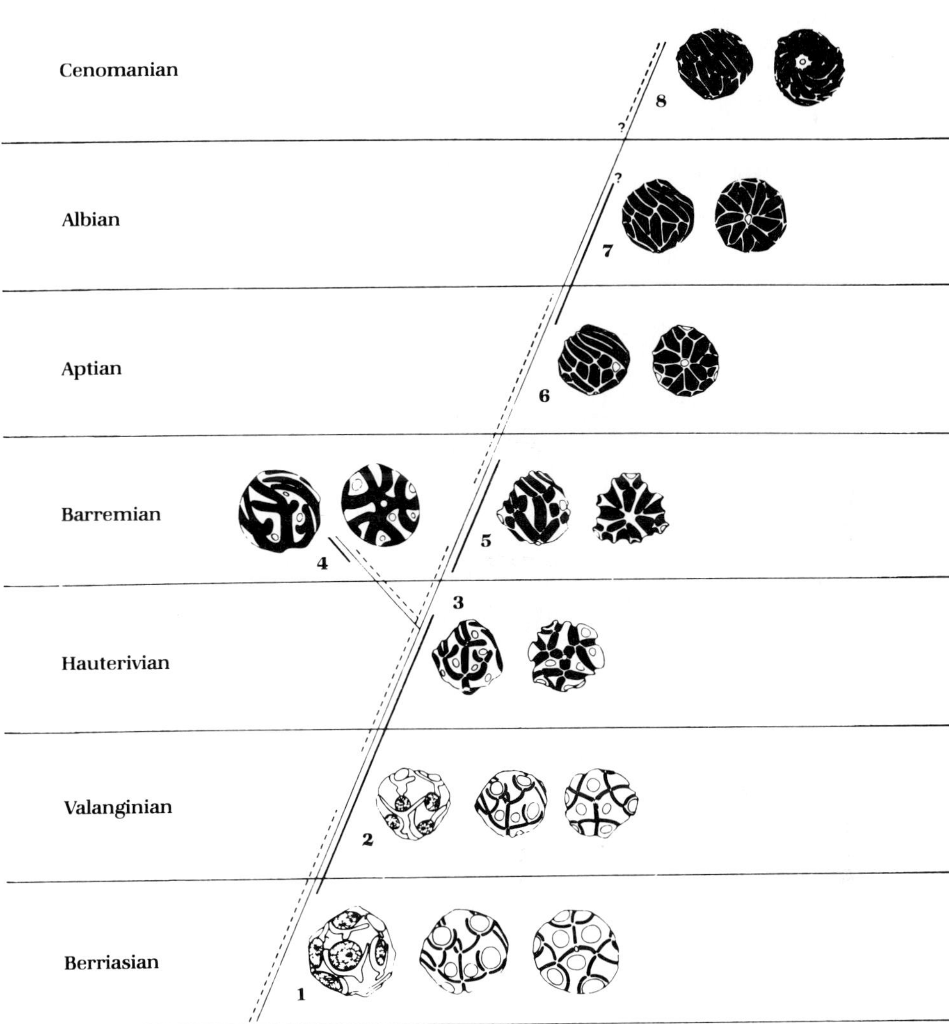

Figure 11.31

Interpretation of *Perimneste-Atopochara* lineage in the Lower and Middle Cretaceous.
1,2. Left to right shows surface of utricle with antheridia, same with etched surface, and base. **1.** *Perimneste horrida.* **2.** *P. micrandra* Grambast. **3 – 8.** Side and base of utricles.
3. *P. ancora* Grambast. **4.** *P. vidua* Grambast. **5.** *Atopochara trivolvis triquetra* Grambast.
6. *A. trivolvis trivolvis.* **7.** *A. restricta* Grambast. **8.** *A. multivolvis* Peck. Thin solid line represents lineage axis; alternate parallel heavy and broken lines indicate range of the populations of the illustrated grades. From Grambast, 1974.

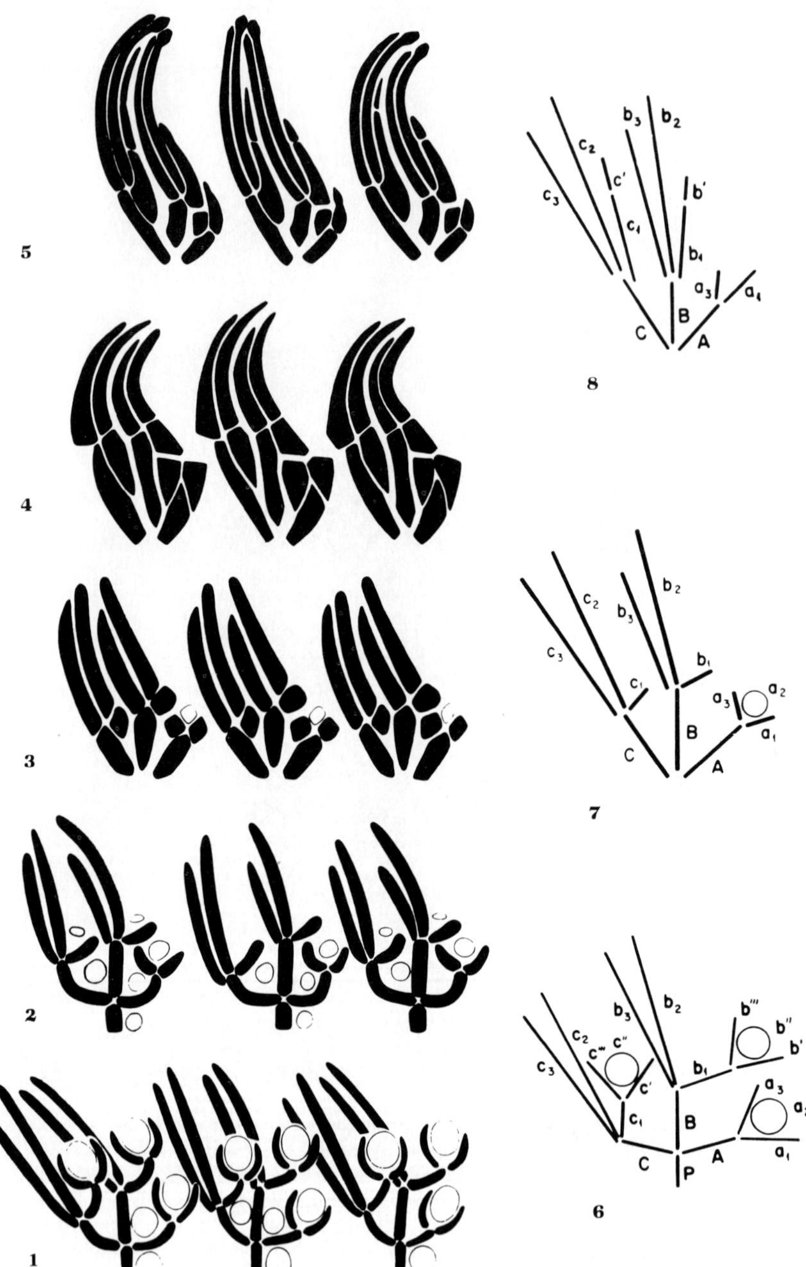

5
4
3
2
1
8
b₃ b₂
c₂
c₃ c'
c₁
b'
b₄
a₃ a₄
C B
A
7
c₂ b₂
b₃
c₃
b₁
c₁
a₃ a₂
B a₁
C
A
6
b₃ b₂
c₂ c''' c''
c₃ c'''' c''
c'
b₁
c₁
b''' b''
b'
a₃ a₂
B
C P A a₁

Figure 11.32 (*facing page*)
Structure of utricle in *Perimneste-Atopochara* sequence shown in Figure 11.31. **1.** Primitive utricle constructed of three series of elongate and ramifying cells, bearing antheridia (open circles) in the forks of the ramifications; *P. horrida* and *P. micranda*. **2.** Progressive decrease in size and number of antheridia; *P. ancora* and *P. vidua*. **3.** Antheridia persisting only on right-hand ramification of each series; *A. trivolvis*. **4,5.** All antheridia lost, other ramifications condensed. **4.** *A. restricta*. **5.** *A. multivolvis*. **6 – 8.** Diagrams of one of three cell series at different evolutionary stages corresponding to species at left, showing progressive loss of antheridia and higher order ramifications. P, original cell of a series; A,B,C, first order ramifications; a_{1-3}, b_{1-3}, c_{1-3}, second order, and b'-b''', c'-c''', third order ramifications, arising respectively from A, B, and C. From Grambast, 1974.

Another evolutionary sequence in the Clavatoraceae was described for the Late Cretaceous (Maastrichtian) *Septorella* (Grambast, 1974). The utricle shows no sign of regression in the various species, but is exceptionally large in the latest species, *S. ultima* Grambast (see Figure 11.33). The upper sterile cells are closely appressed at the apex to cover the germinative pore, leading to the suggestion that it may have been nonfunctional, and that the populations were maintained solely by vegetative multiplication. In the absence of sexuality, the reduced genetic variability prevented adaptation to the changing environmental conditions at the end of the Cretaceous and led to the extinction of *Septorella*.

The Raskyellaceae also appeared in the Mesozoic, although near its close. Gyrogonites of this family have five sinistrally enrolled cells and five apical cells that form an opercular cover. When this cover is shed at germination, a petaloid opening remains (see Figure 11.34). Important in the early Tertiary, the Raskyellaceae disappeared within the Oligocene.

The last family of the charophytes, the Characeae, also appeared within the Mesozoic, perhaps in the Triassic, as indicated by the already mentioned possible presence of *Sphaerochara*. Other genera of the subfamily Charoideae appeared in the Late Jurassic and Cretaceous (see Figures 11.35 and 11.36). True Characeae have a gyrogonite of 5 spiralling cells that join at the summit; a simple basal plate and 5 large more or less persistant coronal cells rarely may be calcified in the subfamily Charoideae, whereas the Nitelloideae have a multiple basal plate and 10 small uncalcified coronal cells in two superposed series. A sparse record of the Nitelloideae in the Jurassic (see Figure 11.37) is based on rare silicified oosporangia, as these structures are not calcified in living species.

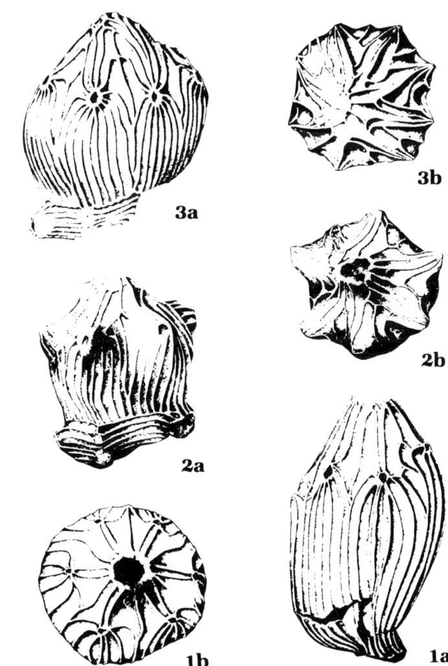

Figure 11.33
Last three successive species of Clavatoraceae in Upper Cretaceous; a, lateral, and b, apical views. **1.** *Septorella campylopoda* Grambast. **2.** *S. brachycera*. **3.** *S. ultima* Grambast, the last species, lacking an apical pore. 1,2, ×28; 3, ×14. From Grambast, 1974.

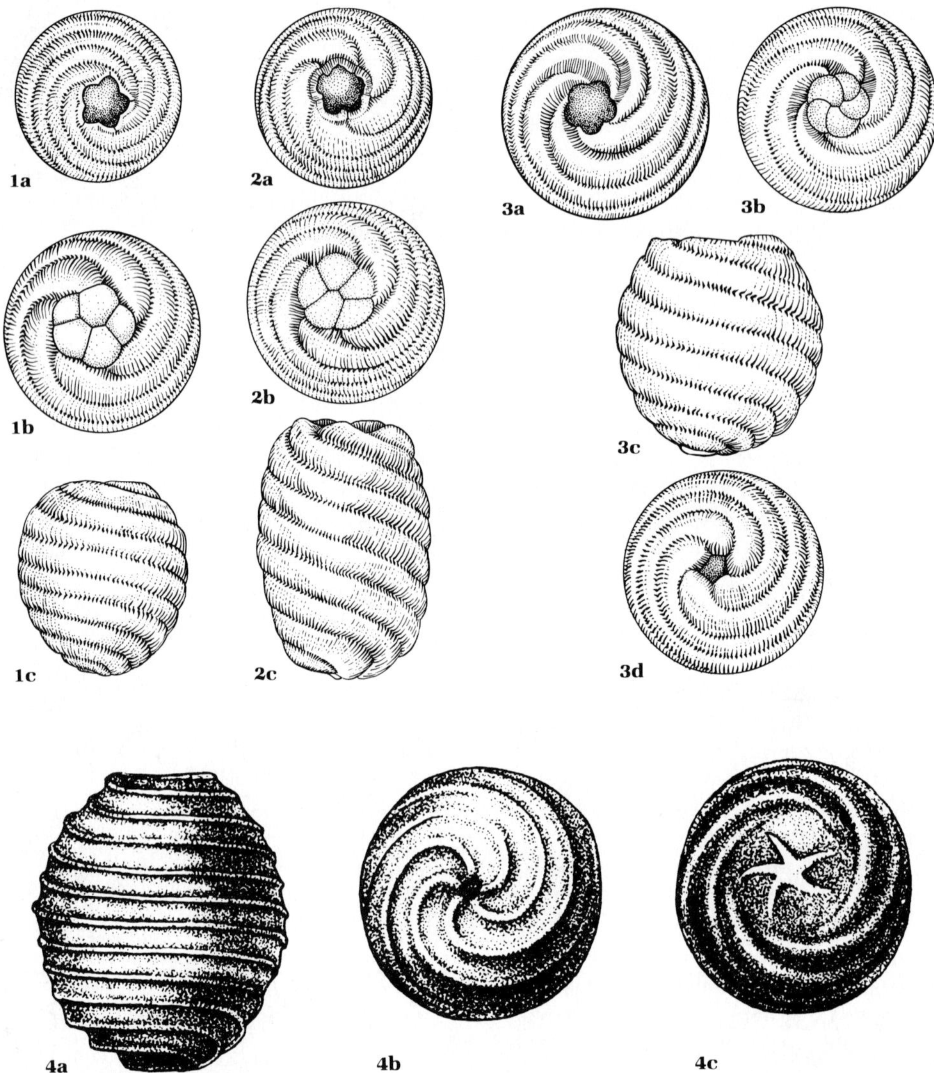

Figure 11.34

Raskyellaceae, showing a, top view after loss of opercular cells; b, top view with opercular cells in place; c, side view; d, base, unless otherwise indicated. **1.** *Raskyella pecki* L. & N. Grambast subsp. *meridionale* L. Grambast, Eocene, northwest Sahara, ×30, redrawn from Grambast, 1960b. **2.** *Rantzieniella nitida* Grambast, upper Oligocene, France, ×40. **3.** *Saportanella maslovi* Grambast, Upper Cretaceous, France, ×30. 2,3, redrawn from Grambast, 1962b. **4.** *Jurella abshirica* Kyansep-Romashkina, Middle Jurassic, southeast Fergana, central Asia; 4a, side; 4b, top; 4c, base; ×90, from Kyansep-Romashkina, 1974.

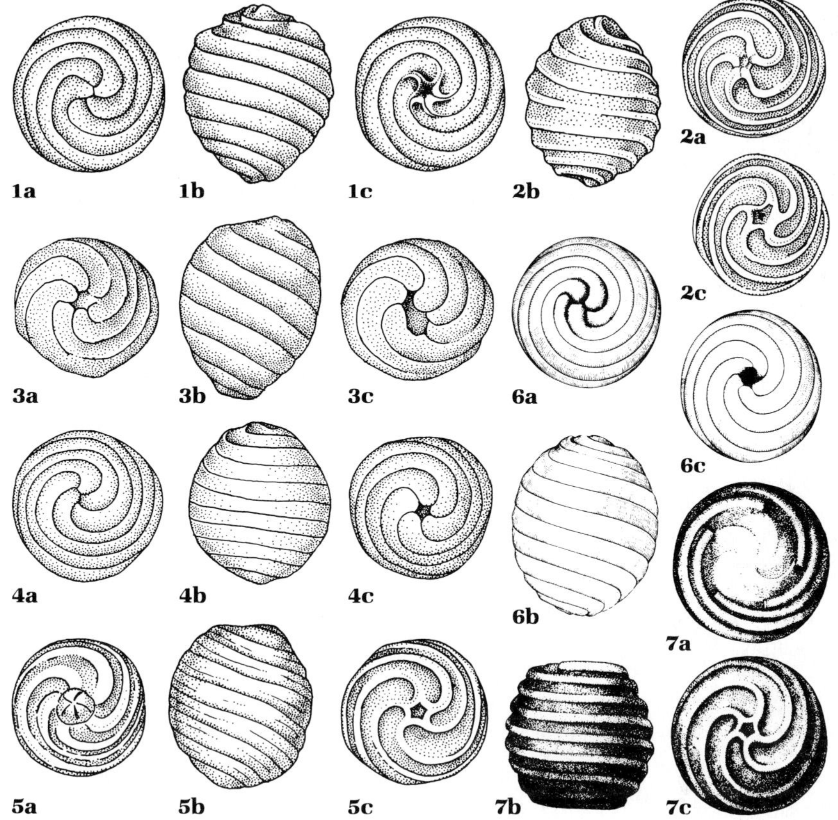

Figure 11.35

Mesozoic Characeae, showing a, top; b, side; c, base. **1.** *Amblyochara begudiana* Grambast, Upper Cretaceous, France, ×23, redrawn from Grambast, 1962b. **2.** *Mesochara symmetrica* (Peck) Grambast, Lower Cretaceous, South Dakota, ×47, drawn from Peck, 1957. **3.** *Obtusochara prima* Mädler, Upper Jurassic, Germany, ×78, redrawn from Mädler, 1953a. **4.** *Peckisphaera verticillata* (Peck) Grambast, Upper Jurassic, Montana, ×43. **5.** *Aclistochara bransoni* Peck, Upper Jurassic, Wyoming, ×43. 4,5, drawn from Peck, 1957. **6.** *Retusochara macrocarpa* Grambast, Upper Cretaceous, France, ×15, from Grambast, 1971. **7.** *Mongolichara deplanata* Kyansep-Romashkina, Upper Cretaceous, Mongolia, ×40, from Kyansep-Romashkina, 1975.

Figure 11.36

Charoideae. **1.** *Dughiella bacillaris* Feist-Castel, Paleocene, France, top, side, and base, ×32, reproduced with permission from Feist-Castel, 1975, *Bull. Soc. Géol. Fr.,* sér. 7, v. 17, p. 88−97. **2.** *Gobichara deserta* Karczewska & Ziembińska-Tworzydło, Paleogene, Mongolia, top, side, and base, ×80, from Karczewska and Ziembińska-Tworzydło, 1972. **3.** *Krassavinella lagenalis* (Straub) Feist-Castel, upper Oligocene, Switzerland, top, base, and side, ×80, from Feist-Castel, 1977c. **4.** *Nothochara apiculata* Musacchio, Upper Cretaceous, Argentina, top, side, and base, ×80, and enlargement of apex in oblique side view, ×120, from Musacchio, 1973. **5.** *Nemegtichara prima* Karczewska & Ziembińska-Tworzydło, Paleogene, Mongolia, top and side, ×80, from Karczewska and Ziembińska-Tworzydło, 1972. **6,7.** *Pseudoharrisichara,* Upper Cretaceous, Argentina, ×54, from Musacchio, 1973. **6.** *P. walpurgica* Musacchio, side and base. **7.** *P. tenuis* Musacchio, top.

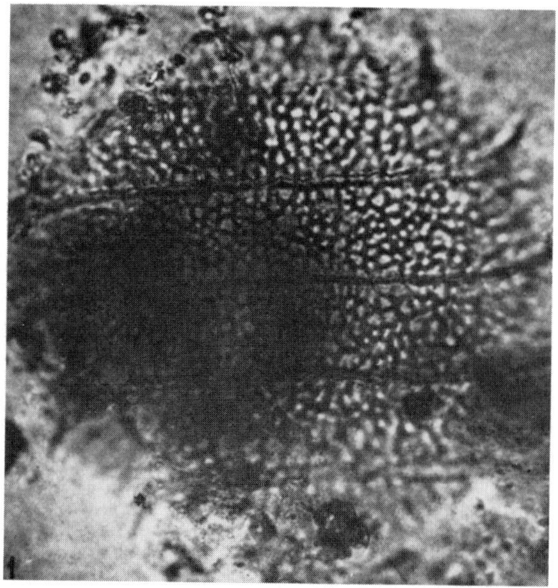

Figure 11.37
Nitellites sahnii Horn af Rantzien, Middle to Upper Jurassic, India. Lateral view of silicified, noncalcareous oosporangium, with reticulate membrane, ×550. From Horn af Rantzien, 1957.

Cenozoic Charophytes

Following the Late Cretaceous and early Paleocene decline, a new charophyte radiation began in the late Paleocene (late Thanetian) and reached a climax in the Eocene and Oligocene. These early Tertiary forms are quite different from post-Oligocene ones, including both the last Raskyellaceae and many new genera of Characeae (see Figure 11.38) that were at places abundant and useful index species, although restricted to the Paleogene. Eocene and Oligocene rocks include some 19 genera, but diversity declined rapidly to 10 genera in the Miocene and only 6 or 7 presently living, excluding the subgeneric taxa that are based solely on vegetative characters that are not preservable. Post-Miocene species represent relatively simple, long-lasting types. In eastern North America, 7 genera, 12 species, and a variety were recognized in Pleistocene glacial and postglacial deposits (Daily, 1961).

Many living species are defined on vegetative characters that are not preserved, whereas nearly all fossil records are based on calcified gyrogonites, and a small number on secondarily silicified ones. However, the Pleistocene record of the noncalcified Nitelloideae is based on preserved organic remains of the oosporangia. No indication is given in the fossils as to the presence or absence of the double row of coronal cells characteristic of the Nitelloideae, hence they have been referred to the form genus *Nitellites,* rather than to *Nitella* itself. The compressed form and even the membrane ornamentation closely resemble living *Nitella.*

General Evolutionary Trends

Among the observable general evolutionary trends in charophytes is the acquisition of the ability to secrete calcium carbonate. This ability may have arisen repeatedly in unrelated lineages. Early evolution had progressed to a high degree of specialization prior to the appearance of calcification in Devonian Charophytes. Similarly, some modern Characeae deposit lime, whereas other congeneric species do

1a
1b
1c
2a
2b
2c
3a
3b
3c
4a
4b
4c
5a
5b
5c
6a
7a
7b
8a
8b
8c
6b
9a
9b
10a
10b
11a
11b
11c
11d
12
13a
13b
14a
14b
15a
15b
15c
16a
16b
16c

not. The Nitelloideae are mostly noncalcified (see Figure 11.39).

Structural diversification was greatest in the oldest charophytes. *Chovanella* may be the most primitive form, having vertical units of a single cell, and most similar to the ancestor that produced the Sycidiaceae with its numerous vertically aligned cells, and the Trochiliscaceae with spiralling cells. True charophytes also may have arisen independently from a primitive type that had no sterile cells around the female gametocyst, hence no gyrogonite. Possibly the Umbellinaceae may be similar to such a primitive form (Grambast, 1974).

The number of cells in the gyrogonite decreased in the later Paleozoic, from the numerous vertically aligned cells to the spiralling and hence fewer cells per lime shell. Coiling of the cells in the Paleozoic Trochiliscales was to the right, and the basic number of cells ranged between seven and ten, commonly nine. Sinistral spiralling also appeared early (*Eochara*), probably in an entirely different lineage. The cell numbers were variable in *Eochara*, reduced in the Pennsylvanian to five (*Stomochara*), or rarely six (*Palaeochara*). All later charophytes have only five spiral cells (disregarding an occasional aberrant specimen). Fewer cells were

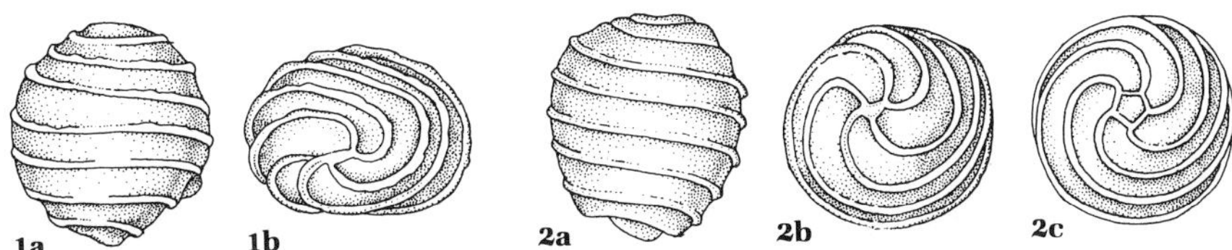

Figure 11.39
Gyrogonites of Nitelloideae. **1.** *Nitella clavata* Kützing, living, side and top views of uncalcified gyrogonite, ×50. **2.** *Tolypella nidifica*, living, side, top, and base of uncalcified gyrogonite, ×48. Both redrawn from Maslov, 1963b.

formed by transverse partitions in younger charophyte gyrogonites, evolving from *Sycidium* with many horizontal partitions, to those with spiral cells and a corona of two rows of cells (Nitelloideae), and finally a corona of a single row of cells (Charoideae and others).

Early Paleozoic species showed an increase in gyrogonite size from the Devonian to Mississippian, but the Charales do not follow this trend, early Tertiary ones having larger lime shells than modern ones.

The oosporangia of the oldest species had a simple open apical pore for access of the male gametes, as in other algae. Modern types have a closed apex, to which access is gained through intercellular slits, a more recent evolutionary development.

An early division of the Characeae led to the Charoideae and Nitelloideae by the Middle Jurassic, the two subfamilies then showing a parallel development, although the former has a far more extensive fossil record. Trends in this family are from a primitive uniform structure to one with differentiated fertile and sterile whorls, and from a general distribution of the gametangia on the plant to their localization on fertile whorls. Commonly both oogonia and antheridia are present on the same plant (monoecious). Some species with dioecious populations represent genetic strains with only half the usual number of chromosomes. Diploidy or polyploidy also has arisen repeatedly. *Chara* has 6 or 7 chromosomes, and *Nitella* basically has 6, but some species may have 12, 14, 18, and even 34. The various numbers probably have resulted from a combination of ploidy and loss of a chromosome, from an original 7.

CHAROPHYTE CLASSIFICATION AND SYSTEMATIC RELATIONSHIPS

Although a relatively small group, the charophytes are structurally quite unlike any other plants, hence generally are regarded as a separate division of the plant kingdom, the Charophyta (or Charophycophyta), although some recognize only an order Charales of the Chlorophyceae or Chlorophyta (green algae), the latter variously being given class or divisional rank.

The exposed nature of the reproductive organs is one of the few characters that separates algae from higher plants. These organs are protected by an outer sheath in the charophytes, hence are intermediate in character between green algae and bryophytes. Similarities to the green algae include the presence of green plastids, the production of starch, and the normally haploid plant. The green dasyclad algae were once regarded as the closest relatives, but charophytes have a phragmoplast type of cell division, unlike most green algae. The earliest known charophytes were already highly differentiated, and their similarities to the green algae are largely those shared by all green plants. The complex charophyte gametangia are suggestive of the archegonia of the bryophytes.

Early classifications and taxa were based on thallus characters, but as these may be quite variable, they do not offer reliable criteria. Gametangial characters were shown by Horn af Rantzien (1959b) to be more useful in modern classifications, hence are equally valid for fossil taxa. These features are indicated in the classification shown in Table 11.3.

Table 11.3
Classification of the charophytes.

Division Charophyta Migula 1890

I. Position uncertain

1. Family Umbellinaceae (Loeblich & Tappan 1962) Loeblich & Tappan 1974

Syn.: Subfamily Umbellinae Fursenko 1959; Subfamily Umbellininae Loeblich & Tappan 1962; Family Umbellaceae Poyarkov 1965; Subfamily Umbellidae Berchenko 1974; Subfamily Biumbellidae Berchenko 1974.

Globular to flask-shaped with simple cavity; wall double, with inner microcrystalline layer and outer radial fibrous layer; basal slit or pore may be present; apical opening or pore may have lid.

Middle Devonian to Lower Mississippian.

Archaelagena Howchin 1888; *Biumbella* Mamet 1970; *Costatumbella* Berchenko 1974; *Elenia* Poyarkov 1965; *Eoumbella* Platonov 1974; *Lagenumbella* Mamet 1970; *Planoumbella* Platonov 1974; ?*Plavskina* Reitlinger 1960; *Protoumbella* Mamet 1970; *Pseudoumbella* Berchenko 1971; *Quasiumbella* Poyarkov 1965; *Quasiumbelloides* Berchenko 1971; *Spinumbella* Platonov 1974; *Umbellina* Loeblich & Tappan 1961 (syn.: *Umbella* Maslov 1955, non d'Orbigny 1841 nec Scudder 1882).

II. Class Sycidiphyceae Langer 1976

A. Order Sycidiales Mädler 1952

1. Family Sycidiaceae Karpinsky 1906.

Gyrogonite consists of numerous cells arranged in 12 to 18 vertical series; wall of laminated calcite perforated by radial pores; apical and basal pores present.

?Upper Silurian, Middle Devonian to Lower Mississippian.

?*Kusjaella* Chuvashov 1973; *Sycidium* Sandberger 1849.

2. Family Chovanellaceae Grambast 1962

Gyrogonite consists of vertically aligned cells, not subdivided.

Middle to Upper Devonian.

Chovanella Reitlinger & Yartseva 1958.

III. Class Charophyceae G. M. Smith 1938

A. Order Trochiliscales Mädler 1952

1. Family Trochiliscaceae Karpinsky 1906

Gyrogonite composed of dextrally spiralling cells, of varied number, but more than six.

Lower Devonian to Lower Mississippian.

Moellerina Ulrich 1886 (syn.: *Trochiliscus* Karpinsky 1906; *T.* (*Eutrochiliscus*) Croft 1954; *Weikkoella* Summerson 1958); *Karpinskya* (Croft 1952) Grambast 1962 (syn.: *Trochiliscus* (*K.*) Croft 1952).

B. Order Charales Lindley 1836

1. Family Eocharaceae Grambast 1959

Gyrogonite with numerous sinistrally spiralling cells, of variable number, but more than six; with apical pore.

Middle Devonian.

Eochara Choquette 1956.

2. Family Palaeocharaceae Pia 1927

Gyrogonite with six sinistrally spiralling cells. As occasional aberrant four- and six-celled gyrogonites occur in the Characeae, the Palaeocharaceae and Porocharaceae may be synonymous.

Pennsylvanian.

Palaeochara Bell 1922.

3. Family Porocharaceae Grambast 1962

Gyrogonite without utricle, with five sinistrally spiralling cells and free apical pore.

a. Subfamily Porocharoideae Grambast 1961

Syn.: Stomocharoideae Saydakovskiy 1968

Gyrogonite with truncate, rounded, or conical apical zone, but no distinct neck.

Pennsylvanian; Triassic to Upper Cretaceous (Maastrichtian).

Altochara Saydakovskiy 1968; *Auerbachichara* Kiselevskiy & Saydakovskiy 1967; *Cuneatochara* Saydakovskiy 1962; *Horniella* Shaykin 1966; *Latochara* Mädler 1955; *Porochara* Mädler 1955; *Stenochara* Grambast 1962 (syn.: *Praechara* Horn af Rantzien 1954 non Birina 1948); *Stomochara* Grambast 1961 (syn.: *Catillochara* Peck & Eyer 1963); *Strobilochara* Grambast 1975.

b. Subfamily Stellatocharoideae Grambast 1962

Syn.: Maslovicharoideae Saydakovskiy 1966

Gyrogonites with apical zone stretched into a neck; with truncate extremity.

Permian to Lower Cretaceous (Albian).

Clavatorites Horn af Rantzien 1954; *Stellatochara* Horn af Rantzien 1954 (syn.: *Leonardosia* Sommer 1954; *Maslovichara* Saydakovskiy 1962).

(continued)

4. Family Clavatoraceae Pia 1927

Gyrogonite surrounded by utricle, an external covering of vegetative elements; apical zone stretched into neck that opens into pore at summit.

Upper Jurassic to Upper Cretaceous.

Ascidiella Grambast 1966; *Atopochara* Peck 1938; *Clavator* Reid & Groves 1916; *Clypeator* Grambast 1962; *Dictyoclavator* Grambast 1966; *Diectochara* Musacchio 1971; *Echinochara* Peck 1957; *Embergerella* Grambast 1969; *Flabellochara* Grambast 1959; *Globator* Grambast 1966; *Lucernella* Grambast & Lorch 1968; *Nodosoclavator* Maslov 1961; *Perimneste* Harris 1939; *Pseudoglobator* Grambast 1969; *Septorella* Grambast 1962; *Triclypella* Grambast 1969.

5. Family Raskyellaceae (N. Grambast & L. Grambast 1955) L. Grambast 1957

Syn.: Subfamily Raskyelloideae N. & L. Grambast 1955

Gyrogonite with five sinistrally spiralling cells in contact with oospore, and five apical cells forming an operculum that leaves a roseate opening after being shed.

Middle Jurassic; Upper Cretaceous (Maastrichtian) to upper Oligocene (lower Aquitanian).

Jurella Kyansep-Romashkina 1974; *Rantzieniella* L. Grambast 1962; *Raskyella* L. & N. Grambast 1954; *Saportanella* Grambast 1962.

6. Family Characeae Richard ex C. Agardh 1824

Syn.: Family Lagynophoraceae Stache 1889; Family Nitellopsidaceae Krassavina 1971

Oogonium and gyrogonite of five sinistrally spiralling cells joined at the summit.

a. Subfamily Charoideae (Leonhardi 1863) Robinson 1906

Oogonium with single oosphere that after calcification forms a gyrogonite of five spiral cells joined at simple basal plate; five large coronal cells, more or less persistent, rarely calcified.

Triassic to Holocene; fresh and brackish water.

Aclistochara Peck 1937; *Amblyochara* Grambast 1962; *Chara* Linnaeus 1753 (syn.: *Characias* Rafinesque 1815; *Charites* Horn af Rantzien 1959; *Charopsis* Kützing 1843; *Croftiella* Horn af Rantzien 1959; *Grambastichara* Horn af Rantzien 1959; *Hornichara* Maslov 1963; *Protochara*

Womersley & Ophel 1947); *Dughiella* Feist-Castel 1975; *Gobichara* Karczewska & Ziembińska-Tworzydło 1972; *Grambastiella* Massieux & Tambareau 1979; *Grovsichara* Horn af Rantzien 1959; *Gyrogona* Lamarck 1804 (syn.: *Gyrogonites* Lamarck 1804, non Pia 1927; *Brachychara* L. & N. Grambast 1954; *Brevichara* Horn af Rantzien 1956); *Harrisichara* Grambast 1957; *Kosmogyra* Stache 1889; *Krassavinella* Feist-Castel 1977; *Lamprothamnium* J. Groves 1916 (syn.: *Lamprothamnus* Braun 1882, non Hiern 1877); *Lychnothamnites* Maslov 1966; *Lychnothamnus* (Ruprecht 1845) Leonhardi 1863 (syn.: subgenus *Chara* (*Lychnothamnus*) Ruprecht); *Maedleriella* Grambast 1957; *Mesochara* Grambast 1962; *Microchara* Grambast 1959; *Mongolichara* Kyansep-Romashkina 1975; *Nitellopsis* Hy 1889 (syn.: *Tolypellopsis* Migula 1898); *N.* (*Campaniella*) (L. Grambast 1972) Grambast & Soulié-Märsche 1972; *N.* (*Microstomella*) Feist-Castel 1977; *N.* (*Tectochara*) (L. & N. Grambast 1954) Grambast 1972 (syn.: *Tectochara* (*Sulcosphaera*) Maslov 1966); *Nemegtichara* Karczewska & Ziembińska-Tworzydło 1972; *Nodosochara* Mädler 1955; *Nothochara* Musacchio 1973; *Obtusochara* Mädler 1952; *Peckichara* Grambast 1957; *Peckisphaera* Grambast 1962; *Platychara* Grambast 1962; *Pseudoharrisichara* Musacchio 1973; *Psilochara* Grambast 1959; *Retusochara* Grambast 1971; *Rhabdochara* Mädler 1955; *Sphaerochara* Mädler 1952 (syn.: *Maedlerisphaera* Horn af Rantzien 1959; *Raskyaechara* Horn af Rantzien 1959); *Stephanochara* Grambast 1959.

b. Subfamily Nitelloideae (Ganterer 1847) Braun ex Migula 1890

Oogonium with multiple cells (usually three) in oosphere, producing calcified gyrogonite; spiral cells joined at composite basal plate; ten small nonpersistent coronal cells in two superposed series.

Jurassic; Pleistocene, Holocene.

Nitella Agardh 1824 (syn.: *Leiacina* Rafinesque 1825; *Charina* Filarszky 1937); *Nitellites* Horn af Rantzien 1957; *Tolypella* (Braun 1849) Leonhardi 1863 (syn.: *Nitella* (*T.*) Braun).

Modern *Chara* is divided on the basis of vegetative characters into the subgenera C. (*Chara*) and C. (*Charopsis*), and *Nitella* into the three subgenera, N. (*Nitella*), N. (*Hyella*), and N. (*Tieffallenia*). These are not recognizable in fossil forms. Fossilized vegetative parts, not associated with gyrogonites, have been referred to *Charaxis* Harris 1939 (syn.: *Characeites* Pia 1927 non Tuzson 1913). *Chariella* Birina 1948 and *Praechara* Birina 1948, based on thin sections, are unrecognizable, as are *Characeites* Tuzson 1913 and *Anomalochara* Maslov 1961. Genera based on incompletely calcified gyrogonites and thus not definitely assignable are *Cristatella* and *Kosmogyrella* Stache 1889 and *Kosmogyrina* Mädler 1952. Questionably a charophyte is *Palaeonitella* (Kidston & Lang 1921) Pia 1927. The Paleocene *Lagynophora* Stache 1889, known only from thin sections, consists of incrusted gyrogonites of a characean, similar to such encrustations of the Eocene *Microchara vestita* Castel, but the identity of Stache's material is uncertain (Bignot and Grambast, 1969).

REFERENCES

Bell, W. A., A new genus of Characeae and new Merostomata from the Coal Measures of Nova Scotia. *Trans. R. Soc. Can.*, ser. 3, v. 16, p. 159–167, 1 pl., 1922.

Berchenko, O. I., Charophyta, p. 115–122, pls. 81–83. *In* D. E. Ayzenverg, *Atlas fauny Turneyskikh otlozheniy Donetskogo Basseyna (s opisaniem novykh vidov)*. Kiev: Akad. Nauk Ukrainskoy SSR, Inst. Geol. Nauk, 1971.

Bignot, Gérard, and Louis Grambast, Sur la position stratigraphique et les charophytes de la Formation de Kozina (Slovénie, Yougoslavie). *C. R. Hebd. Séanc. Acad. Sci., Paris*, v. 269, p. 689–692, pl. 1, 1969.

Birina, L. M., Novye vidy izvestkovykh vodorosley i foraminifer pogranichnykh sloev Devona i Karbona [New species of calcareous algae and foraminifera from strata at the Devonian-Carboniferous boundary]. *Sov. Geol.*, Sbornik 28, p. 154–159, 2 pls., 1948.

Brady, H. B., *A monograph of the Carboniferous and Permian Foraminifera (the genus* Fusulina *excepted)*. London: Palaeontogr. Soc., 166 p., 12 pls., 1876.

Brazhnikova, N. E., and L. F. Rostovtseva, Foraminifery, p. 9–42, pls. 1–27. In *Fauna Nizov Turne (Zony C$_1^i$a) Donetskogo Basseyna*. Kiev: Akad. Nauk Ukrainskoy SSR, Inst. Geol. Nauk, "Naukova Dumka," 1966.

Bruckner, Werner, and Julius v. Pia, Characeenreste im unteren Teil der Zementsteinschichten (oberer Malm) der Griesstock-Decke am Klausenpass (Kt. Uri). *Eclog. Geol. Helv.*, v. 28, p. 115–121, pl. 10, text-figs. 1–4, 1935.

Bykova, E. V., and E. N. Polenova, Foraminifery, radiolyarii i ostrakody devona volgo-ural'skoy oblasti [Foraminifera, radiolaria and ostracoda from the Devonian of the Volga-Ural region]. *Trudy Vses. Neft. Nauchno-Issled. Geol.-Razv. Inst.*, n. s., vyp. 87, p. 5–190, pls. 1–25, 1955.

Castel, Monique, Charophytes de l'Oligocène supérieur de Marseille. *Bull. Soc. Géol. Fr.*, sér. 7, v. 9 (1967), p. 514–519, pls. 19–21, figs. 1–3, 1968.

Castel, Monique, and Louis Grambast, Charophytes de l'Éocène des Corbières. *Bull. Soc. Géol. Fr.*, sér. 7, v. 11, p. 936–943, pls. 30–32, figs. 1–5, 1969.

Choquette, G. B., A new Devonian charophyte. *J. Paleont.*, v. 30, p. 1371–1374, 1 fig., 1956.

Chuvashov, B. I., Foraminifery i vodorosli iz verkhnedevonskikh otlozheniy zapadnogo sklona Srednego i Yuzhnogo Urala [Foraminifera and algae from Upper Devonian deposits on the western slope of the Central and Southern Urals]. *Trudy Inst. Geol., Ural Filial, Akad. Nauk SSSR*, vyp. 74, p. 3–153, 29 pls., 1965.

Chuvashov, B. I., Novye Devonskie vodorosli Urala [New Devonian algae of the Urals]. *Trudy Inst. Geol. Geokhim., Ural. Nauch. Tsentr., Akad. Nauk SSSR*, vyp. 99, *Sbornik po Voprosam Stratig.*, no. 18, p. 28–47, pls. 1–5, figs. 1, 2, 1973.

Conil, Raphaël, and Maurice Lys, Matériaux pour l'étude micropaléontologique du Dinantien de la Belgique et de la France (Avesnois). Première partie: algues et Foraminifères. *Mem. Inst. Géol. Univ. Louvain*, v. 23, p. 1–287, pls. 1–42, 1964.

Conkin, J. E., B. M. Conkin, G. S. Gregory, and A. T. Hotchkiss, Revision of the charophyte genus *Moellerina* Ulrich 1886 and suppression of the genus *Karpinskya* (Croft, 1952) Grambast, 1962. *Stud. Paleont. Stratigr. Univ. Louisville*, no. 3, p. i–iii, 1–18, pls. 1, 2, 1974.

Conkin, J. E., B. M. Conkin, Takashi Sawa, and J. M. Kern, Middle Devonian *Moellerina greenei* Zone and suppression of the genus *Weikkoella* Summerson, 1958. *Micropaleontology*, v. 16, p. 399–406, pl. 1, fig. 1, 1970.

Conkin, J. E., Takashi Sawa, R. G. Coy, and A. M. Salmon, The charophyte genus *Sycidium* in the Upper

Devonian of Ohio. *Micropaleontology*, v. 18, p. 74—80, pl. 1, fig. 1, 1972.

Croft, W. N., A new *Trochiliscus* (Charophyta) from the Downtonian of Podolia. *Bull. Br. Mus. Nat. Hist., Geology*, v. 1, p. 189—220, 2 pls., 7 text-figs., 1952.

Daily, F. K., Glacial and post-glacial charophytes from New York and Indiana. *Butler Univ. Bot. Stud.*, no. 14, p. 39—72, pls. 1—5, 1961.

Daily, F. K., *Lamprothamnium* in America. *J. Phycol.*, v. 3, p. 201—207, figs. 1—46, 1967.

Daily, F. K., Fossil charophytes from South Dakota and Iowa with a review of a modern species. *Am. Midl. Nat.*, v. 84, p. 365—375, figs. 1—26, 1970.

Daily, F. K., Lime shells of *Nitellopsis obtusa* fossil and extant affinities. *Bull. Torrey Bot. Club*, v. 100, p. 75—78, figs. 1—5, 1973.

Daily, F. K., A note concerning calcium carbonate deposits in charophytes. *Phycologia*, v. 14, p. 331—332, 1975.

Daily, F. K., and J. Wyatt Durham, Miocene Charophytes from Ixtapa, Chiapas, Mexico. *J. Paleont.*, v. 40, p. 1191—1199, 2 figs., 1966.

Dollfuss, G. F., and P. H. Fritel, Catalogue raisonné des characées fossiles du Bassin de Paris. *Bull. Soc. Géol. Fr.*, sér. 4, v. 19, p. 243—264, 23 figs., 1920.

Donze, P., Nouvelles espéces de charophytes dans le niveaux de la limite jurassico-crétacé du Jura, des Alpes-Maritimes et de la Provence. *Bull. Soc. Géol. Fr.*, sér. 6, p. 287—290, pl. 13, 1955.

Ehrenberg, C. G., Fortschreitende Erkenntniss massenhafter mikroskopischer Lebensformen in den untersten silurischen Thonschichter bei Petersburg. *Mber. Preuss. Akad. Wiss. Berlin*, p. 295—311, 1858.

Ehrenberg, C. G., Über die mikroskopischen Lebensformen als Nahrung des Höhlen-Salamanders. *Mber. Preuss. Akad. Wiss. Berlin*, p. 581—601, 1 pl., 1862.

Emberger, Jacques, and Jean Magné, Observations sur les niveaux à charophytes de l'Aptien des Monts des Oulad-Naïl (Atlas Saharien, Algérie). *Bull. Soc. Géol. Fr.*, sér. 6, v. 6, p. 1029—1039, 3 figs., 1957.

Feist-Castel, Monique, Sur les charophytes fossiles du bassin tertiaire d'Alès (Gard). *Geobios*, v. 4, p. 157—172, pls. 10—12, figs. 1—5, 1971.

Feist-Castel, Monique, Observations nouvelles sur la paroi des fructifications chez les charophytes. *Geobios*, v. 6, p. 239—242, pls. 13—15, fig. 1, 1973.

Feist-Castel, Monique, Répartition des charophytes dans le Paléocène et l'Éocène du Bassin d'Aix-en-Provence. *Bull. Soc. Géol. Fr.*, sér. 7, v. 17, p. 88—97, pls. 1—3, 1975.

Feist-Castel, Monique, Evolution of the charophyte floras in the Upper Eocene and Lower Oligocene of the Isle of Wight. *Palaeontology*, v. 20, p. 143—157, pls. 21, 22, fig. 1, 1977a.

Feist-Castel, Monique, Etude floristique et biostratigraphique des Charophytes dans les séries du Paleogène de Provence. *Géologie Méditerranéenne*, v. 4, p. 109—138, pls. 1—5, figs. 1—8, 1977b.

Feist-Castel, Monique, Description du nouveau genre *Krassavinella* (Charophytes, Characeae) et répartition de *K. lagenalis* (Straub) dans l'Oligocène supérieur de la molasse suisse. *Eclog. Geol. Helv.*, v. 70, p. 771—775, 1 pl., 1977c.

Forsberg, Curt, Ecological and physiological studies of charophytes. *Abstr. Uppsala Diss. Sci.*, 53, p. 1—10, 1965.

Fritsch, F. E., *The structure and reproduction of the algae.* Cambridge: University Press, v. 1, xvii + 791 p., 245 figs., 1935.

Glukhovskaya, N. B., Girogonity Semeystva Characeae s anomal'nym kolichestvom spiraley [Gyrogonites of the family Characeae with an anomalous number of spirals]. *Dokl. Akad. Nauk SSSR*, v. 224, p. 691—692, fig. 1, 1975.

Grambast, Louis, Sur le genre *Tectochara*. *C. R. Somm. Séanc. Soc. Géol. Fr.*, p. 113—114, 1956a.

Grambast, Louis, Le genre *Gyrogona* Lmk. (Characeae). *C. R. somm. Séanc. Soc. Géol. Fr.*, p. 278—280, 1956b.

Grambast, Louis, La plaque basale des characées. *C. R. Hebd. Séanc. Acad. Sci., Paris*, v. 242, p. 2585—2587, 8 figs., 1956c.

Grambast, Louis, Sur la déhiscence de l'oospore chez *Chara vulgaris* L. et la systématique de certaines Characeae fossiles. *Revue Gén. Bot.*, v. 63, p. 331—336, 3 figs., 1956d.

Grambast, Louis, Ornamentation de la gyrogonite et systématique chez les charophytes fossiles. *Rev. Gén. Bot.*, v. 64, p. 339—360, pls. 5—8, figs. 1—7, 1957.

Grambast, Louis, Tendances évolutives dans le phylum des Charophytes. *C. R. Hebd. Seanc. Acad. Sci., Paris*, v. 249, p. 557—559, fig., 1959.

Grambast, Louis, *Extension Chronologique des Genres chez les Charoideae.* Paris: Soc. Édit. Technip, p. 3—12, figs. 1—4, 1959 [1960a].

Grambast, Louis, Description et signification stratigraphique de deux charophytes d'origine Saharienne. *Revue Micropaléont.*, v. 2, p. 192—198, pls. 1, 2, text-figs. 1, 2, 1960b.

Grambast, Louis, Remarques sur la systématique et la répartition stratigraphique des Characeae prétertiares. *C. R. Somm. Séanc. Soc. Géol. Fr.*, p. 200—202, 1961.

Grambast, Louis, Sur l'intérêt stratigraphique des charophytes fossiles: exemples d'application au Tertiaire parisien. *C. R. Somm. Séanc. Soc. Géol. Fr.*, p. 207—209, 7 figs., 1962a.

Grambast, Louis, Classification de l'embranchement

des charophytes. *Naturalia Monspel., Sér. Bot.,* fasc. 14, p. 63 – 86, 4 figs., 1962b.

Grambast, Louis, Charophytes du Trias, État des connaissances acquises, intérêt phylogénétique et stratigraphique. *Mém. Bur. Rech. Géol. Min.,* no. 15, p. 567 – 569, 1963.

Grambast, Louis, Sur des charophytes remarquables du Crétacé terminal. *C. R. Hebd. Séanc. Acad. Sci., Paris,* v. 258, p. 643 – 646, 1964a.

Grambast, Louis, Indications fournies par les charophytes pour la stratigraphie du Paléogène. *Mém. Bur. Rech. Géol. Min.,* no. 28, p. 1009 – 1011, 1964b.

Grambast, Louis, Precisions nouvelles sur la phylogénie des charophytes. *Naturalia Monspel., Sér. Bot.,* fasc. 16, p. 71 – 77, fig., 1964c.

Grambast, Louis, État des connaissances acquises sur les charophytes du Crétacé inférieur. *Mém. Bur. Rech. Géol. Min.,* no. 34, p. 577 – 582, 1965.

Grambast, Louis, Structure de l'utricule et phylogénie chez les Clavatoracées. *C. R. Hebd. Séanc. Acad. Sci., Paris,* v. 262, p. 2207 – 2210, pls. 1 – 3, 1966a.

Grambast, Louis, Un nouveau type structural chez les Clavatoracées; son intérêt phylogénétique et stratigraphique. *C. R. Hebd. Séanc. Acad. Sci., Paris,* v. 262, p. 1929 – 1932, figs. 1 – 7, 1966b.

Grambast, Louis, La série évolutive *Perimneste-Atopochara* (Charophytes). *C. R. Hebd. Séanc. Acad. Sci., Paris,* sér. D, v. 264, p. 581 – 584, pls. 1 – 4, figs. a – g, 1967a.

Grambast, Louis, Division Charophyta, p. 216 – 217, figs. 3.1A – 3.1B. *In* W. B. Harland et al., *The fossil record.* London: Geological Society London, 1967b.

Grambast, Louis, Evolution of the utricle in the charophyte genera *Perimneste* Harris and *Atopochara* Peck. *J. Linn. Soc. (Bot.),* v. 61 (384), p. 5 – 11, pls. 1 – 3, figs. 1 – 2, 1968.

Grambast, Louis, La symétrie de l'utricule chez les Clavatoracées et sa signification phylogénétique. *C. R. Hebd. Séanc. Acad. Sci., Paris,* sér. D, v. 269, p. 878 – 881, pls. 1 – 4, 1969.

Grambast, Louis, Origine et évolution des *Clypeator* (Charophytes). *C. R. Hebd. Séanc. Acad. Sci., Paris,* sér. D, v. 271, p. 1964 – 1967, pls. 1 – 4, 1970.

Grambast, Louis, Remarques phylogénétiques et biochronologiques sur les *Septorella* du Crétacé terminal de Provence et les Charophytes associées. *Paléobiologie Continentale,* v. 2 (2), p. 1 – 38, pls. 1 – 29, figs. 1 – 23, 1971.

Grambast, Louis, Etude sur les charophytes tertiaires d'Europe occidentale I – Genre *Tectochara. Paléobiologie Continentale,* v. 3(2), p. 1 – 30, pls. 1 – 8, figs. 1 – 11, 1972a.

Grambast, Louis, Principes de l'utilisation stratigraphique des charophytes. Applications au Paléogène d'Europe occidentale. *Mém. Bur. Rech. Geol. Min.,* no. 77, p. 319 – 328, 1972b.

Grambast, Louis, Phylogeny of the Charophyta. *Taxon,* v. 23, p. 463 – 481, figs. 1 – 10, 1974.

Grambast, Louis, Charophytes du Crétacé Supérieur de la region de Cuenca. *1er Symposium sobre el Cretacio de la Cordillera Iberica, Cuenca, 9 – 12 Sept. 1974,* p. 67 – 83, pls. 1 – 6, figs. 1 – 12, 1975.

Grambast, Louis, and Nicole Grambast, Sur la position systématique de quelques charophytes Tertiaires. *Revue Gén. Bot.,* v. 61, p. 665 – 671, figs. 1a – c, 1954a.

Grambast, Louis, and Nicole Grambast, Sur deux espèces de charophytes du Bartonien du Bassin de Paris. *C. R. Somm. Séanc. Soc. Géol. Fr.,* p. 95 – 97, 1 fig., 1954b.

Grambast, Louis, and J. Lorch, Une flore de charophytes du Crétacé Inférieur du proche-Orient. *Naturalia Monspel., Ser. Bot.,* fasc. 19, p. 47 – 56, pls. 1 – 4, figs. 1 – 4, 1968.

Grambast, Louis, and Philippe Paul, Observations nouvelles sur la flore de charophytes du Stampien du Bassin de Paris. *Bull. Soc. Géol. Fr.,* ser. 7, v. 7, p. 239 – 247, 2 pls., 4 text-figs. (1965) 1966.

Grambast, Louis, and Ingeborg Soulié-Märsche, Sur l'ancienneté et la diversification des *Nitellopsis* (Charophytes). *Paléobiologie Continentale,* v. 3(3), p. 1 – 14, 1972.

Grambast, Nicole, and Louis Grambast, Les Raskyelloideae, sous-famille fossile des Characeae. *C. R. Hebd. Séanc. Acad. Sci., Paris,* v. 240, p. 999 – 1001, figs. 1 – 4, 1955.

Groves, James, Fossil charophyte fruits from Texas. *Am. J. Sci.,* ser. 5, v. 10, p. 12 – 14, 3 figs. 1925.

Groves, James, Pars 19: Charophyta, p. 1 – 74. *In* W. Jongmans, ed., *Fossilium Catalogus II: Plantae.* Berlin: W. Junk, 1933.

Groves, James, and G. R. Bullock-Webster, *The British Charophyta,* v. 1, *Nitelleae.* Ray Soc. Publs., p. 1 – 141, 20 pls., 25 text-figs., 1920.

Groves, James, and G. R. Bullock-Webster, *The British Charophyta,* v. 2, *Chareae.* Ray Soc. Publs., p. 1 – 129, pls. 21 – 45, 31 text-figs., 1924.

Hacquaert, A. L., Notes sur les genres *Sycidium* et *Trochiliscus. Bull. Mus. R. Hist. Nat. Belg.,* v. 8, p. 1 – 22, 10 figs., 1932.

Harris, T. M., *British Purbeck Charophyta.* London: British Museum (Natural History), 83 p., 17 pls., 16 text-figs., 1939.

Horn af Rantzien, Henning, On the fossil Charophyta of Latin America. *Svensk Bot. Tidskr.,* v. 45, p. 658 – 677, 1951.

Horn af Rantzien, Henning, Revisions of some Pliocene charophyte gyrogonites. *Bot. Notiser,* v. 107, p. 1 – 33, 8 text-figs., 1954a.

Horn af Rantzien, Henning, Middle Triassic Charophyta of south Sweden. *Op. Bot. Soc. Bot. Lund.,* v. 1, p. 5−81, 7 pls., 3 figs., 1954b.

Horn af Rantzien, Henning, Morphological terminology relating to female charophyte gametangia and fructifications. *Bot. Notiser,* v. 109, p. 212−259, figs. 1a−h, 1956a.

Horn af Rantzien, Henning, An annotated check-list of genera of fossil Charophyta. *Micropaleontology,* v. 2, p. 243−256, 1956b.

Horn af Rantzien, Henning, Nitellaceous charophyte gyrogonites in the Rajmahal Series (upper Gondwana) of India with notes on the flora and stratigraphy. *Stockh. Contr. Geol.,* v. 1, p. 1−29, pls. 1−3, text-figs. 1−14, 1957.

Horn af Rantzien, Henning, Morphological types and organ-genera of Tertiary charophyte fructifications. *Stockh. Contr. Geol.,* v. 4, p. 45−197, pls. 1−21, tables 1−24, 1959a.

Horn af Rantzien, Henning, Comparative studies of some modern, Cenozoic and Mesozoic charophyte fructifications. *Stockh. Contr. Geol.,* v. 5, p. 1−17, 1959b.

Horn af Rantzien, Henning, Recent charophyte fructifications and their relations to fossil charophyte gyrogonites. *Ark. Bot.,* ser. 2, v. 4, p. 165−325, pls. 1−19, 1959c.

Horn af Rantzien, Henning, and Louis Grambast, Some questions concerning Recent and fossil charophyte morphology and nomenclature. *Stockh. Contr. Geol.,* v. 9, p. 135−144, 1962.

Iyengar, M. O. P., *Nitella terrestris* sp. nov., a terrestrial charophyte from South India. *Bull. Bot. Soc. Beng.,* v. 12, p. 85−90, pls. 1, 2, 1960.

Karczewska, Jadwiga, and Maria Ziembińska-Tworzydło, Lower Tertiary Charophyta from the Nemegt Basin, Gobi Desert. *Palaeont. Pol.,* no. 27, p. 51−81, pls. 7−27, figs. 1−14, 1972.

Karpinskiy, A. P., O Trokhiliskakh: Die Trochilisken. *Trudy Geol. Kom.,* n. ser., vyp. 27, p. 1−166, 1906.

Kasaki, Hideo, The Charophyta from the lakes of Japan. *J. Hattori Bot. Lab.,* no. 27, p. 217−314, 15 pls., 8 figs., 10 maps, 1964.

Kiselevskiy, F. Yu., Novye dannye o nizhnetriasovykh kharofitakh Prikaspiyskoy vpadiny [New data on the Lower Triassic charophytes of the Precaspian basin]. *Vop. Geol. Yuzh. Ural i Povolzh'ya, Saratovsk Univ.,* vyp. 4, p. 37−44, 1967.

Kozur, H., and P. Reinhardt, Charophyten aus dem Muschelkalk und dem unteren Keuper Mecklenburgs und Thüringens. *Mber. Dt. Akad. Wiss. Berl.,* v. 11, p. 369−386, pls. 1, 2, figs. 1−3, 1969.

Kyansep-Romashkina, N. P., Znachenie kharovykh vodorosley dlya stratigrafii Mezozoyskikh otlozheniy Fergany i paleolimnologicheskikh rekonstruktsiy [Significance of charophyte algae for stratigraphy of Mesozoic deposits of Fergana and paleolimnological reconstruction], p. 21−37, pls. 1−3, In *Problemy Issledovaniya Drevnikh ozer Evrazii.* Leningrad: Akad. Nauk SSSR, Inst., Ozerovedeniya, 1974.

Kyansep-Romashkina, N. P., Nekotorye pozdneyurskie i melovye kharofity Mongolii [Some Late Jurassic and Cretaceous charophytes of Mongolia], p. 181−204, pls. 1−6. In *Iskopaemaya fauna i flora Mongolii,* Sovmestnaya Sovetsko-Mongol'skaya Paleontologicheskaya Ekspeditsiya, *Trudy,* vyp. 2. Moscow: "Nauka," 1975.

Lamarck, J. B., Suite des mémoires sur les fossiles des environs de Paris. *Annls. Mus. Hist. Nat. Paris,* v. 5, p. 349−357, 1804.

Lamarck, J. B., Explication des planches relatives aux coquilles fossiles des environs de Paris. *Annls. Mus. Hist. Nat. Paris,* v. 9, p. 236−240, pls. 15−18, 1807.

Langangen, Anders, Ecology and distribution of Norwegian charophytes. *Norw. J. Bot.,* v. 21, p. 31−52, figs. 1−27, 1974.

Langer, W., Neufunde von *Sycidium* G. Sandberger (nova class., Charophyta?) aus dem Devon der Eifel. *Paläont. Z.,* v. 50, p. 209−221, pls. 24, 25, figs. 1−3, 1976.

Levy, Alain, and Gabriel Lucas, Remarques sur le rôle des charophytes dans les milieux margino-littoraux. *C. R. Hebd. Séanc. Acad. Sci., Paris,* sér. D, p. 2527−2530, 4 figs., 1971.

Loeblich, A. R., Jr., and Helen Tappan, Suprageneric classification of the Rhizopodea. *J. Paleont.,* v. 35, p. 245−330, 1961.

Loeblich, A. R., Jr., and Helen Tappan, *Treatise on invertebrate paleontology,* Pt. C, Protista 2, Sarcodina chiefly "Thecamoebians" and Foraminiferida. Lawrence, Kans.: Geol. Soc. Am. and Univ. Kansas Press, v. 1, xxxi + 511 p., 399 figs., v. 2, 390 p., 254 figs., 1964.

Loeblich, A. R., Jr., and Helen Tappan, Recent advances in the classification of the Foraminiferida, p. 1−53, 1 fig. *In* R. H. Hedley and C. G. Adams, *Foraminifera,* v. 1. London: Academic Press, 1974.

Lokke, D. H., and J. F. Van Sant, Upper Pennsylvanian Charophyta from Kansas. *J. Paleont.,* v. 40, p. 971−976, 2 figs., 1966.

Mädler, Karl, Charophyten aus dem Nordwestdeutschen Kimmeridge. *Geol. Jb.,* v. 67, p. 1−46, pls. A, B, text-figs. 1−8, (1952) 1953a.

Mädler, Karl, Fossil Charophytes als Zeitmarken. *Erdöl Kohle,* v. 6, p. 63−65, 1953b.

Mädler, Karl, Ein neues system de fossilen Charophyten. *Flora, Jena,* v. 140, p. 474−484, 17 figs., 1953c.

Mädler, Karl, Die taxionomischen Prinzipien bei der Beurteilung fossiler Charophyten. *Paläont. Z.,* v. 29, p. 103−108, 1955a.

Mädler, Karl, Zur Taxionomie der tertiären Charophyten. *Geol. Jb.,* v. 70, p. 265—328, pls. 23—26, 1955b.

Mädler, Karl, Fossil charophytes, their evolution, taxonomy and stratigraphy. *J. Palaeont. Soc. India,* v. 2, p. 42—47, 2 figs., 1957.

Magniez, Guy, Pierre Rat, and Henri Tintant, Découverte d'oogones de charophytes dans le Bathonien marin près de Dijon. *C. R. Hebd. Séanc. Acad. Sci., Paris,* v. 250, p. 1692—1694, 1 fig., 1960.

Mamet, B. L., Sur les Umbellaceae. *Can. J. Earth Sci.,* v. 7, p. 1164—1171, pl. 1, 1970.

Maslov, V. P., Materialy k poznaniyu iskopaemykh vodorosley SSSR, Iskopaemye Khary—znachenie, anatomiya i metodika iz izucheniya (na materiale iz Kirgizii) [Materials for investigation of fossil algae of the USSR. Fossil *Chara*—their importance, anatomy, and the methods of their study (on material from Kirghiz)]. *Byull. Mosk. Obshch. Ispyt. Prir., Otdel Geol.,* v. 22, p. 73—90, text-figs. 1—15, 1947.

Maslov, V. P., Iskopaemye izvestkovye vodorosli SSSR [Fossil calcareous algae of the USSR]. *Trudy Inst. Geol. Nauk, Mosk.,* vyp. 160, p. 1—301, 1956.

Maslov, V. P., K voprosu o klassifikatsii i filogenii kharofitov [A contribution to the problem of classification and phylogenesis of charophytes]. *Dokl. Akad. Nauk SSSR,* v. 133, p. 678—680, 1 fig., 1957.

Maslov, V. P., Ne yavlyayutsya li sitsidii i khovanelly utrikulami kharofitov? *Dokl. Akad. Nauk SSSR,* v. 138, p. 677—680, 1 fig., 1961. [Are not *Sycidium* and *Chovanella* the utricles of charophytes? *Dokl. Acad. Sci. USSR,* Earth Sci. Sect., AGI Transl., v. 138, p. 669—671, figs. a—m, 1962].

Maslov, V. P., Tip Charophyta. Luchnitsy, ili kharovye vodorosli [Phylum Charophyta, Stoneworts, or charophyte algae], p. 224—242, figs. 1—27. *In* Yu. A. Orlov, ed., *Osnovy paleontologii, Vodorosli, mokhoobraznye, psilofitovye, plaunovidnye, chlenistostebel' nye, paporotniki.* Moscow: Akad. Nauk SSSR, 1963a.

Maslov, V. P., Vvedenie v izuchenie iskopaemykh kharovykh vodorosley. [Introduction to the study of fossil charophyte algae]. *Trudy Inst. Geol. Nauk, Mosk.,* vyp. 82, p. 1—104, pls. 1, 2, text-figs. 1—42, 1963b.

Maslov, V. P., Nakhodka izvestkovoy koronki i stroeyenie vershiny organov plodonosheniya iskopaemykh kharofitov. *Dokl. Akad. Nauk SSSR,* v. 149, p. 954—956, 2 figs., 1963c. [The discovery of the calcareous crown and the apical structure of the fruit-bearing organs of fossil charophytes. *Dokl. Acad. Sci. USSR,* Earth Sci. Sect., AGI Transl., v. 149, p. 196—198, 2 figs., 1965].

Maslov, V. P., Prorastanie oospory iskopaemykh kharofitov i novyy organ-rod. *Dokl. Akad. Nauk SSSR,* v. 152, p. 443—445, text-figs. 1, 2, 1963d. [The growth of oospores in fossil charophytes and a new organ-genus. *Dokl. Acad. Sci. USSR,* Earth Sci. Sect., AGI Trans., v. 152, p. 206—207, 2 figs., 1965].

Maslov, V. P., Nekotorye Kaynozoyskie Kharofity Yuga SSSR i metodika ikh izucheniya. [Certain Cenozoic charophytes in the south of the USSR, and methods for their study], p. 10—92, pls. 1—11, In Iskopaemye Kharofity SSSR, *Trudy Inst. Geol. Nauk,* vyp. 143, 1966.

Massieux, M., and Y. Tambareau, Charophytes Thanétiennes et infra-Ilerdiennes des Pyrénées Centrales. *Revue Micropaléont.,* v. 21, p. 140—148, 2 pls., 3 figs., [1978] 1979.

Migula, W., Die Characeen Deutschlands, Oesterreichs und der Schweiz, xv + 765 p., 149 figs. In *L. Rabenhorst's Kryptogamen-Flora von Deutschland, Oesterreich und der Schweiz,* v. 5. Leipzig: Akademische Verlagsgesellschaft, 1897.

Miklukho-Maklay, A. D., Nekotorye Devonskie vodorosli sredney Azii i drugikh oblastey SSSR i ikh porodoobrazuyushchee i paleogeograficheskoe znachenie. *Dokl. Akad. Nauk SSSR,* v. 138, p. 655—658, 1961. [Certain Devonian algae of central Asia and of other regions of the USSR and their rock-forming and paleogeographic importance. *Dokl. Acad. Sci. USSR,* Earth Sci. Sect., AGI Transl., v. 138, p. 541—543, 1962].

Movshovich, E. V., and L. Ya. Saydakovs'kiy, Baskunchats'ki vidkladi nizhn'ogo Triasu pivnichnoskhidnoy okrayni Donbasu [Baskunchatsky horizon of the lower Triassic of the northeastern Donets borderland]. *Geol. Zh.,* v. 26, p. 52—59, 3 figs., 1966.

Musacchio, E. A., Charophytas de la Formacion la Amarga (Cretacico Inferior) Provincia de Neuquen, Argentina. *Revta. Mus. La Plata,* n. ser., v. 6 (*Paleontologia* no. 37), p. 19—38, pls. 1—3, figs. 1—4, 1971.

Musacchio, E. A., Charophytas y ostracodos no marinos del Grupo Neuquen (Cretacico Superior) en algunos afloramientos de la provincias de Rio Negro y Neuquen, Republica Argentina. *Revta. Mus. La Plata,* n. ser., v. 8 (*Paleontologia* no. 48), p. 1—32, pls. 1—7, figs. 1—3, 1973.

Nordstedt, Otto, De Algis et Characeis, 4. Über die Hartschale der Characeenfrüchte. *Acta Univ. Lund.,* v. 25, p. 2—17, 1889.

Nötzold, T., Die Präparation von Gyrogoniten und kalkigen Charophyten-Oogonien aus festen Kalksteinen. *Mber. Dt. Akad. Wiss. Berl.,* v. 7, p. 216—221, 2 pls., 1965.

Olsen, Siguard, Danish Charophyta, chorological, ecological, and biological investigations. *Biol. Skr.,* v. 3, p. 1—240, 2 pls., 44 text-figs., 1944.

Pander, C. H., *Monographie der fossilen Fische des silurischen Systems der Russisch-Baltischen Gouvernements.* St. Petersburg: K. Akad. Wiss., x + 91 p., 7 pls., 10 figs. of text pl., 1856.

Papenfuss, G. F., Classification of the algae, p. 115—225. *In* E. L. Kessel, *A century of progress in the natural sciences.* San Francisco: Calif. Acad. Sci., 1955.

Papp, A., and N. Manolessos, Charophytenreste aus dem Jungtertiär Griechenlands. *Annls. Géol. Pays Hell.,* v. 5, p. 88—92, pl. 22, 1953.

Peck, R. E., The North American trochiliscids, Paleozoic Charophyta. *J. Paleont.*, v. 8, p. 83—119, 5 pls., 2 text-figs., 1934a.

Peck, R. E., Late Paleozoic and early Mesozoic Charophyta. *Am. J. Sci.*, ser. 5, v. 27, p. 49—55, 1 fig., 1934b.

Peck, R. E., Structural trends of the Trochiliscaceae. *J. Paleont.*, v. 10, p. 764—768, text-figs. 1a—h, 1936.

Peck, R. E., Morrison Charophyta from Wyoming. *J. Paleont.*, v. 11, p. 83—90, pl. 14, 1937.

Peck, R. E., A new family of Charophyta from the Lower Cretaceous of Texas. *J. Paleont.*, v. 12, p. 173—176, 1 pl., 1 fig., 1938.

Peck, R. E., Lower Cretaceous Rocky Mountain non-marine microfossils. *J. Paleont.*, v. 15, p. 285—304, pls. 42—44, 1941.

Peck, R. E., Fossil Charophyta. *Am. Midl. Nat.*, v. 36, p. 275—278, 1 fig., 1946.

Peck, R. E., Fossil charophytes. *Bot. Rev.*, v. 19, p. 209—227, 1953.

Peck, R. E., North American Mesozoic Charophyta. *Prof. Pap. U.S. Geol. Surv.*, 294-A, p. 1—44, pls. 1—8, text-figs. 1—7, 1957.

Peck, R. E., On the systematic position of the umbellids. *J. Paleont.*, v. 48, p. 409—414, fig. 1, 1974.

Peck, R. E., and J. A. Eyer, Representatives of *Chovanella*, a Devonian charophyte in North America. *Micropaleontology*, v. 9, p. 97—100, pl. 1, 1963a.

Peck, R. E., and J. A. Eyer, Pennsylvanian, Permian and Triassic Charophyta of North America. *J. Paleont.*, v. 37, p. 834—844, pls. 100, 101, text-fig. 1, 1963b.

Peck, R. E., and J. A. Eyer, Pennsylvanian, Permian and Triassic Charophyta of North America—a correction. *J. Paleont.*, v. 38, p. 426, 1964.

Peck, R. E., and G. A. Morales, The Devonian and Lower Mississippian charophytes of North America. *Micropaleontology*, v. 12, p. 303—324, pls. 1—4, 1966.

Peck, R. E., and C. C. Reker, Cretaceous and lower Cenozoic Charophyta from Peru. *Am. Mus. Novit.*, no. 1369, p. 1—6, 1 pl., 1947.

Peck, R. E., and C. C. Reker, Eocene Charophyta from North America. *J. Paleont.*, v. 22, p. 85—90, pl. 21, 1948.

Pia, J., *Pflanzen als Gesteinbildner.* Berlin: Gebrüder Borntraeger, 355 p., 1926.

Platonov, V. A., K sistematike Umbell (Kharovye vodorosli) [On the systematic of *Umbella* (Charophyte algae)]. *Paleont. Zh.*, 1974(1), p. 101—111, pl. 9, 1974 [Systematics of the Umbellaceae (Charophyta). *Paleont. J.*, v. 8, p. 94—103, pl. 9, AGI Transl. 1974].

Podles'kiy, V. I., Charophyta pivdennozakhidnoi URSR [Charophytes of the southwest UkSSR]. *Zh. Inst. Bot. Kyyiv*, p. 65—69, 1935.

Poyarkov, B. V., Pervyy kollokvium po izucheniyu iskopaemykh kharovykh vodorosley i ikh znacheniya dlya stratigrafii [The first colloquium on the investigation of fossil chara algae and their significance for stratigraphy]. *Paleont. Zh.*, 1964(1), p. 143—144, 1964.

Poyarkov, B. V., O sistematicheskom polozhenii Umbell. *Dokl. Akad. Nauk SSSR*, v. 163, p. 728—730, figs. 1—3, 1965. [Status of *Umbella. Dokl. Acad. Sci. USSR*, Earth Sci. Sect., AGI Transl., v. 163, p. 210—212, figs. 1—3, 1966].

Poyarkov, B. V., Devonskie kharofity Tyan'-Shanya [Devonian charophytes of Tien-Shan]. *Trudy Inst. Geol. Nauk, Mosk.*, vyp. 143, p. 161—200, pls. 1, 2, figs. 1—10, 1966.

Proctor, V. W., The nature of charophyte species. *Phycologia*, v. 14, p. 97—113, figs. 1—7, 1975.

Rasky, K., Fossil Charophyta. *XX Int. Geol. Congr.*, Resumenes de los trabajos presentados, p. 123, 1956.

Reid, Clement, and James Groves, The Charophyta of the lower Headon Beds of Hordle [Hordwell] Cliffs (South Hampshire). *Q. Jl. Geol. Soc. Lond.*, v. 77, p. 175—192, pls. 4—6, 1921.

Reinhardt, P., Charophyten aus dem Unterkeuper Thuringens. *Geologie*, v. 12, p. 224—229, 2 pls., 1963.

Reytlinger, E. A., Kharakteristika Ozerskikh i Khovanskikh sloev po mikroskopicheskim organicheskim ostatkam [Characteristic microscopic organic remains of the Ozerki and Khovanshchina beds]. *Trudy Inst. Geol. Nauk, Mosk.*, vyp. 14, p. 136—177, pls. 1—3, figs. 1—5, 1960.

Reytlinger, E. A., Ob Umbellakh Evropeyskoy chasti SSSR i Zakavkaz'ya [On the Umbellas in the European part of the USSR and Transcaucasia]. *Trudy Inst. Geol. Nauk, Mosk.*, vyp. 143, p. 213—220, 1 pl., 1966.

Reytlinger, E. A., and M. V. Yartseva, Novye karofity verkhnefamenskikh otlozheniy Russkoy Platformy [New charophytes from the upper Fammenian deposits of the Russian Platform]. *Dokl. Akad. Nauk SSSR*, v. 123, p. 1113—1116, 1 pl., 1958.

Ross, C. A., Population study of charophyte species, Morrison Formation, Colorado. *J. Paleont.*, v. 34, p. 717—726, pl. 92, 10 text-figs., 1960.

Samoylova, R. B., Pervaya nakhodka trokhiliskov podroda *Karpinskya* Croft v Devonskikh otlozheniyakh Russkoy Platformy. *Dokl. Akad. Nauk SSSR*, v. 139, p. 206—207, 1 pl., 1961. [The first record of *Trochiliscus* of the subgenus *Karpinskya* Croft in Devonian deposits of the Russian Platform. *Dokl. Acad. Sci. USSR*, Earth Sci. Sect., AGI Transl., v. 139, p. 844—846, 1 pl., 1963.

Sandberger, Guido, Eine neue Polypen-Gattung *Sycidium* aus der Eifel. *Neues Jb. Min. Geol. Paläont.*, v. 20, p. 671—672, pl. 8B, 1849.

Saydakovskiy, L. Ya., Kharofity iz triasovykh pestrotsvetov Bol'shogo Donbassa [Charophytes from the

Triassic variegated rocks of the Great Donets Basin]. *Dokl. Akad. Nauk SSSR,* v. 145, p. 1141—1144, figs. 1—18, 1962.

Saydakovskiy, L. Ya., Biostratigrafiya triasovykh otlozheniy Yuga Russkoy platformy [Biostratigraphy of Triassic deposits in the south of the Russian Platform], p. 93—144, pls. 1—4, *In* Iskopaemye Kharofity SSSR, *Trudy Inst. Geol. Nauk,* vyp. 143, 1966a.

Saydakovskiy, L. Ya., Kharofity iz verkhnepermskikh otlozheniy Yuga Russkoy platformy [Charophytes from the upper Permian deposits of the south of the Russian Platform], p. 145—153. *In* Iskopaemye Kharofity SSSR, *Trudy Inst. Geol. Nauk,* vyp. 143, 1966b.

Saydakovskiy, L. Ya., Kharofity iz Triasa prikaspiyskoy vpadiny [Charophyta from the Triassic of the Caspian Depression]. *Paleont. Zh.,* 1968(2), p. 95—110, pls. 15, 16, 1968. [*Paleont. J.,* AGI Transl., p. 234—247, 1968].

Shaykin, I. M., Characeae verkhnego Karbona Donbassa [Characeae of the upper Carboniferous of the Donets], p. 154—160, 1 pl., 2 figs. *In* Iskopaemye Kharofity SSSR, *Trudy Inst. Geol. Nauk,* vyp. 143, 1966.

Sommer, F. W., Contribuição à paleofitografia do Paraná, p. 175—194, pls. 15—20. *In* F. W. Lange, ed., *Paleontologia do Paraná, Curitiba Comis. Comemor. Cent. Paraná,* 1954.

Stache, G., Die Liburnische Stufe und deren Grenz-Horizonte. *Abh. Geol. Bundesanst., Wien,* v. 13, p. 1—170, pls. 1—5, 1889.

Tappan, Helen, and A. R. Loeblich, Jr., Smaller protistan evidence and explanation of the Permian-Triassic crisis, p. 465—480, figs. 1—5. In A. Logan and L. V. Hills, *The Permian and Triassic Systems and their mutual boundary.* Calgary: Can. Soc. Petrol. Geol., Mem. 2, 1973.

Wang, Shui, Tertiary Charophyta from Chaidamu (Tsaidam) Basin, Qinghai (Chinghai) Province. *Acta Palaeont. Sin.,* v. 9, p. 183—219, pls. 1—7, 1961.

Womersley, H. B. S., and I. L. Ophel, *Protochara,* a new genus of Characeae from western Australia. *Trans. R. Soc. S. Aust.,* v. 71, p. 311—317, 2 figs., 1947.

Wood, R. D., Stability and zonation of Characeae. *Ecology,* v. 31, p. 642—647, figs. 1—4, 1950.

Wood, R. D., The Characeae. *Bot. Rev.,* v. 18, p. 317—353, 1952.

Wood, R. D., Gametangial constants of extant Charophyta for use in micropaleobotany. *J. Paleont.,* v. 33, p. 186—194, 1959.

Wood, R. D., New combinations and taxa in the revision of Characeae. *Taxon,* v. 11, p. 7—25, 1962.

Wood, R. D., A synopsis of the Characeae. *Bull. Torrey Bot. Club,* v. 91, p. 35—46, 1964.

Wood, R. D., and Kozo Imahori, *A revision of the Characeae, Second Part, Iconograph of the Characeae.* Weinheim: J. Cramer, xv + 7 p., 394 pls., 1964.

Wood, R. D., and Kozo Imahori, *A revision of the Characeae, First Part, Monograph of the Characeae.* Weinheim: J. Cramer, xxiv + 904 pp., 8 pls., 28 figs., 1965.

GLOSSARY

acicule: V-shaped hollow spicule at anterior pole of dinoflagellate *Plectodinium,* composition unknown.

acritarch: Acritarcha, phytoplankton fossils of unknown exact affinity; the resistant covering and presence of various excystment mechanisms indicate that they are resting cysts.

acrobase: lines separating nonthecate dinoflagellate epicone surface when silver-stained into apical acromere and adjacent prosomere.

acromere: apical region of nonthecate dinoflagellate, delineated by acrobase line when silver-stained.

acronematic: flagella lacking mastigonemes, cylindrical in section, also termed **whiplash** flagella.

actines: individual branches of the siliceous spicule of the ebriophyceans.

akinete: resting spore, with thickened wall, in blue-green algae.

alar canal: elongate passage through raised keel of diatom valve, opening at one end into raphe canal and at other end into frustule cavity (in *Surirella*).

ambitus: outline of dinoflagellate cyst as viewed from dorsal or ventral side.

amphiesma: cell covering of dinoflagellates, consisting of innermost pellicular layer, thecal layer, vesicular layer, and outer or plasma membrane.

amphitrichate: bacteria with one or more flagella at each pole.

anisopolar: charophyte lime shells with greatest diameter not at equator.

antapical plates: one or more plates at posterior of dinoflagellate theca.

anterior intercalary plates: one to three plates lying between apical and precingular series on dorsal side of dinoflagellate theca.

antheridium: male gametangium.

antiligula: short extension from intercalary band of diatom frustule, pointing in abvalvar direction to close gap in adjacent bands.

apex: part of acritarch vesicle at which excystment opening occurs; anterior end of flagellate cells; growing tip of algal thallus.

apical area: hyaline area at poles of pennate diatom frustule.

apical axis: sagittal or lengthwise axis of pennate diatom frustule.

apical pit: depression surrounded by four projections at anterior end of prasinophyte motile cell, from which arise the four scaly flagella.

apical plates: anteriormost plate or series of plates in dinoflagellate theca.

apical ring: in silicoflagellates, smaller skeletal ring nearest cell periphery, and attached to the main or basal ring by supporting bars, as in *Distephanus.*

aplanospores: statospores, or resting cysts, as in the Chrysophyta.

apochlorotic: algae that lack photosynthetic pigments (some dinoflagellates, euglenids, diatoms).

"

archaeomonadite: siliceous deposit composed in large part of the cysts of the marine chrysophytes or archaeomonads.

archeopyle: dorsal or apical polygonal opening in dinoflagellate cyst, left by removal of operculum at excystment; commonly represents only certain paraplates, hence is evidence of orientation of the cyst.

areolae: subcircular to angular chamberlike structures within the diatom frustule wall, usually covered by thin velum at surface.

armored dinoflagellates: dinoflagellates with cellulosic theca consisting of series of plates.

aspondyl: lateral branches randomly placed on the main axis.

asterolith: star or rosettelike ortholithid coccolith, as *Discoaster*, each ray formed of a single calcite crystal with C-axis perpendicular to plane of disk.

ATP: adenosine triphosphate, a compound involved in energy transformation in metabolism; consists of one molecule each of adenine and D-ribose, and three molecules of phosphoric acid.

atrichate: bacteria lacking flagella, moving only by Brownian movement or current action.

autocyst: dinoflagellate cyst with single wall; also termed **autoblast.**

autophragm: single-layered wall of dinoflagellate cyst.

autotomy: self-pruning of spines by dinoflagellates to maintain symmetry and in response to changed water conditions.

autotrophs: organisms that obtain energy from sunlight by photosynthesis (**phototrophs**) or from chemical oxidation-reduction (**chemotrophs**).

auxiliary cell: specialized cell of carpogonium of some red algae, into which zygote nucleus passes after fertilization.

auxospores: special rejuvenescent cells of diatoms formed during sexual reproductive process, allowing clone to regain maximum cell size.

awns: setae, elongate projections from valve margin of diatom *Chaetoceros*.

axial plastids: plastids that are centrally located within the cell.

axoneme: central part of flagellum, containing the 9+2 fibrils (ring of nine paired fibrils surrounding central pair).

axoskeleton: Y-shaped spicule with bifurcated tips in the dinoflagellate *Dicroerisma;* composition unknown, but insoluble in acetic acid.

bacillus: cylindrical or rodlike bacterial cell.

bacteriophage: filterable virus that infects bacteria and transfers genetic material by transduction.

baculate: echinate surface sculpture, with bluntly terminated cylindrical spinules.

basal plug: calcified nodal cell at base of charophyte lime shell; also termed **basal plate.**

basal pore: rounded to pentagonal pore at base of charophyte lime shell that was previously occupied by a nodal cell.

basal ring: larger or only (*Mesocena*) circular or polygonal ring of the silicoflagellate skeleton; during reproductive cell fission, daughter cell basal ring is produced nearest cell center, against basal ring of mother cell.

bioluminescence: production of cold light by organism, as by the enzyme luciferase oxidizing the substrate luciferin; common in marine dinoflagellates.

bisporangium: sporangium of red algae that produces two spores.

boghead coal: algal coal, formed by *Botryococcus* (Chlorococcales).

branches: laterals of unlimited growth in the charophytes.

branchlets: laterals of limited growth in charophytes.

bulbils: small shoots that arise from nodal cells of charophytes and may give rise to new plants vegetatively after fragmentation.

bullula: void between well-separated areolae in diatom frustule wall that does not open at either surface of wall.

cage: basal structure of charophyte lime shell formed by a downward extension of the intercellular ridges.

calyptra: outer layer of flocculent material covering exterior of some dinoflagellate cysts, readily destroyed.

calyptrolith: holococcoliths with shape of proximally open cap.

canal raphe: raphe of diatom frustule with tubular canal at inner margin separated from cell interior by fibulae.

caneolith: discolithlike coccolith with lath-filled central area and petaloid upper and lower rims.

cap cells: terminal cells on filaments of corallinacean red algae that do not divide, filament development resulting from division of cell beneath the cap cell.

capsule: mucilaginous slime layer surrounding a bacterial cell. In dinoflagellates, may refer to inner body of cavate cyst or endocyst.

carpogonium: female gametangium of red algae.

carposporangium: spore-bearing structure of diploid carposporophyte of red algae, generally produces a single **carpospore**, although some species produce four; more than one carposporangium may arise from carposporophyte, hence numerous carpospores result from a single zygote.

carposporophyte: carpospore-producing generation of red algae; diploid, may be parasitic on female gametophyte after fertilization, or a distinct plant.

catenate: chain-forming; dinoflagellate cells that remain attached at least temporarily after binary fission; attachment commonly by cytoplasmic connection from apical pore of posterior cell to distal end of girdle of next anterior cell of chain.

cavate cyst: dinoflagellate cyst with two or more distinct wall layers forming separate endocyst and pericyst.

central area: hyaline area at center of diatom valve.

central nodule: thickened central region of pennate diatom valve, interrupting the raphe.

central pore: expansion of raphe fissure at central nodule of pennate diatom valve.

centric diatoms: diatoms with basically radial symmetry of the frustule.

centroplasm: DNA region of blue-green algae; see **nucleoplasm.**

ceratolith: horseshoe- or wishbone-shaped coccolith, each a single calcite crystal; *Ceratolithus* has a single large ceratolith per cell.

chasmolithic: living in fissures or cracks in rocks.

chorate cyst: dinoflagellate cyst with prominent processes and crests; has less evident paratabulation than other dinoflagellate cysts.

choristospory: definite reproductive organs for sporangia, as in some advanced Dasycladaceae.

cincture: **girdle** of diatom, consisting of joined connecting bands of the two valves.

cingular plates: series of plates in the girdle of dinoflagellate theca.

cingulum: girdle, a commonly oblique transverse groove on the dinoflagellate motile cell, in which lies the transverse flagellum.

cisternae: swollen saclike structures that comprise the Golgi body in the cell; in haptophytes the Golgi cisternae are the site of coccolith and scale formation.

clades: branches at ends of actines and connecting adjacent actines in the ebriophycean skeleton.

cladospory: lateral branches function as sporangia and gametangia, in some dasyclad algae.

closterite: coallike rock consisting largely of remains of desmids.

coccolithosomes: electron-dense granules produced by Golgi apparatus of coccolithophorids, fuse to form the granular organic matrix on which the coccolith is calcified.

coccoliths: calcified plates forming the outer cover of the haptophyte coccolithophorids.

coccus: ellipsoidal or spherical cell form of bacteria.

coenobium: colony of flagellate cells, as of *Volvox.*

collar: carina, flangelike projection from surface of diatom valve.

conceptacle: spore-bearing structure just beneath surface in coralline red algae; one conceptacle may contain many sporangia, each (usually) with four tetraspores.

conchocelis: free-living or perforating filamentous growth phase, alternating with a leafy thallus phase in some bangiophycid red algae, previously regarded as a separate genus, *Conchocelis.*

conjugation: joining of two cells for transfer of genetic material from one to the other.

connecting band: hooplike ring at edge of diatom valve; that of epivalve is the larger **epicingulum,** that of hypovalve is **hypocingulum.**

coorongite: peat equivalent of the boghead coal formed by *Botryococcus* (Chlorococcales).

copula: intercalary bands of diatom frustule separated from valve by the valvocopulae.

corona: ringlike series of linking spines on surface of diatom valve that join similar projections of adjacent cell of chain.

coronular cells: small terminal cells at end of spiralling cells, at apex of charophyte sporophydium; usually five, but two superposed series of five each in the Nitelloideae.

corrugate: with irregular ridges.

cortex: short-celled filaments that arise from axial medulla and arch upward to the surface epithallium, in erect coralline red algae.

costate: with narrow, parallel ridges, separated by wider interridge areas.

craticula: thick siliceous bars within valve of some pennate diatoms.

cribrilith: discolith coccolith with many central perforations.

cribrum: regularly perforated velum that covers areola in diatom frustule wall.

cricolith: elliptical ringlike coccolith.

cristate: with simple low flangelike crests that may anastomose to appear reticulate.

cruciform rays: meromes, radial skeletal elements of the Receptaculitales, with main shaft that ends distally in four radiating branches perpendicular to the shaft.

cryptopylome: circular opening produced by loss of portion of acritarch vesicle wall at excystment but not preformed as is true pylome.

crystallolith: holococcolith produced by motile form of *Coccolithus pelagicus,* with minute rhombohedra on surface of organic scale.

cyanelle: blue-green algal endosymbiont.

cyanophycin granules: structured granules, spherical to polyhedral and peripherally located; serve for nitrogen storage in blue-green algae.

cyclolith: ringlike coccolith.

cyclopyle: circular preformed opening for excystment of acritarch, may be bordered by collar, or elevated on neck, closed by operculum prior to excystment.

cyrtoliths: basketlike coccoliths with central spine.

cytostome: region of cell with less firm wall, utilized to ingest particles, as by holozoic Euglenophyta.

depauperating division: repeated division of dinoflagellate cells resulting in production of gametes that resemble adult vegetative cell other than in size.

dextrogyre: lines of interference figure in polarized light curve to right as traced out from center of coccolith in distal view.

dinokaryon: mesokaryon nucleus of dinoflagellates and euglenids with continuously condensed and visible chromosomes that appear as beadlike strands containing fibrillar DNA in archlike whorls.

diplobiontic: life cycle with alternate haploid and diploid free-living stages that may or may not be morphologically alike.

diploid: nuclei with doubled number of chromosomes, may consist of a short stage in the life cycle, representing only the zygote, may alternate with an approximately equivalent haploid phase, or may be the major phase of the organism.

discobolocysts: ejectile organelles that release a disc at the end of an elongate thin thread in chrysophytes such as *Ochromonas.*

discolith: discoid coccolith, with raised rim, may have central spike.

ecdysis: shedding of the theca by armored dinoflagellates for cell division or as a response to stress; although the protoplast escapes through an opening in a specific region, this is not an archeopyle, which is an opening in the resting cyst.

echinate: with small thornlike spinules on surface.

ectophragm: veillike reticulum of filmy material, or trabecular network external to the periphragm of some dinoflagellate cysts.

elevation: raised area of diatom frustule surface, with similar structure as remainder of surface.

endedaphic: soil algae living within the soil.

endocoel: cavity of dinoflagellate cyst, bounded by endophragm.

endocyst: inner body of dinoflagellate cyst, constructed of endophragm; also termed **endoblast, endocorpus.**

endolithic: living within the rock, boring into calcium carbonate, or living between sand grains.

endophragm: inner wall layer of dinoflagellate cyst, may be closely appressed to periphragm; inner wall layer of acritarch vesicle.

endospore: in bacteria, spore in which only a part of the original cell is utilized; in blue-green algae, many rounded spores that germinate immediately, in contrast to resting spores; in dinoflagellates, innermost cyst layer, first to dissolve at germination.

endospory: sporangia borne within main axis as in Dasycladaceae.

epedaphic: soil algae living at the surface.

epicone: anterior portion of nonthecate dinoflagellate.

epicyst: epitract, anterior part of dinoflagellate cyst, equivalent to epitheca of motile cell.

epipelic: benthic species that live on silt grains although not firmly attached, as part of the interstitial microflora.

epipsammon: benthic species that are attached to sand grains, as part of the interstitial flora.

epithallium: outer layer of cells of coralline red algae.

epitheca: anterior portion of dinoflagellate theca, terminating at the median cingulum; larger half of the diatom frustule, overlaps the hypotheca, consists of flattened valve and attached connecting band.

epitract: epicyst, anterior part of dinoflagellate cyst, equivalent to epitheca of motile cell.

epityche: broad flap produced by arched rupture of wall of acritarch vesicle at excystment, allowing flap to fold out like that of an envelope.

epivalve: older and larger valve of diatom frustule.

euspondyl: lateral branches distinctly arranged in whorls on main axis.

eutrophic: water of high nutrient level, poorly oxygenated.

exospore: thin outer layer of dinoflagellate cyst, acid-resistant, probably equivalent to cyst that is fossilized.

exuviation: process of shedding a few old plates of dinoflagellate theca to replace them with thinner new ones.

eyespot: stigma of dinoflagellates, a collection of osmiophilic pigmented granules located near flagellar bases; also similar structures in some euglenids and chlorophytes, and in the haptophytes *Pavlova* and *Platychrysis.*

false branching: rapid growth and cell division in blue-green algal filament causing it to buckle and break through the sheath, separating into two ends that continue to divide as distinct trichomes; filament also may break at a heterocyst to produce false branching.

fascia: central hyaline area extending transapically across diatom frustule.

fibulae: silica arches that span the raphe fissure in some diatoms, such as *Cylindrotheca.*

filament: elongate series of cells.

flagellar markings: sculpture in ventral region of dinoflagellate cyst indicating position of flagellar pores in motile cell.

fossulate: surface with closely spaced, elevated granules, separated by irregular grooves, resulting in an incised reticulum.

foveolate: with closely arranged pits, the depressions of greater diameter than inter-pit area.

frustule: siliceous bipartite diatom cell wall.

gametophyte: gamete-producing generation, may be entirely haploid in red algae.

gemmate: surface sculpture with small spherical projections, constricted at attachment.

genicula: uncalcified cells, or **nodes,** of articulated coralline red algae.

genome: complete genetic complement of a cell.

genophore: chromatin granule in bacteria, containing the DNA, but not surrounded by a membrane; also termed **nucleoid.**

girdle: cingulum, a commonly oblique transverse constriction or groove on the dinoflagellate motile cell, in which lies the transverse flagellum; **cincture** or joined connecting bands of the diatom frustule.

girdle lamellae: plastids of some algae in which some thylakoids ring the plastid periphery, enclosing other parallel aligned thylakoids.

Golgi body: organelle consisting of stacks of swollen saclike **cisternae;** site of scale and coccolith formation in haptophytes.

gonidia: enlarged cells at posterior of *Volvox* colony that divide vegetatively to form new daughter colonies.

Gram stain: staining procedure for bacteria, indicating a distinctive cytological character; Gram-positive forms retain a violet stain, but Gram-negative ones are decolorized by alcohol or acetone.

granulate: tiny granules on wall surface.

gyrogonite: lime shell, commonly used to refer to fossil oosporangium of charophyte.

hair cell: trichocyte, much elongated external cell in red algae, lacks nucleus when mature.

half-bands: incomplete girdle bands of diatom girdle or cincture, encircling about half the girdle or less.

haplobiontic: life cycle with only one free-living stage, as opposed to diplobiontic.

haploid: single copy of chromosomes in nucleus, characterizing one phase of the life cycle, of varied importance in different groups; may include only gametes, or may be an entire vegetative generation alternating with a similar diploid generation.

haptonema: flagellumlike appendage in Haptophyta, may be straight or coiled, may serve for temporary attachment; has fewer microtubules than do flagella and these are not paired; also lacks the central pair of microtubules of the 9 + 2 flagella.

helicolith: discolithlike coccolith with spiralling marginal flange.

heliolithae: coccoliths constructed of many tiny crystals, commonly with rotary arrangement.

heterococcoliths: coccoliths produced in vesicles within the cell of the coccolithophorid, deposited on organic matrix, and showing major modification of the crystal form.

heterocyst: enlarged cell with thickened wall, polar pore and plug; contains fewer pigments than vegetative cells and has no phycobilisomes; probable site of nitrogen fixation in blue-green algae.

heterotrophs: organisms that obtain energy by ingesting preformed organic matter.

holococcoliths: coccoliths developed at surface of organic scales, with little modification of crystal form of the rhombohedra or hexagonal calcite prisms.

holoplankton: planktonic organisms that lack a benthic stage, hence may occur in the open ocean.

hormogonia: segments of blue-green algal filaments produced by fragmentation at random or at separation disks; capable of movement.

horns: large hollow projections of dinoflagellate theca, and similar hollow expansions of dinoflagellate cyst; commonly apical and antapical in position, less common at other locations; in diatoms, exceptionally long and narrow elevation on frustule surface.

hyaline field: area of diatom frustule not perforated by areolae or punctae.

hyalosome: hyaline, refractive part of ocellus of dinoflagellates, containing crystalline lens covered by cornea.

hypnoid stage: oval cell with external membrane, but lacking theca and flagella, and moving by amoeboid locomotion, as a temporary phase in life history of some dinoflagellates.

hypnozygote: resting cyst zygote.

hypocaust: continuous space within wall of diatom frustule between well-separated areolae.

hypocone: posterior portion of nonthecate dinoflagellate.

hypocyst: hypotract, posterior part of dinoflagellate cyst, equivalent to hypotheca of motile cell.

hypolithic: soil algae living at soil interface on undersurface of stones.

hypothallium: inner part of encrusting thallus of red algae, lowermost filaments subparallel to substrate, may lack plastids.

hypotheca: posterior portion of dinoflagellate theca, posterior to median cingulum; smaller half of the diatom frustule, overlapped by epitheca, consists of flattened valve and attached connecting band.

hypotract: hypocyst, posterior part of dinoflagellate cyst, equivalent to hypotheca of motile cell.

hypovalve: younger and smaller valve of diatom frustule.

imbrication: overlap of radial elements of coccolith; appears to be in same direction as viewed from either distal or proximal side, in contrast to precession.

incomplete open bands: girdle bands of diatom frustule that encircle more than half of girdle, but are not complete rings.

inner body: endocyst or **capsule,** formed by endophragm that is separated from periphragm of dinoflagellate cyst.

intercalary bands: girdle bands proximal to valve of diatom frustule; the intercalary band or two half bands adjacent to valve form valvocopula, next bands are copulae, and distal elements are pleurae; **megacytic band** or region of growth of individual plates of dinoflagellate theca at their margins.

intergenicula: calcified segments of articulated coralline red algae, alternating with uncalcified genicula.

internodal cells: elongate biconvex cells that alternate with the shorter nodal cells in charophytes; nodal cells may continue to elongate but do not divide further.

interstitial flora: microflora living in interstitial water in shore sands and bottom muds.

isopolar: charophyte lime shells with greatest diameter at equator, separating two like halves.

isthmus: constriction or bridge between two semicells of the desmid algae.

jointive elements: crystals forming coccolith are wedge-shaped or petaloid and not overlapping, in contrast to imbricated elements.

knot-stage: dinoflagellate nucleus with large irregular mass of chromosomes in late stage of cyclosis, just prior to first meiotic division in sexual reproduction.

Krebs cycle: citric acid cycle, one of major ATP-producing enzyme systems in metabolism; converts pyruvic acid to CO_2 in presence of O_2, releasing energy as ATP molecules; cycle first described by Sir Hans Krebs in 1937.

kuckersite: *torbanite*, coal stage of boghead coal formed by *Botryococcus* (Chlorococcales).

labiate process: rimoportula, pore or tube through diatom valve that may project from surface, constricted internally but not at exterior.

laevigate: smooth unornamented surface.

laevogyre: lines of interference figure in polarized light curving to left as traced outwards from center of coccolith in distal view.

left valve: one half of theca of Prorocentrales, possibly equivalent to epitheca of other dinoflagellates; flagella emerge from pore in apical indentation of left valve.

ligula: short extension pointing in advalvar direction from intercalary bands of diatom frustule, may fill gap in adjacent bands.

lime shell: calcified oosporangium of charophyte; **gyrogonite** of fossils.

linking spines: projections from surface of diatom frustule that connect with similar projections from adjacent cell in a chain.

lists: flangelike projections arising at margin of thecal plates of dinoflagellates, particularly prominent at margins of girdle and sulcus in the Dinophysiales, and may be supported by thickened spines.

longitudinal flagellum: in dinoflagellates, occupies the sulcus and extends to posterior, contains 9 + 2 axoneme, and is bordered by two rows of short fine hairs.

longitudinal groove: sulcus on ventral side of dinoflagellate cell, connecting the two ends of the transverse cingulum, and generally extending beyond to the cell's posterior.

lopadolith: basketlike coccoliths, opening distally.

lophotrichate: bacteria with tuft of flagella at each pole.

lorica: flasklike cell wall, as in *Trachelomonas* (Euglenophyta); in ebriophyceans, an inflated chamberlike structure at anterior or nuclear pole, with a large terminal opening.

lower windows: openings left between opisthoclade branches of the ebriophycean skeleton.

loxolith ring: coccolith with outer wall of steeply inclined laths resembling twisted barrel staves; the left-handed imbrication resembles threading of left-handed screw, hence appears same from either side of coccolith.

macrococcolith: larger coccoliths of cells with two sizes of otherwise identical coccoliths, probably formed by mature cell, in contrast to micrococcoliths formed in early development.

marginate cyst: dinoflagellate cyst with processes and crests lateral in position but rare or absent on dorsal and ventral surfaces.

mastigoneme: fine hairs on one or both sides of flagella, visible with electron microscopy.

median rupture: equatorial excystment rupture, resulting in division of the acritarch vesicle into two equal halves.

median windows: openings left between the mesoclade and actine bars of the ebriophycean skeleton.

medulla: axial region of elongate cells in coralline red algae.

megacytic band: region of growth of individual plates of dinoflagellate theca at their margins.

melanosome: heavily pigmented cuplike inner part of dinoflagellate ocellus.

membranate cyst: dinoflagellate cyst with membrane that distally connects adjacent processes.

meromes: cruciform rays, radial skeletal elements of the Receptaculitales, with main shaft that ends distally in four radiating branches perpendicular to the shaft.

meroplankton: planktonic organisms that produce benthic resting spores, hence generally restricted to near-shore or shallow water environment.

mesobase: silver-stained lines dividing nonthecate dinoflagellate surface at posterior margin of girdle region.

mesoclades: connecting bars between divergent branches of a single proclade or opisthoclade of the ebriophycean skeleton.

mesocyst: rare median body of dinoflagellate cyst formed of mesophragm; also termed **mesoblast.**

mesokaryon: dinokaryon nucleus of dinoflagellates and euglenids with continuously condensed and visible chromosomes.

mesomere: silver-stained region in area of the cingulum of nonthecate dinoflagellate, between the silver-stained prosobase and mesobase.

mesophragm: median wall layer, between periphragm and endophragm of a few dinoflagellate cysts; also median wall layer reported in some acritarchs.

mesospore: thin wall layer of dinoflagellate cyst, just within the exospore, readily disintegrates, may form slime that expands to burst cyst at germination.

metaspondyl: lateral branches regularly spaced in tufts on main axis.

micrococcolith: small coccoliths of identical morphology to normal-sized ones, may be those formed during earliest development of cell.

microcysts: small cysts produced on fruiting structures of Myxobacterales.

μg-atoms per liter: measure of concentration of dissolved material in seawater, expressed as mass per liter (in grams or micrograms) divided by atomic weight of the material.

miospores: fossil spores and sporelike bodies smaller than 0.20 mm; may include microspores, megaspores, prepollen, and pollen grains.

mitochondria: organelles surrounded by double membrane, site of energy release in respiration of eucaryotic cells.

monosporangia: sporangia in red algae that produce only one spore.

monospores: single spores produced by red algae.

monotrichate: bacteria with a single simple flagellum.

nannoconid: barrellike globular, conical, cylindrical, or pyriform nannofossil with central canal surrounded by wedges of calcite arranged in low spiral; probably a coccolithophorid.

nannoplankton: very small plankton, passes through 60-μm net or smaller; may be restricted to plankton less than 2 μm in size, with that from 2 to 20 μm termed **ultramicroplankton.**

nematocysts: cnidocysts, more complex ejective organelle than trichocyst, with posterior chamber containing spiral filament and external capsule with trigger, and a tripartite structure termed the **taeniocyst;** present in a few dinoflagellates.

nodal cells: short discoidal cells that alternate with longer internodal cells in charophytes; lateral branches arise by division from the nodal cells.

nodes: uncalcified cells or **genicula** in articulated coralline red algae.

nuclear cyclosis: rotation of nuclear contents during chromosomal pairing at beginning of meiosis in dinoflagellates.

nucleoid: small DNA-bearing particle in bacteria, lacking an enclosing membrane; also termed **genophore.**

nucleoplasm: DNA-region forming a colorless central body in cells of blue-green algae; lacking an enclosing membrane; also termed **centroplasm.**

occluded process: in diatoms, a distally constricted hollow tube through frustule wall.

ocellus: complex photoreceptive organelle in some dinoflagellates, with distinct lens, pigment mass, and sensory core; in diatoms, a siliceous plate with thickened structureless rim pierced by closely spaced holes.

oligotrophic: water of limited nutrient level, may be highly oxygenated.

oncolites: laminated blue-green algal balls, formed like stromatolites, but rolling rather than stationary during growth.

oogonium: female gametangium, produces egg cell.

oosporangium: sporophydium that after fertilization of egg cell of charophyte becomes calcified as a lime shell or gyrogonite.

operculum: portion of dinoflagellate cyst that is released to form the archeopyle at excystment, may consist of one or more pieces representing one or more paraplates and may be freed or remain attached at one side; also circular portion of wall freed to form pylome at excystment of acritarch.

opisthoclades: clades or branches directed backward from the ends of the actines of the ebriophycean skeleton.

opisthomere: region of hypocone of non-thecate dinoflagellate, as divided by the silver-staining network, bounded by mesobase at posterior margin of girdle.

ortholithae: coccoliths consisting of a single or very few crystals, may have radial symmetry, as in asteroliths.

pandasutural zone: wide and transversely striated band between paraplates of dinoflagellate cyst, corresponding to intercalary bands formed by growth at margins of motile cell thecal plates.

paracingulum: region of dinoflagellate cyst that corresponds to girdle or cingulum of motile cell theca; indicated by differential development or absence of ridges and processes.

paraplates: areas of dinoflagellate cyst, equivalent to plates of various series in the dinoflagellate theca.

parasutures: linear features on dinoflagellate cyst whose sculpture indicates approximate position of the true sutures between plates of the motile cell theca; cyst does not separate at these parasutures, except for those producing the excystment opening or archeopyle.

paratabulation: pseudotabulation or **reflected tabulation** of dinoflagellate cyst, indicating by surface sculpture the position of the plate series of the motile cell.

parietal plastids: plastids located adjacent to the outer wall of the cell.

passage pores: pores that connect adjacent loculate areolae within frustule wall of diatom.

pedicel cells: cells at base of charophyte sporophydium, support the sporostegium that surrounds the oogonium.

pellicular layer: plasmalemma; innermost continuous layer of amphiesma of some dinoflagellates; resistant composition suggests that it may be source of the vegetative cyst.

pennate diatoms: diatoms with basically bilateral symmetry of the frustule; includes many with central slit or raphe.

pentalith: ortholithid coccolith with five pieces, each consisting of a single calcite crystal, oriented 72° apart; cleavage plane of crystal is in plane of pentalith, optic axes are oblique to principal axis of pentalith.

pentasters: siliceous internal stellate spicules of the dinoflagellate Actiniscaceae.

pericoel: space left between endophragm and periphragm where these are distinctly separated, hence space left between endocyst and pericyst of cavate dinoflagellate cyst.

pericyst: dinoflagellate with distinctly separated wall layers; outer part is the pericyst constructed of periphragm; also termed **periblast, pericorpus.**

perinuclear capsule: perforated capsule surrounding nuclear region in some dinoflagellates; apparently a modification of endoplasmic reticulum.

periphragm: outermost wall layer of dinoflagellate cyst, carrying the surface sculpture, flanges, and processes; also outer wall layer of acritarch vesicle.

periplekton: special labiate process on diatom valve, with distal extensions that attach to similar structures on adjacent frustules of colony.

perispore: delicate outermost layer of resting cyst of some dinoflagellates, probably remnant of the motile cell theca.

perithallium: medial part of encrusting red algal thallus, of filaments that arch upward from the underlying hypothallium to become perpendicular to substrate.

peritrichate: bacteria with surface covered by many flagella.

perizonium: silicified structure formed within zygote membrane of pennate diatom, within which new initial cell is produced.

pervalvar axis: axis connecting midpoints of valves of diatom frustule.

phototaxis: movement as a reaction to a light stimulus.

phragmoplast: cell division with open persistent nuclear spindles and microtubules algined perpendicular to plane of division; found in higher plants and some Prasinophyta and Chlorophyta, as well as Charophyta.

phycobilisomes: small granules at outer surface of thylakoids containing aggregates of phycobiliprotein pigments, in blue-green algae and red algae.

phycoma: nonmotile stage produced by prasinophytes, cystlike, but enlarges and cell contents divide to produce many nuclei, plastids, and pyrenoids; eventually releases quadriflagellate motile cells.

phycoplast: type of cell division in some Prasinophyta and Chlorophyta, in which nuclear envelope of dividing cell is pierced at poles by microtubules, forming closed spindle; microtubules lie in plane of cell division.

pinnae: first subdivisions from the trunk of the processes of acritarch vesicle.

pinnulae: second subdivision of processes of acritarch vesicle; divisions or branches from the pinnae.

pit connections: opening in wall between successive cells of red algae, filled by a biconvex membrane-bound plug.

placolith: coccolith in form of two shields connected by column.

planozygote: motile zygote.

plasmalemma: plasma membrane, surrounding the cytoplasm of a cell.

plasma membrane: outermost continuous membrane of the dinoflagellate, external to theca of armored species.

plasmodesmata: intercellular connections resulting from incomplete closure of cross walls in filaments of blue-green algae.

pleurae: distal intercalary bands of diatom frustule, separated from valve by valvocopulae and copulae.

plug: organic material closing the pore of a chrysophyte cyst; also material deposited at proximal end of processes of acritarchs, preventing communication with vesicle interior.

polar nodule: thickened area at raphe ends of pennate diatom valve.

polar pore: expansion of raphe fissure at the polar nodules of pennate diatom valve.

polyhedral bodies: angular bodies in region of nucleoplasm of blue-green algal cell, carboxysomes, store enzyme involved in CO_2 fixation.

polyphosphate bodies: small colorless bodies near center of blue-green algal cell involved in storage of phosphorus.

polysporangia: sporangia in red algae that produce more than the usual four spores.

pore: simple opening in cyst of chrysophytes, filled by organic plug.

pore plate: additional preapical plate at apex of some dinoflagellates, ringlike plate at tip of apical horn of *Ceratium*.

portals: windows or openings between bars of the silicoflagellate skeleton.

portulae: holes in inner raphe canal wall in diatom frustule.

postcingular plates: series of plates of dinoflagellate theca just posterior to the girdle.

posterior intercalary plates: additional plates inserted between postcingular and antapical plate series of dinoflagellate theca.

preapical plates: additional series of plates anterior to apical series in theca of some dinoflagellates; may include a pore plate and a canal plate.

precession: deviation from a radial direction of sutures between elements (crystals) of a coccolith; in clockwise precession, suture deviates to right in crossing from center to circumference, viewed from distal side; in contrast to imbrication, precession appears reversed in direction when viewed from proximal side.

precingular plates: plates of dinoflagellate theca just anterior to the girdle.

pre-perizonium: silicified series of rings developed about the auxospore of some centric diatoms, within which the new initial cell is formed.

prismatolith: polygonal prisms, may be solid or axially perforate, that form the wall of *Thoracosphaera*, a probable coccolithophorid.

procaryote: Procaryota or Prokaryota, organisms lacking membrane-bounded nuclei, mitochondria, plastids, or other organelles.

processes: hollow or solid projections from the surface of the acritarch vesicle; hollow ones may connect with main vesicle cavity, or cavity may be separated by a plug at the proximal end of the process.

proclades: anteriorly directed clades, or branches, from ends of upper actines of the ebriophycean skeleton.

prod: tentacle, an elongate projection from the cell of some dinoflagellates used in locomotion (*Noctiluca*) or attachment to a host (*Protoodinium*).

proplastids: nonfunctional and much reduced plastids in some nonpigmented algae, lacking the thylakoid lamellae, but with homogeneous matrix, and bounded by double membrane.

prosobase: line dividing epicone surface of nonthecate dinoflagellates, visible with silver staining at anterior margin of cingulum.

prosomere: region of epicone of nonthecate dinoflagellate separating the silver-staining acrobase and prosobase lines.

protonemata: small horizontally growing plant that develops from germinating zygospore of charophyte and produces axilary shoot to give rise to new erect plant.

proximate cyst: dinoflagellate cyst that closely corresponds to shape of motile cell, commonly with distinct paratabulation.

pseudocellus: field on diatom cell wall with much smaller areolae than elsewhere.

pseudocyst: nonmotile cell in which coccoliths have become laterally fused, as in *Ochrosphaera*.

pseudonodule: marginal indented structure on diatom valve, may be areolate or operculate.

pseudopylome: short, thick-walled cylindrical projection at aboral end of acritarch vesicle, does not have an opening, and may be an aborted or nonfunctional, incompleted pylome.

pseudoraphe: axial area or **hyaline area** along apical axis of some pennate diatoms that lack true raphe on one or both valves.

pseudosepta: internal costae projecting from inner surface of frustule into cell cavity of diatom.

pterate cyst: dinoflagellate cyst with winglike equatorial flange.

punctate: surface with scattered shallow pits.

pusule: saclike vacuole in dinoflagellates, usually one associated with each flagellar pore, membrane-enclosed, contain fluid, function uncertain.

pylome: circular opening for excystment of acritarch; relatively large, and may be produced on a neck or bordered by a collarlike rim; filled by an operculum prior to excystment; also termed **cyclopyle.** Other excystment openings, such as epityche or rupture, are herein distinguished from the pylome.

pyrenoids: dense, refractive, proteinaceous bodies associated with plastids of algae.

raphe: longitudinal slit along apical axis of pennate diatom valve.

reservoir: invaginated anterior end of cell of *Euglena*, from which flagella arise.

réseau argentophile: silver-staining network of nonthecate dinoflagellates, delineating the vesicles beneath the plasma membrane.

rhabde: posterior axial ray or actine of the triaene ebriophycean skeleton.

rhodolite: algal ball, consisting of concentric layers of red algae and associated organisms; preferably termed **rhodolith.**

rhodolith: algal ball, consisting of concentric layers of crustose coralline red algae and associated organisms; also known as **rhodolite** or **rhodoid.**

ribosomes: small particles of protein and ribonucleic acid that are the site of protein synthesis.

right valve: one half of theca of Prorocentrales dinoflagellates, probably equivalent to hypotheca of other dinoflagellates in view of location of flagellar pores.

rimoportula: labiate process, pore or tube through diatom valve that may project from surface, constricted internally but not at exterior.

RNA: ribonucleic acid. Present in cell cytoplasm and nucleolus; contains phosphoric acid, D-ribose, adenine, guanine, cytosine, and uracil.

rota: one or radial bars across areola in diatom frustule wall.

rugulate: surface with fine, irregular wrinkles.

rupture: simple method of acritarch excystment, terminal or lateral in position and apparently not indicated by preformed thinner region of wall; slit usually does not completely free a portion of the wall, but equatorial rupture may split vesicle into two equal halves.

sagittal fission suture: longitudinal division of theca of Dinophysiales dinoflagellate, separating right and left valve at cell division.

sarcina (pl.: **sarcinae**): cubical packets of bacterial cells.

scapholith: diamond-shaped coccolith with central area of parallel laths.

schizospheres: globular nannofossils constructed of two hemispherical to caplike valves; superficially resemble diatom resting spore, but of calcite with distinctive tetraradial crystal arrangement; possibly a resting stage of a haptophyte.

secondary pit connections: pit connections formed by partial dissolution of the intervening wall between cells of adjacent red algal filaments.

segmented bands: partial girdle bands of diatom frustule, with two or more segments separated by gaps in one circumference of the cell.

semicells: two mirror image cell halves of desmid algae.

setae: awns, elongate projections from valve margin of diatom, may be very elongate and may bifurcate (*Chaetoceros, Bacteriastrum*).

sheath: mucilaginous slime layer surrounding a bacterial or blue-green algal filament.

shield: single plate of a placolith coccolith; two shields are joined by a column at their centers.

shield cells: subtriangular sterile outer cells, usually eight in number, of the charophyte antheridium.

silicalemma: three-layered organic membrane, site of silica deposition in diatoms.

silicoflagellite: siliceous rock composed largely of skeletons of silicoflagellates.

skolate cyst: dinoflagellate cyst with tubular processes.

sori: pits in surface of red algal thallus in which tetrasporangia may be produced.

sphenolith: coccolith with prismatic base of radial elements surmounted by a cone.

spiral cells: elongate sterile cells of sporostegium that surround the oogonium of charophyte; calcified after fertilization.

spirillum: bacterial cell with form of spiral rod.

sporangium: structure in which spores are produced.

sporophydium: female gametangium of charophytes, consisting of oogonium, sporostegium, and pedicel cells.

sporophyte: spore-producing generation, generally diploid.

sporopollenin: acid-resistant, high molecular weight material like that of pollen and spores of higher plants, very similar to material of dinoflagellate and acritarch cysts.

sporostegium: nodal cell, spiral enveloping cells, and coronular cells that surround oogonium of charophytes.

stalk: localized slime layer or iron-encrusted material supporting a single bacterial cell or ramified to support numerous cells.

statospores: aplanospores, or resting cysts, as in the Chrysophyta; in diatoms statospore is bivalved, although differing from frustule.

stauros: transapical expansion of the central nodule of pennate diatom valve that may extend nearly to valve margin.

stephanolith: hollow polygonal or cylindrical coccolith with radiating spokes that may project as spines beyond ring.

stichonematic: flagella with one or two rows of numerous fine hairs or mastigonemes along the margin.

stigma: eyespot of dinoflagellates and euglenids, consisting of pigmented granules; in diatoms, perforation of basal siliceous layer of frustule wall near central nodule.

stomatocerque: arcuate projection over pore from one side of chrysophyte cyst.

striated strand: shorter straight strand of protein enclosed in flagellar sheath with the axoneme of the dinoflagellate transverse flagellum.

stromatoids: sets of laminae or successive algal mats that compose stromatolites.

stromatolite: laminated organo-sedimentary structure produced by algal mats that entrap detrital sediment, commonly by blue-green algae, but less frequently by bacteria.

structured granules: cyanophycin granules, spherical to polyhedral and peripherally located; serve for nitrogen storage in blue-green algae.

strutted process: also termed **apiculus, fultoportula,** or **strutted tubulus;** consists of tube through diatom valve with arched internal supports and surrounded by pores or chambers; may serve for extrusion of mucilage.

sulcal plates: plates of ventral area or sulcus of dinoflagellate theca, commonly five to seven in number, but as few as four to as many as ten are known.

sulcus: longitudinal groove on ventral side of dinoflagellate motile cell, connecting the two ends of the transverse cingulum, and generally extending beyond to the cell's posterior.

synclades: connecting bars between distal ends of adjacent proclades or opisthoclades of the ebriophycean skeleton.

tasmanite: white coal, composed of massive numbers of prasinophyte *Tasmanites,* particularly in late Paleozoic and Mesozoic.

tentacle: prod, an elongate projection from the cell of some dinoflagellates in locomotion (*Noctiluca*) or attachment to a host (*Protoodinium*).

tetrasporangia: spore-bearing structures of tetrasporophyte generation of red algae.

tetrasporophyte: tetraspore-producing generation of red algae; results from germination of a carpospore.

thallus: sheet of cells; many-celled leaflike frond.

theca: cell wall of dinoflagellate motile cell, consists of cellulose plates of characteristic number and position.

thecal layer: layer of dinoflagellate amphiesma containing thecal plates.

thylakoid: lamellae consisting of two membranes joined at the ends to form a closed flattened sac; contains photosynthetic pigments and is site of photosynthesis; lamellae of fixed numbers of thylakoids characterize the plastids of different algal divisions.

torbanite: kuckersite, coal stage of boghead coal formed by *Botryococcus* (Chlorococcales).

trabeculae: thin cylindrical projections, apparently solid, that repeatedly branch and anastomose, forming complex arches, as on surface of some acritarch and dinoflagellate cysts.

trabeculate cyst: dinoflagellate cyst with processes interconnected by means of rodlike or tubular trabeculae.

tract: fossil cyst of dinoflagellate.

transapical axis: transverse axis, perpendicular to apical and pervalvar axes of diatom frustule.

transduction: transfer of genetic material in bacteria by parasitic filterable virus or bacteriophage.

transfer connection: specialized pit connection in red alga *Polysiphonia;* intracellular in nature, and may play a role in intercellular nutrient transport.

transformation: genetic recombination method in some bacteria, in which DNA molecules may move from cell to cell through the solution in which they are growing.

transverse flagellum: in dinoflagellates, lies within the girdle or cingulum, has 9 + 2 fibrils in axoneme, enclosed in flagellar sheath that is expanded at one side and encloses a striated strand; single row of fine hairs or mastigonemes at one margin.

triaene: initial four-rayed or tridentlike spicule in ebriophyceans such as *Hermesinum.*

trichocyst: fusiform ejectile organelles of dinoflagellates, discharge a threadlike filament.

trichocyst pores: perforations in thecal plates of armored dinoflagellates through which trichocysts are discharged.

trichocyte: hair cell of red algae, much elongated external cell that lacks a nucleus when mature.

trichogyne: elongate tip of carpogonium of red algae, may be mucilage-covered and receives the spermatium.

trichome: chains of cells of bacteria or blue-green algae.

triode: initial triradial spicule in ebriophyceans such as *Ebria.*

true branching: intermediate cell of blue-green algal trichome divides in plane of trichome axis, continued division producing a branch.

trunk: basal part of complexly branched process of acritarch vesicle.

ultramicroplankton: plankton between nannoplankton and microplankton in size, thus between 2 and 20 μm.

upper windows: openings left between proclade branches of the ebriophycean skeleton.

utricle: outer covering of vegetative branches of charophyte Clavatoraceae surrounding the gyrogonite.

valval plane: plane of division between valves of diatom frustule.

valve: part of bipartite cell wall of diatoms; each half consists of more or less flattened plate (**valve face**), bending at edge to form flange (**valve mantle**); a separate connecting band attaches to valve mantle and meets or overlaps a similar connecting band from opposite valve. Inherited larger valve is **epivalve;** younger and smaller valve is **hypovalve.**

valve face: flattened plate comprising major part of diatom frustule, may bend at periphery to form valve mantle or flange, adjacent to connecting band.

valve mantle: marginal flange at edge of diatom valve.

valvocopula: intercalary band adjacent to valve of diatom frustule.

vellum: tubular or skirtlike extension of outer wall layer of acritarch vesicle.

velum: thin perforated silica layer across areola in diatom frustule wall, plate may have one or numerous pores, or may consist only of a bar (rota) or numerous marginal projections (vola) into areola.

vesicle: main body of acritarch (preferred over use of *test, shell,* or *central body,* which may have alternate meanings); probably a cyst.

vesicular layer: of dinoflagellate amphiesma or cell cover; vesicles contain the thecal plates of armored species, are hollow in unarmored species.

vibrio: elongate, curved bacterial cell of a half turn.

vola: velum consisting of many separate elements projecting inward from margin of areola in wall of diatom frustule.

windows: portals, openings between bars of the silicoflagellate skeleton.

zooxanthellae: endosymbiotic dinoflagellates, common in coelenterates, protozoans, and molluscs.

zygolith: coccolith in form of elliptical ring with arched crossbar.

INDEX

Page numbers in italics refer to illustrations; those in bold-face refer to definition of terms, to location in systematic classification tables (genera), or to page where authorship is given (for species); other pages refer to text.

Abacella, **890**
Abbott, B. C., 317
Abbott, W. H., 645, 647
Abé, T. H., 263, 264, 269, 270, 306
Abedinium, **380**
Abundacapsa, **81**
Acaenolithus, **782**
Acanthaulax, **392**
Acanthochonia, **892**
Acanthochonia barrandei, 854, *854*
Acanthochrysis, 507
Acanthodiacrodium, 149, **208**
Acanthodiacrodium divisum, **152**
Acanthodiacrodium numerosum, **186**, *186*
Acanthodiacrodium selectum, **186**, *186*
Acanthodiacrodium ubui, **176**
Acanthodiscus, **657**
Acanthodiscus antarcticus, **639**
Acanthoeca, **526**
Acanthoecaceae, 526
Acanthoecopsis, **526**
Acanthogonyaulax, 264, 373
Acanthoica, 697, 723, **784**

Acanthoicaceae, 784
Acanthoiceae, 784
Acanthomorphitae, 207
Acanthosolenia, **781**
Acanthosphaera, **208**
Acanthosphaeridium, 512, **528**
Acanthosphaeridium reticulatum, **514**, *514*
Acetabularia, 185, 279, 734, 826, 828, 834, 863, 864, 881, **894**
Acetabularia cyst, 827
Acetabularia acetabulum, 825, 826, 827, 862, 863
Acetabularia antillana, 831, 864
Acetabularia chiavonica, **864**
Acetabularia major, 863, **864**
Acetabularia mediterranea, 826
Acetabularia schenckii, **864**
Acetabulariaceae, 894
Achilleodinium biformoides, 413, *413*
Achnanthaceae, 658
Achnanthales, 658
Achnanthes, 577, 579, 607, 613, 622, 626, 628, 653, **658**
Achnanthes exigua, **616**

Achnanthes longipes, **578**, *578*
Achomosphaera, 358, **404**
Achomosphaera alcicornu, **406**, *406*
Achomosphaera ramulifera, **406**, *406*
Achomosphaera ramulifera v. perforata, **406**, *406*
Achradina, 255, 256, 338, **378**
Achradina pulchra, 254, 256, **256**
Achradina pulchra f. sulcata, 256
Achromatiaceae, 39
Achromobacteriaceae, 40
Acicularia, 864, **894**
acicule, 254, **964**
Aciculella, **892**
Aciculites, **208**
Ackman, R. G., 812
Aclistochara, 926, 933, **956**
Aclistochara bransoni, 949, *949*
aclistocharoid gyrogonite, 925, 926
Acriora, **208**
acritarch, 351, **964**
Acritarcha, 148, 148–224
  aggregates, *151*
  clusters, 149
  colonies, 149

Acritarcha (*continued*)
  classification, 205
  contamination, 199
  ecology
    proximity to shore, 200
    salinity, 204
    temperature, 199
    water depth, 199, 202
  evolution, 197
  genera, 208
  genera, rate of description, 4
  generic diversity, 184
  geologic distribution, 185
  morphology, 149, 167
    apex, **167**
    polar axis, 167
  paleoecology, 199–204
  phylogenetic trends, 197
  planktonic habit, 199
  preservation, 204
  reworking, 199
  size, 94, 149
  species diversity, *187*
  suprageneric taxa, 206
  systematic position, 181
  vesicle, 149
  vesicle, processes, 160
  vesicle shape, 149
  wall composition, 153
  wall preservation, 154
  wall sculpture, 154, *157*
acrobase, **258**, 258, **964**
Acrochaetiaceae, 139
Acrochaetium, 124, **139**
acromere, **258**, *258*, **964**
acronematic, **248**, **964**
Acropora, 126
Acroporella, **892**
Acrum, **208**
Acrotylaceae, 142
Actinella, **658**
Actinella punctata, **582**, *582*
actines, ebridian skeleton, **465**, **964**
Actiniscaceae, 250, 254, 299, 378, 386
Actiniscus, 250, 251, 252, 253, 254, 256,
    343, 358, 367, **378**, **386**, 479, 546,
    559
Actiniscus elegans, **358**, *358*
Actiniscus pentasterias, **251**, *251*, 252,
    252, *358*
Actiniscus sirius, **252**
Actiniscus tetrasterias, **252**
Actinochloris, **888**
Actinoclava, 637
Actinocyclus, 598, 605, 639, **656**
Actinocyclus curvatulus, **651**
Actinocyclus divisus, **624**, 651
Actinocyclus ehrenbergii, **650**, *650*
Actinocyclus ingens, **644**, 645
Actinocyclus normanii, **591**
Actinocyclus ochotensis, **651**
Actinocyclus octonarius, **583**, *583*, 591
Actinocyclus oculatus, **644**
Actinodiscaceae, 656
Actinodiscus, **656**
Actinomyces, **42**
Actinomycetaceae, 42
Actinomycetales, 42

actinomycetes, 31
Actinoplanaceae, 42
Actinoporella, 868, **892**
Actinoporella podolica, **870**, *870*
Actinoptychus, 598, **656**
Actinoptychus adriaticus v.
    pumila, **641**, *641*
Actinoptychus gruendleri, **645**
Actinoptychus heliopelta, **647**
Actinoptychus packi, **639**, *639*
Actinoptychus senarius, **591**, *592*
Actinoptychus splendens v. *ha-*
    lionyx, **633**, *633*
Actinoptychus undulatus, **599**
Actinosphaera, **783**
Actinotheca, **409**
Actinothecaceae, 409
Actinotodissus, 160, **208**
Actinotodissus longitaleosus, **150**, *150*,
    163, *163*
Actinotophasis, **208**
Actinotophasis complurilata, **194**, *194*
Actinozygaceae, 779
Actinozygus, **779**
Actipilion, 152, **208**
Actipilion druggii, **153**, *153*, 157
Acuturris, **778**
Adamczak, E., 79
Adamiker, D., 752
Adams, I., 692, 712, *712*
Adams, T. D., 734
Adara, **208**
Adarve, **208**
Addison, R. F. (Ackman et al.), 812
Adelseck, C. G., Jr., 769
Adenoïdes, **374**
adenosine triphosphate, 16
Adey, W. H., 123, 126, 127, 140
Adinomonadaceae, 368
Adinomonas, **368**
Adnatosphaeridiaceae, 424
Adnatosphaeridium, **424**
Adnatosphaeridium reticulense, **361**,
    362, *362*
Adnatosphaeridium robustum, **426**,
    426
Adnatosphaeridium vittatum, **360**,
    360, **362**, *362*
Adorfia, **212**
Adornofusa, **208**
aerobic bacteria, oxygen require-
    ment, 31
aerobic metabolism, efficiency, 29
aerobic respiration, 30
Aethesolithon, **140**
Aethesolithon grandis, **119**, *119*
Afghanopolia, **893**
Africanosphaeroides, **81**
Afzelius, B. A., 236
Agalmatoaster, **778**
agar, 110
Agardhiella, **142**
Agardhiellopsis, **142**
Agardhiellopsis cretacea, **134**, *134*
Agelopoulos, J., *413*
Agmenellum, 44, 45, 81
Agmenellum quadruplicatum, **44**, 51
Ahmadjian, V., 836, 843

Ahmuellerella, **779**
Ahmuellerella phaseolus, **697**, **699**,
    699
Ahmuellerellaceae, 779
Ahnfeltia, 110, **142**
Aireiana, **419**
Aireiana verrucosa, **421**, *421*
Ajakmalajsoria, **892**
Akanthochloris, **498**
Akanthochloris bacillifera, **492**, *492*
akinete, 46, 48, *49*, **964**
Akinina, D. K., 306
Aksiyote, J. (Bonotto et al.), 826
alar canal, **581**, **964**
Alasphaera, **386**
Alatocladia, **141**
Albertaporella, **892**
Alberti, G., 171, 403, 427
Albertia, **392**
Aldanella, 77, 83
Aldorfia, 296, **392**
Aldorfia aldorfensis, **283**, *283*, 297
Aldorfia deflandrei, **413**, *413*
Aldrich, D. V., 319, 326, 341
Alfordia, **779**
Alfordiaceae, 779
algal balls, 127
algal blooms, blue-green algae, 53
algal biscuits, 59
algal boring, depth limits, 60
algal mats, 58
algal peat, 57
algal symbionts, productivity, 126
Algirosphaera, **780**
Alisocysta ornata, **398**, *398*
Alisphaera, **783**
Alisum, **208**
Alldredge, A. L., 316
Allemann, F., 738, 740, 741, *741*, 776
Allemannites, **781**
Allen, J. R., 232, 234, 236, 237, 238
Allen, J. R. (Roberts et al.), 235, 236
Allen, M. B., 138
Allen, W. E., 621, 623, 641
Allison, C. W., 74
Alliumella, **208**
Allman, G. J., 272, 348
Allsopp, A., 128, 137, 138
Almassy, R. J. (Dickerson et al.), 30
α-spores, red algae, 116
Alterbia, **392**
Altmann, R., 89
Altochara, 933, **955**
Altochara continua, **941**, *941*
Alvearium, 735, **780**
Alvearium dorsetense, **737**, *737*
alveolar membrane, 260
Amazonites, **208**
Amaurolithus, 706, 750, **778**
Amaurolithus amplificus, **761**, 766
Amaurolithus primus, **761**, 766
Amaurolithus tricorniculatus, **755**,
    755, **761**, 766
Ambistria, 641, **657**
ambitus, **284**, **964**
Amblyochara, 933, **956**
Amblyochara begudiana, **949**, *949*
Ames, B. N., 31

Amgaella, **892**
Amgaelleae, **892**
amoeboflagellates, 138
Amoebophrya, 310, 311, **382**
Amoebophrya ceratii, **310**
Amoebophryaceae, **382**
Ammodochiaceae, **480**
Ammodochiidae, **480**
Ammodochiinae, **480**
Ammodochium, 256, 465, 469, 472, 476, **480**
Ammodochium ampulla, **467**, *467*, *471*
Ammodochium danicum, **469**, *469*
Ammodochium prismaticum paradoxum, **469**, *469*
Ammodochium pyramidale, **485**, *485*
Ammodochium rectangulare, **469**, *469*, *471*, *485*, *550*
Ammonidium, **208**
Ammonidium uncinum, **195**, *195*
Ammonidium waldronensis, **155**
Amorfia, **142**
Amosopollis, **208**
Amphicampa, **657**
Amphidiadema, 293, **392**
Amphidinium, 313, 315, 317, 319, 337, 338, 347, 348, **378**
Amphidinium carterae, **233**, 234, 237, 239, 241, 244, 258, 306, 320, 331, 344
Amphidinium herdmanii, **231**, 244, 259, 347
Amphidinium klebsii, **232**, 233, 237, 239, 244, 316
Amphidinium pellucidum, **309**, 347
Amphidoma, **373**
amphiesma, **256**, *260*, **964**
Amphilithopyxis, **528**
Amphilithopyxis aenigmatica, **528**, *528*
Amphipentas, **656**
Amphipleura, 613, 622, **658**
Amphipleura pellucida, **585**
Amphipleura rutilans, **571**, **572**, 607
Amphiprora, 613, 624, **658**
Amphiroa, 117, **141**
Amphiroa prefragilissima, 119, *119*
Amphiroa rigida, **119**, *119*
Amphiroideae, **141**
Amphiscoleps, **314**
Amphisolenia, 269, 306, 311, 324, 344, **371**
Amphisoleniaceae, 371
Amphispongia, **891**
Amphispongia oblonga, **856**, *856*
Amphispongieae, **891**
Amphitetras, **656**
Amphitholus, 253, 254, **374**
Amphitholus elegans, ̄253, 374, *374*
amphitrichate, **16**, **964**
Amphizygus, **780**
Amphora, 600, 613, **659**
Amphora montana, **610**
Amphora ovalis, **611**
Amphorosphaeridium, **419**
Amphorula, **419**
ampulla, 247
Anabaena, 45, 49, 53, 54, 55, 58, 63, **82**

Anabaena cylindrica, **44**, 49, 52, 54
Anabaena doliolum, **52**
Anabaena flos-aquae, **58**
Anabaenaceae, **82**
Anabaenidium, 74, **82**
Anabaenopsis, **63**
Anacanthoica, 723, **784**
Anacystis, **91**
Anacystis nidulans, 46, 51
Anadyomenaceae, **890**
Anadyomene, **890**
anaerobes, facultative, 19
anaerobes, obligate, 19
anaerobic respiration, fermenting, 22
Anatolipora, **892**
Anaulus, 641, **657**
Anaulus subantarcticus, **639**
Anchicodium, **890**
Andalusiella, **392**
Anderson, G. C., 616, 618
Anderson, O. R., 315
Anderson, T. F., 770
Andrews, G. W., 639, 641, *641*, 643, 648, 648, 649
Andri, E., 734
Andriella, **785**
Andrieu, B., 507, 509
Andrusoporella, **892**
Anfractuosoporella, **894**
Angioporella, **894**
Angst, E. C., 623, 641
Angularia, **208**
Angulisphaerina, **208**
Angulofenestrellithus, **780**
Angulolithina, **778**
Anguloplanina, **208**
Animikiea, 69, **82**
Animikiea septata, **70**, *70*
Anisonema, **898**
anisopolar gyrogonite, **924**, *924*, **964**
anisopolarity index, **924**
Anisoporella, **892**
Anisostomum, **507**
Annellus, **656**
Annulatopsophosphaera, **208**
Annulithus arkellii, **735**, 749
Annulodiscus, **818**
Anodonta, 845
Anomalochara, 927, **957**
Anomalochara polymorpha, **929**, *929*
Anomaloides, **891**
Anomaloplaisium, **208**
Anomaloplaisium lumariacuspis, **155**, *157*, *157*
Anomoeoneis, 600, 641, **658**
Anonymous, 581
Anoplosolenia, **781**
Anoplosolenia brasiliensis, **703**, *703*, *703*
Anorthoneis, **658**
antapical archeopyle, 298
antapical plate series, **263**, **964**
antapical process, acritarch, 167
anterior intercalary plate series, **263**, **964**
Anthatractus, **208**
antheridium, 117, *125*, **917**, **964**
Anthopleura elegantissima, 313

Anthosphaera, **778**, **783**
Anthosphaera robusta, **721**
Anthosphaeridium, **419**
Anthracoporella, **892**
Anthracoporellopsis, **892**
Antia, N. J., 84, 497
Antia, N. J. (Duff et al.), 620
antiligula, diatom frustule, **557**, **964**
Antithamnion, **142**
Antruejadina, **208**
Aoujgalia, **142**
Aoujgalia variabilis, **134**, *134*
Aoujgaliaceae, 142
Aoujgaliida, 133
Apedinella, **525**
Apel, K. (Schweiger et al.), 863, 864
Apelt, G., 613
Apertapetra, **777**
apex, of gyrogonite, **924**, **964**
Aphanizomenon, 54, 58, **82**
Aphanizomenon flos-aquae, **46**
Aphanocapsa, 46, 59, **81**
Aphanocapsites, **81**
Aphanothece, 59, 62
Aphroditicodium, **890**
apical archeopyle, 295, 296
apical area, **577**, **964**
apical axis, **576**, **964**
apical pit, **804**, **964**
apical plate series, **263**, **964**
apical platelet, 263
apical processes, acritarch, 167
apical ring, **539**, **964**
apical structures of gyrogonite, *925*
apiculus, **975**
Apidium, **891**
Apiroschadon, **822**
Apistonema, 684, 713, 714, **777**
aplanospores, **494**, 828, **964**, **975**
apochlorotic, 29, 51, **964**
Apodiniaceae, 380
Apophoretella, **83**
Apstein, C., 335
Aptea, **424**
Aptea anaphrissa, **362**
Aptea polymorpha, **427**, *427*
Apteodiniaceae, 395
Apteodinium, 361, **395**
Apteodinium granulatum, **396**, *396*
Apteodinium spiridoides, **283**, *283*
Aquadulcum, **430**
Aquadulcum myalupensis, **365**, *365*
Aquadulcum pikei, **365**, *365*
Arabicodium, **890**
Arabisphaera, **818**
Arachnochloris, **498**
Arachnoidiscus, **656**
Arachnoidiscus ehrenbergii, **583**, *583*, *585*
Arachnoidiscus ornatus, **569**, *569*
aragonite, 110
Aranidium, **208**
Araucarioxylon, 836
Arbusculidium, 176, **208**
Archaelagena, **955**
Archaelagena howchiniana, **934**
Archaeobatophora, 868, **893**
Archaeocatastomum, **507**

Archaeocladus, 893
Archaeocorynbostomum, 507
Archaeodiscina, 208
Archaeofavosina, 208
Archaeohystrichosphaeridium, 209
Archaeolithophyllum, 140
Archaeolithothamnium, 126, 140
Archaeolithothamnium lugeoni, 136, 136
Archaeolithothamnium mega-sporum, 136, 136
Archaeomassulina, 208
archaeomonadite, 517, 965
Archaeomonadaceae, 512, 528
Archaeomonadopsis, 512, 528
Archaeomonadopsis elegante, 515, 515, 516
Archaeomonadopsis frenguellii, 528, 528
Archaeomonas, 512, 528
Archaeomonas angulosa, 516, 516
Archaeomonas chiarugii, 514, 514, 528, 528
Archaeomonas cratera, 514, 514, 516
Archaeomonas cretacea, 515, 515
Archaeomonas membranosa, 515, 515
Archaeomonas paucispinosa, 514, 514
Archaeomonas vermiculosa, 515, 515
Archaeonema, 74, 82
Archaeonema longicellulare, 74, 74
Archaeoperisaccus, 208
Archaeopertusina, 208
Archaeopsophosphaera, 208
Archaeopteris, 843, 844
Archaeorestis, 69
Archaeorestis schreiberensis, 73, 73
Archaeosacculina, 208
Archaeosphaeridium, 512, 528
Archaeosphaeridium australensis, 516, 516
Archaeosphaeridium dangeardianum, 528, 528
Archaeosphaeridium dumitricae, 516, 516
Archaeosphaeridium tasmaniae, 516, 516
Archaeosphaerodiniopsaceae, 371
Archaeosphaerodiniopsis, 371
Archaeosphaerodiniopsis verrucosa, 370, 370
Archaeosphaeroides, 81
Archaeosphaeroides barbertonensis, 68, 68
Archaeotectatum, 395
Archaeothrix, 82
Archaeotrichion, 42, 69
Archaias, 835
Archamphiroa, 141
Archangiaceae, 39
archeopyle, 227, 288, 293, 295, 965
archeopyle, miscellaneous types, 298
archeopyle types, 296
Archezoa, 39
Aremoricanium, 149, 178, 197, 208
Aremoricanium cylindrosum, 178
Aremoricanium decoratum, 150, 150, 181
Aremoricanium rigaudi, 198, 198

Aremoricanium simplex, 198, 198
Aremoricanium squarrosum, 198, 198
areolae, 581, 583, 965
Areoligera, 295, 358, 424
Areoligera cassicula, 362, 362
Areoligera senonensis, 361
Areoligeraceae, 424
Areosphaeridium, 409
Arkhangelskiella, 782
Arkhangelskiella ethmopora, 745, 748
Arkhangelskiellaceae, 782
Arkhangelskielloideae, 782
Arkonia, 208
Arkonites, 208
armored dinoflagellates, 227, 230, 965
Arpyloraceae, 386
Arpylorus, 351, 356, 386
Arpylorus antiquus, 352, 353
Arthrobacter, 31, 41
Arthrocardia, 141
Arthrodesmis, 885, 886, 894
Arthrodiniales, 378
Artisphaeridium, 512, 528
Artisphaeridium fragile, 515, 515
Asaul, Z. I., 897
Ascidiella, 933, 956
Ascidiella iberica, 928, 928
Ascodinium, 298, 389
Ascodinium acrophorum, 390, 390
Ascodinium ovale, 390, 390
Ascophyllum, 613
Ascostomocystis, 152, 178, 208
Ascostomocystis hydria, 153, 153, 181
asexual reproduction
    bacteria, 17
    Chlorophyta, 828
    Chrysophyta, 504
    diatoms, 595
    dinoflagellates, 320
    euglenids, 896
    haptophytes, 710
    Prasinophyta, 806
    red algae, 116, 123
    silicoflagellates, 542
    Xanthophyta, 494
Asketopalla, 178, 185, 187, 208
Asketopalla formosula, 157, 157, 170, 178, 178, 181
Asparagopsis, 139
Aspidolithus, 782
Aspidorhabdus, 784
aspondyl, 868, 965
Assipetra, 778
Astasia, 896, 898
Asterionella, 607, 626, 634, 658
Asterionella formosa, 614
Asterionella japonica, 618, 620
Asterocytis, 116, 126, 139
Asterodinium, 254, 378
Asterodinium spinosum, 255, 255
Asterolampra, 639, 656
Asterolampra marylandica, 624
Asterolampraceae, 656
asterolith, 706, 707, 965
Asteromonas, 818
Asteromphalus, 585, 587, 639, 656
Asteromphalus imbricatus, 624
Asteromphalus robustus, 632

Asteromphalus roperianus, 605
Asterostomum, 507
Astrionis, 779
Astrocysta, 389
Ataxiodinium, 404
Athiorhodaceae, 42
Atlanticellodinium, 382
Atlanticellodinium tregouboffi, 310
atmospheric oxygen, 95
Atopochara, 933, 941, 944, 956
Atopochara multivolvis, 945, 945, 947
Atopochara restricta, 945, 945, 947
Atopochara trivolvis, 928, 928, 944, 945, 947
Atopochara trivolvis triquetra, 945, 945
ATP, 16, 965
Atractophora, 139
atrichate, 16, 965
Attheya, 657
Attritasporites, 208
Aubry, M. P., 729, 733, 733, 734
Audomonas, 525
Audouinella, 139
Auerbachichara, 933, 955
Auerbachichara saidakovskyi, 941, 941
Aulacodiscus, 637, 656
Aulacodiscus margaritaceus, 569, 569
Aulacodiscus reticulatus, 585
Auliscaceae, 656
Auliscus, 591, 622, 653, 656
Auliscus caelatus, 590, 590, 641
Aureodinium, 374
Aureodinium pigmentosum, 237, 241, 244, 315, 374
Aureosphaera, 208
Aureotesta, 208
Auricula, 659
Auriculaceae, 659
Auriculopsis, 657
Australiella, 392
autoblast, 284, 965
autocyst, 284, 965
autogamy, diatoms, 599
autophragm, 281, 284, 965
autotomy, dinoflagellates, 327, 328, 965
autotrophs, 18, 308, 965
auxiliary cell, red algae, 117, 965
auxospores, diatoms, 596, 599, 965
auxospore formation, 596, 598
Avrainvillea, 890
Avrainvilleopsis, 890
awns, diatoms, 585, 964, 965
Awramik, S. M., 94
Ax, P., 613
axial area, 577, 974
axial plastids, 965
axis, red algal thallus, 117
Axisphaeridium, 149, 178, 185, 208
axoneme, 248, 249, 965
Axopodorhabdus, 735, 781
Axopodorhabdus cylindratus, 735, 737, 749
axoskeleton, 254, 965
Azomonas, 40
Azotobacter, 14, 22, 40

Azotobacteraceae, 40
Azyrtalia, 77, **83**

Bass-Becking, L. G. M., 110, 112
Baccinellula, **894**
Baccinellula cosmarioides, **886**
Bachmann, A., 541, *545*
Bachmann, A. (Ichikawa et al.), *545*
Bachmannocena, 539, **559**
Bacillaceae, 16, 41, 91
Bacillaria, 579, 607, 610, **659**
Bacillariophyceae, 654
Bacillariophyta, 491, 567, 654, 655
   See also diatoms
bacillus, 11, 12, **965**
Bacillus, 14, 16, 24, 30, 31, **41**, 93,
   234
Bacinella, **890**
Bacisphaeridium, **208**
bacteria, 10−42
   biochemical evolution, 31
   cell, 11
   cell division, 11
   cell form, 12
   classification, 38
   chromosome, 11, 15
   distribution, 23
   dormancy, 37
   endospore, 17
   evolution, 28, 37−38
   fission, rate of, 17
   flagellar motion, 16
   flagellar structure, *15*
   fossils, 11, 24, 31
      Paleozoic, 35−37
      Precambrian, 32, 37−38
      Tertiary, 36, 37
   geologic agents, 24
   penetrating rock, 37
   primitive organisms, 28, 31
   purple nonsulfur, 14
   respiratory enzymes, 14
   secondarily simplified orga-
      nisms, 29
   sheath composition, 13
   utilized as food, 21
   viable in salt, 37
Bacteriaceae, 40
bacteria, abundance, 21, 22
bacterial decomposition, 21
bacterial respiration, 16, 20
Bacteriastraceae, 657
Bacteriastrum, 559, 597, 600, 605, 607,
   613, 625, **657**, 974
Bacteriastrum delicatulum, **585**, *587*,
   603
Bacteriastrum hyalinum, **597**, *598*, *599*,
   601, **605**, *609*
Bacteriastrum solitarium, **605**
bacteriochlorophylls, 19
bacteriophage, **18**, *18*, **965**
Bacteriophyta, 38, 39
Bacterium calcis, **28**
Bacteroidaceae, 41
Bacterosira, **655**
Bacterosira fragilis, **624**
baculate, 155, **965**
Badour, S. S., 596

Bailey, J. W., 463, 535
Bainbridge, R., 621
Baiomeniscus, **208**
Baiomeniscus camarus, **190**, *190*
Baker, H., 225
Balamuth, W. (Honigberg et al.), 226
Balaniger, **778**
Balch, W. E. (Fox et al.), 31
Balderston, W. L., 314
Balech, E., 262, 264, 270, 287, *325*, 345,
   346, 347, 351
balkhashite, 841
Ballantine, D., *307*, *307*, 317
Balme, B. E., 176, 204, 838
Balmeella, **208**
Baltisphaera, **211**
Baltisphaeridiaceae, 207
Baltisphaeridium, 149, 151, 152, 153,
   167, 173, 178, 184, 185, 190, 200,
   202, 206, **208**
Baltisphaeridium accinctum, **189**, *189*
Baltisphaeridium digitatum, **200**
Baltisphaeridium longispinosum, **152**
Baltisphaeridium multipilosum, **152**
Baltisphaeridium pachyacanthum, **152**
Baltisphaeridium perclarum, **150**, *150*
Baltisphaeritae, 207
Bambuites, **208**
banded iron formations, 69
Bangia, 108, 112, 114, 116, 126, 137,
   139
Bangia phase, 114
Bangiaceae, 139
Bangiales, 114, 139
Bangiophycidae, 94, 108, 112, 113, 116,
   138
   morphologic phases, 113
   multicellular, 114
   spores, 116
   sexual reproduction, 116
Bangiopsis, **139**
Bankia, 881
Banner, F. T., 733, 734, *734*
Banse, K., 616, 618
Barakella, **208**
Barathrisphaeridium, **208**
Barber, H. G., 649
Barghoorn, E. S., 32, *35*, 68, 69, 70, 71,
   73, 74, 74, 80, 94, 95, 128, 129
Barghoorn, E. S. (Awramik et al.), 94
Barghoorn, E. S. (Knoll et al.), 74
Barghoorn, E. S. (Meinschein et
   al.), 35
Barghoorn, E. S. (Schopf et al.), 32,
   34, *34*
Bargoniella, 309
Barker, H. A., 308, 326, 344
Barker, I. W., 637
Barnard, T. (Rood et al.), 735, 737, *738*,
   776
Barnes, B., 699, 703, 745, 764
Barnes, J., 127
Barringtonella, **780**
Barron, J. A., 639, 645, *651*, *651*, 652
Barta-Calmus, S., *877*
basal cage of gyrogonite, *925*
basal plate, 965
basal plug, **924**, **965**

basal pole, gyrogonite, 924
basal pore, gyrogonite, **924**, **965**
basal ring, silicoflagellates, **539**, **965**
Baschnagel, R. A., 182, 843, *883*, 886,
   887
Batiacasphaera, **395**
Batiacasphaera compta, **290**, *290*
Batinevia, **141**
Batioladinium, **403**
Batophora, 834, 860, 864, 868, 881, **893**
Batophora oerstedii, **869**, *869*
Batophoreae, 893
Batrachospermaceae, 139
Batrachospermum, **139**
Bauld, J. (Walter et al.), 28
Bavlinella, **208**
Baxteriopsis, **656**
Bazin, M. J., 51
Bé, A. W. H., 315, *681*, *699*, *706*, *711*,
   719, *720*, *720*
Bé, A. W. H. (McIntyre et al.), 720, 721,
   751
Beam, C. A., 235, 326, 329
Beckspringia, **82**
Beggiatoa, 22, 27, 29, 30, 39, 55, 80
Beggiatoaceae, 39
Beggiatoales, 24, 39
Beijerinckia, **40**
Bein, A., 733, 734
Beklemishev, K. V., 609
Belcher, J. H., 807, 841
Bellamy, D. J. (Barnes et al.), 127
Bellerochea, 598, **657**
Belodiniaceae, 395
Belodinium, **395**
Belodinium dysculum, **398**, *398*
Belyaeva, T. V., 628, 629
Belzungia, **893**
Benedek, P. N., 283, 430, 433
Benetorus, **656**
Bennett, A. G., 607
Bennocyclus, **782**
Benson, A. A. (Patton et al.), 339
Berchenko, O. I., 932, 934
Beresella, 868, **892**
Beresella erecta, **868**, *868*
Bereselleae, 892
Berge, G., 719
Berger, S. (Schweiger et al.), *863*, *864*
Berger, W. H., 769
Berggren, W. A., 751
Bergh, R. S., 227, *323*, *323*
Bergonia, **656**
Berkeleya, **658**
Berkeleya hyalina, **585**, *585*
Berkeleya scopulorum, **585**, *585*
Berman, T. (Gressel et al.), 137, 234
Bernard, F., 725
Bernardinium, **378**
Bernardosphaera, **784**
Bernier, P., 881
Berry, W. B. N., 201
Bertaud, W. S., 581
Bertholdia, **141**
Bertrand, C. E., 841
β-spores, red algae, 116
Bevocastria, **83**, **890**
Bevocastria conglobata, **847**, *847*

Beyers, R. J., 314
Biantholithus, **779**
Biantholithus sparsus, 763
Bibby, B. T., 315, 334
Bicarinellum, **386**
bicavate cysts, 286
Bicoeca, **518**
Bicorium, **141**
Bicosta, **525**
Biddulphia, 572, 587, 598, 599, 637, **657**
Biddulphia aurita, **622**, 624, *650*
Biddulphia biddulphiana, **585**
Biddulphia levis, **607**
Biddulphia pulchella, **572, 587, 606**
Biddulphia rhombus, **598**
Biddulphia sinensis, **621**
Biddulphia sparsepunctata, **639**
Biddulphia weissflogii, **624**
Biddulphiaceae, 657
Biddulphiales, 657
Bidiscus, **782**
Bidiscus ignotus, **741**, *741*
Biecheler, B., 253, 254, *258*, 269, 309, 330, 334
Biechelerella, **386**
Bien, E., 37
Bigeminococcus, 74, **81**
Biggley, W. H. (Swift et al.), 316
Bignot, G., 926, 957
Bija, **140**
Bikosoeca, 504, **518**
Bikosoeca crystallina, **505**, *505*
Bikosoeca epiphytica, **505**, *505*
Bikosoeca eurystoma, **505**, *505*
Bikosoecaceae, 518
Bikosoecophyceae, 490, 518
biliproteins, 108
binary fission, 17
binucleate dinoflagellates, 312
Biocatenoides, **41**, 69
Biocatenoides sphaerula, **73**, *73*
bioluminescence, **316, 965**
bioluminescent dinoflagellates, 316
Bion, **181**
Biorbifera, **403**
biostratigraphic zonation
  calcareous nannoplankton, 748, 749, 762, 763, 766, 767
  Charophyta, 933
  diatoms, 639, 644–647
  dinoflagellates, 361, 362
  silicoflagellates, 549, 550, 551
Bipedinomonas, **818**
Bipodorhabdus, **781**
Birkelundia, **783**
Birkelundia staurion, **757**, 762
Birle, J. D. (Schopf et al.), *24, 26, 36*
Bisacculoides, **81**
Bisalputra, T., 257, 261
Bisalputra, T. (Antia et al.), 497
Biscutaceae, 782
Biscutum, **782**
Biscutum ellipticum, **741**, *741*
bisporangium, **124**, *965*
Bisset, K. A., 29
Bitectatodinium, **395**
Bitectatodinium tepikiense, 283, **299**, 301

Bitterli, P., 204
Biumbella, **955**
Biumbella brazhnikovae, **934**, *934*
Biumbellidae, 955
Blacic, J. M., 74, 94, 128, *130*
Black, M., 697, *699*, 703, 703, 705, 706, 707, 708, 737, 738, 743, 745, 746, 752, 754, 756, 761, 764, 776, 782
Blackburn, K. B., 841
Blackites, **784**
Blackites amplus, **705**, *705*
Blackman, V. H., 726
Blackwelder, P. L., 692
Blanchard-Babillot, C., 237
Blastodiniaceae, 380
Blastodiniales, 380
Blastodinium, 311, **380**
Blecker, H. H. (Holton et al.), 46
Bleecken, S., 15
Blepharocysta, 346, **377**
blue-green algae, 43–83
  abundance, 57
  blooms, 47
  cell, 43
  freshwater, 57
  genetic recombination, 51
  growth, 46
  hot springs, 58
  metabolism, 53
  mineral requirements, 54
  mobility, 52
  mutations, 51
  nitrogen deficiency, 49
  nonmineralized fossils, 66
  packets of cells, 48
  pigments, 45
  polyploidy, 44
  rate of reproduction, 46
  sheath, 43
  symbionts, 55
  trichome, 47
  wall, 13, 43, *43*
Blunt, Sister M. X. (Odum et al.), 831
Boalch, G. T., 808
Boalch, G. T., (Parke et al.), 808, *810*
Bock, W., 77, 881
Boderia, 138
Boekel, N. M. van, *817*
Boekelovia, **519**
Bogdanov, Yu. A., 617, 752
bog iron ores, 24
boghead coal, 496, 832, 965
Bogorad, L., 88, 91
Bogorovia, **658**
Bogorovia veniamini, **646**, 647
Bogutchania, **81**
Bogutschanophycus, **890**
Böhm, A., 345
Boichenko, E. A., 617
Bolboforma, 509, *509*, **528**
Bolboforma rotunda, **509**, *509*
Bolboforma spinosa, **509**, *509*
Bold, H. C., 497, 499, 837, 882
Boldia, **139**
Boldiaceae, 139
Boletuvelum, **781**
Boleyn, B. J., 609, 632
Bolivina, 635
Bolli, H. M., 733, 734, 734

Boltovskoy, A., 295, 321, 323, 334
Bomolithus, **779**
Bonadonna, F. P., 639
Bond, T. A., 649
Bonet, F., 734
Bonetocardiella, **785**
Boney, A. D., 124, 540, 546, 714, 721
Bonnemaisonia, 112, 124, 130, **139**
Bonnemaisonia asparagoides, **116**
Bonnemaisonia hamifera, **116, 131**
Bonnemaisonia nootkana, **116**
Bonnemaisoniaceae, 113, 139
Bonotto, S., 825, 826
Boodlea, **834, 891**
Boodleaceae, 891
Boodleopsis, **890**
Borge, O., 839, 887
Borgert, A., 233, 330
Borgert, A. H. C., 535
boring algae, 59
boring algae, fossil, 65
Bornetella, 862, 869, **893**
Bornetella capitata, **862**, *862*
Bornetelloideae, 893
Bornetia, **142**
Borowitzka, M. A., 112
Borrelia, **40**
Borsetti, A. M., 718
Borzinema, **83**
Borzinemataceae, 83
Bosellini, A., 127
Bosporella, **480**
Bossea, **141**
Bossiella, **141**
Bossiella californica, **121**, *121*
Bossiella chiloensis, **122**, *122*
Bossiella plumosa, **122**, *122*
Bostrychia, **142**
Botrydium, 493, 495, **498**
Botryocladia, **142**
Botryococcaceae, 841, 889
Botryococcus, 77, 149, 204, 496, 825, 832, 833, 841, 842, **889**, 965, 966, 970, 975
  geologic record, 841
Bouck, G. B., 247
Boucot, A. J., 201, 202, 204
Boudreaux, J. E., 776
Boueina, **890**
Bouquaheux, F. (Cachon et al.), 233, 309, *310*
Bourrelly, P., 58, 124, 265, 348, 368, 499, 518, 525, 649, 654, 836, 837, 838
Bovee, E. C. (Honigberg et al.), 226
Boyer, C. S., 639, 641, 654
Braarud, T., 342, 343, 344, 609, 713, 717, 723, 726, 770
Braarudosphaera, 5, 706, 745, 770, 771, 773, 774, **779**
Braarudosphaera bigelowii, **708**, *708*, 745, 747
Braarudosphaera discula, **747**
Braarudosphaera magnei, **747**
Braarudosphaera rosa, **747**, *770*, 771
Braarudosphaera chalk, 770
Braarudosphaeraceae, 778
Brachiolithus, **777**
Brachychara, **956**

Brachydiniaceae, 254, 378
Brachydinium, 254, **378**
Brachydinium capitatum, **255**, *255*
Bracteacoccus, 836, **888**
Bradbury, J. P., 649
Bradley, W. H., 36, 37, 77, 510, 835, 837, 839, *841*, 841, 842, *844*, 845, 884, 896, *897*
Brady, H. B., 934
Bramlette, M. N., 5, 632, 636, 678, 725, 726, 745, 748, *761*, 768, 772
Bramletteius, **784**
Bramletteius serraculoides, **757**, 762
branches, Charophyta, **915**, **965**
branchlets, Charophyta, **915**, **965**
Brazhnikova, N. E., *934*
Brazilea, **208**
Brebissonia, **658**
Breed, R. S., 38
Brenneckella, 725
Brevichara, **956**
brevicharoid apical structure, gyrogonite, **925**, *925*
Brigger, A. L., 587, *588*, *589*, 639
Brigger, A. L. (Hanna et al.), 639
Brightwellia, 637, **656**
Brito, I. M., 205, *817*, 824
Brochopsophosphaera, **208**
Brock, T. D., 23, 28, 58
Brock, T. D. (Mosser et al.), 835
Brock, T. D. (Walter et al.), 28
Broeckella, **893**
Broinsonia, 769, **782**
Broinsonia dentata, **692**, *692*
Broinsonia gammation, **741**, 748
Broinsonia lacunosa, 748
Broinsonia parca, **745**, *745*, 748
Brokaw, C. J., 250
bromine, in red algae, 109
Brongersma-Sanders, M., 619
Bronikovsky, N., 344, 345, 346
Bronnimann, P., 843
Brooks, J., 68, 69
Broomea, **386**
Broomea ramosa, **389**, *389*
Brosius, M., 204
Brown, R. M., 63, 92, 93, 684, 688, 837
Bruce, D. E. (Duff et al.), 620
brucite, 110
Brunia, **656**
Bryopsidaceae, 890
Bryopsidophyceae, 890
Bryopsis, 826, 834, **890**
Buchanan, R. E., 38
Buchanan, R. J., 324, 326
Buedingiisphaeridium, **208**
Buetow, D. E., 91, 92
Buisonjé, P. H. de, 769
Bujak, J. P., *357*, 358, *361*, *389*, 390
Bukry, D., 512, 539, *550*, *551*, *559*, 697, 706, 721, 741, 743, 745, *745*, 748, 750, 752, *752*, 756, 764, 766, 770, 771, 776
Bukryaster, **780**
Bukrylithus, **780**
bulbils, **915**, *917*, **965**
Bulbodinium, **392**
Bullasphaera, 74
bullula, diatom frustule, **585**, **965**

Bumilleria, 493
Bunt, J. S., 616
Burckle, L. H., 604, 629, 639, *644*, 649
Burgess, J. D., 358
Burklew, M. A. (Donnelly et al.), 341
Burmann, G., 154
Burns, D. A., 719, 769
Burrows, A., 714, 721
Bursa, A. S., 252, 253, 335, 752
Bursaria hirundinella, 227
Burton, H., 825
Busby, W. F., 619
Busby, W. F. (Lewin et al.), 572, 595, 618, 619
Bütschli, O., 229, 244, 261, 265, 272, 293, 364, 366, 379
Bybell, L. M., *757*, 770
Bykova, E. V., 932, *934*
Byrnes, J. G., 856, 857, *858*

Cabioch, J., 117, 140
Cabrieropora, **892**
Cachon, J., 233, 244, 309, *310*, 310, 311
Cachon, J. (Hollande et al.), 232, 367
Cachon, J. (Sournia et al.), 368
Cachon, M., 309, 310, 311
Cachon, M. (Cachon et al.), 233, 244, 309, 310
Cachon, M. (Hollande et al.), 232, 367
Cachon, M. (Sournia et al.), 368
Cachon-Enjumet, M., 31
Cachon-Enjumet, M. (Hollande et al.), 367
Cachonina, 264, 297, 301, 324, **374**
Cachonina niei, **231**, 232, 233, 237, 238, 239, *241*, 245, 257, 265, 295, 306, 321
Cadosina, **785**
Cadosinella, **785**
Cadosinopsis, **785**
cage, gyrogonite, **924**, **965**
Caiacorymbifer, **208**
Cairns, J., 15
Calathella, **891**
Calathieae, 891
Calathium, **891**
Calathium egerodae, **856**, *856*
calcareous algae, distribution, 831
calcareous deposits, green algae, 828
calcareous tubes, blue green algal, 65
calcareous nannoplankton, 678 – 803
  Cenozoic species, 745
  classification, 776
  evolution, 771
  extinctions, 771
  genera, rate of description, *5*
  generic diversity, 773
  Mesozoic species, 728
  Paleozoic species, 727
  species diversity, 772, *774*, *775*
Calciarcus, **777**
Calcicalathina, **780**
Calcicalathina oblongata, **735**, 738, 749
Calcicarpinum, **386**
Calcidiscus, **783**
Calcidiscus formosus, **757**, 762, 771
Calcidiscus gammation, **771**
Calcidiscus leptoporus, 703, **705**, *705*, 717, 719

Calcidiscus macintyrei, **761**, 764, 766
Calcidiscus margaritae, **757**, 762
Calcidiscus robustus, **757**, 763
calcification
  Charophyta, 919
  Dasycladales and Receptaculitales, 859
  red algae, 110
Calcifolium, 848
Calcigonellum, **386**
Calcioconus, **784**
Calciodinellaceae, 279, 302, 386
Calciodinellum, 279, 280, 299, 383, **386**
Calciodinellum operosum, **280**, *280*, 387, 747
Calciogranellum, **386**, 747
Calciopappus, **781**
Calciopappus caudatus, **721**, 723
Calciopilleus, **780**
Calciopilleus obscurus, **764**, *764*
Calciosolenia, 703, **781**
Calciosolenia sinuosa, **703**, *703*
Calciosoleniaceae, 781
Calcipterellum, **386**
Calcisphaera lemoni, **937**
Calcisphaera robusta, **937**
Calcisphaerellum, **386**
Calcisphaerula, **785**
Calcisphaerulidae, 785
calcispheres, 3, 937
calcite, 110
calcium carbonate deposition, 24
  bacteria, 28
    blue-green algae, 55
    dissolution, blue-green algae, 59
calcium phosphate deposition, 24
calcium requirement for motility, 52
Calculites, **778**
Calculites obscurus, **745**
Caledonidinium, **395**
Calef, G. W., 62
Calicipedinium, 253, **386**
Calicipedinium quadripes, **387**, *387*
Caligodinium, **386**
Caligodinium amiculum, **290**, *290*, 389
Callaiosphaeridium, **419**
Callaiosphaeridium asymmetricum, **419**, *419*
Calliacantha, **525**
Calliacantha natans, **526**, *526*
Calliarthron, **141**
Callipsygma, **890**
Callithamnion, **142**
Callithamniopsis, **892**
Callithamniopsis fruticosa, **864**
Callophyllis, **141**
Caloglossa, **142**
Calolithus, **783**
Caloneis, **658**
Calosiphonia, **141**
Calosiphoniaceae, 141
Calothrix, 44, 45, 62, **83**
Calvert, S. E., 630, 632, 634, 635, 636
Calycomonas, **518**
Calycomonas pascheri, **519**, *519*
Calycomonas vangoorii, **519**, *519*
Calyculaceae, 779
Calyculus, **779**

Calyptothrix, 74, **82**
calyptra, **281, 965**
calyptrolith, **694,** *700,* **965**
Calyptrolithus, **778**
Calyptrosphaera, 684, **778**
Calyptrosphaera oblonga, **694**
Calyptrosphaera sphaeroidea, **692**
Calyptrosphaeraceae, 777
Cambroporella, **892**
Cambroporelleae, 892
Campaniella, 933, **956**
Campanoeca, 504, **526**
Campanoeca dilatata, **505,** *505*
Campbell, A. S., 559
Campbelliella, 881, **893**
Campenia, 812, **818**
Campylodiscus, **659**
Campylodiscus clypeus, **582,** *582*
Campyloneis, **658**
Campylosira, **658**
Campylosphaera, **784**
Campylosphaera eodela, **757,** *763*
Campylostylus, **658**
canal plate, 265
canal raphe, **579, 965**
Candelasphaeridium, **208**
Candida, 95
Cane, R. F., 833
caneolith, **697,** *700,* **965**
Caneosphaera, **783**
Cann, J. P. (Pennick et al.), 806, *806,*
    807
Canningia, **395**
Canningia attadalica, **361**
Canningia scabrosa, **403,** *403*
Canningiaceae, 395
Canninginopsaceae, 424
Canninginopsis, **424**
Canninginopsis denticulata, **426,** *426*
Cannopilus, 549, 553, **559**
Cannopilus calyptra, **541**
Cannopilus depressus, **553,** *553*
Cannopilus hemisphaericus, **549,** *550,*
    559
Cannopilus hexacantha, **541**
Cannopilus major, **550,** *552*
Cannosphaeropsis, 290, 361
Cannosphaeropsitaceae, 424
Cantabrica, **208**
cap cells, **113, 966**
Capon, B., 306, 348
Capsosiraceae, 83
capsule, bacteria, 12, **13, 966**
    dinoflagellates, **283, 966, 969**
Caratini, C., 745
carbon cycle, 20
carbon fixation, rate, 21
Carduifolia, 253, 386, **483**
Carinolithus, **779,** *780*
Carmichael, W. W., 58
Carminella, **208**
Carminella maplewoodensis, **171,** *174*
Carminellidae, 208
Carnegia, 507, **528**
Carnegia armata v. *uncinata,* **508,** *508*
Carnegia cristata v. *ornatissima,* **508,**
    508
Carnegia forcipata, **529,** *529*

Carnegia operculata, **508,** *508*
Carpathoporella, **892**
Carpelan, L. H., 474
Carpenter, E. J., 54
Carpenter, J. H. (Seliger et al.), 316
Carpenterella, **894**
Carpistomiosphaera, **785**
Carpodinium, **392**
Carpodinium granulatum, **395,** *395*
carpogonium, **117, 125, 966**
carposporangial conceptacle, *121, 123*
carposporangium, *111,* **123, 966**
carpospore, **115, 123, 966**
carposporophyte, **116, 123,** *125,* **966**
Carr, J. F., 49, 54
Carr, N. G., 51, 91, 128
Carré, D., 313
carrageenan, 110
Carter, H. J., 229, 272
Carteria, 835, 882, **888**
Carterius, 836
Caryosphaeroides, 69, 74, 81, 838, **888**
Caryosphaeroides pristina, **838,** *838*
Casselton, P. J., 493, 497
Cassie, V., 570, 581, 847
Cassiopeia, 314, 315
Castenholz, R. W., 52
Castracane, A. F., 623
Catastomum, 507
Catellaria, **893**
Catena, **891**
catenate dinoflagellates, **325, 966**
Catenella, **142**
Catenula, **658**
Caterella, **783**
Cati, F., 718
Catillochara, **955**
Catillopsis, **392**
Catillopsis abdita, **357,** *357*
Catillus, **658**
Catinaster, **778**
Catinaster calyculus, **761,** *767*
Catinaster coalitus, **761,** *761,* 767
Catinella, **81**
Cattell, S. A., 254, *255*
Caudiculophycus, 74, **83**
Caudiculophycus rivularioides, **77,** *77*
Caudwell, C., 845
Caulerpa, 826, 834, 848, **891**
Caulerpa prolifera, **852,** *852*
Caulerpa sertularioides, **848**
Caulerpaceae, 891
Caulerpales, 890
    fossils, 848
    morphology, 847
Caulobacter, 14, **40**
Caulobacteraceae, 40
cavate cyst, **284,** *286, 286,* 287, **966**
Cavalier-Smith, T., 88, *89,* 90, 137
cavernous travertine, *57*
Cayeuxia, **83,** 847
cell asymmetry, relationship to move-
    ment, 304
cell buoyancy, dinoflagellates, 306
cell division
    blue-green algae, 46
    dinoflagellates, 323
cell movement, dinoflagellates, *304*

Haptophyta, 683
cell size, Precambrian, 94
cell size, Rhodophyta, 107
cell structure
    Acritarcha, 149
    bacteria, 11
    Charophyta, 915
    Chlorophyta, 825
    Chrysophyta, 498
    coccolithophorids, 679
    Cyanophyta, 43
    diatoms, 568
    dinoflagellates, 229
    ebridians, 464
    Euglenophyta, 895
    Eustigmatophyceae, 497
    Haptophyta, 679
    Prasinophyta, 805
    Rhodophyta, 107
    silicoflagellates, 535
    Xanthophyta, 491, 496
cell tetrads, 94
cell wall
    algae, 13
    bacteria, *13,* 13
    Charophyta, 915
    Chlorophyta, 826
    Cyanophyta, 52
    diatoms, 572
    dinoflagellates, 256
    fungi, 13
    Haptophyta, 684
    Rhodophyta, 110
    Xanthophyta, 493
Celtiberium, **208**
Cenozoic
    calcareous nannoplankton, 745
    charophytes, 951
    coccolith diversity, *774*
    diatom biostratigraphy, 644
    dinoflagellates, 358
    Discoaster diversity, *775*
Centosphaera, 747, **777**
central area, **577, 966**
central body, acritarch, 149
central nodule, **577, 966**
central pore, **579, 966**
Centrales, 568
centric diatoms, **568, 966**
Centrobacillariophyceae, 655
Centrococcales, 887
Centrodiniaceae, 371
Centrodinium, 263, **371**
Centrodinium elongatum, **370,** *370*
Centromonadales, 887
Centronella, **658**
Centropalmellales, 887
centroplasm, blue-green algae, **44,**
    966, 971
Centroplastophyceae, 887
Centroporus, **656**
Centrostomatales, 887
Centrothricales, 887
Čepek, P., 745, 748
Cepekiella, **783**
Cephaleuros, 843, **889**
Cephalia, **208**
Cephalophytarion, 74, **82**

Cephalophytarion grande, **77**, *77*
Cepillum, **208**
Ceramiaceae, 142
Ceramiales, 113, 124, 142
Ceramium, **142**
Cerataulina, 601, 613, **657**
Cerataulus, 639, **656**
Cerataulus smithii, 598
Ceratiaceae, 233, 371
Ceratiopsis, **392**
Ceratium, 250, 254, 265, 272, 273, 274,
   275, 276, 284, 288, 302, 306, 308,
   317, 319, 324, 325, 326, 327, 334,
   335, 337, 338, 343, 344, 345, 346,
   351, 354, 367, **371**, 383, 424, 973
Ceratium arcticum, **345**
Ceratium breve, **344**
Ceratium californiense, 326
Ceratium candelabrum, **344**
Ceratium contrarium, **370**, *370*
Ceratium cornutum, **273**, 275, 335, *336*
Ceratium digitatum, **344**
Ceratium divaricatum, **370**, *370*
Ceratium extensum, **345**
Ceratium furca, **237**, 244, 273, 306, 345
Ceratium fusus, **229**, *229*, 273, 317,
   330, 345
Ceratium gravidum, **344**
Ceratium hirundinella, 227, **230**, *230,*
   232, 233, 244, 259, 265, 272, 274,
   275, 293, 308, 330, 335, 337, 348
Ceratium hirundinella f. *reticula-*
   tum, **274**, *274*
Ceratium horridum, **328**, *328,* 330,
   335, *336,* 345
Ceratium incisum, **344**
Ceratium macroceros, **326**, *325,* 345
Ceratium massiliense, 325, **326**, *328,*
   345
Ceratium ostenfeldii, 326
Ceratium pavillardii, **325**
Ceratium tetraceros, **272**, *272*
Ceratium trichoceros, **328**, *328,* 344
Ceratium tripos, **229**, *229,* 260, **261**,
   273, 304, 323, 324, 326, 327, 330,
   344
Ceratium tripos v. *subsalsum,* **233**
Ceratium vultur, **325**, *325,* 345, 346
Ceratocoryaceae, 371
Ceratocorys, 264, 265, 324, 325, 345,
   371
Ceratocorys horrida, 246, **345**
Ceratocystidiopsis, **424**
Ceratocystidiopsis simplex, **427**, *427*
ceratolith, **706**, *707,* **966**
Ceratolithaceae, 778
Ceratolithina, **778**
Ceratolithoides, **778**
Ceratolithoides aculeus, **745**, 748
Ceratolithus, 706, 745, 750, 776, **778**,
   966
Ceratolithus acutus, **761**, 766
Ceratolithus cristatus, 706, **708**, *708,*
   719, 751, 761, 766, 770
Ceratolithus rugosus, **761**, *764,* 766
Ceratoneis, **658**
Cerionites, **891**
Cernichiari, E., 314

Cerodinium, **392**
Cestodiscus muhinae, **646**
Chabiosphaera, **208**
Chadefaud, M., 86, 87, 804, 887
Chaetangiaceae, 139
Chaetoceraceae, 657
Chaetoceros, 62, 311, 573, 585, 598,
   600, 601, 604, 607, 609, 613, 619,
   622, 623, 625, 634, 639, **657**, 965,
   974
Chaetoceros atlanticus, **621**
Chaetoceros borealis, **613**
Chaetoceros compressus, **601**, 623
Chaetoceros curvisetus, **619**
Chaetoceros debilis, **621**
Chaetoceros decipiens, **568**, 613
Chaetoceros dichaeta, **623**
Chaetoceros furcellatus, **623**, 624
Chaetoceros galvestonensis, **568**, **615**
Chaetoceros gracilis, **618**
Chaetoceros laciniosus, **621**
Chaetoceros laevis, **623**
Chaetoceros lauderi, **609**
Chaetoceros neglectus, **623**
Chaetoceros pseudocurvisetum, **601**
Chaetoceros radicans, **621**
Chaetomorpha, 826, **890**
Chaetophora, **889**
Chaetophoraceae, 889
Chaetophoraceae, fossils, 845
Chaetophorales, 889
Chaetophorites, 65, 83, 845, 889
Chaetophorites gomontoides, **845**
Chaetophorites tertiarius, **845**
Chaetotyphla, 517
chain formation, dinoflagellates, 325
Chalkopyxis, **518**
Chamaesiphon, 55, **82**
Chamaesiphonaceae, 82
Chamaesiphonales, 58, 82
Chalmasia, **894**
Champia, **142**
Champiaceae, 142
Chandler, C. D. (Stewart et al.), 805
Chandler, P. T. (Patton et al.), 339
Chandra, A., 841
Chapman, G. B., 839
Chara, 589, 848, 915, 923, 924, 925,
   929, 932, 933, 937, 939, 954, **956**
Chara canescens, **930**, *930*
Chara fragilis, **930**
Chara galioides, **930**
Chara globularis, **915**, *925,* 930
Chara globularis v. *aspera,* **931**
Chara hispida, **925**, *925,* 930
Chara intermedia, **918**, *918*
Chara minutissima, 924
Chara tomentosa, **931**
Chara vulgaris, 915, **916**, *916,* 929, 930,
   931
Chara lakes, 930
Characeae, 956
Characeites, **957**
Characias, **956**
Characiopsis, **498**
Characiopsis gracilis, **495**, *495*
Characium, 836, **888**
Charales, 955

Charaxis, **957**
Charcotia, **656**
Charcotia actinochilus, **624**
Charcotia oculoides, **624**
Charcotia symbolophorus, **624**
Chariella, **957**
Charina, **956**
Charites, **956**
charoid apical structure, gyrogonite, *925,* 925
Charoideae, 956
Charophyceae, 955
Charophyta, 913−963
   antheridium, 917
   calcification, 919
   cell wall, 915
   Cenozoic, 951
   classification, 954
   development of lime shell, 921
   ecology, 927
   evolution, 939
   evolutionary trends, 951
   fossils, 931
   gametangia, 915
   genera, occurrence, *933*
   lime requirements, 930
   Mesozoic, 941
   oosporangium, 919
   oxygenation limitation, 930
   Paleozoic, 937
   pH, 930
   plant, 915
   plastids, 915
   polarity indices, 924
   reproduction, 915
   salinity limits, 930
   species diversity, end Paleozoic, 942
   sporophydium, 919
   utricle, 926
   water depth, 927
Charopsis, **956**
Chase, D. (Margulis et al.), 80, 93
Chasea, **657**
Chasea ornata, **639**
chasmolithic, **57**, **966**
Chatangiella, **392**
Chatangiella ditissima, **357**, *357*
Chatangiella scheii, **357**, *357*
Chatangiellaceae, 389
Chatton, E., 226, 247, 311, 329, 367,
   368, *379*
Chave, K. E., 110
Cheilosporum, **141**
Cheleutochroa, **208**
chemolithotroph, 20
chemoorganotroph, 20
chemosynthesis, 18
chemotroph, **18**, **965**
chemotrophic metabolism, 20
Chen, L. C. M., 131
Cheng, J. Y. (Antia et al.), 497
Cheremisinova, E. A., 639
Chiarugi, A., 850
Chiasmolithus, 770, 776, **783**
Chiasmolithus bidens, **757**, *759,* 763
Chiasmolithus danicus, **757**, *757,* 763
Chiasmolithus gigas, **757**, *757,* 763
Chiasmolithus grandis, **757**, *759,* 762

Chiasmolithus oamaruensis, **757,** *759,* 762

Chiasmolithus solitus, **757,** *759*

Chiastozygaceae, 780

Chiastozygus, 703, **780**

Chiastozygus initialis, **745,** 748

Chiastozygus irregularis, **745,** 748

Chiastozygus litterarius, **741,** *741,* 748

Chihara, M., 116, 117, 140, 807

Chiharaea, **141**

Chinianella, **894**

Chiphragmalithus, **780**

Chiropteridium, **424**

Chiropteridium dispersum, **361**

Chitinozoa, 3

Chlamydiales, 42

Chlamydobacteriales, 39

Chlamydomonadaceae, 888

Chlamydomonadales, 888
  morphology, 834
  reproduction, 834

Chlamydomonas, 805, 826, 828, *830,* 833, 834, *835,* 835, 882, 888
  fossil, 835

Chlamydomonas hedleyi, **835**

Chlamydomonas nivalis, **835**

Chlamydomonopsis, 69, 81

Chlamydophorella, **409**

Chlamydophorella grossa, **411,** 411

Chlorallantus, **498**

Chlorallantus oblongus, **492,** 492

Chloramoebales, 498

Chloraster, **818**

Chlorella, 234, 619, 620, 825, 826, 827, 829, 833, 836, 838, 888

Chlorella vulgaris, **836**

Chlorellaceae, 888
  fossils, 838
  reproduction, 838

Chlorellopsis, 838, **888**

Chloridella, 493

chlorine, in red algae, 109

Chlorobacteriaceae, 42

Chlorobiaceae, 42

Chlorobium, **42,** 92

chlorobium chlorophyll, 19

Chlorobotrys, 185, 497, **498**

Chlorobotrys regularis, **497,** 497

Chlorochytrium, 836, **888**

Chlorocladus, **893**

Chlorococcaceae, 888
  fossils, 837
  reproduction, 837

Chlorococcales, 888
  in lichens, 836
  morphology, 836

Chlorococcum, 836, 837, **888**

Chlorodesmis, **890**

Chlorogloea, 45, 53, 81

Chlorohydra, endosymbionts of, 314

Chloromeson, **498**

Chloromeson agile, **495,** 495

chloromonad, 240

chlorophyll a, 19, 53
  in blue-green algae, 45
  degradation products, 35

Chlorophyceae, 888

Chlorophyta, 804, 825—894
  cell wall, 826
  classification, 887
  depth distribution, 834
  flagella, 826
  growth requirements, 833
  life history, 833
  morphology, 825
  nucleus, 826
  pigments, and plastids, 825
  reproduction, 828
  rock forming, 831
  sexual reproduction, 829, *830*

Chlorosarcinaceae, 889

Chlorosarcinales, 889

Chlorosarcinia, **889**

Chlorotylites, 845, **889**

Chlorotylites berryi, **844,** *844*

Chlorotylium, 845, **889,** 889

Choanoeca, **525**

choanoflagellates, 525

choanoflagellate collar, 504

Chodat, R., 507

Chomotriletes, **208**

Chondria, **142**

Chondrides, **142**

Chondriellaceae, 141

Chondrus, 110, **142**

chorate cyst, dinoflagellates, **284, 285,** 286, **966**

Choreocolacaceae, 141

Choreonema, **140**

choristospory, **868, 966**

Chovanella, 933, 936, 937, 953, **955**

Chovanella kovalevi, **937,** *937*

Chovanellaceae, 955

Christenhuss, R., 581

Christensen, T., 518, 525, 776, 887

Chromatiaceae, 42

Chromatium, **42,** 92

Chromophyta, 491

chromosome number in dinoflagellates, 233

Chromulina, 506, 507, 510, **518,** 518

Chromulina pascheri, **520,** *520*

Chromulinales, 518

Chroococcaceae, 70, 74, 77, 81

Chroococcales, 45, 53, 58, 63, 80, 81

Chroococcidiaceae, 82

Chroococcus, 46, 48, 57, 80, 81, 94

Chroodactylon, 139

Chroothece, 138

Chrysamoeba, 501, **519**

Chrysapion, **519**

Chrysapionales, 519

Chrysapsis, 185, **518**

Chrysapsis yserensis, **520,** *520*

Chrysastrella, 507, **528**

Chrysastrella incerta, **529,** *529*

Chrysobotriella, **519**

Chrysocampanula, **784**

Chrysocapsales, 518

Chrysochromulina, 735, 782, **784**

Chrysochromulina adriatica, **684,** *685*

Chrysochromulina brevifilum, **684**

Chrysochromulina camella, **684**

Chrysochromulina kappa, **681,** 683, 684, 715

Chrysochromulina parva, **725**

Chrysococcus, 499, 510, **518**

Chrysococcus dokidophorus, **520,** *520*

Chrysococcus radians, **520,** *520*

Chrysolykos, 499, **519**

chrysomonad, 138, 510
  cyst abundance, *517*

Chrysonema, 714, **777**

Chrysophyceae, 477, 490, 518, 554

Chrysophyta, 138, 477, 490, 491, 499—529
  cell, 499
  classification, 518
  cysts, 506
  cyst genera and families, 528
  cytology, 501
  flagella, 501
  geologic occurrence, 510
  lorica, 499
  nutrition, 504
  occurrence, 510
  reproduction, 504

Chrysopyxis, **518**

Chrysosaccales, 518

Chrysosphaera, **518,** 714

Chrysosphaerella, **519**

Chrysostomataceae, 507, 512, 528

Chrysostomum, 507, **528**

Chrysostomum simplex, **529,** 529

Chrysothylakion, **519**

Chrysotila, 684, 714, 715, **777**

Chrysotila lamellosa, **683,** 683, 716

Chrysotila stipitata, **716,** *716*

Chuaria, **818**

Churchill, D. M., 204, 364, *365,* 401, 838

Chuttecloska, **208**

Chuttecloska athyrma, **171,** *174*

Chuvashov, B. I., 932, 938

Chytriodiniaceae, 233, 380

Chytriodiniales, 233, 380

Chytriodinium, 309, **380**

Chytroeisphaeridia, **395**

Chytroeisphaeridia cariacoensis, **299,** 403

Chytroeisphaeridia simplicia, **299**

Cienkowski, L., 506

Ciesielski, P. F., *551,* 552

Cilioflagellata, 227, 248

Cinctactiniscus, 253, **386**

Cinctactiniscus cinctus, **387,** *387*

cincture, diatom, **576, 966,** *968*

cingular list, dinoflagellates, 271

cingular plate series, **263, 966**

cingulum, 226, **227,** 263, **966,** *968*

Circulisporites, 176, 190, 204, **209**

Circulisporites parvus, **183,** *183*

cisternae of Golgi body, Haptophyta, **688, 966,** *968*

Cistula, **658**

Citharistes, 311, 371

Citharistes regius, **370,** *370*

clades, ebridian skeleton, **465, 966**

Cladocephalus, **890**

Cladogramma, **656**

Cladophora, 55, 826, 831, 833, 846, **890**

Cladophoraceae, 890

Cladophorales, 890

morphology and life cycle, 846
Cladophorites, 846, **890**
Cladopyxidaceae, 371
Cladopyxidium, 295, **395**
Cladopyxidium saeptum, **401**, *401*
Cladopyxis, **371**
Cladopyxis brachiolata, **272**, *372*
cladospory, **868**, **966**
Claparède, E., 227, 229, 334
Clarke, B. (Parke et al.), 681, 683, 684, 715, 725
Clarke, K. J., 248, *500*, 500, *501, 503*
Clarke, K. J. (Pennick et al.), *806*, 806, 807
Clarke, R. F. A. (Evitt et al.), 298, *355*
classification
  Acritarcha, 205
  bacteria, 38
  calcareous nannoplankton, 776
  Charophyta, 954
  Chlorophyta, 887
  Chrysophyta, 518
  diatoms, 654
  dinoflagellates, cyst fossils, 383
  dinoflagellates, living, 368
  Ebriophyceae, 477
  Euglenophyta, 897
  Haptophyta, 776
  Prasinophyta, 816
  Rhodophyta, 138
  silicoflagellate, 554
  Xanthophyta, 498
Clastidiaceae, 82
Clathrium, *549*
Clathrium reticulare, **559**, *559*
Clathroctenocystis, **395**
Clathrolithus, 770, **778**
Clathrolithus spinosus, **752**
Clathromorphum, **141**
Clathromorphum parcum, **123**, *123*
Clathropyxidella, 549, *559*
Clathropyxidella similis, **556**, *556*
Claus, G., 314
Clavaphysoporella, **892**
Clavaporella, **892**
Clavator, 933, **956**
Clavator reidii, **928**, *928*
Clavatoraceae, *956*
  evolution, 947
Clavatorites, 933, **955**
Clavatorites hoellvicensis, **941**, *941*
Clavicula, **658**
Clavodiscoaster, **778**
Clavosphaera, **784**
Cleistorhabdus, **781**
Cleistosphaeridiaceae, 419
Cleistosphaeridium, **419**
Cleistosphaeridium huguoniotii, **422**, *422*
Cleistosphaeridium polypes, **361**
Cleithronetrum, **208**
Cleithronetrum cancellatum, **159**, *159*
Clendenning, K. A., 342, 346, 347
Clericia, **528**
Clericia ansulata, **508**, *508*
Clericia aureolata, **529**, *529*
Clericia colligera, **529**, *529*

Clericia elegantissima, **529**, *529*
Clericia frenguellii, **529**, *529*
Clericia irregolare, **529**, *529*
Clericia pulcherrima, **508**, *508*
Clericia retiforme, **529**, *529*
Clericia zanoni, **529**, *529*
Cleve-Euler, A., 637
Clibeca, **890**
Clifton, H. F., 833
Climacodium, **657**
Climacosphenia, **658**
Clinorhabdus, **779**
Clocchiatti, M., *705*, 754, *755*
Clonophycus, **138**
Closteridium, **888**
Closterieae, 894
closterite, **833**, **966**
Closterium, 828, 833, 884, 885, *885*, **894**
Clostridium, 17, 22, 29, *30*, **41**, 92
Cloud, P. E., 10, *32*, 65, 68, 69, 74, 80, 94, *129*
Cloud, P. E. (Licari et al.), 74
Clypeator, 933, **956**
Clypeator corrugatus, **928**, *928*
Clypeina, **892**
cnidocysts, **247**, **971**
Coccidiniales, 382
Coccidinium, **382**
Coccobotrys, 843, **889**
Coccochrysis, **777**
coccoliths, **679**, 684, **966**
  abundance in rock, 761
  Cenozoic biostratigraphy, 745
  crystal structure, 694
  deposits, 761
  diversity in Cenozoic, 774
  in fecal pellets, 769
  formation, 687, 688
  geologic occurrence, 727, 735
  imbrication, 705
  Mesozoic biostratigraphy, 748
  mineralogy, 692, *692*
  morphology, 694
  Neogene biostratigraphy, 766
  organic matrix, 684
  oozes, 761
  Paleogene latitudinal zonation, 771
  rate of sinking, 769
  shape, 700
  structure, 705
Coccolithaceae, 783
Coccolithales, 782
coccolithid coccolith, *705*
Coccolithidae, 783
Coccolithinales, 782
Coccolithoideae, 783
Coccolithophora, **783**
Coccolithophora cretacea, **782**
Coccolithophoraceae, 783
Coccolithophorales, 776, 782
Coccolithophorea, 776
coccolithophores, 510
  ecology, 715
  floral zones, 720
  heterotrophy, 725
  latitudinal distribution, 716
  life cycle, 710

  light requirements, 723
  motile cell, 679
  oxygen tolerance, 724
  paleoecology, 769
  salinity effects, 721
  temperature limitations, 719
  vitamin requirements, 724
  water depth, 715
Coccolithophorida, 782
Coccolithophoridaceae, 776
Coccolithophorides, 776
Coccolithophoriinae, 776
Coccolithophorineae, 776
Coccolithophyceae, 776
coccolithosomes, **689**, *690*, **966**
Coccolithus, 684, 694, 777, **783**
Coccolithus miopelagicus, **761**, 767
Coccolithus pelagicus, **683**, 687, 688, 689, 691, 692, 694, 695, 710, *711*, 711, 712, 719, 721, 725, 770, 771, **967**
Coccomonas, 836, **888**
Coccomyxa, 836, **889**
Coccomyxaceae, 889
Cocconeis, 577, 579, 600, 613, 649, 653, **658**
Cocconeis ceticola, **607**
Cocconeis placentula v. *lineata*, **649**
Coccosphaera, **783**
Coccosphaerales, 782
coccosphere, 679
coccus, *11, 12*, **966**
Cochlodinium, 227, 317, **378**
Cochlodinium geminatum, **319**
Cochlodinium strangulatum, **379**, *379*
Cockbain, A. E., 65
Codiaceae, 890
Codiales, 846, 890
Codiolum, 843, **889**
Codiolophyceae, 843, 889
Codium, 826, 834, 846, 847, **890**
Codium fragile, 846, *846*, 847
Codium magnum, **847**
Codomier, L. (Feldmann et al.), 850
Codomonas, **518**
Codonia, **430**
Codoniella, **430**
Codoniella campanulata, **435**, *435*
Codonocladium, **525**
Codonomonas, **518**
Codonosiga, **504**
Codosiga, **525**
Codosigaceae, 525
Coelastraceae, 839, 889
Coelastrum, 185, 839, **889**
Coelastrum verrucosum, **841**, *841*
Coeloceratioides, **891**
Coelosphaeridium, **891**
Coenobiodiscus, 607, **655**
Coenobiodiscus muriformis, **608**, *608*
coenobium, 828, **966**
Cohen, C. L. D., 727
Cohen, N. (Gressel et al.), 137, 234
Cohen, S. S., 91
Cohen, Y., 53, 59
Colacium, 895, **898**
Cole, K., 116
Cole, S. A., 770
Coleobacter, **39**

Coleochaete, 805, **894**
Coleochaetaceae, 894
Coleochaetales, 894
collar, 587, **966**
Collard, P. A. (Hand et al.), 306
Collenia, 64, 66
Collier, A., 568, 615
Collins, G. B., *589*
Colom, G., 726, 729, 843
Colomisphaera, 733, **785**
Colvillea, **783**
Comasphaeridium, **209**, 364
Combaz, A., 149, *195,* 206, 810, 814, *841*
combination archeopyles, 296, 297
Cometodinium, **409**
Cometodinium obscurum, **416,** *416*
Comparodinium, **424**
Compositosphaeridiaceae, 409
Compositosphaeridium, **409**
Compositus, **656**
Compsopogon, 114, 139
Compsopogonaceae, 139
Compsopogonales, 139
Comptaluta, **209**
Comptaluta pirellum, **155,** *159*
Concentricystes, 818
Concentrites, **209**
conceptacle, 111, 122, **123, 966**
Conchocelis, 114, 126, **139, 966**
conchocelis phase, *112, 112,* **114,** *115, 116,* **966**
  fossils, 132
conchospores, *115*
Conejoconus, **785**
Confervites, **82**
Congoites, **209**
Conil, R., *937*
Conipora, **893**
Coniporeae, **893**
Coniporella, **893**
Coniporella clavaeformis, **872,** *872*
Conjugales, **882**
  See also Zygnematales
Conjugatophyceae, **887**
conjugation, **966**
  bacterial, 17
Conkin, B. M. (Conkin et al.), 924, 939
Conkin, J. E., 924, 939
connecting band, *573,* **966**
Connexia, **892**
Conococcolithus, **783**
Conococcolithus panis, **757,** *763*
Conodictyum, **893**
Conodictyum striatum, **872,** *872*
Conosphaeridium, **409**
Conover, S. A. M., 472, 547
Consinocodium, **890**
Conti, S. F. (Gant et al.), *109, 110*
Contortothrix, 74, **82**
Conulina, **893**
Conusphaera, **779**
Convoluta, 812, 818
Convoluta roscoffensis, **613**
Conrad, W., 349, 350, 507, 508, 517, *519, 520, 523, 525, 526, 897*
Conradidium, **209**
Conradiella, **519**

Constrictosphaerina, **209**
Conway, E., *115,* 116
Cookson, I. C., 290, 293, 362, 385, 389, *390, 391, 395, 396, 398, 403, 408, 411, 415, 416, 421, 426, 427, 428, 433, 435, 819, 821, 836, 839*
Cooksonella, **818**
Cooksoniella, **392**
Cooksoniellaceae, 389
Coolia, **374**
Coombs, J., 573, 618
Coombs, J. (Healey et al.), 571
Coombs, J. (Lauritis et al.), 573, 611
Cooper, A. H. (Wilson et al.), 133, 142
coorongite, **832,** *833,* **841, 966**
Copeland, H. F., 85
coprolites, 36
  microorganisms in fossils, 835
Coptolithus, **783**
copula, **576, 966**
Corallina, 108, 112, 117, 119, 124, 141
Corallina cossmannii, **119,** *119*
Corallina officinalis v. *chilensis,* *111*
Corallina vancouveriensis, **119,** *119*
Corallinaceae, 110, 112, 113, 123, 135, 140
coralline algae
  articulated, 117
  encrusting, 117
Corallinoideae, 141
Corannulus, **783**
Corbisema, 539, 549, **559**
Corbisema apiculata, *551,* **552**
Corbisema bimucronata, *551,* **552**
Corbisema disymmetrica, **542,** *542*
Corbisema geometrica, **542,** *542,* 556
Corbisema hastata, *551,* **552,** *552*
Corbisema hexacantha, *551,* 552
Corbisema sphaericus, **559,** *559*
Corbisema spinosa, *551,* **552**
Corbisema triacantha, *550,* 551, **552**
Corbisemaceae, 559
Corcontochrysis, **784**
Cordosphaeridiaceae, 419
Cordosphaeridium, 293, 364, **419**
Cordosphaeridium funiculatum, **419,** *419*
Cordosphaeridium inodes, **281,** *419*
Corethron, 625, **655**
Corethronaceae, 655
Corisphaera, **778**
Corliss, J. O. (Honigberg et al.), 226
Cornell, W. C., 508, 512, *515, 516, 517, 517,* 552
Cornell, W. C. (Tappan et al.), 190
Cornua, 549, 553, 554, **559**
Cornua aculeifera, **560,** *560*
Cornua trifurcata, **560,** *560*
Cornuaceae, 559
Cornplanktona, **209**
Cornutosphaera, **209**
Corollithion, 735, **780**
Corollithion exiguum, **745,** *745,* 748
Corollithion geometricum, 735, *738,* 749
Corollithion signum, **745,** *745,* 748
corona, **587, 966**
Coronaster, **779**

Coronifera, **419**
Coronifera oceanica, **421,** *421*
Coronifera striolata, **416,** *416*
Coroniferaceae, 419
Coronocyclus, **783**
Coronocyclus prionon, **705,** *705*
Coronosphaera, **783**
coronula, 919
coronular cell, **919, 967**
Corrudinium, **392**
Corrudinium incompositum, **383,** *383, 395*
corrugate, **155, 967**
cortex, **117, 967**
Corymbellus, **784,** *784*
Corymbostomum, 507
Corynebacterium, 31, **41**
Corynebacteriaceae, 41
Coryphidium, **209**
Coryphomorphitae, 208
Coscinoconus, **894**
Coscinodiscaceae, 655
Coscinodiscus, 572, 591, 605, 609, 613, 623, 634, 637, 639, 646, **656,** 725
Coscinodiscus africanus, **624**
Coscinodiscus asteromphalus, **568**
Coscinodiscus centralis, **609**
Coscinodiscus concinnus, **597,** *597,* 605
Coscinodiscus crenulatus, **624**
Coscinodiscus curvatulus, **624,** *650*
Coscinodiscus elliptiporus, **644**
Coscinodiscus excavatus v. *quadriocellatus,* **646**
Coscinodiscus excentricus, **650,** *650*
Coscinodiscus gigas v. *diorama,* **652,** *652*
Coscinodiscus granii, **605**
Coscinodiscus insignis, **644**
Coscinodiscus kolbei, **644**
Coscinodiscus kurzii, **645**
Coscinodiscus lentiginosus, **624,** *625,* 644
Coscinodiscus lewisianus, **644,** 645
Coscinodiscus lineatus, **605,** *625,* 651
Coscinodiscus lineatus v. *leptopus,* **650,** *650*
Coscinodiscus marginatus, **644,** 645, 651
Coscinodiscus nodulifer, **604, 624,** *624,* 625, 651, 652
Coscinodiscus oblongus, **647**
Coscinodiscus oculus-iridis, **583,** *583,* **633,** *633*
Coscinodiscus plicatus, **641,** *641,* 647
Coscinodiscus praenitidus, **647**
Coscinodiscus radiatus, **624,** 651
Coscinodiscus vigilans, **646, 647**
Coscinodiscus yabei, **644,** 645
Coscinosira, **656**
Cosmarieae, **894**
Cosmarium, *885,* 886, *886,* 887, **894**
Cosmarium pachydermum, **886**
Cosmarium salinum, **885**
Cosmiodiscus, 639, **656**
Cosmocladium, **894**
Costa, L. I., 358
Costacentrum, **782**

costate, **157, 967**
Costatilobus, **209**
Costatosphaerina, **209**
Costatumbella, **955**
Coste, B., 547
Courteville, H., *416*
Courvoisiella, **889**
Courvoisiella ctenomorpha, 843, **844,**
844
Couté, A., 265
Couté, A. (Feldmann et al.), 850
Cox, E. J., 585
Cox, E. R., 237, 240, 311
Cox, E. R. (Tomas et al.), 311
Cox, R. L., 284
Cramer, F. H., 151, 152, 154, 167, *173,*
173, 190, 199, 200, 201, 205
Crameria, **210**
Craniopsidae, 480
Craniopsis, 476, **480**
Craniopsis minor, **469,** *469*
Craniopsis octo, **483,** *483*
Craspedodinium, **389**
Craspedodiscus, 637, **656**
Craspedodiscus coscinodiscus, **646**
Craspedolithus, **783**
Craspedomonadales, 525
Craspedomonadophyceae, 490, 525
Craspedomonadophycidae, 525
Craspedophyceae, 525
Craspedoporus, **656**
Craspedoporus elegans, **605**
Craspedotella, **380**
Crassapontosphaera, **780**
Crassiangulina, **209**
Crassisphaeridium, **209**
Crassosphaera, **818**
Crassosphaera concinna, **821,** *821*
Crassosphaera stellulata, **821,** *821*
Crassosphaeridae, 818
craticula, 587, **967**
Crawford, R. M., 231, 233, 237, *241,*
241, 242, 242, 244, 250, 257, 258,
259, 265, 308, 570, 581, 585, *587,*
587, 594, 606, 896
Creberlumectum, 204, **430**
Creberlumectum tinglewoodense, **365,**
365
Crenalithus, **777**
Crenalithus doronicoides, **761,** 766
Crenothrix, 24, **40**
Crenotrichaceae, 40
Crepidolithaceae, 780
Crepidolithus, 735, **780**
Crepidolithus crassus, 735, **736,** *736,*
749
crests, dinoflagellate cysts, **288,** *289*
Cretaceous CCD, 773
Cretadiscus, **781**
Cretarhabdaceae, 781
Cretarhabdella, **781**
Cretarhabdus, **781**
Cretarhabdus angustiforatus, **741,** 749
Cretaturbella, **779**
Cretosiphon, **891**
Cretosiphon caulerpoides, **848**
Cribricatillus, **782**
cribrilith, **697,** *700,* **967**

Cribrocentrum, **783**
Cribrocentrum reticulatum, **771**
Cribrocorona gallica, **743,** *743*
Cribroperidinium, **392**
Cribroperidinium edwardsii, **395,** *395*
Cribroperidinium muderongense, **395,**
395
Cribrosphaera, **781,** *781*
Cribrosphaerella, **781**
Cribrosphaerella ehrenbergii, **692,** *692*
Cribrosphaerella murrayi, **746,** *746*
Cribrosphaerella primitiva, **735**
Cribrosphaerelloideae, **781,** *781*
Cribrosphaeroideae, 781
cribrum, **585, 967**
cricolith coccolith, *686, 700,* **703, 967**
Cricolithus, **728, 783**
Cricosphaera, 684, 714, 721, **777**
Cricosphaera carterae, **679,** 681, 684,
686, 689, 690, 692, 703, 710, *713,*
714, 723
Cricosphaera elongata, **679,** 710, 721
Cricosphaera roscoffensis, **679,** 721
Cridland, A. A., 275
Crinalium, **82**
Crinolina, **526**
cristate, **157, 967**
Cristatella, **957**
Crockett, L. J. (Lee et al.), 835
Croft, W. N., 77, 79
Croftiella, 925, **956**
Cronquist, A., 87
Cros, P., 872
Cross, A.T., 275
cruciate tetraspores, *124*
Crucicribrum, **782**
Cruciellipsis, **781**
Cruciellipsis chiastus, 748
Cruciellipsis cuvillieri, **738,** *738,* 749
cruciform rays, **856, 967,** 970
Crucilithus, **783**
Cruciplacolithus, **783**
Cruciplacolithus tenuis, **757,** *757,* 763,
775
Crucirhabdus, **779**
Crucirhabdus primulus, **735,** *735,*
749
Cruoria, **141**
Cruoriaceae, 141
Cruoriopsis, **141**
crustose coralline red algae, 126
Cryptarchaeodinium, **392**
Crypthecodiniaceae, 233, 371
Crypthecodinium, 309, 320, 337, **371**
Crypthecodinium cohnii, **232,** *232,233,*
234, 235, 236, 237, 244, 273, 319,
326, 329, 330, 344
Cryptonemia, 132, **141**
Cryptonemiaceae, 141
Cryptonemiales, 110, 128, 140
cryptopylome, 176, **967**
Cryptosphaera, **209**
Cryptostromatium, **209**
cryptosuture, 176
Cryptozoon, 64, 66
crystallolith coccolith, **692, 694, 967**
Crystallolithus, 692, 694, **783**

Crystallolithus hyalinus, 689, **691,** *691,*
695, 711, 712
Ctenidodiniaceae, 403
Ctenidodinium, 283, 290, **404**
Ctenidodinium continuum, **290**
Ctenidodinium elegantulum, **362**
Ctenidodinium panneum, **362**
Ctenidodinium schizoblatum, **403,** *403*
Cuatrifolia, **209**
Cummings, R. H., *134*
Cumulosphaera, 69, **81**
Cuneatochara, 933, **955**
Cuneatochara acuminata, **941,** *941*
Cuneiphycus, **141**
Cupp, E. E., 613, 621, 623, 641
Cushing, D. H. (Hendey et al.), 585
cyanelle, **62, 967**
Cyanidium, **138**
Cyanobacteria, 42, 43, 81
Cyanocyta, **63**
Cyanoderma, **139**
Cyanonema, 74, **82**
cyanophage viruses, 51
Cyanophanaceae, 82
Cyanophanon, **82**
Cyanophora, **63**
Cyanophyceae, 43
cyanophycin granule, *43, 46, 50,* **967,**
975
Cyanophyta, 42 – 83
evolution, 80
wall, 13
Cyanospira, **92**
Cyanospira aeruginosa, **63**
Cyatholithus, **783**
Cyclagelosphaera, 703, **783**
Cyclagelosphaera margereli, **736,** *736*
Cyclapophysis, **409**
Cyclicargolithus, **783**
Cyclicargolithus floridanus, **761,** *762,*
771
Cyclicargolithus reticulatus, **757,** *762*
Cyclocalyptra, **778**
Cyclococcolithina, **783**
Cyclococcolithoideae, 783
Cyclococcolithus, **783**
Cyclocrineae, **891**
Cyclocrinites, **891**
Cyclocrinus, **891**
Cyclocrinus porosus, **859,** *859*
cyclolith coccolith, 703, **967**
Cyclolithella, **780**
Cyclolithella annulus, **719**
Cyclolithus, **780**
Cyclonephelium, 358, **424**
Cyclonephelium pastielsi, **362,** *362*
Cycloplacolithella, **783**
Cycloplacolithus, **783**
Cyclophora, **658**
Cyclopsiella, 178, **209**
Cyclopsiella vieta, **178,** *178*
cyclopyle, **178, 964, 967**
Cyclotella, 572, 583, 597, 607, 613, 620,
626, **655**
Cyclotella cryptica, **568,** 572, 596, 619
Cyclotella meneghiniana, **598**
Cylindralithus, **780**
Cylindralithus laffittei, **746,** *746*

Cyclindrocystis, **894**
Cylindroporella, **892**
Cylindropyxis, **655**
Cylindropyxis profunda, **568**
Cylindrospermum, 49, 53, **83**
Cylindrotheca, 572, 579, 618, **659**, 968
Cylindrotheca closterium v. *cali-
   fornica,* **618**
Cylindrotheca fusiformis, **568**, 571,
   579, 581, 596, 618, 619, 620
Cymatiogalea, 151, 187, **209**
Cymatiomorpha, **209**
Cymatiosphaera, 187, 199, 808, 809,
   821, **822**
Cymatiosphaeraceae, 207, 818
Cymatiosphaeropsis, **822**
Cymatodiscus, **656**
Cymatoneis, **658**
Cymatopleura, 626, **659**
Cymatosira, **658**
Cymatosira biharensis, **645**
Cymatotheca, **656**
Cymbella, 600, 605, 626, 628, **659**
Cymbella cistula, **587**, *589*
Cymbella ehrenbergii, **569**, *569*
Cymbella tumida, **648**, *648*
Cymbellaceae, 659
Cymbellonitzschia, **659**
Cymbodinium, **380**
Cymbosphaeridium, **209**
Cymbosphaeridium bikidium, **167**
Cymbosphaeridium pilar, **151**, 176,
   178
Cymopolia, 861, **894**
Cymopolia barbata, **861**, *861*
Cymopolia paktia, **880**, *880*
cyrtolith coccolith, **697**, **967**
Cyrtophora, 510, **525**
cysts
   bacteria, 16
   Chlorophyta (*Acetabularia*), 827
   Chrysophyta, 506
   dinoflagellates, 271
      inclusions, 293
      morphology, 284
      types, 284
      wall composition, 278
      wall structure, 281
      wall surface, *290, 290*
   Prasinophyta
      composition, 809
      development, 808
      wall structure, 809
   Xanthophyceae, 496
cytochromes, **29**
cytochrome sequences, 93, 95
Cytophagaceae, 39
Cytophagales, 39
cytoplasmic membrane
   bacteria, 13
   dinoflagellates, *257*
cytosine, 31
cytostome, **896**, **967**

Dactyliosolen, 613, 637, **655**
Dactyliosolen antarcticus, **577**, *577*
Dactylofusa, 200, **210**

Dactylopora, **893**
Dactylopora pauperata, **876**, *876*
Dactyloporeae, 893
Daday, E. von, 374
Daillydium, **210**
Daily, F. K., 932, 951
Daily, W. A., 58, 81
Daktylethra, 770, **778**
Daktylethra punctulata, **697**, *697*, **752**
Dale, B., 275, 276, 279, 283, 284, 293,
   298, *301, 302*, 329, 334, 337, 351,
   364, 383, 385, 424
Dale, B. (Wall et al.), 275, 278, 279,
   279, 280, 289, 290, *301*, 364, *396*,
   398
Dalmatella, 59
Dal Pont, G. (Grant et al.), 617, 618
Daly, O. (Siegel et al.), 69
dancing groups, 331, 332
Danea, **404**
Danea californica, **406**, *406*
Dangeard, L., 726, 838
Daniels, C. H. von, 509
Dapcodinium, **395**
Dapcodinium priscum, 353, **401**, *401*
Darby, D. G., 94, 129
Darko, D. A., 761
Darland, G. K., 23, 58
Darley, W. M., 597, 598, 605, 607, 614,
   618
Darley, W. M. (Coombs et al.), 573
Darley, W. M. (Loeblich et al.), 607,
   608
Das, S. C. (Venkataraman et al.), 572
Dasya, **142**
Dasyaceae, 142
Dasycladaceae, 892
Dasycladales, 891
   cysts, 862
   evolution, *869*
   fossils, 864
   morphology, 860
Dasycladeae, 893
Dasycladoideae, 892
Dasycladus, 826, **893**
Dasycladus clavaeformis, **860**, *860,
   860*, 861
Dasydiacrodium, **210**
Dasyhapsis, **822**
Dasypilula, **210**
Dasypilula compacta, **193**, *193*
Dasyporella, **892**
Dasyporelleae, 892
Dateriocradus, **210**
Daturellina, 881, **894**
Davenport, D. (Hand et al.), 306
Davey, R. J., 199, 202, 204, 284, *395,
   401, 406, 408, 415, 416, 423, 427, 429*
Davidson, S. E., 273, 275, 278, 285, 351
Davis, C. A., 884
Davis, E. F., 636
Davis, J. S., 839
Davis, J. T. (Steidinger et al.), 308, 341
Dawes, C. J., 237
Dawson, G. M., 725
Dawson, E. Y., 136, 597, 848
Dawson, P. A., 571
Dayhoff, M. O., 92, 95

Dayman, J., 725
Deane, H., 350
Deflandrastrum, 839, **888**
Deflandrastrum colonnae, **841**, *841*
Deflandrastrum millepiedi, **841**, *841*
Deflandrastrum speciosum, **841**, *841*
Deflandre, G., 154, 171, *173*, 181, 182,
   205, 248, 278, 279, 281, *281*, 283,
   286, 293, 294, 349, 350, 351, *355*,
   383, 387, 406, 415, 416, 423, 427, 435,
   463, 465, 467, 469, 471, 477, 479,
   482, 483, 484, 487, 512, 517, 528,
   539, 541, 545, 556, 559, 560, 630,
   637, 726, 727, 729, 776, 841, 896, 897
Deflandre, M., 205
Deflandrea, 5, 276, 281, 283, 284, 286,
   287, 288, 290, 293, 294, 295, 296,
   301, 302, 355, 358, 383, **392**
Deflandrea medcalfii, **357**, *357*
Deflandrea heterophlycta, **361**
Deflandrea phosphoritica, **357**, *357*
Deflandrea speciosa, **361**
Deflandreaceae, 389
Deflandreia, 507, **528**
Deflandreia longispina, **529**, *529*
Deflandriaceae, 781
Deflandrius, 741, **782**
Deflandrius columnatus, **741**, *746*, 748
Deflandrius cretaceus, **741**, *741*
Deflandryocha, 549, **559**
Deflandryocha cymbiformis, **542**, *542,
   559*
Deflandryocha spathulata, **559**, *559*
dehydrogenation, 20
Delage, Y., 243, 272, 302
Delesseria, **142**
Delesseriaceae, 142
Delesserites, 132, **142**
De Ley, J., 11, 31, 37
Delphineis, **658**
Delphineis angustata, **643**, *643*
Delphineis penelliptica, **647**
Demorhethium, **210**
Demorhethium lappaceum, **194**, *194*
Dendractis, **83**
denitrifying bacteria, 22
Denticula, 600, **659**, 639
Denticula antarctica, **644**, 645, 646
Denticula hustedtii, **644**
Denticula hyalina, **647**
Denticula kamtschatica, **644**
Denticula lauta, 587, **588**, *588*, 644, 645
Denticula nicobarica, **645**, 646
Denticula ovata, **647**
Denticula penelliptica, **647**
Denticula seminae, **587**, *587*, 587, 624,
   644, 650
Denticula tenuis, **581**, *581*
Denton, E. J., 341
deoxyribonucleic acid, 11
depauperating division, 331, **967**
Depeche, F., 733, *733*
depth of water, acritarchs, 202
depth distribution, Chlorophyta, 834
depth zonation, algae, 126
Derbesia, 826, 827, 829, 850, 882, **891**
Derbesiaceae, 891
Derbesiales, 891

reproduction, 850
Dermatolithon, **141**
Dermatophilaceae, 42
Derwalker, J. C. van (Clifton et
  al.), *833*
Derxia, **40**
Desikachary, T. V., *572, 581*
Desmarella, **525**
Desmatractum, 185, 838, **888**
Desmidiaceae, 894
desmids, 884
  sexual reproduction, 885
  vegetative reproduction, 885
Desmidium, **894**
Desmidium cylindricum, **885**
Desmobacteriales, 39
Desmocapsa, **368**
Desmocapsaceae, 368
Desmocapsales, 368
Desmomastigaceae, 368
Desmomastix, **368**
Desmogonium, 626, **658**
Desulfovibrio, 20, 22, 30, **40**, 93
Desulfovibrio desulfuricans, **27**
Detonula, **655**, 601
Detonula confervacea, **593**, *593,* 601
Detonula pumila, **585**, *585*
Dettmann, M. E., *819*
Deunff, J., 151, 154, 181, 182, 187, 190
Deunffia, 171, 200, **210**
Deusilites, **210**
Deutschlandia, **784**
Deutschlandiaceae, 784
Devonian-Mississippian microplankton
  diversity, *196*
dextrogyre coccoliths, **694, 967**
Diacanthum, **392**
Diacanthum filapicatum, **292**, *292*
Diacrocanthidium, **430**
Diacrocanthidium echinulatum, **435**,
  435
Diacrodiaceae, 207
Diacromorphitae, 207
Diacronema, **784**
Diademopetra, **783**
Diadorhombus, **780**
Diadorhombus rectus, **735**, 749
Diadozygus, **780**
Diadozygus dorsetense, 735, *738,* 749
Diaphanoeca, **526**
Diaphanoeca grandis, **511**, *511*
Diaphorochroa, **210**
Diatoma, **658**
Diatomaceae, 657
diatomaceous sediments and diatomi-
  tes
  accumulation, 634
  diagenesis, 636
  distribution, *628, 629, 630*
Diatomella, **658**
diatoms, 567−677
  abundance in sediments, 628
  auxospore, *597*
  auxospore development, *599*
  biostratigraphy
    Cenozoic zonation, 644
    Upper Cretaceous zonation, 639
  cell, 568

cell size, 595
classification, 654
colonies, 607
distribution, 520
  in fresh water, 626
ecology, 615
evolution, 652
frustule, 572
  composition, 572
  morphology, *573, 575, 576*
girdle, 576
  half-bands, **576**
  incomplete open bands, **576**
habitat, 605
heterotrophy, 613
movement, 610, *612*
nucleus, 568
paleotemperature indicators, 650,
  650, 651, 652
parasites, 613
photosynthesis, 614
pigments, 571
plastids, 571
productivity, 614
pyrenoids, 571
raphe system, *581*
  modifications, *582*
  morphology, *579*
rate of sinking, 609
sexual reproduction, 596
  centric, female gamete, 598
  centric, male gamete, 597
  pennate, 599
sheaths, stalks, 572
size distribution, *595*
size range, 603
species diversity, *653*
statospores, 600
systematic position, 568
vegetative reproduction, 595
wall, fine structure, 581
Diazomatolithus, **783**
Diazomatolithus lehmanii, **735, 736**,
  736, 749
Dichadogonyaulacaceae, 403
Dichadogonyaulax, **404**
Dichotomosiphon, **890**
Dichotomosiphonaceae, 890
Dichothrix, 59, **83**
Dichothrix fucicola, **54**
Dickerson, R. E., *30*
Dicladia, 463, *476, 479,* **480**
Dicladia clathrata, 463, 479, *484*
Dicommopalla, 200, **210**
Dicommopalla macadamii, **157**, *178,*
  178
Diconodinium, **392**
Diconodinium caulleryi, **353**
Diconodinium multispinum, **391**, *391*
Dicranemaceae, 142
Dicraspedella, **525**
Dicrateria, 715, **777**
Dicrodiacrodium, **210**
Dicroerisma, 254, **378**, 965
Dicroerisma psilonereiella, **255**, *255*
Dictyocha, 251, 463, 474, 512, 535, 539,
  541, 542, 546, 549, 552, 554, **559**
Dictyocha aculeata, **549**, *550, 551*

Dictyocha aspera, **542**, *542, 550,* 552
Dictyocha aspera v. *pygmaea,* **549**, *551*
Dictyocha ausonia, **545**, *545*
Dictyocha bimucronata, 551, 552
Dictyocha boliviensis, **549**, *550,* 551
Dictyocha deflandrei, 551, **552**, *552*
Dictyocha epiodon, **549**, *550*
Dictyocha fibula, **536**, *536,* 537, 538,
  541, *541,* 541, 542, 546, 547, 550,
  552, *554*
Dictyocha fibula v. *aculeata,* **549**, *550*
Dictyocha fibula v. *aspera,* 550, 552
Dictyocha fibula v. *brevispina,* **545**,
  545
Dictyocha fibula v. *octogona,* 550
Dictyocha fibula v. *pentagona,* **541**
Dictyocha fischeri, 552
Dictyocha frenguellii, 551, 552
Dictyocha hexacantha, **541**, *551,* 552
Dictyocha medusa, 551, **552**
Dictyocha navicula, 546
Dictyocha naviculoidea, 551, 552
Dictyocha perlaevis, 550, 552
Dictyocha pseudofibula, **549**, *550,* 551,
  552
Dictyocha quadralta, **559**, *559*
Dictyocha quadria, **542**, *542,* 545, 551,
  552
Dictyocha rhombica, 550, 552
Dictyocha stapedia, **554**, *554*
Dictyocha subarctios, 550
Dictyocha transitoria, 551, **552**
Dictyocha tripartita, 463
Dictyocha/Distephanus ratios, 552
Dictyochaceae, 554
Dictyochales, 554
Dictyochroa, **822**
Dictyochroa ovalis, **822**, *822*
Dictyoclavator, 933, **956**
Dictyoclavator fieri, **943**, *943*
Dictyococcites, **777**
Dictyococcites abisectus, **757**, 762
Dictyococcites bisectus, **761**, 762, 771
Dictyococcites hesslandii, **771**
Dictyocrinus, **891**
Dictyolithus, **781**
Dictyoneis, **658**
Dictyopyxidia, **403**
Dictyopyxis, **403**
Dictyosphaera, **210**
Dictyosphaeria, 826, **891**
Dictyosphaeridium, **210**, 822
Dictyotidium, **822**
Didymosphenia, **658**
Diectochara, 933, **956**
Diectochara andica, **928**, *928*
Diersche, V., 130, *131*
Diexallophasis, 206, **210**
Diexallophasis denticulata, **166**, *166*
Díez de Cramer, M. C., 151, 152, 154,
  167, 200, 205
Digitella, **893**
Dikimdinella, 77, **83**
Dilatisphaera, **210**
Dimastigobolus, 171, **210**
Dimastigobolus longifilum, **173**, *173*
Dimeregramma, **658**
Dimidiata, **658**

Dimorphosiphon, 848, **890**

Dimorphosiphon rectangularis, **851,** *851*

Dimorphostroma, **83**

Dimroth, E., 80, 95

Dinamoeba, **381**

Dinamoebidiaceae, 380

Dinamoebidiales, 380

Dinamoebidium, **381**

Dinastridium, **380**

Dinastridium sexangulare, **381,** *381*

Dinetromorphitae, **208**

Dingman, R., 649

Dingodinium, **392**

Dingodinium cerviculum, **391,** *391*

Dinobryaceae, 519

Dinobryon, 138, 499, 506, 510, **519**

Di Nocera, S., 728, 733

Dinocloniaceae, 381

Dinoclonium, **381**

Dinococcales, 380

Dinoflagellata, 229

dinoflagellates, 225—435

  amphiesma, 256, *259*

  bioluminescence, 316

  cell, 229

  cell movement, 302

  cell orientation, 306

  chain formation, 325

  circadian rhythms, 237

  classification, 364

    of fossil cyst taxa, 386

    of living taxa, 368

  cysts, 227, 271, 275

    flagellar markings, 290

    history of study, 272

    morphology, 284

    and motile cell theca, 273, 299, 301, 302

    processes, horns, and crests, 288

    rate of description, genera, 4

    in sediments, 275

    types, 284

    wall composition, 272, 278

    wall structure, 281

  cytoplasm, 231

  division rates, 319

  fission, 319

  flagella, 247, 249

  distribution, 342

    Atlantic, 345

    fresh water, 348

    interstitial flora, 347

    Pacific, 345

    tidepools, 348

    tropics, 346

  ecology, 338

    nutrient requirements, 345

    physical conditions, 343

    salinity, 344

    temperature tolerance, 344

  generic diversity, *357*

  geologic occurrence, 351

    biostratigraphic zonation, 361

    Cenozoic fossils, 358

    Mesozoic fossils, 353

    nonmarine fossils, 364

    Paleozoic fossils, 352

  growth, 326

  nucleus, chromosomes, and DNA, 231—237

  nutrition, 308

    autotrophs, 308

    auxotrophs, 308

    endosymbionts, 240

    host symbionts, 311

    myxotrophs, 308

    parasites, 308, 309

    saprophytes, 309

  phylogenetic relationships, 364

  physiology, 302

  plastids and pigments, 237, 239

  productivity, 338

    blooms in fresh water, 348

    red tides, 339

  pyrenoids, 240, *241*

  ribosomal RNA, 234

  reproduction

    of *Noctiluca,* 335

    nuclear cyclosis and knot stage, 330

    sexual, 329

    vegetative, 319

  species diversity, *352*

  theca, 258

    comparison to cyst, 273

    composition, 261

    lists, 271

    modification by exuviation, 327

    ornamentation, 265

    plate overlap, 265, 267

    plate synthesis, *260*

    scales, 248

    structure of, *260*

    tabulation, 261

  toxins, 317

Dinogymniaceae, 386

Dinogymnium, 292, 293, 298, 302, 354, **386**

Dinogymnium acuminatum, **355,** *355*

Dinogymnium euclaense, **361**

Dinogymnium longicorne, **298**

Dinogymnium nelsonense, **298**

Dinogymnium sibiricum, **298**

Dinogymnium strombomorphum, **355,** *355*

Dinogymnium westralium, **355,** *355*

dinokaryon nucleus, **227,** 231, **967, 971**

  in ebridians, 464

Dinophyceae, 233, 368

Dinophysiaceae, 371

Dinophysiales, 261, 301, 351, 368, 369, 386

  division, 324

  tabulation, 269, *270*

Dinophysis, 271, *301,* 306, 319, 324, 338, 343, 344, 346, 353, **371**

Dinophysis acuta, **258**

Dinophysis caudata, **325**

Dinophysis diegensis, **301,** *301*

Dinophysis fortii, **270,** *270*

Dinophysis odiosa, **271,** *271*

Dinopterygium, **409**

Dinopterygium cladoides, **411,** *411*

Dinosphaera, **371**

Dinosphaera palustris, **372,** *372*

Dinosphaeraceae, 371

Dinotrichaceae, 381

Dinotrichales, 381

Dinothrix, **381**

dioecious, 917

Diornatosphaera, **208**

Diornatosphaeridae, 207

Dioxya, **392**

Diphyes, **419**

Diphyes colligerum, **361,** **421,** *421*

diplobiontic, **829, 967**

Diploeca, **525**

Diploeca costata, **526,** *526*

Diplofusa, **424**

Diplofusa gearlensis, **435,** *435*

diploid, 16, 335, **967**

Diplomorphaceae, 380

Diploneis, **658**

Diploneis bombus, **575,** *575*

Diploneis ovalis, **611**

Diplonema, **83**

Diplonemataceae, 83

Diplopeltopsis, 302

Diplopora, 868, **893**

Diplopora annulatissima, **871,** *871, 872*

Diplopora helvetica, **870,** *870*

Diplopora tubispora, **869,** *869*

Diploporeae, 892

Diploporundus, 881, **893**

Diplopsalis, 302, **374**

Diplopsalopsis, 302, **374**

Diplopsalopsis orbicularis, **301,** 301

Diplotesta, **424**

Diplotestaceae, 424

Diplotestidae, 208

Diplotheca, **526**

Diporina, 838

disaster species, 5

Discoaster, 706, 745, 751, 770, 771, **778,** 965

  crystallographic structure, 707, 708

  diversity in Cenozoic, *775*

Discoaster abisectus, 767

Discoaster adamanteus, **706, 707,** *707, 708*

Discoaster aster, **752**

Discoaster asymmetricus, **756,** *756,* 761, 766

Discoaster barbadiensis, **754,** *754,* **757,** 762

Discoaster bellus, **761,** 767

Discoaster berggrenii, **761,** 766

Discoaster bifax, **757,** 762

Discoaster binodosus, **752,** *752,* **757,** *763*

Discoaster brouweri, **751,** 752, 756, 761, 766

Discoaster calcaris, **755,** *755,* **761,** 767

Discoaster challengeri, **752,** 755

Discoaster deflandrei, **761,** 767

Discoaster diastypus, **757,** *763*

Discoaster dilatus, **756,** *756*

Discoaster divaricatus, **756,** *756*

Discoaster druggii, **755,** *755,* **761,** 767

Discoaster elegans, **754,** *754*

Discoaster exilis, **755,** *755,* 761, 767

Discoaster gemmeus, **752**, *752*, *757*, *763*
Discoaster hamatus, **761**, *767*
Discoaster kugleri, **755**, *755*, *761*, *767*
Discoaster lodoensis, **754**, *754*, *757*, *763*
Discoaster mirus, **757**, *763*
Discoaster mohleri, **757**, *763*
Discoaster multiradiatus, **752**, *752*, *757*, *763*
Discoaster neohamatus, **761**, *766*
Discoaster neorectus, **761**, *766*
Discoaster nobilis, **757**, *763*
Discoaster obtusus, **706**, *707*, *707*, *708*
Discoaster pentaradiatus, **756**, *756*, *761*, *766*
Discoaster perplexus, **770**
Discoaster planktonicus, **751**, *752*
Discoaster quinqueramus, **755**, *755*, *761*, *767*
Discoaster saipanensis, **754**, *754*, **757**, *762*
Discoaster strictus, **757**, *763*
Discoaster sublodoensis, **754**, *754*, *757*, *763*
Discoaster surculus, **756**, *756*, *761*, *766*
Discoaster tamalis, **761**, *766*
Discoaster tani, **757**, *762*, *763*
Discoaster tani nodifer, **754**, *754*, **757**, *763*
Discoasteraceae, 778
Discoasterales, 778
Discoasteridae, 778
Discoasteroides, **778**
Discoasteroides kuepperi, 752, **757**, *763*
Discoasteromonas, 752
Discoasteromonas calciferus, **752**
discobolocysts, **501**, **967**
Discoidella, **818**
Discoidella hannae, **819**, *819*
discolith, **697**, *700*, **967**
Discolithina, 694, 770, **780**
Discolithina japonica, **770**
Discolithina macropora, **752**
Discolithus, 728, **780**
Discorhabdaceae, 782
Discorhabdus, **782**
Discorhabdus jungii, **735**, 749
Discorhabdus patulus, **736**, *736*
Discorhabdus tubus, **735**, *737*, 749
Discosphaera, 721, **784**
Discosphaera tubifera, 697, 770, **703**, *703*, 717, 719
Discoturbella, **778**
Disectispora, **210**
Disparifusa, 176, **210**
Disparifusa hystricosa, **155**, *157*, *157*, *159*, **175**, *175*
Disphaeromorphitae, **208**
Dissocladella, **893**
Dissocladelleae, 893
Dissodinium, **377**
Dissodinium pseudocalani, **309**
Dissodium, 320, 344, **374**
Dissodium lenticulum, **272**, *334*
Distanov, U. G., 547, 548
Distatodinium ellipticum, **416**, *416*

Distephanaceae, 559
Distephanus, 535, 539, 542, 549, 554, **559**, *559*, 964
Distephanus corona, **545**, *545*
Distephanus crux, 550, 551, **552**, *554*
Distephanus crux v. *longispina,* 550, 552
Distephanus fibula v. *octogona,* **552**
Distephanus heptacanthus, **549**
Distephanus longispina, 550, **552**
Distephanus mesophthalmus, 550, **552**
Distephanus octacanthus, 550, 551, **552**
Distephanus octangulatus, **549**, *550*, *551*
Distephanus octogonus, **549**, *551*
Distephanus octonarius, **540**, *545*, *546*, *547*, **549**, *550*, *551*
Distephanus pentagonus, **552**
Distephanus quinquangellus, **552**
Distephanus schauinslandii, **542**, *542*, *550*, *552*
Distephanus speculum, **536**, *536*, *540*, *541*, *545*, *546*, *549*, *550*, *551*, *552*
Distephanus speculum v. *pentagonus,* 550, *551*, **552**
Distephanus speculum f. *pseudofibula,* 550, 552
Distephanus speculum septenarius, **541**
Distephanus speculum regularis, **546**, *546*
Distephanus stauracanthus, 550, **552**
Distephanus subarctios, **549**
Distichoplax, 140, **141**
Distigma, **898**
Distigmopsis, **898**
Ditripodiaceae, 480
Ditripodiidae, 480
Ditripodium, 476, **480**
Ditripodium elephantinum, **469**, *469*, *484*
Ditripodium fenestratum, **467**, *467*
Ditripodium latum, **469**, *469*, *484*
Ditylum, 637, **657**
Ditylum brightwellii, **596**, *601*, *603*, *603*, *604*
Diversispina, **419**
Diversocallis, **139**
Divietipellis, **210**
Diwald, K., 275, 332, 332
Dixon, H. H., 685, 688, 697, 726
Dixon, P. S., 107, 117, 124, 128, 137, 138, 714, 776
DNA, 11
  in dinoflagellates, 232, 234
  in mitochondria, 85
  percent G+C, bacteria, 31
  percent G+C, blue-green algae, 44, 46
  percent G+C, diatoms, 568
  in plastids, 85
Dobunniella, **893**
Docidium, **894**
Dodekapodorhabdus, **781**
Dodge, J. D., 230, 231, 232, 233, 234, 236, 237, 240, 241, 241, 242, 242,

244, 245, 248, 249, 250, 257, 258, 259, 265, 308, 311, 315, 334, 374, 896
Dodge, J. D. (Crawford et al.), 258
Dodson, A. N., 618
Doemel, W. N., 28
Domasia, 149, 171, 200, 202, 204, **210**
Domasia elongata, **200**, *200*
Domasia trispinosa, **201**, *201*
Domasiella, **210**
Dombrowski, H., 37
Dominopolia, **210**
Donahue, J. G., 639, 650, 651
Donezella, **142**, 868, **892**
Donkinia, **658**
Donnell, J. R. (Tourtelot et al.), 814
Donnelly, P. V., 341
Dorigan, J. L., 689
Dorocysta, **409**
Dorsennidium, **212**
Dossetia lacera, **603**, *603*
double skeletons, silicoflagellates, 539
Dougherty, C. C., 138
Douglas, R. G. (Bukry et al.), 770
Downie, C., 149, 154, 167, 171, 181, 182, 187, 190, 202, 205, 206, 281, 284, 285, 286, 295, 297, 350, 351, 352, 357, 358, 362, 364, 368, 383, 385, 391, 392, 411, 429
Downiea, **210**
Dracodinium, **392**
Dragesco, J., 309, 334, 347, 347
Drebes, G., 309, 597, 598, 599, 599, 600, 601, 605
Drepanotheca, **658**
Drew, E. A. (Barnes et al.), 127
Drew, G. H., 28
Drew, K. M., 116
Droop, M. R., 344, 620, 724, 725
Drouet, F., 58, 81
Drugg, W. S., 173, 178, 181, 289, 290, 290, 351, 357, 362, 383, 385, 389, 392, 395, 396, 401, 404, 406, 408, 416, 419, 421, 422, 424, 426
Druggidium, **392**
Druggidium apicopaucicum, **396**, *396*
Drum, R. W., 571, 607, 610, 611, 612
Druridgea, **655**
Duboscquella, 311, 382
Duboscquella tintinnicola, **311**
Dubosquellaceae, 382
Dürr, G., 258, 260, 265, 267, 321, 322, 322, 324, 327
Duff, D. C. B., 620
Dughiella, 933, **956**
Dughiella bacillaris, **950**, *950*
Duke, E. L., 612
Dumitrică, P., 253, 387, 483, 539, *541*, *542*, 542, 550, 556
Dumontia, **140**
Dumontiaceae, 140
Dunaliella, 311, 834, 835, **888**
Dunaliellaceae, 888
Duncan, P. M., 65, 67
Duosphaeridium, **419**
Duosphaeridium rugosum, **421**, *421*
Durand Delga, M., 843
Durham, J. W., 932
Durrieu, L., 726

Duthie, H. C., 886
Duve, C. de, 90
Duvernaysphaera, **822**, 822, 824
Duvernaysphaera tenuimarginata, **822**, 824
Dvinella, 868, 892
Dvinella (Trinodella) bifurcata, **868**, 868
Dweltz, N. E., 572
Dwivedi, J. N., 843, 882, 884

Eatonicysta, **409**
Eatonicysta ursulae, **411**, 411
Ebria, 463, 464, 465, 472, 474, 476, 476, 479, **483**, 976
Ebria tripartita, **463**, 465, 472, 473, 474, 487
Ebriaceae, 480
Ebriacidae, 480
Ebriales, 381, 479, 480, 535
ebridians, 463—489
    cell, 464
    classification, 477, 480
    ecology, 472
    evolutionary trends, 476
    flagella, 464
    fossil record, 474
    geologic range of genera, 476
    hypersilicification, 469
    living organism, 464
    rate of description, genera, 4
    reproduction, 472
    skeletal lorica, 469
    skeleton, 465
    surface ornamentation, skelton, 465, 469
Ebriella, **483**
Ebriida, 477, 479
Ebriidae, 480
Ebriidales, 479
Ebriideae, 477
Ebrinula, 476, **483**
Ebrinula paradoxa, **487**, 487
Ebrioidae, 480
Ebriophyceae, 381, 479, 480
Ebriopsidae, 480
Ebriopsis, 463, 469, 472, 476, **480**
Ebriopsis antiqua, **463**, 471, 480, 480, 482, 550
Ebriopsis aplanata, **482**, 482
Ebriopsis crenulata, **472**, 479
ecdysis, 258, 260, 265, 295, 319, 320, **967**
Echaninia, 77, 81
Echlin, P., 13, 14, 80, 92
echinate, 155, **967**
Echinochara, 933, **956**
Echinochara spinosa, **928**, 928
Eckert, R., 316, 317
Ecmelostoiba, **210**
ecologic pioneers, blue-green algae, 57
ecologic succession, patterns, 7
ecology and distribution
    Acritarcha, 199
    bacteria, 23
    blue-green algae, 57
    Charophyta, 927

Chrysophyta, 510
coccolithophores, 715, 769
Cyanophyta, 57
diatoms, 605, 615, 620
dinoflagellates, 338
ebridians, 472
euglenids, 896
Haptophyta, 715, 769
Prasinophyta, 812
Rhodophyta, 126
silicoflagellates, 546, 552
Xanthophyta, 496
ecosystem, 6
Ecthymabrachion, **210**
Ecthymapalla, **210**
Ecthymapalla echinata, **160**, 160
ectophragm, 281, **967**
Ectothiorhodospira, **42**, 92
Ectothiorhodospira mobilis, **85**
Ectypolopus, **210**
Ectypolopus elimatus, **193**, 193
Eddy, S., 348
Edelstein, T. (Chen et al.), 131
Edelsteinia, **892**
Edhorn, A. S., 69, 94
Edromorphida, 207
Edwards, A. R., 752, 757, 761, 764
Edwards, M. R. (Gant et al.), 109, 110
Edwards, P., 125
Egmontodinium, **392**
Ehlen, J. (Orr et al.), 639
Ehlers, E. G. (Schopf et al.), 24, 26, 36
Ehlersospongia, **891**
Ehrenberg, C. G., 24, 227, 229, 251, 275, 281, 293, 349, 385, 463, 479, 484, 512, 559, 630, 641, 654, 725, 937
Ehrlich, A., 641
Eiffellithaceae, 779
Eiffellithales, 779
Eiffellithus, **779**
Eiffellithus augustus, **743**, 743
Eiffellithus eximius, **745**, 746, 748
Eiffellithus turriseiffelii, **741**, 741, 748
Eisenack, A., 149, 153, 160, 173, 178, 181, 182, 185, 205, 206, 243, 276, 278, 286, 288, 290, 293, 294, 351, 352, 368, 383, 389, 390, 391, 395, 396, 398, 403, 406, 408, 411, 426, 428, 433, 435, 812, 815, 819, 836
Eisenack, A. (Cookson et al.), 385, 390, 391, 403, 427
Eisenackia, **395**
Eisenackia scrobiculata, **398**, 398
Eisenackidium, **210**
Elaeophyton, 841
Elektoriskos, 160, **210**
Elektoriskos aurora, **160**, 160
Elenia, **955**
Elenia aculeata, **934**, 934
Elenia famena, **934**, 934
elevation, **585**, 967
Eliasum, **210**
Elliott, G. F., 135, 851, 854, 864, 866, 869, 875
Ellipsagelosphaera, 694, **783**
Ellipsagelosphaera britannica, 738, 749, 770

Ellipsagelosphaera communis, 735, **738**, 738, 749, 768
Ellipsagelosphaera lucasi, **769**
Ellipsagelosphaeraceae, 782
Ellipsaletes, **210**
Ellipsochiastus quadriserratus, **741**, 749
Ellipsoidictyum, **395**
Ellipsoidinium, **395**
Ellipsoidion, **498**
Ellipsoidomorphida, 207
Ellipsolithus, **784**
Ellipsolithus macellus, **757**, 757, 763
Ellipsoplacolithus, **777**
Ellisiella, **526**
Ellobiophyceae, 381
Elson, K. G. R., 620
Emberger, L., 496, 868
Embergerella, 933, **956**
Embergerella cruciata, **943**, 943
Emerson, R., 382
Emerson, S. U., 15
Emmetrocysta, **419**
Emmetrocysta urnaformis, **421**, 421
Emig, W. H., 55, 496, 833
Emiliania, **777**
Emiliania annula, **761**, 764, 766
Emiliania huxleyi, 679, **681**, 681, 687, 688, 689, 692, 699, 710, 712, 716, 717, 719, 720, 721, 723, 724, 761, 764, 766, 769
    cold water phenotype, 681
    warm water phenotype, 681
Emiliania ovata, **761**, 766
Emslandia, **395**
Encrusta, 83
endedaphic, **57**, 967
Endictya, 637, **655**
endoblast, **284**, 967
Endoceratiaceae, 424
Endoceratium, **424**
Endocladia, **140**
Endocladiaceae, 140
endocoel, **283**, 967
endocorpus, **284**, 967
endocyst, **284**, 286, **967**, 969
Endodinium, 313, **380**
Endodinium chattonii, **232**, 233, 241, 244, 314, 315
endolithic algae, **57**, 62, 836, **967**
Endonema, 82
Endonemataceae, 82
endophragm, 152, **281**, 284, **967**
Endoscriniaceae, 430
Endoscrinium, **430**
endospore, **16**, 17, 48, 278, **967**
endospory, 868, **967**
endosymbionts
    blue-green algae, 62
    dinoflagellates, 313
endosymbiotic origin
    of mitochondria, 91
    of plastids, 91
Engelhardt, D. W., 364
Energlynia, **404**
Ensiculifera, 301, 302, 327, **374**
Ensiculifera loeblichii, **239**
Ensiculifera mexicana, **279**, 302, 386

Enterobacteriaceae, 40, 58
Enteromorpha, 834, **889**
Entodesmis, **519**
Entogonia, 587, *588*, **657**
Entogonia davyana v. *orbicularis*, **589**, *589*
Entomosigma, **378**
Entomosigma peridinioides, **379**, *379*
Entomosigmaceae, 378
Entophysalidaceae, 81
Entophysalis, 81
Entophysalites, 81
Entopyla, **658**
Entosphaeroides, 69, **82**
Entosphaeroides amplus, **70**, *70*
Entz, G., 233, 274, 275, 330, 335
Eoastrion, 34, **40**
Eoastrion bifurcatum, **35**, *35*
Eoastrion simplex, **35**, *35*
Eobacterium, **41**
Eobacterium isolatum, **32**, *32*
Eochara, 919, 933, 953, **955**
Eochara wickendeni, **941**, *941*
Eocharaceae, 955
Eocladopyxis, **419**
Eodasycladus, **893**
Eodasycladus ogilviae, **872**, *872*
Eodinia, **403**
Eodinia pachytheca, **283**, *283*, 290, *403*
Eodiniaceae, 403
Eodiscoaster, **779**
Eoentophysalis, 69, **81**
Eogoniolina, **893**
Eomicrhystridium, **210**
Eomycetopsis, 69
Eomycetopsis filiformis, **73**, *73*
Eomycetopsis robusta, **77**, *77*
Eosphaera, 94, **138**, 835
Eosphaera tyleri, **128**, *129*
Eosynechococcus, 69, **81**
Eotetrahedrion princeps, 128, **130**, *130*
Eothrix, 843
Eoumbella, **955**
Eoumbella ollaria, **934**, *934*
Eovelebitella, 868, **893**
Eovelebitella occitanica, **870**, *870*
Eovolvox, 835, **888**
Eovolvox silesiensis, **835**, *835*
Eozoon, 10
Eozygion, 69, 74, **81**
epedaphic, **57**, *968*
Epelidosphaeridia, **403**
Ephelopalla, **210**
epicingulum, **576**, *966*
epicone, 226, **230**
epicyst, 284, **968**
epicystal archeopyle, *296*, 297
epigenetic sulfur, 27
Epimastopora, **891**
Epimastopora malaysiana, **854**, *854*
epipelic, **621**, *968*
Epiphytaceae, 141
epiphytic diatoms, 607
Epiphyton, 141
Epiplosphaera, **419**
Epipolaia, 841
epipsammon, **621**, *968*

Epipyxis, **500**
Epistacheoides, **142**
epithallium, **117**, *968*
epitheca, **230**, 263, 297, **573**, *968*
Epithelion russicum, **639**
Epithemia, 626, 628, **659**
Epithemia cistula lunaris, **581**, *581*
Epithemia turgida, **582**, *582*
Epithemiaceae, 659
epitract, **284**, 297, *968*
epityche, **176**, *968*
epivalve, **576**, *968*, *976*
Epivalvia, 83
Eppley, R. W., 239, 306, 342, 596
Eprolithaceae, 780
Eprolithus, **780**
Eprolithus floralis, **740**, *740*, *741*
equatorial process, acritarchs, 167
equatorial rupture, 176
Erdtman, G., 275, 350, 839
Eremosphaera, 841, 842, **889**
Eren, J., 240, 348
Ericanthea, **210**
Ericsonia, **784**
Ericsonia subdisticha, **757**, *762*
Ericsonia subpertusa, **771**, *771*
Erikania, **430**
Ernst, W. G., 636
Erythrocladia, **139**
Erythropeltidaceae, 139
Erythropsis, **380**
Erythropsidinium, 231, 243, **380**
Erythropsidinium agile, **250**
Erythropsidinium pavillardii, **243**, *243*, 310
Erythrotrichia, 116, 126, **139**
Escherichia, 11, 30, **41**, 93
Escherichia coli, 14, 15, 17, 18, 31, 37, 44, 45, 51, 234
Esgia, **783**
Estiastra, **210**
Estiastra stellata, **201**, *201*
Ethelia, **140**, *141*
Ethmodiscus, 576, 628, 630, **655**
Ethmodiscus rex, **568**, 609, 624, 628, *629*, *631*
Ethmorhabdoideae, 781
Ethmorhabdus, **781**
Ethmosphaeridium, **210**
Euastrum, 887, **894**
Eubacteria, 16, 24, 40
Eubacteriales, 11, 40
Eubacterium, **42**
Eucampia, 622, **657**
Eucampia balaustium, **624**
Eucapsis, 69, 80
eucaryotes, origin of, 84—96
  autogenous theories, 85
    cluster-clone hypothesis, 88
    coenocyte hypothesis, 87
    invagination of surface hypothesis, 86, *87*
    phagocytosis hypothesis, 88, *89*
    surface protrusion hypothesis, 86, *86*
  endosymbiotic theory, 89
  time of origin, 93, 95
eucaryotes, Precambrian, 93

Eucladinium, **392**
Eucommiidites minor, **361**
Eudorina, 835
Eudorina elegans, **835**, *835*
Euglena, 91, 92, 234, 826, 895, 896, **898**, 964
Euglena gracilis, **91**, 896
Euglena mesnili, **895**, *895*
Euglenales, 898
Euglenamorpha, **898**
Euglenamorphales, **898**
Euglenophyta, 804, 895, **898**
  cell, 895
  classification, 897
  fossils, 896
  nutrition, 896
  occurrence, 896
  pigments, 896
  reproduction, 896
Eugonophyllum, **890**
Eunaviculopsis, **559**
Eunaviculopsis navicula, 550, 551, **552**, *559*
Eunotia, 579, 596, 600, 607, 626, **658**
Eunotiaceae, 658
Eunotiopsis, **657**
Eunotogramma, **657**
Eunotogramma weissei, **587**, *587*, 588
Euphausia pacifica, parasite of, 382
Eupodiscaceae, 656
Eupodiscales, 655
Eupodiscus, **656**
Eupoikilofusa, 200, **210**
Eurhabdus, **778**
Eurydinium, **392**
Eurydinium ingrami, **391**, *391*
euspondyl, **868**, *968*
Eustigmatophyceae, 490, 491, 496, 498
  cell, 497
Eusyracosphaera, **783**
Eutreptia, **898**
Eutreptiales, 898
Eutreptiella, **898**
Eutrochiliscus, **955**
eutrophic lakes, 930
eutrophic water, 604, *968*
eutrophication, 46
Evensen, D. L., 593
Evitt, W. R., 148, 149, 155, 171, 182, 205, 230, 243, 265, 272, 273, 275, 276, *276*, 278, 281, 284, 285, 286, 288, 293, 294, 295, 296, 297, 298, 323, 335, 351, 353, *353*, 354, 355, 364, 383, 385, 389, 401, 411, 427, 428, 430, 839
Evitt, W. R. (Downie et al.), 149, 205
Evittia, **210**
Evittia cymosa, **157**, *157*, 195
Evittodinium, **392**
evolution
  Acritarcha, 197
  bacteria, 28
  blue-green algae, 80
  Charophyta, 939, 951
  diatoms, 652
  dinoflagellates, 364
  ebridians, 476
  Haptophyta, 771

evolution (*continued*)
 Rhodophyta, 136
 silicoflagellates, 553
evolutionary succession, 6, **7**
Exanthemachrysis, 715, **784**
Excultibrachium, **210**
Excultibrachium concinnum, **167**, *167*
excystment openings
 acritarchs, 173
 chrysomonads, 506
 dinoflagellates, 293
 Prasinophyta, *808, 810, 812*
Exilisphaeridium, **210**
Exochoderma, **210**
Exochosphaeridiaceae, 409
Exochosphaeridium, **409**
*Exochosphaeridium pseudhystrichodi-
 nium*, **416**, *416*
exospore, **278, 968**
external metabolites, blue-green algae, 55
extinctions
 acritarchs, 190
 calcareous nannoplankton, 771
 dinoflagellates, 358
exuviation, dinoflagellates, **327, 968**
Exvotarisella, **892**
Eyachia, **430**
Eyer, J. A., 921, 941
eyespots, **968, 975**
 dinoflagellates, 242
 euglenids, 895
 Eustigmatophyceae, 497, *499*
 Haptophyta, **681**
 Xanthophyceae, 493
Eyrea, **430**
Ezo, **140**

facultative anaerobes, 22
Faegri, K., 154
Fairchild, T. R., 843
false branching, **47**, *48*, **68, 968**
Falsebria, 476, 480, 482
Falsebria ambigua, **482**, *482*
Falsebria ambigua f. *coralloidea*, **482**,
 482
Famintzen, A., 89
Fanesella, **893**
Fanesella dolomitica, **872**, *872*
Fankboner, P. V., 313, 315
Farinacci, A., 881
fascia, **587, 968**
Fasciculithaceae, 779
Fasciculithus, **779**
Fasciculithus tympaniformis, 752, **757**,
 763
Favocentrum, **781**
Favososphaera, **210**
Favososphaeridium, **210**
Favre, J., 881
Fay, M., 130, 131
Fay, P., 49
Fay, P. (Fogg et al.), 45, 46, 51, 57, 92
fecal pellets, 36
 coccoliths in, 769
Feist-Castel, M., 921, *922, 950, 953*
Feldman, J., 850
Felluga, B. (Bonotta et al.), 826

female conceptacles, 121
Fenestrella, **656**
Fenical, W., 109
Fenner, J., 639, 641, 645, 647
Fenwick, M. G., 62
Ferganella, **893**
Ferganelleae, 893
fermentation, 20
fermentation efficiency, 29
Fermoria, **818**
Ferrimonilis, 39
Ferrobacillus, 24, 41
Feys, R., 841
Fibradinium, **395**
Fibradinium annetorpense, **398**, *398*
Fibrocysta lappacea, **289**, *289, 290, 416*
Fibrocysta radiata, **416**, *416*
fibulae, **579, 968**
filament, **12, 43, 968**
Filamentella, **83**
Filfilan, S. A., 232
Filiconstrictosus, 74, **82**
Filipescu, M. G., 547, 548
Filisphaeridium, **210**
Filodinium hovassei, **309**
filterable virus, 18
Fimbriaglomerella, 152, 176, **210**
Findlay, M., 316
Fine, K. E., 240, 279, 311
Fiore, M., 848
Fischer, A., 828, 884
Fischer, A. G., 774, 775
Fischer, J. C., 872
Fisher, D. W., 181, 182
Fisher, M. J., 353, *357, 389, 390*
Fisherites, **891**
fission, transverse, 11
Flabellaria, **890**
Flabellites, **781**
Flabellochara, 933, **956**
Flabellochara harrisii, **928**, *928*
flagella, origin of, 93
flagellar markings, dinoflagellate
 cysts, **290**, *292*, **385, 968**
flagellar structure, bacteria, **12, 15**, *15*
flagellar structure, eucaryotes, **15**, *15*
 Chlorophyta, 826
 Chrysophyta, 501, 510
 dinoflagellates, 247, 249
 movement of, 250, 302
 ebridians, 464
 Haptophyta, 683
 Prasinophyta, 805
 Xanthophyta, 493, 497, *499*
Fliche, P., 77
Flintiella, **138**
Florentinia, **419**
Florentinia clavigera, **423**, *423*
Florentinia deanei, **423**, *423*
Florentinia laciniata, **423**, *423*
Florentiniaceae, 419
Florideophycidae, 107, 108, 112, 113,
 116, 139
 life history, 124
 morphologic phases, 116 – 125
 thallus, 116
flotational adaptations, diatoms, 607
Florisphaera, **784**

Florisphaera profunda, **717**
Florisphaeridium, **210**
Flügel, E., 761
Fogg, G. E., 45, 46, 51, 57, 92, 614, 615,
 618, 620, 833, 841
Fogg, G. E. (Fay et al.), 49
Foliactiniscus, 253, **386**
Foliactiniscus folia, **358**, *358, 387*
Folk, H. (Brown et al.), 684
Folk, R. L. (Fairchild et al.), 843
food chains, 5, 21
Forchheimer, S., 697, *741, 745, 746,
 776*
Ford, J. E., 725
Ford, T. D., 154
Foreliella, 845, **889**
Forti, A., 637
Fosliella, **140**
fossil record
 Acritarcha, 185
 bacteria, 31
 blue-green algae (Cyanophyta), 63
 Charophyta, 931, *933*
 Chlorophyta, 831, 833
 dasyclads, 864
 Chrysophyta, 510
 diatoms, 636
 dinoflagellates, 349, 351
 abundance, 349
 ebridians, 474, 476
 Euglenophyta, 896
 Haptophyta
 calcareous nannoplankton, 725,
 727
 fossil holococcoliths, 752
 Nannoconus, 731
 Prasinophyta, 813
 Rhodophyta, 128
 silicoflagellates, 547
 umbellinaceans, 932
fossulate, **155, 968**
Foster, J. H., *550, 551*
Fott, B., 226, 258, 502, 506, *507*
Foucher, J. C., 403, 406, 408, 413, 415,
 416, 419, 421, 422, 423, 433, 435
Fournier, R. O., 725
Fourstonella, **142**
Fourstonella fusiformis, **133**, *133*
Foveofusa, **210**
foveolate, **155, 968**
Fox, G. E., 31
Fragilaria, 572, 601, 607, 609, 622, 624,
 626, 641, 649, **658**
Fragilaria brevistriata, **649**
Fragilaria crotonensis, **649**
Fragilariales, 657
Fragilariella, **658**
Fragilariopsis, **659**
Fragilariopsis antarctica, **624**
Fragilariopsis curta, **624**
Fragilariopsis cylindrus, **624**
Fragilariopsis kerguelensis, **625**
Fragilariopsis sublinearis, **616**
Fragilidium, 265, **378**
Franke, W. W., 684
Franke, W. W. (Brown et al.), 684
Frankea, **210**
Frankiaceae, 42

Franz, H. E., 761
Frederica, 869, 893
Frederica villiersi, 877, 877
Frenguelli, G., 623, 637, 639, 648, 649
Frenguelli, J., 467, 471, 529, 541, 559, 641
freshwater occurrence
  blue-green algae, 57
  Chrysophyta, 510
  coccolithophores, 721
  diatoms, 626
  dinoflagellates, 348
  fossil diatoms, 641
  fossil dinoflagellates, 365
Fresnel-Morange, J., 714, 747, 777
Freudenthal, H. D., 315
Frey, D. G., 490
Freytet, P., 65
Frickia, 658
Friedmann, I., 57, 836
Friedlander, C. G. I. (Dombrowski et al.), 37
Fries, M., 276
Fritsch, F. E., 649, 654, 804, 836, 862, 913
Fritschiella, 889
Fromea, 395
Fromea amphora, 403, 403
Fromeaceae, 395
frustule, diatoms, 586, 572, 968
Frustulia, 626, 658
Fryxell, G. A., 581, 592
Fryxell, G. A. (Gaarder et al.), 512, 512
Fucus, 371, 496, 831
Fueloepia, 210
Fütterer, D., 279, 280, 708, 747
Fuge, D. P. (Long et al.), 637
Fuji, N. (Ichikawa et al.), 545
Fuller, G. (Patton et al.), 339
fultoportula, 975
fungal hyphae, 94
Furcatolithus, 778
Furcellaria, 141
Furcellariaceae, 141
Furcoporella, 894
Fusilites, 176, 210
Fusomorphida, 207

Gaarder, K. R., 512, 512, 613, 695, 717, 723, 725, 750, 750, 752, 782
Gaarder, K. R. (Braarud et al.), 726
Gaarderella, 782
Gabriel, M. L., 88
Galaxaura, 139, 139
Galliher, E. W., 110, 112
Gallionella, 13, 24, 34, 36, 40
Gallionella pyritica, 36, 36
Gallois, R. W., 770
gametangia, Charophyta, 915
gamete formation, diatoms, 597
gametes, dinoflagellates, 337
gametophyte, 116, 968
Gantt, E., 109, 110
Gardner, J. V., 629
Gardodinium, 409
Gardodinium trabeculosum, 413, 413
Garrett, P., 59
Garstang, W., 604

Gartner, S., Jr., 697, 706, 726, 727, 728, 745, 750, 751, 752, 757, 759, 761, 766, 770, 776
Gartnerago, 782
Gartnerago obliquum, 745, 746, 748
Garwood, E. J., 66, 847
Garwoodia, 83, 890
Garwoodia gregaria, 66, 66, 847, 847
Garwoodiaceae, 890
gas vesicles, 46
Gasse, F., 581
Gaudsmith, J. T., 237
Gawlik, S. R., 828, 839
Gayral, P., 714, 747, 777
Geehan, G. W. (Adelseck et al.), 769
Geiselodinium, 389, 392
Geitler, L., 92, 596, 597
Gelidiaceae, 140
Gelidium, 108, 110, 140
Gemeinhardt, K., 541, 545, 546, 559, 623
Gemeridella, 890
Geminella, 843, 889
gemmate, 157, 968
generic diversity, through geologic time
  Acritarcha, 184
  calcareous nannoplankton, 773
  Charophyta, 933
  dinoflagellates, 357
  ebridians, 476
  silicoflagellates, 547, 549
genetic transfer, bacteria, 17
genetic transfer, blue-green algae, 51
genicula, 111, 117, 968, 971
Genicularia, 882
genome, 968
genophore, bacterial, 12, 14, 17, 43, 968, 971
Gentile, R., 727, 728
geologic occurrence. *See* fossil record
George, E. A., 77, 79
Gephyria, 658
Gephyrocapsa, 777
Gephyrocapsa caribbeanica, 761, 766, 770
Gephyrocapsa ericsonii, 719
Gephyrocapsa oceanica, 710, 717, 719, 761, 764, 766, 769, 770
Gephyrocapsaceae, 777
Gephyrorhabdus, 781
Geppella, 890
Gerlach, E., 433
Gerloff, J., 512
German, T. N., 205
germination, of charophyte zygospore, 927, 929
Geroch, S., 637
Geron, 171, 210
Geron guerillerus, 173, 173
Geronidae, 208
Gerrath, J. F., 349
Gessner, F., 649
Gessnerium, 317, 373
Gessnerium acatenellum, 318, 339, 340
Gessnerium catenellum, 232, 236, 265, 310, 317, 326, 339
Gessnerium fraterculum, 325, 325

Gessnerium monilatum, 317, 317, 319, 326
Gessnerium tamarense, 232, 233, 236, 237, 244, 317, 318, 339
Getaeia, 139
Gharagozlou, I., 93
Gibbons, N. E., 38
Gibbs, S. P., 108
Gigartina, 110, 142
Gigartinaceae, 142
Gigartinales, 135, 141
Ginginodinium, 290, 389
Ginginodinium ornatum, 385, 385
Ginsburg, R. N., 127, 275
Ginsburg, R. N. (Logan et al.), 65
Girvanella, 65, 83
Girvanella problematica, 66, 66
girdle, diatoms, 576, 966, 968
girdle, dinoflagellate, 226, 227, 263, 966, 968
girdle lamellae, 493, 968
girdle lists, 271
Gitmez, G. U., 398
Gittleson, S. M., 304, 306
Gladius, 637, 656
Glaessner, M. F., 65
Glaphrocysta, 424
Glaphrocysta ordinata, 362, 362
Glaphrocysta retiintexta, 362, 362
Glaucocystis, 62, 63, 92, 836, 889
Glaucocystis nostochinearum, 62
Glauert, A. M., 14
Glaukolithus, 780
Glenn, D. A., 604
Glenobotrydion, 69, 74, 81, 128, 138, 838, 888
Glenobotrydion aenigmatis, 77, 77
Glenobotrydion majorinum, 73, 73
Glenodiniaceae, 374
Glenodiniopsaceae, 371
Glenodiniopsis, 371
Glenodinium, 325, 374
Glenodinium danicum, 274
Glenodinium pulvisculus, 330
Glezer, Z. I., 539, 547, 548, 637, 639, 658
gliding motion
  blue-green algae, 52
  diatoms, 610
  red algae, 113
global ecosystem, 7
Globator, 933, 956
Globator trochiliscoides, 928, 928
Globigerinoides sacculifer, symbionts of, 315
Globochaete, 843
Globophycus, 69, 74, 81, 838, 888
Globophycus rugosum, 77, 77
Globuliferoporella, 891
Globuloella, 83
Gloeocapsa, 46, 54, 57, 81
Gloeocapsa alpicola, 81
Gloeocapsomorpha, 77, 149, 841
Gloeocapsomorpha prisca, 79, 79, 200
Gloeochaete, 92
Gloeocystis, 888
Gloeocystis oxfordiensis, 836
Gloeodiniopsis, 74, 81

Gloeodiniopsis lamellosa, **77,** *77*
Gloeodiniaceae, 381
Gloeodiniales, 381
Gloeodinium, **381**
Gloeothamnion, 714
Gloeotrichata, 83
Gloeotrichia, 83
Gloioconis, 81
Gloiosiphonia, 140
Gloiosiphoniaceae, 140
Glomodinium, **386**
Glorioptychus, **656**
Glottimorpha, **210**
Glukhovskaya, N. B., 939
Glyphanodinium, **395**
Glyphanodinium facetum, **401,** *401*
Glyphodesmis, **658**
Gobichara, 933, **956**
Gobichara deserta, **950,** *950*
Gocht, H., 278, 279, 281, 283, 285, 288,
    290, *290,* 292, *292, 293, 294, 297*
    301, 351, 353, 355, 383, 403, 415,
    433, 837
Gochtia, **210**
Godjics, M. (Honigberg et al.), 226
Goksøyr, J., 87, 89, 90
Goksuella, **868, 892**
Gold, K., *511*
Golgi body, Haptophyta, 681, **968**
Golovenkina, N. I., 648
Golubic, S., 44, 62, 68, 74, 80, 94
Golubic, S. (Awramik et al.), 94
Golubic, S. (Knoll et al.), 74, *74*
Gombos, A. M., Jr., 639, 641, 645, 647
Gomontiella, **82**
Gomontiellaceae, 82
Gomphocymbella, **658**
Gomphoneis, **659**
Gomphonema, 572, 607, 610, 624, 628,
    649, **659**
Gomphonema olivaceum, **572**
Gomphonema parvula, **571,** *571*
Gomphonemaceae, 659
Gomphonitzschia, **659**
Gonamophyton, 77, **81**
gonidia, **828,** 834, **968**
Goniodoma, **373**
Goniodoma acuminatum, 272
Goniodoma pseudogonyaulax, **330**
Goniolina, 859, **893**
Goniolina geometrica, **874,** *874*
Goniolinites, 881, **894**
Goniolinites hemisphaerica, **881**
Goniolithaceae, 779
Goniolithon, 110, **140**
Goniolithus, **779**
Goniolithus fluckigeri, **757,** *757*
Goniolopadion, **210**
Goniophyllum, 67
Goniosphaeridium, 152, 184, 206, **210**
Goniosphaeridium balticum, **152**
Goniosphaeridium polygonale, **160**
Goniothecium, 600, **657**
Goniothecium tenue, **645**
Goniotrichaceae, 139
Goniotrichales, 138
Goniotrichopsis, **139**
Goniotrichum, **139**

Gonyaulacaceae, 233, 239, 371
Gonyaulacysta, 283, 290, 292, 294, 296,
    392
Gonyaulacysta cassidata, **395,** *395*
Gonyaulacysta jurassica, **362**
Gonyaulacysta wetzelii, **293,** *294,* **395,**
    395
Gonyaulacystaceae, 392
Gonyaulax, 231, 264, 265, 274, 275, 286,
    299, 302, 316, 317, 319, 320, 324,
    326, 337, 338, 339, 340, 341, 343,
    344, 345, 348, 368, 371, **373,** 383,
    385, 395, 404
Gonyaulax catenata, **317,** 325
Gonyaulax digitalis, **276, 283,** 284, 293,
    299, 317, 364, 364, 404
Gonyaulax grindleyi, **237,** 239, 274,
    283, 299, 302, 350, 424
Gonyaulax hyalina, **317,** 346
Gonyaulax pacifica, **254,** 346
Gonyaulax polyedra, **228,** *228,* 234,
    236, 237, 239, 244, 247, 258, 263,
    264, 265, 266, 267, 275, 295, 299,
    301, 306, 316, *316,* 317, 319, 321,
    322, *322,* 323, 324, 327, 334, 339,
    342, 350, 424
Gonyaulax polygramma, **233,** 317, 324,
    346
Gonyaulax scrippsae, **299,** 404
Gonyaulax series, **326**
Gonyaulax sphaeroidea, **317,** 319
Gonyaulax spinifera, **272,** *283,* 299,
    301, 317, 325, 337, 346, 385, 404
Gordon, R., 612
Gordon, R. O. (Schopf et al.), 32,
    34
Goreau, T. F., 314, 831
Gorgonisphaeridium, **210**
Gorgonisphaeridium ohioense, **203,**
    203
Gorgonisphaeridium plerispino-
    sum, **160,** *160*
Gorham, P. R., 58
Górka, H., 202, 745, *822*
Goryniella, **210**
Gossleriella, **656**
Gourret, P., 366
Gracilaria, 110, **141**
Gracilariaceae, 141
Gracilisphaeridium, **210**
Gracilisphaeridium encantador, **165,**
    165
Graham, A., 852, *858*
Graham, H. W., 258, 261, 263, 264,
    325, 326, 344, 345, 346, 367
Gram reaction, **14, 968**
Gram-negative bacteria, 14
    reproduction, 17
    wall structure, 13, *13*
Gram-positive bacteria, 14, 80
    reproduction, 17
    wall structure, 13, *14*
Grambast, L., 924, 925, 926, 927, *928,*
    932, *936, 942, 943, 944,* 944, *945,*
    947, *947,* 948, *949, 953,* 953, 957
Grambastichara, 925, 939, **956**
Grambastiella, **956**
Grammatophora, 622, **658**

Grammatophora marina, **569,** *569,*
    577, 606, 653
Gran, H. H., 607, 623, 641, 770
Granodiscus, **818**
Granomarginata, **210**
Grant, B. R., 617, 618
Grant, W. M., 639, 648, 649
Grantarhabdus, **781**
granulate, **155, 968**
Grateloupia, **141**
Gray, J., 202, 204, 837, 839, 841
Greber, C. (Feys et al.), 841
Grebespora, **210**
Green, B. R., 826
Green, J. C., 681, *683,* 683, 714, 715,
    716, 776
green algae. *See* Chlorophyta
green bacteria, 19
Gregory, G. S. (Conkin et al.), 924, 939
Grell, K. G., 226, 231
Gressel, J., 137, 234
Greuet, C., 231, 243, *243,* 244, 247, *247,*
    310
Greuet, C. (Cachon et al.), 244
Grice, G. D., 62
Griffithsia, 108, 110, **142**
Griffithsia pacifica, 137
Griphoporella, **893**
Groentvedia, **655**
Gromia, 309
Gross, F., 335, 597, 601, *603,* 603, 604,
    609
Grove, E., 637
Grovesichara, 925, 933, **956**
Grovesichara distorta, **953,** *953*
Grün, W., 729, 738, 740, 741, *741,* 776
Grüss, J., 843
Gruner, J. W. (Cloud et al.), 69
Grunowiella, **658**
Grypania, **139**
Grypania spiralis, **132,** *132*
guanine + cytosine (G + C) per-
    cent, 92
Gümbel, C. W., 761
Guillard, R. R. L., 571, 596, 604, 605,
    614, 615, 616, 619, 621, 623, 632,
    655
Guillard, R. L. (Wall et al.), 275, 279,
    280
Guinardia, **657**
Guinardia flaccida, **598**
Gunflintia, 69, 80, 83
Gunflintia grandis, **71,** *71*
Gunflintia minuta, **71,** *71*
Guttatisphaeridium, **210**
Guttatisphaeridium pandum, **203,** *203*
Gutwinskiella, **656**
Gyalorhethium, **210**
Gymnaster, 252, **378**
Gymnocodiaceae, 133, 135, 139
Gymnocodium, **139**
Gymnodiniaceae, 233, 239, 294, 378
Gymnodiniales, 233, 351, 368, 378, 386
Gymnodinium, 231, 286, 292, 294, 306,
    308, 310, 313, 315, 317, 319, 332,
    337, 338, 341, 342, 347, 348, 350,
    366, **378,** 380, 479, 510
Gymnodinium aeruginosum, **239**

Gymnodinium arenicolus, **347**
Gymnodinium breve, **317,** 318, 341, 344
Gymnodinium catenatum, **326**
Gymnodinium coeruleum, **239**
Gymnodinium dodgei, **233**
Gymnodinium flavum, **317**
Gymnodinium fucorum, 273
Gymnodinium fungiforme, **308**
Gymnodinium fuscum, **231,** 244
Gymnodinium hippocastanum, **275**
Gymnodinium micrum, **236,** 240, 241, 244, 245, 249, 249
Gymnodinium nelsonii, **234,** 244, 247, 339
Gymnodinium pigmentosum, 241
Gymnodinium pseudopalustre, **319,** 329, 331, 334
Gymnodinium rubrum, **335**
Gymnodinium simplex, **229,** 230, 237, 239, 241
Gymnodinium splendens, **239,** 319
Gymnodinium veneficum, **240,** 366
Gymnodinium vitiligo, 233, 240
Gymnodinium vorticella, **272**
Gymnodinium zachariasii, **233**
Gymnogongrus, **142**
Gymnophlaeaceae, 141
Gymnosclerotaceae, 254
Gyratosphaerina, **210**
Gyrodinium, 231, 306, 307, 309, 337, 338, 341, 342, 344, 348, 371, **378,** 380, 479
Gyrodinium aureolum, **341,** 342
Gyrodinium dorsum, **237,** 239, 307
Gyrodinium pavillardii, **258,** 258, 309, 330
Gyrodinium resplendens, **234**
Gyrodinium rubrum, 335
Gyrodinium vorax, **309**
Gyrodiscoaster, 754, **778**
Gyrodiscus, **656**
Gyrogona, 919, 924, 925, 926, 931, 932, 933, **956**
Gyrogona medicaginula, **925,** 925
gyrogonite, charophyte, **919,** 968, **970**
  development, *921*
  wall structure, *920*
Gyrogonites, 919, **956**
Gyrogonites medicaginula, 932
Gyroporella, 868, **893**
Gyrosigma, 626, **658**

Haapala, O. K., 235
Habib, D., *396,* 413, *819*
Hadziella, 881, **894**
Hadziina, 881, **894**
Haeckel, E., 343, *545*
Haematococcaceae, 888
Haematococcus, 55, 833, 835, **888**
Hafniasphaera, **404**
Hagen, H., 69, *129*
Hagen, H. (Cloud et al.), 69
Hagen, J. (Lee et al.), 835
hair cell, red algae, **113, 968, 976**
Hajós, M., 479, 508, 512, *514, 515,* 637, *639,* 639
half-bands, diatom frustule, **968**

Halfen, L. N., 52
Halicalyptra, **559,** *559*
Halicalyptra virginica, **559,** *559*
Halicki, P. J. (Coombs et al.), 573, 618
Halicoryne, 826, *862,* 862, **894**
Halicystis, 827, 829, 850, **891**
Halim, Y., 338, 474
Halimeda, 112, 826, 831, 832, 834, 848, 890
Halimeda gracilis, *832,* 832
Haliptylon, **141**
Hall, J. B., 29, 31, 85, 91
Hall, R. P. (Honigberg et al.), 226
Halldal, P., *696*
Halobacteriaceae, 40
Halopappaceae, 784
Halopappus, **784**
Halophoridia, **395**
Halosphaera, 185, 805, 808, 812, 814, 818
Halosphaera, parasites of, 309
Halosphaera viridis, **812**
Halosphaeraceae, 818
Halosphaeropsis, **818**
Halymenia, **141**
Halythrix, 69, 74, **85**
Hamulusella, **894**
Hand, C., 314
Hand, W. G., 306, 307
Hanna, G. D., *556, 559,* 560, 633, 637, 639, 648, 649
Hannaea, **658**
Hannaites, **542,** *545*
Hantschel, W., 839
Hantzschia, 605, 626, **659**
Hantzschia virgata v. *intermedia,* **581,** *581*
Hapalophloea, **139**
Hapalosiphon, 46, 47, **83**
Haplobacteriacei, 40
haplobiontic, *829,* **968**
Haplocystia, **818**
Haplocystia pellucida, **811,** *811*
Haplohermesinum, **480**
haploid, **16, 969**
Haplozooaceae, 380
Happey, C. M. (Crawford et al.), 258
Hapsidaulax, **424**
Hapsidopalla, **210**
haptonema, **679,** *679,* **969**
  function of, 683
Haptophyceae, 491, 776
Haptophyta, 491, 678–803
  cell, 679
  cell movement, 683
  classification, 776
  coccoliths, 684
  ecology, 715
  eyespots, 681
  flagella, 683
  fossil record, 725, 727
  Golgi body, 681
  haptonema, 683
  life cycle, 710
  nucleus, 679
  plastids, 681
  pigments, 681
  reproduction, 710

reticular body, 688
scales, 679, 681, 684, *685,* 691
Haq, B. ul, 464, *708,* 726, *754,* 761, 764, 770, 771, 774, 774, 775, 775, 776
Harada, K. (Wall et al.), 278, 279, 289, *290, 301, 396, 398*
Harbour, D. S. (Parke et al.), 808, *810*
Hargraves, P. E., 465, 474, *587,* 601, 603
Harker, S. D., 351
Harland, R., 204, 344, 353, 358, 364, 383, *390*
Harlanjohnsonella, **893**
Harlanjohnsonella annulata, **875,** *875*
Harmon, W. M. (Jahn et al.), 250, 304
Harper, J. F., 611, *612,* 612
Harper, M. A., 610, 611, *612,* 612
Harris, T. M., *928*
Harris, W. K., 364
Harrisichara, 922, 926, 933, **956**
Harrisichara vasiformis, **953,** *953*
Hart, T. J., 340
Hartog-Adams, I. den, 808
Harveyella, **141**
Haskell, T. R., 353
Hasle, G. R., 581, *582,* 585, *587,* 592, *593,* 613, 653, 655, 719, 725
Hasle, G. R. (Gaarder et al.), *512,* 512
Hasle, G. R. (Ross et al.), *587,* 589, 592, *594*
Haslingfieldia, **780**
Hastigerina pelagica, symbionts of, 315, 316
Hastings, J. W., 317, 319
Hattin, D. E., 761, 769
Hauge, H. V., 275, 276, 335, 337
Haworth, E. Y., 649
Hay, W. W., 745, 748, 776
Hay, W. W. (Rood et al.), 735, *737, 738,* 776
Hayaster, **778**
Hayella, **783**
Hayella situliformis, **757,** 762
Hayesites, **780**
Hayesites albiensis, **745,** *745,* 748
Haynaldia, **657**
Hays, J. D., *629,* 751
Haxo, F. T. (Jeffrey et al.), 239, 240
Healey, F. P., 571
Heath, I. B., 597, 598, 607
Hebecysta, 353, **392**
Hebecysta brevicornuta, **357,** *357*
Heck, S. E. van, 776
Hedlund, R. W., 297
Hedstroemia, 83, **890**
Heibergella, 353, 367, **386**
Heibergella asymmetrica, **389,** *389*
Heimdal, B. R., 342, 581, *750,* 750, 752
Heimiella, **777**
Helenea, **781**
Helenoideae, 781
Helgolandinium, 334, 337, **378**
Helgolandinium subglobosum, **320,** 377
Heliaktis, **519**
helicolith coccolith, **703, 969**
Helicolithus, **779**
Helicolithus trabeculatus, **692,** *692*

Heliconema, 74, **82**
Heliconema australiense, **77,** *77*
Helicopontosphaera, **781**
Helicosphaera, 776, **781**
Helicosphaera ampliaperta, **761,** *764,*
  767
Helicosphaera carteri, **703,** *703, 703,*
  719, 761, 767
Helicosphaera compacta, **757,** *762*
Helicosphaera hyalina, **718,** *718*
Helicosphaera reticulata, **757,** *761, 762*
Helicosphaera sellii, **761**
Helicosphaeraceae, 780
Heliodinium, **404**
Heliodiscoaster, 754, **778**
heliolithae, **694,** 969
Heliolithaceae, 778
Heliolithus, **778**
Heliolithus kleinpellii, 752, **757,** *763*
Heliolithus riedeli, **752,** *752, 757, 763*
Heliopeltaceae, 656
Heliorthus, **780**
Helios, **822**
Heliospermopsis, **822**
Helladosphaera, **778**
Hellebust, J. A., 339
Helmcke, J. G., 572, 581
Helminthocladia, **139**
Helminthocladiaceae, 139
Helminthopsis, **658**
Helminthora, **139**
Helosphaeridium, **210**
Hemer, D. O., 203, 838
Hemiaulaceae, 657
Hemiaulus, 510, 585, 601, 607, 613,
  623, 637, **657**
Hemiaulus capitatus, **591,** *594*
Hemiaulus haitensis, **589,** *589*
Hemiaulus inaequilaterus, **647**
Hemiaulus incisus, **647**
Hemiaulus includens, **592,** *592*
Hemiaulus polycystinorum, **646**
Hemiaulus polymorphus, **585,** *587*
Hemibaltisphaeridium, 173, **210**
Hemicystodinium, 302, 371, **419**
Hemicystodinium zoharyi, **299**
Hemideunffia, **210**
Hemidinium, 347, 348, **371, 381**
Hemidinium nasutum, **372,** *372*
Hemidiscaceae, 656
Hemidiscoaster, **778**
Hemidiscus, 639, **656**
Hemidiscus cuneiformis 591, 624, 625,
  632, 650, 651
Hemidiscus hardmannianus, **604**
Hemidiscus karstenii, 645
Hemiplacophora, **419**
Hemipodorhabdus, **781**
Hemisphaera, **785**
Hemisphaerium, **210**
Hemistomiosphaera, 733
Hemmingsen, B. B. (Lauritis et
  al.), 571, 605
Hemplectostomum, 507
Hendey, N. I., 568, 571, 581, 585, 610,
  621, 622, 623, 625, 629, 631, 641,
  654, 655
Hendey, N. I. (Hanna et al.), 639

Hendeya, **656**
Henry, J. L., 187
Hensen, V., 329, 335
Hentschel, E., 255, 338
Herakella, **893**
Hercotheca, **657**
Herdman, E. C., 347, 622
Herendeenia, **392**
Herkomorphitae, **207**
Herman, E. M., 237
Hermesinaceae, 480
Hermesinella, 464, 474, 476, 476, **480**
Hermesinella transversa, **482,** *482*
Hermesinidae, 480
Hermesinopsis, 476, **480**
Hermesinopsis caulleryi, **482,** *482*
Hermesinum, 256, 464, 465, 472, 474,
  476, 476, 477, 479, **480,** 976
  cell division, 473
  skeletal terminology, 467
Hermesinum adriaticum, **464,** *464,*
  465, 468, 469, 474, 474
Hermesinum geminum, **465,** *468*
Hermesinum longispinosum, **480,** *480*
Hermesinum platense, **474**
Hermesinum praesidis, **466,** *466*
Hermesinum schulzii, **465,** *469, 469*
Hérouard, E., 243, 272, 302
Heslertonia, **404**
Heslertoniaceae, 404
Heteraulacaceae, 373
Heteraulacacysta, **404**
Heteraulacacysta campanula, **297, 404,**
  404, 408
Heteraulacacystaceae, 404
Heteraulacus, 265, 320, 324, 337, **373**
Heteraulacus polyedricus, **272,** *272,*
  346
Heterocapsa triquetra, 330
Heterocapsales, 498
Heterochloridales, 498
Heterochloris, **498**
Heterococcales, 498
heterococcoliths, **689,** 694, 695, 969
Heterococcus, **498**
heterocyst, 48, **49,** *49, 49, 50,* 53, 68,
  969
Heteroderma, **140**
Heterodiacrodeae, 207
Heterodictyon, **656**
Heterodiniaceae, 373
Heterodinium, 263, 265, 338, 346, **373**
Heterodinium scrippsi, **373,** *373*
heterogamy, dinoflagellates, 334
Heterogloeales, 498
Heterokontae, 491
Heteromarginatus **780, 782**
Heteromastix, **818**
Heteromorphidion, **210**
Hetermorophidion echinopillum, **195,**
  195
Heteronema, **898**
Heteronematales, 898
Heteroporella, **894**
Heteroporella lepina, **880,** *880, 881*
Heterorhabdus, **781**
Heteroschisma, **371**
Heterosiphonia, 124, **142**

Heterosphaeridium, **419**
heterothallous, **331**
Heterotrichales, 498
heterotrophy, 18, **969**
  in coccolithophores, 725
  in dinoflagellates, 308
Heurck, H. van, 623, 641, 654
Hexabactron, **891**
Hexagonifera, **389**
Hexagoniferaceae, 389
Hexalithus, **779**
Hexangulolithus, **779**
Hexapodorhabdus, **781**
Hexasterias, **818**
Heyneckia, **783**
Hibberd, D. J., 491, *493, 493,* 494, 496,
  497, 499, 501, *504,* 506, 681, 715, 776
Hidasia, **822**
Hikorocodium, **890**
Hildenbrandia, 124, **140,** 140
Hildenbrandiaceae, 140
Hill, M. E., III, 692, 729, 738, 740, 741,
  745, 747, 749, 769
Hilliard, D. K., 505
Hills, L. V., 355, 389
Himes, M., 235, 326, 329
Hispanaediscus, **210**
Histiocysta, **403**
Histiocysta palla, **403,** *403*
Histioneis, 311, **371**
Hoch, G. (Bunt et al.), 616
Hochuli, P., 509
Hoffman, N., 776
Hoffmeister, W. S., 839, *841*
Hofmann, H. J., 34, 44, 65, 68, 69, 73,
  74, 94
Hollande, A., 93, 232, 313, 367
Hollandella, **382**
Holmes, R. W., 587, *588, 589,* 597, 605,
  618, 619, 639
Holmes, R. W. (Eppley et al.), 596
Holm-Hansen, O., 62
Holm-Hansen, O. (Coombs et al.), 573,
  618
Holm-Hansen, O. (Eppley et al.), 306
holococcolith, **689,** 694, 695, *697,* **969**
  fossils, 752
Holodiscolithus, **778**
Holodiscolithus macroporus, **697,** *697*
holoplankton, **343,** 969
Holothuriadeigma, **210**
Holton, R. W., 46
Holttum, R. E., 1
Holosporella, **893**
Homodiacrodeae, 207
Homodiacromorphitae, 207
Homophora, **659**
homothallous, **331**
Homotrybliaceae, 409
Homotryblium, 302, **419**
Homotryblium plectilum, **419,** *419*
Homozygosphaera, **778**
Homozygosphaera tholifera, **695**
Homozygosphaera triarcha, **695, 696,**
  696
Homozygosphaera wettsteinii, **695**
Honigberg, B. M., 226
Honjo, S., *706, 710, 716, 718, 723,* 769

Hooper, S. N. (Ackman et al.), 812
Hopkins, J. T., 610, 611, 612
Hormatonema paulocellulare, **62**
hormogonia, **47, 969**
Hormosporites, 843, **889**
Horn af Rantzien, H., 913, 919, *920*, 924, 926, *941*, 941, *951*, *953*, 954
horns, dinoflagellate cyst, **288, 969**
horns, diatom frustule, **585, 969**
Hornibrookina, **783**
Hornichara, **956**
Horniella, 933, **955**
Horniella robertsi, 941, *941*
Horodiscus, **656**
Horodiscus rugosus, **639**
Horologinella, **395**
Horologinella spinosa, **403**, *403*
Horowitz, A., 353
hot springs
  bacteria, 28
  blue-green algae, 58
  stromatolites, 28
Hotchkiss, A. T. (Conkin et al.), 924, 939
Hotchkiss, S. A. (Gittleson et al.), 304, 306
Houghton, J. A., 51
Hovasse, R., *464*, 464, *468*, 469, *473*, 477, 479, *483*, 484, 545
Hovassebria, 469, 472, 476, **480**
Hovassebria brevispinosa, **482**, *482*
Howe, M. A., *844*, 845
Howland, G. P., 137
Hsü, K. J., 770
Huber, G., 275, 293, 335
Hudson, J. D., 847
Hufford, T. L., *589*
Humboldt, A. von, 349
Hungarodiscus, **818**
Huroniospora, 69, 74, 81, 94, 128, **138**
Huroniospora microreticulata, **70**, *70*, **128**, *129*
Huroniospora macroreticulata, **70**, *70*
Huroniospora psilata, **70**, *70*
Hussain, M. A. (Downie et al.), 358
Hustedt, F., 623, 649, 654
Hutner, S. H., 725
Huxley, T. H., 725
hyaline area, **577, 974**
hyaline field, **585, 969**
Hyalobryon, 500
Hyalodiscus, 622, **655**
Hyalophysa, **818**
hyalosome, **243, 969**
Hyalotheca, **894**
Hydra, 836
Hydrodictyaceae, 838, 839, **888**
Hydrodictyolites, 839, **888**
Hydrodictyon, 827, 828, 839, **889**
hydrogen ion
  bacteria, 23
  blue-green algae, 54
  charophytes, 930
  diatoms, 616
Hydrogenomonas, 20
Hydrolithon, **140**
Hydrolithon iranicum, **135**, *135*
Hydrosera, **657**

Hydrosilicon, **659**
Hydrosirella, **655**
Hydrurus, 510, **518**
Hyella, 59, 62, **82**, 956
Hyella caespitosa, **62**
Hyellaceae, 82
Hymenoclonium serpens, **116**
Hymenomonadaceae, 777
Hymenomonadales, 777
Hymenomonas, 684, 714, 719, 721, 724, 725, **777**
Hymenomonas coronata, **681**
Hymenomonas pringsheimii, **683**, 684, 688, 689, 692, 713
Hymenomonas roseola, **723**, *723*
Hymenophacoïdes, 149
hypersilicification, ebridians, 469
Hyphomicrobiaceae, 40
Hyphomicrobiales, 40
Hyphomicrobium, **40**
Hypneaceae, 141
hypnoid stage, 276, **969**
hypnozygote, **227, 329**, 337, **969**
hypocaust, **585, 969**
hypocingulum, **576, 966**
hypocone, 226, 230, **969**
hypocyst, 167, **284, 969**
hypolithic, **57, 969**
hypothallium, **117, 969**
hypothallus, *119*
hypotheca, **969**
  diatoms, 573, **969**
  dinoflagellates, 230, 263, **969**
hypotract, **284, 969**
hypovalve, **576, 969, 976**
hyrax, 507, 539
Hystrichodiniaceae, 404
Hystrichodinium, **404**
Hystrichodinium pulchrum, **405**, *405*
Hystrichogonyaulax, **355**
Hystrichogonyaulax cladophora, **362**
Hystrichokolpoma, 349, **409**
Hystrichokolpoma bulbosum, **413**, *413*
Hystrichokolpoma cinctum, **413**, *413*
Hystrichokolpomaceae, 409
Hystrichophyta, 148, 181, 205, 816
Hystrichosphaera, 148, 167, 285, **404**
Hystrichosphaeridiaceae, 409
Hystrichosphaeridium, 178, 294, 295, 302, 349, 358, **409**
Hystrichosphaeridium choanophorum, **361**
Hystrichosphaeridium difficile, **415**, *415*
Hystrichosphaeridium diploporum, **178**
Hystrichosphaeridium truncigerum, **361**
Hystrichosphaeridium tubiferum, **415**, *415*
Hystrichosphaerina schindewolfi, **361**
Hystrichosphaeropsis, **430**
Hystrichosphaeropsis complanata, **433**, *433*
Hystrichosphaeropsis quasicribrata, **433**, *433*
hystrichospheres, 148

Hystrichostrogylon, **404**
Hystrichotriangulatum, **212**

Iberosphaeridium, **210**
Ichikawa, W., *545*, 639
Ignatiades, L., 547
Ikebea, **658**
Ilselithina, **783**
Imahori, K., 913, 930
Imantonia, 683, 715, **777**
imbrication, coccoliths, **703, 969**
Impagidinium, **395, 404**
Impagidinium alectrolophum, **395**, *395*
Imperiella, **893**
Impletosphaeridium, **419**
Impletosphaeridium densicomatum, **423**, *423*
Impletosphaeridium rugosum, **423**, *423*
impoverishing division, dinoflagellates, 335
Impluviculus, 178, **210**
Impluviculus miloni, **151**
Incisoria, **658**
incomplete open bands, diatom frustule, **969**
Incurvatina, **822**
Inderites, **818**
Indopolia, **894**
Indumentalithus, **780**
Induoglobus, **210**
Induoglobus latipenniscus, **193**, *193*
Induropilarius, **210**
Ingle, R. M., 341
Ingle, R. M. (Donnelly et al.), 341
Ingram, B. S. (Cookson et al.), *385*, *390*, *391*, *403*, *427*
inner body, **283, 969**
Inopinatella, **892**
Inopinatella lawsoni, **864**, *866*
intercalary archeopyle, dinoflagellates, 295, *296*
intercalary bands
  diatoms, **576, 969**
  dinoflagellates, **258**, *326*, *327*, **969**
intergenicula, *111*, 117, *122*, **969**
Intermurella, **892**
internal skeleton, dinoflagellates, 250
internodal cells, **915, 969**
interstitial flora, dinoflagellates, 347, **969**
intussusception, 258
Inversidinium, 298, **389**
iodine, in red algae, 109
iridescent cells, red algae, 113
iron-depositing bacteria, 24
Isabelia, **392**
Isabelidinium, **392**
Isabelidinium acuminatum, **391**, *391*
Isabelidinium amphiatum, **357**
Isabelidinium microarmum, **357**, *357*
Ischadites, 852, 856, *858*, **891**
  living position, *858*
Ischadites hemisphericus, **856**, *857*
Ischadites murchisoni, **854**, *854*
Ischyrospongia, 133, 142
Isenberg, H. D., 687
Isochrysidaceae, 777

Isochrysidales, 777
Isochrysis, 684, 715, **777**
Isochrysis galbana, **679**
isogamy
  Chlorophyta, 830
  dinoflagellates, 330
isopolar, **924**, *924*, **970**
Isthmia, 576, 637, **657**
Isthmia nervosa, *572, 574,* 597, 598
Isthmolithus, **780**
Isthmolithus recurvus, *757, 761,* 762
isthmus, desmids, **884**, **970**
Ivanovia, **890**
Iversen, J., 154
Iyengar, M. O. P., 927
Iyengariella, **83**
Izhella, **83**

Jackson, G. D., 69, *73,* 94
Jackson, T. A., 34
Jacobson, S. R., 200
Jacutiella, **892**
Jacutiella aciculata, **865,** *865*
Jafar, S. A., 776
Jahn, T. L., 250, *304,* 304
Jania, **141**
Jania guamensis, **119,** *119*
Jansaella, **892**
Jansonius, J., 151, 352, 353
Jansonius, J. (Staplin et al.), 160, 173,
    181, 205
Jardiné, S., *195, 405, 428*
Jarosch, R., 16, 52, 610, 611
Jatuliana, 77, **83**
Javor, B. J., 74
Jayaswal, M. C. (Donnelly et al.), 341
Jeffrey, S. W., 239, 240
Jendrzejewski, J. P., 552
Jerković, L., 465, 539, 542, *548, 559,*
    581, 587
Jodotella, **894**
Jørgensen, B. B. (Cohen et al.), 53
Jørgensen, E. G., 596, 620, 632, 635
Johannes, R. E., 23
Johannesbaptista, **82**
Johansen, H. W., 117, *119, 121, 123,*
    124, 138, 140
John, P., 91
Johnson, J. H., *119, 135, 136, 355, 389,*
    831, 832, 832
jointive elements, coccoliths, **703,**
    970
Jolley, J. W., Jr. (Murphy et al.), 341
Jollos, V., 273, 293
Jones, B. M., 316
Jones, D. J. (Barnes et al.), 127
Jones, R. F., 126
Jousé, A. P., 628, 641, 651
Jousea, **658**
Jowett, R. (Parke et al.), 808, *810*
Joyce, E. A., Jr., 319, 340
Jubatasphaera, **210**
Jubatasphaera dispariluma, *193, 193*
Jurella, **933, 956**
Jurella abshirica, **948,** *948*
Jusella, **386**
Jux, U., 152, 184, 206, 281, *283,* 295,
    301, 302, 810, 812, 814

Kaever, M., *880*
Kakabekia, 69, **81**
Kakabekia barghoorniana, **69**
Kakabekia umbellata, **73,** *73*
Kalan, E. B. (Patton et al.), 339
Kalkowsky, E., 58
Kalley, J. P., 257, 261
Kalley, J. P. (Antia et al.), 497
Kallosphaeridium, **403**
Kallymenia, **141**
Kallymeniaceae, 141
Kalyptea, 281, **386**
Kalyptea stegasta, **389,** *389*
Kamaena, **892**
Kamaenella, 868, **892**
Kamptner, E., 692, 726
Kamptnerioideae, 782
Kamptnerius, **782**
Kamptnerius magnificus, **741,** *743,* 748
Kanaya, T., 634, 637, 639, 650
Karczewska, J., *950*
Kareliana, 77, **81**
Karpinskya, 921, 924, 933, 938, **955**
Karpinskya laticostata, **939,** *939*
Karpinskya oscolensis, **923,** *923*
Karpinskiy, A. P., 937
Karreria, **894**
Karsten, G., 623
Katahiraia, **659**
Katavella, **141**
Katodinium, 311, 348, **374**
Katodinium rotundatum, **231,** *232,*
    233, 237, 244, *259,* 306
Kaufman, J. W., 110
Kazakhstanelia, **892**
Kaźmierczak, J., 138, **835,** *835*
Keck, A., 542
Keega, 140
Kellerman, K. F., 28
Kelly, M. G., 317
Kenleyia, **395**
Kenleyia lophophora, **396,** *396*
Kentrodiscus, **657**
Kentrodiscus armatus, **639**
Kentrosiga, **525**
Kenyon, C. N., 45
Kephyrion, **518**
Kephyrion petasatum, **520,** *520*
Kephyriopsis, **506**
Kephyriopsis cincta, **507,** *507*
Keppenodinium, 311, **382**
Keratophora, **657**
Kern, J. M. (Conkin et al.), 939
Kesling, R., 852, *858*
Keupp, H., 735, 747
Khalili, M. (Adams et al.), 734
Khmeleva, N. N., 339
Khyen, N. T., 51
Kidston, R., 77
Kidstoniella, **83**
Kidstoniella fritschii, **79,** *79*
Kildinella, **818**
Kilham, P., 604, 615, 621, 623
Kimberley, M. M., 80, 95
Kimor, B., 345
Kimura, M., 95
King, K., Jr., 573, 636
Kiselev, I. A., 226

Kisselevia, **392**
Kisselevia coleothrypta, **392,** *392*
Kisseleviella, **656**
Kitakamiania, **890**
Kittonia, 637, **656**
Kjellström, G., 152, 286, 287, 368
Klaveness, D., 679, 688, 689, 692, 713,
    724
Klebs, G., 227, 229, 248, 897
Klebshormidium, **889**
Klebsiella, **41**
Klein, R. M., 87
Kleinig, H. (Brown et al.), 684
Kleithriasphaeridium, **419**
Kleithriasphaeridium readei, **423,** *423*
Klement, K. W., 294, 368, 430
Kling, S. A. (Bukry et al.), 770
Kloppstech, K. (Schweiger et al.), 863,
    864
Klotz, L. C. (Allen et al.), 232, 234, 236,
    237, 238, *343*
Klotz, L. C. (Roberts et al.), 44, 235,
    236
Klumpp, B., 294, 413, 726
Knauer, J., 843
Knoll, A. H., 68, 74, *74,* 94, 95, 128
knot-stage, dinoflagellate nu-
    cleus, **330,** *330,* **970**
Koczyia, **780**
Koczyia fimbriata, **757,** *757*
Kofoid, C. A., 226, *226,* 230, 244, 247,
    254, 256, 261, *263,* 304, 305, 306,
    308, 310, 321, 324, 325, *325,* 326,
    327, 328, 335, 337, 339, 348, 365,
    366, 370, 372, 373
Kofoidinium, 311, **380**
Kofoidinium splendens, **311**
Kofoidopsis, 352
Koizumi, I., 634, 639, 644, 650
Kolbe, R. W., 581, 617, 621, 623, 625,
    630, 634
Kolkwitz, R., 510
Kolkwitziella, **275**
Kolkwitziella salebrosa, **274,** *276*
Kolkwitziellaceae, **274**
Kolkwitziellales, **274**
Komar, V. A., 65
Komewuia, **386**
Komia, 133, **142,** *142*
Komura, S., 639
Koninckopora, **891**
Koninckoporoides, **891**
Konishi, K., 881
Konzalová, M., 841
Kopetdagaria, **893**
Kopetdagarieae, 893
Kopetzky-Rechtperg, O., 884
Koppen, J. D., 626
Korde, K. B., *865*
Koriba, K., 839
Kornmann, V., 843
Kosmogyra, 933, **956**
Kosmogyra superba, **953,** *953*
Kosmogyrella, **957**
Kosmogyrina, **957**
Koths, K. E., 45
Kowallik, K. (Hendey et al.), 568
Kowallik, K. (Manton et al.), 568

Kozhova, O. M., 348
Kozlova, O. G., 623, 624, 625, 629, 632, 641
Kozlova, O. G. (Jousé et al.), 628, 641, 651
Kozlova, O. G. (Zhuze et al.), 623, 624
Kozloviella, 656
Kozur, H., 941
Kozyrenko, T. F., 639
Krasheninnikov, V. (Bukry et al.), 770
Krassavinella, 956
Krassavinella lagenalis, 950, *950*
Krauskopf, K. B., 632
Krebs cycle, 30, 970
Kreger, D. R., 683, 827
Kremp, G. O. W., 154
Krieger, W., *525,* 581
Krimodiscus, 818
Krishnania, 818
Kristiansen, J., 500, *505,* 518
Krivan-Hutter, E., 841
Krumbein, W. E., 59
Krutzsch, W., 364
Krylov, I. N., 65
Ktenodiscus, 637
Kubai, D. F., 233, 234, 236, 237
kuckersite, 832, 970, 975
Kudo, R. R. (Honigberg et al.), 226
Kühn, R. (Dombrowski et al.), 37
Kuenzler, E. J. (Odum et al.), 831
Kulik, E. L., *868*
Kumar, H. D., 51
Kury, W. (Jones et al.), 126
Kusjaella, 938, 955
Kvaleya, 141
Kyansep-Romashkina, N. P., *948, 949*
Kyliniella, 139
Kymalithon, 141
Kyrtuthrix, 59, *60,* 83

Labeyrie, L., 651
labiate process, diatoms, 591, 964, 970
Lachmann, J., 227, 229, 334
Lackey, J. B., 342, 346, 347, 348
Laciniadinium, 392
Laciniadinium orbiculatum, 391, *391*
Lacrymasphaera, 778
Lactobacillaceae, 41
Lacunalites, 210
Lacunalites sphaericus, 203, *203*
Lacunopsophosphaera, 210
laevigate surface, 155, 970
laevogyre coccolith, 694, 970
Laffittius, 782
Lageniastrum, 842, 889
Lagenumbella, 955
Lagenumbella lageniformis, 934, *934*
Lagerheim, G., 836, 839, 887
Lagynion, 519
Lagynophora, 926, 957
Lagynophora liburnica, 926, *926*
Lagynophoraceae, 956
lagynophoroid gyrogonite, 926
Laird, H. C., 182
lake balls, 55, 63
lake biscuits, 63
LaLou, C., 28
Lamarck, J. B., 931, 932

Lamella, 658
Lampropedia, 14, 41
Lamprothamnium, 933, 956
Lamprothamnium macropogon, 953, *953*
Lamprothamnium papulosum, 922, *922, 930,* 931
Lamprothamnus, 956
Lanceoforma, 138, 139, 142
Lanceoforma striata, 132, *132*
Lancettopsis, 812, 813, 818
Lancicula, 890
Lancicula alta, 850, *850*
Landon, M. (Jahn et al.), 250, 304
Lang, N. J., 44, 45
Lang, W. H., 77
Langangen, A., 927, 930
Lange, F. W. (Combaz et al.), 206
Lange, W., 62
Langer, W., *938,* 938
Langiella, 83
Lankester, E. R., 261
Lanterna, 395
Lanternithus, 770, 778
Lanternithus minutus, 697, *697,* 752
Lanternosphaeridium, 409
Lanveocia, 210
Lapideacassis, 780
Larkum, A. W. D. (Borowitzka et al.), 112
Larvaria, 869, 894
Larvaria encrinula, 879, *879*
Larvaria limbata, 877, *877, 879*
laterally linked hemispheroids, stromatolites, 64
Laterilites, 210
Laticarinina pauperata, perforating algae in, *60, 60*
Latifascia, 270, 369, 371
Latisphaera, 74, 889
Latochara, 933, 955
Latochara collina, 941, *941*
Laubenfels, M. W. de, 852
Lauderia, 655
Lauritis, J. A., 571, 573, 605, 611
Lauritis, J. A. (Coombs et al.), 573
Lauritzen, O., 65
Lauterborn, R., *63*
Lavine, L. S. (Isenberg et al.), 687
Leadbeater, B. S. C., 244, 245, 248, *249,* 250, *501, 511,* 512, 518, *521,* 525, *526, 527,* 681, 684, *685,* 714, 725
LeBlanc, R., 601
Lebour, M. V., 226, 230, 242, 265, 269, 344, 345, 623, 641
Lecalia, 784
Lecaniella, 210, 430
Lecaniellaceae, 430
LeCorre, C., 154
Lee, J. J., 835
Lee, K. W., 497, *499*
Lee, R. E., 92, *112,* 137
Leech, R. M., 91
Leedale, G. F., 491, *493,* 493, 494, 496, 497, 499, *691,* 692, 710, 714, 895, 897
Lefèvre, M., 240, 261, 262, 350
Lefort, F., 714, 747
Leiacina, 956

Leioaletes, 210
Leiodiscina, 210
Leiofusa, 160, 199, 200, 206, 207, 210
Leiofusa granulacutis, *157, 157,* 163
Leiofusa rhikne, 201, *201*
Leiofusidaceae, 207
Leiofusidae, 207
Leioligotriletes, 210
Leiomarginata, 151, 210
Leiopsophosphaera, 210
Leiosphaeraceae, 818
Leiosphaeridaceae, 207
Leiosphaeridae, 207
Leiosphaeridia, 149, 178, 204, 206, 496, 812, *813, 814,* 818
Leiosphaeridia oelandica, 812
Leiosphaeridia voighti, 812
Leiosphaeridiaceae, 207, 818
Leiovalia, 178, 210
Leipokatium, 389
Leitgeb, H., 864
Lejeune-Carpentier, M., *287,* 287, 288, *293, 294, 294,* 349, 350, *385*
Lejeunia, 392
Lejeunia kozlowskii, 392, *392*
Lemanea, 124, 132, 139
Lemaneaceae, 139
Lemmermann, E., 226, 273, 348, 464, 477, 535
Lemoine, M., 134, 140, *872*
Lemoinella, 894
Lenoble, A., 897
Lentin, J. K., 287, 288, 299, 368
Lentin, J. K. (Evitt et al.), 284, 295
Lentin, J. K. (Millioud et al.), 351
Leonardosia, 955
Leoniella, 210
Lepidodendron, parasites on spores of, 842
Lepidodiscus, 656
Lepidolites, 891
Leptocylindraceae, 655
Leptocylindrus, 600, 655
Leptocylindrus danicus, 603, *603,* 623
Leptodinium, 395
Leptodinium aculeatum, 395, *395*
Leptodinium hyalodermopse, 395, *395*
Leptodiscus, 380
Leptophyllus, 380
Leptophytum, 141
Leptoscaphos, 658
Leptosira, 843, 889
Leptotrichaceae, 39
Leptothrix, 40
Leucotrichaceae, 39
Leucothrix, 39
Leudugeria, 657
Leudugeria janischii, 641, *641*
Levine, N. D. (Honigberg et al.), 226
Levring, T., 834
Lewin, D. (Smith et al.), 339
Lewin, J. C., 571, 572, 595, 596, 605, 610, 613, 614, 615, 617, 618, 619, 620, 632, 828, 831
Lewin, J. C. (Reimann et al.), 573, 579, 581, 610
Lewin, R. A., 84, 92, 613, 614, 725, 814, 833

Lewin, R. A. (Lewin et al.), 605, 610
Liagora, 112, 124, **139**, *139*
Licari, G. R., 69, 74
Licari, G. R. (Cloud et al.), 74, 94
lichens, 55, 63
Licmophora, 607, **658**
Licmophora hyalina, **613**
Licmosphenia, **658**
Liebisch, W., 611
light-dependent motility, blue-green algae, 53
Lignier, O., 836
ligula, **576, 970**
Liguori, V. R. (Gold et al.), 511
Likanella, **893**
lime availability, charophytes, 930
lime deposition
    bacteria, 28
    Chlorophyta, 828, 831
        Dasycladales, 860
    red algae, 110
lime shell, Charophyta, **919, 923, 968, 970**
Lin, H., 114
Lindemann, E., 274, 275, 276
Ling, H. Y., 465, 479, 539, 542, *542, 550*
Lingulasphaera, **419**
Lingulasphaera spinula, **289,** *289, 290,* 422
Lingulodiniaceae, 424
Lingulodinium, 295, 299, **424,** *424*
Lingulodinium machaerophorum, **299,** *301, 334*
Lingulodinium siculum, **424,** *424*
Lingulodinium solarum, **424,** *424*
linking spines, diatoms, **587, 591, 970**
Linoporella, **893**
Linoporelleae, 893
Lipkin, Y. (Friedmann et al.), 57, 836
Lipman, C. B., 28
Lipps, J. H., 538, 542, 547, 708, 764, 776
Liradiscus, **656**
Lisitsyn, A. P., 617, 628, 630, 634, 635
Lissodinium, **377**
Lister, T. R., 167, 173, 176, 182, 206
lists, dinoflagellate, **271, 970**
Litanaia, **890**
Litharchaeocystis, **528**
Litharchaeocystis glabra, **528,** *528*
Lithastrinaceae, 780
Lithastrinus, 728, **780**
Litheusphaerella, 512, **528**
Litheusphaerella cretacea, **515,** *515*
Litheusphaerella spectabilis, **528,** *528*
Lithocaulon, **140**
Lithocodium, **890**
Lithodesmium, 598, **657**
Lithodesmium cornigerum, **633,** *633*
Lithodesmium minusculum, **633,** *633*
Lithodesmium undulatum, **568,** *599*
Lithodinia, **395**
Lithodinia bulloidea, **401,** *401*
Lithodinia valensii, **401,** *401*
Lithoperidiniaceae, 386
Lithoperidinium, **386**
Lithophylloideae, 140
Lithophyllum, 112, 117, 126, **140**

Lithophyllum quadrangulum welschii, **135,** *135*
Lithoporella, 126, **140**
Lithostromatiaceae, 779
Lithostromation, **779**
Lithostromationidae, 779
Lithothamnium, 117, 126, 140, **141**
Lithothamnium bofillii, **119,** *119*
Lithothamnium ridge, 126
Lithothrix, **141**
lithotrophs, **18**
lithotrophic metabolism, 19
Lithraphidites, **781**
Lithraphidites acutus, **741,** *741, 745,* 748
Lithraphidites bollii, **735,** *738, 749*
Lithraphidites carniolensis, *738, 741*
Lithraphidites quadratus, *743,* **745,** 748
Lithuropyxis, **528**
Lithuropyxis barbadensis, **528,** *528*
Litopora, **894**
Litosphaeridium, **409**
Litosphaeridium conispinum, **416,** *416*
Litosphaeridium flosculus, **416,** *416*
Littler, M. M., 126
Lobelia-lakes, 930
Locker, S., *556, 559, 559*
Loeblich, A. R., Jr., 4, 5, 6, 7, *131,* 150, 155, *157, 159, 160, 160, 162, 163, 165, 167, 173, 174, 175, 176, 178, 178, 181, 182, 184, 187, 188, 189, 190, 193, 194, 195, 197, 197, 198, 201,* 202, 230, 233, 271, 293, 302, 330, 339, 351, 352, 355, 357, 358, 367, 368, 370, 383, 385, 404, 408, 419, 421, 424, 476, 477, 548, 549, *549,* 613, 618, 630, 653, 654, 771, 772, *772, 773, 776,* 805, 816, 939, 942
Loeblich, A. R., Jr. (Honigberg et al.), 226
Loeblich, A. R., Jr. (Loeblich III et al.), 539, 541, 549
Loeblich, A. R., III, *228,* 230, 233, 234, 235, 236, 240, 241, 245, 246, 248, 256, *257, 258, 261, 264, 265, 266, 267, 268, 269, 269, 271, 278, 279,* 288, 293, 295, 302, 308, 311, *312, 313,* 313, 315, 317, *318,* 319, 320, 321, 324, 327, 329, 330, 339, *340,* 344, 351, 366, 368, *369, 370,* 474, 479, 539, 541, 549, 568, 571, 607, *608, 614, 681, 686*
Loeblich, A. R., III (Allen et al.), *232, 234, 236, 237, 238, 343*
Loeblich, A. R., III (Patton et al.), 339
Loeblich, A. R., III (Roberts et al.), 44, 235, 236
Loeblich, L. A., *317, 318, 340,* 568, 571, 614, 681
Loeblich, L. A. (Loeblich III et al.), 539, 541, 549
Loftus, M. (Seliger et al.), 316
Logan, B. W., 65
Lohman, K. E., 567, 632, 636, 639, 641, 648, 649
Lohmann, G. P., 770, 771
Lohmann, G. P. (Haq et al.), 770

Lohmann, H., 181, 254, 255, *256,* 273, 510, 678, 726
Lohmannella, **382**
Lohmannia, **382**
Lohmannosphaera, **780**
Lombard, A., 843
Lombard, E. H., 306, 348
Lombardia, 843
Lombardia arachnoidea, **843**
Lomentaria, **142**
Lomentarites, **142**
Long, J. A., 637
longitudinal flagellum, **247, 248, 250, 970**
longitudinal groove, **227, 970, 975**
lopadolith coccolith, **697,** *700,* **970**
Lophodiacrodium, **208**
Lophodictyotidium, **822**
Lophodiniaceae, 253, 292, 374
Lophodinium, 351, **374,** *374*
Lophodinium dadayi, **374,** *374*
Lophodinium polylophum, **374,** *374*
Lophodolithus, **780**
Lophominuscula, **210**
Lophorytidodiacrodium, **208**
Lophosphaeridium, *190,* **210**
Lophotheca, **658**
lophotrichate, *12, 16,* **970**
Lorch, J., 944
Loreau, J.-P. (Feldmann et al.), 850
lorica, **970**
    Chrysophyta, 499
    ebridian, 469
    euglenid, 895
Loring, D. M. (Dombrowski et al.), 37
Lotharingiaceae, 782
Lotharingius, **783**
Lowenstam, H. A., 831
lower windows, ebridian skeleton, **465, 970**
loxolith ring, coccolith, **703, 970**
loxolithid coccolith, *705*
Loxolithus, 703, **780**
Lucernella, 933, **956**
Lucernella ampullacea, **944,** *944*
Lucianorhabdus, **778**
luciferin, 316
Ludvik, L., 258
Luehndea, **404**
Lund, J. W. G., 619
Lunulidia, **211**
Lurquin P. (Bonotto et al.), 825, 826
Lusata, **211**
Lusatia, **211**
Lutetianella, **559**
Luxadinium, **392**
Lychnothamnites, **956**
Lychnothamnites narynensis, **953,** *953*
Lychnothamnus, 915, 933, **956**
Lychnothamnus barbatus, **953,** *953*
Lyngbya, 44, 49, 51, 55, 57, 59, **82,** 833
Lyramula, 549, 554, **559**
Lyramula furcula, *550, 551,* **552, 560,** *560*
Lyramulaceae, 559
Lys, M., *937*
lysis, bacterial, 18

Lysvaella, **141**
Lythgoe, J. N. (Barnes et al.), 127

MacAdam, R. B., *150, 197, 198*
McAndrews, J. H., 284, 364
McCarthy, J. J. (Eppley et al.), 342
McCollum, D. W., 641, 644, 646
McCready, R. G. L., 37
Macdonald-Pfitzer hypothesis, *595, 595*
McElroy, W. D., 316
McElroy, W. D. (Seliger et al.), 316
McHugh, J. L., 568, 628, 629
McIntire, C. D., 621
McIntyre, A., *681, 699, 701, 703, 706, 708, 711, 719, 720, 720, 721, 723, 751, 755, 769*
McIntyre, D. J., *357, 385, 391, 392*
Macintyre, I. G., 126, 127, 140
McIntyre, R., 769
McLachlan, J., 344
McLachlan, J. (Chen et al.), *131*
McLaughlin, J. J. A., 313
McLaughlin, P. J., 95
McLaughlin, R. B., 604
McLean, D. M., 295, 298, *401*
MacLean, J. L. (Worth et al.), 317, 341
McPherson, L. M., 465, 479, *542, 542*
McQuade, A. B., 137, 138
MacRaild, G. N., 850, *853*
macrococcoliths, **706,** 706, **970**
macrogamete, dinoflagellate, *336*
Macroporella, **893**
Macroptycha, **211**
Macrora, 254, 512, **519, 559**
Macrora barbadensis, **512,** *512*
Macrora stella, **512,** *512*
Maculatasporites, 810, **818**
Maculatasporites amplus, **811,** *811*
Maculatasporites minimus, **811,** *811*
Maculosphaera, 74, **889**
Madgwick, J. (Grant et al.), 617, 618
Maduradinium, **389**
Maduradinium pentagonum, **385,** *385*
Mädler, K. A., 181, 205, 812, 816, *949*
Maedleriella, 926, 933, **956**
Maedleriella mangenoti, **925,** *925*
Maedlerisphaera, **956**
Magloire, L., *195*
Magrum, L. J. (Fox et al.), 31
Mahadevan, C., 843
Mahler, H. R., 86, 90, 91
Mahnken, C. V. W. (Clifton et al.), 833
Makarova, I. V., 639, 641
malformed coccoliths, 710
Mallomonas, 501, 504, 512, **519,** 654
Mallomonas harrisae, **502,** *502,* **512,** *512*
Mallomonas leboimei, **502,** *502*
Mallomonas litomesa, **525,** *525*
Mallomonas lychenensis, **525,** *525*
Mallomonas teilingii, **525,** *525*
Mallomonas tonsurata, **525,** *525*
Mamet, B. L., 133, 142, 932, *936*
Mametella, **142**
Mamillata, **211**
Mancodinium, **404**
Mandel, M., 37

Mandelli, E. F., 240, 615
Mandra, H., 552
Mandra, Y. T., 552, *545*
Mangin, L. A., 273
Manivit, H., 741, 748
Manivitella, **783**
Manivitella pemmatoidea, **692,** *692*
Mann, A., *581,* 581, 623, 629, 641
Mann, D. G., 581
Mantell, G. A., 350
Manton, I., 512, 525, 568, 597, 681, 683, *691,* 692, 710, 714
Manton, I. (Parke et al.), 681, 683, 684, 715, 725
Manum, S., 276, *357, 385, 391, 411, 415, 427, 821*
Mapletoft, H., 334
Maranhites, **818**
Maranhites brasiliensis, **817,** *817*
Margalef, R., 604, 625
marginate cyst, dinoflagellate, **286, 970**
Marginisporum, **141**
Margolatus, **783**
Margosphaera, **211**
Margulis, L., 80, 89, *90,* 93, 366
marine dinoflagellates, 342
Marinella, **890**
Markali, J., *695, 696,* 706
Markali, J. (Braarud et al.), 726
Markali, J. (Gaarder et al.), *723*
Markalius, **783**
Markalius astroporus, **757,** 763, 771, 774
Markalius inversus, **754,** *754,* **757**
Marpolia, **82**
Marrocanium, **211**
Marsh, J. A., 126
Marshall, S. M., *536, 537, 538*
Marszalek, D. S., 828, 831, 864
Marthasteritaceae, 777
Marthasterites, 715, **777**
Marthasterites contortus, **757,** 763
Marthasterites furcatus, **745,** *745,* **745,** 748
Marthasterites tribrachiatus, **752,** *752,* 757, 763
Martin, F., 152
Martin, T. C., 43
Martini, E., 465, 471, 482, *542, 542, 550, 551,* 745, 752, 757, 766, 772
Martiniaster, **779**
Masek, J., 841
Maser, M. D. (Schopf et al.), *32, 34*
Maslov, V. P., 65, 77, *850, 868, 881, 921, 923, 924, 925, 928, 929, 932, 937, 941, 953*
Maslovella, **783**
Maslovichara, **955**
Maslovicharoideae, 955
Maslovina, **890**
Massalski, A., 493
Massartia, **374**
Mastigocladaceae, 48, 83
Mastigocladopsis, **83**
Mastigocladopsidaceae, 83
Mastigocladus, 53, **83**
Mastigocoleus, 59, **83**

Mastigocoleus testarum, 60, **62**
mastigoneme, **248, 970**
Mastigophora, 477, 554
Mastogloia, 649, 652, **658**
Mastogonia, **656**
Mastophoroideae, 140
Mastophora, **140, 891**
Mastopora, **891**
Mattox, K. R. (Stewart et al.), 805
Maturodinium, **395**
Matvienko, O. M., 518
Matzenauer, L., 346
May, F. E., 285, 292, 351
Maynard, N. G., 623, 625
Mazza, A., (Bonotto et al.), 825, 826
Meakin, S. H., 637
Mecsekia, **211**
Mediales, 568
median rupture, acritarch excystment, **176, 970**
median windows, ebridian skeleton, **465, 970**
medulla, **117, 970**
megacytic band, dinoflagellates, **326, 969, 970**
Megasacculina, **211**
Megasphaeromorphida, **208**
Mehlisphaeridium, 190, **211**
Mehlisphaeridium fibratum, **183,** *183*
Mehta, S. C. (Venkataraman et al.), 572
Meijer, J. J. de, 132
Meinschein, W. G., 35
Meiourogonyaulax, **395**
Mejerella, **892**
Mejerella ramosa, **865,** *865*
melanosome, **244, 970**
Melasmatosphaera, 69, 81
Melikeriopalla, 157, **822**
Melikeriopalla amydra, **809,** *809*
Melobesia, **141**
Melobesioideae, 140
Melosira, 576, 587, 596, *597,* 597, 598, 601, 604, 607, 626, 628, 637, 641, 649, **655**
Melosira ambigua, **585**
Melosira architecturalis, **647**
Melosira granulata, **570,** *570,* **648,** 648, 649
Melosira italica, **568**
Melosira moniliformis, **583,** *585,* 606
Melosira nummuloides, **587,** *587,* 587, *594,* 616
Melosira sulcata, **604**
Melosira varians, **570,** *570,* 573, 598, 599
Melosiraceae, 655
membranate cyst, dinoflagellate, **286, 970**
Membranilarnacia, 290, **409**
Membranilarnaciaceae, 409
Membranolimbus, **211**
Membranosphaera, **403**
Meminella, **894**
Menneria, **818**
Mereschkowsky, C., 89
Meretrosulus, **656**
Meridion, **658**
Meringosphaera, 510, **518**

Meringosphaera mediterranea, **521**, *521*
Merismopedia, 81
meromes, receptaculitid, **856**, 857, 967, **970**
meroplankton, **343, 970**
Merrill, J. A., 350
mesobase, **258**, *258*, **970**
mesoblast, **284, 971**
Mesocaryota, 225, 231
mesocaryote, 231, 237
Mesocena, 539, 542, *549*, 553, **559**, 965
Mesocena apiculata, **545**, *545*, 551, 552
Mesocena circulus, 551, **552**, 557
Mesocena circulus v. *apiculata*, *550*, **552**, *557*
Mesocena diodon, **542**, *542*, 551, 552
Mesocena elliptica, **542**, *542*, 549, *550*
Mesocena hexagona, **539**, *539*, 557
Mesocena nodulifera, **545**, *545*
Mesocena quadrangula, **549**, *550*
Mesochara, **933, 956**
Mesochara symmetrica, **949**, *949*
mesoclades, **465, 971**
mesocyst, **284, 971**
mesokaryon nucleus, **227, 967, 971**
Mesolithon, **141**
mesomere, **258**, *258*, **971**
mesophragm, acritarch, 152
mesophragm, dinoflagellate cyst, **281**, **284, 971**
Mesophyllum, 126, **141**
Mesophyllum guamense, **136**, *136*
Mesophyllum ishigakiensis, **136**, *136*
Mesoporos, **369**
Mesoprotista, 138
mesospore, **278, 971**
Mesotaeniaceae, 894
Mesotaenium, **894**
Messikommer, E., 887
metabolism
   bacteria, 18
   blue-green algae, 53
   Chrysophyta, 504
   coccolithophores, 725
   diatoms, 613
   dinoflagellates, 308
   euglenids, 896
   Haptophyta, 725
Metacyclina, 881, **894**
Metagoniolithoideae, 141
Metagoniolithon, 141
Metaleiofusa, **211**
Metallogenium, **40**
Metamastophora, **140**
Metapeyssonnelia, **140**
Metasolenopora, **140**
metaspondyl, **868, 971**
Methanobacteriaceae, 41
Methanobacterium, 31, **41**
Methanomonadaceae, 40
Methanomonas, 20, **40**
Methanosarcina, 31
Methanospirillum, 31
Methylomonadaceae, 40
Meunier, A., 274, 275, 293, 613, 623, 641
Meunier, V., 339

Meyer, K. I., 887
Michaelsarsia, **784**
Michaelsarsia elegans, **710**
Michener, J. R., 372
Micrampulla, 512, **528**
Micrampulla parvula, **528**, *528*
Micrantholithus, 747, 770, **779**
Micrantholithus altus, **757**, *757*
Micrantholithus hoschulzi, **745**, *745*, 749
Micrantholithus obtusus, **741**
Micrantholithus pinguis, **761**, *761*
Micrasterias, 828, 884, 885, **894**
Micrhystridium, 151, 160, 167, 171, 173, *183*, 185, 187, 190, 197, 199, 202, 204, 206, **211**, 364
Micrhystridium arachnoides, **151**
Micrhystridium minutispinum, **199**
Micrhystridium variabile, **199**
Microchaetaceae, 83
Microchaete, 83
Microchara, 926, *926*, 933, **956**
Microchara hystrix, **953**, *953*
Microchara vestita, **957**
Micrococcaceae, 41
micrococcolith, **706, 971**
Micrococcus, 30, **41**, 80
Micrococcus lignitum, **21**, *21*
Microcoleus, 57, 59
Microconcentrica, **211**
Microcystis, 58, 81
Microcystis aeruginosa, **44**, 46
microcysts, **16, 971**
Microdiniaceae, 395
Microdinium, **395**
Microdinium ornatum, **398**, *398*
Microdistephanus, **559**
microenvironments, 23
microgamete, dinoflagellate, *336*
Microglena, **519**
μg-atoms per liter, **971**
Micromarsupium, 469, 472, 476, **480**
Micromarsupium anceps, *482*
Micromarsupium anceps curticannum, **471**, *471*
Micromarsupium rostovense, **471**, *471*, *482*
Micromonas, **818**
Micromonosporaceae, 42
microplankton, abundance, *196*
microplankton, diversity, *196*
Micropodiscus, **655**
Microrhabdulaceae, 781
Microrhabdulidae, 781
Microrhabdulinus, **781**
Microrhabduloidus, **781**
Microrhabdulus, **781**
Microrhabdulus belgicus, **746**, *746*
Microsiphona, **655**
Microspora, **889**
Microspora pachyderma, **845**
Microsporaceae, 845, 889
Microsporales, 889
Microstaurus, **781**
Microstomella, **956**
microtubules, *260*
  in dinoflagellates, 324

microtubule organizing centers (MTOC), 89, 93
Micula, **778**
Micula concava, 748
Micula decussata, **745**, *745*
Micula murus, **745**, 748, 774
Micula staurophora, 741, **745**, 748
Mignot, J. P., 232
Migula, W., 914, 916, 917, 918, 929, 931
Mikkelsen, N., 632, 761
Miklukho-Maklay, A. D., 932
Milanović, M., 871
Miliola (Holococcus?) panderi, 937
Millaria, 81
Millepora, 126, 314
Miller, B. T., 465, 474
Miller, J. D. A., 826
Miller, U., 581, 639
Millington, W. F., 828, 839
Millioud, M. E., 351, *428*
Millioud, M. E. (Evitt et al.), 284, 295
Mills, F. W., 654
Milon, Y., 846
Minas, H.-J. (Coste et al.), 547
Minidiscus, **655**
miospores, **204, 971**
Miravetesina, **781**
Misceomarginatus, **780**
Mischococcales, 498
Mitcheldeania, 83, **890**
mitochondria, 14, *110*, **971**
   DNA in, 85
   endosymbiotic origin, 91
Mitosia, **780**
Mitrolithus, **780**
Mizzia, 832, 868, 881, 893
Mizzia velebitana, **870**, *870*
Mizziella, **890**
Moberly, R., Jr., 110, 831
Moellerina, 933, 937, 938, 939, **955**
Moellerina greenei, **939**, *939*
Moesiodinium, **392**
Moestrup, Ø., 496
Mohler, H. P., 776
Moiseeva, A. I., 639
Mollicutes, 42
Monallantus, **498**
Monas, 504, 518, **519**
Monaster, 253, 374
Monaster rete, 374, **374**
Moner, J. G., 839
Monera, 1, 14, 38, 39
Mongolichara, 933, **956**
Mongolichara deplanata, **949**, *949*
Mongoloporella, **891**
Mongoloporella rozovi, **854**, *854*
Mongoloporellaceae, 891
Monobrachia, **657**
Monodus, 493, **498**
monoecious, 915
Monomarginatus, **780**
Monosiga, **525**
Monosphaeritae, 207
monosporangia, **124**, *124*, **971**
monospores, 114, *115*, 116, **971**
Monostroma, **890**
Monostromataceae, 890
monotrichate, 12, **16, 971**

Montgomery, M. (Mapletoft et al.), 334
Montiella, **894**
Monty, C. L. V., 58, 59, 80
Moodie, R. L., 37
Moore, W. W., 621
Moorman, M. A., 74
Morales, G. A., 938, *939*
Morania, **83**
Moravamminida, 133
Morbey, S. J., 353, *354*
Morbey, S. J. (Harland et al.), *390*
Morellet, J., 832
Morellet, L., 832
Morelletina, **894**
Morelletpora, **893**
Morelletporeae, 893
Morgenroth, P., *173*, 353, *395, 396, 398, 401, 406, 411, 413, 415, 416, 419, 421, 422, 423, 426, 429, 430, 435, 815*
Morkallacysta, **389**
Morland, H., 605
Morris, I., *13, 14,* 80
Mosser, A. G. (Mosser et al.), 835
Mosser, J. L., 835
Mougeotia, **884**
Mountjoy, E. W., 74
Moyeria, **211**
mucilage, blue-green algae, 43, 44
Muderongia, **424**
Muderongia mcwhaei, **428**, *428*
Muderongiaceae, 424
Muelleriella, **655**
Muelleriopsis, **655**
Muhina, V. V. *See* Mukhina, V. V.
Muir, M. D., 74
Muir, M. D. (Brooks et al.), 68, 69
Muir, M. D. (Oehler et al.), 94, 128
Muiradinium, **430**
Muiradinium dorsispirale, **365**, *365*
Muiriella, **395**
Muiriella plioplax, **401**, *401*
Mukhina, V. V., 623, 624, 632
Mukhina, V. V. (Zhuze et al.), 623, 624, 628, 641, 651
Müller, A. H., 852
Müller, C., *542, 542, 551, 755, 756, 761, 764*
Muller, J., 275
Müller, O., 637
Müller, O. F., 227
multiaxial filament, red algae, 117
Multiplicisphaeridium, *171*, **211**
Multiplicisphaeridium bifurcatum, **163**, *163*
Multiplicisphaeridium corallinum, **200**
Multiplicisphaeridium digitatum, **160**
Multiplicisphaeridium fisheri, **169**, *169*
Multiplicisphaeridium mergaeferum, **167**, *167*
Multispinula, **389**, *389*
Munier-Chalmas, E., 864
Munieria, **893**
Murandavia, *77,* **81**
Muraticavea, **822**
Muraticavea enteichia, **821**, *821*
Muratodinium, **395**
Muratodinium fimbriatum, **289**, *289, 290, 290, 396*

Murphy, A., 568, 615
Murphy, E. B., 341
Murphy, L. S., 605, 655
Murray, G., 726
Murrayella, **374**
Musacchio, E. A., *928, 950*
Muscatine, L., 314
Muscatine, L. (Smith et al.), 339
Mussill, M., 16
Mutabilimorpha, **208**
mutation rates, bacteria, 11
Mycacanthacoccus cellaris, **128**
Mycelites, 65
Mychodeaceae, 141
Mychota, 38, 39
Mycobacteriaceae, 42
Mycobacterium, 31, **42**
Mycoplasma, 11, **42**
Mycoplasmatales, 42
Myers, G. E., 37
Myrmecia, **888**
Myrmekioporella, **893**
Myxobacterales, 16, 39
Myxococcaceae, 39
Myxococcoides, 69, 74, **81**
Myxococcoides minor, **77**, *77*
Myxococcoides muricata, **74**, *74*
Myxodinium, 308, **380**
Myxodinium pipiens, **233**, *309, 310*
Myxophyceae, 43, 81
Myxosarcina, **82**
myxospores, 39
myxotrophic dinoflagellates, 308

Naccaria, **139**
Naccariaceae, 139
Naevisphaeridium, **211**
Nagappa, Y., 841
Nagy, L. A., 69
Naias, 607
Nakajima, T., *572*
Naninfula, **778**, *784*
Naninfula deflandrei, **697**, *697*, **761**, *761*
Nannoceratopsaceae, 386
Nannoceratopsiella, 352
Nannoceratopsis, *294, 301, 301,* 351, 353, 354, **386**
Nannoceratopsis gracilis, **301**, *301, 355*
nannoconid, 728, **729**, *971*
Nannoconaceae, 779
Nannoconidaceae, 779
Nannoconidae, 779
Nannoconus, 728, **779**
  geologic occurrence, 731
Nannoconus abundans, **729**, *729*
Nannoconus bermudezi, **731**, *731*
Nannoconus boneti, **731**, *731*
Nannoconus bronnimanni, **731**, *731*
Nannoconus bucheri, **731**, *731*
Nannoconus carniolensis, **731**, *731*
Nannoconus colomii, **731**, *731,* 749
Nannoconus dauvillieri, **731**, *731*
Nannoconus dolomiticus, **731**, *731*
Nannoconus elongatus, **729**, *729,* 731
Nannoconus elongatus cylindrus, **731**, *731*
Nannoconus globulus, **731**, *731*
Nannoconus kamptneri, **731**, *731*

Nannoconus minutus, **731**, *731*
Nannoconus multicadus, **729**, *729, 731*
Nannoconus planus, **731**, *731*
Nannoconus robustus, **731**, *731*
Nannoconus steinmannii, **729**, *729, 731*
Nannoconus truittii, **729**, *729,* 731
Nannoconus wassalli, **731**, *731*
Nannocorbis, **780**
Nannocorbis challengeri, **764**, *764*
Nannopatina, **785**
nannoplankton, **678**, *971*
Nannotetraster, **778**
Nannotetrina, **780**
Nannotetrina alata, **757**, *763*
Nannotetrina quadrata, **757**, *762, 763*
Nannoturbella, **778**
Nanococcus, **81**
Nanocyclopia, **211**
Nanocyclopia aspratilis, **193**, *193*
Nanoneis, 579, **658**
Nanopora anglica, **866**, *866*
Nass, S., 87
Natarajan, K. V. (Venkataraman et al.), 572
Nathan, H. (Seigel et al.), 69
Naumova, S. N., 182, 205
Navicula, 572, 587, 600, 612, 613, 622, 624, 626, 628, 641, 652, **658**
Navicula cuspidata, **571**, *587*
Navicula hennedyi, **633**, *633*
Navicula pelliculosa, **568**, 571, 572, 573, 596, 613, 617, 619
Navicula praetexta, **633**, *633*
Navicula radiosa, **610**
Navicula spectabilis, **578**, *578, 587*
Naviculaceae, 658
Naviculales, 658
Naviculopsis, 549, 554, **559**
Naviculopsis biapiculata, 551, **552**, *552,* **559**, *559*
Naviculopsis constricta, 551, **552**
Naviculopsis foliacea, 550, 551, **552**
Naviculopsis lata, 550, 551, **552**
Naviculopsis minor, 551, **552**
Naviculopsis quadrata, 551, **552**
Naviculopsis regularis, 551, **552**
Naviculopsis robusta, 551, **552**
Naviculopsis trispinosa, 551, **552**, **559**, *559*
Navifusa, 206, **211**
Navifusa ancepsipuncta, **157**, *157*
Navisolenia, **781**
Neale, J. W., 383
Necrobroomea, **403**
Necrobroomea jaegeri, **403**, *403*
Necrobroomea longicornuta, **403**, *403*
Neevea, 116, **139**
Neidium, 626, **658**
Neisseriaceae, 41
Nelcanella, 77, **83**
Nelchinopsaceae, 392
Nelchinopsis, **392**
Nelchinopsis kostromiensis, **390**, *390*
Nelsoniella, **392**
Nelsoniella tuberculata, **390**, *390*
Nelsoniellaceae, 392
Nemaliales, 110, 112, 124, 139

Nemalion, 117, 126, 132, **139**
Nematochrysis, **519**, 714
Nematochrysopsis, **525**
nematocyst, 231, 244, **247**, *247*, 971
Nematodinium, 243, 247, **380**
Nematosphaeropsis, 273, 299, **404**
Nematosphaeropsis balcombiana, 301,
  301, **406**
Nematosphaeropsis labyrinthea, **299**
Nematosphaeropsis philippoti, **406**,
  406
Nematosphaeropsis pusulosa, **426**, *426*
Nemegtichara, 933, **956**
Nemegtichara prima, **950**, *950*
Neoanchicodium, **890**
Neochiastozygus, **780**
Neococcolithes, **780**
Neogoniolithon, 126, **140**
Neogyroporella, **893**
Neoischadites, **892**
Neomacroporella, **893**
Neomeridoideae, 893
Neomeris, 860, *861, 862, 864, 869,* **894**
Neomeris annulata, **860**
Neomizzia, **893**
Neopolyporolithon, **141**
Neopolyporolithon reclinatum, **123**,
  123
Neosolenopora, **140**
Neosphaera, **783**
Neostreptotheca, **657**
Neoveryhachium, **211**
Neoveryhachium carminae, 200, **201**,
  201
Nephrolithus, **781**
Nephrolithus frequens, **745**, 748
Nephroselmidaceae, 818
Nephroselmis, **818**
Neresheimeria, **382**
Netrelytraceae, 386
Netrelytron, 281, **386**
Netromorphitae, **207**
Netzel, H., 258, *260,* 265, 267, 288, 321,
  322, *322,* 324, 327
Neurospora, 95
Nevidia, 149, **211**
Nevo, Z., 261
Newman, K. R., 839
Newton, E. T., 814, *816*
Nicholls, K. H., 240
Nichols, H. W., 113, *114,* 115, 126, 218
Nidulites, **891**
Niel, G. B. van, 9, 29, 80, 85, 93
Niklas, K. J., 843, *844*
Nipkow, F., 275, 293, 335
Nipponophycus, **141**
Nitecki, M. H., 852, *856,* 856, *857,* 857,
  858, 868
Nitella, 915, 921, 924, 929, 930, 933,
  951, 954, **956**
Nitella clavata, **953**, *953*
Nitella flexilis, **914**, *914,* 930
Nitella furcata mucronata, 931
Nitella gracilis, **914**, *914,* 930
Nitella syncarpa, **931**
Nitella terrestris, **927**
Nitella translucens, **930**
Nitellites, 933, 951, **956**

Nitellites sahnii, **951**, *951*
Nitelloideae, 956
Nitellopsidaceae, 956
Nitellopsis, 925, 933, **956**
Nitellopsis obtusa, **921**, *923,* 931
Nitellopsis (Campaniella) helic-
  teres, **953**, *953*
Nitellopsis (Microstomella) apten-
  sis, **953**, *953*
Nitellopsis (Tectochara) merianii, **920**,
  920, 924, 925, 953
Nitellopsis (Tectochara) pa-
  laeohungarica, **924**
nitrate-reducing bacteria, 22
nitrate respiration, 30, 31
nitrifying bacteria, 22
Nitrobacter, 20, 22, **41**
Nitrobacteraceae, 41
nitrogen cycle, 22
nitrogen fixation, 22
  anaerobic, 53
  bacteria, 54
  blue-green algae, 53, 54
Nitrosomonas, 20, 22, **41**
Nitzsch, C. L., 654
Nitzschia, 572, 579, 600, 605, 613, 617,
  619, 620, 621, 622, 623, 625, 626,
  641, 645, **659**
Nitzschia alba, **571**, 573, 605, 611, 613,
  619
Nitzschia angularis, **571**
Nitzschia barkleyi, **623**
Nitzschia bicapitata, **624**
Nitzschia calva, **648**, *648*
Nitzschia curta, **623**
Nitzschia cylindrus, **623**
Nitzschia delicatissima, **621**, *621*
Nitzschia denticula, **626**, *626*
Nitzschia fossilis, **645**
Nitzschia interfrigidaria, **644**
Nitzschia interrupta, **624**
Nitzschia jouseae, **644**, *644*
Nitzschia kerguelensis, **644**
Nitzschia leucosigma, **613**
Nitzschia linearis, **573**, 611, 632
Nitzschia maleinterpretaria, **647**, *647*
Nitzschia marina, **624**, *625,* 651
Nitzschia miocenica, **644**
Nitzschia porteri, **644**
Nitzschia praeinterfrigidaria, **644**
Nitzschia princeps, **633**, *633*
Nitzschia pseudonana, **623**
Nitzschia pusilla, **647**
Nitzschia putrida, **571**, 613
Nitzschia reinholdii, **645**
Nitzschia seriata, **609**
Nitzschia sicula, **624**
Nitzschiaceae, 659
Nival, P., 253, 255, *256,* 256, 338, 547
Nival, P. (Coste et al.), 547
Nocardiaceae, 42
Nockolds, C. E. (Borowitzka et al.), 112
Noctiluca, 225, 226, 227, 231, 236, 237,
  250, 311, 313, 317, 329, 335, 337,
  339, 340, 344, 347, **380**, 973, 975
Noctiluca miliaris, **226**, *226,* 311, 316,
  317, 337
Noctiluca scintillans, **317**, *337*

Noctilucaceae, 380
Noctilucales, 380
nodal cells, charophytes, **915**, 919,
  971
nodes, coralline algae, **117**, **971**
Nodosella, **780**
Nodosochara, 925, 933, **956**
Nodosochara clivulata, **953**, *953*
Nodosoclavator, 933, **956**
Nodosoclavator nodosus, **928**, *928*
Nodularites, **83**
Noël, D., 735, 736, 737, 738, 745, 768,
  776
Noellithina, **782**
Nooden, L. D., 234
Nordli, E., 275, 344, 717
Nordli, E. (Braarud et al.), 726
Nordli, O., 342
Norstedt, O., 926
Noremia, 816, **818**
Noricysta, 353, 367, **389**
Noricysta fimbriata, **390**, *390*
Norris, G., 284, 295, 297, 299, 364, 368,
  403
Norris, R. E., 124, 538, 541, 546, 579,
  607, 706, 750, 751, 776, 806
Nostoc, 44, *50,* 52, 53, 57, 63, **83**
Nostoc fritschii, **91**
Nostoc muscorum, **91**
Nostocaceae, 53, 71, 74, 82
Nostocales, 47, 49, 53, 58, 59, 63, 80,
  82
Nostochopsidaceae, 83
Nostocopsis, 74, **83**
Nothochara, 933, **956**
Nothochara apiculata, **950**, *950*
Nothooidium, **211**
Nothyocha, 549, **559**
Nothyocha insolita, **559**, *559*
Novantiella, **892**
Nowak, W., 733
Nucellohystrichosphaera, **211**
Nucellosphaeridium, **211**
nuclear cyclosis, dinoflagellates, 330,
  971
nucleoid, **12**, 14, *17,* **968**, **971**
nucleolus, *109*
nucleoplasm, blue-green algae, **44**,
  966, **971**
nucleotide sequences, 92
nucleus
  Clorophyta, 826
  dinoflagellates, 231−237
  euglenids, 894
  origin of, 85, 90
  Rhodophyta, 108
Nuia, **890**
nutrition
  Chrysophyta, 504
  dinoflagellates, 308, 345
  euglenids, 896
Nygreen, P. W., *203,* 838

Obconicophycus, 74, **82**
Oberlies, F., 32, *32,* 34, 35
Oblea, **374**
Obtusochara, 933, **956**
Obtusochara prima, **949**, *949*

Ocampo-Friedmann, R., 57
Ocampo-Paus, R. (Friedmann et al.), 57, 836
Occisucysta, **395**
occluded process, diatoms, **591, 971**
ocellus, dinoflagellate, 231, 242, **243**, *243*, **971**
ocellus, diatoms, **587**, *589, 590*, **971**
Ochetodinium, **404**
Ochromonadaceae, 519
Ochromonadales, 518
Ochromonas, 501, 504, **519**, 967
Ochromonas crenata, **523**, *523*
Ochromonas tuberculata, **504**, *504*, **506**
Ochrosphaera, 138, 506, 684, 714, 747, **777**, 974
Ochrosphaera neapolitana, **683**, *713*, 714, 747
Ochrosphaera verrucosa, **713**, *714*
Ochrosphaeraceae, 777
Ocridoligotriletes, 818
Ocridosphaeridium, **211**
Octactis, 549, **559**
Octactis bioctonaria, **556**, *556*
Octactis pulchra, **556**, *556*
Octaplata, **211**
Octocyclus, **781**
Octoedryxium, **211**
Octogoniella, 838, **888**
Octogonium, **209**
Octolithites, **779**
Octopodorhabdus, **781**
Odontella, **626**
Odontella regia, **599**
Odontochitina, 288, 302, **424**
Odontochitina costata, **427**, *427*
Odontochitina operculata, **361**
Odontochitiniaceae, 424
Odontochitinopsis, **424**
Odontochitinopsis molesta, **427**, *427*
Odontomorpha, **211**
Odontotropis, 639, **657**
Odontotropis klavensii, **647**
Odum, E. P., 831
Oedogoniaceae, 894
Oedogoniales, 894
    fossil, 882
    morphology, 882
Oedogoniites, **894**, 896
Oedogoniites palanensis, 882
Oedogonites, **882**
Oedogoniophyceae, 894
Oedogonium, 55, 828, 833, 882, **894**
Oehler, D. Z., 68, 74, 94
Oehler, D. Z. (Oehler et al.), 94, 128
Oehler, D. Z. (Walter et al.), 94, 96, 130, *132*
Oehler, J. H., 35, 94, 128
Oehler, J. H. (Walter et al.), 94, 96, 130, *132*
Oesterlin, R. (Wong et al.), 317
Oestrupia, **658**
Ogawa, R. E., 49, 54
Ohta, T., 95
Oikopleura, 254, 310
Okada, H., 706, 710, 716, *718*
Okedonia, **659**

Okkolithus, **778**
Okuno, H., 581, 648
Oligopogonium, **211**
Oligoporella, **893**
Oligosphaeridium, **409**
Oligosphaeridium asterigerum, **415**, *415*
Oligosphaeridium complex, **415**, *415*
Oligosphaeridium prolixispino-sum, **415**, *415*
Oligosphaeridium pulcherrimum, **361**
oligotrophic lakes, 930
oligotrophic water, **604, 971**
Olisthodiscus, **519**
Ollaria, **893**
Olpidium, 613
Olsen, S., 930
Omanodinium, **389**, *389*
Omatia, **409**
Omphalotheca, **600**
Oncobyrsella, **82**
oncolites, *63*, 64, *64*, 127, **971**
Onondagella, **211**
Oocardium, 885, **894**
Oocardium stratum, **886**, *886*
Oocystaceae, 889
    fossils, 842
    morphology, 841
Oocystis, **889**
Oodiniaceae, 380
Oodnadattia, **409**
oogonia, **598, 919**
Ooidaceae, 207
Oodinium, 309, 310, **380**
Ooidium, 149, 160, **211**
Ooidium rossicum, **157**, *162*
Ooidomorphida, 207
Oolithotus, **783**
Oolithotus antillarum, **770**
Oolithotus fragilis, **715**, 716, 717, 719, 721, 724, 725
Oomorphitae, 207
oosporangium, charophyte, **919, 971**
oospore membrane, charophyte, **926**
    sculpture of, 927
Opephora, 657, **658**
Operculites, 178, **211**
Operculodinium, 295, 299, 349, **424**, *424*
Operculodinium centrocarpum, *283*, **299**, *302*, 424
Operculodinium erinaceum, **423**, *423*
Operculodinium israelianum, **299**
Operculodinium placitum, **424**, *424*
Operculodinium psilatum, **299**
Operculodinium tiara, **294**, *422*
operculum, 178, *293*, **971**
Ophiaster, **784**
Ophiaster hydroideus, **721**, *723*
Ophiobolidae, 208
Ophiobolus, 171, **212**
Ophiocytium, 491, 493, 494, **498**
Ophiocytium gracilipes, **492**, *492*
Ophiocytium majus, **493**, *493*
opisthoclades, ebridian skeleton, **465**, 971
opisthomere, **258**, *258*, **972**
Oppenheimer, C. H., 27

Oppilatala, **211**
Oppilatala vulgaris, **193**, *193*
Orastrum, **778**
Ordovicidium, 152, **211**
Ordovicidium elegantulum, **167**, *167*, *189*
organelles
    fossilized, 94
    origin of, 86
organic matrix, coccoliths, **684**
organic matrix, diatom frustule, **572**
organic matter, in solution, 21
organotrophs, **18**, 20
origin of eucaryotes, theories
    autogenous origin, 85
    endosymbiosis, 89
    evolutionary, 85
    membrane invaginations, 85
    phagocytosis, 88
    serial symbiosis, 89
Orlando, H. A., 623, 641
Ornithocercaceae, **371**
Ornithocercus, 306, 311, 324, 325, 326, 344, 346, **371**
Ornithocercus magnificus, **271**, *271*
Ornithocercus splendidus, **346**
Orioporella, **893**
Orr, W. N., 639
ortholithae, coccoliths, **694, 706**, *707*, **972**
Orthorhabdus, **778**
Orthosphaeridium, 149, 160, 176, *197*, *197*, 206, **211**
Orthosphaeridium insculptum, **165**, *165*
Orthosphaeridium vibrissiferum, **157**, *157, 188*
Orthozygus, **778**
Orthozygus aureus, **697**, *697*, **752**
Orthriosiphon, **890**
Orthriosiphonoides, **890**
Ortonella, 83, 847, **890**
Ortonella furcata, **66**, *66*
Orygmatosphaeridium, **818**
Oscillatoria, 29, 32, 45, 47, 54, 55, 57, 59, 62, 69, 80, **82**, 833
Oscillatoria amoena, **44**
Oscillatoria animalis, **52**
Oscillatoria limnetica, **53**, 59
Oscillatoria limosa, **59**
Oscillatoria princeps, **43**, 52
Oscillatoria rubescens, **46**
Oscillatoriaceae, 45, 49, 53, 62, 70, *77*, 82
Oscillatoriites, **82**
Oscillatoriopsis, 74, **82**
Oscillatoriopsis obtusa, **77**, *77*
Osorio-Tafall, B. F., 374
Ostenfeld, C. H., 273, 726, 808, 812
Ostreopsidaceae, 374
Ostreopsis, **374**
Ostreopsis siamensis, **373**, *373*
Ottavianus, **778**
Outesia, 507, **528**
Outesia perlata, **529**, *529*
Outesia robusta, **508**, *508*
Outesia yberiensis v. *reticulato-spinosa*, **508**, *508*

Outka, D. E., 684, 689, *690*, 714
Overstreet, K. A. (Donnelly et al.), 341
Ovnia, **822**
Ovocalcisporites, **139**
Ovocalcisporites multidissaeptus, **130**, *131*
Ovoidinium, 295, 389
Ovulites, 848, 891
Ovulites elongatus, **848**, *848*
Ovulites margaritula, **848**, *848*
Ovulites oehlerti, **848**, *848*
Ovum, 181
Owens, J. (Hastings et al.), 317
Owens, O. van H. (Bunt et al.), 616
oxalic acid, 60
oxygen respiration, *30, 31*, 85
oxygen limitations, blue-green algae, 55
oxygen tolerance, coccolithophores, 724
oxygenation, charophytes, 930
oxygenic atmosphere, 95
Oxyphysis, 324, **371**
Oxyphysis oxytoxoides, **370**, *370*
Oxyrrhis, 319, 337, 366, **378**
Oxyrrhis marina, **231**, *233*, *237*, 244, 248, 250, *259*, 308, 334, 344, 724
Oxytoxaceae, 374
Oxytoxum, 338, 374, *374*
Oxytoxum scolopax, **377**, *377*
Ozotobrachion, 160, **211**
Ozotobrachion dicros, **193**, *193*
Ozotobrachion pulvinus, **195**, *195*

Paasche, E., 113, 128, 687, 688, 706, 713, 719, 720, 721, 724
Paasche, E. (Eppley et al.), 596
Pachysoeca, **525**
Pachysphaera, 152, 206, 496, 805, 808, 809, 812, 814, 816, 818
Pachysphaera pelagica, **808**
Pachysphaeridium, **211**
Padan, E. (Cohen et al.), 53
Paddock, T. B. B., *581*, 581, *582*, *626*, 689, 714
Pagodaporella, 869, **893**
Palacanthus, **211**
Palaeachlya, 65, **83**
Palaeachlya perforans, **67**, *67*
Palaeoanacystis, 69, 74, **81**
Palaeoanacystis vulgaris, **77**, *77*
Palaeoberesella, **892**
Palaeobereselleae, 892
Palaeobion, **211**
Palaeochara, 919, 933, 939, 953, **955**
Palaeochara acadica, **941**, *941*
Palaeocharaceae, 955
Palaeocladus, **893**
Palaeocodium, **891**
Palaeocodium saharianum, **850**, *850*
Palaeocryptidium, **211**
Palaeocystodinium, **392**
Palaeocystodinium incertum, **392**, *392*
Palaeodasycladus, **893**
Palaeodasycladus mediterraneus, **874**, *874*
Palaeogeminella, 843, **889**
Palaeohystrichophora, **430**

Palaeohystrichophora infusorioides, **433**, *433*
Palaeolyngbya, 74, **82**
Palaeolyngbya barghoorniana, **77**, *77*,
Palaeomicrocoleus, **82**
Palaeomicrocystis, **81**
Palaeonitella, **957**
Palaeonostoc, **83**
Palaeopede, 65, **83**
Palaeoperidiniaceae, 383, 389
Palaeoperidinium, 294, 349, 352, 383, **389**
Palaeoperidinium pyrophorum, **295**, *361*, *385*
Palaeophalacroma, 265, 346, **371**
Palaeophalacroma verrucosa, **372**, *372*
Palaeophyllum, **141**
Palaeopleurocapsa, **82**
Palaeopleurocapsa wopfnerii, **74**, *74*
Palaeopontosphaera, **782**
Palaeopontosphaera dubia, **735**, *737*, *749*
Palaeoporella, **891**
Palaeorivularia, **83**
Palaeoscytonema, 69, **83**
Palaeosiphonella, 74, 846, **890**
Palaeospiralis, 69, **82**
Palaeospirulina, 69, **82**
Palaeospongia, **892**
Palaeostomocystis, **430**
Palaeostomocystis reticulata, **435**, *435*
Palaeotetradinium, 385, **430**
Palaeotetradinium silicorum, **435**, *435*
Palaeothamnium, **141**
Palambages, 838, **888**
Palambages morulosa, **837**, *837*
Palamphimorphium, **211**
Palanaea, **211**
Paleoclosterium, 886, **894**
Paleoclosterium leptum, **887**, *887*
Paleococcolithus missouriensis, **727**, *728*, *727*
Paleodictyon, **839**
paleoecology, acritarchs, 199
paleoecology, coccolithophores, 769
paleolimnology, diatoms, 649
Paleooedogonium, 882, **894**
Paleooedogonium micrum, **883**, *883*
Paleopedicystus, **211**
Paleopikea, **140**
Paleosiphonia, **140**
paleotemperatures
  coccolithophores, 769
  diatoms, 650
  silicoflagellates, 552
Paleothamnion, **142**
Paleozoic occurrence
  acritarchs, 185
  charophytes, 937
  coccoliths, 727
  dinoflagellates, 352
Palinosphaera, **785**
Palmellaceae, 888
Palmellococcus, 836, **889**
Palmeria, **656**
Palmnickia, **409**
Palmogloea, 836, **888**
Palmophylloites, **888**

Palmophylloites europaea, **836**
Palusphaera, **784**
pandasutural zone, dinoflagellate cyst, **288**, *972*
Pander, C. H., 937
Pankratz, H. S., 571, 585, 610
Panomnimella, 77, **83**
Pansart, J. (Combaz et al.), 206
Pantocsekia, **657**
pantonematic flagella, 895
Papenfuss, G. F., 80, 138, 226, 477, 499, 654, 913
Papillomembrana, 864, **892**
Papillomembraneae, 892
Papinochium, **211**
Pappomonas, **784**
Papposphaera, **784**
Paracapsa, 55, **81**
Paracarnegia, **507**
Parachaetetes, **140**
Parachrysostomum, **507**
paracingulum, dinoflagellate cyst, **284**, *972*
Paraclericia, **507**
Paracoccus, **30**
Paracoccus denitrificans, **91**
paraconformities, Cretaceous, 772
Paradeflandreia, **507**
Paradella, **891**
Paradictyocha, 549, 554, 559
Paradictyocha dumitricae, *551*, **552**
Paradictyocha polyactis, **556**, *556*
Paragarwoodia, **890**
Parahistioneis, 311, **371**
Parakamaena, **892**
Paralecaniella, **430**
Paralia, **655**
paralytic shellfish poisoning, 317
Parammodochium, **480**
Paranetrelytron, 281, **386**, *389*
Paraoutesia, **507**
Paraphysomonas, 500, 519
Paraphysomonas cylicophora, **501**, *501*
Paraphysomonas corbidifera, **501**, *501*
paraplates, dinoflagellate cyst, **972**
Pararchaeomonas, 512, *515*, **528**
Pararchaeomonas colligera, **528**, *528*
parasites, dinoflagellate, 309
Parastacheia, **142**
Parastomiosphaera, 733, **785**
parasutures, dinoflagellate cysts, **287**, *972*
paratabulation, dinoflagellate cyst, **287**, *972*
Parathranium, 469, 472, 476, 479, **480**, *480*
Parathranium biclathratum, **479**
Parathranium intermedium, **471**, *471*
Parathranium quadrispinum, **484**, *484*
Parathranium tenuipes, **469**, *469*, *471*, *484*
Parebria, **480**
Parebriopsis, 476, **480**
Parebriopsis fallax, **483**, *483*
Pareodinia, 281, 383, **386**
Pareodinia ceratophora, **281**, *389*
Pareodinia evittii, **389**, *389*
Pareodiniaceae, 386

Parhabdolithus, 735, **780**
Parhabdolithus embergeri, **735**, *738,* 749
Parhabdolithus liasicus, **735**, 749
Parhabdolithus marthae, **735**, 749
parietal plastids, **116, 972**
Parietoplastophyceae, 887
Parke, M., 307, 518, 681, *683,* 683, 684, *692, 712,* 712, *713,* 713, 714, 715, 716, 725, 776, *806,* 806, 808, *810*
Parker, B. C., 136, 828
Parkerella, **894**
Parrocelia, **377**
Partitiofilum, 74, **82**
Parvicorbicula, **526**
Parvicorbicula campaniformis, **527**, 527
Parvicorbicula spinifera, **527**, 527
Parvocavatus, **389**
Pascal, M. (Feys et al.), 841
Pasceolus, **891**
Pascher, A., *381,* 491, *492, 494, 495, 497,* 518, 836
Pascherella, **518**
Pascherinema, **82**
Pascherinemataceae, 82
passage pores, diatom frustule, **585, 972**
Pastiels, A., 278
Patrick, R., 607, 626, 628, 649
Patton, S., 339
Paul, P., *925*
Paulinella, symbionts of, 63, 92, 311
Paulinella chromatophora, 63
Paulsen, O., 273, 274, 275, 329
Paulsenella, 311
Paulsenella chaetoceratis, **613**
Pavillard, J., 326, 623
Pavillardia, **380**
Pavillardia tentaculifera, **250**
Pavillardinium, 265, **374**
Pavillardinium kofoidii, **254**, 256
Pavlova, **784**, *968*
Pavlova gyrans, **681**, 683
Pavlova lutheri, **679**
Pavlova mesolychnon, **681**
Pavlovaceae, 784
Pavlovales, 784
Pavlovskya, **211**
Pearse, V. B., 314
Pearson, B. R., 806
Peck, R. E., 921, *928,* 936, 937, 938, *939, 941,* 941, *949,* 953
Peckichara, 925, 933, **956**
Peckichara varians, **953**, *953*
Peckisphaera, 933, **956**
Peckisphaera verticillata, **949**, *949*
Pediastrum, 149, 204, 828, 829, 839, *841,* 889
   wall of, 839
Pediastrum biradiatum, **839**
Pediastrum boryanum, **839**
Pediastrum kajaites, **841**, *841*
Pediastrites, 839, **889**
pedicel cells, charophyte, **919, 972**
Pedinella, **525**
Pedinella hexacostata, **526**, *526*
Pedinellaceae, 525

Pedinocyclus, **783**
Pedinomonas, 805, **818**
Pedinomonadaceae, 818
Pedinomonadales, 818
Pedobesia, 850, **891**
Pedobesia clavaeformis, **853**, *853*
Pekiskopora, **893**
pellicular layer, dinoflagellate amphiesma, **257**, *257,* 261, **972**
Pelomyxa, 91
Pelonemataceae, 39
Peltacystia, 176, 190, 204, **211**
Peltacystia calvitium, **177**, *177*
Peltacystia venosa, **177**, *177*
Pemma, 770, **779**
Pemma papillatum, **757**, *762*
Pemma rotundum, **761**, *761*
Penicilloides, **891**
Penicillus, 826, 831, 834, **891**
Penieae, **894**
Peniguel, G., *195*
Penium, **894**
Pennales, 568
pennate diatoms, **568, 972**
Pennatibacillariophyceae, 657
Pennick, N. C., 248, *500,* 500, *501, 503,* 806, *806, 807*
Pentadinium, **430**
Pentadinium circumsutum, **406**, *406*
Pentadinium laticinctum, 361, **433**, *433*
Pentadinium laticinctum imaginatum, **433**, *433*
Pentadinium taeniagerum, **433**, *433*
Pentagonum, **389**
pentalith coccolith, **706**, *707,* **972**
Pentaporella, **893**
Pentaster, 770, **779**
pentasters, **250, 972**
Peptococcaceae, 41
Peragallo, H., 623
Peragallo, M., 623
Peranema, 896, **898**
percent G+C, **31**
Perch-Nielsen, K., *465, 468,* 508, 512, *514, 515, 516, 551, 559,* 697, *705,* 708, 726, 741, *743, 745,* 745, *752,* 754, *755, 757,* 761, *761, 764, 766,* 771
Perch-Nielsen, K. (Manivit et al.), *741, 748*
Percival, S. F., *774, 775*
Percivalia, **782**
Percivaliaceae, 782
Percultisphaera, **211**
perforating algae, 59, *60,* 65, 67, 126
perforating fungus, 60
periblast, dinoflagellate cyst, **284, 972**
pericoel, dinoflagellate cyst, **283, 284,** 286, **972**
pericorpus, dinoflagellate cyst, **972**
pericyst, dinoflagellate cyst, **284,** 286, **972**
Peridiniaceae, 233, 239, 374
Peridiniales, 233, 261, 301, 351, 368, 371, 386
   tabulation of, 261
Peridiniopsis, 274, 337, 341, 343, 344, 348, **374**

Peridiniopsis borgei, **274**, *274,* 275
Peridiniopsis cunningtoni, **275**
Peridiniopsis lubiniensiforme, **331**, *332*
Peridinitaceae, 386
Peridinites, 279, 350, 383, **386**
Peridinites maculatum, **387**, *387*
Peridinites oamaruense, **281**, *281*
Peridinium, 229, 231, 232, 264, 265, 272, 274, 275, 276, 278, 283, 286, 288, 297, 299, 301, 302, 320, 321, 337, 340, 348, 373, 374, **377**, 383, 385, 389
Peridinium aciculiferum, **324**
Peridinium balticum, **237**, 239, 311, 320, 366
Peridinium bipes, **229**, *229,* 321
Peridinium cinctum, 227, 230, **237**, 244, 272, 301, 331
Peridinium cinctum f. *ovoplanum,* **329**, 331
Peridinium cinctum f. *westii,* **234**, 240, 242, 261, 348
Peridinium delitiense, **349**
Peridinium depressum, 367
Peridinium elpatiewskyi, **275**
Peridinium foliaceum, **237**, 239, 240, 242, *242,* 244, 259, 272, 311, 312, 366, *377*
Peridinium gargantua, **309**
Peridinium gatunense, **321**, 331
Peridinium gregarium, **306**, 348
Peridinium limbatum, **242**, 276, 288, 297, 299, 301, 367
Peridinium limbatum v. *minnesotense,* **297**
Peridinium nivale, **274**
Peridinium ponticum, **278**, *279, 279*
Peridinium priscum, 349
Peridinium pyrophorum, 349
Peridinium sociale, **330**, 330
Peridinium triquetrum, **234**, *241,* 241, 244, 272, 319, 330, 340, 343, 350
Peridinium uberrimum, **348**
Peridinium volzii, **321**
Peridinium willei, **321**
Peridinium wisconsinense, **240**
Perimneste, 933, 941, **956**
Perimneste ancora, **944**, **945**, *945,* 947
Perimneste horrida, **943**, *943,* **944**, 945, 947
Perimneste micrandra, **945**, *945,* 947
Perimneste vidua, **945**, *945,* 947
Perimneste-Atopochara lineage, *945, 947*
perinuclear capsule, **231, 972**
Periphyllophora, **778**
Periphyllophora mirabilis, 695, **696**, *696*
periphragm, **972**
   acritarch, 152
   dinoflagellate cyst, **281, 284**
periplekton, diatom frustule, **591, 972**
Periptera, 600, **657**
Perisaccus, **818**
perispore, **278**, 320, **972**
Perisseiasphaeridium, **409**

Perisseiasphaeridium pannosum, **415,** *415*
Perissocyclus, **781**
Perissolagonella, **211**
Perissolagonella amsdenii, *193, 193*
perithallium, **117, 972**
perithallus, *119*
Peritrachelina, 770, **778**
Peritrachelina joidesa, **697,** *697,* **752**
peritrichate bacteria, *12,* **16, 972**
perizonium, diatoms, **600, 972**
Perkins, R. D., 62
Permocalculus, **139**
Permoperplexella, **893**
Peronia, **658**
Peroniopsis, **658**
Perret, M.-F., 848
pervalvar axis, **576, 972**
Petalomonas, **898**
Petalosphaera, **778**
Petalosporites, 176, **211**
Petasus, **784**
Peteinosphaeridium, 152, 178, 184, **211**
Peteinosphaeridium brevira-diatum, **178**
Peteinosphaeridium majorfurca-tum, **152**
Peteinosphaeridium trifurcatum, **152**
Peteinosphaeridium velatum, **170,** *170, 181*
Peters, N., 326, 327, *327,* 335, 337
Petraphera, 69, 80, **83**
Petrascula, **893**
Petrasculeae, **893**
Petrikova, M. N. (Beklemishev et al.), 609
Petrocelis, **142**
Petroglossum, *124*
petroleum accumulation, 27
Petrophyton, **140**
Petryk, A. A., 133, 142
Petschoria, **142**
Peyssonelia, 117, **140**
Peyssoneliaceae, 110, 135, 140
Pfiester, L. A., 278, 329, 331
Pfister, R. M. (Gold et al.), *511*
Pflug, H. D., 68
Phacodiscus, **656**
Phacotaceae, **888**
Phacotus, 828, 835, **888**
Phacotus lenticularis, **836**
Phacus, 895, 896, **898**
Phacus caudata, **897,** *897*
Phaeocystaceae, **784**
Phaeocystis, **784**
Phaeocystis pouchetii, **715**
Phaeodactylaceae, 659
Phaeodactylum, 572, 620, **659**
Phaeodactylum tricornutum, **572,** *604, 610, 614, 618, 619*
Phaeogloea, **519**
Phaeoplacales, 519
Phaeothamniales, 519
Phaeothamnion, **519**
phage, 18
Phalacroma, **353**
Phanerodinium, 350

Phanerosphaerops, 74, **81**
Phanulithus, **778**
Pharetronida, 133
Pharus, **778**
Phelodinium, **392**
Phenacocladus, **142**
Pheoclosterium, **211**
Pheoclosterium fuscinulaegerum, **157,** *157*
Pheopolykrikos, **378**
Pheopolykrikos beauchampi, **379,** *379*
Phillips, D. O., 128
Philpott, D. E. (Lewin et al.), 605, 610
Phoberocysta, **424**
Phoberocysta neocomica, 362, **428,** *428*
Phormidium, 29, 51, 54, 80, **82**
phosphorus cycle, 22
phosphorus depletion, 22
Photobacteria, 38, 42
photoorganotrophs, 19
photosynthesis, 687
anoxygenic, 53
bacterial, 19
blue-green algae, 53
evolution of, 30
photosystem I, 53
photosystem II, 53
phototaxis, dinoflagellates, **306, 972**
phototropism, 65
phototrophs, **18, 965**
Phragmonema, **139**
Phragmonemataceae, 139
phragmoplast cell division, **805,** 913, **972**
Phthanoperidiniaceae, 389
Phthanoperidinium, **389**
Phthanoperidinium amoenum, **385,** *385*
phycobilin pigments, 239
phycobilisomes, **45,** *50,* **108,** *109,* **972**
phycocyanin, 53, 108
Phycochromaceae, 81
phycoerythrin, 45, 53, 108
phycoma stage, prasinophytes, 149, **807,** *810,* **972**
Phycopeltis, 843, **889**
Phycopeltis microthyrioides, **845**
phycoplast cell division, **805, 972**
Phyllodictyocha, 549, **559**
Phyllodictyocha recta, **556,** *556*
Phyllodictyocha schulzii, **556,** *556*
Phyllophora, **142**
Phyllophoraceae, 142
Phymatolithon, **141**
Physoporella, **893**
Phytodiniaceae, 239, 380
Phytodiniales, 380
Phytomastigophorea, 477, 554
phytoplankton biomass, 21
phytoplankton diversification, *5,* 7
phytoplankton extinction, 5, 7
Pia, J., 637, 654, *827, 845, 859, 860, 861, 866, 867, 868, 870, 874*
Pianella, **893**
Pickett-Heaps, J. D., 89, 93, 137, 805
Picostella, **211**
Pictonicopila, 149

Pienaar, R. N., 745
Pierce, M. J. (Worth et al.), 317, 341
Pierce, S. T., 822
Piggott, G. H., 91
pigments
bacterial, 19
Cyanophyta, 45
Chlorophyta, 825
Chrysophyta, 504
diatoms, 571
dinoflagellates, 237, 239
Haptophyta, 681
Rhodophyta, 108
Xanthophyta, 493
pigments, light requirements, 108
Pilasporites, **211**
Piliferosphaera, **211**
Piliferosphaera setosa, **201,** *201*
Pillotina calotermitidis, **93**
Pilodea, **139**
pinnae, acritarchs, **167, 973**
pinnulae, acritarchs, **167, 973**
Pinnularia, 571, 583, 624, 626, 628, 641, **658**
Pinnularia torta, **648,** *648*
Pinnularia viridis, **648,** *648*
Pirea, **311**
Pirini Radrizzani, C., *727, 728*
pit connections, red algae, *111,* **112,** *112,* **973**
Pithonella, 279, 733, 734, 747, **785**
Pithonella carteri, **734,** *734*
Pithonella cooki, **734,** *734*
Pithonella edgari, **734,** *734*
Pithonella francadecimae, **734,** *734*
Pithonella ovalis, **733, 734,** *734*
Pithonella quiltyi, **734,** *734*
Pithonella veeversi, **734,** *734*
Pithonellidae, 785
Placklesia, **893**
placolith coccolith, **697,** *700,* **973**
Placozygus, **779**
Plaesiodictyon, **889**
Plagiodiscus costatus, **587**
Plagiogramma, 653, **658**
Plagiogramma nankoorense, **589,** *589*
Plagiostomum, 507
Planinosphaeridium, **404**
Planktonetta atlantica, **310**
Planktoniella, **655**
Planktoniella sol, **621,** *623*
Planoumbella, **955**
Planoumbella costata, **934,** *934*
Planoumbella patella, **934,** *934*
planozygote, **329,** 334, **973**
plant protists
fluctuations in diversity, *5*
geologic occurrence, 2
as rock builders, 3
plasmalemma, **43,** *50,* 260, **261, 973**
plasmodesmata, **47, 973**
plastids
DNA in, 85
endosymbiotic origin, 91
See also pigments
plate symbols, dinoflagellate theca, 262, 263
Platonov, V. A., *934*

Platychara, 933, **956**
Platychara compressa, **953,** *953*
Platychrysidaceae, 784
Platychrysis, 681, **784,** 968
Platymonas, *805,* 805, 806, 807, 812, 818, *818*
Platymonas subcordiformis, 813
Platymorphitae, **208**
Plavskina, **955**
Plavskina piriformis, **934,** *934*
Playford, G., 838
Playford, P. E., 65
Plaziat, J. C., 65
Plectodinium, 254, **378,** 964
Plectodinium nucleovolvatum, **253,** *253,* 367
Plectonema, 51, 59, 83, *83*
Plectonema terebrans, **62**
Pleonosporium, 124
pleurae, diatoms, **576,** 973
Pleurangium, 185
Pleurocapsa, **82**
Pleurocapsaceae, 82
Pleurocapsales, 58, 59, 74, 80, 82
Pleurochloris, **498**
Pleurochrysis, 714, **777**
Pleurasiga, 499, *504,* **526**
Pleurasiga reynoldsii, **527,** *527*
Pleurosigma, 624, **658**
Pleurotaenium, 828, **894**
Pleurozonaria, **785**
Plocamiaceae, 141
Plocamium, **141**
Ploiaria, **657**
plug, chrysophyte cyst, **506,** 973
Plumaria, **142**
Pluriarvalium, **386**
Pocock, S. A. J., 205, 355
Pocock, S. A. J. (Staplin et al.), 160, 173, 181, 205
Pocockia, **389**
Podamphora, 469, 472, 476, **480**
Podamphora elgeri, **471,** *471,* 483
Podamphoraceae, 480
Podamphoropsis, 476, **480**
Podamphoropsis joidesi, **483,** *483*
Podium, 480
Podocystis, **658**
Podolampaceae, 377
Podolampas, **377,** *377*
Podolampas bipes, **377,** *377*
Podoliella, **211**
Podolina, **211**
Podorhabdaceae, 781
Podorhabdales, 781
podorhabdid coccolith, *705*
Podorhabdinales, 781
Podorhabdoideae, 781
Podorhabdus, **781**
Podorhabdus albianus, **741,** *741,* 745, 748
Podorhabdus rahla, **735,** *737,* 749
Podosira, **655**
Poelchau, H. S., 552
Poignant, A. F., 135
Poikilofusa, **211**
Poikiloporella, **893**

polar nodule
of blue-green algal heterocyst, *50*
of diatom, **577,** 973
polar plug, heterocyst of blue-green algae, 49
polar pore, of diatom, **579,** 973
Polenova, E. N., 932, *934*
Polyancistrodorus, 178, 181, 187, **211**
Polyancistrodorus columbariferus, **165,** *165,* 188
Polyangiaceae, 39
Polycladolithus, **778**
Polycostella, **779,** 780
Polycostella beckmannii, **735**
Polycyclolithaceae, 780
Polycyclolithus, **780**
Polydeunffia, **211**
Polyebriopsis, 476
Polyedriella, **498**
Polyedrosphaeridium, **818**
Polyedryxidae, 207
Polyedryxium, **822**
Polygonifera, **389**
Polygonium, 166, *188,* 201, **211**
Polygonomorphitae, **207**
Polygonosphaerites, **892**
polyhedral bodies, blue-green algae, *43,* 46, *50,* **973**
Polyideaceae, 142
Polyides, **142**
Polykrikaceae, 378
Polykrikos, 247, 308, 317, 319, 347, **378,** 479
Polykrikos lebourae, **347**
Polykrikos schwartzii, **247,** *247,* 317, 379
Polymorphocodium, **890**
Polymyxus, **656**
Polynucella, **822**
Polyoeca, **526**
polyphosphate bodies, blue-green algae, 46, **973**
Polyphysa, **894**
Polyplanifer, **211**
polyploidy, blue-green algae, 51, 87
Polypodorhabdus, **781**
Polypodorhabdus escaigii, **735,** *737,* 749
Polyporolithon, **141**
Polysphaeridium, **419**
Polysphaeridium pumilum, **423,** *423*
Polysphaeridium simplex, **283**
Polysphaeritae, **208**
Polysiphonia, 112, 124, 126, **142,** 975
Polysiphonia boldii, **125,** *125*
Polysiphonides, **142**
polysporangia, *124,* 124, **973**
Polystephanephorus, **419**
Polytoma, 835, **888**
Polytomella, 825, **888**
Pomatodinium, **380**
Pomeroy, L. R., 342
Pomotophyllum, **141**
Pomphodiscus, **656**
Pontilithus, **780**
Pontolithina, **780**
Pontosphaera, 721, **780,** *780*
Pontosphaera borealis, **721**

Pontosphaera discopora, 697, **701,** *701*
Pontosphaera syracusana, 697, **701,** *701*
Pontosphaeraceae, 780
Pool, R. R. (Muscatine et al.), 314
Porata, 208
pore, chrysophyte cyst, **506,** 973
pore plate, dinoflagellate theca, 265, 973
pore plug, red algae, 112
porelli, **591**
Poretzkia, **657**
Porites, 314
Porochara, 933, **955**
Porochara abjecta, **941,** *941*
Porocharaceae, 955
Porocharoideae, 955
Porochrysis, **518**
Porodiscus, **656**
Porodiscus interruptus, **605**
Porolithon, 126, 140
poroliths, of *Thoracosphaera,* 706
Porosira, 601, **655**
Porostromata, 83
Porpeia, **657**
Porphyra, 108, 110, 112, 114, 116, 126, 137, **139**
Porphyra leucostica, 112
Porphyra stage, 116
Porphyra umbilicalis, **115,** *115*
Porphyrella, 132, **139**
Porphyridiaceae, 113, 138
gliding movement of, 113
Porphyridiales, 107, 114, 138
Porphyridium, 91, 110, 113, 137, **138**
Porphyridium aerugineum, **109,** *109,* 110, 137
Porphyridium cruentum, 144, 126
Porphyridium purpureum, **113,** *113,* 114, 126, 128
Porphyropsis, 116, **139**
Porpita, 314
Portalites, **211**
portals, silicoflagellate skeleton, **539,** 973, 977
portulae, diatoms, **581,** 973
postcingular plate series, **263,** 973
post-equatorial process, acritarchs, 167
posterior intercalary plates, **263,** *263,* 973
posterior process, acritarchs, 167
Potamodiscus, 512, **519**
Potamodiscus kalbei, **512,** *512*
Potamogeton lakes, 930, 939
Pöthe de Baldis, E. D., *390*
Pouchet, G., 243, 323, 324, 325, 327, 330
Pouchetia, **380**
Poyarkov, B. V., 932, 936
Praechara, **955,** *957*
Praechroococcus, 77, **81**
Praedonezella, **868**
Praeepithemia, **658**
Prakash, A. (Ackman et al.), 812
Prashnowsky, A. A., 32, *32,* 34, 35
Prasinochloris, **818**
Prasinocladaceae, 818
Prasinocladales, 818

Prasinocladus, 807, **818**
Prasinophyceae, 149, 152, 205
Prasinophyta, 804–824
  cell, 804, **805**
  classification, 816
  cyst or phycoma, 807
  cyst composition, 809
  cyst development, 808
  cyst structure, 809
  distribution, 812
  ecology, 812
  flagella, 804
  fossil genera, rate of description, 4
  geologic occurrence, 813
  reproduction, 806
  scales, motile cell, *806,* **806,** 807
Prasiola, 834, **890**
Prasiolaceae, 890
Prasiolales, 890
Praturlon, A., *851, 880, 881*
preapical plates, dinoflagellate
  theca, **973**
Precambrian life, 10
  bacteria, *34*
  eucaryotes, *93*
precession, in coccoliths, **703, 973**
precingular archeopyle, 296, *296*
precingular plate series, dinoflagellate
  theca, **263, 973**
Prediscosphaera, **782**
Prediscosphaera decorata, **782**
Prediscosphaera spinosa, 748
Prediscosphaeraceae, 781
pre-equatorial processes, acri-
  tarch, 167
Preikstas, R. (McIntyre et al.), 751
Preiss, W. V., 65
Preiss, W. V. (Glaessner et al.), 65
Premoli-Silva, I. (Haq et al.), 770
pre-perizonium, diatoms, **599, 973**
Prescott, G. W., 834
pressure tolerance, bacteria, 23
Preston, R. D., 827, 846
Price, C. C., 54
primary pit connections, red al-
  gae, 112
primary productivity, variations, 6
Primicorallina, 864, **892**
Primicorallina trentonensis, **866,** *866*
Primicorallineae, 892
Primorivularia, 69, 83
Pringsheim, E. G., 895
Pringsheim, O., 895
Prins, B., 735
Prins, B. (Manivit et al.), 741, 748
Prinsiaceae, 784
Prinsius, **784**
Prinsius bisulcus, **771**
Prinsius martinii, **771**
Prionodinium, **395**
Priscogalea, 151, 178, *178, 206,* **211**
Priscotheca, **211**
Prismatocystis, **211, 424**
prismatolith coccolith, **706,** *707,* **973**
Prismatomorphitae, **207**
Proaulopora, **892**
Procaryota, 1, 38, 39
procaryote, 9, **973**

diversification of, *30*
  evolution, *30*
  heterotrophs, 84
  photosynthetic procaryotes, 84
processes, acritarchs, **152,** 160, 167,
  973
processes, dinoflagellate cysts, **288,**
  289, **973**
Prochloraceae, 84
Prochlorales, 84
Prochloron, **84**
Prochloron didemni, **84**
Prochlorophyceae, 84
Prochlorophyta, 38, 42, 84
proclades, ebridian skeleton, **465, 973**
Proctor, V. W., 881, 886, 917
prod, dinoflagellate, 226, **250, 973,**
  975
productivity
  blue-green algae, 54
  coccolithophores, 719, 761
  diatoms, 614
  dinoflagellates, 338
Proebria, **483**
proembryo, charophyte, 927
Progonoia, **658**
proheterocysts, blue-green algae, 49
Prolatipatella, **782**
Prolixosphaeridium, **419**
*Prolixosphaeridium mixtispino-
  sum,* 421, *421*
Pronoctiluca, **378**
Pronoctilucaceae, 378
Propionibacteriaceae, 41
proplastids, **896, 973**
Prorocentraceae, 233, 239, 369
Prorocentrales, 48, 261, 301, 351, 369
  division of, 323
  tabulation, 265
Prorocentrum, 229, 232, 239, 240, 306,
  317, 319, 323, 338, 340, 343, 344, 345,
  346, 366, **369**
  apical region, 268, 269
Prorocentrum balticum, **232,** 233, 237,
  240, 244, 343, 344
Prorocentrum cassubicum, **234,** 237,
  239, 269, 369
Prorocentrum marinum, **248,** *248,* 347
Prorocentrum micans, **231,** 232, 233,
  235, 236, 241, 244, 247, 269, 319,
  342
Prorocentrum minimum, **232,** *232,*
  233, **235,** *235,* 241, 244, 248, 259,
  268, 269, *269,* 307, 339
  movement of, 307
Prorocentrum minutum, **301**
Prorocentrum pusillum, **233,** 244
Prorocentrum rhathymum, **248**
Prorocentrum subinerme, **301**
Prorocentrum triestinum, **232,** 233,
  248
Proshkina-Lavrenko, A. I., 548, 623,
  641, 654
prosobase, **258,** *258,* **973**
prosomere, **258,** *258,* **973**
Protaspidaceae, 381
Protaspidales, 381
Protaspis, **381**

Protaspis glans, **382,** *382*
Proterospongia, **525**
Proterotainia, 138, **139**
Proterotainia montana, **130**
Proterotainia neihartensis, **130,** *132*
Proterythropsis, **380**
Proterythropsis crassicaudata, **250**
Proteus, 31, 41
Protoarchaeosacculina, **211**
Protobolella, **818**
Protoceratium reticulatum, 274, 299,
  302, 350
Protoceratium splendens, 274
Protochara, **956**
protochlorophyll, 19
Protoctista, 3
Protodinifer, **378**
Protodiniferidae, 378
Protoellipsoidinium, **404**
protoeucaryote, 86
Protoleiosphaerideae, 207
Protoleiosphaeridium, **818**
protonemata, charophyte, **927, 973**
Protoodiniaceae, 380
Protoodinium, 250, 309, 310, **380,** 973,
  975
Protoperidinium, 263, 275, 301, 317,
  320, 338, 343, 344, 346, 374, **377**
Protoperidinium avellana, **299**
Protoperidinium brochii, **317**
Protoperidinium compressum, **389**
Protoperidinium conicoides, **299**
Protoperidinium conicum, **275,** 317,
  377, 389
Protoperidinium depressum, **244,** 306,
  317
Protoperidinium divergens, **317,** 377
Protoperidinium elegans, **346**
Protoperidinium globulus, **317**
Protoperidinium granii, **274,** 275, 317
Protoperidinium inflatum, **346**
Protoperidinium leonis, **317**
Protoperidinium mediterraneum, **306**
Protoperidinium nudum, **389**
Protoperidinium oceanicum, **301,** *301,*
  316, **317**
Protoperidinium ovatum, **272**
Protoperidinium pallidum, **317**
Protoperidinium pentagonum, **317**
Protoperidinium punctulatum, **317**
Protoperidinium subinerme, **317,** *327,*
  327, 389
Protoperidinium steinii, 317, *343*
Protophyta, 9, 38, 39
Protoraphidaceae, 658
Protoraphis, **658**
Protorivularia, 83
Protosiphon, 836, **888**
Protosiphonaceae, 888
Protospira, 77, **82**
Protospiraceae, 82
Protosphaeridium, **818**
Protospongia, **525**
Prototheca, 836, **888**
Prototrematosphaeridium, **211**
Protoumbella, 936, **955**
Protoumbella saccamminiformis, **934,**
  934

Protovalonia, **891**
Protovalonia ichinzaganense, **854**, *854*
Protovaloniaceae, 891
Provasoli, L., 834, 895
Prowse, G. A., 607
proximate cysts, dinoflagellate, **284**, *294*, **973**
Prymnesiaceae, 784
Prymnesiales, 784
Prymnesiophyceae, 776
Prymnesiophyta, 776
    See also Haptophyta
Prymnesium, **784**
Prymnesium parvum, **681**, *725*
Psaligonyaulax, **430**
Psaligonyaulax deflandrei, **430**, *430*
Psenotopus, **211**
Psenotopus chondrocheus, **157**, *157, 157, 165*
Pseudaethesolithon, **140**
Pseudammodochium, 465, 469, 472, 476, **480**
Pseudammodochium dictyoides, **469**, *469, 487*
Pseudammodochium robustum, **471**, *471, 487*
Pseudammodochium sphericum, **469**, *469*
Pseudamphiprora, **658**
Pseudoaulacodiscus, **657**
Pseudobryopsis, **890**
Pseudobryopsis planktonica, **847**
pseudocellus, diatom frustule, **587**, **974**
Pseudocerataulus, **656**
Pseudoceratiaceae, 424
Pseudoceratium, 294, 302, **424**
Pseudoceratium pelliferum, **428**, *428*
Pseudochaetetes, **140**
Pseudocharaciopsis, 497, 498, **498**
Pseudocharaciopsis texensis, **499**, *499*
Pseudochlorella, **888**
Pseudochlorella pyrenoidosa, **833**
Pseudochlorodesmis, **891**
Pseudoclypeina, **893**
Pseudocodium, **891**
Pseudocodium convolvens, **851**, *851*
Pseudocymopolia, **894**
pseudocyst, **713**, **974**
Pseudodeflandrea, **392**
Pseudodimerogramma, **658**
Pseudodimerogramma elegans, **647**
Pseudodimerogramma filiformis, **647**
Pseudoemiliania, **777**
Pseudoepimastopora, **891**
Pseudoeunotia, 634, **659**
Pseudoeunotia doliolus, **624**, *625, 644, 651*
Pseudoglobator, 933, **956**
Pseudoglobator fourcadei, **944**, *944*
Pseudoharrisichara, 933, **956**
Pseudoharrisichara tenuis, **950**, *950*
Pseudoharrisichara walpurgica, **950**, *950*
Pseudohimantidium, **658**
Pseudokamaena, **892**
Pseudokephyrion, 499, 510, **519**

Pseudokephyrion formosissimum, **523**, *523*
Pseudokomia, **142**
Pseudolithophyllum, **140**
Pseudolithothamnium, **140**
Pseudolithraphidites, **781**
Pseudolunulidia, **818**
Pseudolunulidia imperatrizensis, **817**, *817*
Pseudomasia, 171, **211**
Pseudomesocena, 549, **559**
Pseudomicromarsupium, **559**
Pseudomonadaceae, 40
Pseudomonadales, 40
Pseudomonas, 11, 20, 30, 31, **40**, 91
Pseudomonas aeruginosa, 37
Pseudomonas calcis, **28**
Pseudomonas halocrenaea, 37
Pseudonanopora, **892**
pseudonodule, diatom, **591**, *591*, **974**
Pseudopedinella, **525**
Pseudoperonia, **658**
Pseudopleurococcus, 843, **889**
Pseudopodosira, **656**
Pseudopodosira westii, **639**, *639*
pseudopylome, acritarch, **181**, **974**
Pseudopyxilla, **657**
Pseudopyxilla jouseae, **639**
pseudoraphe, diatom, **577**, **974**
Pseudorocella, 254, 512, 519, **559**, *559*
Pseudorocella corona, **512**, *512*
Pseudorutilaria, **656**
Pseudoschizaea, **818**
Pseudoschizaea circula, **819**, *819*
Pseudoschizaea rubina, **819**, *819*
Pseudoscourfieldia, **818**
pseudosepta, diatom, **587**, **974**
Pseudostacheoides, **142**
Pseudostictodiscus, **656**
Pseudosynedra, **658**
pseudotabulation, **287**, **972**
Pseudotasmanites, **818**
Pseudotetraëdron, 185, **498**
Pseudotetraëdron neglectum, **492**, *492*
Pseudotriquetrorhabdulus, **781**
Pseudoumbella, **955**
Pseudoumbella donbassica, **934**, *934*
Pseudozonosphaeridium, **818**
Psilochara, 933, **956**
Psilochara undulata, **953**, *953*
Psilosphaera, **211**
pterate dinoflagellate cyst, **286**, **974**
pterocavate dinoflagellate cyst, 286
Pterocladia, 110, **140**
Pterococcus, 273
Pterodinium, **404**
Pterodinium aliferum, **406**, *406*
Pteroidea, **210**
Pteromorphitae, 207
Pterosperma, 185, 808, 816, **818**
    life cycle, *810*
Pterosperma labyrinthus, **273**
Pterosperma marginatum, **808**
Pterosperma moebii, **808**
Pterosperma ornatum, **751**
Pterospermataceae, 818
Pterospermatales, 818
Pterospermella, 199, 808, **822**, *824*

Pterospermella dichlidosis, **824**, *824*
Pterospermella reticulata, **808**, *808, 824*
Pterospermellaceae, 207, 822
Pterospermopsidae, 207
Pterospermopsimorpha, **822**
Pterospermopsis, 206, 808
Pterosphaeridia, **822**, *822*
Pterotheca, **656**
Pterotheca aculeifera, **647**
Pterotheca cretacea, **639**, *639*
Pterosphaeridiaceae, 207, 822
Pterotosphaerula, **822**
Ptilotoscolus, **822**
Puiseux-Dao, S., 860
Pulvinomorpha, **211**
Pulvinosphaeridium, 200, **211**
Pulvinosphaeridium parvum, **190**, *190*
Pulvinosphaeridium-Estiastra facies, 200
Pulvinulariaceae, 83
punctate surface, **155**, **974**
purple bacteria, 19
Pustularia, **141**
Pustulariaceae, 141
pusules, dinoflagellates, 231, **244**, *244, 245*, **974**
Pustulisphaeridium, **211**
Puteoscortum, **211**
Pycnoporidium, 852, **891**
pylome, acritarch, **178**, **974**
Pyramidosporites, **211**
Pyramimonadaceae, 818
Pyramimonadales, 818
Pyramimonas, 805, 806, 807, 808, 818
Pyramimonas amylifera, **808**
Pyramimonas grossii, 805, **806**, *806*
Pyramimonas obovata, **806**, *806, 807*
Pyramimonas parkeae, **807**, *807*
pyrenoids, **974**
    of diatoms, 571
    of dinoflagellates, 240, *241*
Pyrgodiscus, 637, **657**
Pyrgupyxis, **656**
Pyrgupyxis eocenica, **647**
Pyrgupyxis prolongata, **646**
Pyrocyclus, **784**
Pyrocystaceae, 377
Pyrocystis, 316, 339, 346, 368, **377**
Pyrocystis acuta, 316, *320, 321*
Pyrocystis elegans, 315, 316
Pyrocystis fusiformis, **316**, *316, 320, 339, 346*
Pyrocystis lunula, **239**, 316, 317, 320
Pyrocystis pseudonoctiluca, **316**, *316, 317*
Pyrodinium, 302, 317, 324, 371, **373**, 409
Pyrodinium bahamense, **237**, 299, 316, 317, 324, 326, 341, 419
Pyrophacaceae, 377
Pyrophacus, 298, 302, **378**, 409
Pyrophacus horologium, **283**
Pyrophacus vancampoae, **298**, *299, 302*
Pyrrhophyta, 225, 237, 386
Pyxidicula, **655**, *729*
Pyxidicula bollensis, 637
Pyxidicula liasica, 637

Pyxidiella, **389**
Pyxilla, 637, **656**
Pyxillaceae, 656
Pyxolithus, **778**

Quadraditum, **211**
Quadratimorpha, **211**
Quadrisporites, **211**
Quadruilobus, **211**
Quadrum, **778**
Quadrum gartneri, 748
Quadrum gothicum, 748
Quadrum trifidum, 748
Quasiumbella, 936, **955**
Quasiumbella pseudorotunda, 934,
   934
Quasiumbella rotunda, **934**, *934*
Quasiumbelloides, **955**
Quasiumbelloides laevis, **934**, *934*
Quinquerhabdus, **779**
Quisquilites, 816, **818**
Quisquilites buckhornensis, **817**, *817*

Raaben, M. E., 65
Radialetes, **211**
Radiodiscoaster, **778**
Radiodiscus, 637
Radiolithus, **780**
Radoičić, R., 881
Radoiciciella, **893**
Rae, P. M. M., 234
Raff, R. A., 86, 90, 91
Ragotzkie, R. A., 342
Ramacia, **39**
Ramanujam, C. G. K., 841
Rampi, L., 370, 512, *528, 529*
Ramsfjell, E., 597, 598
Ramsfjell, E. (Gaarder et al.), 723
Ramus, J., *110, 137*
Rantzieniella, 933, **956**
Rantzieniella nitida, **948**, *948*
Rao, A. R., 843, 882, 896
raphe, diatoms, **577, 974**
Raphidodinium, **430**
Raphidodinium fucatum, **435**, *435*
Raphidodiscus, **658**
Raphidodiscus marylandicus, **641**, *641,*
   647
Rapoport, H. (Wong et al.), 317
Raskyaechara, **956**
Raskyella, 926, 933, **956**
Raskyella pecki meridionale, **948**, *948*
Raskyellaceae, 956
Raskyelloideae, 956
raskyelloid gyrogonite, *925,* 926
Rasul, S. M., 151
rate of division, dinoflagellates, 319
rate of movement, dinoflagellates, 306
rate of sinking, coccoliths, 769
Rattrayella, **656**
Rauff, H., *854,* 856
Raven, P. H., 89, 92
Ray, S. M. (Aldrich et al.), 319, 326
Raynaud, J. F., *405, 428*
Rayns, D. G., 714
Reade, J. B., 349
Receptaculitaceae, 891
Receptaculitales, 852, 891

Receptaculiteae, 891
Receptaculites, *859,* **892**
Receptaculites neptuni, **854**, *854*
red algae. *See* Rhodophyta
red tides or red water, 225
   blue-green algae, 62
   dinoflagellates, 339
reef-building algae, 126
reflected tabulation, 287, **972**
Reid, P. C., 275, 364, 385
Reijnders, L., 87, 91
Reimann, B. E. F., 568, 573, 579, *581,*
   605, 610, 612
Reimann, B. E. (Lewin et al.), 572, 595,
   618, 619
Reimer, C. W., 626, 649
Reinhardt, P., *743, 745, 776, 941, 941*
Reinhardtites, **779**
Reinhardtites anthophorus, **692**, *692*
Reinhold, Th., 639
Reinsch, P. F., 350
Reinschia, 841
Reiss, Z., 733, 734
Reker, C. C., *953*
Renault, B., *21, 37, 77*
Renidinium, **424**
Renidinium membraniferum, **426**, *426*
Repagulum, **784**
reproduction
   bacteria, 16
   blue-green algae, 46
   Charophyta, 915
   Chlorophyta, 828
   Chrysophyta, 504
   diatoms, 595, 597, 599
   dinoflagellates, 319
   ebridians, 472
   euglenids, 896
   Haptophyta, 710
   Prasinophyta, 806
   Rhodophyta, 113 – 125
   silicoflagellates, 542
   Xanthophyta, 494
réseau argentophile, *258, 258,* **974**
reservoir, euglenid, **895, 974**
resting spores
   bacteria, 16
   blue-green algae, 46, 48
   diatoms, 600
Retecapsa, **781**
Retecapsa crenulata, **740**, *740*
Retecapsaceae, 781
Retecapsoideae, 781
Retialetes, **822**
Reticulofenestra, **777**
Reticulofenestra laevis, **757**, *762*
Reticulofenestra pseudoumbilica, 764,
   766, **767**
Reticulofenestra umbilica, **757**, *761,*
   762, 771
Reticulosphaeromorphitae, 207
Retisphaeridium, **211**
Retractomorphitae, 208
Retusochara, 933, **956**
Retusochara macrocarpa, **949**, *949*
Revinotesta, **211**
Reyes-Vasques, G., 622
Reytlinger, E. A., *934, 937*

Rezak, R. (Logan et al.), 65
rhabde, ebridian skeleton, **465, 974**
Rhabdochara, 926, 933, 939, **956**
Rhabdochara stockmansi, **925**, *925*
Rhabdocyclus, **784**
Rhabdolekiskus, **783**
rhabdolith coccolith, *700*
Rhabdolithes, **784**
Rhabdolithina, **779**
Rhabdolithus, **784**
Rhabdomonadales, 898
Rhabdomonas, 896, **898**
Rhabdonema, 622, **658**
Rhabdonema adriaticum, **571**, 596,
   600, 653
Rhabdoniaceae, 142
Rhabdopleura, 140
Rhabdoporella, 864, **892**
Rhabdoporelleae, 892
Rhabdosphaera, 721, 770, **784**
Rhabdosphaera clavigera, 697, **699**,
   699, 717, 719
Rhabdosphaera inflata, **757**, 763
Rhabdosphaera stylifera, 697, **699**,
   699, 719
Rhabdosphaeraceae, 784
Rhabdosphaerales, 784
Rhabdothorax, 750, **784**
Rhabdothorax regalis, **747**, *750, 752*
Rhachosoarium, **211**
Rhaetogonyaulax, 297, 353, 367, **392**
Rhaetogonyaulax arctica, 298, **390**, *390*
Rhaetogonyaulax rhaetica, **390**, *390*
Rhagodiscaceae, 779
Rhagodiscus, **779**
Rhagodiscus angustus, 740, **741**, 748
Rhaphoneis, 657, **658**
Rhaphoneis amphiceros v.
   elongata, **645**
Rhaphoneis angularis, **633**, *633*
Rhaphoneis margaritata, **643**, *643*
Rhicnonema, **41**, 69, 82
Rhipidodesmis, **891**
Rhipocephalus, 831, **891**
Rhiptosocherma, **211**
Rhiptosocherma improcera, **159**, *159*
Rhizobiaceae, 40
Rhizobium, 31, **40**
Rhizochloridales, 498
Rhizochloris, **498**, *498*
Rhizochrysidales, 519
Rhizochrysis, **519**
Rhizoclonium, 826, **890**
Rhizophyllidaceae, 140
Rhizosolenia, 62, 510, 601, 607, 609,
   613, 619, 621, 623, 625, 626, 639,
   657
Rhizosolenia acuminata, **577**, *577,* **621**
Rhizosolenia alata, **599**, 604
Rhizosolenia barboi, **644**
Rhizosolenia bergonii, **624**, 625
Rhizosolenia bulbosa, **647**
Rhizosolenia calcar-avis, **619**
Rhizosolenia curvirostris, **644**
Rhizosolenia fragilissima, **621**
Rhizosolenia hebetata, **600**, **624**
Rhizosolenia hebetata fa. *hi-*
   emalis, **650**

Rhizosolenia hebetata fa. *semi-spina*, **577**, *577*, 598
Rhizosolenia miocenica, **645**
Rhizosolenia norwegica, **647**
Rhizosolenia praebergonii, **644**
Rhizosolenia robusta, **623**
Rhizosolenia setigera, **601**, 609
Rhizosolenia shrubsolei, **599**
Rhizosolenia styliformis, **604**, 625
Rhizosoleniaceae, 656
Rhizosoleniales, 656
Rhodella, 113, 126, 128, 137, **138**
Rhodella maculata, **113**, 128
Rhodobacillaceae, 42
Rhodobacteria, 42
Rhodobacteriaceae, 42
Rhodochaetales, 139
Rhodochaete, 116, **139**
rhodoids, 127, **974**
rhodolite, 127, **974**
rhodolith, **127**, *127*, **974**
Rhodomela, **142**
Rhodomelaceae, 112, 142
Rhodomenites, **142**
Rhodomicrobium, **42**, 91
Rhodophyceae, 138
Rhodophyllidaceae, 142
Rhodophyllis, **142**
Rhodophyta, 107—147, 138
  asexual reproduction, 113
  budding, 113
  carposporophyte, 123
  classification, 138
  distribution, 126
    freshwater, 126
    marine, 126
  evolution, 136
  fossils, 128
  gametophyte, 117
  gliding movement, 113
  growth phases, 114
  life history, 116
  pH range, 126
  pigments, 126
  salinity range, 126
  sheath, 110
  spores, 124
    cruciate tetraspores, 124
    zonate tetraspores, 124
    tetrahedral tetraspores, 124
  storage products, 109
  tetrasporophyte, 123
  unicells, 113
  vitamin requirement, 126
  wall, 110
Rhodosorus, 137, **138**
Rhodosorus marinus, **137**
Rhodospirillaceae, 42
Rhodospirillales, 24, 42
Rhodospirillum, 95
Rhodospora, **138**
Rhodymenia, **142**
Rhodymeniaceae, 142
Rhodymeniales, 113, 142
Rhoicosphenia, 626, **658**
Rhomboaster, **778**
Rhombodella, **403**
Rhombodinium, 294, **392**

Rhombodinium glabrum, **392**, *392*
Rhombogyrus, **780**
Rhombolithion, **780**
Rhombolithion rhombicum, **745**, *745*
Rhopaliophora, 149, **211**
Rhopaliophora foliatilis, **157**
Rhopaliophora impexa, **150**, *150*, 157
Rhopalodia, **659**
Rhynchodiniopsis, 302, **395**
Rhyniella, **83**
ribonucleic acid (RNA), 12
ribosomal RNA, comparison of pro-caryotes and eucaryotes, 31, 91, 93, 95, 137, 234
ribosome granules, *50*
ribosomes, **12**, *12*, *43*, *46*, **974**
Ricard, M., 258, 295, 324
Richard, A., 881
Richelia, 62, 63
Richelia intracellularis, **613**
Rickettsiales, 42
Riculacysta, **424**
Riculasphaera, 176, **211**
Riculasphaera fissa, 171, *174*
Riding, R., 65, 68
Ridley, S. M., 91
Riedel, W. R., 628, 629, 726
Riedelia, 637, **657**
Rietschel, S., 852, *854*, 856, 857, *858*, *859*, 867
Rifeniata, **211**
Rigby, J. K., 649
Rigden, E. J., 326
Riley, L. A., 351
rimoportula, **591**, **970**, **974**
Ripley, G. W. (Hendey et al.), 585
Ris, H., 89, 233, 234, 236, 237
Risatti, J. B., 745
Rissoëllaceae, 141
Rivularia, 55, 59, 69, **83**
Rivulariaceae, 53, 74, 77, 83
Rivularialithus, **83**
Rivularites, **83**
Rizzo, P. J., 234
RNA. *See* ribonucleic acid
rRNA. *See* ribosomal RNA
Roberts, B. S. (Murphy et al.), 341
Roberts, K. (Siegel et al.), 69
Roberts, T. M., 44, 45, 235, 236
Roberts, T. M. (Allen et al.), 232, 234, 236, 237, *238*, 343
Robertson, J. D., *86*, *86*
Rocella, 254, *550*, **559**, *559*, **656**
Rocella gelida, **633**, *633*
Rocella gemma, *551*
Rocella schraderi, **633**, *633*
Roche, A., *21*
Roche, M. B. (McIntyre et al.), 720, 721
rock-forming algae
  blue-green, 63
  chrysomonads, 517
  coccolithophores, 761
  diatoms, 628, 634
  green, 831
  red, 126, 127
  silicoflagellates, 547
rod, bacterial cell form, 11

Rogers, J. N. (Eppley et al.), 342
Rögl, F., 509
Romasphaera, **822**
Romein, A. J. T., 773, 774
Rood, A. P., 735, 737, 738, 776
Roperia, **656**
Roperia tesselata, **591**, 624
Roscoffia, **378**
Ross, R., 581, 583, 587, 588, 589, 591, 592, 594, 639, 653, 654, 655
Rossignol, M., 275, 293, 364, *819*
Rostovtseva, L. F., *934*
Rostroporella, **894**
rota, diatoms, **585**, **974**
Rotelapillus, **780**
Roth, P. H., 766
Roth, P. H. (Adelseck et al.), 769
Rothpletz, A., 181, 637, 692
Rottnestia, **430**
Rottnestia borussica, **430**, *430*
Round, F. E., 596, 655
Roux, A., 142
Rouxia, **659**
Rubina, N. V., 648
Rucinolithus, **780**
Rucinolithus hayii, 748
Rucinolithus irregularis, **735**, 741
Rucinolithus wisei, **735**, *738*
Rugocystis, **212**
Rugubivesiculites rugosus, **361**
Rugulasphaeridium, **212**
rugulate surface, 157, **974**
Rugulidium, **212**
rupture, acritarch excystment, **173**, **974**
Rutilaria, *591*, **656**
Rutilaria radiata, **592**, *592*
Rutilariaceae, 656
Russell, D. J., 607
Russell, F. E., 317, 318
Russellia, **778**
Rutte, E., 65, 77, 836
Ruttnera, **777**
Rylandsia, **656**
Ryther, J. H., 338, 621

Saar, A. du, 637
Saccoma, 843
Saeptodinium, **389**
Saeptodinium eurypylum, **385**, *385*
Saeptodinium hansonianum, **364**, *385*
Sagan, L., 89, 90, 93
sagittal axis, diatom frustule, **576**, **974**
sagittal fission suture, dinoflagellates, **269**, **974**
Saharidia, 178, **212**
Said, A. K. (Adams et al.), 734
Saito, T., 770
Sakkionella, **894**
salinity, effect on
  acritarchs, 204
  charophytes, 930
  coccolithophores, 721
  diatoms, 616
  dinoflagellates, 344
Salopekiella, **893**
Salopekiella velebitana, **871**, *871*
Salpa, 547

Salpicola, 309
Salpingoeca, **525**
Salpingoecaceae, 525
Salpingoporella, 868, **893**
Samarella, **892**
Samarella setosa, **868**, *868*
Sameioneis, 607, **659**
Samlandia, **409**
Samlandia chlamydophora, **411**, *411*
Samoylova, R. B., 923
Sandberger, G., 937
Saportanella, 933, **956**
Saportanella maslovi, **948**, *948*
saprophytes, dinoflagellates, 309
sarcina, **12**, **974**
Sarcina, 80
Sarcinochrysidales, 525
Sarcinochrysis, **525**, 684
Sarcodiaceae, 141
Sarcomastigophora, 477, 554
Sarfert, E. (Bleecken et al.), 15
Sarjeant, W. A. S., 167, 178, 182, 204,
    281, 284, 285, 286, 295, 297, 349,
    350, 351, 352, 353, 358, 364, *365,*
    368, *383, 385, 389, 395, 401, 413,*
    430, 433, 435
Sarjeant, W. A. S. (Downie et al.), 149,
    205
Sarjeant, W. A. S. (Harland et al.), *390*
Sarma, S. R., 843
Sarma, Y. S. R. K., 233
Saturnus, **212**
Saunders, R. P., 341, 604
Saunders, R. S., 202
Savillea, **526**
Savillea parva, **527**, *527*
Savitrinia, **822**
Sawa, T. (Conkin et al.), 939
Sawamuraia, **658**
saxitoxin, dinoflagellates, 318, 319
Saydakovskiy, L. Ya., *941, 941*
scales
    chrysophytes, 500
    dinoflagellates, 248
    haptophytes, 684, *685, 691*
    prasinophytes, 806
Scampanella, **778**
Scampanella cornuta, **697**, *697*
Scandone, P., 728, 733
Scaphita, **212**
Scaphodinium, **380**
scapholith coccolith, 700, **703**, **974**
Scapholithus, 703, 728, **781**
Scaphomorphida, 207
Scaphospinosa, **212**
Scenedesmaceae, 842, 889
Scenedesmus, 619, 829, 836, 843, 889
Sceptroneis, 637, 647, 657, **658**
Sceptroneis caducea, **647**
Sceptroneis grunowii, **639**, *639*
Sceptroneis pupa, **647**
Schatz, A., 895
Schei, B., 721
Schematophora, **419**
Schematophora speciosa, **421**, *421*
Schepelevia, **212**
Schidlowski, M., 32
Schiller, J., 248, 368, 372, 379, 556

Schilleriella, **498**
Schilleriella anuraea, **495**, *495*
Schilling, A. J., 226, 348, 368
Schimperiella, 605, **656**
Schimperiella antarctica, **624**
Schismatosphaeridium, **212**
Schizochlamys, 836, **888**
Schizochlamys haywellensis, 836
Schizocystia, **212**
Schizocystia pilosa, **195**, *195*
Schizodiacrodium, **212**
Schizomorphitae, 208
Schizomycetae, 39
Schizomyceteae, 80
Schizomycetes, 38, 39
Schizomycophyta, 39
Schizophacus, **818**
Schizophacus rugulatus, **819**, *819*
Schizophacus spriggii, **819**, *819*
Schizophyceae, 80, 81
Schizophyta, 9, 38, 39, 81
Schizosphaerella, 728, **785**
Schizosphaerella punctulata, **733**, *733*
Schizosphaerellaceae, 784
schizospheres, 729, **974**
    microstructure of, 733
Schizosporales, 11, 24, 40
Schizosporis, 204, **822**
Schizosporis reticulatus, **822**, *822*
Schizothrichites, 77, **82**
Schizothrichites ordoviciensis, **79**, *79*
Schizothrix, 29, 57, 59, 79, 80, **82**
Schizothrix calcicola, **58**, *59*
Schizymenia, **141**
Schmalz, R. F., 110
Schmidt, A., 623, 641, 654
Schmidt, J., 373
Schmidt, J. A., 397
Schmidt, R. J. (Loeblich III et al.), 248,
    269
Schmitzia, **141**
Schnepf, E., 63, 92, 93
Schopf, J. M., 24, 26, 36, 155
Schopf, J. W., *32, 32, 34,* 35, 68, 69, 71,
    74, 74, 77, 80, 93, 128, *130,* 138, 182,
    825, *838, 838,* 843
Schopf, J. W. (Fairchild et al.), 843
Schopf, J. W. (Meinschein et al.), 35
Schrader, H. J., 579, 581, 582, 629, 637,
    639, 644, 645, 647, 653
Schroderella, **655**
Schütt, E., 243, 248, 252, 253, 256, 272,
    293, 323, 334, 374, 379, 623, 654
Schuettia, **656**
Schultz, M. E., 597, 598
Schulz, P., *560,* 637, 641
Schulz-Baldes, M., 84
Schumann, J., 463
Schwartz, R. M., 92
Schwarz, E. H. L., 713
Schwarz, W., 37
Schweiger, H. G., *863, 864*
Scinaia, **139**
sclerosponges, 133
Sclerothamnium, **141**
Scoliopleura, **658**
Scoliotropis, **658**
Scotobacteria, 38, 39

Scott, J. L., 124
Scott, R. A., 155
Scribroporella, **892**
Scribroporella socialis, **867**, *867*
Scribroporelleae, 892
Scriniocassiaceae, 409
Scriniocassis, **409**
Scriniocassis dictyotus, **413**, *413*
Scriniodinium, 286, 294, 302, **430**
Scriniodinium attadalense, **433**, *433*
Scriniodinium galeritum, **430**, *430*
Scriniodinium luridum, **290**
Scriniodinium subvallare, **433**, *433*
Scrippsiella, 262, 299, 301, 302, 374,
    377, **383**, 777
Scrippsiella sweeneyae, **279**, 299
Scrippsiella trochoidea, **229**, *229,* 231,
    232, 233, 234, 239, 241, 244, 247,
    262, 279, 280, 299, 302, 339, 343,
    344, 386, 750
Scutellomorphitae, 208
Scuticabolaceae, 208
Scuticabolus, 149, 171, 176, **212**
Scuticabolus lapidaris, **171**, **173**, *173*
Scyphosphaera, 770, **780**
Scyphosphaera amphora, 697, **701**,
    701
Scyphosphaera apsteini, 697, **701**, *701*
Scyphosphaeraceae, 780
Scytonema, 57, 59, 62, **83**, *83*
Scytonema myochrous, 59
Scytonemataceae, 53, 74, 83
Sebdenia, **141**
Sebdeniaceae, 141
secondary pit connections, red al-
    gae, 112, **964**
sedimentary iron ores, 24
segmented bands, diatoms, **576**, **974**
Segroves, K. L., 176, *183,* 204, 205, *811,*
    838
Seilacher, A., 839
Selenoides, **892**
Seletonella, **892**
Seletonellaceae, 892
Seletonelleae, 892
Seliger, H. H., 316
Seliger, H. H. (Swift et al.), 316
Sellberg, B., 286, 287
Selliporella, **893**
Semantebria, **483**
semicells, desmids, **884**, **974**
Semina, G. E. (Beklemishev et
    al.), 609
Semihololithus, **778**
Semikhatov, M. A., 65
Senegalinium, **392**
Senispora, **212**
Senn, G., *886*
Senoniasphaera, **389**
Senoniasphaera inornata, **426**, *426*
Senzeillea, **822**
separation disc, blue-green algae, 47,
    47
Septamesocena, 539, 545, 559
Septorella, 933, 947, **956**
Septorella brachycera, **928**, *928,* 947
Septorella campylopoda, **947**, *947*
Septorella ultima, **947**, *947*

serial symbiosis theory, eucaryote origin, *90*
Seribiscutum, **782**
Serratia, 31, **41**
Serraticardia, **141**
Sestrosphaera, **893**
Sestrosphaera liasina, **872**, *872*
setae, diatoms, **585**, **964**, **975**
settling rate, diatoms, 632
Shamray, I. A., 745
Shapiro, J., 54
Sharon, N., 261
Shaw, G. (Brooks et al.), 68, 69
sheath, **974**
  bacteria, **13**
  blue-green algae, 43
  red algae, 110
Shecodium, **891**
Sheridan, R. P., 616
Sherley, J. L., 313, 315
Sherley, J. L. (Loeblich III et al.),
  248, 269
Sheshukova-Poretskaya, V. S., 639
Shestakov, S. V., 51
shield, coccoliths, 697, **975**
shield cells, charophyte, **917**, **975**
Shigaporella, **891**
Shigeru, M., 839
Shilo, M. (Cohen et al.), 53
Shublikodiniaceae, 392
Shublikodinium, **392**
Shukla, V. B., 884
Shumenko, S. I., 745
Shyam, R., 233
Sibiriella, **892**
Sibirielleae, 892
Siderocapsa, 24, **41**
Siderocapsaceae, 41
Siegel, B. Z., 69
Siegel, S. M., 69
Sielicki, M. (Jeffrey et al.), 239, 240
Siesser, W. G., 127
Sigee, D. C., 232
Sigmopollis, **212**
silica deposition
  diatoms, 573
  green algae, 828
  silicoflagellates, 547
silica diagenesis, 636
silica dissolution, 632
silicalemma, diatoms, **573**, **975**
Silicisphaera, **419**
Silicisphaera ferox, **423**, *423*
Silicisphaera tridactylites, **423**, *423*
Silicisphaeraceae, 419
Silicoflagellata, 535, 554
silicoflagellates, 510, 535–566
  abundance, 517, 547
  biostratigraphy, 549
  cell, 535
  classification, 554
  distribution, 546
  ecology, 546
  evolution, 553
  genera, rate of description, *5*
  generic ranges, *549*
  geologic occurrence, 547
  paleotemperatures, 552

reproduction, 542
  skeleton, 538
    double, 539
  species diversity, 548
Silicoflagellatophycidae, 477
Silicoflagellida, 554
silicoflagellite, **548**, **975**
silicoflagellithite, 547
silicoflagellitite, 548
Silva, E. S., 335
Silva, P. C., 654
silver-staining network,
  dinoflagellate, 258
Similicoronilithus, **782**
Simon, M. I. (Emerson et al.), 15
Simonsen, R., *591*, *649*, 655
Simonsiellaceae, 39
Sims, P. A., *581*, *581*, *582*, *583*, 588,
  589, *591*, *592*, *594*, *626*, *653*, 655
Sims, P. A. (Ross et al.), *587*, *589*, *592*,
  594
Simsangia, **212**
Sinclair, C., 44, 54, 55
Singer, G. E., 31
Singh, C., *173*, *391*
Singh, G., 364
Singh, H. N., 52
Singraulipollenites, **212**
Sinha, V., 205
Sinodinium, **371**
Sinophysis, **371**
Sinophysis ebriola, 309, 347
Siphonocladaceae, 891
Siphonocladales, 852, 891
Siphonocladus, 826, **891**
Siphononemataceae, 82
Siphonophycus, 74, **82**
Siphonophycus kestron, **77**, *77*
Siphonotestales, 464, 535, 554
Sirmiodinium, **430**
Sirmiodinium grossi, **435**, *435*
Sitle, P. (Brown et al.), 684
Skeletonema, 604, 605, 613, 618, 619,
  620, 625, 634, **655**
Skeletonema costatum, **568**, *593*, *596*,
  604, 609, 614, 620, 623
Skeletonema potamos, **591**, *593*
Skeletonemaceae, 655
Skogsberg, T., 226, 258, 269, *270*, *271*,
  326, 367
skolate cyst, dinoflagellate, **286**, **975**
Skuja, H., 348, **382**
Skujapelta, 62
Sloan, P. R., 240
Small, J., 654
Smayda, T. J., 345, 346, 609, 610, 616,
  621, 622, 623, 629, 632, 719
Smith, C. C., 726, 748
Smith, D. L., 339
Smith, J. (Long et al.), 637
Smith, N. D., 202
Smith, V. E., 240
Smith, W., 654
Smith, W. D. (Licari et al.), 74
Smithora, **139**
Smolenskiella, **430**
Soaniella, **395**
Sober, H. A., 568

soil algae, 57
soil bacteria, 23
Sokač, B., *871*, *872*
Sol, **212**
Sol radians, **163**, *163*
Solatisphaera, **212**
Solenomeris, **140**
Solenopora, **140**
Solenoporaceae, 132, 135, 140
Solieria, **142**
Solieriaceae, 142
Solisphaeridium, **212**
Solisphaeridium apodasmion, **203**, *203*
Solita, **212**
Sollasitaceae, 782
Sollasites, **782**
Solonoporidium, **891**
Sommer, F. W., 817
Sommerfeld, M. R., 113, *114*, 115, 126,
  128
Sommerfeld, M. R. (Lin et al.), 114
Sorby, H. C., 692, 725, 761
sori, red algae, **123**, **975**
Sournia, A., 59, 62, 254, *255*, 345, 368
Soyer, M. O., 235
Spatulodinium, **380**
Spatulodinium pseudonoctiluca, **250**
species diversity
  acritarchs, *187*
  calcareous nannoplankton, *772*, *774*,
    775
  charophytes, *942*
  diatoms, *653*
  dinoflagellates, *352*
  ebridians, *477*
  silicoflagellates, *548*
Speer, H. L. (Jones et al.), 126
Speetonia, **782**
Speetonia colligata, **740**, *740*
Speetoniaceae, 782
spermatangial conceptacles, red algae, 121
Sphaeripara, **382**
Sphaeriparaceae, 382
Sphaerocalyptra, **778**
Sphaerocalyptra papillifera, 694, **696**,
  696
Sphaerochara, 925, 926, 933, 941, 947,
  956
Sphaerochara ulmensis, **953**, *953*
sphaerocharoid gyrogonite, 925, *925*
Sphaerococcaceae, 141
Sphaerococcides, **141**
Sphaerococcus, **141**
Sphaerocodium, **83**
Sphaerocongregus, 74, **82**
Sphaerocongregus variabilis, **74**
Sphaerodinium, **371**
Sphaerohystrichomorphida, 207
Sphaeromorphida, 207, 208
Sphaeromorphitae, 207
Sphaeronites, **892**
Sphaerophycus, 69, 74, **81**
Sphaerophycus parvum, **73**, *73*, *77*
Sphaeroplea, 185, 830, **890**
Sphaeropleaceae, 890
Sphaeropleales, 890
Sphaeroporalites, **212**

Sphaeropsis, 818
Sphaerosomatites, 181
Sphaerospongia, 892
Sphaerotilaceae, 39
Sphaerotilales, 24, 39
Sphaerotilus, 24, 34, 39
Sphaerotilus catenulatus, 26, 26
Sphaerozosma, 894
Sphagnum bogs, 842, 885
Sphaleromantis, 504, 518
sphenolith coccolith, 706, 707, 975
Sphenolithaceae, 778
Sphenolithidae, 778
Sphenolithus, 745, 778
Sphenolithus belemnos, 761, 764, 767
Sphenolithus ciperoensis, 757, 761, 762
Sphenolithus distentus, 757, 761, 762
Sphenolithus furcatolithoides, 757, 762
Sphenolithus heteromorphus, 761, 761, 767
Sphenolithus neoabies, 761, 766
Sphenolithus predistentus, 761, 762
Sphenolithus pseudoradians, 757, 761, 762
Sphenolithus radians, 757, 763
Sphenomonadales, 898
Sphenomonas, 898
Sphenoradiatus, 783
Spheripollenites, 212
Sphinctoporella, 894
Sphynctolethus, 657
Spiegler, D., 509
Spinidinium, 392
Spinidinium sagittulum, 392, 392
Spinidinium vestitum, 361
Spiniferitaceae, 404
Spiniferites, 5, 148, 153, 285, 294, 295, 299, 302, 349, 350, 358, 404
Spiniferites belerius, 299
Spiniferites bentori, 283, 285, 293, 299, 364
Spiniferites buccinus, 405, 405
Spiniferites bulloideus, 299
Spiniferites cruciformis, 289, 289
Spiniferites elongatus, 299
Spiniferites mirabilis, 299, 301
Spiniferites nodosus, 299
Spiniferites ramosus, 272, 287, 288, 293, 299, 301, 337
Spiniferites ramosus v. *pachydermus,* 299
Spiniferites scabratus, 361
Spinumbella, 955
Spinumbella spinifera, 934, 934
spiral enveloping cells, charophyte gyrogonite, 919, 975
Spiraulaxina, 324, 373
spire, spirillum, bacterial cells, 11, 12, 975
Spirilla, 30
Spirillaceae, 40
Spirillum, 14, 40
Spirillum volutans, 16
Spirochaeta, 40
Spirochaetaceae, 40
Spirochaetales, 40

spirochaete, ectosymbiotic, 93
Spirochrysis, 507
Spirodinium, 334, 378
Spirogyra, 828, 829, 882, 883, 884, 894
Spirogyra wyomingia, 884
Spirogyrites, 884
Spirulina, 59, 82
Spjeldnaes, N., 864
Spolsky, C., 95
Spondylosium, 894
Spongebria, 476, 480
Spongebria marthae, 483, 483
Spongiochloris, 836, 888
Spongiocysta, 818
Spongodinium, 349, 395
Spongodinium delitiense, 293, 396
sporangium, bacterial, 16, 975
spore coat, bacterial, 16, 17
Sporodinium pseudocalani, 309
Sporolithon, 141
sporophydium, Charophyta, 919, 975
sporophyte, 975
sporopollenin, 153, 278, 975
sporostegium, charophyte, 919, 975
Sportelloeca, 526
Spotnitz, A. (Isenberg et al.), 687
Spumiosa, 212
Spumosina, 212
Spurimoyeria, 212
Spyrebria, 483
Squamariaceae, 140
Stache, G., 926, 953
Stacheia, 142
Stacheia marginulinoides, 133, 133
Stacheoides, 142
Stacheoides polytrematoides, 133, 133, 134
stalk, bacterial, 13, 975
Stanier, R. Y., 9, 29, 45, 80, 85, 93
Staphylococcus, 30, 41
Staplin, F. L., 160, 167, 173, 181, 202, 205
Staplinium, 822
Starmach, K., 77, 79
Starr, R. C., 837
statospores, chrysophyte, 494, 964, 975
statospores, diatoms, 600, 975
Staurastrum, 885, 886, 887, 894
Staurastrum enteroxenum, 886
Staurolithites, 703, 779
Staurolithites matalosus, 741, 741, 748
Stauroneis, 613, 626, 641, 658
Staurorhabdus, 779
stauros, diatom frustule, 577, 975
Steidinger, K. A., 252, 253, 308, 317, 318, 319, 339, 340, 341, 343, 344
Steidinger, K. A. (Murphy et al.), 341
Steidinger, K. A. (Tomas et al.), 311
Stein, F. R. von, 227, 229, 261, 272, 273, 274, 293, 302, 323, 329, 330, 335, 364, 370, 372, 377
Stelladinium, 389, 389
Stellasphaeroides, 81
Stellatochara, 933, 941, 955
Stellatochara maedleri, 920, 920
Stellatochara sellingii, 920, 920, 941
Stellatocharoideae, 955

Stelliferidium, 212
Stellinium, 212
Stellinium micropolygonale, 203, 203
Stenixis, 894
Stenixis cosmarioides, 886
Stenocalyx, 506
Stenochara, 933, 955
Stenochara maedleri, 941, 941
Stenogramme, 142
Stenogrammites, 142
Stenoneis, 658
Stenopterobia, 659
Stenozonoligotriletes, 212
Stephanelytraceae, 430
Stephanelytron, 430
Stephanelytron redcliffense, 435, 435
Stephanochara, 933, 956
Stephanochara compta, 953, 953
Stephanodiscus, 596, 604, 655
Stephanodiscus hantzschii, 614, 649
Stephanoeca, 526
Stephanoeca pedicellata, 511, 511
Stephanogonia, 657
stephanolith coccolith, 700, 706, 975
Stephanolithiaceae, 780
Stephanolithion, 735, 780
Stephanolithion bigotii, 735, 737, 749
Stephanolithion hexum, 735, 749
Stephanolithion speciosum, 735, 749
Stephanolithion speciosum v. *octum,* 735, 749
Stephanolithionaceae, 780
Stephanomorphitae, 207
Stephanoporos, 519
Stephanoptera, 818
Stephanopyxis, 598, 601, 637, 655
Stephanopyxis turris, 573, 575, 598, 601, 603
Stephodiniaceae, 430
Stephodinium, 286, 430
Stephodinium coronatum, 429, 429
Stereotestales, 464, 477, 535
Stevens, T. S. (Holton et al.), 46
Stewart, K. D., 805
Stewart, W. D. P., 53
Stewart, W. D. P. (Fay et al.), 49
Stewart, W. D. P. (Fogg et al.), 45, 46, 51, 57, 92
Stichochrysis, 519
Stichococcus, 894
Stichogloea, 525
Stichogloeales, 525
stichonematic flagella, 248, 975
Stichoporella, 893
Stichoporelleae, 893
Stictocyclus, 656
Stictodiscaceae, 656
Stictodiscus, 254, 512, 656
Stictodiscus gelidus, 559
Stictosphaeridium, 212
Stictosporaceae, 141
Stigeoclonium, 845, 845, 889
Stigeoclonium lubricum, 844, 844
stigma
    diatom frustule, 587, 975
    dinoflagellate, 242, 975
Stigonema, 47, 83
Stigonemataceae, 83

Stigonematales, 47, 49, 53, 59, 80, 83
Stiles, D. V. (Schopf et al.), 24, 26, 36
Stipitochloris, **498**
Stipitococcus, **498**
Stipitococcus vas, **492**, *492*
Stipulella, **81**
Stoermer, E. F., 585
Stolley, E., *860*
stomatocerque, chrysophyte
    cyst, **506**, *508*, **975**
Stomiosphaera, 733, **785**
Stomiosphaeraceae, 279, 733, 785
Stomochara, 933, 953, **955**
Stomochara moreyi, **921**, *921*, *941*
Stomocharoideae, 955
Stone, R. J. (Lee et al.), 835
stoneworts, 913
    See also charophytes
storax, 539
Stosch, H. A. von, 258, 275, 278, 319,
    320, 329, 330, 331, *333*, *334*, 335,
    336, 351, 374, 377, 576, 577, *577*,
    596, 597, 603, 619, 653, 654
Stosch, H. A. von (Hendey et al.), 568
Stosch, H. A. von (Manton et al.), 568
Stoschia, **656**
Stover, L. E., 286, 351, 357, 398, 421
Stover, L. E. (Evitt et al.), 284, 295
Strachan, I., 204
Stradner, H., 508, 539, 697, 729, 745,
    752, 755, 756, 761, 764, *764*
Stradneria, **781**
Stradnerius, **777**
Stradnerlithus, **780**
Strel'nikova, N. I., 637
Streptococcaceae, 41
Streptococcus, 29, **41**
Streptomyces, **42**
Streptomycetaceae, 42
Streptotheca, **657**
striated strand, dinoflagellate flagel-
    lum, 248, *249*, **975**
Striatella, **658**
Striatococcolithus, **783**
Striatomarginis, **783**
Striatotheca, **212**
Strickland, J. D. H. (Eppley et al.), 306
Strohbach, G. (Bleecken et al.), 15
Strobilochara, **955**
stromatoids, **975**
Strombidium, 309
Strombomonas, **898**
stromatolites, 28, **975**
    bacterial, 28
    blue-green algal, **58**
    fossil record, 63
stromatoporoids, 133
structured granules, blue-green al-
    gae, **46**, **967**, **975**
strutted process, diatom frus-
    tule, **591**, **975**
strutted tubulus, diatom frustule, **975**
Stubbe, W., 86, 92
Studer, H. P., 831
Sturt, G., 637
Sturtiella, **656**
Stylochromonas, **525**
Stylocodium, **890**

Stylodiniopsis, **212**
styrax, 539
subapical processes, acritarchs, 167
Subsilicea, **658**
Subterraniphyllum, **140**
Subterraniphyllum thomasii, **135**, *135*
Subtetrapedia, **81**
Subtilidinium, **395**
Subtilidinium minutum, **401**, *401*
Subtilisphaera, **389**
Subtilisphaera perlucida, 361
Subtilisphaera trendalli, **385**, *385*
Succodium, **891**
Suchlandt, O., 240, 349
Suessia, 353, **430**
Suessia swabiana, 353, **354**, *354*
Sulcosphaera, **956**
sulcal lists, dinoflagellate, 271
sulcal plates, dinoflagellate, **264**, *264*,
    975
sulcus, dinoflagellate, 226, **227**, **970**,
    975
Sulcatosphaeridium, **212**
sulfur bacteria, 23, 24
sulfur cycle, 22
sulfur deposits
    bacteria, 22, 27
    blue-green algae, 55
Sumatradinium, **392**
Suppiluliumaella, **893**
Suppiluliumaella polyreme, **875**, *875*
Surculosphaeridium, **409**
*Surculosphaeridium longifurca-
    tum*, 361, **415**, *415*
Surirella, 579, 581, 600, 610, 626, **659**,
    964
Surirella elegans, **582**, *582*, 587
Surirella robusta, **581**
Surirellaceae, 659
Surirellales, 659
Svalbardella, **392**
Svalbardella cooksoniae, **357**, *357*
Sverdrupiella, 353, 367, **392**
Sverdrupiella septentrionalis, **357**, *357*
Swafford, J. R. (Lin et al.), 114
Swale, E. M. F., 806
Sweeney, B. M., 237, 247, 311, 317, 319
Sweeney, B. M. (Hastings et al.), 317
Swezy, O., 226, 230, *244*, 247, 308, 310,
    339, 348, 365, *366*
Swift, E. 5th, 316, *320*, *321*, 339, 368,
    629, *631*
Swift, E. J. (Wall et al.), 279, *280*
Swinchatt, J. P., 60
Sycidiaceae, 955
Sycidiales, 955
Sycidiphyceae, 955
Sycidium, 933, 937, 938, 954, **955**
Sycidium clathratum, **938**
Sycidium melo, **937**, *937*
Sycidium reticulatum, **938**, *938*
Sycidium volborthi eifelicum, **938**, *938*
Sydow, B. von, 581
Sylvanidium, 176, **212**
Sylvanidium paucibrachium, **157**,
    157, *175*
Symbiodinium, 313, **380**
symbiosis

blue-green algae, 55, 57
    diatoms, 613
    dinoflagellates, 311, 313
    green procaryote, 84
Symplassosphaeridium, 149, **212**
synclades, ebridian skeleton, **465**, **975**
Syncrypta, 500, **519**
Syncrypta glomerifera, **500**, *500*, 503
Syndendrium, **657**
Syndetocystis, **656**
Syndetoneis, **656**
Syndiniaceae, 233, 382
Syndiniales, 233, 236, 381
Syndiniophyceae, 233, 236, 381
Syndinium, 236, 311, **382**
Syndinium turbo, **232**, *233*
Synechococcus, 28, 43, 45, 62, 63, *63*,
    81, 84, 92, 311
Synechococcus elongatus, **51**
Synechococcus lividus, **58**
Synechocystis, 84
Synedra, 591, 600, 607, 613, 622, 624,
    626, 628, 657, **658**
Synedra hyalina, **571**
Synedra jouseana, **647**
Synedrosphenia, **658**
Synsphaeridium, 149, *151*, **212**, 818
Synsphaeridium gotlandicum, **151**,
    151
Synsphaeroides, **82**
Synura, 501, 504, 512, **519**, 654
Synura adamsii, **525**, *525*
Synuraceae, 519
Syracolithus, **783**
Syracorhabdus, **783**
Syracosphaera, 684, 714, **783**
Syracosphaera nodosa, 718
Syracosphaera pirus, 697, **703**, *703*
Syracosphaera pulchra, 697, **703**, *703*,
    703
Syracosphaera pulchroides, **710**
Syracosphaeraceae, 783
Syracosphaerales, 783
Syringidium, 637, **657**
Systematophora, **419**
Systematophora placacantha, **421**, *421*
Systematophoraceae, 419

Tabellaria, 607, 626, **658**
Tabellaria flocculosa v. *floccu-
    losa*, **626**, *626*
tabulation, dinoflagellate theca, 261
    of Dinophysiales, 269, *270*
    of Peridiniales, 261
    of Prorocentrales, 265
taeniocyst, dinoflagellate, **247**, *247*,
    971
Taeniophora, **419**
Tai, L. S., 258, 269, 270, 271, 326, 366
Tailleur, I. L. (Tourtelot et al.), 814
Takahashi, E., 502, 512, 654
Takahashi, K., 364
Tan Sin Hok, 692, 726, 751, 761
Tanada, T., 615
Tannuolaia, **892**
Tannuolaiaceae, 892
Tanyosphaeridiaceae, 409
Tanyosphaeridium, **409**

Tanyosphaeridium xanthiopyx-
   ides, **416,** *416*
Tapajonites, 816, 818
Tapajonites roxoi, **817,** *817*
Tappan, H., 4, 5, 6, 7, 7, 23, 74, *90, 94,*
   96, 128, 131, 138, 148, *150,* 155, *157,*
   159, 165, 167, 173, 176, *178, 181,*
   182, 184, 187, *188, 189, 190, 197,*
   202, 293, 302, *339, 351, 352, 355,*
   357, 358, 476, 477, 548, 549, 613,
   618, 630, 653, 654, 771, 772, 772,
   773, 773, 776, 939, 942
Tappan, H. (Loeblich III et al.),   539,
   541, 549
Tarsisphaeridium, **818**
Tasmanaceae,   208, 818
tasmanite,   814, **975**
Tasmanites,   5, 149, 152, 153, 178, 206,
   496, 808, 812, 814, *815, 816,* 816,
   818, 975
Tasmanites globulus, **815,** *815*
Tasmanites huronensis, **815,** *815*
Tasmanites martinssoni, **812**
Tasmanites medius, **812**
Tasmanites punctatus, **816,** *816*
Tasmanites verrucosus, **812**
Tasmanititae, **208**
Tauridium, **891**
Taylor, D. L.,   92, 233, 241, 313, 314,
   315, 316
Taylor, F. J. R.,   85, 87, 88, *249, 250,*
   254, 255, 258, 271, 306, 311, 326
Taylor, W. R.,   126, 834
Tectatodinium,   299, **395**
Tectatodinium pellitum,   **299,** *301*
Tectatodinium psilatum,   *290,* **290,** *396,*
   398
Tectitheca, **212**
Tectochara,   925, 932, 933, 939, **956**
Tectochara diluviana,   **921,** *923*
tectocharoid, gyrogonite,   *925,* 925
Tegumentum, **780**
Teichorhabdus, **781**
temperature, effects
   on acritarchs,   199
   on bacteria,   23
   on coccolithophores,   719
   on dinoflagellates,   344
      thecal modifications,   *328*
   on motility of blue-green algae,   52
Temperley, B. N.,   841
Templuma sinica, **864**
Temporina, **212**
Tenarea, **140**
Teneridinium, **392**
tentacle, dinoflagellate,   226, **250, 973,**
   975
Tenua,   349, **403**
Tenuofilum,   74, **82**
Tergestiella, **783**
Tergestielloideae,   783
Termier, G.,   134
Termier, G. (Termier et al.),   133, 142,
   868
Termier, H.,   133, *134,* 142, 868
Ternithrix, **83**
Terpsinoe, **657**
Terquemella,   869, **893**

Terquemella parisiensis, **877,** *877*
Terquemella parvula, **877,** *877*
Ters, M.,   154
Tersella, **894**
Tetracyclus, **658**
Tetracyclus cruciformis, **648,** *648*
Tetracystis, 830, 837, **888**
tetrad cells, Precambrian fossils,   128
Tetradinium, **212, 380**
Tetradinium minus, **381,** *381*
Tetraëdriella, **498**
Tetraedrixium, **212**
Tetraëdron,   828, 837, 838, 842, **888**
Tetraëdron regulare torsum, **837,** *837*
Tetragoniella, **498**
Tetraletes, **212**
Tetralithus,   714, 715, 716, **777**
Tetralithus malticus, **741,** *749*
Tetralithus pyramidus, **745,** *748*
Tetralithus trifidus, **745**
Tetraniveum, **212**
Tetrapodorhabdus, **781**
Tetraporina,   204, **212,** *838*
Tetrapterites, **212**
Tetraselmis, **818**
Tetraspora, **888**
Tetrasporaceae,   888
Tetrasporales,   834, 836, **888**
tetrasporangia, red algae,   **123, 975**
tetrasporangial conceptacles,   *123*
tetrasporophyte, red algae,   116, *125,*
   975
Tett, P. B.,   317
Tettragonis, **892**
Tettragonis orbis, **854,** *854*
Teutloporella, **893**
Teutloporelleae,   893
Thaiporella, **891**
Thalassia testudinum, **622**
Thalassionema,   625, 634, 639, **658**
Thalassionema nitzschioides, **623,**
   624, 634
Thalassiosira,   *593,* 601, 607, 622, 623,
   639, **653**
Thalassiosira angstii, **591**
Thalassiosira antarctica, **616**
Thalassiosira antiqua, **645**
Thalassiosira convexa, **644**
Thalassiosira decipiens, **621,** *624,* 625
Thalassiosira excentrica, **624,** *625,* 632
Thalassiosira fluviatilis, **568,** *572,* 605,
   614, 655
Thalassiosira fraga, **647**
Thalassiosira gracilis, **624, 633,** *633*
Thalassiosira gravida, **585,** *621,* 624,
   625, **651**
Thalassiosira gravida v. fossilis, **645**
Thalassiosira hyalina, **623**
Thalassiosira hyalinopsis, **645**
Thalassiosira irregulata, **647**
Thalassiosira kryophila, **645**
Thalassiosira lineata, **624,** *632*
Thalassiosira nana, **609,** *632*
Thalassiosira nordenskioeldi, **568,**
   621, 624
Thalassiosira oestrupii, **624,** *645,* **651**
Thalassiosira praeconvexa, **644**

Thalassiosira pseudonana, **605,** *621,*
   655
Thalassiosira rotula, **609**
Thalassiosira spinosa, **647**
Thalassiosira spumellaroides, **647**
Thalassiosira zabelinae, **644**
Thalassiosiraceae,   655
Thalassiothrix,   621, 625, 634, 639, **658**
Thalassiothrix antarctica, **624**
Thalassiothrix longissima, **624,** *625*
Thalassiphora, **430**
Thalassiphora dynamica, **429,** *429*
Thalassiphora pelagica, **429,** *429*
Thalassiphora reticulata, **429,** *429*
Thalassiphoraceae,   424
Thalassomyces, **381**
Thalassomyces fagei, **382**
Thalassomycetaceae,   381
Thalassomycetales,   381
Thalassopappus, **784**
Thallochrysis,   185, **518,** 714
Thallochrysidales,   518
thallus,   43, **975**
Thamnastraea, **67**
Tharama, **141**
Thaumatodiscaceae,   656
Thaumatoporella, **893**
theca, dinoflagellates,   **230, 975**
   thecal layer,   257, 258, **975**
   thecal membrane,   260
   thecal plate,   260
   thecal vesicle,   257
   thecal partition,   323
Thecadiniaceae,   378
Thecadinium,   334, 347, 348, **378**
Thecadinium petasatum,   **334,** *334, 347,*
   347
Thibia, **892**
Thienemann, A.,   348
Thierry, J.,   *872*
Thierstein, H. R.,   735, 738, 741, *741,*
   743, 745, *745, 748,* 769, 770, 776
Thiessen, R.,   841
Thiobacillus,   20, 24, 30, **41**
Thiobacillus thiooxidans, **22**
Thiobacillus thioparus, **27**
Thiobacteriaceae,   41
Thiobacterium, **41**
Thioploca,   29, 80
Thiorhodaceae,   42
Thiothrix,   29, *30,* 55, 80
Thomas, J.,   53
Thomas, J. P.,   124
Thomas, W. H.,   618
Thompson, G. C. (Cross et al.),   275
Thompson, R. H.,   275, 335
Thoracosphaera,   5, 279, 280, 706, 714,
   735, 745, 747, 771, 774, **777,** *973*
Thoracosphaera albatrosiana,   279,
   280, *280,* **747, 777**
Thoracosphaera deflandrei,   279, **708,**
   708
Thoracosphaera granifera, **279**
Thoracosphaera heimi,   279, **708,** *708,*
   747, *747*
Thoracosphaera pelagica, **747**
Thoracosphaera saxea, **279**
Thoracosphaera tesserula, **279**

Thoracosphaera tuberosa, **279, 747, 777**
Thoracosphaeraceae, 777
Thornley, M. J., 14
Thorosphaera, **784**
Thorosphaera flabellata, **717**
Thraniinae, 480
Thranium, 476, **480**
Thranium crassipes, **484,** *484*
Throndsen, J., 113, 128, 505, *511*
Thusu, B., 202
thylakoids, *43, 45, 50,* **975**
  blue-green algae, 43, 46
  chrysophytes, 501
  coccolithophores, 681
  diatoms, 571
  dinoflagellates, 237
  green algae, 825
  red algae, 108
  silicoflagellates, 537
  xanthophytes, 493
Thyrsoporella, **893**
Thyrsoporelleae, 893
Thysanoplanta, 77, **82**
Thysanoplantaceae, 82
Thysanoprobolus, **212**
Thysanoprobolus polykion, **194,** *194*
Tiarolithus, **783**
Tieffallenia, 956
Tieghem, Ph. van, 37
Timanella, 864, **892**
Timanellaceae, 892
Timkovich, R. (Dickerson et al.), 30
Timofeev, B. V., 154, 182, 205, 206
Tintinnus, 613
To, L. (Margulis et al.), 80, 93
Tomas, R. N., 236, 237, 240, 311, 326
Tokuyasu, K. (Emerson et al.), 15
Tolypella, 921, 923, 924, 933, **956**
Tolypella hispanica, **930**
Tolypella intricata, **931**
Tolypella nidifica, **931,** *931, 953*
Tolypella nidifica v. glomerata, **931**
Tolypellopsis, **956**
Tolypellopsis stelligera, **917,** *917*
Tolypothrix, *48, 48, 57, 68,* **83**
Toolongia, **409**
Toolongiaceae, 409
Tophoporata, **212**
torbanite, **832,** *841,* **970,** *975*
Tornacia, **212**
Toroformis, **212**
Tourtelot, H. A., 814
Toweius, **784**
Toweius craticulus, **771**
toxins
  blue-green algae, 58
  dinoflagellates, 317
trabeculae, acritarch and dinoflagellate
    cysts, **160,** *975*
trabeculate cyst, dinoflagellate, **286,** *975*
Trachelomonas, 517, 895, 896, *897,* **898,** *970*
Trachelomonas hispida, **897,** *897*
Trachelomonas raciborskii, **897,** *897*
Trachelomonas scabra, **897,** *897*
Trachelomonas woycickii, **897,** *897*

Trachelostomum, 507, **528**
Trachelostomum rampii, **529,** *529*
Trachydiacrodidae, 207
Trachyhystrichosphaera, **212**
Trachyneis, **658**
Trachyoligotriletes, **212**
Trachyrytidodiacrodium, **212**
Trachysphaeridium, **818**
Trachysphenia, **658**
Trachyzonodiacrodium, **212**
tract, **284,** *975*
Trainor, E. R., 597, 598
Trainor, F. R., 843
Tranolithus, **780**
transapical axis, diatom frustule, **576,** *976*
transduction, bacterial, **18,** *976*
transfer connection, red algae, **112,** *976*
transfer RNA, 95
transformation, bacterial, **17,** *975*
transverse axis, diatom frustule, **576,** *975*
transverse flagellum, dinoflagellate, **247, 248,** *976*
Transversopontis, 770, **780**
Transversopontis exilis, **757,** *757*
Travers, A., 546
Travers, M., 546
Traverse, A., 275, 364, 841
travertine, 55, *57,* 833
Trebouxia, **836,** *888*
Trejo, M., 729, *729,* 734
Tremalithus, **777**
Trematoligotriletum, **212,** *818*
Trematophora, **212**
Trematosphaeridium, **212**
Trench, R. K. (Muscatine et al.), 314
Trentepohlia, 826, 843, **889**
Trentepohliaceae, 845, 889
Trentepohliales, 889
Treponema, 40
Tretosestrum, **783**
triaene, ebridian skeleton, **465,** *976*
Triangulina, **212**
Triangumorpha, **212**
Triblastula, 286, **430**
Triblastula utinensis, **430,** *430*
Tribonema, 491, 493, 495, 496, 498
Tribonema intermixtum, **494,** *494*
Tribonematales, 498
Tribrachiatus, **778**
Triceratium, 637, 653, **656**
Triceratium antediluvianum, **583,** *583,* 585
Triceratium barbadense, **646, 647**
Triceratium cinnamomeum, **624**
Triceratium elegans, **641,** *641*
Triceratium shadboltianum, **583,** *583,* 585
trichocysts, 231, **976**
  dinoflagellate, **244,** *246*
  prasinophyte, 806
trichocyst pores, dinoflagellate
    theca, **244,** *245,* **976**
trichocyte, red algae, **113, 968,** *976*
Trichodesmium, *53, 54, 62, 82,* 340
Trichodesmium thiebautii, **62**

Trichodinium, **395**
Trichodinium hirsutum, **396,** *396*
trichogyne, red algae, *117, 125,* **976**
trichomes, blue-green algae, *12, 43, 47,* **976**
Trichosphaeridium, **212**
Triclypella, 933, **956**
Triclypella calcitrapa, **943,** *943*
Tridacna, symbionts of, 314, 315
Trigonium, 587, **657**
Trinacria, **657**
Trinocladus, **893**
Trinodella, 868, **892**
triode, ebridian skeleton, **465,** *976*
Trioperculodinium, **424**
Triploporella, 868, 869, **893**
Triploporelleae, 893
Tripodiidae, 480
Tripodium, **480**
Triporina, 838
Triposolenia, 304, 306, **371**
  movement of, *305*
Triposolenia bicornis, **346**
Triposolenia truncata, **370,** *370*
Triquetrorhabdulaceae, 781
Triquetrorhabdulus, 745, **781**
Triquetrorhabdulus carinatus, **761,** *764,* 767
Triquetrorhabdulus rugosus, **761,** *764,* 766
Trithyrodinium, 295, **392**
Trithyrodinium suspectum, **391,** *391*
Traillliella intricata, **116**
Trochastrites, **778**
Trochiliscaceae, 955
Trochiliscales, 955
Trochiliscus, 938, **955**
Trochiliscus (Karpinskya) oscolensis, **923**
Trochiscia, 55, 833, 836, **888**
Trochoaster, **778, 779**
Trochosira, **656**
Tropidoneis, **658**
Troxel, B. W. (Cloud et al.), 74, 94
Truania, **656**
true branching, blue-green algae, **47,** *976*
Truncatoscaphus, **780**
Truncodiscoaster, **778**
Trunculumarium, **212**
trunk, of acritarch processes, 167, *976*
Tryblioptychus, **656**
Trypanodinium, 311, **382**
Tschudy, R. H., 155
Tsentas, C. I., 62
Tsumura, K., *545, 556, 559*
Tuberculodinium, 378, **409**
Tuberculodinium vancampoae, **298,** *299, 302*
Tubidermodinium, **409**
Tubidermodinium sulcatum, **411,** *411*
Tubiellaceae, 82
Tubirhabdus, **780**
Tubodiscus, **783**
Tubodiscus jurapelagicus, **741,** *749*
Tubotuberella, **430**
Tubularia, **658**

Tubulospina, **212**
Tunisphaeridium, 160, 167, **212**
Tunisphaeridium caudatum, **170,** *170*
Tunisphaeridium concentricum, **155,** *155, 170*
Turbiosphaera, **404**
Turbodiscoaster, **778**
Turkmeniaria, **894**
Turkmeniaria adducta, **881,** *881*
Turner, M. F., 126
Turpin, C. R., 349
Turrispira, **778**
Tuttle, R. C., 319, 329
Tuttle, R. C. (Allen et al.), 236
Tydemania, **891**
Tyler, S. A., 35, 69, 70, 71, 73, 129
Tyligmasoma, 206, **212**
Tylotopalla, **212**
Tylotopalla caelamenicutis, **157,** *157, 166, 190*
Tylotopalla digitifera, **190,** *190*
Tynan, E. J., 517
Tytthodiscus, 816, 818
Tytthodiscus suevicus, **819,** *819*

Udotea, 831, 833, 834, 847, **891**
Udotea minima, **848**
Udotea javensis, **848,** *848*
Udotea palmetta, **848**
Udoteaceae, **890**
Ulothrix, 829, 834, 843, 845, 896, **889**
Ulotricaceae, 843, **889**
Ulotrichales, **889**
ultramicroplankton, **678, 971, 976**
Ulva, 829, 834, 845, 845, **889**
Ulvaceae, 845, **889**
Ulvales, **889**
Umbella, 932, **955**
Umbella lineata, **932**
Umbellaceae, 955
Umbellasphaeridium, **212**
Umbellidae, 955
Umbellina, 932, 936, 937, 937, **955**
Umbellina bella, **934,** *934*
Umbellina lineata, **932, 934,** *934*
Umbellinaceae, 955
umbellinaceans, 913, 932
  compared to charophytes, 936
Umbellinae, 955
Umbellininae, 955
Umbellosphaera, **784**
Umbellosphaera irregularis, **706,** *706,* 717, **718,** *718,* 719, 721, 770
Umbellosphaera tenuis, **706,** *706,* 717 **718,** *718,* 719, 770
Umbilicosphaera, **783**
Umbilicosphaera hulburtiana, **717**
Umbilicosphaera mirabilis, **719**
Umbilicosphaera sibogae, **718,** *718, 719, 769*
Umnova, N. I., 205
Uncinasphaera, **212**
Unella, **892**
Unellium, 149, 173, **212**
Unellium ampullium, **175,** *175, 203*
Ungdarella, **142**
Ungdarellaceae, 133, 135, 142
Unio, 845

Unjaella, **891**
upper windows, ebridian skeleton, **465, 976**
Uragiella, **893**
Uraimella, **140**
Uralella, **892**
uralga, 87
Uralites, **142**
Uraloporella, **868**
Uranovia, 77, 81
Urban, J. B., *817*
Urceolus, **898**
Uroglena, **519**
Uroglena soniaca, **523,** *523*
Ushakova, M. G., 752
Uteria, **893**
Uterieae, **893**
utricle, charophyte, **926,** *947,* **976**
Utricularia, 607
Uva, **891**
Uvatodinium, **392**
Uvatodinium marginatum, **391,** *391*
Uzzell, T., 95

Vachard, D., 848, *870*
Vachard, D. (Termier et al.), 133, 142, 868
Vachey, G., *195*
Vagalapilla, **779**
Vaginella striata, **881**
Valencia, F. G. (Gittleson et al.), 304, 306
Valensi, L., 151, 349, 350
Valensiella, **409**
Valensiella vermiculata, **362**
Valet, G., 860
Vallacerta, 549, **559**
Vallacerta hortonii, **542,** *542, 550, 552, 559*
Vallacertaceae, 559
Vallacertales, 525, 554
Vallenia, **212**
Valonia, 827, **891**
Valoniaceae, **891**
valve, diatom frustule, **573, 976**
  face, **573, 976**
  mantle, **573, 976**
  plane, **576, 976**
valvocopula, diatom frustule, **576, 976**
Van Geel, B., **884**
Vanguestaine, M., 187
Vanheurckia, **658**
Vanhoeffenia, **138**
VanLandingham, S. L., 626, 648, 654
Van Valkenburg, S. D., *536, 537, 538, 541, 546*
Varma, C. P., 140
Vassar, J. M., 126
Vaucheria, 55, 493, 495, 496, *496,* 496, **498,** *833*
Vaucheria antiqua, **496**
Vaucheriales, **498**
Vavososphaeridium, **212**
Vavrdová, M., 173, 205
Vavrdovella, **212**
Veer, J. van der, 683
Veillonellaceae, 41

Vekshina, V. N., 745
Vekshinella, **779**
Vekshinella stradneri, **735,** *738, 749*
Velebitella, **893**
Velella velella, symbionts of, 314, 315
Veliferites, **822**
Veliferites tenuimarginatus, **822**
vellum, acritarch, **171, 976**
velum, diatom frustule, **583, 976**
Venkataraman, G. S., 572
Venrick, E. L., 623, 624
Verbeek, J. W., 741, 745, 748
Verbeek, J. W. (Manivit et al.), 741, 748
Verdier, J. P., *395, 401, 408, 415, 416, 423, 429*
Vermiculithina, **779**
Vermiporella, 864, 867, **892**
Verniory, R., 843
Versimorphida, **208**
Verticillodesmis, 852, **891**
Verticilloporella, **893**
Veryhachium, 149, 151, 152, 153, 160, 171, 185, 187, 190, 197, 199, 202, 204, 206, **212,** 838
Volcani, B. E. (Lauritis et al.), 571, 573, 605, 611
Volcani, B. E. (Lewin et al.), 572, 595, 618, 619
Volcani, B. E. (Reimann et al.), 573, 579, *581,* 610
Volkova, N. A., 128, 151, 181, 182, 185
Vologdin, A. G., 64, 65, *854*
Vologdinella, **892**
Vologdinella fragilis, **865,** *865*
Vologdinelleae, 892
Volvocaceae, **888**
Volvocales, 834, 835, **888**
Volvox, 828, 829, *830,* 834, 835, **888,** 968
Volvox aureus, **829,** *829*
Vorticella, 243
Vorticella cincta, 227
Votta, J. J., 304
Vozzhennikova, T. F., 383, *387*
Vozzhennikovia, **392**
Vuille, J. (Donnelly et al.), 341
Vulcanisphaera, **212**

Wahlquist, C. L., 341
Wailes, G. H., *546, 546*
Waines, R. H. (Wilson et al.), 133, 142
Walcott, C. D., 32
Wall, D., 202, 230, 243, 265, 273, 275, 276, *276,* 278, 279, *279,* 280, 283, 284, *285, 285,* 286, 288, 289, 290, 293, 297, 298, *301,* 302, 320, 321, 329, 334, 335, 337, 344, 351, 364, *364,* 368, 383, 385, *395, 396, 398, 403, 406,* 424, *424,* 428, 810, 814
Wall, J. H., *173, 391*
wall
  acritarchs, 152
  bacteria, 13
  blue-green algae, 43, *43*
  chrysophytes, 498
  diatoms, 572
  dinoflagellates, 256

eucaryotic algae, 43
green algae, 826
prasinophyte cyst, 809
red algae, 110
xanthophytes, 493
Waller, B. A. (Clifton et al.), 833
Wallich, G. C., 725
Wallner, J., *886*, 886
Wallodinium, **424**
Wallodinium bidigitatum, **427**, *427*
Wallodinium krutzschi, **427**, *427*
Walsby, A. E., 52
Walsby, A. E. (Fay et al.), 49
Walsby, A. E. (Fogg et al.), 45, 46, 51, 57, 92
Walter, M. R., 28, 65, 94, 96, 130, *132*
Walter, M. R. (Glaessner et al.), 65
Wanaea, **404**
Wanaea clathrata, **408**, *408*
Waputikia, 132, **142**
Warnowia, 231, 243, 308, **380**
Warnowia fusus, **379**, *379*
Warnowiaceae, 231, 378
Warren, J. S., 275, 364, 435
Watabe, N., 687, 688, *689*, 692, 720
Watabe, N. (Wall et al.), 279, *280*
Waters, J. (Mapletoft et al.), 334
Watznaueria, **783**
Watznaueria barnesae, **738**, *738*, 769
Watznaueria diaphanae, **741**, *749*
Watznaueria oblonga, **741**, *749*
Watznaueriaceae, 782
Ważyńska, H., 182
Weaver, F. M., 552
Webby, D. B., 839
Weber, J. N., 110
Weeksia, **140**
Weeksiaceae, 140
Weikkoella, 938, **955**
Weiser, J. (Honigberg et al.), 226
Weiss, R. E. (Blackwelder et al.), 692
Weissfellner, H. (Isenberg et al.), 687
Wells, P. (Mapletoft et al.), 334
Wendiella, **818**
Wenrich, D. H. (Honigberg et al.), 226
Werner, D., 614
Werner, R. G., 649
West, G. S., 239, 273, 274, 293, 324, 334, 649, 654
West, J. A., 124
Wetherbee, R., 257, 258, *260*, 261, 324, 325, 326, 327
Wetzel, O., 181, 182, 281, 293, 350, 383
Wetzel, W., 349, 392
Wetzeliella, 286, 288, 290, 294, 295, 296, 302, 355, 358, **392**
Wetzeliella articulata, **391**, *391*
Wetzeliella solida, **281**
Wetzeliellaceae, 389
Weyland, H., 886
Whatley, F. R., 91
whiplash flagella, **248**, 826, **964**
White, H. H., 350
White, M. C., 181
Whitney, J. D., 636
Whittle, S. J., 493, 497
Whitton, B. A., 44, 51, 54, 55
Whitton, B. A. (Barnes et al.), 127

Wicander, E. R., 160, *174*, *175*, 182, *193*, *194*, *196*, 202, *203*, 816
Wicander, E. R. (Tappan et al.), 190
Wiesner, M. (Hastings et al.), 317
Wiggins, V. D., 297, *298*, 353, *389*, *390*
Wight, W. W. (Loeblich III et al.), 607, 608
Wigwamma, **777**
Wilbur, K. M., 687, 688, 689, 720
Wilbur, K. M. (Blackwelder et al.), 692
Wilcox, M., 49
Wilcoxon, J. A., 735, *761*
Wille, W., *837*
Williams, D. B., 275, 358, 364
Williams, D. C., 684, 689, *690*, 714
Williams, G. L., 287, 288, 299, *357*, 358, *361*, *362*, 368, *391*, *392*, 406, 411, 415, 429
Williams, G. L. (Downie et al.), 205, 358
Williams, G. L. (Evitt et al.), 284, 295
Williams, G. L. (Millioud et al.), 351
Williams, J., 252, 253, 339, 341, 343, 344
Williams, J. (Donnelly et al.), 341
Williams, J. (Murphy et al.), 341
Williams, J. (Steidinger et al.), 308, 341
Wilson, E. C., 133, 142
Wilson, G. J., *392*
Wilson, L. R., *817*, 839, *841*
Wilson, W. B. (Aldrich et al.), 319, 326
Wilsonastrum, **212**
Wilsonidium, **392**
Wilsonidium echinosuturatum, **392**, *392*
windows, ebridian skeleton, **539**, *973*, **977**
Windsoporella, **893**
Winslow, M. R., 814
Winwaloeusia, **212**
Wise, S. W., Jr., 770
Wiseman, J. D. H., 629, *631*
Withers, N. W., 84
Woese, C. R. (Fox et al.), 31
Wolfarth-Bottermann, K. E., 231
Wolk, C. P., 44
Woloszynskia, 254, 275, 332, 334, 337, 351, **374**
Woloszynskia apiculata, **278**, 319, 329, 332, *333*, 334, 374
Woloszynskia coronata, **242**, *244*, 259
Woloszynskia pascheri, **240**, 334, 348
Woloszynskia reticulata, **334**
Woloszynskia tenuissima, **258**
Woloszynskia tylota, **244**, 334
Womersley, H. B. S., 850, *853*
Wong, J. L., 317
Wood, A., 66, *847*, *866*
Wood, E. J. F., 343, 345, 346, 605, 616, 623, 836
Wood, R. D., 913, 923, 930
Wornardt, W. W., Jr., 542, 635, 639, 649
Worsley, D., 65
Worsley, T. R., 735, 757, 773
Worth, G. K., 317, 341
Wray, J. L., 126, 133
Wright, L., 250

Wright, L. A. (Cloud et al.), 74, 94
Wulff, E., 861
Wyatt, J. T., 43

Xanthidium, 181, 275, 349, 350, 885, **894**
Xanthidium aculeatum, 349
Xanthidium bulbosum, 349
Xanthidium delitiense, 349
Xanthidium furcatum, 272, 349
Xanthidium hirsutum, 349
Xanthidium pilosum, 349
Xanthidium ramosum, 272, 349
Xanthidium tubiferum, 349
Xanthiopyxis, 600, **657**
Xanthiopyxis grantii, **639**, *639*
Xanthophyta, 490, 491–499
cell, 491
cysts, 496
classification, 498
occurrence, 496
plastids and pigments, 493
reproduction, 494
wall, 493
zoospores, 491
Xanthophyceae, 490, 491, 498
Xenascus, **424**
Xenicodinium, **395**
Xenicodinium lubricum, **396**, *396*
Xenikoon, **392**
Xiphophoridiaceae, 404
Xiphophoridium, **409**
Xiphophoridium alatum, **408**, *408*
Xystotheca, **657**

Yamadaea, 141
Yartseva, M. V., *937*
Yashnov, V. A., 812
Yoshidaia, **659**

Zahl, P. A., 313
Zaitzeff, J. B. (Cross et al.), 275
Zaitzeff, J. B. (Orr et al.), 639
Zalessky, M. D., 832, 833
Zanon, V., 507
Zaporella, **891**
Zarillo, G. A., 552
Zedebauer, E., 330
Zetella, 881, **894**
Zeugrhabdotus, **780**
Zeuthen, E., 609
Zhuze, A. P., 623, 624, 628, 637, 639, 644, 646
Ziembińska-Tworzydło, M., *950*
Zimmermann, W., 250, *252*, 252, 367
Zingmark, R. G., 231, 236, 237, 335, 337, *337*
Zittelina, **893**
Zobell, C. E., 9
Zonaletes, **818**
Zonooidium, **211**
Zonosphaeridium, **818**
Zonotrichites, **83**
Zoochlorella, 836
Zooxanthella, 313, **380**
Zooxanthella microadriatica, **234**, 244, 313, 315
Zooxanthellaceae, 380

zooxanthellae, **313, 977**
Zooxanthellales, 380
Zosterosphaera, 69, 74, **81**
Zosterosphaera tripunctata, **77**, *77*
Zygabikodinium, 344, **377**
Zygnema, *883*, **894**
Zygnemaphyceae, 894
Zygnemataceae, 884, 894

Zygnematales, 882, *883*, 894
Zygoceros, **656**
Zygodiscaceae, 780
Zygodiscus, **780**
Zygodiscus spiralis, 748
Zygolithaceae, 780
Zygolithites, 703, **780**
zygolith coccolith, **694, 977**

Zygolithus, **780**
Zygosphaera, **778**
Zygosphaera divergens, **695**
Zygosphaeraceae, 777
Zygosphaerales, 777
Zygostephanos, **779**
Zygrhablithus, 770, **778**
Zygrhablithus bijugatus, **697**, *697*, **752**